Atlas of Functional Anatomy for Regional Anesthesia and Pain Medicine

Miguel Angel Reina
Editor

José Antonio De Andrés • Admir Hadzic
Alberto Prats-Galino • Xavier Sala-Blanch
André A.J. van Zundert
Associate Editors

Atlas of Functional Anatomy for Regional Anesthesia and Pain Medicine

Human Structure, Ultrastructure and 3D Reconstruction Images

Springer

Editor
Miguel Angel Reina, MD, PhD
Department of Clinical Medical Sciences
and Institute of Applied Molecular Medicine
School of Medicine
University of CEU San Pablo
Madrid
Spain

Department of Anesthesiology
Madrid-Montepríncipe University Hospital
Madrid
Spain

Associate Editors
José Antonio De Andrés, MD, PhD
Department of Surgical Specialties
School of Medicine, University of Valencia
Department of Anesthesia, Critical Care,
and Pain Management
General University Hospital
Valencia
Spain

Admir Hadzic, MD, PhD
Department of Anesthesiology
St. Luke's-Roosevelt Hospital Center
New York School of Regional
Anesthesia (NYSORA)
New York, NY
USA

College of Physicians and Surgeons
Columbia University
New York, NY
USA

Alberto Prats-Galino, MD, PhD
Laboratory of Surgical NeuroAnatomy
Human Anatomy and Embryology Unit
School of Medicine, University of Barcelona
Barcelona
Spain

Xavier Sala-Blanch, MD
Human Anatomy and Embryology Unit
School of Medicine, University of Barcelona
Barcelona
Spain

Department of Anesthesiology and Critical Care
Clinic Hospital
Barcelona
Spain

André A.J. van Zundert, MD, PhD, FRCA,
EDRA, FANZCA
Discipline of Anesthesiology
The University of Queensland and Royal
Brisbane and Women's Hospital
Brisbane, Queensland
Australia

ISBN 978-3-319-09521-9 ISBN 978-3-319-09522-6 (eBook)
DOI 10.1007/978-3-319-09522-6
Springer Cham Heidelberg New York Dordrecht London

Library of Congress Control Number: 2014957316

Printed on acid-free paper

Springer is part of Springer Science+Business Media (www.springer.com)

To my children, Yamila, Lucila, and Emmanuel, for their shared interest and for being the source of inspiration in the development of a lifelong project.

To my beloved wife, Emilse, for her enduring love and unconditional support and for transforming each day into a unique experience with the gift of her smile.

Preface

Several excellent books on anatomy and histology comprehensively address features of the entire human body. Some are based on drawings and sketches; others are collections of photographic images obtained from cadavers. In recent decades, not only anesthesiologists with an interest in techniques related to anesthetic blockades and pain treatment, but also orthopedic surgeons and neurosurgeons interested in surgical techniques on the spinal column have searched for answers to their inquiries through the pages of these types of books. The elaboration of atlases on general histology has enabled professionals to assess the role that human tissues play in processes related to the pharmacokinetics and pharmacodynamics of certain drugs such as the analysis of the possible mechanisms leading to several complications.

This Atlas is the result of anatomical and histological research oriented and applied to solve specific questions expressed by practitioners that may arise during clinical practice, with answers not often found in traditional general atlases.

Practitioners who carry out their activities in the operating room during long shifts may have difficulty finding time to research subjects of interest. Often, hypotheses supporting clinical investigations have been based on anatomy and general histology books.

Our inquiring minds prompted us to seek possible answers to problems befalling patients due to regional anesthetic practice in the operating room, leading us to revise in the laboratory basic concepts in anatomy and histology. In several instances, the search for answers with regard to needs in regional anesthesia and related medical specialties has enabled us to support previously published results, but in few occasions the opposite findings obtained in our studies led us to question theoretical considerations from books that had been traditionally accepted over the past century. Such is the case in our revision of the spinal dural sac in humans, where examination of its ultrastructure displayed differences that can no longer support the existence of a real subdural space as it had been described decades ago. Instead, the subdural space is an acquired space, in many cases of artifactual origin, due to manipulation during the process of sample extraction from cadavers. Evidence outlined here, as well as other facts discussed in this Atlas, raises awareness of the need for further critical review of many hypotheses to objectively analyze previous contradictory data.

The authors of chapters in this Atlas are experts in their respective fields of research who present images from their own work, offering relevant anatomical insights of interest to several medical disciplines. In an effort to improve scientific communication between researchers and readers, a great effort has been made to collect numerous recently acquired images in each chapter, with many remaining unpublished at present.

Instead of barely presenting facts in the form of a traditional textbook in which explanations are provided mainly in written form, this Atlas offers a brief introduction in each chapter but prioritizes the illustrating potential of real human images from tissue samples carefully selected to provide results open to inquiry and interpretation by the reader. Specialists may use this work as a source of information according to their own interests.

For this reason, we believe this Atlas closes a gap that existed in the past decade concerning new scientific research technologies of interest to anesthesiologists, orthopedic surgeons, neurosurgeons, and medical practitioners.

However, this Atlas is not intended to replace textbooks on regional anesthesia, techniques in the treatment of pain, or surgical techniques of the spine or peripheral nerves. Each chapter intends to act as complement and facilitate the understanding of chapters of books that do not have this type of graphic material.

The Atlas includes more than 1,600 images. Each image is followed by a short text to aid in its interpretation. The amount of work involved in obtaining these images extends over a period of 25 years.

The different types of images include those on gross anatomy obtained from anatomical dissections, specifically designed to meet the requirements of this Atlas. In addition, transmission and scanning electron microscopy images help to interpret the ultrastructure of tissues that may be of particular interest to professionals for whom this work is intended. Three-dimensional (3D) image reconstruction of magnetic resonance images (MRI) has become increasingly relevant as it can display 3D images of anatomical structures from different perspectives as well as their relationships, significantly meeting not only the diagnostic purposes of radiology departments but also the goals of several medical disciplines such as therapy units and educational departments' surgical units, among others.

The first part of the Atlas presents the macroscopic morphology of peripheral nerves along with their ultrastructure and the types of tissue damage that may be caused due to accidental puncture of these nerves. This part incorporates the analysis of anatomical models on which "in vitro" intraneural injections had been performed.

The second part presents the macroscopic morphology, ultrastructure, and 3D reconstructions obtained from MRIs of human spinal meninges, the nerve roots of the cauda equina, spinal ligaments, and epidural fat. This part includes chapters in which the ultrastructure of tissue damaged during "in vitro" lumbar punctures is examined, adopting anatomical models to illustrate injuries caused by different types of needle tips of routine use during regional anesthesia. Similar models are presented to evaluate the consequences of selective root anesthetic blocks and to assess spinal devices for neurostimulation instruction.

The third part of the Atlas describes devices required in the realization of anesthetic blocks, each displayed and examined in an illustrative manner, including different types of needles and catheters used in central and peripheral blockade and relevant features of epidural filters.

Finally in the fourth part, chapters are organized by methodological aspects relative to techniques applied in the production of images illustrating this Atlas.

The authors have made a significant effort to produce a remarkable collection of scientific images. Their work also reveals its message in part through brief elucidating descriptions that accompany these images. Each of these images contains details that may support or challenge current concepts accepted in our clinical practice. Facts are exposed in this Atlas to serve the purpose of presenting current insights and controversies to readers. It is our hope that reviews of scientific work may help strengthen the foundations of medical knowledge to benefit clinical practice, which has no other aim than caring for patients worldwide.

Madrid, Spain — Miguel Angel Reina
Valencia, Spain — José Antonio De Andrés
New York, NY, USA — Admir Hadzic
Barcelona, Spain — Alberto Prats-Galino
Barcelona, Spain — Xavier Sala-Blanch
Brisbane, QLD, Australia — André A.J. van Zundert

Acknowledgments

I am sincerely grateful for the opportunity given to us by Springer Science+Business Media, LLC. Without the support of all members of their team, it would not have been possible to gather such a number of scientific researchers from different continents who, as a result of their common interest in basic sciences applied to the field of anesthesiology, have produced the work presented throughout the chapters of this Atlas. My deepest gratitude to all researchers involved in this project for their invaluable contribution in providing unique and updated original research material in each chapter, which undoubtedly represents the work of not only one but several lifetimes.

I would like to thank Dr. Andrés López, who has actively collaborated and supported this research for more than 20 years as Head of Department of Anesthesiology HM University Hospitals. I wish to extend my gratitude to all physicians and technical staff at the Madrid-Montepríncipe University Hospital, and the Faculty of Medicine of the Universidad CEU San Pablo, in particular to the technical staff at the Institute of Applied Molecular Medicine.

I would also like to thank my many colleagues who have contributed to this book. I am deeply grateful. I acknowledge in particular doctors, nurses, nursing auxiliaries, technicians, and caretakers of the Mostoles University Hospital, Madrid, where I carried out the first stage of my research.

Electron microscopy works were achieved thanks to the help of Ana Vicente Montaña, Agustín Fernández Larios, María Luisa García Gil, Alfonzo Rodríguez Muñoz, and Alfredo Fernández Larios in the ICTS Electron Microscopy National Center, Complutense University, Madrid.

It is a pleasure to acknowledge the skilled work on human anatomy by professors Alberto Prats-Galino, Xavier Sala-Blanch, and Anna Puigdellívol Sánchez from the University of Barcelona and Anna Carrera at the University of Girona.

Many people were involved in various aspects of this project including text translation and revision (Fabiola Machés). I would like to acknowledge and thank them for their precious time, valuable suggestions, and encouragement during the various phases of the elaboration of this book.

Contents

Part IV Research Techniques

Contributors

Editor

Miguel Angel Reina, MD, PhD Department of Clinical Medical Sciences and Institute of Applied Molecular Medicine, School of Medicine, University of CEU San Pablo, Madrid, Spain

Department of Anesthesiology, Madrid-Montepríncipe University Hospital, Madrid, Spain

Associate Editors

José Antonio De Andrés, MD, PhD Department of Surgical Specialties, School of Medicine, University of Valencia, Valencia, Spain

Department of Anesthesia, Critical Care, and Pain Management, General University Hospital, Valencia, Spain

Admir Hadzic, MD, PhD Department of Anesthesiology, St. Luke's-Roosevelt Hospital Center, New York School of Regional Anesthesia (NYSORA), New York, NY, USA

College of Physicians and Surgeons Columbia University, New York, NY, USA

Alberto Prats-Galino, MD, PhD Laboratory of Surgical NeuroAnatomy (LSNA), Human Anatomy and Embryology Unit, School of Medicine, University of Barcelona, Barcelona, Spain

Xavier Sala-Blanch, MD Human Anatomy and Embryology Unit, School of Medicine, University of Barcelona, Barcelona, Spain

Department of Anesthesiology and Critical Care, Clinic Hospital, Barcelona, Spain

André A. J. van Zundert, MD, PhD, FRCA, EDRA, FANZCA Discipline of Anesthesiology, The University of Queensland and Royal Brisbane and Women's Hospital, Brisbane, QLD, Australia

Contributors

Henning Lykke Andersen, MD, PhD Department of Anaesthesiology, Copenhagen University Hospital, Herlev, Copenhagen, Denmark

Sofie Lykke Andersen, MD Department of Anaesthesiology, Copenhagen University Hospital, Rigshospitalet, Copenhagen, Denmark

Riánsares Arriazu Navarro, Pharm.D. PhD Histology Unit, Department of Basic Medical Sciences and Applied Molecular Medicine Institute, School of Medicine, CEU San Pablo University, Madrid, Spain

Rafael Blanco, MD Anaesthetic Department, Corniche Hospital, Abu Dhabi, Arab Emirates

Anna Carrera, MD, PhD Neuroscience, Embriology, Molecular Oncology and Clinical Anatomy Group (NEOMA), School of Medicine, University of Girona, Girona, Spain

Julio Castedo, MD Department of Clinical Medical Science, School of Medicine, CEU San Pablo University, Madrid, Spain

Department of Radiology, Madrid-Torrelodones University Hospital, Madrid, Spain

Clive Collier, MD, MRCP, FRCA, FANZCA Department of Obstetric Anaesthesia, Prince of Wales Private Hospital, Sydney, NSW, Australia

Emilse Colman Peyrano, BS Research Unit, Anesthesia Reina Colman SLP, Madrid, Spain

Pilar De Diego-Isasa, MD Department of Anesthesiology, Hospital Asepeyo, Madrid, Spain

Oscar De León Casasola, MD Department of Anesthesiology, Roswell Park Cancer Institute, Buffalo, NY, USA

Department of Anesthesiology, University at Buffalo, School of Medicine and Biomedical Sciences, Buffalo, NY, USA

Ana Delgado, Tch Department of Radiology, Complejo Hospitalario de Jaén, Jaén, Spain

Rafael García De Sola, MD, PhD Department of Surgery, School of Medicine, Autonomous University of Madrid, Madrid, Spain

Department of Neurosurgery, La Princesa University Hospital, Madrid, Spain

Concepción Del Olmo, MD Department of Anesthesiology, Hospital Asepeyo, Madrid, Spain

Jorge Diamantopoulos, MD Neurosurgical Department, Madrid-Montepríncipe University Hospital, Madrid-Sanchinarro University Hospital, Madrid, Spain

Martin Dittmann, MD, PhD Department of Anesthesia, Kreiskrankenhaus, Bad Säckingen, Germany

Esther M. Durán Mateos, BS Histology Unit, Department of Basic Medical Sciences and Institute of Applied Molecular Medicine Institute, School of Medicine, CEU San Pablo University, Madrid, Spain

José Manuel Escobar, Tech Department of Radiology, Madrid-Montepríncipe University Hospital, School of Medicine, CEU San Pablo University, Madrid, Spain

Paloma Fernández, BS Histology Unit. Institute of Applied Molecular Medicine, School of Medicine, University of CEU San Pablo, Madrid, Spain

Manuel Fernández Domínguez, MD, PhD Department of Clinical Medical Sciences and Institute of Applied Molecular Medicine, School of Medicine, CEU San Pablo University, Madrid, Spain

Department of Maxillofacial Surgery, Madrid-Montepríncipe University Hospital and Madrid-Sanchinarro University Hospital, Madrid, Spain

Agustín Fernández Larios, Tech Transmission Electron Microscopy Unit, ICTS Electron Microscopy National Center, Complutense University, Madrid, Spain

Alfredo Fernández Larios, Tech Scanning Electron Microscopy Unit, ICTS Electron Microscopy National Center, Complutense University, Madrid, Spain

Maria Luisa García Gil, BS, PhD Transmission Electron Microscopy Unit, ICTS Electron Microscopy National Center, Complutense University, Madrid, Spain

Jose Luís Gonzalez, MD Department of Radiology, Complejo Hospitalario de Jaén, Jaén, Spain

José María Hernández, MD Department of Clinical Medical Sciences, School of Medicine, CEU San Pablo University, Madrid, Spain

Department of Anesthesiology, Madrid-Montepríncipe University Hospital, Madrid, Spain

Juan A. Juanes Méndez, MD, PhD Human Anatomy and Embryology Unit, School of Medicine, University of Salamanca, Salamanca, Spain

Manoj Kumar Karmakar, MD, FRCA, FHKCA, FHKAM Department of Anaesthesia and Intensive Care, The Chinese University of Hong Kong, Prince of Wales Hospital, Hong Kong, China

L. M. Arno Lataster, MSc Department of Anatomy and Embryology, Faculty of Health, Medicine and Life Sciences, Maastricht University, Maastricht, The Netherlands

Philipp Lirk, MD, PhD Department of Anaesthesiology, Academic Medical Center, University of Amsterdam, Amsterdam, The Netherlands

Andrés López, MD Department of Clinical Medical Sciences, School of Medicine, CEU San Pablo University, Madrid, Spain

Department of Anesthesiology, Madrid-Montepríncipe University Hospital, Madrid-Torrelodones University Hospital, Madrid-Sanchinarro University Hospital, Madrid, Spain

Manuel Llusá, MD, PhD Laboratory of Macro-micro Dissection and Surgical Anatomy, Human Anatomy and Embryology Unit, School of Medicine, University of Barcelona, Barcelona, Spain

Orthopaedic and Traumatology Surgery Unit, Vall d'Hebron Hospital, Barcelona, Spain

Fabiola Machés, MD Department of Anesthesiology, Madrid-Montepríncipe University Hospital, Madrid, Spain

Félix Manzarbeitia, MD, PhD Department of Pathology, Fundación Jiménez Díaz University Hospital, School of Medicine, Autonomous University of Madrid, Madrid, Spain

Marija Mavar, MSc Laboratory of Surgical NeuroAnatomy (LSNA), Human Anatomy, and Embryology Unit, School of Medicine, University of Barcelona, Barcelona, Spain

Ritsuko Masuda, MD, PhD Department of Anesthesiology, Tokyo Hospital, Tokai University, Tokyo, Japan

Ana Vicente Montaña, DVM Scanning Electron Microscopy Unit, ICTS Electron Microscopy National Center, Complutense University, Madrid, Spain

María Rosa Morro, MD Human Anatomy and Embryology Unit, School of Medicine, University of Barcelona, Barcelona, Spain

Orthopaedic and Traumatology Surgery Unit, Vall d'Hebron Hospital, Girona, Spain

Amer Mustafa Gondolbeu, MD Laboratory of Macro-Micro Dissection and Surgical Anatomy, Human Anatomy and Embryology Unit, School of Medicine, University of Barcelona, Barcelona, Spain

Orthopaedic and Traumatology Surgery Unit, Arnau de Vilanova Hospital, Lleida, Spain

Anna Oliva, MD Laboratory of Macro-Micro Dissection and Surgical Anatomy, Human Anatomy and Embryology Unit, School of Medicine, University of Barcelona, Barcelona, Spain

Rehabilitation Unit, Mataró Hospital, Mataró, Barcelona, Spain

Teresa Parras, MD, PhD Anaesthetic Department, St. George's Hospital, London, UK

Anna Puigdellívol-Sánchez, MD, PhD Laboratory of Surgical NeuroAnatomy (LSNA), Human Anatomy and Embryology Unit, School of Medicine, University of Barcelona, Barcelona, Spain

Paloma Pulido, MD, PhD Department of Neurosurgery, La Princesa University Hospital, Madrid, Spain

Jaume Pomés, MD, PhD Musculoskeletal Unit, Radiology Department, Imaging Diagnostic Center, Hospital Clinic Barcelona, Barcelona, Spain

Francisco Reina, MD, PhD Neuroscience, Embryology, Molecular Oncology, and Clinical Anatomy Group (NEOMA), School of Medicine, University of Girona, Girona, Spain

Lucila Reina Colman, BS Master Degree Dentistry Department, Rey Juan Carlos University, Madrid, Spain

Teresa Ribalta, MD, PhD Department of Pathology, Clinic Hospital, University of Barcelona, Barcelona, Spain

Alfonso Rodríguez Muñoz, Tech Scanning Electron Microscopy Unit, ICTS Electron Microscopy National Center, Complutense University, Madrid, Spain

Joan San Molina, MD, PhD Neuroscience, Embryology, Molecular Oncology, and Clinical Anatomy Group, School of Medicine, University of Girona, Girona, Spain

Kumiko Tanuma, PhD Department 2nd Anatomy, Nippon Medical School, Shinjuku Vocational College of Acupuncture and Judo Therapy, Ogura Academy, Tokyo, Japan

Ignacio Tardieu, Tch Histology Unit, Institute of Applied Molecular Medicine, School of Medicine, CEU San Pablo University, Madrid, Spain

Jørgen Tranum-Jensen, MD Department of Cellular and Molecular Medicine, School of Medicine, University of Copenhagen, Copenhagen, Denmark

Catherine Vandepitte, MD Department of Anaesthesiology, Catholic University of Leuven, Leuven, Belgium

Ilvana Vuckovic-Hasanbegovic, MD, PhD Department of Anatomy, University of Sarajevo School of Medicine, Sarajevo, Bosnia Herzegovina

Part I

Human Peripheral Nerve

1 Ultrastructure of Myelinated and Unmyelinated Axons

Miguel Angel Reina, Riánsares Arriazu Navarro, and Esther M. Durán Mateos

The macroscopic anatomy of peripheral nerves results from the hierarchic arrangement of nerve fibers formed by microscopic groups of motor and sensory neuronal cytoplasmic elongations known as axons [1–8]. The latter enable conduction of electrical impulses along their plasma membranes, as well as chemically mediated signal transduction throughout cytoplasmic organelles. Most peripheral nerves are mixed nerves containing efferent motor fibers, afferent sensory fibers, and sympathetic fibers. The initial segment of an axon (AIS) originates in the soma and is located between the cell body and the beginning of the myelin sheath. This site is a polarized structure containing proteins such as voltage-gated sodium channels (VGCs), which are responsible for producing the inward ionic flow that generates the action potential at the AIS. VCGs at the AIS, together with the linked spectrin–actin membrane cytoskeleton may function as a diffusion barrier preventing axonal proteins from leaking out of the neuron. Nerve fibers have two types of axons, myelinated and unmyelinated. Each myelinated nerve fiber consists of an axon covered by myelin sheaths produced by Schwann cells alternating with areas called nodes of Ranvier [1–4]. The latter are sites of discontinuity between successive myelin sheaths along the axon [5, 6]. Mechanisms regulating the production and distribution of myelin take place in an extremely tight compartment located between the plasma membranes of both neurons and Schwann cells [7]. An inner layer, or basal membrane, and an outer layer, known as the endoneurium, enclose each myelinated nerve fiber. The axon contains an extremely dense cytoplasm, with an estimated viscosity five times greater than that of water. Inside the cytoplasm are microtubules, neurofilaments, mitochondria, vesicles, cisterns of cytoplasmic reticulum, and lysosomes, whereas ribosomes and Golgi apparatus are present uniquely in the AIS [5, 6]. The external diameter of a myelinated axon measures between 2 and 18 μm, whereas its length varies remarkably, ranging from just a few millimeters to as long as 1 m [5, 6].

Schwann cells wrap around axons at regular intervals, leaving uncovered portions known as Ranvier nodes [7]. The internodal distance is the interval between Ranvier nodes, is occupied by alternating Schwann cells, and measures about 0.4–1.2 mm.

Membrane depolarization and repolarization of myelinated axons during the propagation of action potentials occur at the Ranvier nodes; here, the proportion of VGCs is greater than in other unmyelinated areas.

Unmyelinated axons are not enclosed within multilayered myelin sheaths [8]. Instead, a single Schwann cell appears at the center of axonal groups, emitting cytoplasmic prolongations that separate each of the surrounding unmyelinated axons. Here, groups of six to eight axons are held together and partially covered by simple, nonwrapping prolongations originating in a single Schwann cell. In addition, groups of unmyelinated axons contain bundles of collagen fibers that confer mechanical resistance to the axonal group. The diameter of unmyelinated axons measures about 0.1–3 μm (Figs. 1.1, 1.2, 1.3, 1.4, 1.5, 1.6, 1.7, 1.8, 1.9, 1.10, 1.11, 1.12, 1.13 and 1.14).

M.A. Reina, MD, PhD (✉)
Department of Clinical Medical Sciences and Institute of Applied Molecular Medicine, School of Medicine, University of CEU San Pablo, Madrid, Spain

Department of Anesthesiology, Madrid-Montepríncipe University Hospital, Madrid, Spain
e-mail: miguelangel@perticone.e.telefonica.net

R. Arriazu Navarro, Pharm.D. PhD
Histology Unit, Department of Basic Medical Sciences and Applied Molecular Medicine Institute, School of Medicine, CEU San Pablo University, Madrid, Spain
e-mail: arriazun@ceu.es; riansares.arriazu@gmail.com

E.M. Durán Mateos, BS
Histology Unit, Department of Basic Medical Sciences and Institute of Applied Molecular Medicine Institute, School of Medicine, CEU San Pablo University, Madrid, Spain
e-mail: esthermaria.duranmateos@ceu.es

M.A. Reina et al. (eds.), *Atlas of Functional Anatomy for Regional Anesthesia and Pain Medicine: Human Structure, Ultrastructure and 3D Reconstruction Images*, DOI 10.1007/978-3-319-09522-6_1,

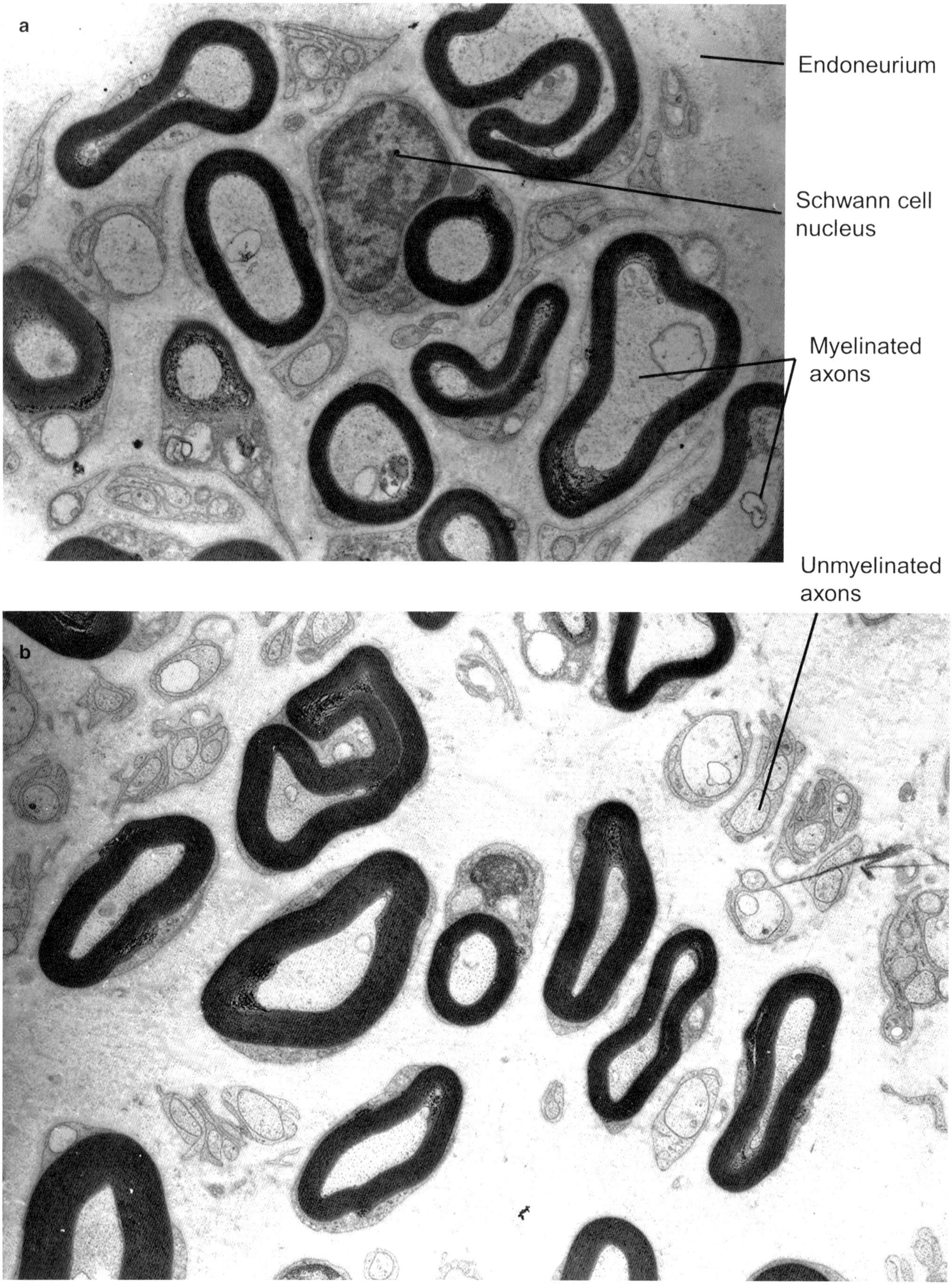

Fig. 1.1 Axons inside fascicles of a sciatic nerve. Transmission electron microscopy, magnification: ×7,000 (**a**); ×3,000 (**b**)

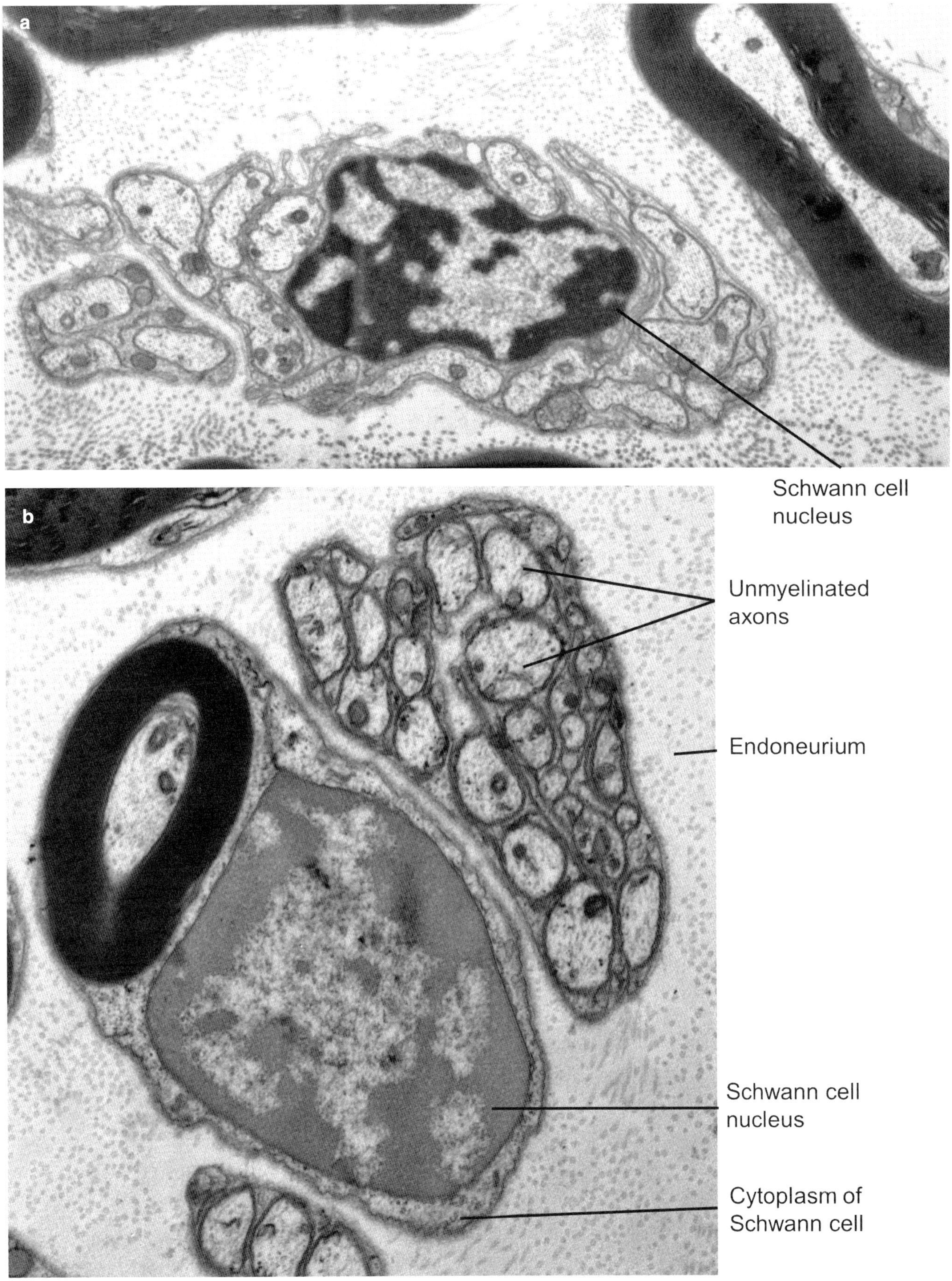

Fig. 1.2 (**a**) Unmyelinated axon of a human nerve rootlet. (**b**) Myelinic axon surrounded by Schwann cells and an unmyelinated axon of a human nerve rootlet. Transmission electron microscopy, magnification: ×15,000 (**a**); ×25,000 (**b**)

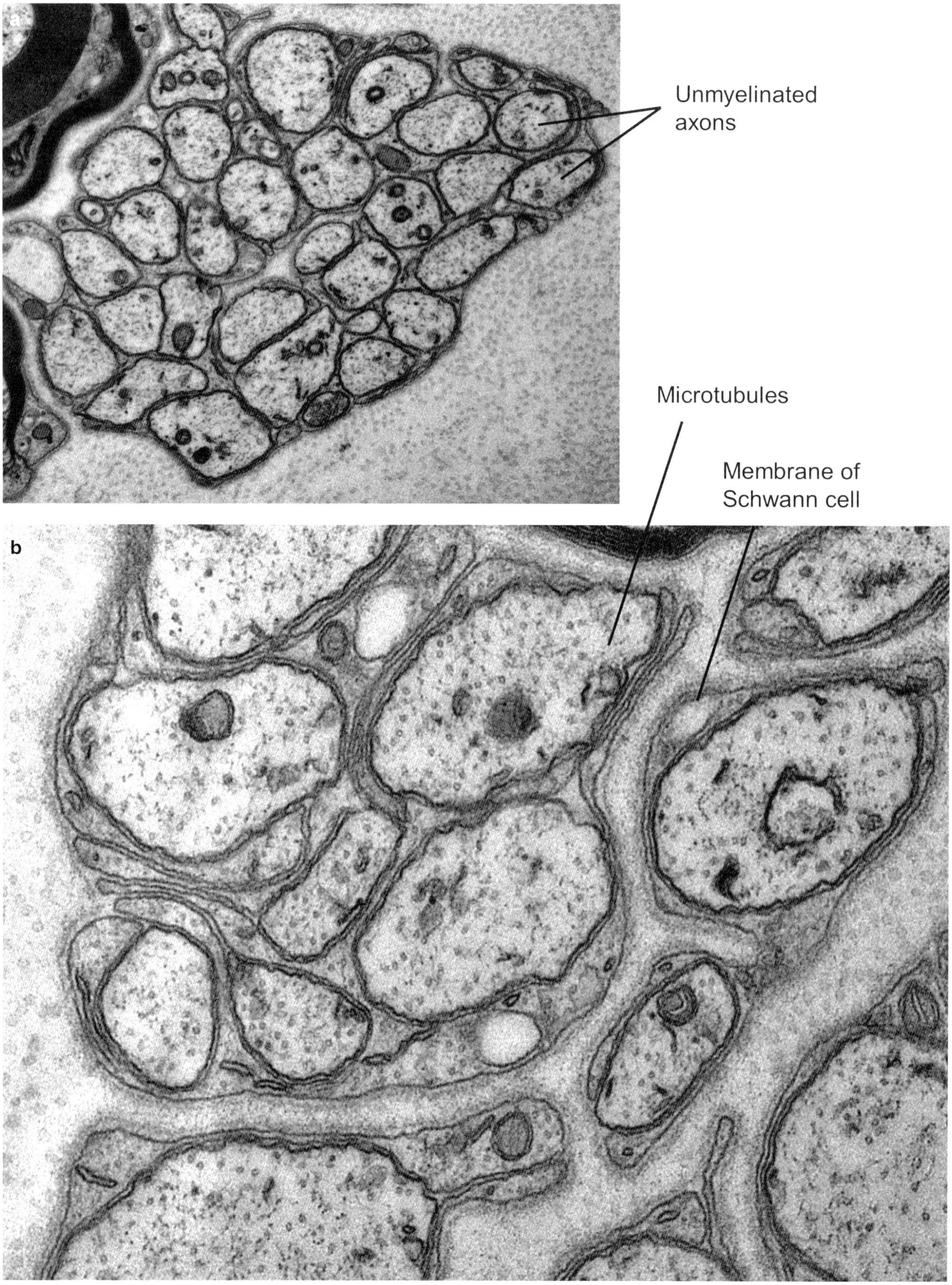

Fig. 1.3 Unmyelinated axon of a human nerve rootlet. Transmission electron microscopy, magnification: ×30,000 (**a**); ×50,000 (**b**)

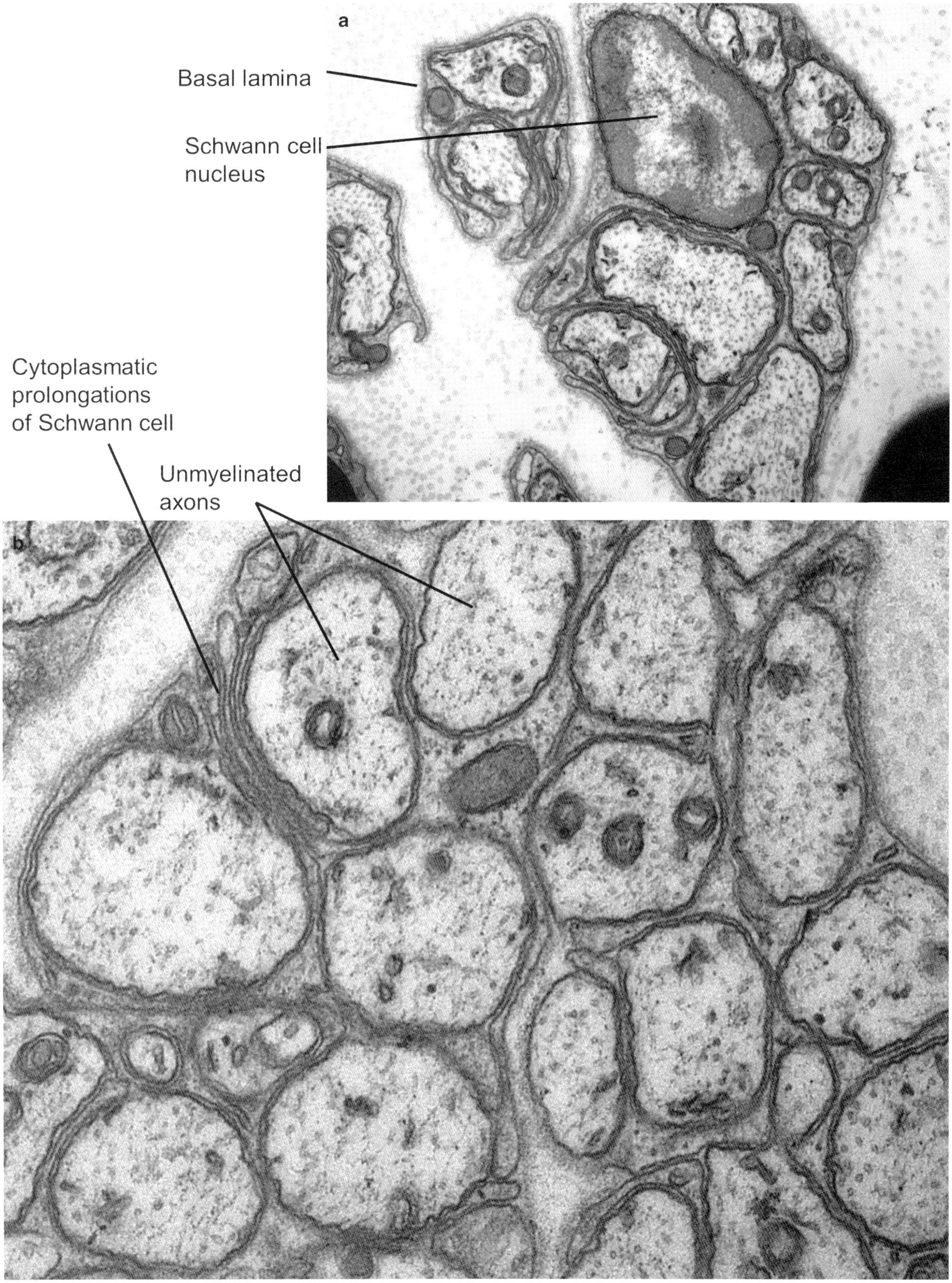

Fig. 1.4 Unmyelinated axon of a human nerve rootlet. Transmission electron microscopy, magnification: ×40,000 (**a**); ×50,000 (**b**)

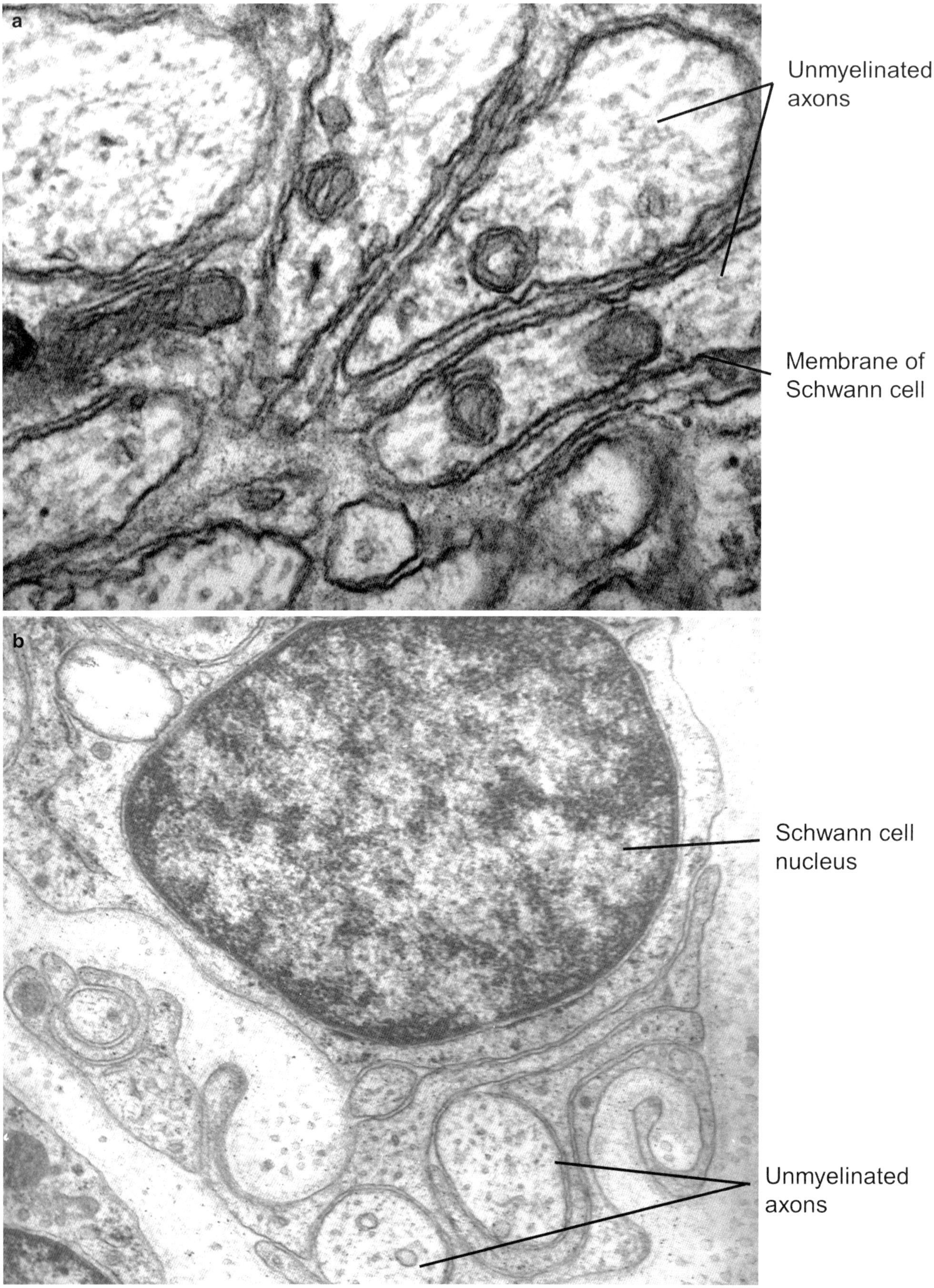

Fig. 1.5 (**a**) Unmyelinated axon of a human nerve rootlet. (**b**) Unmyelinated axon of a sciatic nerve (From De Andrés et al. [4]; with permission) Transmission electron microscopy, magnification: ×100,000 (**a**); ×20,000 (**b**)

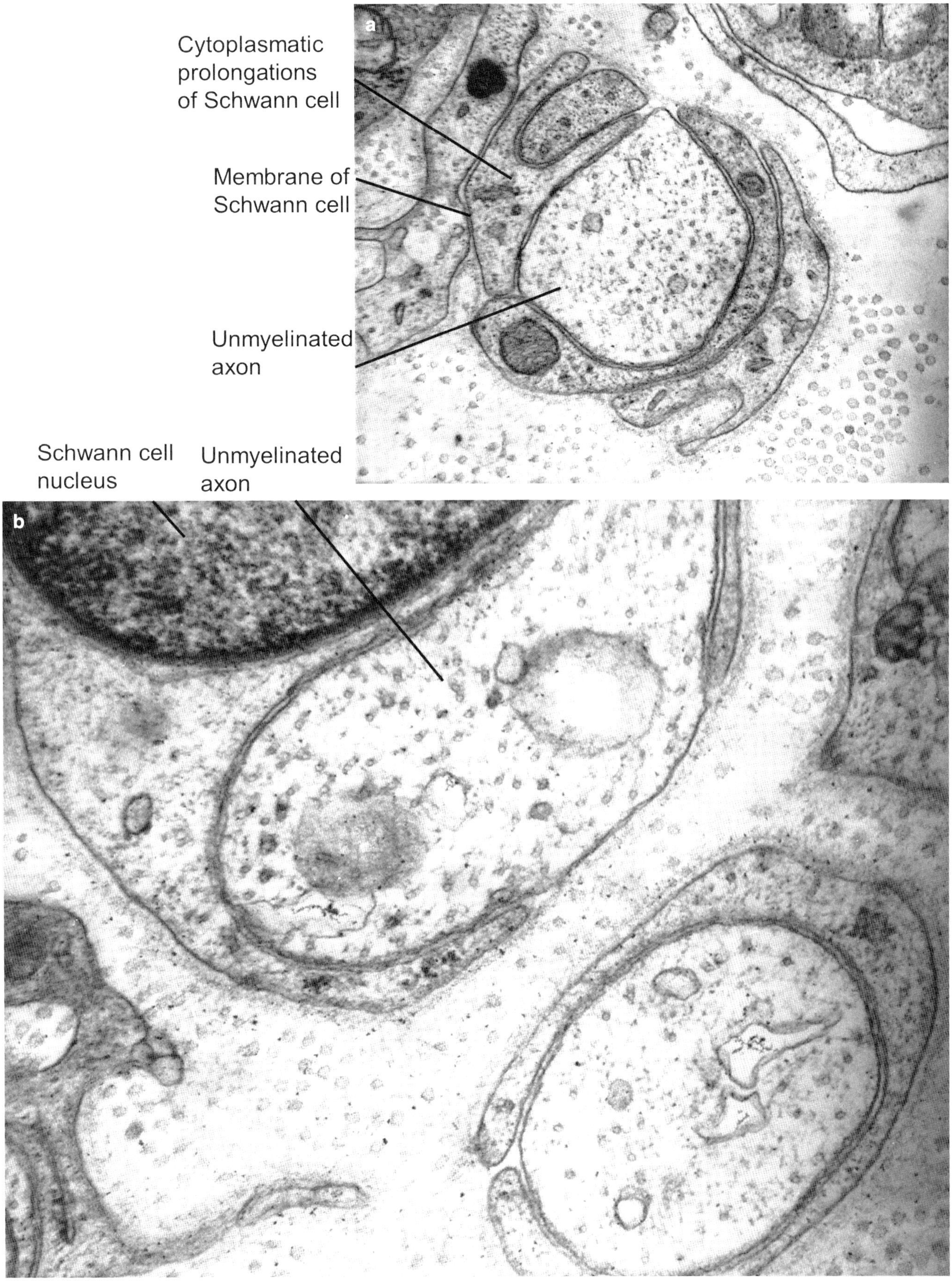

Fig. 1.6 Unmyelinated axon of a sciatic nerve. Transmission electron microscopy, magnification: ×30,000 (**a**); ×30,000 (**b**)

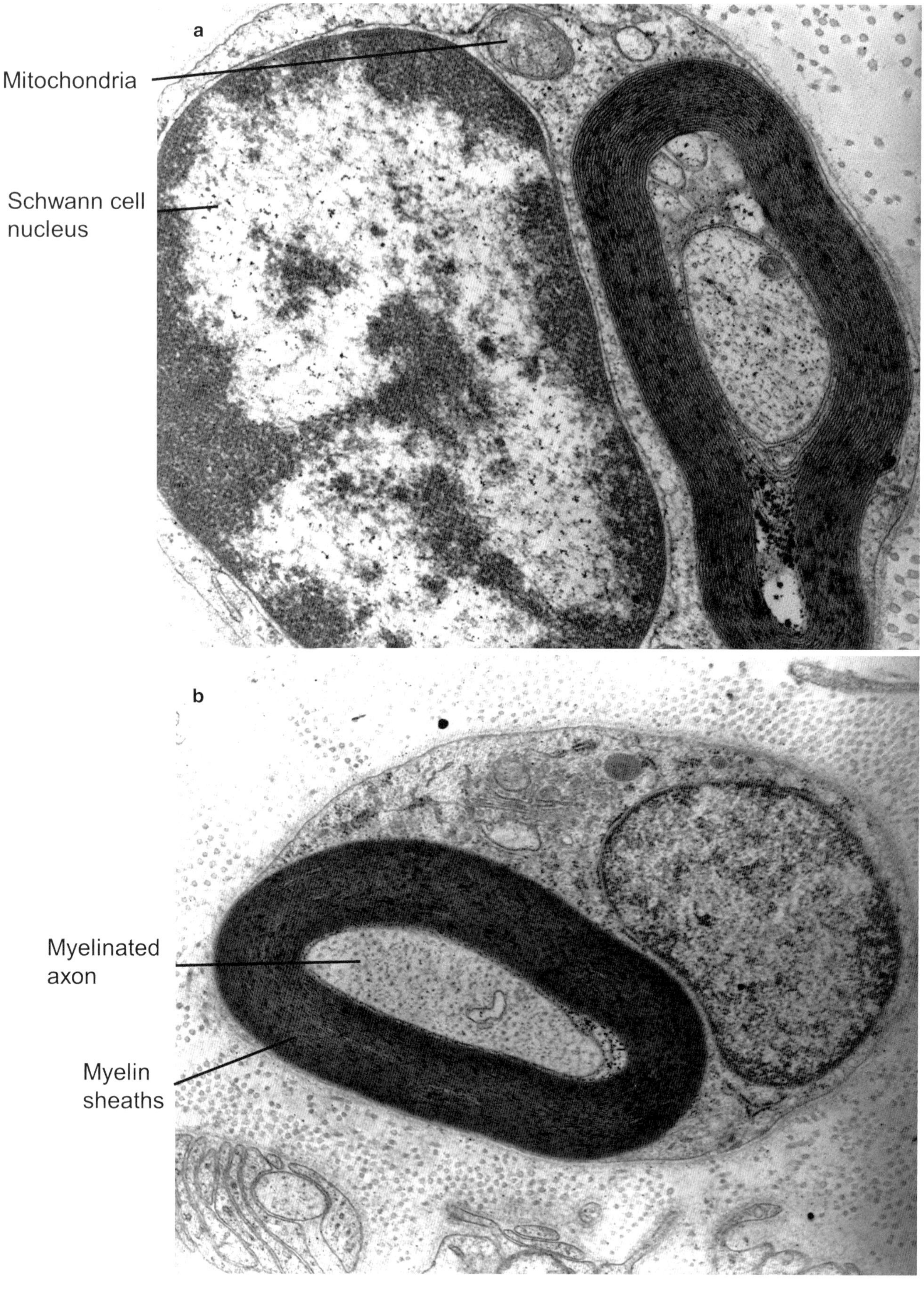

Fig. 1.7 Myelinated axon of a sciatic nerve. Transmission electron microscopy, magnification: ×20,000 (**a**); ×12,000 (**b**) (Panel **b** from Reina et al. [1]; with permission)

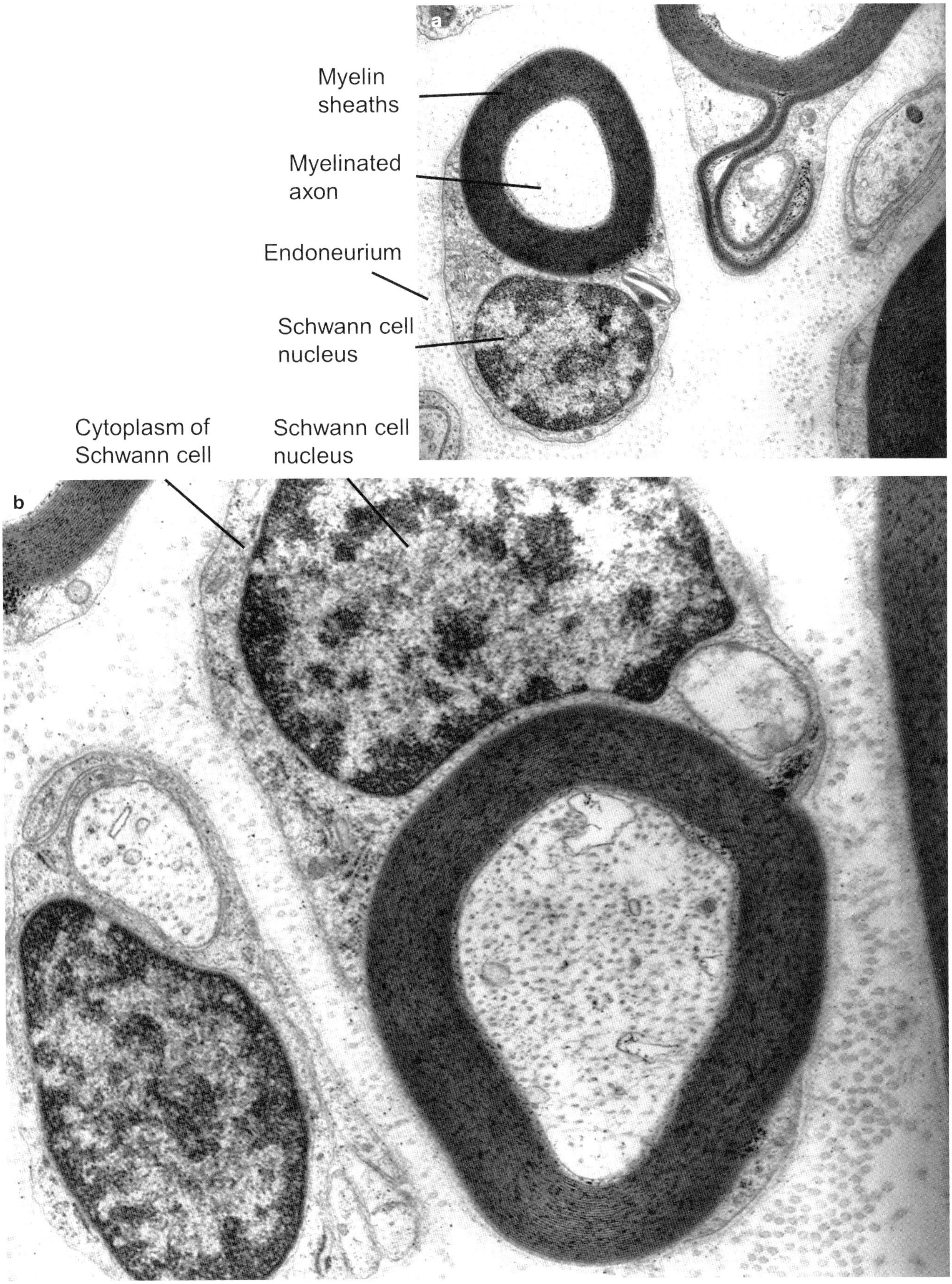

Fig. 1.8 Myelinated axon of a sciatic nerve. Transmission electron microscopy, magnification: ×12,000 (**a**); ×12,000 (**b**)

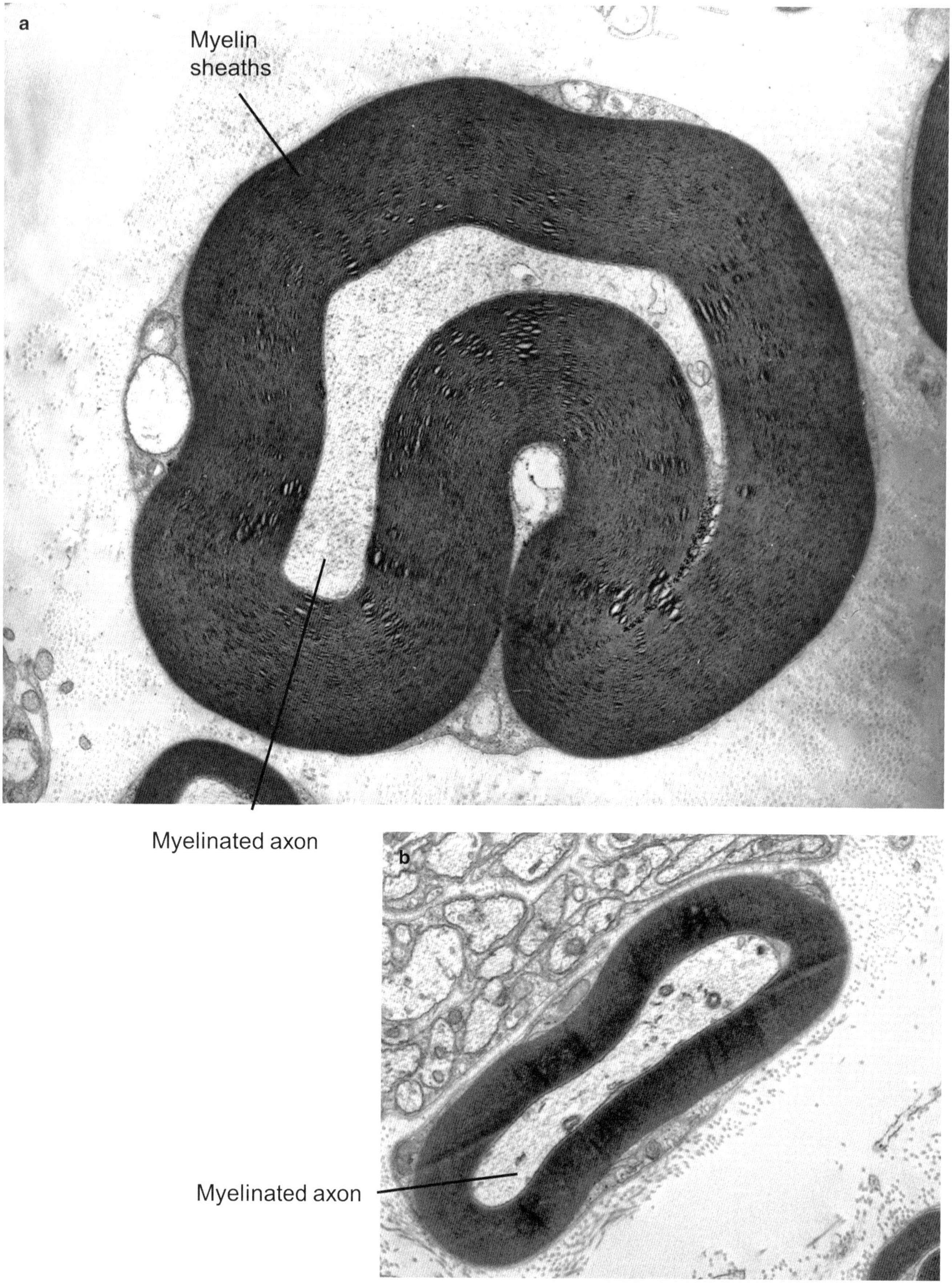

Fig. 1.9 Myelinated axon of a sciatic nerve. Transmission electron microscopy, magnification: ×7,000 (**a**); ×25,000 (**b**)

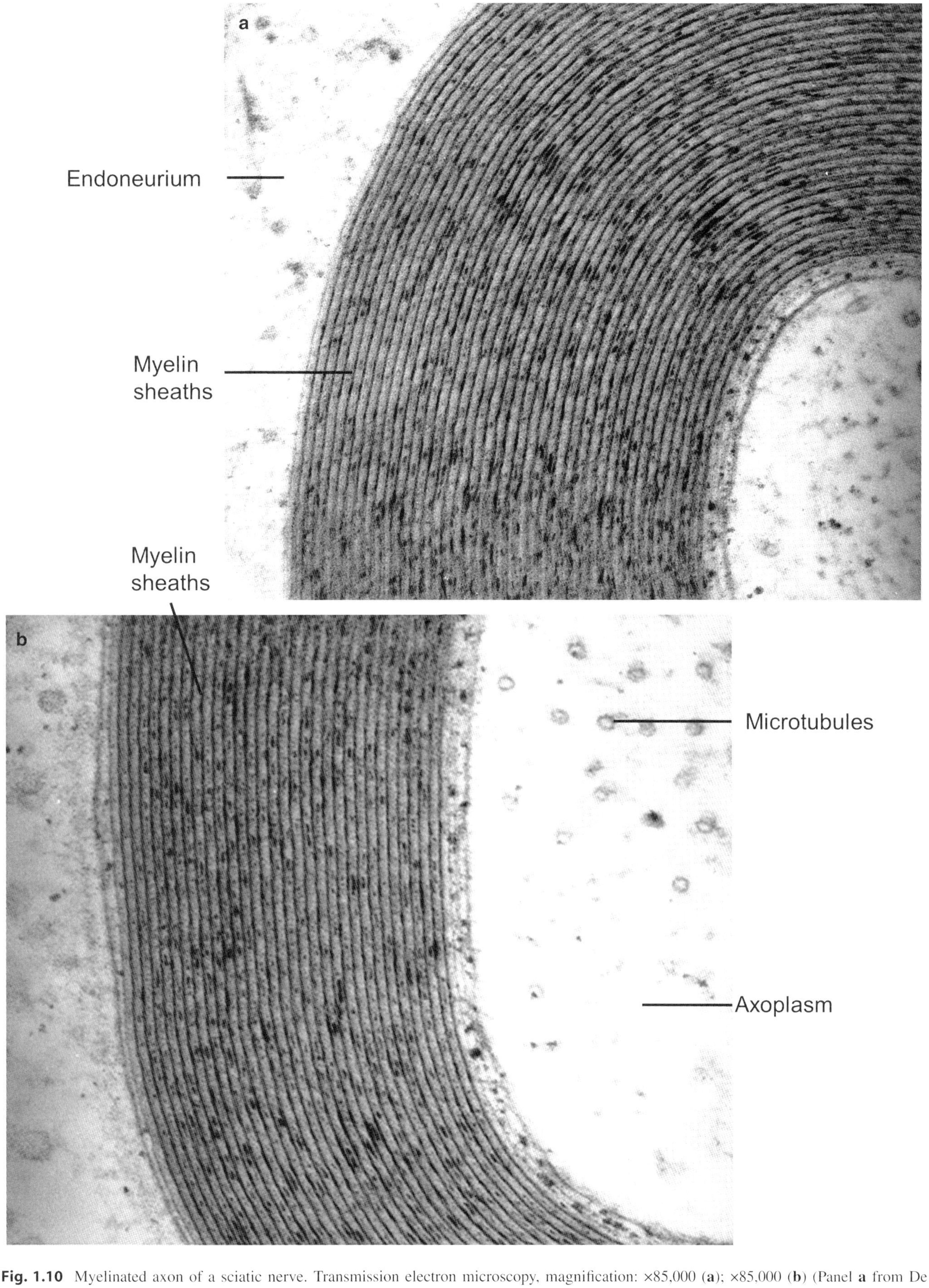

Fig. 1.10 Myelinated axon of a sciatic nerve. Transmission electron microscopy, magnification: ×85,000 (**a**); ×85,000 (**b**) (Panel **a** from De Andrés et al. [4]; with permission)

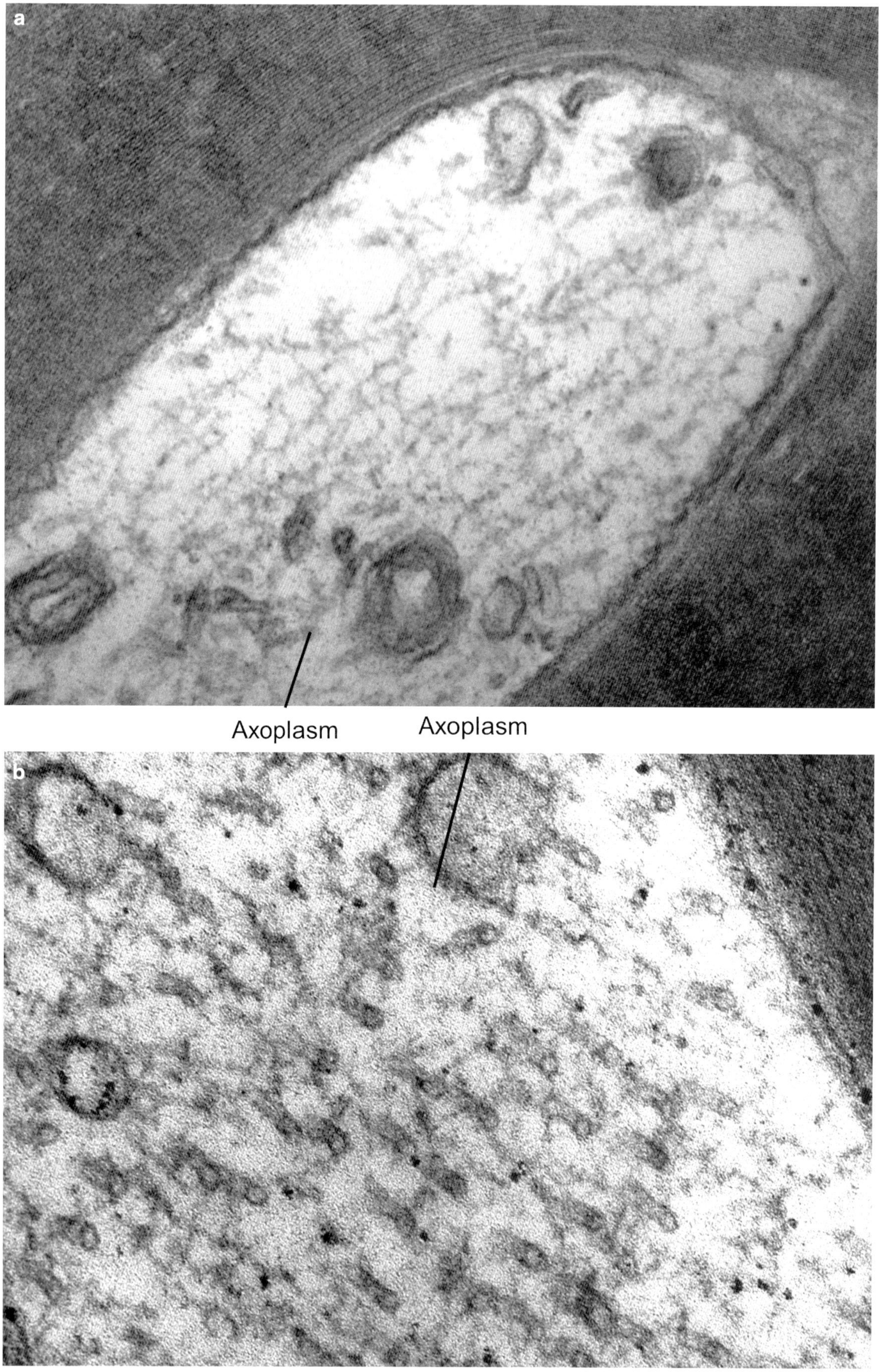

Fig. 1.11 Myelinated axon of a sciatic nerve. Transmission electron microscopy, magnification: ×120,000 (**a**); ×140,000 (**b**)

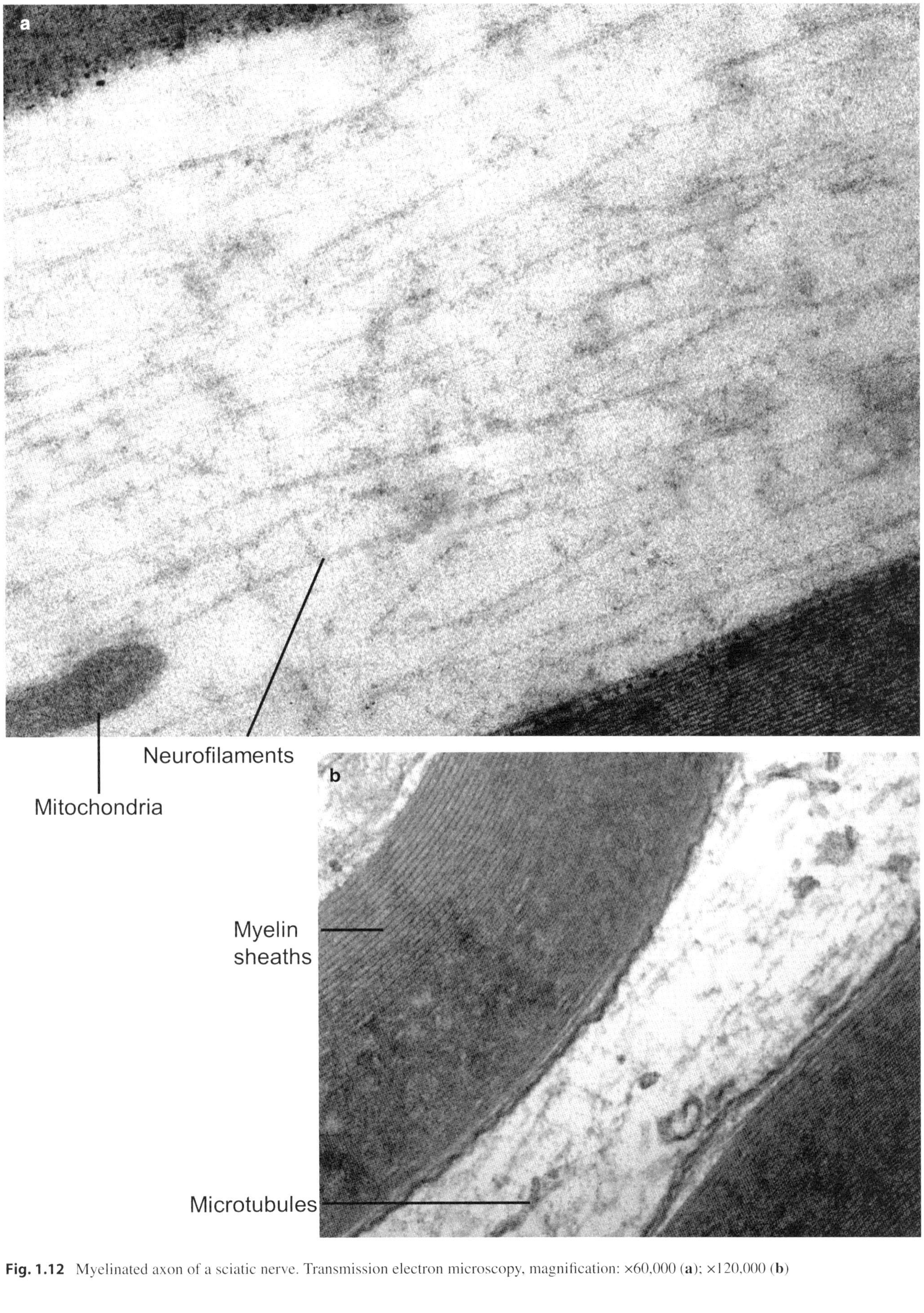

Fig. 1.12 Myelinated axon of a sciatic nerve. Transmission electron microscopy, magnification: ×60,000 (**a**); ×120,000 (**b**)

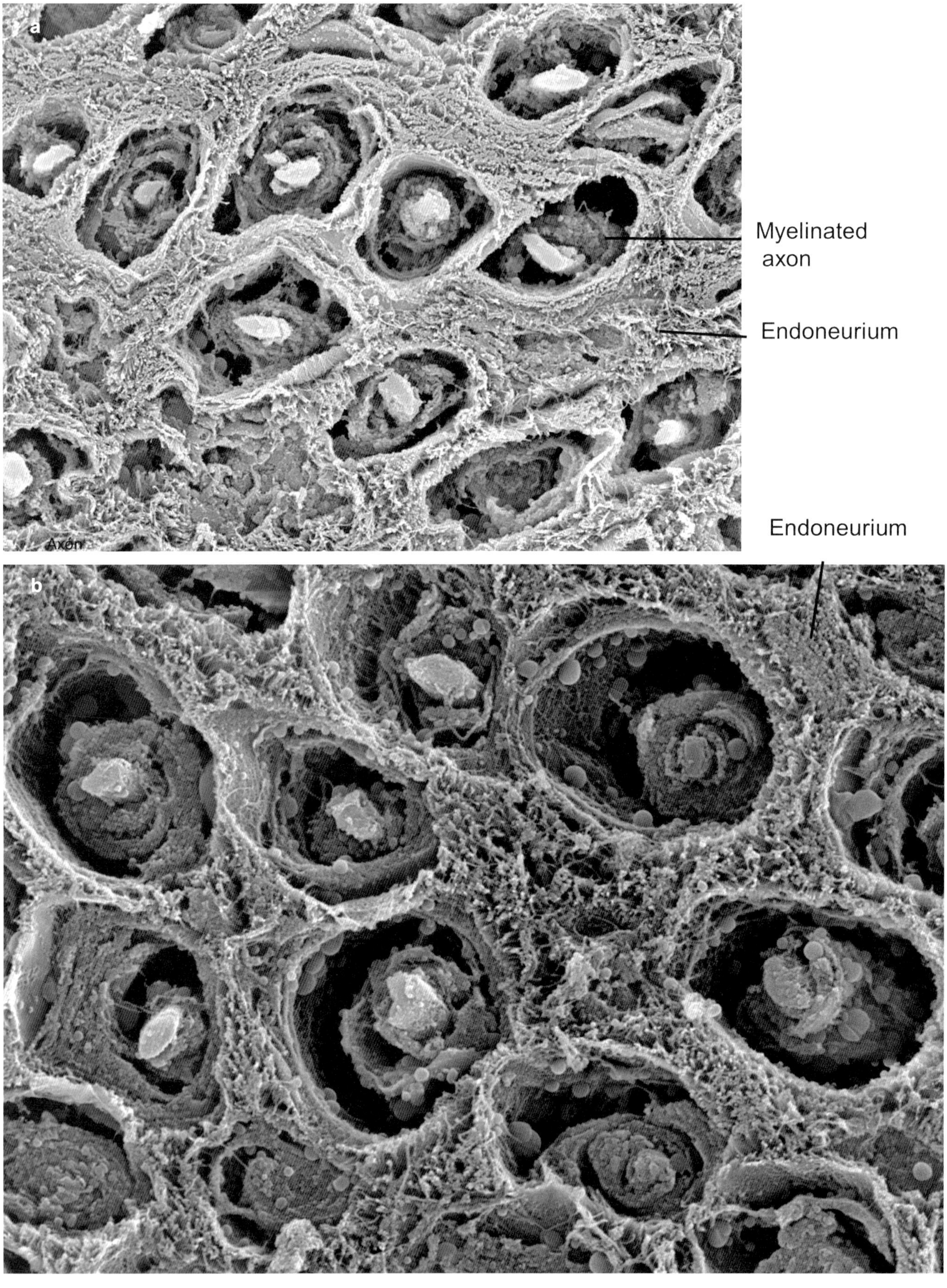

Fig. 1.13 Myelinated axon of a sciatic nerve. Scanning electron microscopy, magnification: ×1,900 (**a**); ×2,000 (**b**)

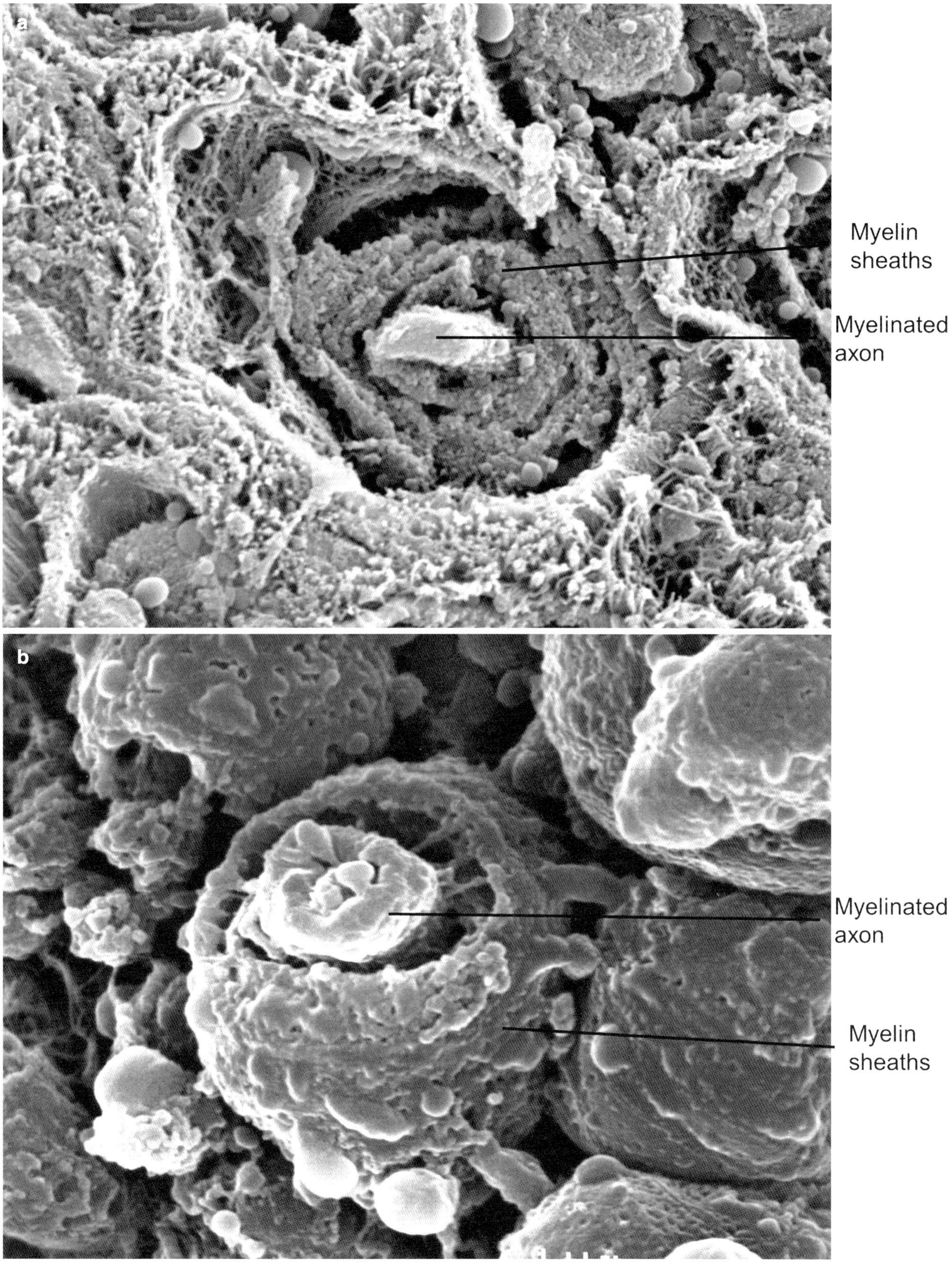

Fig. 1.14 Myelinated axon of a sciatic nerve. Scanning electron microscopy, magnification: ×4,000 (**a**); ×5,000 (**b**)

References

1. Reina MA, Arriazu R, Collier CB, Sala-Blanch X. Histology and electron microscopy of human peripheral nerves of clinical relevance to the practice of nerve blocks. Rev Esp Anestesiol Reanim. 2013;60:552–62.
2. Reina MA, López A, Villanueva MC, De Andrés JA, León GI. Morfología de los nervios periféricos, de sus cubiertas y de su vascularización. Rev Esp Anestesiol Reanim. 2000;47:464–75.
3. Reina MA, De Andres JA, Hernández JM, Arriazu Navarro R, Durán Mateos EM, Prats-Galino A. Successive changes in extraneural structures from the subarachnoid nerve roots to the peripheral nerve, influencing anesthetic block, and treatment of acute postoperative pain. Eur J Pain Suppl. 2011;5:377–85.
4. De Andrés JA, Reina MA, López A, Sala-Blanch X, Prats A. Blocs nerveux périphériques, paresthésies et injections intraneurales. Le Practicien En Anesth Réanim. 2010;14:213–21.
5. Berthold CH, Fraher JP, King RHM, Rydmark M. Microscopic anatomy of the peripheral nervous system. In: Dyck PJ, Thomas PK, editors. Peripheral neuropathy. 4th ed. Philadelphia: Elsevier; 2005. p. 35–91.
6. Peters A, Palay SL, Webster HF. The axon. In: Peters A, Palay SL, Webster HF, editors. The fine structure of the nervous system. 3rd ed. New York: Oxford University Press; 1991. p. 101–39.
7. Kleitman N, Bunge RP. The Schwann cell: morphology and development. In: Waxman SG, Kocsis JD, Stys PK, editors. The axon. New York: Oxford University Press; 1995. p. 97–115.
8. King RHM. Unmyelinated fibers. In: Dyck PJ, Thomas PK, editors. Peripheral neuropathy, vol. 1. 4th ed. Philadelphia: Elsevier; 2005. p. 62–7.

2 Macrophages, Mastocytes, and Plasma Cells

Miguel Angel Reina, Félix Manzarbeitia, and Andrés López

Fibroblasts, macrophages, and mast cells may be present throughout the spinal meninges, within the thickness of the three membranes: dura mater, pia mater, and arachnoid. In addition, these cell types are present inside nerve roots, dorsal root ganglia, and peripheral nerves at the endoneurial level, within the perineurial laminas, interfascicular connective tissue, and epineurium [1, 2].

Fibroblasts present among collagen fibers have an elongated shape, and their nuclei are oblong or elliptic and may contain one or two nucleoli. Mitochondria often are located proximal to the nucleus. The amount of rough endoplasmic reticulum fluctuates throughout inactive and active phases; it is scarce during the former and abundant in the latter. Fibroblasts contribute to the production of collagen fibers and glycoproteins in the extra-fibrillar matrix or ground substance, which fill areas not occupied by collagen fibers.

Macrophages are highly specialized phagocytic cells that play an essential role in immune and regeneration processes, as well as in tissue maintenance, acting as scavengers by removing cellular debris and other particles that reach lysosomal enzymes [3–5]. Macrophages have fusiform or stellate shapes and often are present near small blood vessels and perineurial laminas [3].

Macrophages originate from monocytes in the bone marrow; the latter migrate along the bloodstream and cross through endothelial cells of postcapillary venules to reach connective tissues, where they transform into macrophages [4–8]. These become highly specialized in phagocytic activity, engulfing particles by means of pseudopods or cellular membrane projections that extend and contract. The surface of pseudopods is covered by many microvilli and undulating folds called lamellipodia, where the cytoplasm contains typical heterogenic granules [1, 2].

Among other characteristics, macrophages have an essential lytic function mediated by acid phosphatase and an exceedingly effective phagocytic capability. Macrophages establish special connections with endothelial cells during phagocytic processes, leading to removal of myelin debris [1, 2].

Mast cells are large, 20–30 μm, with a round nucleus, conspicuous granules, and folds of membrane that project from the surface [4]. Mast cells or mastocytes have oval or elliptic shapes and cytoplasmic projections containing vacuoles filled with various types of amines and heparins [9, 10].

Plasma cells are products of B-lymphocyte activation that synthesize and secrete antibodies. These cells have an eccentric nucleus, a well-developed Golgi apparatus, and a cytoplasm packed with dilated rough endoplasmic reticulum (Figs. 2.1, 2.2, 2.3, 2.4, 2.5, 2.6, 2.7, 2.8, 2.9, 2.10, 2.11, 2.12, 2.13, 2.14, 2.15, 2.16) [4].

M.A. Reina, MD, PhD (✉)
Department of Clinical Medical Sciences and Institute of Applied Molecular Medicine, School of Medicine, University of CEU San Pablo, Madrid, Spain

Department of Anesthesiology, Madrid-Montepríncipe University Hospital, Madrid, Spain
e-mail: miguelangel@perticone.e.telefonica.net

F. Manzarbeitia, MD, PhD
Department of Pathology, Fundación Jiménez Díaz University Hospital, School of Medicine, Autonomous University of Madrid, Madrid, Spain
e-mail: FManzarbeitia@fjd.es

A. López, MD
Department of Clinical Medical Sciences, School of Medicine, CEU San Pablo University, Madrid, Spain

Department of Anesthesiology, Madrid-Montepríncipe University Hospital, Madrid-Torrelodones University Hospital, Madrid-Sanchinarro University Hospital, Madrid, Spain
e-mail: alg00009@telefonica.net

M.A. Reina et al. (eds.), *Atlas of Functional Anatomy for Regional Anesthesia and Pain Medicine: Human Structure, Ultrastructure and 3D Reconstruction Images*, DOI 10.1007/978-3-319-09522-6_2,

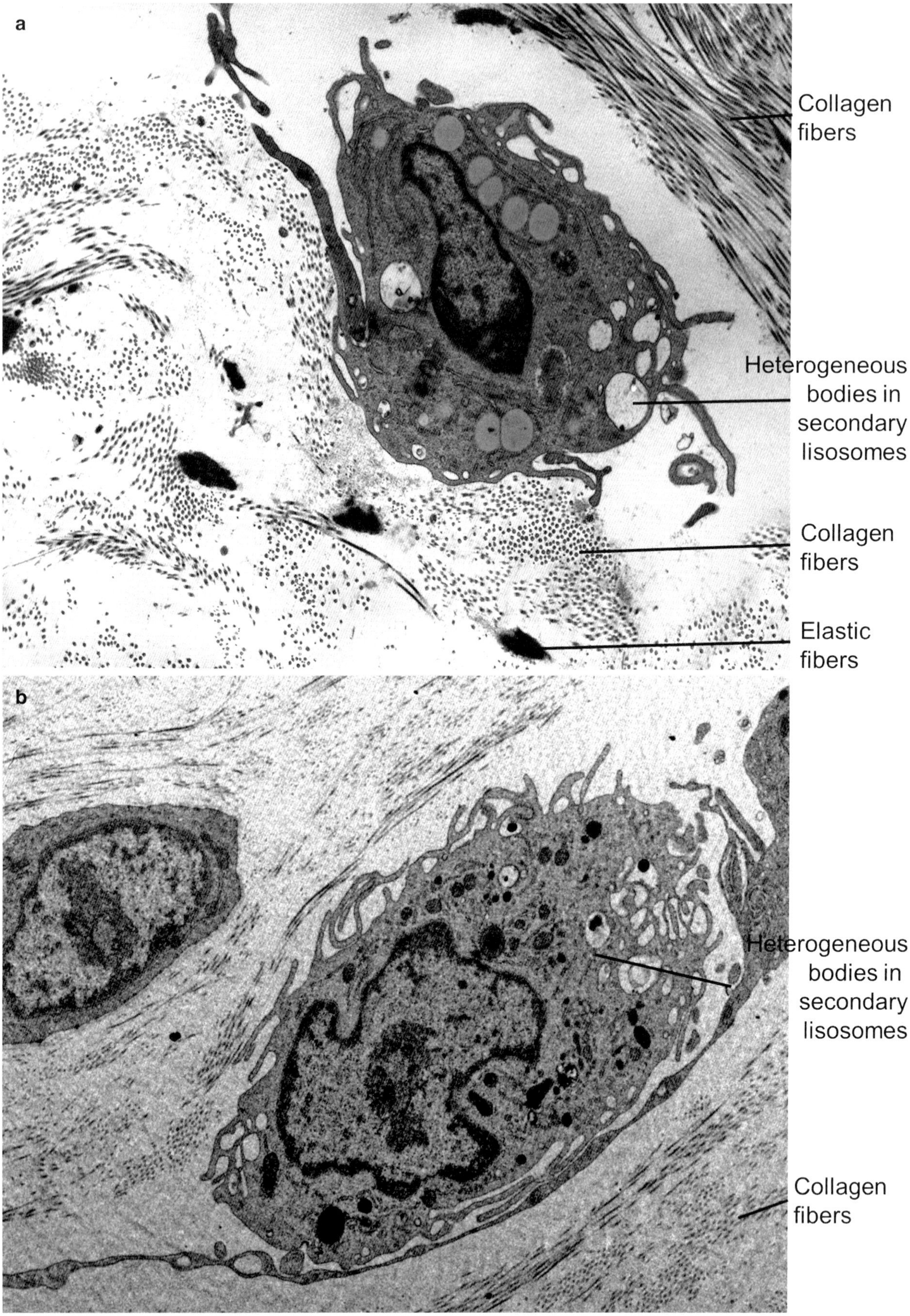

Fig. 2.1 Macrophage (functional form of a histiocyte) within human dura mater. (**a**, **b**) The cell has fine pseudopodia lacking organelles. Cisternae from the rough endoplasmic reticulum, mitochondria, lysosomes, and Golgi complex are identified. Magnification: ×7,000 (Panel **a** from Reina et al. [6]; with permission)

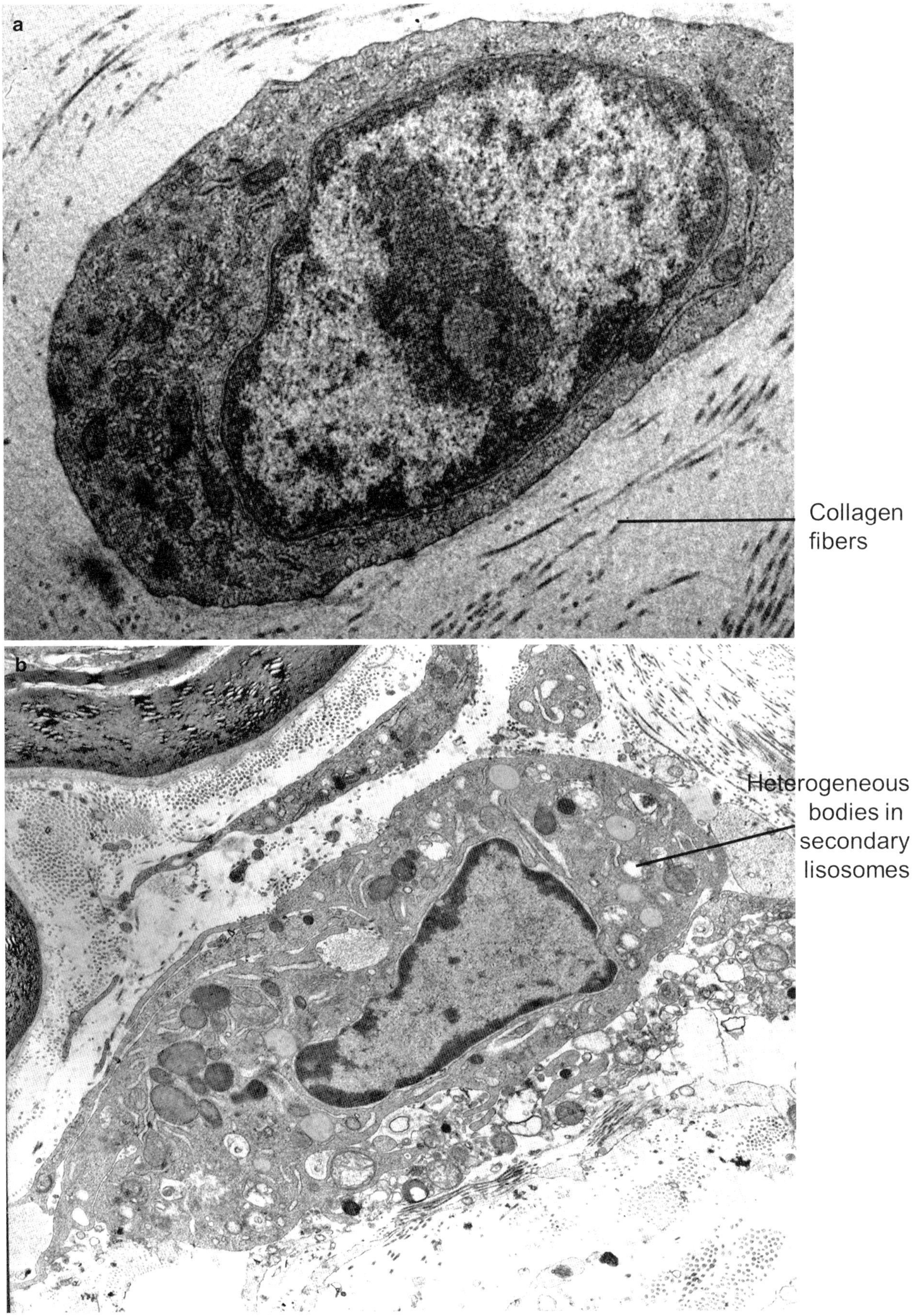

Fig. 2.2 (**a**) A cell similar to a myofibroblast present in human dura mater. (**b**) Macrophage located among nerve roots in the cauda equina. Transmission electron microscopy, magnification: ×15,000 (**a**); ×8,000 (**b**)

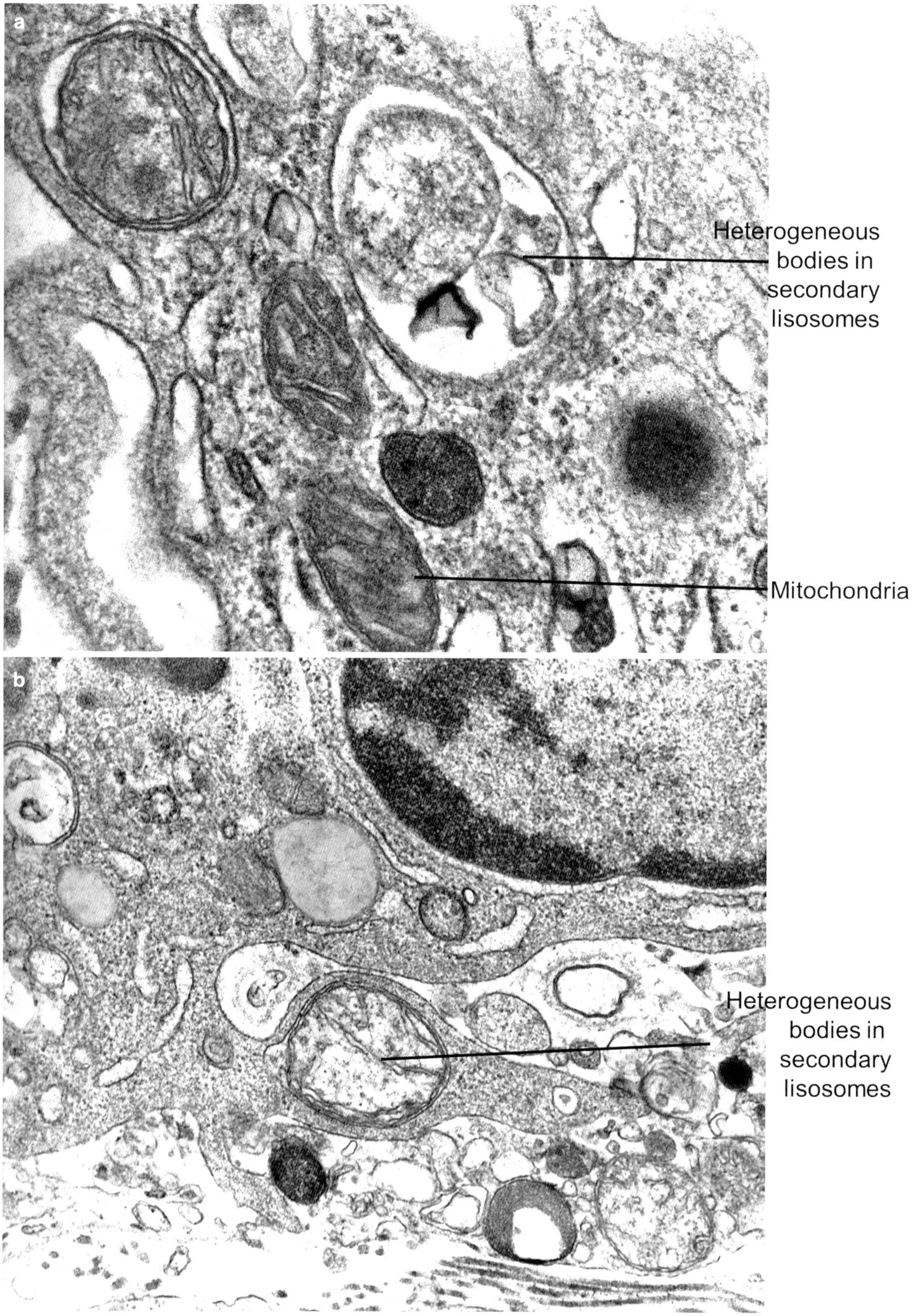

Fig. 2.3 (**a**) View of the cytoplasm of a macrophage located within the pia mater from a nerve root from the cauda equina. Transmission electron microscopy. (**b**) Macrophage cell present in a nerve root from the cauda equina. Electron microscopy (From Reina et al. [8]; with permission) Magnification: ×10,000 (**a**); ×20,000 (**b**)

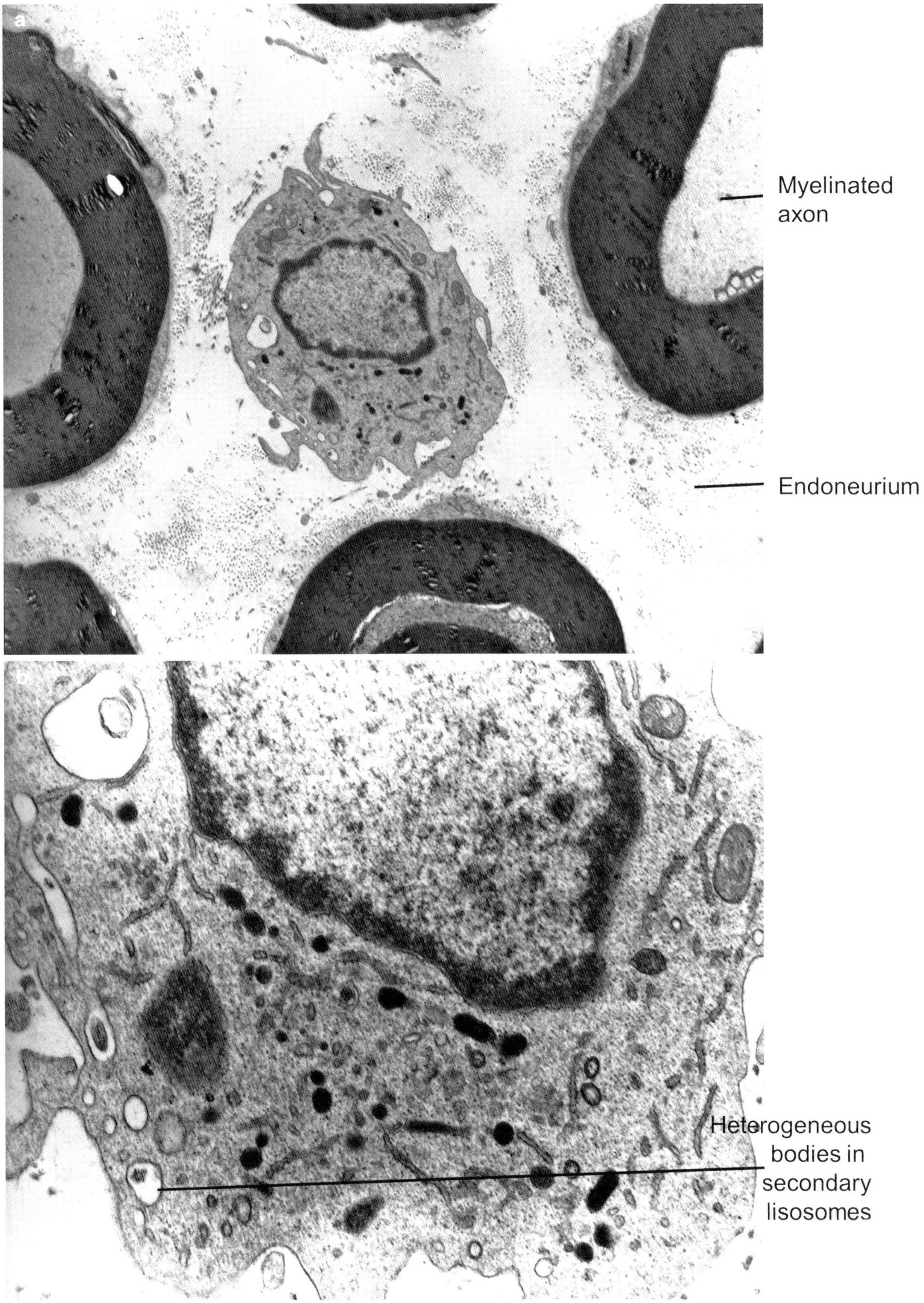

Fig. 2.4 Macrophage in a nerve root from the cauda equina. (**a**) Cytoplasmic pseudopodia are identified. (**b**) The same image at higher magnification. Transmission electron microscopy, magnification: ×4,400 (**a**); ×12,000 (**b**)

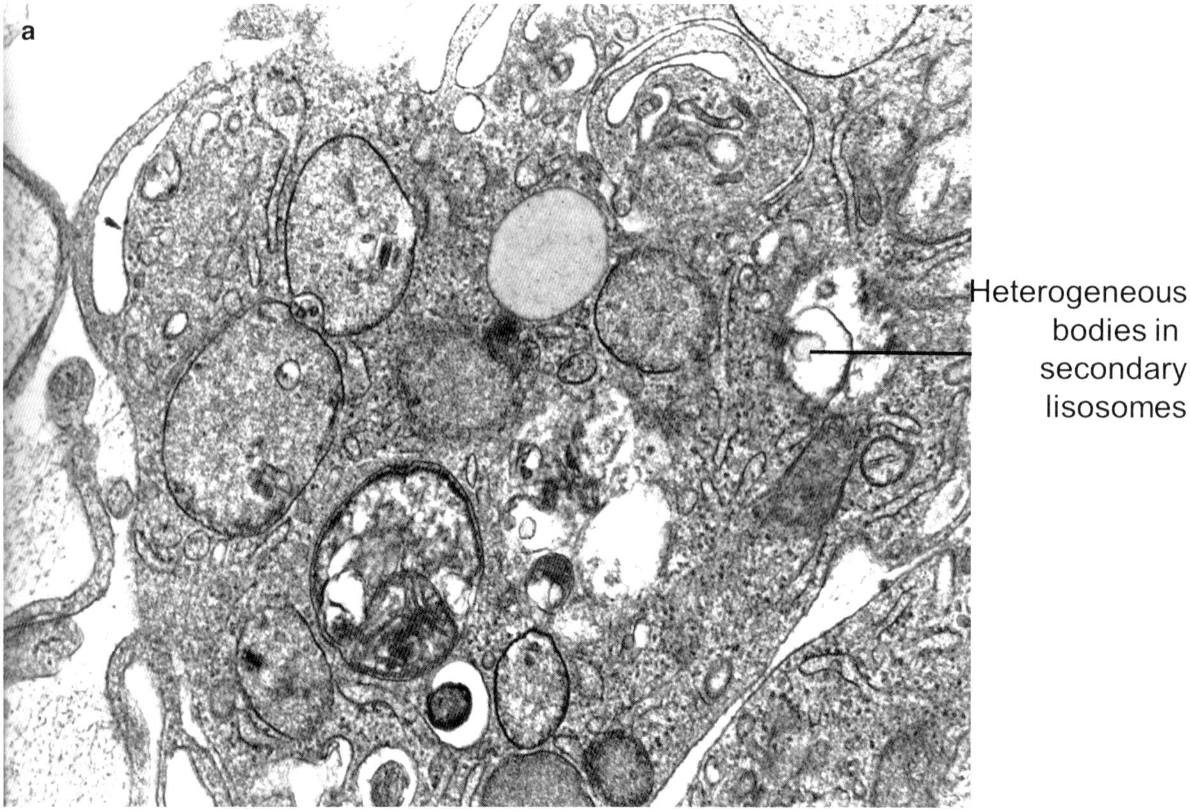

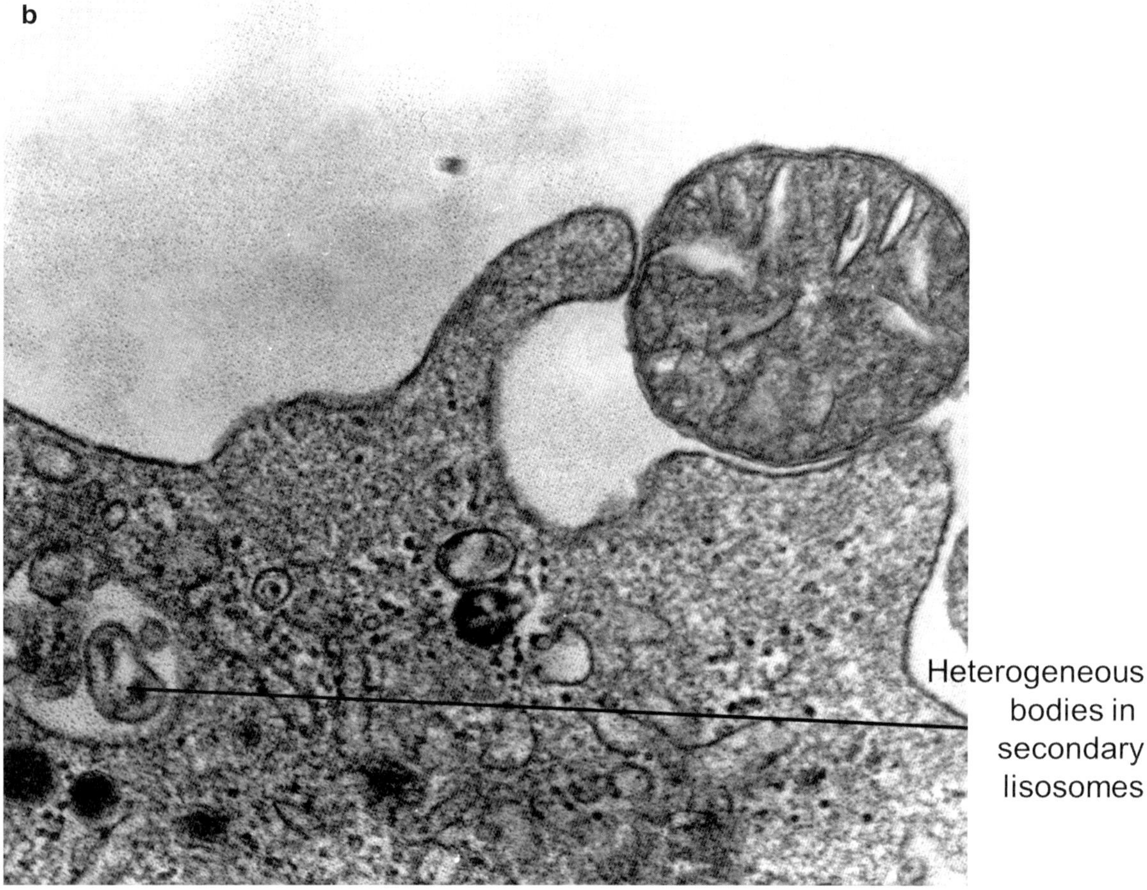

Fig. 2.5 Macrophage in a nerve root from the cauda equina. (**a**) Detail of lysosomes, smooth endoplasmic reticulum, and mitochondria. (**b**) Two pseudopods during phagocytosis of a cell particle (From Reina et al. [8]; with permission) Transmission electron microscopy, magnification: ×20,000 (**a**); ×50,000 (**b**)

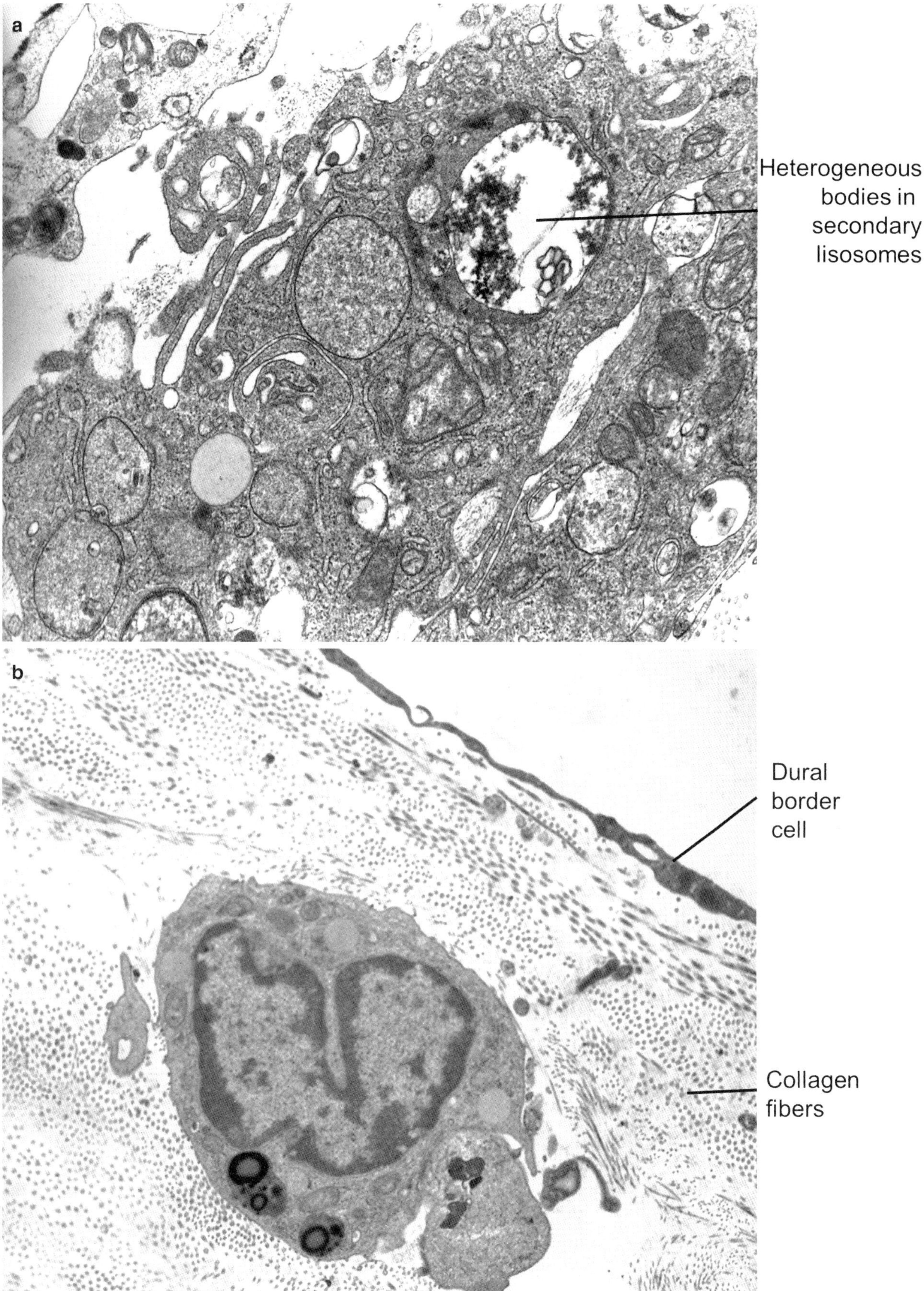

Fig. 2.6 Macrophage. (**a**) Detail of a macrophage in a nerve root from the cauda equina. Lysosomes are identified. (**b**) Macrophage located inside a human spinal subdural compartment. Transmission electron microscopy, magnification: ×12,000 (**a**); ×7,000 (**b**)

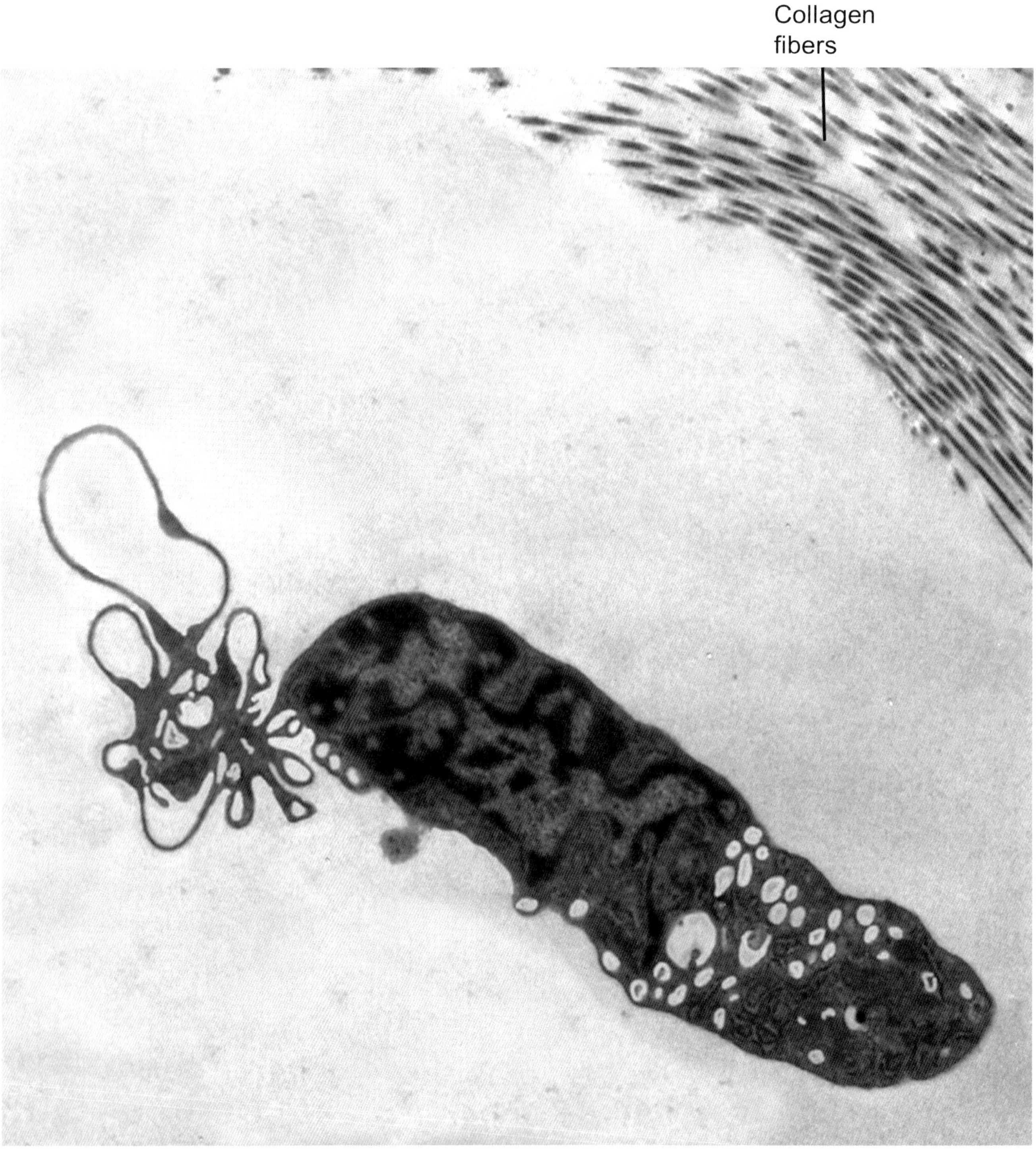

Fig. 2.7 Free cell filled with numerous lysosomes, located inside the human spinal subdural compartment. Transmission electron microscopy, magnification: ×7,000

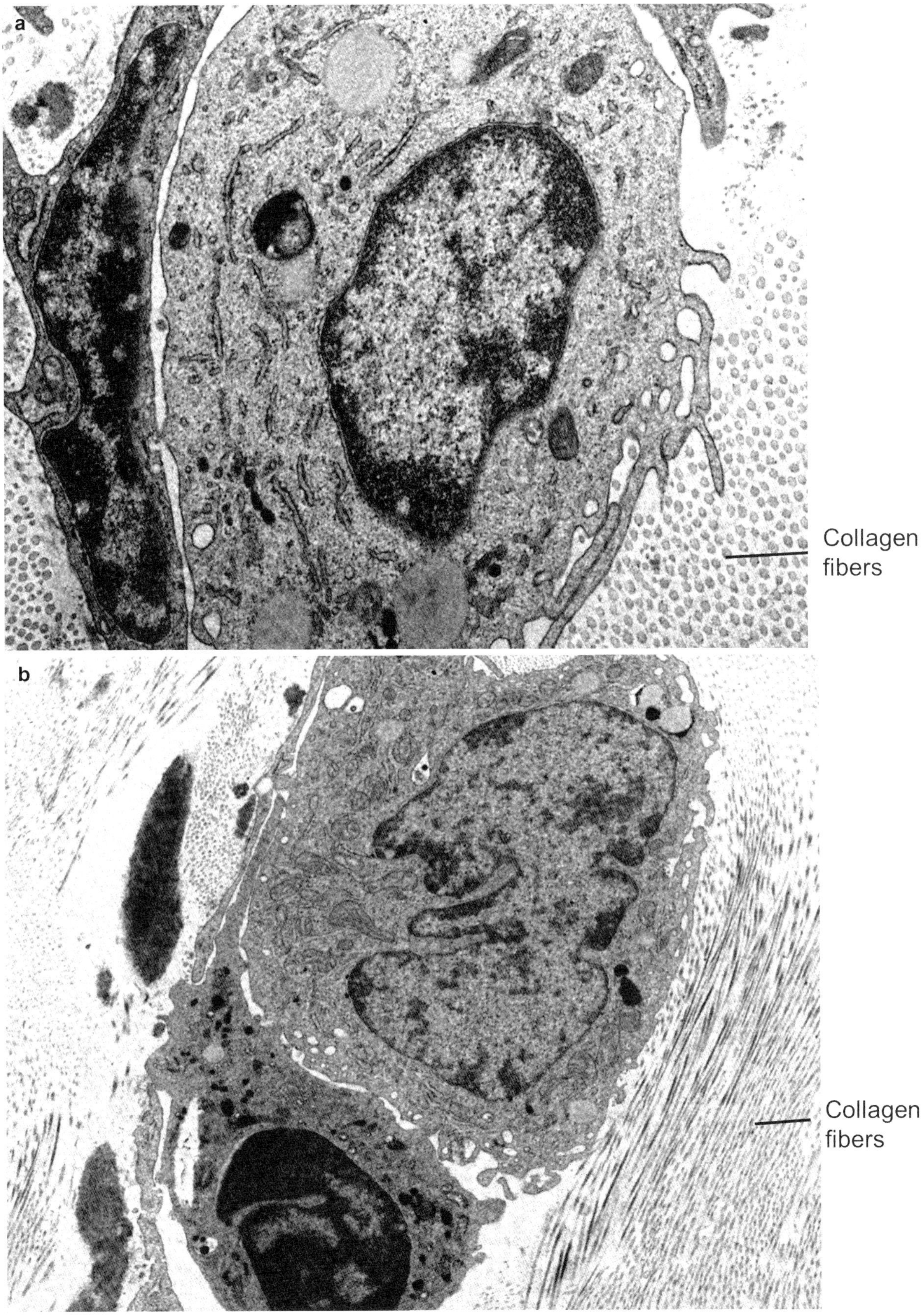

Fig. 2.8 Macrophage located between dural laminas from human spinal dura mater. (**a**) Several elongated pseudopods lacking organelles. (**b**) Two adjacent cells in close contact. Above, a macrophage shows pseudopods. Transmission electron microscopy, magnification: ×12,000 (**a**); ×10,000 (**b**)

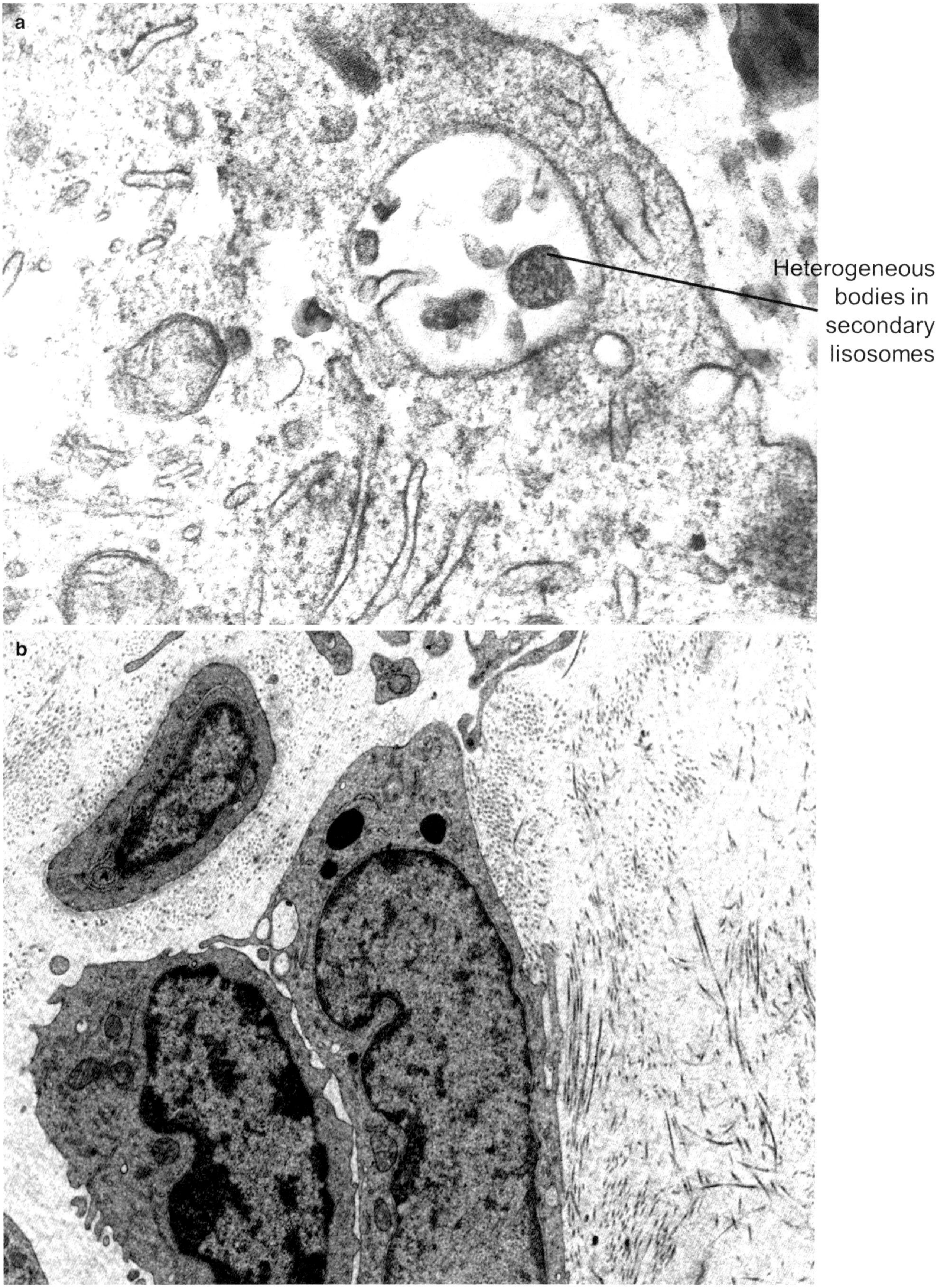

Fig. 2.9 Macrophage and fibroblast. (**a**) Macrophage in a human spinal subdural compartment. (**b**) Two cells showing features of fibroblasts between human spinal dura mater. Transmission electron microscopy, magnification: ×40,000 (**a**); ×12,000 (**b**)

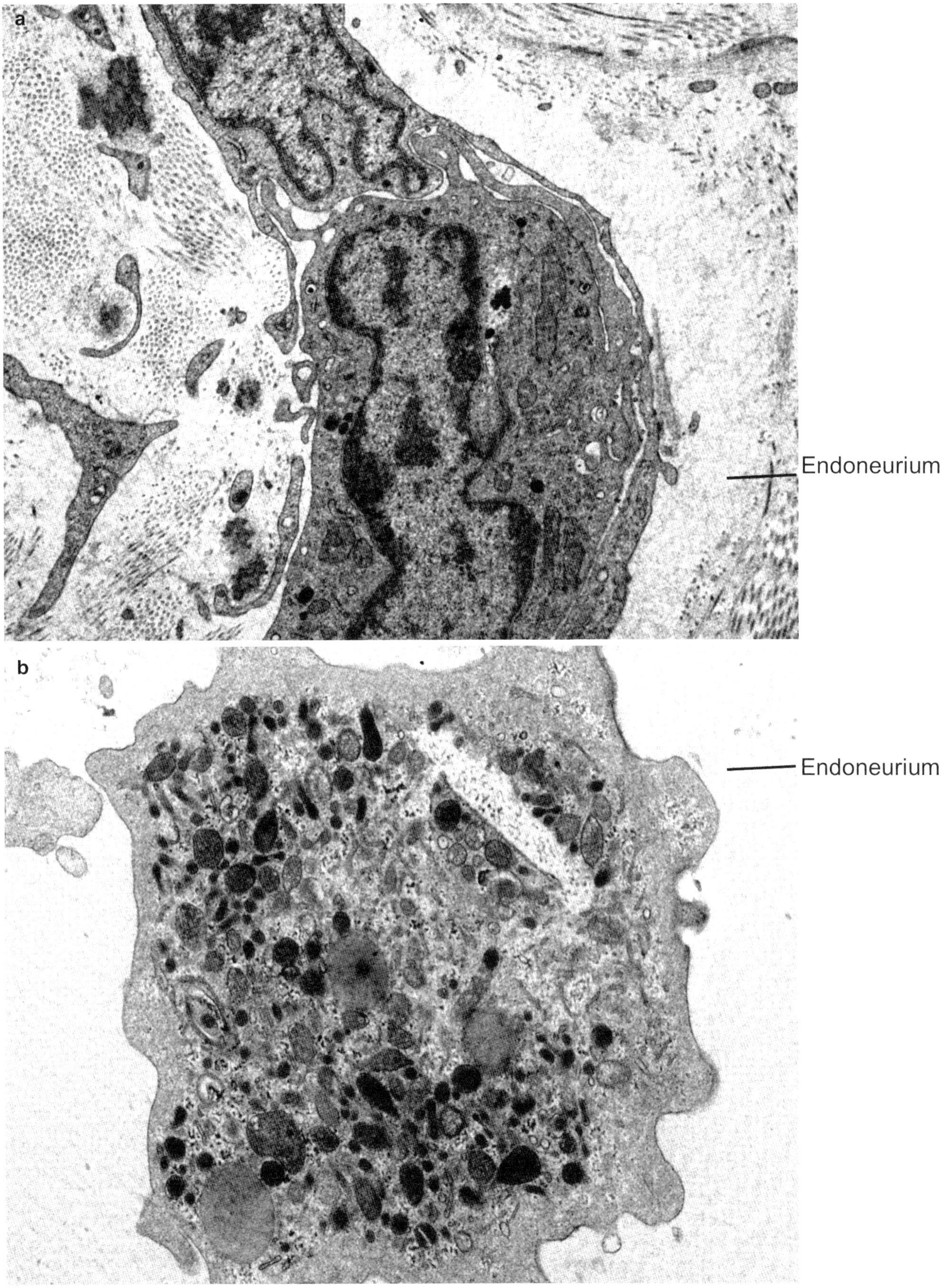

Fig. 2.10 Macrophages and a multinucleated neutrophil. (**a**) Two adhering macrophages (histiocytes) (From Reina et al. [8]; with permission) (**b**) Multinucleated neutrophil in a nerve root from human cauda equina. Transmission electron microscopy, magnification: ×12,000 (**a**); ×10,000 (**b**)

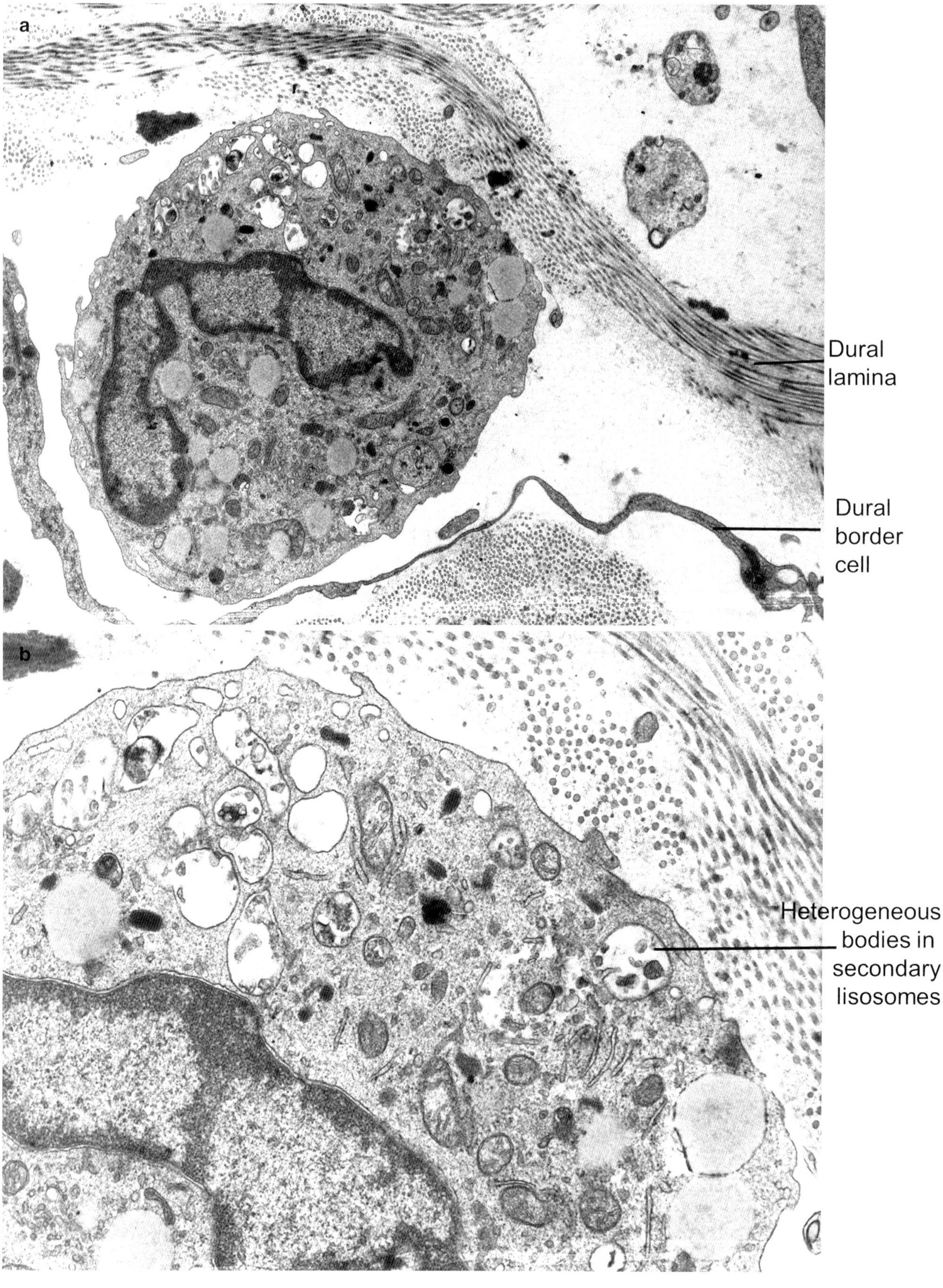

Fig. 2.11 Macrophage located inside a human spinal subdural compartment. (**a**) Macrophage. (**b**) Detail of the same image at higher magnification. Transmission electron microscopy, magnification: ×5,000 (**a**);×12,000 (**b**)

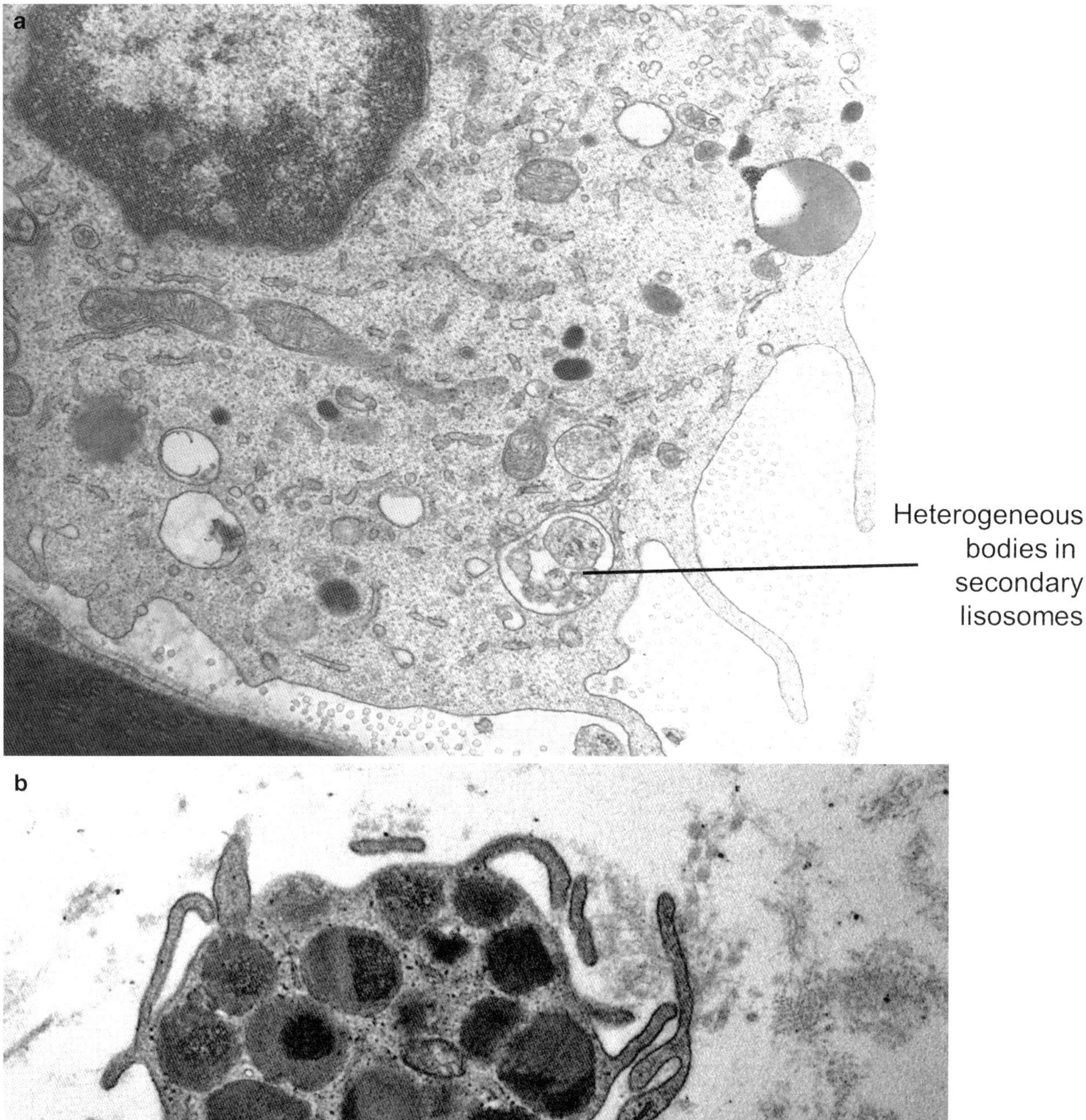

Fig. 2.12 (**a**) Macrophage inside a nerve root from the cauda equina (From Reina et al. [8]; with permission) (**b**) Macrophage located inside a human spinal subdural compartment. Transmission electron microscopy, magnification: ×12,000 (**a**, **b**)

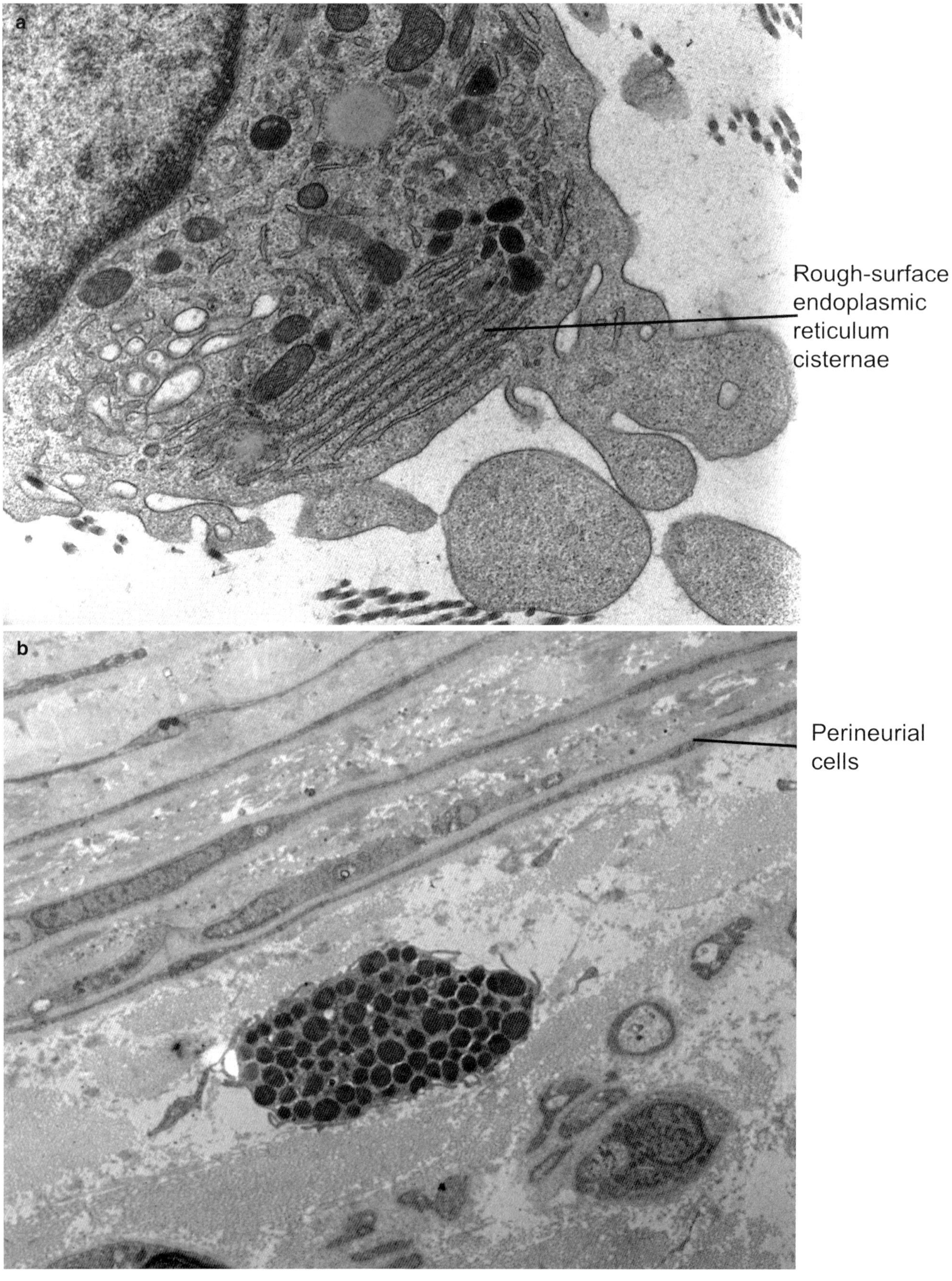

Fig. 2.13 Macrophage and mastocyte. (**a**) Macrophage located inside a human spinal subdural compartment. (**b**) Mastocyte located between perineurial laminas of a sciatic nerve. Transmission electron microscopy, magnification: ×6,000 (**a**); ×3,000 (**b**)

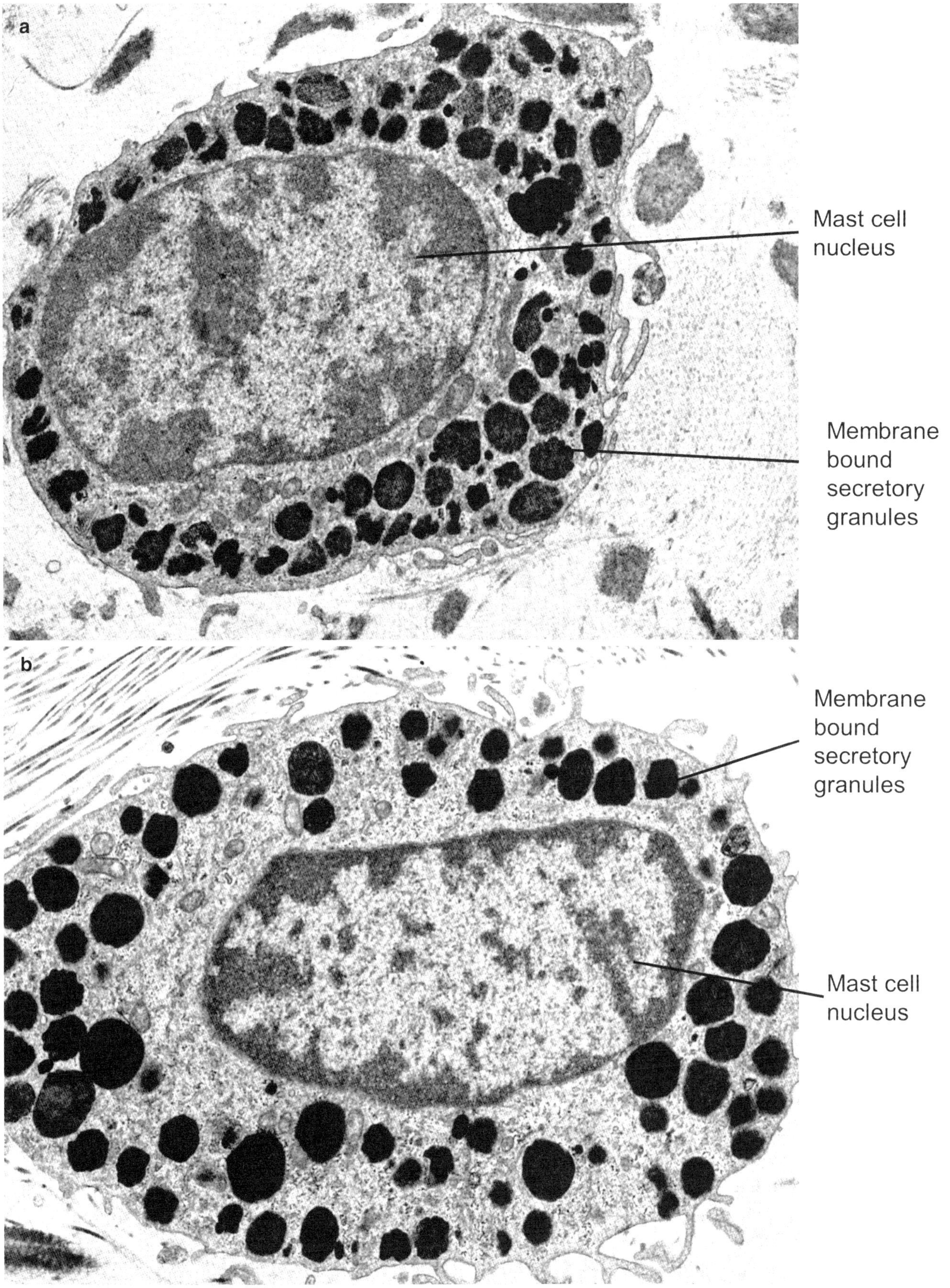

Fig. 2.14 Mastocytes located between laminas of human spinal dura mater. (**a**, **b**) Mastocytes. Transmission electron microscopy, magnification: ×15,000 (**a**, **b**) (Panel **b** from Reina et al. [7]; with permission)

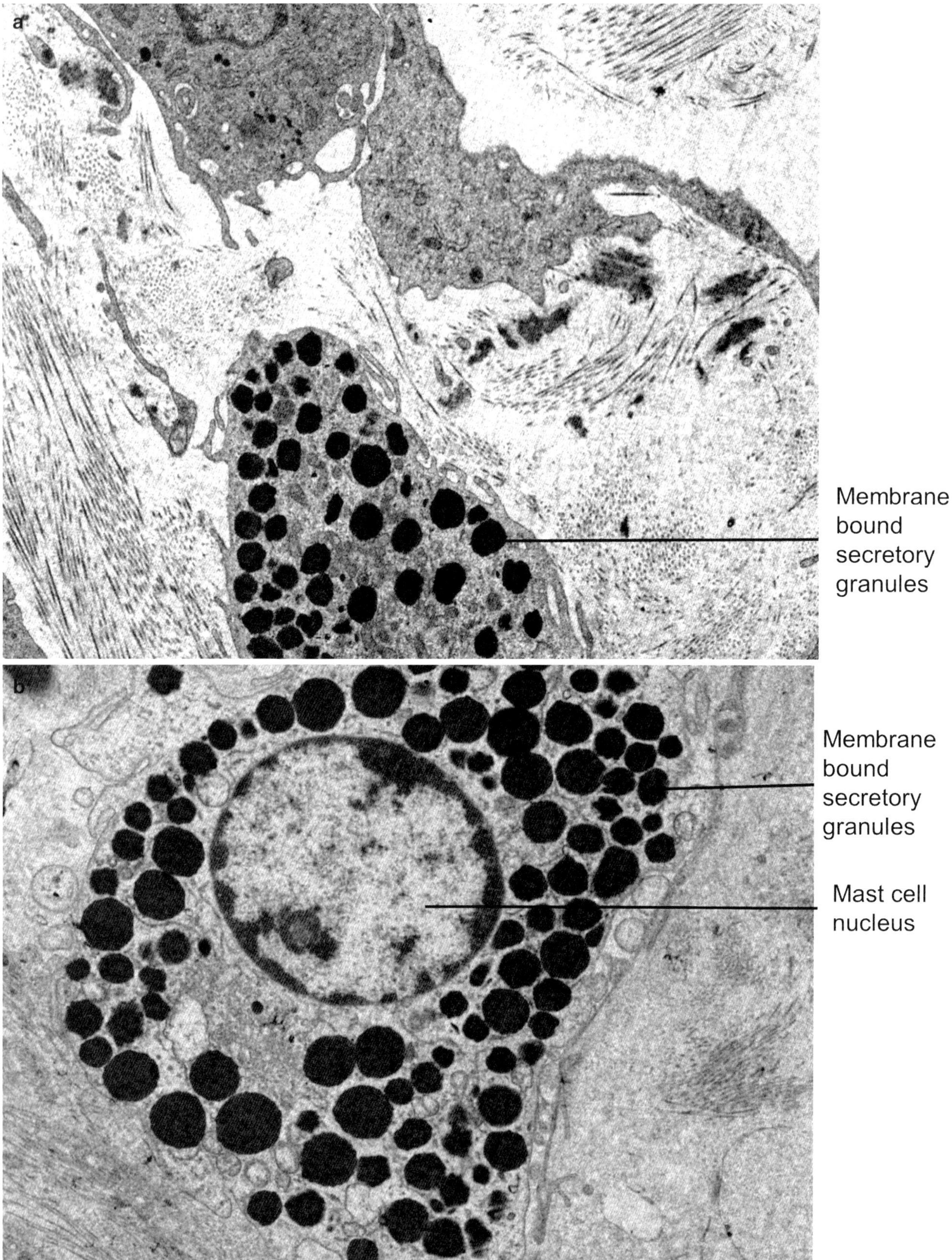

Fig. 2.15 Mastocytes located among laminas of human spinal dura mater. (**a**) Mastocytes and fibroblasts. Characteristic cytoplasmic granules are identified. (**b**) Mastocyte with its nucleolus at the center of the cytoplasm. Transmission electron microscopy, magnification: ×10,000 (**a**, **b**)

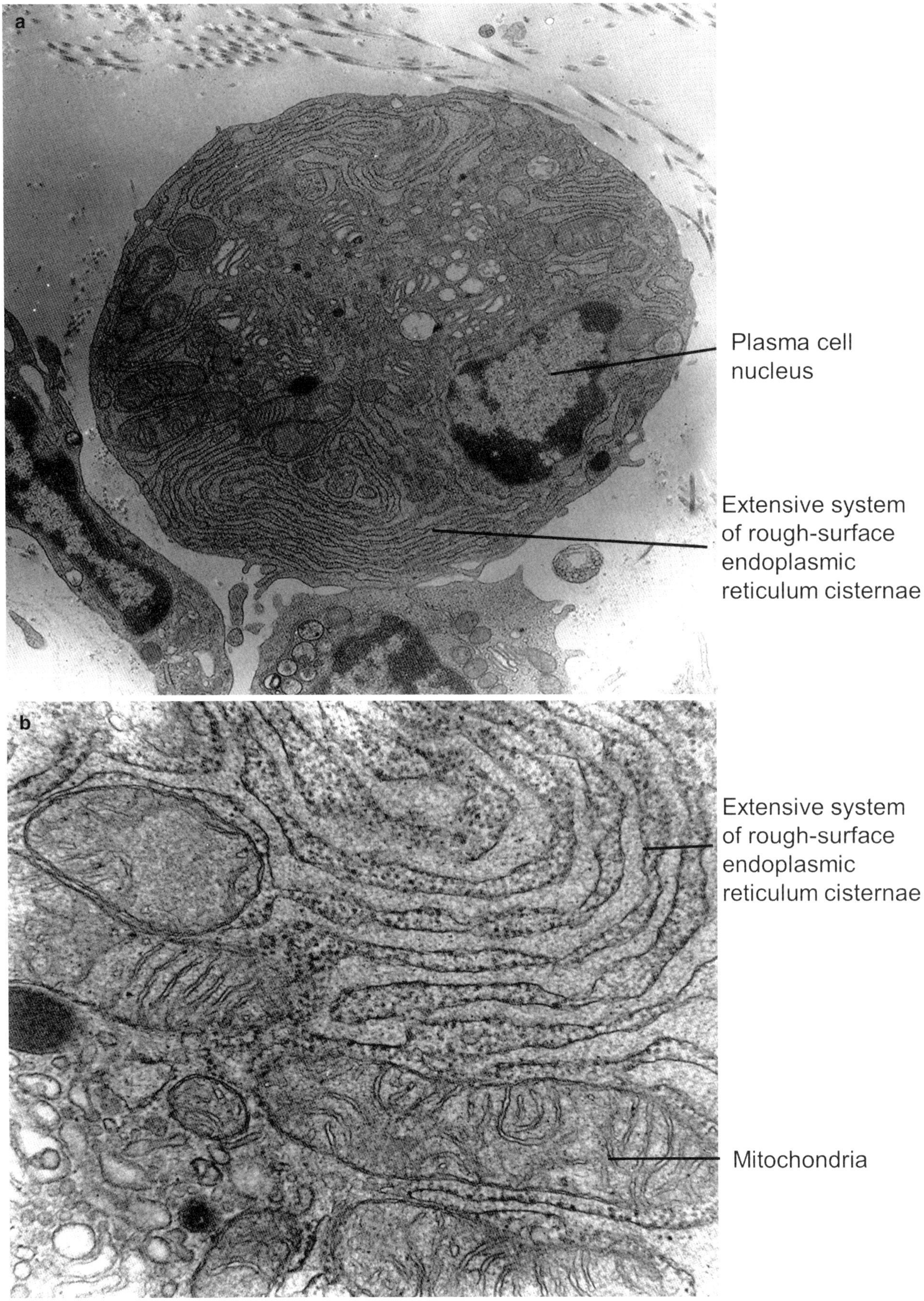

Fig. 2.16 Plasma cell. (**a**) Plasma cell near the dorsal nerve ganglia. The Golgi complex is located in the central and perinuclear areas. (**b**) Detail of several rough endoplasmic reticulum cisternae and mitochondria. Magnification: ×7,000 (**a**); ×30,000 (**b**)

References

1. Reina MA, Arriazu R, Collier CB, Sala-Blanch X. Histology and electron microscopy of human peripheral nerves of clinical relevance to the practice of nerve blocks. Rev Esp Anestesiol Reanim. 2013;60:552–62.
2. Reina MA, López A, Villanueva MC, De Andrés JA, León GI. Morfología de los nervios periféricos, de sus cubiertas y de su vascularización. Rev Esp Anestesiol Reanim. 2000;47:464–75.
3. Cross PC, Mercer KL. Connective tissue: macrophages, mast cells, lymphocytes and plasma cells. In: Cells and tissue ultrastructure. New York: WH Freeman; 1999. p. 74–81.
4. Oldfors A. Macrophages in peripheral nerves. An ultrastructural and enzyme histochemical study on rats. Acta Neuropathol. 1980;49: 43–9.
5. Sacco RE. The mononuclear phagocyte system and lymphocytes. In: Cheville NF, editor. Ultrastructural pathology. 2nd ed. Ames: Wiley-Blackwell; 2009. p. 269–93.
6. Reina MA, De Andrés JA, López A. Subarachnoid and epidural anesthesia. In: Raj P, editor. Textbook of regional anesthesia. New York: Churchill Livingstone; 2002. p. 307–24.
7. Reina MA, Pulido P, López A. The human dural sac. Morphology of the spinal dura-arachnoid membrane. Origin of the spinal subdural space. Rev Arg Anest. 2007;65:167–84.
8. Reina MA, Villanueva MC, López A. Trabecular arachnoid membrane, human spinal pia mater and subarachnoid anesthesia. Rev Arg Anest. 2008;66:111–33.
9. Olsson Y. Mast cells in human peripheral nerve. Acta Neurol Scand. 1971;47:357–68.
10. Monk KR, Wu J, Williams JP, Finney BA, Fitzgerald ME, Filippi MD, Ratner N. Mast cells can contribute to axon-glial dissociation and fibrosis in peripheral nerve. Neuron Glia Biol. 2007;3:233–44.

Ultrastructure of the Endoneurium

3

Miguel Angel Reina, Fabiola Machés, Pilar De Diego-Isasa, and Concepción Del Olmo

The endoneurium is a fine cylindrical lamina that encloses groups of axons with their respective Schwann cells, creating a specific microenvironment around them. Collagen fibers, which are the main constituent of endoneurium, surround both myelinated and unmyelinated axons. Being permeable, the endoneurium does not interfere with the passage of molecules.

Collagen fibers present in the endoneurium are mainly type I, and their characteristic features compare with collagen fibers in the layers of the perineurium, as well as with collagen fibers intermingled with the adipocytes that give shape to interfascicular tissue. Collagen is synthesized by Schwann cells and, to a lesser extent, by fibroblasts. This type of fiber confers strength to units formed by axons and their respective Schwann cells.

Despite their scarcity, fibroblasts may be identified readily inside the endoneurium because of their elongated shape and sharp contours. In nerve fibers, fibroblasts tend to be located in the remaining spaces between axons. Fibroblasts have important distinguishing features, including the absence of a basal membrane and large amounts of ribosomes and rough endoplasmic reticulum inside their cytoplasm.

The characteristic tubular shape of the endoneurium is specifically apparent in images obtained under scanning electron microscopy, in which the multichanneled structure formed by the endoneurium to enclose the axons individually is significantly better appreciated.

Continuous capillaries are found inside the endoneurial compartment. These types of capillaries contribute to the blood–nerve barrier. The inner lining of all blood vessels is formed by a single layer of endothelial cells (Figs. 3.1, 3.2, 3.3, 3.4, 3.5, 3.6, 3.7, 3.8, 3.9, 3.10, 3.11, 3.12, 3.13, 3.14, 3.15, 3.16, 3.17, 3.18, 3.19, 3.20, 3.21, 3.22, and 3.23) [1–7].

M.A. Reina, MD, PhD (✉)
Department of Clinical Medical Sciences and Institute of Applied Molecular Medicine, School of Medicine, University of CEU San Pablo, Madrid, Spain

Department of Anesthesiology, Madrid-Montepríncipe University Hospital, Madrid, Spain
e-mail: miguelangel@perticone.e.telefonica.net

F. Machés, MD
Department of Anesthesiology, Madrid-Montepríncipe University Hospital, Madrid, Spain
e-mail: fabmami@gmail.com

P. De Diego-Isasa, MD • C. Del Olmo, MD
Department of Anesthesiology, Hospital Asepeyo, Madrid, Spain
e-mail: pilar.dediego@gmail.com; delolmo.c@gmail.com

M.A. Reina et al. (eds.), *Atlas of Functional Anatomy for Regional Anesthesia and Pain Medicine: Human Structure, Ultrastructure and 3D Reconstruction Images*, DOI 10.1007/978-3-319-09522-6_3,

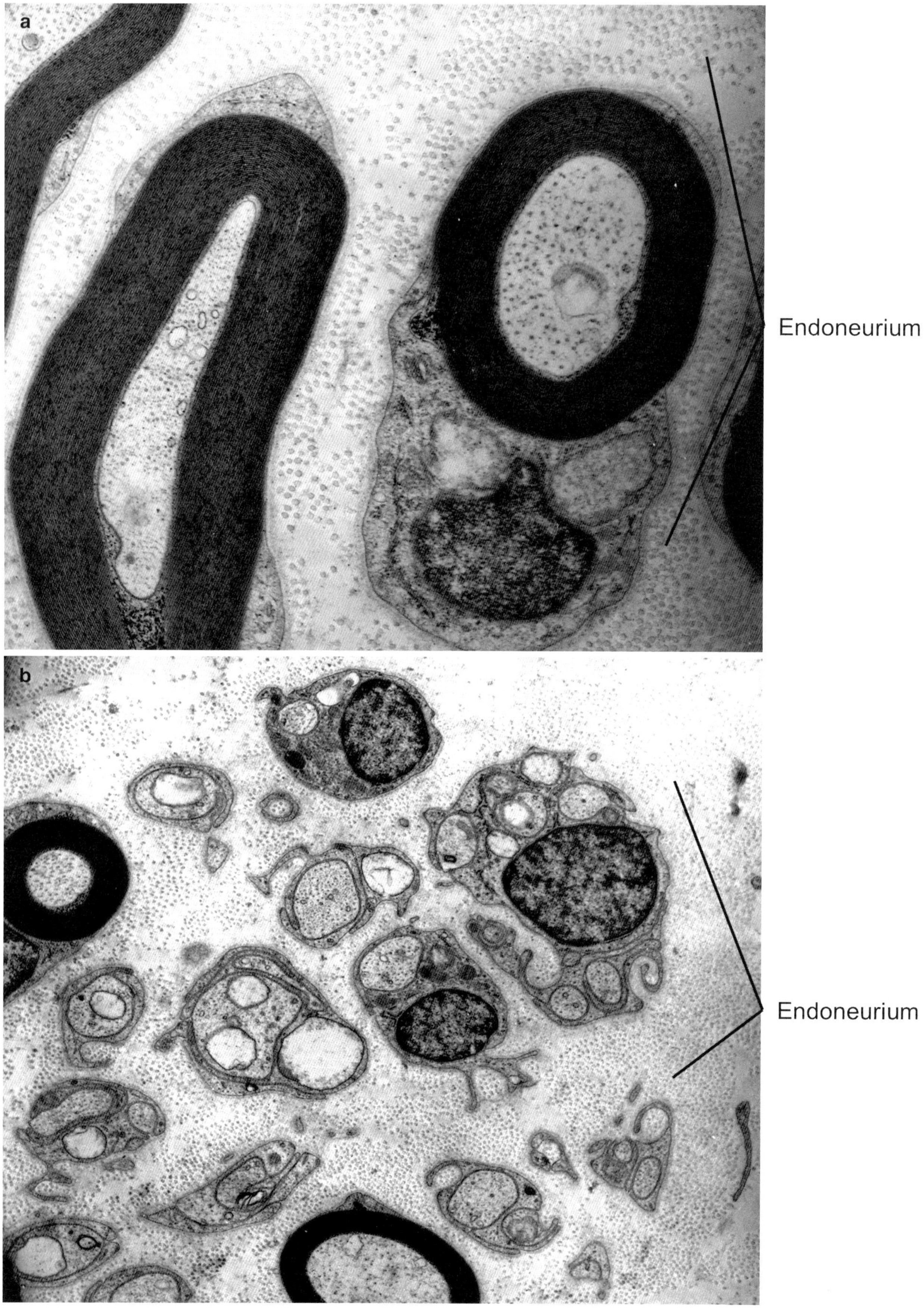

Fig. 3.1 Endoneurium. (**a**, **b**) Myelinated axons enveloped by collagen fibers in a human sciatic nerve. Transmission electron microscopy, magnification: ×12,000 (**a**); ×7,000 (**b**)

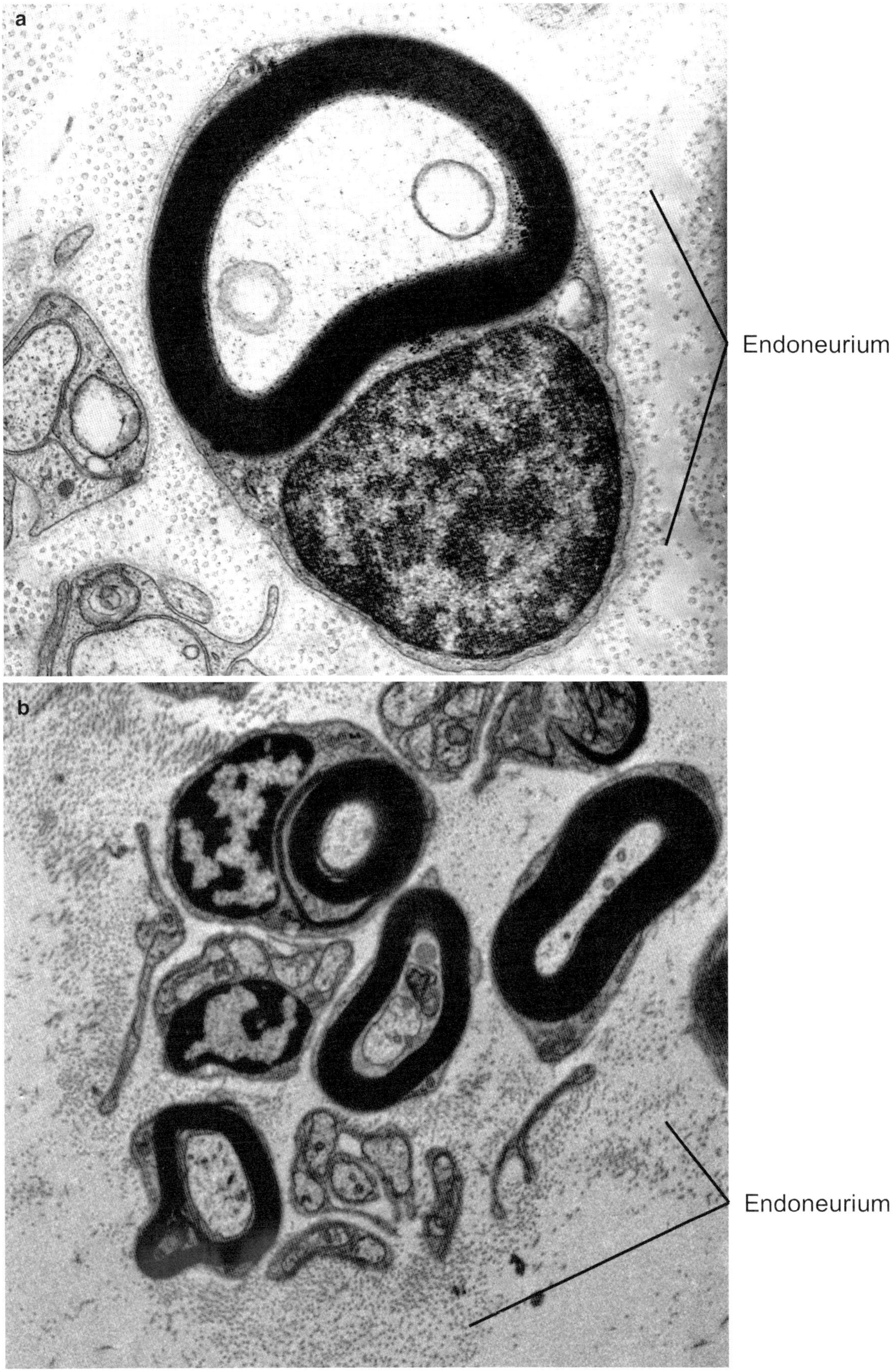

Fig. 3.2 Endoneurium. (**a**, **b**) Myelinated and unmyelinated axons enveloped by endoneurium. This sample was obtained from a human sciatic nerve. Transmission electron microscopy, magnification: ×12,000 (**a**); ×8,000 (**b**)

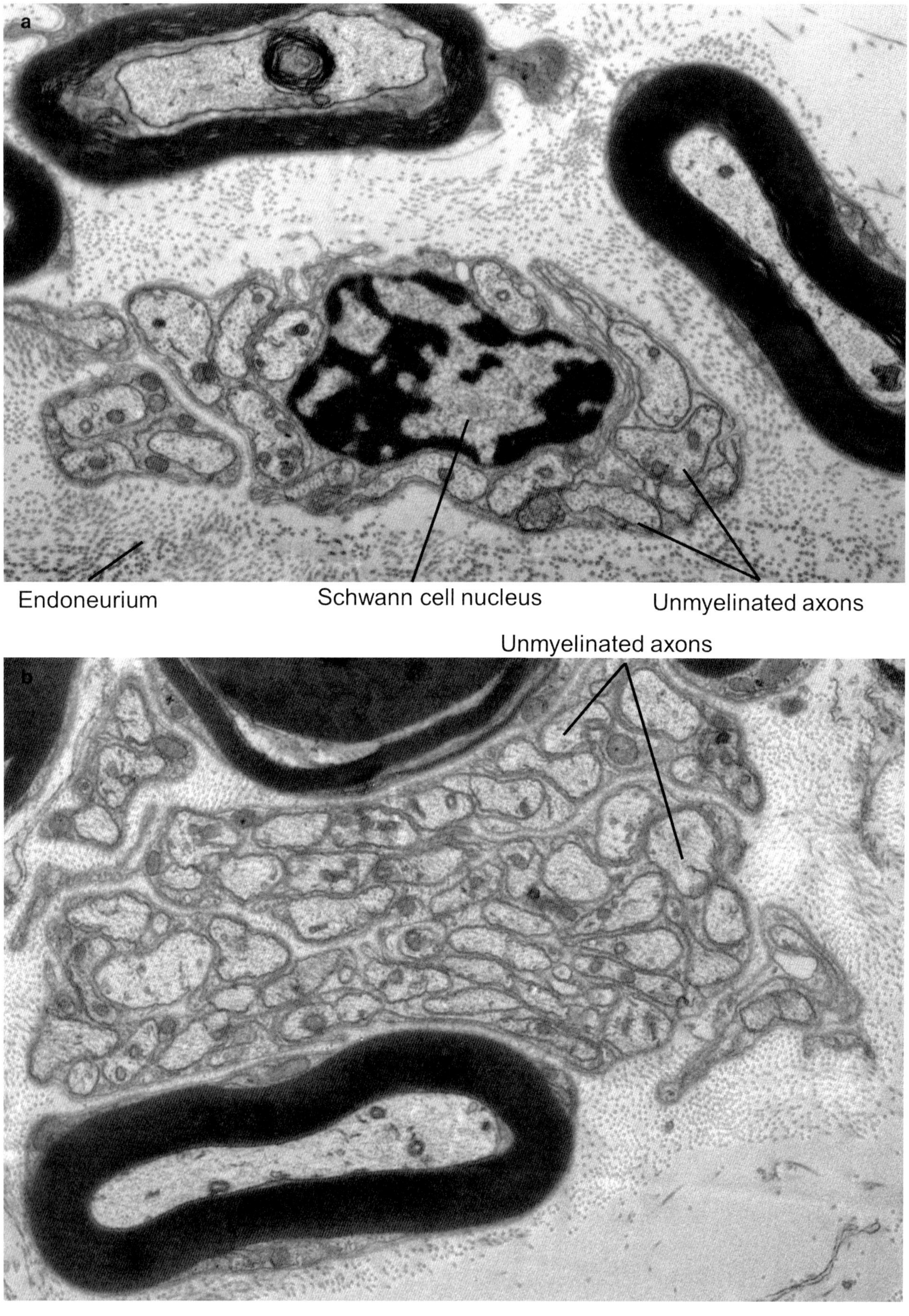

Fig. 3.3 Endoneurium. (**a**, **b**) Myelinated and unmyelinated axons enveloped by endoneurium. The sample was obtained from a human sciatic nerve. Transmission electron microscopy, magnification: ×15,000 (**a**, **b**)

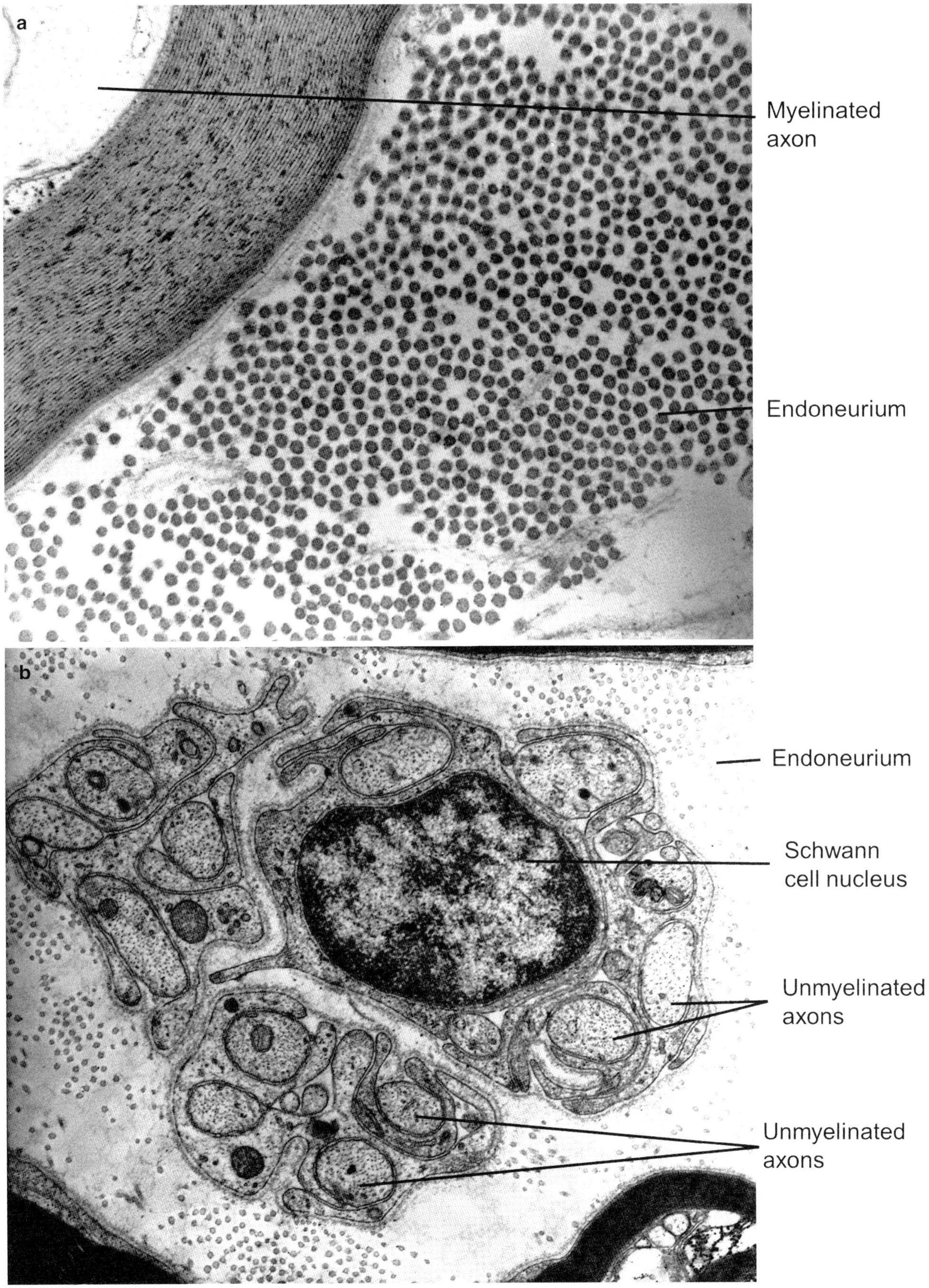

Fig. 3.4 Endoneurium. (**a**) Myelinated axon enveloped by endoneurium. (**b**) Unmyelinated axons enveloped by endoneurium. The sample was obtained from a human sciatic nerve. Transmission electron microscopy, magnification: ×30,000 (**a**); 25,000 (**b**)

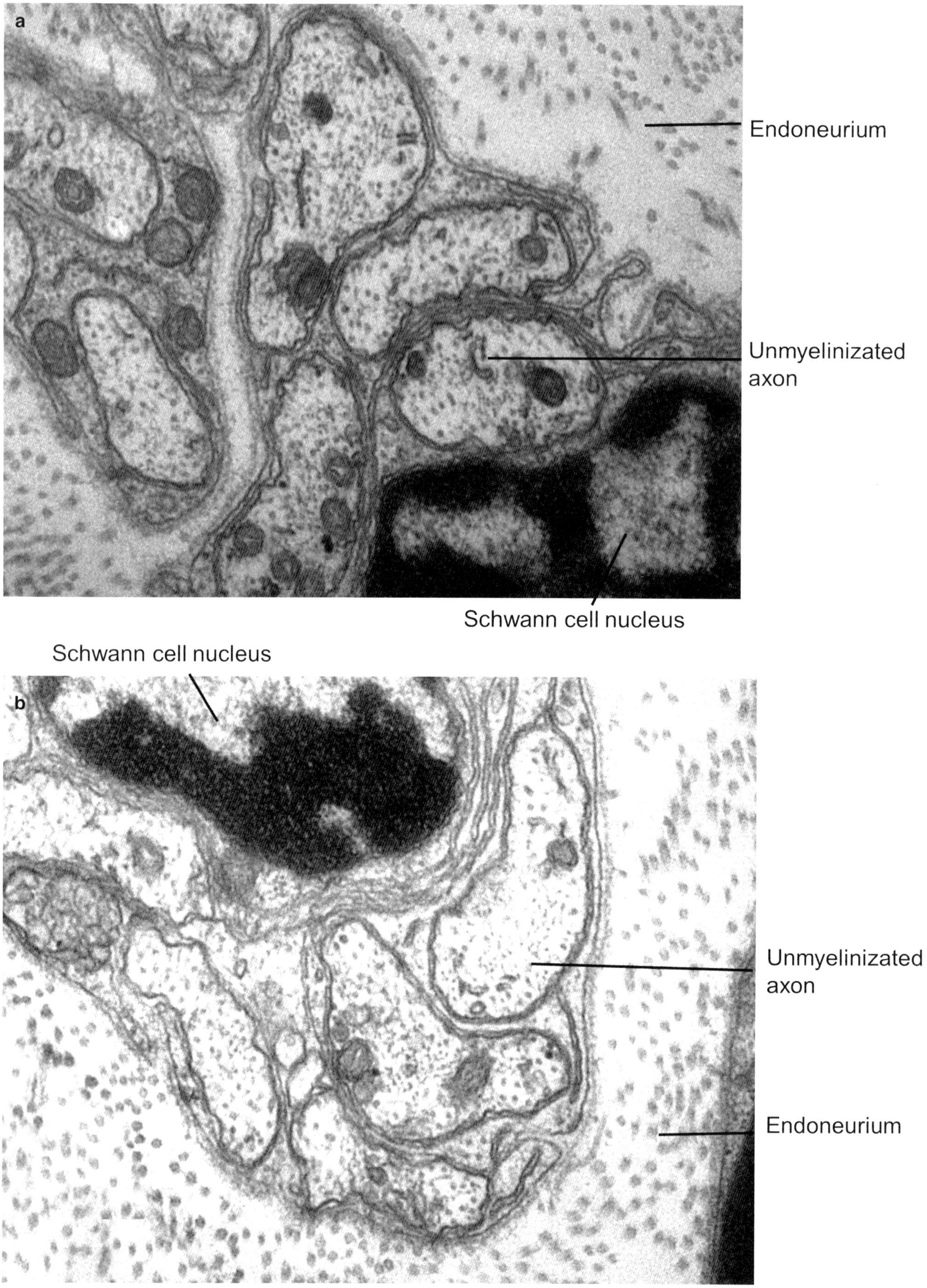

Fig. 3.5 Endoneurium. (**a**, **b**) Unmyelinated axons enveloped by endoneurium. The sample was obtained from a human sciatic nerve. Transmission electron microscopy, magnification: ×50,000 (**a**, **b**)

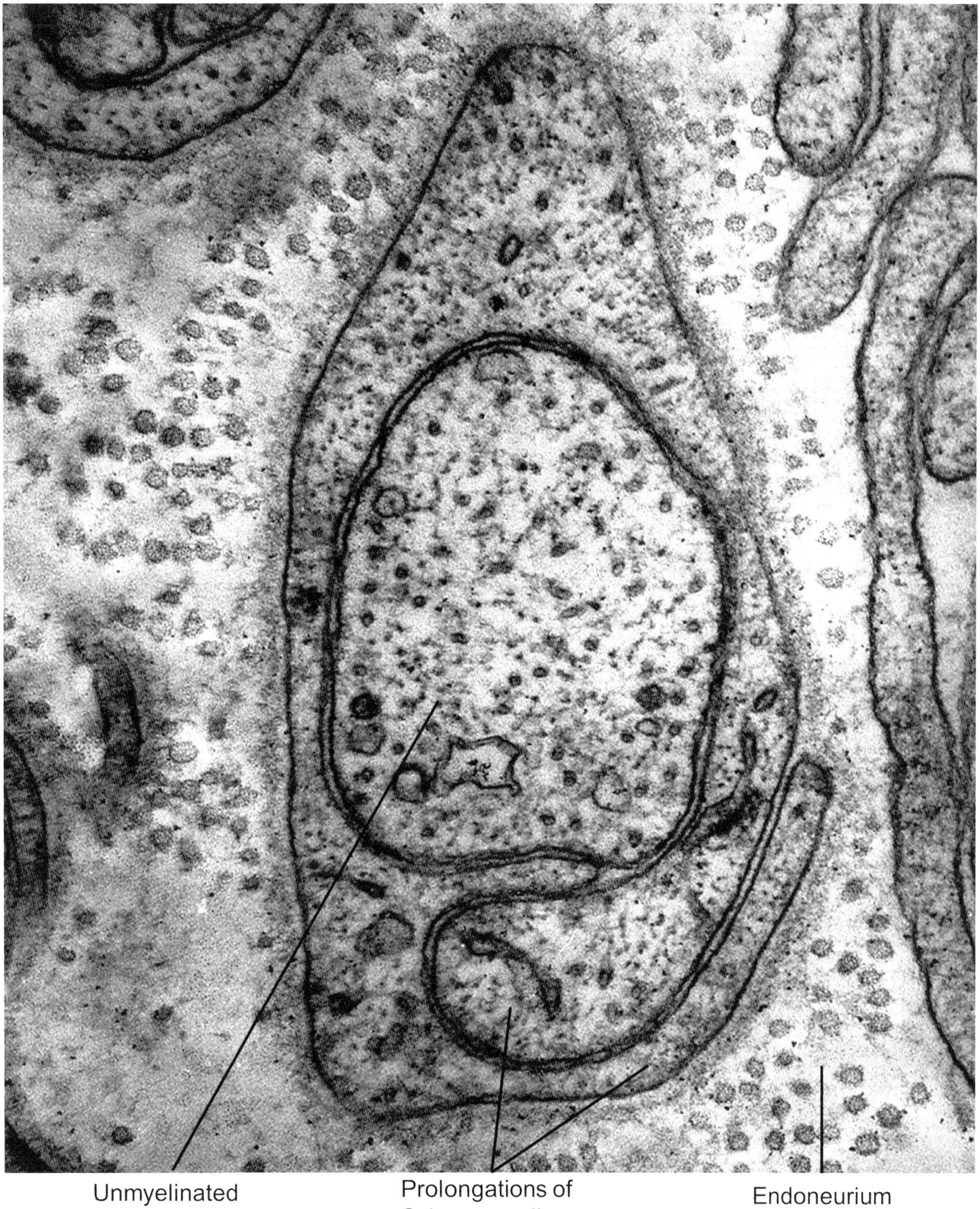

Fig. 3.6 Endoneurium. Unmyelinated axons enveloped by endoneurium. The sample was obtained from a human sciatic nerve. Transmission electron microscopy, magnification: ×150,000

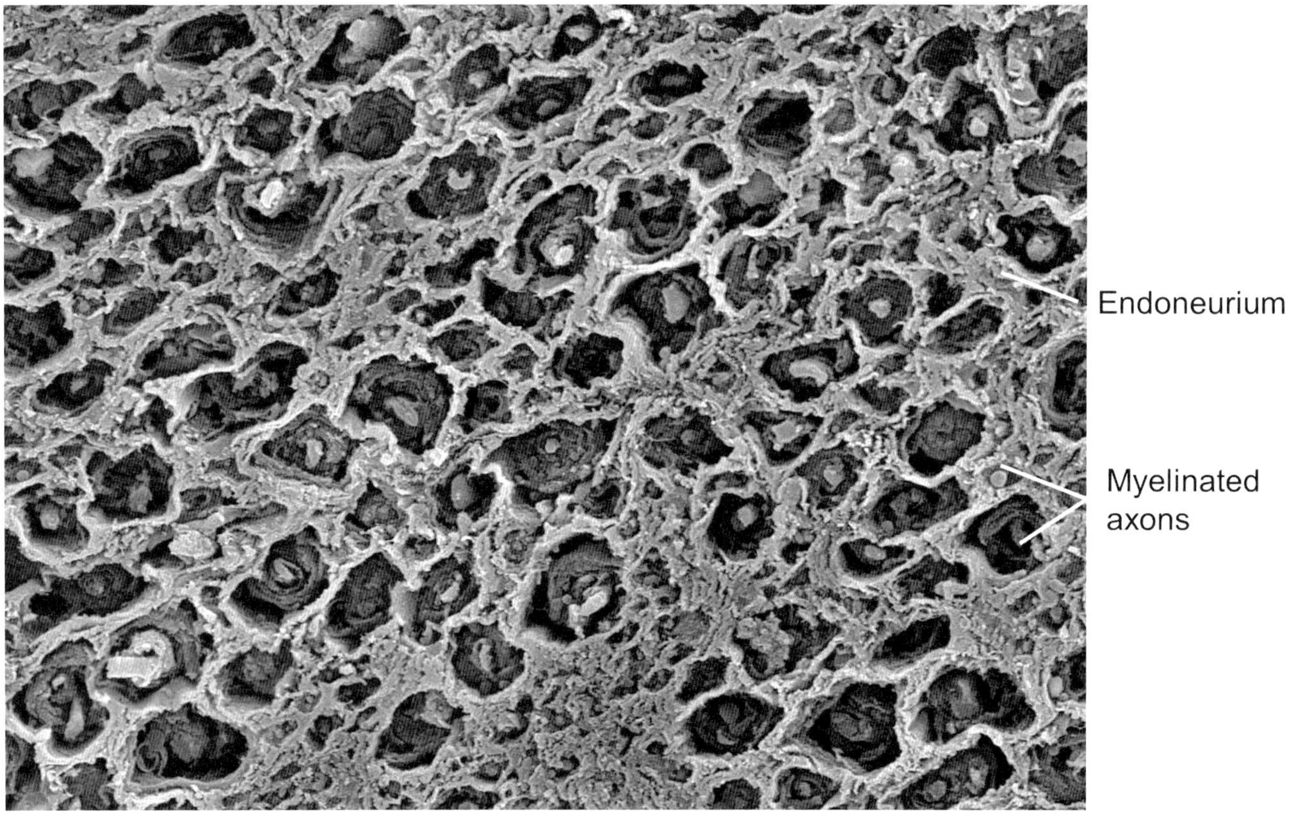

Fig. 3.7 Endoneurium. Myelinated axons enveloped by endoneurium. The sample was obtained from a human sciatic nerve. This technique cannot be used to identify unmyelinated axons. Scanning electron microscopy, magnification: ×1,000

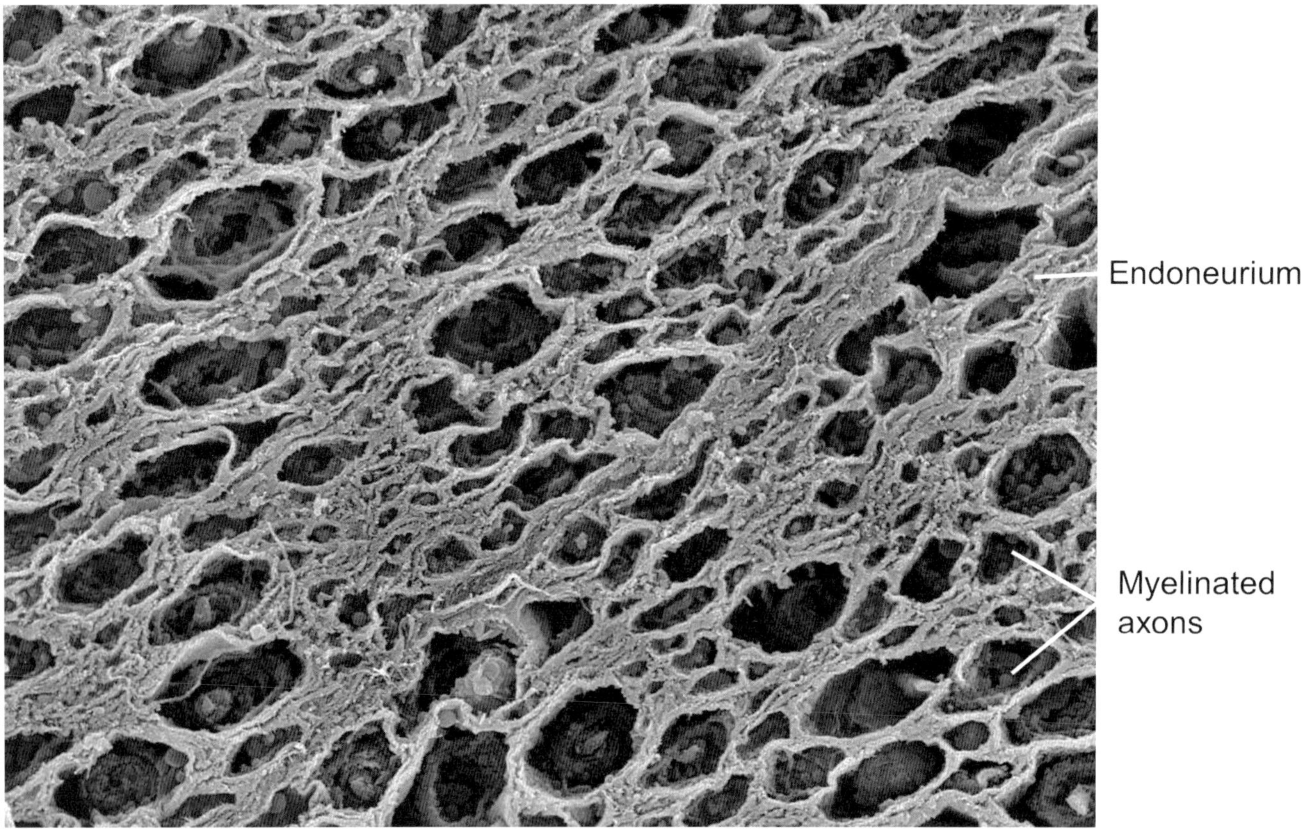

Fig. 3.8 Endoneurium. Myelinated axons enveloped by endoneurium. The sample was obtained from a human sciatic nerve. This technique cannot be used to identify unmyelinated axons. Scanning electron microscopy, magnification: ×1,000

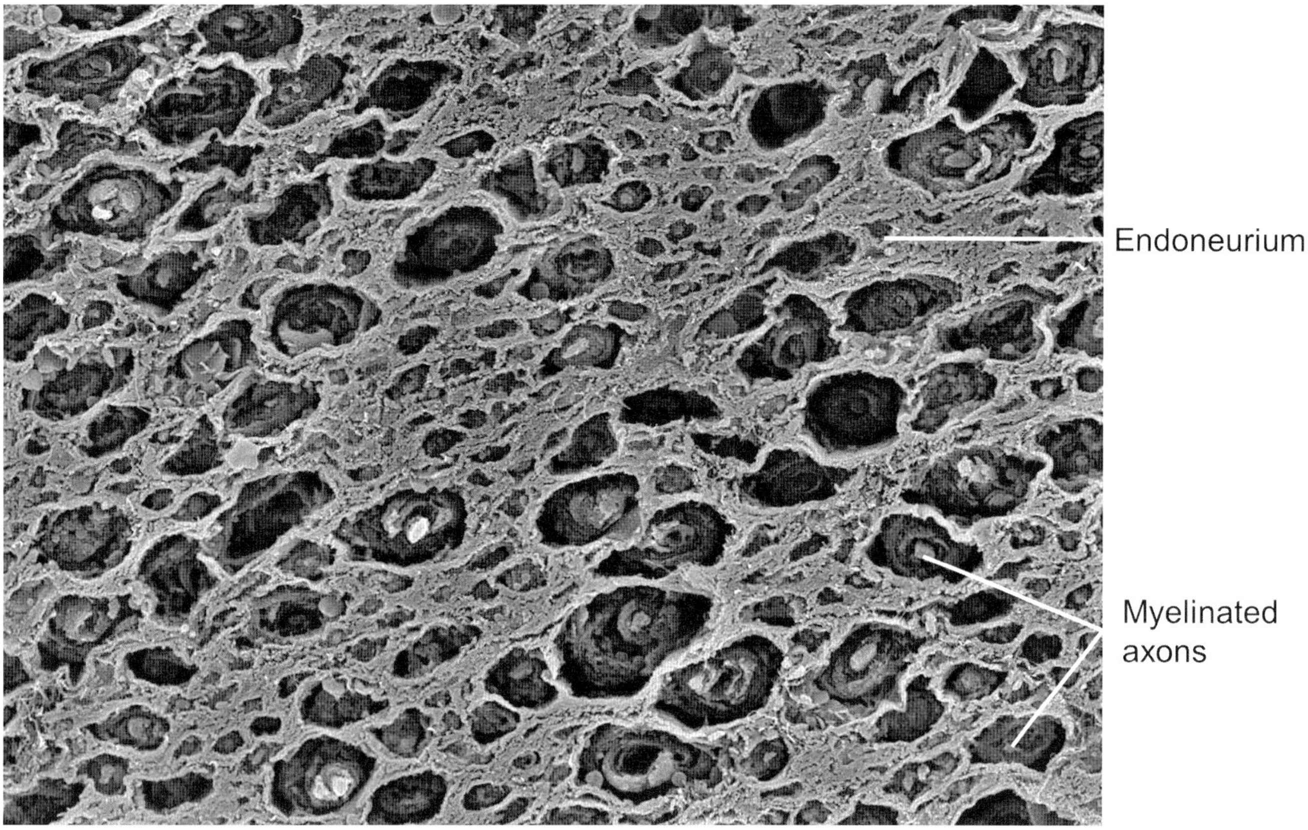

Fig. 3.9 Endoneurium. Myelinated axons enveloped by endoneurium. The sample was obtained from a human sciatic nerve. This technique cannot be used to identify unmyelinated axons. Scanning electron microscopy, magnification: ×1,000

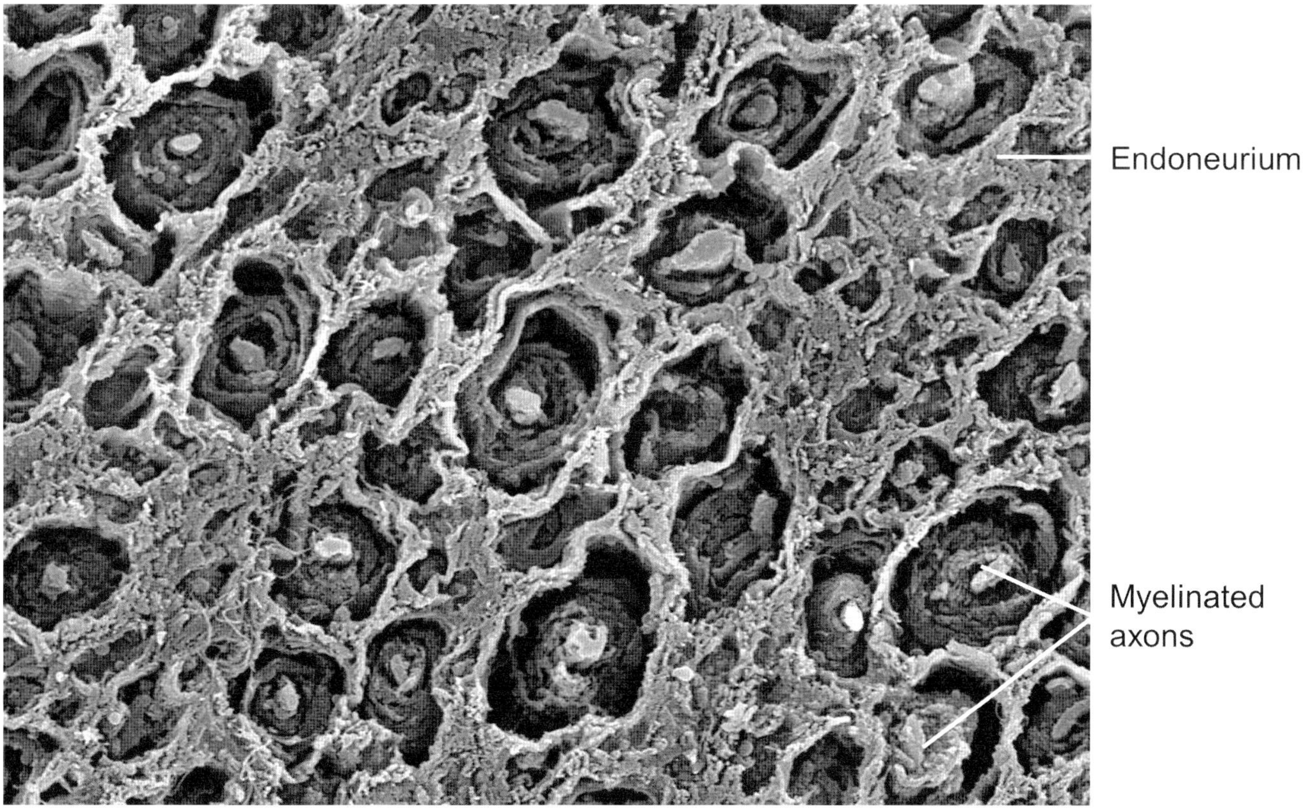

Fig. 3.10 Endoneurium. Myelinated axons enveloped by endoneurium. The sample was obtained from a human sciatic nerve. This technique cannot be used to identify unmyelinated axons. Scanning electron microscopy, magnification: ×1,500

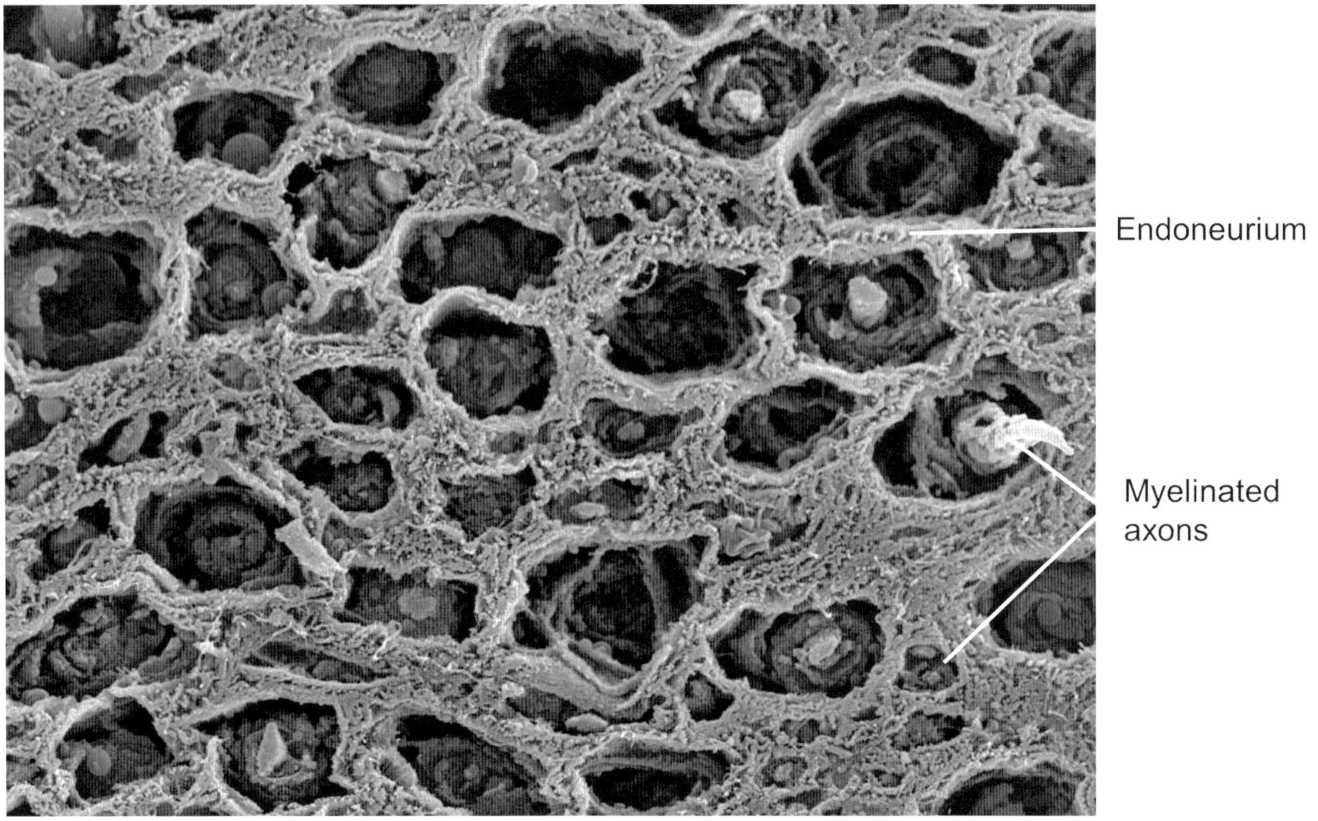

Fig. 3.11 Endoneurium. Myelinated axons enveloped by endoneurium. The sample was obtained from a human sciatic nerve. This technique cannot be used to identify unmyelinated axons. Scanning electron microscopy, magnification: ×1,500

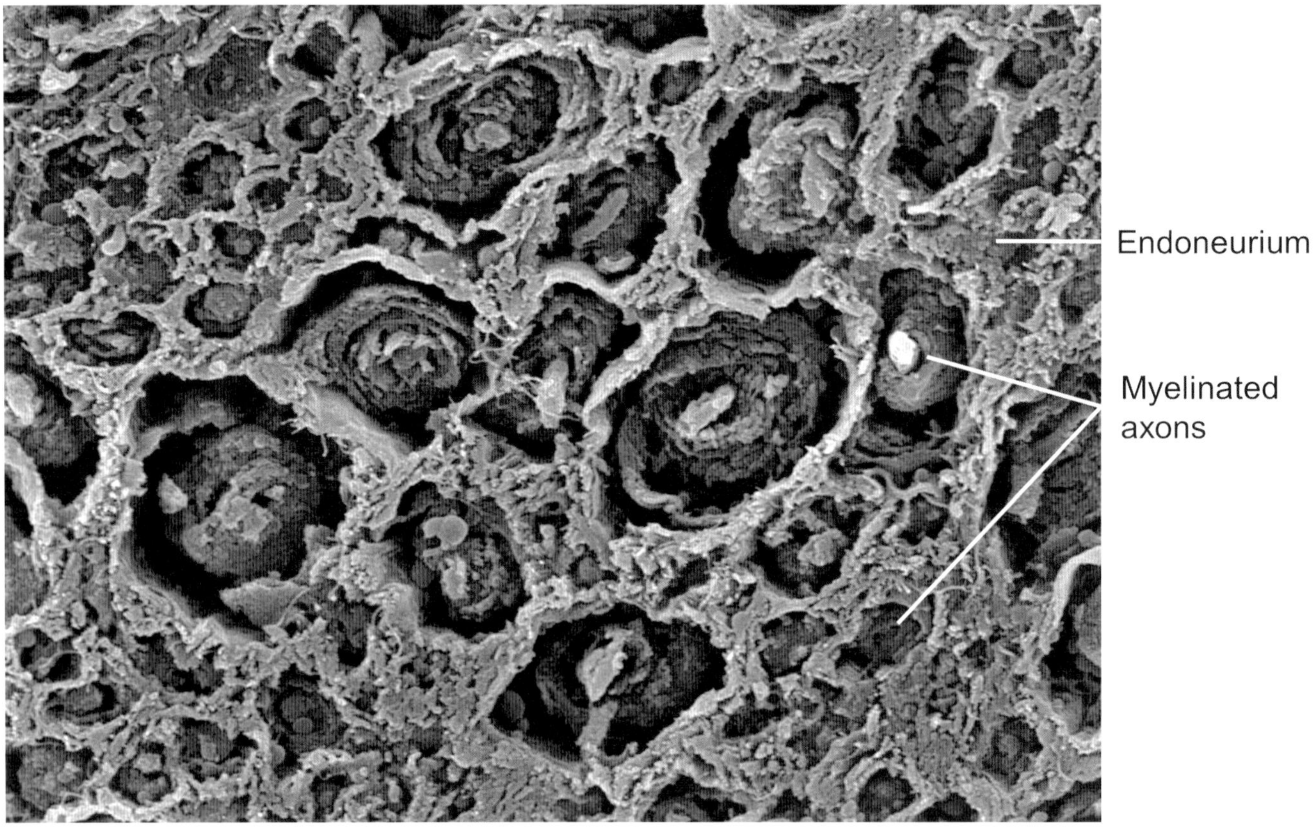

Fig. 3.12 Endoneurium. Myelinated axons enveloped by endoneurium. The sample was obtained from a human sciatic nerve. This technique cannot be used to identify unmyelinated axons. Scanning electron microscopy, magnification: ×2,000

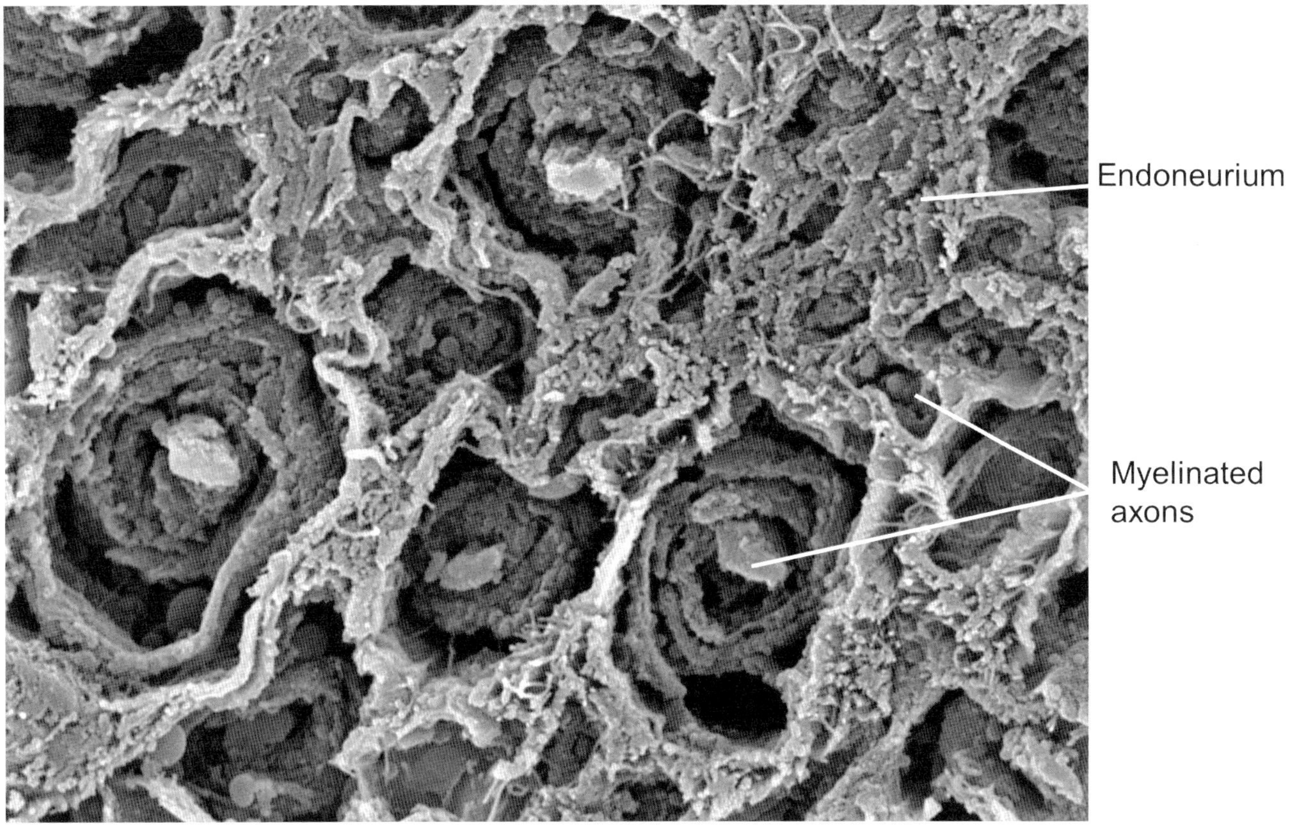

Fig. 3.13 Endoneurium. Myelinated axons enveloped by endoneurium. The sample was obtained from a human sciatic nerve. This technique cannot be used to identify unmyelinated axons. Scanning electron microscopy, magnification: ×3,000

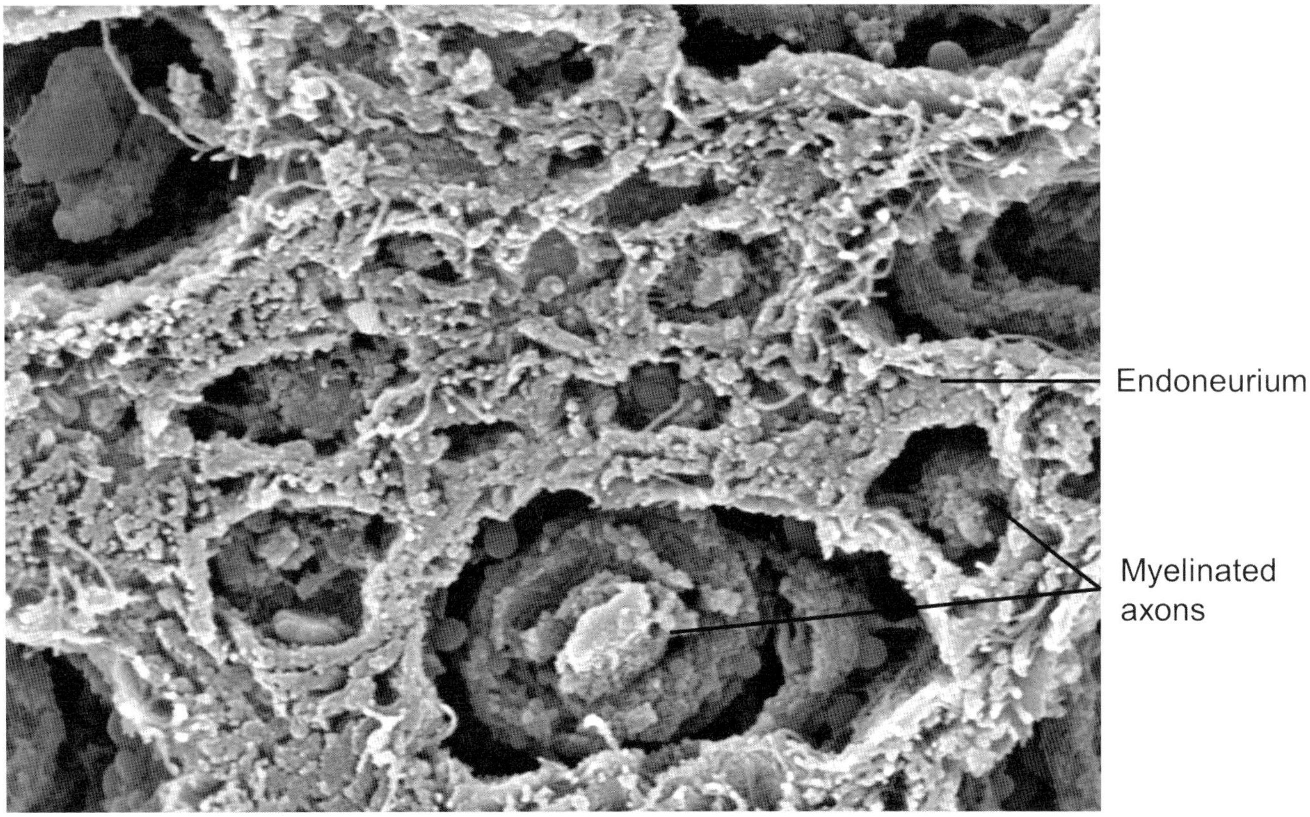

Fig. 3.14 Endoneurium. Myelinated axons enveloped by endoneurium. The sample was obtained from human sciatic nerve. This technique cannot be used to identify unmyelinated axons. Scanning electron microscopy, magnification: ×4,500

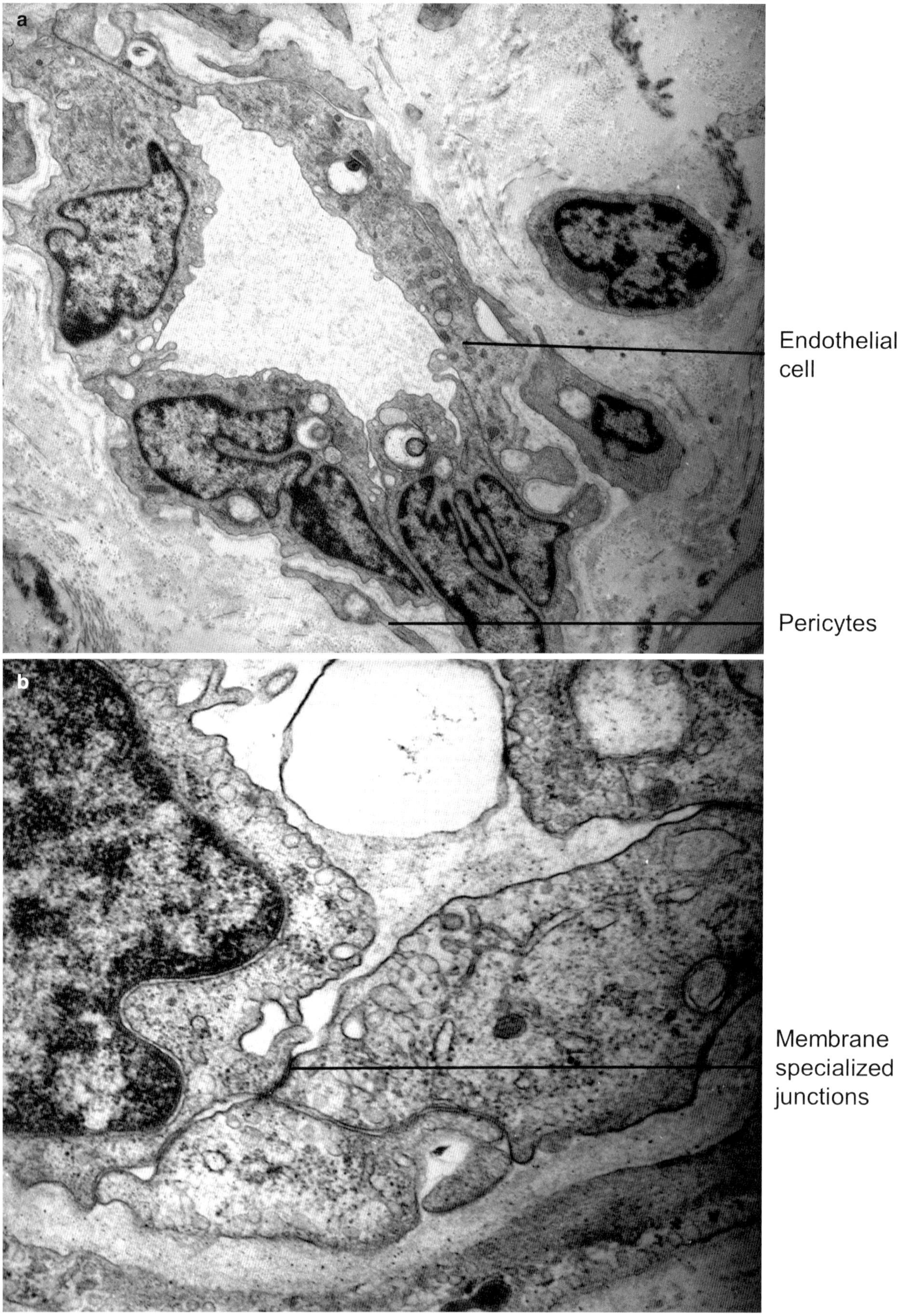

Fig. 3.15 Endoneurium. Endoneural continuous capillaries. Transmission electron microscopy, magnification: ×7,000 (**a**); ×20,000 (**b**)

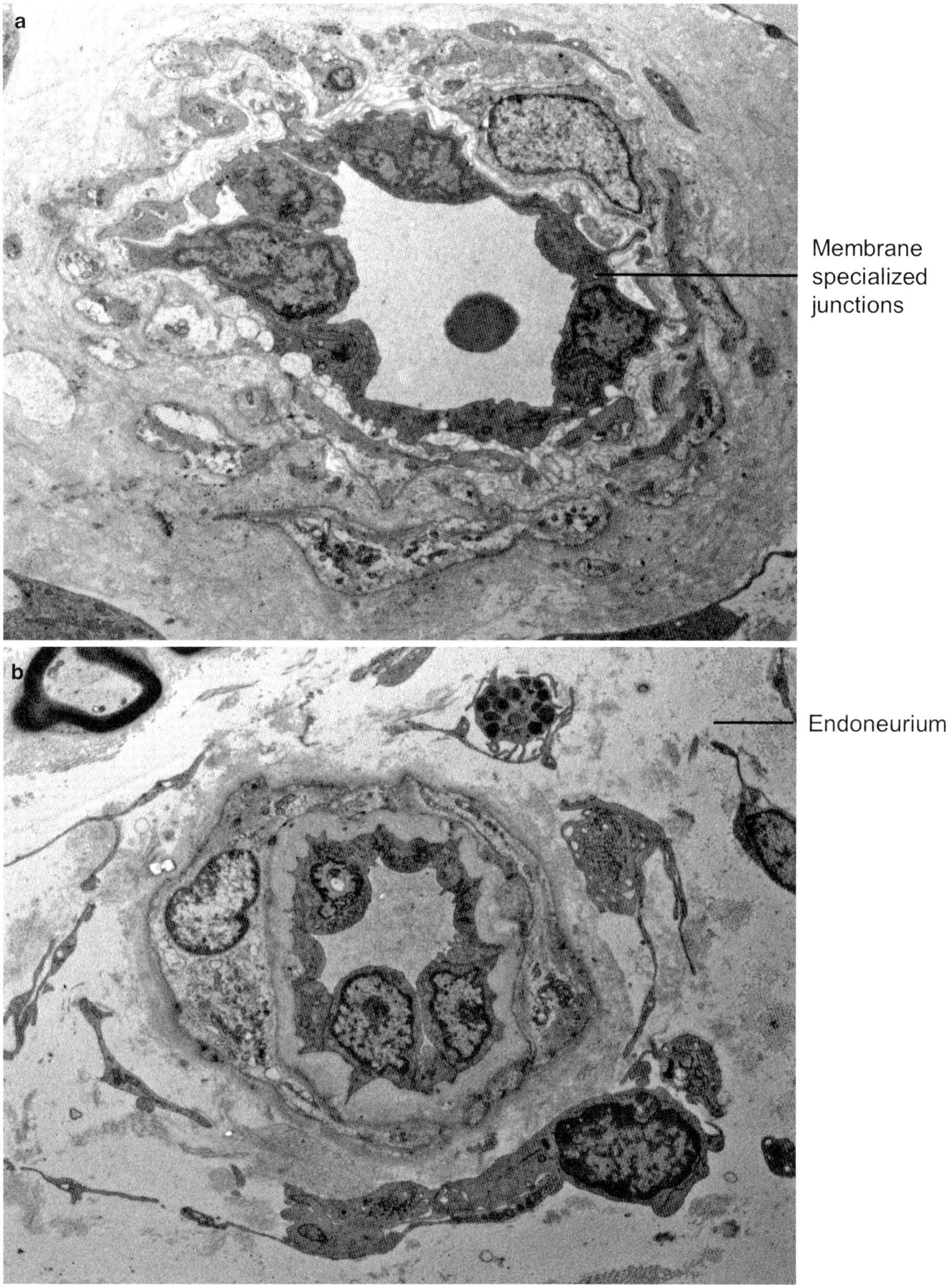

Fig. 3.16 Endoneurium. Endoneural continuous capillaries. Transmission electron microscopy, magnification: ×5,000 (**a**, **b**)

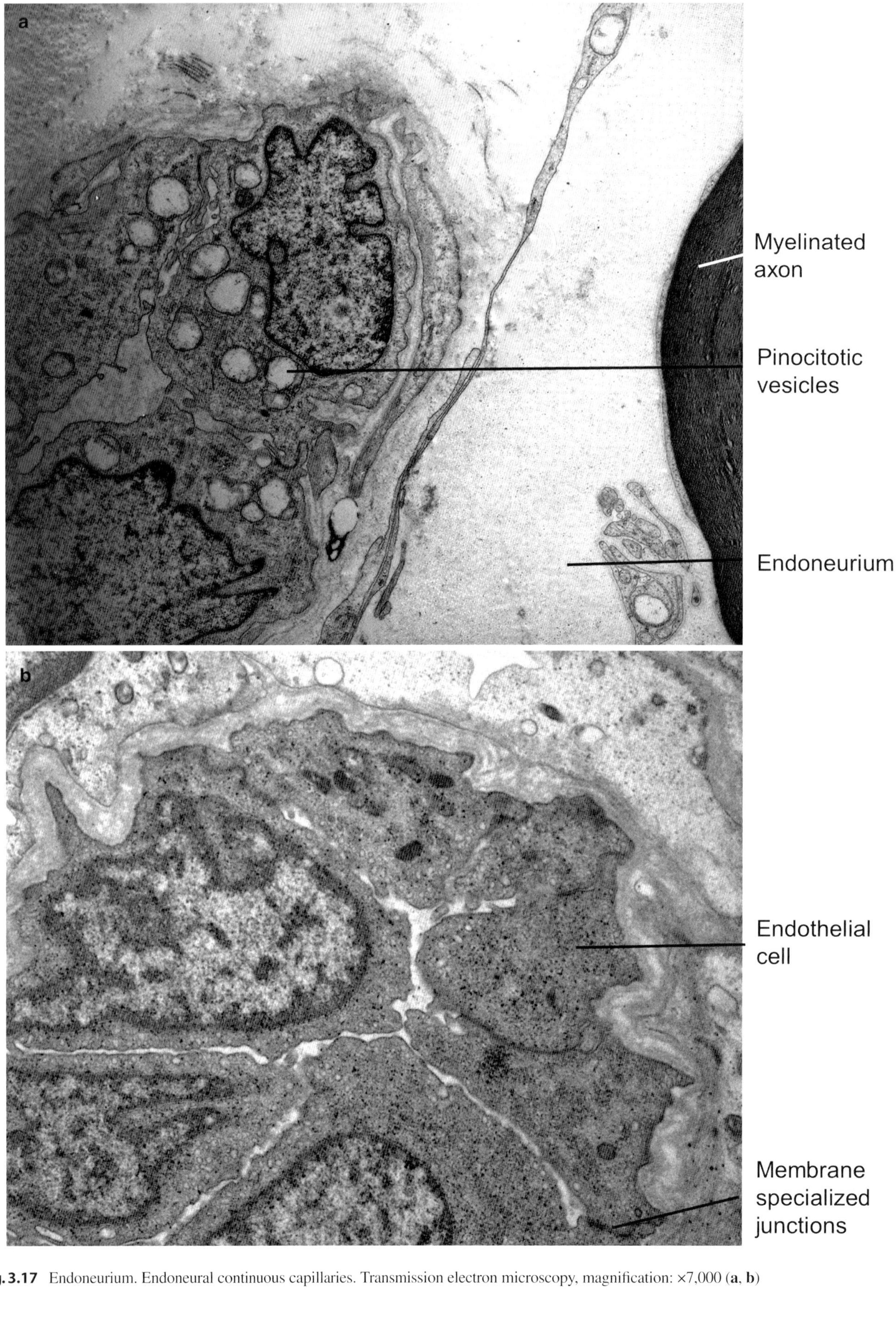

Fig. 3.17 Endoneurium. Endoneural continuous capillaries. Transmission electron microscopy, magnification: ×7,000 (**a**, **b**)

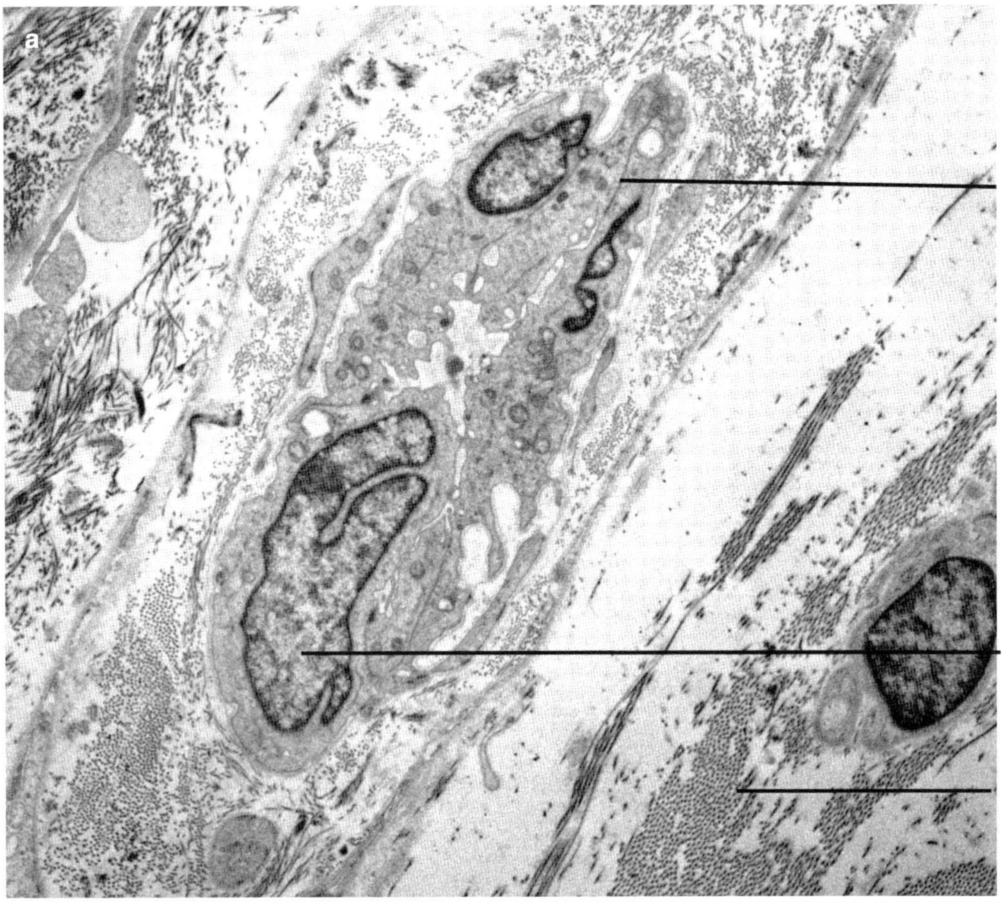

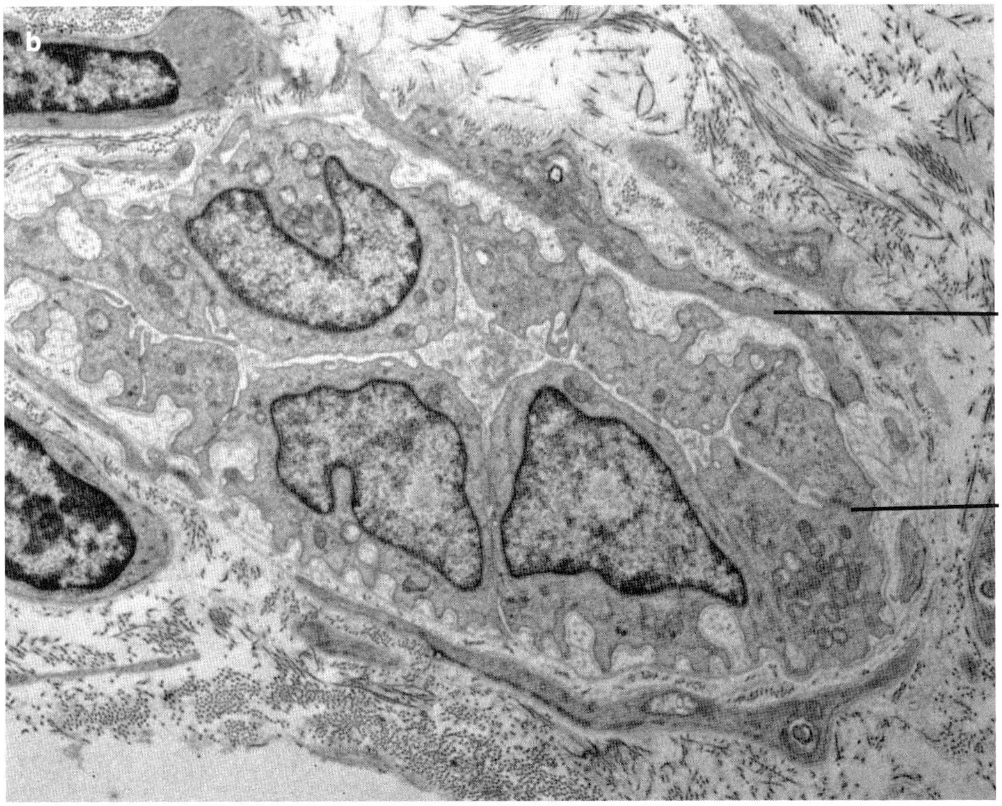

Fig. 3.18 Endoneurium. Endoneural continuous capillaries. Transmission electron microscopy, magnification: ×6,000 (**a**); ×8,000 (**b**)

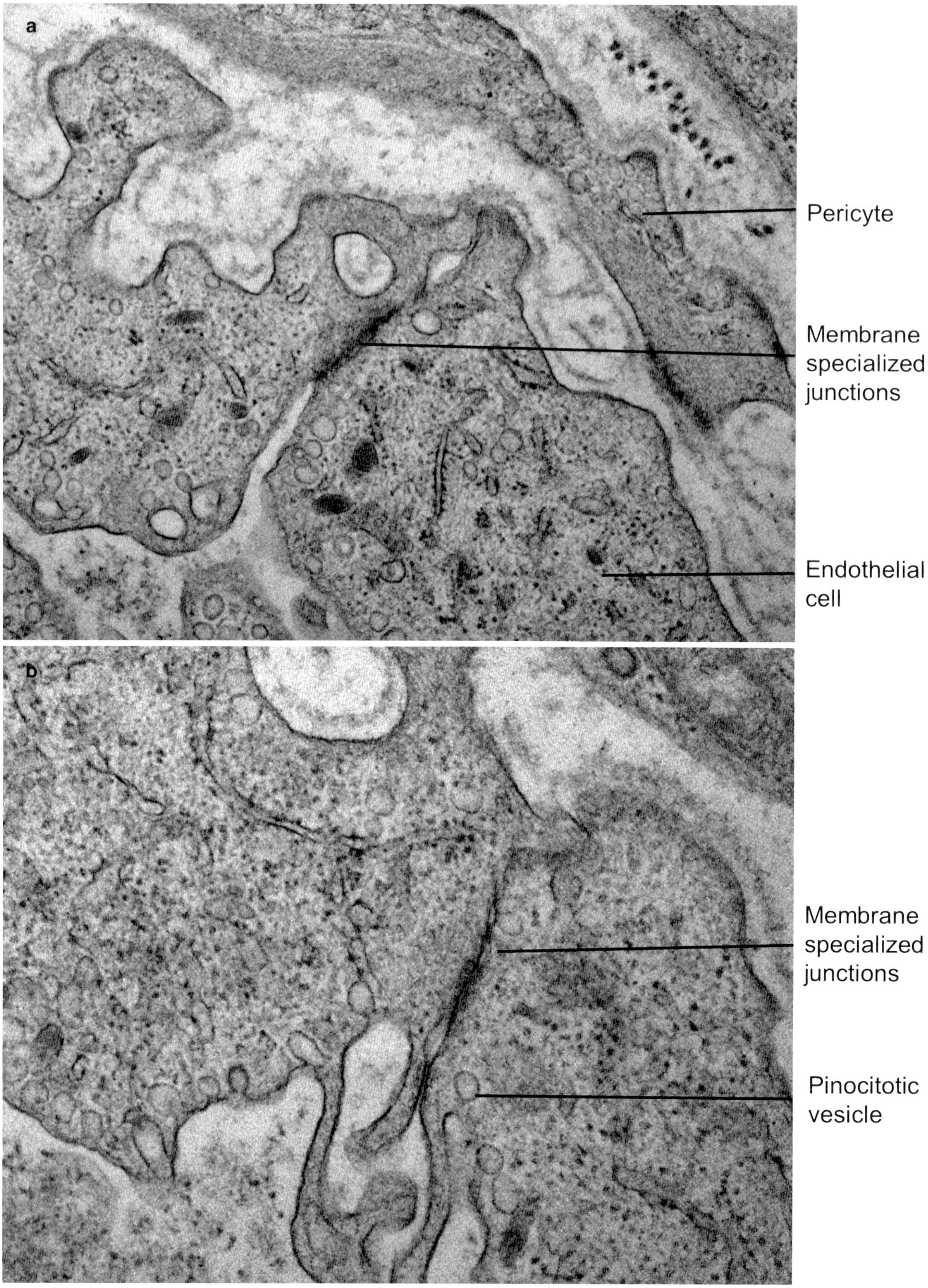

Fig. 3.19 Endoneurium. Endoneural continuous capillaries. Shown are details of the specialized membrane junction between endothelial cells. Transmission electron microscopy, magnification: ×40,000 (**a**); ×60,000 (**b**)

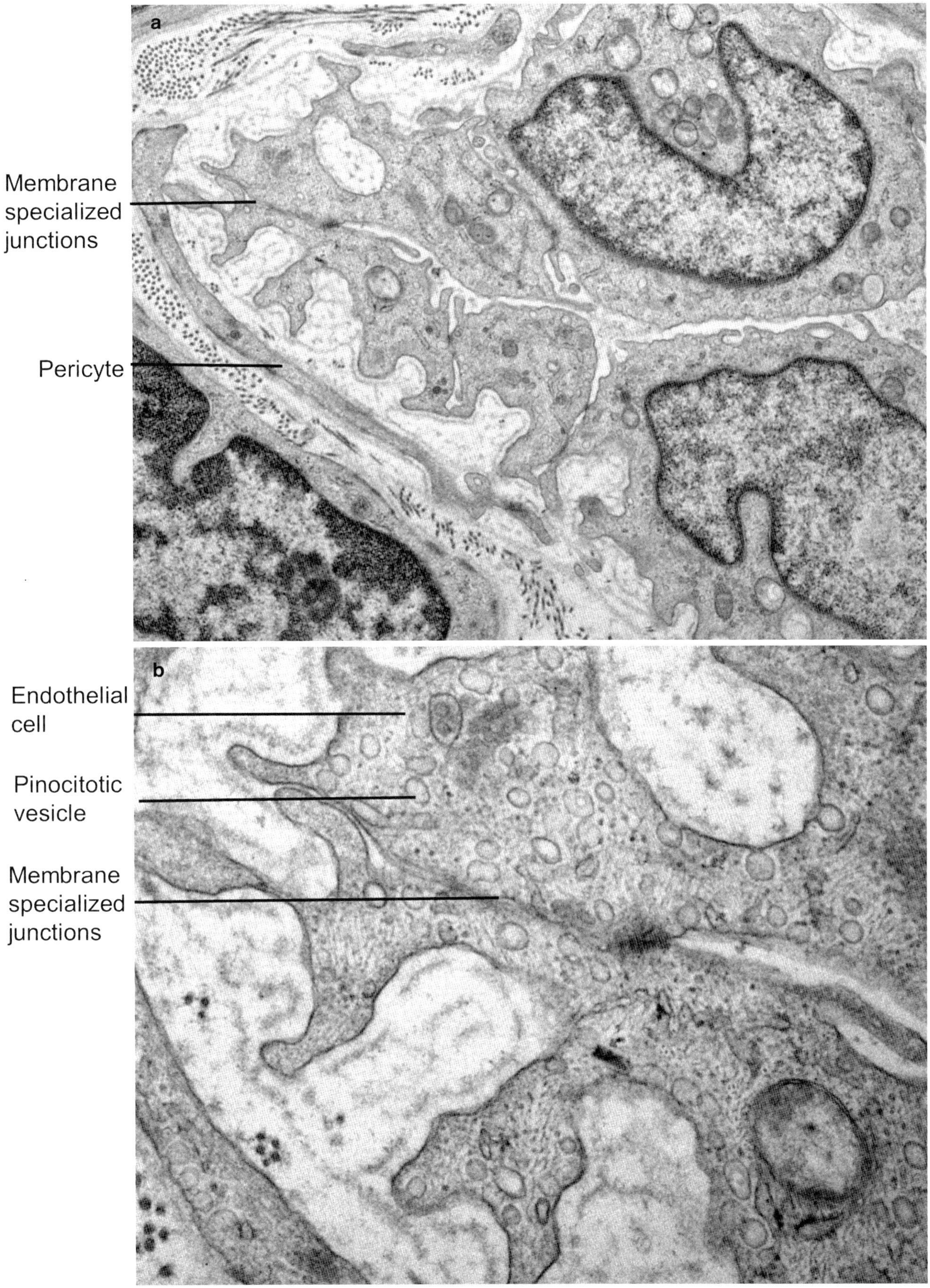

Fig. 3.20 Endoneurium. Endoneural continuous capillaries. Shown are details of the specialized membrane junction between endothelial cells. Transmission electron microscopy, magnification: ×15,000 (**a**); ×50,000 (**b**)

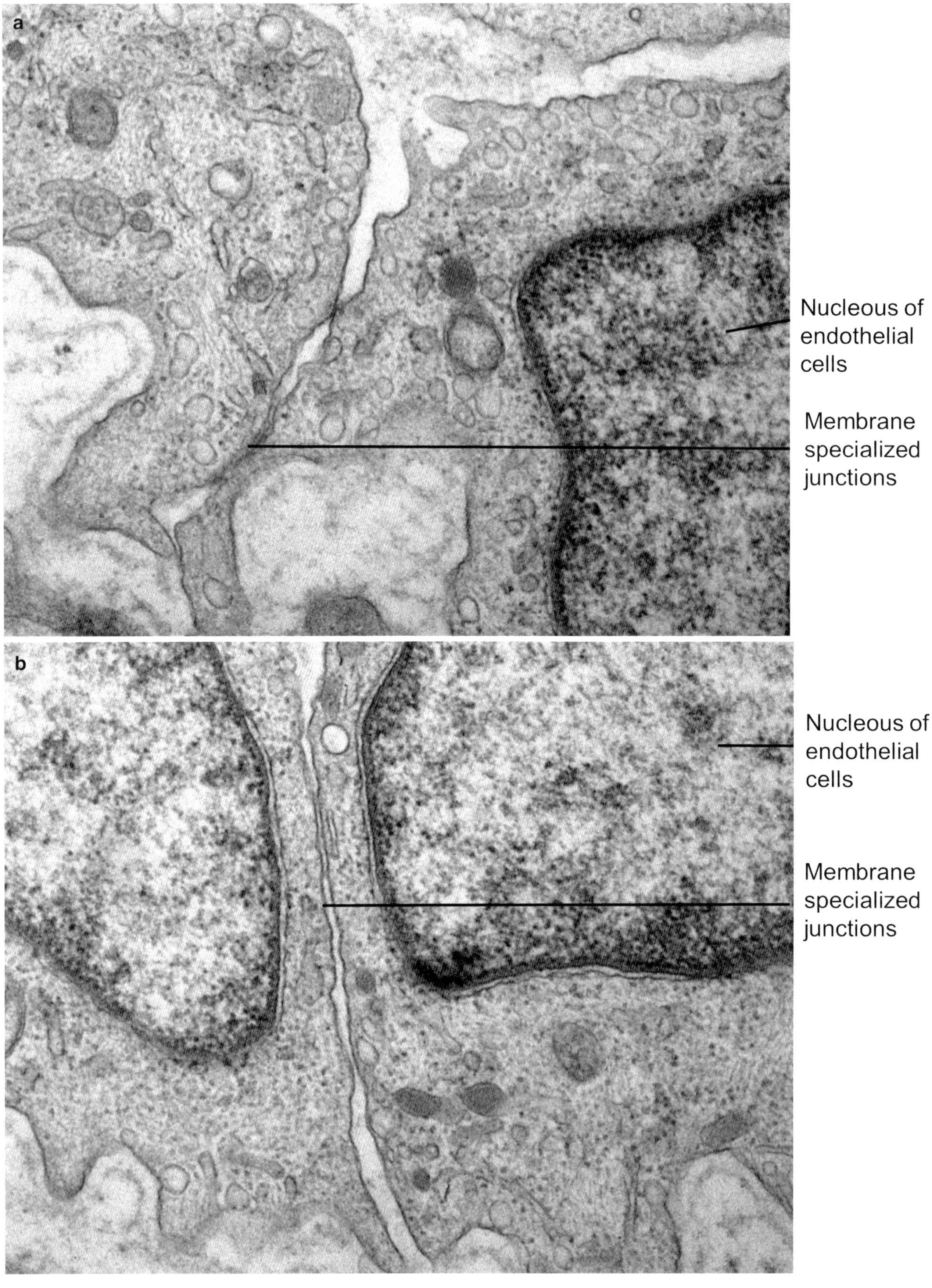

Fig. 3.21 Endoneurium. Endoneural continuous capillaries. Shown are details of the specialized membrane junction between endothelial cells. Transmission electron microscopy, magnification: ×50,000 (**a**, **b**)

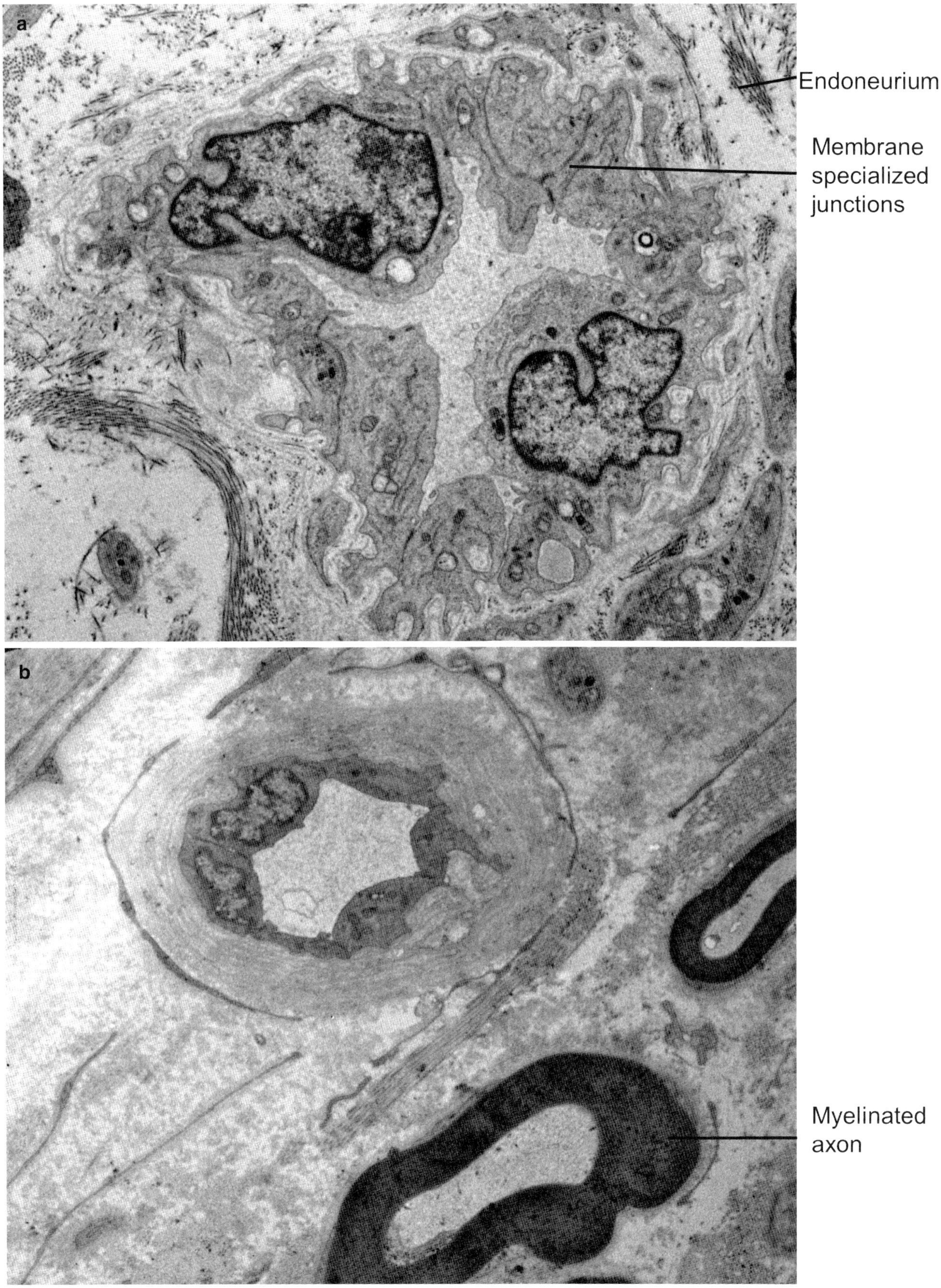

Fig. 3.22 Endoneurium. Endoneural continuous capillaries. Transmission electron microscopy, magnification: ×8,000 (**a**); ×7,000 (**b**)

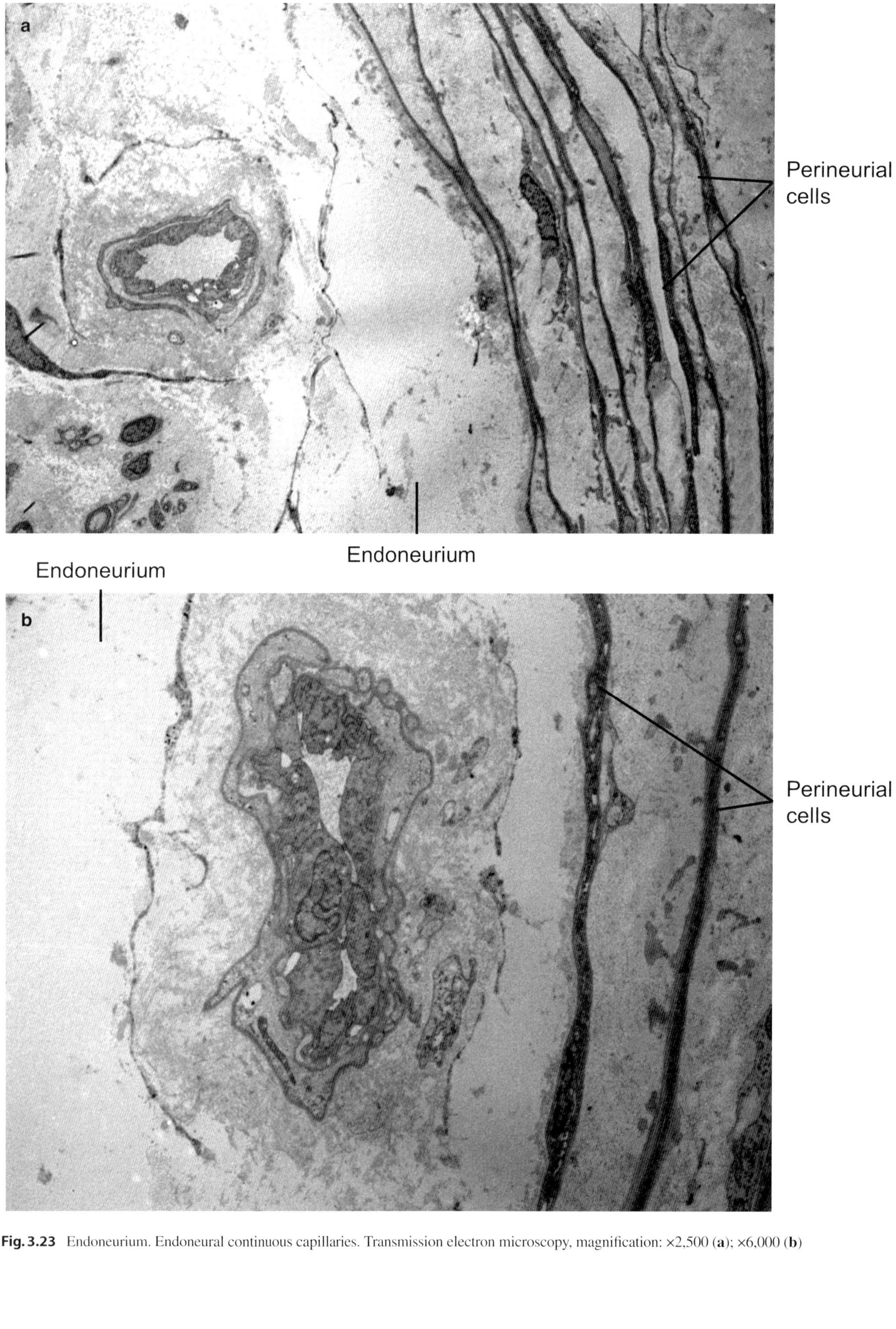

Fig. 3.23 Endoneurium. Endoneural continuous capillaries. Transmission electron microscopy, magnification: ×2,500 (**a**); ×6,000 (**b**)

References

1. Reina MA, Arriazu R, Collier CB, Sala-Blanch X. Histology and electron microscopy of human peripheral nerves of clinical relevance to the practice of nerve blocks. Rev Esp Anestesiol Reanim. 2013;60:552–62.
2. Friede RL, Bischhausen R. The organization of endoneural collagen in peripheral nerves as revealed with the scanning electron microscope. J Neurol Sci. 1978;38:83–9.
3. Ushiki T, Ide C. Three-dimensional organization of the collagen fibrils in the rat sciatic nerve as revealed by transmission and scanning electron microscopy. Cell Tissue Res. 1990;260:175–84.
4. Mizisin AP, Weerasuriya A. Homeostatic regulation of the endoneurial microenvironment during development, aging and in response to trauma, disease and toxic insult. Acta Neuropathol. 2011;121:291–312.
5. Kline DG, Hackett ER, Davis GD, Myers MB. Microcirculation of peripheral nerves. J Neurosurg. 1975;42:114–21.
6. Lundborg G. Structure and function of the intraneural microvessels as related to trauma, edema formation and nerve function. J Bone Joint Surg. 1975;57:938–48.
7. Bush MS, Reid AR, Allt G. Blood-nerve barrier: ultrastructural and endothelial surface charge alterations following nerve crush. Neuropathol Appl Neurobiol. 1993;19:31–40.

Ultrastructure of the Perineurium

4

Miguel Angel Reina, Emilse Colman Peyrano, Jorge Diamantopoulos, and José Antonio De Andrés

The perineurium is composed of multiple concentric single-cell layers enclosing individual nerve fascicles. Each layer has a thickness equivalent to the width of a perineurial cell. Groups of these cells join by means of tight junctions and desmosomes to form layers that function as a barrier against diffusion of particles across them. Perineurial internal layers have more of these specialized unions among perineurial cells, which are proximal to nerve fascicles [1–4].

Perineurial cells have basal laminas constituted by glycoproteins such as fibronectin, as well as heparin sulfate and laminin, and their nuclei are flattened. The cytoplasm is granular and contains few mitochondria and pinocytic vesicles. Measurements of the thickness of the perineurium are related to the number of layers contributing to its shape, which ranges from 8 to 16 perineurial layers. Extracellular matrix, together with collagen fibers and a few fibroblasts, occupies the spaces between perineurial laminas [1–6].

The membrane allows the axons a certain degree of mobility within the nerve fascicles, and it becomes progressively thinner as the number of fascicles increases. Tension exerted on the perineurium is transmitted successively toward the endoneurium, affecting the intracellular pressure in the axon. The perineurium, together with endothelial cells of intrafascicular capillaries, contributes to the blood–nerve barrier (Figs. 4.1, 4.2, 4.3, 4.4, 4.5, 4.6, 4.7, 4.8, 4.9, 4.10, 4.11, 4.12, 4.13, 4.14, 4.15, 4.16, 4.17, 4.18, 4.19, 4.20, 4.21, 4.22, 4.23, 4.24, 4.25, 4.26, 4.27, 4.28, 4.29, 4.30, 4.31, and 4.32) [7–10].

M.A. Reina, MD, PhD (✉)
Department of Clinical Medical Sciences and Institute of Applied Molecular Medicine, School of Medicine, University of CEU San Pablo, Madrid, Spain

Department of Anesthesiology, Madrid-Montepríncipe University Hospital, Madrid, Spain

E. Colman Peyrano, BS
Research Unit, Anesthesia Reina Colman SLP, Madrid, Spain
e-mail: emilsecolpey@yahoo.es

J. Diamantopoulos, MD
Neurosurgical Department, Madrid-Montepríncipe University Hospital, Madrid-Sanchinarro University Hospital, Madrid, Spain
e-mail: jorgediaman@yahoo.com

J.A. Andrés, MD, PhD
Department of Surgical Specialties, School of Medicine, University of Valencia, Valencia, Spain

Department of Anesthesia, Critical Care, and Pain Management, General University Hospital, Valencia, Spain
e-mail: deandres_jos@gva.es

M.A. Reina et al. (eds.), *Atlas of Functional Anatomy for Regional Anesthesia and Pain Medicine: Human Structure, Ultrastructure and 3D Reconstruction Images*, DOI 10.1007/978-3-319-09522-6_4,

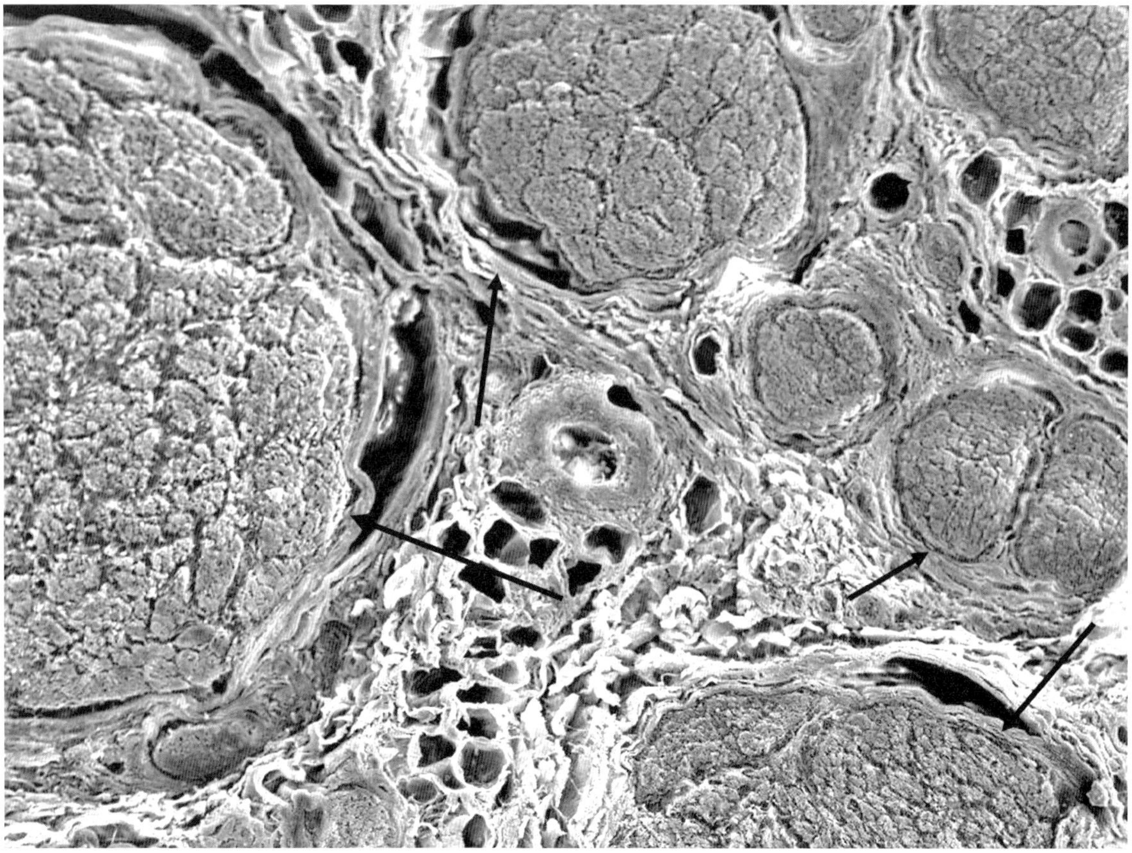

Fig. 4.1 Perineurium. A few nerve fascicles are enclosed in perineurium from a human sciatic nerve. *Arrows* indicate the perineural layer. Scanning electron microscopy, magnification: ×70 (From Reina [3]; with permission)

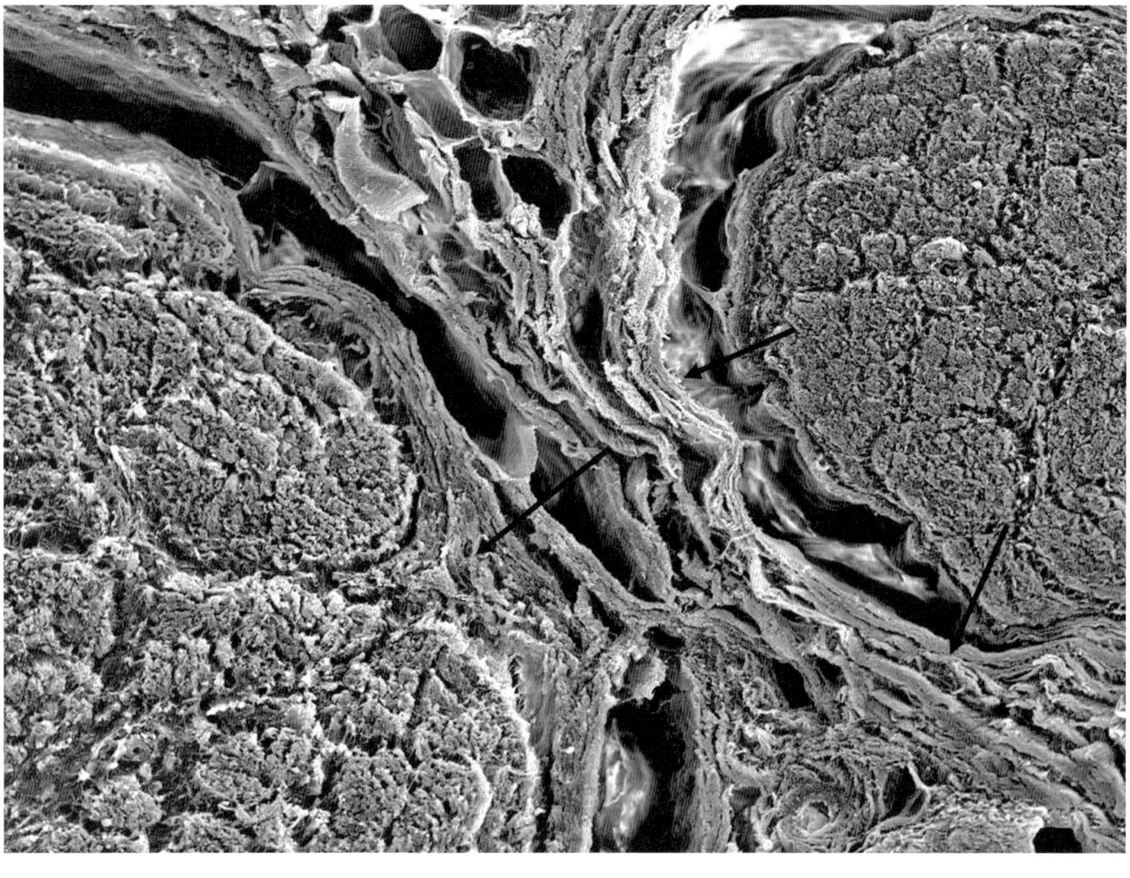

Fig. 4.2 Perineurium. Two fascicles are surrounded by perineurium from a human sciatic nerve. Tridimensional view shows details of the perineural layers (*arrows*). Scanning electron microscopy, magnification: ×150 (From Reina et al. [1]; with permission)

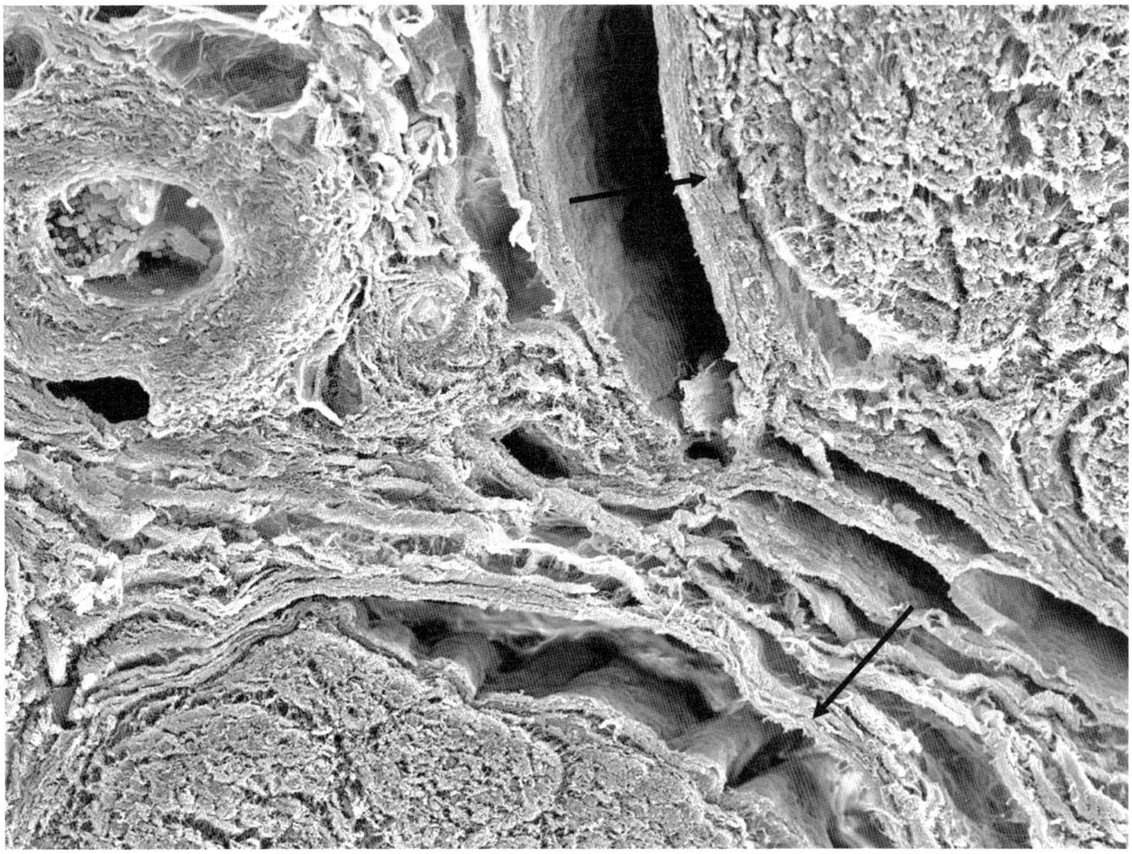

Fig. 4.3 Perineurium. Two fascicles are surrounded by perineurium from a human sciatic nerve. Tridimensional view shows details of the perineural layers (*arrows*). Scanning electron microscopy, magnification: ×200 (From Reina [3]; with permission)

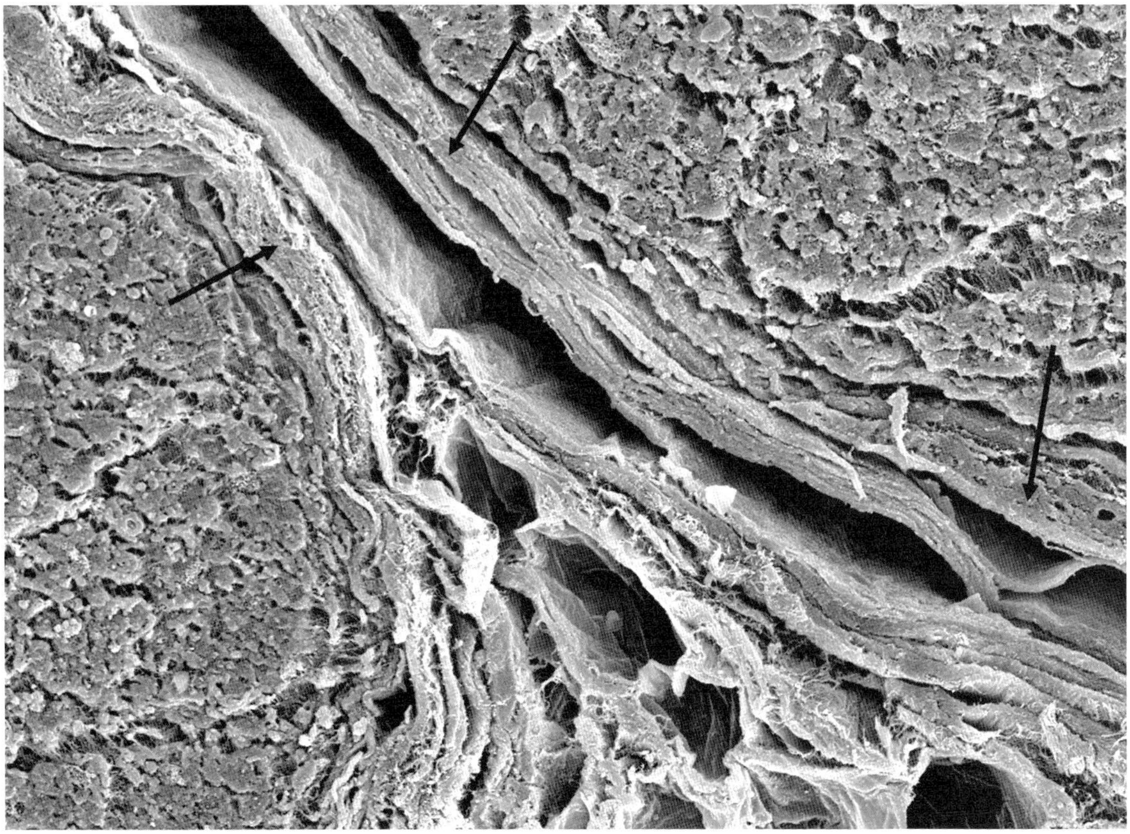

Fig. 4.4 Perineurium. Two fascicles are surrounded by perineurium from a human sciatic nerve. Tridimensional view shows details of the perineural layers (*arrows*). Scanning electron microscopy, magnification: ×300

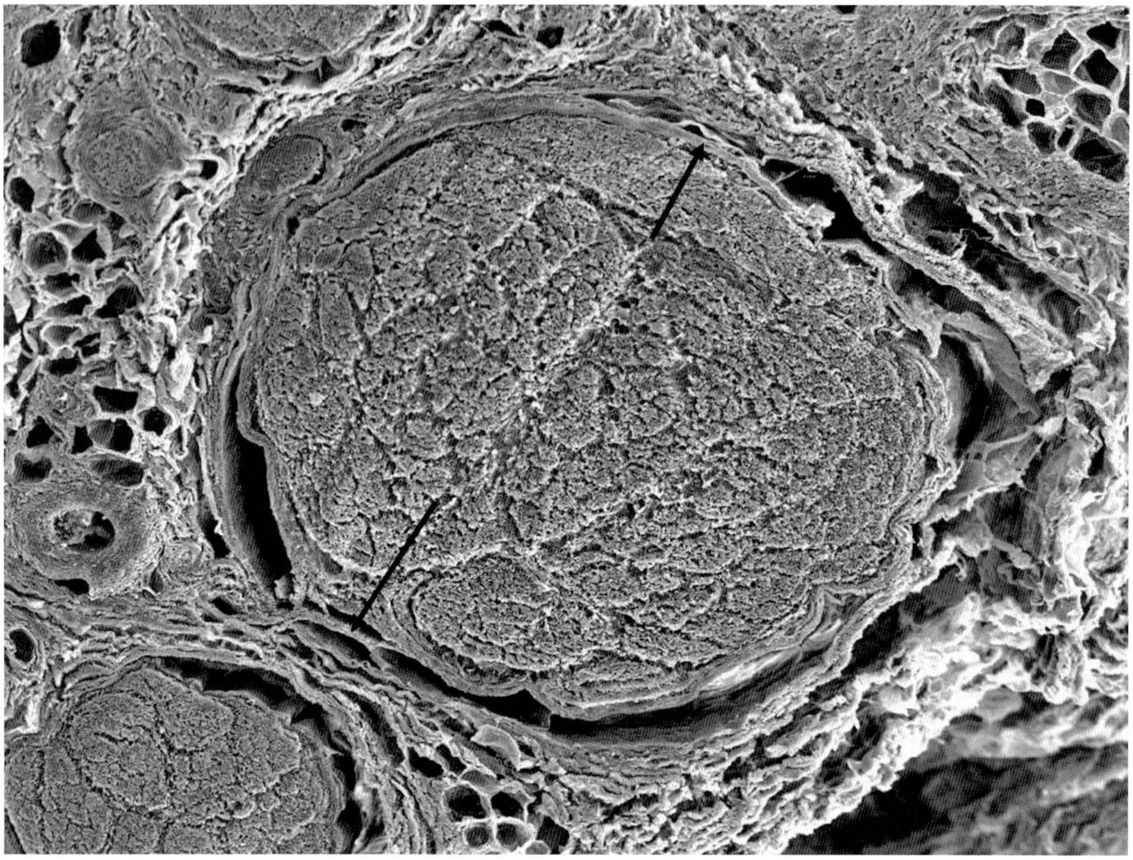

Fig. 4.5 Perineurium. Fascicles are surrounded by perineurium from a human sciatic nerve. Tridimensional view shows details of the perineural layers (*arrows*). Scanning electron microscopy, magnification: ×70

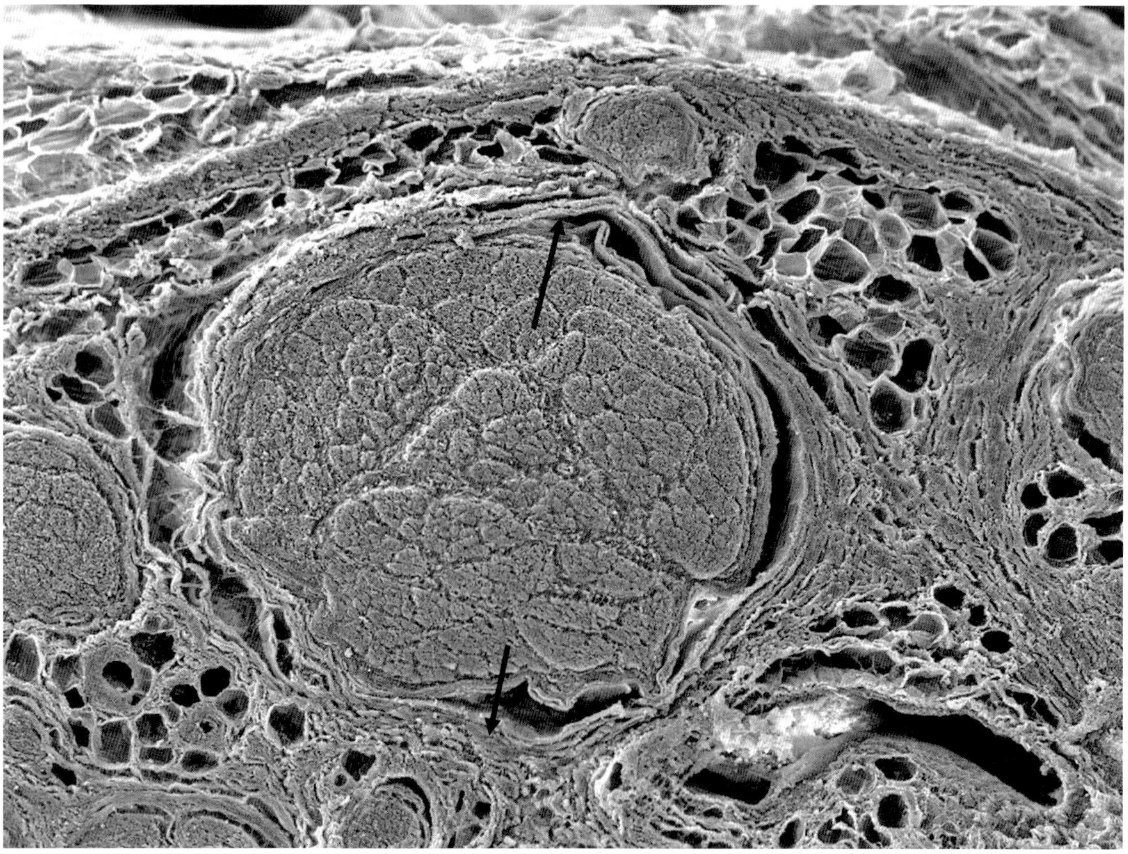

Fig. 4.6 Perineurium. Fascicles are surrounded by perineurium from a human sciatic nerve. *Arrows* indicate a perineural layer. Scanning electron microscopy, magnification: ×70

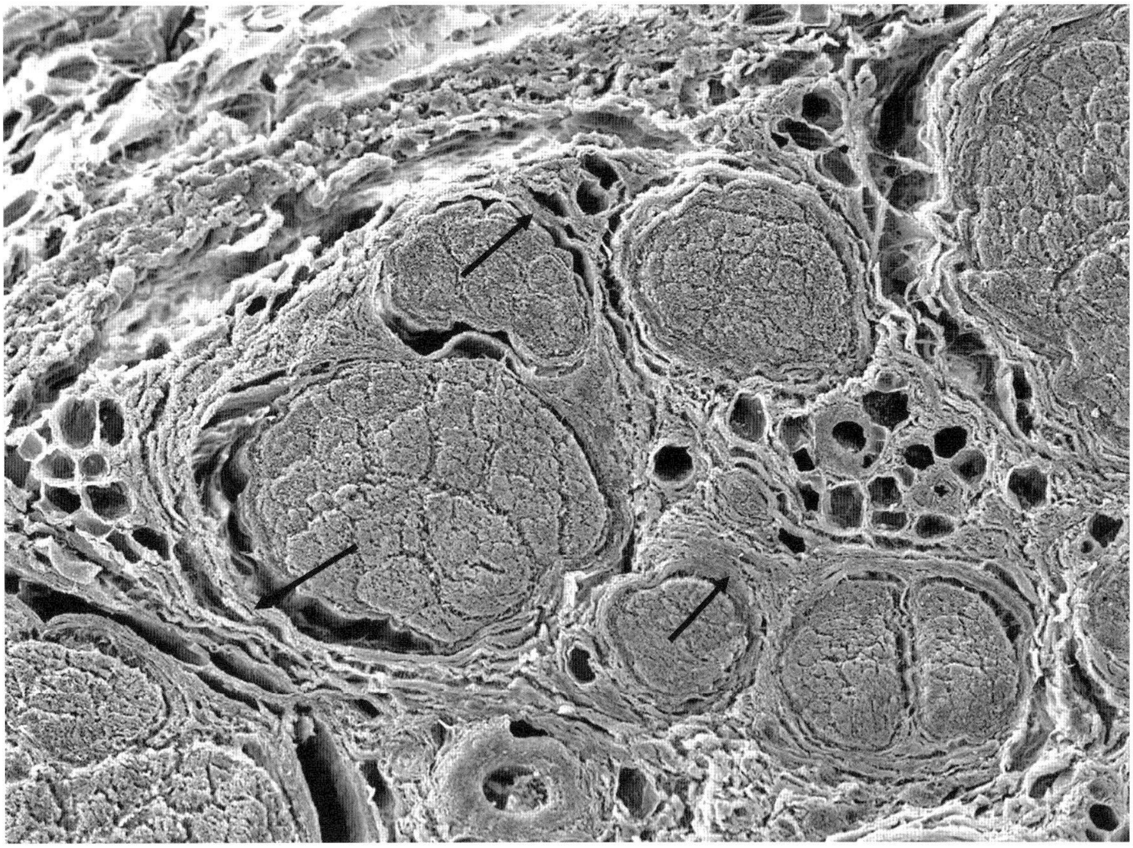

Fig. 4.7 Perineurium. Fascicles are surrounded by perineurium from a human sciatic nerve. *Arrows* point toward a perineural layer. Scanning electron microscopy, magnification: ×75 (From Reina et al. [11]; with permission)

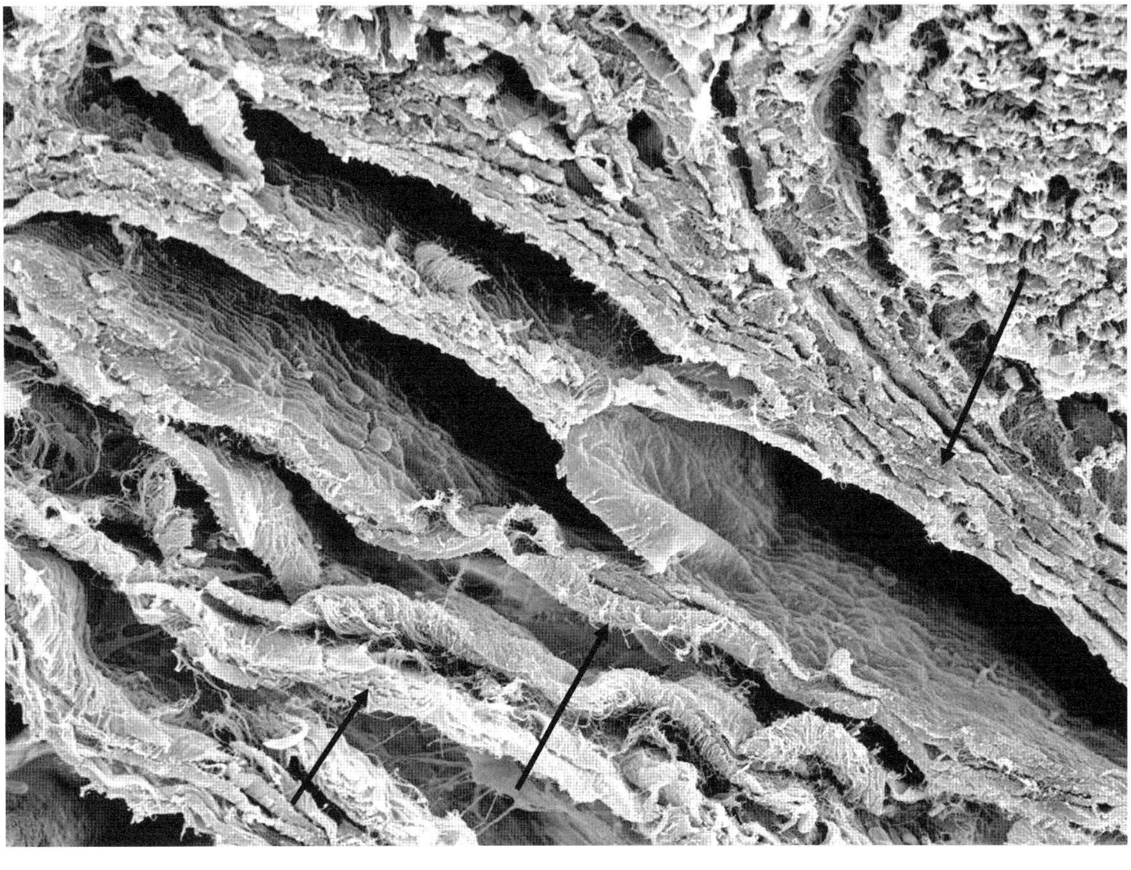

Fig. 4.8 Perineurium. Perineural layers from a human sciatic nerve. Shown is a tridimensional view of the perineural layers (*arrows*). Scanning electron microscopy, magnification: ×500 (From Reina et al. [1]; with permission)

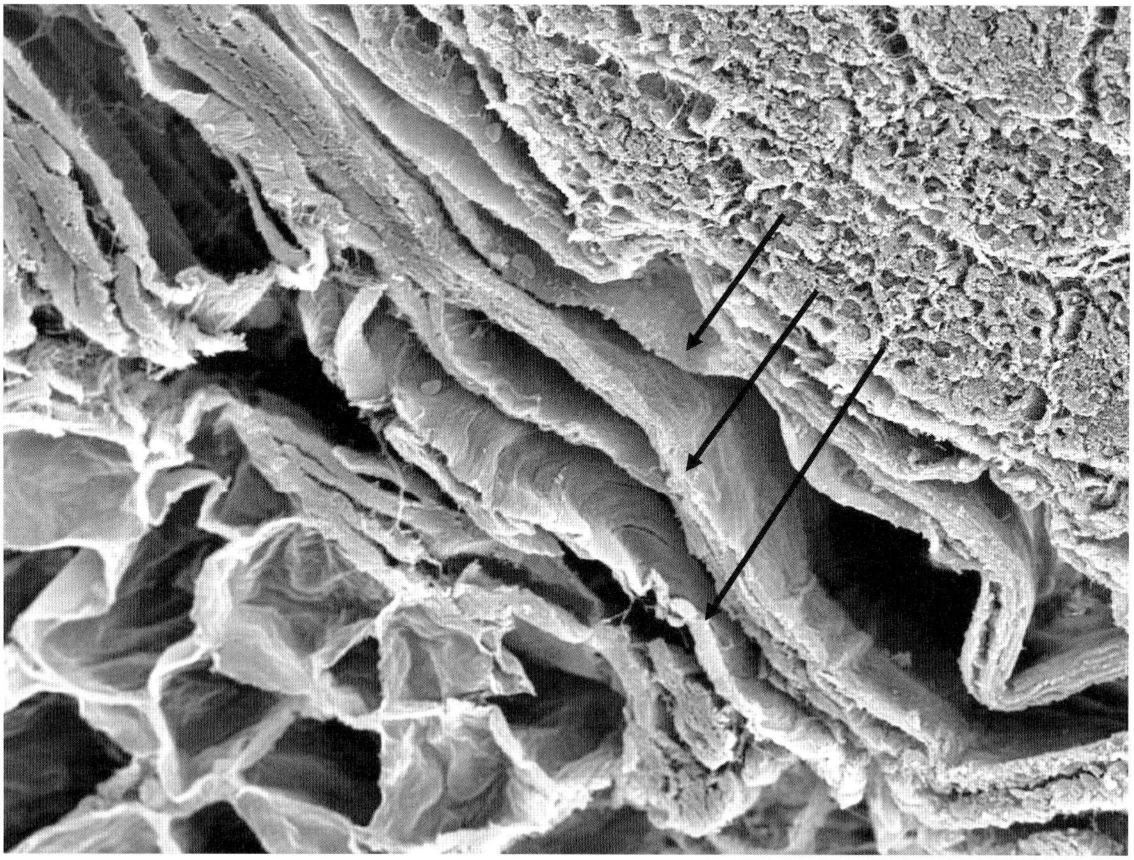

Fig. 4.9 Perineurium. Perineural layers from a human sciatic nerve. Shown is a tridimensional view of the perineural layers (*arrows*). Scanning electron microscopy, magnification: ×350

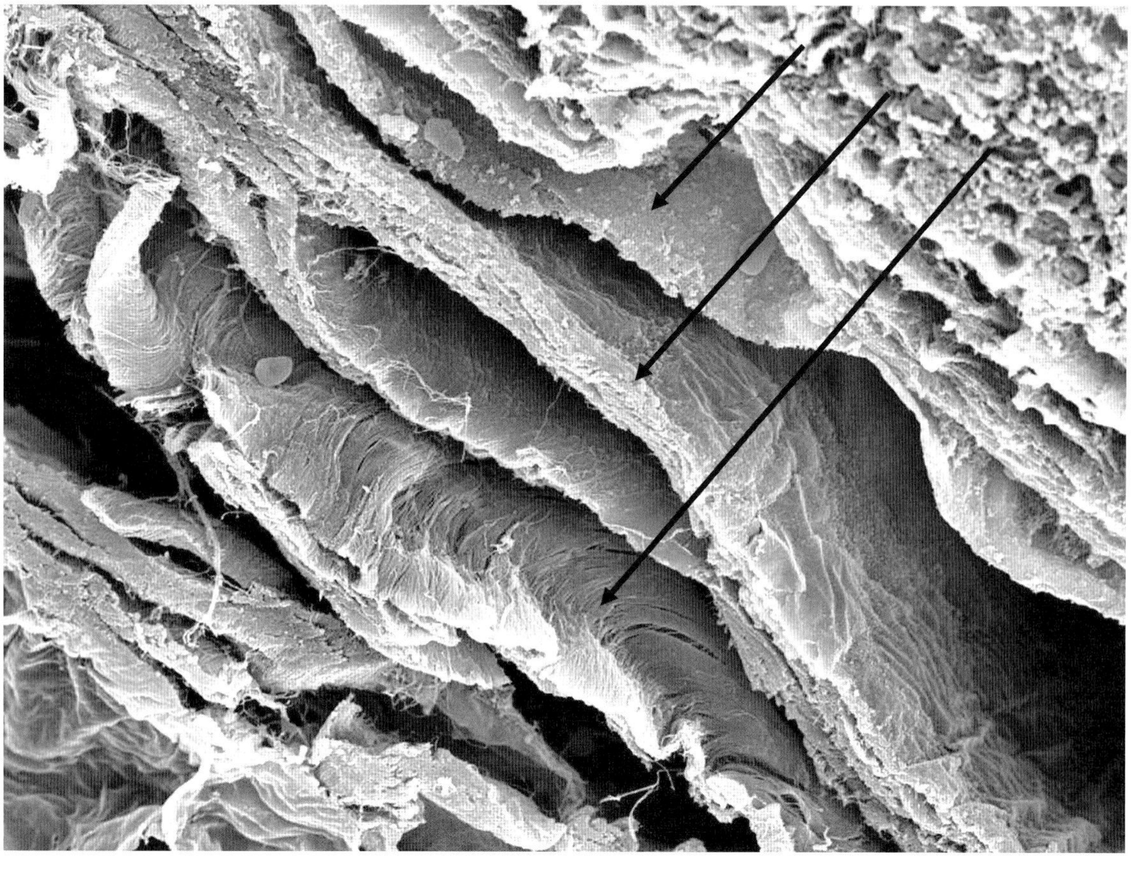

Fig. 4.10 Perineurium. Perineural layers from a human sciatic nerve. Shown is a tridimensional view of the perineural layers (*arrows*). Scanning electron microscopy, magnification: ×650

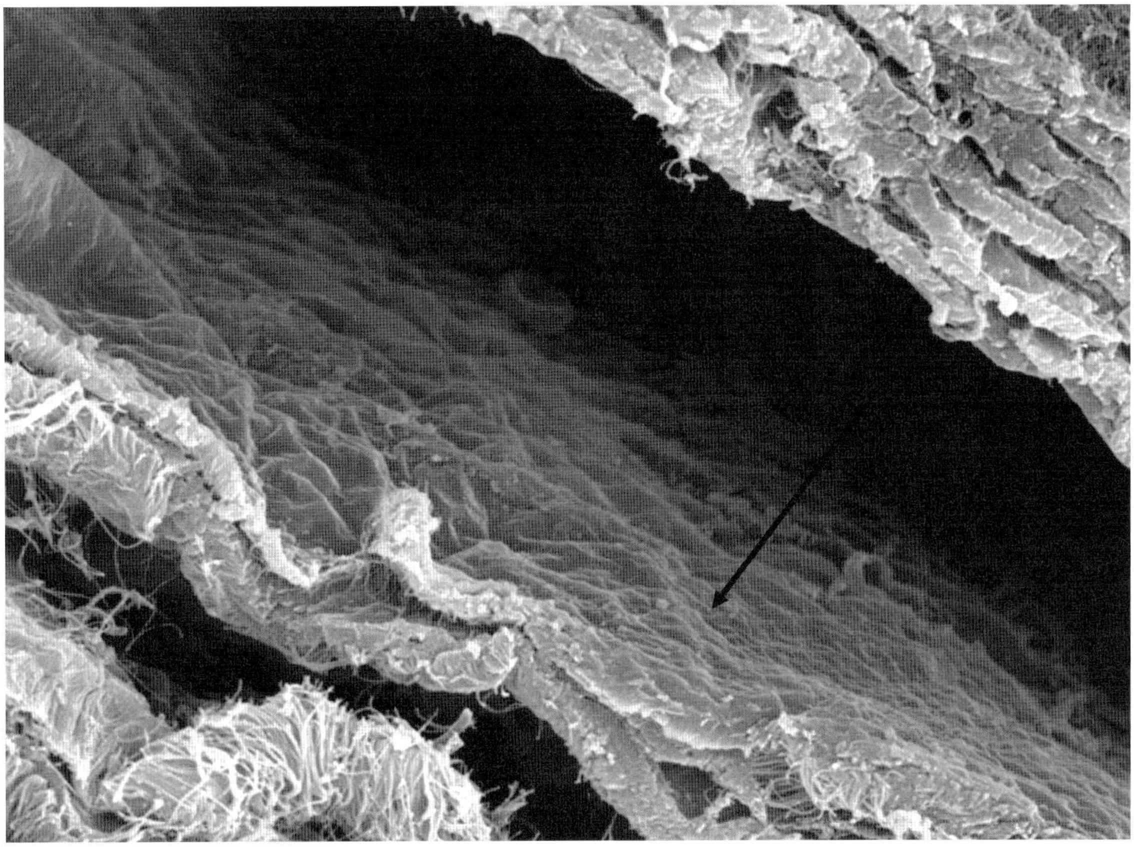

Fig. 4.11 Perineurium. Perineural layers from a human sciatic nerve. Shown is a tridimensional view of the perineural layers (*arrows*). Scanning electron microscopy, magnification: ×1,600

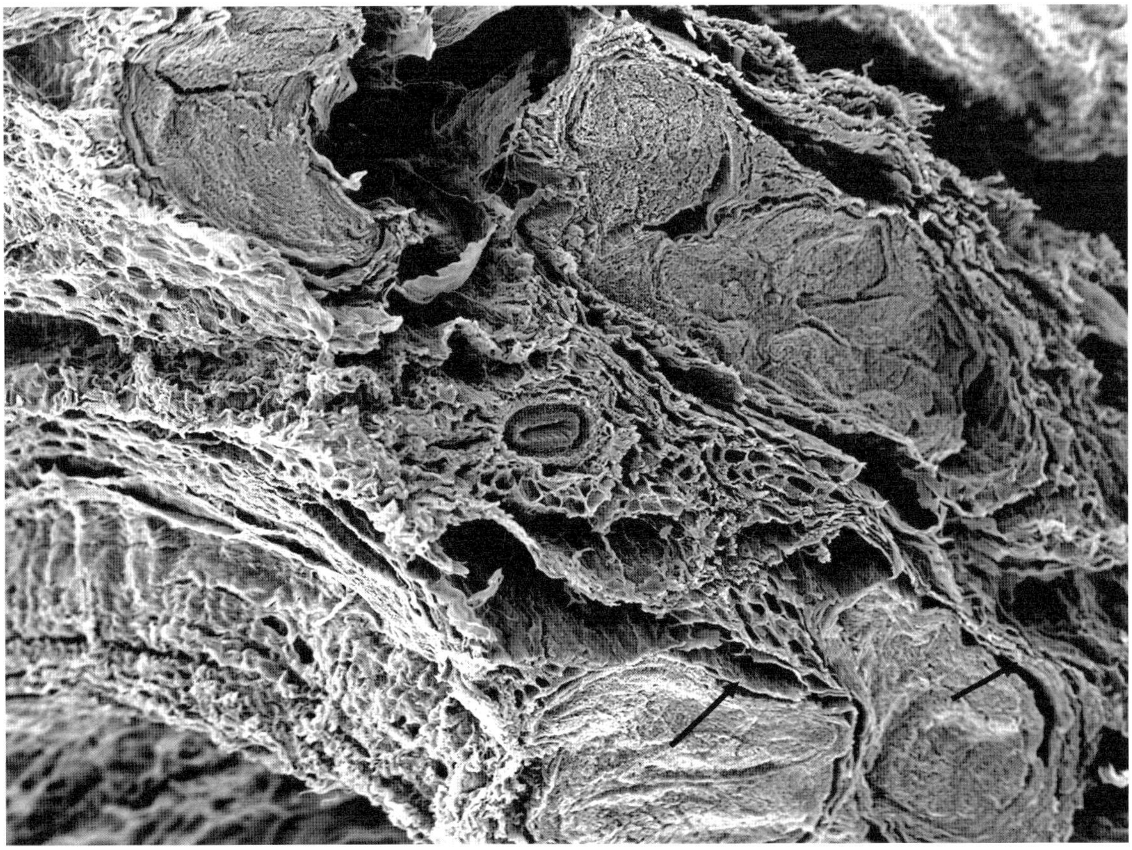

Fig. 4.12 Perineurium. Fascicles are surrounded by perineural layers of human sciatic nerve. Shown is a tridimensional view of perineural layers (*arrows*) and interfascicular tissue. Scanning electron microscopy, magnification: ×50

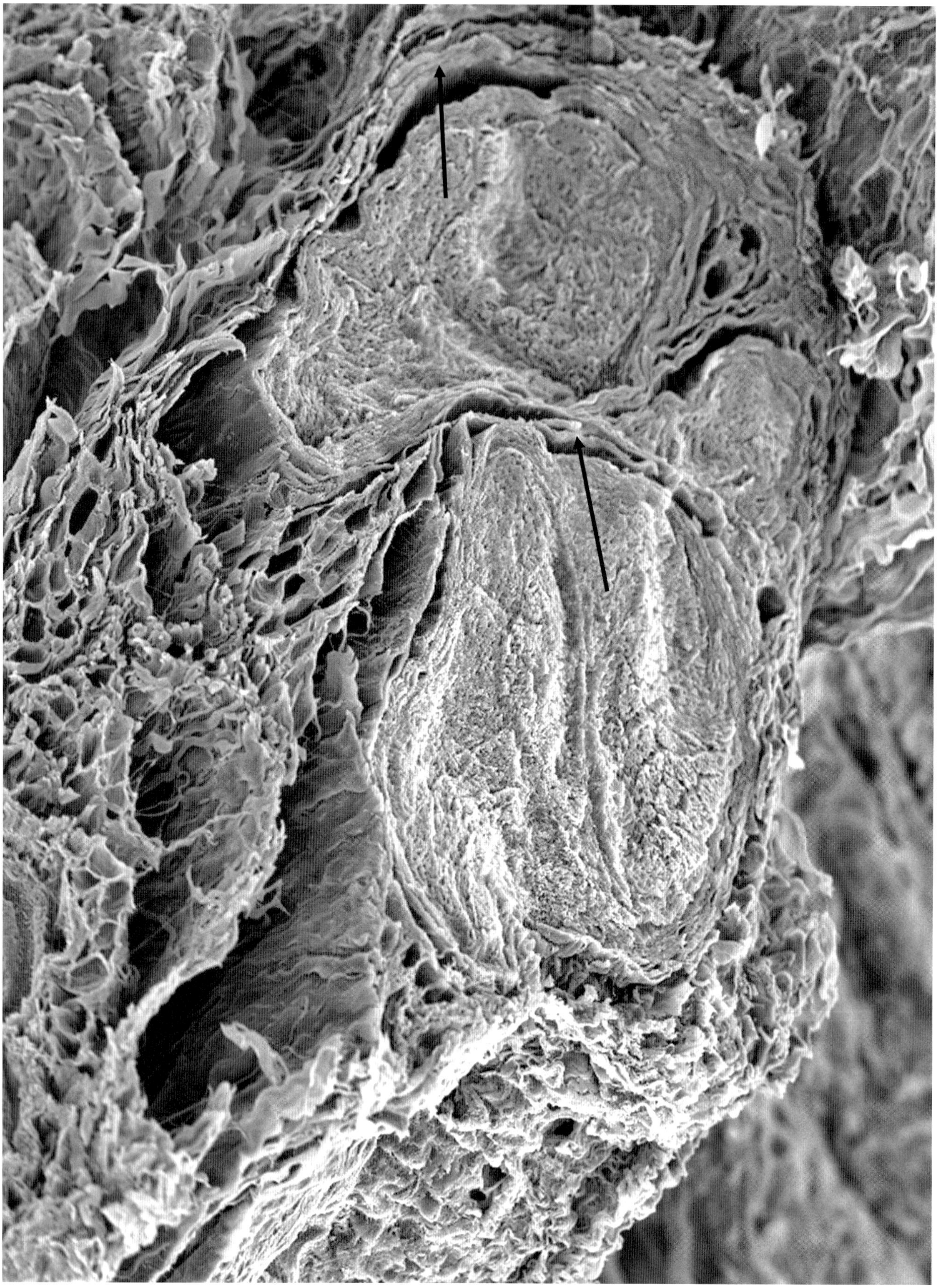

Fig. 4.13 Perineurium. Fascicles are surrounded by perineural layers of human sciatic nerve. Shown is a tridimensional view of the perineural layers (*arrows*) and interfascicular tissue. Scanning electron microscopy, magnification: ×75

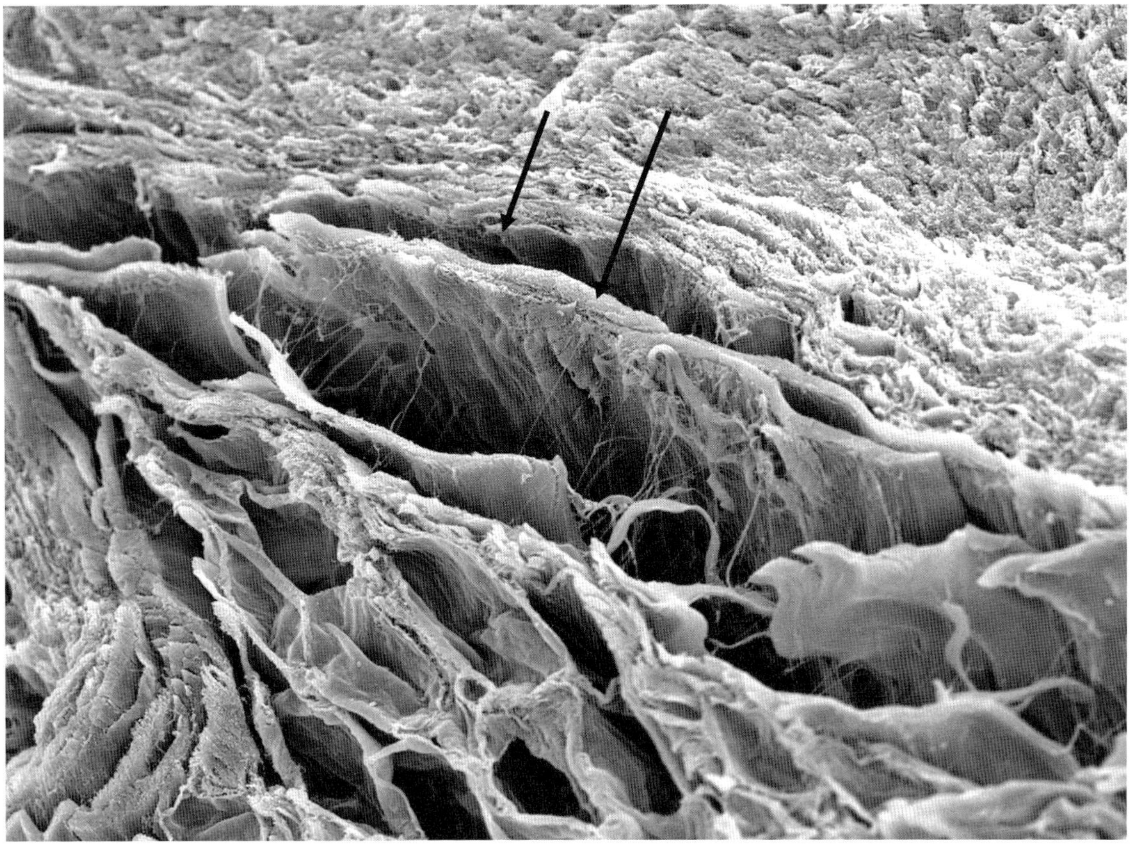

Fig. 4.14 Perineurium. Perineural layers of human sciatic nerve. Shown is a tridimensional view of the perineural layers (*arrows*). Scanning electron microscopy, magnification: ×300

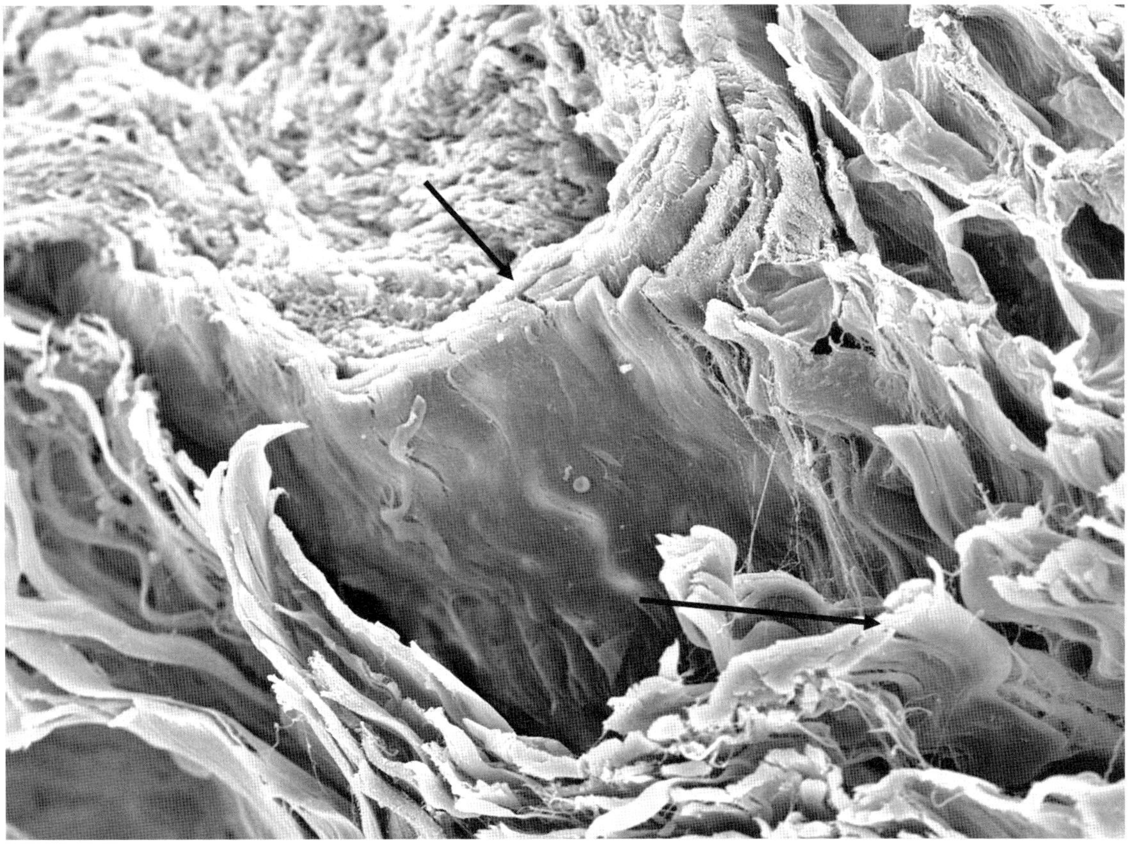

Fig. 4.15 Perineurium. Perineural layers of human sciatic nerve. Shown is a tridimensional view of the perineural layers (*arrows*). Scanning electron microscopy, magnification: ×300

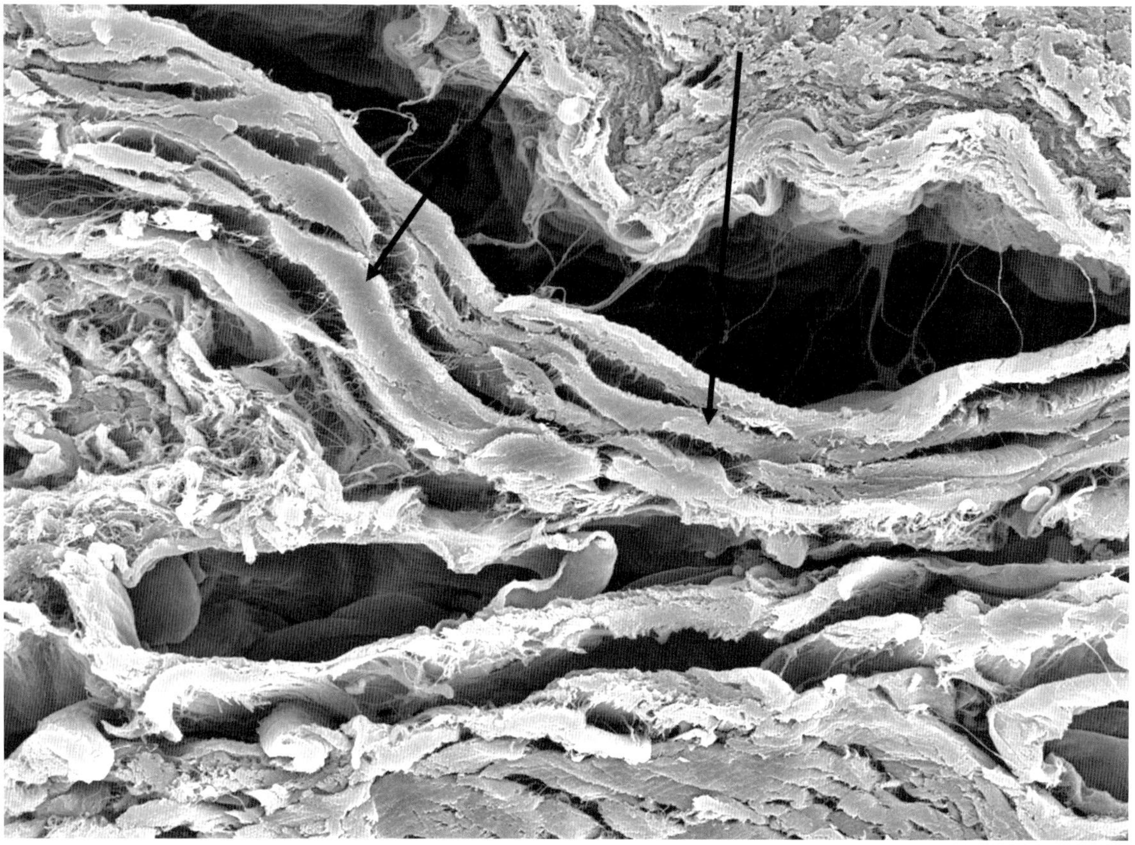

Fig. 4.16 Perineurium. Perineural layers of human sciatic nerve. Shown is a tridimensional view of the perineural layers (*arrows*). Scanning electron microscopy, magnification: ×600

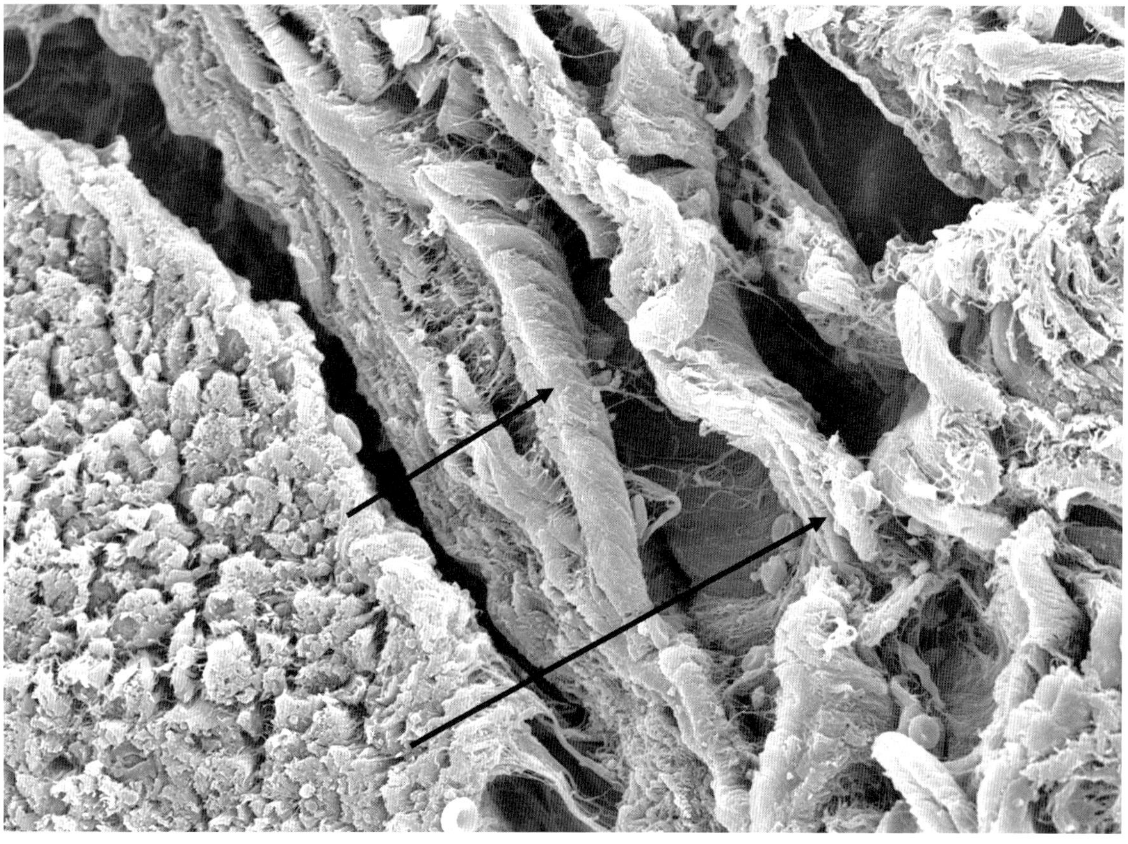

Fig. 4.17 Perineurium. Perineural layers of human sciatic nerve. Shown is a tridimensional view of the perineural layers (*arrows*). Scanning electron microscopy, magnification: ×750

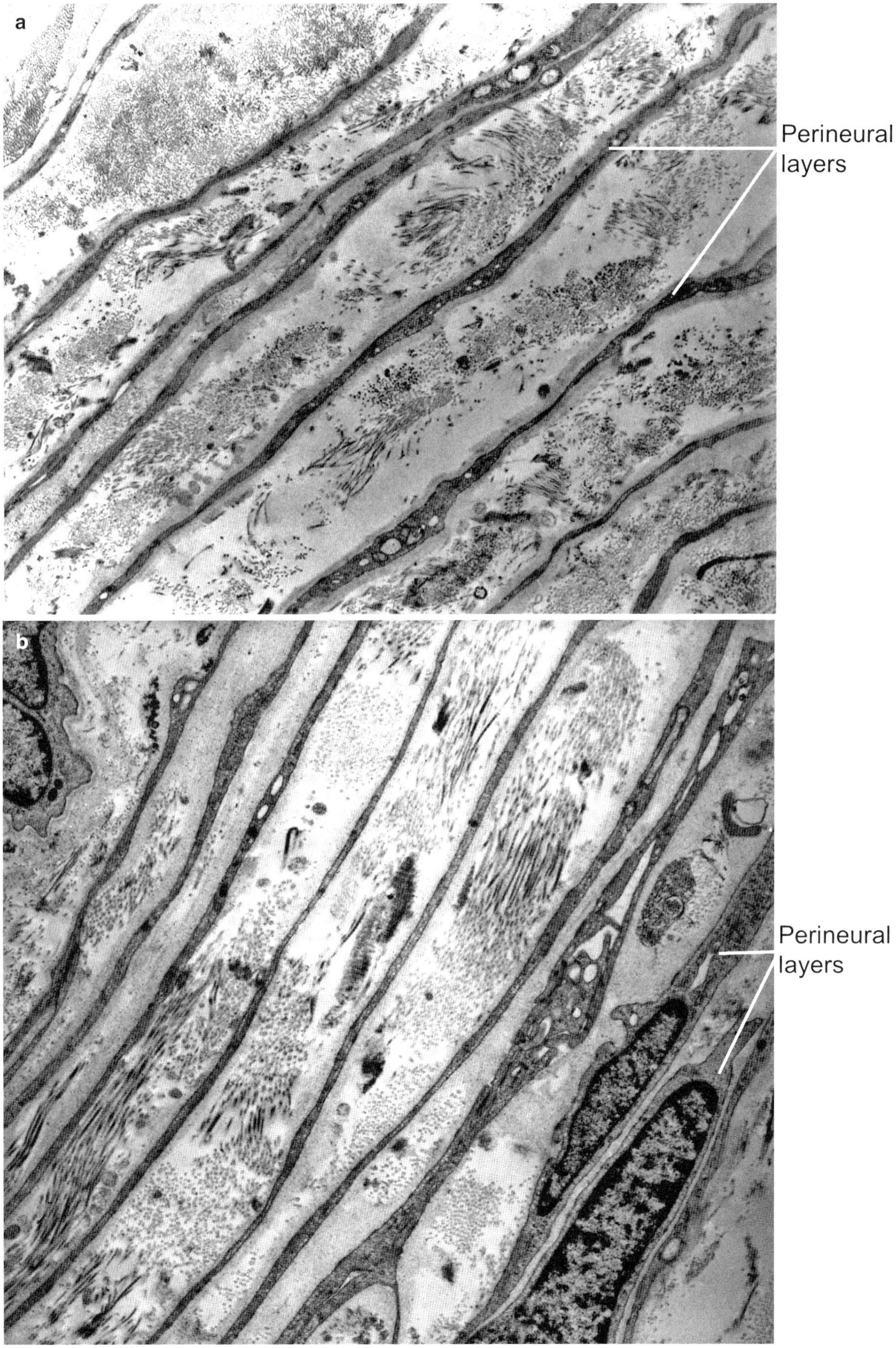

Fig. 4.18 Perineurium. Perineural layers of human sciatic nerve. Transmission electron microscopy, magnification: ×3,000 (**a**); ×7,000 (**b**) (Panel **b** from Reina et al. [9]; with permission)

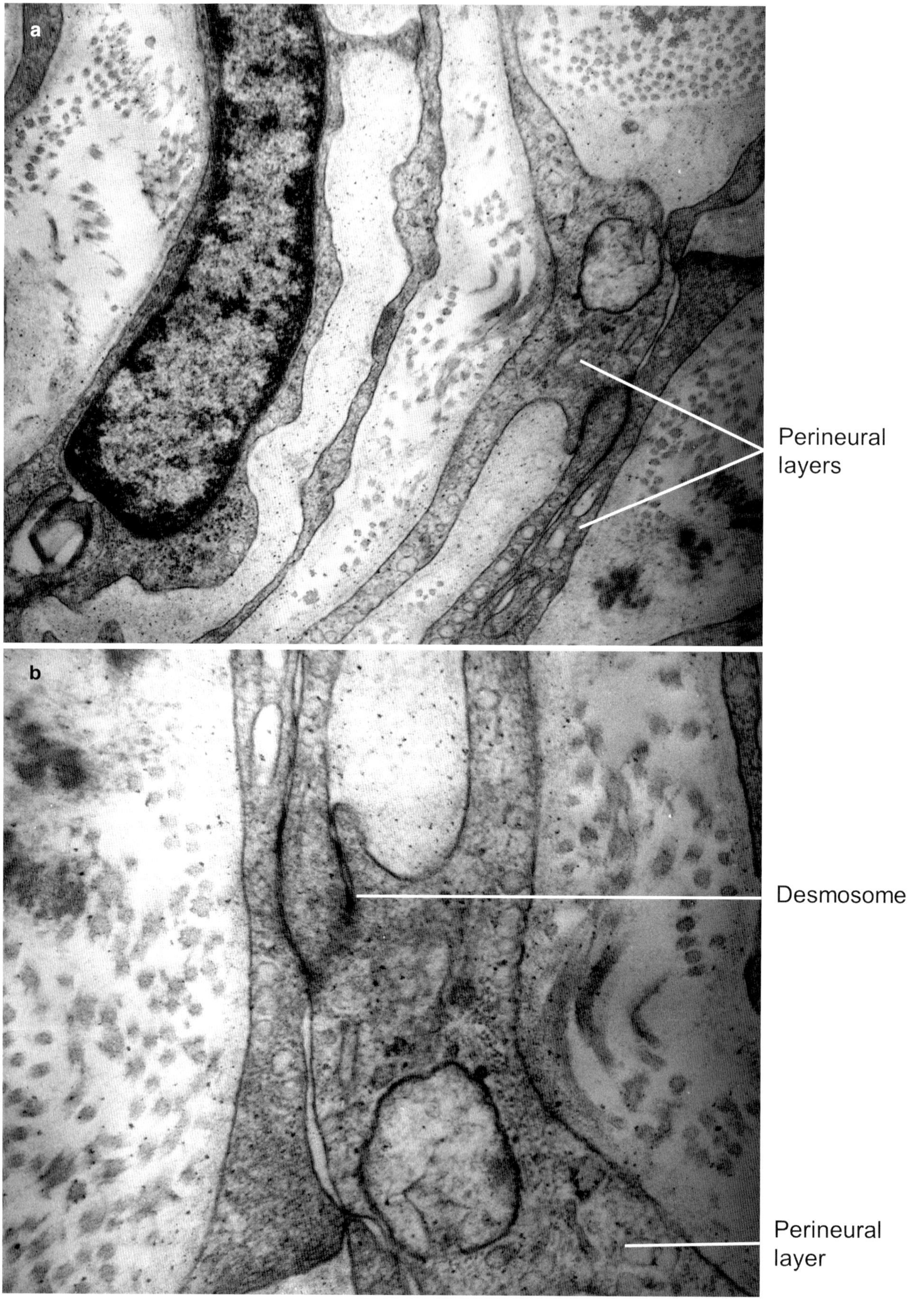

Fig. 4.19 Perineurium. (**a**, **b**) Perineural layers of human sciatic nerve. Transmission electron microscopy, magnification: ×3,000 (**a**); ×1,100 (**b**) (Panel **b** from Reina et al. [9]; with permission)

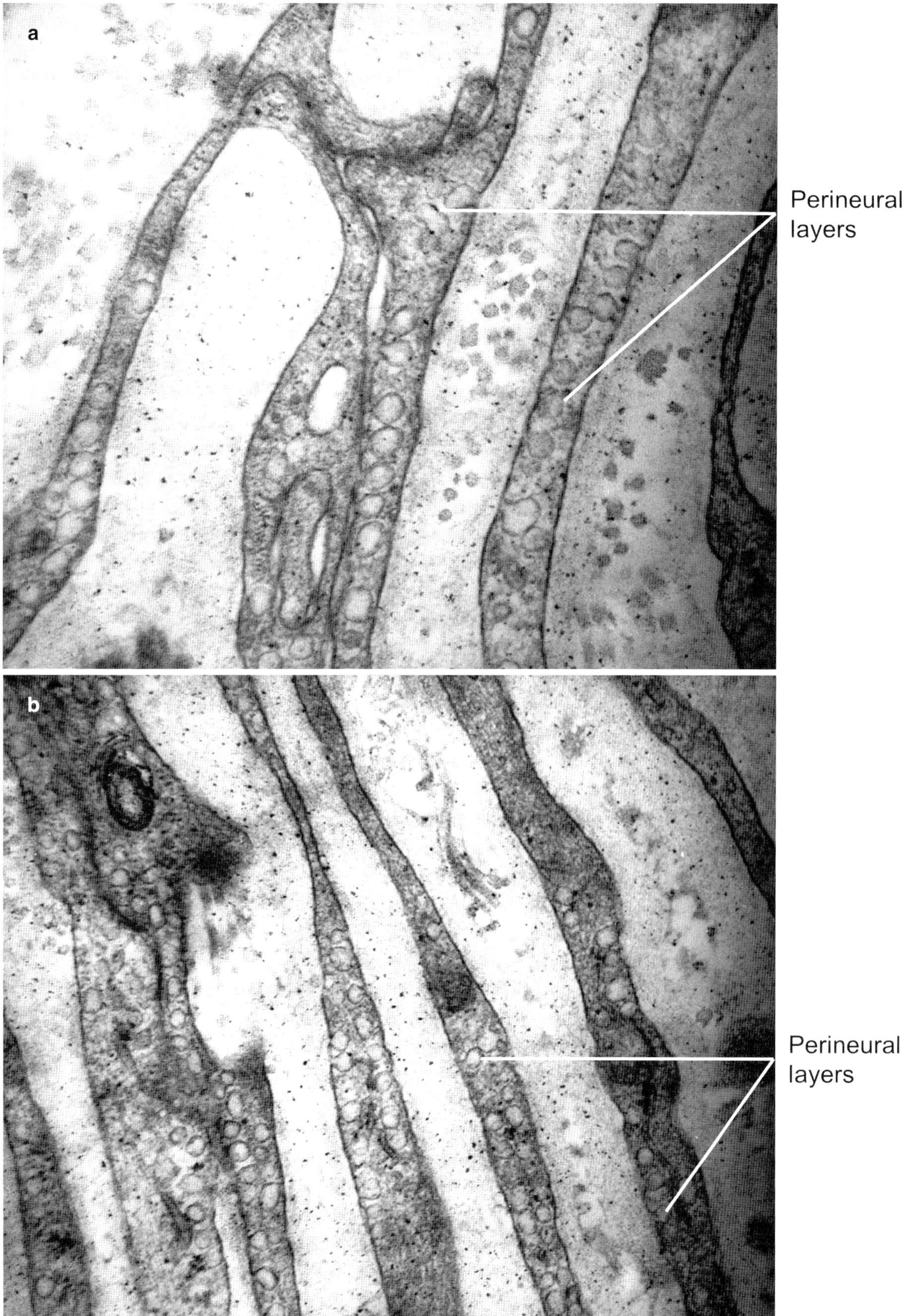

Fig. 4.20 Perineurium. (**a**, **b**) Perineural layers of human sciatic nerve. Specialized membrane junctions are shown in detail. Transmission electron microscopy, magnification: ×20,000 (**a**); ×50,000 (**b**) (Panel **a** from Reina et al. [1]; with permission. Panel **b** from Reina et al. [9]; with permission)

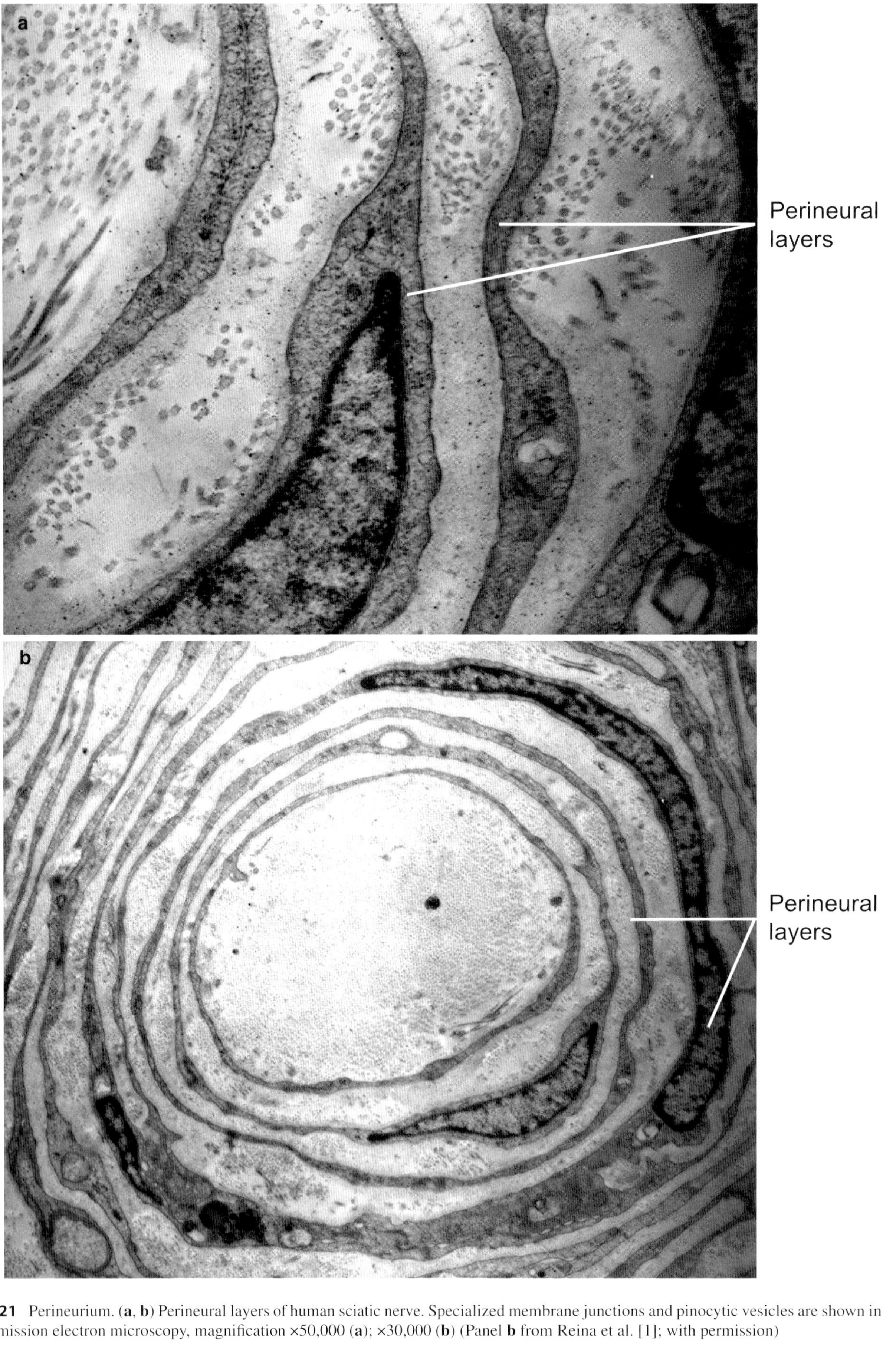

Fig 4.21 Perineurium. (**a**, **b**) Perineural layers of human sciatic nerve. Specialized membrane junctions and pinocytic vesicles are shown in detail. Transmission electron microscopy, magnification ×50,000 (**a**); ×30,000 (**b**) (Panel **b** from Reina et al. [1]; with permission)

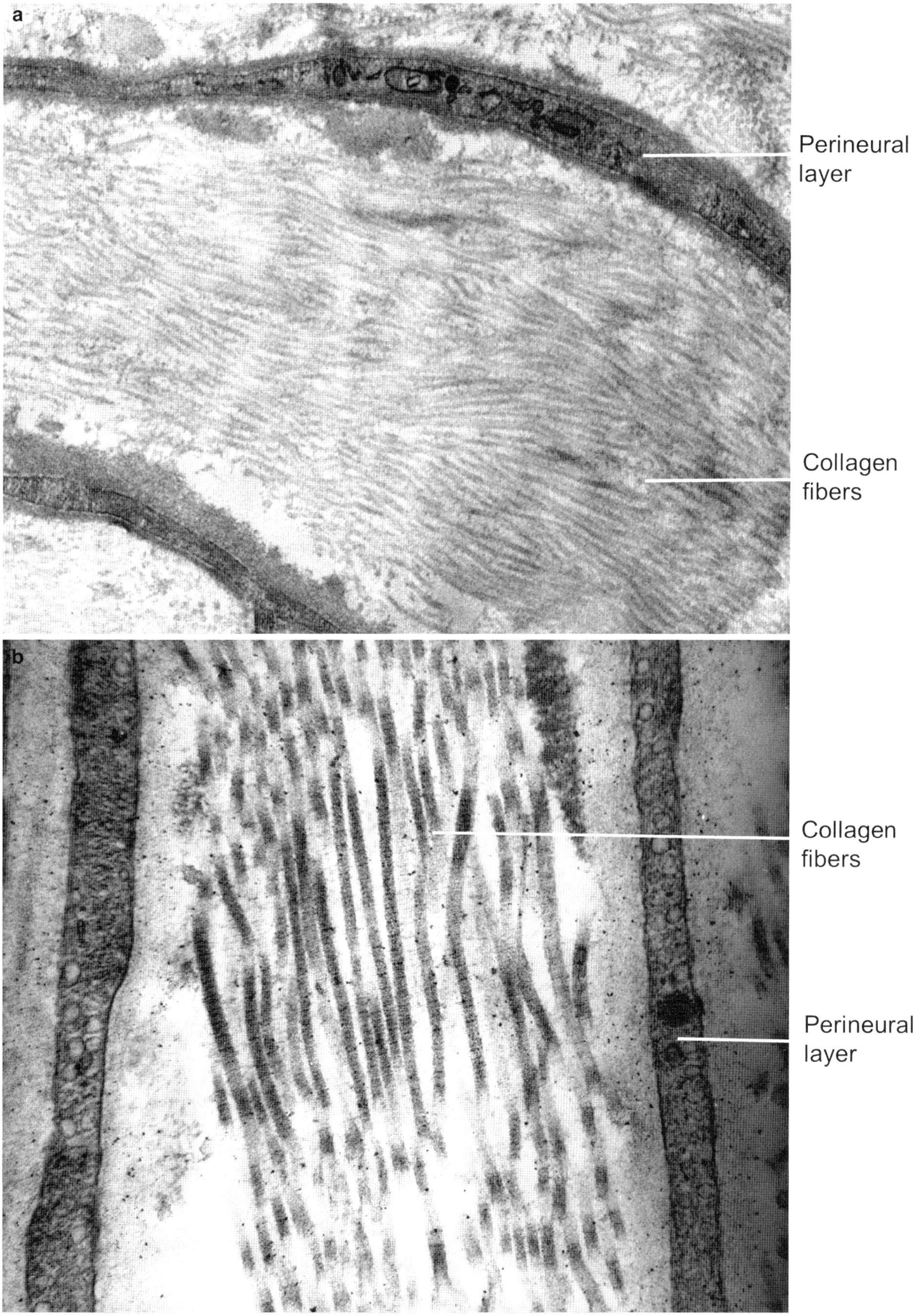

Fig. 4.22 Perineurium. (**a**, **b**) Perineural layers of human sciatic nerve. Transmission electron microscopy, magnification: ×20,000 (**a**); ×7,000 (**b**)

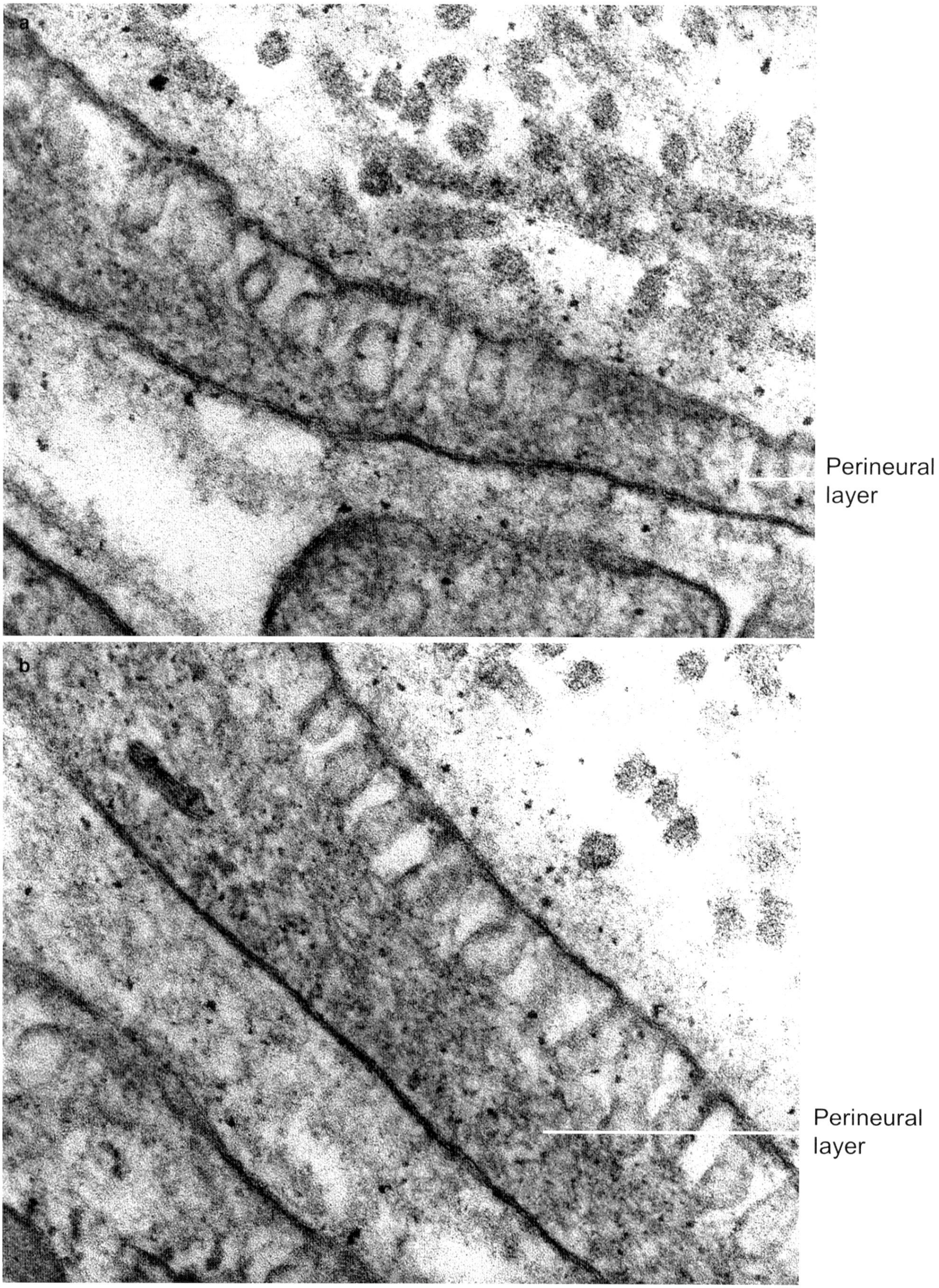

Fig. 4.23 Perineurium. (**a**, **b**) Perineural layers of human sciatic nerve. Collagen fibers between perineural layers are shown in detail. Transmission electron microscopy, magnification: ×25,000 (**a**); ×30,000 (**b**)

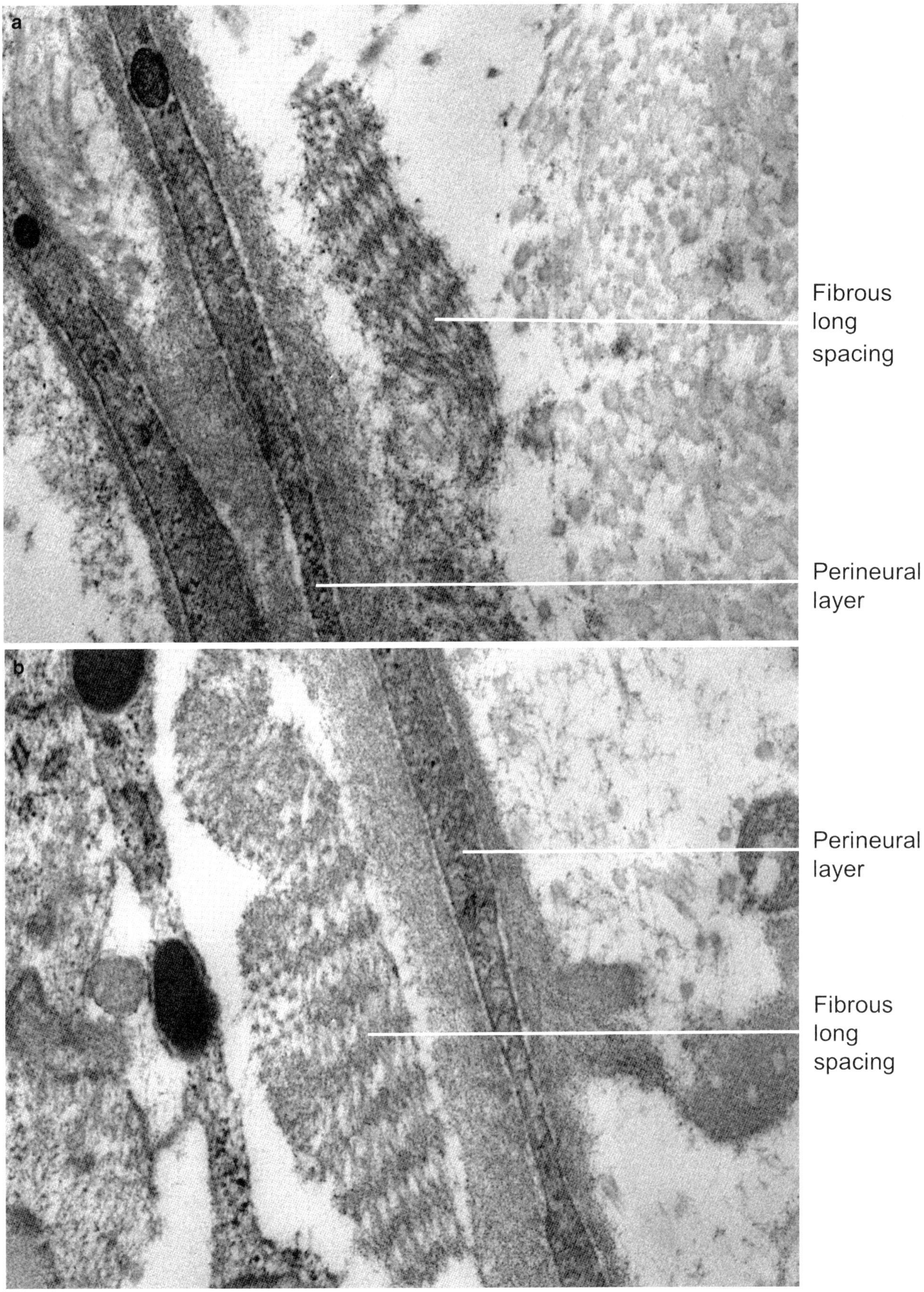

Fig. 4.24 Perineurium. (**a**, **b**) Perineural layers of human sciatic nerve. The pinocytic vesicles are shown in detail. Transmission electron microscopy, magnification: ×80,000 (**a**); ×80,000 (**b**) (Panel **b** from Reina et al. [2]; with permission)

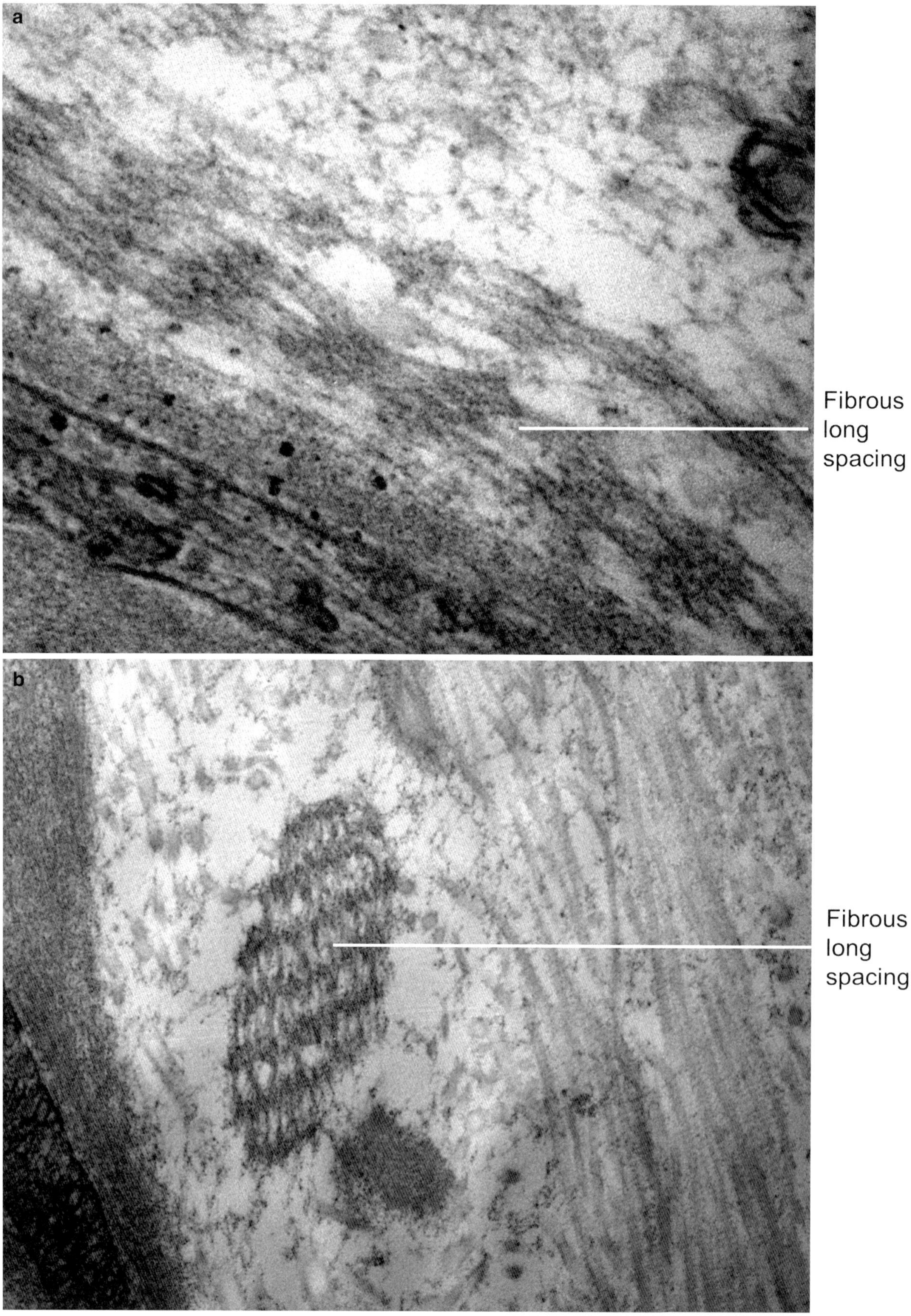

Fig. 4.25 Perineurium. Fibrous long spacing (FLS) collagen is seen among perineural layers of human sciatic nerve. Transmission electron microscopy, magnification: ×60,000 (**a**, **b**)

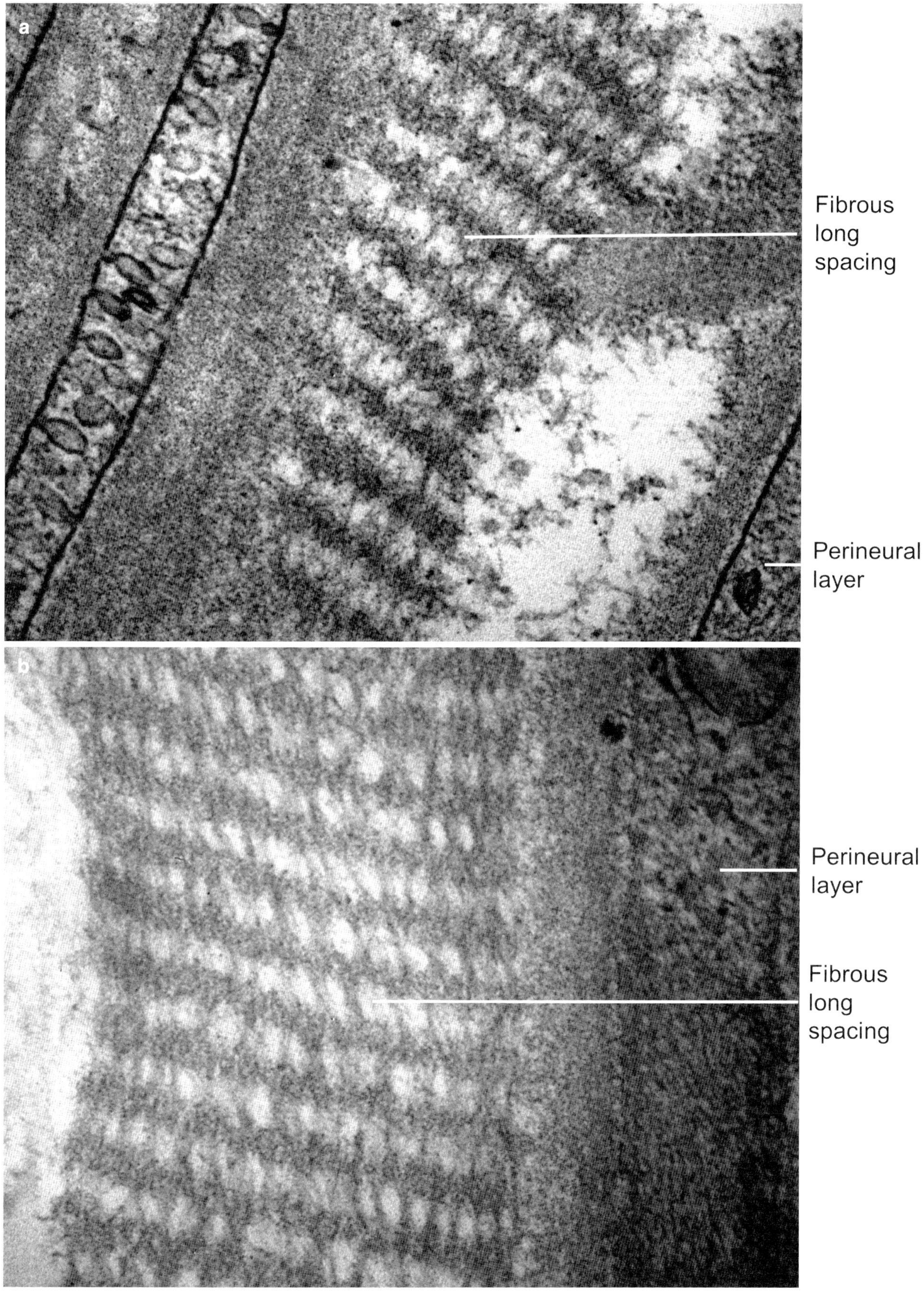

Fig. 4.26 Perineurium. FLS collagen is seen among perineural layers of human sciatic nerve. Transmission electron microscopy, magnification: ×150,000 (**a**); ×60,000 (**b**)

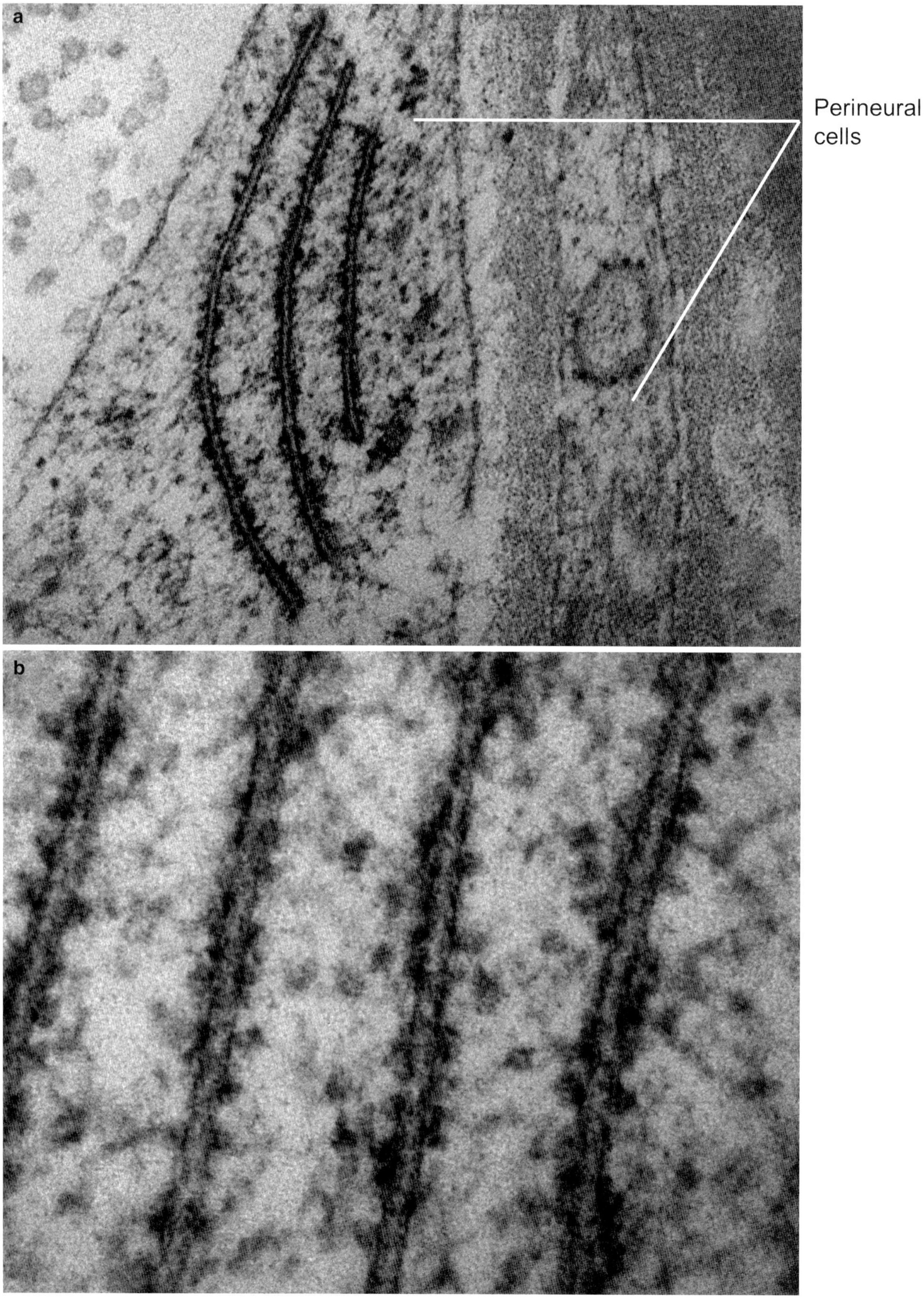

Fig. 4.27 Perineurium. FLS collagen is seen among perineural layers of human sciatic nerve. Transmission electron microscopy, magnification: ×100,000 (**a**); ×120,000 (**b**) (Panel **a** from Reina et al. [1]; with permission)

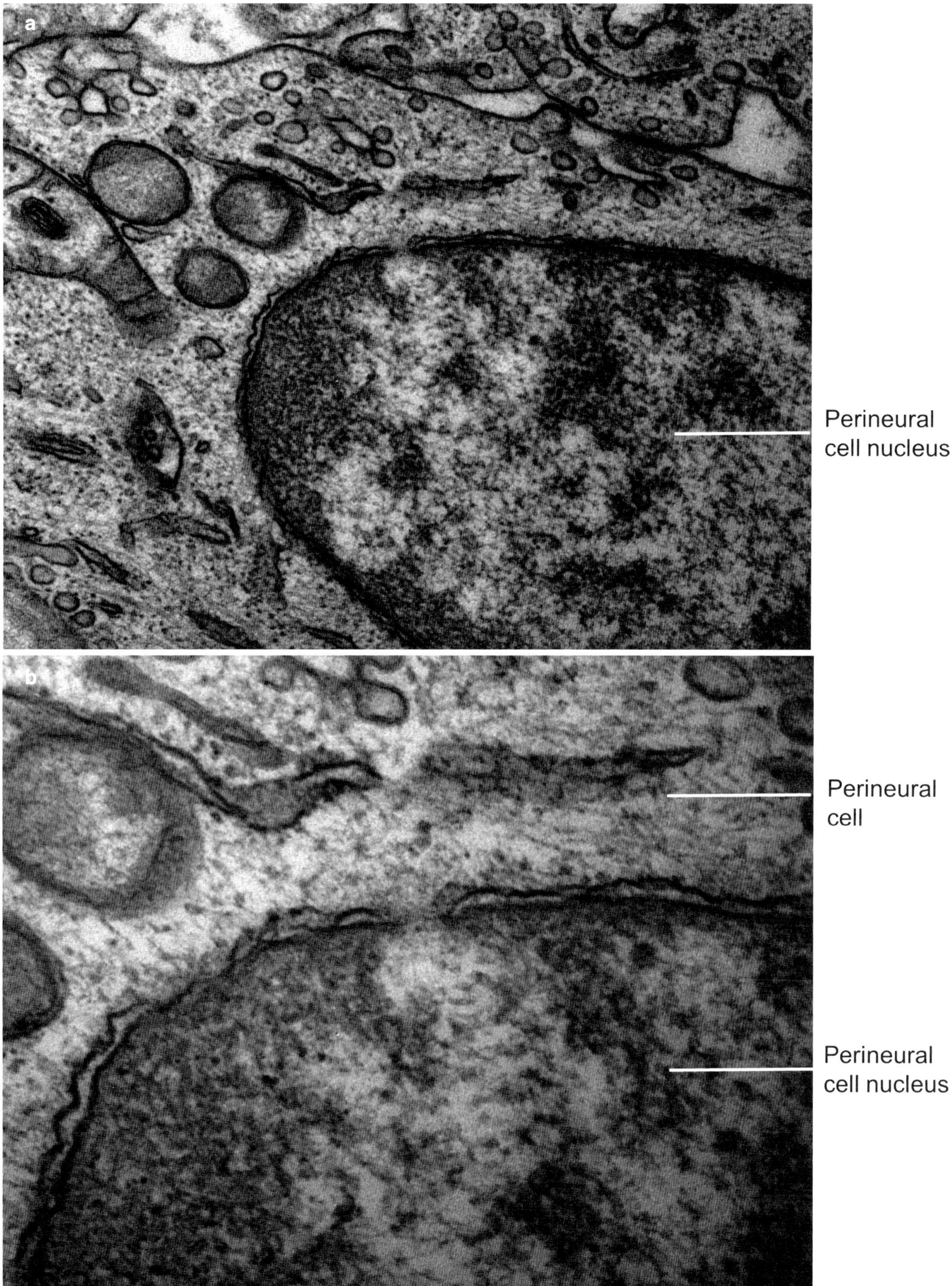

Fig. 4.28 Perineural cell. The rough endoplasmic reticulum is shown in detail. Transmission electron microscopy, magnification: ×100,000 (**a**); ×300,000 (**b**)

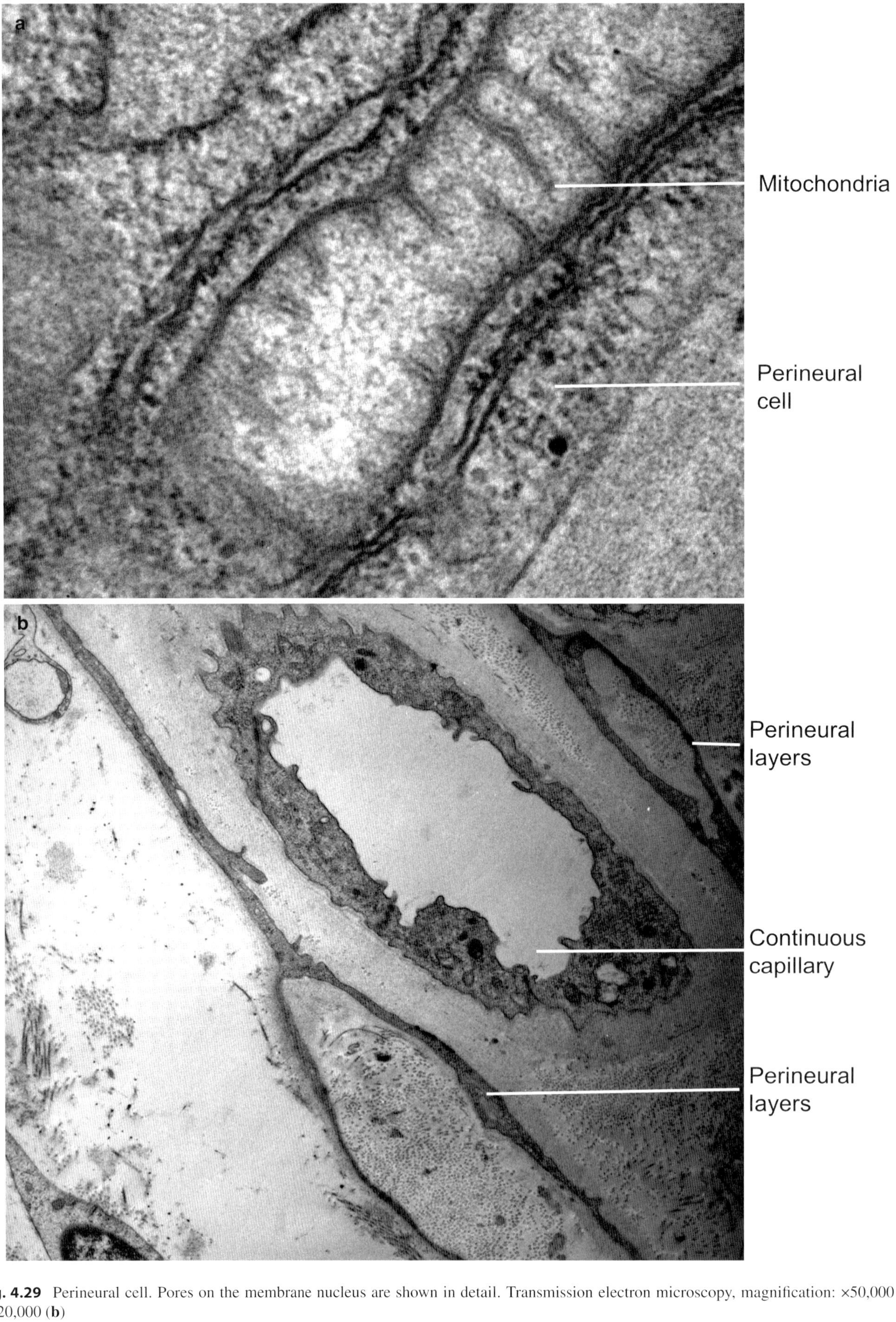

Fig. 4.29 Perineural cell. Pores on the membrane nucleus are shown in detail. Transmission electron microscopy, magnification: ×50,000 (**a**); ×120,000 (**b**)

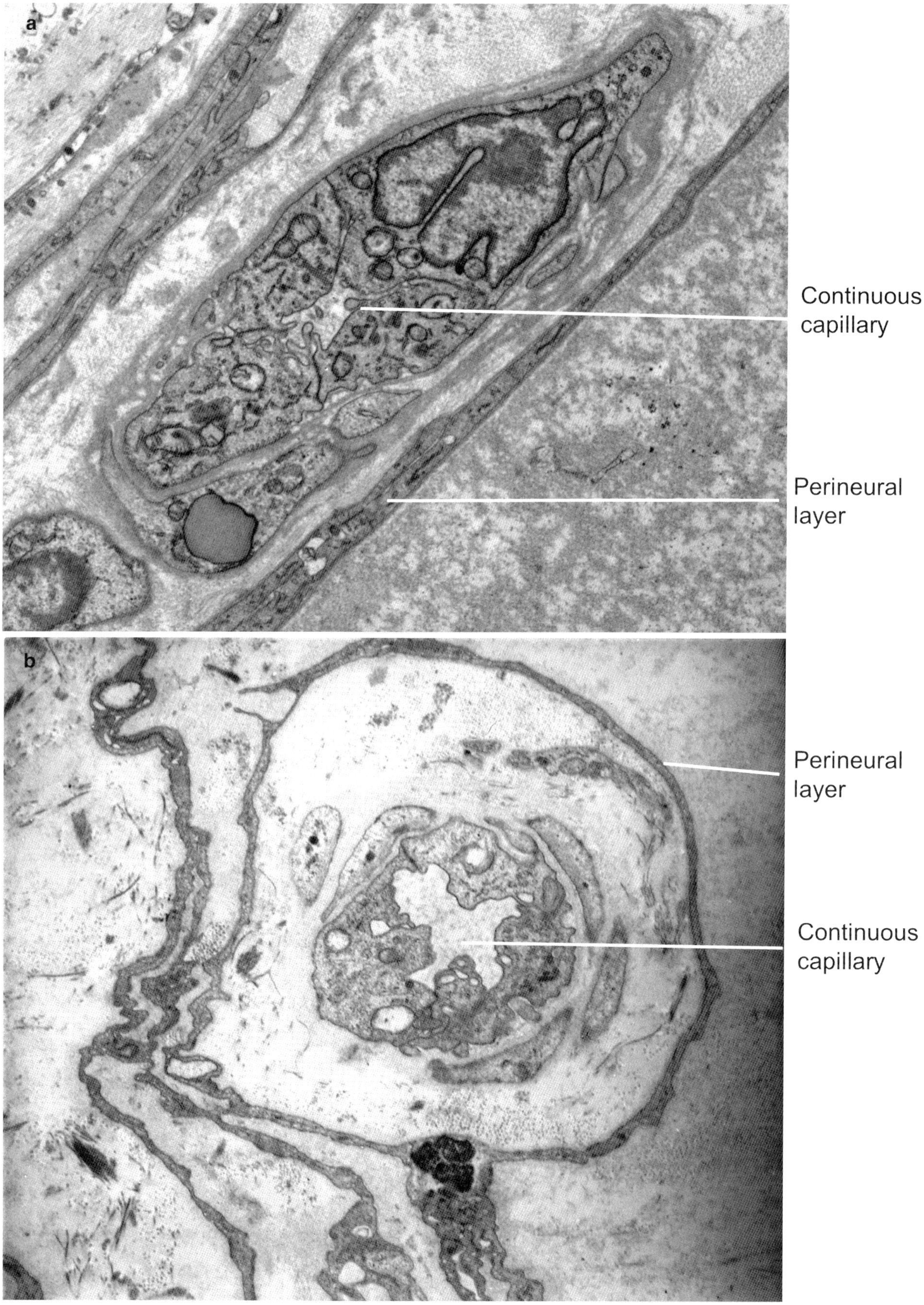

Fig. 4.30 Perineurium. (**a**) Detail of mitochondria within a perineural cell. (**b**) Continuous capillary within perineural layers of human sciatic nerve. Transmission electron microscopy, magnification: ×150,000 (**a**); ×7,000 (**b**)

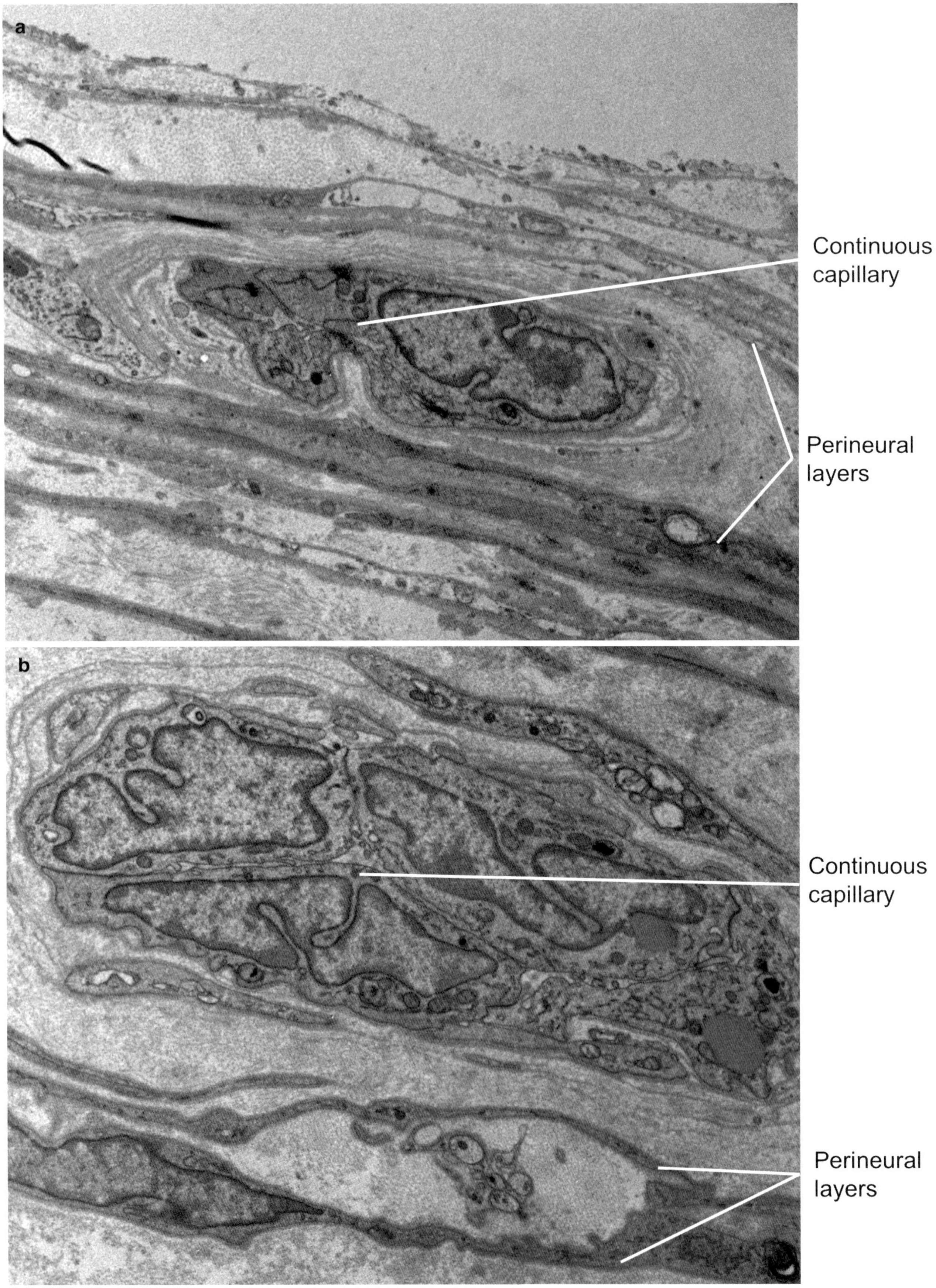

Fig. 4.31 Perineurium. (**a**) Detail of a continuous capillary between perineural cells. (**b**) Detail of a continuous capillary surrounded by the same perineural layer. Transmission electron microscopy, magnification: ×12,000 (**a**); ×7,000 (**b**)

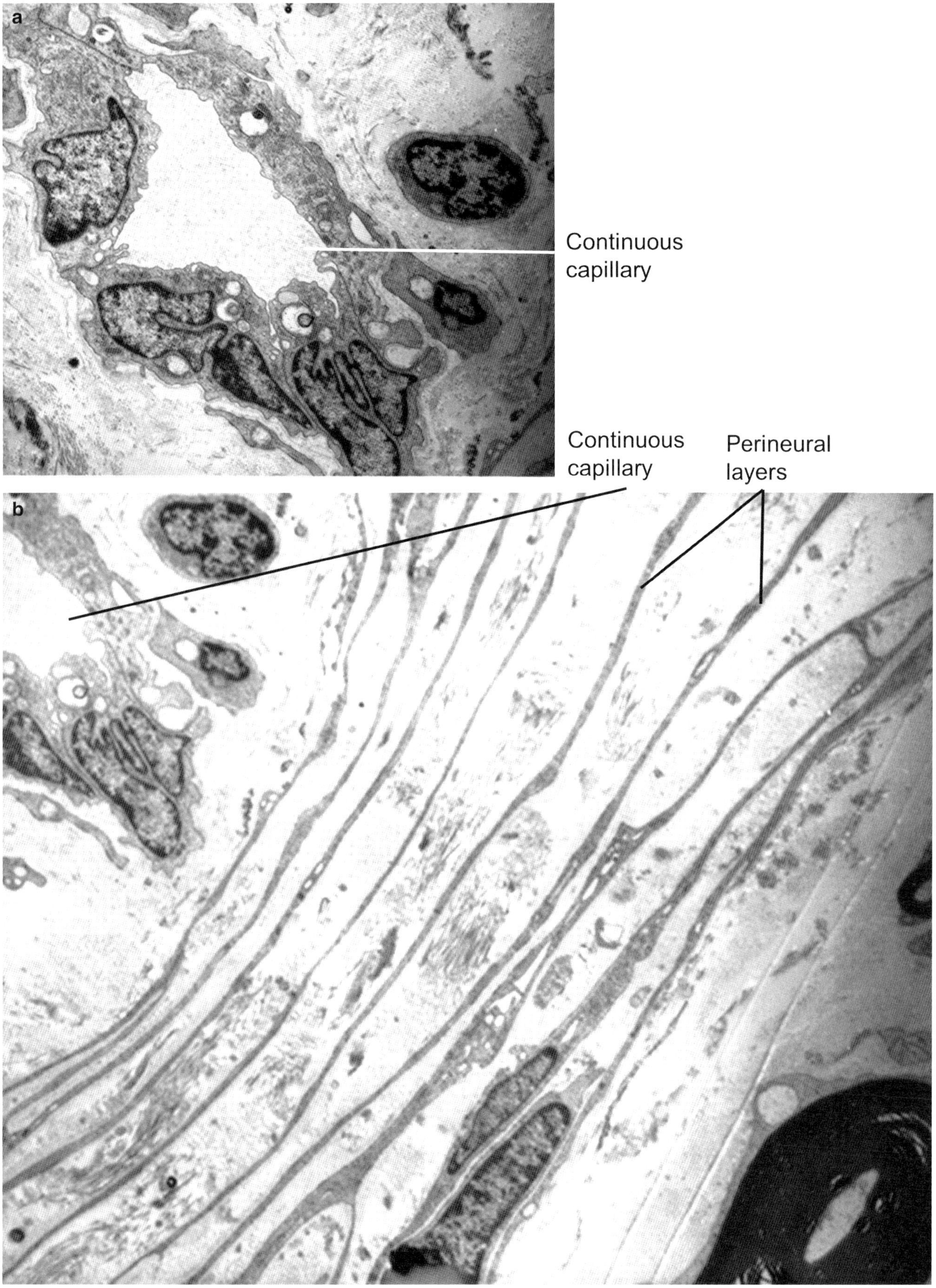

Fig. 4.32 Perineurium. (**a**, **b**) Detail of a continuous capillary between perineural cells. Transmission electron microscopy, magnification: ×8,000 (**a**); ×10,000 (**b**)

References

1. Reina MA, Arriazu R, Collier CB, Sala-Blanch X. Histology and electron microscopy of human peripheral nerves of clinical relevance to the practice of nerve blocks. Rev Esp Anestesiol Reanim. 2013;60:552–62.
2. Reina MA, De Andres JA, Hernández JM, Arriazu Navarro R, Durán Mateos EM, Prats-Galino A. Successive changes in extraneural structures from the subarachnoid nerve roots to the peripheral nerve, influencing anesthetic block, and treatment of acute postoperative pain. Eur J Pain Suppl. 2011;5:377–85.
3. Reina MA. Contribute of electron microscopy in the regional anesthesia. Reg Anesth Pain Med Suppl. 2011;36:E81–5.
4. De Andrés JA, Reina MA, López A, Sala-Blanch X, Prats A. Blocs nerveux périphériques, paresthésies et injections intraneurales. Le Practicien En Anesth Réanim. 2010;14:213–21.
5. Piña-Oviedo S, Ortiz-Hidalgo C. The normal and neoplastic perineurium: a review. Adv Anat Pathol. 2008;15:147–64.
6. Pummi KP, Heape AM, Grenman RA, Peltonen JT, Peltonen SA. Tight junction proteins ZO-1, occludin, and claudins in developing and adult human perineurium. J Histochem Cytochem. 2004;52:1037–46.
7. Olsson Y, Kristensson K. The perineurium as a diffusion barrier to protein tracers following trauma to nerves. Acta Neuropathol. 1973;23:105–11.
8. Soderfeldt B, Olsson Y, Kistensson K. The perineurium as a diffusion barrier to protein tracers in human peripheral nerve. Acta Neuropathol. 1973;25:120–6.
9. Reina MA, López A, De Andrés JA. The blood-nerve barrier in peripheral nerves. Rev Esp Anestesiol Reanim. 2003;50:80–6.
10. Shinowara NL, Michel ME, Rapoport SI. Morphological correlates of permeability in the frog perineurium: vesicles and transcellular channels. Cell Tissue Res. 1982;227:11–22.
11. Reina MA, Wikinski J, Prats A, Machés F. Estructura del nervio periférico [in Spanish]. In: Wikinski J, Reina MA, Bollini C, et al., editors. Diagnóstico, prevensión y tratamiento de las complicaciones neurológicas asociadas con la anestesia regional periférica y central. Buenos Aires: Editorial Médica Panamericana; 2011. p. 71–86.

5 Ultrastructure of the Epineurium

Miguel Angel Reina and Xavier Sala-Blanch

The epineurium is the most external layer of the sheaths comprising the neural connective tissue. This sheath is composed of relatively dense connective tissue wrapping around nerve fascicles. The epineurium contains mainly collagen fibers; fat also is present between the epineurium and nerve fascicles [1–3]. Monofascicular nerves lack epineurium, and around the nerve endings, the epineurium eventually ends. Descriptions of the structures comprising the epineurium have led to controversy [4]. Whereas some authors state that the epineurium constitutes the entire contents of connective tissue extending from each fascicle to the macroscopic limit of the nerve, others distinguish between the portion of collagen enclosing each nerve tissue (epineurium) and connective tissue located among nerve fascicles (interfascicular tissue). In addition, the interfascicular connective tissue may be called *internal epineurium* and the external collagen lamina referred to as *epi-epineurium*. Neural connective tissue adopts a hierarchical structure, which at the level of the sciatic nerve, acquires a distinct complexity. Groups of nerve fascicles inside this nerve are found enclosed in sheaths of epineurium-like connective tissue, which at present, have not been denominated [1–3]. These structures become increasingly intricate as groups of fascicles enclosed in their respective connective fascias join larger groups of nerve bundles, which are compartmented further by layers of neural connective fascia with a morphology to similar to that of the epineurium. Depending on the level at which the sciatic nerve is being examined along its trajectory, the following independent macroscopic structures may be encountered inside this nerve: the tibial and peroneal components, as well as various bundles originating from the first branches of the tibial and peroneal nerves, respectively. A neural connective tissue layer or epineurium (also known as epi-epineurium) wraps around each of these components and ultimately forms the epineurium enclosing the sciatic nerve. However, despite sharing identical morphology, the term *epineurium* conventionally refers to the outermost connective layer, excluding all other connective layers enclosing groups of fascicles. The relevance of such considerations lies in the fact that the epineurium of nerve branches arising from each of the nerve components is a direct prolongation of the neural connective tissue enclosing groups of nerve fascicles. Therefore, damage to these connective layers may affect groups of nerve fascicles constituting entire nerve branches that will arise later in the nerve's trajectory [1–3].

Unlike the perineurium, the epineurium is a permeable structure because of the absence of tight junctions and specialized unions among the scarce number of cells that compose this connective layer (Figs. 5.1, 5.2, 5.3, 5.4, 5.5, 5.6, 5.7, 5.8, 5.9, 5.10, 5.11, and 5.12).

M.A. Reina, MD, PhD (✉)
Department of Clinical Medical Sciences and Institute of Applied Molecular Medicine, School of Medicine, University of CEU San Pablo, Madrid, Spain

Department of Anesthesiology, Madrid-Monteprincipe University Hospital, Madrid, Spain
e-mail: miguelangel@perticone.e.telefonica.net

X. Sala-Blanch, MD
Human Anatomy and Embryology Unit, School of Medicine, University of Barcelona, Barcelona, Spain

Department of Anesthesiology and Critical Care, Clinic Hospital, Barcelona, Spain
e-mail: xavi.sala.blanch@gmail.com

M.A. Reina et al. (eds.), *Atlas of Functional Anatomy for Regional Anesthesia and Pain Medicine: Human Structure, Ultrastructure and 3D Reconstruction Images*, DOI 10.1007/978-3-319-09522-6_5,

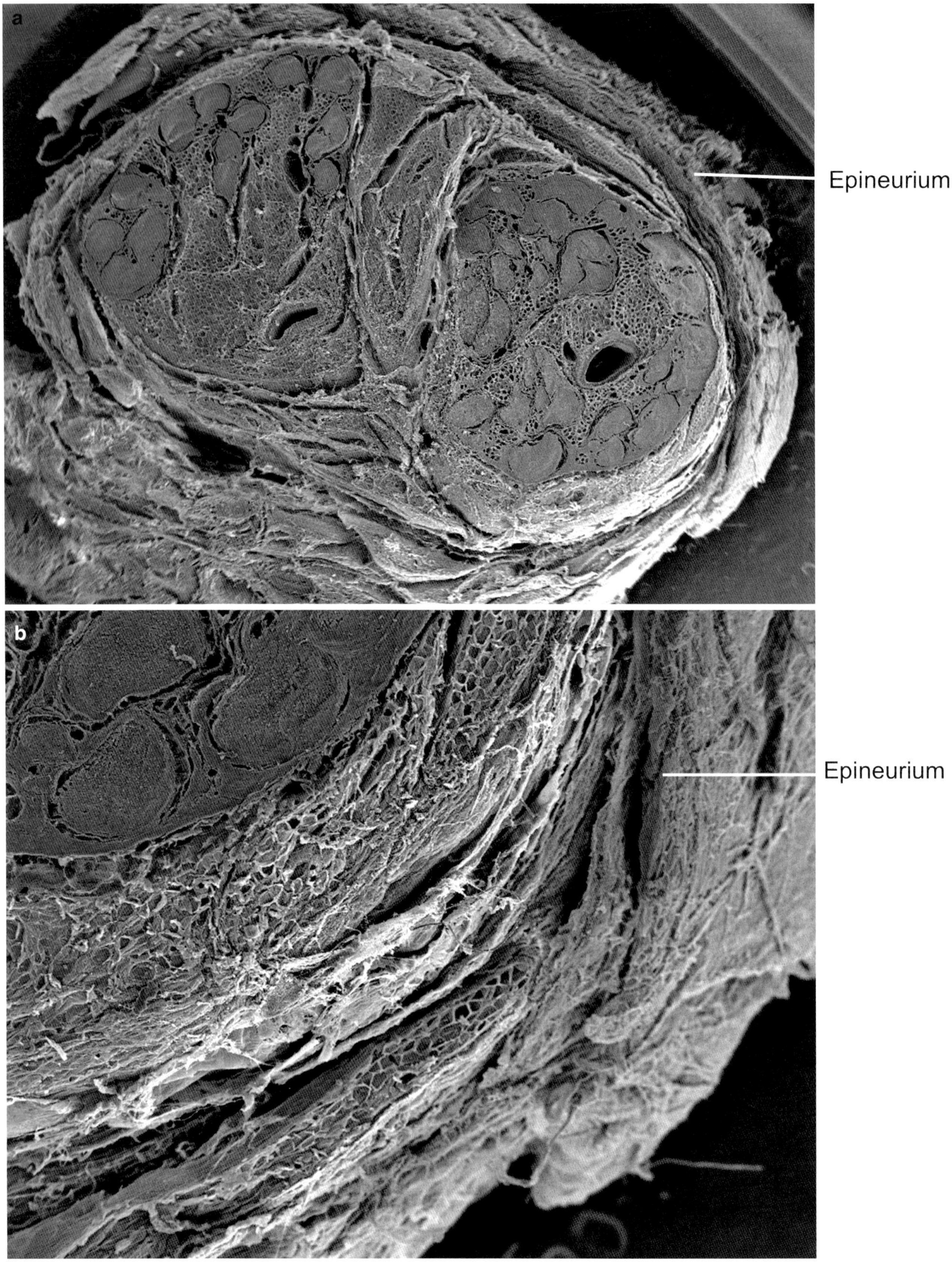

Fig. 5.1 Epineurium. (**a**) Human sciatic nerve. (**b**) Detail of panel **a**. Scanning electron microscopy, magnification: ×10 (**a**); ×40 (**b**)

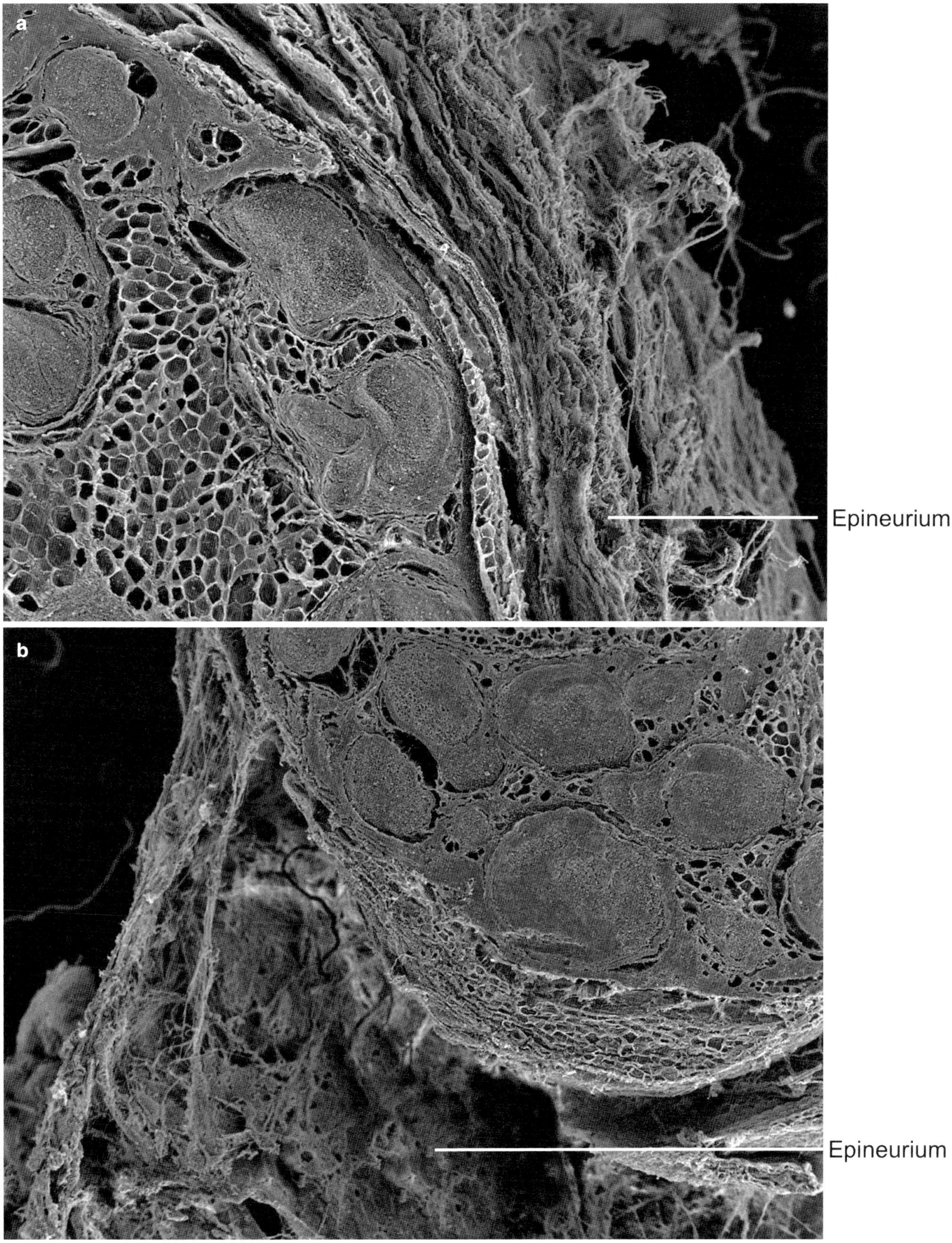

Fig. 5.2 Epineurium. Human tibial nerve. (**a**, **b**) Detail of fascicles, interfascicular tissue, and epineurium. Scanning electron microscopy, magnification: ×40 (**a**); ×30 (**b**)

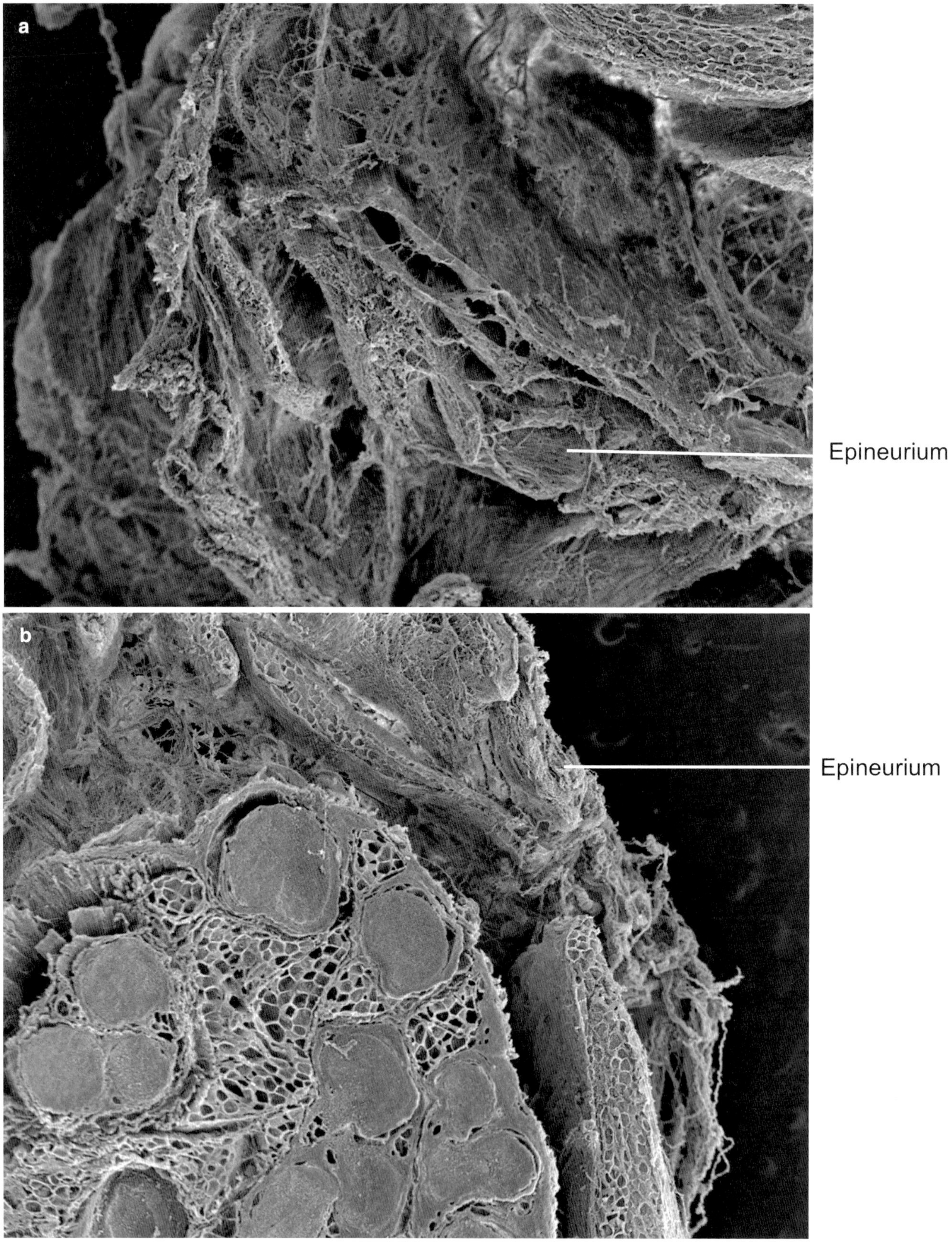

Fig. 5.3 Epineurium. Human tibial nerve. (**a**, **b**) Detail of fascicles, interfascicular tissue, and epineurium. Scanning electron microscopy, magnification: ×30 (**a**, **b**)

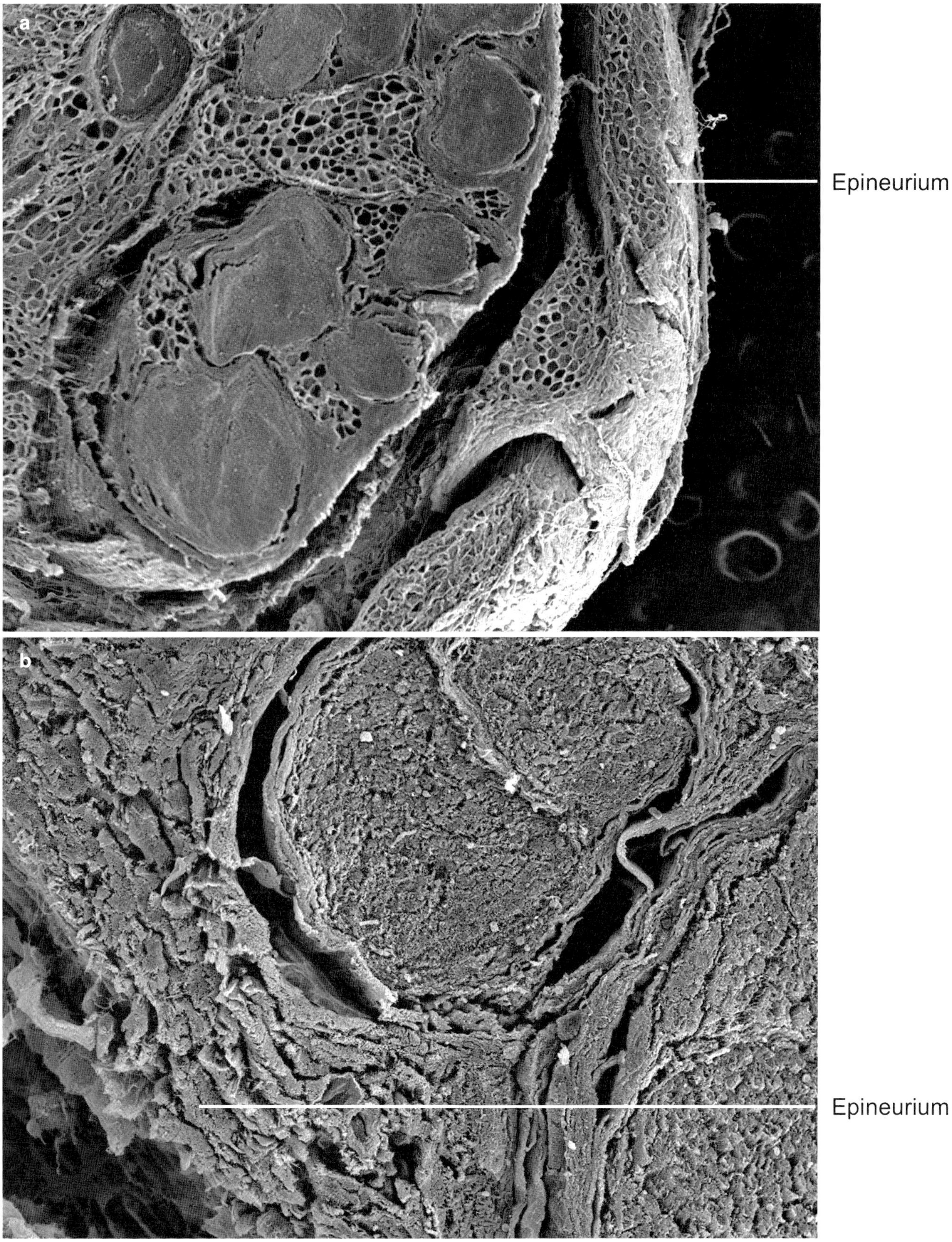

Fig. 5.4 Epineurium. Human tibial nerve. (**a**, **b**) Detail of fascicles, interfascicular tissue, and epineurium. Scanning electron microscopy, magnification: ×30 (**a**); ×200 (**b**)

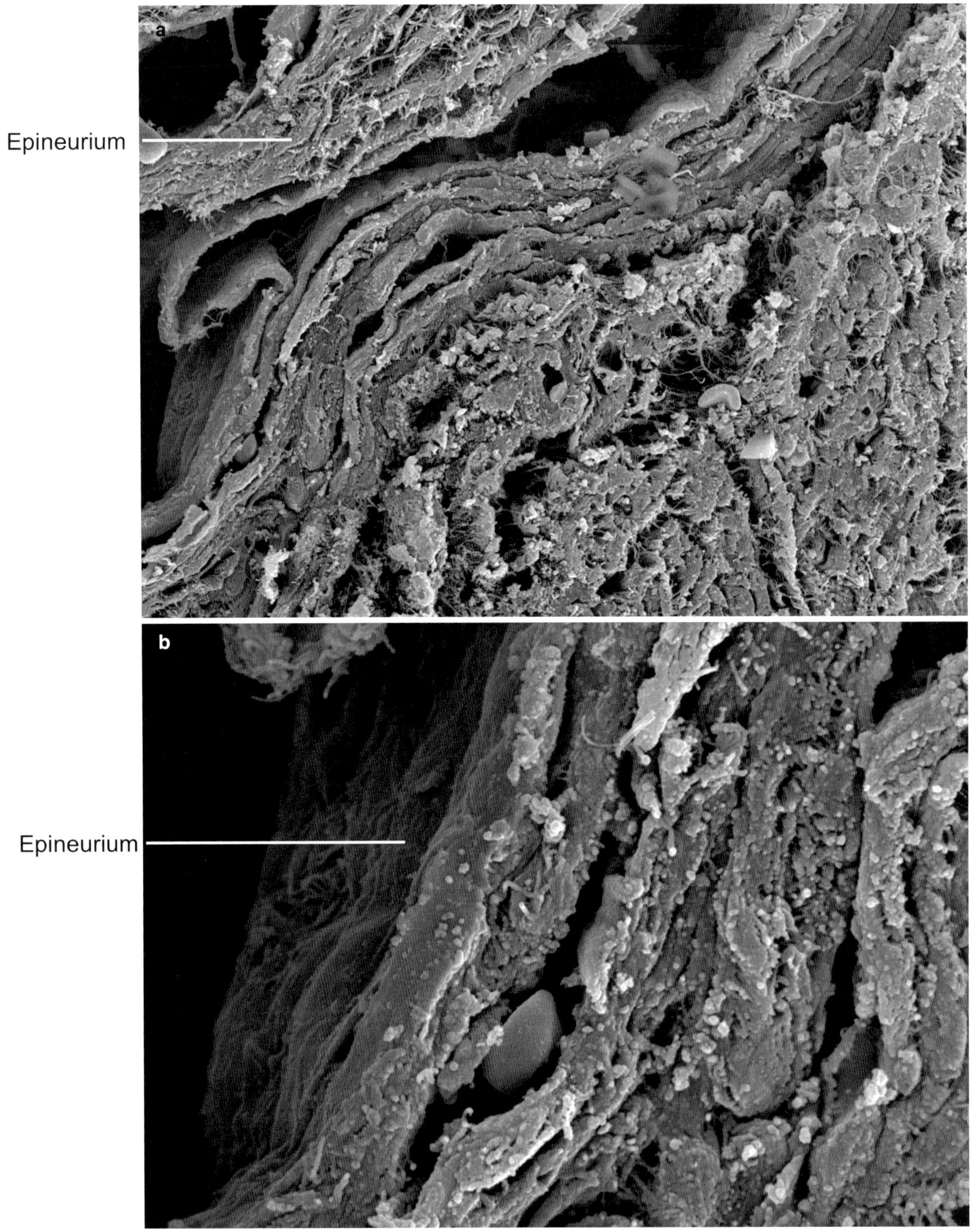

Fig. 5.5 Epineurium. Human tibial nerve. (**a**, **b**) Detail of fascicles, interfascicular tissue, and epineurium. Scanning electron microscopy, magnification: ×1,000 (**a**); ×3,500 (**b**)

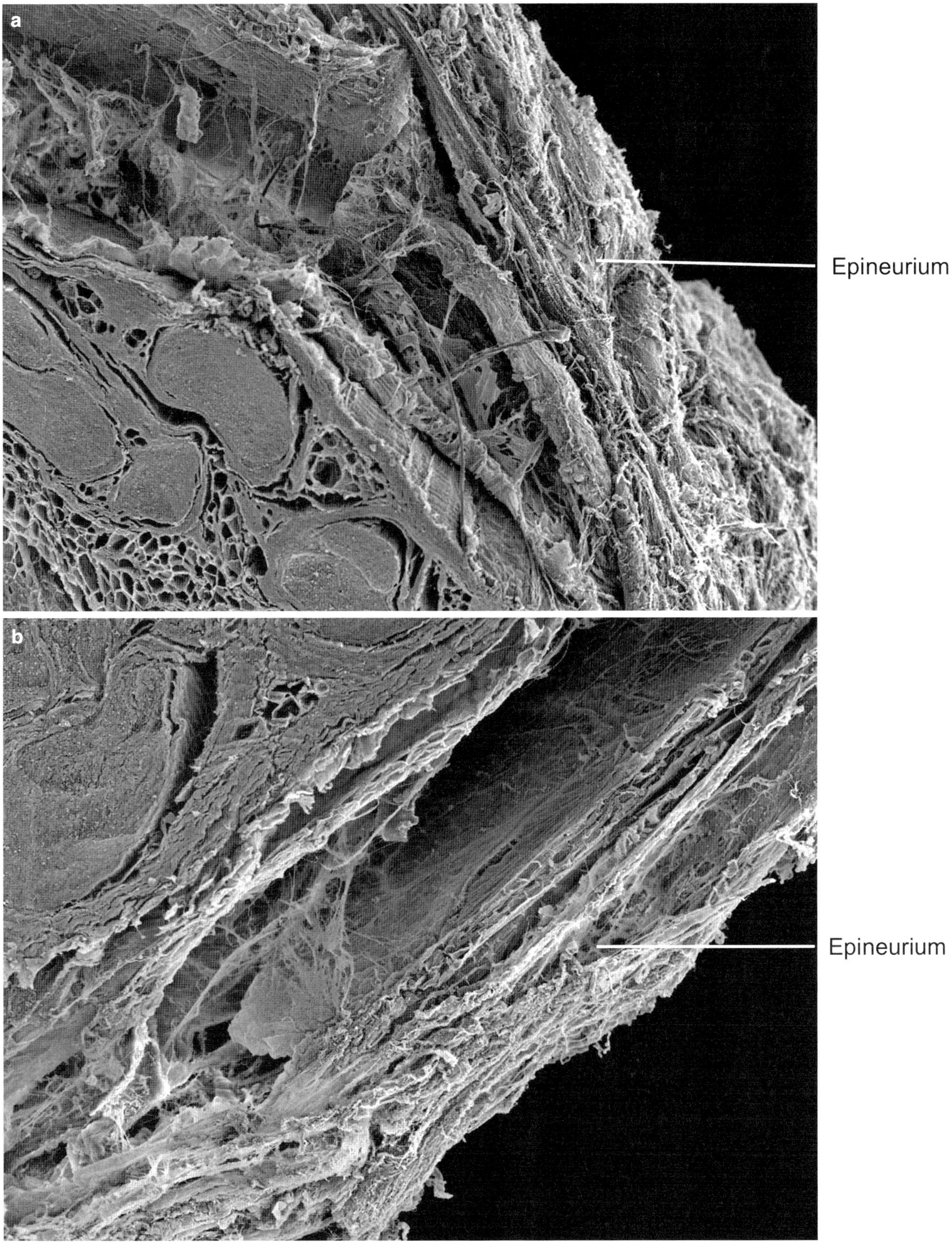

Fig. 5.6 Epineurium. Human tibial nerve. (**a**, **b**) Detail of fascicles, interfascicular tissue, and epineurium. Scanning electron microscopy, magnification: ×40 (**a**); ×60 (**b**)

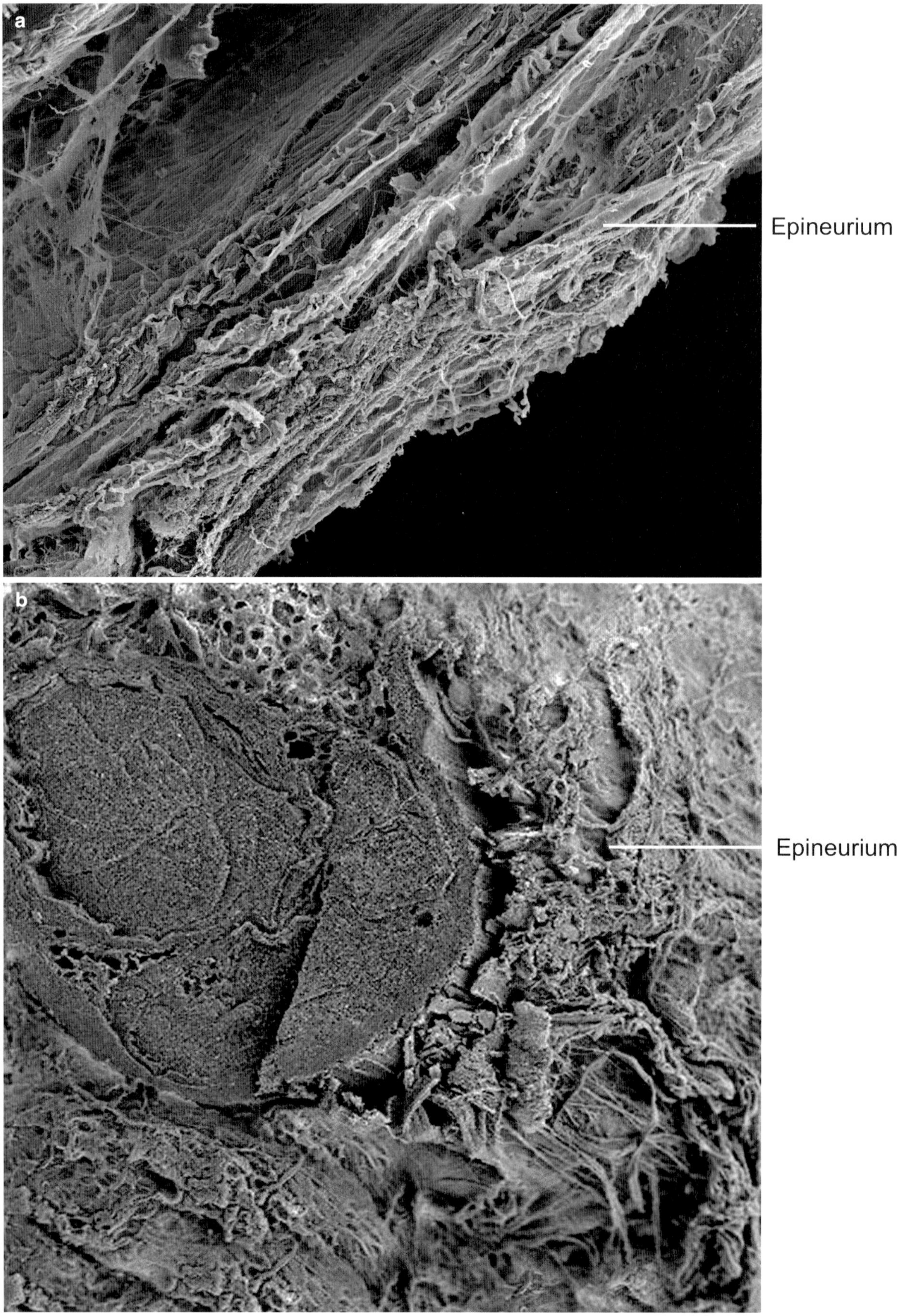

Fig. 5.7 Epineurium. Human tibial nerve. (**a**, **b**) Detail of fascicles, interfascicular tissue and epineurium. Scanning electron microscopy, magnification: ×100 (**a**); ×20 (**b**)

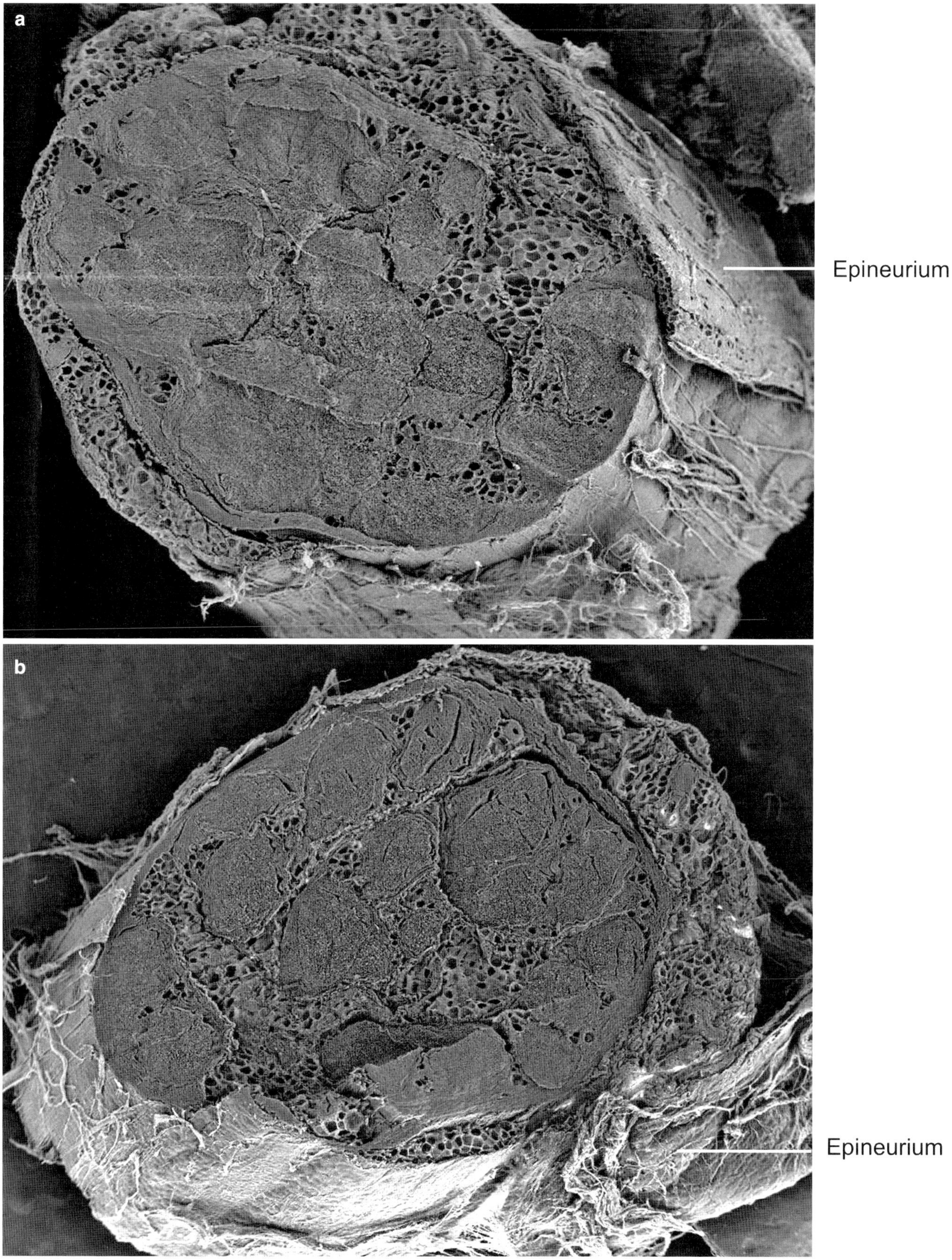

Fig. 5.8 Epineurium. Human tibial nerve. (**a**, **b**) Detail of fascicles, interfascicular tissue, and epineurium. Scanning electron microscopy, magnification: ×20 (**a**); ×20 (**b**) (Panel **b** from Reina et al. [1]; with permission)

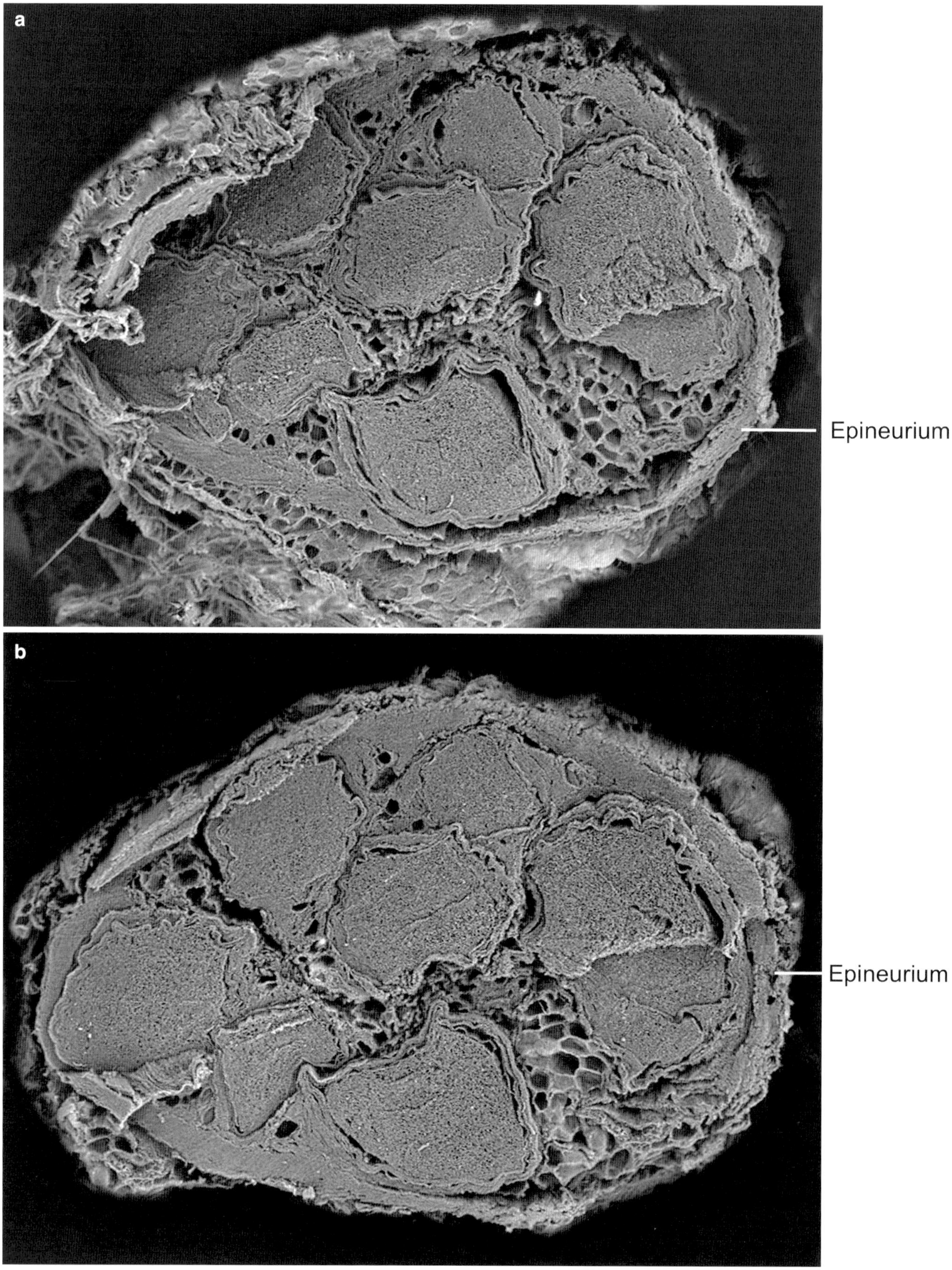

Fig. 5.9 Epineurium. Human tibial nerve. (**a**, **b**) Detail of fascicles, interfascicular tissue, and epineurium. Scanning electron microscopy, magnification: ×40 (**a**, **b**)

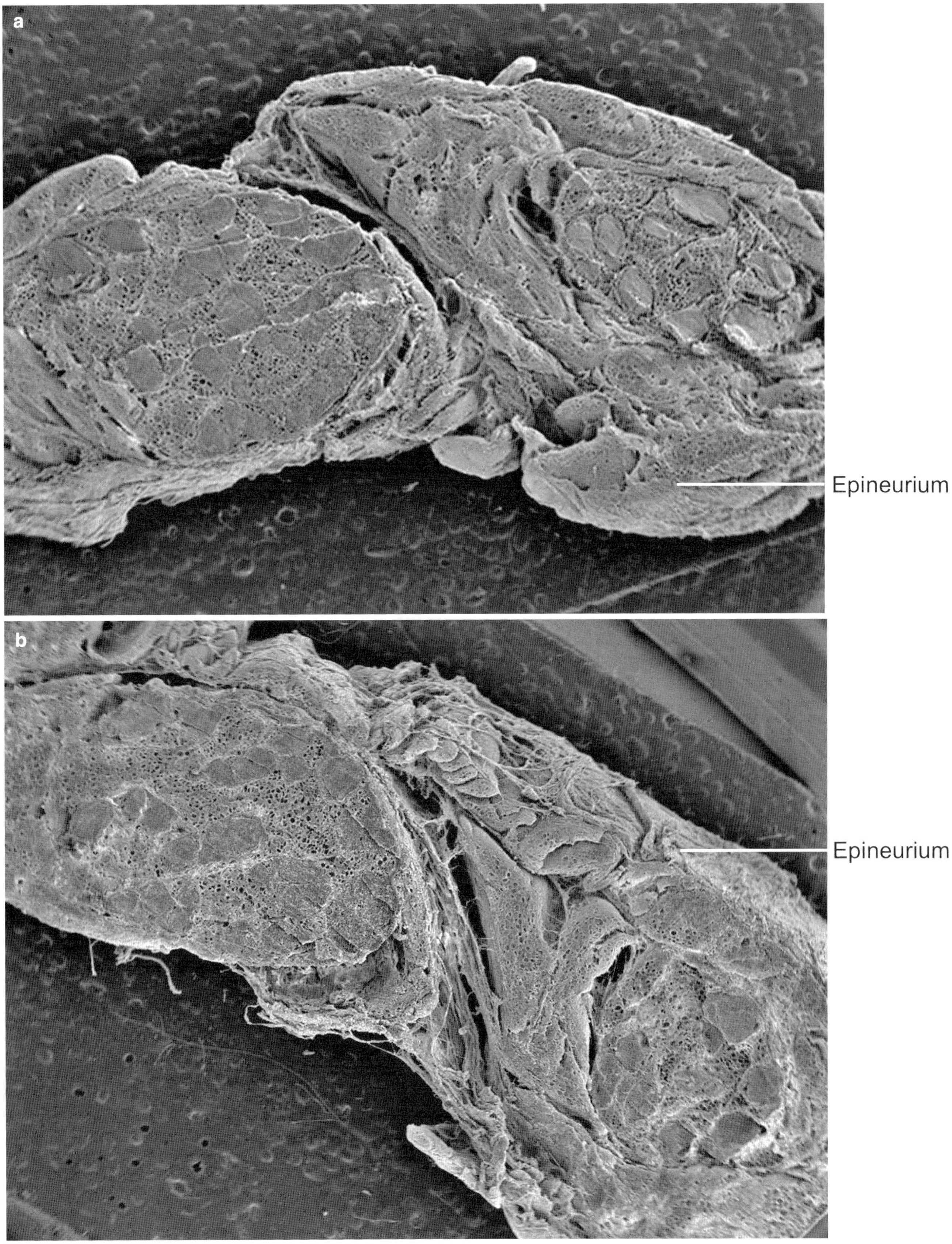

Fig. 5.10 Epineurium. Human sciatic nerve. (**a**, **b**) Detail of fascicles, interfascicular tissue, and epineurium. Scanning electron microscopy, magnification: ×10 (**a**, **b**)

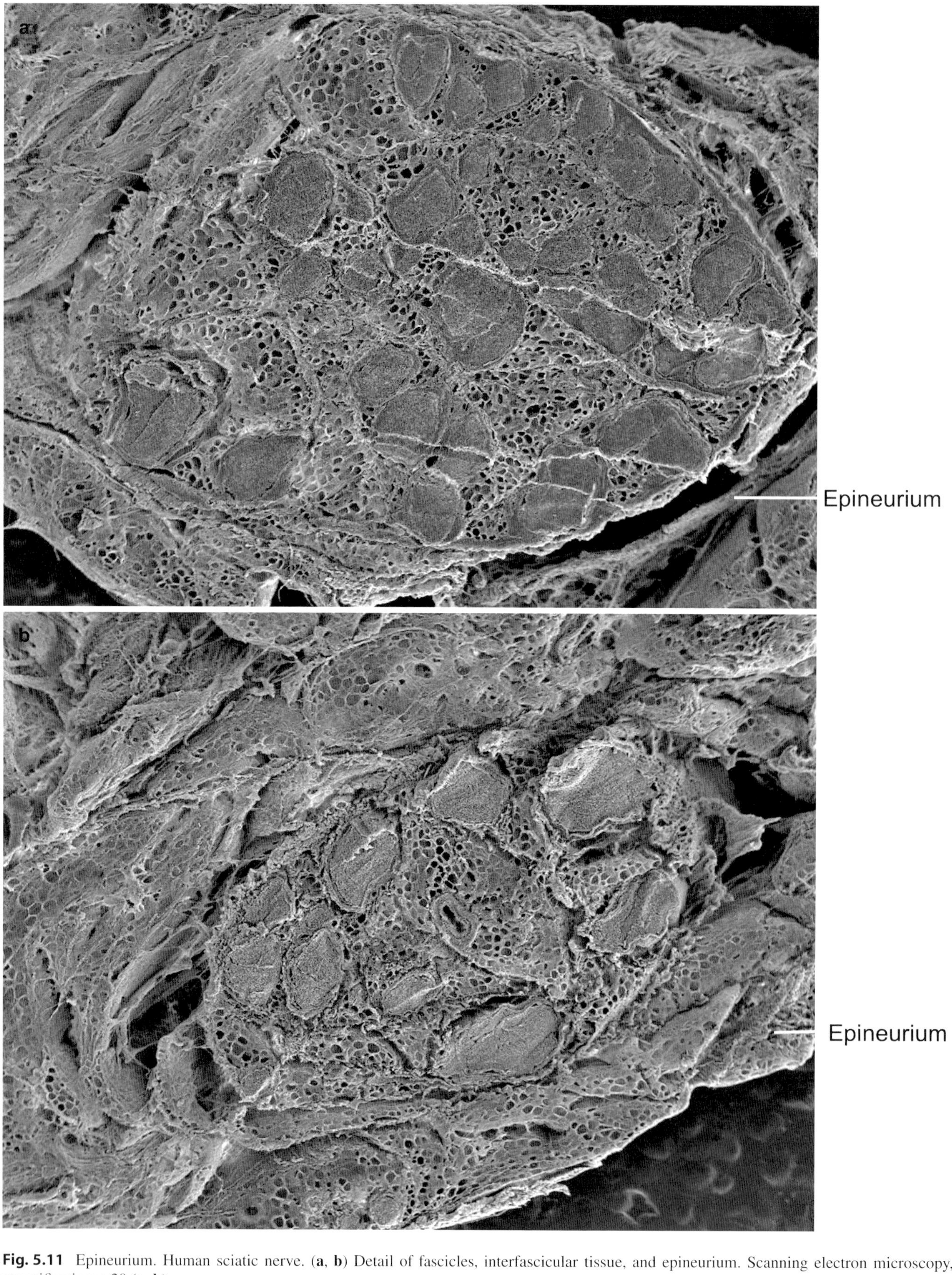

Fig. 5.11 Epineurium. Human sciatic nerve. (**a**, **b**) Detail of fascicles, interfascicular tissue, and epineurium. Scanning electron microscopy, magnification: ×20 (**a**, **b**)

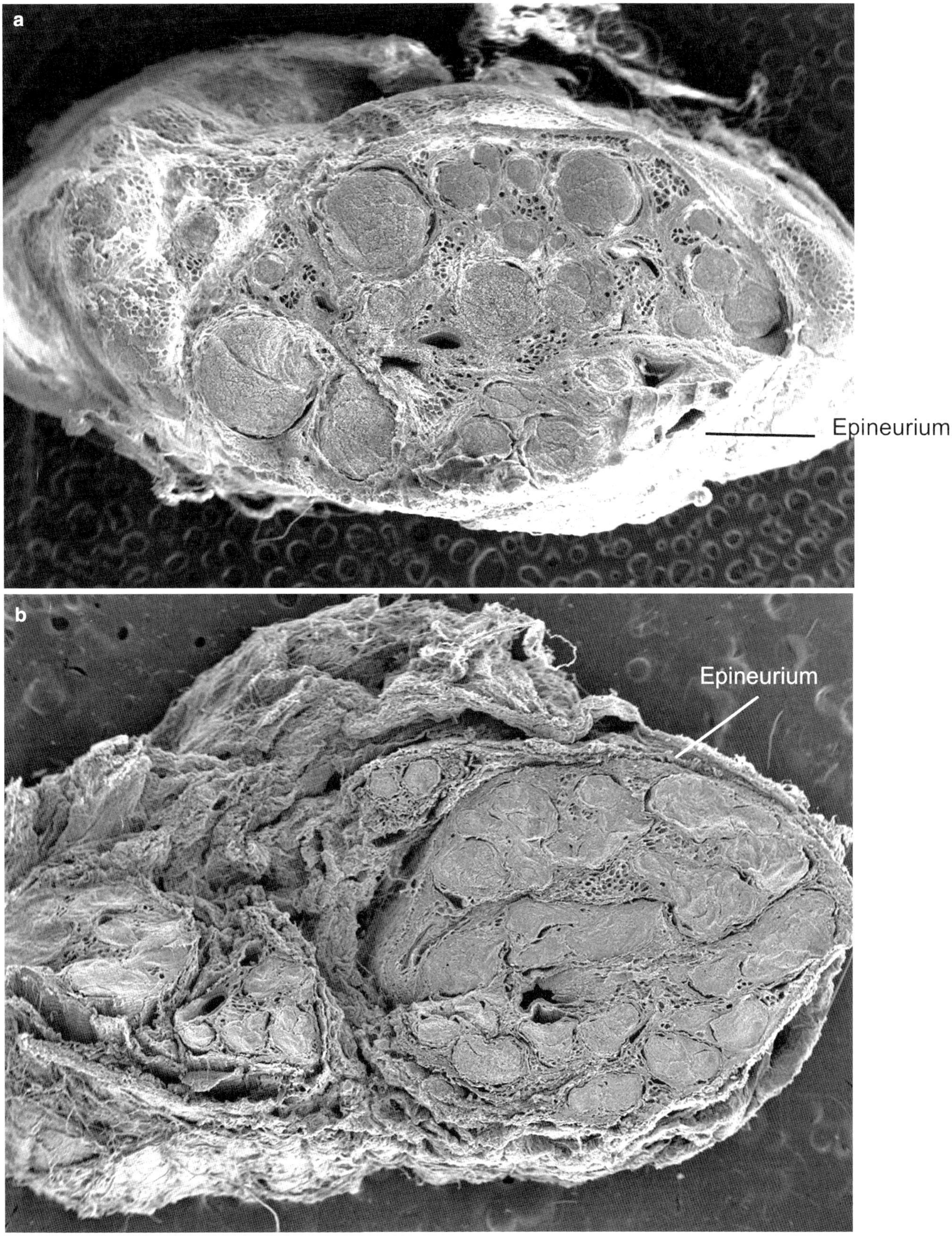

Fig. 5.12 Epineurium. Human sciatic nerve. (**a**, **b**) Detail of fascicles, interfascicular tissue, and epineurium. Scanning electron microscopy, magnification: ×15 (**a**); ×17 (**b**)

References

1. Reina MA, Arriazu R, Collier CB, Sala-Blanch X. Histology and electron microscopy of human peripheral nerves of clinical relevance to the practice of nerve blocks. Rev Esp Anestesiol Reanim. 2013;60:552–62.
2. Reina MA, De Andres JA, Hernández JM, Arriazu Navarro R, Durán Mateos EM, Prats-Galino A. Successive changes in extraneural structures from the subarachnoid nerve roots to the peripheral nerve, influencing anesthetic block, and treatment of acute postoperative pain. Eur J Pain Suppl. 2011;5:377–85.
3. Reina MA, Sala-Blanch X, Fabiola M, Arriazu R, Prats-Galino A. Connective tissues of peripheral nerves. In: Hadzic A, editor. Textbook of regional anesthesia and acute pain management. 2nd Edition. New York: McGraw-Hill Medical; 2015.
4. Sala-Blanch X, Reina MA, Ribalta T, Prats-Galino A. Sciatic nerve structure and nomenclature: epineurium to paraneurium. Is this a new paradigm? Reg Anesth Pain Med. 2013;38:463–5.

6 Origin of the Fascicles and Intraneural Plexus

Miguel Angel Reina, Manuel Fernández Domínguez,
and Ignacio Tardieu

Based on the organization of axons within a peripheral nerve, one might logically think that axons are located more superficially when their pathway is short and more centrally when their trajectory is more extensive. However, this is not the case [1–3].

Within fascicles, axons form intraneural plexuses, in which they take up different positions in different fascicles along their paths. If one made successive cuts alongside the nerve a few millimeters apart, the topographic map of the fascicles would vary from section to section. Sunderland and coworkers [4–7] observed how the topography of fascicles varied 23 times over just 46 mm. This type of distribution, as the intraneural plexus, possibly is related to embryonic development.

The topography of fascicles at a given location in a nerve may be almost, but not entirely, identical between individuals. The difference in fascicle topography along a nerve is explained by the exchange of axons between fascicles. Axon transfer begins in the proximal region of the spinal nerve root cuffs, close to the dural sac (at sites more proximal to dorsal root ganglia) [8, 9].

Spinal nerve roots advance along their subarachnoid course inside special coverings known as nerve root cuffs, which extend up to the beginning of the spinal ganglion; these contain axons from the anterior and posterior roots. Here, the transfer of axons from the anterior and posterior roots begins.

Along the ganglion, communication occurs that facilitates the passage of groups of axons between both anterior and posterior roots at the level of the vertebral foramen. This exchange of axons gives rise to mixed nerve.

The initial portion of mixed nerves contains groups of fascicles and small fascicles that together with larger fascicles, form fascicular interconnections. These interconnections are more frequent among adjacent fascicle groups and absent among fascicles that belong to different groups (Figs. 6.1, 6.2, 6.3, 6.4, 6.5, 6.6, 6.7, 6.8, 6.9, 6.10, 6.11, 6.12, 6.13, 6.14, 6.15, 6.16, 6.17, 6.18, 6.19, 6.20, 6.21, 6.22, 6.23, 6.24, 6.25, and 6.26).

M.A. Reina, MD, PhD (✉)
Department of Clinical Medical Sciences and Institute of Applied Molecular Medicine, School of Medicine, University of CEU San Pablo, Madrid, Spain

Department of Anesthesiology, Madrid-Monteprı́ncipe University Hospital, Madrid, Spain
e-mail: miguelangel@perticone.e.telefonica.net

M. Fernández Domínguez, MD, PhD
Department of Clinical Medical Sciences and Institute of Applied Molecular Medicine, School of Medicine, CEU San Pablo University, Madrid, Spain

Department of Maxillofacial Surgery, Madrid-Montepríncipe University Hospital and Madrid-Sanchinarro University Hospital, Madrid, Spain
e-mail: clinferfun@yahoo.es

I. Tardieu, Tch
Histology Unit, Institute of Applied Molecular Medicine, School of Medicine, CEU San Pablo University, Madrid, Spain
e-mail: ignacio.tardieuchorro@ceu.es

M.A. Reina et al. (eds.), *Atlas of Functional Anatomy for Regional Anesthesia and Pain Medicine: Human Structure, Ultrastructure and 3D Reconstruction Images*, DOI 10.1007/978-3-319-09522-6_6,

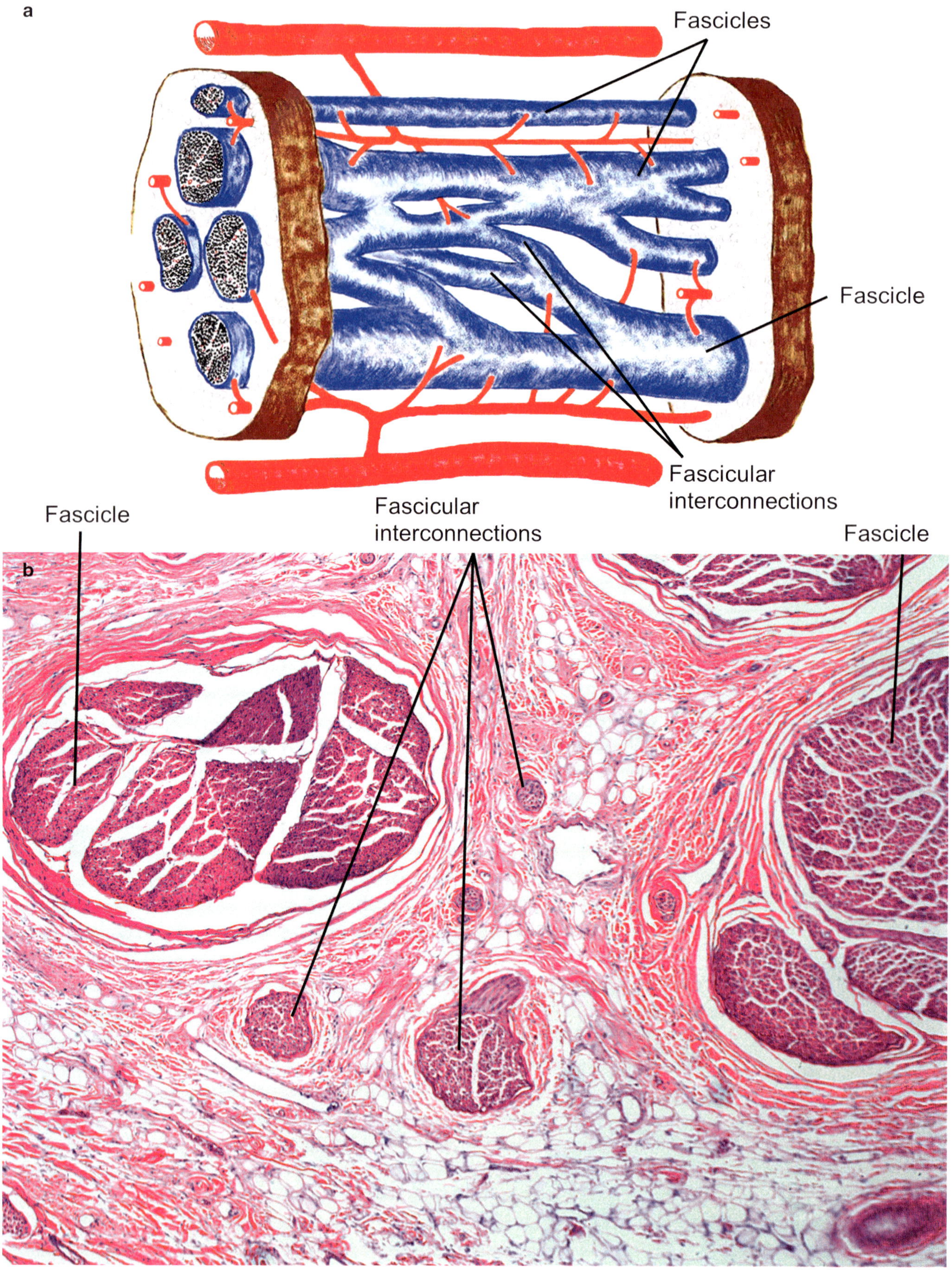

Fig. 6.1 Intraneural plexus. (**a**) Diagram. (**b**) Fascicle interconnections present in Fig. 6.5a (Panel **a** from De Andrés et al. [3]; with permission)

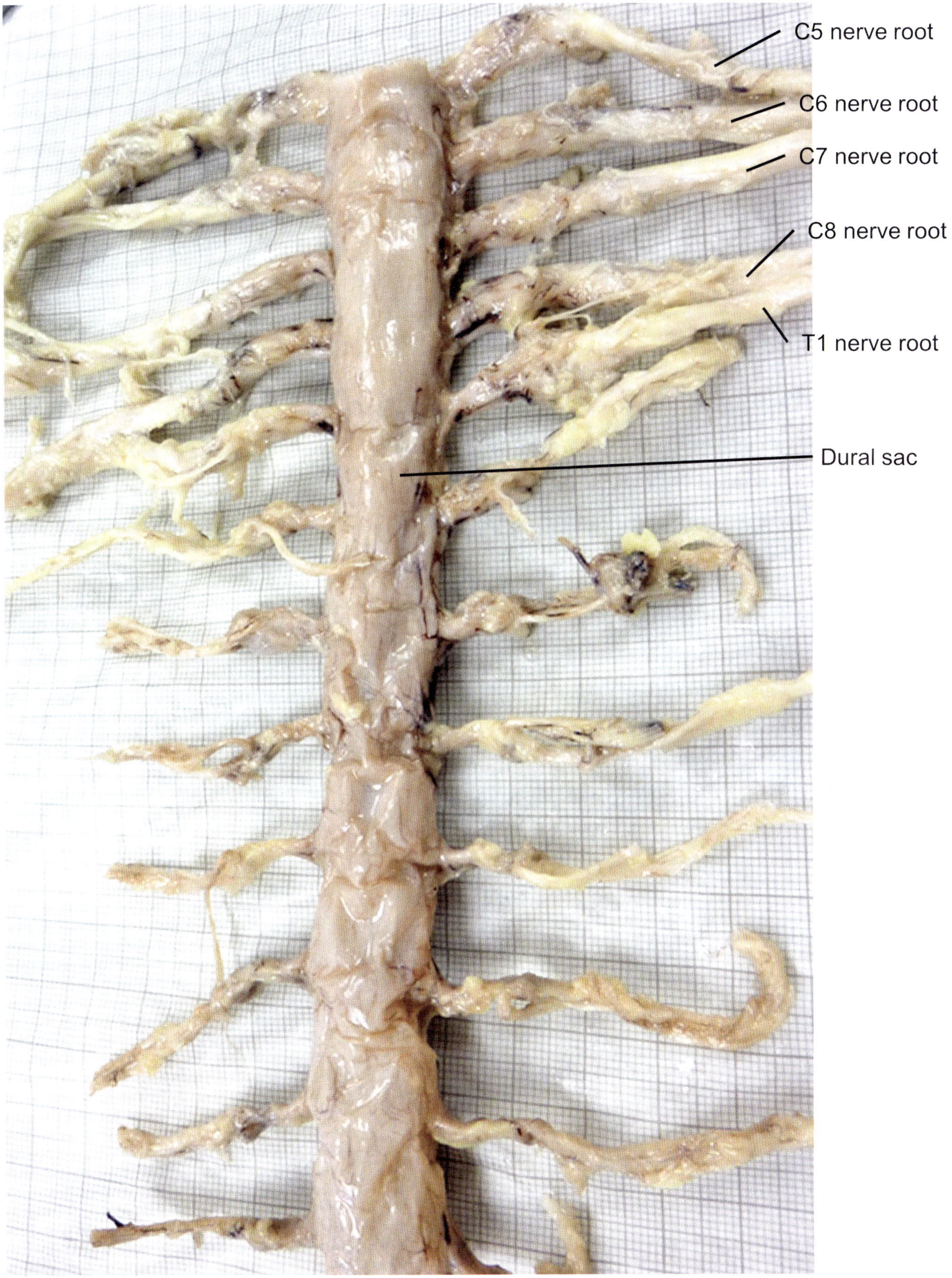

Fig. 6.2 Human dural sac and nerve root cuffs. Cervical roots forming the brachial plexus. Serial cuts from the 6th cervical (C6) nerve root

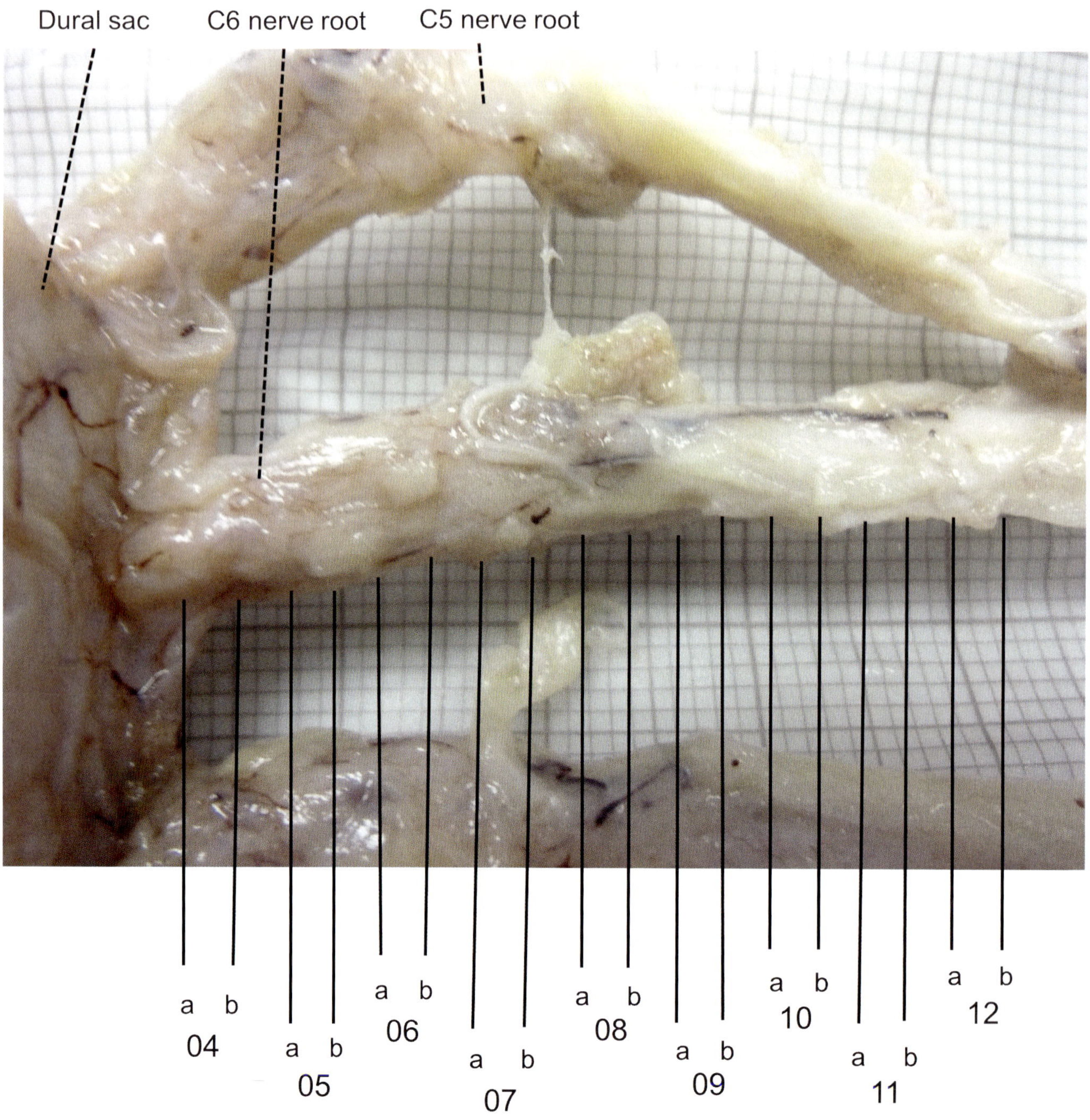

Fig. 6.3 Root cuff from the C6 nerve. Sites where serial cuts were made

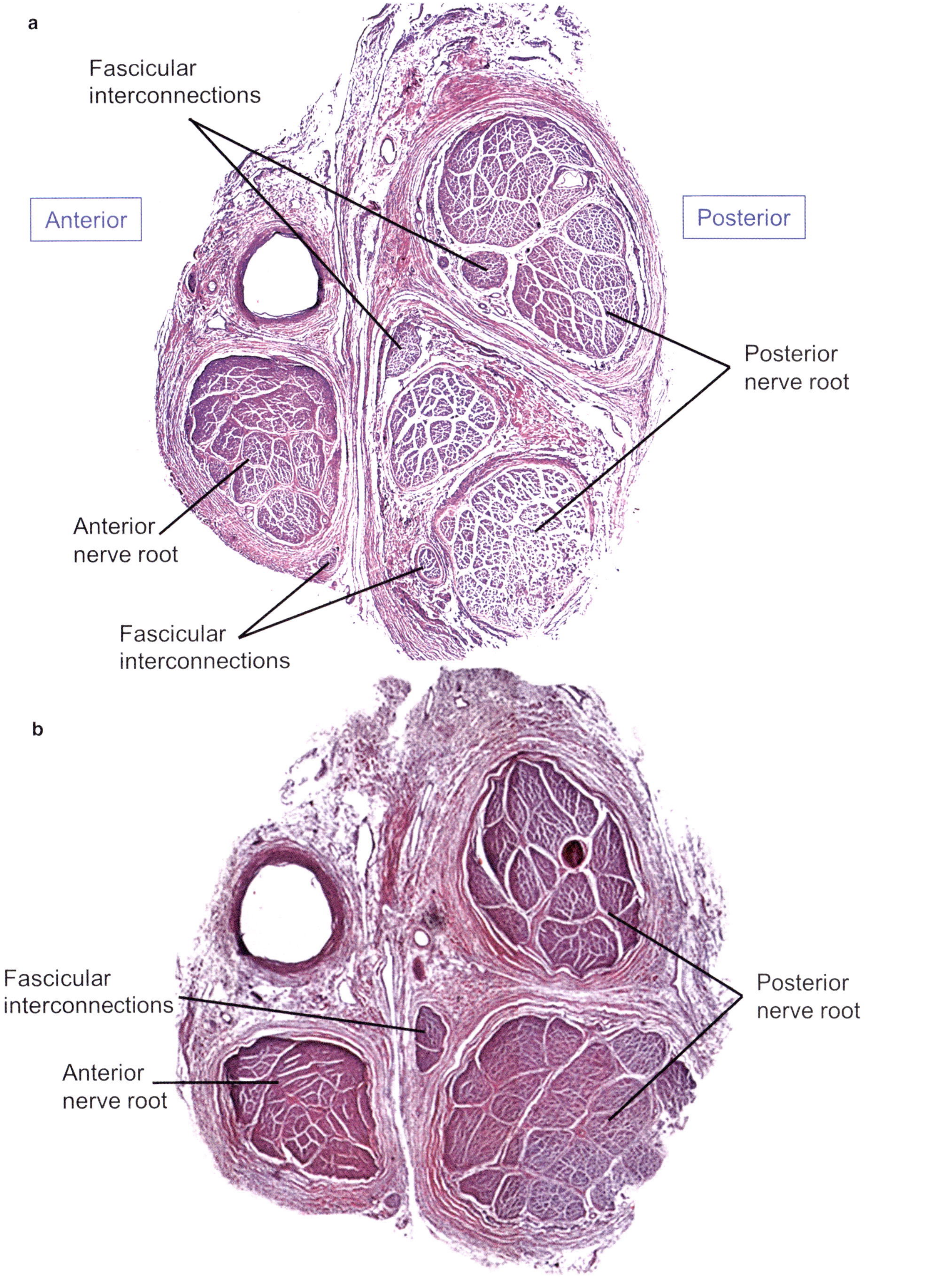

Fig. 6.4 Anterior and posterior nerve roots. Transverse section of C6 nerve at the level shown in Fig. 6.3. (**a**, **b**) First and second cuts proximal to the dural sac. Interconnecting fascicles may be seen. Hematoxylin and eosin (H&E) stain (Panel **a** from Reina et al. [2]; with permission)

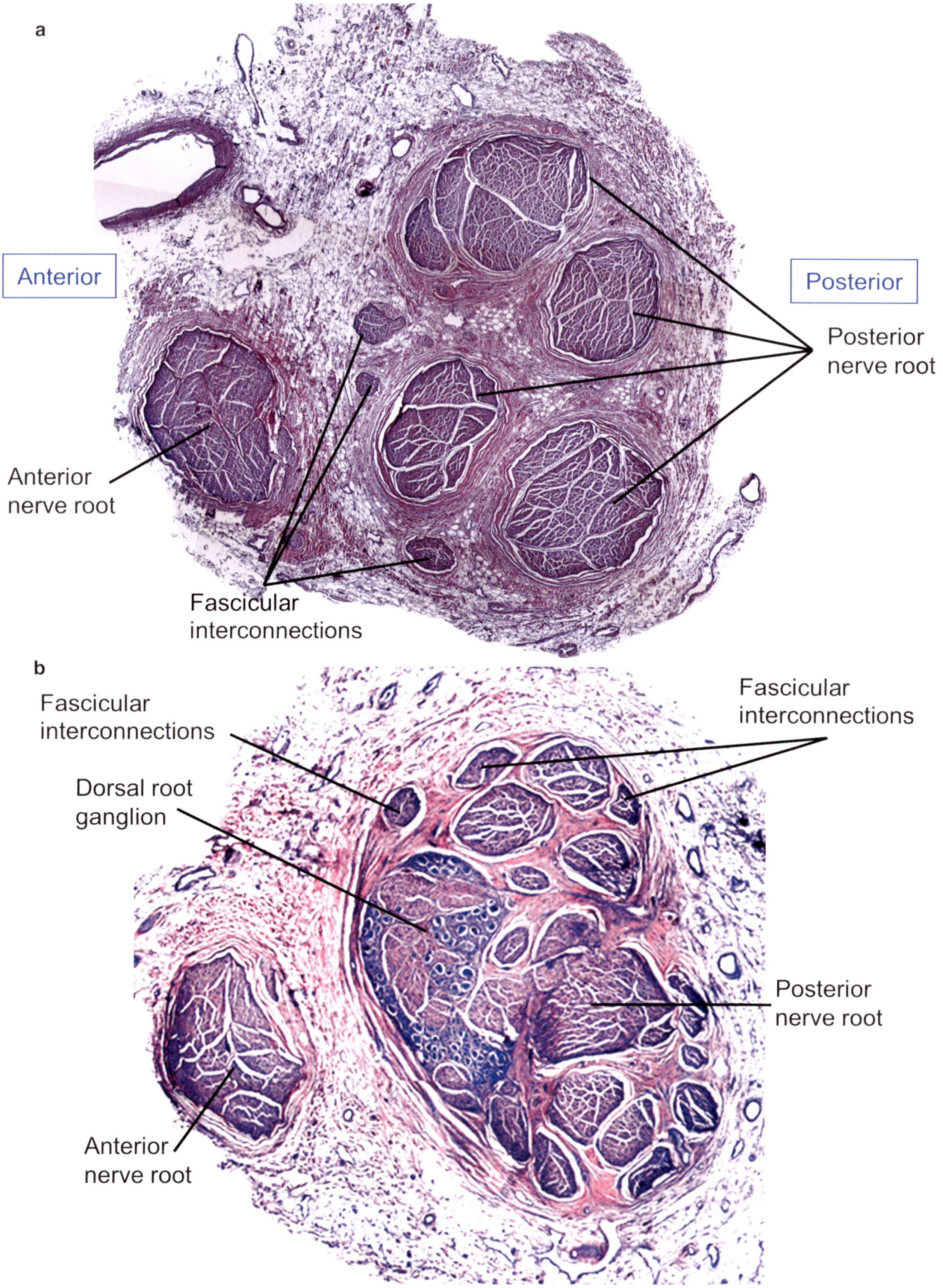

Fig. 6.5 Anterior and posterior nerve roots. Transverse section of C6 nerve. (**a**, **b**) Interconnecting fascicles. (**b**) Minute portion of a dorsal root ganglion (DRG). H&E stain

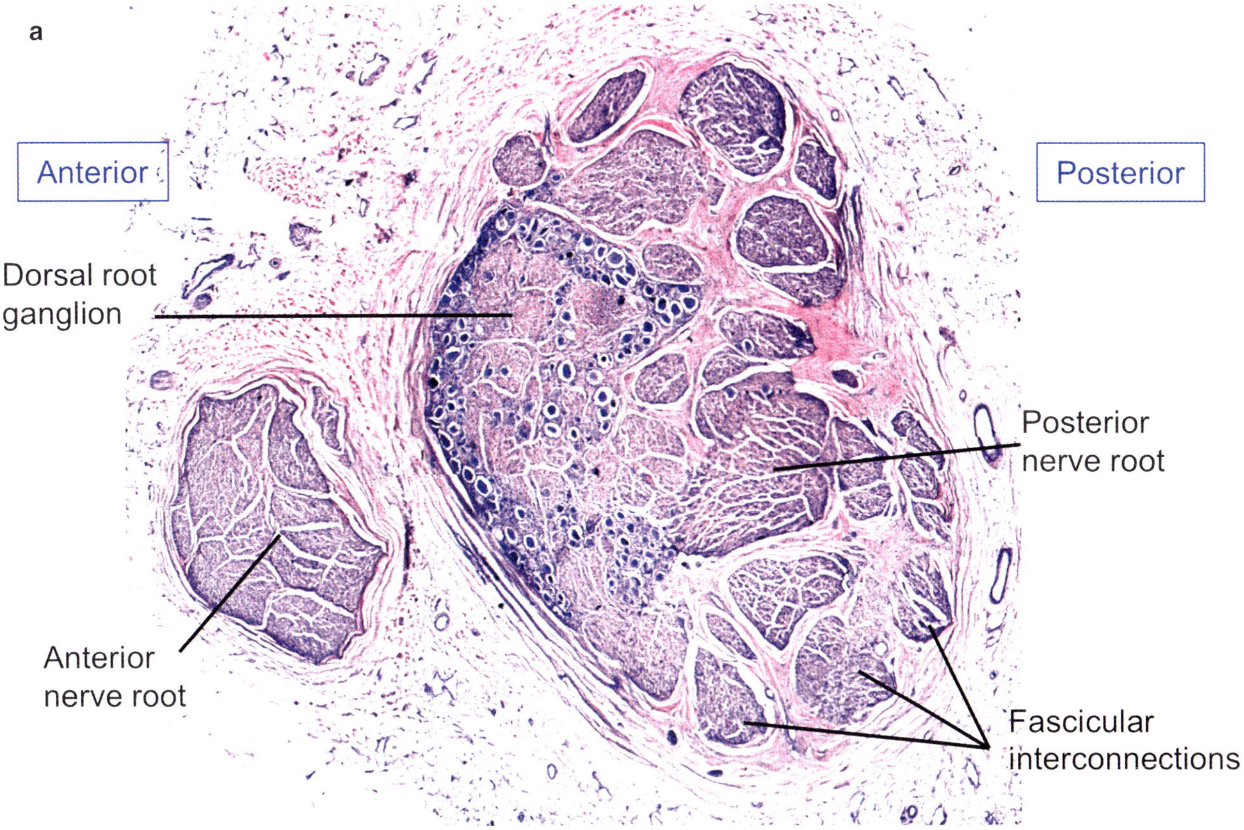

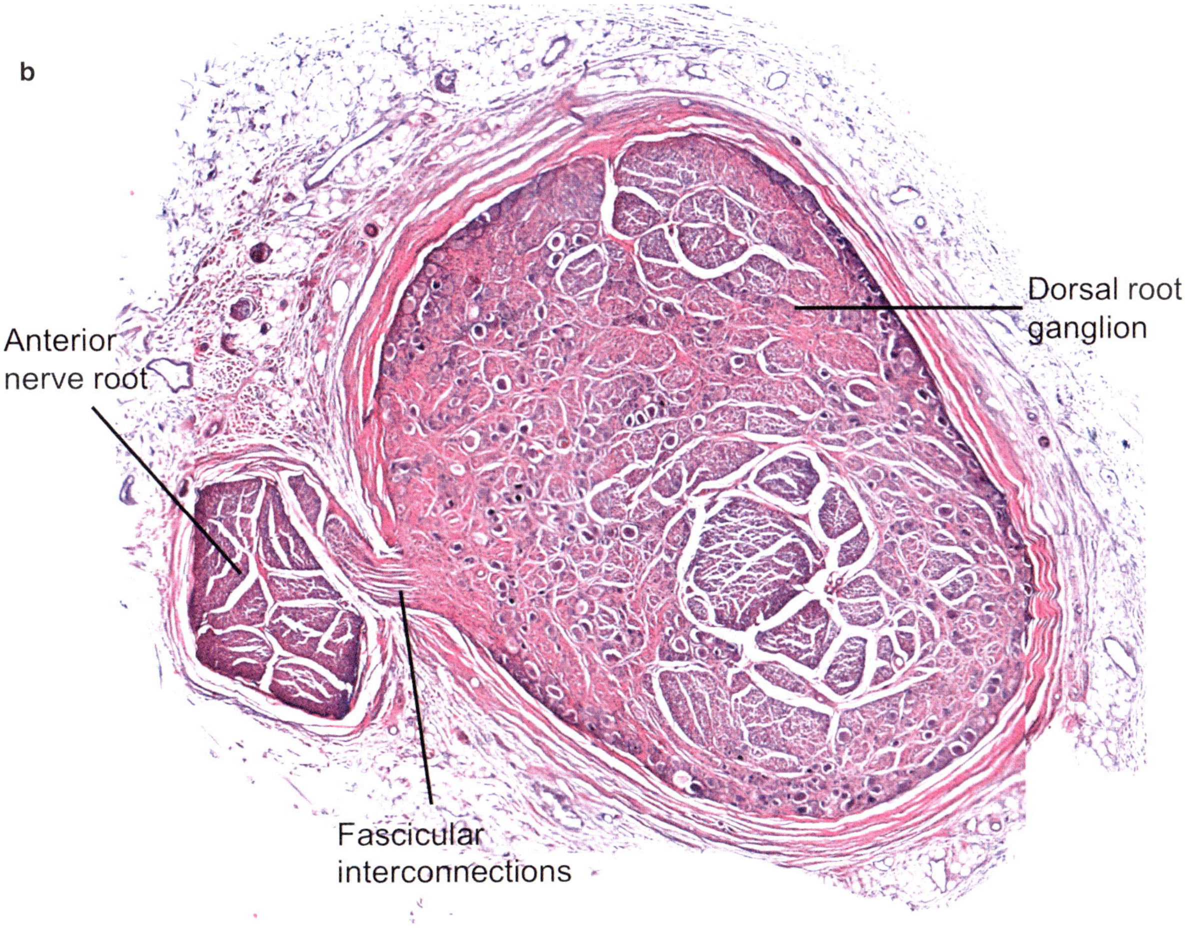

Fig. 6.6 Anterior and posterior nerve roots. Transverse section of C6 nerve. (**a**) A portion of the root ganglion is shown. (**b**) Cut from the DRG and passage of axons between dorsal and anterior nerve roots. H&E stain

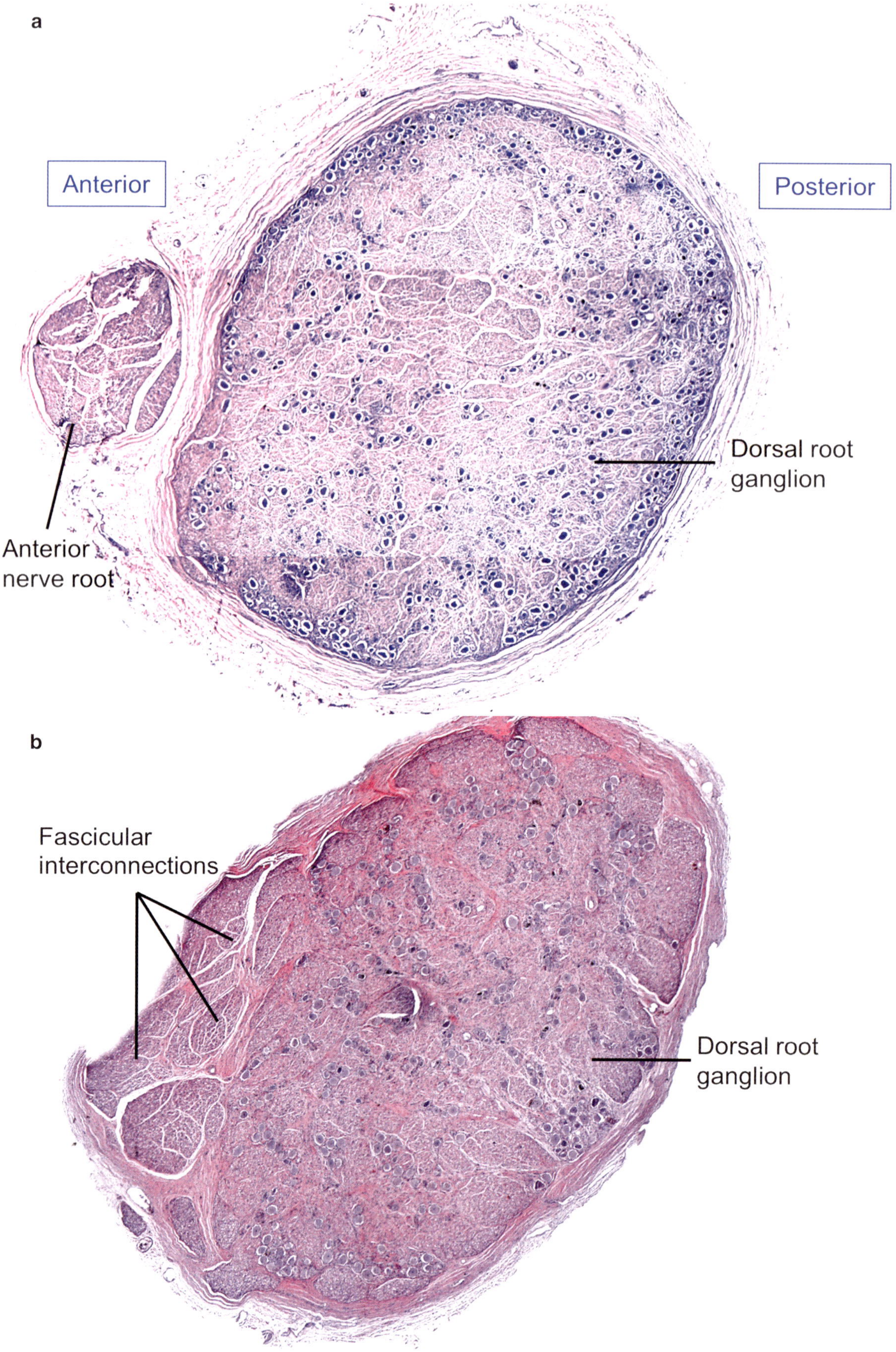

Fig. 6.7 Dorsal root ganglion. Transverse section of C6 nerve. (**a**) Cut at the level of the dorsal nerve ganglion. (**b**) Anterior and posterior roots are not identified. Shown are the nerve fascicles and a portion of the DRG. H&E stain (Panel **a** from Reina et al. [2]; with permission)

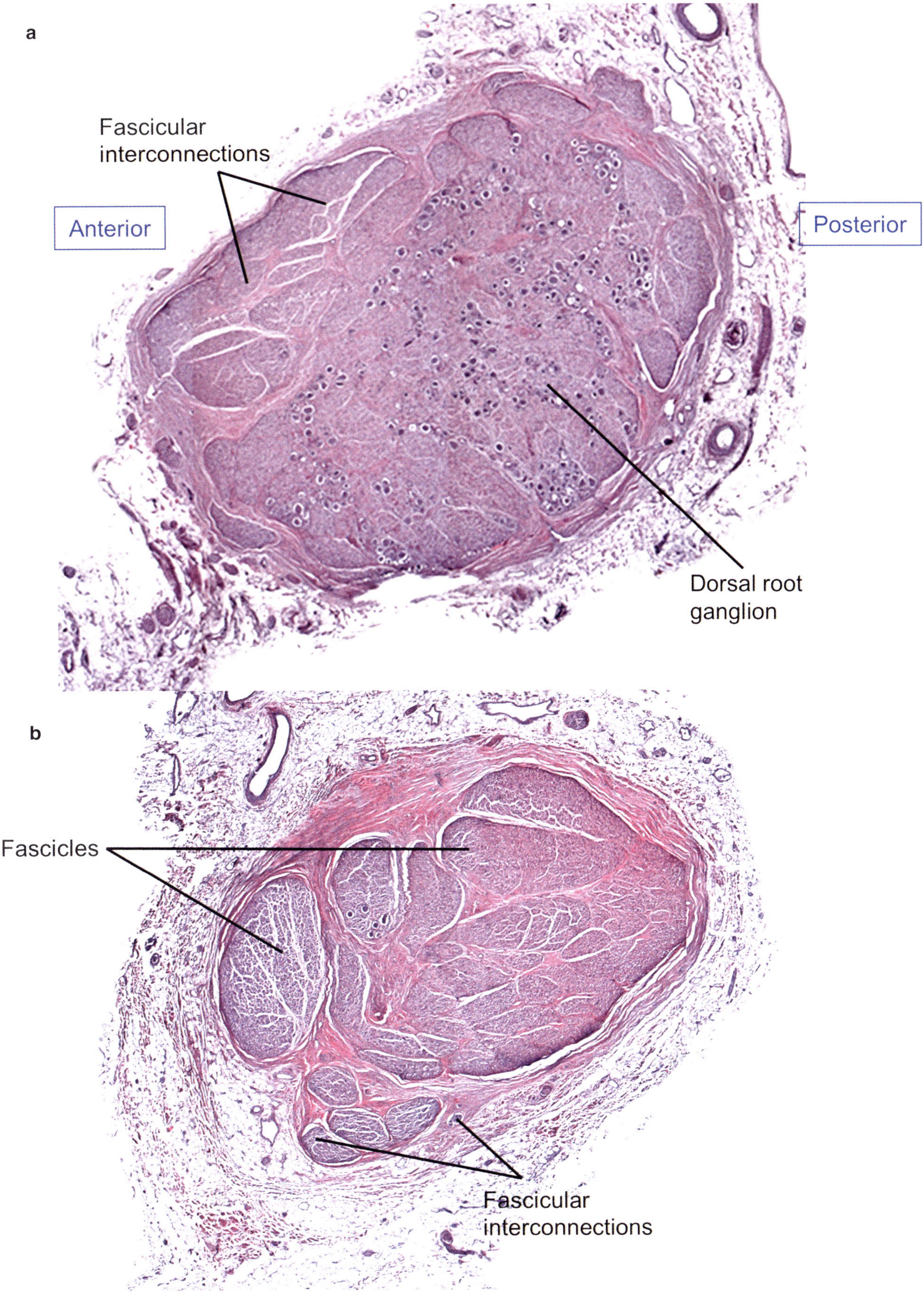

Fig. 6.8 Dorsal root ganglion. Transverse section of C6 nerve. (**a**) DRG and fascicles. (**b**) Nerve fascicles from a mixed peripheral nerve. H&E stain

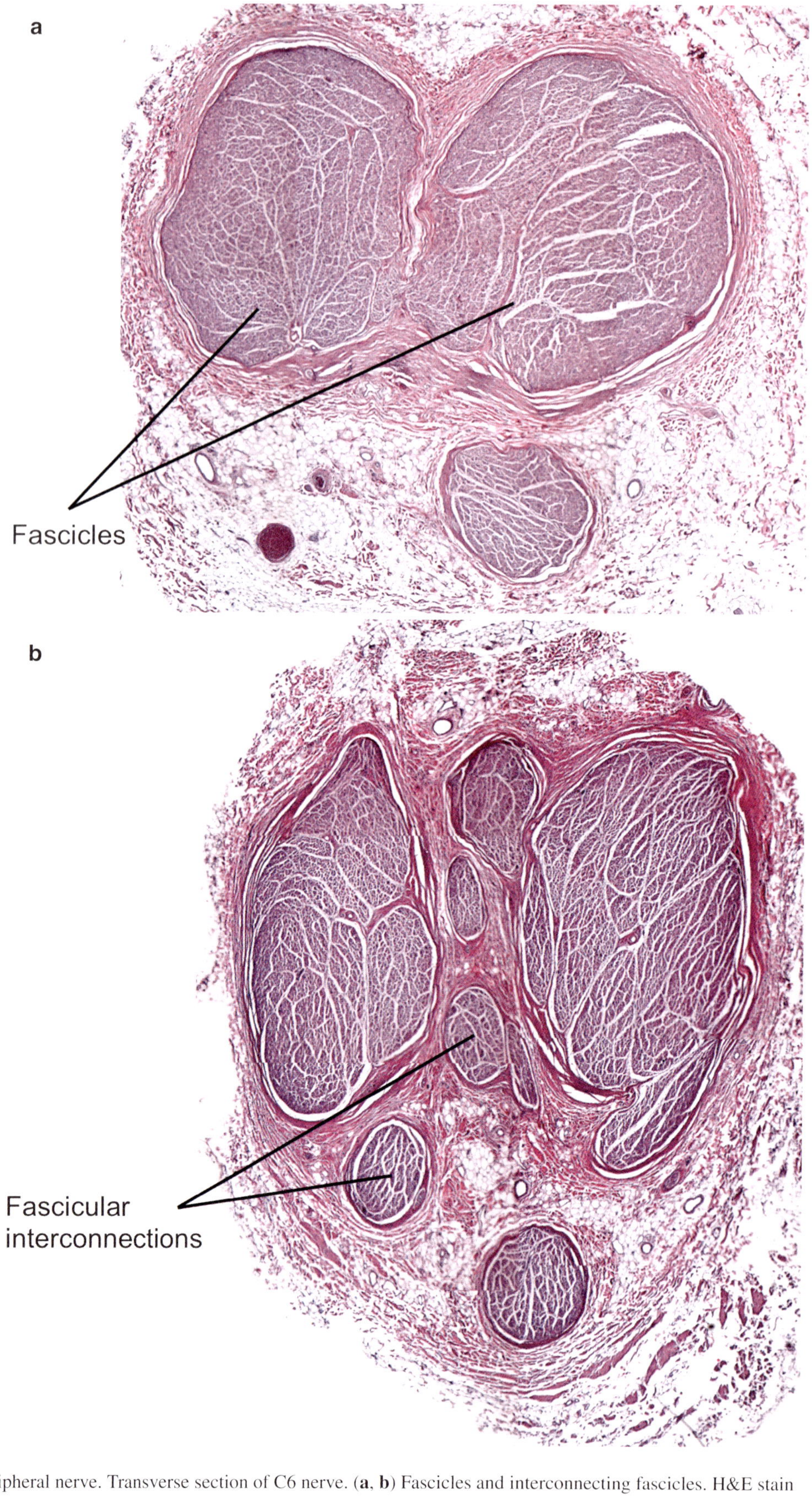

Fig. 6.9 Mixed peripheral nerve. Transverse section of C6 nerve. (**a**, **b**) Fascicles and interconnecting fascicles. H&E stain

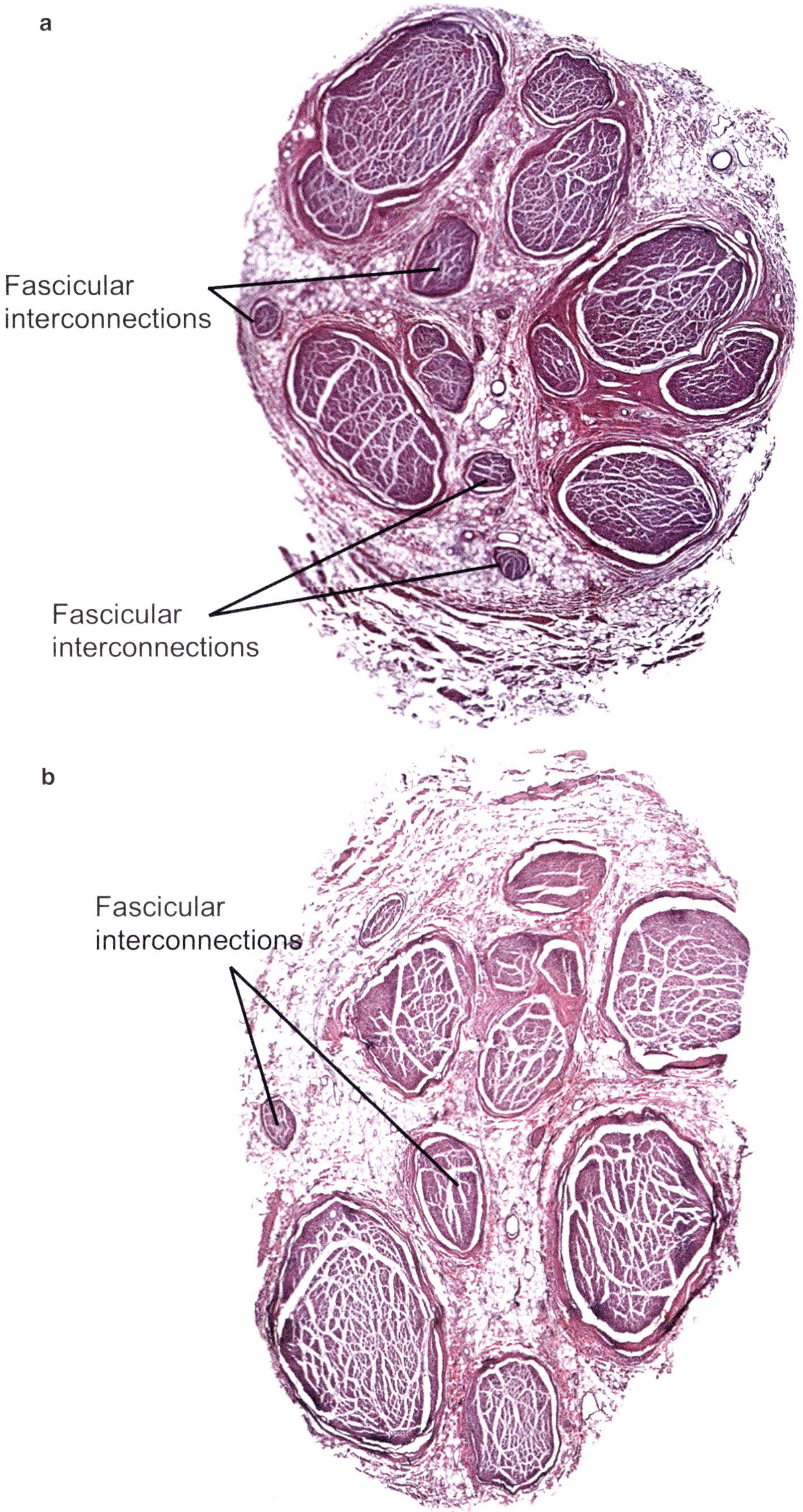

Fig. 6.10 Mixed peripheral nerve. Transverse section of C6 nerve. (**a**, **b**) Fascicles and interconnecting fascicles. H&E stain

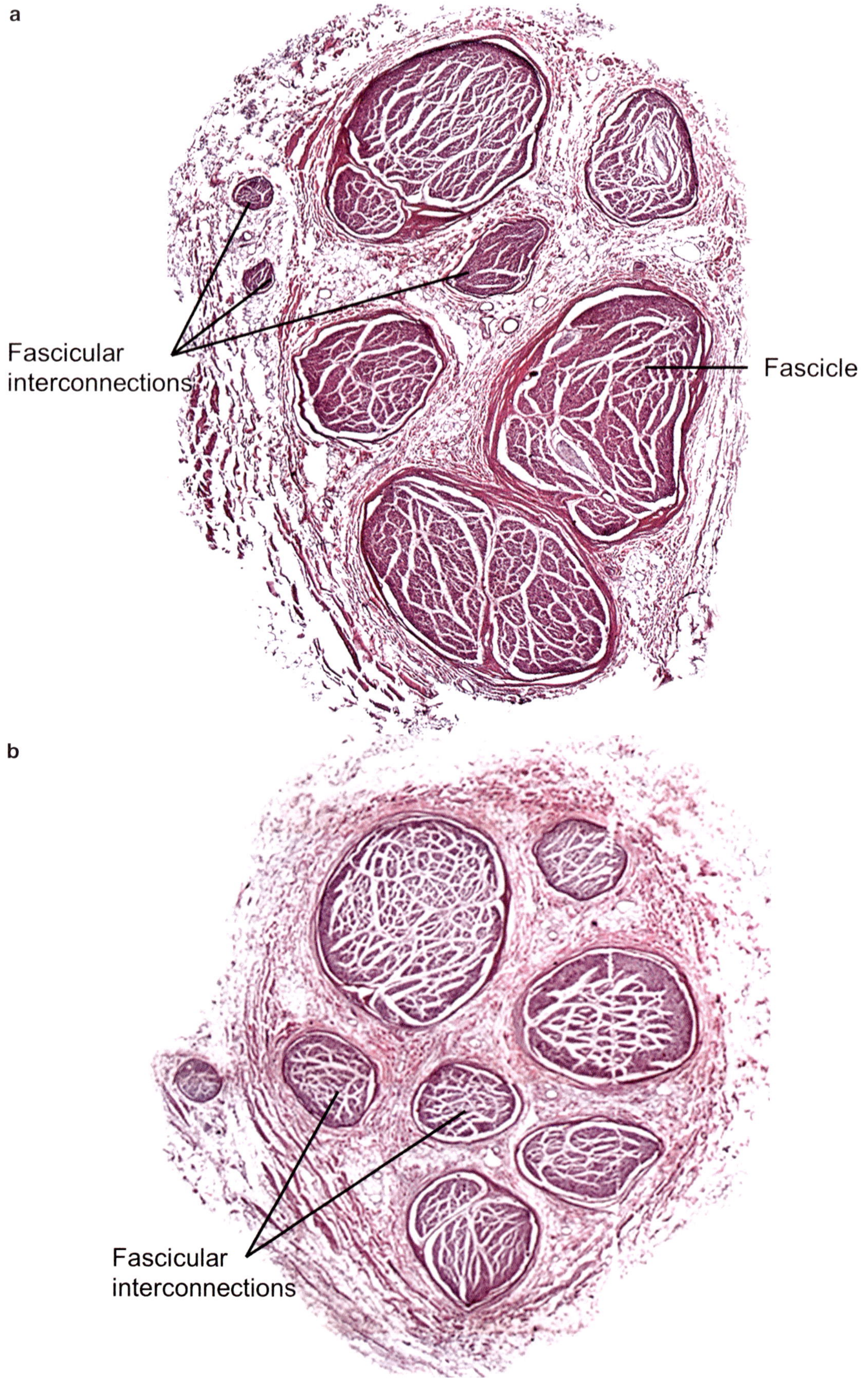

Fig. 6.11 Mixed peripheral nerve. Transverse section of C6 nerve. (**a**, **b**) Fascicles and interconnecting fascicles. H&E stain (Panel **a** from Reina et al. [2]; with permission)

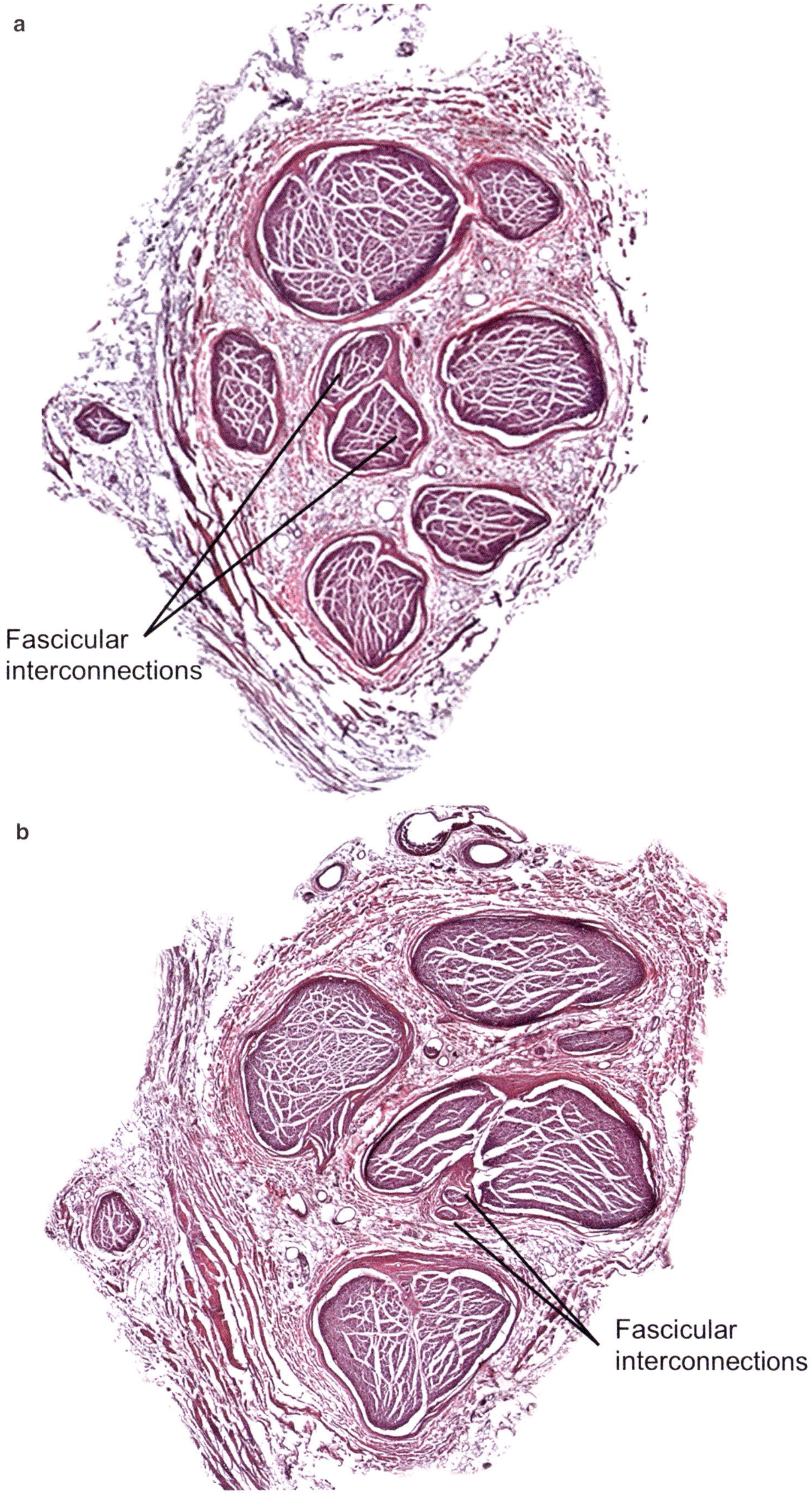

Fig. 6.12 Mixed peripheral nerve. Transverse section of C6 nerve. (**a**, **b**) Fascicles and interconnecting fascicles. H&E stain

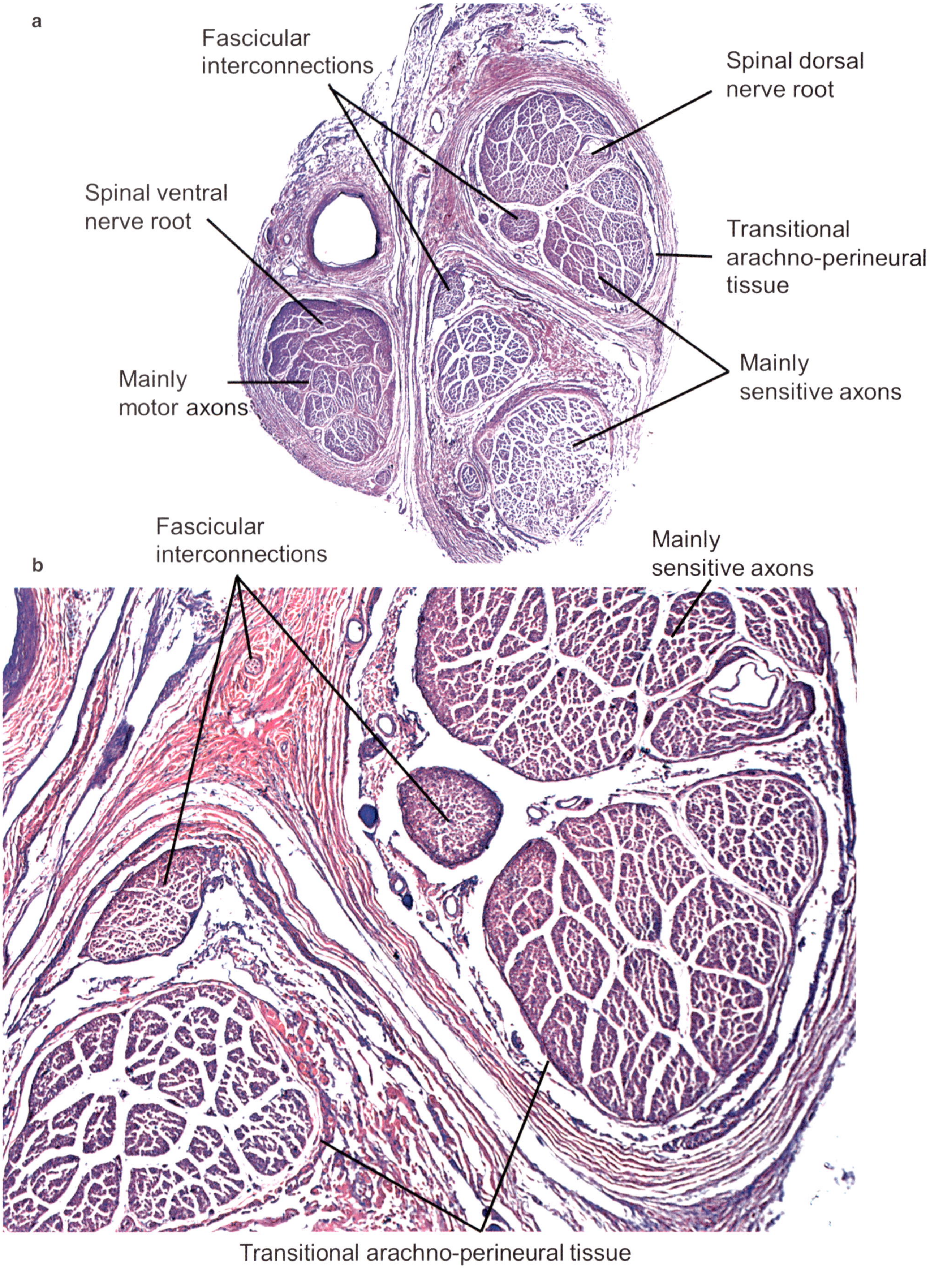

Fig. 6.13 Intraneural plexus. (**a**, **b**) Transverse section of C6 nerve. Details of the interconnecting fascicles in Fig. 6.4a. H&E stain

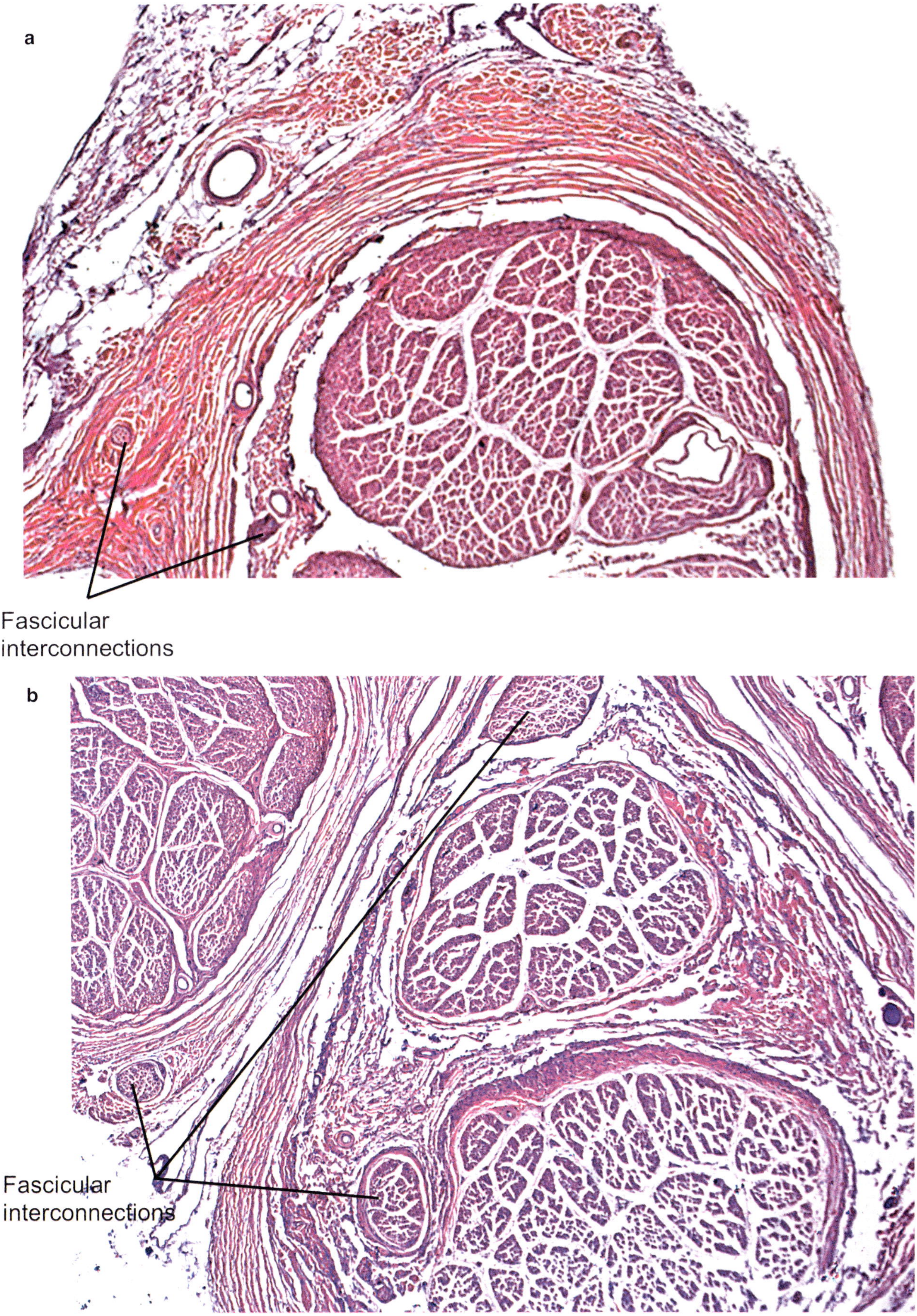

Fig. 6.14 Intraneural plexus. (**a**, **b**) Transverse section of C6 nerve. Details of the interconnecting fascicles in Fig. 6.4a. H&E stain

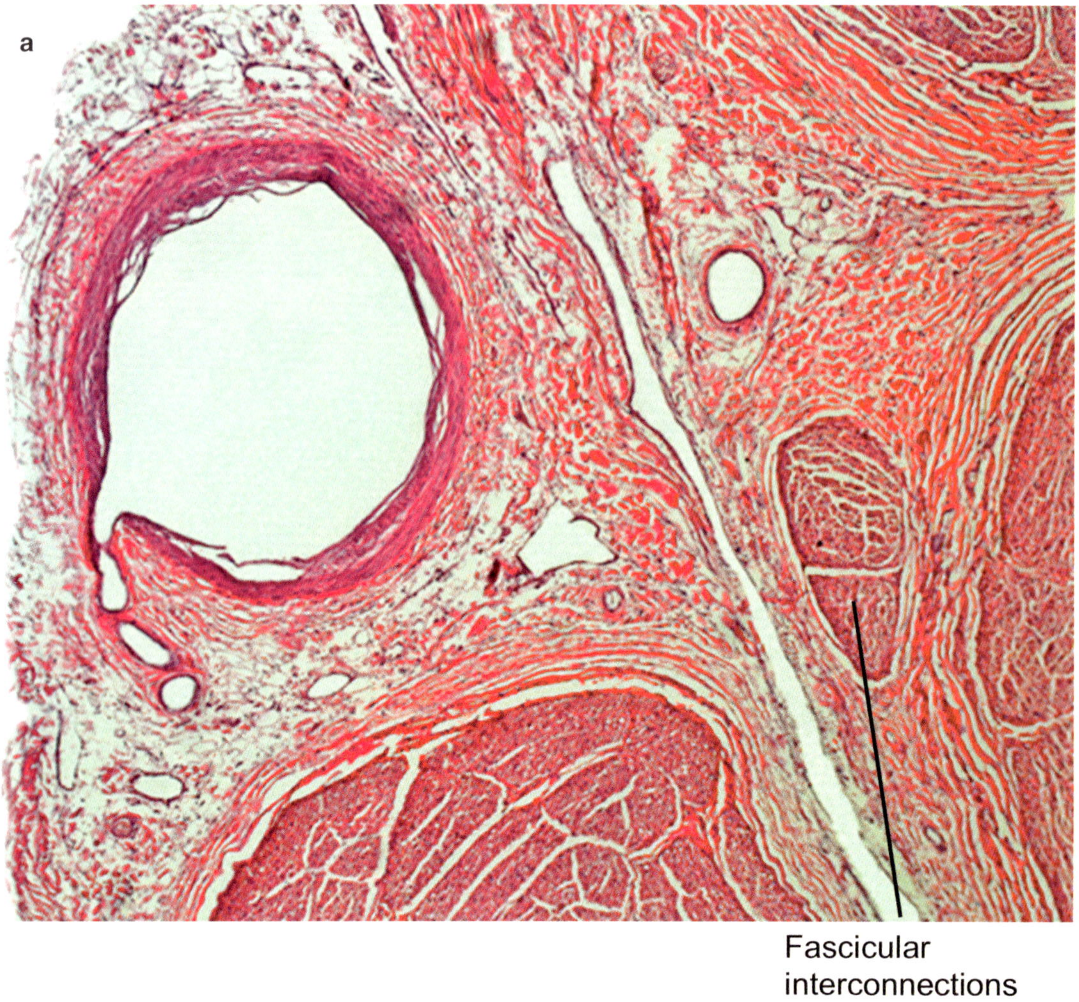

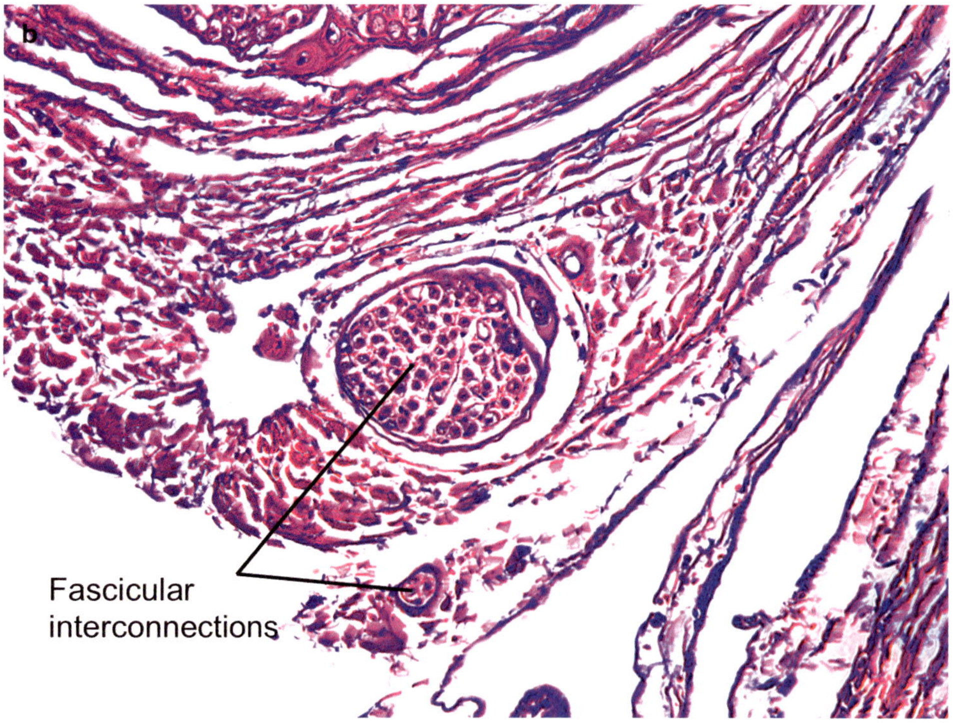

Fig. 6.15 (**a**, **b**) Transverse section of C6 nerve. Details of the interconnecting fascicles in Fig. 6.4b. H&E stain

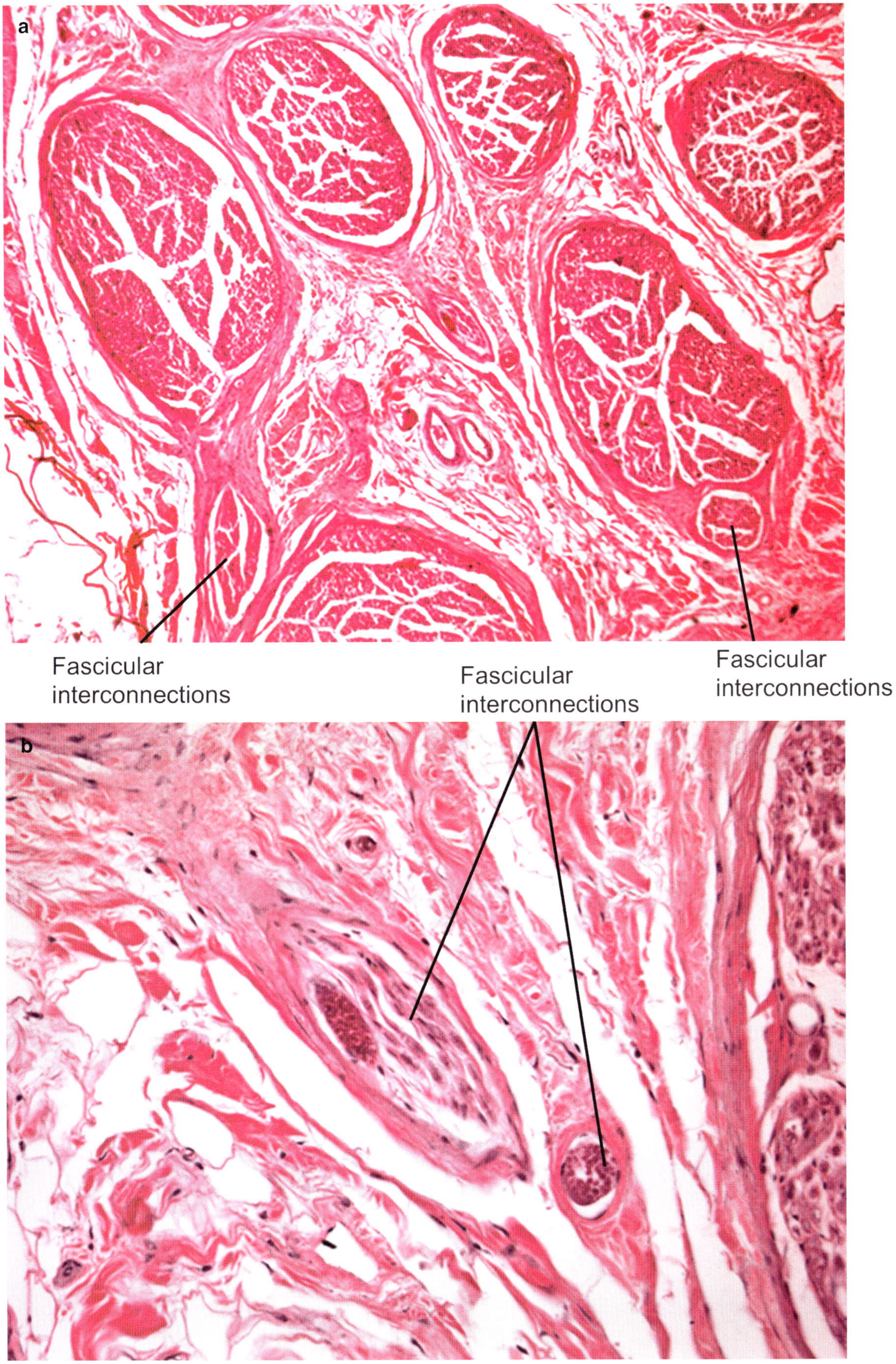

Fig. 6.16 Fascicles and interconnecting fascicles. (**a**, **b**) Detail of interconnecting fascicles at a location similar to that of Fig. 6.5a. H&E stain

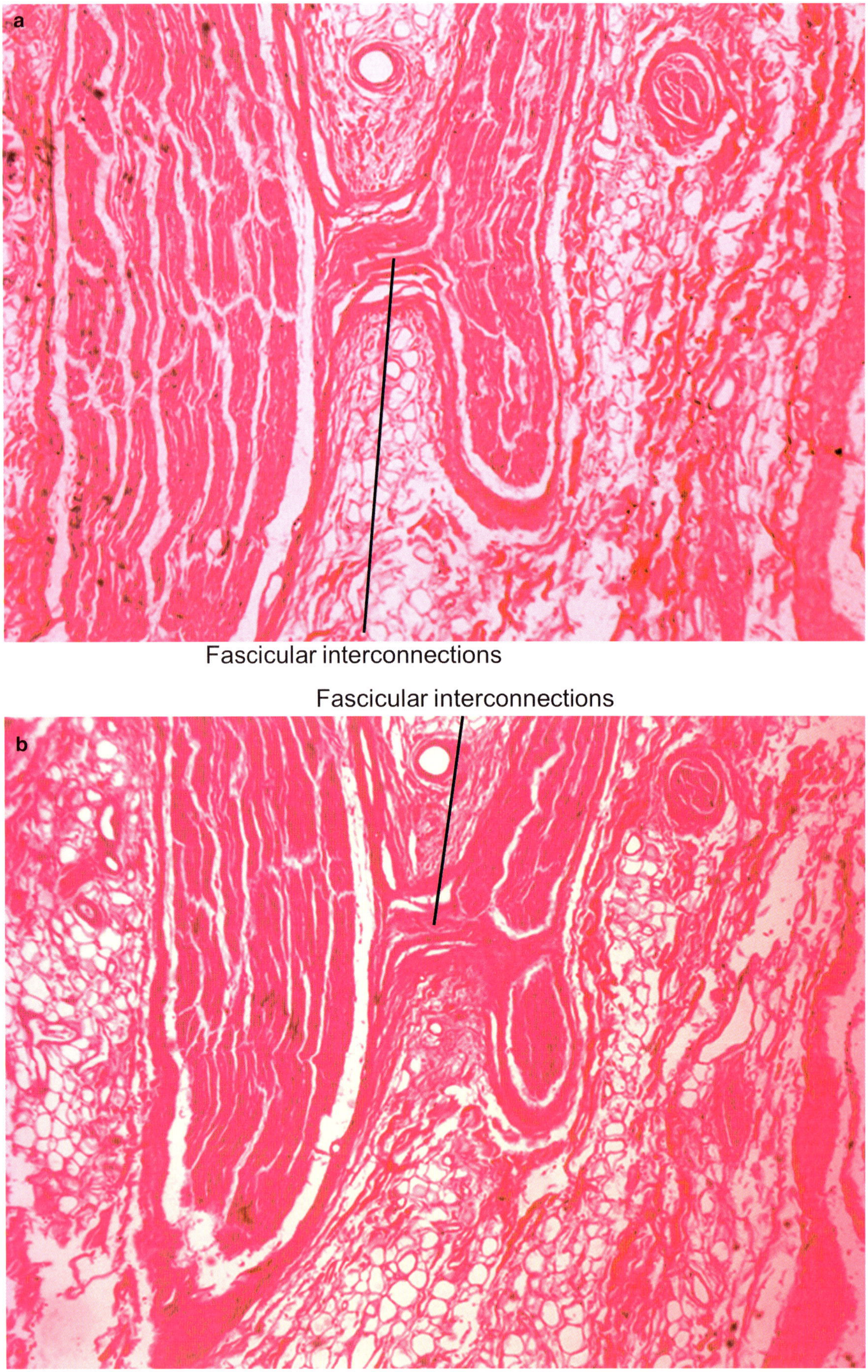

Fig. 6.17 Intraneural plexus. (**a**, **b**) Longitudinal section of C6 nerve. Detail of interconnecting fascicles at a location similar to that of Figs. 6.4 and 6.5. H&E stain

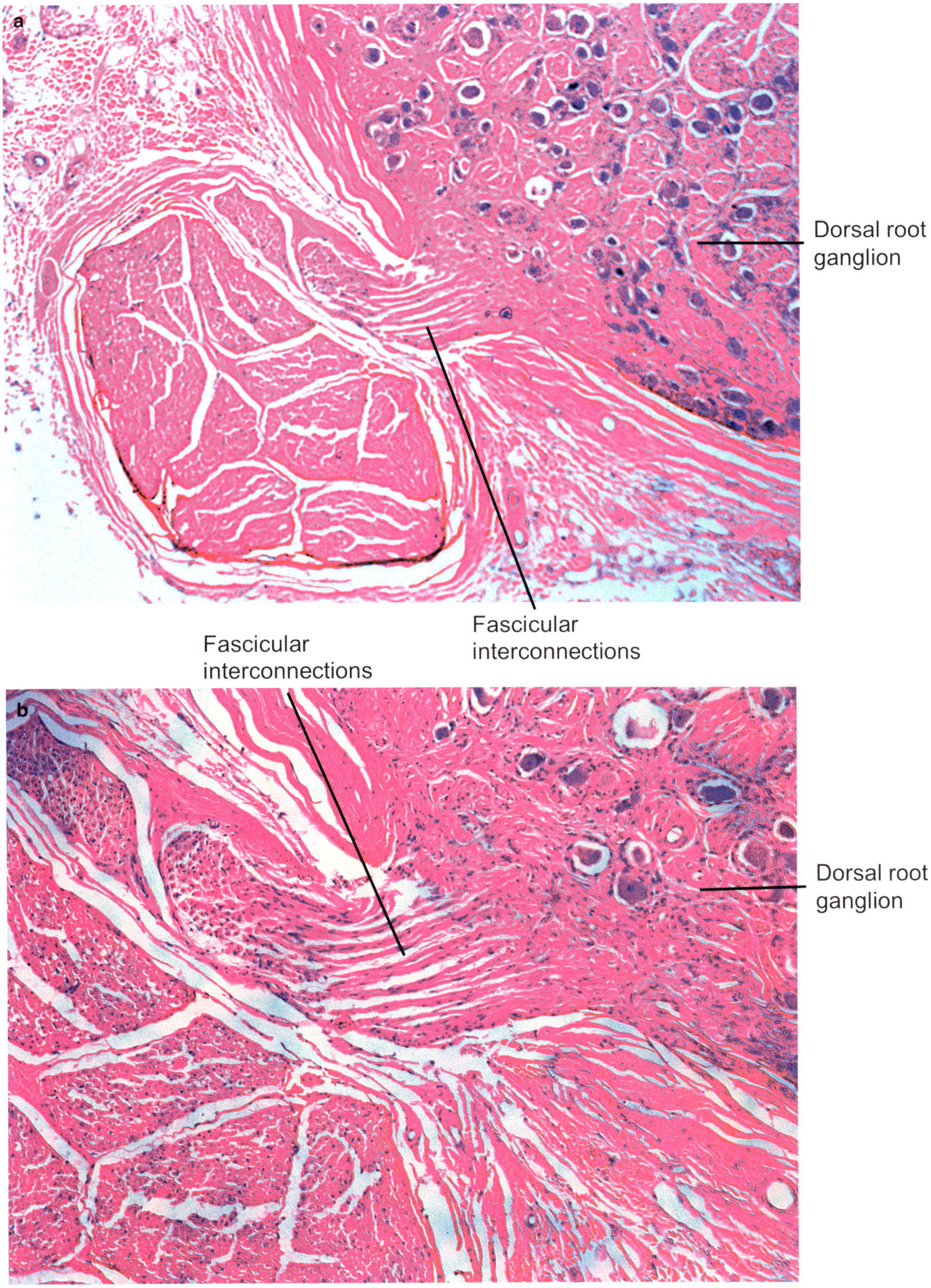

Fig. 6.18 Intraneural plexus. (**a**, **b**) Transverse section of C6 nerve. Details of interconnecting fascicles from Fig. 6.6b. H&E stain (Panel **a** from Reina et al. [1]; with permission)

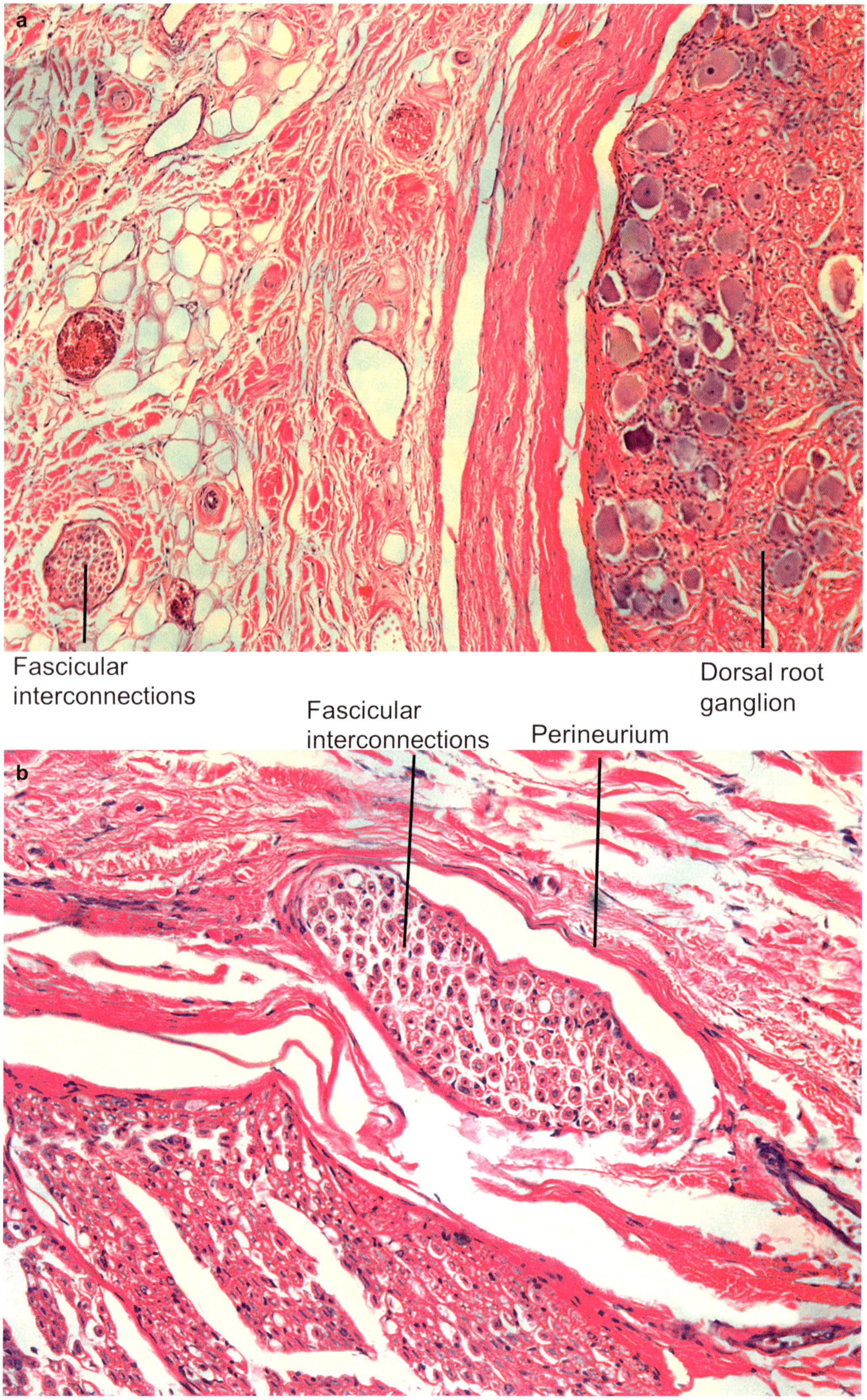

Fig. 6.19 Intraneural plexus. (**a**, **b**) Transverse section of C6 nerve. Details of interconnecting fascicles from (**a**) Fig. 6.6b and (**b**) Fig. 6.5a. H&E stain

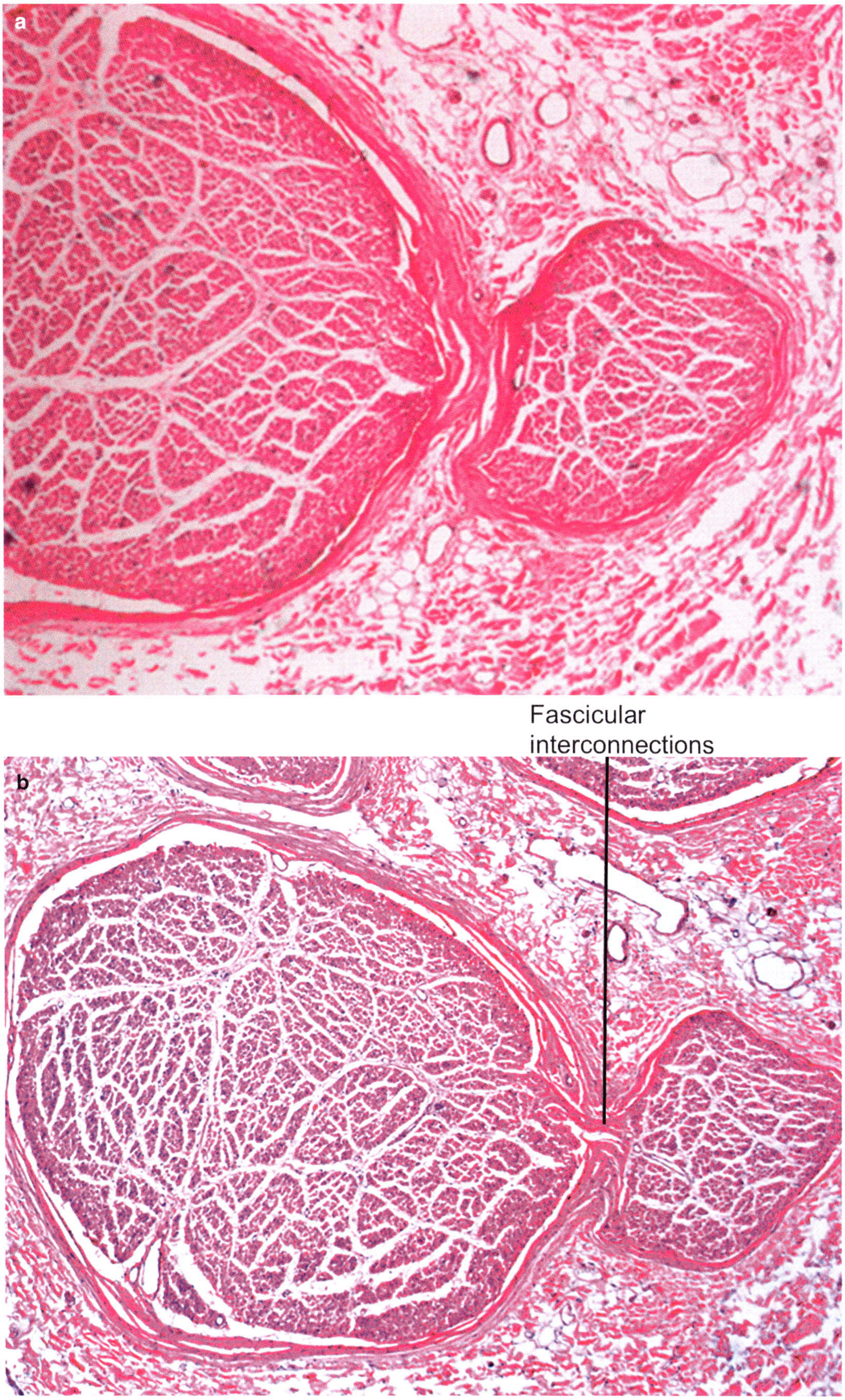

Fig. 6.20 Intraneural plexus. (**a**, **b**) Transverse section of C6 nerve. Details of interconnecting fascicles in a peripheral nerve distal to the DRG. H&E stain (Panel **b** from Reina et al. [1]; with permission)

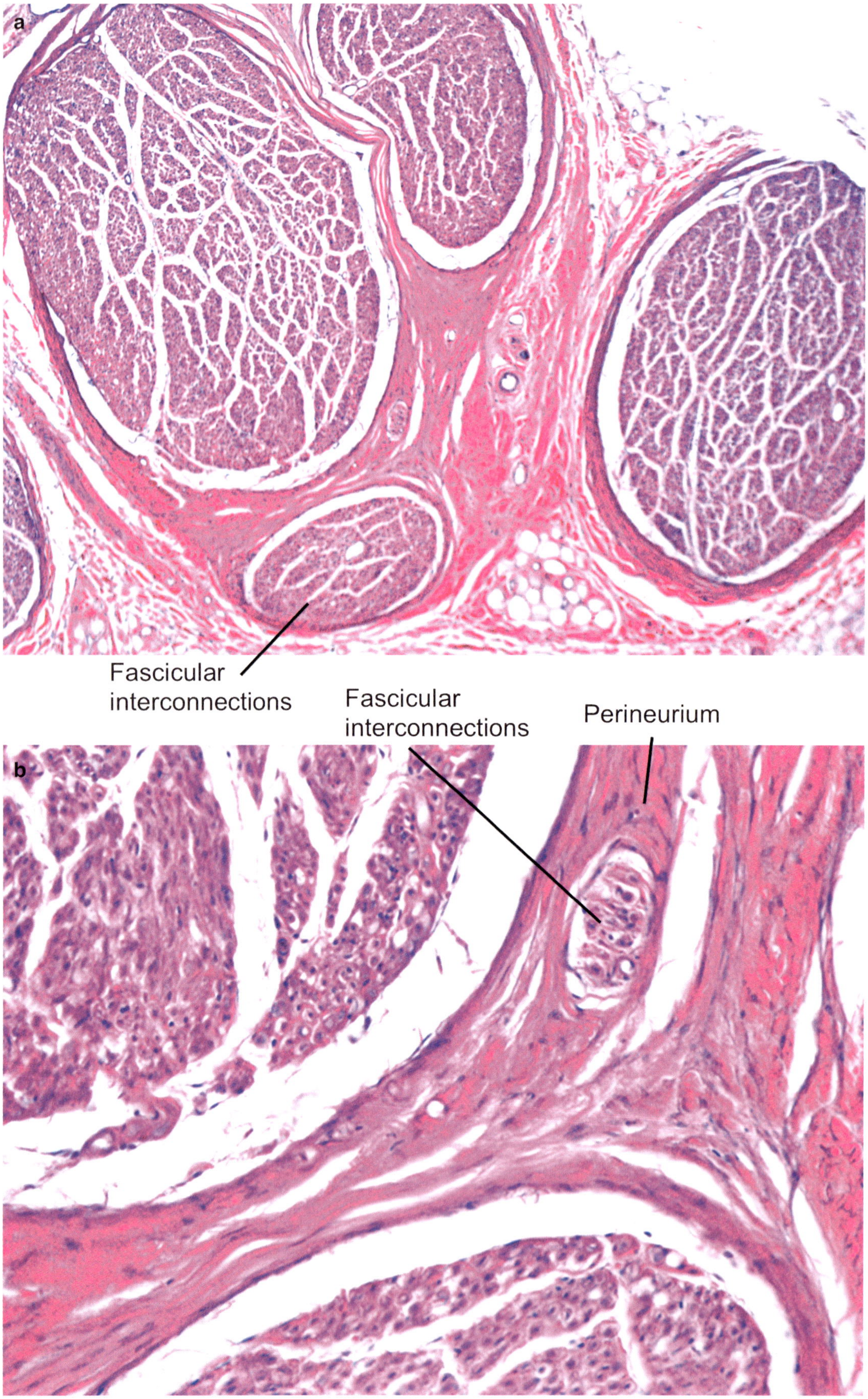

Fig. 6.21 Intraneural plexus. (**a**, **b**) Transverse section of C6 nerve. Details of interconnecting fascicles in a peripheral nerve distal to the DRG. H&E stain

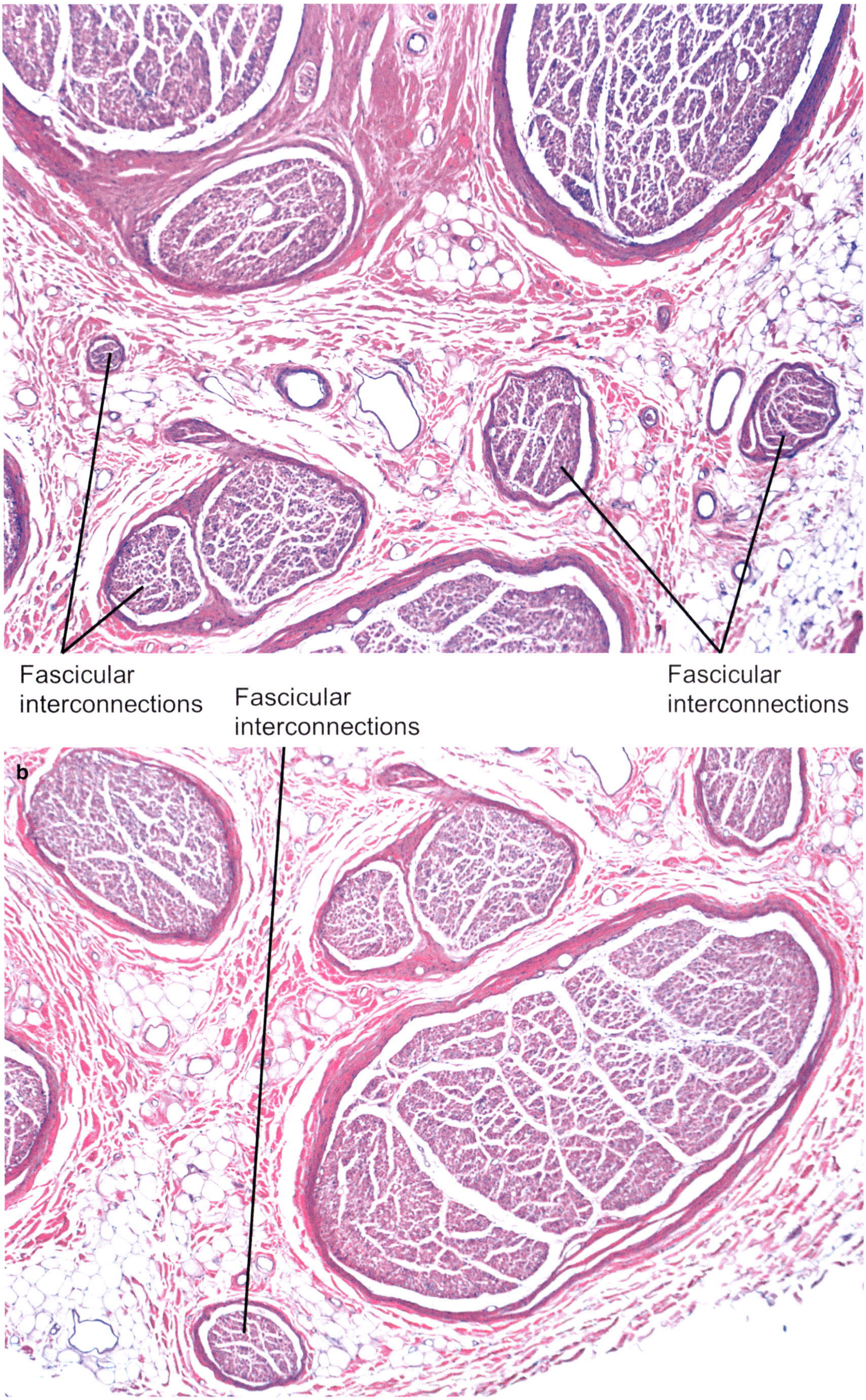

Fig. 6.22 Intraneural plexus. (**a**, **b**) Transverse section of C6 nerve. Details of interconnecting fascicles in a peripheral nerve distal to the DRG. H&E stain (Panel **a** from Reina et al. [1]; with permission)

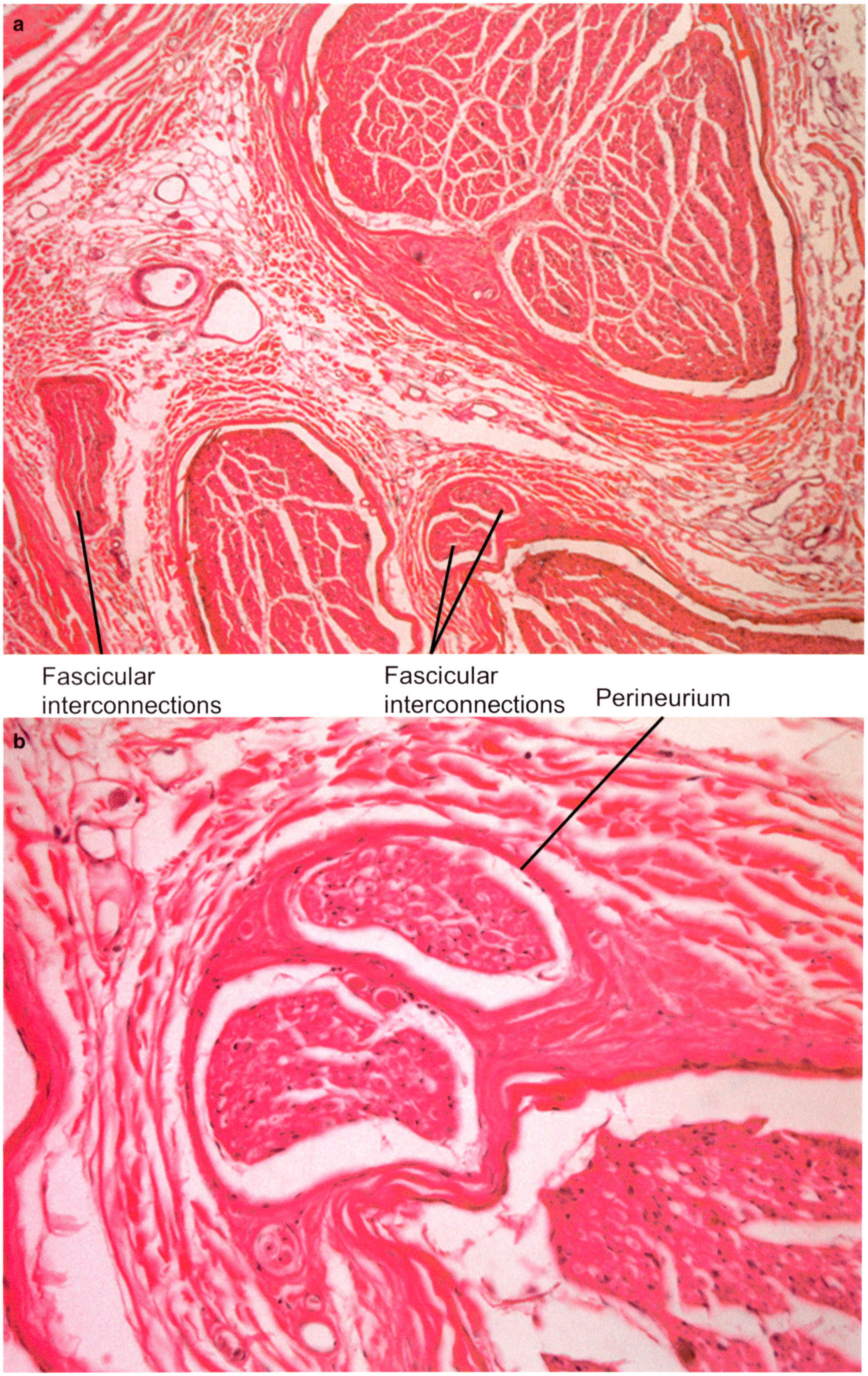

Fig. 6.23 Intraneural plexus. (**a**, **b**) Transverse section of C6 nerve. Details of interconnecting fascicles in a peripheral nerve distal to the DRG. H&E stain

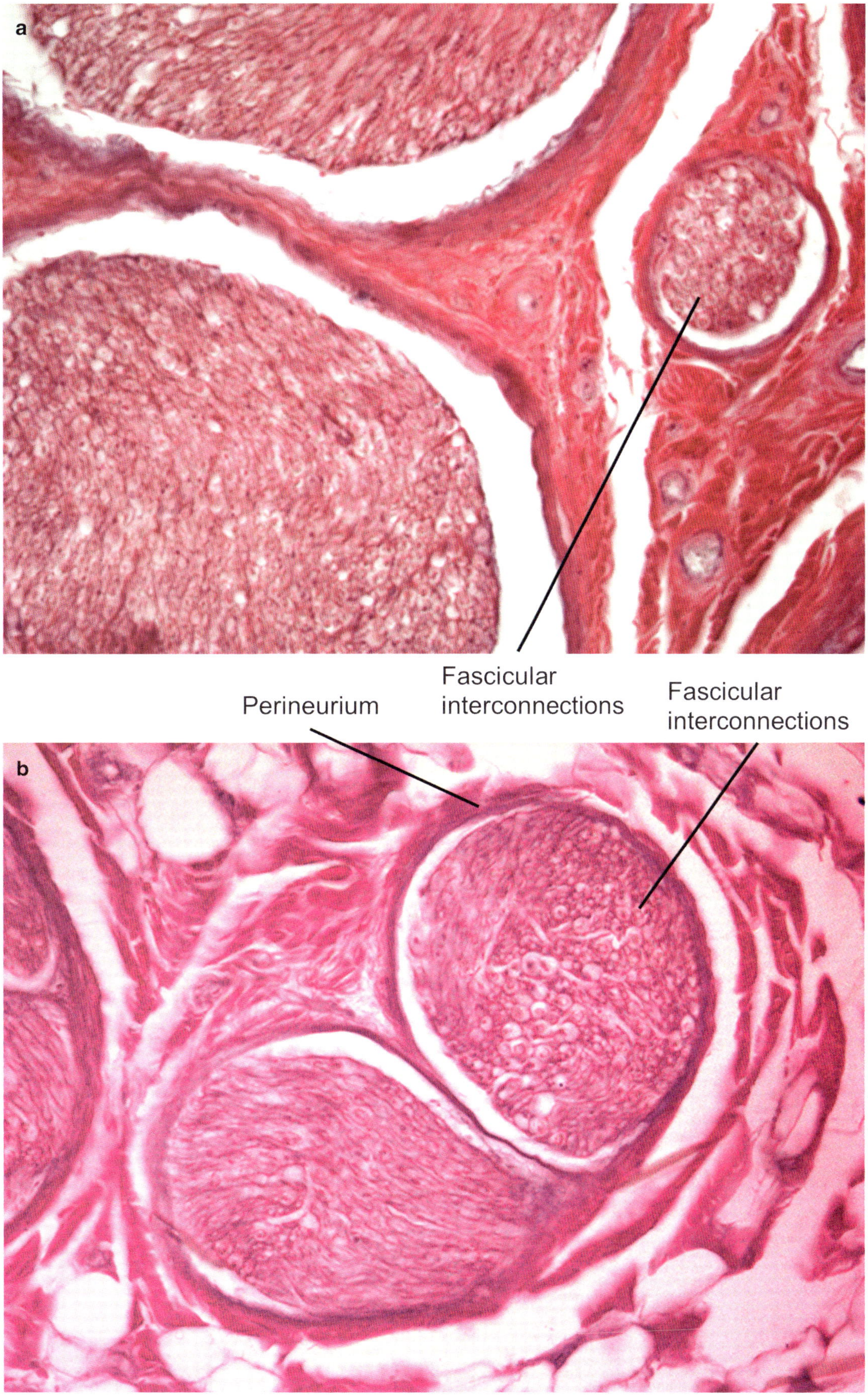

Fig. 6.24 Intraneural plexus. (**a**, **b**) Transverse section of C6 nerve. Details of interconnecting fascicles from the brachial plexus. H&E stain

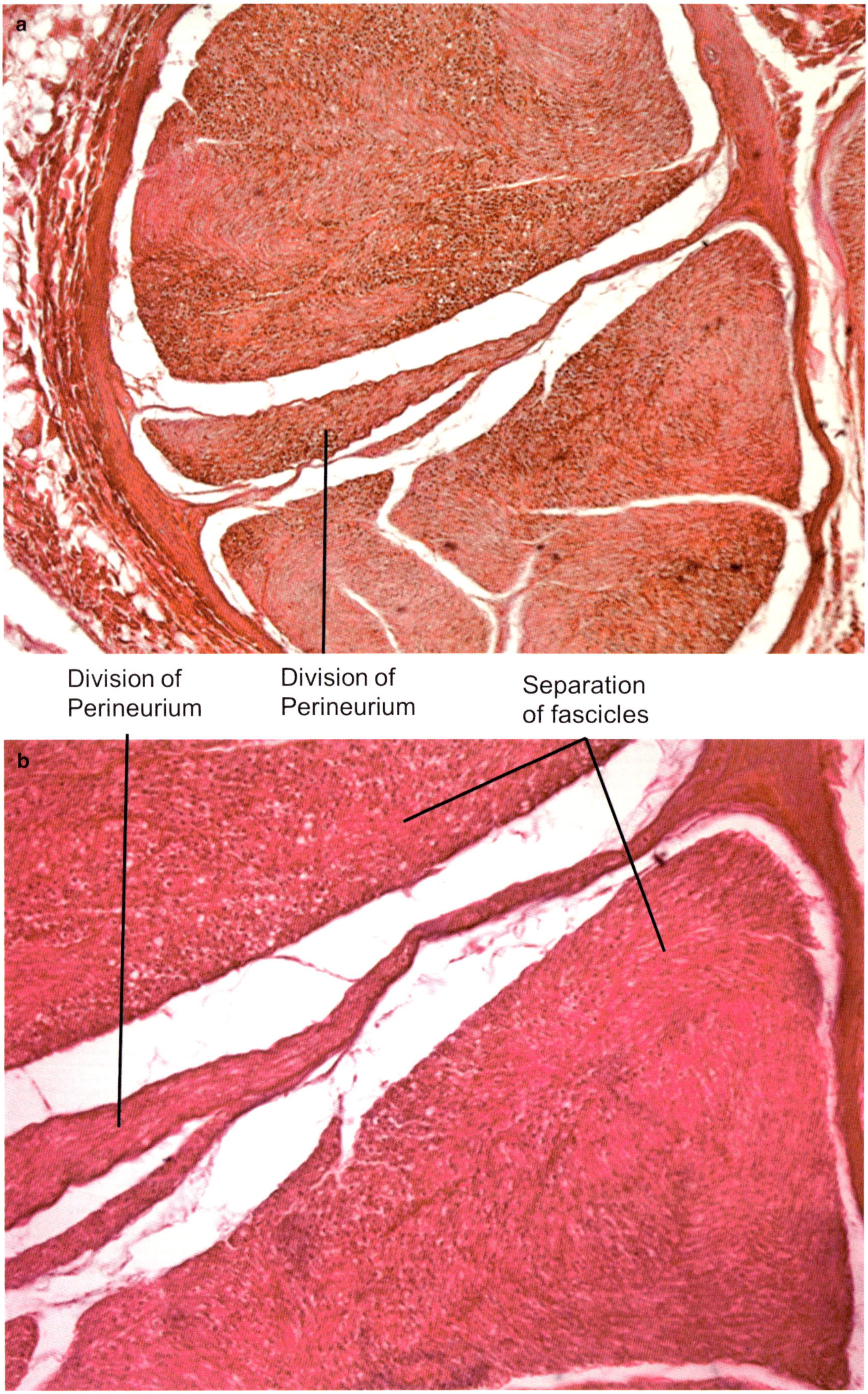

Fig. 6.25 Intraneural plexus. (**a**, **b**) Transverse section of C6 nerve. Details of interconnecting fascicles from the brachial plexus. H&E stain

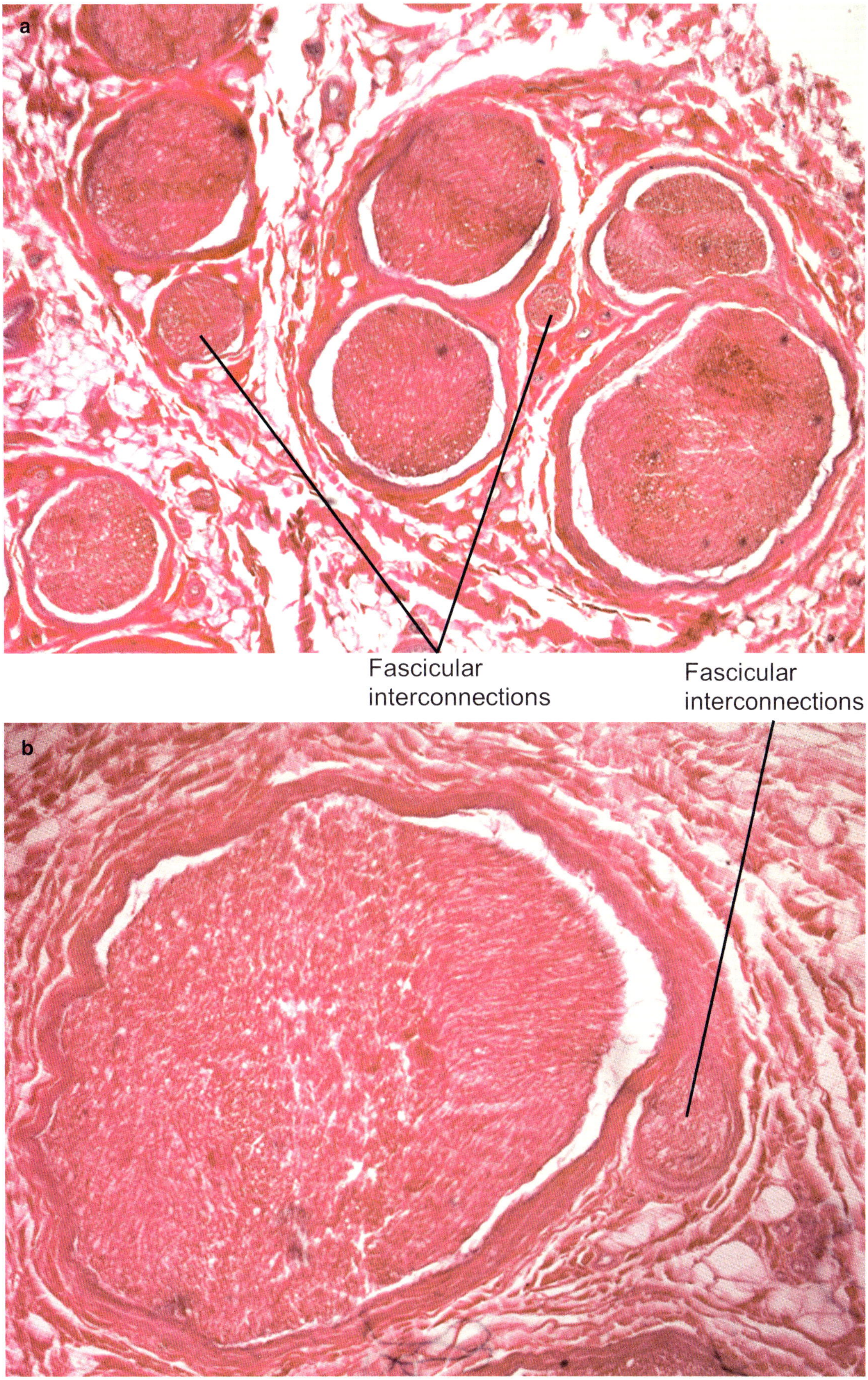

Fig. 6.26 Intraneural plexus. (**a**, **b**) Transverse section of C6 nerve. Details of interconnecting fascicles from the brachial plexus. H&E stain

References

1. Reina MA, Arriazu R, Collier CB, Sala-Blanch X. Histology and electron microscopy of human peripheral nerves of clinical relevance to the practice of nerve blocks. Rev Esp Anestesiol Reanim. 2013;60:552–62.
2. Reina MA, De Andres JA, Hernández JM, Arriazu Navarro R, Durán Mateos EM, Prats-Galino A. Successive changes in extraneural structures from the subarachnoid nerve roots to the peripheral nerve, influencing anesthetic block, and treatment of acute postoperative pain. Eur J Pain Suppl. 2011;5:377–85.
3. De Andrés JA, Reina MA, López A, Sala-Blanch X, Prats A. Blocs nerveux périphériques, paresthésies et injections intraneurales. Le Practicien En Anesth Réanim. 2010;14:213–21.
4. Sunderland S, Marshall RD, Swaney WE. The intraneural topography of the circumflex musculocutaneous and obturator nerves. Brain. 1959;82:116–29.
5. Sunderland S, Ray LJ. The intraneural topography of the sciatic nerve and its popliteal divisions in man. Brain. 1948;71:242–58.
6. Sunderland S. The intraneural topography of the radial, median and ulnar nerves. Brain. 1945;68:243–55.
7. Sunderland S, Bradley KC. The cross-sectional area of peripheral nerve trunks devoted to nerve fibres. Brain. 1949;72:428–49.
8. Yokoyama I. Study on the intraneural topography of the brachial plexus. J Jpn Orthop Assoc. 1989;63:1085–102.
9. Yokoyama I, Matsuda H, Hashimoto T, Kureya S, Shimazu A. Intraneural topography of the brachial plexus. Microscopic dissection in human cadavers. J Jpn Soc Surg Hand. 1986;3:138–45.

Macroscopic View of the Cervical Plexus and Brachial Plexus

7

Anna Carrera, Francisco Reina, Xavier Sala-Blanch, María Rosa Morro, and Amer Mustafa Gondolbeu

The lateral anatomic region of the neck adjacent to the cervical spine contains, among its muscular elements, the cervical plexus and brachial plexus. Nerve plexuses are axon exchange networks that allow peripheral nerves to form from the fibers of two or more consecutive spinal nerves. In this sense, deep to the upper half of the sternocleidomastoid muscle, the spinal nerves from C1 to C4 exchange the axons of their anterior rami in the cervical plexus, giving rise to the peripheral nerves distributed in the anterolateral region of the neck.

Meanwhile, in the interscalene space, the anterior branches of spinal nerves C5 to T1 convene in the brachial plexus. Nerves resulting from this plexus pass under the clavicle and through the axilla to reach the upper extremity. Knowledge of the specific anatomic locations of these neural structures, as well as their function, is essential to the clinical application of anesthetic blocks and to understanding their results (Figs. 7.1, 7.2, 7.3, 7.4, 7.5, 7.6, 7.7, 7.8, 7.9, 7.10, 7.11, 7.12, 7.13, 7.14, 7.15, 7.16, 7.17, 7.18, 7.19, 7.20, 7.21, 7.22, 7.23, 7.24, 7.25, 7.26, 7.27, 7.28, 7.29, 7.30, 7.31, 7.32, and 7.33).

A. Carrera, MD, PhD (✉) • F. Reina, MD, PhD
Neuroscience, Embryology, Molecular Oncology, and Clinical Anatomy Group (NEOMA), School of Medicine, University of Girona, Girona, Spain
e-mail: anna.carrera@udg.edu; francisco.reina@udg.edu

X. Sala-Blanch, MD
Human Anatomy and Embryology Unit, School of Medicine, University of Barcelona, Barcelona, Spain

Department of Anesthesiology and Critical Care Clinic Hospital, Barcelona, Spain
e-mail: xavi.sala.blanch@gmail.com

M.R. Morro, MD
Human Anatomy and Embryology Unit, School of Medicine, University of Barcelona, Barcelona, Spain

Orthopaedic and Traumatology Surgery Unit, Vall d'Hebron Hospital, Barcelona, Spain
e-mail: rosamorro@yahoo.es

A. Mustafa Gondolbeu, MD
Human Anatomy and Embryology Unit, School of Medicine, University of Barcelona, Barcelona, Spain

Orthopaedic and Traumatology Surgery Unit, Arnau de Vilanova Hospital, Lleida, Spain
e-mail: spike.tenma@gmail.com

M.A. Reina et al. (eds.), *Atlas of Functional Anatomy for Regional Anesthesia and Pain Medicine: Human Structure, Ultrastructure and 3D Reconstruction Images*, DOI 10.1007/978-3-319-09522-6_7,

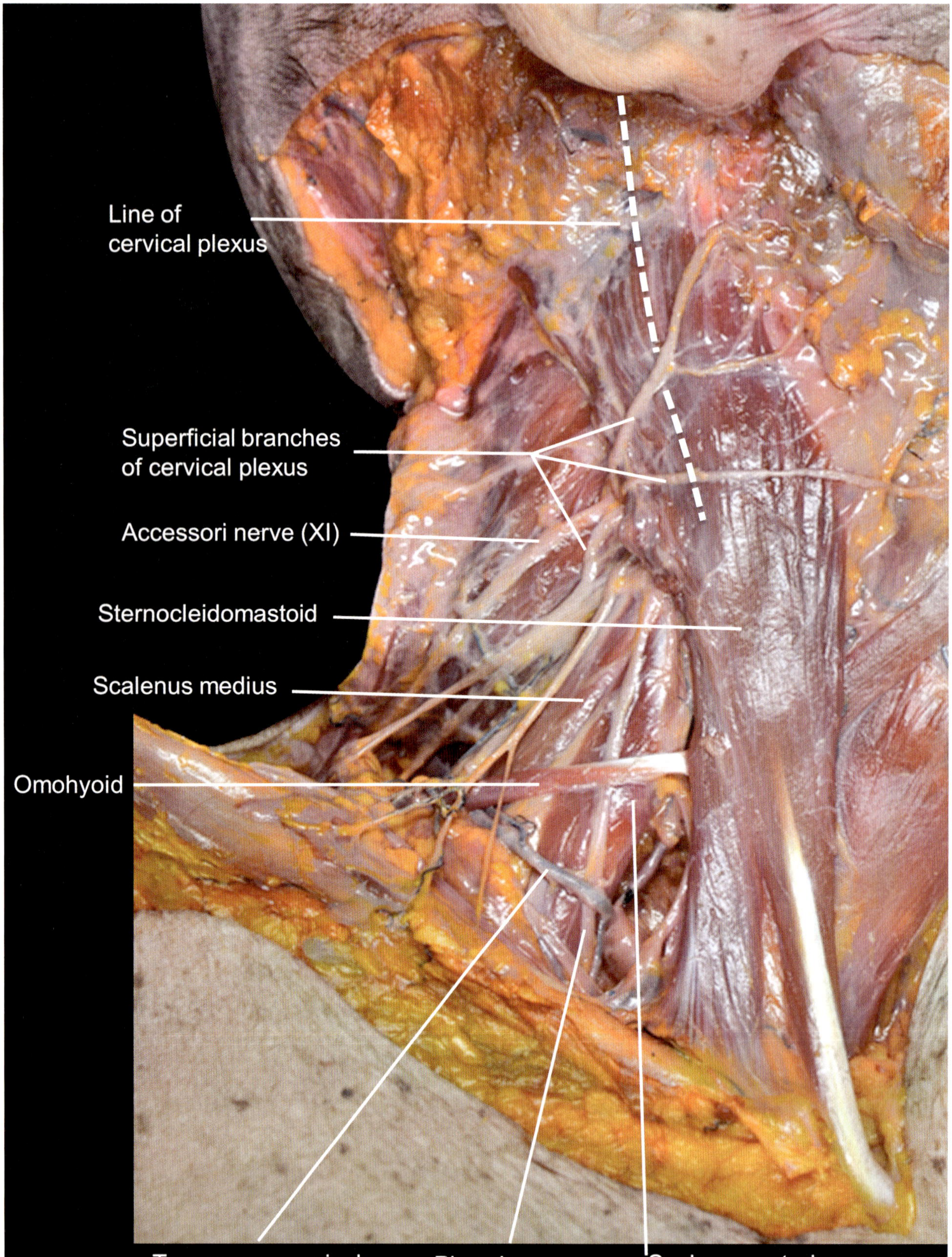

Fig. 7.1 Cervical plexus. The cervical plexus may be visualized as a line under the upper half of the sternocleidomastoid muscle, following the transverse processes of the first cervical vertebrae. Cutaneous branches of this plexus are distributed in the lateral superficial area of the neck

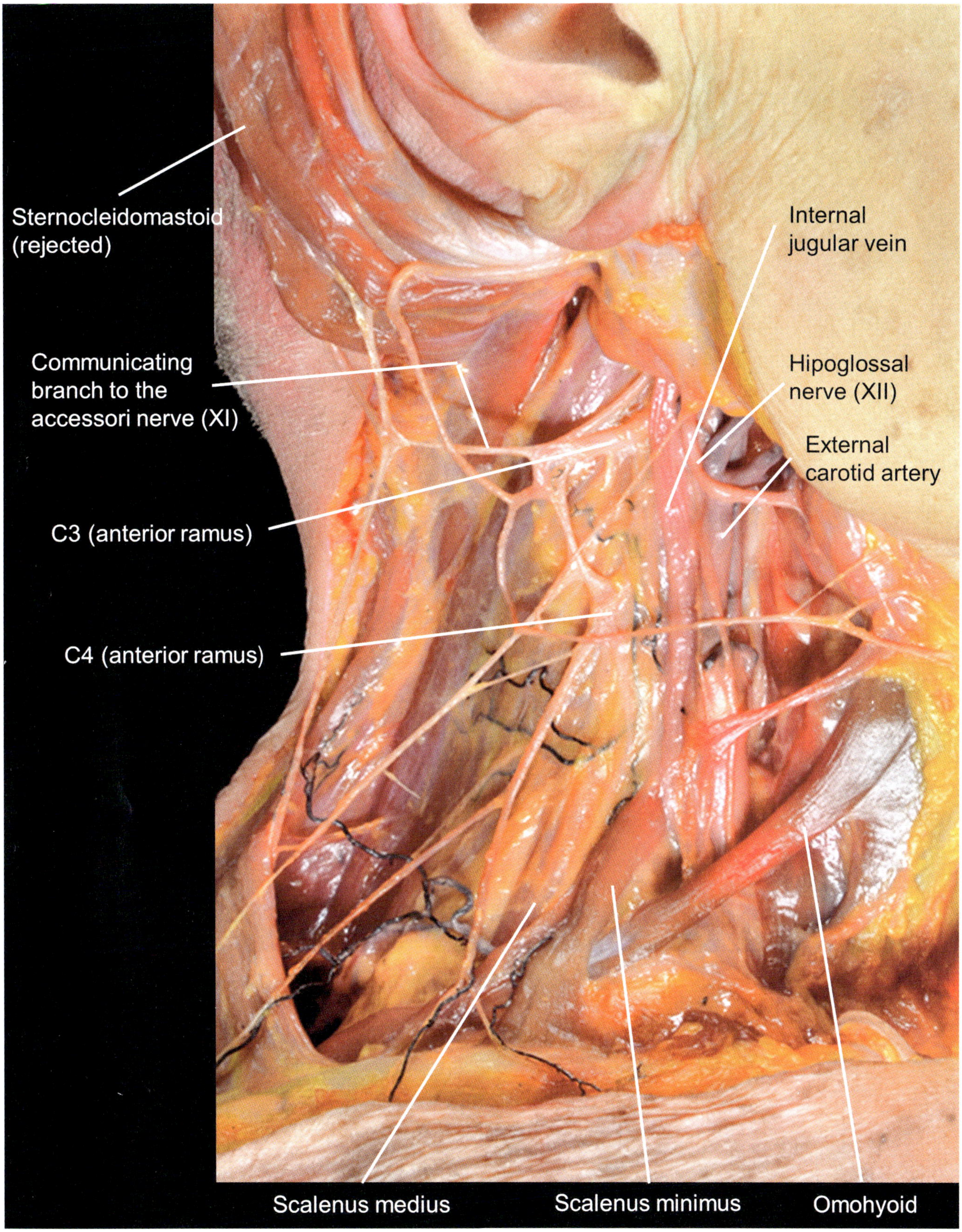

Fig. 7.2 Cervical plexus. Under the sternocleidomastoid muscle, the internal jugular vein is the element of the neck's neurovascular bundle that is most intimately related to the output course of the anterior branches of spinal nerves C1, C2, C3, and C4

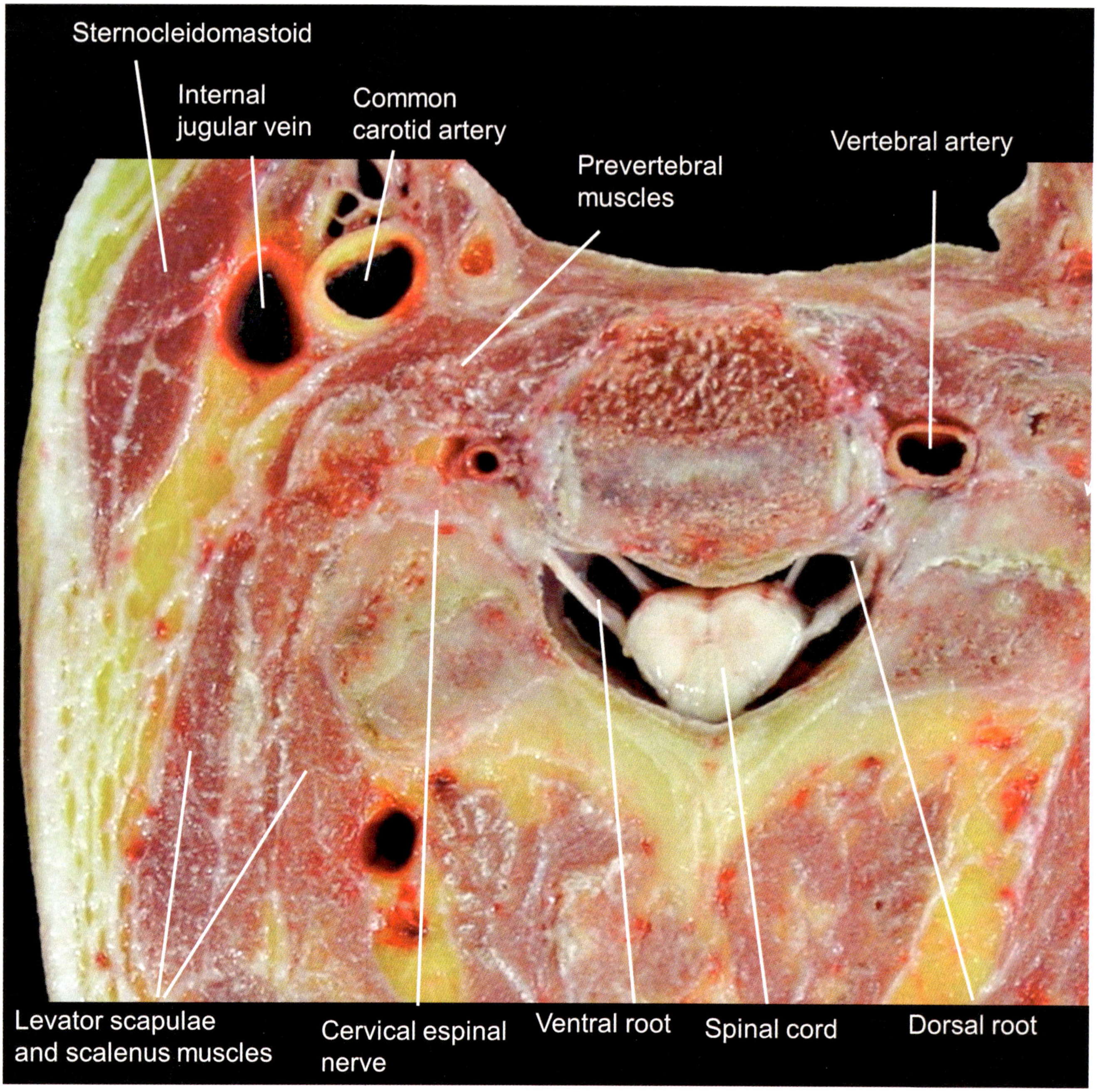

Fig. 7.3 Cervical plexus. In a cross-section of the cervical region, it may be seen that the output of the anterior branches of the spinal nerves forming the cervical plexus is located in the boundary between the insertion of the prevertebral muscles anteriorly and the levator scapulae and scalene muscles posteriorly. On its way through the intervertebral foramen, each spinal nerve is located posterior to the ascending trajectory of the vertebral artery

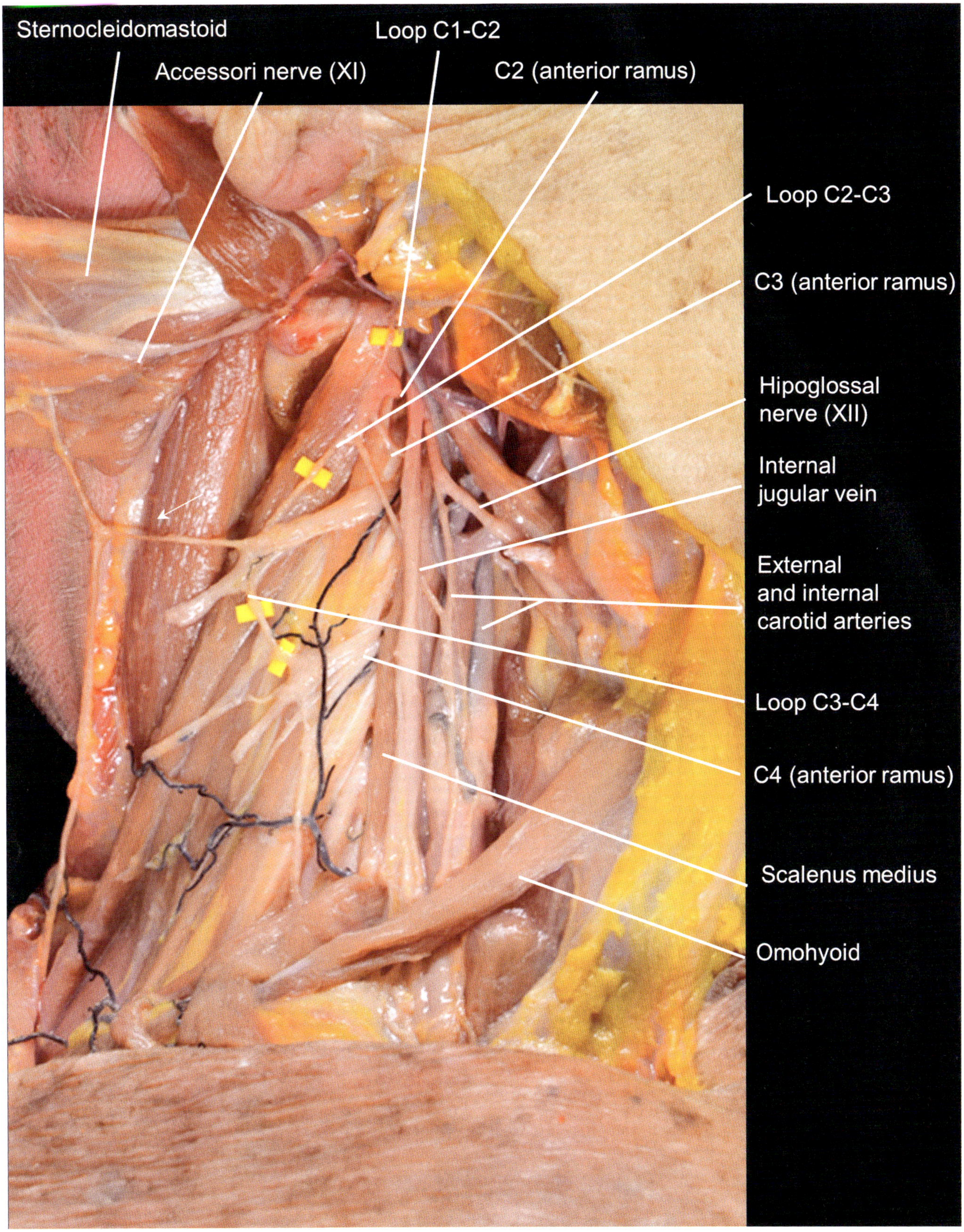

Fig. 7.4 Cervical plexus. The cervical plexus consists of three arches of nervous connection between the spinal nerves that integrate it. Each spinal nerve combines its axons with those of the adjacent spinal nerves, forming the C1–C2 loop, or arch of the atlas (arranged around the first cervical vertebra); the C2–C3 loop, or arch of the axis (surrounding the second cervical vertebra); and the C3–C4 loop

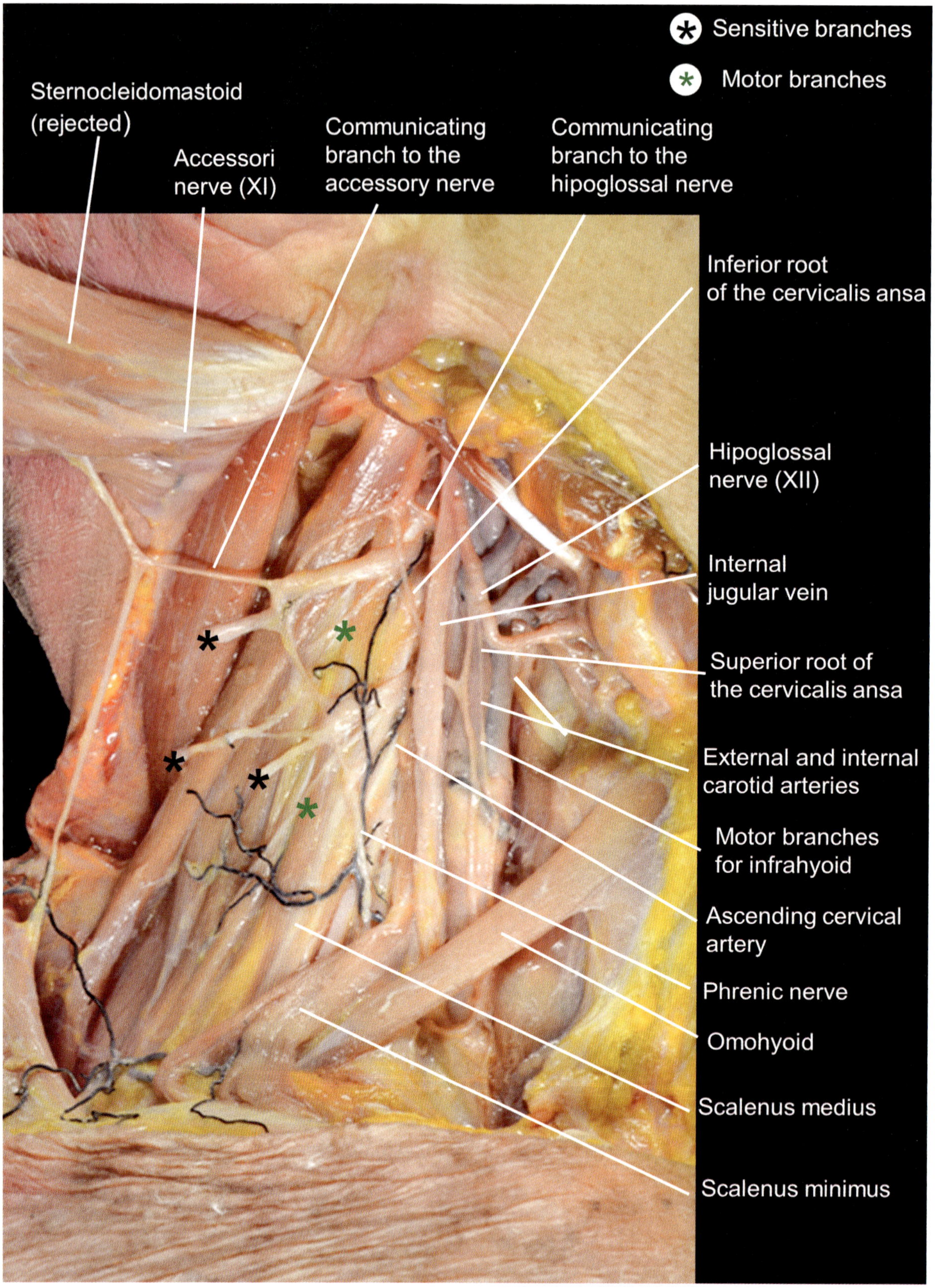

Fig. 7.5 Cervical plexus. From the connections among the C1, C2, C3, and C4 nerves, specific nerves of this plexus are issued: motor nerves that innervate muscles in the region, sensory nerves (cut in this preparation) that will access the superficial plane to be distributed widely in the anterolateral region of the neck, and communicating branches to the sympathetic chain and cranial nerves XI and XII. Connections to the hypoglossal nerve (XII pc) provide part of the motor innervation of this plexus. Thus, the thyrohyoid nerve, the geniohyoid nerve, and the upper root of the cervical loop correspond to C1 and C2 nerve fibers that previously were incorporated into the path of this cranial nerve. Further, the upper root of the cervical loop joins the inferior root formed by branches from C2 and C3. Motor innervation of the sternothyroid, thyrohyoid, and omohyoid muscles is provided by this cervical loop

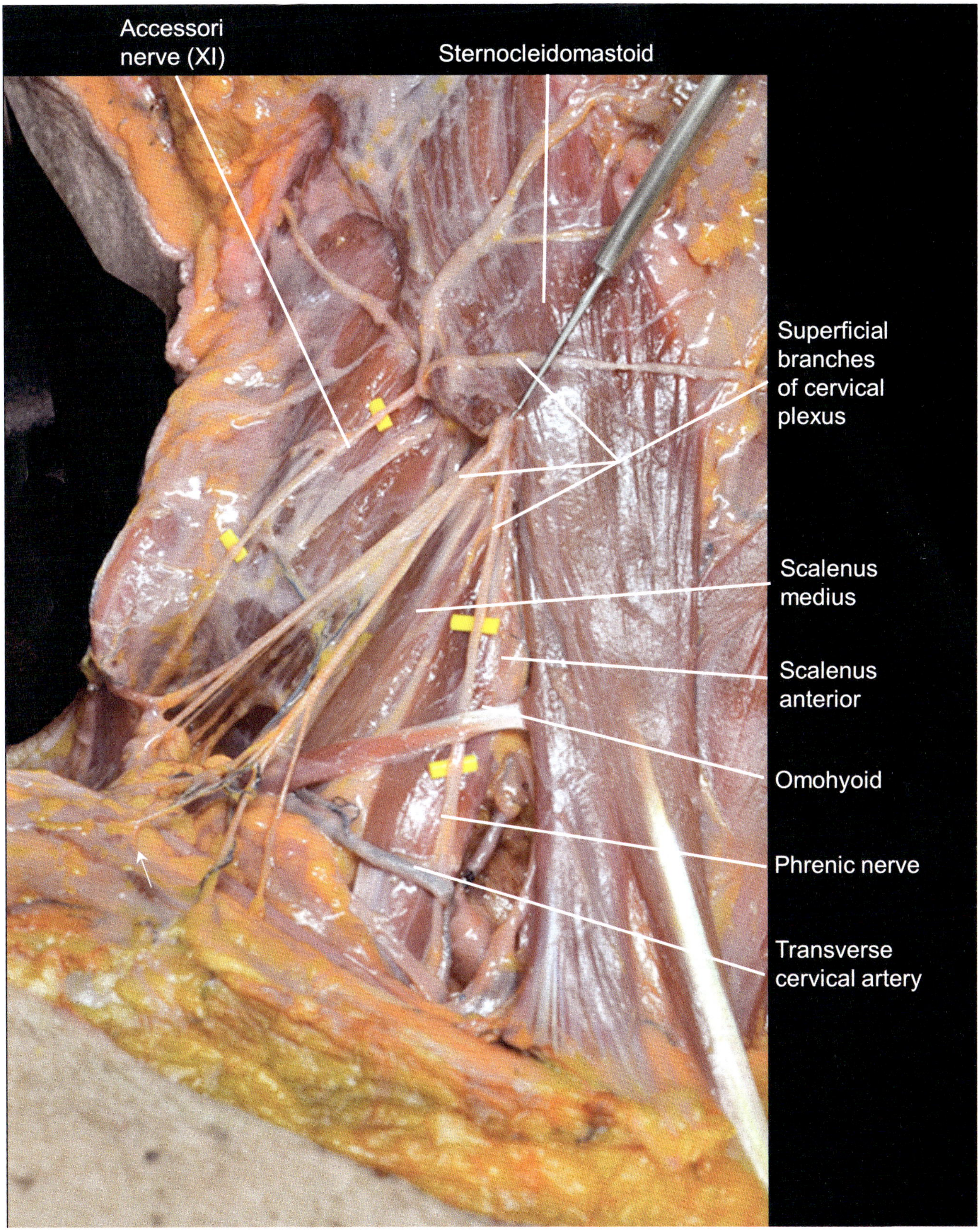

Fig. 7.6 Cervical plexus. One of the main branches of the cervical plexus is the phrenic nerve, which provides motor innervation to the diaphragm and sensory branches to the pericardium and the diaphragmatic and mediastinal pleurae. This nerve originates from C4, with contributions from C3 and C5, and descends along the anterior scalene muscle belly to access the thorax. In the distribution area of the cervical plexus branches, also lies the path of the accessory nerve (cranial nerve XI). This trajectory may be found between the junction of the upper and middle third of the posterior sternocleidomastoid muscle border and the junction point of the lateral and middle third of the anterior border of the trapezius muscle

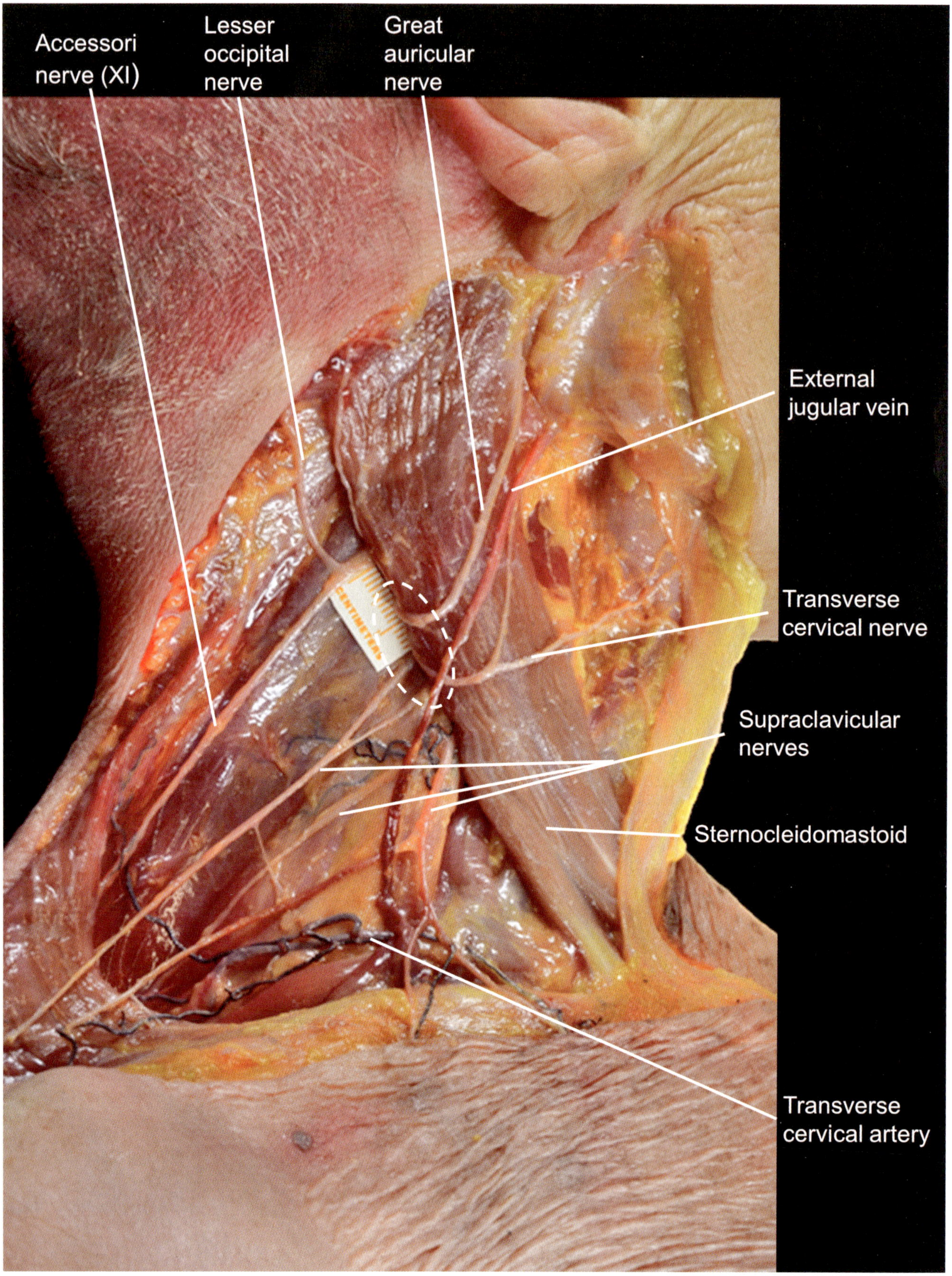

Fig. 7.7 Cervical plexus. Sensory nerves originate in the cervical plexus. Before their subcutaneous distribution, they are gathered in a small area called the nervous neck point or Erb's point. This point (circle) is an anatomic reference for anesthetic blockade of these nerve branches, and it is located at the midpoint of the posterior border of the sternocleidomastoid muscle. Another reference to its location is the path of the external jugular vein; the nerve point is cranial to the crossing point between this vein and the posterior border of the muscle

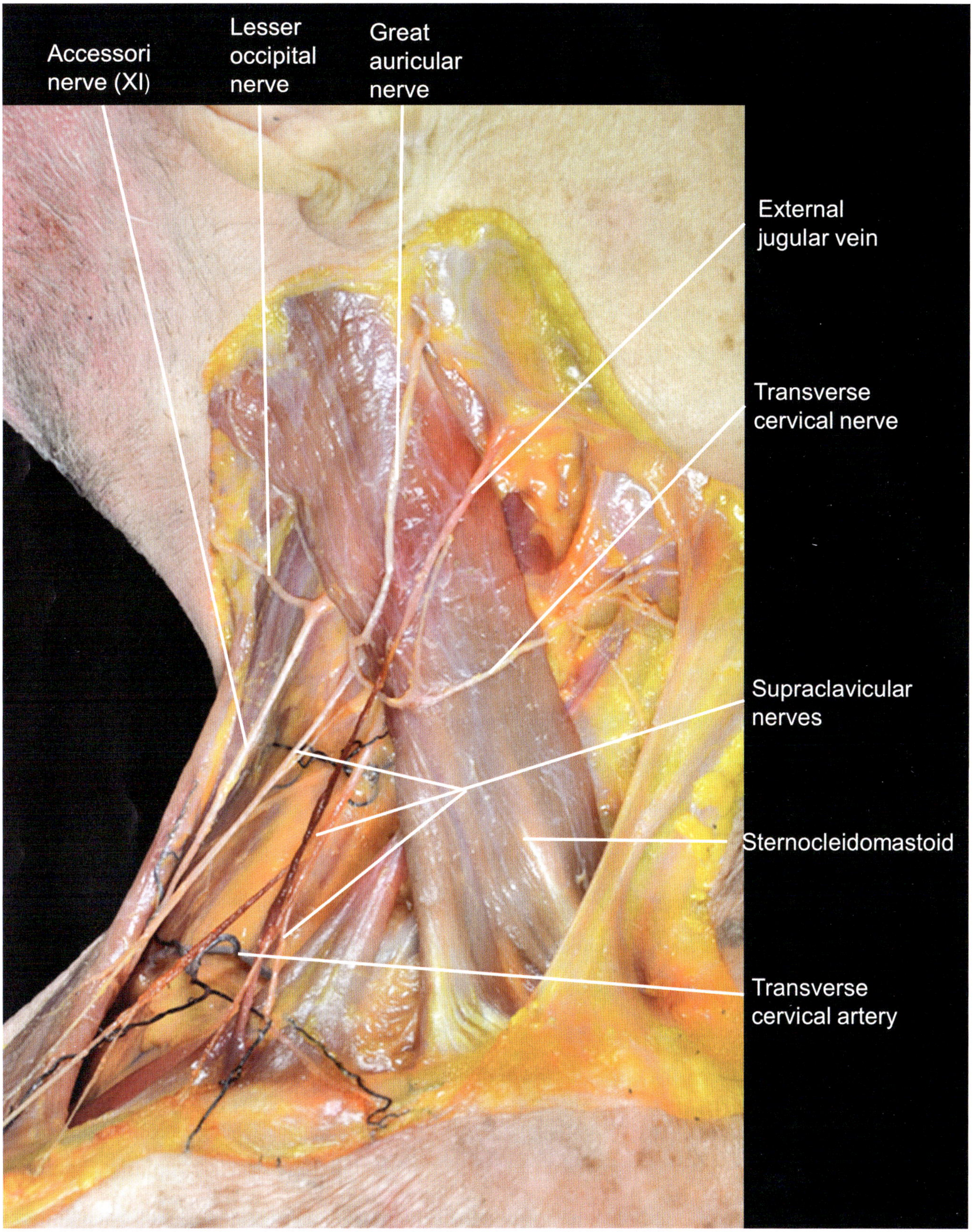

Fig. 7.8 Cervical plexus. From the nervous neck point, the sensory nerves of the cervical plexus radiate to the anterior region of the neck, the supraclavicular area from the sternum to the acromion, the retroauricular occipital zone, the parotid region, and the ear

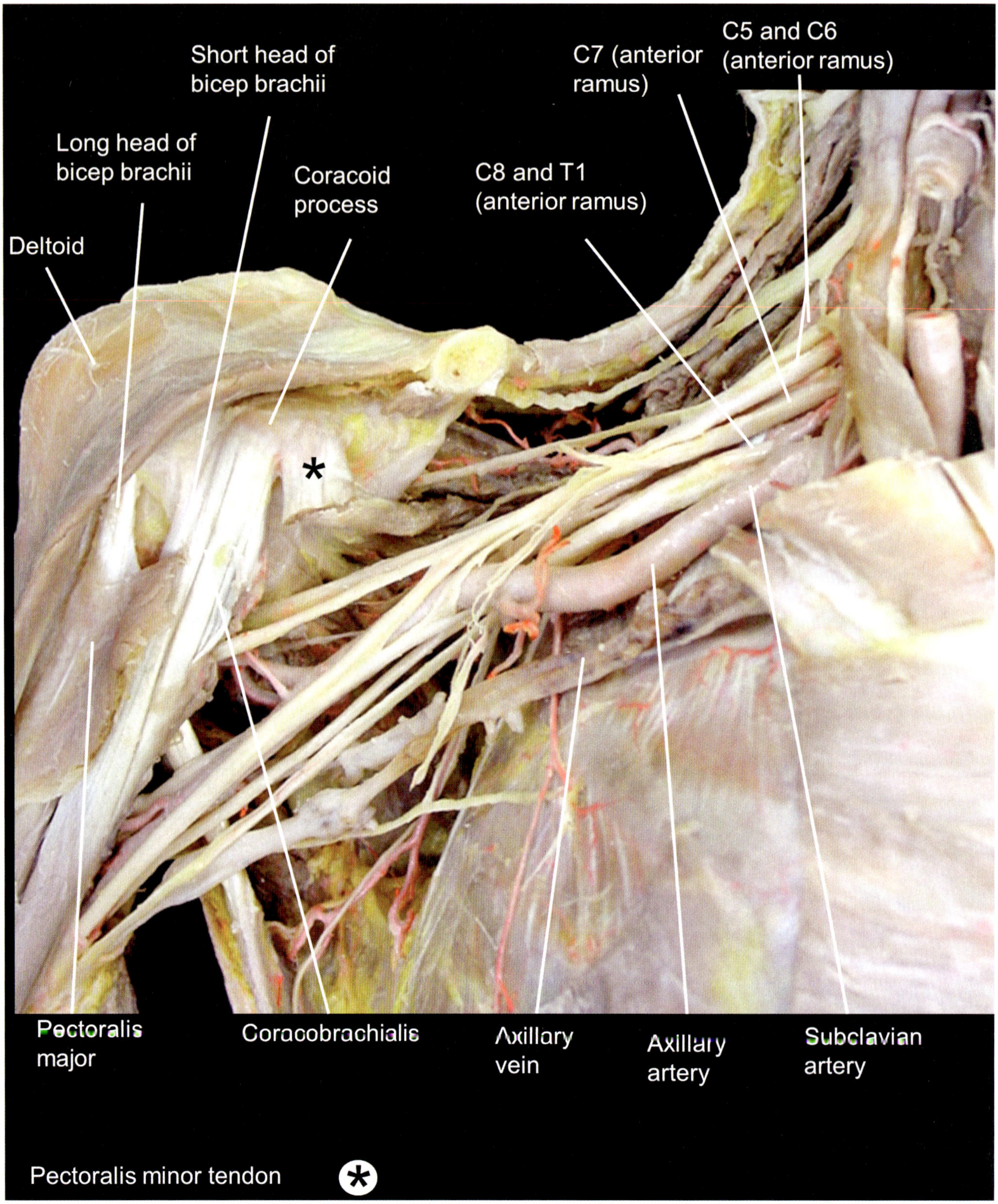

Fig. 7.9 Brachial plexus. The brachial plexus is formed by the anterior branches of the spinal nerves from C5 to T1. From the brachial plexus, the peripheral nerves to the upper extremities originate. From the output of these spinal nerves in the paravertebral line of the lower cervical region, the brachial plexus passes through the interscalene space, the costoclavicular corridor, and the axillary cavity to reach the root of the upper extremity, where its terminal branches are constituted. In this trajectory, the plexus is accompanied by the axillary artery and the subclavian–axillary vein

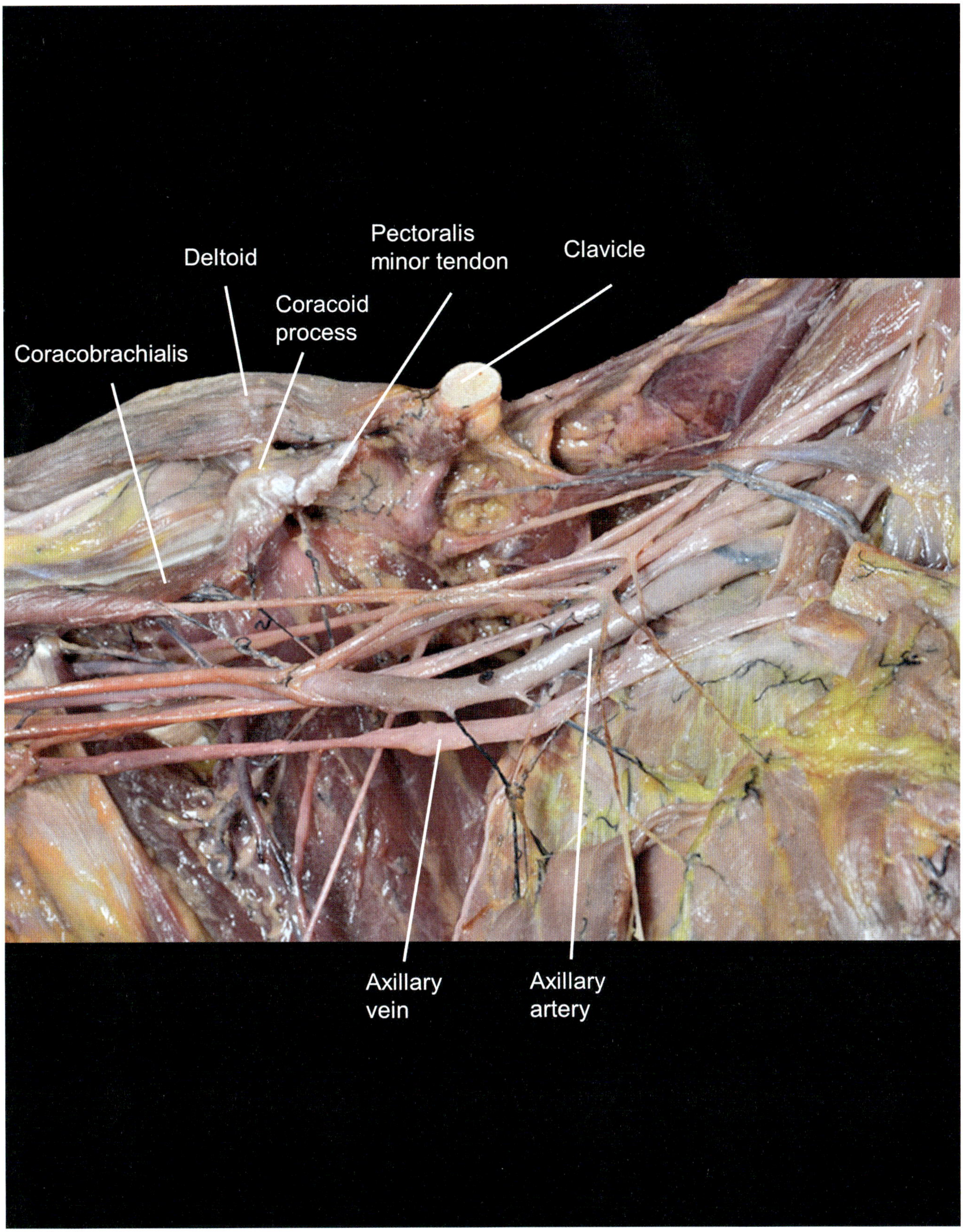

Fig. 7.10 Brachial plexus. The main trajectory of the brachial plexus is inferior, lateral, and posterior to the axillary cavity. It has an overall hourglass shape, expanded at the ends and clustered at its center, coinciding with its passage below the clavicle

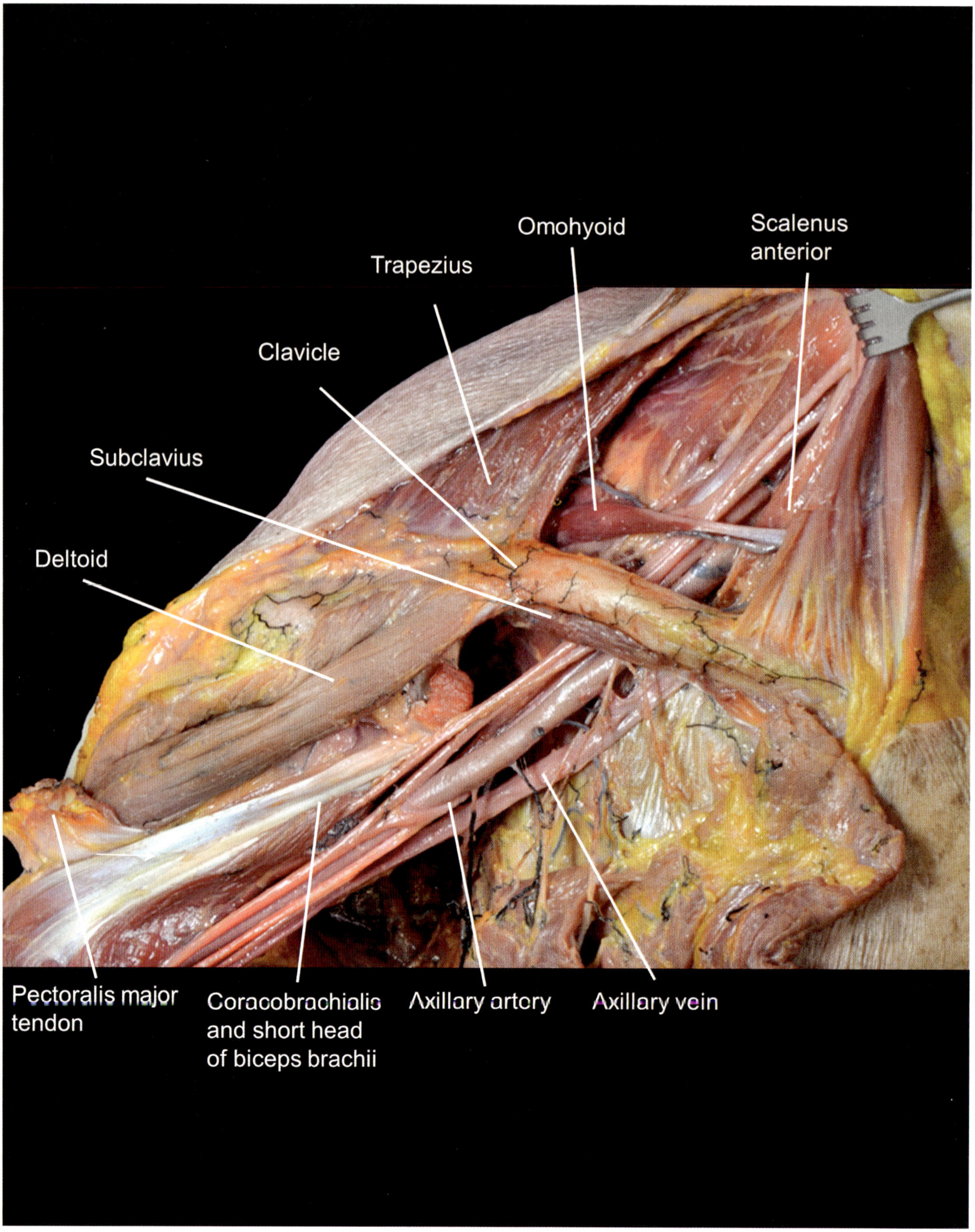

Fig. 7.11 Brachial plexus. The passage of the brachial plexus under the clavicle allows it to divide into two main segments: the supraclavicular plexus and the infraclavicular plexus. Each segment has characteristic anatomic references, specific topographic relations, collateral branches, and functional implications

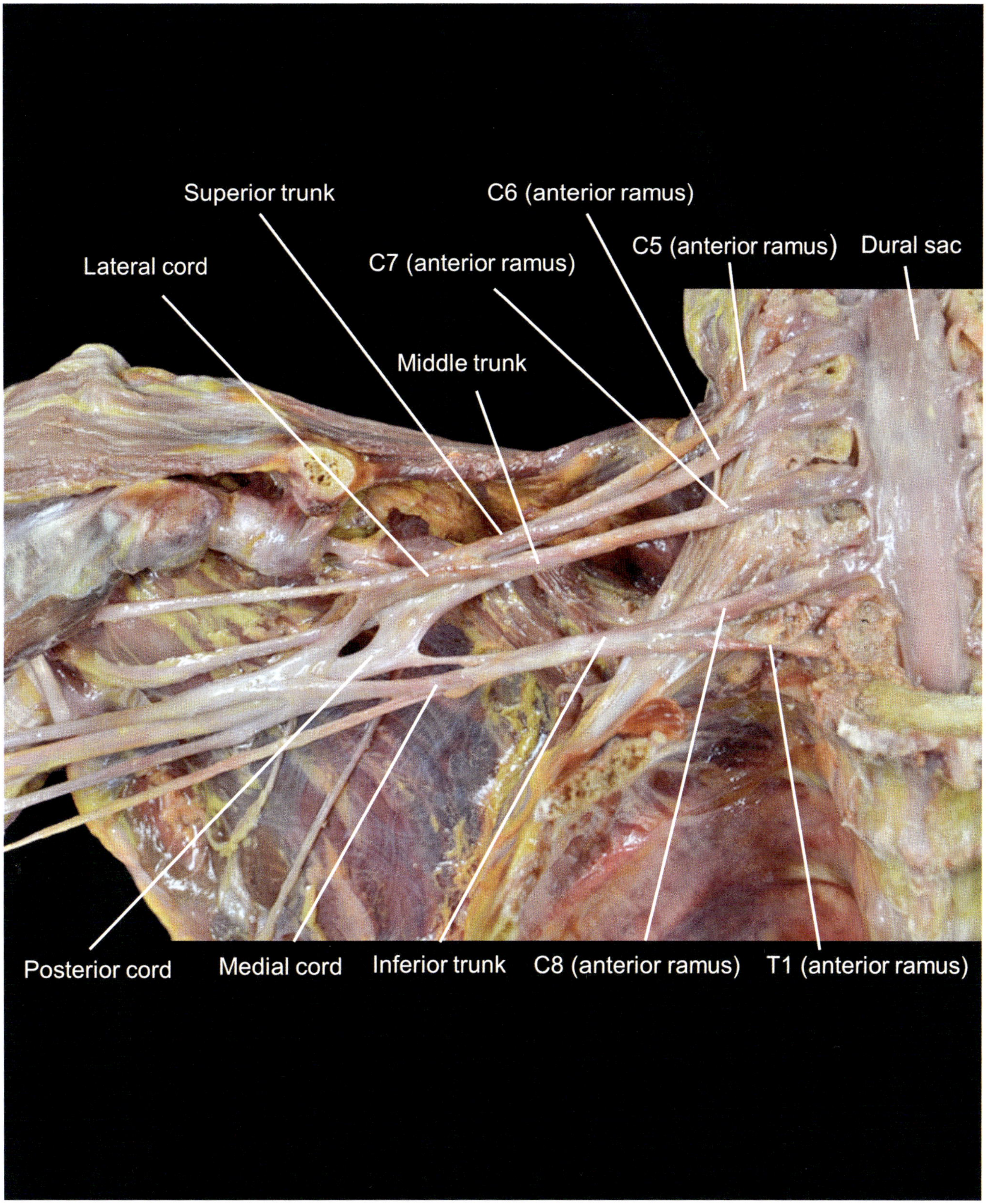

Fig. 7.12 Brachial plexus. The axonal recombination that occurs in the brachial plexus allows sensory and motor information from the spinal nerves that constitute it to be distributed to different peripheral nerves. This exchange follows a pattern that creates a series of defined and recognizable neural elements. Thus, the anterior rami of spinal nerves C5 to T1 combine to form trunks, each of which divides into an anterior and a posterior division. The combination of the anterior divisions on one hand and the posterior divisions on the other hand leads to the formation of brachial plexus fasciculi

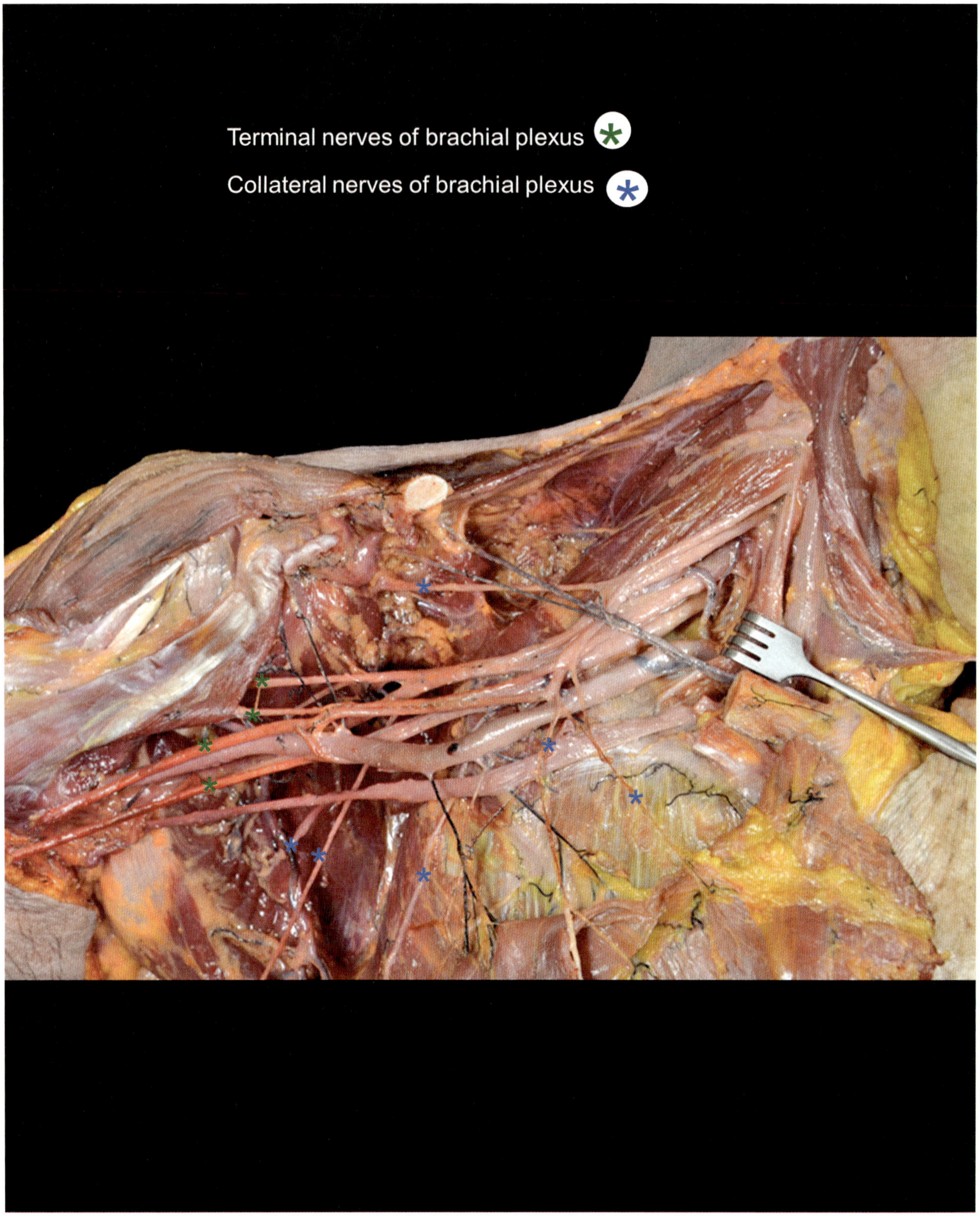

Fig. 7.13 Brachial plexus. From this sequence of brachial plexus elements (anterior branches of the spinal nerves, trunks, divisions, and fasciculi), different collateral nerves originate. These nerves innervate the structures of the upper extremity in the region of the shoulder girdle. Meanwhile, the arm, forearm, and hand regions are innervated by the terminal nerves of the brachial plexus, which are formed by the division and/or combination of brachial plexus fasciculi

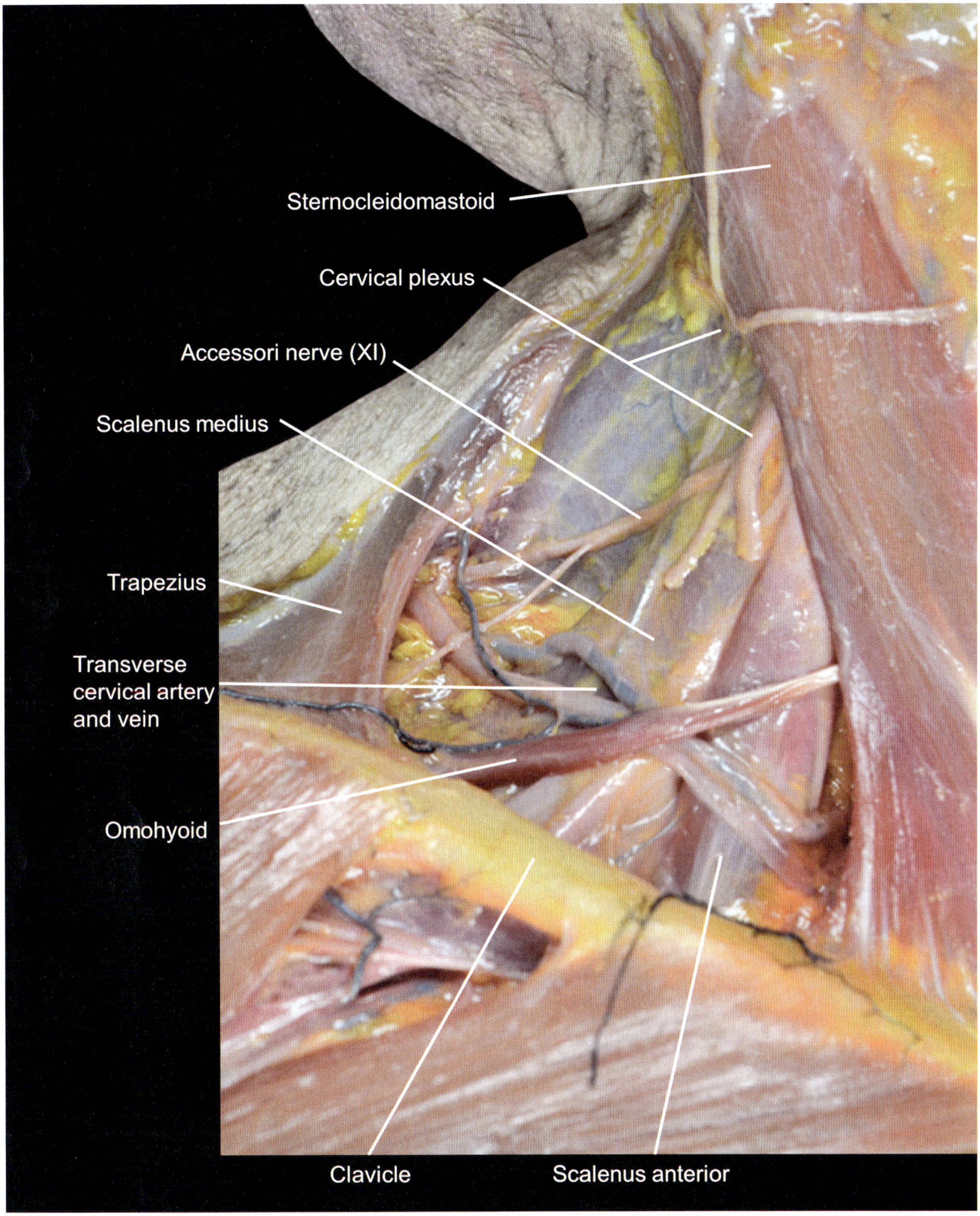

Fig. 7.14 Supraclavicular brachial plexus. The supraclavicular brachial plexus is located topographically in the lateral region of the neck, in the area bounded by the clavicle and the line drawn from the posterior border of the sternocleidomastoid muscle at the level of the cricoid cartilage to the midpoint of the clavicle. The inferior belly of the omohyoid muscle crosses over the brachial plexus. This segment of the brachial plexus corresponds to the anterior branches of the spinal nerves and the trunks formed by their combinations

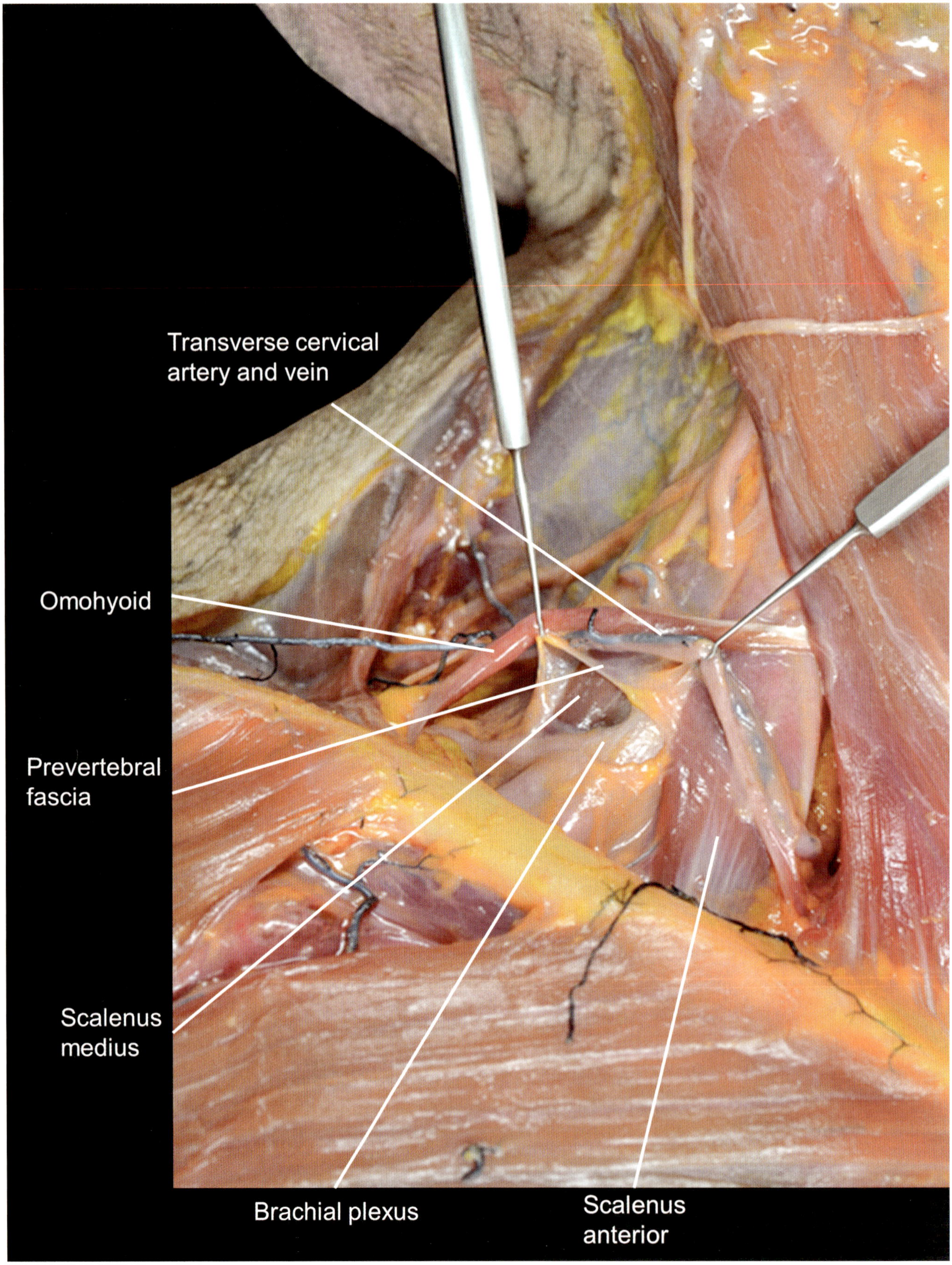

Fig. 7.15 Supraclavicular brachial plexus. All the supraclavicular plexus elements lie beneath the prevertebral (deep cervical) fascia. This fascia forms an anatomic compartment in which an anesthetic agent can flow and bath all the elements contained therein

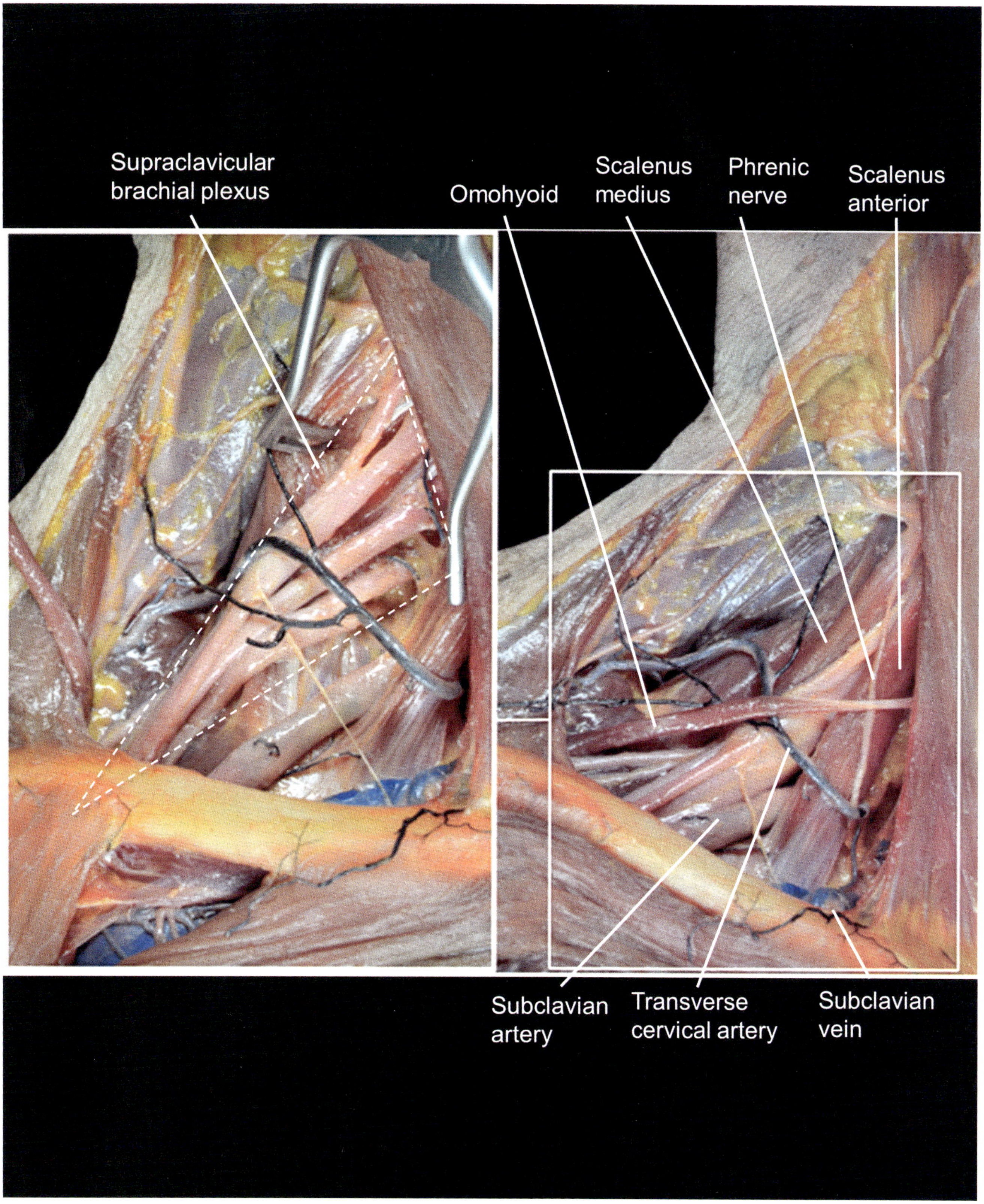

Fig. 7.16 Supraclavicular brachial plexus. The elements of the supraclavicular brachial plexus form a triangular sheet that extends from the cervical paravertebral line to the center of the clavicle. In this lateral region of the neck, the brachial plexus elements exist at two levels. First, the anterior branches of the spinal nerves are found in the cervical paravertebral line, at the output of their intervertebral foramina, between the muscles of the paravertebral channels. Then, the brachial plexus trunks run through the interscalene space between the anterior and middle scalene muscles

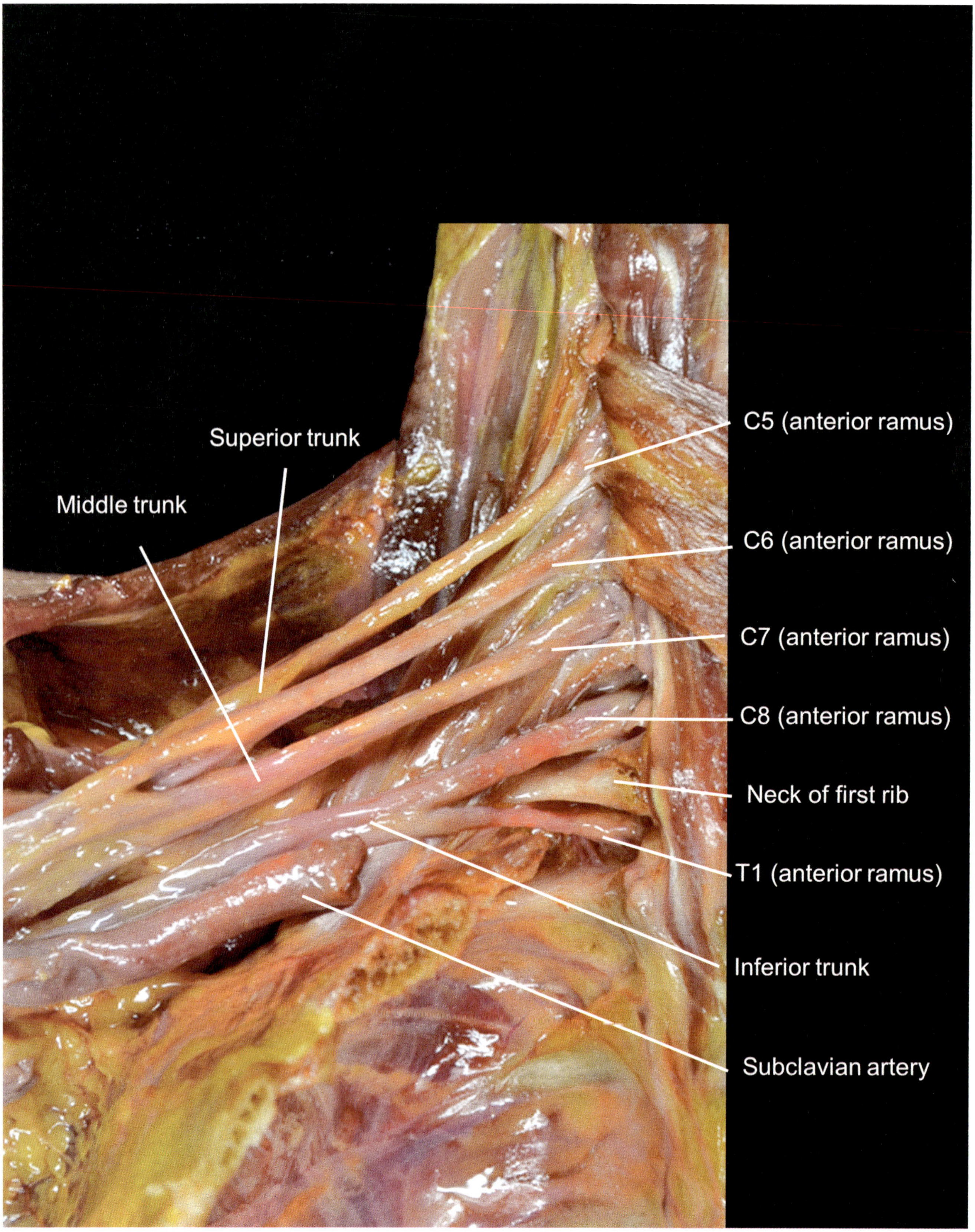

Fig. 7.17 Supraclavicular brachial plexus. The union of the anterior branches of spinal nerves C5, C6, C7, and C8 and most of the ventral branch of spinal nerve T1 forms three trunks. Specifically, the C5 and C6 anterior branches join to form the upper trunk, C7 remains isolated as the middle trunk, and C8 and T1 bind to constitute the lower trunk

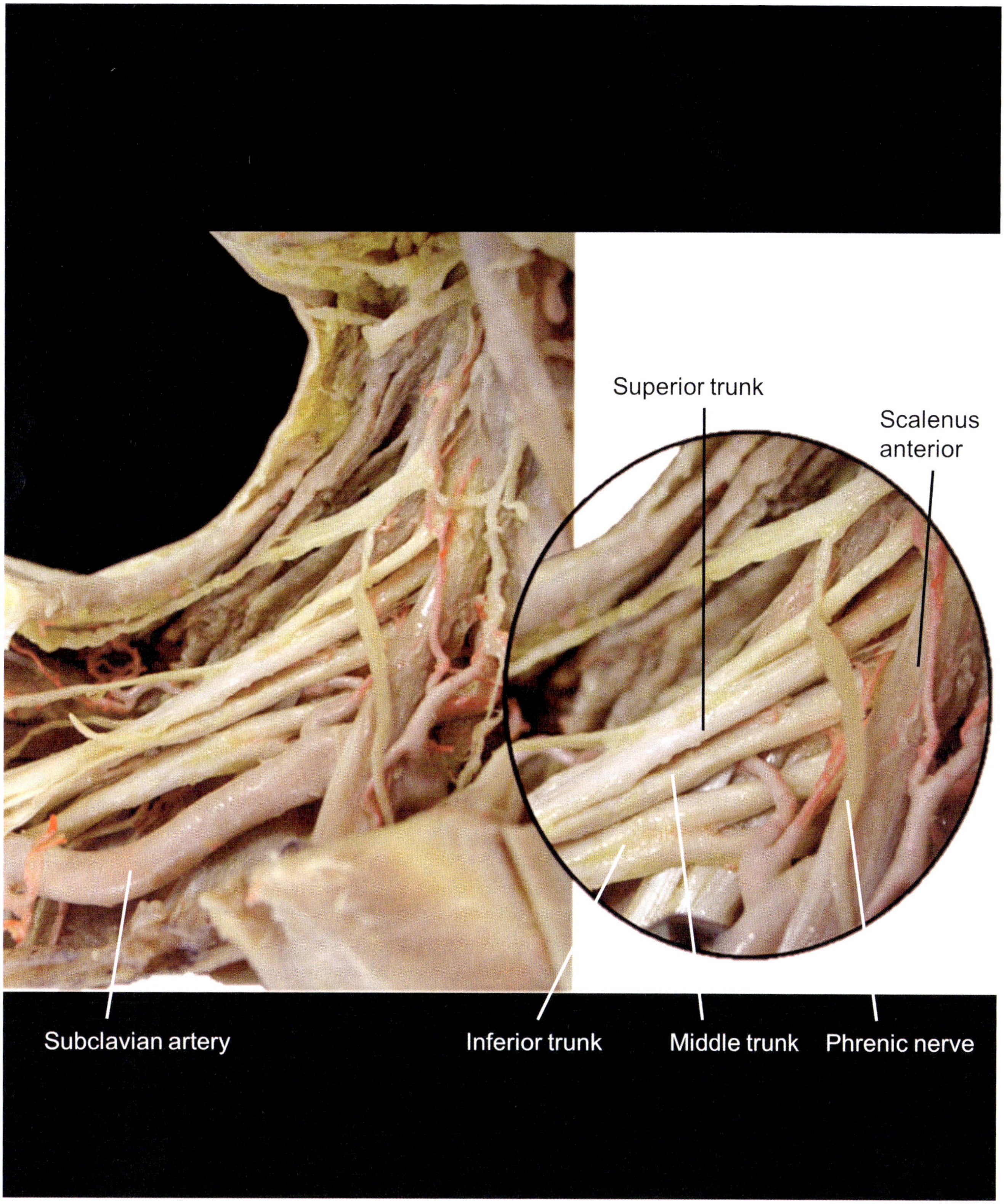

Fig. 7.18 Supraclavicular brachial plexus. The arrangement of the anterior branches of the spinal nerves of the brachial plexus and trunks changes gradually. Whereas the upper trunk and nerve C5 are positioned vertically, the lower trunk adopts a transverse arrangement. The anterior branch of C8 is disposed upward. The close relationship of C8 and the lower trunk with the subclavian artery explains the delayed or even insufficient anesthesia of these elements that sometimes occurs with supraclavicular brachial plexus block. In the cervical paravertebral area, the vertebral artery rises through the cervical transverse processes anterior to each of these nerves. In this space, one must account for the proximity of the paravertebral sympathetic chain and its interconnections to the elements of the plexus

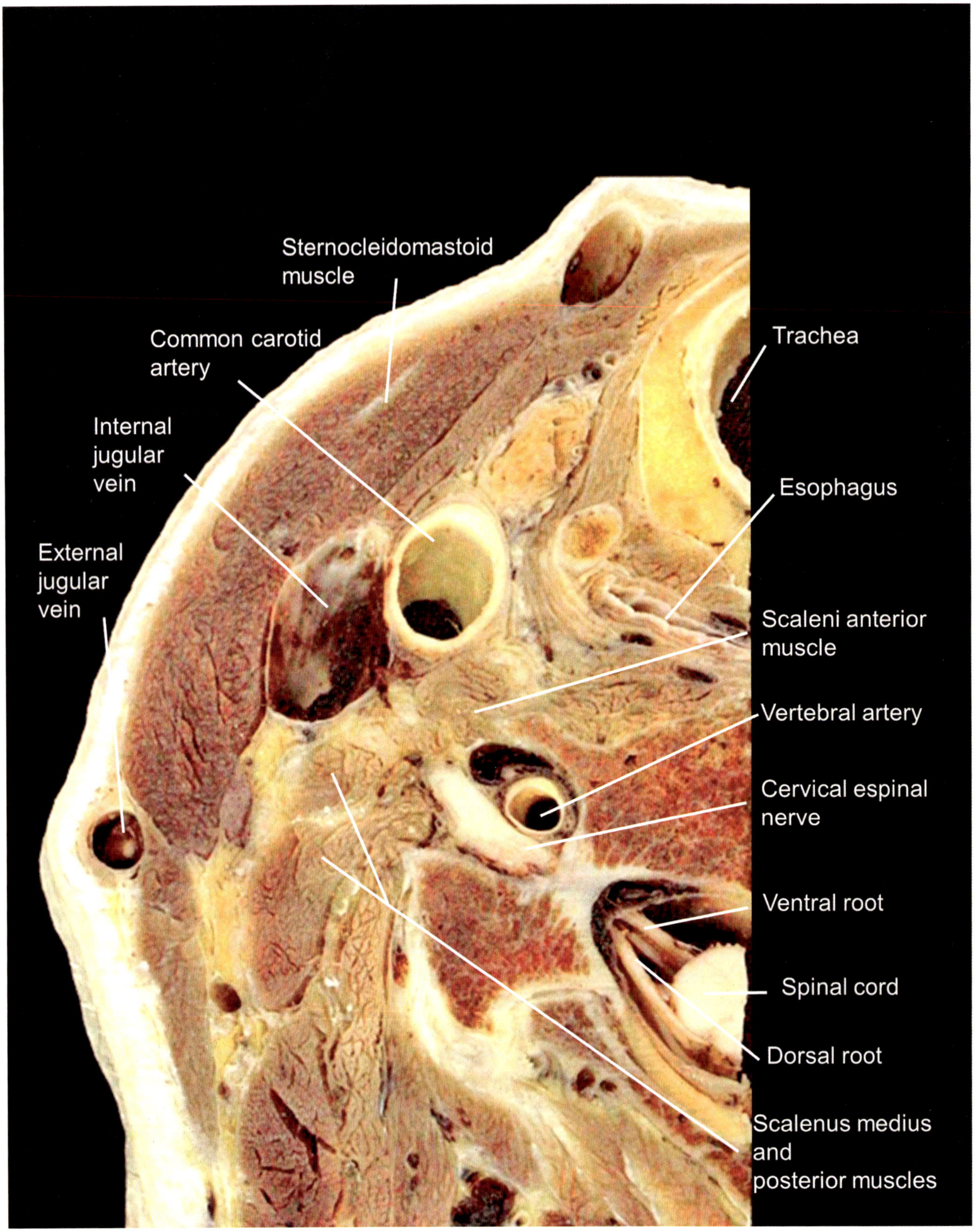

Fig. 7.19 Supraclavicular brachial plexus. This cross-section of a cervical transverse process shows the characteristic arrangement of the spinal nerve anterior branch, between the anterior and posterior tubercles of the transverse process. The relationship between the nerve root and the vertebral artery and vein also may be seen

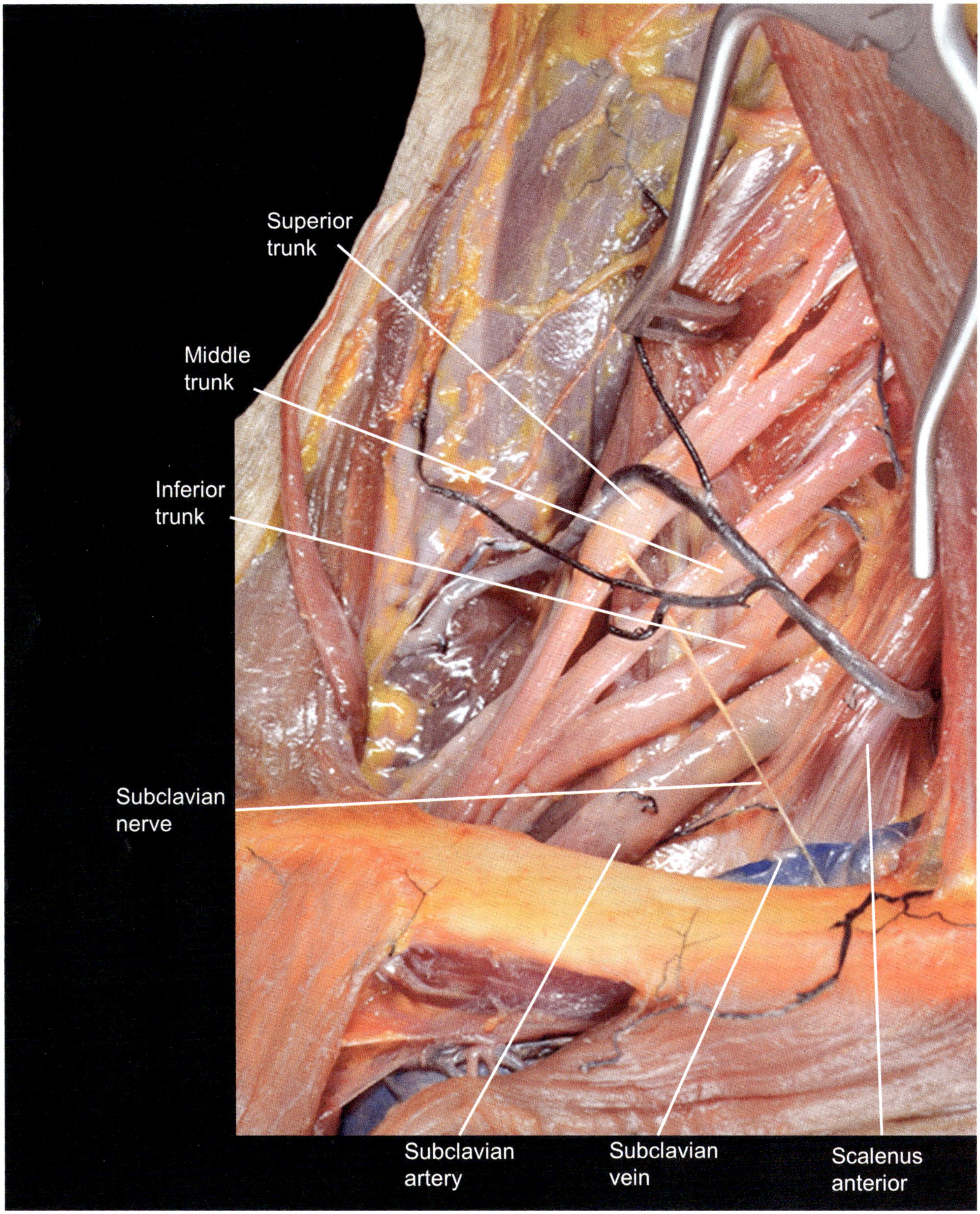

Fig. 7.20 Supraclavicular brachial plexus. The elements of the brachial plexus in the interscalene space are the upper, middle, and lower trunks. Their names refer to the orderly arrangement from cranial to caudal, coinciding with the arrangement of the anterior branches of the spinal nerves at which they originate. The transverse cervical artery runs superficial to the trunks. Caudally, the trunks of the brachial plexus tend to cluster, forming a nerve cord that passes under the clavicle. In the interscalene space, the subclavian artery is in front of the trunks as it passes over the first rib. The subclavian vein is in front of the anterior scalene muscle insertion. It is related to the phrenic nerve, which characteristically is in the anterior face of this muscle

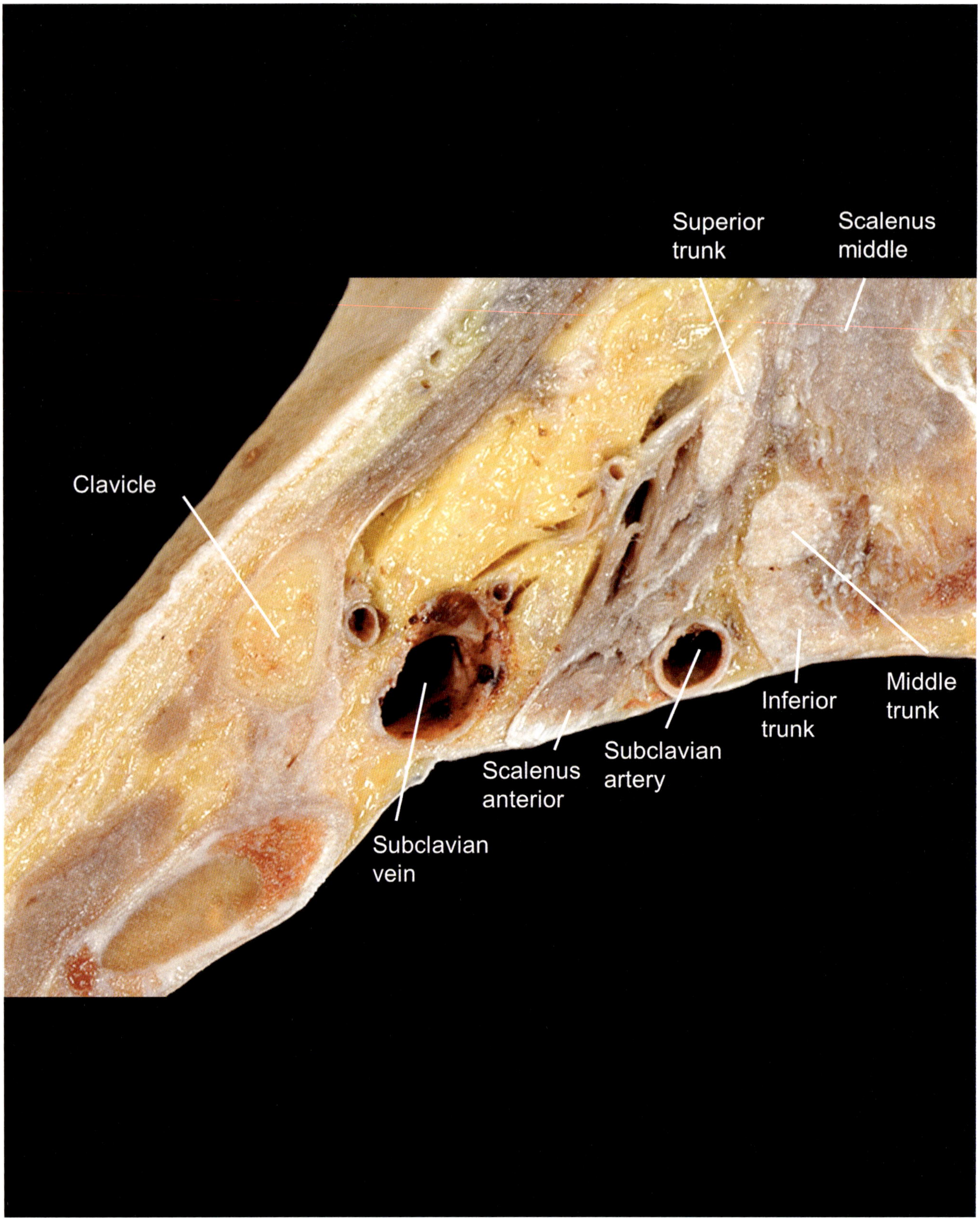

Fig. 7.21 Supraclavicular brachial plexus. This sagittal section through the interscalene space shows how the trunks of the brachial plexus, the subclavian artery, and the subclavian vein are arranged cranial to caudal and posterior to anterior. The trunks of the brachial plexus and the subclavian artery are positioned at the interscalene space, whereas the subclavian vein passes in front of the anterior scalene muscle

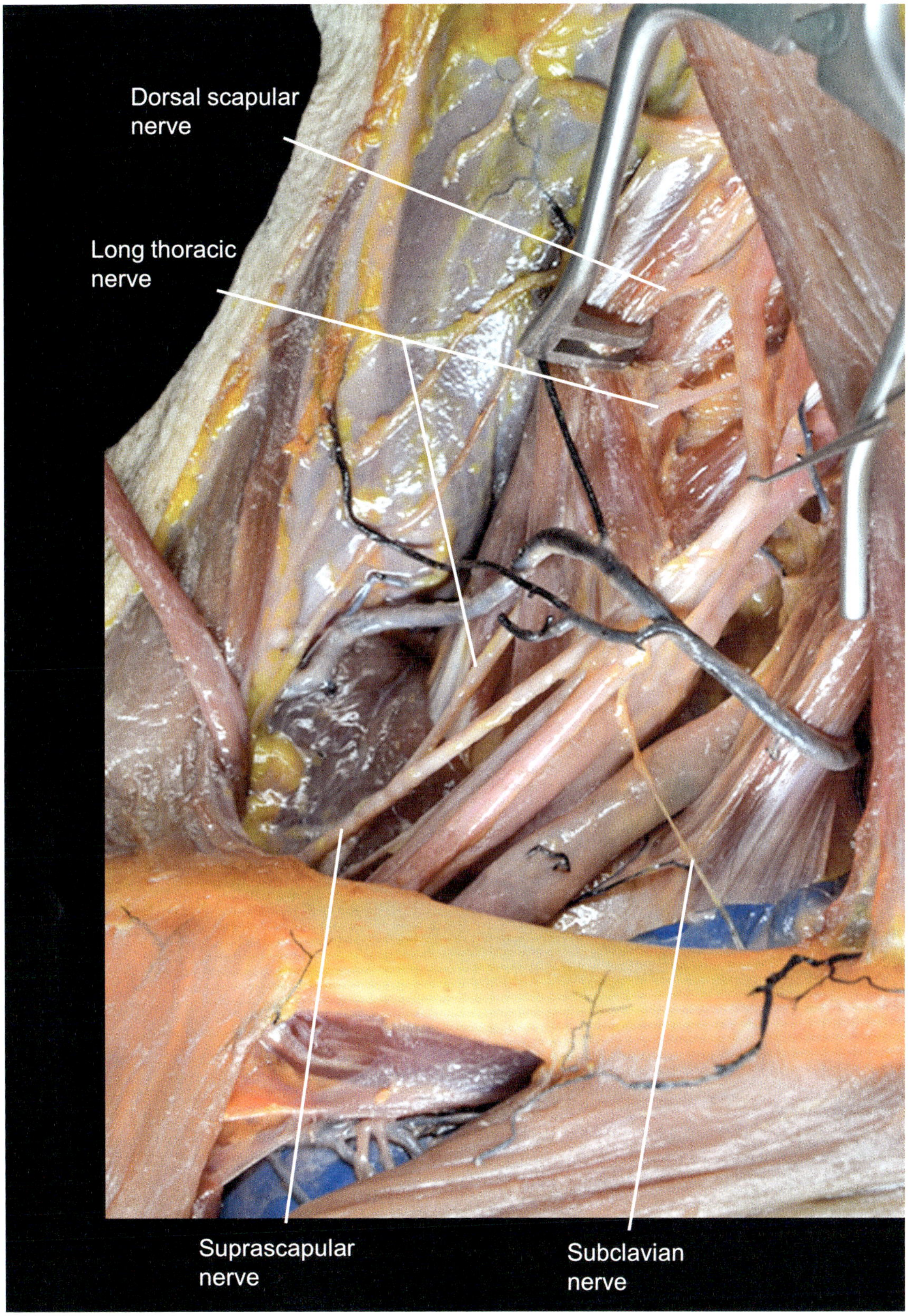

Fig. 7.22 Supraclavicular brachial plexus. Several collateral nerves originate from the supraclavicular brachial plexus. The dorsal scapular nerve arises from the back side of nerve C5. The long thoracic nerve roots originate on the back side of nerves C5, C6, and C7; the suprascapular nerve emerges from the posterior face of the upper trunk. Another collateral branch, the subclavian muscle nerve, arises from the anterior face of the upper trunk. Besides these collateral branches, the supraclavicular plexus begins the innervation of the scalene muscles, longus colli muscle, and connections between the anterior branch of nerve C5 and the phrenic nerve

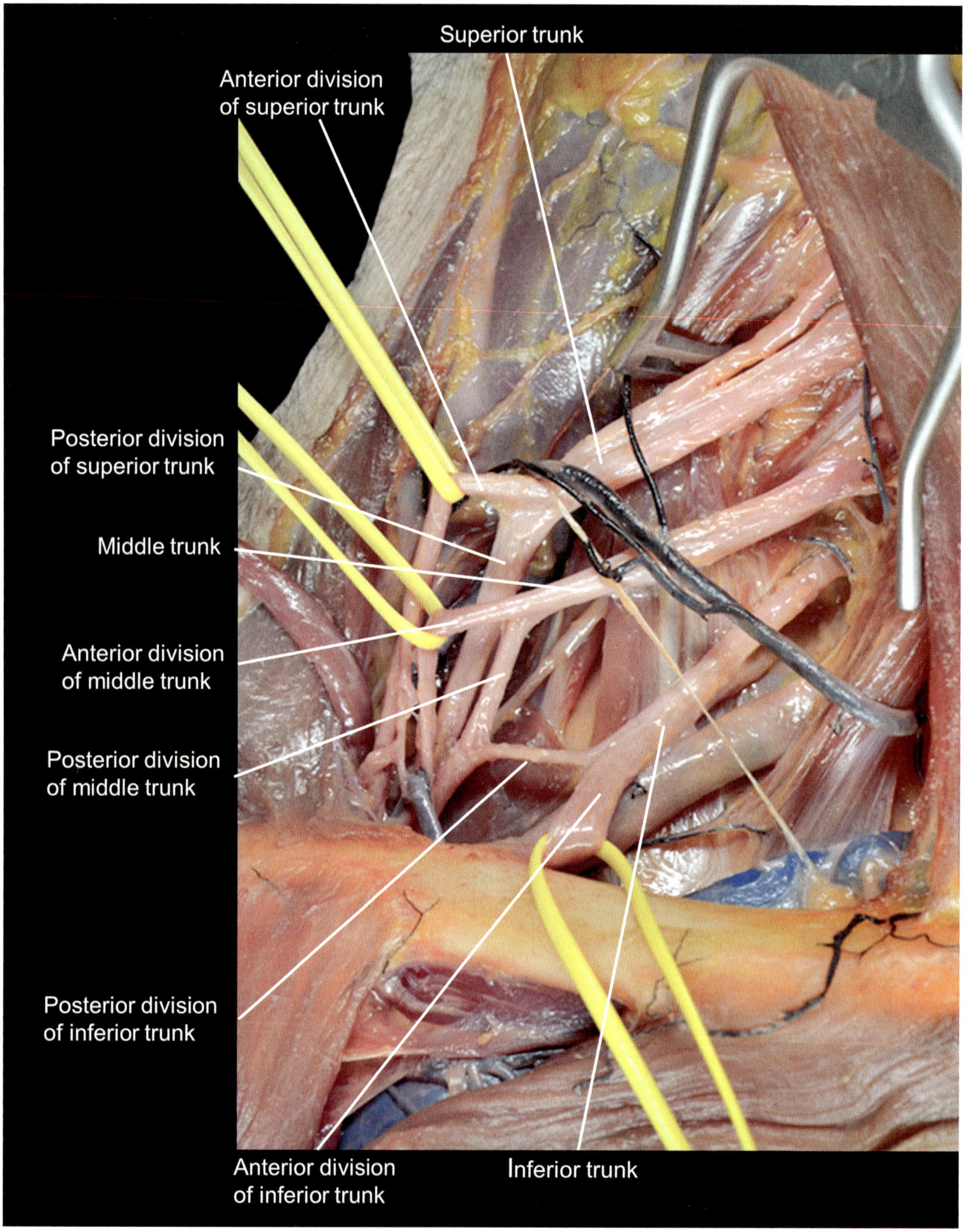

Fig. 7.23 Supraclavicular brachial plexus. Near the costoclavicular corridor, each trunk of the brachial plexus divides into an anterior and a posterior segment. These divisions constitute a functional organization of the plexus. The anterior divisions give rise to nerves distributed on the front face of the upper extremity; therefore, they have a major functional flexor component. Meanwhile, the posterior divisions give rise to nerves toward the posterior face of the limb, with an extensor main functional component

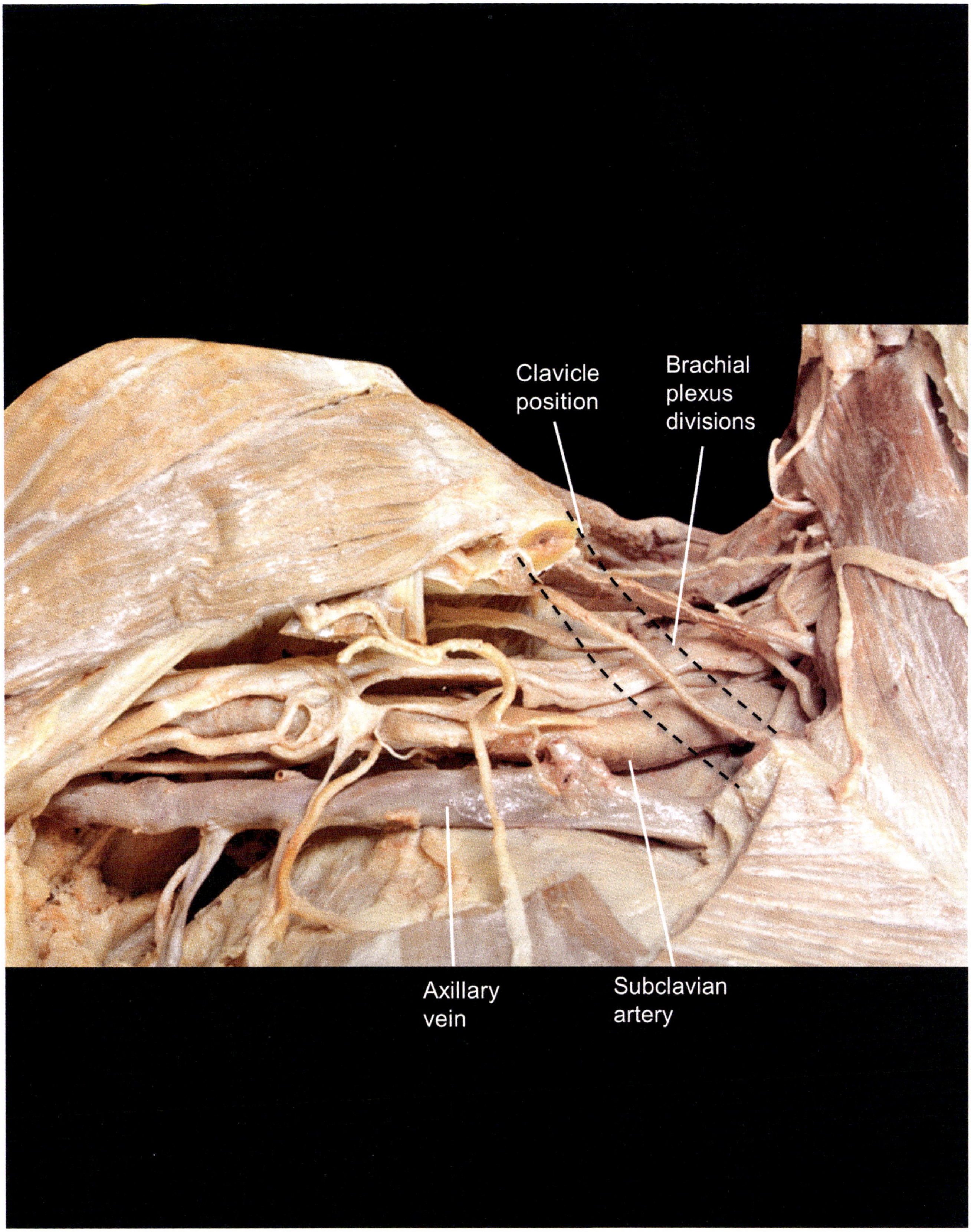

Fig. 7.24 Supraclavicular brachial plexus. Although the brachial plexus is divided into six parts (three anterior and three posterior), these divisions are grouped together to form a single nerve cord positioned at the junction between the lateral third and the two medial thirds of the clavicle, posterior and lateral to the subclavian artery. Thus, the plexus minimizes its size to pass under the clavicle to access the armpit

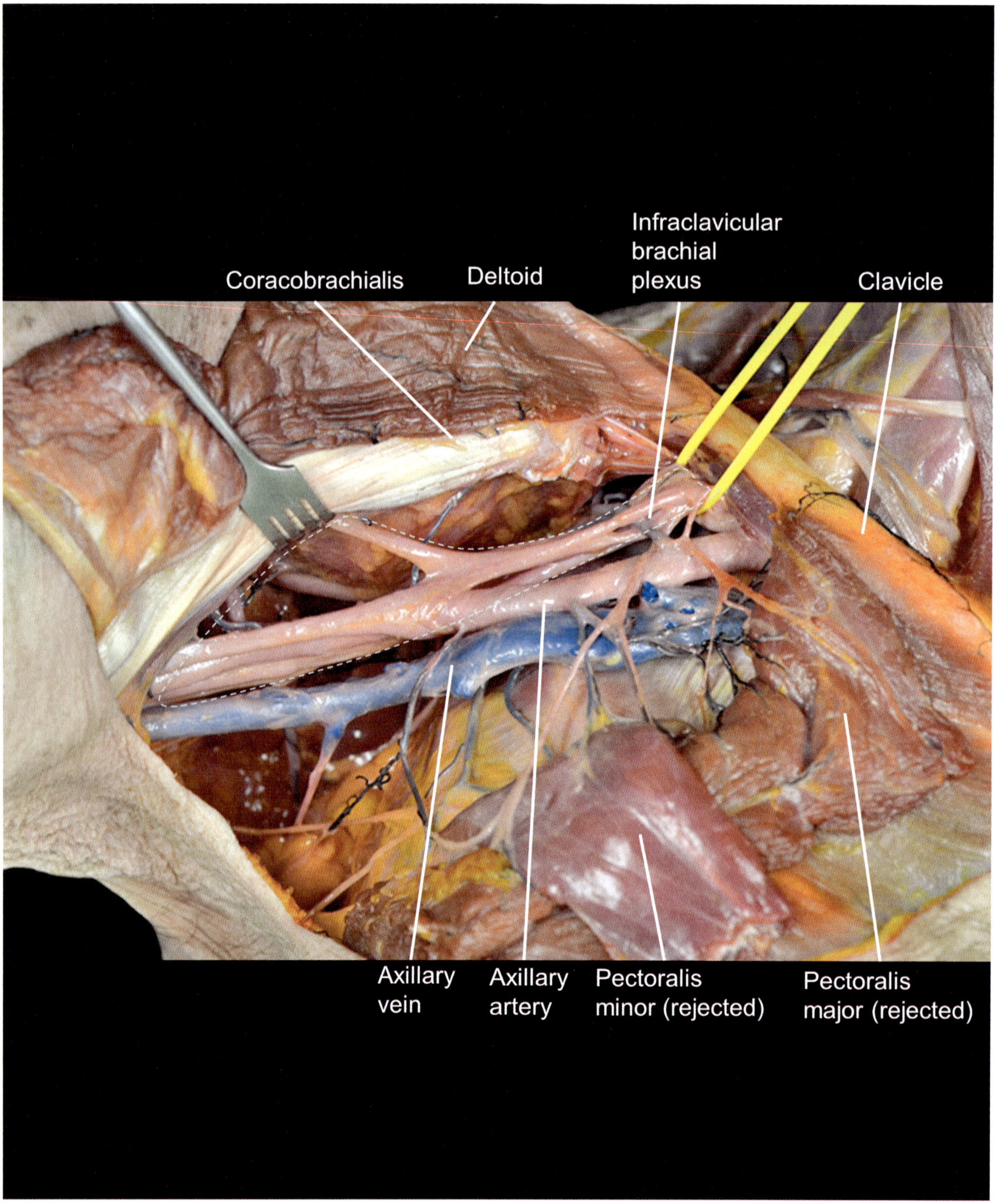

Fig. 7.25 Infraclavicular brachial plexus. The infraclavicular brachial plexus is located in the axillary cavity and consists of cords and terminal nerves. Overall, this portion of the brachial plexus is cone shaped. At its apex, at the entrance of the axillary cavity from the retroclavicular space, its constituent elements remain grouped. Distally, the plexus expands, coinciding with the origin of its terminal branches, which are arranged around the axillary artery. In anatomic position, the infraclavicular brachial plexus has an oblique downward, lateral, and slightly posterior disposition. However, clinically it must be remembered that when the arm is abducted for an anesthetic procedure, this anatomic positions becomes horizontal and the plexus is placed now closer to the skin surface

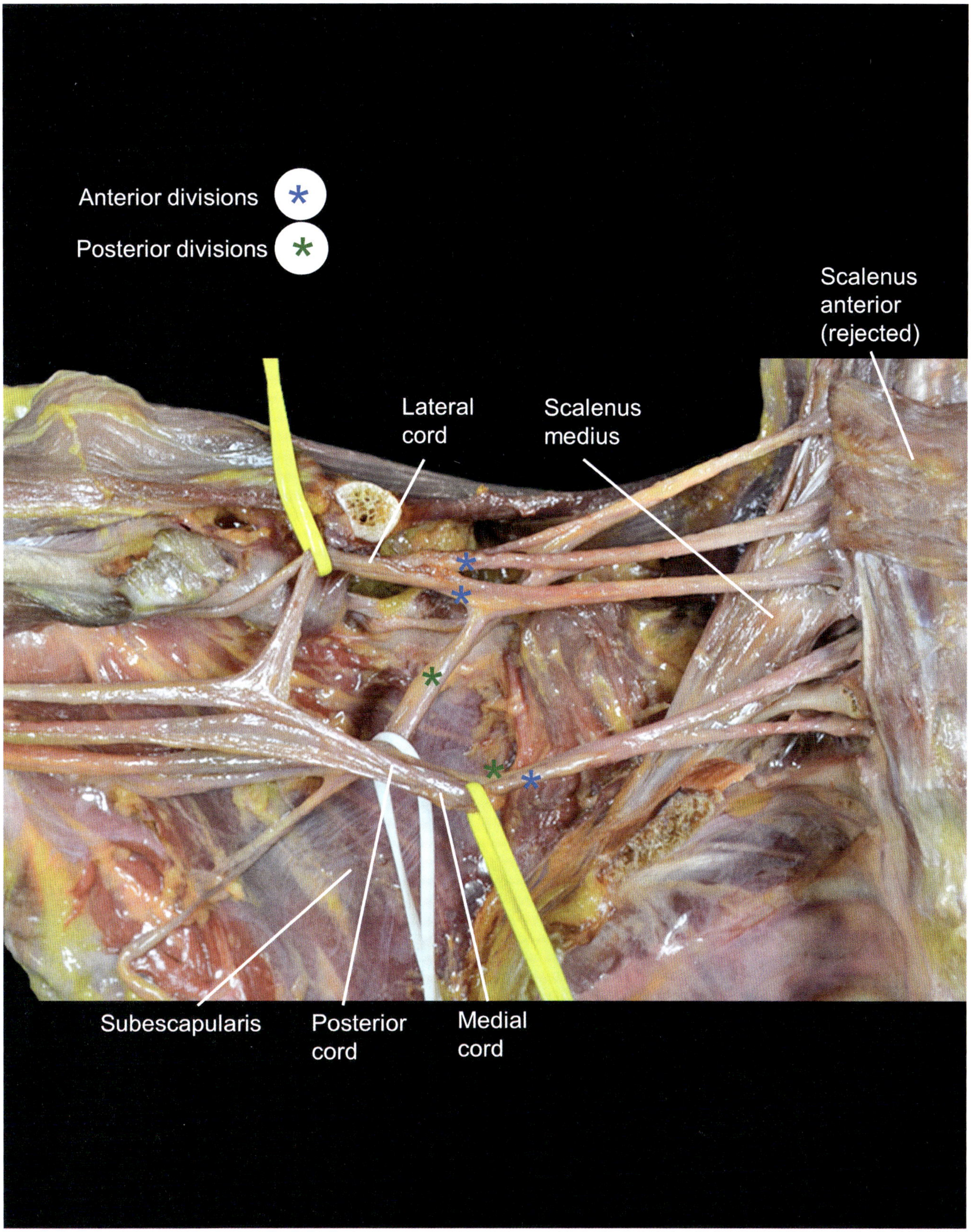

Fig. 7.26 Infraclavicular brachial plexus. In the infraclavicular brachial plexus segment, the anterior and posterior trunk divisions are combined, resulting in cords that give rise to the terminal branches of the plexus. Specifically, the anterior divisions of the upper and middle trunks result in the lateral cord, the anterior division of the lower trunk in the medial cord, and the three posterior divisions in the posterior cord

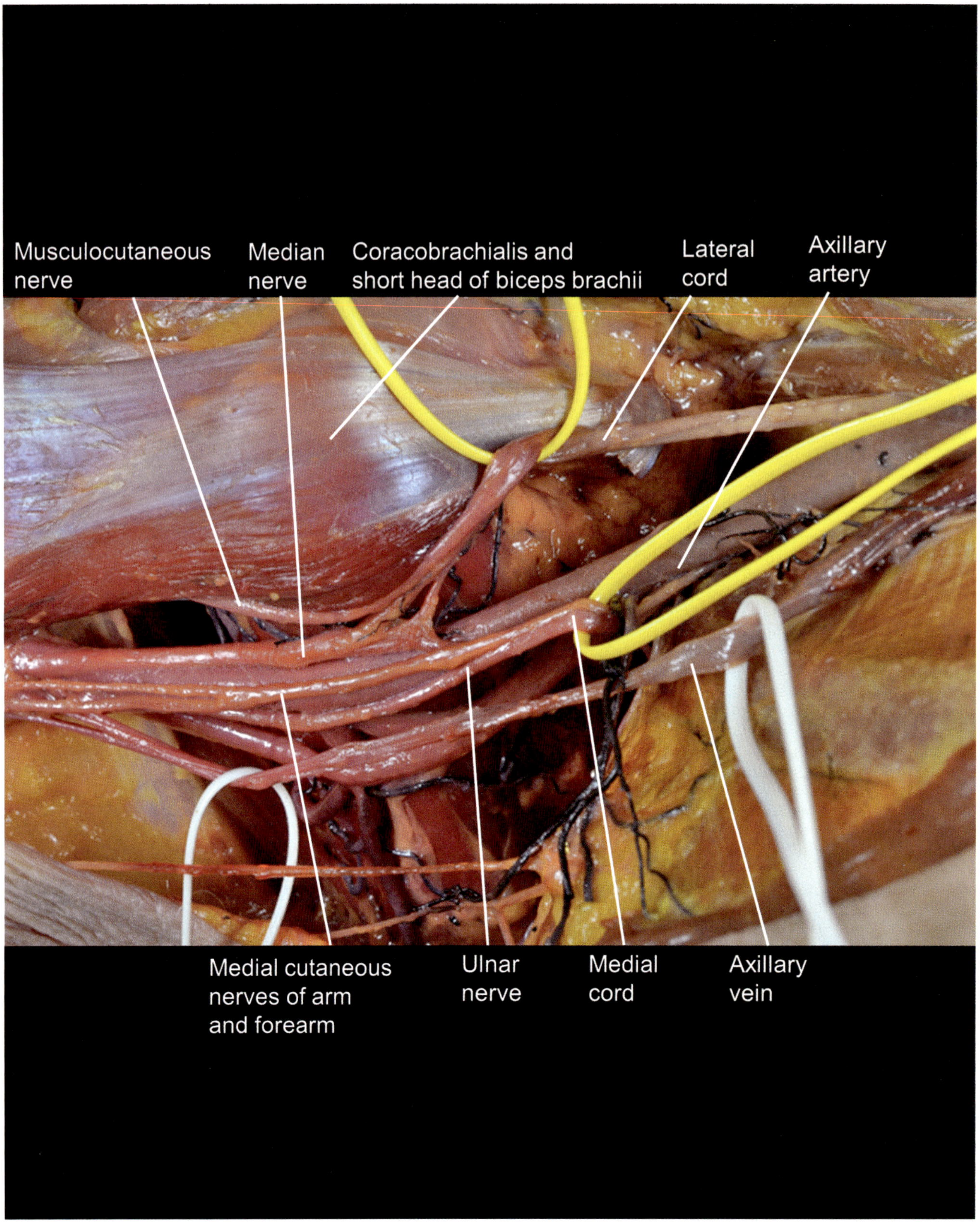

Fig. 7.27 Infraclavicular brachial plexus. The brachial plexus terminal nerves correspond to five mixed nerves—the musculocutaneous, median, ulnar, axillary, and radial nerves—and two exclusively sensory nerves—the medial brachial cutaneous and medial antebrachial cutaneous nerves. From the lateral cord, the musculocutaneous nerve and the lateral root of the median nerve arise; the medial cord divides into the medial root of the median nerve, the ulnar nerve, and the medial brachial and antebrachial cutaneous nerves

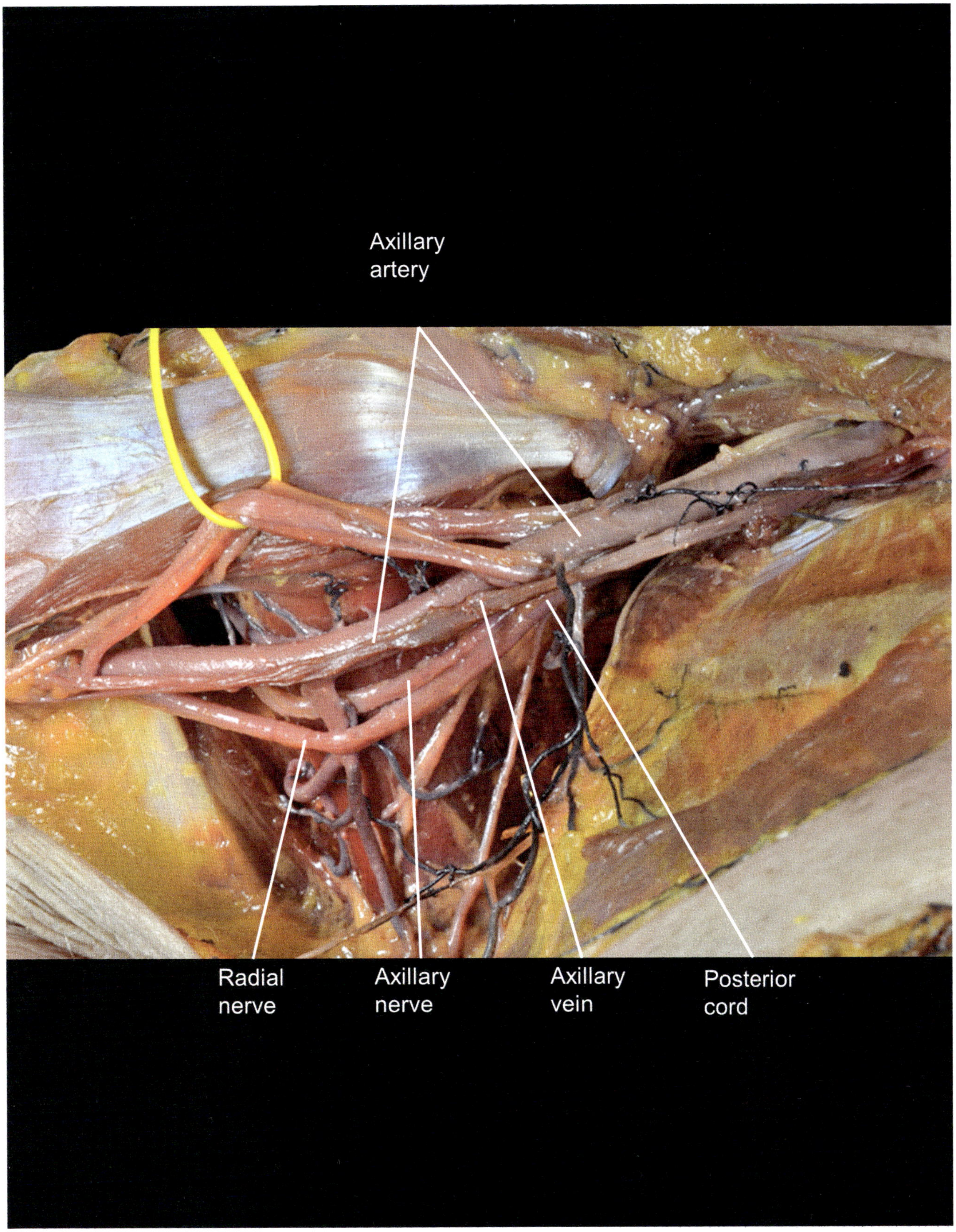

Fig. 7.28 Infraclavicular brachial plexus. The axillary and radial nerves arise from the posterior cord

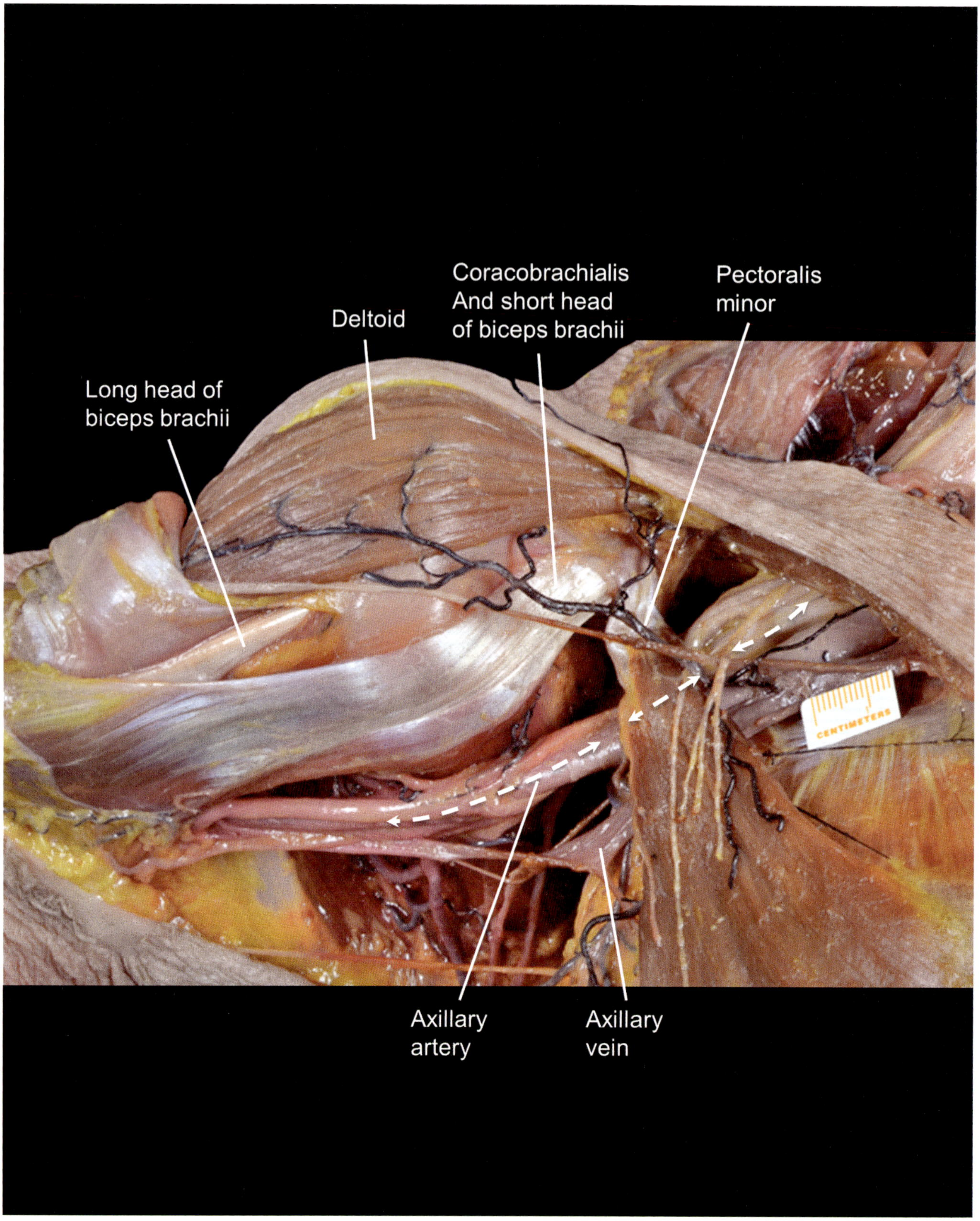

Fig. 7.29 Infraclavicular brachial plexus. In the anterior wall of the axilla, and by using the insertion of the pectoralis minor muscle on the coracoid process of the scapula as a reference, the elements constituting the infraclavicular plexus can be organized. Three regions can be distinguished: a region between the clavicle and the upper margin of the pectoralis minor muscle, a region posterior to this muscle, and a region between the lower edge of the pectoralis muscle and the base of the axilla

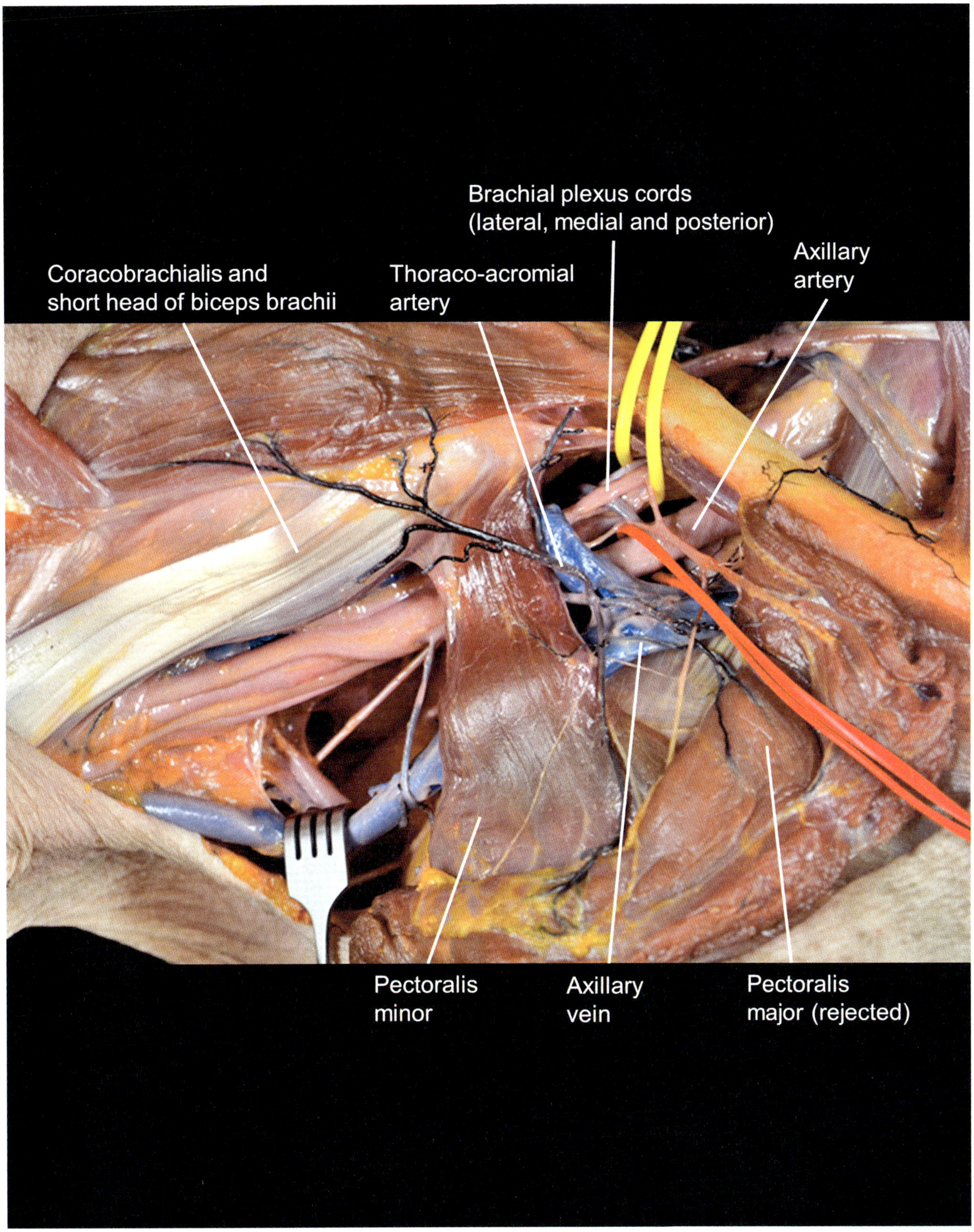

Fig. 7.30 Infraclavicular brachial plexus. Between the clavicle and the upper margin of the pectoralis minor muscle, infraclavicular plexus cords are grouped into a nerve cord located posterolateral to the axillary artery. The axillary vein, meanwhile, is the most medial and anterior element of the axillary neurovascular bundle

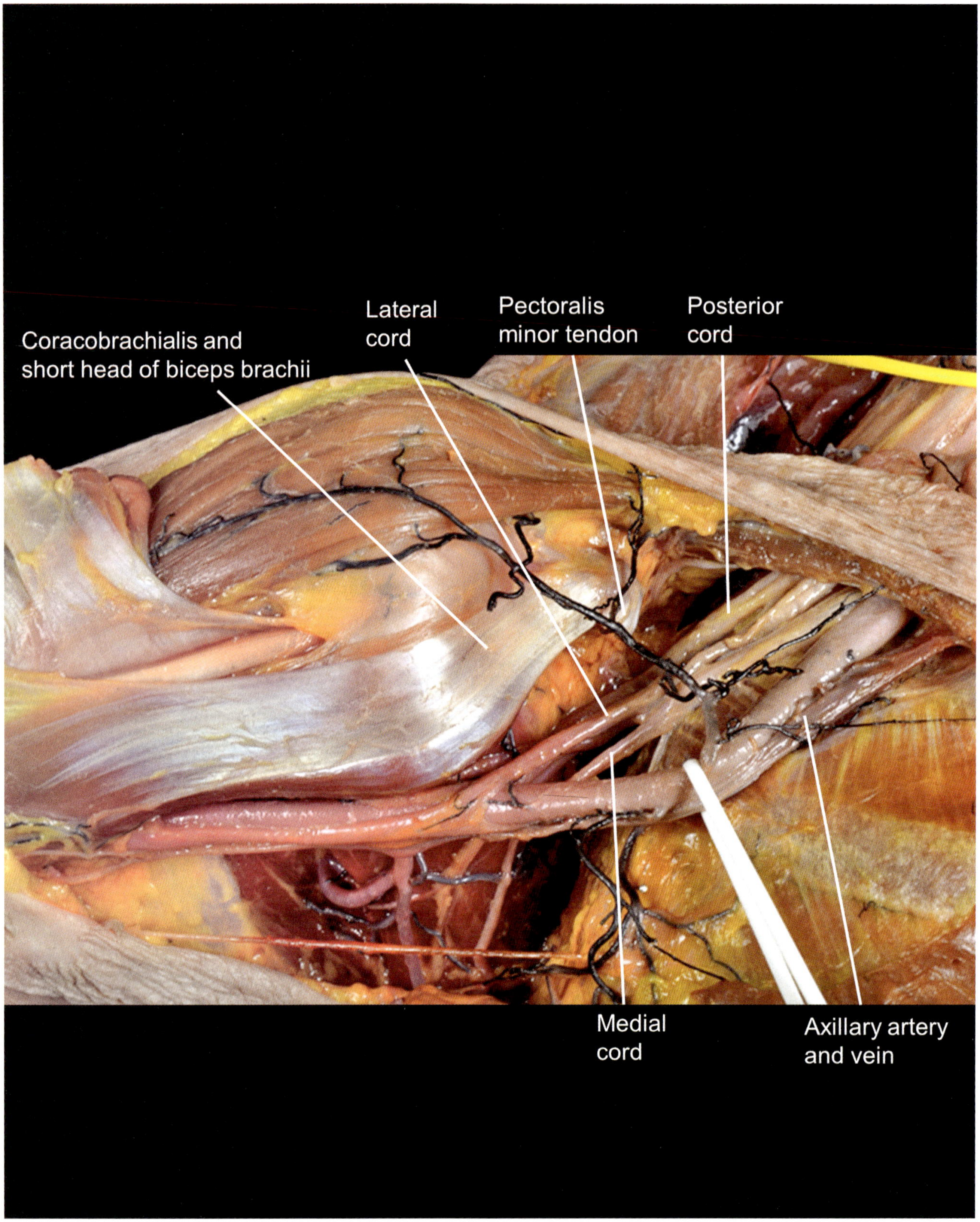

Fig. 7.31 Infraclavicular brachial plexus. Under the pectoralis minor tendon, the brachial plexus cords surround the axillary artery in positions that, in fact, form the underlying nomenclature. Specifically, the lateral cord descends to lie anterolateral to the artery, whereas the medial cord lies obliquely behind the axillary artery and above the posterior cord to a position between the axillary artery and vein, which is the most medial element in this area. The posterior cord then is located behind the artery

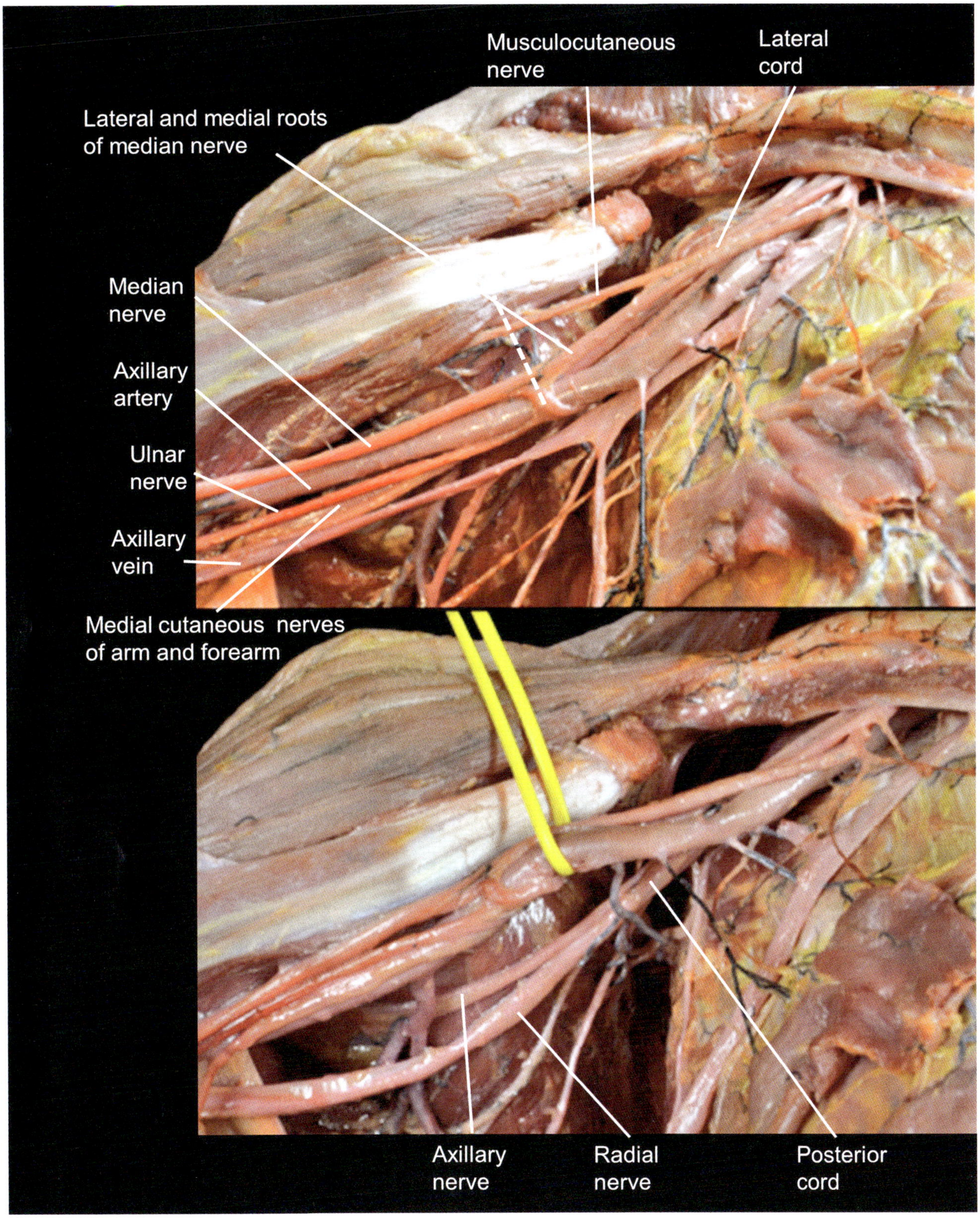

Fig. 7.32 Infraclavicular brachial plexus. At the lower edge of the pectoralis minor muscle, the cords of the brachial plexus become terminal nerves. Given their origin, these nerves are arranged around the axillary artery, occupying a position very similar to that of the cords from which they originated. Thus, the musculocutaneous nerve from the lateral cord is anterolateral to the axillary artery. The median nerve, formed by the union of two components from the lateral and medial cords, is located on the front side of the axillary artery. The ulnar nerve, the medial antebrachial cutaneous nerve, and the medial brachial cutaneous nerve arising from the medial cord lie anteromedial to the artery. Finally, the nerves formed from the posterior cord (the axillary nerve and the radial nerve) are located on the back side of the artery

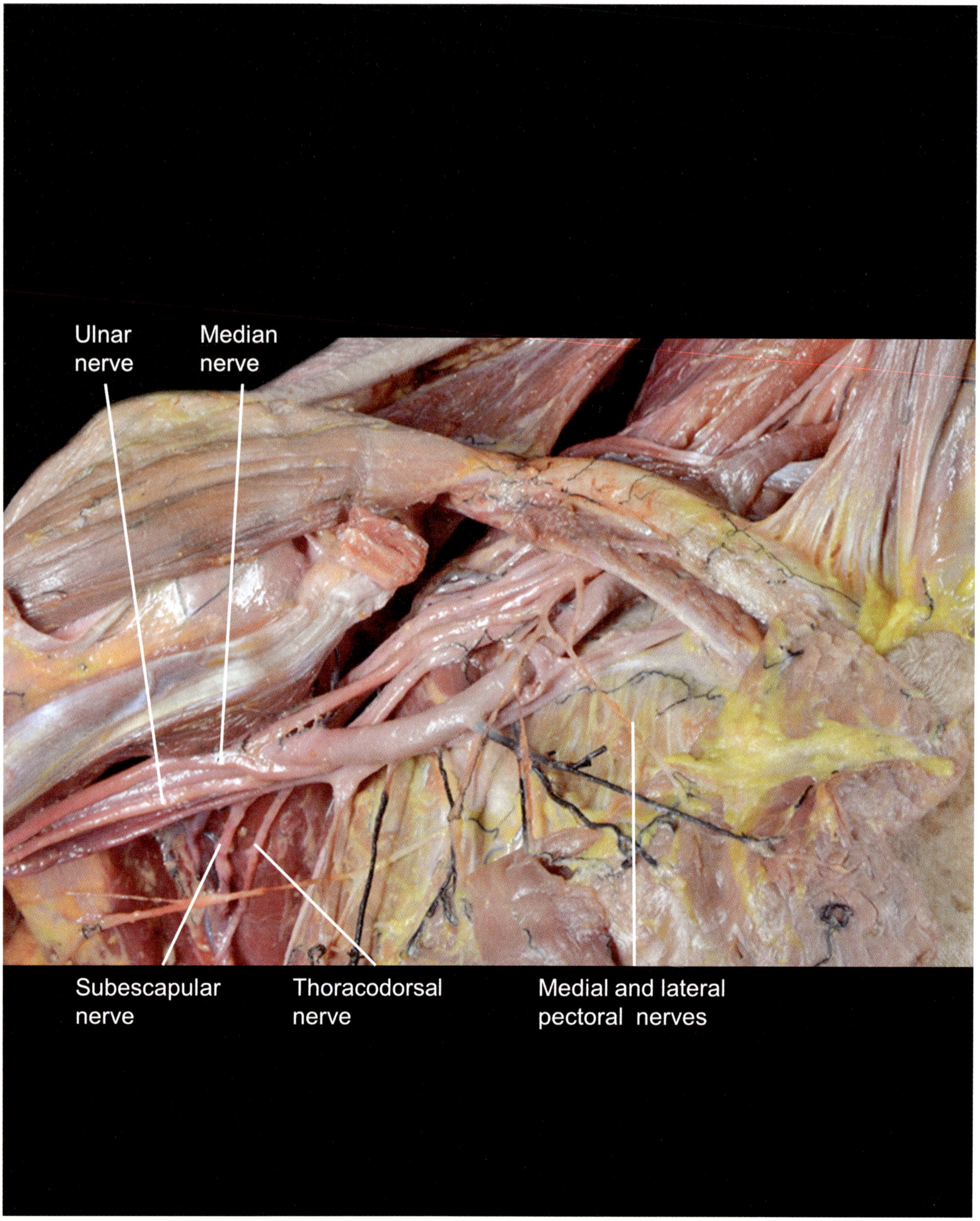

Fig. 7.33 Infraclavicular brachial plexus. Besides the seven terminal nerves already mentioned, the infraclavicular segment of the brachial plexus also gives rise to some collateral nerves. These are the lateral and medial pectoral nerves that arise from the anterior aspect of the lateral and medial cords, respectively, and the upper subscapular, thoracodorsal, and lower subscapular nerves that emerge from the back of the posterior cord, craniocaudally in that order

Cross-Sectional Microscopic Anatomy of the Brachial Plexus and Paraneural Sheaths

8

Miguel Angel Reina and Xavier Sala-Blanch

Ultrasound allows visualization of nerve trunks and branches to guide the position of the needle during anesthetic blockade. Although image resolution of newer ultrasound devices is much better than the resolution of earlier equipment, the invaluable contribution of histological techniques and the relevance of the morphological features they reveal cannot be refuted.

In this regard, details of the type and size of fascicles inside nerves achieved with current ultrasound equipment do not match those obtained by histological methods. There is lack of correlation between ultrasound images and the real anatomy of nerve fascicles. In addition, images of fascicles visualized in a histological section do not match ultrasound images of the same nerve in the same location.

Sectional anatomy aids in the interpretation of images obtained by ultrasound, although it is rare for cross sections to accurately match the orientation of the section produced with the ultrasound probe. At the level of the brachial plexus, ultrasound sections are typically transversal to the direction of the trunks and nerve branches.

The aim of this chapter is to present histological images from real anatomic sections at an anatomical level identical to that of ultrasound images obtained during anesthetic blockade. The resulting images enable accurate identification of microscopic features of anatomical structures identified by ultrasound imaging, as well as structures that are too small to be seen in ultrasound images [1–6]. Sections of this type also allow precise identification of connective tissue enclosing different structures comprising the brachial plexus, including paraneural layers that form several compartments containing fat and small vessels (Figs. 8.1, 8.2, 8.3, 8.4, 8.5, 8.6, 8.7, 8.8, 8.9, 8.10, 8.11, 8.12, 8.13, 8.14, 8.15, 8.16, 8.17, 8.18, 8.19, 8.20, 8.21, 8.22, 8.23, 8.24, 8.25, and 8.26). Figures 8.1–8.23 are images of successive sections from same cadaver, from interscalenic region to axillary region.

M.A. Reina, MD, PhD (✉)
Department of Clinical Medical Sciences and Institute of Applied Molecular Medicine, School of Medicine, University of CEU San Pablo, Madrid, Spain

Department of Anesthesiology, Madrid-Montepríncipe University Hospital, Madrid, Spain
e-mail: miguelangel@perticone.e.telefonica.net

X. Sala-Blanch, MD
Human Anatomy and Embryology Unit, School of Medicine, University of Barcelona, Barcelona, Spain

Department of Anesthesiology and Critical Care, Clinic Hospital, Barcelona, Spain
e-mail: xavi.sala.blanch@gmail.com

M.A. Reina et al. (eds.), *Atlas of Functional Anatomy for Regional Anesthesia and Pain Medicine: Human Structure, Ultrastructure and 3D Reconstruction Images*, DOI 10.1007/978-3-319-09522-6_8,

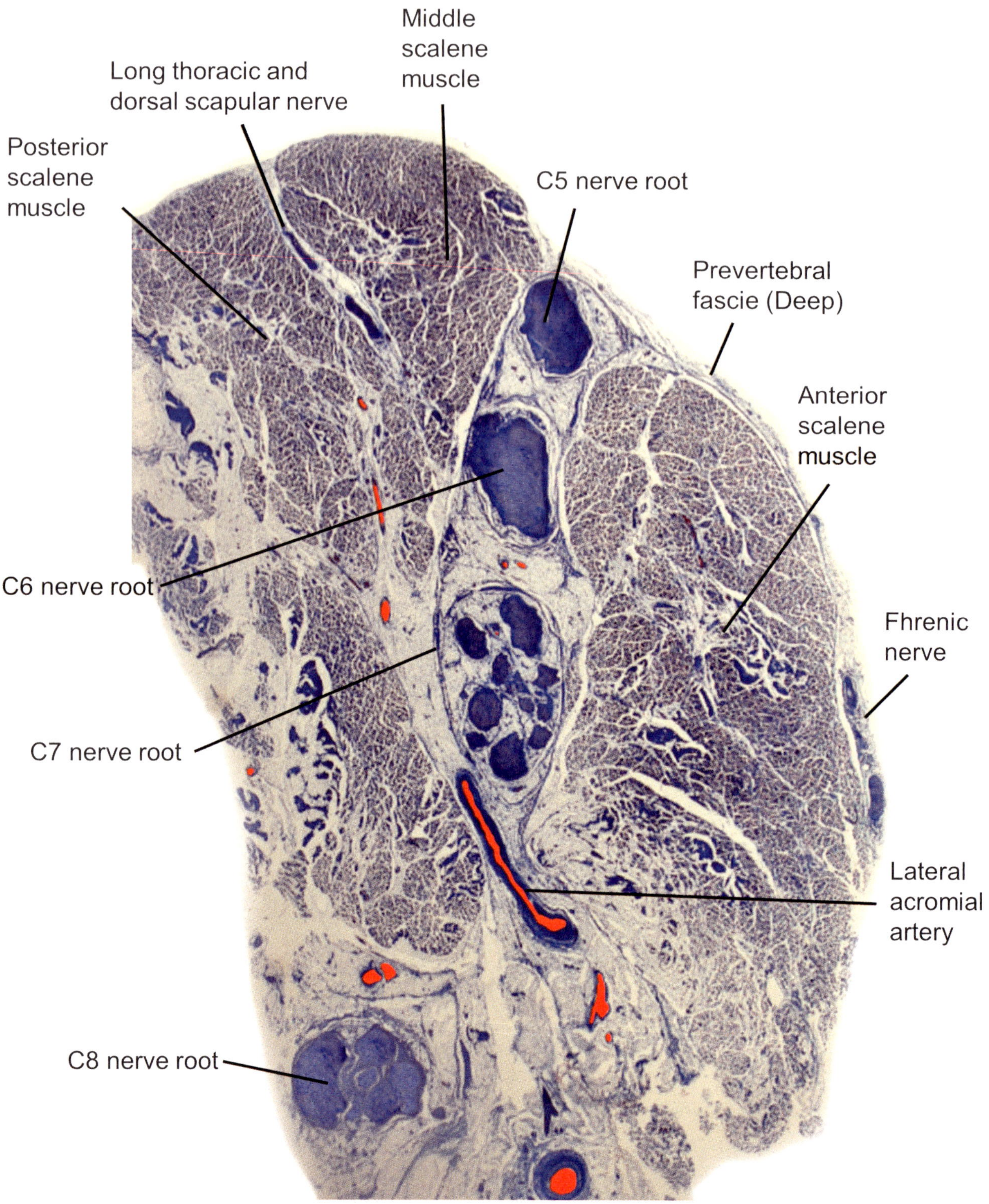

Fig. 8.1 Brachial plexus, interscalenic region. Masson's trichrome stained. First section of a serially sectioned sample

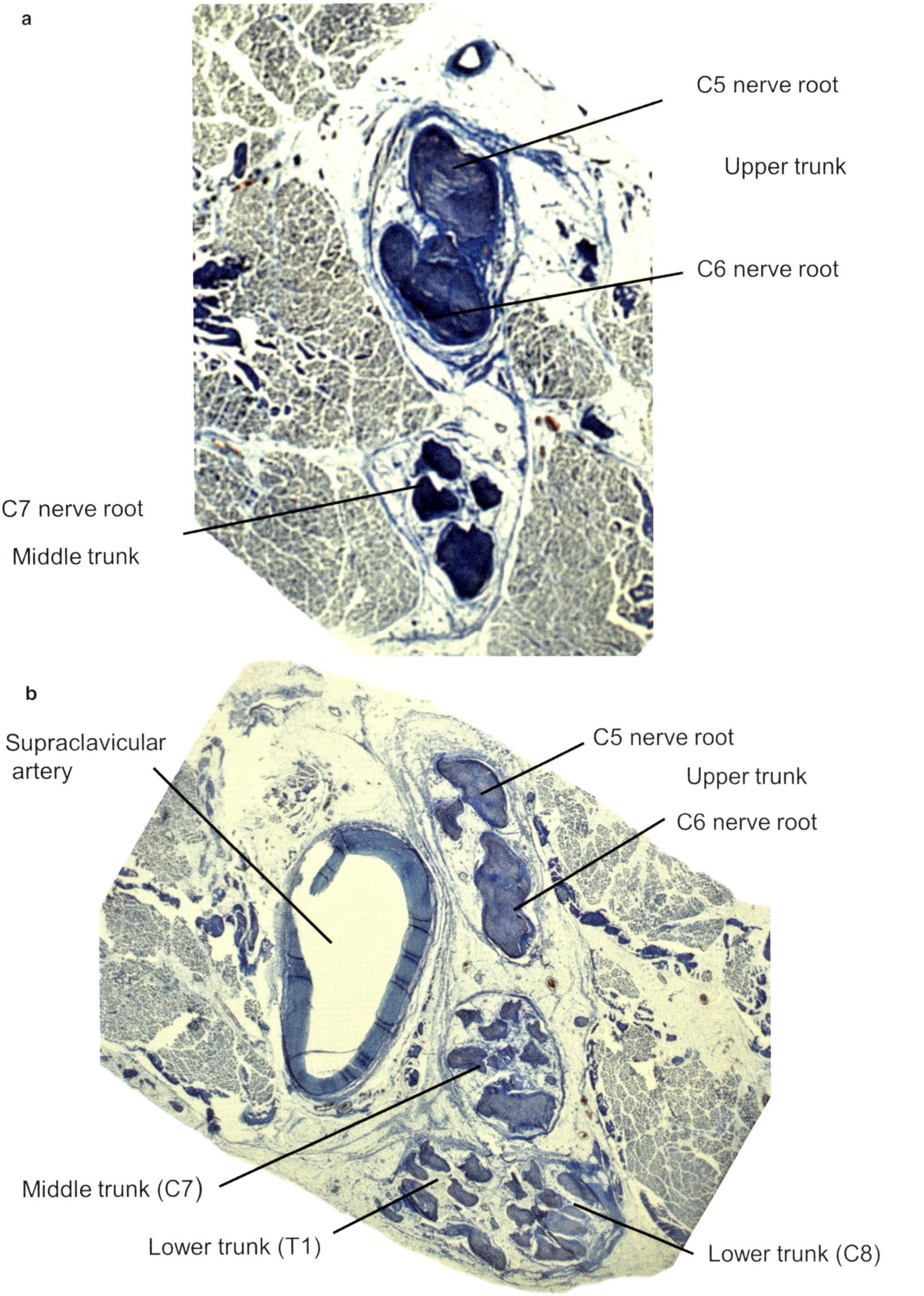

Fig. 8.2 Brachial plexus, interscalenic region. (**a**) and (**b**) are two successive sections. Masson's trichrome stained. Following section of a serially sectioned sample from Figs. 8.1 to 8.23

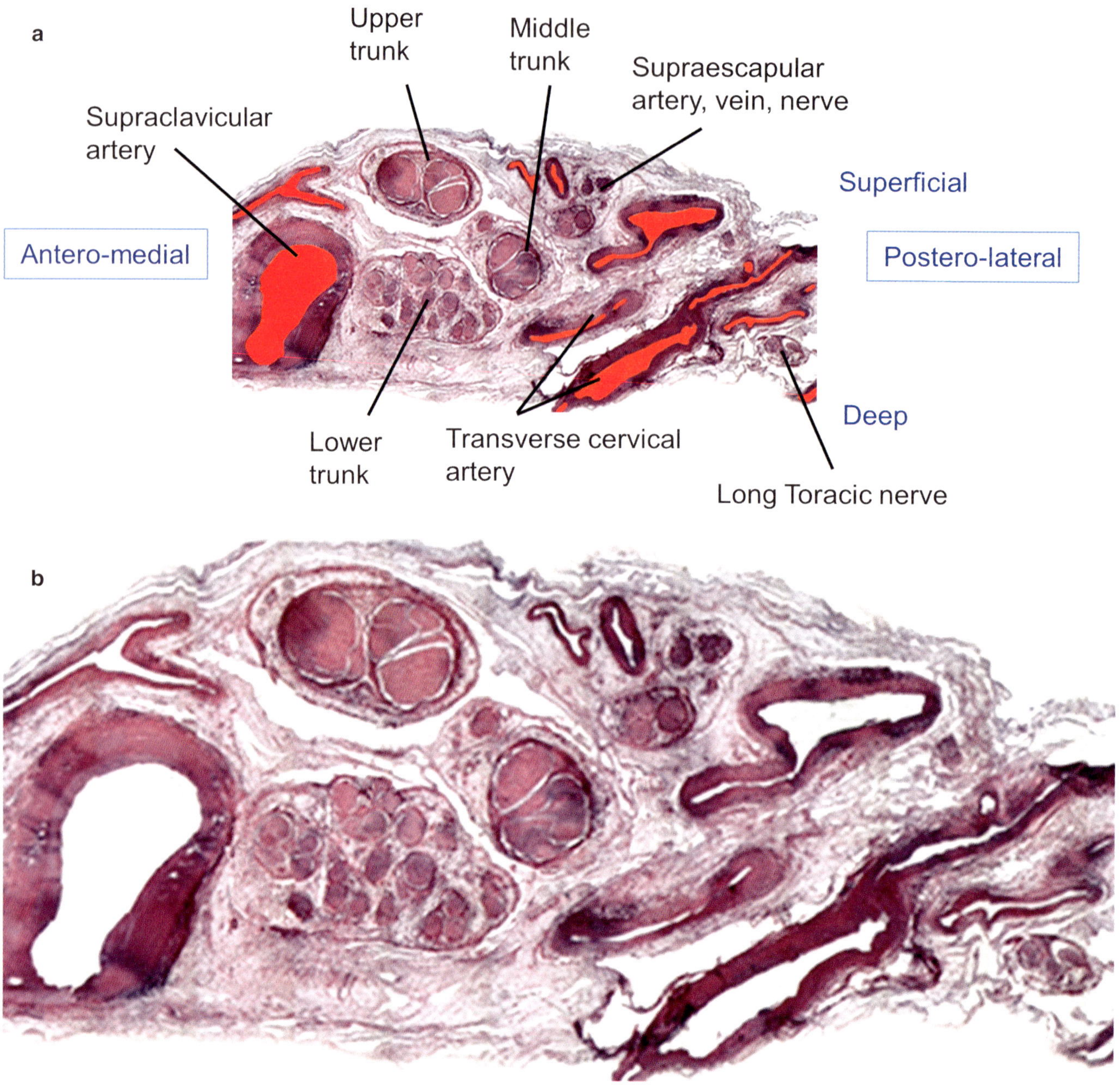

Fig. 8.3 (**a**) Brachial plexus, supraclavicular region. (**b**): (**a**) at more magnification. Hematoxylin-eosin stained. Following section of a serially sectioned sample from Figs. 8.1 to 8.23

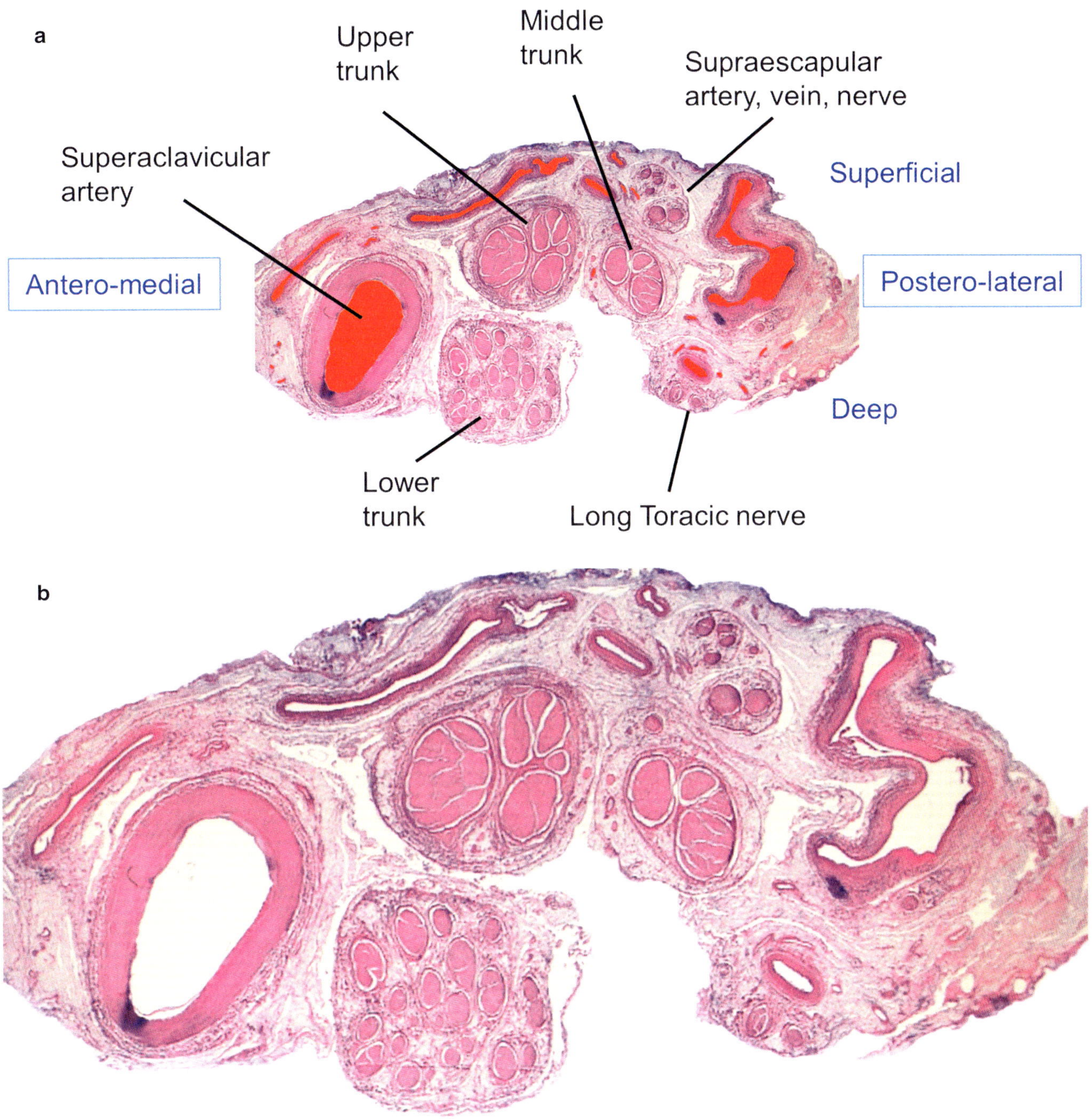

Fig. 8.4 (**a**) Brachial plexus, supraclavicular region. (**b**): (**a**) at more magnification. Hematoxylin-eosin stained. Following section of a serially sectioned sample from Figs. 8.1 to 8.23

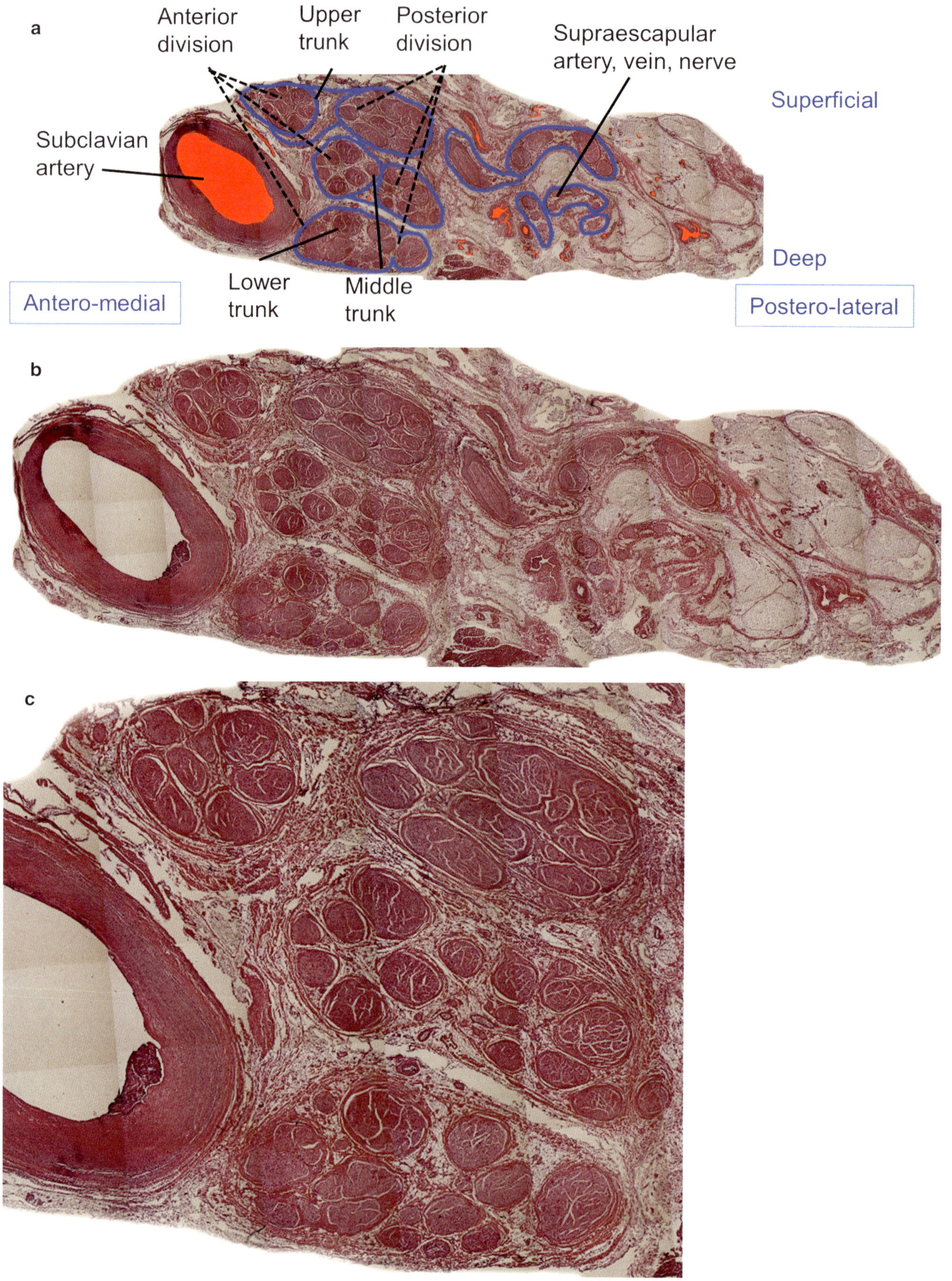

Fig. 8.5 (**a**) Brachial plexus, supraclavicular region. (**b**): (**a**) at more magnification. (**c**) Detail of (**b**). Hematoxylin-eosin stained. Following section of a serially sectioned sample from Figs. 8.1 to 8.23

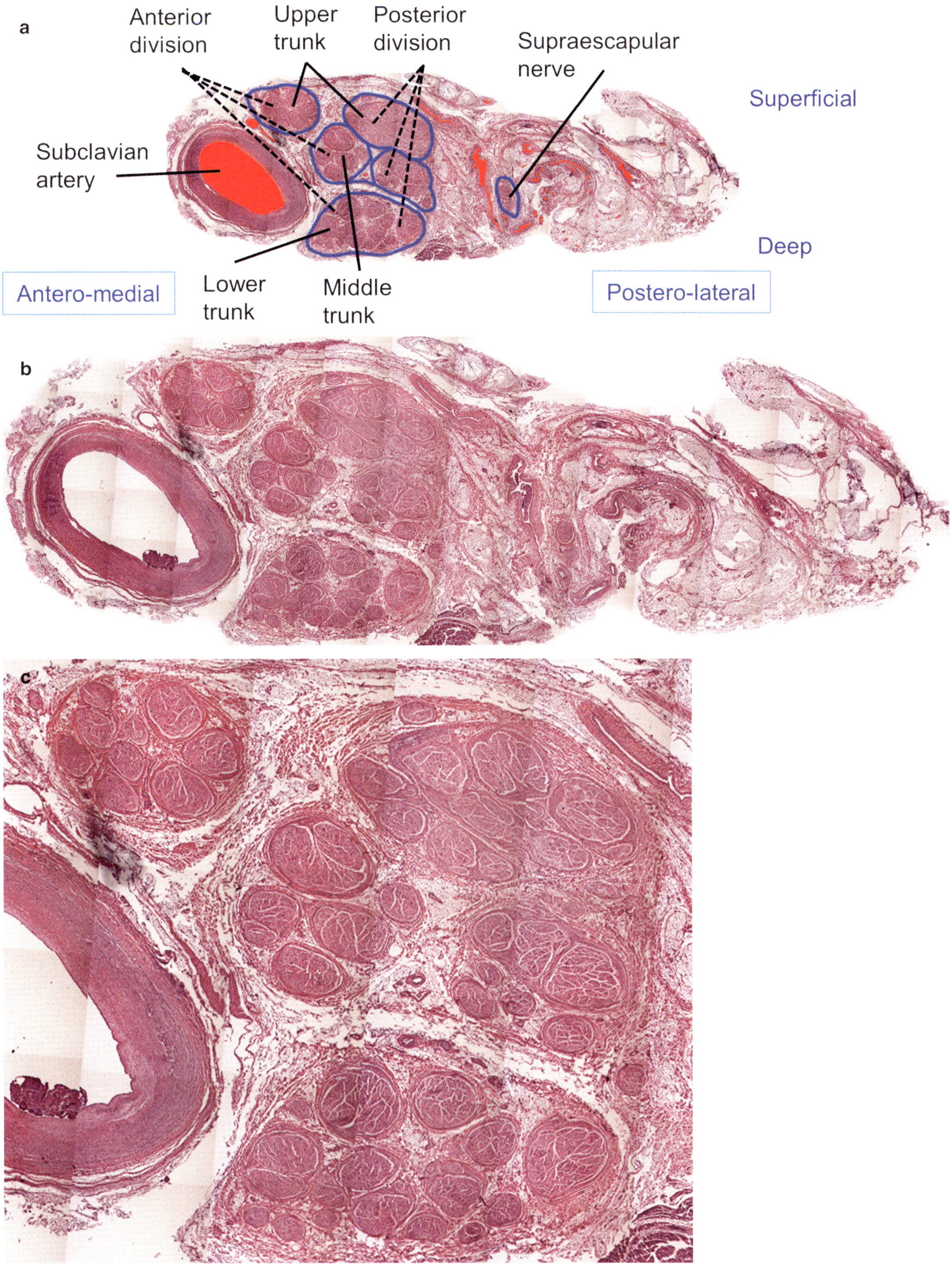

Fig. 8.6 (**a**) Brachial plexus, supraclavicular region at retroclavicular level. (**b**): (**a**) at more magnification. (**c**) Detail of (**b**). Hematoxylin-eosin stained. Following section of a serially sectioned sample from Figs. 8.1 to 8.23

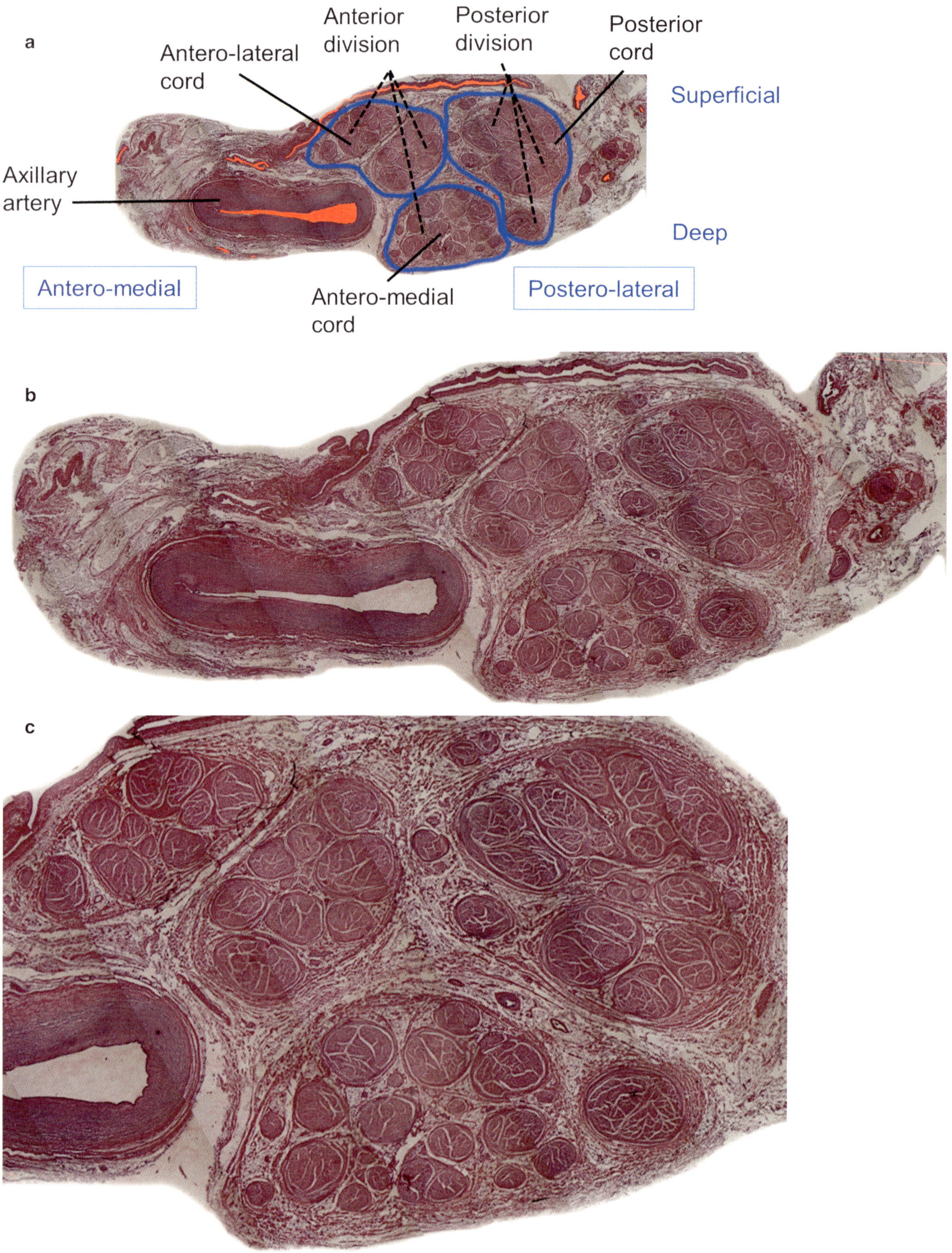

Fig. 8.7 (**a**) Brachial plexus, supraclavicular region at retroclavicular level. (**b**): (**a**) at more magnification. (**c**) Detail of (**b**). Hematoxylin-eosin stained. Following section of a serially sectioned sample from Figs. 8.1 to 8.23

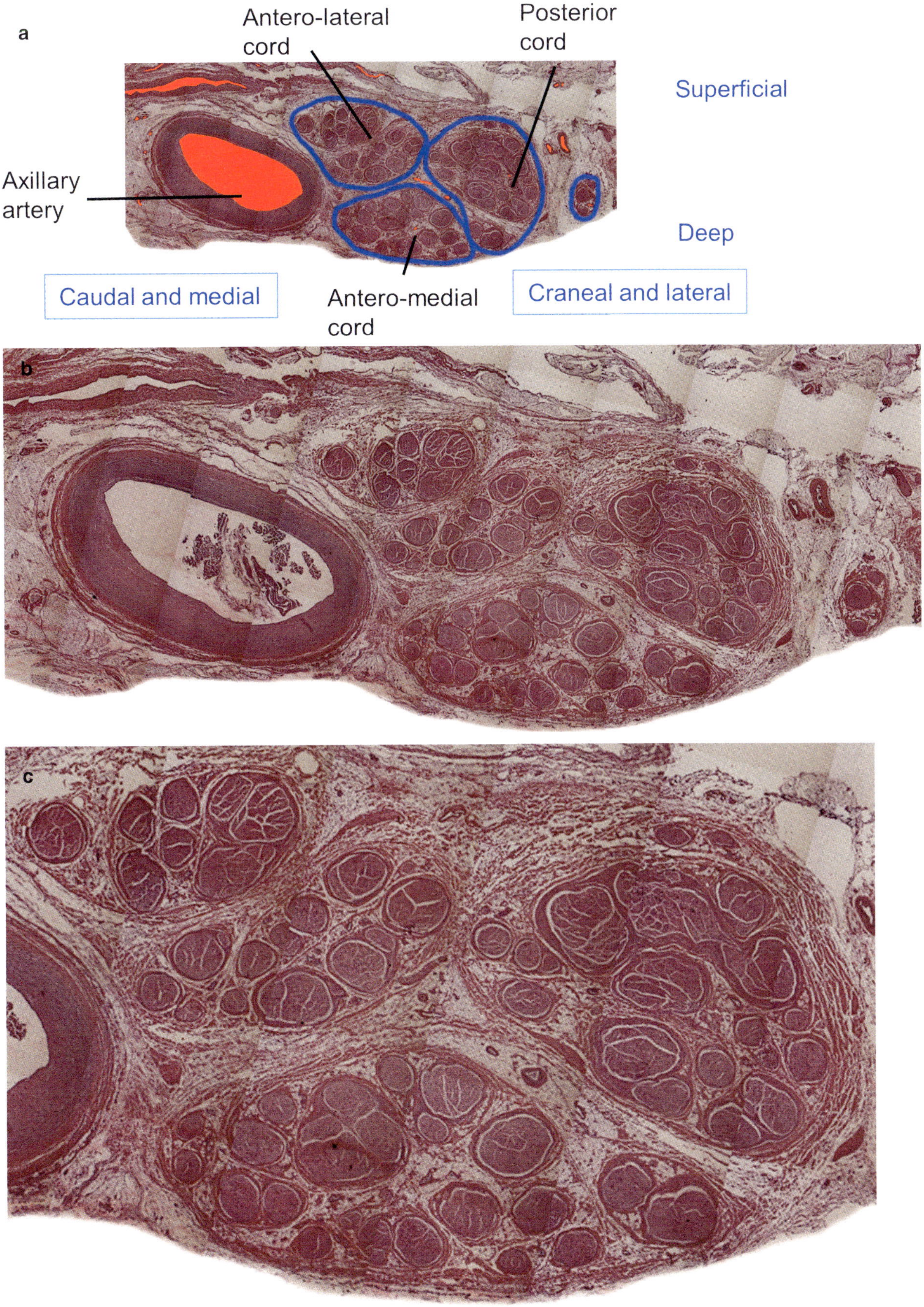

Fig. 8.8 (**a**) Brachial plexus, infraclavicular region. Hematoxylin-eosin stained. (**b**): (**a**) at more magnification. (**c**) Detail of (**b**). Following section of a serially sectioned sample from Figs. 8.1 to 8.23

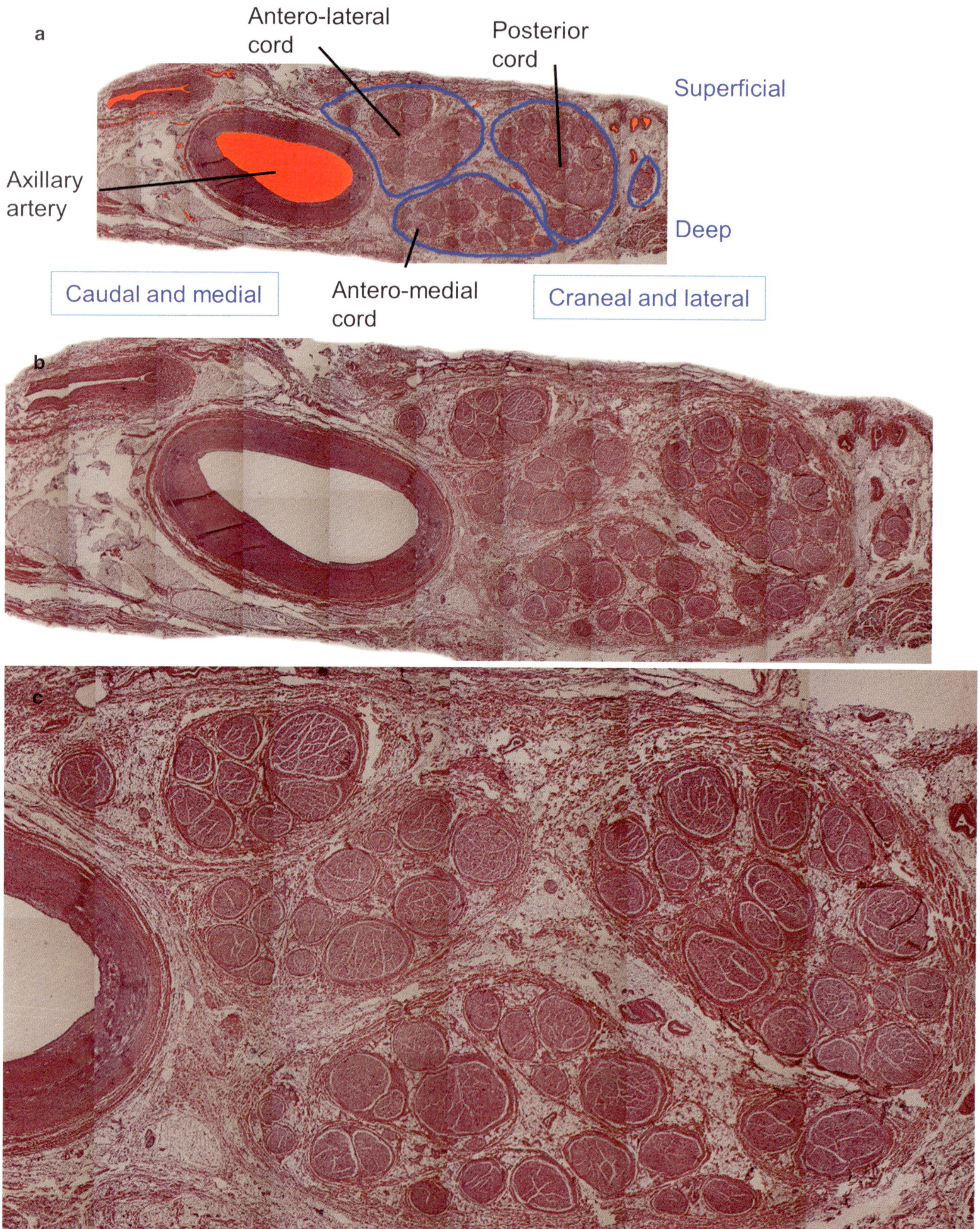

Fig. 8.9 (**a**) Brachial plexus, infraclavicular region at first part level. (**b**): (**a**) at more magnification. (**c**) Detail of (**b**). Hematoxylin-eosin stained. Following section of a serially sectioned sample from Figs. 8.1 to 8.23

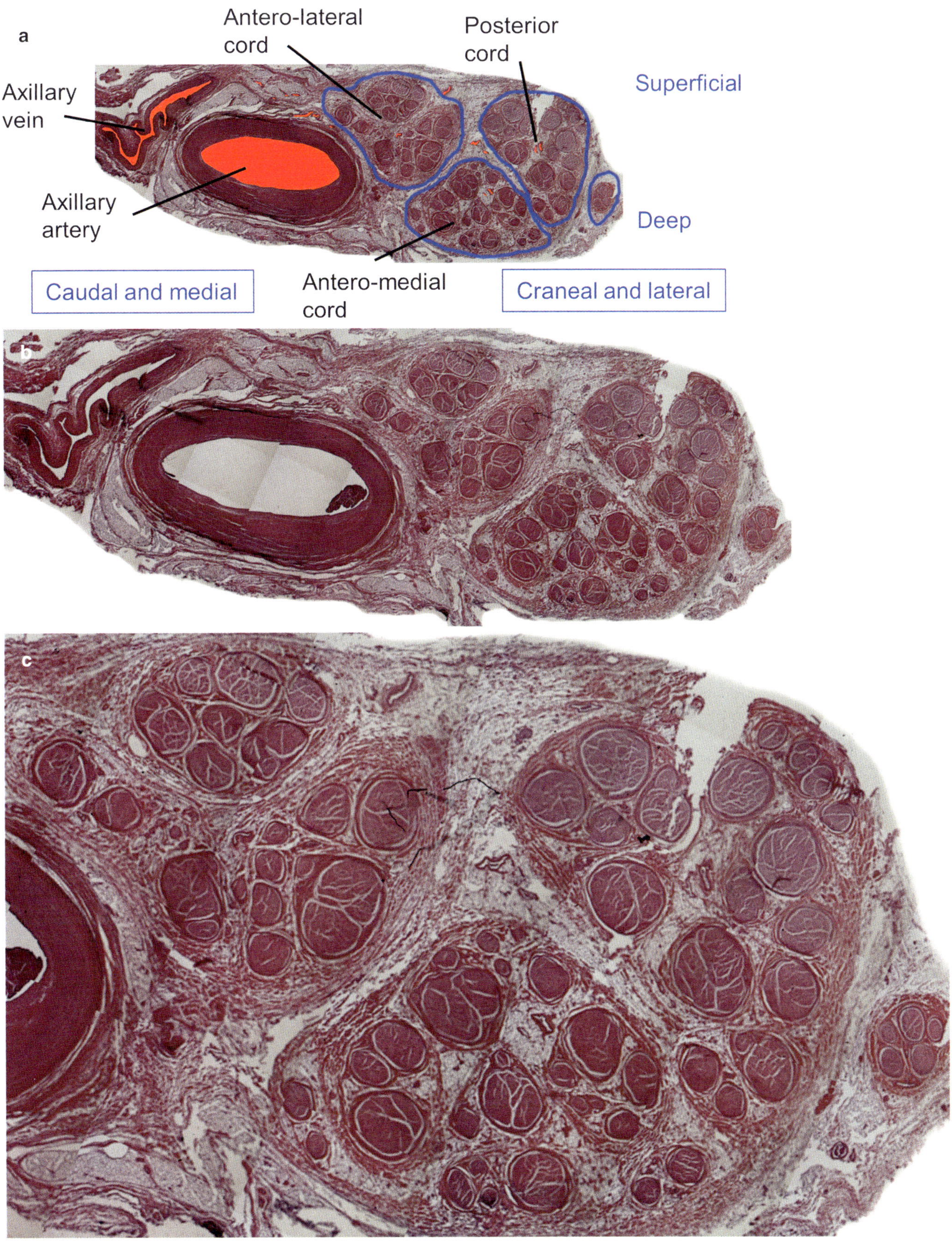

Fig. 8.10 (**a**) Brachial plexus, infraclavicular region at first part level. (**b**): (**a**) at more magnification. (**c**) Detail of (**b**). Hematoxylin-eosin stained. Following section of a serially sectioned sample

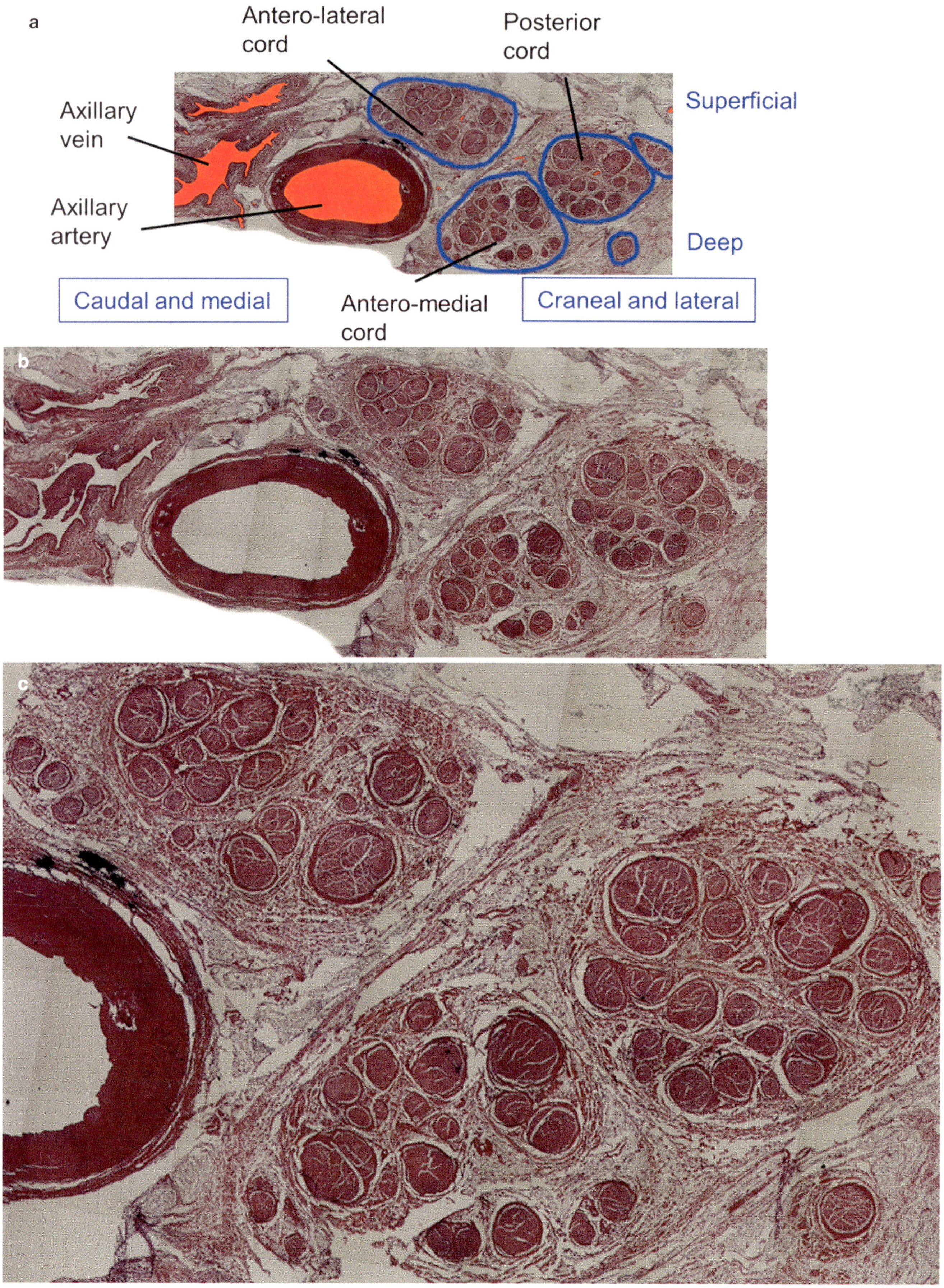

Fig. 8.11 (**a**) Brachial plexus, infraclavicular region at first part level. (**b**): (**a**) at more magnification. (**c**) Detail of (**b**). Hematoxylin-eosin stained. Following section of a serially sectioned sample from Figs. 8.1 to 8.23

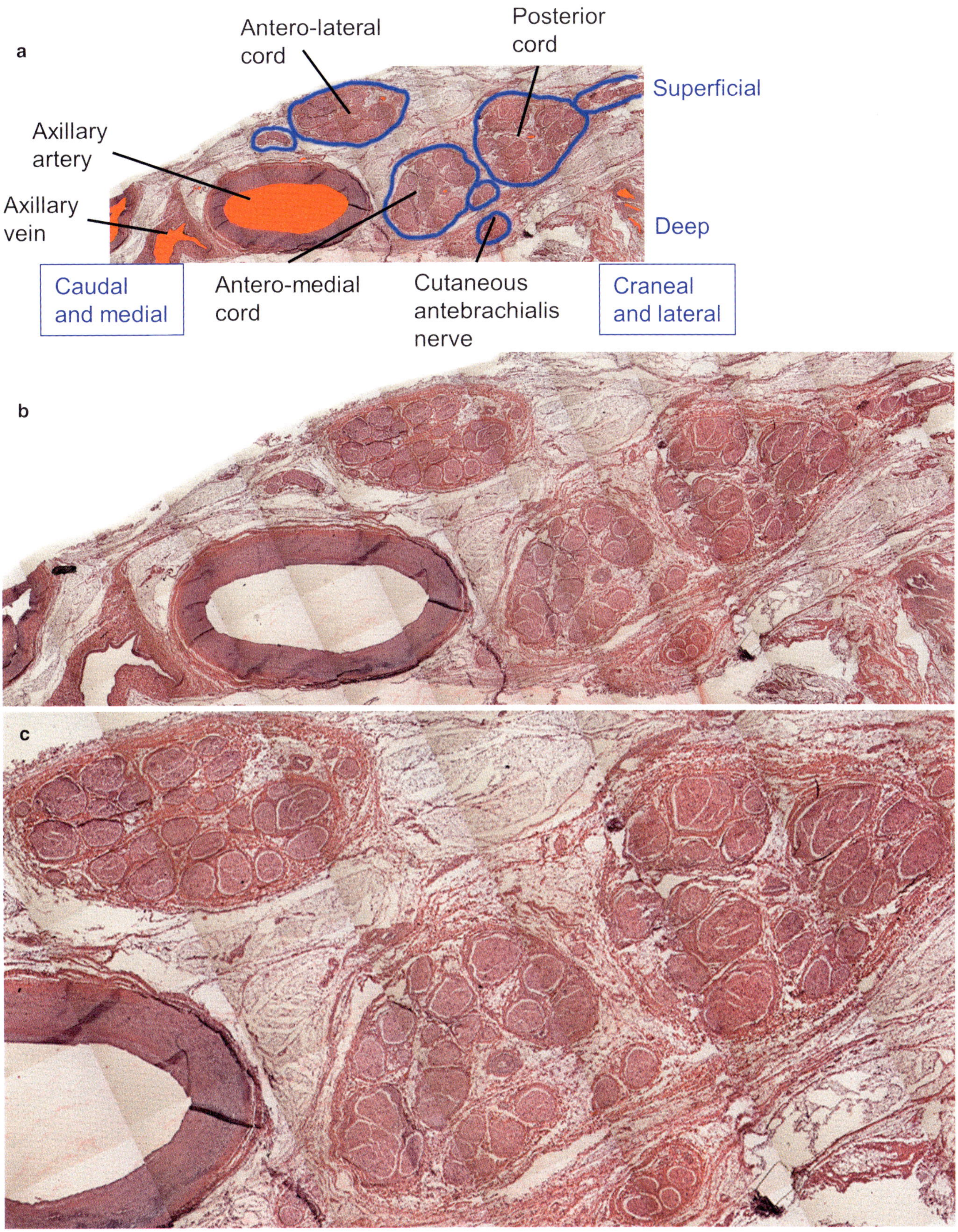

Fig. 8.12 (**a**) Brachial plexus, infraclavicular region at first part level. (**b**): (**a**) at more magnification. (**c**) Detail of (**b**). Hematoxylin-eosin stained. Following section of a serially sectioned sample from Figs. 8.1 to 8.23

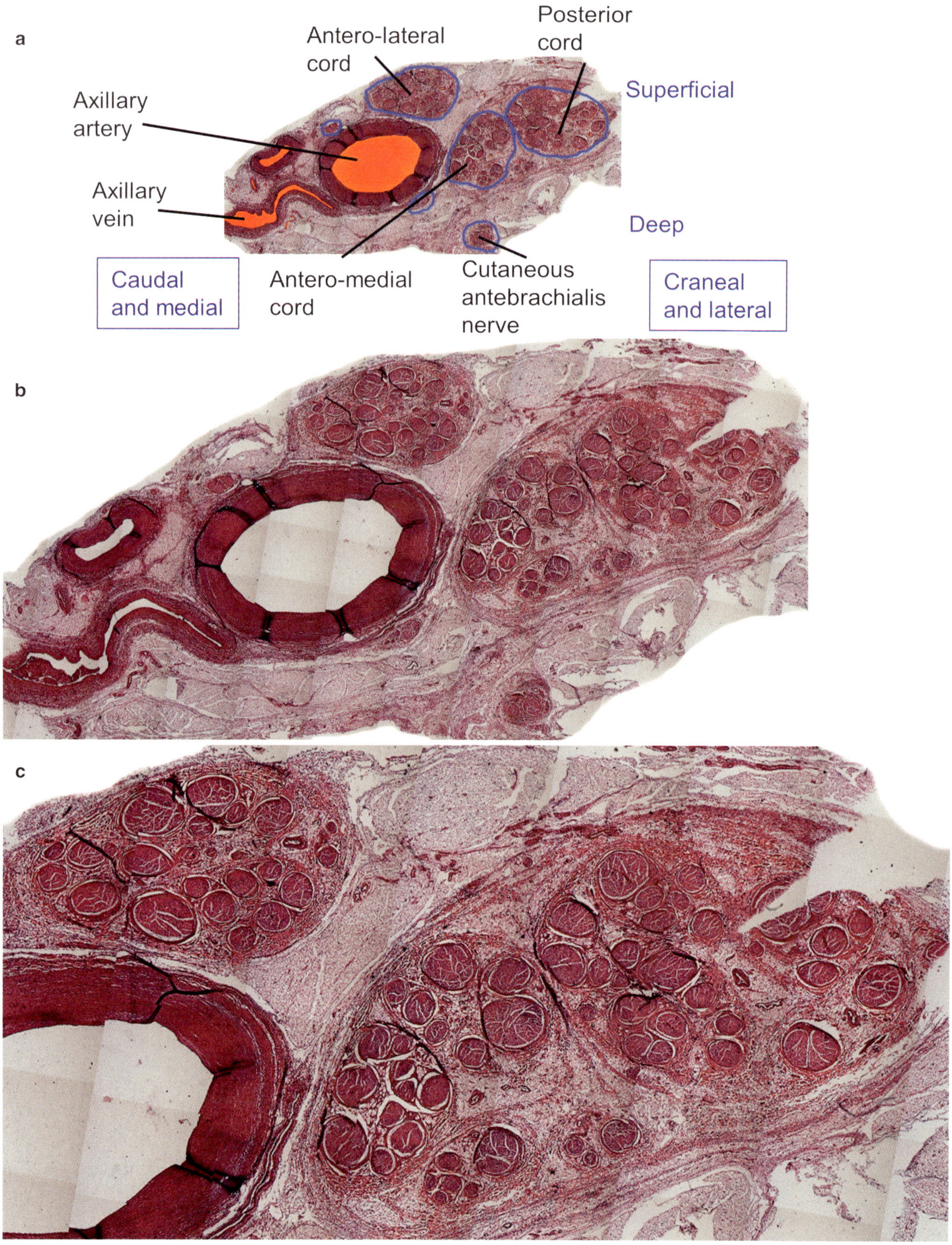

Fig. 8.13 (**a**) Brachial plexus, infraclavicular region at first part level. (**b**): (**a**) at more magnification. (**c**) Detail of (**b**). Hematoxylin-eosin stained. Following section of a serially sectioned sample from Figs. 8.1 to 8.23

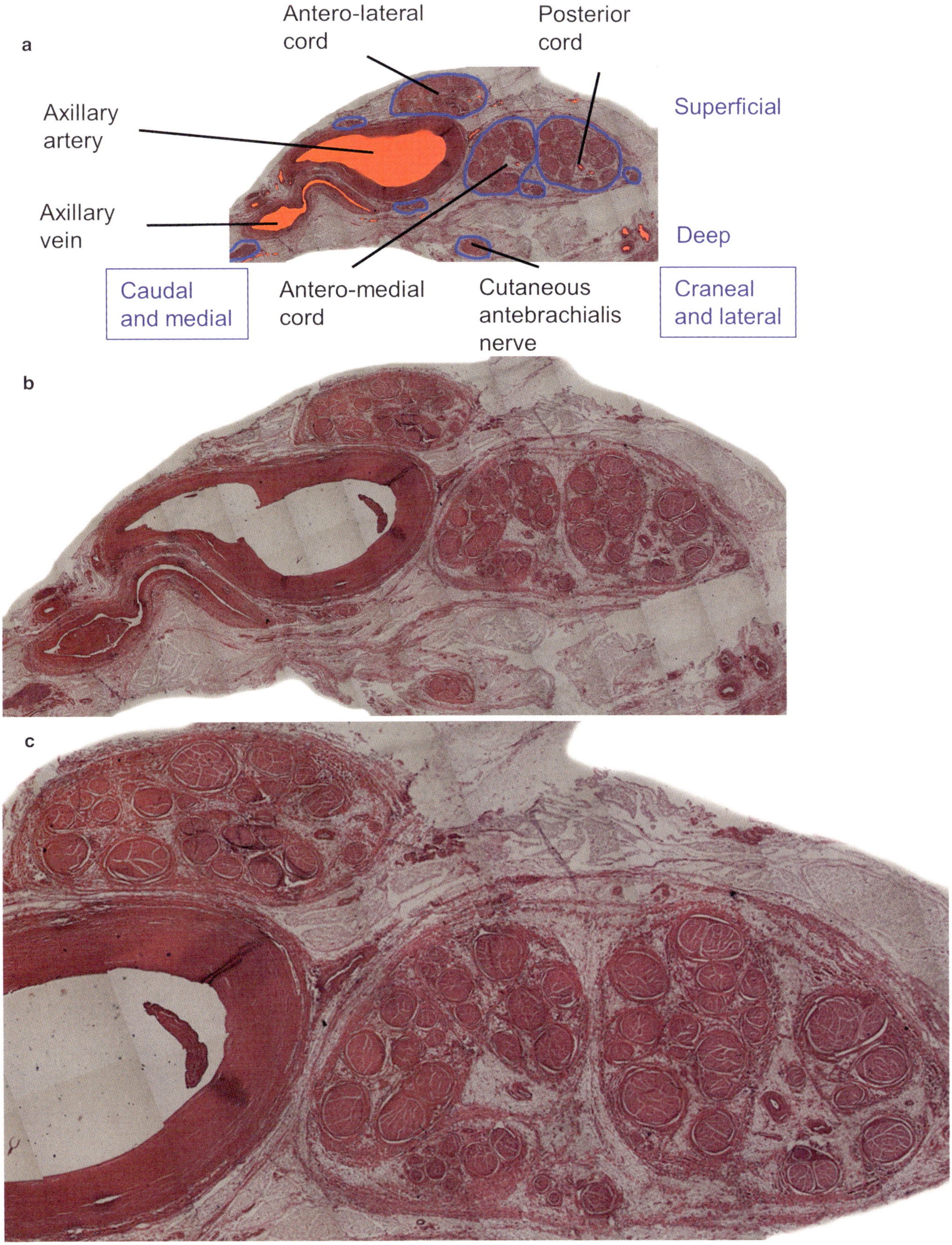

Fig. 8.14 (**a**) Brachial plexus, infraclavicular region at second part level (coracoid). (**b**): (**a**) at more magnification. (**c**) Detail of (**b**). Hematoxylin-eosin stained. Following section of a serially sectioned sample from Figs. 8.1 to 8.23

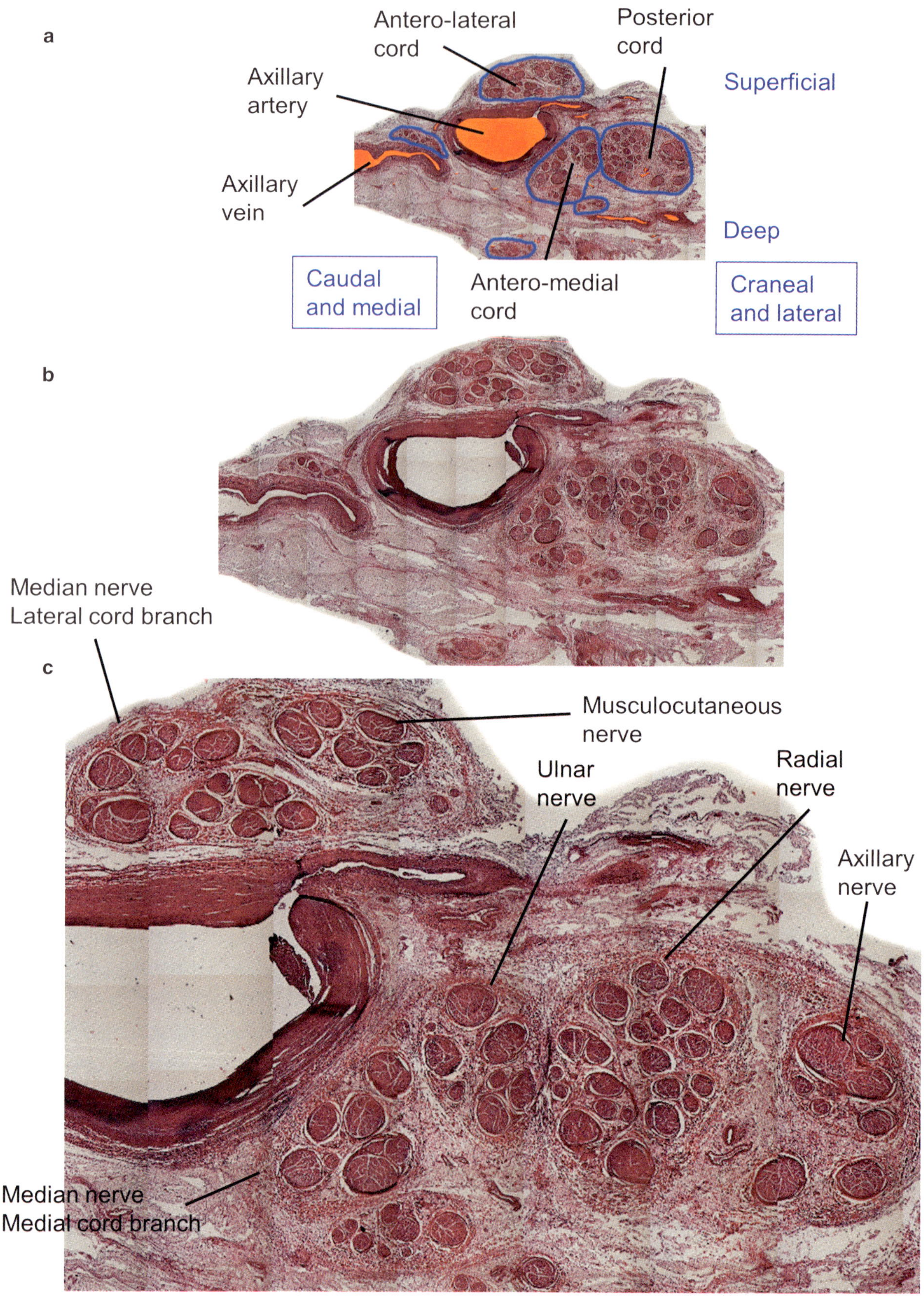

Fig. 8.15 (**a**) Brachial plexus, infraclavicular region at second part level (coracoid). (**b**): (**a**) at more magnification. (**c**) Detail of (**b**). Hematoxylin-eosin stained. Following section of a serially sectioned sample from Figs. 8.1 to 8.23

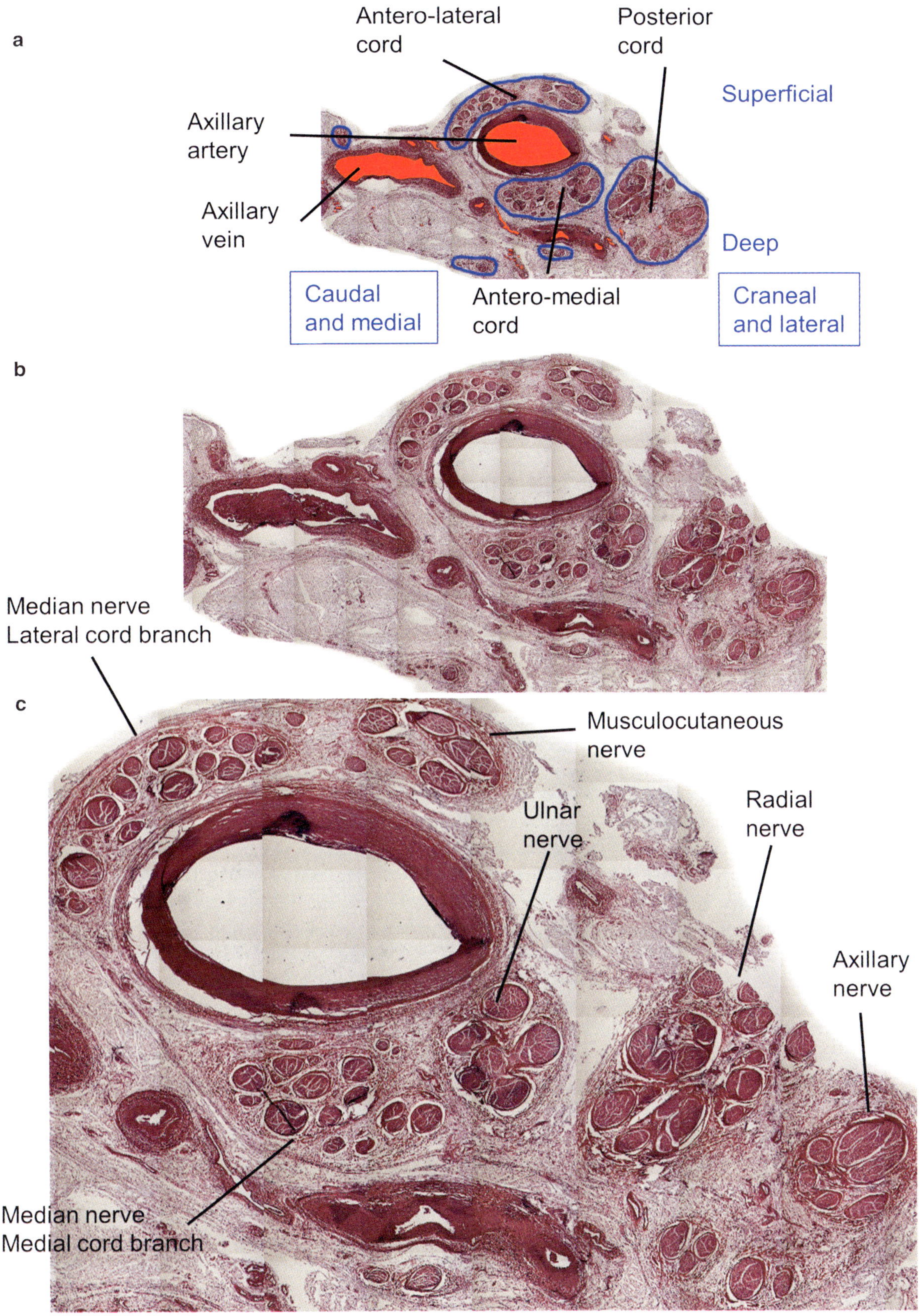

Fig. 8.16 (**a**) Brachial plexus, infraclavicular region at second part level (coracoid). (**b**): (a) at more magnification. (**c**) Detail of (**b**). Hematoxylin-eosin stained. Following section of a serially sectioned sample from Figs. 8.1 to 8.23

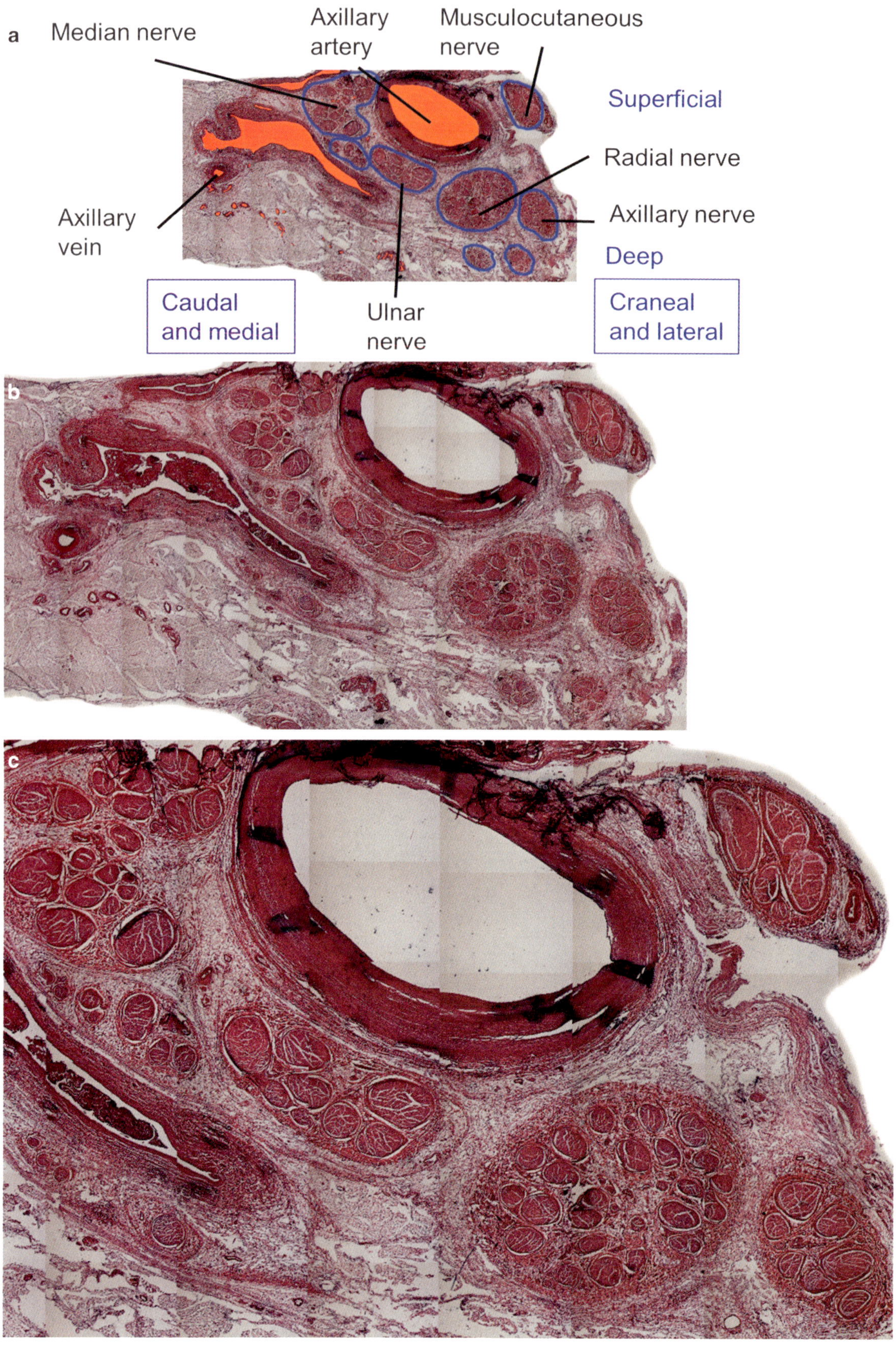

Fig. 8.17 (**a**) Brachial plexus, infraclavicular region at second part level (coracoid). (**b**): (**a**) at more magnification. (**c**) Detail of (**b**). Hematoxylin-eosin stained. Following section of a serially sectioned sample from Figs. 8.1 to 8.23

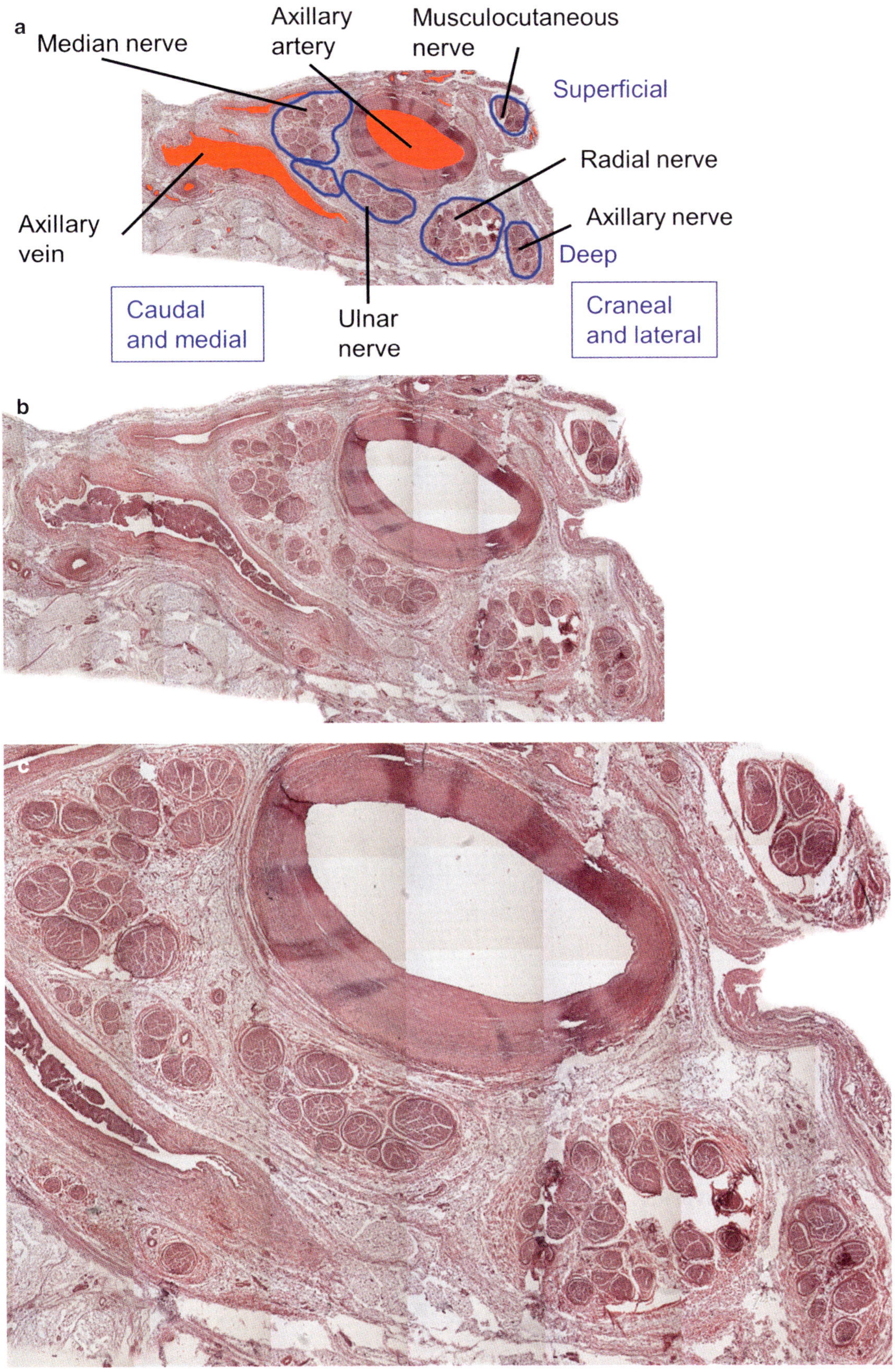

Fig. 8.18 (**a**) Brachial plexus, infraclavicular region at second part level (coracoid). (**b**): (**a**) at more magnification. (**c**) Detail of (**b**). Hematoxylin-eosin stained. Following section of a serially sectioned sample from Figs. 8.1 to 8.23

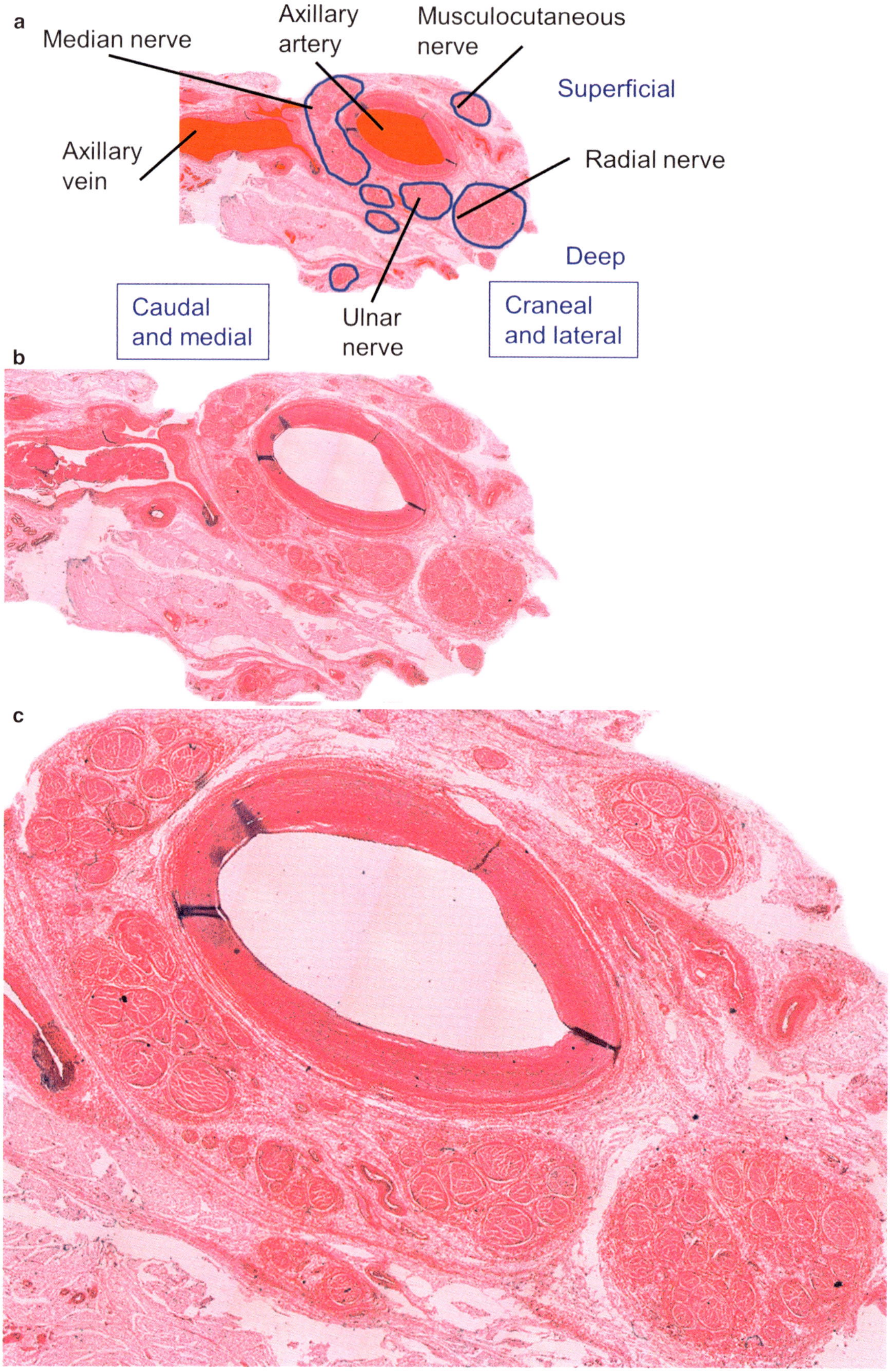

Fig. 8.19 (**a**) Brachial plexus, infraclavicular region at second part level (coracoid). (**b**): (**a**) at more magnification. (**c**) Detail of (**b**). Hematoxylin-eosin stained. Following section of a serially sectioned sample from Figs. 8.1 to 8.23

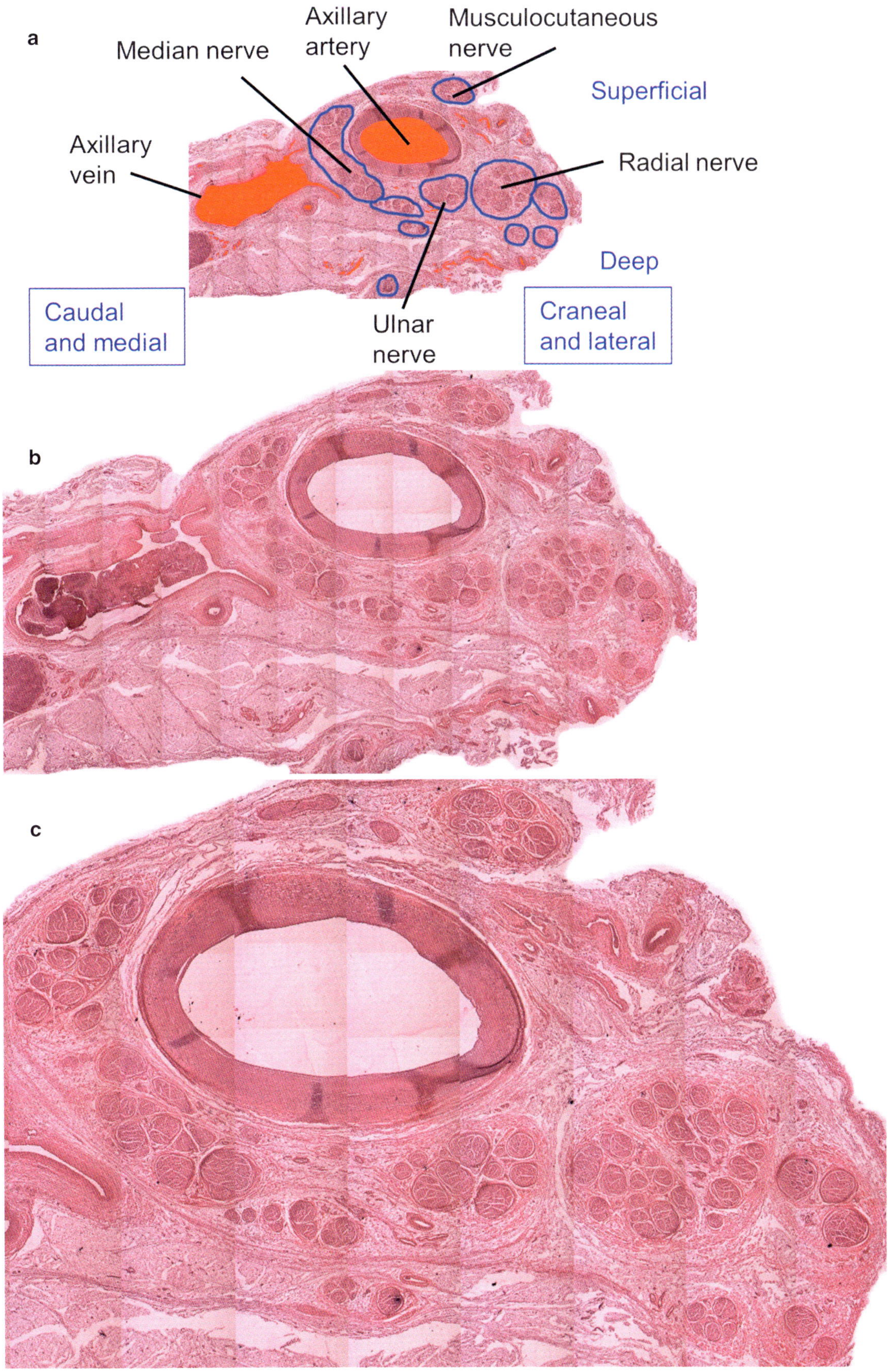

Fig. 8.20 (**a**) Brachial plexus, infraclavicular region at second part level (coracoid). (**b**): (**a**) at more magnification. (**c**) Detail of (**b**). Hematoxylin-eosin stained. Following section of a serially sectioned sample from Figs. 8.1 to 8.23

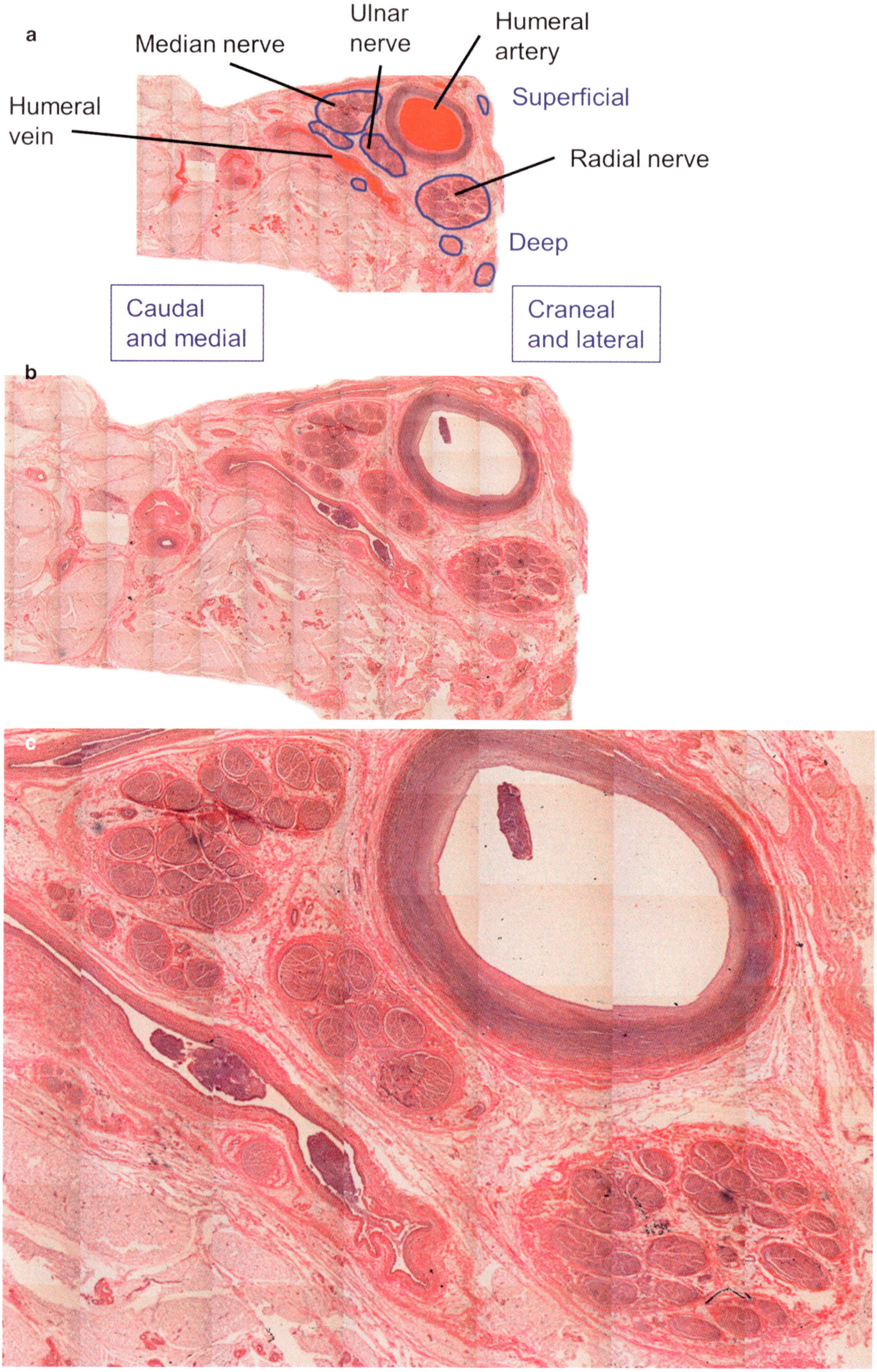

Fig. 8.21 (**a**) Brachial plexus, axillary region. (**b**): (**a**) at more magnification. (**c**) Detail of (**b**). Hematoxylin-eosin stained. Following section of a serially sectioned sample from Figs. 8.1 to 8.23

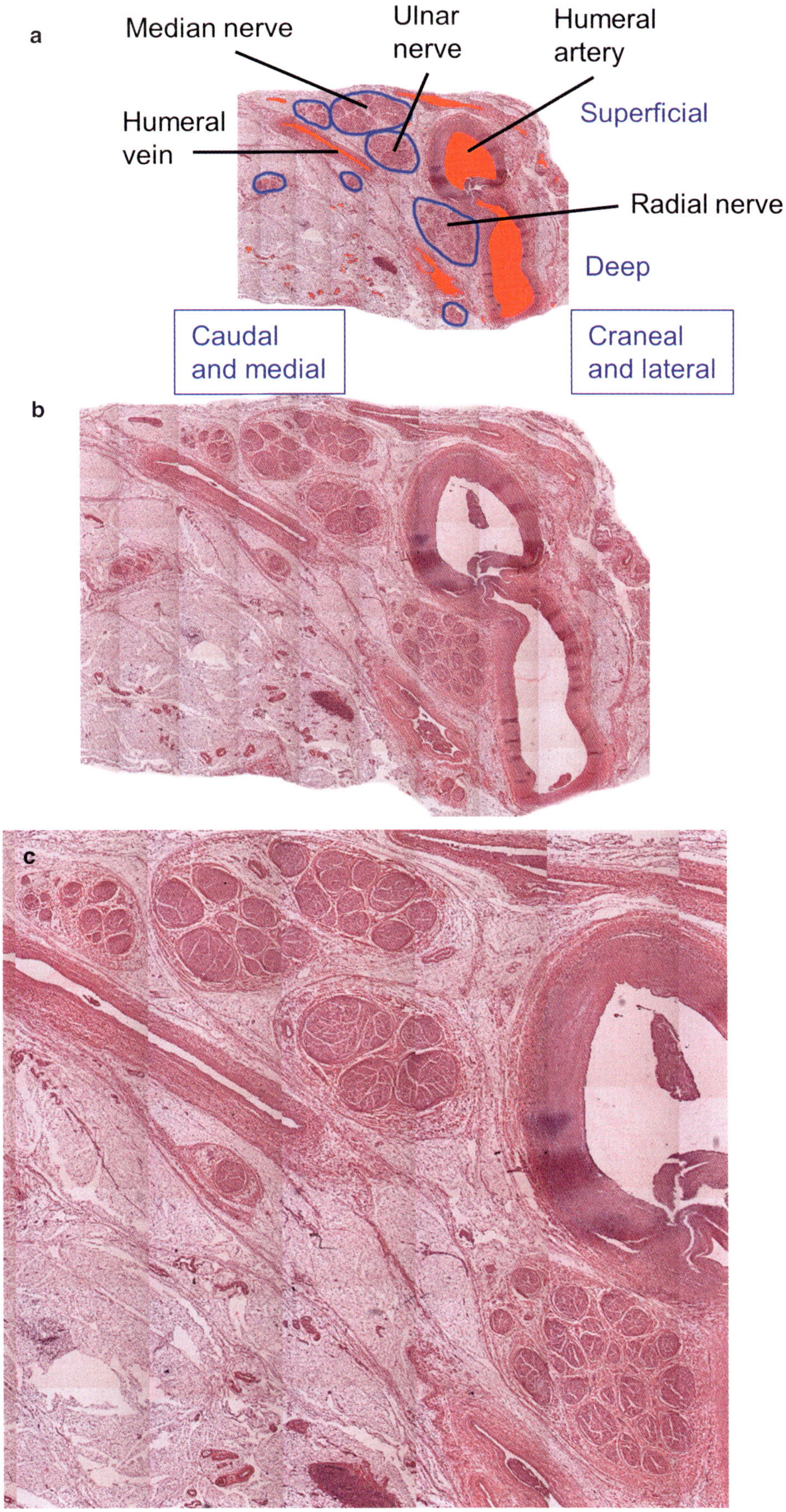

Fig. 8.22 (**a**) Brachial plexus, axillary region. (**b**): (**a**) at more magnification. (**c**) Detail of (**b**). Hematoxylin-eosin stained. Following section of a serially sectioned sample from Figs. 8.1 to 8.23

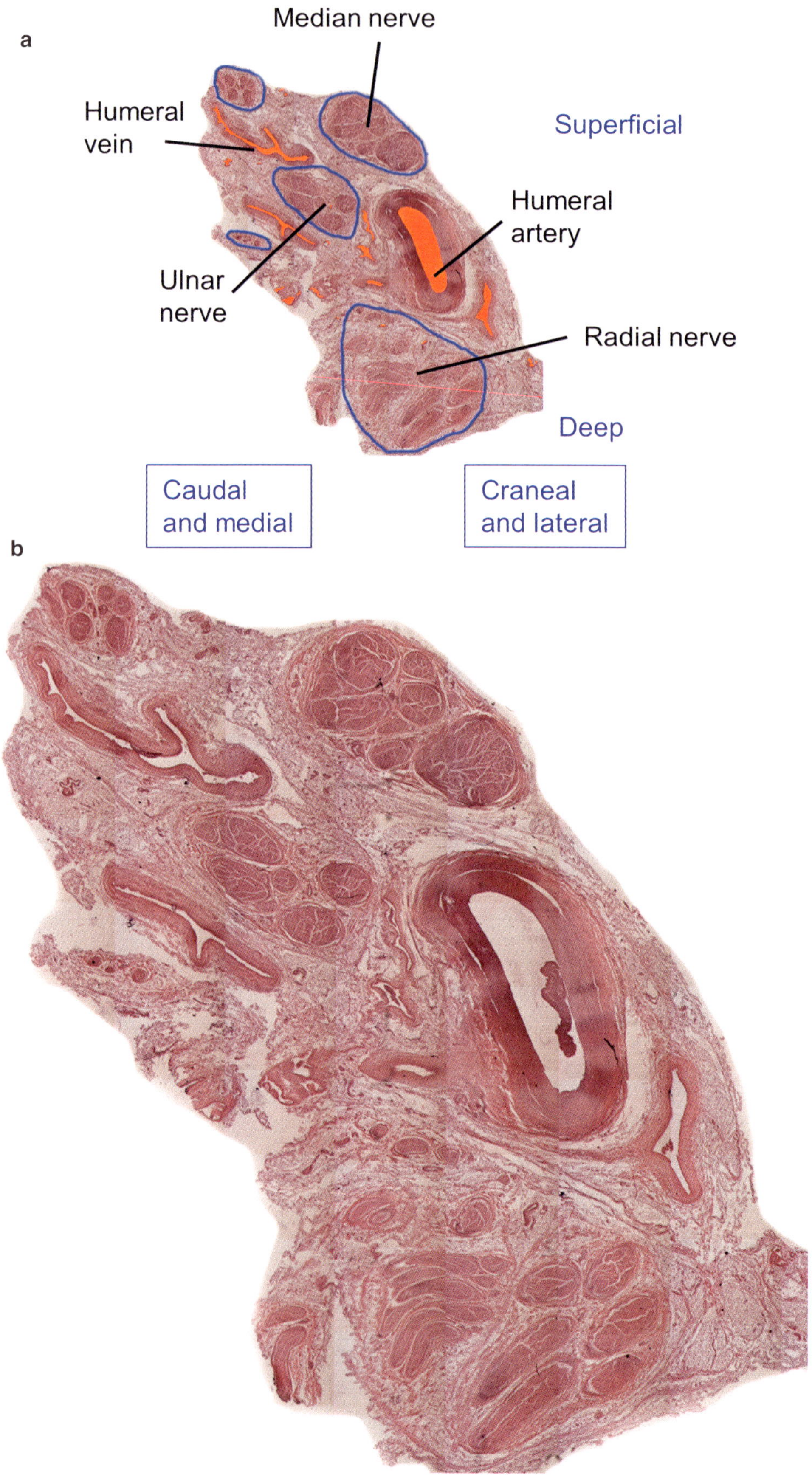

Fig. 8.23 (**a**) Brachial plexus, axillary region. (**b**): (**a**) at more magnification. (**c**) Detail of (**b**). Hematoxylin-eosin stained. Last section of a serially sectioned sample from Figs. 8.1 to 8.23

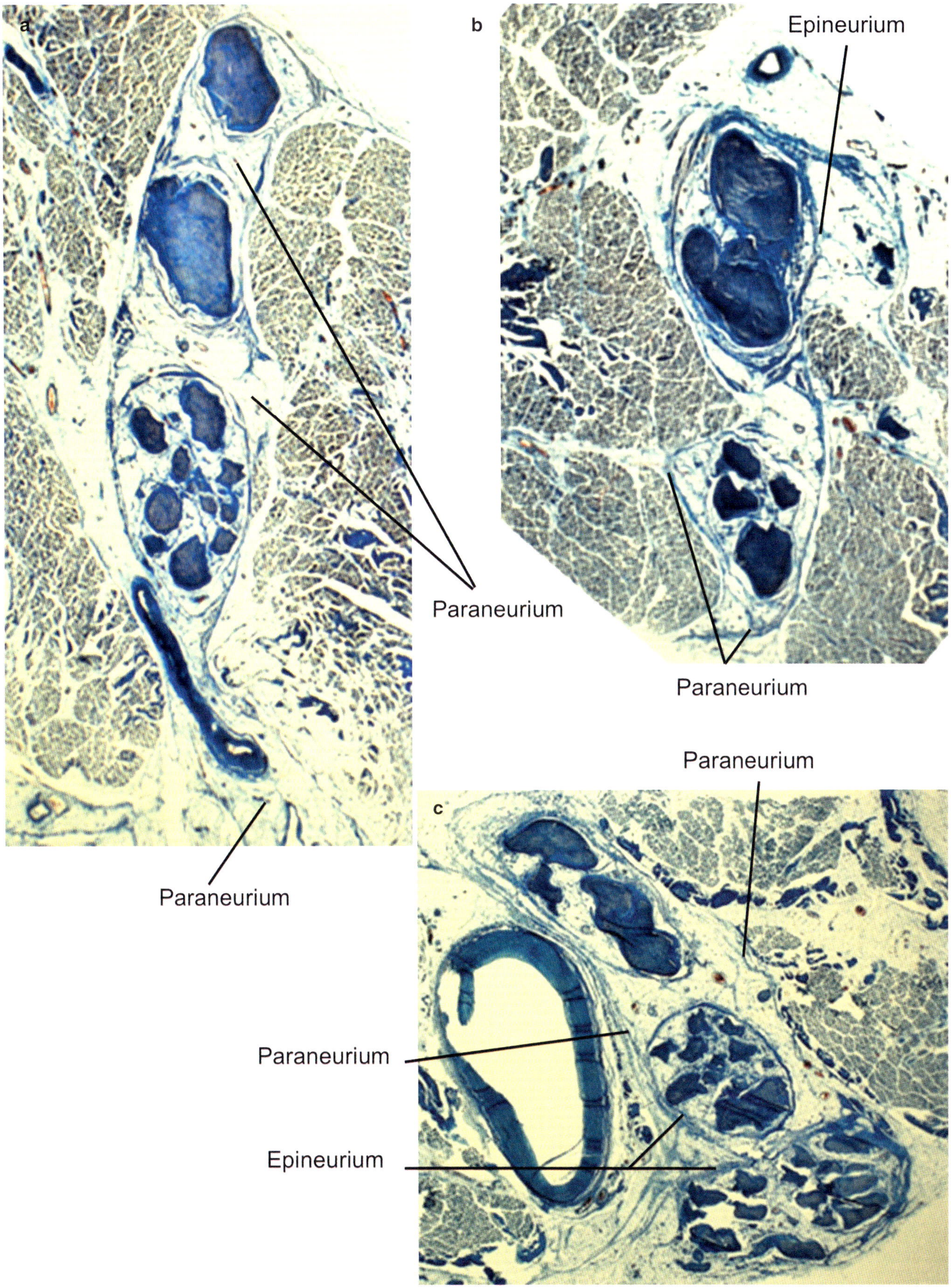

Fig. 8.24 Details of internal structure of nerve root in interscalenic region. (**a**) Detail obtained from Fig. 8.1, (**b**) Detail obtained from Fig. 8.2a, (**c**) Detail obtained from Fig. 8.2b. Masson's trichrome stained

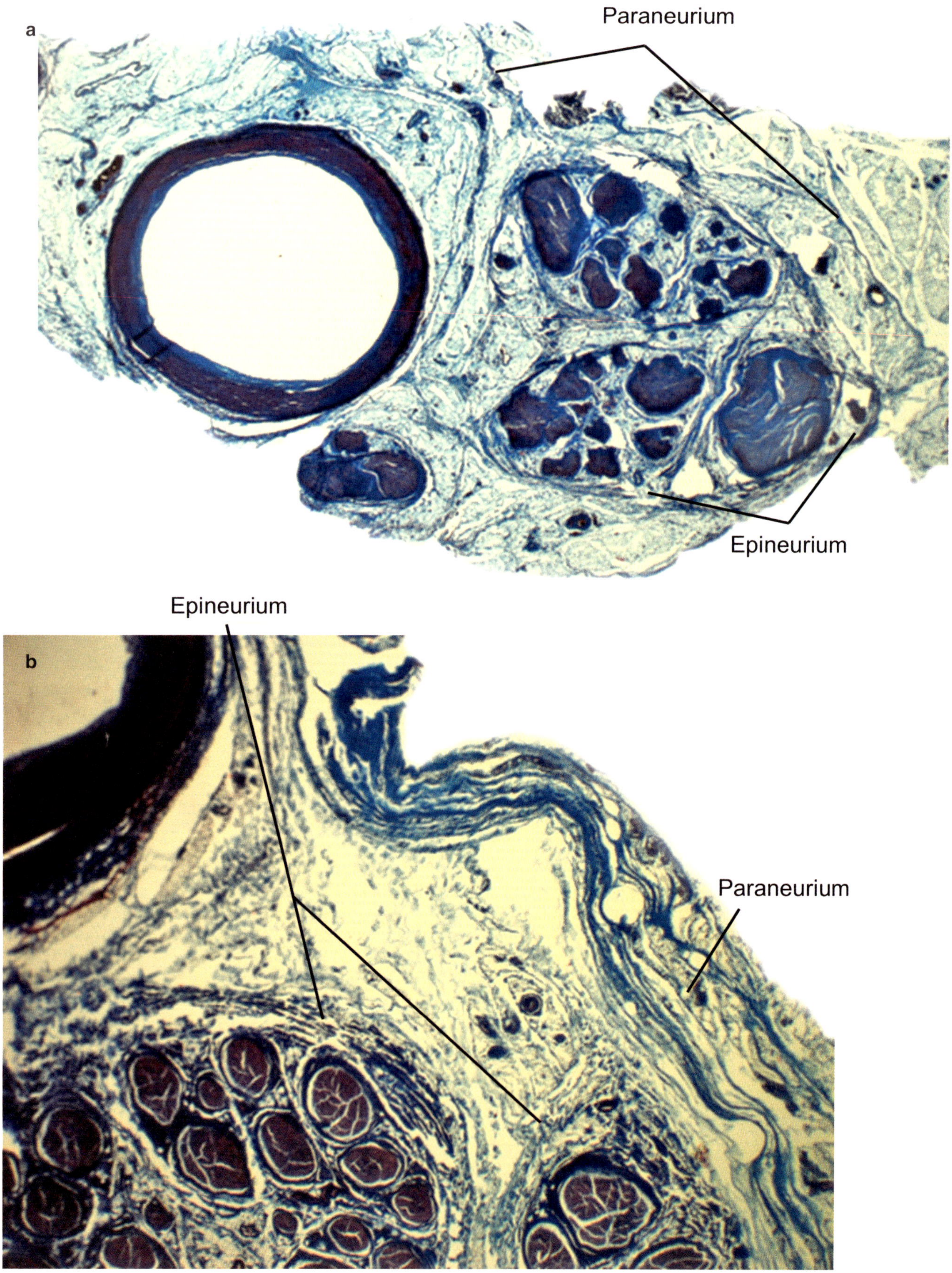

Fig. 8.25 (**a**, **b**) Details of internal structure of nerve trunks in supraclavicular region obtained from different samples. Masson's trichrome stained

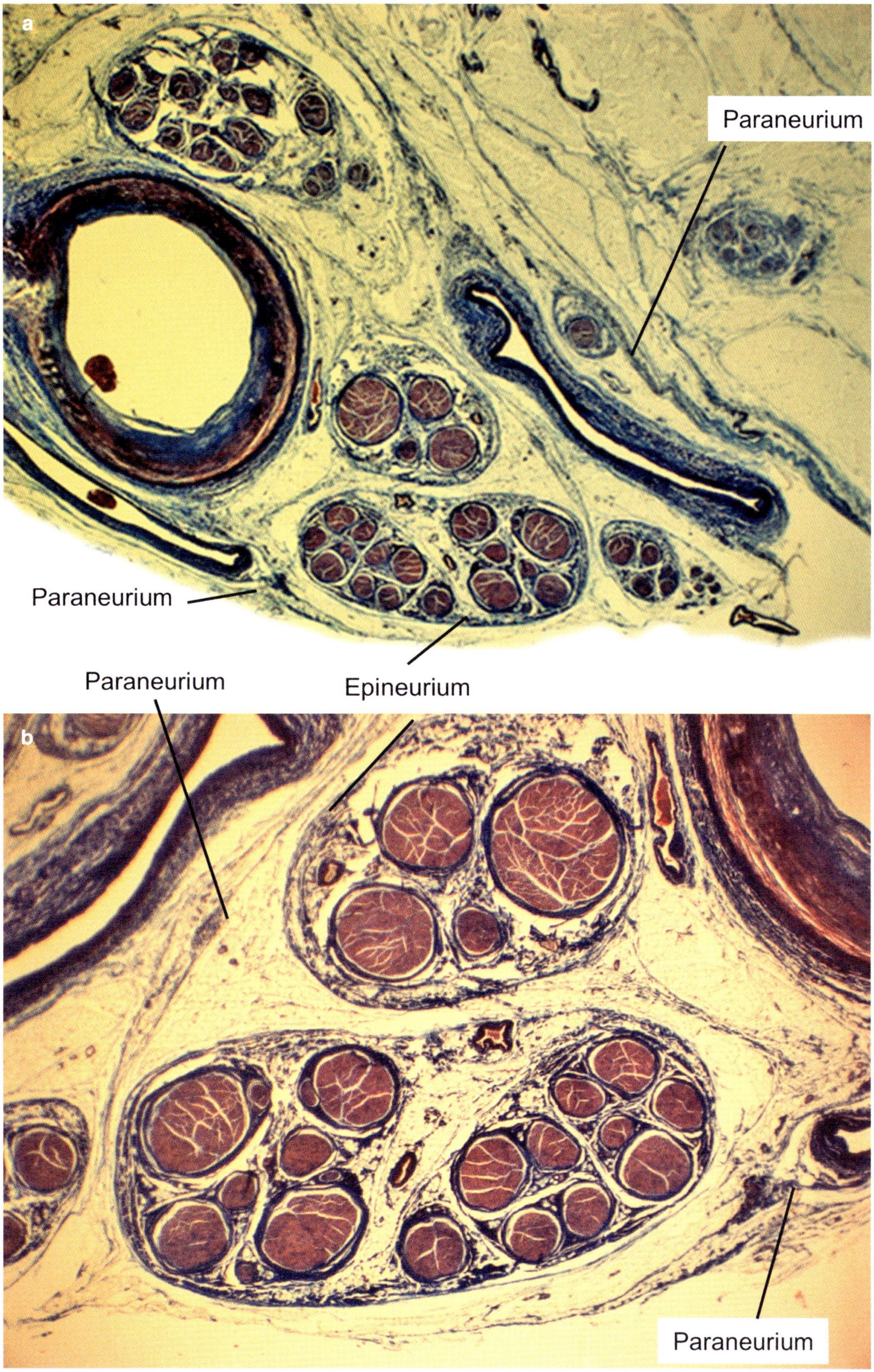

Fig. 8.26 (**a**, **b**) Details of internal structure of nerves in infraclavicular region. (**b**) Detail at more magnification from (**a**). Masson's trichrome stained

References

1. Reina MA, De Andrés JA, Hernández JM, Arriazu Navarro R, Durán Mateos EM, Prats-Galino A. Successive changes in extraneural structures from the subarachnoid nerve roots to the peripheral nerve, influencing anesthetic block, and treatment of acute postoperative pain. Eur J Pain Suppl. 2011;5:377–85.
2. Reina MA, Arriazu R, Collier CB, Sala-Blanch X, Izquierdo L, de Andrés J. Electron microscopy of human peripheral nerves of clinical relevance to the practice of nerve blocks. A structural and ultrastructural review based on original experimental and laboratory data. Rev Esp Anestesiol Reanim. 2013;60:552–62.
3. Yokoyama I. Study on the intraneural topography of the brachial plexus [in Japanese]. Nihon Seikeigeka Gakkai Zasshi. 1989;63:1085–102.
4. Yokoyama I, Matsuda H, Hashimoto T, Kureya S, Shimazu A. Intraneural topography of the brachial plexus. Microscopic dissection in human cadavers. J Jpn Soc Surg Hand. 1986;3:138–45.
5. Sunderland S, Marshall RD, Swaney WE. The intraneural topography of the circumflex musculocutaneous and obturator nerves. Brain. 1959;82:116–29.
6. Sunderland S. The intraneural topography of the radial, median and ulnar nerves. Brain. 1945;68:243–55.

9 Macroscopic View of the Lumbar Plexus and Sacral Plexus

Francisco Reina, Anna Carrera, Manuel Llusá, Anna Oliva, and Joan San Molina

The motor and sensory innervation of the lower limb depends on the anterior branches of the lumbar and sacral spinal nerves. In the lumbar region, the combination of the anterior rami of spinal nerves L1 to L4 forms the lumbar plexus. Its collateral and terminal nerves are distributed through the lower region of the abdominal wall, external genitalia, anterior region of the thigh, and medial cutaneous territory of the leg and foot.

Inside the pelvis and anterior to the sacrum, the ventral rami of nerves S1, S2, and S3 combine with the lumbosacral trunk to form the sacral plexus. The lumbosacral trunk comprises the whole of the anterior division of L5 and part of the anterior branch of the L4 nerve. This plexus provides innervations to most of the structures in the pelvic girdle and lower limb, except those receiving innervations from the lumbar plexus.

The existence of a lumbosacral trunk connecting the two plexuses has led to the concept of the lumbosacral plexus, but it is important to remember that each plexus shows differences in regards to topographic location and distribution, as well as functional characteristics pertaining to its respective neural branches (Figs. 9.1, 9.2, 9.3, 9.4, 9.5, 9.6, 9.7, 9.8, 9.9, 9.10, 9.11, 9.12, 9.13, 9.14, 9.15, 9.16, 9.17, 9.18, 9.19, 9.20, 9.21, and 9.22).

F. Reina, MD, PhD (✉) • A. Carrera, MD, PhD
J. San Molina, MD, PhD
Neuroscience, Embryology, Molecular Oncology, and Clinical Anatomy Group (NEOMA), School of Medicine, University of Girona, Girona, Spain
e-mail: francisco.reina@udg.edu; anna.carrera@udg.edu; joan.san@udg.edu

M. Llusá, MD, PhD
Laboratory of Macro-Micro Dissection and Surgical Anatomy, Human Anatomy and Embryology Unit, School of Medicine, University of Barcelona, Barcelona, Spain

Orthopaedic and Traumatology Surgery Unit, Vall d'Hebron Hospital, Barcelona, Spain
e-mail: mllusa@ub.edu

A. Oliva, MD
Laboratory of Macro-Micro Dissection and Surgical Anatomy, Human Anatomy and Embryology Unit, School of Medicine, University of Barcelona, Barcelona, Spain

Rehabilitation Unit, Mataró Hospital, Mataró, Barcelona, Spain
e-mail: annao.oliva@gmail.com

M.A. Reina et al. (eds.), *Atlas of Functional Anatomy for Regional Anesthesia and Pain Medicine: Human Structure, Ultrastructure and 3D Reconstruction Images*, DOI 10.1007/978-3-319-09522-6_9,

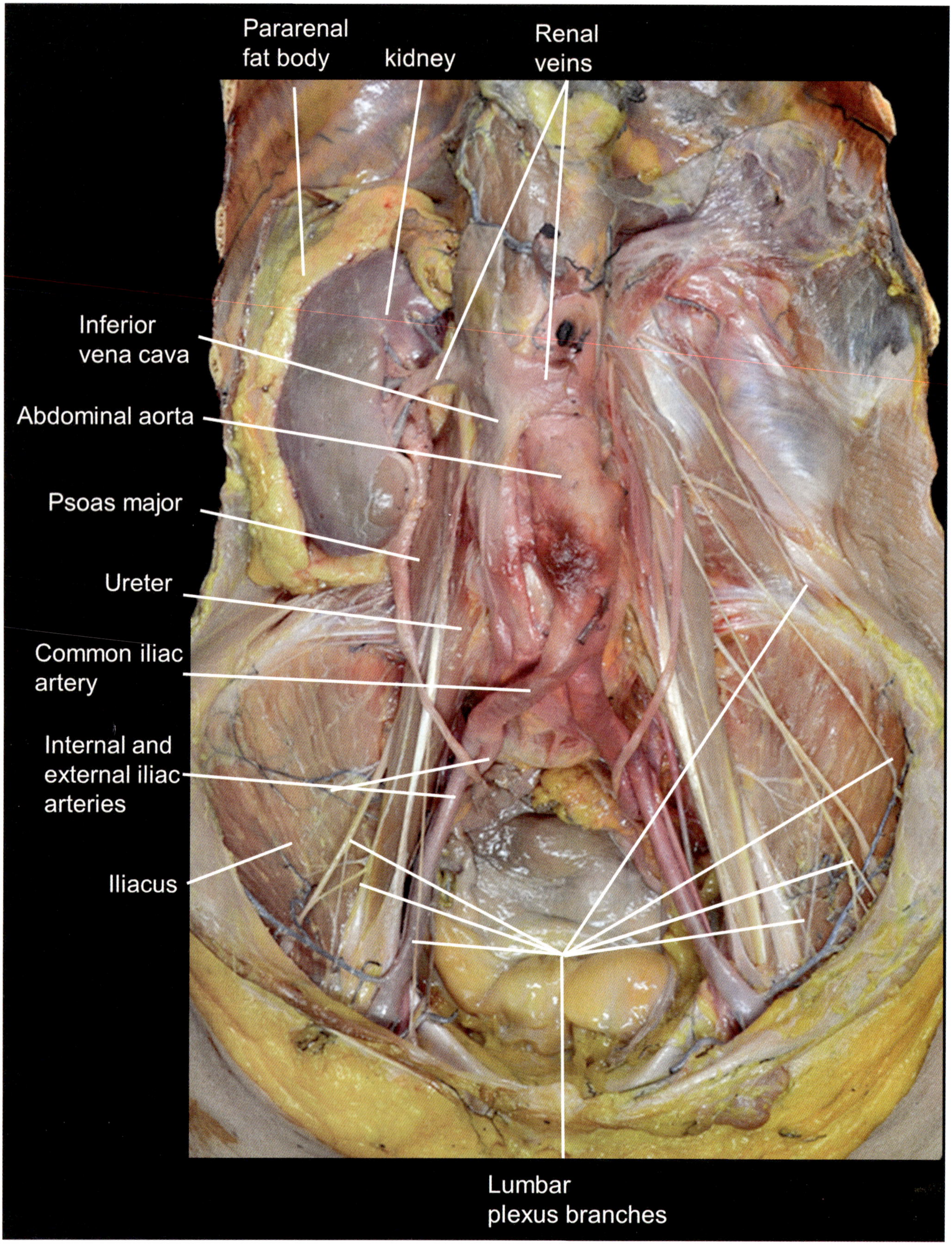

Fig. 9.1 The lumbar plexus is formed inside the psoas major muscle, in the posterior wall of the abdomen. The lumbar ventral rami enter this muscle, where they combine to form the lumbar plexus. Nerves formed from this plexus descend along this muscle, finding their way either across its ventral surface or along its lateral and medial edges. Their distribution is located in different positions relative to the orientation of the psoas major

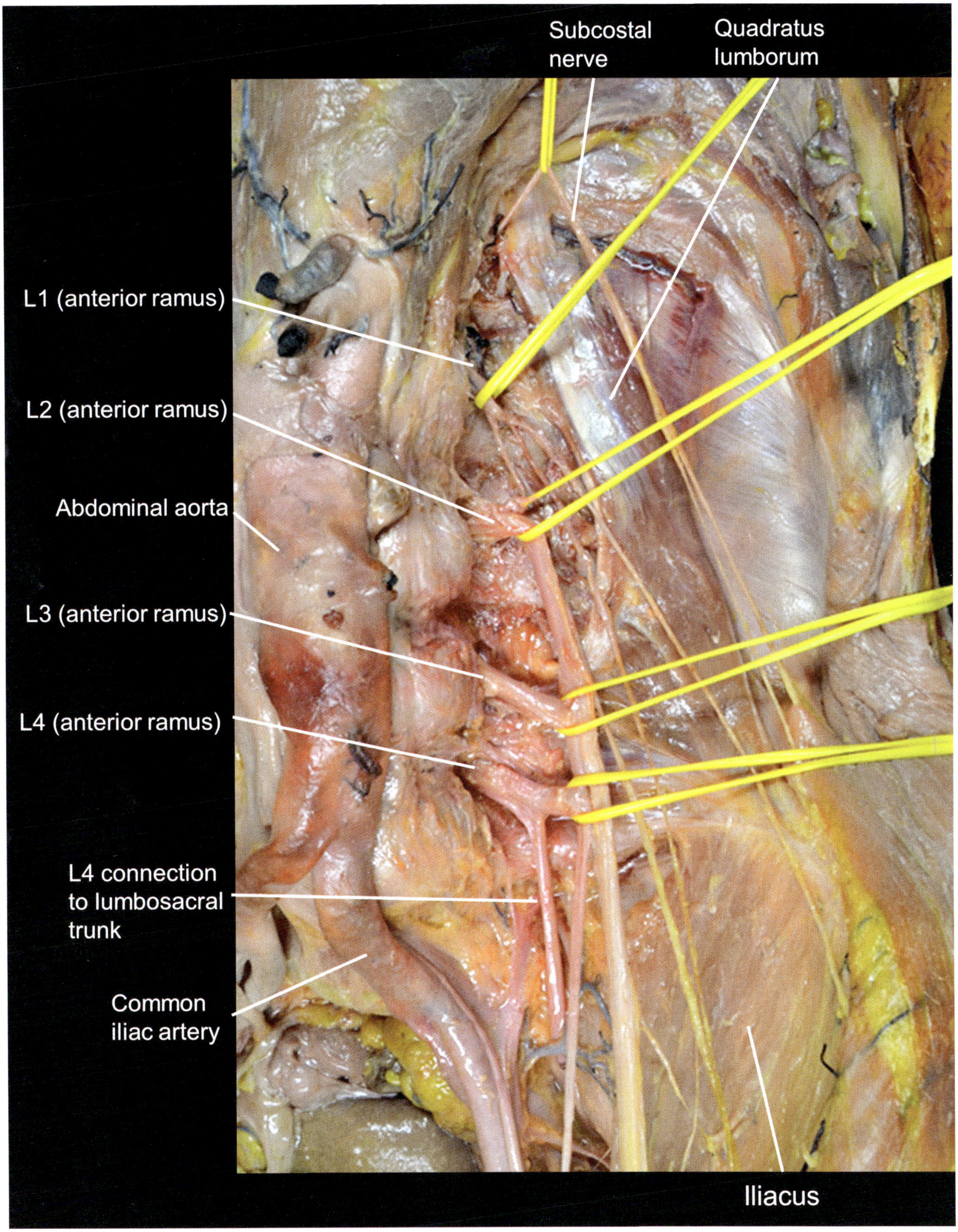

Fig. 9.2 The ventral rami of the L1, L2, L3 spinal nerves and a large part of L4, which constitute the lumbar plexus, are located at the exit of their respective foramina in the lumbar paravertebral region. Often the involvement of the anterior branch of spinal nerve T12 in the formation of this plexus is observed (pre-fixed plexus)

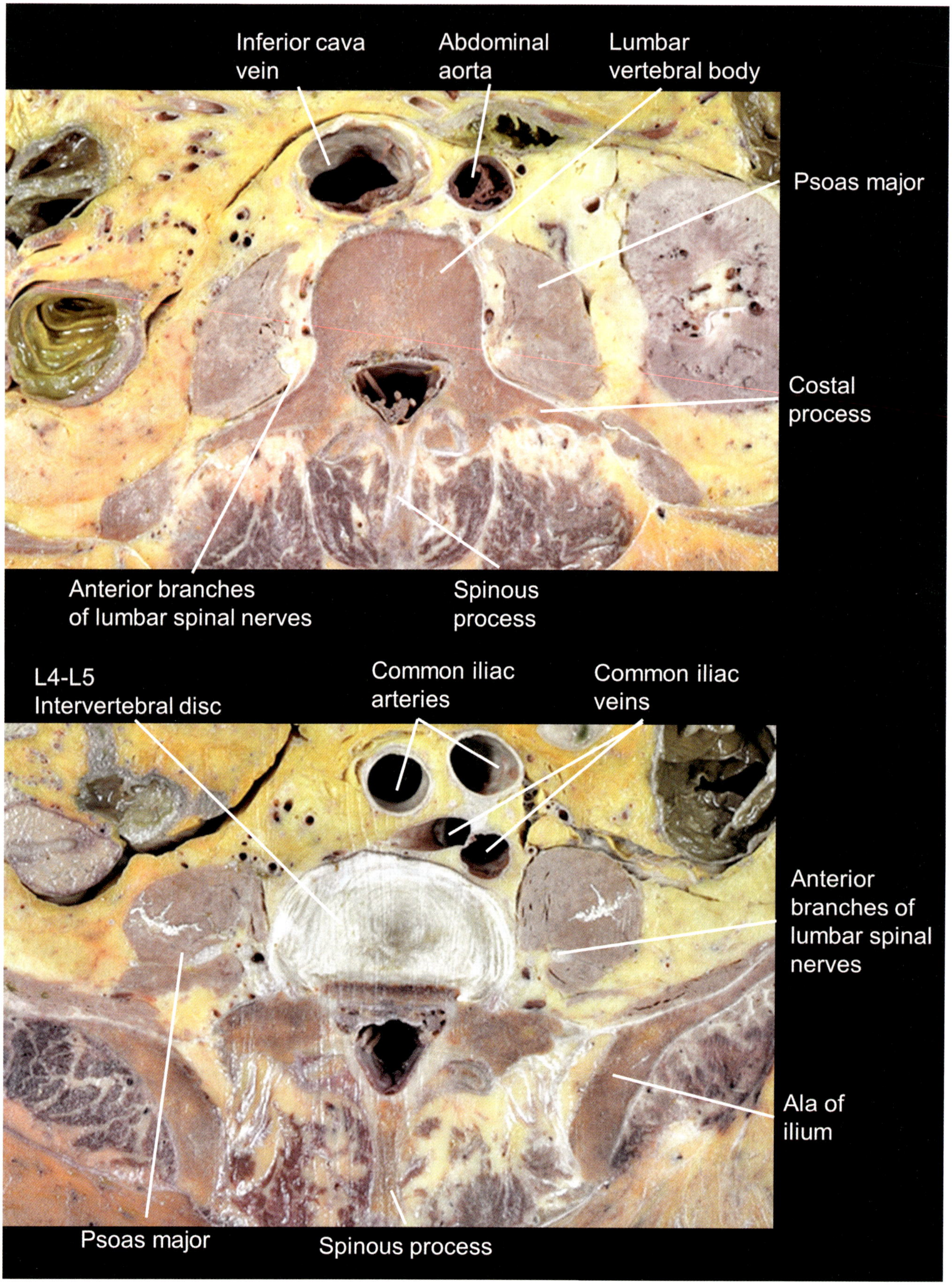

Fig. 9.3 In a cross-section, the lumbar plexus is located in the angle between bodies and transverse processes of the lumbar vertebrae, between the superficial and deep origin of the psoas muscle. In this segment, the lumbar plexus is related to the lumbar arteries and the ascending lumbar vein and is connected to the lumbar sympathetic trunk by rami communicantes

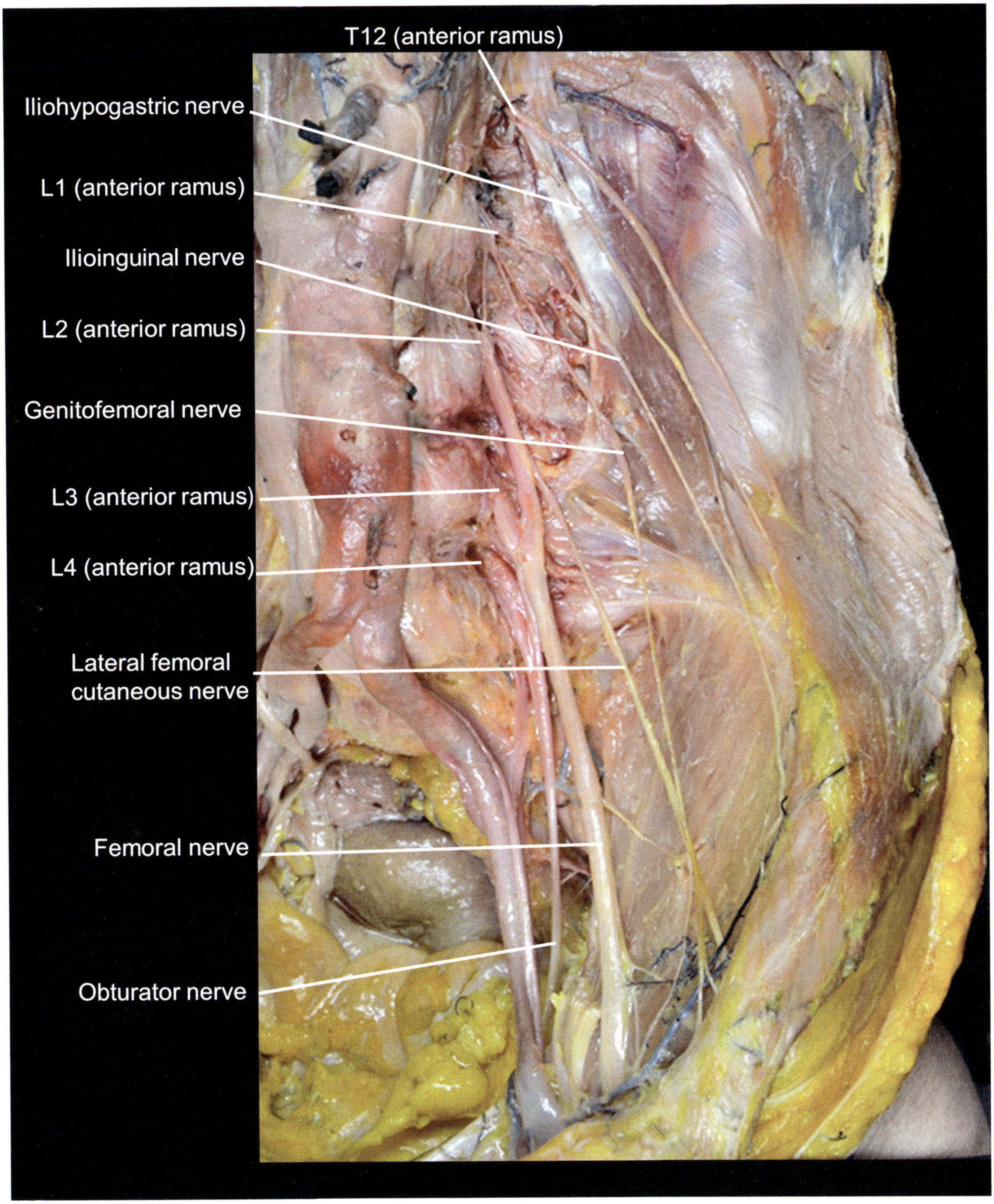

Fig. 9.4 The spinal nerves that form the lumbar plexus follow variable combination patterns, but the arising branches are kept constant. These comprise the iliohypogastric nerve, ilioinguinal nerve, lateral femoral cutaneous nerve, femoral nerve (*not shown*), and obturator nerve

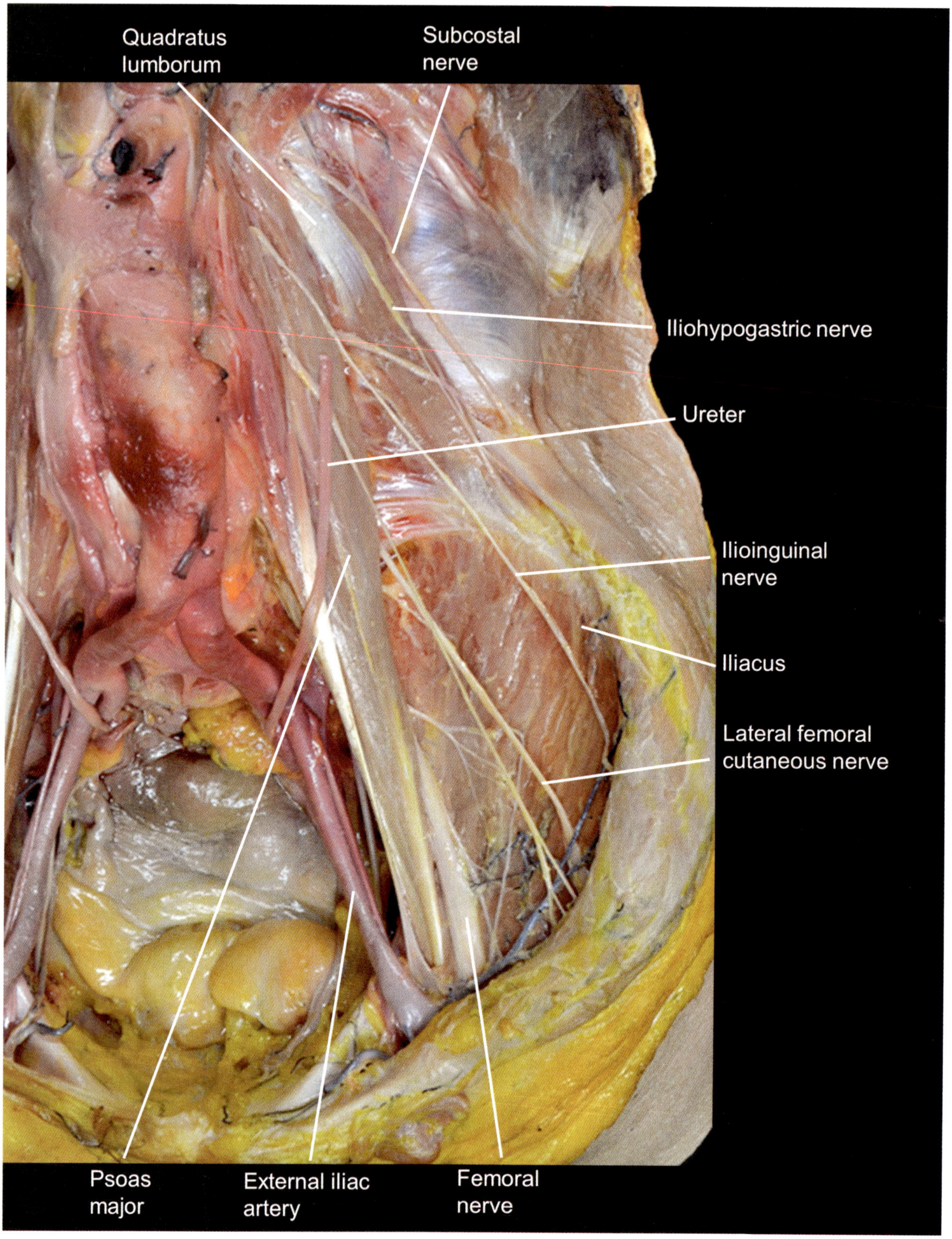

Fig. 9.5 The iliohypogastric, ilioinguinal, lateral femoral cutaneous, and femoral nerves emerging caudally from the lateral side of the psoas major muscle. The iliohypogastric nerve runs behind the kidney and ventral to the quadratus lumborum. It pierces the posterior part of the transversus abdominis above the iliac crest and divides into lateral and anterior cutaneous branches. The ilioinguinal nerve follows a similar course extending towards the iliac crest, where it pierces the transversus and internal oblique muscles. The lateral femoral cutaneous nerve runs obliquely ventral to the iliacus muscle towards the anterior superior iliac spine. The femoral nerve (L4, together with L2 and L3) is the largest branch of the lumbar plexus and is the lowest emerging branch, appearing along the lateral aspect of the psoas major. It descends between the psoas and iliacus muscles. The lumbar plexus also gives direct branches to the quadratus lumborum, psoas major, and psoas minor muscles

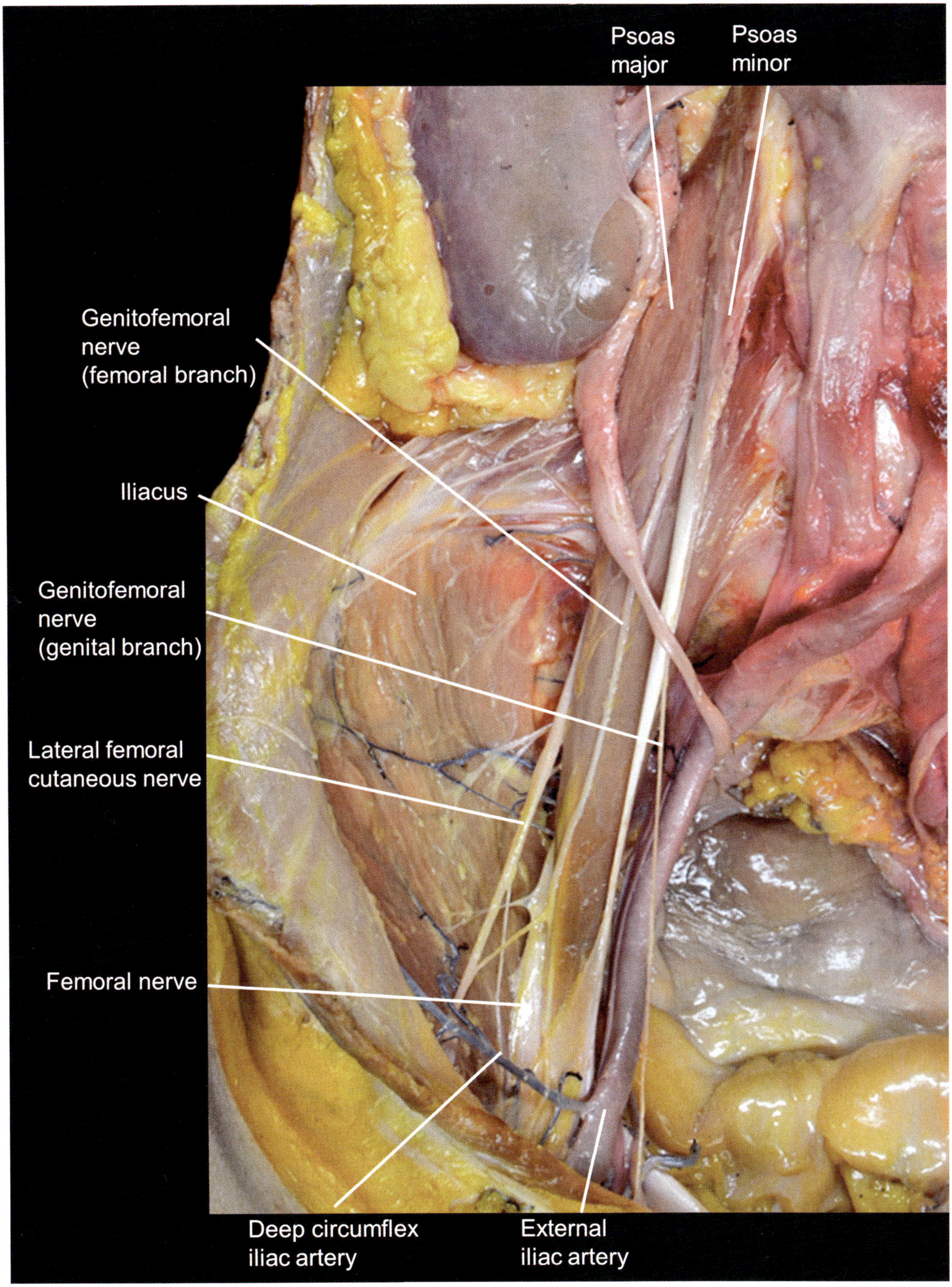

Fig. 9.6 The genitofemoral nerve descends within the psoas major muscle and emerges from its anterior surface. It descends ventrally to the psoas, dividing into genital and femoral branches. Occasionally (as shown in this picture), this division occurs within the psoas muscle, so the branches appear separately at the anterior surface of the muscle

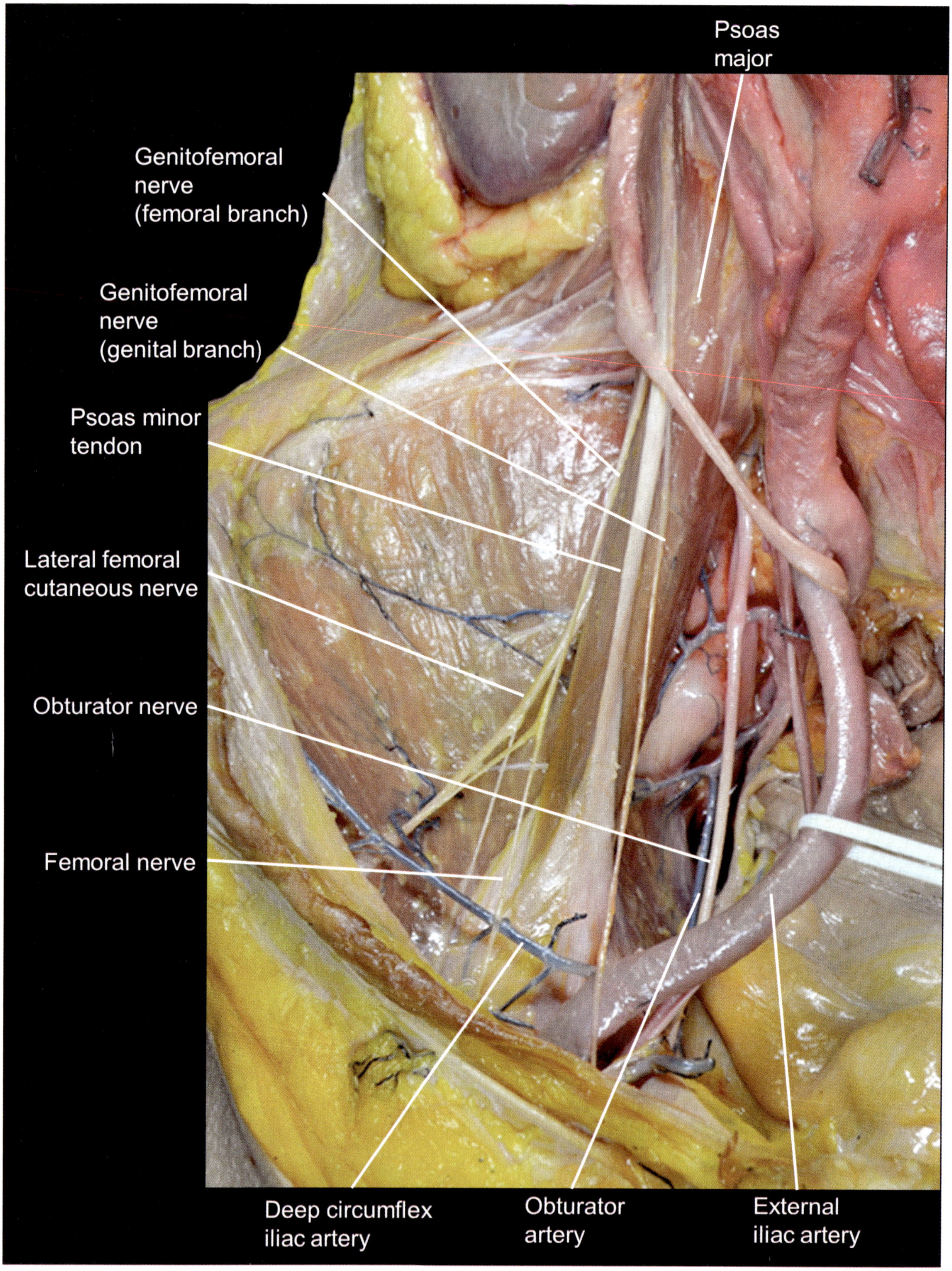

Fig. 9.7 The obturator nerve emerges from the medial aspect of the psoas at the inlet of the pelvis. It runs downwards and forwards along the lateral wall of the lesser pelvis. It is placed caudal to the external iliac artery in the obturator internus muscle fascia. The nerve enters the thigh through the obturator foramen

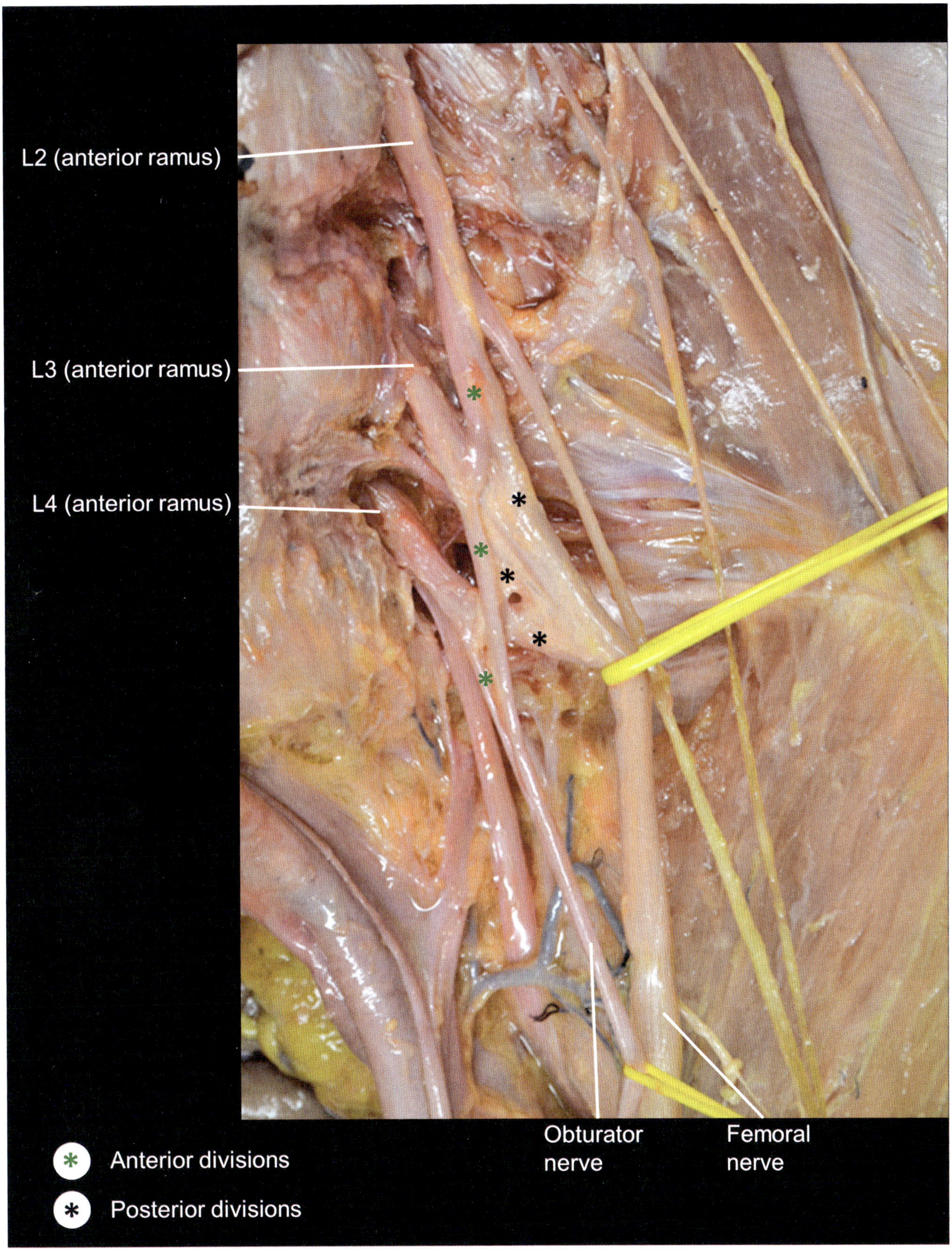

Fig. 9.8 Although it is difficult to demonstrate the formation of anterior and posterior divisions in the structure of the lumbar plexus, some authors state that the ventral rami of the L2, L3, and L4 spinal nerves give an anterior and posterior division for each one. This division represents an essential functional organization of the plexus in its flexor and extensor components, as occurs in the trunks of the brachial plexus. The anterior divisions merge to form the obturator nerve, whereas the posterior divisions give rise to the formation of the femoral nerve

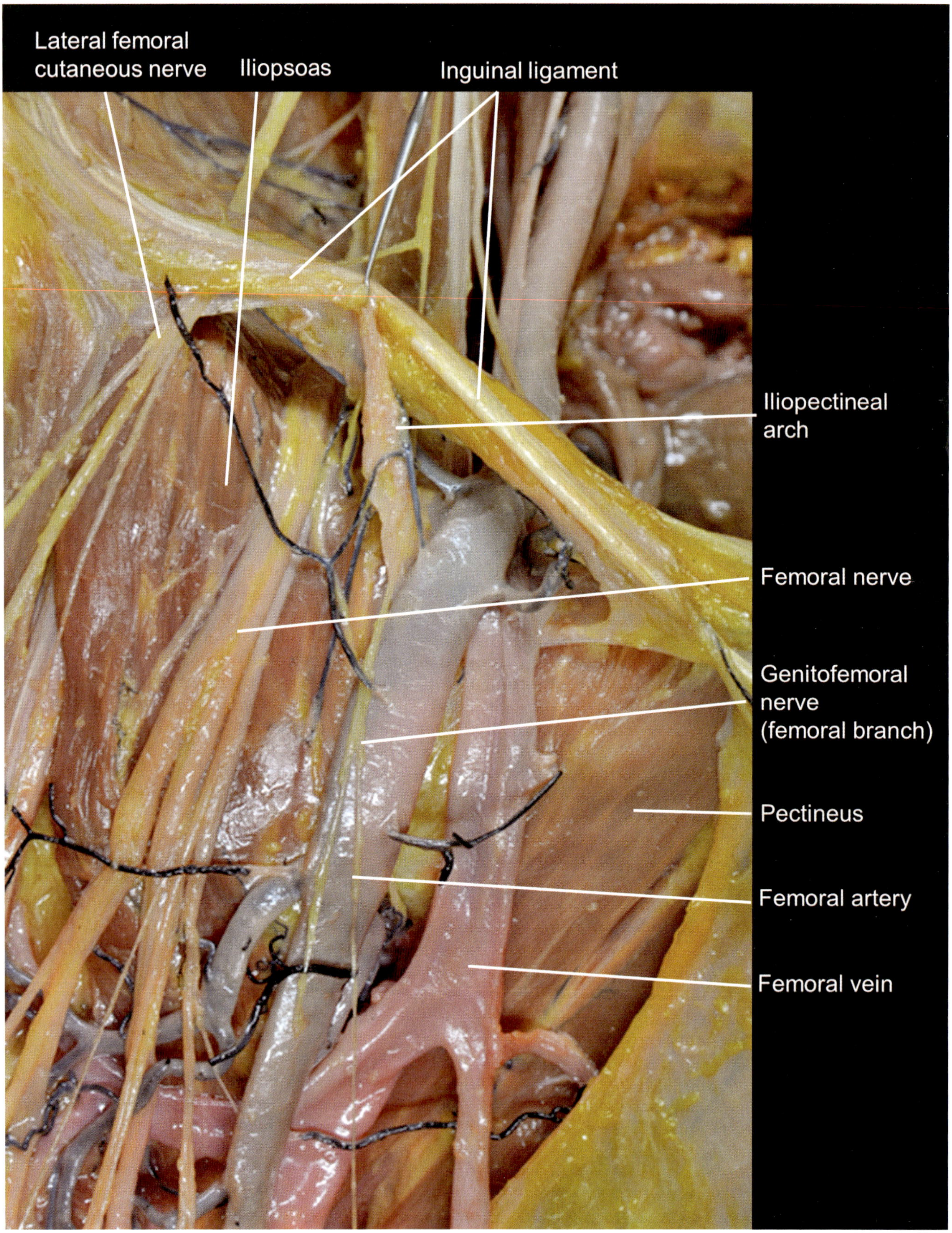

Fig. 9.9 The largest branch of the lumbar plexus is the femoral nerve, which is characteristically flattened. It runs caudally between the iliacus and the psoas, enclosed in its fascia. The femoral nerve enters the thigh behind the inguinal ligament. At this point, the fascia of the psoas thickens to form the iliopectineal arch, separating the femoral nerve from the femoral artery, which is placed medial to the nerve. The femoral vein is the most medial element in the femoral ring

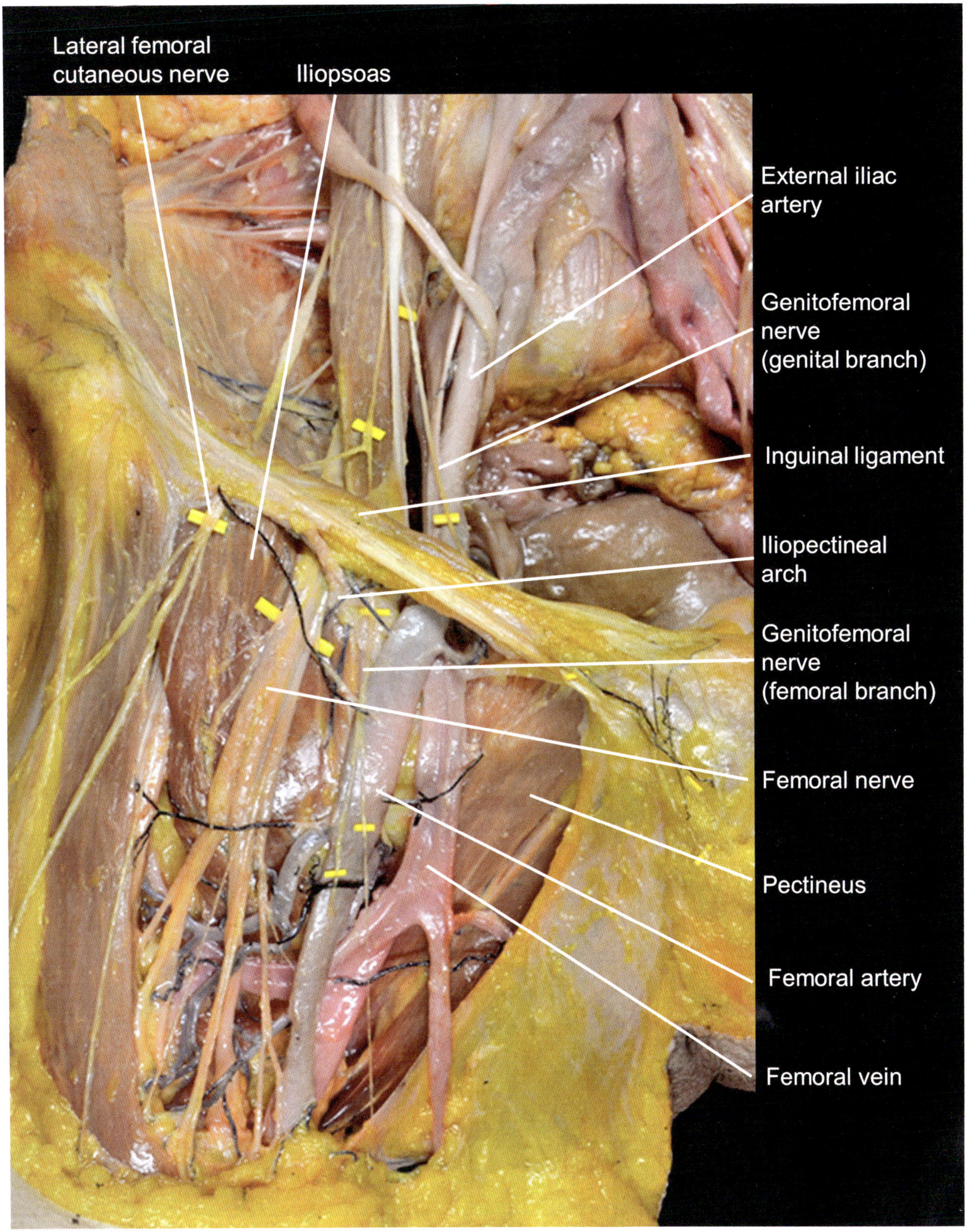

Fig. 9.10 At the femoral triangle, few terminal branches of the lumbar plexus are distributed. In this anatomical region, the femoral nerve gives rise to several motor and sensory branches supplying the quadriceps femoris, the pectineus and sartorius muscles, and the skin of the front and medial side of the thigh, the medial side of the leg, and the hip and knee joints. In addition to the femoral nerve branches, the lateral femoral cutaneous nerve in relation to the anterior superior iliac spine, the femoral branch of the genitofemoral nerve accompanying the femoral vessels, the ilioinguinal nerve, and the genital branch of the genitofemoral nerve emerge from the superficial ring of the inguinal canal

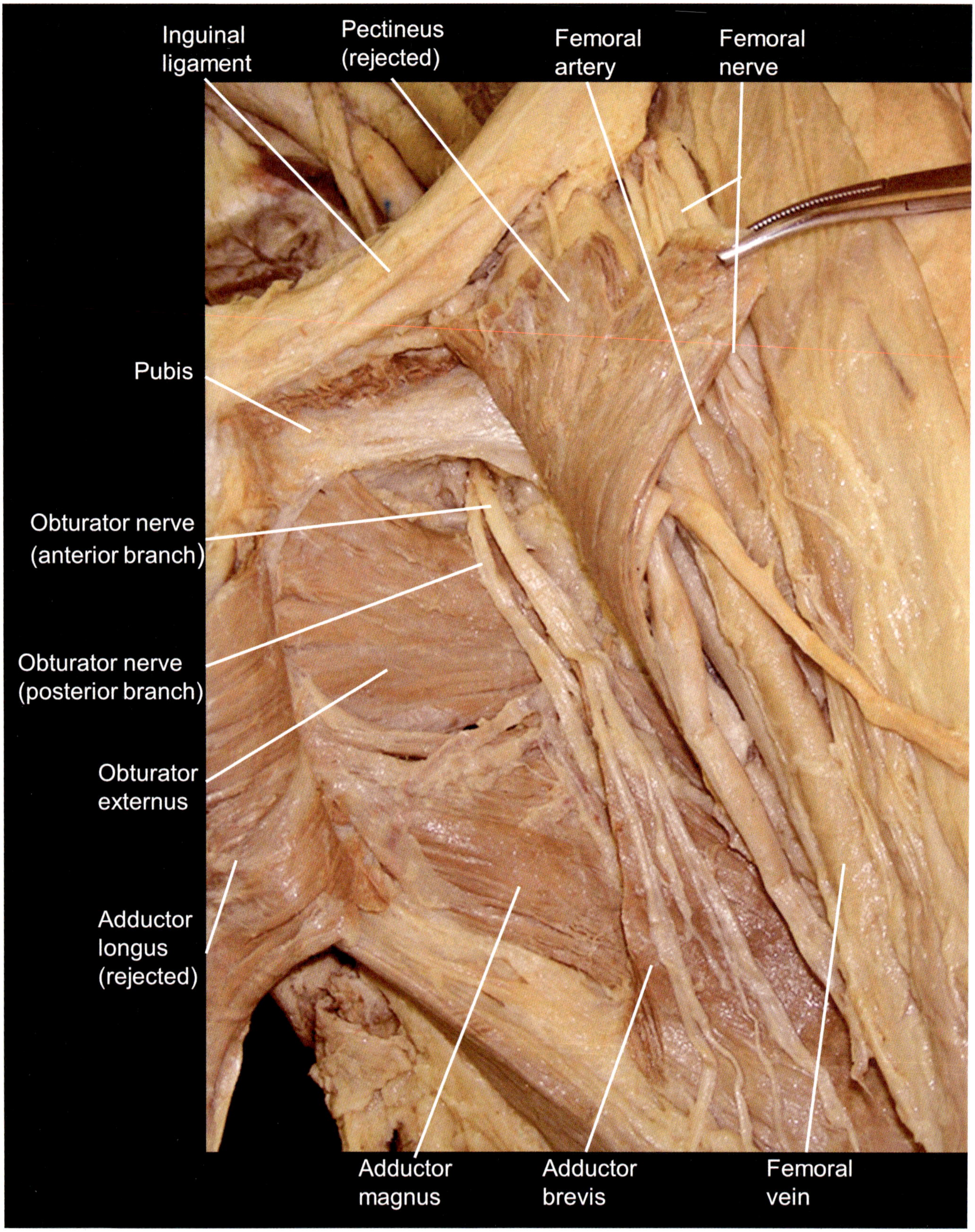

Fig. 9.11 The obturator nerve passes through the obturator groove, where it divides into anterior and posterior branches. These pass through the obturator foramen to reach the thigh, where they are separated by the adductor brevis muscle

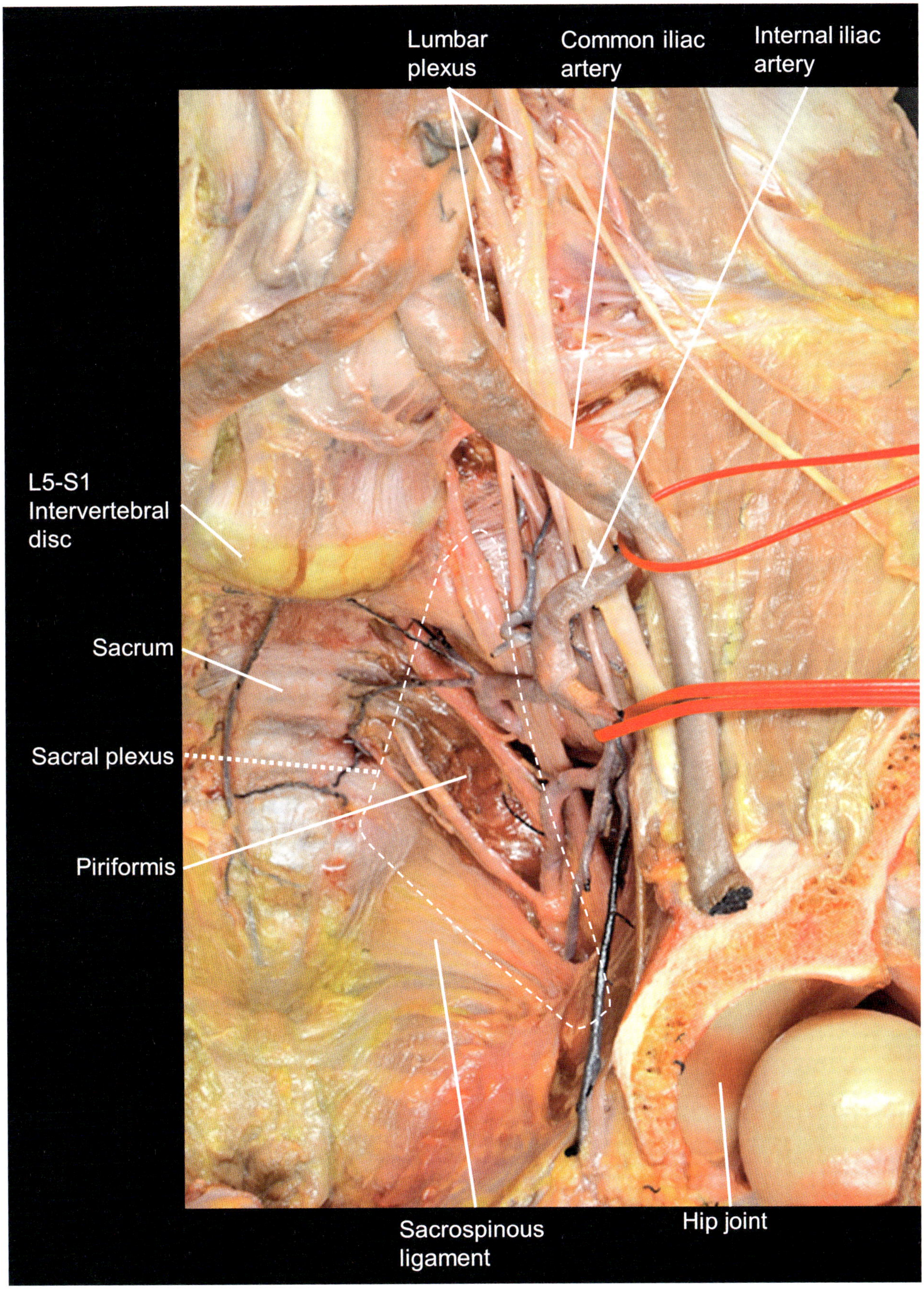

Fig. 9.12 The sacral plexus is formed from the ventral rami of nerves S1, S2, and S3 combining with the lumbosacral trunk. The plexus lies on the dorsal wall of the pelvis, mainly between the piriformis muscle and its fascia. Anterior to that fascia lie the internal iliac vessels and the ureter. The nerves entering the plexus converge to form a triangular band, which passes through the greater sciatic foramen below the piriformis muscle. The sciatic nerve, its main terminal branch, enters the gluteal region below the piriformis, thus continuing the plexus direction from its vertex

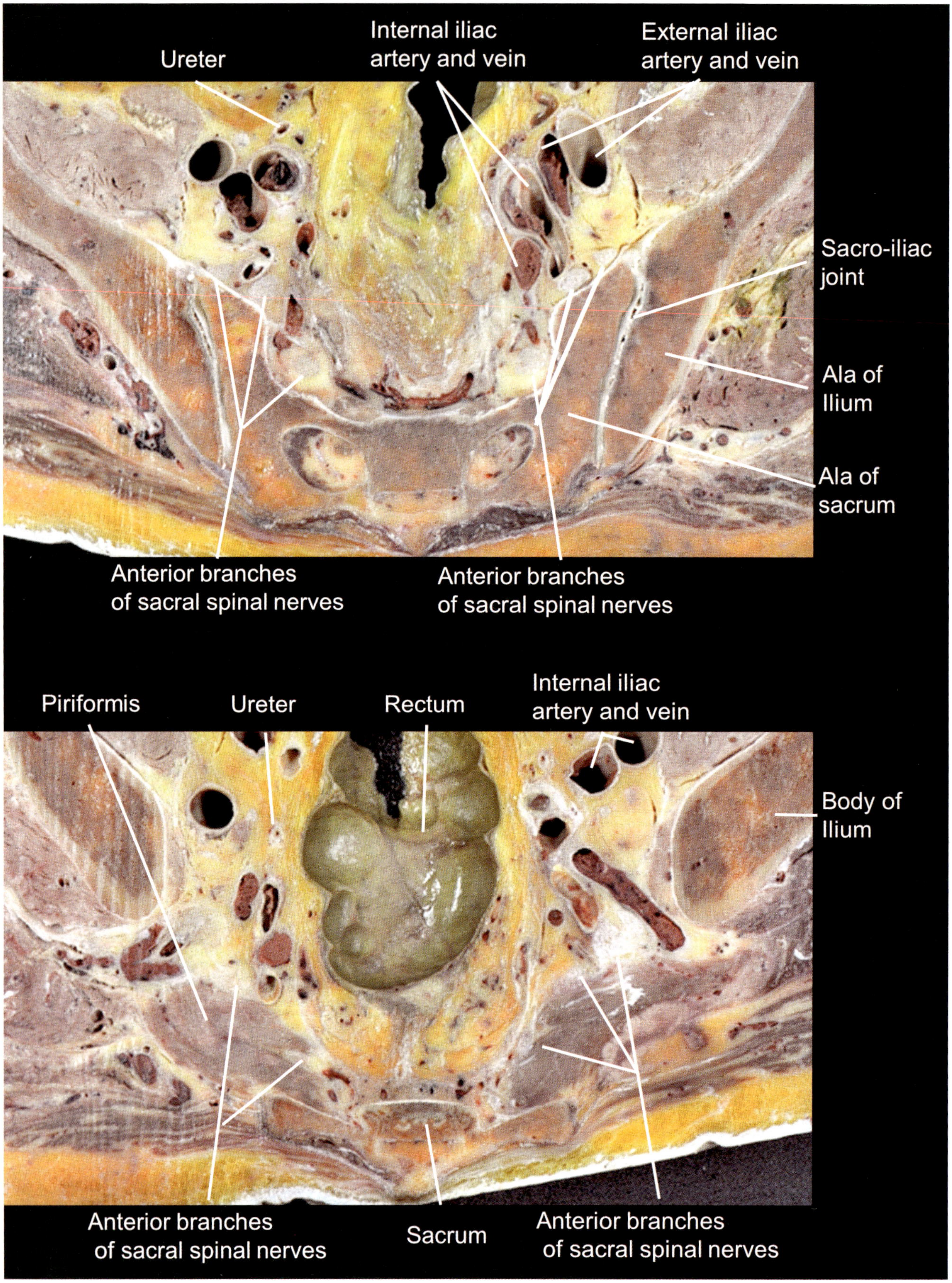

Fig. 9.13 In a pelvis cross-section, it can be observed how the internal iliac vessels, the ureter, and (further anteriorly) the sigmoid colon on the left and coils of ileum on the right lie anterior to the sacral plexus. The sacral plexus is separated from these elements by the parietal pelvic fascia

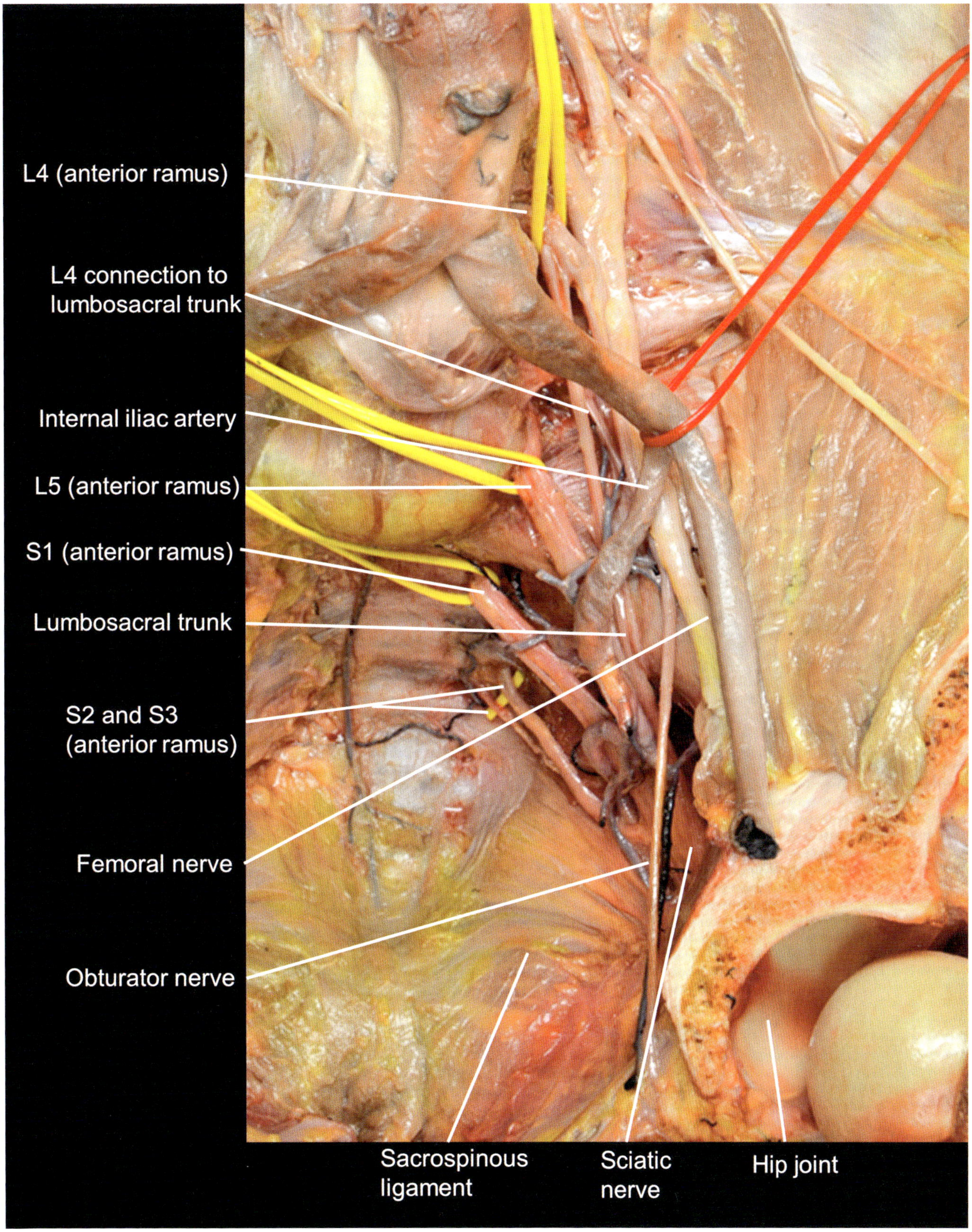

Fig. 9.14 The formation of the sacral plexus is the result of the union of the first three ventral rami of the sacral nerves (S1, S2, S3) with the lumbosacral trunk. Anterior and posterior divisions of the first three sacral rami have been described, but, as in the case of the lumbar plexus, they are difficult to demonstrate

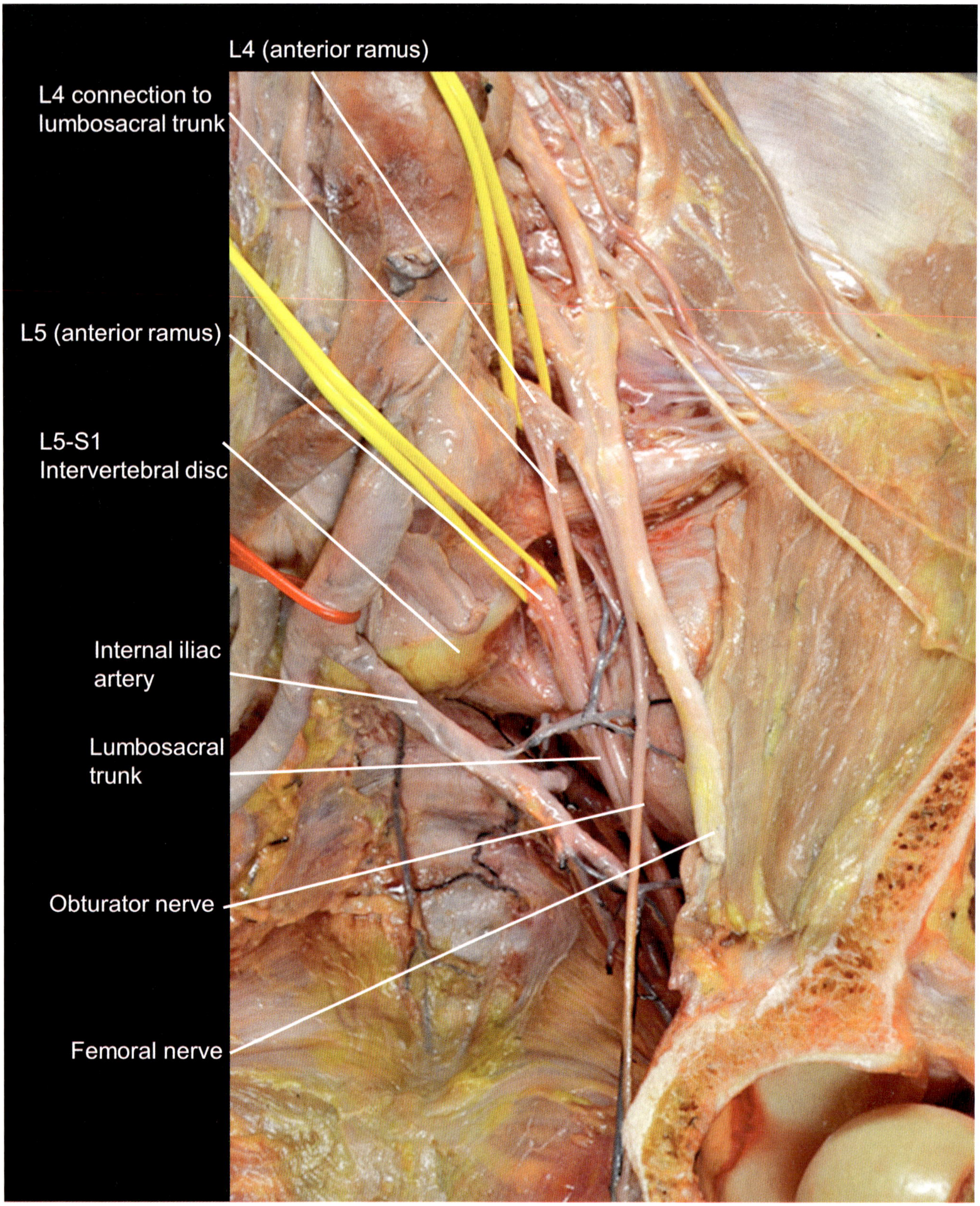

Fig. 9.15 The lumbosacral trunk joins the first, second, and third sacral ventral rami to form the sacral plexus. The lumbosacral trunk is formed when the descending branch of the L4 ventral ramus, emerging from the medial border of the psoas muscle medial to the obturator nerve and posterior to the common iliac vessels, joins the L5 ventral ramus, which has reached the sacroiliac joint by passing over the lateral part of the sacrum. The superior gluteal vessels usually pass between the lumbosacral trunk and the ventral ramus of S1

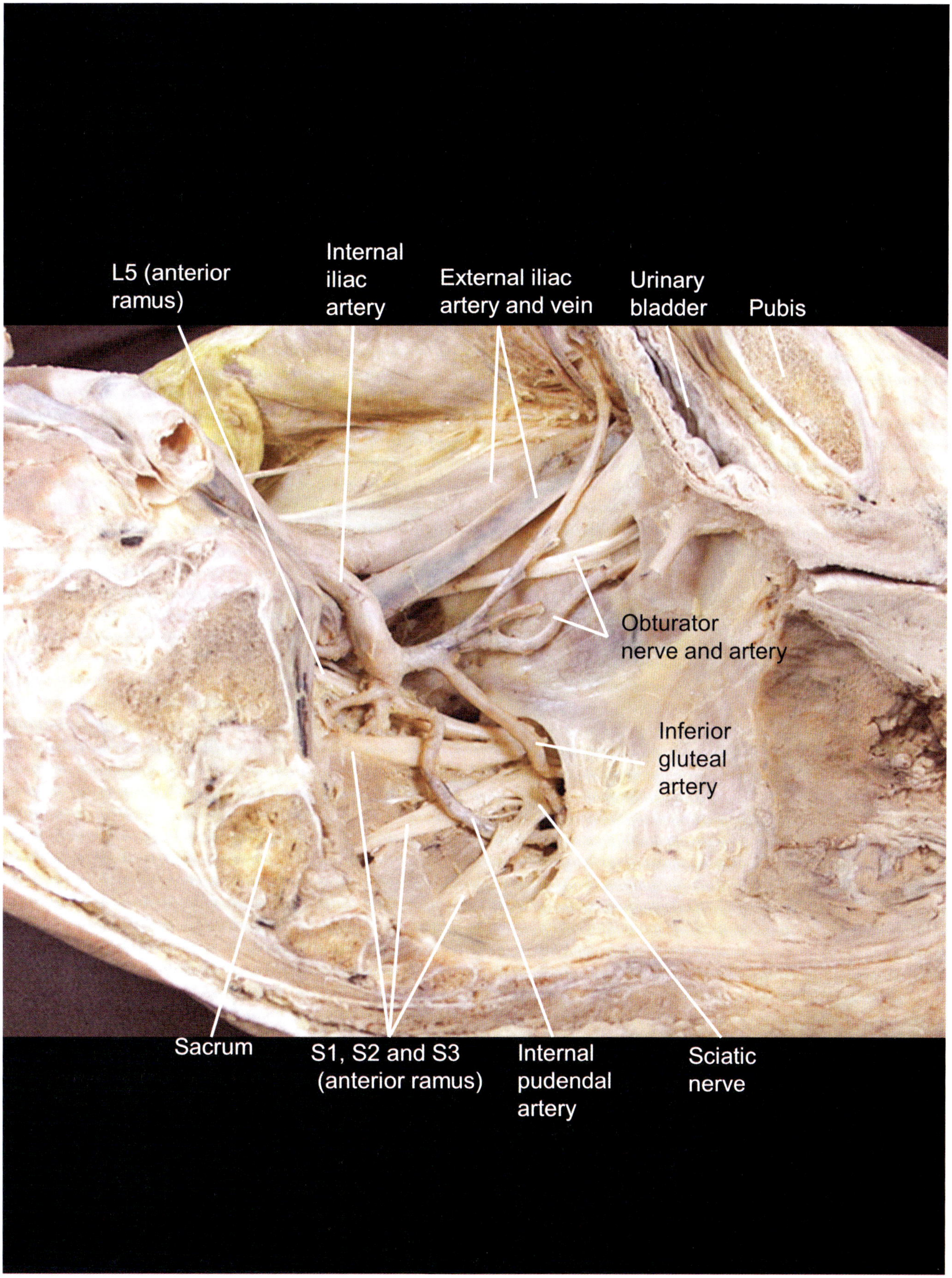

Fig. 9.16 The dorsal and ventral divisions of the sacral plexus converge on the lower part of the greater sciatic foramen to form the sciatic nerve. The dorsal divisions of L4, L5, S1, and S2 form the common peroneal nerve (extensor from a functional point of view). The ventral divisions of L4, L5, S1, S2, and S3 unite to form the tibial nerve (flexor from a functional point of view)

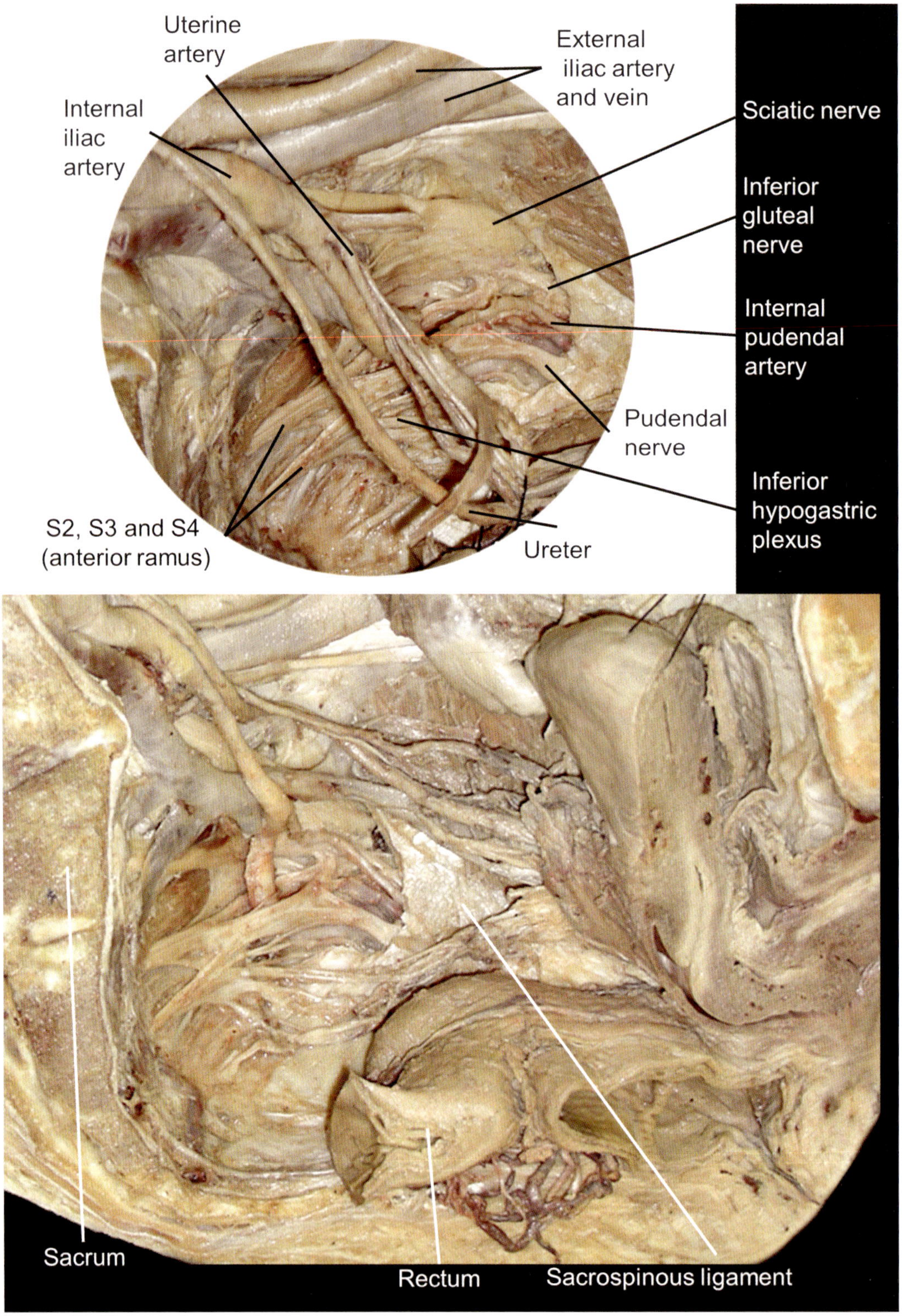

Fig. 9.17 The sacral plexus has a series of collateral branches and a terminal branch, the sciatic nerve. Seven of these branches are distributed to the buttock region and the lower limb: the superior and inferior gluteal nerves, the nerve to the quadratus femoris, the nerve to the obturator internus, the posterior femoral cutaneous nerve, the perforating cutaneous nerve, and the sciatic nerve as terminal branch. The other collateral branches supply structures belonging to the pelvis: the nerve to the piriformis, the nerves to the levator ani and coccygeus muscles, the nerve to the sphincter ani externus, the pelvic splanchnic nerves that contain parasympathetic preganglionic and sensory fibers, and finally, the pudendal nerve. The pudendal nerve (S2, S3, and S4) takes its origin from the lower margin of the sacral plexus. Most of the innervation of the perineum depends on this nerve. The pudendal nerve passes through the greater sciatic foramen, below the piriformis. It crosses back of the ischial spine, where it is located medial to the internal pudendal artery. It then enters the perineum through the lesser sciatic foramen to achieve the pudendal canal in the lateral wall of the ischiorectal fossa. Here, it gives off the inferior rectal nerve and shortly afterwards divides into the perineal nerve and the dorsal nerve of the penis or clitoris

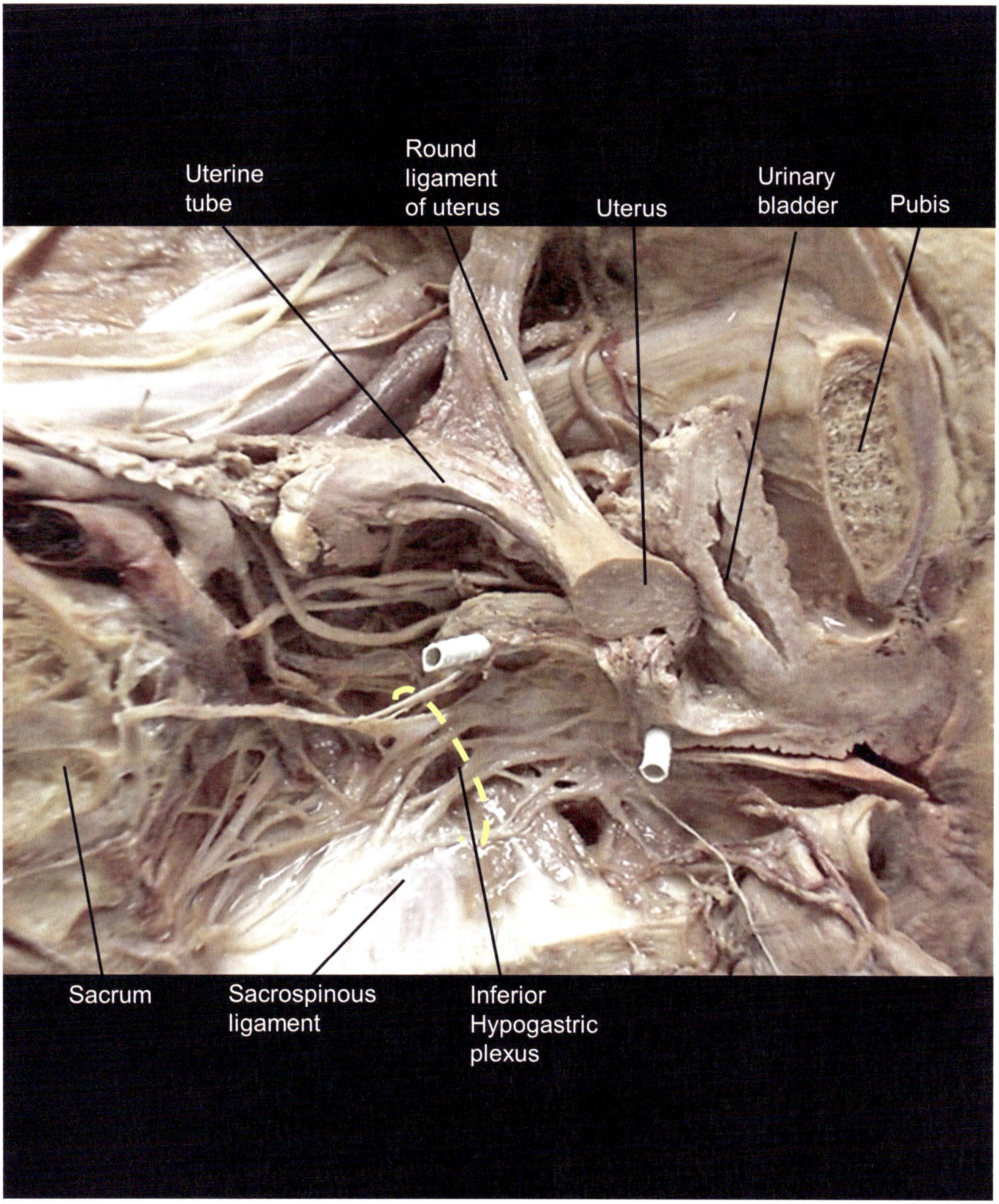

Fig. 9.18 Pelvic splanchnic nerves consist of preganglionic parasympathetic nerve fibers that arise from the second and third sacral ventral rami, or the third and fourth rami. They pass forwards to join the hypogastric nerves, thus forming the inferior hypogastric plexus, through which they supply the urogenital organs and the lower part of the large intestine

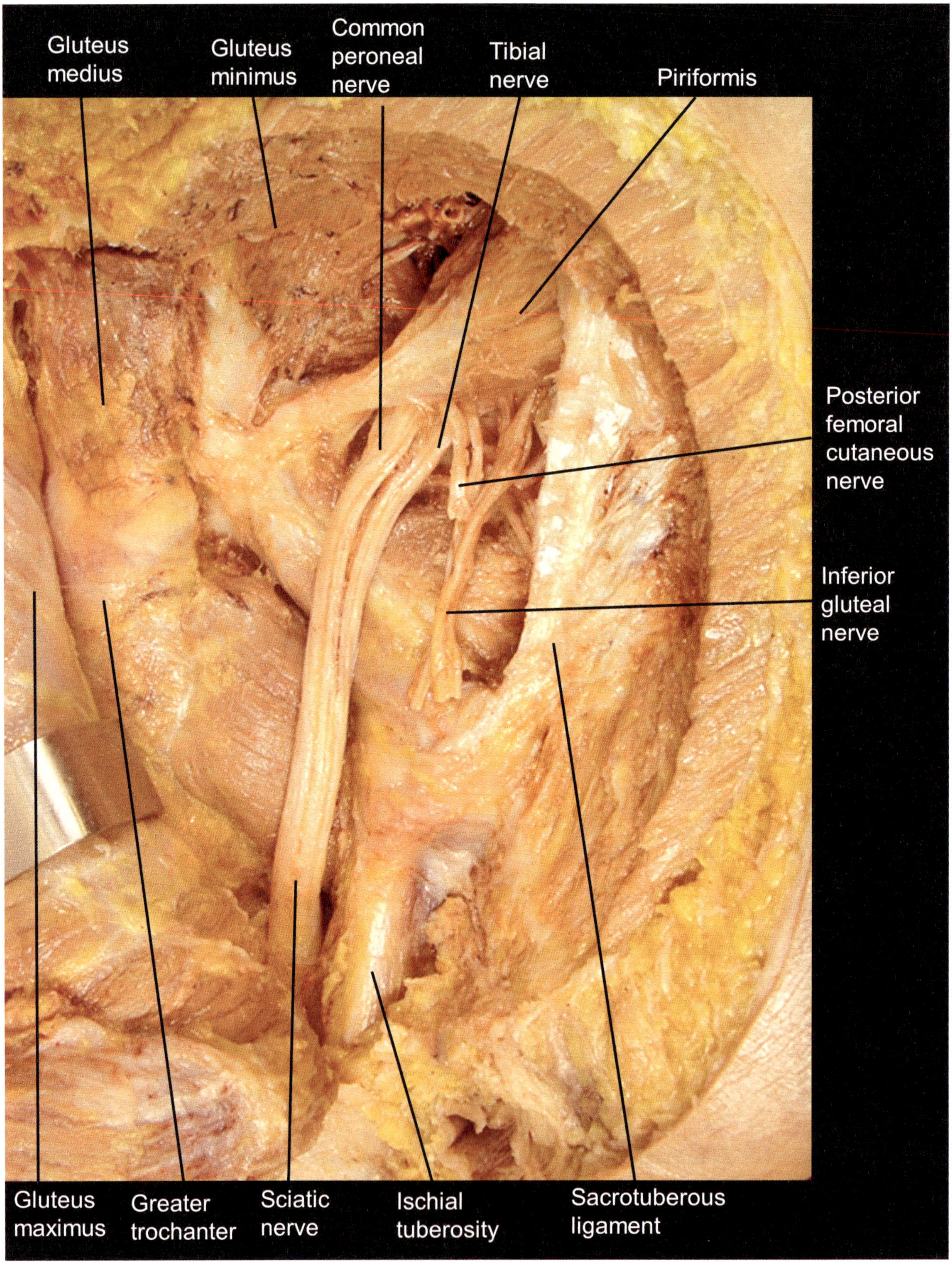

Fig. 9.19 The sciatic nerve emerges from the pelvis to the buttock region through the greater sciatic foramen, below the piriformis and above the sacrospinous ligament. Inferior to the piriformis, it lies deep to the gluteus maximus and descends vertically, posterior to the tendon of the obturator internus and the gemelli. The sciatic nerve consists of a medially placed tibial nerve and its branches to the hamstring muscles, and a laterally placed common peroneal nerve and its branch to the short head of biceps, bound together in a common sheath

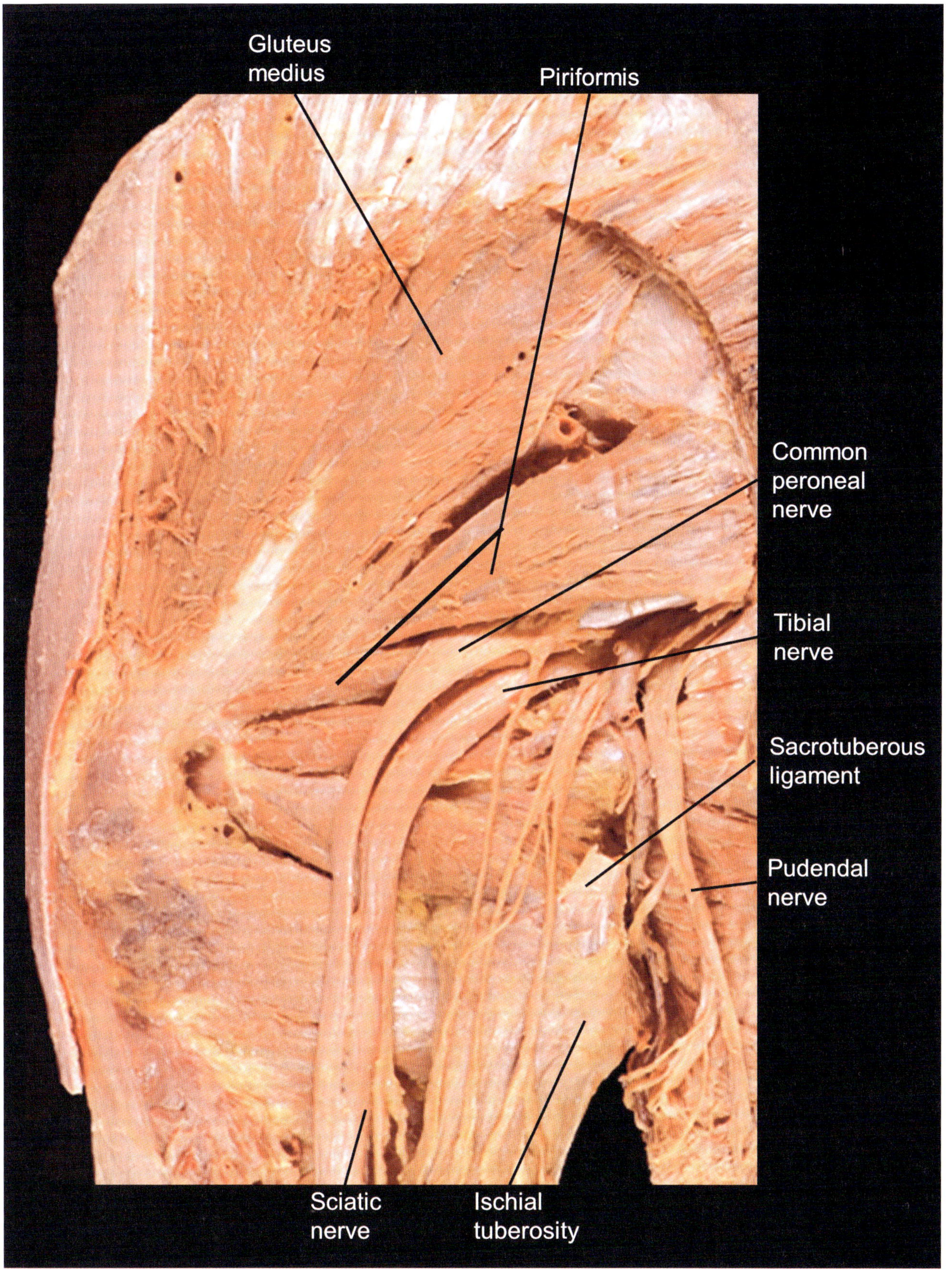

Fig. 9.20 The sciatic nerve consists of a medially placed tibial nerve and its branches to the hamstring muscles, and a laterally placed common peroneal nerve and its branch to the short head of biceps, bound together in a common sheath. In some cases, the tibial and the peroneal components of the sciatic nerve are not bound together, but instead leave the pelvis separately

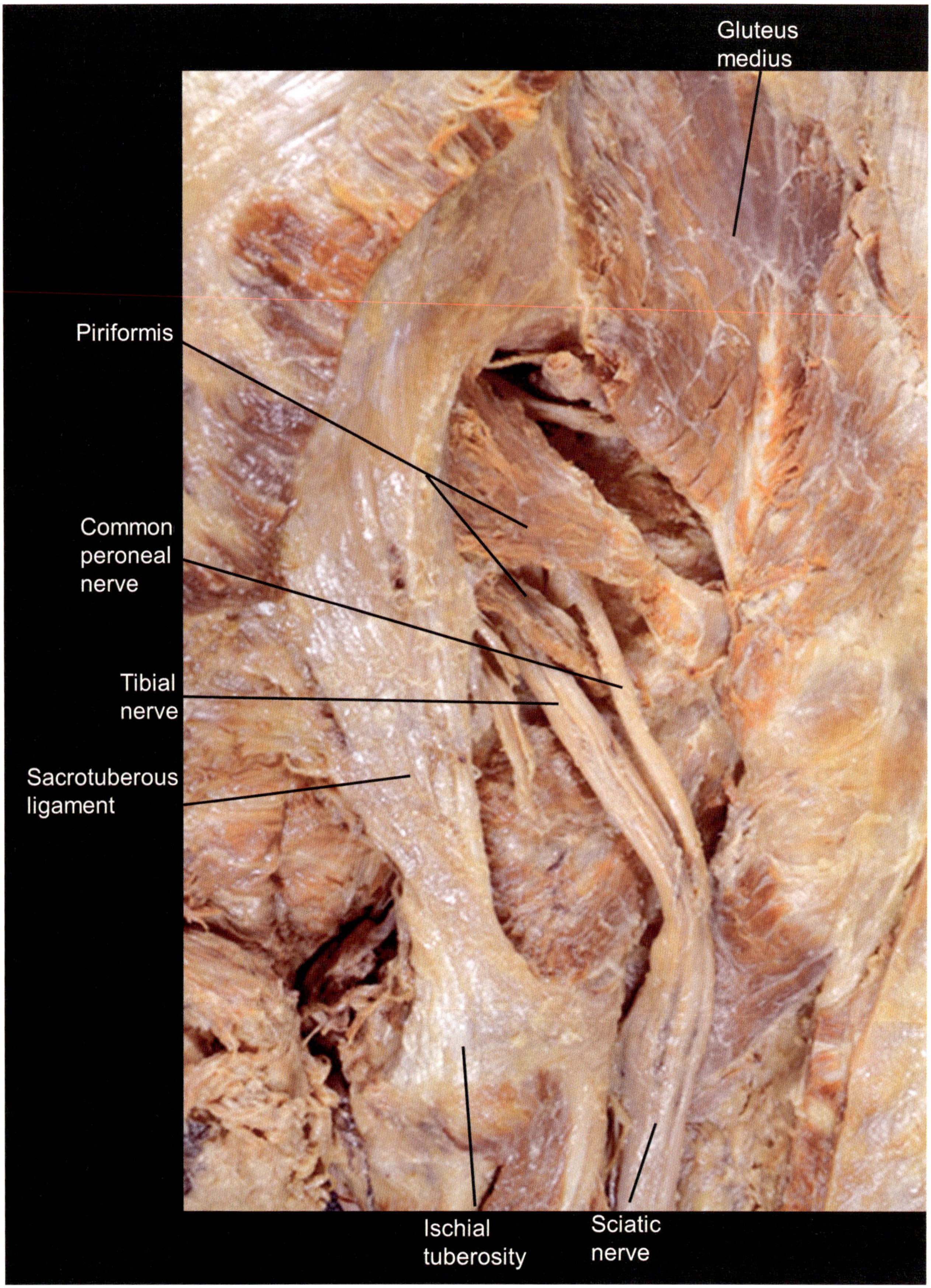

Fig. 9.21 Dissection similar to Fig. 9.20. In these cases, the common peroneal nerve may pierce the piriformis or emerge from the pelvis above it. Anatomical variations of this kind can predispose to certain sciatic nerve compression syndromes

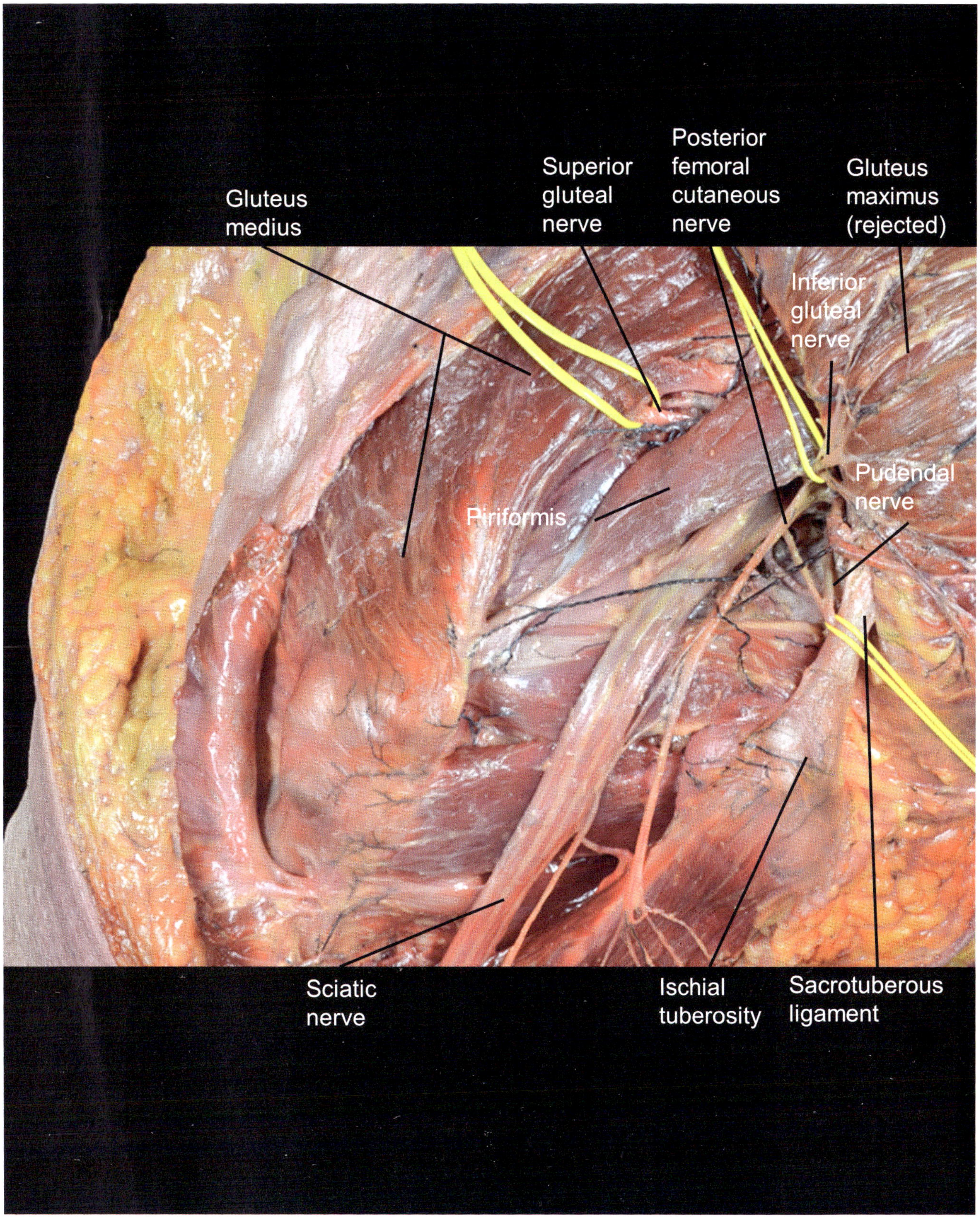

Fig. 9.22 Several branches of the sacral plexus pass to the gluteal region through the superior part of the greater sciatic foramen (above the piriformis muscle) or the inferior part of this foramen (below the piriformis muscle). The superior gluteal nerve passes through the upper portion accompanied by the superior gluteal vessels. These vessels usually pass between the lumbosacral trunk and the ventral ramus of S1. Through the lower portion of the greater sciatic foramen, the sciatic nerve passes accompanied by the inferior gluteal nerve, the posterior femoral cutaneous nerve, and the pudendal nerve. The inferior gluteal and the internal pudendal vessels also run through this foramen. The inferior gluteal vessels pass between the ventral branches of S1 and S2, or S2 and S3. The internal pudendal vessels run between the sciatic and pudendal nerves

Cross-sectional Microscopic Anatomy of the Sciatic Nerve and its Dissected Branches

10

Miguel Angel Reina, Xavier Sala-Blanch and Paloma Fernández

Axons with shorter trajectories are expected to have a more superficial location within the nerves, whereas axons traversing longer distances are expected to have a more central position, but this is not the case in peripheral nerves. Axons constantly fluctuate in their positions along the path of the nerve. The topographic map of the fascicles continually varies. Sunderland and Ray [1] observed that the topography of fascicles varied 23 times in just 46 mm of length.

Variations in topography of the fascicles are due to the exchange of axons between different nerve fascicles. This exchange begins at the proximal end of the spinal dorsal root ganglion, close to the epidural space, and continues along the intervertebral canal [2, 3].

In the subarachnoid space and enclosed by the nerve root cuffs, which extend towards the medial aspect of the spinal dorsal root ganglion, nerve roots primarily contain axons from the anterior and posterior rootlets. These appear clearly separated from one another and have few interconnecting axons [2]. Alongside the spinal dorsal root ganglion, the number of groups of axons increases towards the anterior root, and both anterior and posterior nerve roots are located within the root cuff inside the foramen. At the distal end of the spinal dorsal root ganglion, a fascicular structure identical to that seen in peripheral nerves begins to emerge. The number of fascicles increases along the course of the nerves. In mixed nerves, single fascicles, groups of fascicles, and "small" fascicles may be found, with the latter forming fascicular interconnections [2].

This chapter looks at the sciatic nerve and its branches from three different samples. Figures 10.1, 10.2, 10.3, 10.4, 10.5, 10.6, 10.7, 10.8, 10.9, 10.10, 10.11, 10.12, 10.13, 10.14, 10.15, 10.16, 10.17, 10.18, 10.19, 10.20, 10.21, 10.22, and 10.23 show sequential anatomical sections from proximal to distal locations.

M.A. Reina, MD, PhD (✉)
Department of Clinical Medical Sciences and Institute of Applied Molecular Medicine, School of Medicine, University of CEU San Pablo, Madrid, Spain

Department of Anesthesiology, Madrid-Montepríncipe University Hospital, Madrid, Spain
e-mail: miguelangel@perticone.e.telefonica.net

X. Sala-Blanch, MD
Human Anatomy and Embryology Unit, School of Medicine, University of Barcelona, Barcelona, Spain

Department of Anesthesiology and Critical Care, Clinic Hospital, Barcelona, Spain
e-mail: xavi.sala.blanch@gmail.com

P. Fernández, BS
Histology Unit, Institute of Applied Molecular Medicine, School of Medicine, University of CEU San Pablo, Madrid, Spain
e-mail: paloma.fernandezmartinez@ceu.es

M.A. Reina et al. (eds.), *Atlas of Functional Anatomy for Regional Anesthesia and Pain Medicine: Human Structure, Ultrastructure and 3D Reconstruction Images*, DOI 10.1007/978-3-319-09522-6_10,

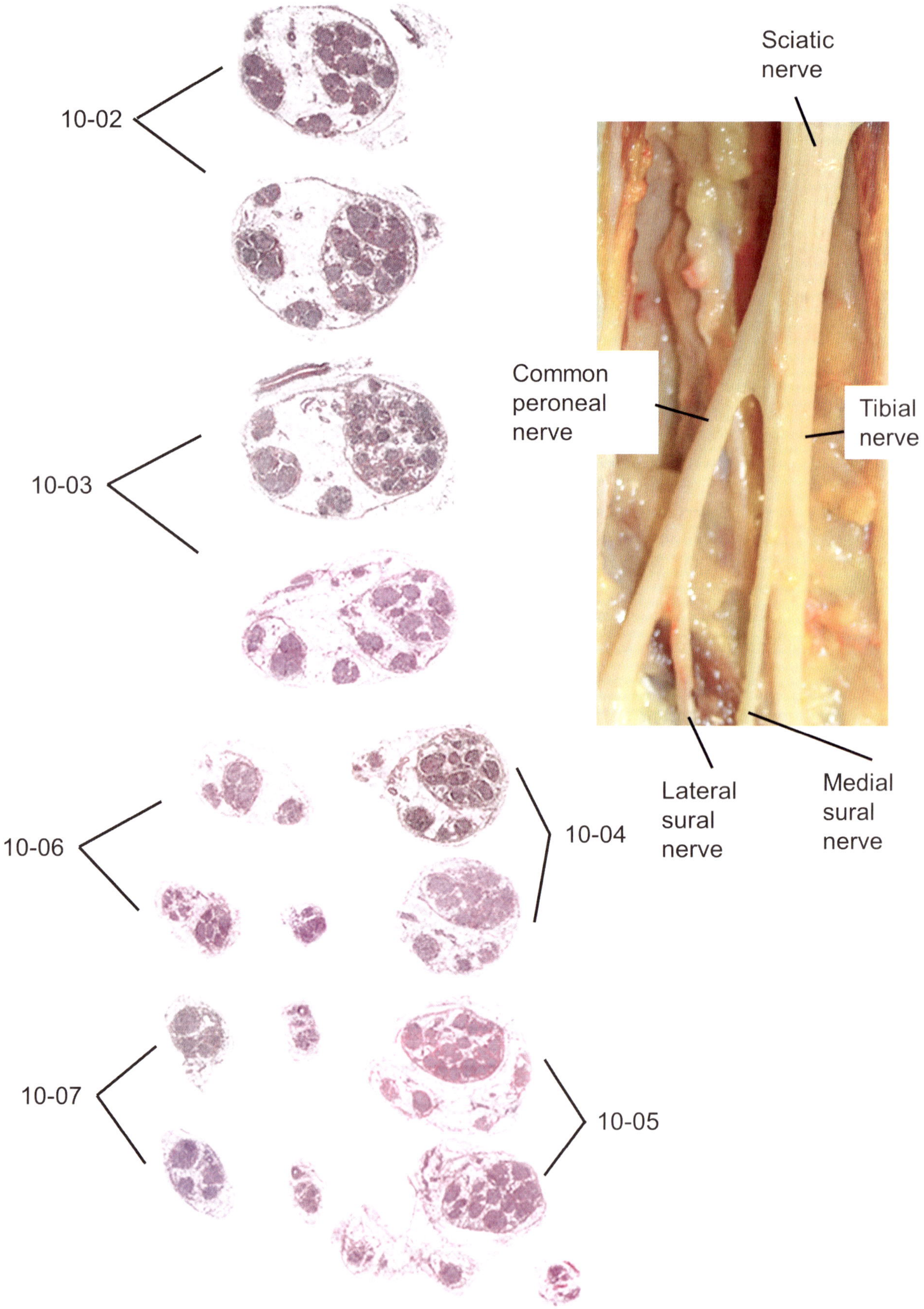

Fig. 10.1 Cross-section from human sciatic nerve and its branches, sample 1. Hematoxylin-eosin stained

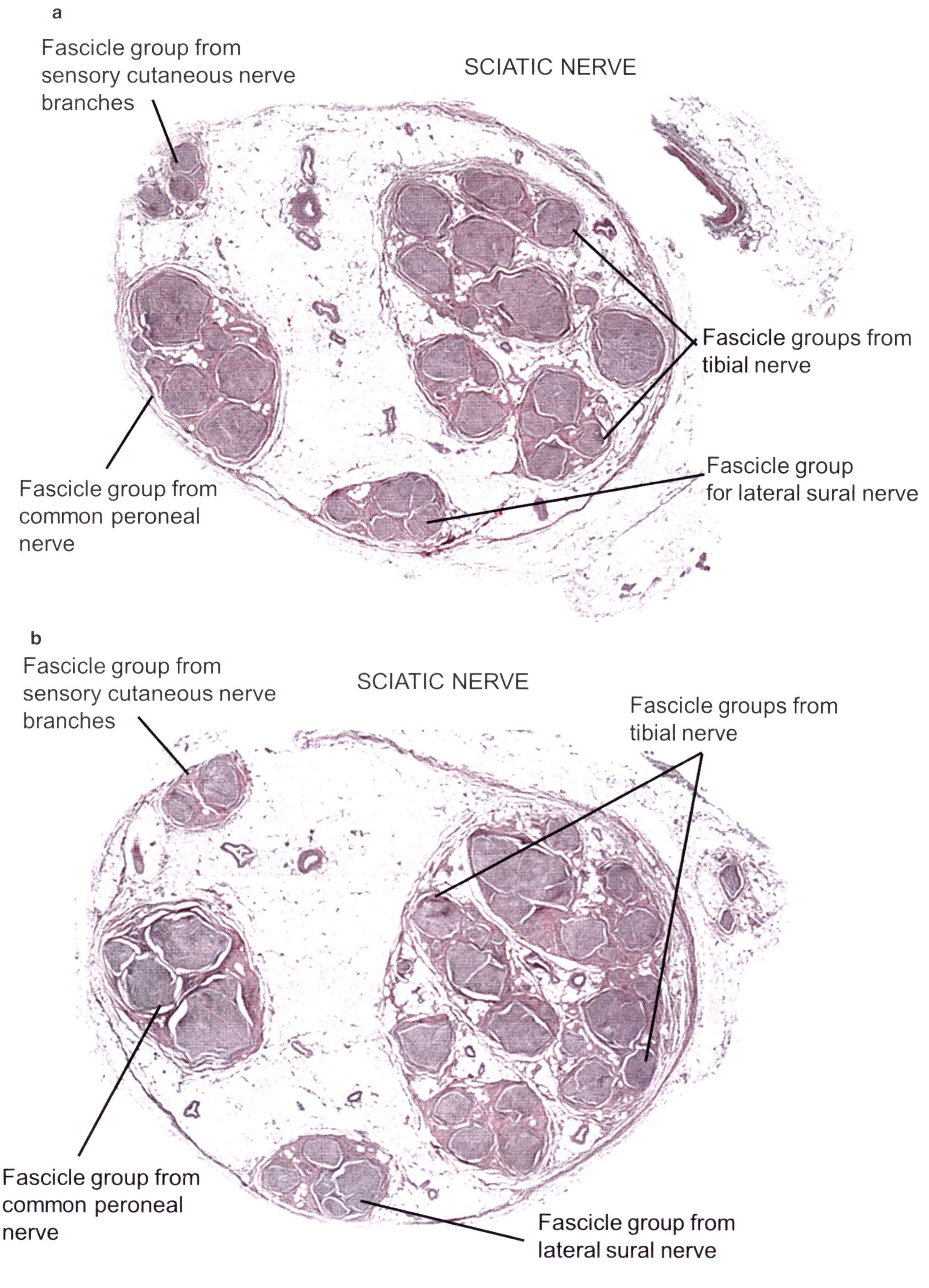

Fig. 10.2 (**a**, **b**) Cross-section from human sciatic nerve. Hematoxylin-eosin stained (**b** *from* Reina et al. [2]; with permission)

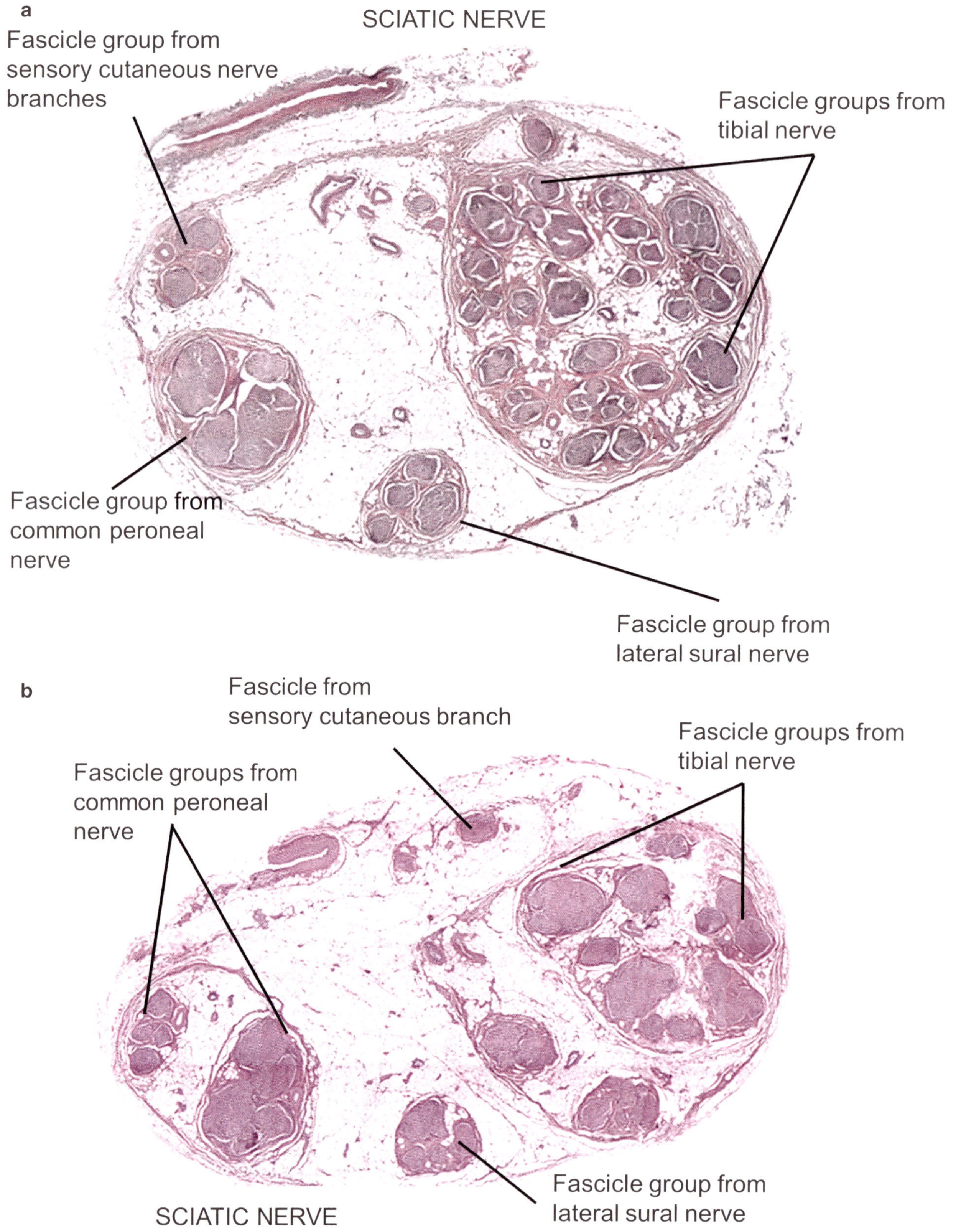

Fig. 10.3 (**a**, **b**) Cross-section from human sciatic nerve. Hematoxylin-eosin stained (**a** *from* Reina et al. [3]; with permission)

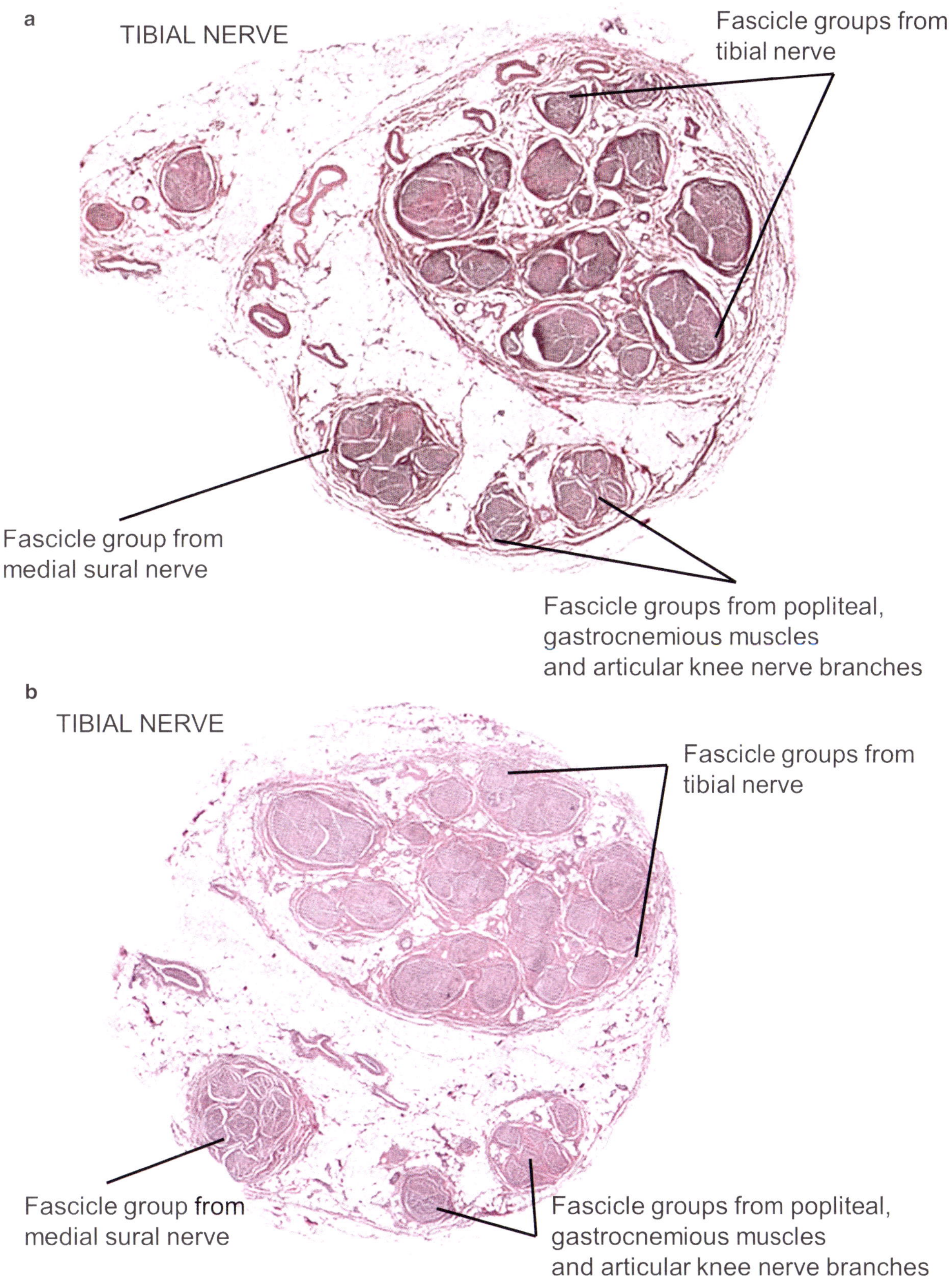

Fig. 10.4 (**a**, **b**) Cross-section from human tibial nerve. Hematoxylin-eosin stained

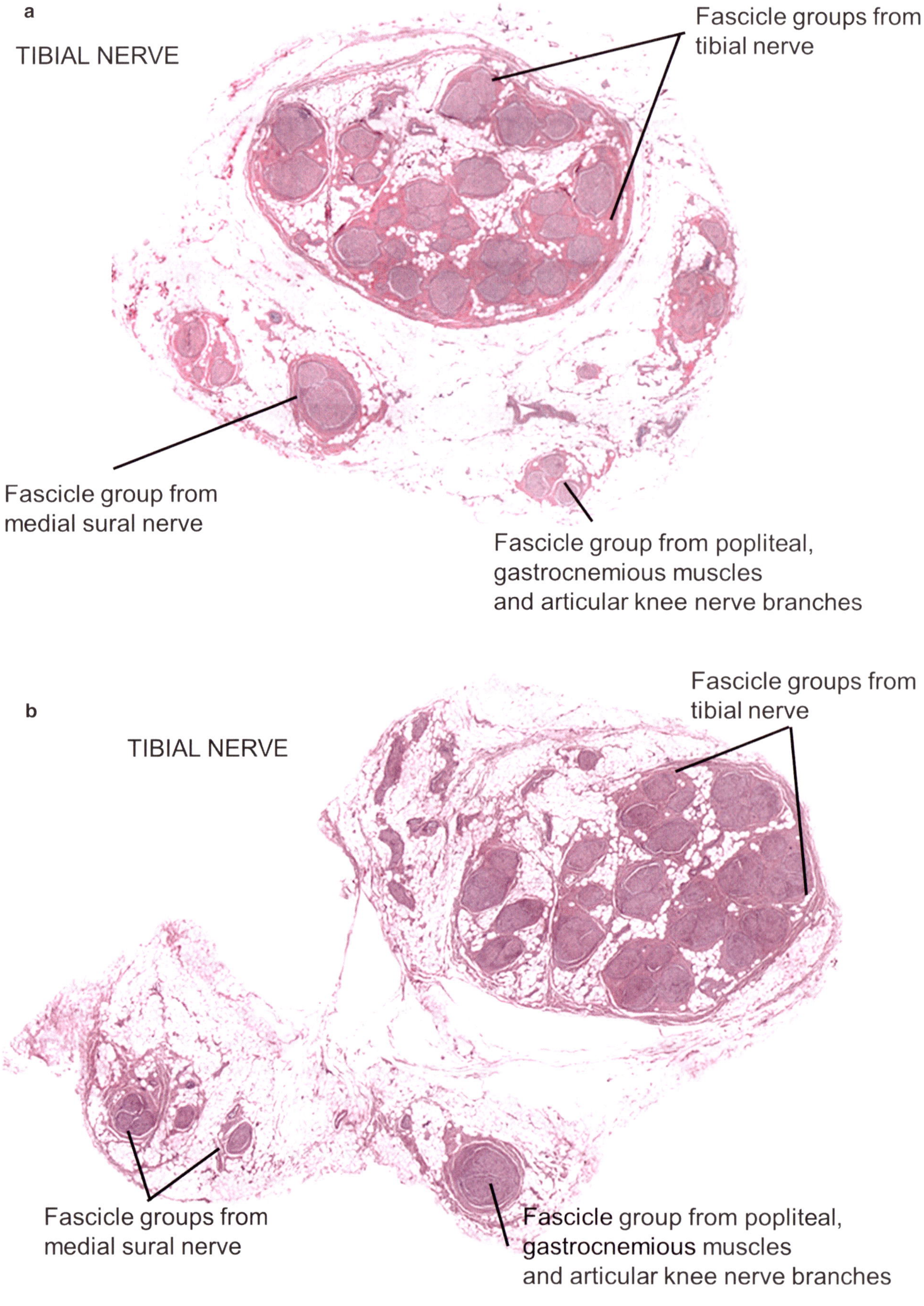

Fig. 10.5 (**a**, **b**) Cross-section from human tibial nerve. Hematoxylin-eosin stained

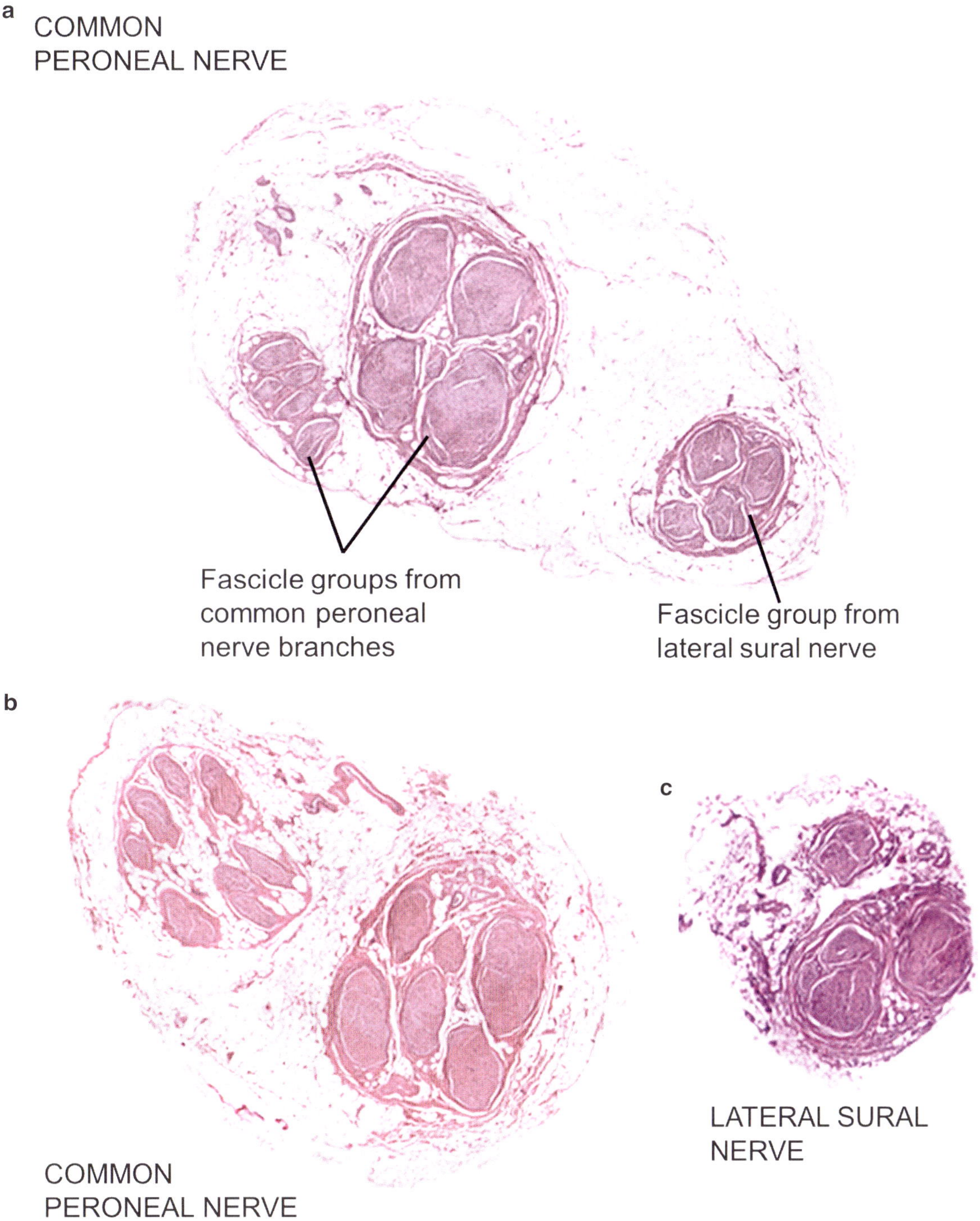

Fig. 10.6 (**a–c**) Cross-section from human common peroneal nerve. Hematoxylin-eosin stained

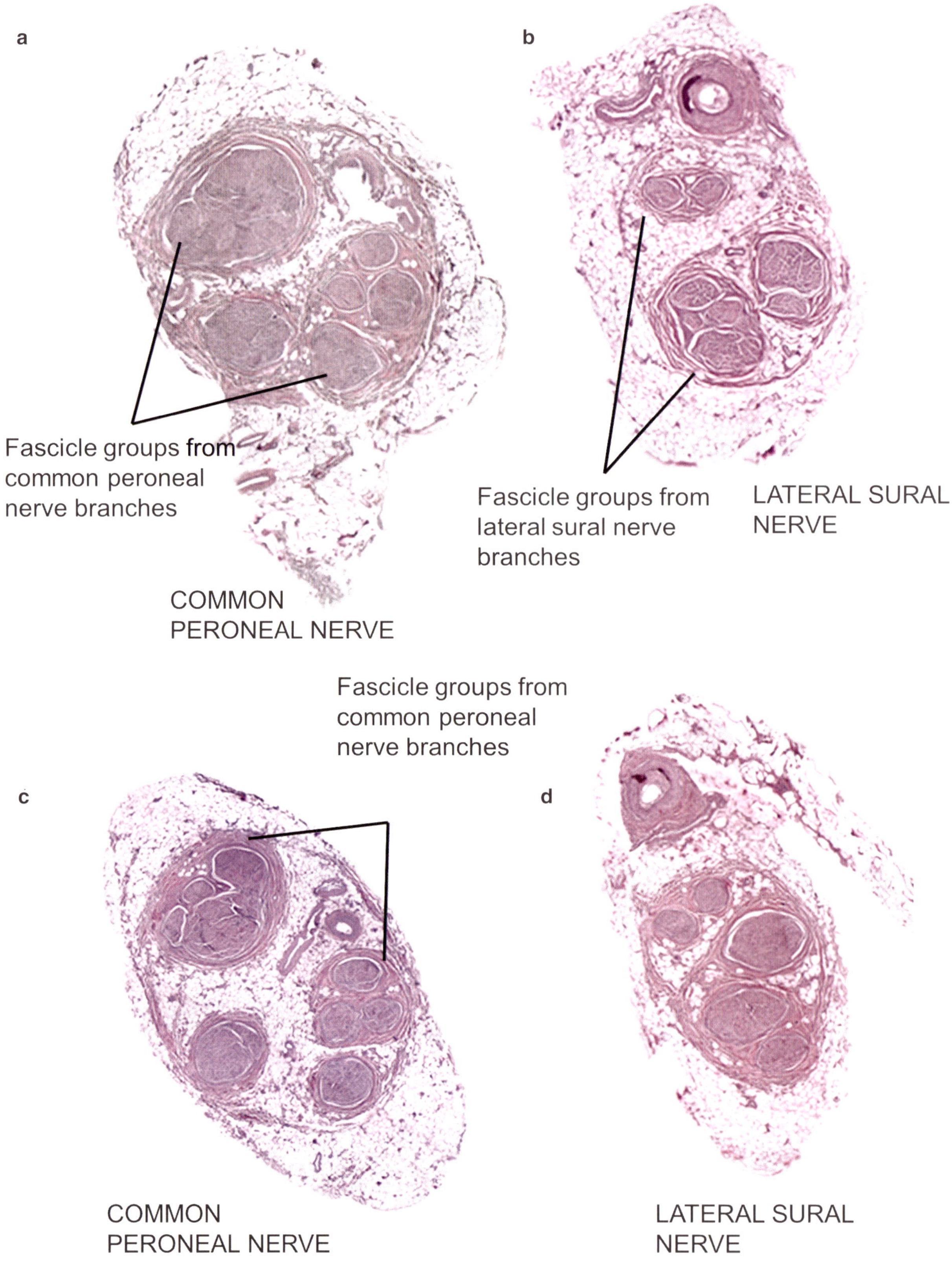

Fig. 10.7 (**a–d**) Cross-section from human common peroneal nerve. Hematoxylin-eosin stained

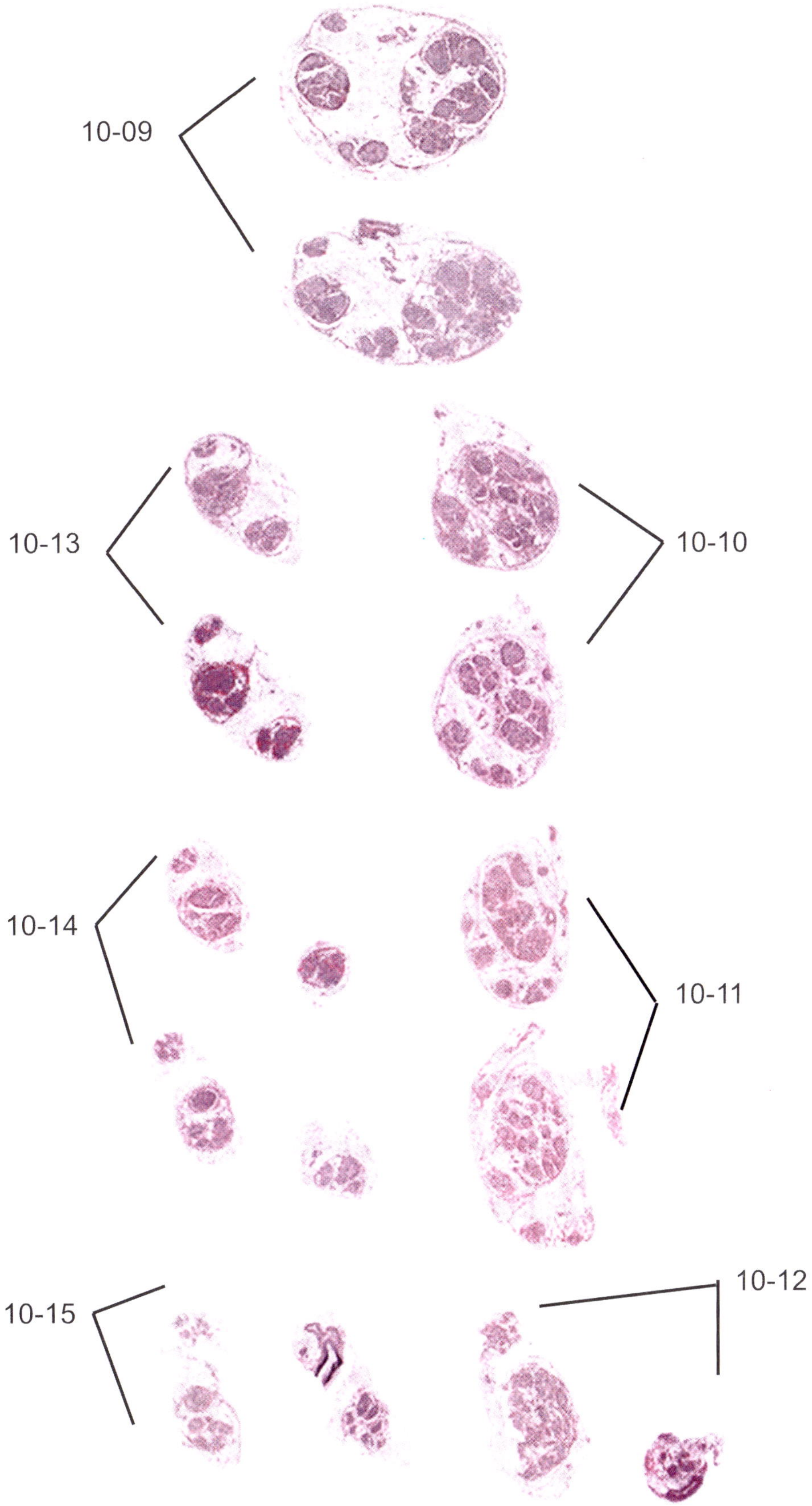

Fig. 10.8 Cross-section from human sciatic nerve and its branches, sample 2. Hematoxylin-eosin stained

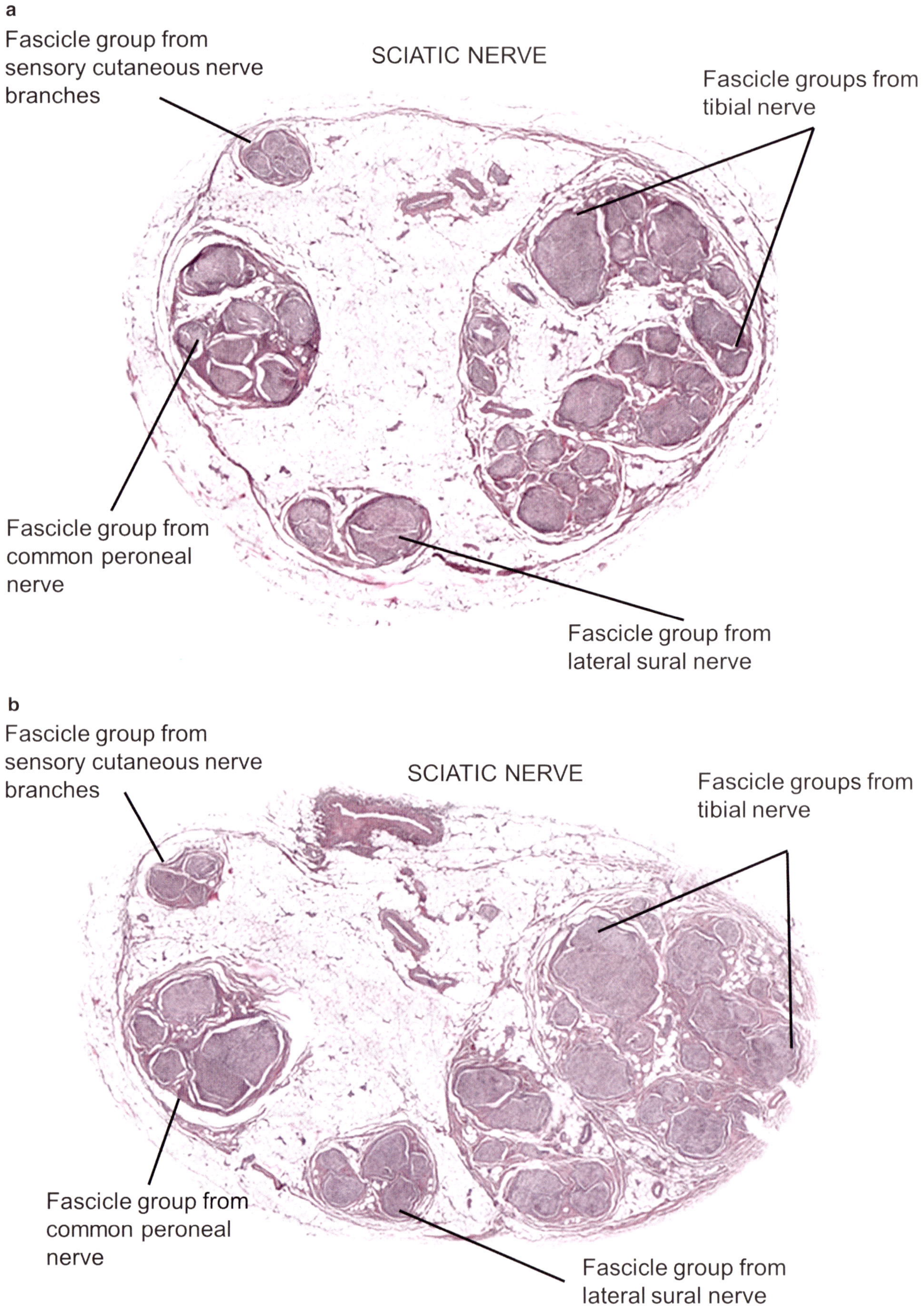

Fig. 10.9 (**a**, **b**) Cross-section from human sciatic nerve. Hematoxylin-eosin stained

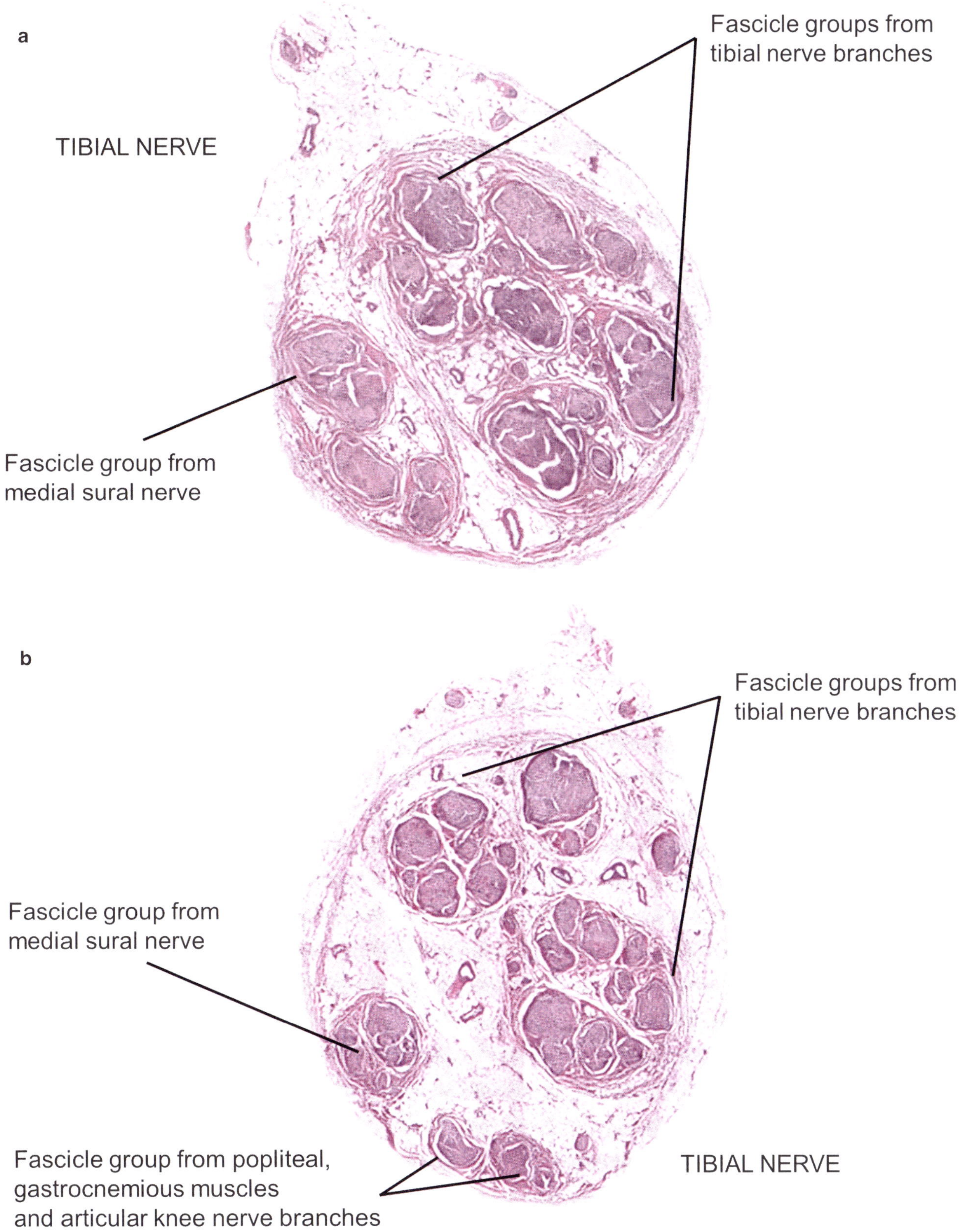

Fig. 10.10 (**a**, **b**) Cross-section from human tibial nerve. Hematoxylin-eosin stained

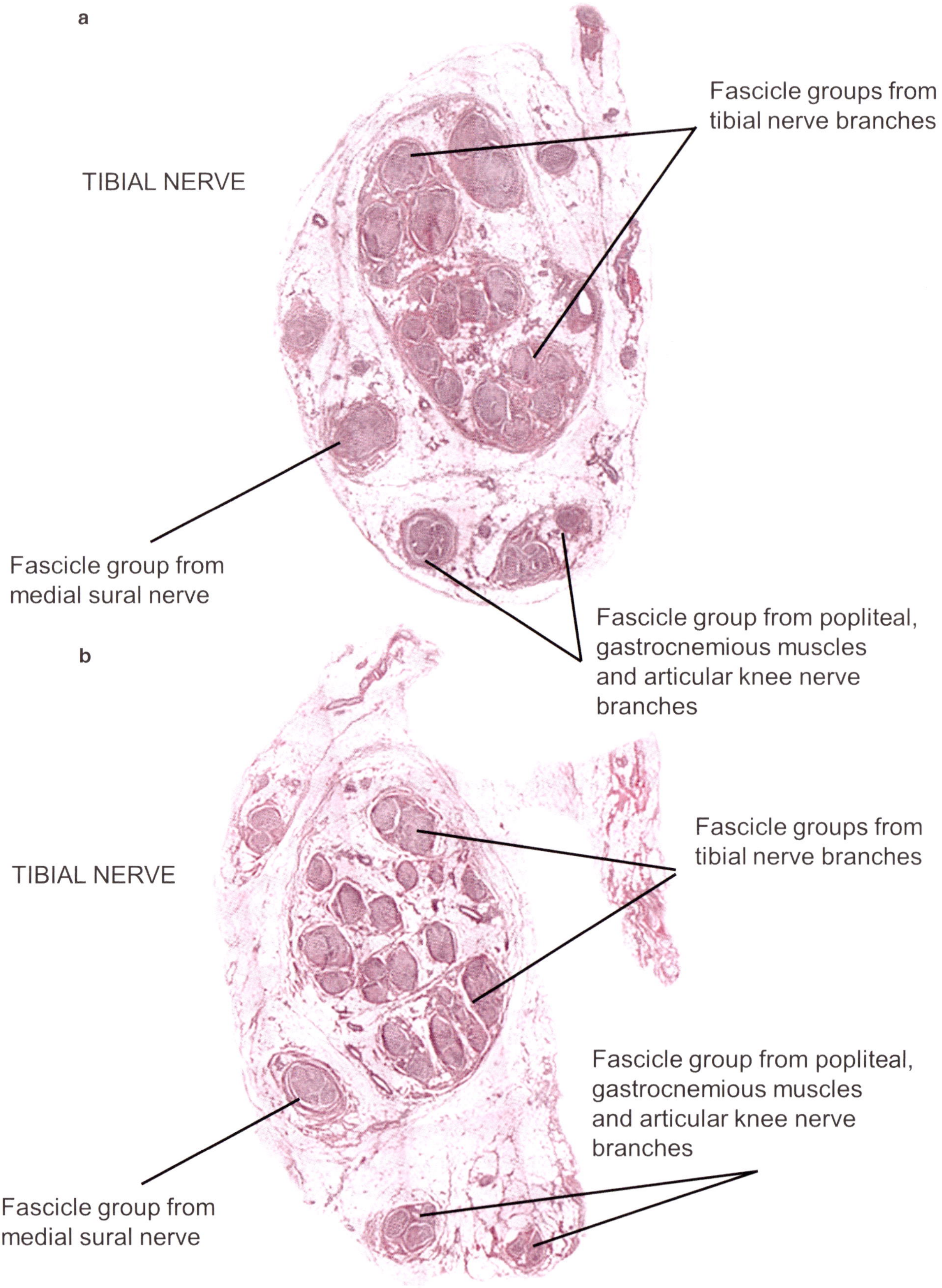

Fig. 10.11 (**a**, **b**) Cross-section from human tibial nerve. Hematoxylin-eosin stained

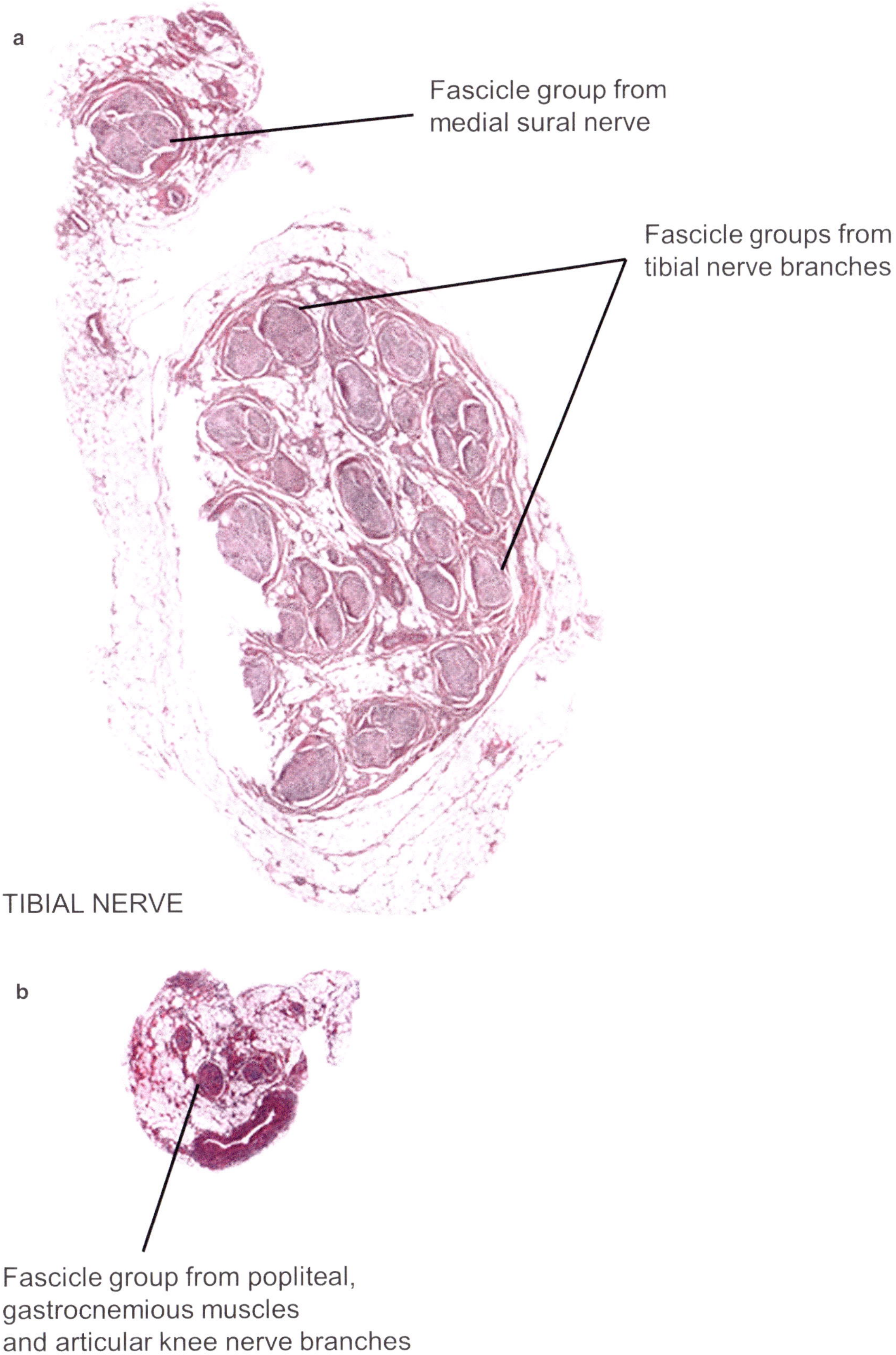

Fig. 10.12 (**a**, **b**) Cross-section from branches of human tibial nerve. Hematoxylin-eosin stained

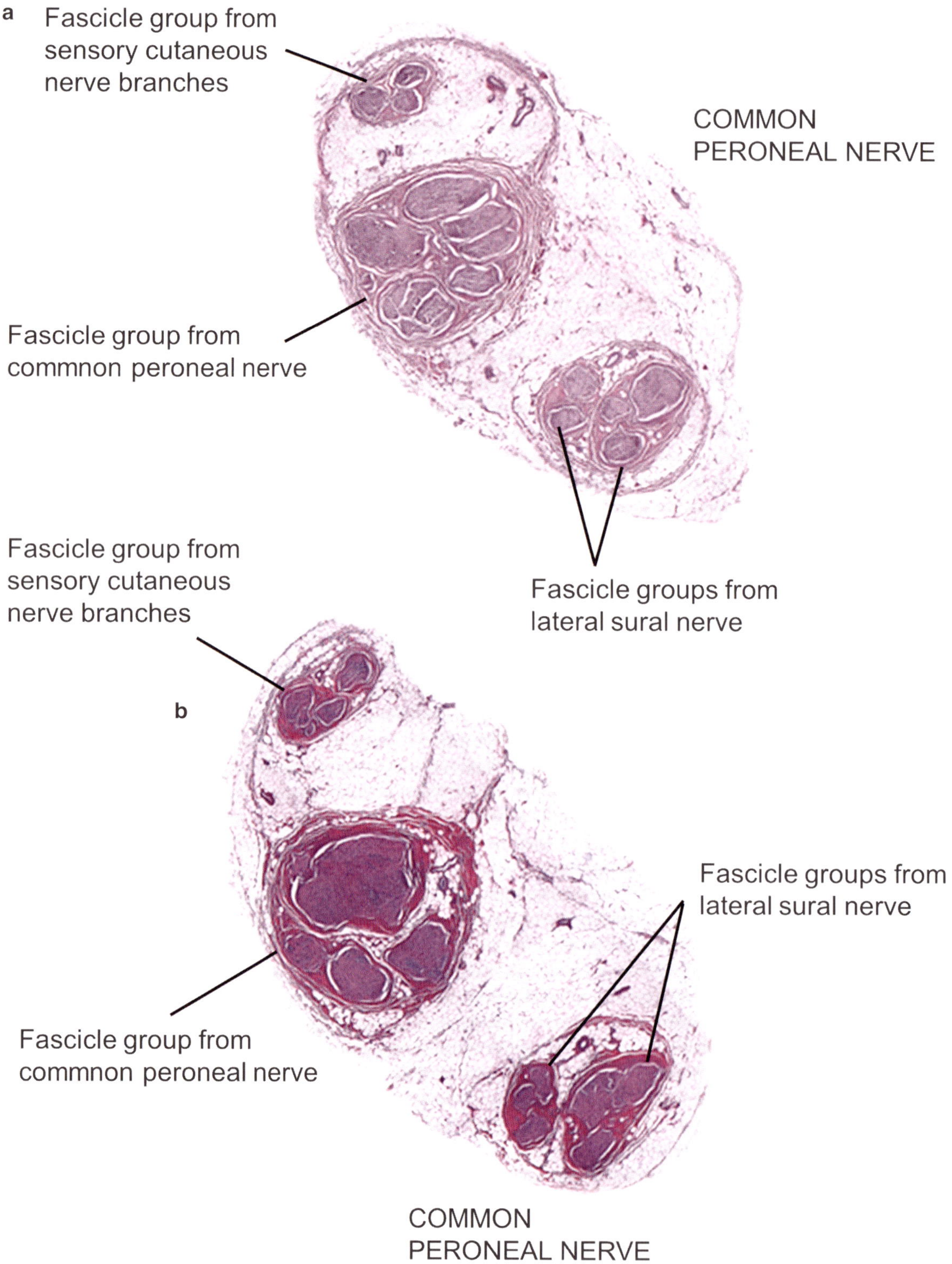

Fig. 10.13 (**a**, **b**) Cross-section from human common peroneal nerve. Hematoxylin-eosin stained

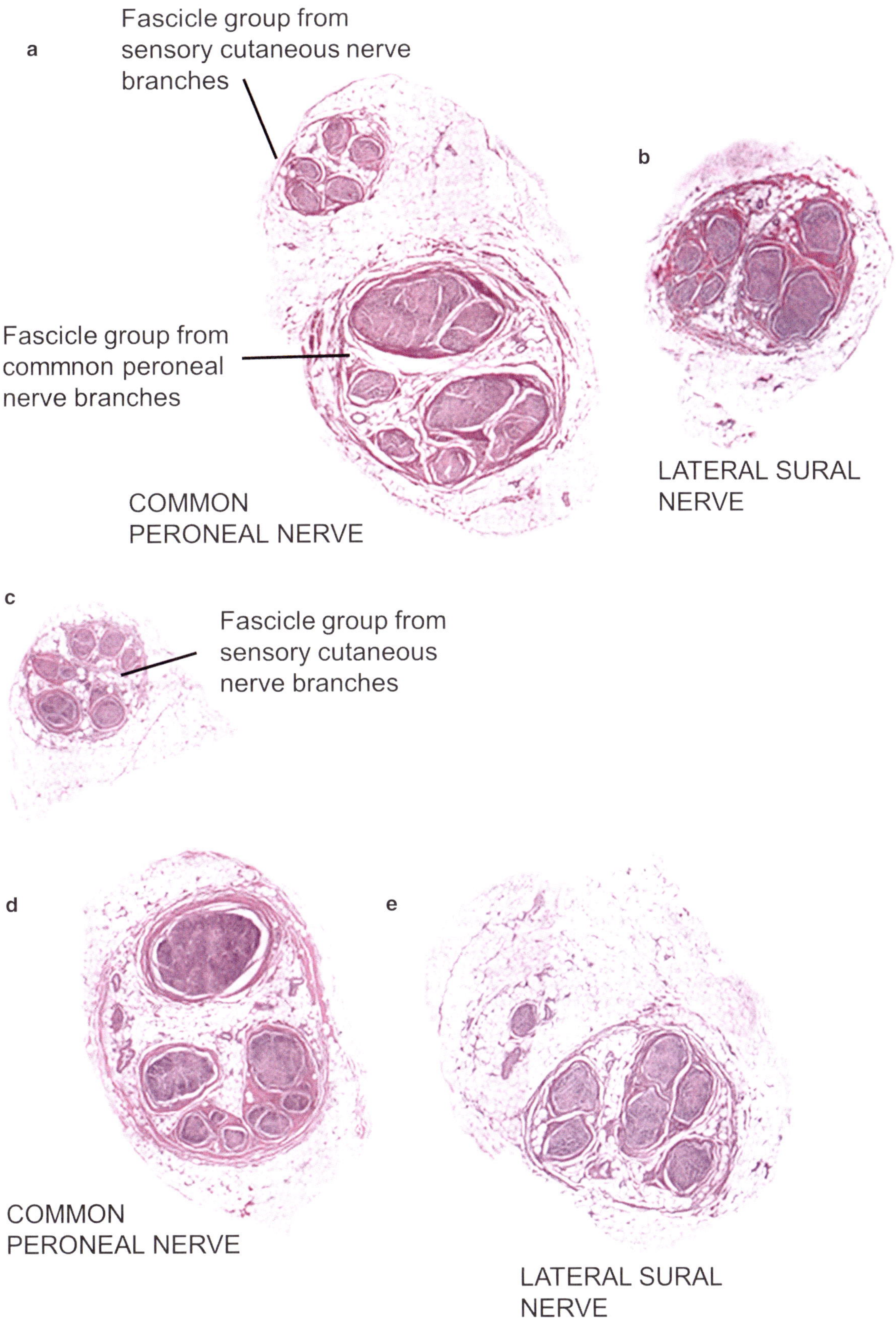

Fig. 10.14 (**a**–**e**) Cross-section from branches of human common peroneal nerve. Hematoxylin-eosin stained

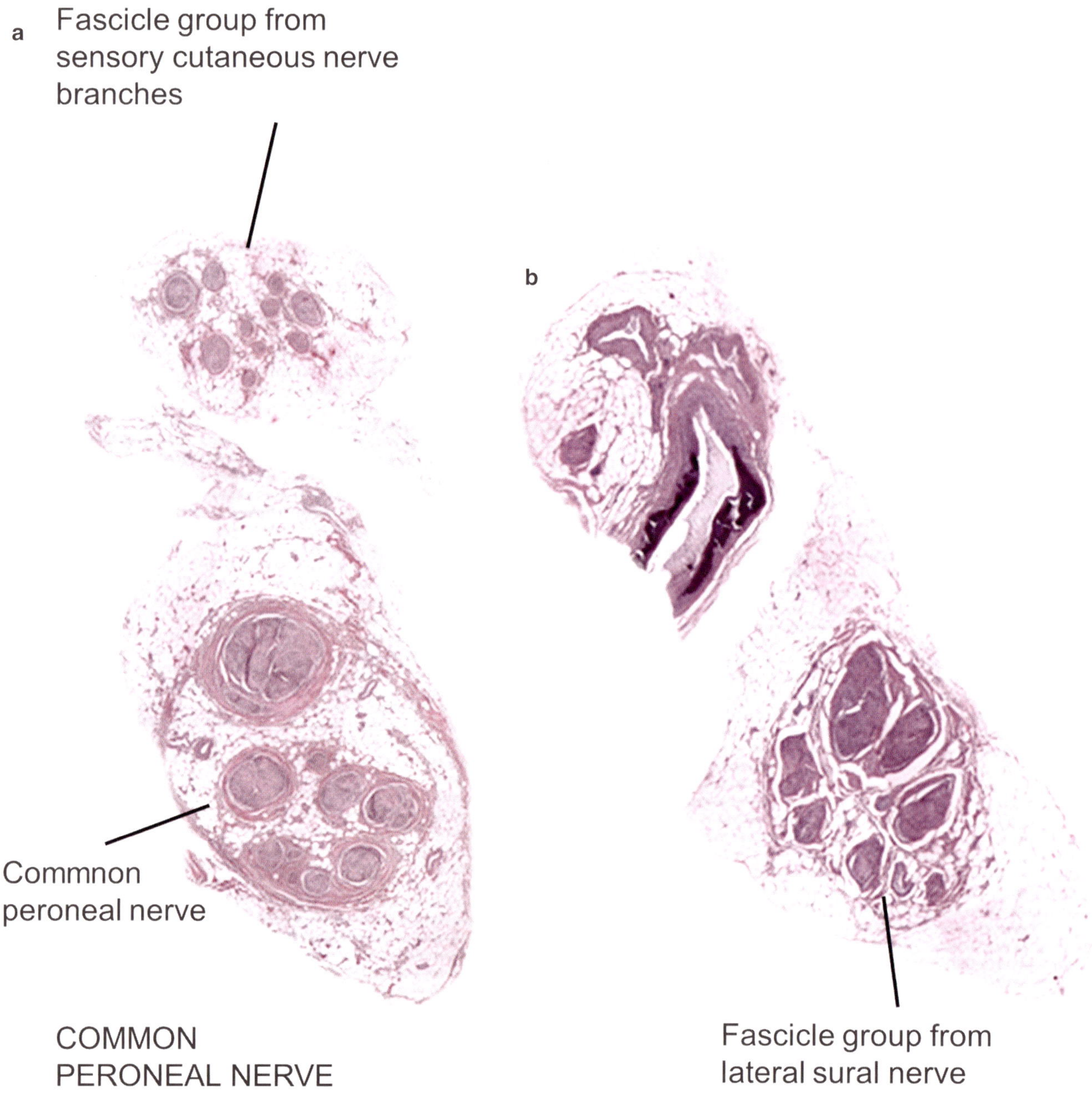

Fig. 10.15 (**a**, **b**) Cross-section from branches of human common peroneal nerve. Hematoxylin-eosin stained

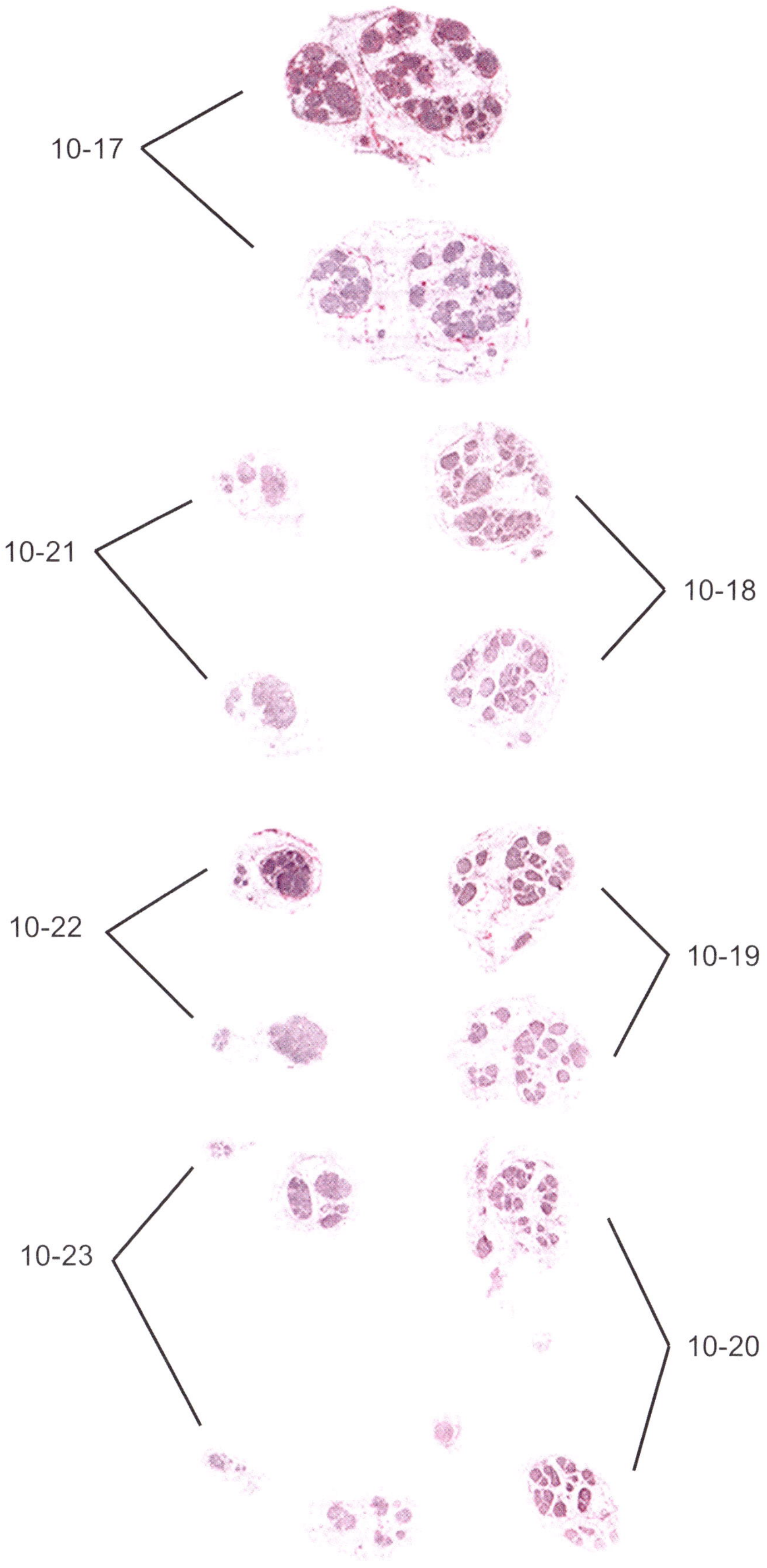

Fig. 10.16 Cross-section from human sciatic nerve and its branches, sample 3. Hematoxylin-eosin stained

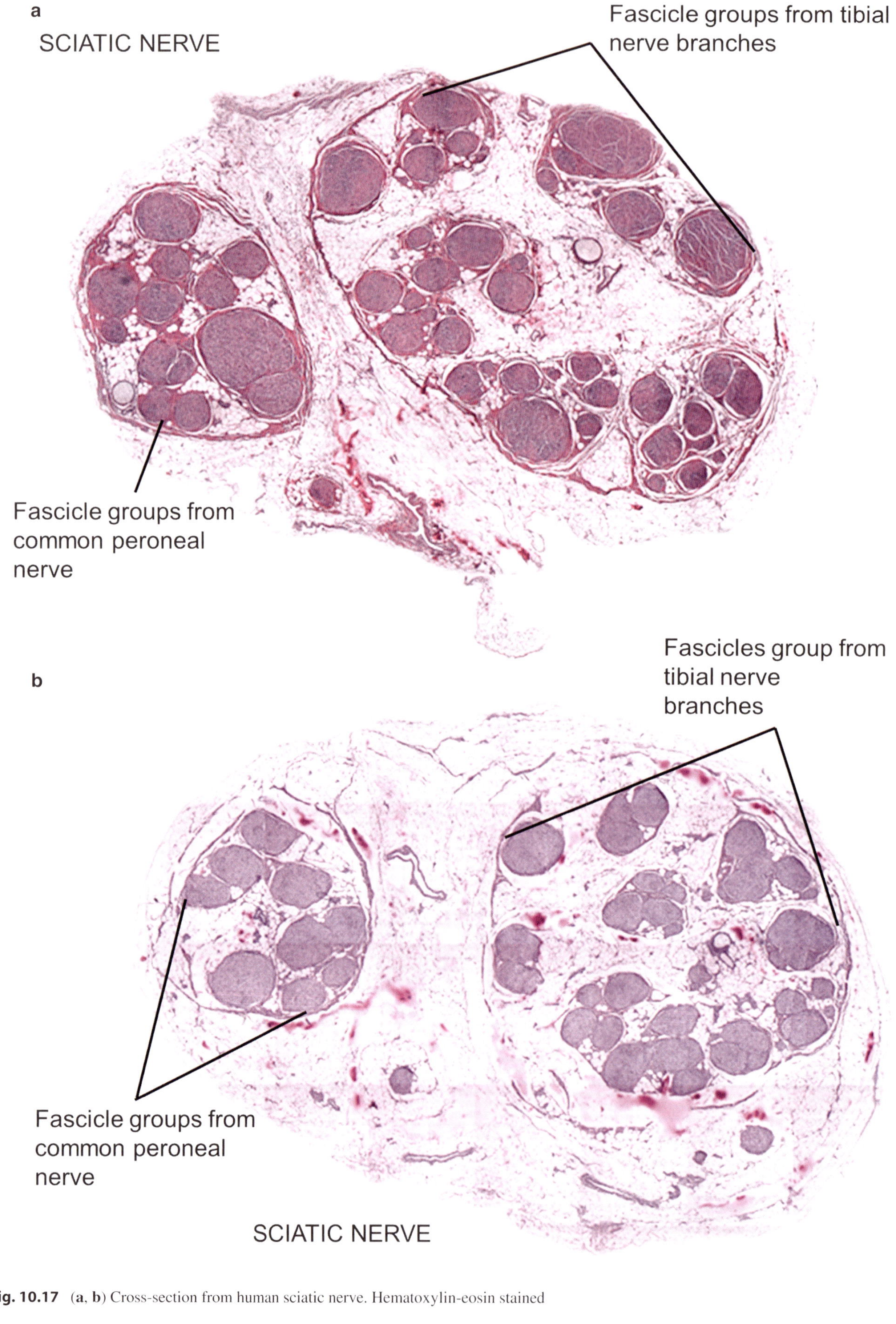

Fig. 10.17 (**a**, **b**) Cross-section from human sciatic nerve. Hematoxylin-eosin stained

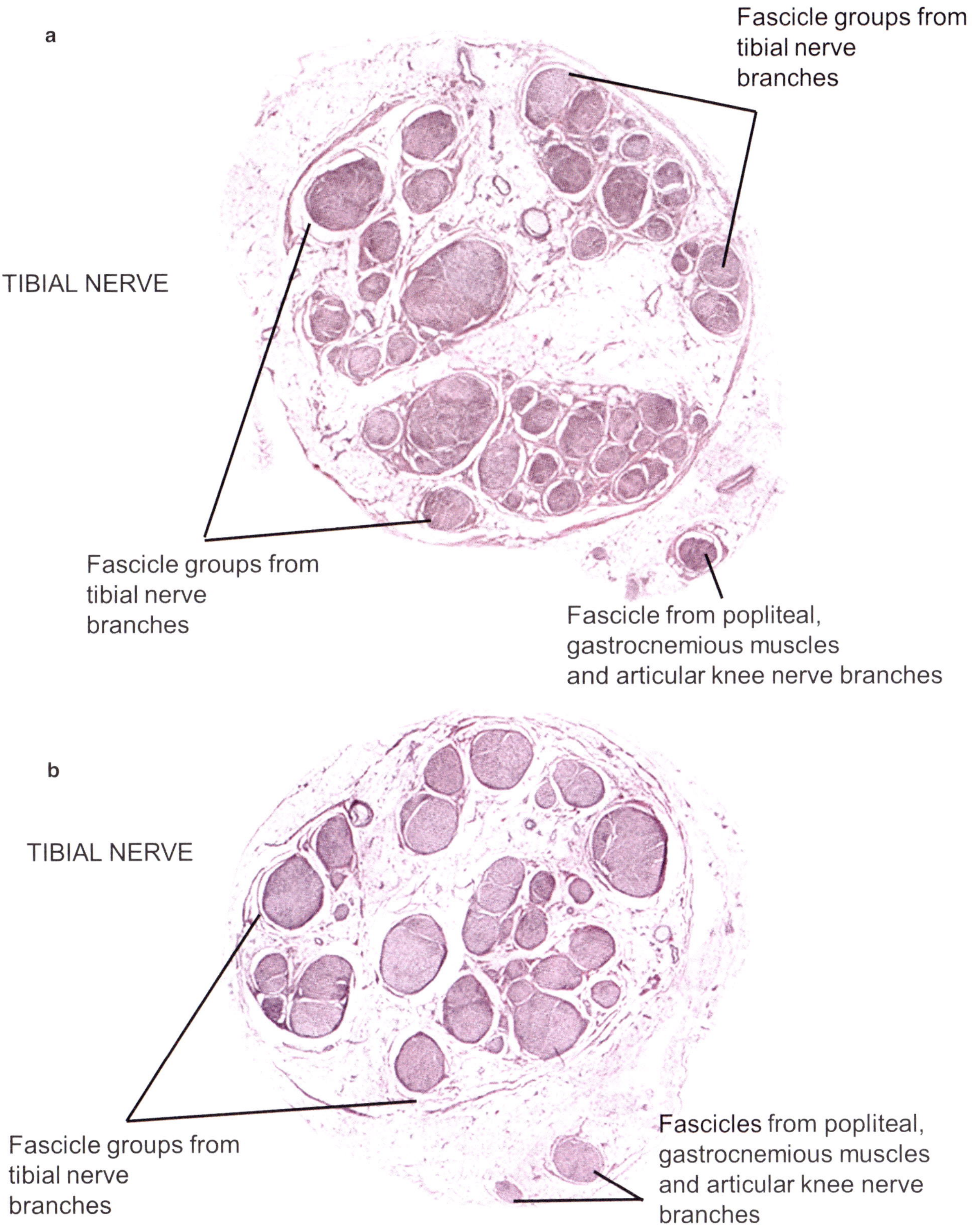

Fig. 10.18 (**a**, **b**) Cross-section from human tibial nerve. Hematoxylin-eosin stained

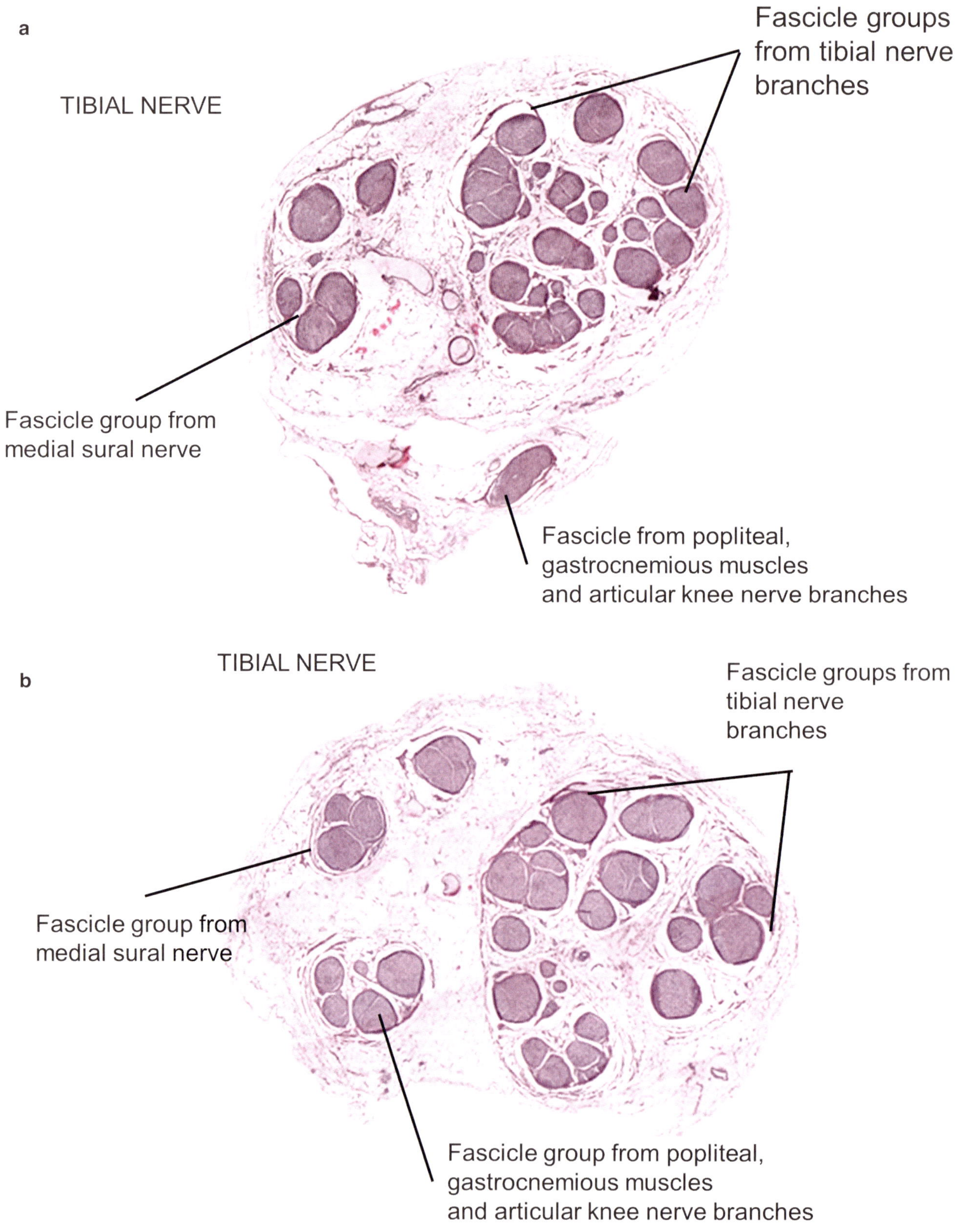

Fig. 10.19 (**a**, **b**) Cross-section human from tibial nerve. Hematoxylin-eosin stained

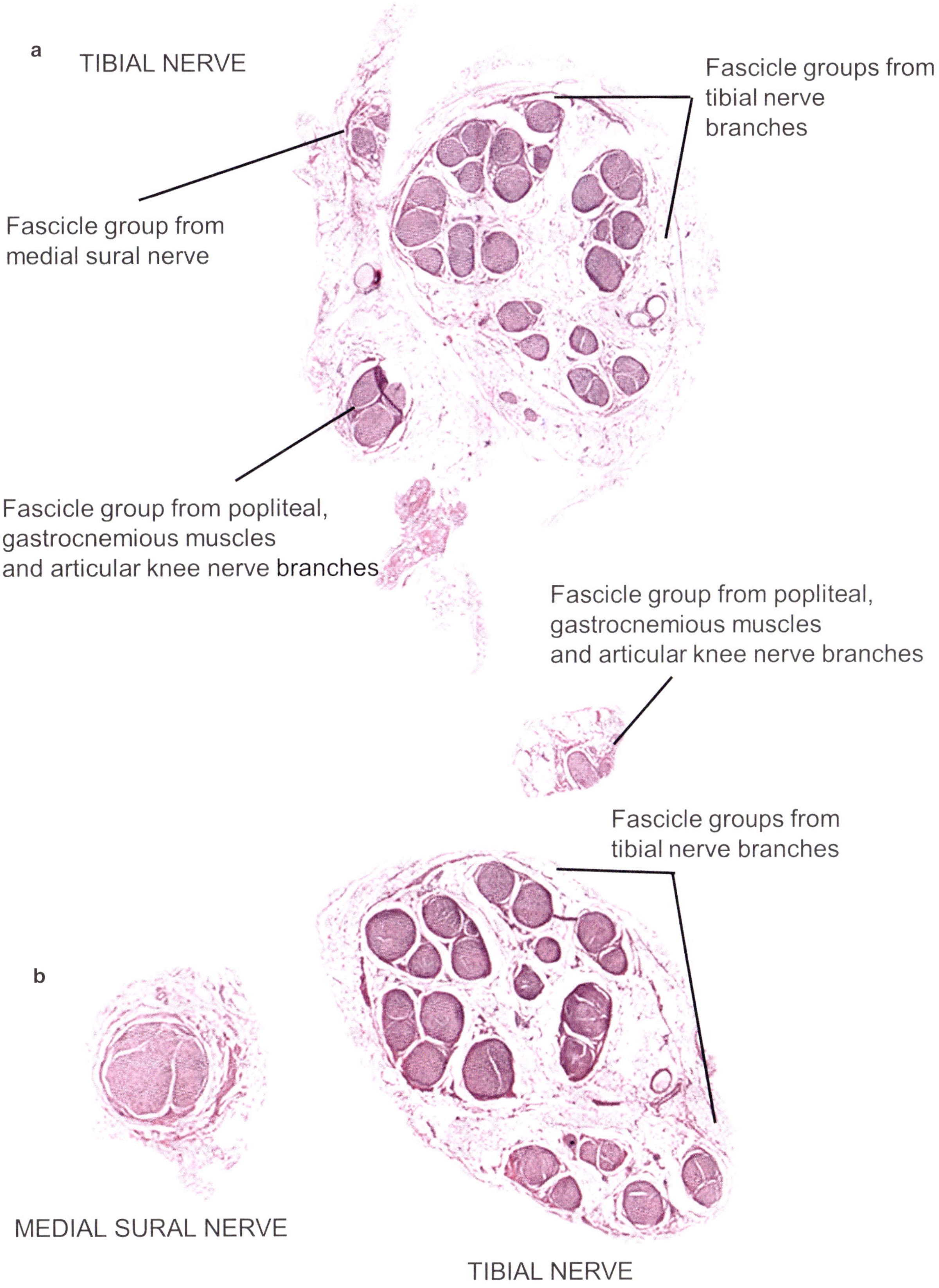

Fig. 10.20 (**a**, **b**) Cross-section of branches of human tibial nerve. Hematoxylin-eosin stained

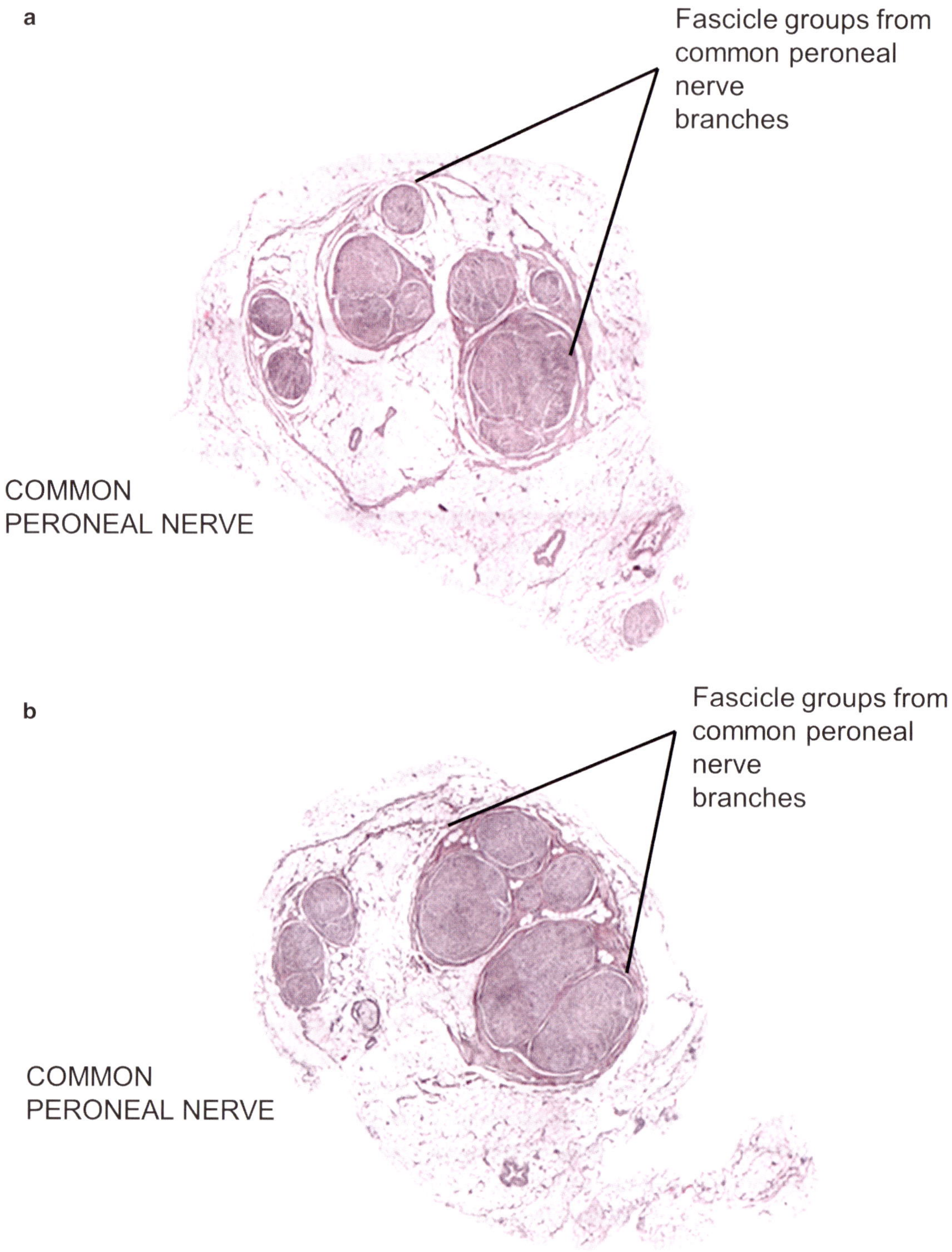

Fig. 10.21 (**a**, **b**) Cross-section human common peroneal nerve. (**a**, **b**) Hematoxylin-eosin stained

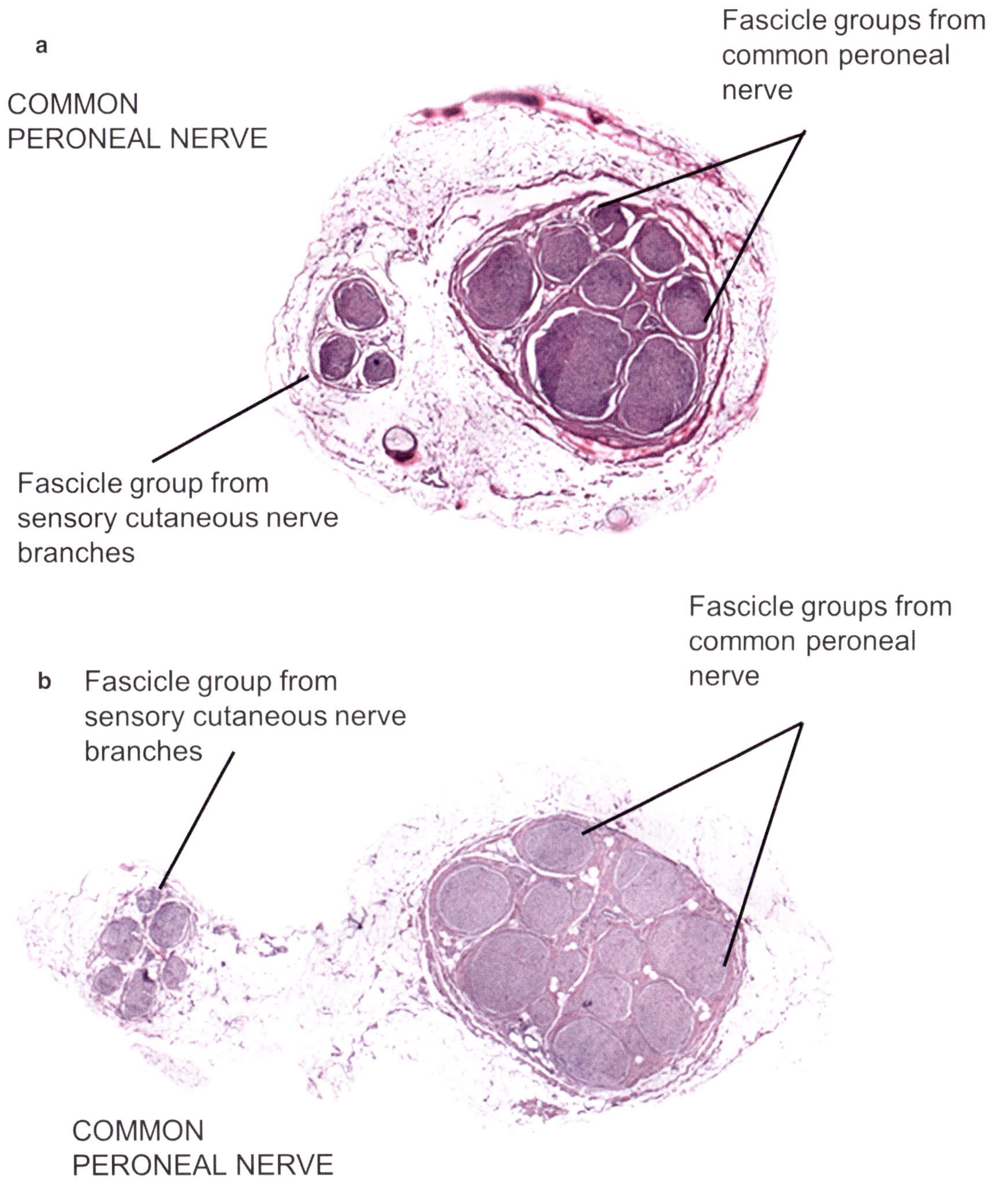

Fig. 10.22 (**a**, **b**) Cross-section from human common peroneal nerve. Hematoxylin-eosin stained

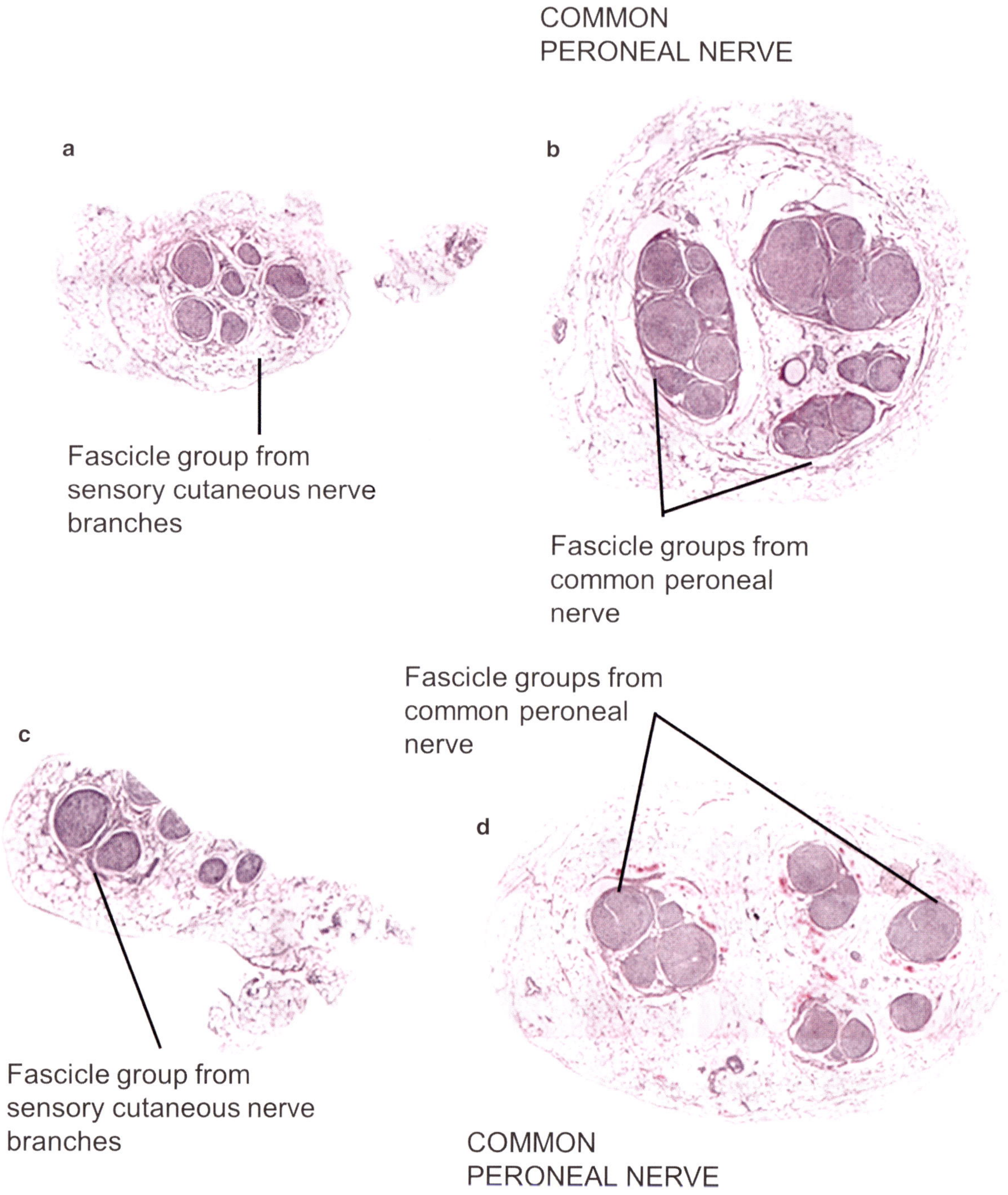

Fig. 10.23 (**a–d**) Cross-section from branches of human common peroneal nerve. Hematoxylin-eosin stained

References

1. Sunderland S, Ray LJ. The intraneural topography of the sciatic nerve and its popliteal divisions in man. Brain. 1948;71:242–73.
2. Reina MA, Arriazu R, Collier CB, Sala-Blanch X, Izquierdo L, de Andrés J. Electron microscopy of human peripheral nerves of clinical relevance to the practice of nerve blocks. A structural and ultrastructural review based on original experimental and laboratory data. Rev Esp Anestesiol Reanim. 2013;60:552–62.
3. Reina MA, De Andrés JA, Hernández JM, Arriazu Navarro R, Durán Mateos EM, Prats-Galino A. Successive changes in extraneural structures from the subarachnoid nerve roots to the peripheral nerve, influencing anesthetic block, and treatment of acute postoperative pain. Eur J Pain Suppl. 2011;5:377–85.

Cross-sectional Microscopic Anatomy of the Sciatic Nerve and Paraneural Sheaths

11

Miguel Angel Reina and Xavier Sala-Blanch

Cross-sectional images from sciatic nerve that include surrounding structures such as fat and muscles have the advantage of revealing features that are lost once these structures are removed from the anatomical section during the dissection process. This type of extensive anatomical tissue section enables identification of the entire nerve structure as well as its neighboring structures, including collagen layers delimiting compartments around the nerve that are filled with fat. Each compartment is enclosed by a layer formed of collagen fibers, identical in nature to the epineurium located outside the nerve. The fat present in these compartments is similar to that found among groups of fascicles. The images in this chapter show sequential anatomical sections of this type, from proximal to distal locations. Collagen layers giving origin to compartments around the nerve may be generally known as *perineurial sheaths*, and compartments filled with fat as *paraneural compartments* [1, 2]. The concentric distribution of paraneural compartments along the nerve varies according to the topographic region being examined. These compartments contain fat, small vessels, and capillaries.

Images from anatomical sections show paraneural sheaths extending towards larger, neighboring blood vessels and groups of muscles. The structure of sheaths of this type is similar to that of fascias enclosing groups of muscle cells. Capillaries are often found along the trajectory of paraneural sheaths. Paraneural sheaths may contribute to the protection of nerves against compression and stretching during articular movement. Paraneural compartments facilitate longitudinal displacement of nerves controlling body movement.

The inspection of images obtained from successive cuts performed at sequential intervals of 1 cm on a single nerve show that these concentric compartments vary in shape, number, thickness, and distribution. Images in this chapter show serial cuts from the sciatic nerve and its neighboring structures, extending from the subgluteal region to the popliteal region. For identification purposes, the femoral region here comprises the extent limited cranially by the subgluteal region and caudally by the popliteal region.

In addition to the term paraneural sheath, other names that may refer to the same anatomical structure include *paraneurium*, common epineural sheath, conjunctiva nervorum, adventitia, or neural fascia (Figs. 11.1, 11.2, 11.3, 11.4, 11.5, 11.6, 11.7, 11.8, 11.9, 11.10, 11.11, 11.12, 11.13, 11.14, 11.15, 11.16, 11.17, 11.18, 11.19, 11.20, 11.21, 11.22, 11.23, 11.24, 11.25, 11.26, 11.27, 11.28, 11.29, 11.30 and 11.31).

M.A. Reina, MD, PhD (✉)
Department of Clinical Medical Sciences and Institute of Applied Molecular Medicine, School of Medicine,
University of CEU San Pablo, Madrid, Spain

Department of Anesthesiology,
Madrid-Montepríncipe University Hospital, Madrid, Spain
e-mail: miguelangel@perticone.e.telefonica.net

X. Sala-Blanch, MD
Human Anatomy and Embryology Unit, School of Medicine,
University of Barcelona, Barcelona, Spain

Department of Anesthesiology and Critical Care, Clinic Hospital,
Barcelona, Spain
e-mail: xavi.sala.blanch@gmail.com

M.A. Reina et al. (eds.), *Atlas of Functional Anatomy for Regional Anesthesia and Pain Medicine: Human Structure, Ultrastructure and 3D Reconstruction Images*, DOI 10.1007/978-3-319-09522-6_11,

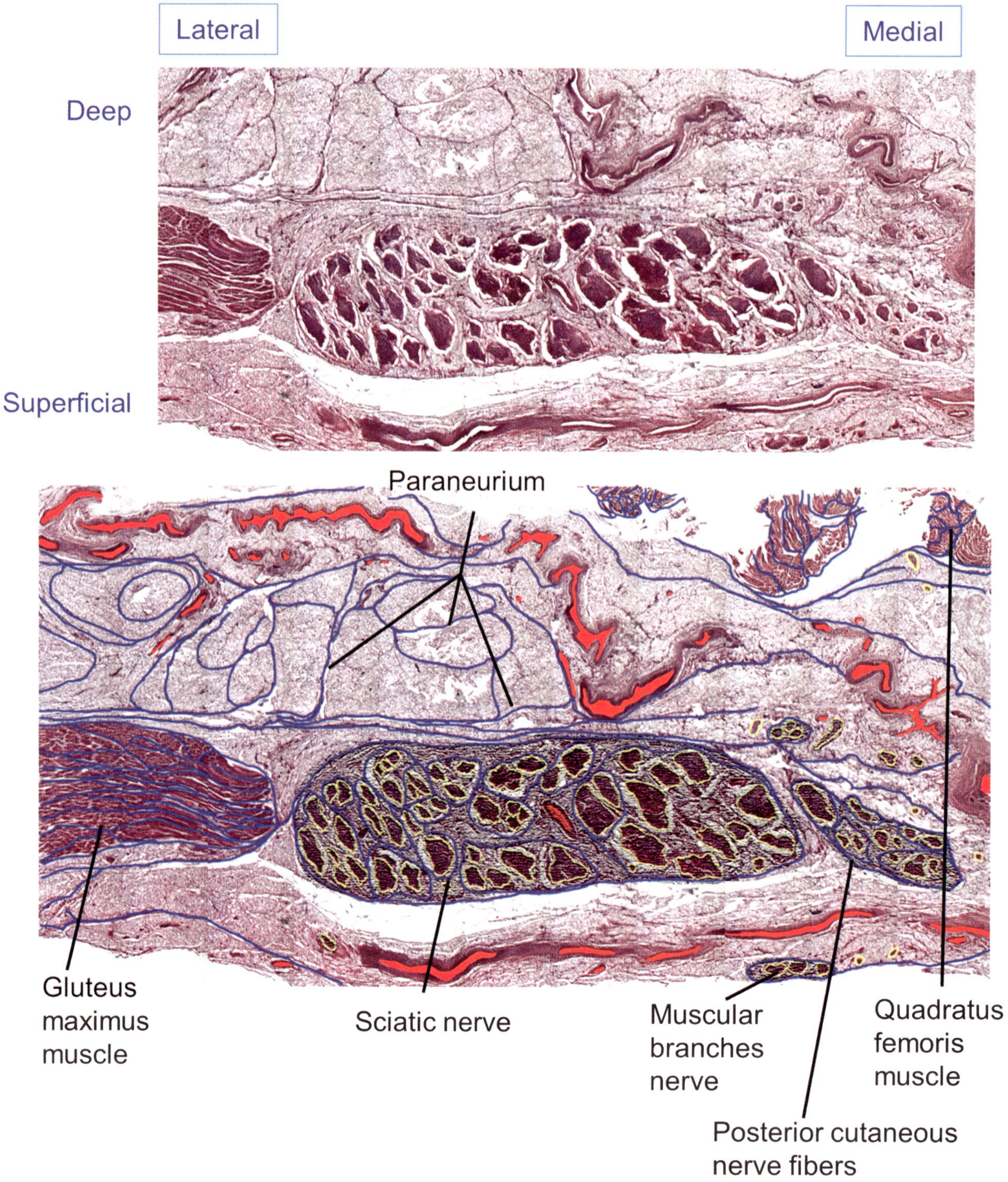

Fig. 11.1 Cross-section of human sciatic nerve and its surrounding tissue. Subgluteal region. Hematoxylin-eosin stained

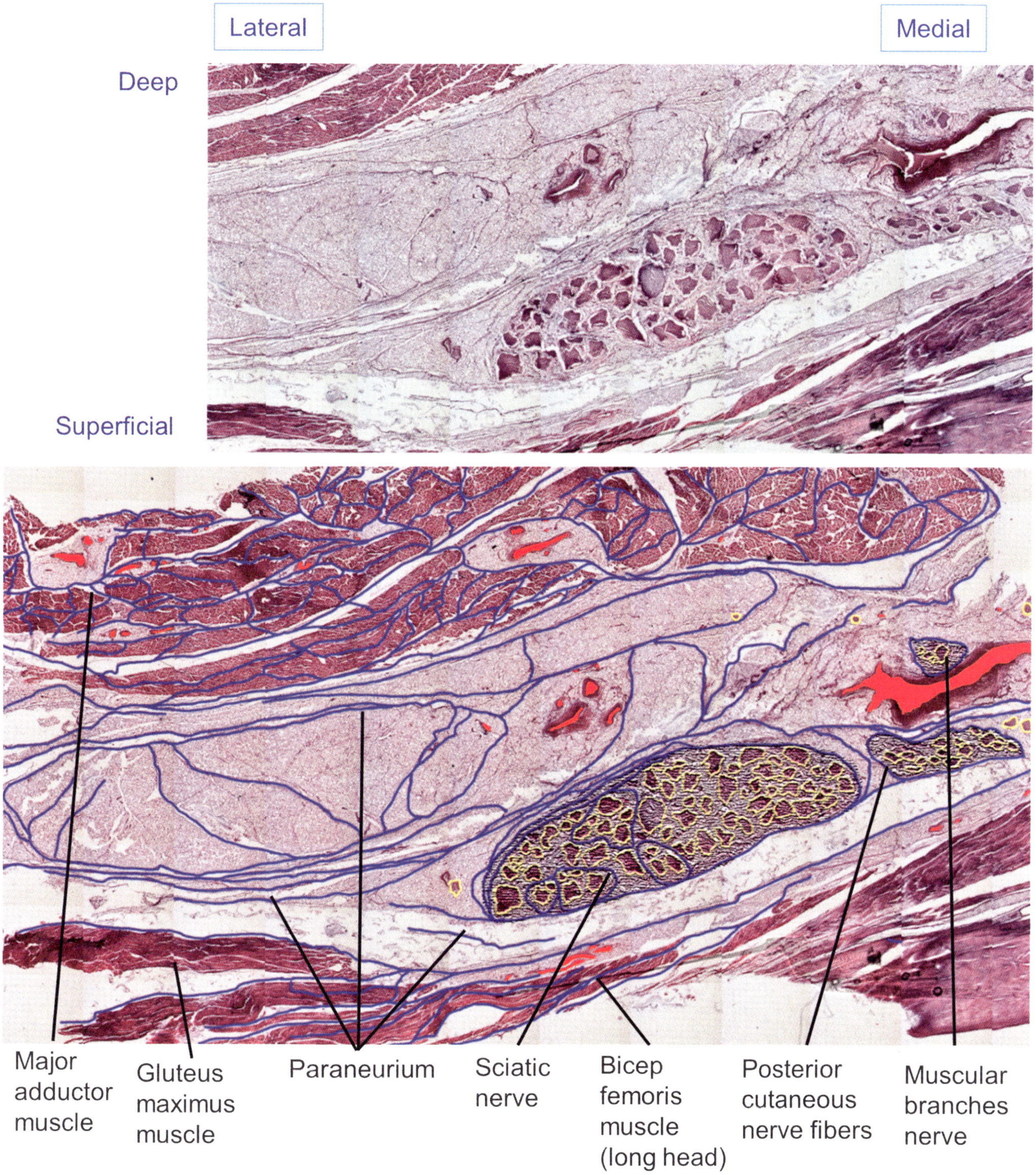

Fig. 11.2 Cross-section of human sciatic nerve and its surrounding tissue. Subgluteal region. Hematoxylin-eosin stained

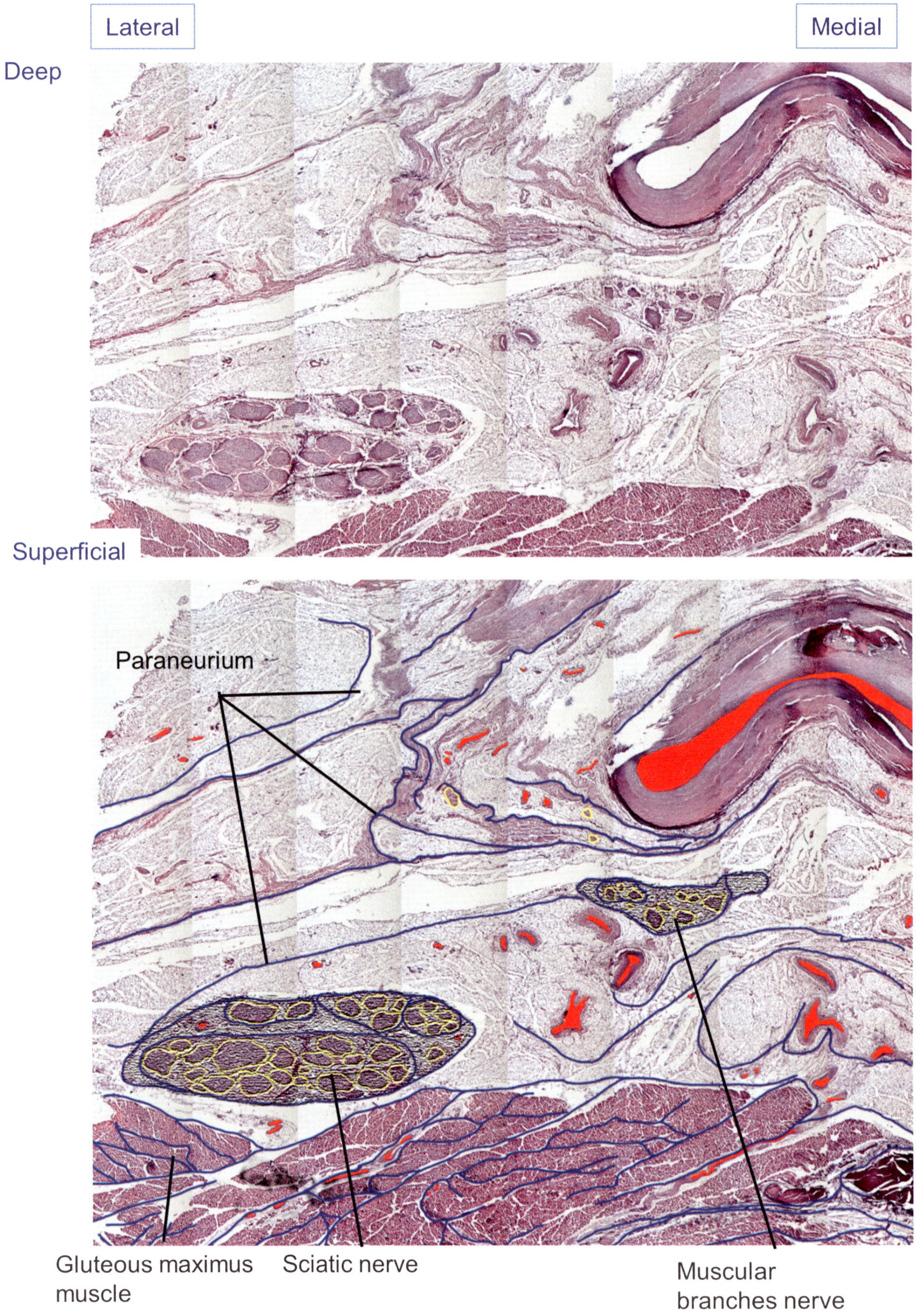

Fig. 11.3 Cross-section of human sciatic nerve and its surrounding tissue. Subgluteal region. Hematoxylin-eosin stained

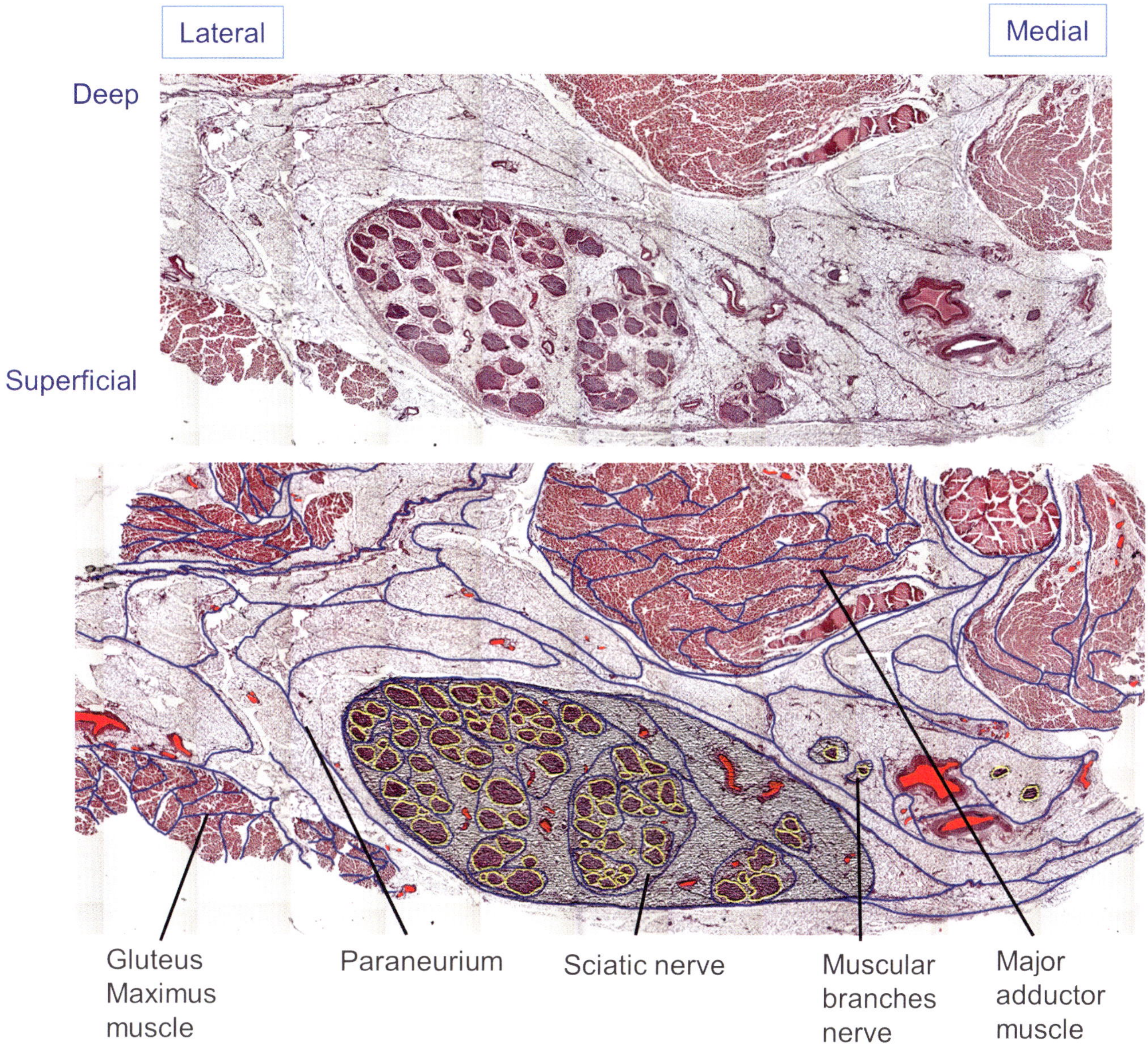

Fig. 11.4 Cross-section of human sciatic nerve and its surrounding tissue. Femoral region. Hematoxylin-eosin stained

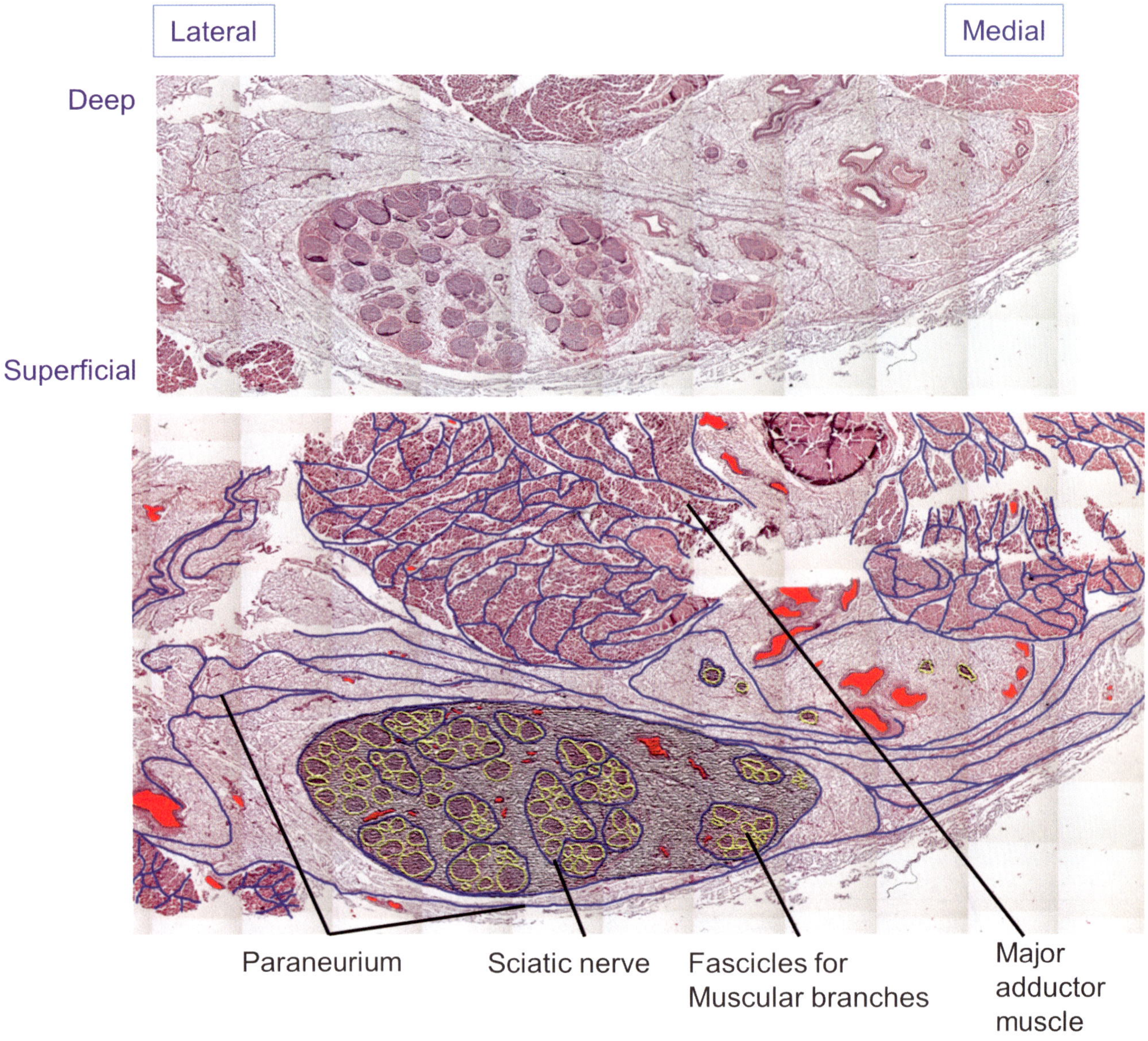

Fig. 11.5 Cross-section of human sciatic nerve and its surrounding tissue. Femoral region. Hematoxylin-eosin stained

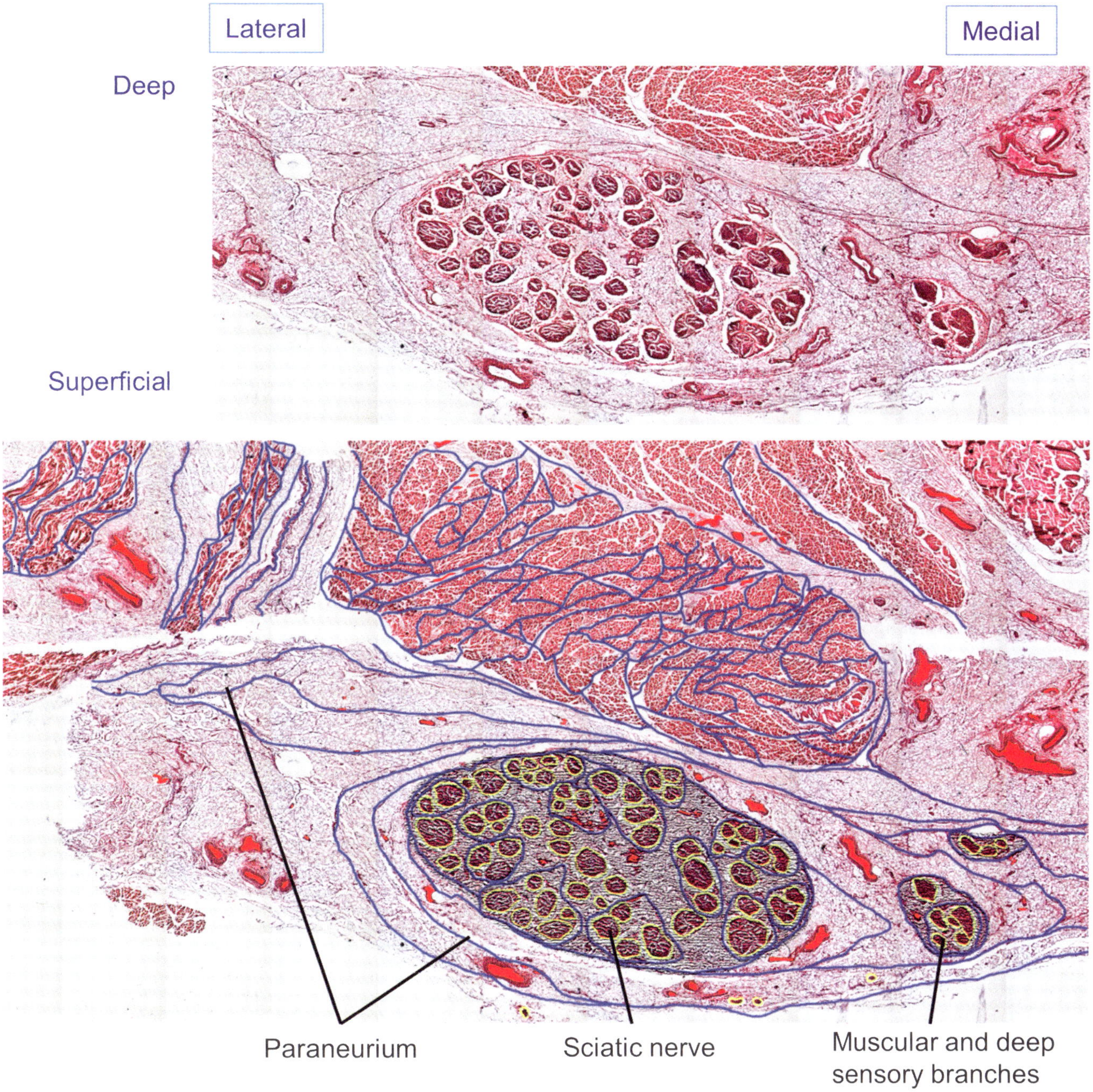

Fig. 11.6 Cross-section of human sciatic nerve and its surrounding tissue. Femoral region. Hematoxylin-eosin stained

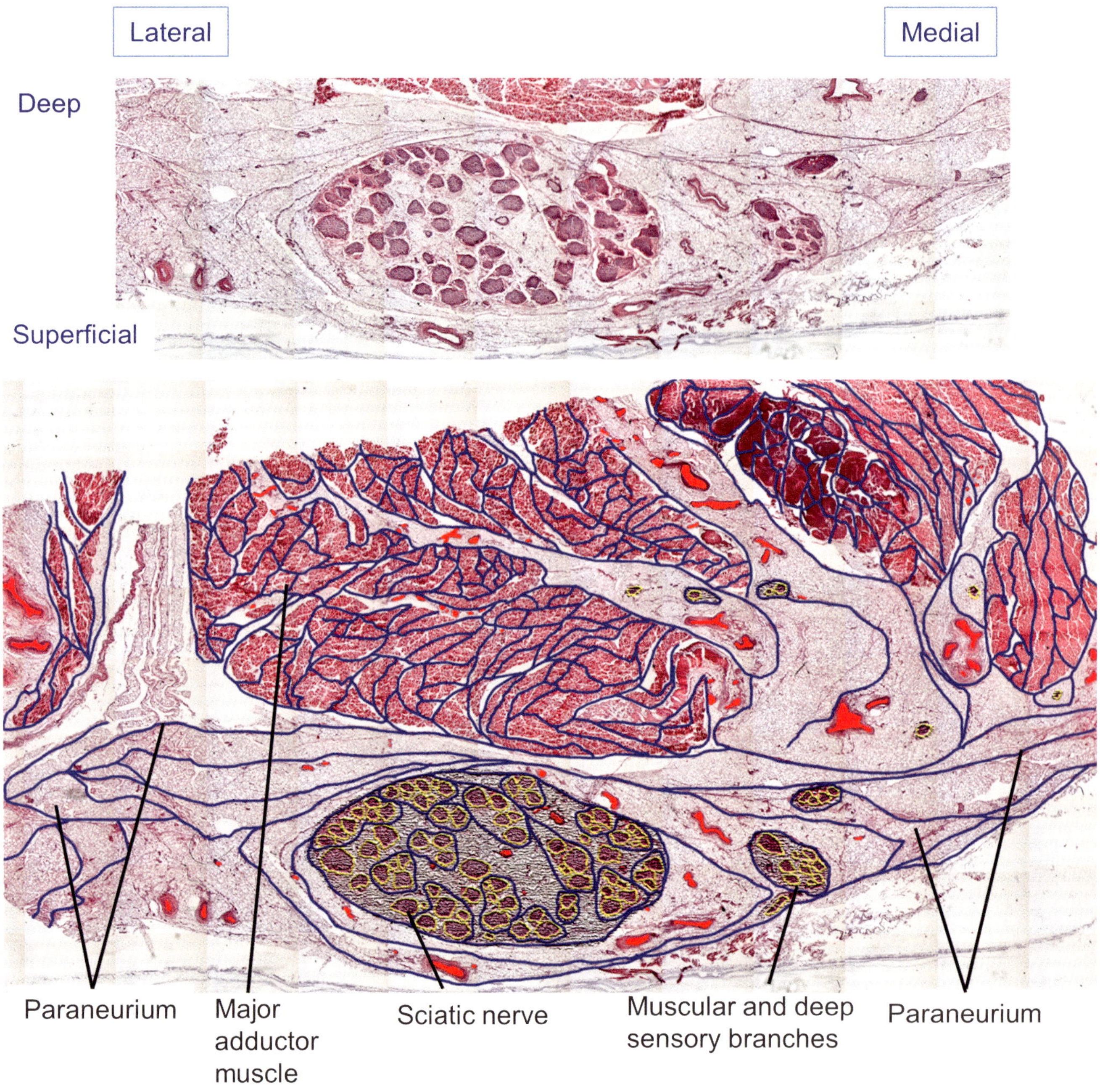

Fig. 11.7 Cross-section of human sciatic nerve and its surrounding tissue. Femoral region. Hematoxylin-eosin stained

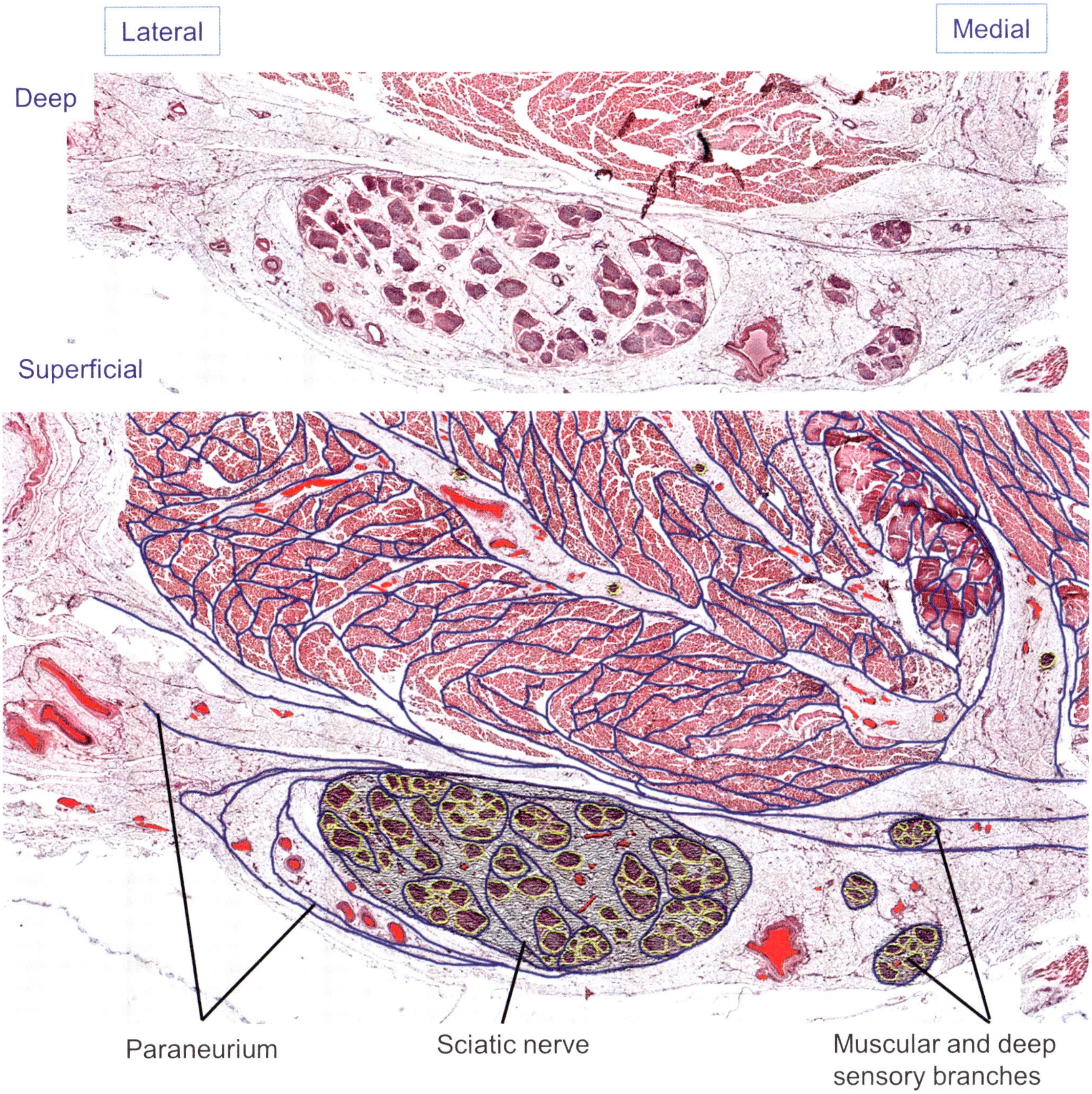

Fig. 11.8 Cross-section of human sciatic nerve and its surrounding tissue. Femoral region. Hematoxylin-eosin stained

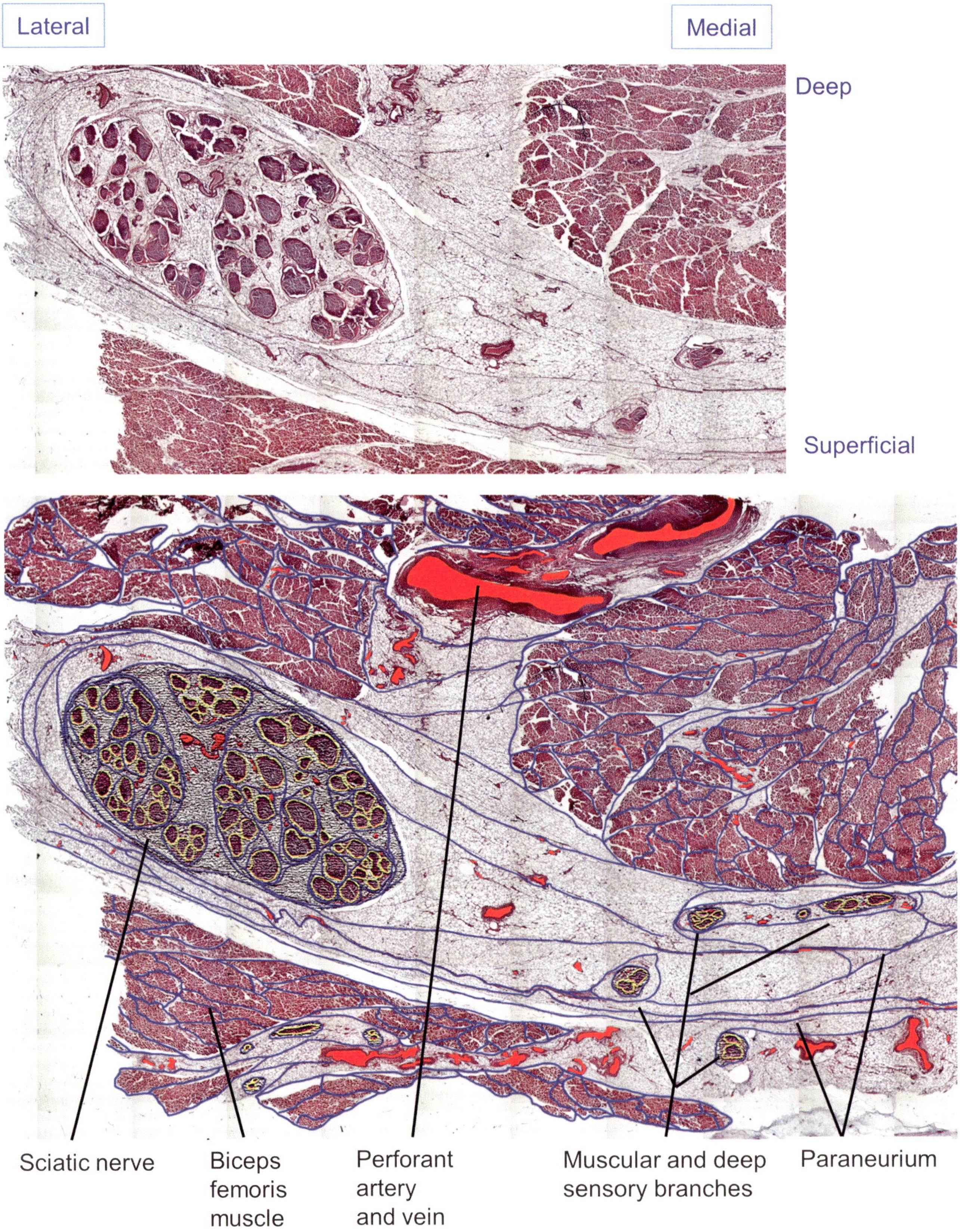

Fig. 11.9 Cross-section of human sciatic nerve and its surrounding tissue. Femoral region. Hematoxylin-eosin stained

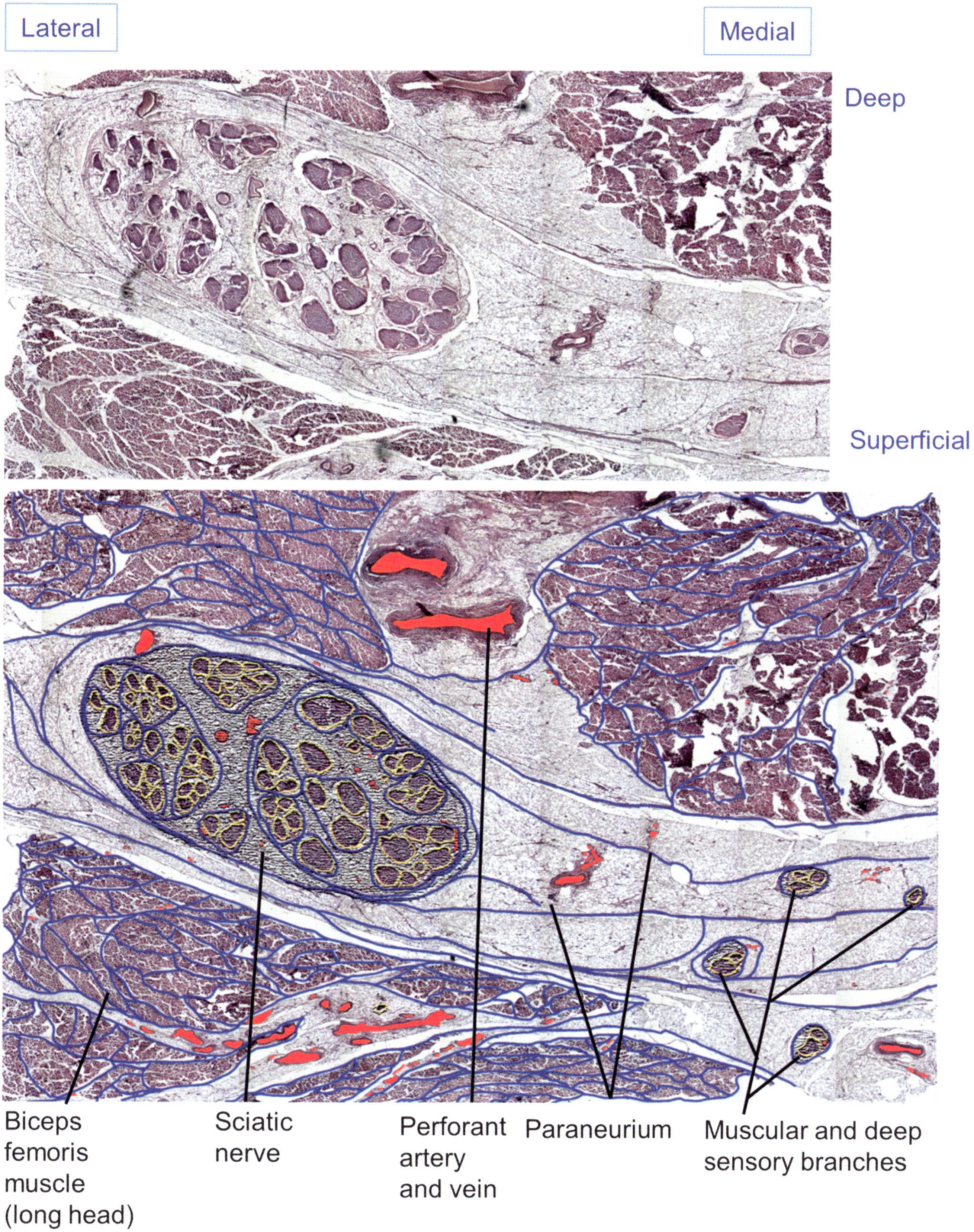

Fig. 11.10 Cross-section of human sciatic nerve and its surrounding tissue. Femoral region. Hematoxylin-eosin stained

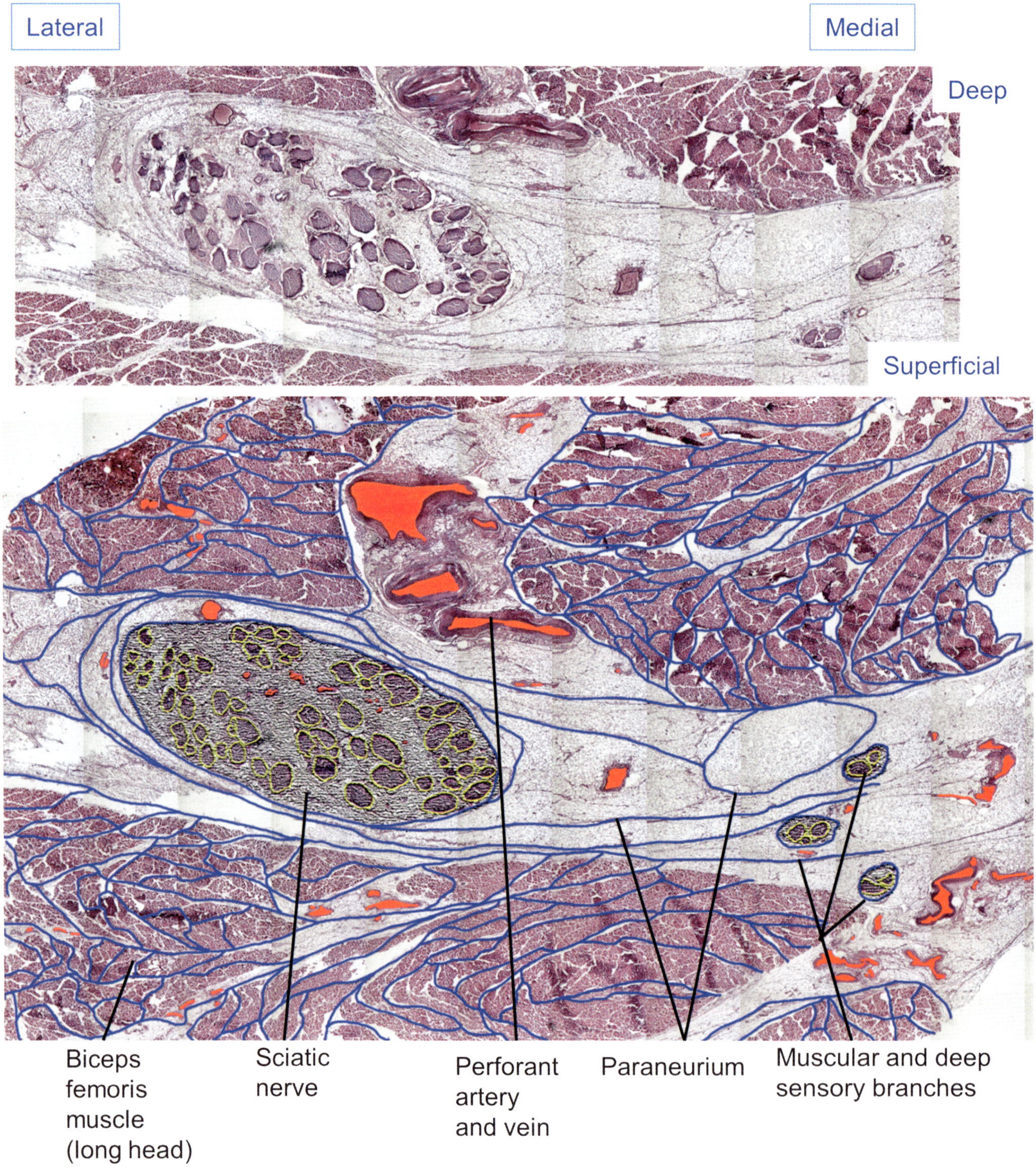

Fig. 11.11 Cross-section of human sciatic nerve and its surrounding tissue. Femoral region. Hematoxylin-eosin stained

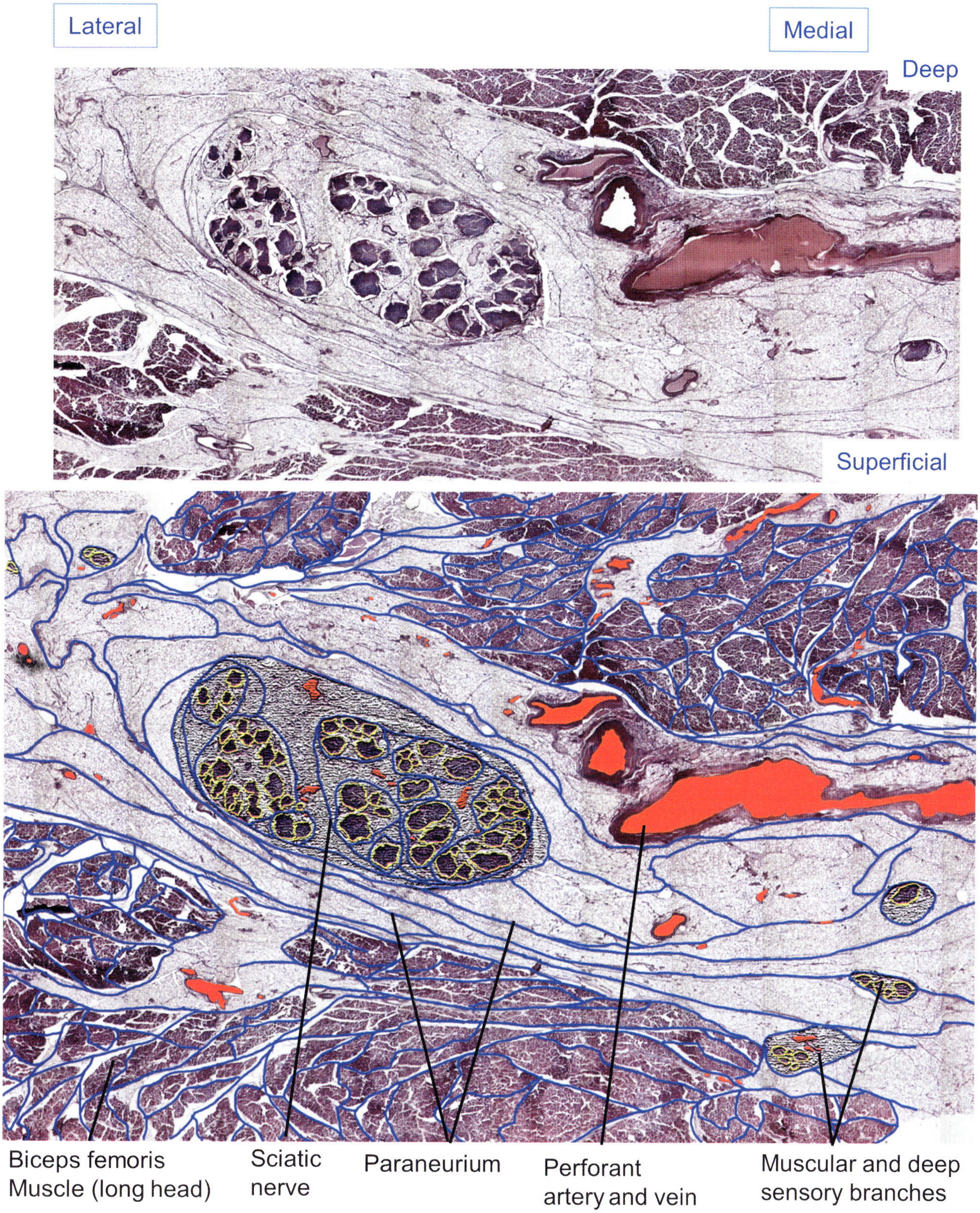

Fig. 11.12 Cross-section of human sciatic nerve and its surrounding tissue. Femoral region. Hematoxylin-eosin stained

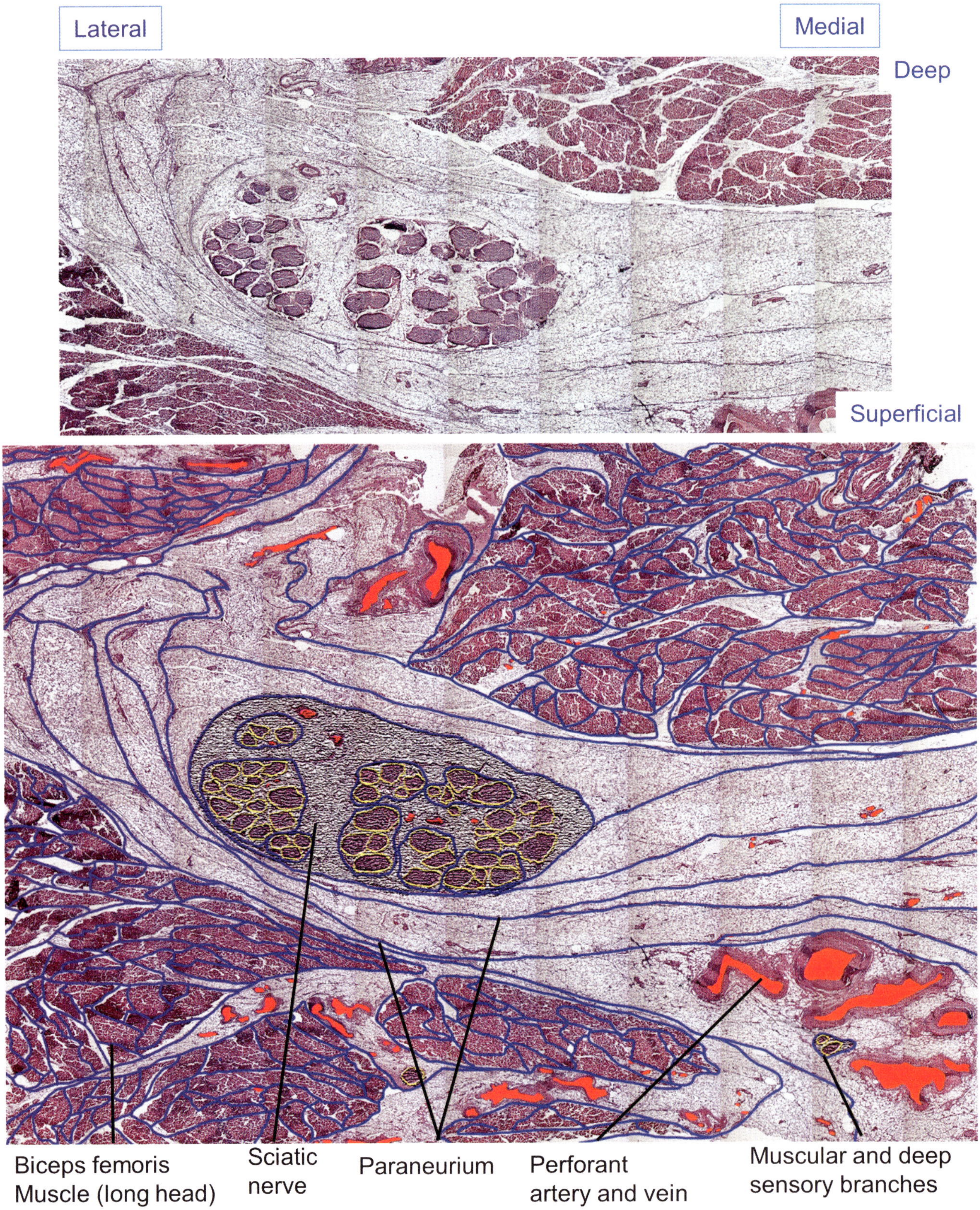

Fig. 11.13 Cross-section of human sciatic nerve and its surrounding tissue. Femoral region. Hematoxylin-eosin stained

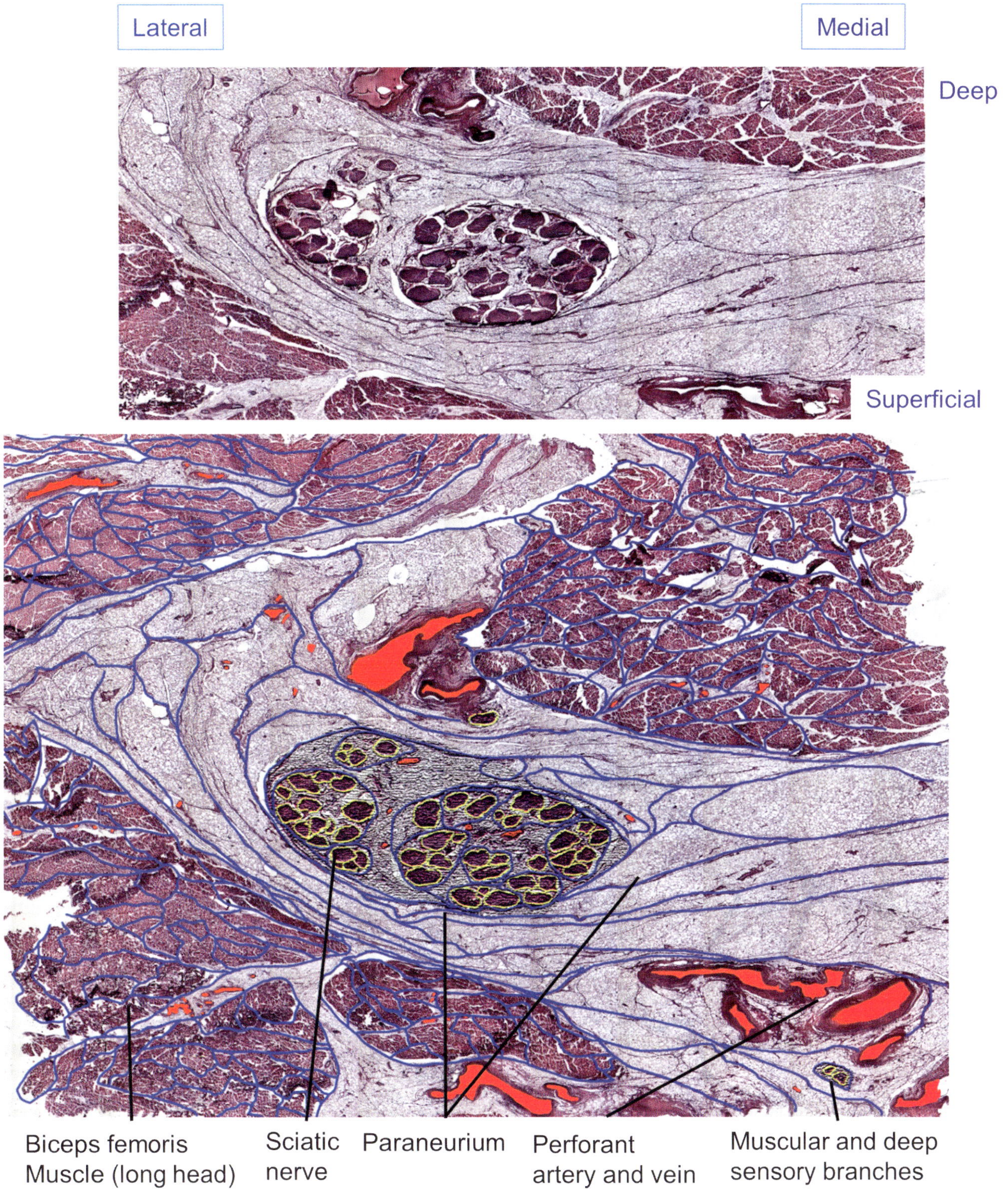

Fig. 11.14 Cross-section of human sciatic nerve and its surrounding tissue. Femoral region. Hematoxylin-eosin stained

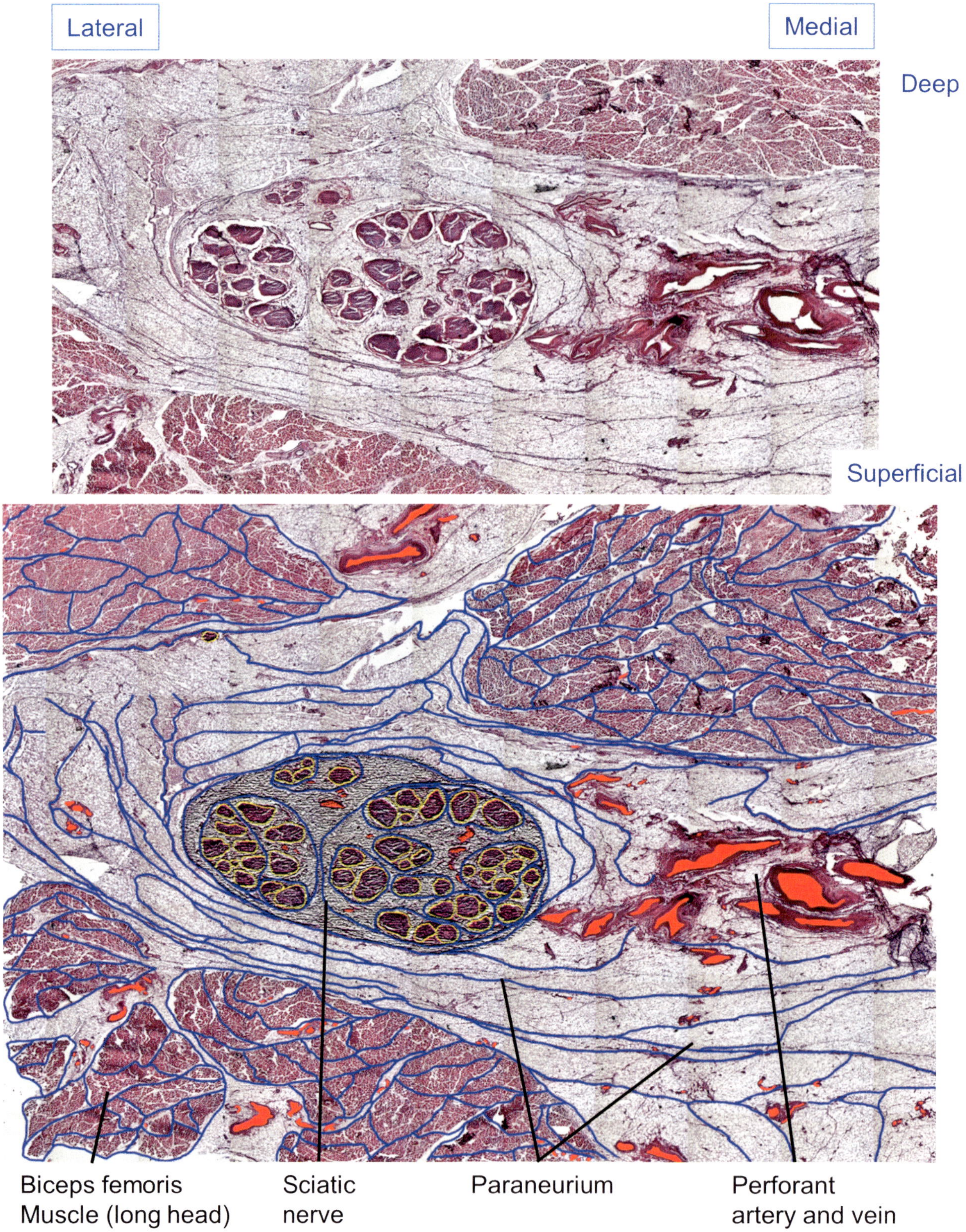

Fig. 11.15 Cross-section of human sciatic nerve and its surrounding tissue. Femoral region. Hematoxylin-eosin stained

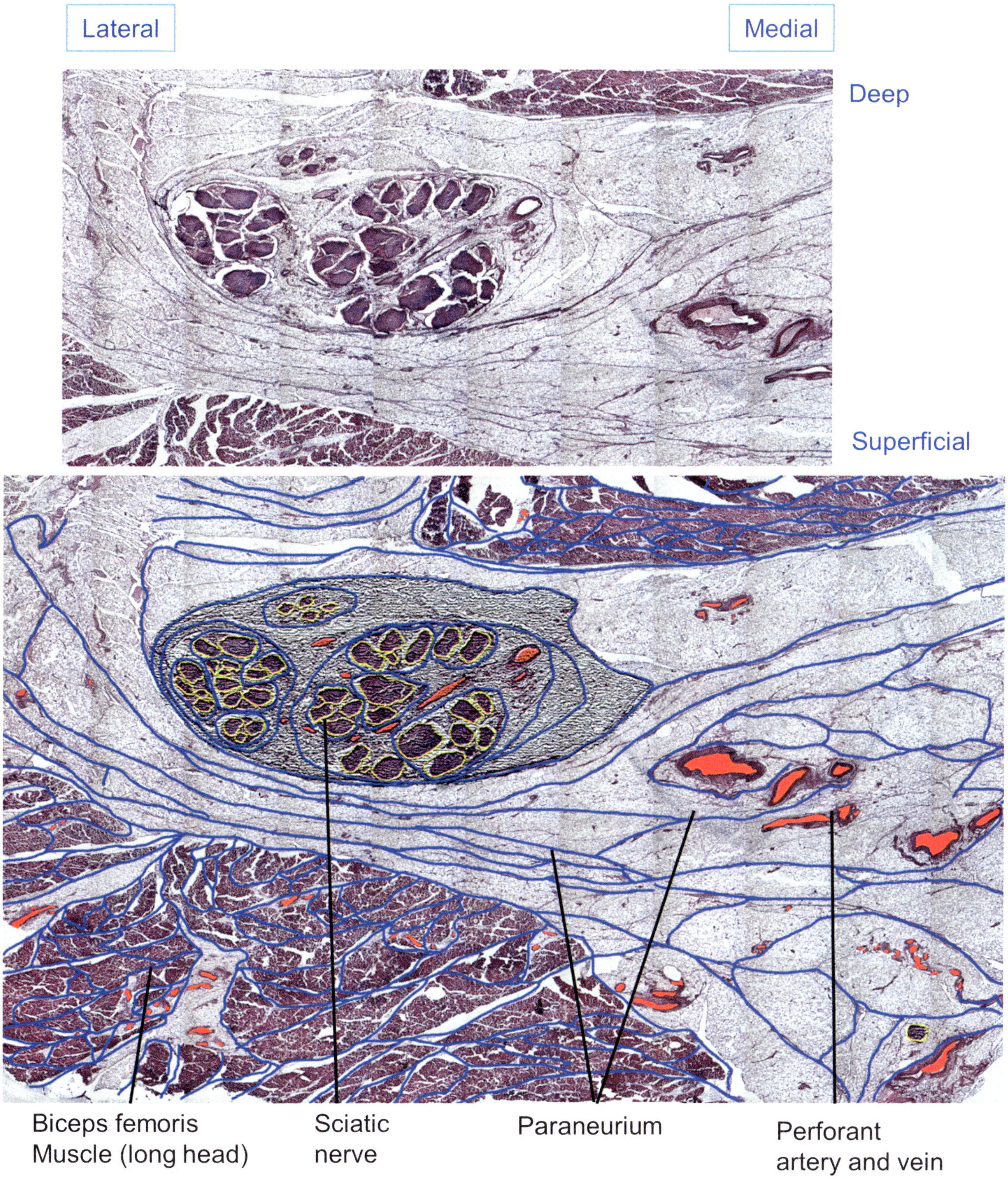

Fig. 11.16 Cross-section of human sciatic nerve and its surrounding tissue. Femoral region. Hematoxylin-eosin stained

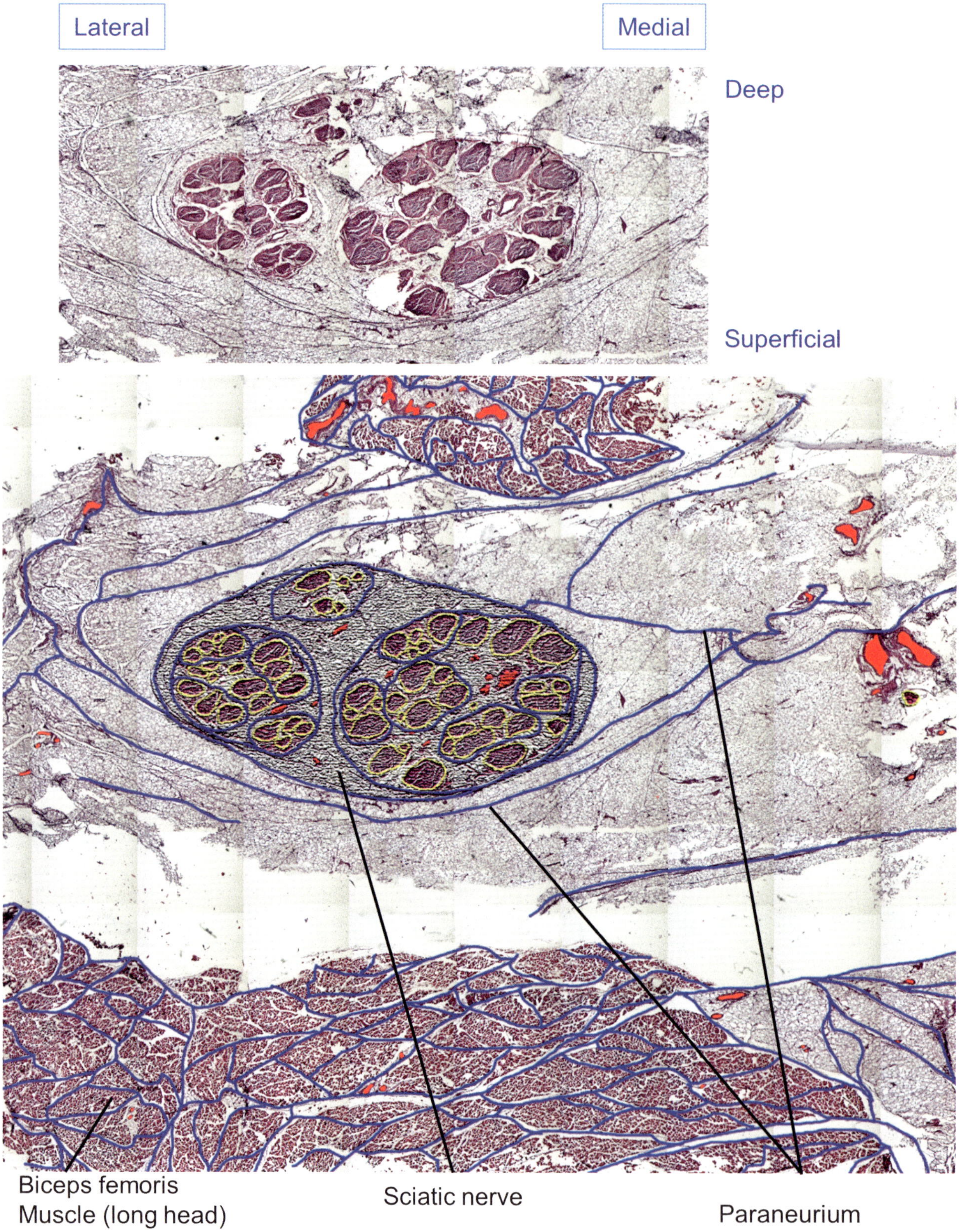

Fig. 11.17 Cross-section of human sciatic nerve and its surrounding tissue. Femoral region. Hematoxylin-eosin stained

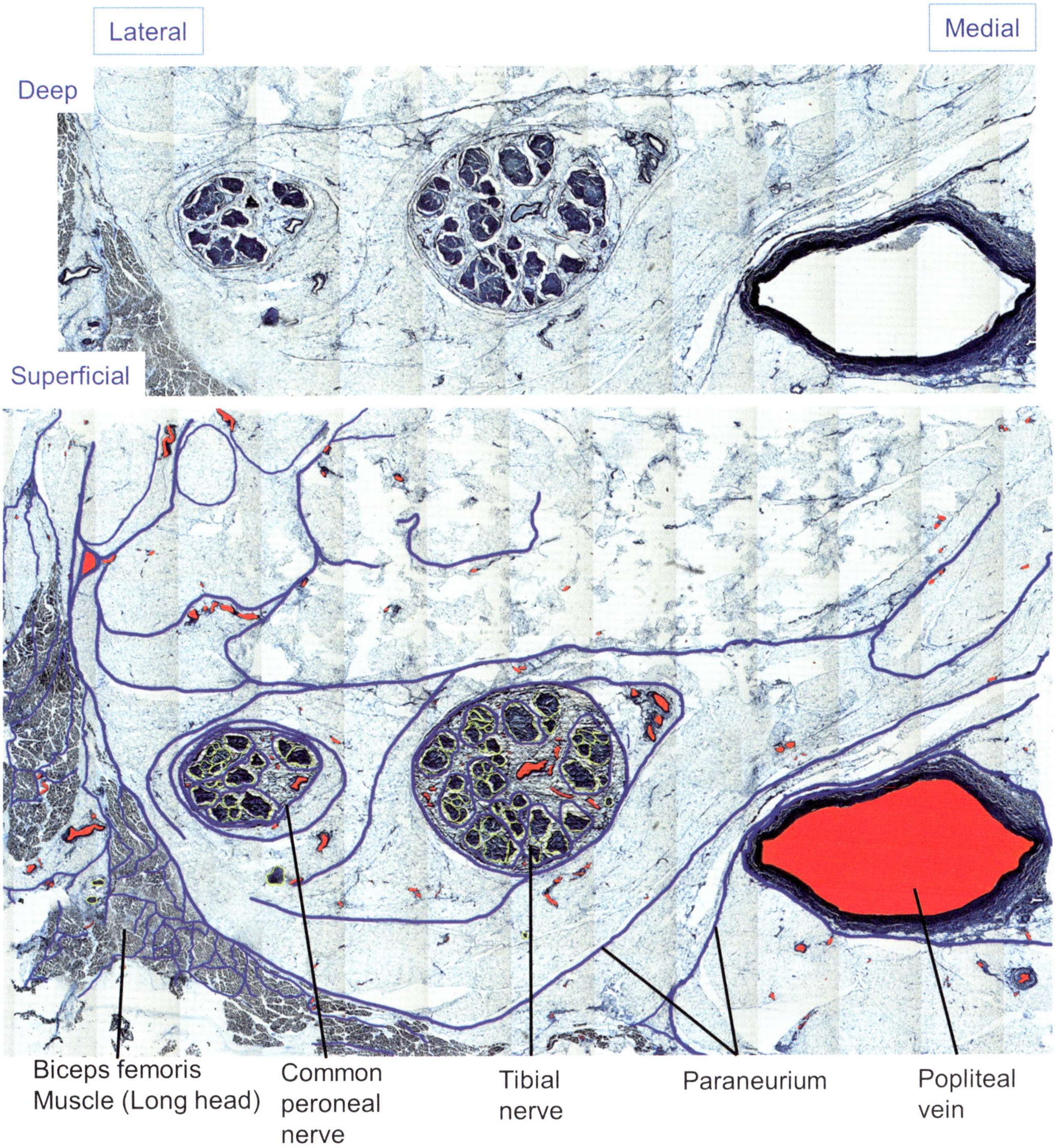

Fig. 11.18 Cross-section of human sciatic nerve and its surrounding tissue. Popliteal region. Hematoxylin-eosin stained

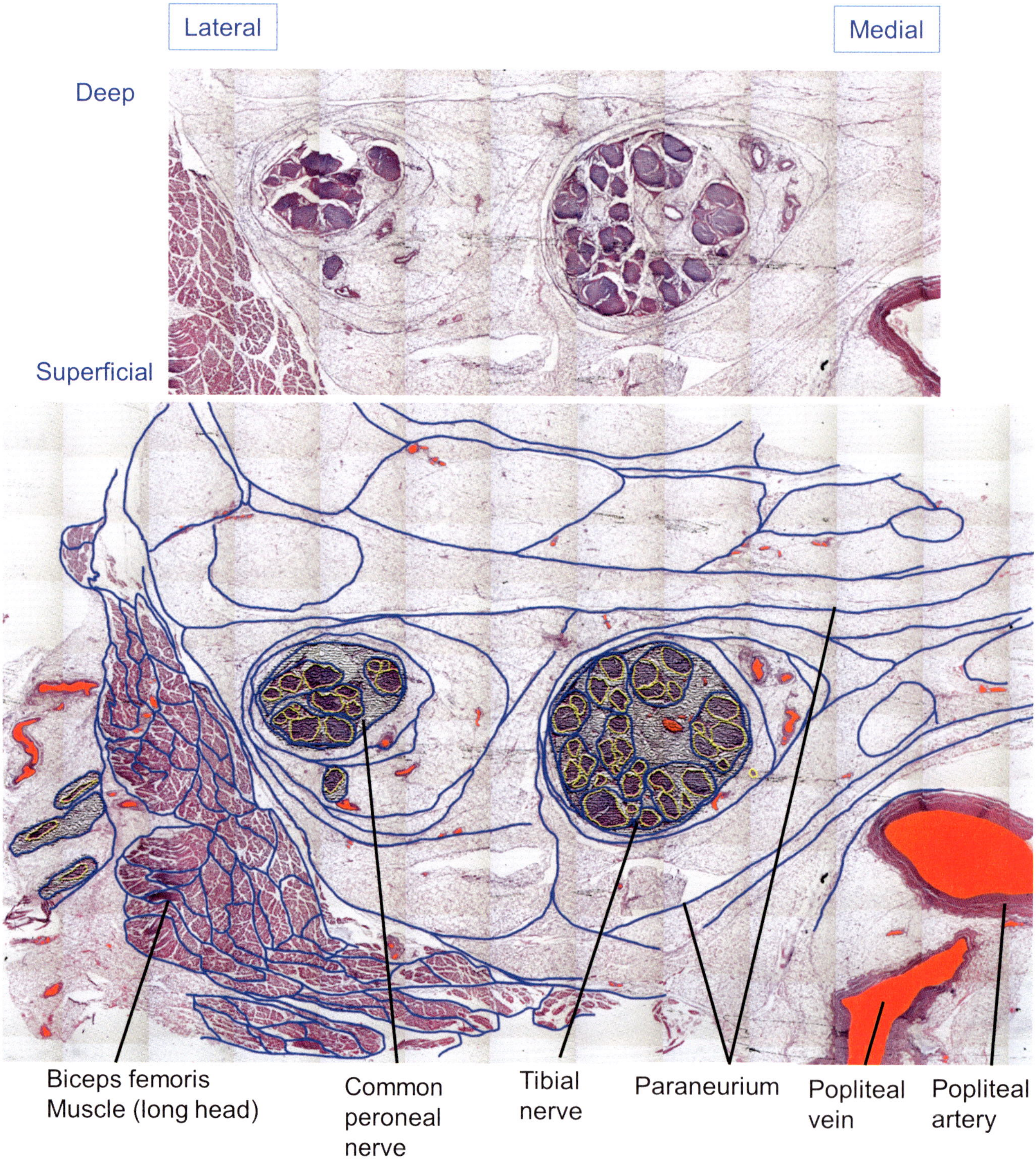

Fig. 11.19 Cross-section of human sciatic nerve and its surrounding tissue. Popliteal region. Hematoxylin-eosin stained

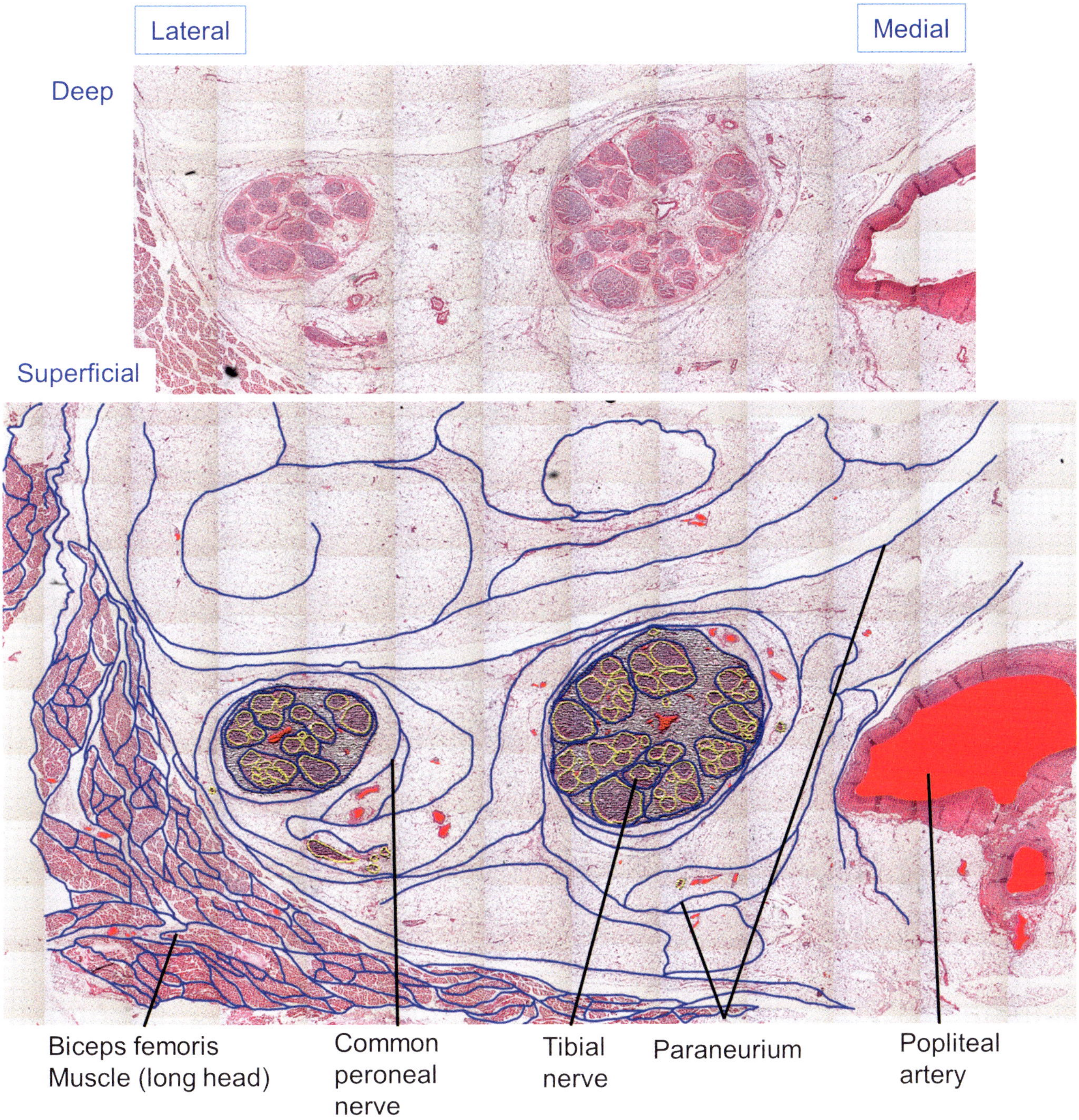

Fig. 11.20 Cross-section of human sciatic nerve and its surrounding tissue. Popliteal region. Hematoxylin-eosin stained

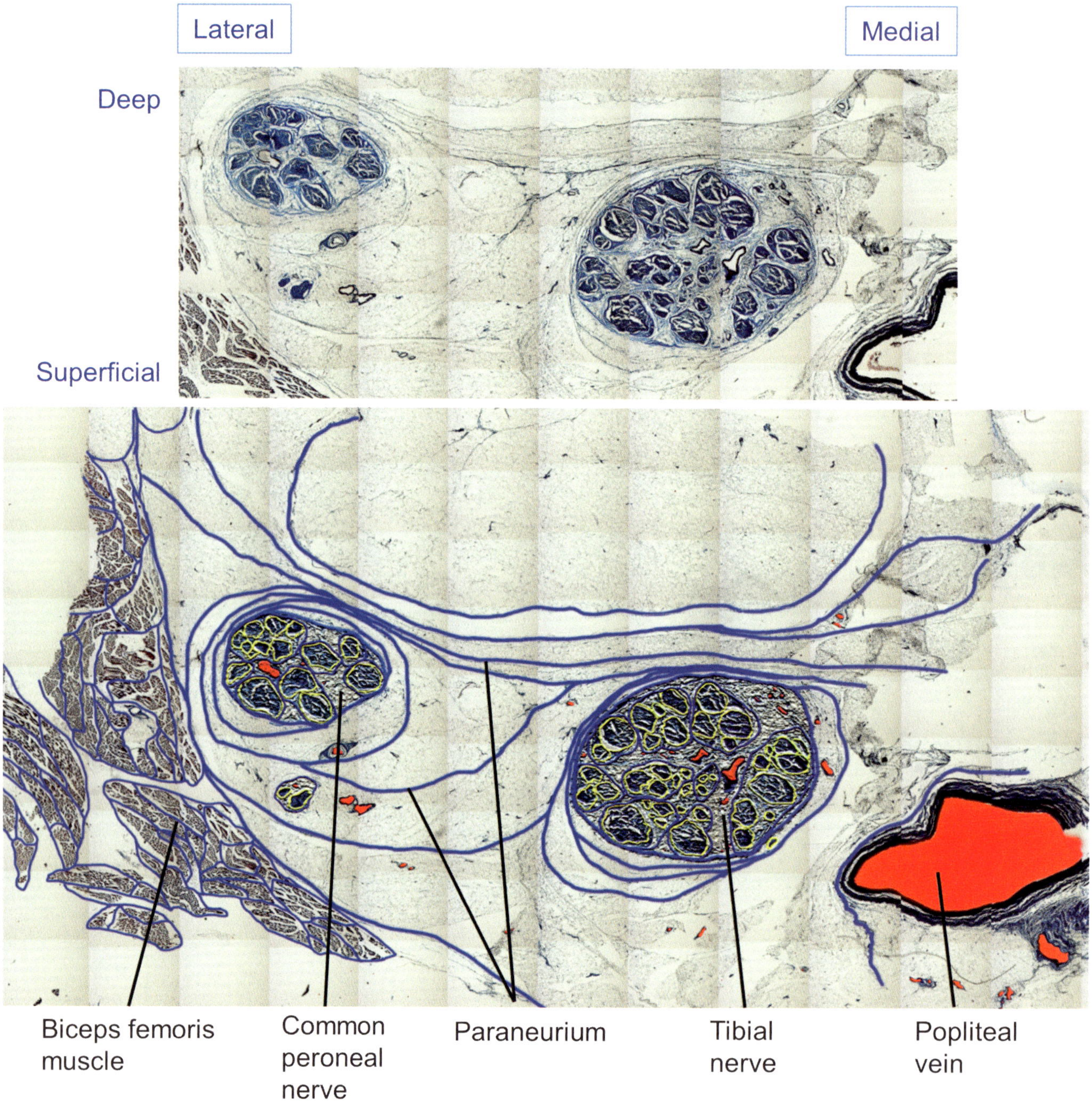

Fig. 11.21 Cross-section of human sciatic nerve and its surrounding tissue. Popliteal region. Hematoxylin-eosin stained

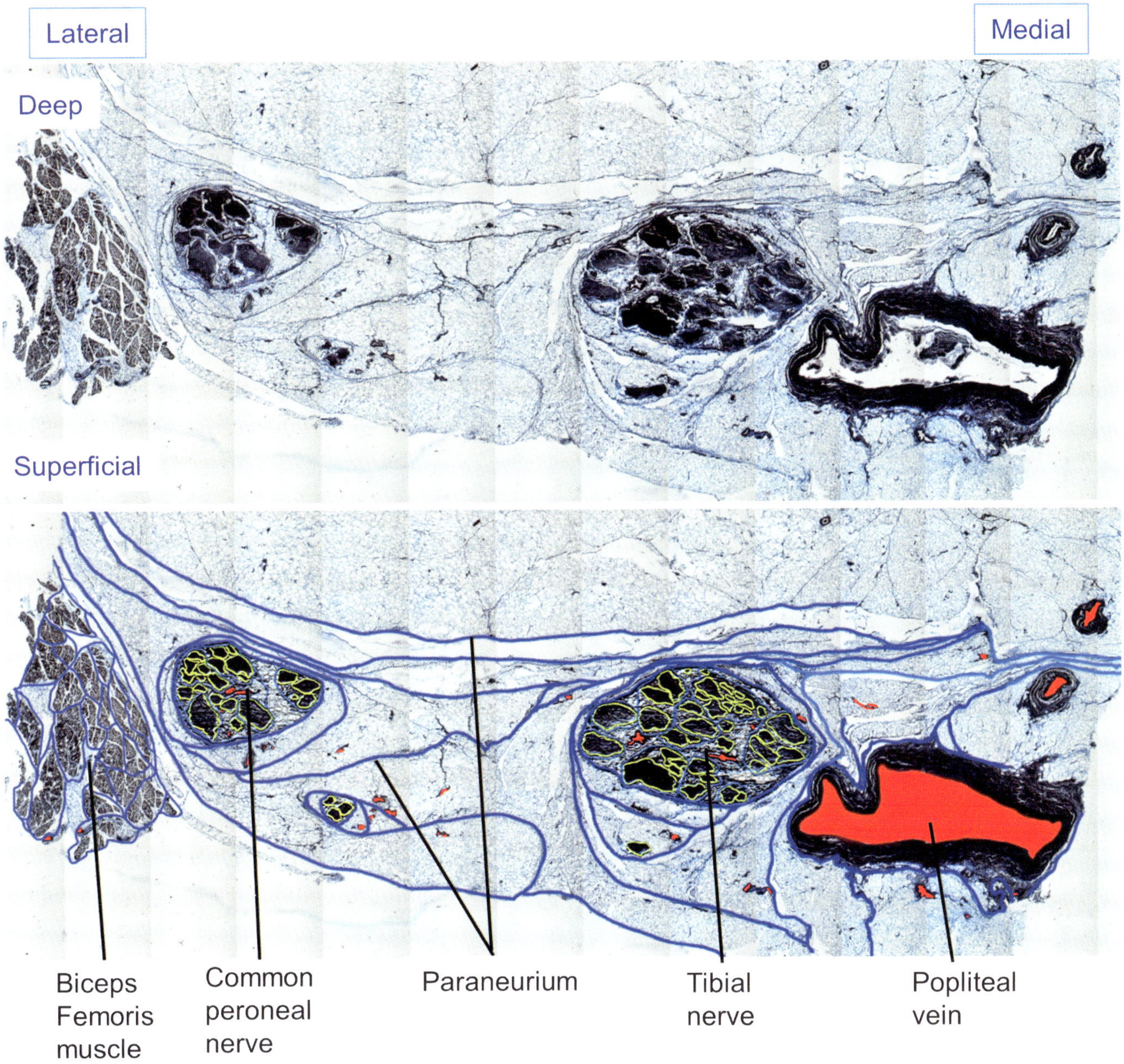

Fig. 11.22 Cross-section of human sciatic nerve and its surrounding tissue. Popliteal region. Hematoxylin-eosin stained

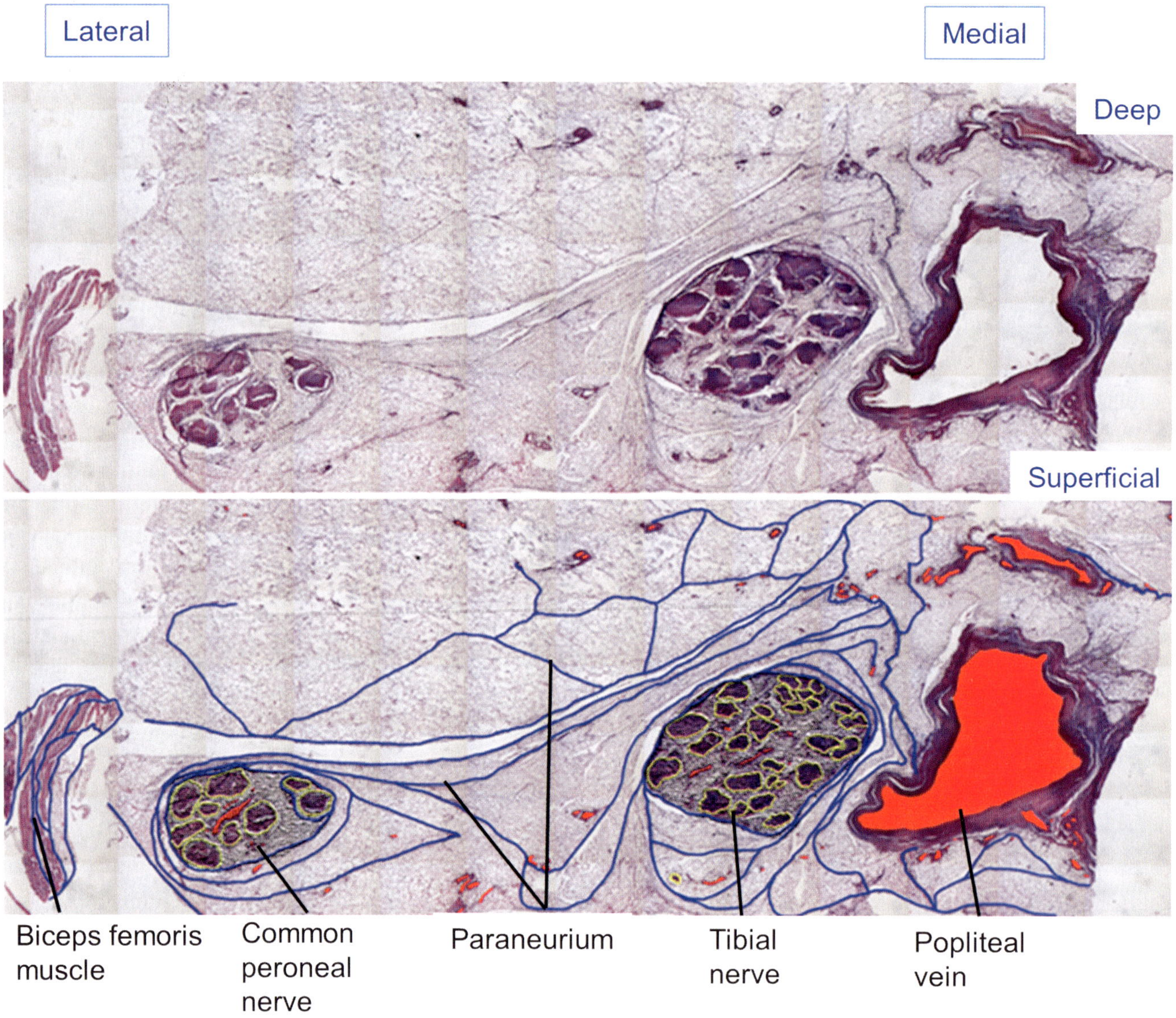

Fig. 11.23 Cross-section of human sciatic nerve and its surrounding tissue. Popliteal region. Hematoxylin-eosin stained

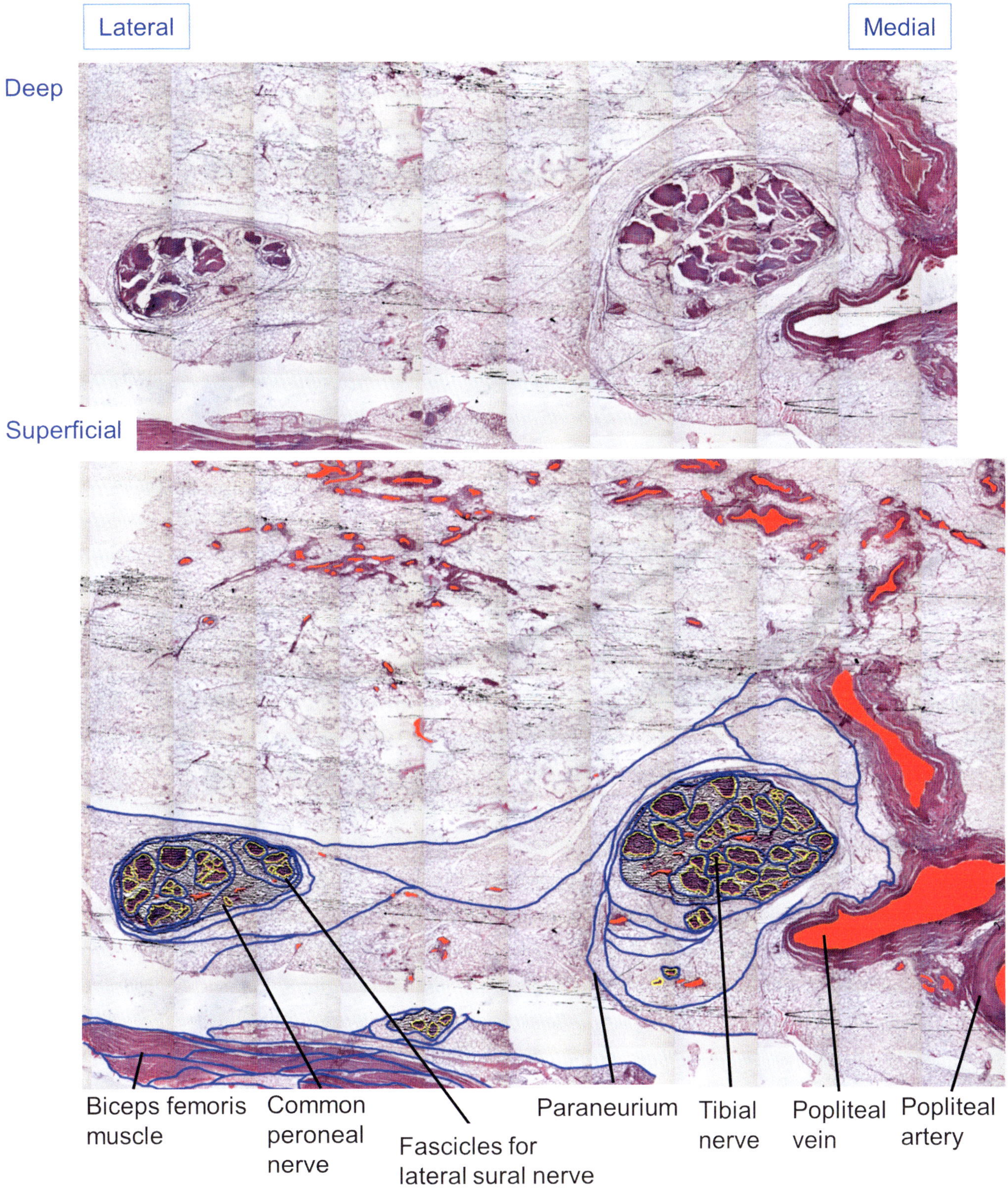

Fig. 11.24 Cross-section of human sciatic nerve and its surrounding tissue. Popliteal region. Hematoxylin-eosin stained

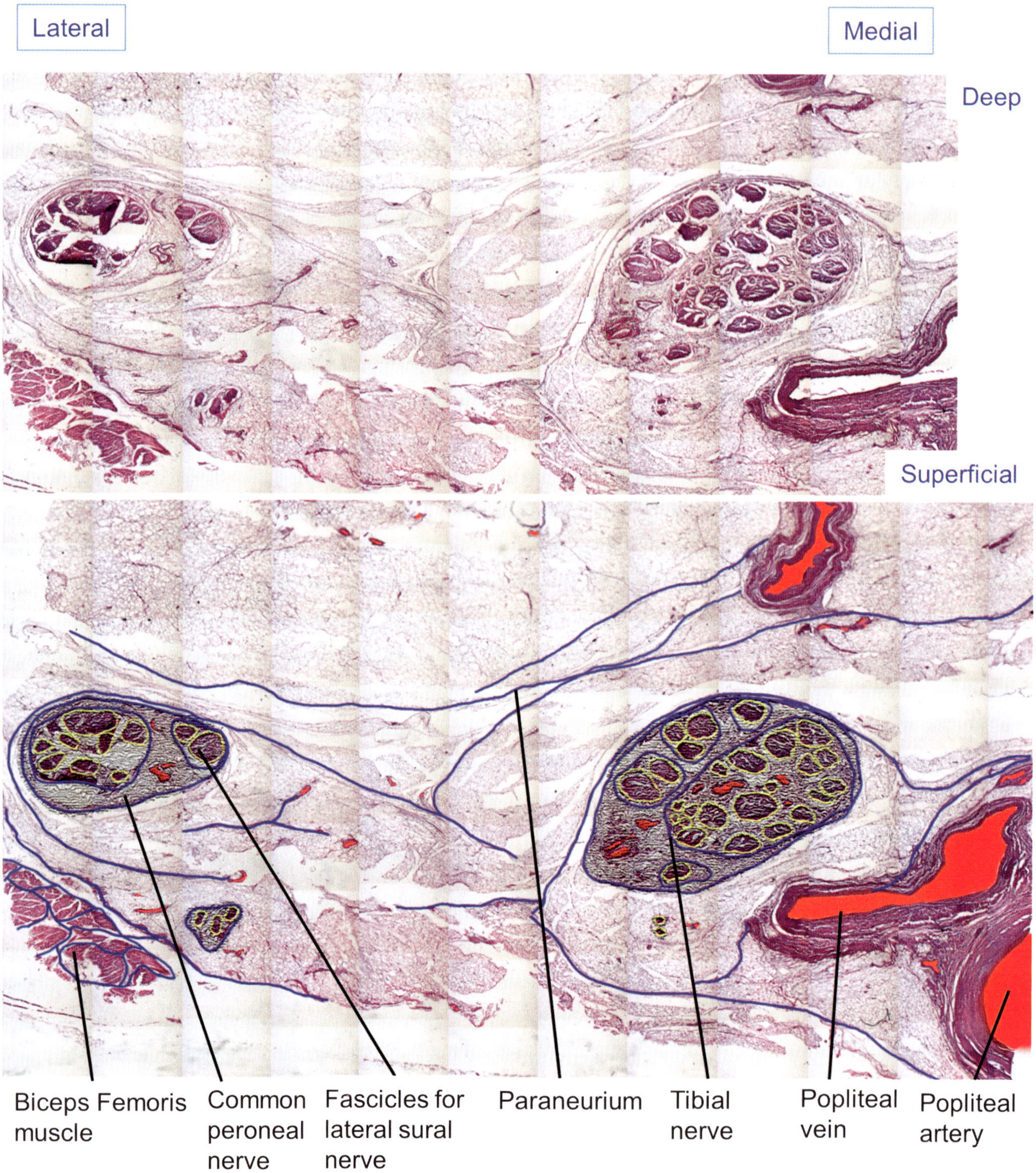

Fig. 11.25 Cross-section of human sciatic nerve and its surrounding tissue. Popliteal region. Hematoxylin-eosin stained

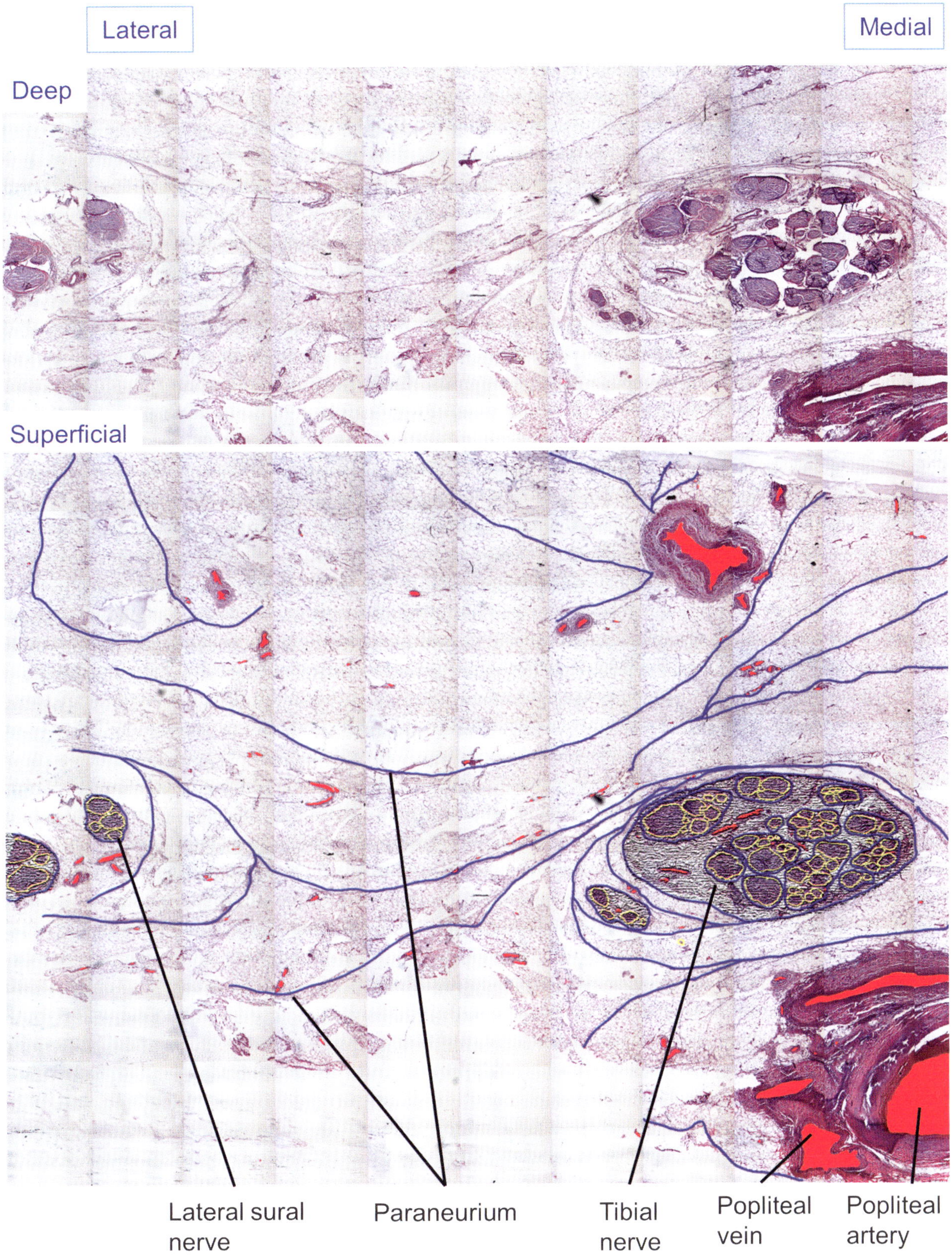

Fig. 11.26 Cross-section of human sciatic nerve and its surrounding tissue. Popliteal region. Hematoxylin-eosin stained

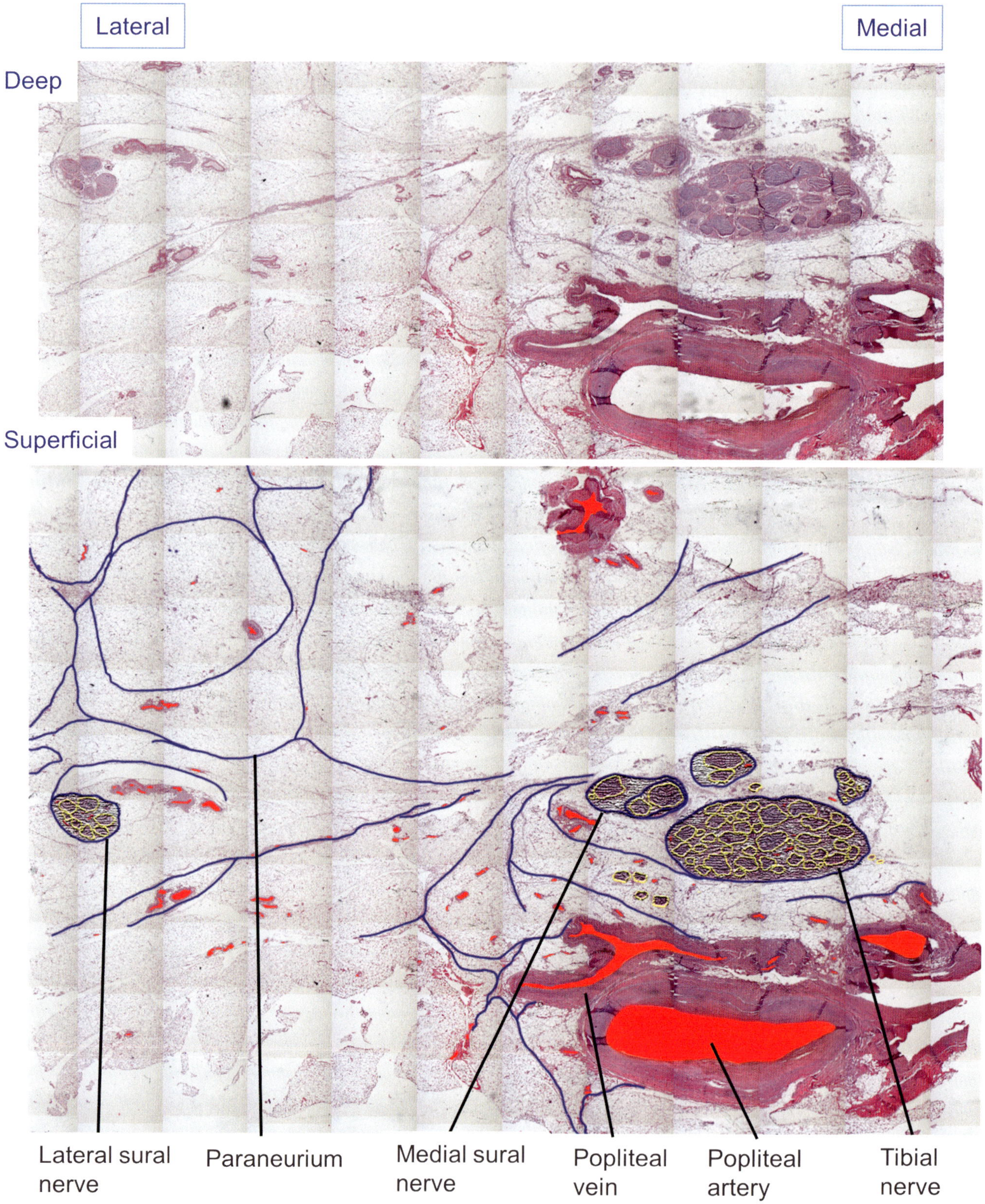

Fig. 11.27 Cross-section of human sciatic nerve and its surrounding tissue. Popliteal region. Hematoxylin-eosin stained

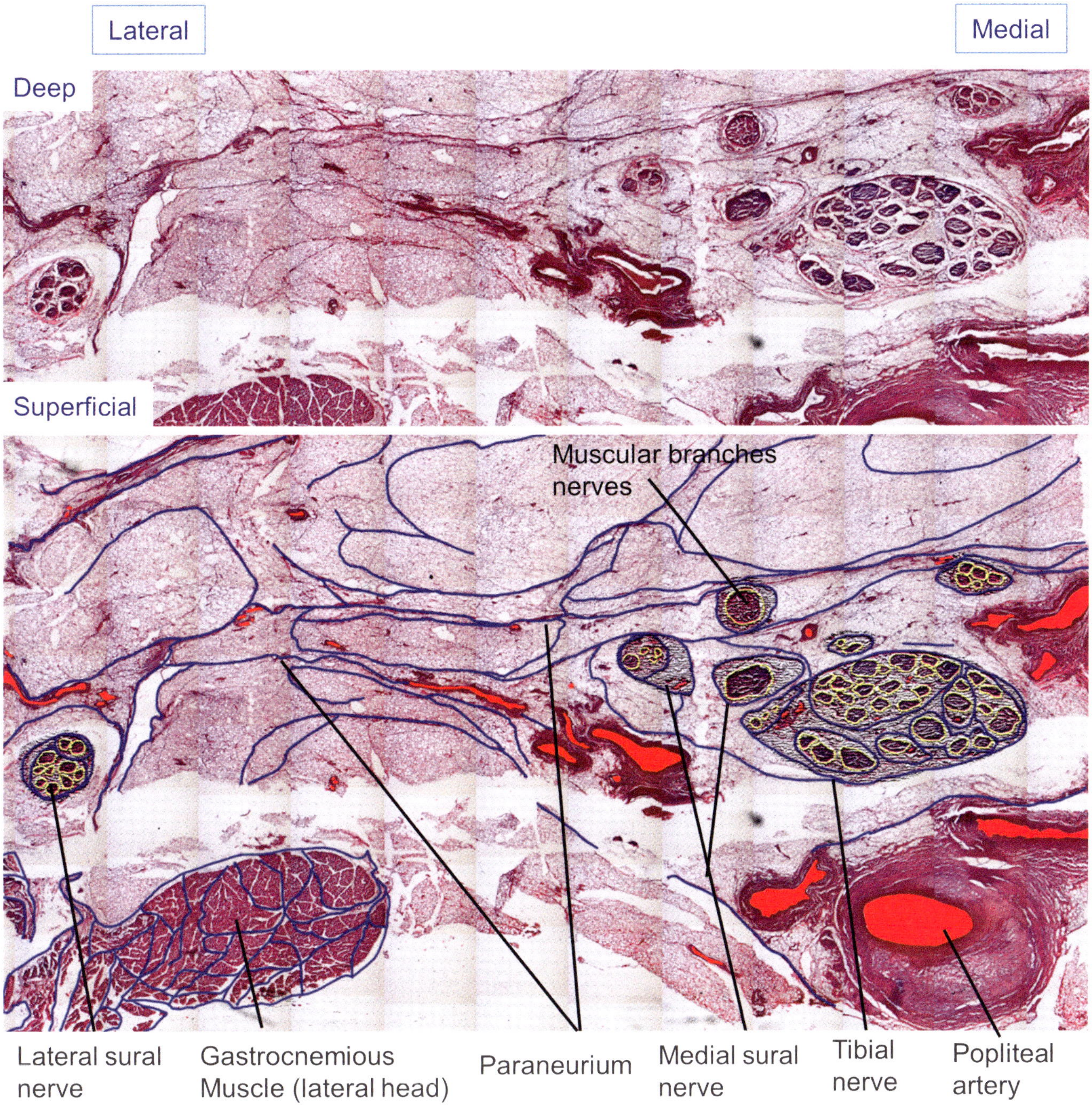

Fig. 11.28 Cross-section of human sciatic nerve and its surrounding tissue. Popliteal region. Hematoxylin-eosin stained

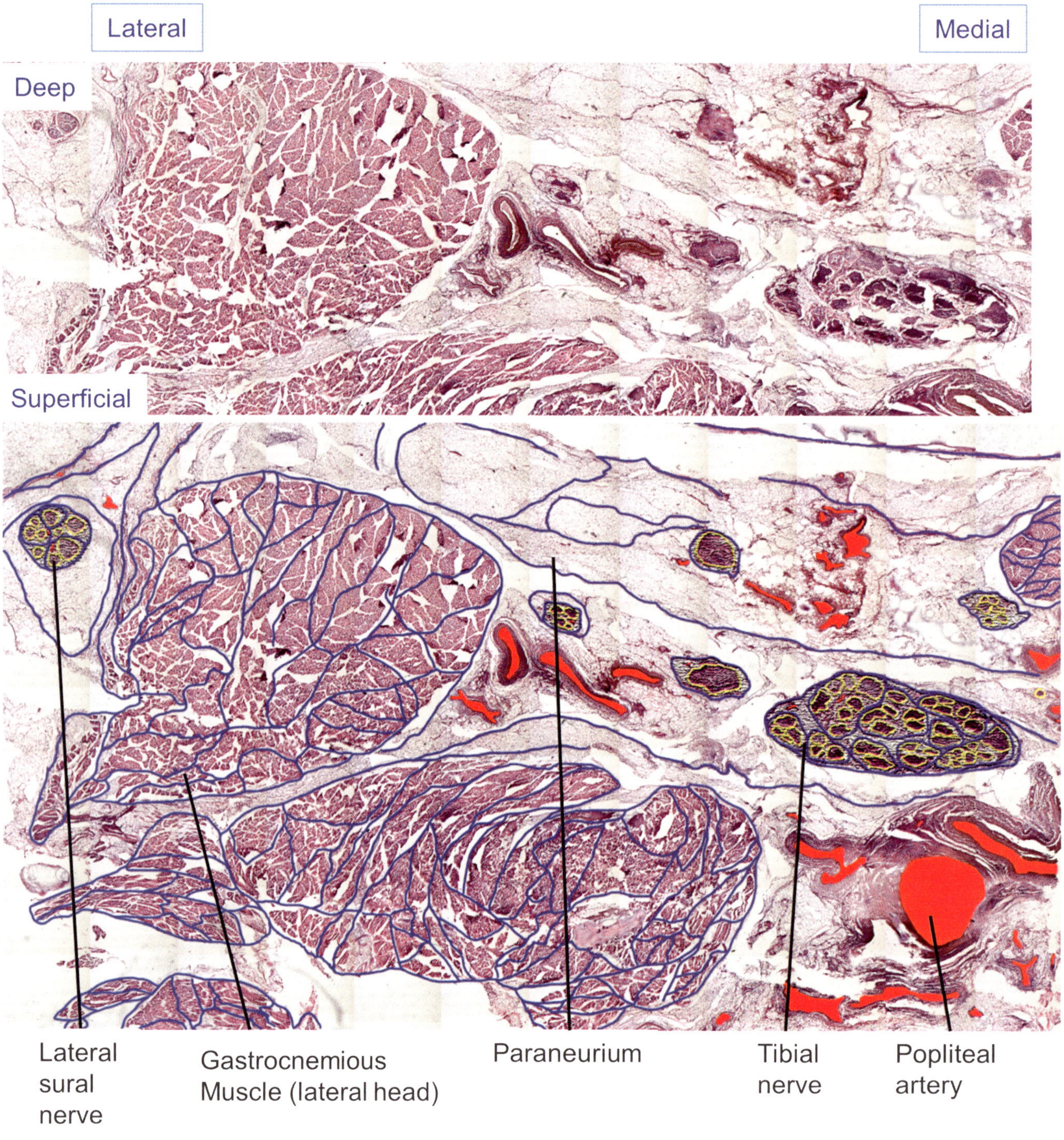

Fig. 11.29 Cross-section of human sciatic nerve and its surrounding tissue. Popliteal region. Hematoxylin-eosin stained

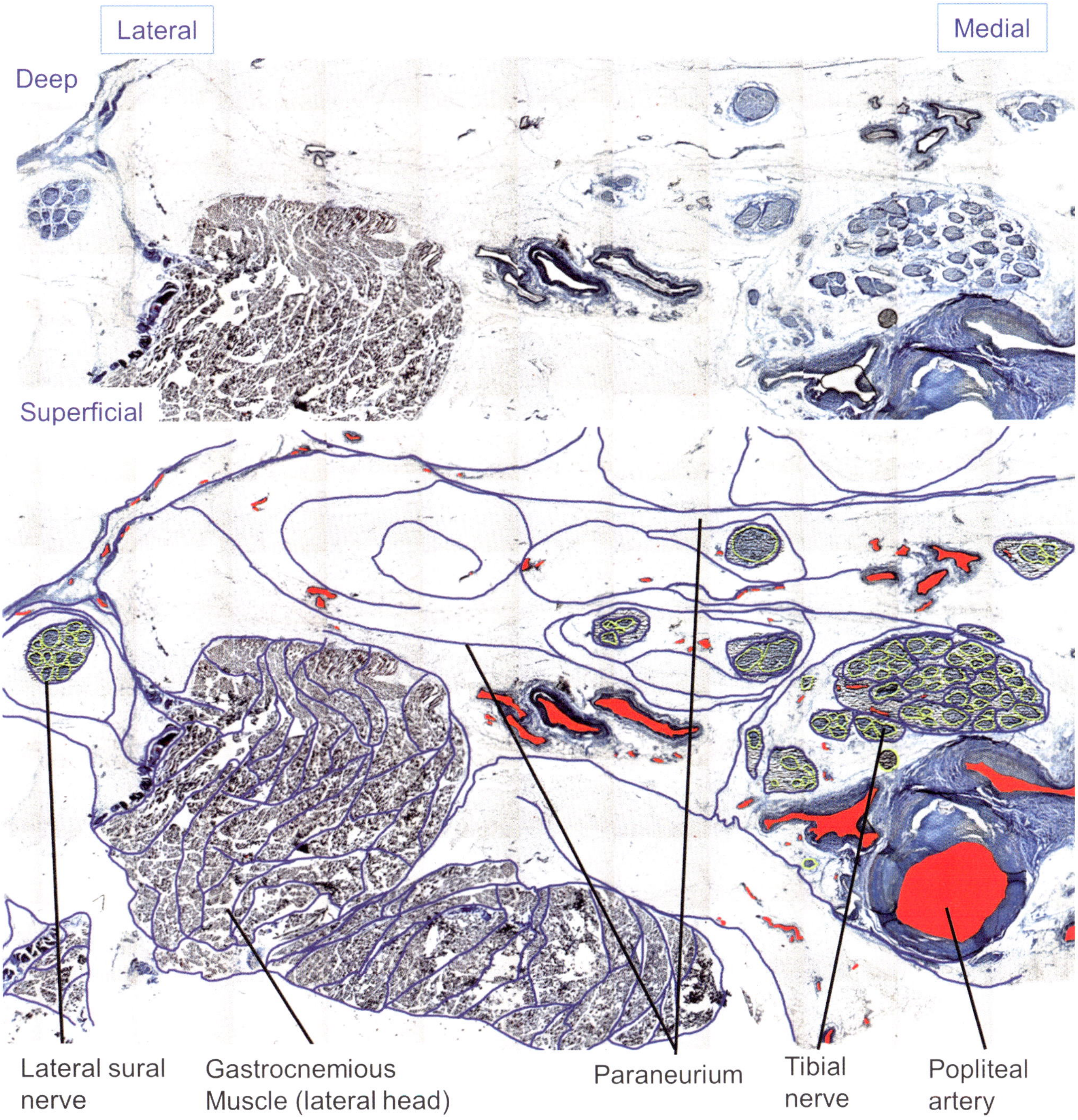

Fig. 11.30 Cross-section of human sciatic nerve and its surrounding tissue. Popliteal region. Hematoxylin-eosin stained

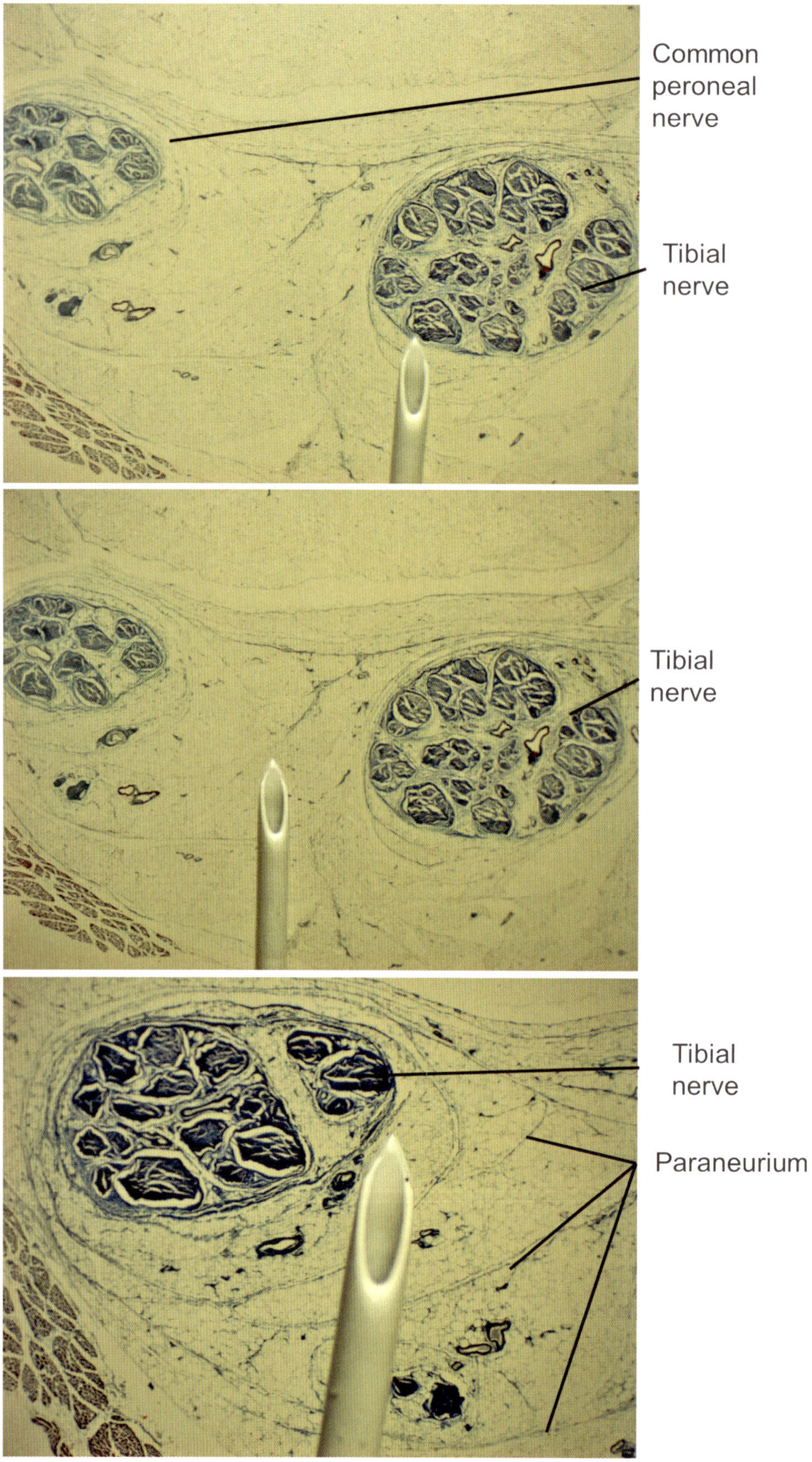

Fig. 11.31 Detail of tip of the needle inside paraneural sheaths enveloping the tibial nerve and common peroneal nerve. Masson trichrome stained

References

1. Reina MA, Arriazu R, Collier CB, Sala-Blanch X, Izquierdo L, de Andrés J. Electron microscopy of human peripheral nerves of clinical relevance to the practice of nerve blocks. A structural and ultrastructural review based on original experimental and laboratory data. Rev Esp Anestesiol Reanim. 2013;60:552–62.
2. Sala-Blanch X, Reina MA, Ribalta T, Prats-Galino A. Sciatic nerve structure and nomenclature: epineurium to paraneurium: Is this a new paradigm? Reg Anesth Pain Med. 2013;38:463–5.

12 Computerized Tomographic Images of Intraneural Injection

Jaume Pomés and Xavier Sala-Blanch

In the early years of regional anesthesia, the nerve block technique was closely related to the injection of cocaine into the nerve, this behavior being known as "intraneural injection." Thus, Victor Pauchet [1], in his book *L'Anesthesie Regionale*, defines what happens after an intended intraneural injection: "…A fusiform enlargement of the nerve ensues that disappears quickly. The injectate diffuses along both sides, that is why an intraneural injection can exit through the branches that leave the nerve close to the injection site…." These signs have recently been reconsidered as signs of intraneural injection.

In specialized anesthesiology literature, the definition of intraneural injection is not clear [2–6], even though the peripheral nerve is a clearly characterized structure, as explained in previous chapters. Intraneural injection must be defined as drug injection inside the epineurium sheath (interfascicular tissue). We can then divide intraneural injection into intrafascicular injection, injection inside the nerve fascicle (with disruption of the perineurium), or intraepineurial-extrafascicular injection, which happens inside the epineurial tissue (interfascicular tissue).

After localizing the nerve with nerve stimulation, computerized tomographic (CT) scanning could reveal that the needle has been placed intraepineurially [7]. CT images show contrast medium inside the nerve. The contrast occupies the interfascicular space, and the catheter inserted through the stimulator needle is positioned inside the nerve too.

The signs of nerve puncture and intraneural injection are:

1. View the needle inside the nerve (Figs. 12.1a, 12.3b, 12.10 and 12.11)
2. Contrast inside the nerve with increment of area (Cross Sectional Area –CSA-) (Figs. 12.1b, 12.3, 12.5, 12.6, 12.7, 12.8, 12.9, 12.10 and 12.11)
3. Air inside the nerve (Figs. 12.3c and 12.9b)
4. Thin halo of contrast medium subepineurally (Figs. 12.1b, 12.3, 12.6, 12.7, 12.8 and 12.9)
5. Separation of nerves and fascicles (and increment of CSA) (Figs. 12.4, 12.5 and 12.7)
6. Concentric layers of contrast (Figs. 12.1b, 12.3, 12.4, 12.6, 12.8 and 12.9)
7. Proximal spread (Figs. 12.3a, 12.5 and 12.7)
8. Distal spread (Figs. 12.3c, 12.5 and 12.7)

J. Pomés, MD, PhD
Musculoskeletal Unit, Radiology Department, Imaging Diagnostic Center, Hospital Clinic Barcelona, Barcelona, Spain
e-mail: jpomes@clinic.ub.es

X. Sala-Blanch, MD (✉)
Human Anatomy and Embryology Unit, School of Medicine, University of Barcelona, Barcelona, Spain

Department of Anesthesiology and Critical Care, Clinic Hospital, Barcelona, Spain
e-mail: xavi.sala.blanch@gmail.com

M.A. Reina et al. (eds.), *Atlas of Functional Anatomy for Regional Anesthesia and Pain Medicine: Human Structure, Ultrastructure and 3D Reconstruction Images*, DOI 10.1007/978-3-319-09522-6_12,

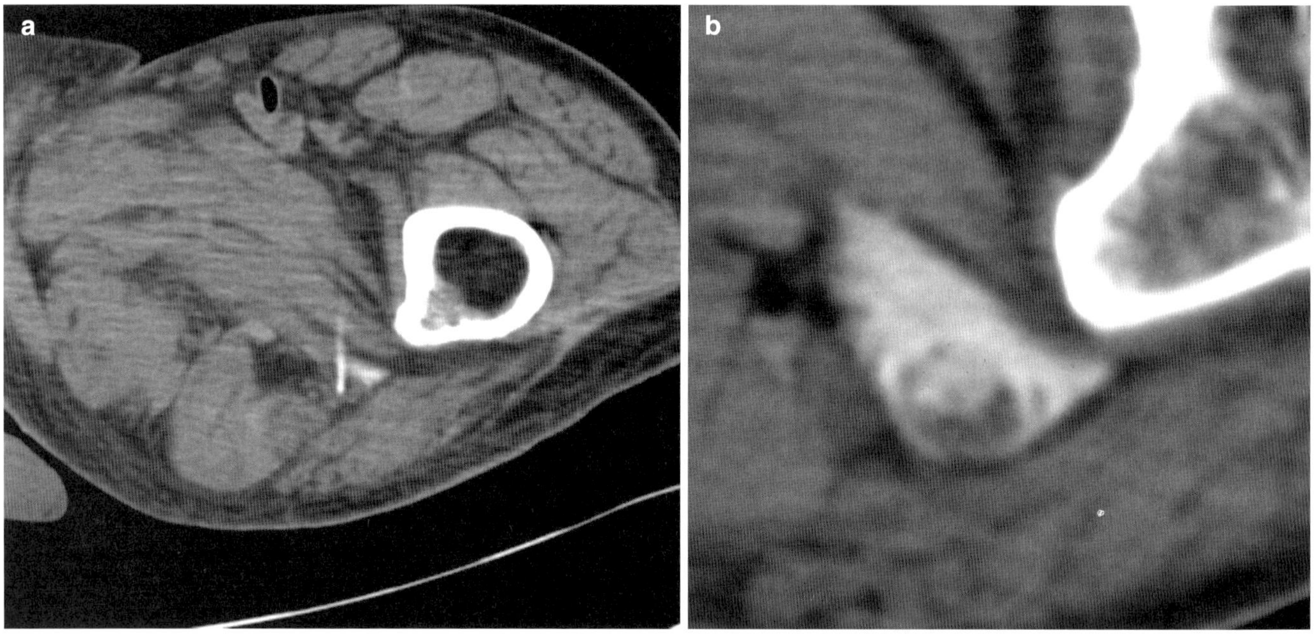

Fig. 12.1 Intraneural injection of 5 mL of contrast medium into the sciatic nerve (anterior approach in cadaver specimen). The needle enters the sciatic nerve (**a**), and the contrast spreads intraneurally and extraneurally (**b**)

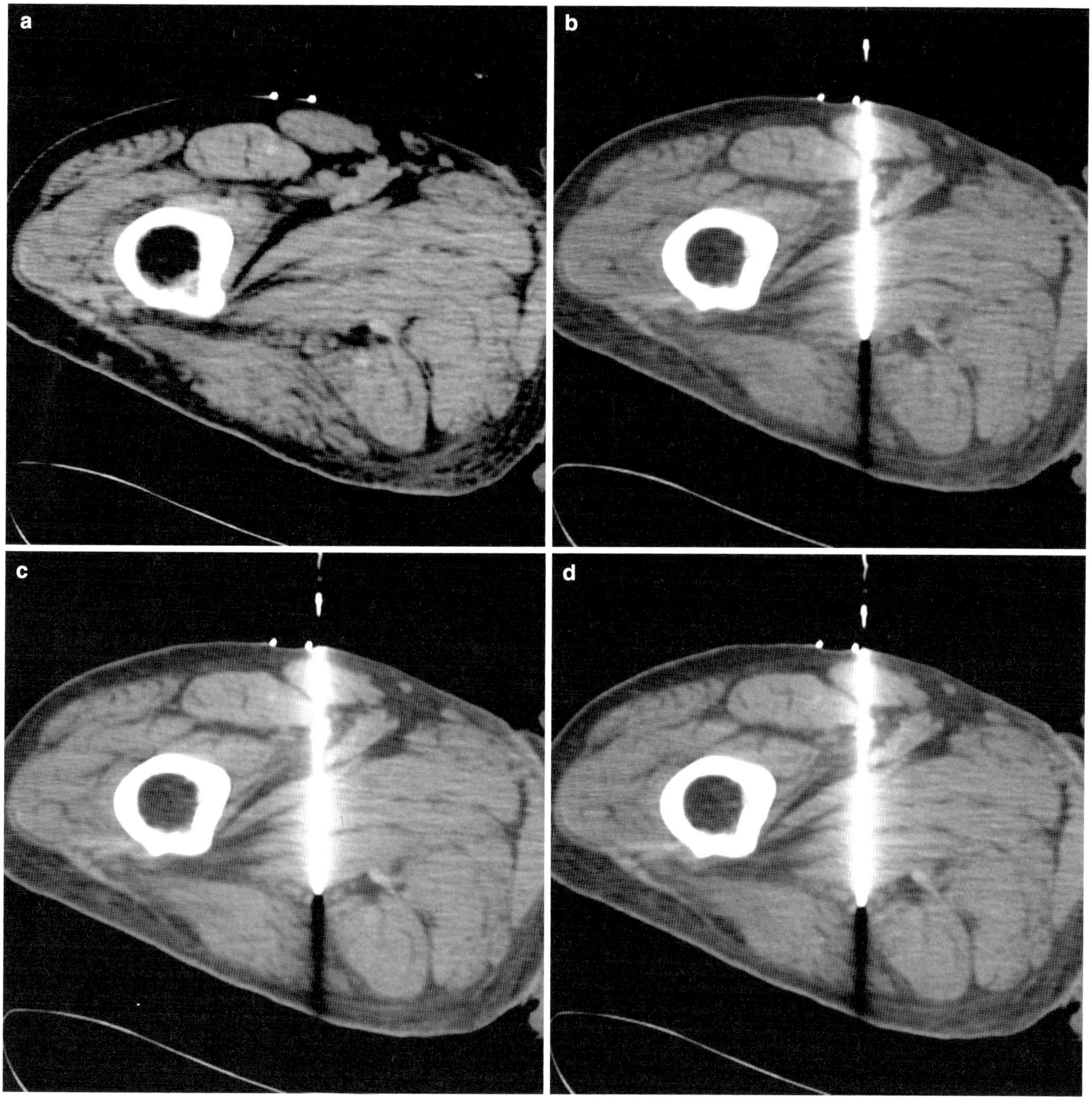

Fig. 12.2 Elastic resistance to puncture of the sciatic nerve (cadaver specimen). CT scan of the thigh at the level of the trochanter minor (**a**). You can see the sciatic nerve between the adductor muscles (Adductor Magnus muscle) and hamstrings (Long hear of biceps muscle and semitendinosus). In figures (**b**), (**c**), and (**d**), we can see the elastic resistance to puncture the nerve. This effect is important to prevent the puncture of the fascicles or group of fascicles (see Chap. 15)

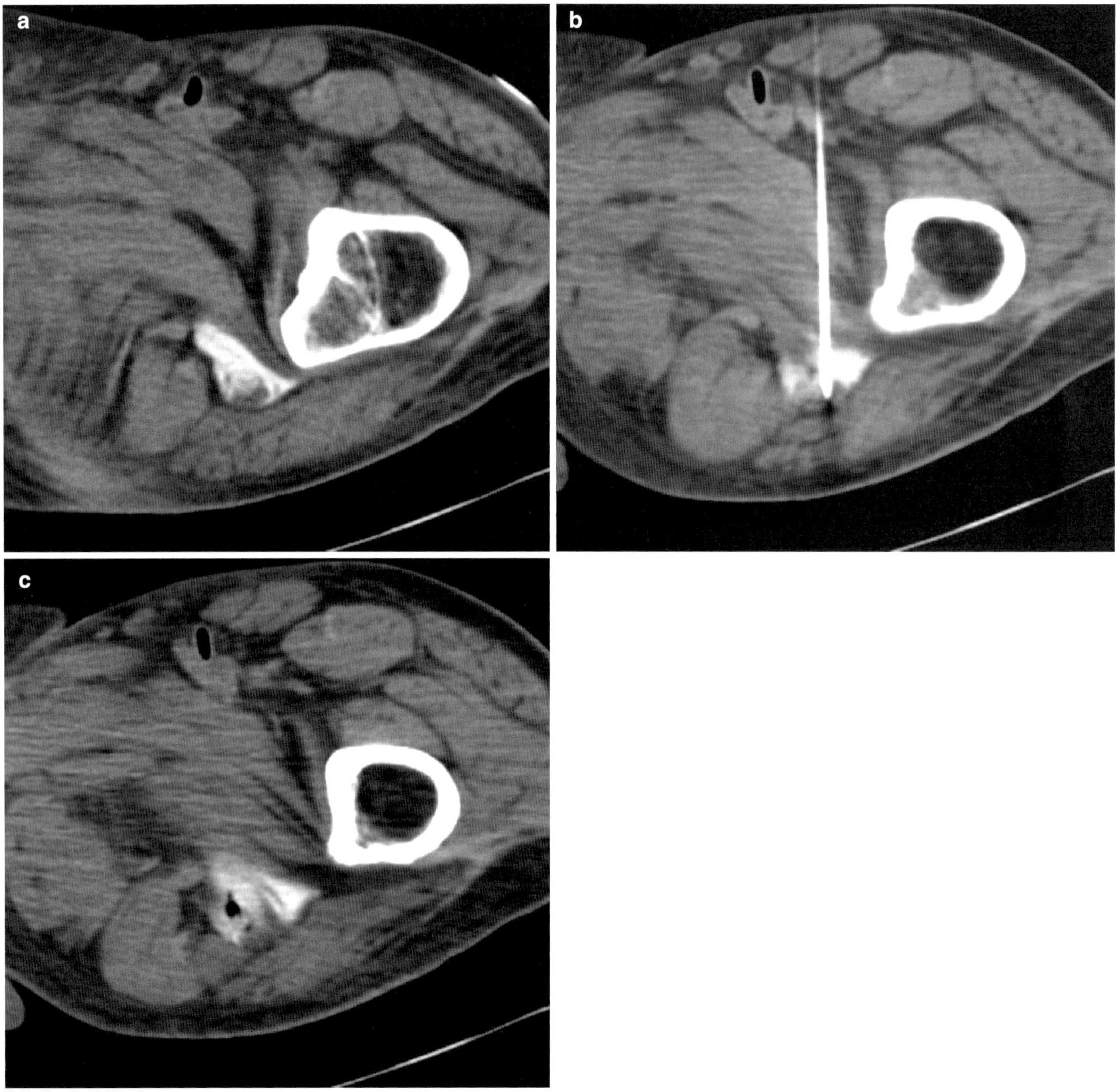

Fig. 12.3 Sciatic nerve intraneural injection (cadaver study). (**a**) Proximal spread. (**b**) View of the needle inside the nerve. (**c**) Distal intraneural spread. (**a**) The subepineurial contrast in the middle of the nerve is surrounded by a thin halo. (**c**) Air can be seen inside the nerve, with an increase of the area of the nerve

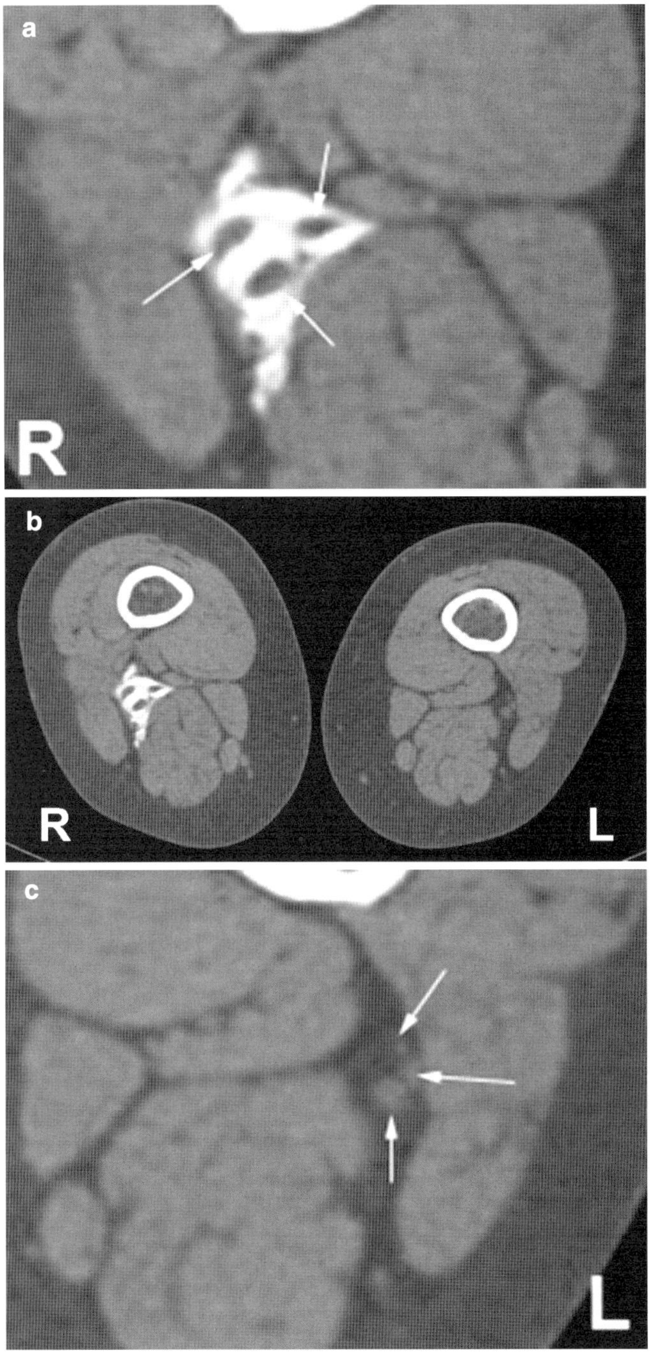

Fig. 12.4 (a–c) Sciatic nerve intraneural injection after popliteal block. View the separation of fascicles of the sciatic nerve, with concentric layers of contrast on the right side (**a**). The section area of the block side (**a**) is greater than the area of the contralateral side (**c**) (*white arrows*)

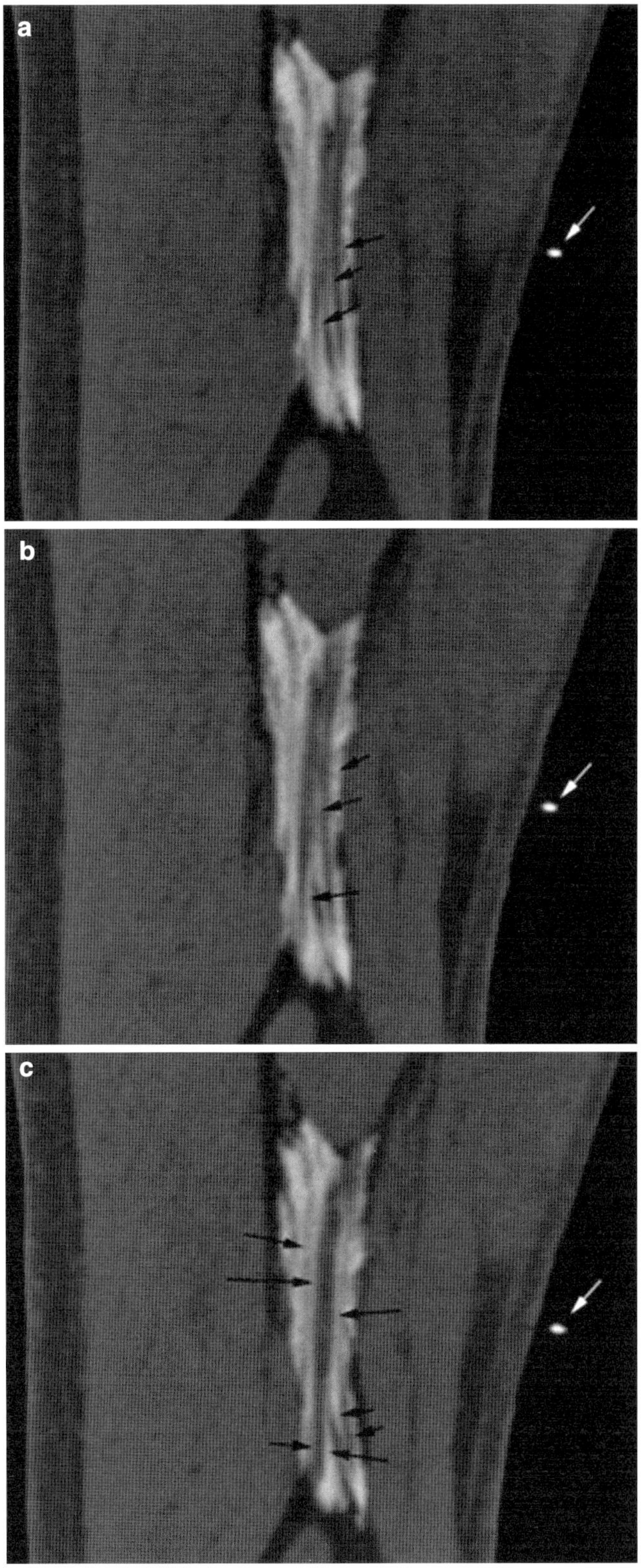

Fig. 12.5 Sciatic nerve intraneural injection after popliteal block. Coronal view in consecutive sections that show the separation of fascicles of the sciatic nerve (*black arrows*), with contrast between the fascicles. *White arrows* are marking the level of injection and showing proximal and distal spread between the fascicles

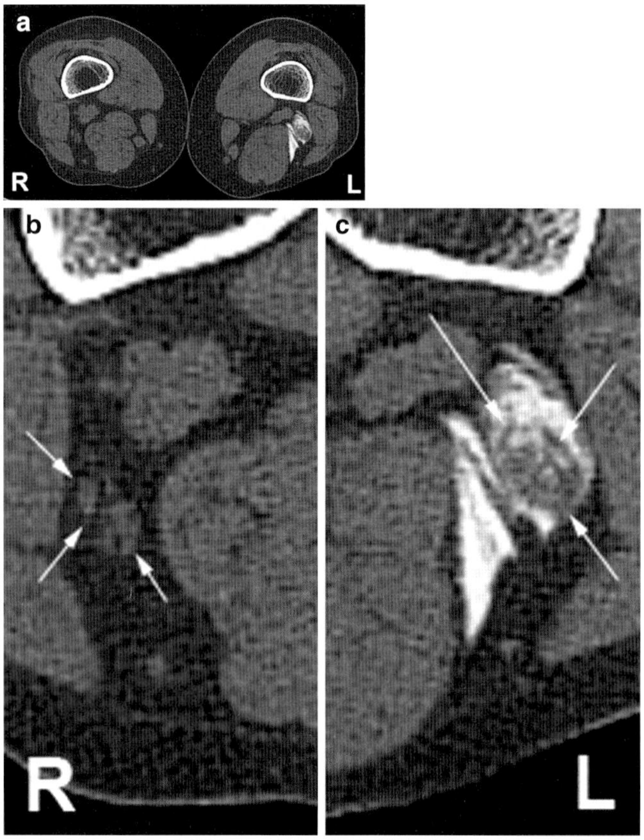

Fig. 12.6 Figure shows the axial view of sciatic nerve after intraneural injection at popliteal level (**a**). Figure **b** shows the right (non injected) side while the figure **c** shows the left (injected) side. On the injected side we can see the separation of fascicles, contrast between the fascicles, a thin halo around the nerve, and increment of nerve area in the short axis. The section area of the block side (**a**) is greater than the area of the contralateral side (**c**) (*white arrows*)

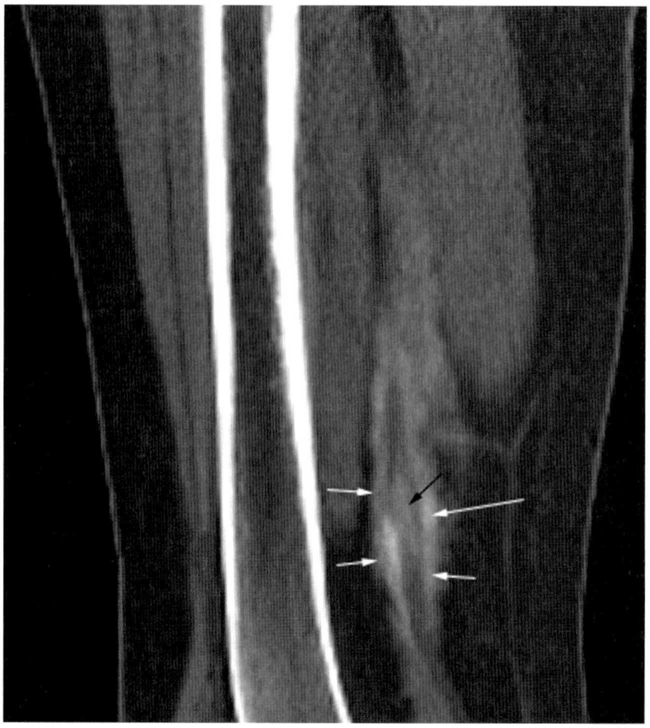

Fig. 12.7 Sagittal view of intraneural injection after sciatic popliteal block. The contrast medium is seen into the nerve, between fascicles (*black arrow*). *White arrows* are showing the limits of the sciatic nerve

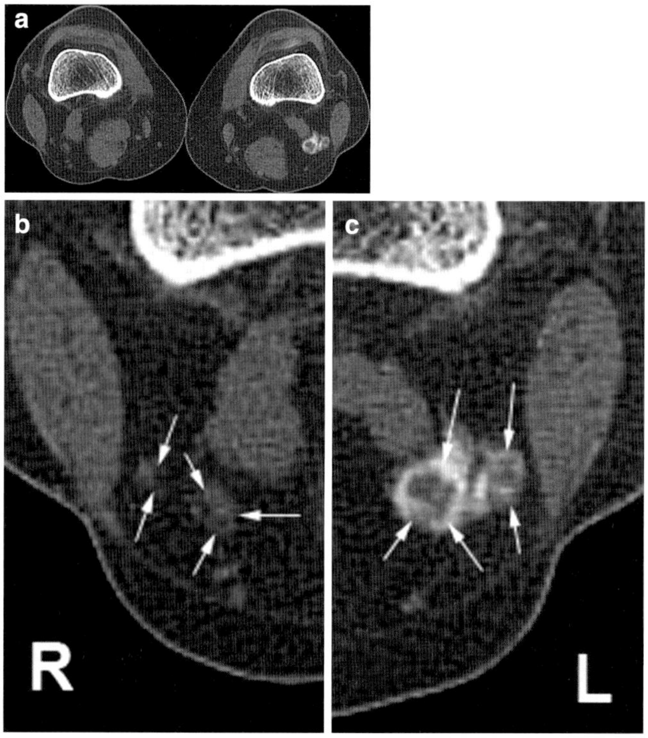

Fig. 12.8 Axial view of intraneural injection after popliteal sciatic nerve block (**a**). Figures **b** and **c** show a detail of the sciatic nerve, in right side (non injected side) (**b**) and left (injected side) (**c**). We can see a thin halo of contrast around tibial and common peroneal nerves at the level of bifurcation. Contrast is present into both nerves with increment of CSA

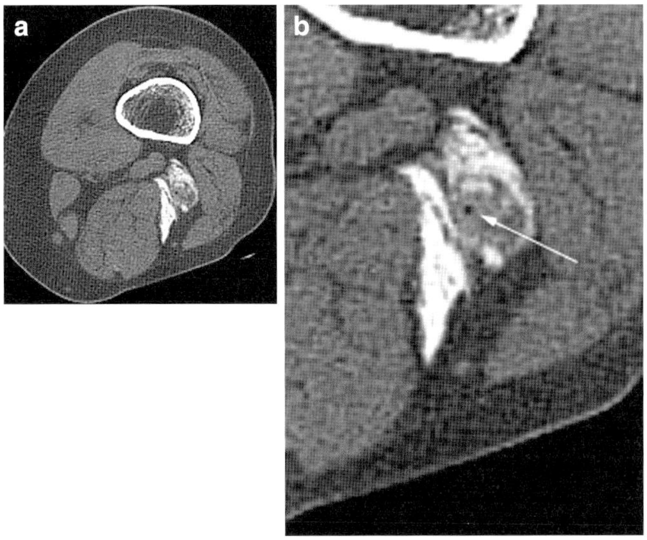

Fig. 12.9 Intraneural injection observed after popliteal sciatic nerve block (**a**). View in detail (**b**) the thin halo of contrast around the nerve at the level of puncture, as well as contrast inside the nerve and a bubble of air (*arrow*)

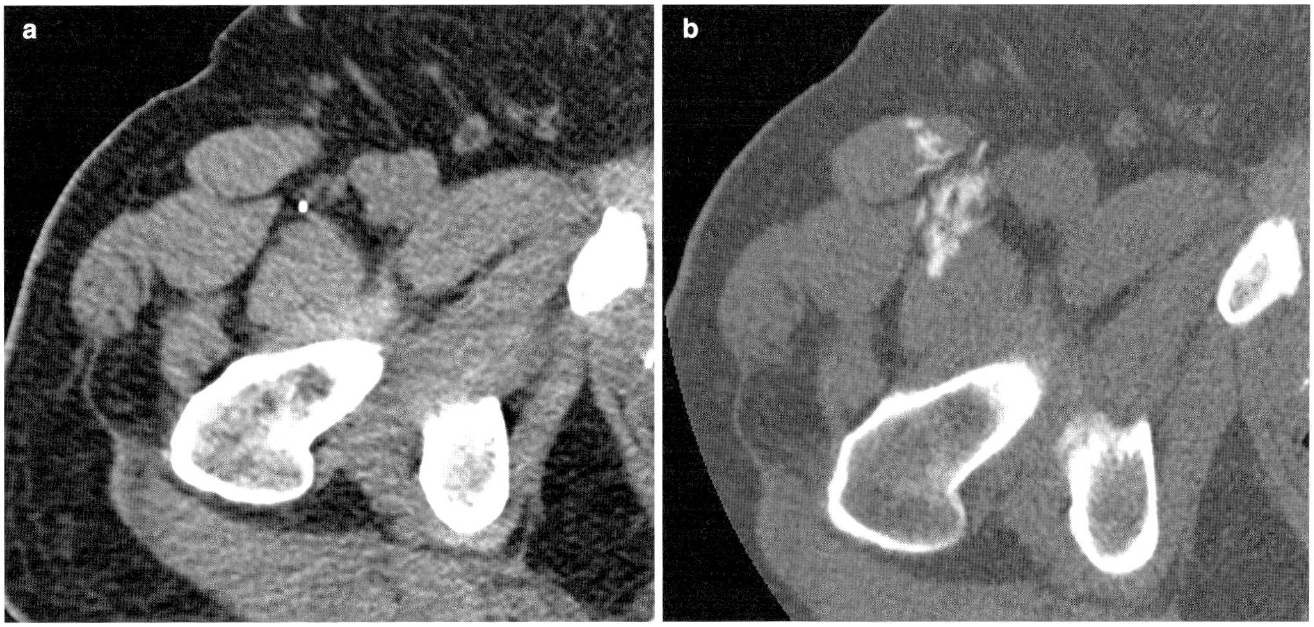

Fig. 12.10 Intraneural injection after continuous femoral nerve block. (**a**) The catheter tip is seen in the femoral nerve (probably subepineurially), with contrast. (**b**) An increment of nerve area, separation of fascicles, and contrast between them

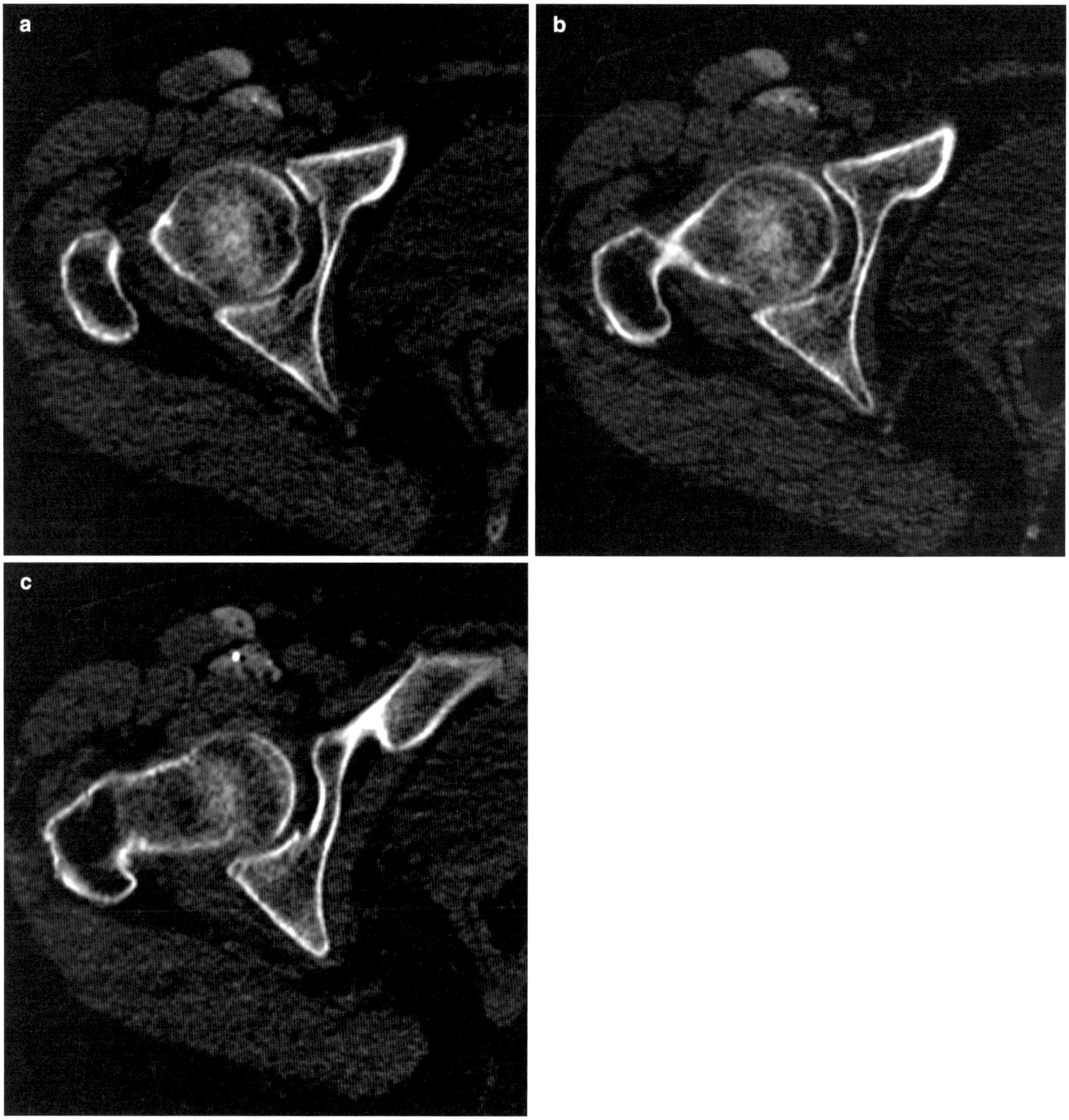

Fig. 12.11 Intraneural injection after continuous femoral nerve block in three consecutive CT scan sections (**a**, **b** and **c**). The catheter is visible in the middle of the femoral nerve (**a**, **b**) and the tip of the catheter subepineurally mor proximal (**c**). In all figures (**a**, **b** and **c**) contrast medium inside the nerve is visible

References

1. Pauchet V, Sourdat P. L'anesthesie regionale. París: Ed Octave Dion; 1914.
2. Hirschel G. Compendio de anestesia local para medicos y estudiantes. Barcelona: Salvat Ed; 1914.
3. Braun H. Local anesthesia. Its scientific basis and practical use. New York: Lea & Febiger; 1914.
4. Labat G. Regional anesthesia. Its technic and clinical application. Philadelphia: W.B. Saunders; 1924.
5. Borgeat A. Regional anesthesia, intraneural injection, and nerve injury: beyond the epineurium. Anesthesiology. 2006;105:647–8.
6. Sala-Blanch X, De Andrés J. Image-guided techniques for peripheral nerve blocks. Curr Opin Anaesthesiol. 2004;17:409–15.
7. Sala-Blanch X, Pomés J, Matute P, Valls-Solé J, Carrera A, Tomás X, García-Diez AI. Intraneural injection during anterior approach for sciatic nerve block. Anesthesiology. 2004;101:1027–30.

13 Ultrasound View of Intraneural Injection

Xavier Sala-Blanch and Jaume Pomés

Presently we can identify signs to help guide the administration of anesthetic drugs in the intraneural tissue. Nevertheless, these signs remain loosely defined, or are too difficult to identify or interpret to be useful in forming the criteria that ensure we have performed an intraneural injection [1–10].

The main disadvantage of imaging techniques is that often they cannot delineate the nerves' limits, much less identify fascicles. Therefore, ultrasound has a low predictive capacity in defining the site of injection and thus, of predicting the possible risk of postblock dysfunction.

Ultrasound signs of puncture and/or intraneural injection are varied and observer dependent. The established criteria should be capable of evaluating the incidence of (1) intraneural injection in non–ultrasound-guided anesthetic techniques (i.e., neurostimulation-guided techniques) and (2) puncture control performed in real time during ultrasound-guided techniques. In both situations, there will be signs of higher or lower specificity:

- Ultrasound-guided regional anesthesia; dynamic signs of neural puncture and injection:
  - Viewing the needle inside the nerve. Neural puncture is an occasional, undesired observation; what usually may be observed is neural injection. The needle tip is controlled and viewed better when an "in-plane" technique performed and when the puncture angle is ≤30°.
  - "Pop" sign. Displacement of the nerve's contact with the needle tip is associated with a loss of resistance due to the needle passing quickly through the nerve because of the elastic resistance of the nerve sheath.
  - Impalement nerve sign. In a short-axis view of a nerve, the nerve moves with lateral movement of the needle.
  - Dynamic view of the area's increase. Nerve distension, as well as fascicle separation (an increase of the neural diameter), is visualized directly at the start of drug injection.
  - Separation of the fascicles and/or fascicular bundles by the injectate.
  - Elastic reduction of the area at cessation of drug injection. This sign suggests elastic distention of the epineurium, which will disappear if there is rupture of the epineurial cover or anesthetic diffusion along the neural structure, as suggested by Pauchet in 1914.
- Static ultrasound signs of neural injection:
  - Increase in the nerve's area and diameter due to the anesthetic's expansive effect at the injection site.
  - Diffusion of the anesthetic, creating a fine halo proximally, distally, and along the nerve branches.
  - Nerve swelling. Change in echogenicity due to an edema effect or to an increase in the aqueous proportion in the tissues. It is correlated with an increased nerve area.
  - Viewing the epineurial sheath after injection. A thin layer of collagen surrounds the nerve.

The signs may be separated into two groups, major and minor signs:

Major signs	Minor signs
Needle inside the nerve	Pop sign
Impalement sign	Halo around the nerve
Increased nerve area	Swelling
View of epineurial sheath	Reduced elastic area
	Proximal diffusion
	Distal diffusion

Intraneural injection diagnosis is the presence of more than one major sign or a major sign with the presence of two or more minor signs.

X. Sala-Blanch, MD (✉)
Human Anatomy and Embryology Unit, School of Medicine, University of Barcelona, Barcelona, Spain

Department of Anesthesiology and Critical Care, Clinic Hospital, Barcelona, Spain
e-mail: xavi.sala.blanch@gmail.com

J. Pomés, MD, PhD
Musculoskeletal Unit, Radiology Department, Imaging Diagnostic Center, Hospital Clinic Barcelona, Barcelona, Spain
e-mail: jpomes@clinic.ub.es

M.A. Reina et al. (eds.), *Atlas of Functional Anatomy for Regional Anesthesia and Pain Medicine: Human Structure, Ultrastructure and 3D Reconstruction Images*, DOI 10.1007/978-3-319-09522-6_13,

Other clinical signs that indicate a probable intraneural injection are:

- Persistent paresthesia before injection
- Paresthesia or clinical symptoms during injection (harm or cold during injection)
- Motor response in a low-current intensity
- Fast-onset block

These clinical signs may be considered minor signs of intraneural injection (Figs. 13.1, 13.2, 13.3, 13.4, 13.5, 13.6, 13.7, 13.8, 13.9, 13.10, 13.11, 13.12, 13.13, 13.14, 13.15, 13.16, 13.17, 13.18, 13.19, 13.20, and 13.21).

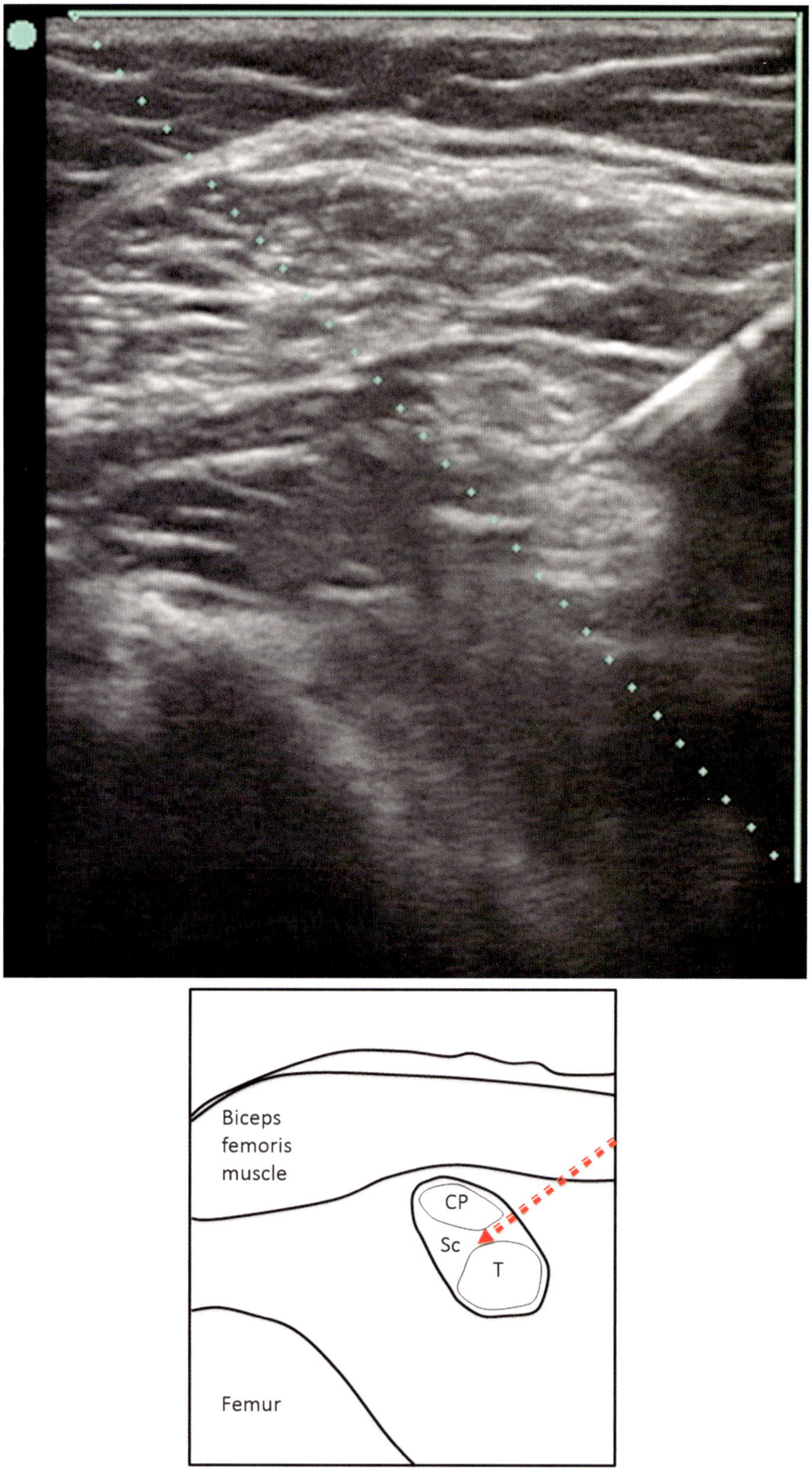

Fig. 13.1 Needle inside the nerve. *CP* common peroneal nerve, *T* tibial nerve, *Sc* sciatic nerve

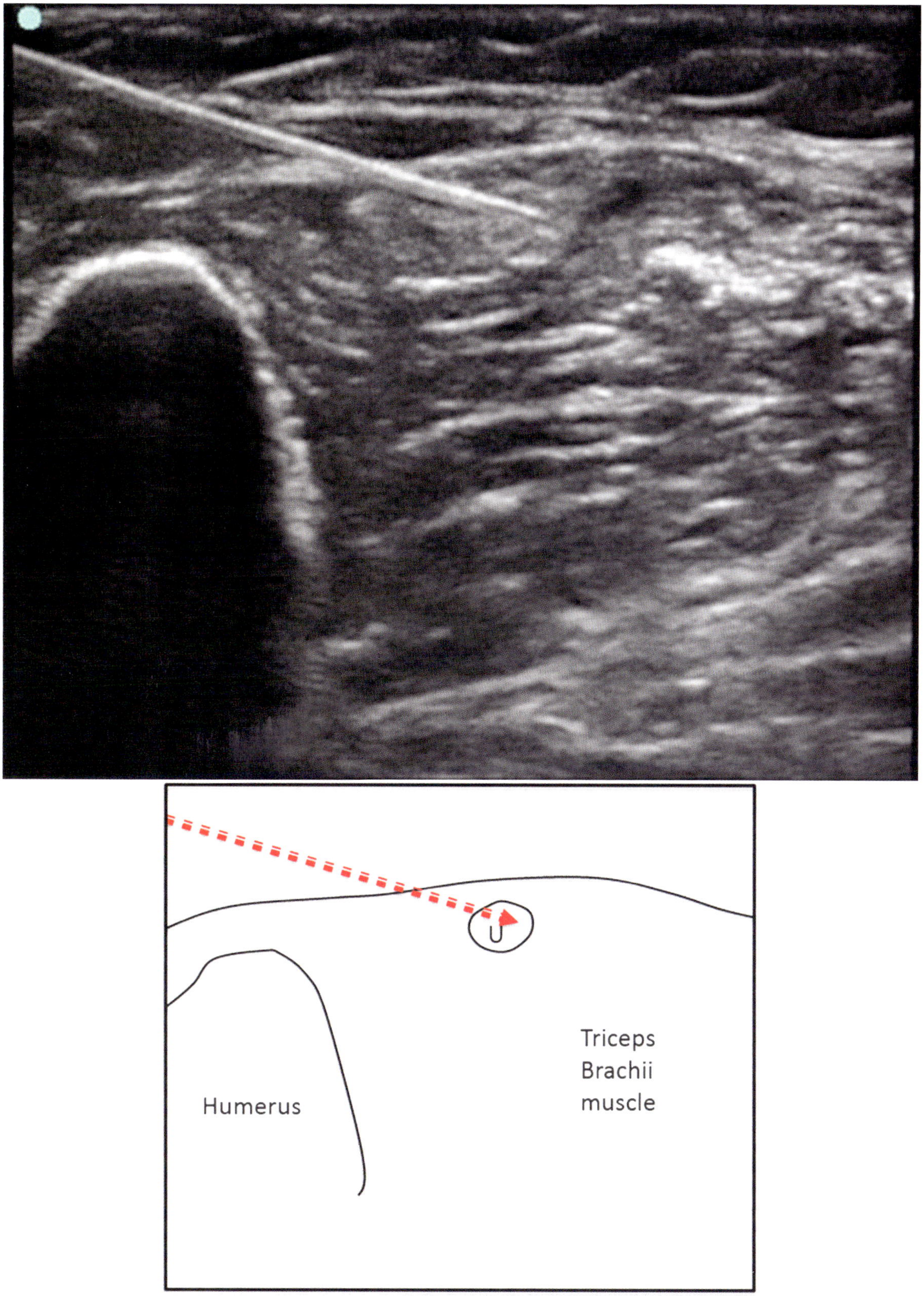

Fig. 13.2 Needle inside the nerve. *U* ulnar nerve

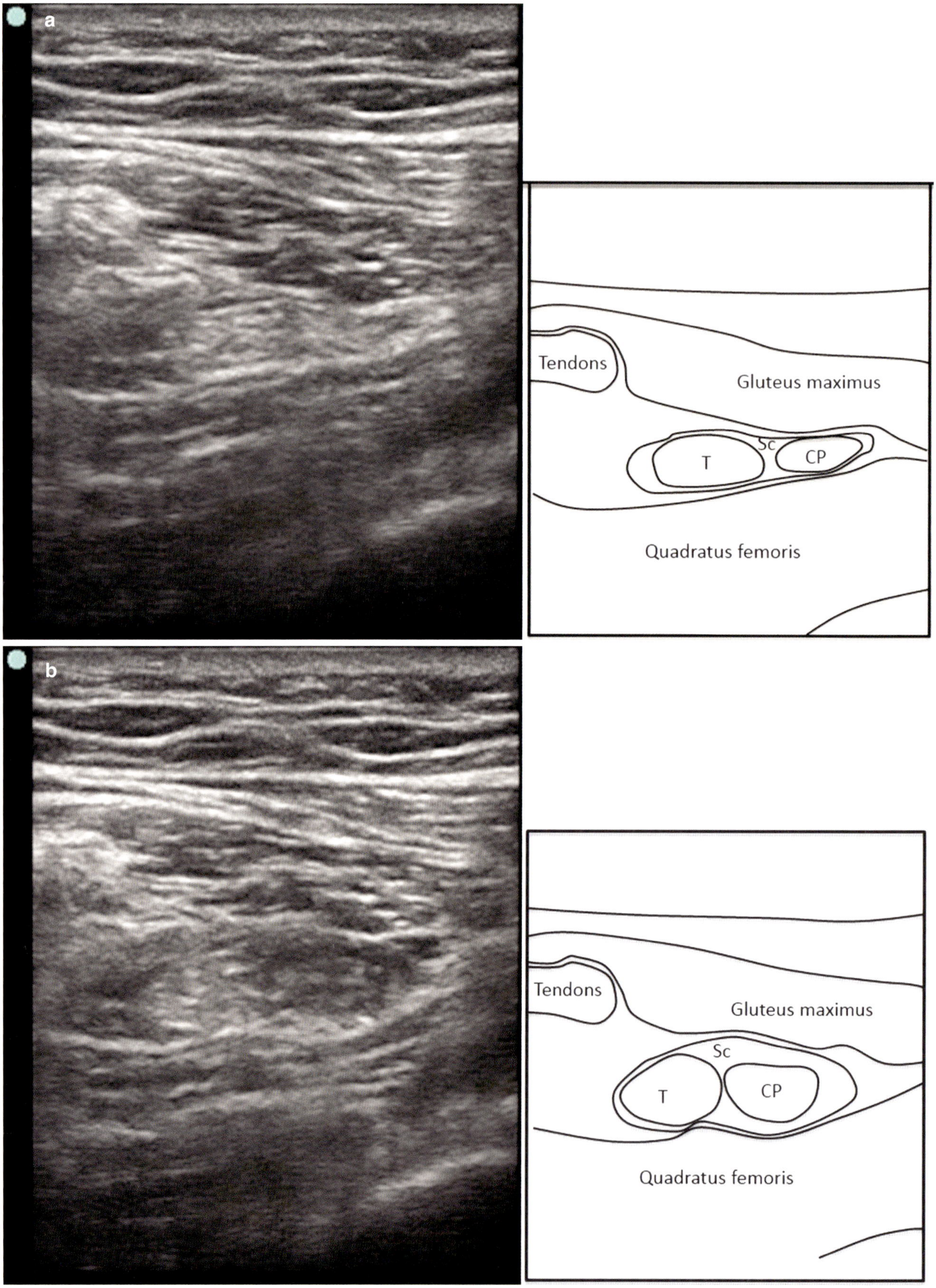

Fig. 13.3 Subgluteal sciatic nerve block before injection (**a**) and after injection (**b**) showing swelling and increase in the cross sectional area. *CP* common peroneal nerve, *T* tibial nerve, *Sc* sciatic nerve

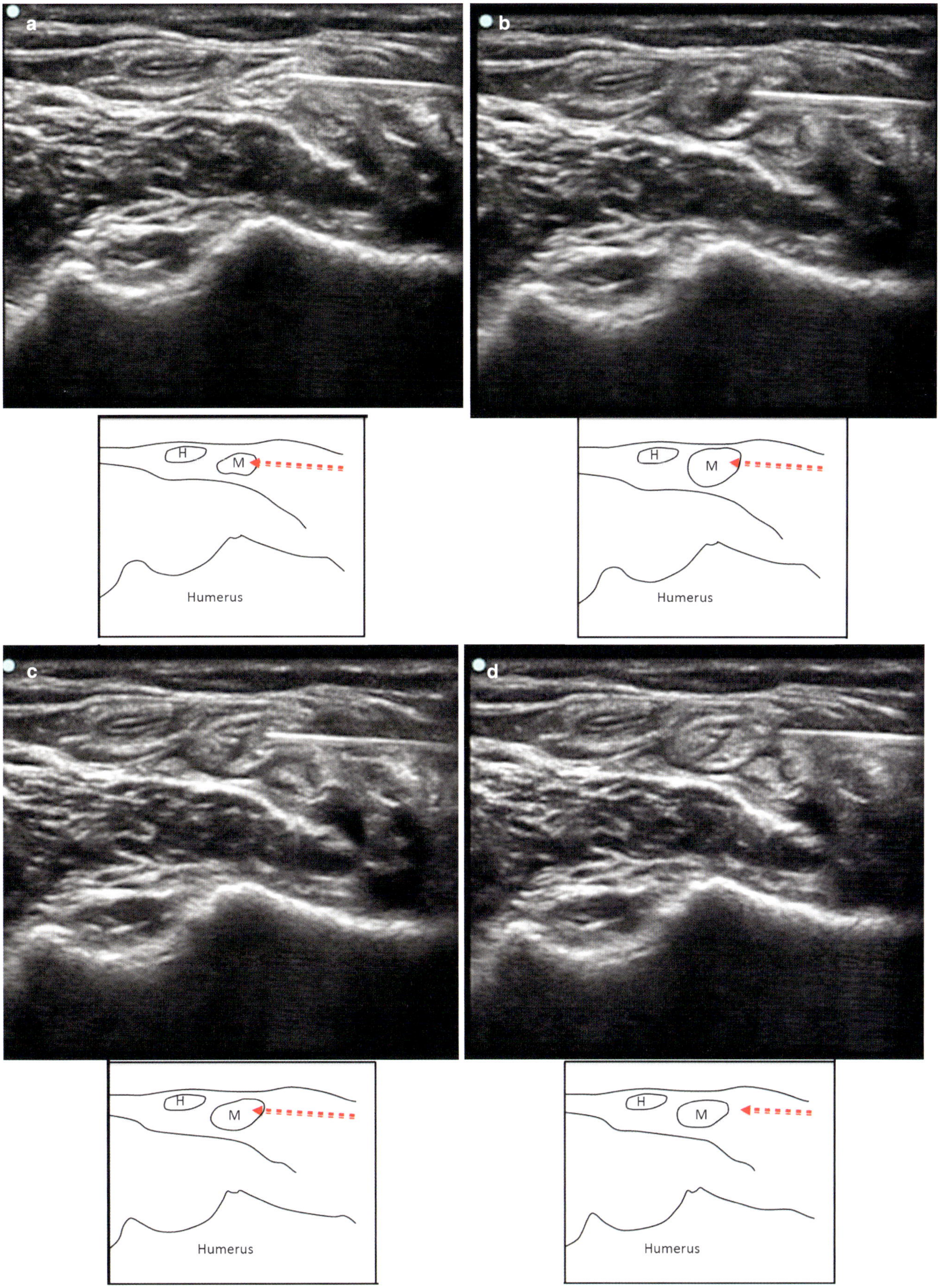

Fig. 13.4 Intraneural needle (**a**–**c**), fascicle spacing (**b**), swelling (**d**), increased area (**b**), and elastic retraction (**b**–**d**). *H* humeral artery, *M* median nerve

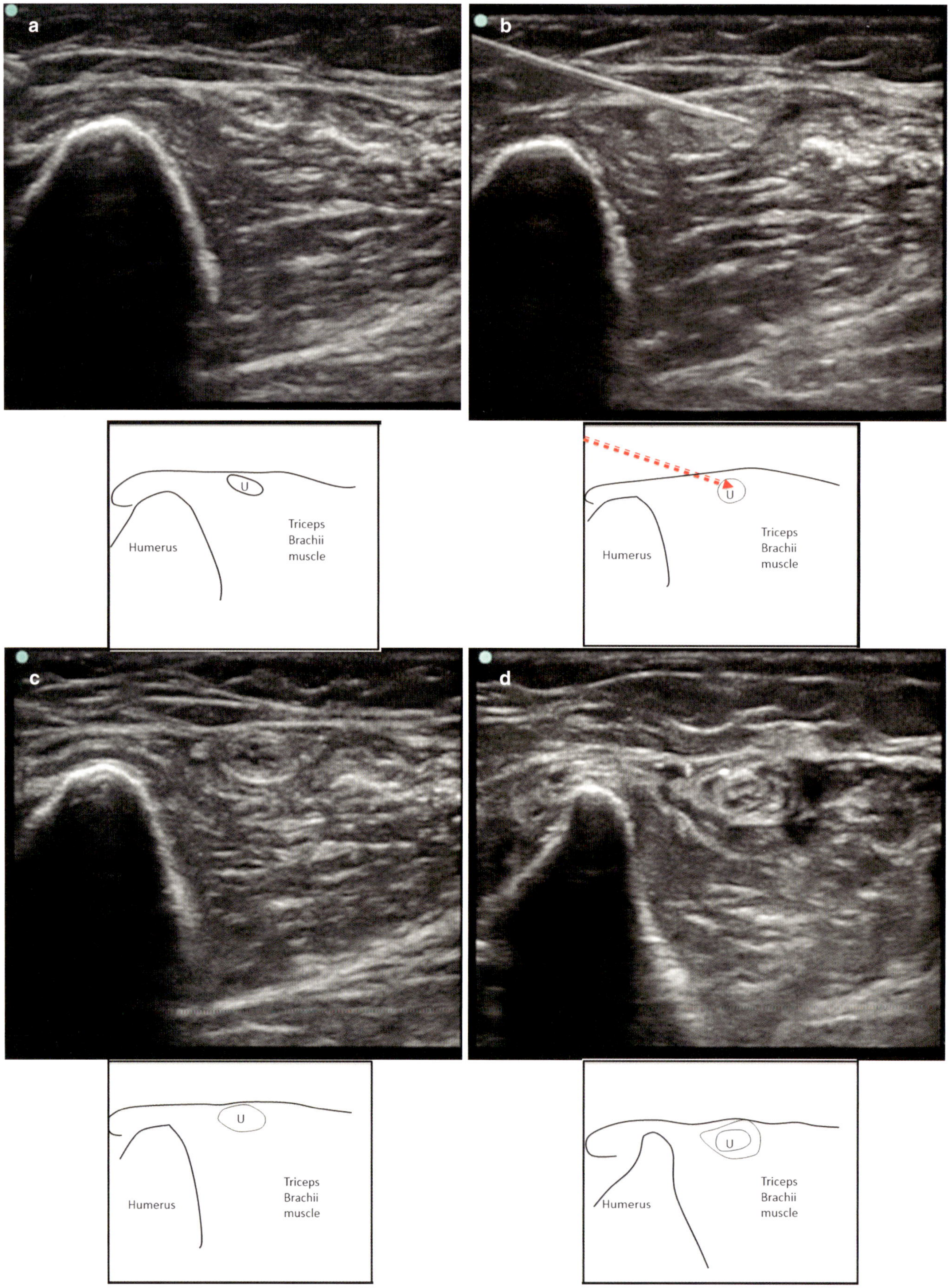

Fig. 13.5 Standard view (ulnar nerve) (**a**), intraneural needle (**b**), swelling (**c**), increased area (**c**), and distal diffusion (**d**). *U* ulnar nerve

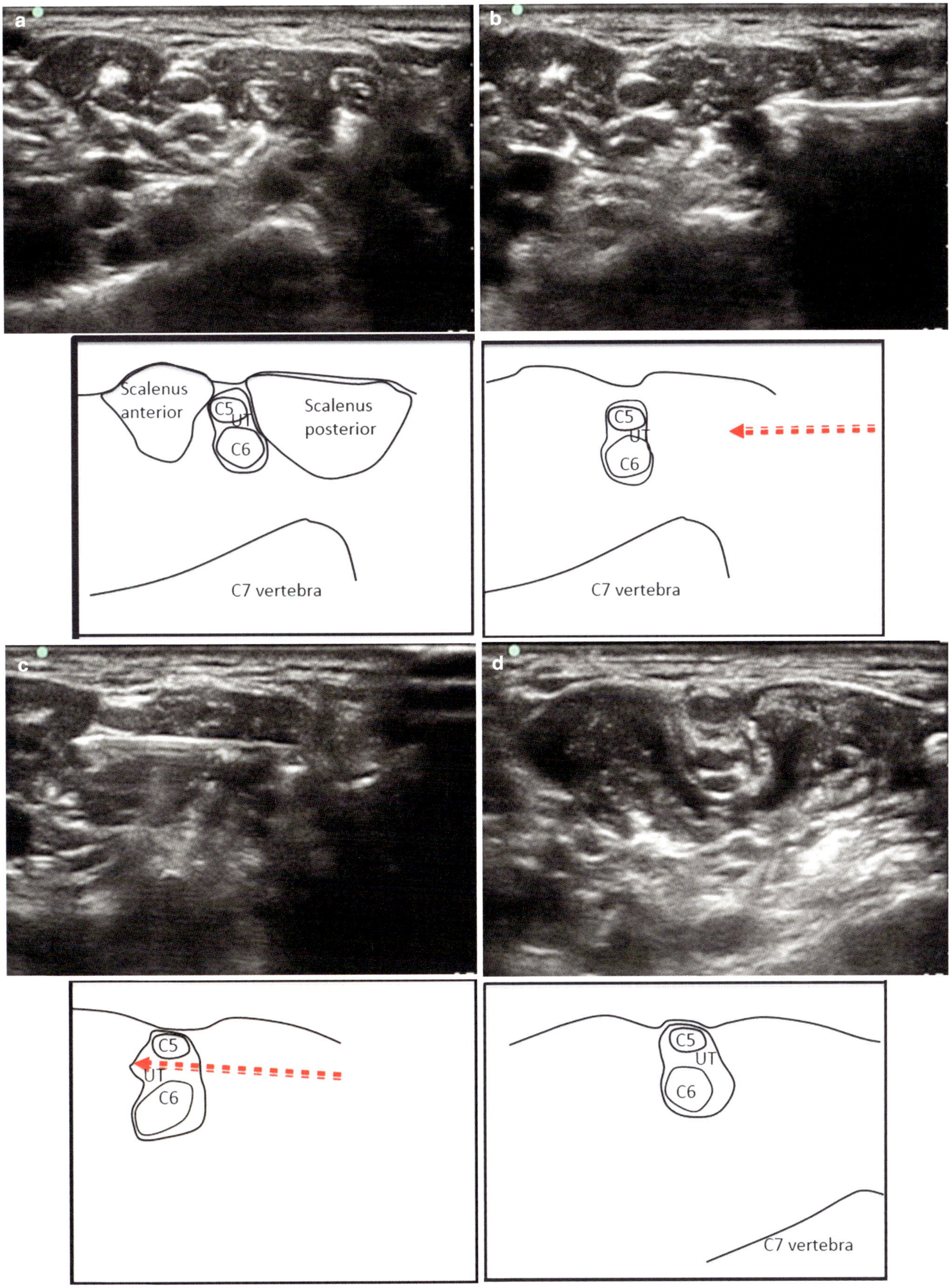

Fig. 13.6 Ultrasound guided interscalen brachial plexus block at the level of sixth cervical root (**a**, **b**). Intraneural needle position in upper trunk (between C5 and C6) (**c**) and posterior area increase of the upper trunk (**d**). *Red dashed* line indicate the position of the needle. *UT* upper trunk, *C5* fifth cervical root, *C6* sixth cervical root, *C7* seventh cervical root

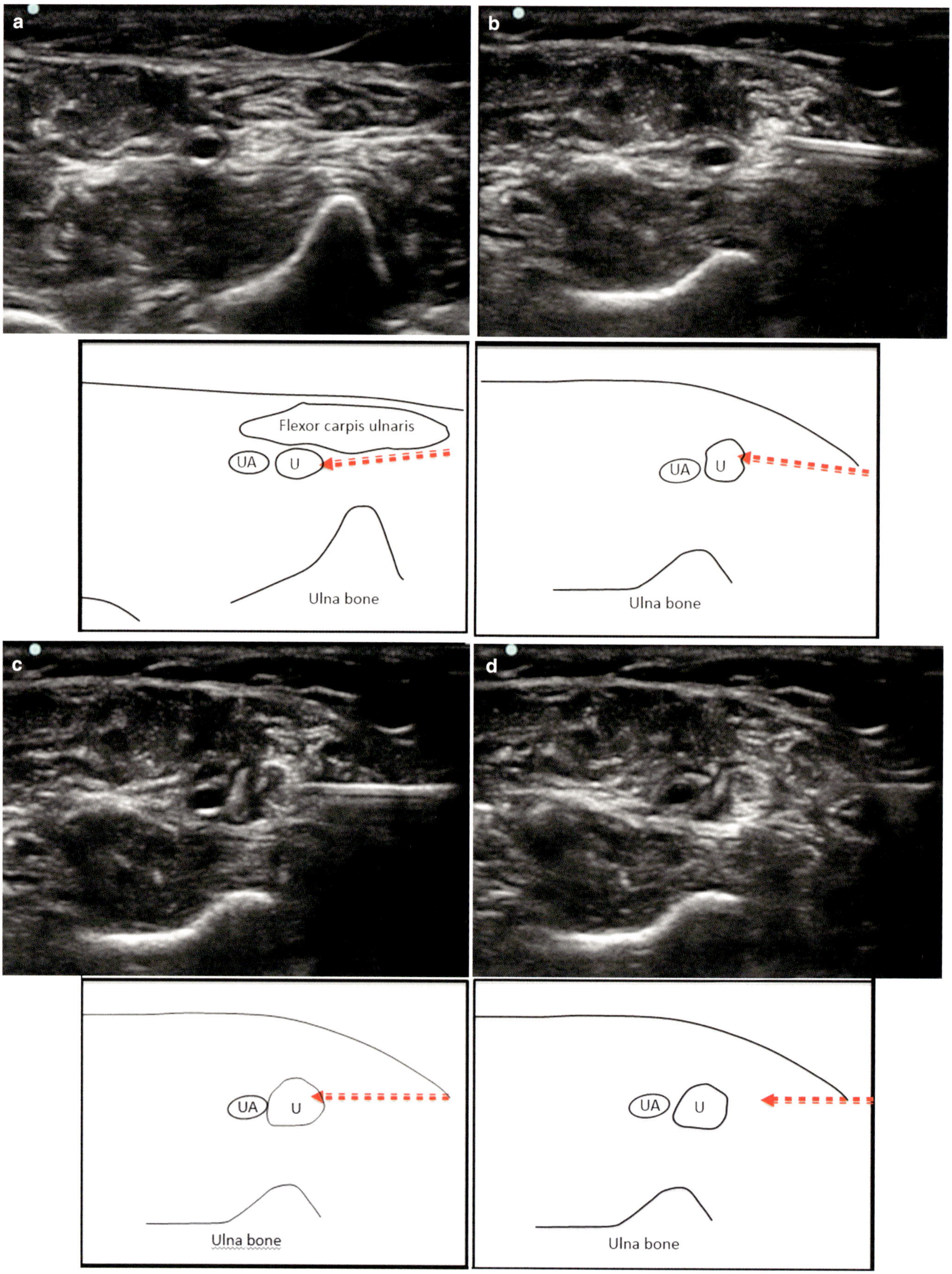

Fig. 13.7 Ultrasound guided ulnar nerve block (**a**) ulnar nerve piercing (**b**) and posterior area increase (**c, d**) and swelling (**d**). *Red dashed* line indicate the position of the needle. *U* ulnar nerve, *UA* ulnar artery

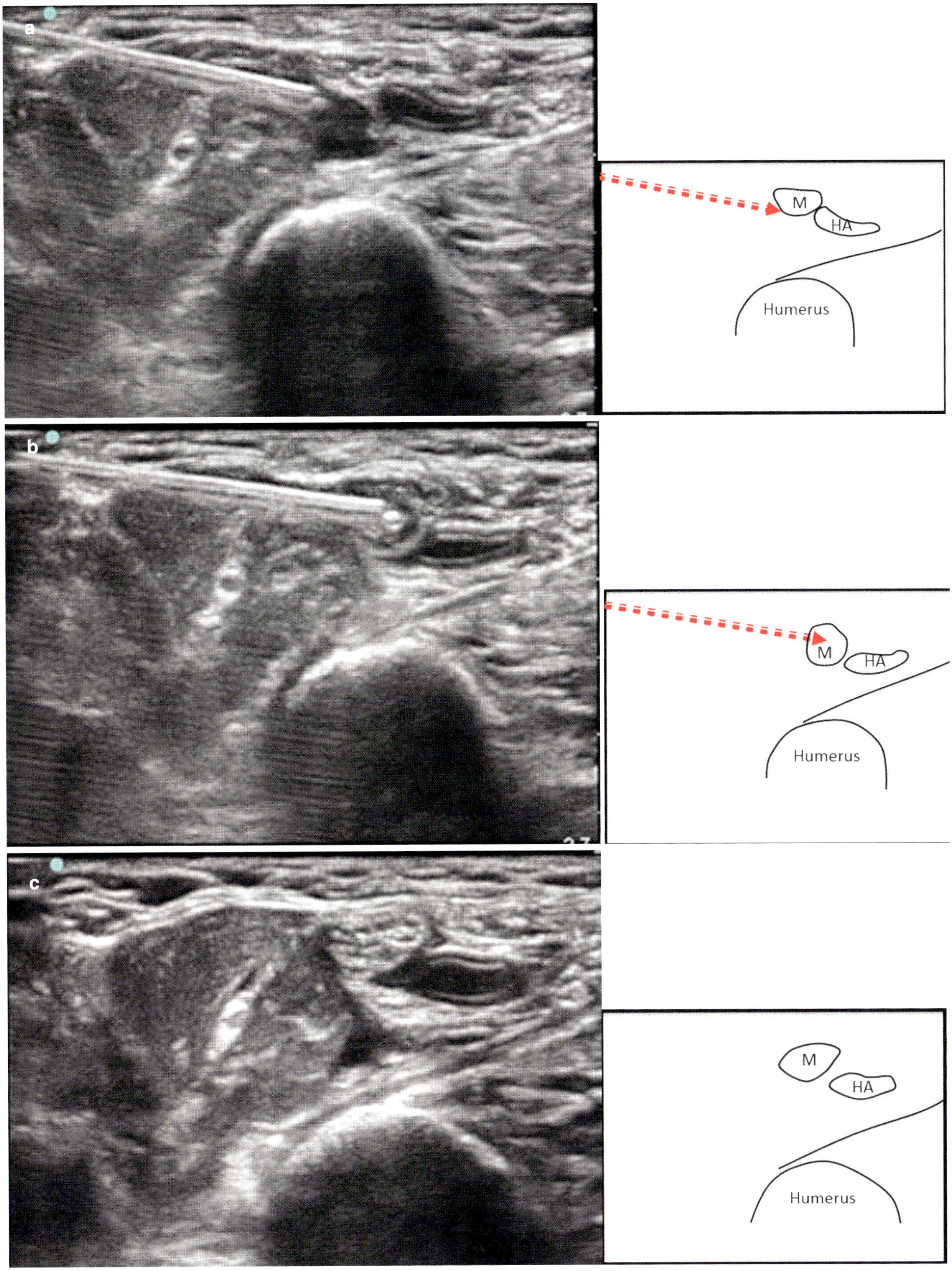

Fig. 13.8 Needle inside the median nerve (**b**) and swelling/area increase (**c**). *Red dashed* line indicate the position of the needle. *M* median nerve, *HA* Humeral artery

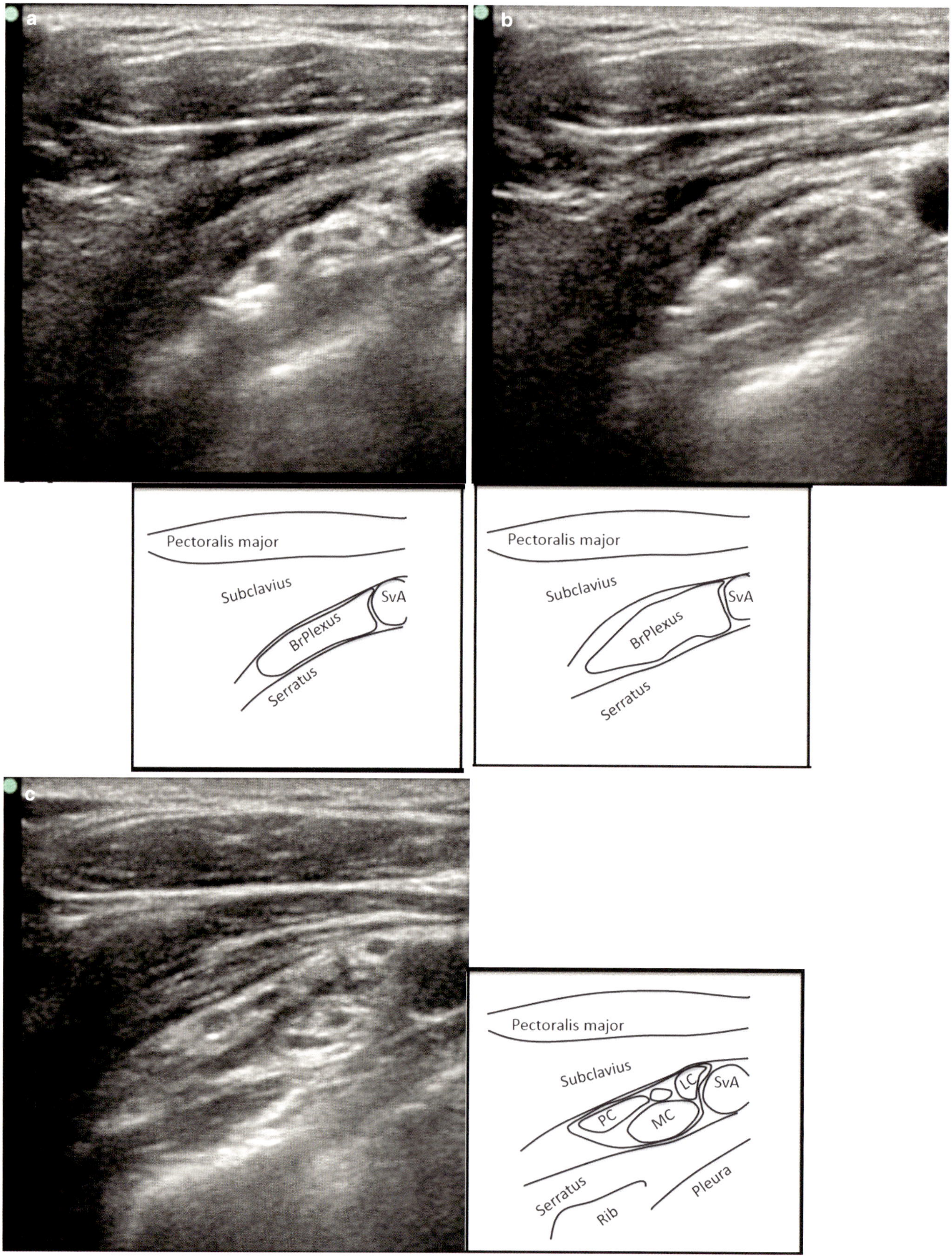

Fig. 13.9 Proximal infraclavicular block (**a**). Area increase, swelling, and fascicular separation (**b**, **c**). *BrPlexus* brachial plexus, *SvA* subclavian artery, *MC* medial cord, *LC* lateral cord, *PC* posterior cord

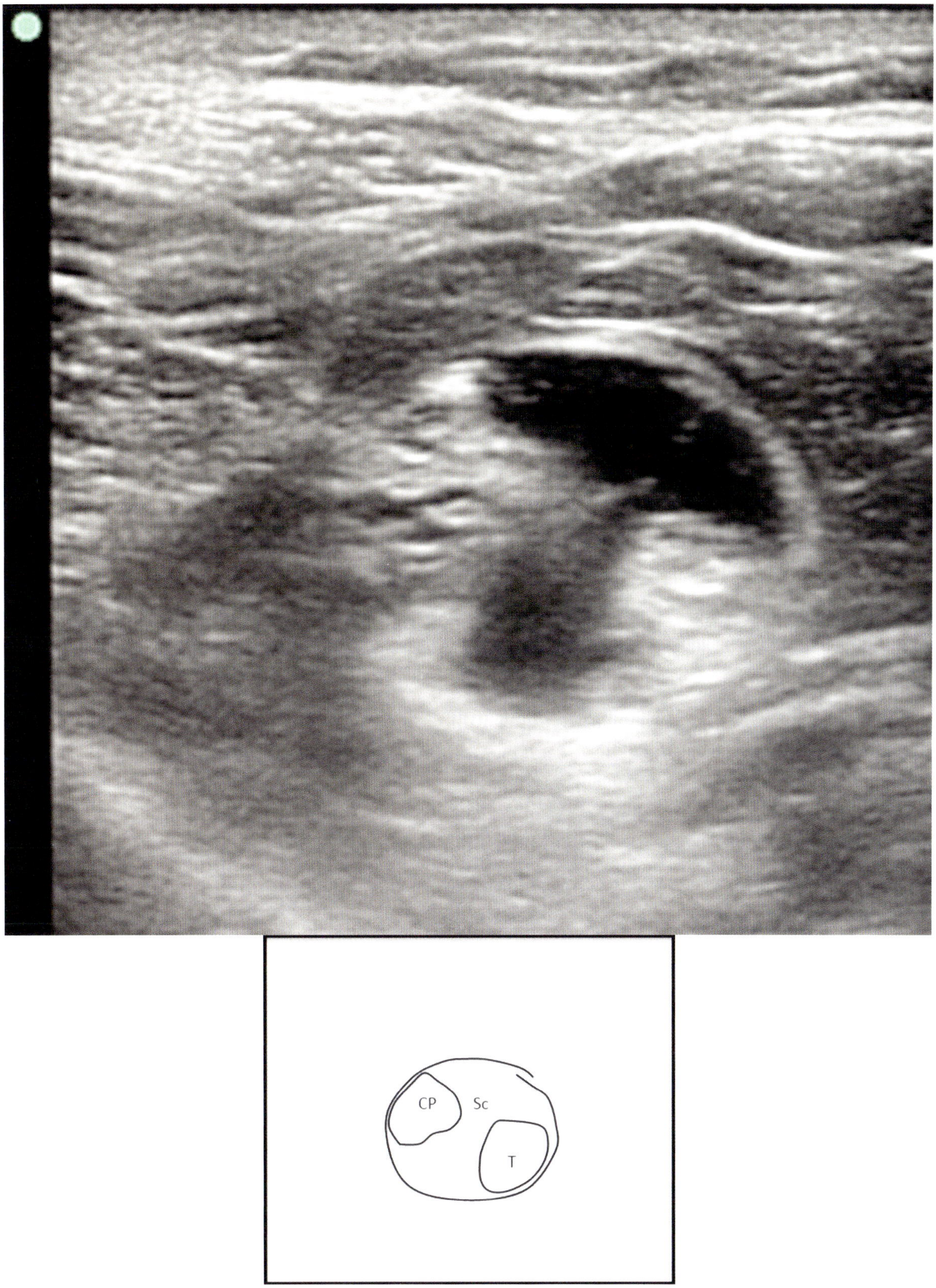

Fig. 13.10 Epineurial view during injection (centripetal spread and elastic reduction of the epineurium sheath) and after injection (sheath view) of sciatic nerve. *CP* common peroneal nerve, *T* tibial nerve, *Sc* sciatic nerve

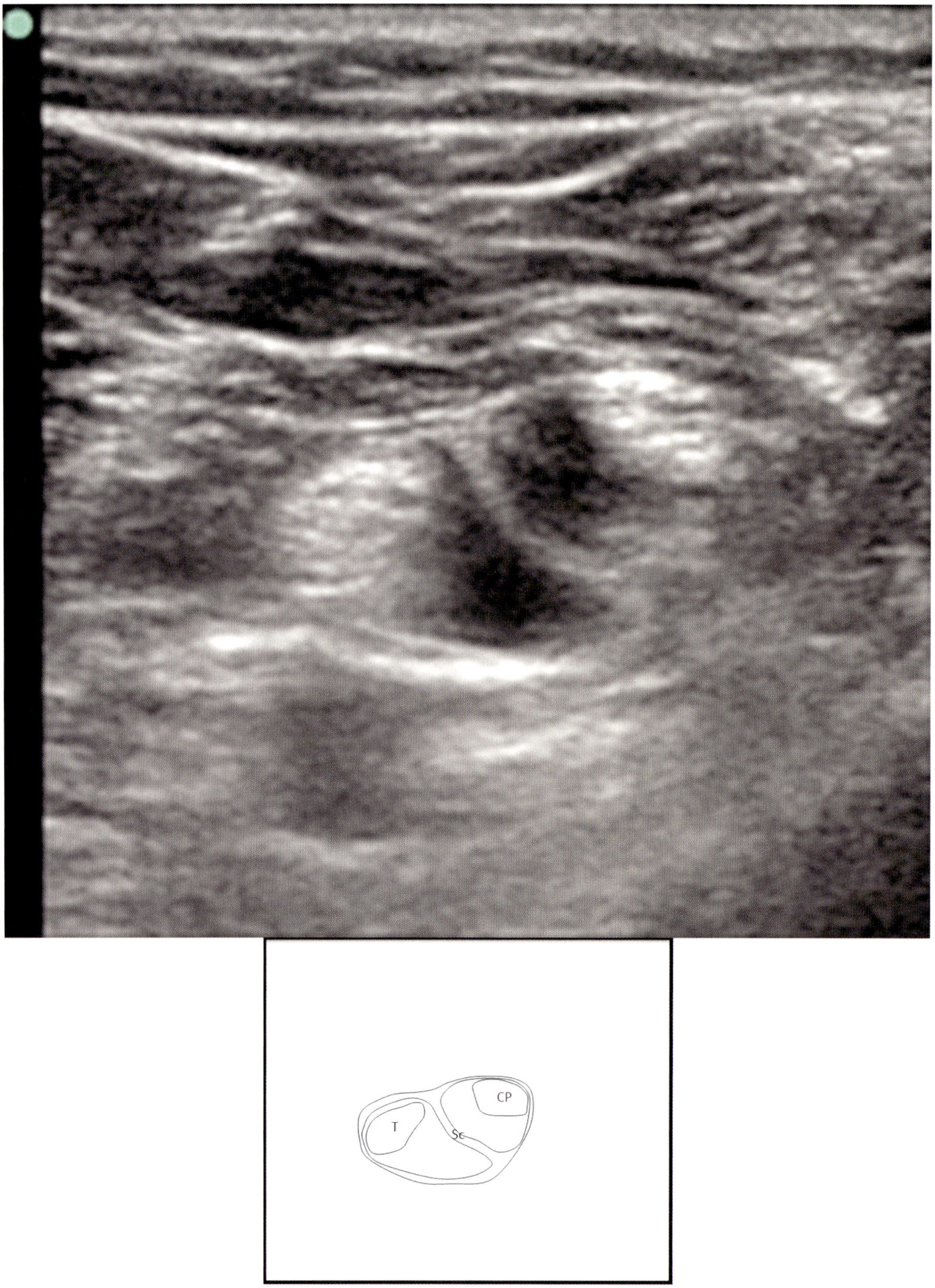

Fig. 13.11 Epineurial view of the tibial and common peroneal nerves at the level of sciatic nerve division. *CP* common peroneal nerve, *T* tibial nerve, *Sc* sciatic nerve

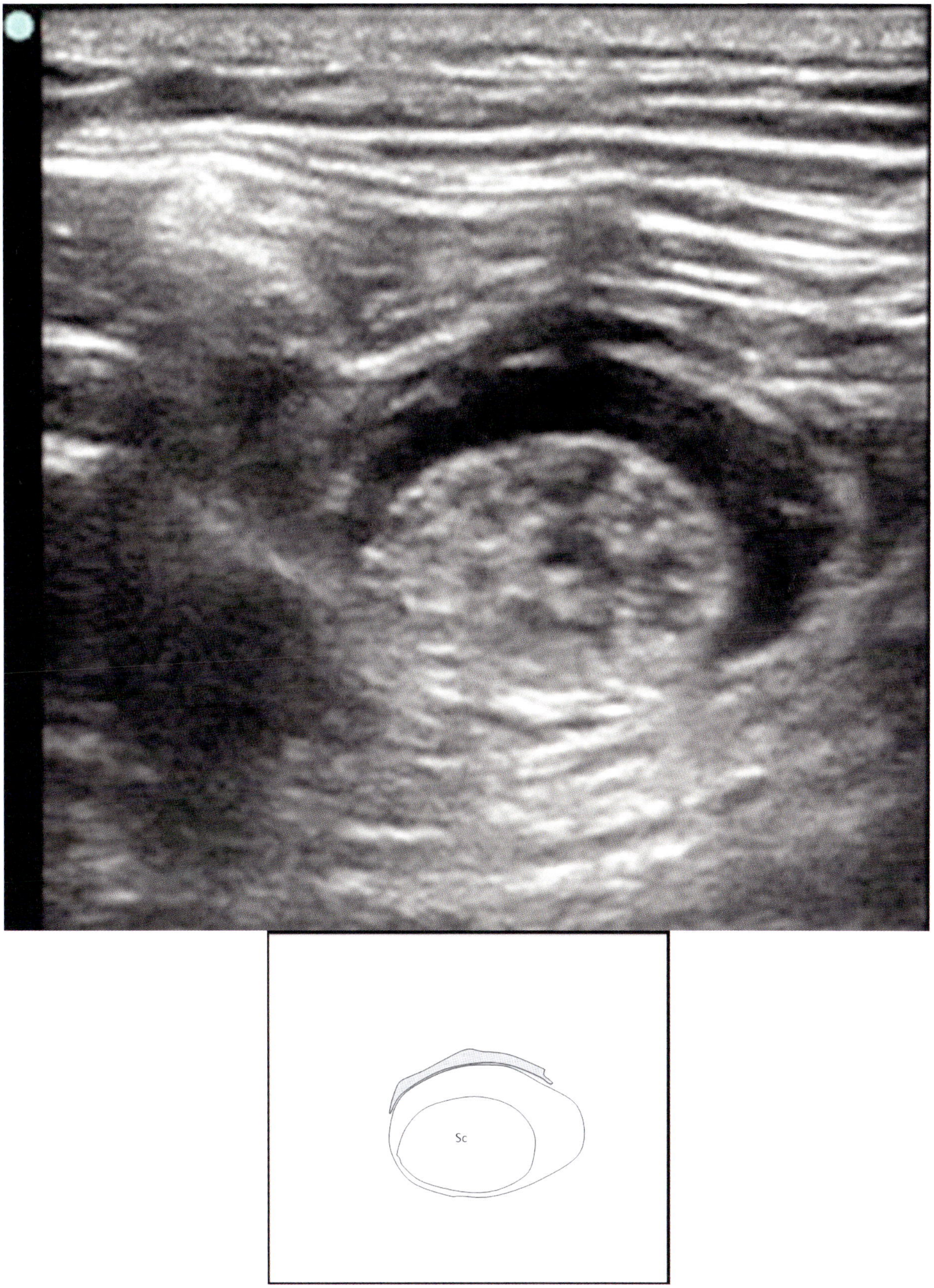

Fig. 13.12 Swelling of the sciatic nerve proximal to the point of injection with epineurial sheath view. *Sc* sciatic nerve

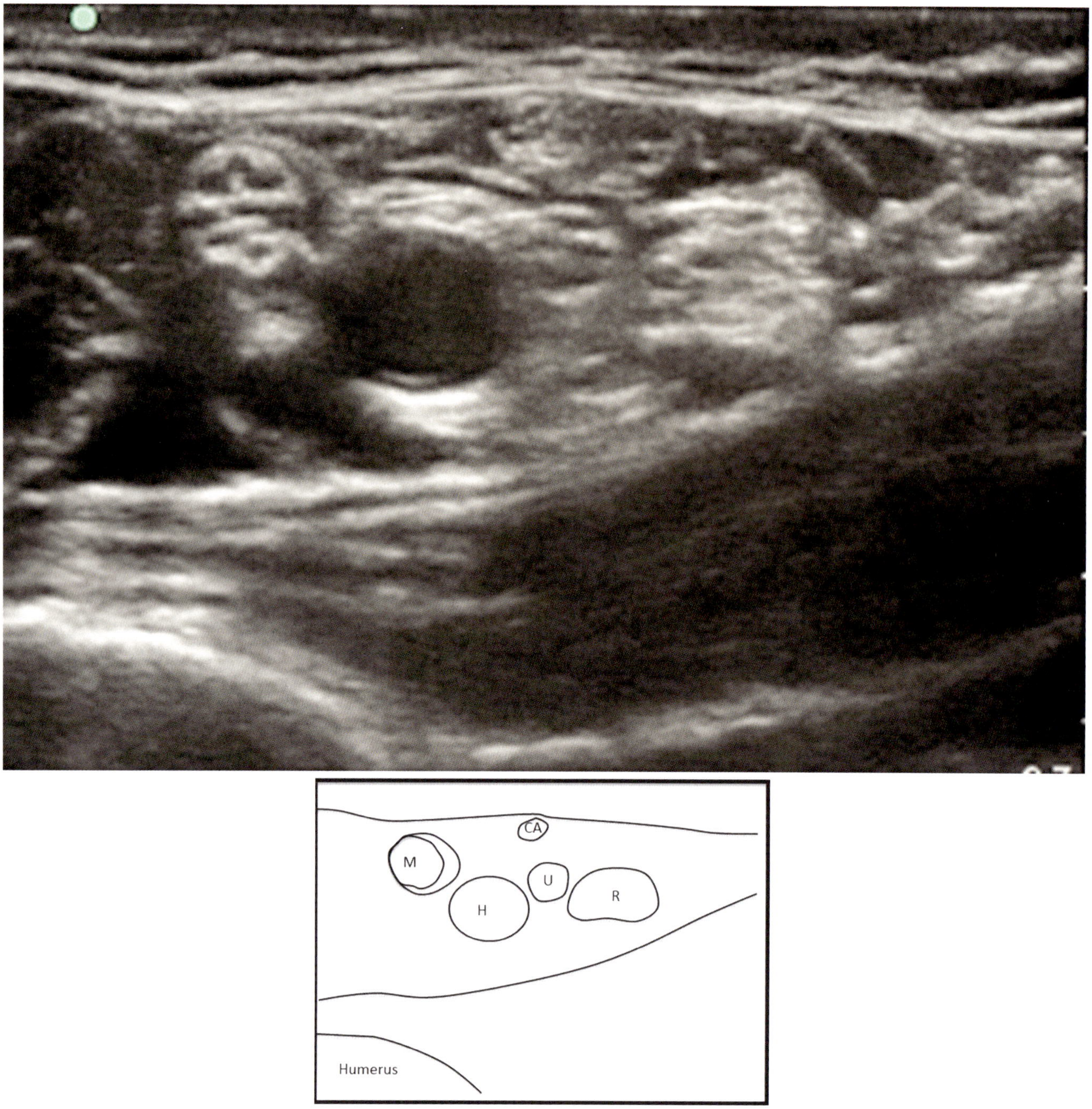

Fig. 13.13 Swelling of the median nerve after axillary block with epineurial sheath view. *H* humeral artery, *M* median nerve, *U* ulnar nerve, *R* radial nerve, *CA* cutaneous antebrachial nerve

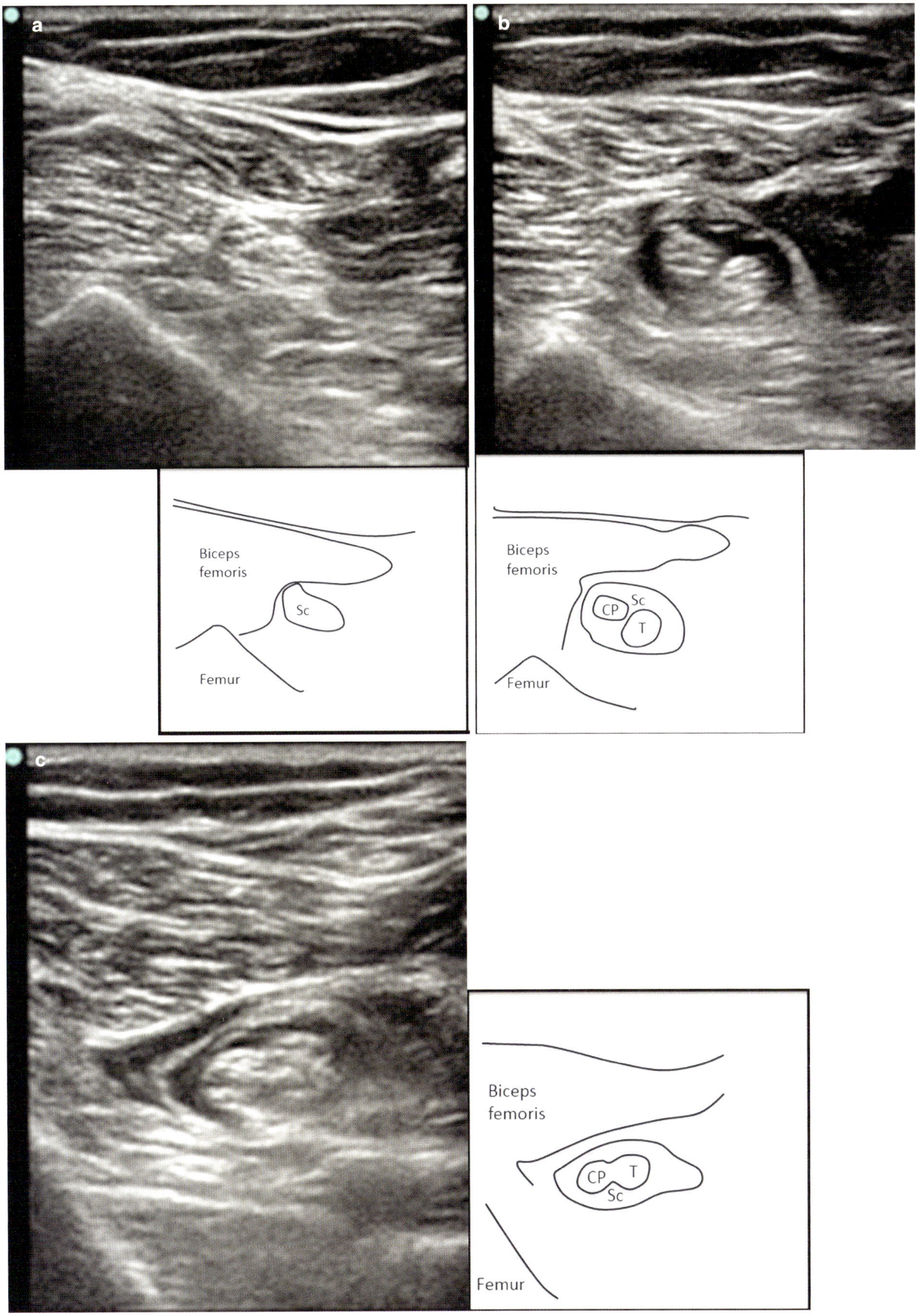

Fig. 13.14 Popliteal sciartic nerve block (**a**). Intraneural sciatic nerve injection with epineurial sheath view at the level of injection (**b**) and proximally (**c**). *CP* common peroneal nerve, *T* tibial nerve, *Sc* sciatic nerve

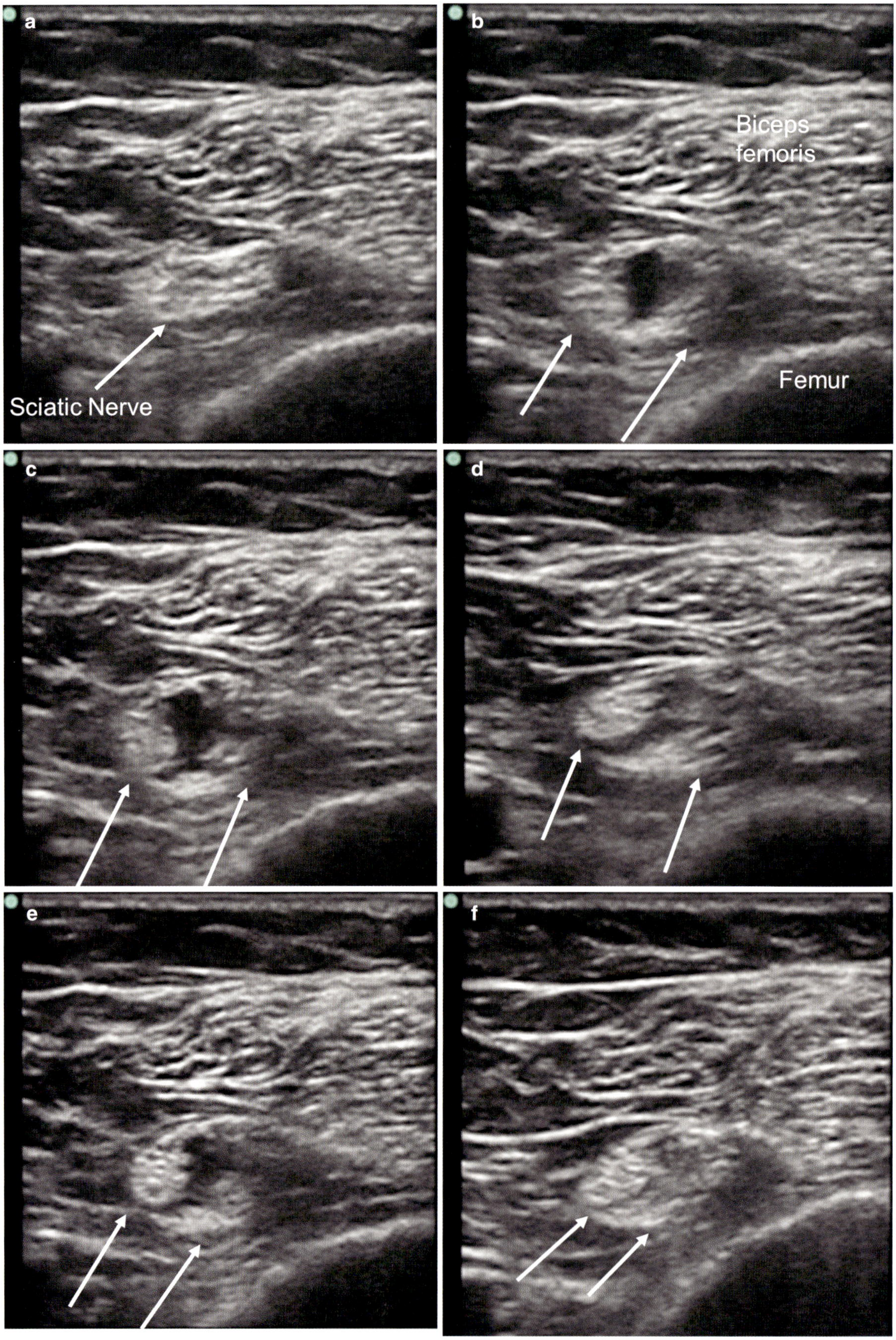

Fig. 13.15 Popliteal block at the level of sciatic nerve division (**a**). *Arrows* are showing the limits of sciatic nerve. Nerve separation during injection (**b**, **c**) and elastic retraction after the injection (**d**–**f**)

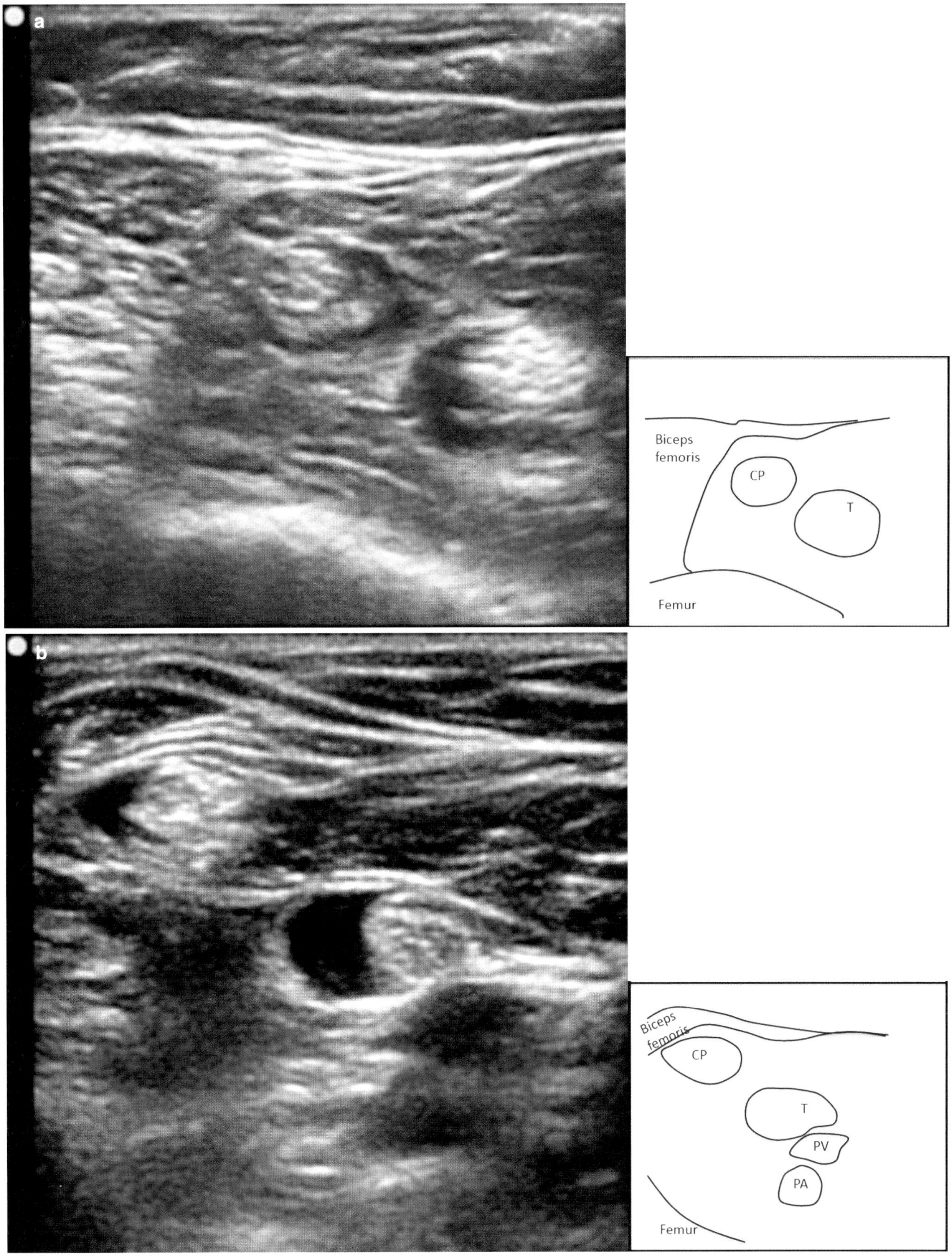

Fig. 13.16 Two examples of distal spread after popliteal sciatic nerve block (**a**, **b**). CP Common peroenal nerve, *T* tibial nerve, *PV* popliteal vein, *PA* popliteal artery

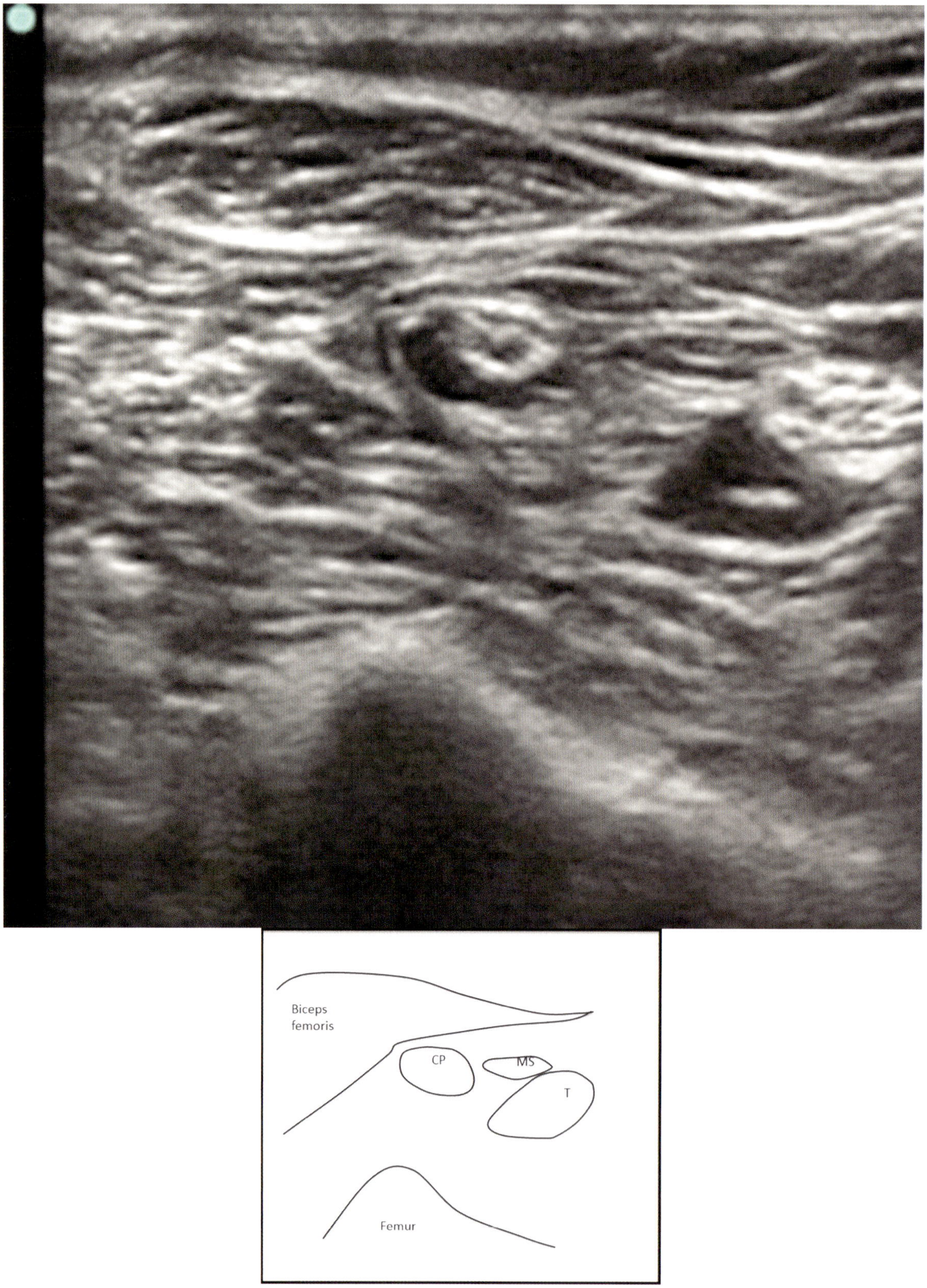

Fig. 13.17 Diffusion along the divisions of the nerve branches. *CP* common peroneal nerve, *T* tibial nerve, *MS* medial sural cutaneous nerve

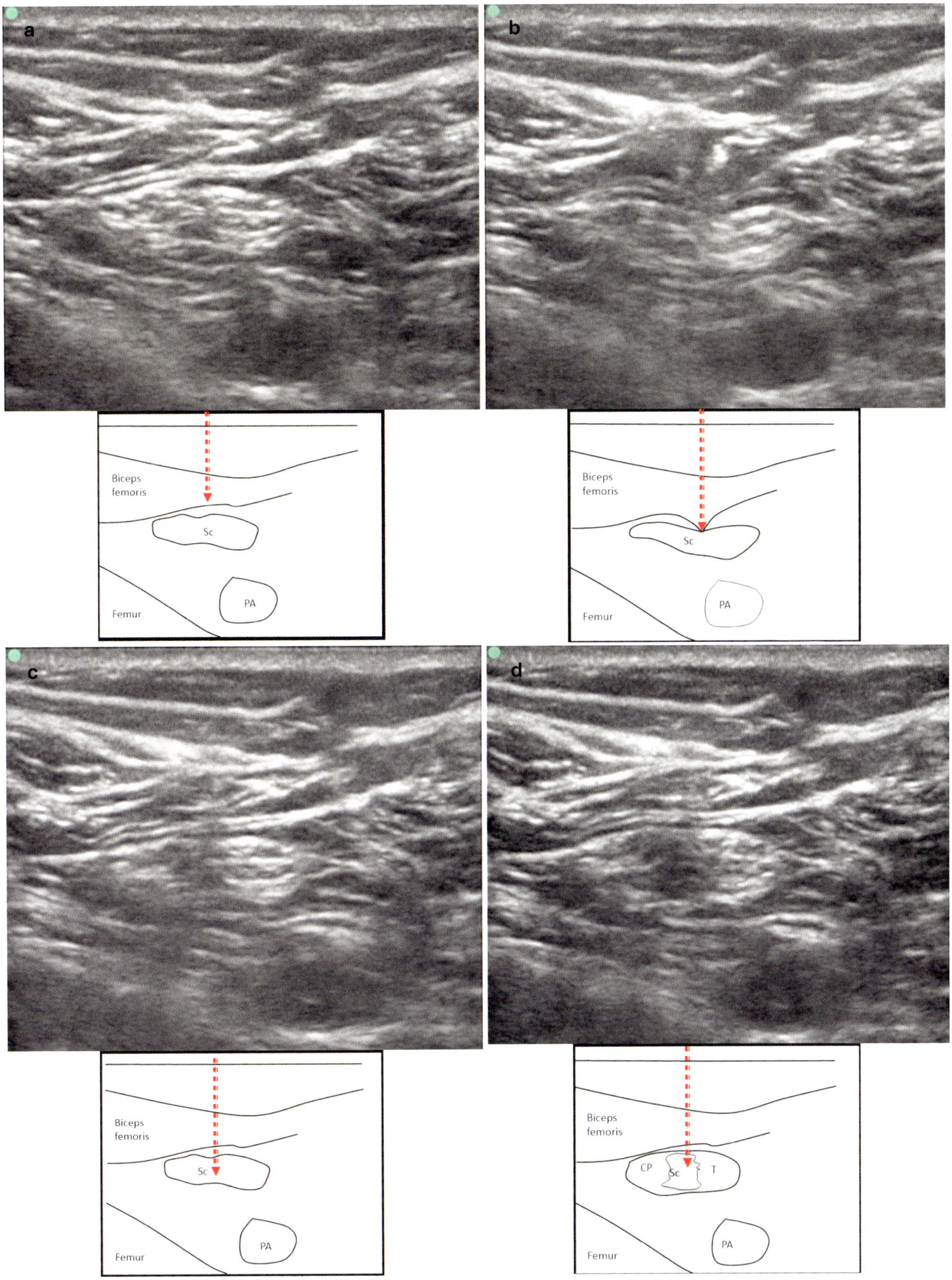

Fig. 13.18 Sheath (fasciae) resistance to piercing. Elastic resistance to puncture (**a**, **b**) after crossing the epineurium (**c**) and the demostration of intraneural spread (**d**). *Red dashed* line indicate the position of the needle. *CP* common peroenal nerve, *T* tibial nerve, *Sc* sciatic nerve, *PA* popliteal artery

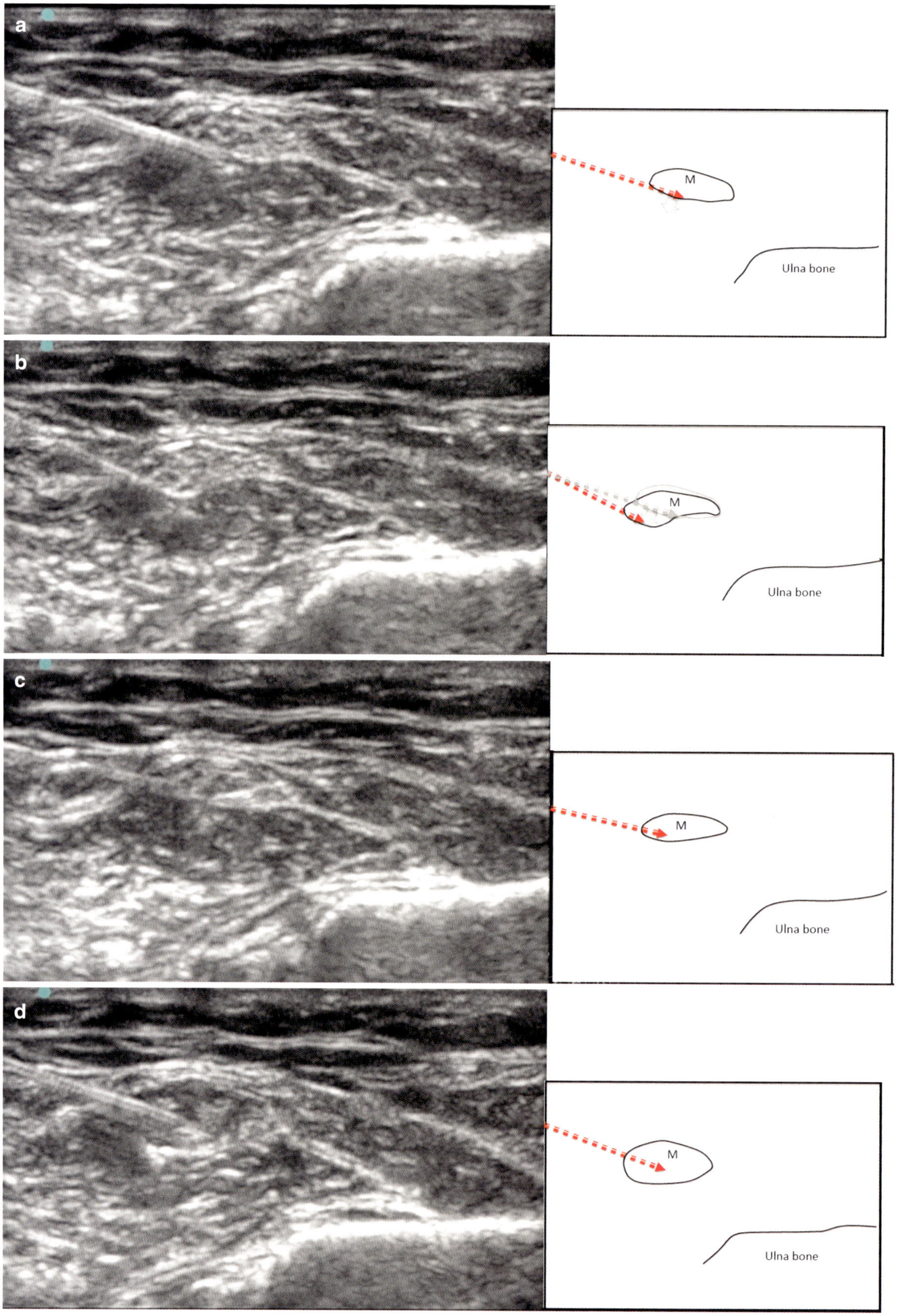

Fig. 13.19 Nerve movement associated with needle movement (piercing the nerve with the needle (**a**-**c**) and the demostration of intraneural spread (**d**). *Red dashed* line indicate the position of the needle. *M* median nerve

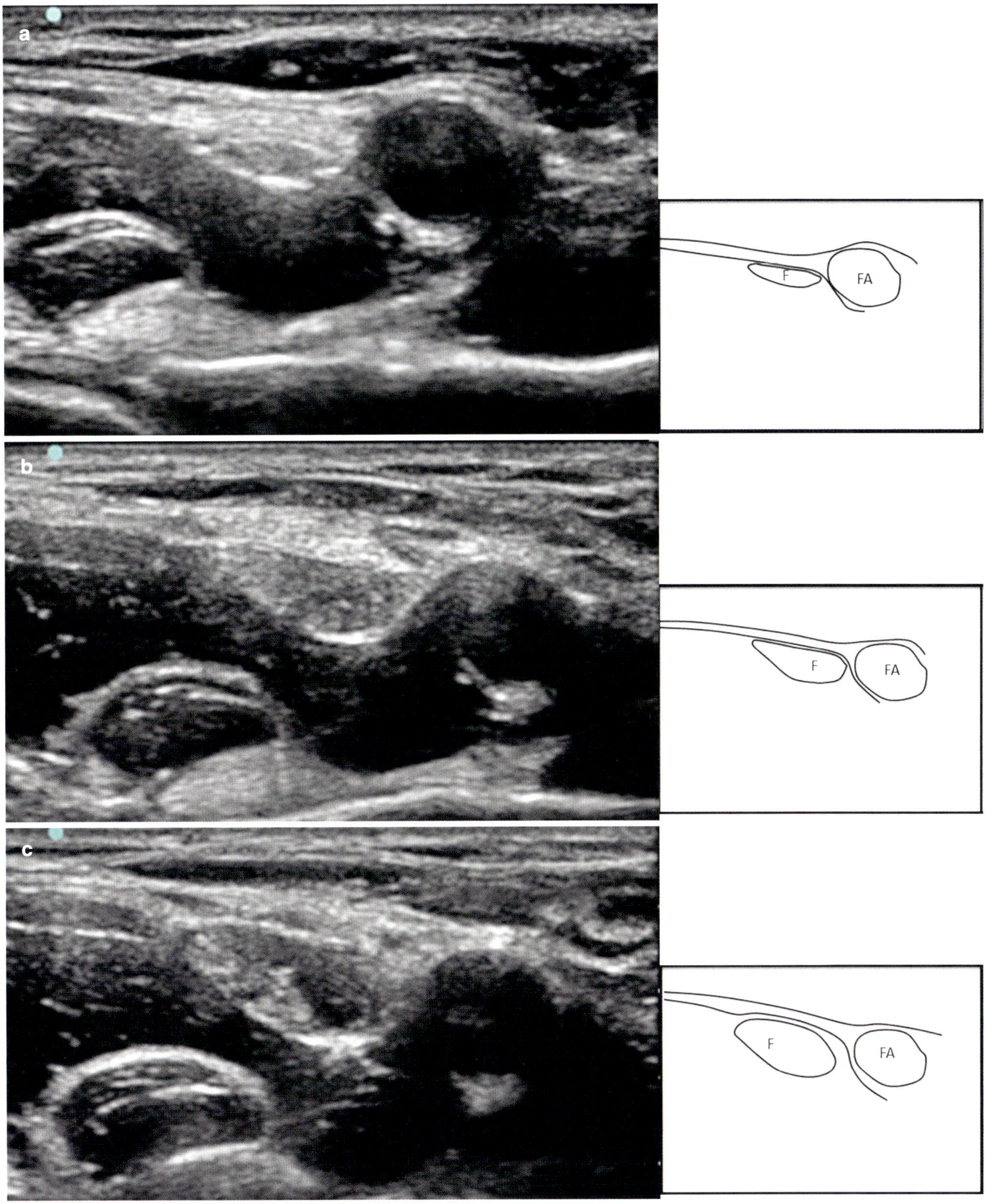

Fig. 13.20 Ultrasound guided femoral nerve block (**a**). Intraneural injection in a femoral nerve, with swelling and area increase (**b**, **c**). *F* femoral nerve, *FA* femoral artery

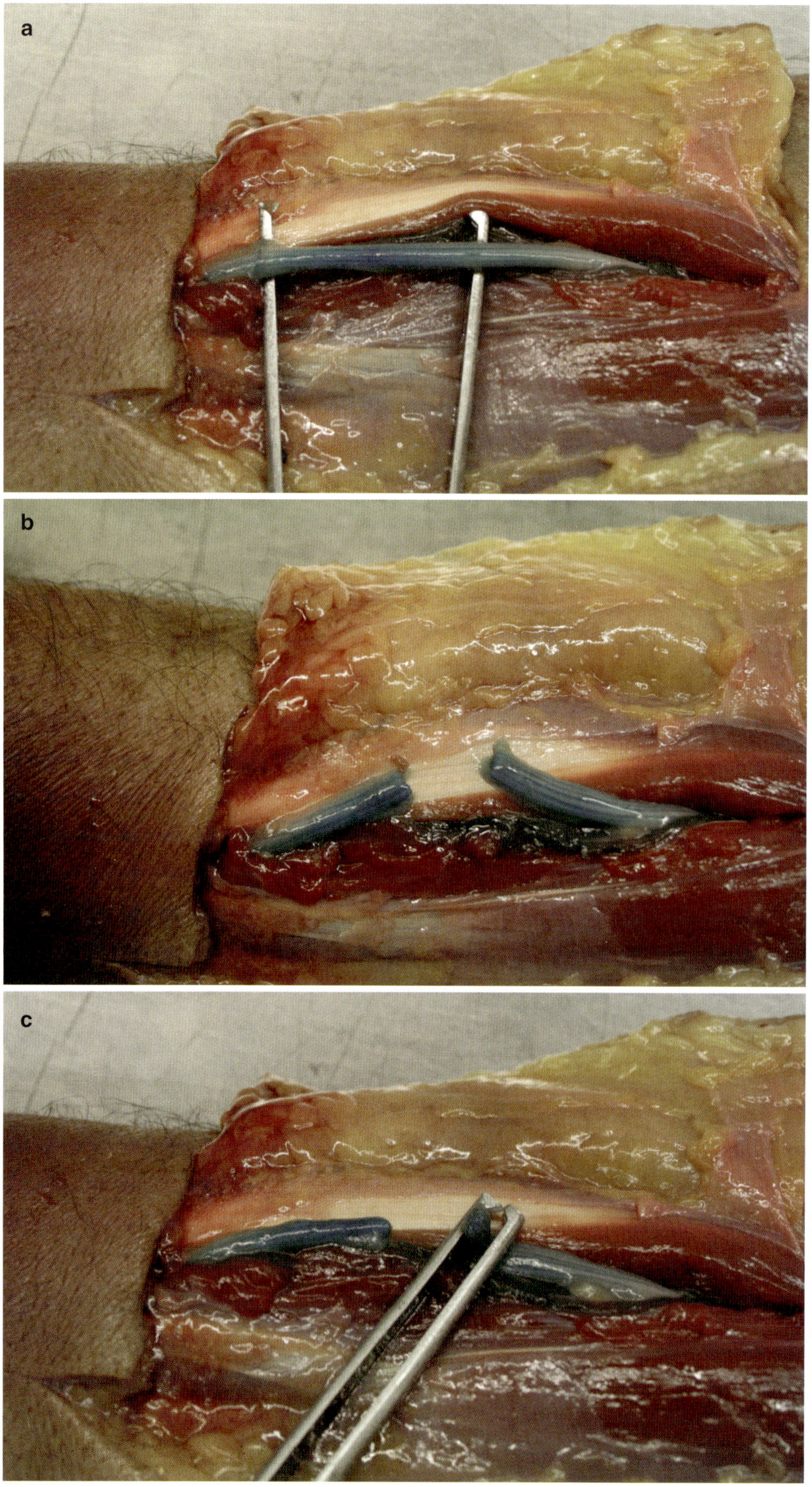

Fig. 13.21 (**a**–**c**) Intraneural injection of 1 mL of methylene blue in an ulnar nerve (cadaver). View the spread along the nerve and the dye inside the nerve

References

1. Ruiz A, Sala-Blanch X, Martinez-Ocón J, Carretero MJ, Sánchez-Etayo G, Hadzic A. Incidence of intraneural needle insertion in ultrasound-guided femoral nerve block: a comparison between the out-of-plane versus the in-plane approach. Rev Esp Anestesiol Reanim. 2014;61:73–7.
2. Sala-Blanch X, Reina MA, Ribalta T, Prats-Galino A. Sciatic nerve structure and nomenclature: epineurium to paraneurium: is this a new paradigm? Reg Anesth Pain Med. 2013;38:463–5.
3. Sala-Blanch X, de Riva N, Carrera A, López AM, Prats A, Hadzic A. Ultrasound-guided popliteal sciatic block with a single injection at the sciatic division results in faster block onset than the classical nerve stimulator technique. Anesth Analg. 2012; 114:1121–7.
4. Sala-Blanch X, López AM, Pomés J, Valls-Sole J, García AI, Hadzic A. No clinical or electrophysiologic evidence of nerve injury after intraneural injection during sciatic popliteal block. Anesthesiology. 2011;115:589–95.
5. Bargalló X, Carrera A, Sala-Blanch X, Santamaría G, Morro R, Llusá M, Gilabert R. Ultrasound-anatomic correlation of the peripheral nerves of the upper limb. Surg Radiol Anat. 2010;32:305–14.
6. Robards C, Hadzic A, Somasundaram L, Iwata T, Gadsden J, Xu D, Sala-Blanch X. Intraneural injection with low-current stimulation during popliteal sciatic nerve block. Anesth Analg. 2009;109:673–7.
7. Sala-Blanch X, Ribalta T, Rivas E, Carrera A, Gaspa A, Reina MA, Hadzic A. Structural injury to the human sciatic nerve after intraneural needle insertion. Reg Anesth Pain Med. 2009;34:201–5.
8. Sala Blanch X, López AM, Carazo J, Hadzic A, Carrera A, Pomés J, Valls-Solé J. Intraneural injection during nerve stimulator-guided sciatic nerve block at the popliteal fossa. Br J Anaesth. 2009;102: 855–61.
9. Hadzic A, Sala-Blanch X, Xu D. Ultrasound guidance may reduce but not eliminate complications of peripheral nerve blocks. Anesthesiology. 2008;108:557–8.
10. Sala Blanch X, Domingo Triadó V, Bargalló X, Carrera A. Intraneural sciatic nerve puncture in the popliteal fossa demonstrated by ultrasound. Rev Esp Anestesiol Reanim. 2006;53:269–70.

Histologic Features of Needle-Nerve and Intraneural Injection Injury as Seen on Light Microscopy

14

Ilvana Vuckovic-Hasanbegovic, Catherine Vandepitte, and Admir Hadzic

Needle-nerve injury and an intrafascicular injection of local anesthetic are relatively uncommon occurrences but may result in devastating complications after peripheral nerve blocks [1, 2]. Numerous case reports and case series have been published on disabling neurologic consequences after intramuscular injection of penicillin intraneurally into the sciatic nerve [1, 3–6]. Likewise, studies in experimental animals have documented neurotoxic effects of intraneural injections of local anesthetics [7, 8]. An inadvertent intrafascicular injection may lead to a variety of nerve injuries, some of which may result in permanent neurologic injury [9]. Intrafascicular injection of local anesthetic agents, in particular, leads to at least some degree of nerve damage, typically characterized by demyelination and Wallerian degeneration [10]. Of note, the most severe injury occurs closest to the injection site, whereas the extent of injury diminishes reciprocally with increasing distance from the injection site [10].

Several factors may influence the severity and consequences of needle-nerve or injection injury. The type of needle tip, needle size, insertion site, angle of insertion, pressure achieved during injection, and type and dose of medication injected (toxicity) all may affect or combine to produce various degrees of nerve injury [8–10]. The multitude of the aforementioned factors accounts for the sometimes conflicting reports in the literature regarding the incidence and magnitude of nerve injury with anesthetic injection [8, 11, 12]. The effects of intrafascicular and extrafascicular injection of numerous medications and local anesthetic solutions in common use have been well studied. Most injectates cause nerve injury when injected intrafascicularly, whereas extrafascicular injections produce little to no damage [13]. In addition, there appears to be a direct correlation between histologic injury and the concentration of medications (e.g. local anesthetics) and time of exposure to the nerve [14, 15]. After exposure to local anesthetics, rat, porcine, and canine sciatic nerves show infiltration by macrophages and myelin damage [15]. Besides apoptosis, local anesthetics may induce vasospasm of the intraneural blood vessels with a consequent decrease in the neural blood flow and nerve ischemia.

High injection pressure also may contribute to the extent of neurologic injury. Orebaugh et al. [16] reported that intrafascicular injections at the level of the brachial plexus root in fresh human cadavers invariably requires a high injection pressure (>15 psi) to initiate the injection. After an intraneural injection of local anesthetic in experimental animals, evidence of histologic and clinical nerve injury was reported, primarily after injections with high opening injection pressures, suggesting that high opening injection pressure may serve as an indicator of injection into fascicles or other tissues with reduced compliance.

Peripheral nerve injury typically is classified into five degrees, as originally described by Sunderland [17]. First-degree injury, or neurapraxia, involves segmental demyelination. Second-degree injury, or axonotmesis, involves injury to both the axon and the myelin without injury to the endoneurial tissue. Third-degree injury involves the axon, myelin, and endoneurium but leaves the perineurium intact. Fourth-degree injury involves the axon, myelin, endoneurium, and perineurium, resulting in derangement of the fascicular

I. Vuckovic-Hasanbegovic, MD, PhD (✉)
Department of Anatomy, School of Medicine, University of Sarajevo, Sarajevo, Bosnia Herzegovina
e-mail: ilvana.hasanbegovic@gmail.com, ilvana2810@hotmail.com

C. Vandepitte, MD
Department of Anesthesiology, ZOL Genk, Catholic University of Leuven, Leuven, Belgium
e-mail: belgette@gmail.com

A. Hadzic, MD, PhD
Department of Anesthesiology, St. Luke's-Roosevelt Hospital Center, New York School of Regional Anesthesia (NYSORA), New York, NY, USA

College of Physicians and Surgeons Columbia University, New York, NY, USA
e-mail: admir@nysora.com

M.A. Reina et al. (eds.), *Atlas of Functional Anatomy for Regional Anesthesia and Pain Medicine: Human Structure, Ultrastructure and 3D Reconstruction Images*, DOI 10.1007/978-3-319-09522-6_14,

architecture by scar tissue with the epineurium left unaffected. Fifth-degree injuries occur when the entire nerve fascicle is transected.

Knowledge of peripheral nerve histology is crucial for understanding the mechanisms of nerve injury. Nerve fibers, the basic structural and functional units of peripheral nerves, are made up of axons and Schwann cells. Individual nerve fibers are enveloped in a delicate layer of connective tissue called endoneurium. Many nerve fibers collect to form a fascicle. Fascicles are bound by the perineurium, a connective tissue sheath consisting of several layers of perineurial cells. The connective tissue surrounding each fascicle is called the internal epineurium, whereas the connective tissue surrounding the entire nerve is called external epineurium, composed primarily of collagen and elastic fibers. The mesoneurium, the loose areolar tissue surrounding the outside of the nerve, extends from the epineurium to the surrounding tissue.

Connective tissue of peripheral nerves also may be classified as high-compliance, expansible connective tissue, or epineurium, and low-compliance connective tissue, or perineurium and the intraperineural contents (fascicles), in which the perineural cells are densely packed. This is the reason injection inside perineurium requires more force to initiate (greater opening pressure) [12, 16].

After an intraneural injection, connective tissue within the epineurium and perifascicular regions typically demonstrates hypercellularity of the mononuclear inflammatory process type, with an increased number of macrophages, lymphocytes, and plasma cells. Blood vessels may show signs of hyperemia and extravasation in some areas.

Intrafascicularly, axons may appear swollen, whereas Schwann cells may be enlarged with hyperchromatic nuclei compared with the nerve fibers located centrally, which show no evidence of injury. With intrafascicular injections, a variable degree of injury to Schwann cells, blood vessels, perineurium, and epineurium may be seen. The perineurium often demonstrates division of lamellas with disintegration at the place of needle injury and loss of demarcation toward the surrounding perifascicular connective tissue and the closest axons. The axons of fascicles in which local anesthetics are injected show damage in general, although subperineural injury often is worse than injury to more centrally located axons [10]. Some axons demonstrate progressive axolysis up to complete disintegration, resulting in residual structureless tissue on light microscopy. In other cases, the myelin sheath shows increased acidophilia and occasionally loss of normal structure and transformation into thin hyperacidophilic structureless rings. Schwann cells may appear hypertrophic, and some protrude noticeably into the interstitium, giving the impression that the fiber structure has separated. The epineurium also shows hypercellularity of the mononuclear inflammatory process type, with an increased number of macrophages, lymphocytes, and plasma cells. Edema, a sign of acute trauma, usually is noticed first in the subperineural region, along with a lack of interstitial collagen matrix, and may be identified as an area of "lost structure." Edema may be seen in the endoneural space and perivascular region around blood vessels and, in the most serious cases, in the interstitial space between nerve fibers.

In summary, needle–nerve contact, needle-nerve injury, and intraneural injection of local anesthetic result in highly variable degrees of nerve injury, depending on many anatomic, physical, pharmacologic, and hydrodynamic factors. Consequently, the light microscopy expression of these injuries varies greatly, and sometimes it is a challenge to distinguish the actual injury from the artifacts incurred during the processing of histologic samples. The figures refer to common histologic features of intact peripheral nerves and nerves that have been exposed to various types of needle-nerve injury and intraneural injection of local anesthetics (Figs. 14.1, 14.2, 14.3, 14.4, 14.5, 14.6, and 14.7).

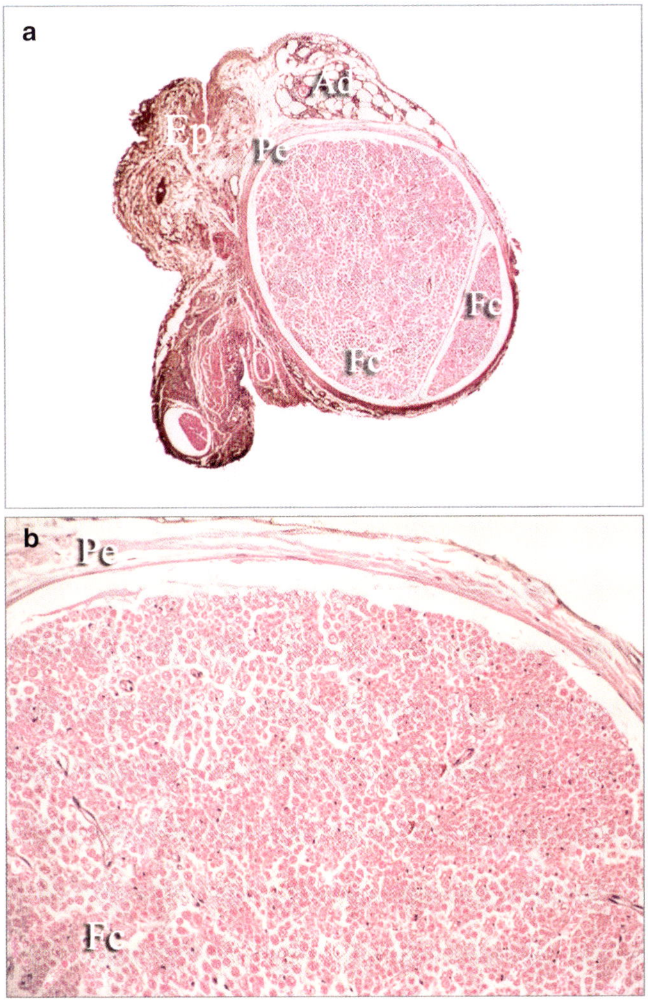

Fig. 14.1 Typical histologic appearance of a rat peripheral nerve. (**a**) Connective tissue of the epineurium is unbalanced in terms of collagen richness, with partly altered tinctorial properties and areal-located adipocytes. The structures of the perineurium and nerve fibers are regular. (**b**) Detail of panel **a**. Typical laminar structure of the perineurium and a delicate fascicular architecture without lesions. *Ad* adipocyte, *Ep* epineurium, *Fc* fascicle, *Pe* perineurium. Hematoxylin and eosin (H&E) stain. Magnification: ×40 (**a**); ×100 (**b**)

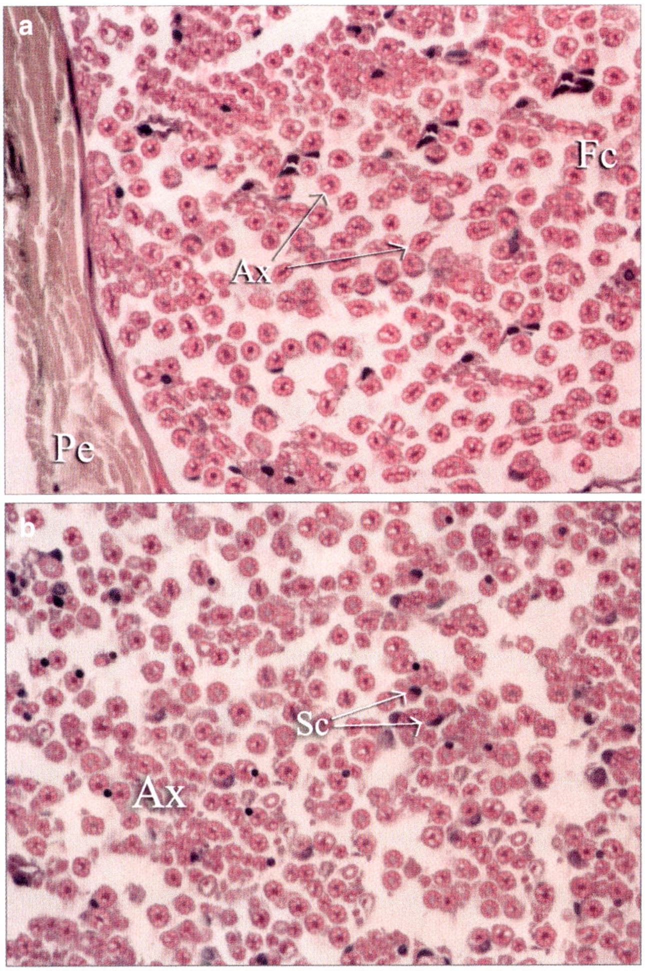

Fig. 14.2 Typical histologic appearance of a rat peripheral nerve. (**a**) Magnified image with intact perineurium and axons. (**b**) Intraneural, extrafascicular application of 2 % lidocaine. Centrally located nerve fibers show no signs of degenerative changes. Schwann cell nuclei are prominent. *Ax* axon, *Fc* fascicle, *Pe* perineurium, *Sc* Schwann cell. H&E stain. Magnification: ×400 (**a**, **b**)

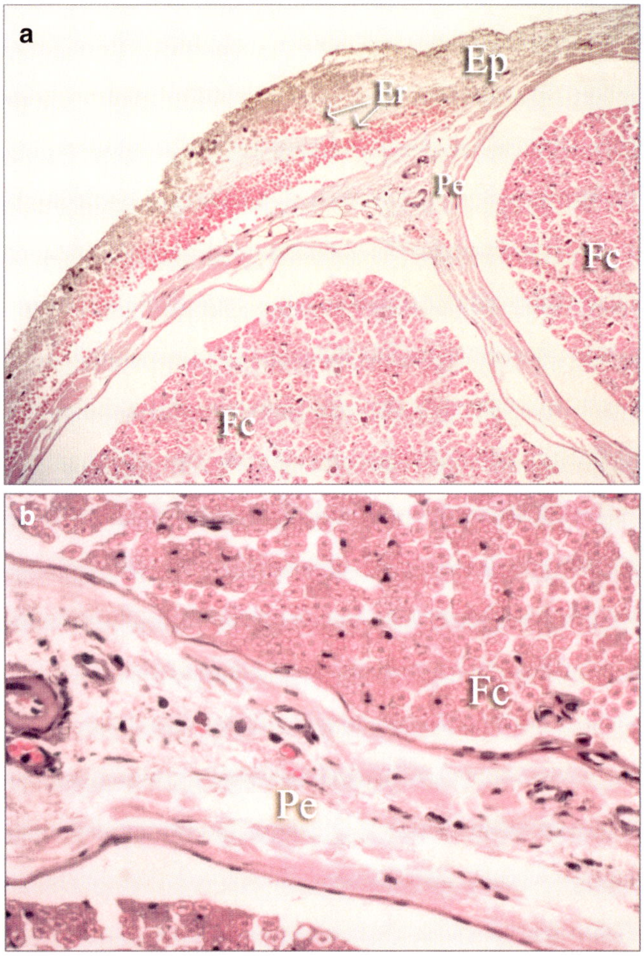

Fig. 14.3 Intraneural extrafascicular application of 0.75 % ropivacaine followed by pressure. (**a**) Epineurium infused with erythrocytes. The perineurium is preserved, as is the axonal structure intrafascicularly. (**b**) Detail of panel **a**. *Ep* epineurium, *Er* erythrocytes, *Fc* fascicle, *Pe* perineurium. H&E stain. Magnification: ×100 (**a**); ×400 (**b**)

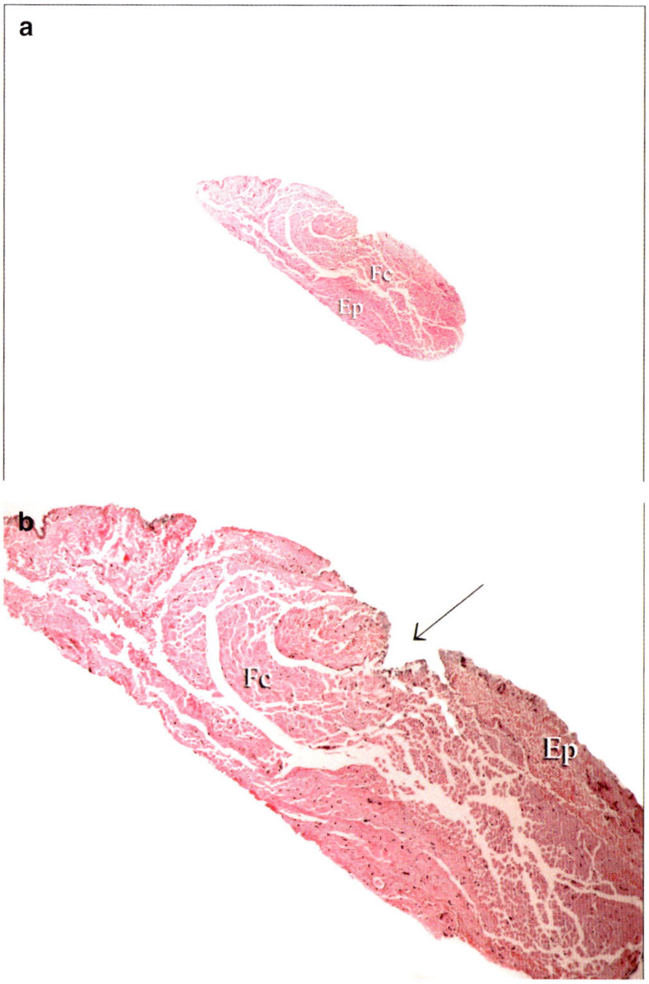

Fig. 14.4 Intraneural intrafascicular application of saline. (**a**) Section of rat sciatic nerve composed of a single nerve fascicle. (**b**) Detail of panel **a**. Rupture of epineural connective tissue (*arrow*) is noted, with a loss of structural space intrafascicularly. *Ep* epineurium, *Fc* fascicle. H&E stain. Magnification: ×10 (**a**); ×40 (**b**)

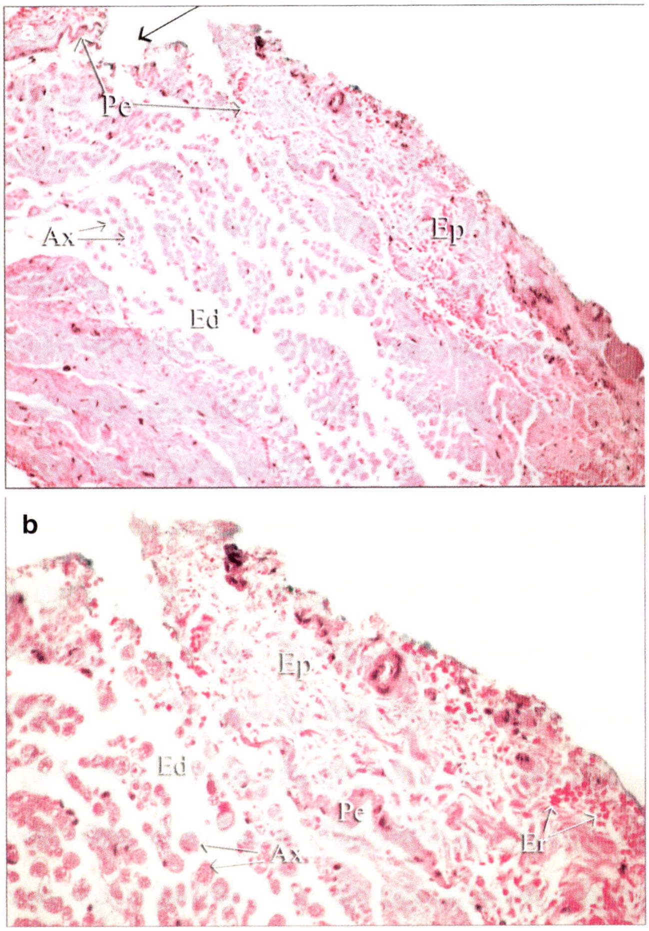

Fig. 14.5 Intraneural intrafascicular application of H&E stain. (**a**) Visible injury to the epineurium and perineurium, probably corresponding to the site of needle penetration (*thick arrow*). Diffuse axonal injury is present. (**b**) Detail of panel **a**. Intraneural intrafascicular application of saline with high (>20 psi) injection pressure. Collagen fibers of epineurium are disintegrated, and there is extravasation of erythrocytes. The perineurium is ruptured. Axons show evidence of nerve injury, with increased interaxonal distances and edema. *Ax* axon, *Ed* edema, *Ep* epineurium, *Er* erythrocyte, *Pe* perineurium. Magnification: ×100 (**a**); ×250 (**b**)

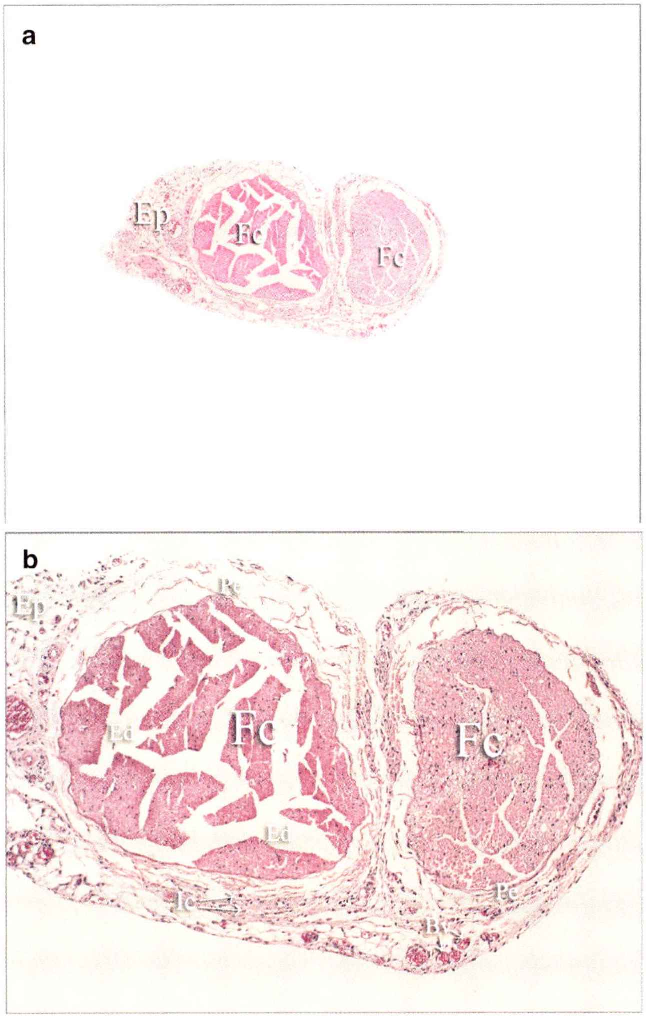

Fig. 14.6 Intraneural intrafascicular application of 2 % lidocaine with high (>20 psi) injection pressure. (**a**) Rat sciatic nerve composed of two nerve fasciculi. (**b**) Detail of panel **a**. The epineurium is edematous, with infiltration of inflammatory elements and hyperemic blood vessels. The perineurium appears delaminated. Subperineural edema is present in the deeper area of the fasciculus. *Bv* blood vessels, *Ed* edema, *Ep* epineurium, *Fc* fascicle, *Ic* inflammatory cells, *Pe* perineurium. H&E stain. Magnification: ×10 (**a**); ×40 (**b**)

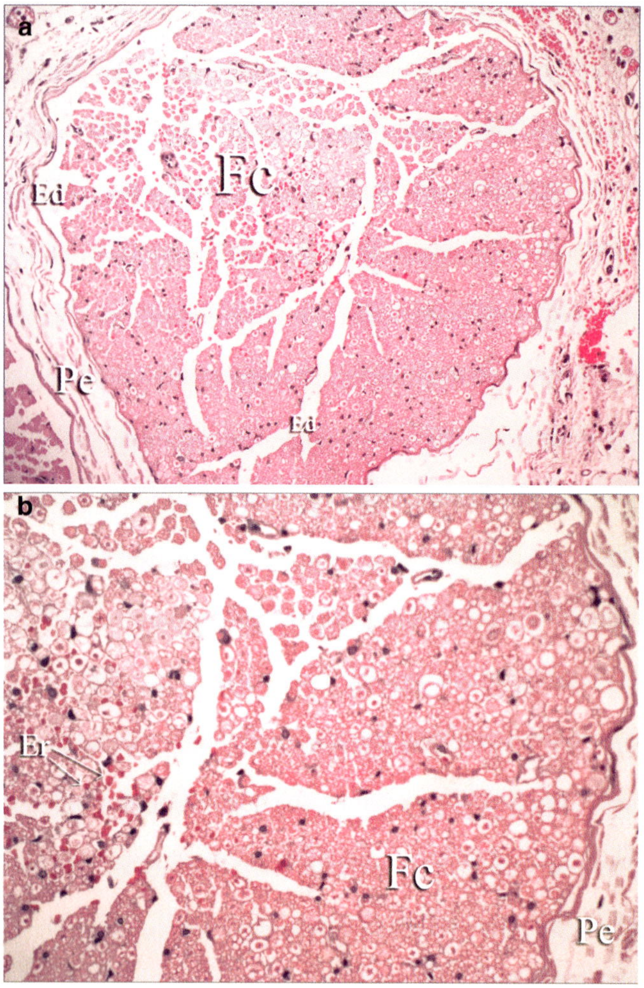

Fig. 14.7 Intraneural intrafascicular application of 2 % lidocaine with high (>20 psi) injection pressure. (**a**) The perineurium shows disintegration and separation of lamellas. Intrafascicularly, diffuse nerve fiber damage is present. (**b**) Detail of panel **a**. The axons show varying degrees of diffuse injury, ranging from changes in their tinctorial properties and position within the axon fibers to axolysis, unequal fiber diameter, and complete disintegration. Extravasally located erythrocytes are visible in the intrafascicular area. *Er* erythrocytes, *Fc* fascicle, *Pe* perineurium. H&E stain. Magnification: ×100 (**a**); ×250 (**b**)

References

1. Lenters TR, Davies J, Matsen 3rd FA. The types and severity of complications associated with interscalene brachial plexus block anesthesia: local and national evidence. J Shoulder Elbow Surg. 2007;16:379–87.
2. Shah S, Hadzic A, Vloka JD, Cafferty MS, Moucha CS, Santos AC. Neurologic complication after anterior sciatic nerve block. Anesth Analg. 2005;100:1515–7.
3. Fatunde OJ, Familusi JB. Injection-induced sciatic nerve injury in Nigerian children. Cent Afr J Med. 2001;47:35–8.
4. Villarejo FJ, Pascual AM. Injection injury of the sciatic nerve (370 cases). Childs Nerv Syst. 1993;9:229–32.
5. Ramtahal J, Ramlakhan S, Singh K. Sciatic nerve injury following intramuscular injection: a case report and review of the literature. J Neurosci Nurs. 2006;38:238–40.
6. Senes FM, Campus R, Becchetti F, Catena N. Sciatic nerve injection palsy in the child: early microsurgical treatment and long-term results. Microsurgery. 2009;29:443–8.
7. Gentili F, Hudson AR, Hunter D, Kline DG. Nerve injection injury with local anesthetic agents: a light and electron microscopic, fluorescent microscopic, and horseradish peroxidase study. Neurosurgery. 1980;6:263–72.
8. Whitlock EL, Brenner MJ, Fox IK, Moradzadeh A, Hunter DA, Mackinnon SE. Ropivacaine-induced peripheral nerve injection injury in the rodent model. Anesth Analg. 2010;111:214–20.
9. Selander D, Brattsand R, Lundborg G, Nordborg C, Olsson Y. Local anesthetics: importance of mode of application, concentration and adrenaline for the appearance of nerve lesions. An experimental study of axonal degeneration and barrier damage after intrafascicular injection or topical application of bupivacaine (Marcain). Acta Anaesthesiol Scand. 1979;23:127–36.
10. Farber SJ, Saheb-Al-Zamani M, Zieske L, Laurido-Soto O, Bery A, Hunter D, et al. Peripheral nerve injury after local anesthetic injection. Anesth Analg. 2013;117:731–9.
11. Bouaziz H, Iohom G, Estèbe JP, Campana WM, Myers RR. Effects of levobupivacaine and ropivacaine on rat sciatic nerve blood flow. Br J Anaesth. 2005;95:696–700.
12. Hadzic A, Dilberovic F, Shah S, Kulenovic A, Kapur E, Zaciragic A, et al. Combination of intraneural injection and high injection pressure leads to fascicular injury and neurologic deficits in dogs. Reg Anesth Pain Med. 2004;29:417–23.
13. Mackinnon SE, Hudson AR, Llamas F, Dellon AL, Kline DG, Hunter DA. Peripheral nerve injury by chymopapain injection. J Neurosurg. 1984;61:1–8.
14. Perez-Castro R, Patel S, Garavito-Aguilar ZV, Rosenberg A, Recio-Pinto E, Zhang J, et al. Cytotoxicity of local anesthetics in human neuronal cells. Anesth Analg. 2009;108:997–1007.
15. Yang S, Abrahams MS, Hurn PD, Grafe MR, Kirsch JR. Local anesthetic Schwann cell toxicity is time and concentration dependent. Reg Anesth Pain Med. 2011;36:441–51.
16. Orebaugh SL, Mukalel JJ, Krediet AC, Weimer J, Filip P, McFadden K, Bigeleisen PE. Brachial plexus root injection in a human cadaver model: injectate distribution and effects on the neuraxis. Reg Anesth Pain Med. 2012;37:525–9.
17. Sunderland S. The anatomy and physiology of nerve injury. Muscle Nerve. 1990;13:771–84.

Structure of Nerve Lesions After "In Vitro" Punctures

15

Xavier Sala-Blanch, Miguel Angel Reina, Teresa Ribalta, Admir Hadzic, and Anna Carrera

The neurologic consequences of an intraneural needle insertion and injection of local anesthetic depend on many factors, such as the design of the needle tip, the injection force, and the chemical structure and concentration of the injectate. However, the anatomic specifics of a given nerve and how its organization can protect it against external injury may be the most significant factor determining the likelihood of injury [1, 2]. Finally, any preexisting pathology of the peripheral nerve also contributes to the risk of postblock neurologic symptoms.

The most important factors determining the vulnerability of a given nerve to an intraneural needle insertion or injection of local anesthetic probably are the ratio of connective (epineural) tissue to axonal tissue (fascicles) and the size of the nerve fascicles or fascicular bundles. The less connective tissue and the larger the fascicles (such as at the root level), the greater the chances an advancing intraneural needle will enter any impeding fascicles. The mechanisms of injection injury include (1) direct needle damage and (2) intraneural hydraulic distention (injection) of the nerve and disruption of the communication among fascicles (interfascicular plexus) and nutrient vessels. Finally, the anatomic organization of the nerve vasculature (vasa nervorum) also plays a role as injury or injection pressure may lead to an intraneural hematoma, nerve ischemia from flow disruption to a watershed area (as occurs in the brain and spinal cord). Pressure of the injected local anesthetic also may add to the ischemia by compressing capillaries and venules. Direct chemical toxicity from the local anesthetic itself or from additives such as epinephrine likely is enhanced in the presence of mechanical trauma to the nerve fascicles and disruption of the perineurium.

The absence of neurologic injury with intraneural injections during sciatic popliteal block in various nerve block models is at apparent odds with conventional thinking that intraneural injection invariably leads to nerve injury. However, the physical location of the injectate (intrafascicular versus peri- or extrafascicular) is the key factor determining whether neurologic injury will result from an intraneural injection. The low incidence of complications after intraneural injections with sciatic popliteal block likely is a result of the anatomic characteristics of the sciatic nerve within the popliteal fossa. The connective tissue of the sciatic nerve comprises up to 80 % of the cross-sectional area of the nerve at this level, thus favoring the path of the needle through the path of lesser resistance (adipose tissue) rather than passing into and through the more densely packed fascicles. For instance, needles deliberately inserted into sciatic nerves

X. Sala-Blanch, MD (✉)
Human Anatomy and Embryology Unit, School of Medicine, University of Barcelona, Barcelona, Spain

Department of Anesthesiology and Critical Care, Clinic Hospital, Barcelona, Spain
e-mail: xavi.sala.blanch@gmail.com

M.A. Reina, MD, PhD
Department of Clinical Medical Sciences and Institute of Applied Molecular Medicine, School of Medicine, University of CEU San Pablo, Madrid, Spain

Department of Anesthesiology, Madrid-Monteprı́ncipe University Hospital, Madrid, Spain
e-mail: miguelangel@perticone.e.telefonica.net

T. Ribalta, MD, PhD
Department of Pathology, Clinic Hospital, University of Barcelona, Barcelona, Spain
e-mail: tribalta@clinic.ub.es

A. Hadzic, MD, PhD
Department of Anesthesiology, St. Luke's-Roosevelt Hospital Center, New York School of Regional Anesthesia (NYSORA), New York, NY, USA

College of Physicians and Surgeons Columbia University, New York, NY, USA
e-mail: admir@nysora.com

A. Carrera, MD, PhD
Neuroscience, Embriology, Molecular Oncology and Clinical Anatomy Group (NEOMA), School of Medicine, University of Girona, Girona, Spain
e-mail: anna.carrera@udg.edu

M.A. Reina et al. (eds.), *Atlas of Functional Anatomy for Regional Anesthesia and Pain Medicine: Human Structure, Ultrastructure and 3D Reconstruction Images*, DOI 10.1007/978-3-319-09522-6_15,

are more likely to pass between rather than through the fascicles. Two recent studies in animal models suggest that low-pressure sciatic nerve injection is associated with interfascicular rather than intrafascicular injection. Indeed, resistance to injection was reported as "normal" in all studies with intraneural injections resulting in no neurologic complications. Motor response during low–current intensity nerve stimulation, high injection pressure, and pain on injection were present simultaneously in several patients with suspected intraneural intrafascicular injections. Because none of the monitoring methods is failsafe with regard to causing nerve injury, their use in combination is suggested for routine monitoring and documentation to reduce the risk of an intraneural intrafascicular injection.

Because an intraneural injection is much more likely to occur in the nonconducting connective tissue elements of the peripheral nerve than in the axon-containing fascicle(s), it is not surprising that long-term nerve injury is rare during peripheral nerve blocks. In fact, it appears that the long-established practice of low–current intensity nerve stimulator guidance during peripheral nerve blocks without ultrasound often results in subepineural placement of the needle and an intraneural but extrafascicular injection with a low incidence of permanent neurologic complications. This is because a motor response to nerve stimulation at low current intensities (0.3–0.5 mA) often is accomplished only after the needle tip enters the epineural layer (of the sciatic nerve).

The absence of neurologic complications reported in a recent small series, however, should not lead one to assume that the indiscriminate practice of intraneural injection in all peripheral nerve block models is a safe and recommended practice. Injection within an epineural or tissue sheath enveloping a major plexus (e.g., brachial plexus) and the popliteal sciatic nerve currently is recommended. However, intraneural injection in other peripheral nerves may be associated with an unacceptable risk of complications and cannot be recommended. This is because these nerves have a less defined epineural layer or sheath and a greater ratio of nerve to connective tissue, making it much more likely that an intraneural injection will result in disruptive injury to the fascicles. Fortunately, the smaller proportion of the connective tissue in these nerves allows perineural injections to be effective, even with small volumes of local anesthetic.

In summary, intraneural, extrafascicular injection of local anesthetics may be a common occurrence with low–current intensity nerve stimulator–guided popliteal sciatic nerve blocks. Such injections appear to be associated with a high success rate and a low risk of neurologic injury, provided that injections are not delivered when the motor response is <0.3 mA (0.1 ms), the injection pressure is low (<15 psi), and/or the injection is asymptomatic. Therefore, based on the available data, it may be suggested that the goal of ultrasound-guided blocks of the brachial plexus and popliteal sciatic nerve should be similar to that of nerve stimulation: injection under the outer epineural layer (sheath) but outside the fascicles or fascicular bundles. In contrast, no data exist to suggest that injections into roots (including interscalene blocks) or intraneural injections in distal peripheral nerves are equally safe, nor is this an established practice; therefore, it cannot be recommended. Because none of the methods of nerve localization is failsafe, a multimodal combination of ultrasound monitoring, nerve stimulation, and assessment of resistance to injection is recommended. To decrease the risk of intrafascicular injection, local anesthetic should not be injected when the needle is inserted into fascicles or roots of a nerve plexus (as seen on ultrasound), motor response is present with low–current nerve stimulation (≤0.3 mA; 0.1 ms), or pain or high resistance to injection (≥15 psi) is present (Figs. 15.1, 15.2, 15.3, 15.4, 15.5, 15.6, 15.7, 15.8, 15.9, 15.10, 15.11, 15.12, 15.13, 15.14, 15.15, 15.16, 15.17, 15.18, 15.19, 15.20, 15.21, and 15.22) (Table 15.1).

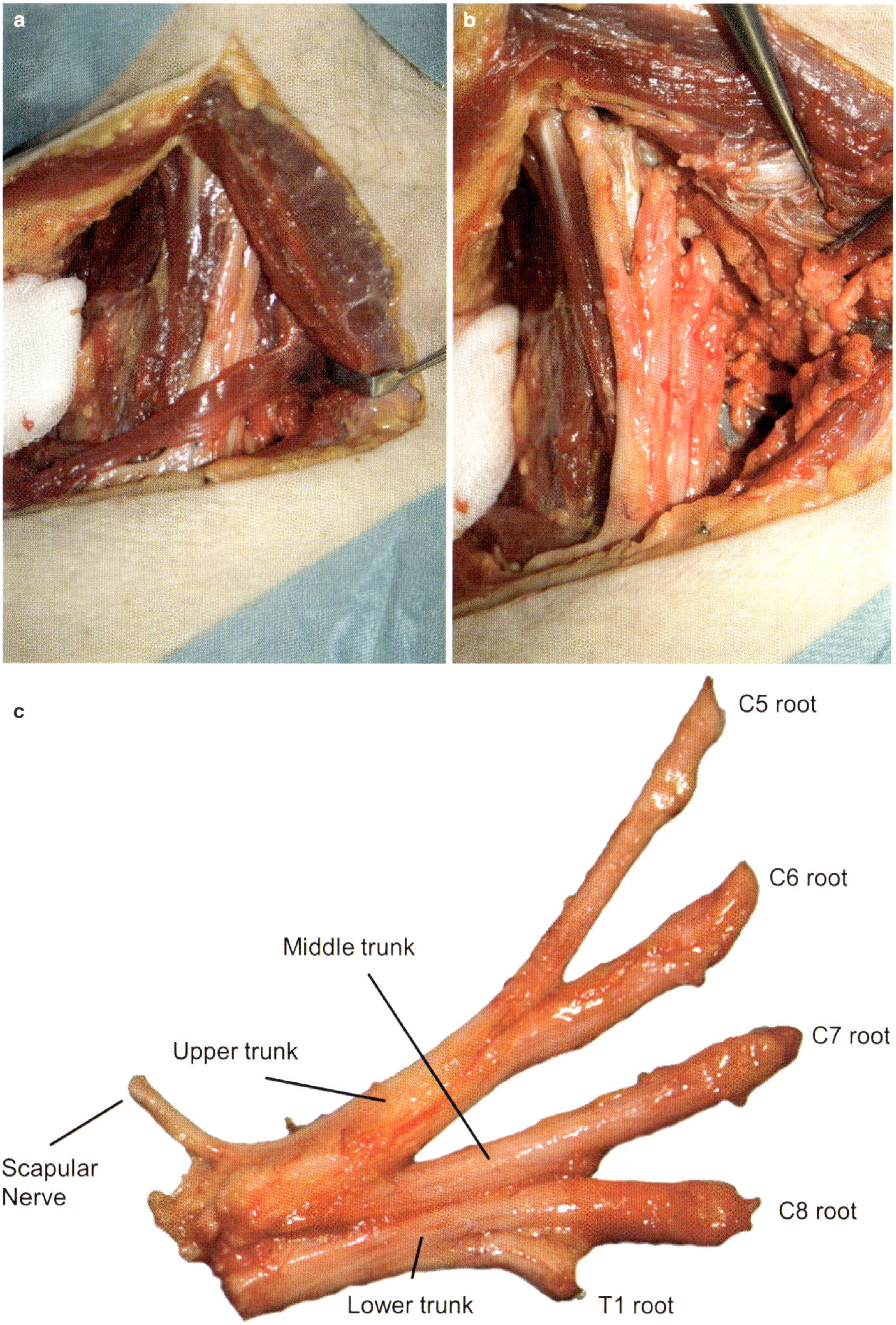

Fig. 15.1 (**a–c**) Dissection of supraclavicular brachial plexus from the roots (anterior divisions of spinal nerves C5 to T1) to divisions of the trunks (upper [*UT*], middle [*MT*], and lower [*LT*]) with the origin of the scapular nerve (*ScN*)

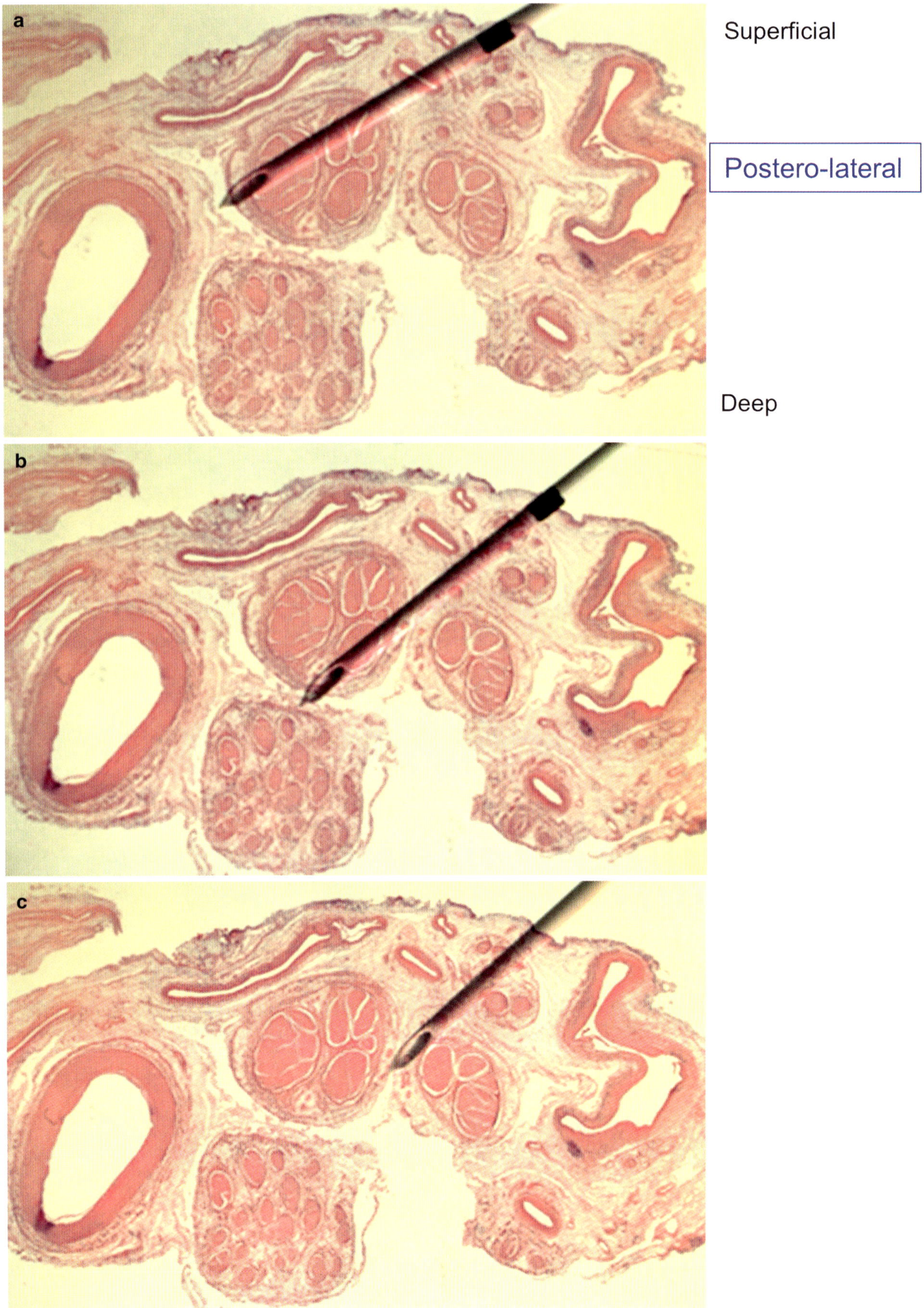

Fig. 15.2 (**a**–**c**) Overlap of the needle in the supraclavicular brachial plexus. If the neurostimulation needle pierces the nerve, based on the static images, the needle injures the fascicles. Observe the difference in size between the needle and fascicles

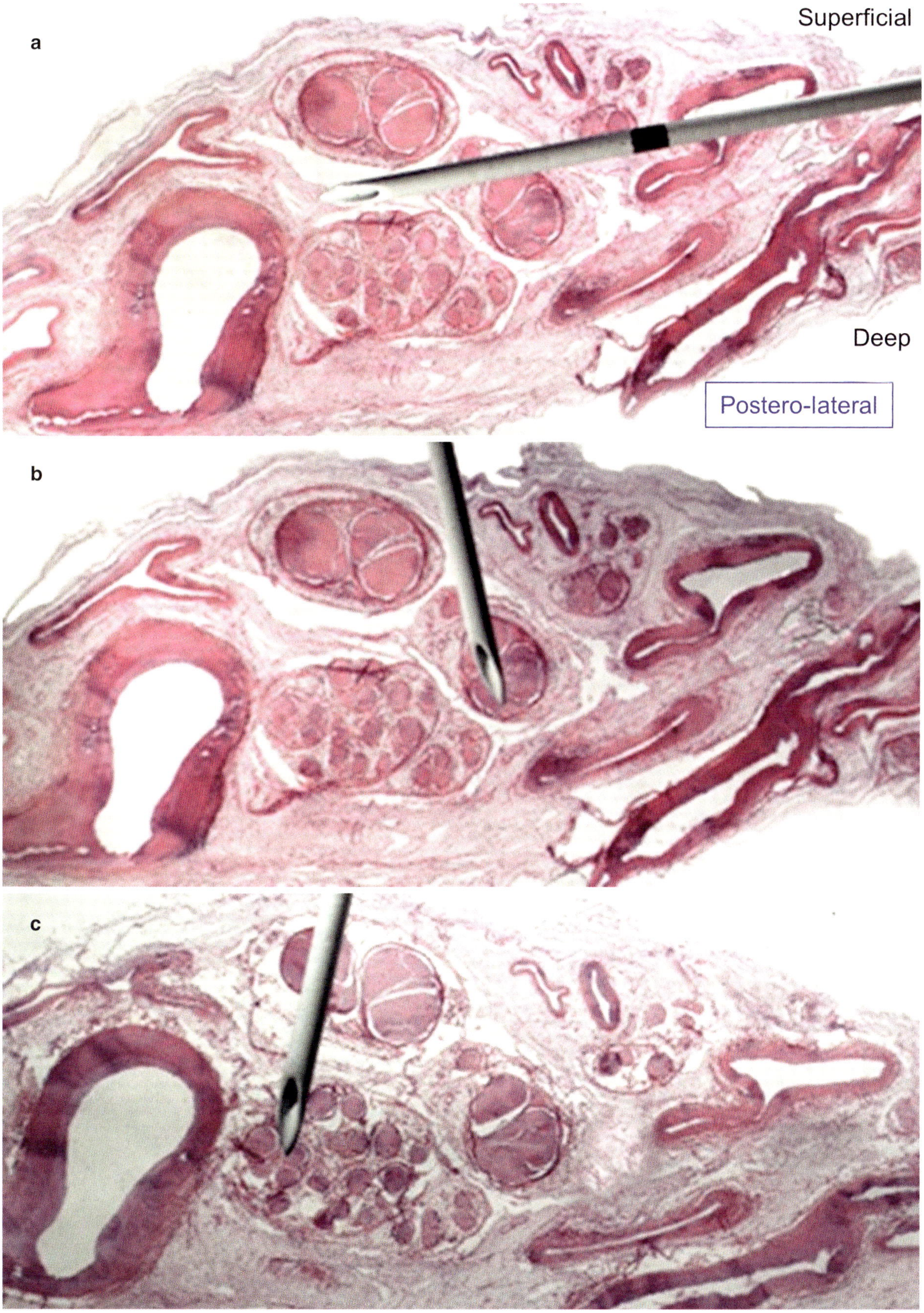

Fig. 15.3 (**a**–**c**) Overlap of the needle in the supraclavicular brachial plexus. If the neurostimulation needle pierces the nerve, based on the static images, the needle injures the fascicles. Observe the difference in size between the needle and fascicles. Otherwise, if the needle were in the connective tissue between trunks or divisions, the risk of fascicular injury would decrease significantly

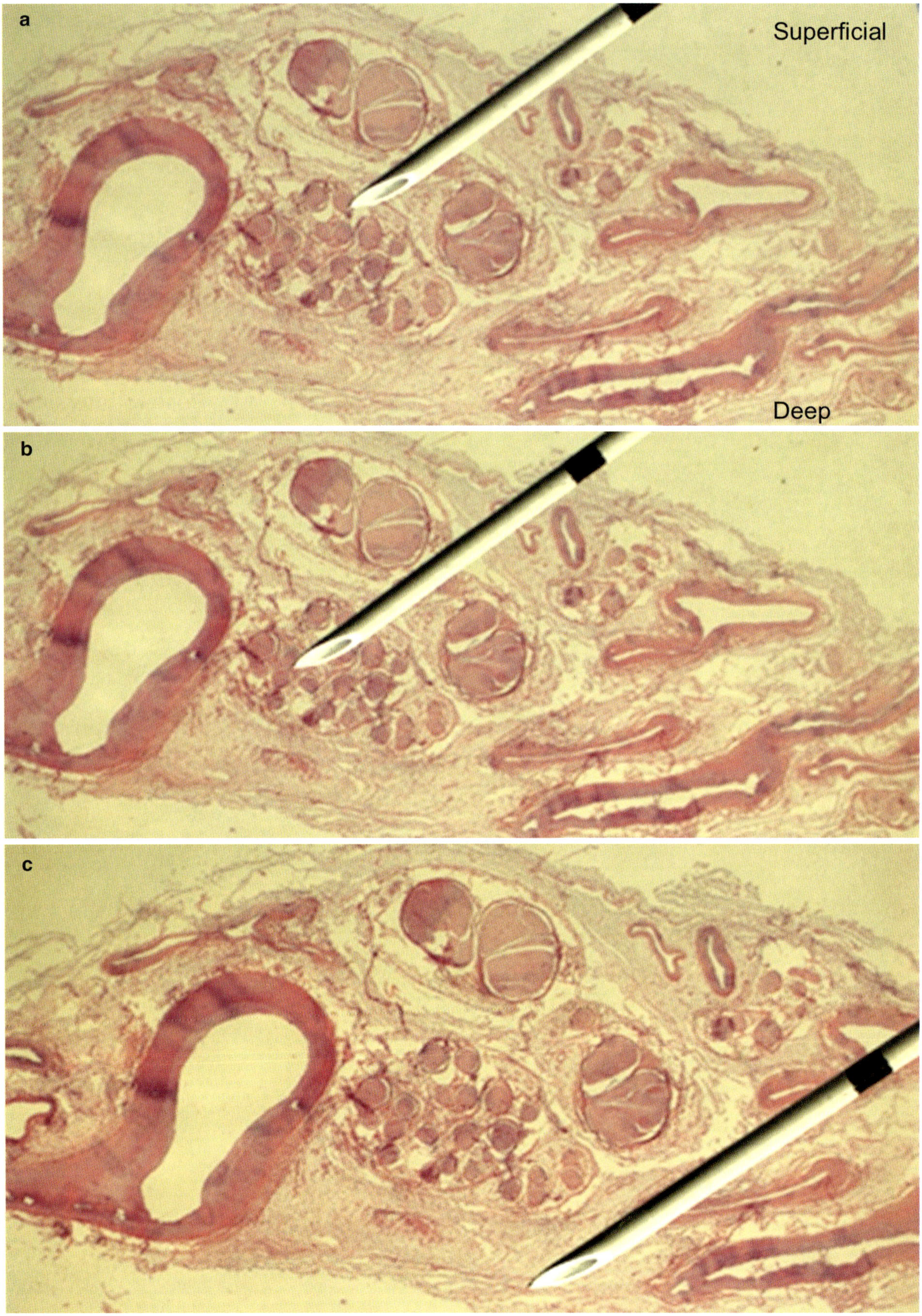

Fig. 15.4 (**a–c**) Overlap of the needle in the supraclavicular brachial plexus. If the neuro-stimulation (NRS) needle pierces the nerve, as compared with the static images, the needle injures the fascicles

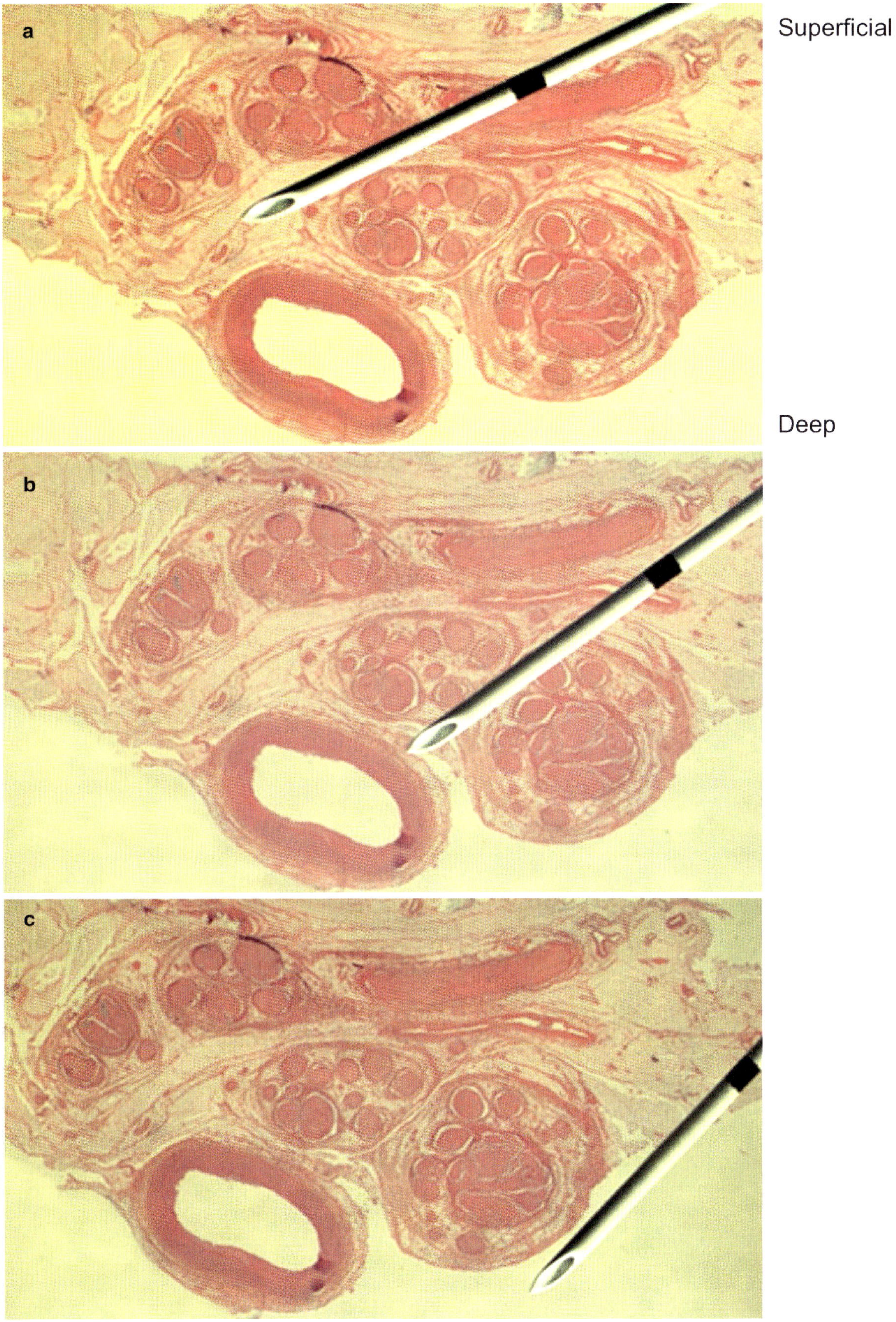

Fig. 15.5 (**a**–**c**) Overlap of the needle in the infraclavicular brachial plexus at coracoid level. If the neurostimulation needle pierces the nerve, as compared with the static images, the needle injures the fascicles

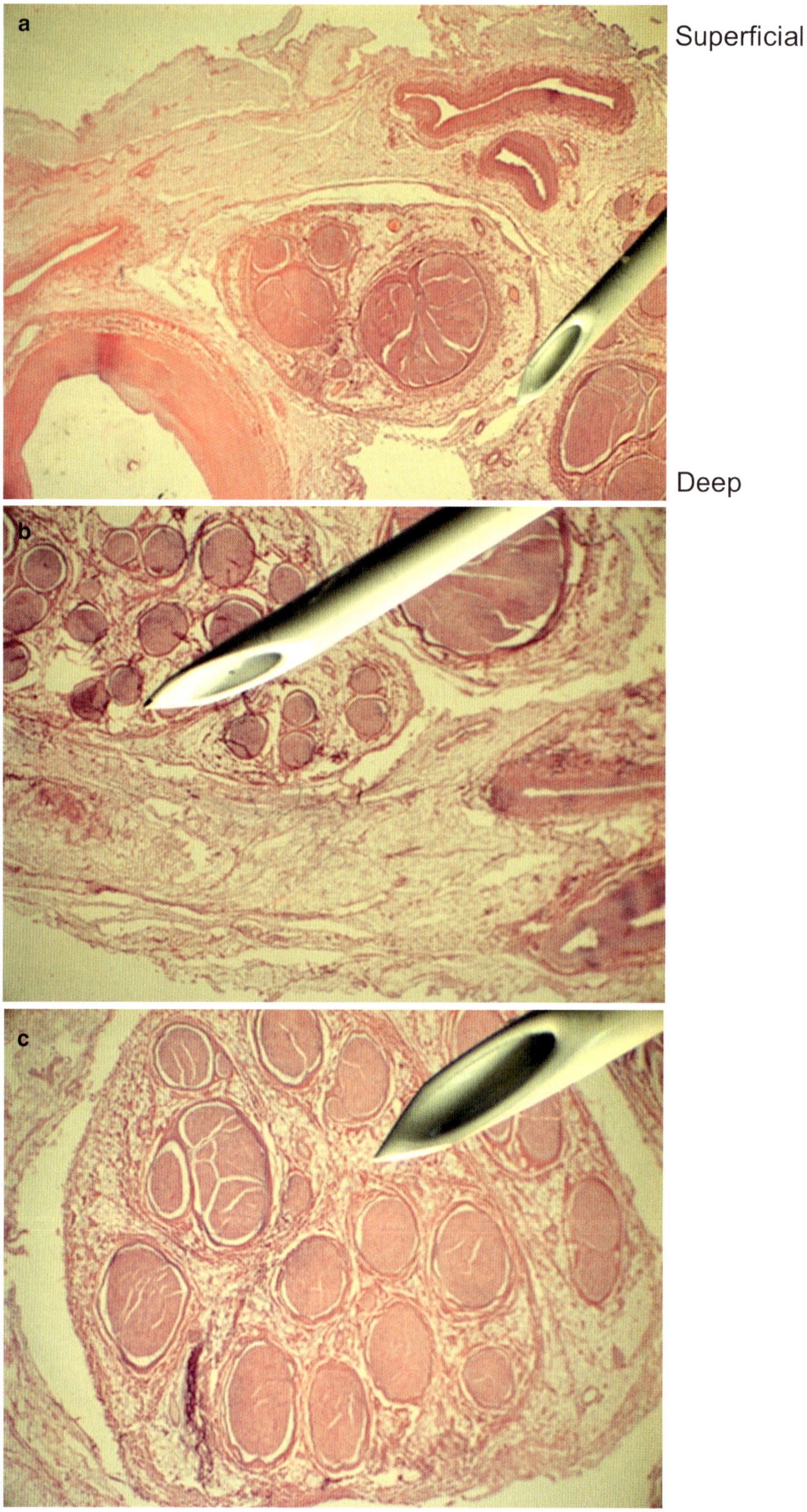

Fig. 15.6 (**a**–**c**) Overlap of the needle in the supraclavicular brachial plexus. If the neurostimulation needle pierces the nerve, as compared with the static images, the needle injures the fascicles

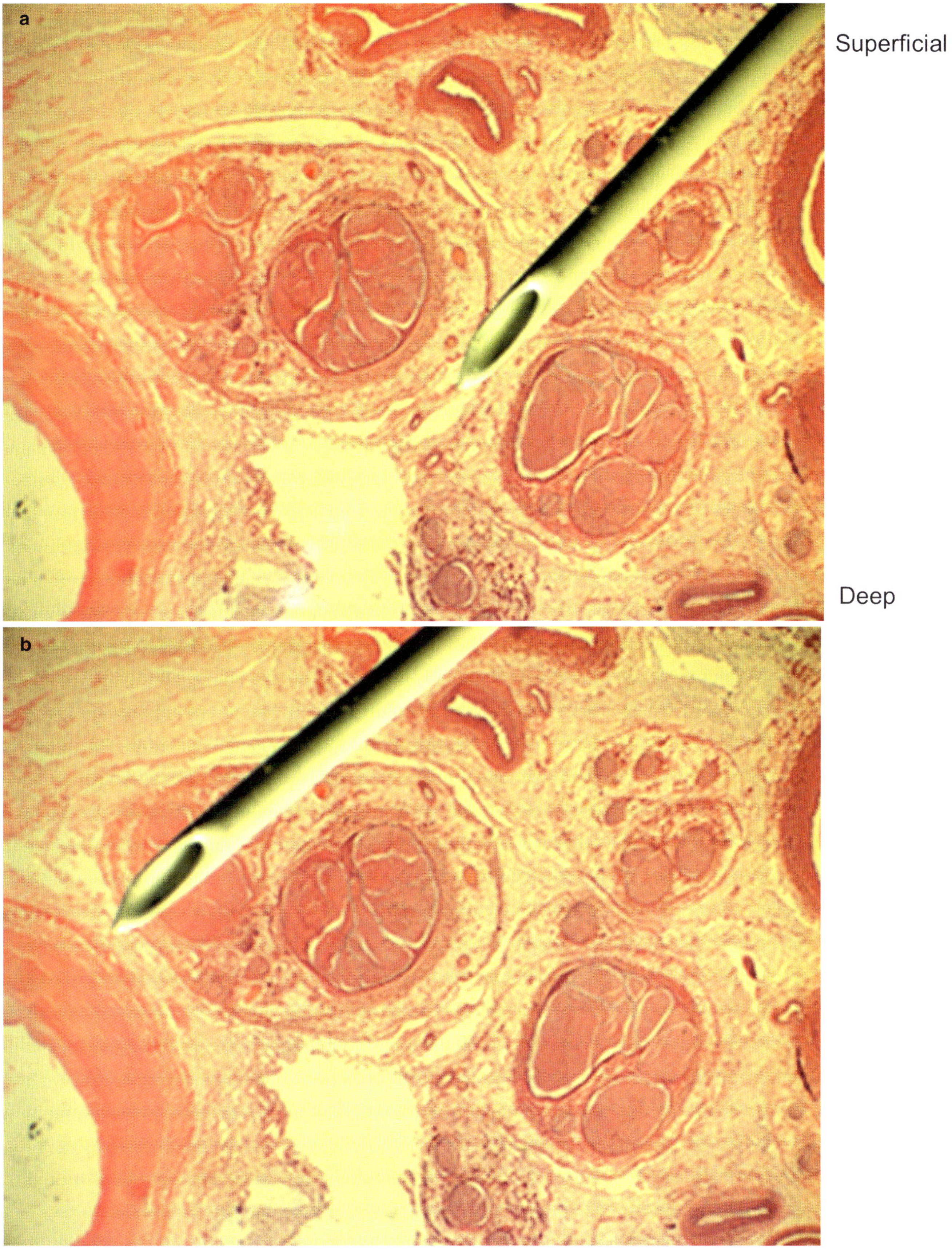

Fig. 15.7 (**a**, **b**) Overlap of the needle in the supraclavicular brachial plexus. If the neurostimulation needle pierces the nerve, as compared with the static images, the needle injures the fascicles. Observe the difference in size between the needle and fascicles

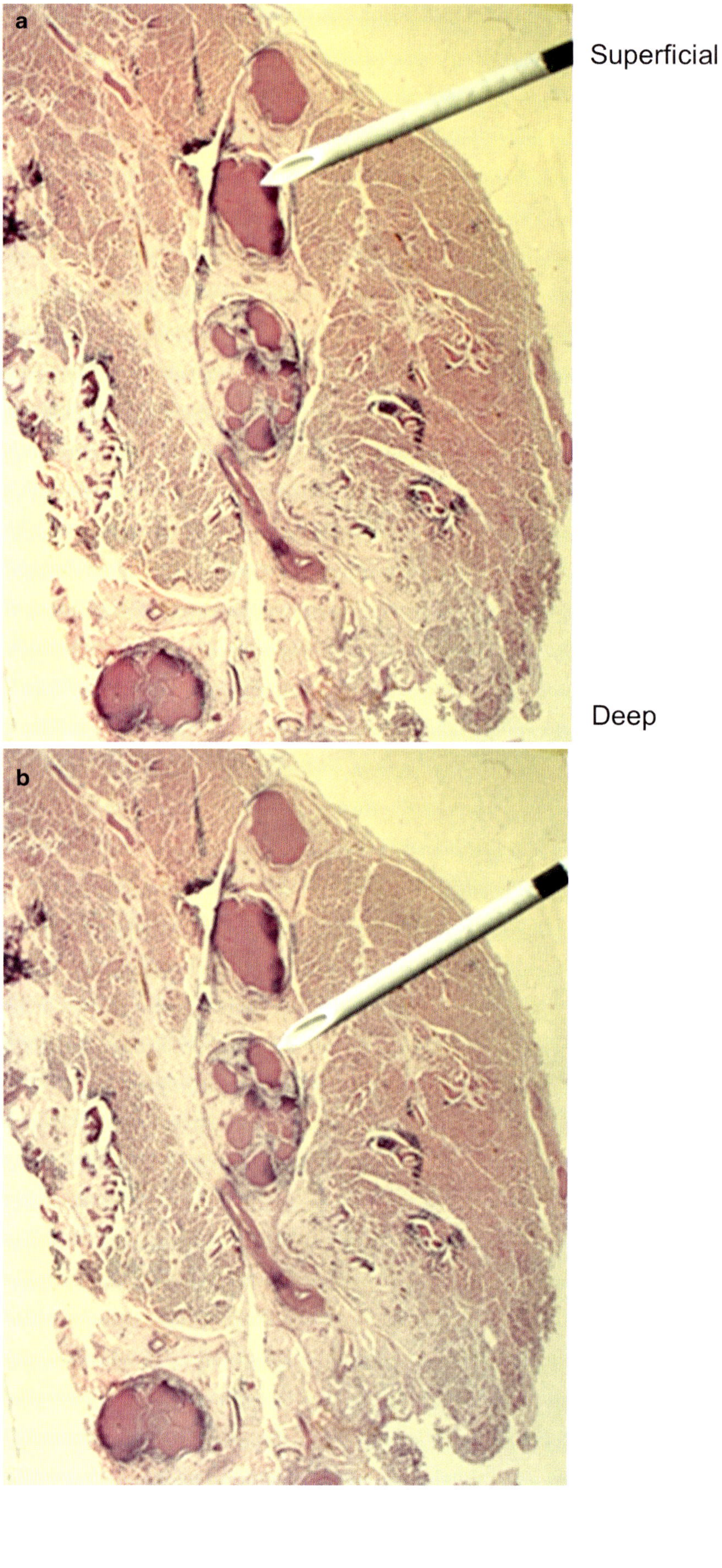

Fig. 15.8 (**a**, **b**) Overlap of the needle in the interscalene brachial plexus. If the neurostimulation needle pierces the nerve, as compared with the static images, the needle injures the fascicles. Observe the difference in size between the needle and fascicles

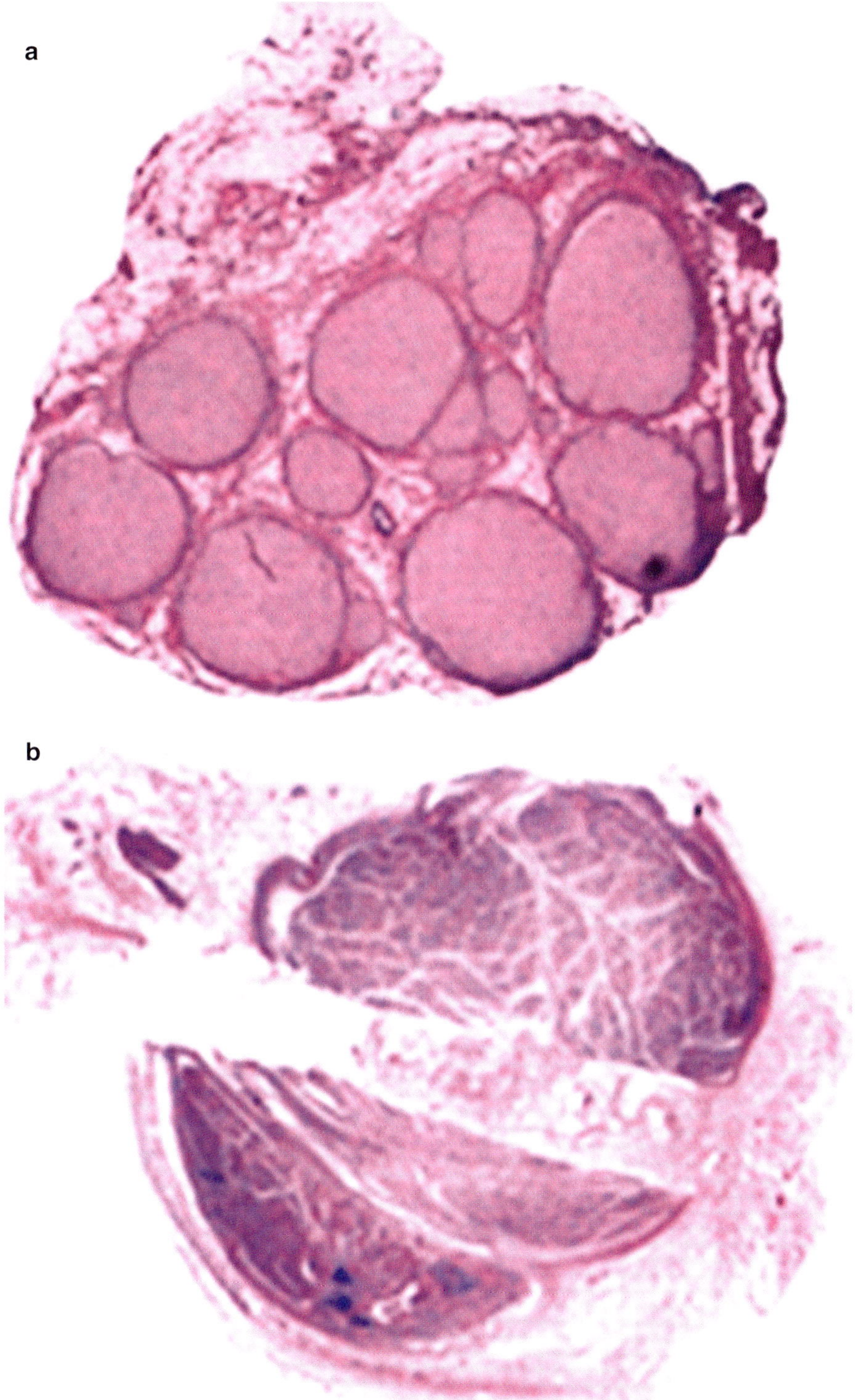

Fig. 15.9 Characteristics of the C8 nerve root. There are many compact fascicles (**a**) and a monofascicular nerve root (**b**). Note the significant damage from the needle (**b**)

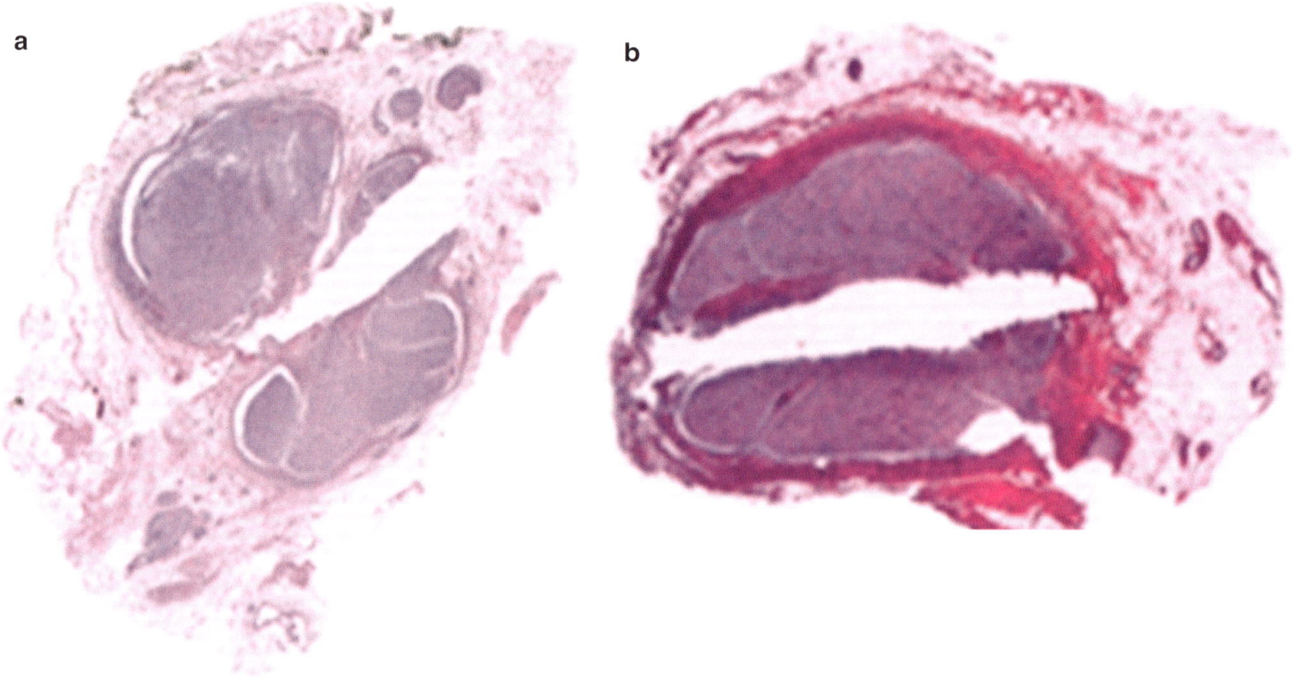

Fig. 15.10 Characteristics of the C6 nerve root. There are few compact fascicles (**a**) and a monofascicular nerve root (**b**). Note the significant damage from the needle (**b**) and the decrease in injury when the nerve has many fascicles (**a**)

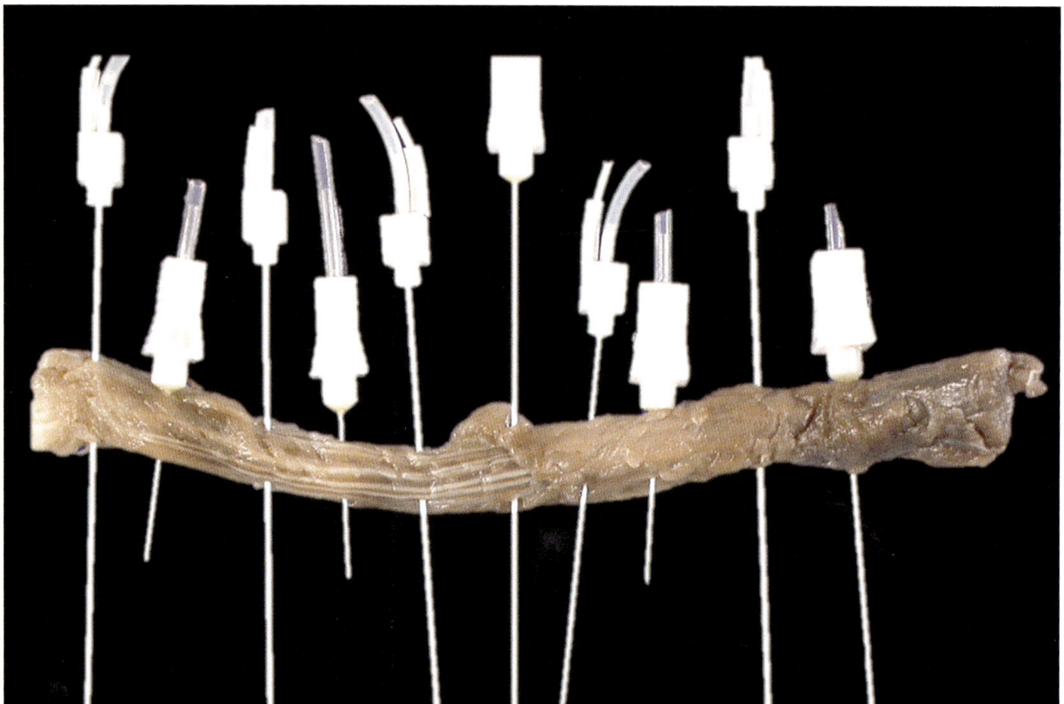

Fig. 15.11 Introduction of 22-gauge NRS needles in a fresh human sciatic nerve laterally (common peroneal) to medially (tibial component)

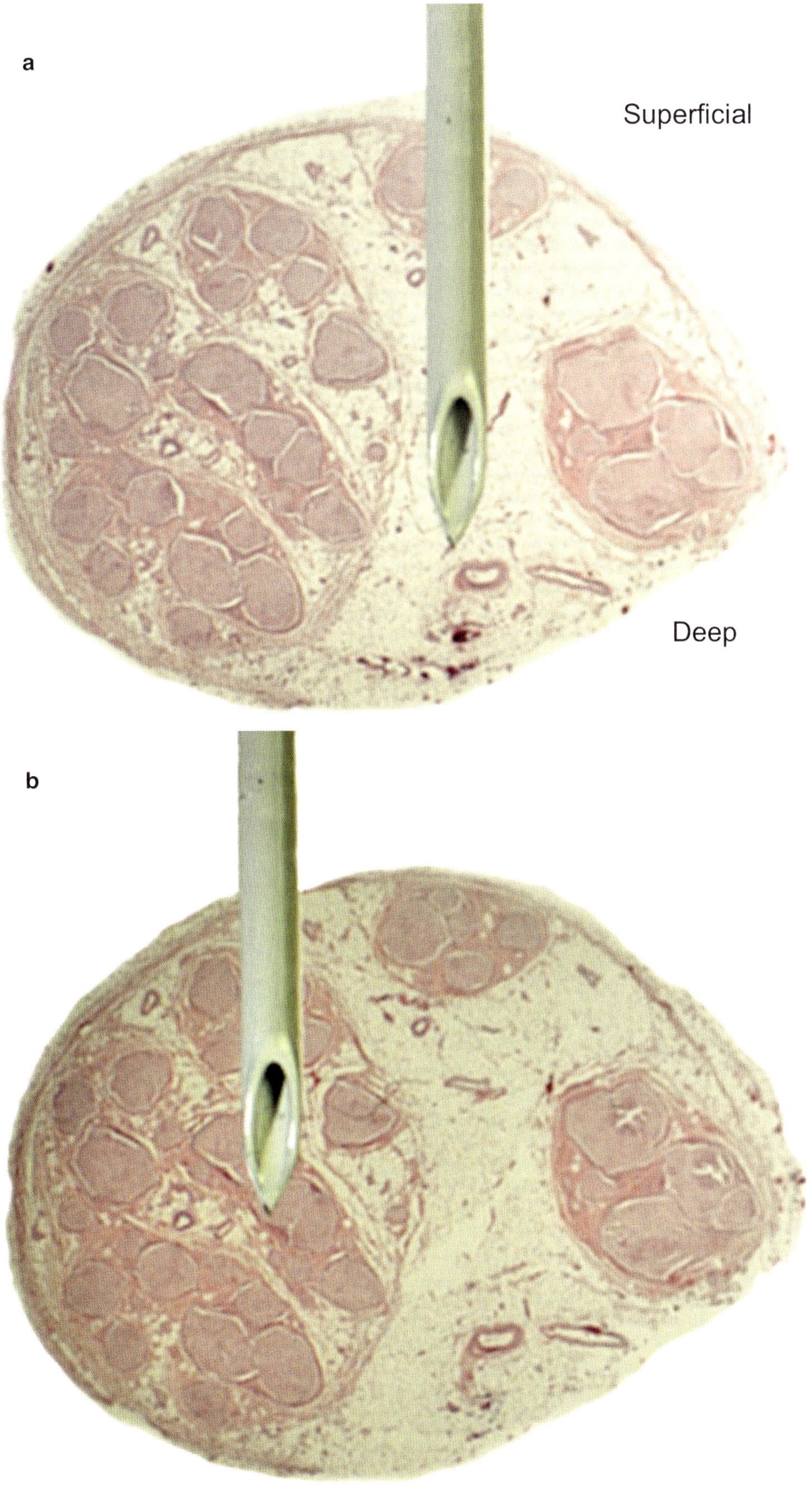

Fig. 15.12 (**a**, **b**) Overlap of the needle in the sciatic nerve, posterior approach. If the neurostimulation needle pierces the nerve, based on the static images, the needle injures the fascicles. Observe the difference in size between the needle and fascicles. **a** If there is a large amount of epineural tissue, the risk of fascicular injury is reduced

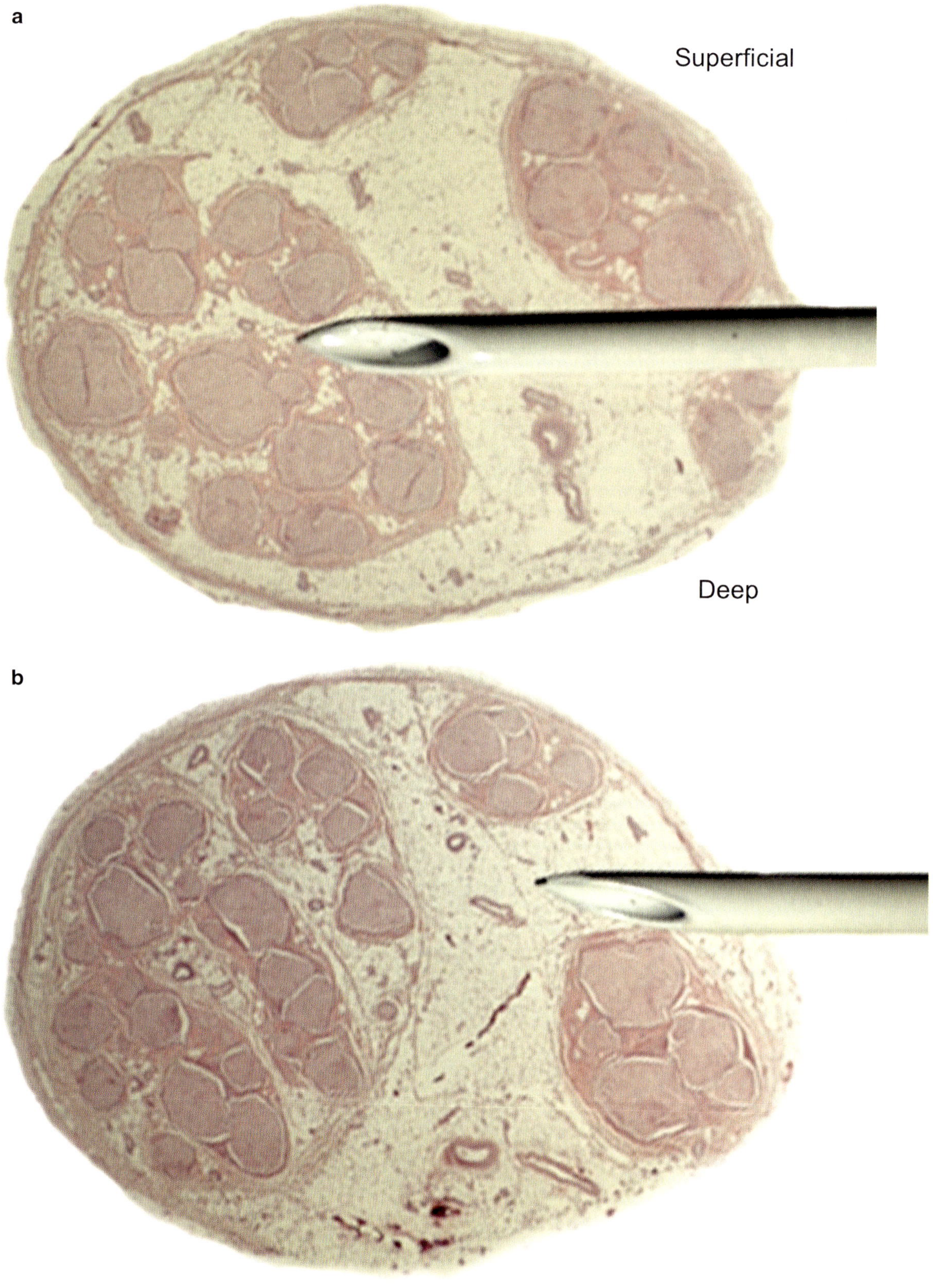

Fig. 15.13 (**a**, **b**) Overlap of the needle in the sciatic nerve, lateral approach. If the neurostimulation needle pierces the nerve, based on the static images, the needle injures the fascicles. Observe the difference in size between the needle and fascicles. (**b**) If a large amount of epineural tissue is present, the risk of fascicular injury is reduced

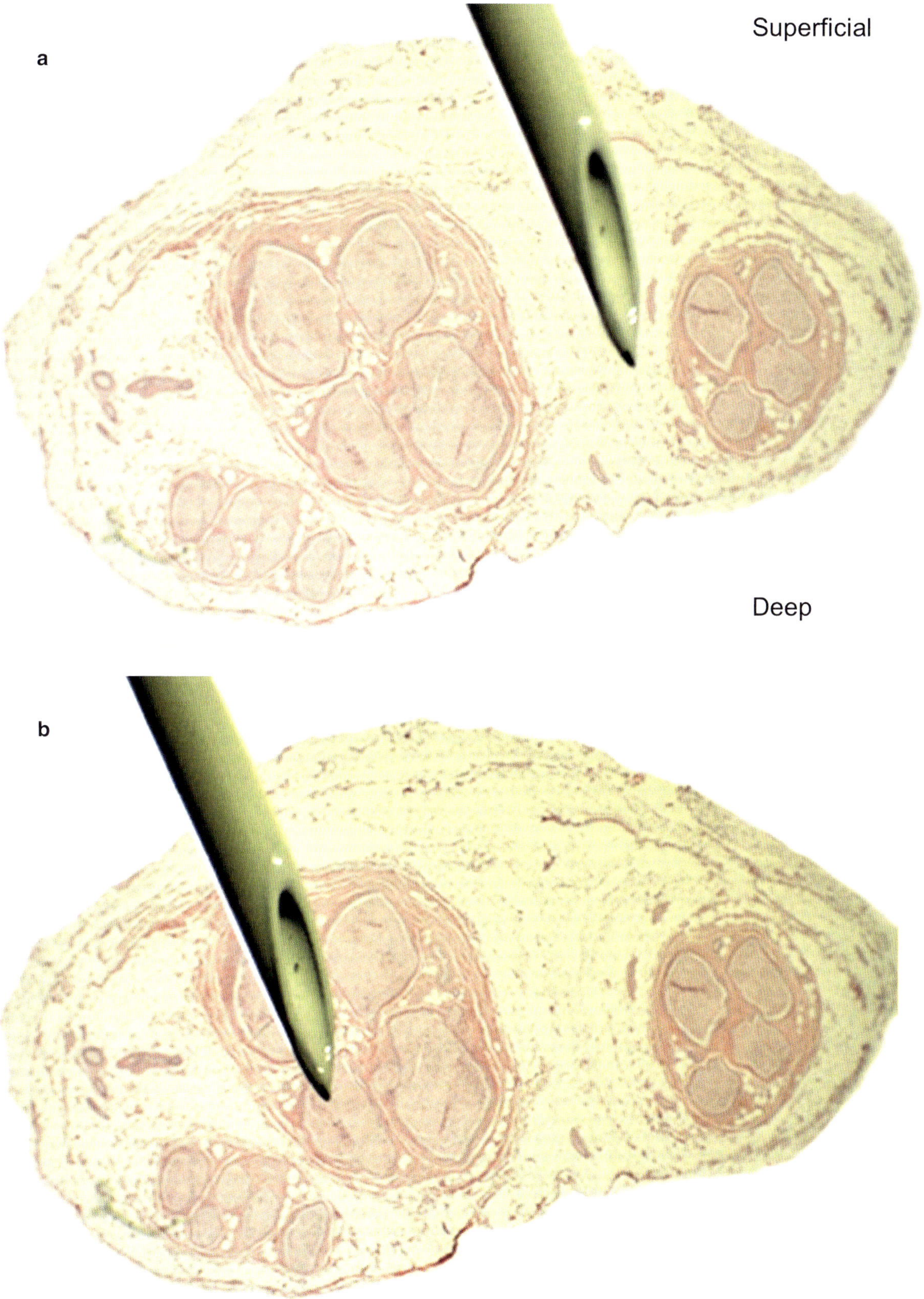

Fig. 15.14 (**a**, **b**) Overlap of the needle in the peroneal (fibular) nerve. If the neurostimulation needle pierces the nerve, based on the static images, the needle injures the fascicles. Observe the difference in size between the needle and fascicles. (**a**) If a large amount of epineural tissue is present, the risk of fascicular injury is reduced

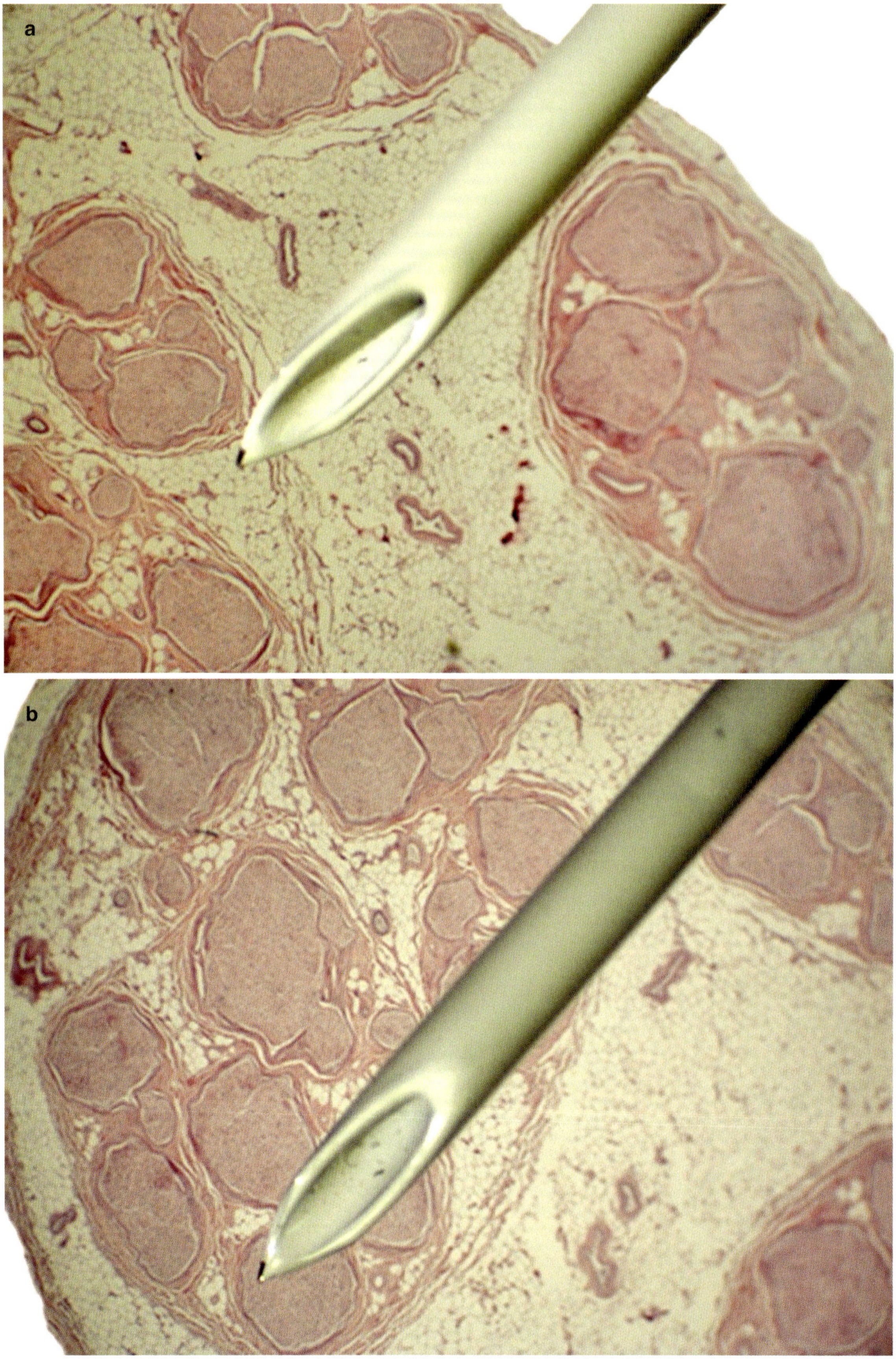

Fig. 15.15 (**a**, **b**) Overlap of the needle in the sciatic nerve, posterior approach. Details are shown at increased magnification. Observe the difference in size between the needle and fascicles

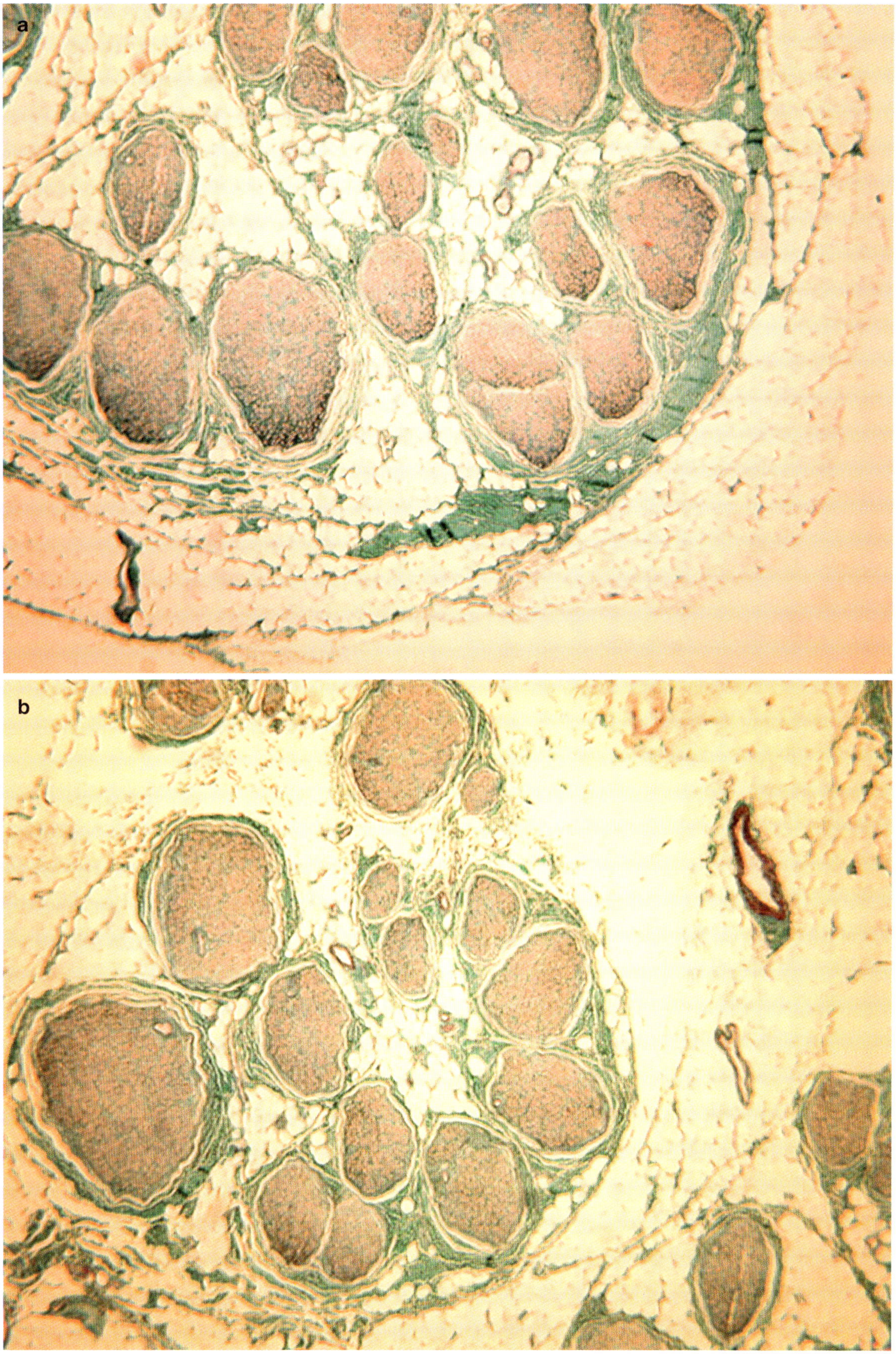

Fig. 15.16 (**a**, **b**) A large amount of fat tissue surrounding the group of fascicles reduces the risk of fascicular injury and facilitates needle withdrawal, preventing damage to the fascicles or group of fascicles

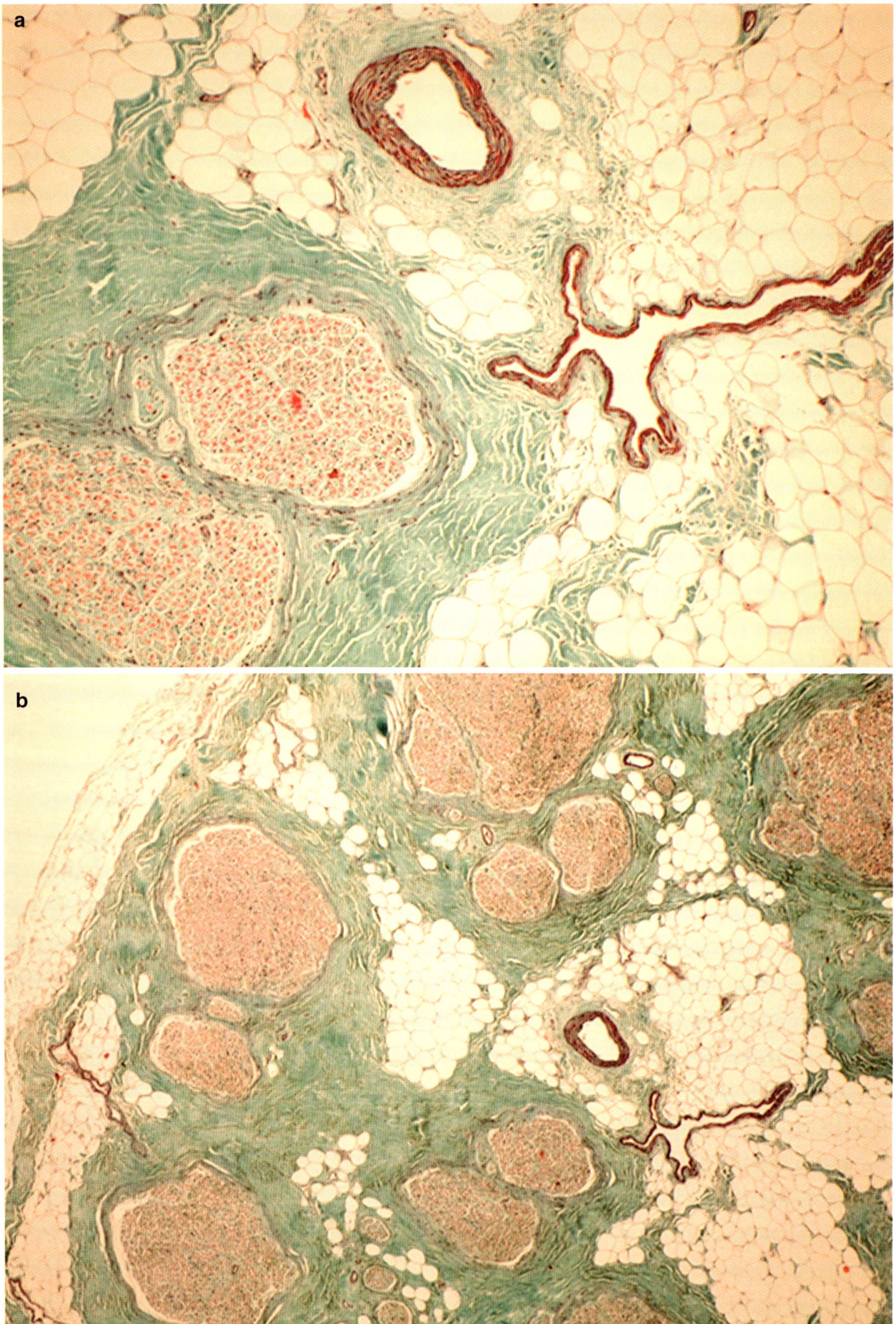

Fig. 15.17 (**a**, **b**) The content of connective tissue inside the sciatic nerve differs among nerves and among areas. A large amount of connective tissue reduces tissue compliance (increases the open pressure), placing the epineural vessels at risk

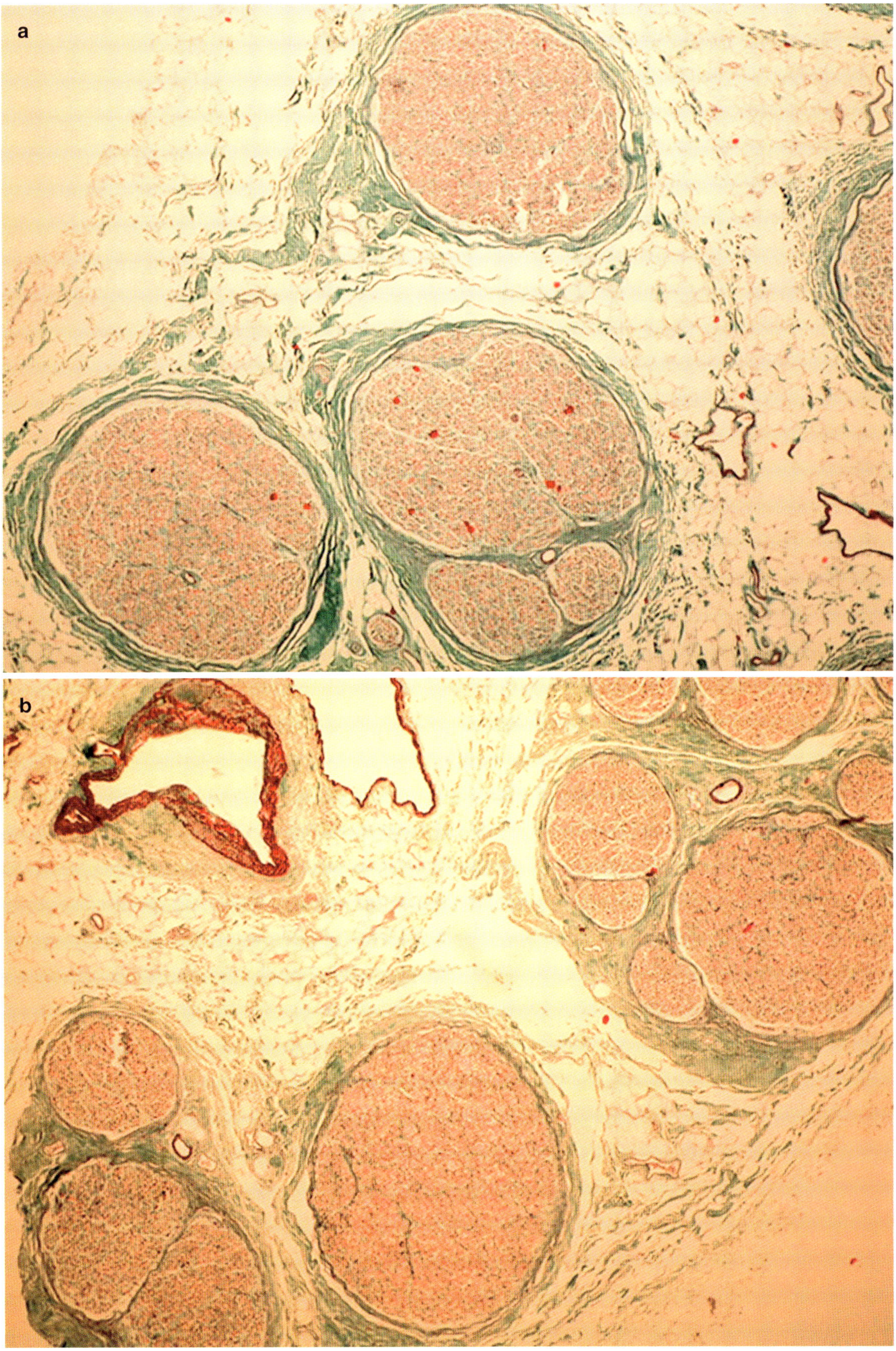

Fig. 15.18 (**a**, **b**) The content of connective tissue inside the sciatic nerve differs among nerves and among different areas. A large amount of connective tissue reduces tissue compliance (increases the open pressure), placing the epineural vessels at risk

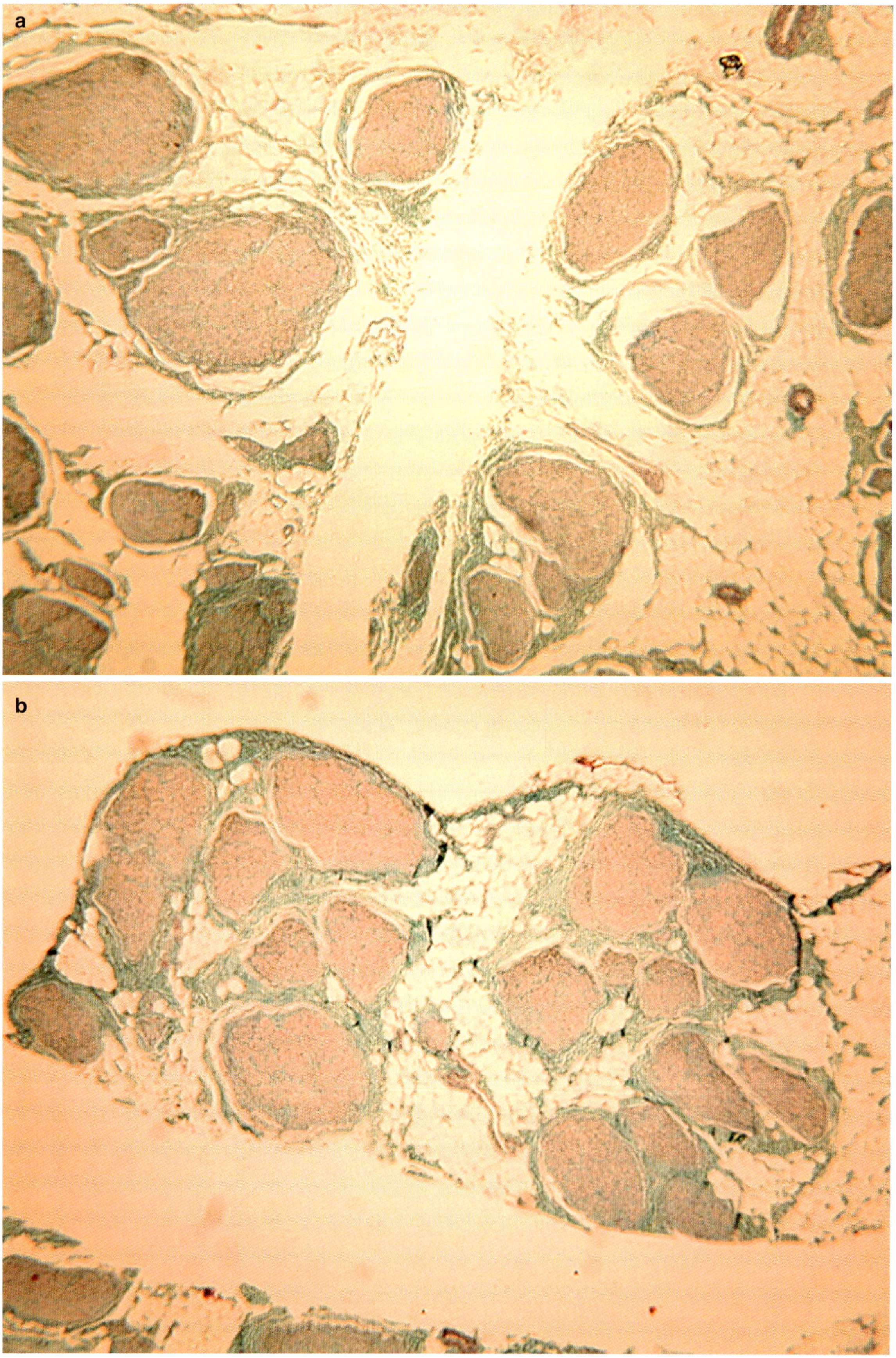

Fig. 15.19 (**a**, **b**) If the needle transfixes the sciatic nerve, which has a large amount of epineural tissue (>80 % of the area of section), it usually can be withdrawn without damaging the fascicles (<3 % of cases result in injury)

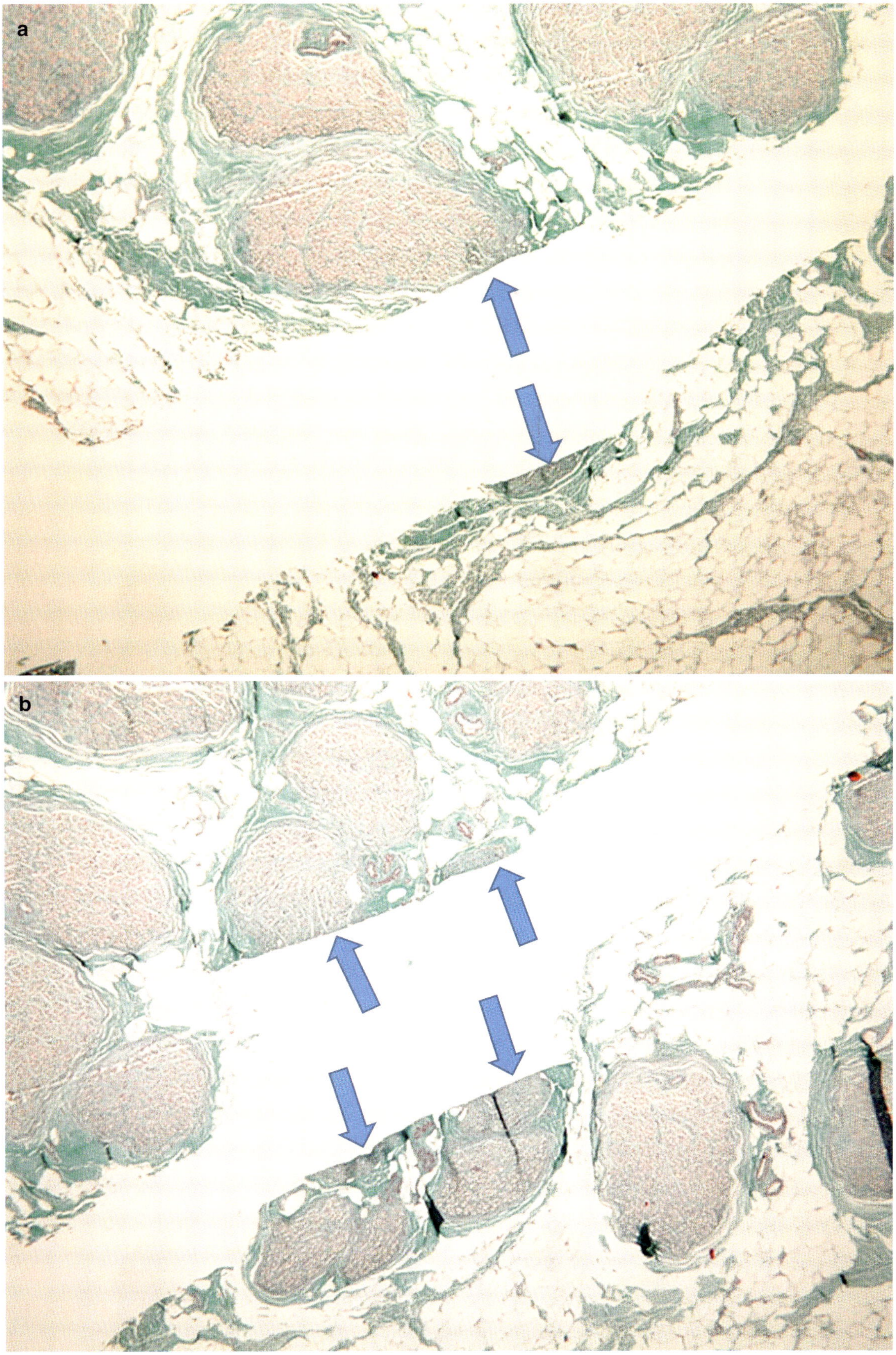

Fig. 15.20 (**a**, **b**) If the needle transfixes the sciatic nerve, injury to the marginal fascicles sometimes occurs (*blue arrows*)

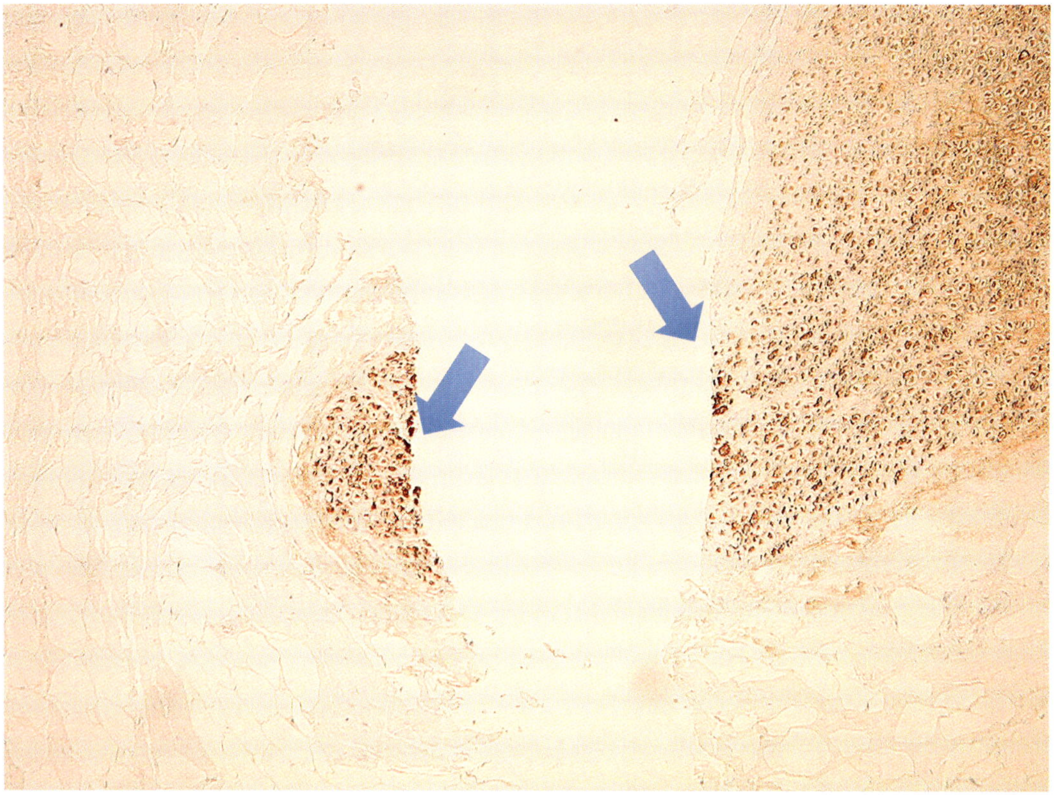

Fig. 15.21 Needle injury to a fascicle (*blue arrows*)

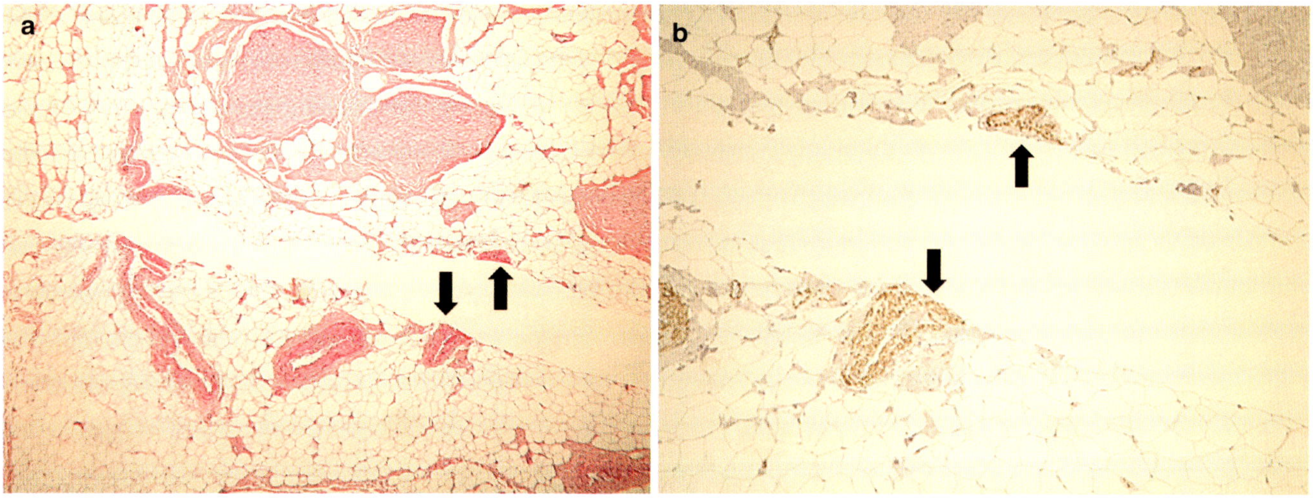

Fig. 15.22 Marginal epineural vessel injury (*black arrows*) is showing in **a** and a detail of the same injury in **b**

Table 15.1 Characteristics of intraneural injection

Route	Paresthesia?	NRS (motor response)	Pressure	Volume distribution	Risk of postoperative dysfunction
Extraneural	No	No response or high intensity (>1 mA)	Low (<5 psi)	Erratic	Very low
Epineural					
Interbranch	Uncommon	No response or high intensity (>1 mA)	Low (<5 psi)	Good distribution in a single injection; distribution along the nerve	Very low
Subepineural	Possible, uncommon	<1 mA (frequently <0.5 mA)	Low (<5 psi)	Thin halo around nerve; epineural sheath visible; increased area of nerve and distribution along nerve	Very low
Intraepineural	Frequent	<0.5 mA (frequently <0.3 mA)	Low (5–10 psi)	Swelling; increased area and distribution along nerve	Low
Intra–fascicle group	Very common	<0.3 mA	High (10–15 psi)	Swelling; increased area and distribution along nerve	High
Intrafascicular	Yes (always)	<0.1 mA	High (>15 psi)		Very high

References

1. Sala-Blanch X, Vandepitte C, Laur J, Horan P, Xu D, Reina MA, Hadzic A. A practical review of perineural versus intraneural injections: a call for standard nomenclature. Int Anesthesiol Clin. 2011;49:1–12.
2. Sala-Blanch X. Nerve injury. How can risk be minimized? Reg Anesth Pain Med. 2013;38:E8–11.

Scanning Electron Microscopy View of In Vitro Intraneural Injections

16

Miguel Angel Reina and Xavier Sala-Blanch

Scanning electron microscopy (SEM) technique allows identification of features from anatomical structures that are traversed by the tip of the needle during an accidental dural puncture [1–3]. Lesions to axons enclosed in injured fascicles may occur in conjunction with damage to blood vessels within the intrafascicular tissue [4]. As a result of blood vessel tissue repair and hematoma reabsorption, variable degrees of fibrosis may affect areas in the proximity of axons. The study of distortions in "used" needle tips, lesions caused to nerve fascicles, and the diameters of blood vessels may aid in the understanding of the mechanisms leading to tissue damage and repair after accidental dural puncture (Figs. 16.1, 16.2, 16.3, 16.4, 16.5, 16.6, 16.7, 16.8, 16.9, 16.10, 16.11, 16.12, 16.13, 16.14, 16.15, 16.16 and 16.17).

M.A. Reina, MD, PhD (✉)
Department of Clinical Medical Sciences and Institute of Applied Molecular Medicine, School of Medicine, University of CEU San Pablo, Madrid, Spain

Department of Anesthesiology, Madrid-Montepríncipe University Hospital, Madrid, Spain
e-mail: miguelangel@perticone.e.telefonica.net

X. Sala-Blanch, MD
Human Anatomy and Embryology Unit, School of Medicine, University of Barcelona, Barcelona, Spain

Department of Anesthesiology and Critical Care, Clinic Hospital, Barcelona, Spain
e-mail: xavi.sala.blanch@gmail.com

M.A. Reina et al. (eds.), *Atlas of Functional Anatomy for Regional Anesthesia and Pain Medicine: Human Structure, Ultrastructure and 3D Reconstruction Images*, DOI 10.1007/978-3-319-09522-6_16,

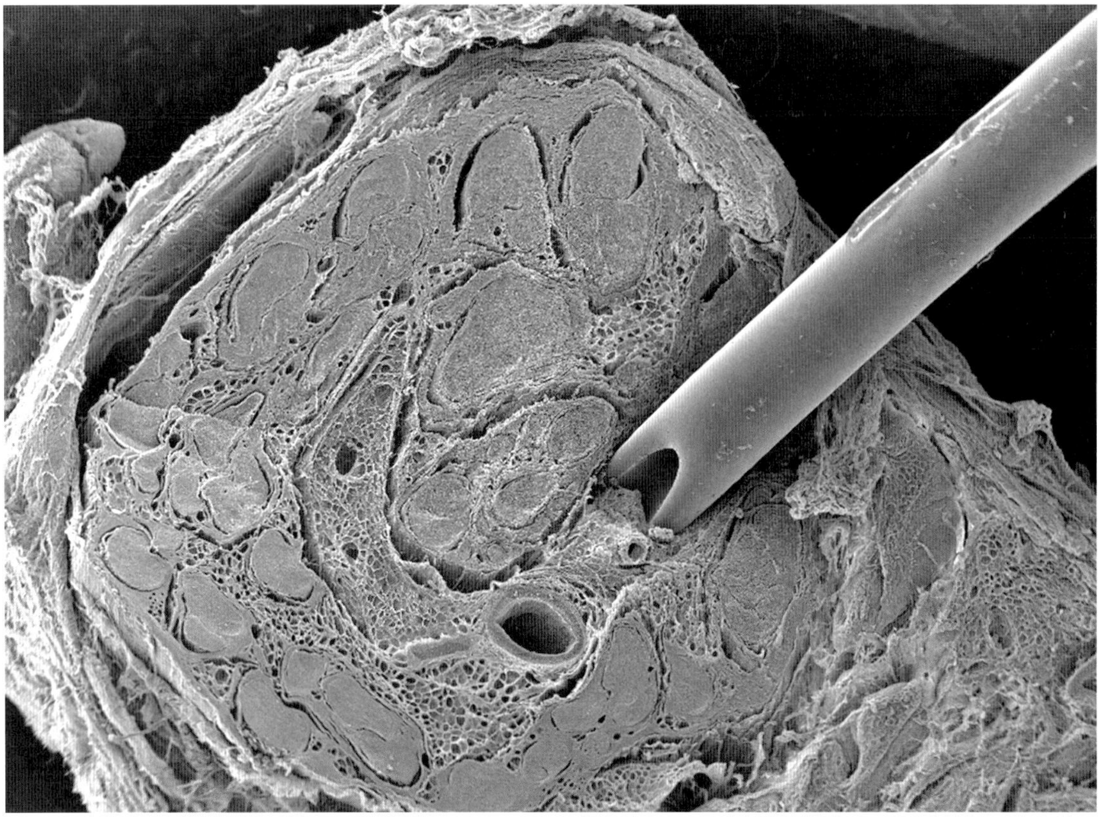

Fig. 16.1 Human tibial nerve. In vitro puncture of nerve with neurostimulation needle. Comparative view. Scanning electron microscopy. Magnification ×15

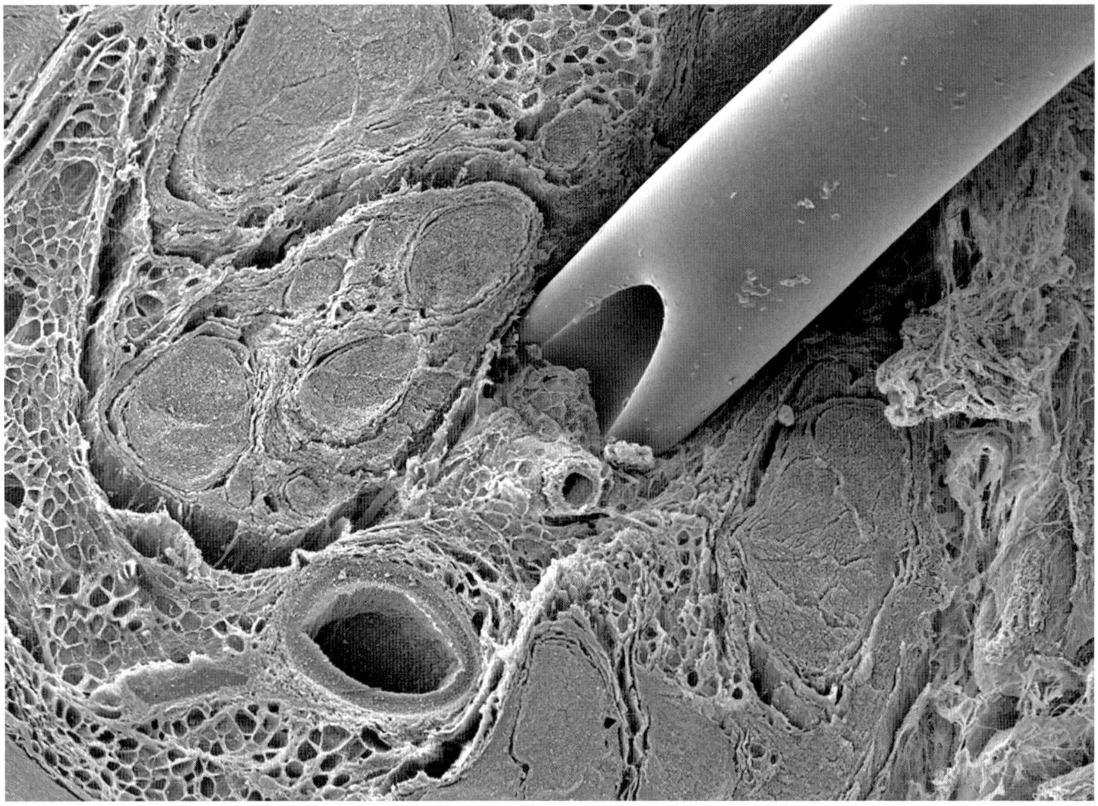

Fig. 16.2 Human tibial nerve. In vitro puncture of nerve with neurostimulation needle. Detail at higher magnification of Fig. 16.1. Scanning electron microscopy. Magnification ×30

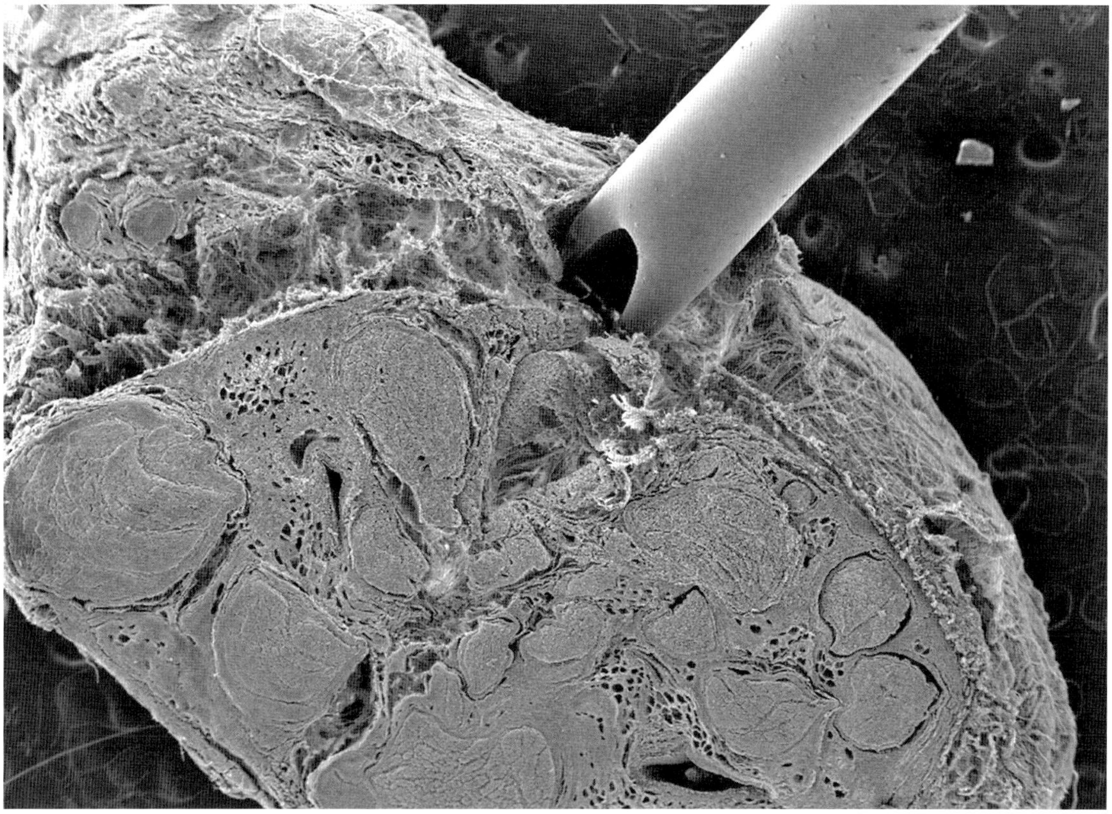

Fig. 16.3 Human tibial nerve. In vitro puncture of nerve with neurostimulation needle. Comparative view. Scanning electron microscopy. Magnification ×20

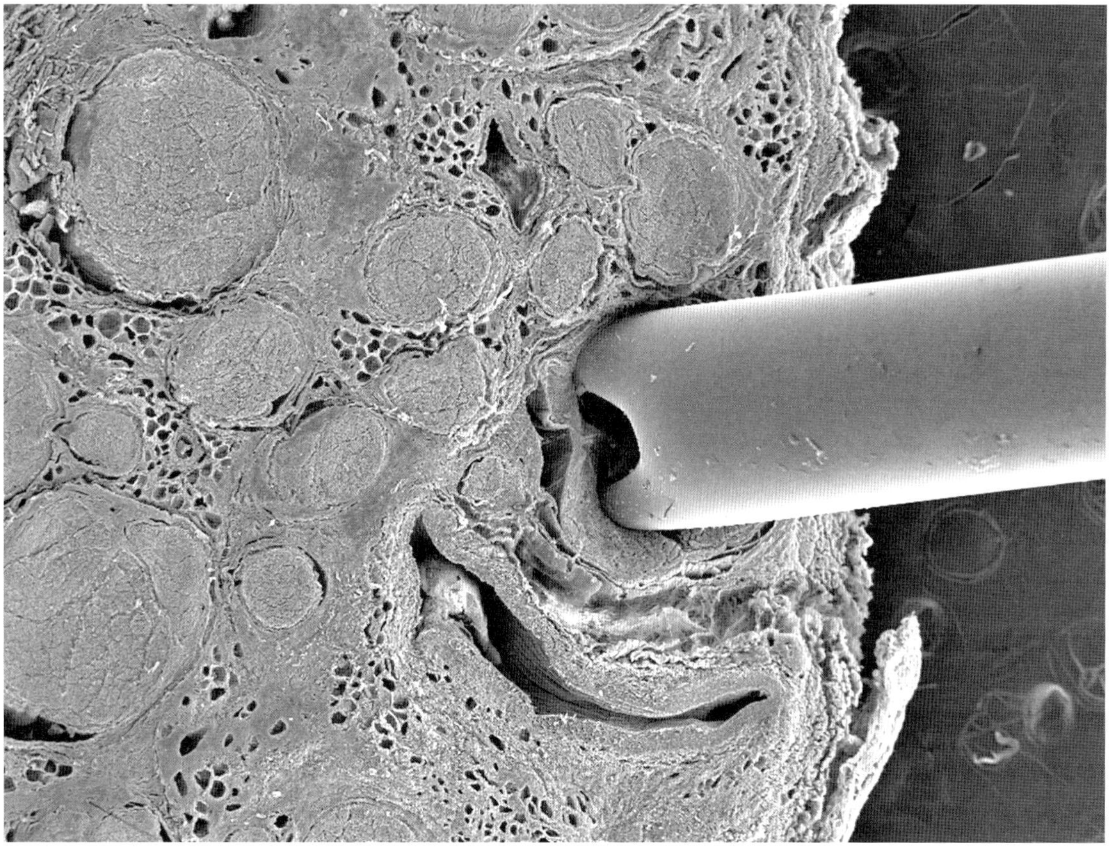

Fig. 16.4 Human tibial nerve. In vitro puncture of nerve with neurostimulation needle. Fascicles can be identified. Scanning electron microscopy. Magnification ×30

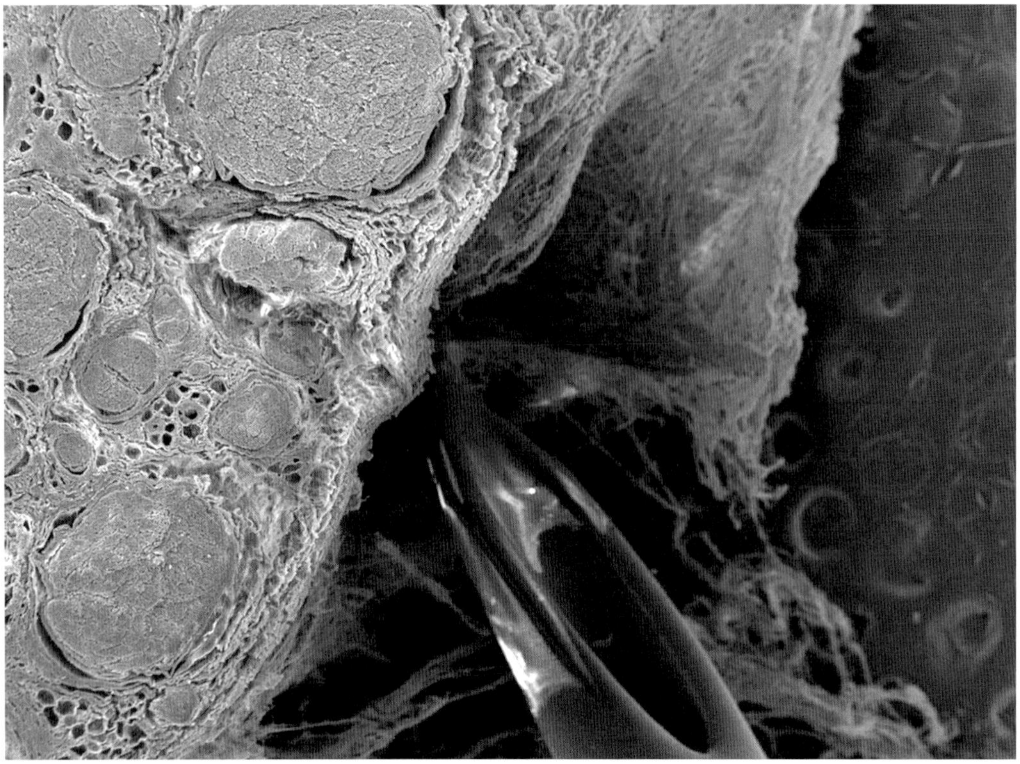

Fig. 16.5 Human tibial nerve. In vitro puncture of nerve with neurostimulation needle. Tip of the needle in contact with the external surface of the nerve. Scanning electron microscopy. Magnification ×30

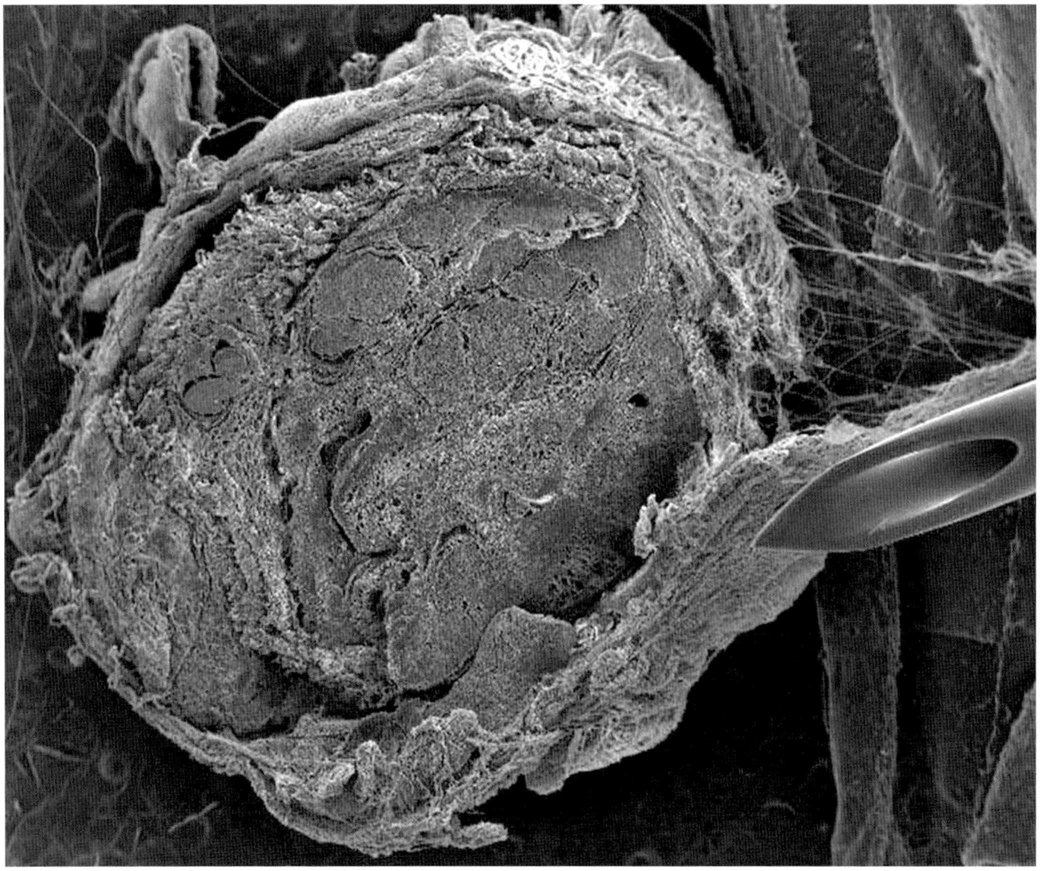

Fig. 16.6 Human tibial nerve. In vitro puncture of nerve with neurostimulation needle. Comparative view. Scanning electron microscopy. Magnification ×10

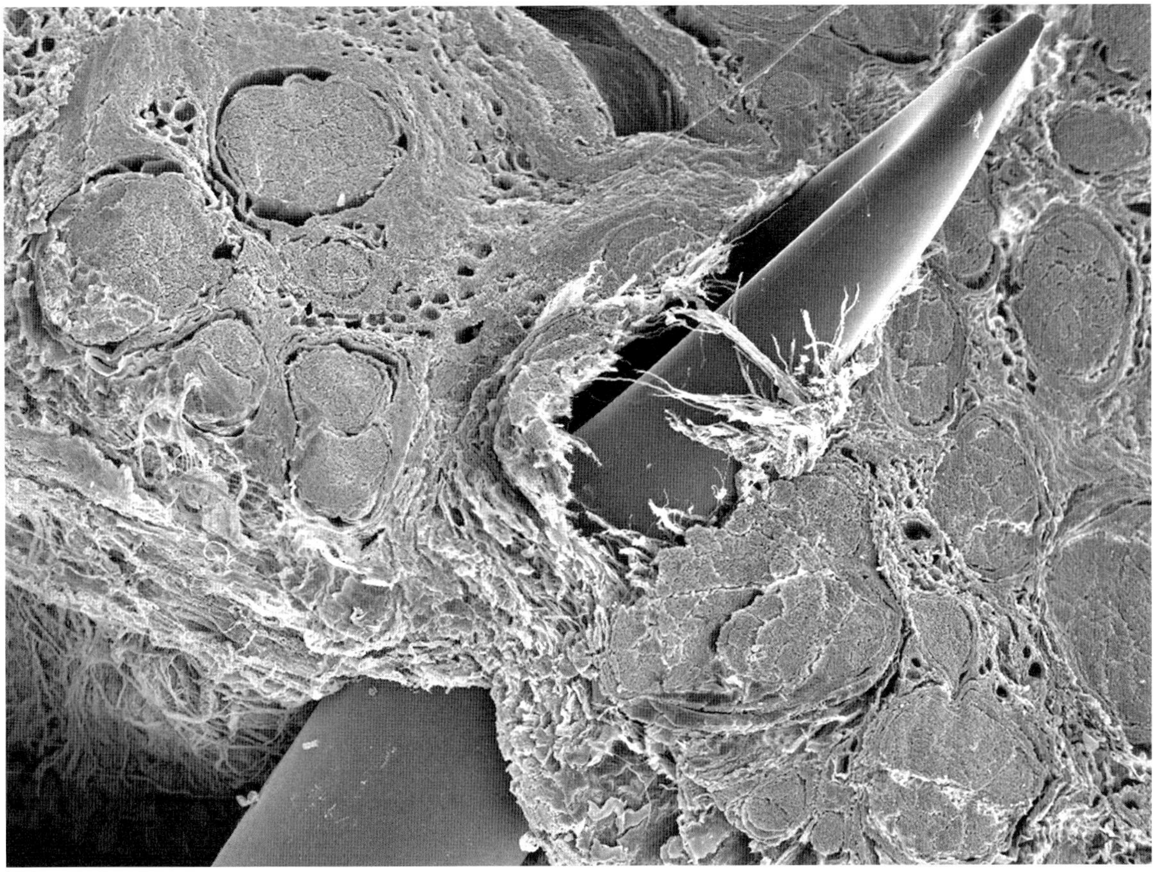

Fig. 16.7 Human tibial nerve. In vitro puncture of nerve with neurostimulation needle. Needle traversing the fascicles. Scanning electron microscopy. Magnification ×35 (From De Andrés et al. [1], with permission)

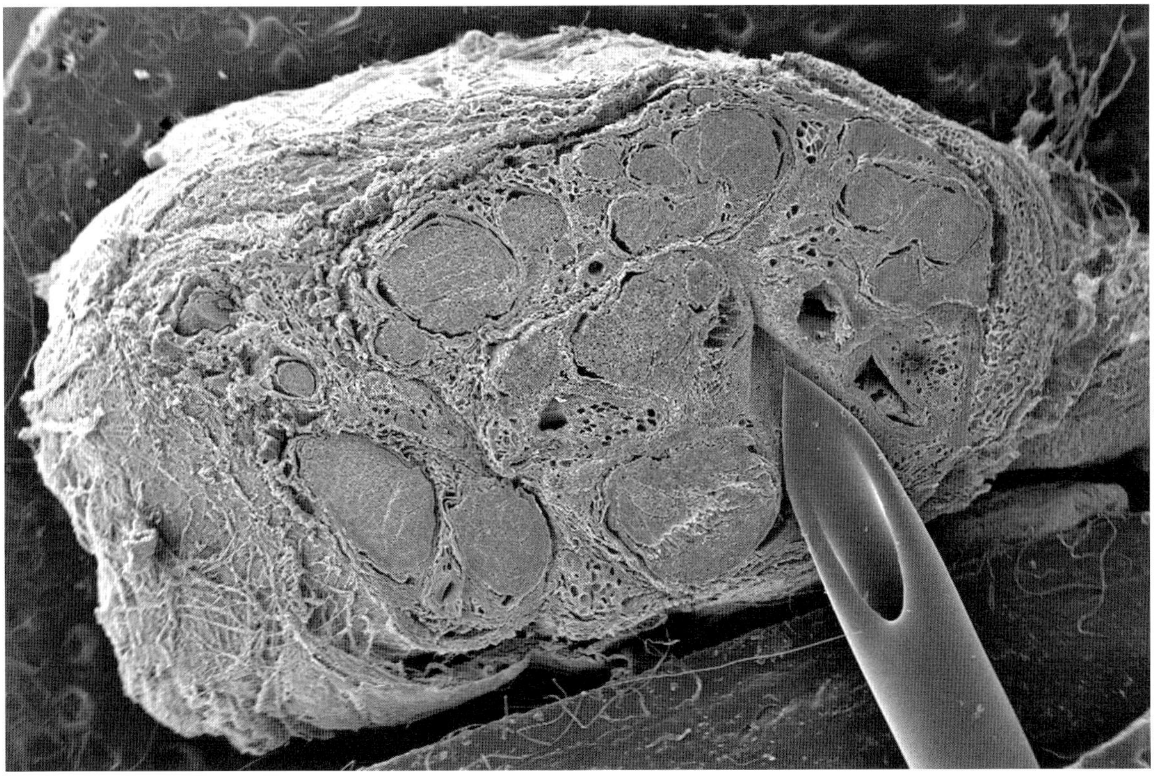

Fig. 16.8 Human tibial nerve. In vitro puncture of nerve with neurostimulation needle. View illustrates comparative sizes. Scanning electron microscopy. Magnification ×15

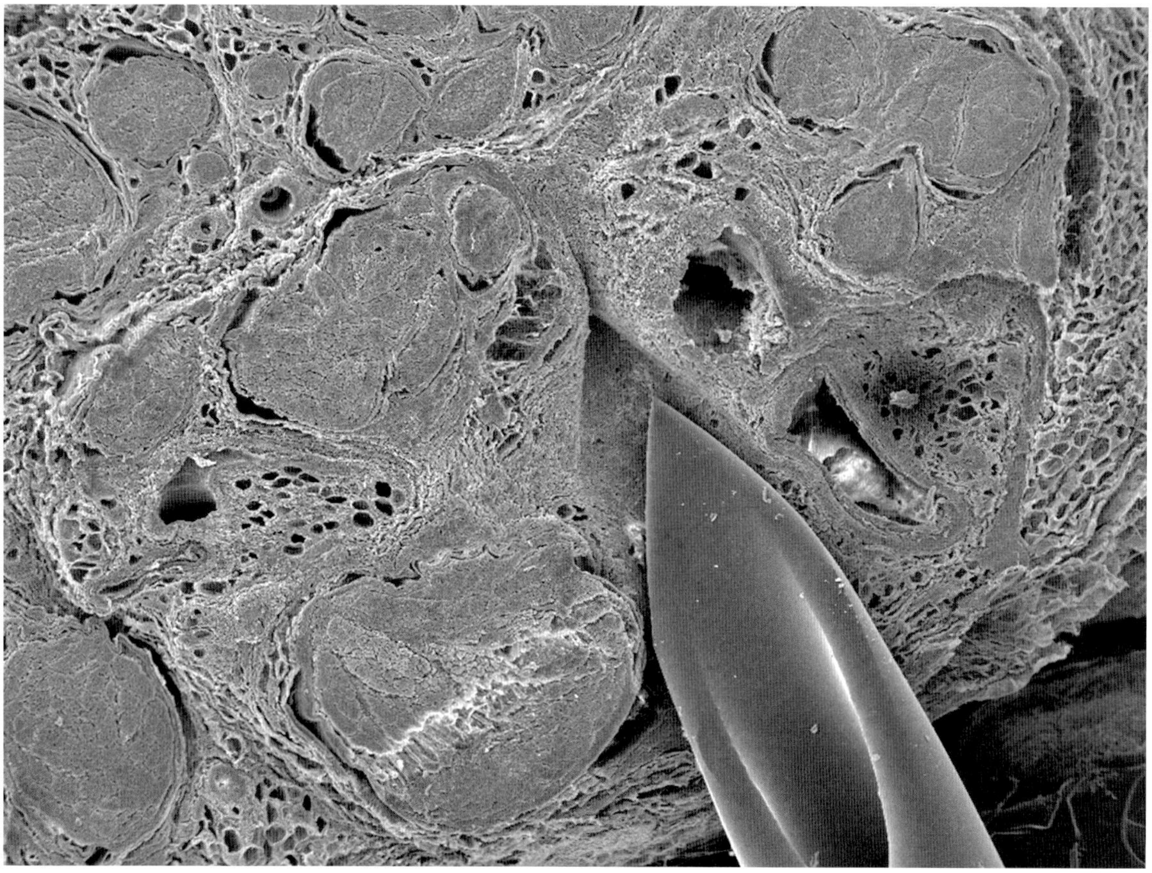

Fig. 16.9 Human tibial nerve. In vitro puncture of nerve with neurostimulation needle. Detail at higher magnification of Fig. 16.8 Scanning electron microscopy. Magnification ×30

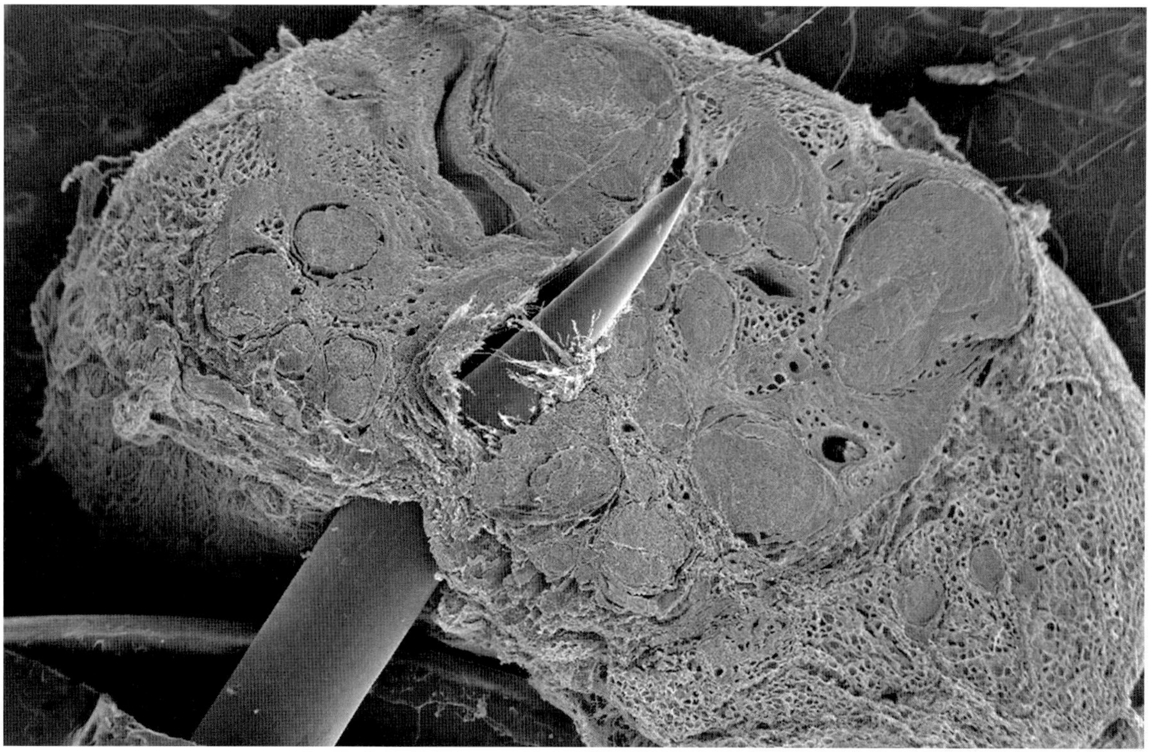

Fig. 16.10 Human sciatic nerve. In vitro puncture of nerve with neurostimulation needle. Needle traversing the fascicles. Scanning electron microscopy. Magnification ×20

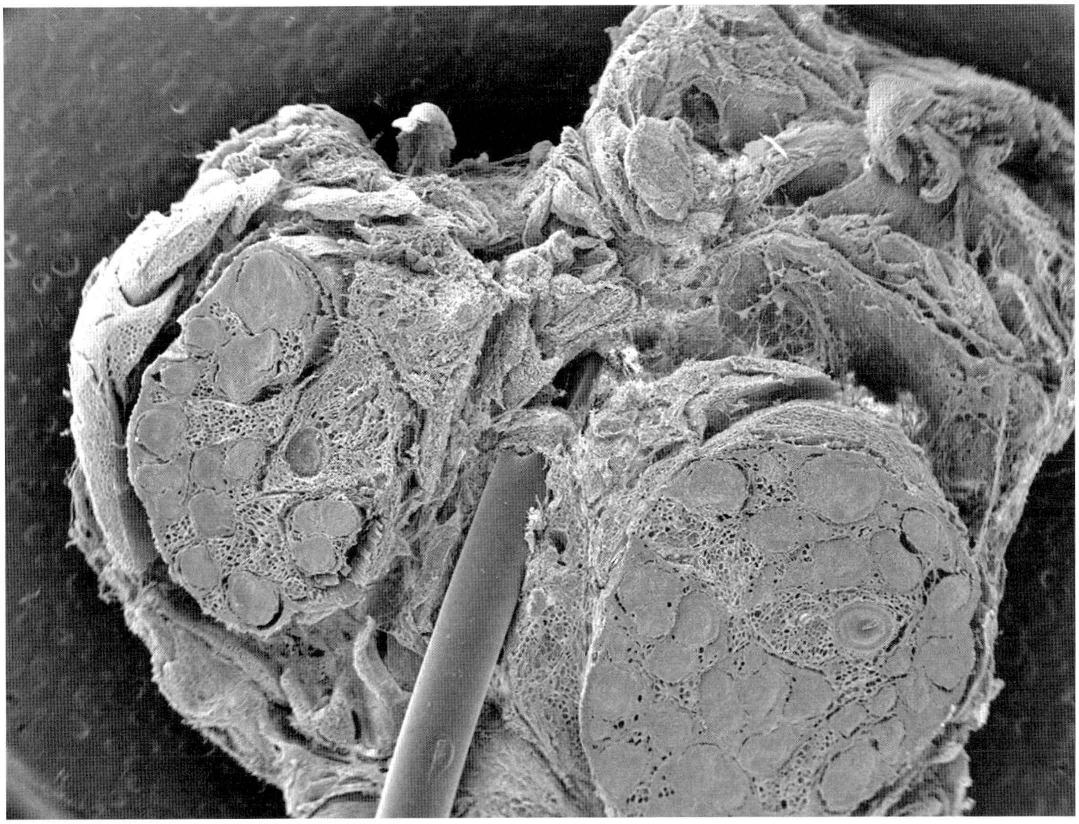

Fig. 16.11 Human sciatic nerve. In vitro puncture of nerve with neurostimulation needle. Needle inserted into the interfascicular fat. Tibial and peroneal components of the sciatic nerve can be identified on the sides of the needle. Scanning electron microscopy. Magnification ×10

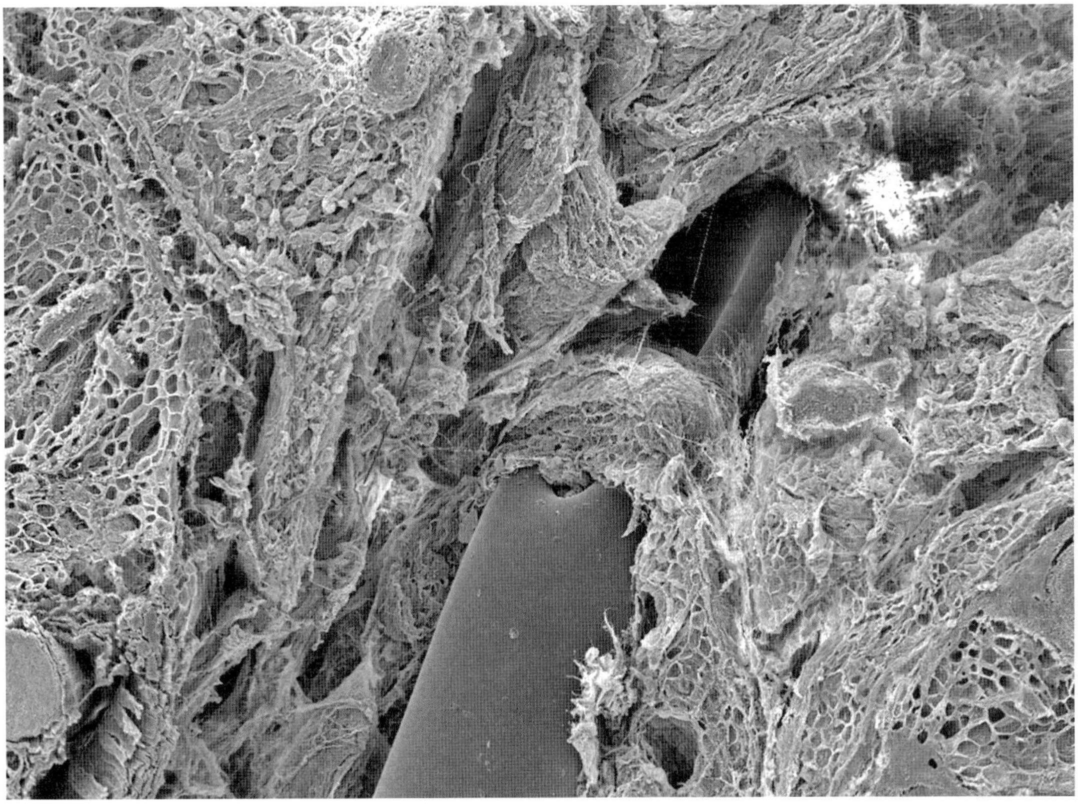

Fig. 16.12 Human sciatic nerve. In vitro puncture of nerve with neurostimulation needle. Detail at higher magnification of Fig. 16.11. Scanning electron microscopy. Magnification ×30

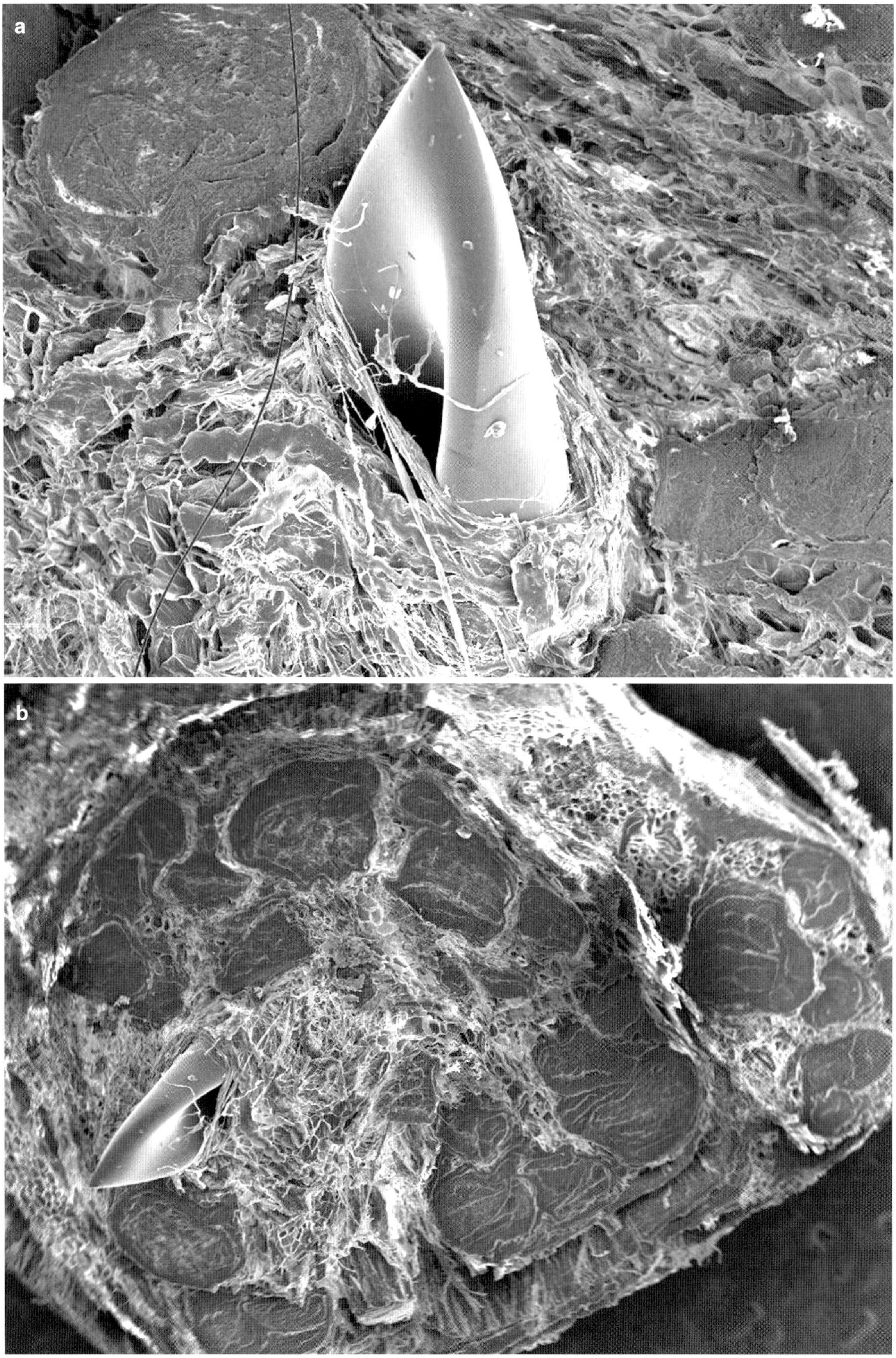

Fig. 16.13 Human tibial nerve. In vitro puncture of nerve with neurostimulation needle. Needle traversing the fascicles viewed from the exit point. (**a**) Detail at higher magnification of **b**. Scanning electron microscopy. Magnification: **a**, ×50; **b**, ×20

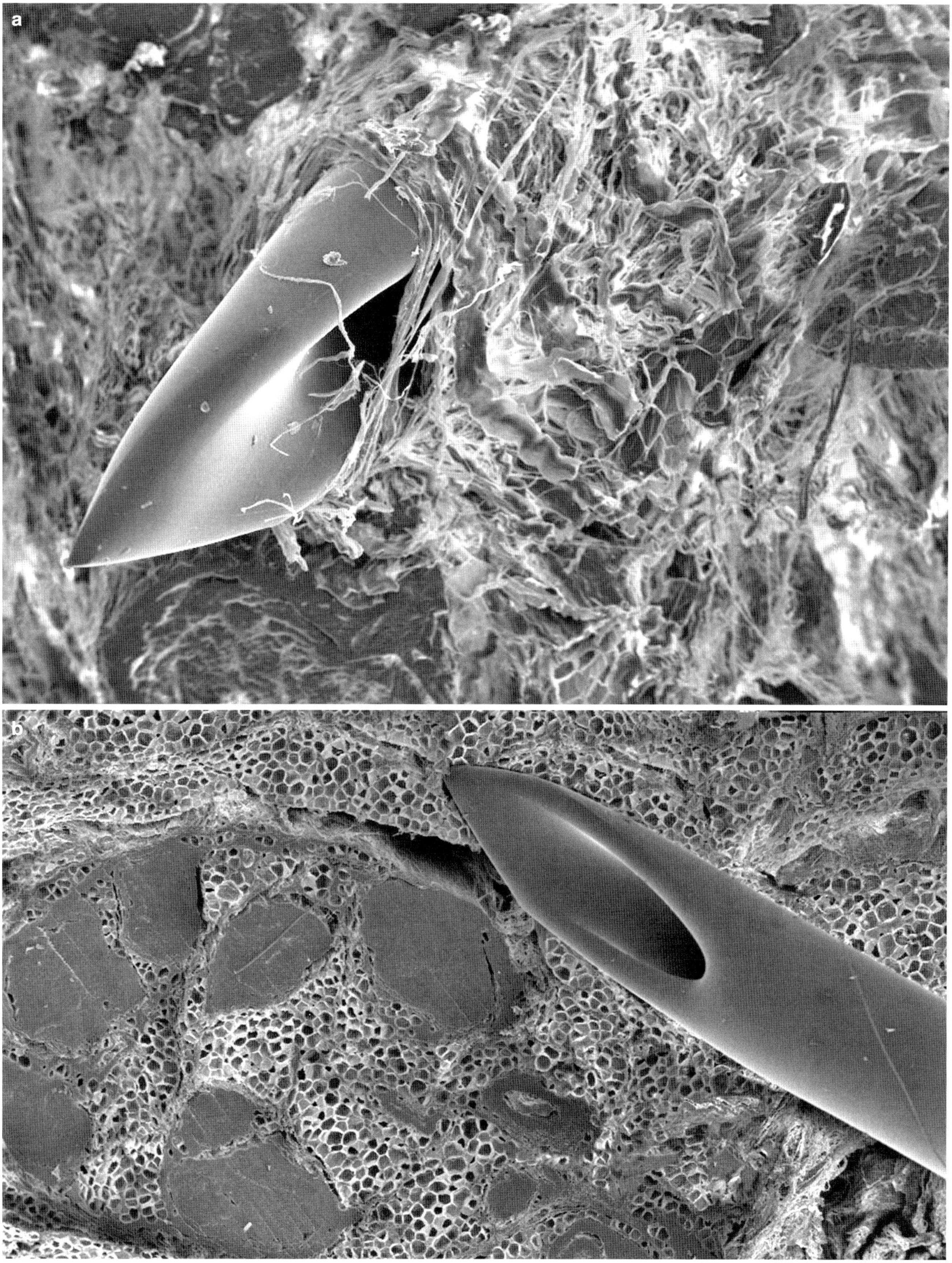

Fig. 16.14 Human tibial nerve. In vitro puncture of nerve with neurostimulation needle. Comparative view. Scanning electron microscopy. Magnification: **a**, ×50; **b,** ×22

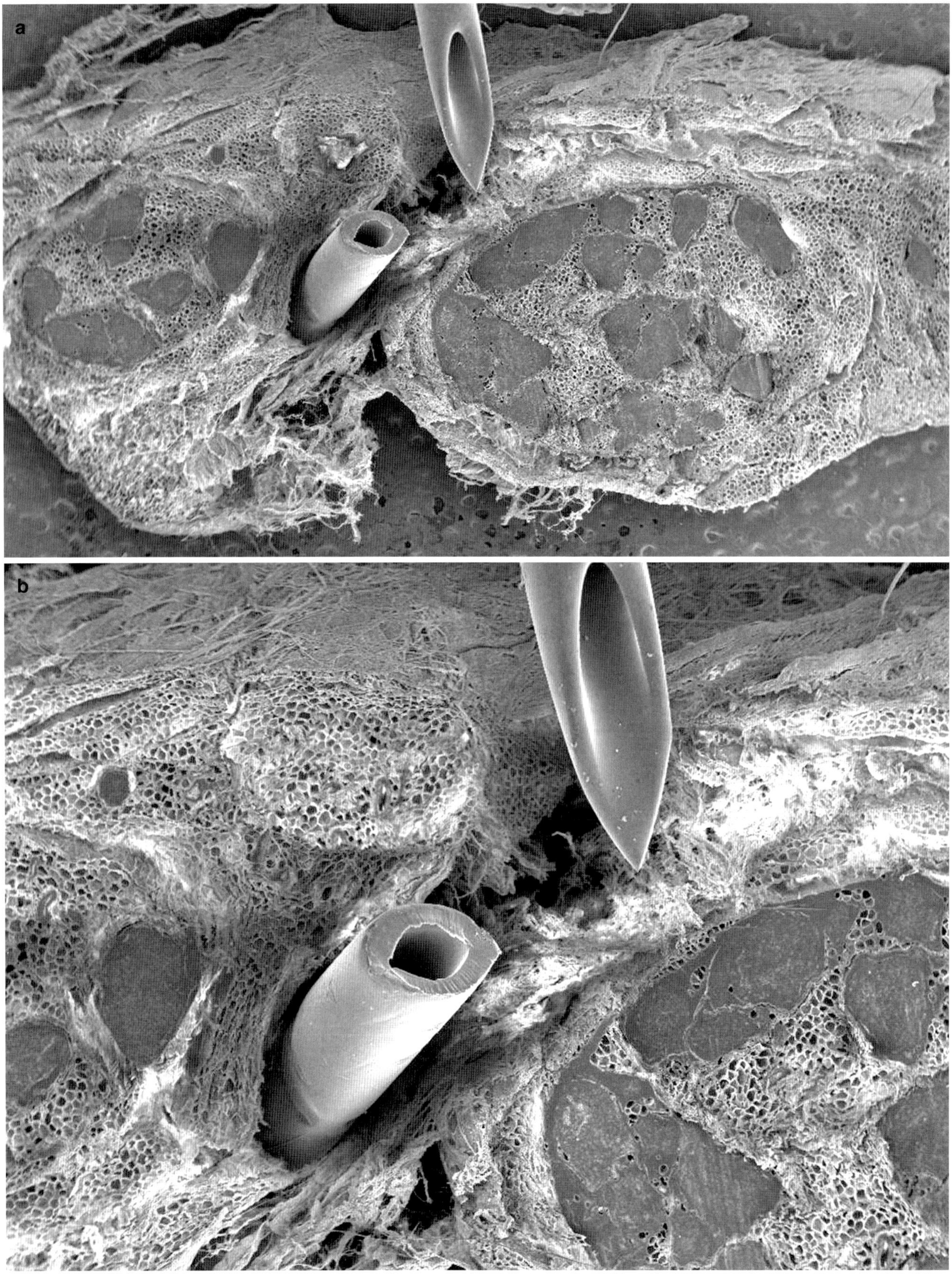

Fig. 16.15 Human sciatic nerve. In vitro intraneural placement of peripheral catheter. Scanning electron microscopy. Magnification: **a**, ×10; **b**, ×20 (Part **b** from De Andrés et al. [1], with permission)

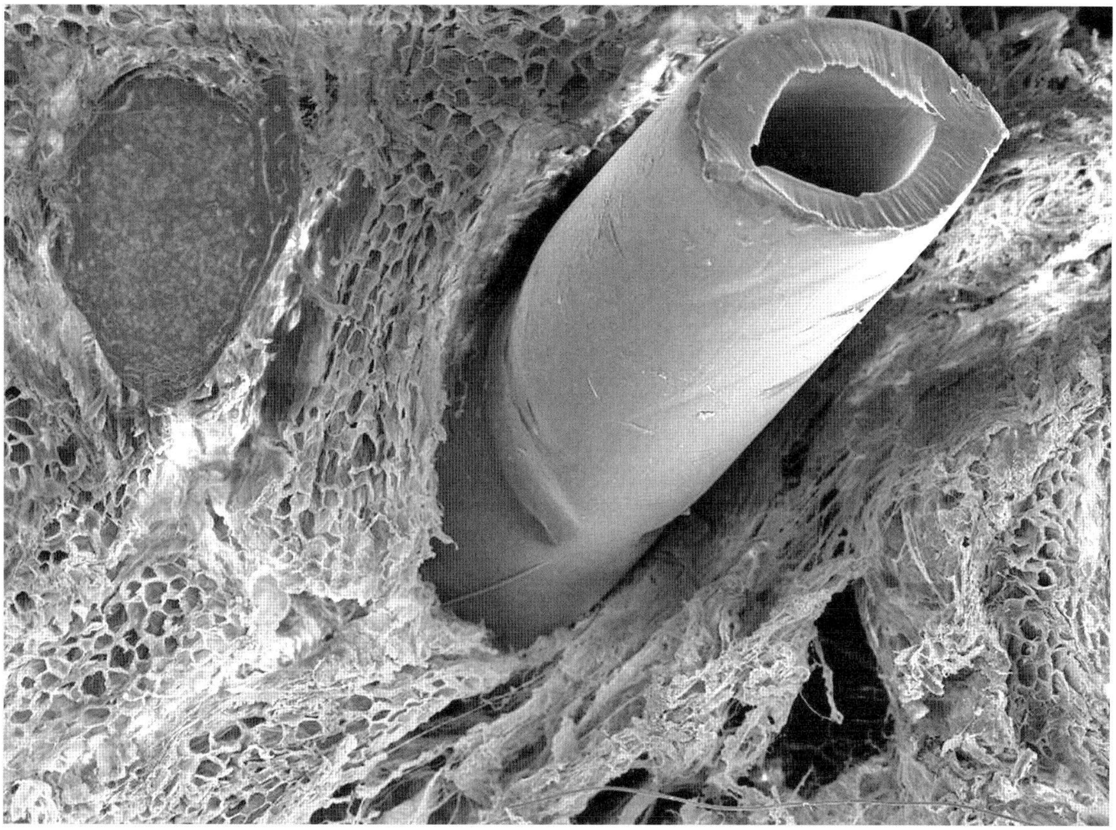

Fig. 16.16 Human sciatic nerve. In vitro intraneural placement of peripheral catheter. Scanning electron microscopy. Magnification ×40

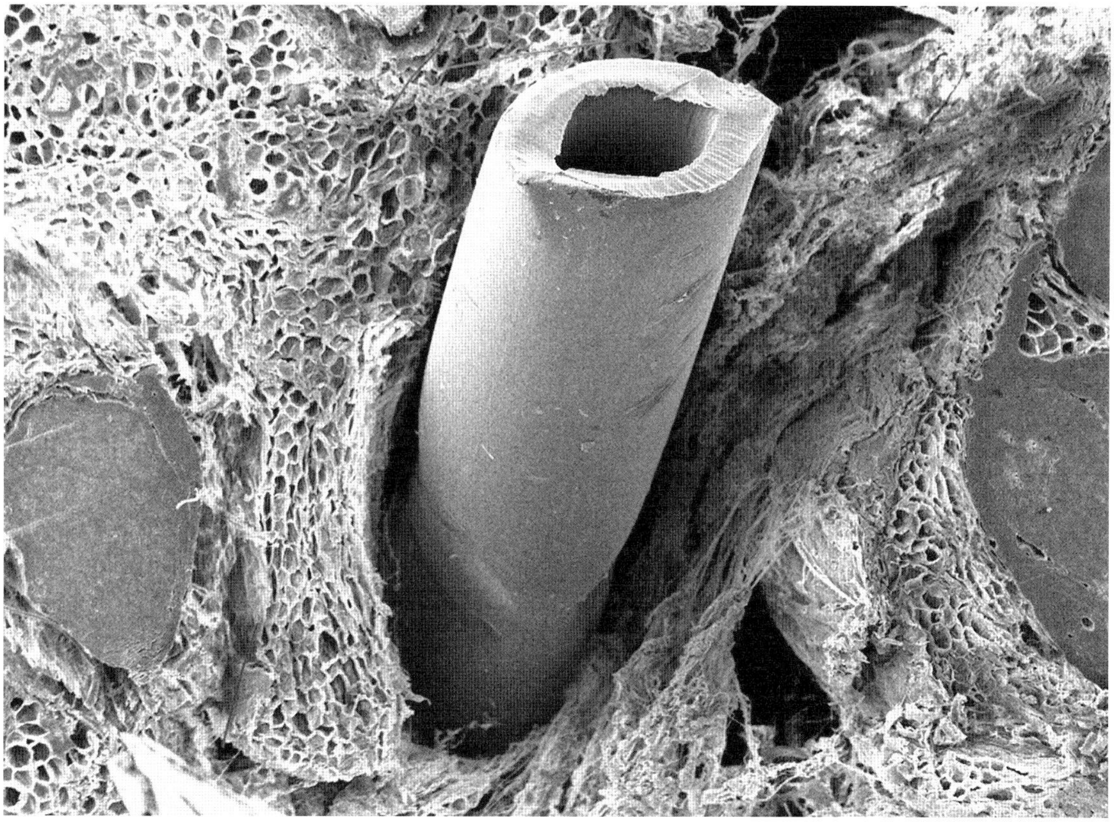

Fig. 16.17 Human sciatic nerve. In vitro intraneural placement of peripheral catheter. Scanning electron microscopy. Magnification ×35

References

1. De Andrés JA, Reina MA, López A, Sala-Blanch X, Prats A. Blocs nerveux périphériques, paresthésies et injections intraneurales. Le Practicien en Anesthésie Réanimation. 2010;14:213–21.
2. Reina MA, López A, De Andrés JA. Adipose tissue within peripheral nerves. Study of the human sciatic nerve. Rev Esp Anestesiol Reanim. 2002;49:397–402.
3. Reina MA, López A, De Andrés JA, Machés F. Possibility of nerve lesions related to peripheral nerve blocks. A study of the human sciatic nerve using different needles. Rev Esp Anestesiol Reanim. 2003;50:274–83.
4. Sala-Blanch X, Ribalta T, Rivas E, Carrera A, Gaspa A, Reina MA, Hadzic A. Structural injury to the human sciatic nerve after intraneural needle insertion. Reg Anesth Pain Med. 2009;34:201–5.

17 Injection of Dye Inside the Paraneural Sheath of the Sciatic Nerve in the Popliteal Fossa

Henning Lykke Andersen, Sofie Lykke Andersen, and Jørgen Tranum-Jensen

The connective tissue around the sciatic nerve in the popliteal fossa forms a sheath-like structure that has a significant influence on the spread of an injection of local anesthetic around the sciatic nerve [1, 2].

The paraneural sheath around the sciatic nerve is a structural part of the connective tissue in the popliteal fossa, to which it is loosely attached. It is multilayered, about 0.5 mm thick, transparent, fragile, and tightly fitted around the sciatic nerve [3]. The sheath creates a virtual cleavage (or fluid conduit) between the paraneural tissue and the epineurium of the sciatic nerve [1, 3]. As the sciatic nerve divides in the popliteal fossa, the paraneural sheath divides as well to surround each of the tibial and common peroneal nerves. During dissection, the sheath is loosened easily from the paraneural tissue and is easy to distinguish macroscopically from the epineurium by manipulation. In some individuals, the sheath is visible on ultrasound at the sciatic nerve division and in some individuals, over the subtle cleft on the surface of the nerve between the two components, where a small amount of fat may be seen under the sheath. Several thin, fascia-like connective tissue layers traverse the popliteal fossa; they also may be seen ultrasonographically. During an injection in the virtual cleavage, the injectate will spread readily inside the paraneural sheath along the surface of the nerve while the sheath expands and eventually ruptures.

Several approaches may be used to place an ultrasound-guided injection inside the paraneural sheath. One is to penetrate the sheath at the sciatic nerve division by using an in-plane, or out-of-plane, short-axis needle approach; another is to penetrate the sheath at the 12 o'clock position in the cleft on the nerve surface by using an in-plane short-axis approach just proximal to the division, where the nerve often is easier to visualize ultrasonically. While the needle tip is pressed gently against the paraneural sheath, a few test injections are made to determine whether the needle tip has penetrated the sheath and entered the cleavage. Often, a loss of resistance may be felt and seen ultrasonically. Once the needle tip is placed correctly, the injectate presses the nerve away from the needle tip while the expansion of the sheath forms a hypoechoic cleavage on the surface of the nerve and between the nerve components. The paraneural sheath then may be seen as a sharp hyperechoic, ballooning boundary of the cleavage with the surrounding tissue. If the transducer is moved distally, the injectate can be seen spreading around each of the nerve branches. This way, the injectate spreads inside the paraneural sheath in the cleavage along the sciatic nerve while leaving the nerve structure intact.

On a microscopic cross-section of the sciatic nerve, the paraneural sheath may be difficult to determine, unless dye has been injected to distinguish the sheath from the epineurium before the preparation of the tissue sample (Figs. 17.1, 17.2, 17.3, 17.4, 17.5, 17.6, 17.7, 17.8, 17.9, 17.10, 17.11, 17.12, 17.13, 17.14, 17.15, 17.16, 17.17, 17.18, 17.19, 17.20, 17.21, 17.22, 17.23, 17.24, 17.25, and 17.26.

H.L. Andersen, MD, PhD (✉)
Department of Anaesthesiology,
Copenhagen University Hospital, Herlev, Copenhagen, Denmark
e-mail: henning.lykke.andersen@regionh.dk

S.L. Andersen, MD
Department of Anaesthesiology,
Copenhagen University Hospital, Rigshospitalet,
Copenhagen, Denmark
e-mail: sofie.lykke.andersen@regionh.dk

J. Tranum-Jensen, MD
Department of Cellular and Molecular Medicine,
School of Medicine, University of Copenhagen,
Copenhagen, Denmark
e-mail: jtj@sund.ku.dk

M.A. Reina et al. (eds.), *Atlas of Functional Anatomy for Regional Anesthesia and Pain Medicine: Human Structure, Ultrastructure and 3D Reconstruction Images*, DOI 10.1007/978-3-319-09522-6_17,

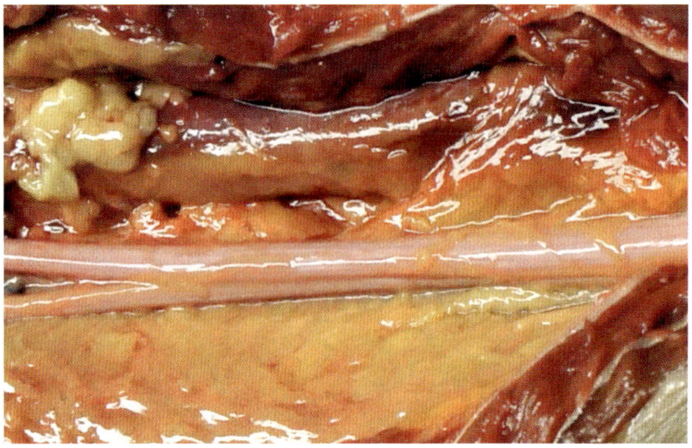

Fig. 17.1 Gross dissection of the popliteal fossa showing the intact sciatic nerve tightly surrounded by the paraneural sheath

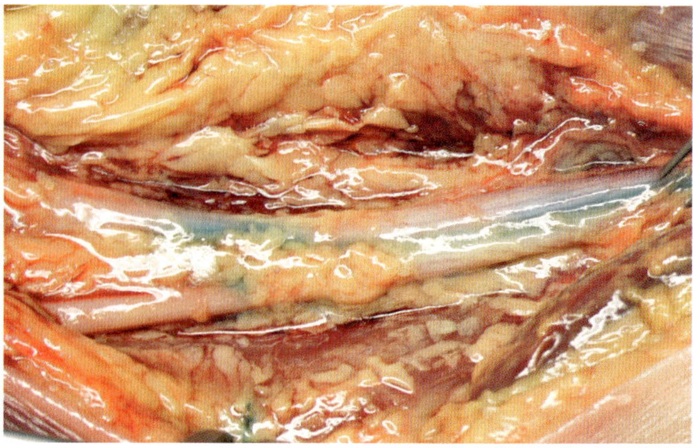

Fig. 17.2 Before the dissection of this nerve, an ultrasound-guided injection of water with a drop of blue dye was made inside the paraneural sheath near the nerve division

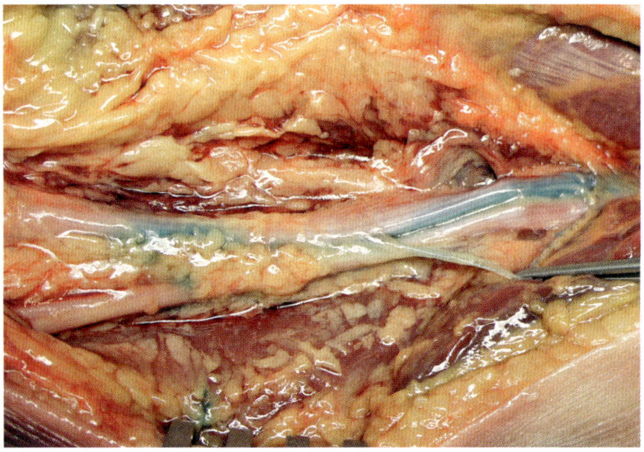

Fig. 17.3 The paraneural sheath is cut open. A forceps holds the sheath aside to expose the sciatic nerve

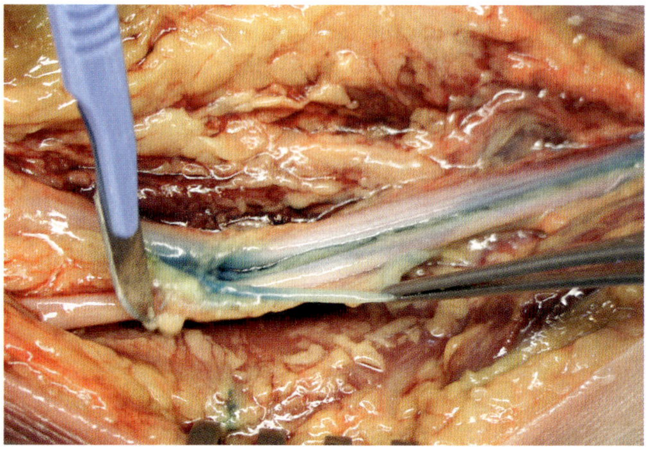

Fig. 17.4 The paraneural sheath is cut open distally to the division of the nerve and sheath

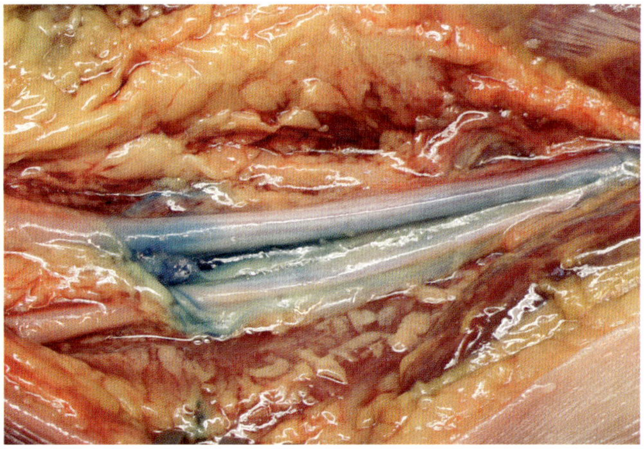

Fig. 17.5 This sciatic nerve actually comprises two separate nerve components colored by blue dye and enveloped in a common paraneural sheath

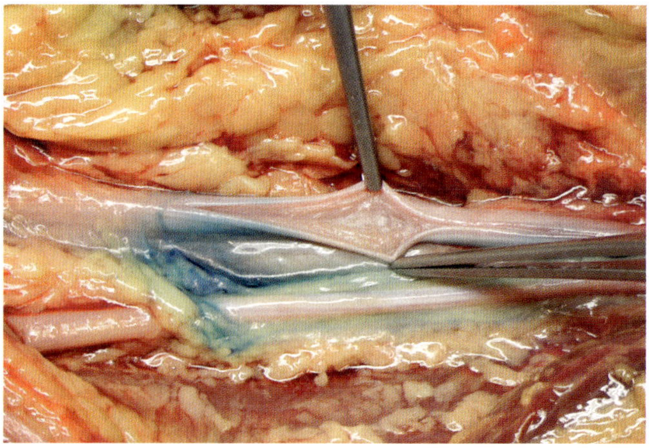

Fig. 17.6 The tibial nerve component is cut open through the epineurium to show that the injection was not placed intraneurally

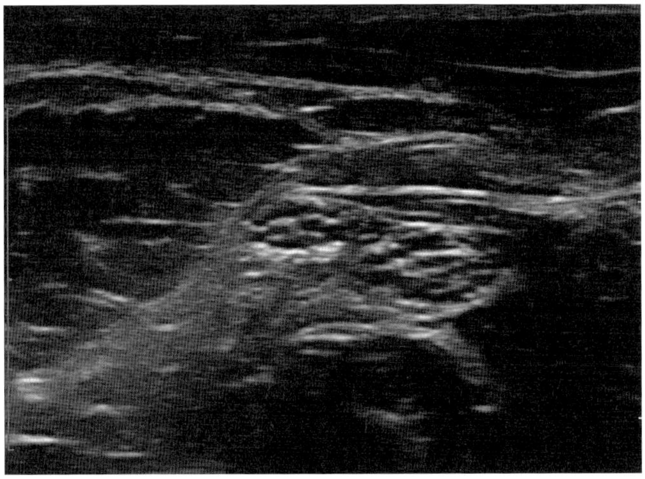

Fig. 17.7 Ultrasound image of the sciatic nerve shown in Figs. 17.2, 17.3, 17.4, 17.5 and 17.6, before injection

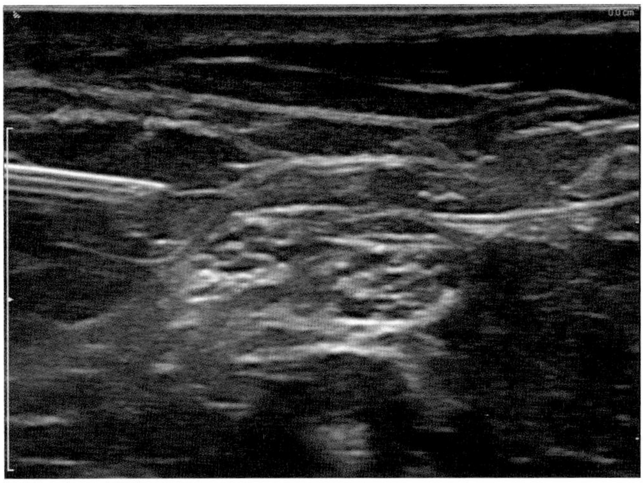

Fig. 17.8 A needle is inserted laterally through the biceps muscle

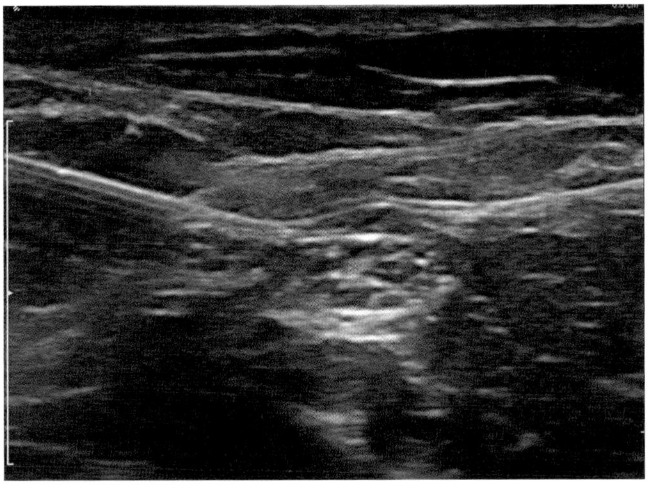

Fig. 17.9 The needle is directed toward the nerve between the two nerve components.

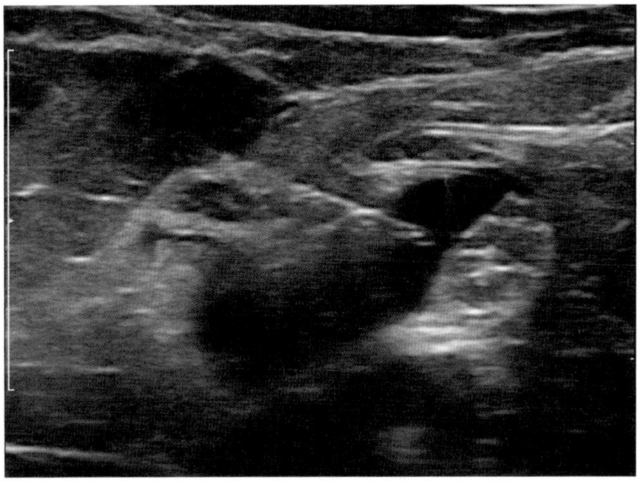

Fig. 17.10 The needle tip has penetrated the paraneural sheath, and dye is injected between the nerve components

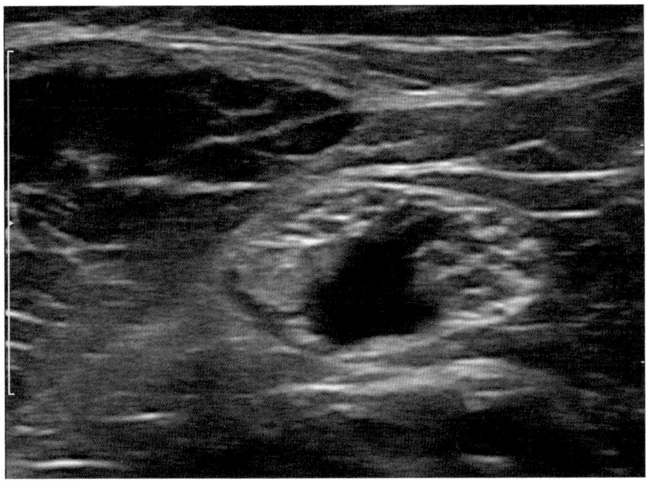

Fig. 17.11 Dye between the nerve components has expanded the paraneural sheath, which now can be seen clearly now

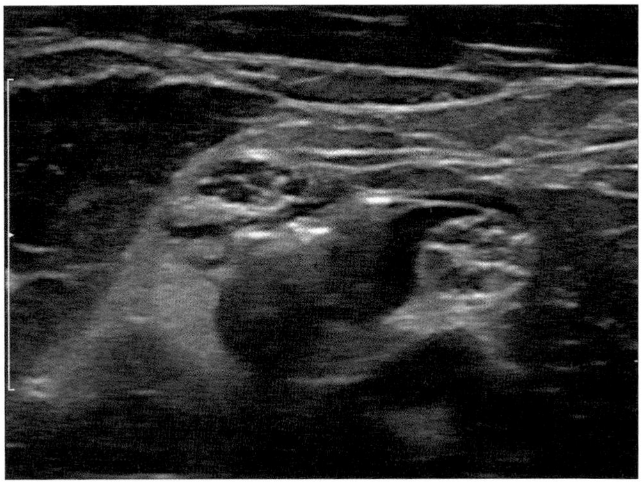

Fig. 17.12 The dye may be traced distally inside the paraneural sheath around each of the tibial and common peroneal nerves

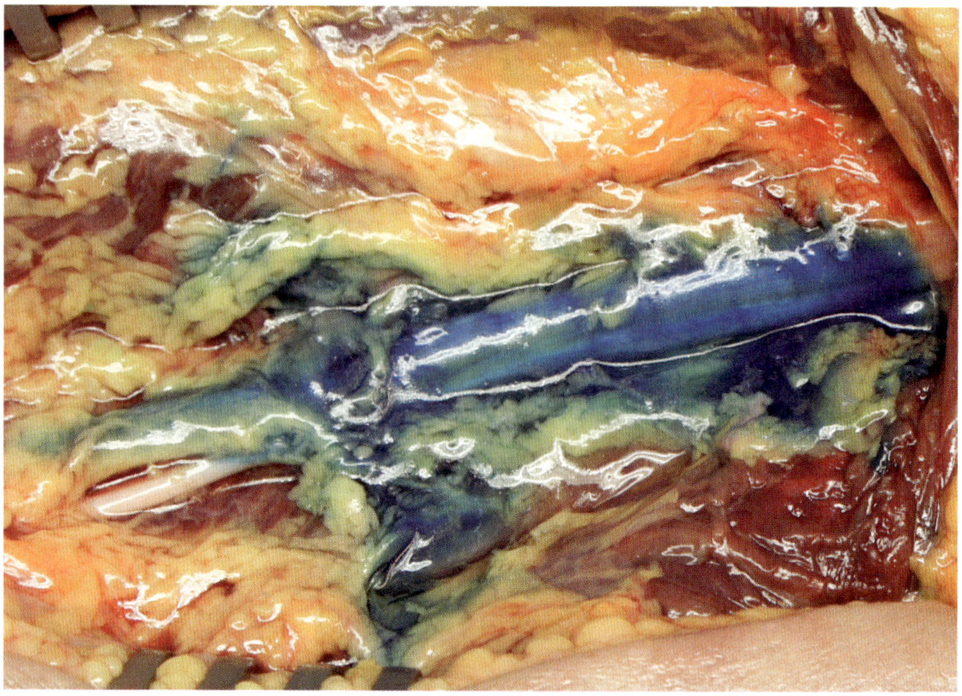

Fig. 17.13 Gross dissection of the popliteal fossa after an ultrasound-guided injection of water with blue dye is placed inside the paraneural sheath near the nerve division. Through the transparent expanded paraneural sheath, the sciatic nerve can be seen floating in the blue watery dye. Some dye also has leaked through the sheath near the injection site (From Andersen et al. [3]; with permission)

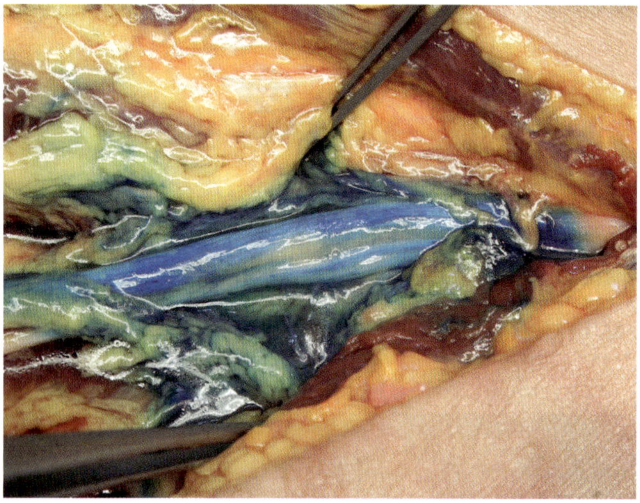

Fig. 17.14 The paraneural sheath is cut open, and a forceps holds aside the sheath to expose the blue-colored epineurium of the sciatic nerve. This sciatic nerve is one solid nerve consisting of two main nerve components. A fat-containing cleft on the surface of the nerve can be seen under the sheath, indicating the two nerve components (From Andersen et al. [3]; with permission)

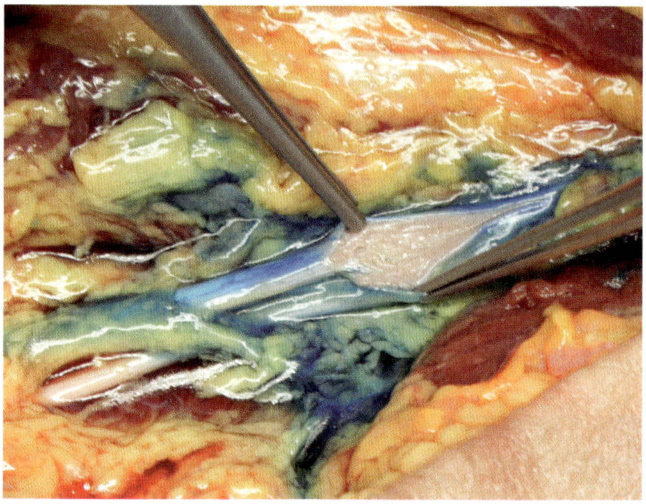

Fig. 17.15 The epineurium is cut open to show that the dye was not injected intraneurally. A forceps holds aside the superficial epineurium to expose the inside of the sciatic nerve (From Andersen et al. [3]; with permission)

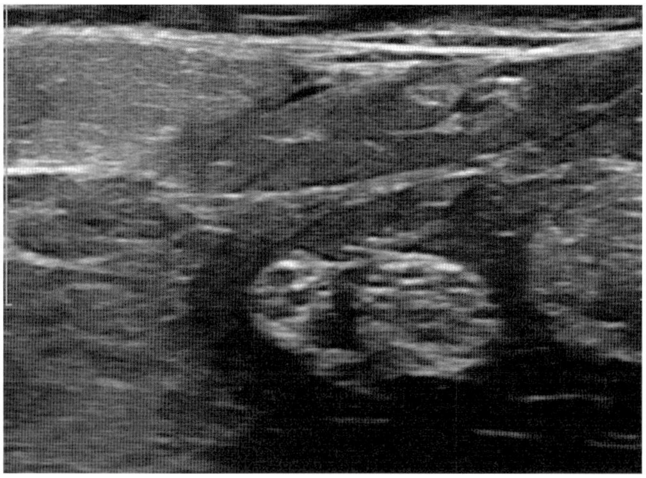

Fig. 17.16 Ultrasound image of the sciatic nerve in Figs. 17.13, 17.14 and 17.15. Showing the division and the paraneural sheath

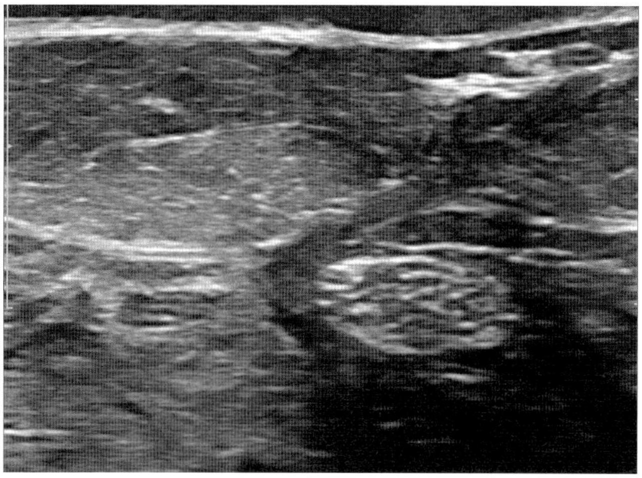

Fig. 17.17 The same sciatic nerve about 1 in. proximal to the nerve division before the injection

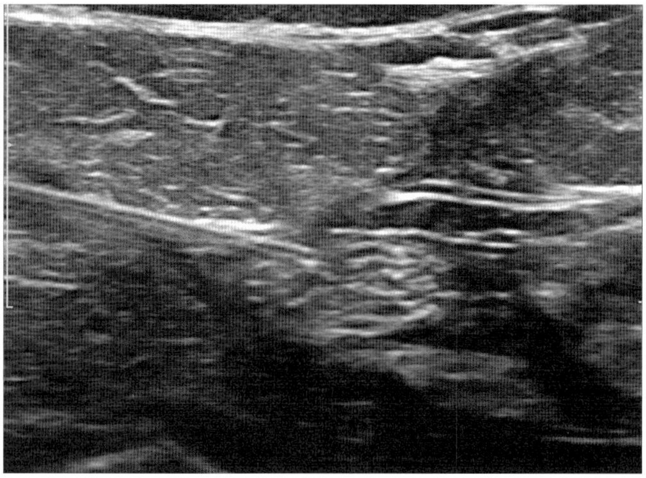

Fig. 17.18 A needle is inserted laterally through the biceps muscle and aimed at the subtle cleft between the two nerve components

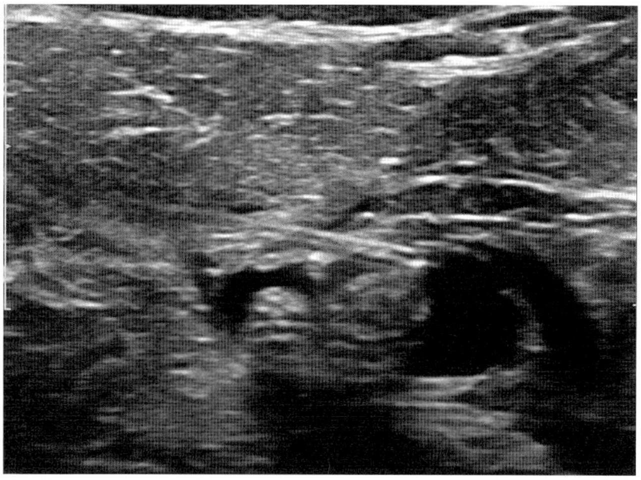

Fig. 17.19 The needle has penetrated the paraneural sheath, and dye is injected. The sheath has expanded and leaked some dye

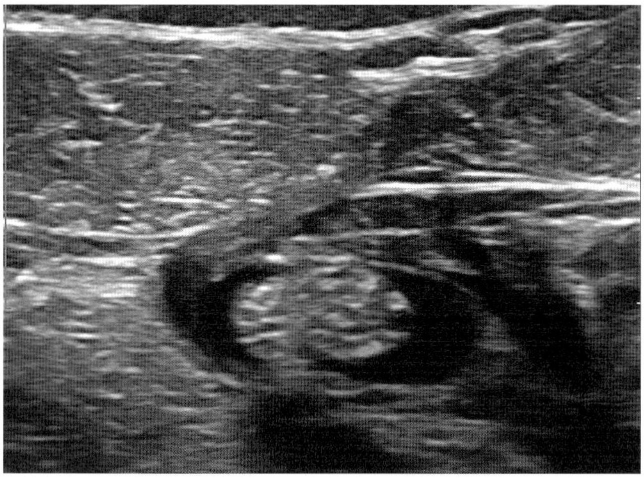

Fig. 17.20 The hypoechoic dye surrounding the sciatic nerve has expanded the paraneural sheath, which now can be seen clearly now

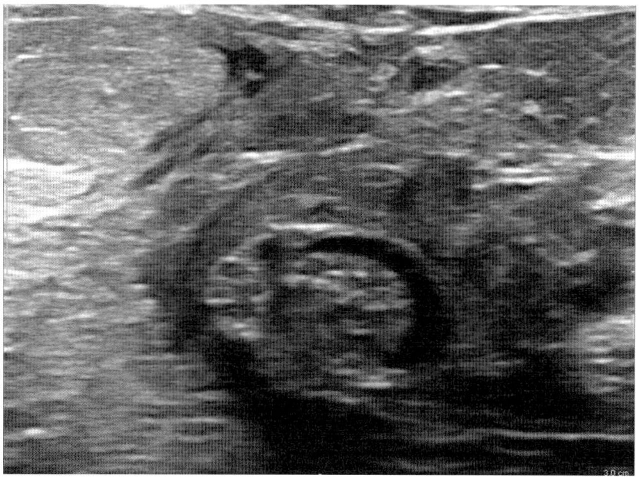

Fig. 17.21 The dye can be traced distally to the division inside the paraneural sheath, mainly around the tibial nerve

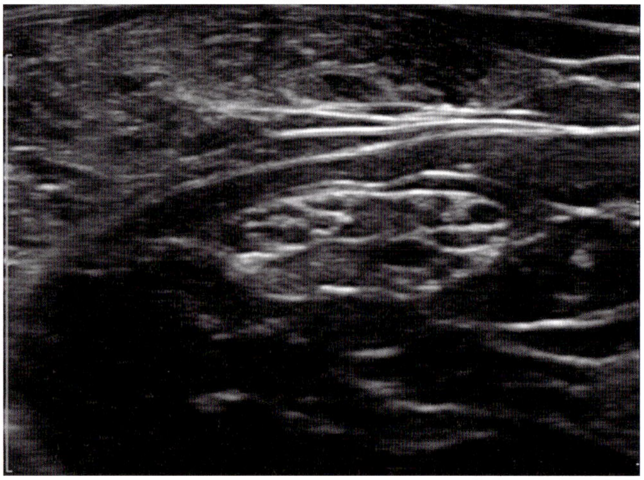

Fig. 17.22 An ultrasound image of the sciatic nerve at the division. The paraneural sheath can be seen bridging the two nerve components.

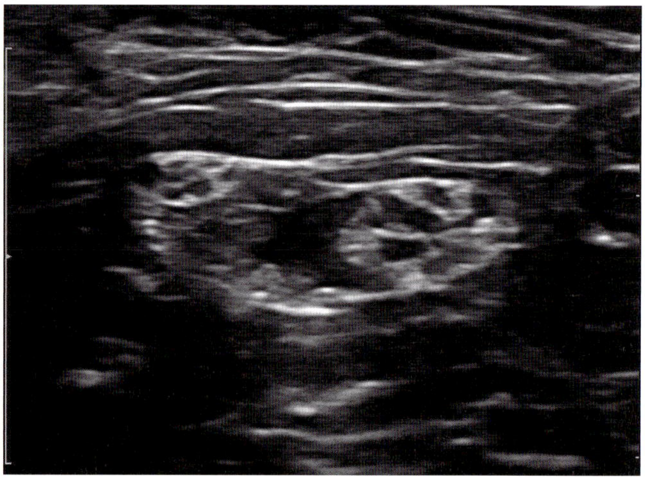

Fig. 17.23 An injection of dye has expanded the paraneural sheath between the two nerve components

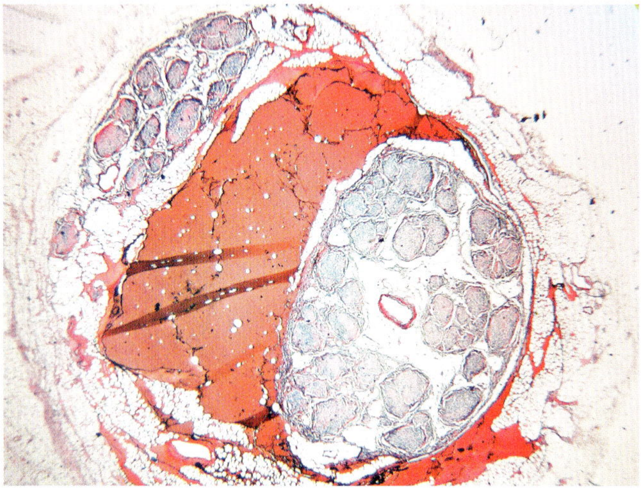

Fig 17.24 A histologic cross-section of the sciatic nerve. The paraneural sheath can be seen surrounding the red dye and the two nerve components

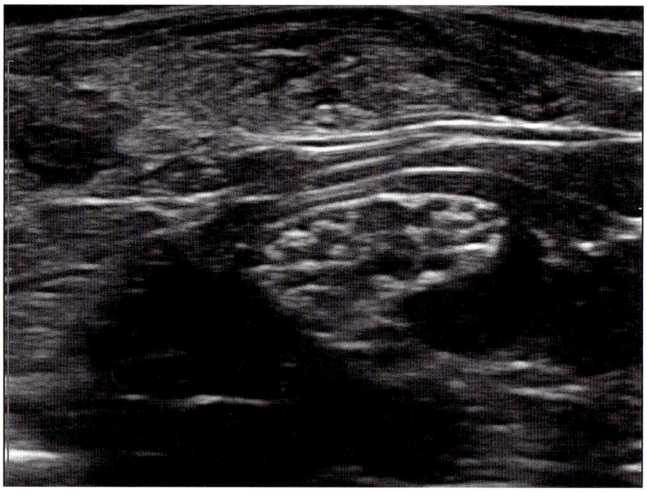

Fig. 17.25 (**a**) The injection of dye also has created a hypoechoic semilunar cleavage inside the paraneural sheath around the sciatic nerve

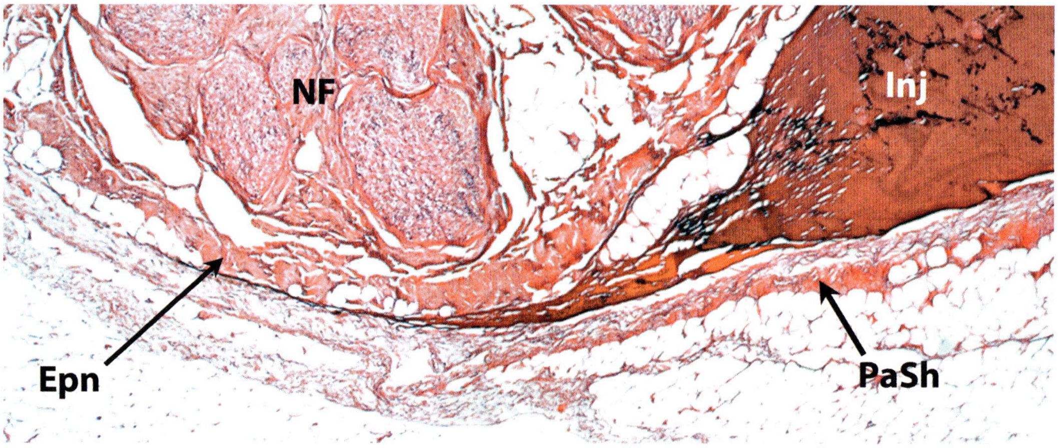

Fig. 17.26 Close-up of a histologic cross-section after dye injection along the sciatic nerve containing bundles of nerve fascicle (*NF*). The injected dye (*Inj*) has made a cleavage between the epineurium (*Epn*) and the paraneural sheath (*PaSh*). Note the difference in tissue structure

References

1. Vloka JD, Hadzić A, Lesser JB, Kitain E, Geatz H, April EW, et al. A common epineural sheath for the nerves in the popliteal fossa and its possible implications for sciatic nerve block. Anesth Analg. 1997;84:387–90.
2. Vloka JD, Hadzić A, April E, Thys DM. The division of the sciatic nerve in the popliteal fossa: anatomical implications for popliteal nerve blockade. Anesth Analg. 2001;92:215–7.
3. Andersen HL, Andersen SL, Tranum-Jensen J. Injection inside the paraneural sheath of the sciatic nerve: direct comparison among ultrasound imaging, macroscopic anatomy, and histologic analysis. Reg Anesth Pain Med. 2012;37:410–4.

High-Definition and Three-Dimensional Volumetric Ultrasound Imaging of the Sciatic Nerve

18

Manoj Kumar Karmakar

The sciatic nerve, the largest (thickest) nerve of the body, is formed by the union of the anterior primary rami of the spinal nerves of L4, L5, S1, S2, S3, and S4 (lumbosacral plexus). The sacral plexus lies deep within the pelvis between the piriformis muscle posteriorly and the pelvis fascia anteriorly. After exiting the pelvis through the greater sciatic foramen, the sciatic nerve enters the subgluteal space between the gluteus maximus muscle posteriorly and the quadratus femoris muscle anteriorly. Thereafter it descends in an intermuscular fascial plane on the posterior aspect of the thigh, lying deep to the semitendinosus and biceps femoris muscles, to about its lower third, where it bifurcates into its two branches, the tibial and common peroneal (fibular) nerves. This bifurcation is variable and can take place at any point between the origin of the sciatic nerve at the sacral plexus and the lower third of the thigh, or at a variable distance from the popliteal crease. The tibial and common peroneal nerves may also arise separately from the sacral plexus. In this chapter, the sonoanatomy of the sciatic nerve from its exit from the pelvis (parasacral location) to its bifurcation at the popliteal fossa is demonstrated using high-definition and three-dimensional ultrasound imaging. The sonoanatomy of the paraneural connective tissue sheath and the perineural fascial compartments (*ie*, the subepimyseal perineural and subparaneural compartment) at the popliteal fossa are also demonstrated using ultrasound images acquired during sciatic nerve block at the popliteal fossa (Figs. 18.1, 18.2, 18.3, 18.4, 18.5, 18.6, 18.7, 18.8, 18.9, 18.10, 18.11, 18.12, 18.13, 18.14, 18.15, 18.16, 18.17, 18.18, 18.19, 18.20, 18.21, 18.22, 18.23, 18.24, 18.25, 18.26, 18.27, 18.28, 18.29, 18.30, 18.31, 18.32, 18.33, 18.34, 18.35, 18.36, 18.37, 18.38, 18.39, 18.40, 18.41, 18.42, 18.43, 18.44, and 18.45) [1–4].

M.K. Karmakar, MD, FRCA, FHKCA, FHKAM
Department of Anaesthesia and Intensive Care, The Chinese University of Hong Kong, Prince of Wales Hospital, Hong Kong, China
e-mail: karmakar@cuhk.edu.hk

M.A. Reina et al. (eds.), *Atlas of Functional Anatomy for Regional Anesthesia and Pain Medicine: Human Structure, Ultrastructure and 3D Reconstruction Images*, DOI 10.1007/978-3-319-09522-6_18,

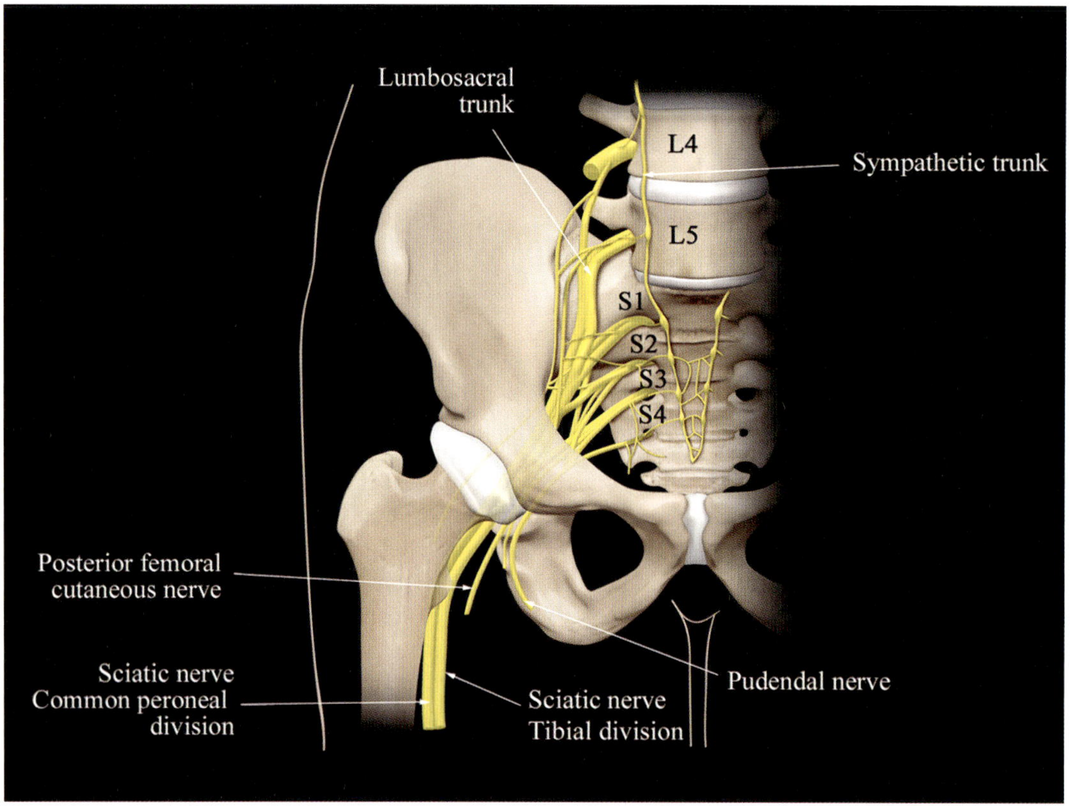

Fig. 18.1 Anatomical illustration showing the formation of the lumbosacral plexus by the union of the anterior primary rami of the *L4, L5, S1, S2, S3*, and *S4* spinal nerves

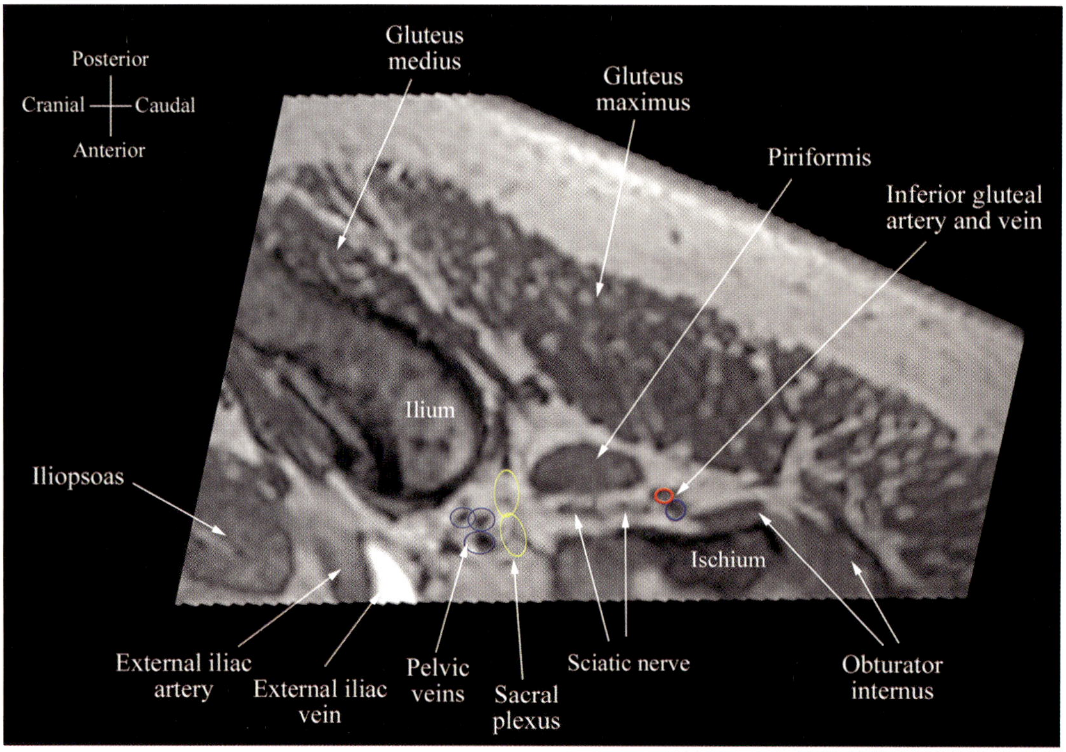

Fig. 18.2 Rendered sagittal oblique MRI image from the level of the greater sciatic foramen, demonstrating the sacral plexus and the parasacral relations of the sciatic nerve. Note that the sciatic nerve exits the pelvis through the greater sciatic foramen lying deep to the piriformis muscle

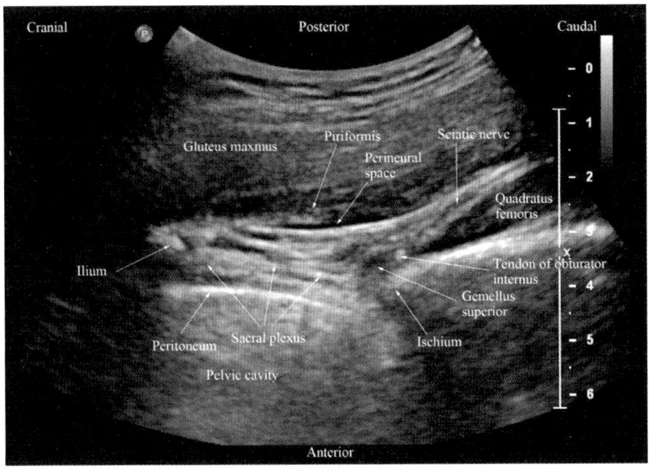

Fig. 18.3 Sagittal sonogram at the level of the greater sciatic foramen (parasacral scan) showing the sacral plexus and the sciatic nerve as it exits the pelvis to enter the infra-piriformis fossa. Note the close relationship of the sacral plexus to the pelvic peritoneum (hyperechoic), and the hypoechoic perineural space deep to the piriformis muscle and posterior to the sciatic nerve

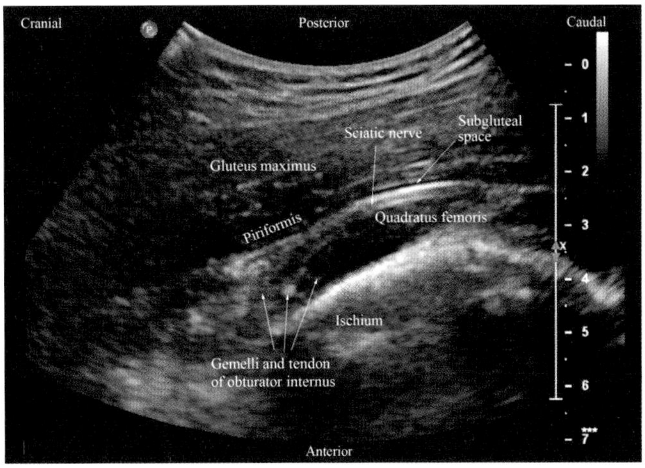

Fig. 18.4 Sagittal sonogram showing the sciatic nerve as it exits the greater sciatic foramen between the piriformis muscle and the gemelli muscles and the tendon of the obturator internus. Note the hypoechoic perineural space between the sciatic nerve and the piriformis muscle posteriorly. The sciatic nerve is also seen to continue distally to enter the subgluteal space between the gluteus maximus posteriorly and the quadratus femoris anteriorly

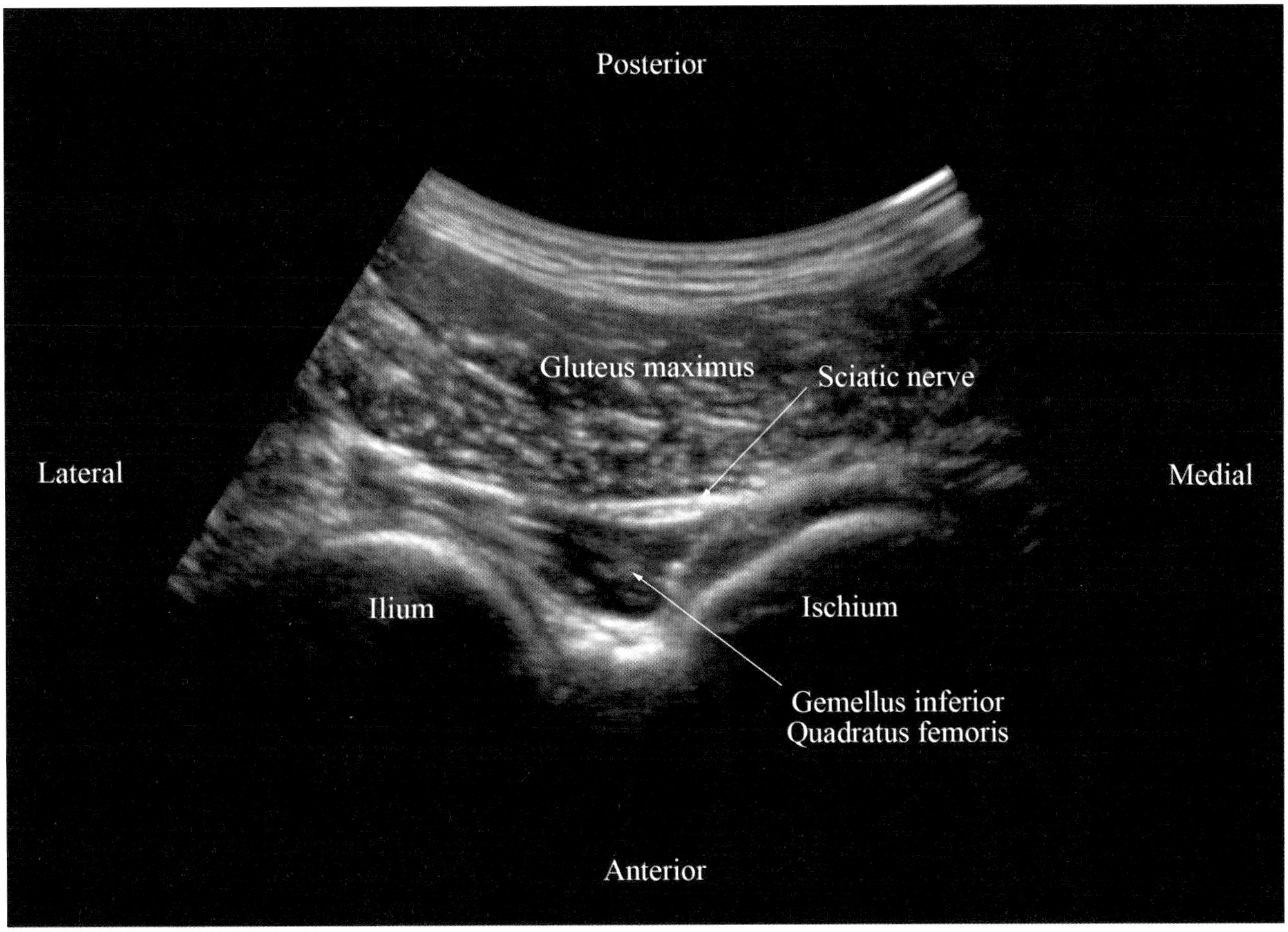

Fig. 18.5 Transverse sonogram of the gluteal region showing the sciatic nerve between the gluteus maximus and the gemelli muscles, immediately distal to the greater sciatic foramen (parasacral position)

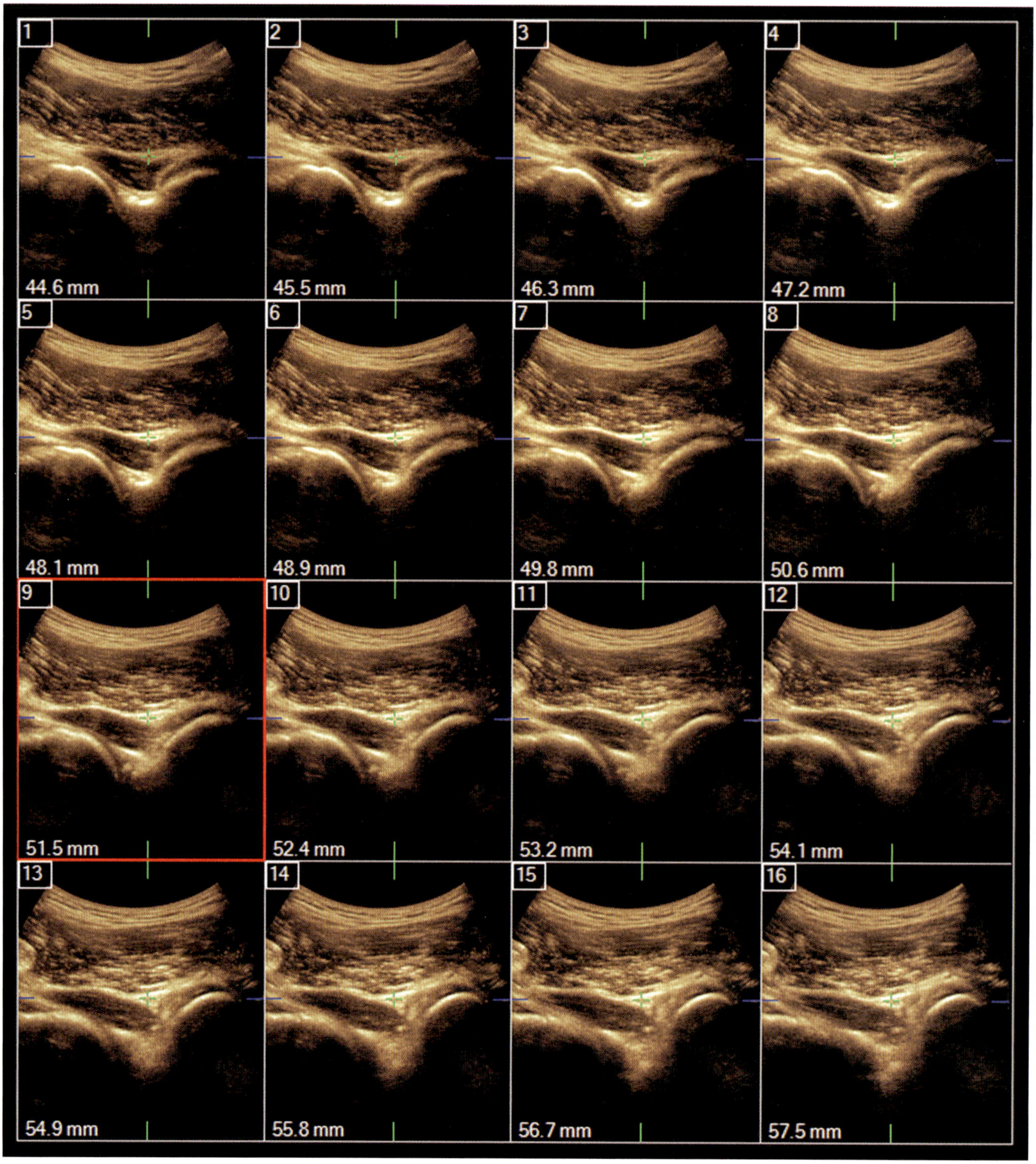

Fig. 18.6 A transverse iSlice display of the sciatic nerve, rendered from a three-dimensional (3D) ultrasound volume of the parasacral region, in color (*sepia tone*) mode. In this figure, 16 contiguous transverse cuts of the acquired sciatic nerve volume (0.8–0.9 mm apart) are displayed. The iSlice mode allows display of the acquired volume similar to a CT scan, and this can be done in all three planes (transverse, sagittal, and coronal). By varying the thickness of the slice, one can accurately delineate changes in the anatomy of the volume as one navigates through it. The distance covered by a single sweep with the volume linear array transducer used (VL13, 13–5 MHz, (Philips Healthcare, Andover, MA, USA) is 42 mm

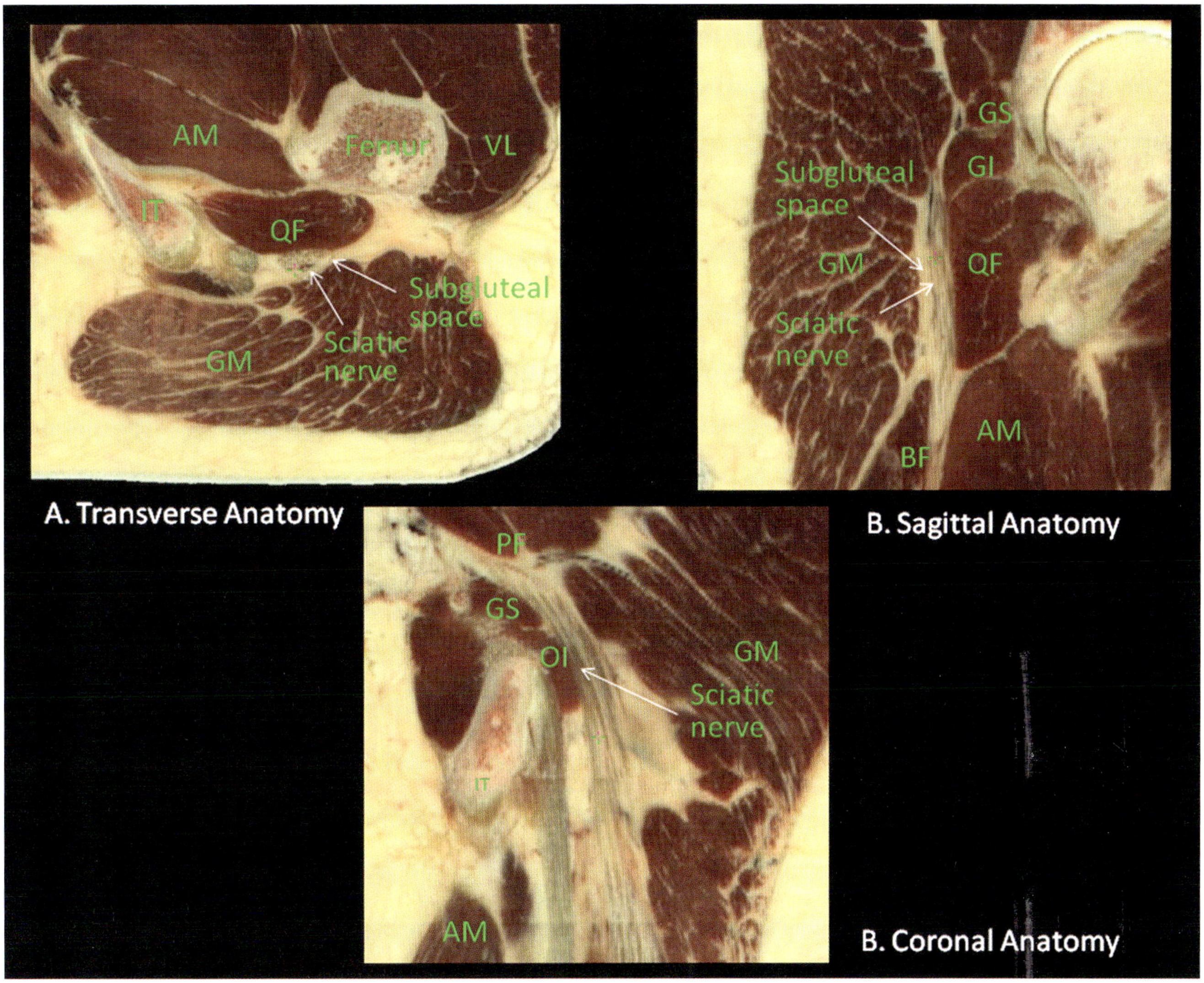

Fig. 18.7 Multiplanar anatomy (rendered from the Visible Human data set) of the sciatic nerve at the subgluteal space. The reference marker (*green crosshair*) has been placed over the sciatic nerve in the transverse view and its corresponding position in the sagittal and coronal images can be seen. *AM* adductor magnus, *BF* biceps femoris, *GI* gemellus inferior, *GM* gluteus maximus, *GS* gemellus superior, *IT* ischial tuberosity, *OI* obturator internus, *PF* piriformis, *QF* quadratus femoris, *VL* vastus lateralis (Courtesy of the Visible Human Server at Ecole Polytechnique Fédérale de Lausanne, Visible Human Visualization Software (http://visiblehuman.epfl.ch)

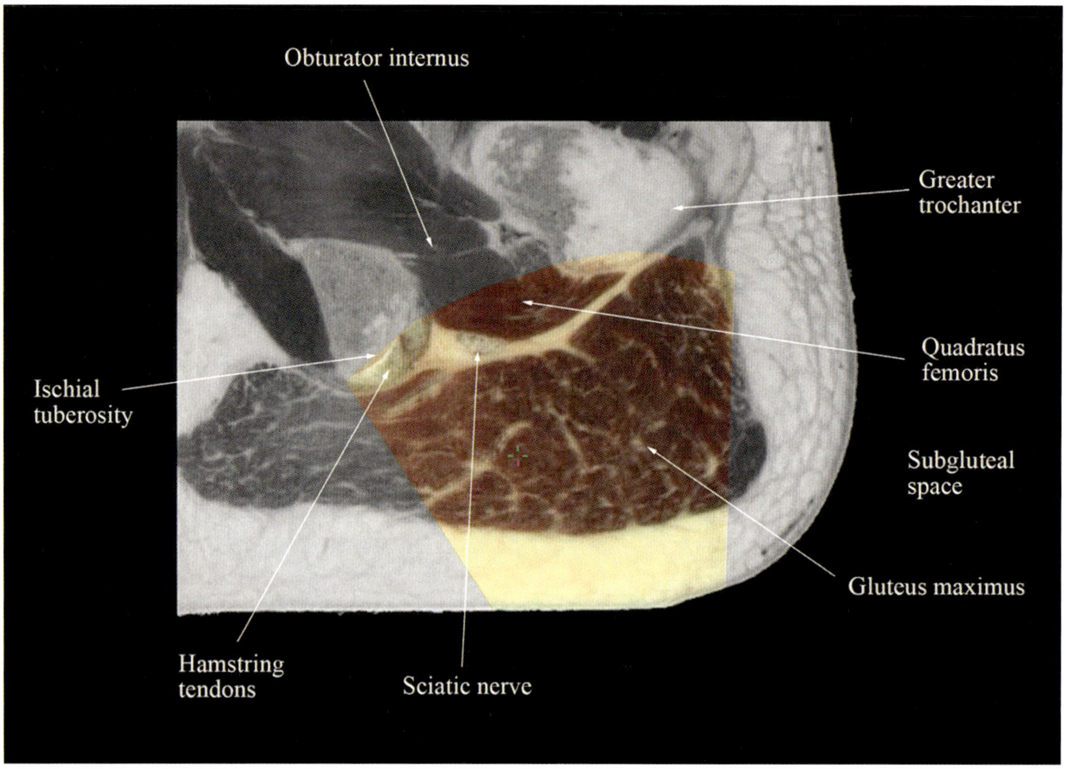

Fig. 18.8 Figure highlighting the anatomical structures that are insonated during a transverse ultrasound scan for the sciatic nerve at the subgluteal space, between the greater trochanter and ischial tuberosity. The plane of the ultrasound beam has been superimposed over the anatomy imaged (rendered from the Visible Human data set) (Courtesy of the Visible Human Server at Ecole Polytechnique Fédérale de Lausanne, Visible Human Visualization Software (http://visiblehuman.epfl.ch)

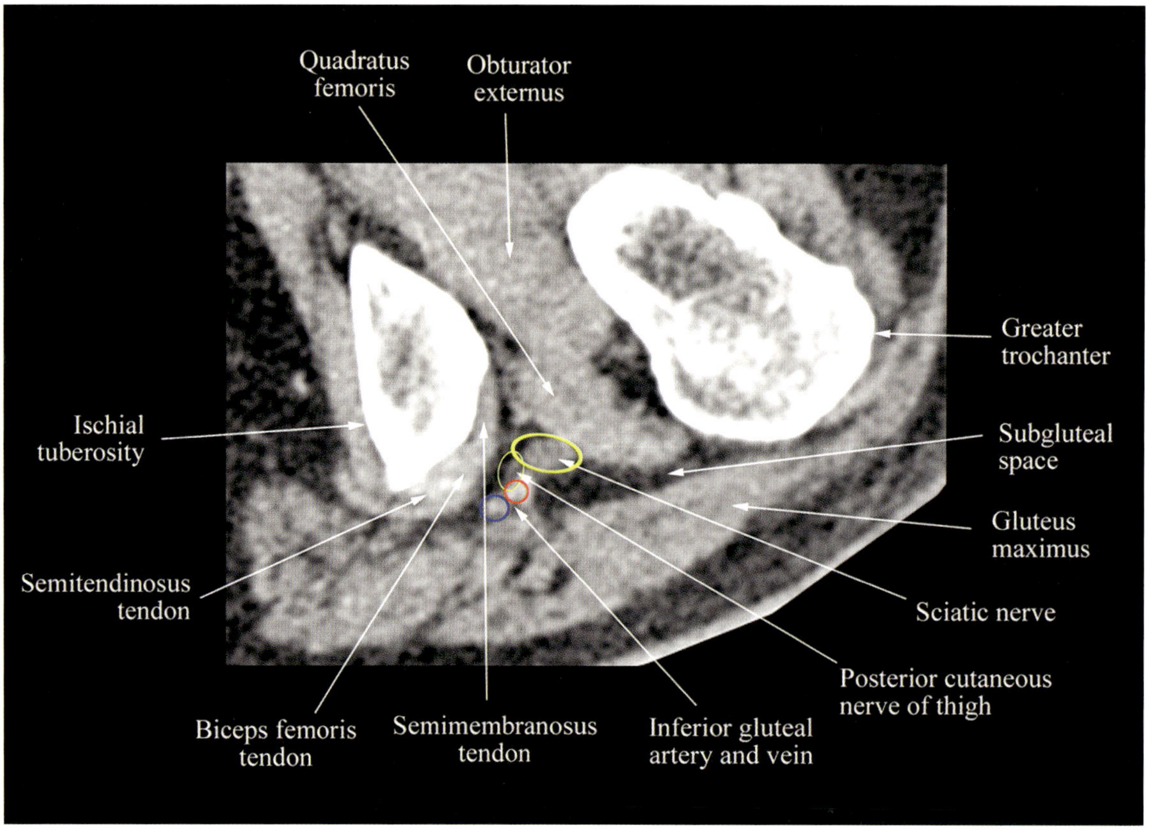

Fig. 18.9 Transverse (axial) CT image showing the subgluteal space between the gluteus maximus muscle posteriorly and the quadratus femoris muscle anteriorly, at the level of the greater trochanter and ischial tuberosity

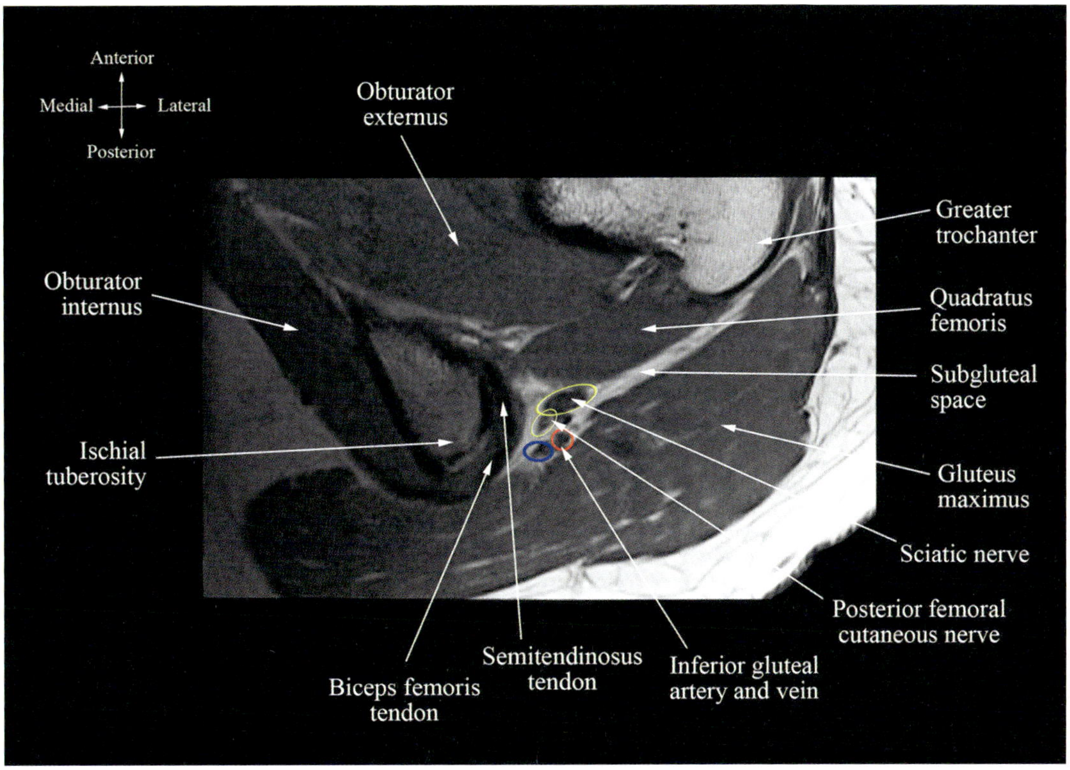

Fig. 18.10 Transverse (axial) MRI scan demonstrating the subgluteal space between the gluteus maximus muscle posteriorly and the quadratus muscle anteriorly, at the level of the greater trochanter and ischial tuberosity. Note the tendons of the semitendinosus and biceps femoris at the medial end of the subgluteal space

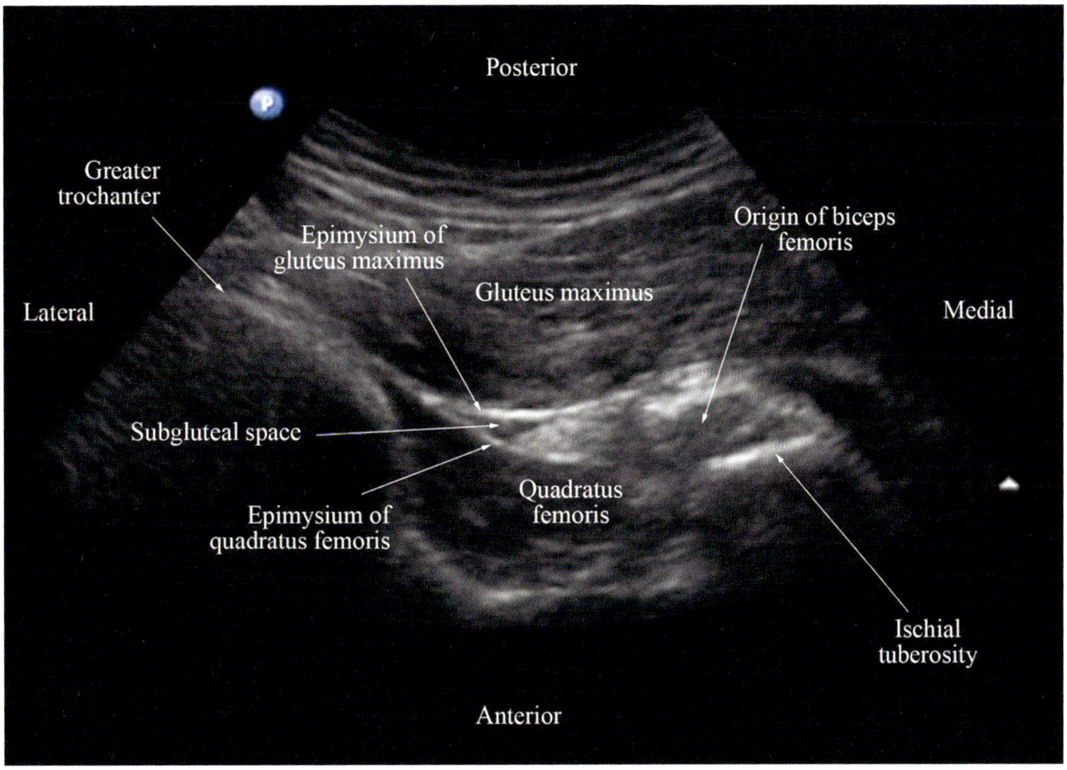

Fig. 18.11 Transverse sonogram of the gluteal region between the greater trochanter and ischial tuberosity, demonstrating the hypoechoic subgluteal space between the hyperechoic epimysium of the gluteus maximus muscle posteriorly and the quadratus femoris muscle anteriorly. The sciatic nerve is seen as a hyperechoic nodule in the medial aspect of the subgluteal space. Also note the origin of the tendon of the biceps femoris from the ischial tuberosity

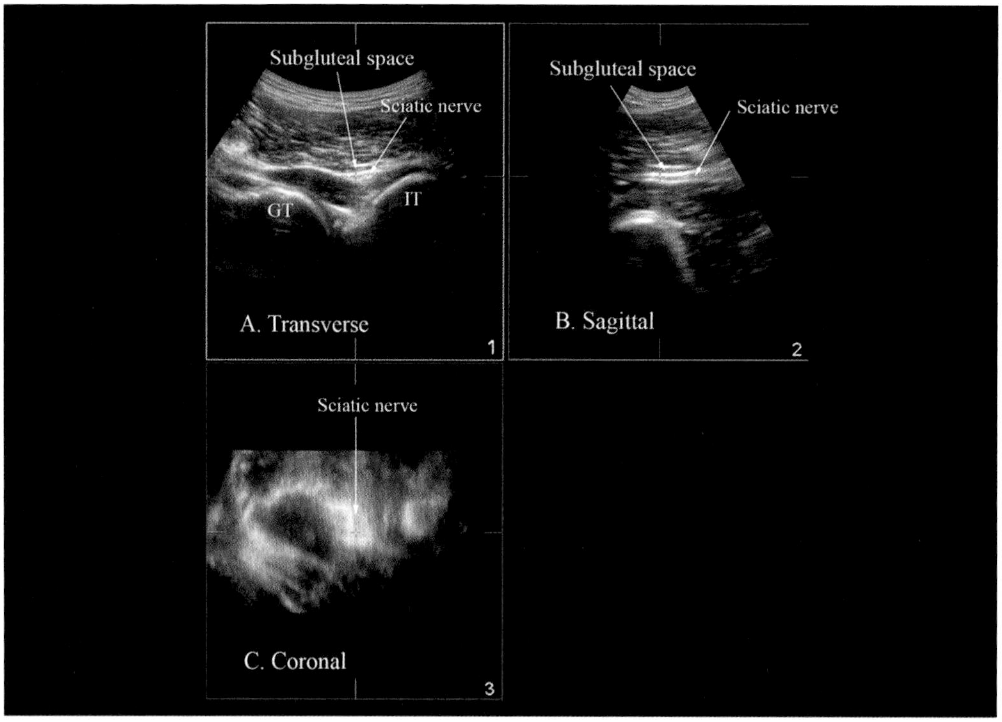

Fig. 18.12 A multiplanar 3D ultrasound image of the sciatic nerve at the subgluteal space between the greater trochanter and ischial tuberosity. The "reference maker" (*green crosshair*) has been placed over the sciatic nerve, and corresponding views of the sciatic nerve in the (**a**) transverse, (**b**) sagittal, and (**c**) coronal planes can be seen. Also note the hypoechoic subgluteal space in the transverse and sagittal sonograms. *GT* greater trochanter, *IT* ischial tuberosity

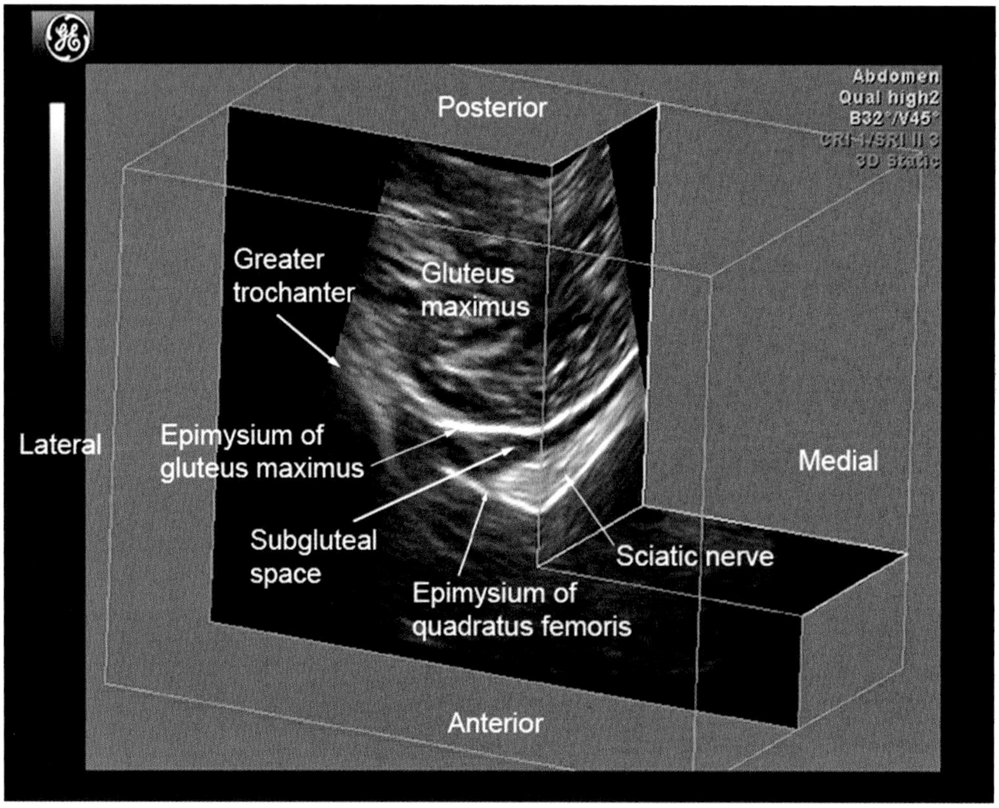

Fig. 18.13 A multiplanar (niche view) 3D ultrasound image of the sciatic nerve between the epimysium of the gluteus maximus and the quadratus femoris muscle at the subgluteal space

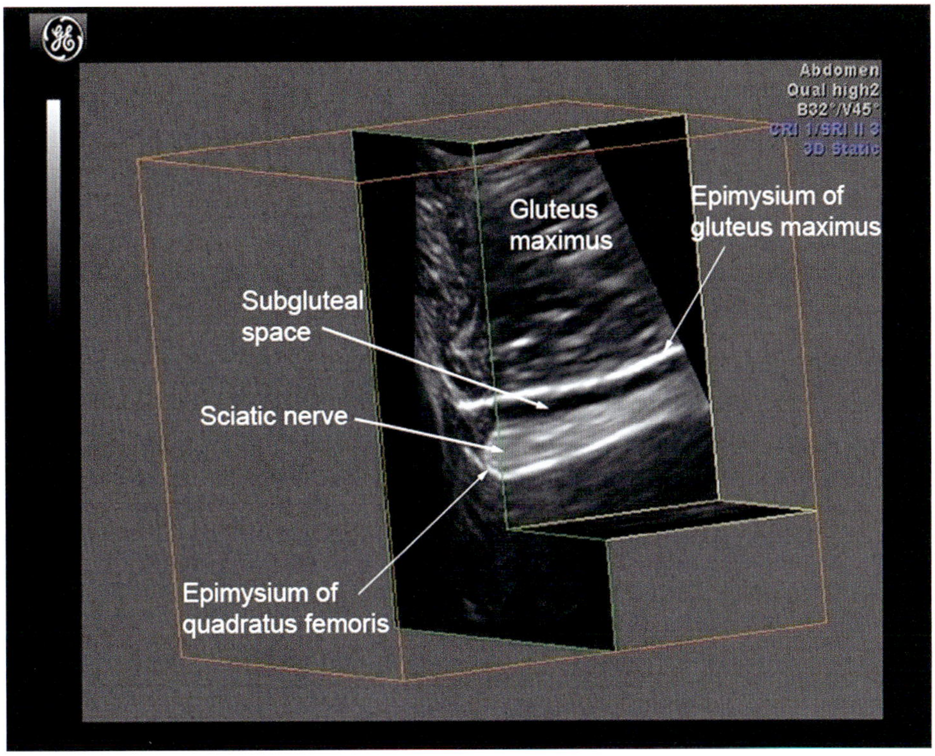

Fig. 18.14 A multiplanar (niche view) 3D ultrasound image of the sciatic nerve at the subgluteal space showing the cranial extension of the perineural space

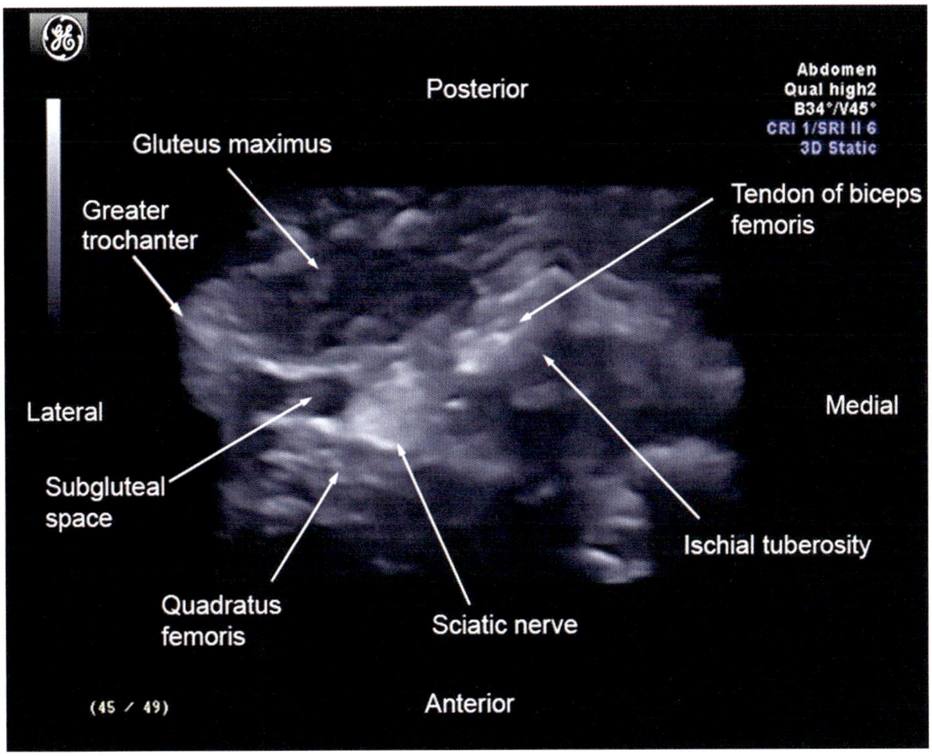

Fig. 18.15 A rendered 3D ultrasound image of the sciatic nerve at the subgluteal space such that the front and left surfaces of the 3D volume are displayed

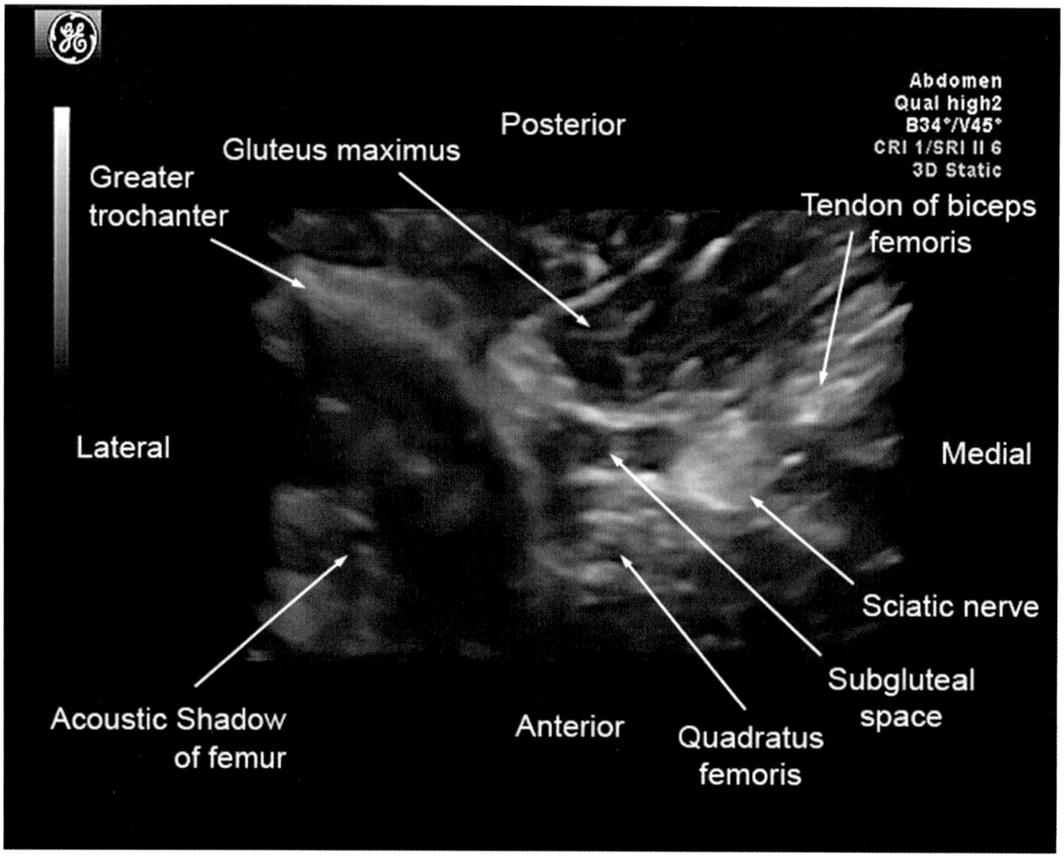

Fig. 18.16 A rendered 3D ultrasound image of the sciatic nerve at the subgluteal space such that the front and right surfaces of the 3D volume are displayed

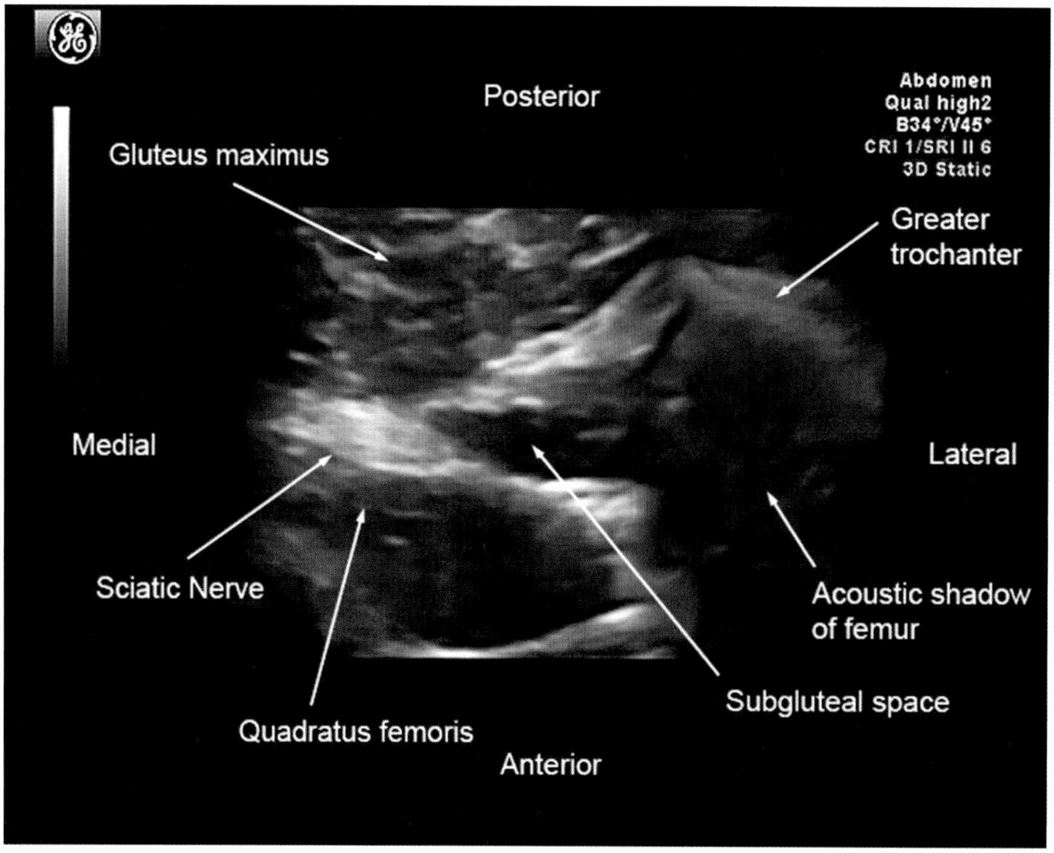

Fig. 18.17 A rendered 3D ultrasound image of the sciatic nerve at the subgluteal space such that the back and right surfaces of the 3D volume are displayed

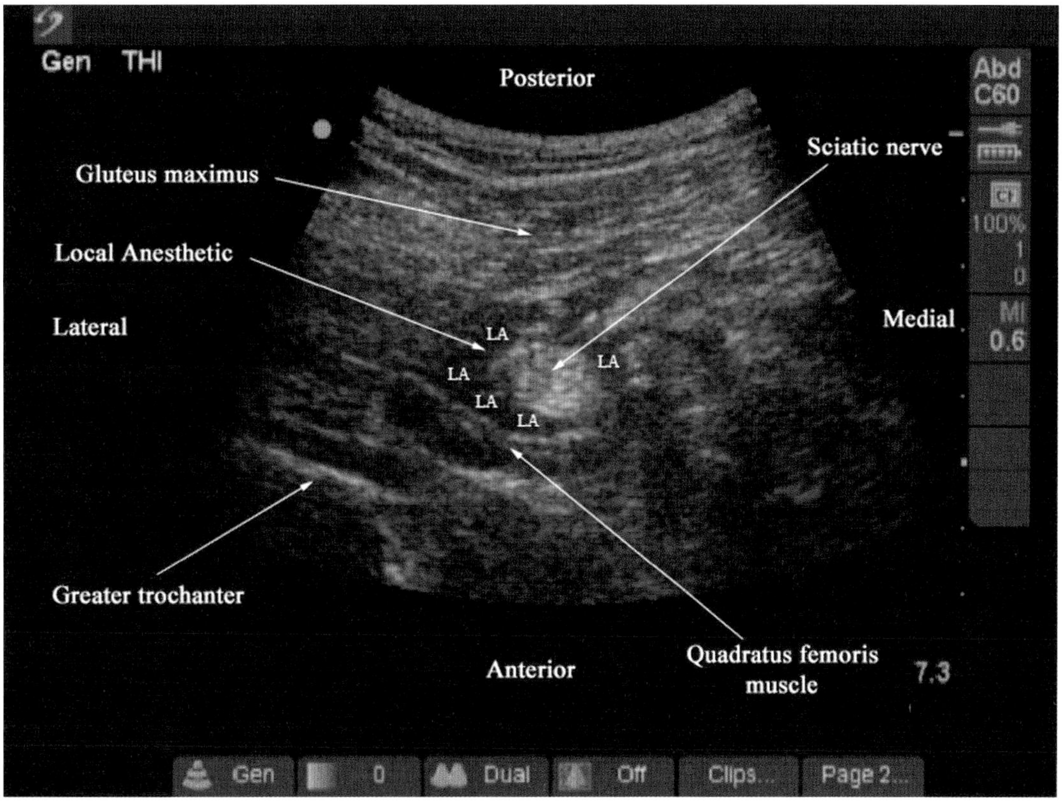

Fig. 18.18 Transverse sonogram of the sciatic nerve at the subgluteal space after an ultrasound-guided sciatic nerve block at the subgluteal space. Note the distension of the subgluteal space by the local anesthetic (*LA*)

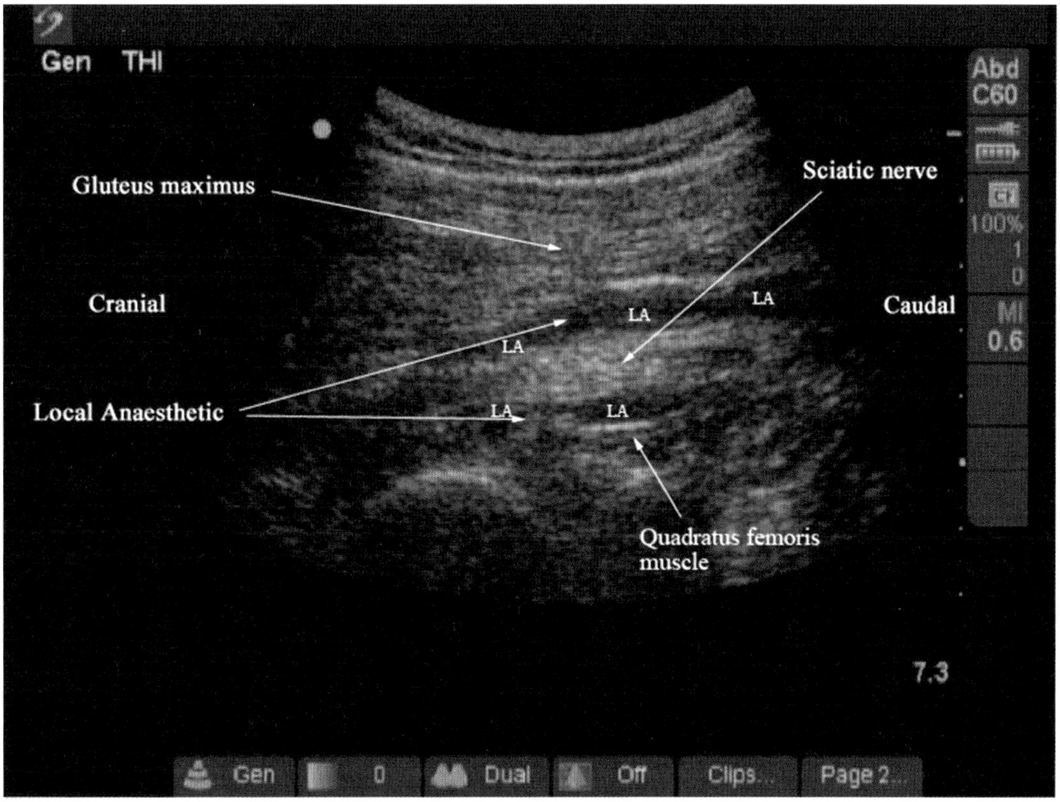

Fig. 18.19 Sagittal sonogram of the sciatic nerve at the subgluteal space after a sciatic nerve block at the subgluteal space. Note the cranio-caudal spread of local anesthetic (*LA*) within the subgluteal (perineural) space

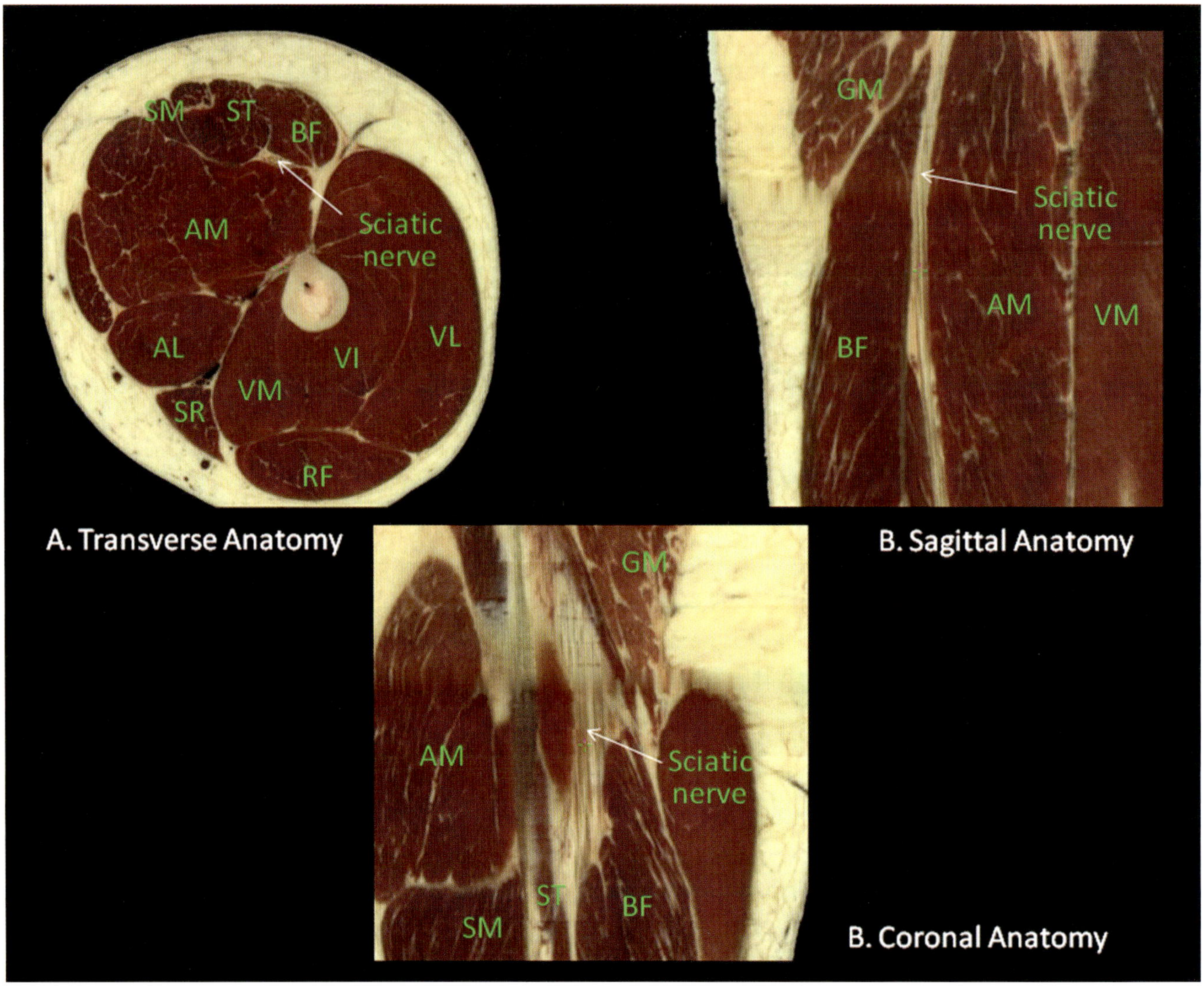

Fig. 18.20 Multiplanar anatomy (rendered from the Visible Human data set) of the sciatic nerve at the mid thigh. *AL* adductor longus, *AM* adductor magnus, *BF* biceps femoris, *GM* gluteus maximus, *RF* rectus femoris, *SM* semimembranosus, *SR* sartorius, *ST* semitendinosus, *VI* vastus intermedialis, *VL* vastus lateralis, *VM* vastus medialis (Courtesy of the Visible Human Server at Ecole Polytechnique Fédérale de Lausanne, Visible Human Visualization Software (http://visiblehuman.epfl.ch)

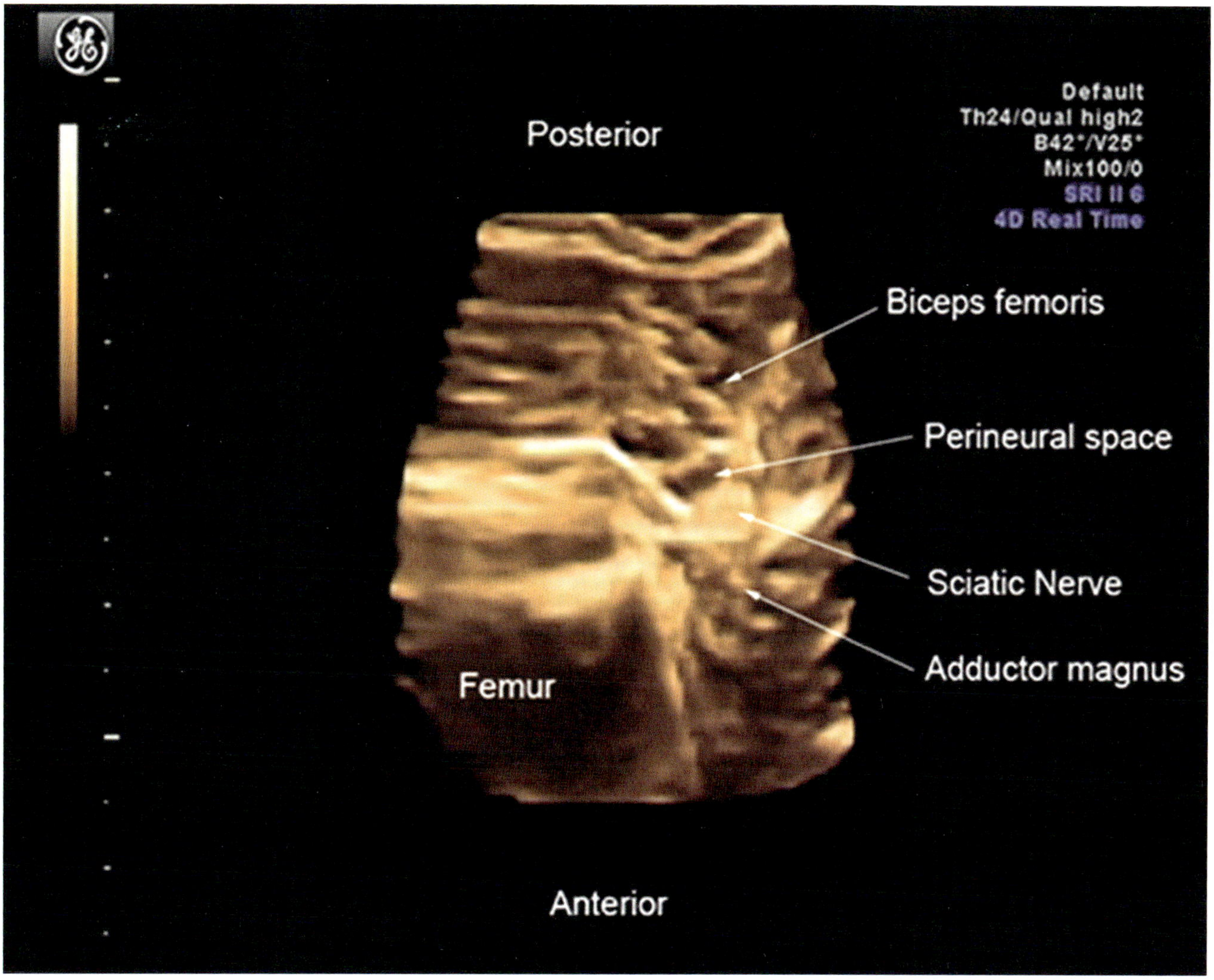

Fig. 18.21 A rendered 3D ultrasound image of the sciatic nerve at the mid thigh. The front and right surfaces of the 3D volume are displayed. Note the hypoechoic perineural space posterior to the sciatic nerve in this image

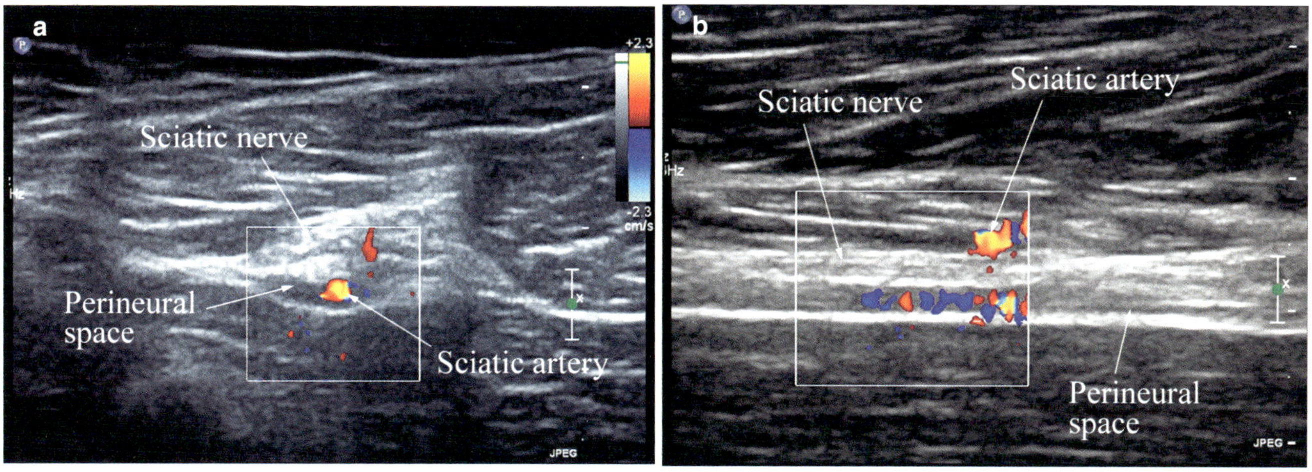

Fig. 18.22 Transverse (**a**) and sagittal (**b**) sonogram of the sciatic nerve at the mid thigh and the Doppler signals of the blood vessels within the hypoechoic perineural space

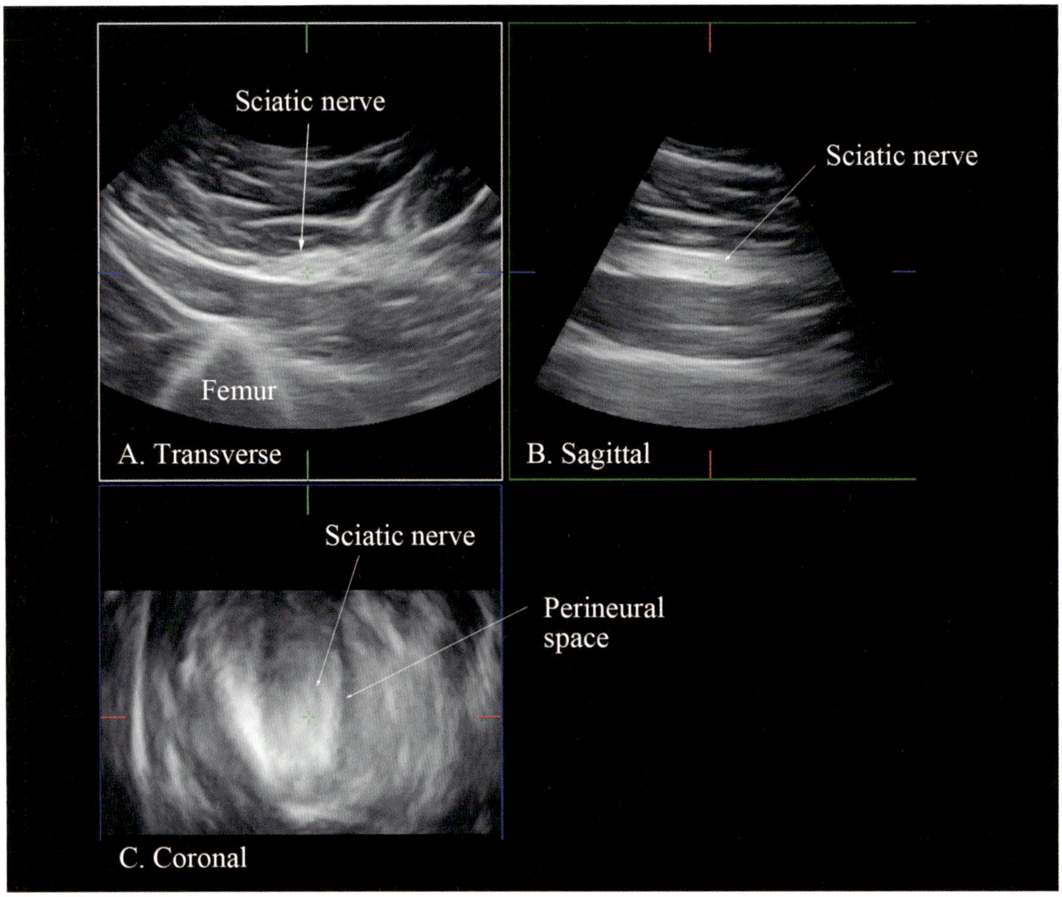

Fig. 18.23 A multiplanar 3D ultrasound image of the sciatic nerve at the mid thigh with the reference marker (*green crosshair*) placed over the sciatic nerve

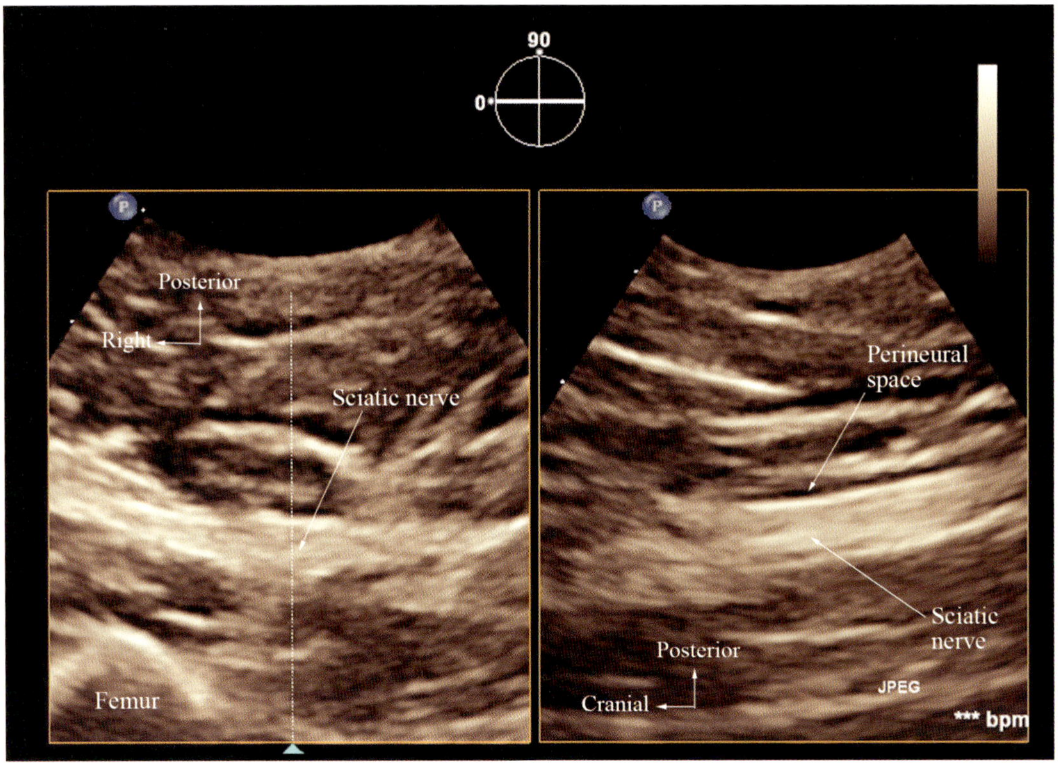

Fig. 18.24 A biplanar 3D ultrasound image of the sciatic nerve at the mid thigh, showing both the transverse and sagittal views. The sagittal plane is at 90° to the transverse axis

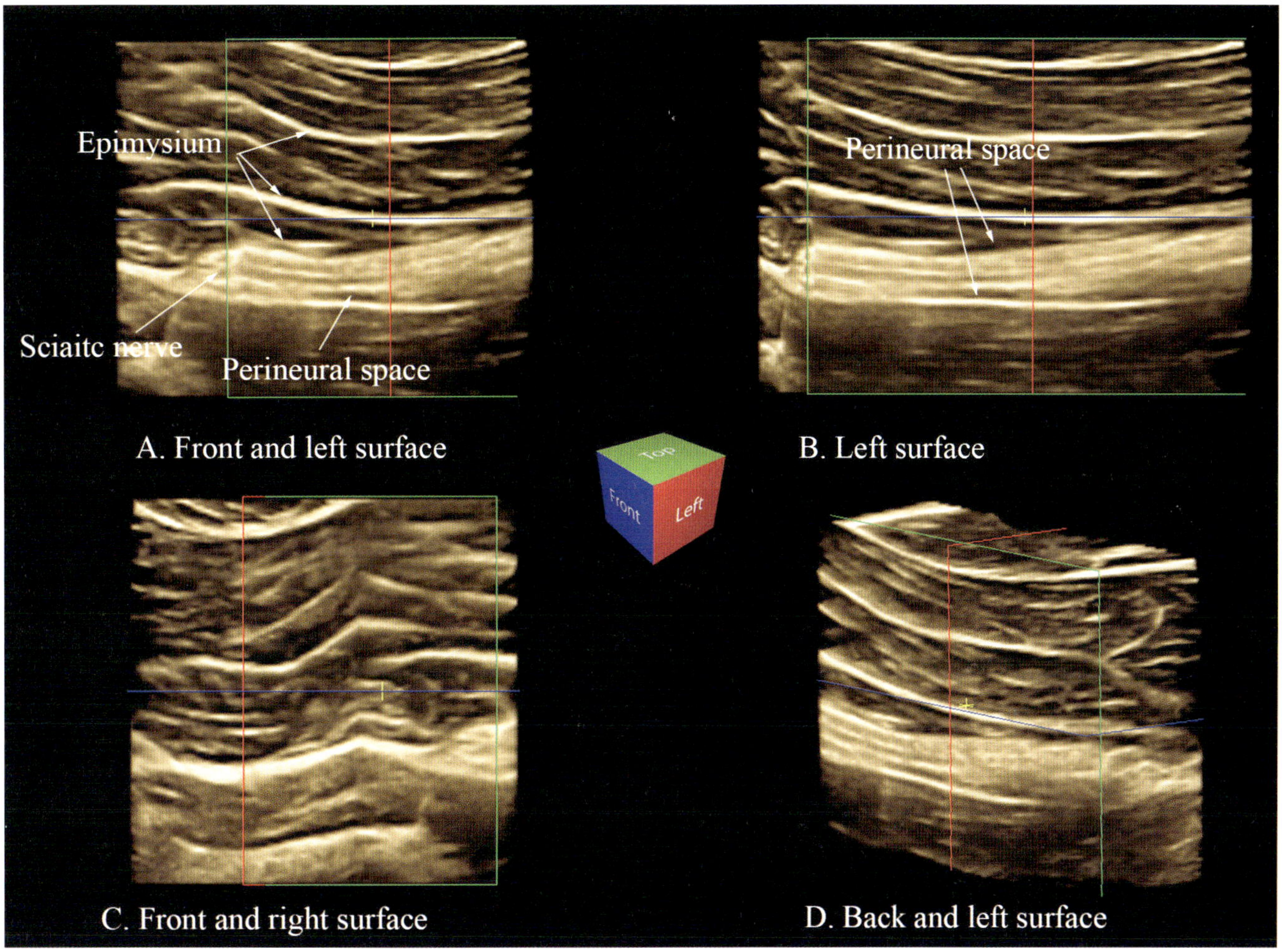

Fig. 18.25 A rendered 3D ultrasound image of the sciatic nerve at the lower thigh, showing the various surfaces of the acquired 3D volume

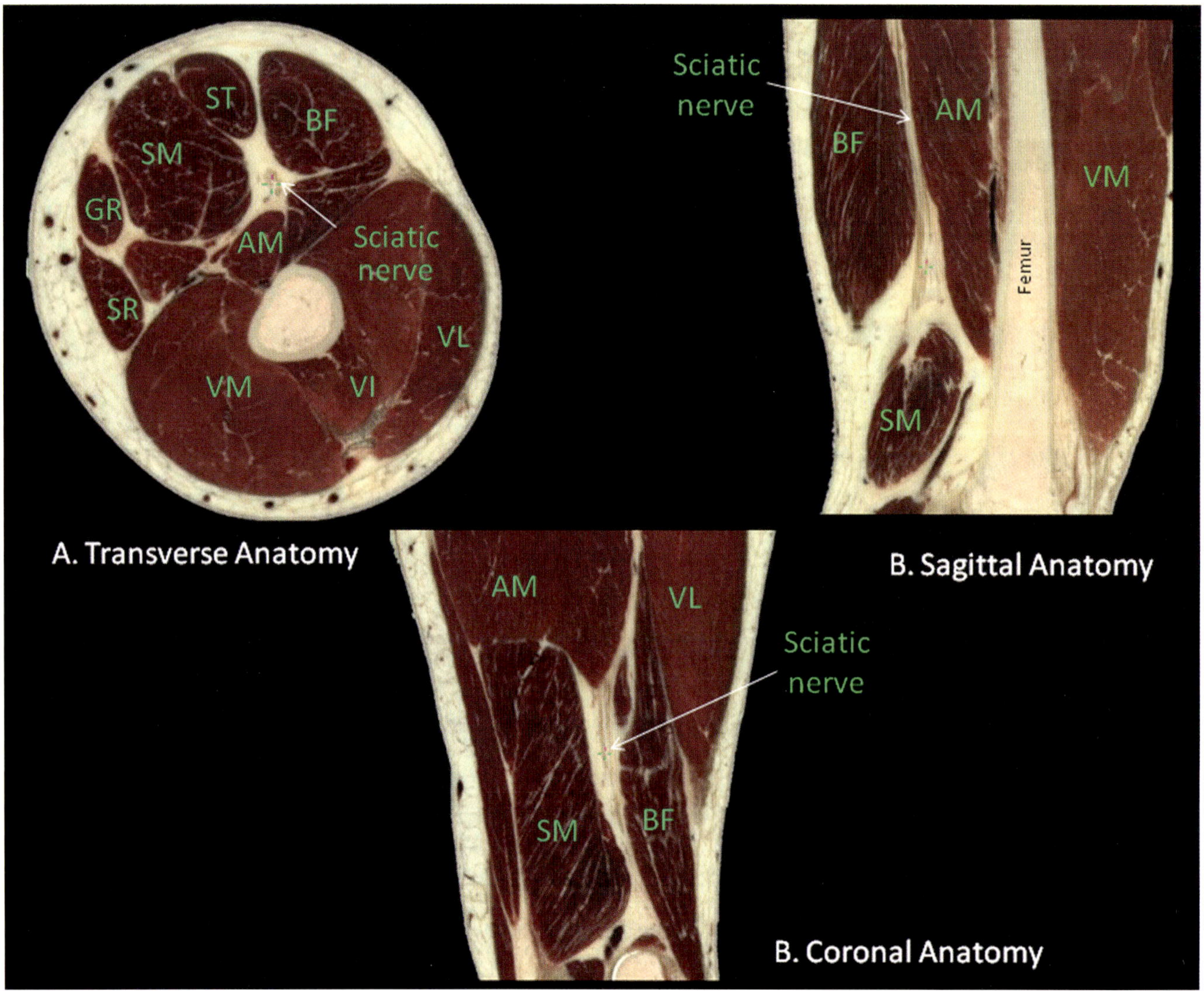

Fig. 18.26 Multiplanar anatomy (rendered from the Visible Human data set) of the sciatic nerve at or close to the apex of the popliteal fossa. *AM* adductor magnus, *BF* biceps femoris, *GR* gracilis, *SM* semimembranosus, *SR* sartorius, *ST* semitendinosus, *VI* vastus intermedialis, *VL* vastus lateralis, *VM* vastus medialis (Courtesy of the Visible Human Server at Ecole Polytechnique Fédérale de Lausanne, Visible Human Visualization Software (http://visiblehuman.epfl.ch)

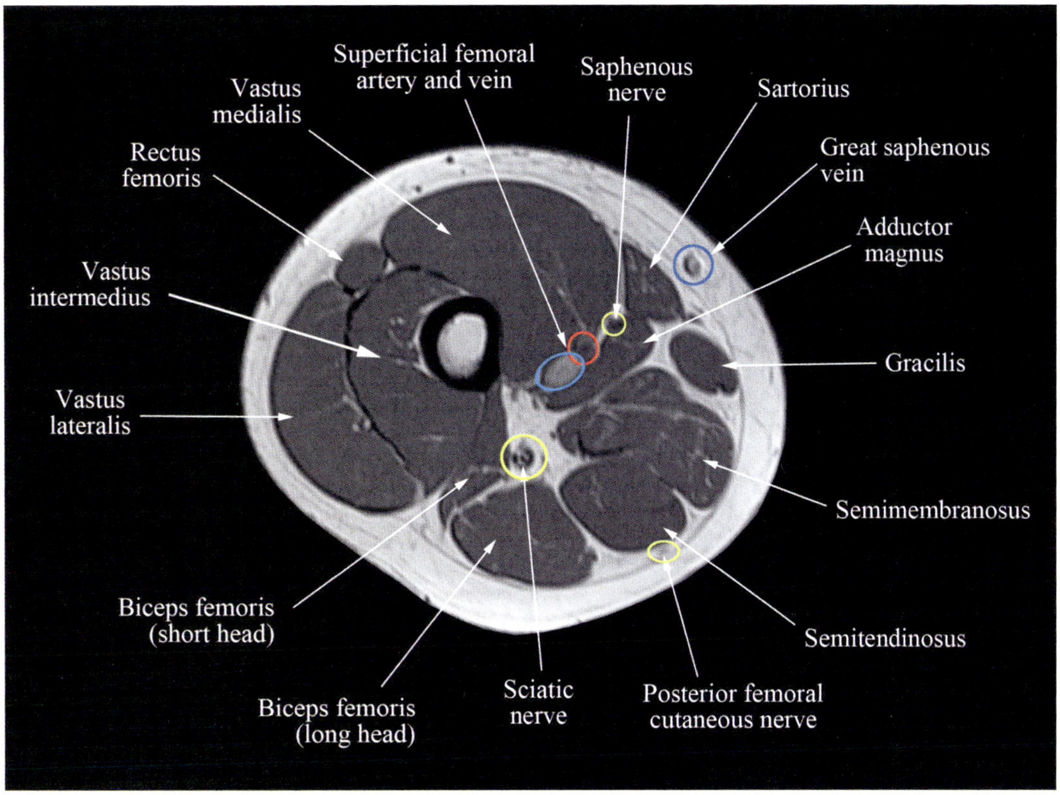

Fig. 18.27 Transverse (axial) MRI of the lower thigh showing the relations of the sciatic nerve before its division into the tibial and common peroneal nerve. The perineural space is clearly delineated and filled with hyperdense fat

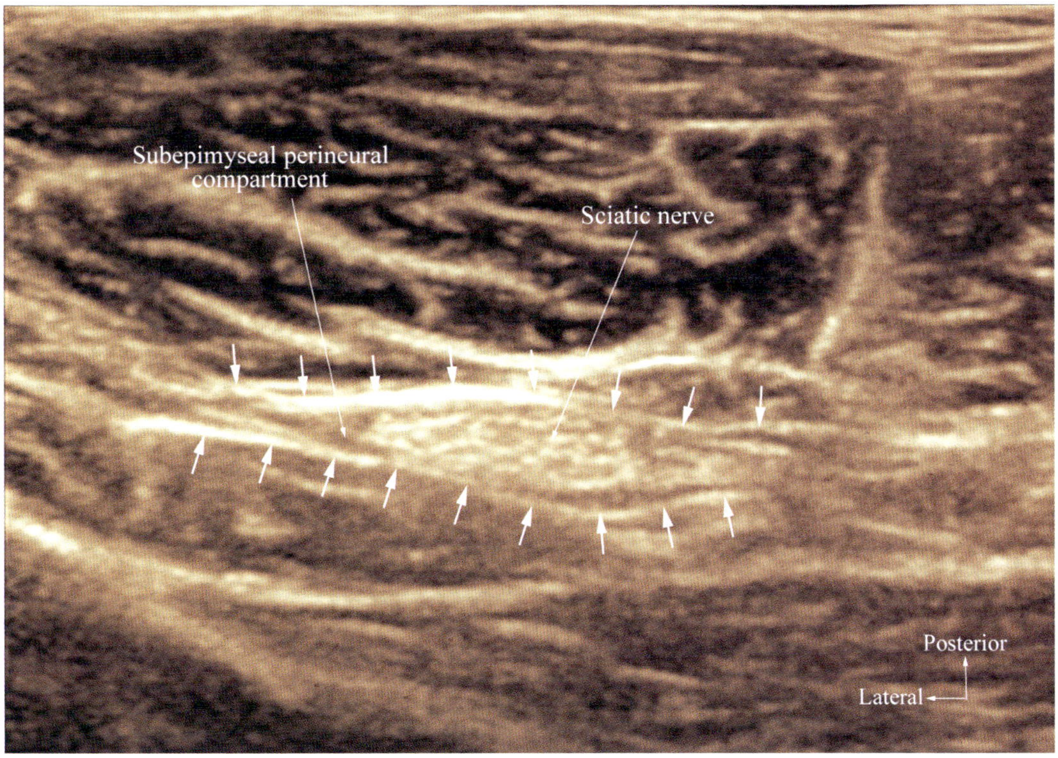

Fig. 18.28 Transverse sonogram of the sciatic nerve from just above the apex of the popliteal fossa in color (*sepia*) mode. The sciatic nerve is seen as an oval, hyperechoic structure within a narrow hypoechoic space (perineural space) between the hyperechoic epimysium (*short white arrows*) of the adjacent muscles

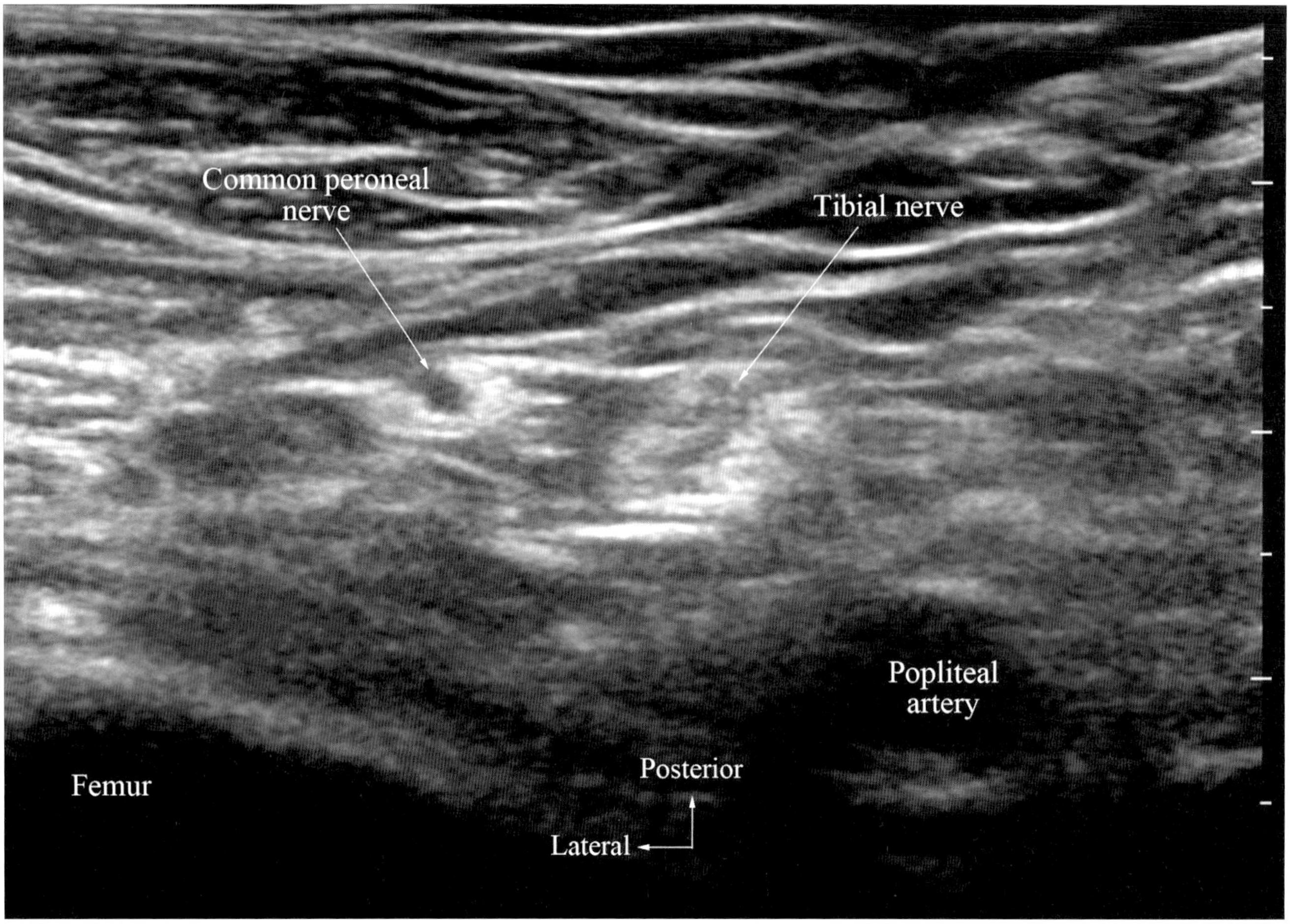

Fig. 18.29 Transverse sonogram of the sciatic nerve at its bifurcation into the common peroneal and tibial nerves

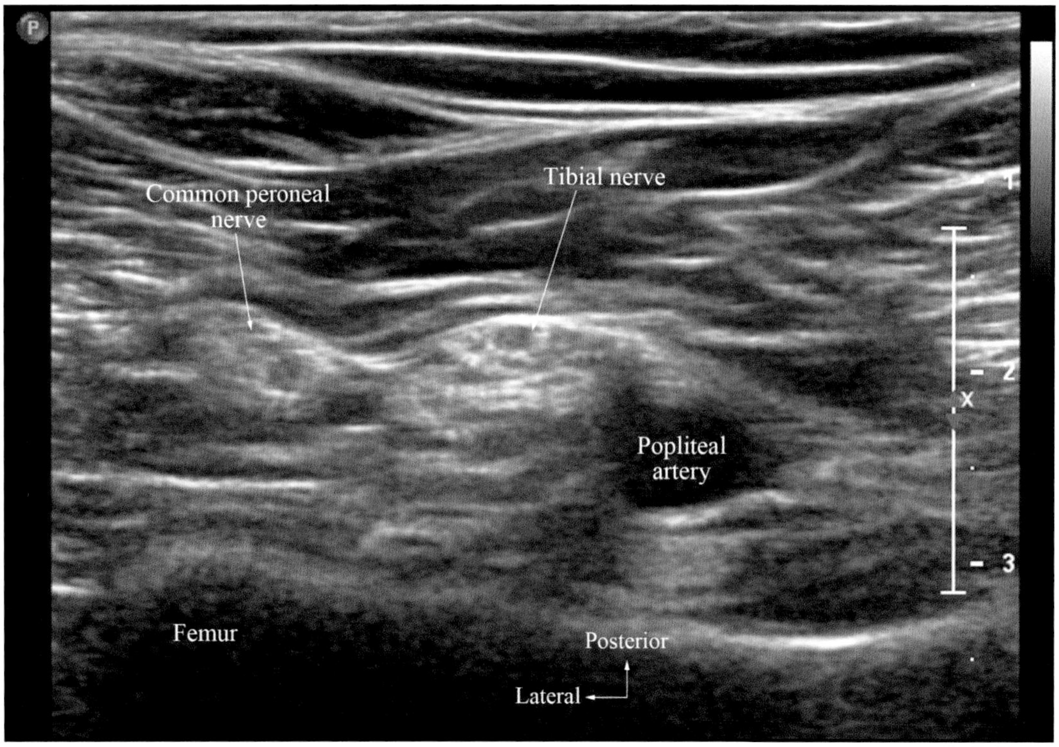

Fig. 18.30 Transverse sonogram of the sciatic nerve distal to its bifurcation into the common peroneal and tibial nerves

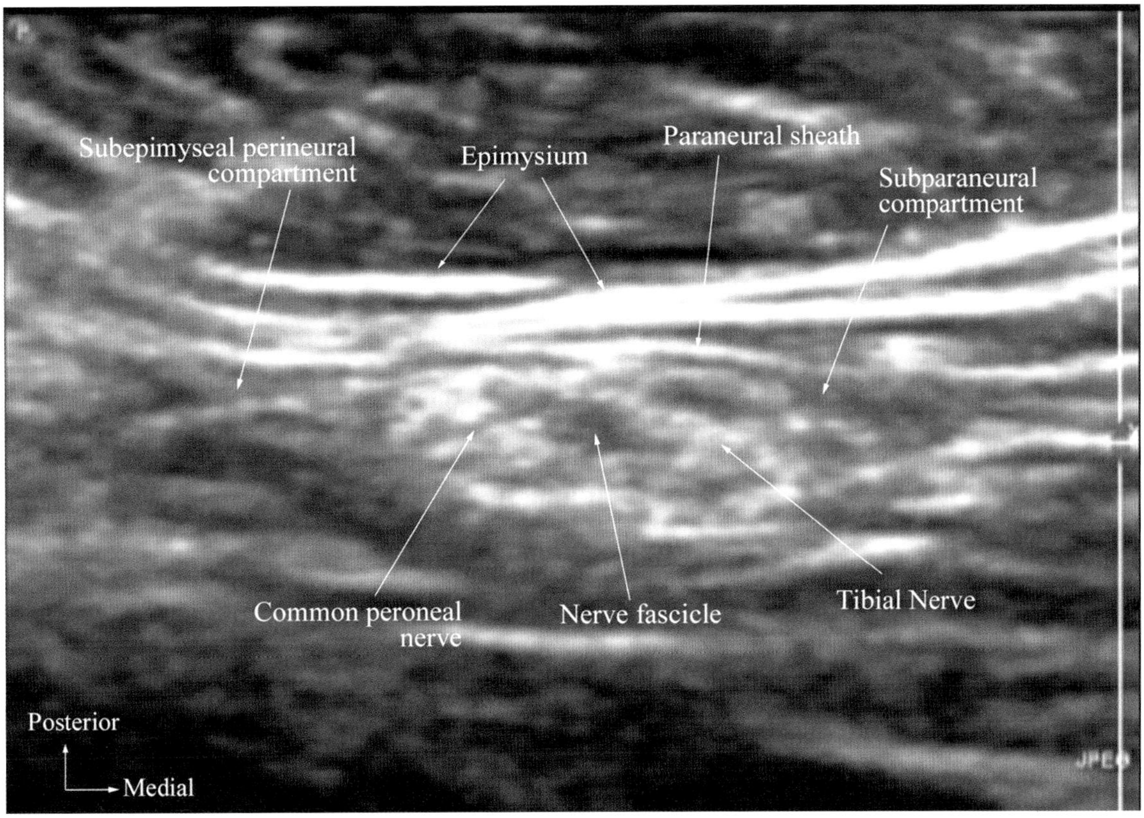

Fig. 18.31 High-definition transverse sonogram of the sciatic nerve at the popliteal fossa just above its bifurcation into the common peroneal and tibial nerves. The paraneural sheath is clearly delineated posterior to the sciatic nerve

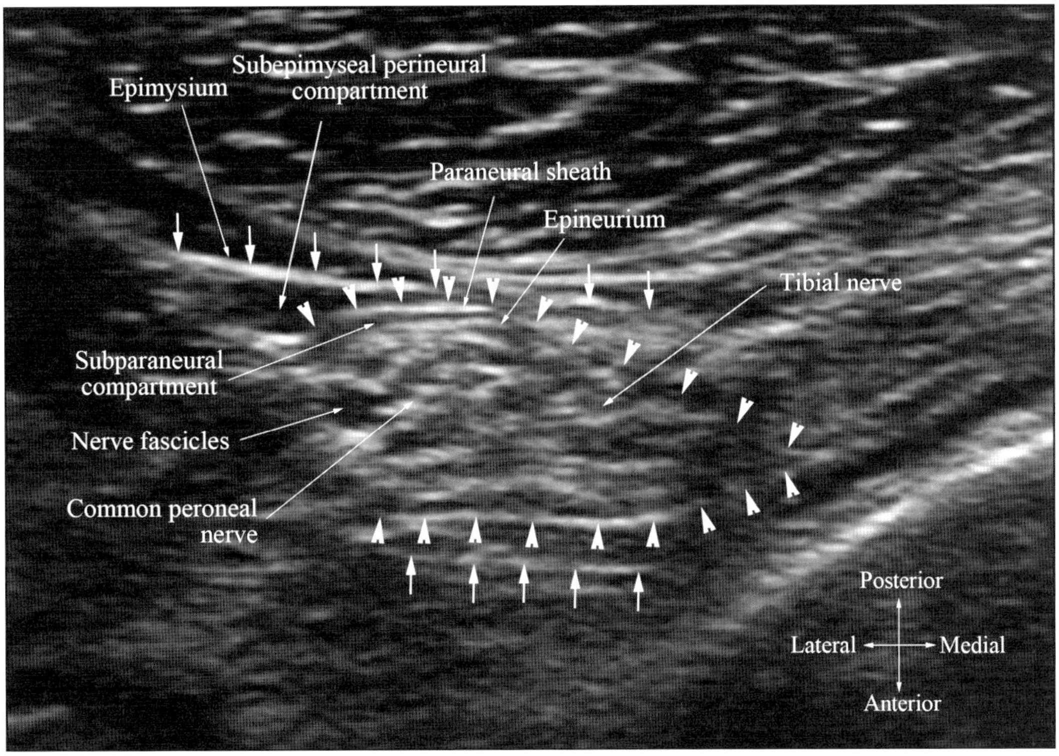

Fig. 18.32 High-definition (and zoomed) transverse sonogram of the sciatic nerve at the level of its bifurcation into the tibial and common peroneal nerves at the popliteal fossa. The paraneural sheath (*white arrowheads*) appears hyperechoic, is very closely related to the sciatic nerve, and is interposed between the epimysium (*short white arrows*) of the surrounding muscles and the outer surface of the sciatic nerve (epineurium). The subepimyseal (perineural) compartment is seen as a hypoechoic area between the epimysium and the paraneural sheath; the subparaneural compartment is the hypoechoic area between the paraneural sheath and the epineurium

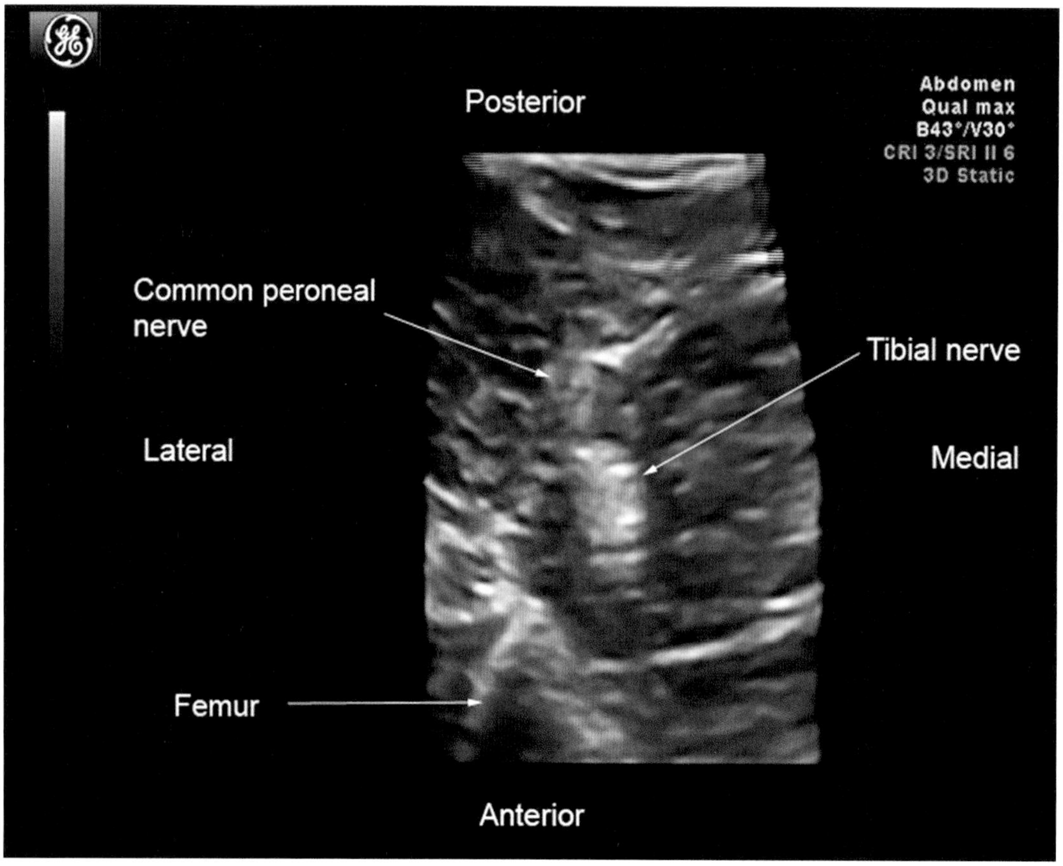

Fig. 18.33 A rendered 3D ultrasound image of the sciatic nerve in the lower thigh. Note the two divisions of the sciatic nerve (the common peroneal and tibial nerves)

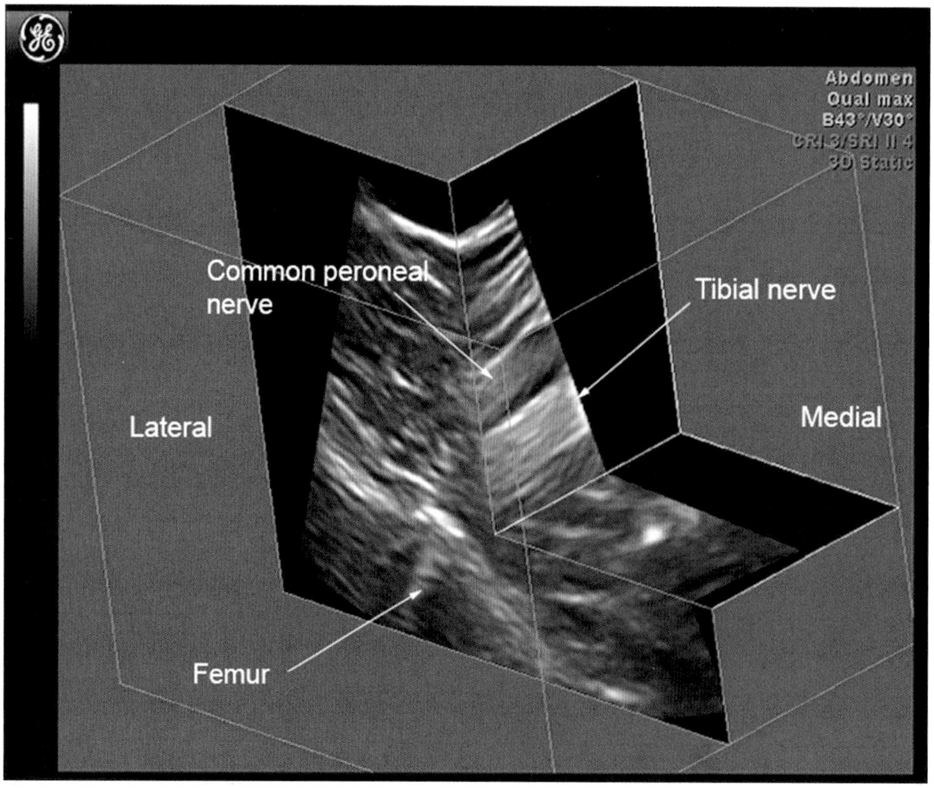

Fig. 18.34 A multiplanar (niche view) 3D ultrasound image of the sciatic nerve at the lower thigh, showing the two divisions of the sciatic nerve (common peroneal and tibial nerves). The hypoechoic perineural space is also seen in this multiplanar view

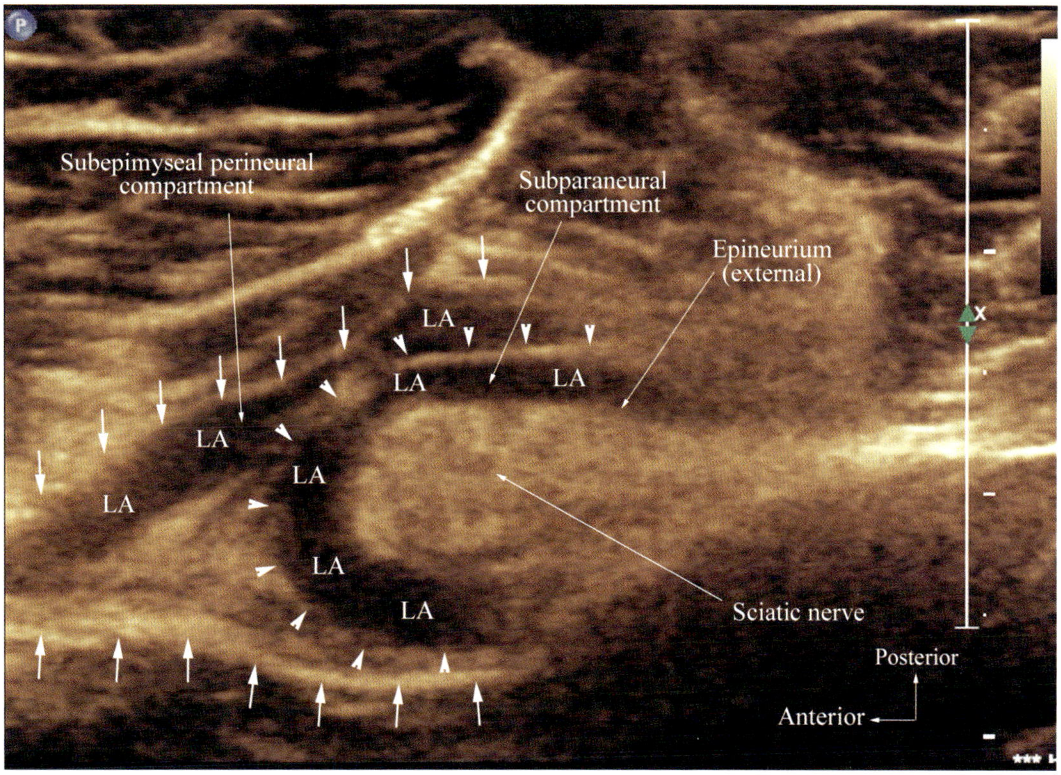

Fig. 18.35 High-definition transverse sonogram of the sciatic nerve at the lower thigh, above its bifurcation, after a subparaneural sciatic nerve block. Note the collection of local anesthetic deep to the paraneural sheath (*white arrowheads*) in the subparaneural compartment. The *short white arrow* indicates the epimysium of the surrounding muscles

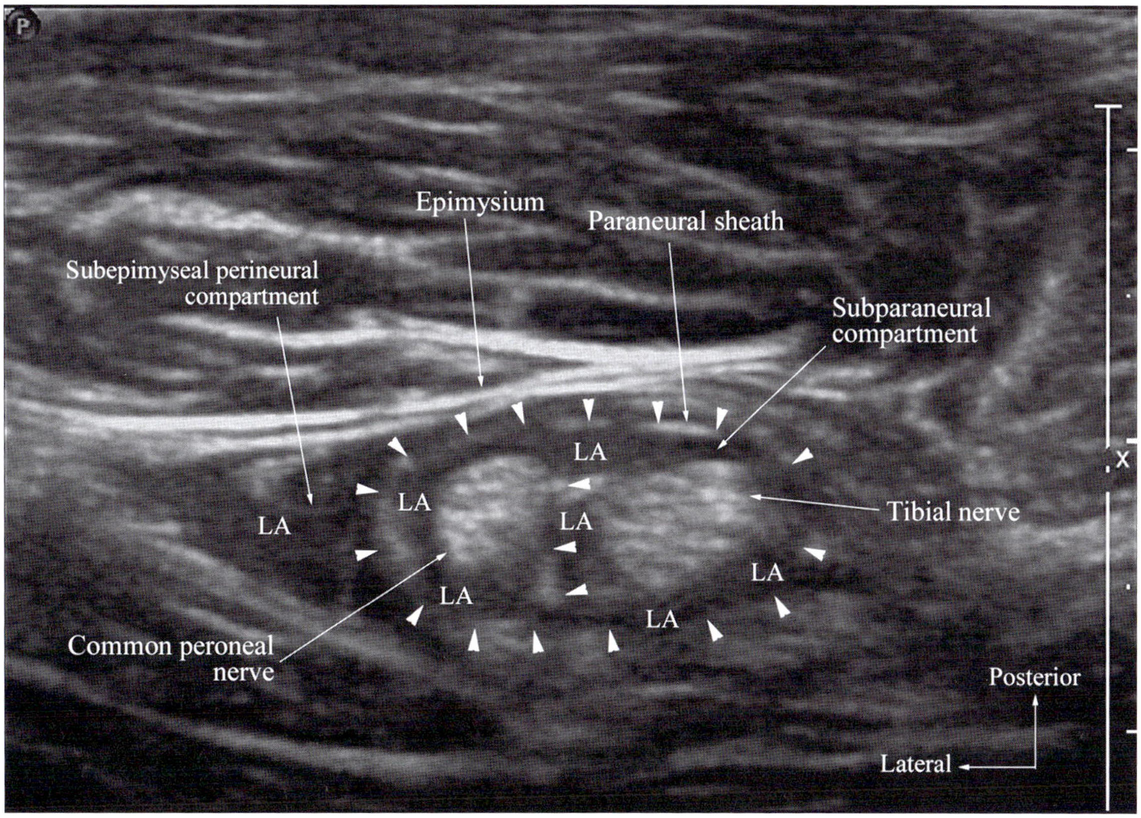

Fig. 18.36 Transverse sonograms demonstrating the spread of local anesthetic (*LA*) in both the subepimyseal and subparaneural compartments surrounding the sciatic nerve after an ultrasound-guided popliteal block of the sciatic nerve above its bifurcation. The paraneural sheath (*white arrowheads*) is interposed between the epimysium of the surrounding muscles and the outer surface (epineurium) of the sciatic nerve

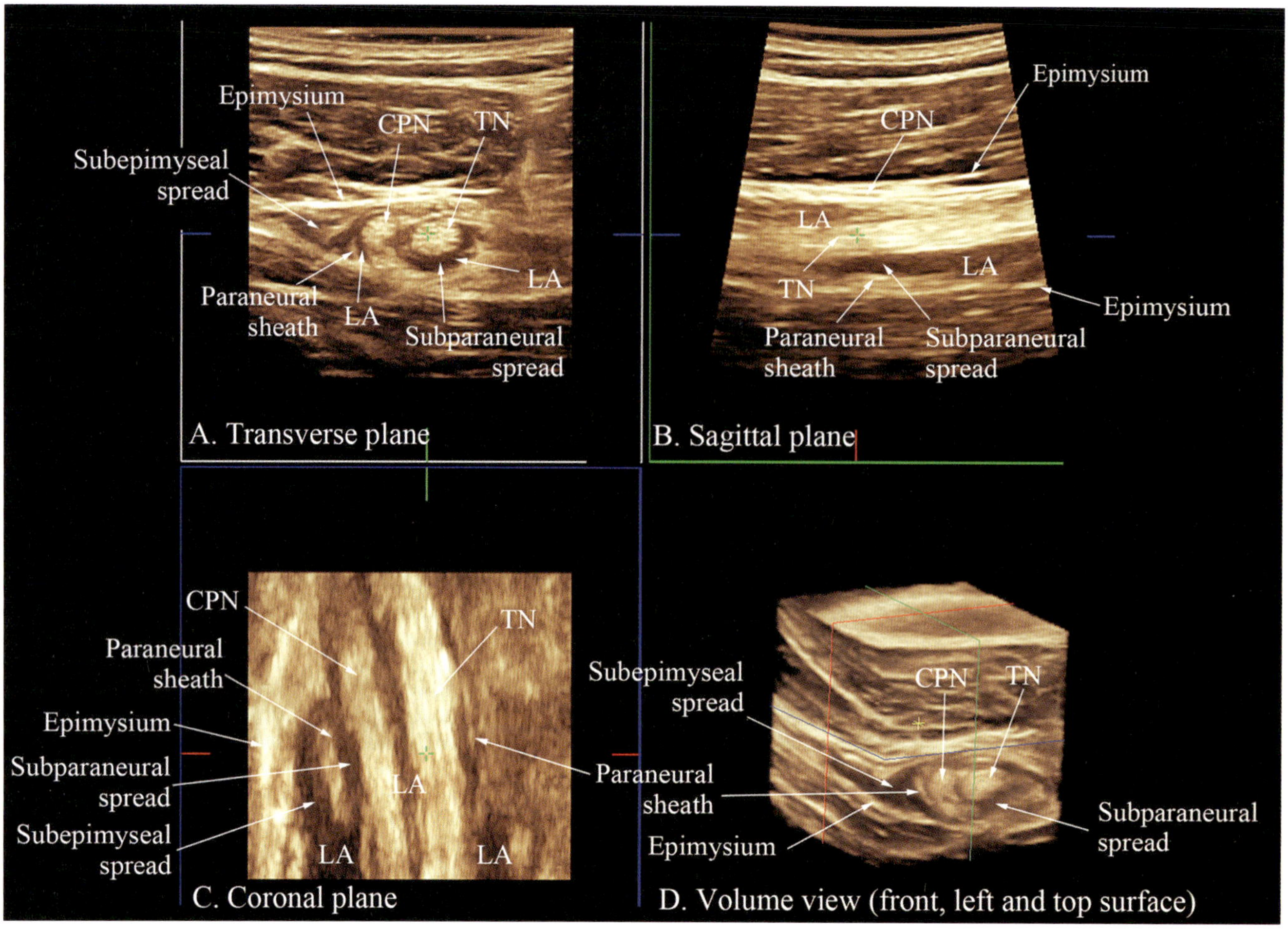

Fig. 18.37 A multiplanar 3D ultrasound image demonstrating subparaneural spread of the local anesthetic (*LA*) after a popliteal sciatic nerve block. The reference marker has been placed over the tibial nerve (*TN*). A rendered 3D volume demonstrating the front, right, and top surfaces is presented in **d**. *CPN* common peroneal nerve

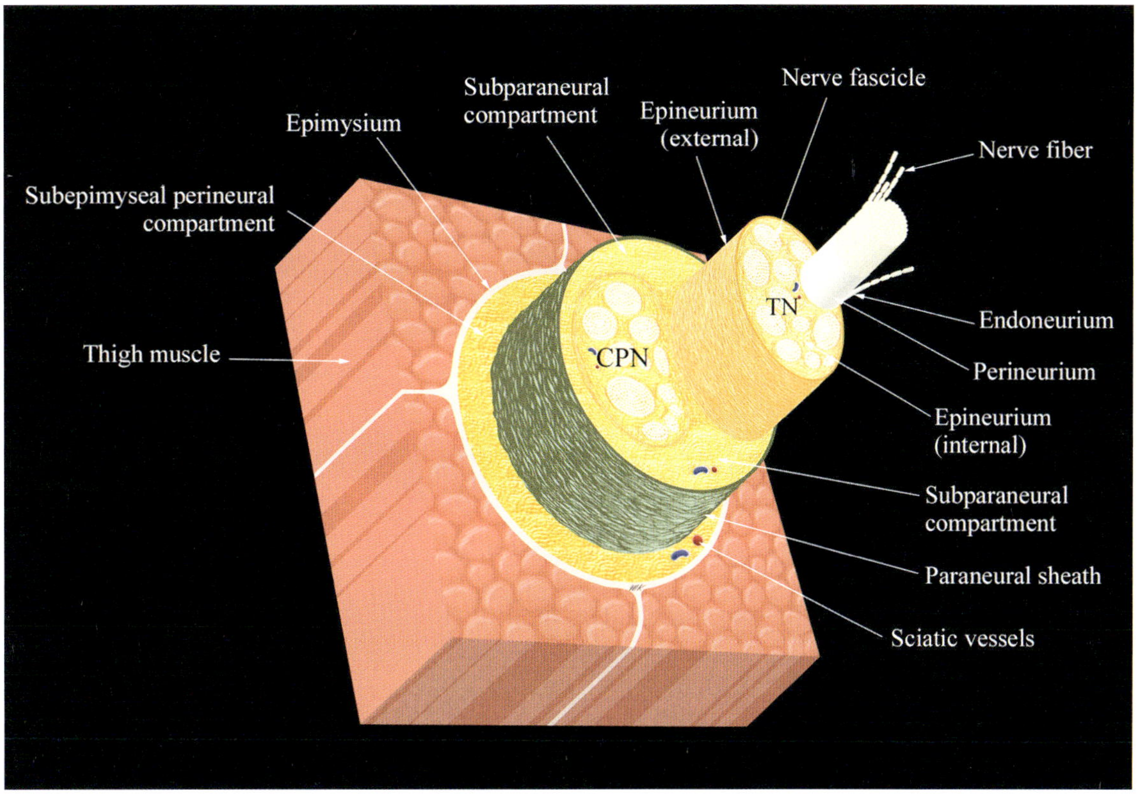

Fig. 18.38 Schematic diagram demonstrating the functional anatomy of the sciatic nerve, paraneural sheath, and the fascial compartments surrounding the sciatic nerve above its bifurcation into the common peroneal nerve (*CPN*) and tibial nerve (*TN*) at the popliteal fossa

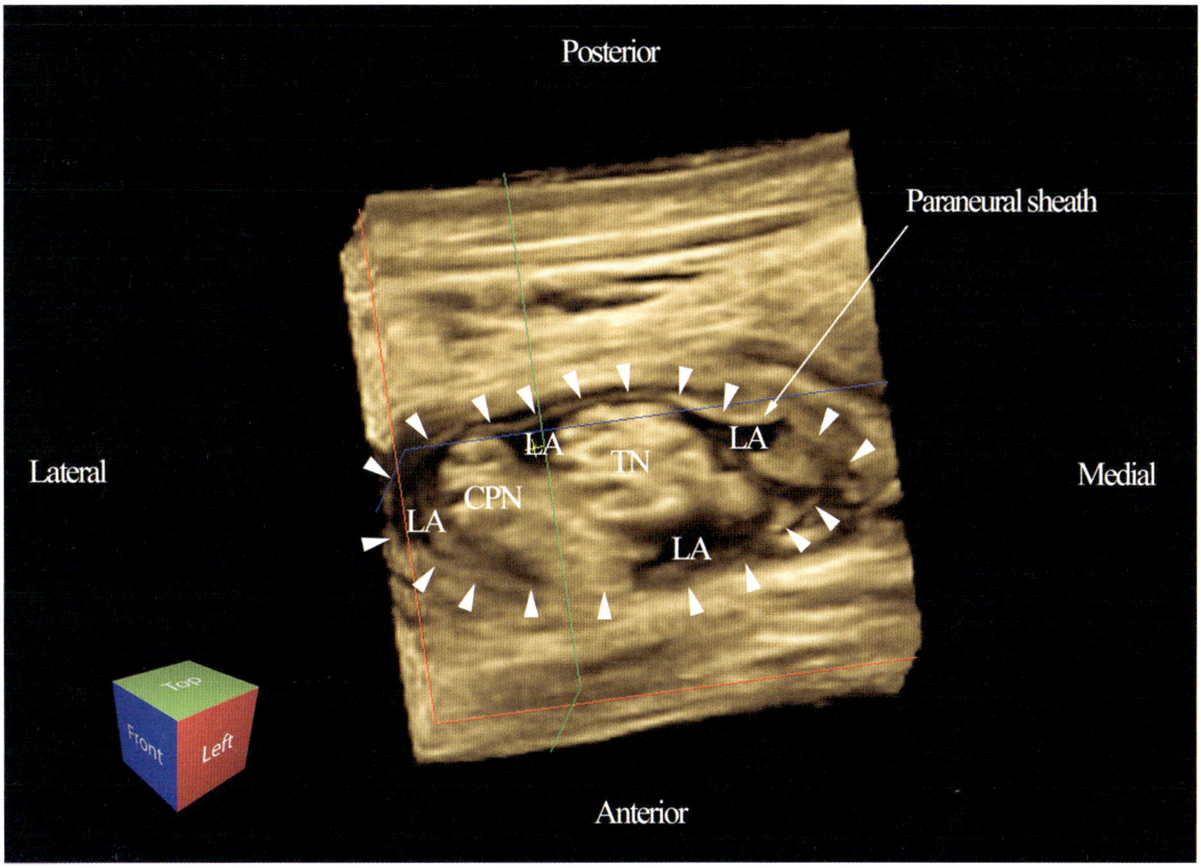

Fig. 18.39 A rendered 3D ultrasound image of the sciatic nerve distal to its bifurcation, after a popliteal sciatic nerve block, demonstrating subparaneural spread of local anesthetic (*LA*). The *white arrowheads* indicate the paraneural sheath. *CPN* common peroneal nerve, *TN* tibial nerve

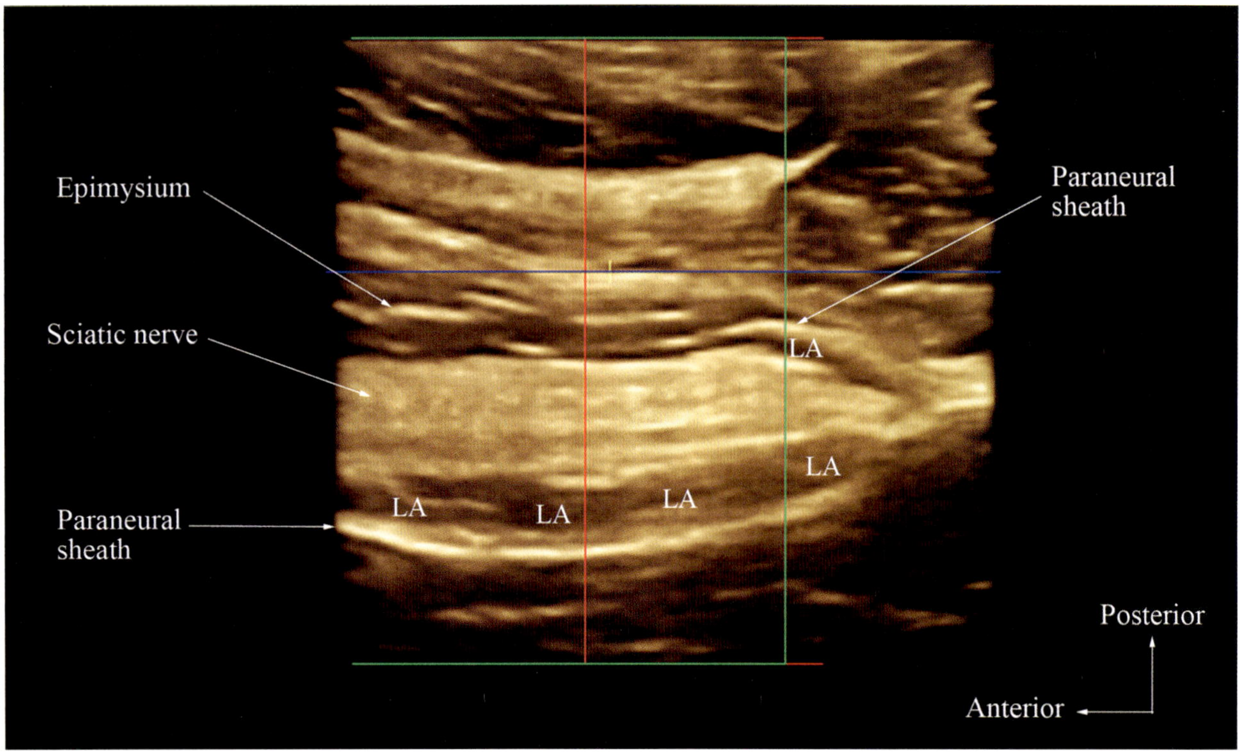

Fig. 18.40 A rendered 3D ultrasound image of the sciatic nerve at the popliteal fossa after a popliteal sciatic nerve block, displaying the front and right surfaces of the 3D volume. Note the spread of local anesthetic deep to the paraneural sheath *LA* local anesthetic

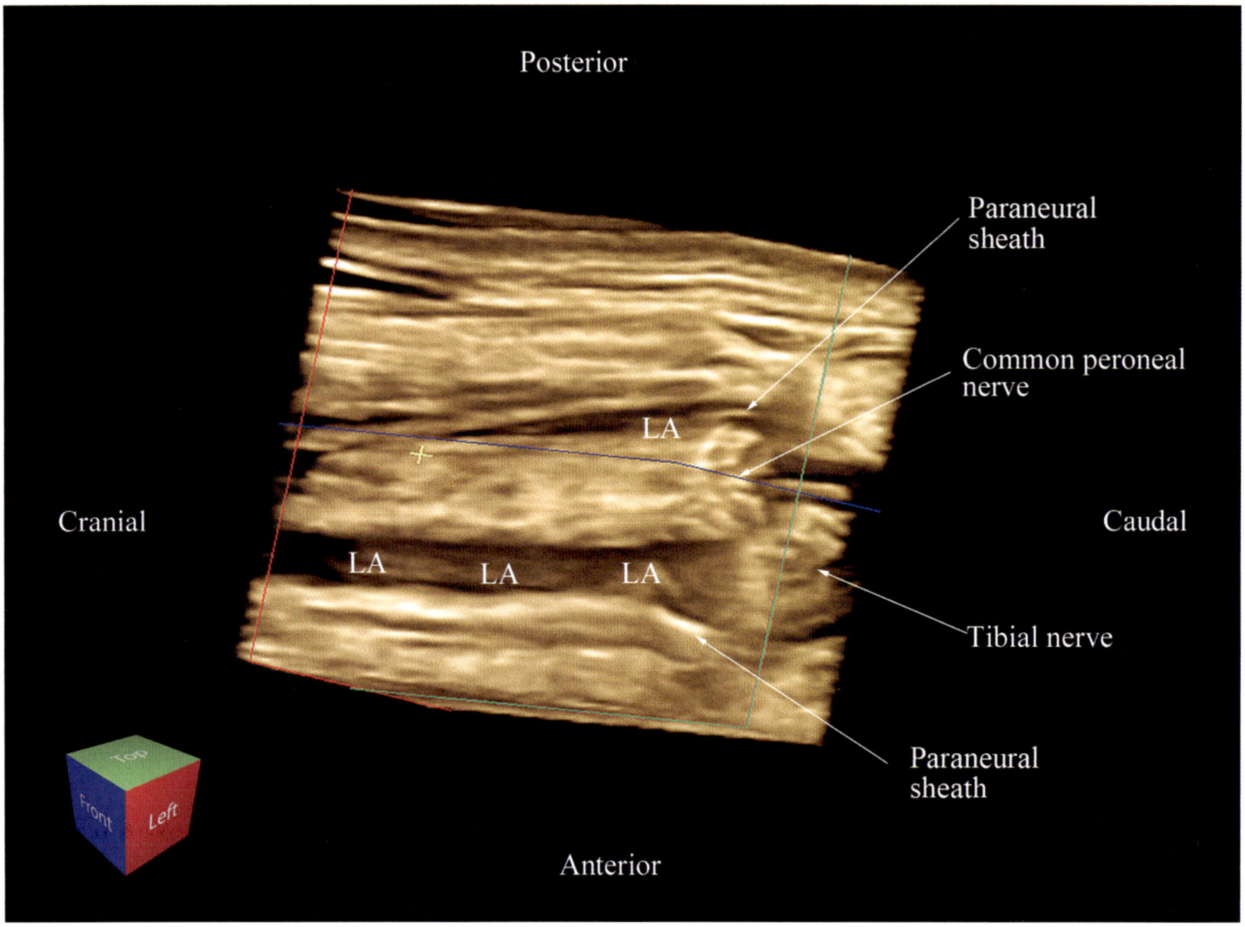

Fig. 18.41 A rendered 3D ultrasound image of the sciatic nerve at the lower thigh after a popliteal sciatic nerve block. The front and right surfaces of the 3D volume are displayed. *LA* local anesthetic

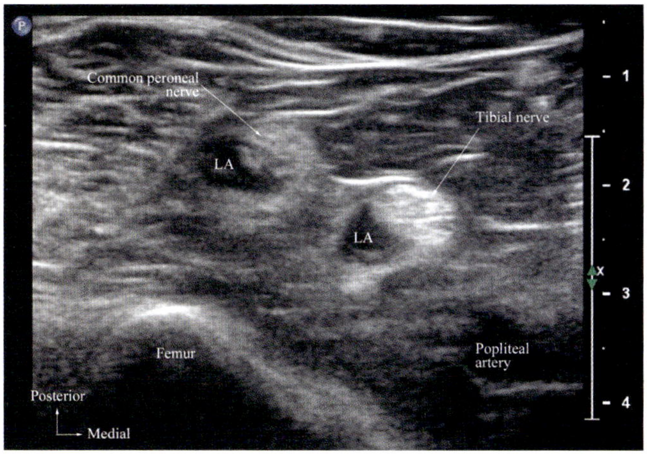

Fig. 18.42 Transverse sonogram of the sciatic nerve distal to its bifurcation, after a popliteal sciatic nerve block, demonstrating distal spread of local anesthetic (*LA*) to the common peroneal and tibial nerves ("twin-halo sign")

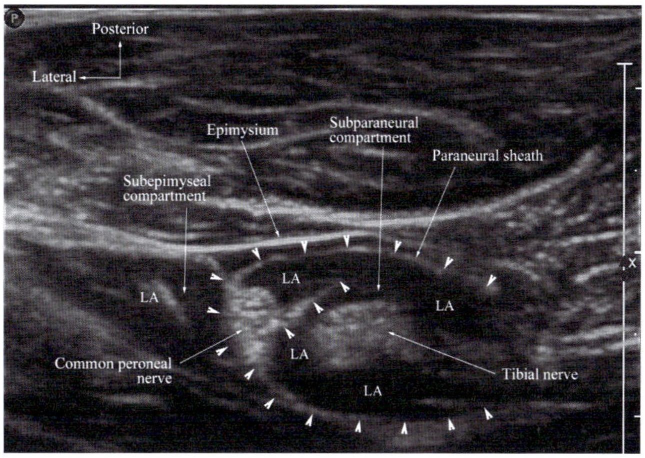

Fig. 18.43 High-definition transverse sonogram of the sciatic nerve after a popliteal sciatic nerve block, demonstrating spread of local anesthetic (*LA*) in the subepimyseal perineural and subparaneural compartments. The individual paraneural sheath and subparaneural compartment of the common peroneal and tibial nerves are clearly delineated. The *white arrowheads* indicate the paraneural sheath

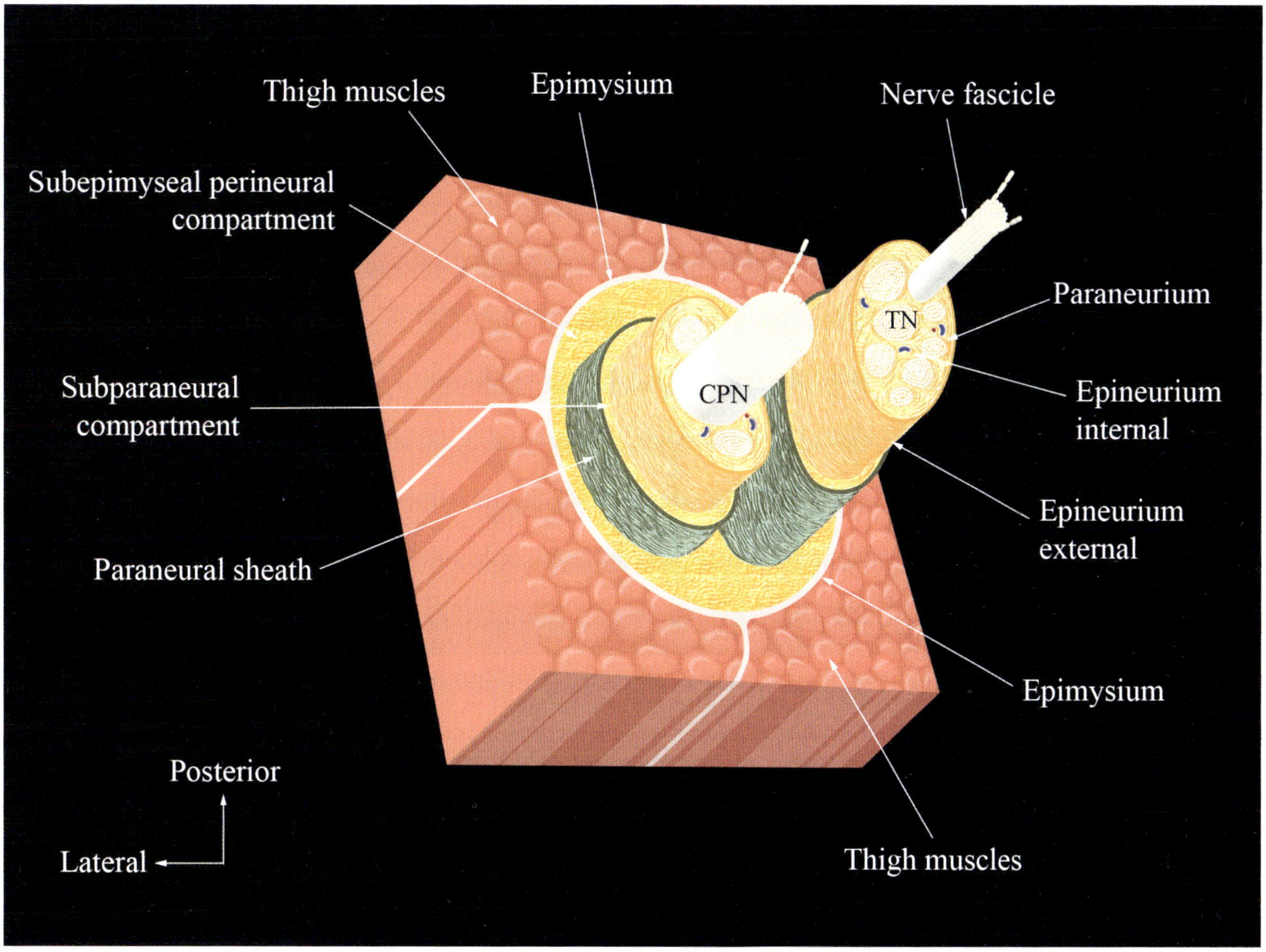

Fig. 18.44 Schematic diagram demonstrating the functional anatomy of the sciatic nerve, the paraneural sheath, and the fascial compartments surrounding the sciatic nerve distal to its bifurcation into the common peroneal nerve (*CPN*) and tibial nerve (*TN*) at the popliteal fossa. Note that the CPN and TN have individual paraneural sheaths and subparaneural compartments

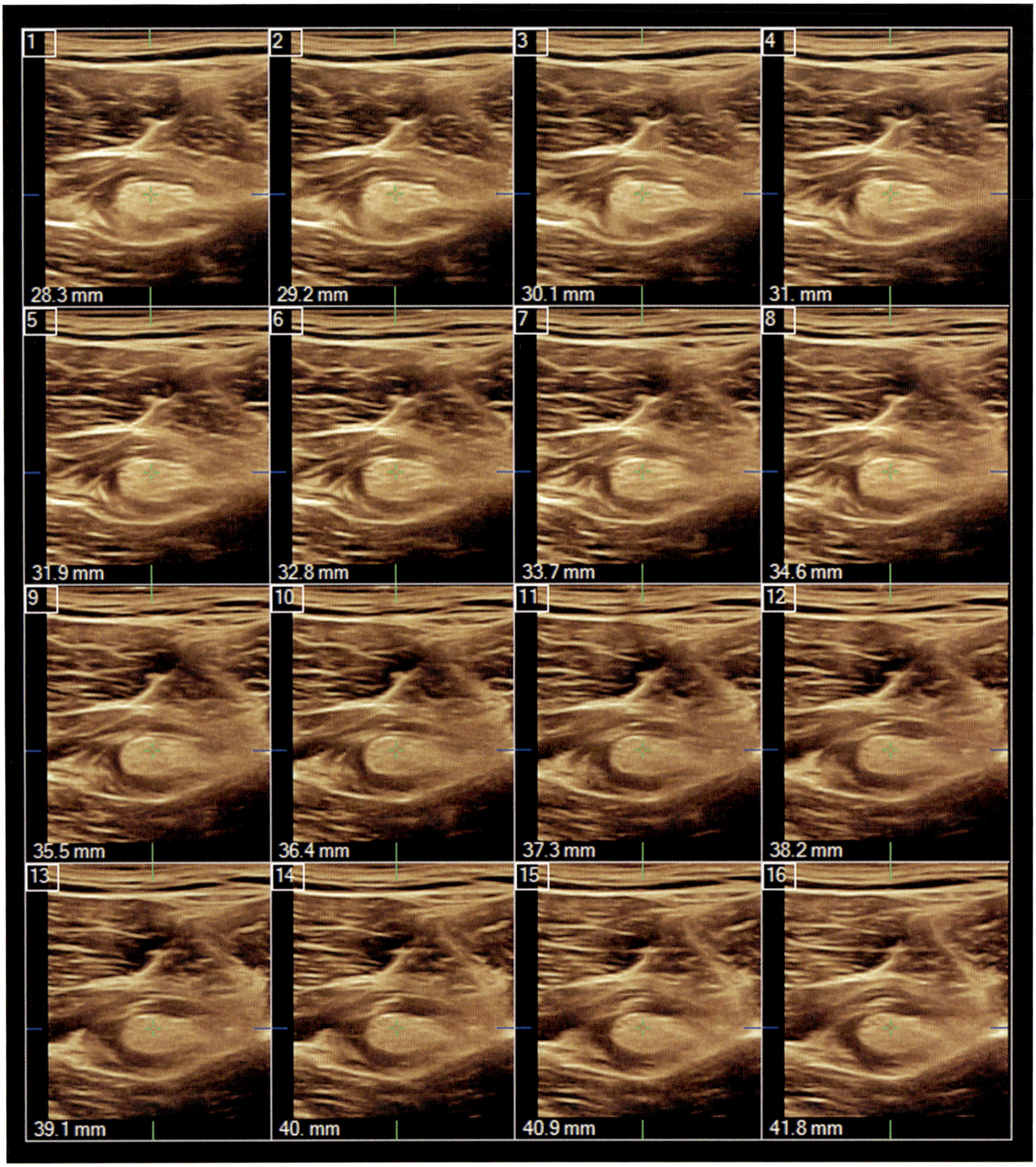

Fig. 18.45 A transverse iSlice display of the sciatic nerve, rendered from a 3D ultrasound volume of the popliteal fossa, after a popliteal sciatic nerve block in color (*sepia tone*). In this figure, 16 contiguous transverse cuts of the acquired volume (0.9 mm apart) are displayed. Note the spread of local anesthetic relative to the paraneural sheath

Acknowledgments The anatomic images (Figs. 18.7, 18.8, 18.20, and 18.26) are courtesy of the Visible Human Server at Ecole Polytechnique Fédérale de Lausanne, Visible Human Visualization Software (http://visiblehuman.epfl.ch). All other figures were reproduced with kind permission from http://www.aic.cuhk.edu.hk/usgraweb.

References

1. Andersen HL, Andersen SL, Tranum-Jensen J. Injection inside the paraneural sheath of the sciatic nerve: direct comparison among ultrasound imaging, macroscopic anatomy, and histologic analysis. Reg Anesth Pain Med. 2012;37:410–4.
2. Chan VW, Nova H, Abbas S, McCartney CJ, Perlas A, Xu DQ. Ultrasound examination and localization of the sciatic nerve: a volunteer study. Anesthesiology. 2006;104:309–14.
3. Karmakar MK, Kwok WH, Ho AM, Tsang K, Chui PT, Gin T. Ultrasound-guided sciatic nerve block: description of a new approach at the subgluteal space. Br J Anaesth. 2007;98:390–5.
4. Karmakar MK, Shariat AN, Pangthipampai P, Chen J. High-definition ultrasound imaging defines the paraneural sheath and the fascial compartments surrounding the sciatic nerve at the popliteal fossa. Reg Anesth Pain Med. 2013;38:447–51.

Part II

Component of the Spinal Canal

19 Spinal Dural Sac, Nerve Root Cuffs, Rootlets, and Nerve Roots

Miguel Angel Reina, Anna Oliva, Anna Carrera, Jorge Diamantopoulos, and Alberto Prats-Galino

Nerve rootlets leave the spinal cord at the anterolateral and posterolateral sulci. Anterior rootlets contain predominantly efferent fibers from the anterior horn and carry motor signals to voluntary muscles. At thoracic and upper lumbar regions, they also carry preganglionic sympathetic fibers from the lateral horns. Posterior rootlets are prolongations of pseudounipolar nerve cells located at dorsal root ganglions (DRGs). Rootlets join to form the anterior roots (from 6 to 8 anterior rootlets) and posterior roots (from 8 to 10 posterior rootlets) [1–5].

This chapter describes the different features of rootlets and nerve roots (anterior and posterior) present on the dorsal and ventral surfaces of the spinal cord. It also highlights the characteristics of the dural sac at different levels of the spinal cord and the differences arising between the ventral and dorsal surfaces in contact with vertebral bodies and discs and between their dorsal surfaces adjacent to the vertebral laminas and ligamentum flavum [1–5]. Other contrasting features are the varying length and diameter of the dural cuffs and dorsal root ganglia along the spinal column (Figs. 19.1, 19.2, 19.3, 19.4, 19.5, 19.6, 19.7, 19.8, 19.9, 19.10, 19.11, 19.12, 19.13, 19.14, 19.15, 19.16, 19.17, 19.18, 19.19, 19.20, 19.21, 19.22, 19.23, 19.24, 19.25, 19.26, 19.27, 19.28, 19.29, 19.30, and 19.31).

M.A. Reina, MD, PhD (✉)
Department of Clinical Medical Sciences and Institute of Applied Molecular Medicine, School of Medicine, University of CEU San Pablo, Madrid, Spain

Department of Anesthesiology, Madrid-Montepríncipe University Hospital, Madrid, Spain
e-mail: miguelangel@perticone.e.telefonica.net

A. Oliva, MD
Laboratory of Macro-Micro Dissection and Surgical Anatomy, Human Anatomy and Embryology Unit, School of Medicine, University of Barcelona, Barcelona, Spain

Rehabilitation Unit, Mataró Hospital, Mataró, Barcelona, Spain
e-mail: annao.oliva@gmail.com

A. Carrera, MD, PhD
Neuroscience, Embryology, Molecular Oncology and Clinical Anatomy Group (NEOMA), Faculty of Medicine, University of Girona, Girona, Spain
e-mail: anna.carrera@udg.edu

J. Diamantopoulos, MD
Neurosurgical Department, Madrid-Montepríncipe University Hospital, Madrid-Sanchinarro University Hospital, Madrid, Spain
e-mail: jorgediaman@yahoo.com

A. Prats-Galino, MD, PhD
Laboratory of Surgical NeuroAnatomy, Human Anatomy and Embryology Unit, School of Medicine, University of Barcelona, Barcelona, Spain
e-mail: aprats@ub.edu

M.A. Reina et al. (eds.), *Atlas of Functional Anatomy for Regional Anesthesia and Pain Medicine: Human Structure, Ultrastructure and 3D Reconstruction Images*, DOI 10.1007/978-3-319-09522-6_19,

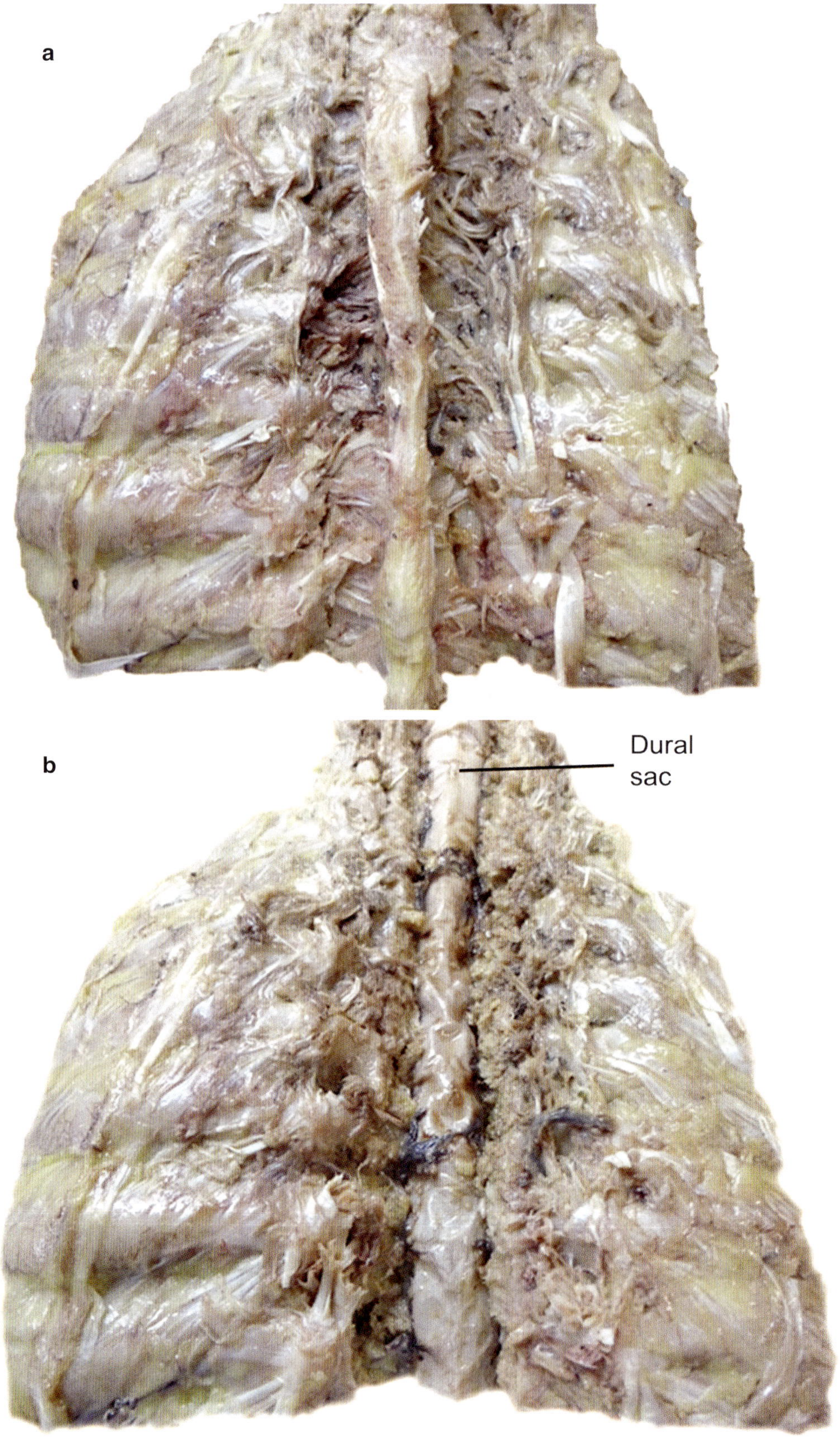

Fig. 19.1 Dissection of human dural sac. (**a**) The spinal ligaments have been partially dissected. (**b**) Detail of dural sac and epidural vein

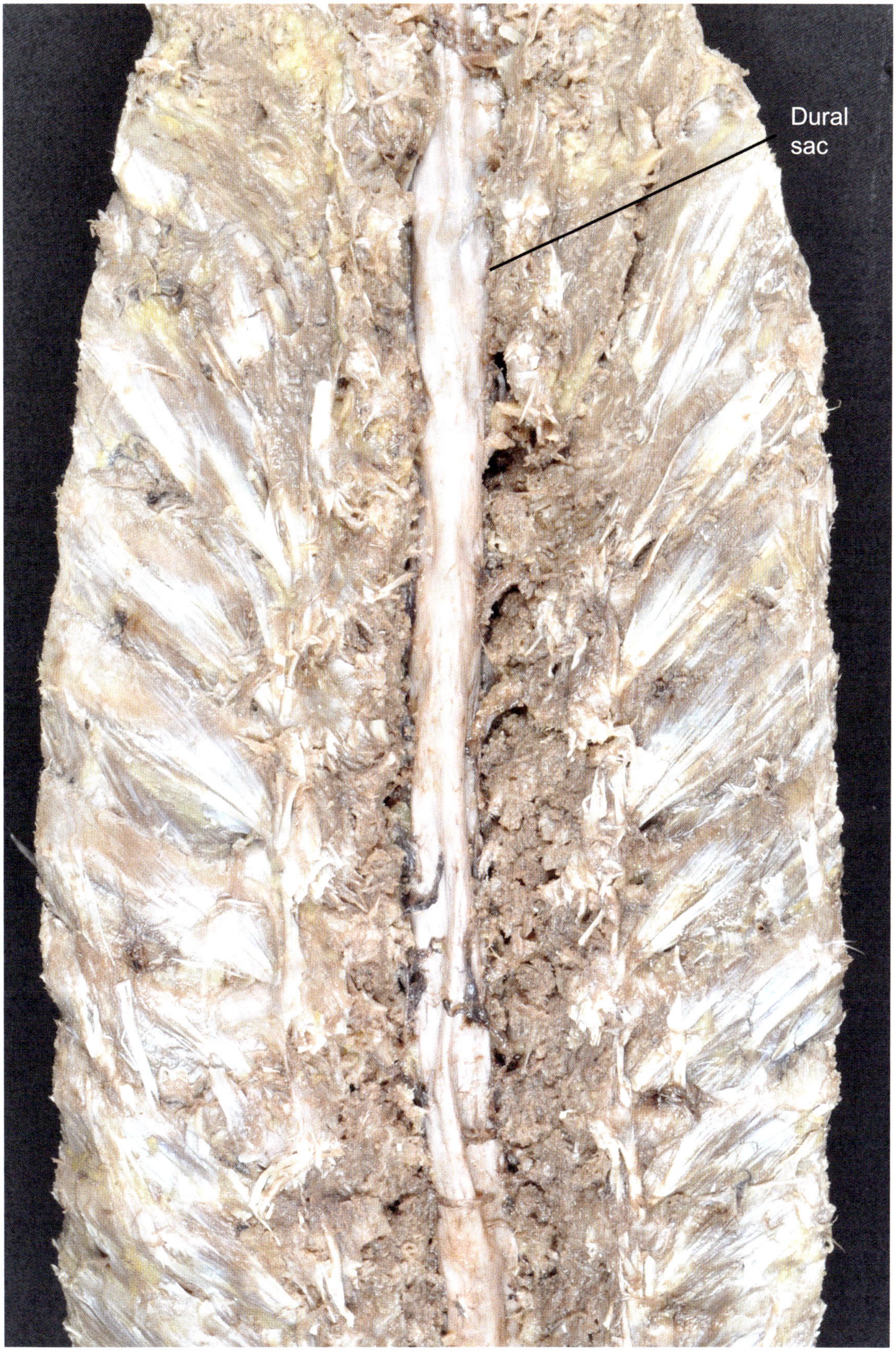

Fig. 19.2 Complete dissection of human dural sac

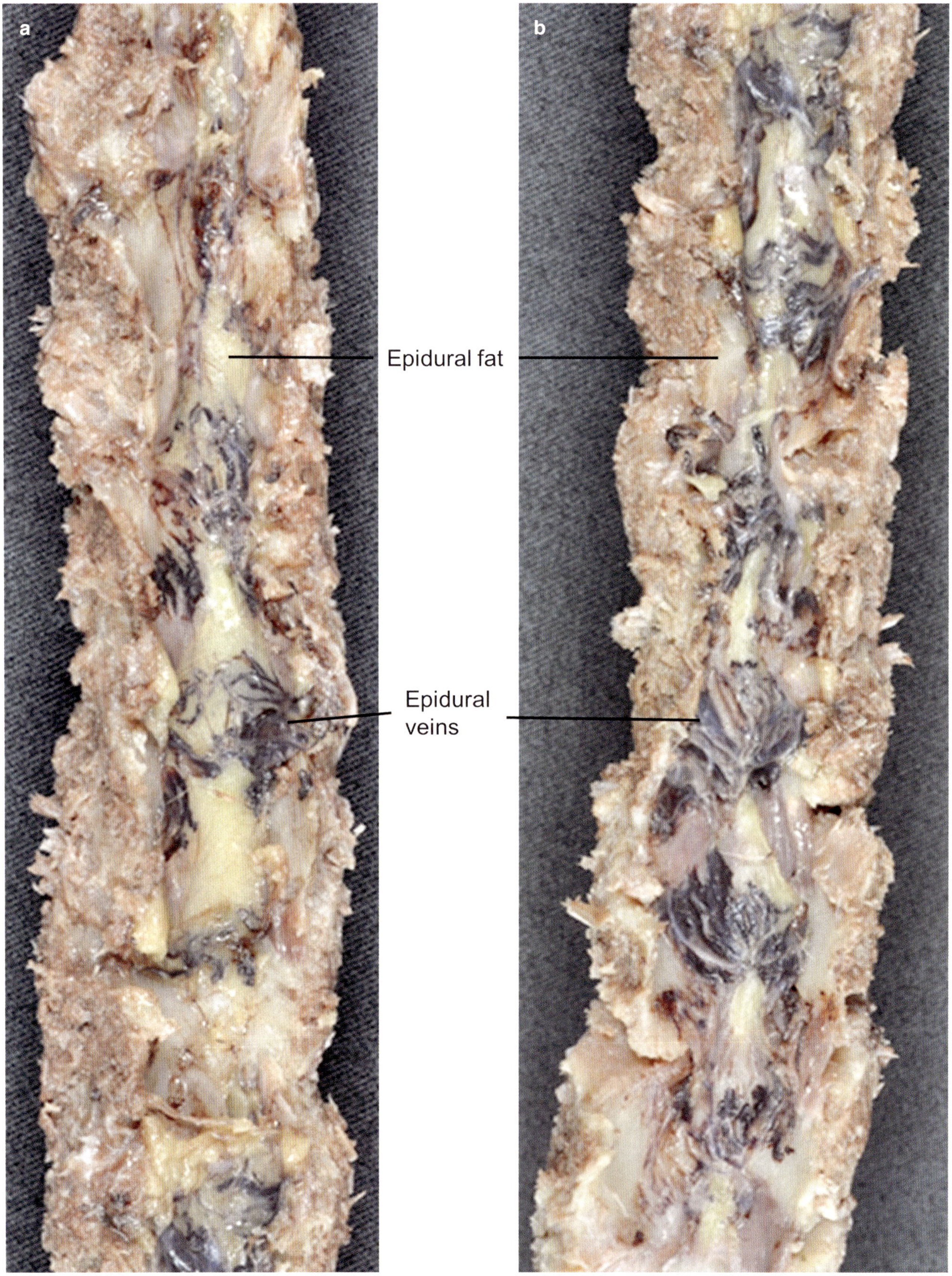

Fig. 19.3 The image shows the posterior arch from within the spinal canal once the spinous processes has been removed. Epidural veins and epidural fat can be identified. (**a**) lumbar region, (**b**) lower thoracic region

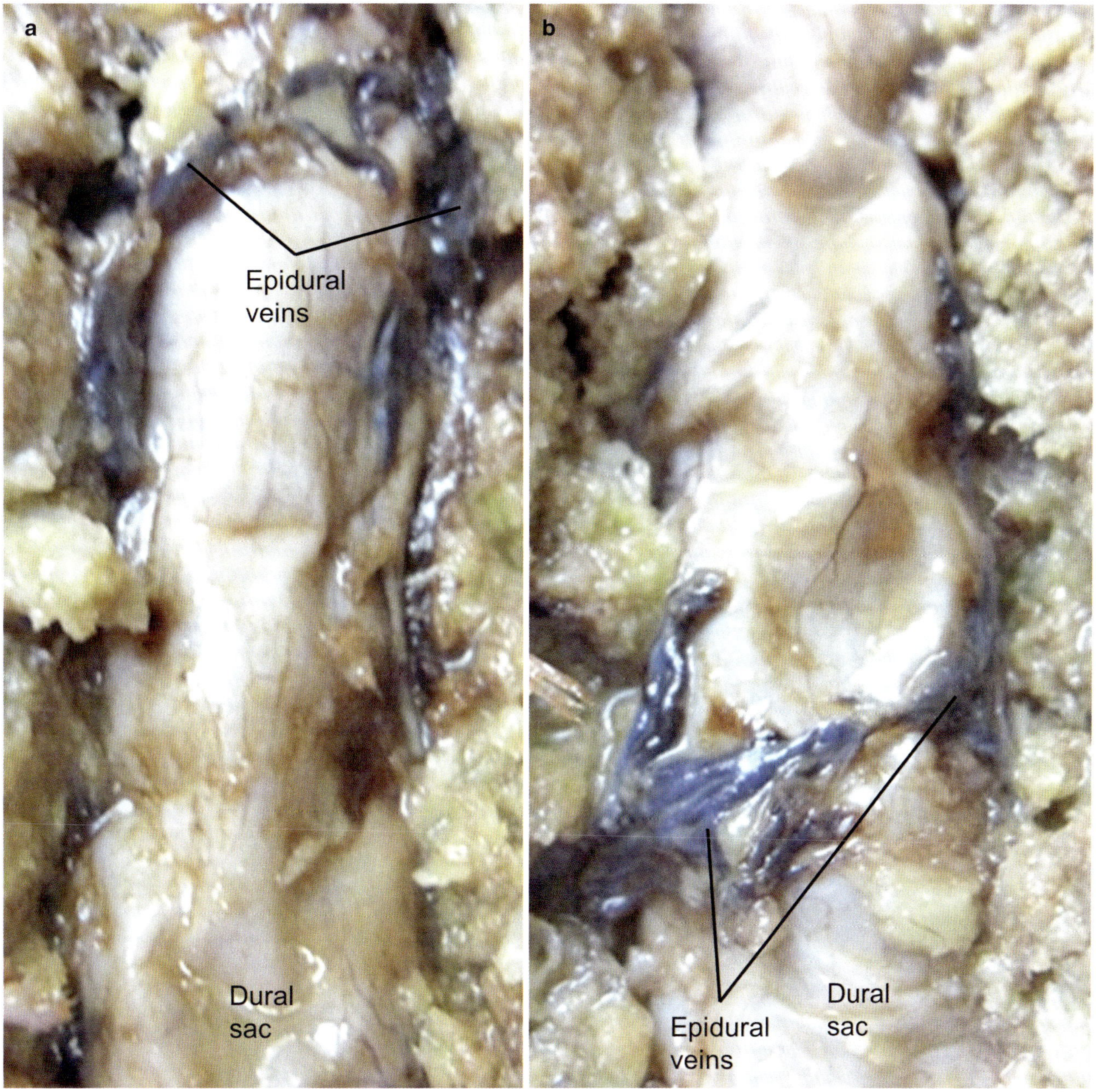

Fig. 19.4 Epidural veins in the posterior epidural space at lumbar region. (**a**, **b**) images obtained from different samples

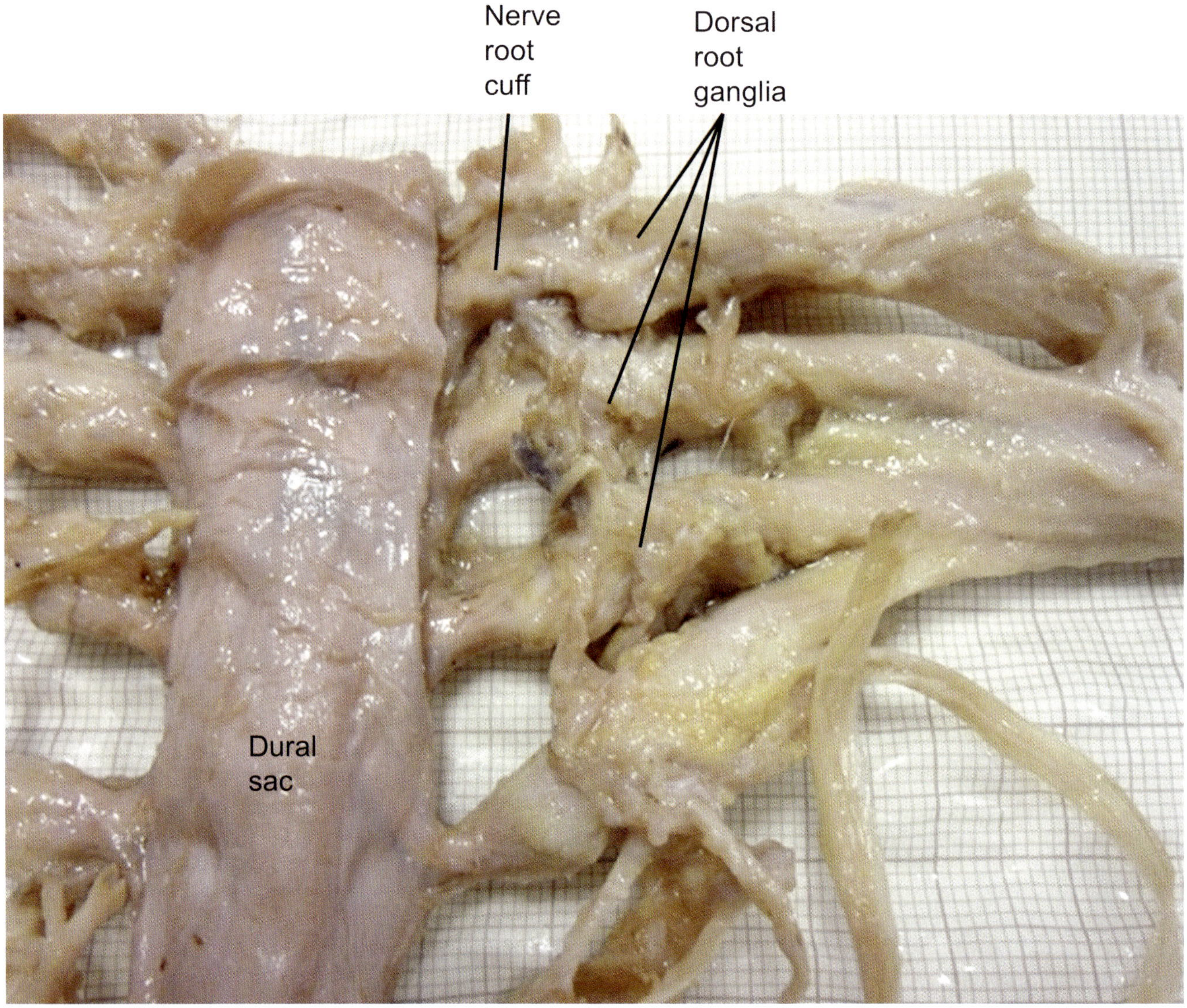

Fig. 19.5 Detail of the dorsal surface of the nerve root cuff at the cervical level and dural sac

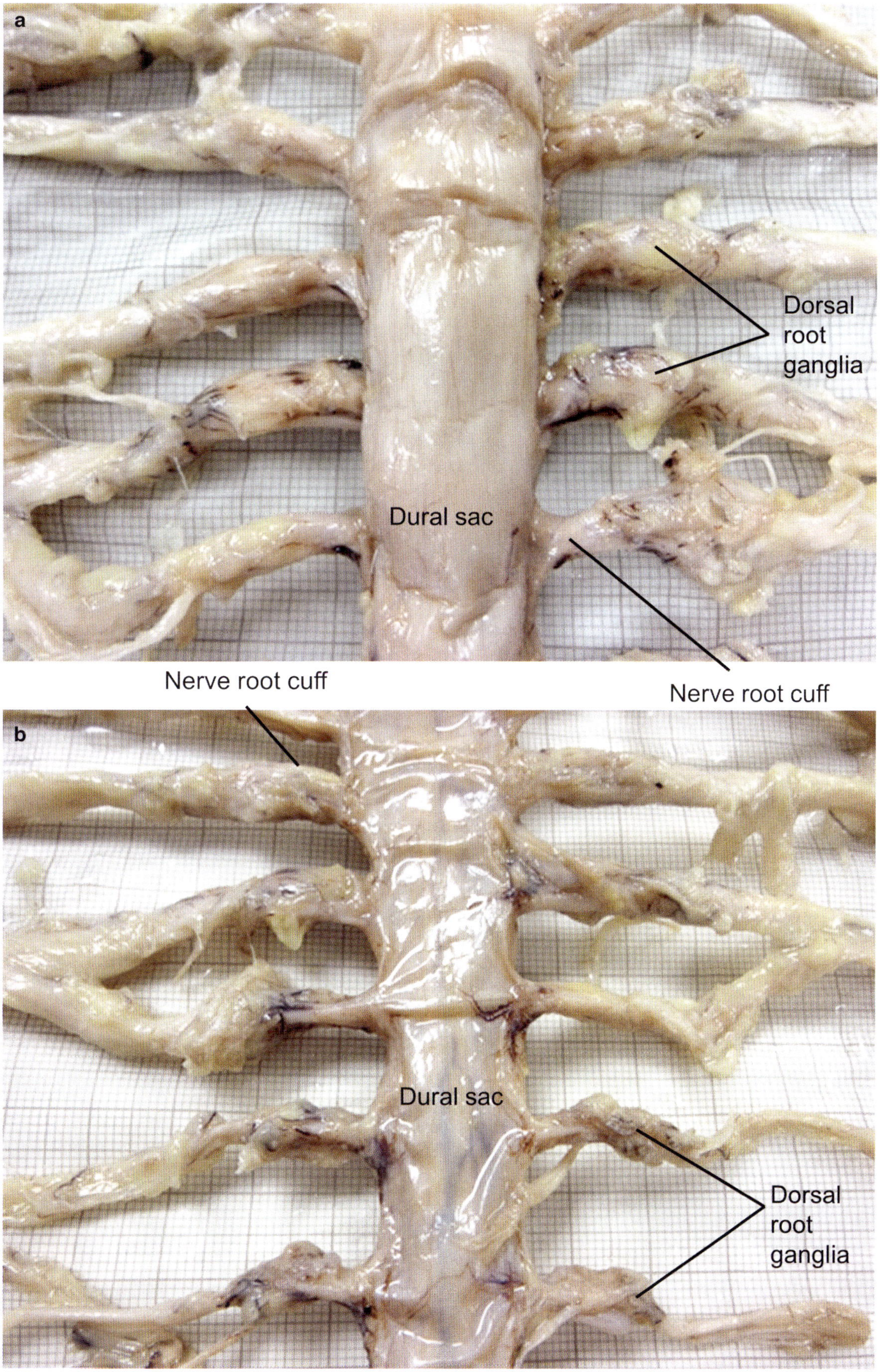

Fig. 19.6 Detail of the nerve root cuff and dural sac at cervical level. (**a**) Dorsal surface. (**b**) Ventral surface (From Reina [1]; with permission)

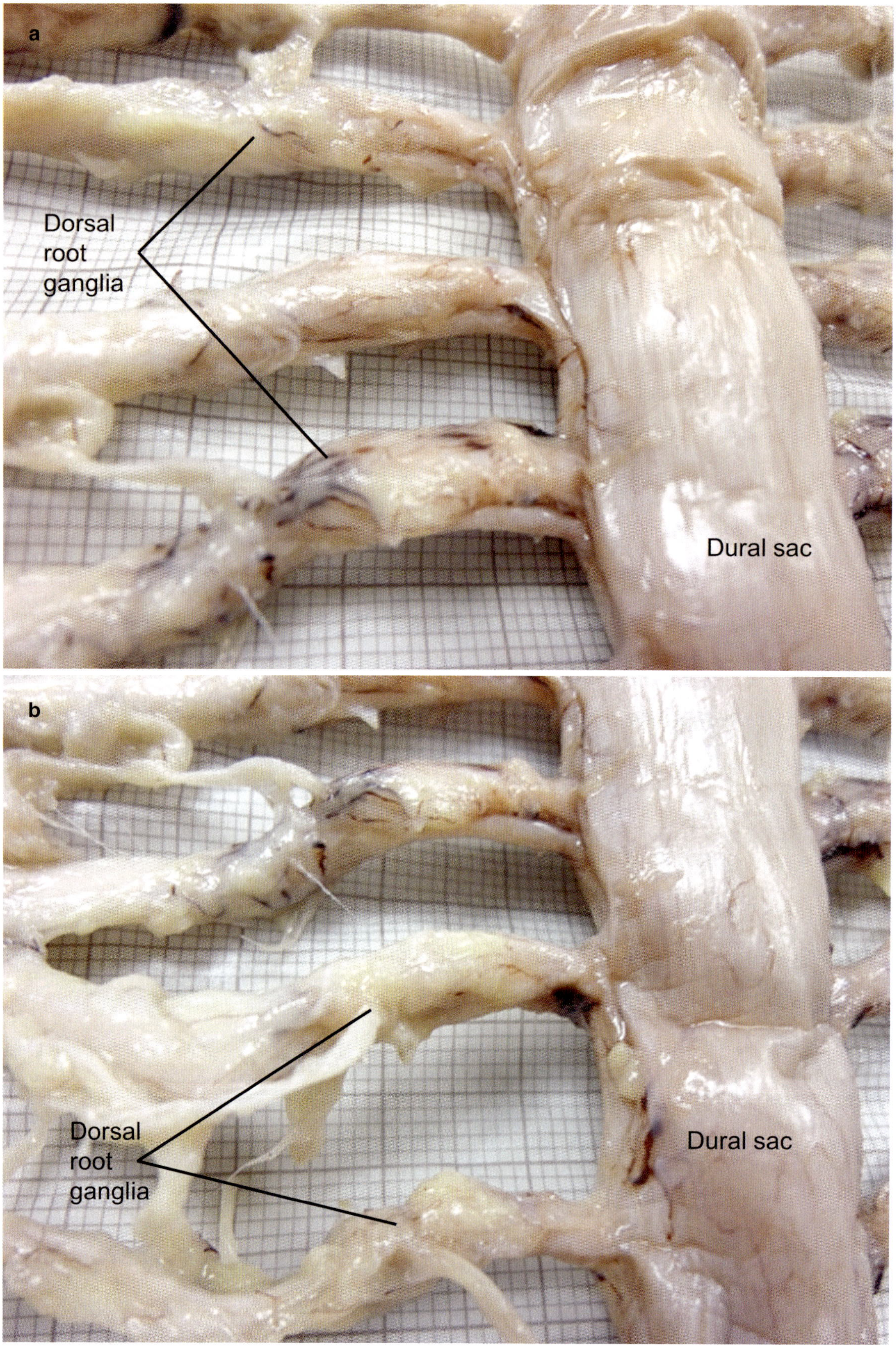

Fig. 19.7 Details of dorsal dural sac and nerve root cuff (**a**) at the cervical region (**b**) cervico-thoracic region

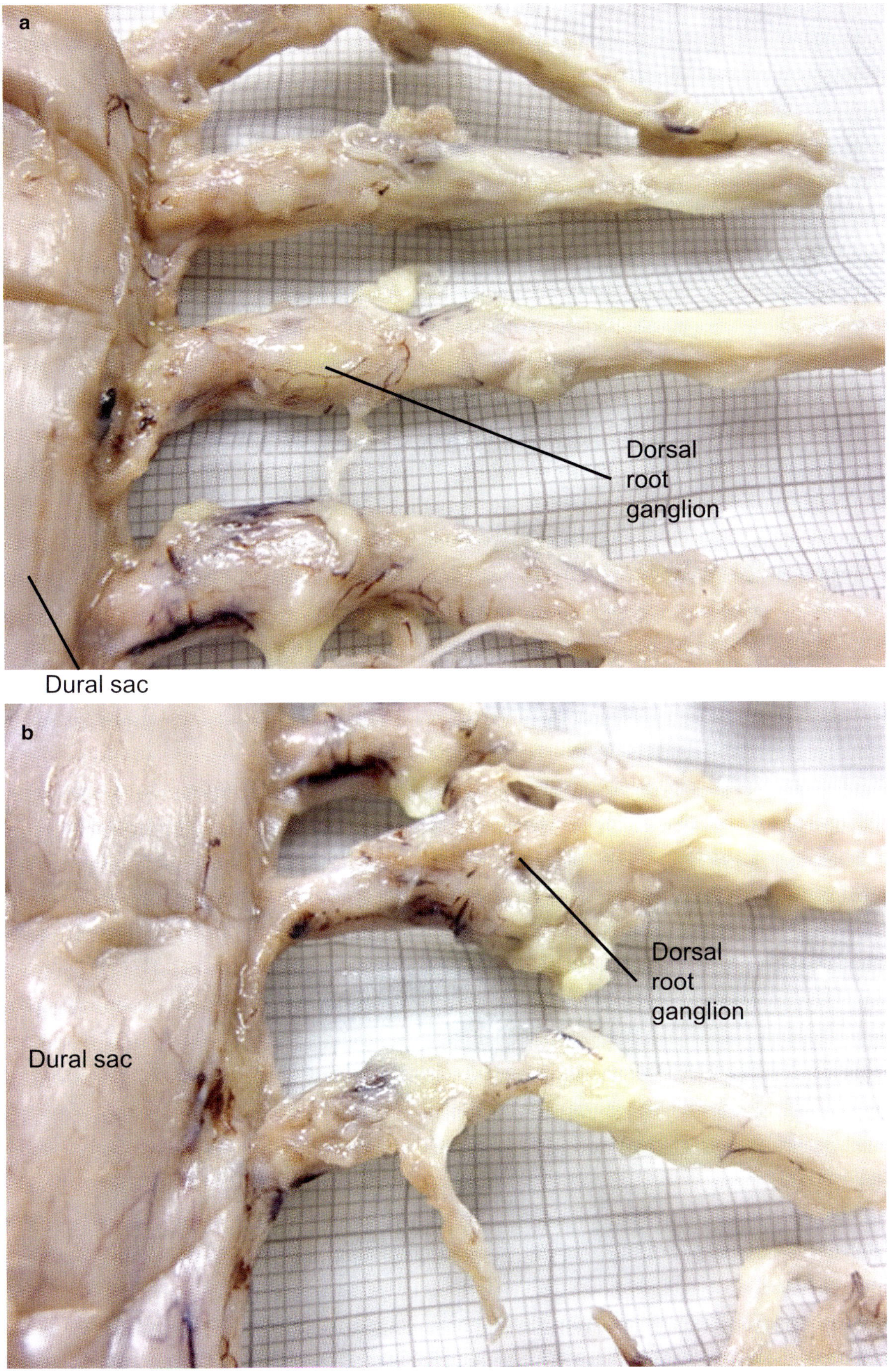

Fig. 19.8 Detail of dorsal surface from dural sac and nerve root cuff. (**a**) Cervical level. (**b**) Thoracic level

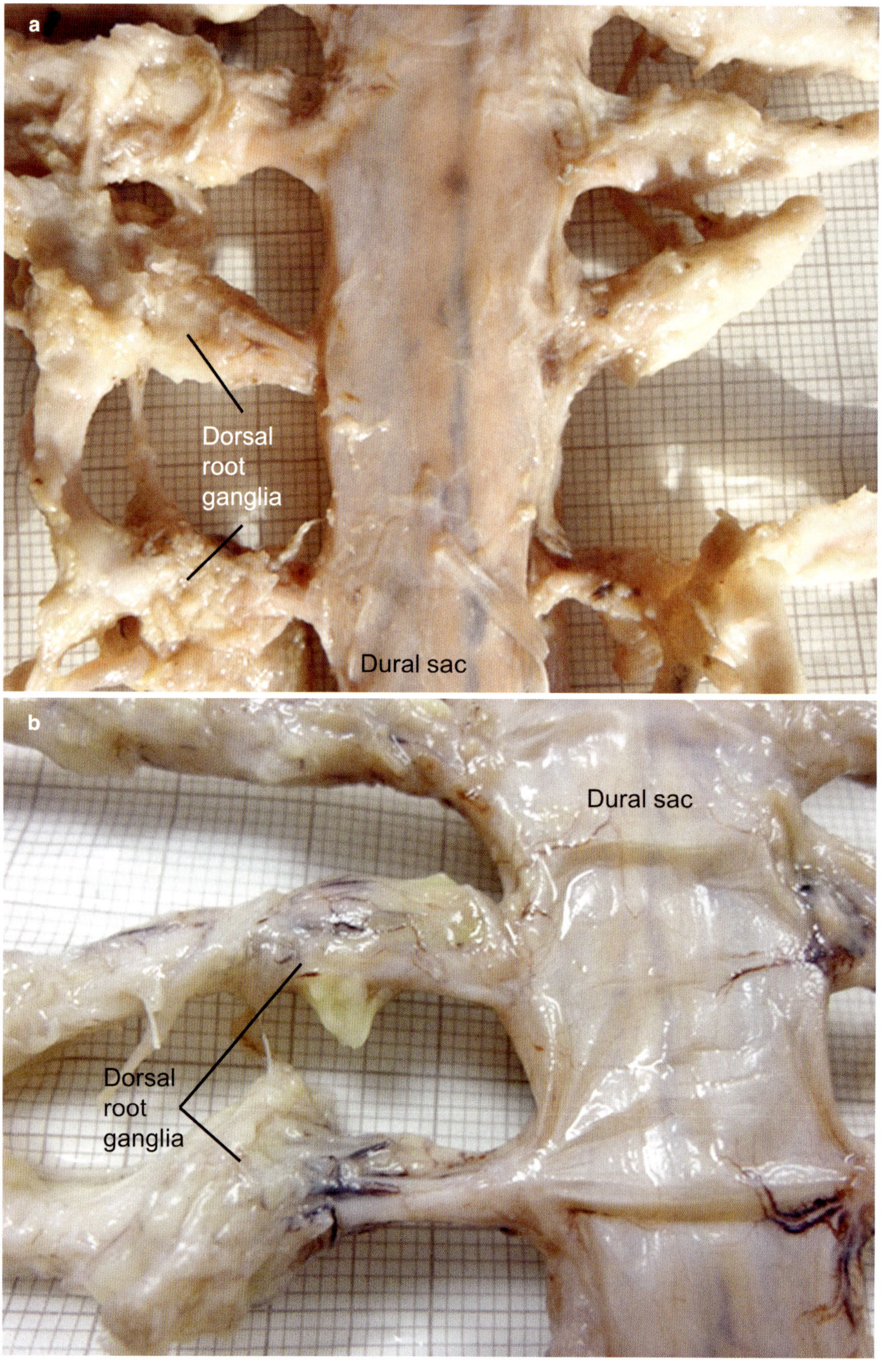

Fig. 19.9 (**a**, **b**) Detail of the ventral surface of dural sac and nerve root cuff at cervical level from different cadavers (From Reina et al. [3]; with permission)

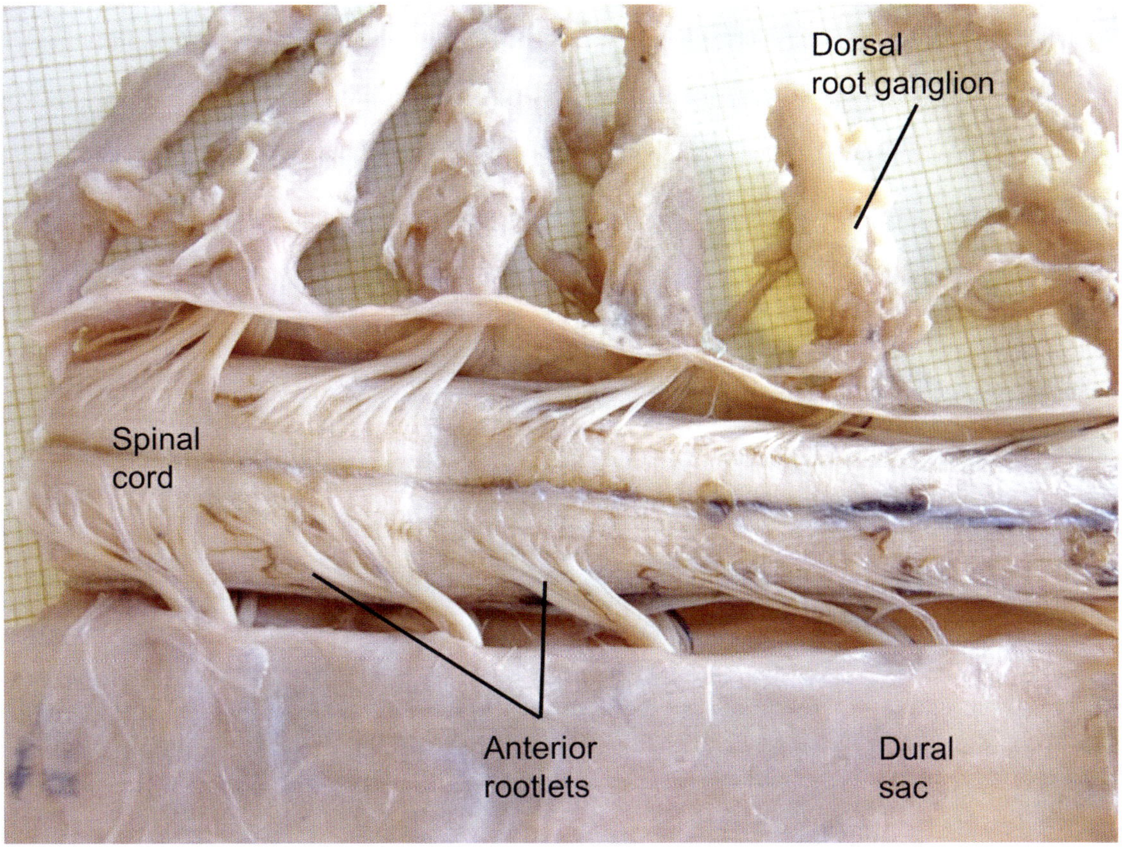

Fig. 19.10 Detail of the spinal cord and rootlets at the cervical and higher thoracic levels (anterior surface)

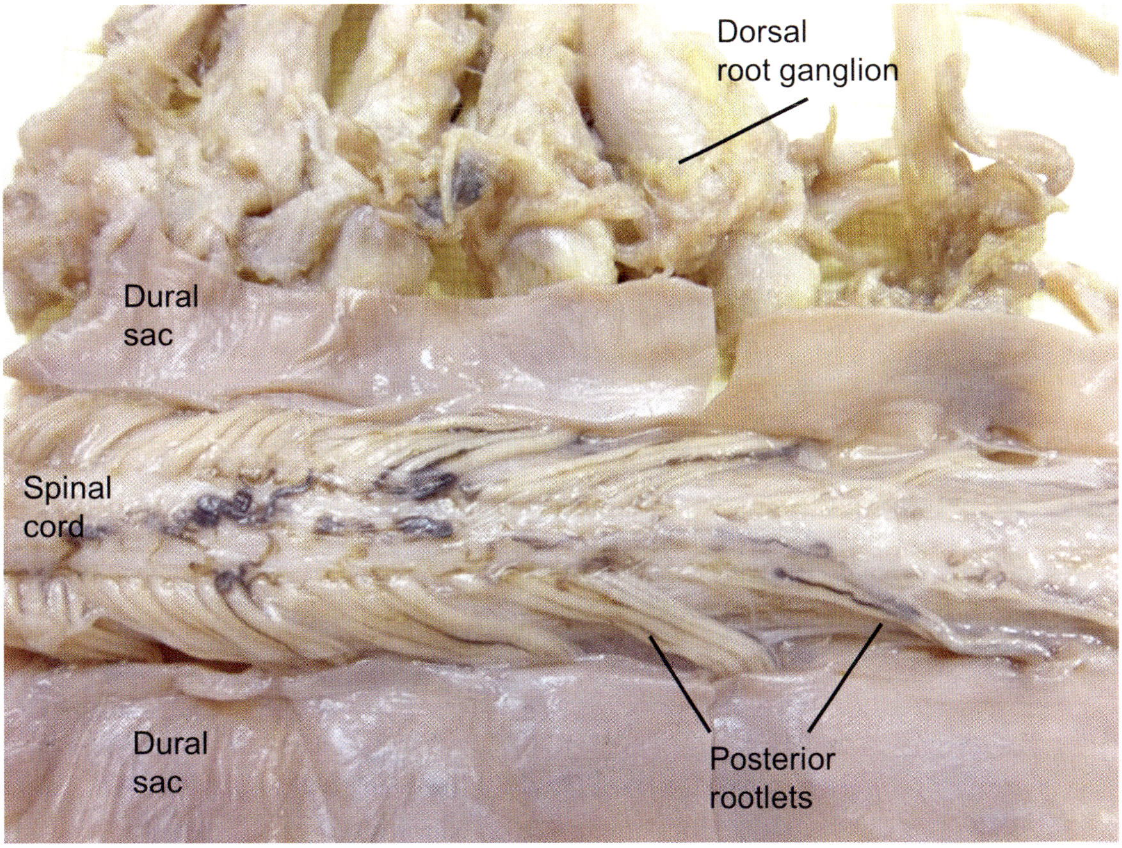

Fig. 19.11 Detail of the spinal cord and rootlets at the cervical level (posterior surface)

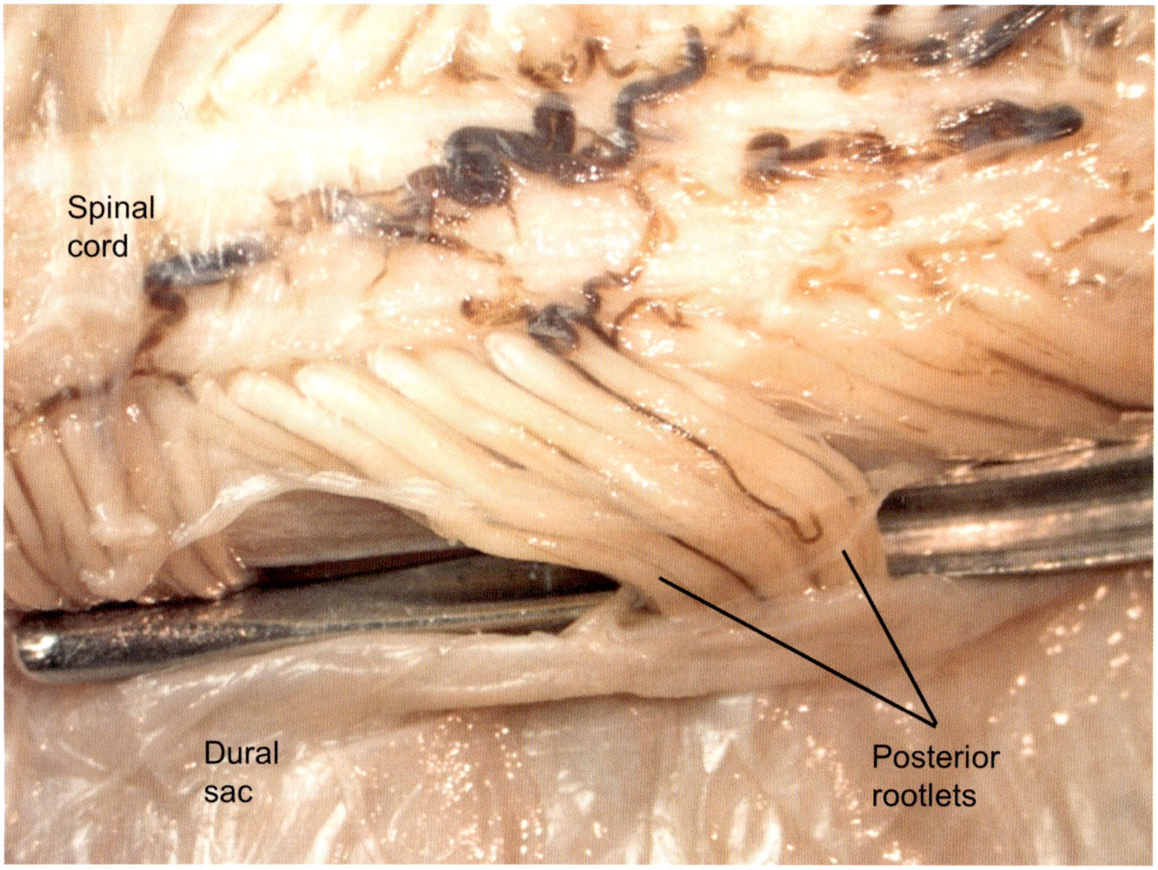

Fig. 19.12 Detail of the spinal cord and rootlets at the cervical level (posterior surface)

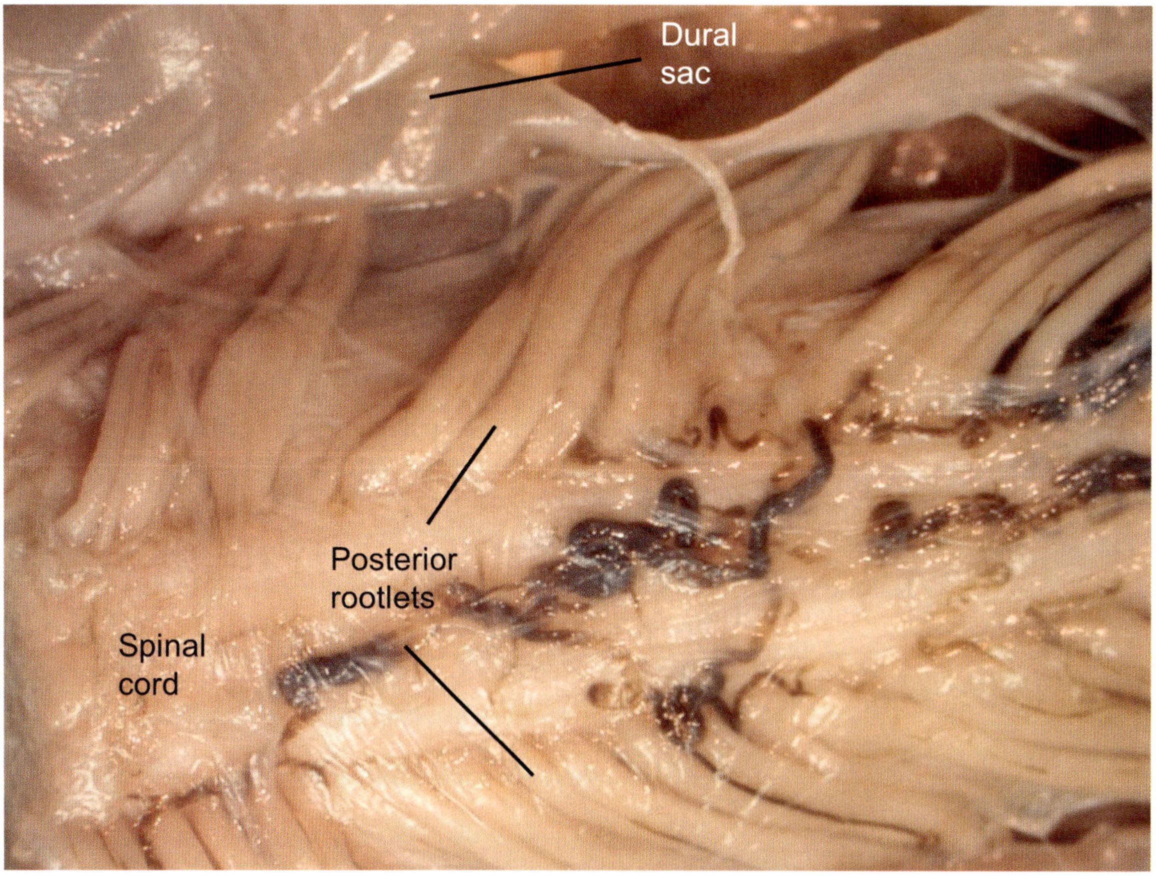

Fig. 19.13 Detail of the spinal cord and rootlets at the cervical level (posterior surface)

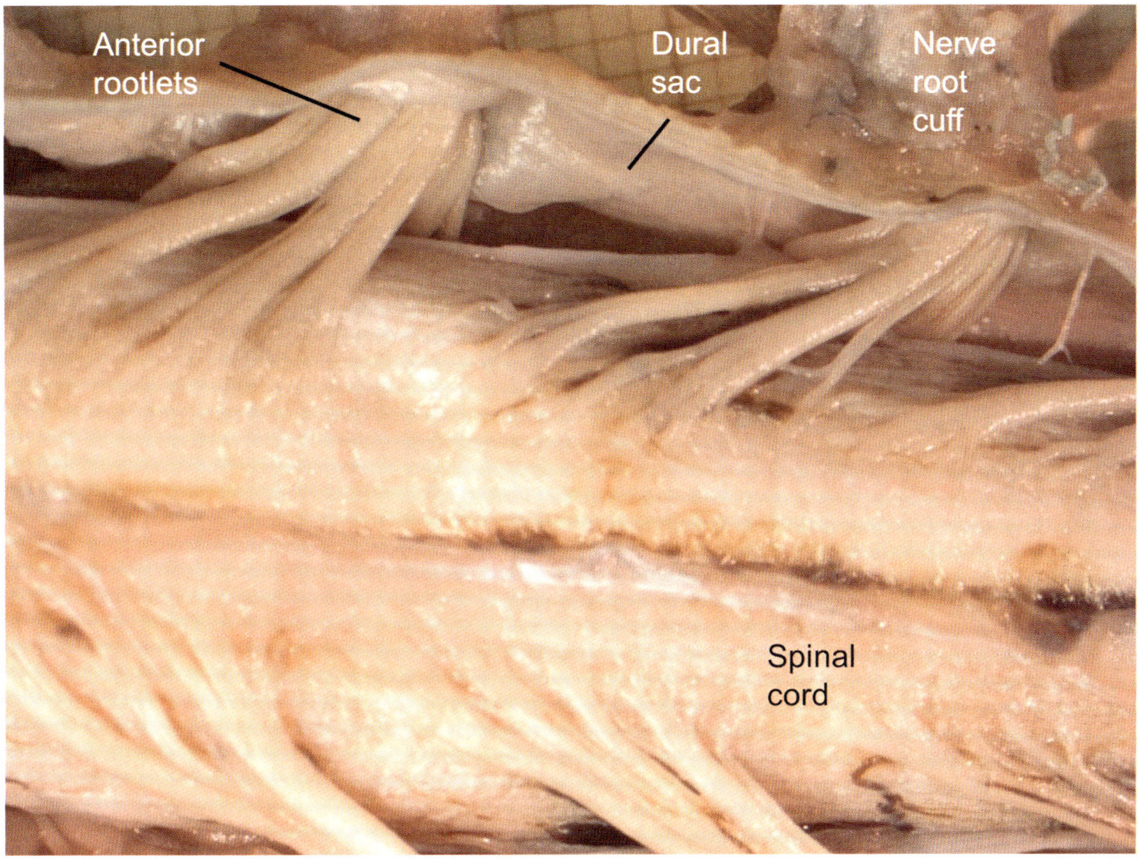

Fig. 19.14 Detail of the spinal cord and rootlets at the cervical level (anterior surface)

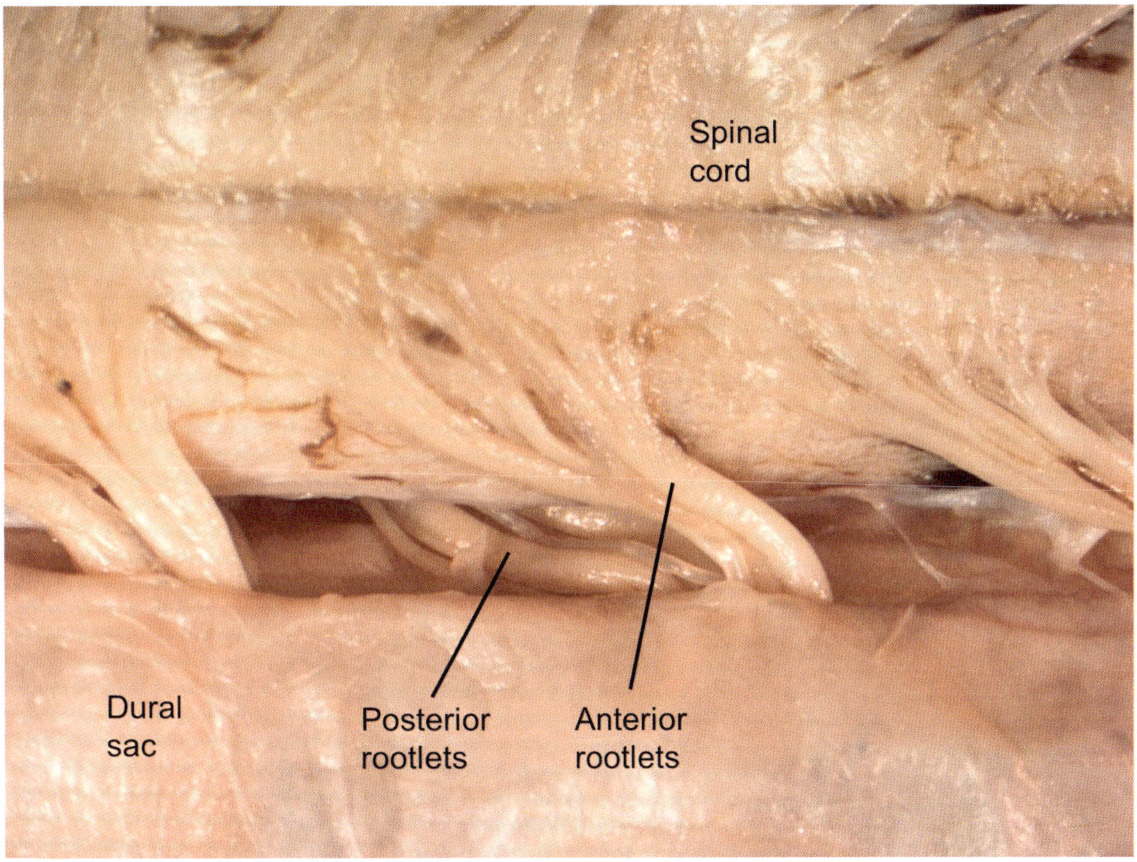

Fig. 19.15 Detail of the spinal cord and rootlets at the cervical level (anterior surface)

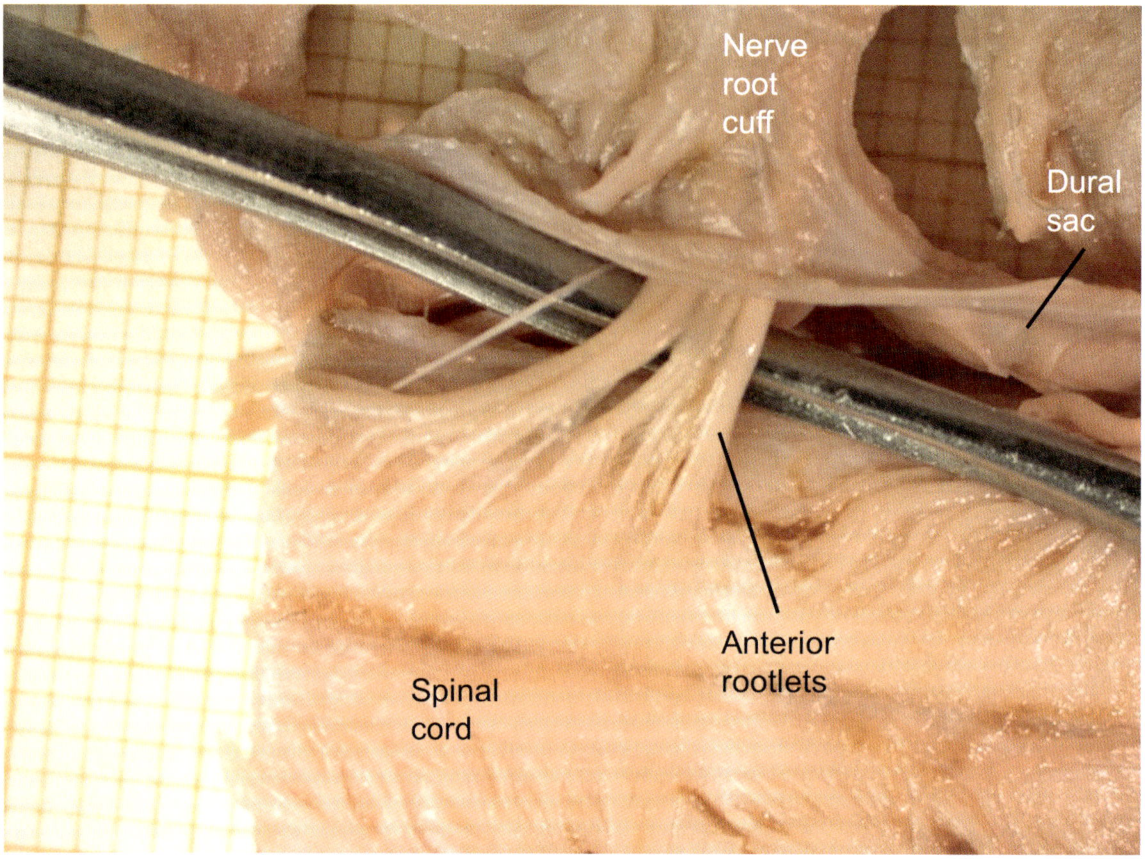

Fig. 19.16 Detail of the spinal cord and rootlets at the cervical level (anterior surface)

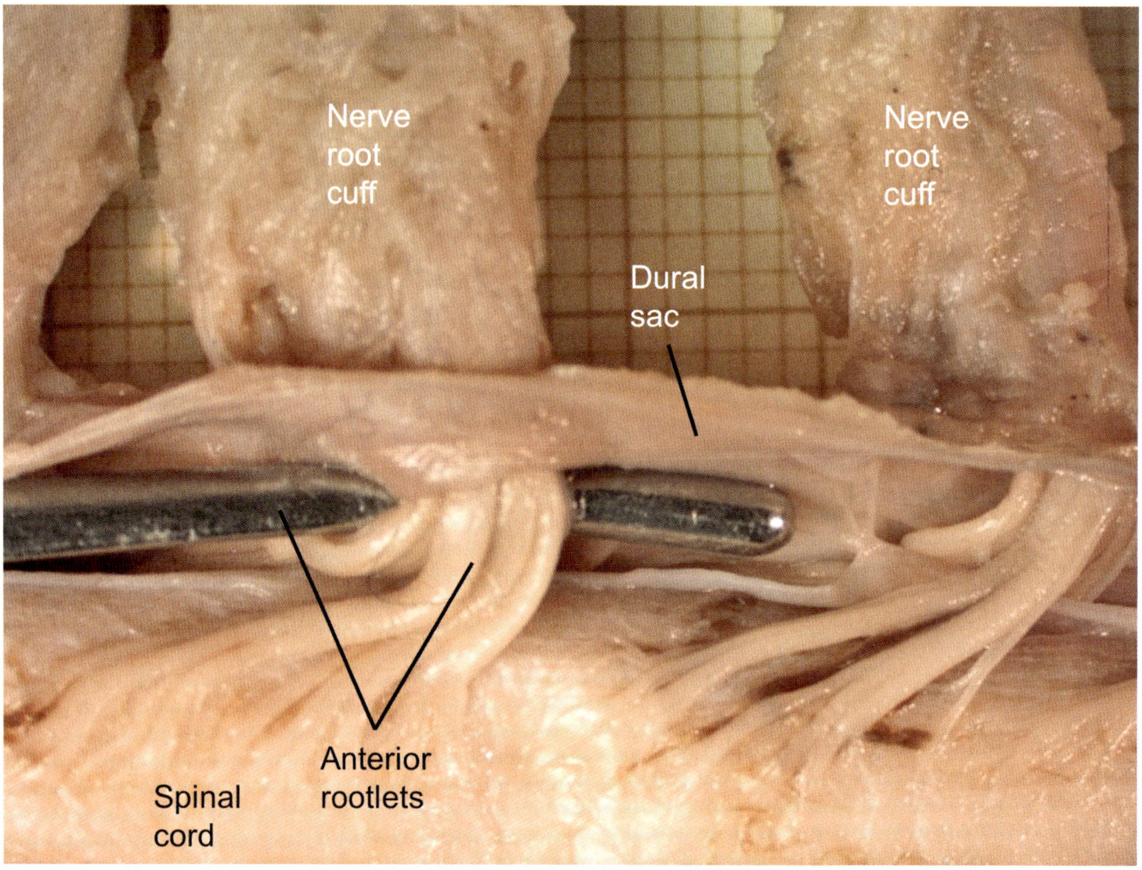

Fig. 19.17 Detail of the rootlets in Fig. 19.16

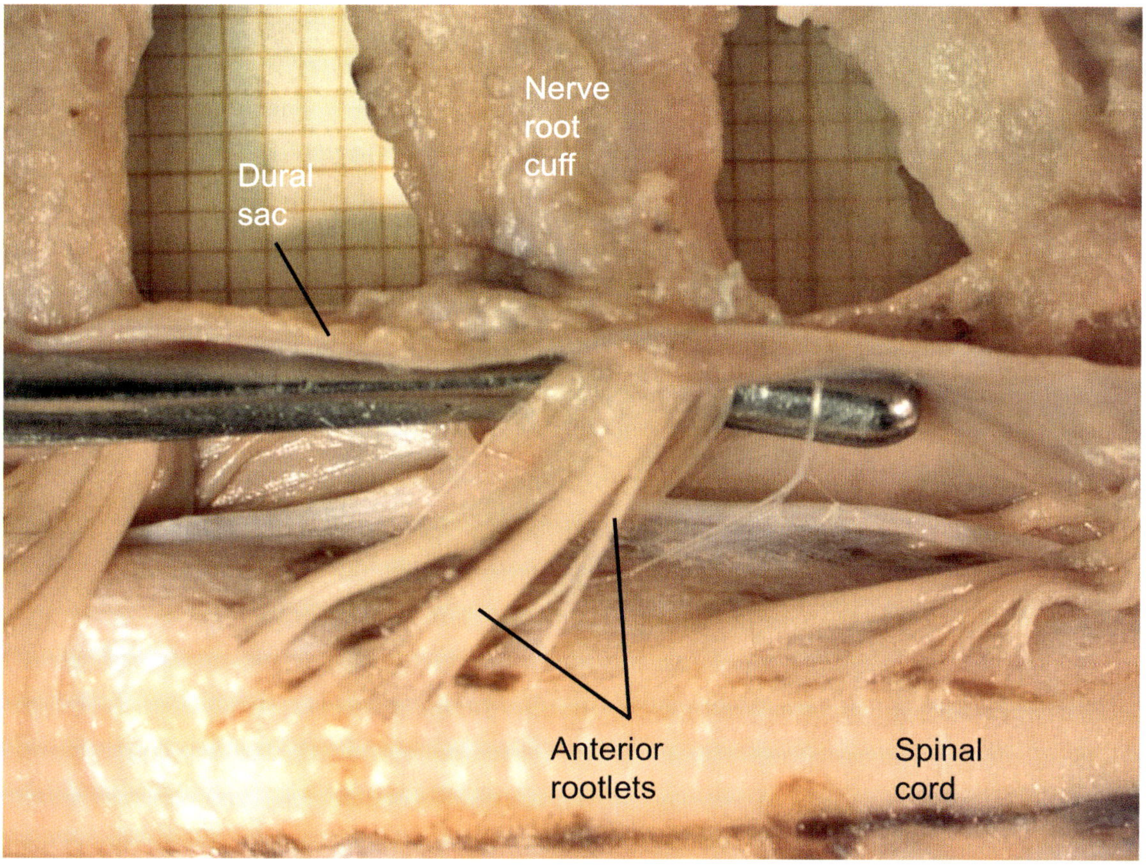

Fig. 19.18 Detail of the rootlets in Fig. 19.16

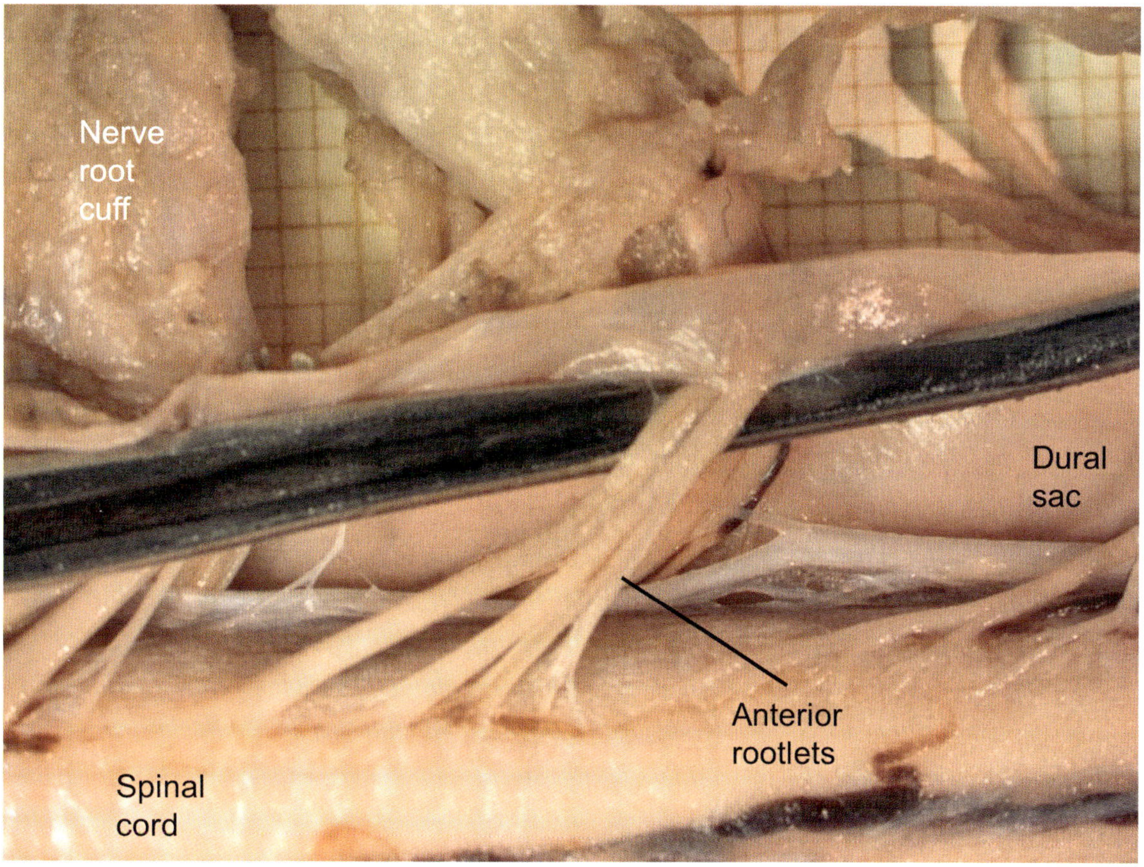

Fig. 19.19 Detail of the rootlets in Fig. 19.16

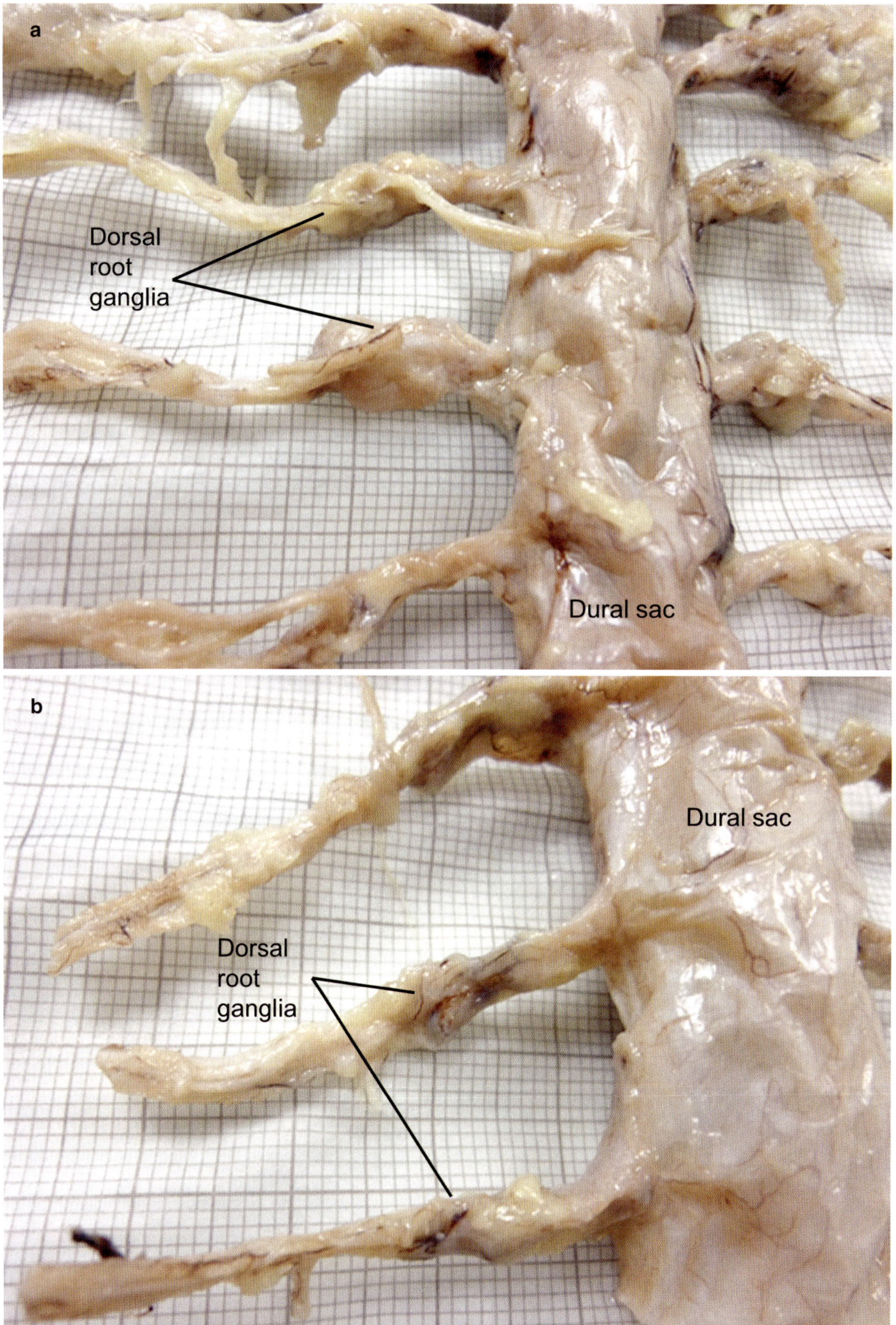

Fig. 19.20 Human dural sac and nerve root cuffs. Posterior view at thoracic level (**a**, **b**) different view of same sample

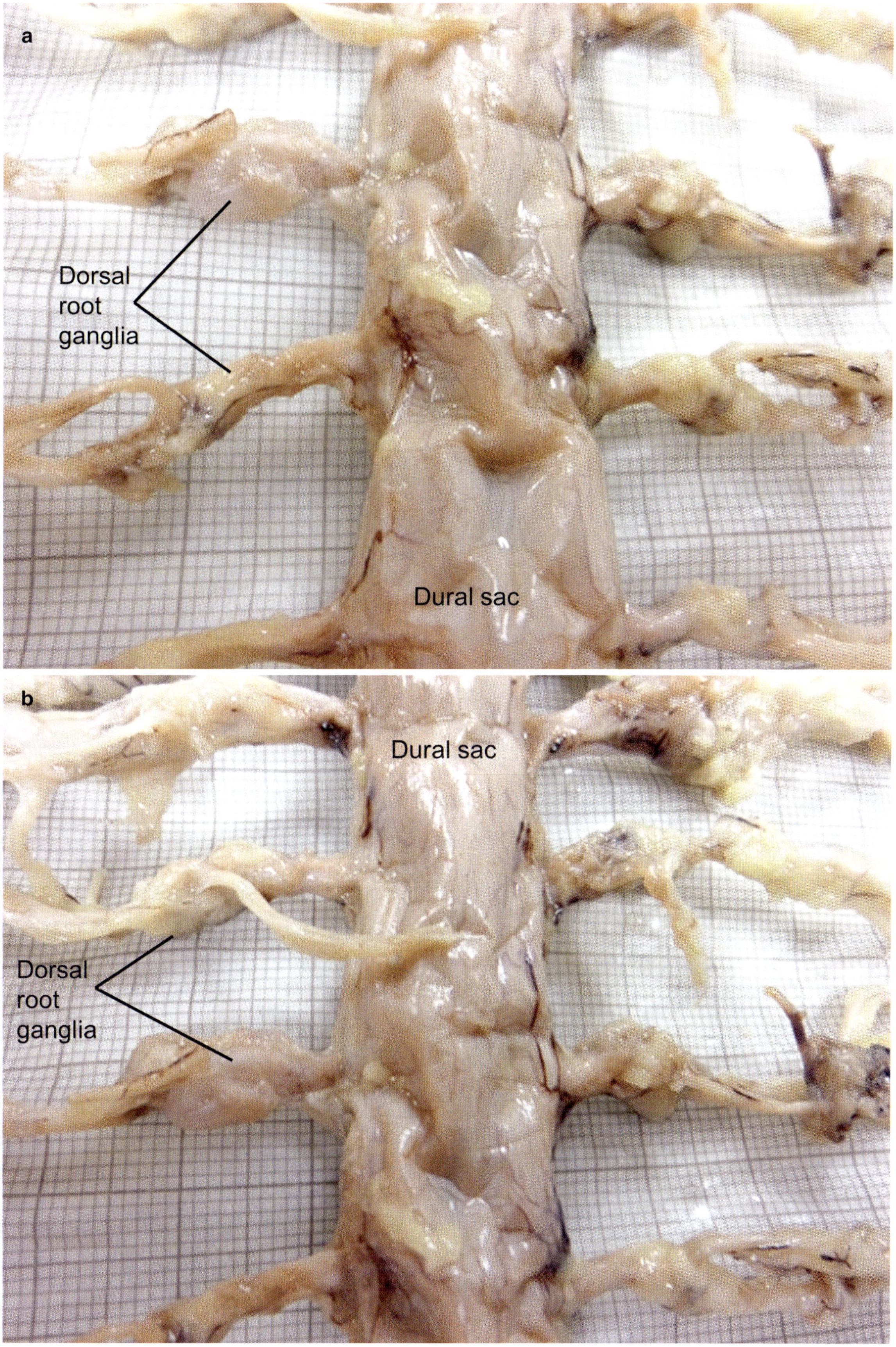

Fig. 19.21 Human dural sac and nerve root cuffs. Posterior view at thoracic level (**a**, **b**) different view of same sample

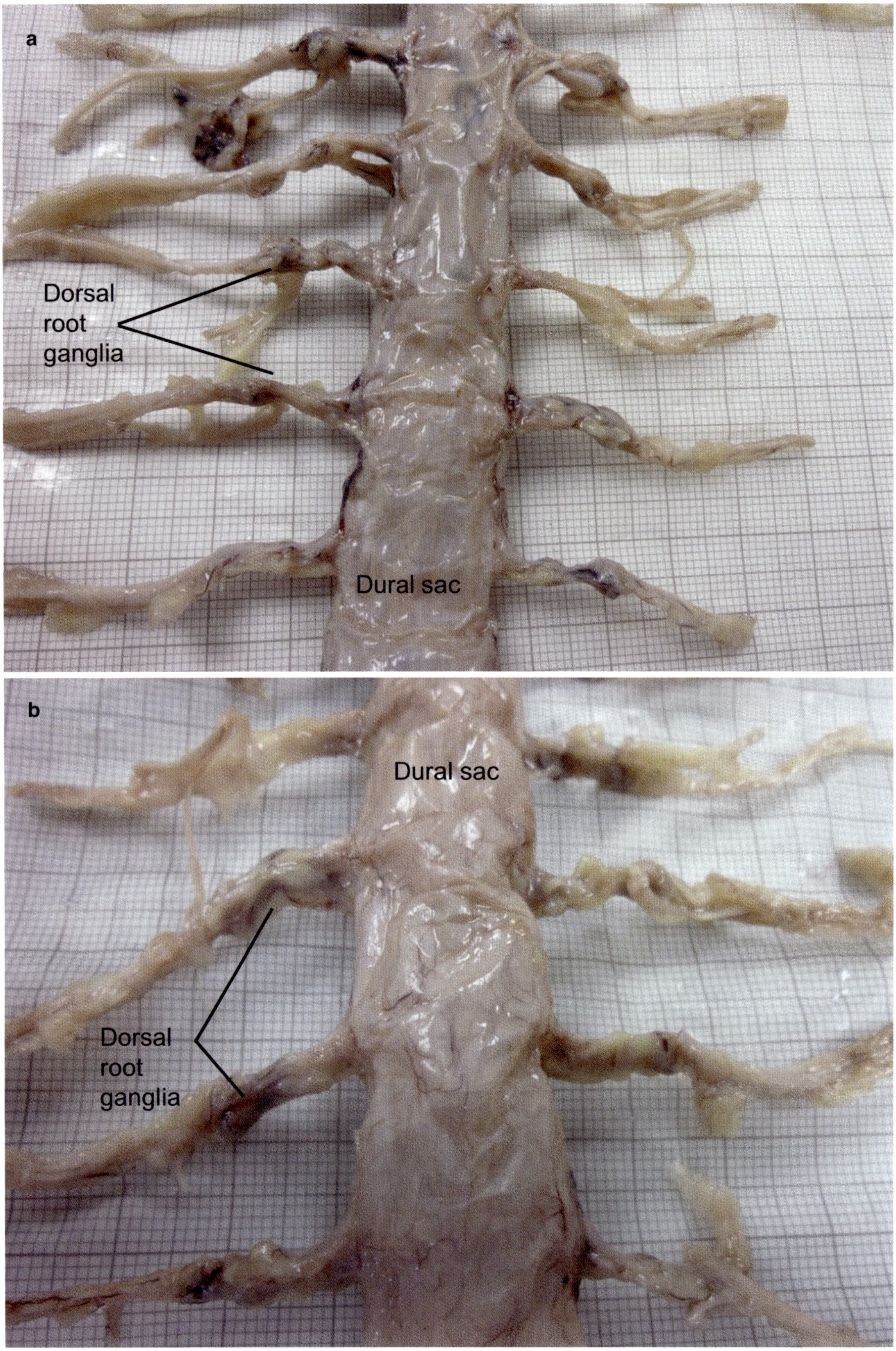

Fig. 19.22 Human dural sac and nerve root cuffs. Anterior view at thoracic level (**a**, **b**) different view of same sample

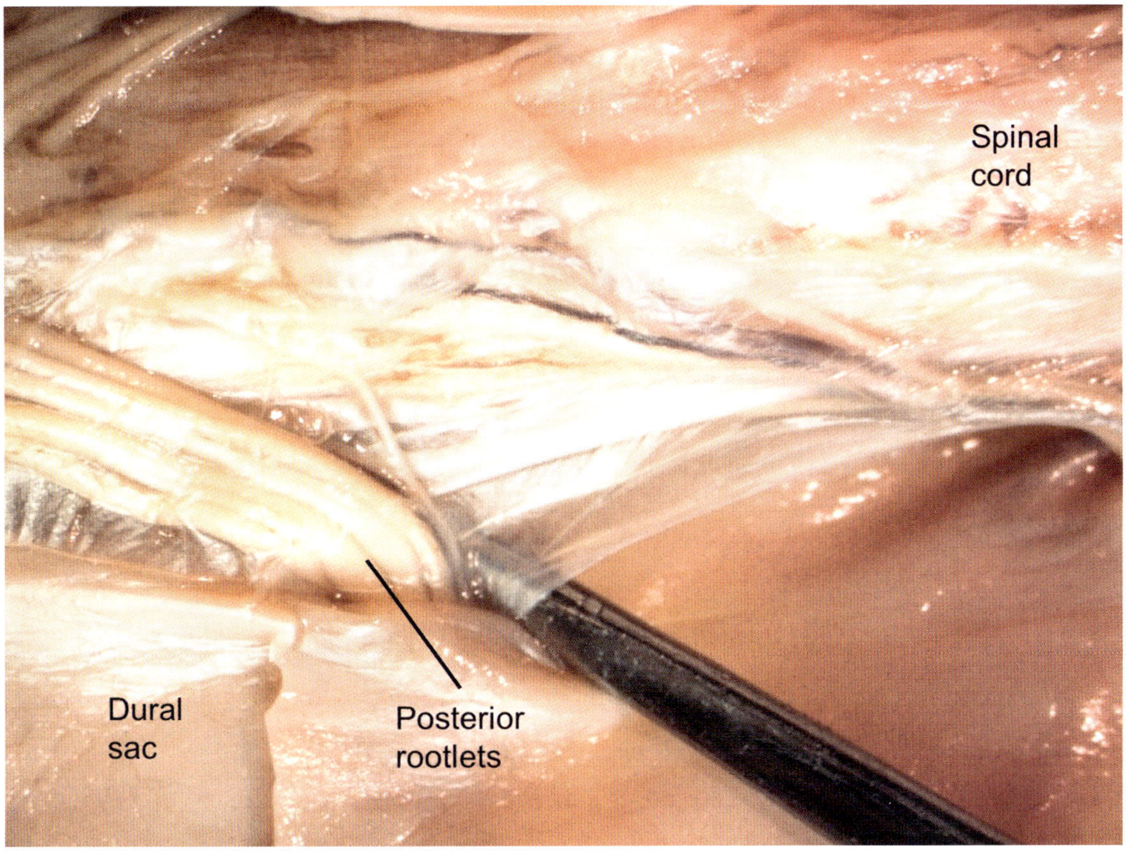

Fig. 19.23 Detail of the spinal cord and rootlets at the thoracic level (posterior surface)

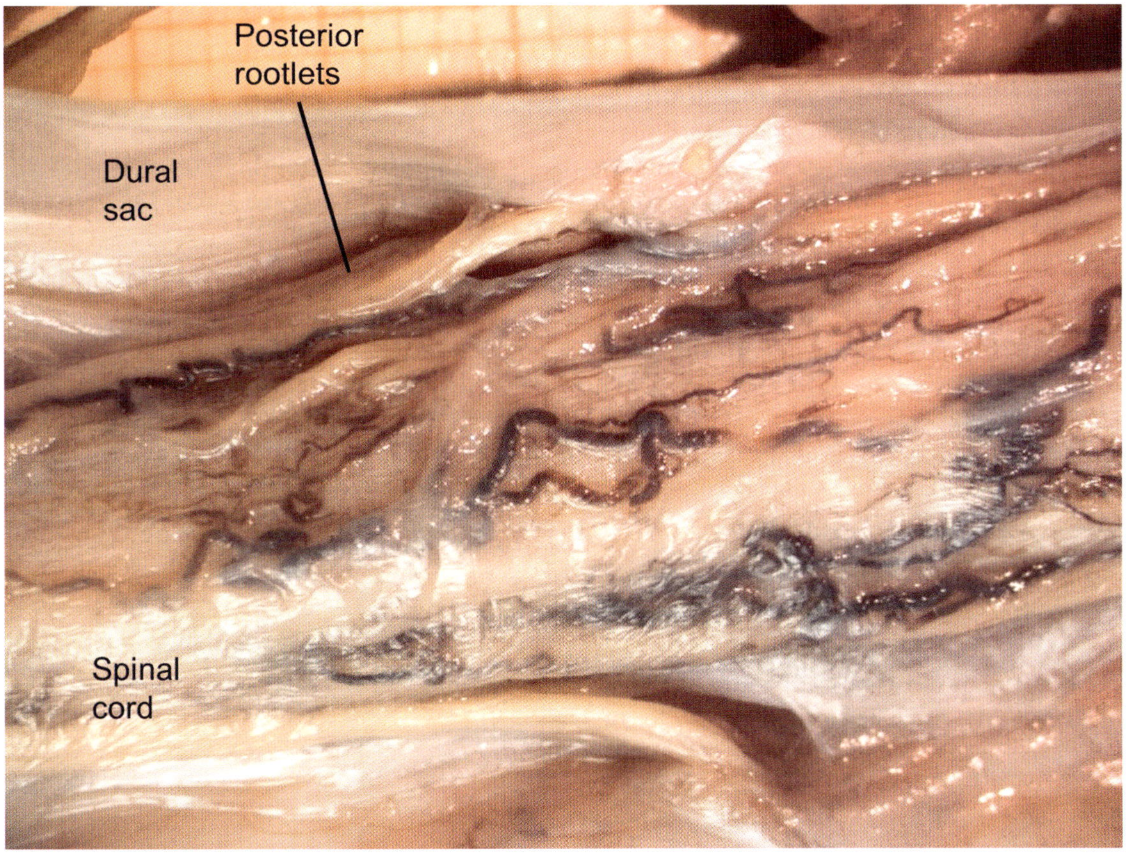

Fig. 19.24 Detail of the spinal cord and rootlets at the thoracic level (posterior surface)

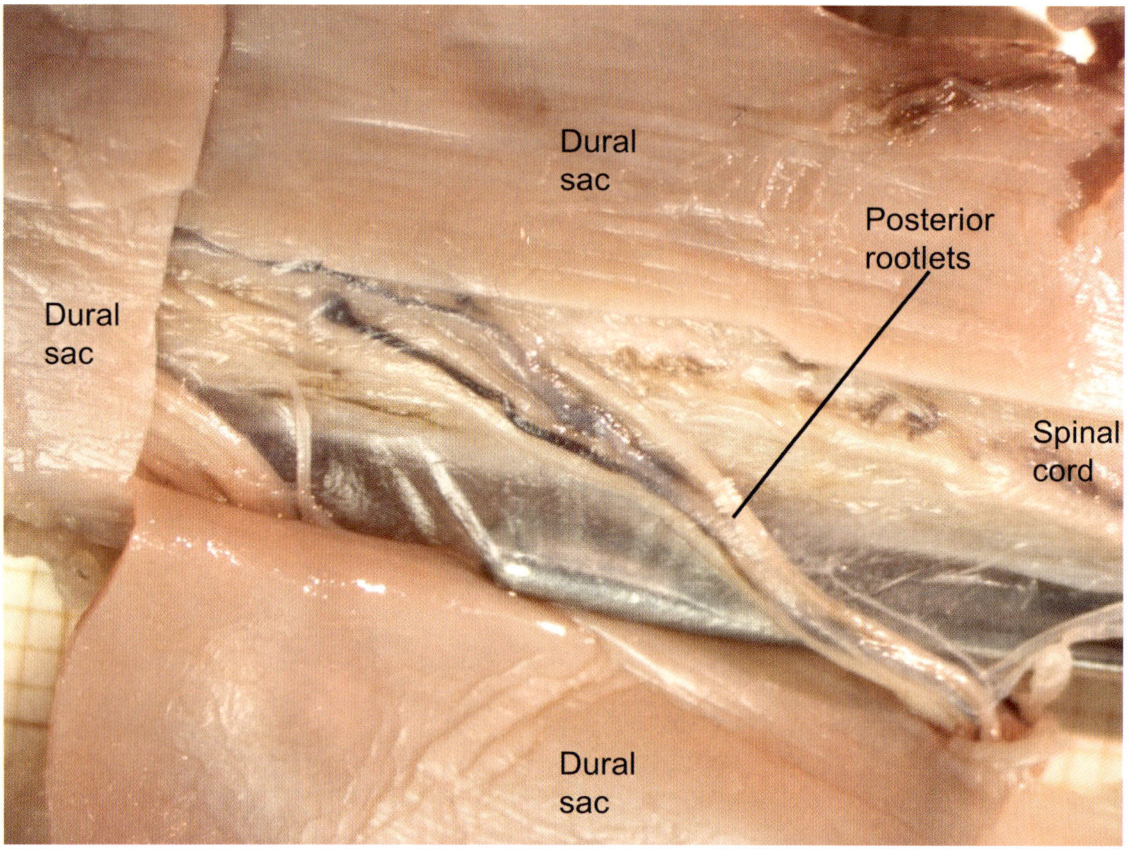

Fig. 19.25 Detail of the posterior surface of the spinal cord and rootlets at the thoracic level

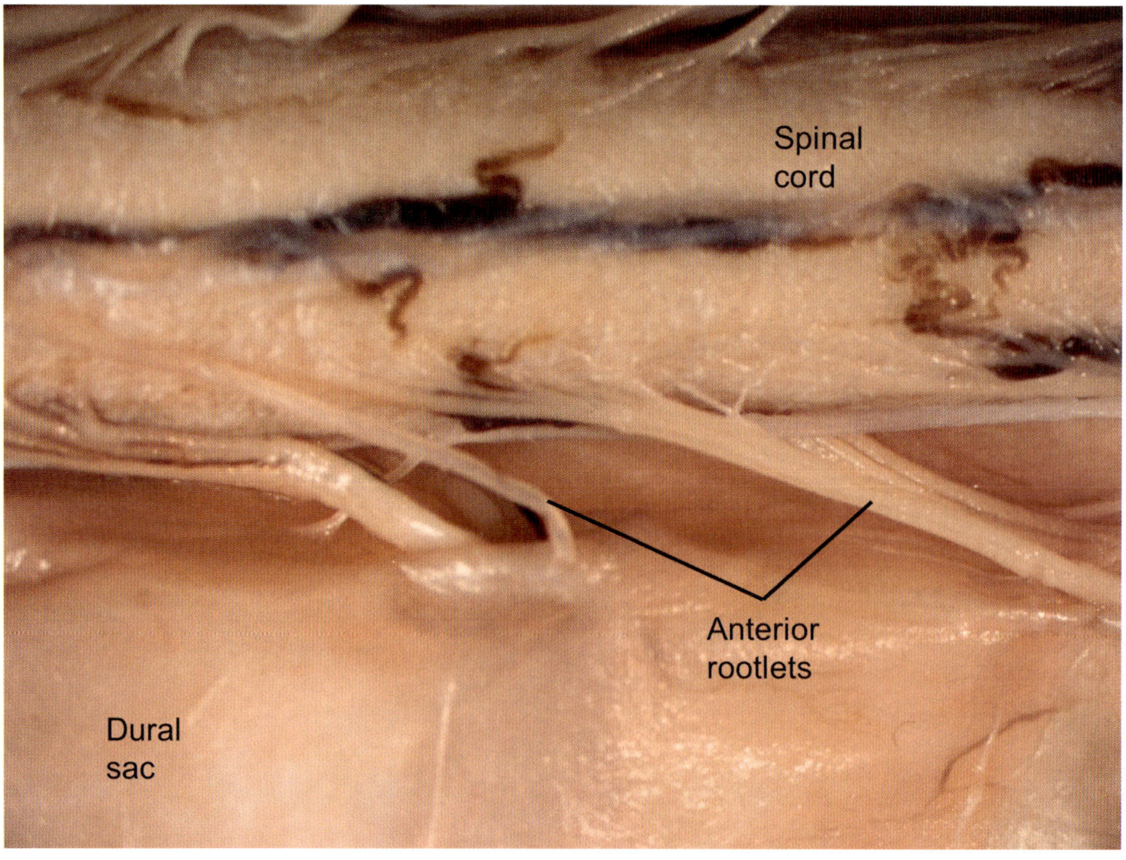

Fig. 19.26 Detail of the anterior surface of the spinal cord and rootlets at the thoracic level

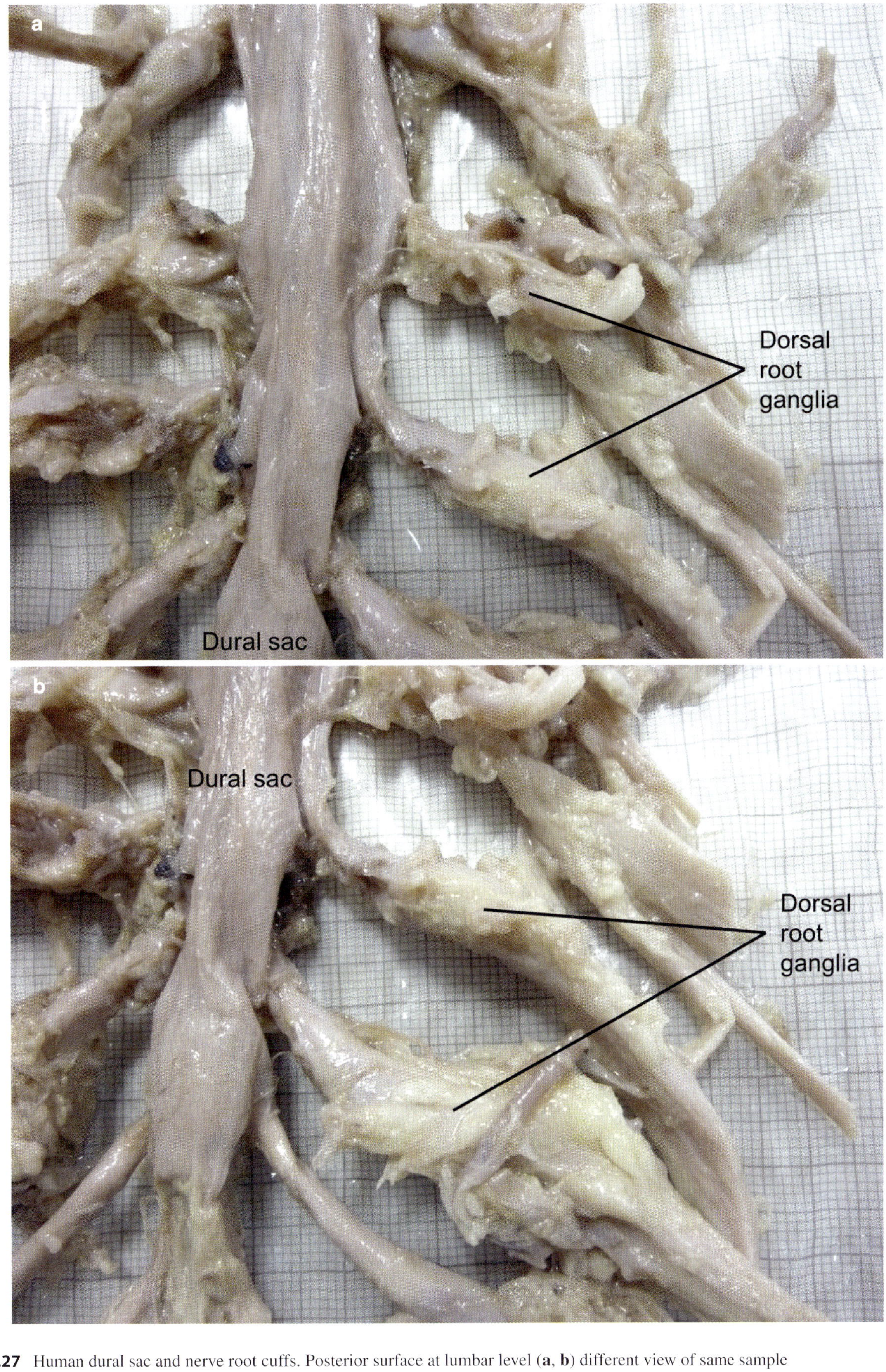

Fig. 19.27 Human dural sac and nerve root cuffs. Posterior surface at lumbar level (**a**, **b**) different view of same sample

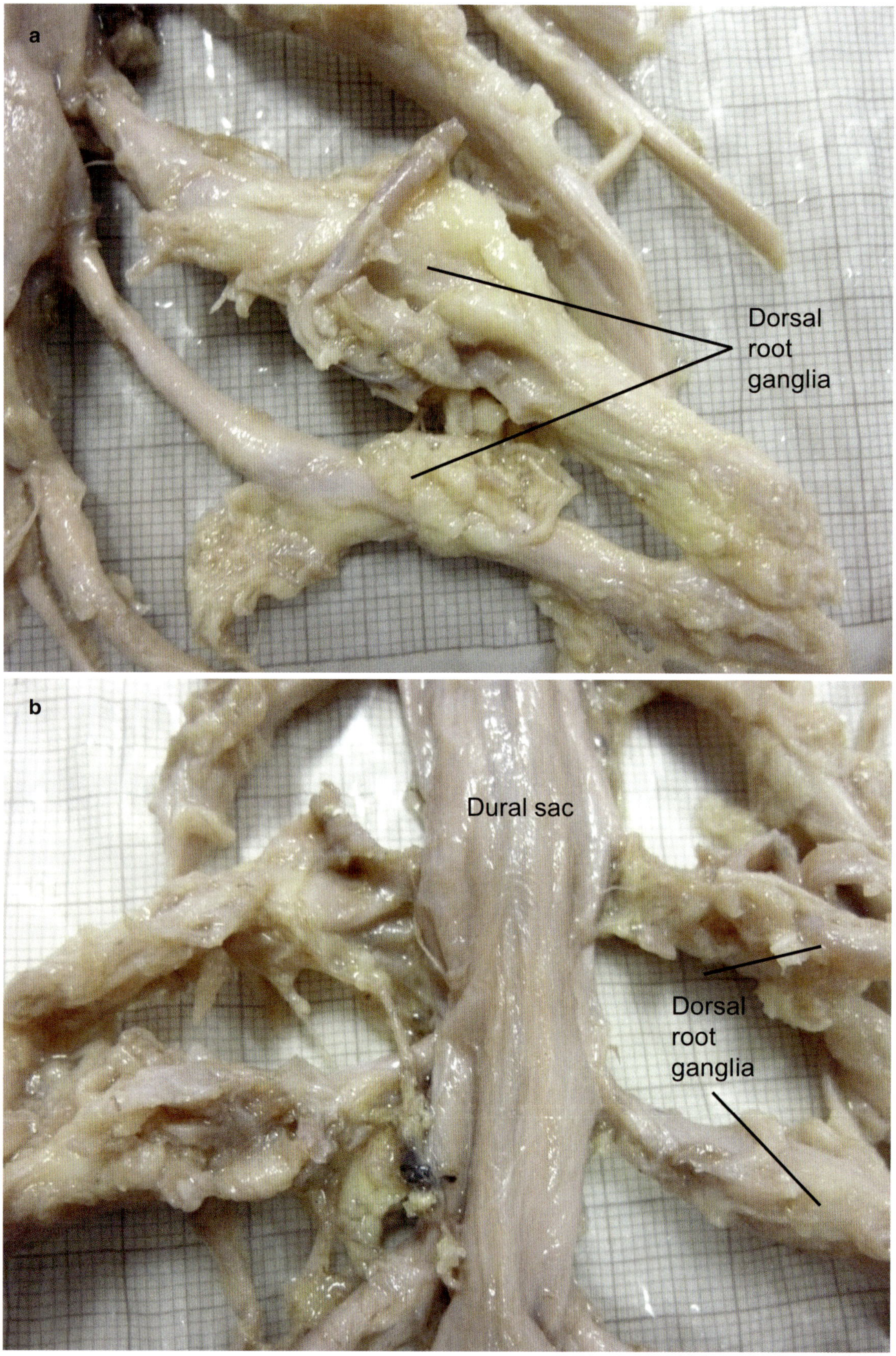

Fig. 19.28 Human dural sac and nerve root cuffs. Posterior surface at lumbar level (**a**, **b**) different view of same sample

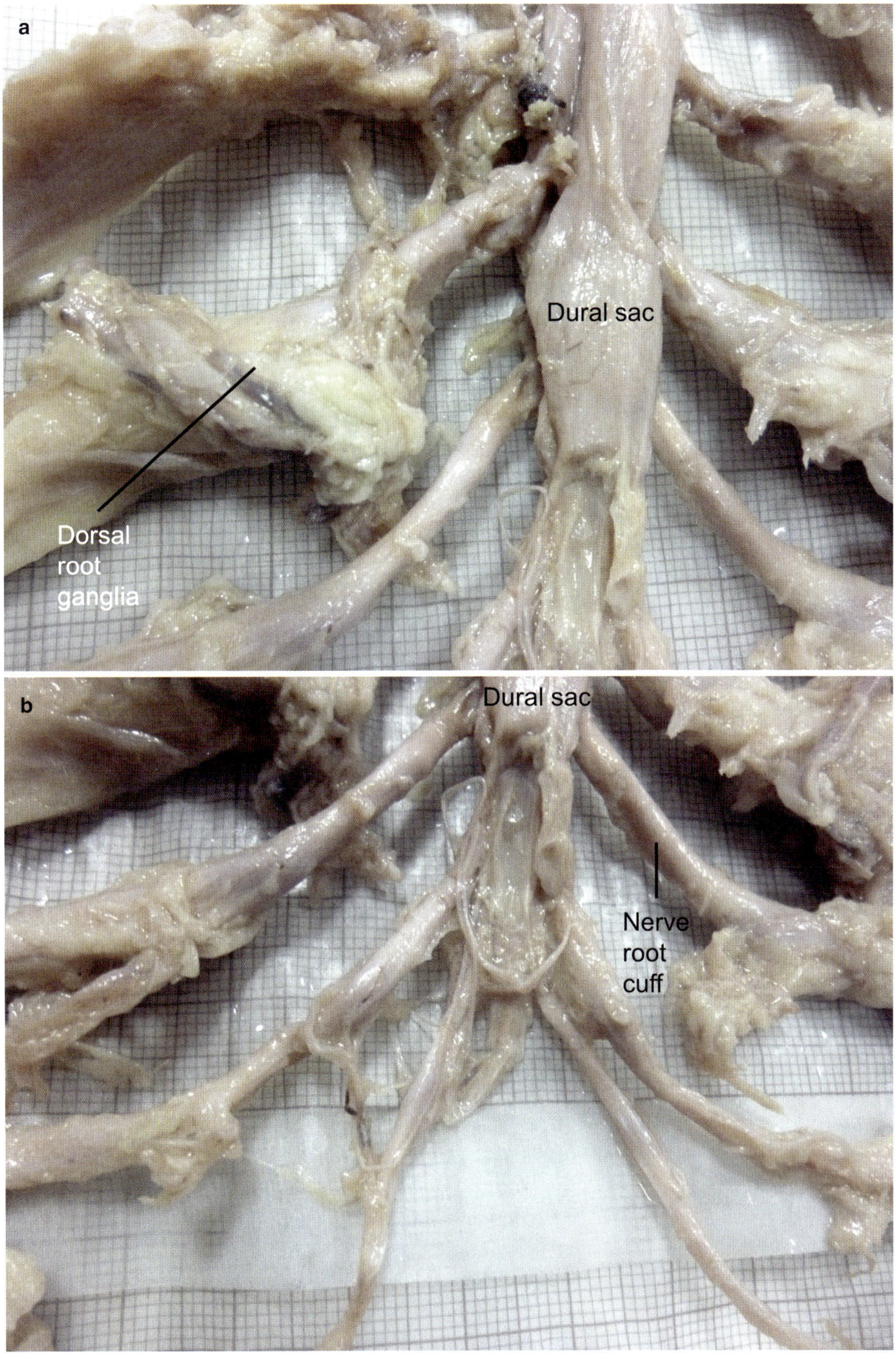

Fig. 19.29 Human dural sac and nerve root cuffs. Posterior surface at lumbar level (**a**, **b**) different view of same sample

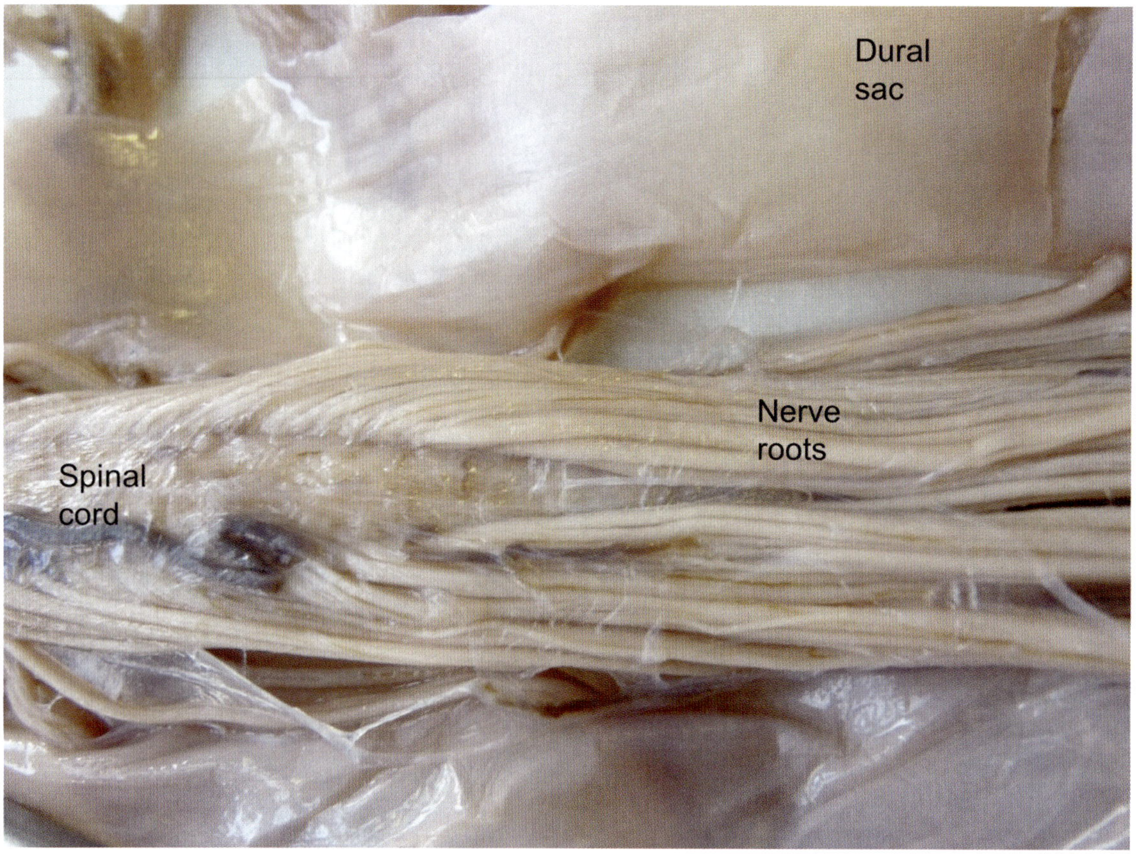

Fig. 19.30 Details of the posterior surface of a nerve root from the cauda equina at the lumbar level

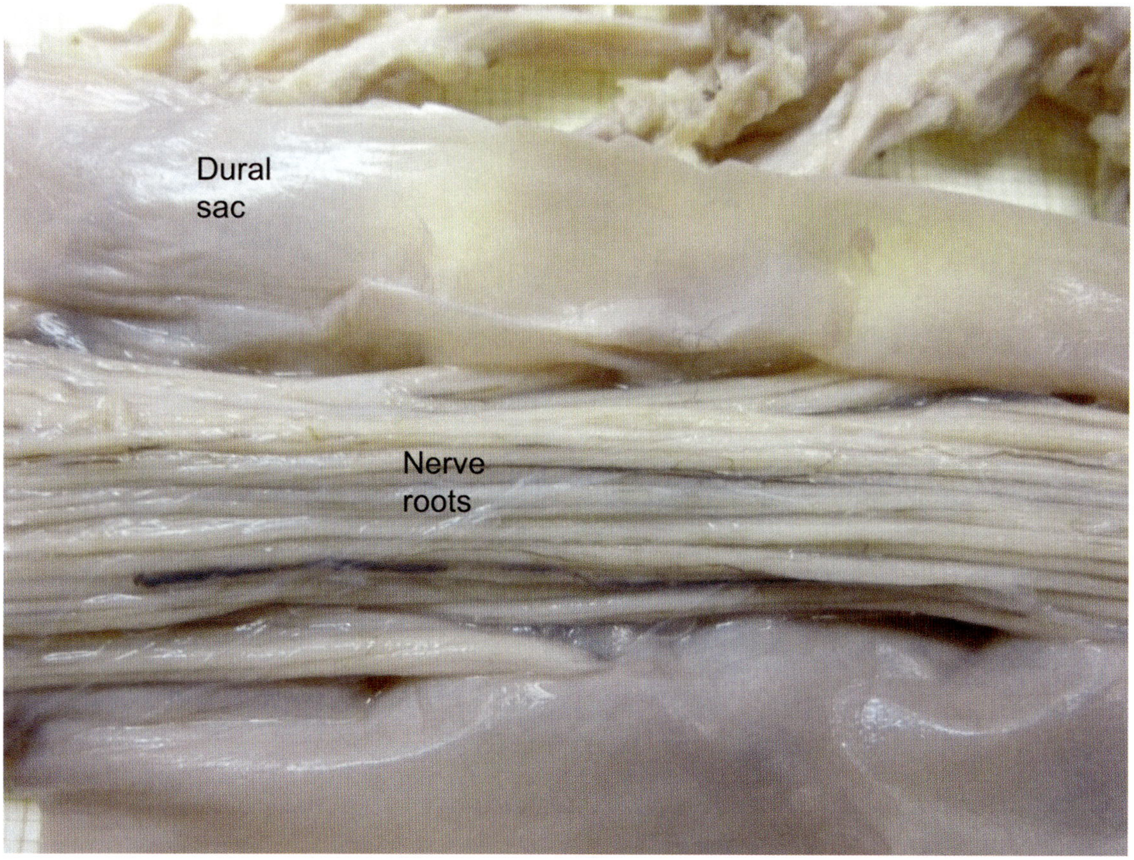

Fig. 19.31 Details of the posterior surface of a nerve root from the cauda equina at the lumbar level

References

1. Reina MA. Contributions of electron microscopy in regional anesthesia. Reg Anesth Pain Med Suppl. 2011;36:E81–5.
2. Reina MA, De Andrés JA, Hernández JM, Arriazu Navarro R, Durán Mateos EM, Prats-Galino A. Successive changes in extraneural structures from the subarachnoid nerve roots to the peripheral nerve, influencing anesthetic block, and treatment of acute postoperative pain. Eur J Pain Suppl. 2011;5:377–85.
3. Reina MA, De Andrés JA, Hernández JM, Prats-Galino A, Machés F, Peláez J. Transforaminal or translaminar approach for dorsal root ganglion and dorsal nerve root. Anatomical reason for technique decision. Eur J Pain Suppl. 2010;4:287–97.
4. Reina MA, Oliva A, Carrera A, Durán Mateos EM, Diamantopoulos Fernández J, Arriazu Navarro R. Origin of postdural puncture headache. Cir May Amb. 2011;16:72–84.
5. Hernández JM, Reina MA, Prats-Galino A, De Andrés JA. Morphological study of nerve root and types of needles used in transforaminal injections. Reg Anesth Pain Med. 2011;36:278–81.

Ultrastructure of Spinal Dura Mater

20

Miguel Angel Reina, Andrés López, Martin Dittmann, and José Antonio De Andrés

Dura mater is the outermost meningeal membrane enclosing the brain and spinal cord. At the spinal level, the dura mater is composed of collagen fibers and about 80 concentric dural laminas measuring approximately 250–400 μm [1–6]. Each dural lamina has a thickness of 5 μm and is formed by thinner laminas (approximately 8–12 subunits). Collagen fibers are the laminas' main component. These collagen fibers are oriented in various directions among the subunits that form each dural lamina, but they are always within the concentric plane of each dural lamina [1–10]. The fibers do not traverse different dural laminas. Each collagen fiber measures approximately 0.1 μm and presents a smooth surface. Elastic fibers are also present in lesser proportion. These fibers measure 2 μm in diameter and have a rough surface. Looking at the dural sac from within the epidural space, the external surface of the outer dural lamina can be seen. In these dural laminas, the fibers do not run longitudinal and parallel to the axis of the vertebral column, as was stated in previous descriptions. At the level of nerve root cuffs, the dura mater is thinner, measuring approximately 80 μm in thickness (Figs. 20.1, 20.2, 20.3, 20.4, 20.5, 20.6, 20.7, 20.8, 20.9, 20.10, 20.11, 20.12, 20.13, 20.14, 20.15, 20.16, 20.17, 20.18, 20.19, 20.20, 20.21, 20.22, and 20.23).

M.A. Reina, MD, PhD (✉)
Department of Clinical Medical Sciences and Institute of Applied Molecular Medicine, School of Medicine, University of CEU San Pablo, Madrid, Spain

Department of Anesthesiology, Madrid-Montepríncipe University Hospital, Madrid, Spain
e-mail: miguelangel@perticone.e.telefonica.net

A. López, MD
Department of Clinical Medical Sciences, School of Medicine, CEU San Pablo University, Madrid, Spain

Department of Anesthesiology, Madrid-Montepríncipe University Hospital, Madrid-Torrelodones University Hospital, Madrid-Sanchinarro University Hospital, Madrid, Spain
e-mail: alg00009@telefonica.net

M. Dittmann, MD, PhD
Department of Anesthesia, Kreiskrankenhaus, Bad Säckingen, Germany
e-mail: mcdittmann@gmail.com

J.A. De Andrés, MD, PhD
Department of Surgical Specialties, School of Medicine, University of Valencia, Valencia, Spain

Department of Anesthesia, Critical Care, and Pain Management, General University Hospital, Valencia, Spain
e-mail: deandres_jos@gva.es

M.A. Reina et al. (eds.), *Atlas of Functional Anatomy for Regional Anesthesia and Pain Medicine: Human Structure, Ultrastructure and 3D Reconstruction Images*, DOI 10.1007/978-3-319-09522-6_20,

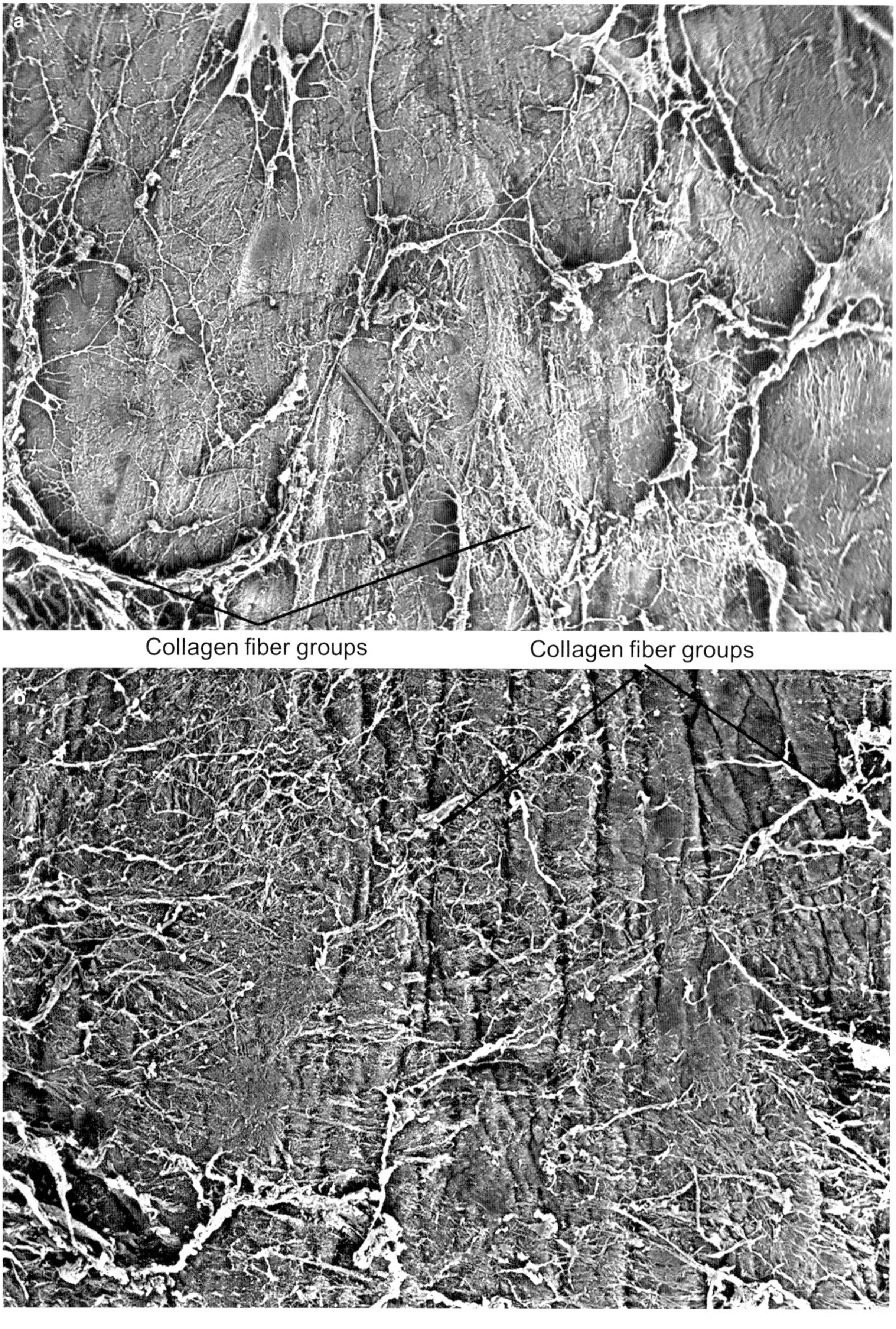

Fig. 20.1 Human spinal dura mater. External surface. Scanning electron microscopy. (**a**) Magnification ×12 (From Reina et al. [1]; with permission.) (**b**) Magnification ×50 (From Reina et al. [6]; with permission)

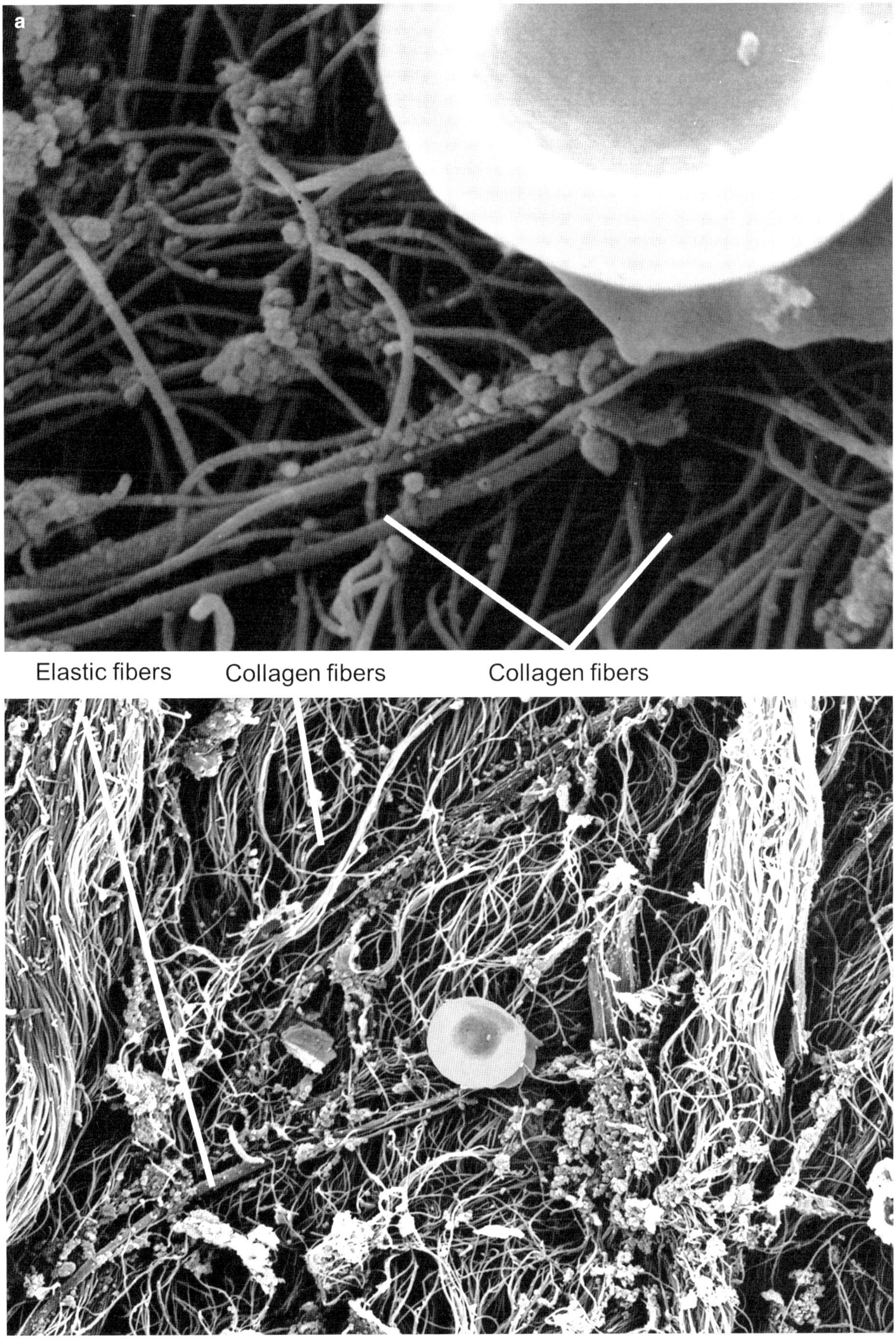

Fig. 20.2 Human spinal dura mater. External surface. Scanning electron microscopy. Magnification: (**a**) ×11,000; (**b**) ×2,000

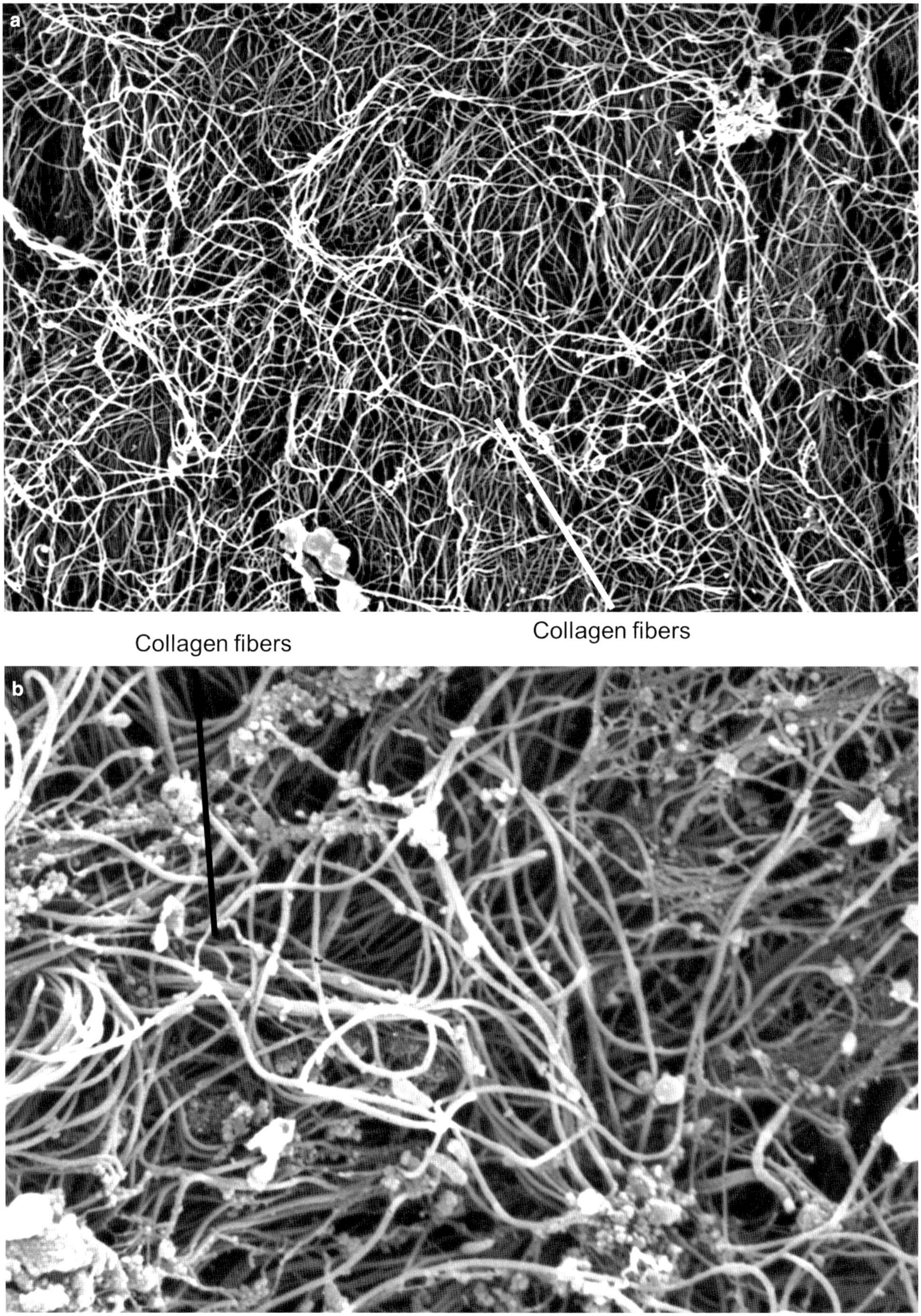

Fig. 20.3 Human spinal dura mater. External surface: Detail of the orientation of collagen fibers. Scanning electron microscopy. (**a**) Magnification ×2,000. (**b**) Magnification ×6,500 (**b** from Dittmann et al. [2]; with permission)

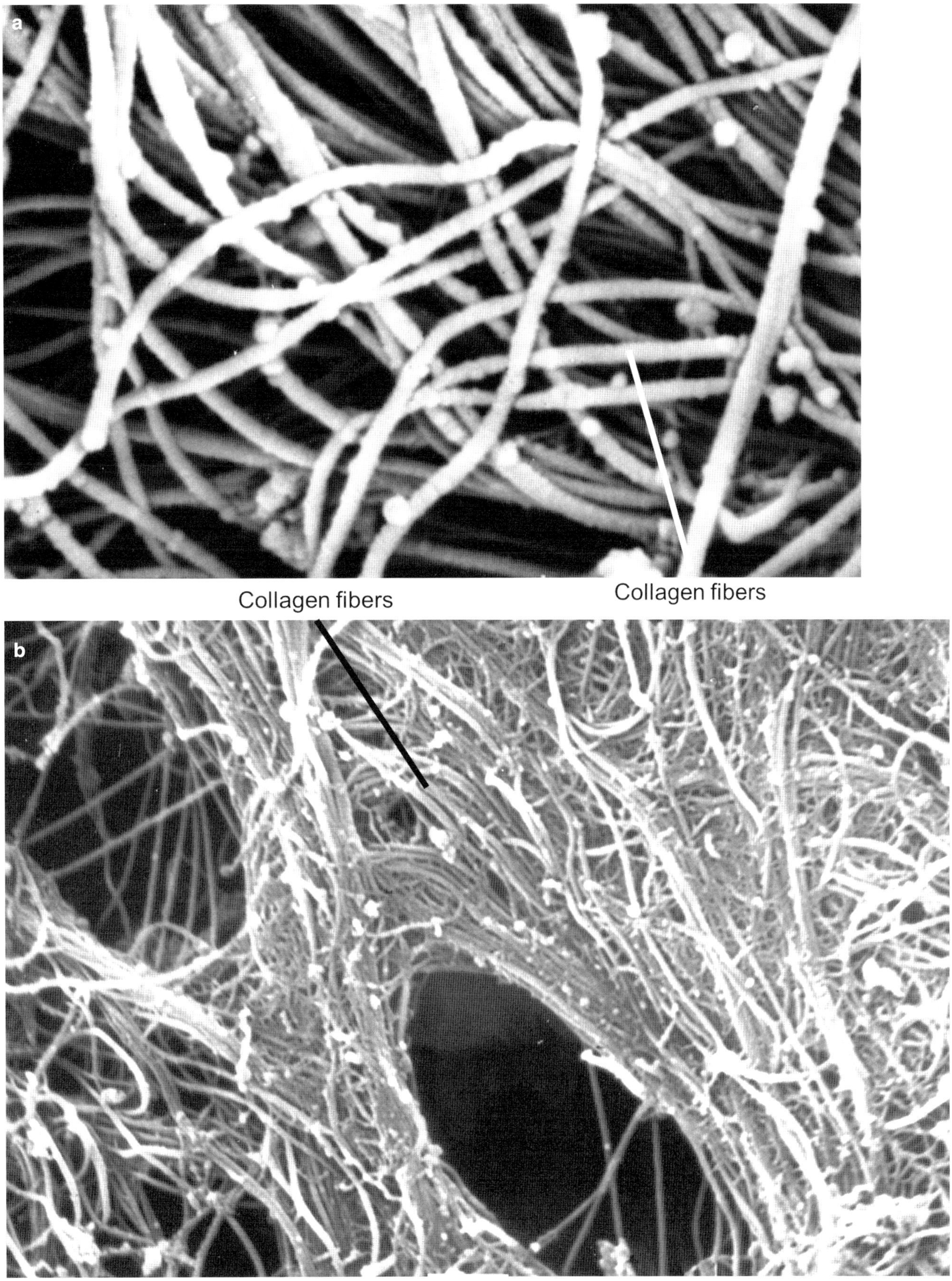

Fig. 20.4 Human spinal dura mater. External surface: Detail of the orientation of collagen fibers. Scanning electron microscopy. (**a**) Magnification ×20,000; (**b**) Magnification ×8,000 (**a**, **b** from Reina et al. [3]; with permission)

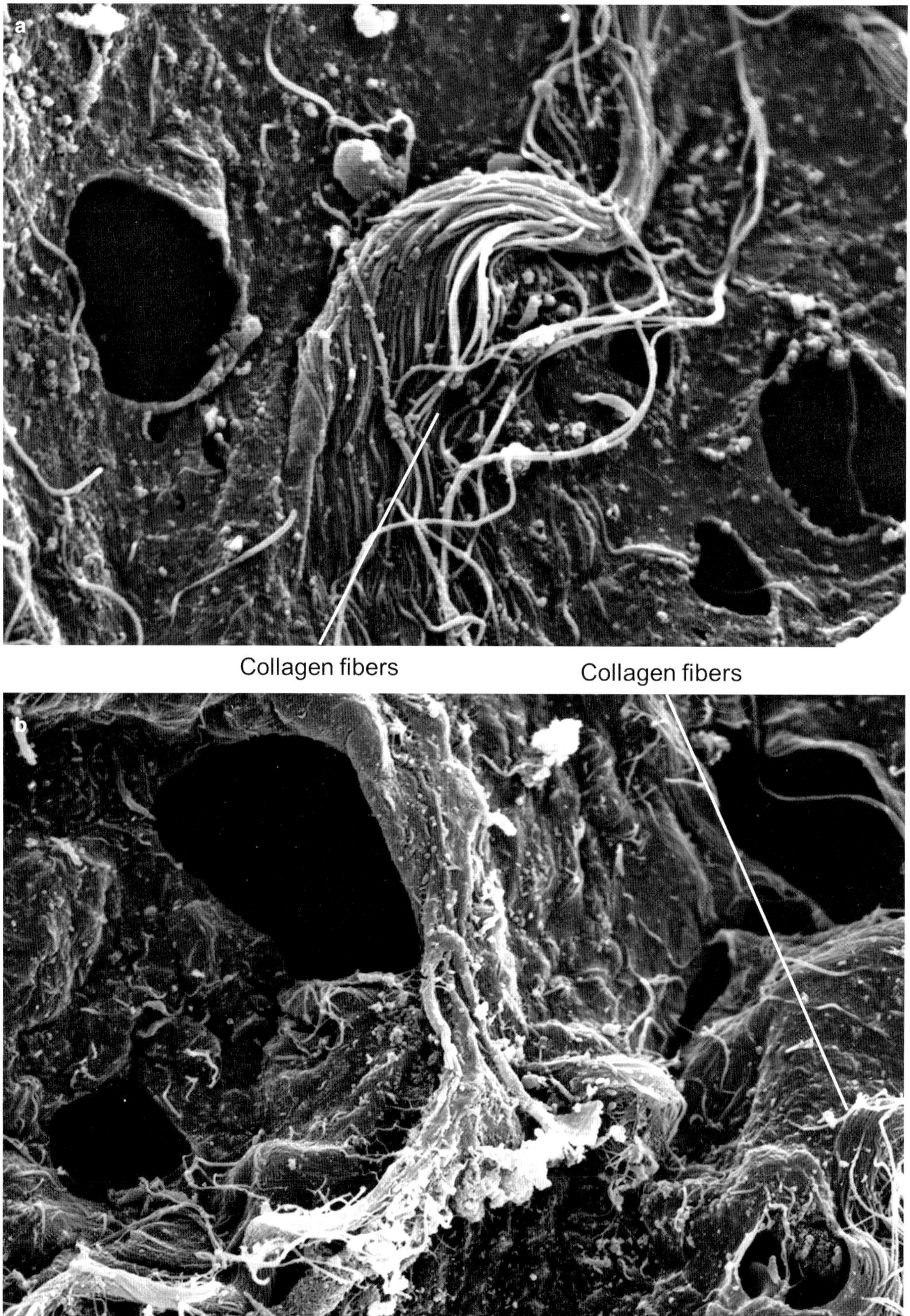

Fig. 20.5 Human spinal dura mater. External surface: Compact layer of dural fibers in detail. Scanning electron microscopy. Magnification: (**a**) ×45,000; (**b**) ×2,200 (**a** from Reina et al. [3]; with permission)

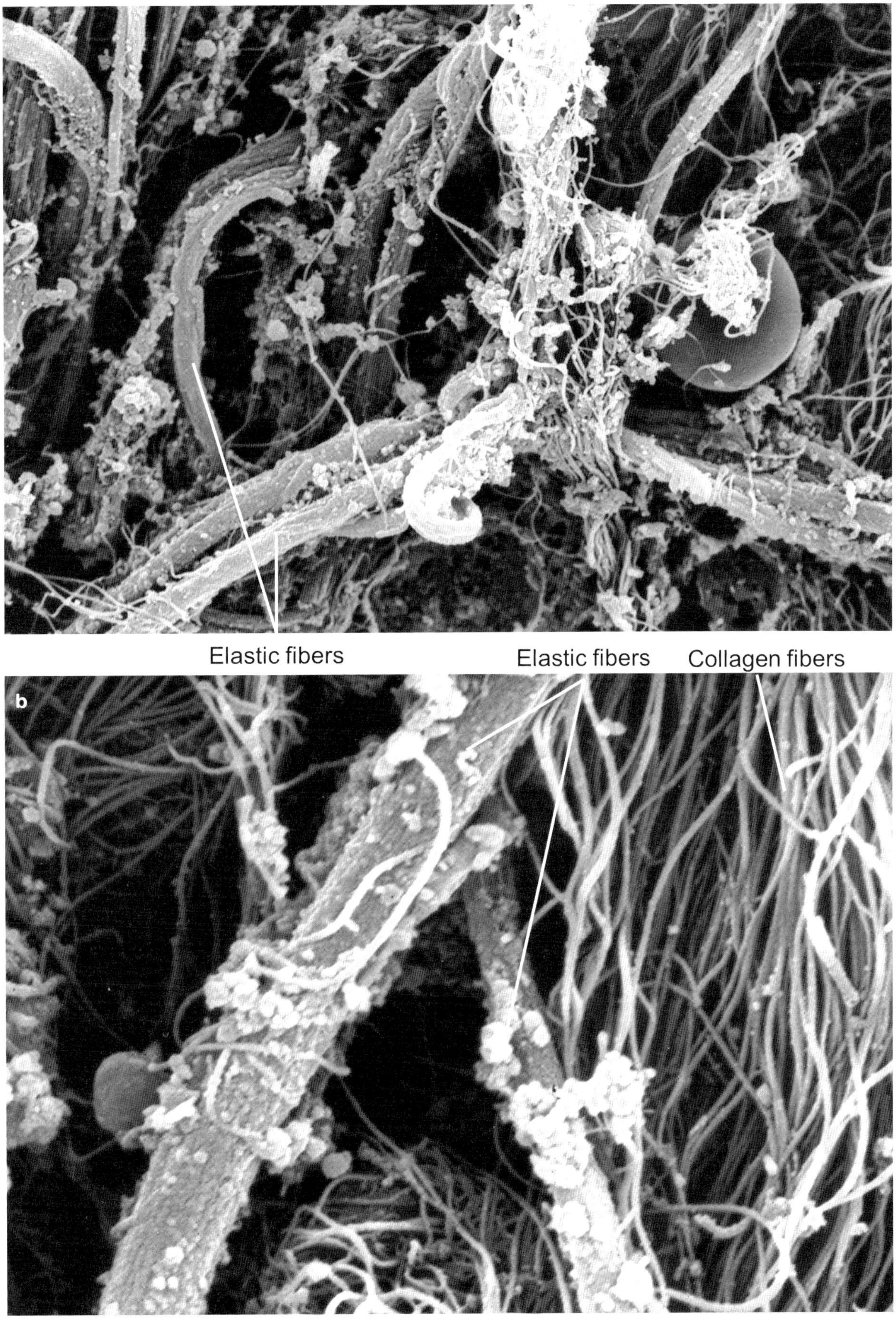

Fig. 20.6 Human spinal dura mater. External surface: Detail of elastic fibers. Scanning electron microscopy. (**a**) Magnification ×3,000 (From Dittmann et al. [2]; with permission). (**b**) Magnification ×6,500 (From Reina et al. [9]; with permission)

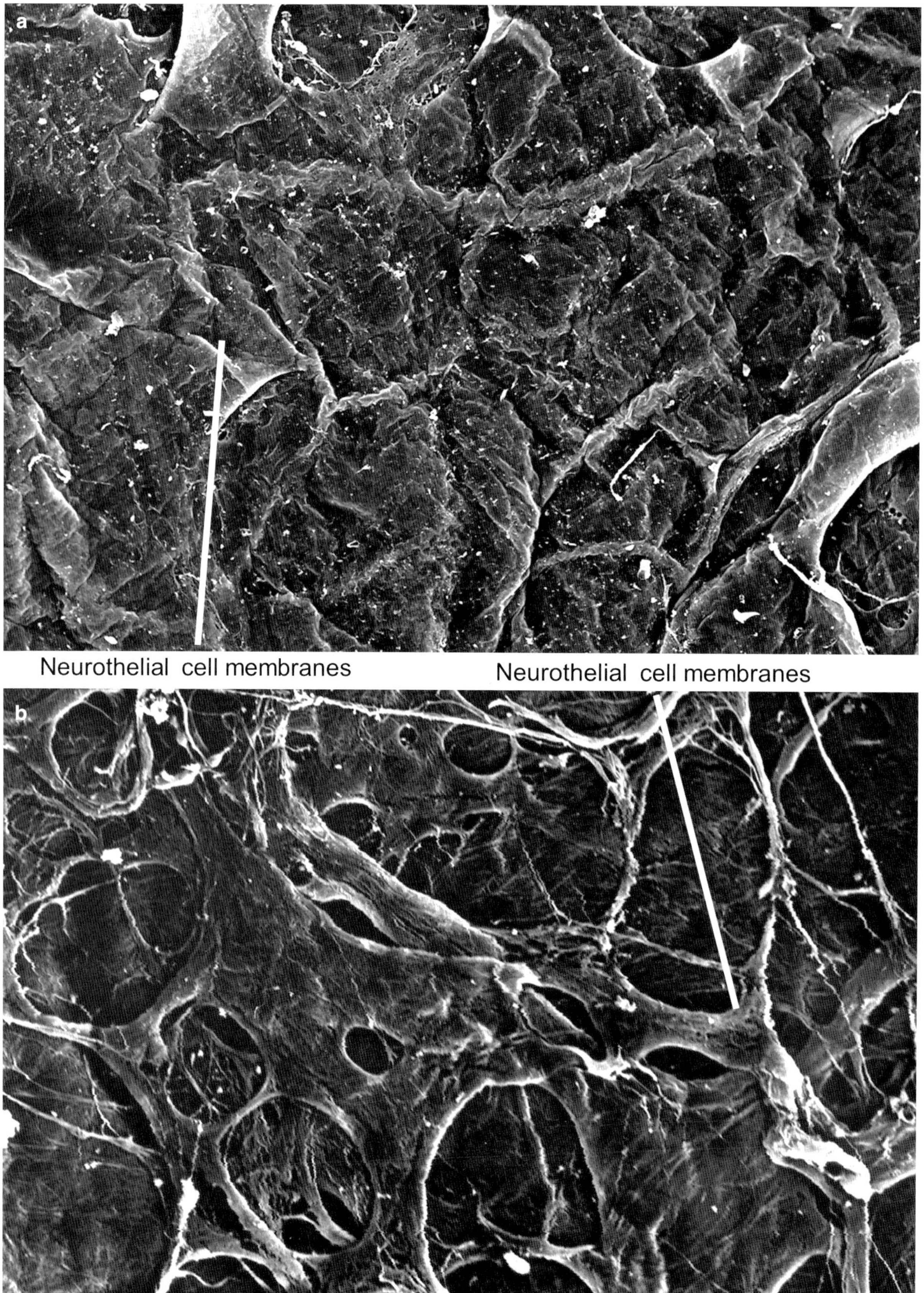

Fig. 20.7 Human spinal dura mater. Internal surface of dura mater adjacent to subdural compartment. Scanning electron microscopy. Magnification: (**a**) ×100; (**b**) ×1,100 (**b** from Reina et al. [1]; with permission)

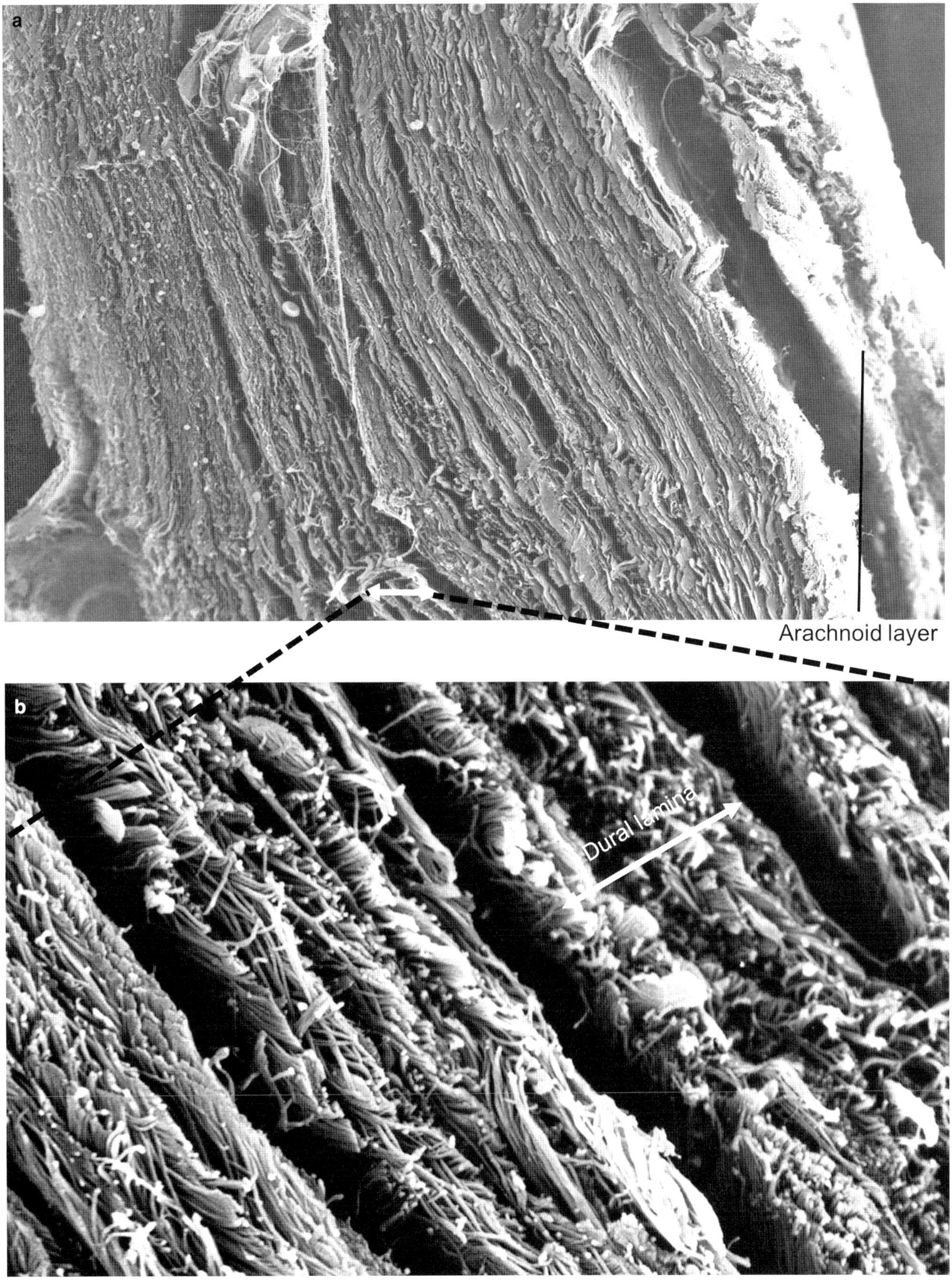

Fig. 20.8 Human spinal dura mater. (**a**) Full thickness (magnification ×300). (**b**) Detail of thickness of four dural laminas (magnification ×4,000). Scanning electron microscopy (**a** from Reina et al. [1]; **b** from Reina et al. [4]; both with permission)

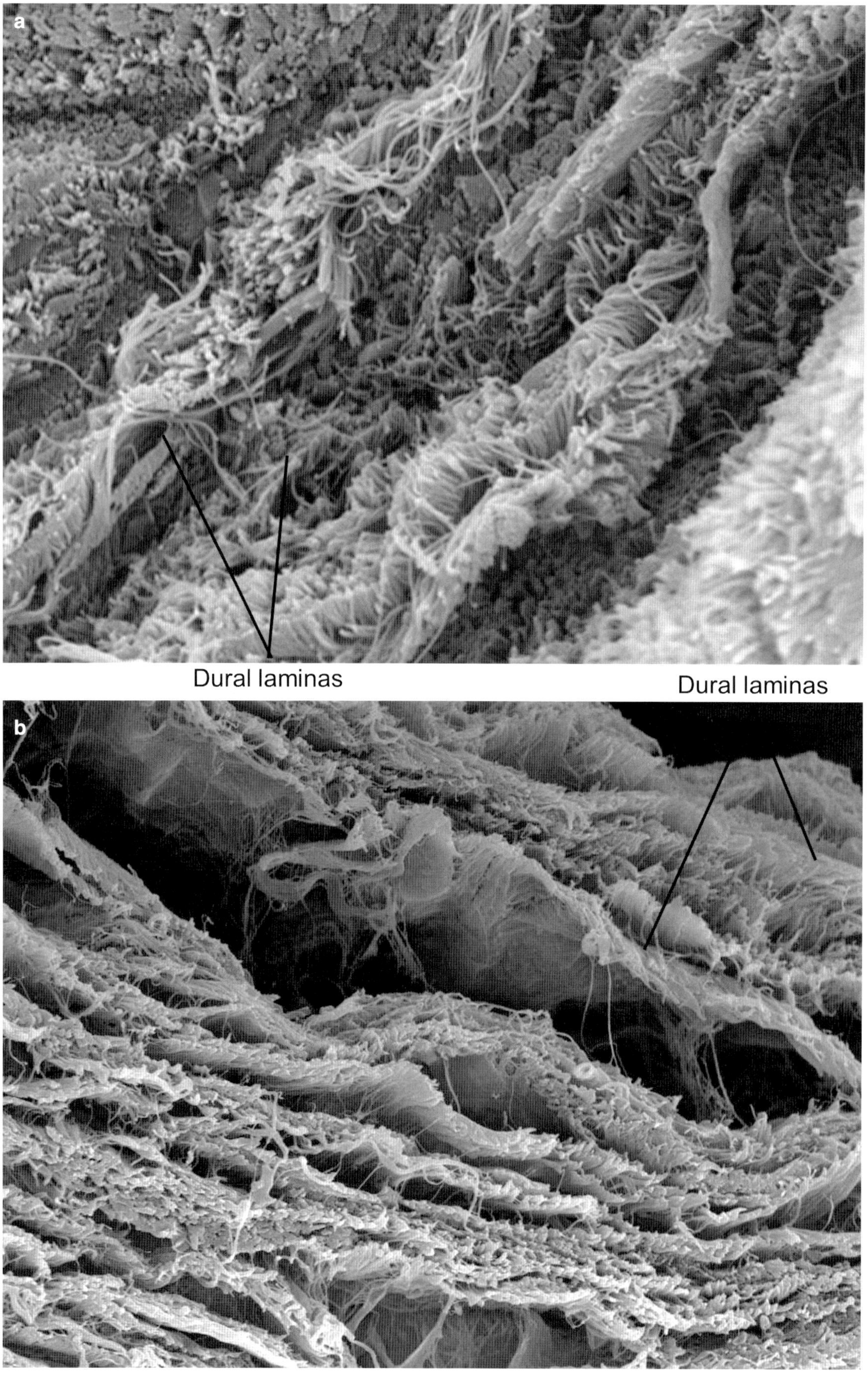

Fig. 20.9 Human spinal dura mater. Partial thickness. Detail of dural laminas. Scanning electron microscopy. Magnification: (**a**) ×4,000; (**b**) ×1,000

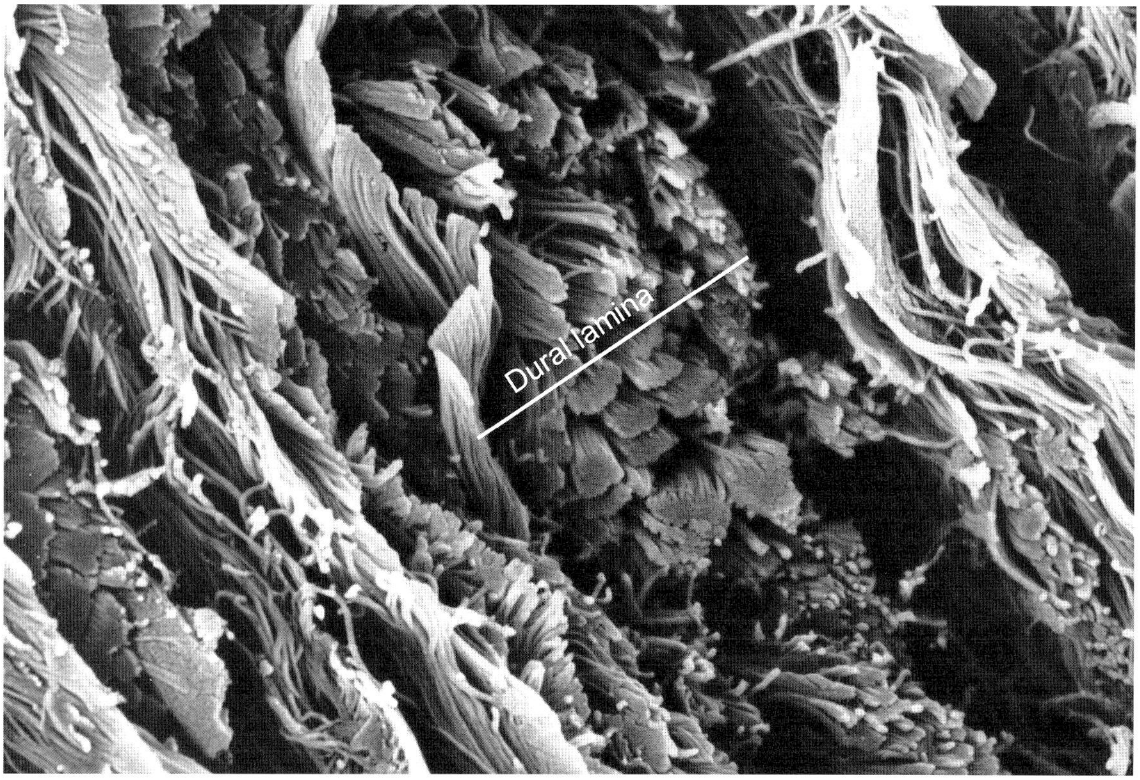

Fig. 20.10 Human spinal dura mater. Partial thickness. Detail of dural laminas. Scanning electron microscopy. Magnification ×5,500 (From Reina et al. [1]; with permission)

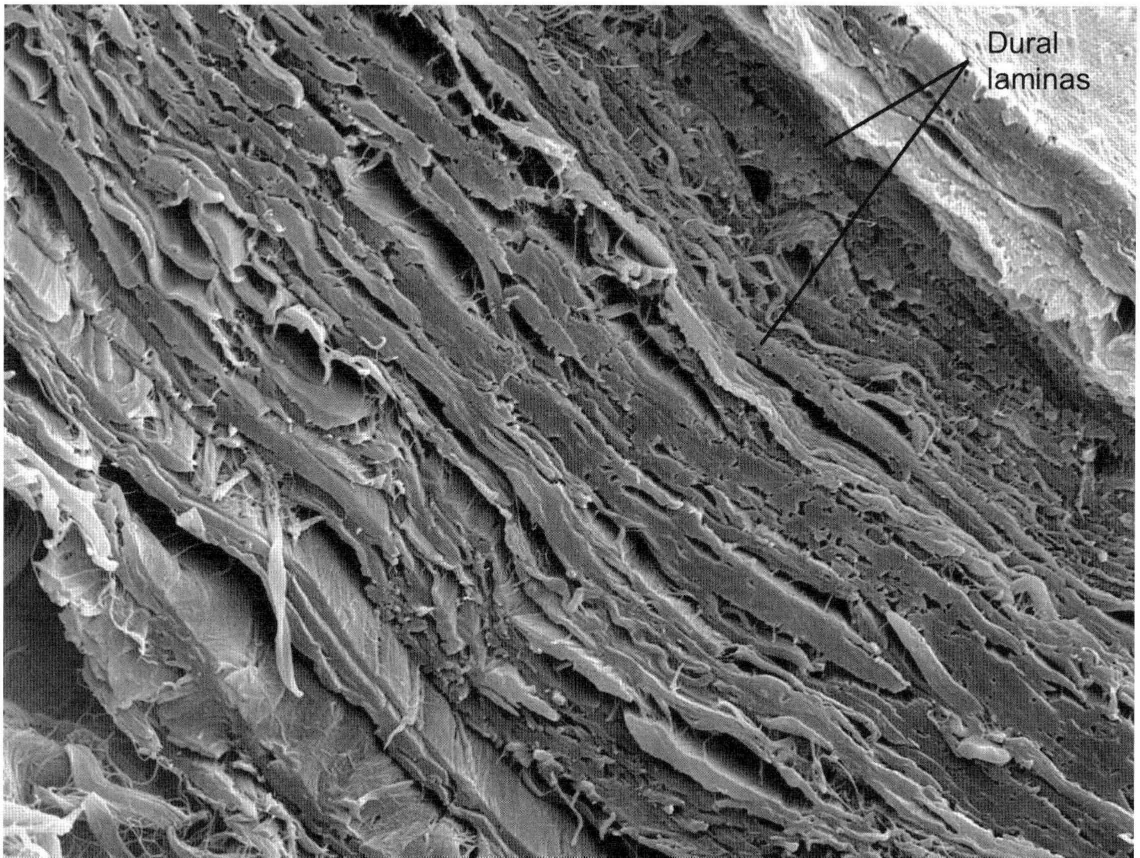

Fig. 20.11 Human spinal dura mater. Partial thickness. Detail of dural laminas. Scanning electron microscopy. Magnification ×1,000 (From De Andrés et al. [10]; with permission)

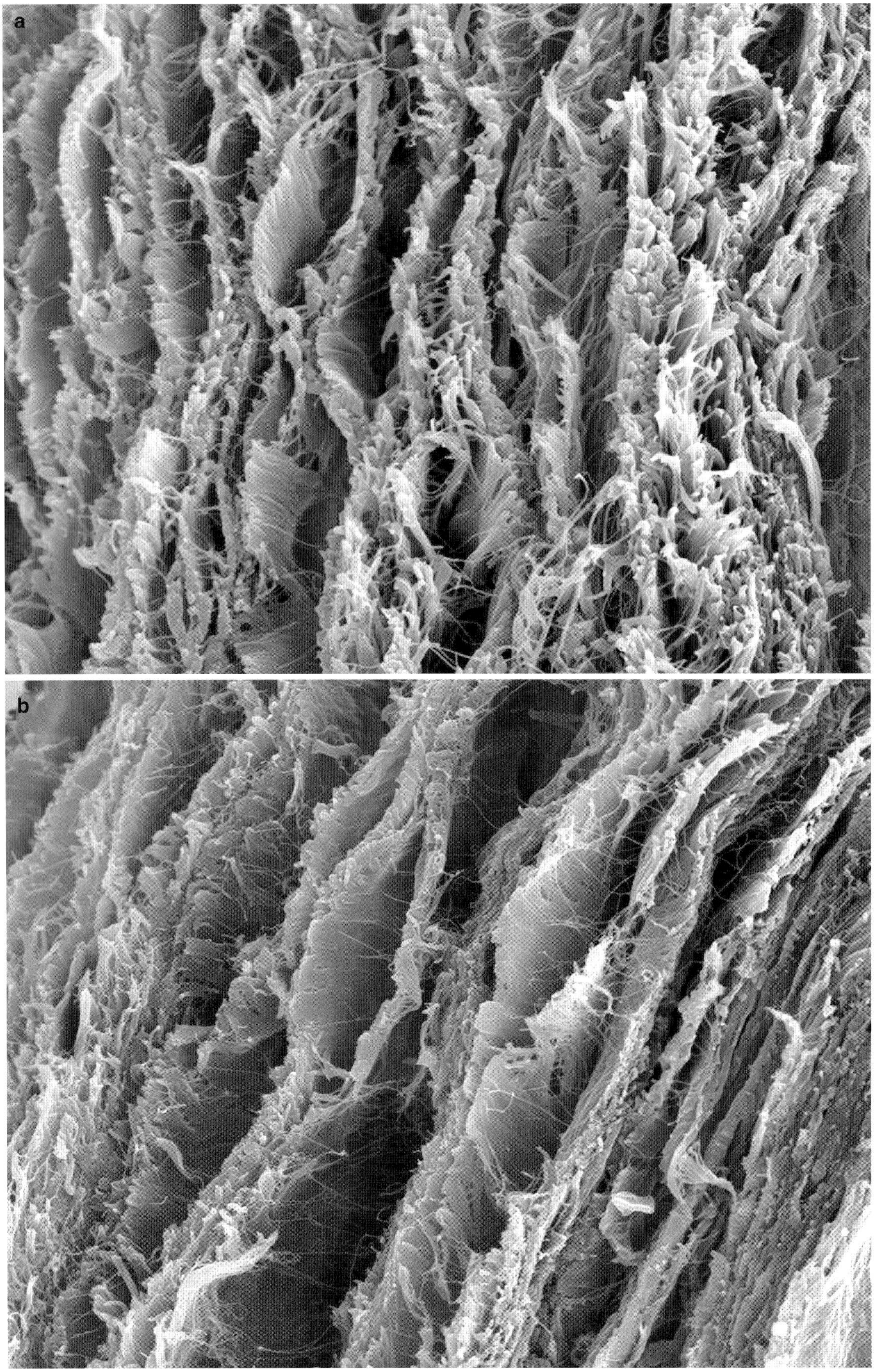

Fig. 20.12 Human spinal dura mater. Partial thickness. Detail of dural laminas. Scanning electron microscopy. Magnification: (**a**) ×1,500; (**b**) ×1,000

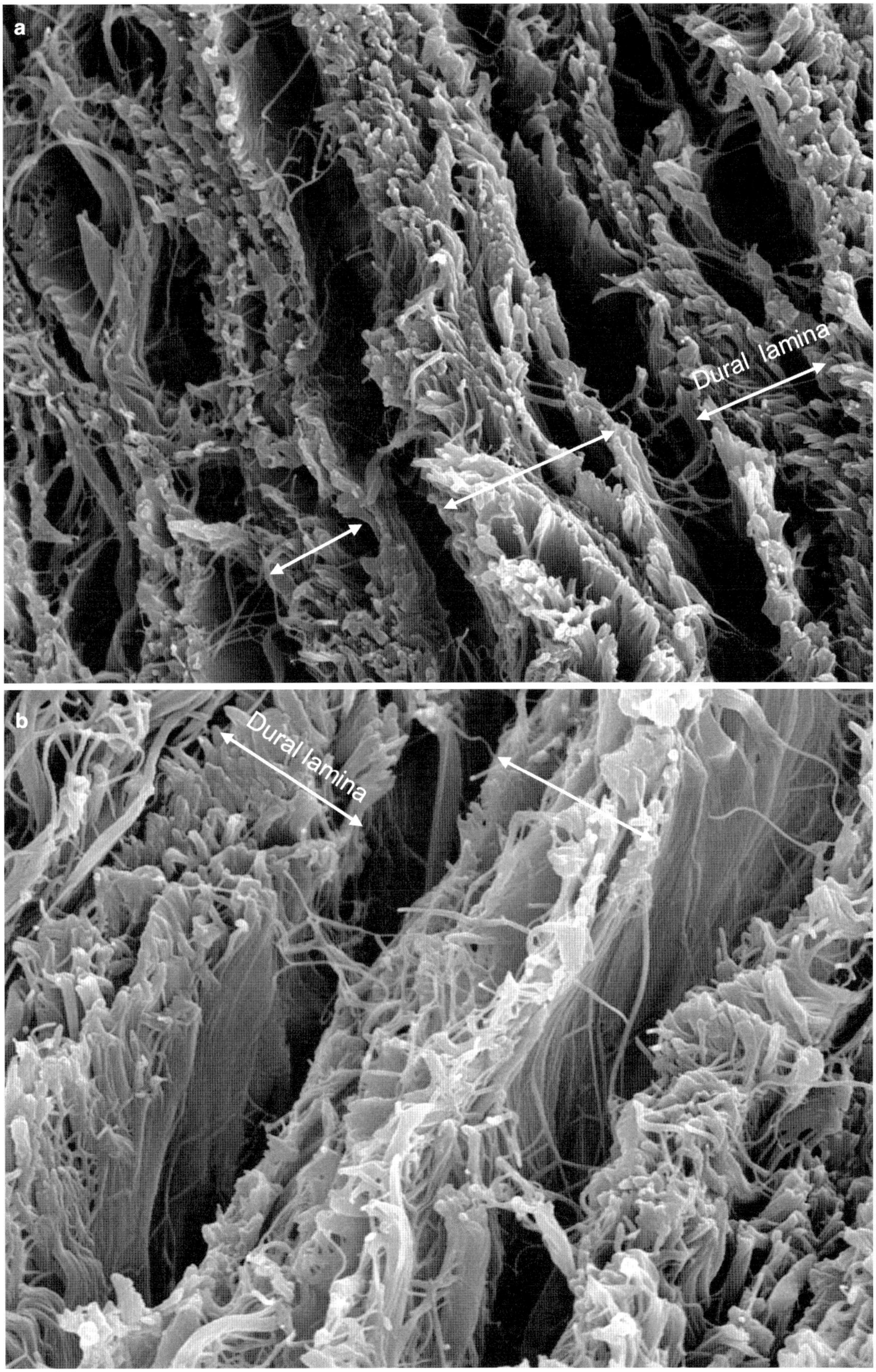

Fig. 20.13 Human spinal dura mater. Partial thickness. Detail of dural laminas. Scanning electron microscopy. Magnification: (**a**) ×2,000; (**b**) ×3,000

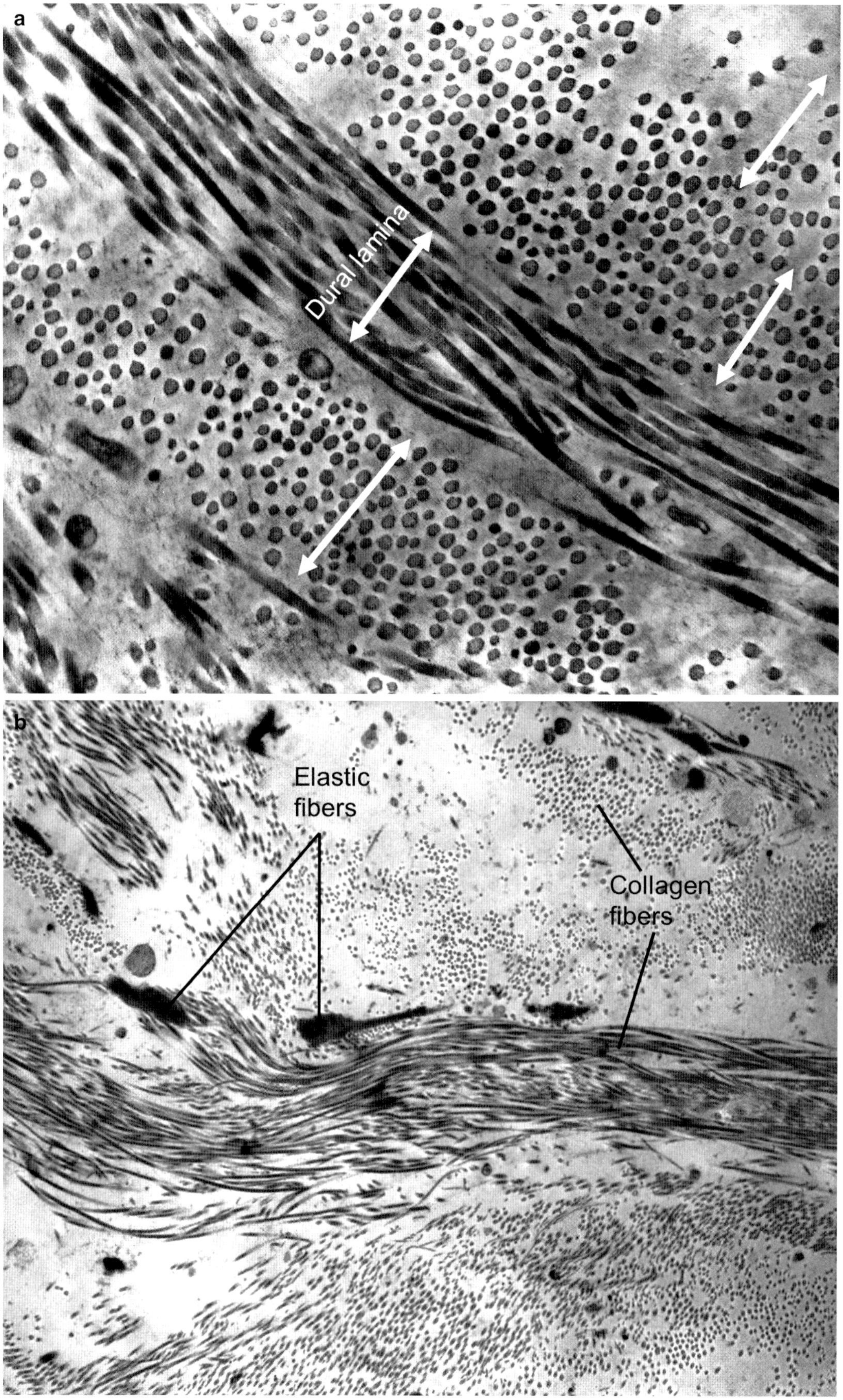

Fig. 20.14 Human spinal dura mater. Partial thickness. Detail of the different orientations of collagen fibers within dural laminas. Transmission electron microscopy. Magnification: (**a**) ×20,000; (**b**) ×7,000 (**a** from Reina et al. [7]; with permission)

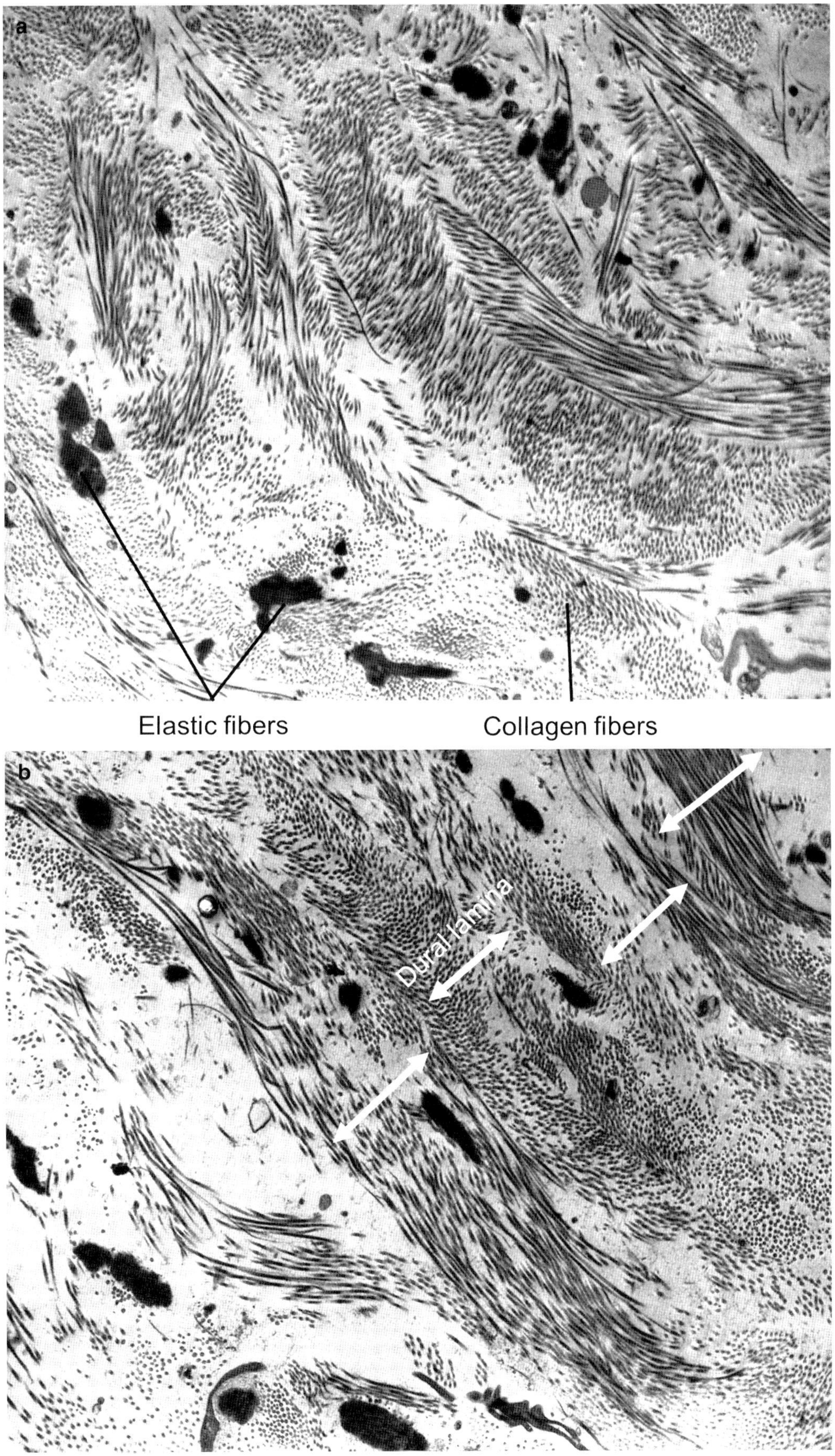

Fig. 20.15 Human spinal dura mater. Partial thickness. Detail of the different orientations of collagen fibers within dural laminas. Collagen and elastic fibers appear sectioned. Transmission electron microscopy. Magnification: (**a**, **b**) ×7,000

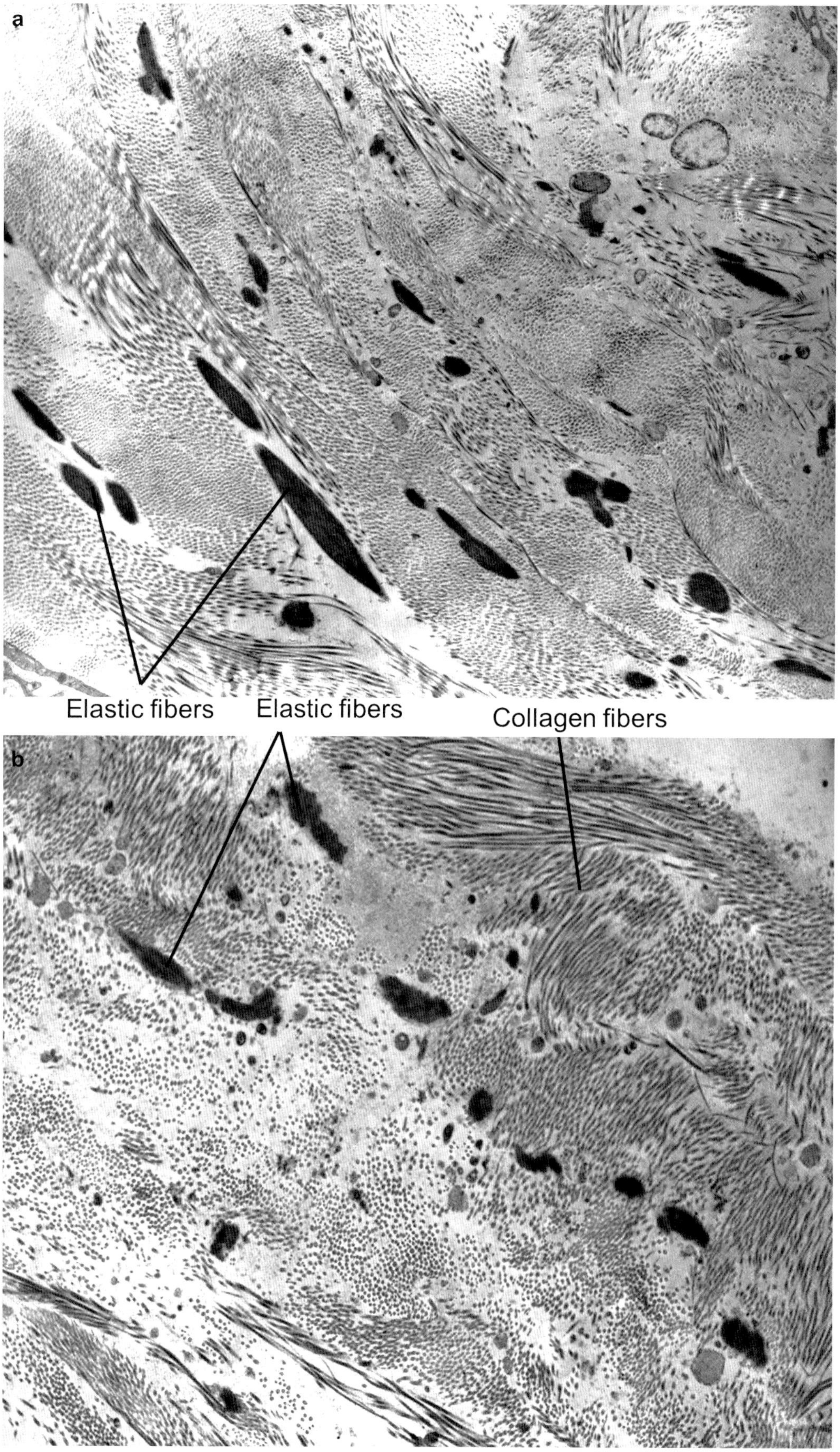

Fig. 20.16 Human spinal dura mater. Partial thickness. Detail of the different orientations of collagen fibers within dural laminas. Collagen and elastic fibers appear sectioned. Transmission electron microscopy. Magnification: (**a**) ×3,000; (**b**) ×7,000

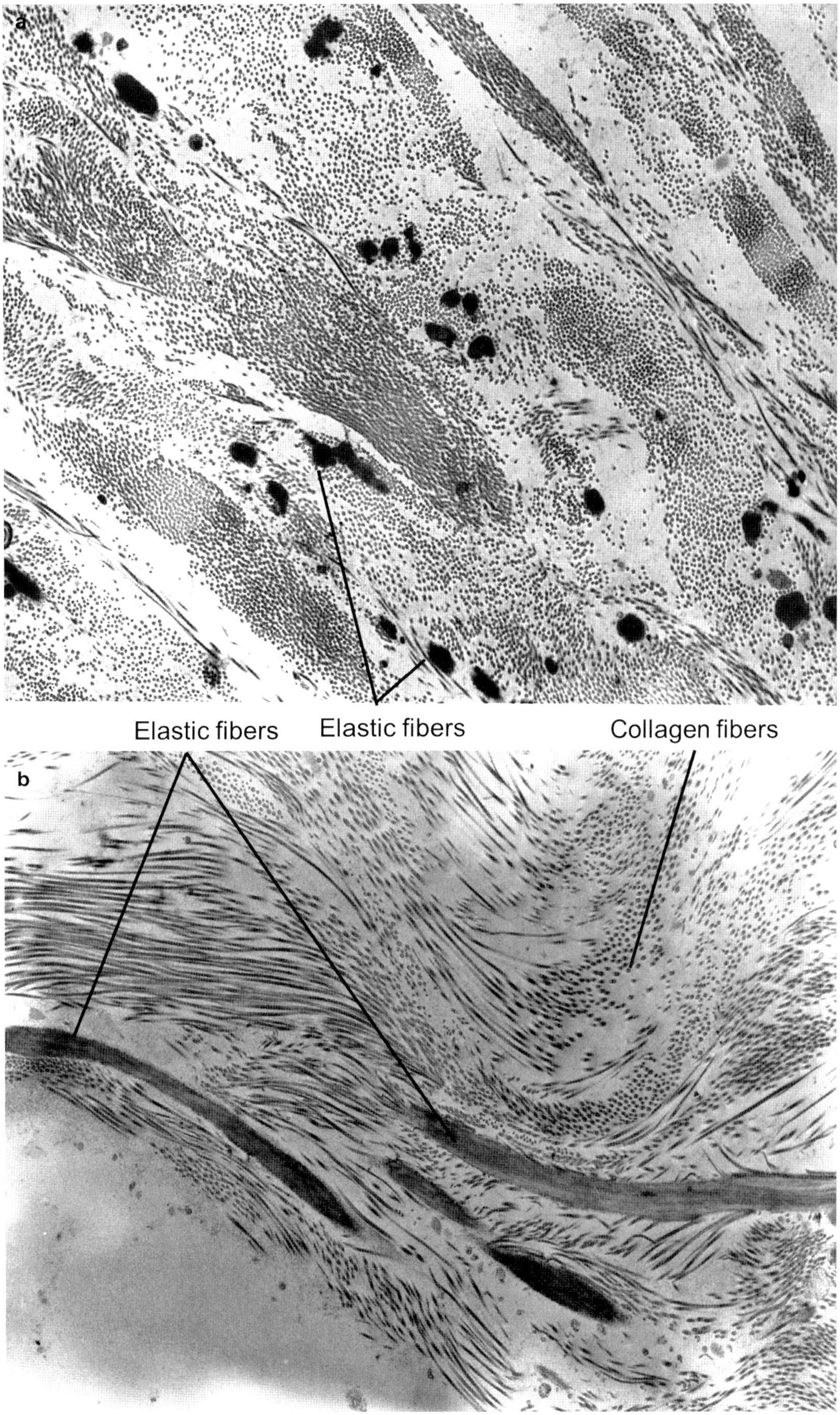

Fig. 20.17 Human spinal dura mater. Partial thickness. Detail of the different directions of collagen fibers within dural laminas. Collagen and elastic fibers appear sectioned. Transmission electron microscopy. Magnification: (**a**, **b**) ×7,000

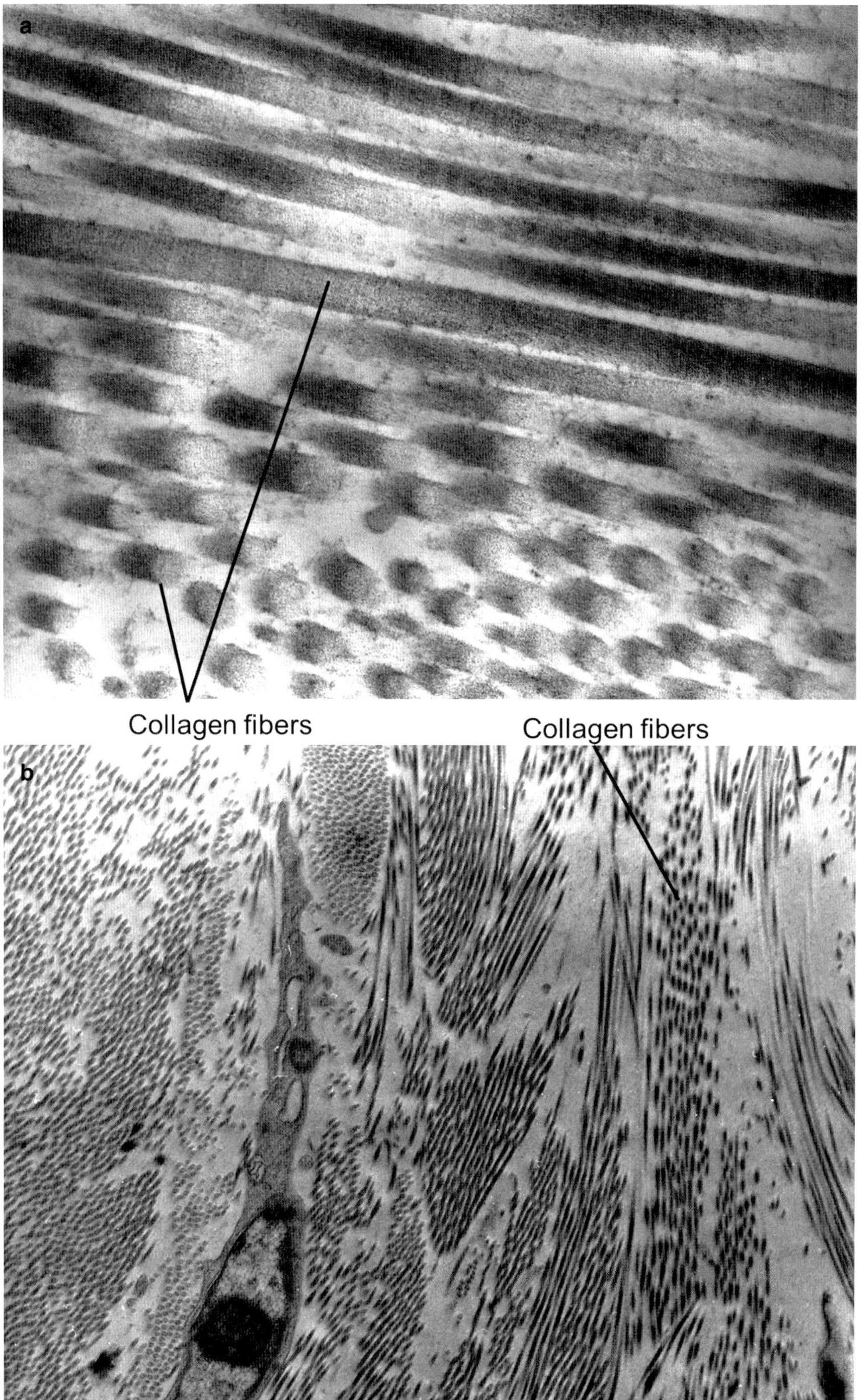

Fig. 20.18 Human spinal dura mater. Detail of collagen fibers. Transmission electron microscopy. Magnification: (**a**) ×80,000; (**b**) ×7,000 (**a** from Reina et al. [6]; with permission)

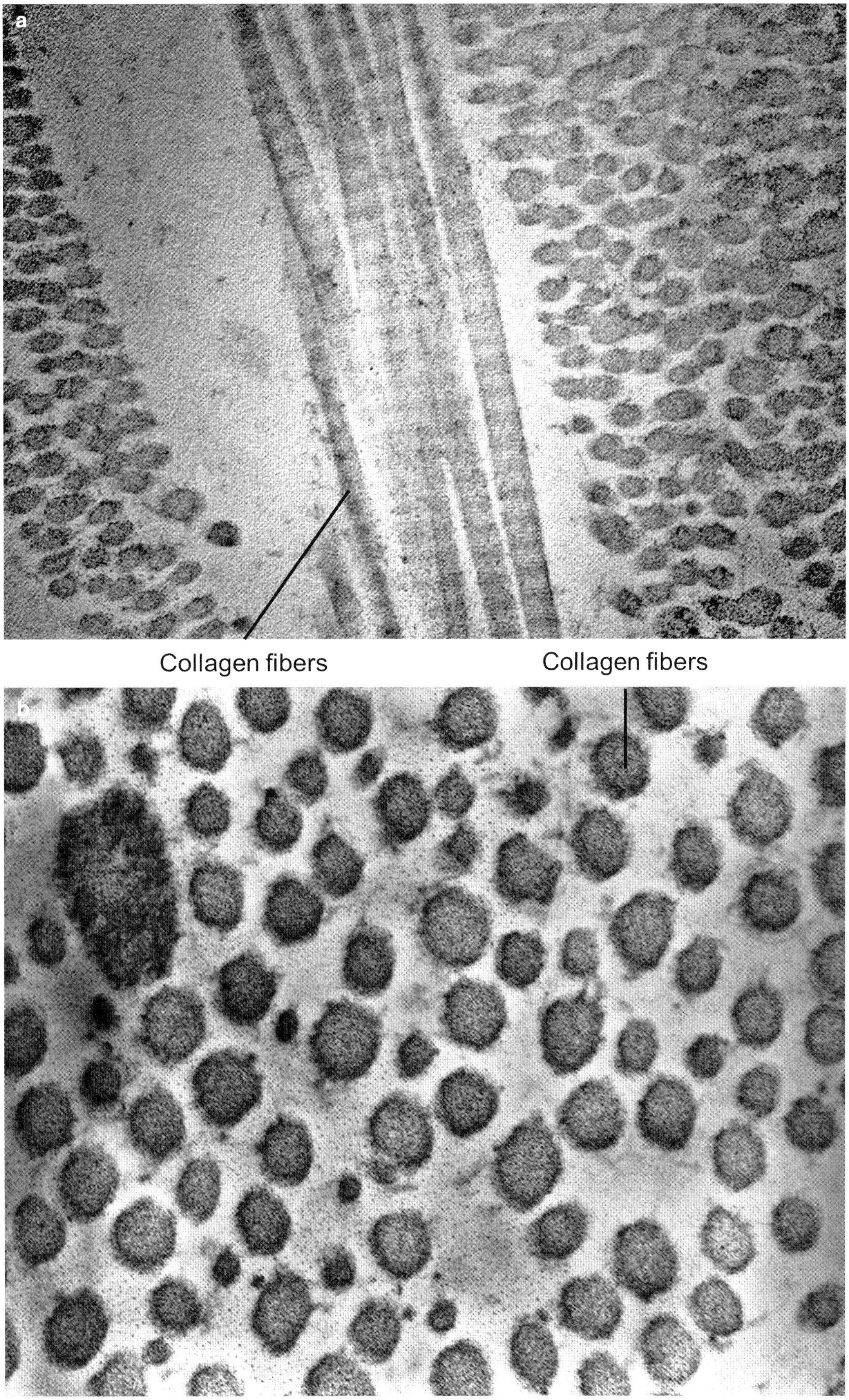

Fig. 20.19 Human spinal dura mater. Detail of collagen fibers. Transmission electron microscopy. Magnification: (**a**, **b**) ×80,000

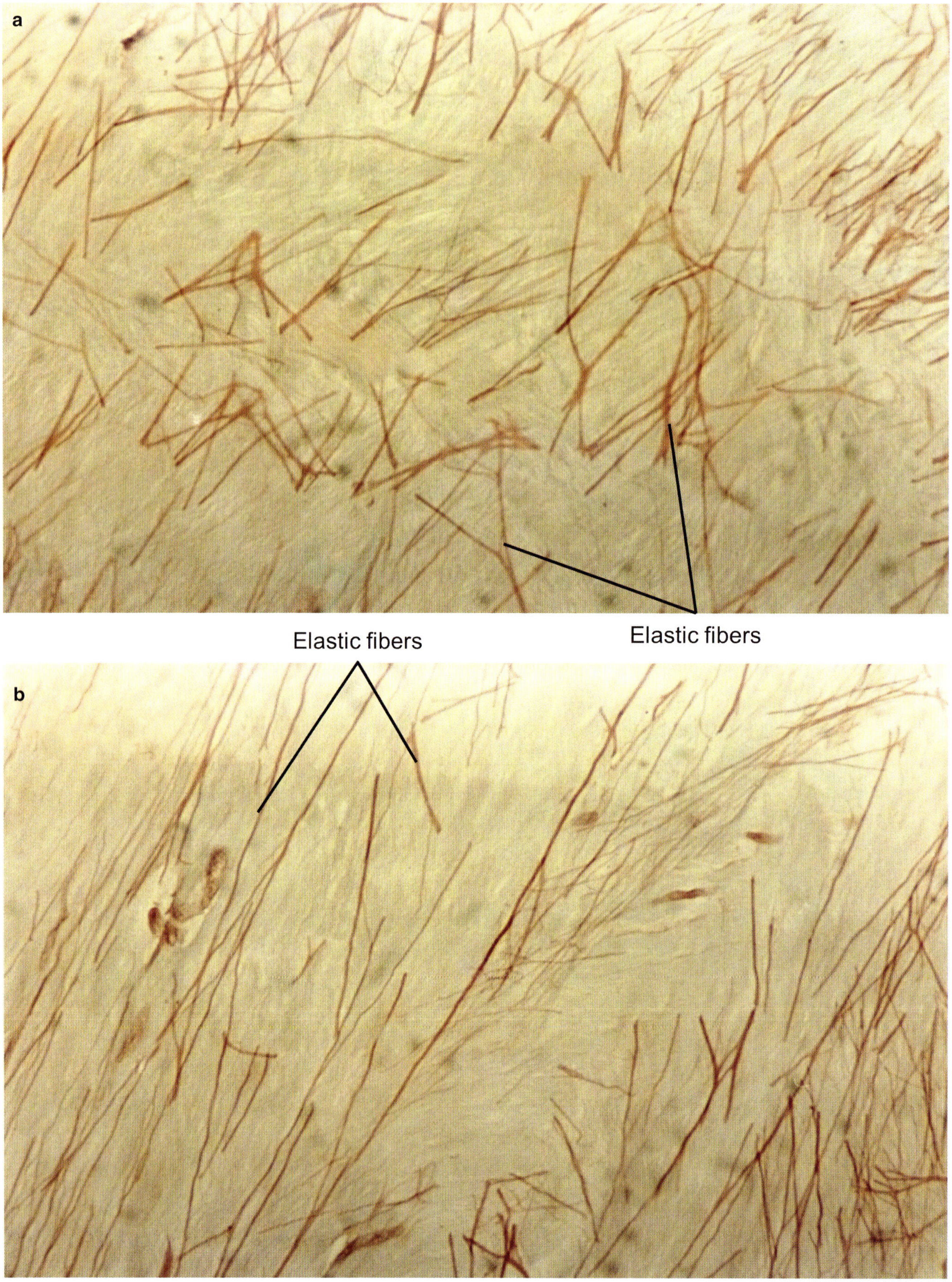

Fig. 20.20 Human spinal dura mater. Elastic fibers identified with orcein stain technique. Fibers appear in different directions. (**a**, **b**) images obtained from different samples

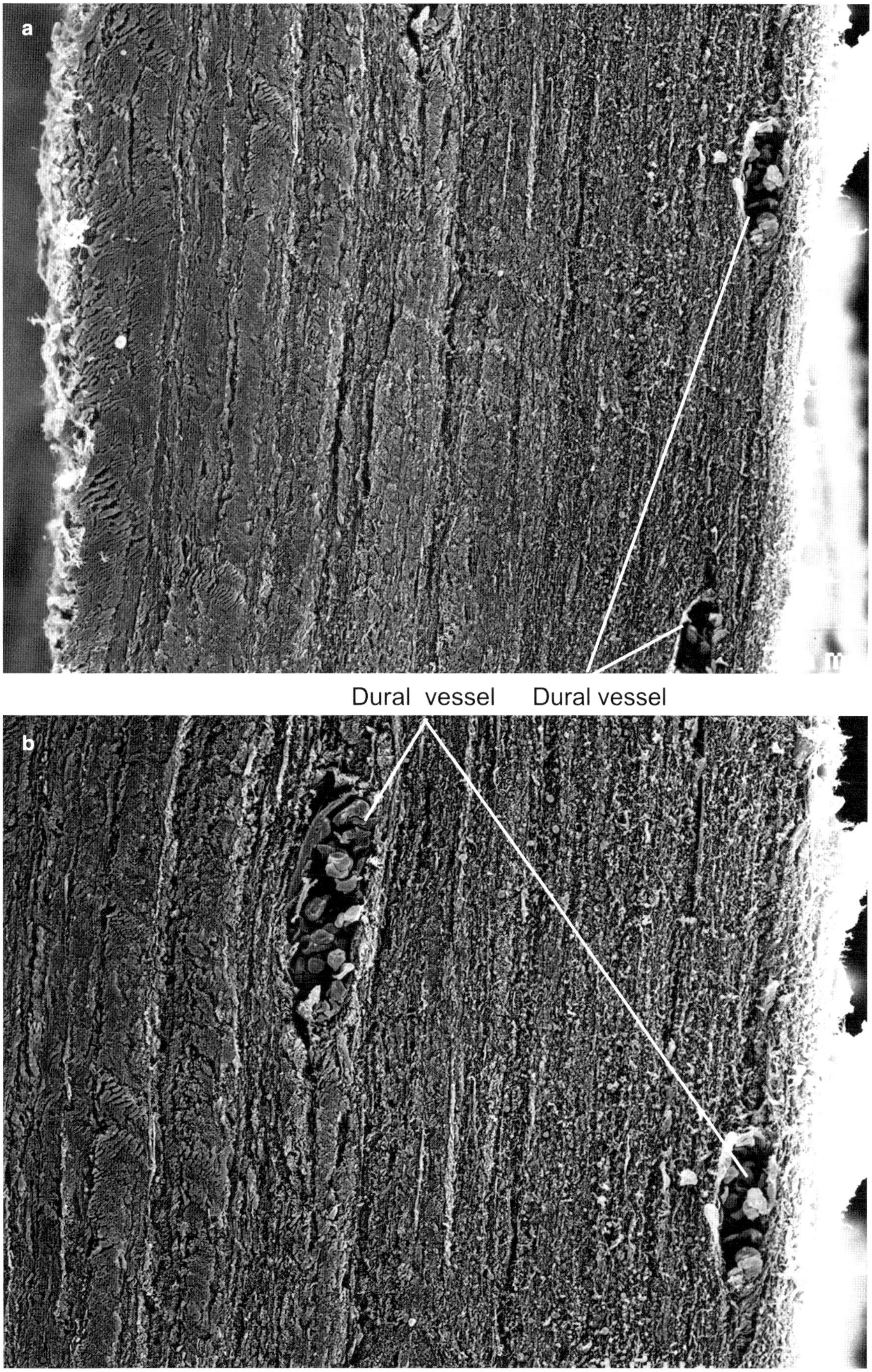

Fig. 20.21 Human spinal dura mater. Transversal section of vessels within the thickness of the dura mater. Scanning electron microscopy. Magnification: (**a**) ×330; (**b**) ×400 (**b** from Reina et al. [8]; with permission)

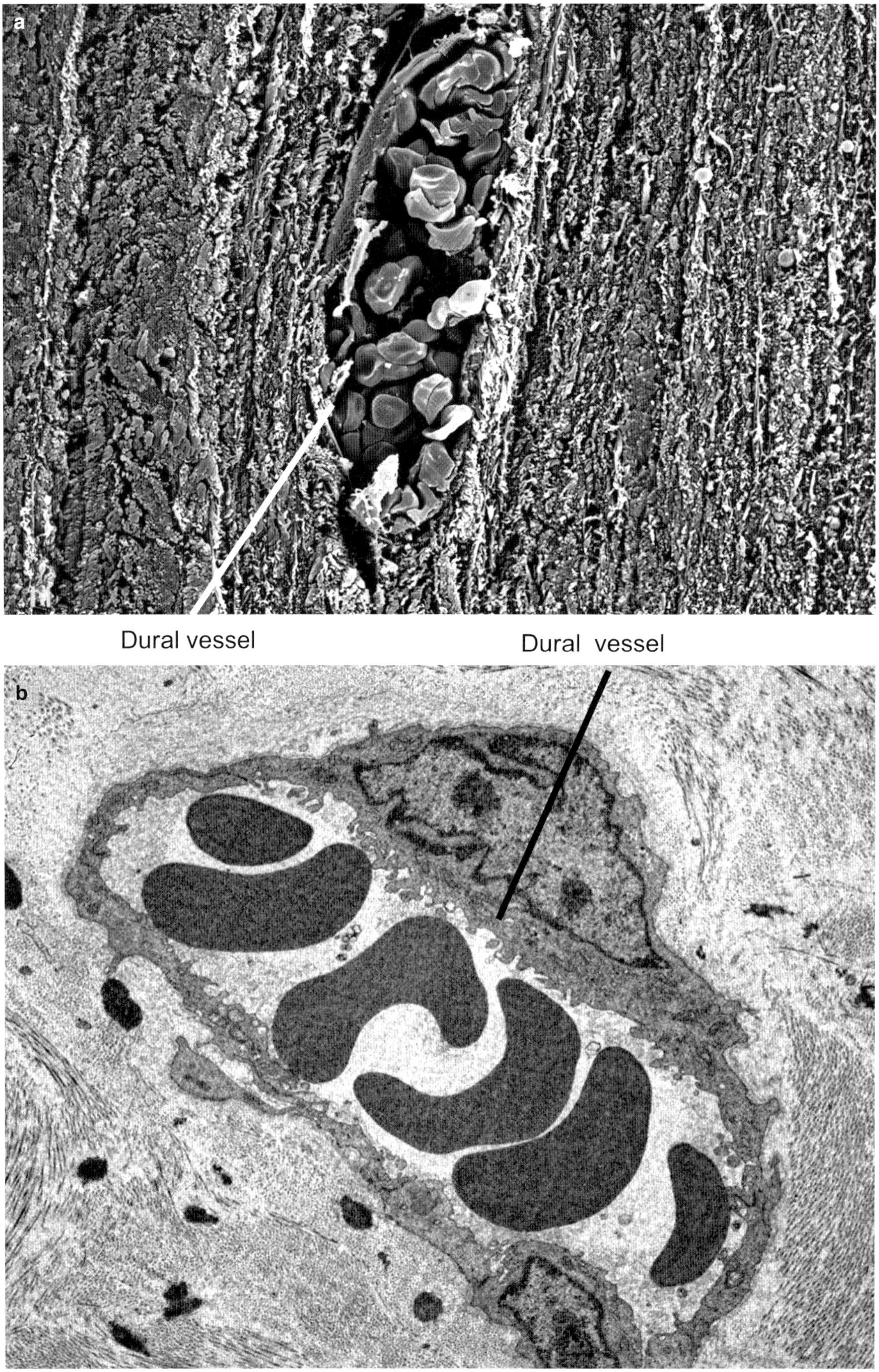

Fig. 20.22 Human spinal dura mater. Detail of transversal section of vessels within the thickness of dura mater. (**a**) Scanning electron microscopy; magnification ×1,000. (**b**) Transmission electron microscopy; magnification ×6,000 (**a** from Reina et al. [8]; with permission)

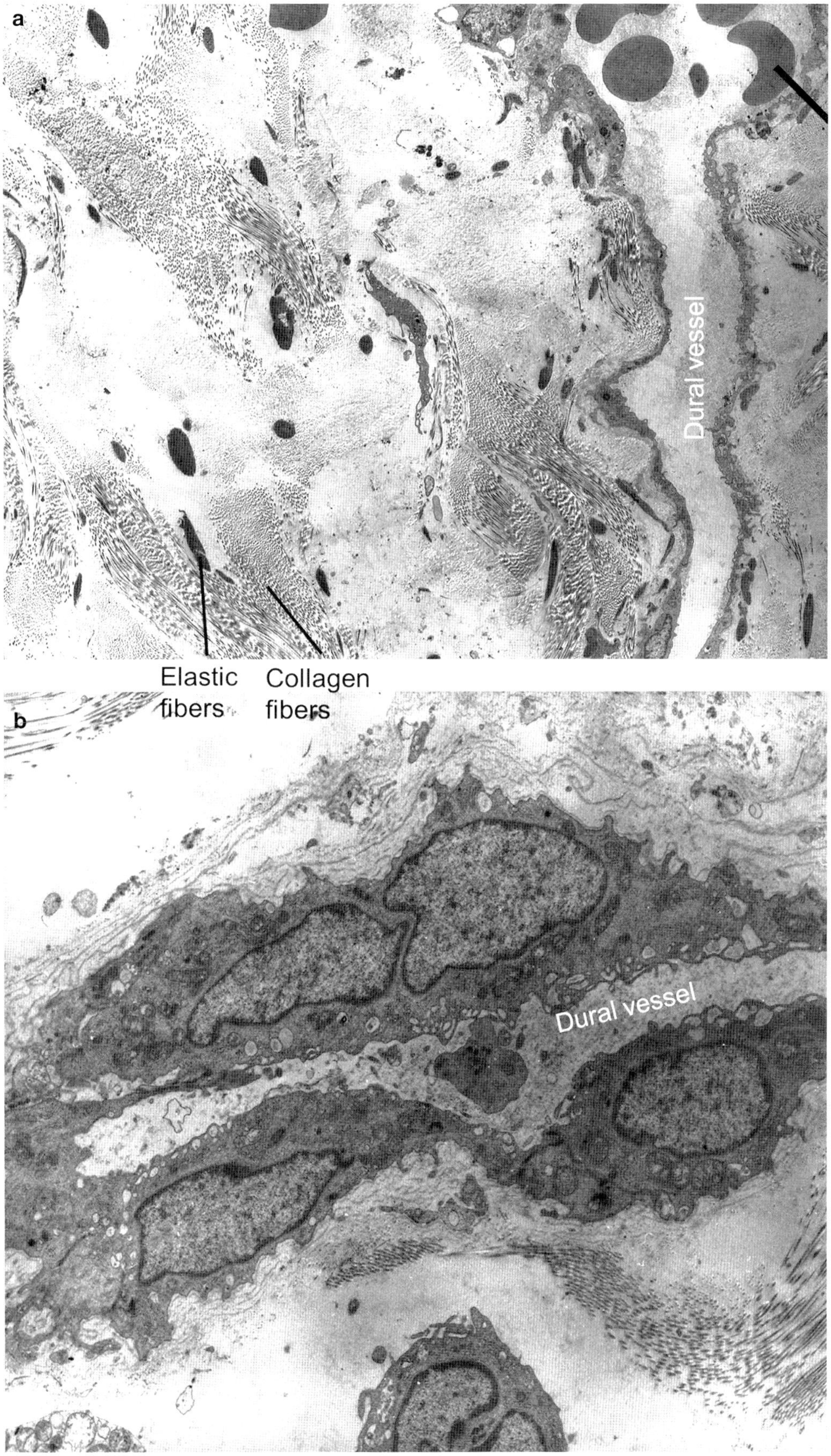

Fig. 20.23 Human spinal dura mater. Detail of longitudinal section of vessels within the thickness of the dura mater. Transmission electron microscopy. Magnification: (**a**) ×1,100; (**b**) ×3,000

References

1. Reina MA, Dittmann M, López A, van Zundert A. New perspectives in the microscopic structure of human dura mater in the dorso lumbar region. Reg Anesth. 1997;22:161–6.
2. Dittmann M, Reina MA, López Garcia A. Neue Ergebnisse bei der Darstellung der Dura mater spinalis mittels Rasterelektronenmikroskopie. Anaesthesist. 1998;47:409–13.
3. Reina MA, López A, Dittmann M, De Andrés JA. Analysis of the external and internal surface of human dura mater with scanning electron microscopy. Rev Esp Anestesiol Reanim. 1996;43:130–4.
4. Reina MA, López A, Dittmann M, De Andrés JA. Structural analysis of the thickness of human dura mater with scanning electron microscopy. Rev Esp Anestesiol Reanim. 1996;43:135–7.
5. Reina MA, López A, De Andrés JA. Thickness variation of the dural sac. Rev Esp Anestesiol Reanim. 1999;46:344–9.
6. Reina MA, Pulido P, López A. The human dural sac. Morphology of the spinal dura-arachnoid membrane. Origin of the spinal subdural space. Rev Arg Anestesiol. 2007;65:167–84.
7. Reina MA, De Andrés JA, López A. Subarachnoid and epidural anesthesia. In: Raj P, editor. Textbook of regional anesthesia. New York: Churchill Livingstone; 2002. p. 307–24.
8. Reina MA, López A, Maches F, De Andrés JA. Origin of spinal subdural haematomas. Rev Esp Anestesiol Reanim. 2004;51:240–6.
9. Reina MA, Oliva A, Carrera A, Durán Mateos EM, Diamantopoulos Fernández J, Arriazu Navarro R. Origin of post dural puncture headache. Cir May Amb. 2011;16:72–84.
10. De Andrés JA, Reina MA, Hernández JM, Carrera A, Oliva A, Prats-Galino A. Role of spinal anatomical structures for neuromodulation. Reg Anesth Pain Med Suppl. 2011;36:E130–8.

Ultrastructure of the Spinal Arachnoid Layer

21

Miguel Angel Reina, Paloma Pulido, and Rafael García De Sola

In the past, the arachnoid layer had been defined as a fine membrane in contact with (but not adhering to) the internal surface of the dura mater, and the subdural space had been described as a virtual space between the arachnoid layer and the dura mater, but this concept has since been modified. (*See* Chaps. 26 and 27). The arachnoid layer, a semipermeable membrane, exerts a barrier effect against the passage of substances across it. The thickness of the arachnoid measures about 35–40 μm, and it is composed of cells strongly bonded by specific membrane junctions [1–4]. The intercellular space is thought to contain collagen fibers that reinforce this structure in order to increase its mechanical resistance. From the outer towards the inner aspect of this membrane, four differentiated structures can be observed:

- The outermost area is occupied by neurothelial cells, also known as dural border cells of the subdural compartment, which become the limits of "acquired subdural spaces" as the process of tearing extends across this layer.
- Collagen fibers oriented in various directions compose the second distinctive area, occupying about 40–50 % of the total thickness of the arachnoid.
- A basal membrane limits a third area filled with arachnoid cells that exert a barrier effect against the passage of substances. This layer comprises four or five cellular planes. These cells are light-colored and elongated, with a thickness of about 1.5–2 μm. The nucleus of these cells is large (1 μm wide by 7–9 μm); in their cytoplasm, the matrix contains fine reticular filaments oriented in different directions, endoplasm, numerous vesicles, and some mitochondria and lysosomes, with a few heterochromatin aggregates. The intercellular space is minimal in areas exerting barrier effect, forming narrow clefts of about 0.02–0.03 μm. Plasma membranes of neighboring cells join together by means of different types of specialized unions, such as tight junctions and desmosomes. The intercellular space lacks collagen, elastic fibers, or microfibrils.
- The innermost area lies in direct contact with cerebrospinal fluid (CSF) and is formed by the reticular arachnoid, where arachnoid cells limit the subarachnoid space [5]. The intercellular space among these cells gradually increases owing to the absence of specialized membrane junctions, giving rise to small intercellular gaps that constitute the trabecular arachnoid extending across the subarachnoid space (Figs. 21.1, 21.2, 21.3, 21.4, 21.5, 21.6, 21.7, 21.8, 21.9, 21.10, 21.11, 21.12, 21.13, 21.14, 21.15, 21.16, 21.17, 21.18, 21.19, 21.20, 21.21, 21.22, 21.23, and 21.24).

M.A. Reina, MD, PhD (✉)
Department of Clinical Medical Sciences and Institute of Applied Molecular Medicine, School of Medicine, University of CEU San Pablo, Madrid, Spain

Department of Anesthesiology, Madrid-Montepríncipe University Hospital, Madrid, Spain
e-mail: miguelangel@perticone.e.telefonica.net

P. Pulido, MD, PhD
Department of Neurosurgery, La Princesa University Hospital, Madrid, Spain
e-mail: paloma.pulido@gmail.com

R.G. De Sola, MD, PhD
Department of Surgery, School of Medicine, Autonomous University of Madrid, Madrid, Spain

Department of Neurosurgery, La Princesa University Hospital, Madrid, Spain
e-mail: rgsola@neurorgs.com

M.A. Reina et al. (eds.), *Atlas of Functional Anatomy for Regional Anesthesia and Pain Medicine: Human Structure, Ultrastructure and 3D Reconstruction Images*, DOI 10.1007/978-3-319-09522-6_21,

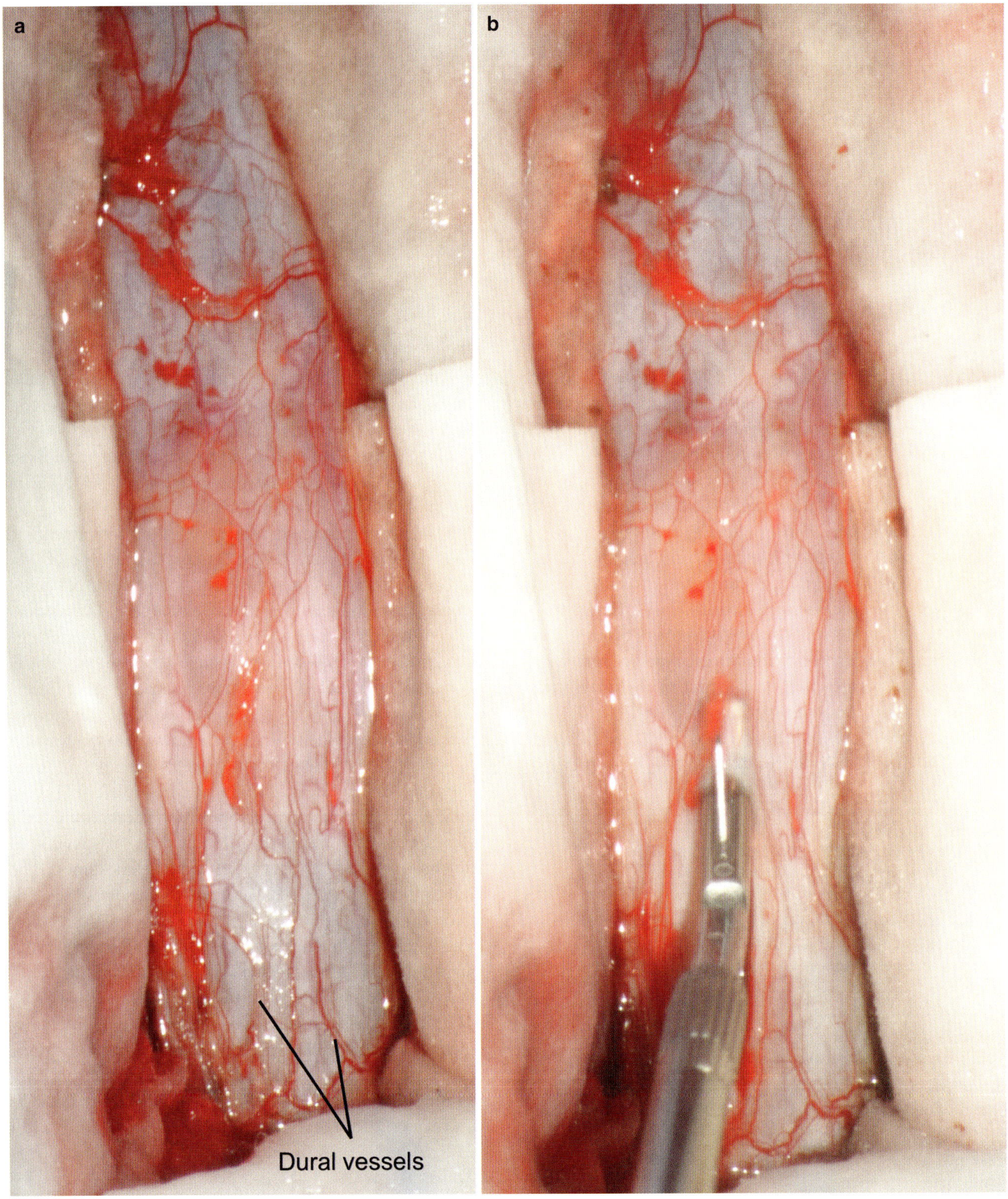

Fig. 21.1 Dissection of human dural sac during neurosurgical intervention (**b** from Reina et al. [1]; with permission)

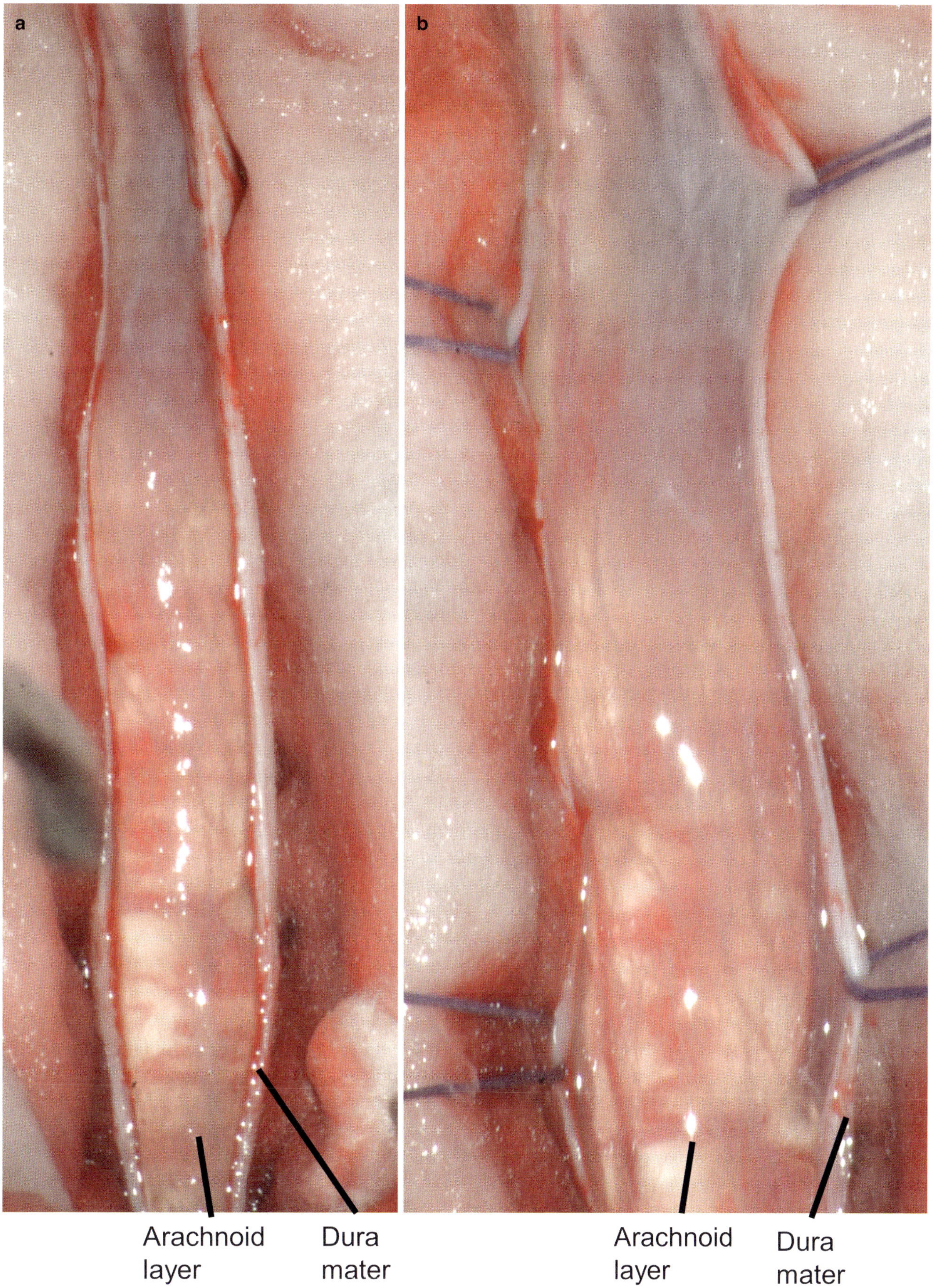

Fig. 21.2 View of human spinal arachnoid layer during neurosurgical intervention (**b** from Reina et al. [1]; with permission)

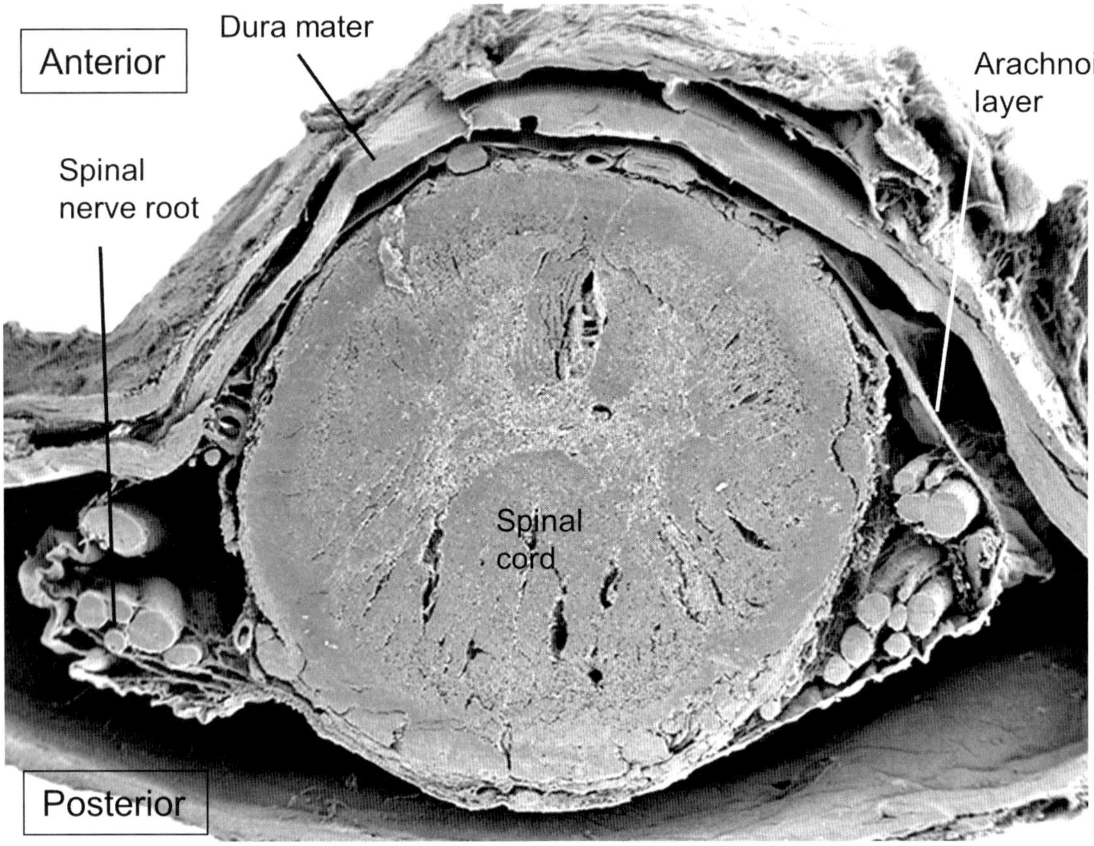

Fig. 21.3 Spinal arachnoid layer. Transverse section of human spinal dural sac and spinal cord. Scanning electron microscopy. Magnification × 10 (From De Andrés et al. [5]; with permission)

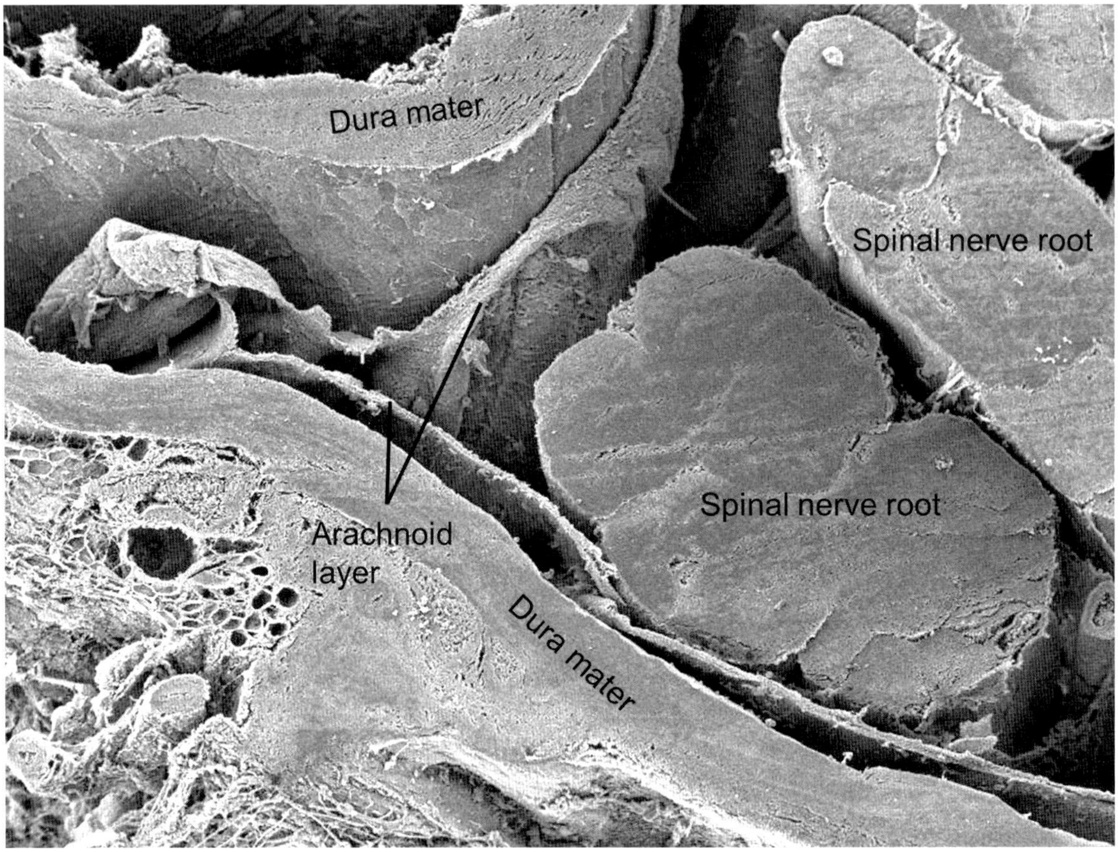

Fig. 21.4 Spinal arachnoid layer. Transverse section of human spinal dural sac and spinal nerve root at lumbar level. Detail of arachnoid layer. Scanning electron microscopy. Magnification × 40

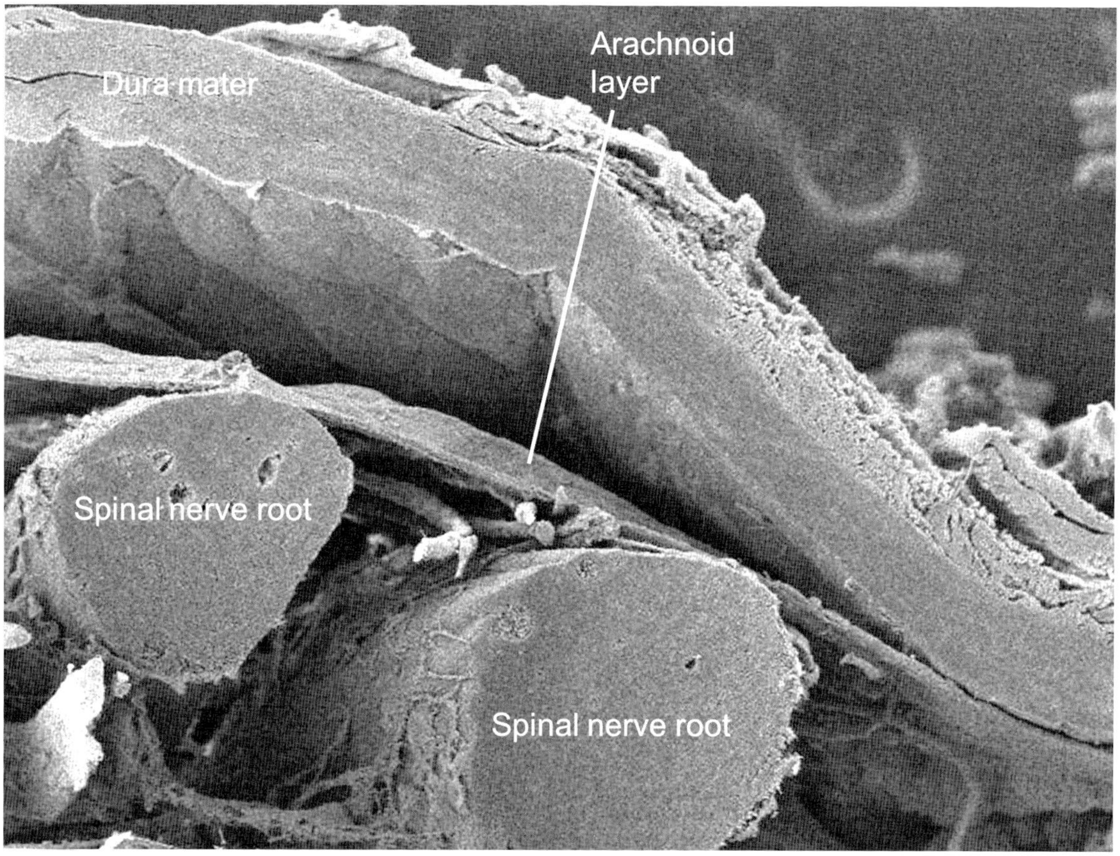

Fig. 21.5 Spinal arachnoid layer. Transverse section of human spinal dural sac and spinal nerve root at lumbar level. Detail of arachnoid layer. Scanning electron microscopy. Magnification × 40

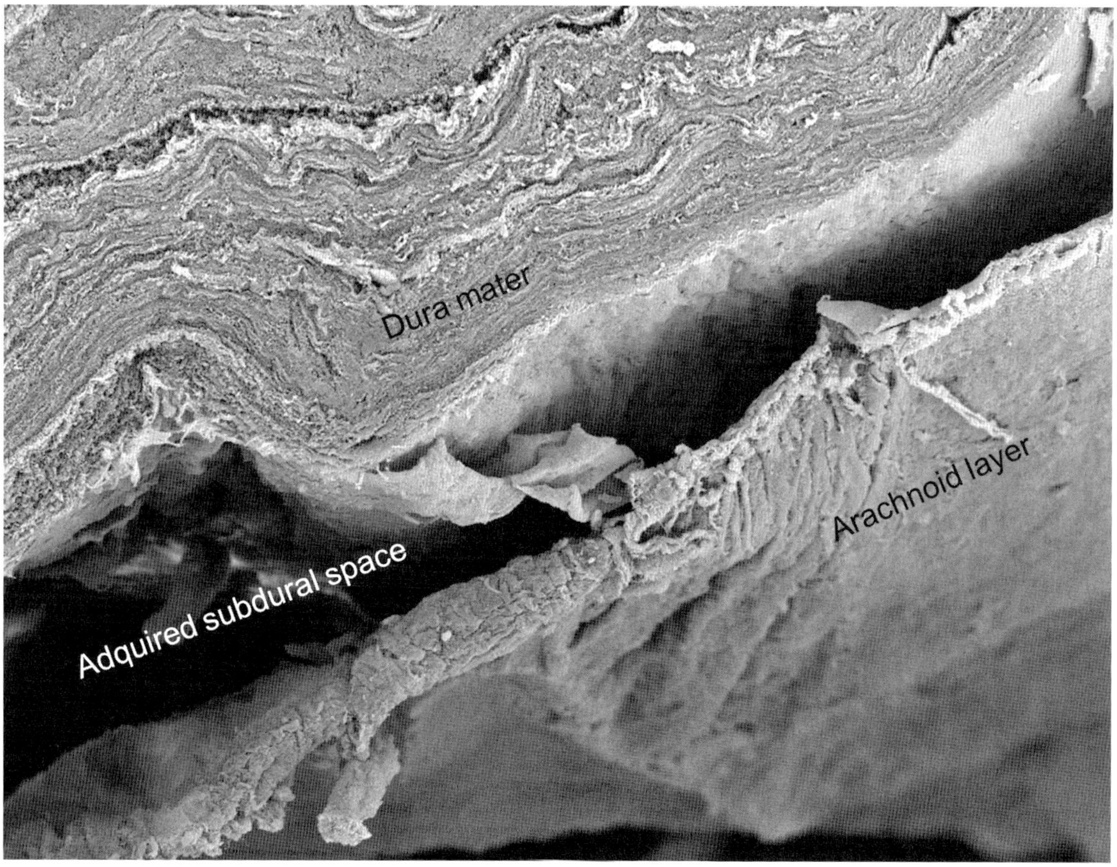

Fig. 21.6 Spinal arachnoid layer. Transverse section of human spinal dural sac and spinal nerve root at lumbar level. Detail of arachnoid layer. Scanning electron microscopy. Magnification × 170 (From Reina et al. [4]; with permission)

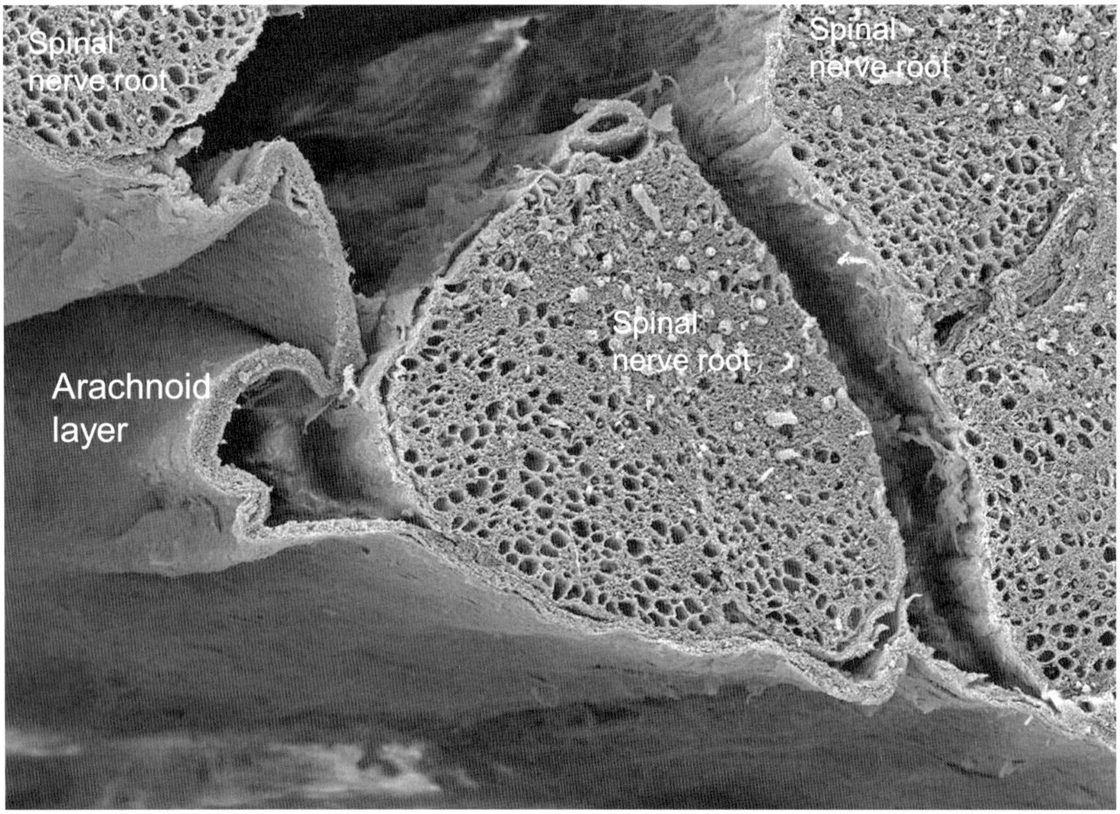

Fig. 21.7 Spinal arachnoid layer. Transverse section of human spinal dural sac and spinal nerve root at lumbar level. Detail of arachnoid layer. Scanning electron microscopy. Magnification × 150

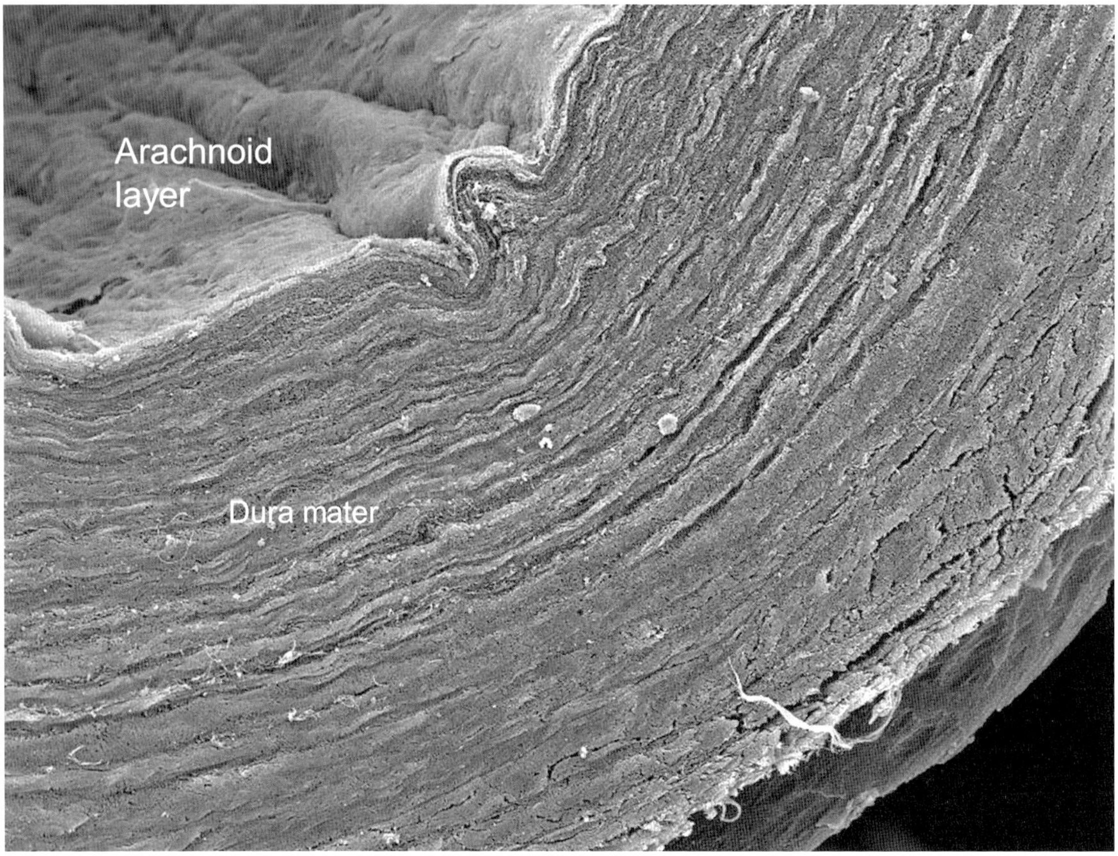

Fig. 21.8 Spinal arachnoid layer. Transverse section of human spinal dural sac at lumbar level. Detail of arachnoid layer. Scanning electron microscopy. Magnification × 300

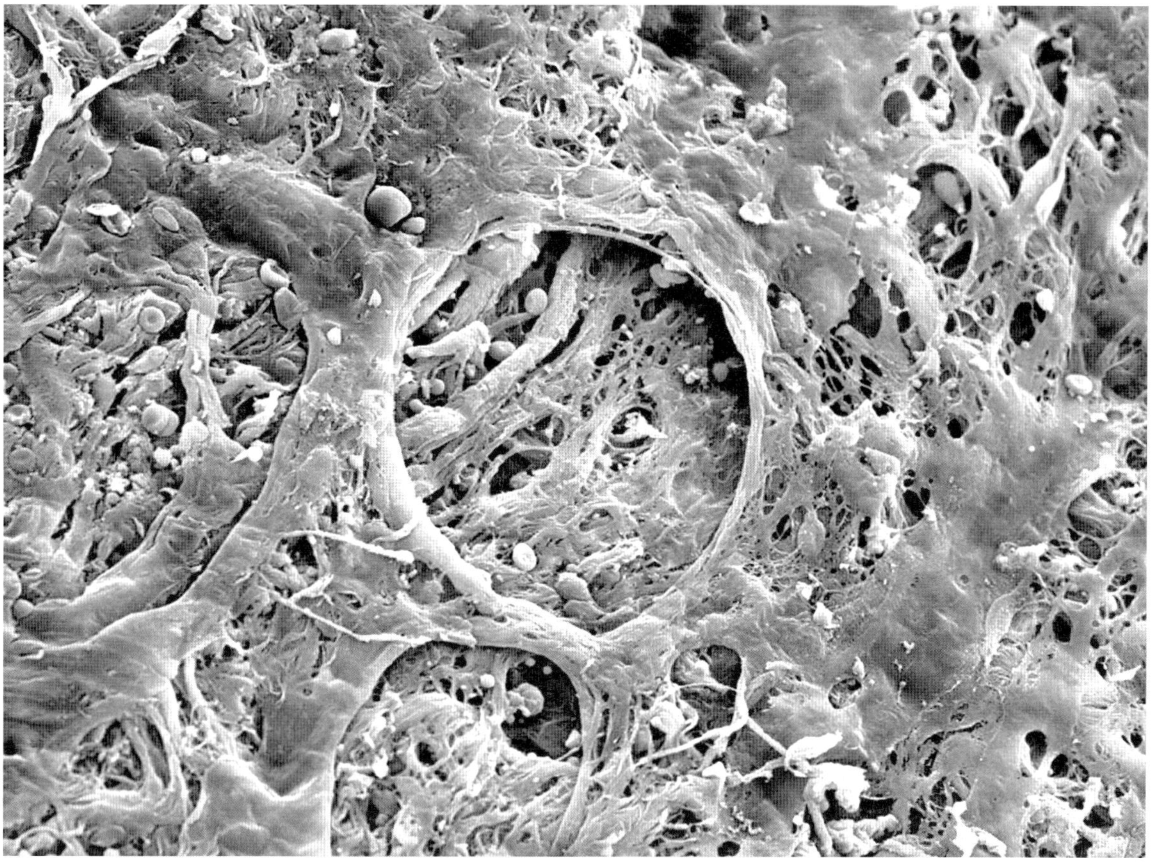

Fig. 21.9 Spinal arachnoid layer. Internal surface of arachnoid layer (internal surface of human spinal dural sac). Scanning electron microscopy. Magnification × 500

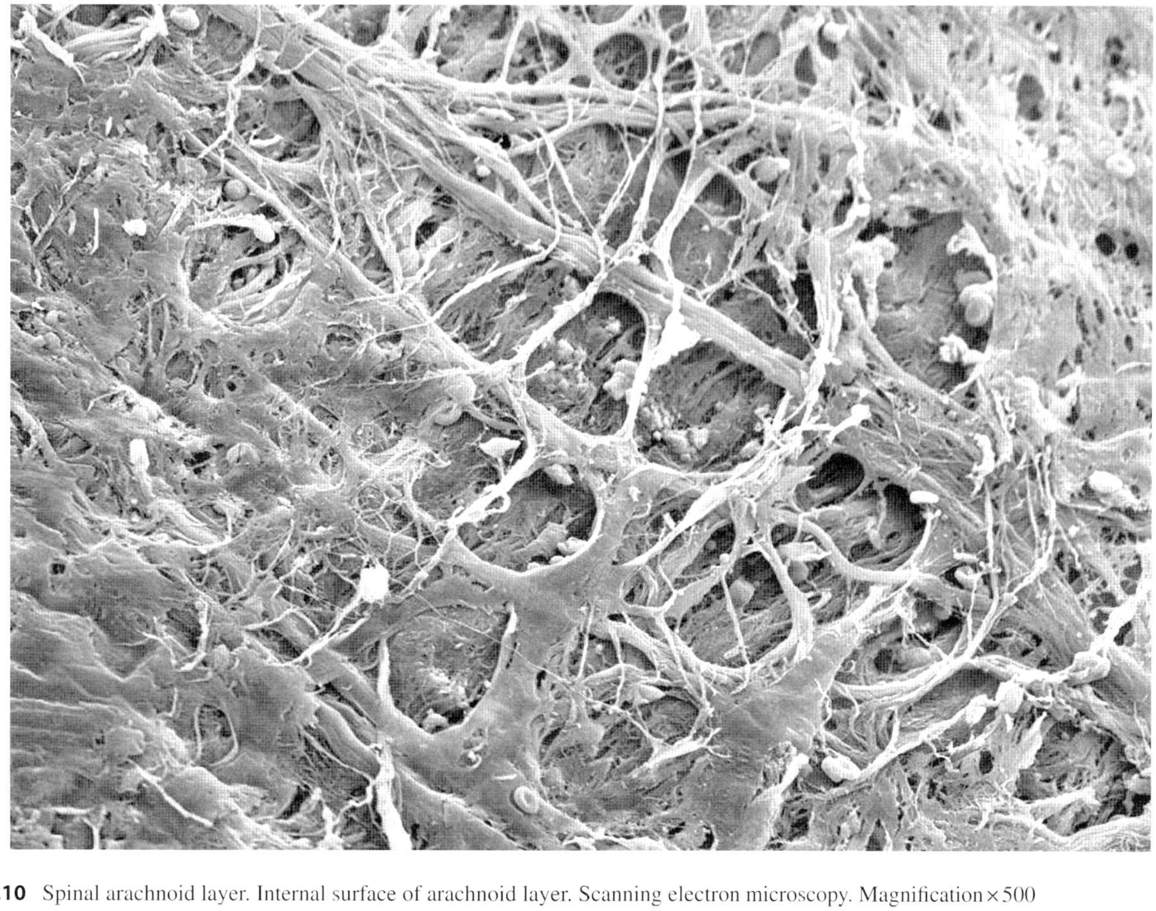

Fig. 21.10 Spinal arachnoid layer. Internal surface of arachnoid layer. Scanning electron microscopy. Magnification × 500

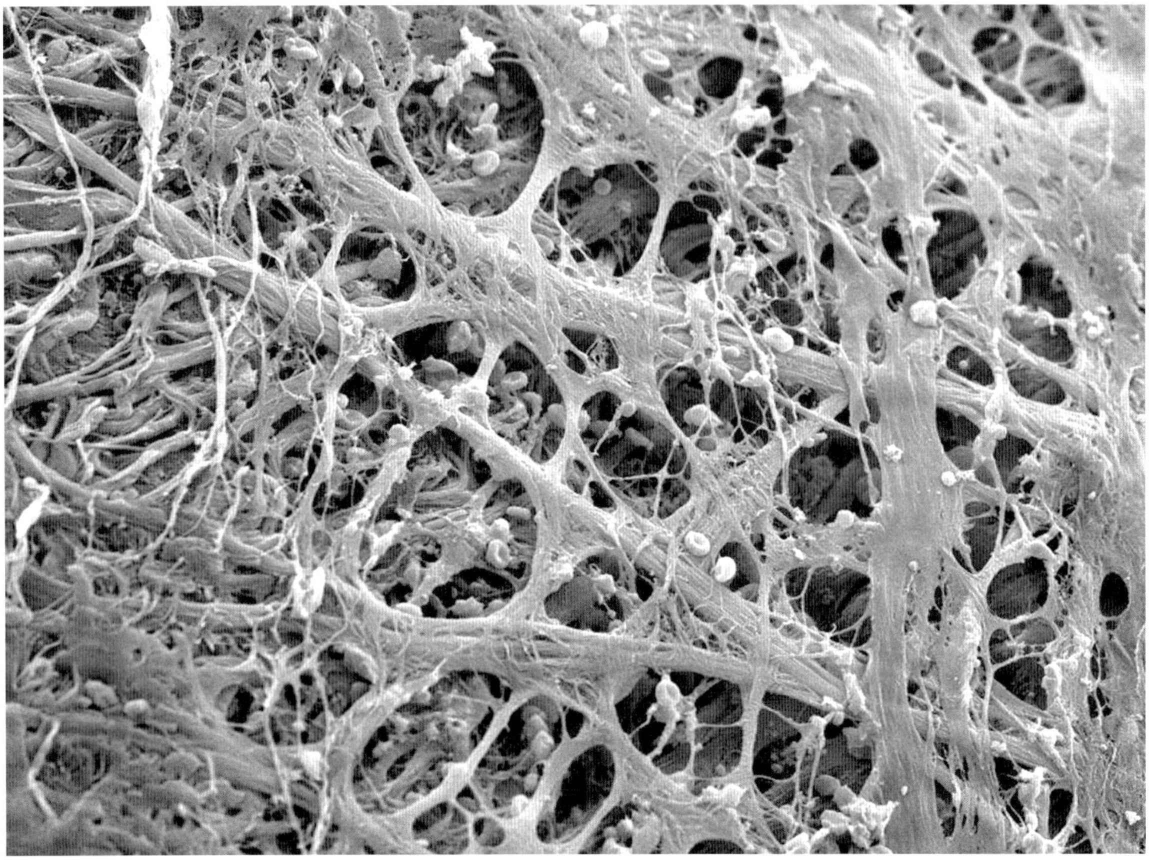

Fig. 21.11 Spinal arachnoid layer. Internal surface of arachnoid layer. Scanning electron microscopy. Magnification × 500

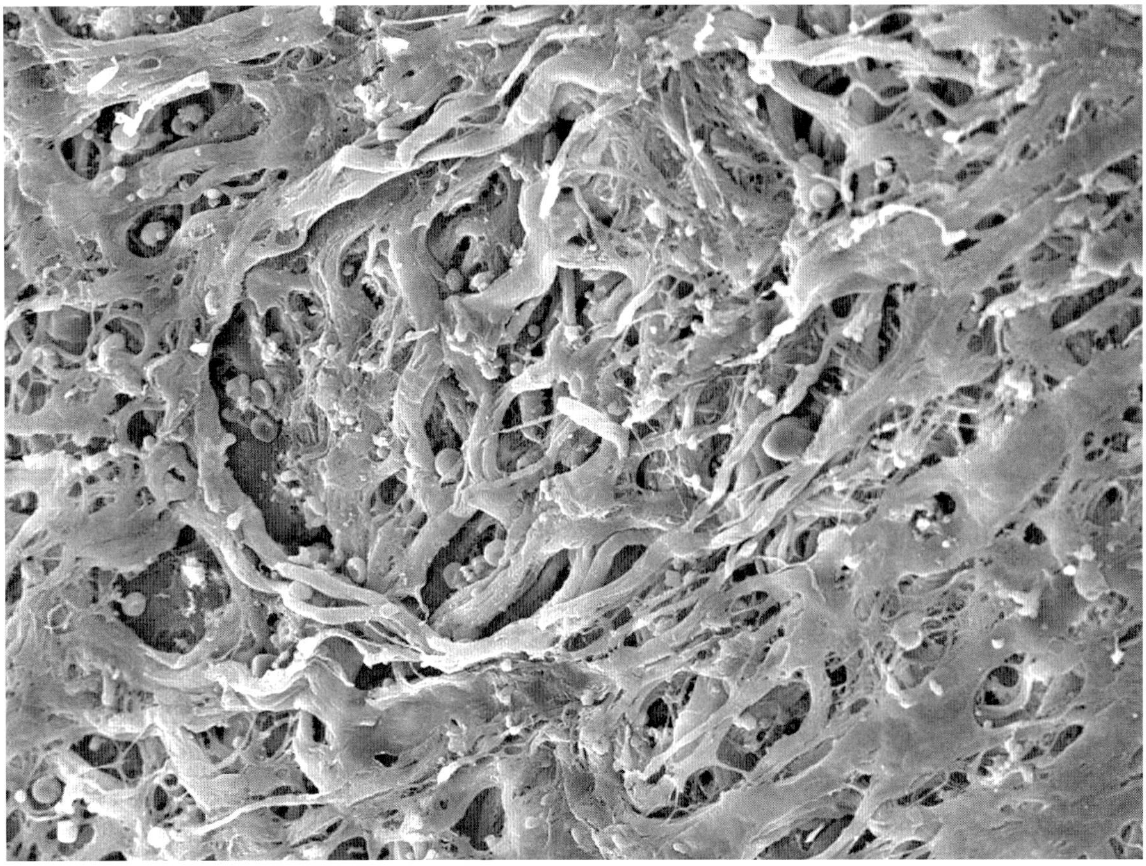

Fig. 21.12 Spinal arachnoid layer. Internal surface of arachnoid layer. Scanning electron microscopy. Magnification × 500

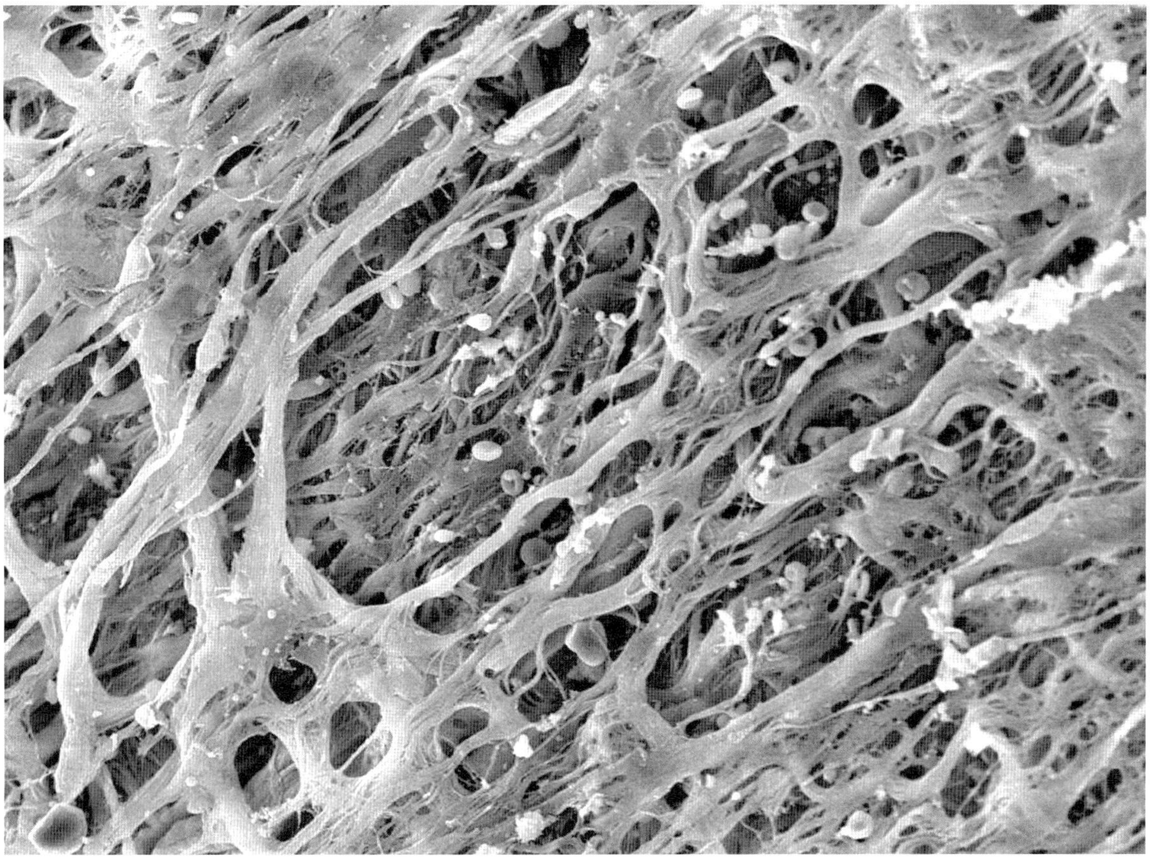

Fig. 21.13 Spinal arachnoid layer. Internal surface of arachnoid layer. Scanning electron microscopy. Magnification × 500

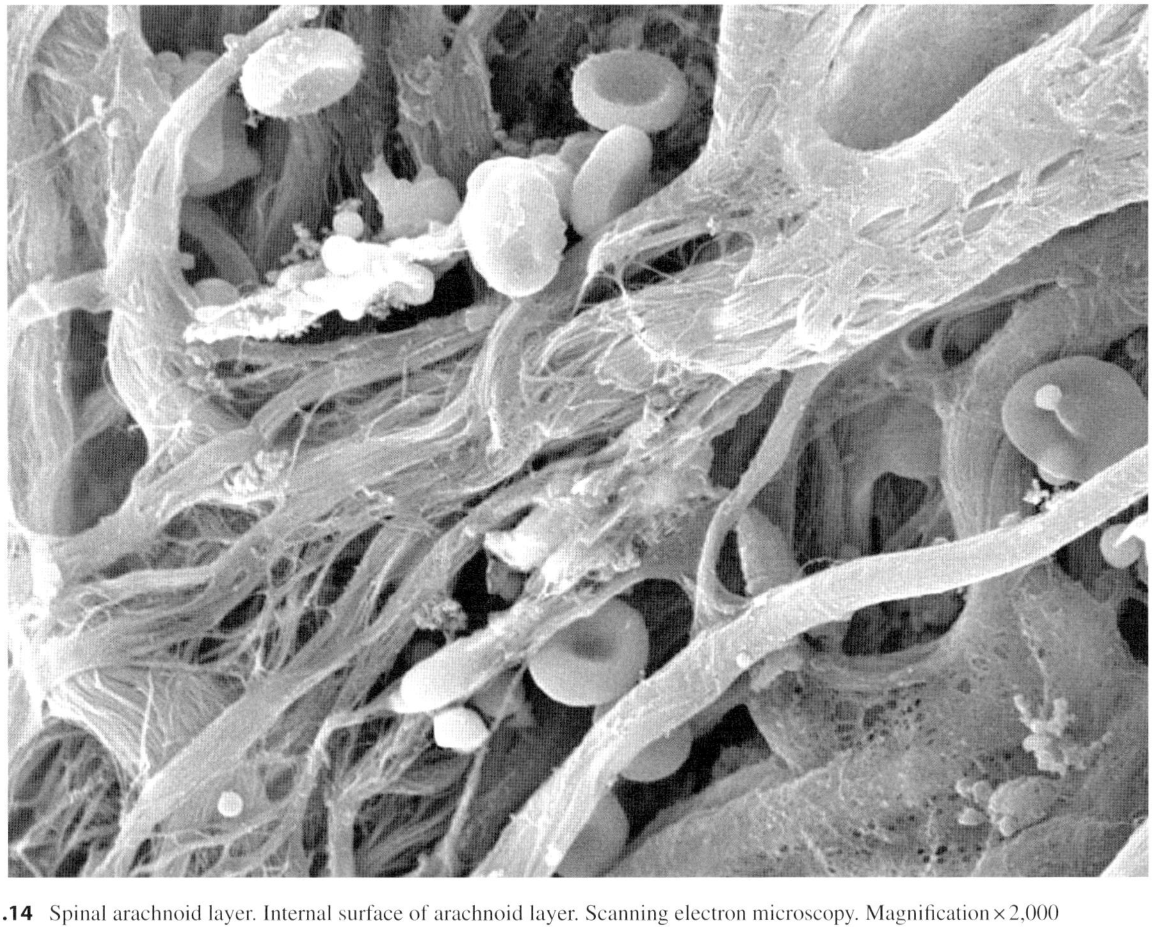

Fig. 21.14 Spinal arachnoid layer. Internal surface of arachnoid layer. Scanning electron microscopy. Magnification × 2,000

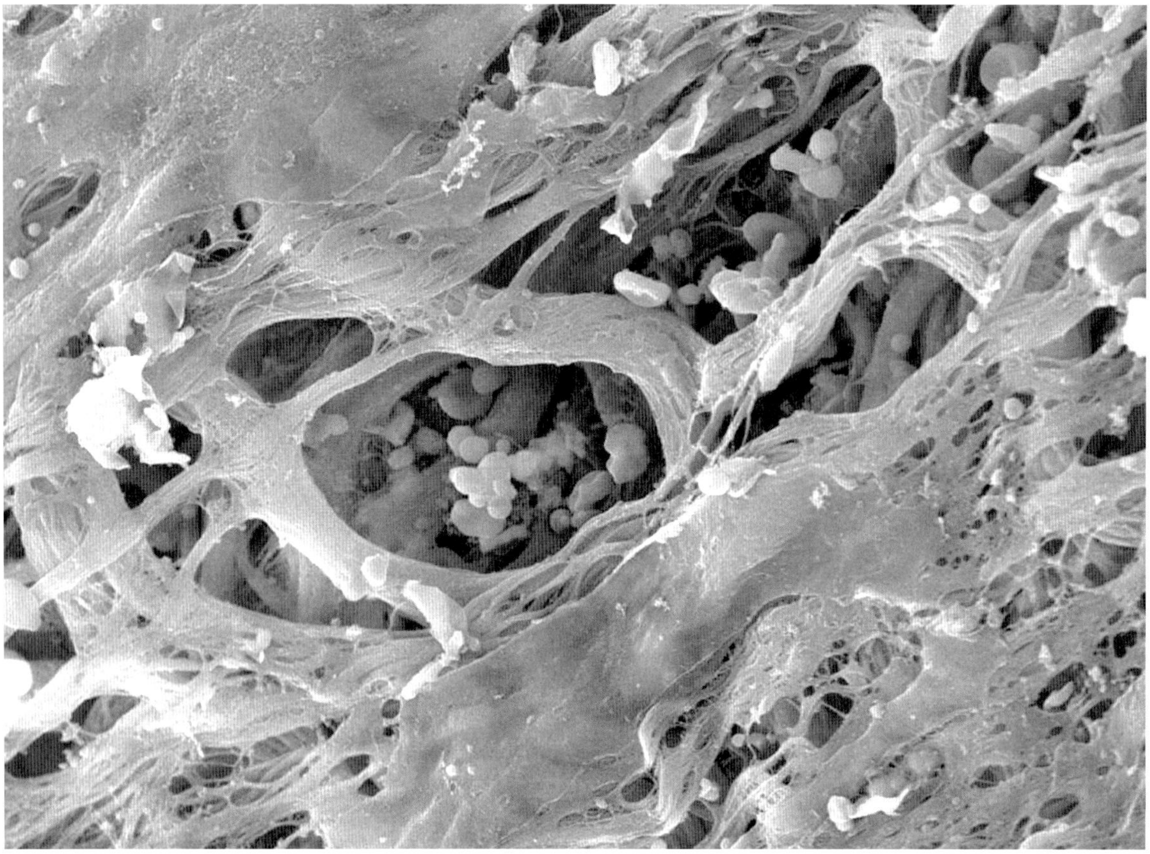

Fig. 21.15 Spinal arachnoid layer. Internal surface of arachnoid layer. Scanning electron microscopy. Magnification × 1,000

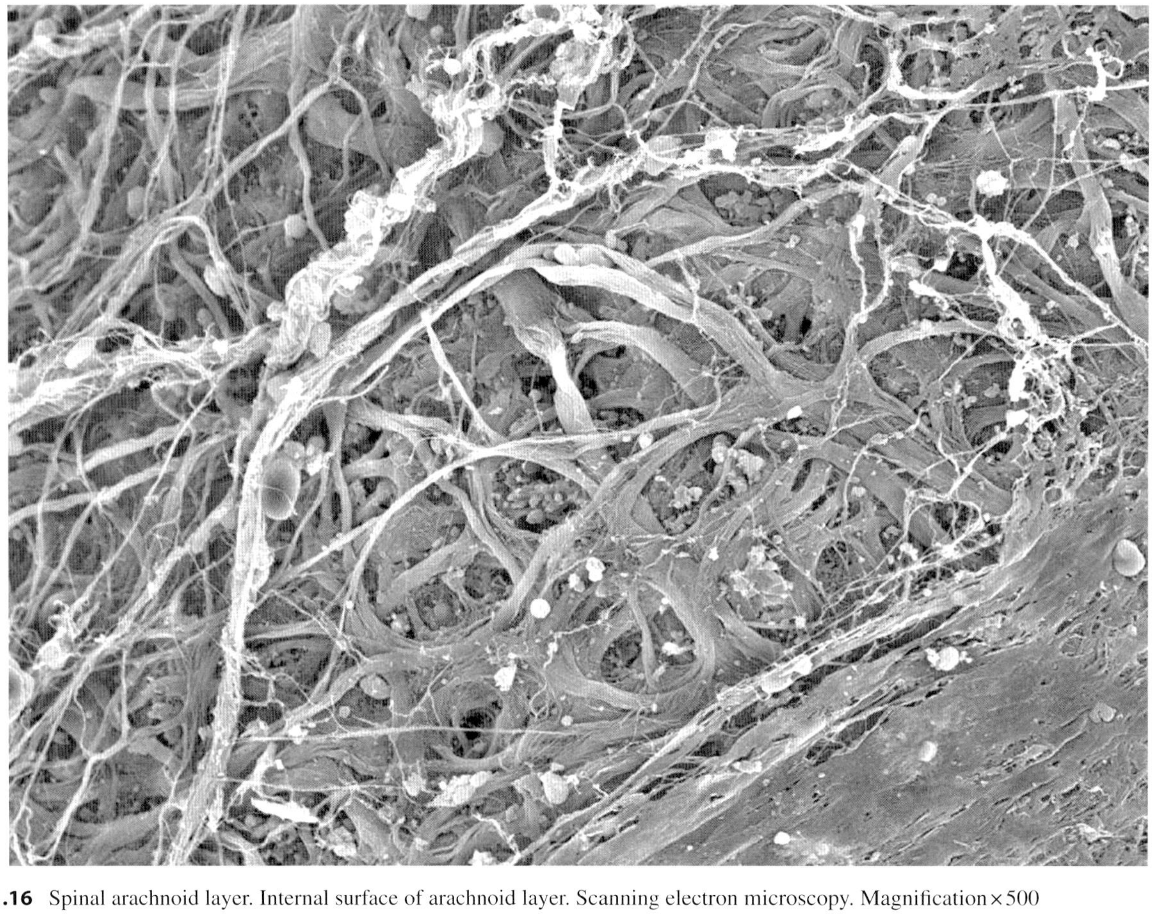

Fig. 21.16 Spinal arachnoid layer. Internal surface of arachnoid layer. Scanning electron microscopy. Magnification × 500

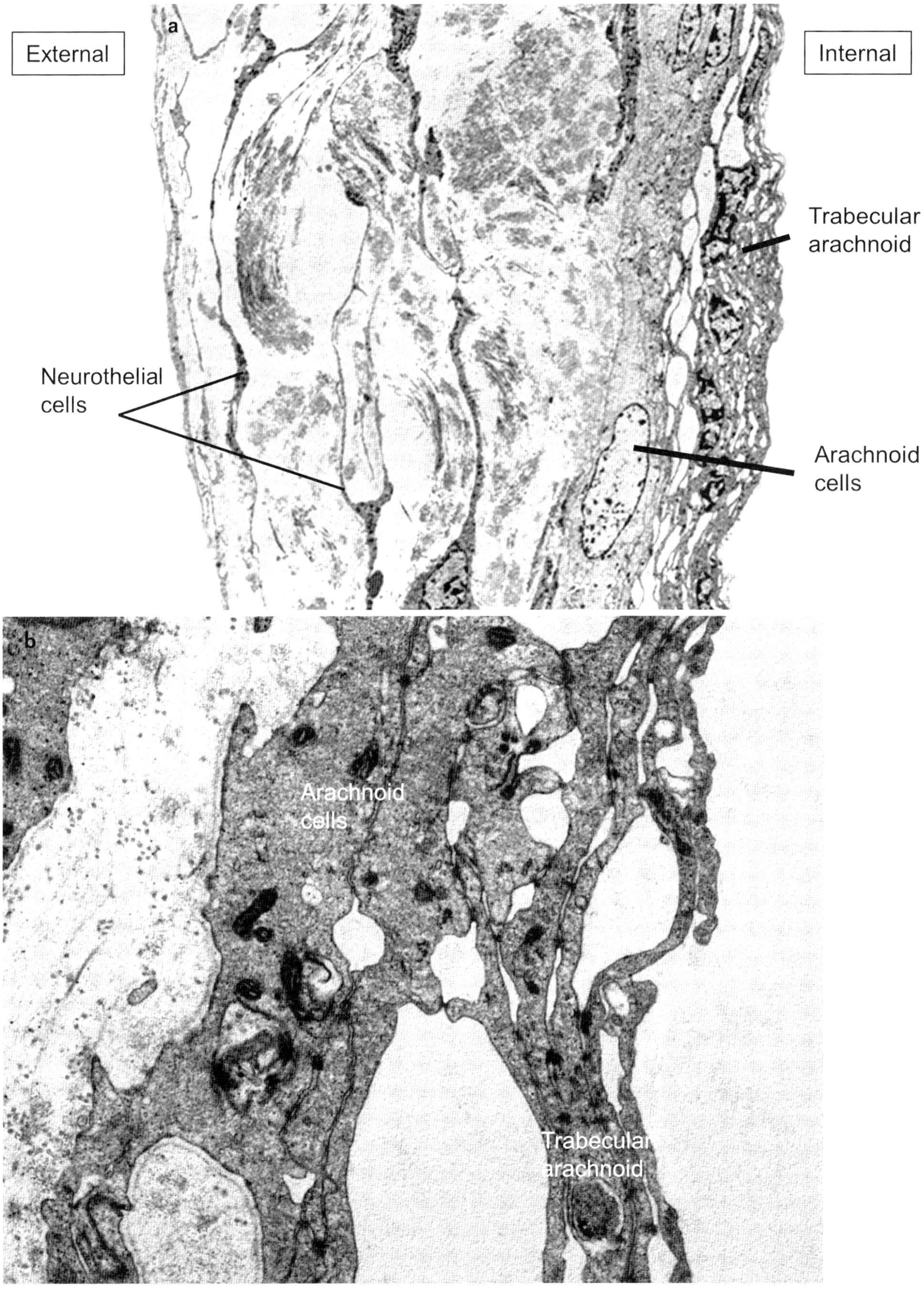

Fig. 21.17 Spinal arachnoid layer. (**a**), Total thickness of the spinal arachnoid layer (magnification × 3,000). (**b**) Detail of (**a**) (magnification × 25,000). Transmission electron microscopy

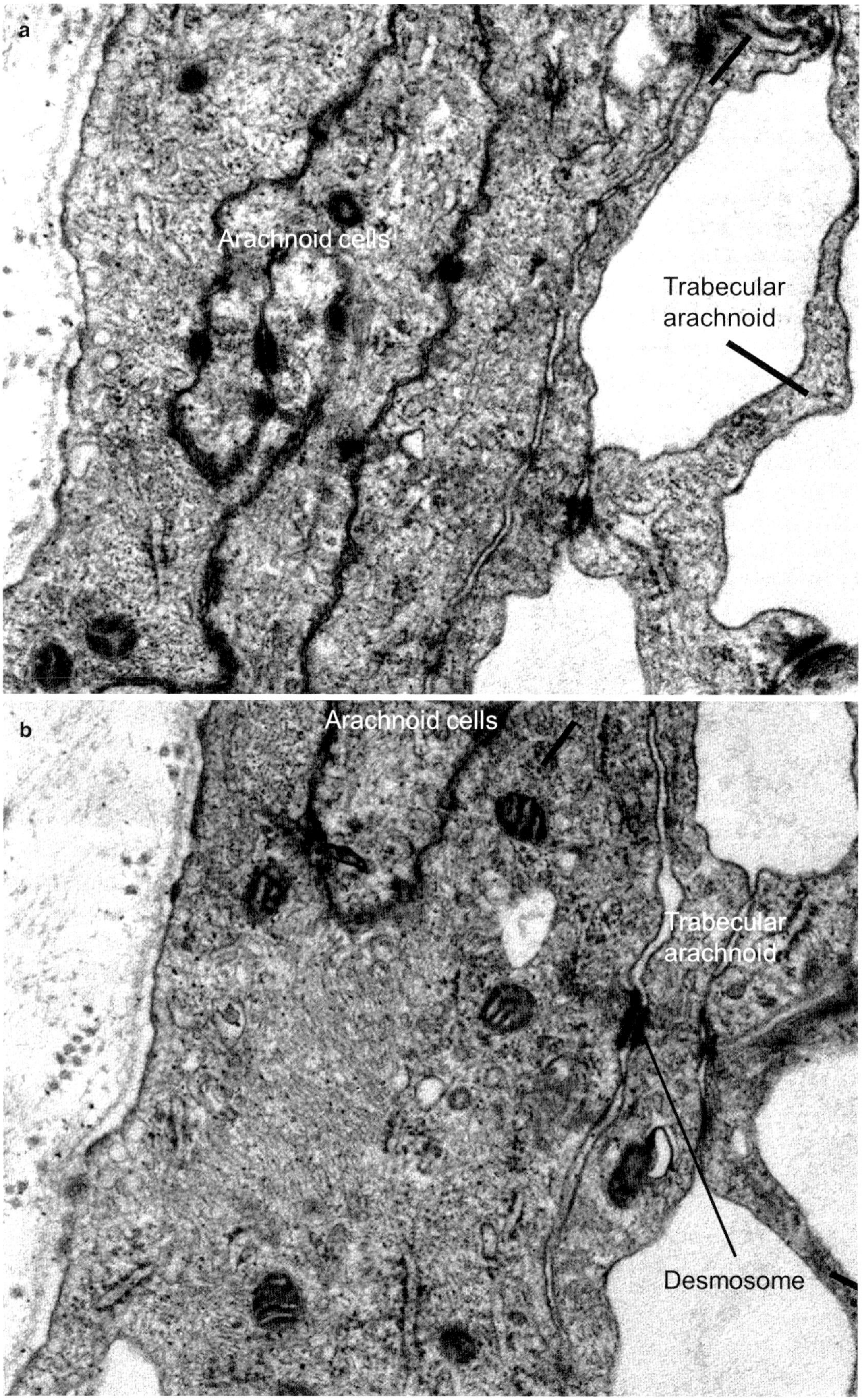

Fig. 21.18 Spinal arachnoid layer, partial thickness. Detail of arachnoid cells and specialized membrane junctions. Transmission electron microscopy. **a** and **b**, Magnification × 50,000

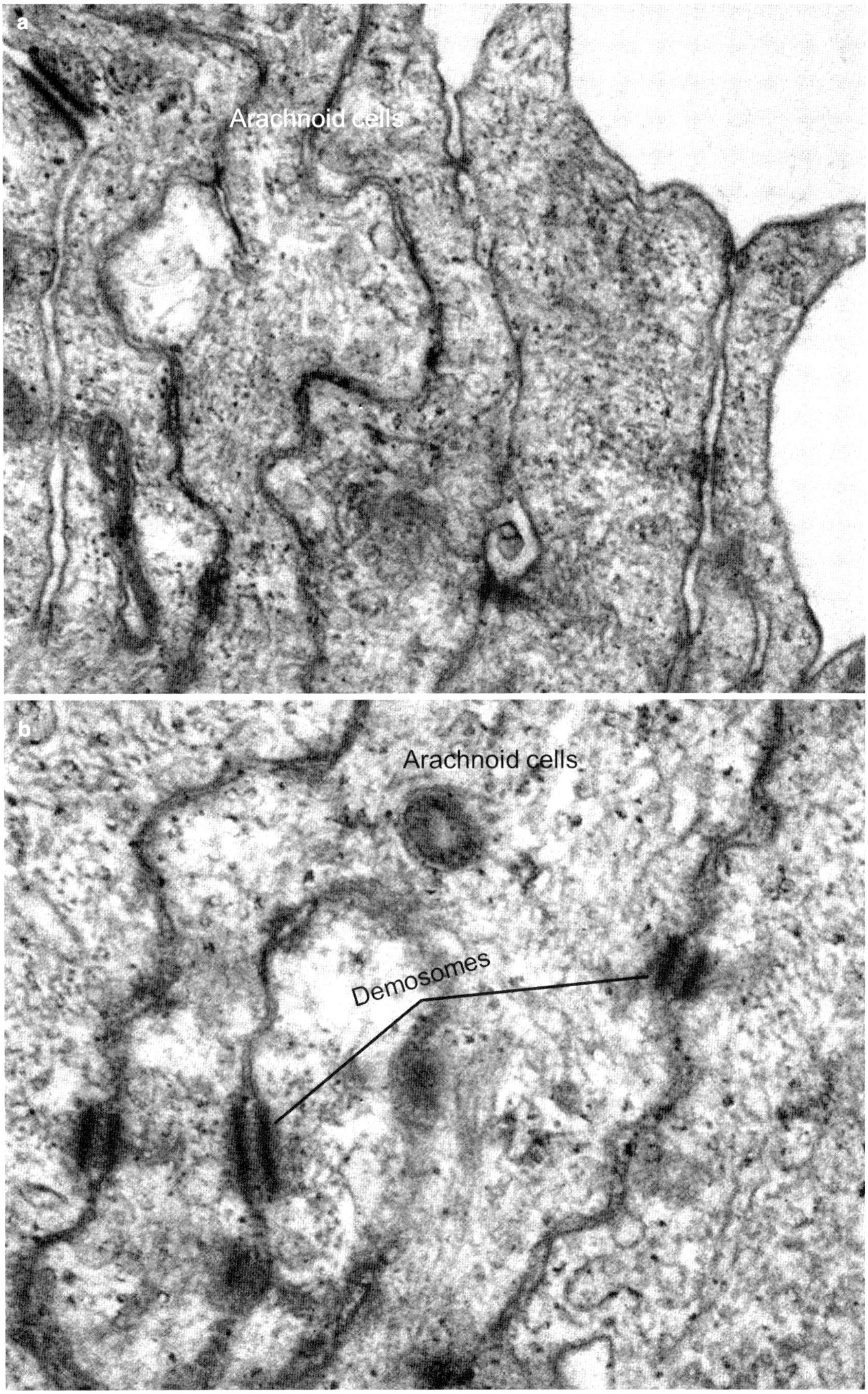

Fig. 21.19 Spinal arachnoid layer, partial thickness. Detail of arachnoid cells and specialized membrane junctions. Transmission electron microscopy. Magnification: **a**, ×80,000; **b**, ×120,000

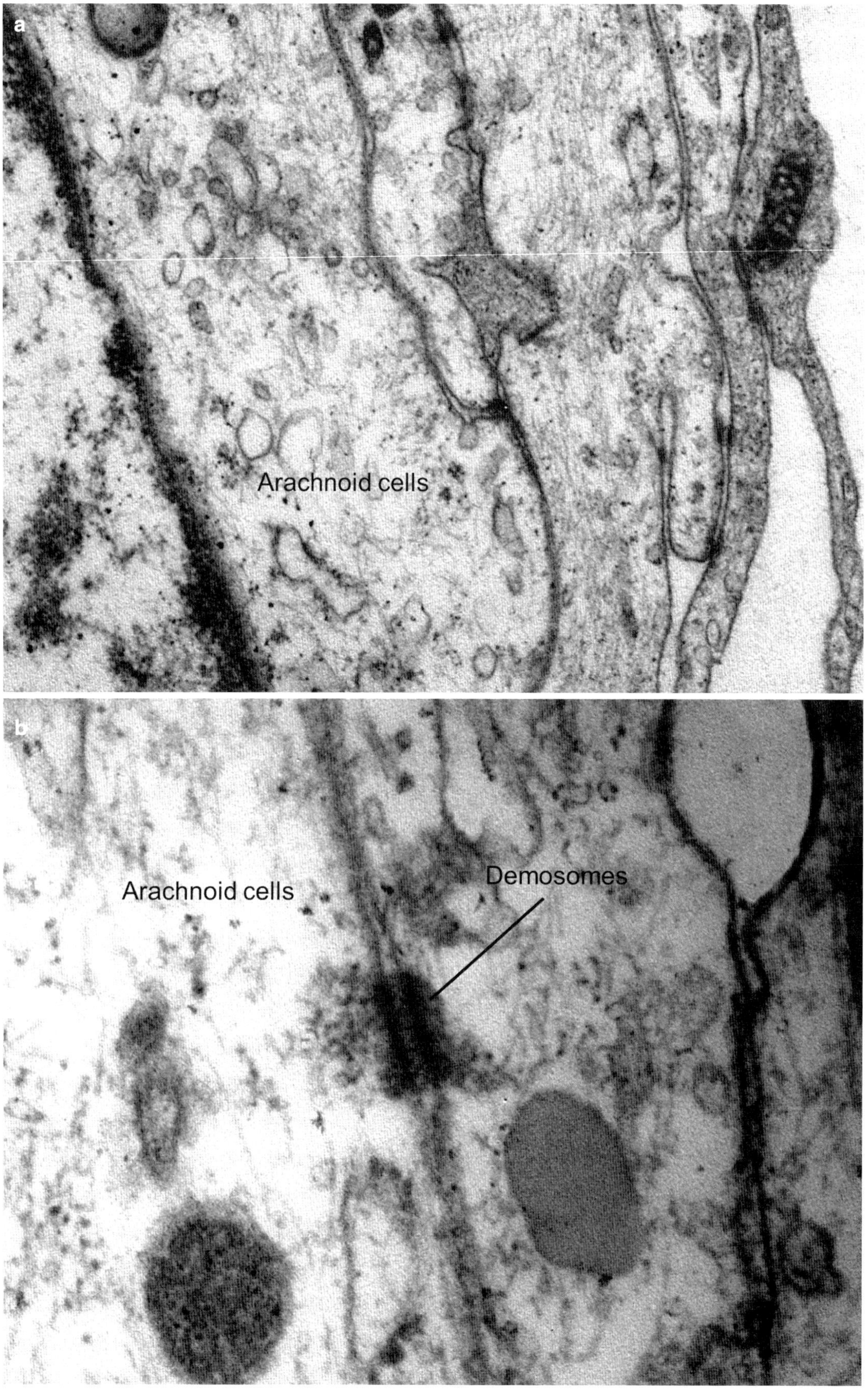

Fig. 21.20 Spinal arachnoid layer, partial thickness. Detail of arachnoid cells and specialized membrane junctions. Transmission electron microscopy. Magnification: **a**, ×50,000; **b**, ×150,000 (**b** from Reina et al. [6]; with permission)

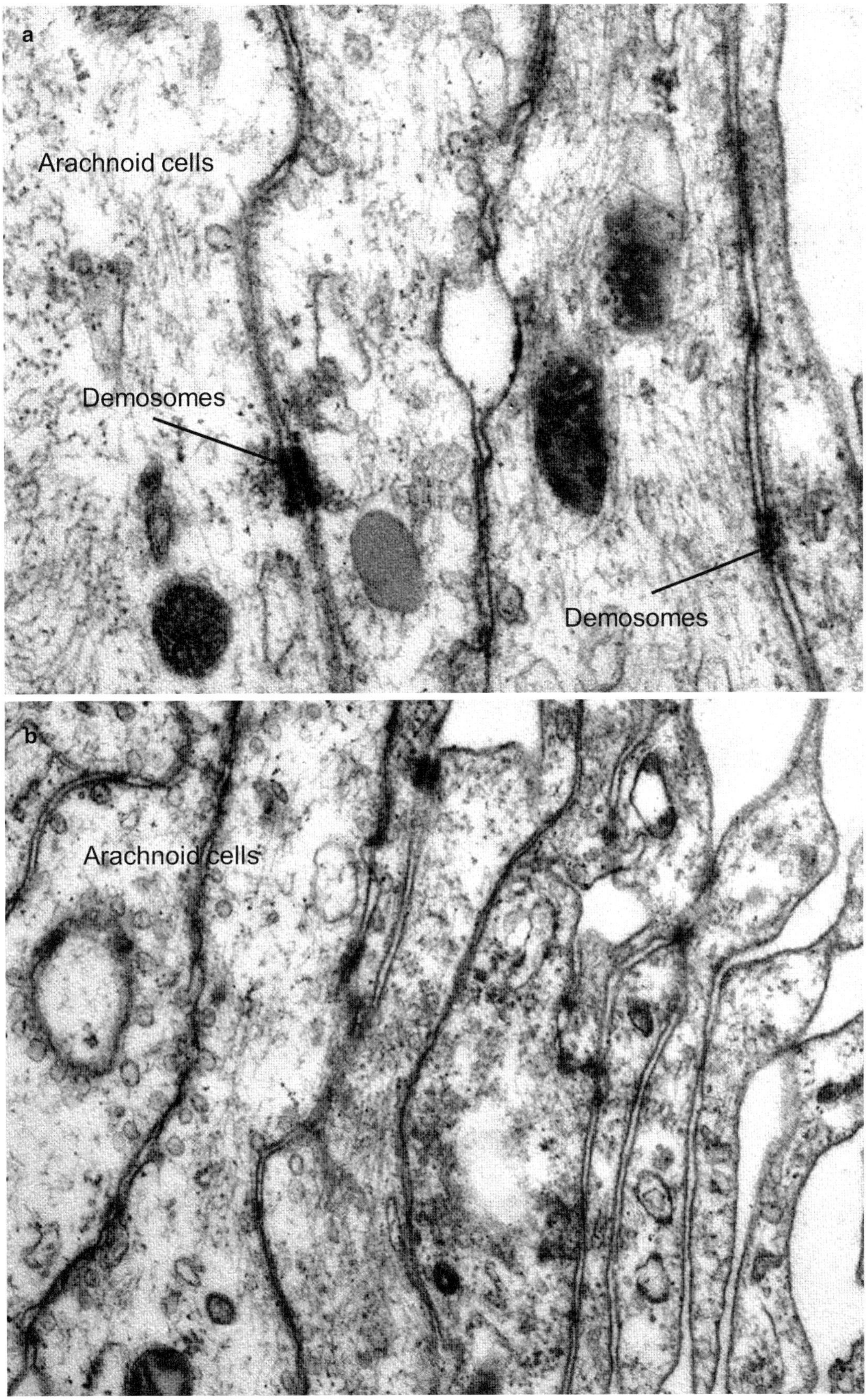

Fig. 21.21 Spinal arachnoid layer, partial thickness. Detail of arachnoid cells and specialized membrane junctions. Transmission electron microscopy. Magnification: **a**, ×80,000; **b**, ×50,000 (**a** from De Andrés et al. [3]; with permission)

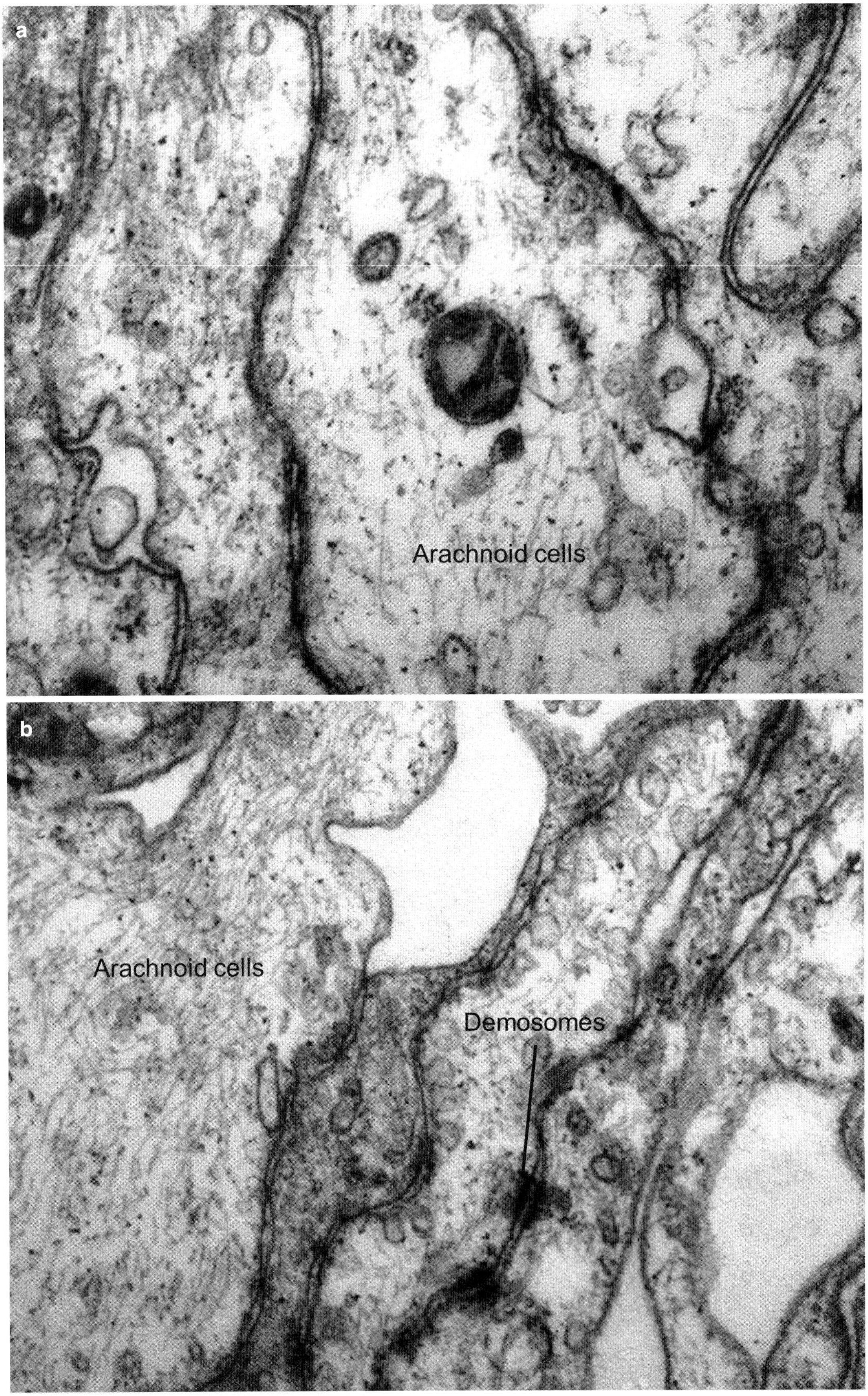

Fig. 21.22 Spinal arachnoid layer, partial thickness. Detail of arachnoid cells and specialized membrane junctions. Transmission electron microscopy. **a** and **b**, Magnification × 80,000

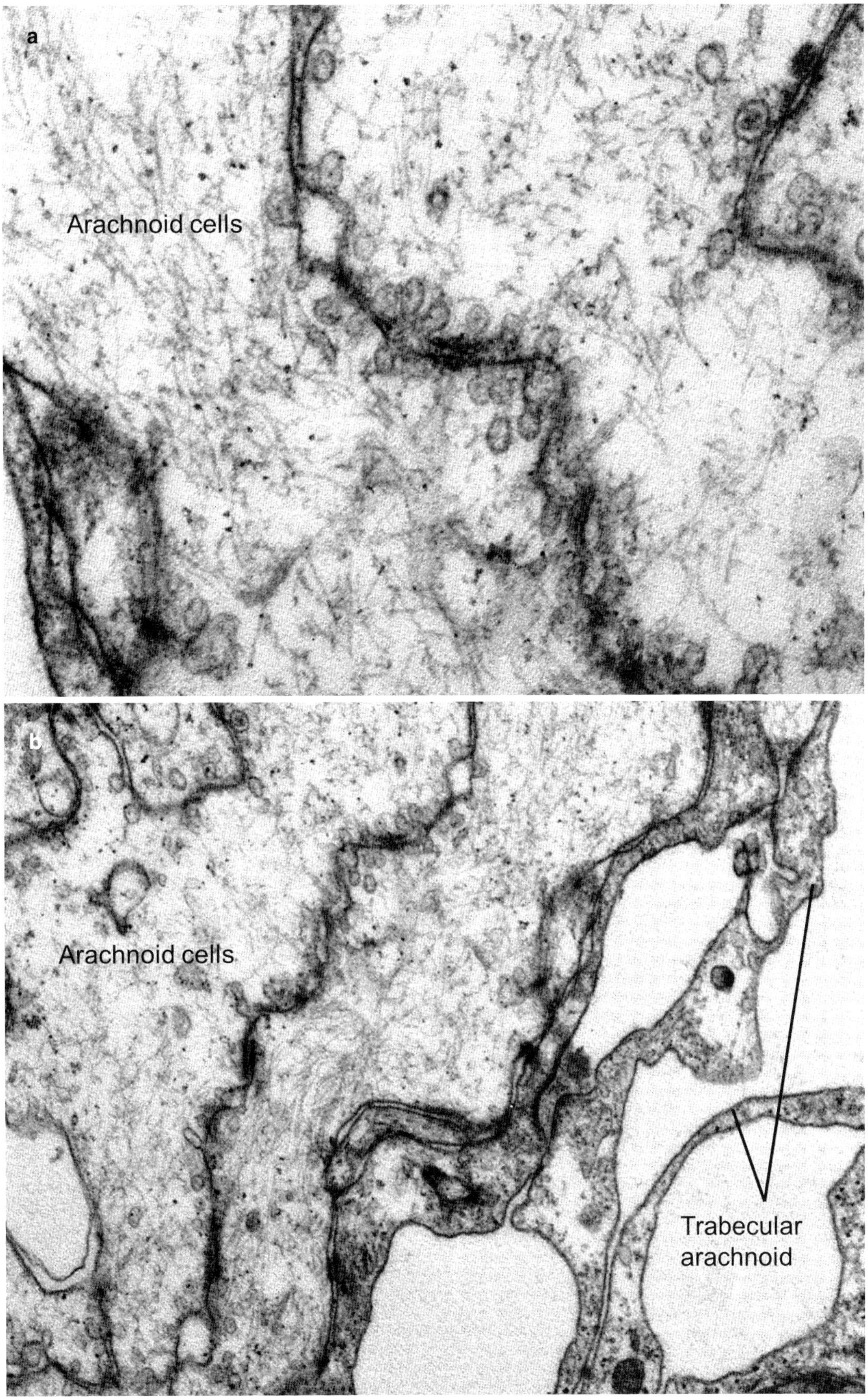

Fig. 21.23 Spinal arachnoid layer, partial thickness. Detail of arachnoid cells and specialized membrane junctions. Transmission electron microscopy. Magnification: **a**, ×80,000; **b**, ×40,000

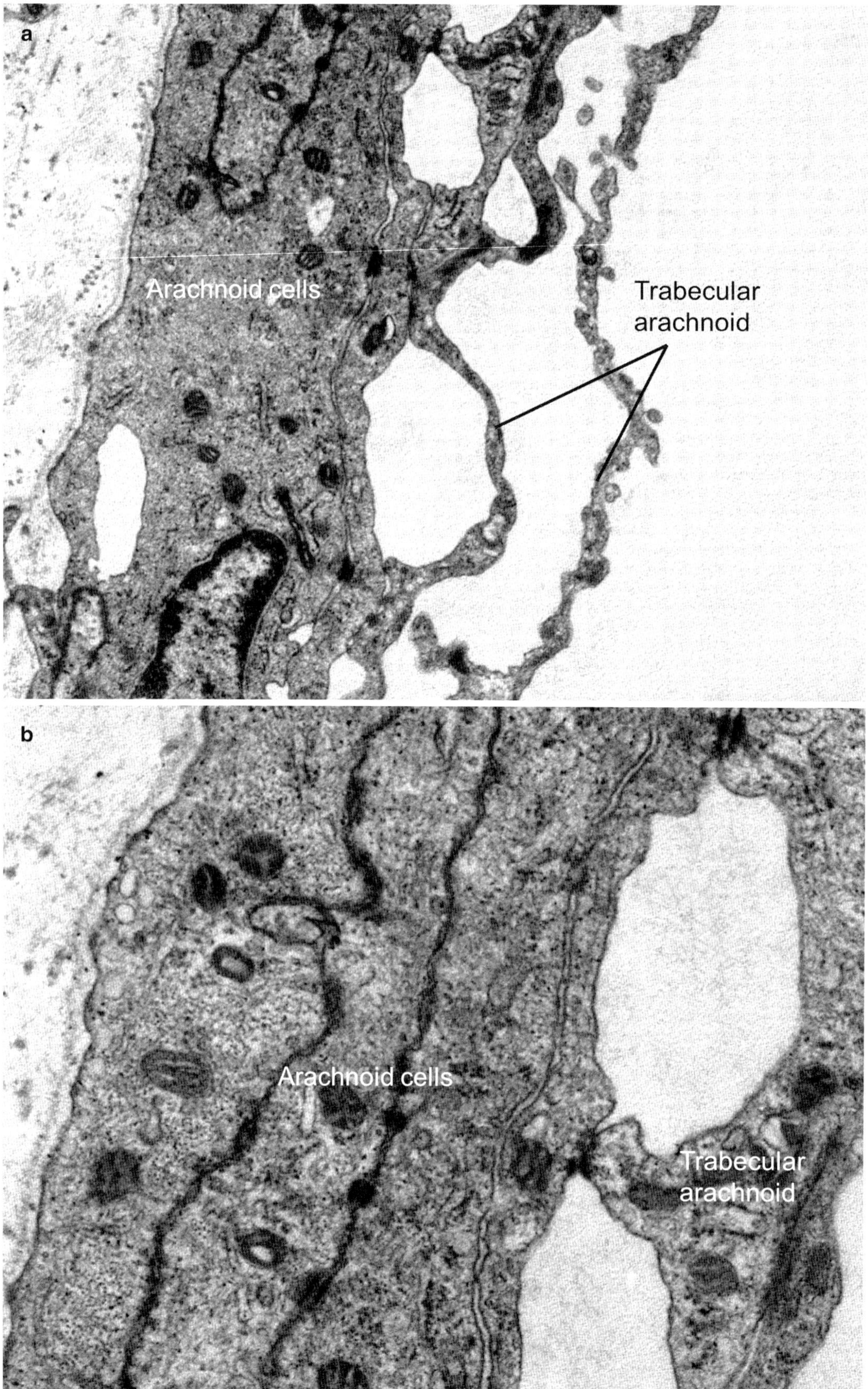

Fig. 21.24 Spinal arachnoid layer, partial thickness. Detail of arachnoid cells and specialized membrane junctions. Transmission electron microscopy. Magnification: **a**, ×25,000; **b**, ×50,000

References

1. Reina MA, Prats-Galino A, Sola RG, Puigdellívol-Sánchez A, Arriazu Navarro R, De Andrés JA. Structure of the arachnoid layer of the human spinal meninges: a barrier that regulates dural sac permeability. Rev Esp Anestesiol Reanim. 2010;57:486–92.
2. Reina MA, Pulido P, López A. The human dural sac. Morphology of the spinal dura-arachnoid membrane. Origin of the spinal subdural space. Rev Arg Anestesiol. 2007;65:167–84.
3. De Andrés JA, Reina MA, Prats A. Epidural space and regional anaesthesia. Eur J Pain Suppl. 2009;3:55–63.
4. Reina MA, De Andrés JA, Hernández JM, Arriazu Navarro R, Durán Mateos EM, Prats-Galino A. Successive changes in extraneural structures from the subarachnoid nerve roots to the peripheral nerve, influencing anesthetic block, and treatment of acute postoperative pain. Eur J Pain Suppl. 2011;5:377–85.
5. De Andrés JA, Reina MA, Hernández JM, Carrera A, Oliva A, Prats-Galino A. Role of spinal anatomical structures for neuromodulation. Reg Anesth Pain Med. 2011;36:E130–8.
6. Reina MA, Oliva A, Carrera A, Durán Mateos EM, Diamantopoulos Fernández J, Arriazu Navarro R. Origin of post dural puncture headache. Cir May Amb. 2011;16:72–84.

Three-Dimensional Reconstruction of Spinal Dural Sac

22

Alberto Prats-Galino, Miguel Angel Reina, Marija Mavar, and Anna Puigdellívol-Sánchez

This chapter describes the spatial disposition and topographical relationships of the dural sac using different three-dimensional (3D) models generated from magnetic resonance imaging (MRI) of the lumbar region. The procedures used to obtain these models are specified in Chap. 47.

The shape of the dural sac resembles an irregular cylinder, with a slight curvature of anterior convexity corresponding to the lumbar lordosis. In T1-weighted images, it appears as a hypointense area occupying the central part of the vertebral canal. In T2 MRI sequences, the dark image of the sac is replaced by the hyperintense signal generated by the cerebrospinal fluid (CSF) contained inside. On both sides, the dural sac gives rise to extensions known as dural sleeves, which surround the spinal nerve roots, extending up to their exit through the intervertebral foramina. From this point onwards, the outer sheaths of the spinal nerves replace the dura mater. In the lumbar region, the dorsal root ganglion and spinal nerve are located on the upper part of the foramen, near the vertebral pedicle (Figs. 22.1, 22.2, 22.3, 22.4, 22.5, 22.6, 22.7, 22.8, and 22.9).

The intervertebral foramen is compartmentalized by collagenous condensations called transforaminal ligaments. Increases in the thickness of these structures may lead to nerve root compression, a circumstance that has been suggested as a possible cause of radicular pain [1]. The arrangement of these ligaments is difficult to identify by imaging techniques, but it has been simulated from anatomical descriptions (Fig. 22.10) [2, 3].

A. Prats-Galino, MD, PhD (✉)
M. Mavar, MSc • A. Puigdellívol-Sánchez, MD, PhD
Laboratory of Surgical NeuroAnatomy (LSNA),
Human Anatomy and Embryology Unit, School of Medicine,
University of Barcelona, Barcelona, Spain
e-mail: aprats@ub.edu; marija.mavar@ub.edu; apuigdellivol@ub.edu

M.A. Reina, MD, PhD
Department of Clinical Medical Sciences and Institute of Applied Molecular Medicine, School of Medicine,
University of CEU San Pablo, Madrid, Spain

Department of Anesthesiology, Madrid-Montepríncipe University Hospital, Madrid, Spain
e-mail: miguelangel@perticone.e.telefonica.net

M.A. Reina et al. (eds.), *Atlas of Functional Anatomy for Regional Anesthesia and Pain Medicine: Human Structure, Ultrastructure and 3D Reconstruction Images*, DOI 10.1007/978-3-319-09522-6_22,

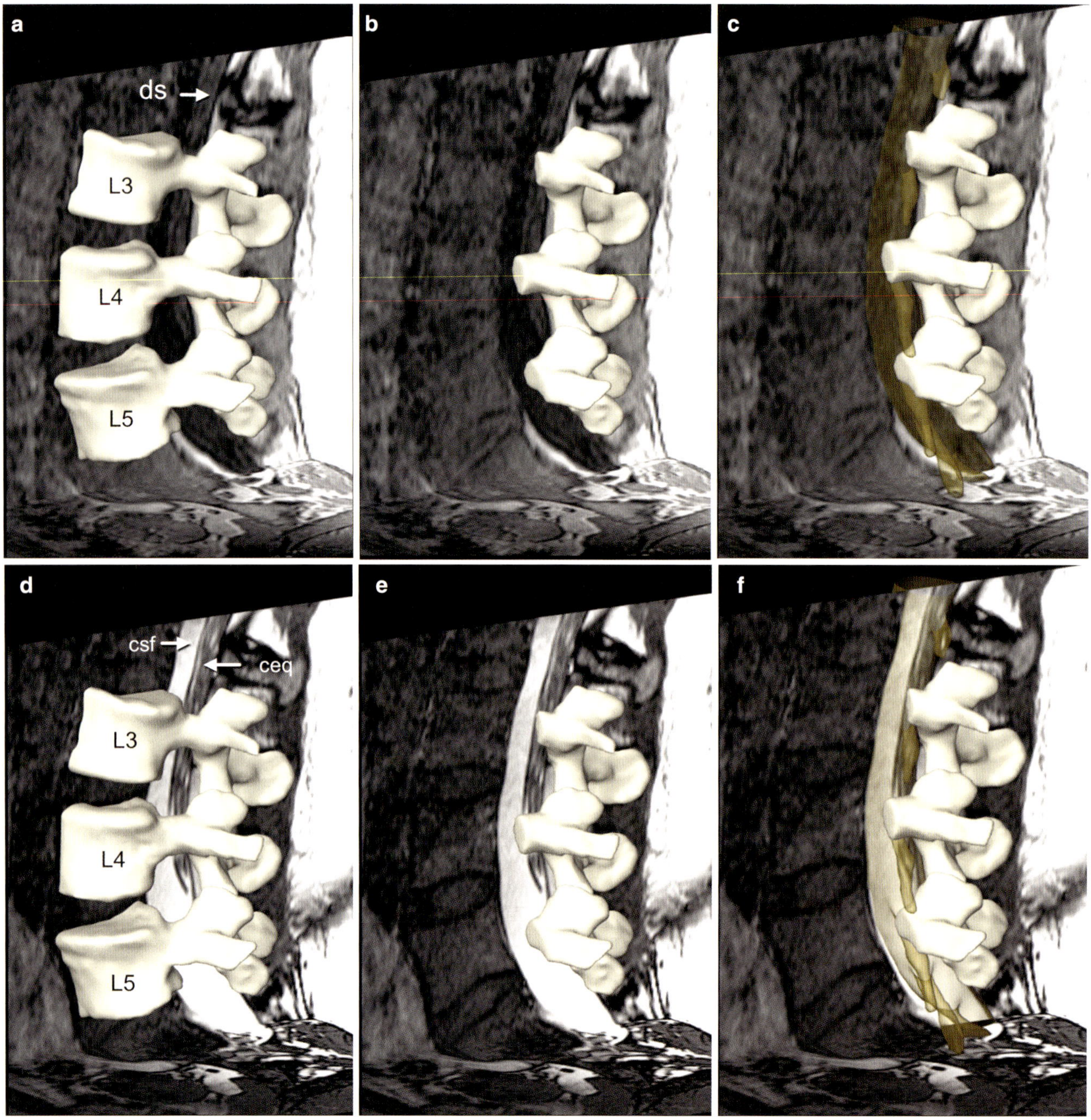

Fig. 22.1 (**a**–**c**) The dural sac (*DS*) appears on T1-weighted MR images as a dark structure that occupies the central region of the vertebral canal. (**d**–**f**) In contrast, T2-weighted images allow visualization of its contents: hyperintense cerebrospinal fluid (*CSF*) and hypointense nerve roots of the cauda equina (*CEQ*). In (**c, f**) the 3D model of the dural sac has been made partially transparent. *L3*—*L3* vertebra, *L4*—*L4* vertebra, *L5*—*L5* vertebra

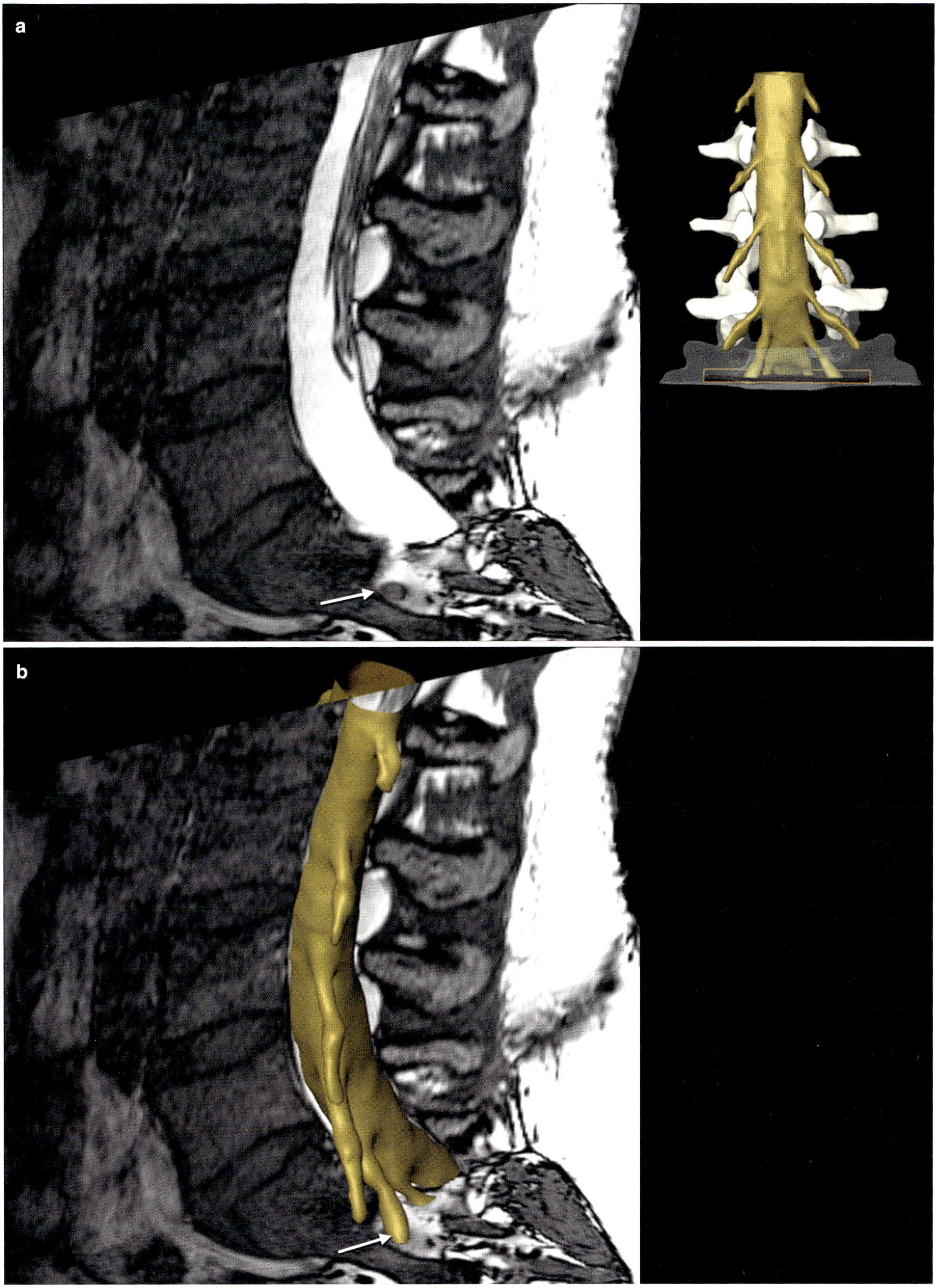

Fig. 22.2 (**a**) Axial section of T2-weighted image of a dural sleeve through the S1 dorsal root ganglion (*arrow*). The top-left image shows the cranio-caudal level of the section plane. (**b**) The reference point for the sagittal section passes approximately through the center of the dural sac (*arrow*) 3D reconstruction of dorsal root ganglion

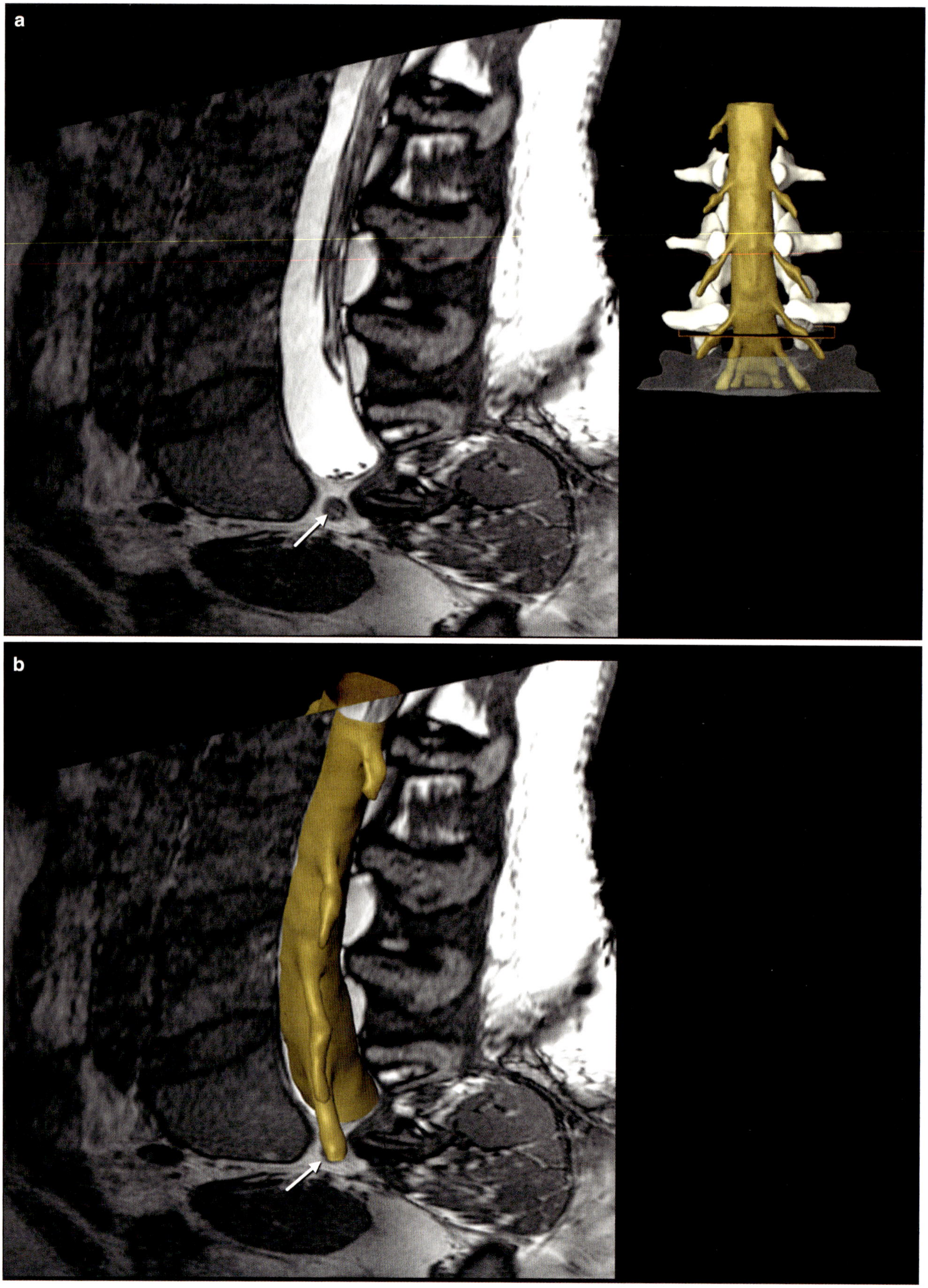

Fig. 22.3 (**a**) Axial section of T2-weighted image of a dural sleeve through the L5 dorsal root ganglion (*arrow*). The image at the top shows the cranio-caudal level of the section plane, located at the L5/S1 intervertebral foramen. (**b**) The reference point for the sagittal section passes approximately through the center of the dural sac (*arrow*) 3D reconstruction of dorsal root ganglion

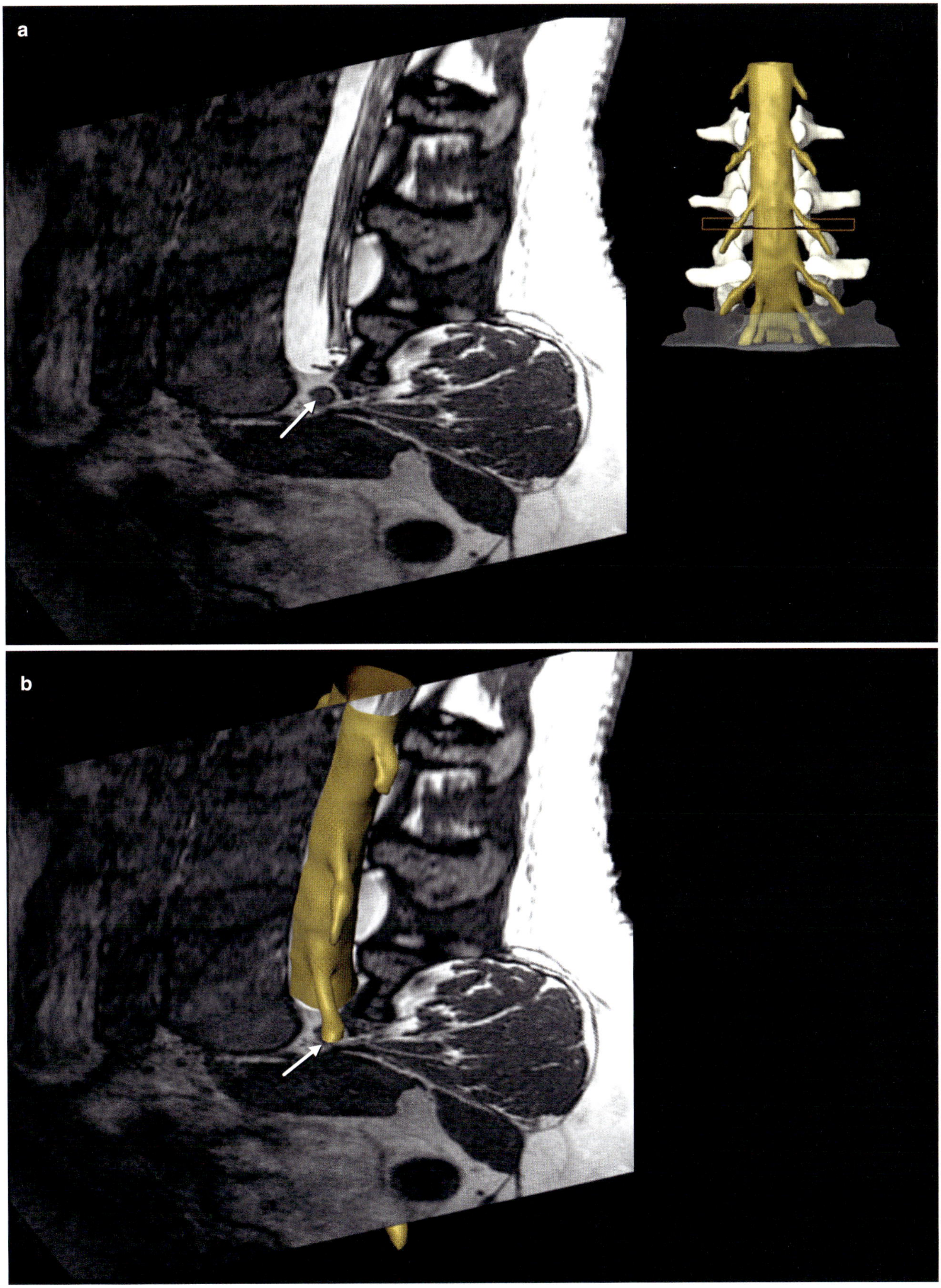

Fig. 22.4 (**a**) Axial section of T2-weighted image of a dural sleeve through the L4 dorsal root ganglion (*arrow*). The top image shows the cranio-caudal level of the section plane, located at the L4/L5 intervertebral foramen. (**b**) The reference point for the sagittal section passes approximately through the center of the dural sac (*arrow*) 3D reconstruction of dorsal root ganglion

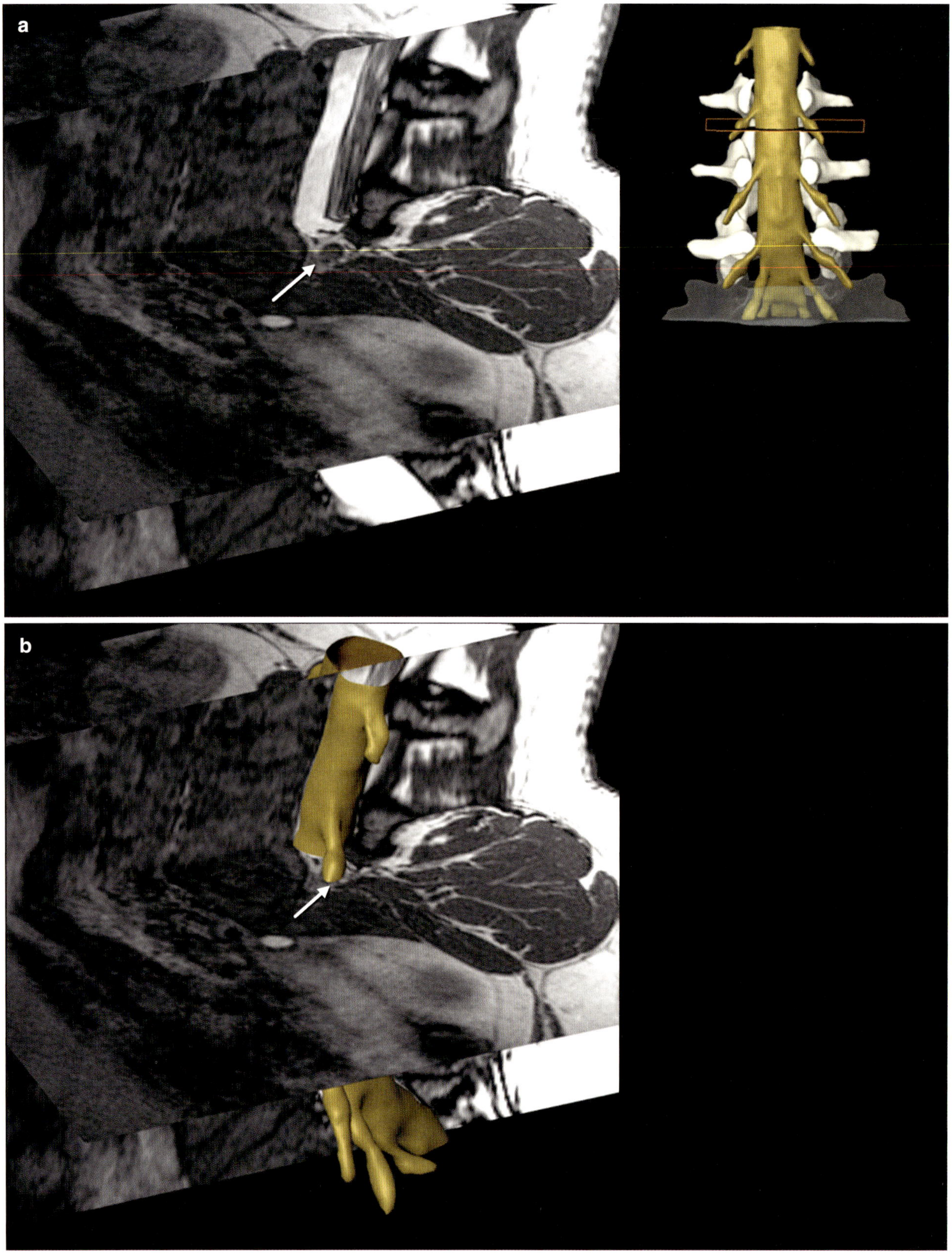

Fig. 22.5 (**a**) Axial section of T2-weighted image of a dural sleeve through the L3 dorsal root ganglion (*arrow*). The top image shows the cranio-caudal level of the section plane, located at the L3/L4 intervertebral foramen. (**b**) The reference point for the sagittal section passes approximately through the center of the dural sac (*arrow*) 3D reconstruction of dorsal root ganglion

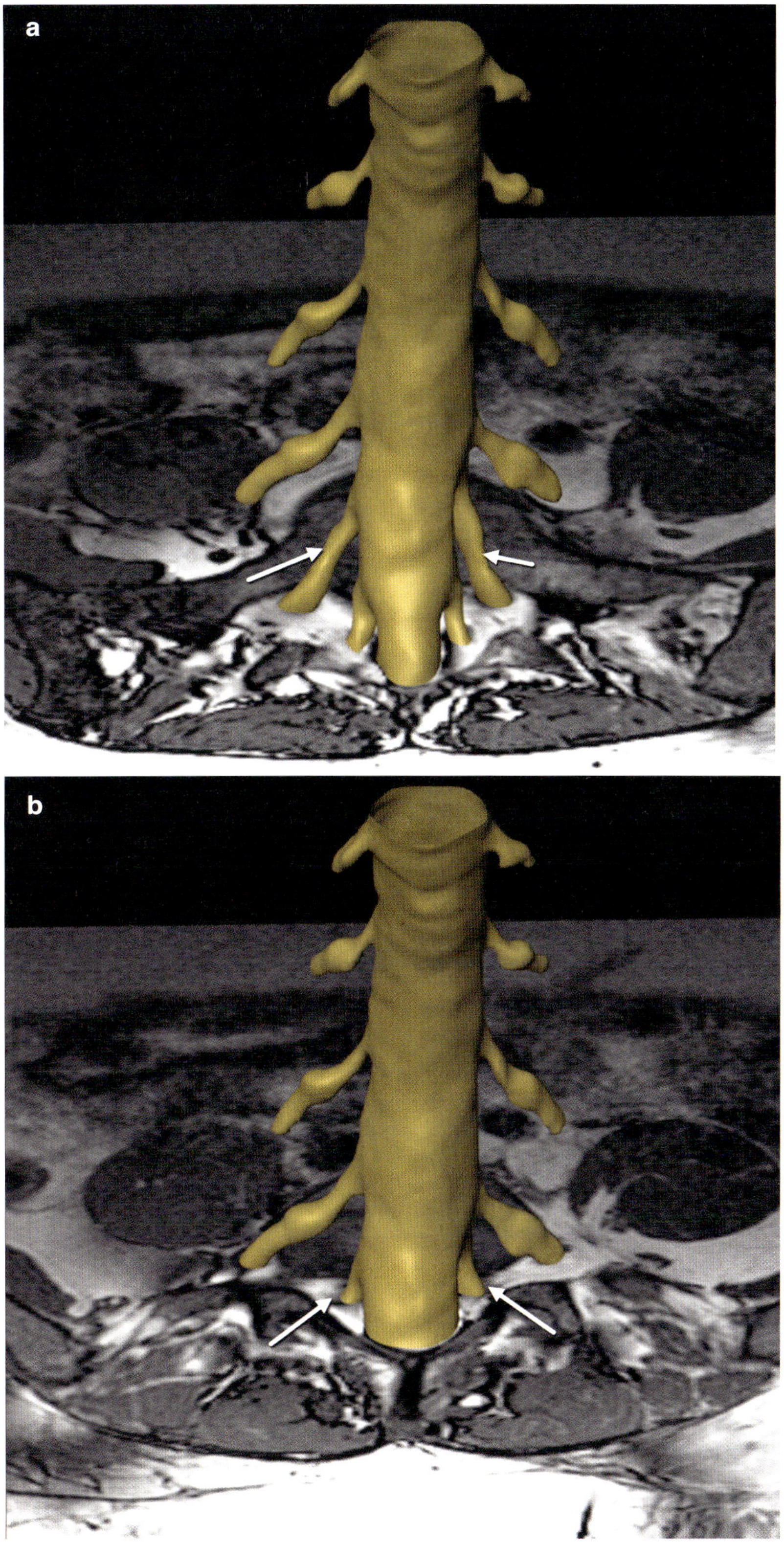

Fig. 22.6 Dorsal view of the dural sac. The reference point for the axial section of a T2-weighted image is located more caudally in (**a**) than in (**b**), at the level of the S1 vertebra. In (**b**), the section passes through the L5/S1 intervertebral disc. The S1 nerve roots are indicated by *white arrows*

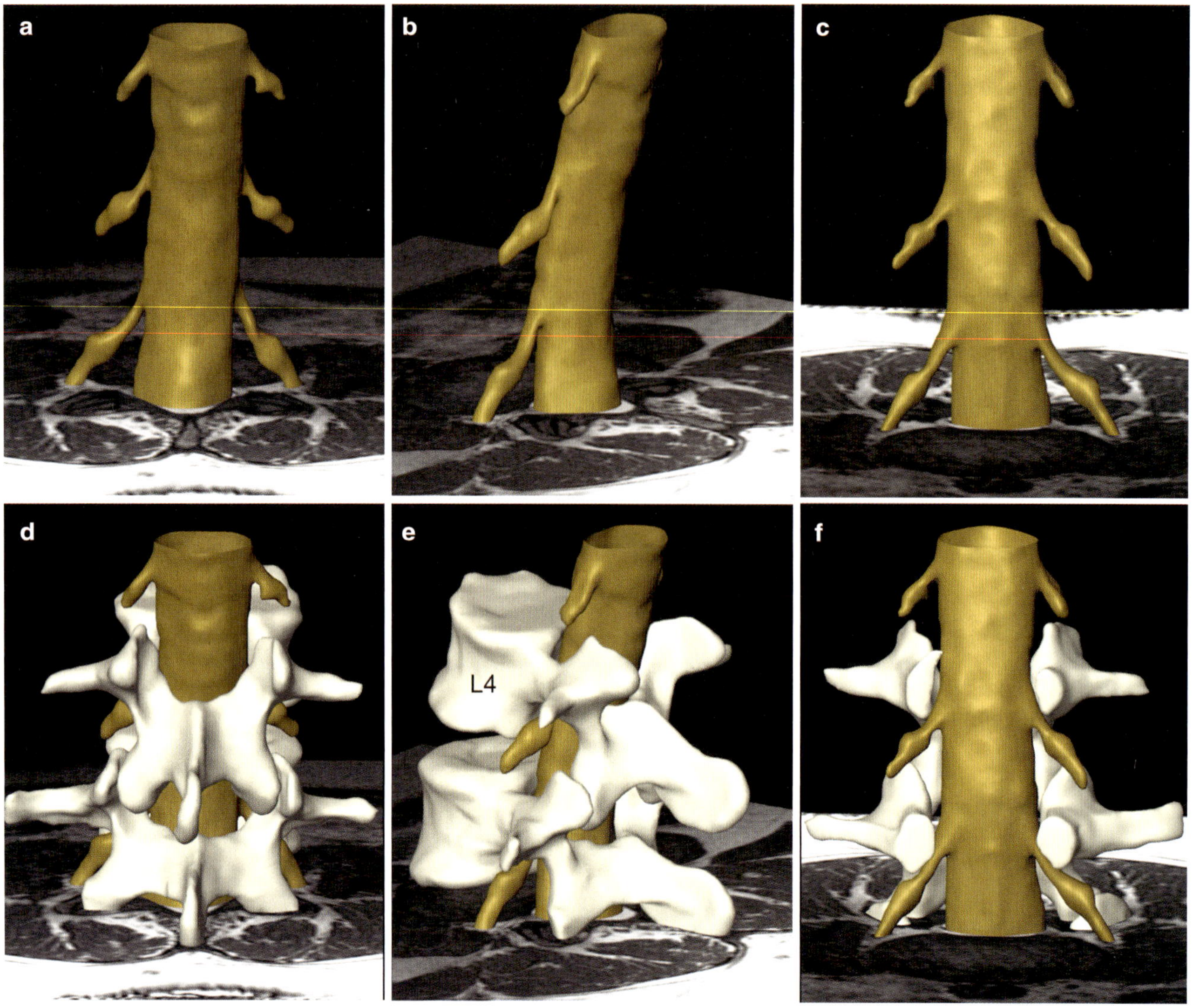

Fig. 22.7 Dorsal (*left column*, **a**, **d**), dorsolateral (*middle column*, **b**, **e**) and ventral (*right column*, **c**, **f**) views of the dural sac. The reference point for the axial section of a T2-weighted image is located at the level of the postganglionic segment of the L5 nerve root, after its exit through the intervertebral foramen. In (**f**), the vertebral bodies of the L3 and L4 model have been removed to enable visualization of the dural sac. *L4*—*L4* vertebra

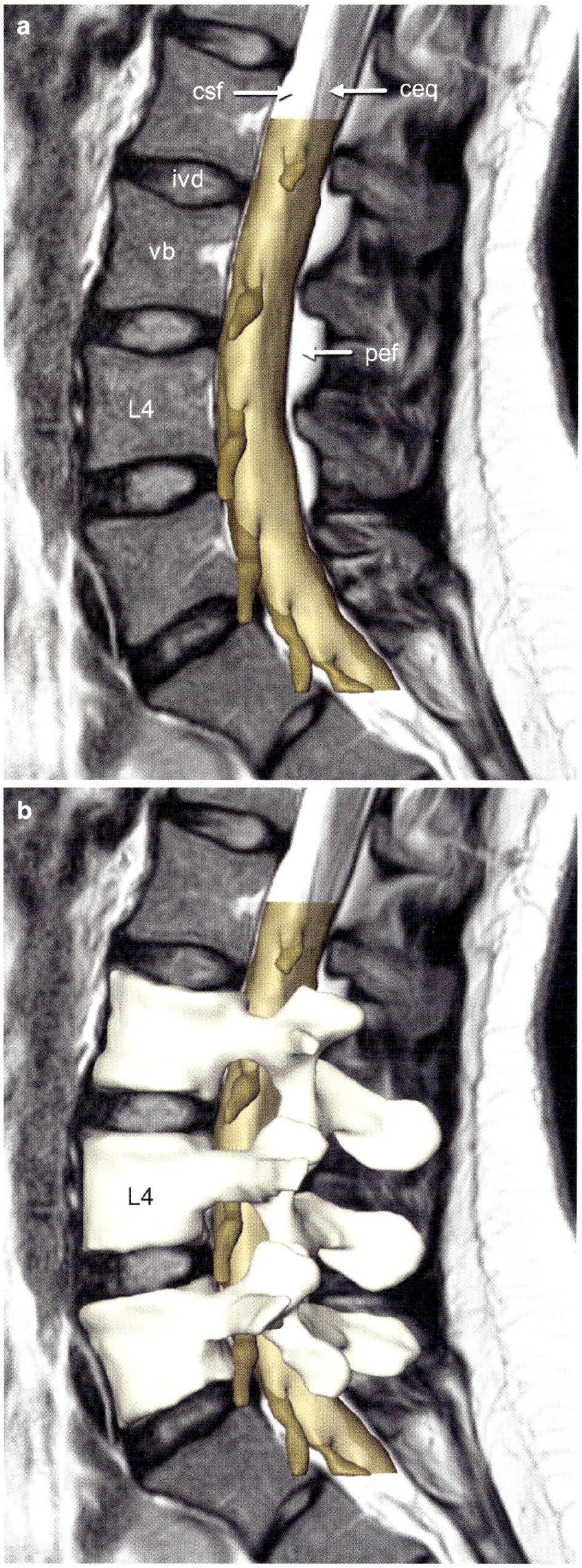

Fig. 22.8 Relationships of the dural sac with the vertebral bodies (*vb*) and intervertebral discs (*ivd*) anteriorly, and with the posterior epidural fat (*pef*) posteriorly. (**a**) The 3D model of the dural sac is represented as partially transparent in a lateral view. (**b**) At the lumbar region, the dorsal root ganglion and spinal nerve roots tend to occupy the upper part of the intervertebral foramen. The reference point for the sagittal section corresponds to a T2-weighted image. *ceq* cauda equina, *csf* cerebrospinal fluid

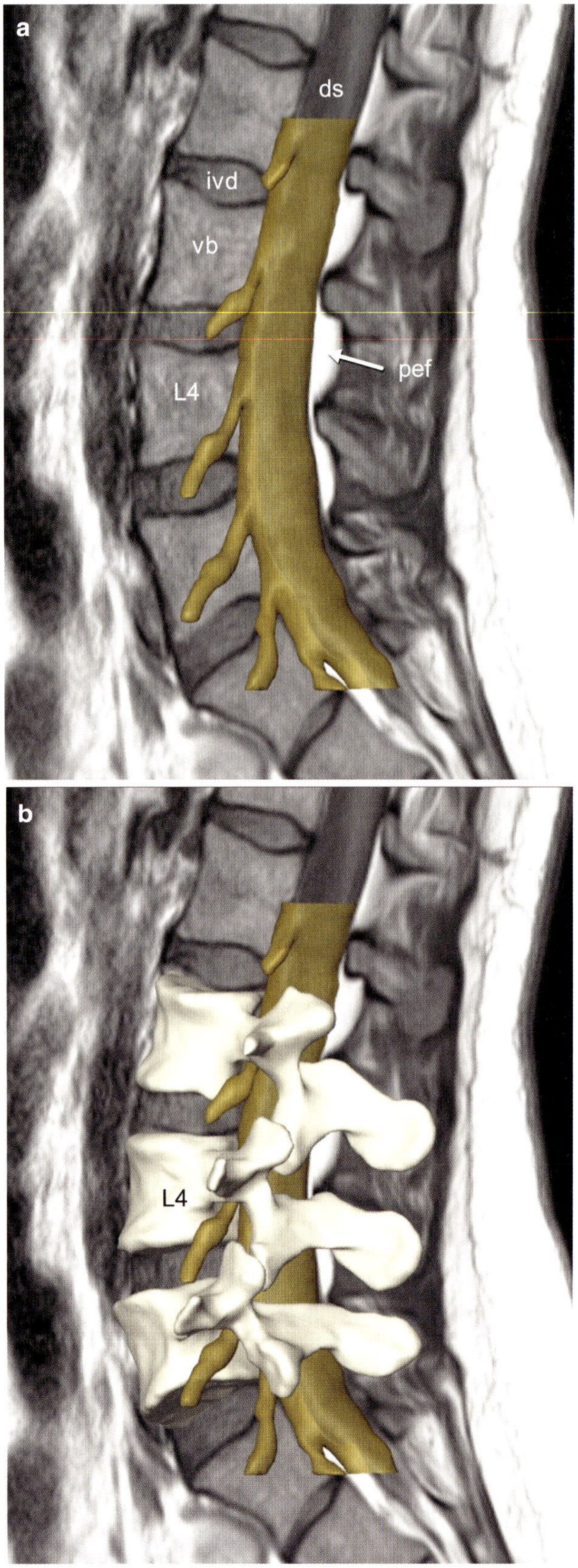

Fig. 22.9 Relationships of the dural sac (*ds*) with the posterior epidural fat (*pef*). In this case, the 3D model of the dural sac is represented in a dorsolateral view, and the reference point for the sagittal section corresponds to a T1-weighted image. *ivd* intervertebral disc, *L4*—*L4* vertebra, *vb* vertebral body. (**a**) 3D reconstruction of dural sac (**b**) 3D reconstruction including dural sac and lumbar vertebrae

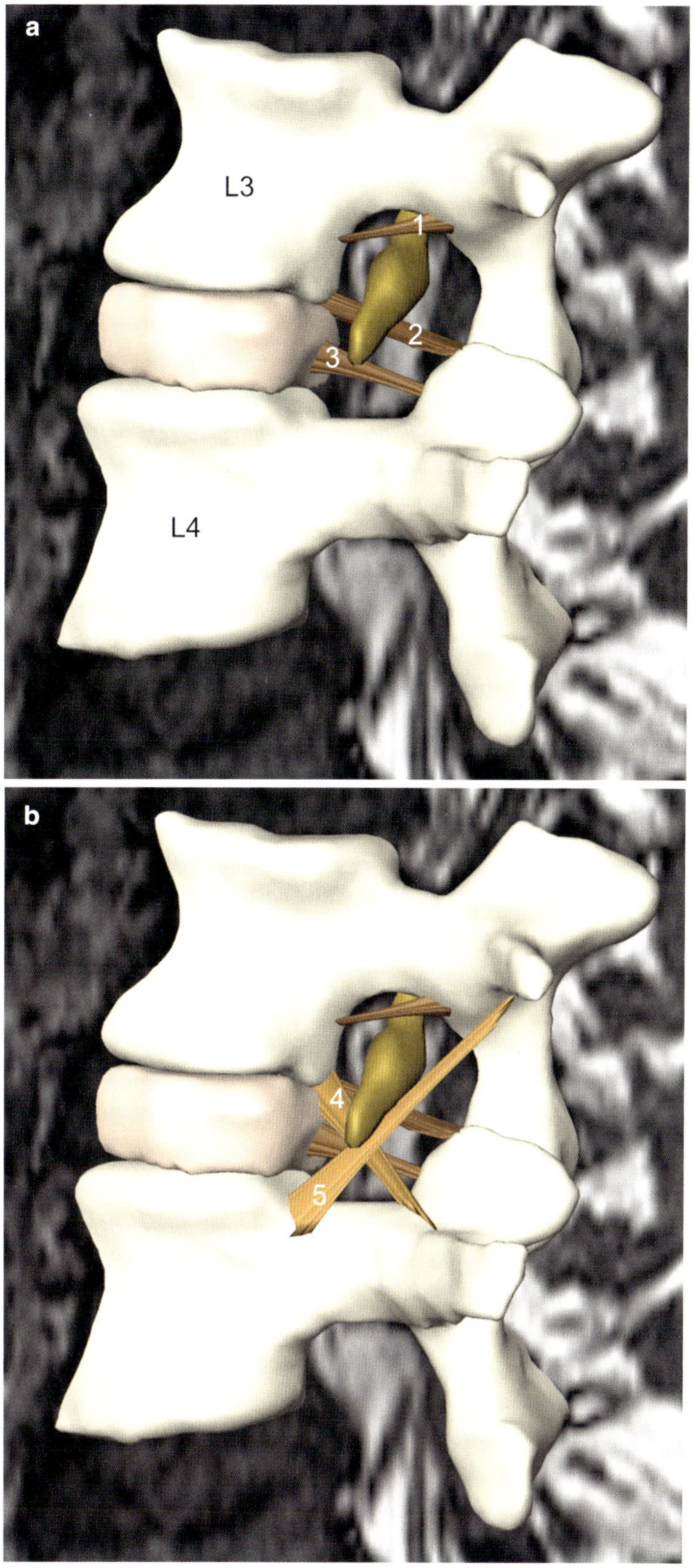

Fig. 22.10 Representation of the main transforaminal ligaments, a group of collagen condensations that compartmentalize the intervertebral foramen. Some of these ligaments have close topographical relationship with the distal dural sleeve enclosing the spinal nerve roots. Their thickening could compress the nerve root. The disposition of the transforaminal ligaments is difficult to identify by imaging techniques, and has been simulated from anatomical descriptions. The ligaments tend to be arranged in deep (*2*) and middle (*1*, *3*) layers (**a**), and superficial (*4*, *5*) layers (**b**). The reference point for the sagittal section corresponds to a T2-weighted image passing through the internal margin of the intervertebral foramen. *1* oblique superior transforaminal ligament, *2* mid-transforaminal ligament, *3* oblique inferior transforaminal ligament, *4* superior corporotransverse ligament, *5* inferior corporotransverse ligament

References

1. Akdemir G. Thoracic and lumbar intraforaminal ligaments. J Neurosurg Spine. 2010;13:351–5.
2. Amonoo-Kuofi HS, El-Badawi MG, Fatani JA, Butt MM. Ligaments associated with lumbar intervertebral foramina. 1. L1 to L4. J Anat. 1988;156:177–83.
3. Min JH, Kang SH, Lee JB, Cho TH, Suh JG. Anatomic analysis of the transforaminal ligament in the lumbar intervertebral foramen. Neurosurgery. 2005;57(1 Suppl):37–41.

Three-Dimensional Reconstruction of Spinal Epidural Fat

23

Alberto Prats-Galino, Juan A. Juanes Méndez,
Miguel Angel Reina, and José Antonio De Andrés

Magnetic resonance imaging is a particularly useful tool that enables the study of ultrastructural aspects of epidural fat [1, 2]. Fat is the main component of the epidural space in the dural sac and protects its contents against the effects of lashing, deceleration, and rotational forces inflicted on the vertebral column. Since epidural fat does not have a semifluid nature, it contributes to the shape of the epidural space in the spinal canal and is grouped in packs inside the epidural space by pedicles containing vessels [3–5]. This epidural fat is a lipophilic deposit of interest in neuraxial anesthetic techniques [6, 7].

Lumbar epidural fat has a metameric and discontinuous topography. It is mainly located in the posterior aspect of the epidural space.

The epidural space is in part real and in part virtual [3–5]. The epidural fat occupies the real epidural space. Three-dimensional (3D) images of epidural fat represent this real space [1, 2].

The epidural fat extends in the craniocaudal direction from the lower aspect of one vertebral lamina to the upper portion of the lamina of the next caudal vertebra and, in the lateral direction, toward the point where articular facets and ligamentum flavum meet. It also fills the space between vertebral arches and intervertebral foramina, wrapping the nerve root cuffs. The epidural fat does not usually adhere to these structures, allowing mobility of the dura within the vertebral canal. However, small amounts of epidural fat covered by a thin connective membrane do adhere to the posterior midline by a vascular pedicle entering the epidural fat at the plane where the right and left portions of the ligamentum flavum meet [5].

The volume of the posterior epidural fat deposits increases caudally, from L_{1-2} to L_{4-5}. Epidural fat in the posterior aspect of the epidural space is uniformly distributed, whereas in the lateral areas fat is compartmented in groups of cells by a net of connective tissue that extends from the intervertebral foramen to the posterior longitudinal ligament. Adjacent to each vertebral disc, connective tissue fills the space left between the posterior longitudinal ligament and each vertebral disc. In the anterior aspect of the epidural space, epidural veins are often found with dura mater attaching to vertebral discs [3].

Epidural fat has different distributions in the cervical, thoracic, lumbar, and sacral areas. This distribution remains constant along each vertebral level. At the cervical level, fat is absent or almost inexistent in the anterior and lateral aspects of the epidural space. In the posterior aspect, a small deposit is sometimes seen. At thoracic levels, the epidural fat forms a broad posterior band with "indentations." It is thicker near the intervertebral disc and less prominent around the middle section of the vertebral bodies and close to the base of the spinous processes. Extending from the upper toward the middle thoracic levels (T_{1-7}), epidural fat is continuous and the indentations are more evident. In the lower thoracic region (T_{8-12}), the distribution of epidural fat becomes patchy. At lumbar levels, the epidural fat in the anterior and posterior

A. Prats-Galino, MD, PhD (✉)
Laboratory of Surgical NeuroAnatomy (LSNA), Human Anatomy and Embryology Unit, School of Medicine,
University of Barcelona, Barcelona, Spain
e-mail: aprats@ub.edu

J.A. Juanes Méndez, MD, PhD
Human Anatomy and Embryology Unit,
School of Medicine, University of Salamanca, Salamanca, Spain
e-mail: jajm@usal.es

M.A. Reina, MD, PhD
Department of Clinical Medical Sciences and Institute of Applied Molecular Medicine, School of Medicine,
University of CEU San Pablo, Madrid, Spain

Department of Anesthesiology, Madrid-Montepríncipe University Hospital, Madrid, Spain
e-mail: miguelangel@perticone.e.telefonica.net

J.A. De Andrés, MD, PhD
Department of Surgical Specialties, School of Medicine,
University of Valencia, Valencia, Spain

Department of Anesthesia, Critical Care and Pain Management,
General University Hospital, Valencia, Spain
e-mail: deandres_jos@gva.es

M.A. Reina et al. (eds.), *Atlas of Functional Anatomy for Regional Anesthesia and Pain Medicine: Human Structure, Ultrastructure and 3D Reconstruction Images*, DOI 10.1007/978-3-319-09522-6_23,

aspects of the epidural space forms two unconnected structures. The posterior epidural fat acquires its greatest volume in caudal lumbar levels [5].

Images of lumbar vertebrae and epidural fat illustrate how the tip of the epidural needle crosses a real epidural space during the performance of epidural anesthetic techniques. On the other hand, the virtual epidural space is placed between the dural sac and vertebral lamina without interposing fat (Figs. 23.1, 23.2, 23.3, 23.4, 23.5, 23.6, 23.7, 23.8, 23.9, 23.10, and 23.11).

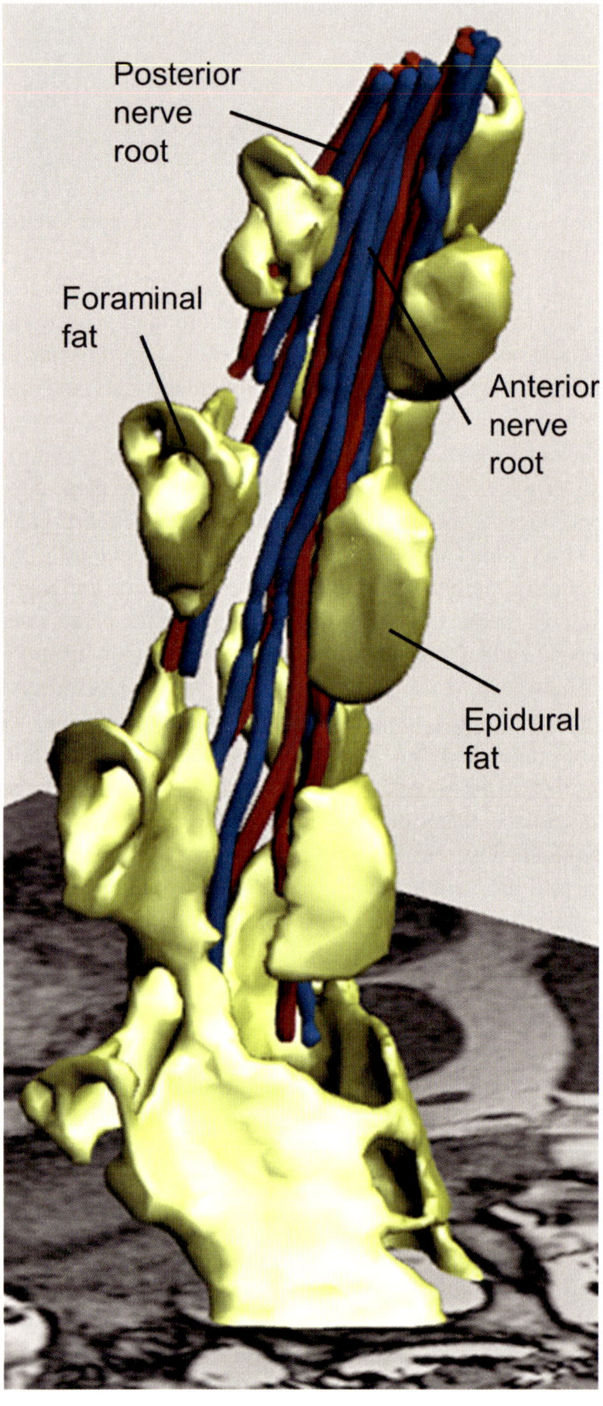

Fig. 23.1 Three-dimensional reconstruction of human spinal epidural fat. Anterior nerve roots (*red*), posterior nerve roots (*blue*), posterior and lateral epidural fat

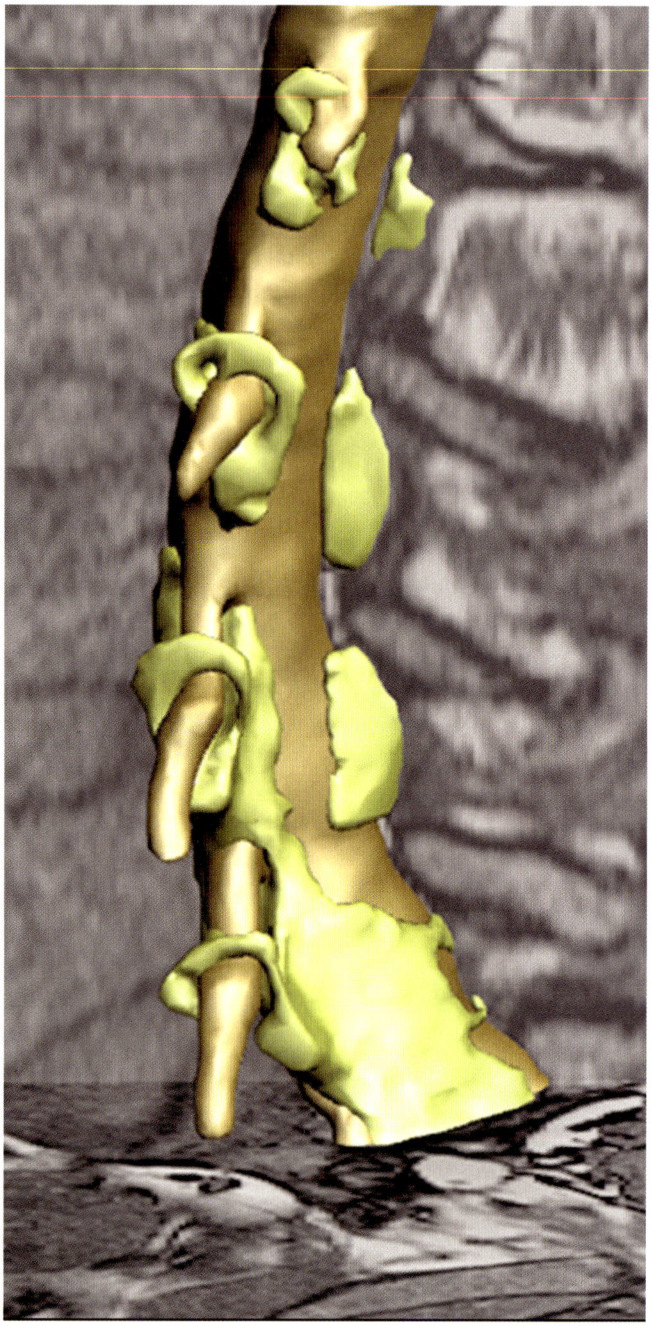

Fig. 23.2 Three-dimensional reconstruction of human spinal epidural fat. Lateral view of spinal epidural fat superimposed on its magnetic resonance image

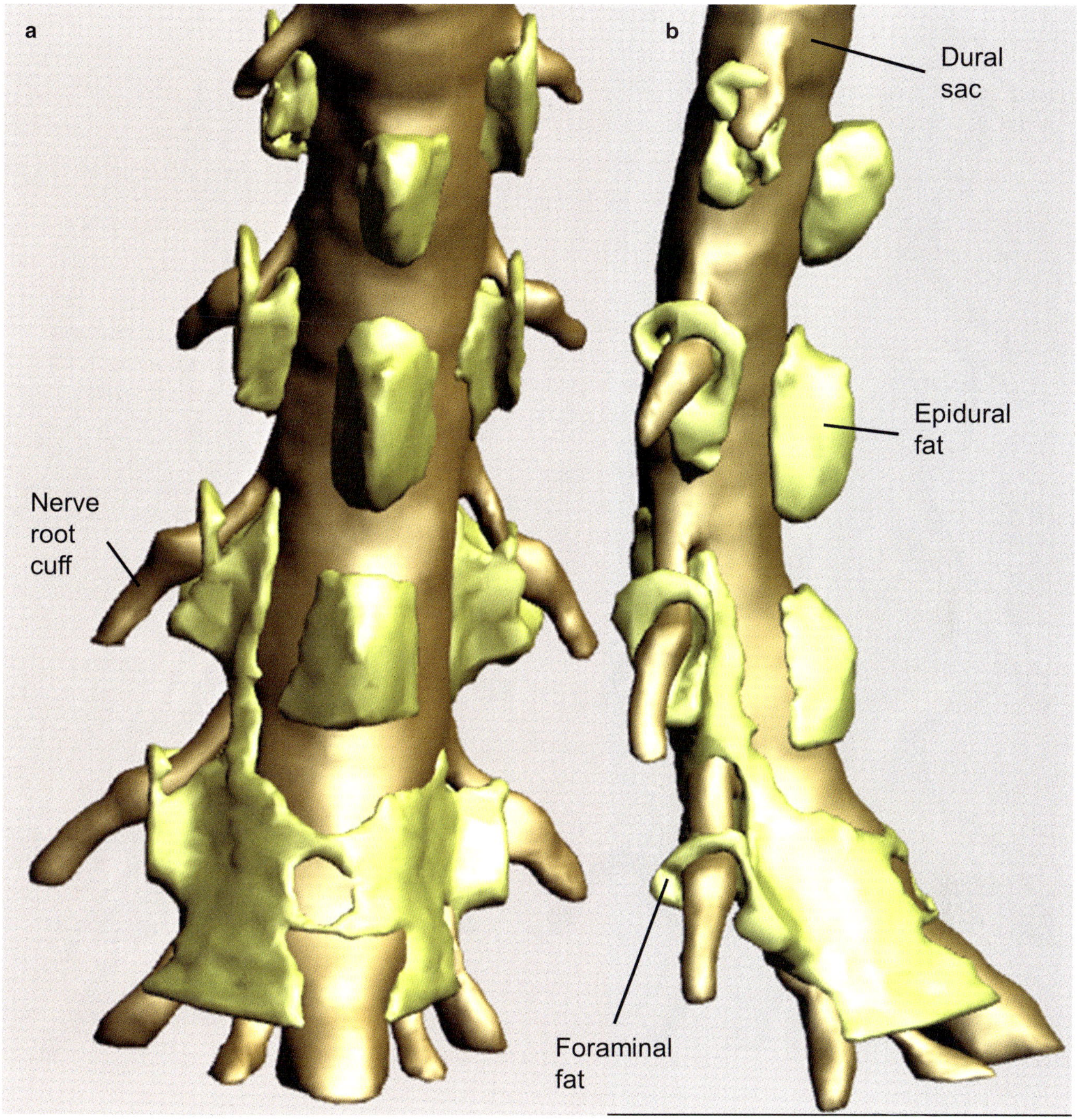

Fig. 23.3 Three-dimensional reconstruction of human spinal epidural fat. Posterior (**a**) and lateral (**b**) view of spinal epidural fat

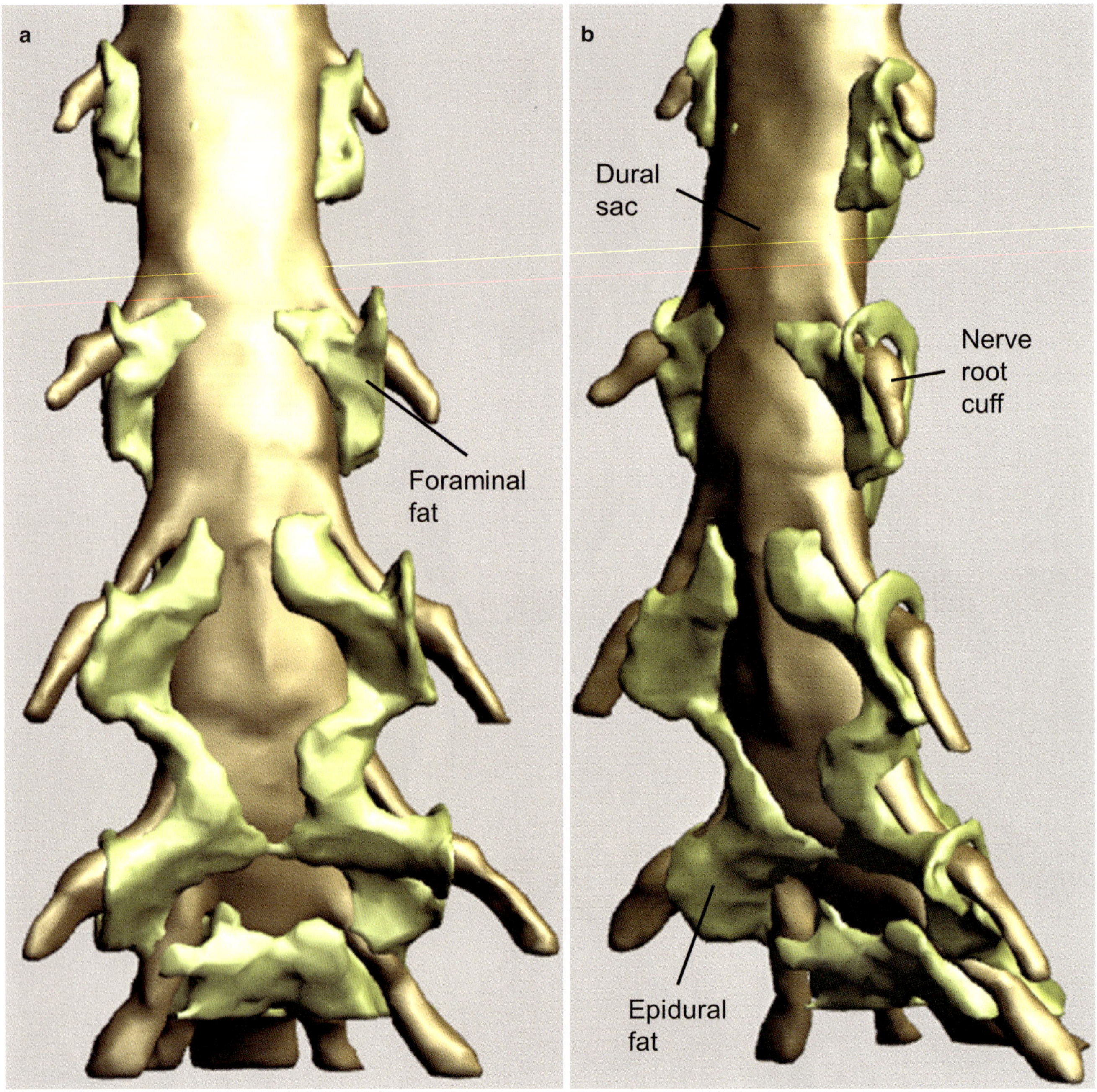

Fig. 23.4 Three-dimensional reconstruction of human spinal epidural fat. (**a**, **b**) Anterior view of spinal epidural fat

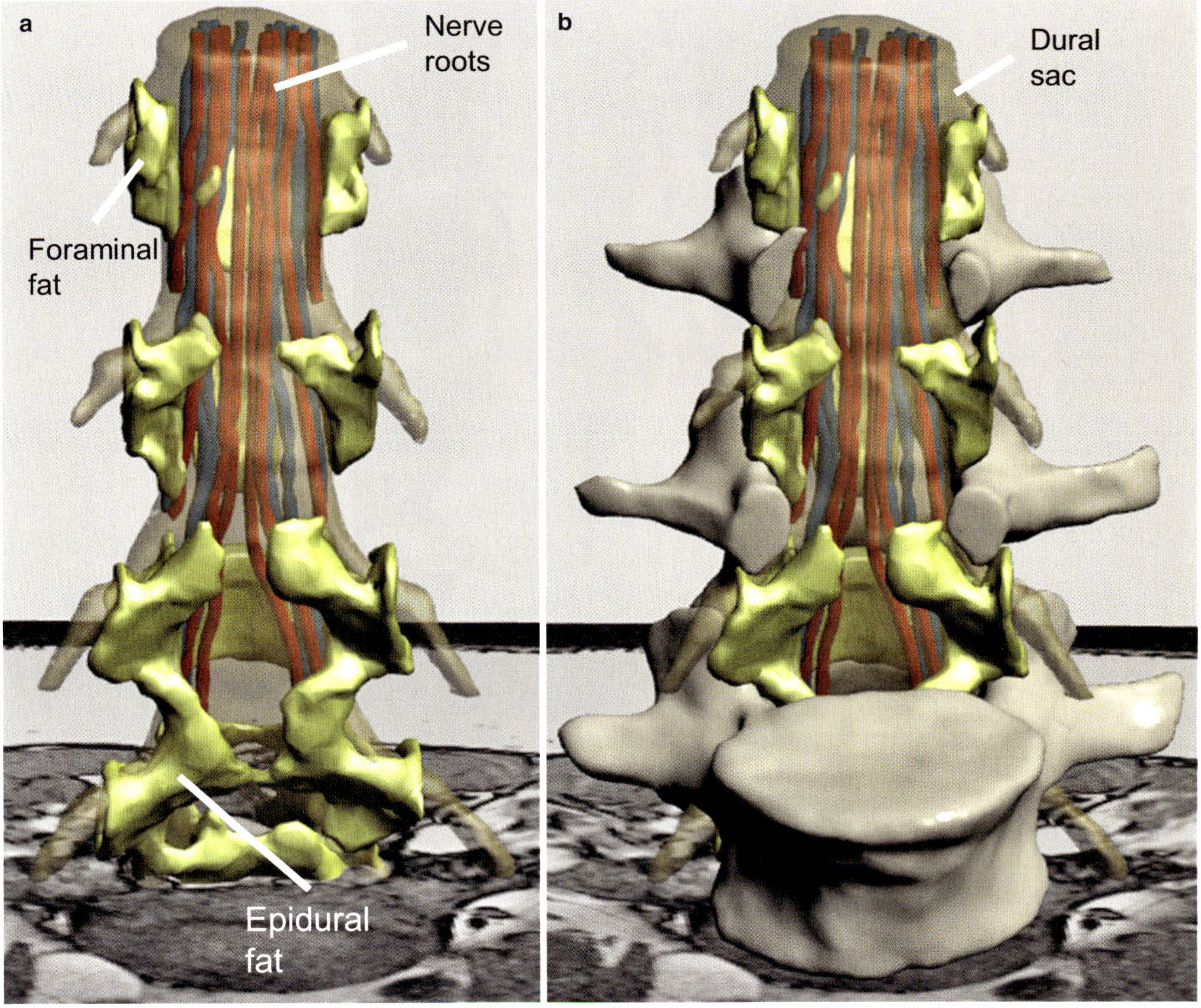

Fig. 23.5 Three-dimensional reconstruction of human spinal epidural fat. (**a**, **b**) Anterior view of spinal epidural fat and nerve roots through translucent dural sac

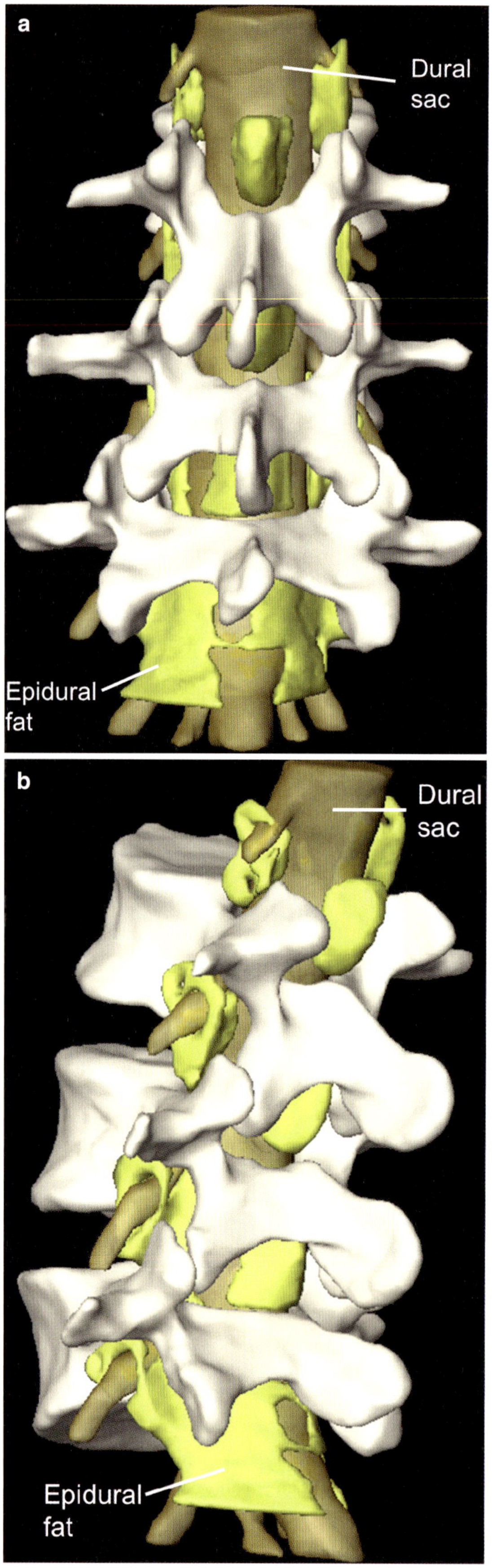

Fig. 23.6 Three-dimensional reconstruction of human spinal epidural fat. Posterior (**a**) and lateral (**b**) view of epidural fat from interlaminar foramen

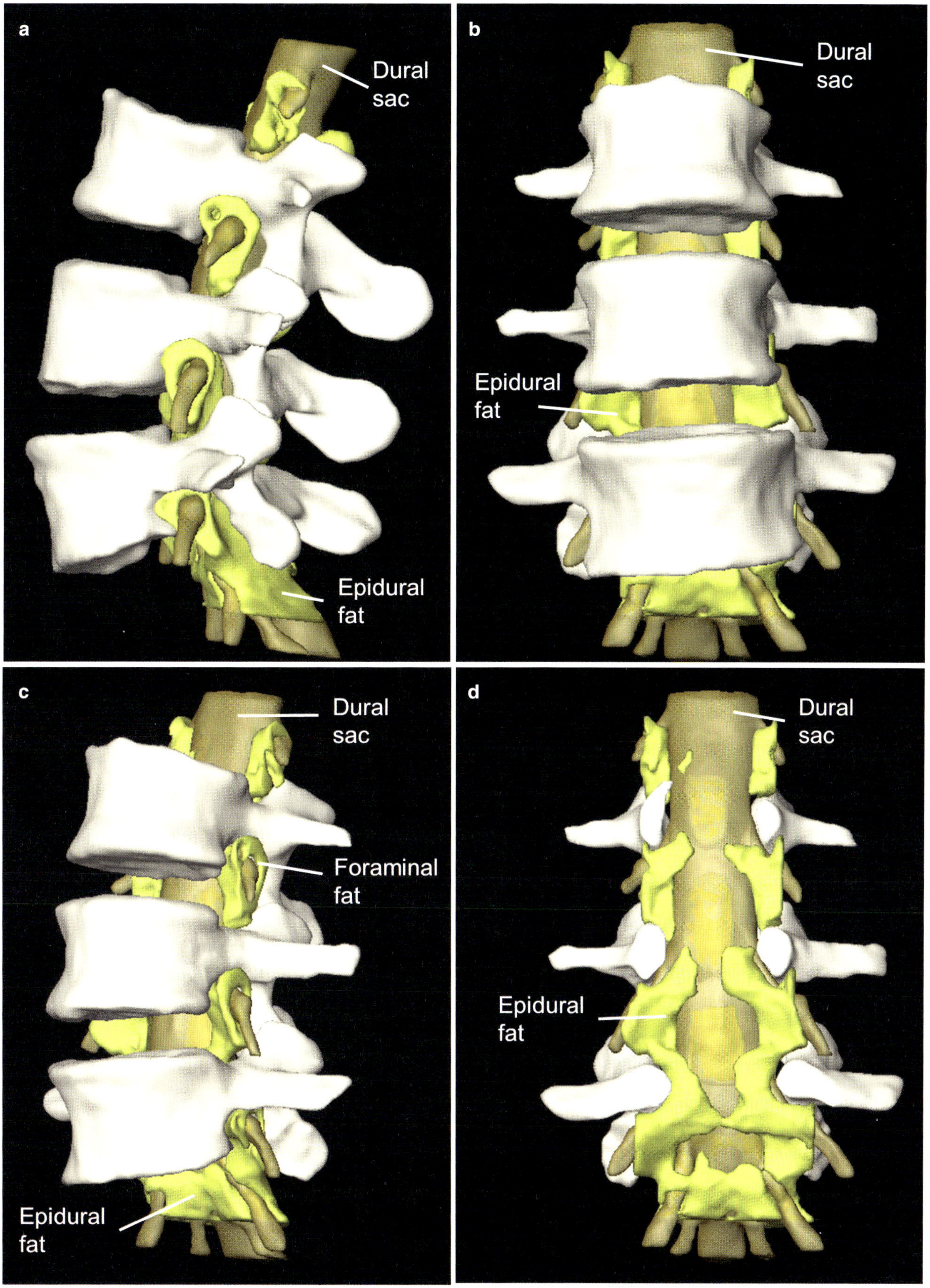

Fig. 23.7 Three-dimensional reconstruction of human spinal epidural fat. Lateral (**a**) anterior (**b**) and oblique (**c**) view of epidural fat among vertebrae. Anterior (**d**) complete view of epidural fat without vertebrae

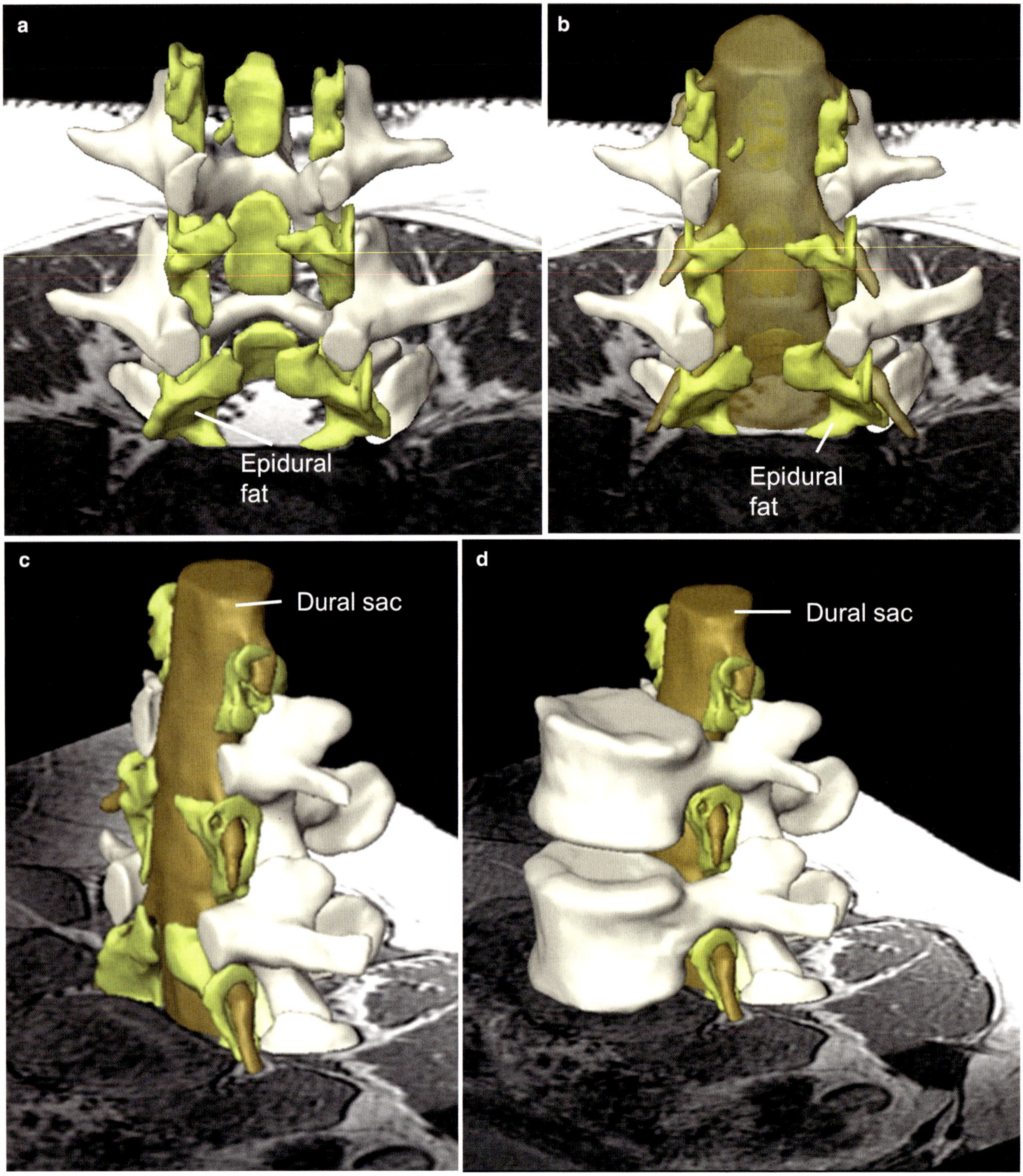

Fig. 23.8 Three-dimensional reconstruction of human spinal epidural fat. Anterior (**a**, **b**) and lateral (**c**, **d**) view of epidural fat and foraminal fat into intervertebral foramen

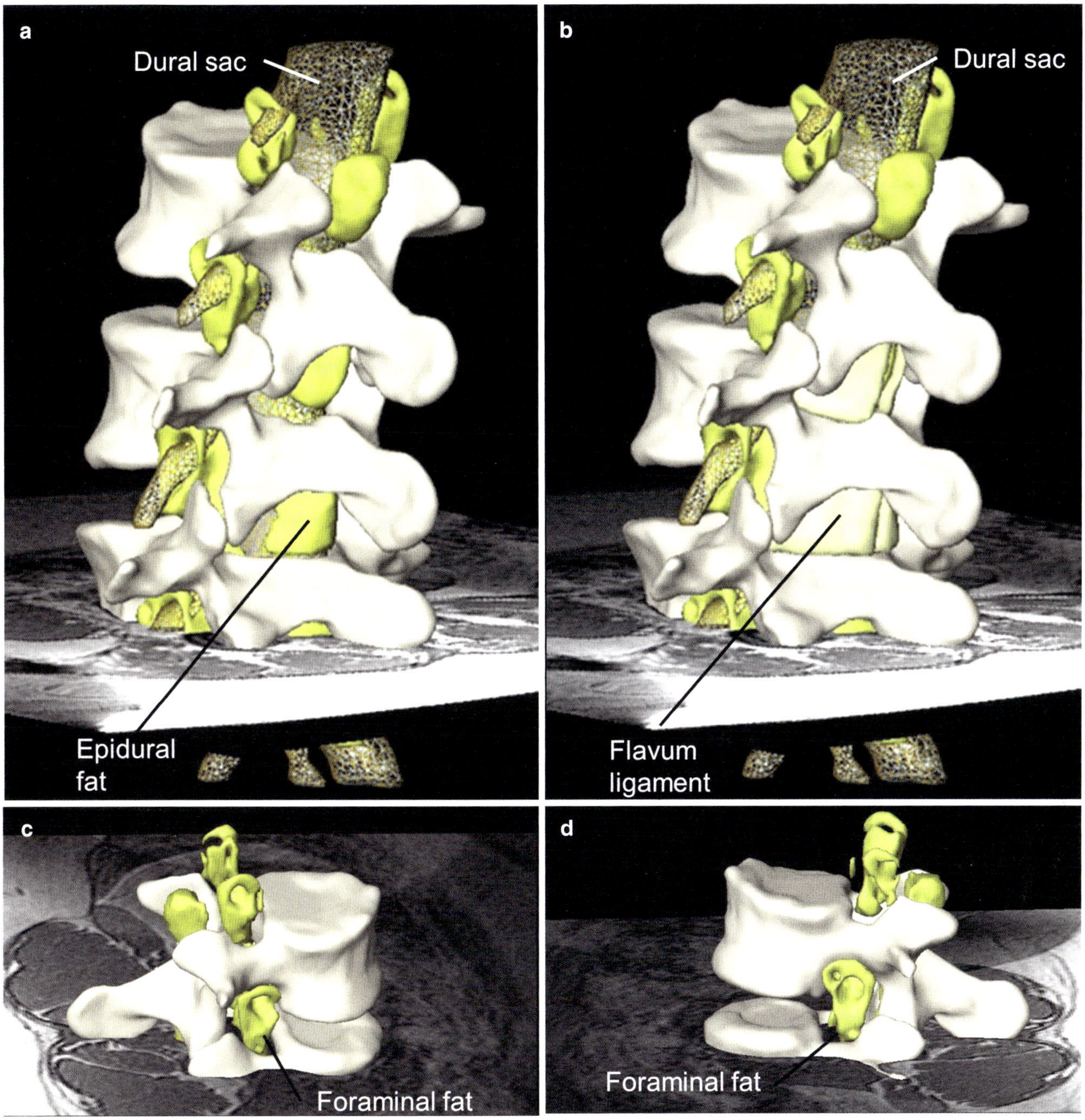

Fig. 23.9 Three-dimensional reconstruction of human spinal epidural fat. Epidural fat (**a**, **b**) and foraminal fat (**c**, **d**) view through lumbar vertebrae

Fig. 23.10 Three-dimensional reconstruction of human spinal epidural fat. Epidural fat through translucent lumbar vertebrae

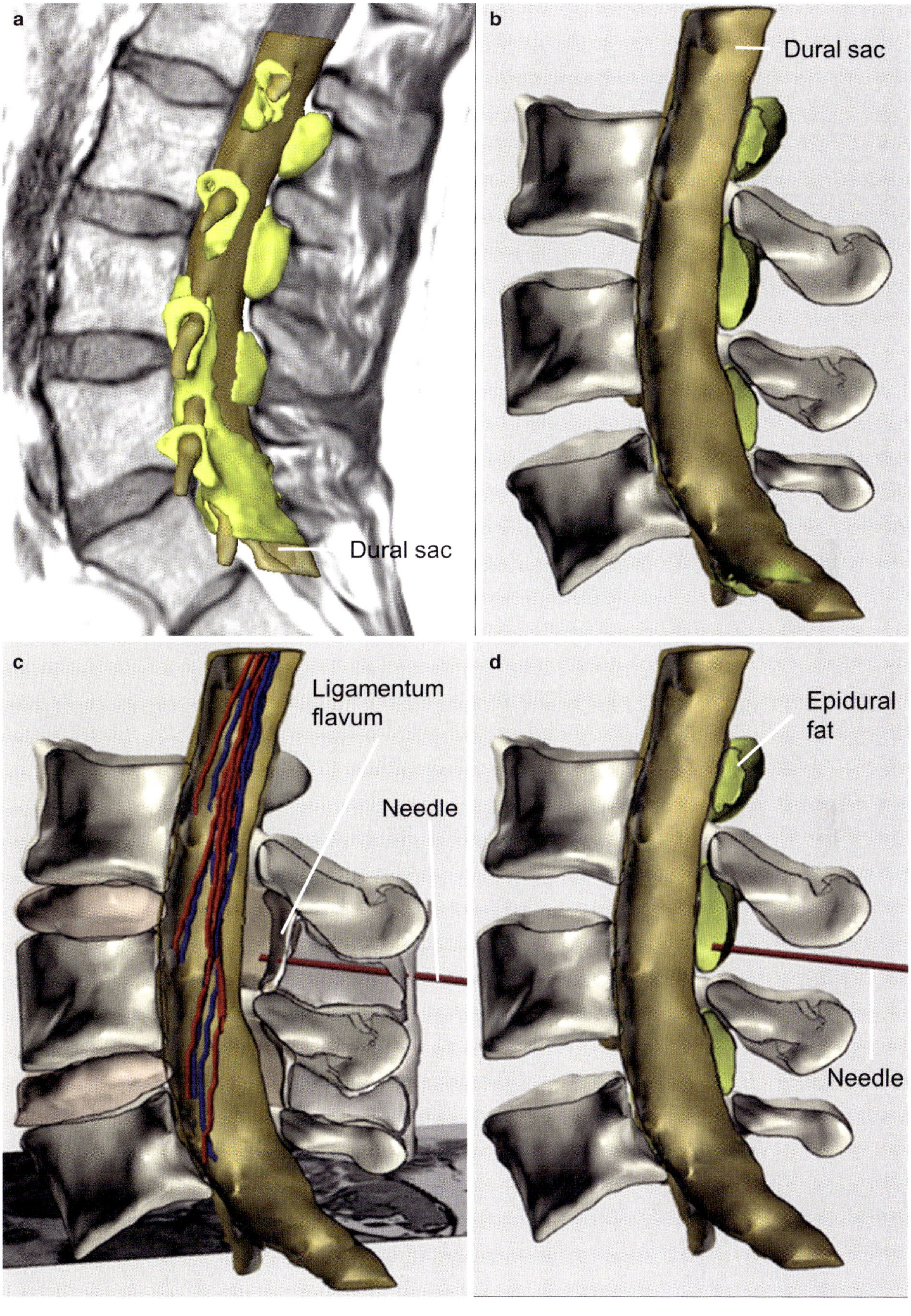

Fig. 23.11 Three-dimensional reconstruction of sagittal sections of human spinal epidural fat. (**a**) Lateral view; (**b–d**) Sagittal sections

References

1. Prats-Galino A, Mavar M, Reina MA, Puigdellívol-Sánchez A, San-Molina J, De Andrés JA. Three-dimensional interactive model of lumbar spinal structures. Anaesthesia. 2014;69:521.
2. Puigdellívol-Sánchez A, Prats-Galino A, Reina MA, Machés F, Hernández JM, De Andrés J, et al. Tridimensional magnetic resonance image of structures enclosed in the spinal canal relevant to anesthetists and estimation of the lumbosacral CSF volume. Acta Anaesthesiol Belg. 2011;62:37–45.
3. Reina MA, Franco CD, López A, De Andrés JA, van Zundert A. Clinical implications of epidural fat in the spinal canal. A scanning electron microscopic study. Acta Anaesthesiol Belg. 2009;60:7–17.
4. Reina MA, Pulido P, Castedo J, Villanueva MC, López A, De Andrés JA, et al. Epidural fat in various diseases: contribution of magnetic resonance imaging and potential implications for neuro axial anesthesia. Rev Esp Anestesiol Reanim. 2007;54:173–83.
5. Reina MA, Pulido P, Castedo J, Villanueva MC, López A, Sola RG. Characteristics and distribution of normal human epidural fat [in Spanish]. Rev Esp Anestesiol Reanim. 2006;53:363–72.
6. Reina MA, De Andrés JA, Hernández JM, Fernández Dominguez M, Juanes JA, Prats-Galino A. Spinal anatomy relevant to intrathecal drug delivery. Reg Anesth Pain Med Suppl. 2012;37:E145–9.
7. De Andrés JA, Reina MA, Machés F, De Sola RG, Oliva A, Prats-Galino A. Epidural fat: considerations for minimally invasive spinal injection and surgical therapies. JNR. 2011;1(S1):45–53.

24 Ultrastructure of Human Spinal Trabecular Arachnoid

Miguel Angel Reina, Andrés López, and José Antonio De Andrés

The arachnoid trabecular layer, located internally to the arachnoid layer, is composed of fibers that form the spider web–like trabecular structure found in the subarachnoid space and in the adventitial layer of blood vessels [1–6]. The arachnoid trabeculae give shape to tubular structures (arachnoid sheaths) for each nerve root and for the spinal cord. Some of these arachnoid trabeculae extend to the pia mater [1–3]. The trabecular arachnoid is an extremely fragile structure, which can easily be damaged. Because of its fragility, this membrane may be destroyed during dissection and manipulation, so that it is not frequently seen or systematically described [7]. Trabecular arachnoid limits nerve root movement to a certain extent, holding each root in its position within the dural sac and in relation to other nerve roots (Figs. 24.1, 24.2, 24.3, 24.4, 24.5, 24.6, 24.7, 24.8, 24.9, 24.10, 24.11, 24.12, 24.13, 24.14, 24.15, 24.16, 24.17, 24.18, 24.19, 24.20, 24.21, 24.22, 24.23, 24.24, 24.25, and 24.26) [4–6].

M.A. Reina, MD, PhD (✉)
Department of Clinical Medical Sciences and Institute of Applied Molecular Medicine, School of Medicine, University of CEU San Pablo, Madrid, Spain

Department of Anesthesiology, Madrid-Montepríncipe University Hospital, Madrid, Spain
e-mail: miguelangel@perticone.e.telefonica.net

A. López, MD
Department of Clinical Medical Sciences, School of Medicine, CEU San Pablo University, Madrid, Spain

Department of Anesthesiology, Madrid-Montepríncipe University Hospital, Madrid-Torrelodones University Hospital, Madrid-Sanchinarro University Hospital, Madrid, Spain
e-mail: alg00009@telefonica.net

J.A. De Andrés, MD, PhD
Department of Surgical Specialties, School of Medicine, University of Valencia, Valencia, Spain

Department of Anesthesia, Critical Care, and Pain Management, General University Hospital, Valencia, Spain
e-mail: deandres_jos@gva.es

M.A. Reina et al. (eds.), *Atlas of Functional Anatomy for Regional Anesthesia and Pain Medicine: Human Structure, Ultrastructure and 3D Reconstruction Images*, DOI 10.1007/978-3-319-09522-6_24,

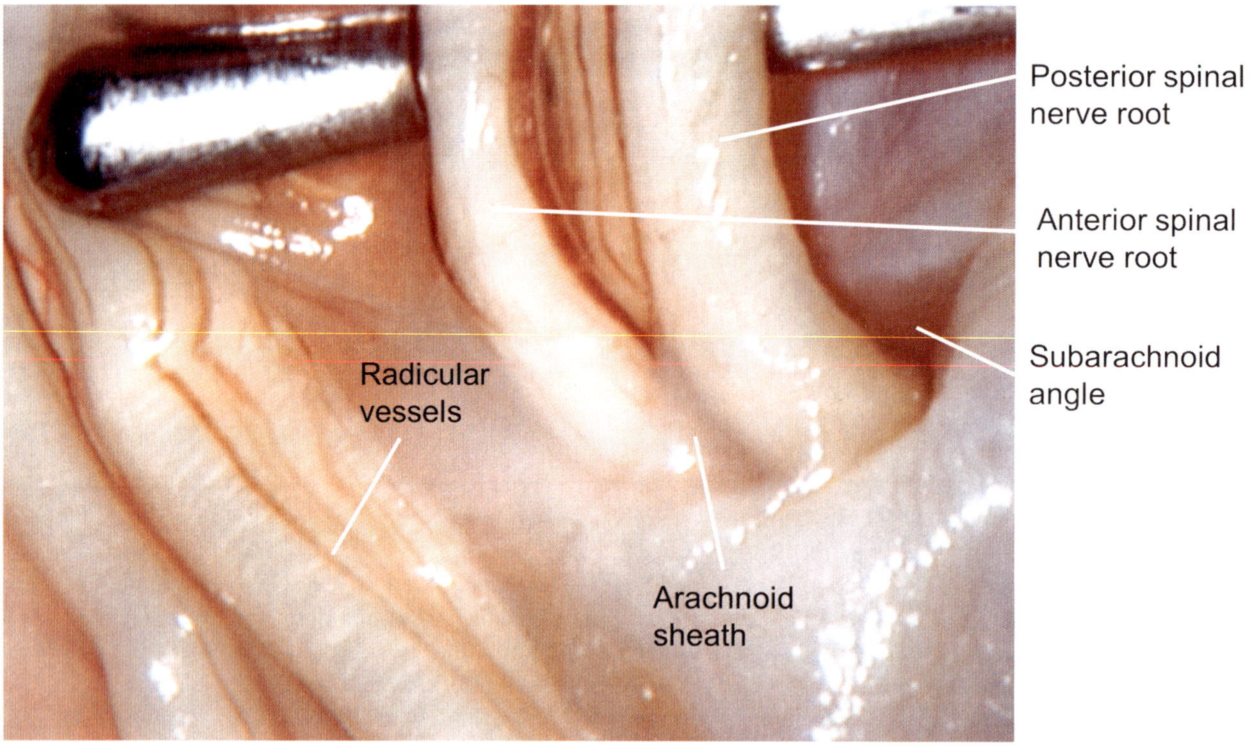

Fig. 24.1 Trabecular arachnoid. Detail of anterior and posterior spinal nerve root and its arachnoid sheath (From Reina et al. [3]; with permission)

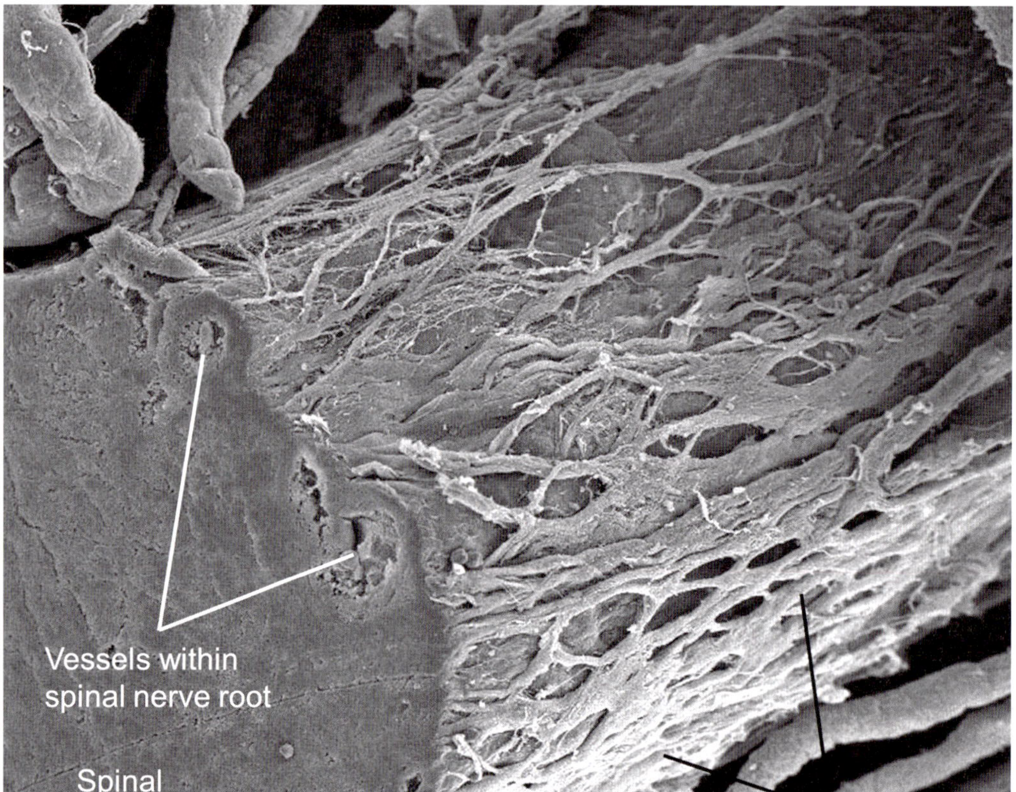

Fig. 24.2 Trabecular arachnoid. Detail of trabecular arachnoid around spinal nerve root. Scanning electron microscopy. Magnification × 120

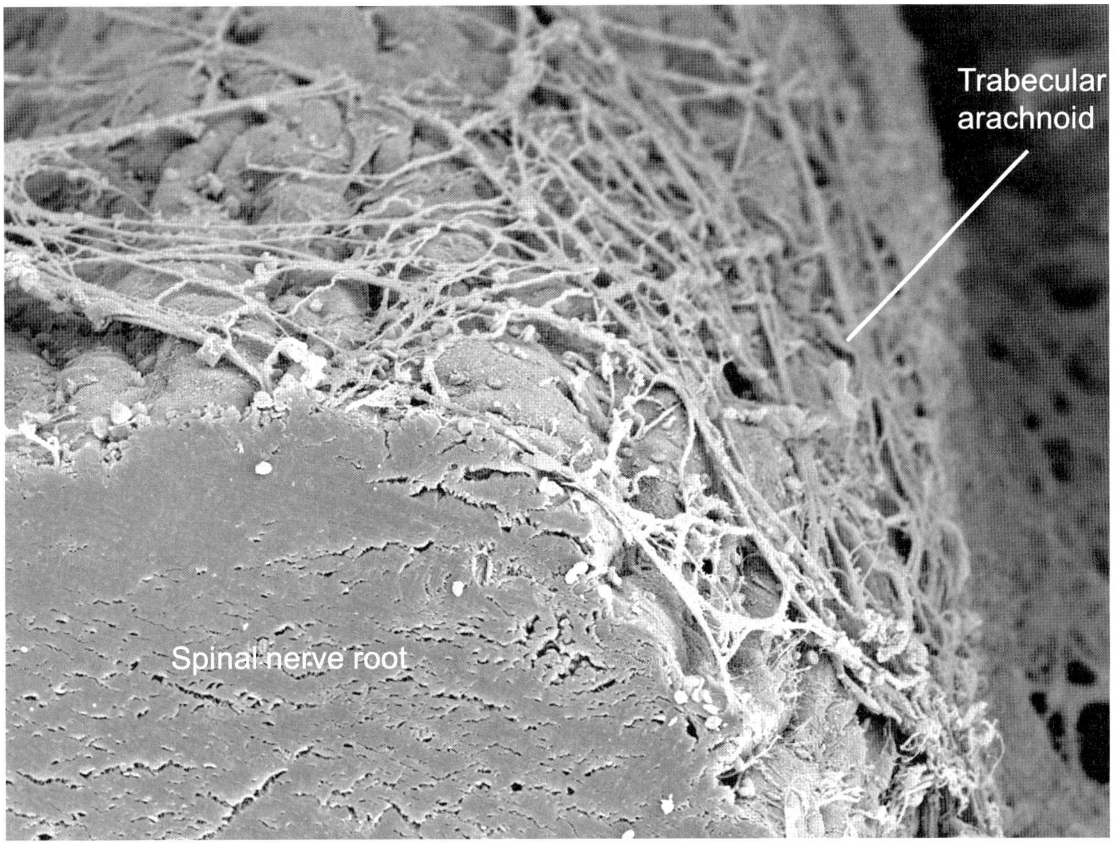

Fig. 24.3 Trabecular arachnoid. Detail of trabecular arachnoid around spinal nerve root. Scanning electron microscopy. Magnification × 250

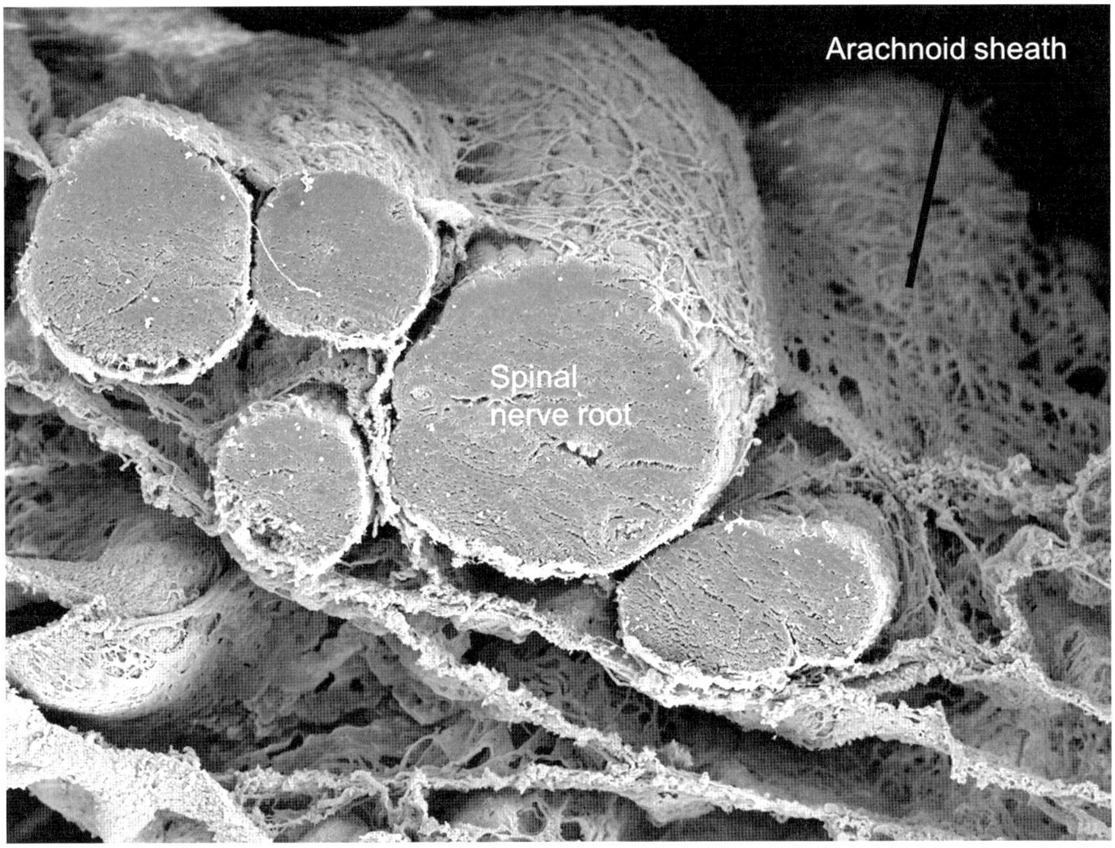

Fig. 24.4 Trabecular arachnoid. Detail of five spinal nerve roots enveloped by arachnoid sheath. Scanning electron microscopy. Magnification × 75

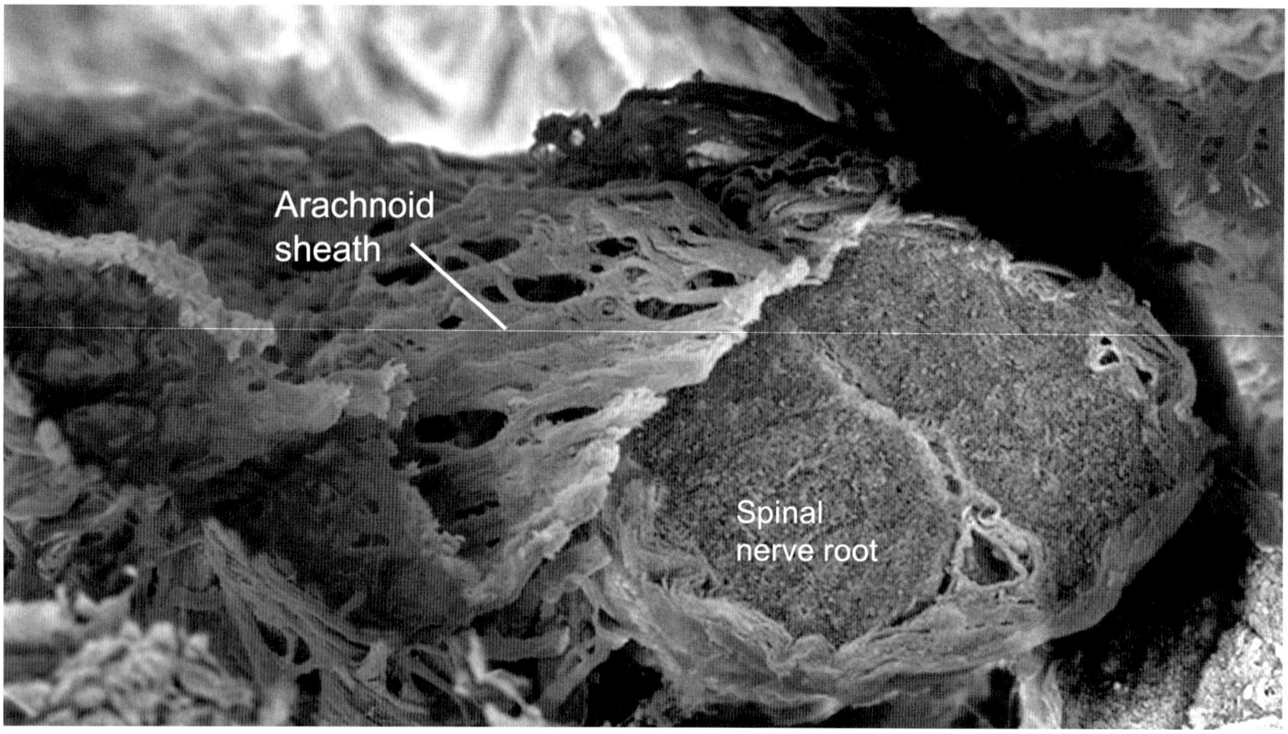

Fig. 24.5 Trabecular arachnoid. Detail of spinal nerve root and arachnoid sheath. Scanning electron microscopy. Magnification × 80

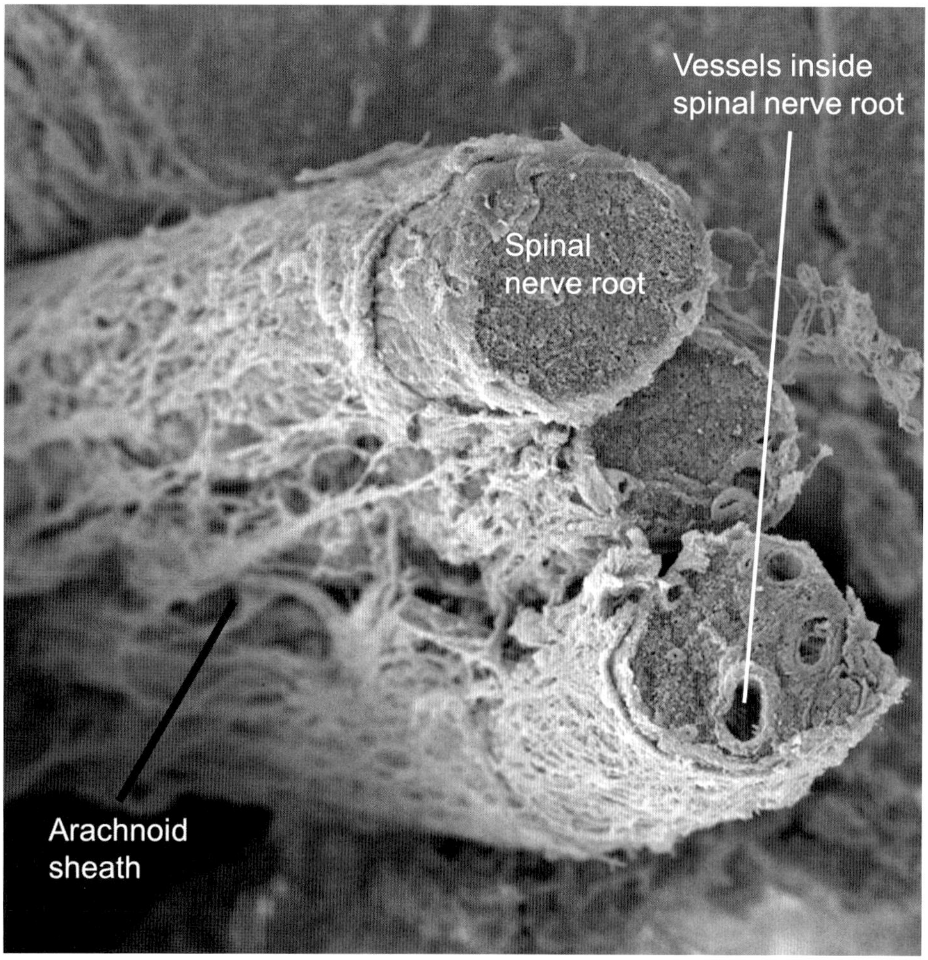

Fig. 24.6 Trabecular arachnoid. Details of three spinal nerve roots with arachnoid sheaths. Scanning electron microscopy. Magnification × 50

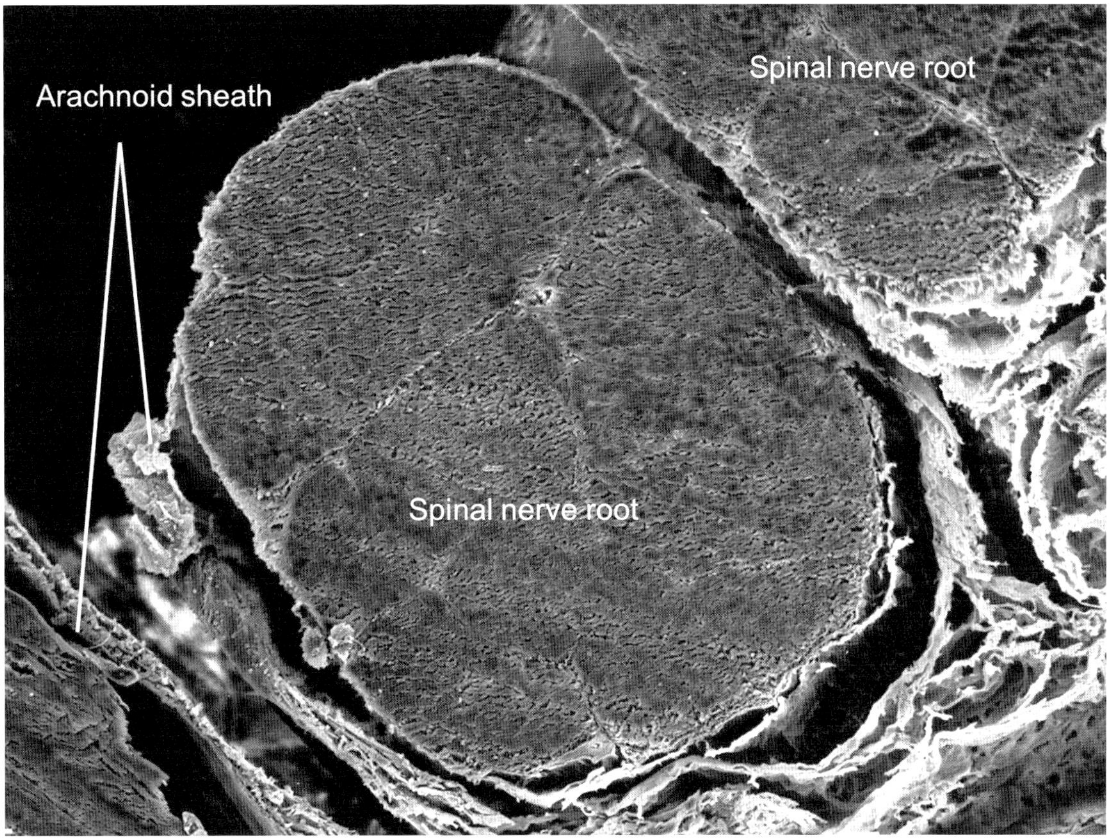

Fig. 24.7 Trabecular arachnoid. Detail of spinal nerve roots and their arachnoid sheaths. Scanning electron microscopy. Magnification × 60 (From Reina et al. [3]; with permission)

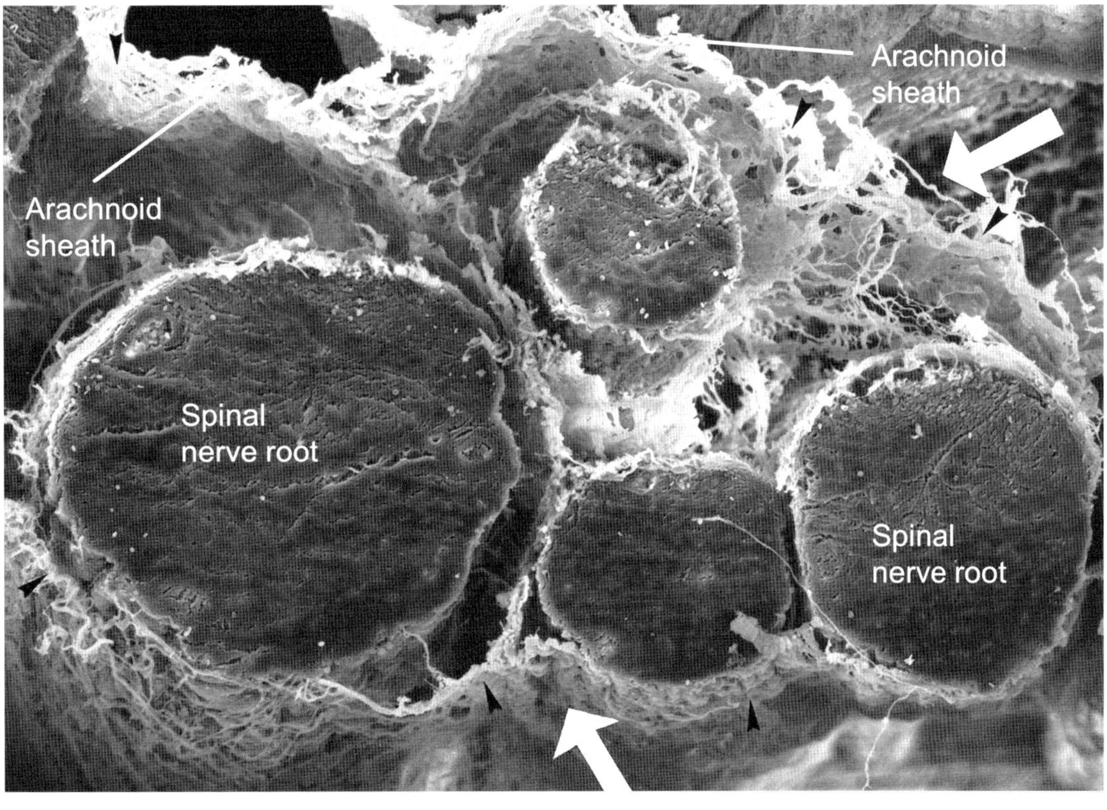

Fig. 24.8 Trabecular arachnoid. Detail of four spinal nerve roots with their arachnoid sheaths. Scanning electron microscopy. Magnification × 100 (From Reina et al. [1]; with permission)

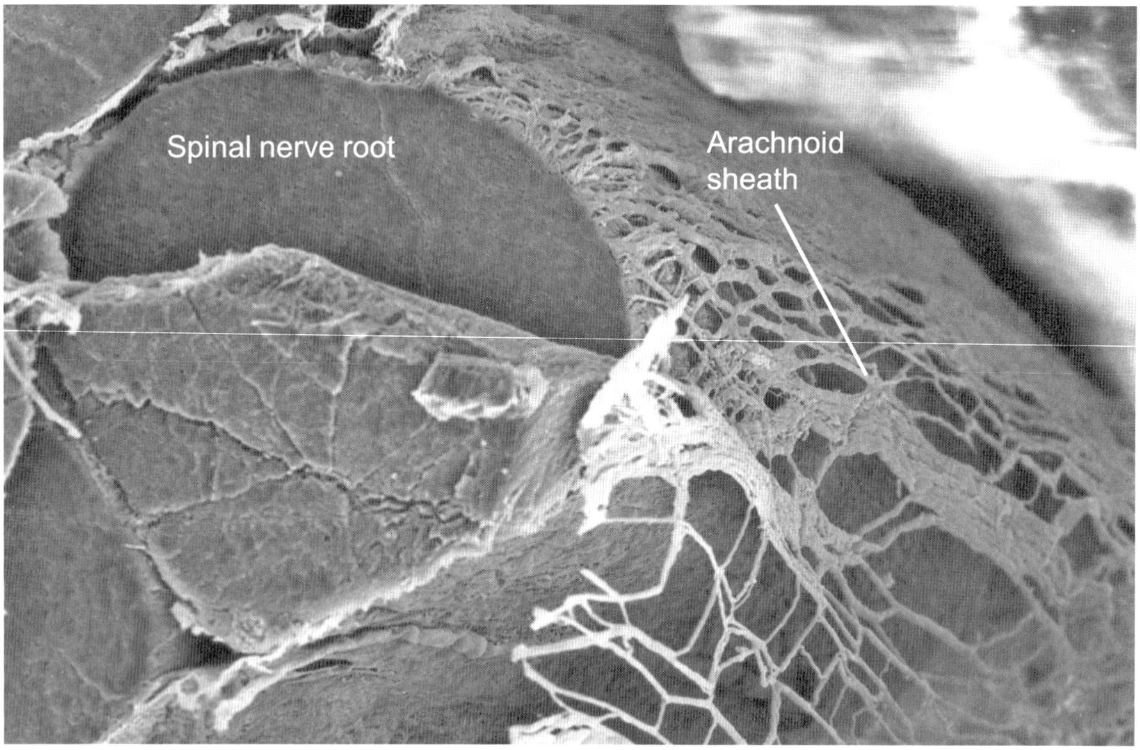

Fig. 24.9 Trabecular arachnoid. Detail of an arachnoid sheath. Scanning electron microscopy. Magnification × 80 (From Reina et al. [6]; with permission)

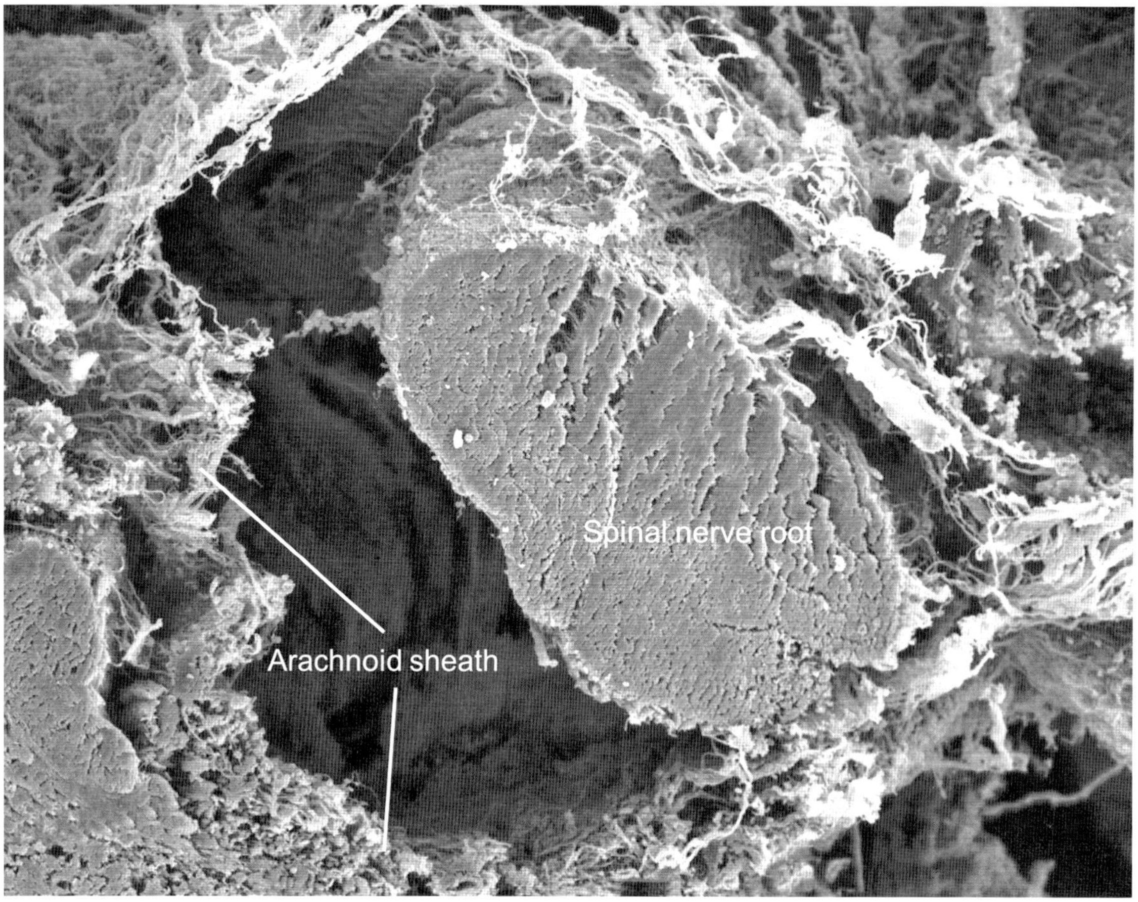

Fig. 24.10 Trabecular arachnoid. Detail of a spinal nerve root with its arachnoid sheath. Scanning electron microscopy. Magnification × 100 (From Reina et al. [3]; with permission)

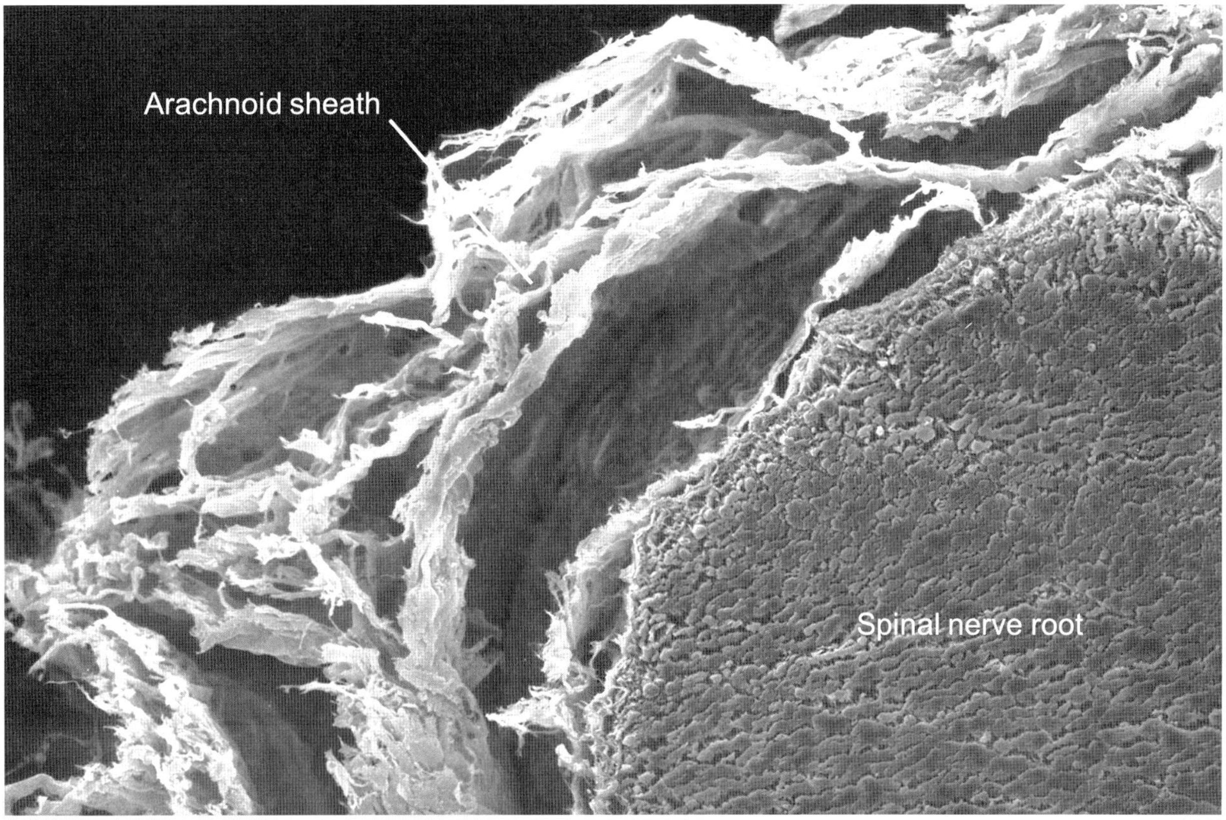

Fig. 24.11 Trabecular arachnoid. Detail of an arachnoid sheath. Scanning electron microscopy. Magnification × 150 (From Reina et al. [1]; with permission)

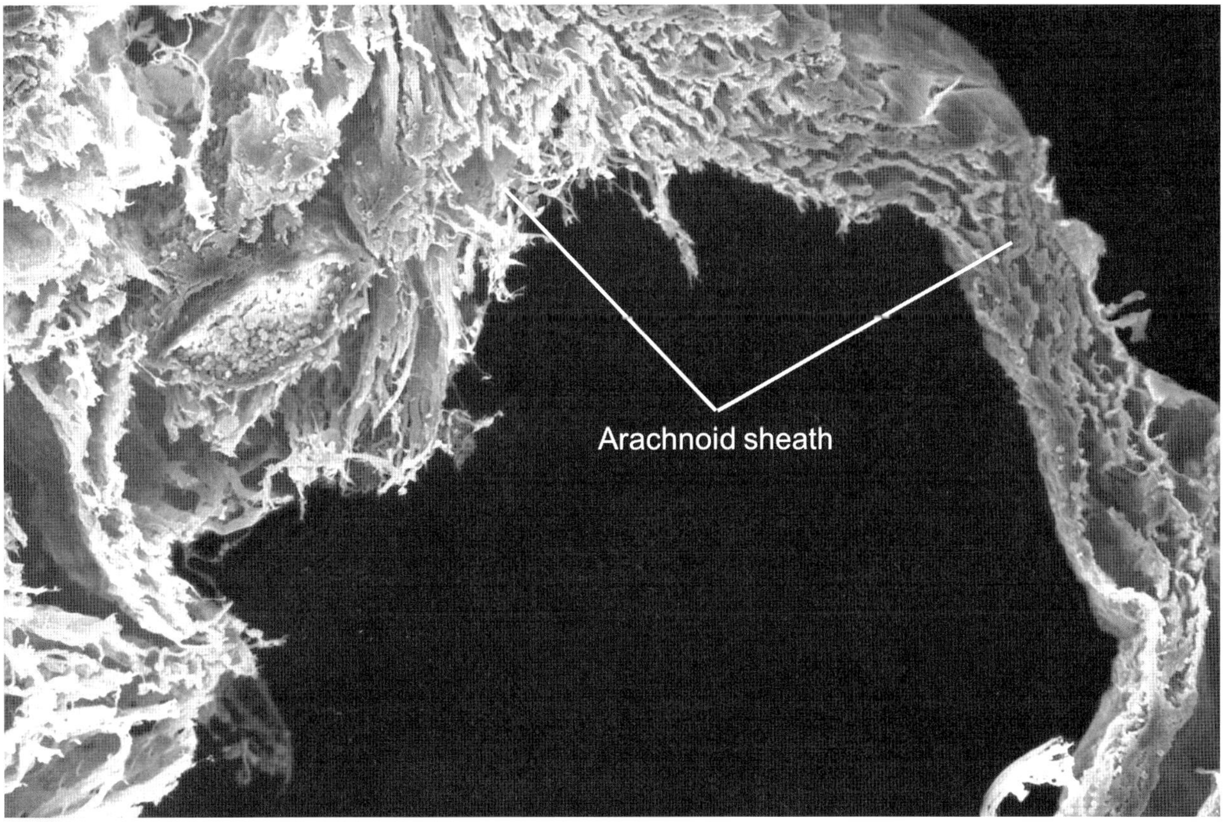

Fig. 24.12 Trabecular arachnoid. Detail of an arachnoid sheath without spinal nerve root. Scanning electron microscopy. Magnification × 100 (From Reina et al. [6]; with permission)

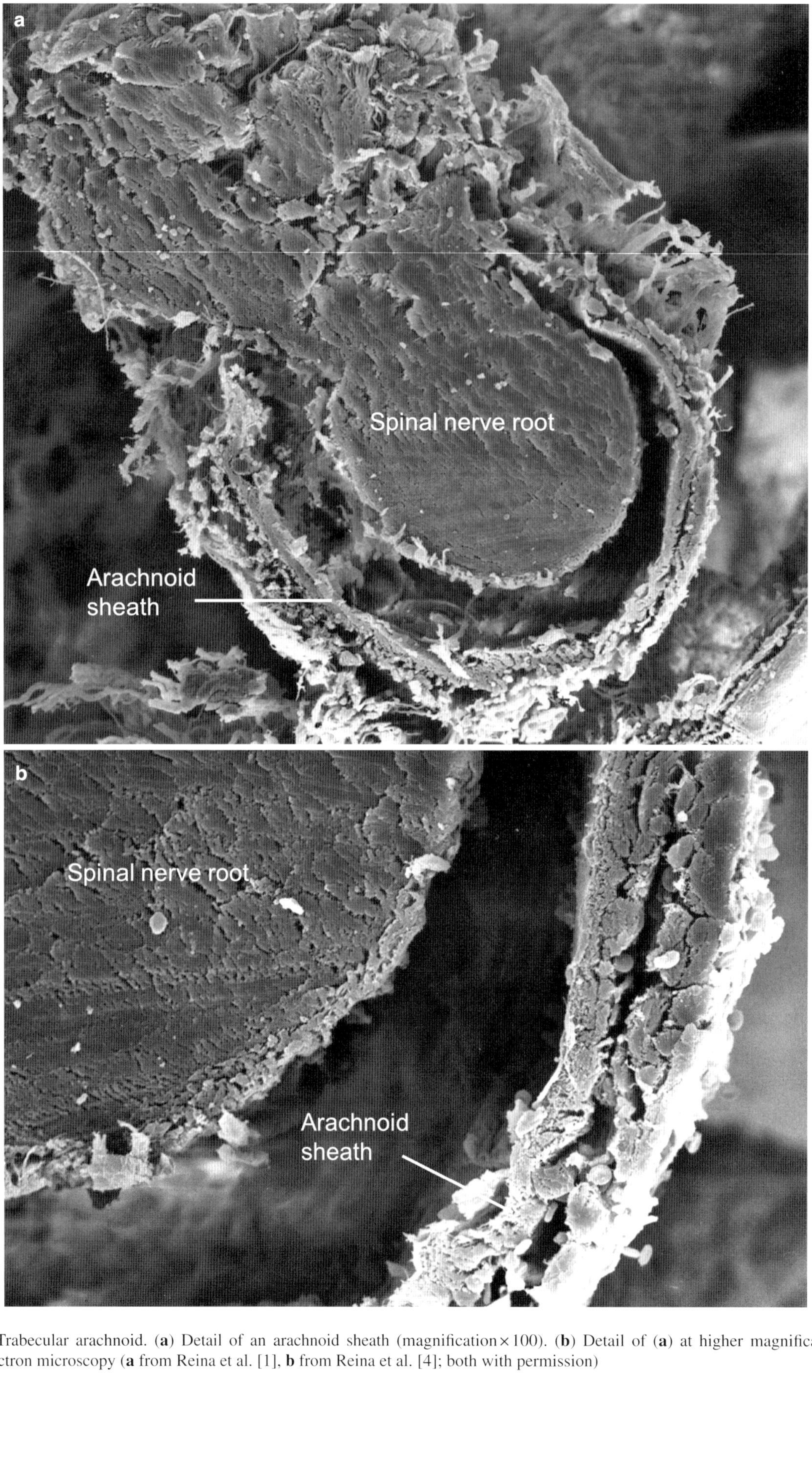

Fig. 24.13 Trabecular arachnoid. (**a**) Detail of an arachnoid sheath (magnification × 100). (**b**) Detail of (**a**) at higher magnification (×400). Scanning electron microscopy (**a** from Reina et al. [1], **b** from Reina et al. [4]; both with permission)

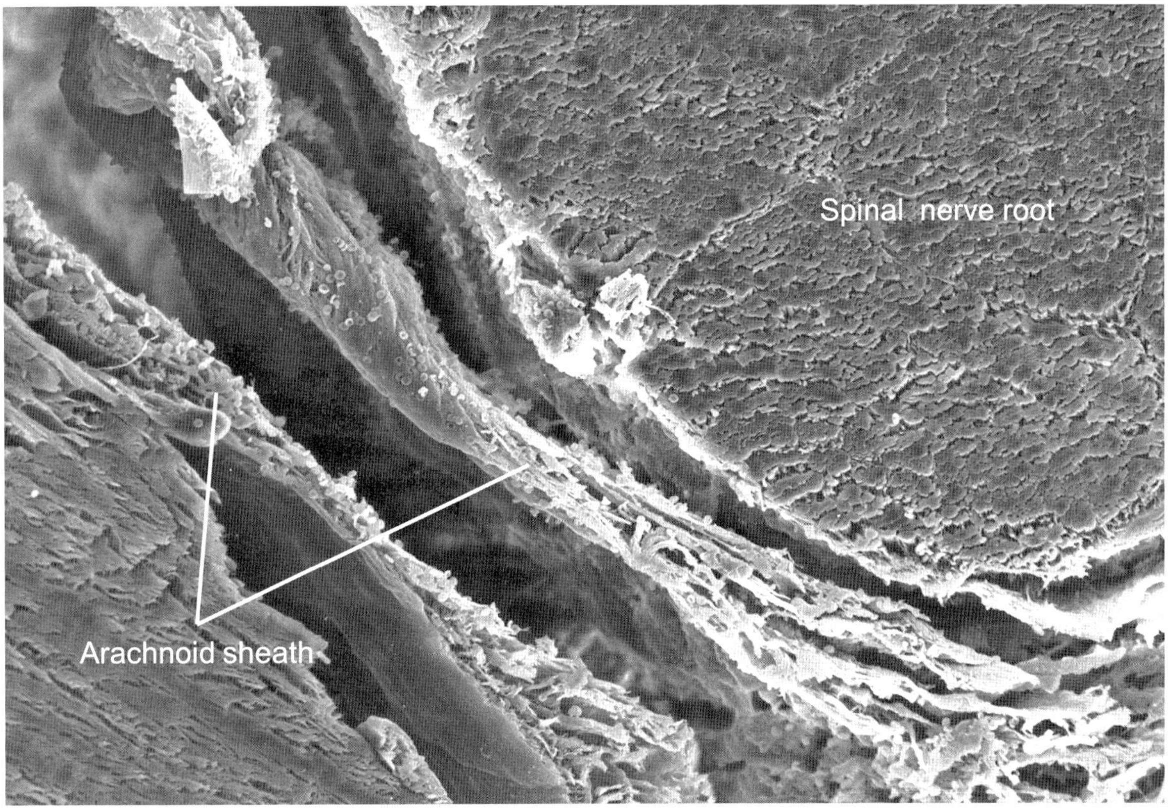

Fig. 24.14 Trabecular arachnoid. Detail of an arachnoid sheath. Scanning electron microscopy. Magnification × 150 (From Reina et al. [6]; with permission)

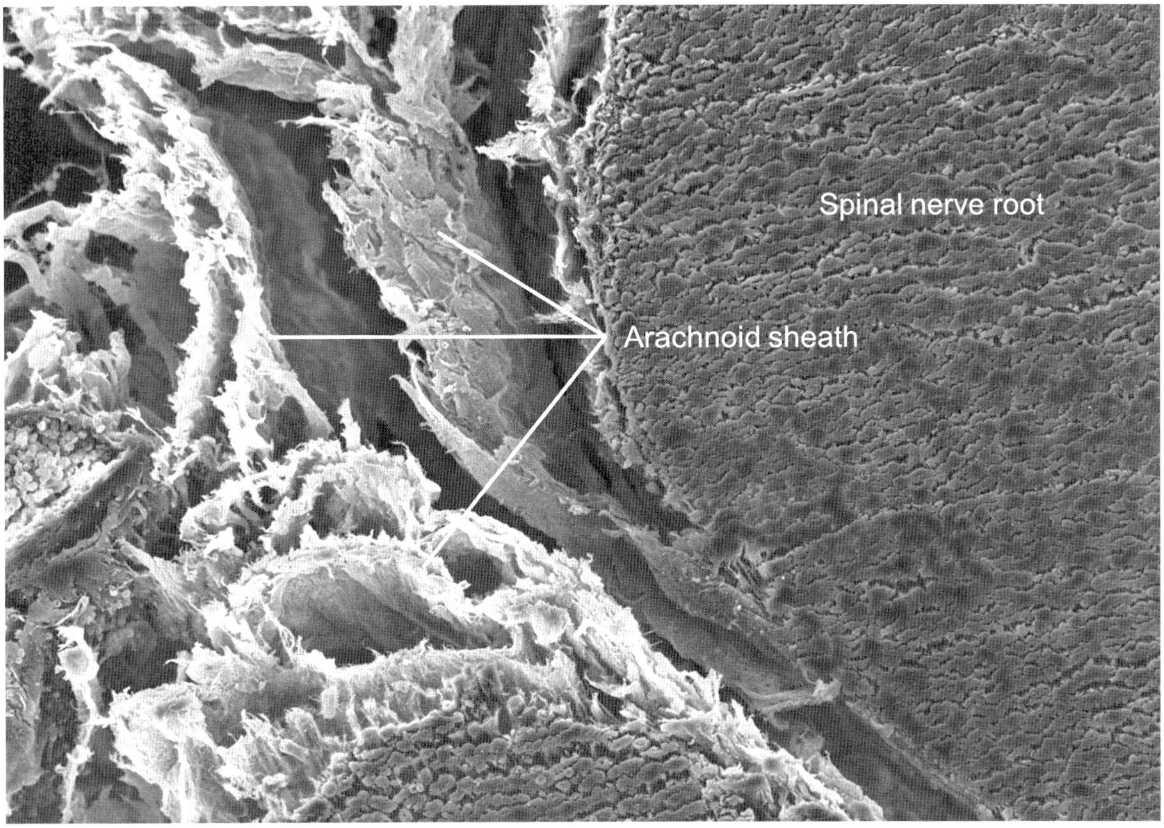

Fig. 24.15 Trabecular arachnoid. Detail of an arachnoid sheath. Scanning electron microscopy. Magnification × 150 (From Reina et al. [6]; with permission)

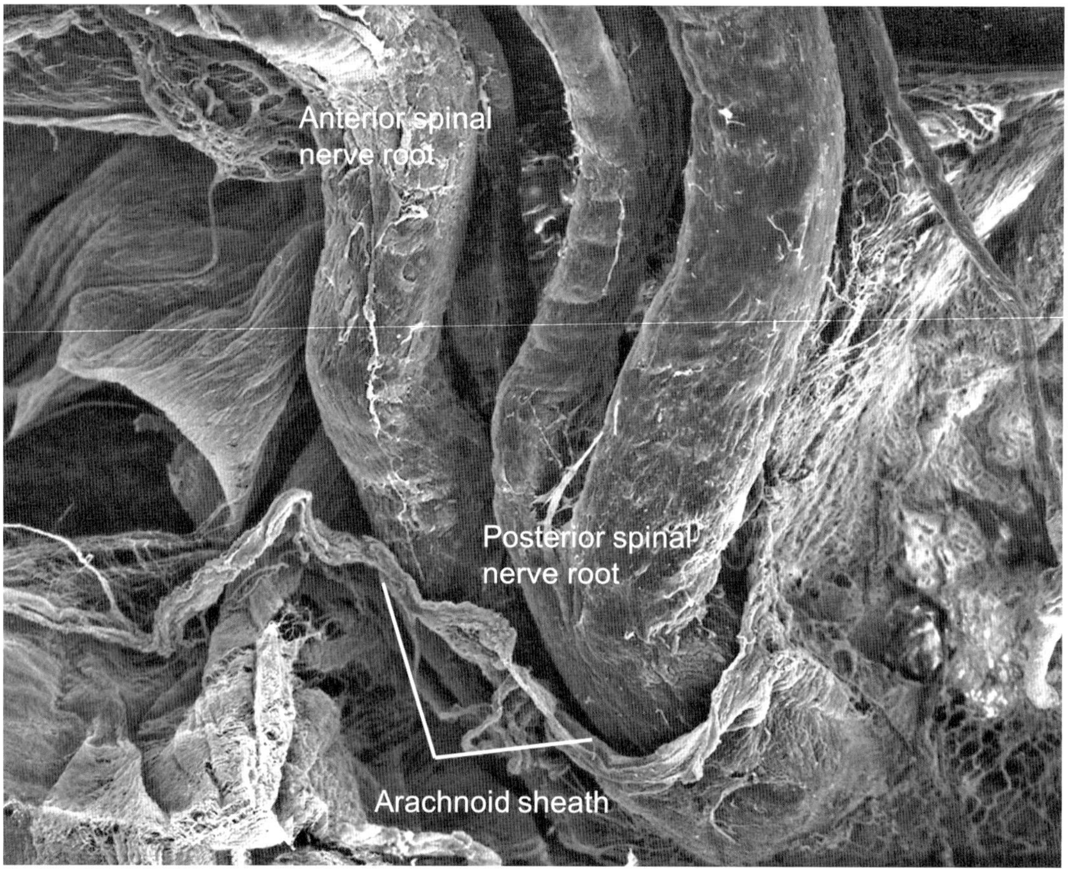

Fig. 24.16 Trabecular arachnoid. Detail of anterior and posterior spinal nerve root enclosed by its arachnoid sheath. Scanning electron microscopy. Magnification × 30 (From Reina et al. [1]; with permission)

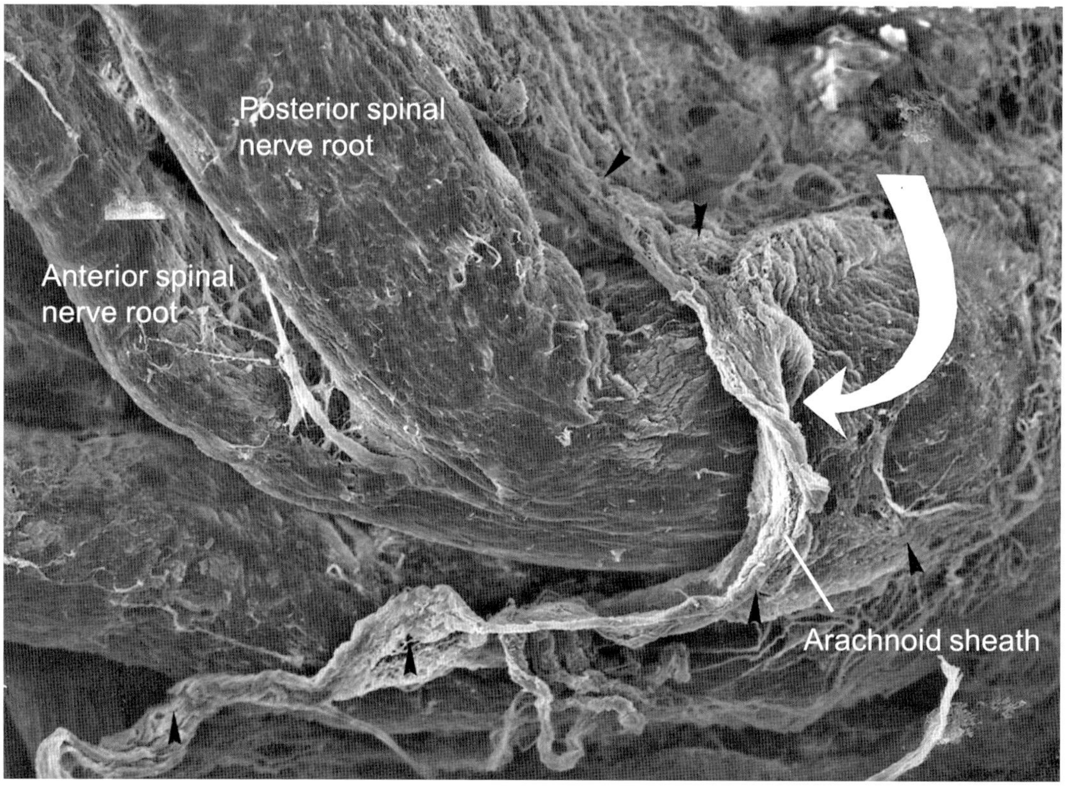

Fig. 24.17 Trabecular arachnoid. Detail of Fig. 24.16 at higher magnification (×55). Scanning electron microscopy (From Reina et al. [1]; with permission)

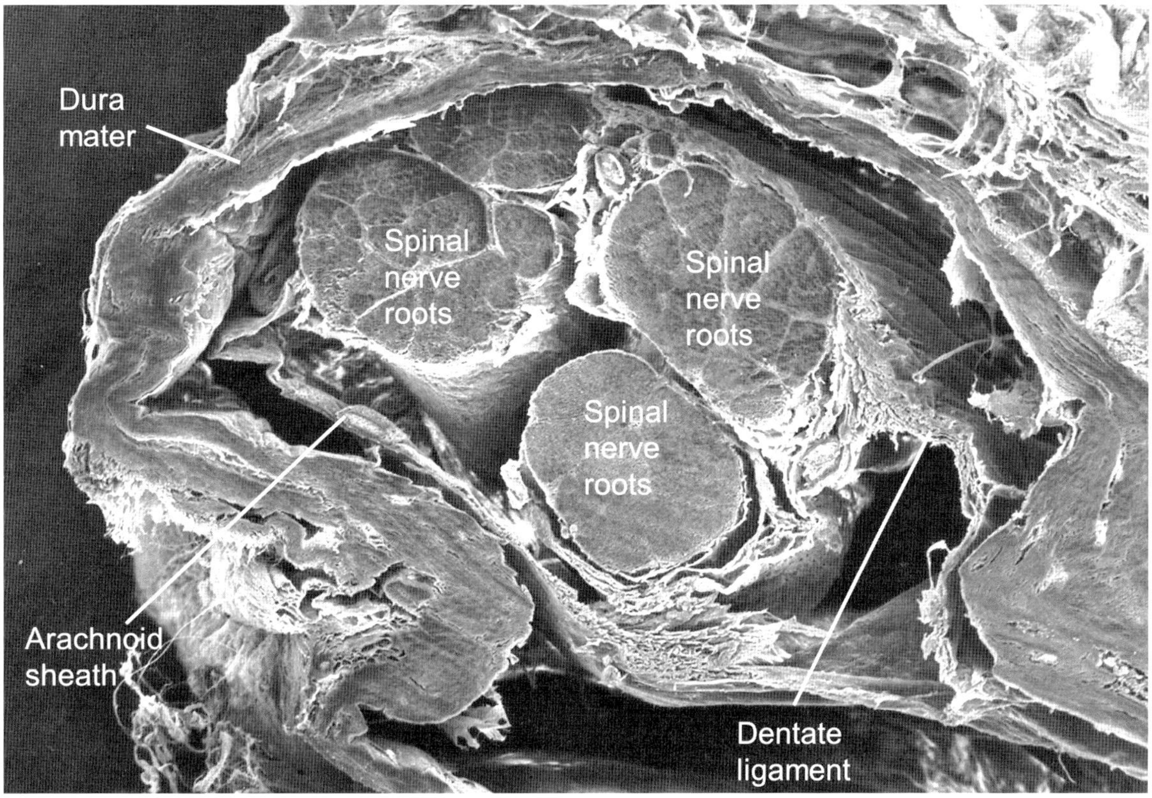

Fig. 24.18 Trabecular arachnoid. Detail of four spinal nerve roots with their arachnoid sheaths. Scanning electron microscopy. Magnification × 20

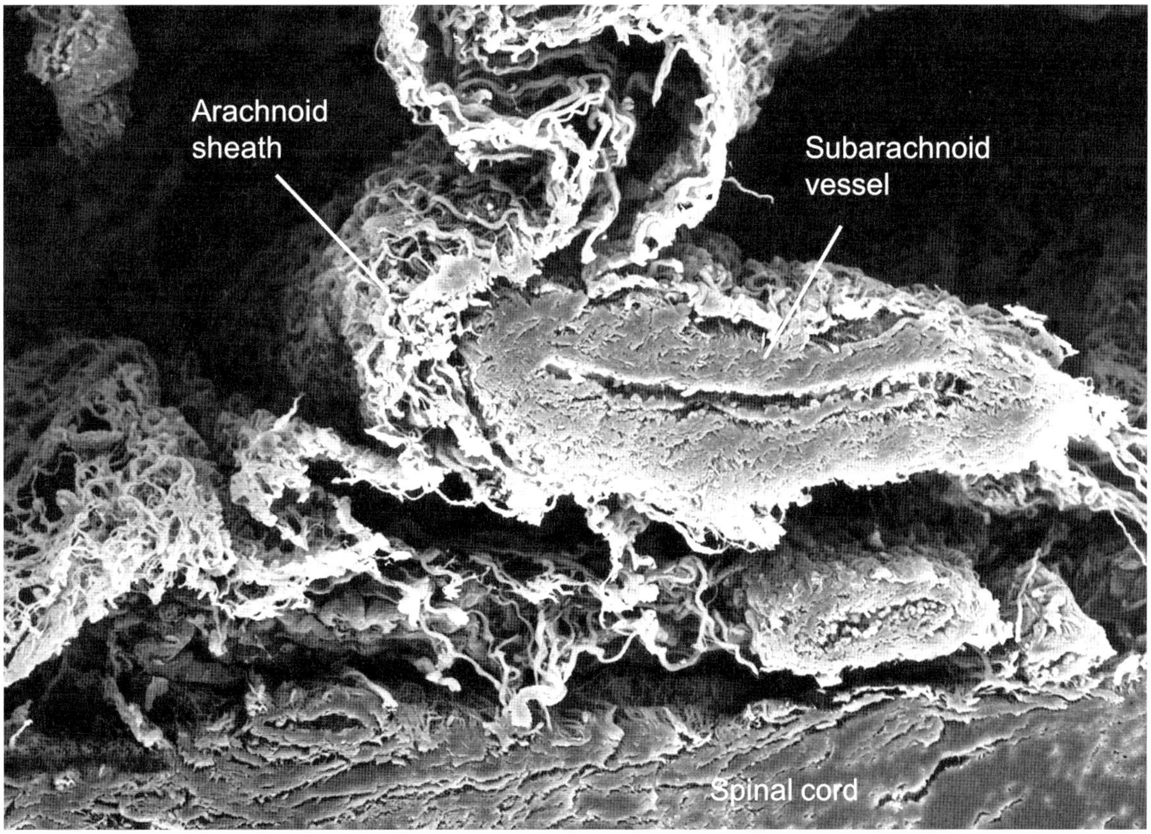

Fig. 24.19 Trabecular arachnoid. Vessel within subarachnoid space surrounded by arachnoid trabecular structure. Scanning electron microscopy. Magnification × 120 (From Reina et al. [4]; with permission)

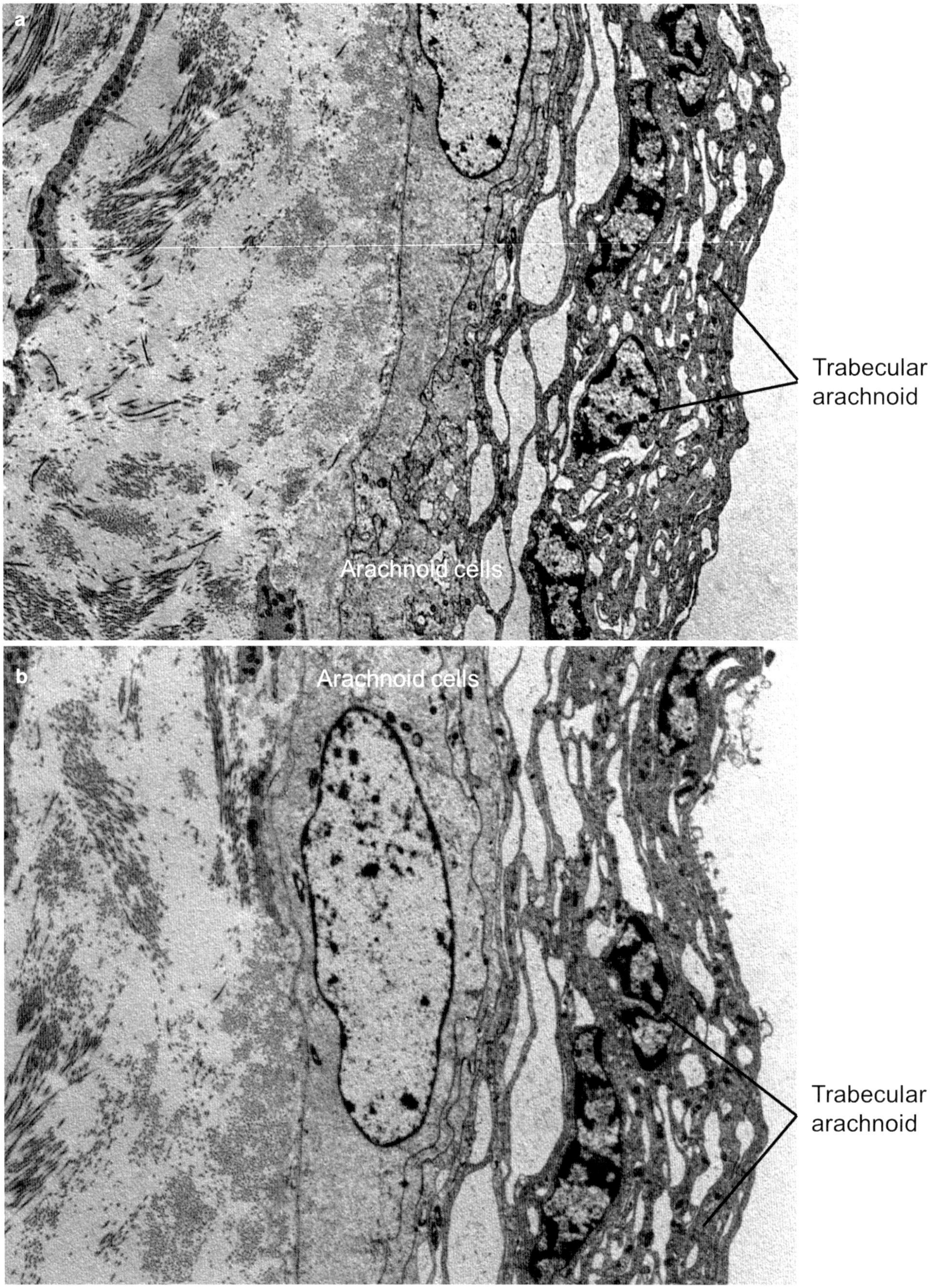

Fig. 24.20 Trabecular arachnoid. (**a**, **b**) Inner portion of trabecular arachnoid close to arachnoid layer. Transmission electron microscopy. Magnification: **a**, ×6,000; **b**, ×8,000

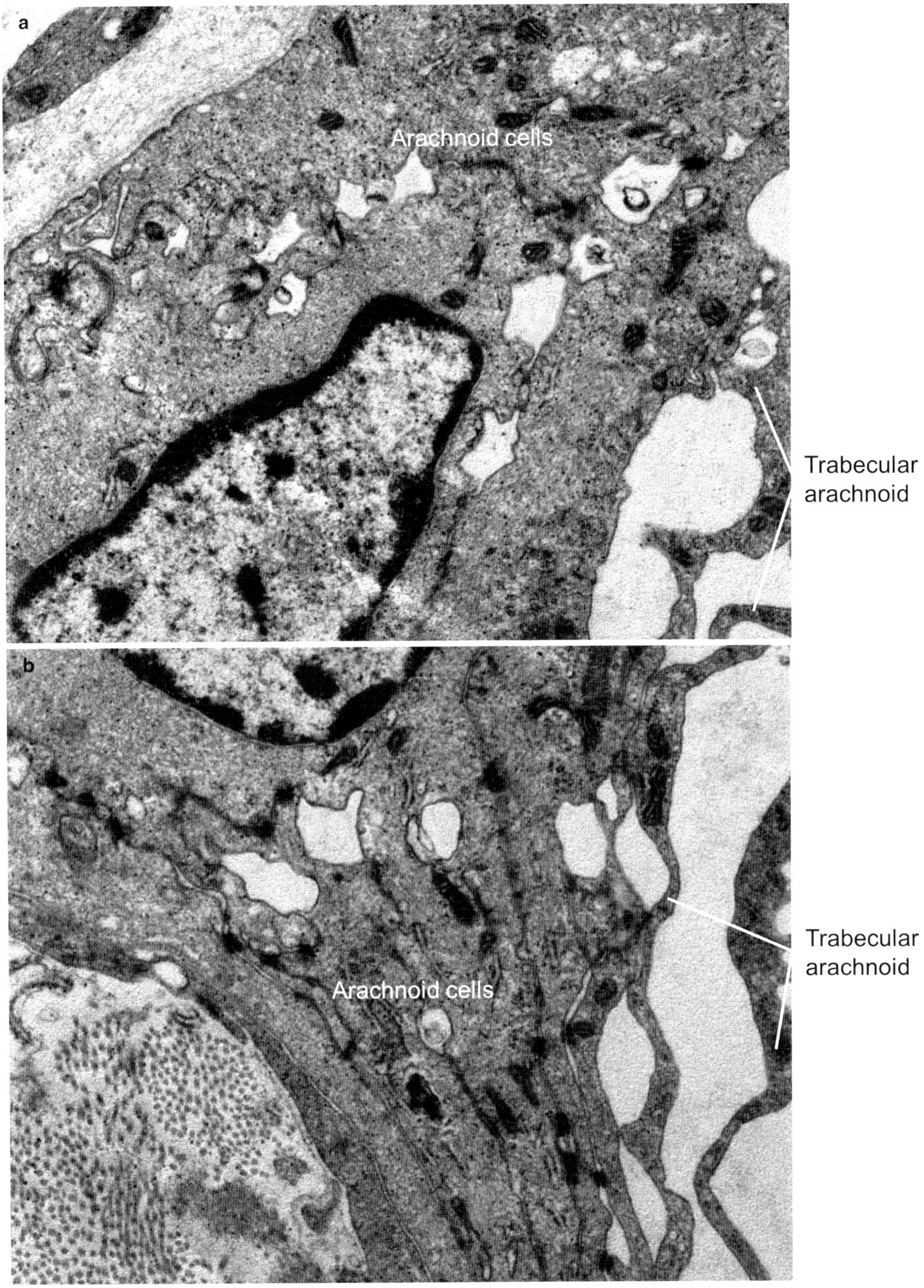

Fig. 24.21 Trabecular arachnoid. (**a**, **b**) Detail of inner portion of trabecular arachnoid close to arachnoid layer (magnification × 25,000). Transmission electron microscopy

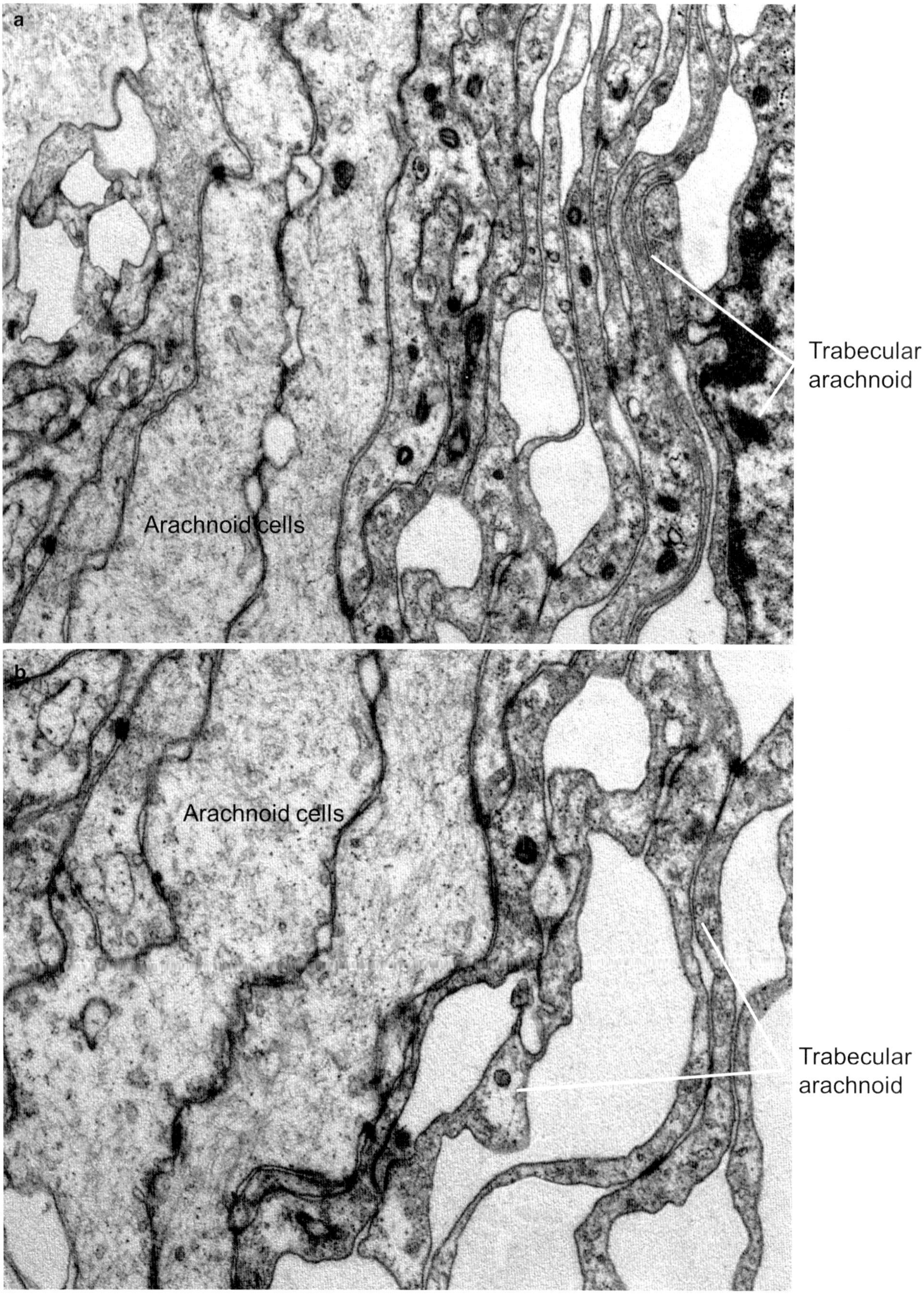

Fig. 24.22 Trabecular arachnoid. (**a**, **b**) Detail of inner portion of trabecular arachnoid close to arachnoid layer. Transmission electron microscopy. Magnification: **a**, ×25,000, **b**, ×30,000

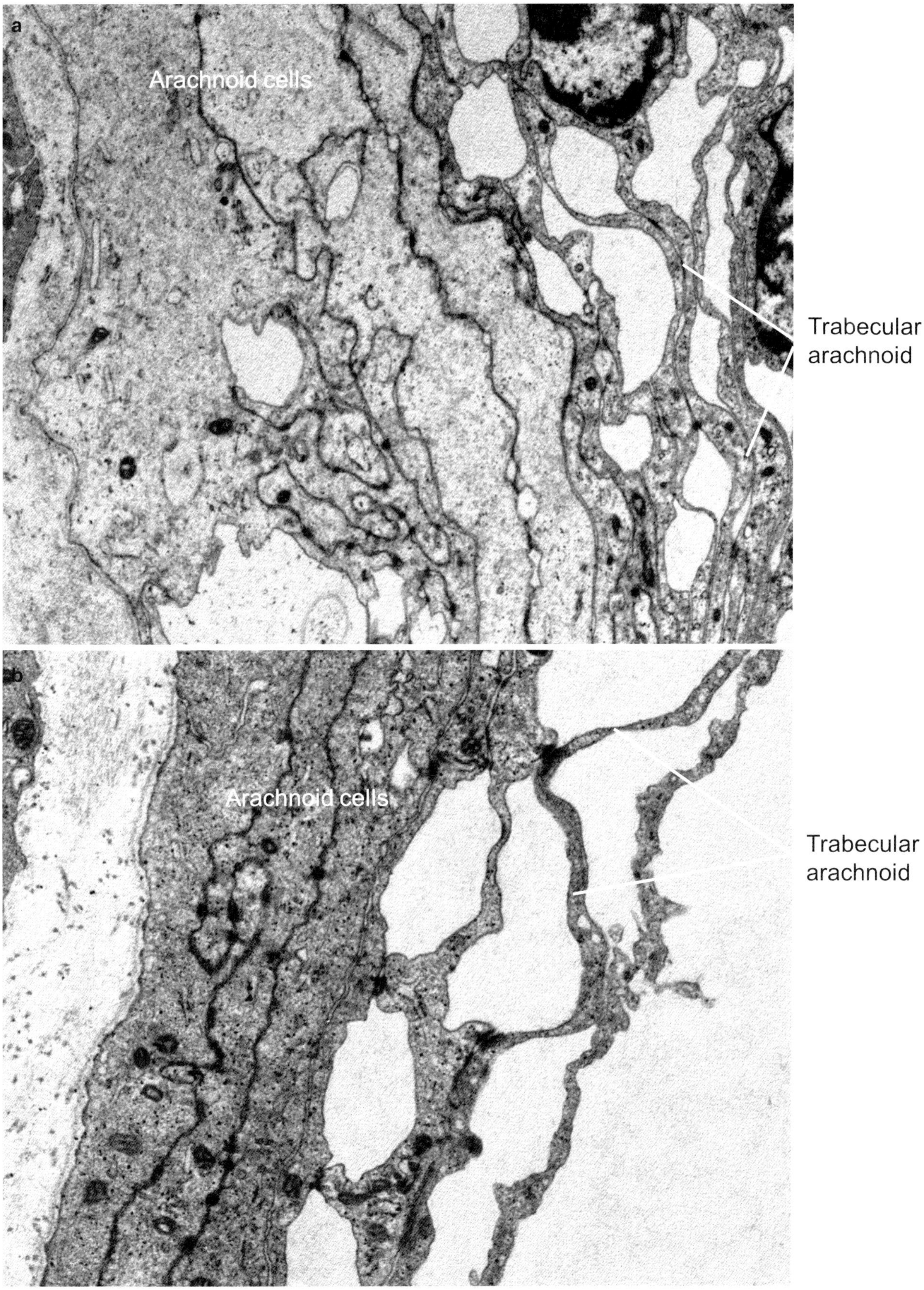

Fig. 24.23 Trabecular arachnoid. (**a**, **b**) Detail of inner portion of trabecular arachnoid close to arachnoid layer. Transmission electron microscopy. Magnification: **a**, ×15,000; **b**, ×25,000 (**b** from Reina et al. [7]; with permission)

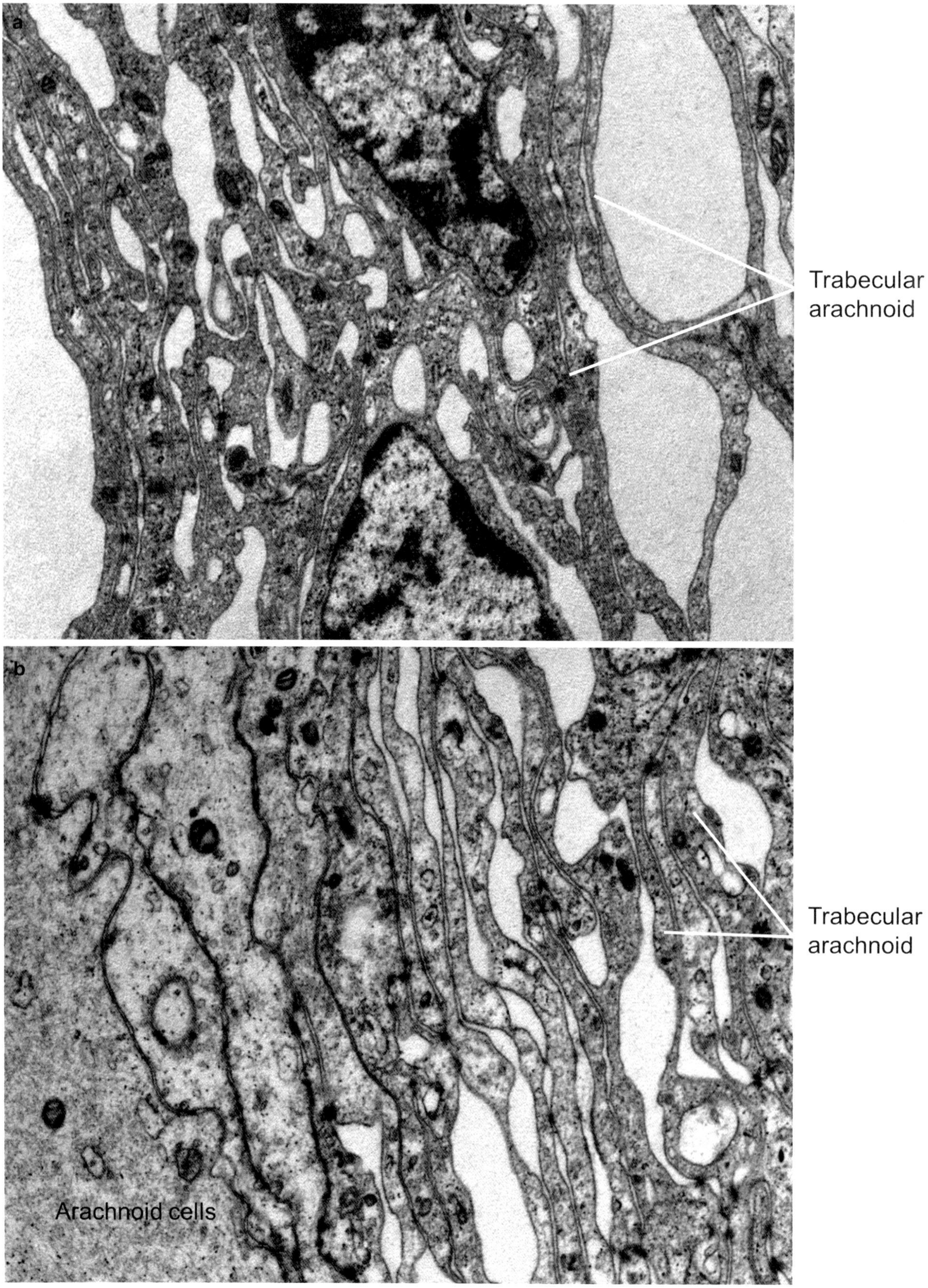

Fig. 24.24 Trabecular arachnoid. (**a**, **b**) Detail of inner portion of trabecular arachnoid close to arachnoid layer. Transmission electron microscopy. Magnification: **a**, ×20,000; **b**, ×25,000

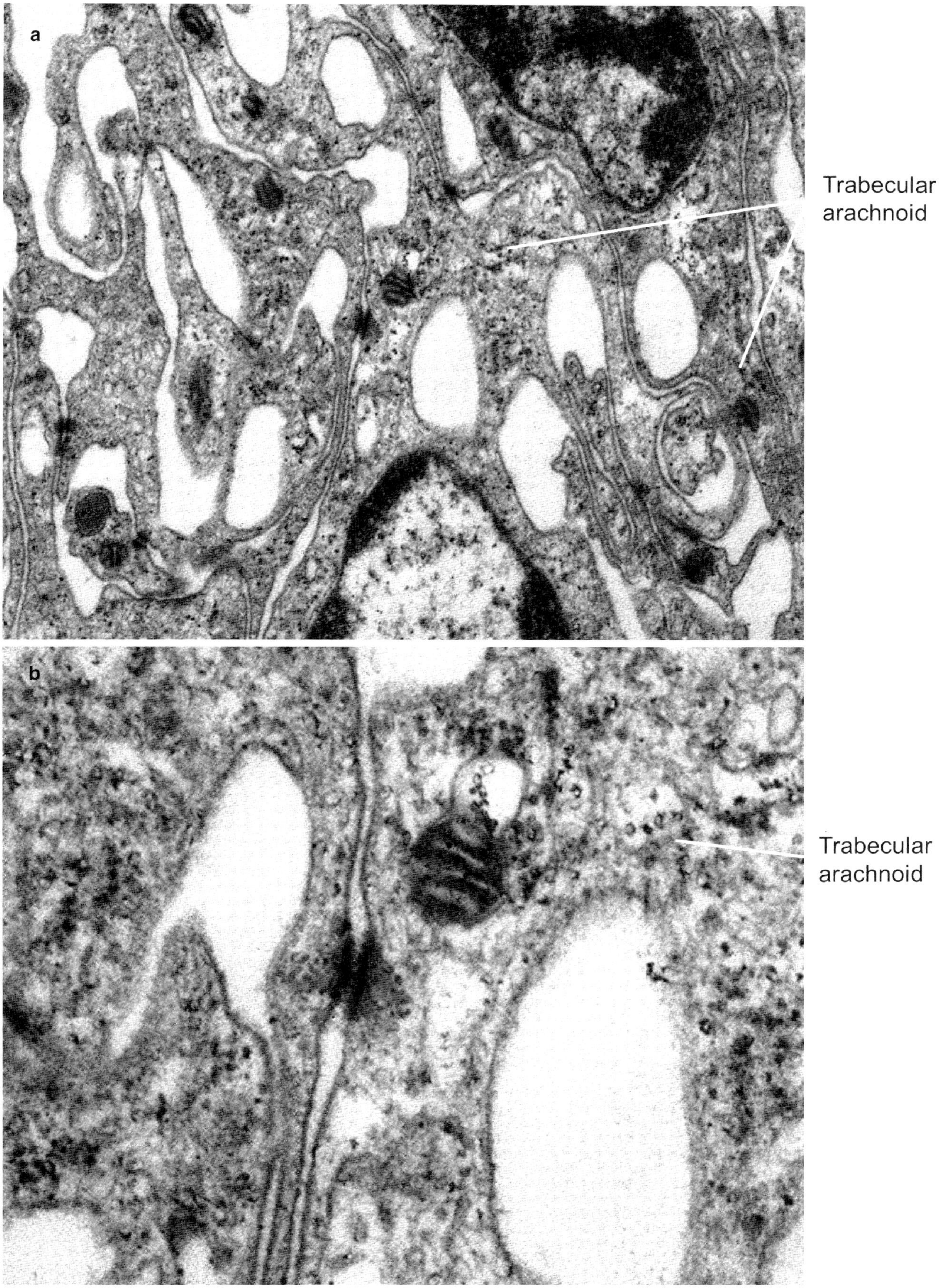

Fig. 24.25 Trabecular arachnoid. (**a**, **b**) Detail of trabecular arachnoid. Transmission electron microscopy. Magnification: **a**, ×40,000; **b**, ×120,000 (**a** from Reina et al. [7]; with permission)

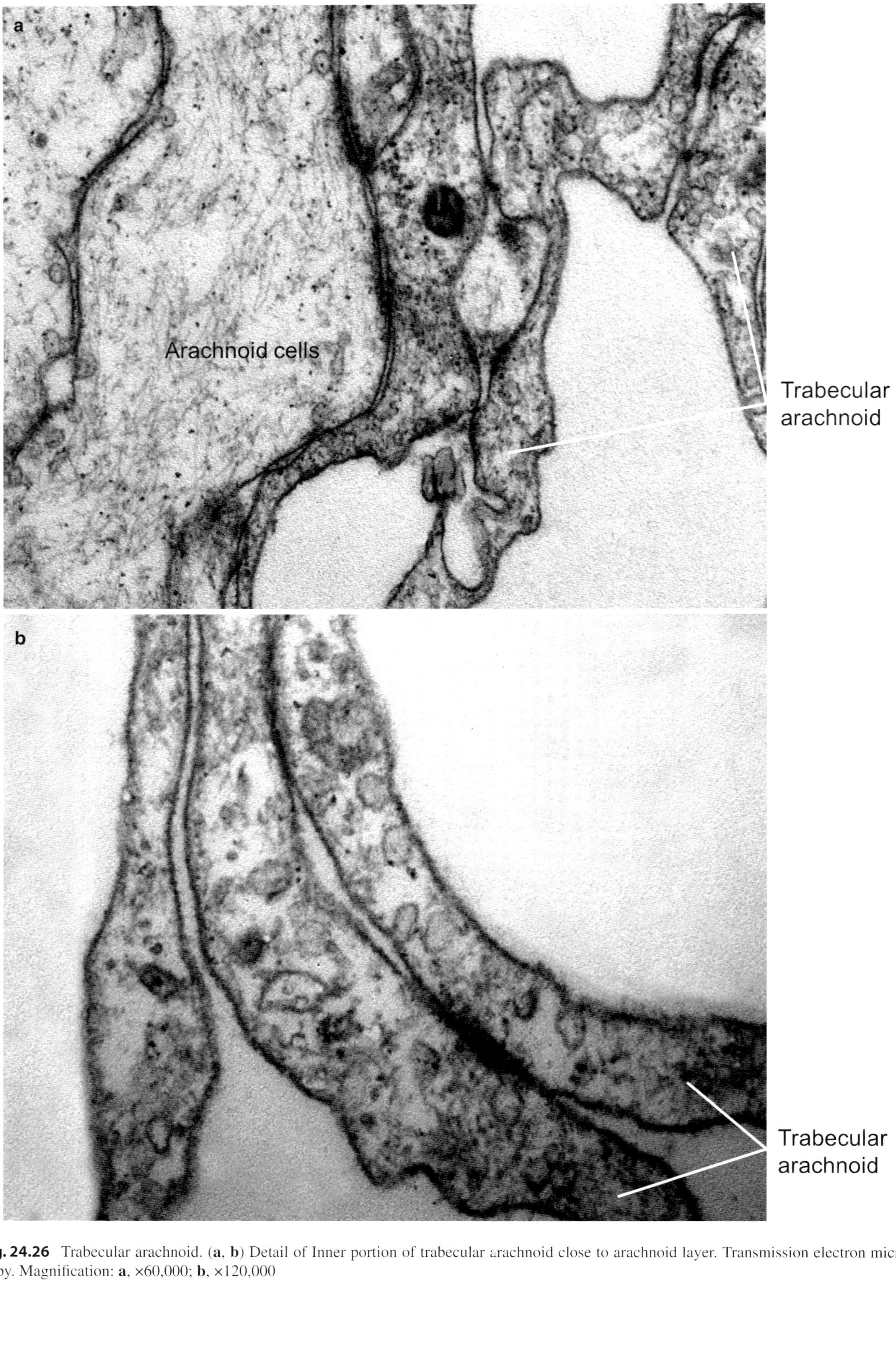

Fig. 24.26 Trabecular arachnoid. (**a**, **b**) Detail of Inner portion of trabecular arachnoid close to arachnoid layer. Transmission electron microscopy. Magnification: **a**, ×60,000; **b**, ×120,000

References

1. Reina MA, López A, De Andrés JA. Hypothesis concerning the anatomical basis of cauda equina syndrome and transient nerve root irritation after spinal anesthesia. Rev Esp Anestesiol Reanim. 1999;46:99–105.
2. Reina MA, Machés F, López A, De Andrés JA. The ultrastructure of the spinal arachnoid in humans and its impact on spinal anesthesia, cauda equina syndrome and transient neurological syndrome. Tech Reg Anesth Pain Management. 2008;12:153–60.
3. Reina MA, De Andrés JA, López A. Anatomía: implicaciones en la práctica de las técnicas de conducción espinal. In: Torres LM, editor. Tratado de anestesia y reanimación. Madrid: Arán; 2001. p. 1135–55.
4. Reina MA, De Andrés JA, López A. Subarachnoid and epidural anesthesia. In: Raj P, editor. Textbook of regional anesthesia. New York: Churchill Livingstone; 2002. p. 307–24.
5. Reina MA, López A, Machés F, de Leon-Casasola O, De Andrés JA. Electron microscopy and the expansion of regional anesthesia knowledge. Tech Reg Anesth Pain Management. 2002;6:165–71.
6. Reina MA, Villanueva MC, López A. Human spinal trabecular arachnoid, pia mater and spinal anesthesia. Rev Arg Anestesiol. 2008;66:111–33.
7. Reina MA, Prats-Galino A, Sola RG, Puigdellívol-Sánchez A, Arriazu Navarro R, De Andrés JA. Structure of the arachnoid layer of the human spinal meninges: a barrier that regulates dural sac permeability. Rev Esp Anestesiol Reanim. 2010;57:486–92.

Ultrastructure of Spinal Pia Mater

25

Fabiola Machés, Miguel Angel Reina, and Oscar De León Casasola

Pia mater is the innermost of the three meningeal membranes, closely surrounding the brain, the spinal cord, and the portion of spinal nerve roots extending towards their exit across the dura mater. A cellular layer and a subpial compartment shape the structure of the pia mater. The cellular layer is made up of flat, overlapping pial cells with amorphous, fundamental substance interposed among them [1–6]. The surface of the cellular plane appears smooth and bright under scanning electron microscopy (SEM). Transmission electron microscopy shows that the cellular component is about three to five pial cells thick at the medullary level and two to three cells thick at the level of the nerve roots. The subpial compartment is limited by a pial cellular layer and a basal membrane and is filled with collagen fibers oriented in various directions [1–6].

Differences can be observed between pia mater at the level of the medullary cone and pia mater from nerve roots. The thickness of both the cellular layer and the subpial compartment is larger at the level of the spinal medulla, becoming thinner towards the medullary cone and the nerve roots (SEM). The thickness of the subpial compartment is more variable than that of the pial cellular layer [1, 2].

At the level of the medullary cone, there are perforations or fenestrations over the entire surface of the cellular layer of the pia mater, distributed in variable proportions. These fenestrations have circular, ovoid, or elliptic shapes (SEM). At the nerve root level, the pia mater also presents similar fenestrations, but these are smaller (Figs. 25.1, 25.2, 25.3, 25.4, 25.5, 25.6, 25.7, 25.8, 25.9, 25.10, 25.11, 25.12, 25.13, 25.14, 25.15, 25.16, 25.17, 25.18, 25.19, 25.20, 25.21, and 25.22) [1, 2].

F. Machés, MD (✉)
Department of Anesthesiology, Madrid-Montepríncipe
University Hospital, Madrid, Spain
e-mail: fabmami@gmail.com

M.A. Reina, MD, PhD
Department of Anesthesiology, Madrid-Montepríncipe
University Hospital, Madrid, Spain

Department of Clinical Medical Sciences
and Institute of Applied Molecular Medicine,
School of Medicine, University of CEU San Pablo, Madrid, Spain
e-mail: miguelangel@perticone.e.telefonica.net

O. De León Casasola, MD
Department of Anesthesiology, Roswell Park Cancer Institute,
Buffalo, NY, USA

Department of Anesthesiology, University at Buffalo,
School of Medicine and Biomedical Sciences, Buffalo, NY, USA
e-mail: oscar.deleon@roswellpark.org

M.A. Reina et al. (eds.), *Atlas of Functional Anatomy for Regional Anesthesia and Pain Medicine: Human Structure, Ultrastructure and 3D Reconstruction Images*, DOI 10.1007/978-3-319-09522-6_25,

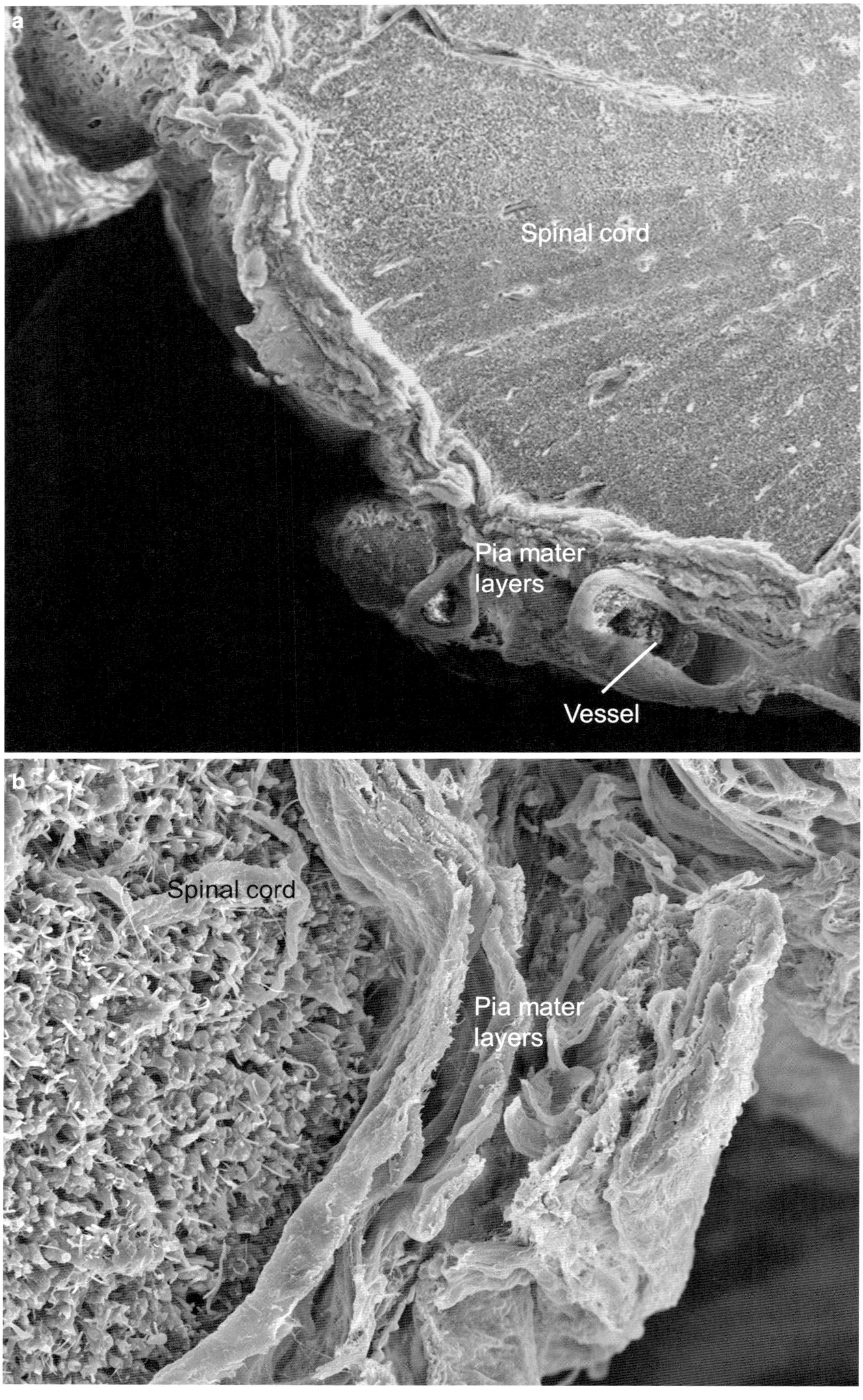

Fig. 25.1 Pia mater. (**a**, **b**) Human spinal pia mater at spinal cord level. Scanning electron microscopy. Magnification: **a**, ×30; **b**, ×300

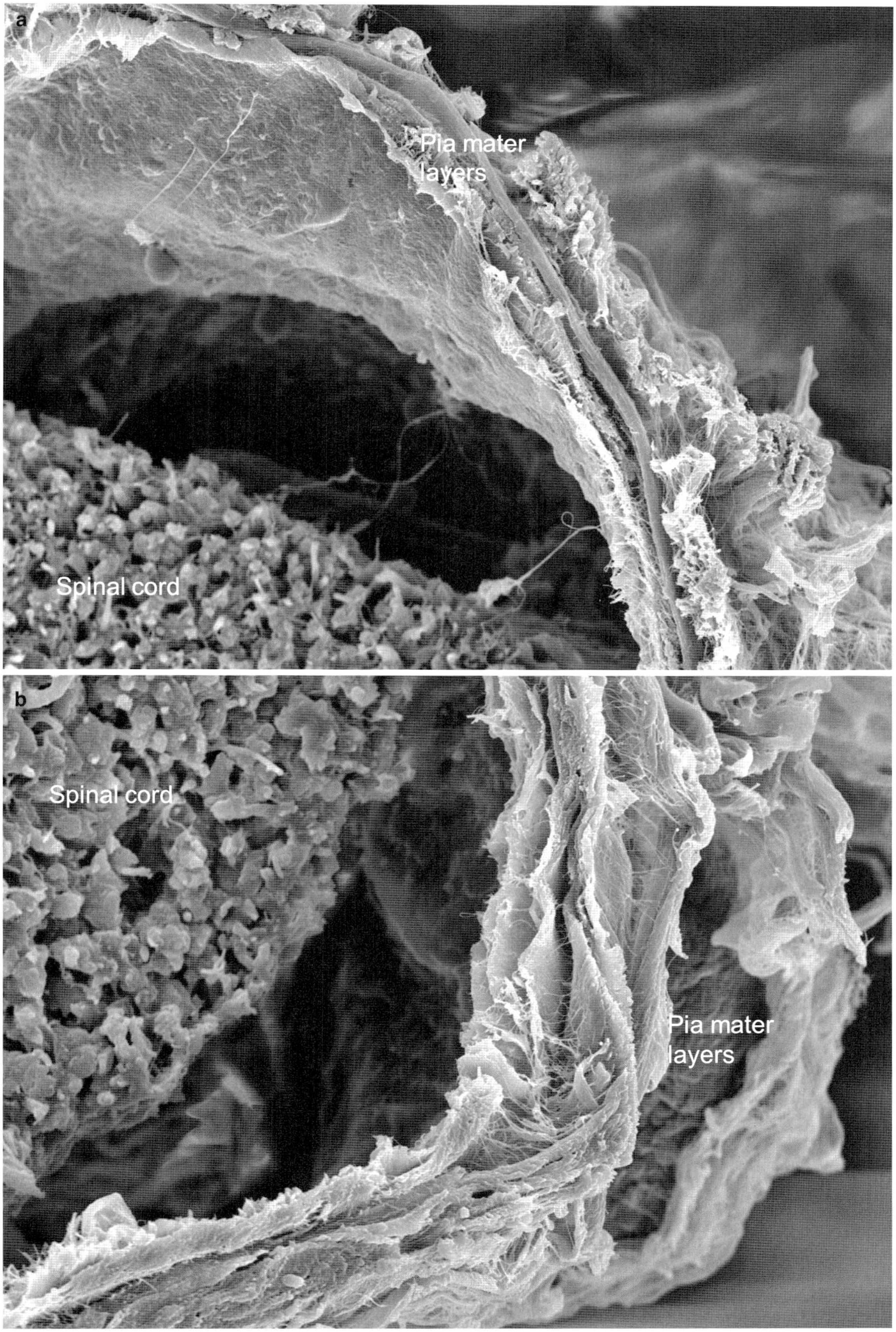

Fig. 25.2 Pia mater. Scanning electron microscopy. Human spinal pia mater at spinal cord level. Magnification: **a**, ×400; **b**, ×500

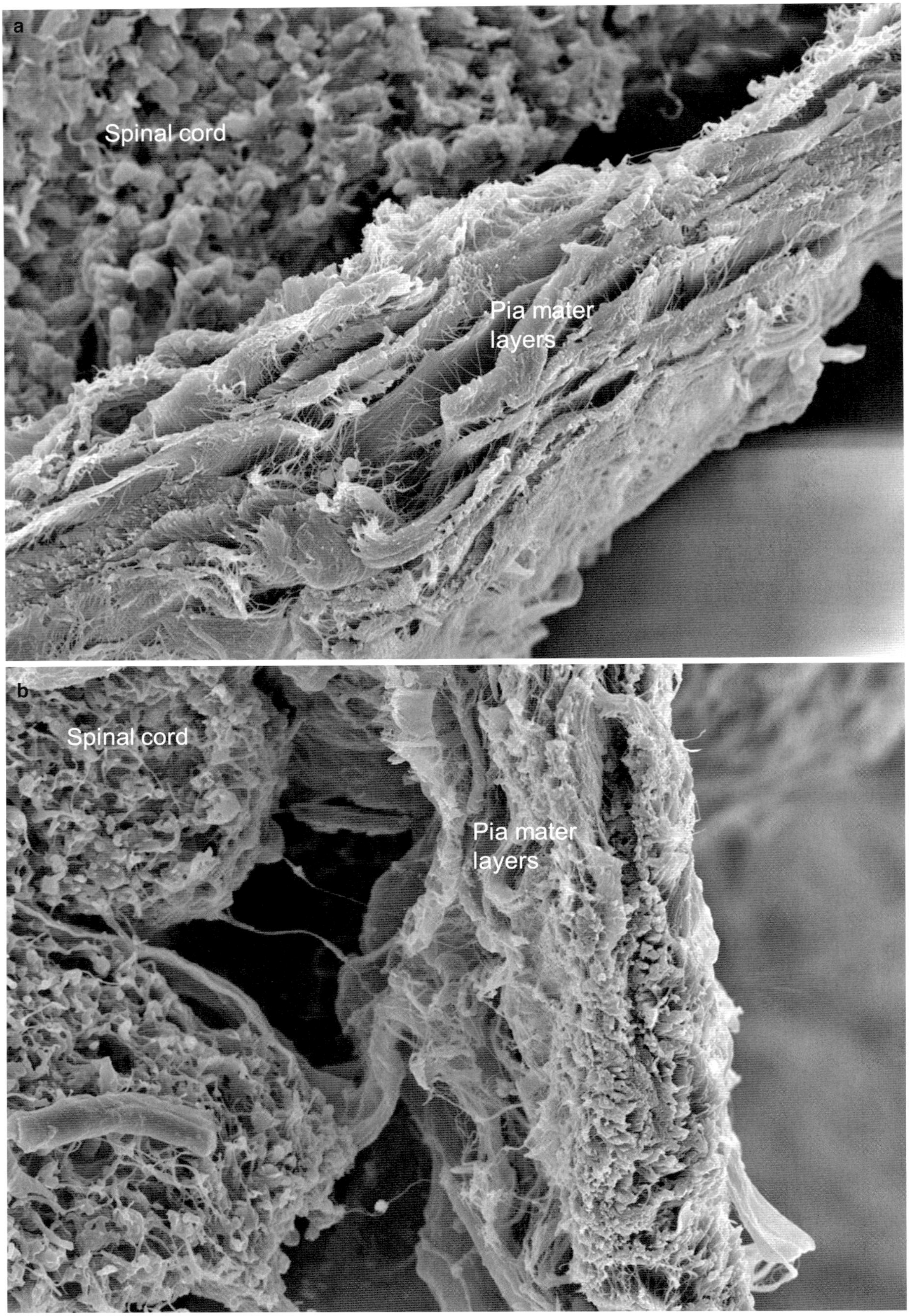

Fig. 25.3 Pia mater. Human spinal pia mater at spinal cord level. Scanning electron microscopy. Magnification: **a**, ×500; **b**, ×500

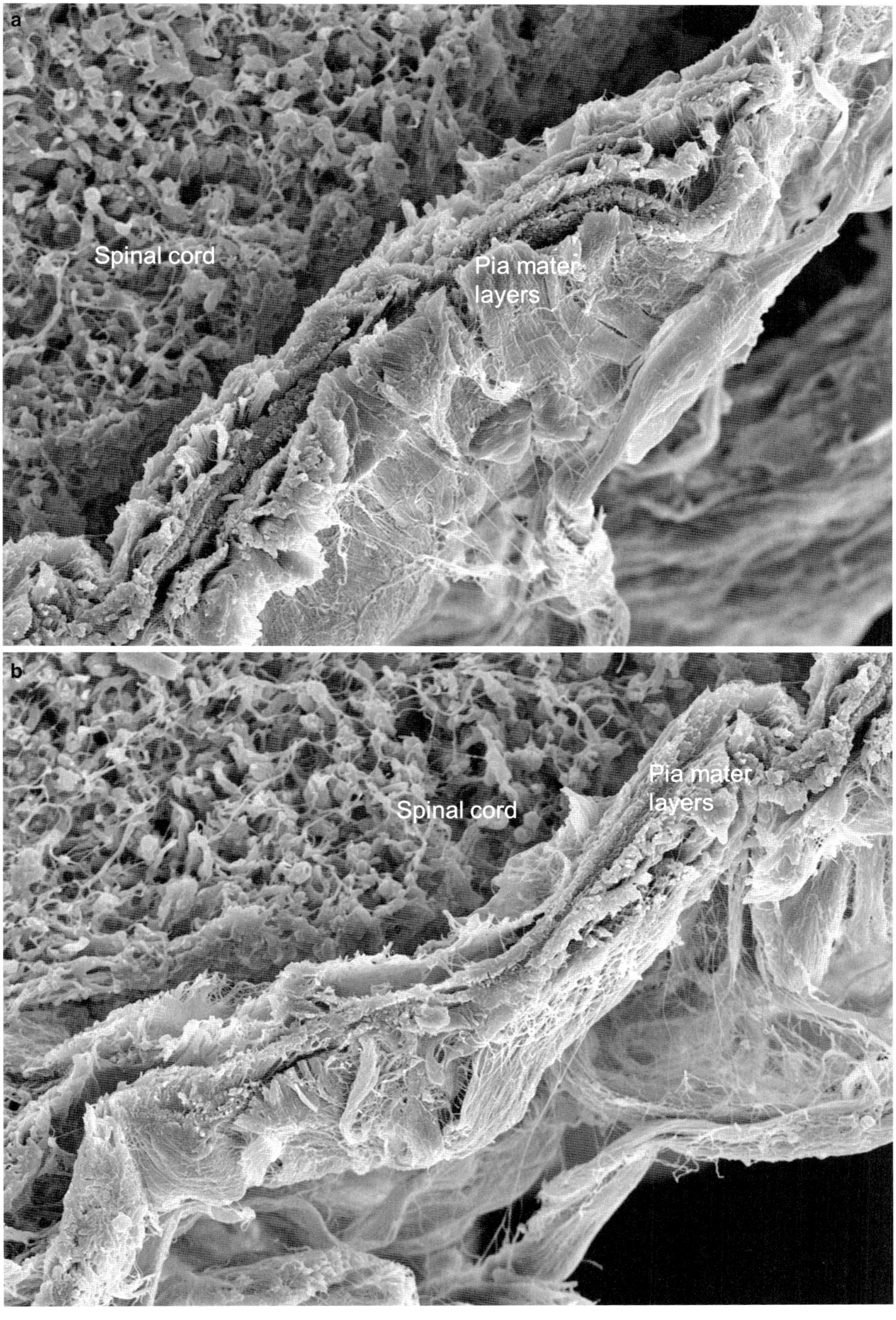

Fig. 25.4 Pia mater. Human spinal pia mater at spinal cord level. Scanning electron microscopy. Magnification: **a**, ×500; **b**, ×500

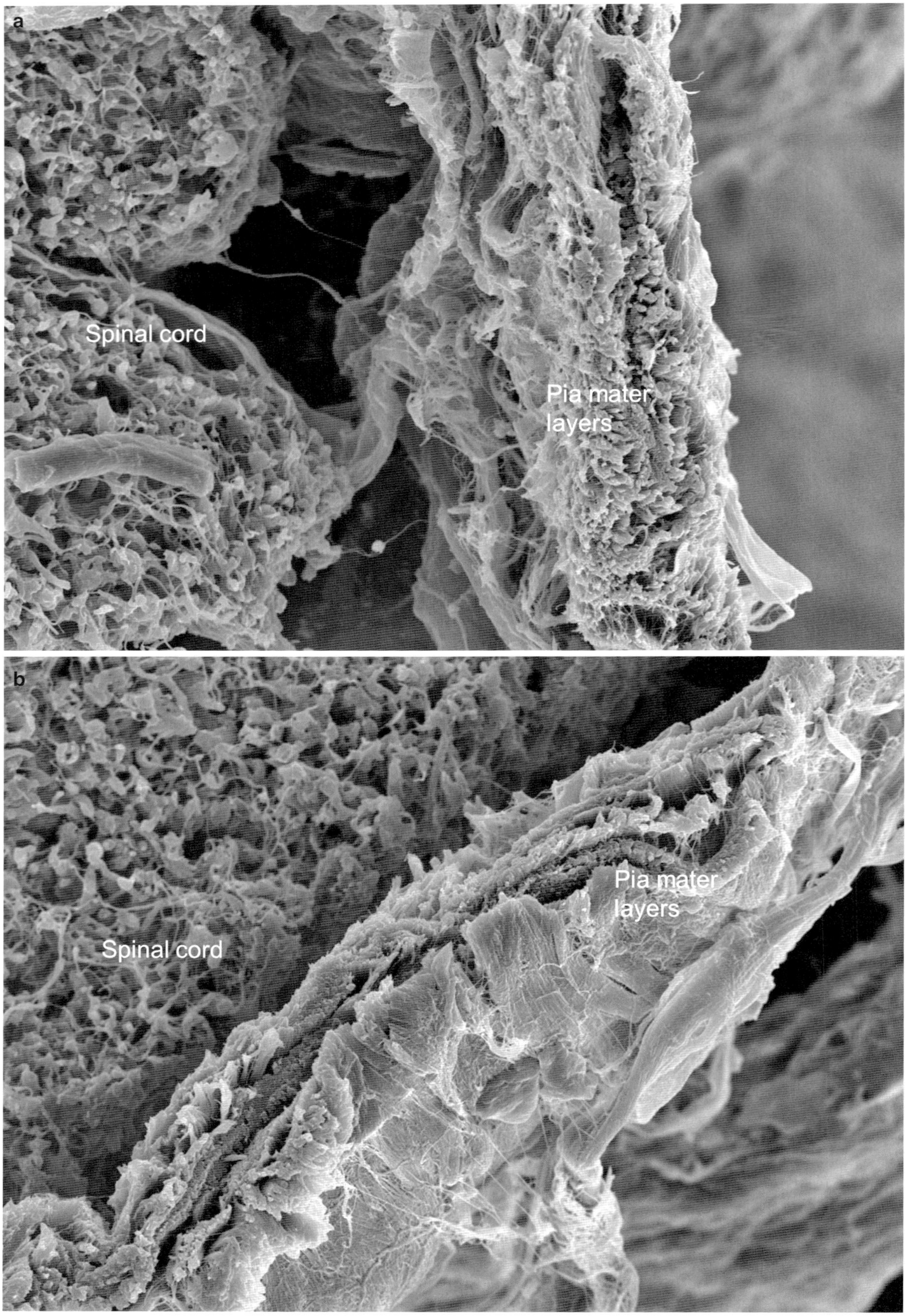

Fig. 25.5 Pia mater. Human spinal pia mater at spinal cord level. Scanning electron microscopy. Magnification: **a**, ×500; **b**, ×500

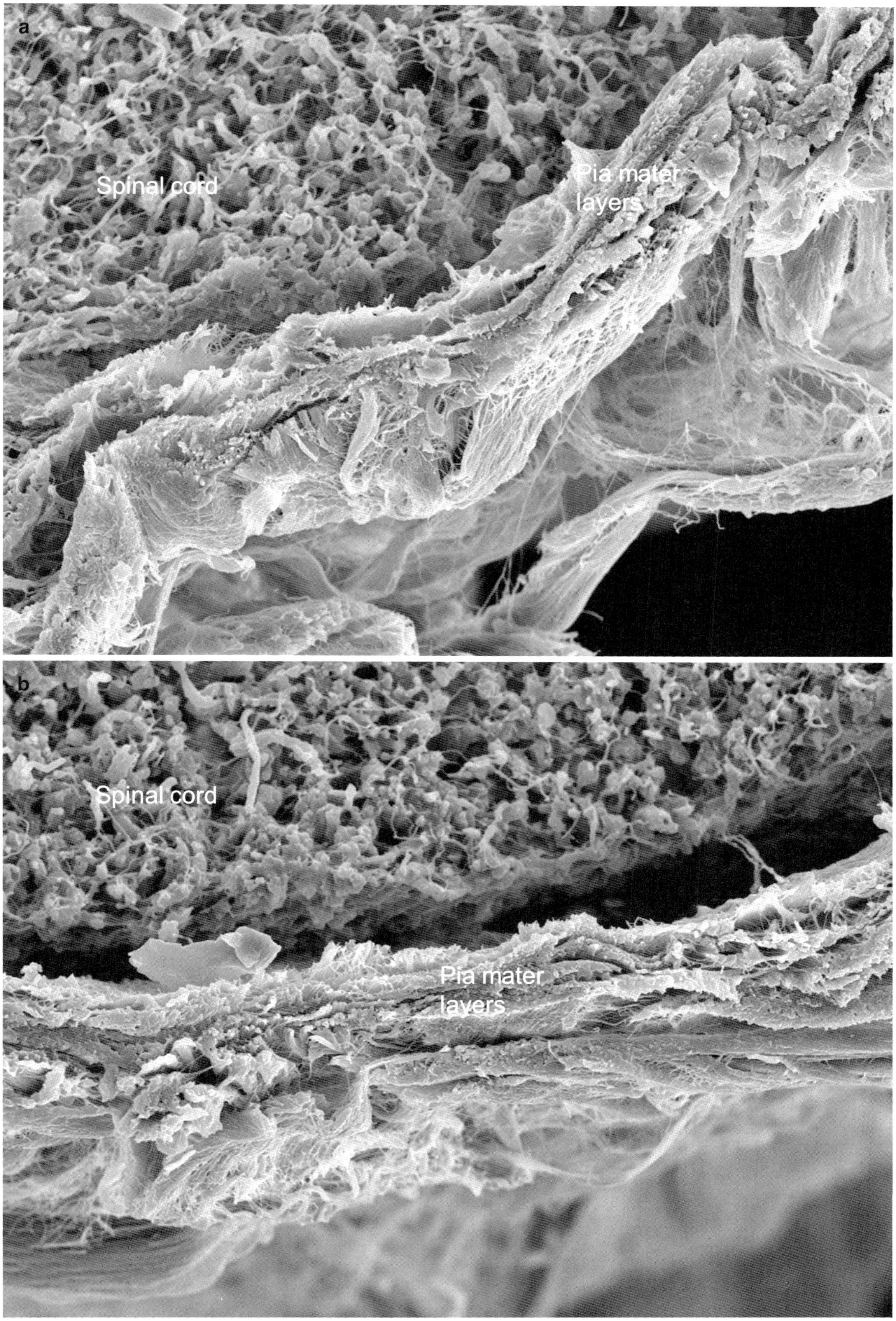

Fig. 25.6 Pia mater. Human spinal pia mater at spinal cord level. Scanning electron microscopy. Magnification: **a**, ×500; **b**, ×500

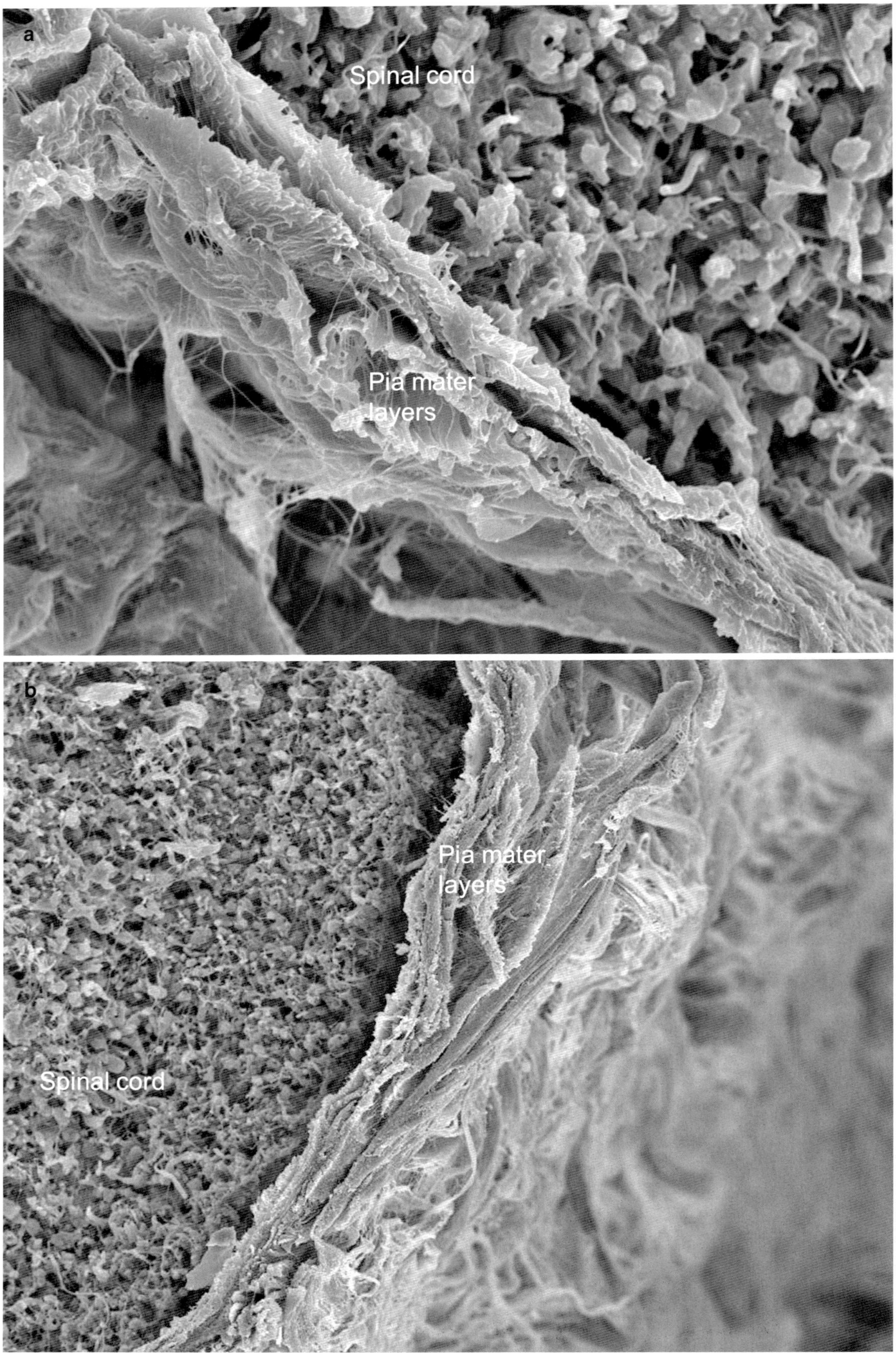

Fig. 25.7 Pia mater. Human spinal pia mater at spinal cord level. Scanning electron microscopy. Magnification: **a**, ×800; **b**, ×250

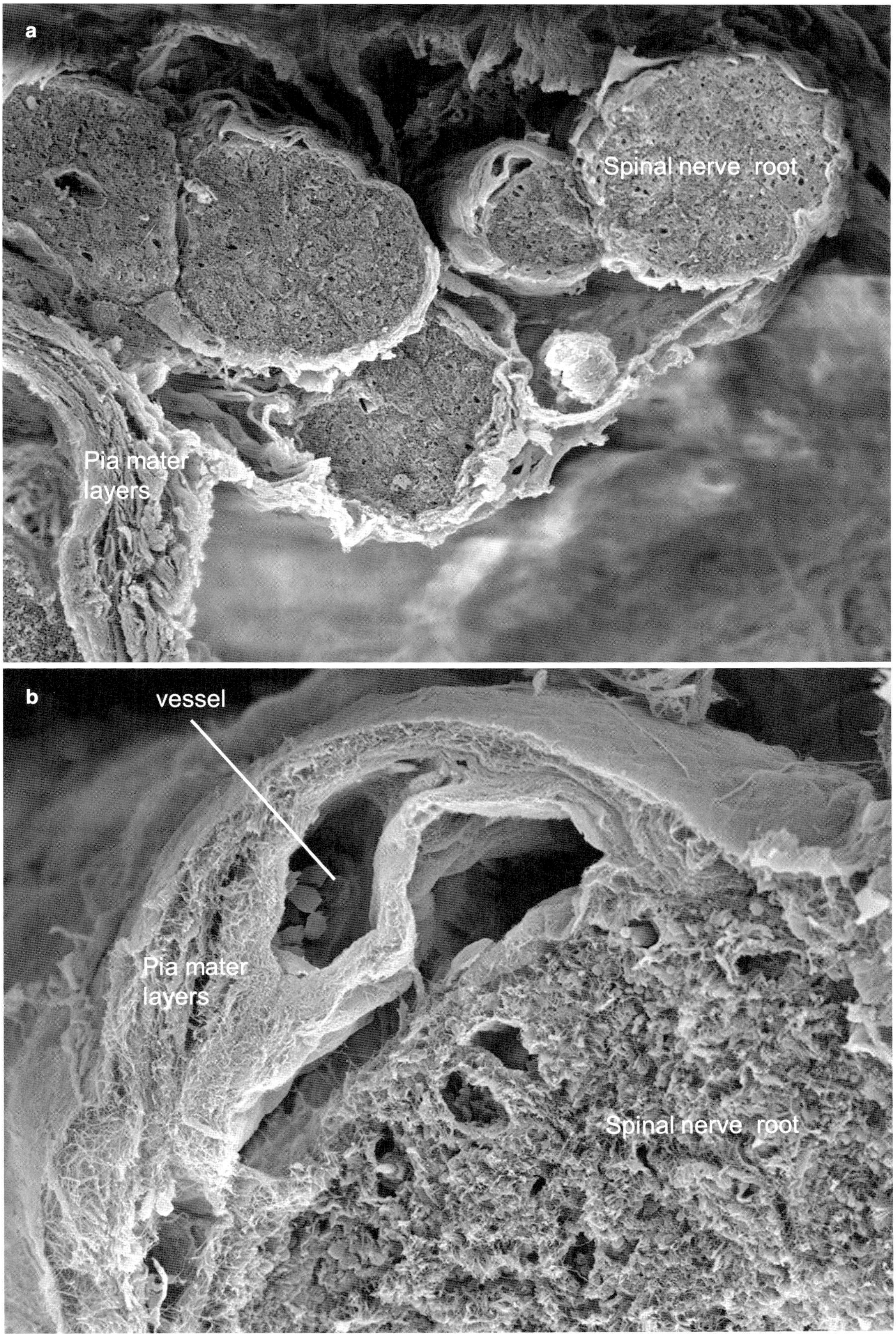

Fig. 25.8 Pia mater. Human spinal pia mater at spinal nerve root level. Scanning electron microscopy. Magnification: **a**, ×90; **b**, ×700

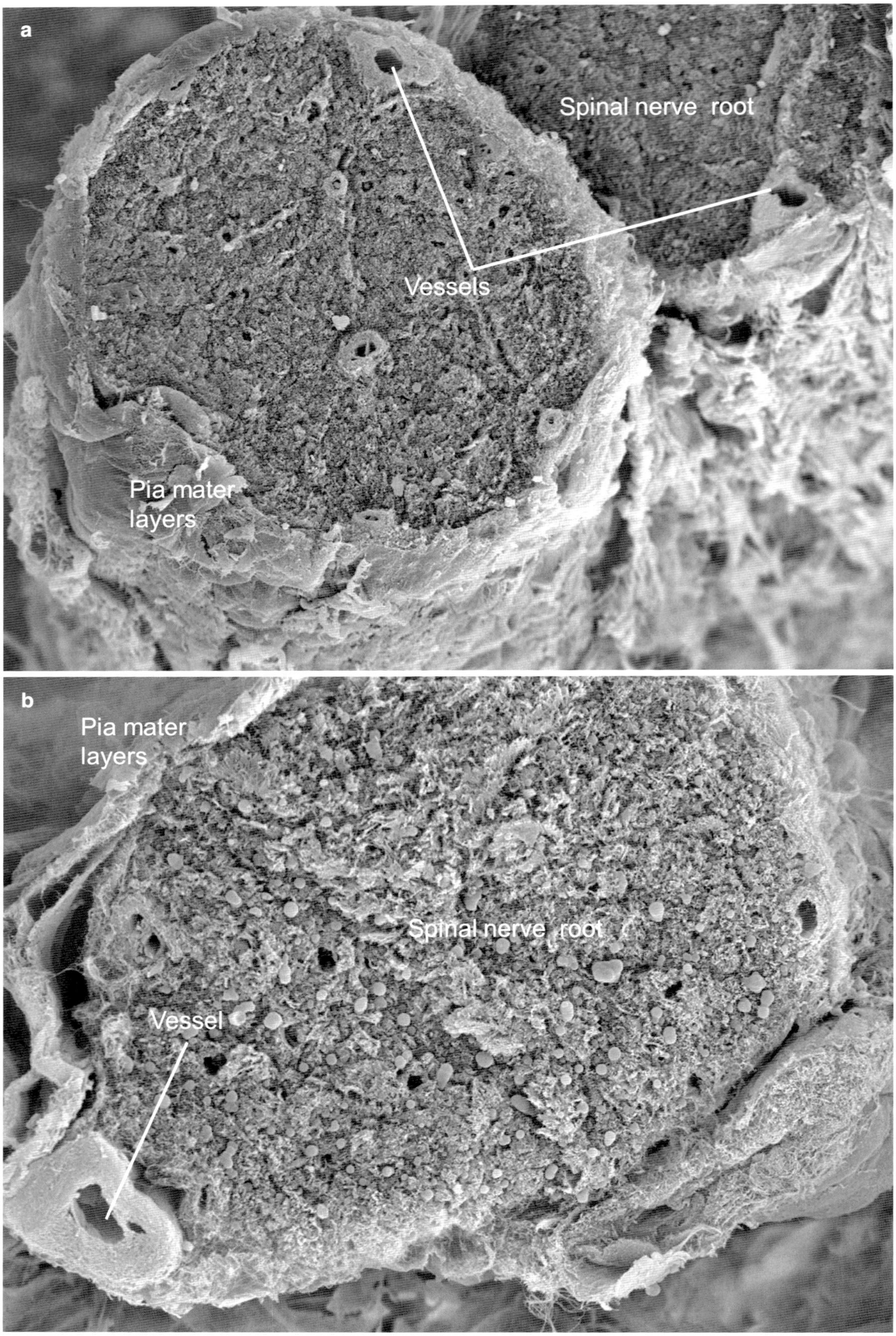

Fig. 25.9 Pia mater. Human spinal pia mater at spinal nerve root level. Scanning electron microscopy. Magnification: **a**, ×200; **b**, ×400

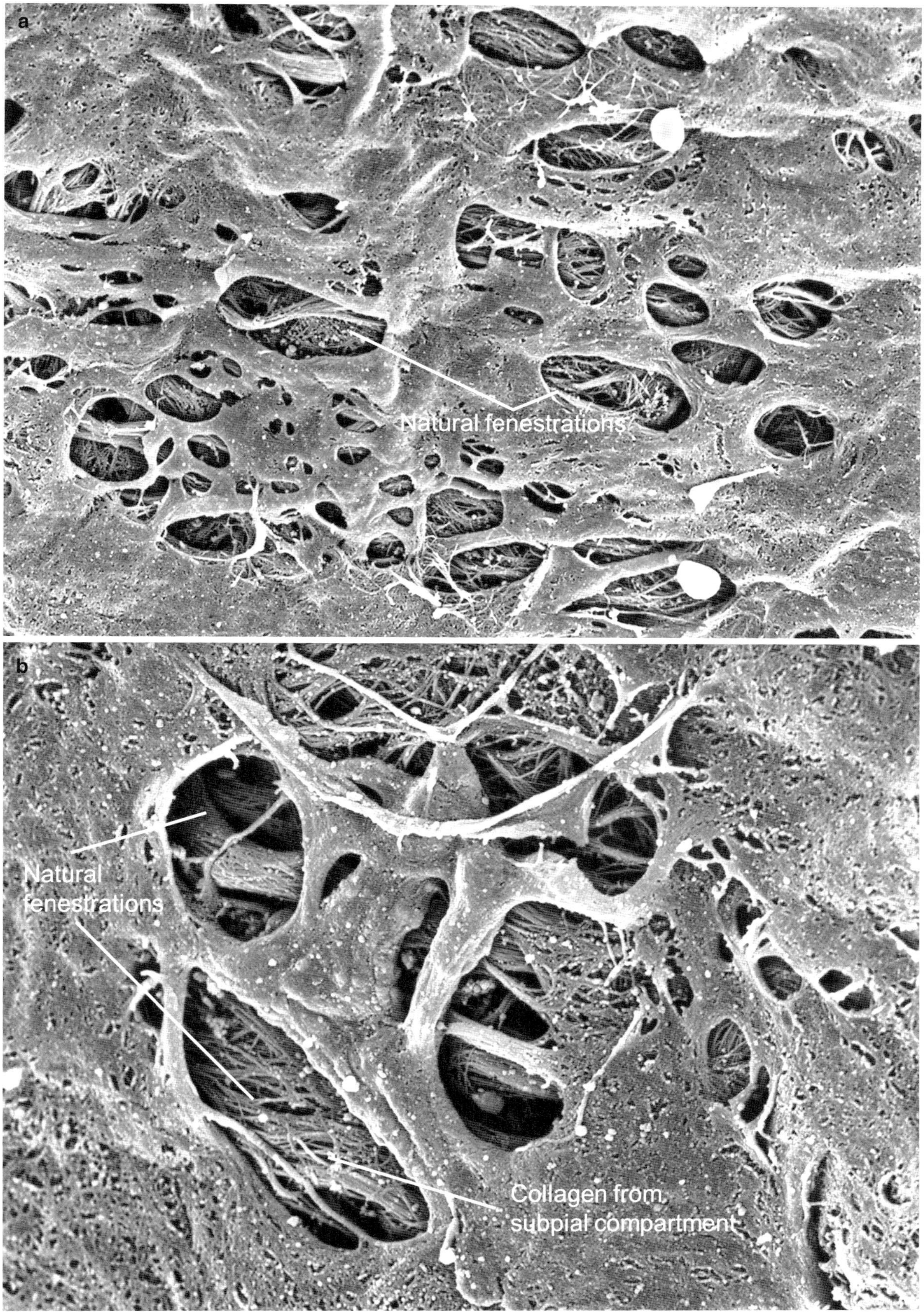

Fig. 25.10 Pia mater. Natural fenestration within human spinal pia mater at spinal cord level. Scanning electron microscopy. Magnification: **a**, ×1,000; **b**, ×2,000 (From Reina et al. [2]; with permission)

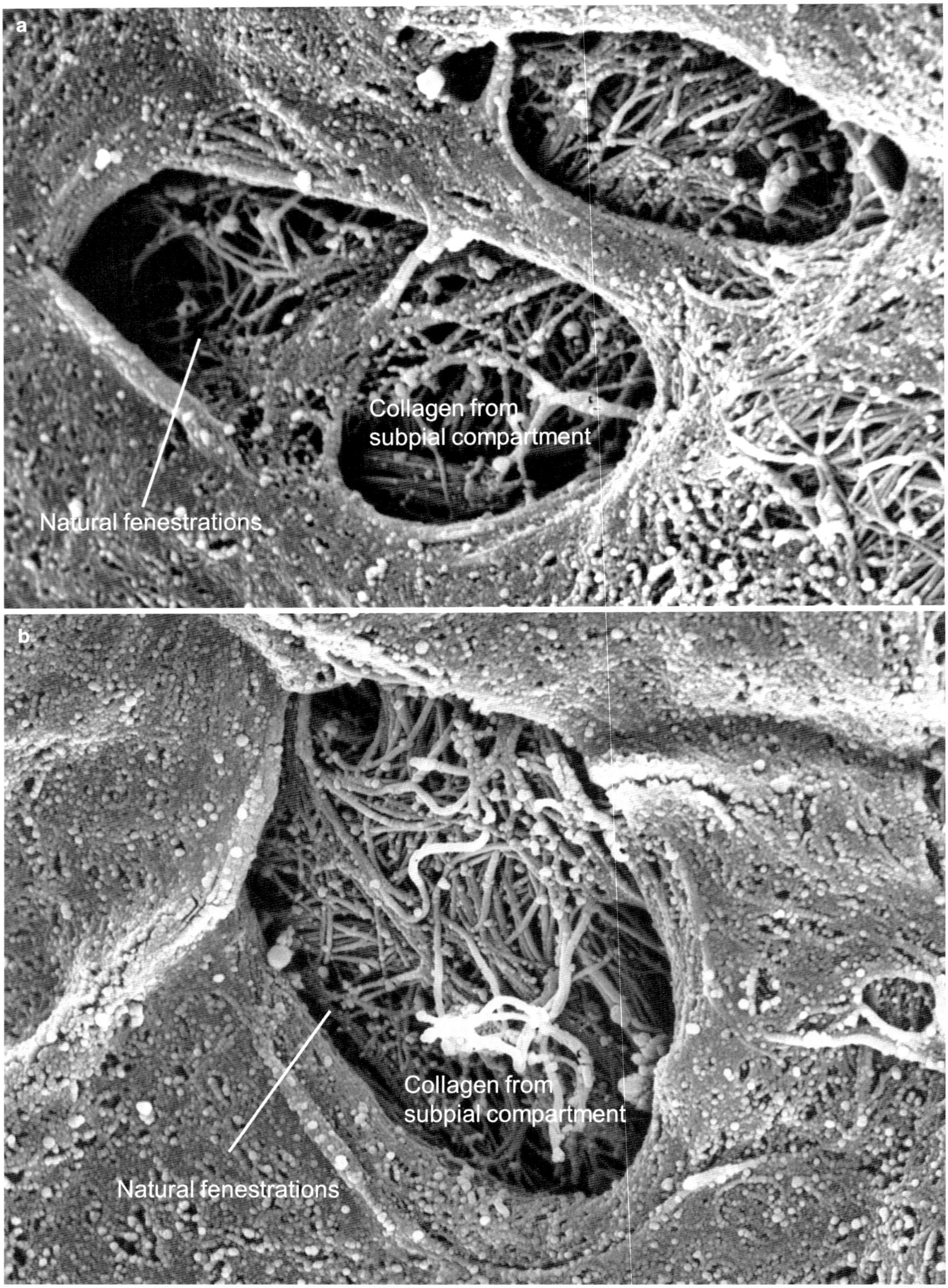

Fig. 25.11 Pia mater. Natural fenestration within human spinal pia mater at spinal cord level. Scanning electron microscopy. Magnification: **a**, ×4,500; **b**, ×1,500 (**a** from Reina et al. [1]; **b** from Reina et al. [2]; both with permission)

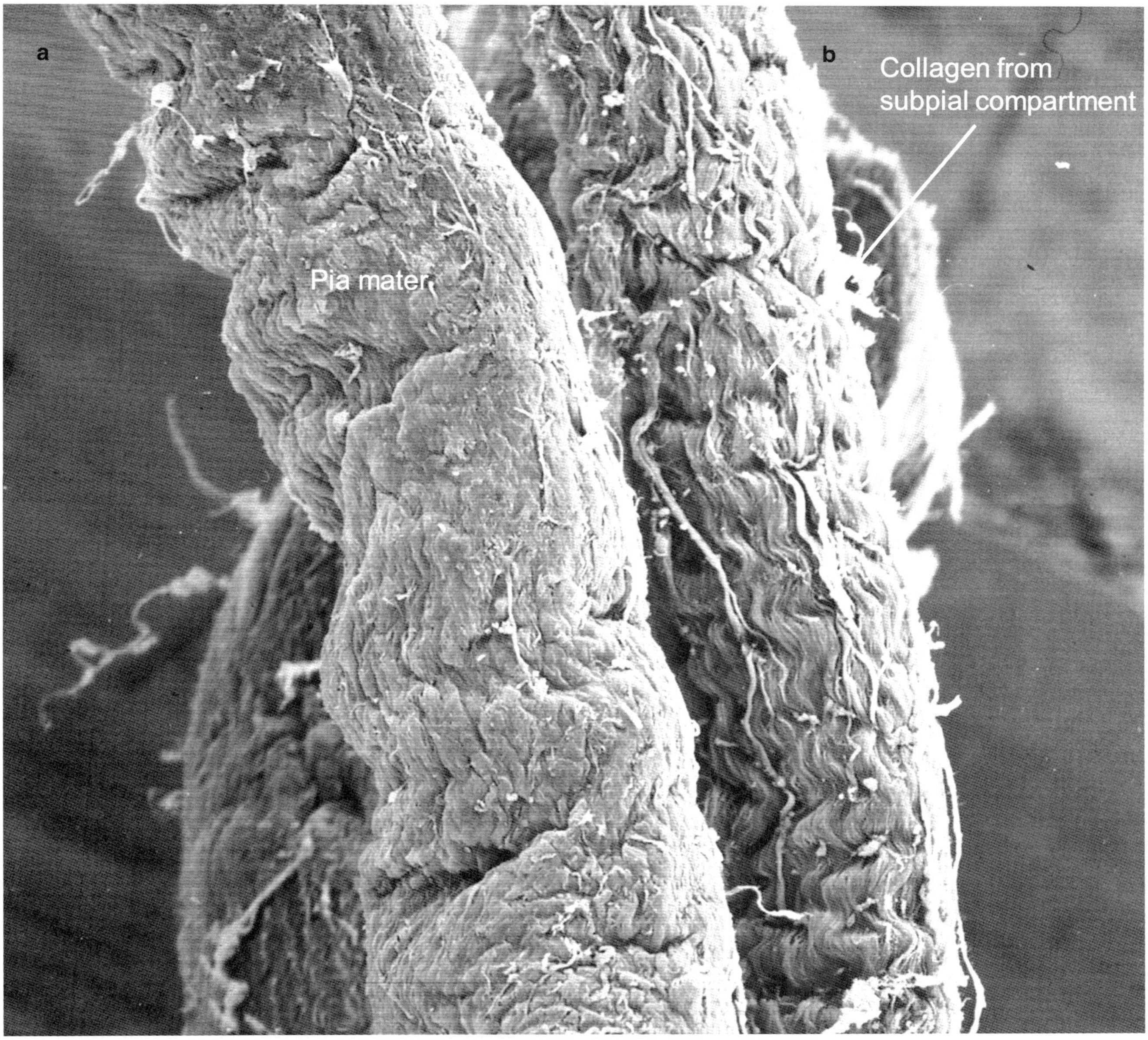

Fig. 25.12 Pia mater. (**a**) Spinal nerve root with pia mater layer. (**b**) Spinal nerve root without pia mater layer. Scanning electron microscopy. Magnification × 100 (From Reina et al. [6]; with permission)

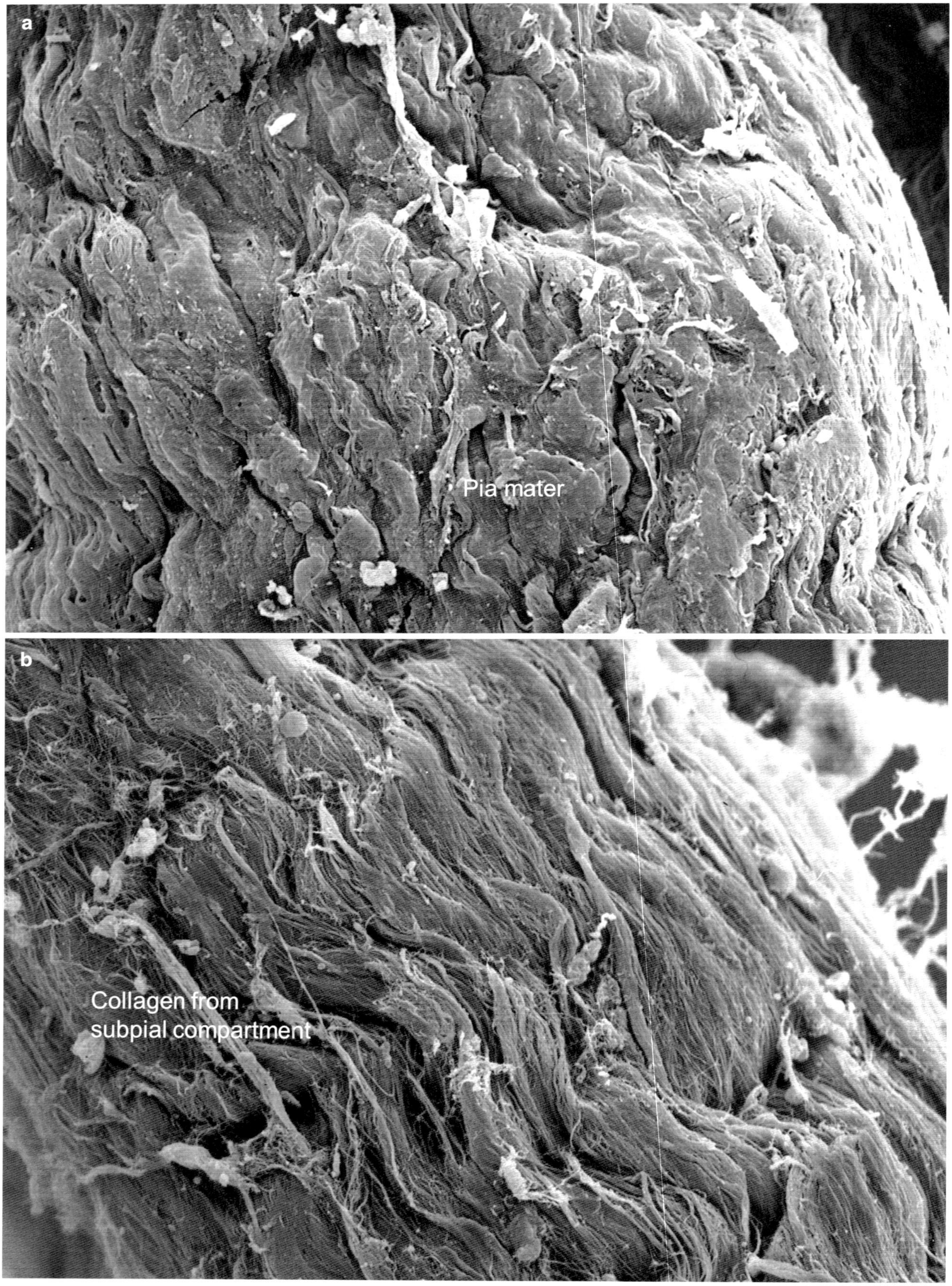

Fig. 25.13 Pia mater. Detail of Fig. 25.12 at higher magnification (×500). (**a**) Spinal nerve root with pia mater layer. (**b**) Spinal nerve root without pia mater layer. Scanning electron microscopy (From Reina et al. [6]; with permission)

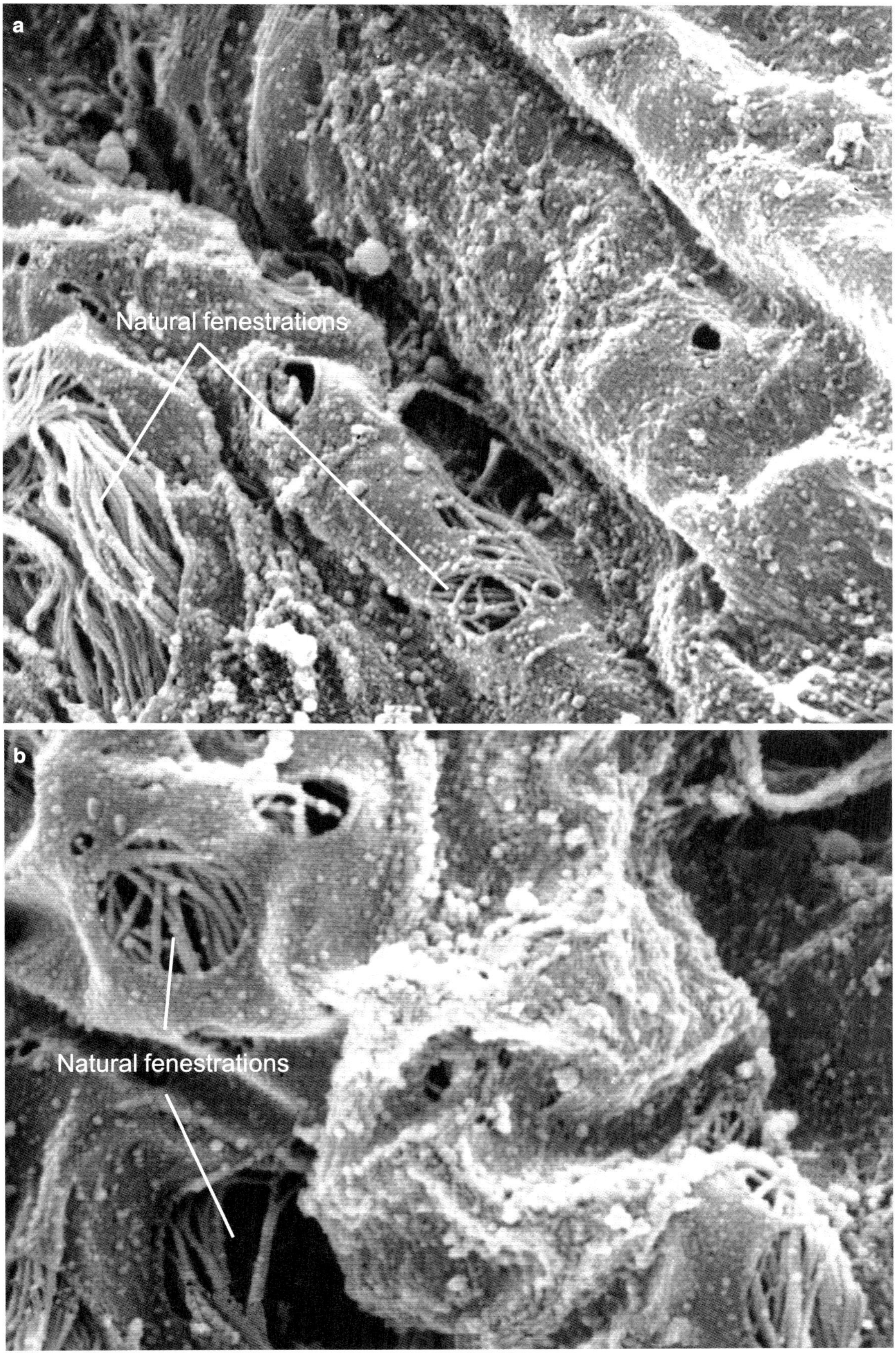

Fig. 25.14 Pia mater. Natural fenestration within human spinal pia mater at spinal nerve root level. Scanning electron microscopy. Magnification: **a**, ×6,000; **b**, ×10,000 (**a** from Reina et al. [6]; **b** from Reina et al. [1]; both with permission)

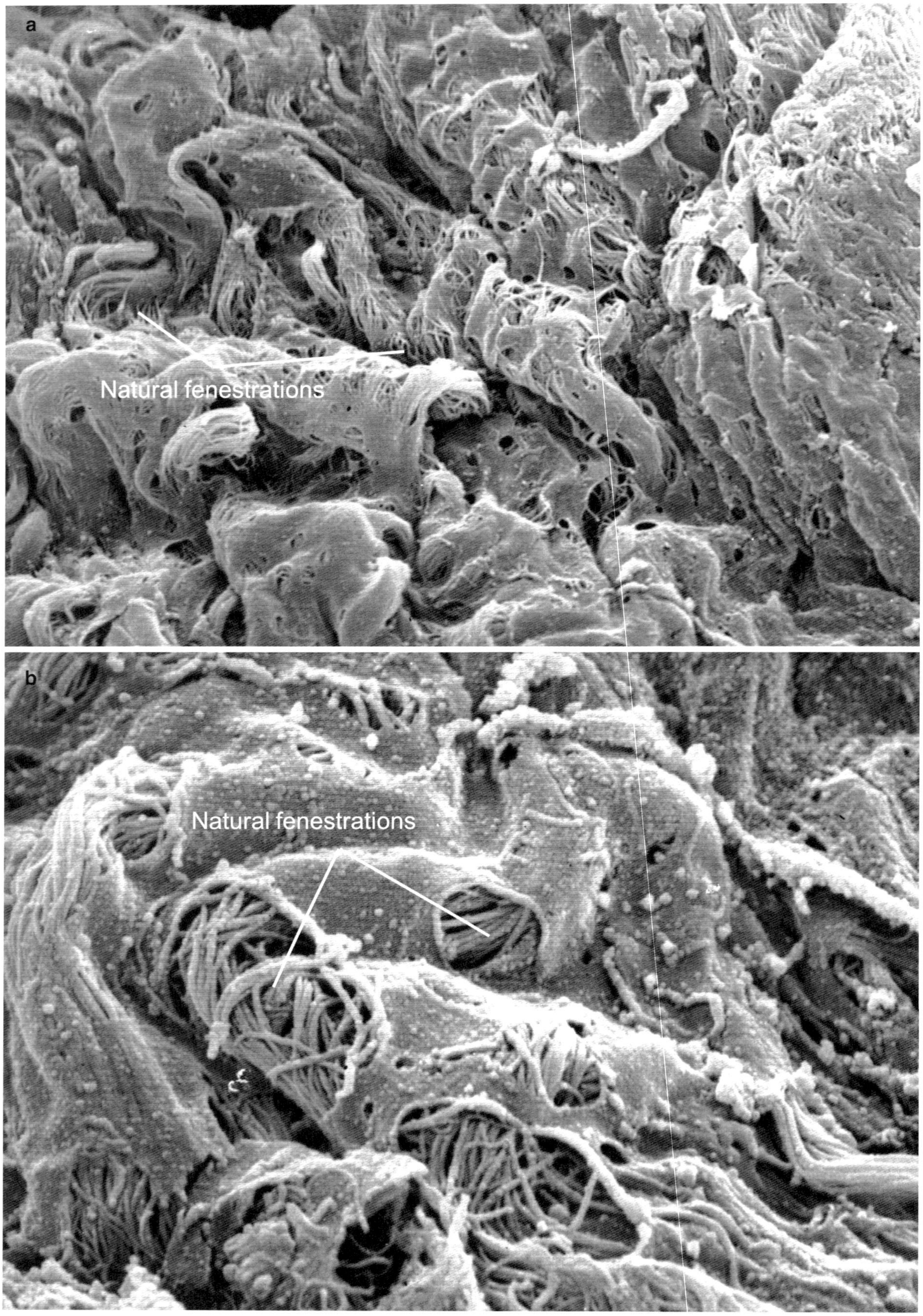

Fig. 25.15 Pia mater. Natural fenestration within human spinal pia mater at spinal nerve root level. Scanning electron microscopy. Magnification: **a**, ×2,000; **b**, ×6,000 (From Reina et al. [6]; with permission)

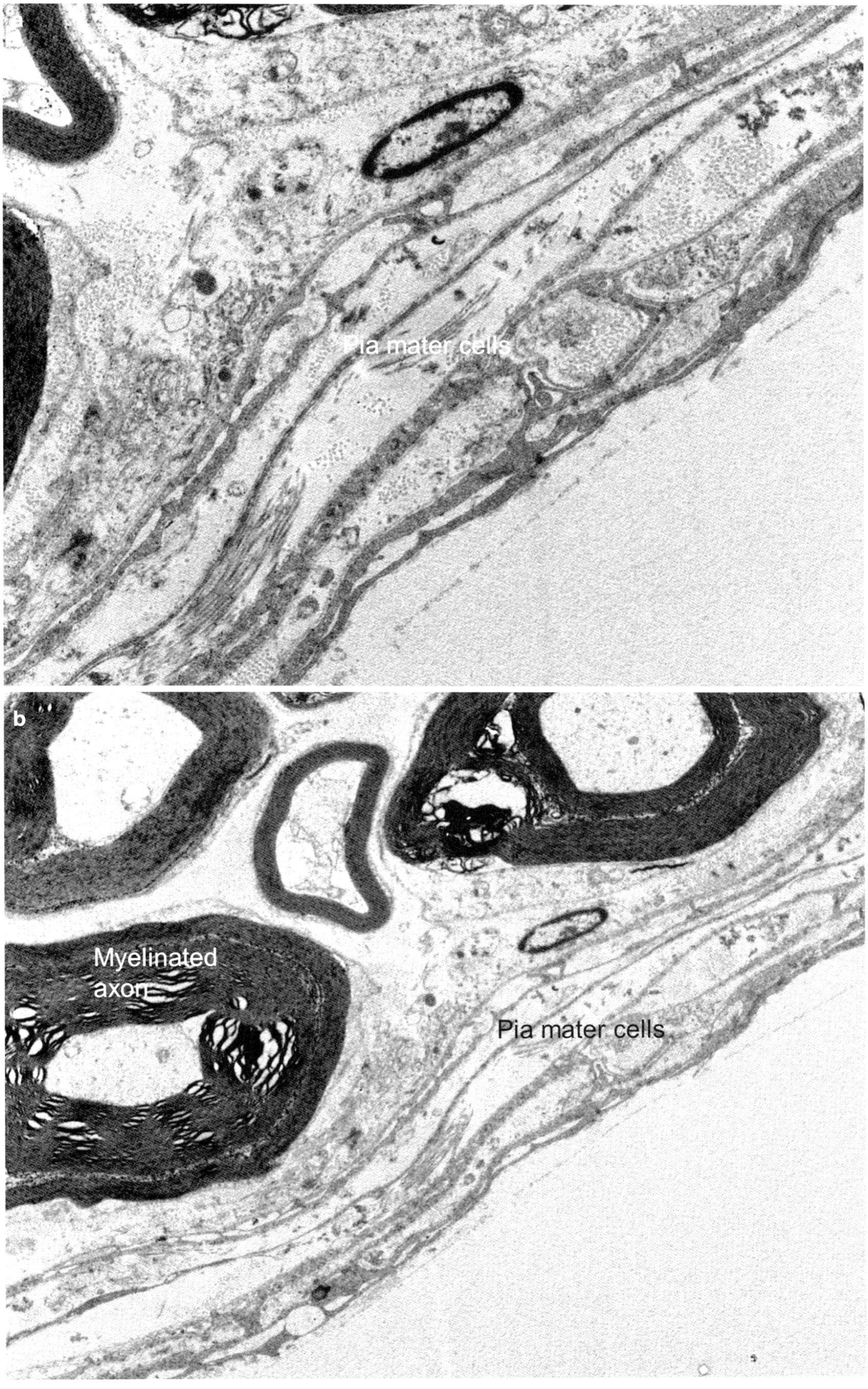

Fig. 25.16 Pia mater. Human pia mater layer at spinal nerve root level. Transmission electron microscopy. Magnification: **a**, ×12,000; **b**, ×6,000

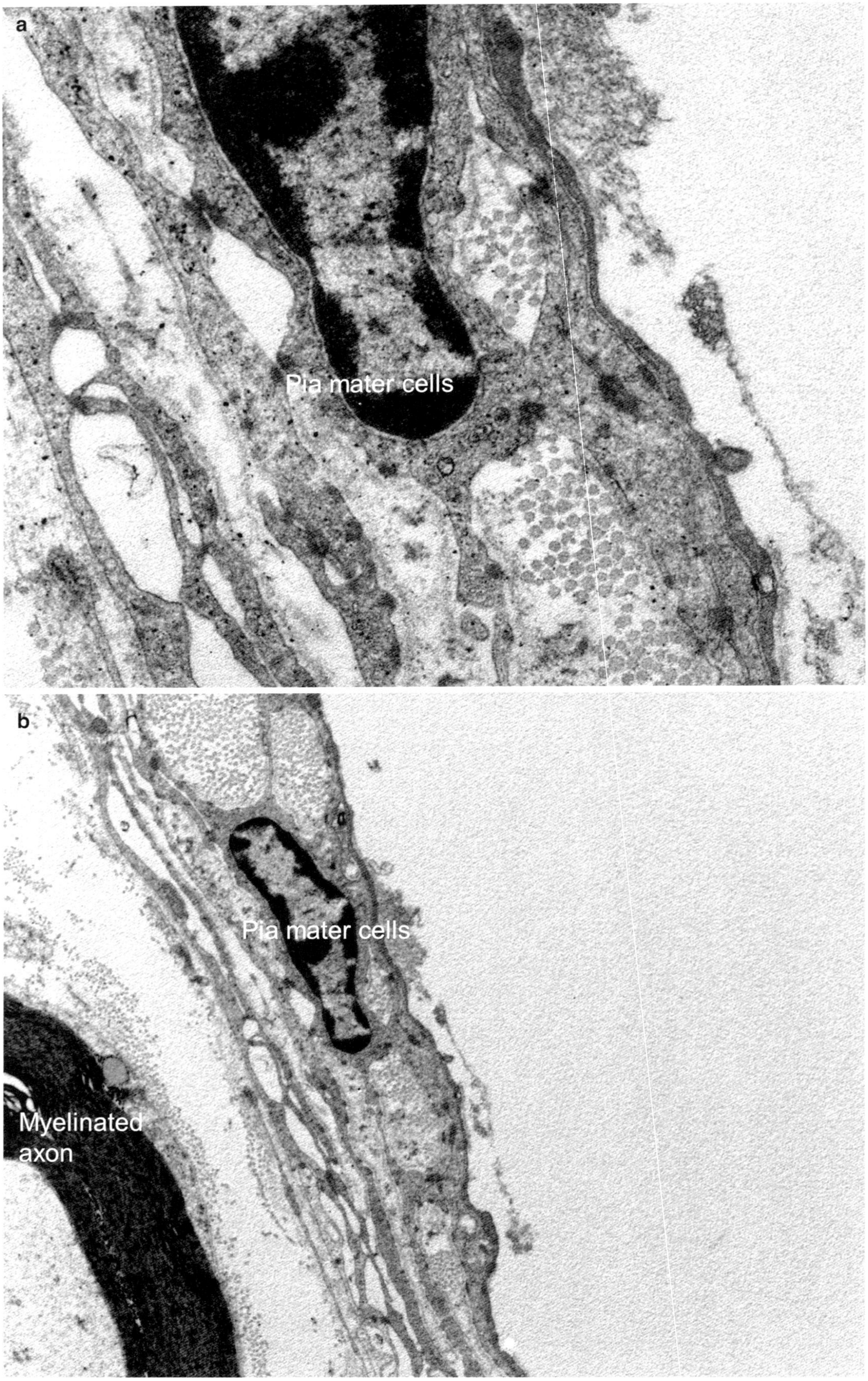

Fig. 25.17 Pia mater. Human pia mater layer at spinal nerve root level. Detail of pial cells. Transmission electron microscopy. Magnification: **a**, ×40,000; **b**, ×12,000 (**b** from Reina et al. [5]; with permission)

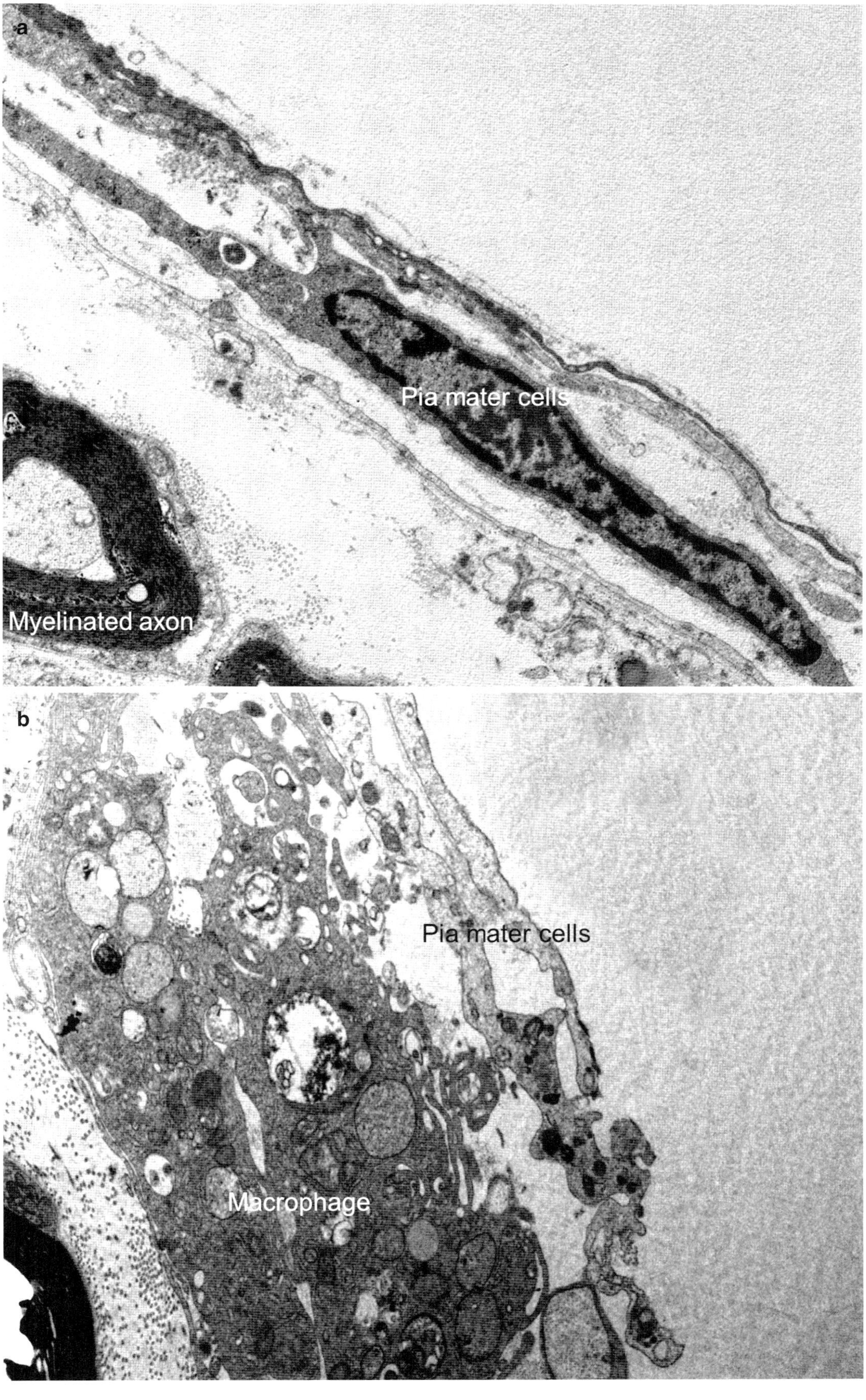

Fig. 25.18 Pia mater. Human pia mater layer at spinal nerve root level. Detail of pial cells. Transmission electron microscopy. Magnification: **a**, ×12,000; **b**, ×7,000 (**b** from Reina et al. [6]; with permission)

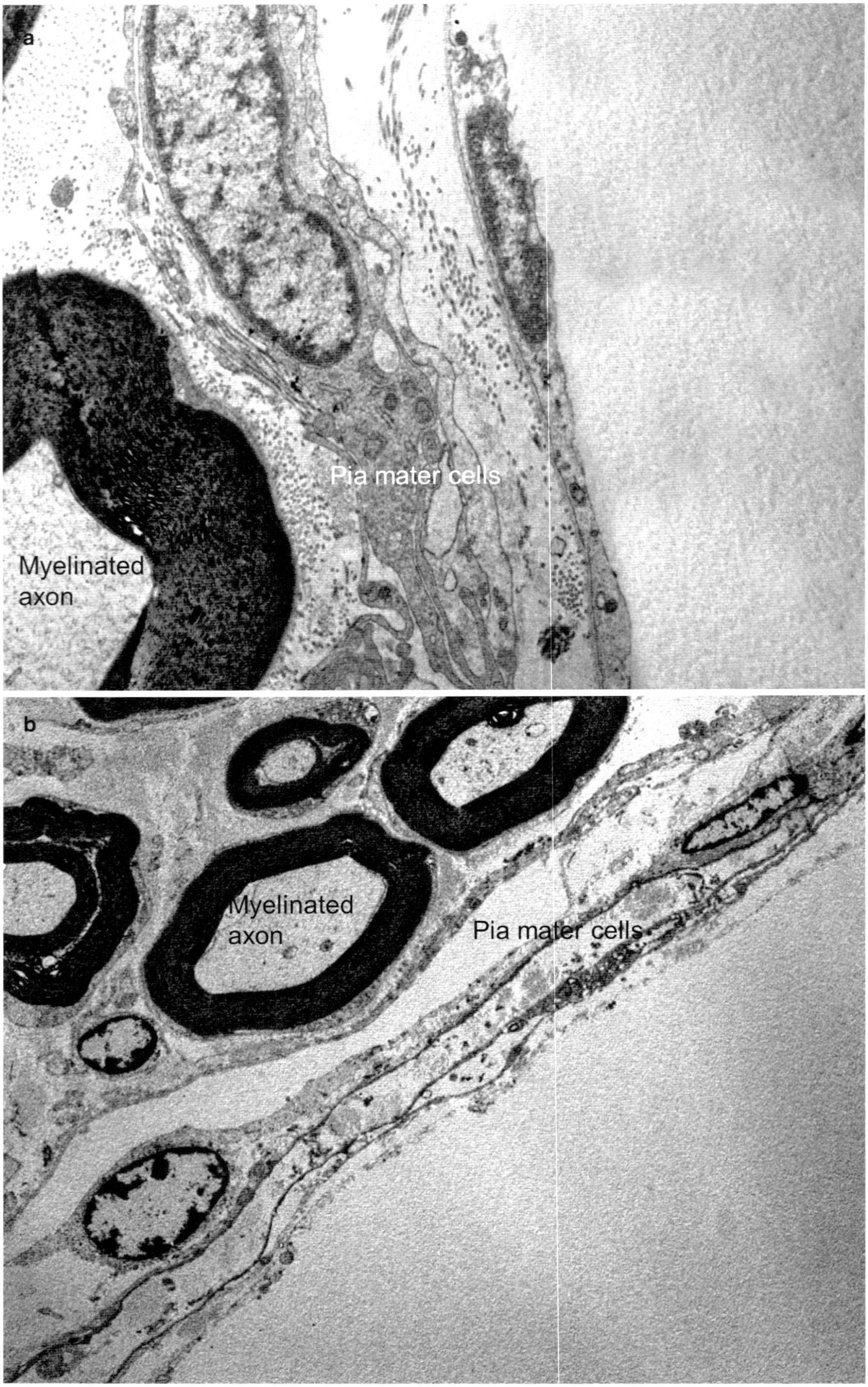

Fig. 25.19 Pia mater. Human pia mater layer at spinal nerve root level. Detail of pial cells. Transmission electron microscopy. Magnification: **a**, ×12,000; **b**, ×4,000 (From Reina et al. [6]; with permission)

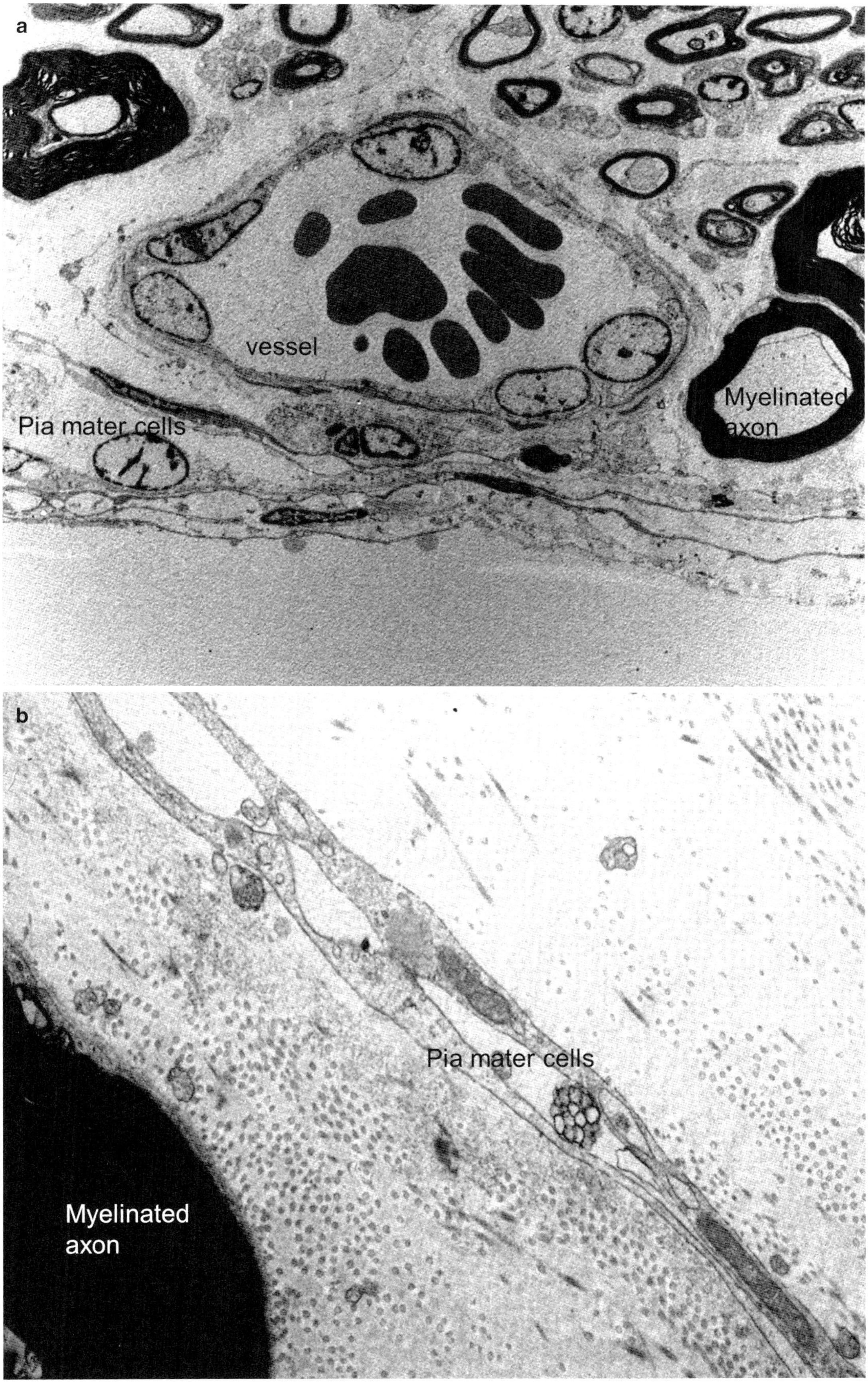

Fig. 25.20 Pia mater. Human pia mater layer at spinal nerve root level. Detail of pial cells. Transmission electron microscopy. Magnification: **a**, ×2,500; **b**, ×12,000

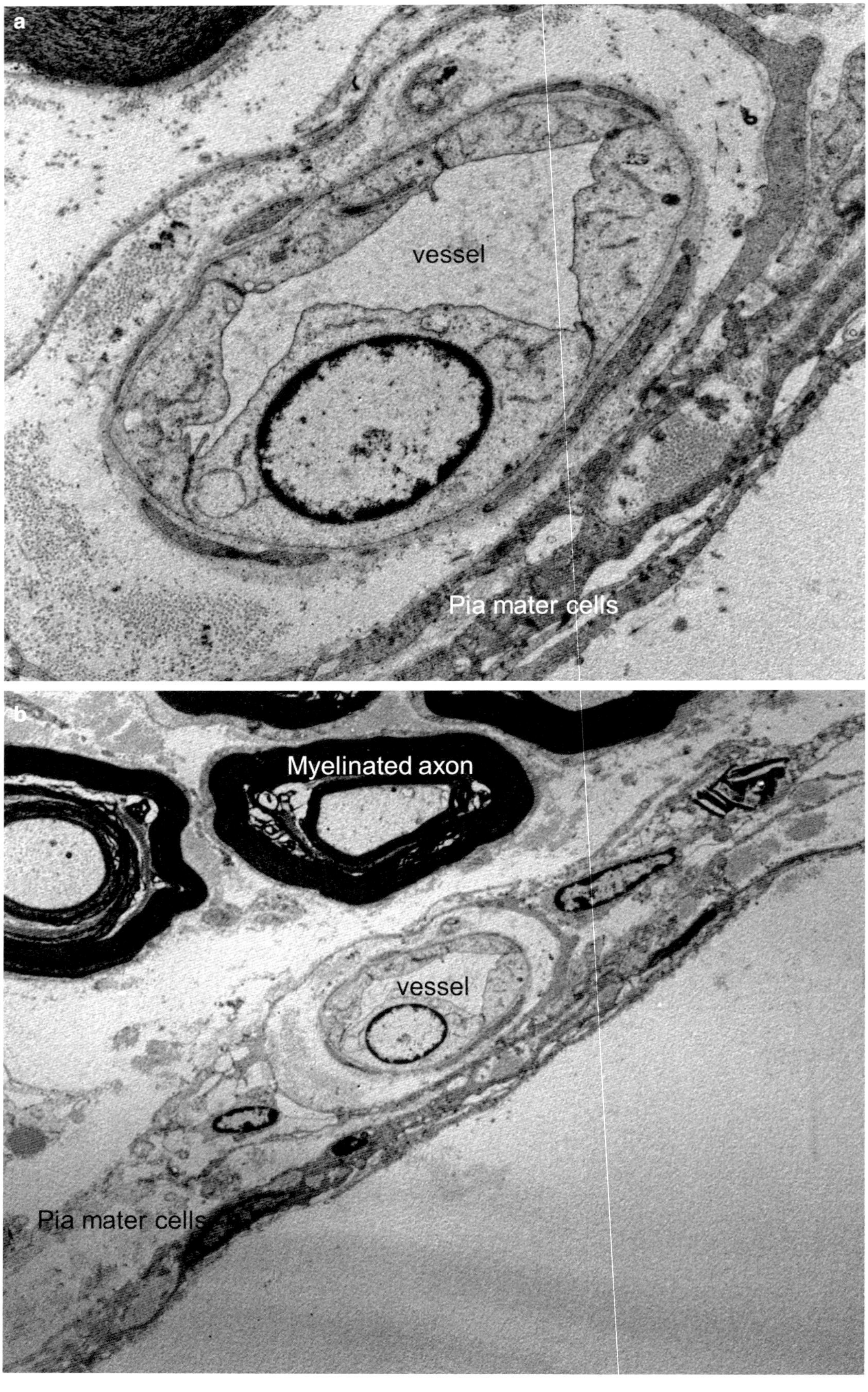

Fig. 25.21 Pia mater. Human pia mater layer at spinal nerve root level. Detail of pial cells. Transmission electron microscopy. Magnification: **a**, ×12,000; **b**, ×4,000

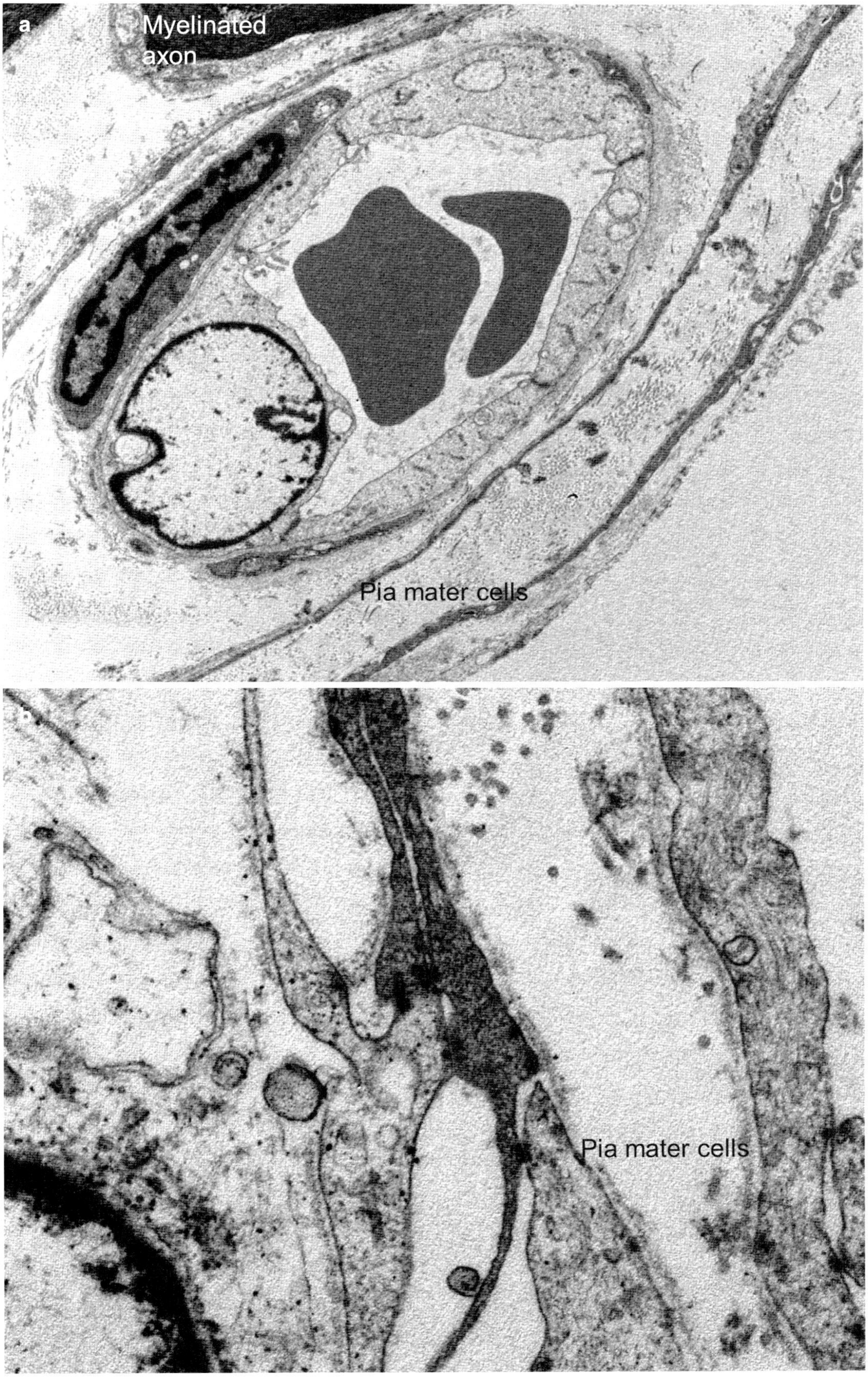

Fig. 25.22 Pia mater. Human pia mater layer at spinal nerve root level. Detail of pial cells. Transmission electron microscopy. Magnification: **a**, ×8,000; **b**, ×50,000

References

1. Reina MA, De León Casasola O, Villanueva MC, López A, Machés F, De Andrés JA. Ultrastructural findings in human spinal pia mater in relation to subarachnoid anesthesia. Anesth Analg. 2004;98: 1479–85.
2. Reina MA, López García A, de Andrés JA. Anatomical description of a natural perforation present in the human lumbar pia mater. Rev Esp Anestesiol Reanim. 1998;45:4–7.
3. Reina MA, De Andrés JA, López A. Anatomía: implicaciones en la práctica de las técnicas de conducción espinal. In: Torres LM, editor. Tratado de anestesia y reanimación. Madrid: Arán; 2001. p. 1135–55.
4. Reina MA, De Andrés JA, López A. Subarachnoid and epidural anesthesia. In: Raj P, editor. Textbook of regional anesthesia. New York: Churchill Livingstone; 2002. p. 307–24.
5. Reina MA, López A, Palácio F, Fornet I. Anestesia del neuroeje. In: Basora Macaya M, Colomina Soler MJ, editors. Anestesia en cirugía ortopédica y en traumatología. Madrid: Editorial Médica Panamericana; 2011. p. 46–58.
6. Reina MA, Villanueva MC, López A. Human spinal trabecular arachnoid, pia mater and spinal anesthesia. Rev Arg Anestesiol. 2008;66:111–33.

26 Ultrastructure of Spinal Subdural Compartment: Origin of Spinal Subdural Space

Miguel Angel Reina, Oscar De León Casasola, and Andrés López

The study of the structure of the subdural compartment is relevant in the performance of neuraxial blockade and in the understanding of the origin of cranial and spinal subdural hematomas associated with cerebrospinal fluid hypotension.

In contrast to classic descriptions, ultrastructural studies demonstrate that there is not a space between the dura mater and the arachnoid lamina. The subdural space is occupied by tissue composed of specialized cells known as neurothelial cells or dural border cells [1–8]. These cells are elongated, with branched extensions. Neurothelial cells are located between the last dural lamina and the barrier arachnoid layer. When tearing occurs within the thickness of the subdural compartment, small fissures convene into larger ones, opening up the subdural space. Intercellular spaces between neurothelial cells are most susceptible to tearing, and cellular fragments may be seen outside torn neurothelial cells. Low cohesion forces between neurothelial cells facilitate the widening of a minimal fissure to yield a "genuine subdural space." Other, smaller fissures can be seen parallel to the main space, and are called secondary subdural spaces. Tearing of the neurothelial cells gives rise to a true subdural space that would not be present otherwise. Therefore, the term "virtual subdural space" is incorrect.

In the samples examined, the subdural compartment is located between the inner part of the dura mater and the external part of the arachnoid lamina. Approximately two to six neurothelial-cell laminas are structured in concentric layers adjacent to the dura, whereas about 10–14 layers are seen adjacent to the arachnoid. Few specialized membrane junctions (desmosomes and tight junctions) are present among neighboring neurothelial cells (Figs. 26.1, 26.2, 26.3, 26.4, 26.5, 26.6, 26.7, 26.8, 26.9, 26.10, 26.11, 26.12, 26.13, 26.14, 26.15, 26.16, 26.17, 26.18, 26.19, 26.20, 26.21, 26.22, 26.23, and 26.24) [1–9].

M.A. Reina, MD, PhD (✉)
Department of Clinical Medical Sciences and Institute of Applied Molecular Medicine, School of Medicine, University of CEU San Pablo, Madrid, Spain

Department of Anesthesiology, Madrid-Montepríncipe University Hospital, Madrid, Spain
e-mail: miguelangel@perticone.e.telefonica.net

O. De León Casasola, MD
Department of Anesthesiology, Roswell Park Cancer Institute, Buffalo, NY, USA

Department of Anesthesiology, University at Buffalo, School of Medicine and Biomedical Sciences, Buffalo, NY, USA
e-mail: oscar.deleon@roswellpark.org

A. López, MD
Department of Clinical Medical Sciences, School of Medicine, CEU San Pablo University, Madrid, Spain

Department of Anesthesiology, Madrid-Montepríncipe University Hospital, Madrid-Torrelodones University Hospital, Madrid-Sanchinarro University Hospital, Madrid, Spain
e-mail: alg00009@telefonica.net

M.A. Reina et al. (eds.), *Atlas of Functional Anatomy for Regional Anesthesia and Pain Medicine: Human Structure, Ultrastructure and 3D Reconstruction Images*, DOI 10.1007/978-3-319-09522-6_26,

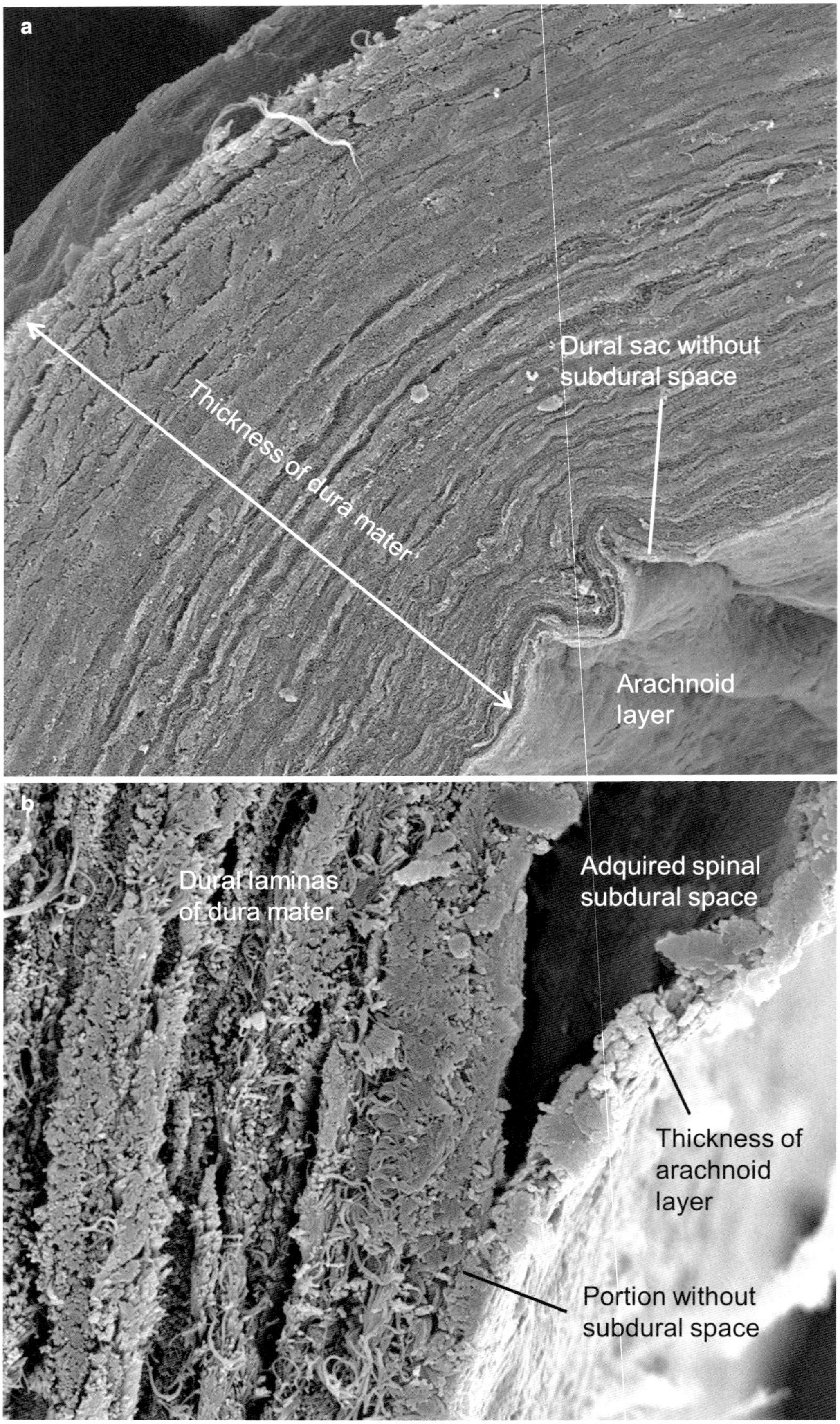

Fig. 26.1 Acquired spinal subdural space. Scanning electron microscopy. Magnification: (**a**) ×300; (**b**) ×1,500 (**a** From Reina et al. [5], **b** From Reina et al. [1]; both with permission)

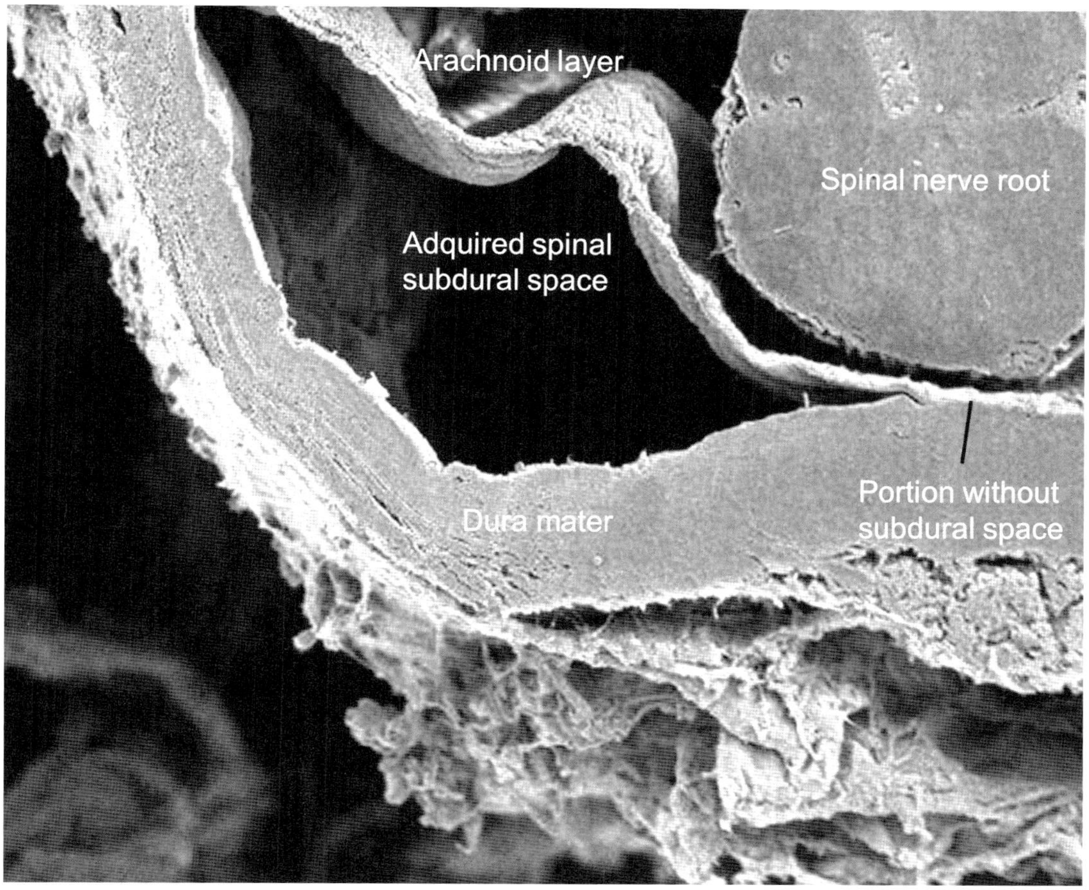

Fig. 26.2 Acquired spinal subdural space. Scanning electron microscopy. Magnification: ×40 (From Reina et al. [1]; with permission)

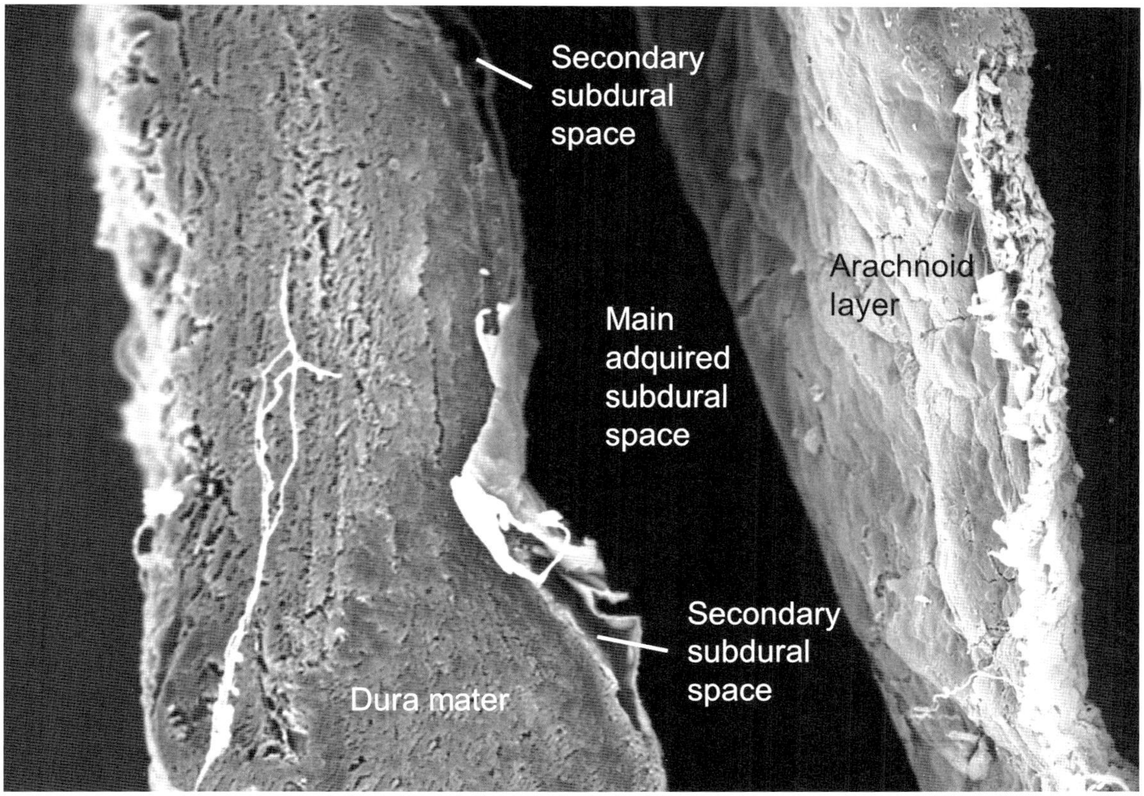

Fig. 26.3 Acquired spinal subdural space. Scanning electron microscopy. Magnification: ×180 (From Reina et al. [2]; with permission)

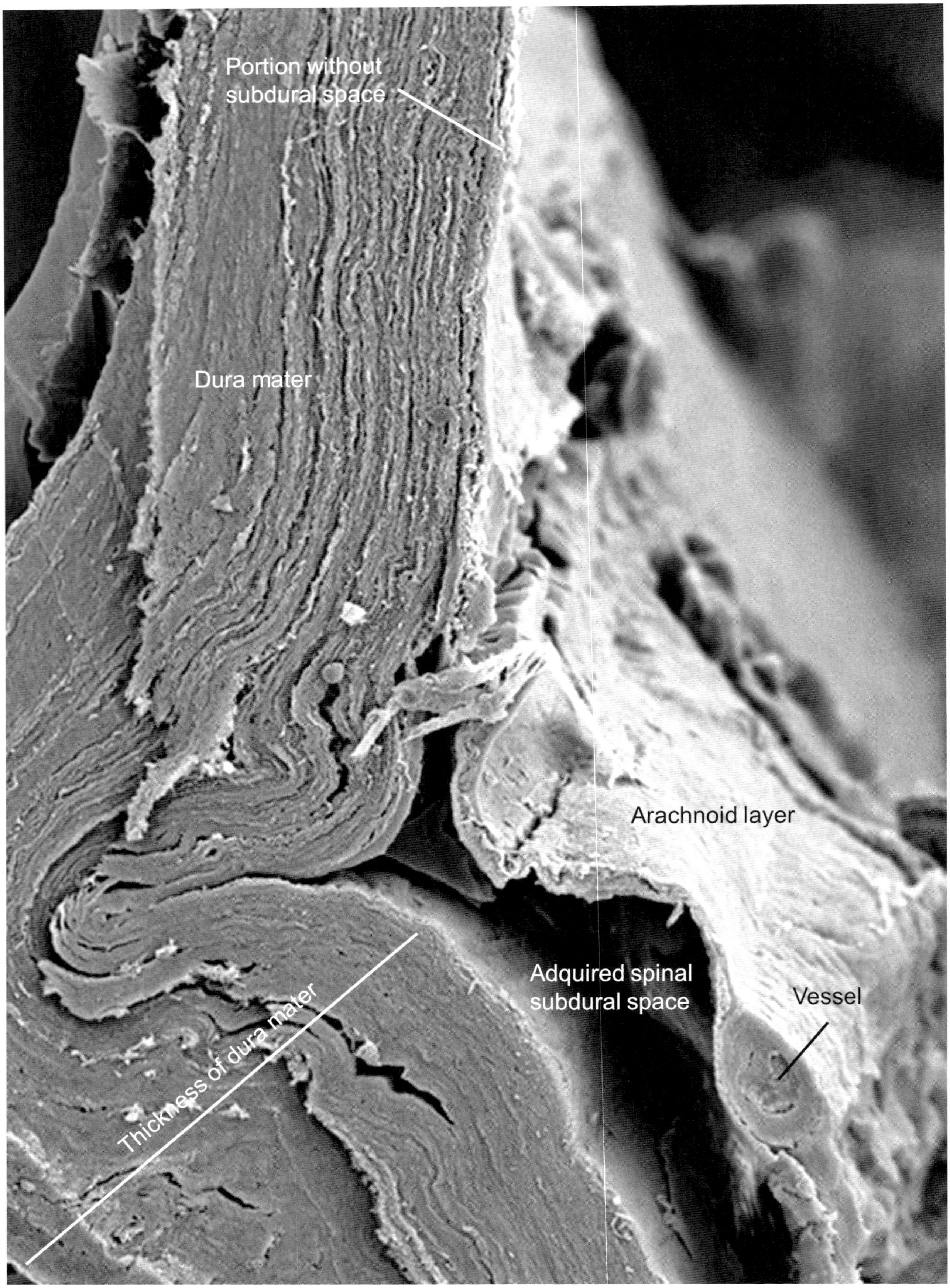

Fig. 26.4 Acquired spinal subdural space. Scanning electron microscopy. Magnification × 90 (From Collier et al. [9]; with permission)

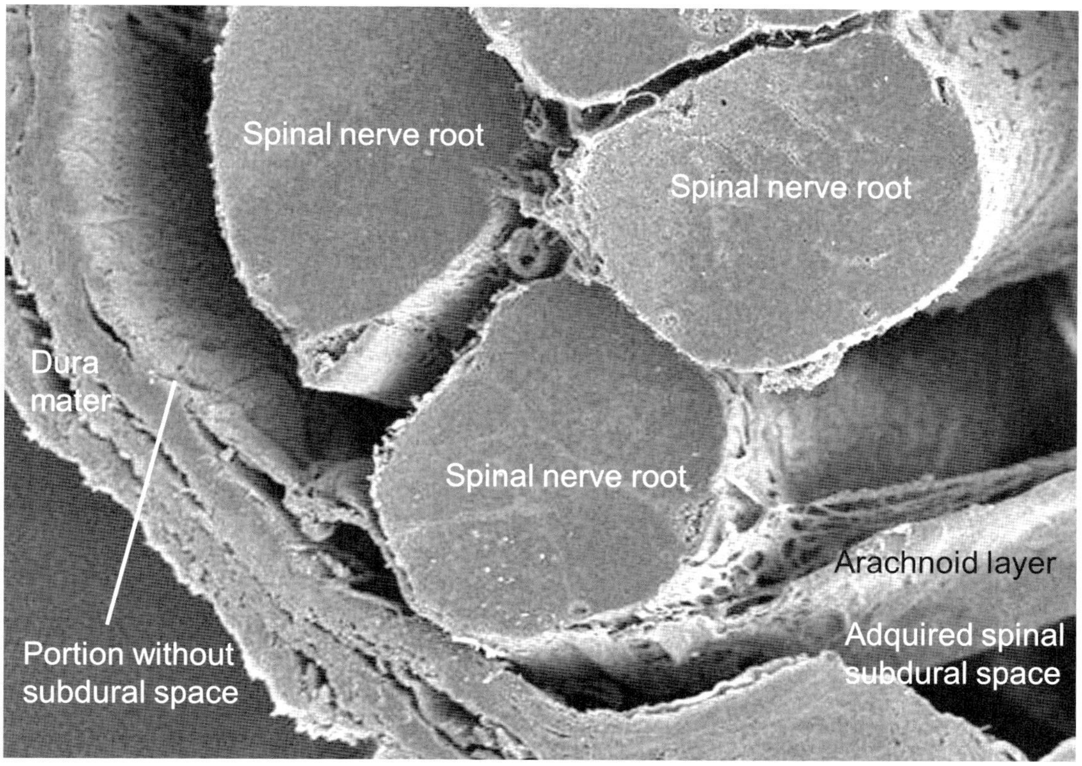

Fig. 26.5 Acquired spinal subdural space. Scanning electron microscopy. Magnification: ×40

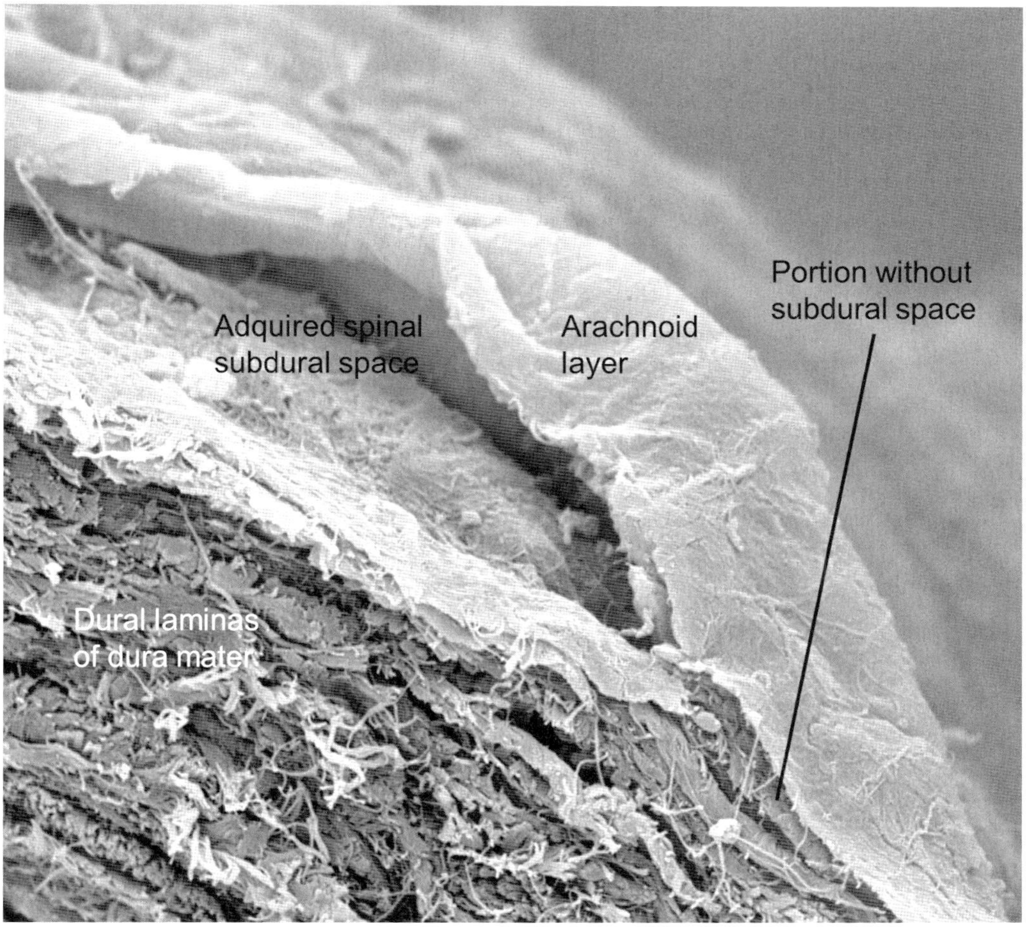

Fig. 26.6 Acquired spinal subdural space. Scanning electron microscopy. Magnification: ×1,600

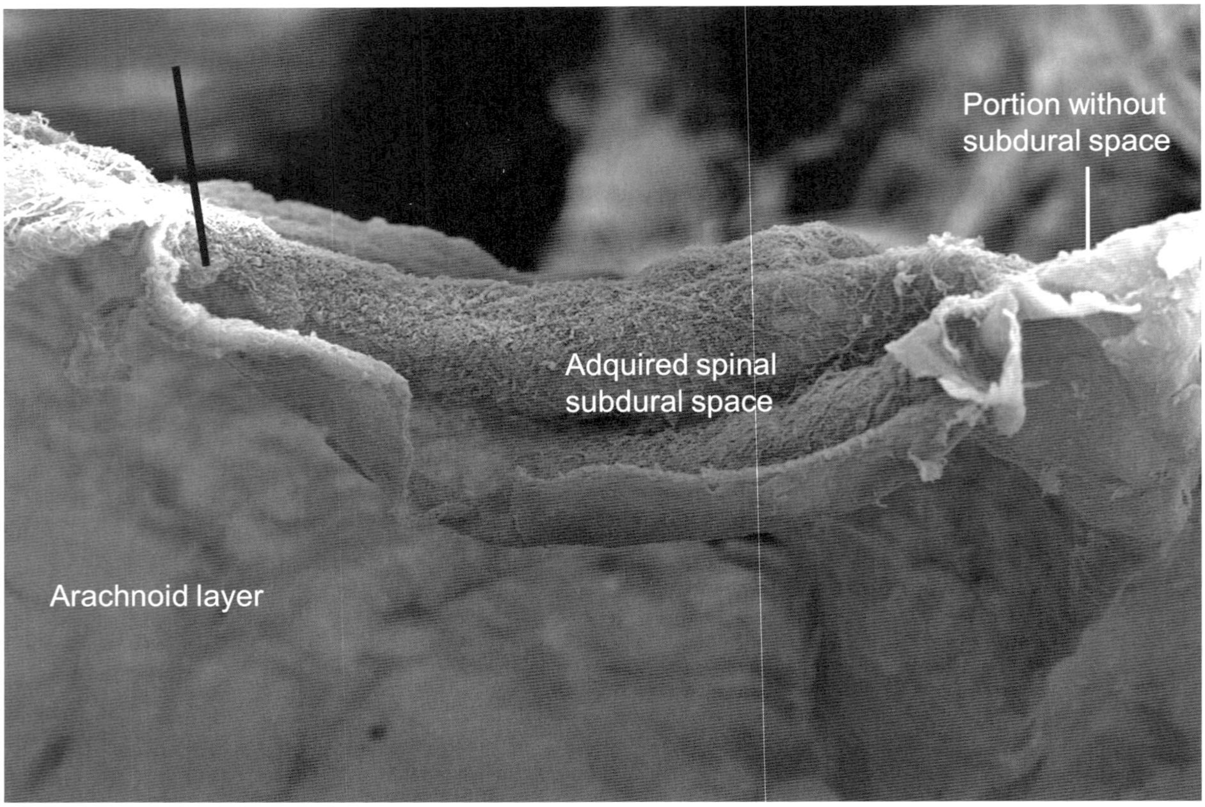

Fig. 26.7 Acquired spinal subdural space. Scanning electron microscopy. Magnification: ×100

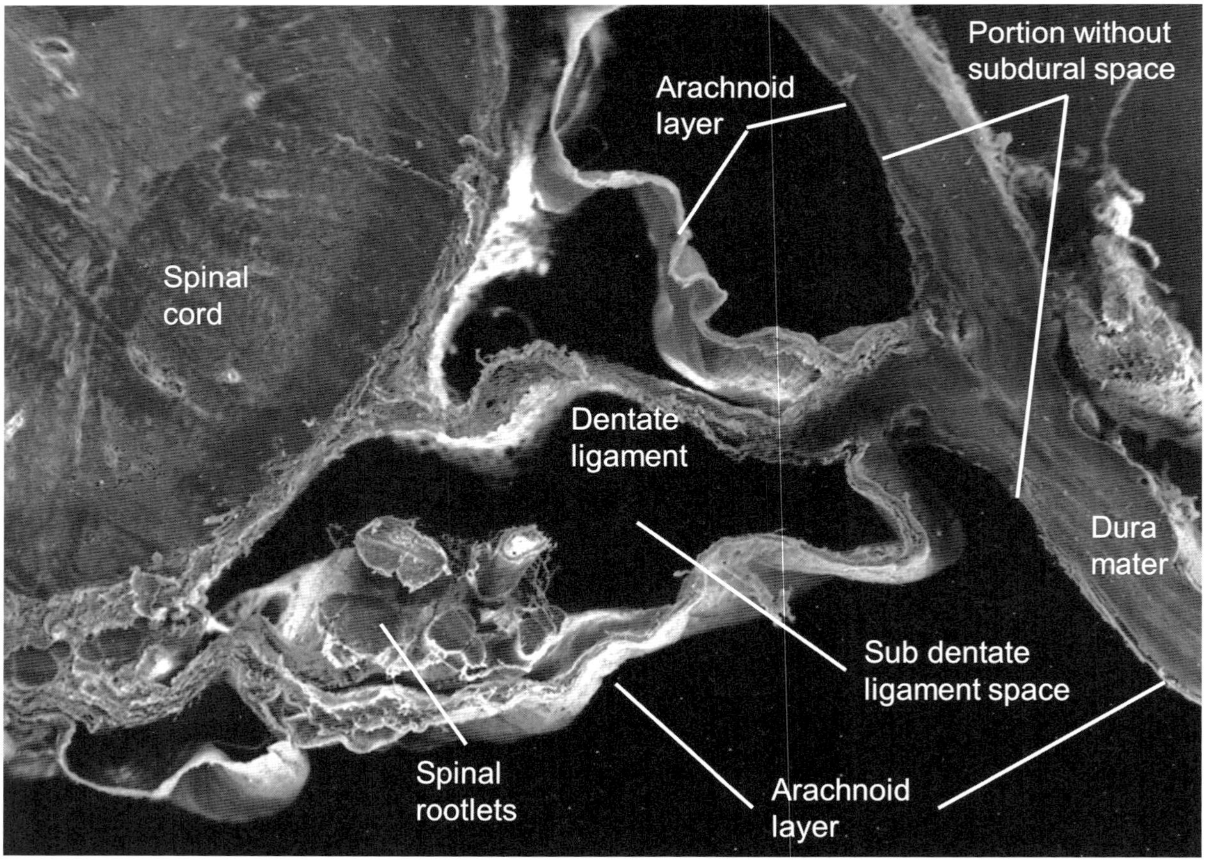

Fig. 26.8 Acquired spinal subdural space. Scanning electron microscopy. Magnification: ×12

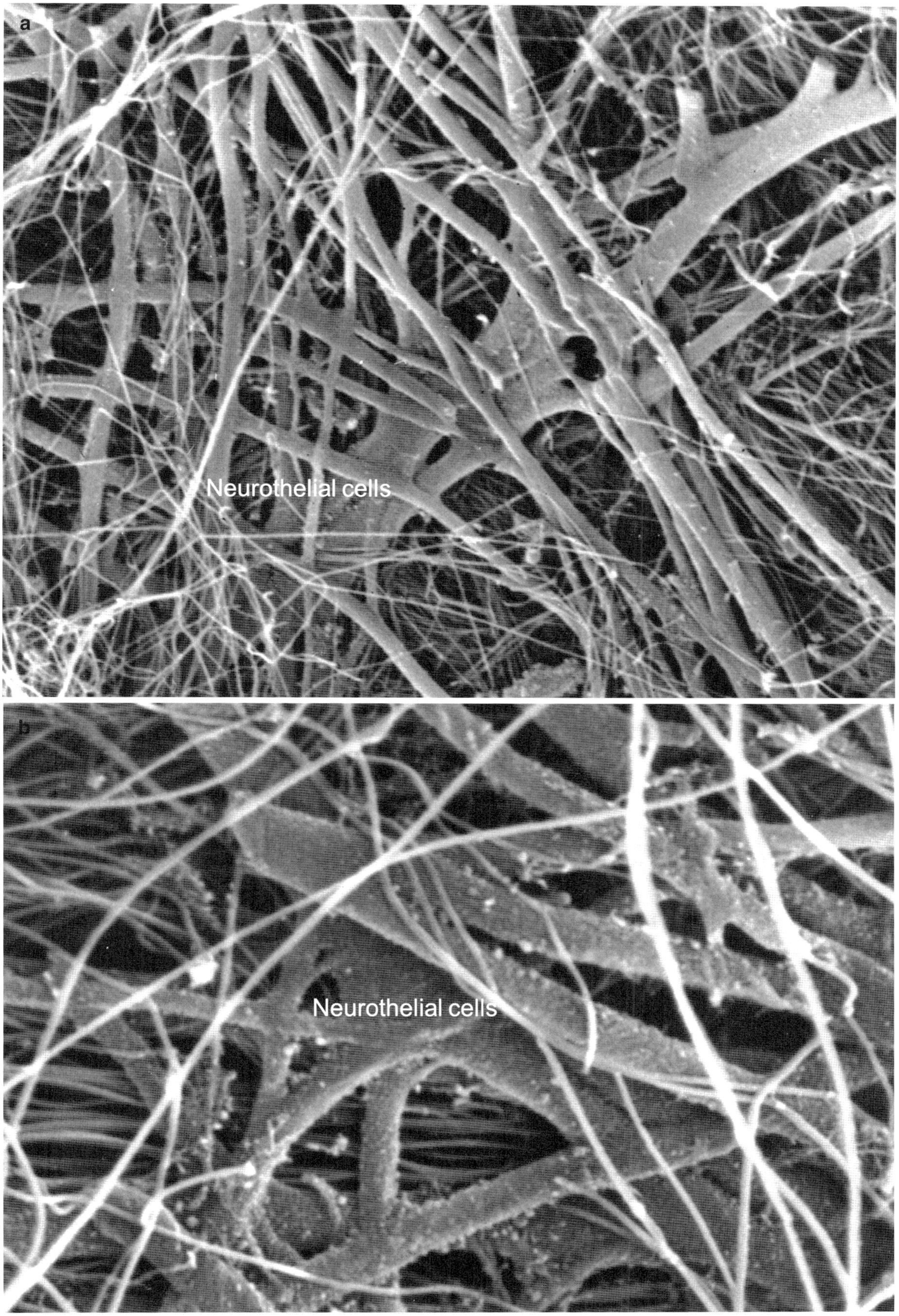

Fig. 26.9 Acquired spinal subdural space. Scanning electron microscopy. Magnification: (**a**) ×3,000; (**b**) ×6,000

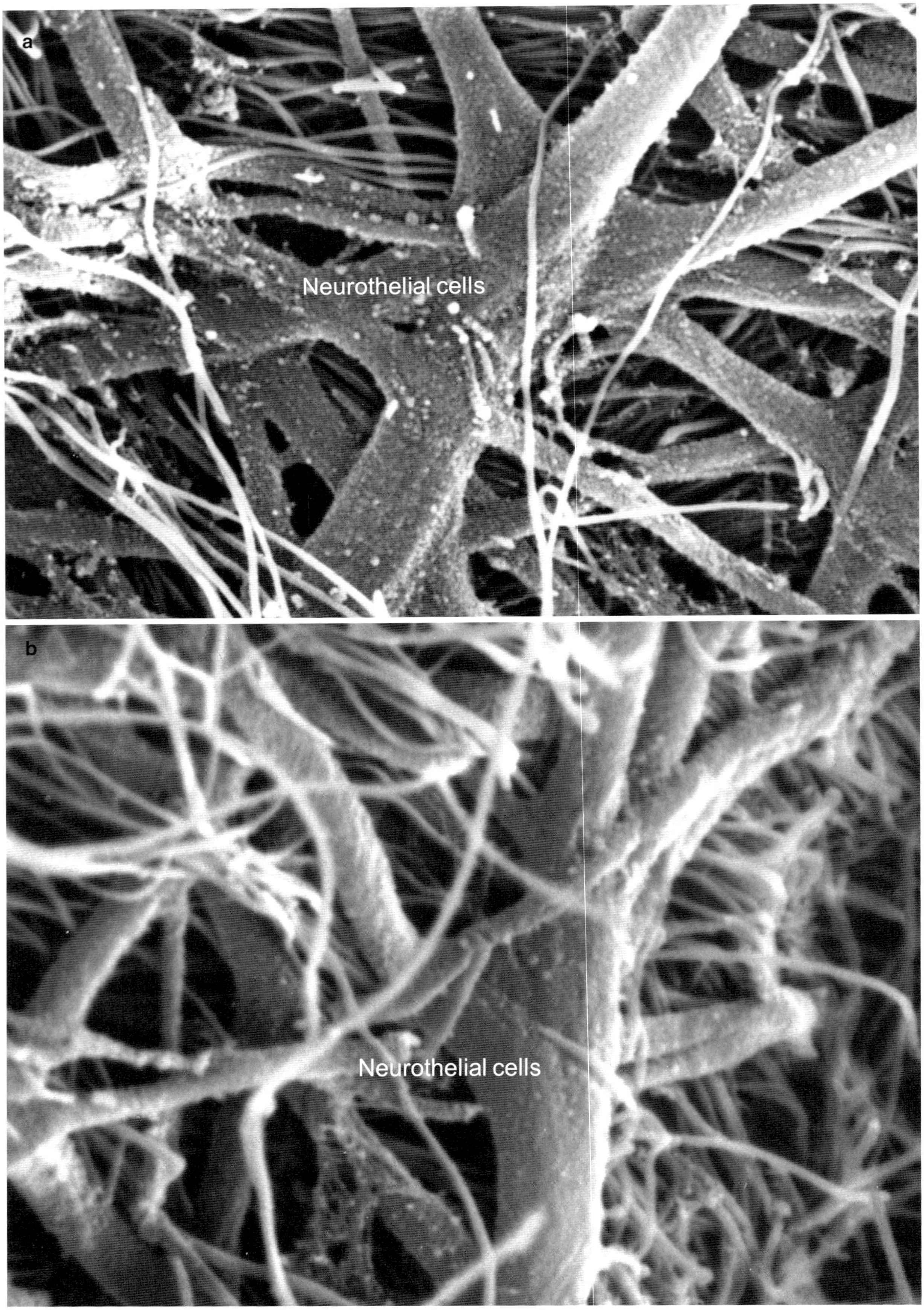

Fig. 26.10 Acquired spinal subdural space. Scanning electron microscopy. Magnification: (**a**) ×10,000; (**b**) ×15,000 (**b** from Reina et al. [2]; with permission)

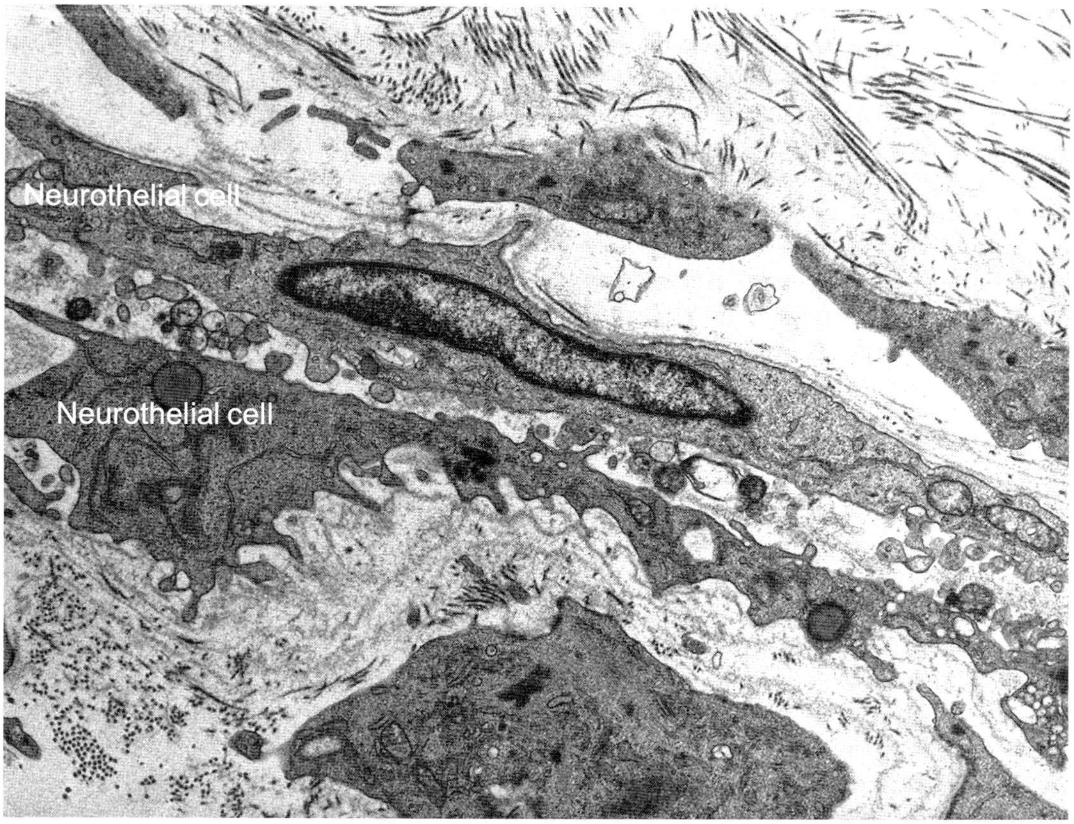

Fig. 26.11 Acquired spinal subdural space. Transmission electron microscopy. Magnification × 5,000

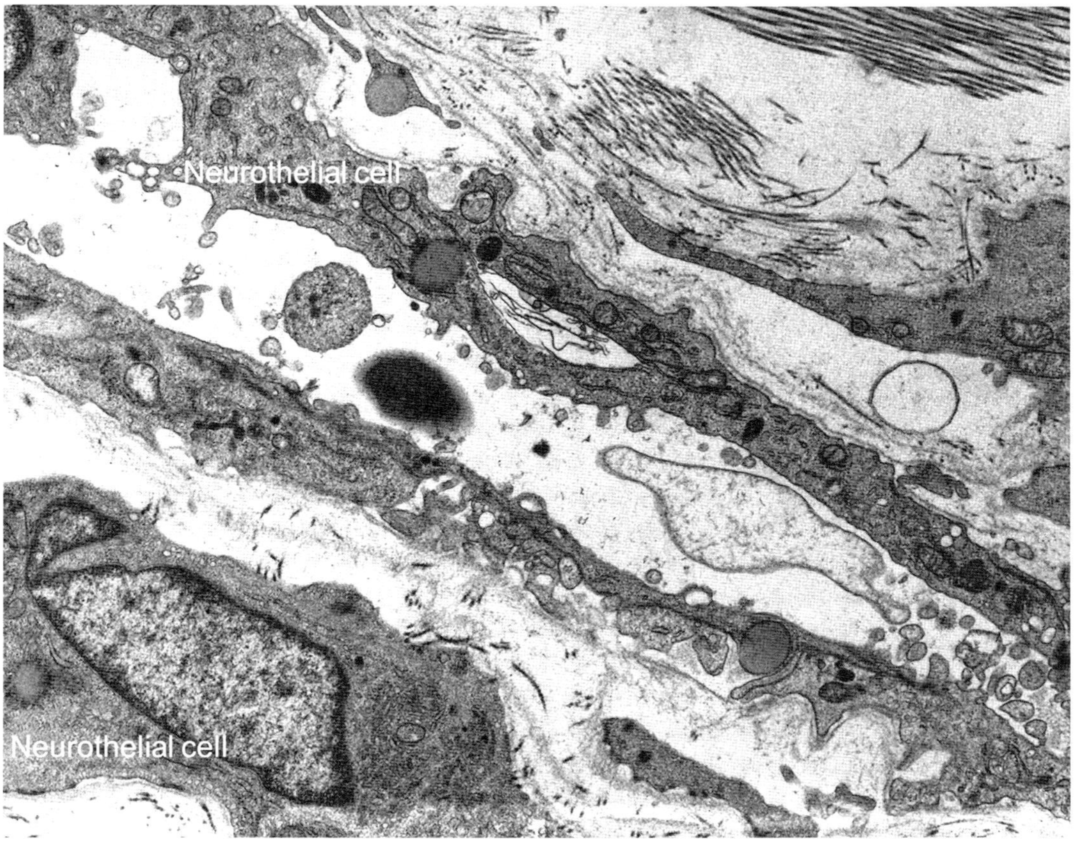

Fig. 26.12 Acquired spinal subdural space. Transmission electron microscopy. Magnification × 5,000 (From Reina et al. [1]; with permission)

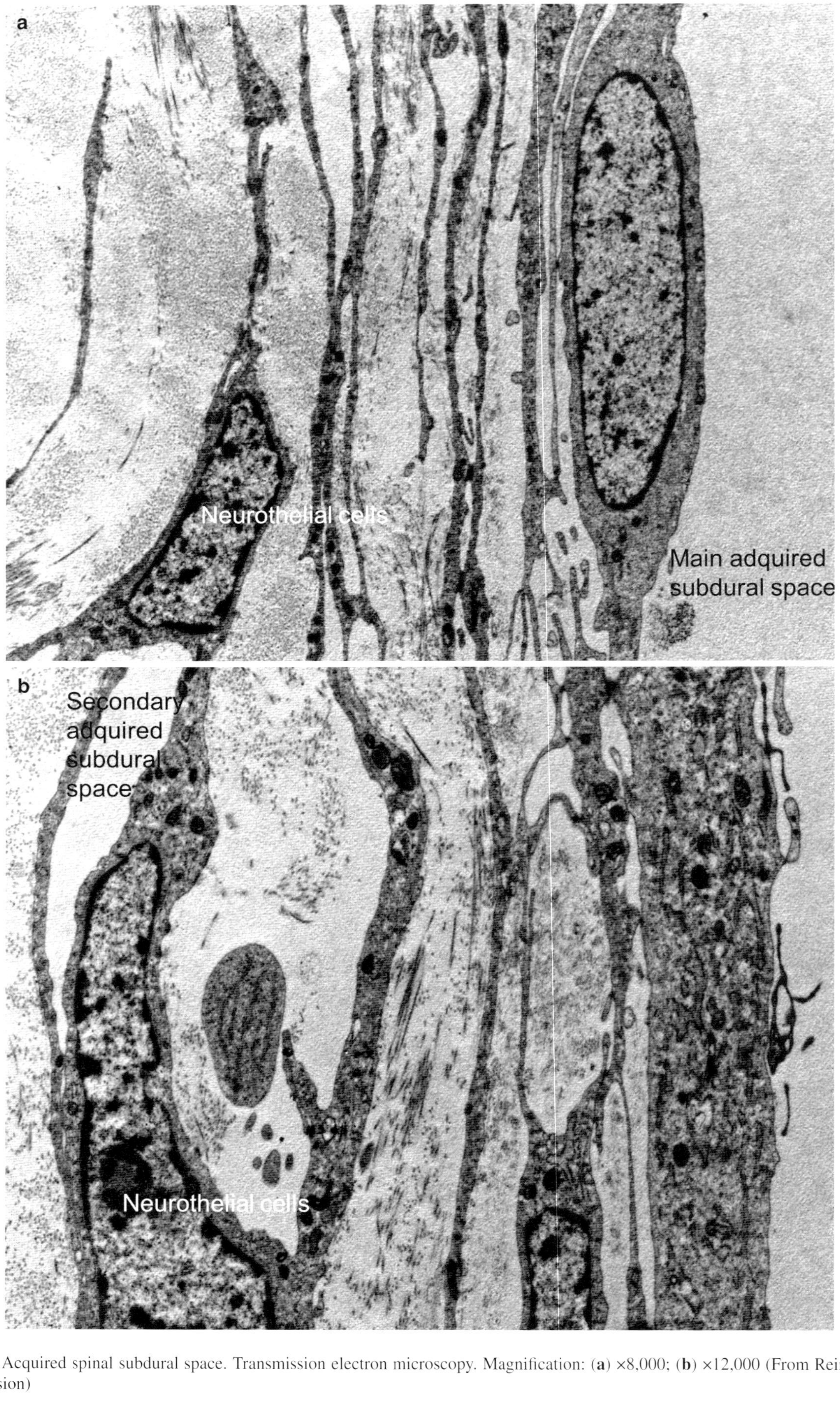

Fig. 26.13 Acquired spinal subdural space. Transmission electron microscopy. Magnification: (**a**) ×8,000; (**b**) ×12,000 (From Reina et al. [1]; with permission)

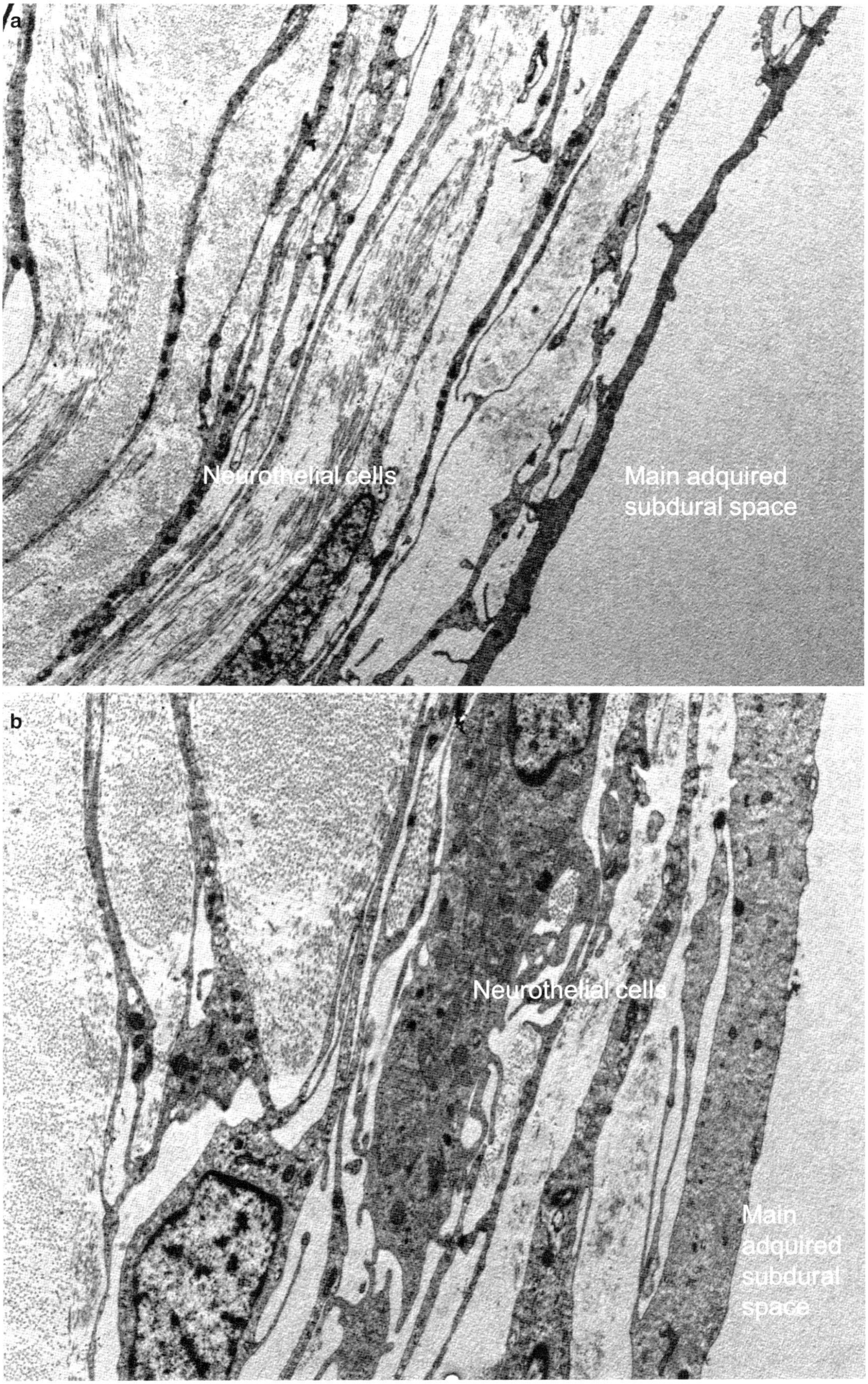

Fig. 26.14 Acquired spinal subdural space. Transmission electron microscopy. Magnification: (**a**) ×6,000; (**b**) ×10,000 (From Reina et al. [1]; with permission)

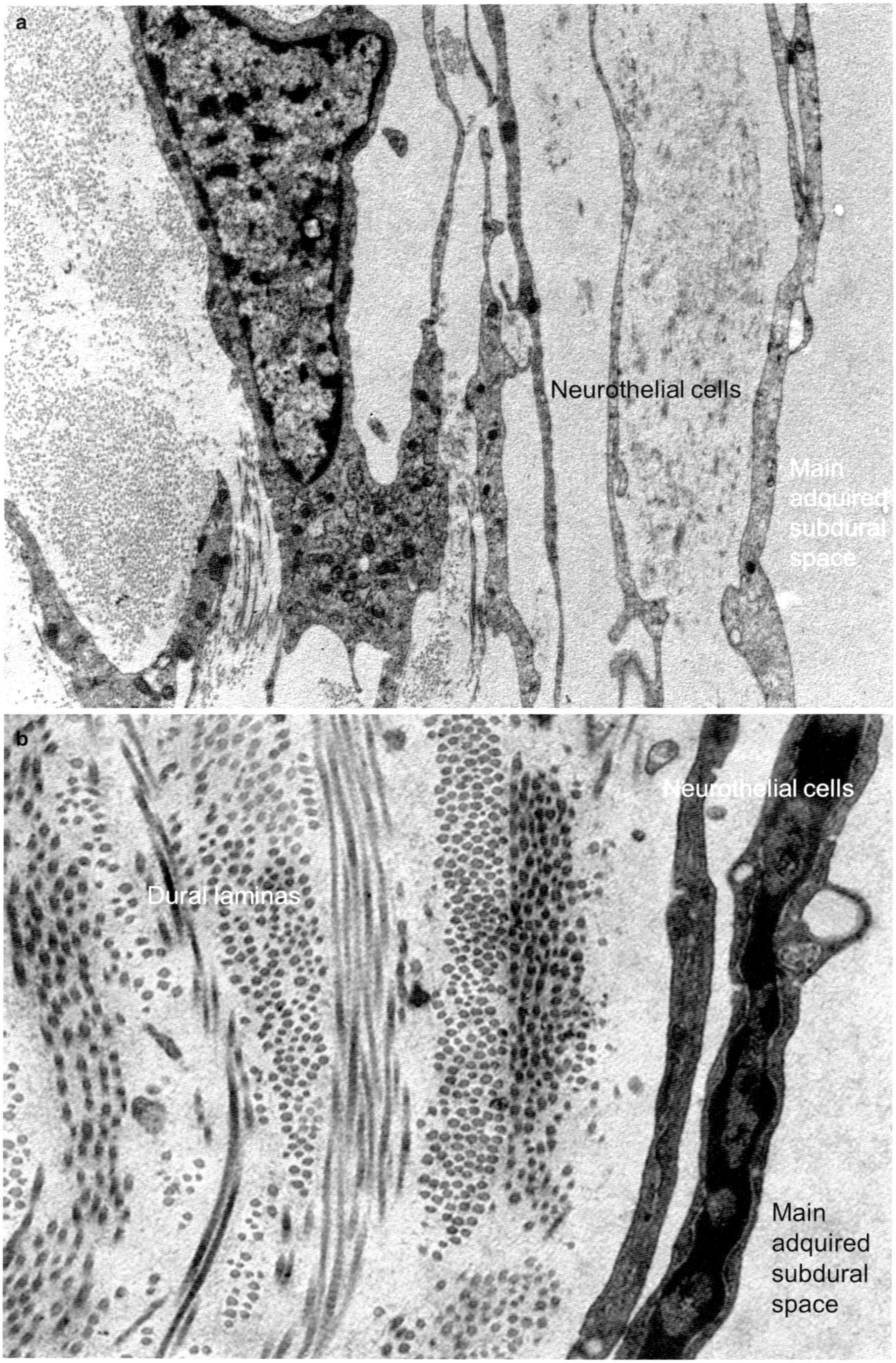

Fig. 26.15 Acquired spinal subdural space. Transmission electron microscopy. Magnification: (**a**) ×10,000; (**b**) ×25,000

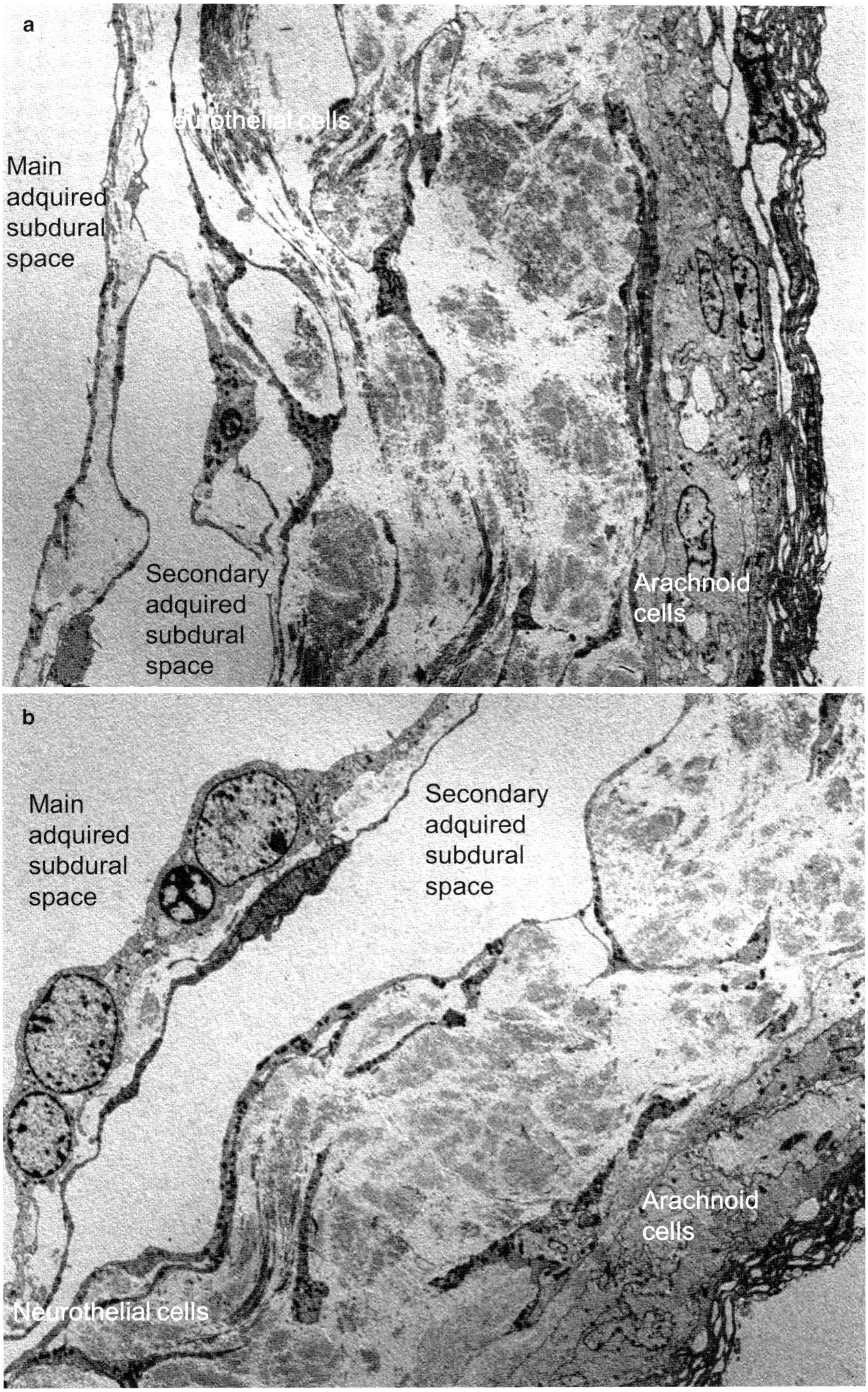

Fig. 26.16 Acquired spinal subdural space. Transmission electron microscopy. Magnification: (**a**) ×3,000; (**b**) ×3,000

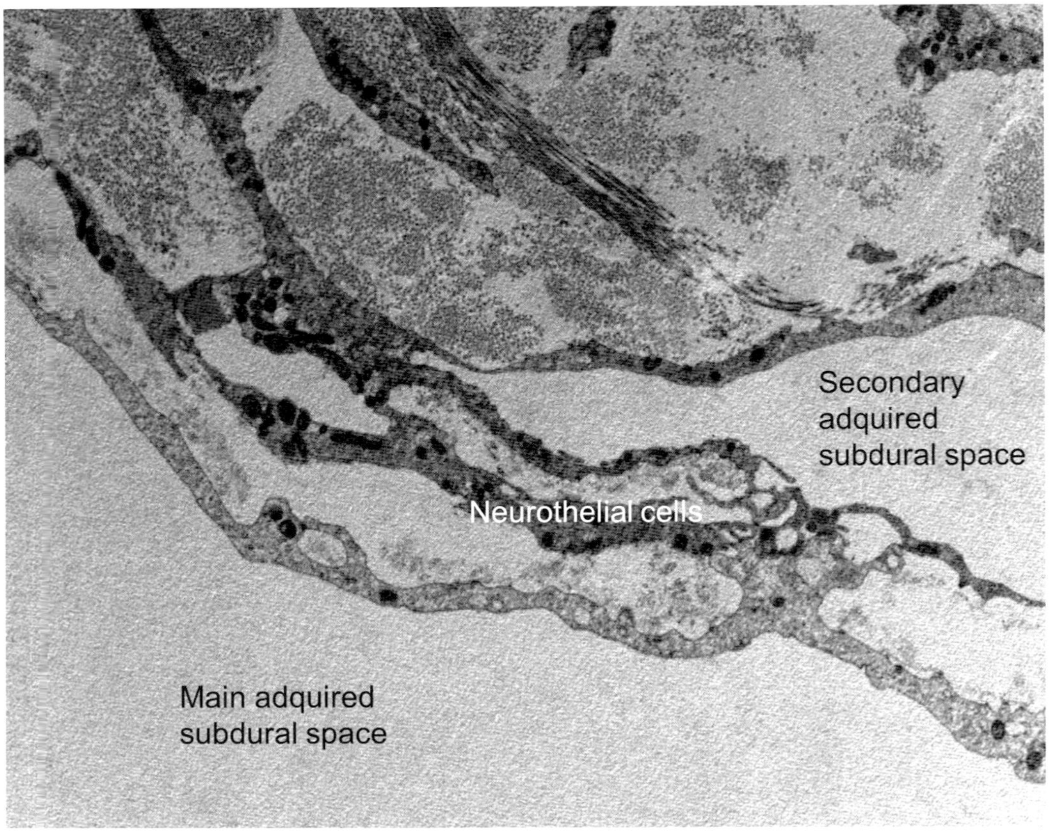

Fig. 26.17 Acquired spinal subdural space. Transmission electron microscopy. Magnification: ×5,000

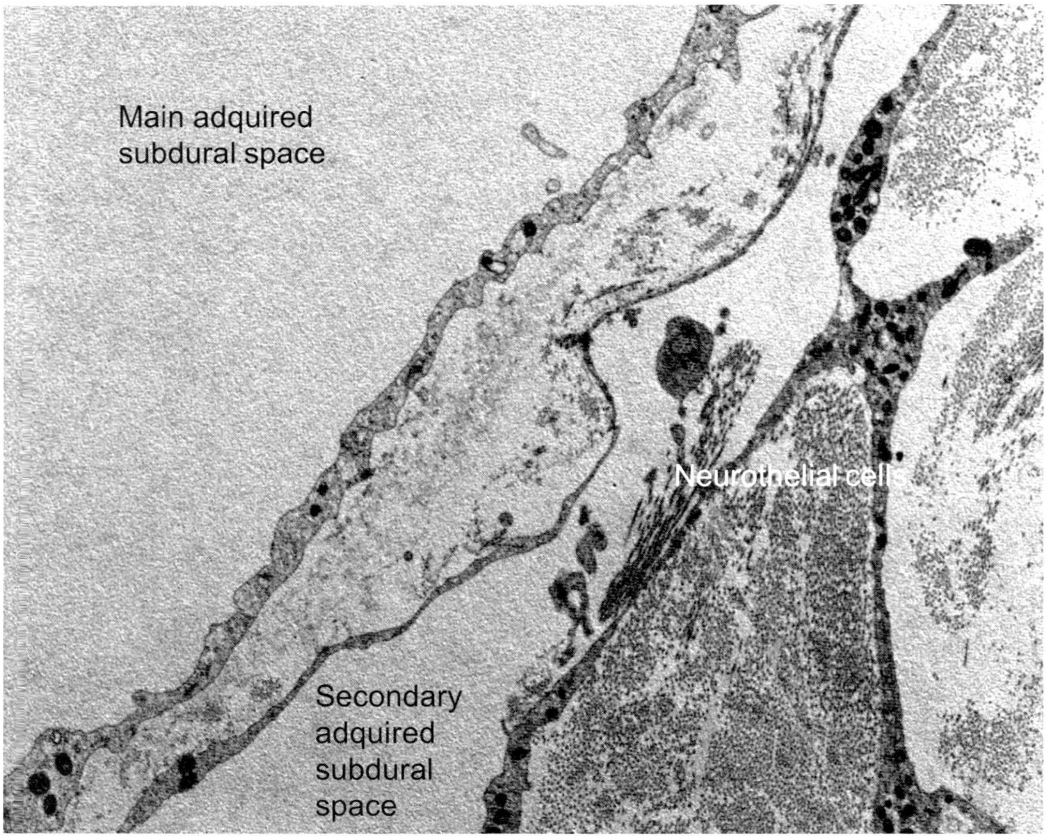

Fig. 26.18 Acquired spinal subdural space. Transmission electron microscopy. Magnification: ×15,000

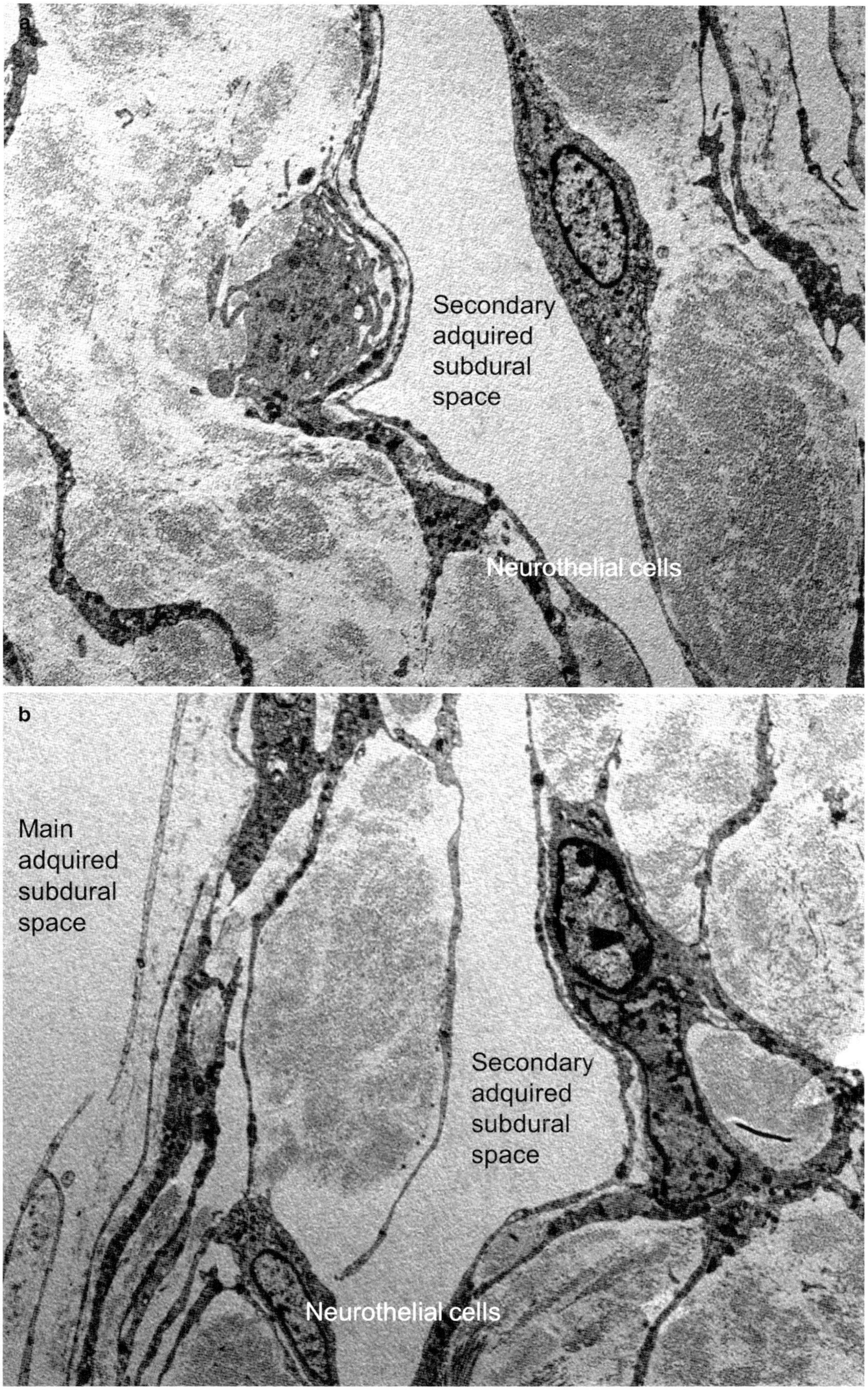

Fig. 26.19 Acquired spinal subdural space. Transmission electron microscopy. Magnification: (**a**) ×5,000; (**b**) ×5,000

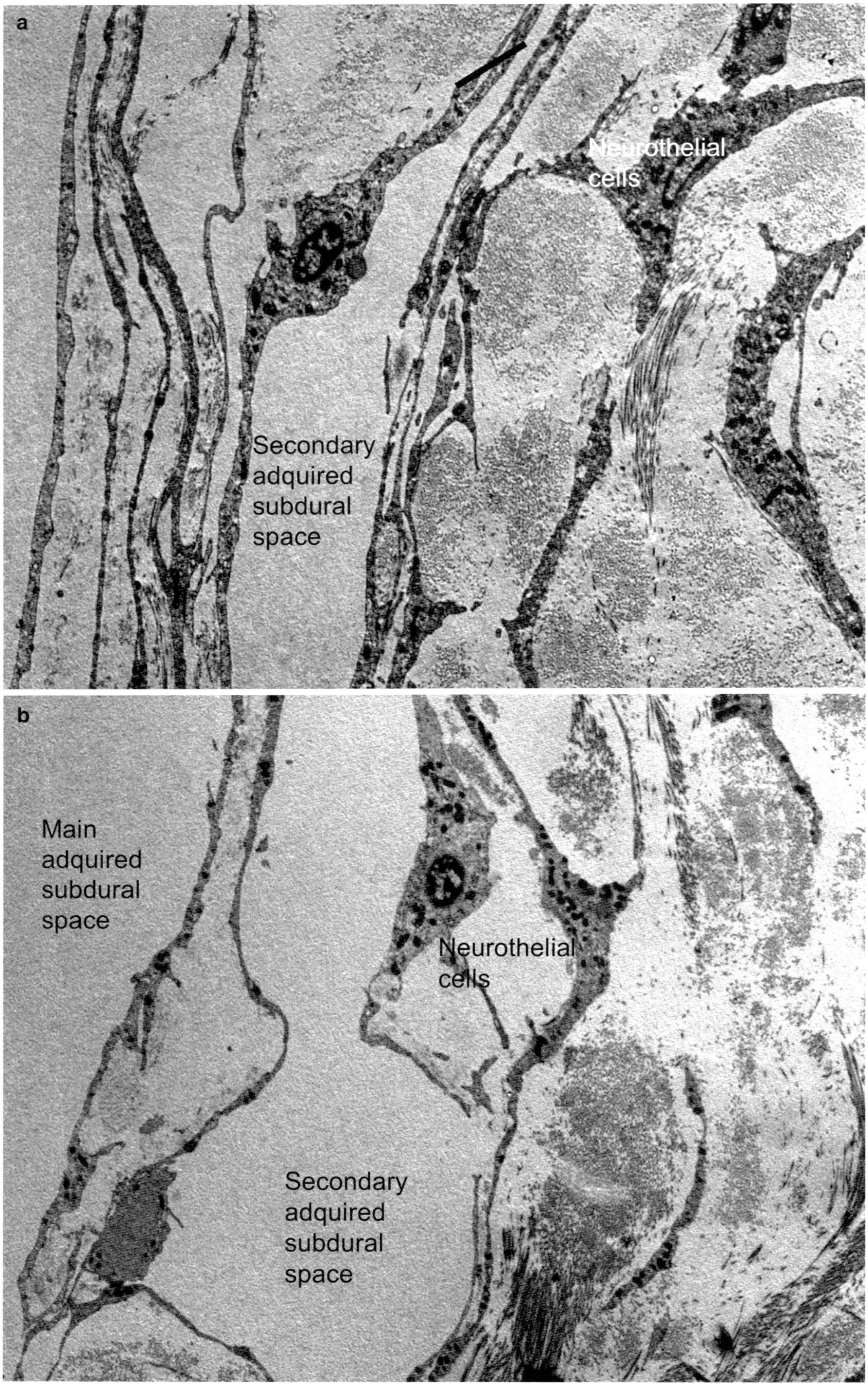

Fig. 26.20 Acquired spinal subdural space. Transmission electron microscopy. Magnification: (**a**) ×6,000; (**b**) ×5,000

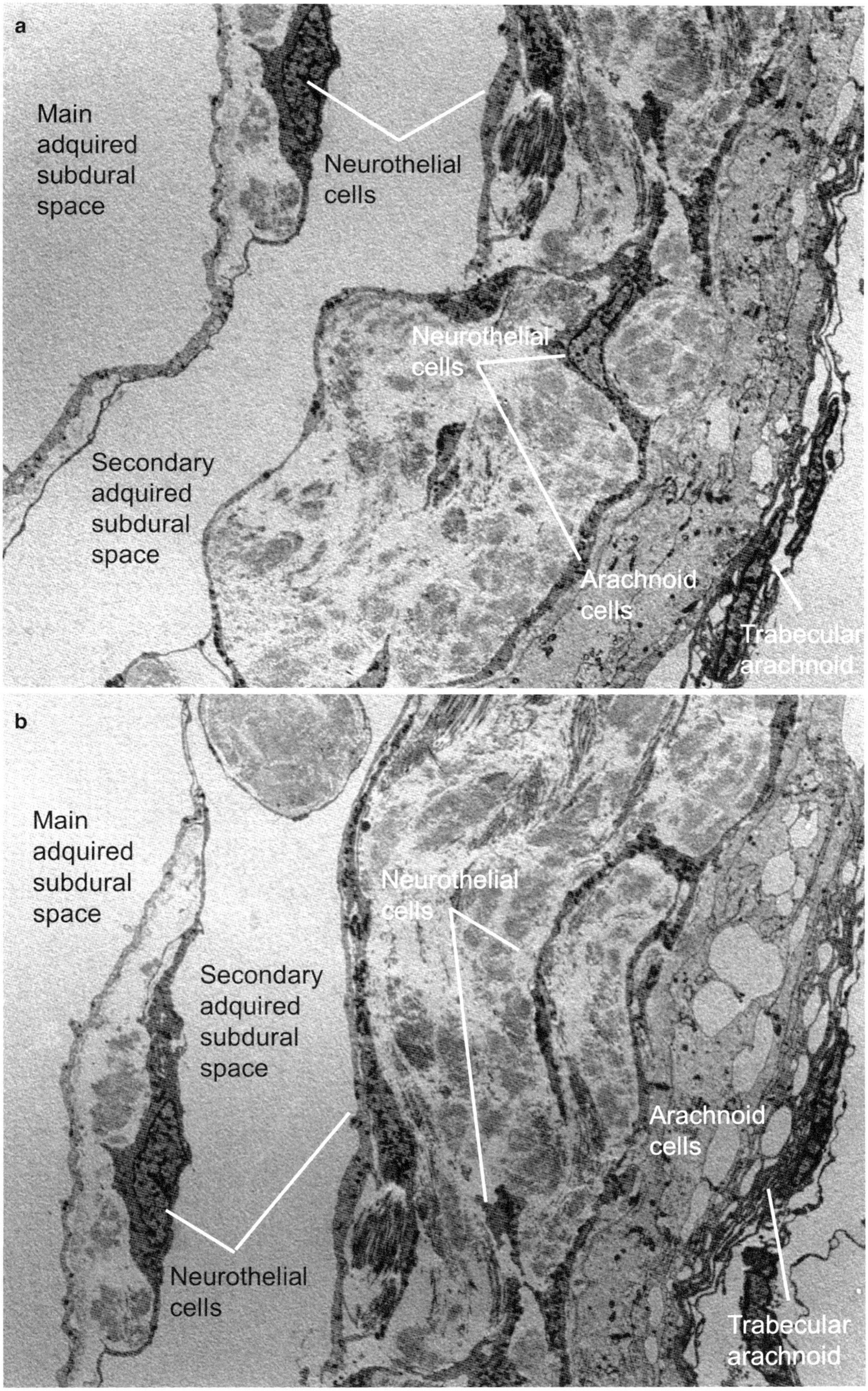

Fig. 26.21 Acquired spinal subdural space. Transmission electron microscopy. Magnification: (**a**) ×3,000; (**b**) ×3,000

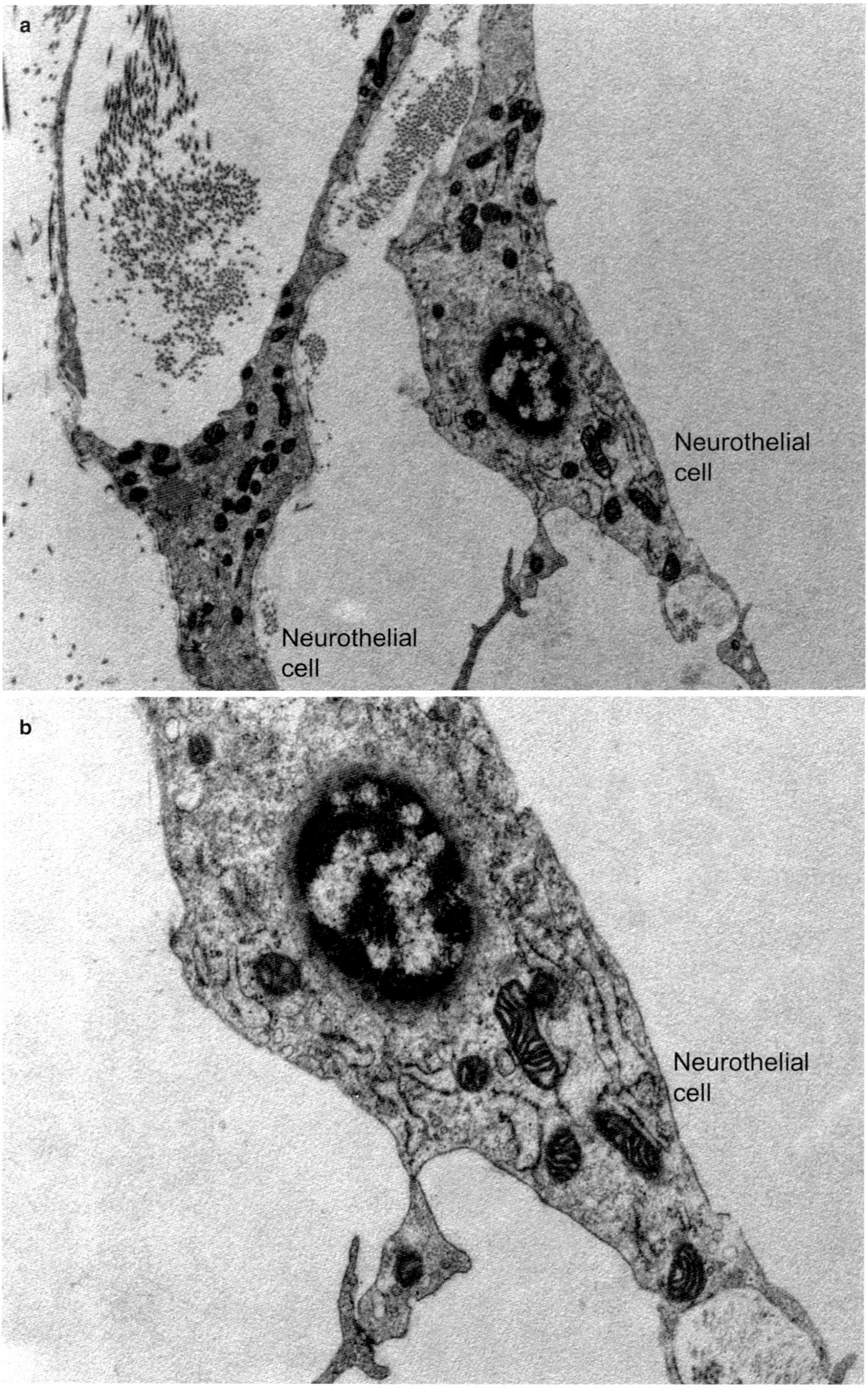

Fig. 26.22 Acquired spinal subdural space. Transmission electron microscopy. Magnification: (**a**) ×12,000; (**b**) ×25,000

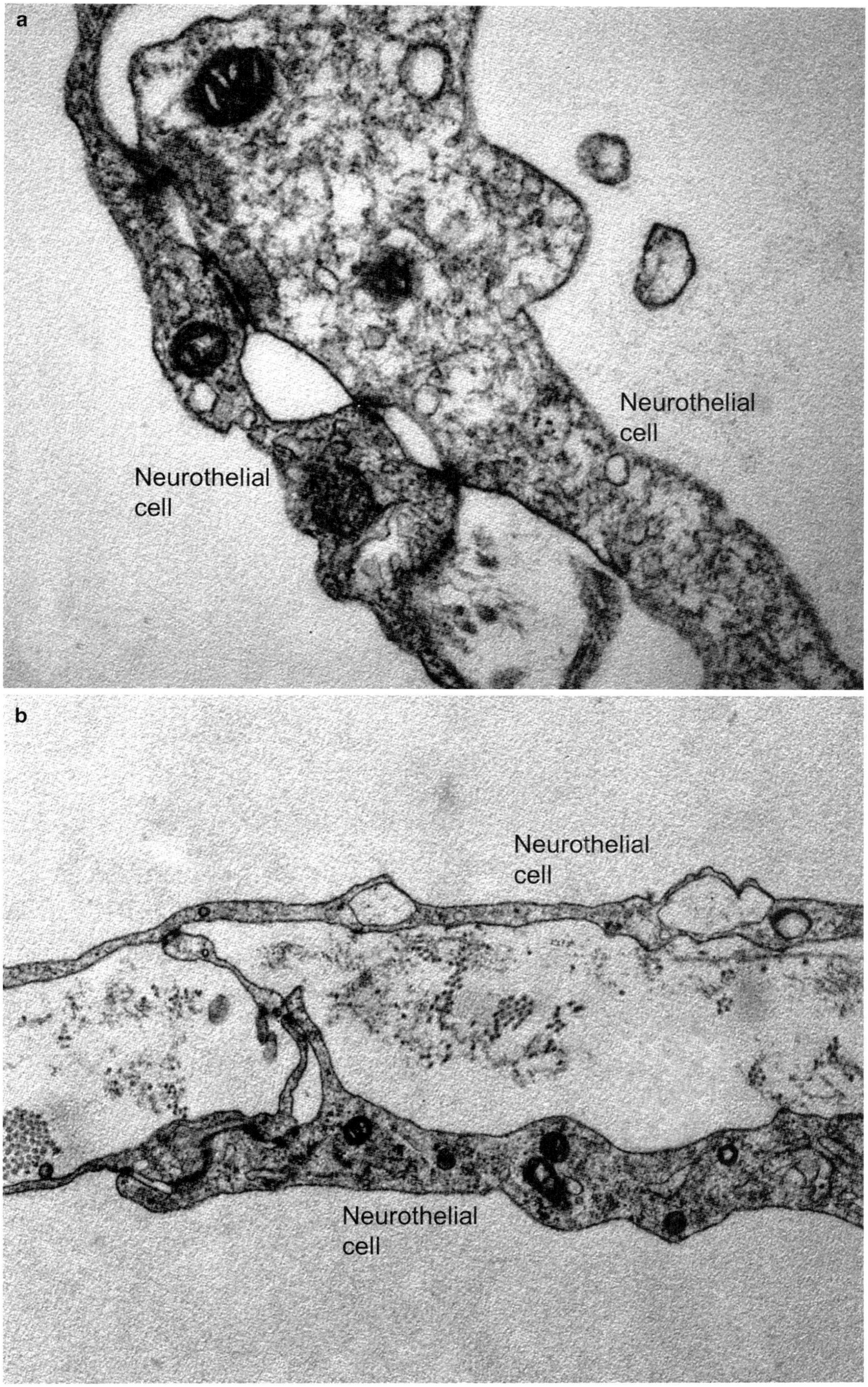

Fig. 26.23 Acquired spinal subdural space. Transmission electron microscopy. Magnification: (**a**) ×60,000; (**b**) ×25,000

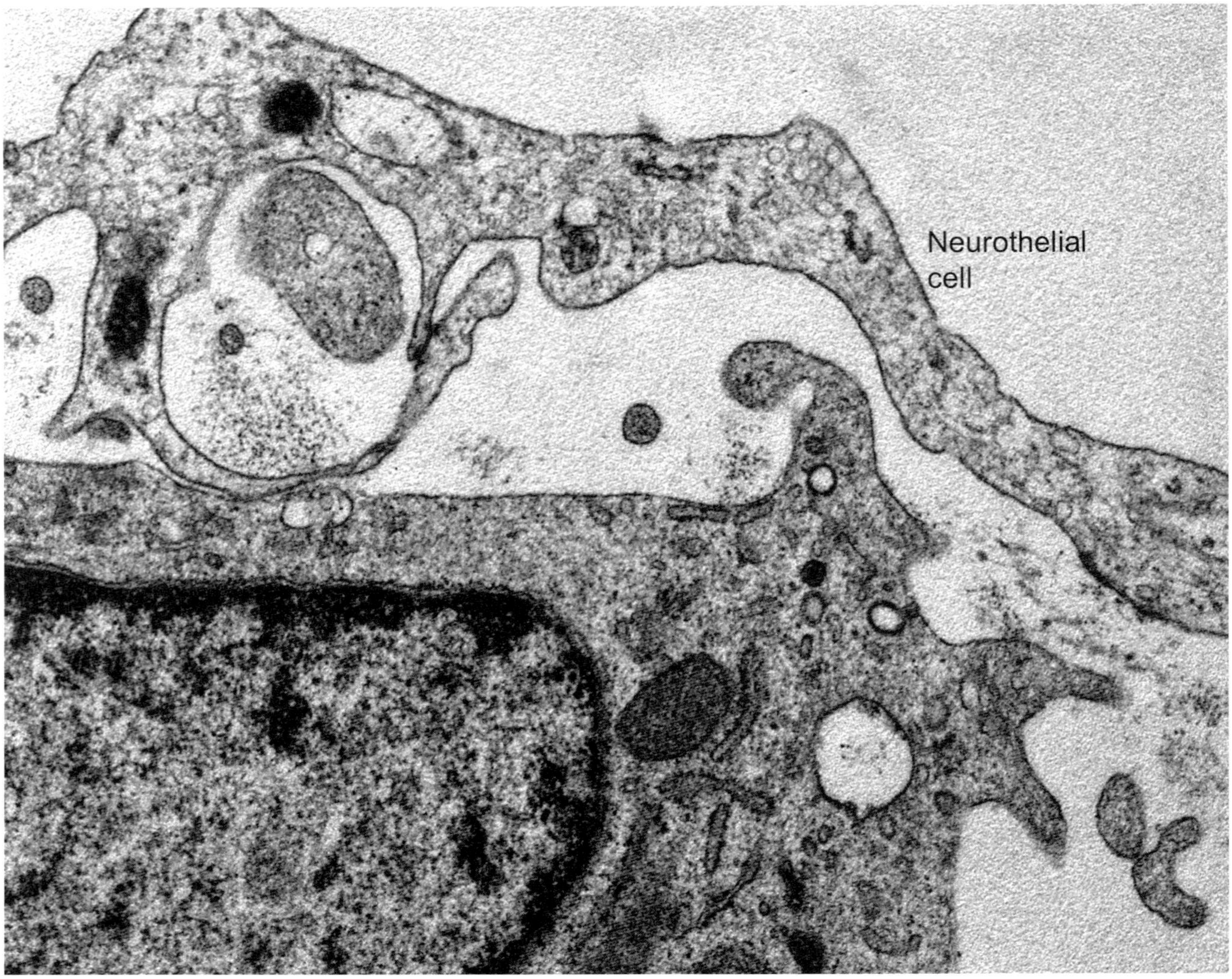

Fig. 26.24 Acquired spinal subdural space. Transmission electron microscopy. Magnification: ×30,000

References

1. Reina MA, De Leon Casasola O, López A, De Andrés JA, Mora M, Fernández A. The origin of the spinal subdural space: ultrastructure findings. Anesth Analg. 2002;94:991–5.
2. Reina MA, López García A, de Andrés JA, Villanueva MC, Cortés L. Does the subdural space exist? Rev Esp Anestesiol Reanim. 1998;45:367–76.
3. Reina MA, Pulido P, López A. The human dural sac. Morphology of the spinal dura-arachnoid membrane. Origin of the spinal subdural space. Rev Arg Anestesiol. 2007;65:167–84.
4. Prats A, Carrera A, Oliva A, Lusá M, Reina MA. Anatomia aplicada a las técnicas neuroaxiales. In: Basora Macaya M, Colomina Soler MJ, editors. Anestesia en cirugía ortopédica y en traumatología. Madrid: Editorial Médica Panamericana; 2011. p. 33–45.
5. Reina MA, López A, Palácio F, Fornet I. Anestesia del neuroeje. In: Basora Macaya M, Colomina Soler MJ, editors. Anestesia en cirugía ortopédica y en traumatología. Madrid: Editorial Médica Panamericana; 2011. p. 46–58.
6. Reina MA, Collier CB, Prats-Galino A, Puigdellívol-Sánchez A, Machés F, De Andrés JA. Unintentional subdural placement of epidural catheters during attempted epidural anesthesia: an anatomic study of spinal subdural compartment. Reg Anesth Pain Med. 2011;36:537–41.
7. Reina MA, De Andrés JA, López A. Anatomía: implicaciones en la práctica de las técnicas de conducción espinal. In: Torres LM, editor. Tratado de anestesia y reanimación. Madrid: Arán; 2001. p. 1135–55.
8. Reina MA, De Andrés JA, López A. Subarachnoid and epidural anesthesia. In: Raj P, editor. Textbook of regional anesthesia. New York: Churchill Livingstone; 2002. p. 307–24.
9. Collier CB, Reina MA. Does an intradural space really exist? Int J Obstet Anesth. 2010;20:89–90.

Unintentional Subdural and Intradural Placement of Epidural Catheters

27

Clive B. Collier and Miguel Angel Reina

Unintentional subdural placement of an epidural catheter is a known complication leading to atypical onset and distribution of epidural blockade. Inadvertent placement of the epidural catheter between the lamina dura and arachnoid is feasible once the tip of the Tuohy needle or the epidural catheter pierces the dura mater.

A study that supported this hypothesis examined whether it was possible to place a 20G catheter (diameter, 0.7 mm) in that space [1]. The characteristics of the thin and fragile arachnoid laminae help to explain unintentional subdural catheter placement, which may occur during the process of needle introduction [2–4]. The orifice produced by the needle or catheter may aid spread along the subdural compartment in the dural sac.

Unintentional intradural (within the dura) placement of an epidural catheter can also occur during the attempted performance of an epidural block in clinical practice (Figs. 27.1, 27.2, 27.3, 27.4, 27.5, 27.6, 27.7, 27.8, 27.9, 27.10, and 27.11) [5].

C.B. Collier, MD, MRCP, FRCA, FANZCA (✉)
Department of Obstetric Anaesthesia,
Prince of Wales Private Hospital, Sydney, NSW, Australia
e-mail: nfi1@pacific.net.au

M.A. Reina, MD, PhD
Department of Clinical Medical Sciences and Institute of Applied Molecular Medicine, School of Medicine, University of CEU San Pablo, Madrid, Spain

Department of Anesthesiology, Madrid-Montepríncipe University Hospital, Madrid, Spain
e-mail: miguelangel@perticone.e.telefonica.net

M.A. Reina et al. (eds.), *Atlas of Functional Anatomy for Regional Anesthesia and Pain Medicine: Human Structure, Ultrastructure and 3D Reconstruction Images*, DOI 10.1007/978-3-319-09522-6_27,

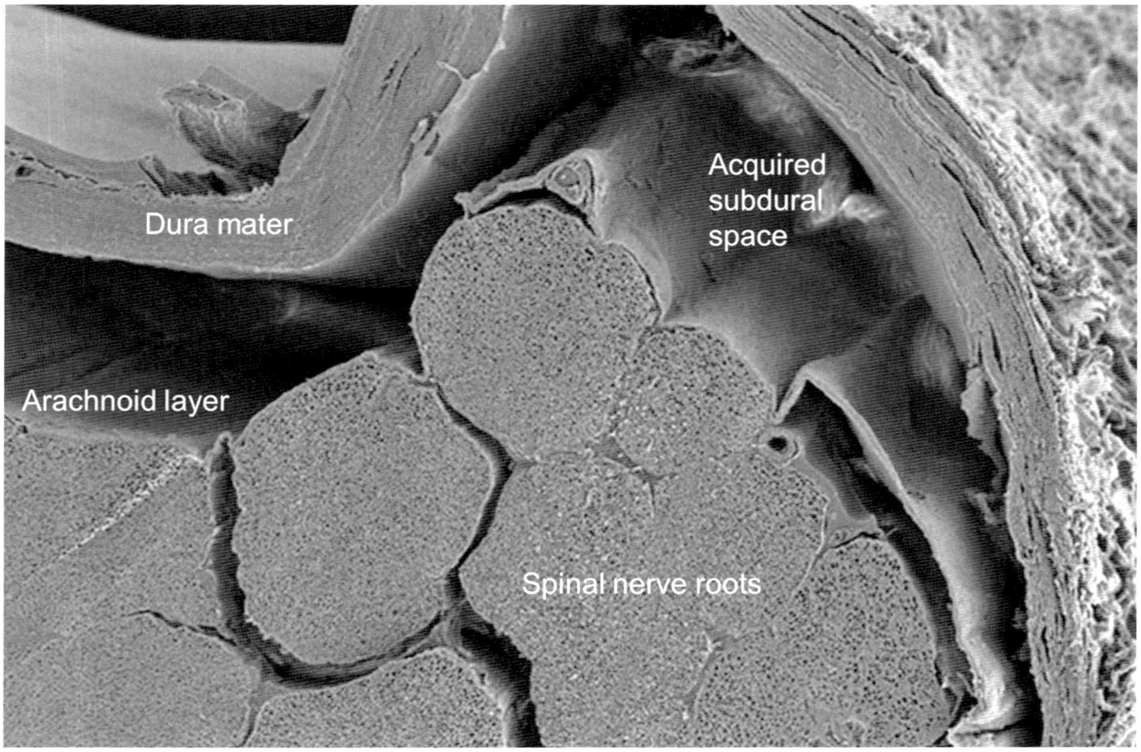

Fig. 27.1 Human spinal acquired subdural space. Scanning electron microscopy. Magnification × 33

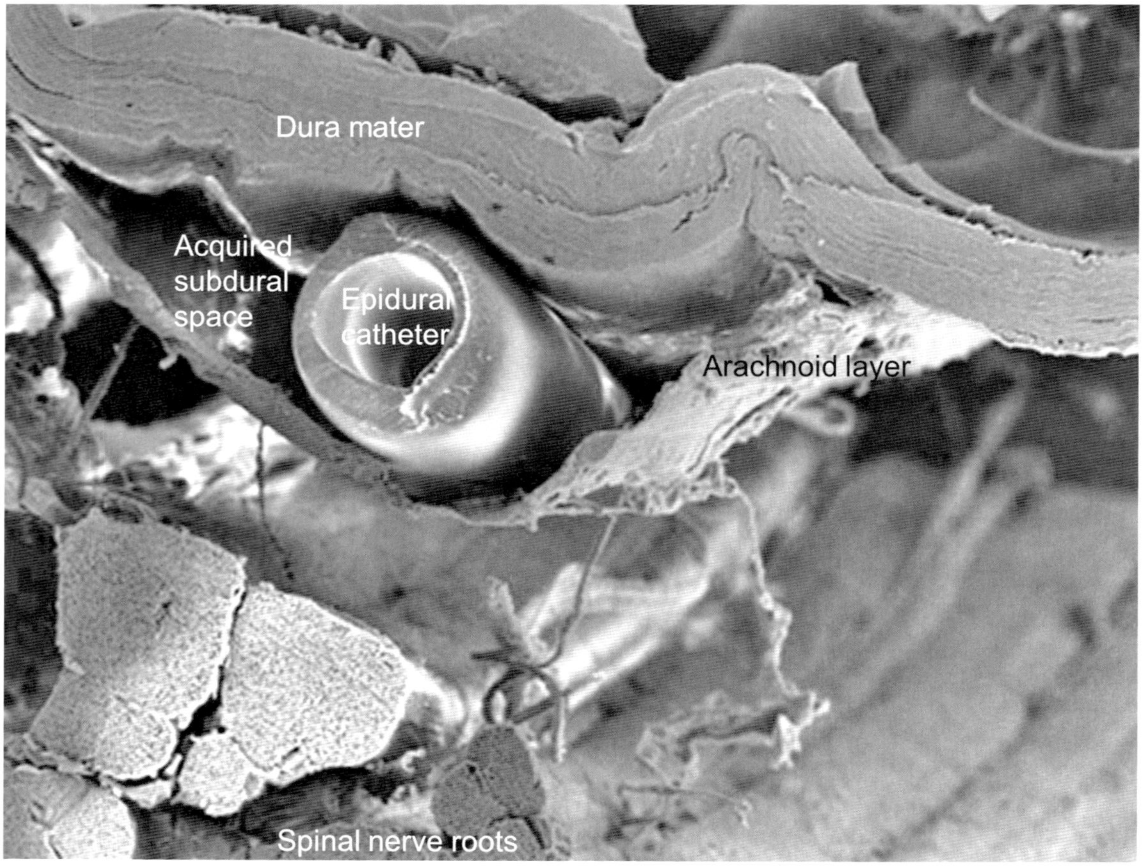

Fig. 27.2 Human spinal acquired subdural space. Scanning electron microscopy. Magnification × 20 (From Reina et al. [1]; with permission)

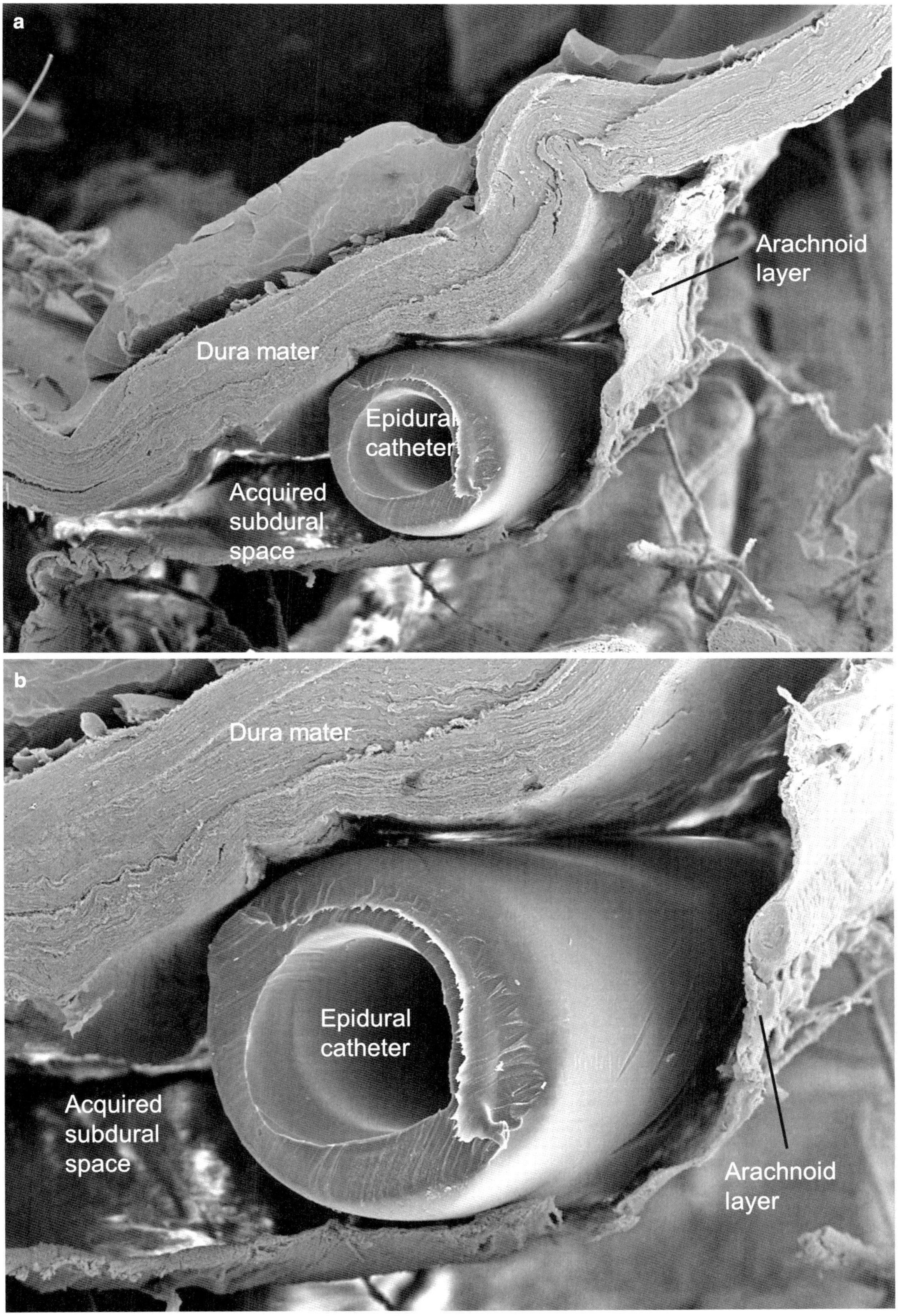

Fig. 27.3 Human spinal acquired subdural space. Scanning electron microscopy. Magnification: (**a**) ×30; (**b**) ×60 (**b** From Reina et al. [1]; with permission)

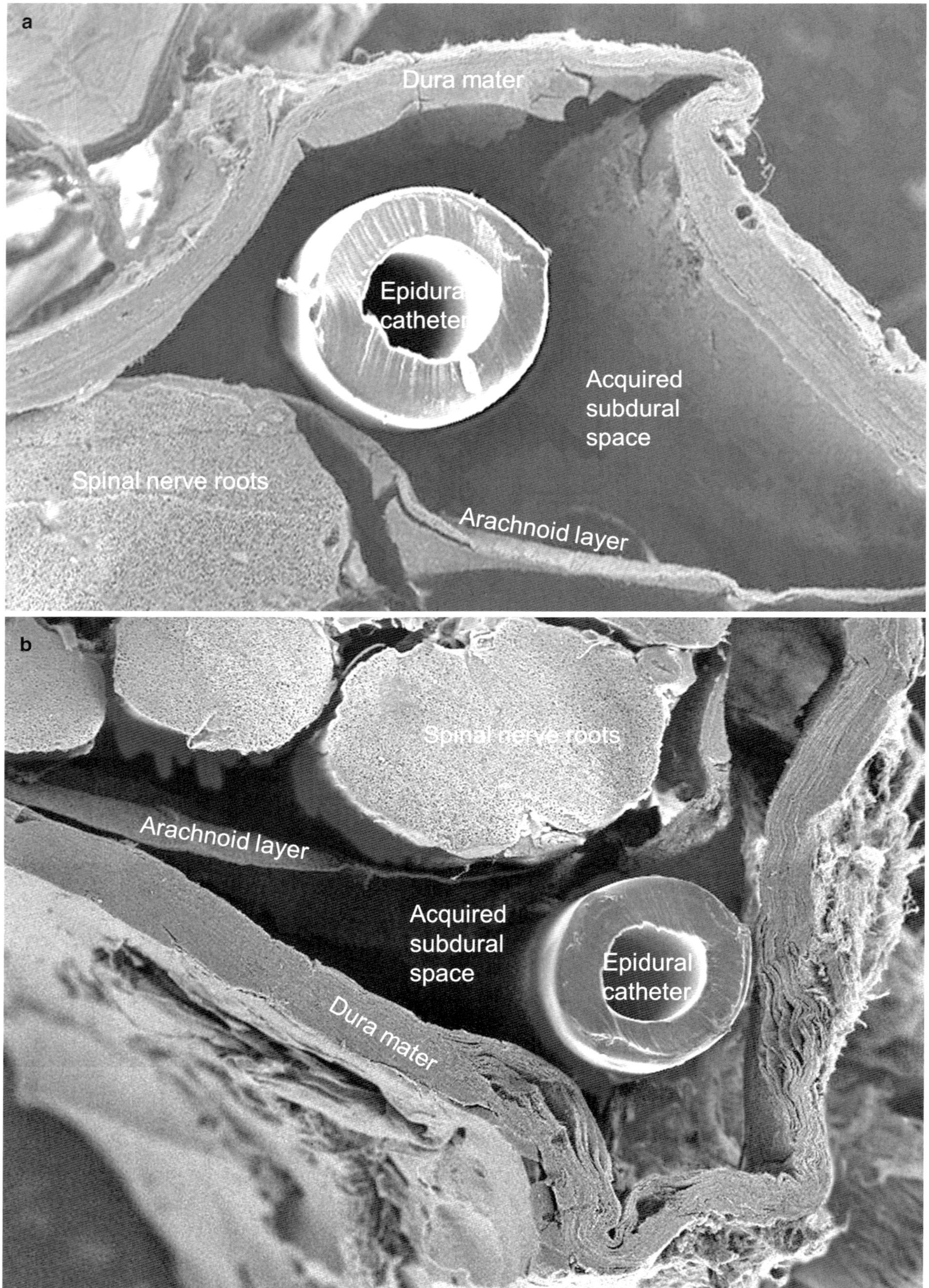

Fig. 27.4 Human spinal acquired subdural space. Scanning electron microscopy. Magnification: (**a**) ×25; (**b**) ×25

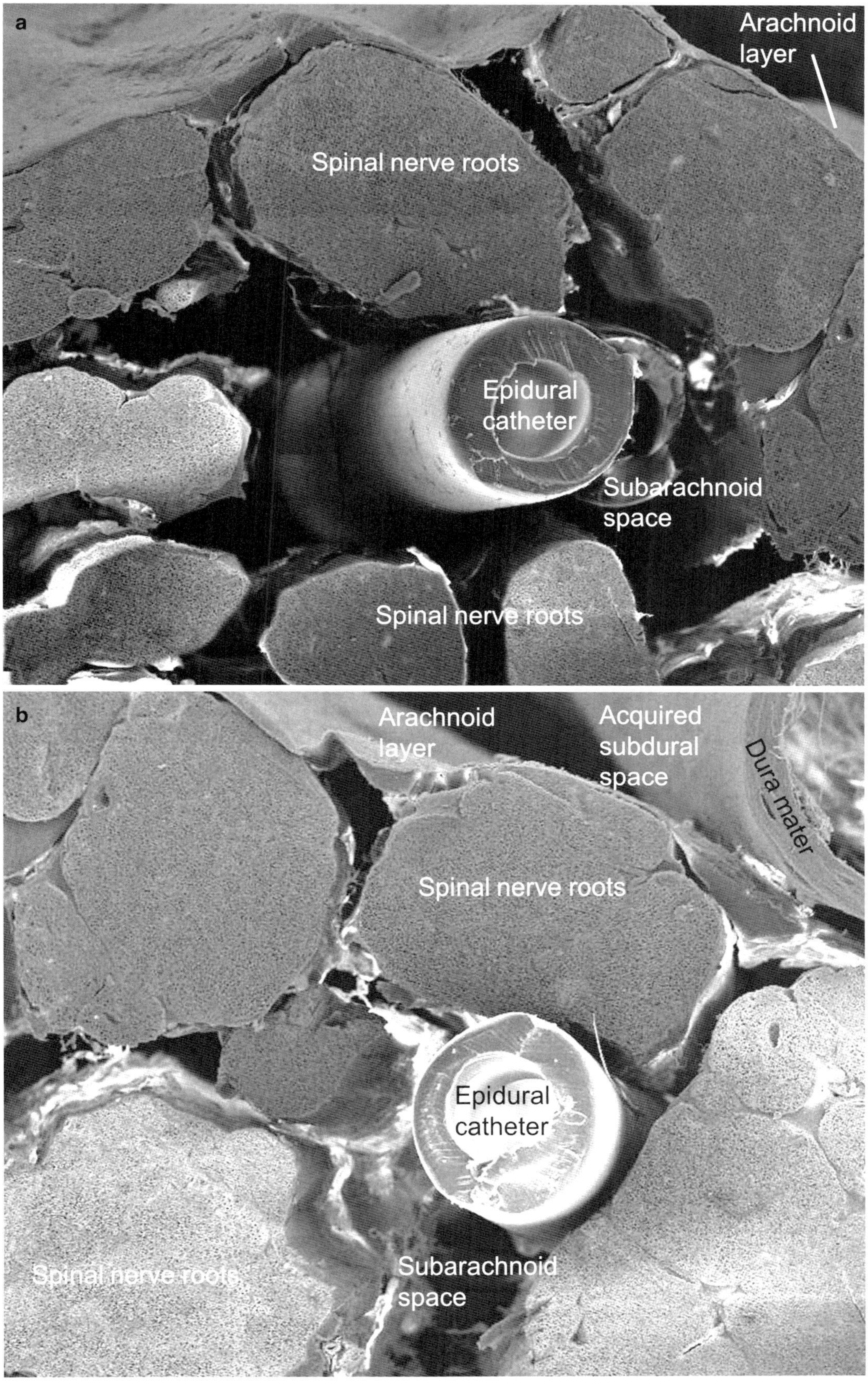

Fig. 27.5 Human spinal acquired subdural space. Scanning electron microscopy. Magnification: (**a**) ×25; (**b**) ×25

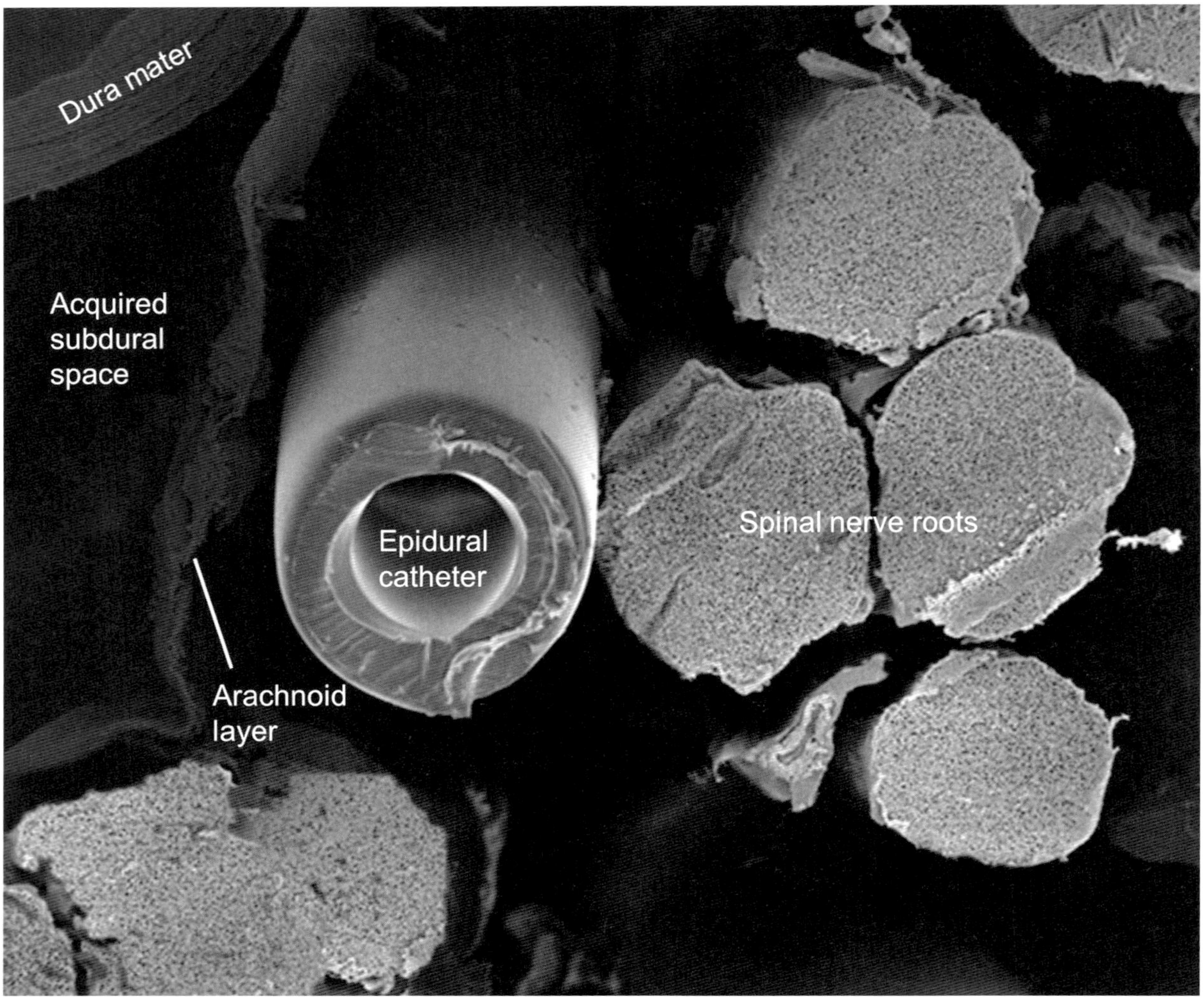

Fig. 27.6 Human spinal acquired subdural space. Scanning electron microscopy. Magnification × 35 (From Reina et al. [1]; with permission)

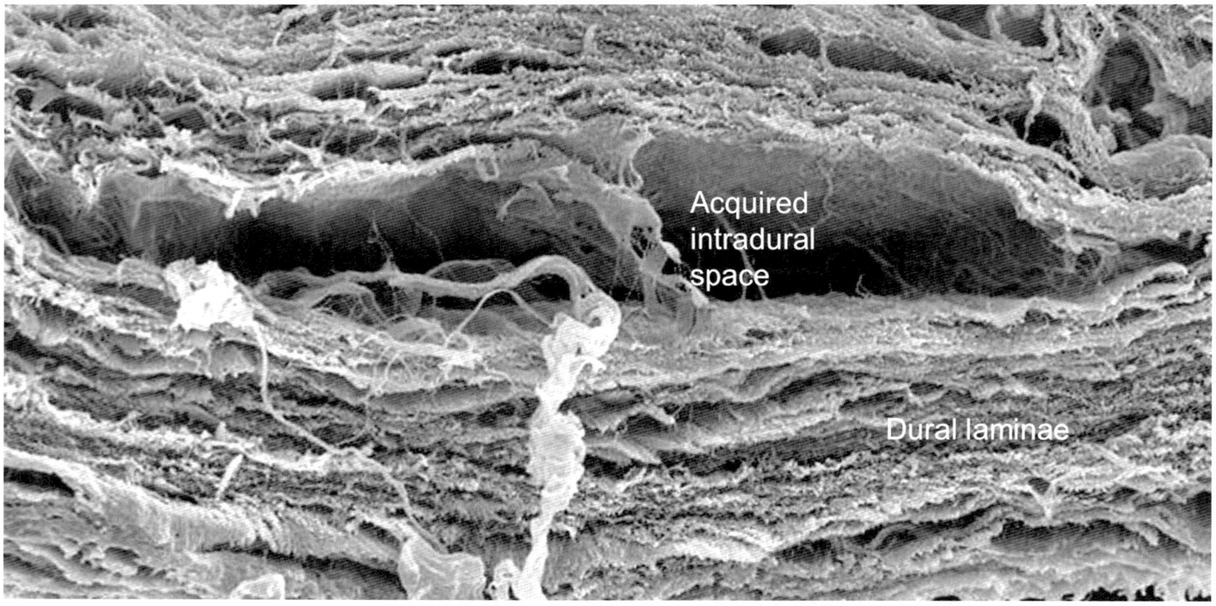

Fig. 27.7 Human spinal acquired subdural space. Scanning electron microscopy. Magnification × 300

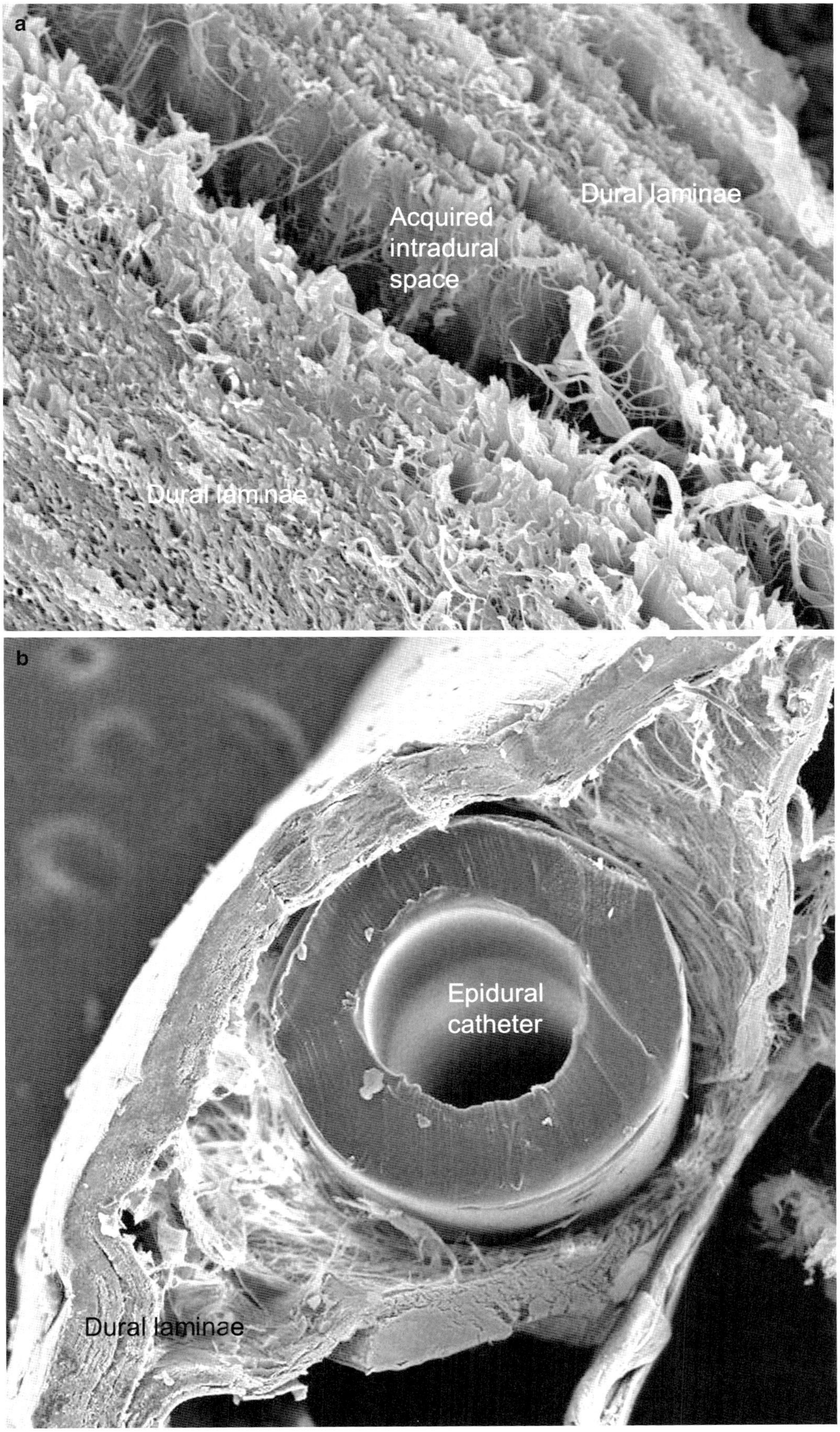

Fig. 27.8 Human spinal acquired subdural space. Scanning electron microscopy. Magnification: (**a**) ×1,000; (**b**) ×50 (**b** From Collier et al. [5]; with permission)

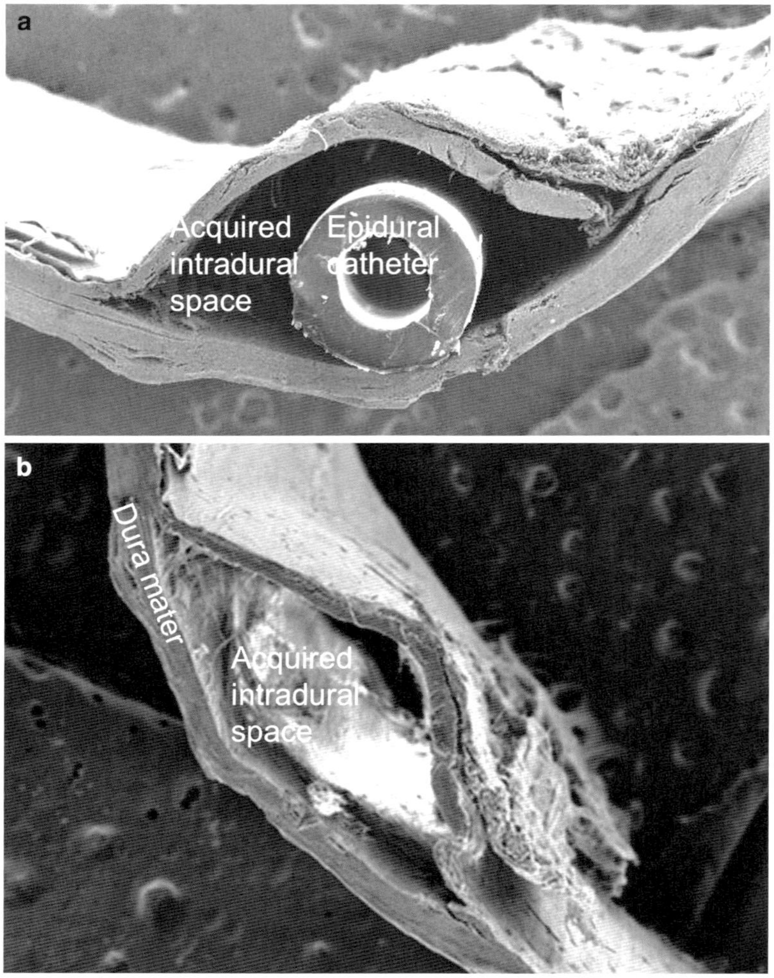

Fig. 27.9 Human spinal acquired subdural space. Scanning electron microscopy. Magnification: (**a**) ×25; (**b**) ×20 (**a** From Collier et al. [5]; with permission)

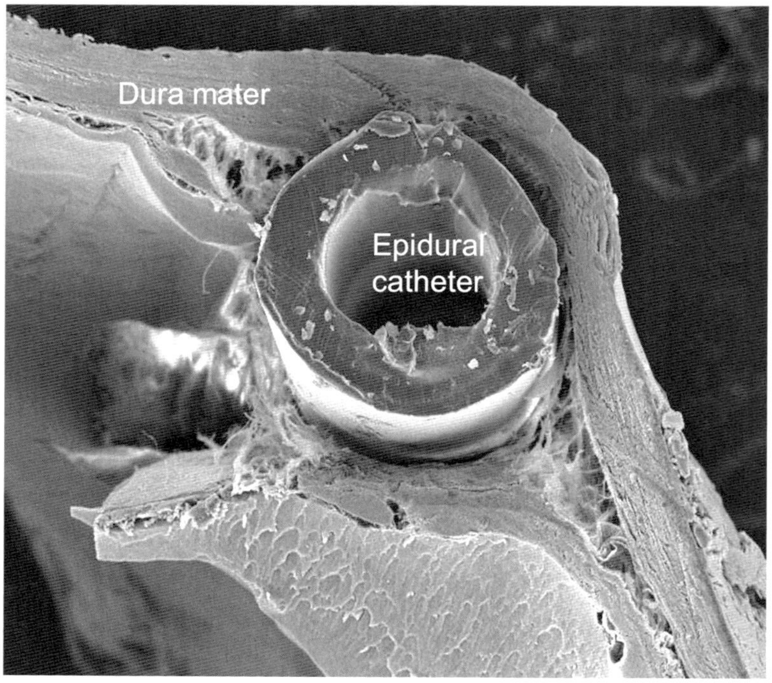

Fig. 27.10 Human spinal acquired subdural space. Scanning electron microscopy. Magnification × 40

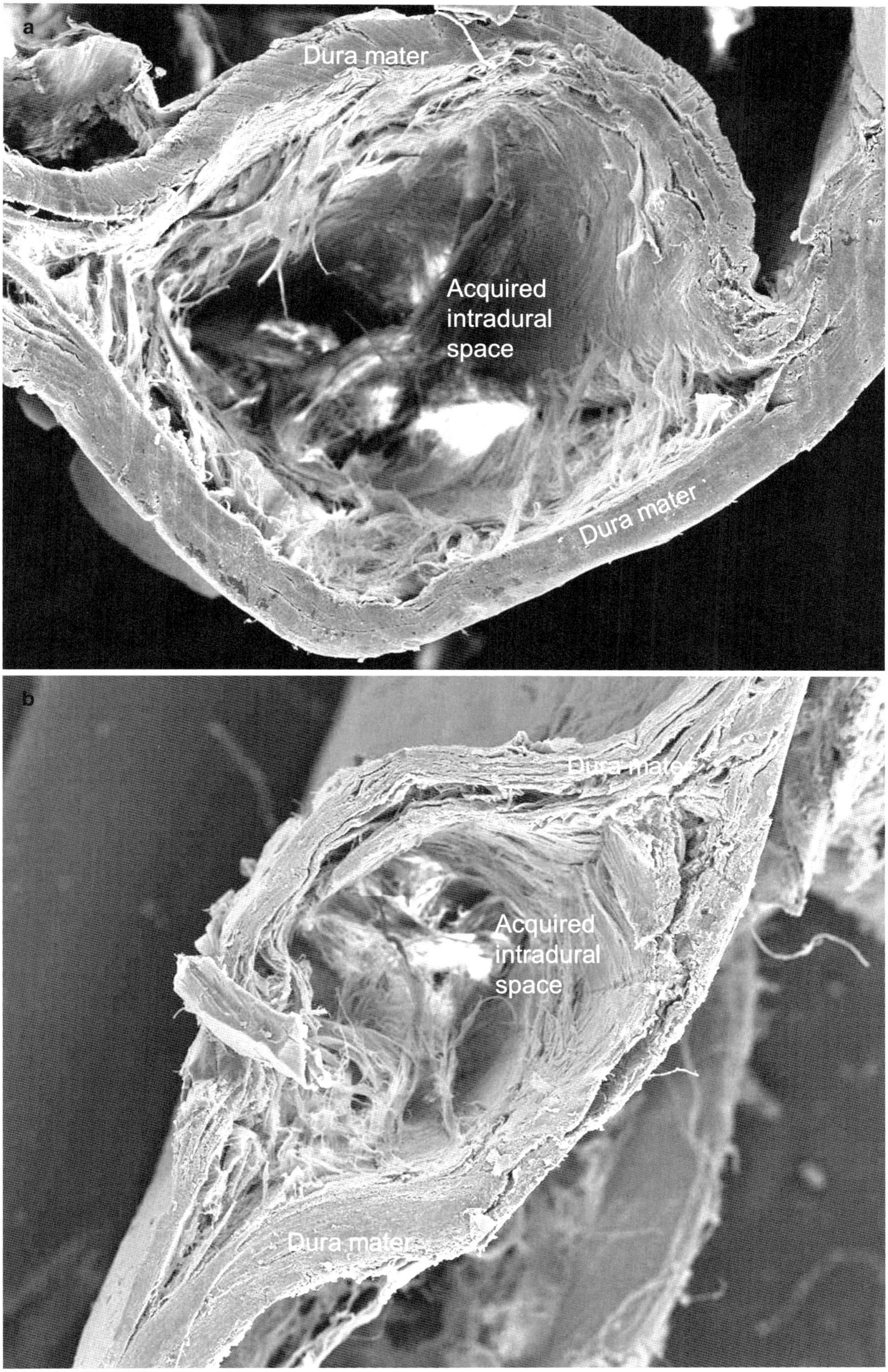

Fig. 27.11 Human spinal acquired subdural space. Scanning electron microscopy. Magnification. (**a**) ×75; (**b**) ×45 (**a** From Collier et al. [5]; with permission)

References

1. Reina MA, Collier CB, Prats-Galino A, Puigdellívol-Sánchez A, Machés F, De Andrés JA. Unintentional subdural placement of epidural catheters during attempted epidural anesthesia: an anatomic study of spinal subdural compartment. Reg Anesth Pain Med. 2011;36:537–41.
2. Reina MA, Pulido P, López A. The human dural sac. Morphology of the spinal dura-arachnoid membrane. Origin of the spinal subdural space. Rev Arg Anestesiol. 2007;65:167–84.
3. Reina MA, De Leon Casasola O, López A, De Andrés JA, Mora M, Fernández A. The origin of the spinal subdural space: ultrastructure findings. Anesth Analg. 2002;94:991–5.
4. Reina MA, López García A, De Andrés JA, Villanueva MC, Cortés L. Does the subdural space exist? Rev Esp Anestesiol Reanim. 1998;45:367–76.
5. Collier CB, Reina MA, Prats-Galino A, Machés F. An anatomical study of the intradural space. Anaesth Intensive Care. 2011;39:1038–42.

28 Ultrastructure of Human Spinal Nerve Roots

Miguel Angel Reina, Lucila Reina Colman, and Alberto Prats-Galino

Motor and sensorial nerve roots are independent structures originating from 8 to 10 sensorial nerve rootlets that emerge from the posterolateral region of the spinal cord and from 6 to 8 motor rootlets located anterolateral to the spinal cord [1–4]. The number of rootlets may increase in roots present at the level of the brachial plexus. The number of rootlets decreases to two to three at the sacral level and to one at the coccygeal level, giving shape to single nerve roots. Rootlet size varies, with the anterior rootlets being finer than the posterior ones. The diameter of nerve roots ranges between 0.7 and 0.3 mm. Compared with the diameter of an anterior rootlet, that of a posterior rootlet is approximately 1.5 times larger in the thoracic region, three times larger at the lumbar level, and three times larger in the cervical region [2].

Nerve roots and rootlets are immersed in cerebrospinal fluid inside the subarachnoid space, with the pia mater as sole protectant and a permeable layer enclosing the axons at this level [3–5]. Blood vessels of various sizes are found along the surface of the nerve roots, usually oriented longitudinally or traversing across the thickness of the roots (Figs. 28.1, 28.2, 28.3, 28.4, 28.5, 28.6, 28.7, 28.8, 28.9, 28.10, 28.11, 28.12, 28.13, 28.14, 28.15, 28.16, 28.17, 28.18, 28.19, 28.20, 28.21, 28.22, 28.23, 28.24, 28.25, 28.26, 28.27, 28.28, 28.29, and 28.30).

M.A. Reina, MD, PhD (✉)
Department of Clinical Medical Sciences and
Institute of Applied Molecular Medicine, School of Medicine,
University of CEU San Pablo, Madrid, Spain

Department of Anesthesiology, Madrid-Montepríncipe
University Hospital, Madrid, Spain
e-mail: miguelangel@perticone.e.telefonica.net

L. Reina Colman, BS Master Degree
Dentistry Department, Rey Juan Carlos University, Madrid, Spain
e-mail: lucilasstu@hotmail.com

A. Prats-Galino, MD, PhD
Laboratory of Surgical NeuroAnatomy (LSNA),
Human Anatomy and Embryology Unit, School of Medicine,
University of Barcelona, Barcelona, Spain
e-mail: aprats@ub.edu

M.A. Reina et al. (eds.), *Atlas of Functional Anatomy for Regional Anesthesia and Pain Medicine: Human Structure, Ultrastructure and 3D Reconstruction Images*, DOI 10.1007/978-3-319-09522-6_28,

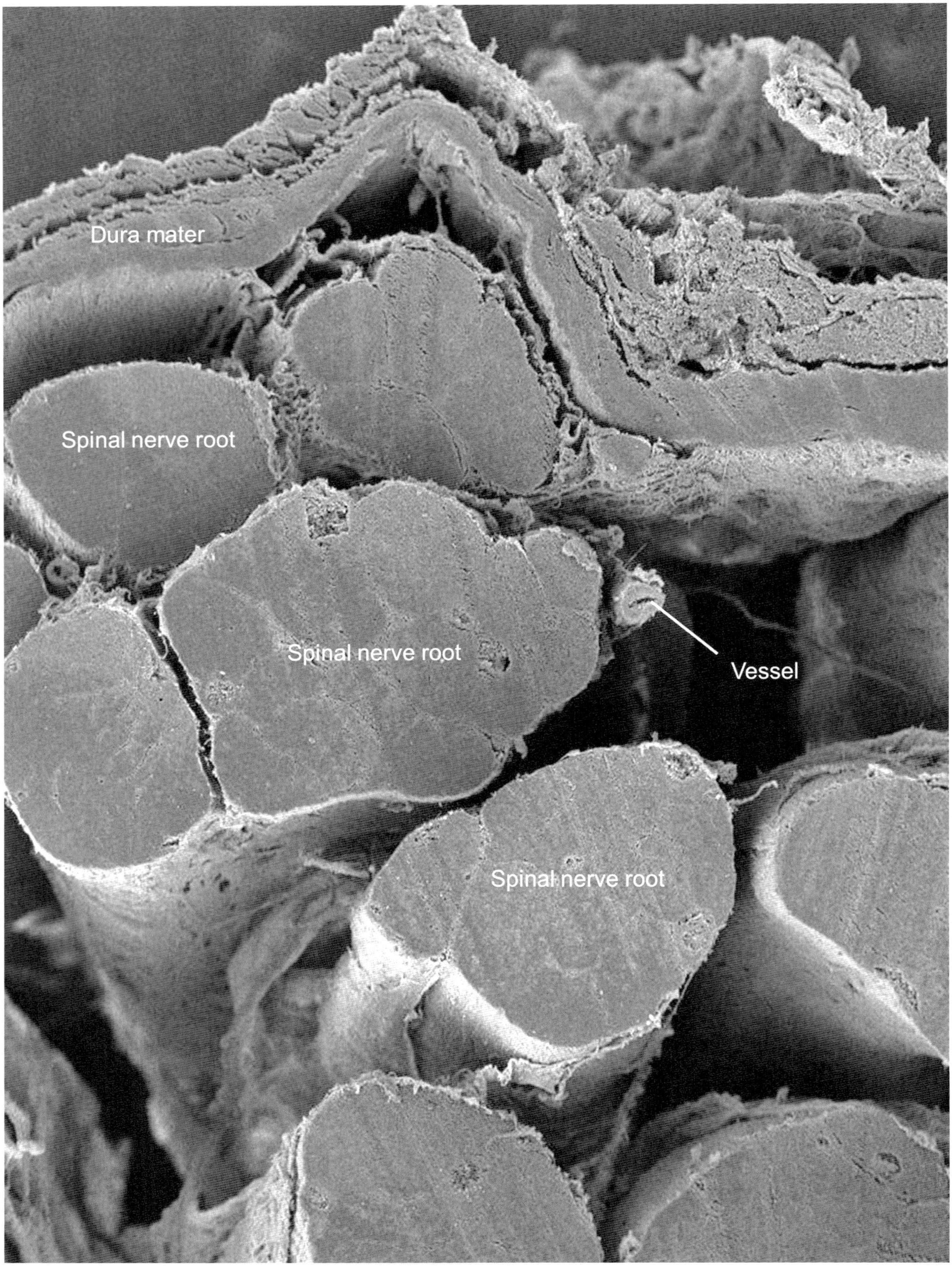

Fig. 28.1 Human spinal nerve root. Scanning electron microscopy; magnification: ×40 (From De Andres et al. [5]; with permission)

Fig. 28.2 Human spinal nerve root. Scanning electron microscopy; magnification: ×40 (**a**, **b**)

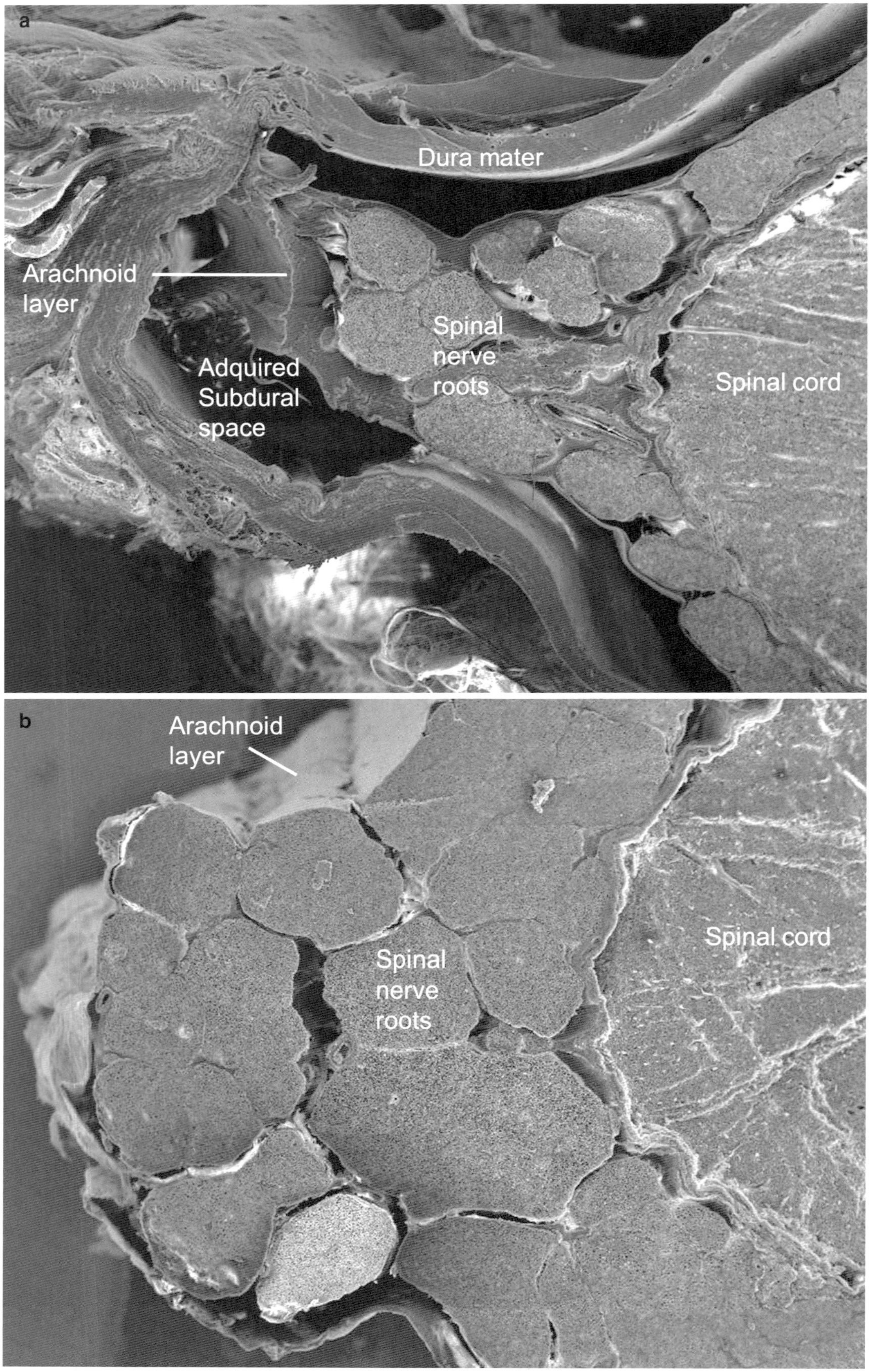

Fig. 28.3 Human spinal nerve root. Scanning electron microscopy; magnification: ×20 (**a**), ×22 (**b**)

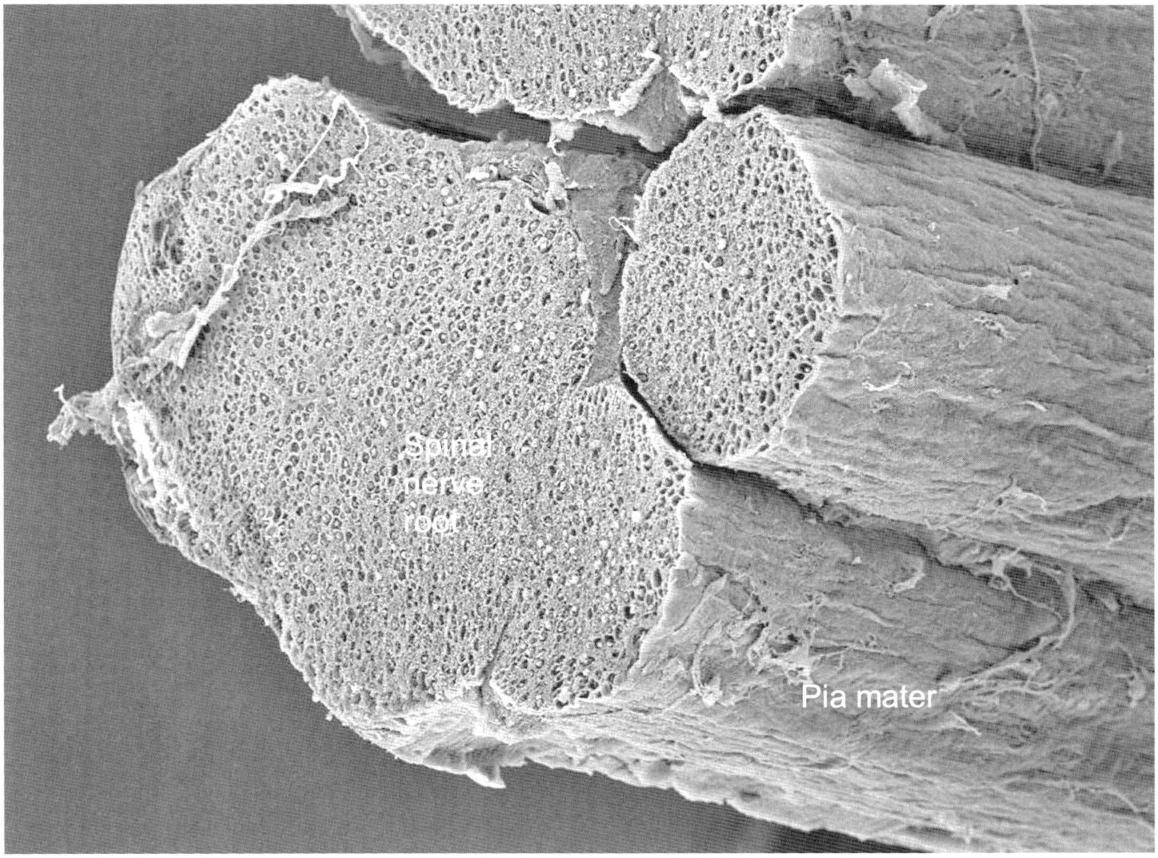

Fig. 28.4 Human spinal nerve root. Scanning electron microscopy; magnification: ×100

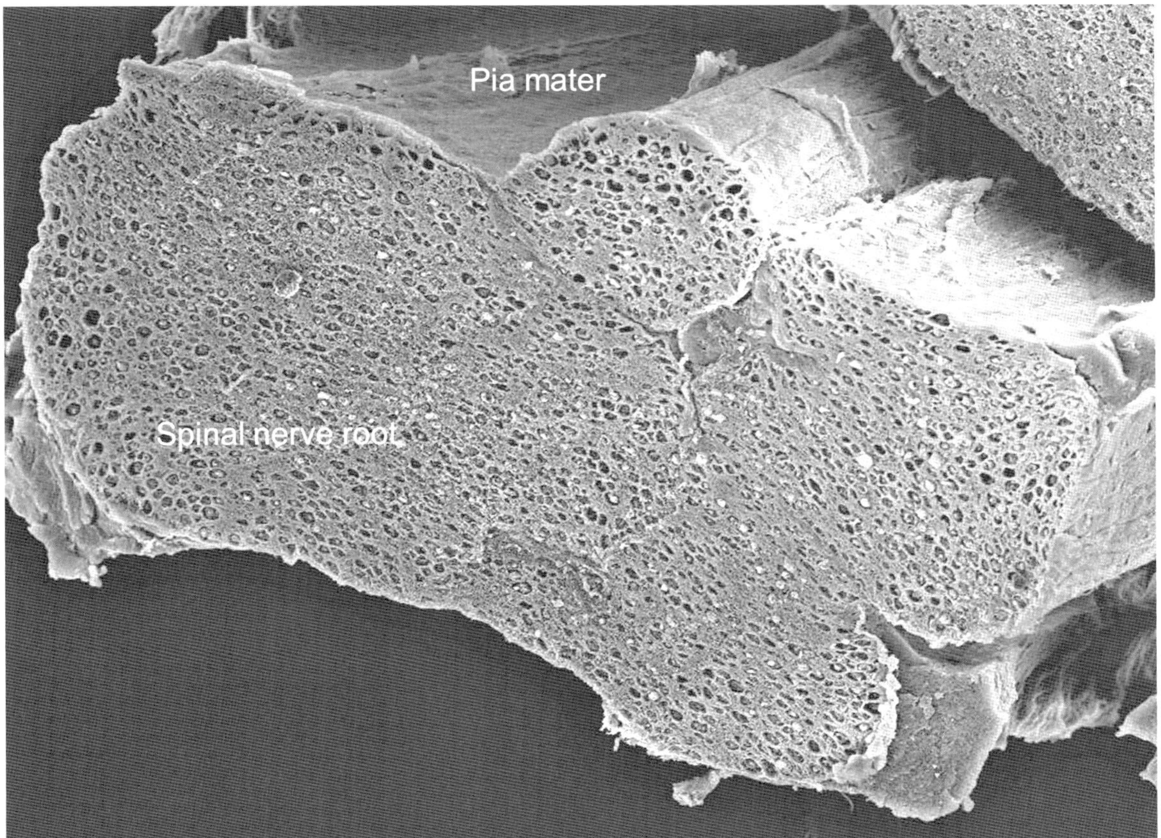

Fig. 28.5 Human spinal nerve root. Scanning electron microscopy; magnification: ×120

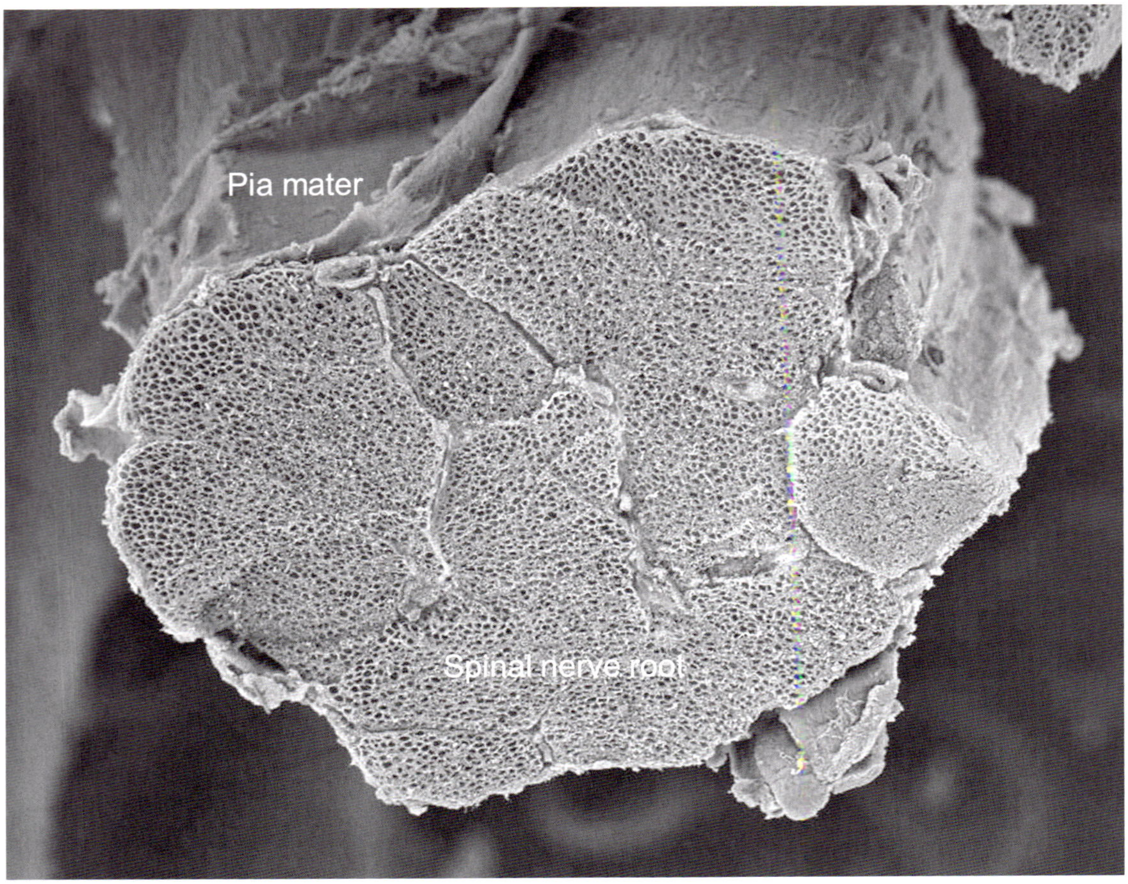

Fig. 28.6 Human spinal nerve root. Scanning electron microscopy; magnification: ×50

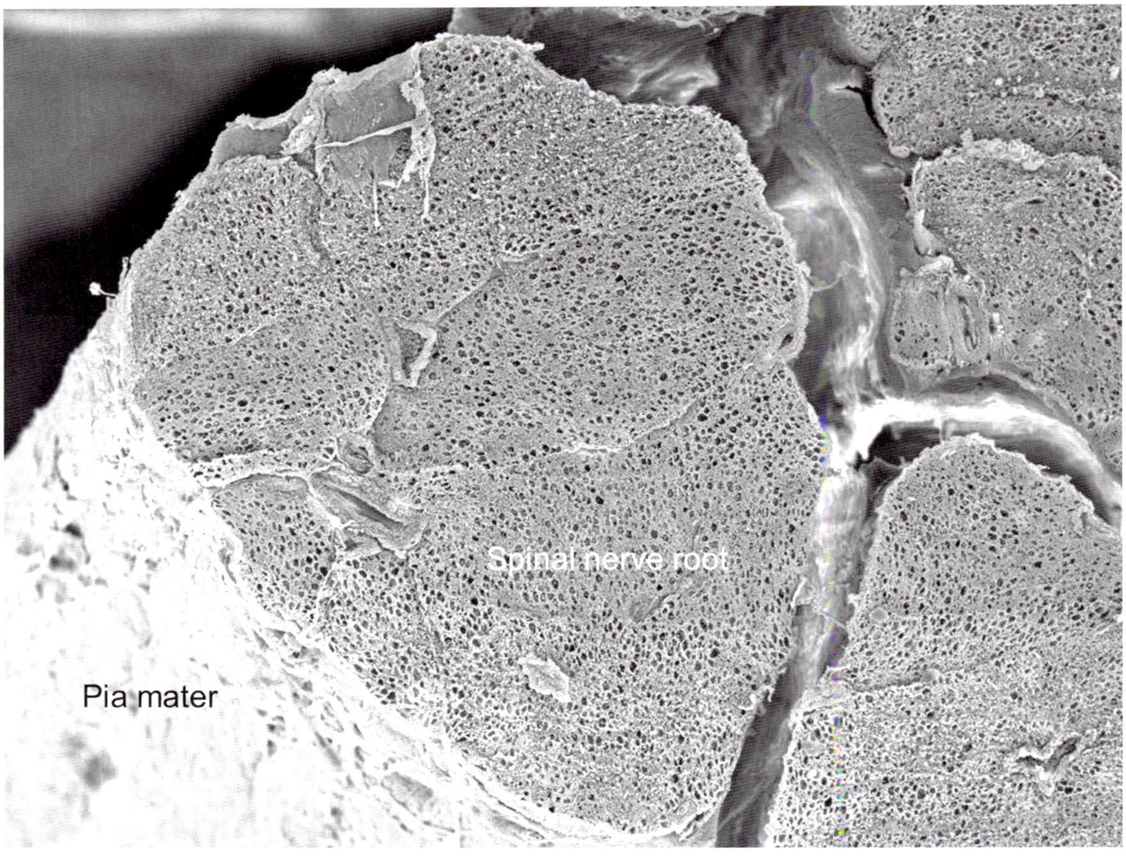

Fig. 28.7 Human spinal nerve root. Scanning electron microscopy; magnification: ×75 (From Reina et al. [1]; with permission)

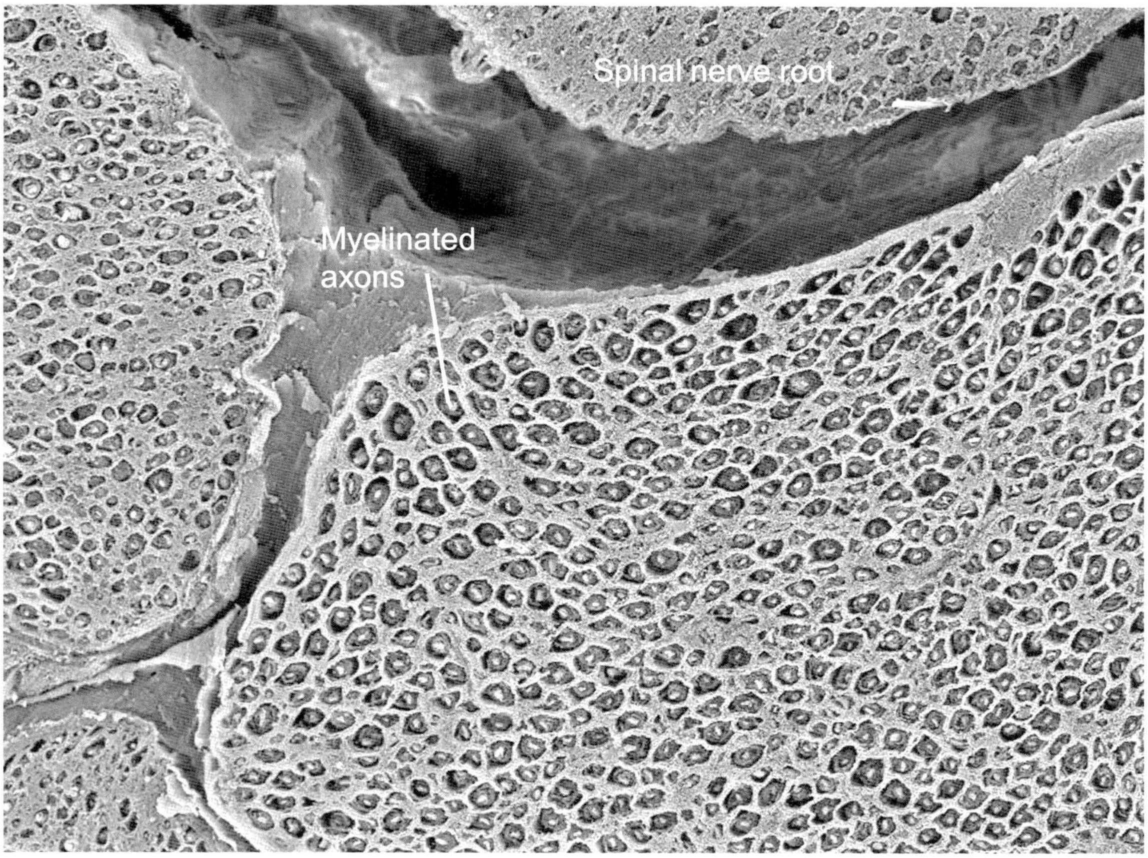

Fig. 28.8 Human spinal nerve root. Scanning electron microscopy; magnification: ×200

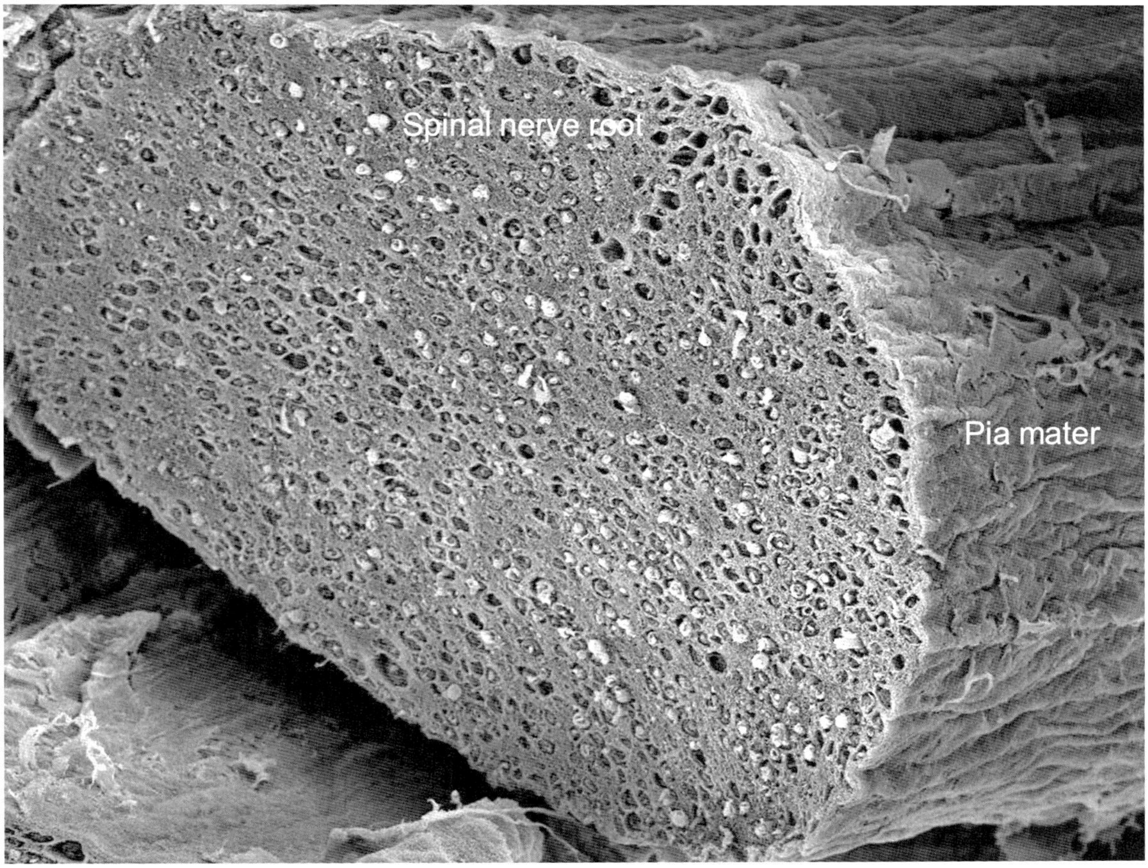

Fig. 28.9 Human spinal nerve root. Scanning electron microscopy; magnification: ×200

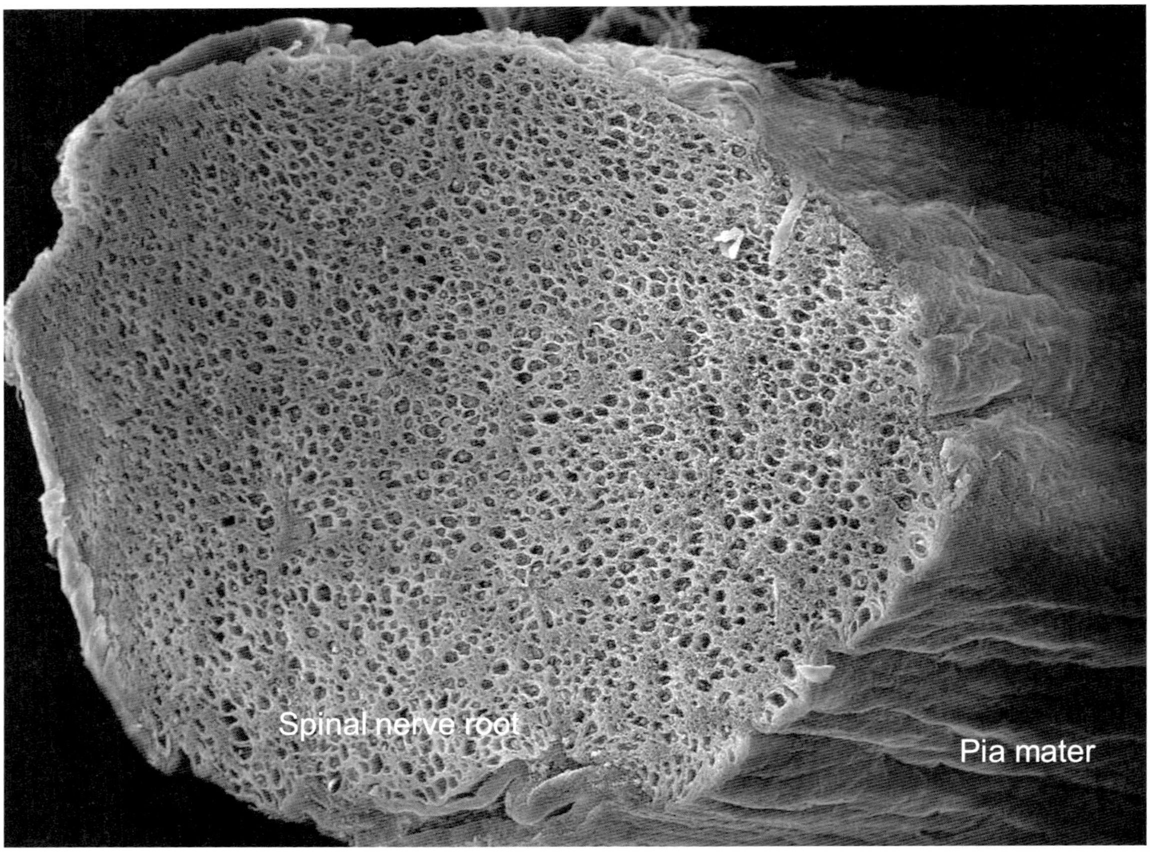

Fig. 28.10 Human spinal nerve root. Scanning electron microscopy; magnification: ×140 (From Reina et al. [4]; with permission)

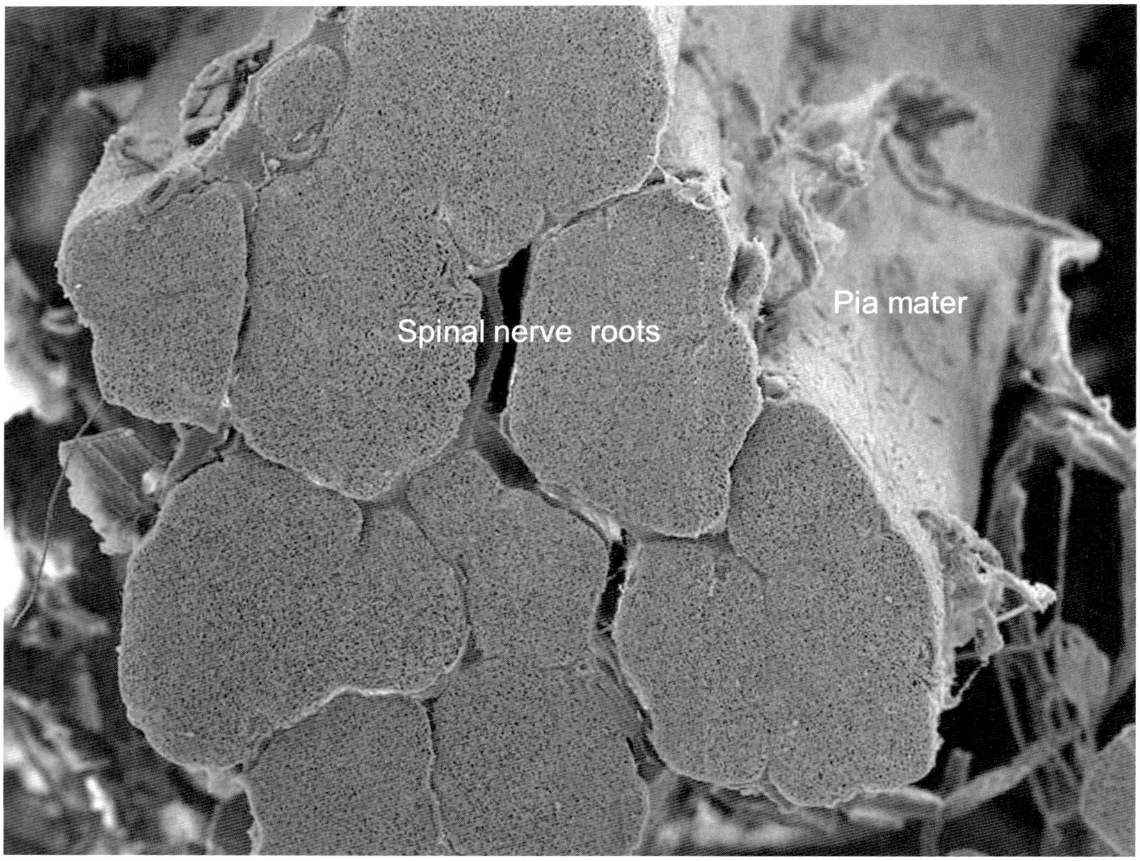

Fig. 28.11 Human spinal nerve root. Scanning electron microscopy; magnification: ×30

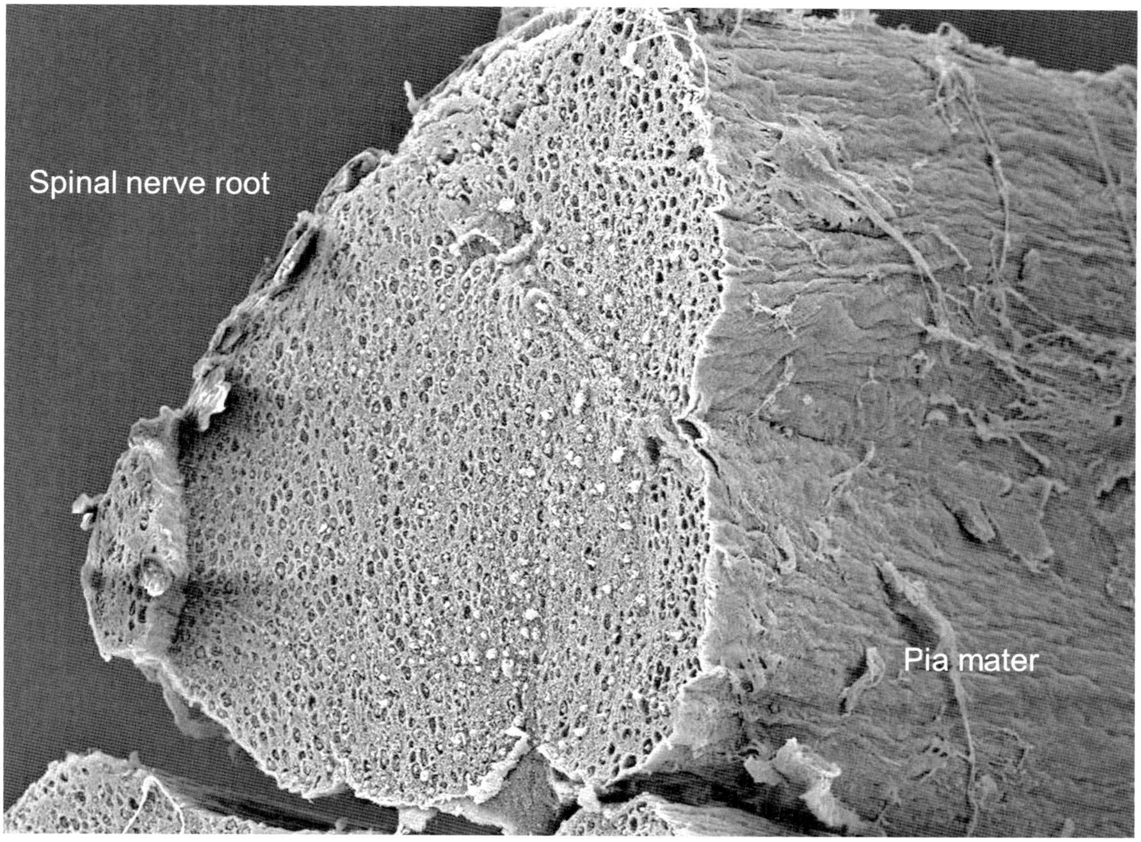

Fig. 28.12 Human spinal nerve root. Scanning electron microscopy; magnification: ×120 (From Prats et al. [3]; with permission)

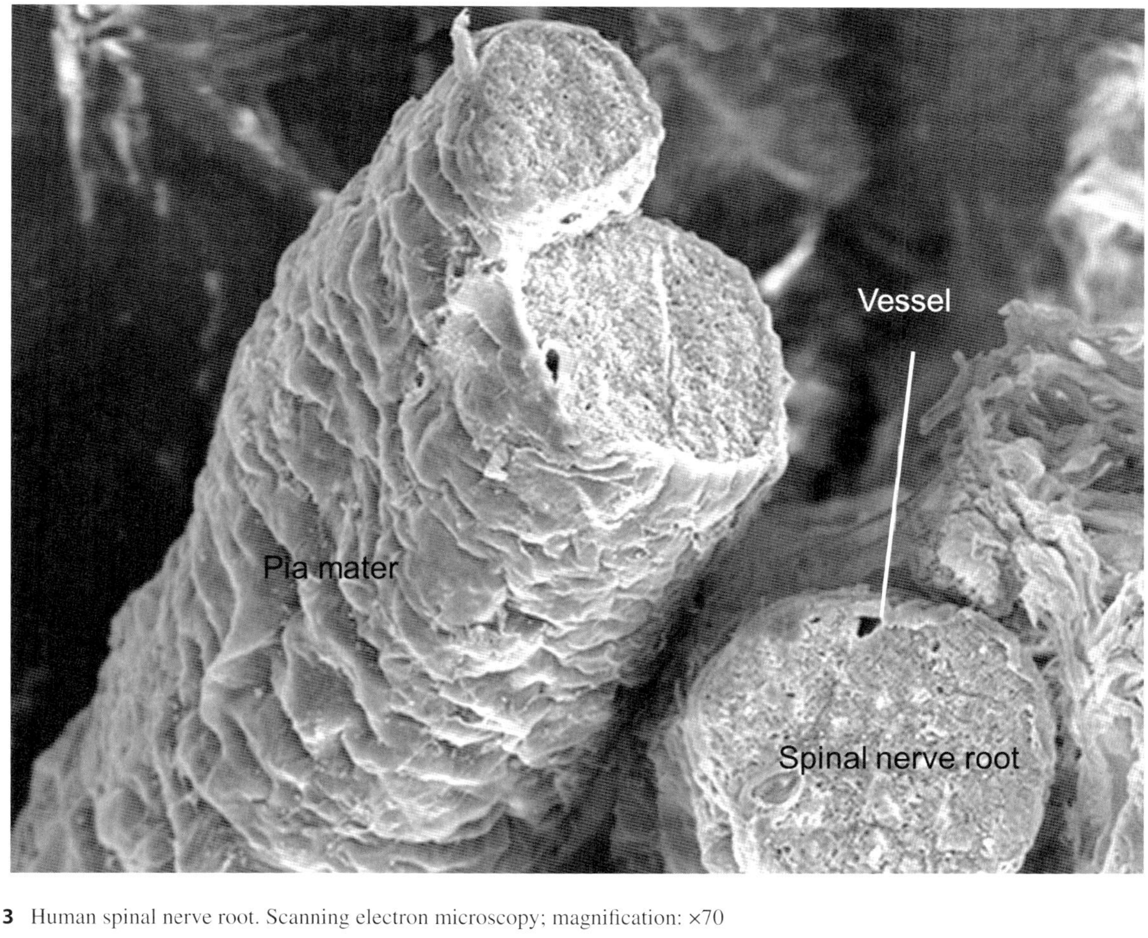

Fig. 28.13 Human spinal nerve root. Scanning electron microscopy; magnification: ×70

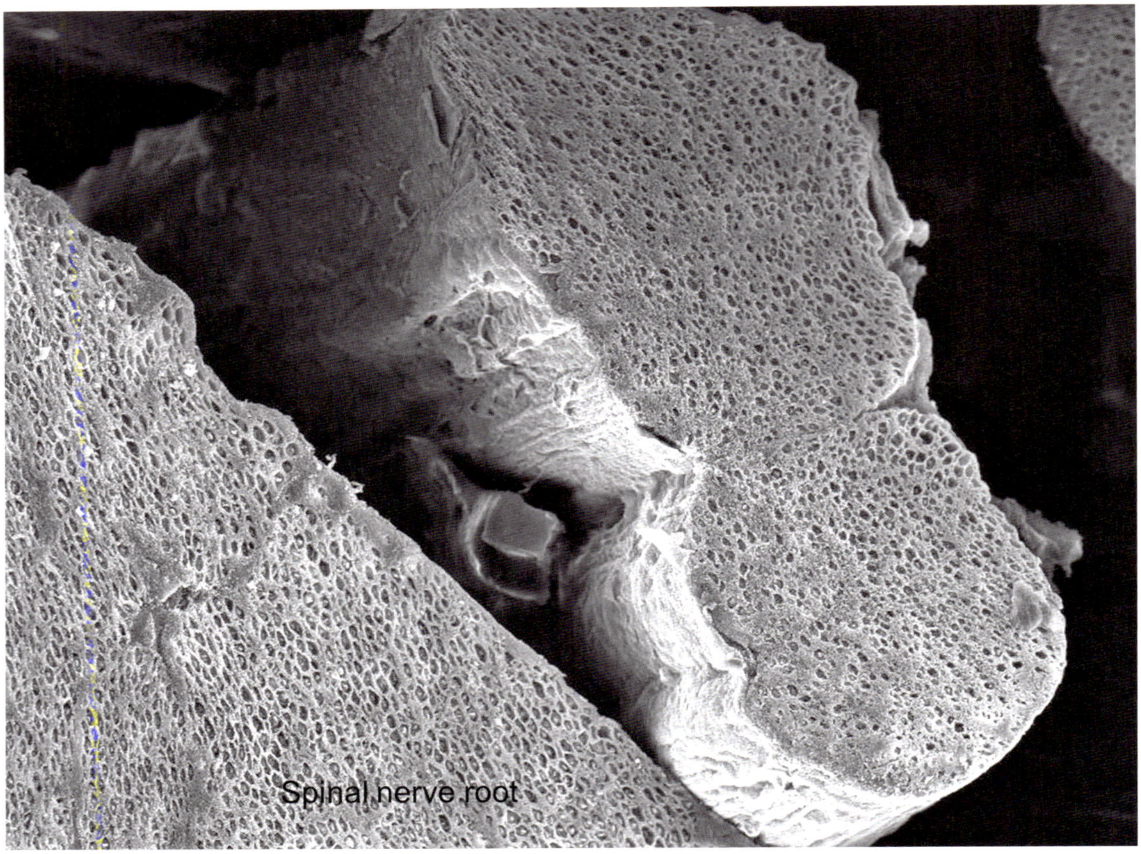

Fig. 28.14 Human spinal nerve root. Scanning electron microscopy; magnification: ×100

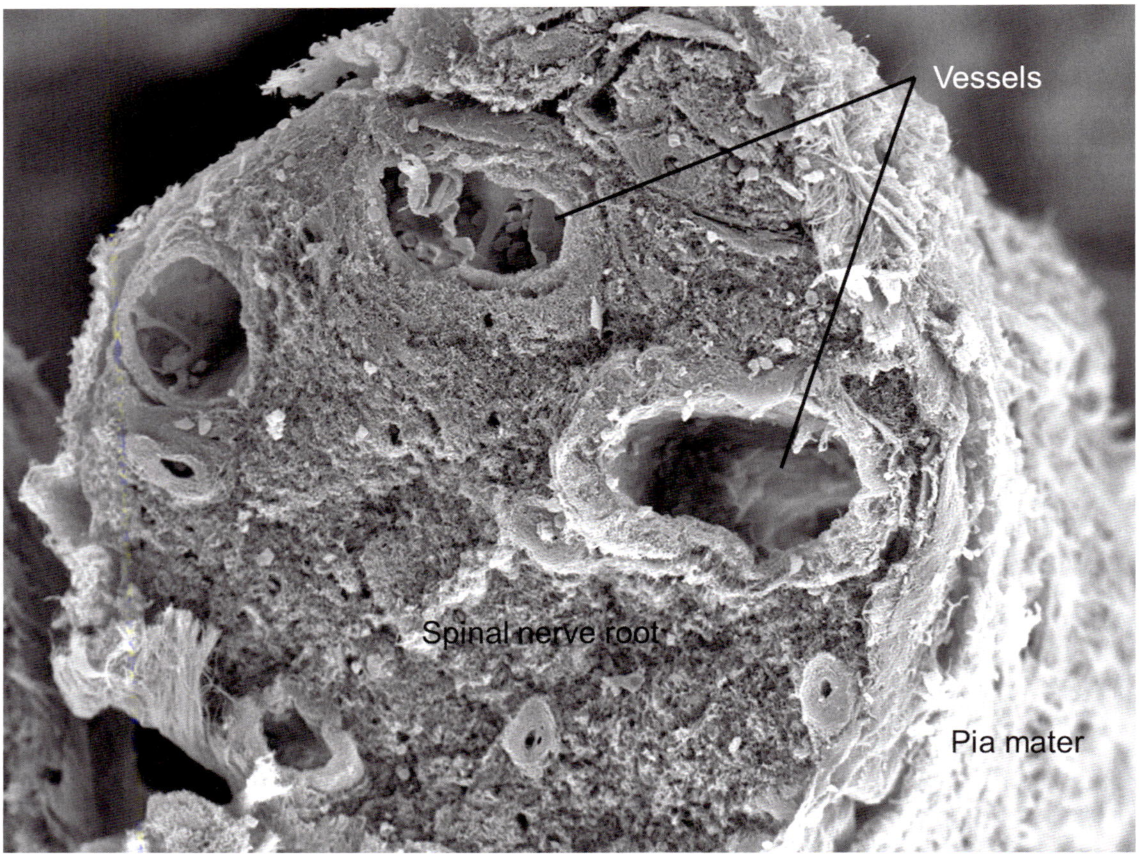

Fig. 28.15 Human spinal nerve root. Scanning electron microscopy; magnification: ×270

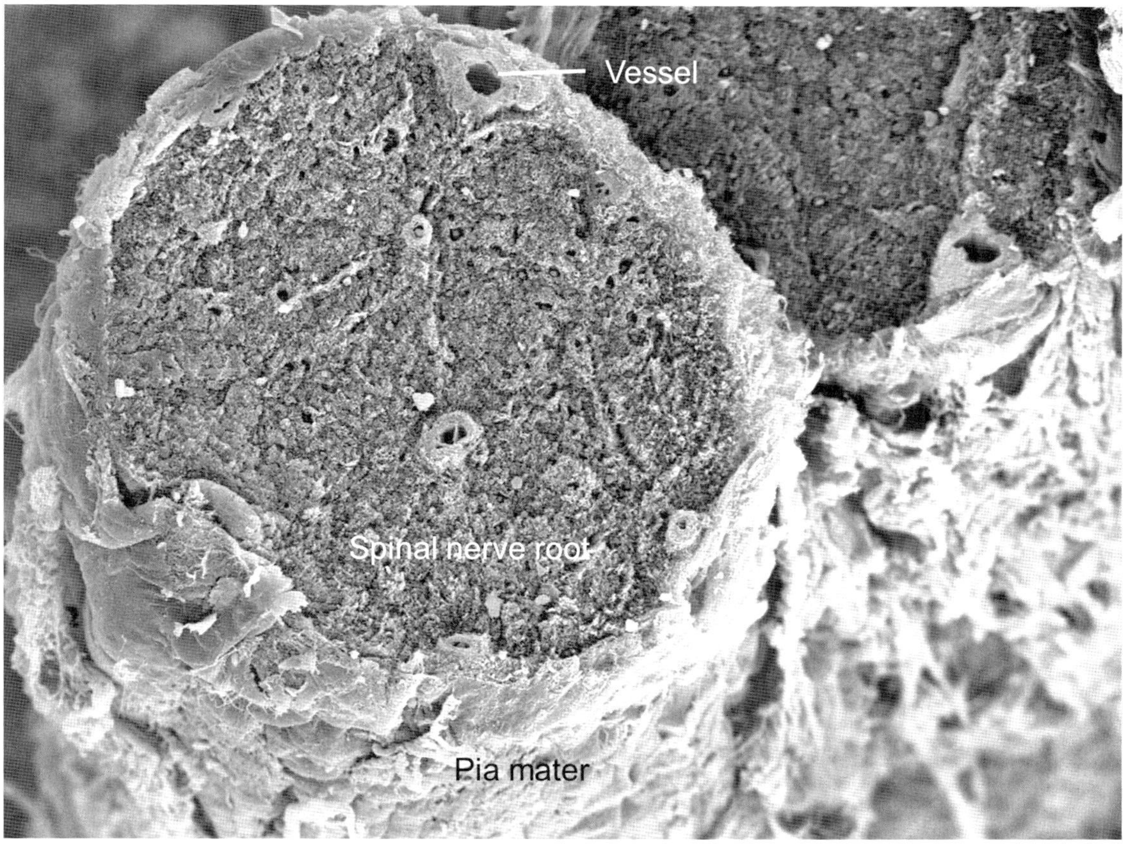

Fig. 28.16 Human spinal nerve root. Scanning electron microscopy; magnification: ×200

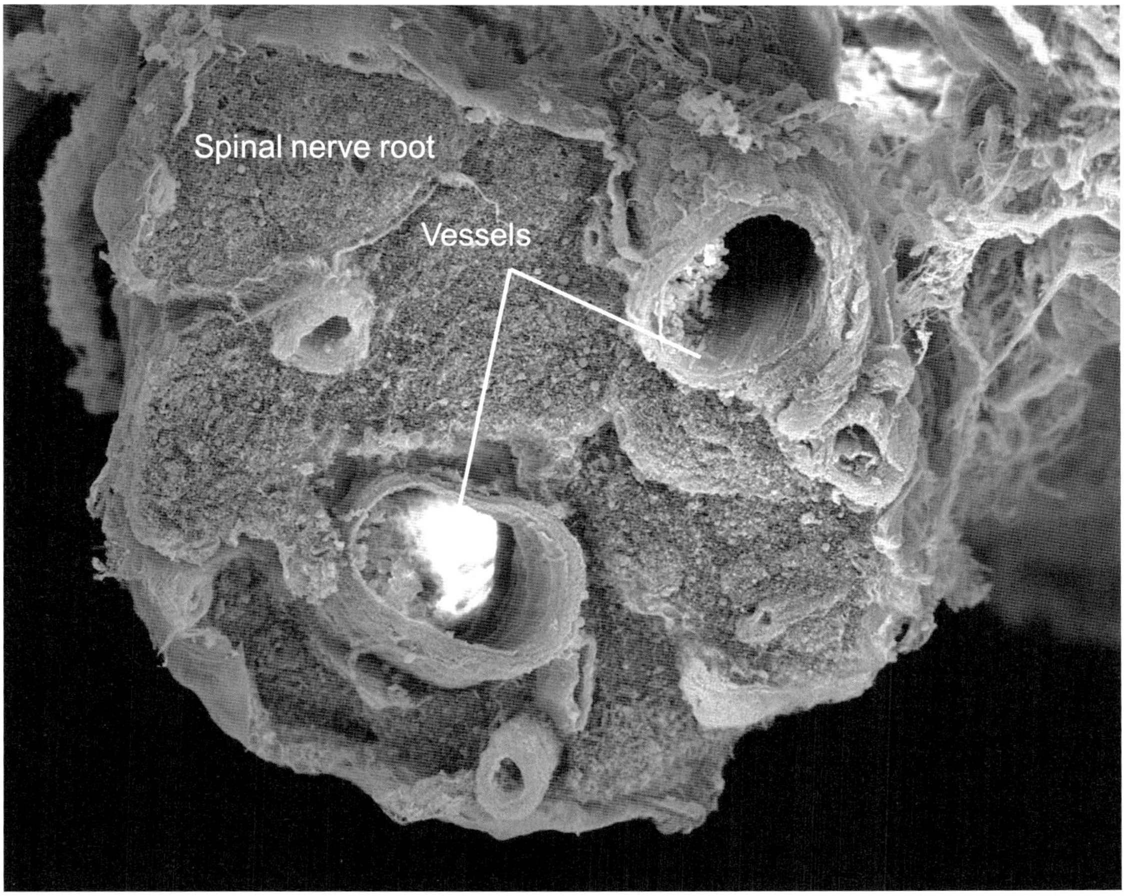

Fig. 28.17 Human spinal nerve root. Scanning electron microscopy; magnification: ×170

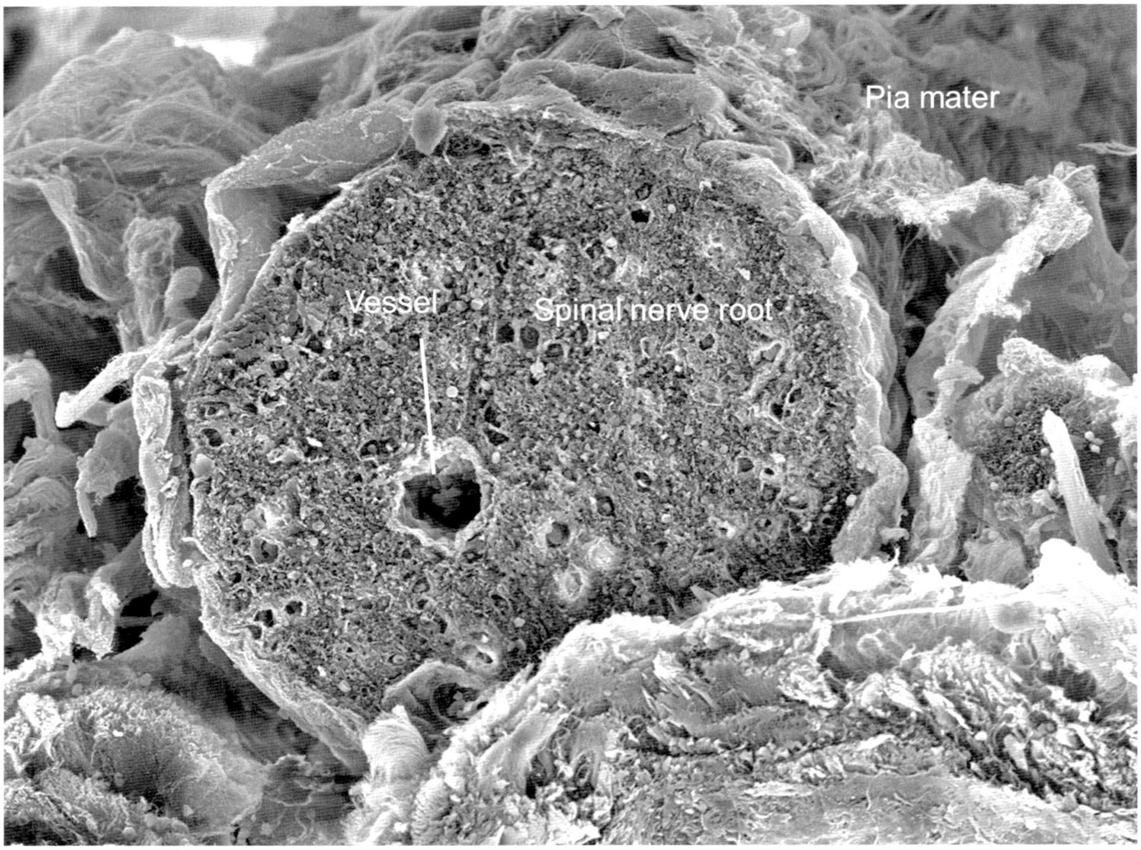

Fig. 28.18 Human spinal nerve root. Scanning electron microscopy; magnification: ×200

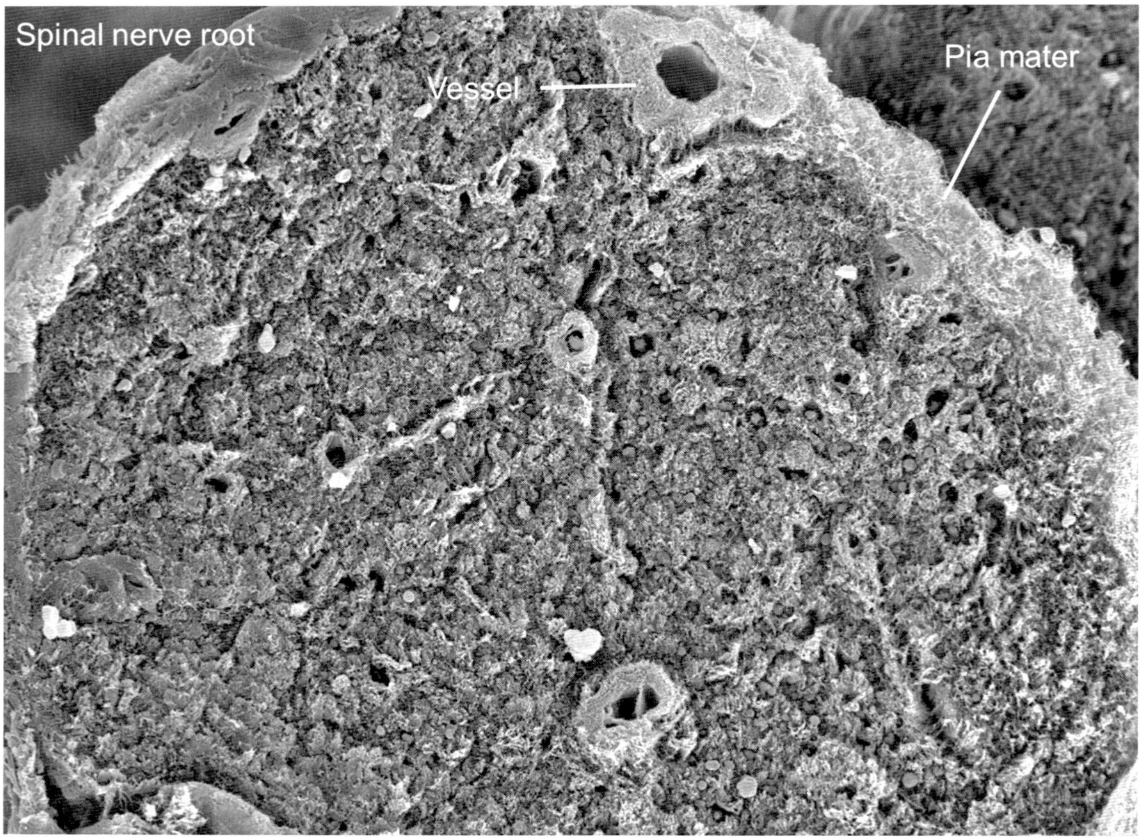

Fig. 28.19 Human spinal nerve root. Scanning electron microscopy; magnification: ×350

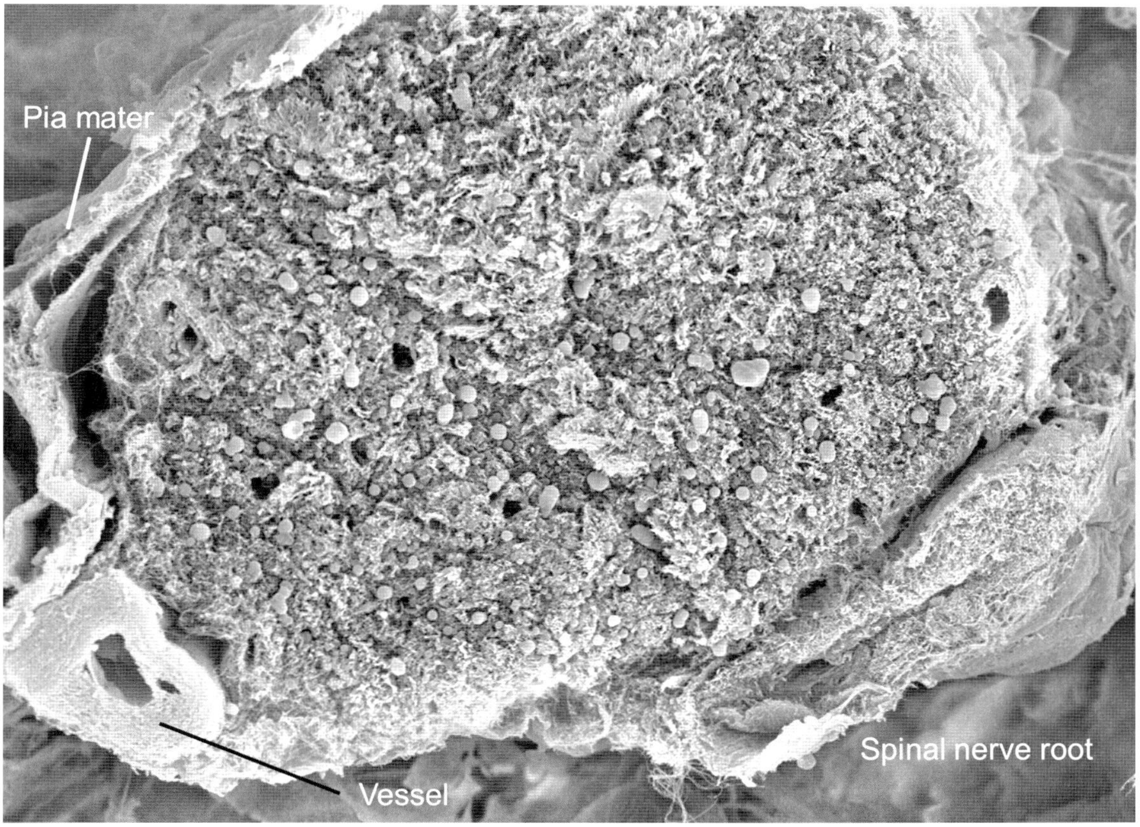

Fig. 28.20 Human spinal nerve root. Scanning electron microscopy; magnification: ×400

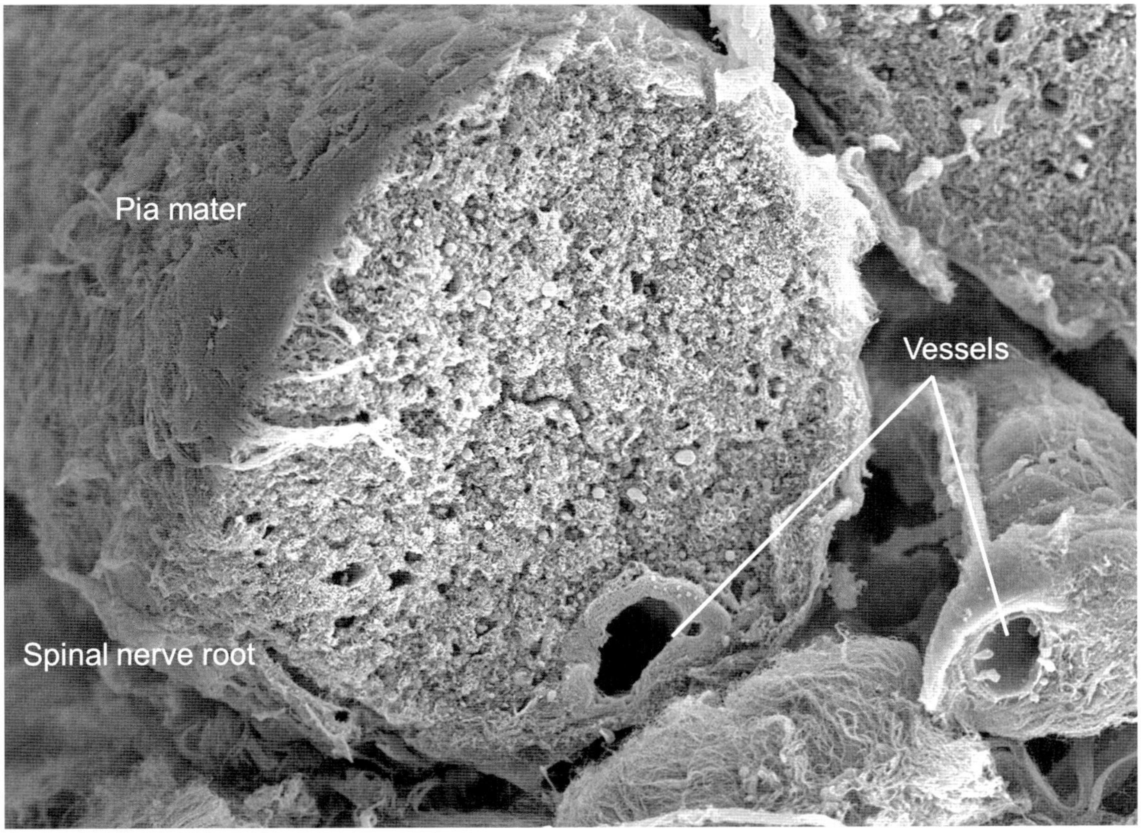

Fig. 28.21 Human spinal nerve root. Scanning electron microscopy; magnification: ×300

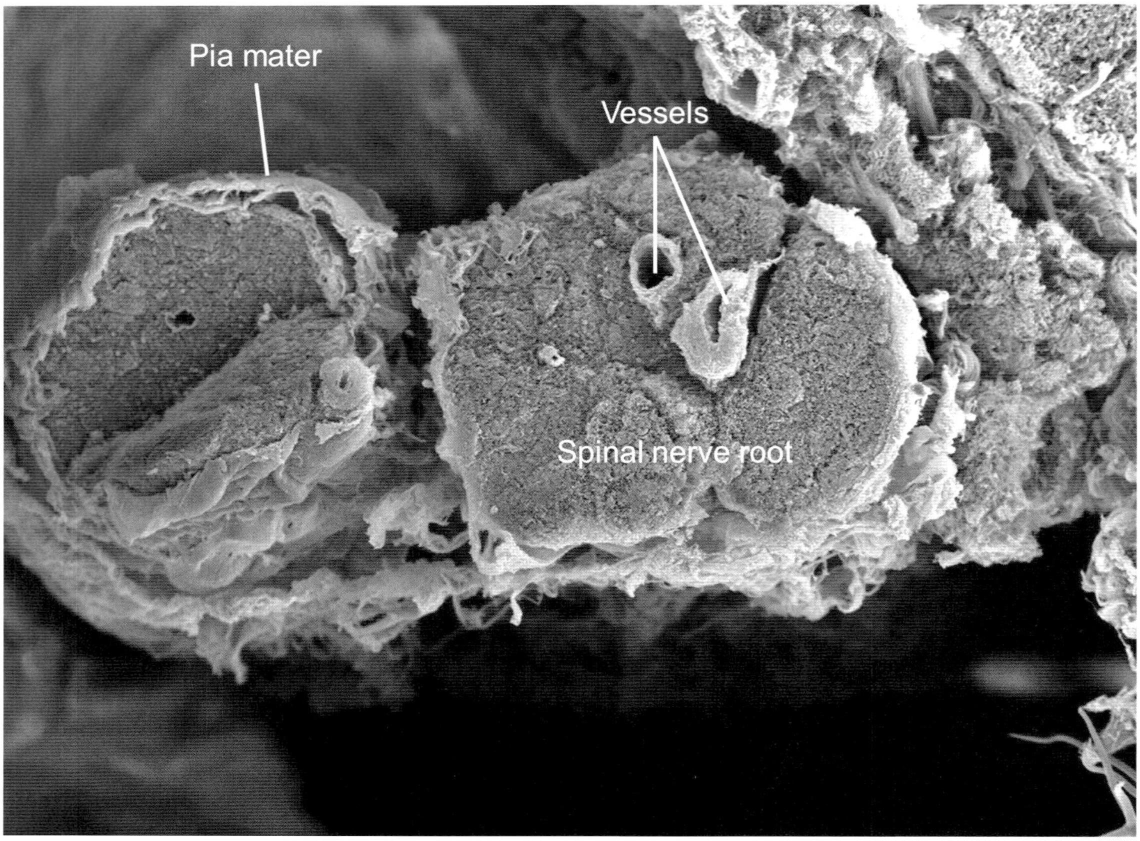

Fig. 28.22 Human spinal nerve root. Scanning electron microscopy; magnification: ×100

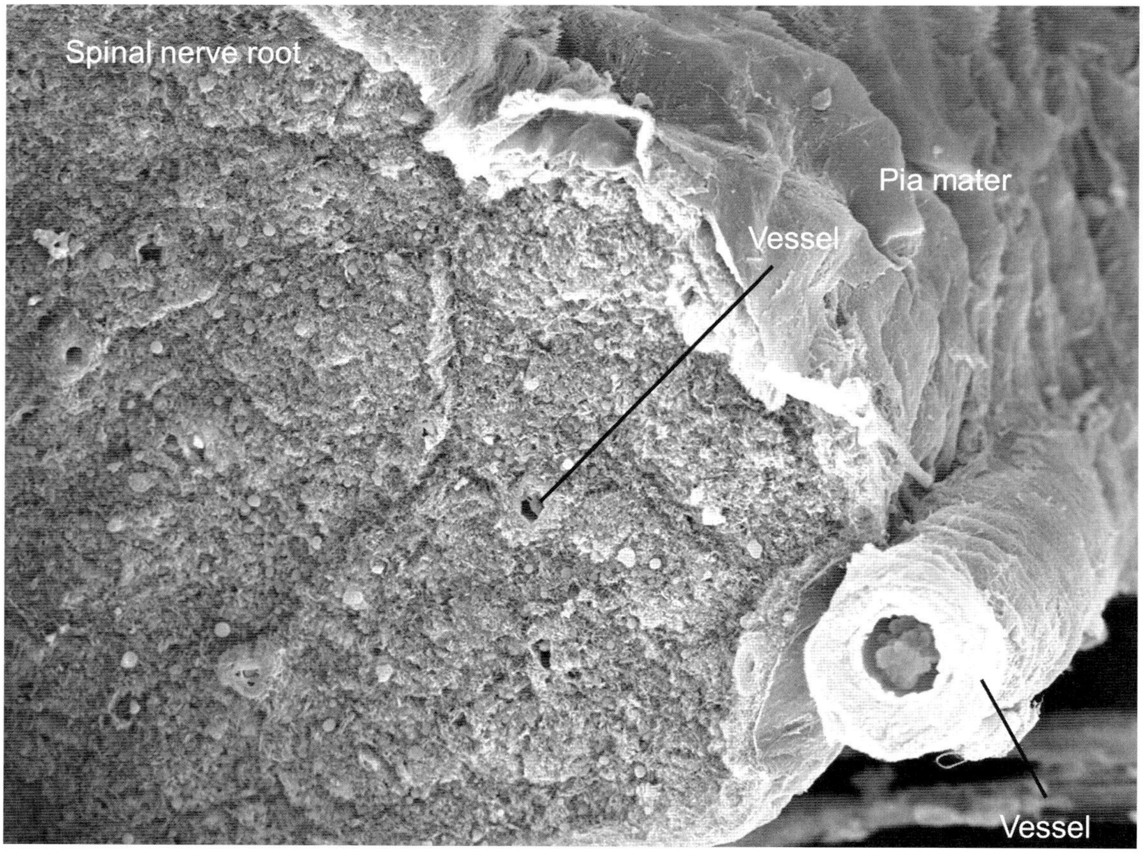

Fig. 28.23 Human spinal nerve root. Scanning electron microscopy; magnification: ×300

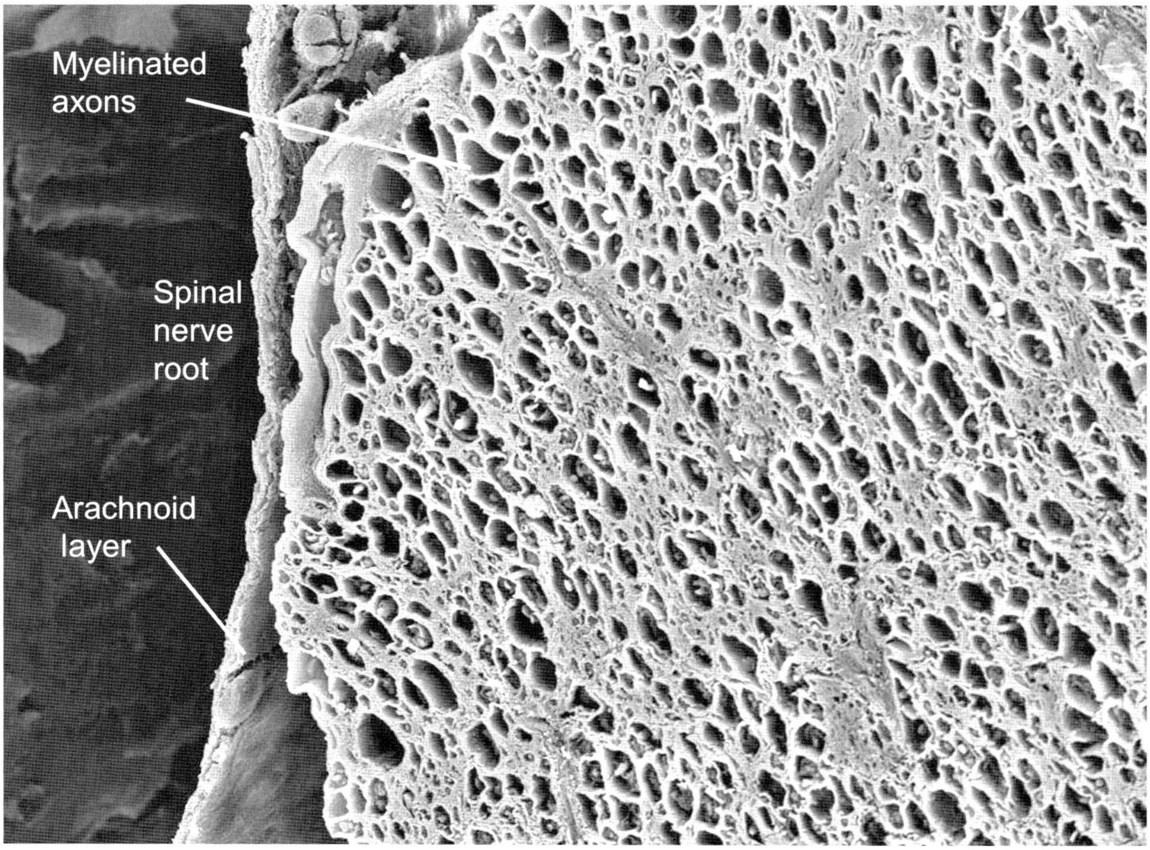

Fig. 28.24 Human spinal nerve root. Detail of spinal nerve root at higher magnification. Scanning electron microscopy; magnification: ×300

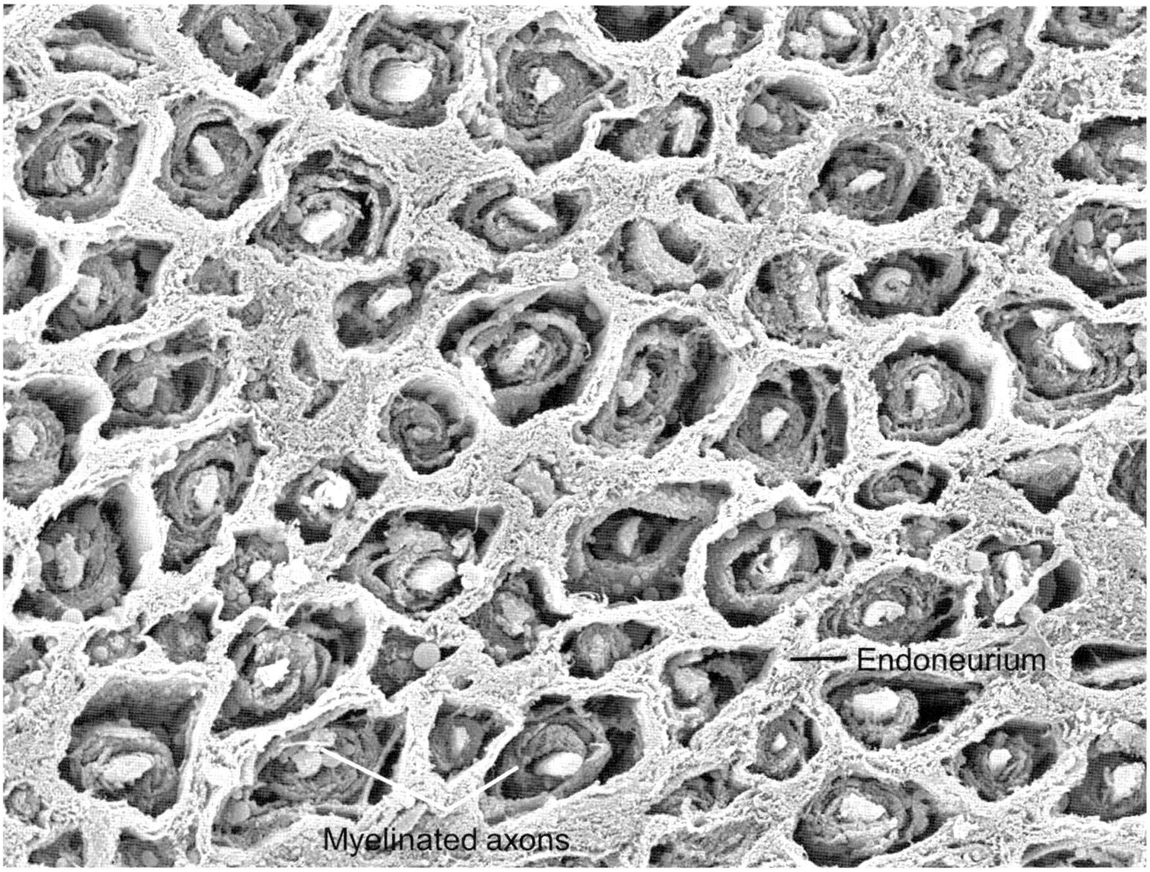

Fig. 28.25 Human spinal nerve root. Detail of spinal nerve root at higher magnification. Scanning electron microscopy; magnification: ×1,000

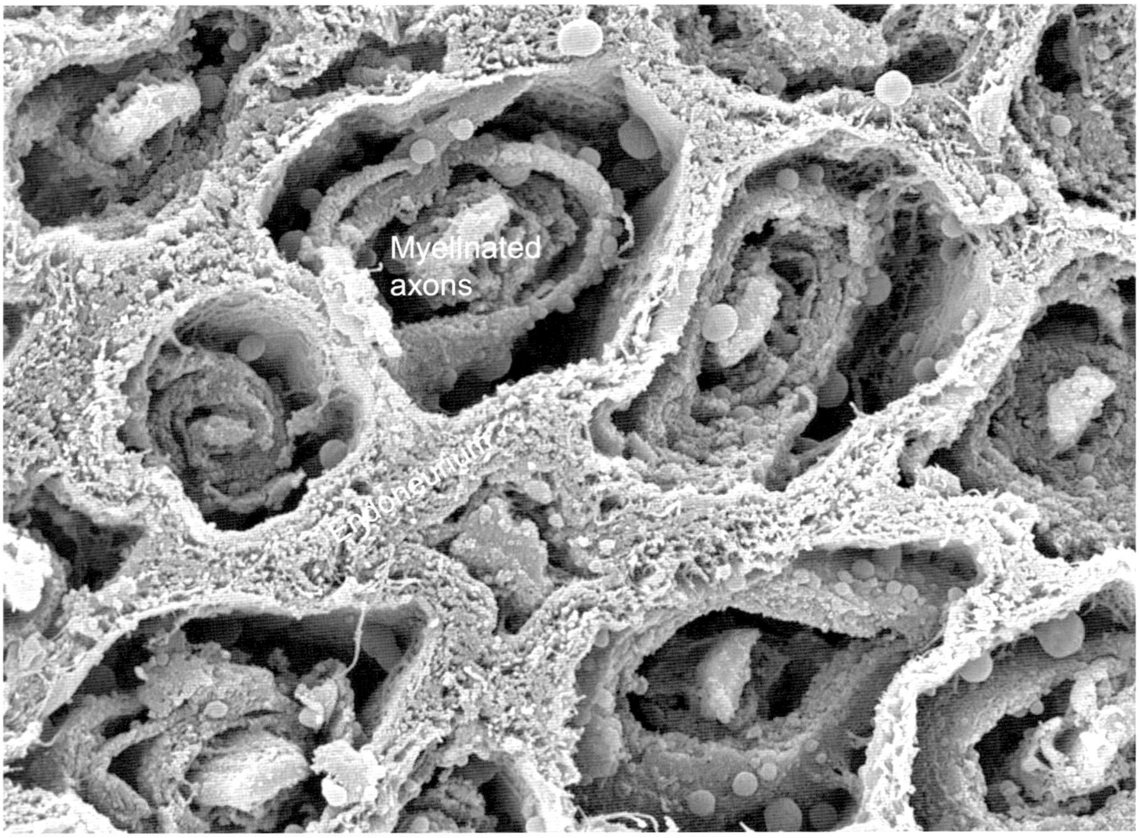

Fig. 28.26 Human spinal nerve root. Detail of spinal nerve root at higher magnification. Scanning electron microscopy; magnification: ×2,500

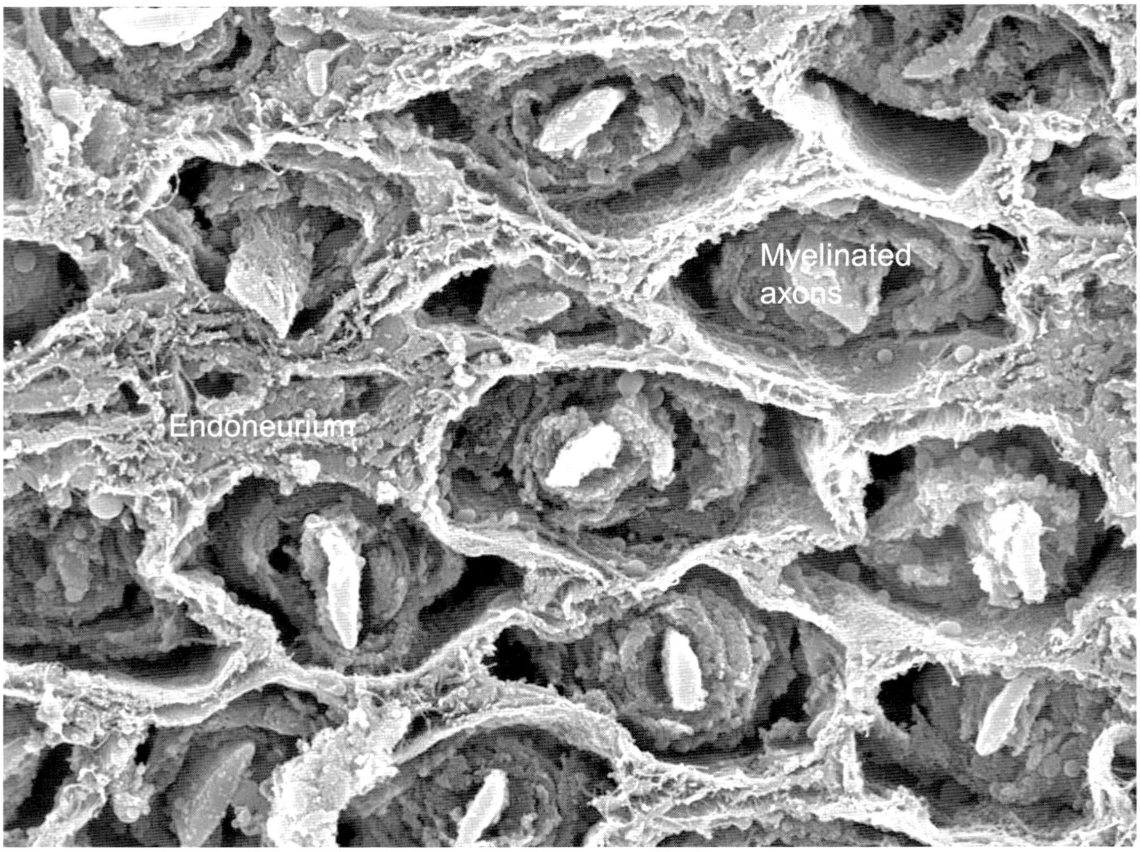

Fig. 28.27 Human spinal nerve root. Detail of spinal nerve root at higher magnification. Scanning electron microscopy; magnification: ×2,000

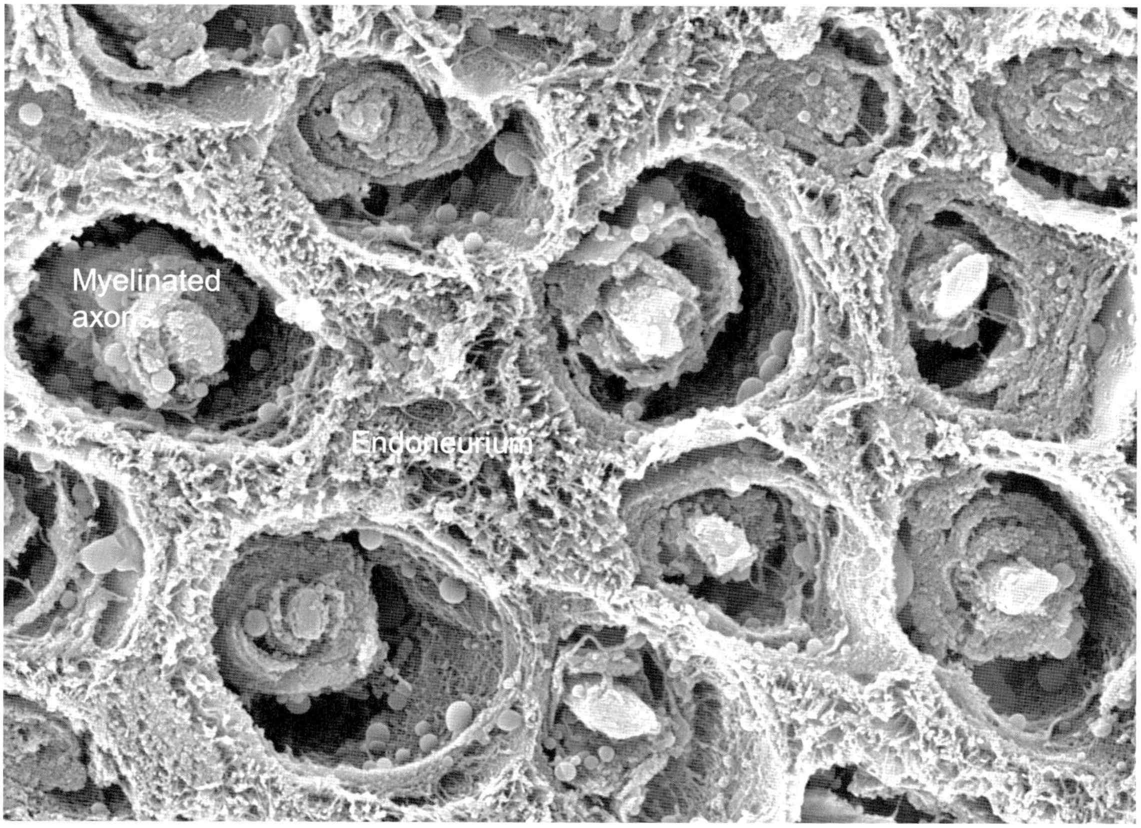

Fig. 28.28 Human spinal nerve root. Detail of spinal nerve root at higher magnification. Scanning electron microscopy; magnification: ×2,000 (From Reina et al. [1]; with permission)

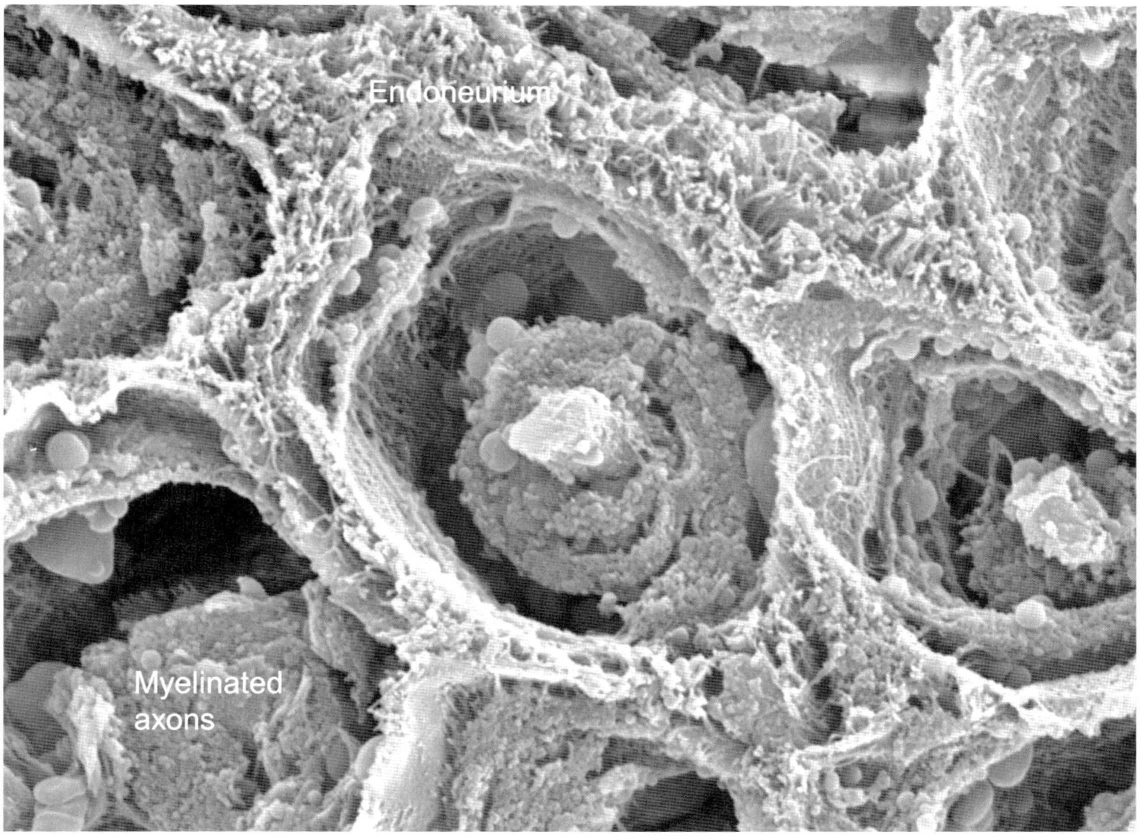

Fig. 28.29 Human spinal nerve root. Detail of spinal nerve root at higher magnification. Scanning electron microscopy; magnification: ×3,300

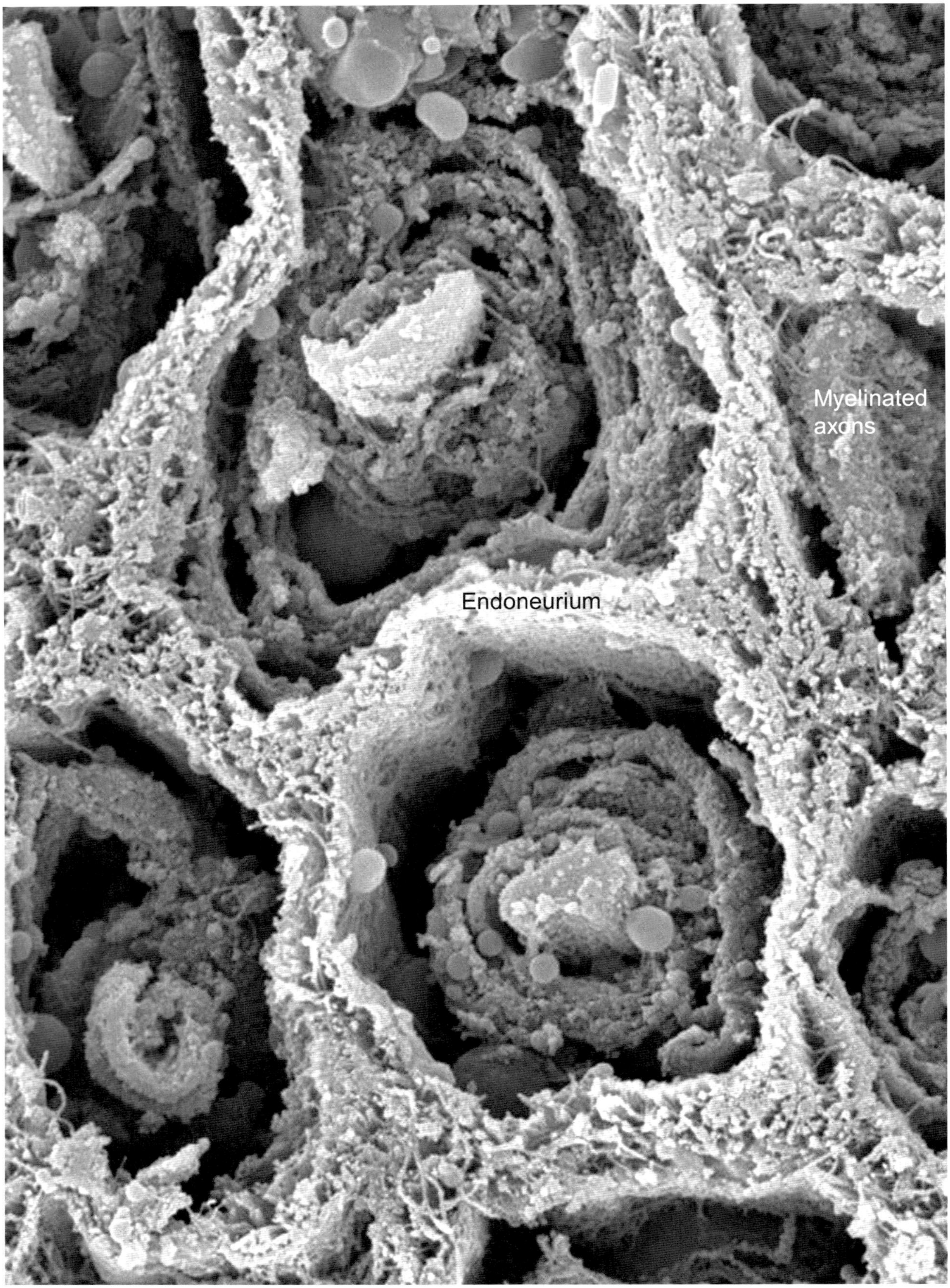

Fig. 28.30 Human spinal nerve root. Detail of spinal nerve root at higher magnification. Scanning electron microscopy; magnification: ×3,500

References

1. Reina MA, De Andres J, Hernández JM, Navarro RA, Pastor J, Prats-Galino A. Looking for the development of paresthesias in the subarachnoid and epidural anaesthesia. A clinical and anatomical analysis. Reg Anesth Pain Med. 2011;36(Suppl):E17–22.
2. Reina MA, De Andres JA, Hernández JM, Arriazu Navarro R, Durán Mateos EM, Prats-Galino A. Successive changes in extraneural structures from the subarachnoid nerve roots to the peripheral nerve, influencing anesthetic block, and treatment of acute postoperative pain. Eur J Pain Suppl. 2011;5:377–85.
3. Prats A, Carrera A, Oliva A, Lusá M, Reina MA. Anatomia aplicada a las técnicas neuroaxiales. In: Basora Macaya M, Colomina Soler MJ, editors. Anestesia en cirugía ortopédica y en traumatología. Madrid: Editorial Médica Panamericana; 2011. p. 33–45.
4. Reina MA, López A, Palácio F, Fornet I. Anestesia del neuroeje. In: Basora Macaya M, Colomina Soler MJ, editors. Anestesia en cirugía ortopédica y en traumatología. Madrid: Editorial Médica Panamericana; 2011. p. 46–58.
5. De Andres JA, Reina MA, Hernández JM, Carrera A, Oliva A, Prats-Galino A. Role of spinal anatomical structures for neuromodulation. Reg Anesth Pain Med Suppl. 2011;36:E130–8.

Three-Dimensional Reconstruction of Cauda Equina Nerve Roots

29

Alberto Prats-Galino, Joan San Molina, Miguel Angel Reina, and Juan A. Juanes Méndez

Nerve roots of the cauda equina adopt specific positions in relation to their neighboring nerve roots within the dural sac along their respective trajectories, and these tend to remain constant for each disposition of the vertebral column. Slight variations in the position of nerve roots within the cauda equina may be observed among patients. The processing of images of transversal sections of the dural sac and its contents in samples obtained from frozen bodies, as well as of magnetic resonance images from patients in supine position, enabled the production of maps called topograms of roots at each vertebral level along the lumbar vertebral column.

Nerve topograms aid in the understanding of the typical positions adopted by nerves inside the dural sac. They also illustrate how the tips of epidural and spinal needles are more likely to encounter different nerve roots during lumbar puncture, according to the vertebral level at which the lumbar puncture is performed [1].

Images in this chapter illustrate schematically various possible positions adopted by the roots as seen on examination of images from samples obtained from several bodies. In addition, there are images obtained from a three-dimensional (3D) model reconstruction [2].

At lumbar level, the nerve roots have a certain degree of mobility within the dural sac, although nerve displacements are restricted by the arachnoid trabecular ligament. At the thoracic level, various ligaments such as the posterior and posterolateral ligaments help to limit the displacement of the spinal cord (Figs. 29.1, 29.2, 29.3, 29.4, 29.5, 29.6, 29.7, 29.8, and 29.9).

A. Prats-Galino, MD, PhD (✉)
Laboratory of Surgical NeuroAnatomy (LSNA), Human Anatomy and Embryology Unit, School of Medicine, University of Barcelona, Barcelona, Spain
e-mail: aprats@ub.edu

J. San Molina, MD, PhD
Neuroscience, Embryology, Molecular Oncology, and Clinical Anatomy Group, School of Medicine, University of Girona, Girona, Spain
e-mail: joan.san@udg.edu

M.A. Reina, MD, PhD
Department of Clinical Medical Sciences and Institute of Applied Molecular Medicine, School of Medicine, University of CEU San Pablo, Madrid, Spain

Department of Anesthesiology, Madrid-Montepríncipe University Hospital, Madrid, Spain
e-mail: miguelangel@perticone.e.telefonica.net

J.A. Juanes Méndez, MD, PhD
Human Anatomy and Embryology Unit, School of Medicine, University of Salamanca, Salamanca, Spain
e-mail: jajm@usal.es

M.A. Reina et al. (eds.), *Atlas of Functional Anatomy for Regional Anesthesia and Pain Medicine: Human Structure, Ultrastructure and 3D Reconstruction Images*, DOI 10.1007/978-3-319-09522-6_29,

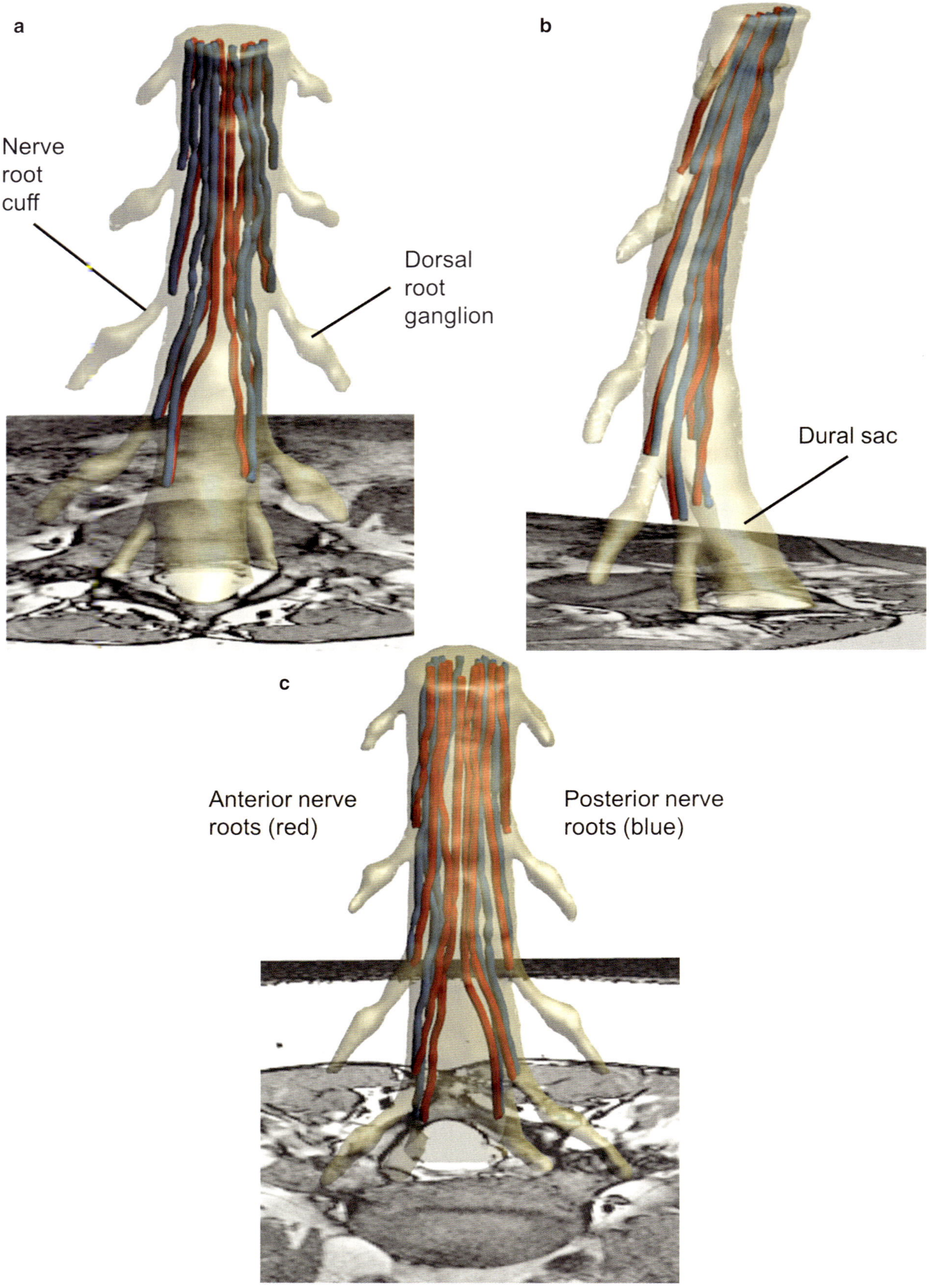

Fig. 29.1 Anterior (motor) nerve roots (*red*) and posterior (sensory) nerve roots (*blue*), as seen in posterior view (**a**), lateral view (**b**), and anterior view (**c**). Nerve roots within the dural sac are identified

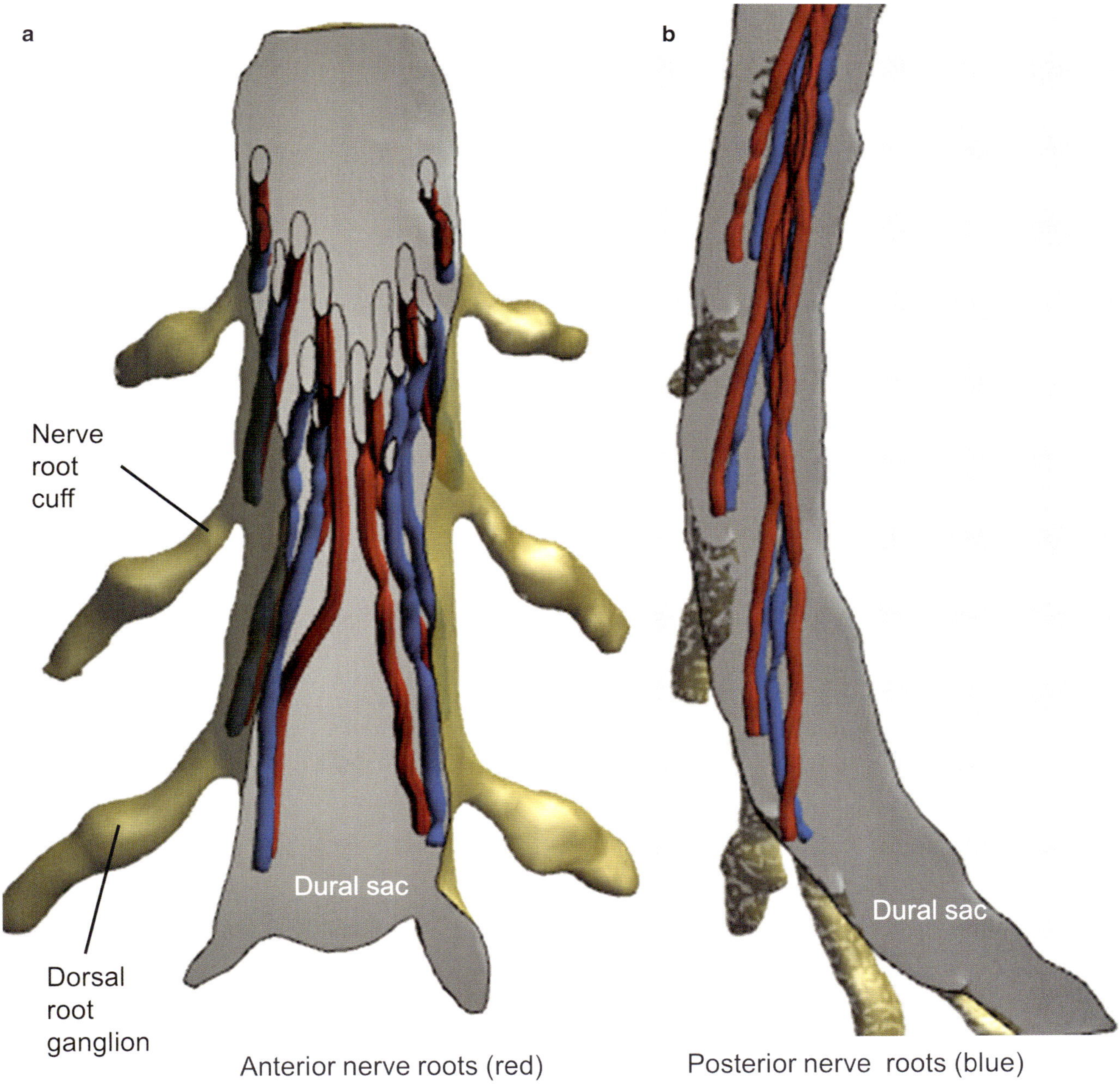

Fig. 29.2 Anterior nerve roots (*red*) and posterior nerve roots (*blue*) within the dural sac, sectioned longitudinally, as seen in posterior view (**a**) and lateral view (**b**)

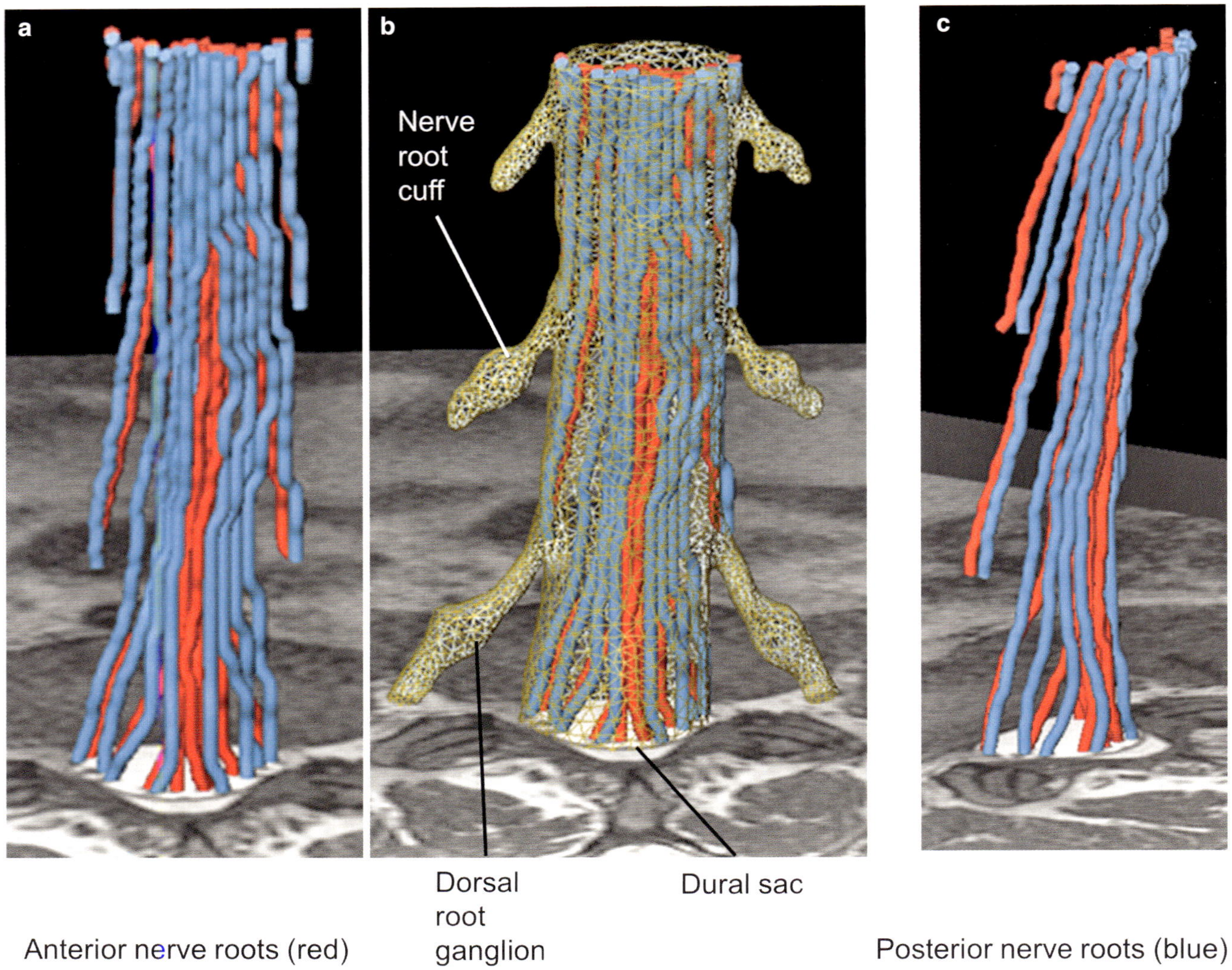

Fig. 29.3 Three-dimensional (3D) reconstruction of anterior roots (*red*) and posterior roots (*blue*). (**a**) Posterior view without dural sac. (**b**) Posterior view with dural sac. (**c**) Lateral view. (**a**, **b**) demonstrate that not all dorsal roots may be located specifically behind the anterior nerve roots; as in the central region of the cauda equina, few anterior nerve roots are identified (*red*)

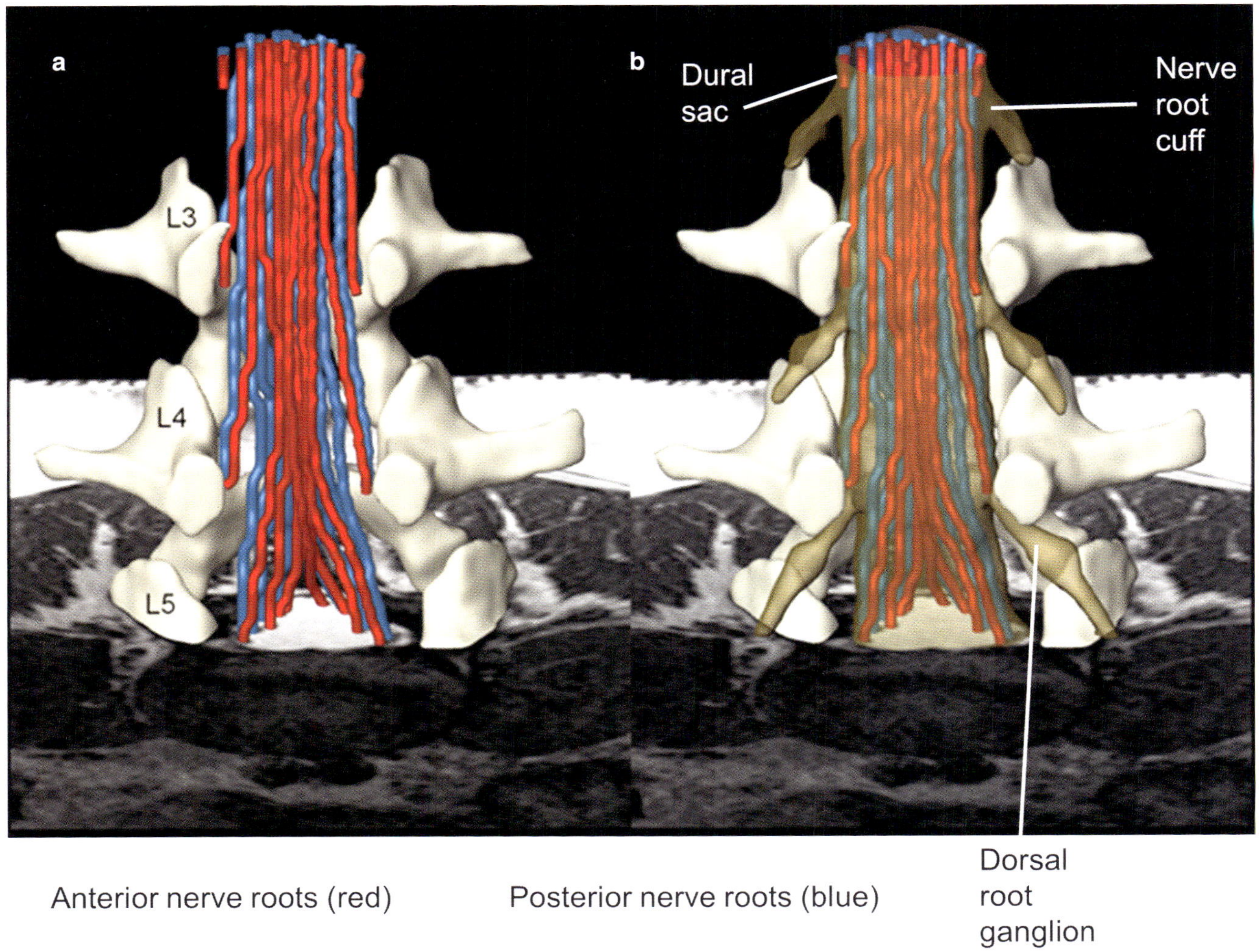

Fig. 29.4 Anterior view of the distribution of dorsal roots (*red*) and posterior roots (*blue*). (**a**) Nerve roots without dural sac. (**b**) Nerve roots with dural sac

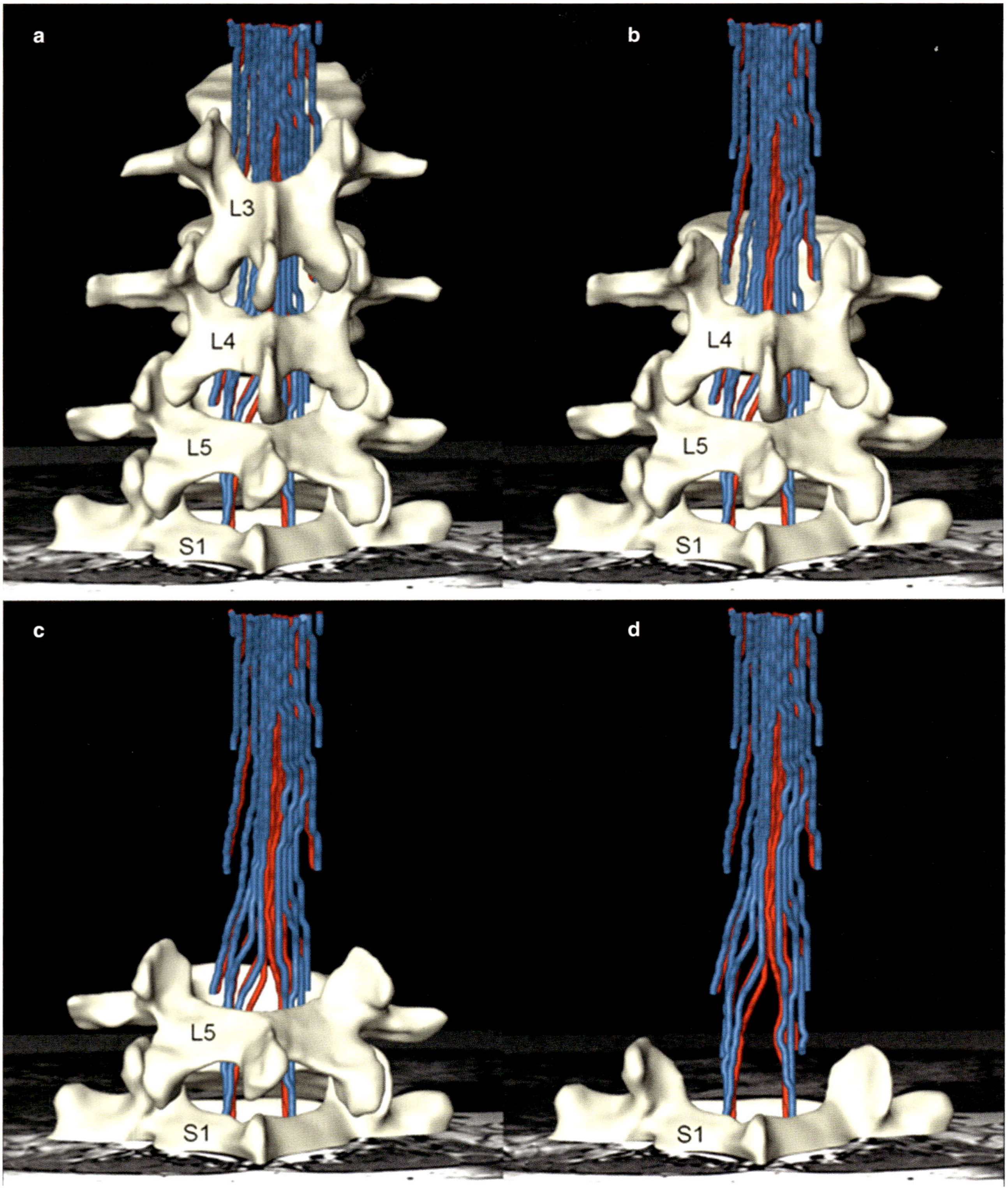

Fig. 29.5 (**a–d**) Nerve roots within the spinal canal, without the dural sac. The lumbar vertebra has been removed to allow the position of the nerve roots at the level of the interlaminar foramina to be viewed in greater detail

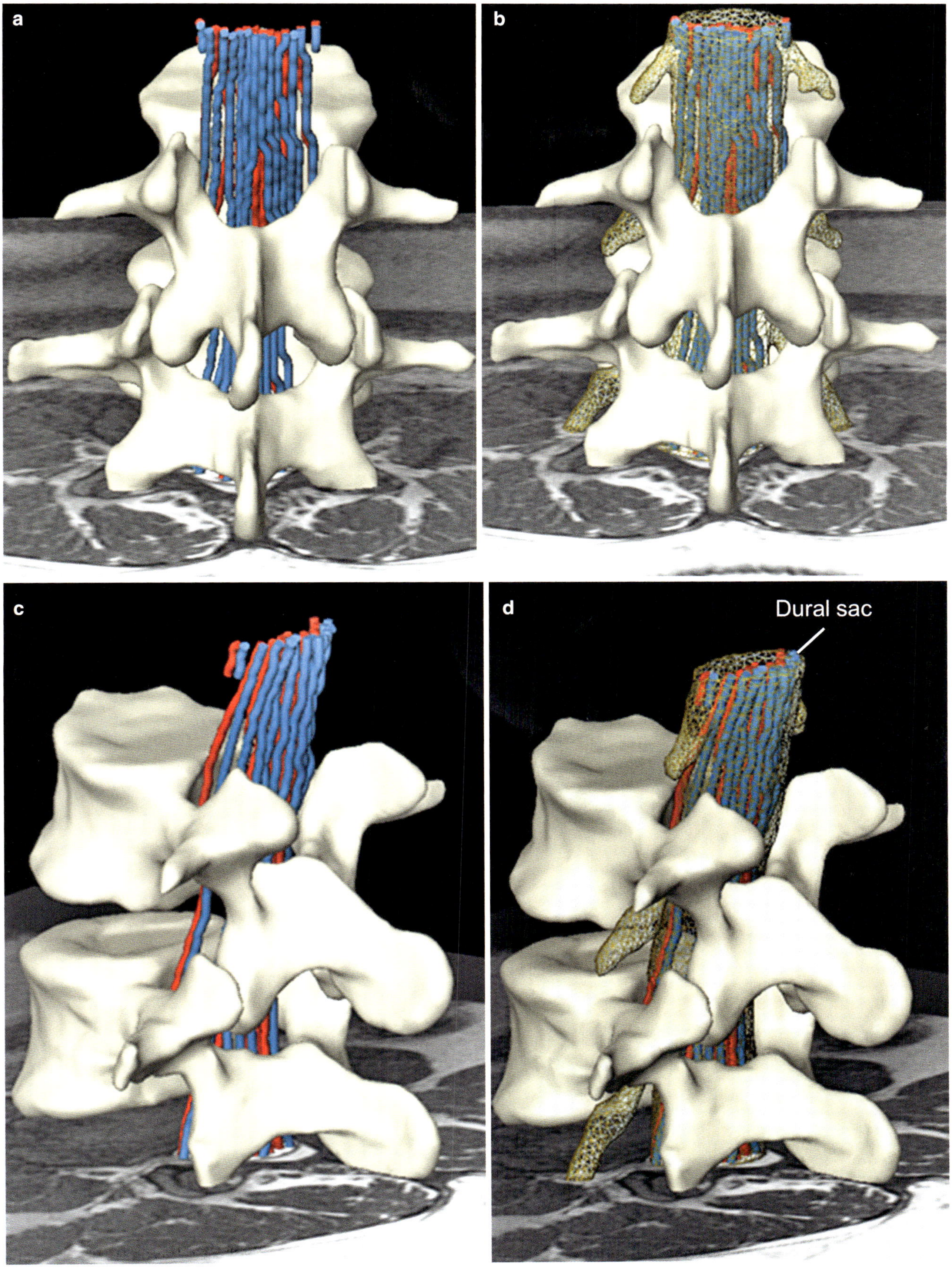

Fig. 29.6 Anterior and posterior nerve roots. (**a**) Dorsal view of the roots without the dural sac. (**b**) Dorsal view with the dural sac. (**c**) Lateral view without the dural sac. (**d**) Lateral view with the dural sac

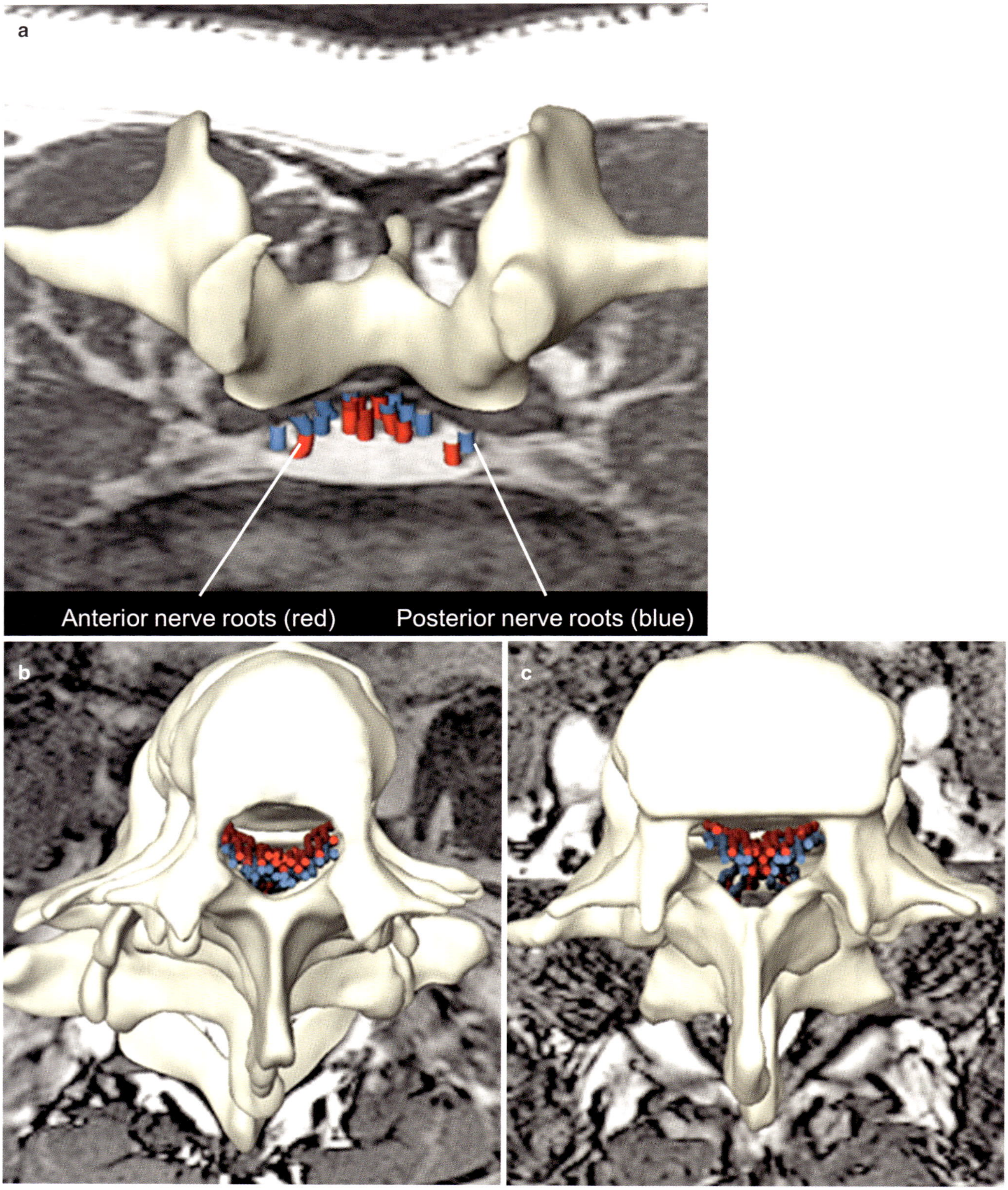

Fig. 29.7 (**a**–**c**) Cross-section images of anterior nerve roots (*red*) and posterior nerve roots (*blue*) at different vertebral levels

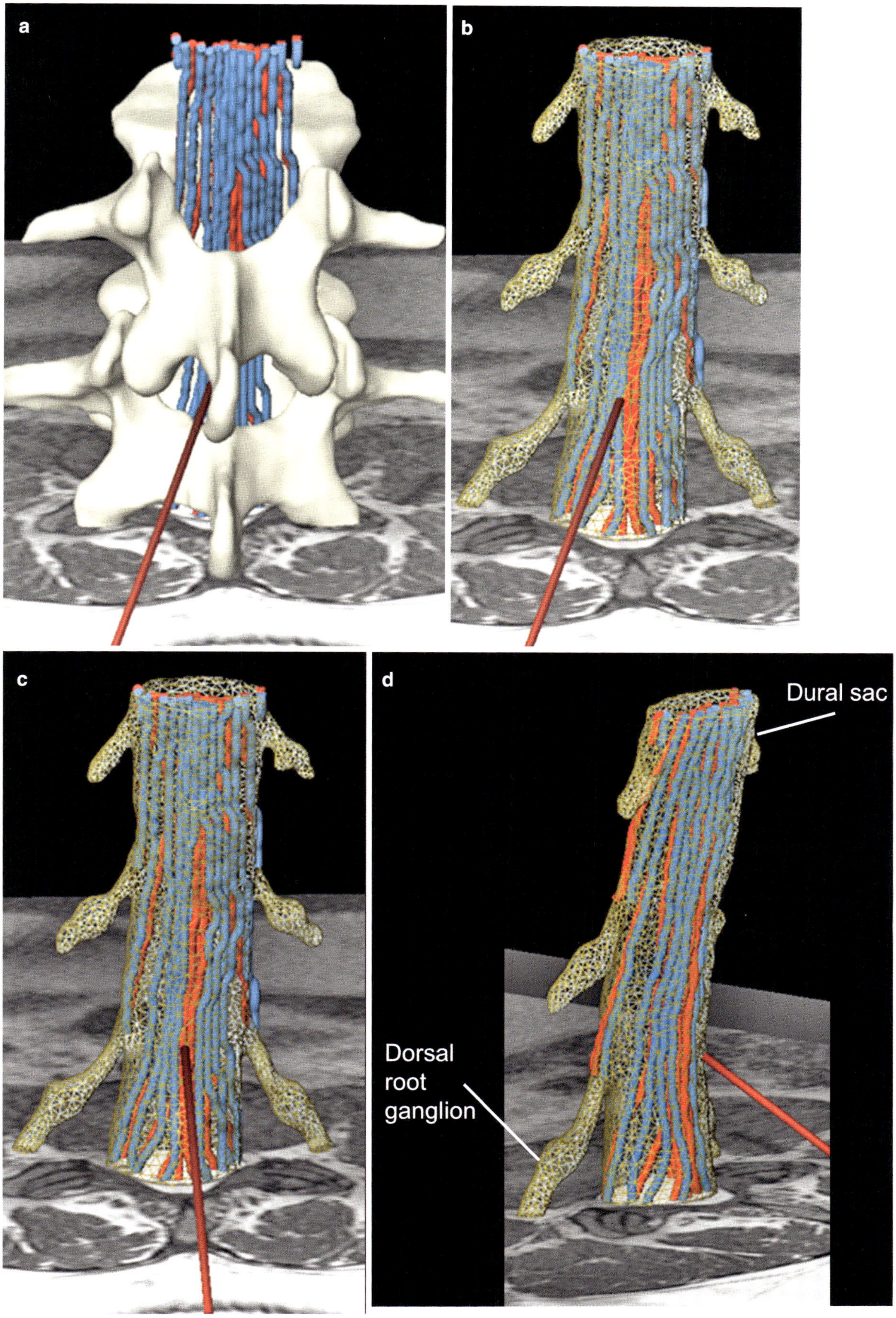

Fig. 29.8 Simulation of lumbar puncture, illustrating the likelihood of causing nerve injury with the tip of the needle at the level of anterior and posterior nerve roots. (**a**, **b**) Lumbar puncture approach along an oblique plane (paramedian approach). (**c**, **d**) Lumbar puncture approach along a mid-sagittal plane (medial approach)

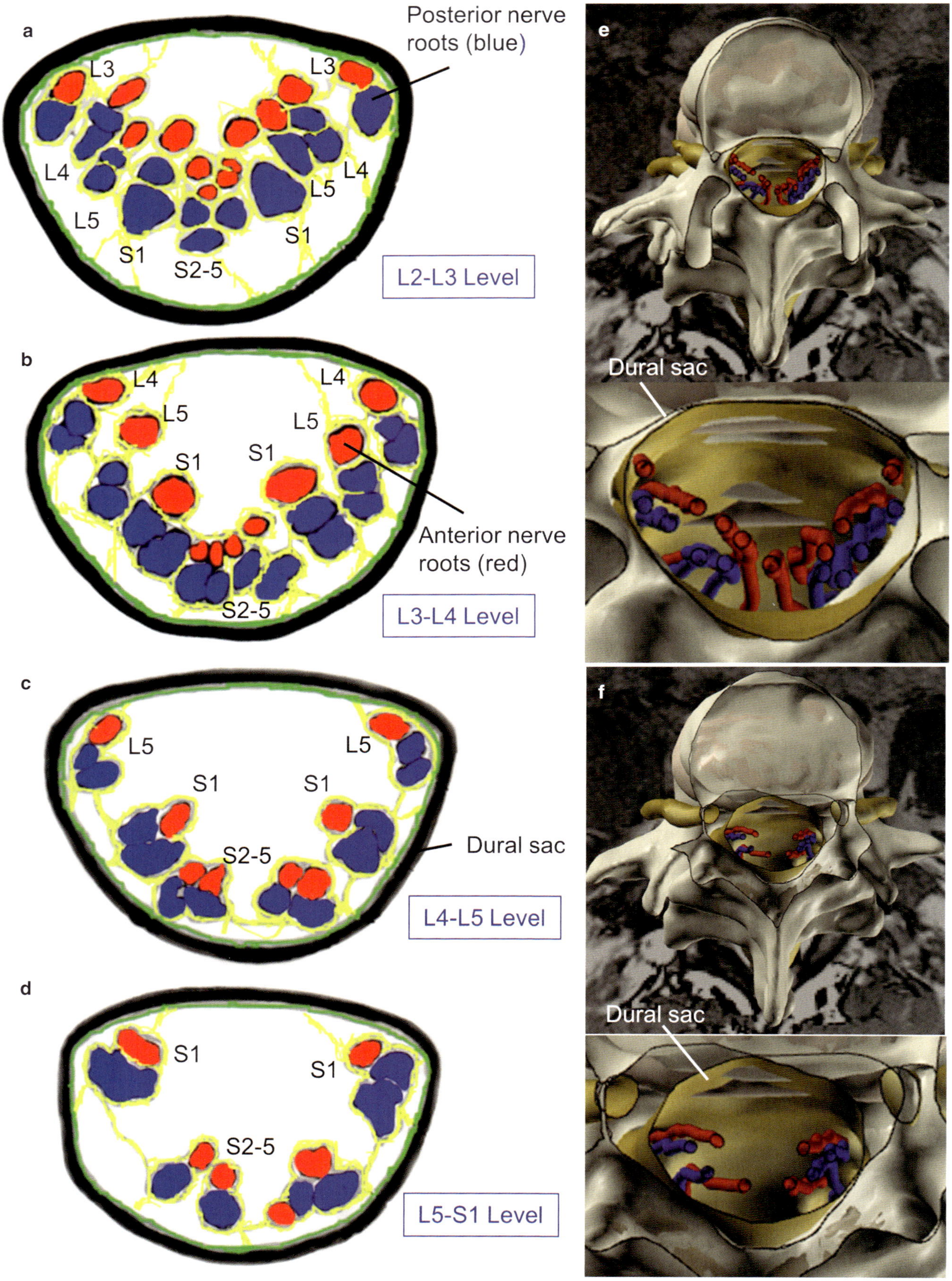

Fig. 29.9 Topograms of nerve roots at different intervertebral levels: (**a**) L2-L3; (**b**) L3-L4; (**c**) L4-L5; (**d**) L5-S1. (**e**) Section of 3D model reconstruction obtained from a patient at lumbar region (**e**) and sacral region (**f**)

References

1. Reina MA, De Andrés JA, Hernández JM, Arriazu Navarro R, Pastor J, Prats-Galino A. Looking for the development of paresthesias in the subarachnoid and epidural anesthesia. A clinical and anatomical analysis. Reg Anesth Pain Med Suppl. 2011;36:E17–22.
2. Prats-Galino A, Mavar M, Reina MA, Puigdellívol-Sánchez A, San-Molina J, De Andrés JA. Three-dimensional interactive model of lumbar spinal structures. Anaesthesia. 2014;69:521.

30 Spinal Nerve Root Lesions After "In Vitro" Needle Puncture

Miguel Angel Reina, Fabiola Machés, and Andrés López

Lumbar punctures often are associated with unintentional paresthesias. The tip of the needle in contact with the spinal nerve root initiates depolarization of the axons, which the patient describes as electric shocks in the dermatomes innervated. The paresthesia is triggered by a mechanical stimulus along the axon [1–3].

Most paresthesias are caused by puncturing of the cauda equina nerves rather than by inadvertent injection within the conus medullaris [4].

Paresthesias due to a traumatic lesion of the nerve have different consequences depending on the size of needle, the type of needle tip, and the depth of the lesion in the nerve root. The damage probably is greater when the roots are exposed to stretching or if the needle is introduced with excessive force. Because direct vision of structures within the spinal canal is not feasible, precise evaluation of the degree of damage to the nerves presently relies solely on clinical evaluation. In theory and based on "in vitro" studies, nerve penetration is possible with 25-gauge or even smaller needles [1, 2].

Nerve roots are located around the posterior area less frequently [5, 6]. This anatomic consideration might be of interest during lumbar puncture procedures, because the spinal needle may advance a few millimeters inside the dural sac without increasing the risk of nerve root damage and paresthesias.

During lumbar puncture, the needle enters the dural sac and progresses a few millimeters before reaching the surface of the nerve roots. As the needle advances, it may contact the nerves located toward the posterior aspect of the dural sac, which would correspond to centrally located nerves if the needle followed a sagittal plane. Otherwise, as the needle is diverted from the sagittal plane, it might pierce nerves lateral to the spinal cord. In patients without spine pathology, paresthesias may be caused by inadequate positioning of the patient during lumbar puncture, which often is associated with inadvertent rotation of the spinal column.

After nerve root damage, it is difficult to determine the number of axons affected as well as the mechanisms of injury, namely compression or sectioning of the nerves, the latter of which is less frequent. Because there are vessels (small capillaries, arteries, or veins) inside as well as on the surface of nerve roots, the likelihood of intraneural hematomas should be considered. The process of reabsorption, inflammatory cascade, and repair of nerve lesions may lead to local fibrosis and, therefore, chronic alteration of conduction in some axons.

Besides paresthesias, other alterations are possible during inadvertent intraneural drug injection, such as transient radicular irritation syndrome, if the tip of the needle is not withdrawn 2–3 mm inside the arachnoid sheath after a paresthesia has been produced. In Chap. 24, the arachnoid sheath around nerve roots is explained in detail (Figs. 30.1, 30.2, 30.3, 30.4, 30.5, 30.6, 30.7, 30.8, 30.9, 30.10, 30.11, 30.12, 30.13, 30.14, 30.15, 30.16, 30.17, 30.18, 30.19, 30.20, 30.21, 30.22, 30.23, 30.24, 30.25, 30.26, 30.27, 30.28, 30.29, 30.30, 30.31, 30.32, 30.33, 30.34, 30.35, 30.36, and 30.37) [7].

M.A. Reina, MD, PhD (✉)
Department of Clinical Medical Sciences and Institute of Applied Molecular Medicine, School of Medicine, University of CEU San Pablo, Madrid, Spain

Department of Anesthesiology, Madrid-Montepríncipe University Hospital, Madrid, Spain
e-mail: miguelangel@perticone.e.telefonica.net

F. Machés, MD
Department of Anesthesiology, Madrid-Montepríncipe University Hospital, Madrid, Spain
e-mail: fabmami@gmail.com

A. López, MD
Department of Clinical Medical Sciences, School of Medicine, CEU San Pablo University, Madrid, Spain

Department of Anesthesiology, Madrid-Montepríncipe University Hospital, Madrid-Torrelodones University Hospital, Madrid-Sanchinarro University Hospital, Madrid, Spain
e-mail: alg00009@telefonica.net

M.A. Reina et al. (eds.), *Atlas of Functional Anatomy for Regional Anesthesia and Pain Medicine: Human Structure, Ultrastructure and 3D Reconstruction Images*, DOI 10.1007/978-3-319-09522-6_30,

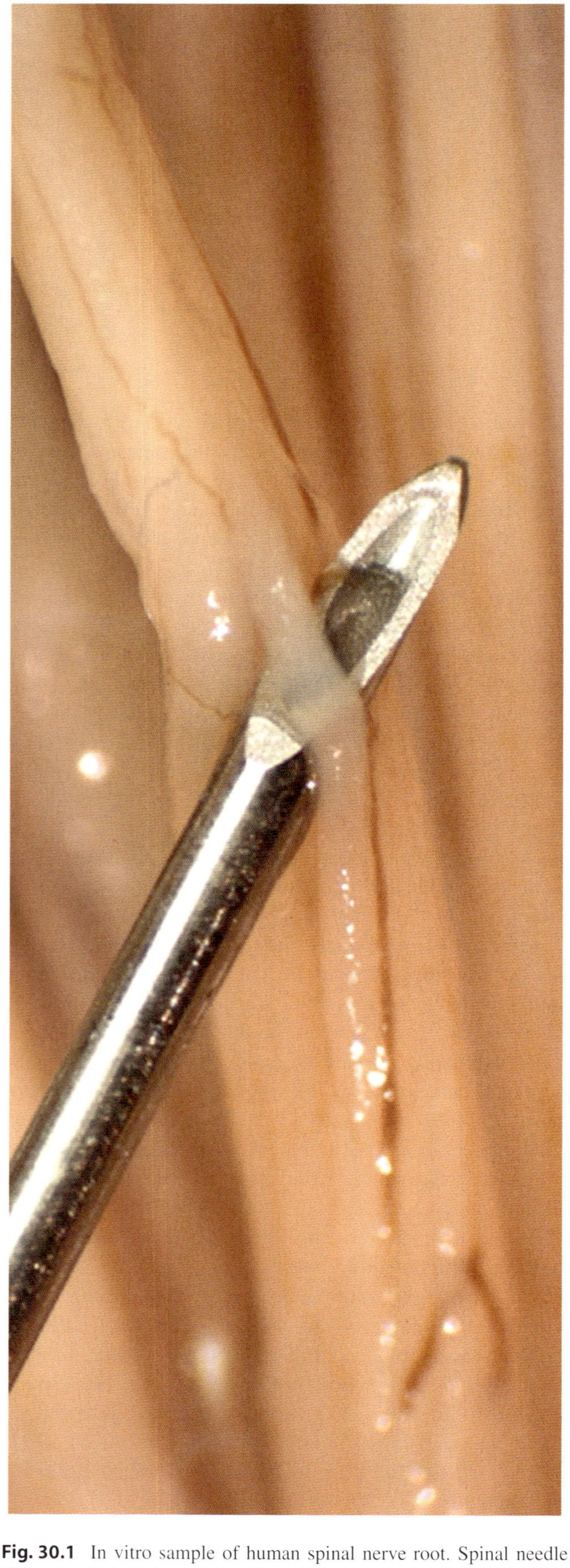

Fig. 30.1 In vitro sample of human spinal nerve root. Spinal needle (22-gauge Quincke) traversing the spinal nerve root

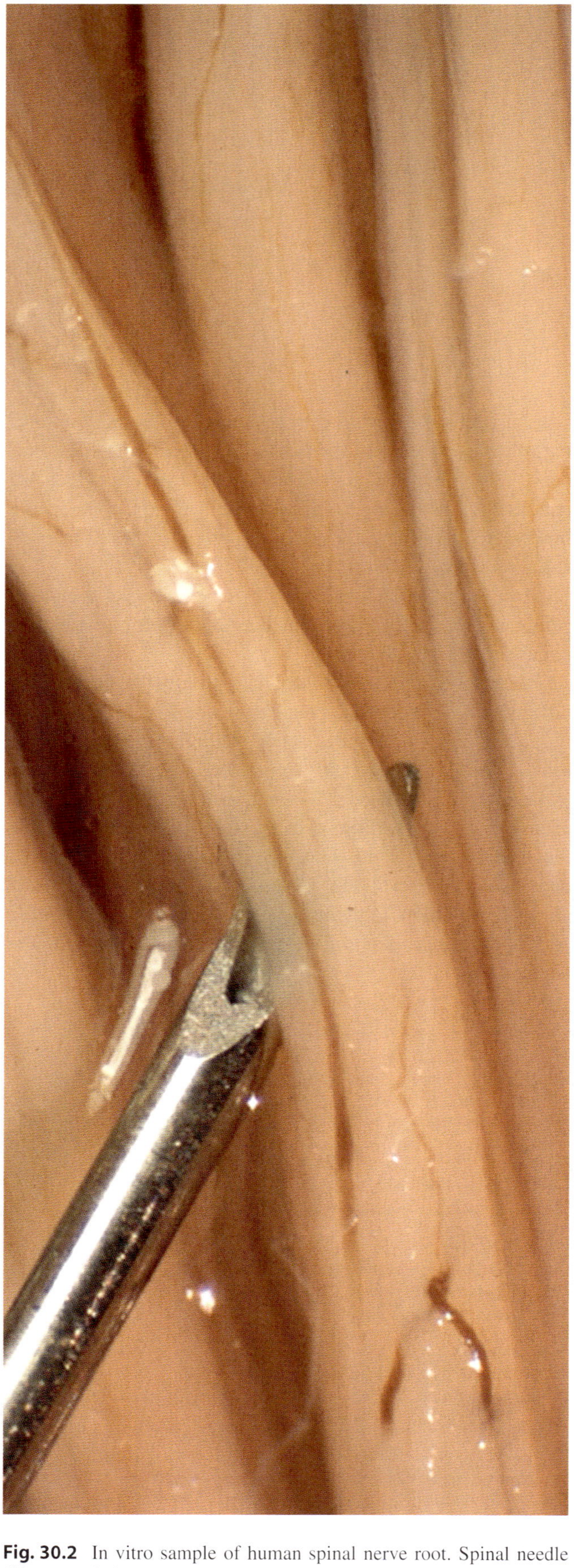

Fig. 30.2 In vitro sample of human spinal nerve root. Spinal needle (22-gauge Quincke) traversing the spinal nerve root

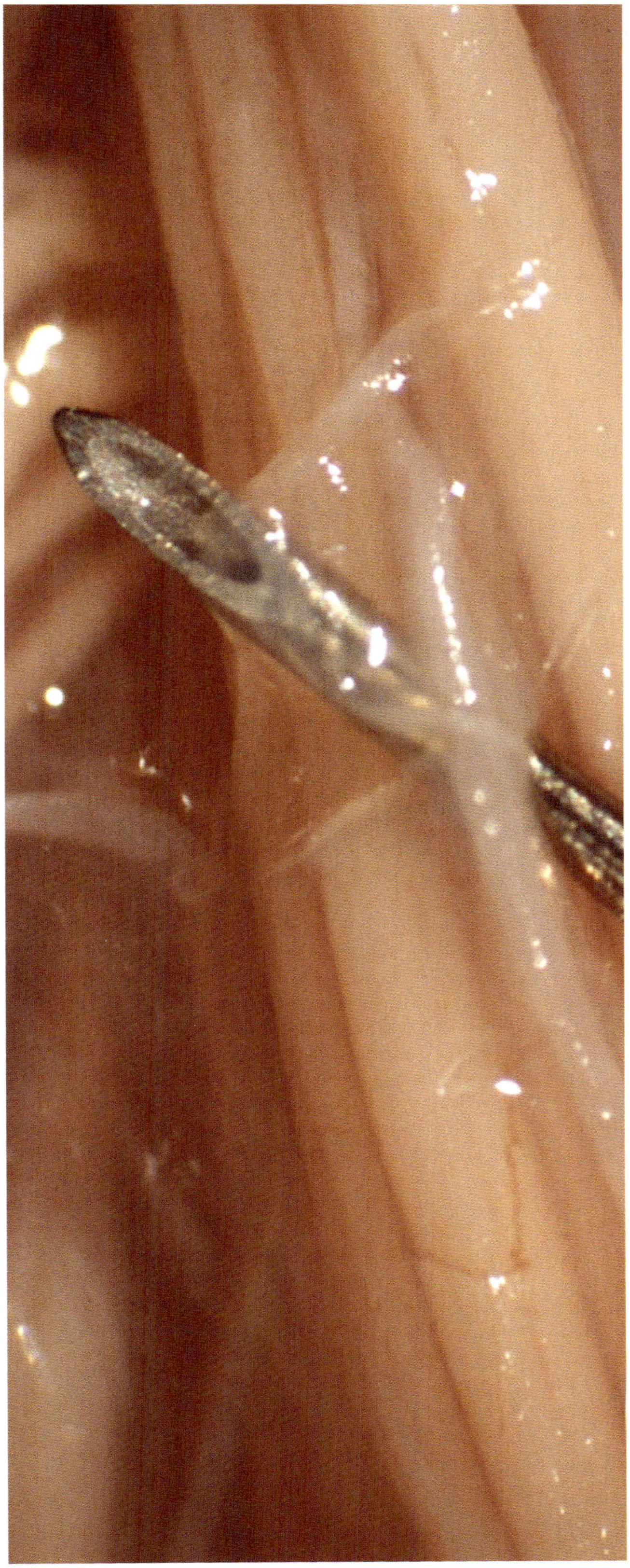

Fig. 30.3 In vitro sample of human spinal nerve root. Spinal needle (22-gauge Quincke) traversing the spinal nerve root

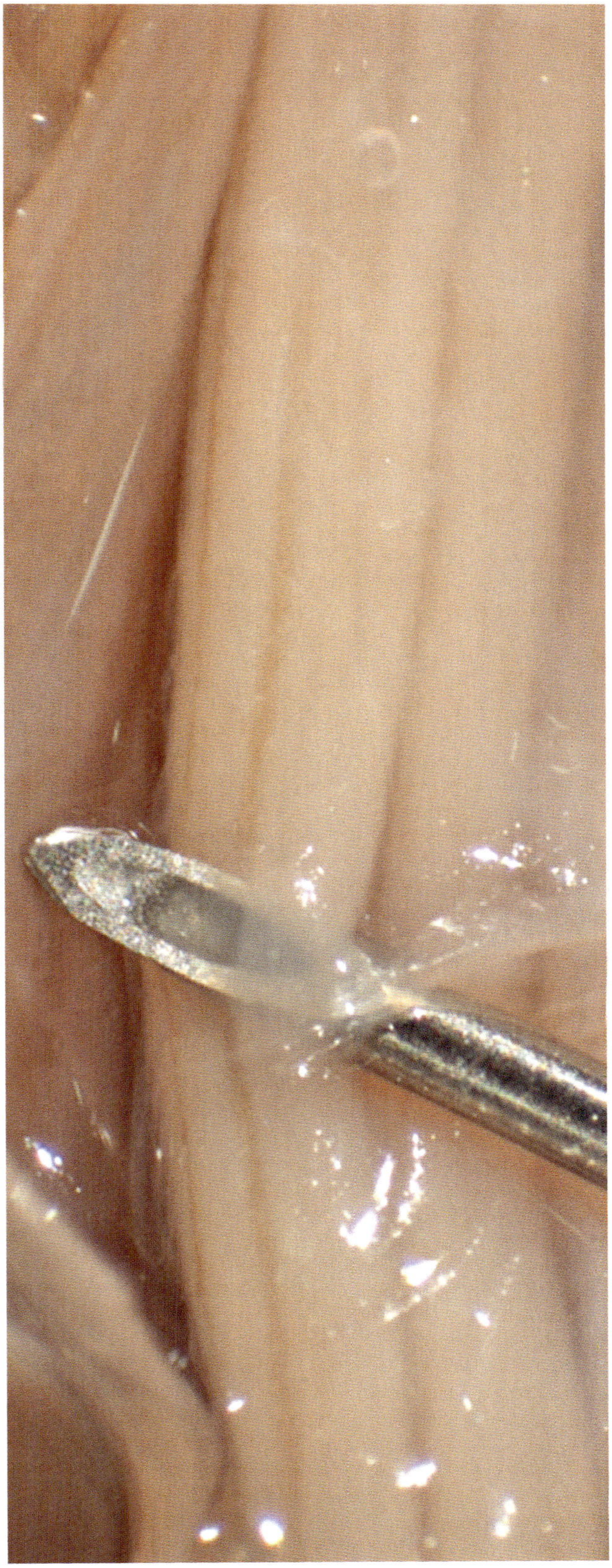

Fig. 30.4 In vitro sample of human spinal nerve root. Spinal needle (22-gauge Quincke) traversing the spinal nerve root

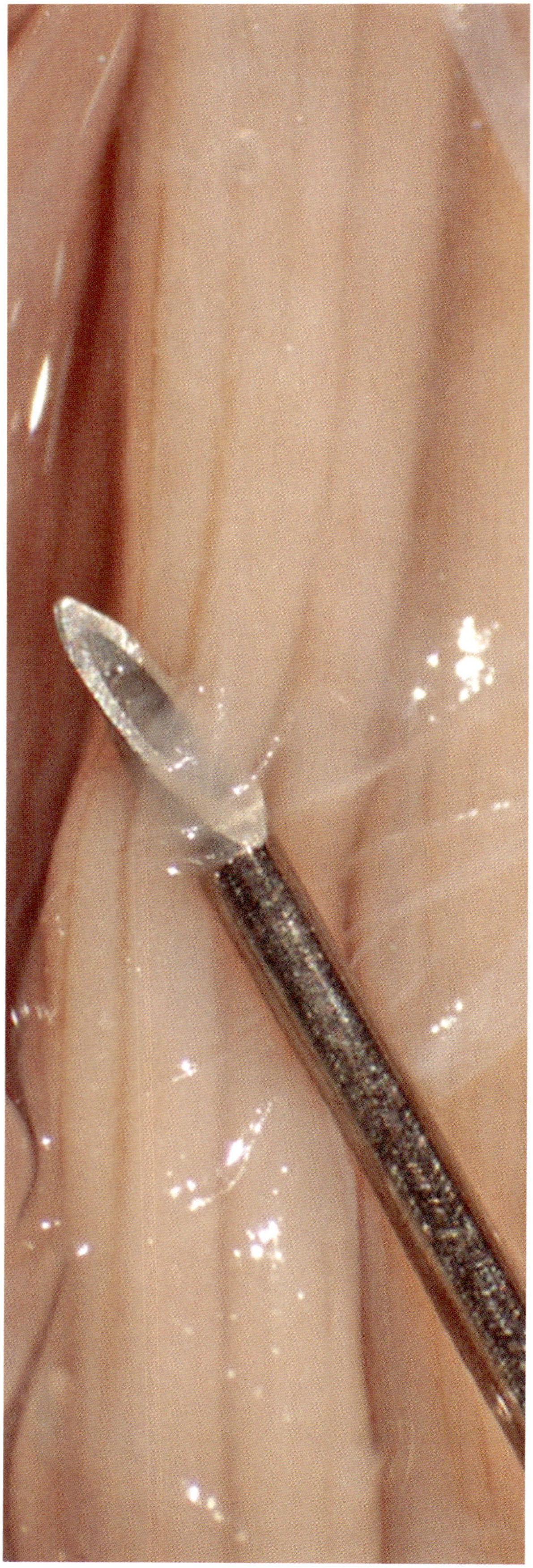

Fig. 30.5 In vitro sample of human spinal nerve root. Spinal needle (22-gauge Quincke) traversing the spinal nerve root

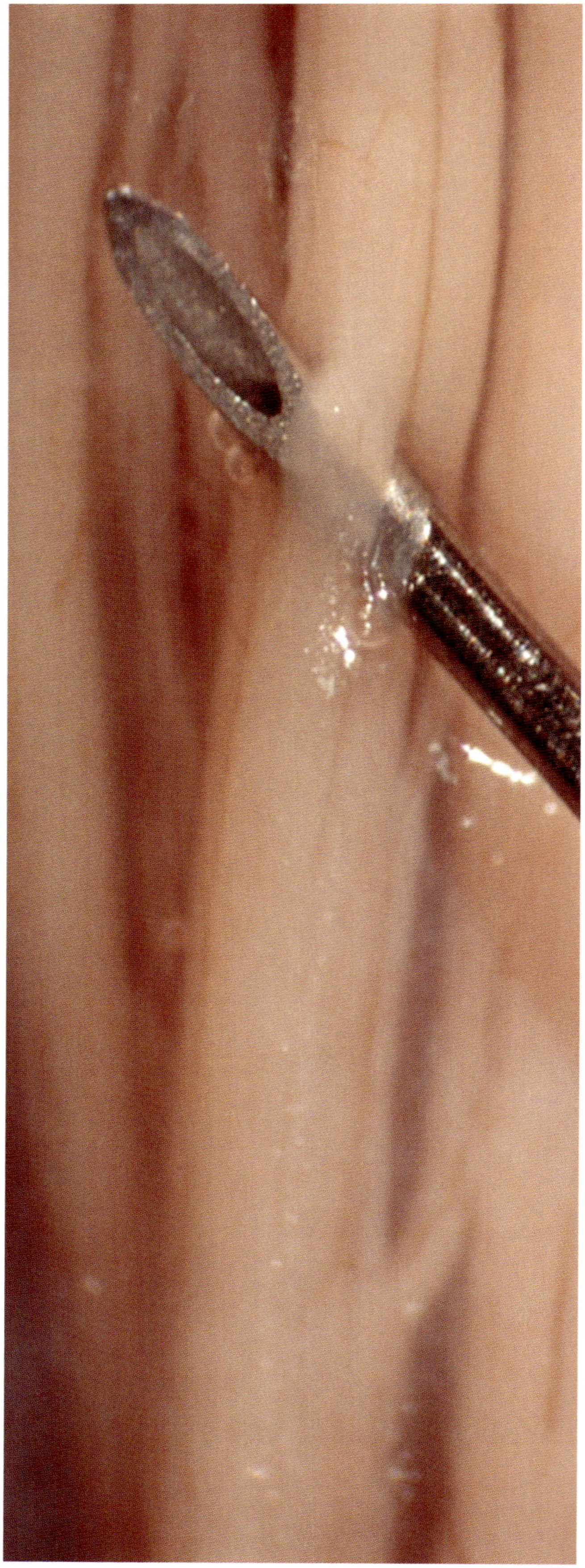

Fig. 30.6 In vitro sample of human spinal nerve root. Spinal needle (22-gauge Quincke) traversing the spinal nerve root

Fig. 30.7 In vitro sample of human spinal nerve root. Spinal needle (22-gauge Quincke) traversing the spinal nerve root

Fig. 30.8 In vitro sample of human spinal nerve root. Spinal needle (22-gauge Quincke) traversing the spinal nerve root

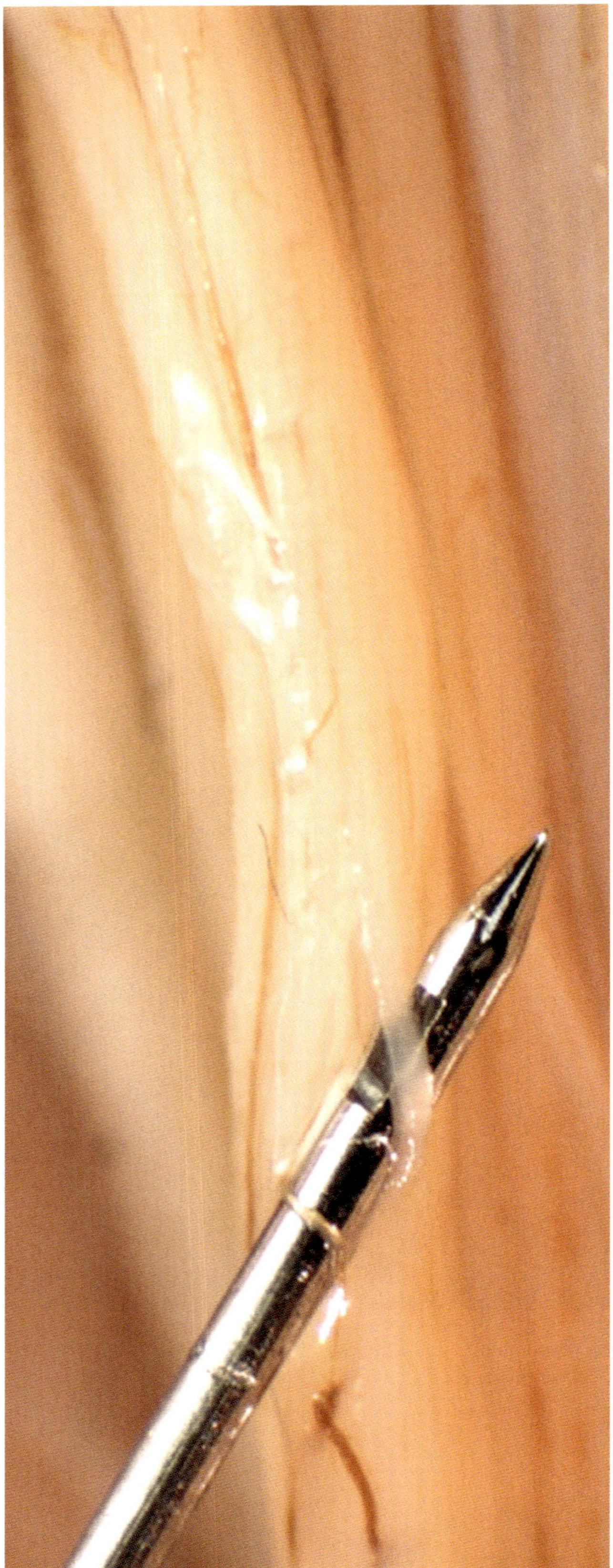

Fig. 30.9 In vitro sample of human spinal nerve root. Spinal needle (27-gauge Whitacre) traversing the spinal nerve root

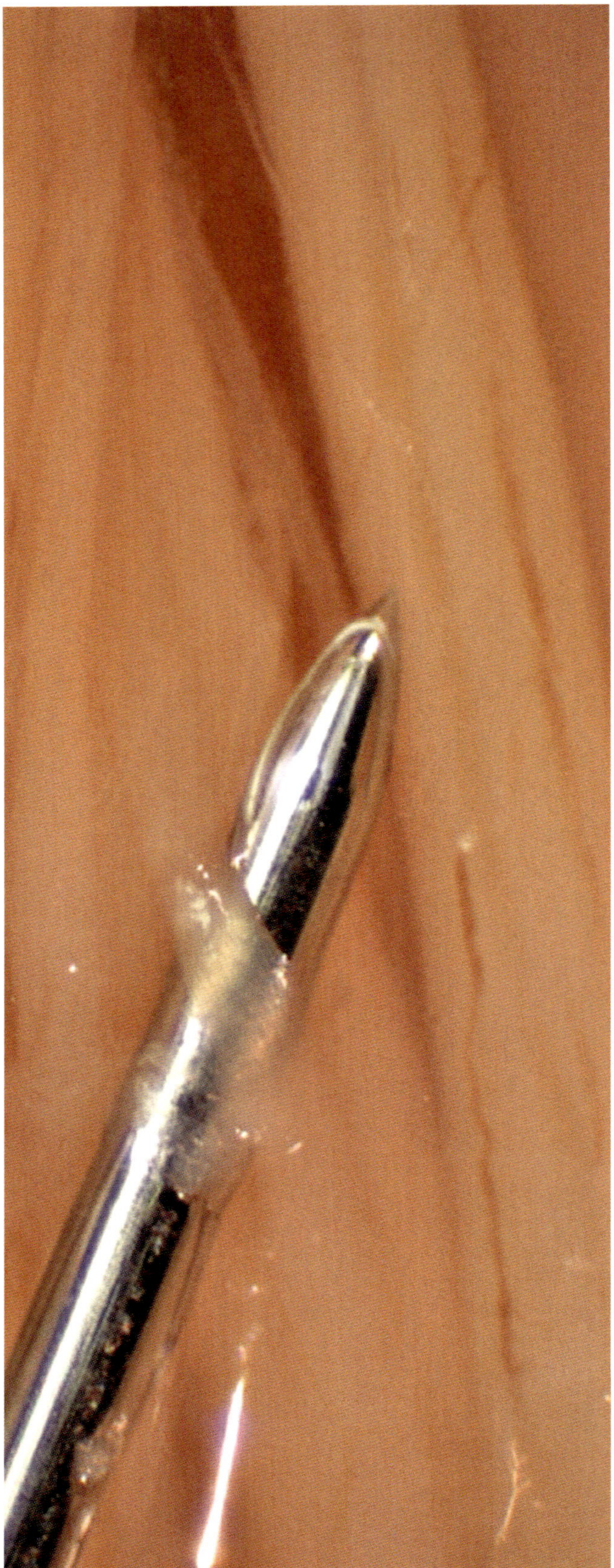

Fig. 30.10 In vitro sample of human spinal nerve root. Spinal needle (25-gauge Whitacre) traversing the spinal nerve root

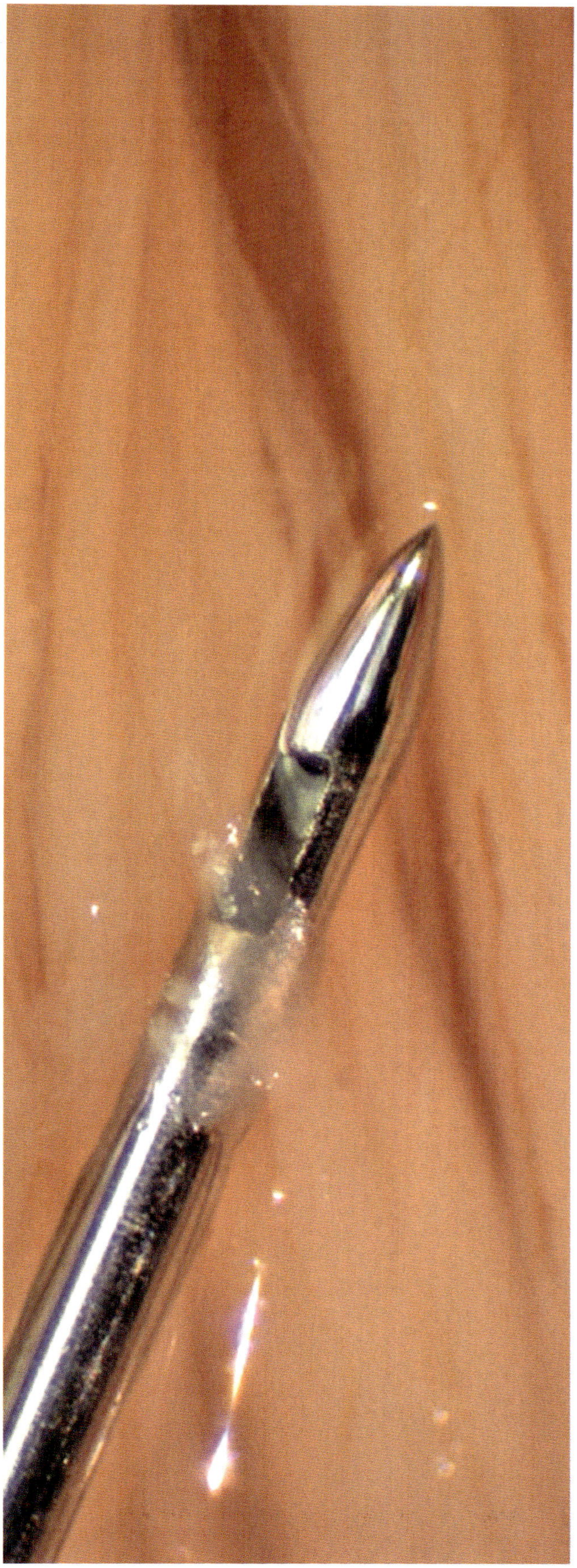

Fig. 30.11 In vitro sample of human spinal nerve root. Spinal needle (25-gauge Whitacre) traversing the spinal nerve root

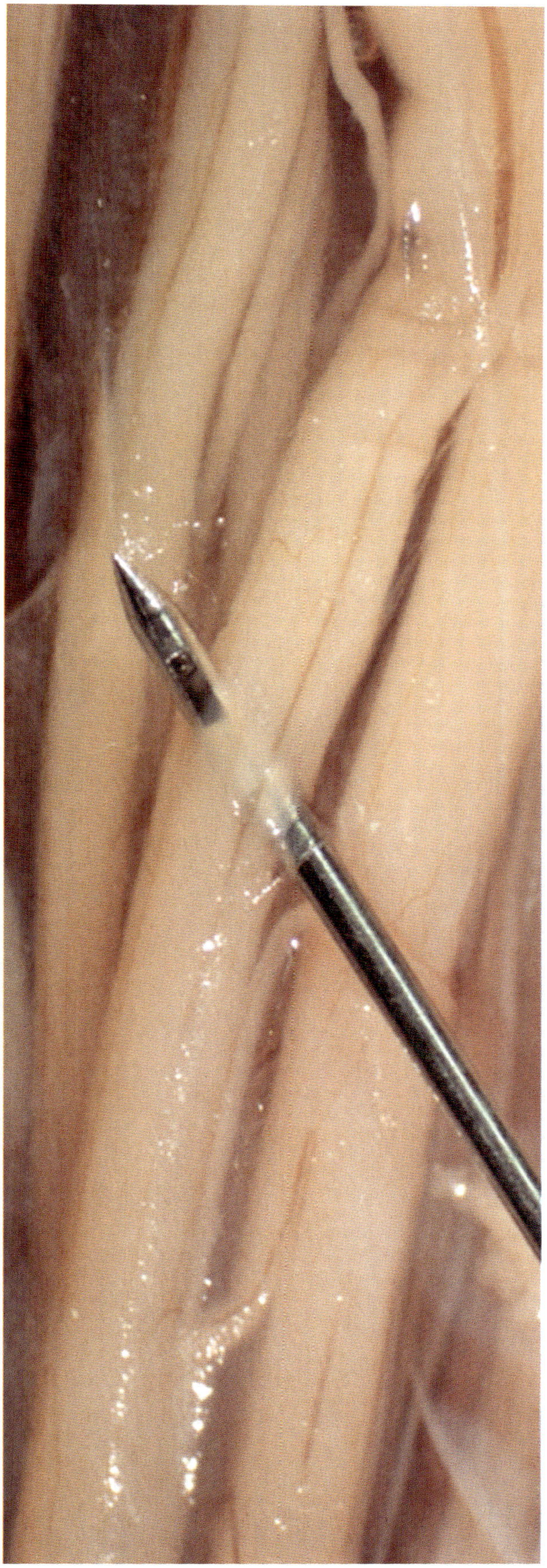

Fig. 30.12 In vitro sample of human spinal nerve root. Spinal needle (27-gauge Whitacre) traversing the spinal nerve root

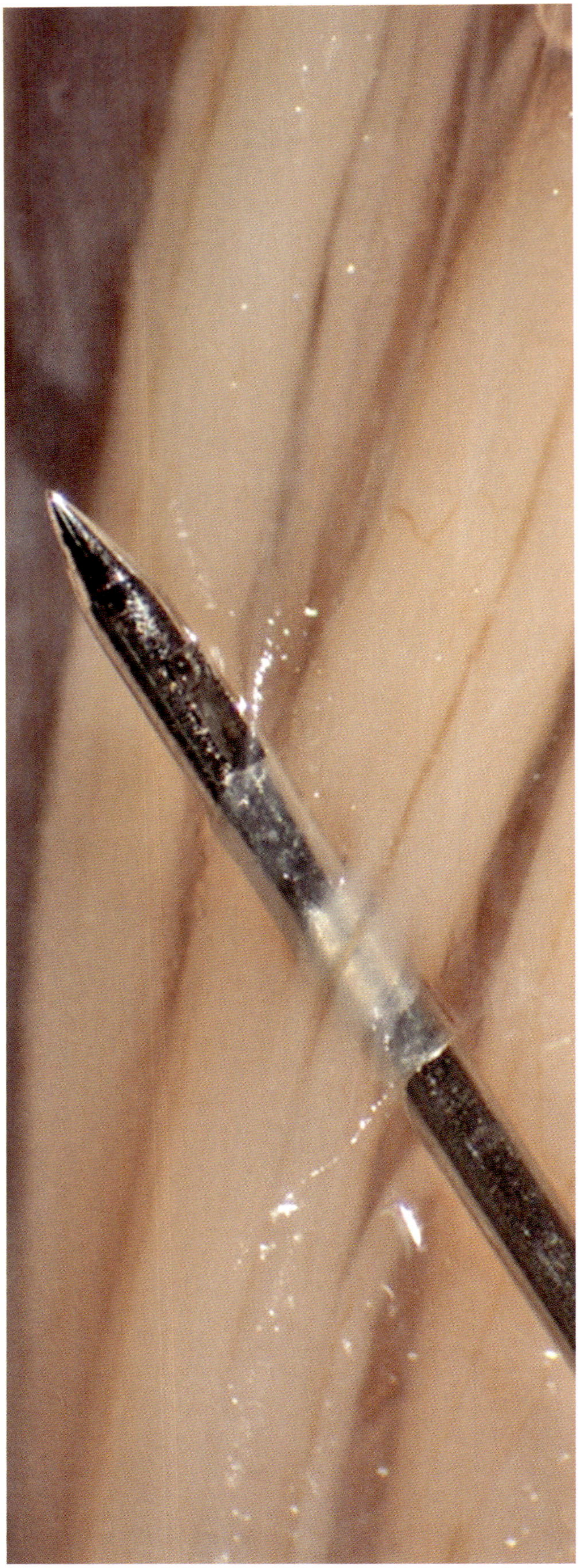

Fig. 30.13 In vitro sample of human spinal nerve root. Spinal needle (25-gauge Whitacre) traversing the spinal nerve root

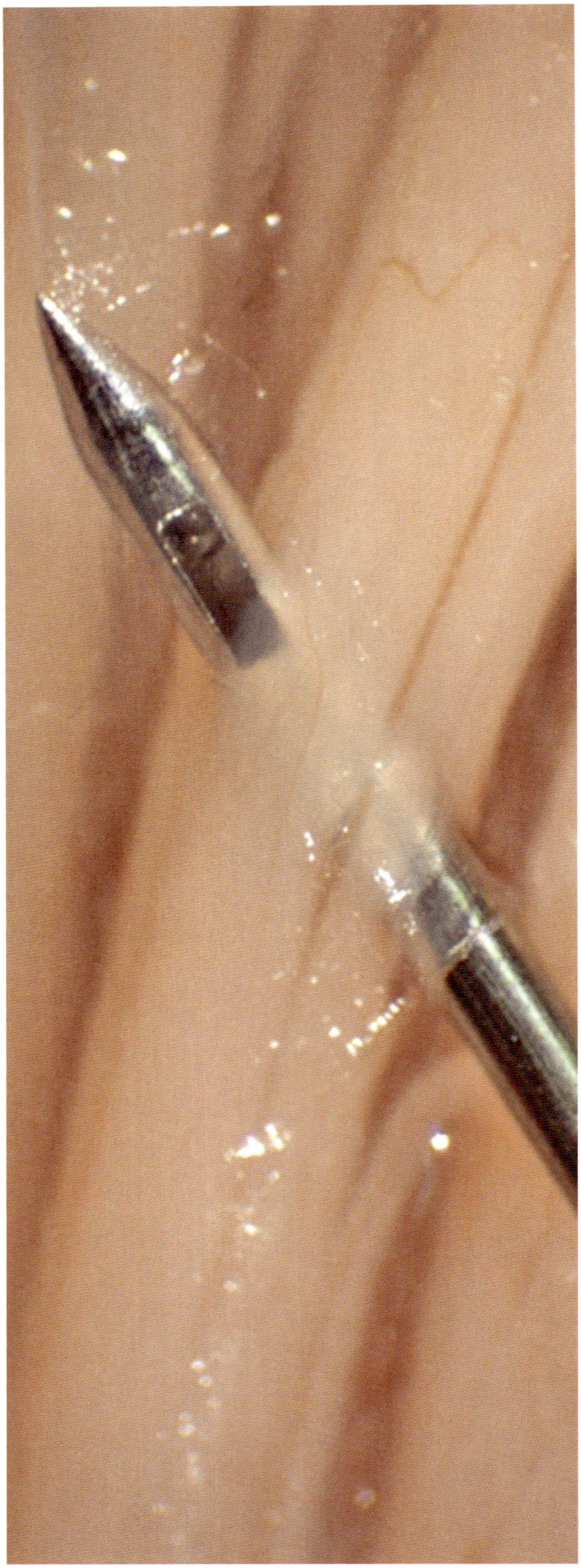

Fig. 30.14 In vitro sample of human spinal nerve root. Spinal needle (25-gauge Whitacre) traversing the spinal nerve root

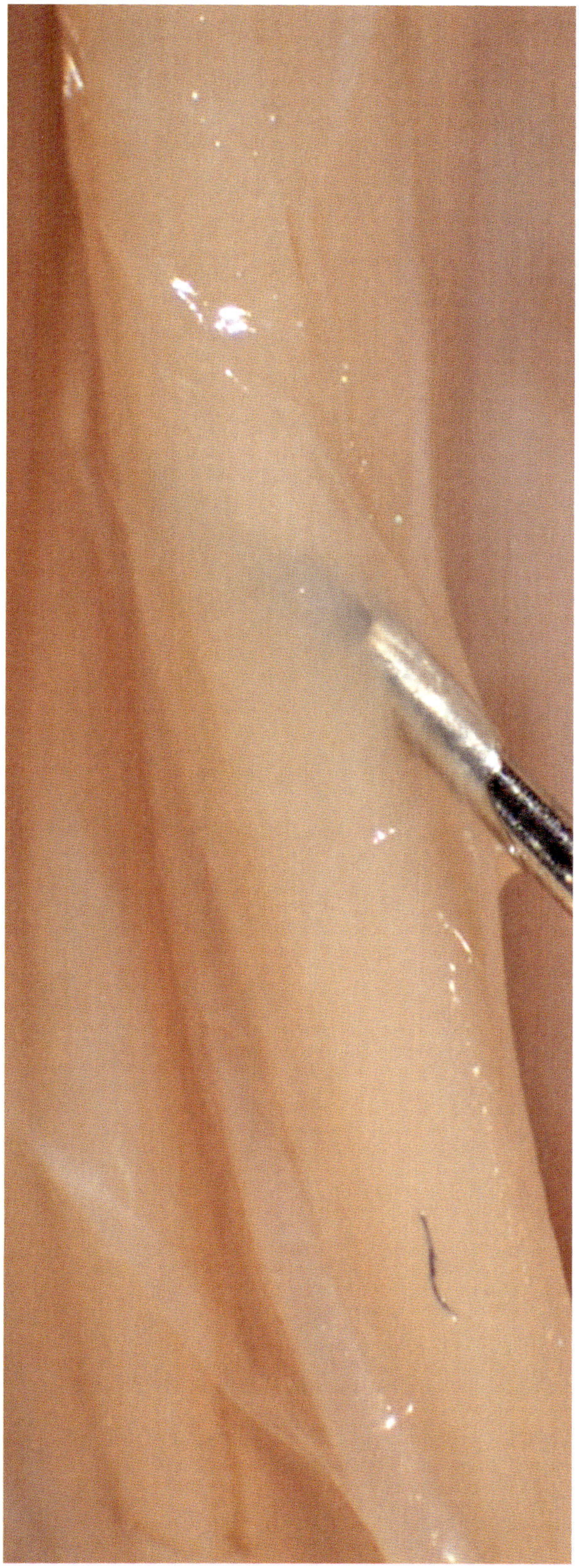

Fig. 30.15 In vitro sample of human spinal nerve root. Spinal needle (27-gauge Whitacre) traversing the spinal nerve root

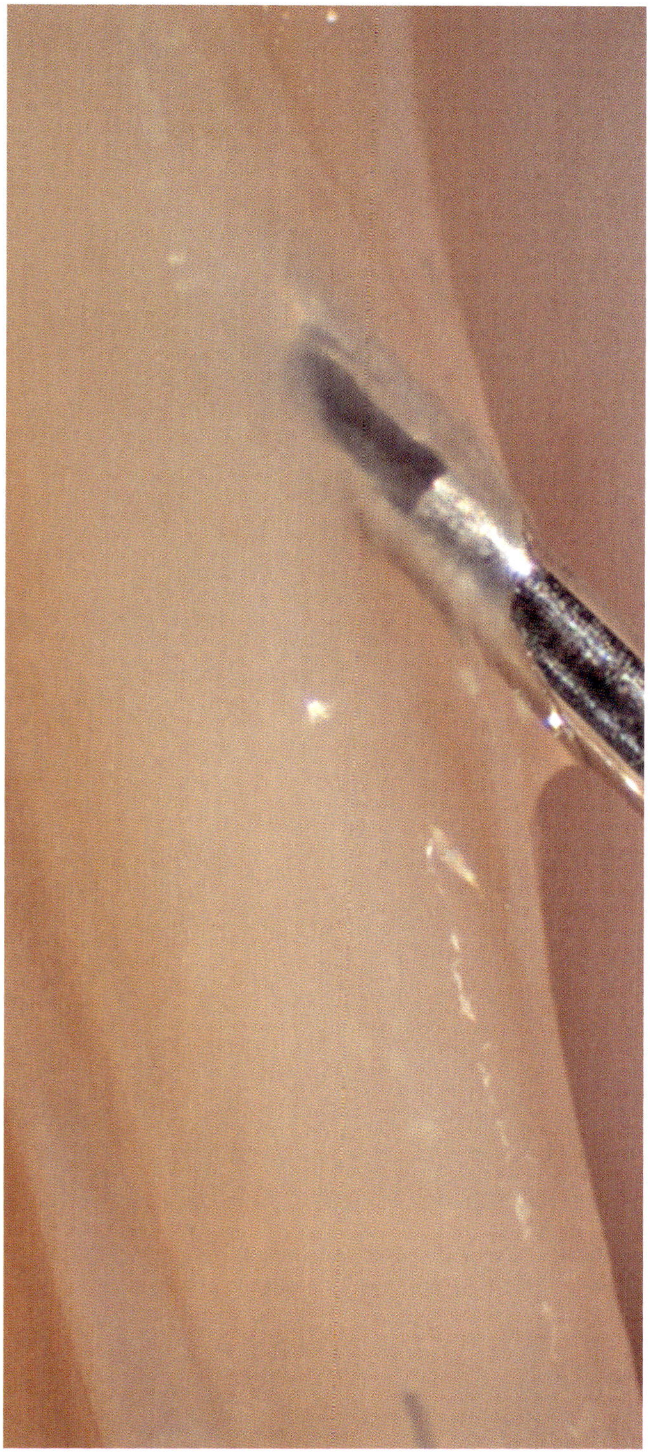

Fig. 30.16 In vitro sample of human spinal nerve root. Spinal needle (25-gauge Whitacre) traversing the spinal nerve root

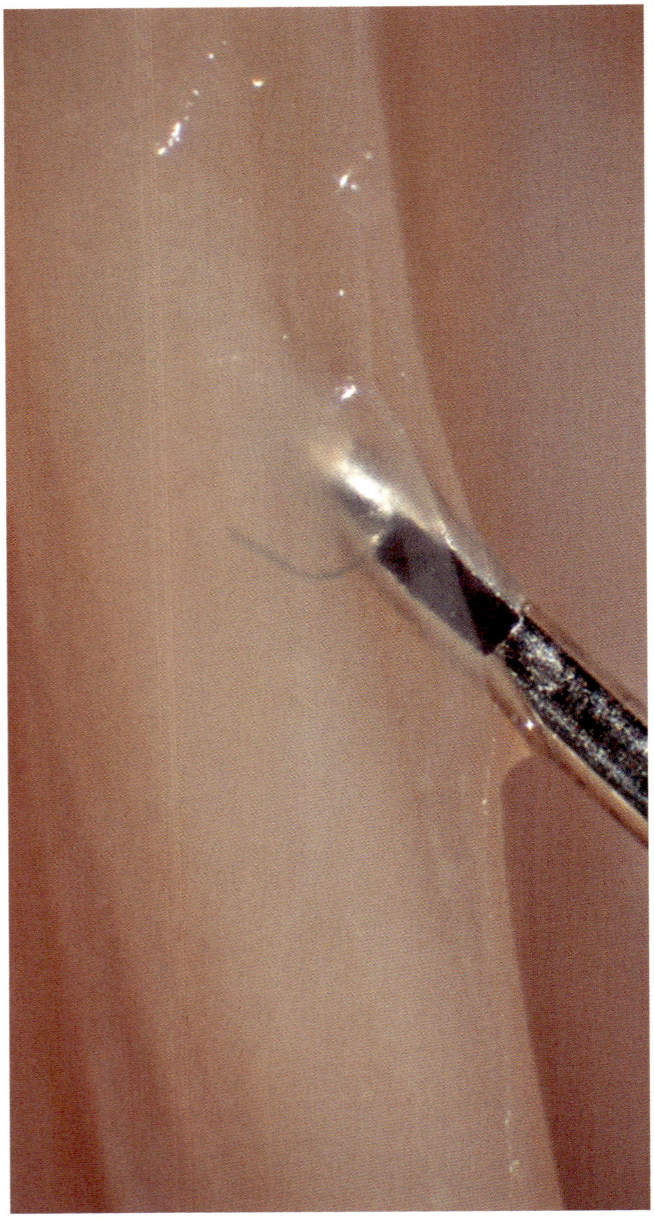

Fig. 30.17 In vitro sample of human spinal nerve root. Spinal needle (25-gauge Whitacre) traversing the spinal nerve root

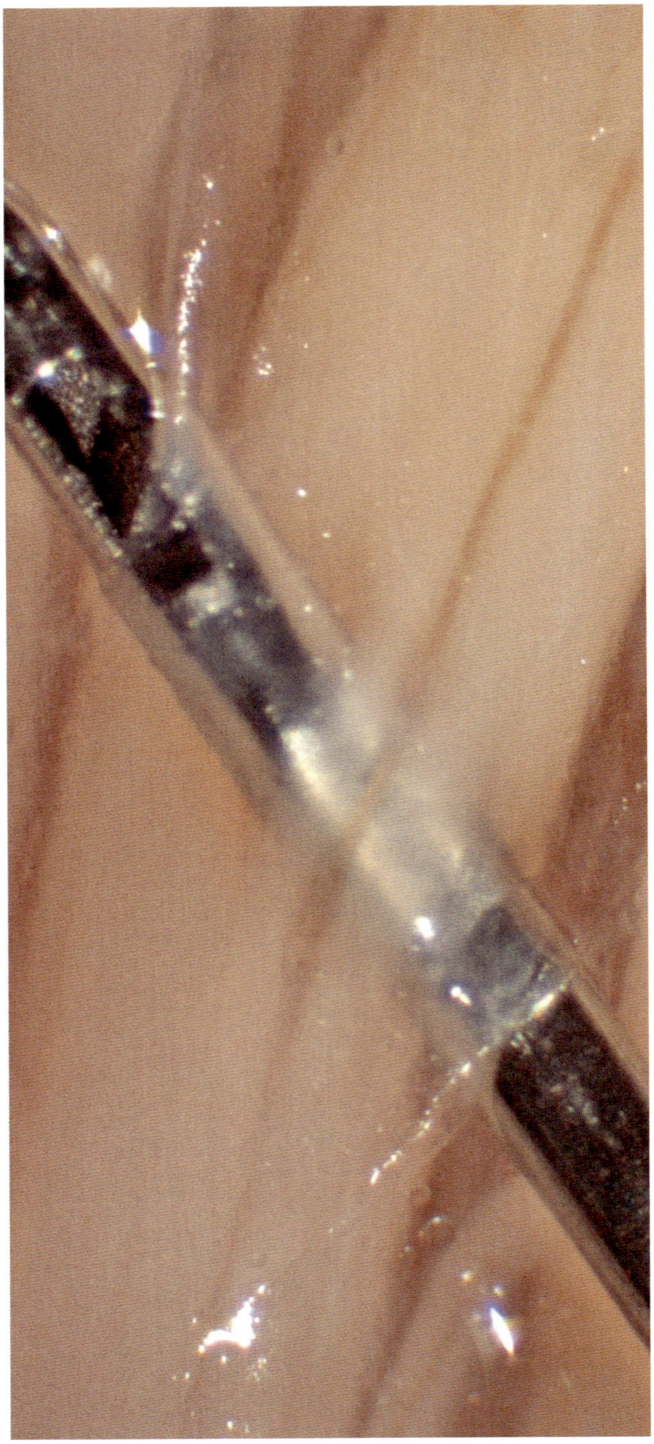

Fig. 30.18 In vitro sample of human spinal nerve root. Spinal needle (25-gauge Whitacre) traversing the spinal nerve root

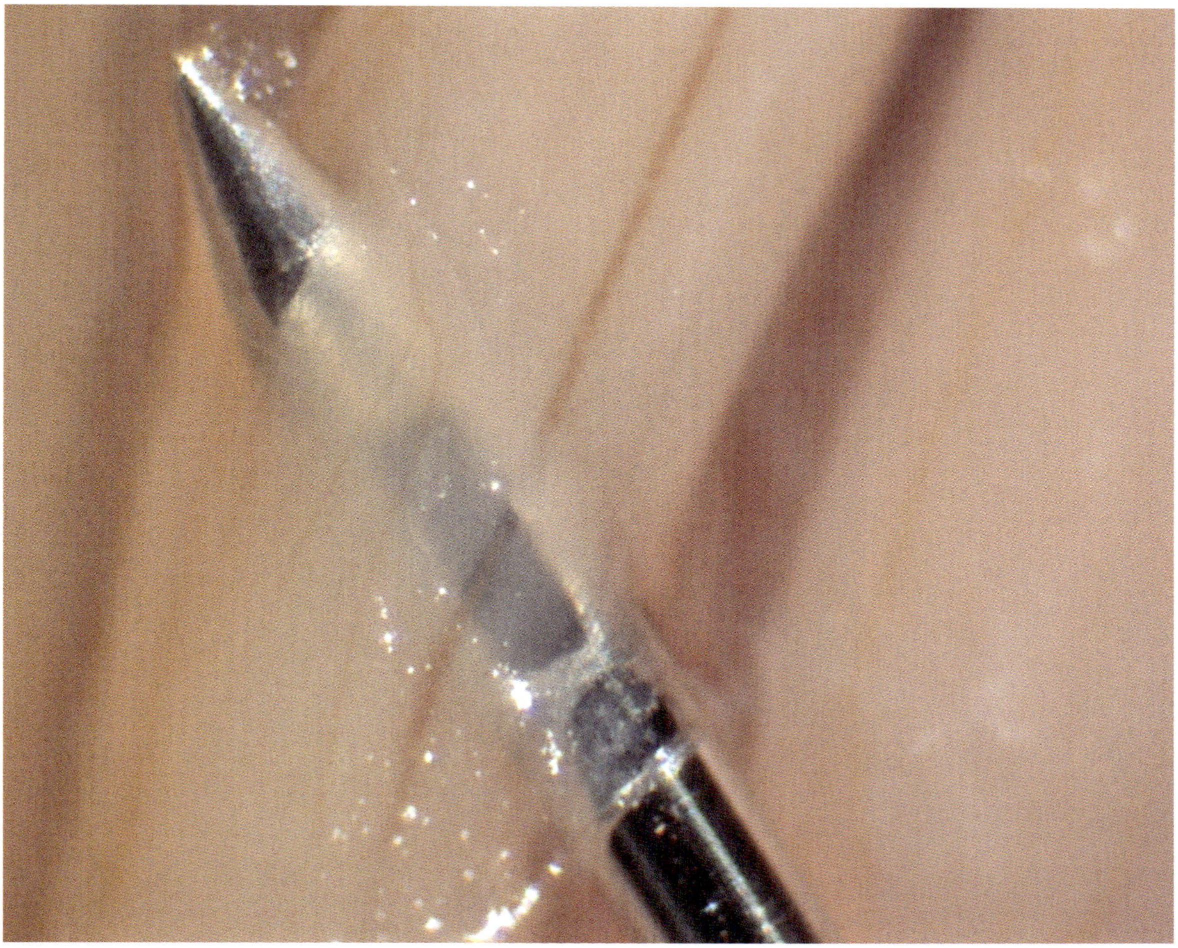

Fig. 30.19 In vitro sample of human spinal nerve root. Spinal needle (25-gauge Whitacre) traversing the spinal nerve root

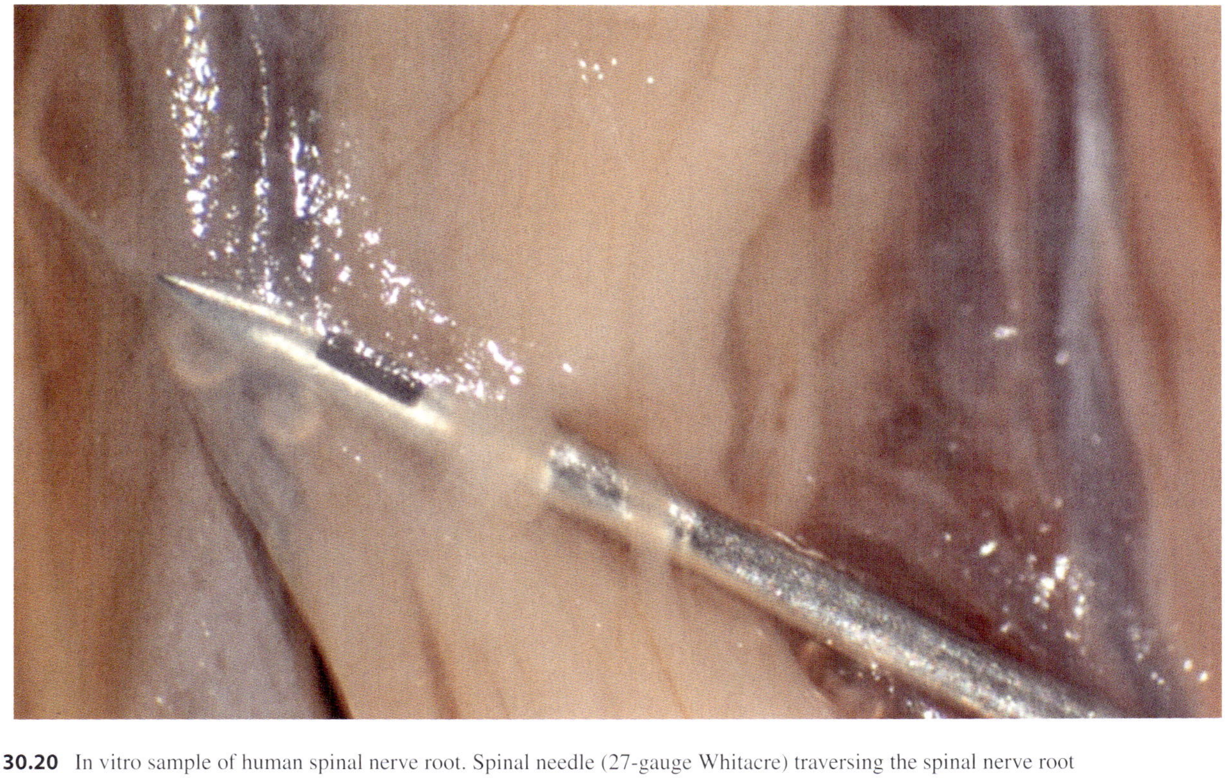

Fig. 30.20 In vitro sample of human spinal nerve root. Spinal needle (27-gauge Whitacre) traversing the spinal nerve root

Fig. 30.21 In vitro sample of human spinal nerve root. Spinal needle (22-gauge) traversing the spinal nerve root. Scanning electron microscopy; magnification: ×40

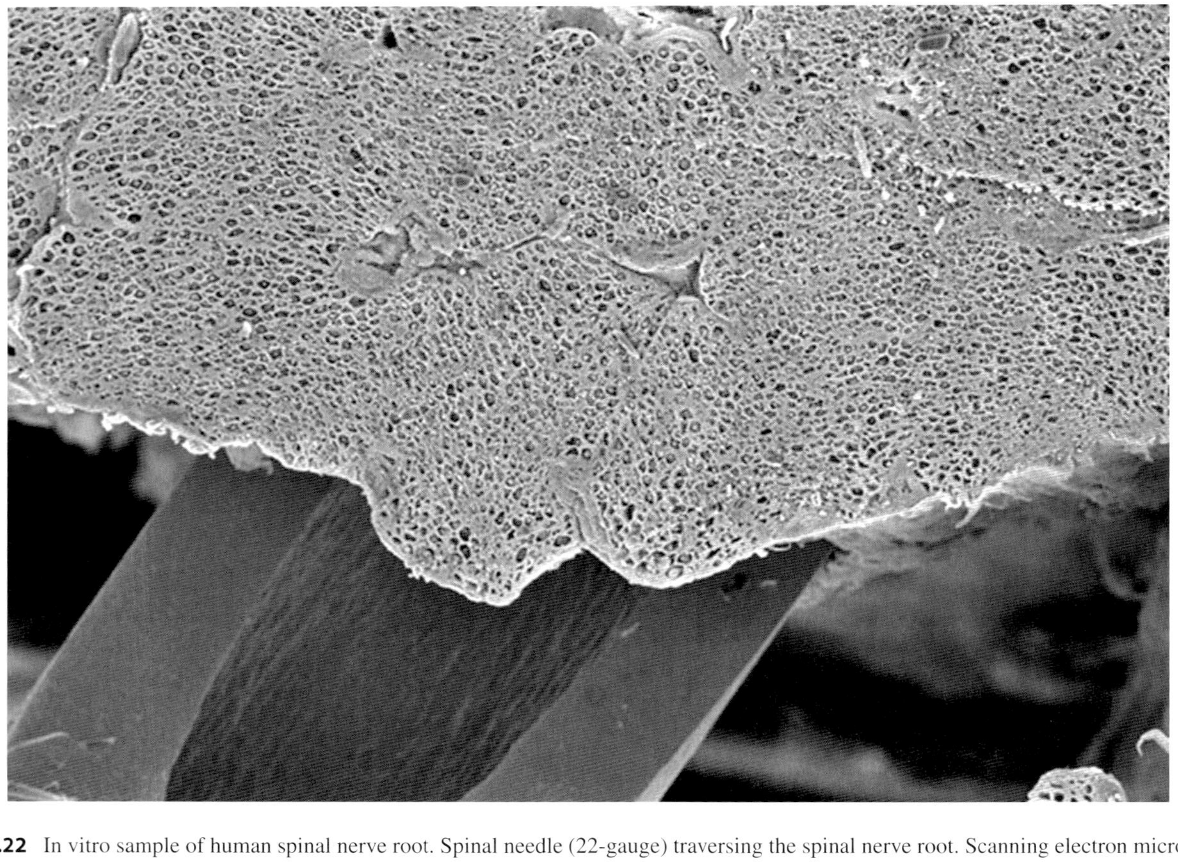

Fig. 30.22 In vitro sample of human spinal nerve root. Spinal needle (22-gauge) traversing the spinal nerve root. Scanning electron microscopy; magnification: ×80

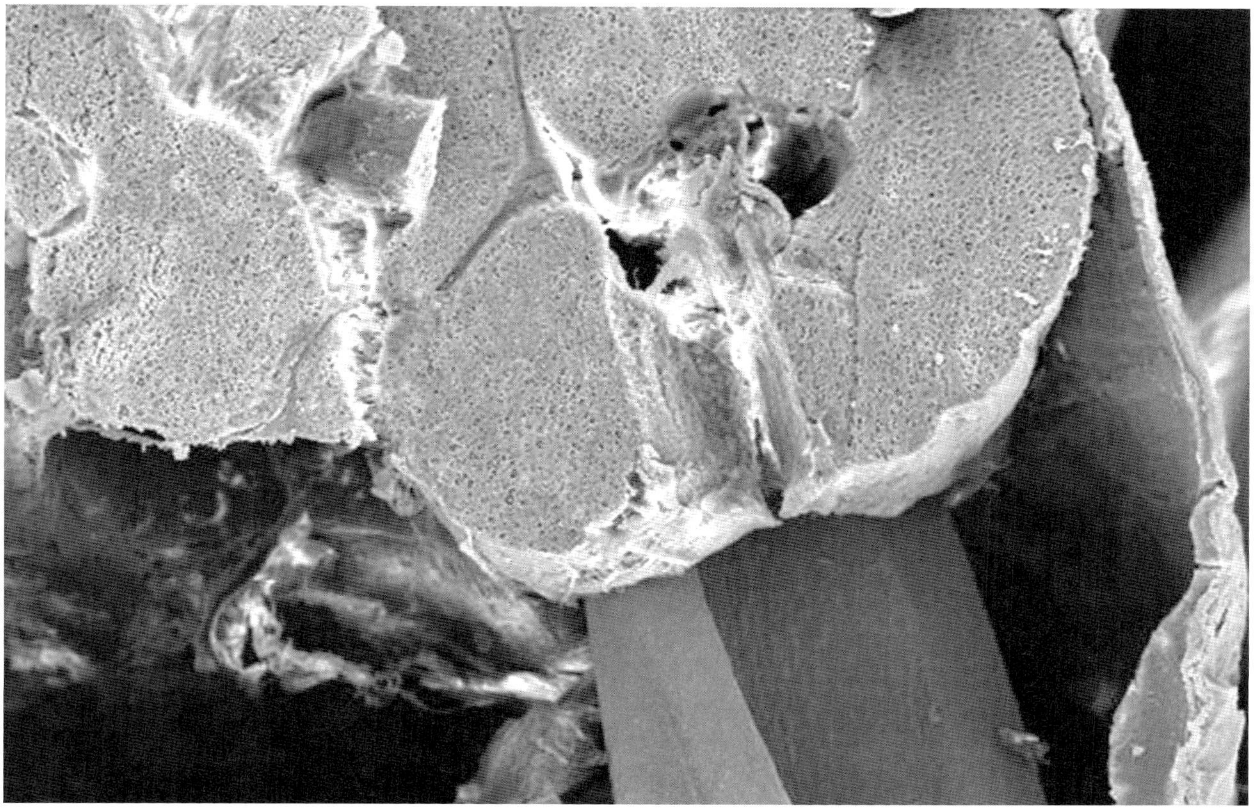

Fig. 30.23 In vitro sample of human spinal nerve root. Spinal needle (22-gauge) traversing the spinal nerve root. Scanning electron microscopy; magnification: ×40

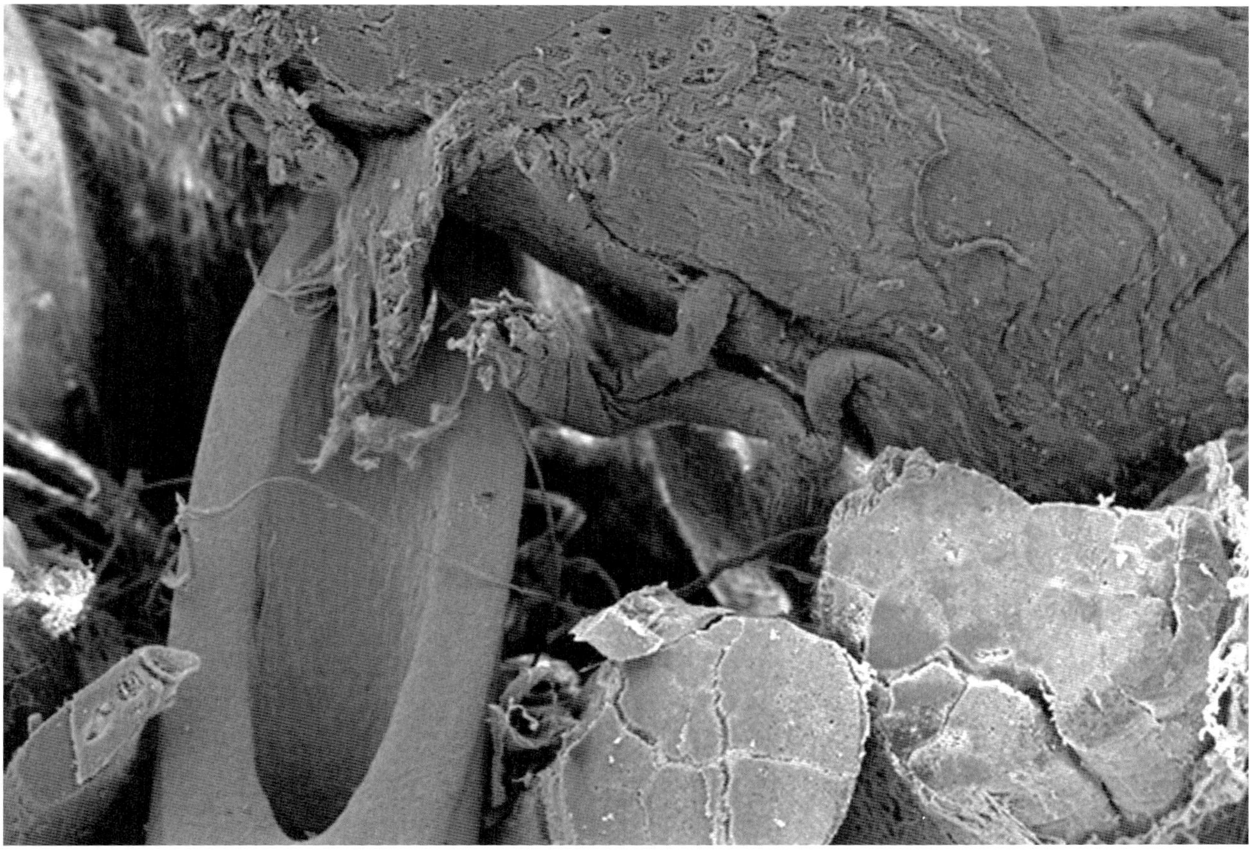

Fig. 30.24 In vitro sample of human spinal nerve root. Spinal needle (22-gauge) traversing the spinal nerve root. Scanning electron microscopy; magnification: ×30

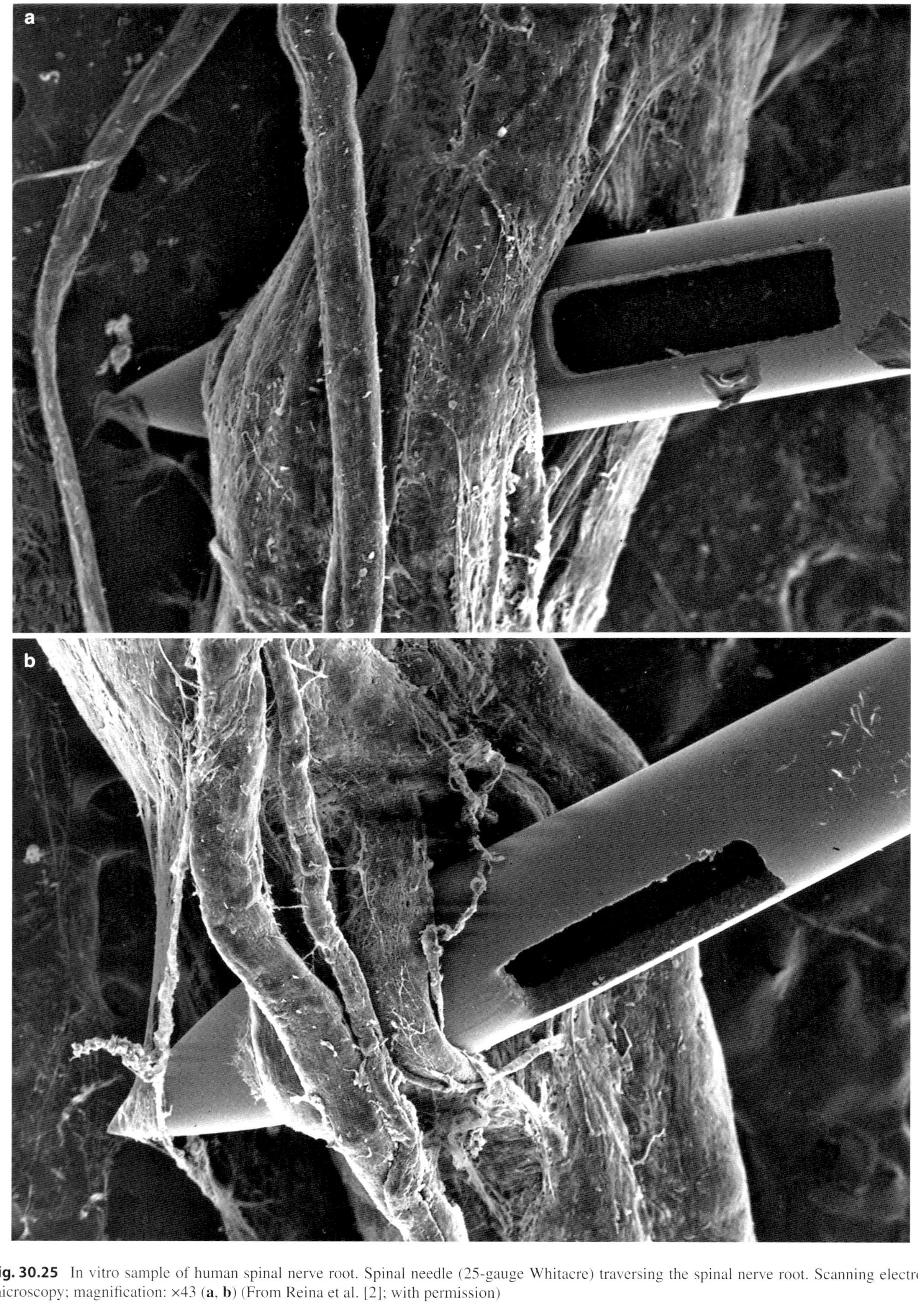

Fig. 30.25 In vitro sample of human spinal nerve root. Spinal needle (25-gauge Whitacre) traversing the spinal nerve root. Scanning electron microscopy; magnification: ×43 (**a**, **b**) (From Reina et al. [2]; with permission)

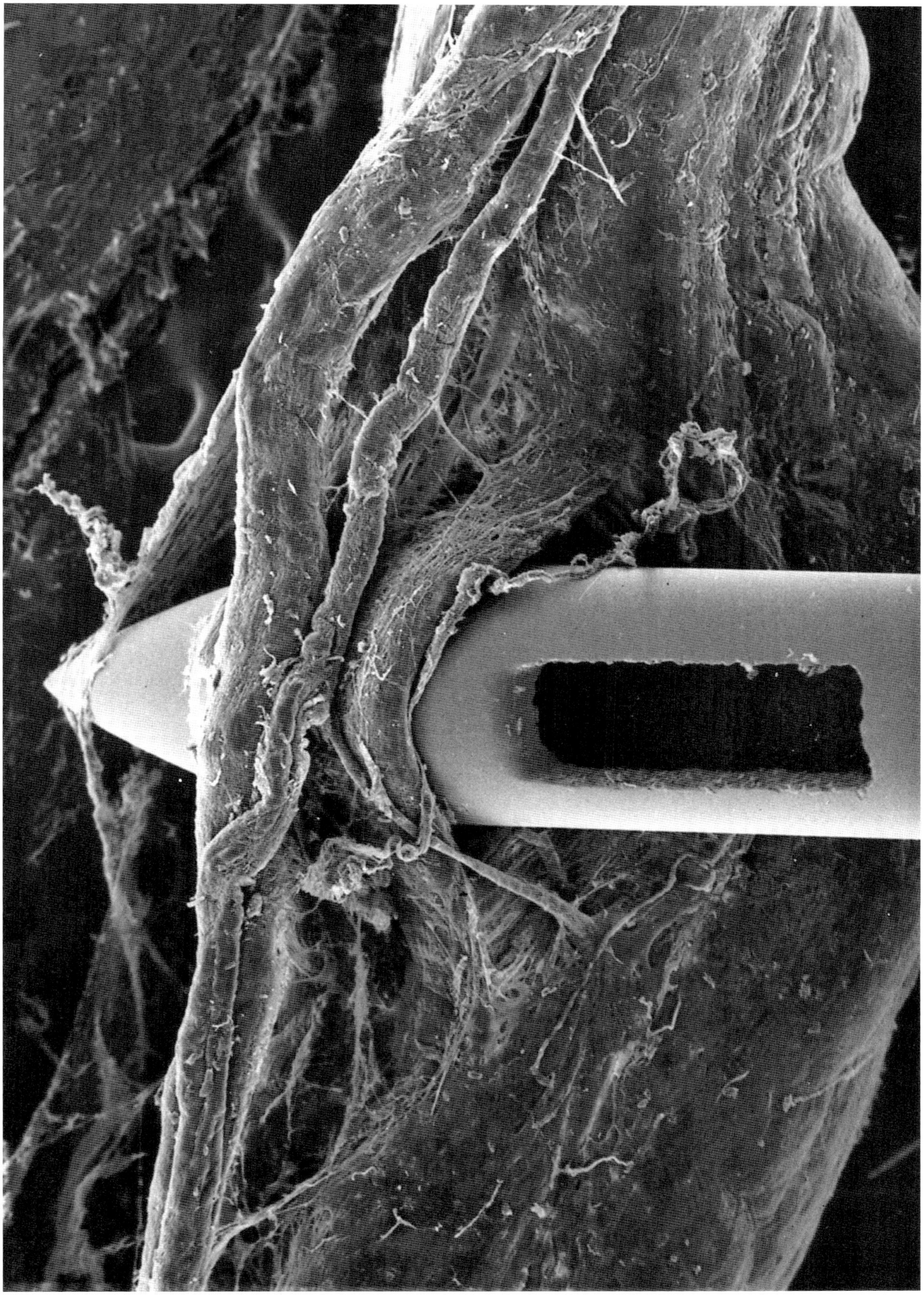

Fig. 30.26 In vitro sample of human spinal nerve root. Spinal needle (25-gauge Whitacre) traversing the spinal nerve root. Scanning electron microscopy; magnification: ×43

Fig. 30.27 In vitro sample of human spinal nerve root. Spinal needle (25-gauge Whitacre) traversing the spinal nerve root. Scanning electron microscopy; magnification: ×50

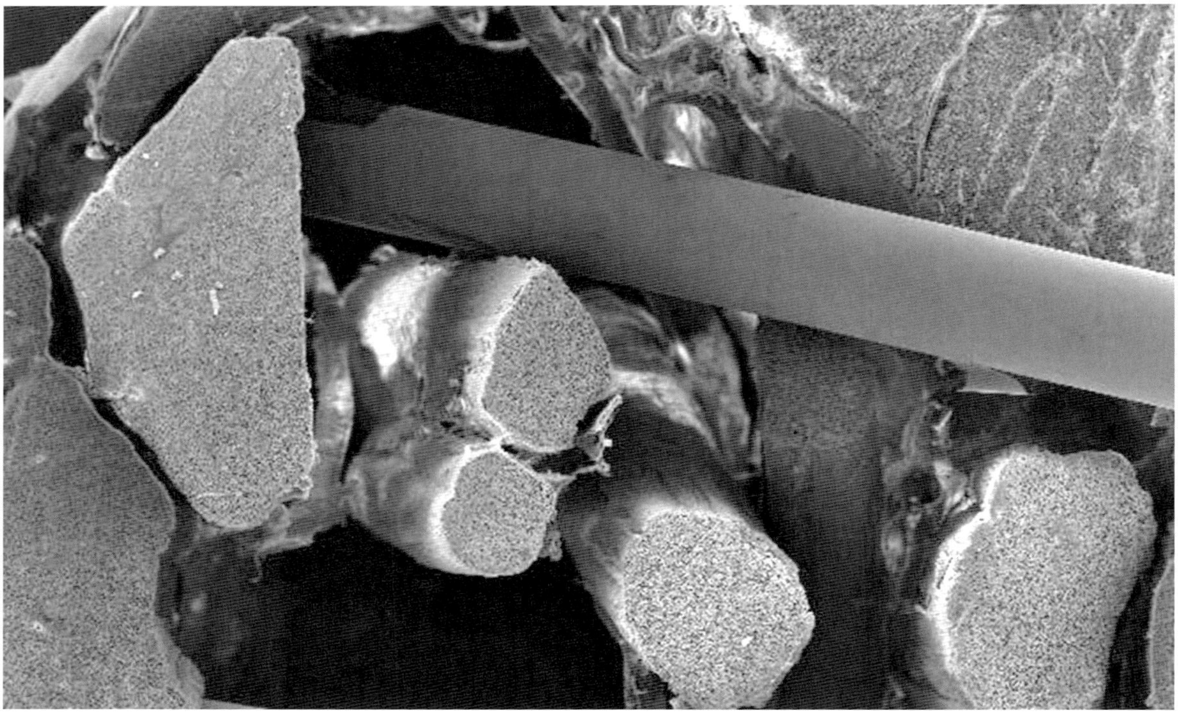

Fig. 30.28 In vitro sample of human spinal nerve root. Spinal needle (25-gauge Whitacre) traversing the spinal nerve root. Scanning electron microscopy; magnification: ×20

Fig. 30.29 In vitro sample of human spinal nerve root. Spinal needle (25-gauge Whitacre) traversing the spinal nerve root. Scanning electron microscopy; magnification: ×90

Fig. 30.30 In vitro sample of human spinal nerve root. Spinal needle (25-gauge Whitacre) traversing the spinal nerve root. Scanning electron microscopy; magnification: ×100

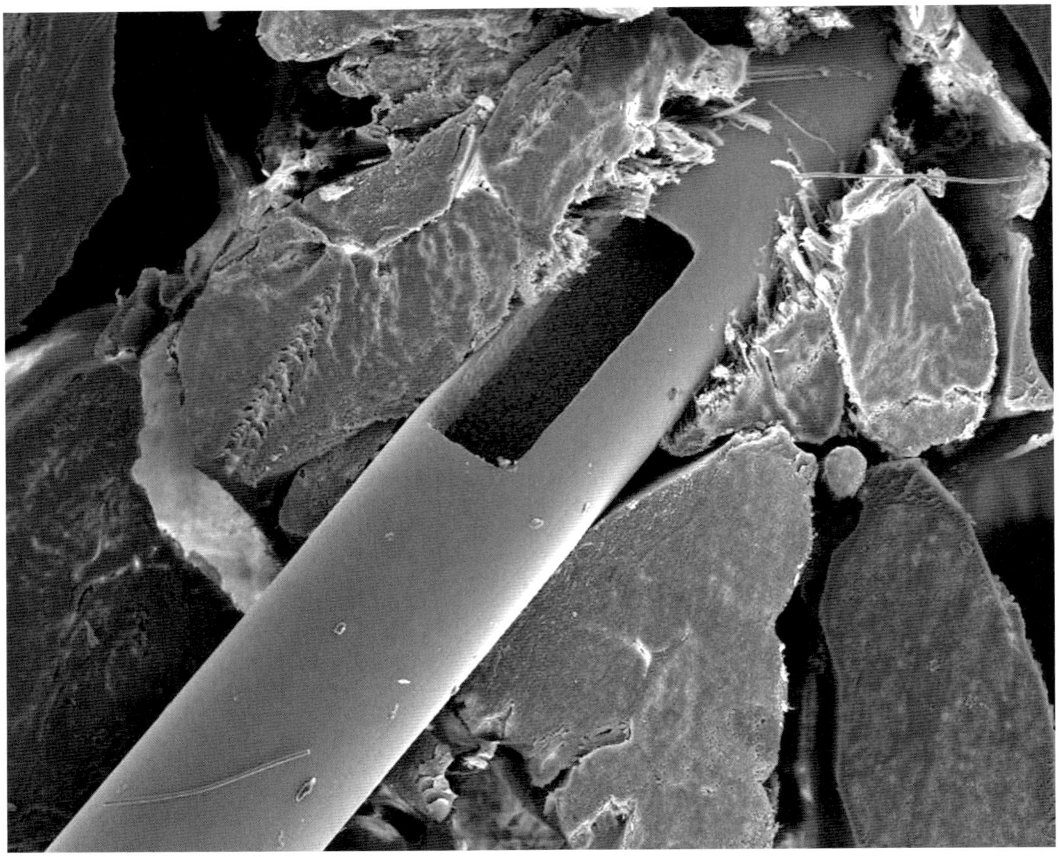

Fig. 30.31 In vitro sample of human spinal nerve root. Spinal needle (25-gauge Whitacre) traversing the spinal nerve root. Scanning electron microscopy; magnification: ×50

Fig. 30.32 In vitro sample of human spinal nerve root. Spinal needle (25-gauge Whitacre) traversing the spinal nerve root. Scanning electron microscopy; magnification: ×30

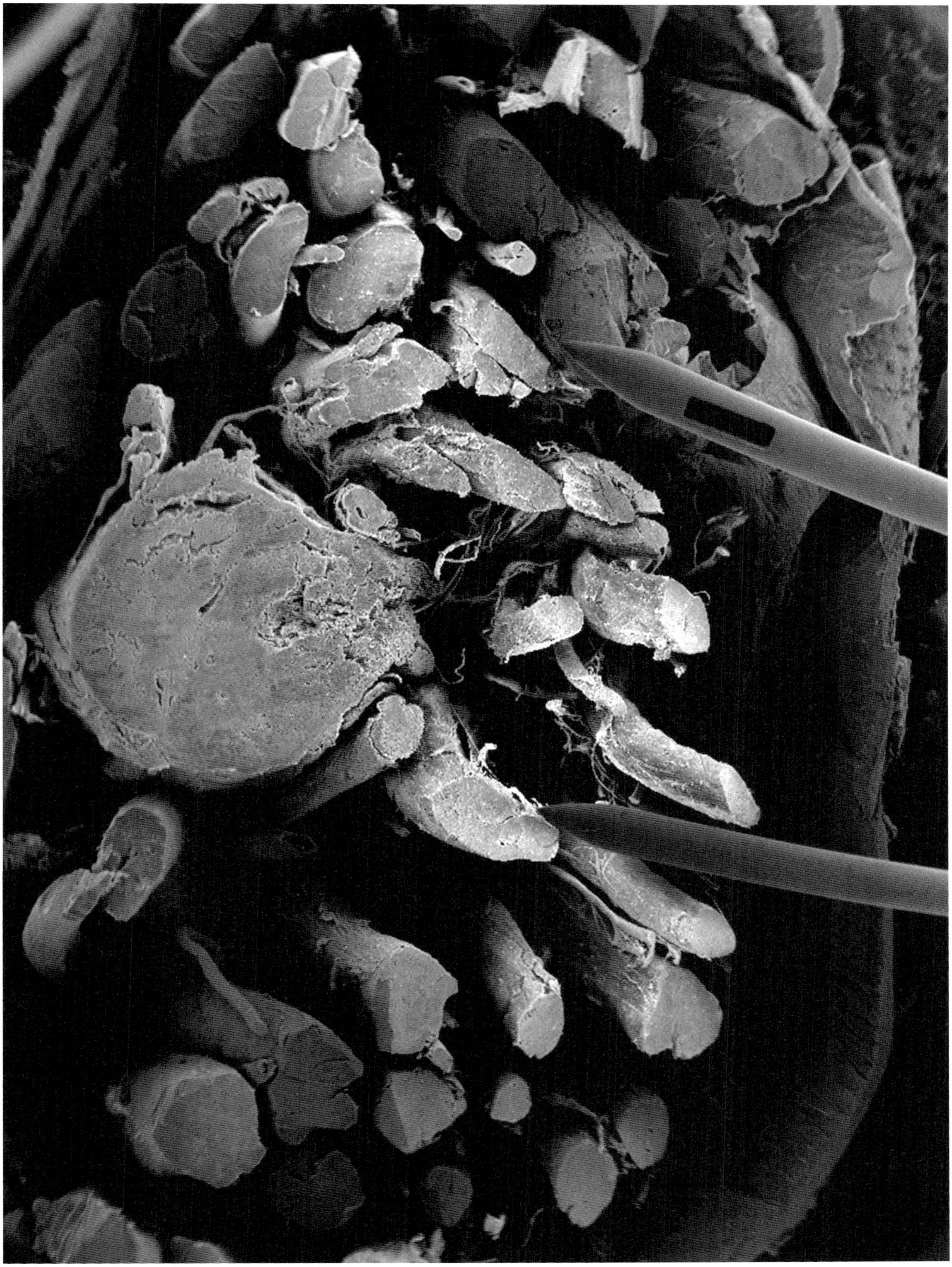

Fig. 30.33 In vitro sample of human spinal nerve root. Spinal needle (25-gauge Whitacre) traversing the spinal nerve root. Scanning electron microscopy; magnification: ×10 (From Reina et al. [1]; with permission)

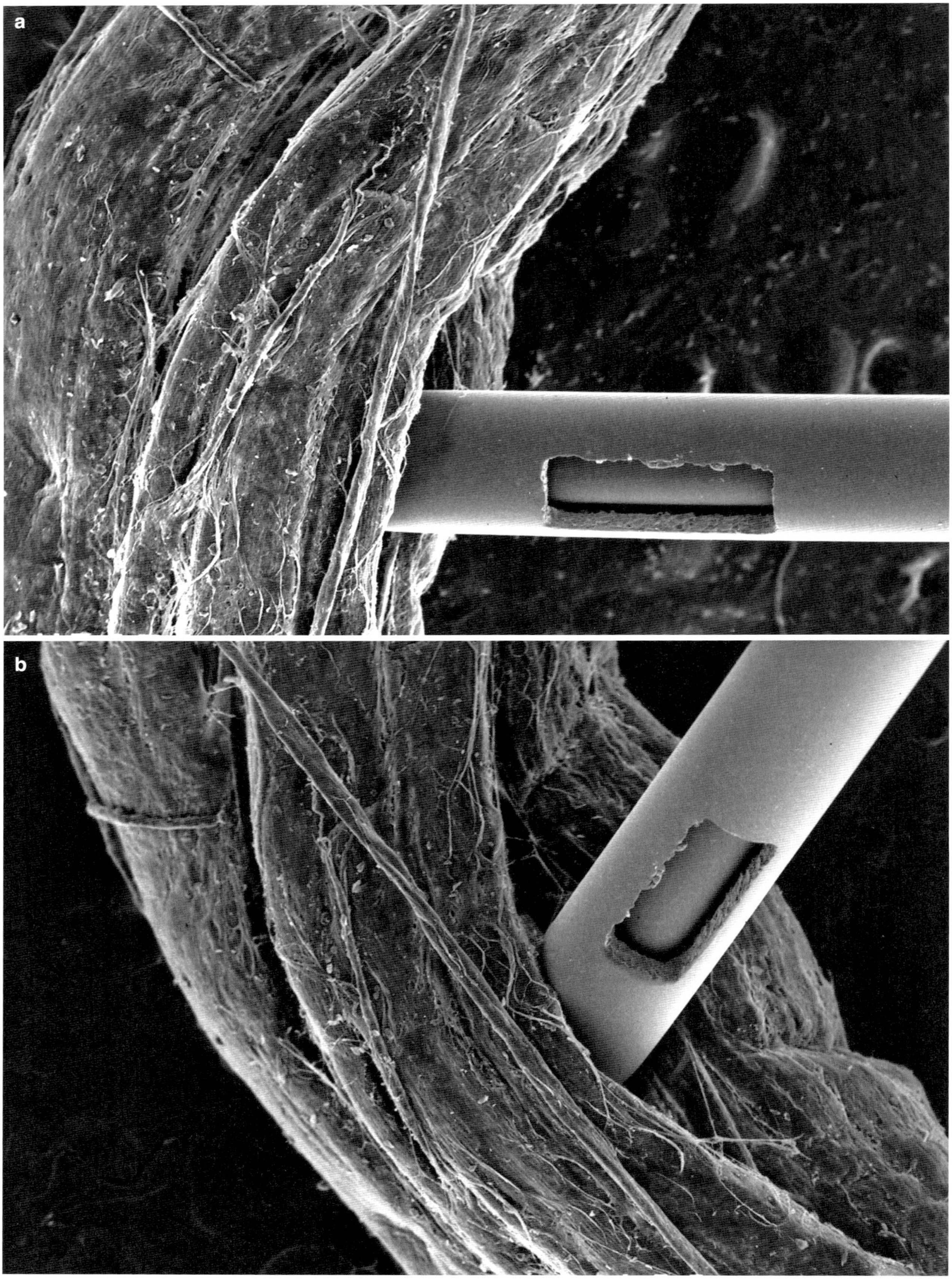

Fig. 30.34 In vitro sample of human spinal nerve root. Spinal needle (27-gauge Whitacre) traversing the spinal nerve root. Scanning electron microscopy; magnification: ×43 (**a**, **b**) (From Palacio et al. [3]; with permission.)

Fig. 30.35 In vitro sample of human spinal nerve root. Spinal needle (27-gauge Whitacre) traversing the spinal nerve root. Scanning electron microscopy; magnification: ×40

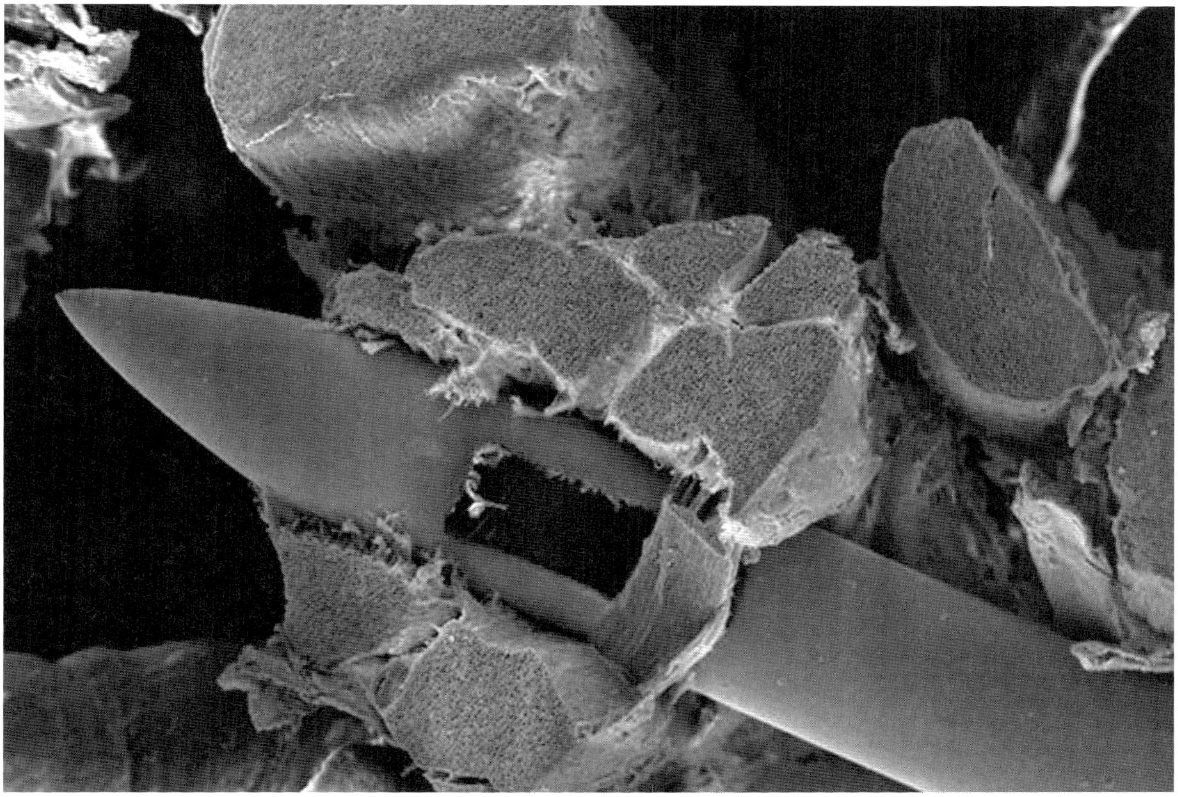

Fig. 30.36 In vitro sample of human spinal nerve root. Spinal needle (27-gauge Whitacre) traversing the spinal nerve root. Scanning electron microscopy; magnification: ×30

Fig. 30.37 In vitro sample of human spinal nerve root. Spinal needle (27-gauge Whitacre) traversing the spinal nerve root. Scanning electron microscopy; magnification: ×30

References

1. Reina MA, De Andres J, Hernández JM, Navarro RA, Pastor J, Prats-Galino A. Looking for the development of paresthesias in the subarachnoid and epidural anaesthesia. A clinical and anatomical analysis. Reg Anesth Pain Med. 2011;36(Suppl):E17–22.
2. Reina MA, Lopez A, Villanueva MC, De Andrés JA. Possibility of cauda equina nerve root damage from lumbar punctures performed with 25-gauge Quincke and Whitacre needles. Rev Esp Anestesiol Reanim. 2005;52:267–75.
3. Palacio F, Reina MA, Fornet I, López A. Parestesias con diferentes técnicas de bloqueo subaracnoideo con una aguja 27-G. Rev Esp Anestesiol Reanim. 2007;54:529–36.
4. Reynolds F. Damage to the conus medullaris following spinal anaesthesia. Anaesthesia. 2001;56:238–47.
5. Wall EJ, Cohen MS, Massie JB, Rydevik B, Garfin SR. Cauda equine anatomy. I: Intrathecal nerve root organization. Spine. 1990;15:1244–7.
6. Cohen MS, Wall EJ, Kerber CW, Abitbol JJ, Garfin SR. The anatomy of the cauda equina on CT scans and MRI. J Bone Joint Surg Br. 1991;73:381–4.
7. Reina MA, Machés F, López A, De Andrés JA. The ultrastructure of the spinal arachnoid in humans and its impact on spinal anesthesia, cauda equina syndrome and transient neurological syndrome. Tech Reg Anesth Pain Manag. 2008;12:153–60.

31 Nerve Root Cuff Lesions After "In Vitro" Needle Puncture and Model of "In Vitro" Nerve Stimuli Caused by Epidural Catheters

Miguel Angel Reina and Emilse Colman Peyrano

The tip of a spinal or epidural needle may stimulate a nerve or even cause nerve lesions, leading to paresthesias during the performance of spinal or epidural anesthetic blockade [1–3]. Similarly, the tip of an epidural catheter may affect nerves, causing paresthesias after continuous anesthetic blockade techniques, in which the catheter may be placed adjacent to the nerve for prolonged periods.

The outcome of unwanted nerve stimulation or lesions varies according to the location at which these nerves are affected, as well as other factors. Therefore, injury caused by the needle to nerve roots inside the dural sac bears no resemblance to damage to motor or sensorial roots at sites where these remain grouped inside nerve root cuffs.

Patients experience paresthesias during anesthetic blockade techniques because of either unwanted nerve stimuli or nerve injury. Nerve root cuff lesions are possible if the needle is directed laterally along the dural sac and advanced toward the anterolateral area of the epidural space in which the nerve root cuffs are located [3]. In this situation, there is no cerebrospinal fluid leakage [1]. Nerve root cuff lesions may be caused by deflection of the epidural or spinal needle toward lateral areas of the epidural space or by rotation of the spine. The degree of damage to the nerves determines the severity of the neurologic effects, ranging from minor paresthesias to permanent neurologic damage.

The diameter of nerve roots inside the dural sac ranges from 0.5 to 1 mm. Both anterior and posterior nerve roots appear enclosed in nerve root cuffs, which are about 6–8 mm thick. Grouping of nerve roots inside the dural sac is less compact; however, displacement inside the nerve root cuffs is limited [2].

Factors such as the size and limited mobility of the nerve root cuffs may influence the degree of damage to nerve structures such as nerve root cuffs.

Nerve stimuli caused by epidural catheters are produced by compression of the catheter tip against nerve structures. They probably occur with anterolateral positioning of the catheter tip within the epidural space. However, they do not produce neural damage (Figs. 31.1, 31.2, 31.3, 31.4, 31.5, 31.6, 31.7, 31.8, 31.9, 31.10, 31.11, 31.12, 31.13, 31.14, 31.15, 31.16, 31.17, 31.18, 31.19, 31.20, 31.21, 31.22, 31.23, and 31.24).

M.A. Reina, MD, PhD (✉)
Department of Clinical Medical Sciences and Institute of Applied Molecular Medicine, School of Medicine, University of CEU San Pablo, Madrid, Spain

Department of Anesthesiology, Madrid-Monteprı́ncipe University Hospital, Madrid, Spain
e-mail: miguelangel@perticone.e.telefonica.net

E. Colman Peyrano, BS
Research Unit, Anesthesia Reina Colman SLP, Madrid, Spain
e-mail: emilsecolpey@yahoo.es

M.A. Reina et al. (eds.), *Atlas of Functional Anatomy for Regional Anesthesia and Pain Medicine: Human Structure, Ultrastructure and 3D Reconstruction Images*, DOI 10.1007/978-3-319-09522-6_31,

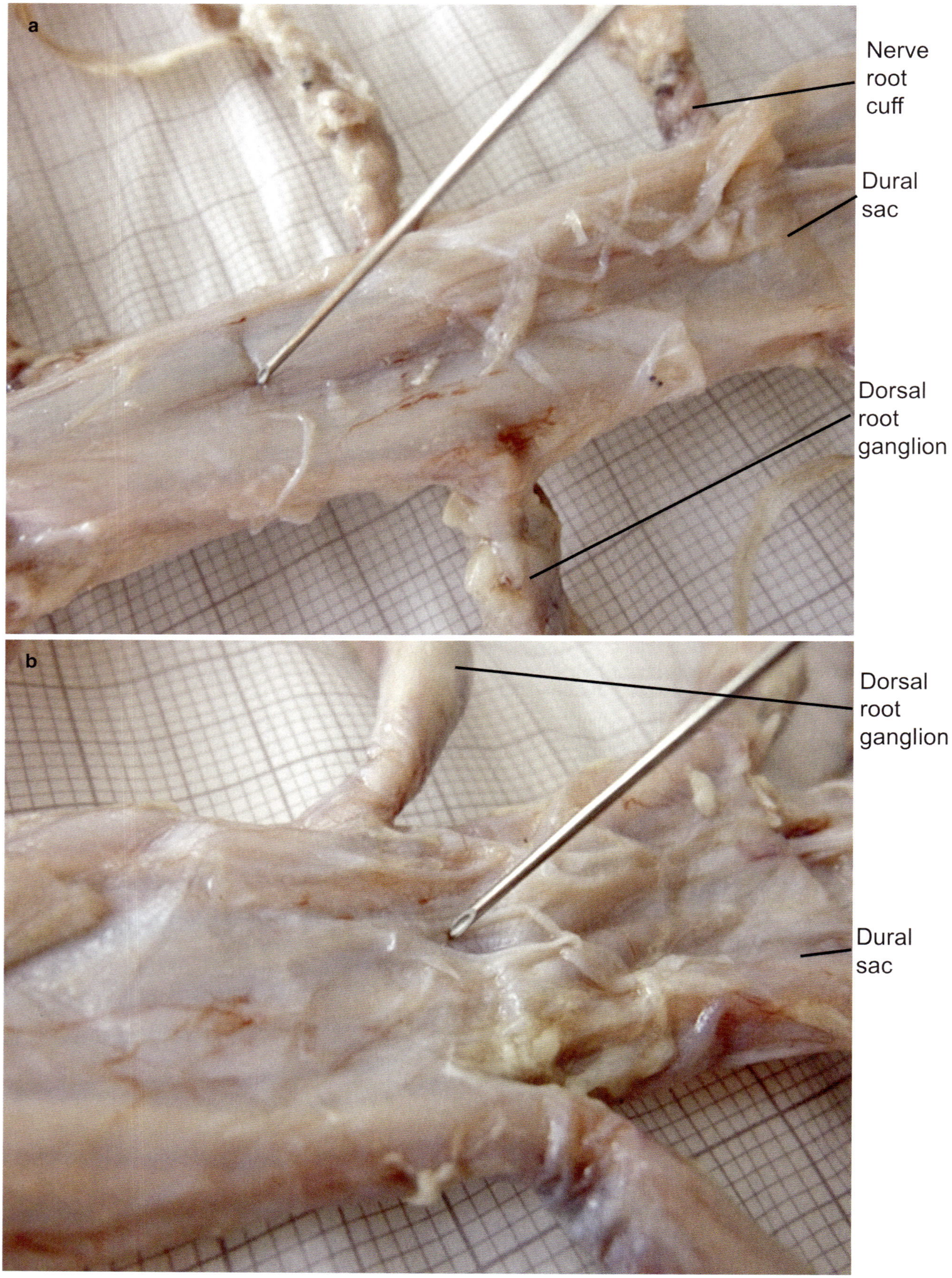

Fig. 31.1 In vitro sample of human dural sac. Spinal needle (22-gauge Quincke) contacting the dural sac. (**a**, **b**) Images obtained from different samples

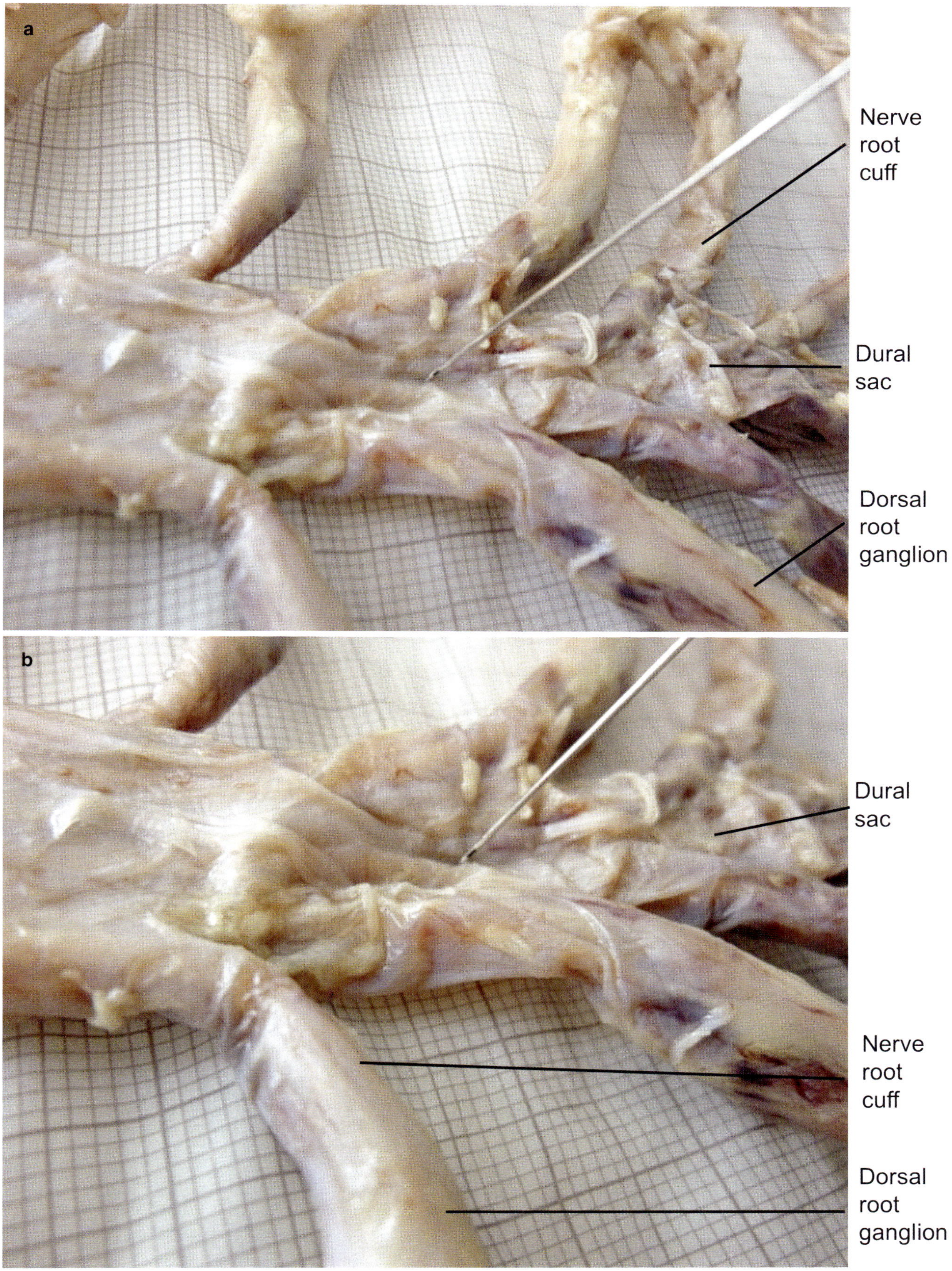

Fig. 31.2 In vitro sample of human dural sac. Spinal needle (25-gauge Whitacre) contacting the dural sac. (**a**, **b**) Images obtained from same sample at different magnification

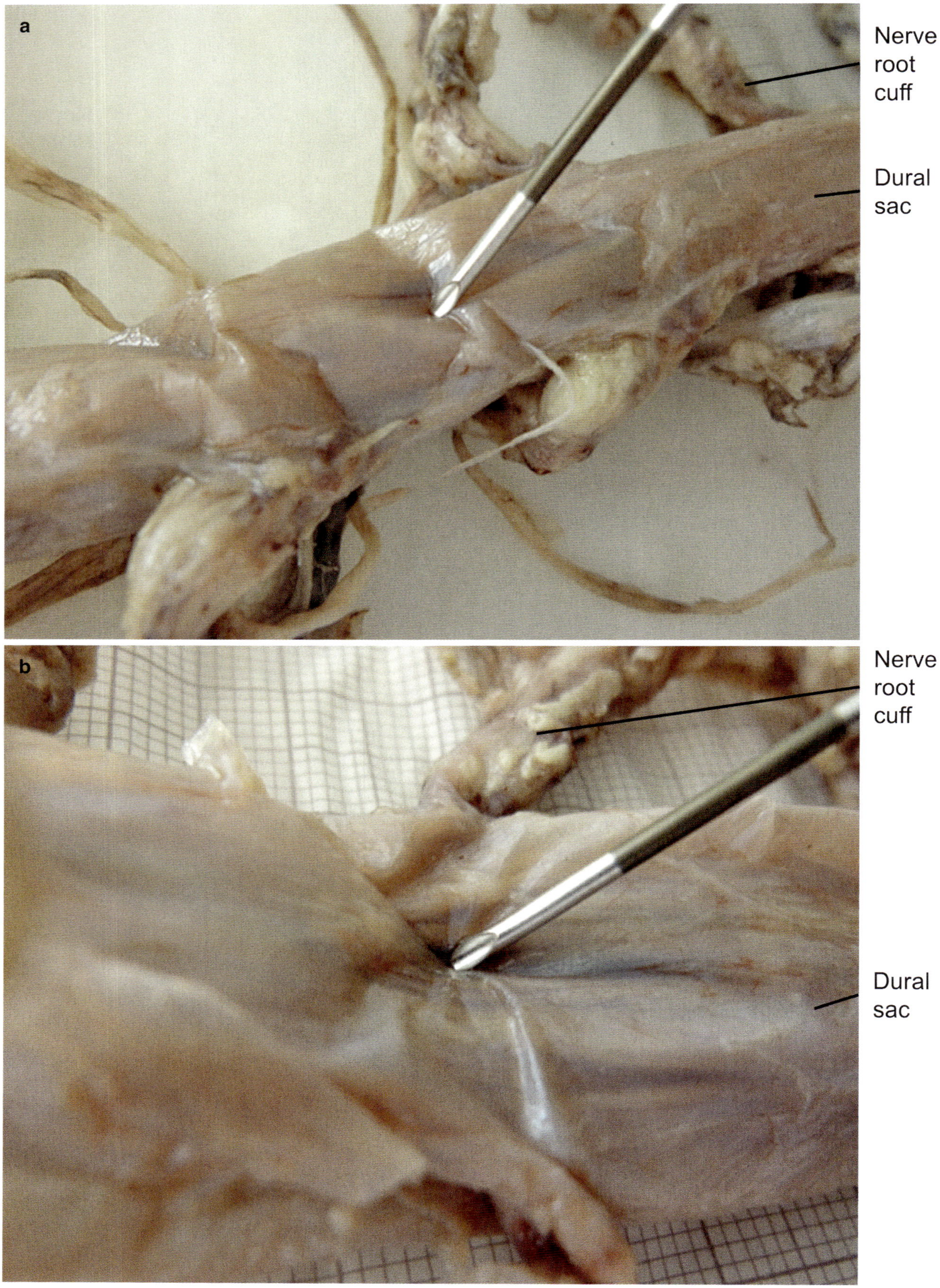

Fig. 31.3 In vitro sample of human dural sac. Epidural needle (18-gauge Tuohy) contacting the dural sac. (**a**, **b**) Images obtained from same sample

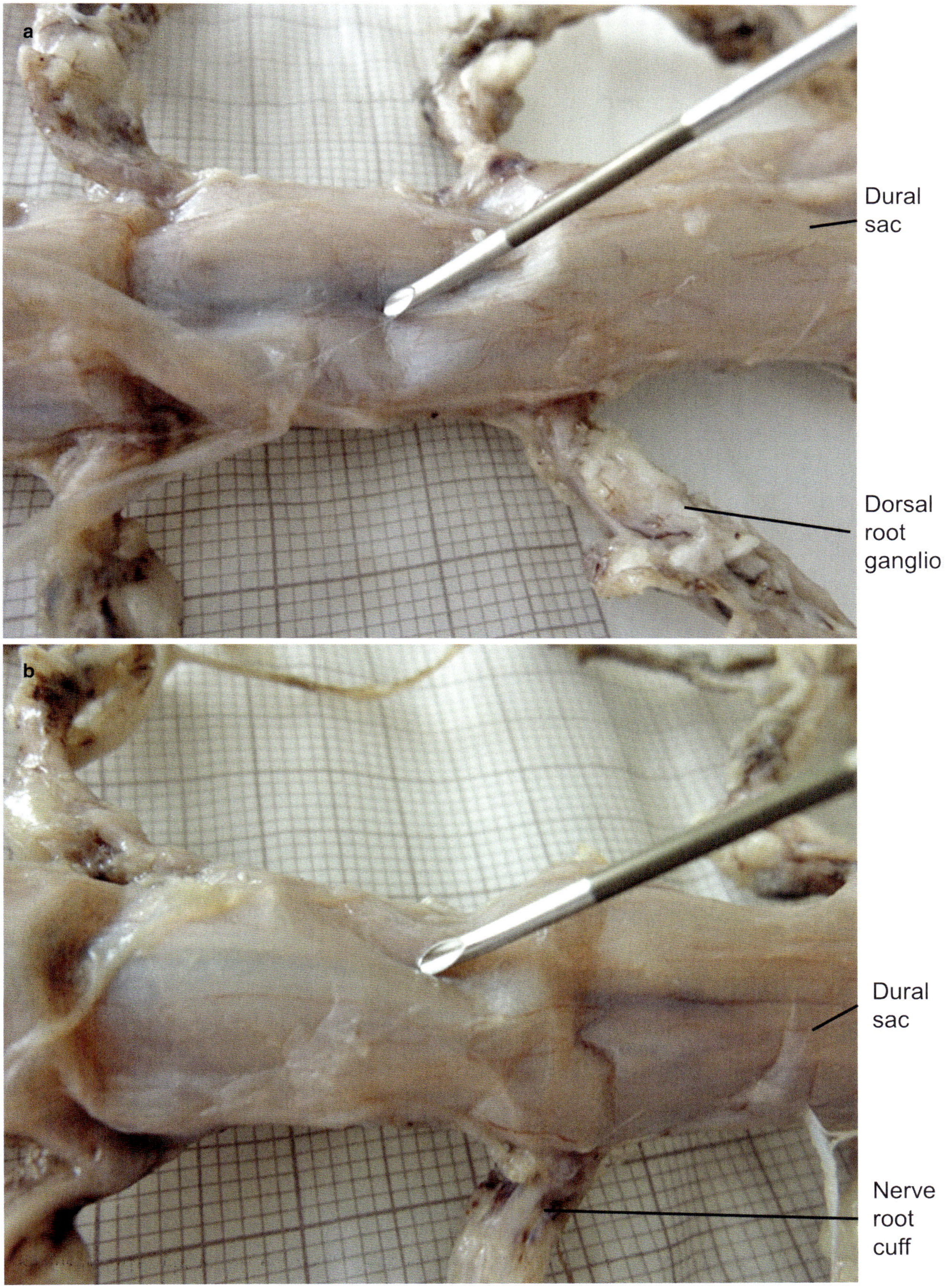

Fig. 31.4 In vitro sample of human dural sac. Epidural needle (18-gauge Tuohy) contacting the dural sac. (**a**, **b**) Images obtained from same sample

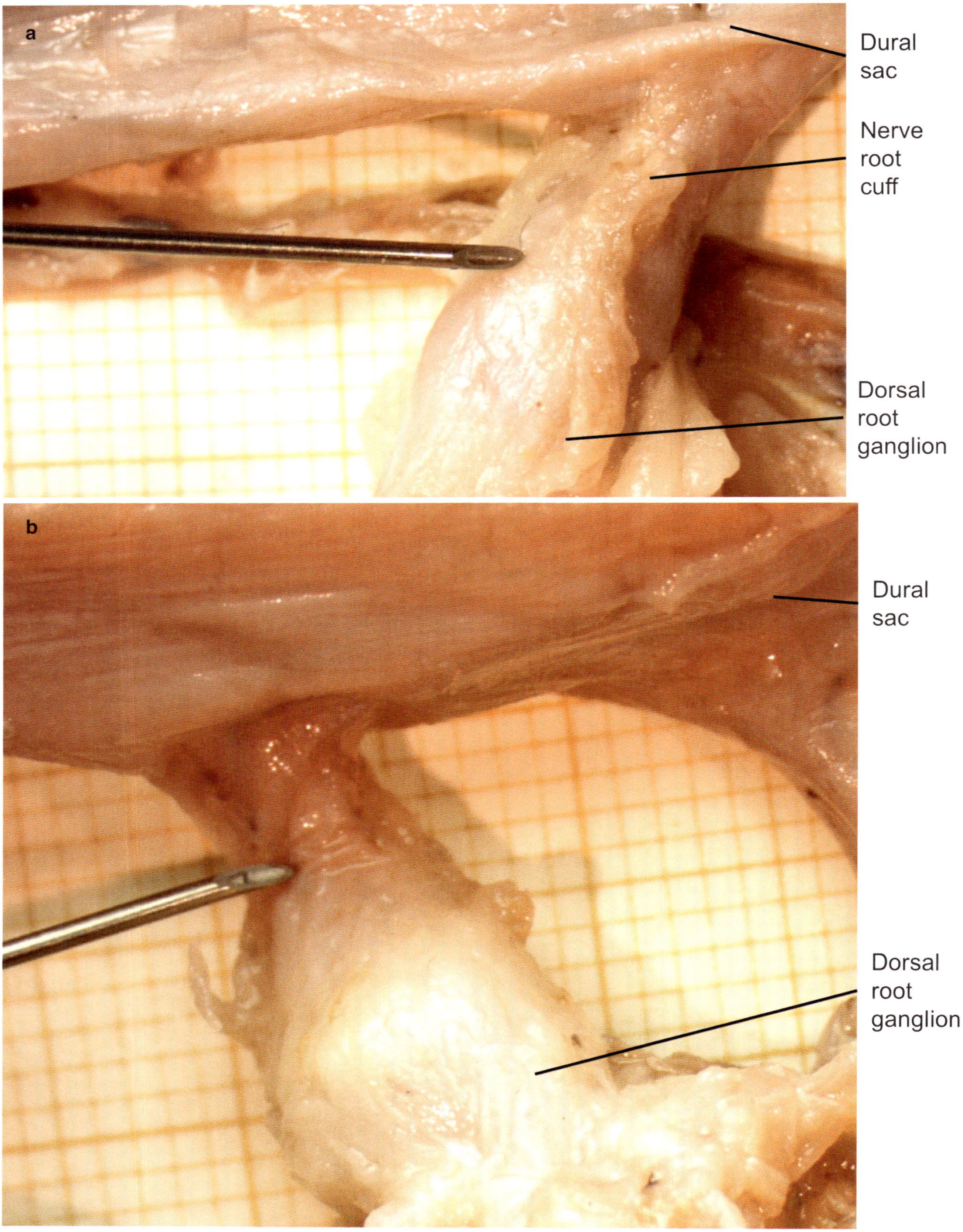

Fig. 31.5 In vitro sample of human dural sac. Spinal needle (22-gauge Quincke) contacting the nerve root cuff. (**a**, **b**) Images obtained from different samples

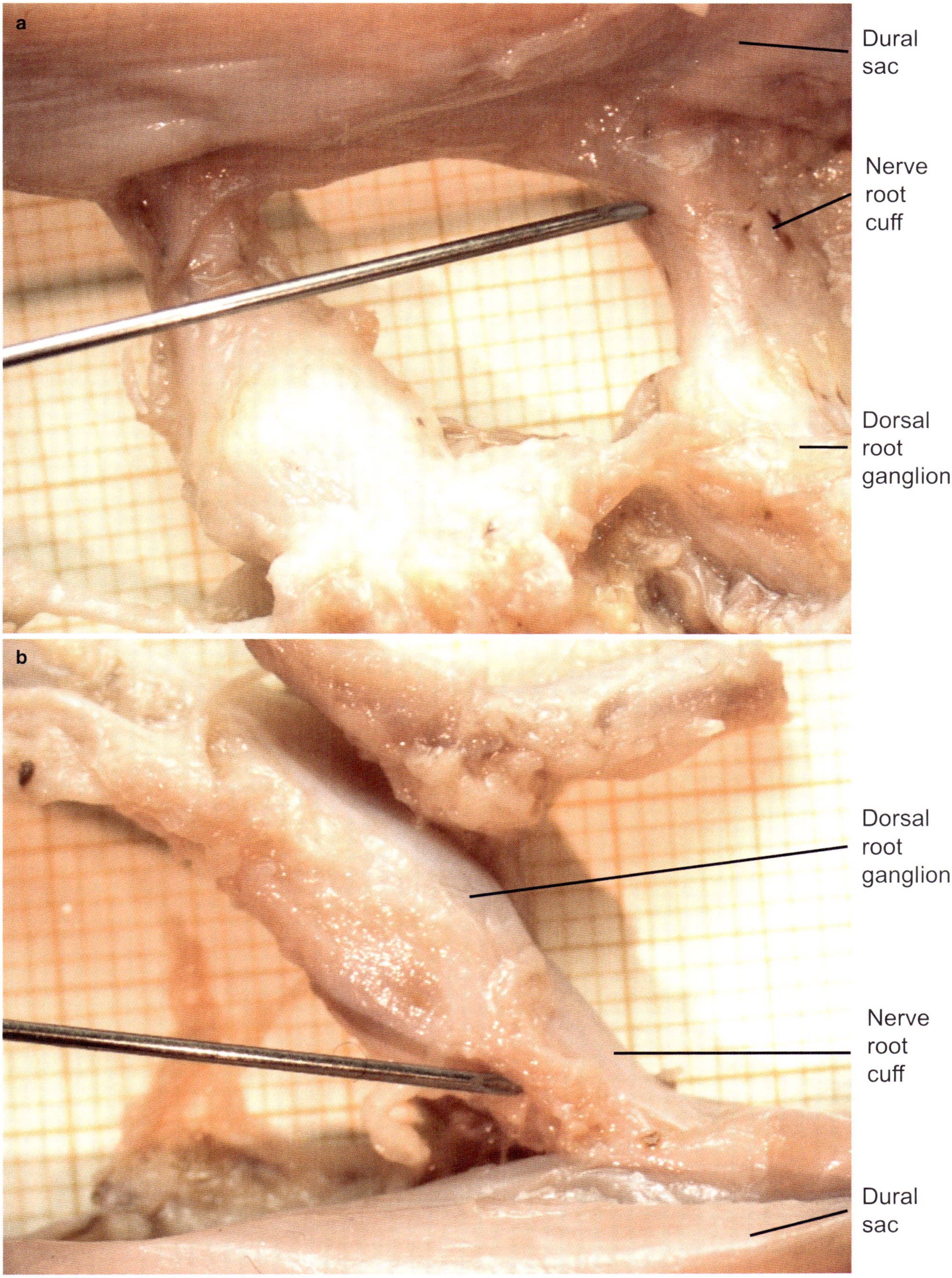

Fig. 31.6 In vitro sample of human dural sac. Spinal needle (22-gauge Quincke) contacting the nerve root cuff. (**a**, **b**) Images obtained from different samples

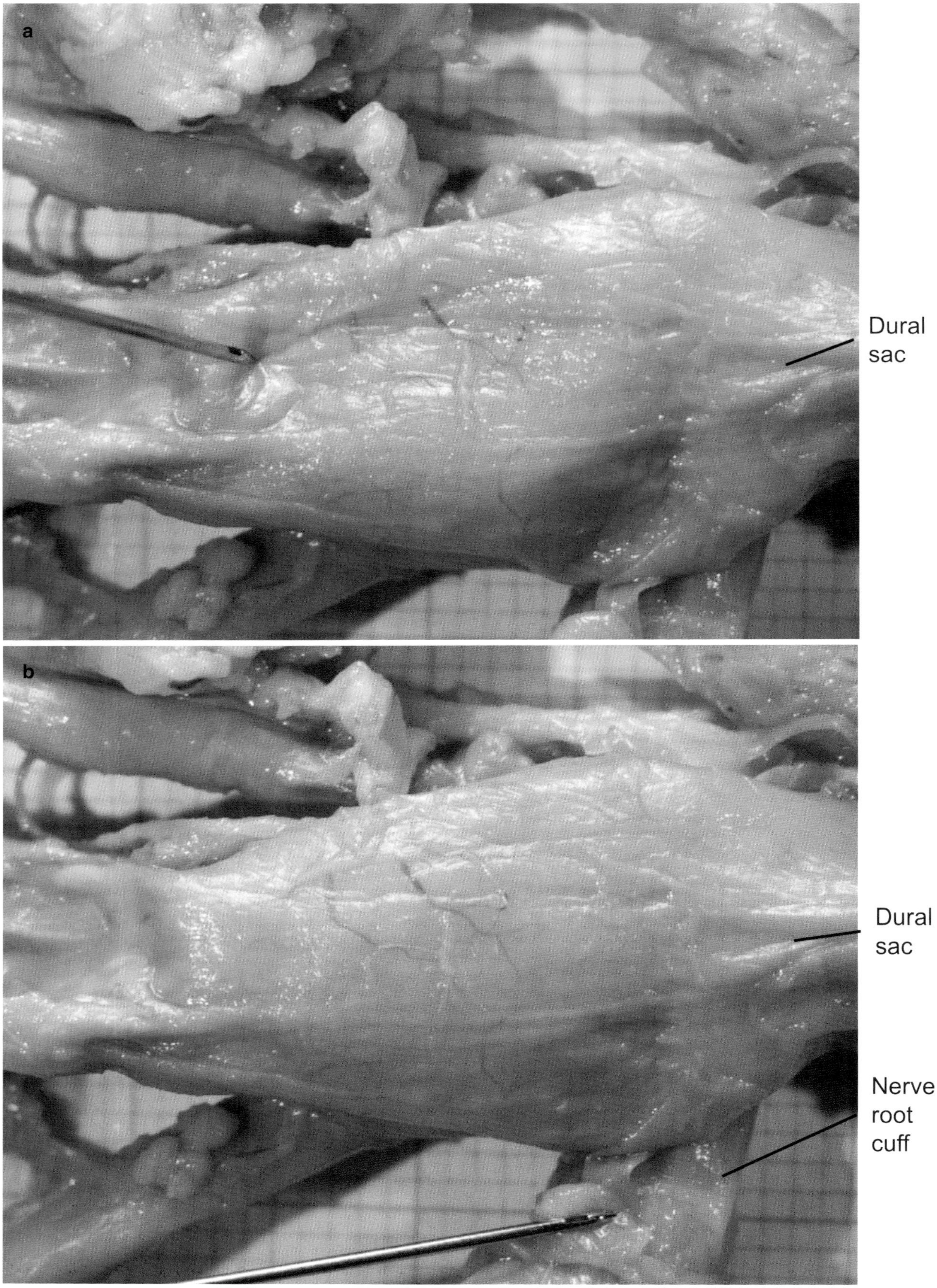

Fig. 31.7 In vitro sample of human dural sac. Spinal needle (25-gauge Whitacre) contacting the nerve root cuff. (**a**, **b**) Images obtained from same sample

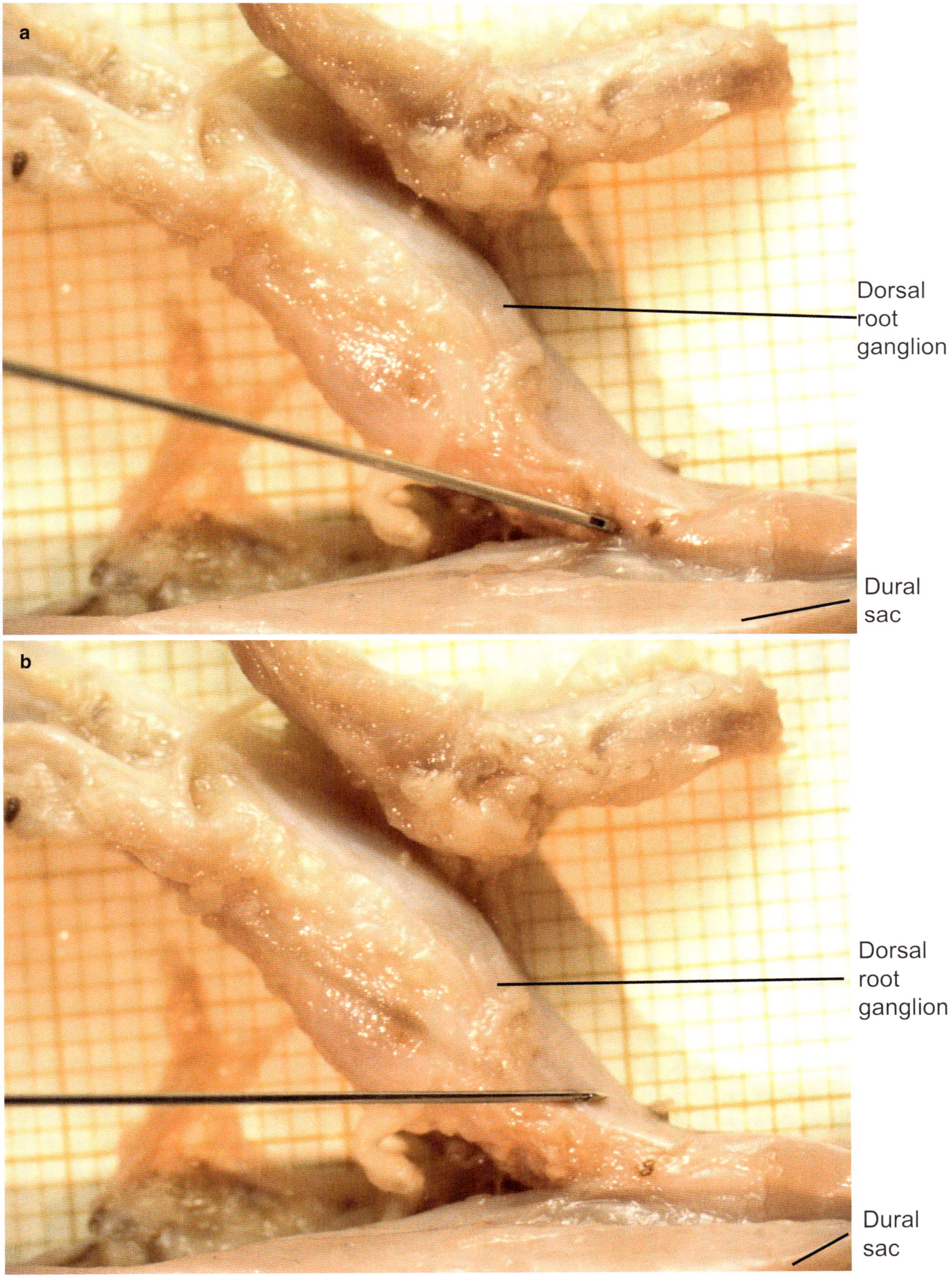

Fig. 31.8 In vitro sample of human dural sac. Spinal needle (25-gauge Whitacre) contacting the nerve root cuff. (**a**, **b**) Images obtained from same sample

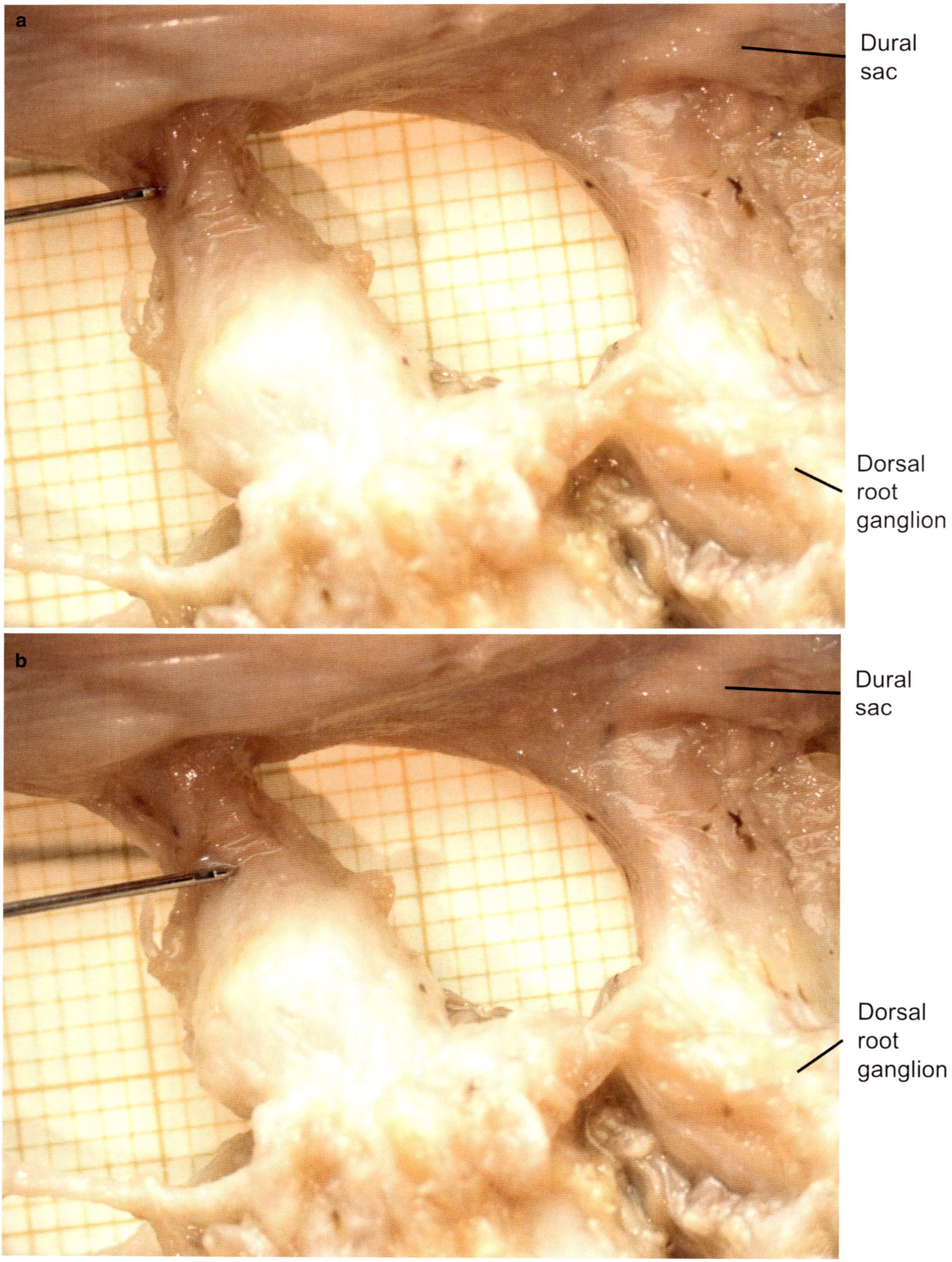

Fig. 31.9 In vitro sample of human dural sac. Spinal needle (25-gauge Whitacre) contacting the nerve root cuff. (**a**, **b**) Images obtained from same sample

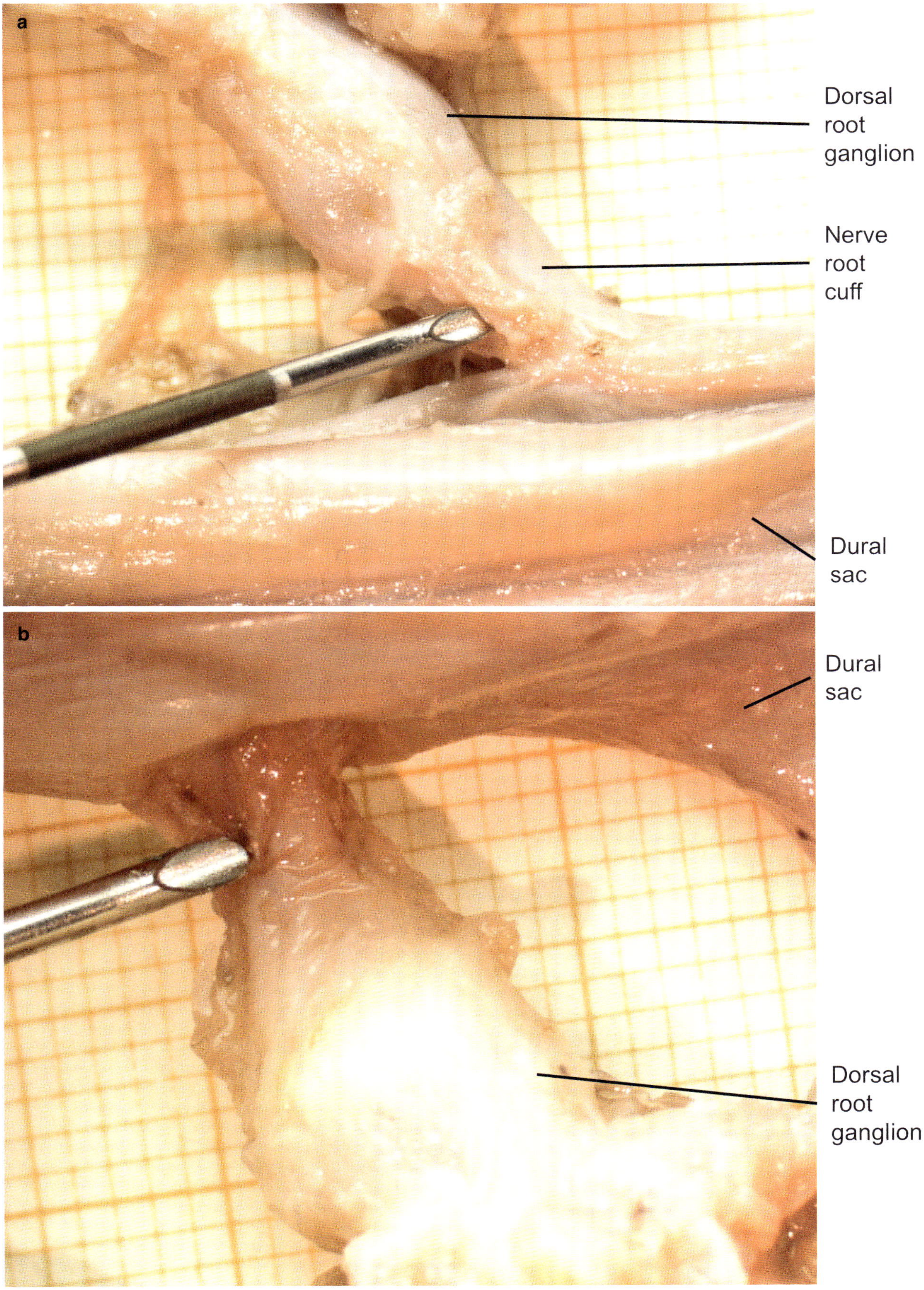

Fig. 31.10 In vitro sample of human dural sac. Epidural needle (18-gauge Tuohy) contacting the nerve root cuff. (**a**, **b**) Images obtained from different samples

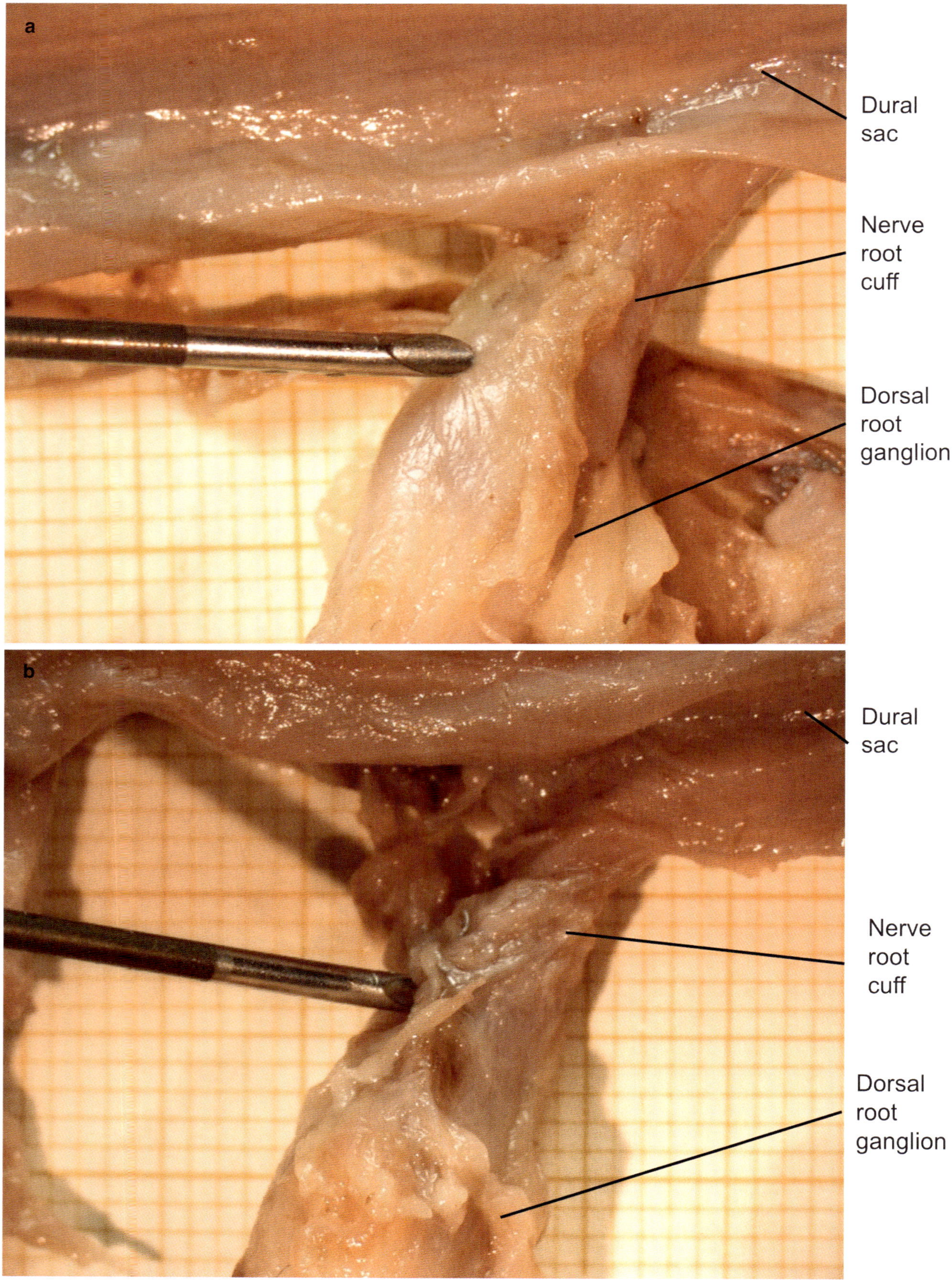

Fig. 31.11 In vitro sample of human dural sac. Epidural needle (18-gauge Tuohy) contacting the nerve root cuff. (**a**, **b**) Images obtained from different samples

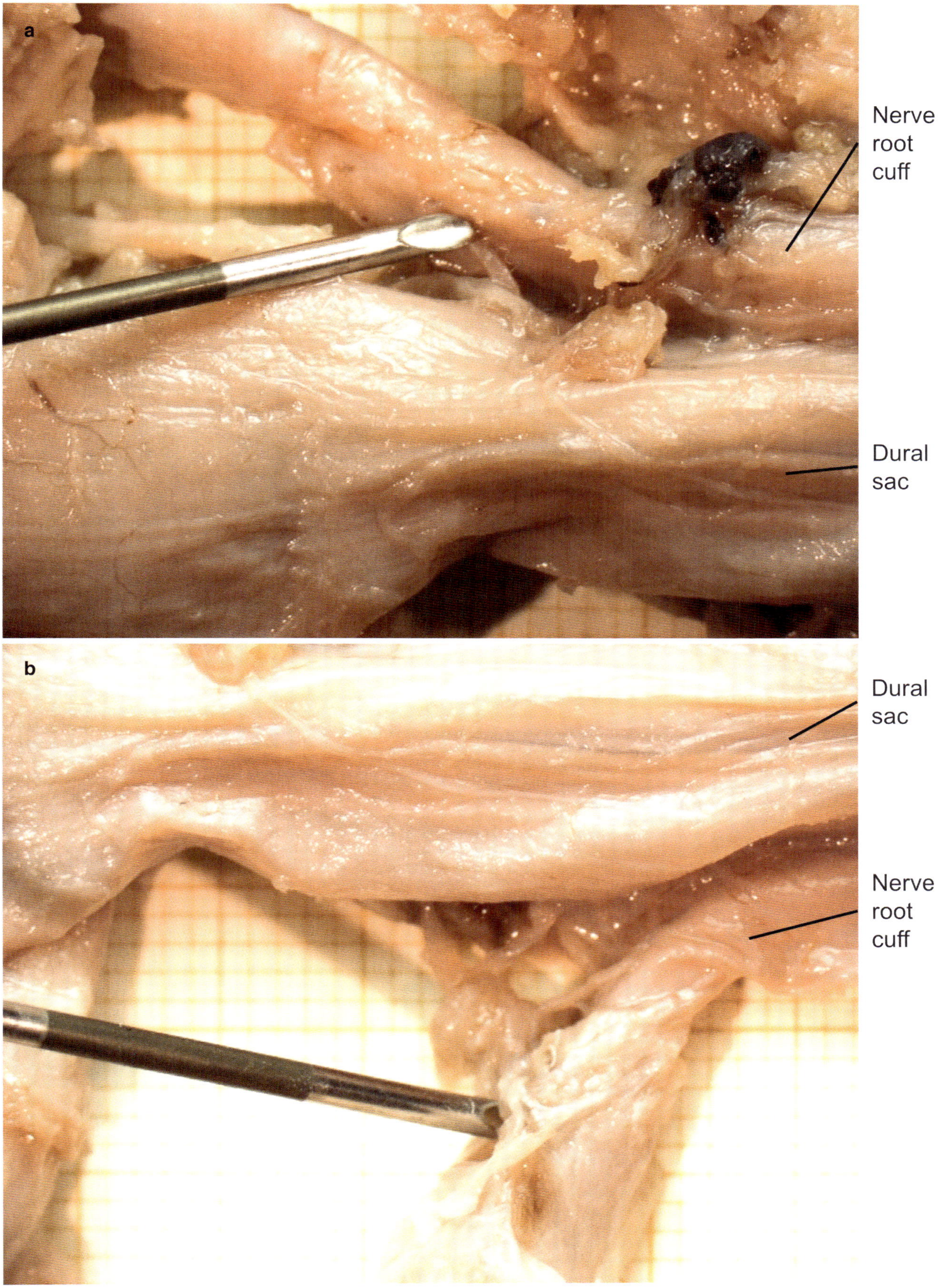

Fig. 31.12 In vitro sample of human dural sac. Epidural needle (18-gauge Tuohy) contacting the nerve root cuff. (**a**, **b**) Images obtained from different samples

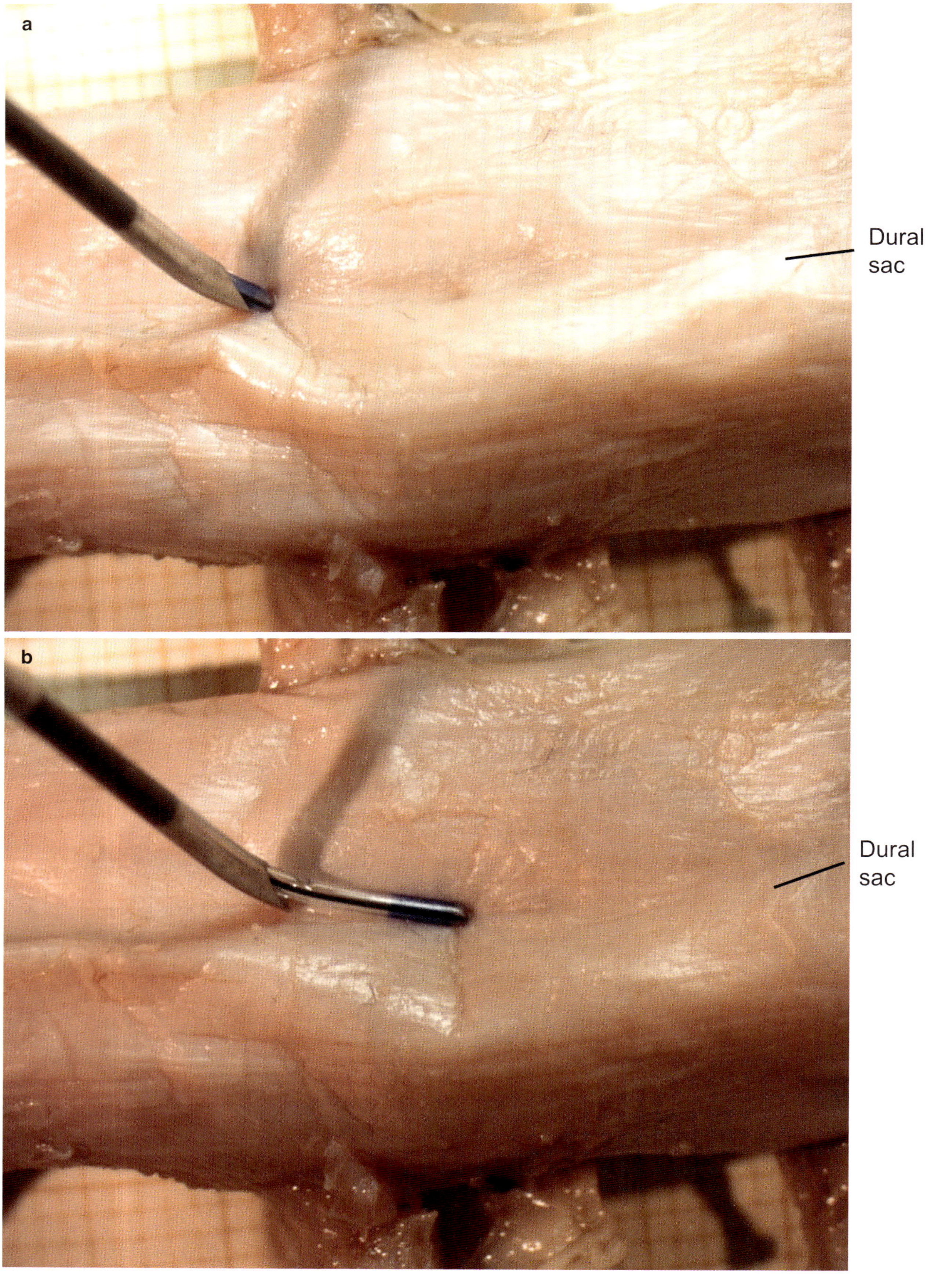

Fig. 31.13 In vitro sample of human dural sac. Epidural catheter (20-gauge with multiorifice, closed-end orifice) contacting the dural sac. (**a**, **b**) Images obtained from same sample

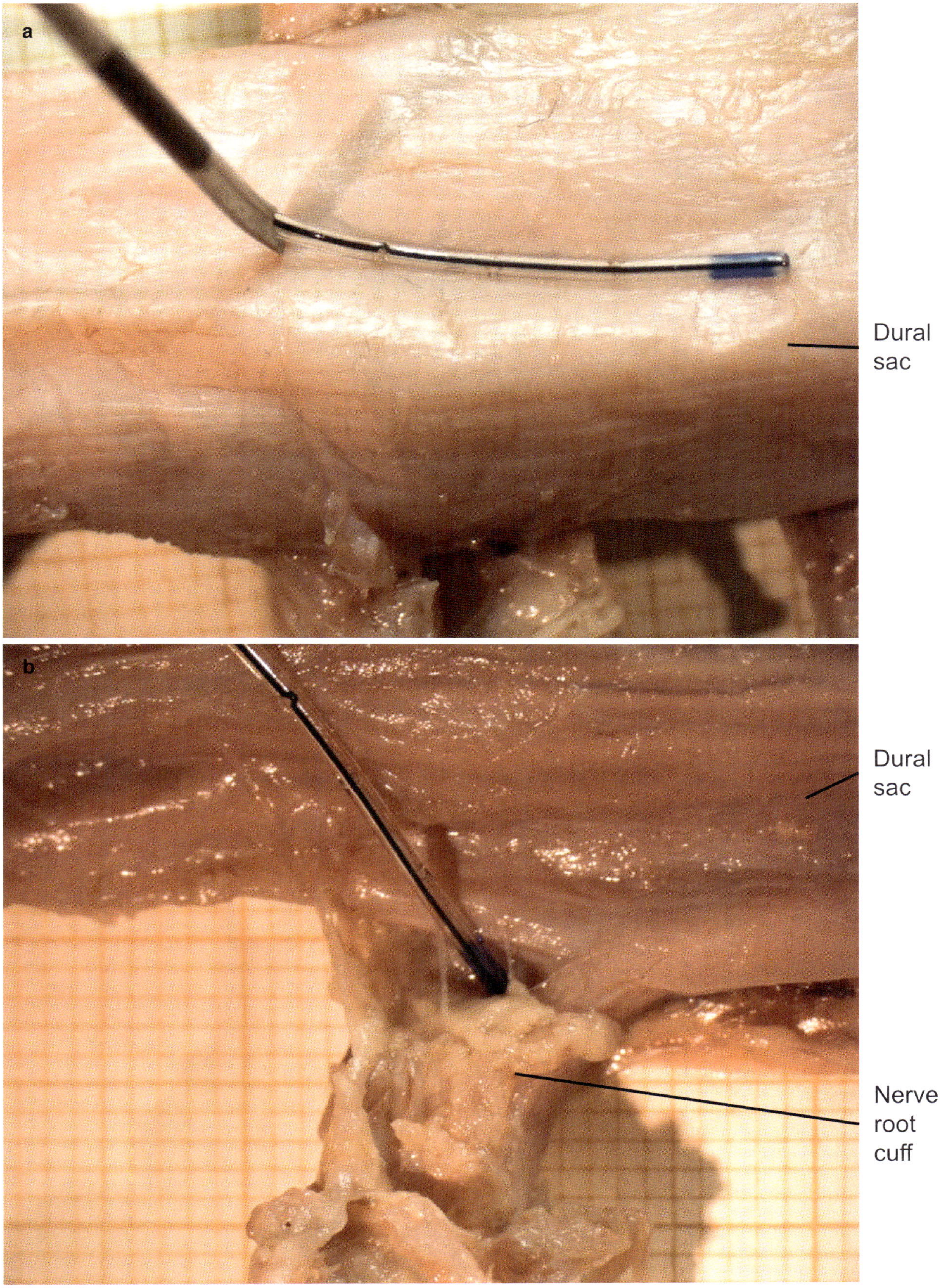

Fig. 31.14 In vitro sample of human dural sac. Epidural catheter (20-gauge with multiorifice, closed-end orifice) contacting the dural sac (**a**) and nerve root cuff (**b**). (**a**, **b**) Images obtained from different samples

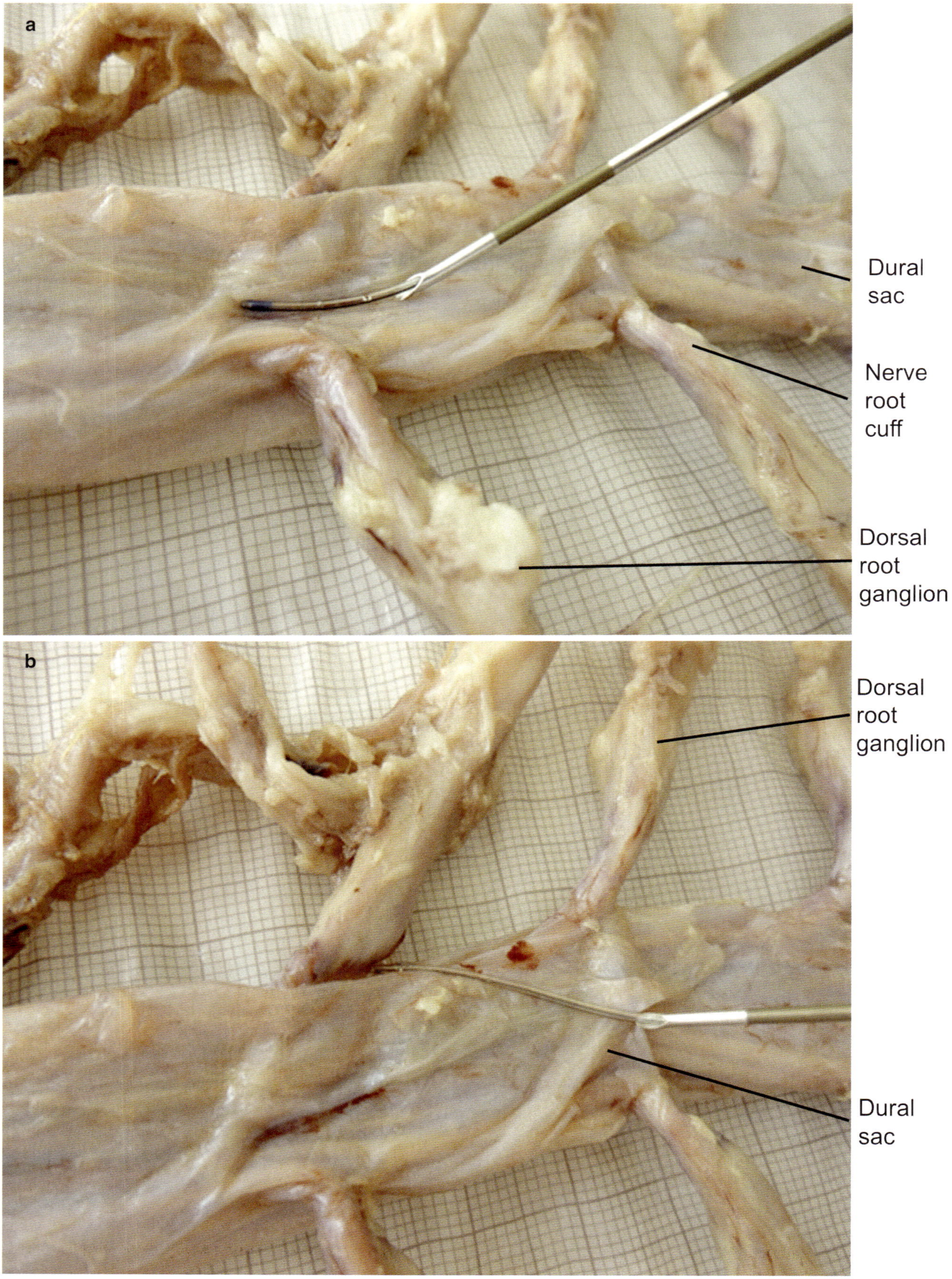

Fig. 31.15 In vitro sample of human dural sac. Epidural catheter (20-gauge with multiorifice, closed-end orifice) contacting the dural sac. (**a**, **b**) Images obtained from same sample

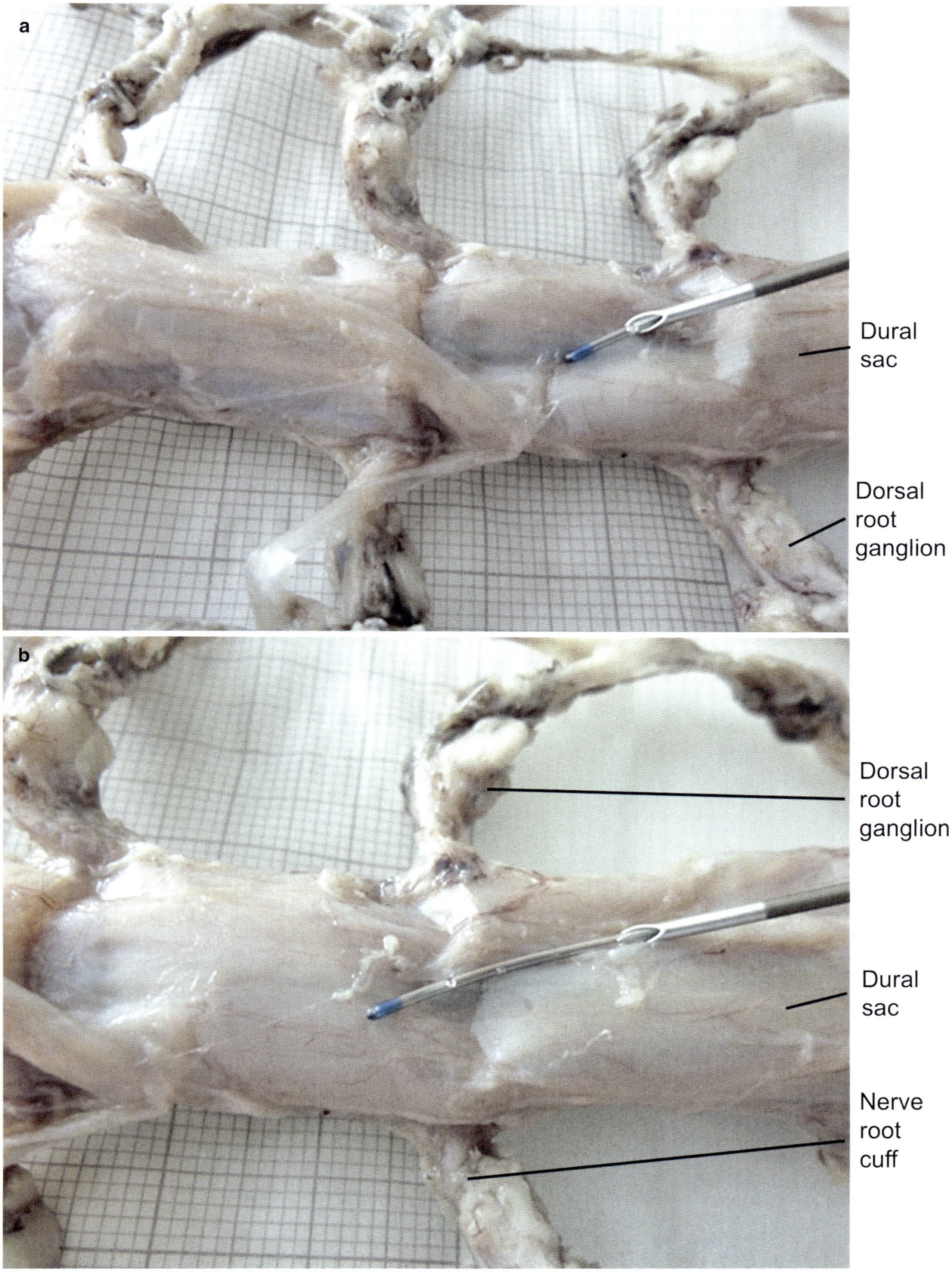

Fig. 31.16 In vitro sample of human dural sac. Epidural catheter (20-gauge with multiorifice, closed-end orifice) contacting the dural sac. (a, b) Images obtained from same sample

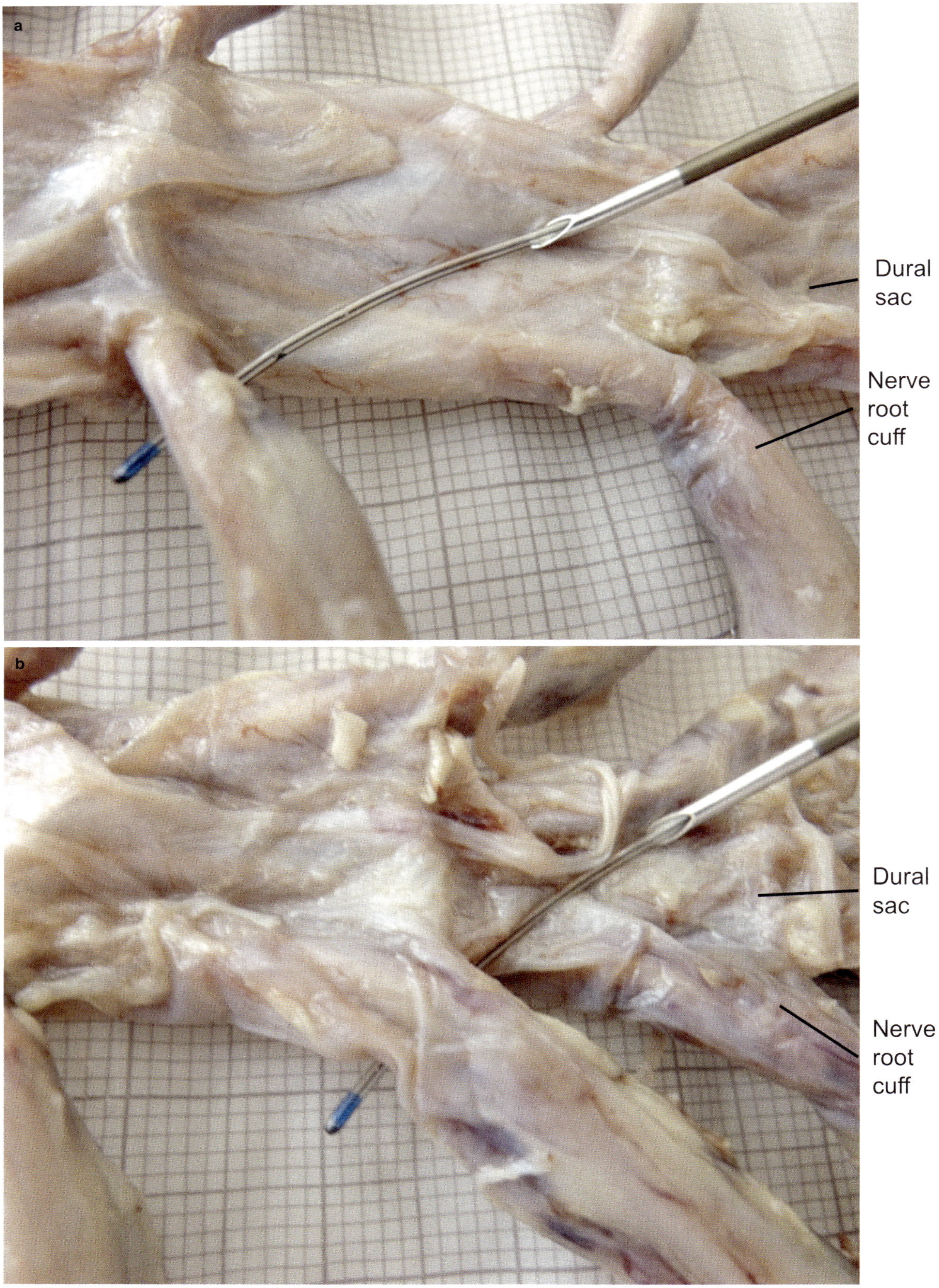

Fig. 31.17 In vitro sample of human dural sac. Epidural catheter (20-gauge with multiorifice, closed-end orifice) contacting the nerve root cuff. (**a**, **b**) Images obtained from same sample

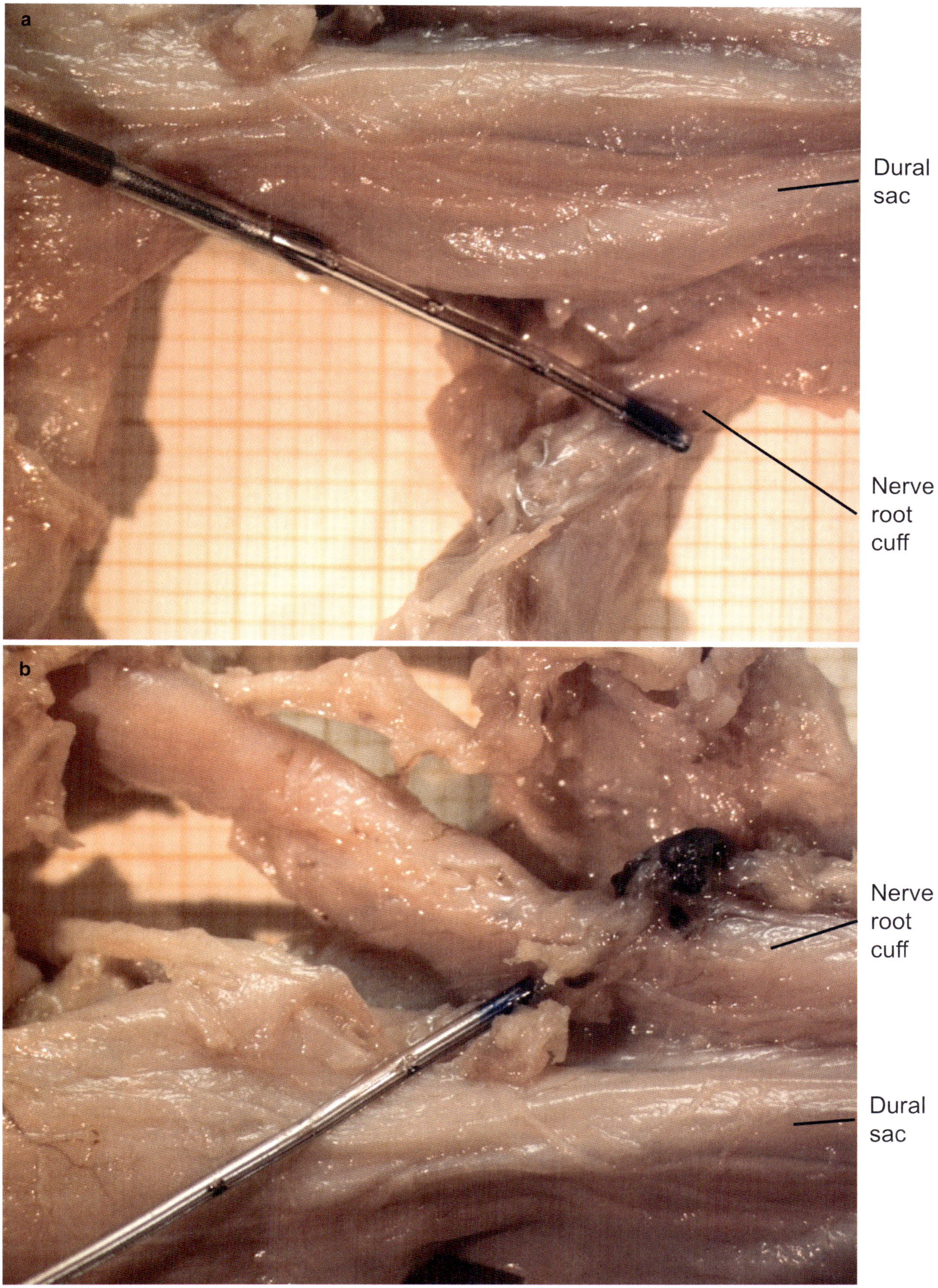

Fig. 31.18 In vitro sample of human dural sac. Epidural catheter (20-gauge with multiorifice, closed-end orifice) contacting the nerve root cuff. (**a**, **b**) Images obtained from different samples

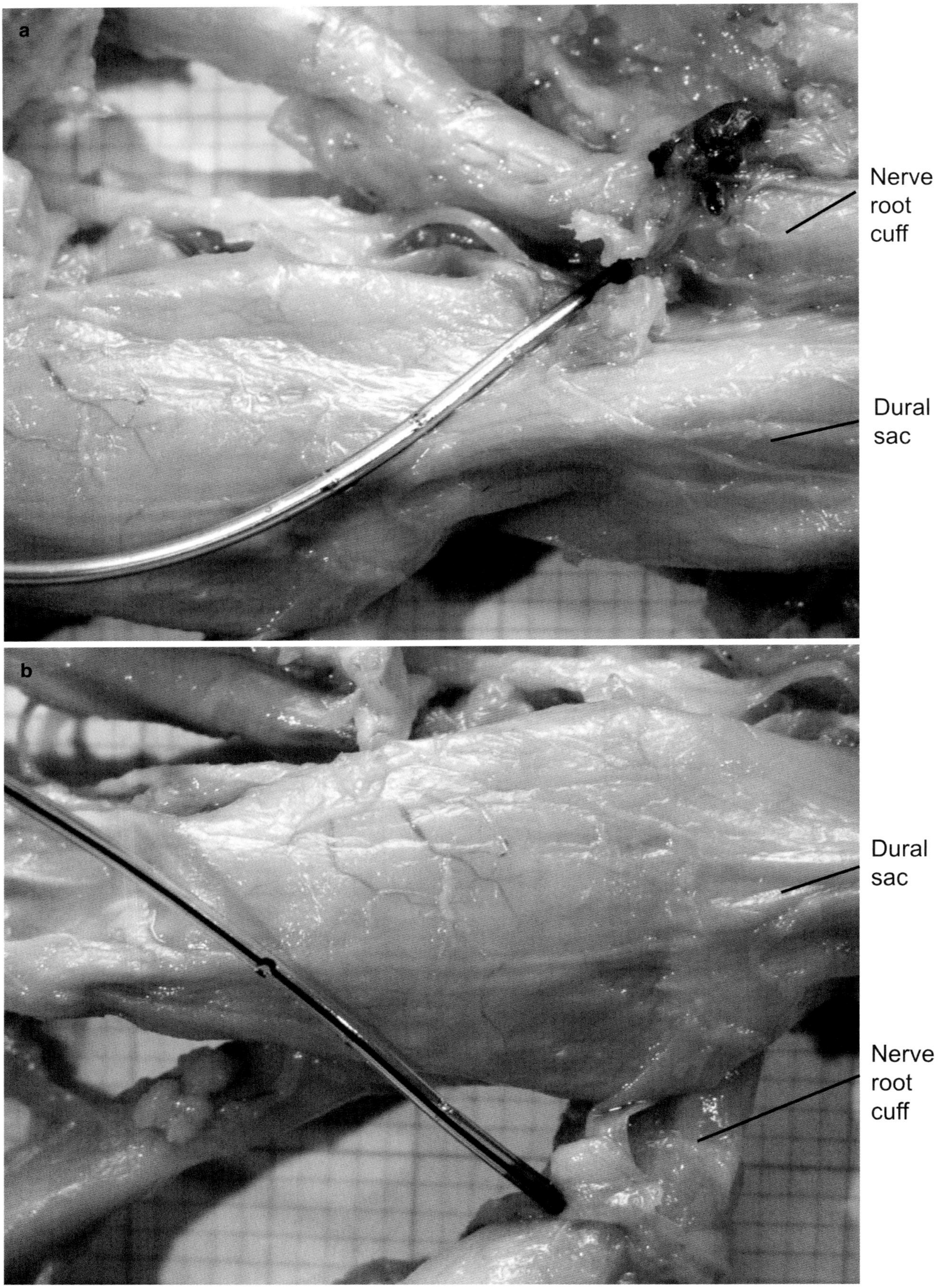

Fig. 31.19 In vitro sample of human dural sac. Epidural catheter (20-gauge with multiorifice, closed-end orifice) contacting the nerve root cuff. (**a**, **b**) Images obtained from same sample

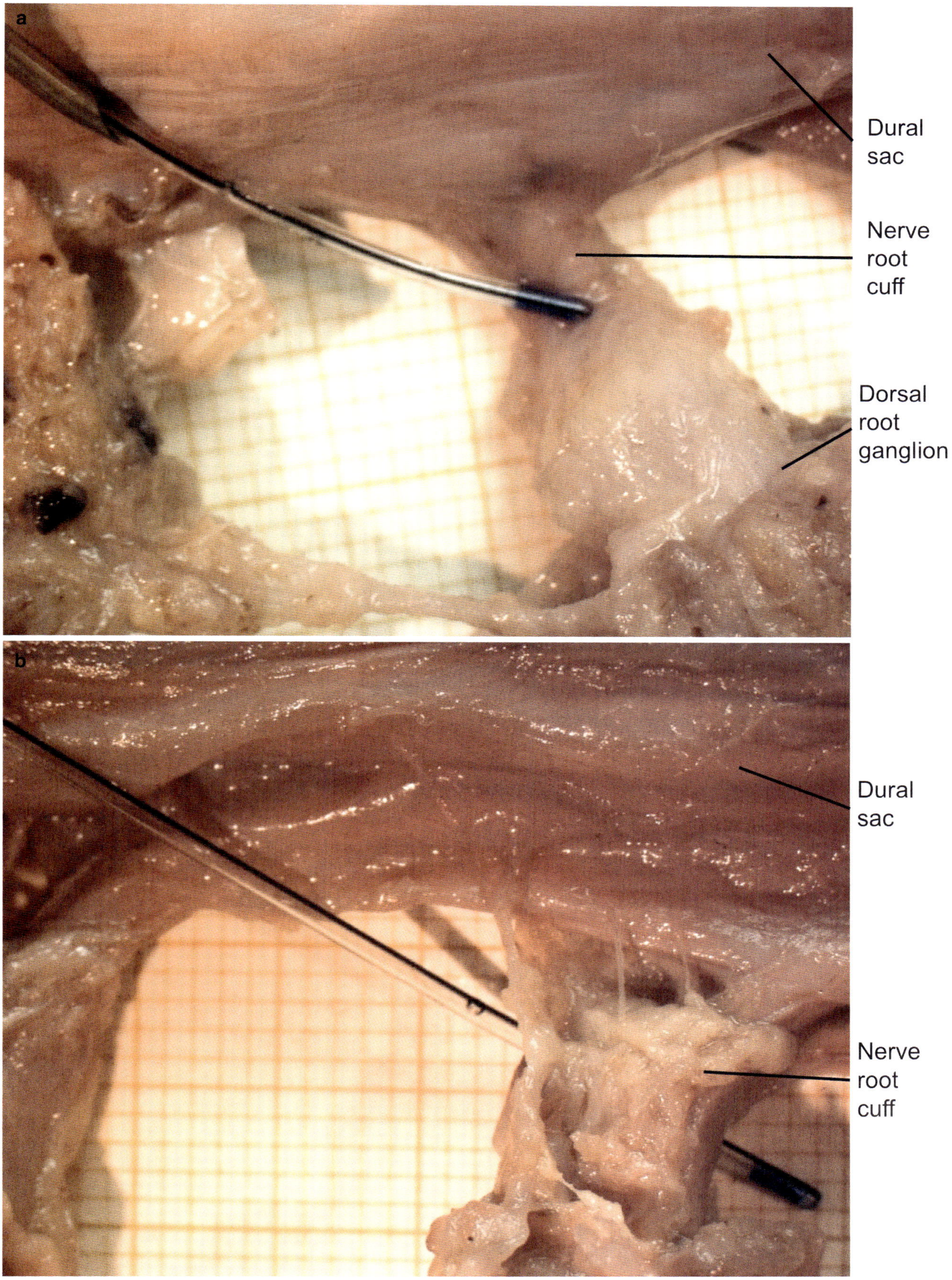

Fig. 31.20 In vitro sample of human dural sac. Epidural catheter (20-gauge with multiorifice, closed-end orifice) contacting the nerve root cuff. (**a**, **b**) Images obtained from different samples

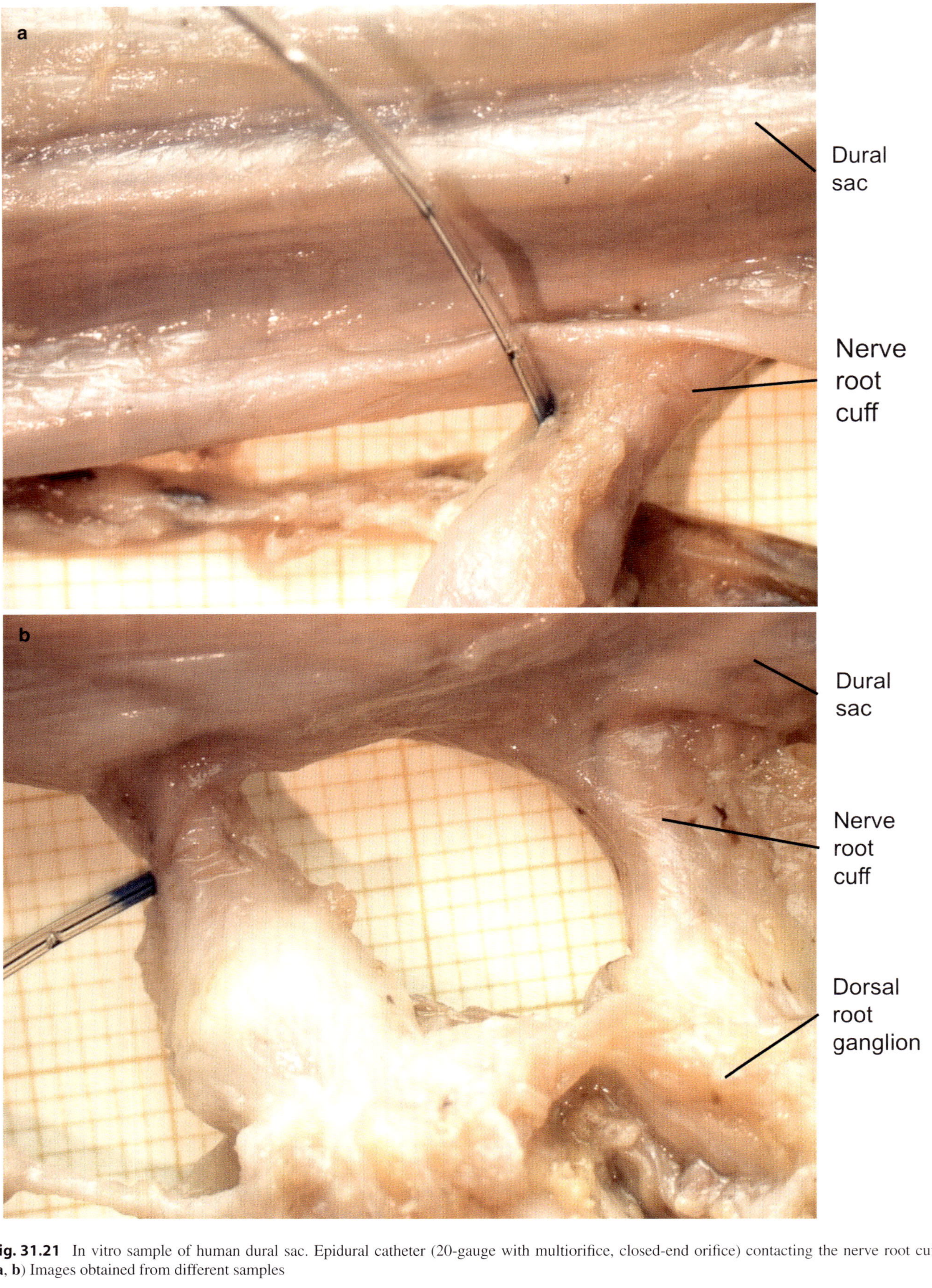

Fig. 31.21 In vitro sample of human dural sac. Epidural catheter (20-gauge with multiorifice, closed-end orifice) contacting the nerve root cuff. (**a**, **b**) Images obtained from different samples

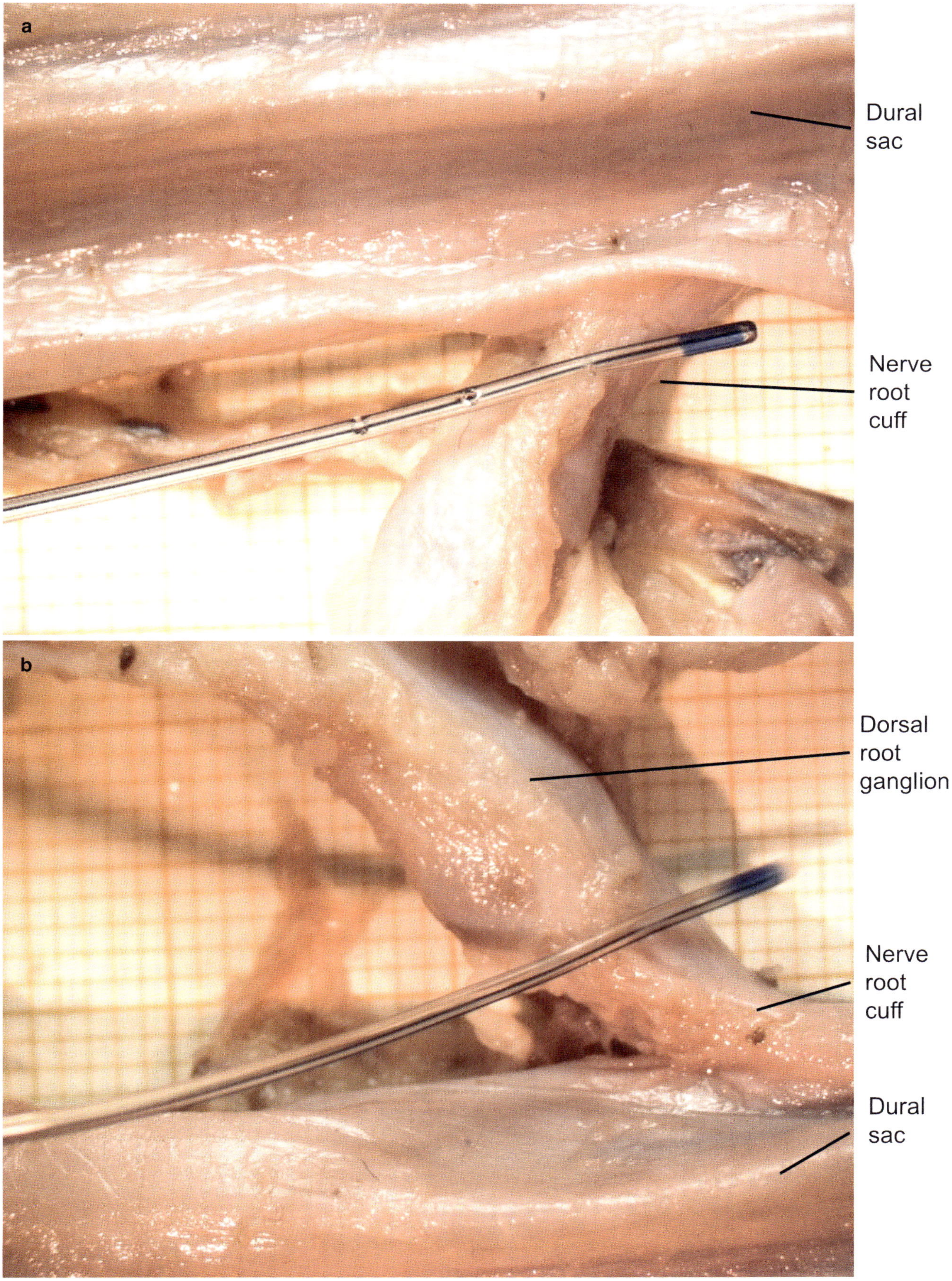

Fig. 31.22 In vitro sample of human dural sac. Epidural catheter (20-gauge with multiorifice, closed-end orifice) contacting the nerve root cuff. (**a**, **b**) Images obtained from different samples

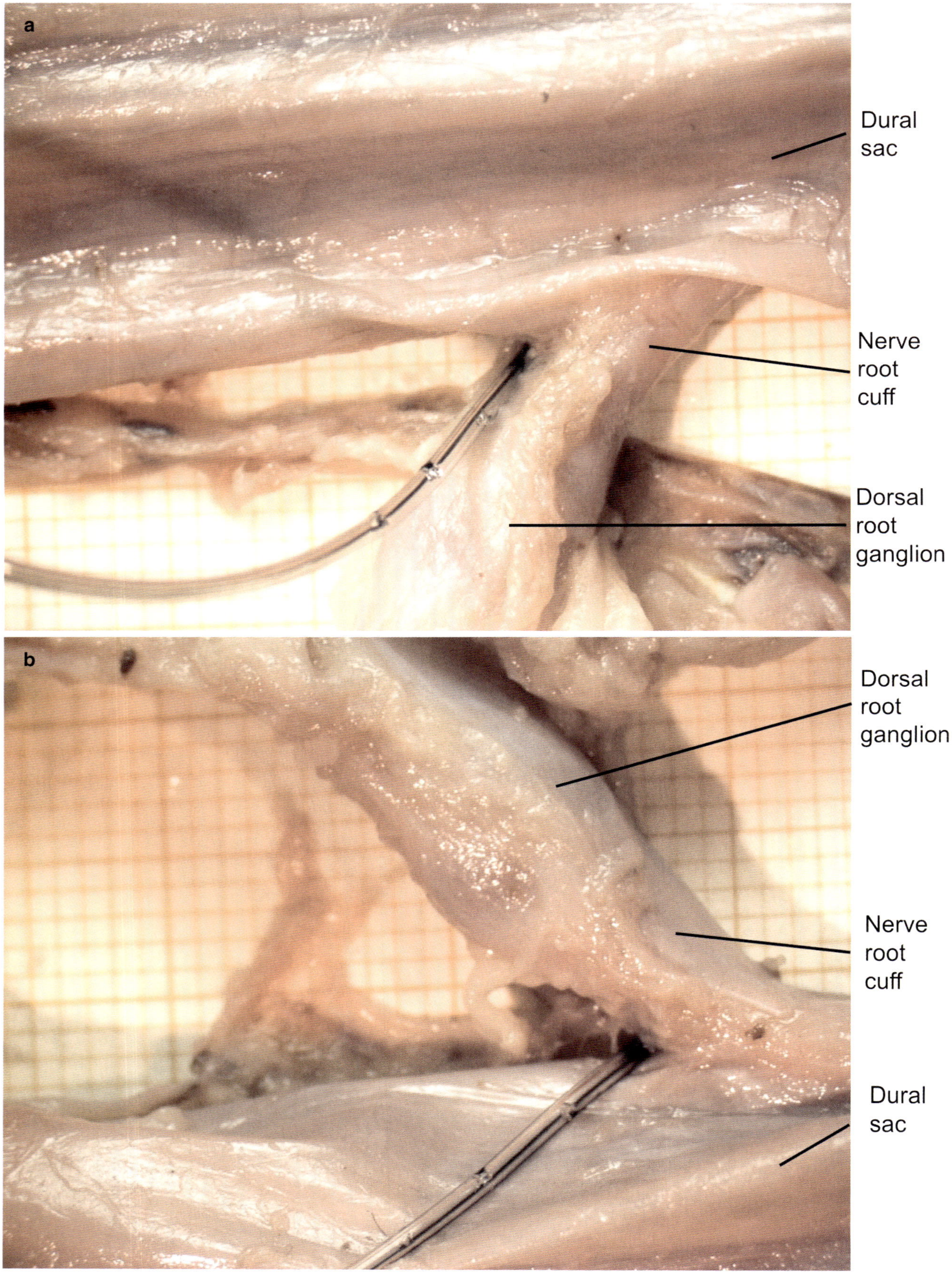

Fig. 31.23 In vitro sample of human dural sac. Epidural catheter (20-gauge with multiorifice, closed-end orifice) contacting the nerve root cuff. (**a**, **b**) Images obtained from different samples

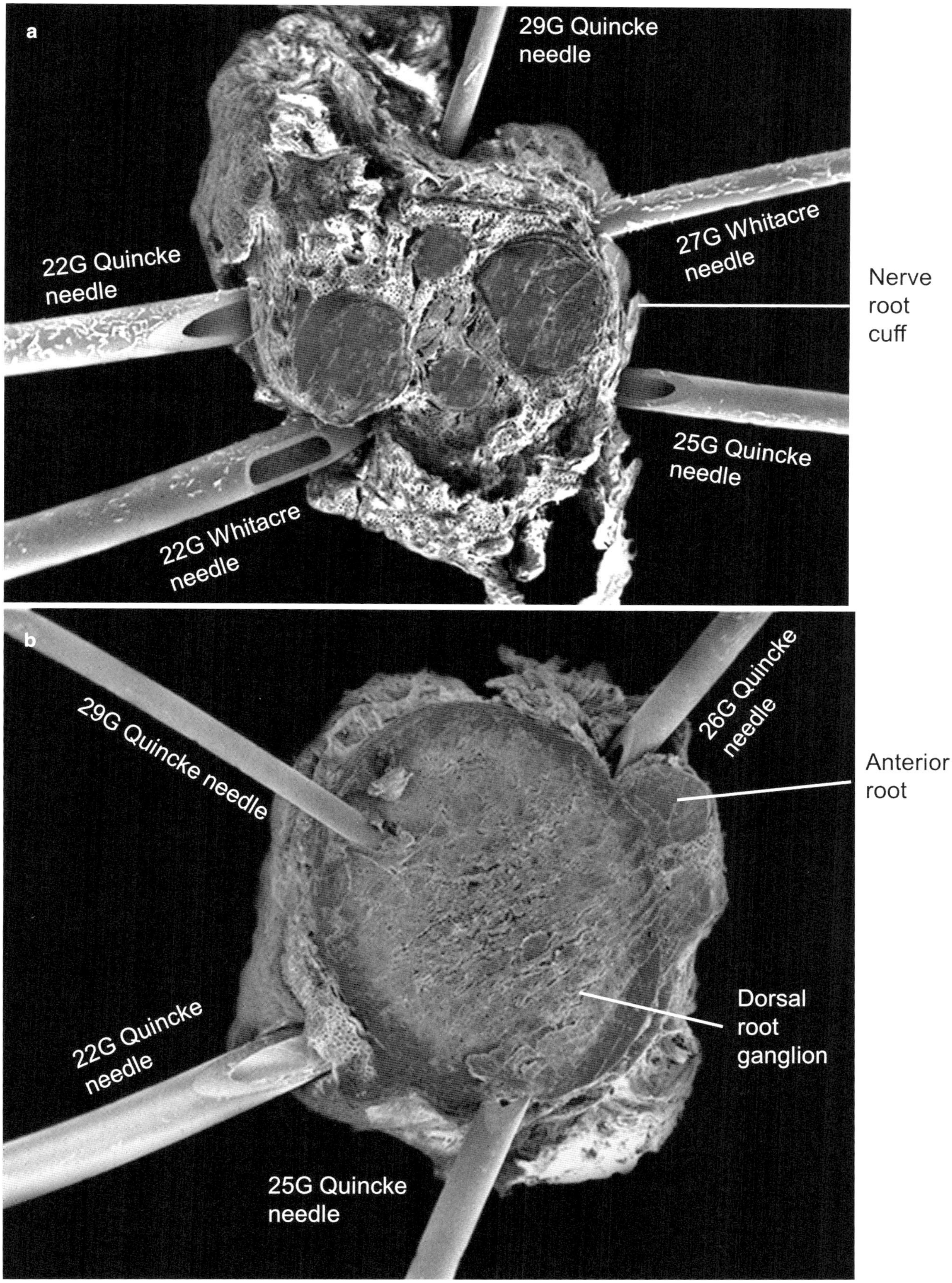

Fig. 31.24 In vitro sample of nerve root cuff. Different types of spinal needles traversing the nerve root cuff. (**a**, **b**) Images obtained from different samples. Scanning electron microscopy; magnification: ×10 (**a**, **b**) (Panel **a** from Reina et al. [1]; with permission. Panel (**b**) from Reina et al. [3]; with permission)

References

1. Reina MA, De Andres J, Hernández JM, Navarro RA, Pastor J, Prats-Galino A. Looking for the development of paresthesias in the subarachnoid and epidural anaesthesia. A clinical and anatomical analysis. Reg Anesth Pain Med. 2011;36(Suppl):E17–22.
2. Reina MA, Villanueva MC, Machés F, Carrera A, López A, De Andrés JA. Ultrastructure of human spinal root cuffs in lumbar spine. Anesth Analg. 2008;106:339–44.
3. Reina MA, Palacio F, Pastor J, Fernández P. Las parestesias en la anestesia regional. In: Wikinski J, Reina MA, Bollini C, et al., editors. Diagnóstico, prevensión y tratamiento de las complicaciones neurológicas asociadas con la anestesia regional periférica y central. Buenos Aires: Editorial Médica Panamericana; 2011. p. 205–18.

32 Ligamentum Flavum and Related Spinal Ligaments

Alberto Prats-Galino, Marija Mavar, Miguel Angel Reina, Anna Puigdellívol-Sánchez, and Joan San Molina

In this chapter, the spatial disposition and topographical relationships of the ligamentum flavum and other ligaments relevant to spinal anesthesia are described. To this end, different three-dimensional (3D) models of MRI of the lumbar region have been used (Figs. 32.1 and 32.2). Details regarding the procedures followed to obtain these models are found in Chap. 47.

The so-called medial approach (Fig. 32.3) is a common procedure used in spinal blocks. At lumbar level, the L3–L4 space is often used, although other spaces may also be used. After passing through the skin and subcutaneous tissue, the needle should move through the supraspinous and interspinous ligaments and finally the ligamentum flavum, reaching the posterior epidural fat.

In the paramedial approach (Fig. 32.4), the needle is placed about 2 cm from the midline and directed medially and cranially to reach the interlaminar space. Once the ligamentum flavum is obliquely crossed, the anesthetic solution can be injected in either the epidural space (epidural anesthesia) or the subarachnoid space (subarachnoid or spinal anesthesia).

A. Prats-Galino, MD, PhD (✉)
M. Mavar, MSc • A. Puigdellívol-Sánchez, MD, PhD
Laboratory of Surgical NeuroAnatomy (LSNA),
Human Anatomy and Embryology Unit,
School of Medicine, University of Barcelona, Barcelona, Spain
e-mail: aprats@ub.edu; marija.mavar@ub.edu; apuigdellivol@ub.edu

M.A. Reina, MD, PhD
Department of Clinical Medical Sciences and Institute of Applied Molecular Medicine, School of Medicine, University of CEU San Pablo, Madrid, Spain

Department of Anesthesiology, Madrid-Montepríncipe University Hospital, Madrid, Spain
e-mail: miguelangel@perticone.e.telefonica.net

J. San Molina, MD, PhD
Neuroscience, Embryology, Molecular Oncology, and Clinical Anatomy Group, School of Medicine, University of Girona, Girona, Spain
e-mail: joan.san@udg.edu

M.A. Reina et al. (eds.), *Atlas of Functional Anatomy for Regional Anesthesia and Pain Medicine: Human Structure, Ultrastructure and 3D Reconstruction Images*, DOI 10.1007/978-3-319-09522-6_32,

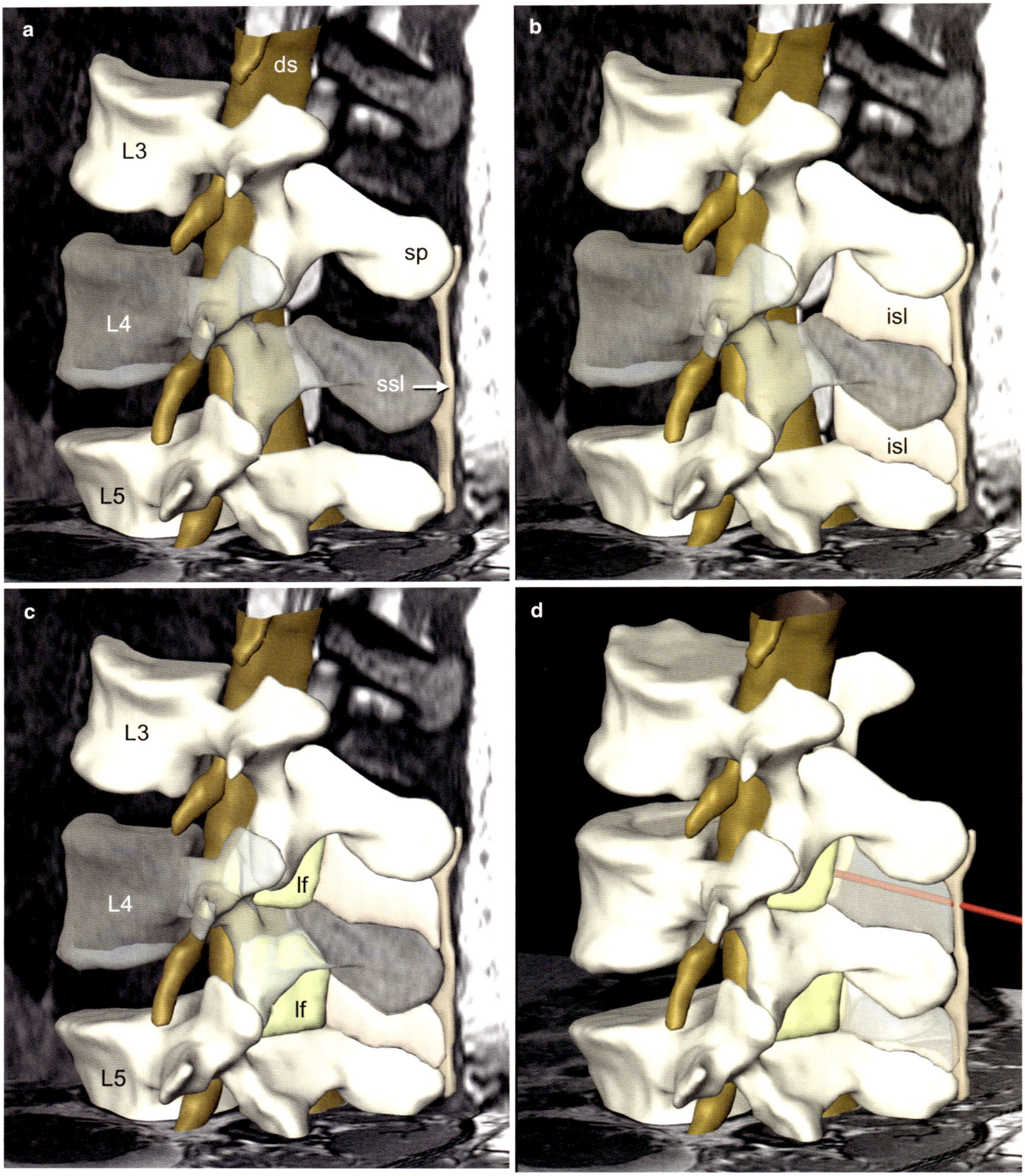

Fig. 32.1 Posterior ligaments of interest in lumbar spinal blockade. Different ligaments may be traversed to reach the spinal canal: (**a**) The supraspinal ligament (*ssl*), a slender structure that extends between the tips of adjacent spinous processes (*sp*); (**b**) The interspinal ligament (*isl*), a band of collagen fibers occupying the interspinal space; (**c**) The ligamentum flavum (*lf*), the Latin name of which refers to its *yellowish* color due to its high content of elastic fibers. (**d**) The trajectory of a needle is visualized in this model by applying a partial transparent effect to the interspinal ligament. It is generally accepted that the sign of loss of resistance results from traversing the ligamentum flavum, and aids in locating the epidural space. The three-dimensional (3D) models of the lumbar spine are built from axial (**a**–**d**) and sagittal (**a**–**c**) T2-weighted reference images. *ds* dural sac, *L3* L3 vertebra, *L4* L4 vertebra, *L5* L5 vertebra

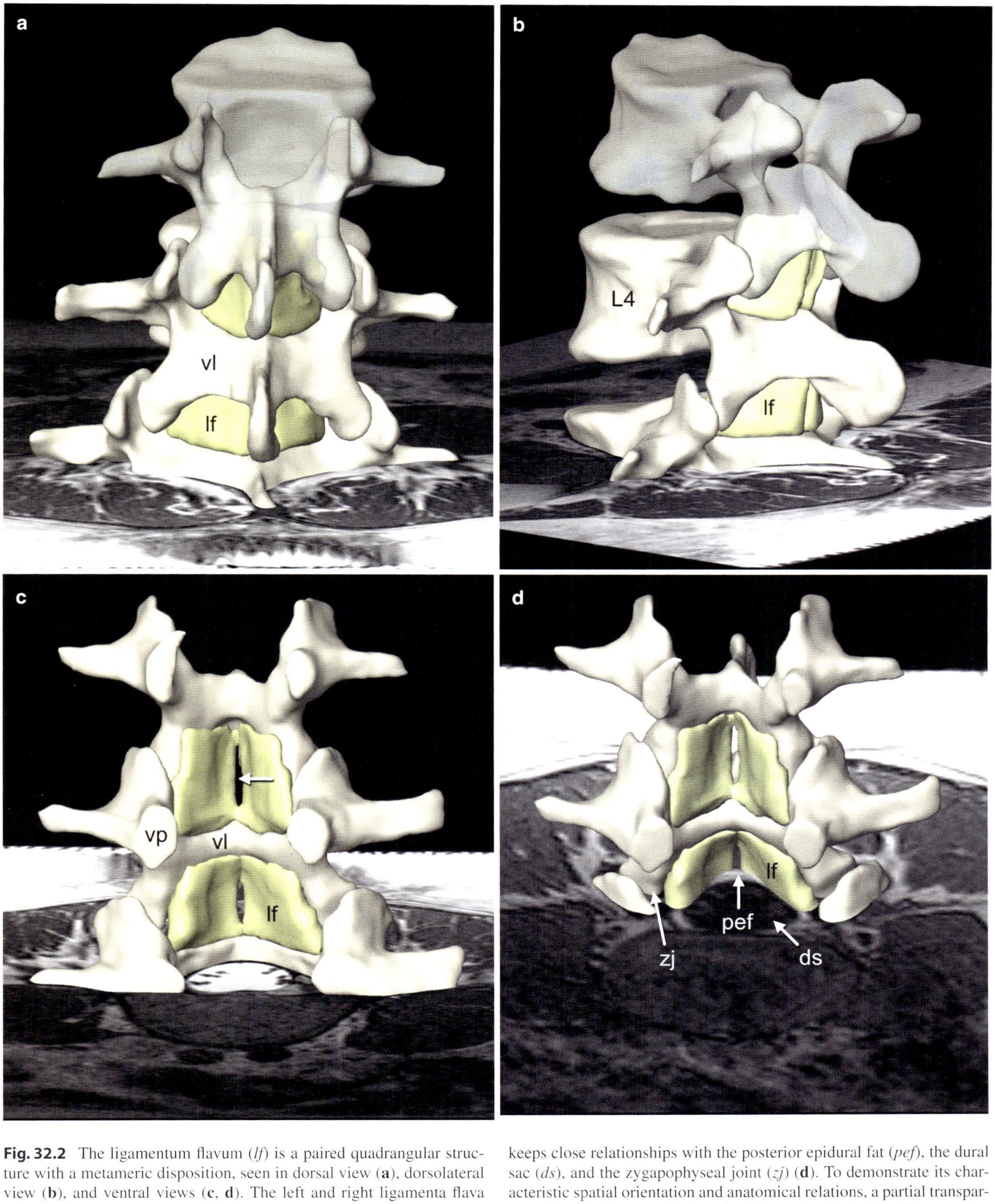

Fig. 32.2 The ligamentum flavum (*lf*) is a paired quadrangular structure with a metameric disposition, seen in dorsal view (**a**), dorsolateral view (**b**), and ventral views (**c**, **d**). The left and right ligamenta flava tend to fuse in the middle, but a gap may persist at some vertebral levels, as represented in this 3D model (*arrow*, **c**). This ligament is attached cranially to the inner surface of the vertebral lamina (*vl*) of the contiguous superior vertebra, as well as to the outer surface of the lamina from the contiguous inferior vertebra. It occupies the interlaminar space and keeps close relationships with the posterior epidural fat (*pef*), the dural sac (*ds*), and the zygapophyseal joint (*zj*) (**d**). To demonstrate its characteristic spatial orientation and anatomical relations, a partial transparent effect has been applied to the L3 vertebra in (**a**, **b**). The posterior vertebral arch is shown by cutting the model through the vertebral pedicles (*vp*), and a T2-weighted axial image (**a**–**c**) or a T1-weighted axial image (**d**) is used as reference. *L4* L4 vertebra, *ss* subarachnoid space

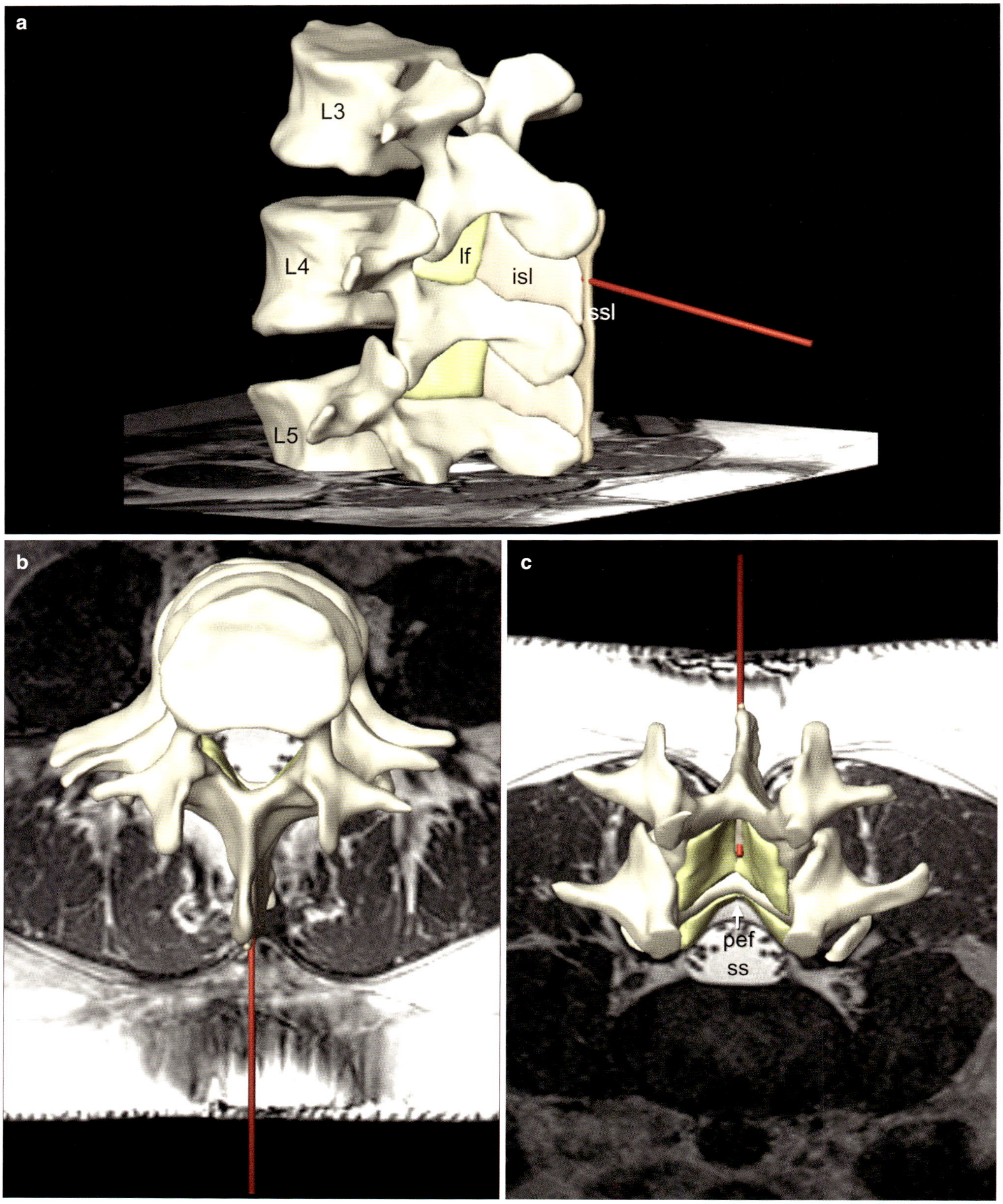

Fig. 32.3 Medial approach used in spinal blockade. The *red cylinder* represents the trajectory of the needle in dorsolateral (**a**), superior (**b**), and ventrocranial views (**c**). This approach enables the identification of the tips of two adjacent spinous processes, usually from L3 and L4 vertebrae, so that the needle can be directed perpendicular to the skin through the supraspinous ligament (*ssl*) and interspinous ligament (*isl*), and then across the ligamentum flavum, to reach the posterior epidural fat (*pef*). The presence of a gap between the ligamentum flavum on both sides may explain a false loss-of-resistance sign, which is used to locate the epidural space in some patients. A T2-weighted axial image is used as reference. *L3* L3 vertebra, *L4* L4 vertebra, *L5* L5 vertebra. *ss* subarachnoid space, *lf* ligamentum flavum

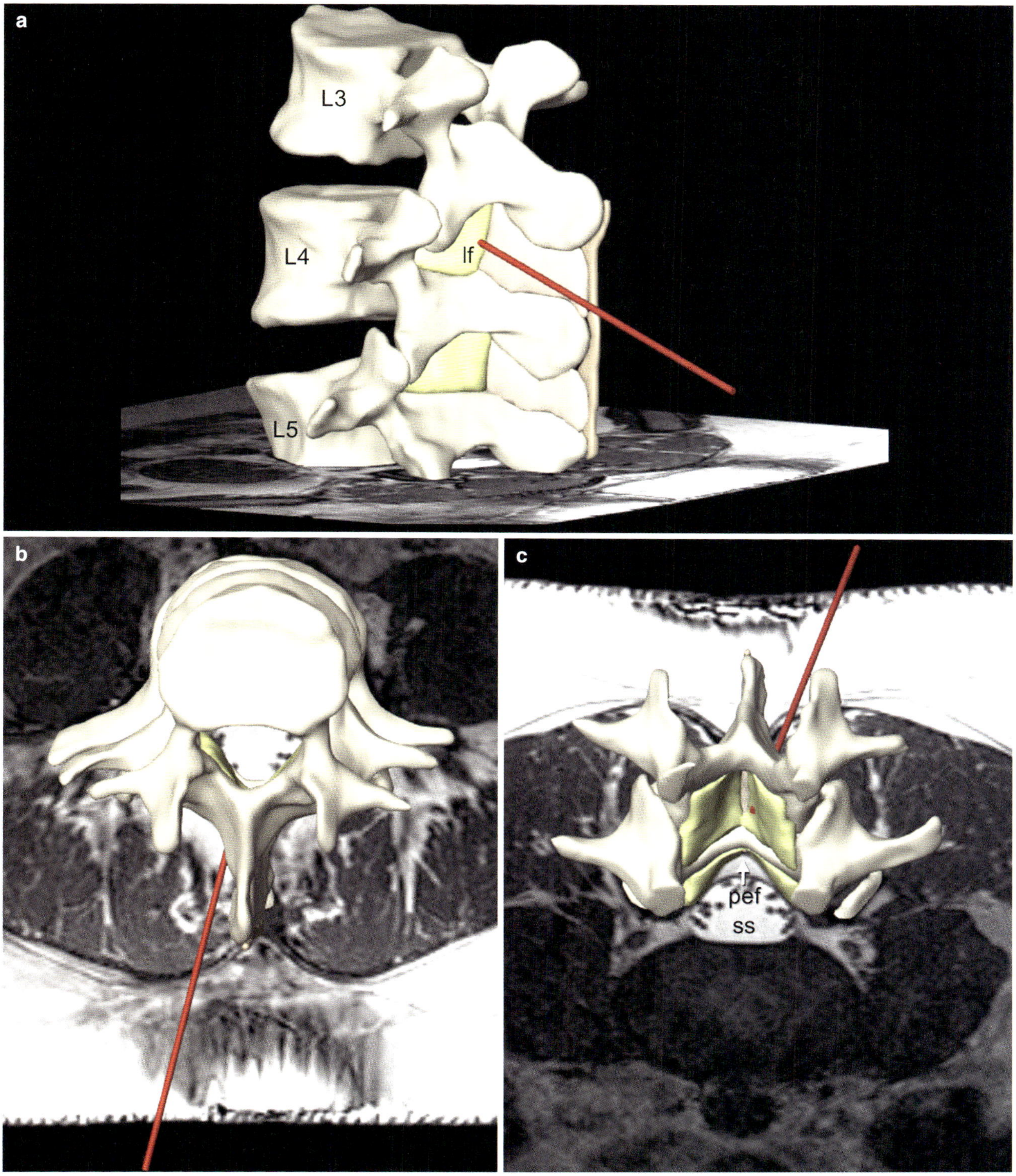

Fig. 32.4 Paramedial approach used in spinal blockade. The *red cylinder* represents the trajectory of the needle in dorsolateral (**a**), superior (**b**), and ventrocranial views (**c**). In this approach, the entry point on the skin is located about 2 cm from the midline. The needle is obliquely inserted in a laterocaudal to mediocranial direction to reach the interlaminar space immediately above the identified spinal process. Once the ligamentum flavum (*lf*) is pierced, the anesthetic solution may be injected into the posterior fat of the epidural space (*pef*) (epidural technique) or in the cerebrospinal fluid of the subarachnoid space (*ss*) (subarachnoid technique). A T2-weighted axial image is used as reference. *L3* L3 vertebra, *L4* L4 vertebra, *L5* L5 vertebra

The Ligamentum Flavum

33

Philipp Lirk

Classically, anesthesiologists identify the epidural space using the "loss of resistance" technique [1]. This method relies on the distinctive drop in resistance against needle advancement and fluid injection as the epidural needle is advanced through the ligamentum flavum [2]. Knowledge of anatomy and of variations of the ligamentum flavum is essential in understanding the loss of resistance technique and its potential anatomic pitfalls.

The ligamentum flavum is one of the main ligaments supporting the static and dynamic functions of the vertebral column. In contrast to the collagenous supraspinous and interspinal ligaments, the ligamentum flavum consists mainly of yellow, elastic fibers. (*Flavus* is Latin for "yellow.") The ligamentum flavum inserts on the superior edge and posterosuperior surface of caudad vertebrae, and the inferior edge and anteroinferior surface of cranial vertebrae [3]. When the spine is flexed, the ligamentum flavum is stretched and stores mechanical energy, which is regained upon extension of the spinal column.

Embryologically, the ligamentum flavum is a paired structure that fuses in the midline [4]. It is formed after the 12th gestational week and increases in thickness from the 15th gestational week [5]. The degree of midline fusion varies between individuals and across vertebral heights. In general, midline gaps appear to be most frequent in the cervical and high thoracic region, where as many as 75 % of ligaments may incompletely fuse [2, 6, 7]. Therefore, when the needle is in the precise midline, the ligamentum flavum is not always perfectly perceptible as a barrier to the epidural space [2]. Conversely, resistance may be increased or passage impeded when the ligamentum flavum is calcified or ossified, as occurs in advanced age [8, 9].

Posteriorly, the ligamentum flavum is bordered by the spinous muscles and (in the midline) by the interspinal ligament; anteriorly, it is bordered by a mixture of fatty and very loose connective tissue (Figs. 33.1, 33.2, 33.3, 33.4, 33.5, 33.6, 33.7, 33.8, 33.9, and 33.10) [10].

P. Lirk, MD, PhD
Department of Anaesthesiology, Academic Medical Center, University of Amsterdam, Amsterdam, The Netherlands
e-mail: p.lirk@amc.uva.nl

M.A. Reina et al. (eds.), *Atlas of Functional Anatomy for Regional Anesthesia and Pain Medicine: Human Structure, Ultrastructure and 3D Reconstruction Images*, DOI 10.1007/978-3-319-09522-6_33,

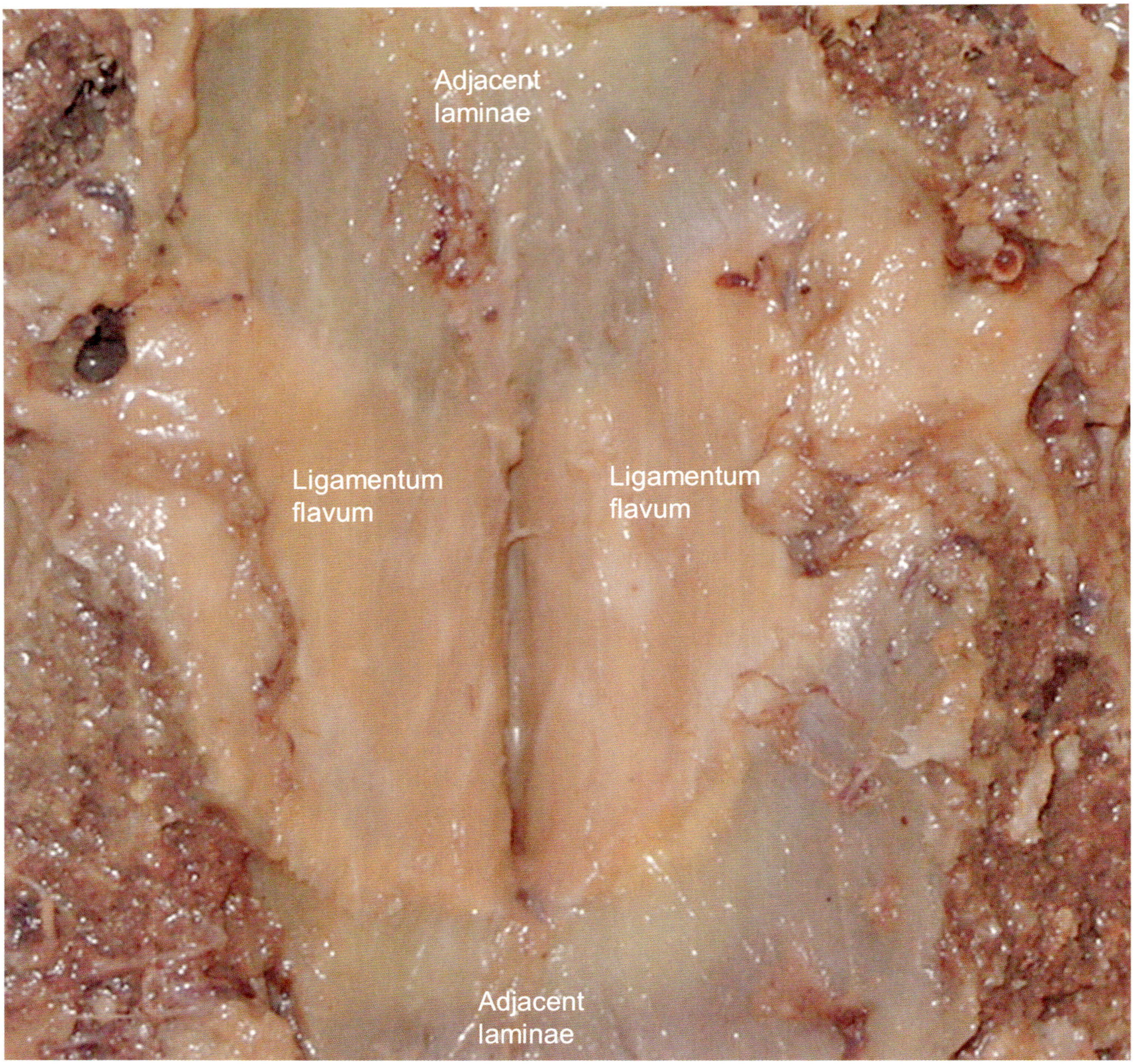

Fig. 33.1 Ligamentum flavum. Anatomy of the paired ligamentum flavum spanning out between adjacent laminae. The view is from the epidural space. In the midline, a gap is apparent between the two ligamenta (Photo taken at the Department of Anatomy, Innsbruck Medical University, Austria. Courtesy of L. Kirchmair, MD, and Prof. B. Moriggl, MD, FIACA)

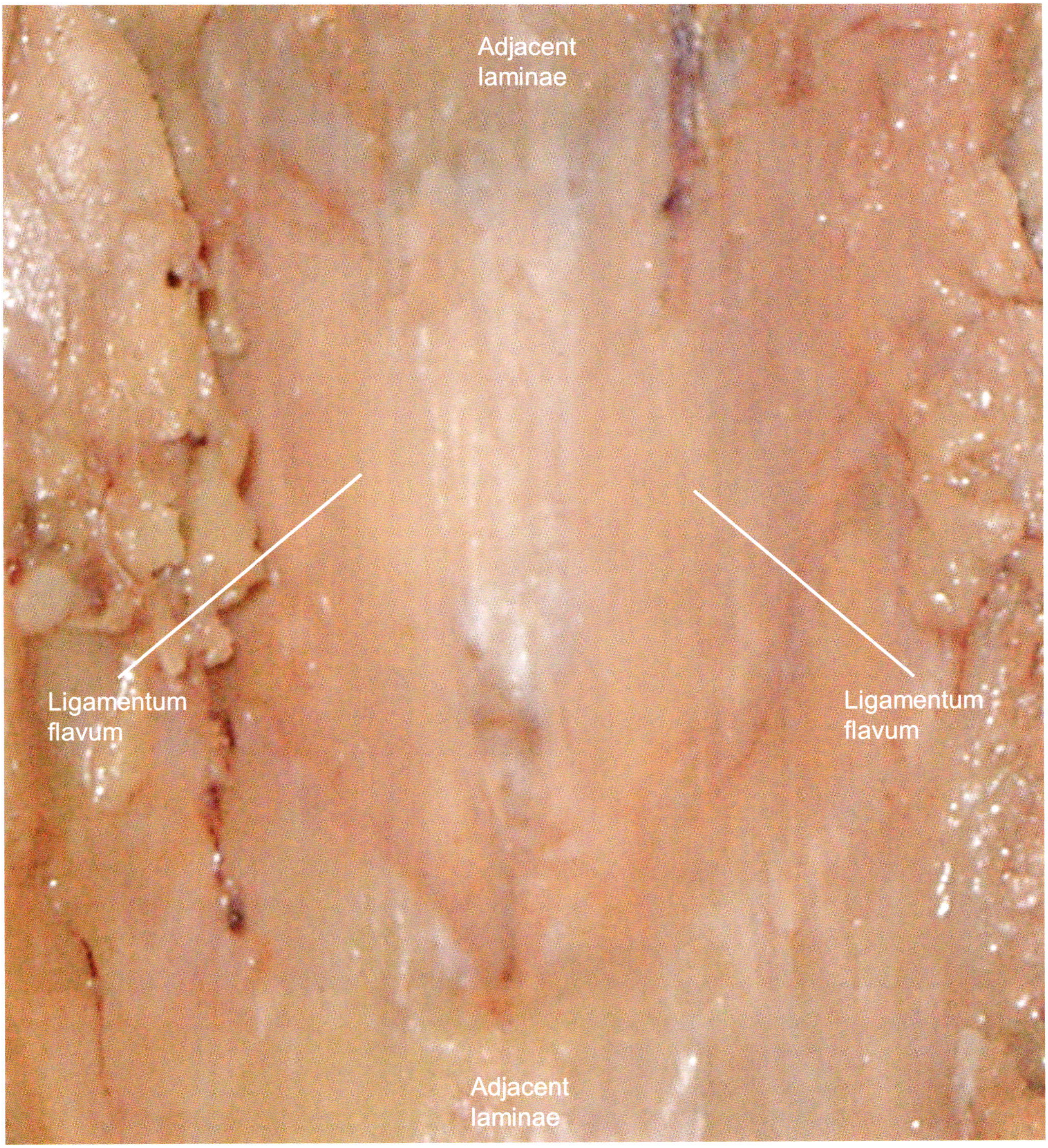

Fig. 33.2 Ligamentum flavum. Anatomy of the ligamentum flavum between adjacent laminae. The view is from the epidural space. The two ligamenta flava are completely fused in the midline (Photo taken at the Department of Anatomy, Innsbruck Medical University, Austria. Courtesy of L. Kirchmair, MD, and Prof. B. Moriggl, MD, FIACA)

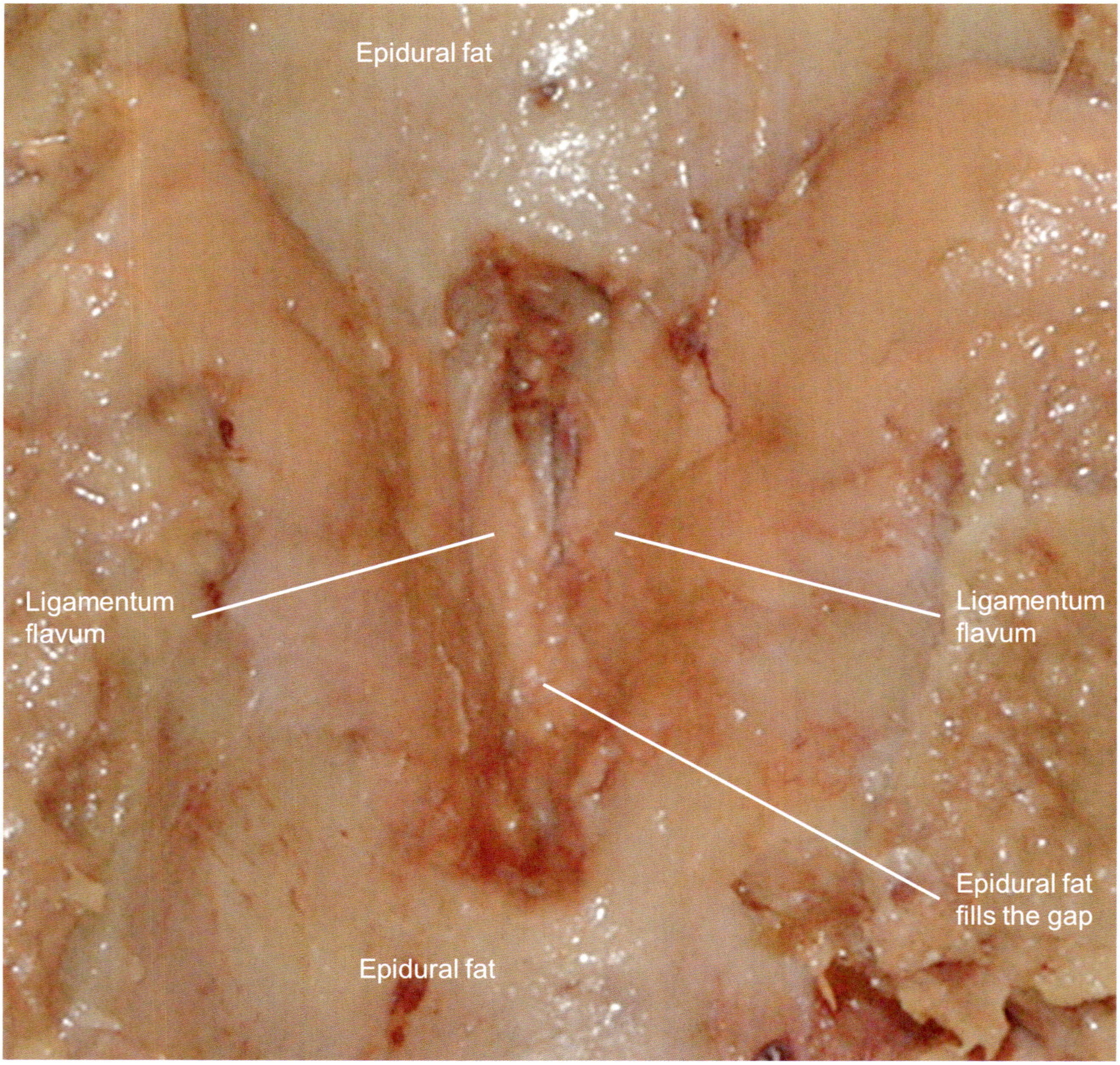

Fig. 33.3 Ligamentum flavum. Anatomy of the paired ligamentum flavum spanning out between adjacent laminae. The view is from the epidural space. Epidural fat fills the gap between the two ligamenta flava. Note the epidural fat, which is compartmentalized and fills substantial portions of the posterior epidural space (Photo taken at the Department of Anatomy, Innsbruck Medical University, Austria. Courtesy of L. Kirchmair, MD, and Prof. B. Moriggl, MD, FIACA)

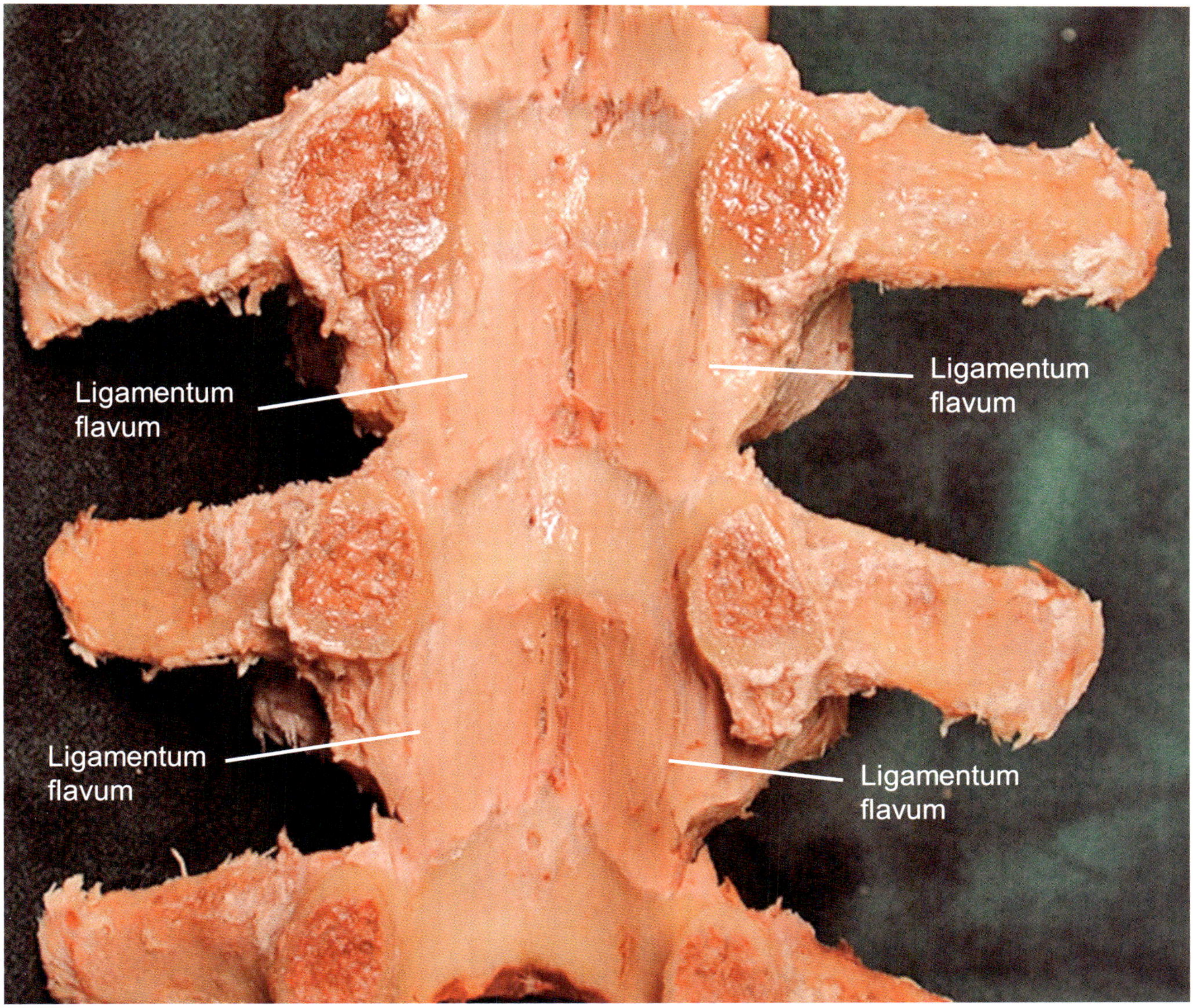

Fig. 33.4 Ligamentum flavum. Anatomy of the ligamentum flavum viewed from the epidural side (Photo taken at the Department of Anatomy, Innsbruck Medical University, Austria. Courtesy of L. Kirchmair, MD, and Prof. B. Moriggl, MD, FIACA)

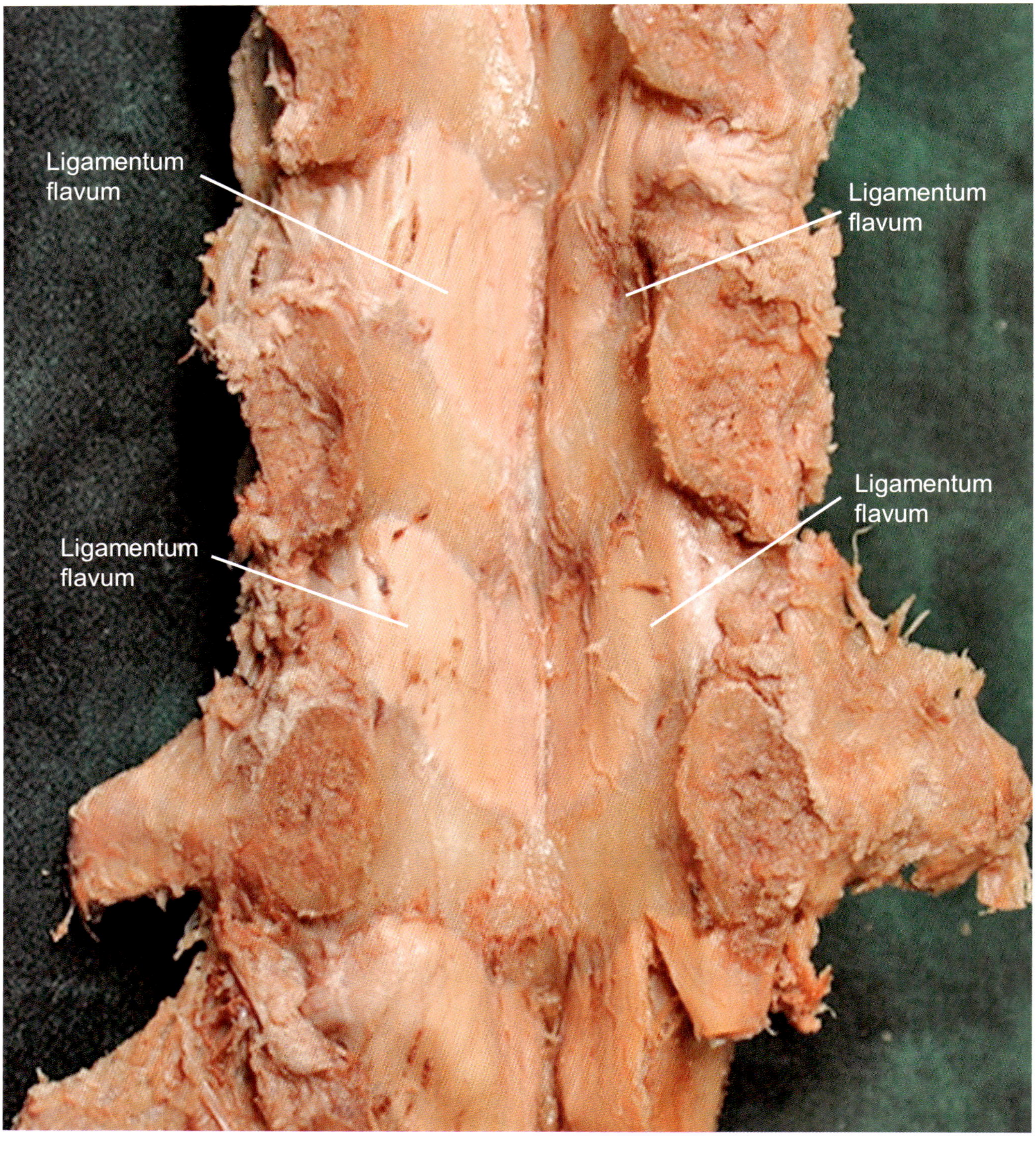

Fig. 33.5 Ligamentum flavum. Anatomy of the ligamentum flavum viewed from the epidural side (Photo taken at the Department of Anatomy, Innsbruck Medical University, Austria. Courtesy of L. Kirchmair, MD, and Prof. B. Moriggl, MD, FIACA)

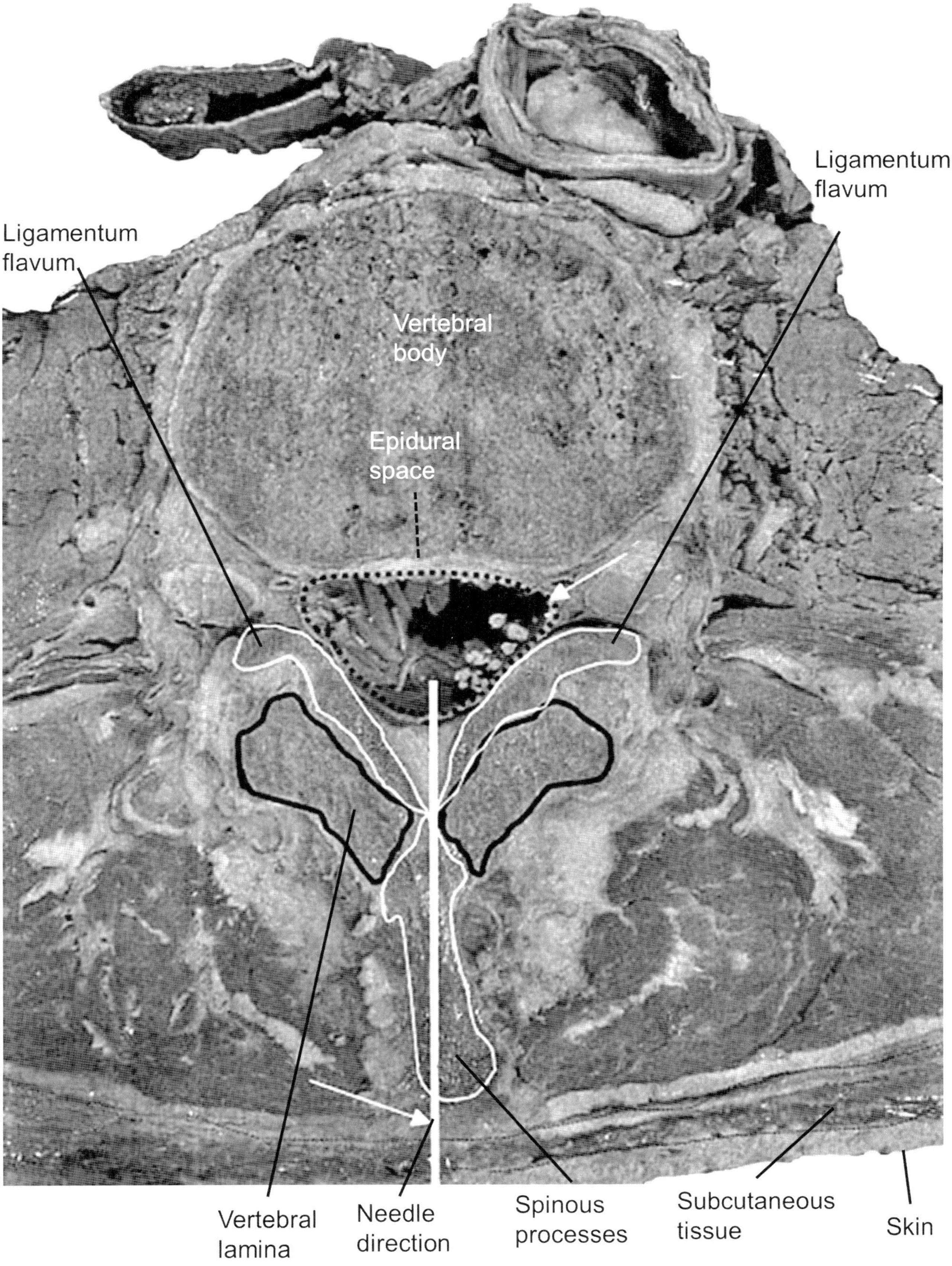

Fig. 33.6 Ligamentum flavum. Horizontal cross-section at the lumbar level detailing the tissues traversed during epidural or spinal puncture. In the figure can be identified skin, subcutaneous tissue, spinous processes joined by the supraspinous and interspinal ligaments, the lamina vertebrae, the ligamentum flavum (here with a very subtle midline gap), epidural space, and vertebral body. The *white line* indicates the needle direction for a perfect midline puncture (From Boon et al. [8]; with permission)

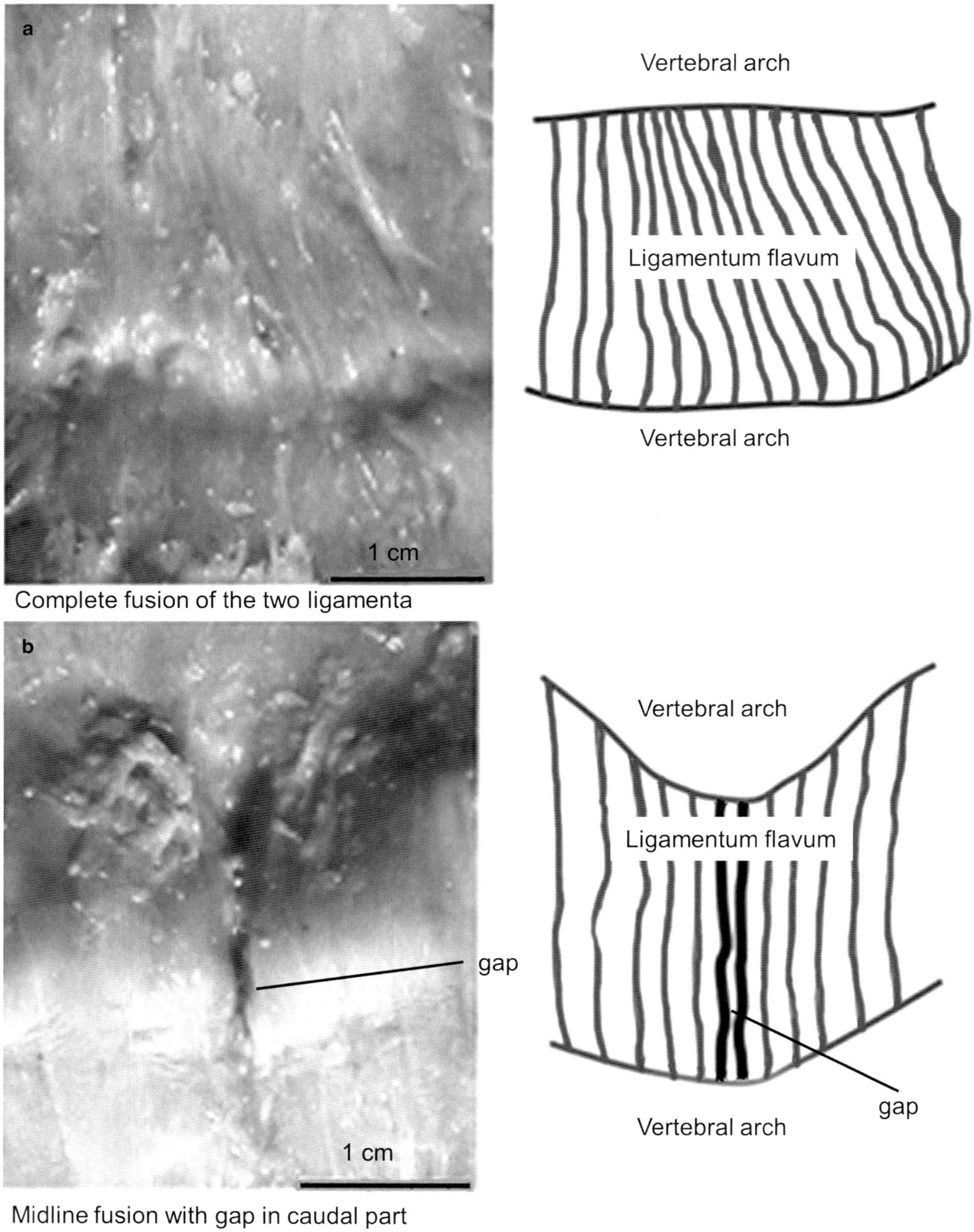

Fig. 33.7 Anatomic variations in ligamentum flavum anatomy. The ligamenta flava are seen from the epidural space, featuring complete fusion of the two ligamenta (**a**), and continuous gaps along the entire height (**b**) (From Lirk et al. [2]; with permission)

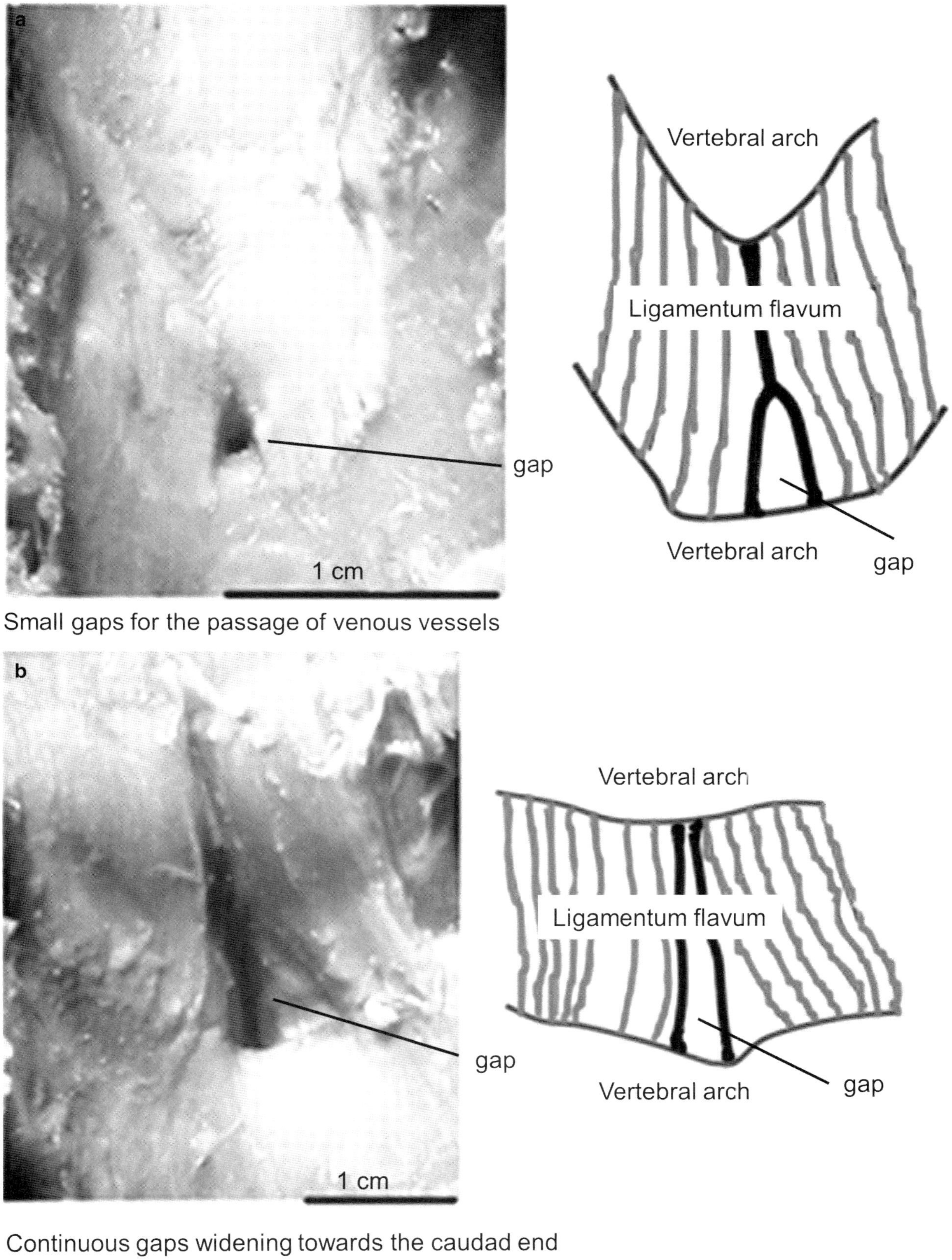

Fig. 33.8 Anatomic variations in ligamentum flavum anatomy. The ligamenta flava are seen from the epidural space. (**a**) Small gaps for the passage of venous vessels connecting the posterior external and posterior internal venous plexuses. (**b**) Continuous gaps (From Lirk et al. [2]; with permission)

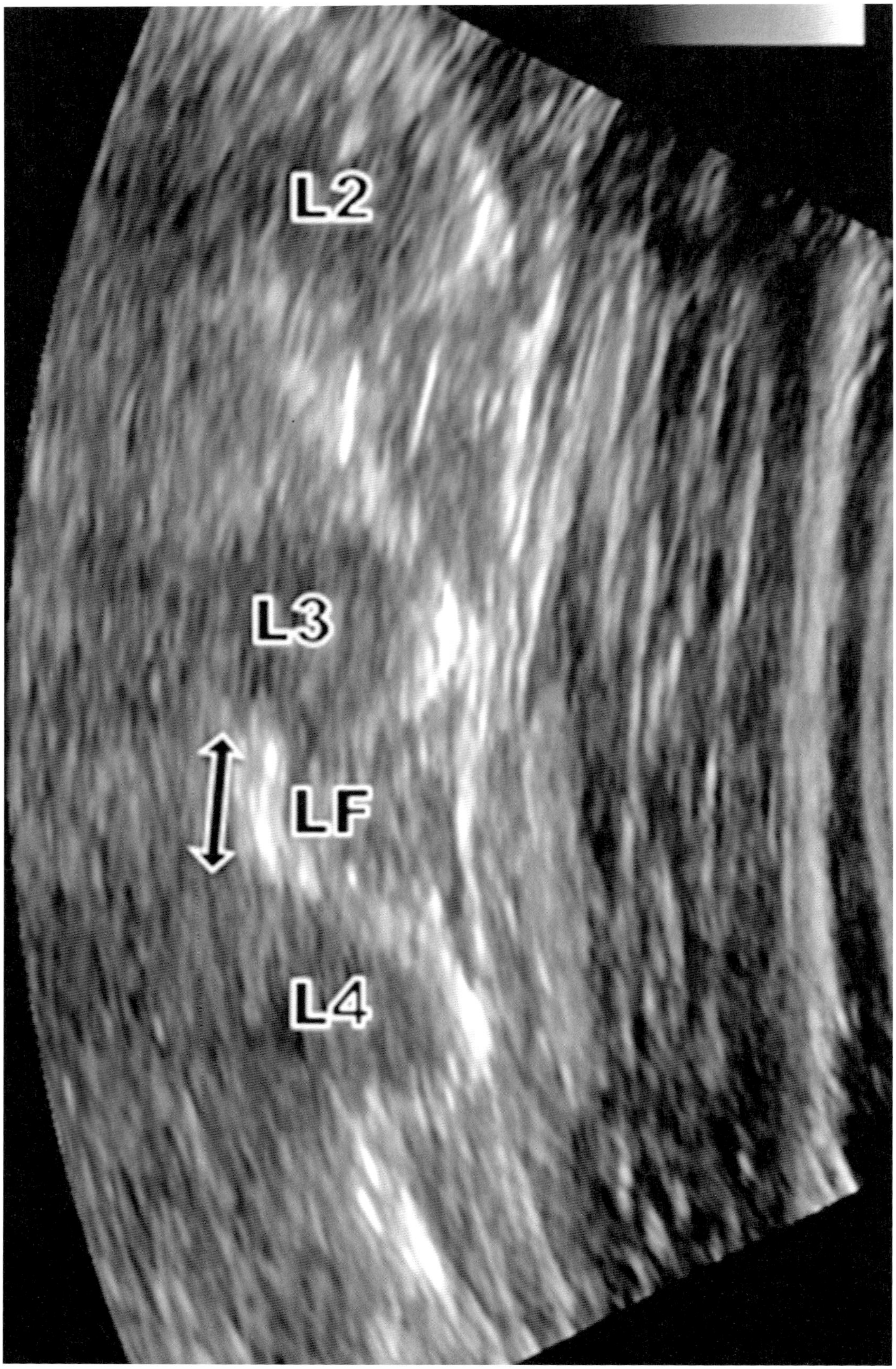

Fig. 33.9 Sonoanatomic representation of the human ligamentum flavum (*LF*) in parturients is seen as a bright, echogenic line between the vertebral laminae (From Jones et al. [11]; with permission)

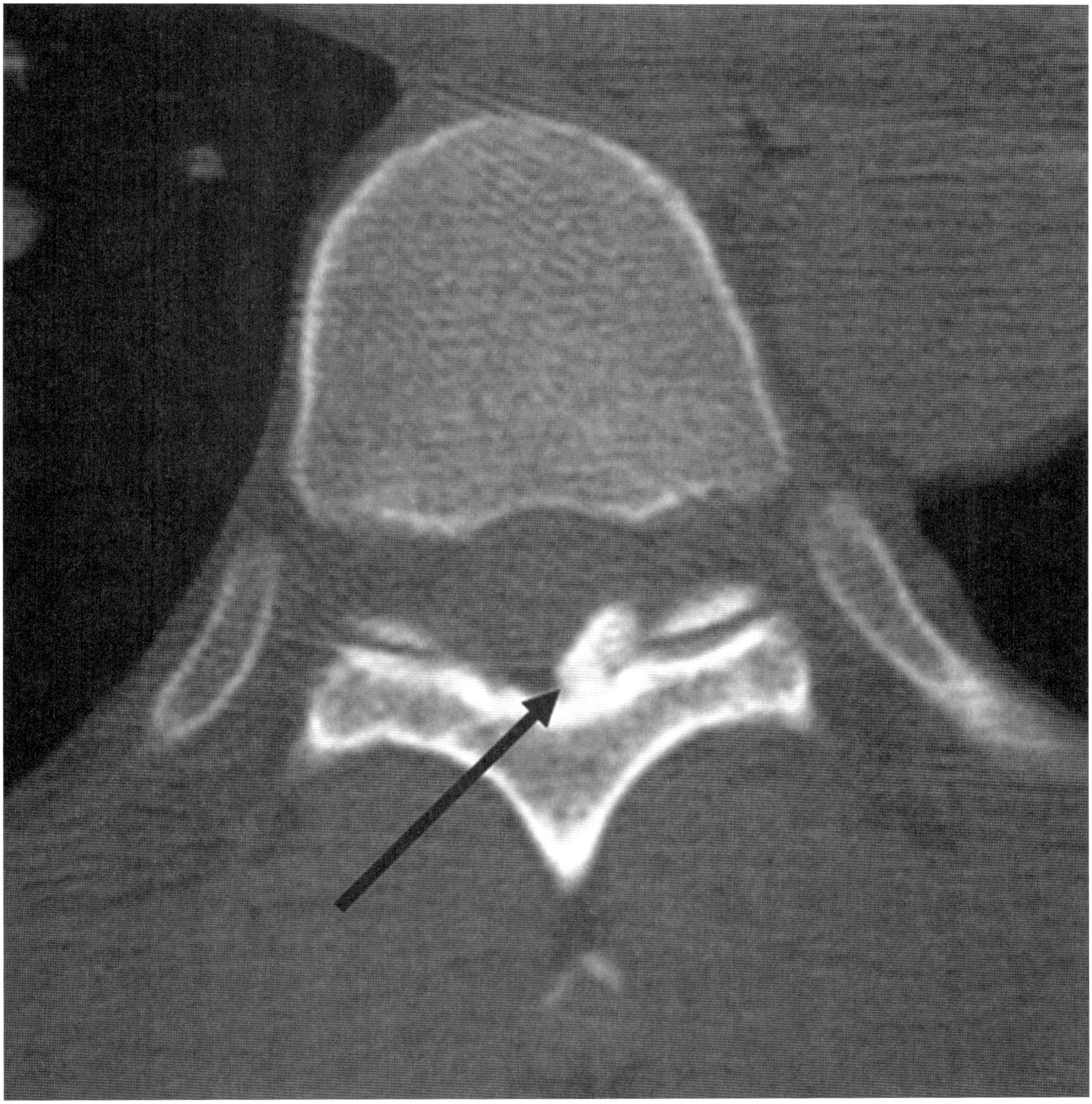

Fig. 33.10 In advanced age, calcification (isolated regions of calcification within the ligamentum flavum) or ossification (bone forming in continuity with the lamina) may appear. The prevalence increases with increasing age, and is highest in the low thoracic region [9]. The *arrow* indicates bone formation at the ligamentum flavum (From Guo et al. [9]; with permission)

References

1. Wantman A, Hancox N, Howell PR. Techniques for identifying the epidural space: a survey of practice amongst anaesthetists in the UK. Anaesthesia. 2006;61:370–5.
2. Lirk P, Kolbitsch C, Putz G, Colvin J, Colvin HP, Lorenz I, et al. Cervical and high thoracic ligamentum flavum frequently fails to fuse in the midline. Anesthesiology. 2003;99:1387–90.
3. Olszewski AD, Yaszemski MJ, White 3rd AA. The anatomy of the human lumbar ligamentum flavum. New observations and their surgical importance. Spine (Phila Pa 1976). 1996;21:2307–12.
4. Zarzur E. Anatomic studies of the human ligamentum flavum. Anesth Analg. 1984;63:499–502.
5. Misawa H, Ohtsuka K, Nakata K, Kinoshita H. Embryological study of the spinal ligaments in human fetuses. J Spinal Disord. 1994;7:495–8.
6. Lirk P, Colvin J, Steger B, Colvin HP, Keller C, Rieder J, et al. Incidence of lower thoracic ligamentum flavum midline gaps. Br J Anaesth. 2005;94:852–5.
7. Lirk P, Moriggl B, Colvin J, Keller C, Kirchmair L, Rieder J, et al. The incidence of lumbar ligamentum flavum midline gaps. Anesth Analg. 2004;98:1178–80.
8. Boon JM, Abrahams PH, Meiring JH, Welch T. Lumbar puncture: anatomical review of a clinical skill. Clin Anat. 2004;17:544–53.
9. Guo JJ, Luk KD, Karppinen J, Yang H, Cheung KM. Prevalence, distribution, and morphology of ossification of the ligamentum flavum: a population study of one thousand seven hundred thirty-six magnetic resonance imaging scans. Spine (Phila Pa 1976). 2010;35:51–6.
10. Ramsey RH. The anatomy of the ligamenta flava. Clin Orthop Relat Res. 1966;44:129–40.
11. Jones AR, Carle C, Columb M. Effect of table tilt on ligamentum flavum length measured using ultrasonography in pregnant women. Anaesthesia. 2013;68:27–30.

34 Subarachnoid (Intrathecal) Ligaments

Ritsuko Masuda and Kumiko Tanuma

Spinal ligaments can be divided into two categories: (1) spinal ligaments [1–6], which are thick and fibrous bands that stabilize the vertebrae; and (2) subarachnoid (intrathecal) ligaments [7–9], which are thin filaments that anchor the spinal cord, spinal roots, and nerves to the subarachnoid space. Spinal ligaments stabilize the vertebral column by connecting vertebral bodies, discs, spinous processes, and laminae of adjacent vertebrae and transverse processes [1]. They also prevent disc herniation and injury due to hyperextension or hyperflexion of the spine. Intrathecal structures anchor the spinal cord and spinal roots. Several narrow fibrous or membranous structures in the subarachnoid space cover and anchor the spinal cord, spinal roots, and nerves. These include the denticulate ligament, the dorsal arachnoid septum, the dorsolateral arachnoid septum, and the filum terminale (Figs. 34.1, 34.2, 34.3, 34.4, 34.5, 34.6, 34.7, 34.8, 34.9, 34.10, 34.11, 34.12, 34.13, 34.14, 34.15, 34.16, 34.17, 34.18, and 34.19).

R. Masuda, MD PhD (✉)
Department of Anesthesiology,
Tokyo Hospital, Tokai University, Tokyo, Japan
e-mail: ritsu@tok.u-tokai.ac.jp; rinet-m@ga2.so-net.ne.jp

K. Tanuma, PhD
Department 2nd Anatomy, Nippon Medical School,
Shinjuki Vocational Collage of Acupuncture and Judo Therapy,
Ogura Academy, Shinjuku Vocational College of Acupuncture
and Judo Therapy, Ogura Academy, Tokyo, Japan
e-mail: tanuma@ssjs.ac.jp

M.A. Reina et al. (eds.), *Atlas of Functional Anatomy for Regional Anesthesia and Pain Medicine: Human Structure, Ultrastructure and 3D Reconstruction Images*, DOI 10.1007/978-3-319-09522-6_34,

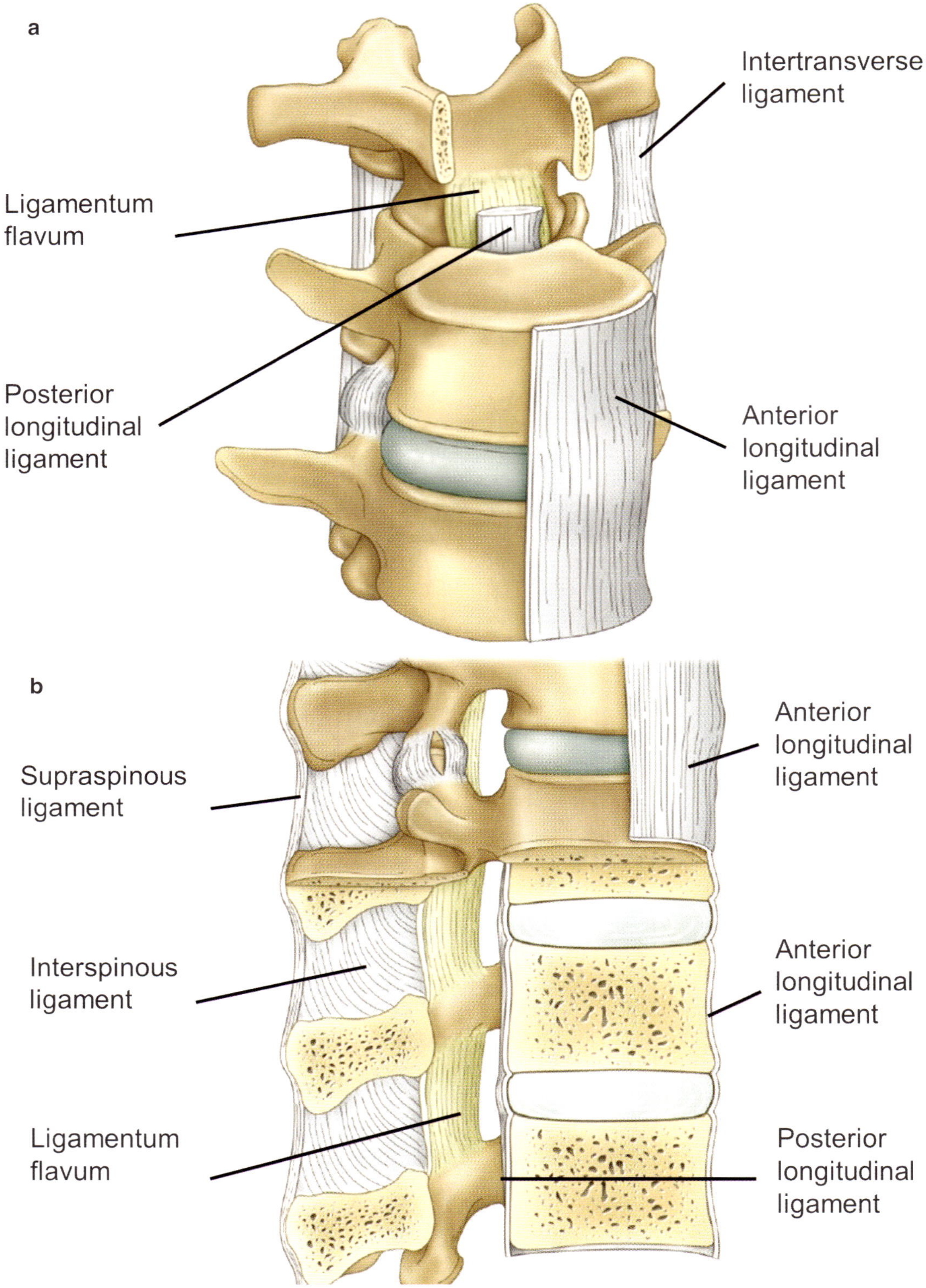

Fig. 34.1 (**a**, **b**) Overview of spinal ligaments of the vertebral column

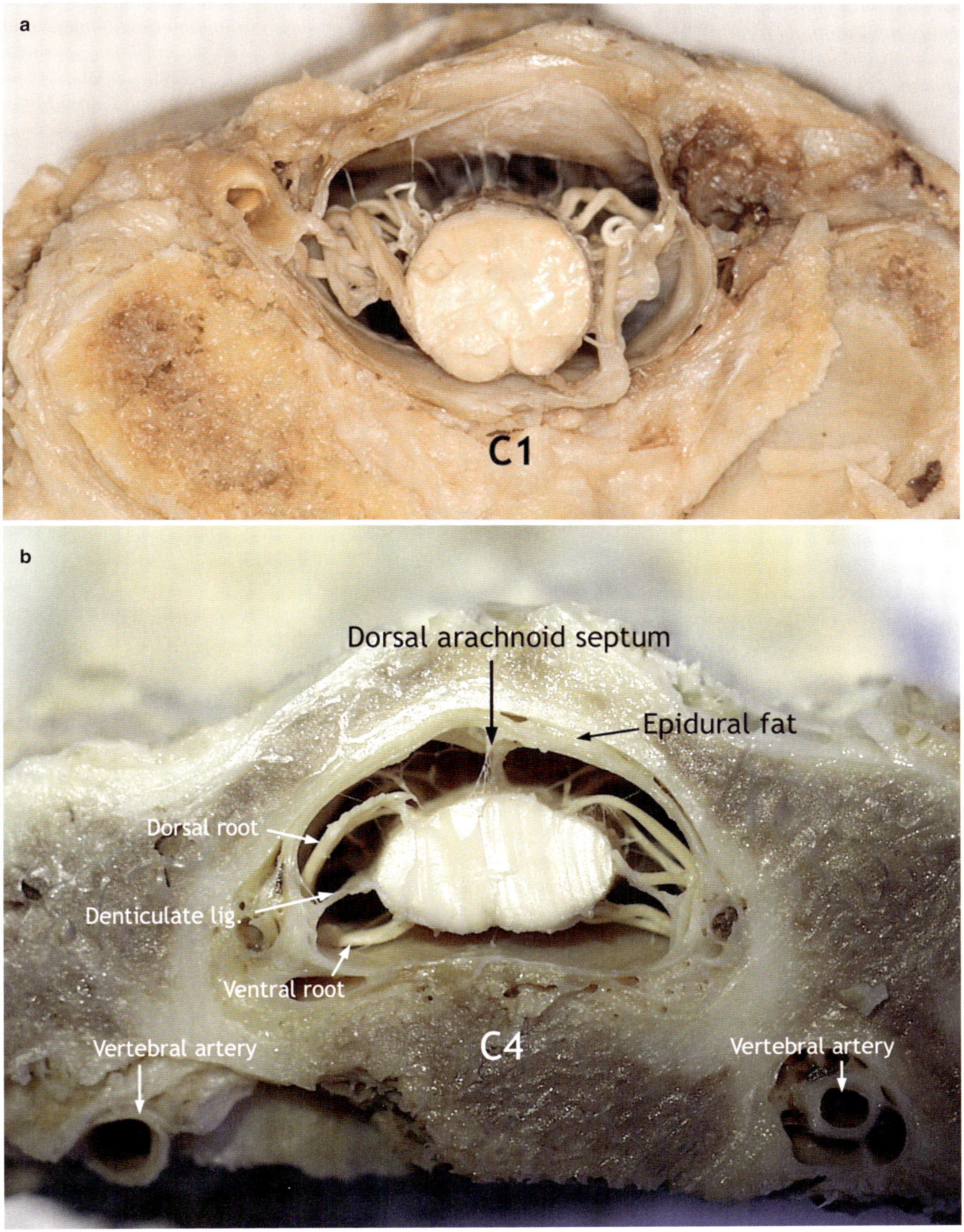

Fig. 34.2 (**a**, **b**) Transverse section of the spinal canal at the C1 and C4 level. The cervical spinal canal is enlarged laterally because of the large cervical nerve roots that supply the upper limbs. The denticulate ligaments are evident on the lateral aspects of the cord, and a thick dorsal arachnoid septum is also present

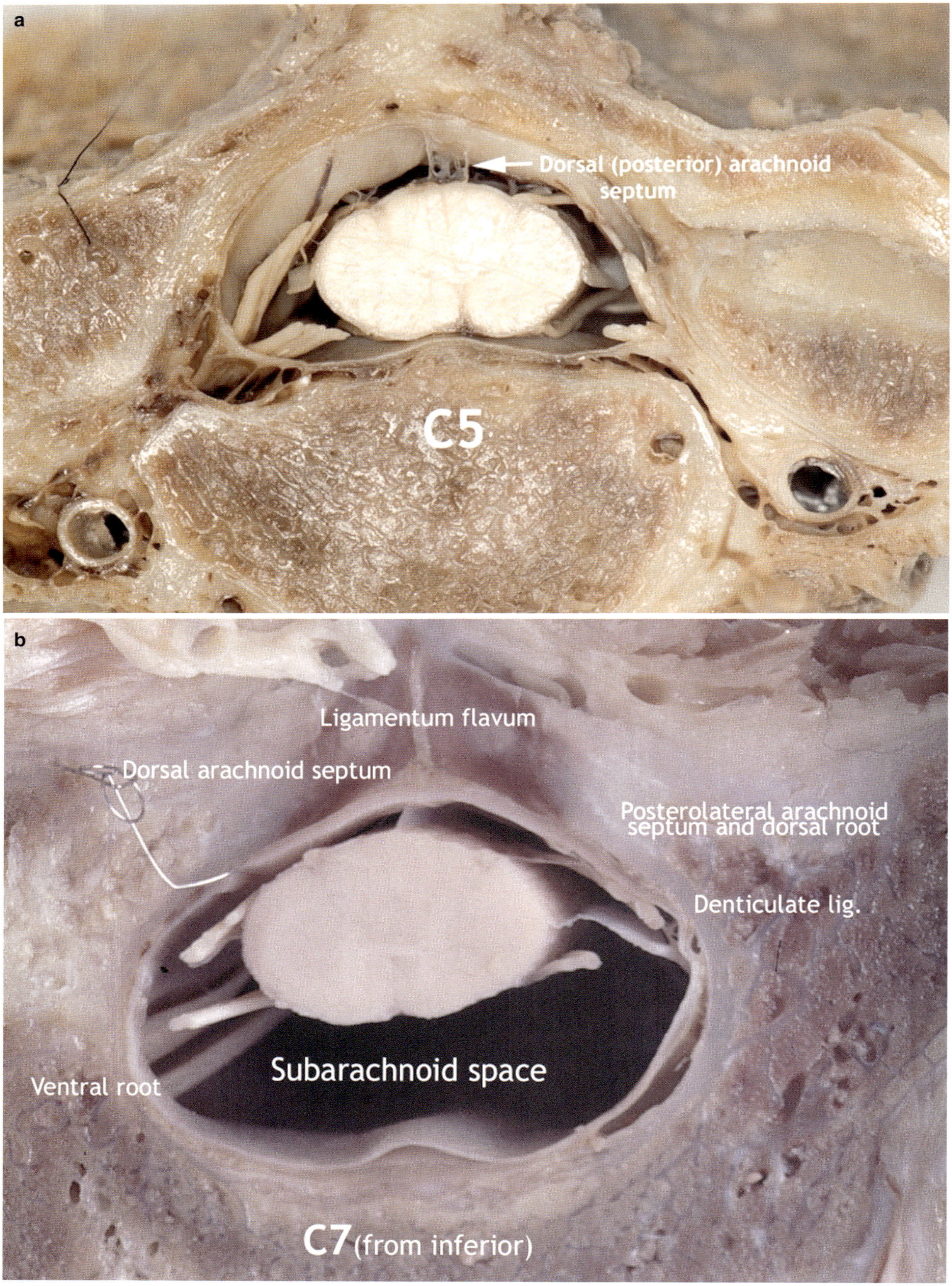

Fig. 34.3 (**a**, **b**) Transverse section of the spinal canal at the C5 and C7 level. The spinal canal becomes rounder at the lower cervical level (*C7*). The denticulate ligament, the dorsal arachnoid septum, and the dorsolateral arachnoid septum cover the dorsal and ventral roots

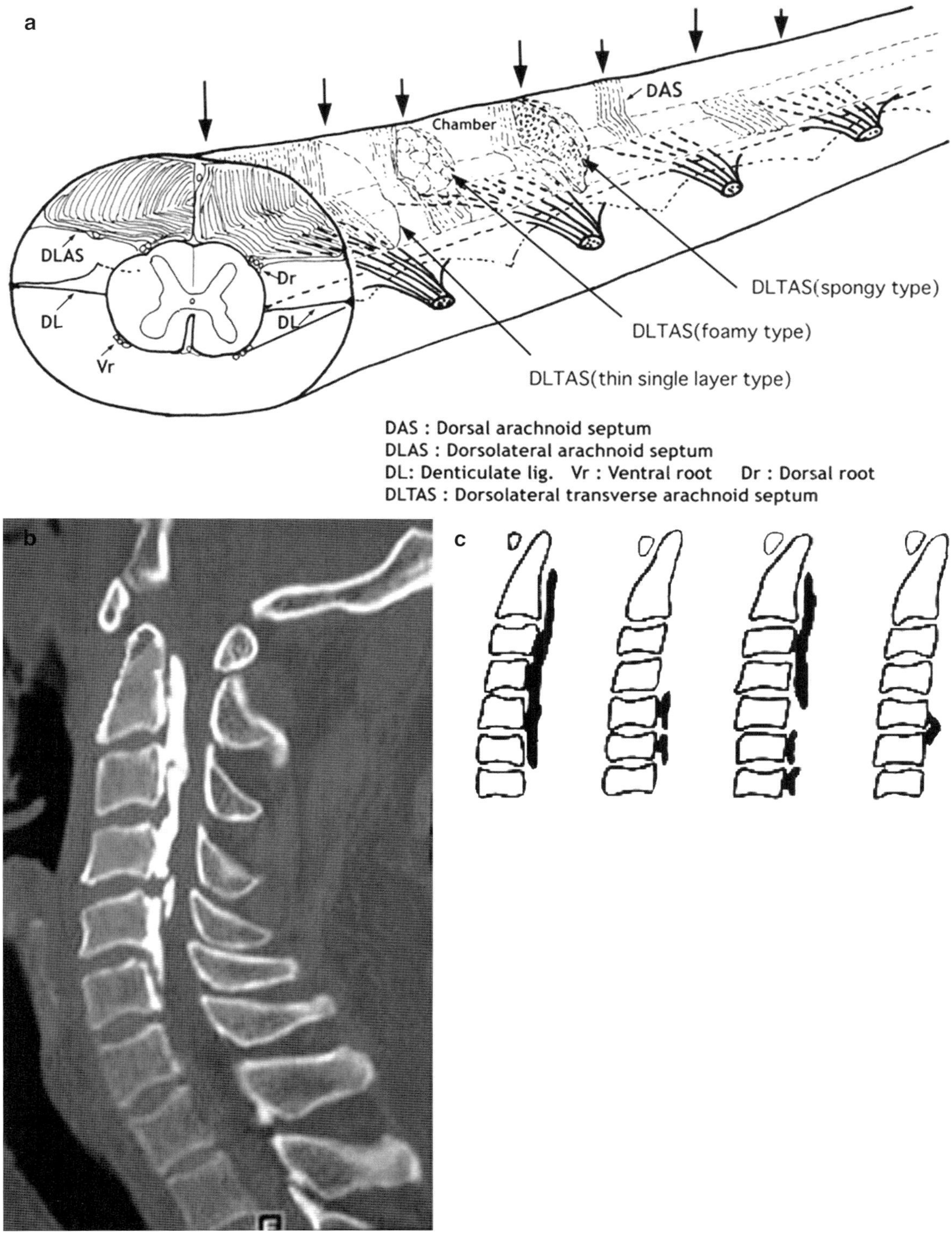

Fig. 34.4 (**a**) Intrathecal structures anchoring the spinal cord and spinal roots. Several narrow, fibrous or membranous structures in the subarachnoid space cover and anchor the spinal cord, spinal roots, and nerves. These include the denticulate ligament (*DL*), the dorsal arachnoid septum (*DAS*), the dorsolateral arachnoid septum (*DLAS*), and the filum terminale (*FT*). The DL, referred to as an intrathecal ligament, is a representative intrathecal structure that anchors the spinal cord and roots [7]. The DL forms tent like structures on both sides of the spinal cord between the ventral and dorsal spinal roots and stabilizes the spinal cord. Each DL extends from the craniovertebral junction to T12, and features 18–20 triangular extensions that attach to the dura at their apices [8]. The DAS is also present in the midline, but it is less conspicuous than the denticulate ligament [9]. It is not a continuous sheet, but rather a series of broad leaflets derived from the intermediate leptomeningeal layer [9]. The DLAS arises as a reflection of the intermediate layer [9]. It is smaller than the DAS and is occasionally present at the cervical and thoracic level. This intermediate layer spreads over the surface of the cord and is connected to blood vessels, nerve roots, and pia mater by fine trabeculae. Thin membranes cross the DAS vertically at regular intervals, from the upper to the lower thoracic level. We have termed these membranes the dorsolateral transverse arachnoid septa (*DLTAS*). They divide the bilateral thoracic subarachnoid columns into small segmental chambers. (**b**, **c**) Pathological forms of spinal ligaments. Ossification of the posterior longitudinal ligament (OPLL) is a hyperostotic condition of the spine associated with spinal canal stenosis and cervical myelopathy. (**b**) MRI of a spine with OPLL. (**c**) Schema showing four types of OPLL [6]

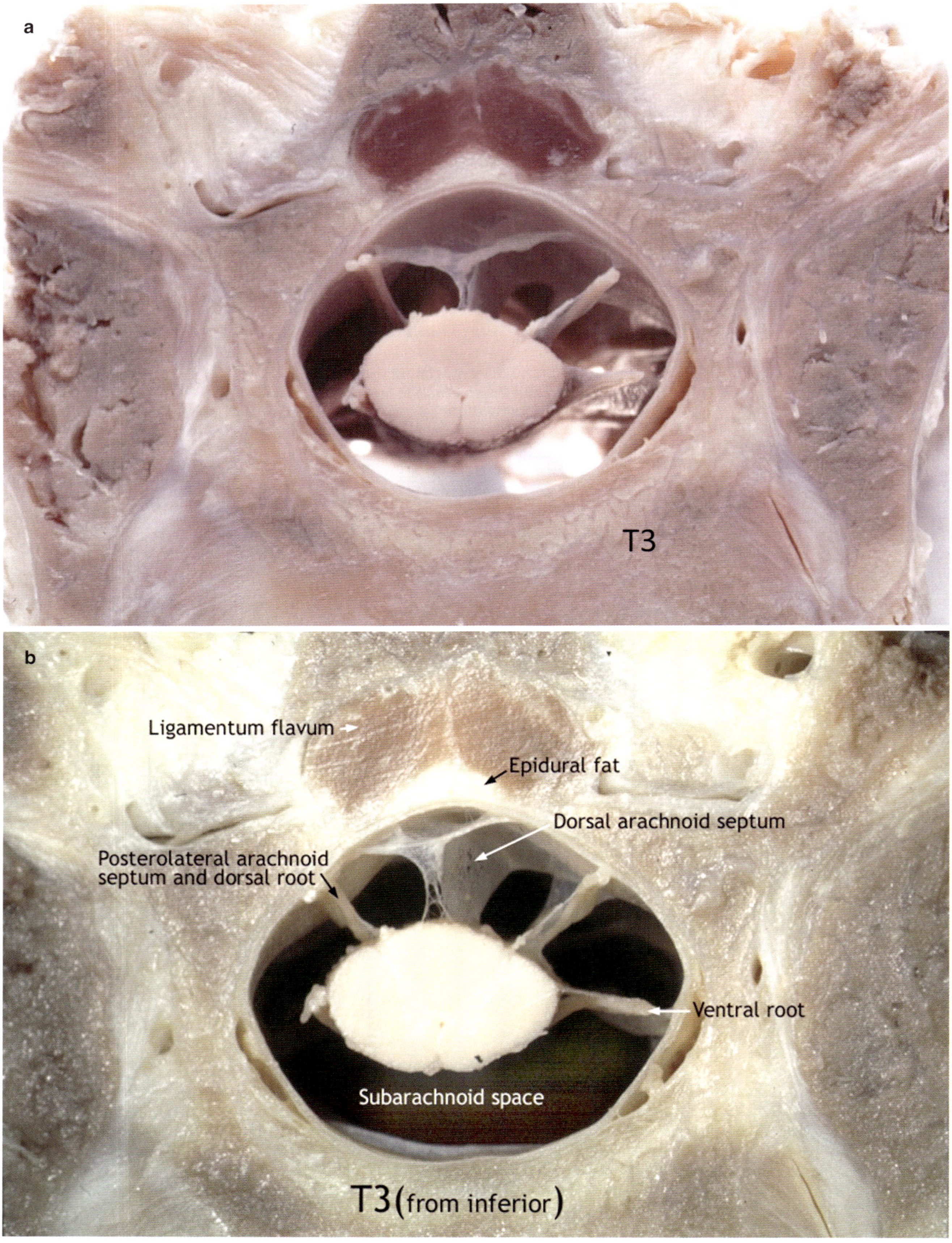

Fig. 34.5 (**a**, **b**) Transverse section cf the spinal canal at the T3 level

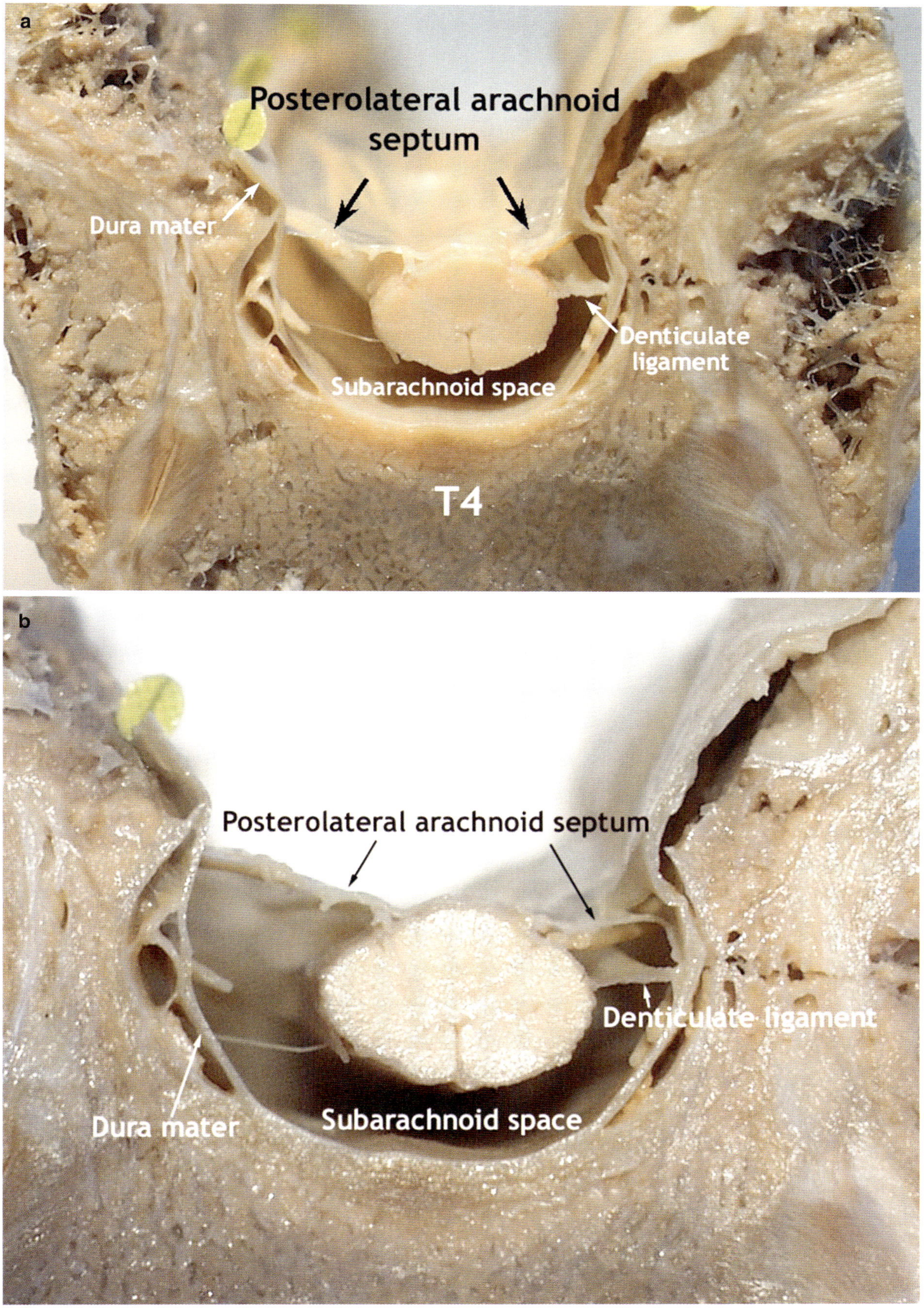

Fig. 34.6 (**a**, **b**) Transverse section of the spinal canal at the T4 level

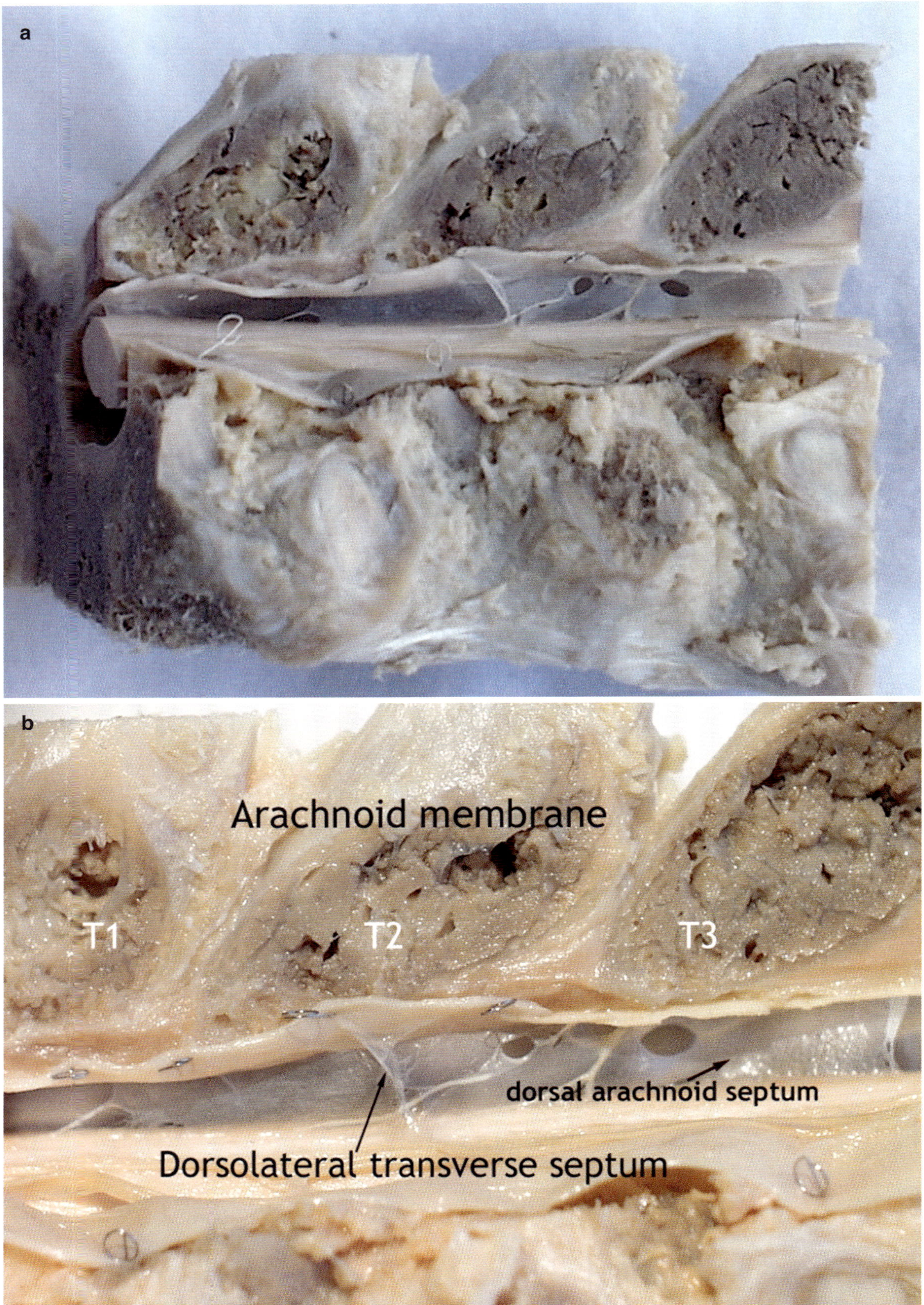

Fig. 34.7 (**a**, **b**) Sagittal section of the spinal canal in the upper thoracic region (*T1–T3*). The *black arrows* indicate the dorsal arachnoid septum and the dorsolateral transverse arachnoid septum. The dorsolateral transverse arachnoid septum is an intermediate layer of leptomeninges that form a discontinuous series of dorsal ligaments attaching the spinal cord to the arachnoid mater

a

b

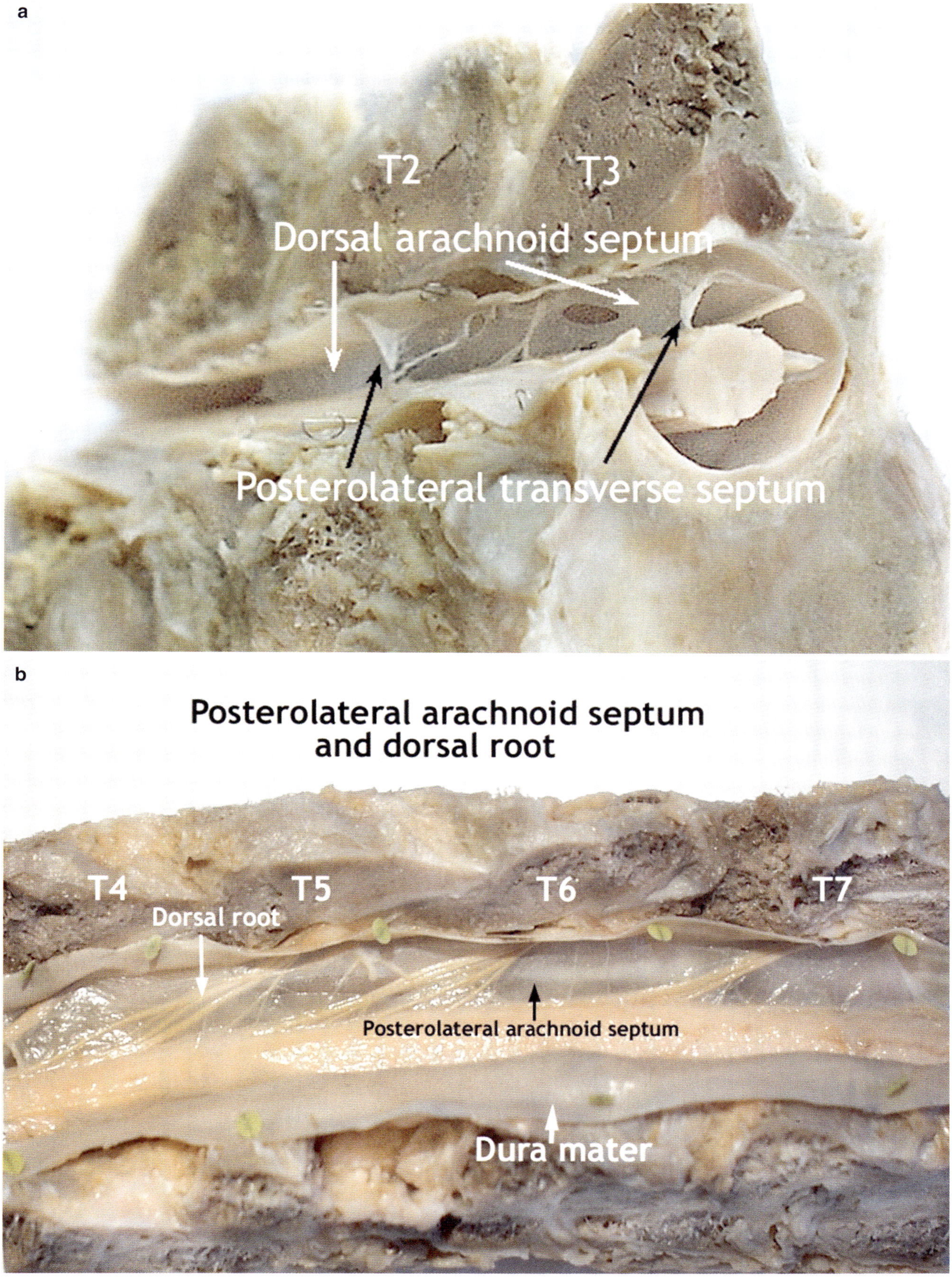

Fig. 34.8 Oblique section of the thoracic spinal canal from T2 to T3 (**a**) and T4 to T7 (**b**). The dorsal arachnoid septum and several dorsolateral (posterolateral) transverse arachnoid septa of the thin-layer type are exposed. The thin layers are concentrated in the dorsal region of the spinal canal

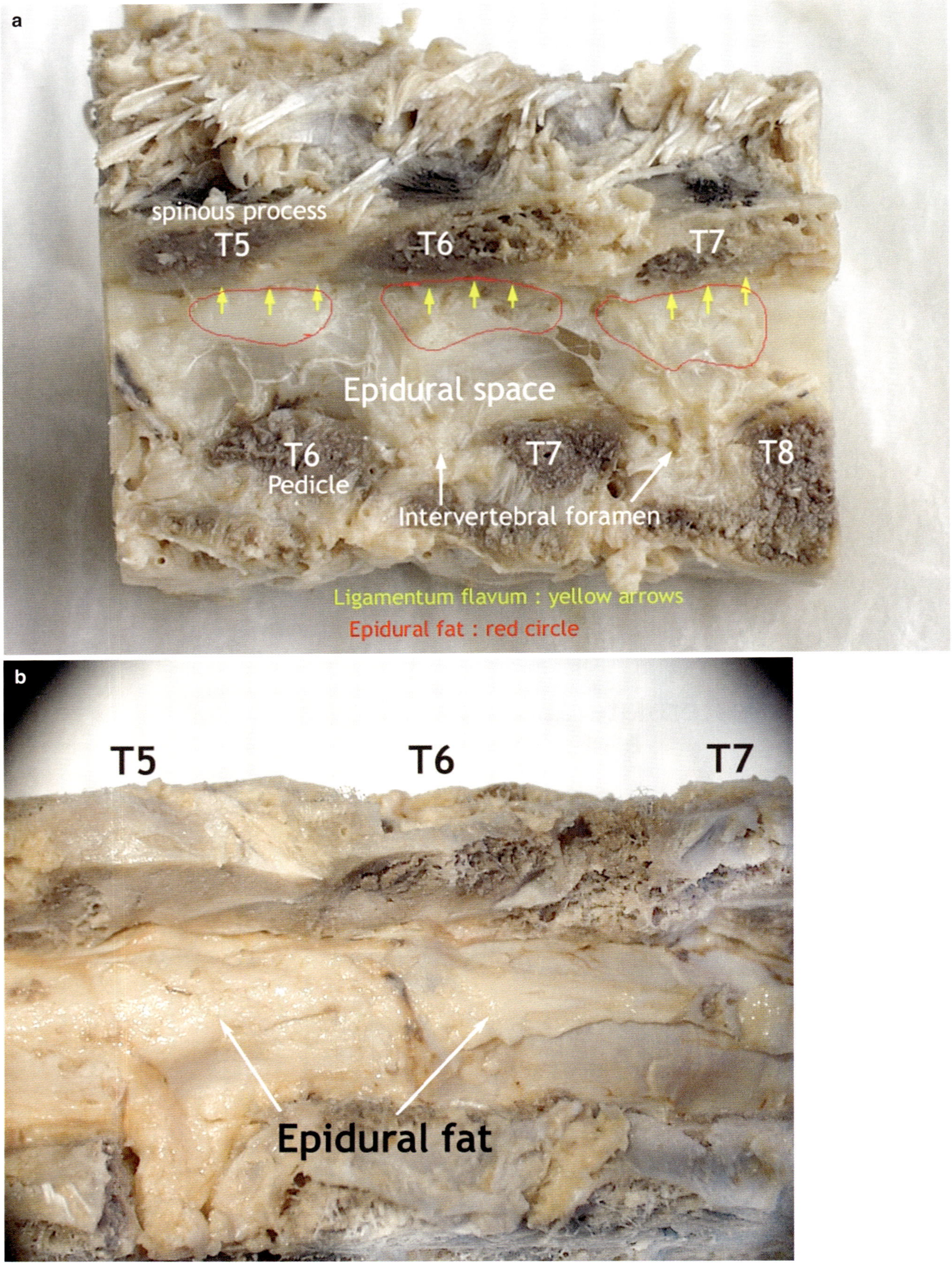

Fig. 34.9 (**a**) Sagittal view of ligamenta flava at the thoracic level. The *red circles* indicate regions of epidural fat, and the *yellow arrows* indicate a ligamentum flavum. (**b**) Epidural fat at the T5 to T7 level

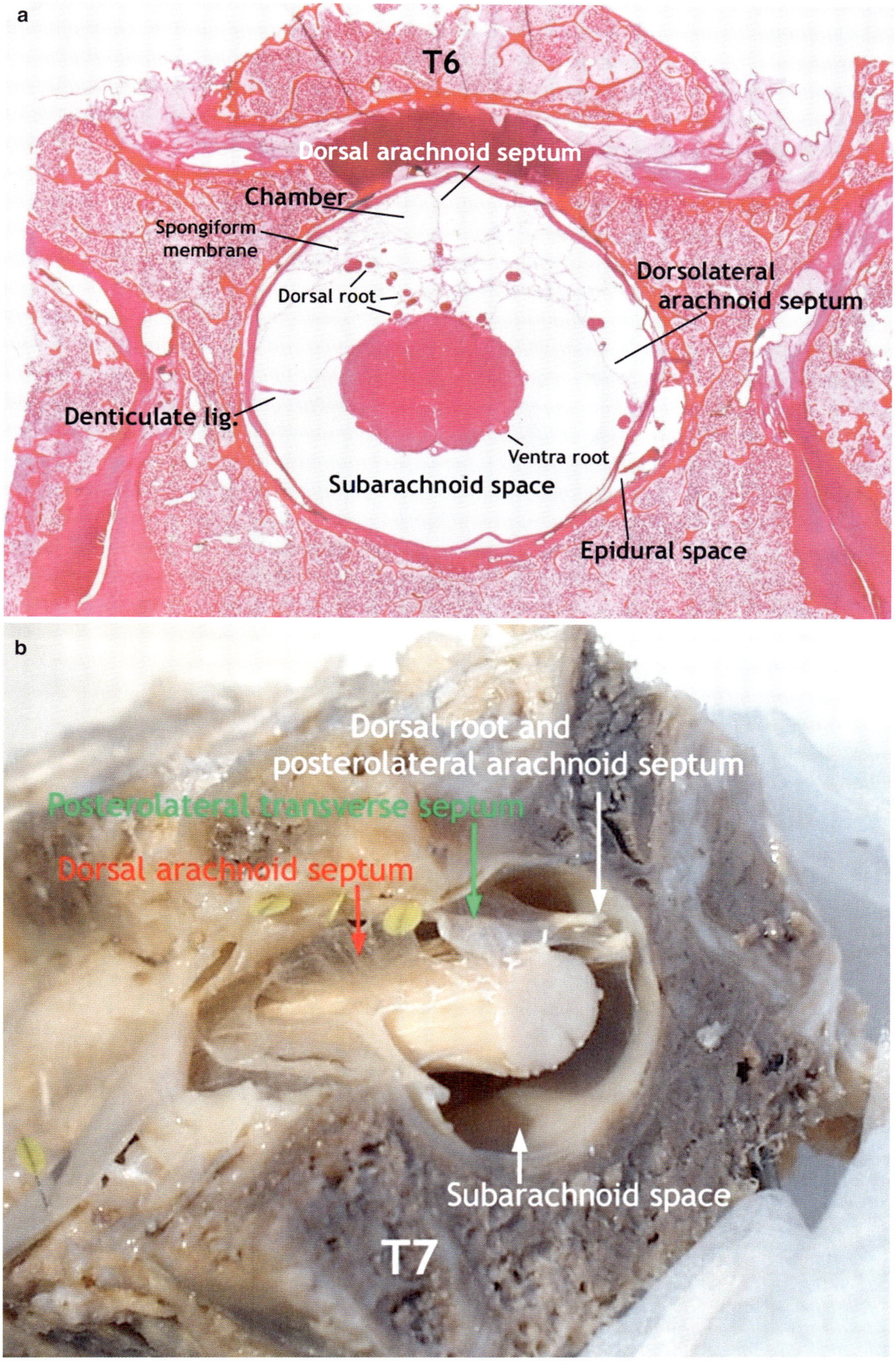

Fig. 34.10 (**a**) Photomicrograph of the T6 spinal cord after hematoxylin and eosin (H & E) staining. The dorsal arachnoid septum, the dorsolateral arachnoid septum, and the dorsolateral transverse arachnoid septum are evident in transverse sections of the thoracic vertebral canal histologically represented as arachnoid membrane. The denticulate ligament is covered by the pia mater. (**b**) Oblique view of a transverse section of the thoracic spinal canal from the T7 level

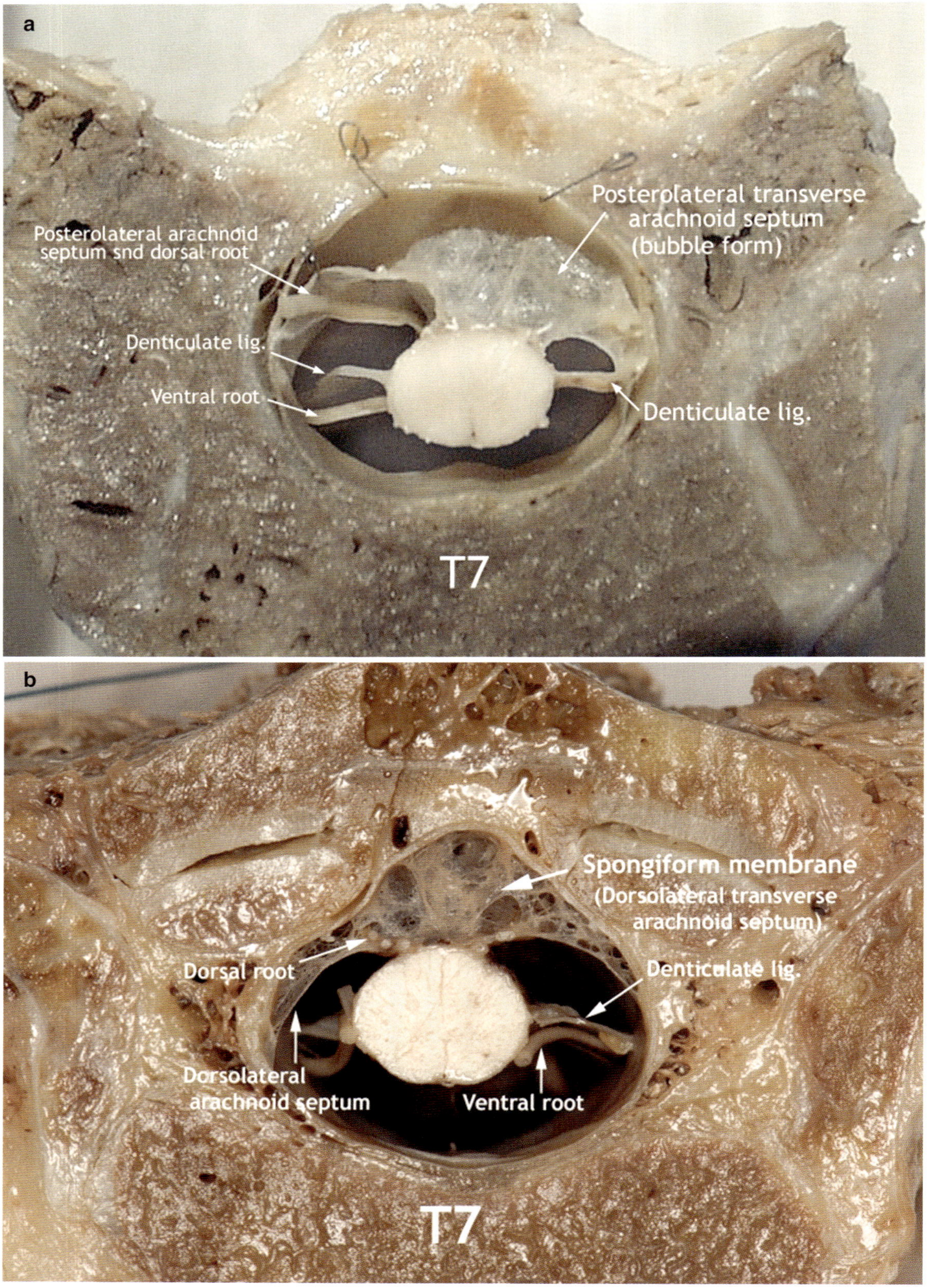

Fig. 34.11 (**a**, **b**) Transverse section of the spinal canal at the T7 level, showing various types of dorsolateral transverse arachnoid septa (DLTAS). Some are thin structures, consisting of a single thin layer or a spongy or foamy film. Various types of spongy or foamy DLTAS vertically cross the dorsal arachnoid septum and the dorsolateral arachnoid septum at regular intervals, dividing the bilateral thoracic subarachnoid columns into segmental chambers. DLTAS are histologically composed of arachnoid tissue, and can be a single or bubble layer type as well as a spongiform membrane type. DLTAS is an intermediate layer of leptomeninges that forms a discontinuous series of dorsal ligaments attaching the spinal cord to the arachnoid mater. The intermediate layer is concentrated in the dorsal half of the subarachnoid space. The intermediate layer serves as a buffer and divides the subarachnoid space into chambers to absorb the flow of cerebrospinal fluid in the spinal canal. These structures may impede the diffusion of agents administered by an intrathecal drug delivery system for the management of chronic pain or cancer pain

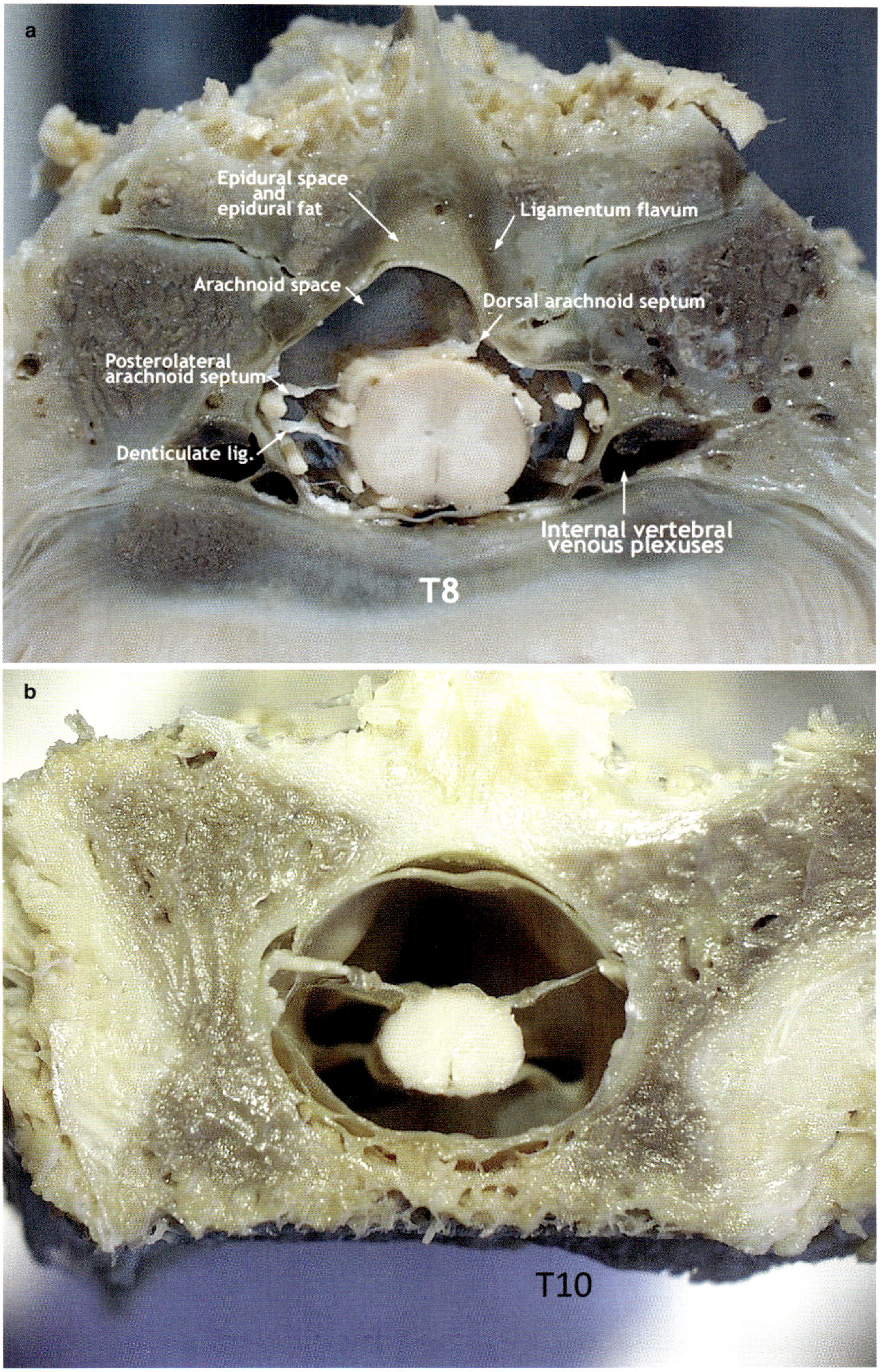

Fig. 34.12 (**a**, **b**) Transverse section of the spinal canal at the T8 and T10 levels. The ligamentum flavum, epidural space and epidural fat, the dorsal arachnoid septum, dorsolateral arachnoid septum (the posterolateral arachnoid septum), and the denticulate ligament are all evident

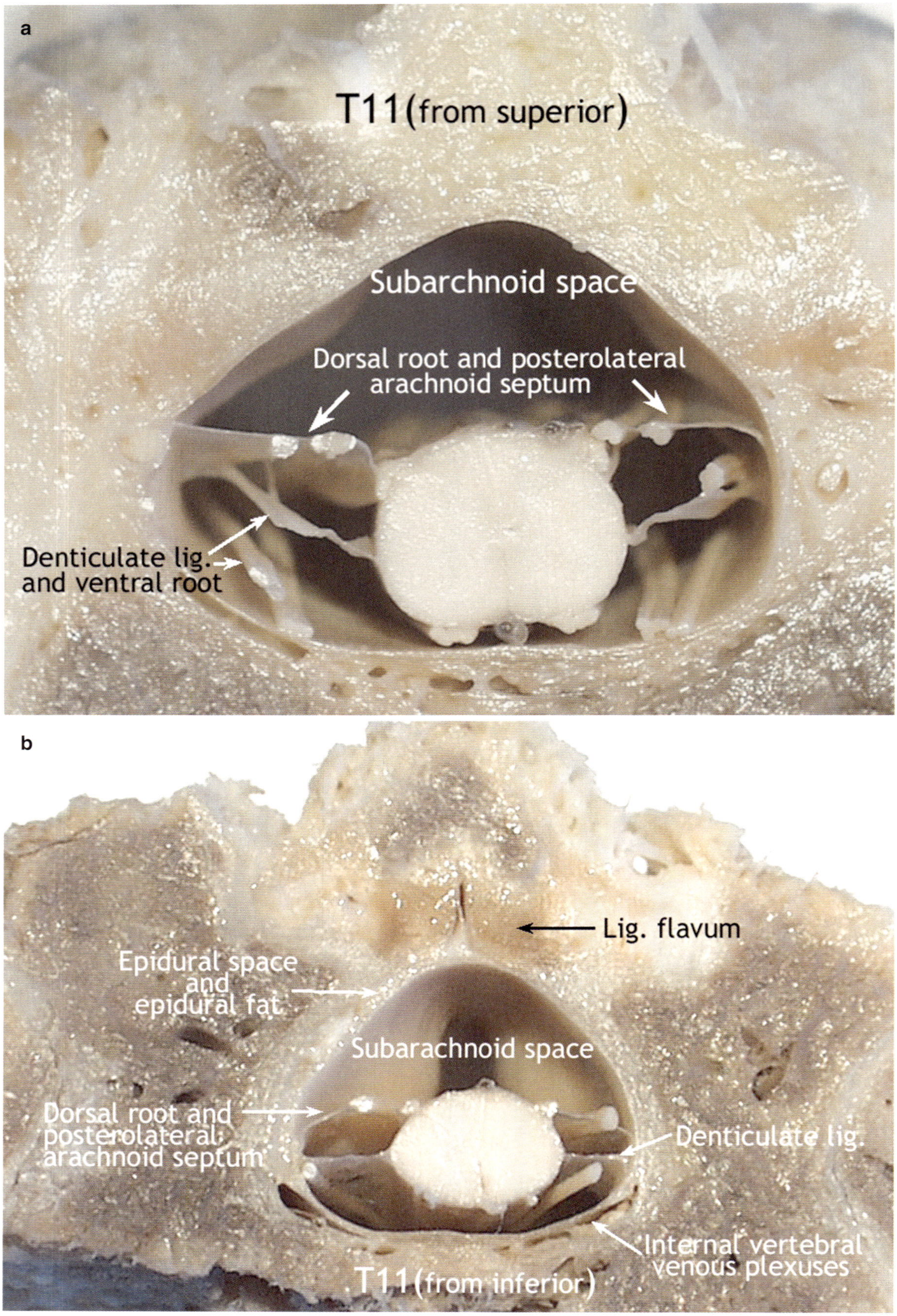

Fig. 34.13 Transverse section of the spinal canal at the T11 level. (**a**) View of canal from the superior aspect. (**b**) View from the inferior aspect. This specimen lacks a dorsal arachnoid septum

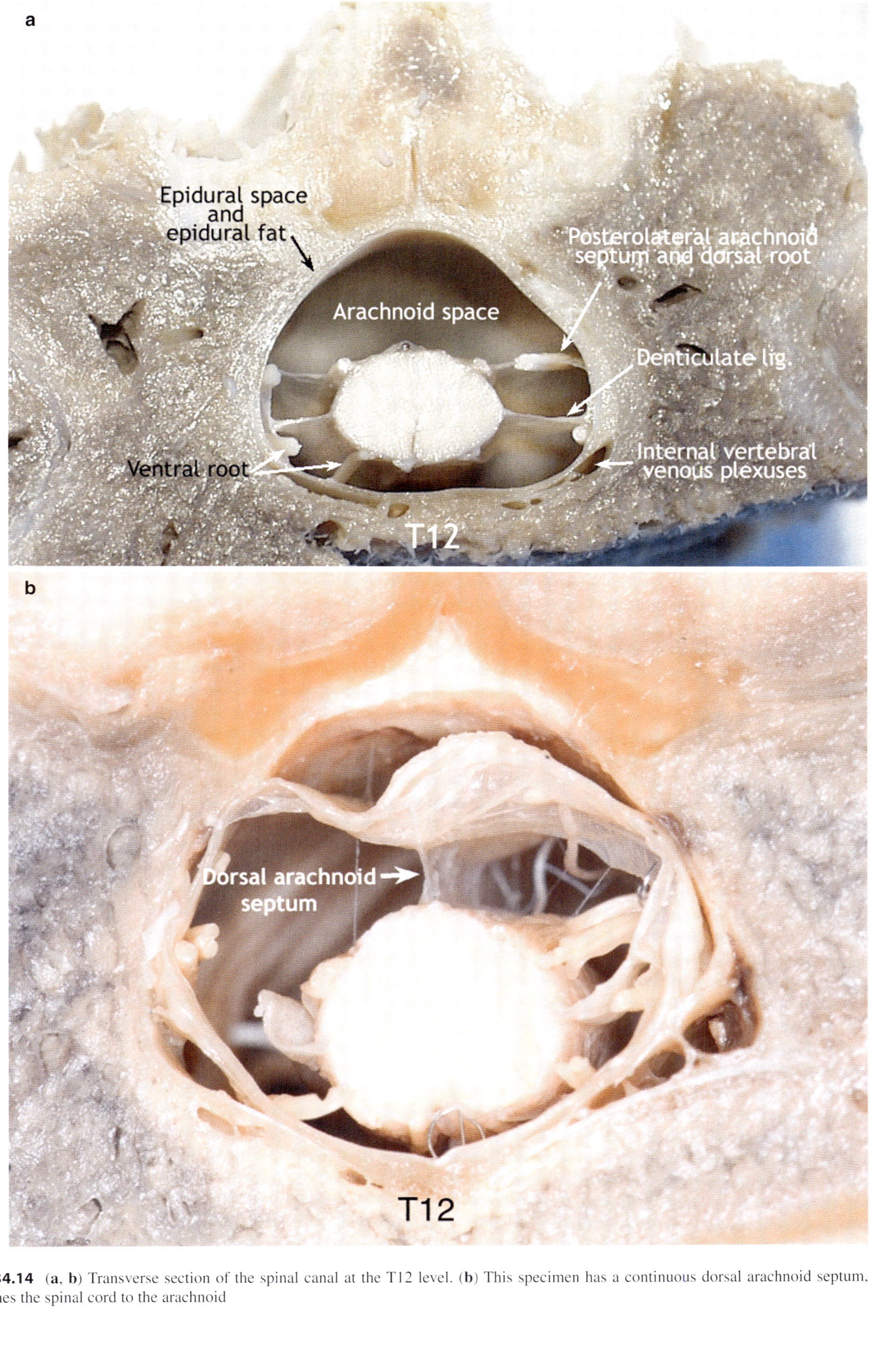

Fig. 34.14 (**a**, **b**) Transverse section of the spinal canal at the T12 level. (**b**) This specimen has a continuous dorsal arachnoid septum, which attaches the spinal cord to the arachnoid

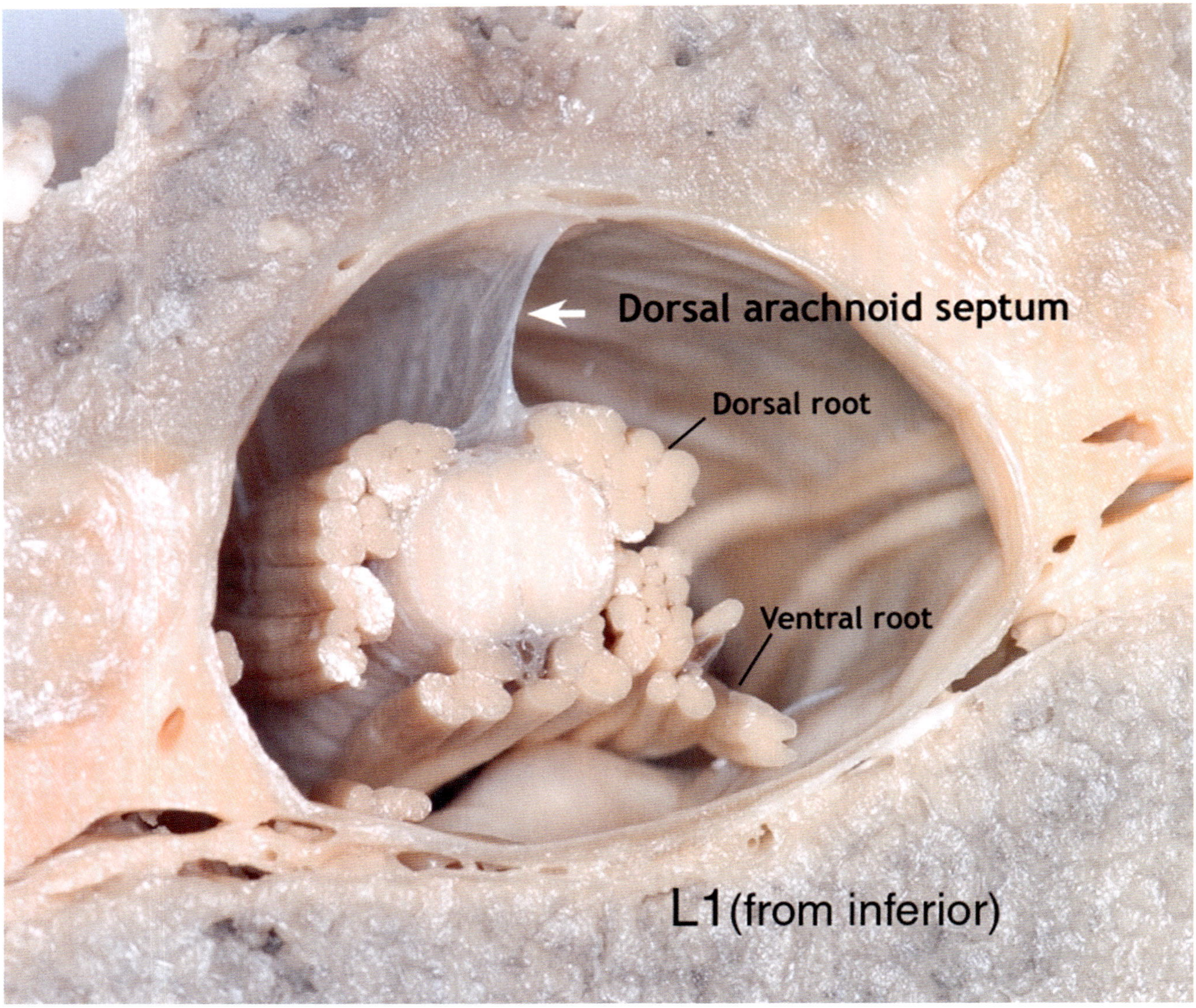

Fig. 34.15 Transverse section of the spinal canal at the L1 level. The spinal cord dwindles into the conus medullaris carrying lumbar spinal nerves at the L1 level. This specimen has a thick and continuous dorsal arachnoid septum in the posterior region

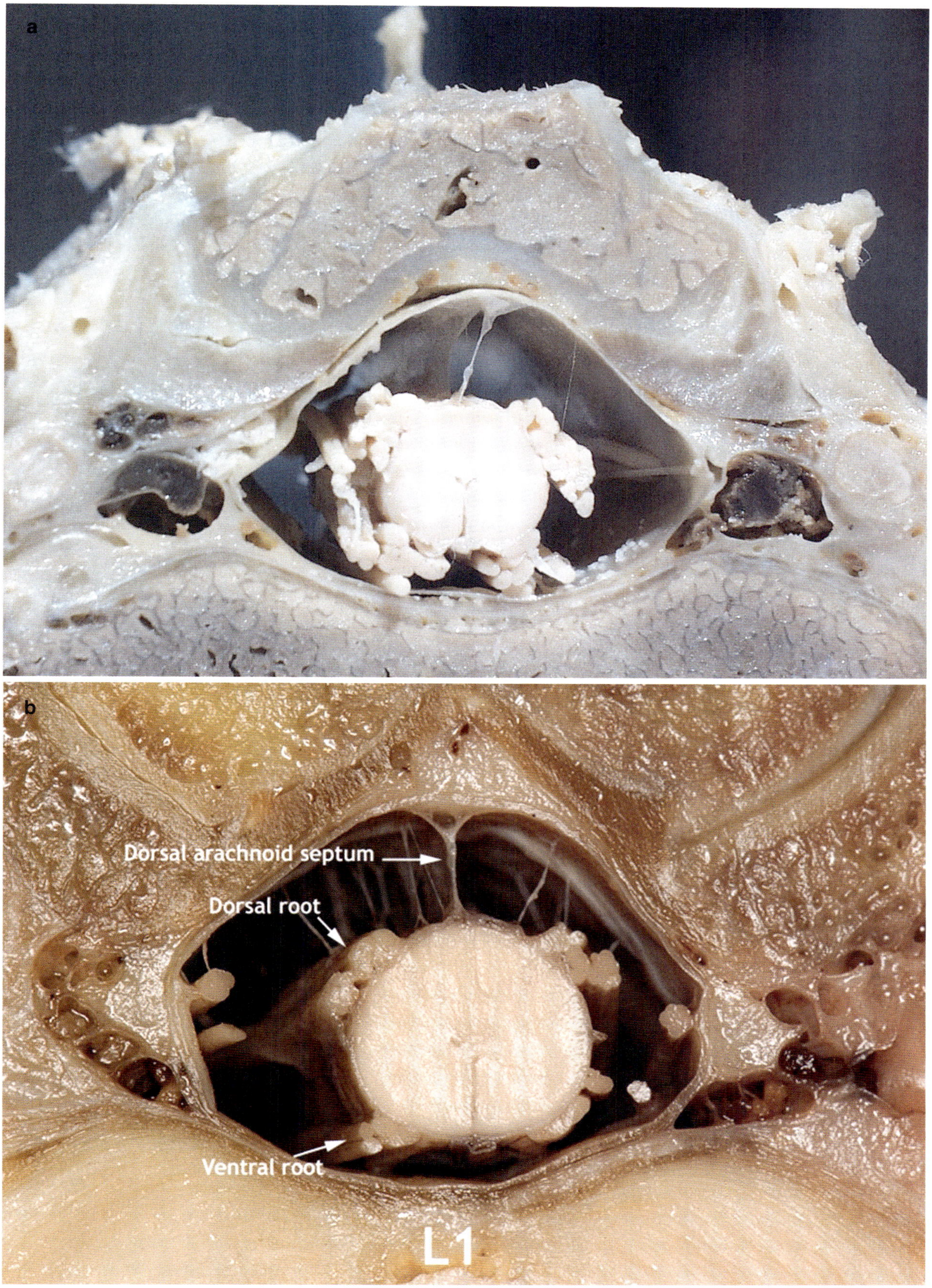

Fig. 34.16 (**a**, **b**) Transverse section of the spinal canal at the L1 level. (**a**) Transverse section of the spinal canal at the L1 level in another specimen. This specimen also has a continuous dorsal arachnoid septum and bilateral basal portions of the denticulate ligament. (**b**) The spinal cord is large at the L1 level in this specimen, but the level at which the spinal cord ends varies considerably between individuals. In 40 % of individuals, it ends below the L1 and L2 level, and occasionally it ends as low as the disc between the L2 and L3 vertebrae [1]

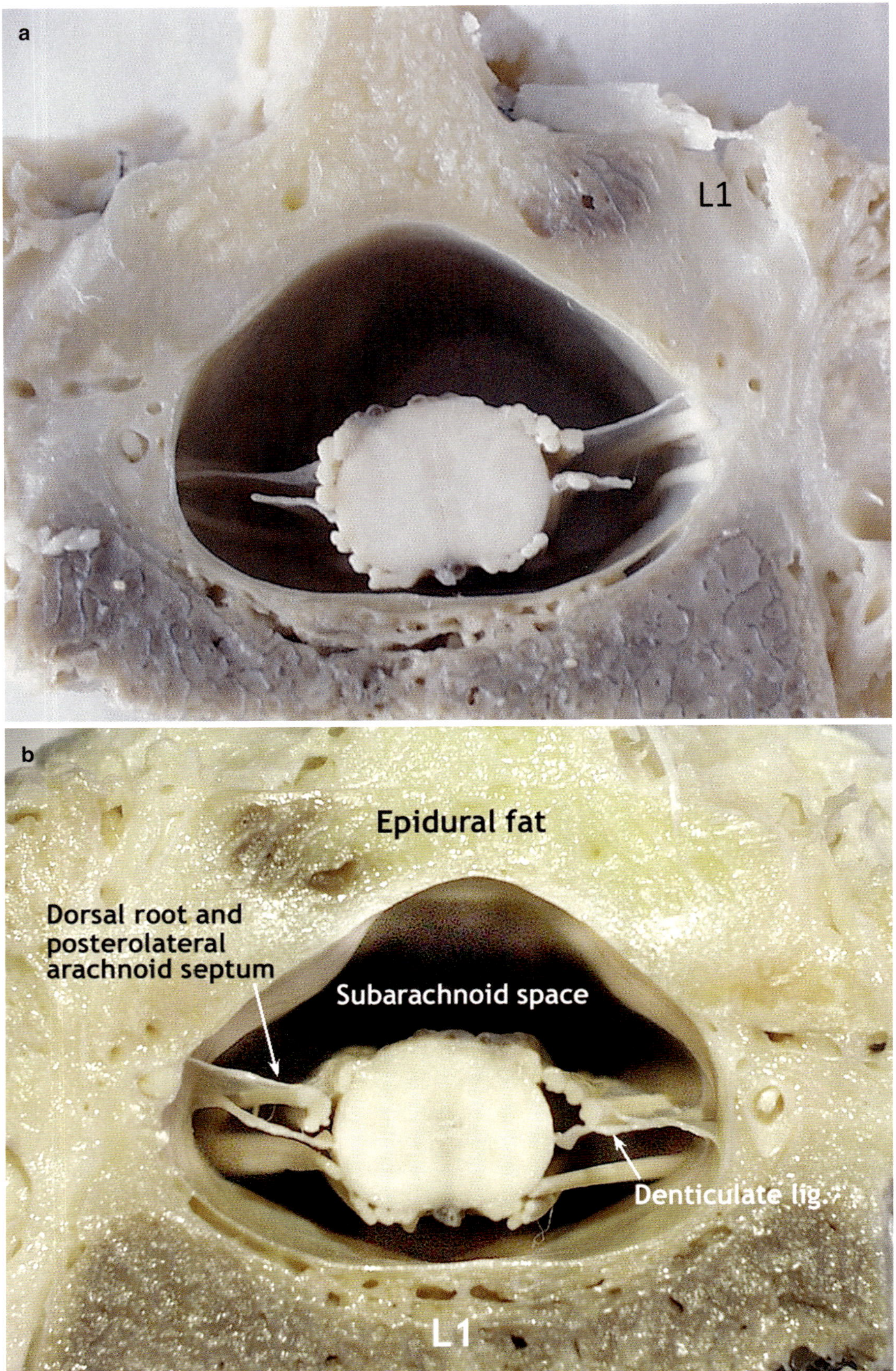

Fig. 34.17 (**a**, **b**) Transverse section of the spinal canal at the L1 level

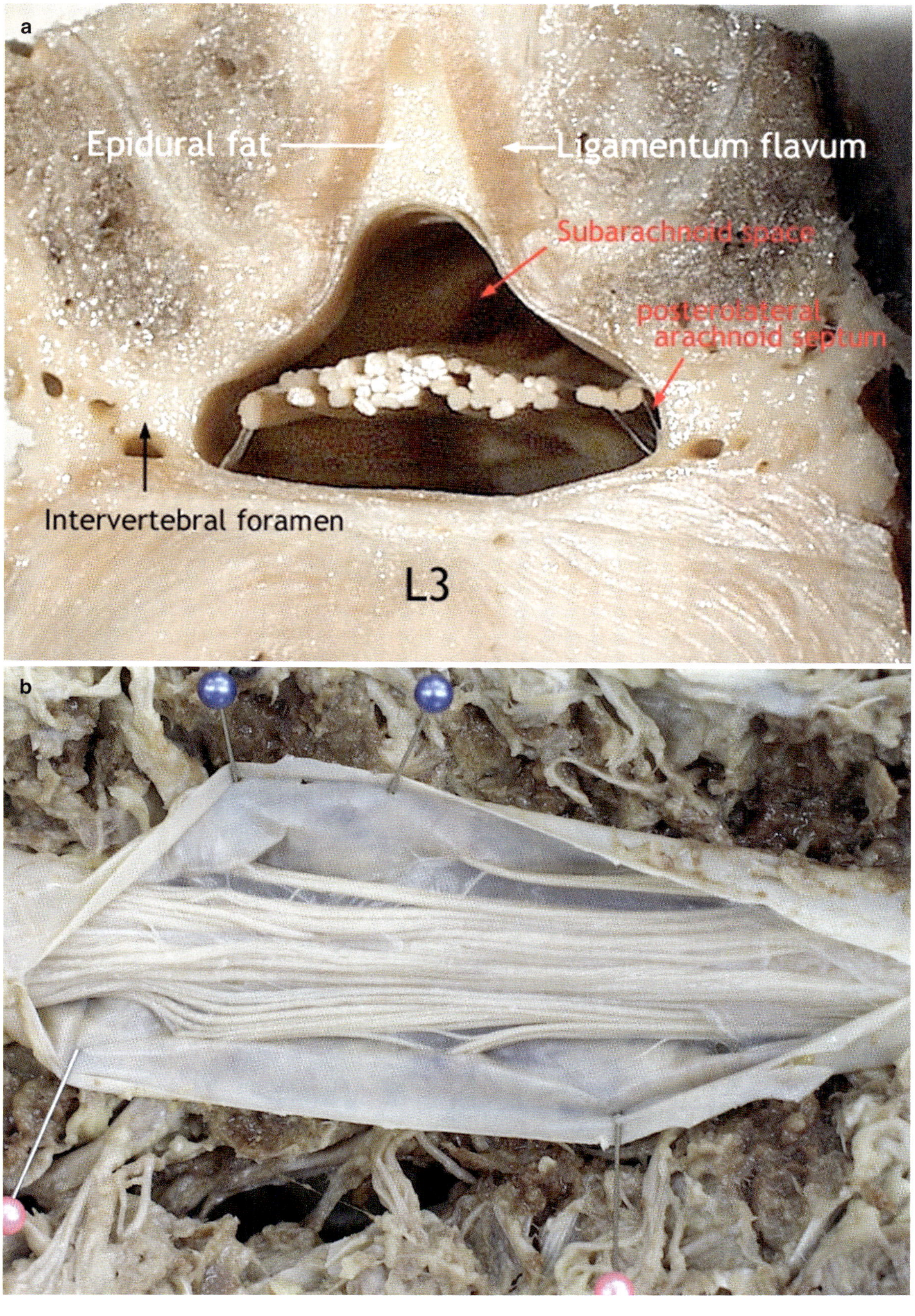

Fig. 34.18 (**a**) Transverse section of the spinal canal at the L3 level and the dural sac dissected at lumbar level. The lumbar intrathecal ligaments are seen in the cauda equina. In the adult vertebral canal, the denticulate ligaments extend laterally from the lateral aspect of the spinal cord to the dura, where they attach [7]. The last of the denticulate ligaments is a narrow, oblique band that descends laterally from the conus medullaris and lies between the exiting T12 and L1 spinal nerves [1]. Kershner and Binhammer [7] showed that intrathecal lumbar ligaments randomly bind the dorsal nerve roots of the cauda equina to the dura, and this is histologically considered to be a fetal development of the denticulate ligament. We have also observed these ligamentous bands in cadavers in our laboratory. (**b**) Longitudinal section of the dural sac

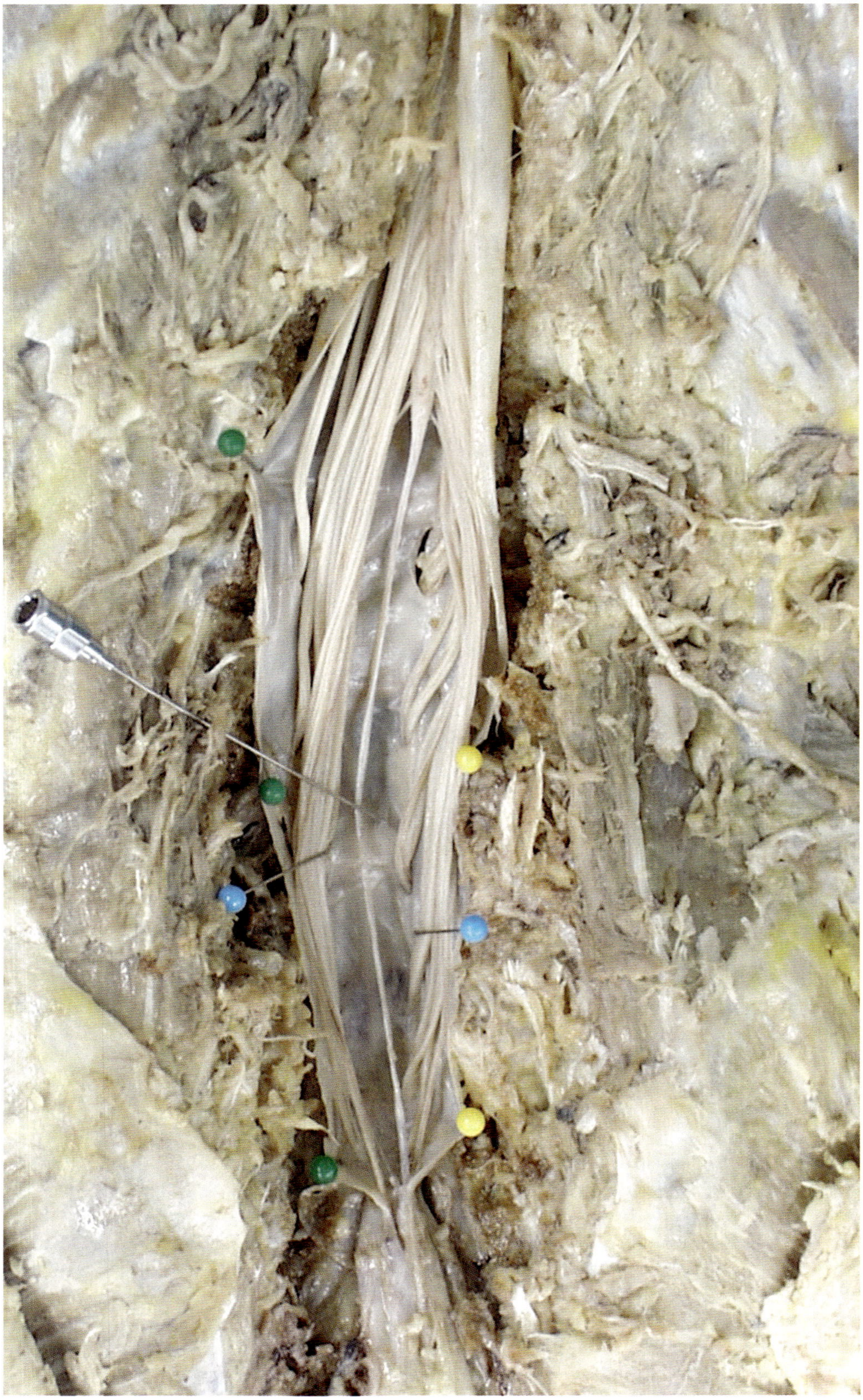

Fig. 34.19 The filum terminale (FT). The FT is a fibrous structure that extends from the conus medullaris of the spinal cord to the coccyx. The upper three quarters of the FT (the FT internum) is a filament from the apex of the conus medullaris that continues within the extension of the dural and arachnoid meninges and reaches the caudal border of the second sacral vertebra. The lower part of the FT (the FT externum) fuses with the investing dura mater and descends to the dorsum of the first coccygeal vertebral segment [1]. Under normal conditions, the FT provides loose support to the spinal cord, anchoring it to the coccyx, and does not cause pathological symptoms. Incomplete involution of the embryonal spinal cord may cause tight FT syndrome, which causes urinary dysfunction, lumbago, and other neurogenic symptoms. Tight FT syndrome is associated with spinal cord tethering and an abnormally positioned conus medullaris below L2–3. The image shown is from a cadaver

References

1. Newell RLM. Spinal cord and spinal nerves: gross anatomy. In: Standring S, editor. Gray's anatomy: the anatomical basis of clinical practice. 40th ed. New York: Churchill Livingstone; 2008. p. 749–62.
2. Hogan QH. Lumbar epidural anatomy. A new look by cryomicrotome section. Anesthesiology. 1991;75:767–75.
3. Lirk P, Kolbitsch C, Putz G, Colvin J, Colvin HP, Lorenz I, et al. Cervical and high thoracic ligamentum flavum frequently fails to fuse in the midline. Anesthesiology. 2003;99:1387–90.
4. Lirk P, Moriggl B, Colvin J, Keller C, Kirchmair L, Rieder J, et al. The incidence of lumbar ligamentum flavum midline gaps. Anesth Analg. 2004;98:1178–80.
5. Lirk P, Colvin J, Steger B, Colvin HP, Keller C, Rieder J, et al. Incidence of lower thoracic ligamentum flavum midline gaps. Br J Anaesth. 2005;94:852–5.
6. Matsunaga S, Sakou T. Ossification of the posterior longitudinal ligament of the cervical spine: etiology and natural history. Spine. 2012;37:E309–14.
7. Kershner DE, Binhammer RT. Lumbar intrathecal ligaments. Clin Anat. 2002;15:82–7.
8. Ceylan D, Tatarli N, Abdullaev T, Seker A, Yildiz SD, Keles E, et al. The denticulate ligament: anatomical properties, functional and clinical significance. Acta Neurochir (Wien). 2012;154:1229–34.
9. Nicholas DS, Weller RO. The fine anatomy of the human spinal meninges. A light and scanning electron microscopy study. J Neurosurg. 1988;69:276–82.

35 Displacement of the Nerve Roots of Cauda Equina in Different Positions

Teresa Parras, Alberto Prats-Galino, Rafael Blanco, Ana Delgado, and José Luís González

Nerve roots adopt relatively fixed positions inside the cauda equina. However, they may experience slight displacements following postural changes of the spine. Magnetic resonance images are obtained from patients in the supine position. This circumstance explains the predominant posterior location of the nerve roots within the cauda equina observed in such images. It is certain that postural changes of the lumbar column lead to displacements of the nerve roots within the dural sac. If the human body lies in the lateral position, the dural shifts laterally toward the lower region of the sac. If the human body lies with the thighs flexed against the abdomen, the nerve roots are displaced forward. In magnetic resonance studies, it is convenient to select an adequate acquisition configuration that enables volumetric reconstruction of each nerve root respectively. Otherwise, the nerve roots will form part of a group of nerve roots, rendering individual volumetric reconstruction impossible. Isometric voxels smaller than 0.6 mm are a feature in optimized software of high-field equipment that improves the quality of nerve root images (Figs. 35.1, 35.2, 35.3, 35.4, and 35.5).

T. Parras, MD, PhD (✉)
Anaesthetic Department, St. George's Hospital, London, UK
e-mail: tparrasmaldonado@gmail.com

A. Prats-Galino, MD, PhD
Laboratory of Surgical NeuroAnatomy (LSNA), Human Anatomy and Embryology Unit, School of Medicine, University of Barcelona, Barcelona, Spain
e-mail: aprats@ub.edu

R. Blanco, MD
Anaesthetic Department, Corniche Hospital, Abu Dhabi, Arab Emirates
e-mail: rafablanco@me.com

A. Delgado, Tch • J.L. González, MD
Department of Radiology, Complejo Hospitalario de Jaén, Jaén, Spain
e-mail: andelbel@hotmail.com; jlgmontane@hotmail.com

M.A. Reina et al. (eds.), *Atlas of Functional Anatomy for Regional Anesthesia and Pain Medicine: Human Structure, Ultrastructure and 3D Reconstruction Images*, DOI 10.1007/978-3-319-09522-6_35,

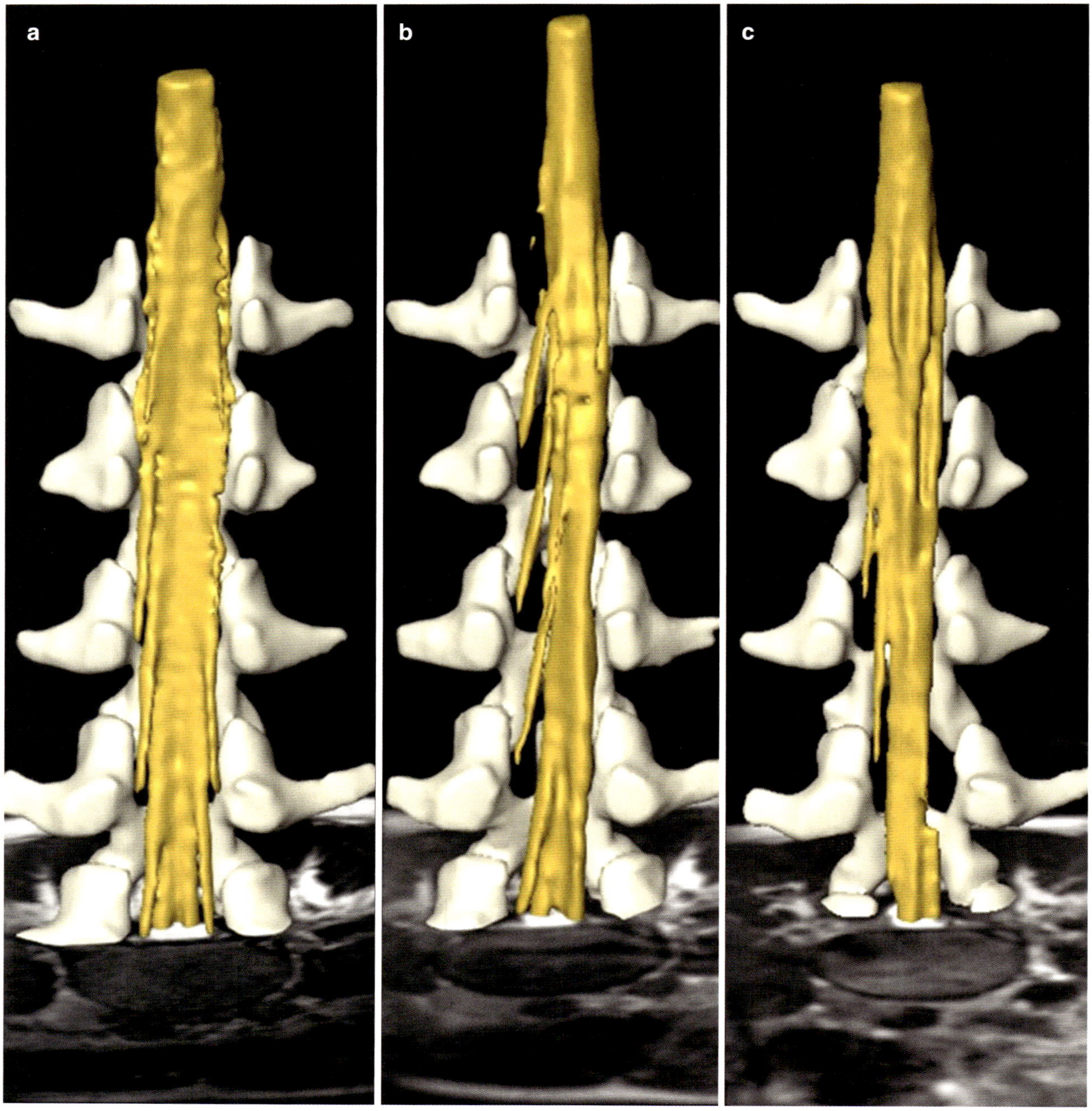

Fig. 35.1 Three-dimensional reconstruction of human nerve root of cauda equina at lumbar spine level. Patient in supine position (**a**), lateral position (**b**) and prone position (**c**). Anterior view

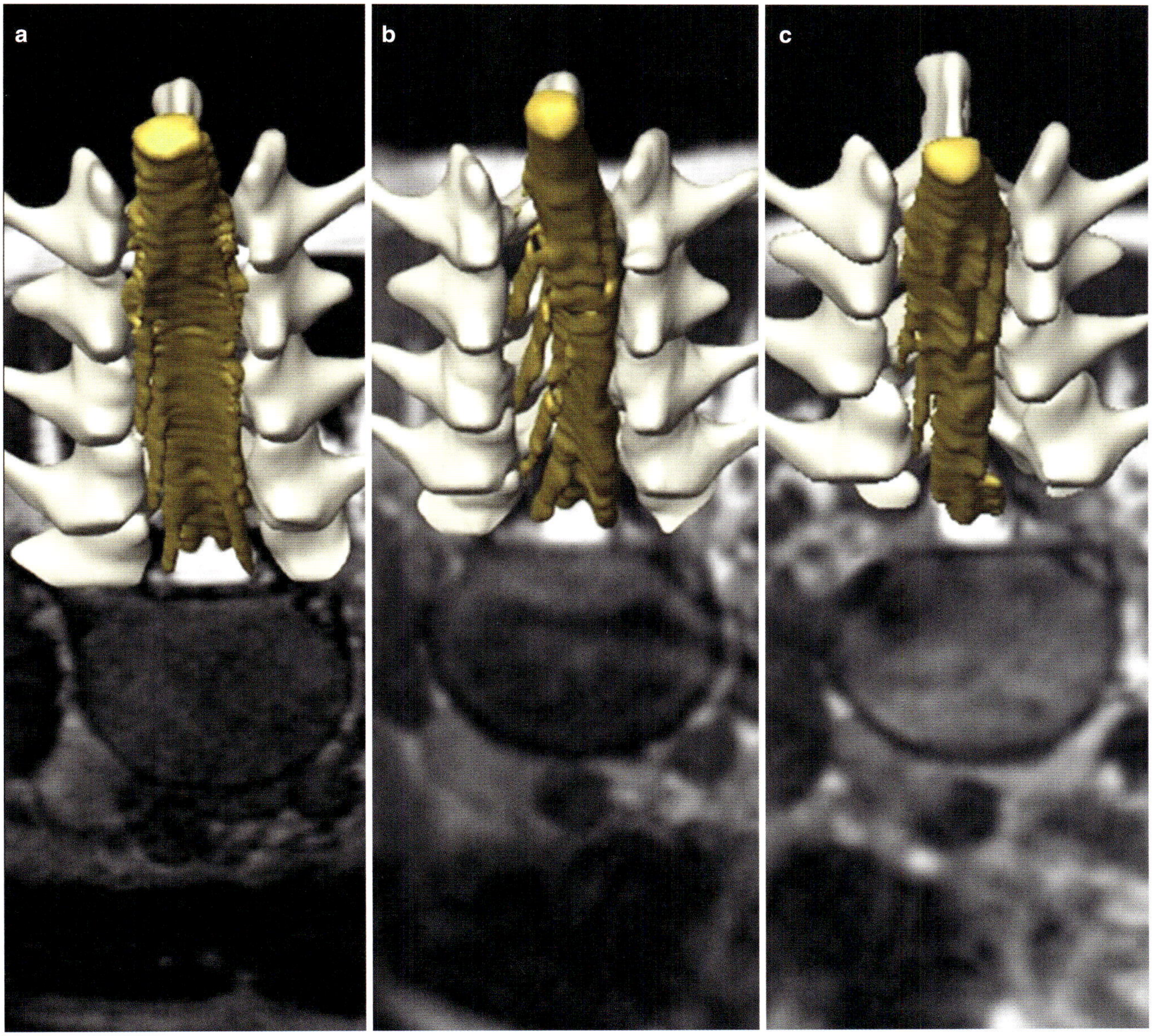

Fig. 35.2 Three-dimensional reconstruction of human nerve root of cauda equina at lumbar spine level. Patient in supine position (**a**), lateral position (**b**), and prone position (**c**). Oblique view

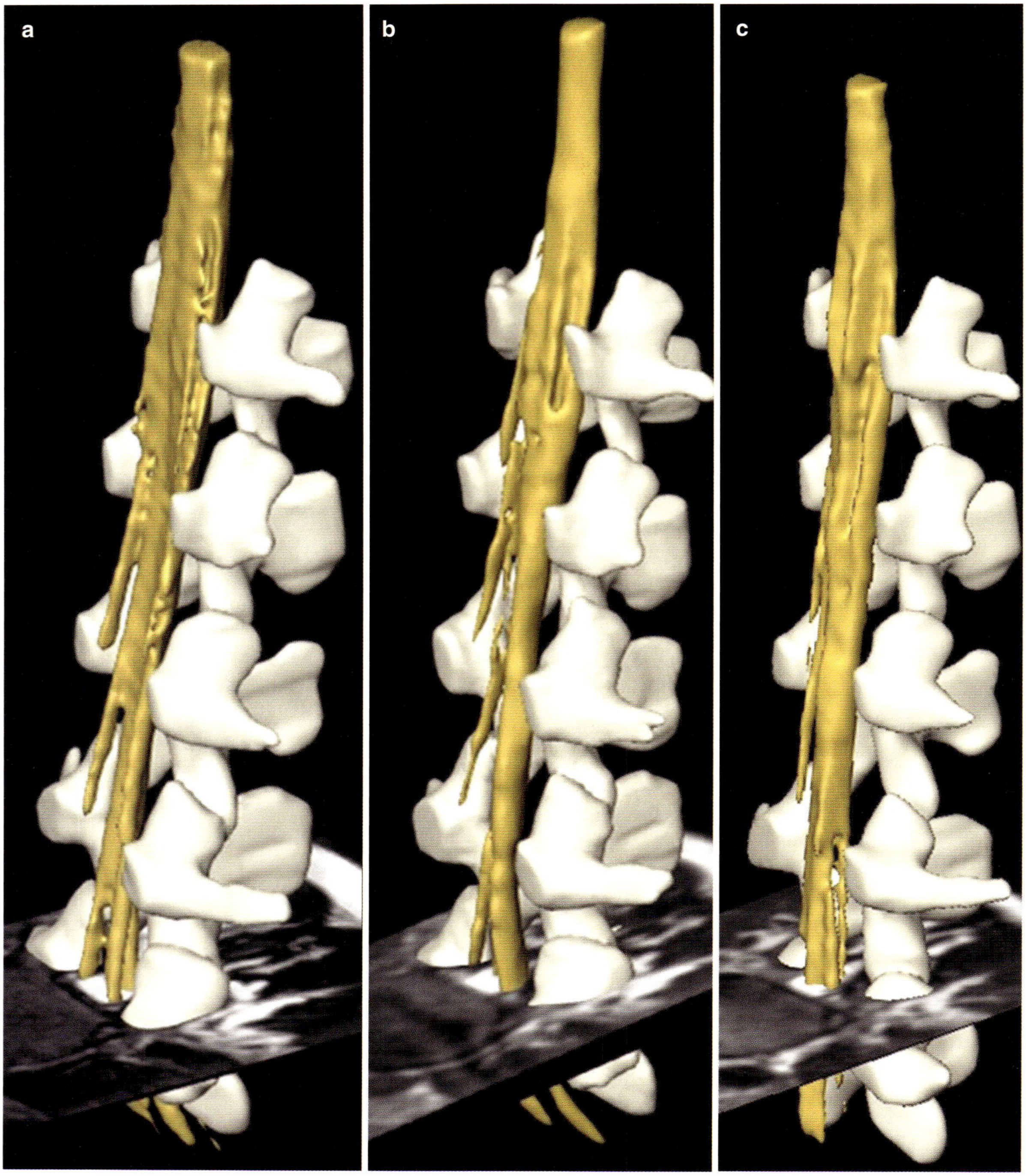

Fig. 35.3 Three-dimensional reconstruction of human nerve root of cauda equina at lumbar spine level. Patient in supine position (**a**), lateral position (**b**), and prone position (**c**). Lateral view

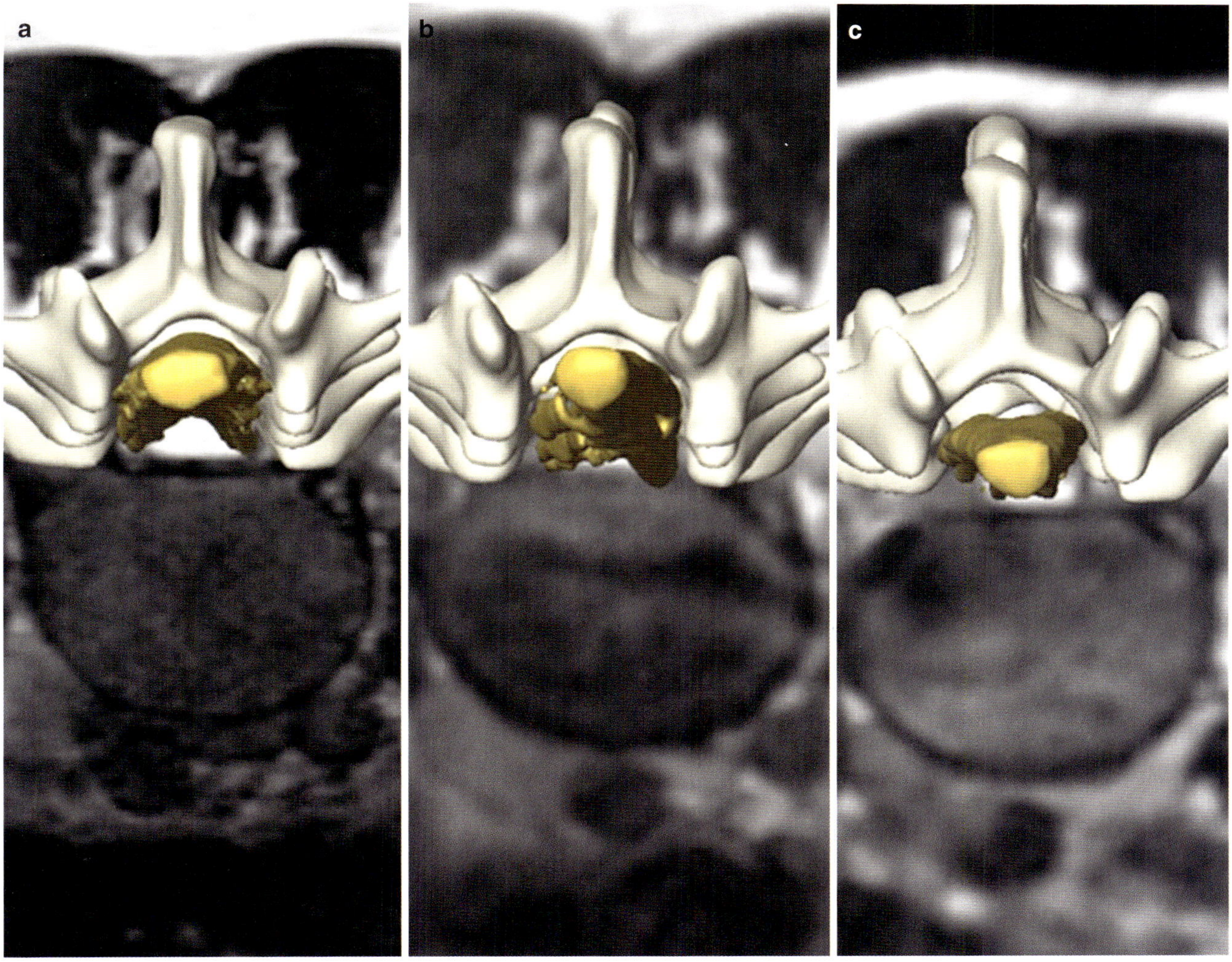

Fig. 35.4 Three-dimensional reconstruction of human nerve root of cauda equina at lumbar spine level. Patient in supine position (**a**), lateral position (**b**), and prone position (**c**). Superior view

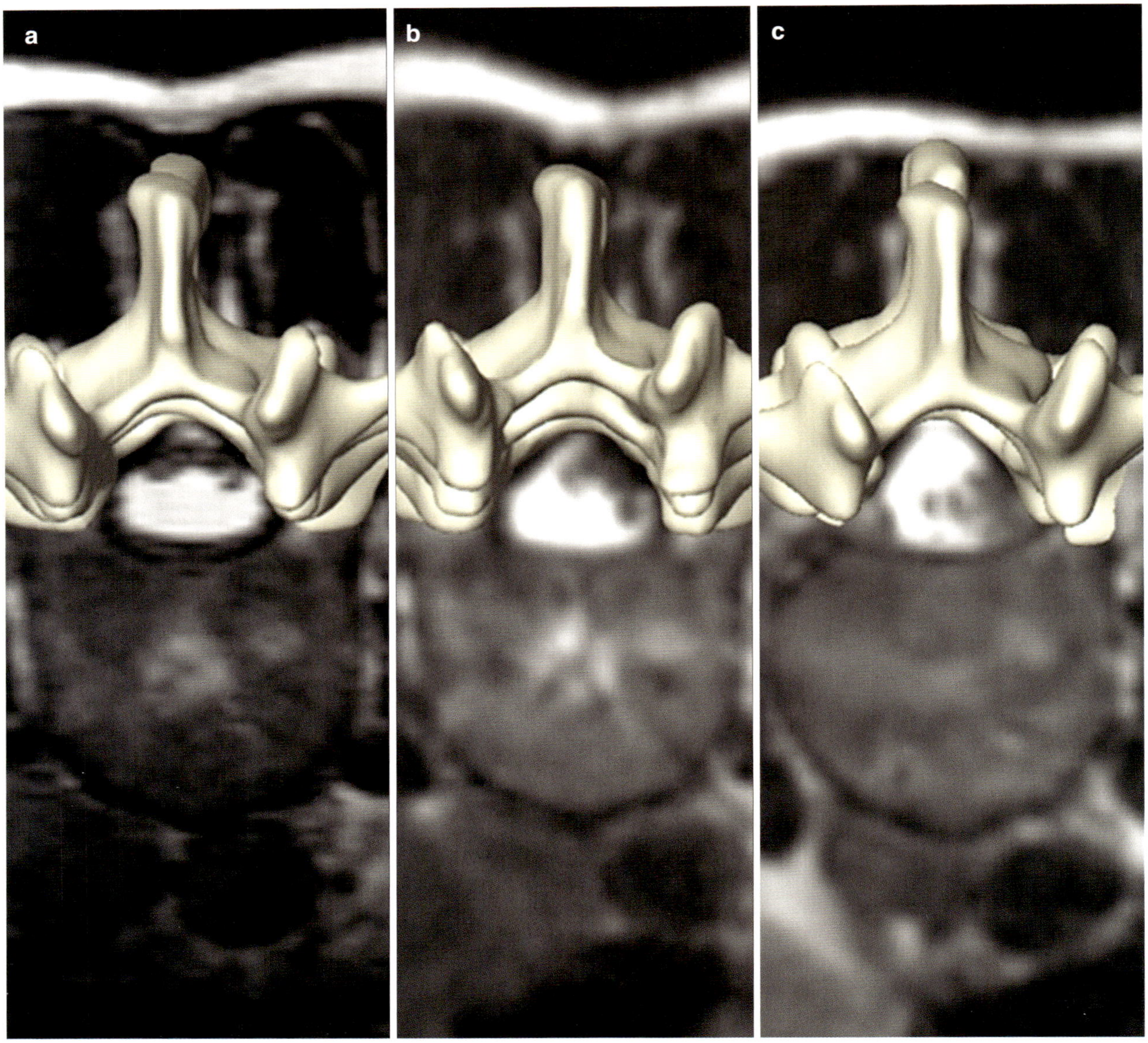

Fig. 35.5 Human nerve root of cauda equina at lumbar spine level. Patient in supine position (**a**), lateral position (**b**), and prone position (**c**). Superior view

36 Anatomy of the Thoracic Spinal Canal in Different Postures: An MRI Investigation

L.M. Arno Lataster and André A.J. van Zundert

The following images were garnered largely from studies using functional MRI (fMRI) techniques to determine the anatomic positions of the spinal canal (e.g., spinal cord, thecal tissue) in various postures and to identify possible implications from different patient positions for neuraxial anesthetic practice. Additionally, this chapter includes many anatomic slices of the spine. This chapter presents an overview of the functional aspects of the tissue architecture, especially of the epidural space and thecal tissues, and the position of the neural tissue.

In the past, high spinal anesthesia was used for craniotomies, and before the introduction of MRI, neurologists and radiologists performed subarachnoid myelographic injections at mainly cervical (and occasionally thoracic) levels. Subarachnoid punctures still are used for relief of cancer pain at the thoracic (subarachnoid catheter) and cervical (cordotomy) levels. The same MRI techniques that replaced radiographic myelography have been used to investigate the position of the spinal cord and surrounding structures (e.g., dura mater, spinal processes) in different postures.

The spinal cord follows the straightest line through the imposed geometry of the spine. Accordingly, there is relatively more posterior separation of the cord and surrounding thecal tissue at midthoracic levels (e.g. T5–T9) in the apex of the thoracic kyphosis. Placing a patient in a position that accentuates the thoracic curvature of the spine (i.e., sitting head-down) increases the posterior separation of the spinal cord and dural sheath at thoracic levels.

The images presented here and the corresponding publications are associated with studies using fMRI-obtained measurements with no clinical component. We can only speculate on their implications for clinical practice. However, it is known that epidural attempts may result in accidental puncture of the dura mater, at both the thoracic and lumbar spaces, with the risk of unintentional damage to the spinal cord. The relatively wide posterior subdural space at the thoracic level potentially allows the Tuohy needle to be halted before it contacts the pia mater or penetrates the spinal tissue. The incidence of serious neurologic complications from thoracic epidural procedures is far lower than that of inadvertent dural tap, and there may be benefits from the practice of spinal blockades at higher levels, although this awaits more clinical investigation and documentation. Because of considerable anatomic variation among patients, and the current lack of a clear correlation between patient characteristics and the distances of pertinent structures of the spinal canal, these findings should be interpreted cautiously (Figs. 36.1, 36.2, 36.3, 36.4, 36.5, 36.6, 36.7, 36.8, 36.9, 36.10, 36.11, 36.12, 36.13, 36.14, 36.15, 36.16, and 36.17).

L.M.A. Lataster, MSc (✉)
Department of Anatomy and Embryology,
Faculty of Health, Medicine and Life Sciences,
Maastricht University, Maastricht, The Netherlands
e-mail: a.lataster@maastrichtuniversity.nl

A.A.J. van Zundert, MD, PhD, FRCA, EDRA, FANZCA
Discipline of Anesthesiology, The University of Queensland
and Royal Brisbane and Women's Hospital,
Brisbane, QLD, Australia
e-mail: vanzundertandre@gmail.com

M.A. Reina et al. (eds.), *Atlas of Functional Anatomy for Regional Anesthesia and Pain Medicine: Human Structure, Ultrastructure and 3D Reconstruction Images*, DOI 10.1007/978-3-319-09522-6_36,

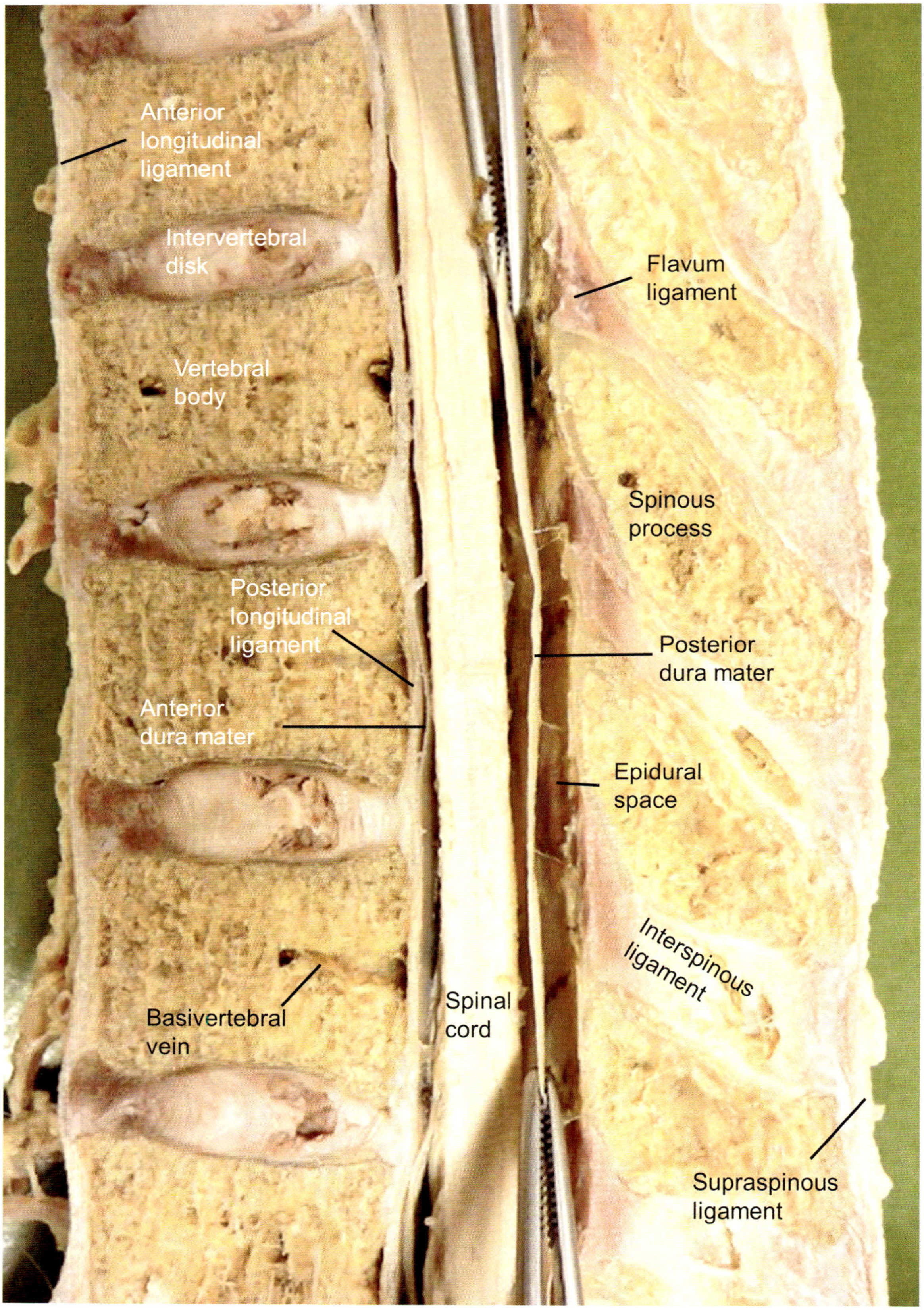

Fig. 36.1 Median section of a part of the human thoracic spinal canal, with the dura held slightly apart from the posterior boundary of the vertebral arches. Often, it is not possible to easily perceive the actual separation of the neural tissue from the thecal tissue because of the process of the dissection. As such, the spinal cord usually is depicted as equally displaced from the anterior and posterior boundaries along the whole length of the spinal column. Many illustrations depict the spinal cord this way, which is incorrect. Anteriorly, the dura mater is attached more to the posterior longitudinal vertebral body and loosely at the middle, where the basivertebral veins run, and fused with the intervertebral disk. Please note that the epidural space is filled with loose connective tissue, fat, and epidural veins throughout. In this section, the fat has been rinsed away and the connective tissue disrupted; however, the MRI figures (Figs. 36.6 and 36.7) clearly show the position of the fat

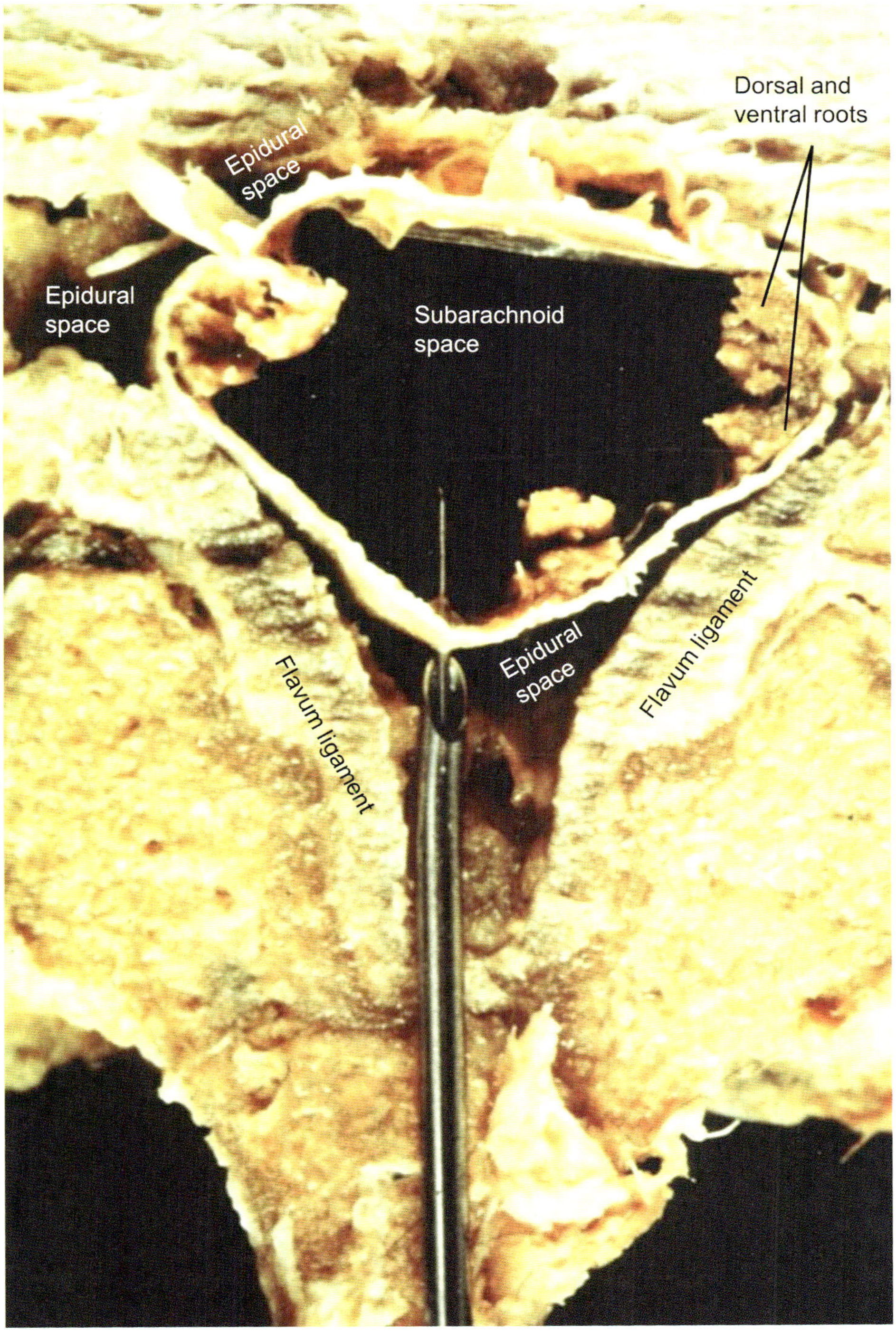

Fig. 36.2 Lumbar-level axial section of the combined spinal–epidural technique in situ. It is of utmost clinical importance that the spinal needle not be advanced too far into the subdural space so that mechanical damage is done to the neural tissue by the needle point. At levels caudal to the cauda equina, the neural tissue may move upon contact with the needle, thus avoiding damage; however, this is not the case above the termination of the spinal cord. Clearly, the Tuohy needle (16-gauge) is much larger than the spinal needle (27-gauge). Hence, the punctures made in the dura and arachnoid mater by the respective needles will differ substantially, which may result in post–dural puncture headache; large-gauge epidural needles should not puncture the dura mater if possible

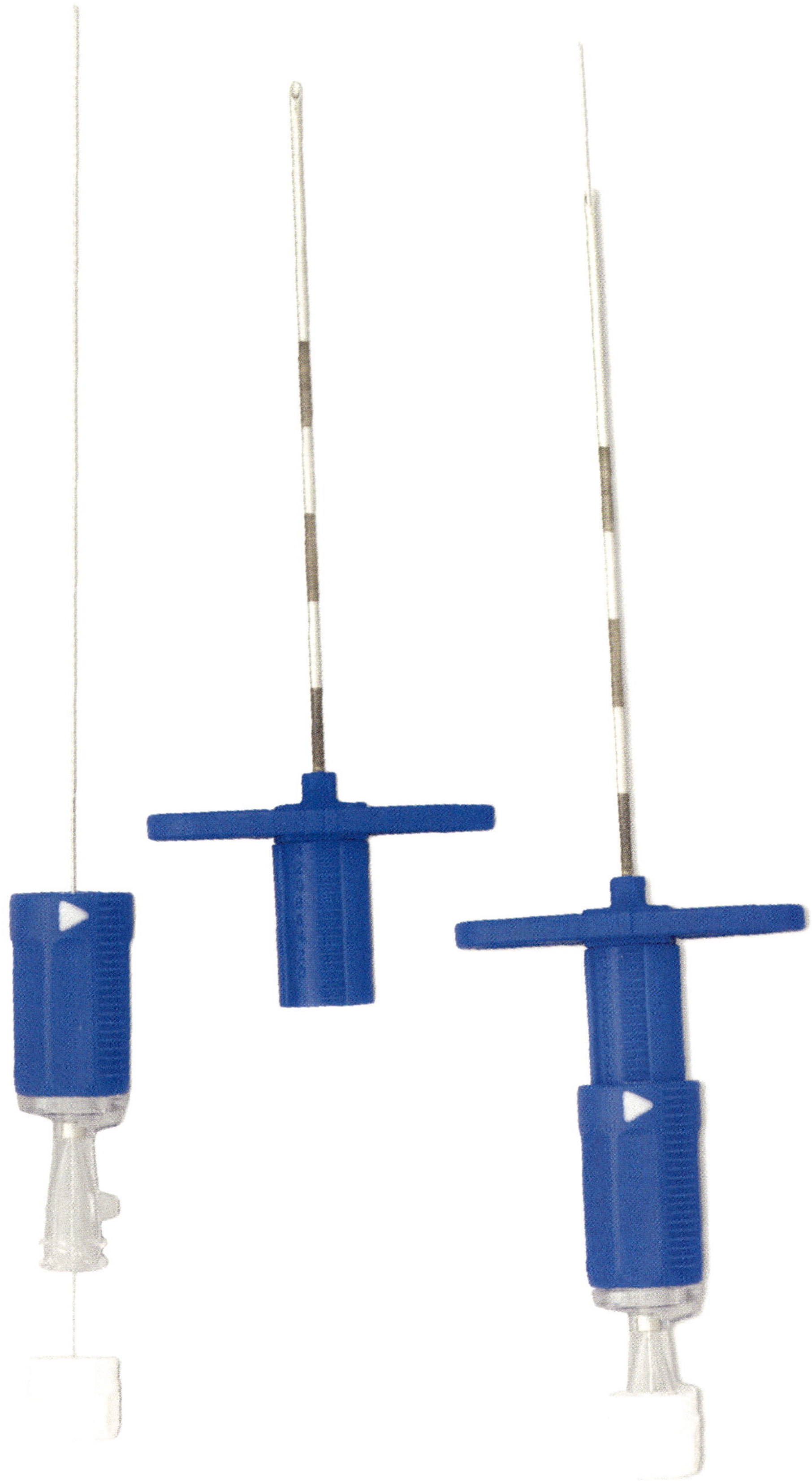

Fig. 36.3 Spinal needle (*top*), epidural needle (*center*), and spinal needle inserted through the cannula of the epidural needle (*bottom*). The needles include centimeter distance markings, as well as a locking hub design, to ensure that the clinician is aware of the distance the spinal needle advances past the end of the epidural needle and that this may be fixed once the dura is breached. CSEcure Combined Spinal Epidural System, Portex (Reproduced courtesy of Smiths Medical, Ashford, Kent, United Kingdom)

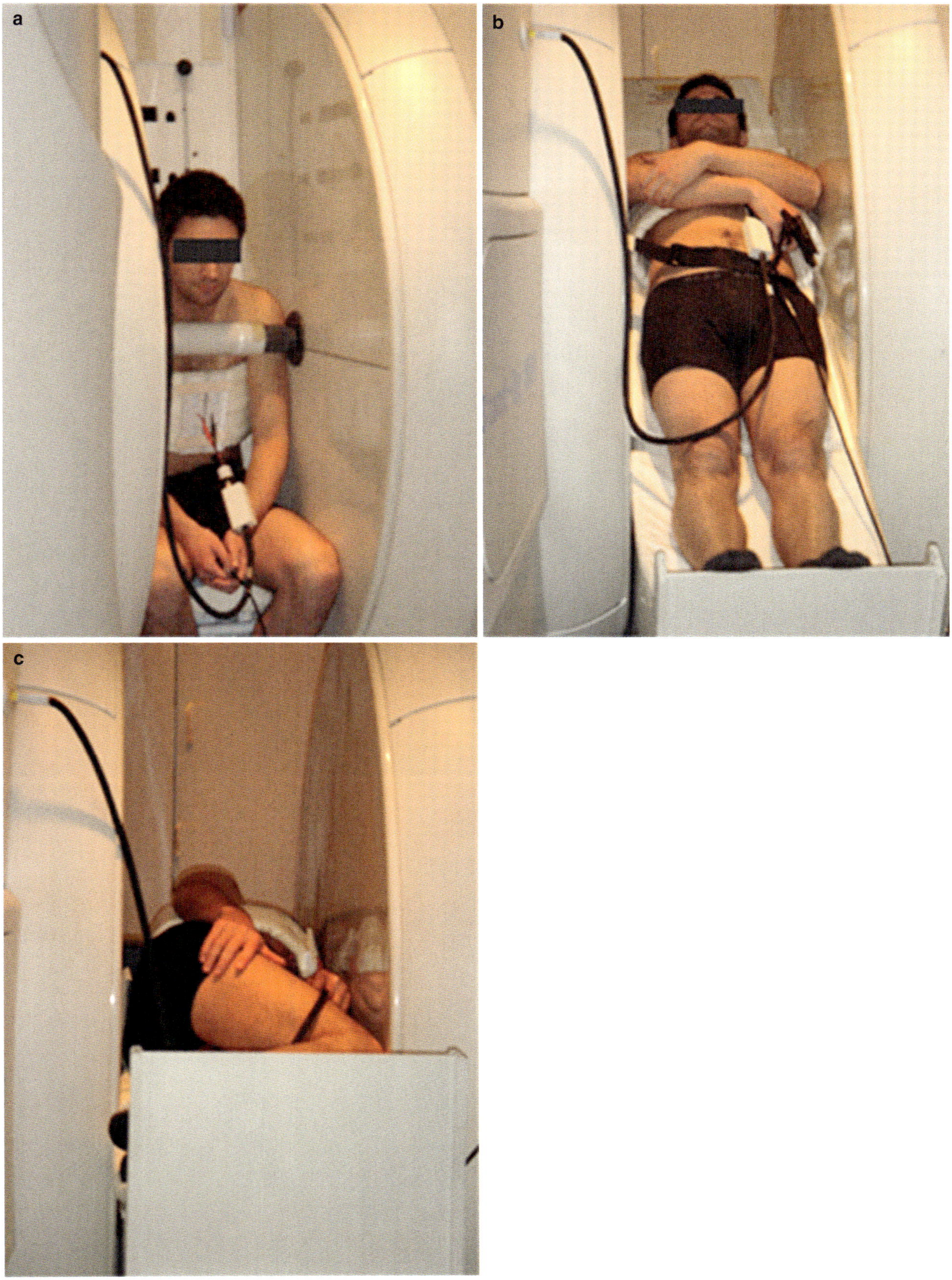

Fig. 36.4 Volunteer positioned ((**a**) sitting, (**b**) supine, and (**c**) laterally recumbent) in a Fonar Upright Multi-Position MRI scanner (Fonar Corporation, Melville, NY) with an antenna attached around the thorax. The volunteers were not limited by curvature of their backs in the sitting position, and they had more curved-back and flexed-head (head-down) positions during the measurements than the straight-backed position illustrated in this figure. fMRI scans are effective for imaging volunteers in postures resembling those in clinical practice during insertion of epidural and spinal needles (From Lee et al. [1]; with permission)

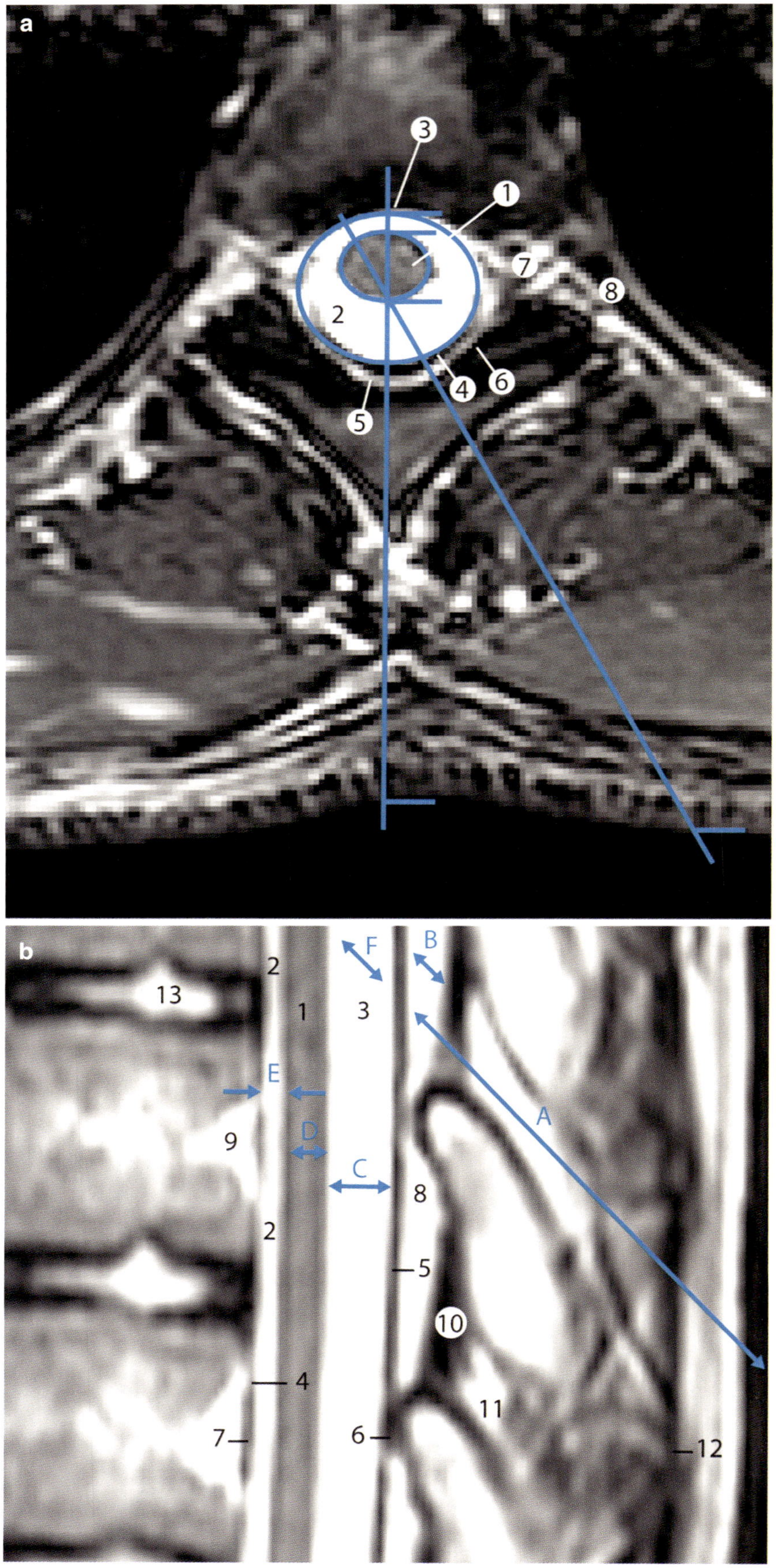

Fig. 36.5 (**a**) Segmentation of the axial MRI images. This MRI image is from the actual dataset and shows the spinal cord and dural sac modeled using an ellipsoid approximation, whereas the measurements were taken automatically. The anterior dura mater is bordered by the posterior longitudinal ligament. *1* Spinal cord; *2* subarachnoid space with cerebrospinal fluid; *3* posterior longitudinal ligament/anterior dura mater; *4* posterior dura mater; *5* posterior epidural space; *6* flavum ligament; *7* neuroforamen, lateral dural space, sleeve region; *8* costa. (**b**) Magnified median MRI of the thoracic spine of a volunteer in the supine position. The *arrows* indicate the pertinent distances for neuraxial anesthesia. *A* skin to dura, *B* epidural space, *C* posterior dura to cord, *D* cord diameter, *E* anterior dura to cord, and *F* posterior dura to cord in needle path. Note that the craniocaudal slope of the spinal processes (analogous to roofing tiles) at the thoracic level increases the separation of the dura from the spinal cord relative to a horizontal approach. *1* Spinal cord, *2* anterior arachnoid space, *3* posterior arachnoid space, *4* anterior dura mater, *5* posterior dura mater, *6* posterior dura mater next to vertebral arch, *7*, posterior longitudinal ligament, *8* posterior epidural space (fat, connective tissue, veins), *9* basivertebral veins(s), *10* flavum ligament, *11* interspinous ligament, *12* supraspinous ligament, *13* intervertebral disk (Panel **a** from Lee et al. [1]; with permission)

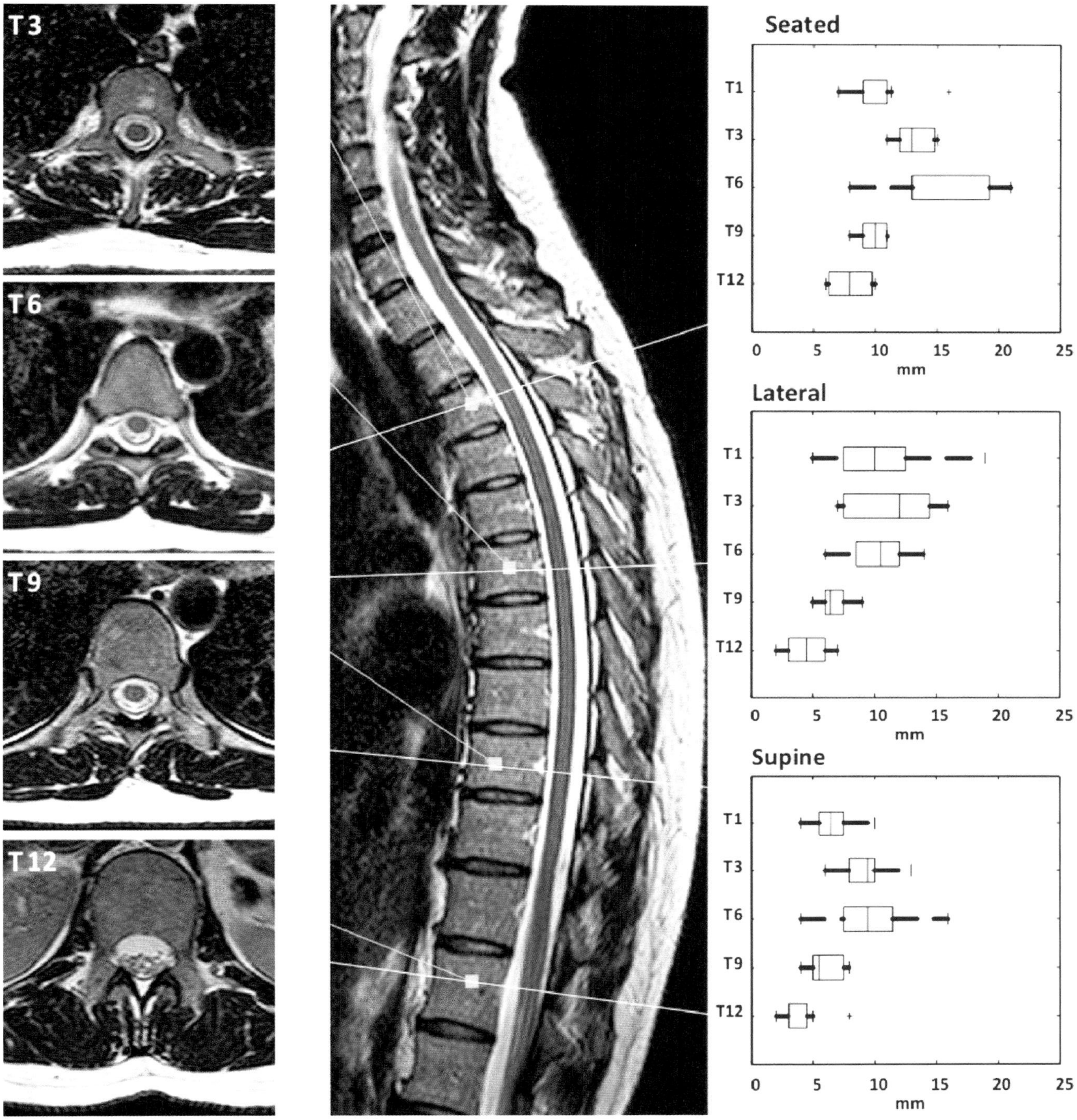

Fig. 36.6 Box plots (median, lower, and upper quartiles; range of the data; and eventual outliers indicated) of midline measurements from the axial dataset showing the measured distances (millimeters) between the dura mater and spinal cord posteriorly. To the left are magnetic resonance images showing the position of the spinal cord at various interspaces in a volunteer in the supine position (T1 image not included for layout considerations). Note the relatively large distance between the posterior aspect of the spinal cord and posterior dura mater, especially considering the actual approach of the epidural–spinal needles midline between the processes of the spinal cord (i.e., roof-tile effect). This distance is larger than the horizontal distance between the ligamentum flavum and dura mater (From Lee et al. [1]; with permission)

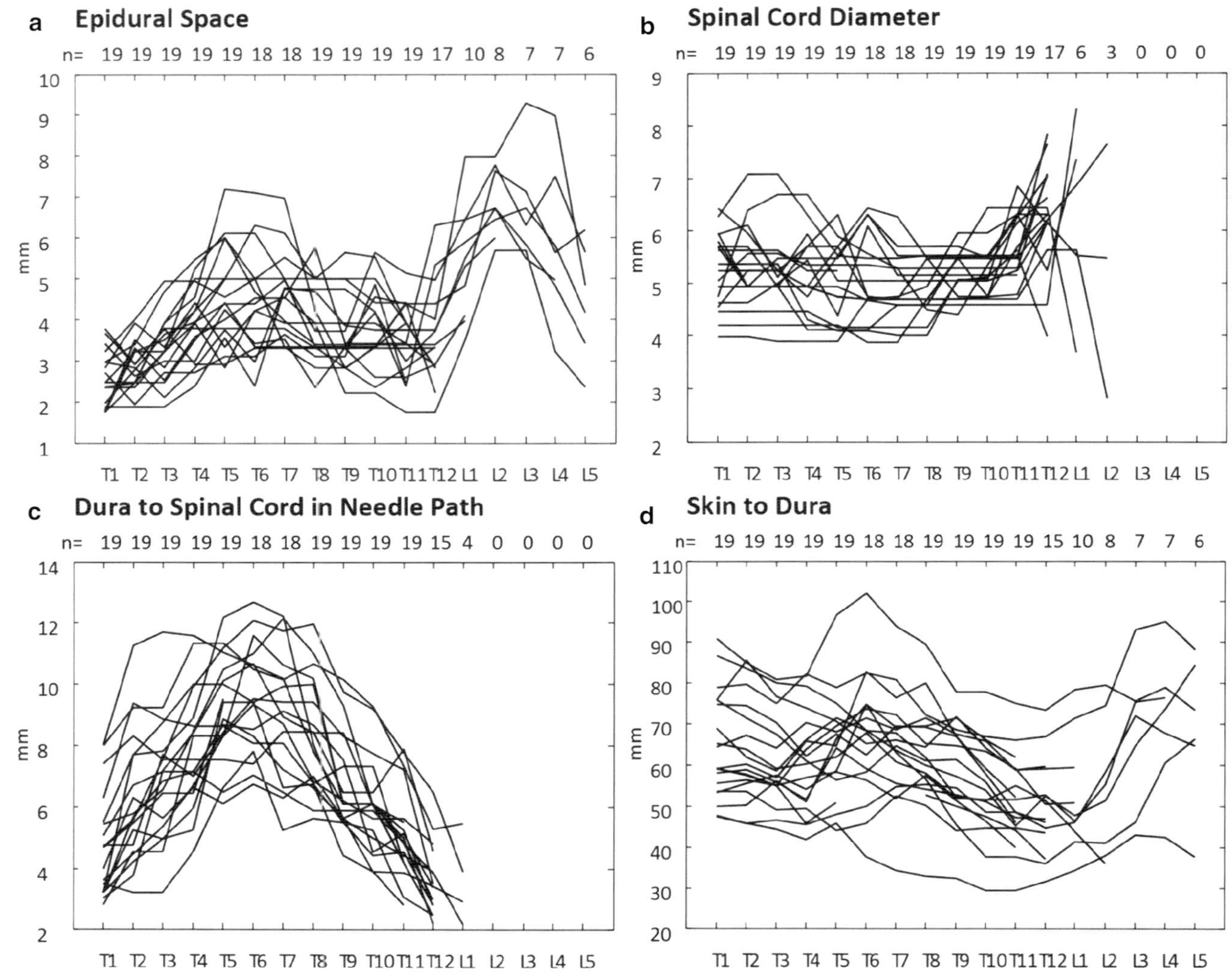

Fig. 36.7 Line plots of measured distances between various anatomic structures (a: Epidural space, b: Spinal cord diameter, c: Dura to spinal cord in needle path, d: Skin to dura) in the thoracic and lumbar spine for respective patients (distances are in millimeters, and the number of patients in each group is indicated above the respective plots). The plots demonstrate the increase in epidural space at lumbar levels. Additionally, the plot once again shows a clear relationship between the vertebral column level and the anterior–posterior positioning of the spinal cord—with the posterior distance maximized in the apex of the thoracic curve (From Lee et al. [2]; with permission)

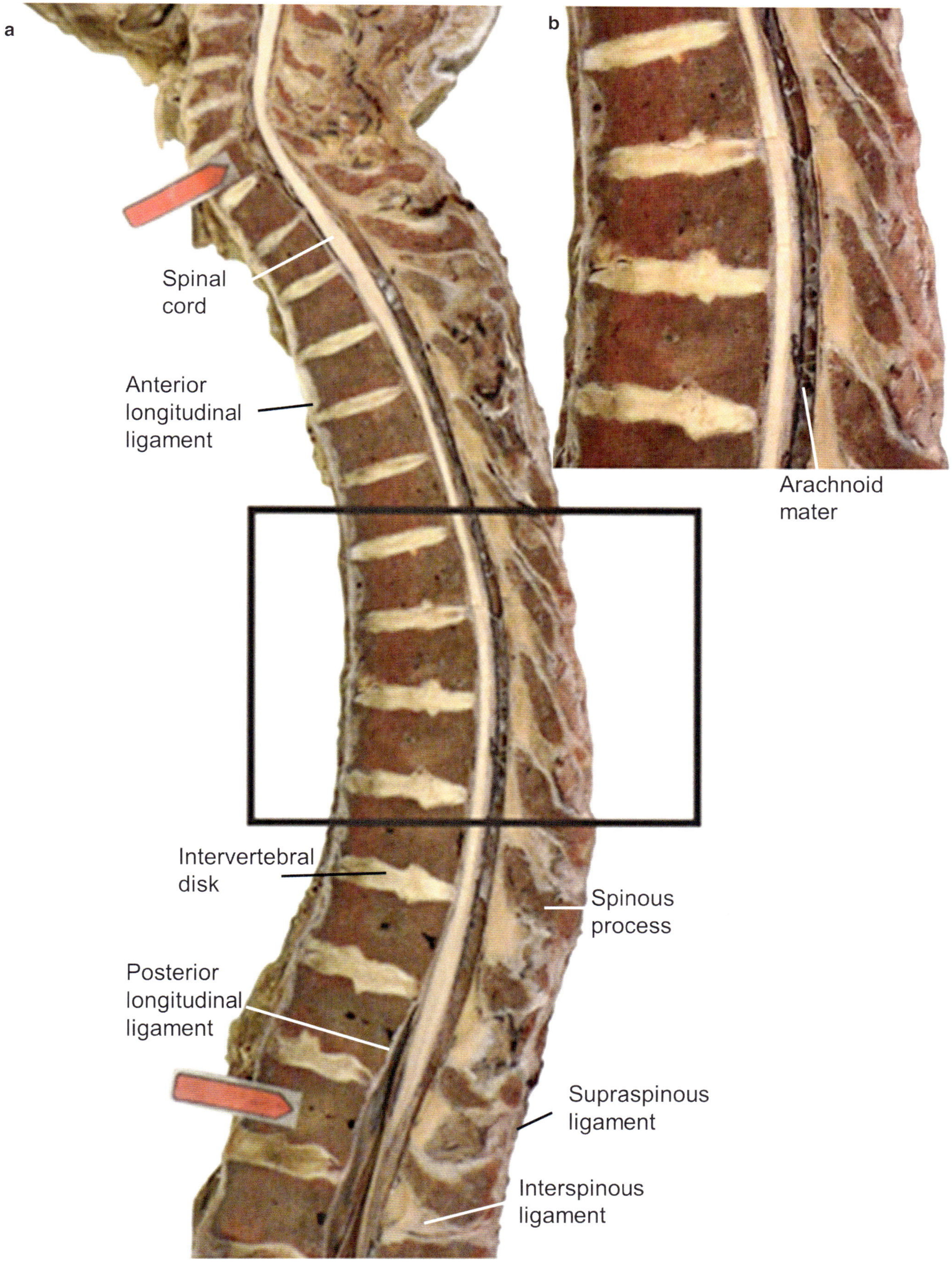

Fig. 36.8 (**a**) Anatomic median is sufficient section of the spinal cord in a human cadaver (*arrows* indicate C7 and L1, respectively). (**b**) Detail is indicated by frame in (**a**). Note that the epidural and arachnoid mater are still intact in this section (From Lee et al. [1]; with permission)

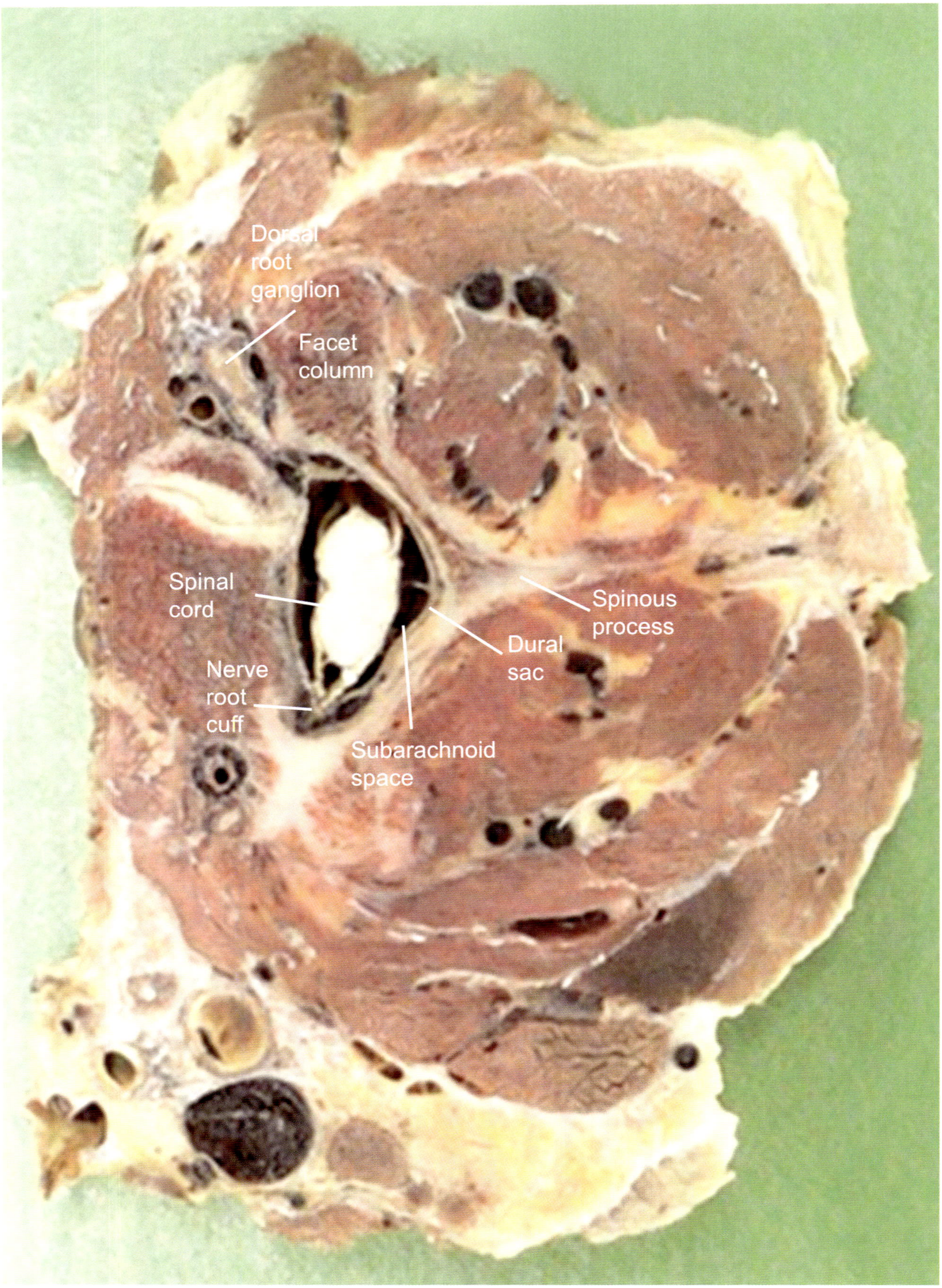

Fig. 36.9 Anatomic axial midcervical section of the spinal canal

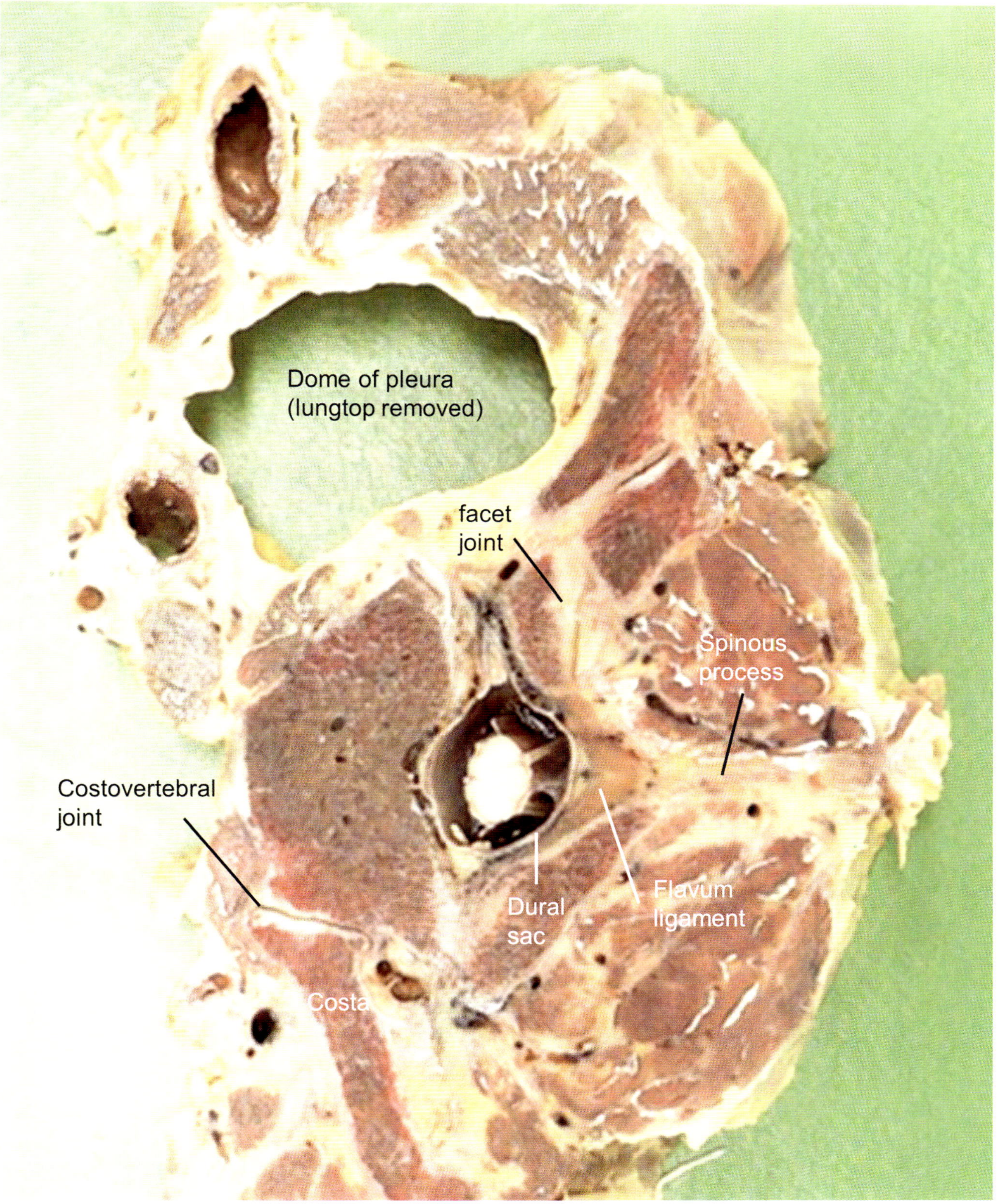

Fig. 36.10 Anatomic axial high thoracic (dome of pleura shown, top of lung removed) section of the spinal canal

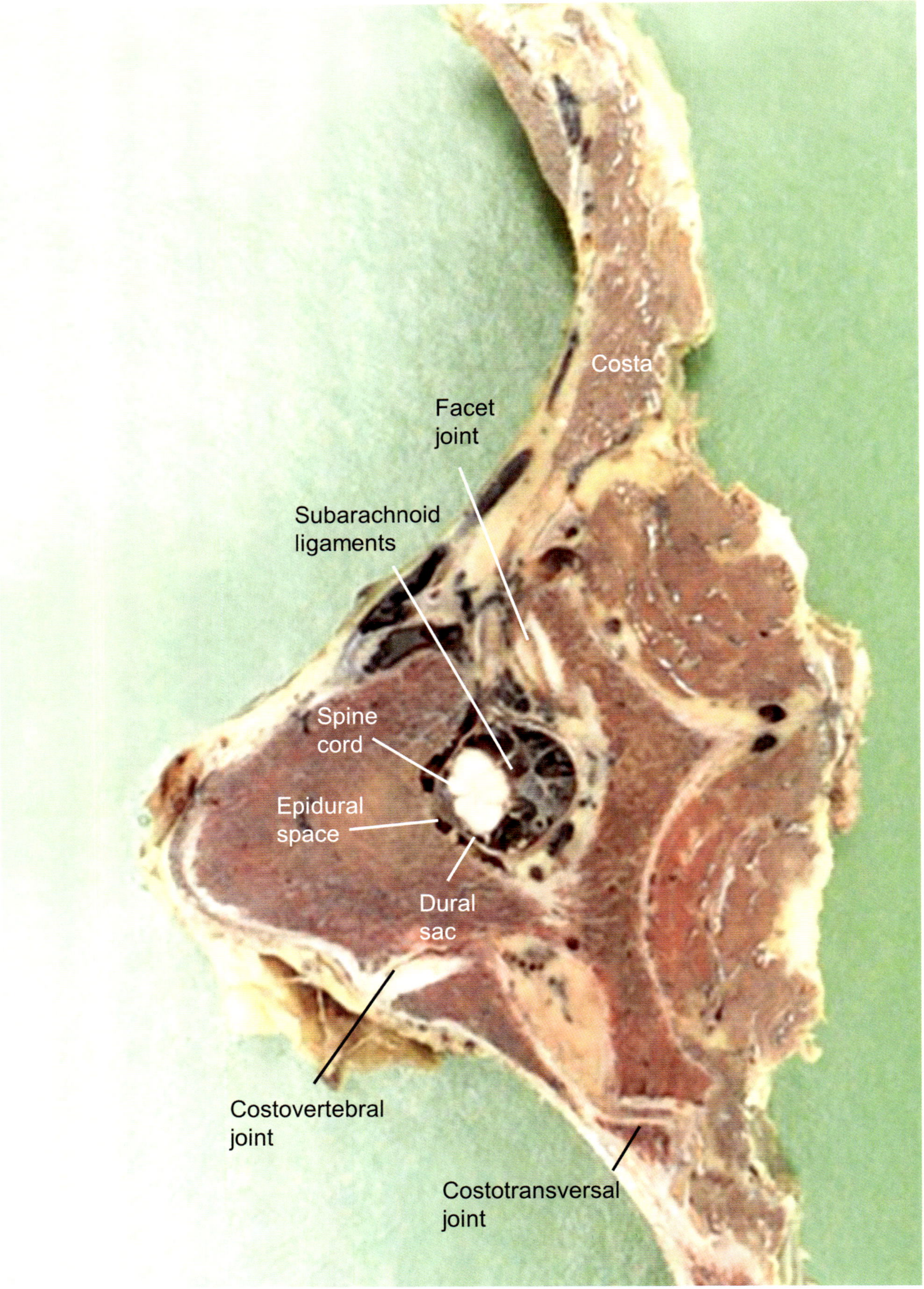

Fig. 36.11 Anatomic axial section at T3–T4 of the spinal canal. Note the anterior position of the spinal cord within its thecal boundaries

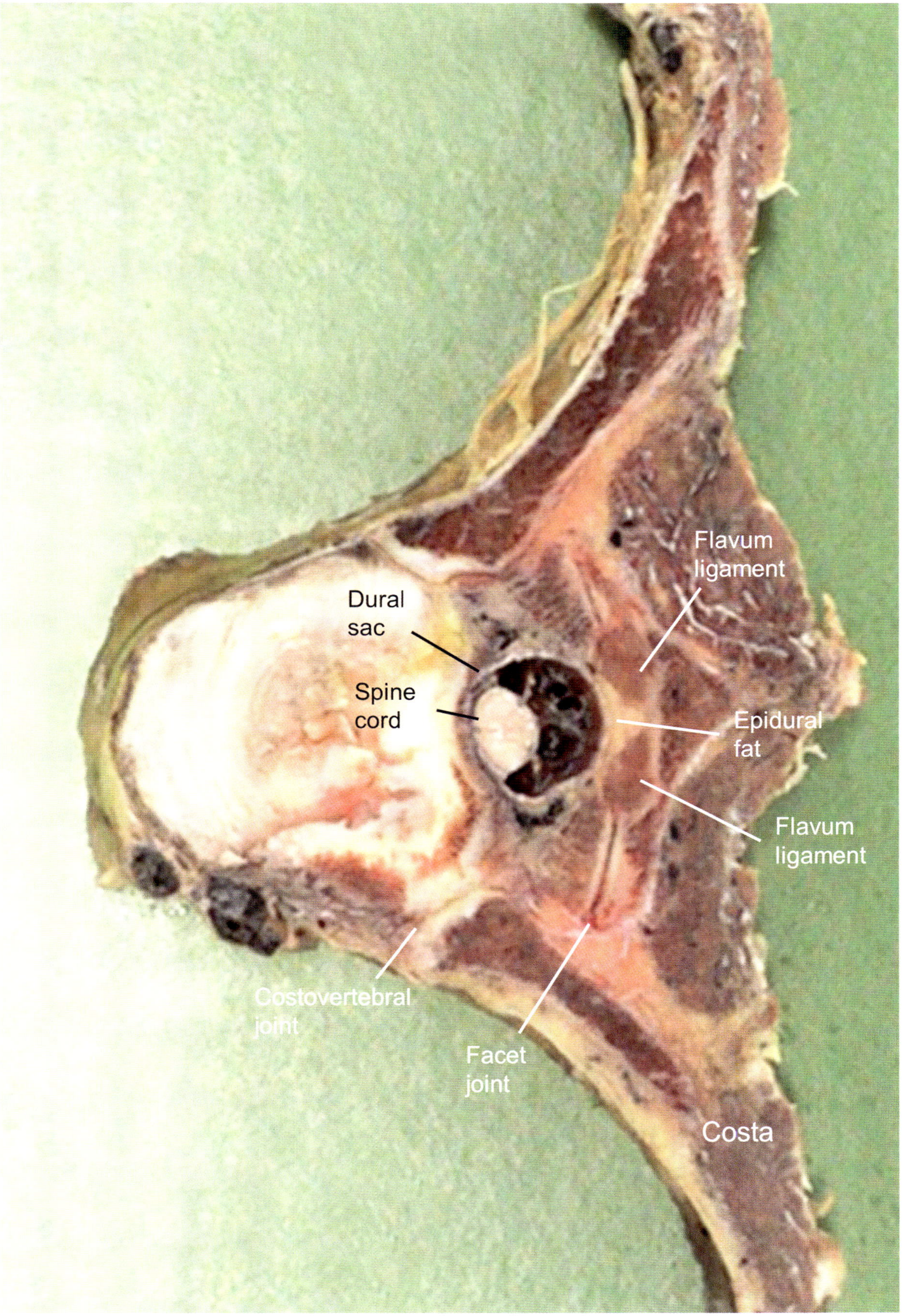

Fig. 36.12 Anatomic axial midthoracic section of the spinal canal. The spinal cord is positioned completely anterior within the imposed geometry of the spinal canal, effectively following the thoracic kyphosis

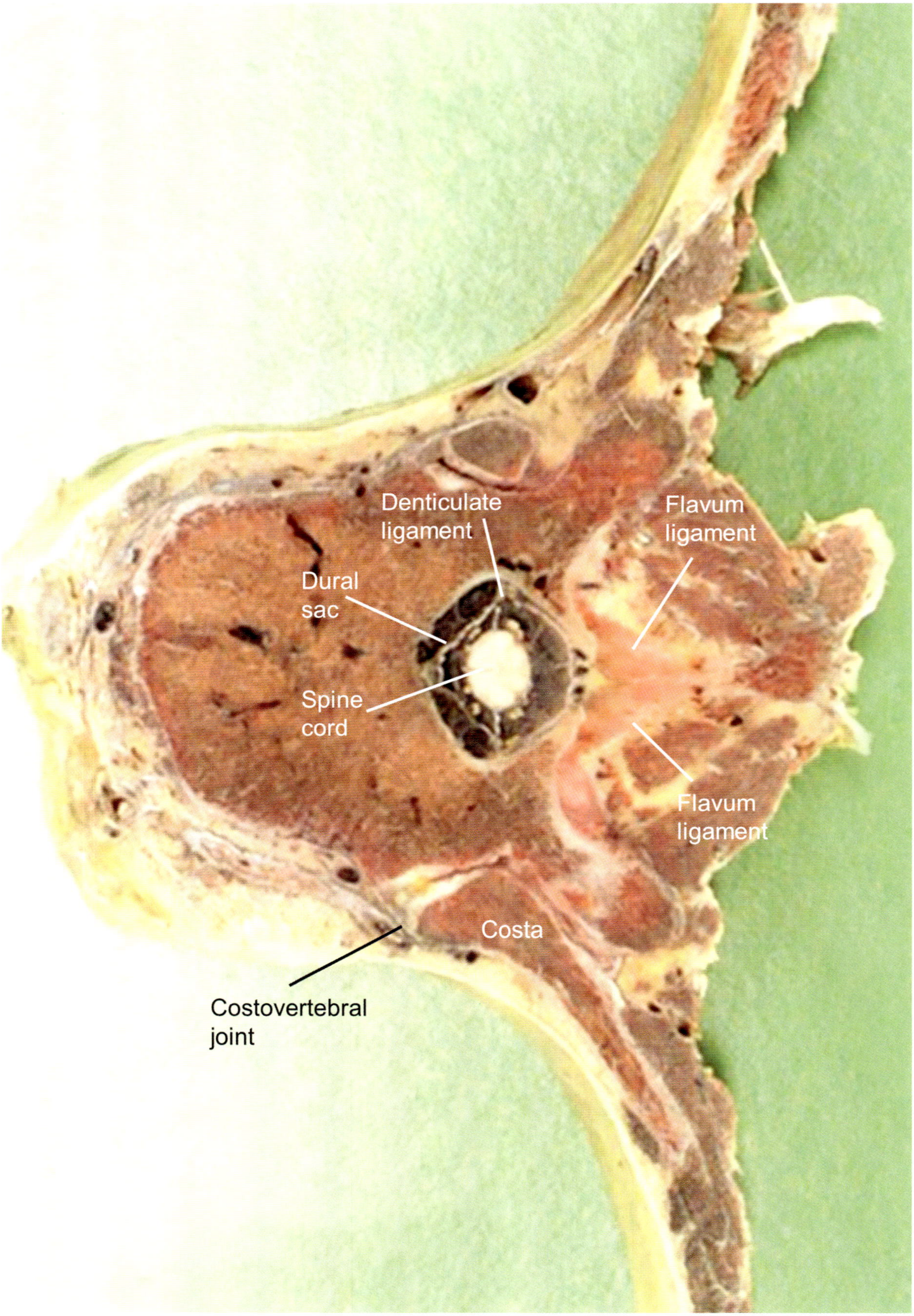

Fig. 36.13 Anatomic axial section of the low thoracic spinal canal

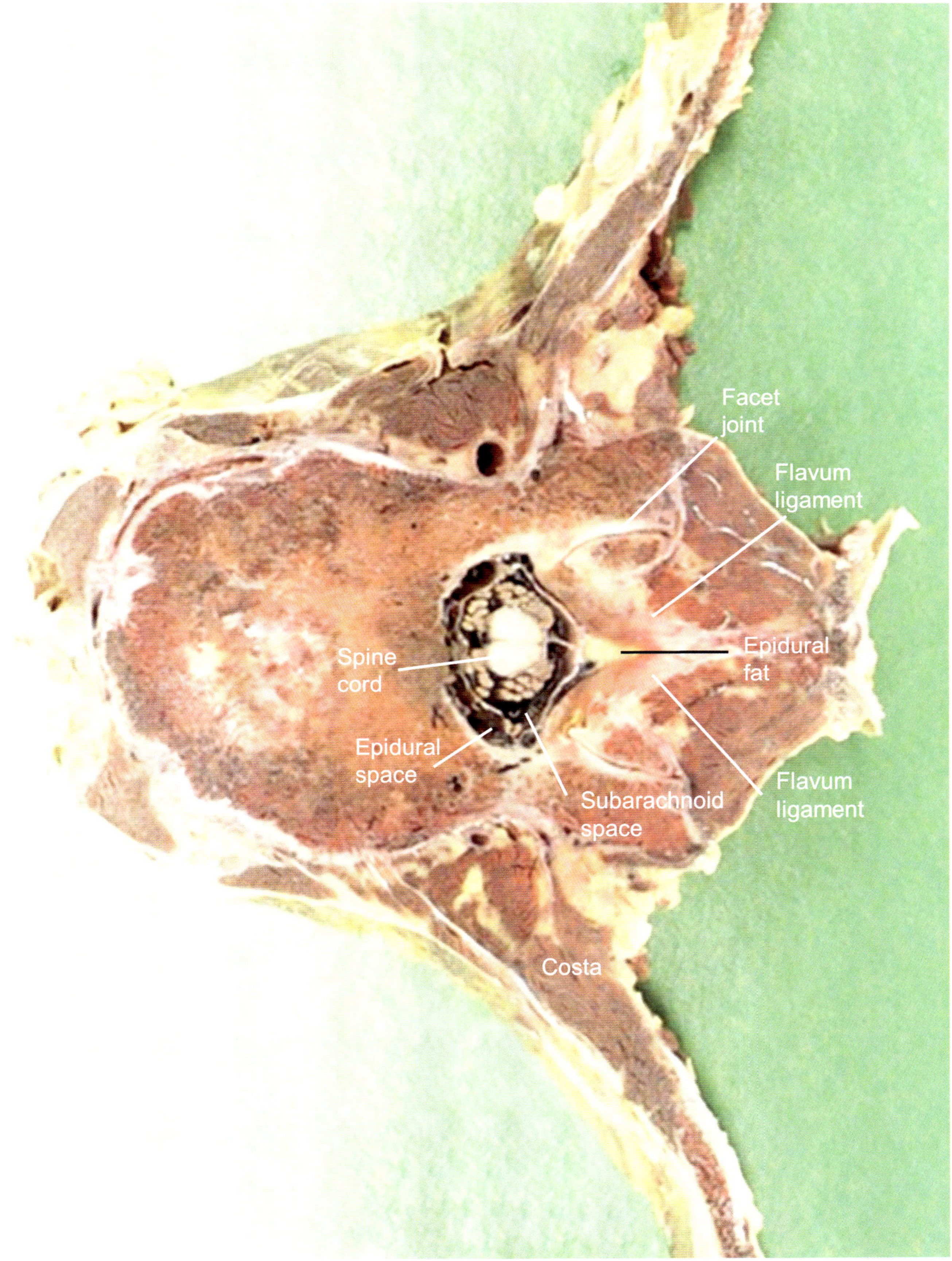

Fig. 36.14 Anatomic axial upper thoracolumbar section of the spinal canal. The spinal cord clearly has reached the beginning of the cauda equina, with the nerve fibers starting to separate (conus medullaris)

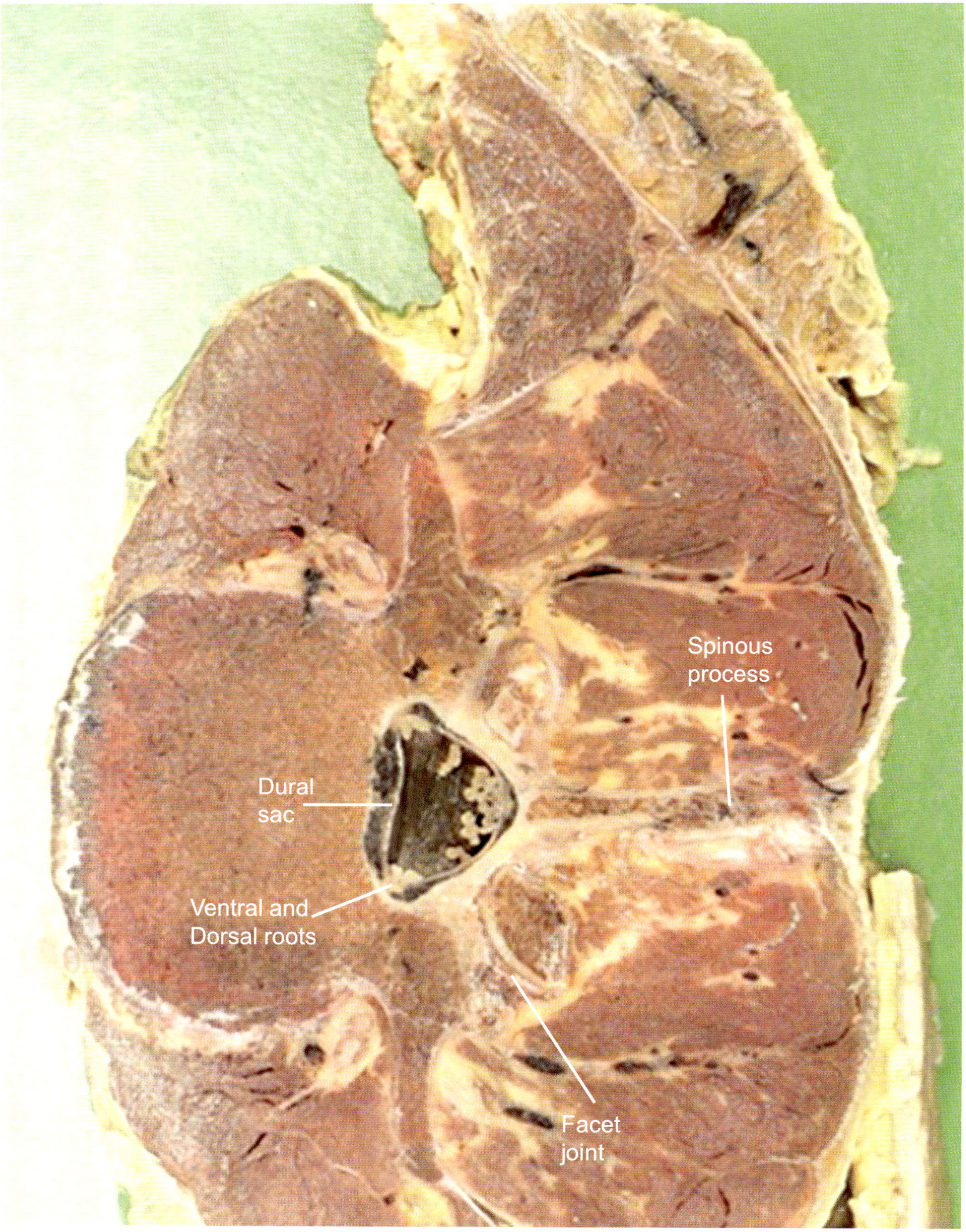

Fig. 36.15 Anatomic axial section at the L3–L4 level. Note the proximity of the cauda equina nerve fibers to the posterior dura mater. This may help explain the incidence of paresthesia during insertion of the spinal needle, with the spinal needle in such direct proximity to the neural tissue upon penetrating the thecal boundary

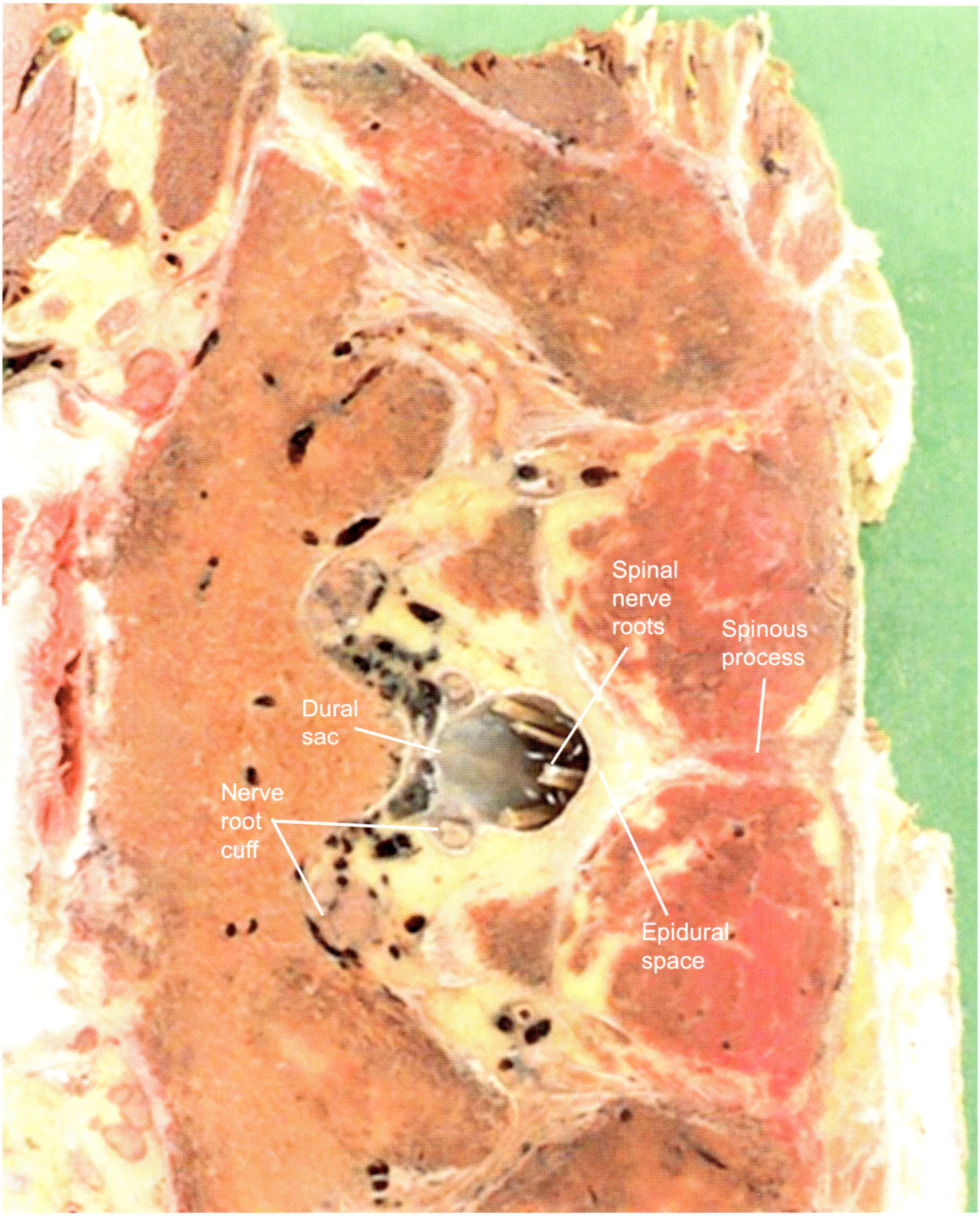

Fig. 36.16 Anatomic axial section at the high sacral level

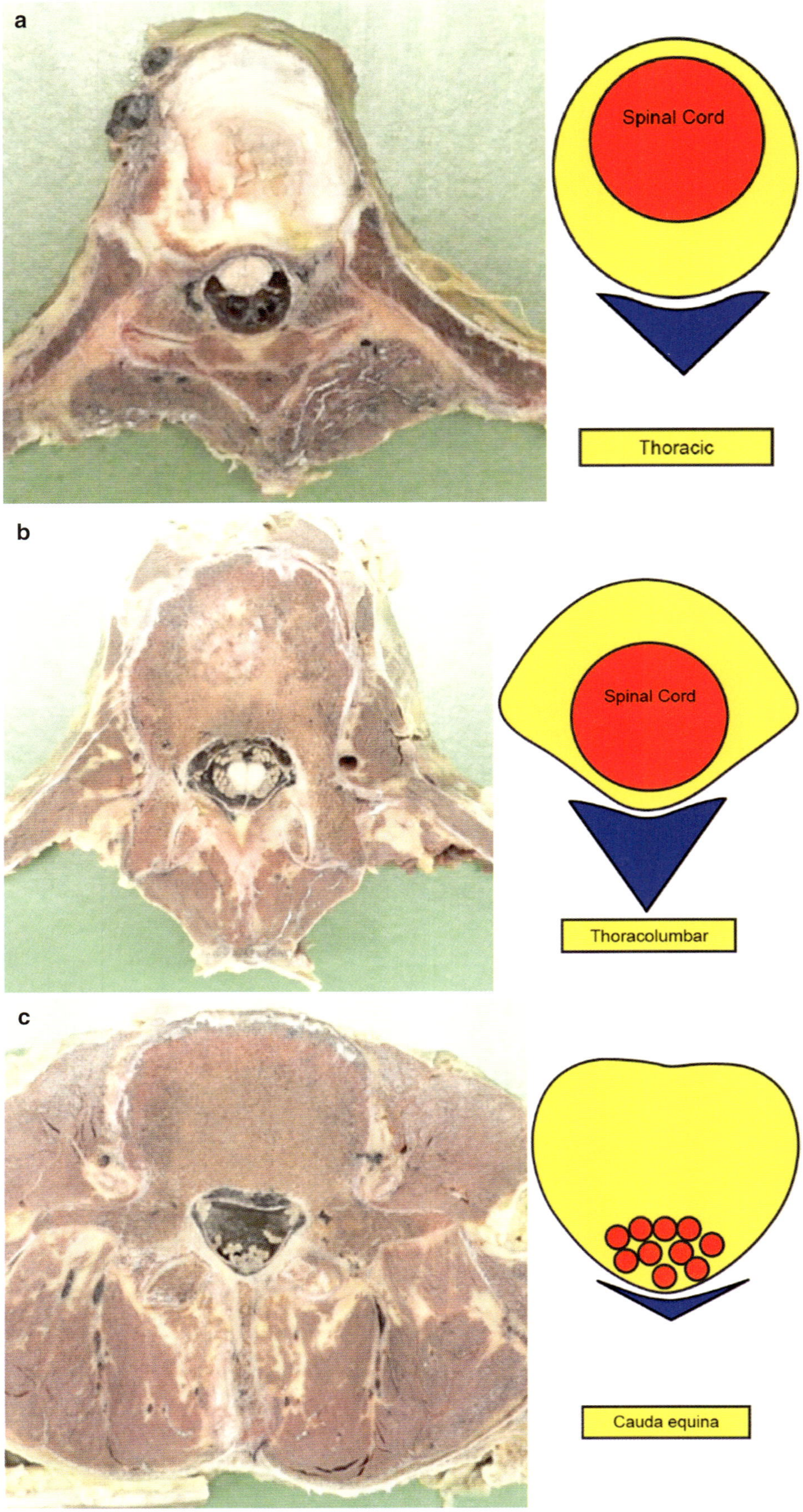

Fig. 36.17 Anatomic axial sections at the (**a**) human midthoracic, (**b**) thoracolumbar, and (**c**) low lumbar interspaces showing the respective relationships of the epidural space (*blue*), the subarachnoid space (*yellow*), and the spinal cord or cauda equina (*red*). Note the distance between the posterior aspect of the spinal cord and the posterior dura mater, which differs substantially according to level

Acknowledgments The authors thank Mrs. Greet Mommen for adjusting and editing the figures.

References

1. Lee RA, van Zundert AAJ, Botha CP, Lataster LMA, van Zundert TCRV, van der Ham WGJM, Wieringa PA. The anatomy of the thoracic spinal canal in different postures: a magnetic resonance imaging investigation. Reg Anesth Pain Med. 2010;35:364–9.
2. Lee RA, van Zundert AA, Breedveld P, Wondergem JH, Peek D, Wieringa PA. The anatomy of the thoracic spinal canal investigated with magnetic resonance imaging (MRI). Acta Anaesthesiol Belg. 2007;58:163–7.

Three-Dimensional Visualization of Spinal Cerebrospinal Fluid and Cauda Equina Nerve Roots, and Estimation of a Related Vulnerability Ratio

37

Alberto Prats-Galino, Anna Puigdellívol-Sánchez, and José Manuel Escobar

The procedure for obtaining a detailed three-dimensional (3D) reconstruction of the cerebrospinal fluid (CSF) and cauda equina roots inside the lumbosacral dural sac is described in Chap. 47 [1].

Briefly, processed T2-weighted magnetic resonance (MR) images show a gray scale range between 0 and 3,000. A segmentation threshold (around 700) is chosen to separate the spinal nerve roots of the cauda equina and caudal spinal cord segments from the CSF inside the dural sac along the lower levels of the vertebral canal. The resulting root segmentation then is corrected manually.

A vulnerability index [2] relating the cauda equina nerve root volume to the dural sac volume (root volume/[root volume + CSF volume]) may be calculated at the different vertebral levels and in the intervertebral spaces. Vulnerability may be represented on a color scale on which reddish segments correspond to levels with higher vulnerability and yellow-green segments correspond to levels that are safer for performing punctures. Vulnerability increases from caudal to rostral lumbar levels, with high variability among cases (Figs. 37.1, 37.2, 37.3, 37.4, and 37.5) (Table 37.1).

Mean and SD of the vulnerability index, calculated according to the formula roots volume/(CSF volume + roots volume) in blocks of 11 axial slices (7.15 mm high), in the intervertebral zones, in 7 cases. There is high variability among cases

A. Prats-Galino, MD, PhD (✉) • A. Puigdellívol-Sánchez, MD, PhD
Laboratory of Surgical NeuroAnatomy (LSNA),
Human Anatomy, and Embryology Unit,
School of Medicine, University of Barcelona,
Gran Via de les Corts Catalanes, 585, 08007 Barcelona, Spain
e-mail: aprats@ub.edu; apuigdellivol@ub.edu

J.M. Escobar, Tech
Department of Radiology, Madrid-Montepríncipe
University Hospital, School of Medicine,
CEU San Pablo University, Madrid, Spain
e-mail: josemanuelescobar@hospitaldemadrid.com

M.A. Reina et al. (eds.), *Atlas of Functional Anatomy for Regional Anesthesia and Pain Medicine: Human Structure, Ultrastructure and 3D Reconstruction Images*, DOI 10.1007/978-3-319-09522-6_37,

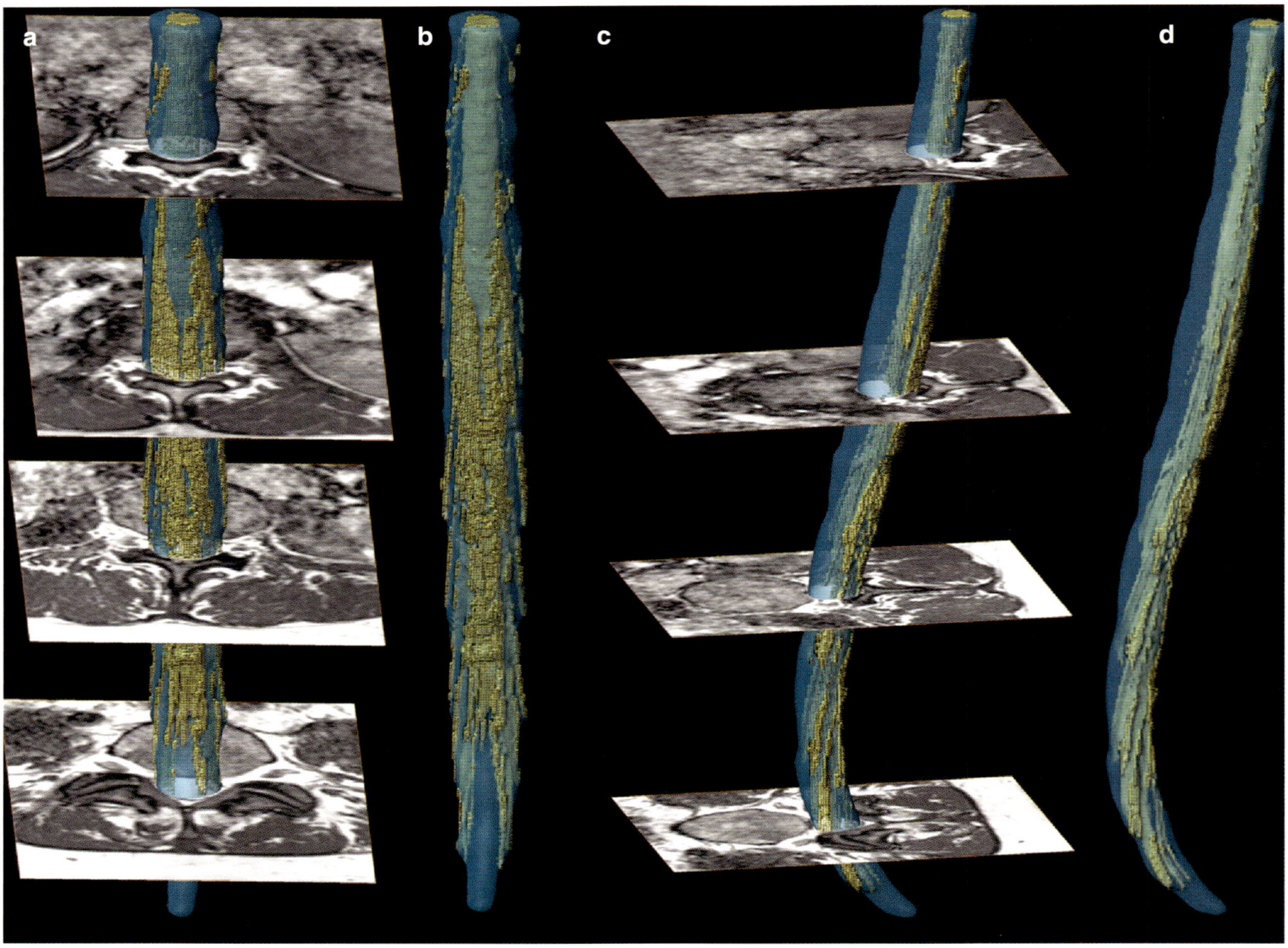

Fig. 37.1 Three-dimensional model of CSF and cauda equina nerve roots from the lowest thoracic levels to the caudal end of the dural sac. Dorsal (**a**, **b**) and lateral (**c**, **d**) views with (**a**, **c**) and without (**b**, **d**) some illustrative axial MR planes. The segmentation applied to the dural sac volume of interest (VOI) allows the separation of the spinal nerve roots of the cauda equina from the CSF inside the dural sac

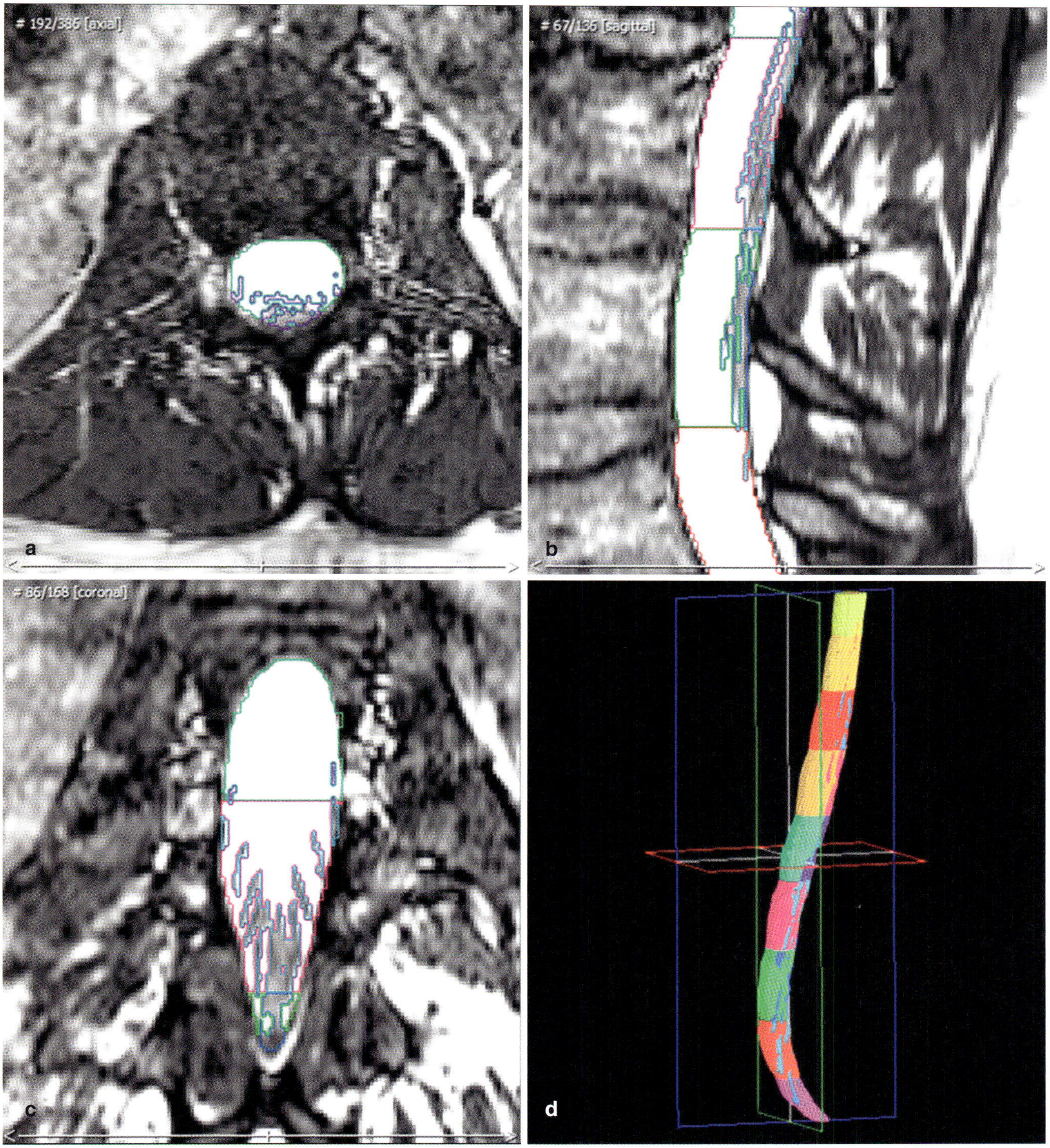

Fig. 37.2 Segmentation of CSF and nerve roots in separate VOIs at each vertebral level. CSF and nerve root VOIs corresponding to the different vertebral levels may be delimited as shown in axial (**a**), sagittal (**b**), coronal (**c**), and 3D model views (**d**), from 10th thoracic vertebra (*upper*) to first sacral vertebra (*lower*). The volume of these segmented VOIs then is calculated by multiplying the number of voxels of the analyzed tissue by the voxel size of the images

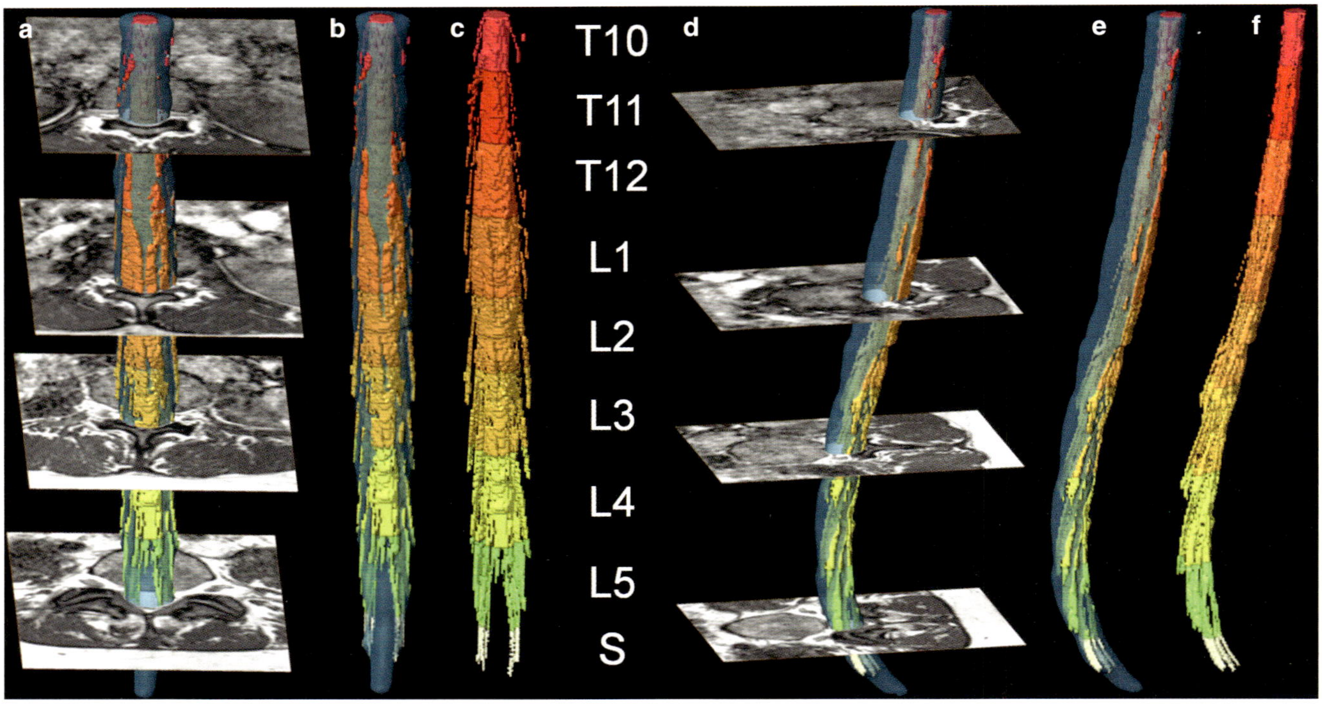

Fig. 37.3 Three-dimensional CSF and cauda equina roots per vertebral level and color-coded vulnerability. Dorsal (**a–c**) and lateral (**d–f**) views, including (**a**, **d**) or not including illustrative MR axial planes with or without (**c**, **f**) CSF. Rostral thoracolumbar vertebral levels (T10–L1) may include the conus medullaris (*red–dark orange*), and a higher proportion of the spinal canal is occupied by cauda equina nerve roots in L2–L4 (*orange–yellow*) than in L5 sacral levels (*green–white*)

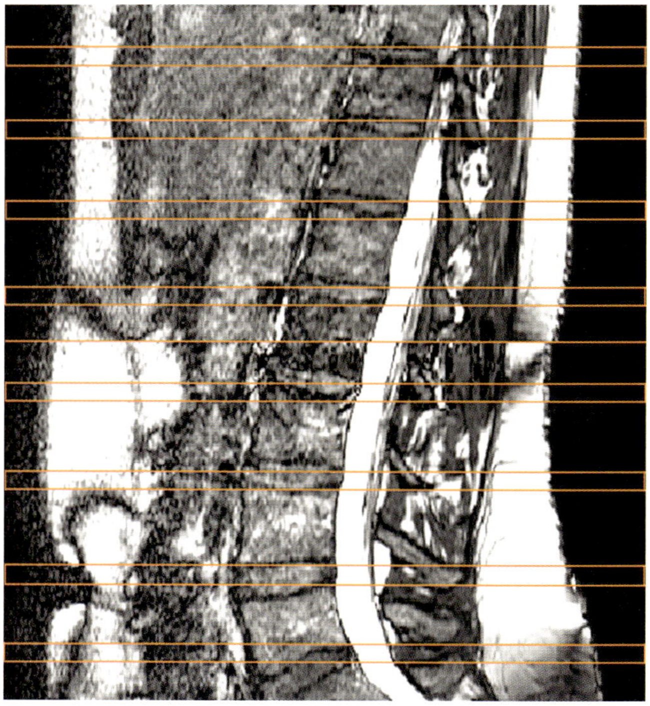

Fig. 37.4 Intervertebral zones. Once the CSF and nerve root VOIs within the dural sac are obtained, volumes and ratios between these two tissues may be calculated exclusively in the intervertebral zones, where punctures may be performed in patients. Here, small blocks of 11 axial slices (7.15 mm high) were analyzed

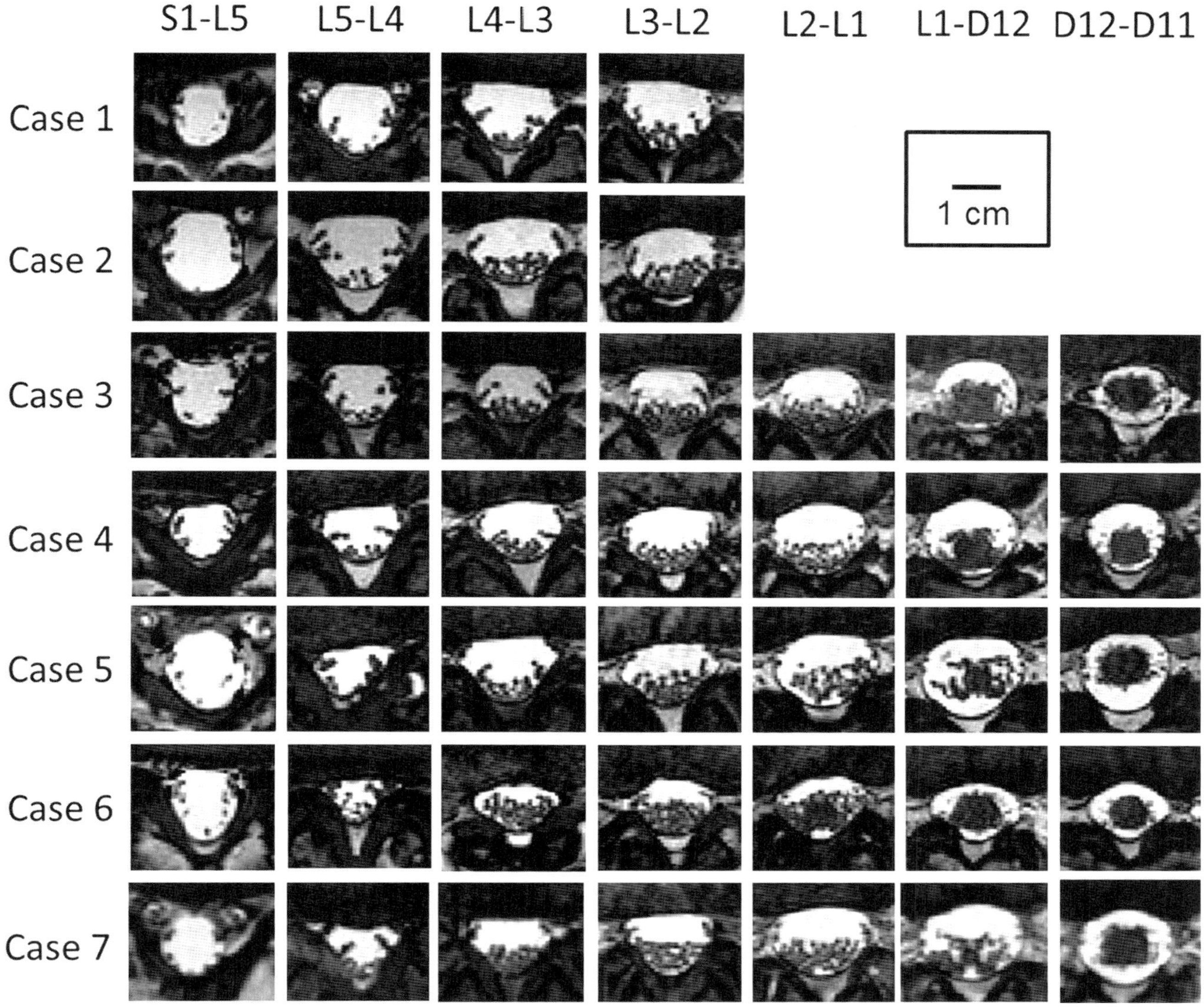

Fig. 37.5 Axial MR planes at the middle zone of the different intervertebral levels in different illustrative cases. There is high variability among cases (Table 37.1)

Table 37.1 Vulnerability ratio in intervertebral levels

	Intervertebral level, %						
	S1–L5	L5–L4	L4–L3	L3–L2	L2–L1	L1–T12	T12–T11
Case							
1	1.4	9.0	21.2	24.9			
2	7.3	20.7	34.9	35.4			
3	8.6	18.2	26.8	39.8	38.3	43.3	48.6
4	10.1	16.2	19.3	21.6	20.8	34.5	32.8
5	2.6	17.7	16.8	19.1	16.1	24.4	31.1
6	4.5	30.7	54.6	22.9	46.2	40.5	31.4
7	6.6	23.4	32.5	33.4	33.0	32.4	35.4
Mean	5.9	19.4	29.4	28.2	30.9	35.0	35.9
SD	3.2	6.7	13.0	8.0	12.4	7.4	7.3
Max.	10.1	30.7	54.6	39.8	46.2	43.3	48.6
Min.	1.4	9.0	16.8	19.1	16.1	24.4	31.1

References

1. Puigdellivol-Sánchez A, Prats-Galino A, Reina MA, Machés F, Hernández JM, De Andrés J, Van Zundert A. Three dimensional magnetic resonance image of structures enclosed in the spinal canal relevant to anesthetists and estimation of the lumbosacral CSF volume. Acta Anaesthesiol Belg. 2011;62:37–45.
2. Prats-Galino A, Reina MA, Puigdellívol-Sánchez A, Juanes Méndez JA, De Andrés JA, Collier CB. Cerebrospinal fluid volume and nerve root vulnerability during lumbar puncture or spinal anaesthesia at different vertebral levels. Anaesth Intensive Care. 2012;40: 643–7.

Ultrastructure of Nerve Root Cuffs: Dura-Epineurium Transition Tissue

38

Miguel Angel Reina, Lucila Reina Colman, and Emilse Colman Peyrano

Nerve root cuffs extend from the dural sac to the dorsal root ganglia and are formed by a cellular component and an intermingling fatty and fibrillar layer [1], similar to interfascicular tissue present in peripheral nerves. This fibrillar layer is mainly composed of collagen fibers, with similar structure to dura mater. As well as the dural laminas present in the dural sac, the dural layer present in nerve root cuffs is permeable and shows features characteristic of transitional tissue, which is found between collagen fibers of spinal dura mater [2] and epineurium of peripheral nerves [3]. Another feature that distinguishes the dural layer of nerve root cuffs from dura mater in the dural sac is the presence of adipose tissue among the dural laminas of the nerve root cuffs.

At the level of the dorsal root ganglia [4–6], the outermost cover resembles epineurium, although the amount of adipose tissue appears to be proportionately reduced (Figs. 38.1, 38.2, 38.3, 38.4, 38.5, 38.6, 38.7, 38.8, 38.9, 38.10, 38.11, 38.12, 38.13, 38.14, and 38.15).

M.A. Reina, MD, PhD (✉)
Department of Clinical Medical Sciences and Institute of Applied Molecular Medicine, School of Medicine, University of CEU San Pablo, Madrid, Spain

Department of Anesthesiology, Madrid-Montepríncipe University Hospital, Madrid, Spain
e-mail: miguelangel@perticone.e.telefonica.net

L. Reina Colman, BS Master Degree
Dentistry Department, Rey Juan Carlos University, Madrid, Spain
e-mail: lucilasstu@hotmail.com

E. Colman Peyrano, BS
Research Unit, Anesthesia Reina Colman SLP, Madrid, Spain
e-mail: emilsecolpey@yahoo.es

M.A. Reina et al. (eds.), *Atlas of Functional Anatomy for Regional Anesthesia and Pain Medicine: Human Structure, Ultrastructure and 3D Reconstruction Images*, DOI 10.1007/978-3-319-09522-6_38,

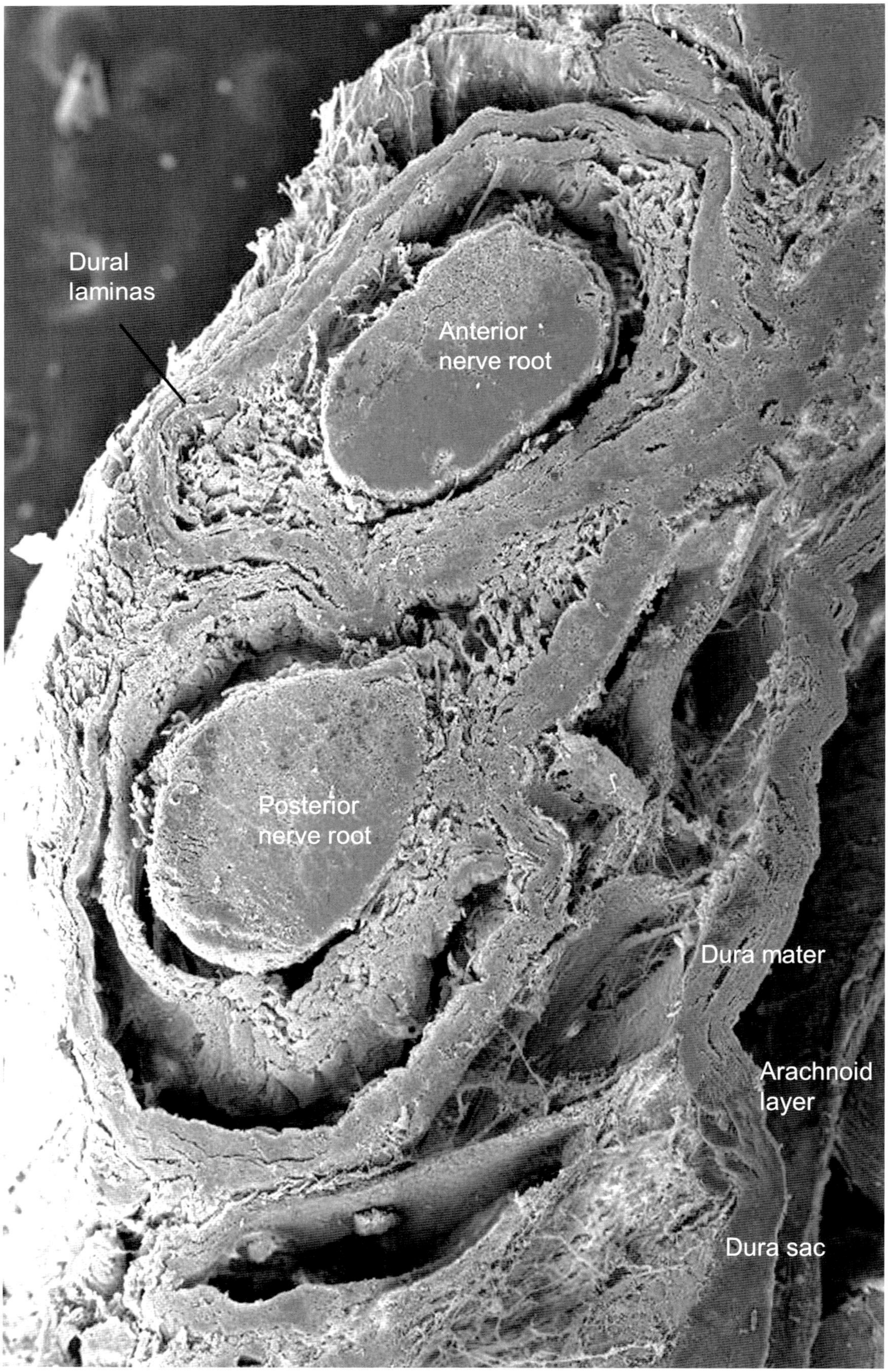

Fig. 38.1 Ultrastructure of human nerve root cuffs. Dura-epineurium transition tissue. Transverse section. Scanning electron microscopy. Magnification ×45 (From Reina et al. [5]; with permission)

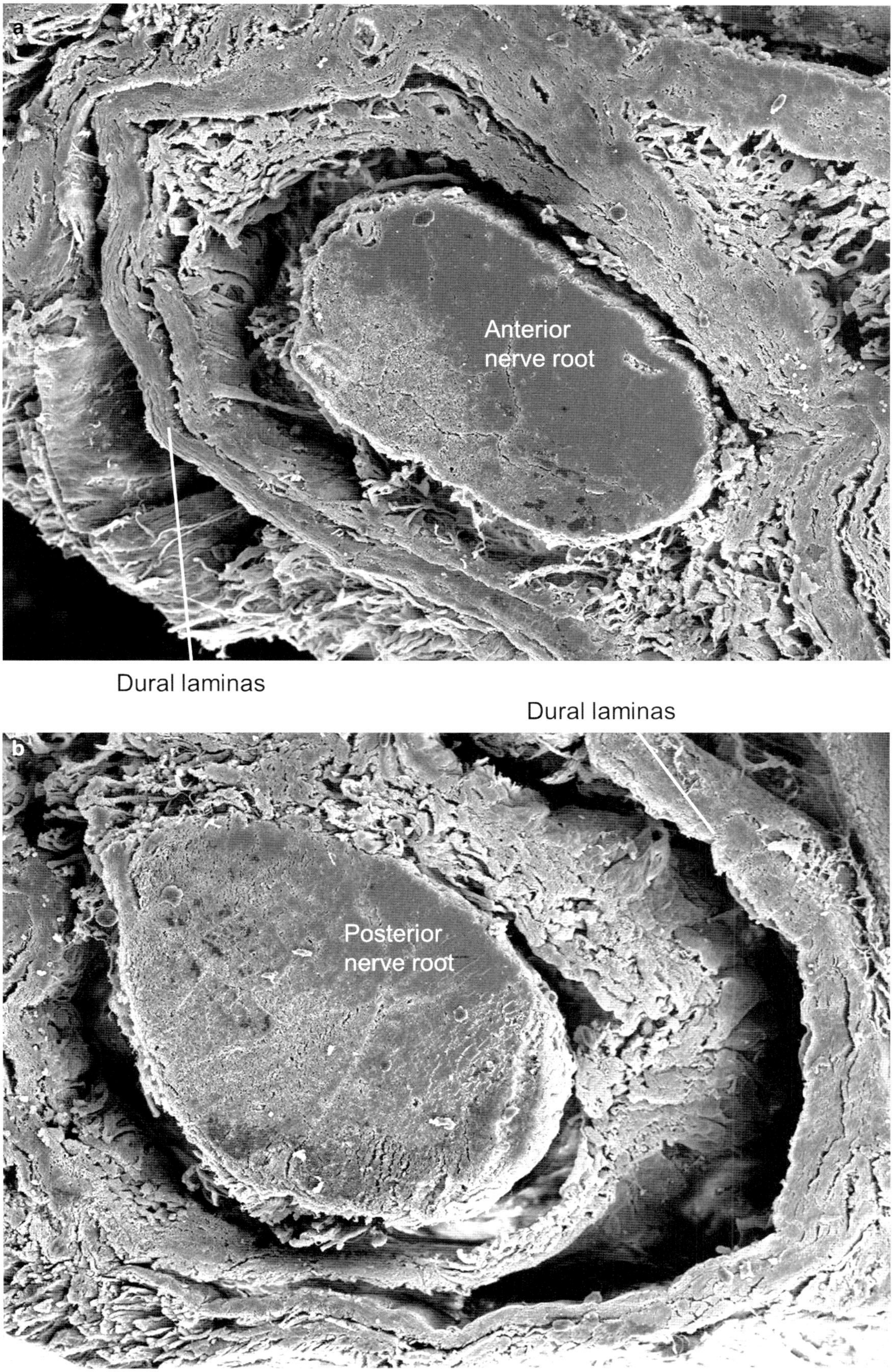

Fig. 38.2 Ultrastructure of human nerve root cuffs. Dura-epineurium transition tissue (details of Fig. 38.1). Transverse section. Scanning electron microscopy. Magnification: **a** ×100; **b** ×100

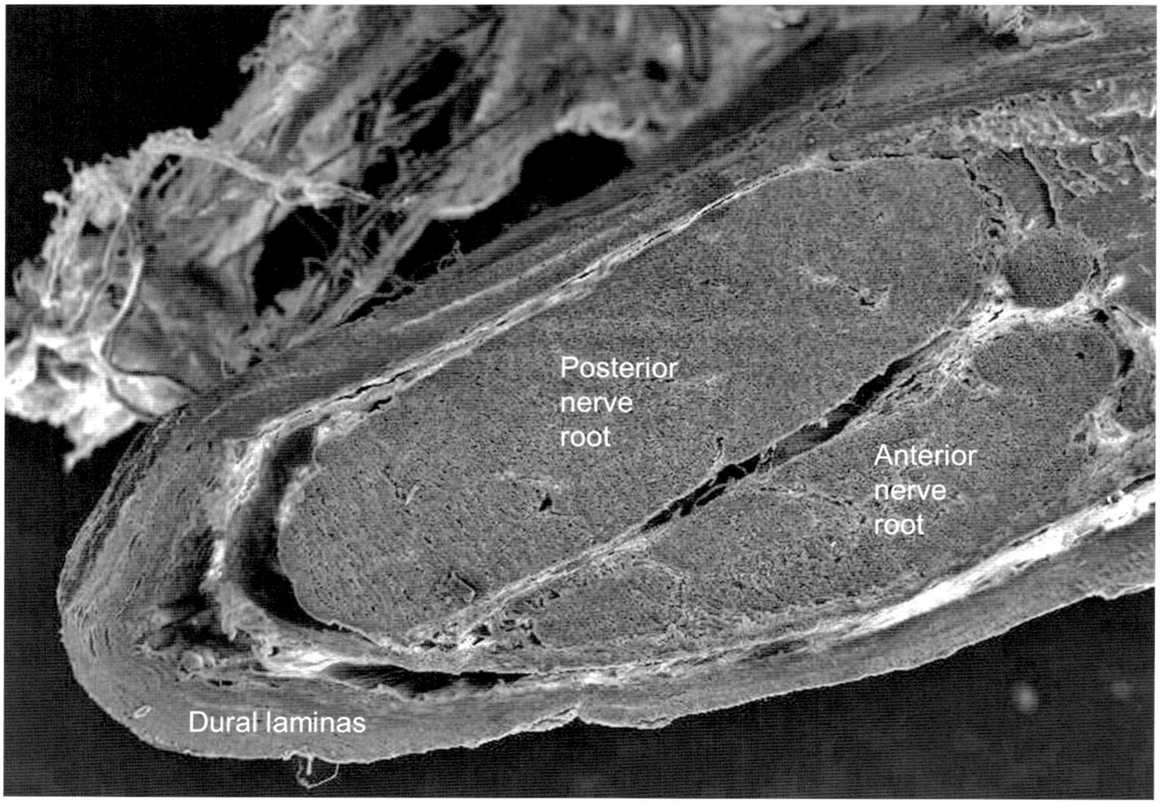

Fig. 38.3 Ultrastructure of human nerve root cuffs. Dura-epineurium transition tissue. Transverse section. Scanning electron microscopy. Magnification ×20 (From Reina et al [6]; with permission)

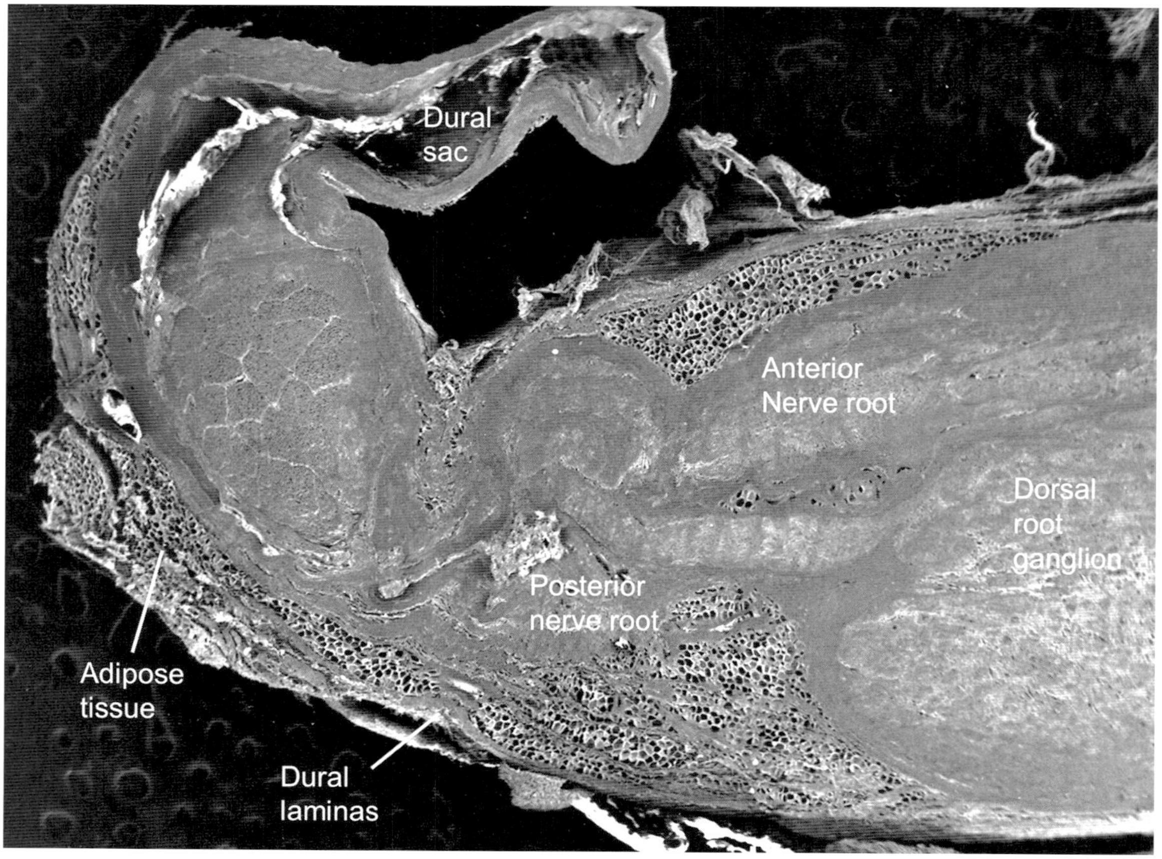

Fig. 38.4 Ultrastructure of human nerve root cuffs. Anterior and posterior nerve root, dorsal nerve ganglion. Longitudinal section. Scanning electron microscopy. Magnification ×12

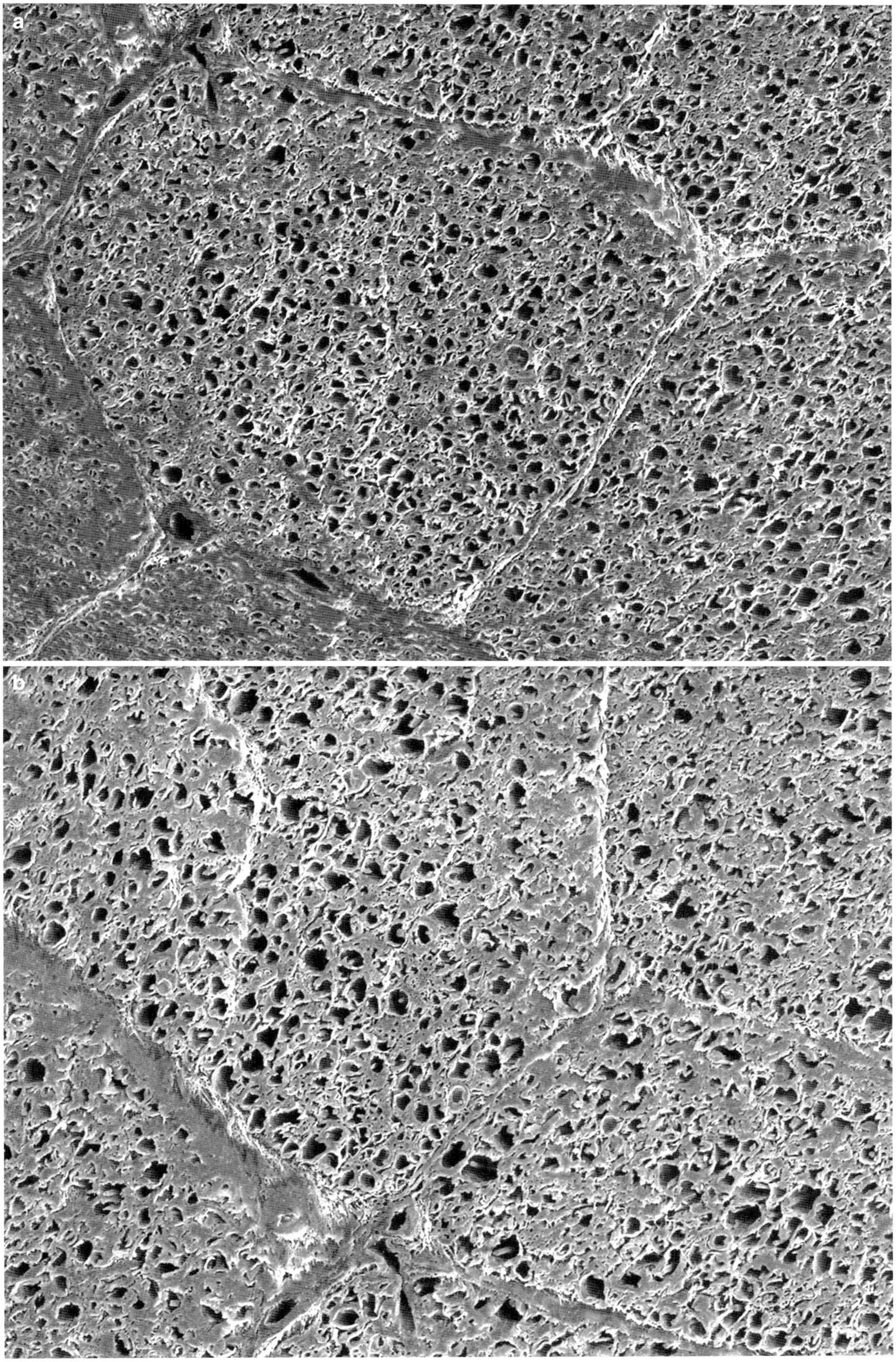

Fig. 38.5 Ultrastructure of human nerve root cuffs. Details of nerve roots. Transverse section. Scanning electron microscopy. Magnification: **a** ×150; **b** ×200

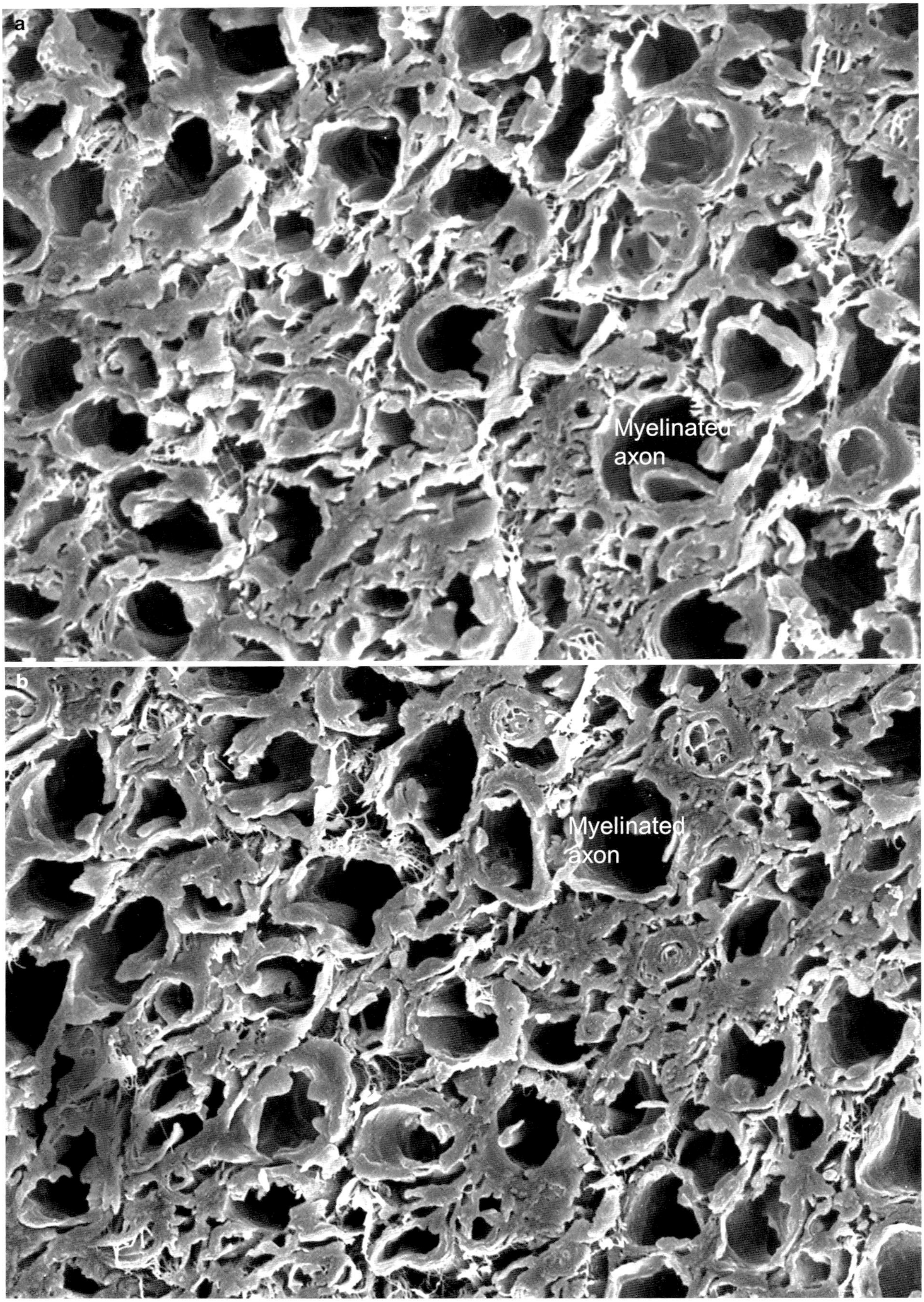

Fig. 38.6 Ultrastructure of human nerve root cuffs. Details of Fig. 38.5 at higher magnification. Scanning electron microscopy. Magnification: **a** ×1,000; **b** ×1,000

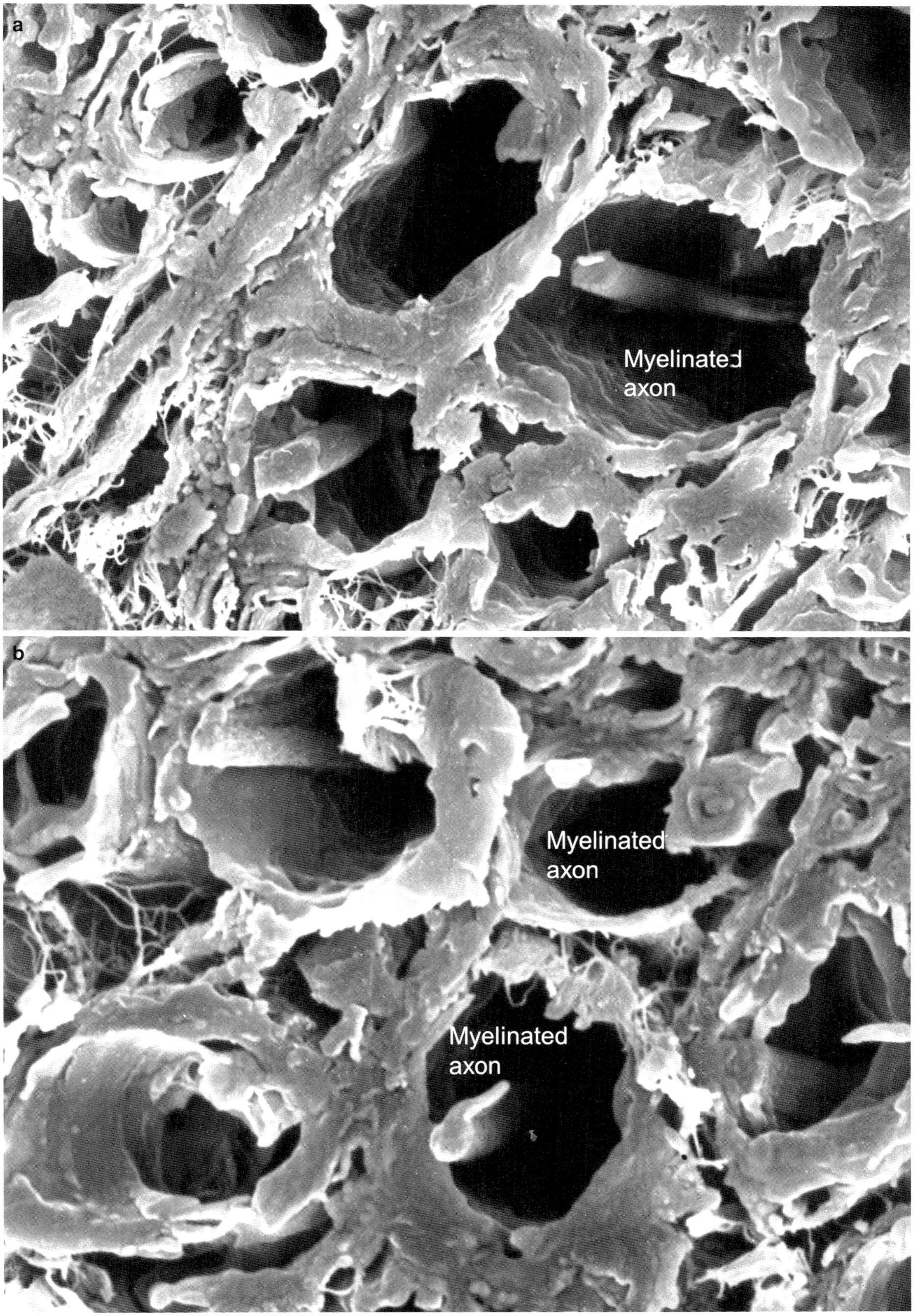

Fig. 38.7 Ultrastructure of human nerve root cuffs. Details of Fig. 38.5 at higher magnification. Scanning electron microscopy. Magnification: **a** ×2,000; **b** ×3,000

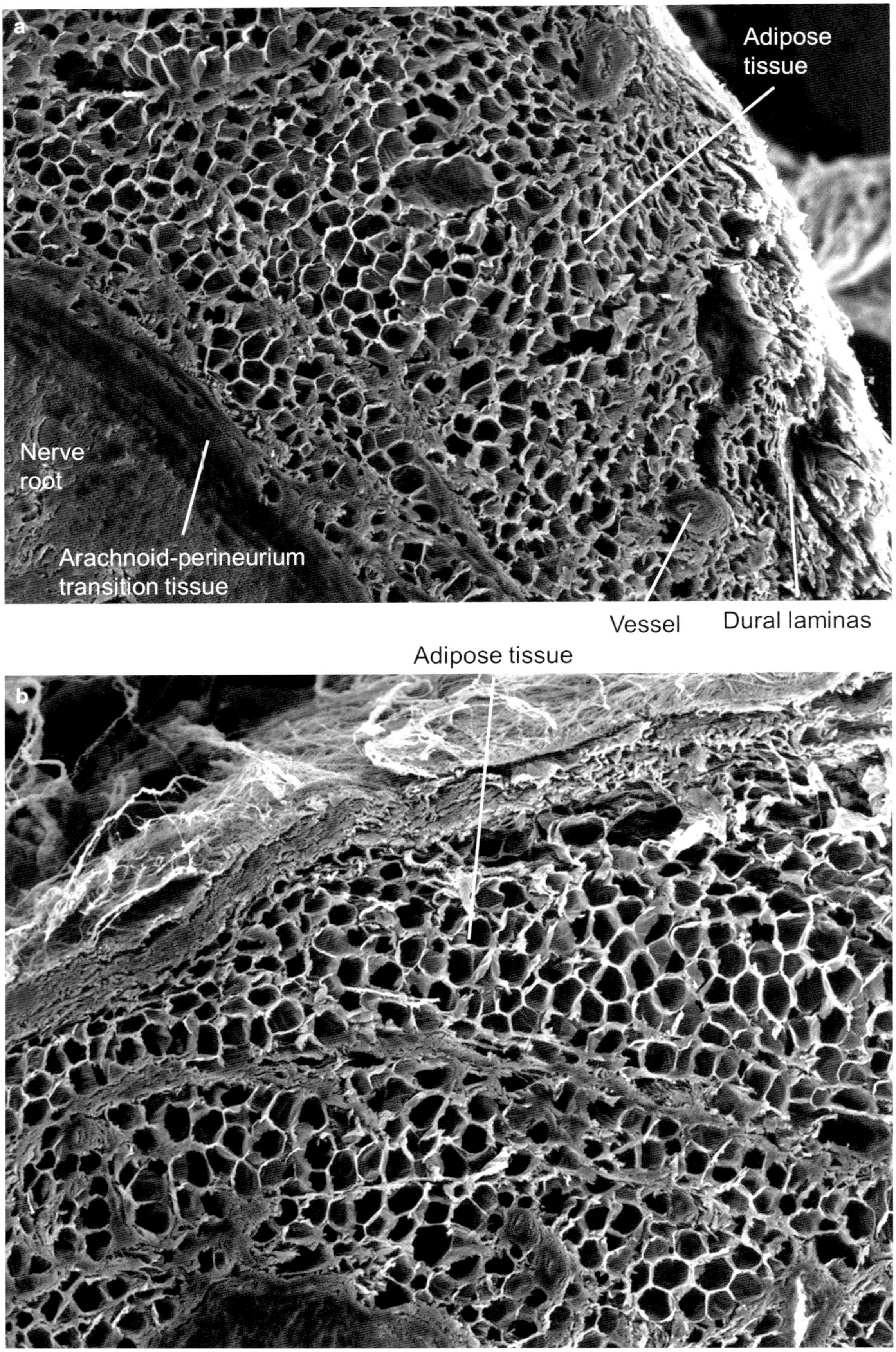

Fig. 38.8 Ultrastructure of human nerve root cuffs. Transverse section. Details of dural laminas and adipose tissue. Scanning electron microscopy. Magnification: **a** ×50; **b** ×50

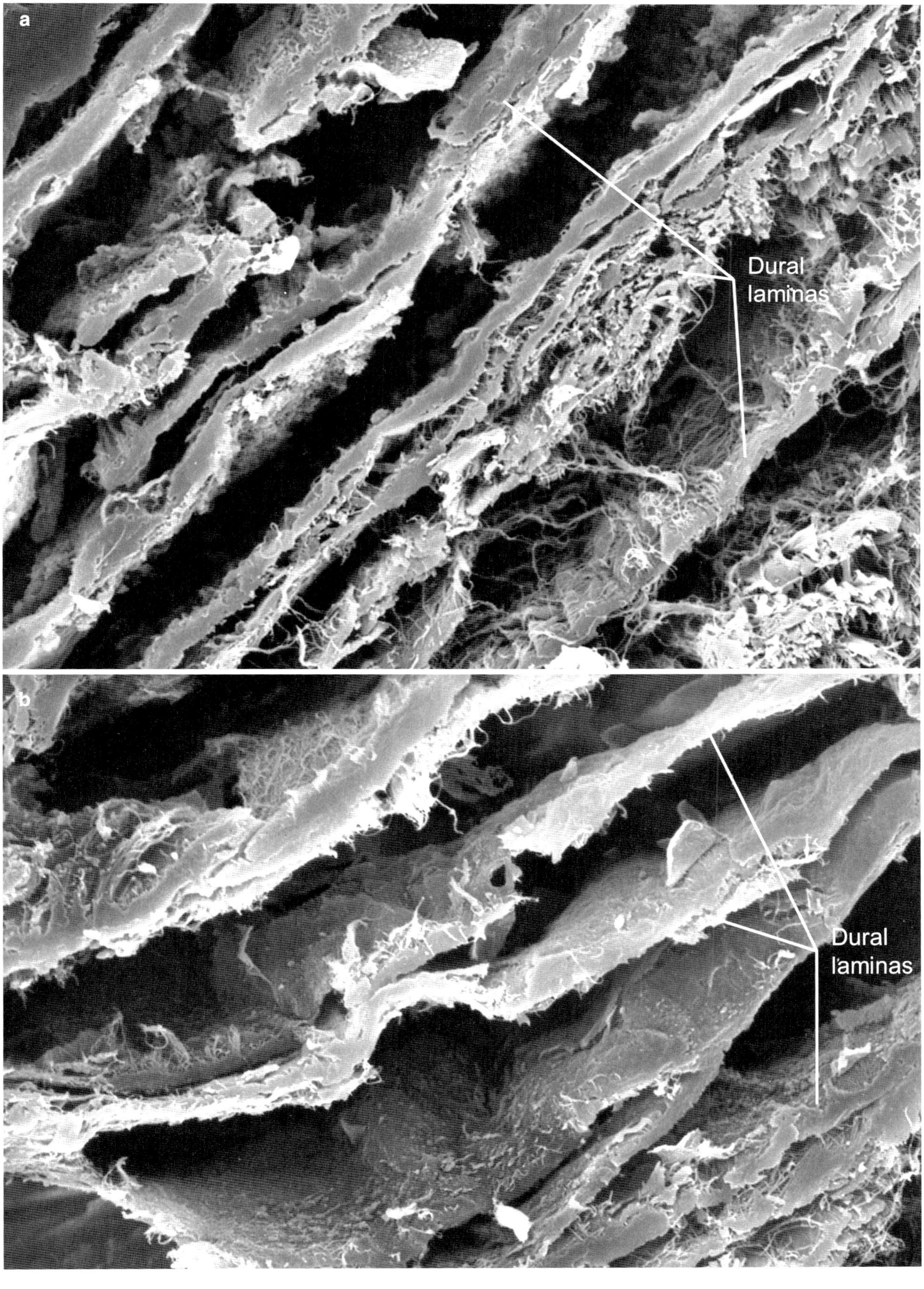

Fig. 38.9 Ultrastructure of human nerve root cuffs. Transverse section. Details of dural laminas. Scanning electron microscopy. Magnification: **a** ×1,000; **b** ×1,000

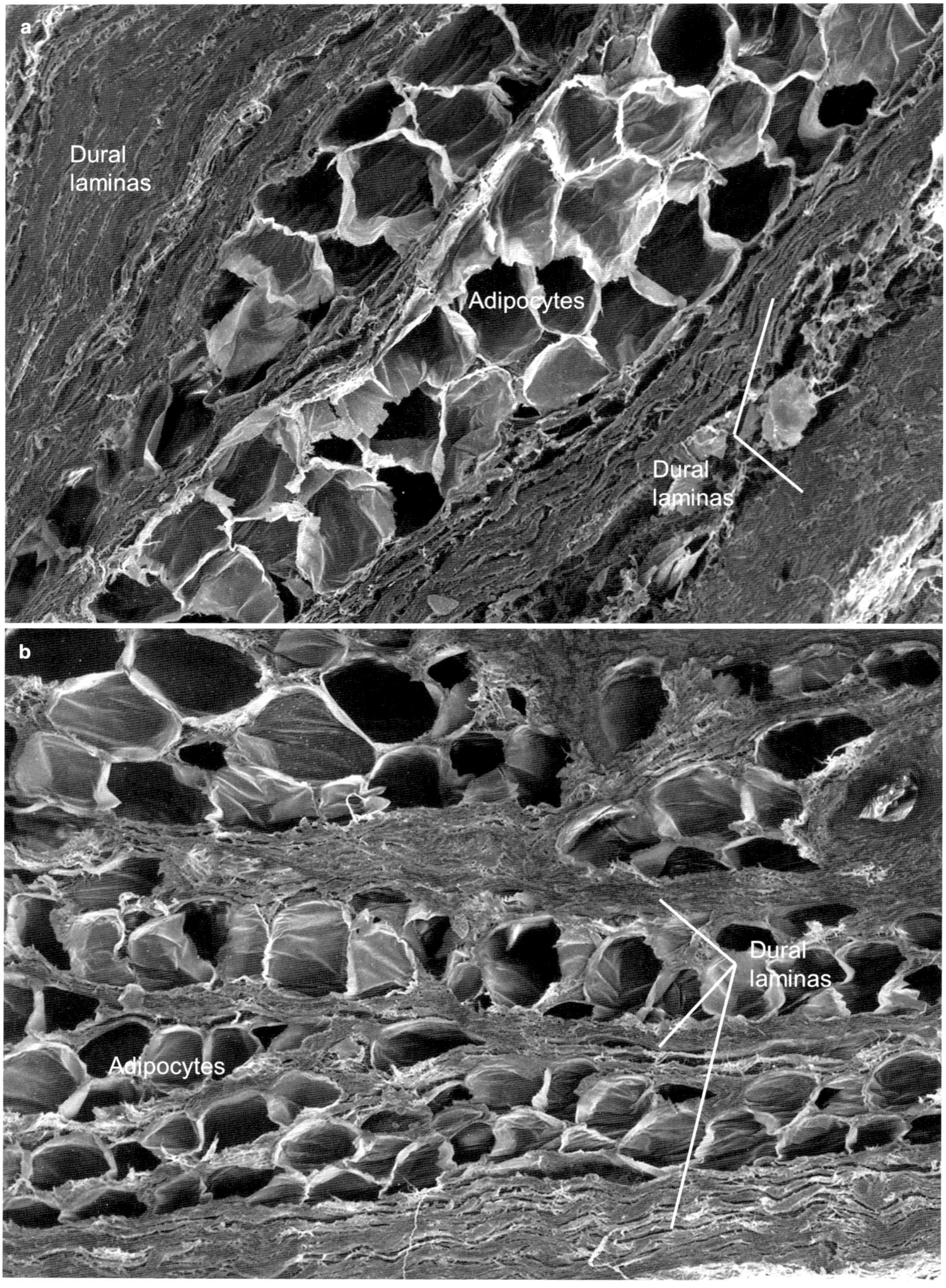

Fig. 38.10 Ultrastructure of human nerve root cuffs. Transverse section. Details of dural laminas and adipose tissue. Scanning electron microscopy. Magnification: **a** ×150; **b** ×140 (From Reina et al. [1]; with permission)

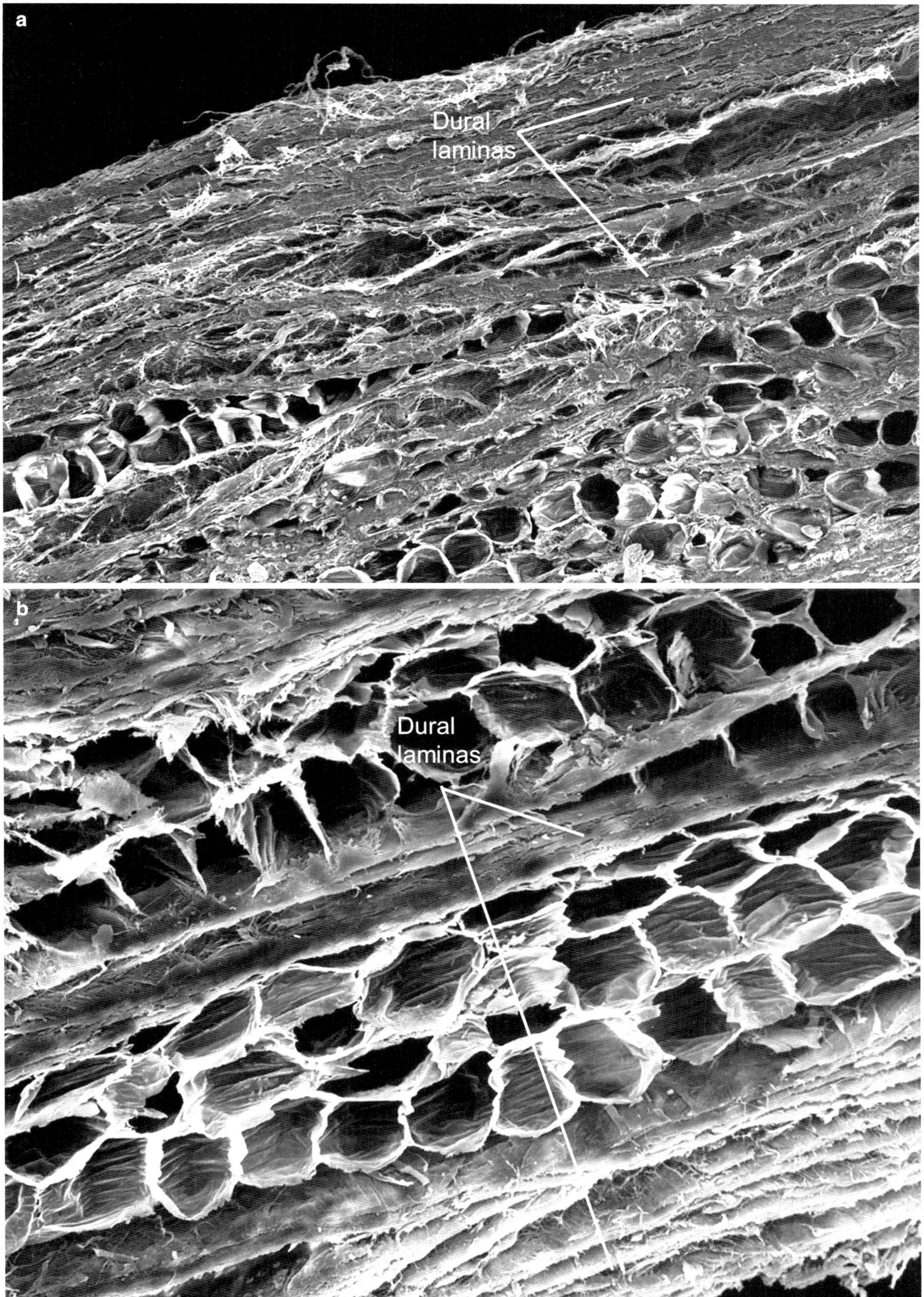

Fig. 38.11 Ultrastructure of human nerve root cuffs. Transverse section. Details of dural laminas and adipose tissue. Scanning electron microscopy. Magnification: **a** ×100; **b** ×150

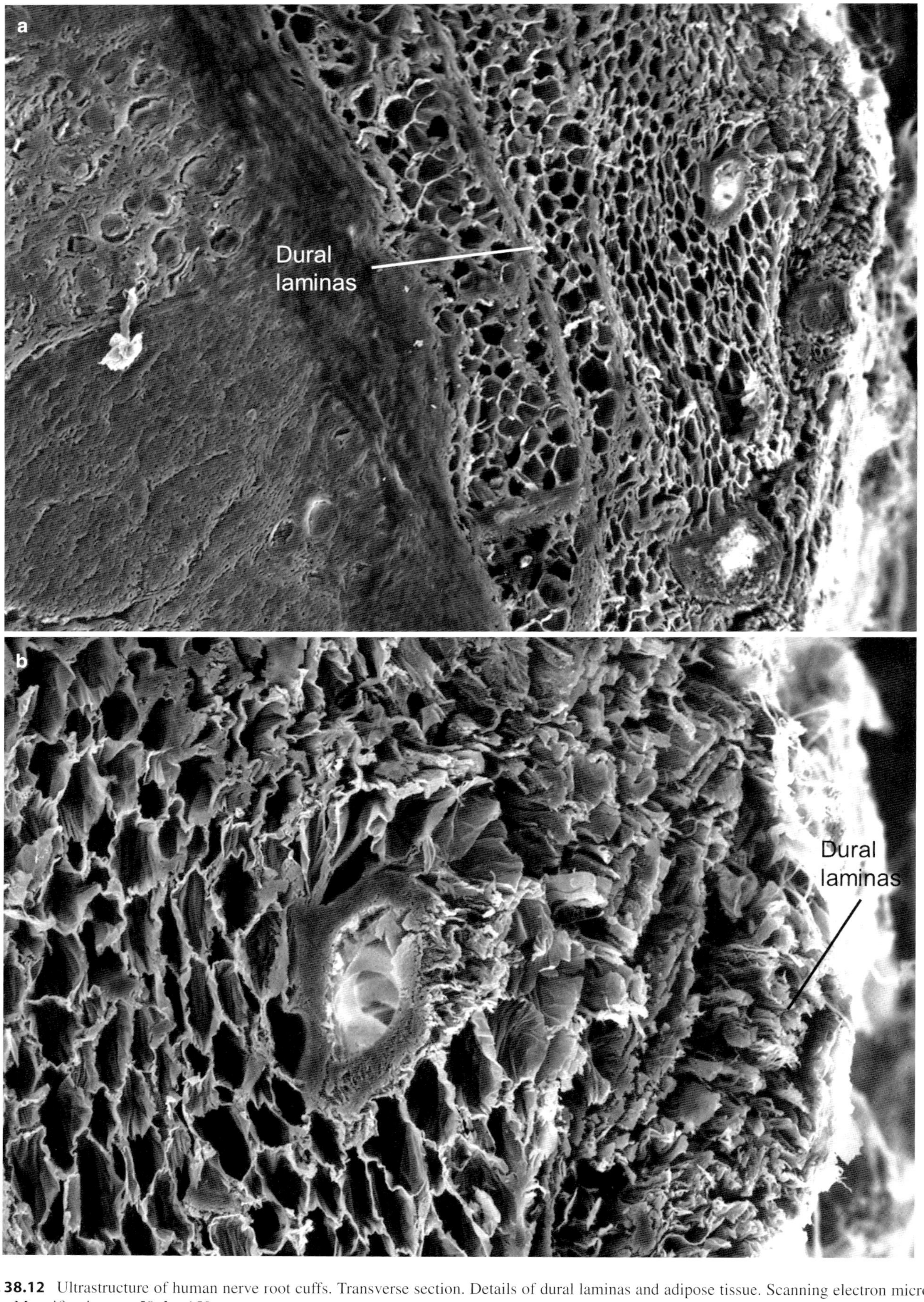

Fig. 38.12 Ultrastructure of human nerve root cuffs. Transverse section. Details of dural laminas and adipose tissue. Scanning electron microscopy. Magnification: **a** ×50; **b** ×150

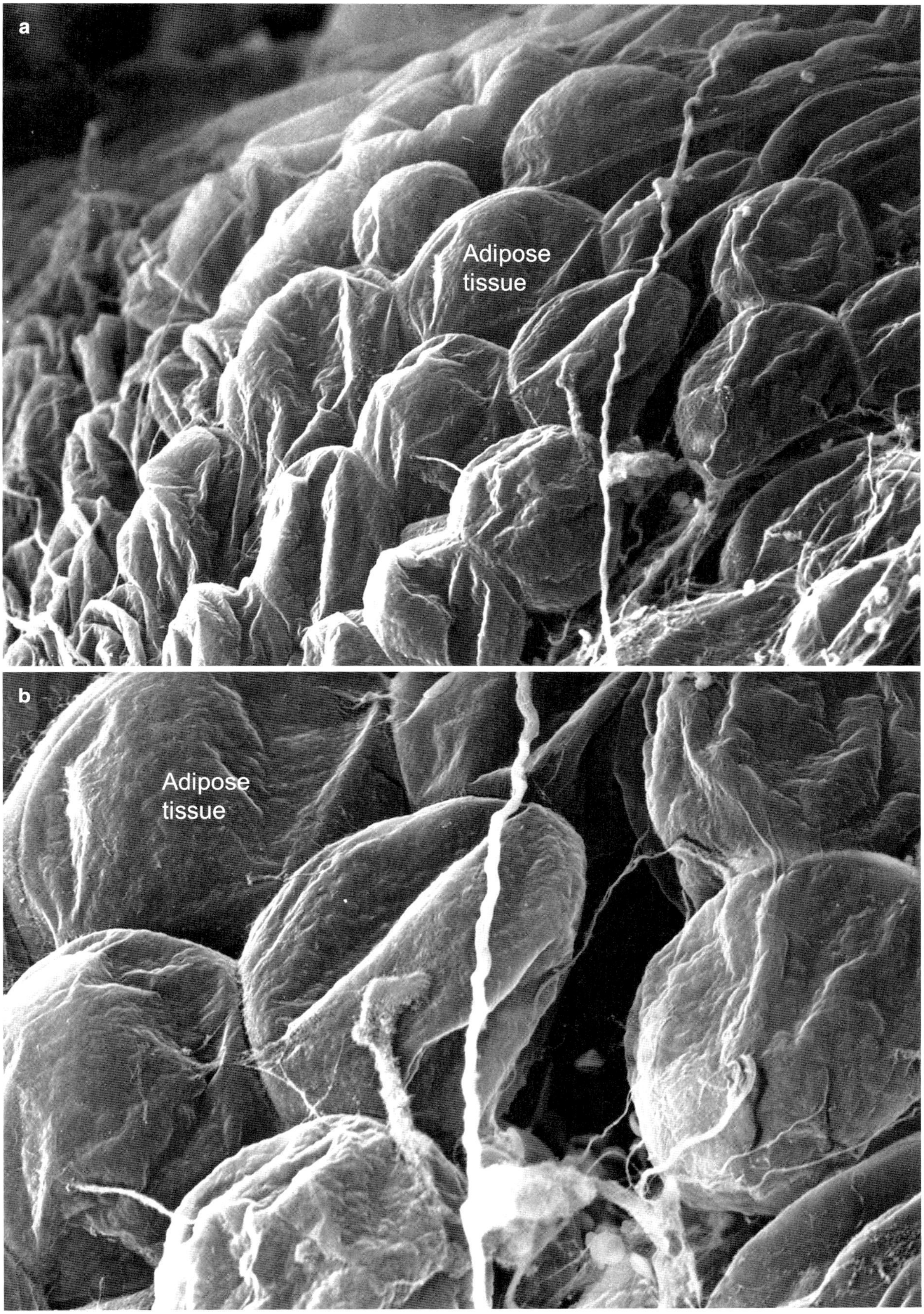

Fig. 38.13 Ultrastructure of human nerve root cuffs. Details of adipose tissue on external surface of nerve root cuff. Scanning electron microscopy. Magnification: **a** ×300; **b** ×700 (From Reina et al. [1]; with permission)

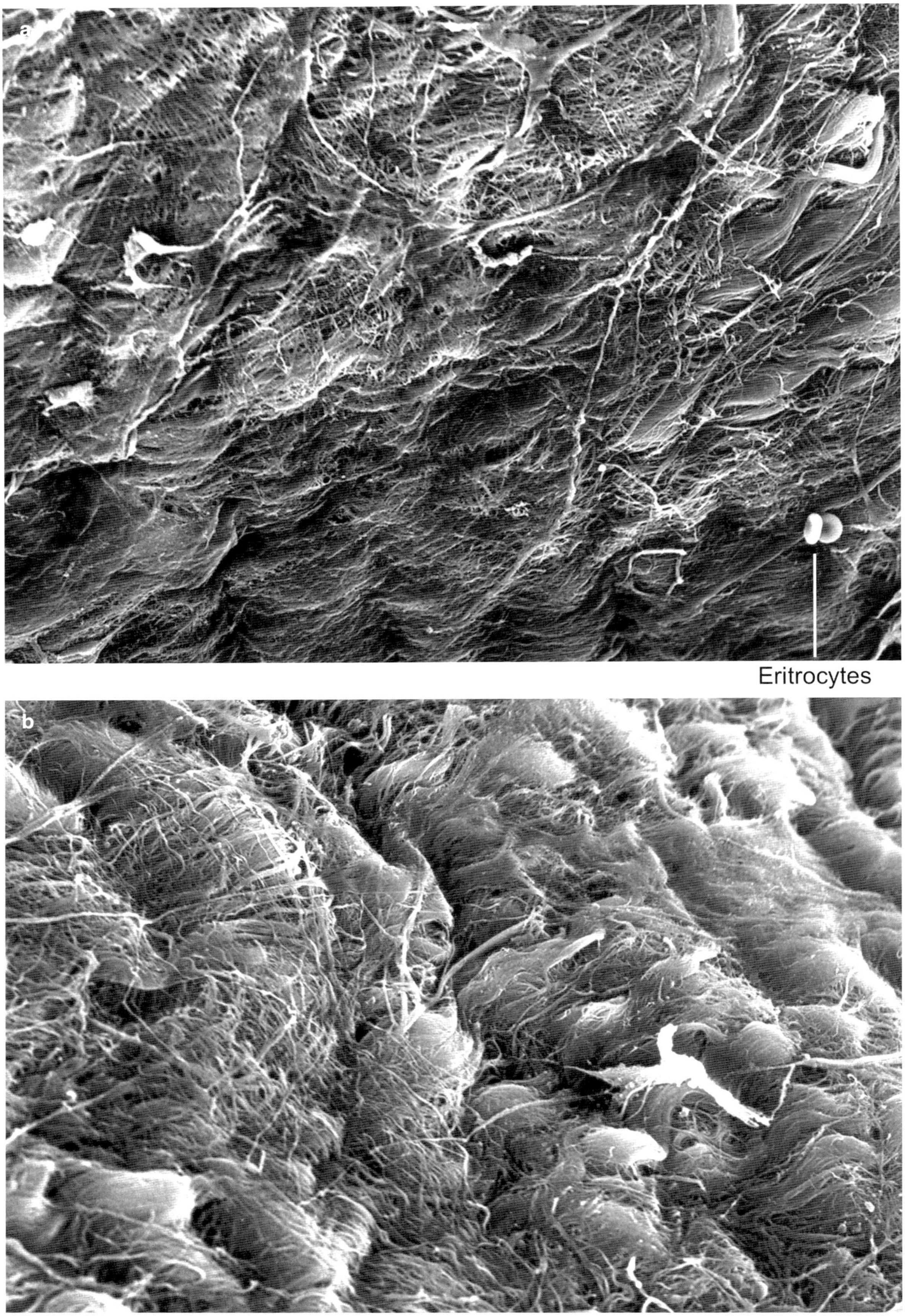

Fig. 38.14 Ultrastructure of human nerve root cuffs. Details of dural laminas on external surface of nerve root cuff. Scanning electron microscopy. Magnification: **a** ×500; **b** ×750

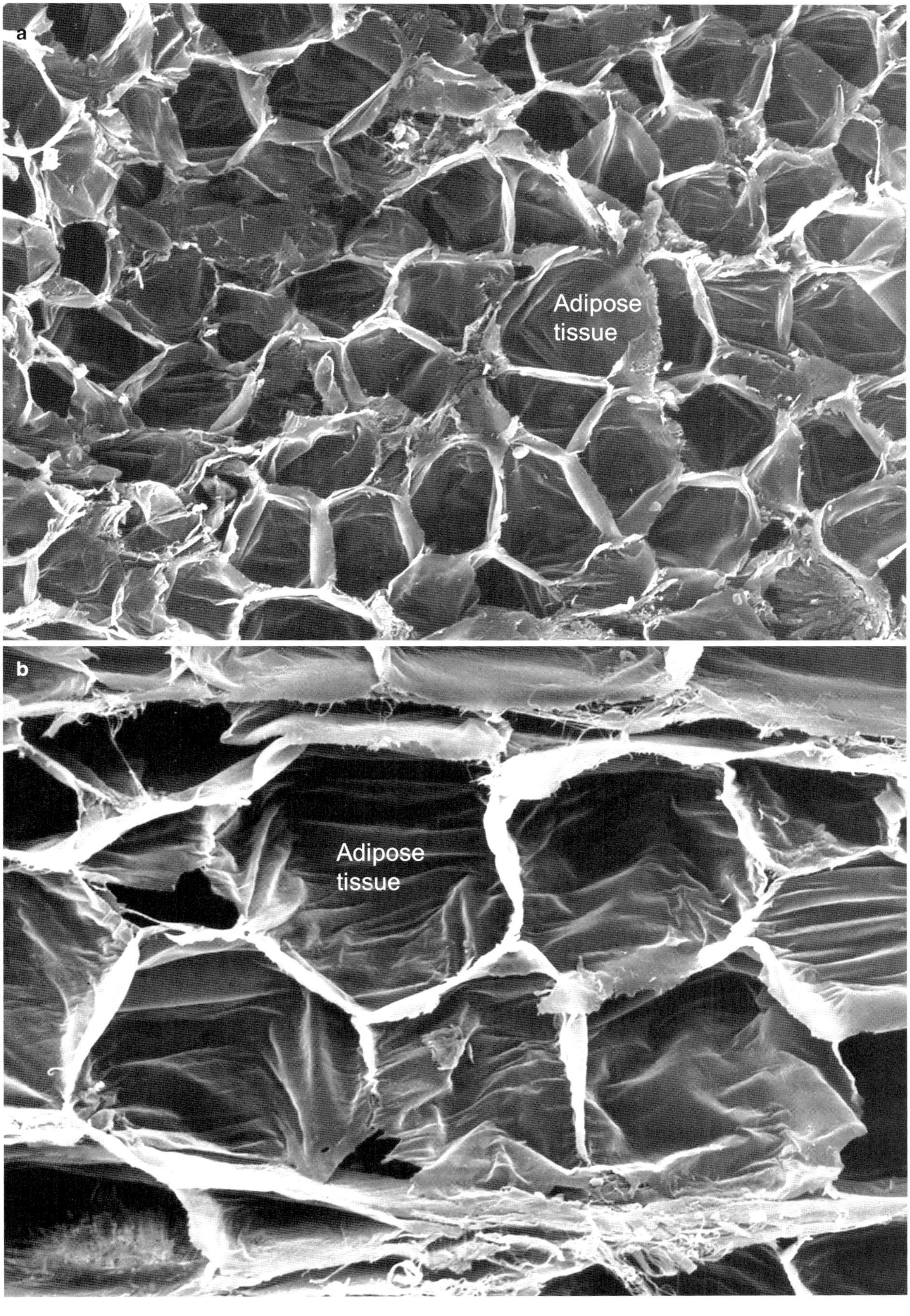

Fig. 38.15 Ultrastructure of human nerve root cuffs. Transverse section. Details of adipose tissue. Scanning electron microscopy. Magnification: **a** ×200; **b** ×400

References

1. Reina MA, Villanueva MC, Machés F, Carrera A, López A, De Andrés JA. The ultrastructure of the human spinal nerve root cuff in the lumbar spine. Anesth Analg. 2008;106:339–44.
2. Reina MA, Dittmann M, López A, van Zundert A. New perspectives in the microscopic structure of human dura mater in the dorso lumbar region. Reg Anesth. 1997;22:161–6.
3. Reina MA, Arriazu R, Collier CB, Sala-Blanch X. Histology and electron microscopy of human peripheral nerves of clinical relevance to the practice of nerve blocks. Rev Esp Anestesiol Reanim. 2013;60:552–62.
4. Reina MA, López A, De Andrés JA. Adipose tissue within peripheral nerves. Study of the human sciatic nerve. Rev Esp Anestesiol Reanim. 2002;49:397–402.
5. Reina MA, De Andrés JA, Hernández JM, Prats-Galino A, Machés F, Peláez J. Transforaminal or translaminar approach for dorsal root ganglion and dorsal nerve root. Anatomical reason for technique decision. Eur J Pain Suppl. 2010;4:287–97.
6. Reina MA, López A, Palácio F, Fornet I. Anestesia del neuroeje. In: Basora Macaya M, Colomina Soler MJ, editors. Anestesia en cirugía ortopédica y en traumatología. Madrid: Editorial Médica Panamericana; 2011. p. 46–58.

39 Ultrastructure of Nerve Root Cuffs: Arachnoid Layer–Perineurium Transition Tissue at Preganglionic, Ganglionic, and Postganglionic Levels

Miguel Angel Reina and Emilse Colman Peyrano

At the level of the spinal epidural space, spinal nerve roots are protected by nerve root cuffs. Nerve root cuffs are lateral prolongations of dura mater and the arachnoid layer that enclose sensory and motor nerve roots crossing the epidural space towards the intervertebral foramen [1].

In the region extending from the dural sac to the dorsal root ganglia, each group of axons is surrounded by a cellular layer, stratified and joined together by tight junctions and desmosomes. The composition of the cellular layer does not resemble that of perineurium. Rather, it is a transitional arachno-perineurial tissue formed by elongated cells that are stratified and oriented longitudinally to the axon groups [1].

They are thick, with a number of flat cells ranging between 14 and 20. These cells have cytoplasmic prolongations that encroach upon neighboring cells, leaving very little extracellular space, with a rough appearance and few collagen fibers. Sometimes two concentric layers appear separated by collagen fibers that probably correspond to the inner and outer aspects of the cellular component. This transitional tissue acts as a barrier at this level and is formed by transition cells ranging from pia mater cells [2], dural border cells (neurothelial cells) [3, 4], arachnoid layer [5] and trabecular arachnoid cells [6], to perineurial cells of more distal peripheral nerves [7]. Outside the cellular component is another fatty and fibrillar layer, similar to interfascicular tissue in peripheral nerves.

At the level of the dorsal root ganglia, fascicles appear enclosed in a sheath with cellular components. The morphology of the cellular layers is similar to perineurium, with identical ultrastructure, although the number of concentric cellular layers increases up to about 25–30.

Distal to the dorsal root ganglia, the ultrastructure of connective tissue layers enclosing the peripheral nerves remains constant along their trajectories. Nerve fascicles appear to be surrounded by perineurium formed by about 9–14 single-cell concentric layers (Figs. 39.1, 39.2, 39.3, 39.4, 39.5, 39.6, 39.7, 39.8, 39.9, 39.10, 39.11, 39.12, 39.13, 39.14, 39.15, 39.16, 39.17, 39.18, 39.19, 39.20, 39.21, 39.22, 39.23, 39.24, and 39.25).

M.A. Reina, MD, PhD (✉)
Department of Clinical Medical Sciences and Institute of Applied Molecular Medicine, School of Medicine, University of CEU San Pablo, Madrid, Spain

Department of Anesthesiology, Madrid-Montepríncipe University Hospital, Madrid, Spain
e-mail: miguelangel@perticone.e.telefonica.net

E.C. Peyrano, BS
Research Unit, Anesthesia Reina Colman SLP, Madrid, Spain
e-mail: emilsecolpey@yahoo.es

M.A. Reina et al. (eds.), *Atlas of Functional Anatomy for Regional Anesthesia and Pain Medicine: Human Structure, Ultrastructure and 3D Reconstruction Images*, DOI 10.1007/978-3-319-09522-6_39,

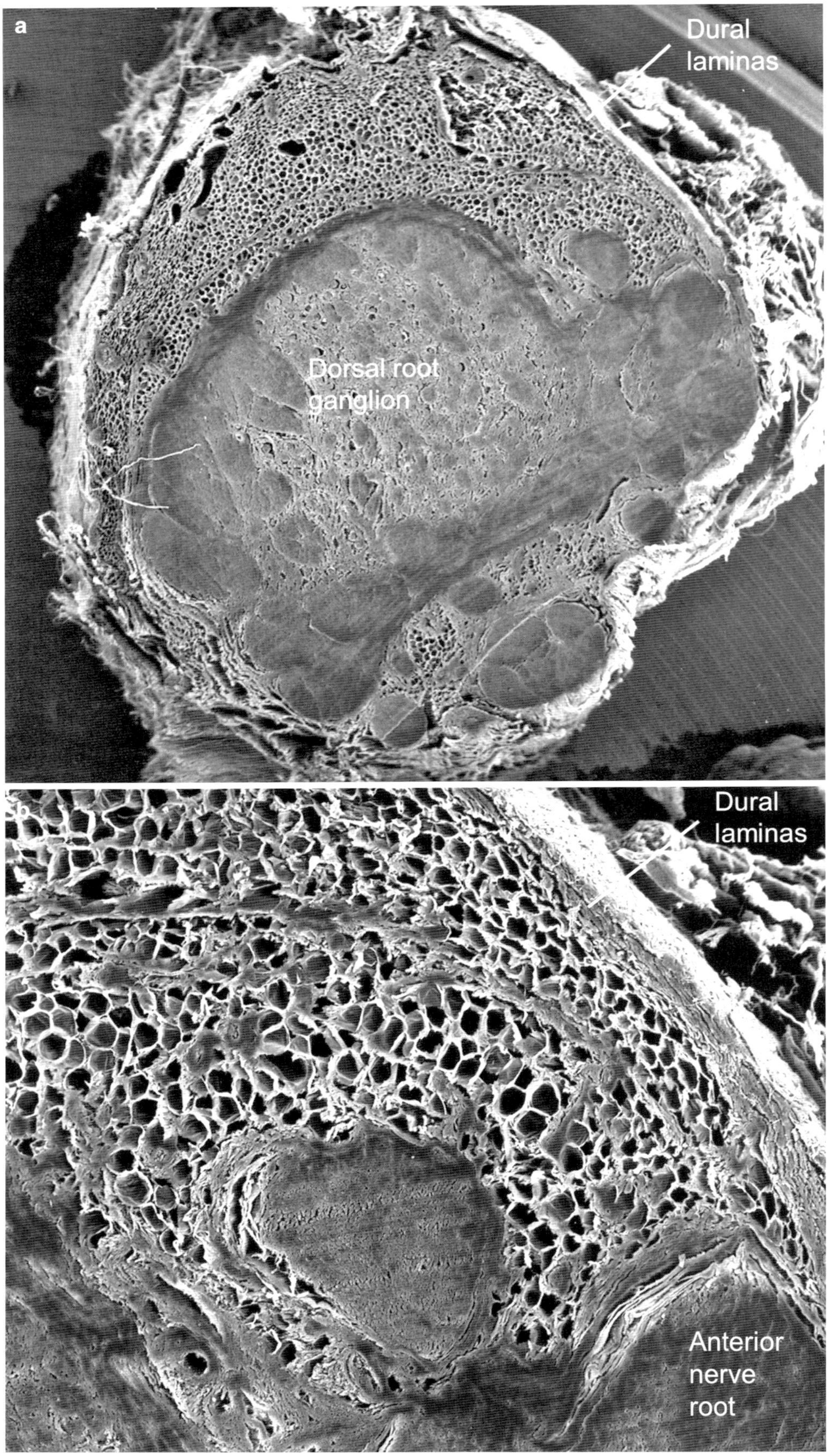

Fig. 39.1 Ultrastructure of human nerve root cuffs at the ganglionic zone. (**a**) Arachnoid layer–perineurium transition tissue. (**b**) Detail of 39.1a. Transverse section. Scanning electron microscopy. Magnification: **a** ×12; **b** ×50 (**a** from Reina et al. [1]; **b** from Reina et al. [8]; both with permission)

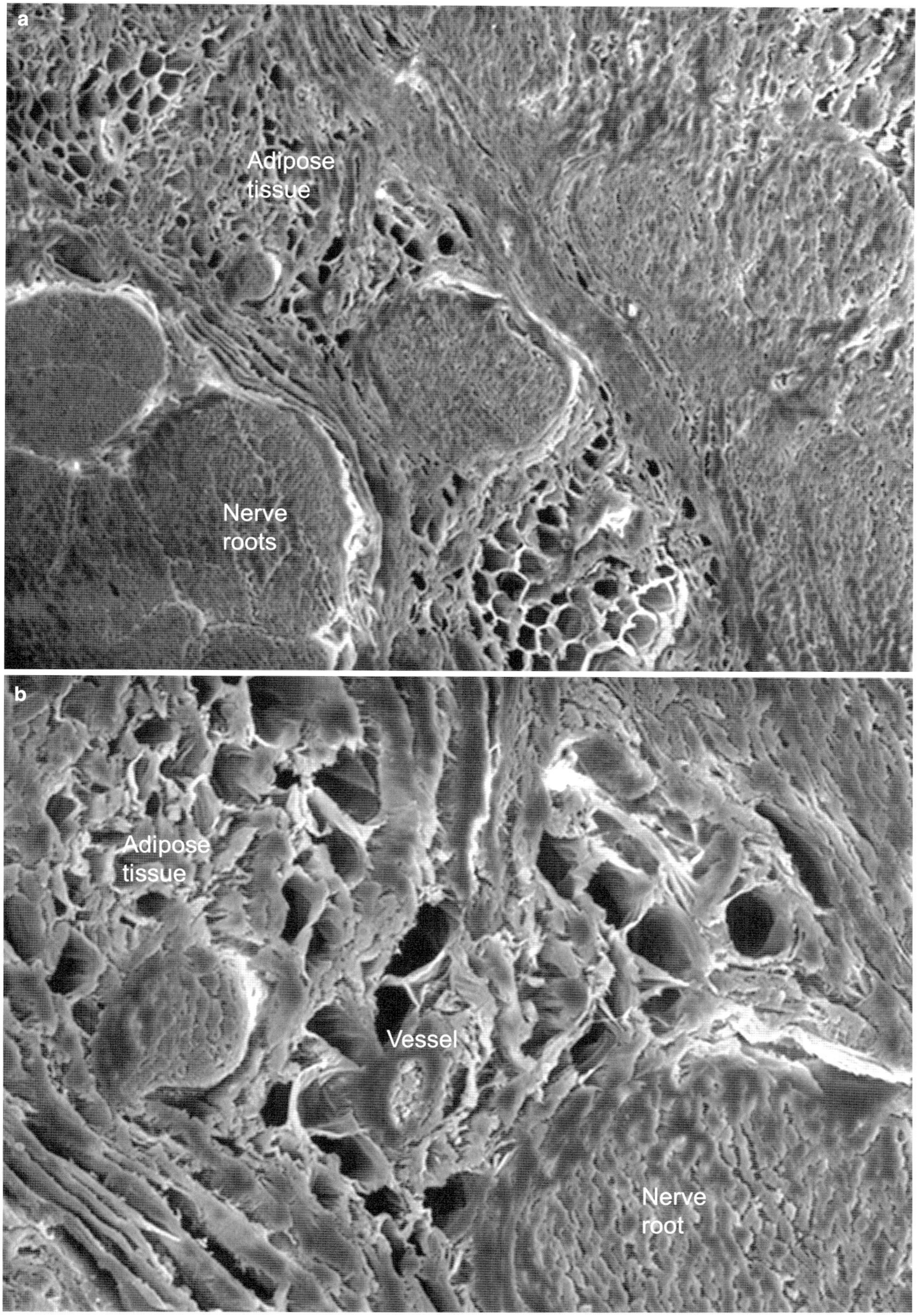

Fig. 39.2 Ultrastructure of human nerve root cuffs at the preganglionic zone. (**a**) Arachnoid layer–perineurium transition tissue. (**b**) Detail of 39.2a. Transverse section. Scanning electron microscopy. Magnification: **a** ×50; **b** ×150

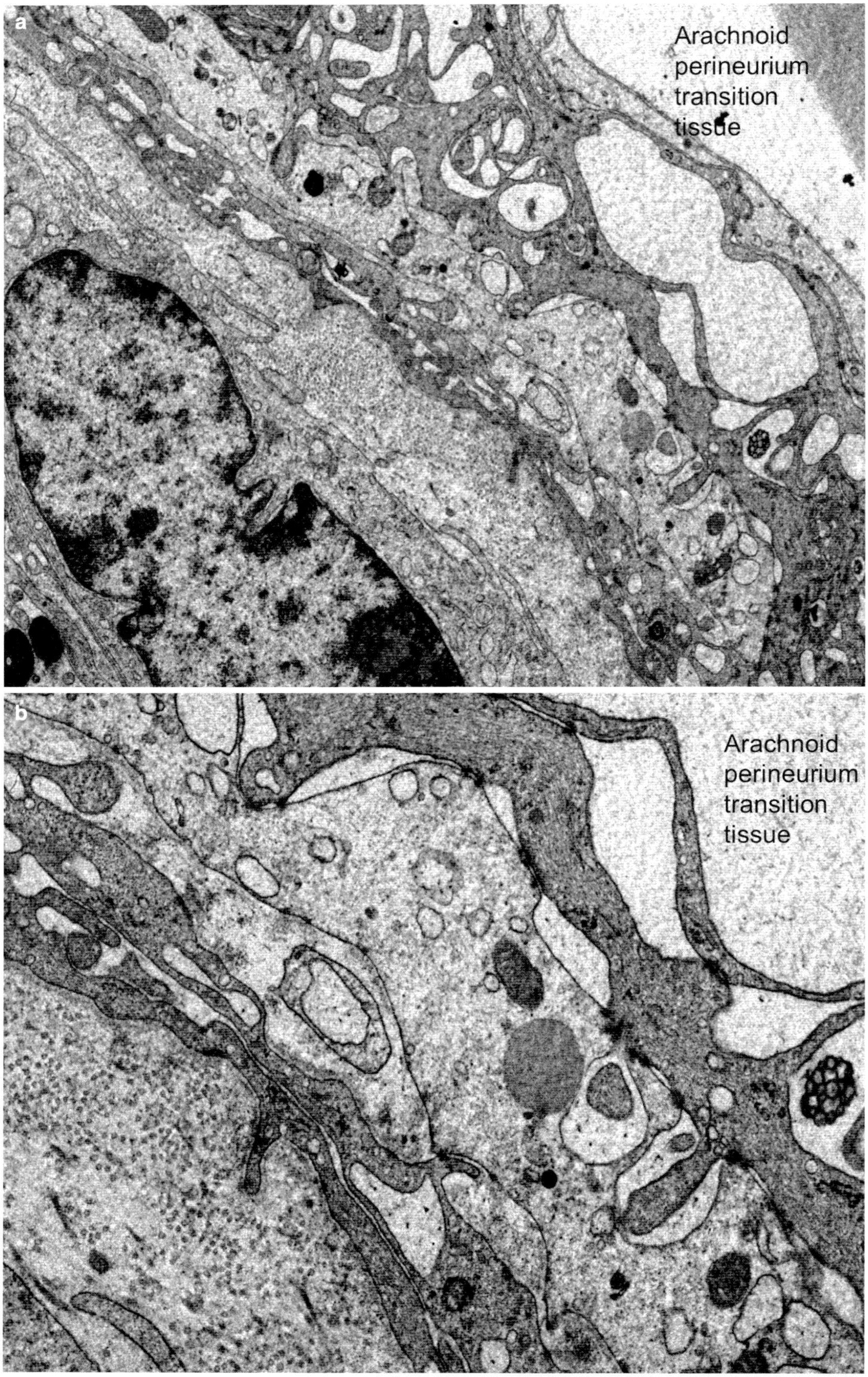

Fig. 39.3 Ultrastructure of human nerve root cuffs at the preganglionic zone. Arachnoid layer–perineurium transition tissue. (**a**, **b**) Images obtained from different samples. Transverse section. Transmission electron microscopy. Magnification: **a** ×10,000; **b** ×3,000 (**b** from Reina et al. [9]; with permission)

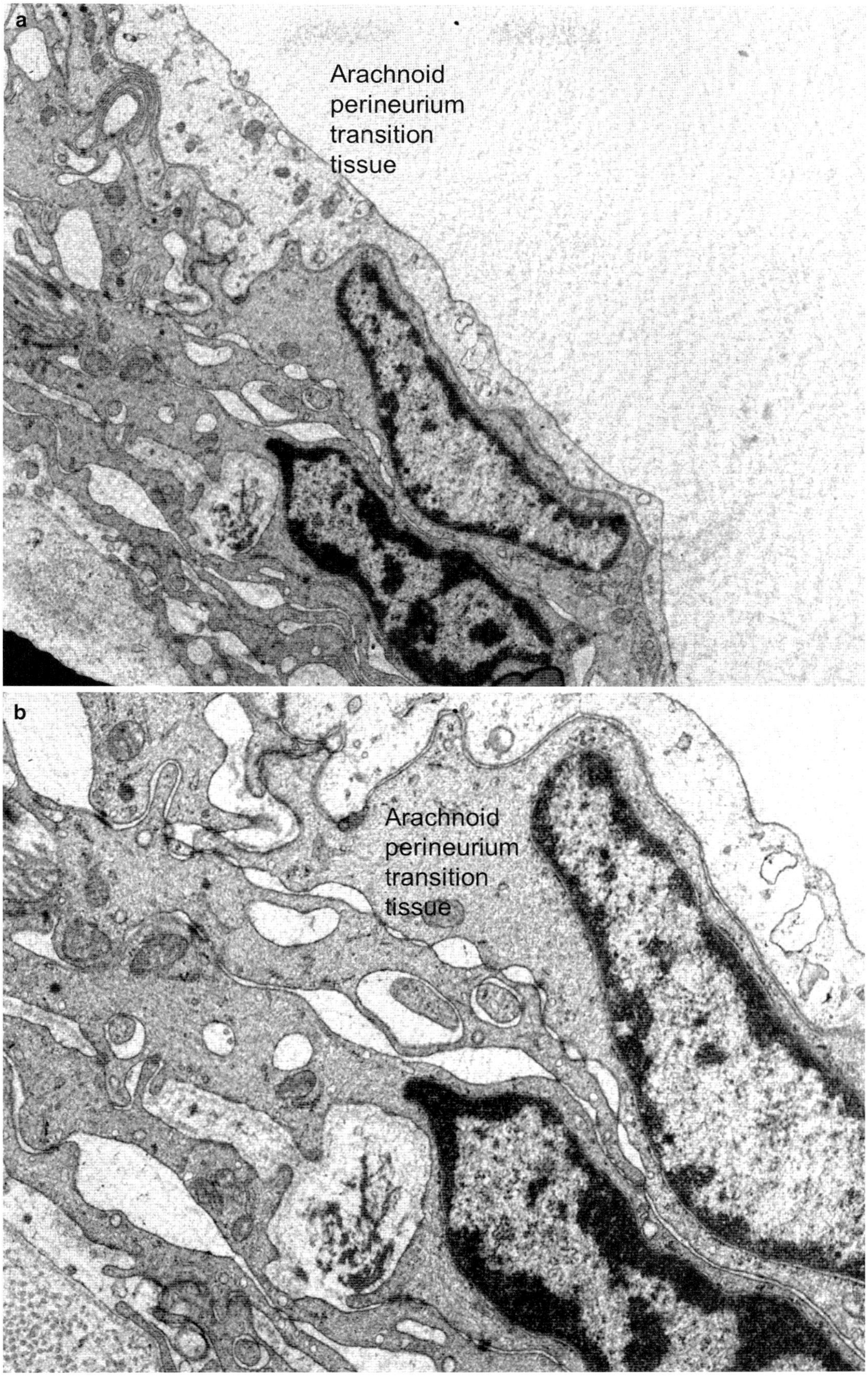

Fig. 39.4 Ultrastructure of human nerve root cuffs at the preganglionic zone. Arachnoid layer–perineurium transition tissue. (**a**, **b**) Images obtained from different samples. Transverse section. Transmission electron microscopy. Magnification: **a** ×10,000; **b** ×20,000 (From Reina et al. [1]; with permission)

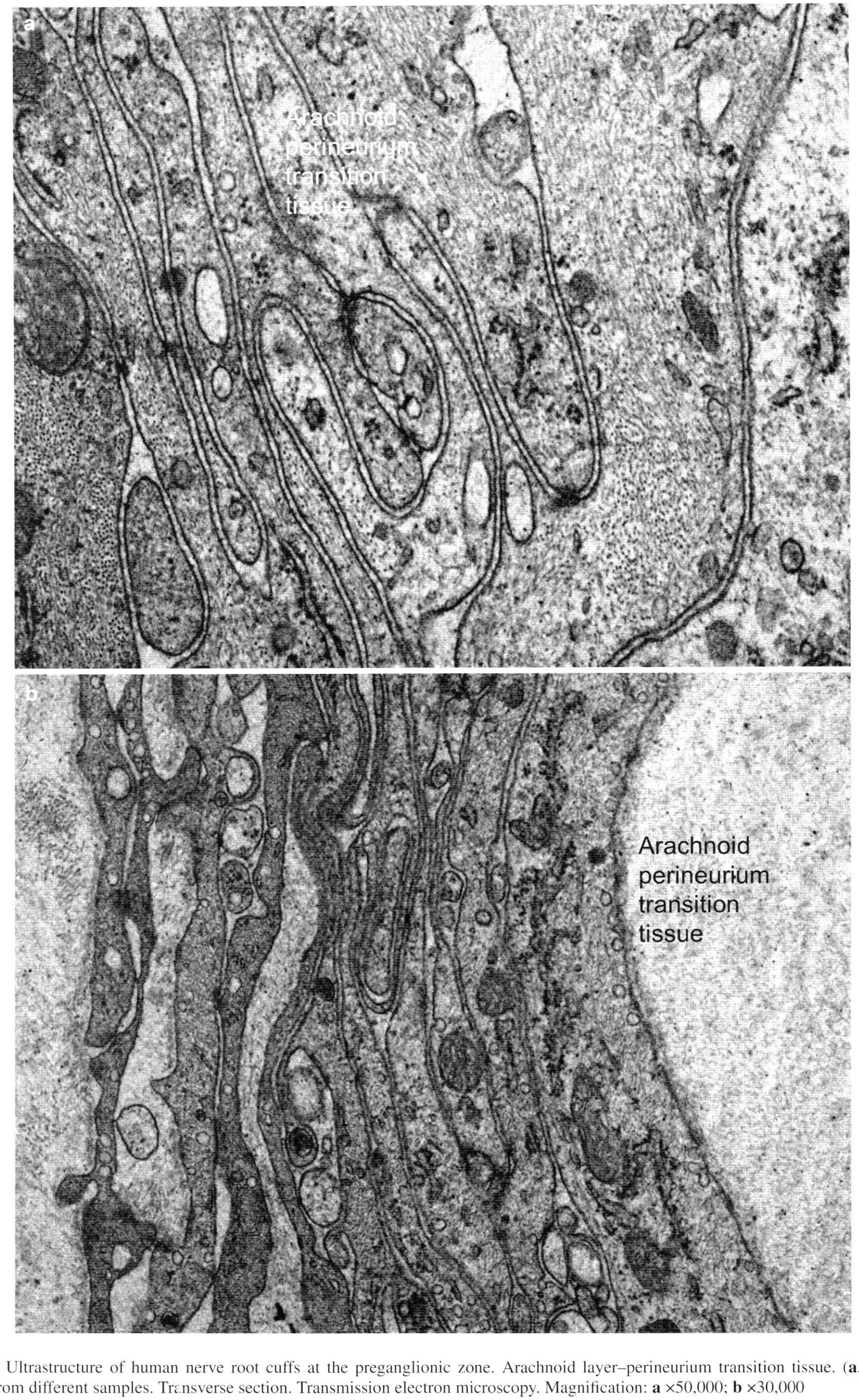

Fig. 39.5 Ultrastructure of human nerve root cuffs at the preganglionic zone. Arachnoid layer–perineurium transition tissue. (**a**, **b**) Images obtained from different samples. Transverse section. Transmission electron microscopy. Magnification: **a** ×50,000; **b** ×30,000

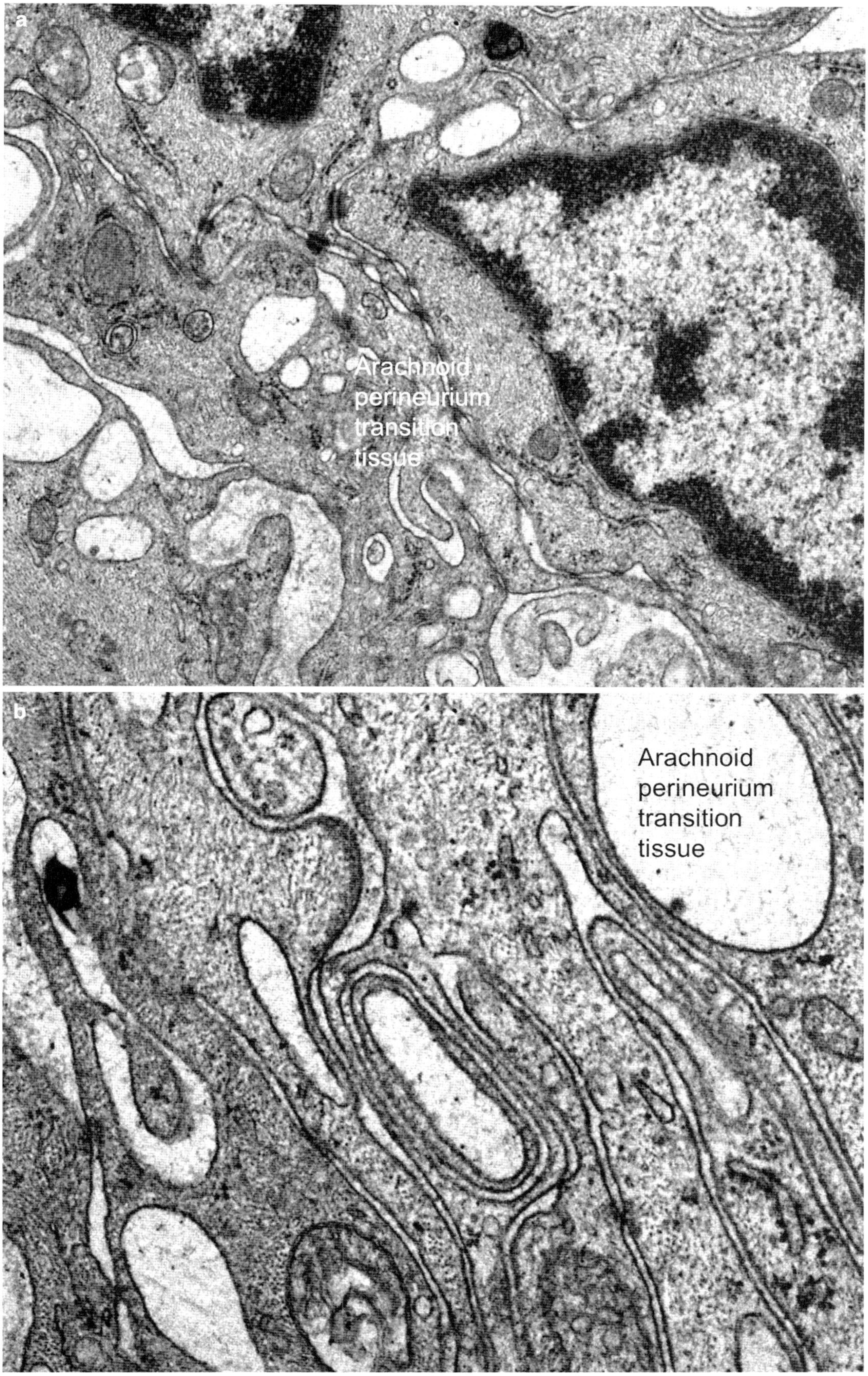

Fig. 39.6 Ultrastructure of human nerve root cuffs at the preganglionic zone. Arachnoid layer–perineurium transition tissue. (**a**, **b**) Images obtained from different samples. Transverse section. Transmission electron microscopy. Magnification: **a** ×30,000; **b** ×60,000

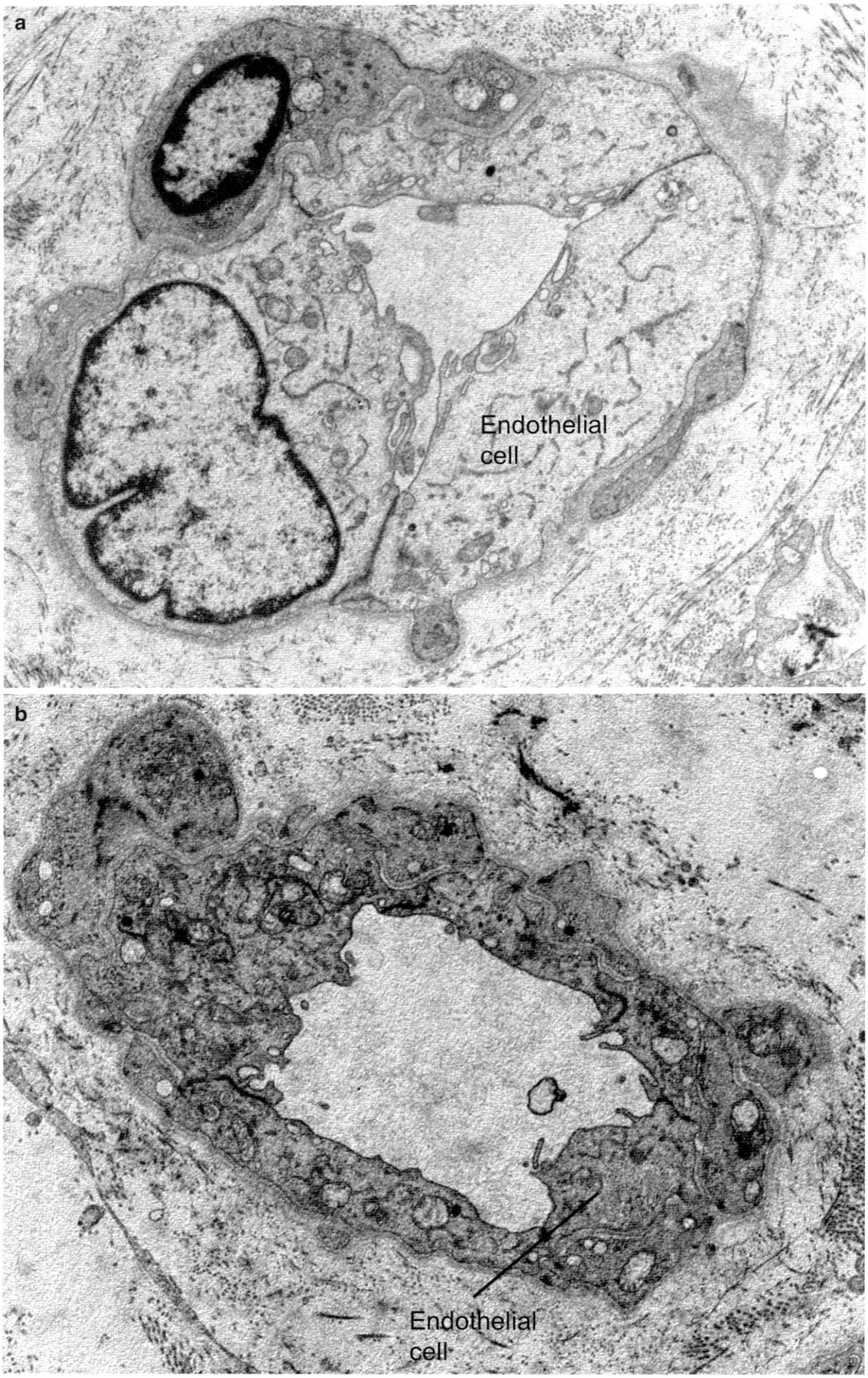

Fig. 39.7 Ultrastructure of human nerve root cuffs at the preganglionic zone. Capillaries within nerve root. (**a**, **b**) Images obtained from different samples. Transverse section. Transmission electron microscopy. Magnification: **a** ×10,000; **b** ×10,000

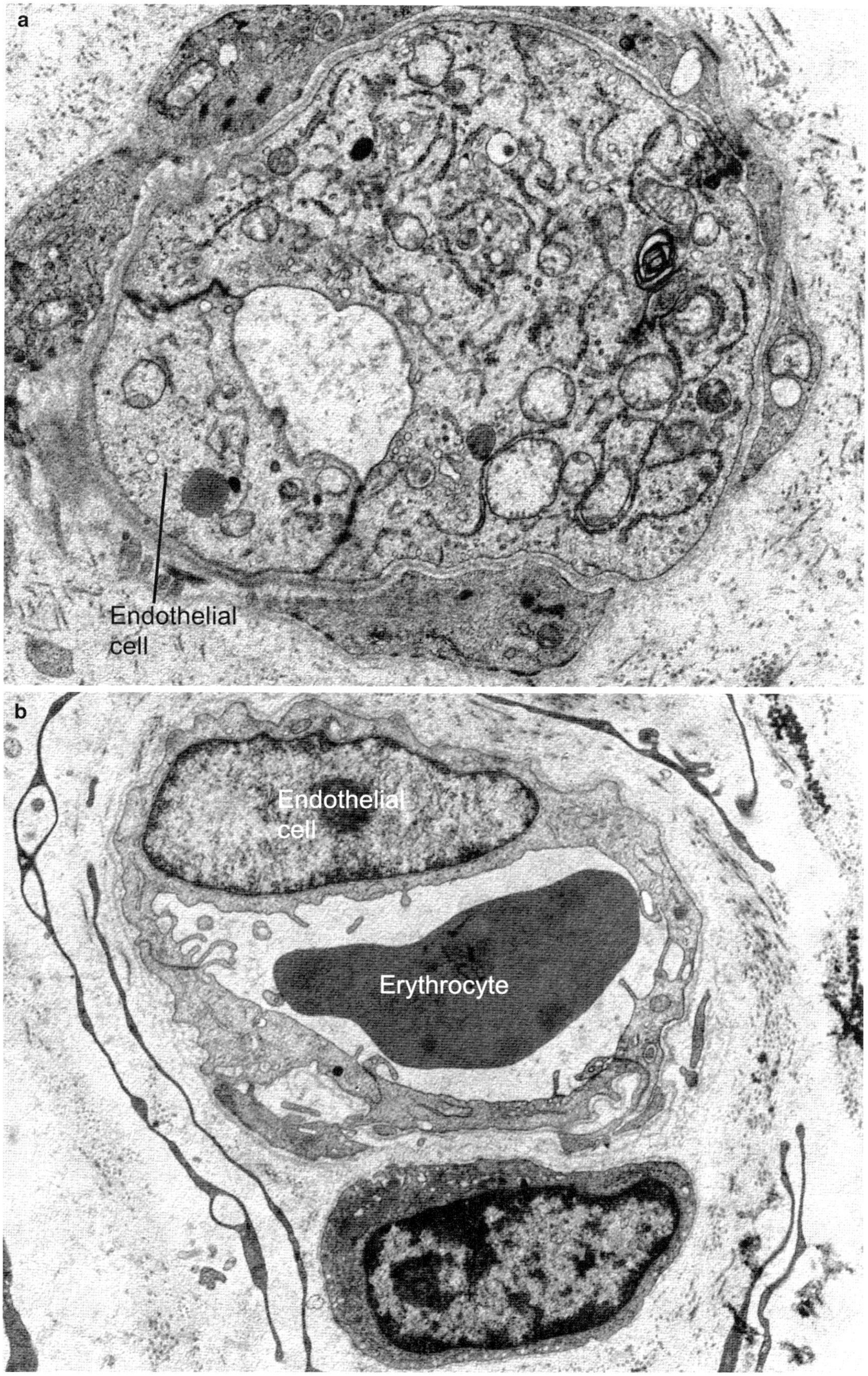

Fig. 39.8 Ultrastructure of human nerve root cuffs at the preganglionic zone. Capillaries within nerve root. (**a**, **b**) Images obtained from different samples. Transverse section. Transmission electron microscopy. Magnification: **a** ×15,000; **b** ×12,000

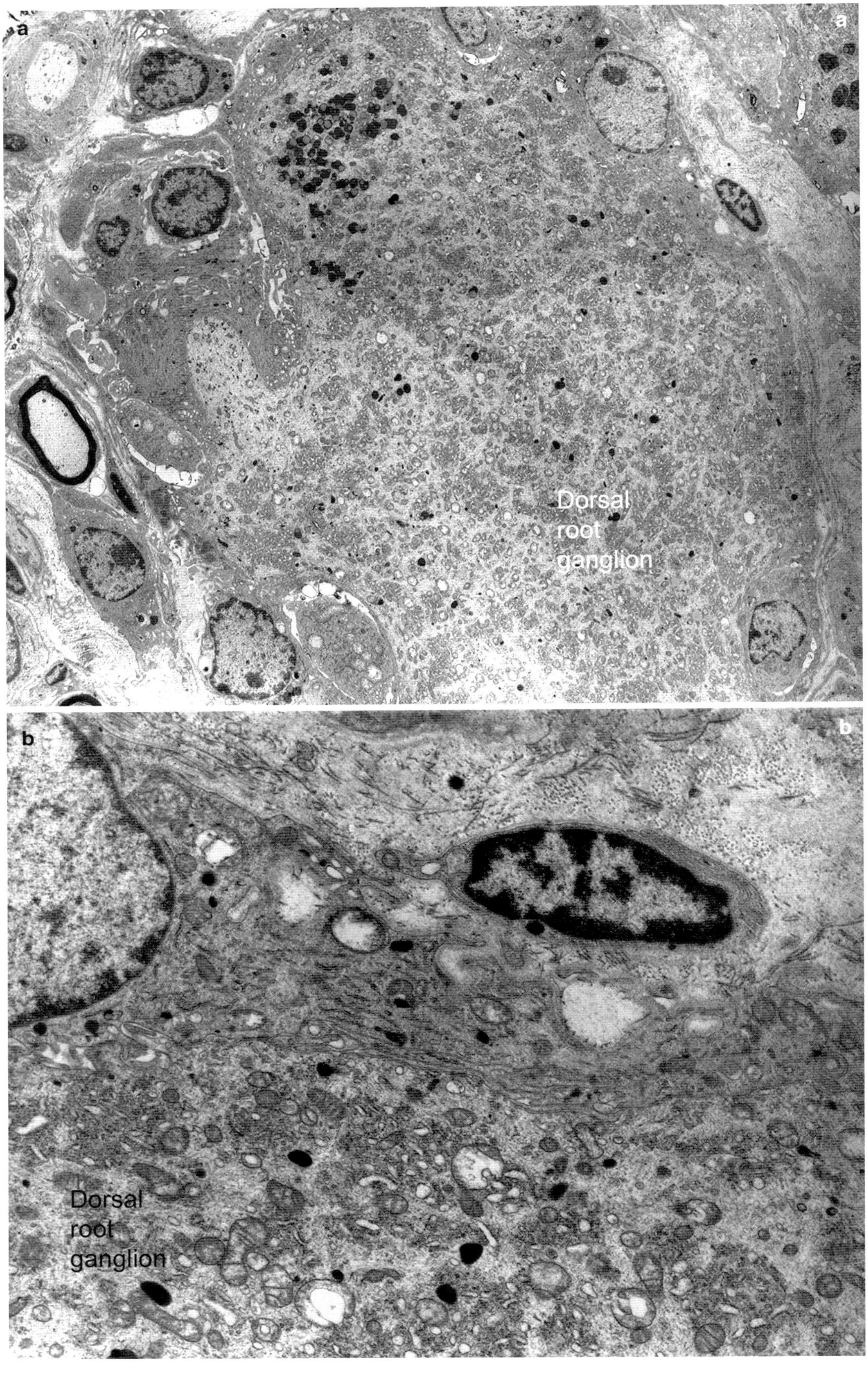

Fig. 39.9 Ultrastructure of human nerve root cuffs at the ganglionic zone. Dorsal root ganglion tissue. (**a**, **b**) Images obtained from different samples. Transverse section. Transmission electron microscopy. Magnification: **a** ×1,100; **b** ×7,000

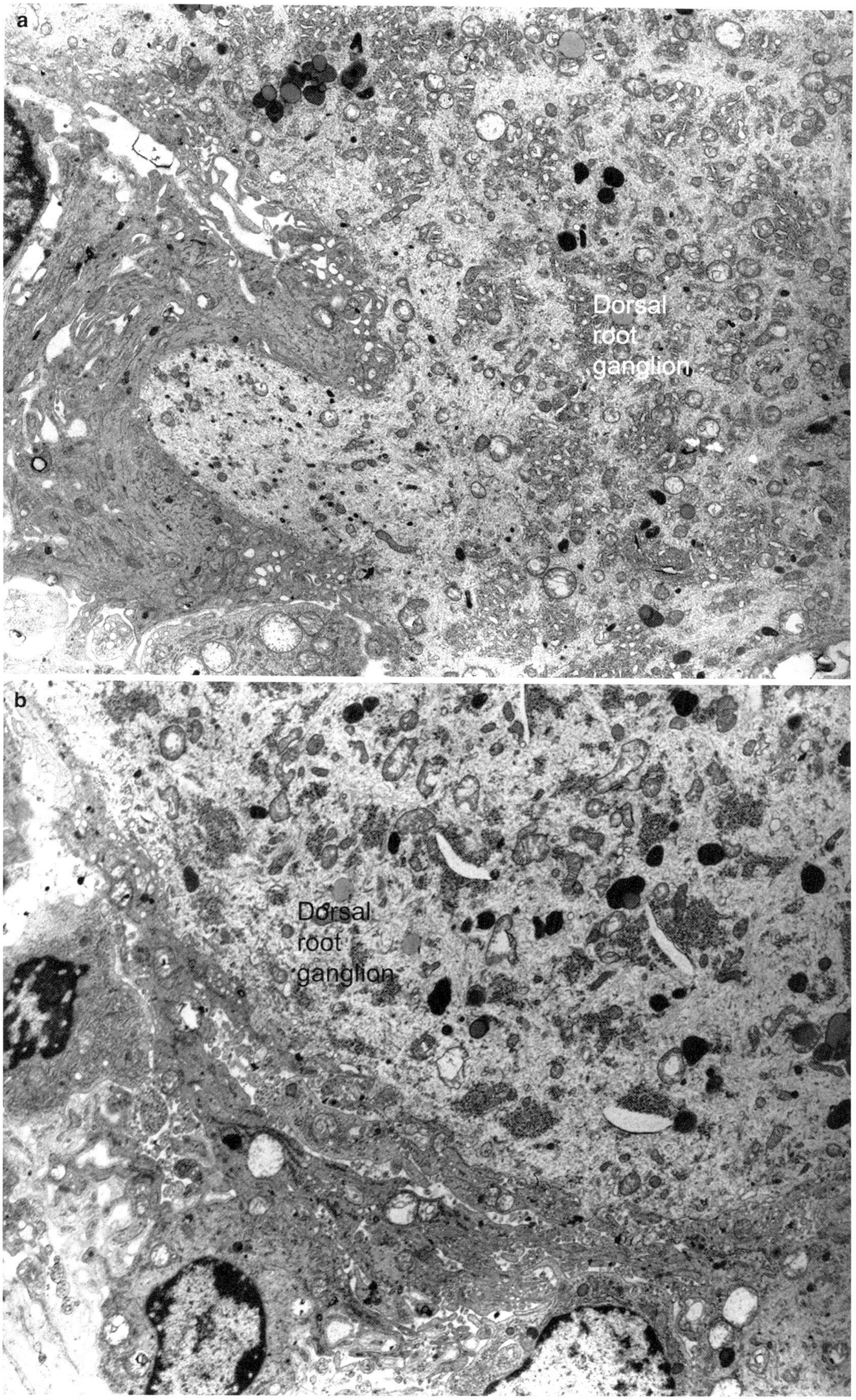

Fig. 39.10 Ultrastructure of human nerve root cuffs at the ganglionic zone. Dorsal root ganglion tissue. (**a**, **b**) Images obtained from different samples. Transverse section. Transmission electron microscopy. Magnification: **a** ×3,000; **b** ×3,000

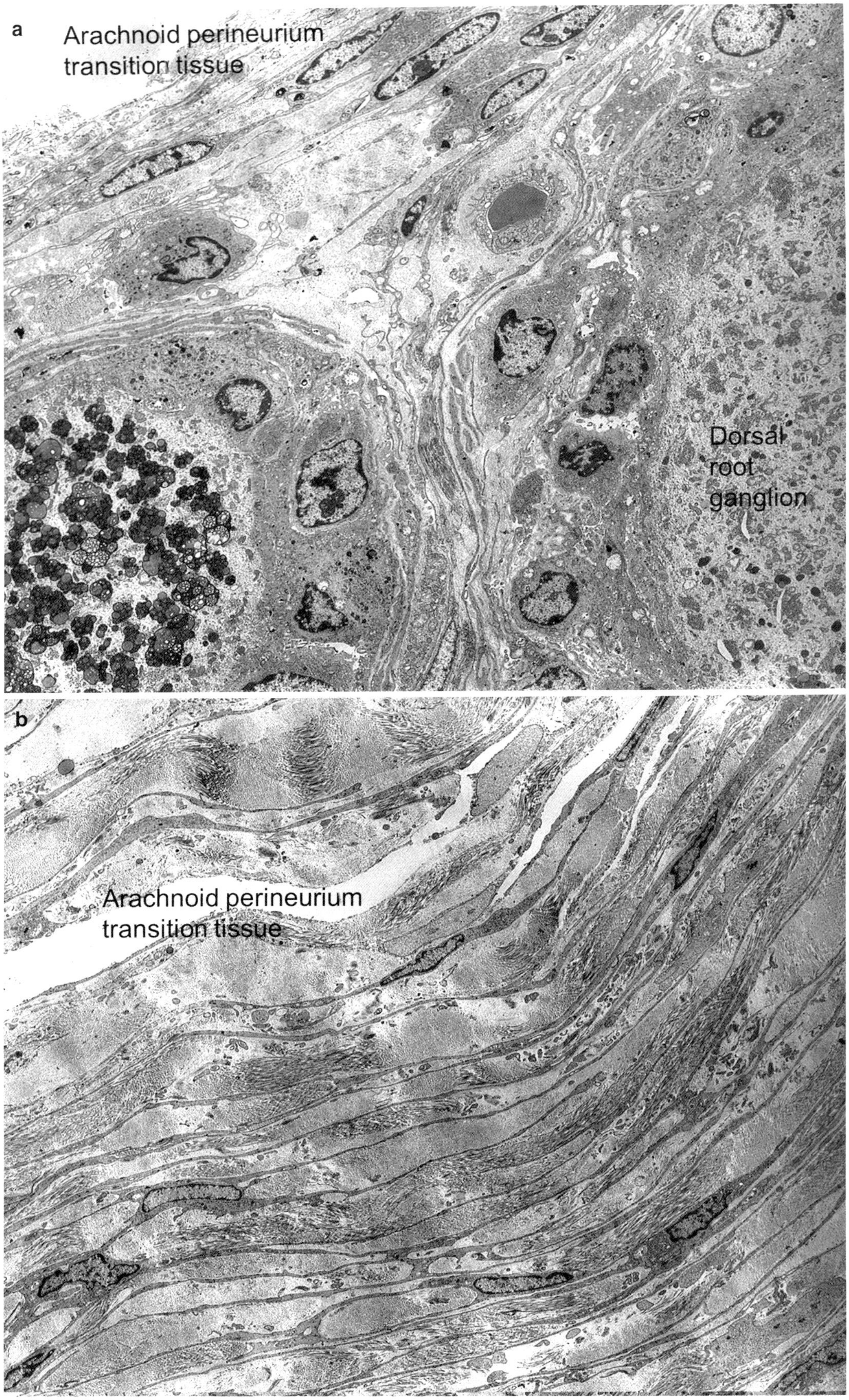

Fig. 39.11 Ultrastructure of human nerve root cuffs at the ganglionic zone. (**a**) Dorsal root ganglion tissue. (**b**) Arachnoid layer–perineurium transition tissue. Transverse section. Transmission electron microscopy. Magnification: **a** ×1,100; **b** ×1,100

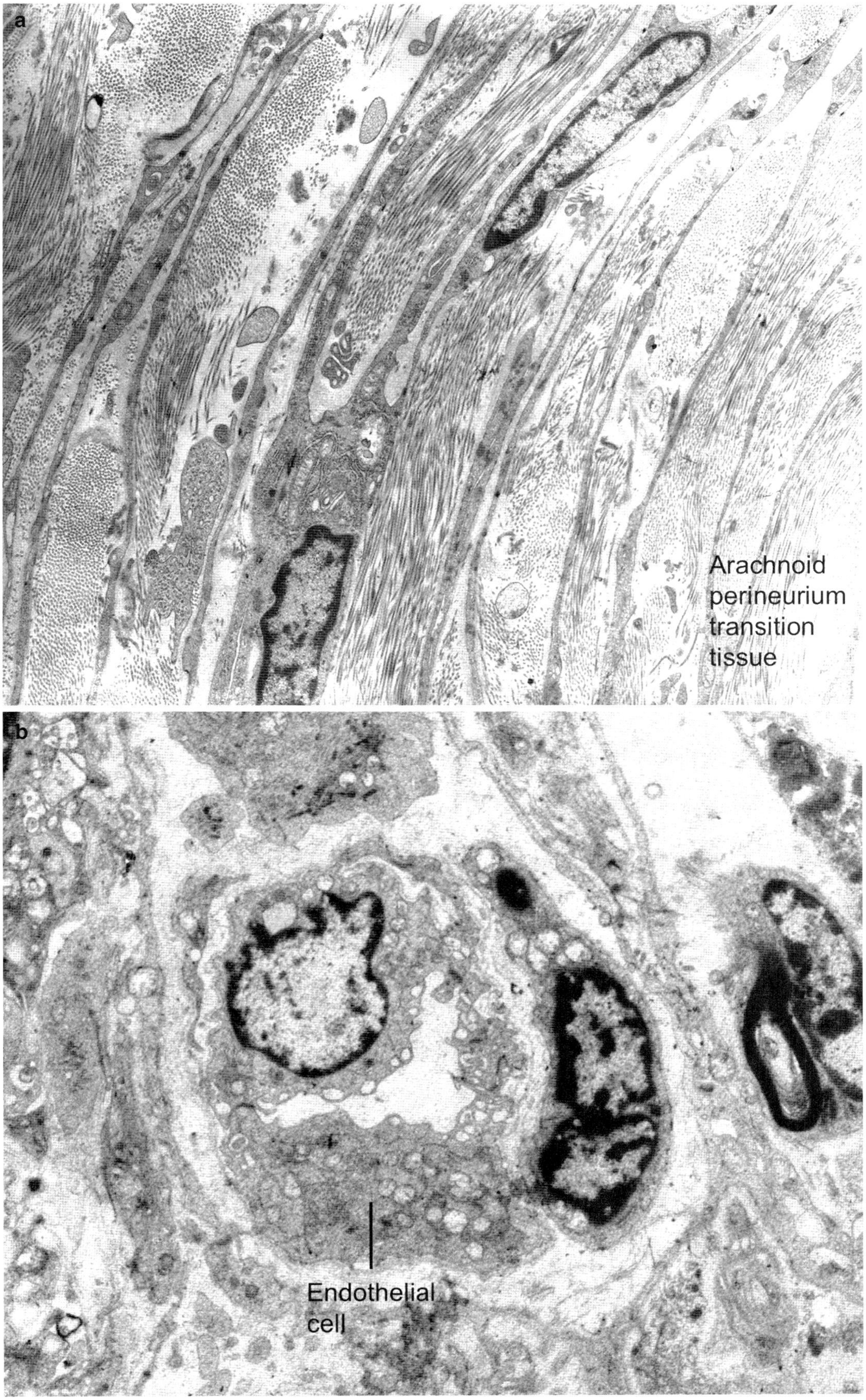

Fig. 39.12 Ultrastructure of human nerve root cuffs at the ganglionic zone. (**a**) Arachnoid layer–perineurium transition tissue. (**b**) Capillary within dorsal root ganglion tissue. Transverse section. Transmission electron microscopy. Magnification: **a** ×4,400; **b** ×8,000 (**a** From Reina et al. [1]; **b** from Reina et al. [8]; both with permission)

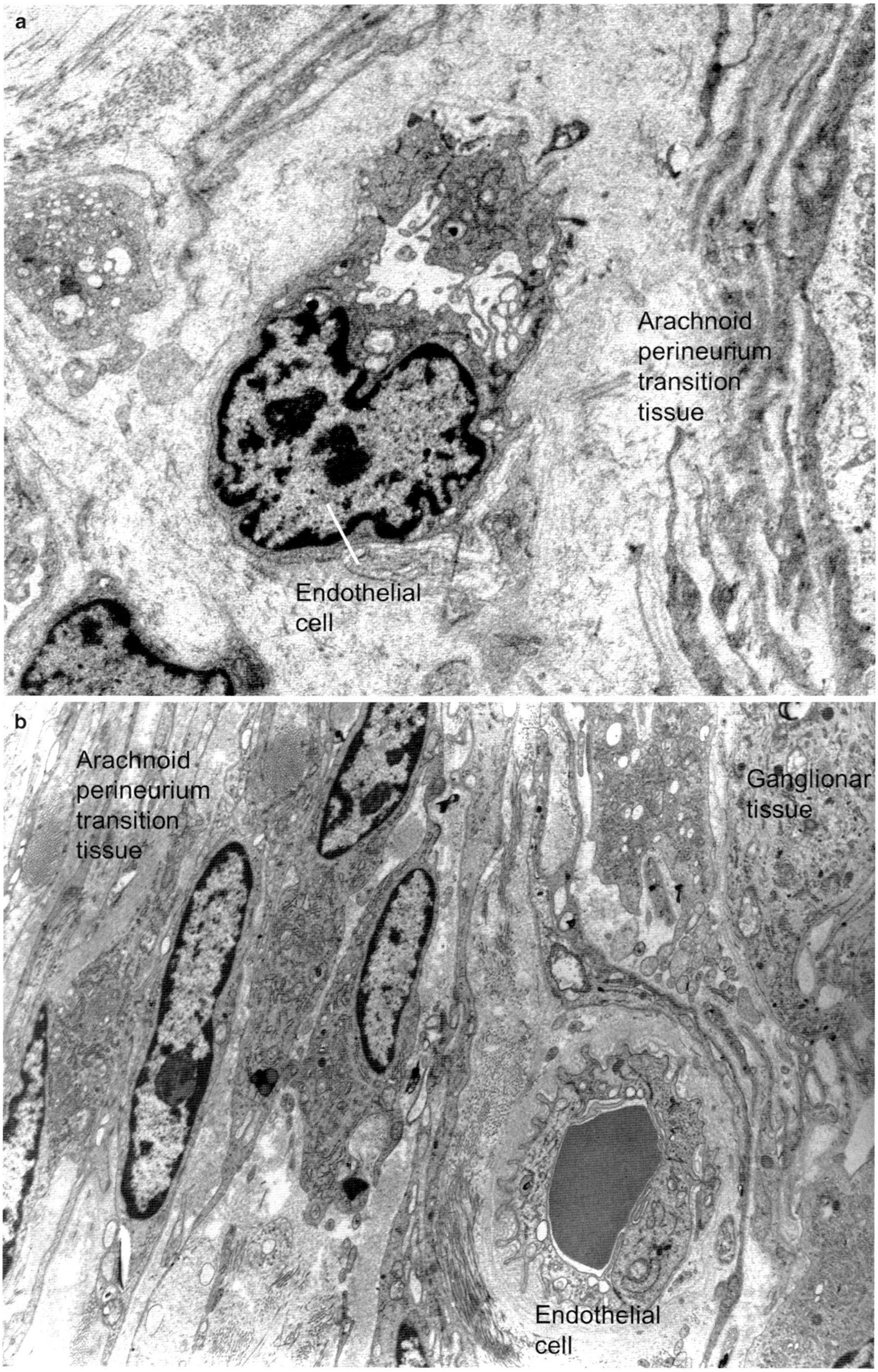

Fig. 39.13 Ultrastructure of human nerve root cuffs at the ganglionic zone. Arachnoid layer–perineurium transition tissue and continuous capillaries with dorsal root ganglion tissue. (**a**, **b**) Images obtained from different samples. Transverse section. Transmission electron microscopy. Magnification: **a** ×3,000; **b** ×3,000

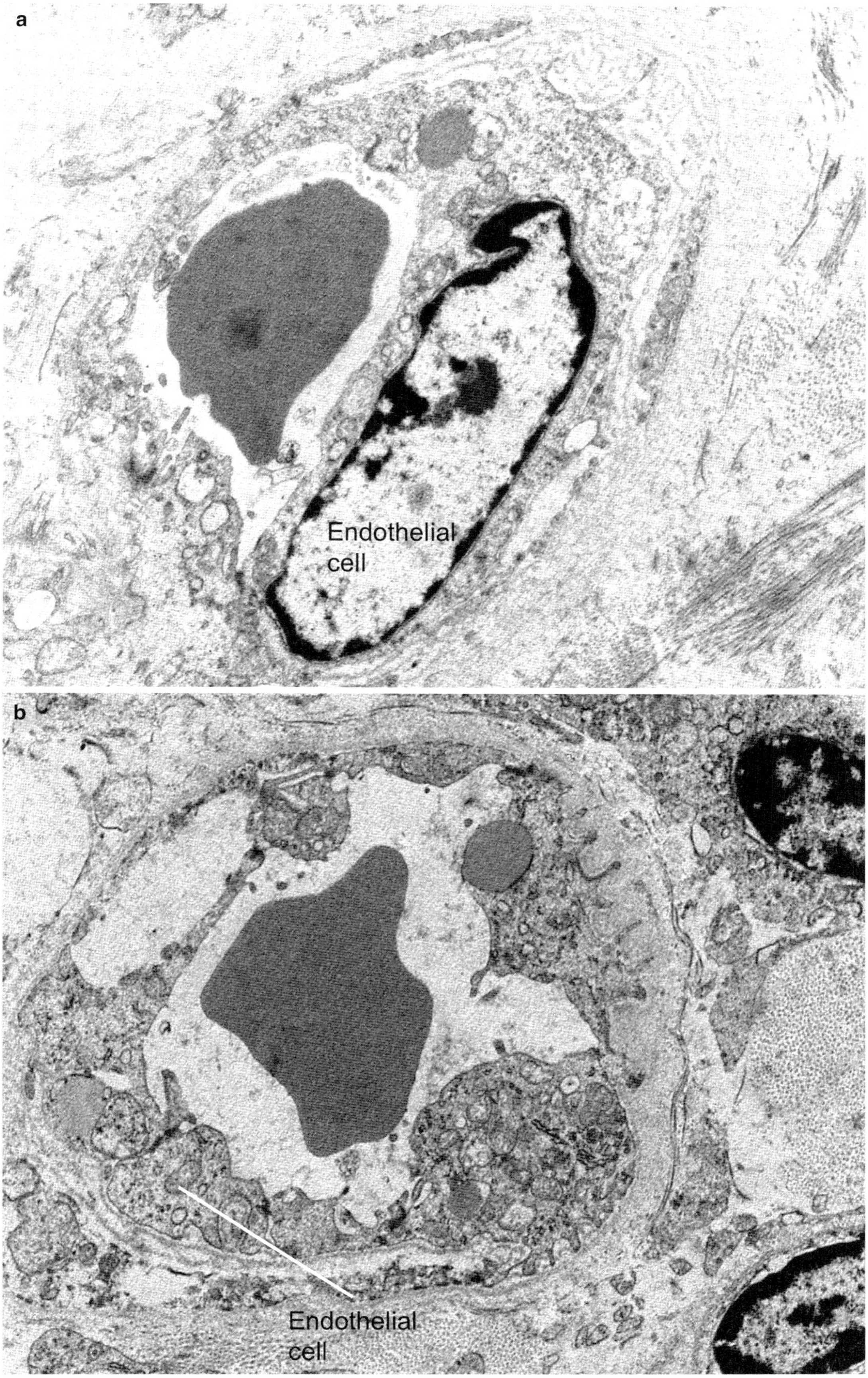

Fig. 39.14 Ultrastructure of human nerve root cuffs at the ganglionic zone. Continuous capillaries within dorsal root ganglion tissue. (**a**, **b**) Images obtained from different samples. Transverse section. Transmission electron microscopy. Magnification: **a** ×12,000; **b** ×12,000

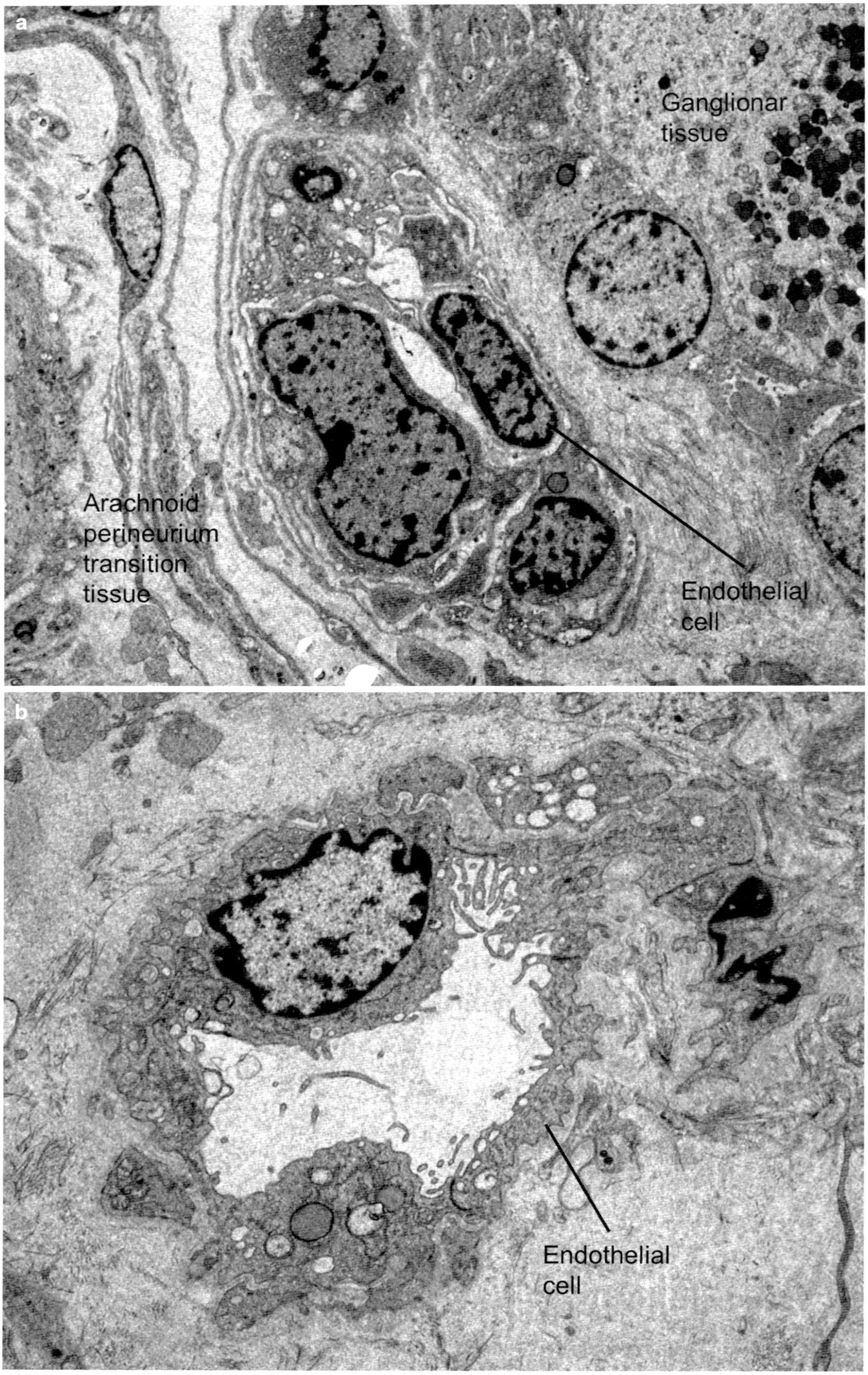

Fig. 39.15 Ultrastructure of human nerve root cuffs at the ganglionic zone. Continuous capillaries within dorsal root ganglion tissue. (**a**, **b**) Images obtained from different samples. Transverse section. Transmission electron microscopy. Magnification: **a** ×4,000; **b** ×8,000

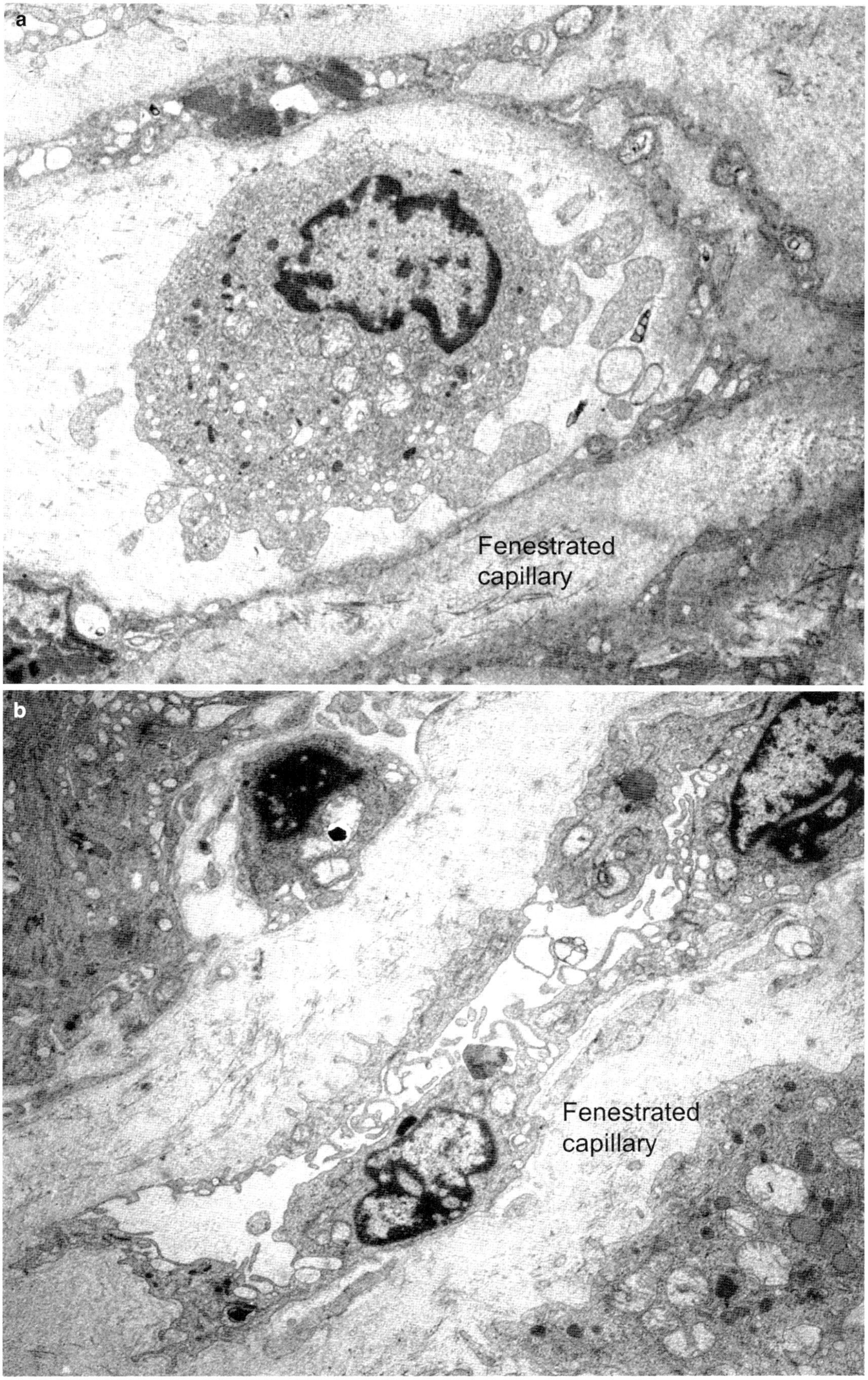

Fig. 39.16 Ultrastructure of human nerve root cuffs at the ganglionic zone. Noncontinuous capillaries (fenestrated capillaries) within dorsal root ganglion tissue. (**a**, **b**) Images obtained from different samples. Transverse section. Transmission electron microscopy. Magnification: **a** ×8,000; **b** ×8,000

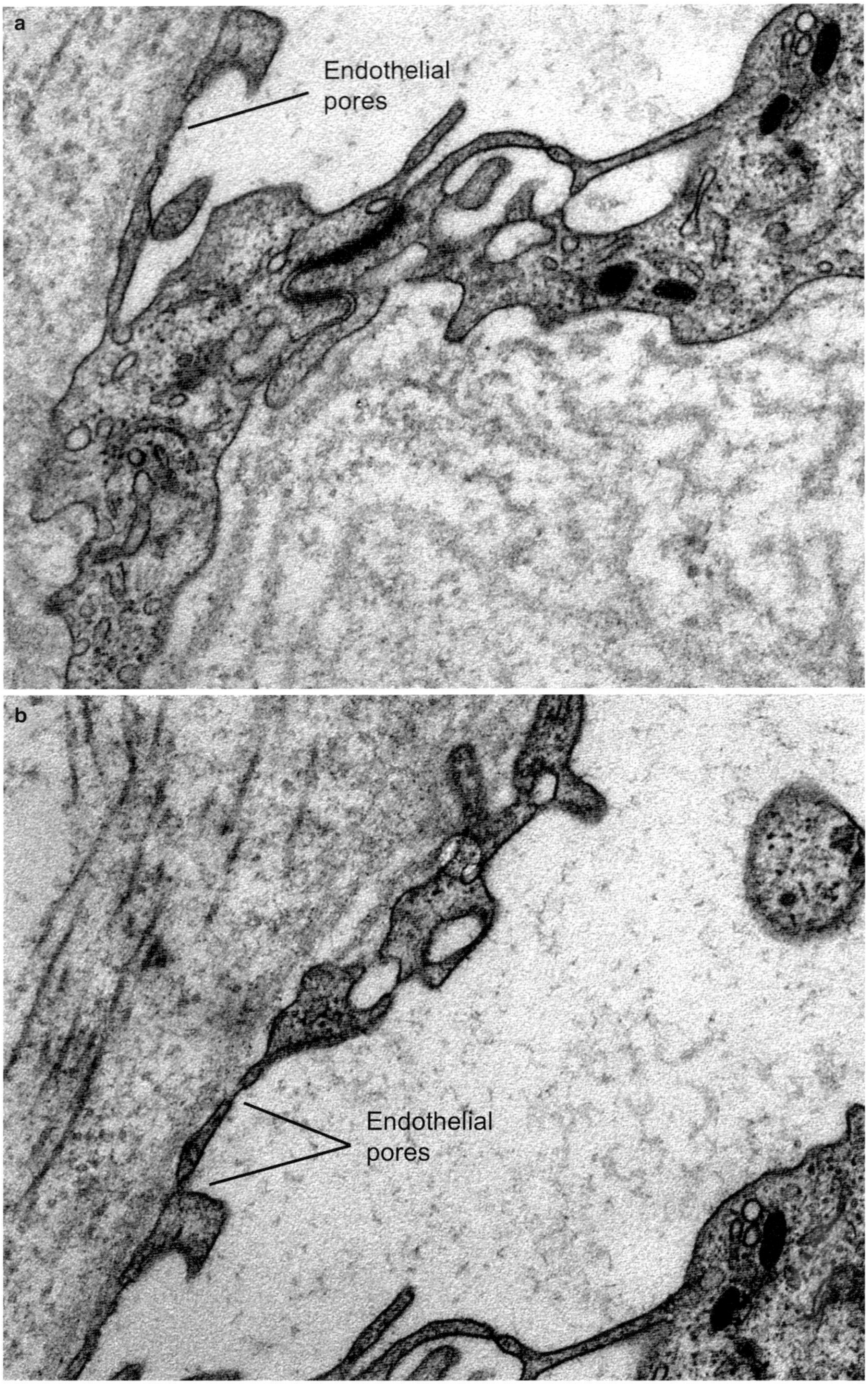

Fig. 39.17 Ultrastructure of human nerve root cuffs at the ganglionic zone. Details of noncontinuous capillaries (fenestrated capillaries) within dorsal root ganglion tissue. (**a**, **b**) Images obtained from different samples. Transverse section. Transmission electron microscopy. Magnification: **a** ×50,000; **b** ×50,000

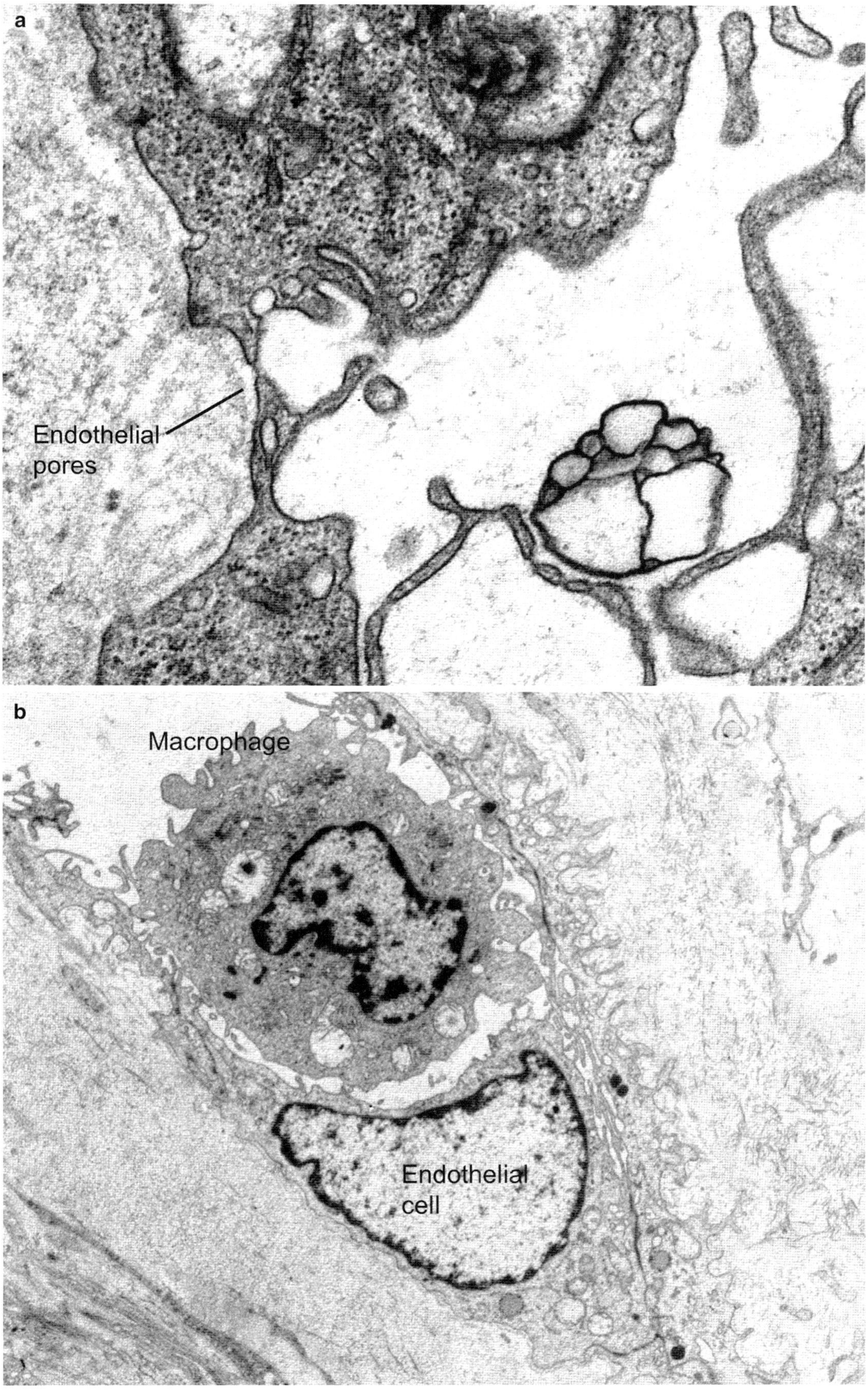

Fig. 39.18 Ultrastructure of human nerve root cuffs at the ganglionic zone. Noncontinuous capillaries (fenestrated capillaries) within dorsal root ganglion tissue. (**a**, **b**) Images obtained from different samples. Transverse section. Transmission electron microscopy. Magnification: **a** ×50,000; **b** ×8,000

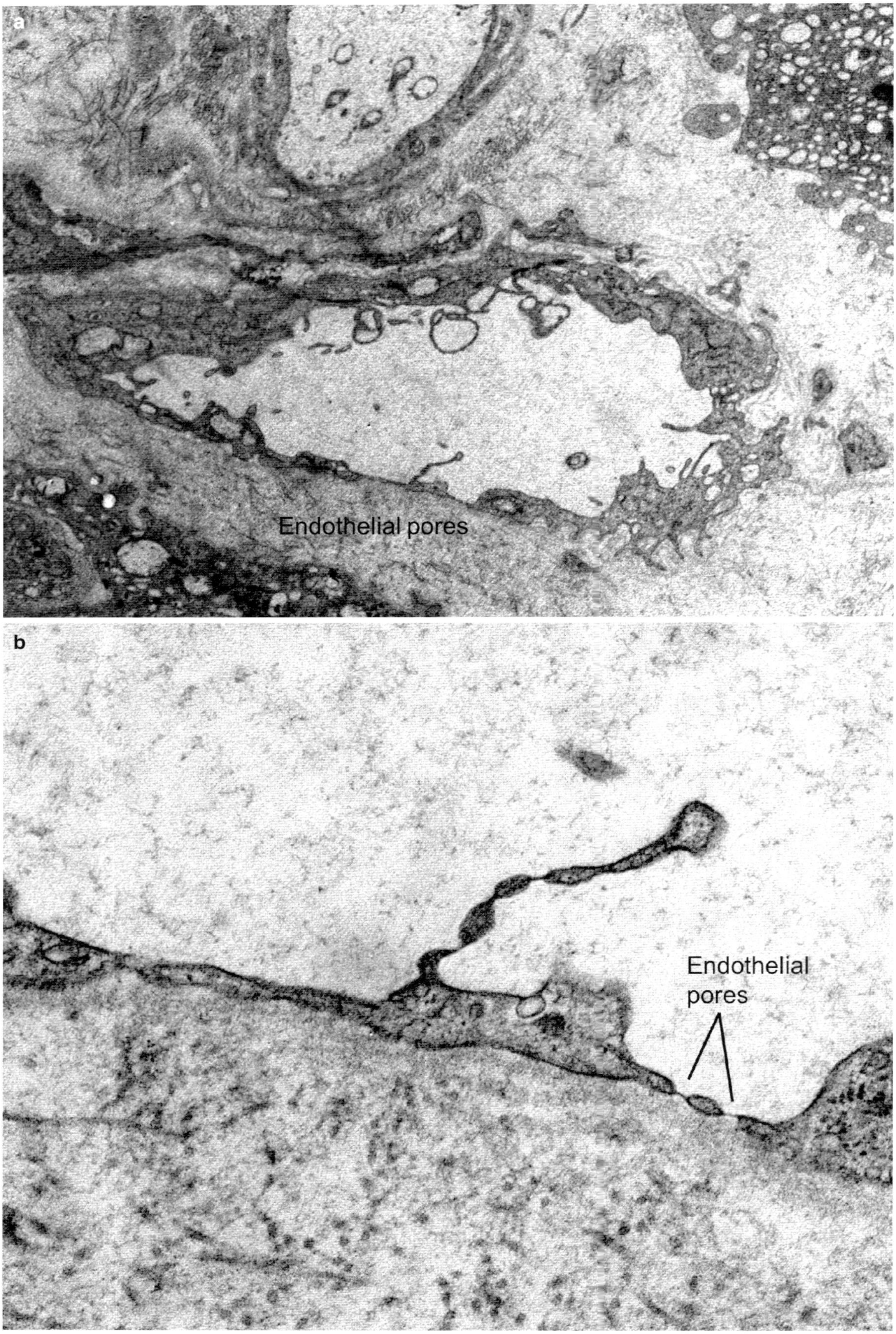

Fig. 39.19 Ultrastructure of human nerve root cuffs at the ganglionic zone. (**a**) Details of noncontinuous capillaries (fenestrated capillaries) within dorsal root ganglion tissue. (**b**) Detail of 39.19a. Transverse section. Transmission electron microscopy. Magnification: **a** ×8,000; **b** ×50,000 (**a** From Reina et al. [1]; **b** From Reina et al. [8]; both with permission)

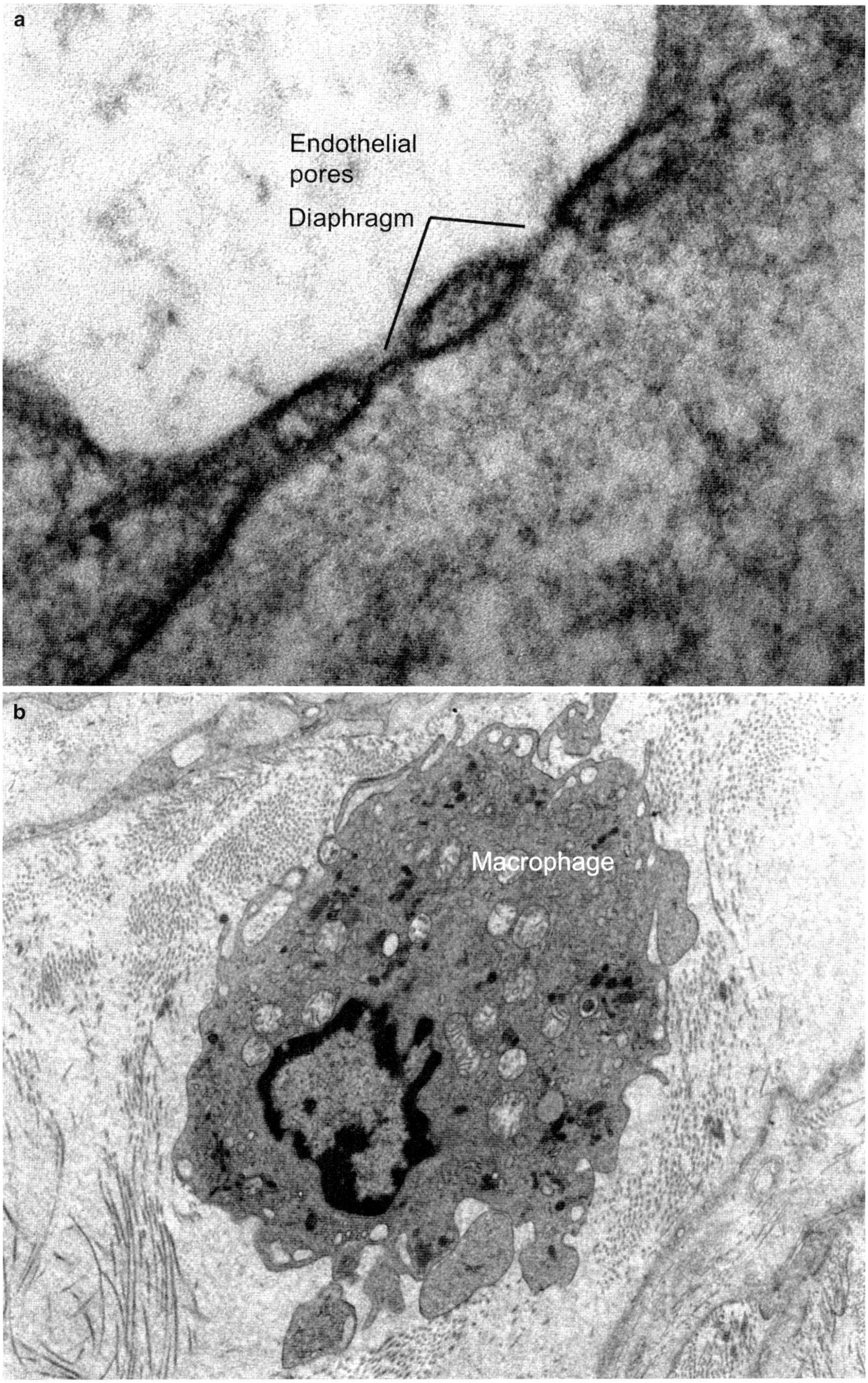

Fig. 39.20 Ultrastructure of human nerve root cuffs at the ganglionic zone. (**a**) Detail of noncontinuous capillaries of 39.19b (fenestrated capillaries) within dorsal root ganglion tissue. (**b**) Macrophage within dorsal root ganglion tissue. Transverse section. Transmission electron microscopy. Magnification: **a** ×200,000; **b** ×12,000

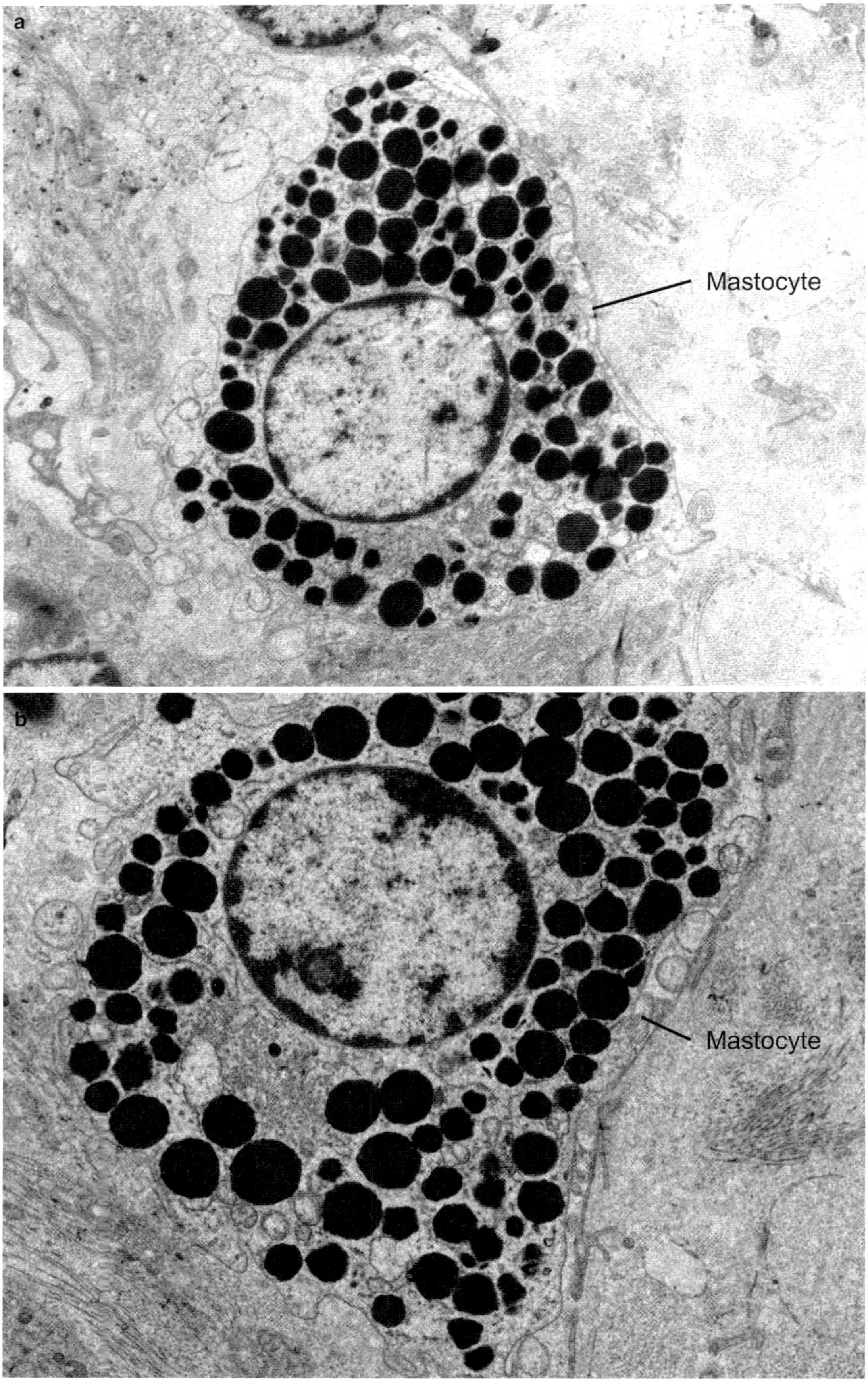

Fig. 39.21 Ultrastructure of human nerve root cuffs at the ganglionic zone. Mastocytes within dorsal root ganglion tissue. (**a**, **b**) Images obtained from different samples. Transverse section. Transmission electron microscopy. Magnification: **a** ×8,000; **b** ×10,000

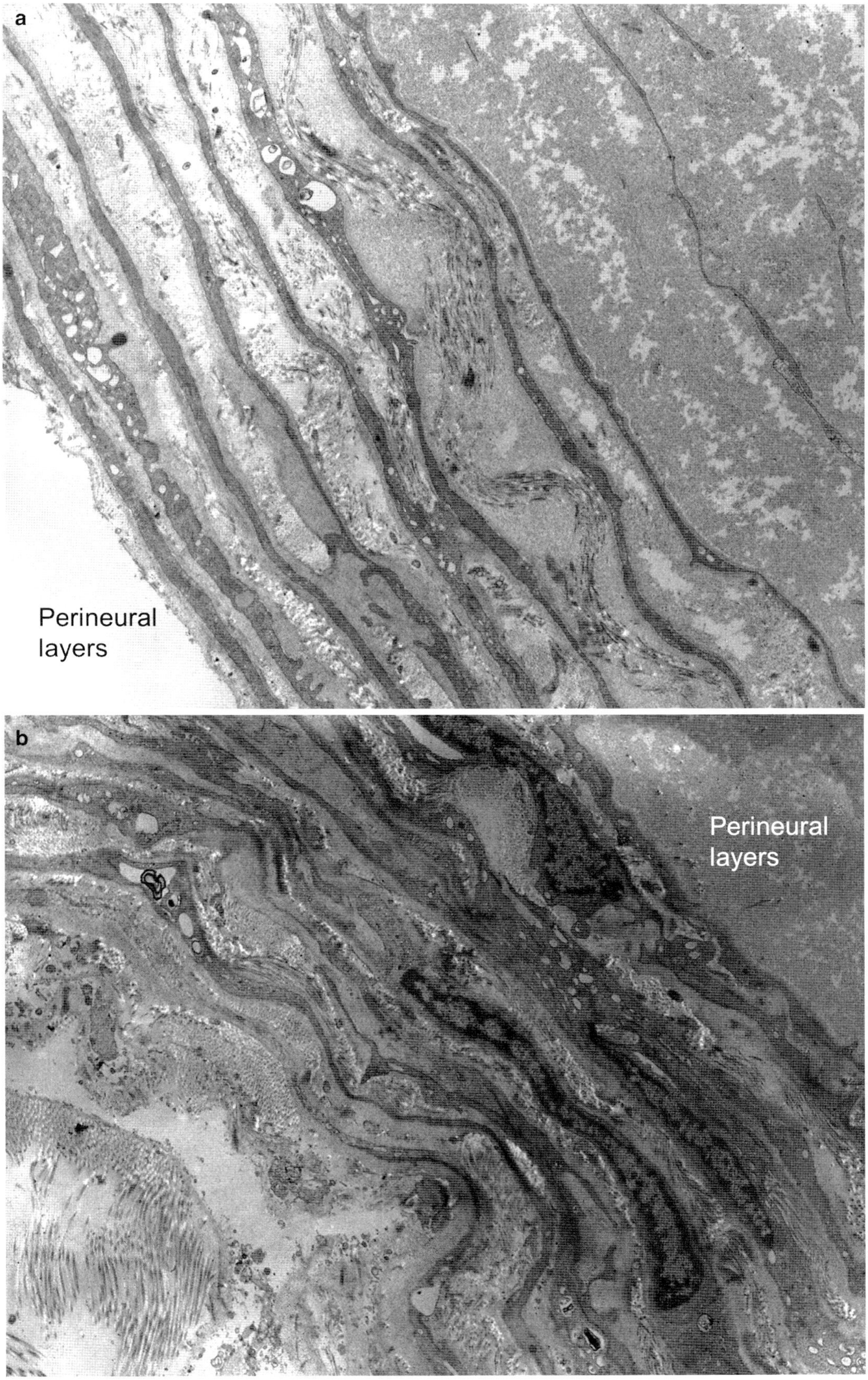

Fig. 39.22 Ultrastructure of human nerve root cuffs at the postganglionic zone. Perineurium layer. (**a**, **b**) Images obtained from different samples. Transverse section. Transmission electron microscopy. Magnification: **a** ×1,100; **b** ×1,100

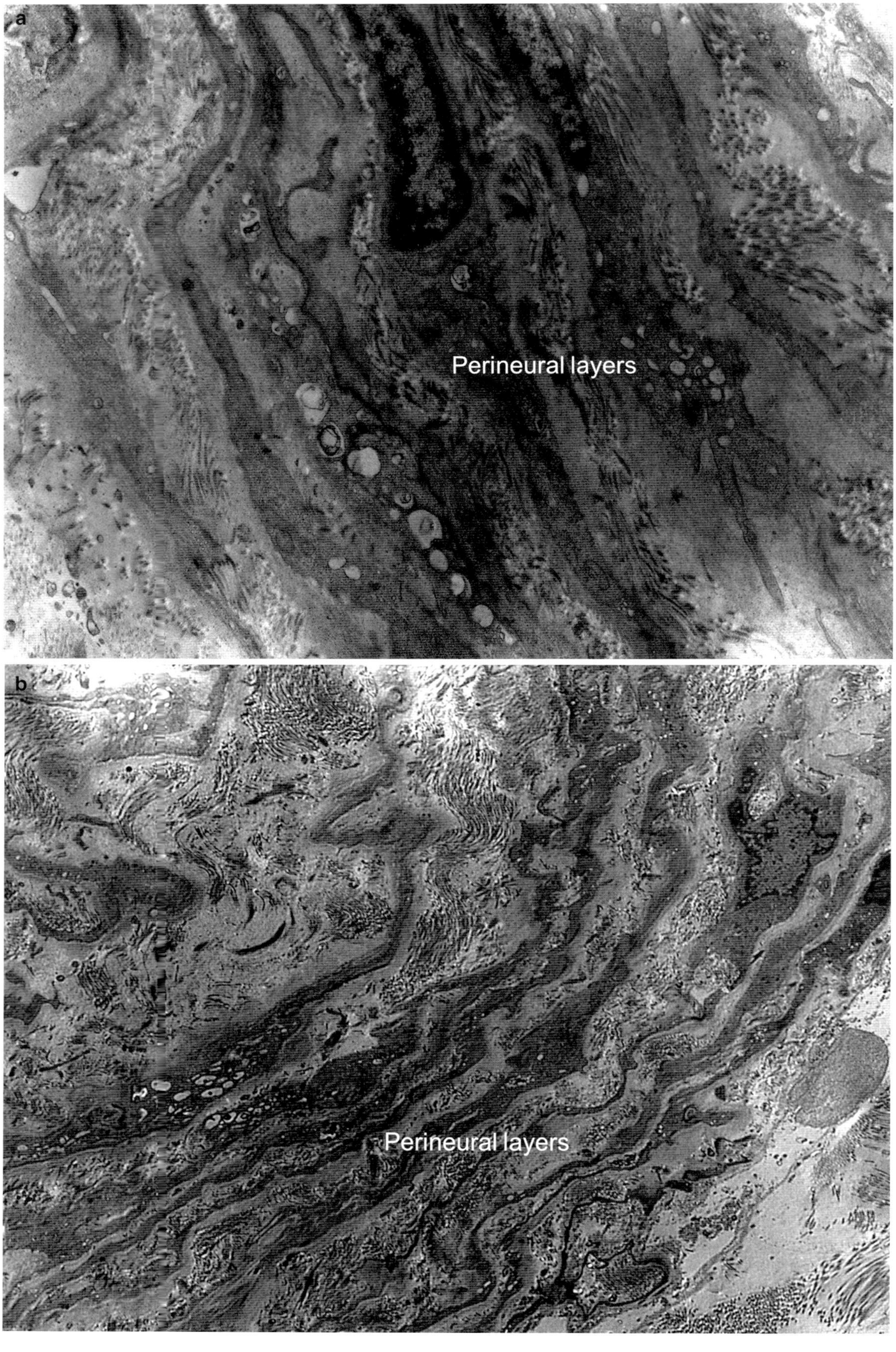

Fig. 39.23 Ultrastructure of human nerve root cuffs at the postganglionic zone. Perineurium layer. (**a**, **b**) Images obtained from different samples. Transverse section. Transmission electron microscopy. Magnification: **a** ×7,000; **b** ×1,100

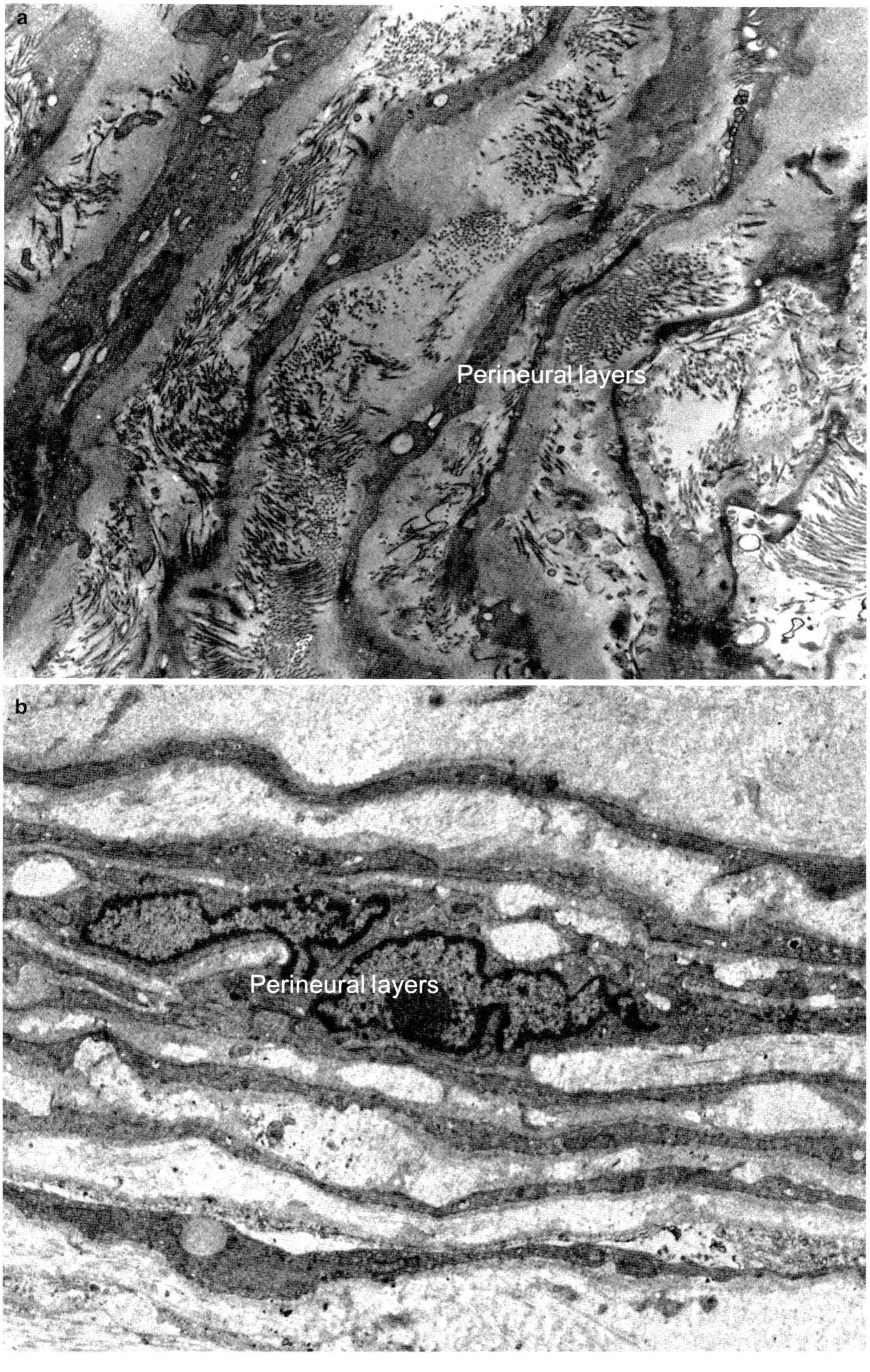

Fig. 39.24 Ultrastructure of human nerve root cuffs at the postganglionic zone. Perineurium layer. (**a**, **b**) Images obtained from different samples. Transverse section. Transmission electron microscopy. Magnification: **a** ×4,400; **b** ×8,000

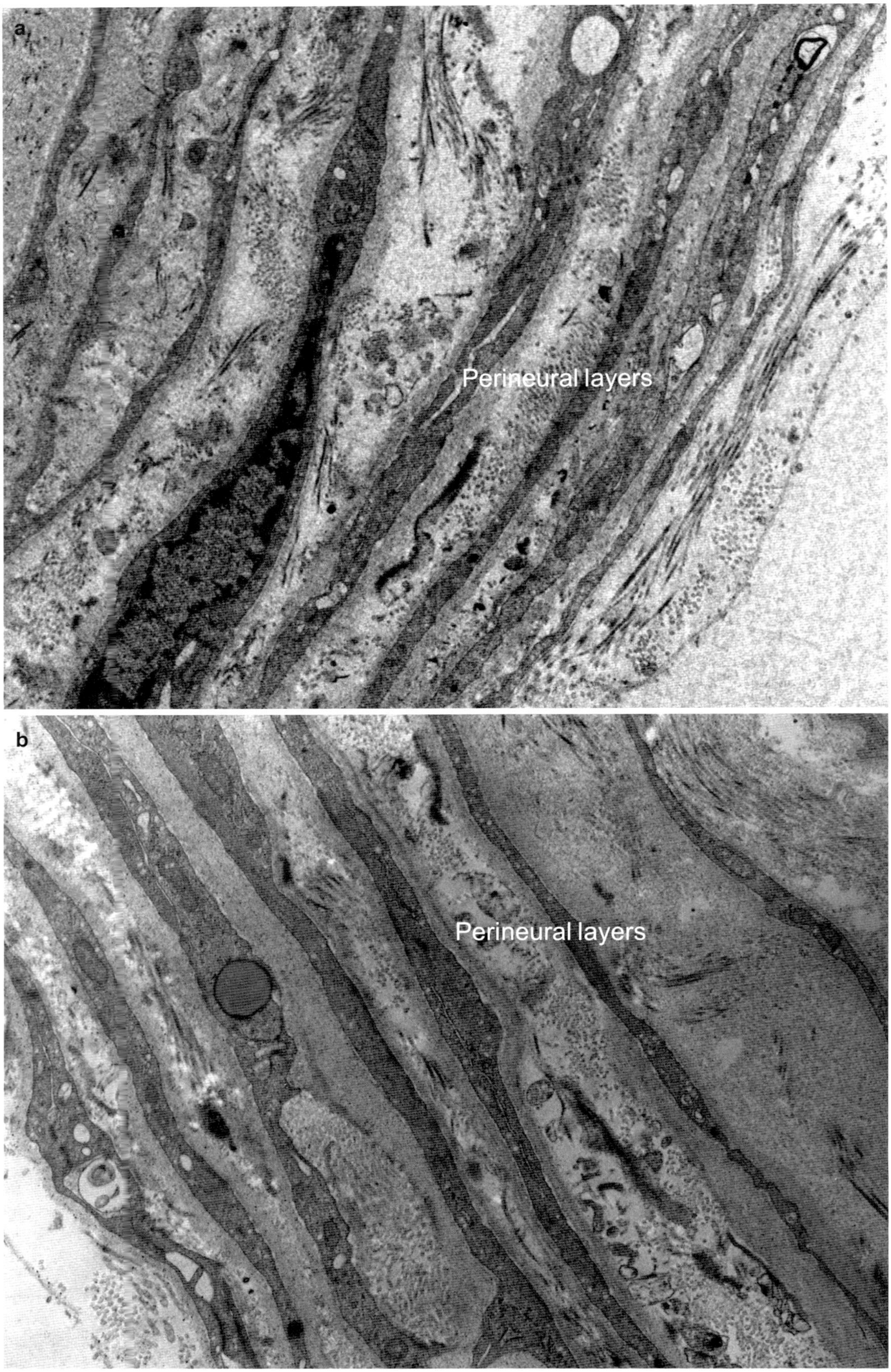

Fig. 39.25 Ultrastructure o human nerve root cuffs at the postganglionic zone. Perineurium layer. (**a**, **b**) Images obtained from different samples.Transverse section. Transmission electron microscopy. Magnification: **a** ×12,000; **b** ×7,000 (**a** From Reina et al. [1]; **b** From Reina et al. [8]; both with permission)

References

1. Reina MA, Villanueva MC, Machés F, Carrera A, López A, De Andrés JA. The ultrastructure of the human spinal nerve root cuff in the lumbar spine. Anesth Analg. 2008;106:339–44.
2. Reina MA, De Leon Casasola O, Villanueva MC, López A, Maches F, De Andrés JA. Ultrastructural findings in human spinal pia mater in relation to subarachnoid anesthesia. Anesth Analg. 2004;98:1479–85.
3. Reina MA, De León Casasola OA, López A, De Andrés JA, Mora M, Fernández A. The origin of the spinal subdural space. Ultrastructure finding. Anesth Analg. 2002;94:991–5.
4. Reina MA, Collier CB, Prats-Galino A, Puigdellívol-Sánchez A, Machés F, De Andrés JA. Accidental subdural placement of epidural catheters during attempted epidural anesthesia. An anatomic study. Reg Anesth Pain Med. 2011;36:537–41.
5. Reina MA, Prats-Galino A, Sola RG, Puigdellívol-Sánchez A, Arriazu Navarro R, De Andrés JÁ. Structure of the arachnoid layer of the human spinal meninges: a barrier that regulates dural sac permeability. Rev Esp Anestesiol Reanim. 2010;57:486–92.
6. Reina MA, Machés F, López A, De Andrés JA. The ultrastructure of the spinal arachnoid in humans and its impact on spinal anesthesia, cauda equina syndrome and transient neurological syndrome. Tech Reg Anesth Pain Manag. 2008;12:153–60.
7. Reina MA, Arriazu R, Collier CB, Sala-Blanch X. Histology and electron microscopy of human peripheral nerves of clinical relevance to the practice of nerve blocks. Rev Esp Anestesiol Reanim. 2013;60:552–62.
8. Reina MA, De Andrés JA, Hernández JM, Prats-Galino A, Machés F, Peláez J. Transforaminal or translaminar approach for dorsal root ganglion and dorsal nerve root. Anatomical reason for technique decision. Eur J Pain Suppl. 2010;4:287–97.
9. Prats A, Carrera A, Oliva A, Lusá M, Reina MA. Anatomia aplicada a las técnicas neuroaxiales. In: Basora Macaya M, Colomina Soler MJ, editors. Anestesia en cirugía ortopédica y en traumatología. Madrid: Editorial Médica Panamericana; 2011. p. 33–45.

Spinal Cord Stimulation

40

José Antonio De Andrés, Miguel Angel Reina, and José María Hernández

Spinal cord stimulation has been used for over 30 years to treat different pain conditions, and has been proved effective in neuropathic or sympathetically mediated pain states [1]. The postulated mechanism of action is selective activation of the large-diameter afferent fibers through electrical stimulation. To obtain stimulation, an array of metal contacts is placed in the dorsal epidural space. A pulse generator supplies the energy, and the contacts are programmed to generate an electric field, which stimulates the axons of dorsal root and dorsal column fibers in the spinal column. The result of the stimulation of dorsal columns is an inhibition of activity in the lateral spinothalamic tract, as well as increased activity in descending antinociceptive pathways.

The electric field generated by the metal contacts is influenced by the morphology of the different structures [2–6] that lay between the contacts and the axons, their thickness, and their electric conductivity [7].

Stimulation of the large, myelinated afferent fibers can occur in four different areas of the spinal cord: the dorsal root, the dorsal root entry zone, the dorsal horn, and the dorsal columns. Stimulation of the dorsal root can be achieved if the electrode is placed laterally in the spinal canal. It may be difficult to differentiate this stimulation from stimulation of the dorsal root entry zone, the dorsal horn, or both [8]. Early recruitment of the segmentary motor fibers (from spread of the current through the cerebrospinal fluid to the anterior roots) associated with sensory paresthesias may be a sign of stimulation of the root filaments. Stimulation of the longitudinal fibers of the dorsal columns is characterized by paresthesias occurring in areas of the body caudal to the level of the electrode; these paresthesias are always ipsilateral to the electrode [8].

Two types of leads are available for spinal cord stimulation: percutaneous leads and paddle leads.

Percutaneous Leads

The percutaneous leads are cylindrical and flexible, showing similar features to catheters. They can be introduced into the spinal column via an epidural needle. The contacts on the leads are also cylindrical and are made of platinum-iridium alloys, although there is a new model with bigger, flat contacts. The lead is made up of an isodiametric polyurethane body containing wires that connect the contacts to a proximal connector, and then to the battery [3].

Percutaneous leads have four or eight contacts; there are models with different distances between contacts. The contacts on the leads are cylindrical and most commonly 3–3.5 mm in length. Multiple contacts have the advantage of recapturing stimulation areas that can be lost in the postoperative phase owing to migration of the lead [3].

J.A. De Andrés, MD, PhD (✉)
Department of Surgical Specialties,
School of Medicine, University of Valencia,
Valencia, Spain

Department of Anesthesia, Critical Care, and Pain Management,
General University Hospital Valencia, Valencia, Spain
e-mail: deandres_jos@gva.es

M.A. Reina, MD, PhD
Department of Clinical Medical Sciences and Institute of Applied Molecular Medicine, School of Medicine,
University of CEU San Pablo, Madrid, Spain

Department of Anesthesiology, Madrid-Montepríncipe University Hospital, Madrid, Spain
e-mail: miguelangel@perticone.e.telefonica.net

J.M. Hernández, MD
Department of Anesthesiology, Madrid-Montepríncipe University Hospital, Madrid, Spain

Department of Clinical Medical Sciences,
School of Medicine, CEU San Pablo University,
Madrid, Spain
e-mail: jhzamora@yahoo.es; jmhernandez@madridsindolor.com

M.A. Reina et al. (eds.), *Atlas of Functional Anatomy for Regional Anesthesia and Pain Medicine: Human Structure, Ultrastructure and 3D Reconstruction Images*, DOI 10.1007/978-3-319-09522-6_40,

Paddle Leads

Modern paddle leads also have multiple contacts, ranging in number from 2 to 20, which are arranged in single or multiple columns. Paddle leads, which are constructed of a flexible silicone, are much wider than percutaneous leads and are flat, with round or rectangular plate contacts placed on one side of the flat portion. Paddle leads have been shown to achieve better technical results than percutaneous leads. The primary reason for the improved stimulation appears to be mechanical, as the flat, relatively rigid shape of the paddle lead tends to compress the pliable dorsal dural sac [3].

Paddle leads require a surgical procedure, such as laminotomy or laminectomy, to introduce them into the epidural space. The types of leads differ in the number of contacts, intercontact center-to-center spacing, contact length, contact width, contact shape, intraspinal lead shape, and (for designs with multiple columns) the relative orientation of the contacts among columns [3].

Images in this chapter show shapes and positioning of contacts from percutaneous and paddle leads in relation to the dorsal dural sac and the distance from the dural sac to neighboring structures such as the spinal cord, nerve roots, and rootlets, including their sizes and locations. To allow visualization of internal structures, the dural sac was removed and percutaneous and paddle leads were placed directly on the surface of the spinal cord, nerve roots, and rootlets (Figs. 40.1, 40.2, 40.3, 40.4, 40.5, 40.6, 40.7, 40.8, 40.9, 40.10, 40.11, 40.12, 40.13, 40.14, and 40.15).

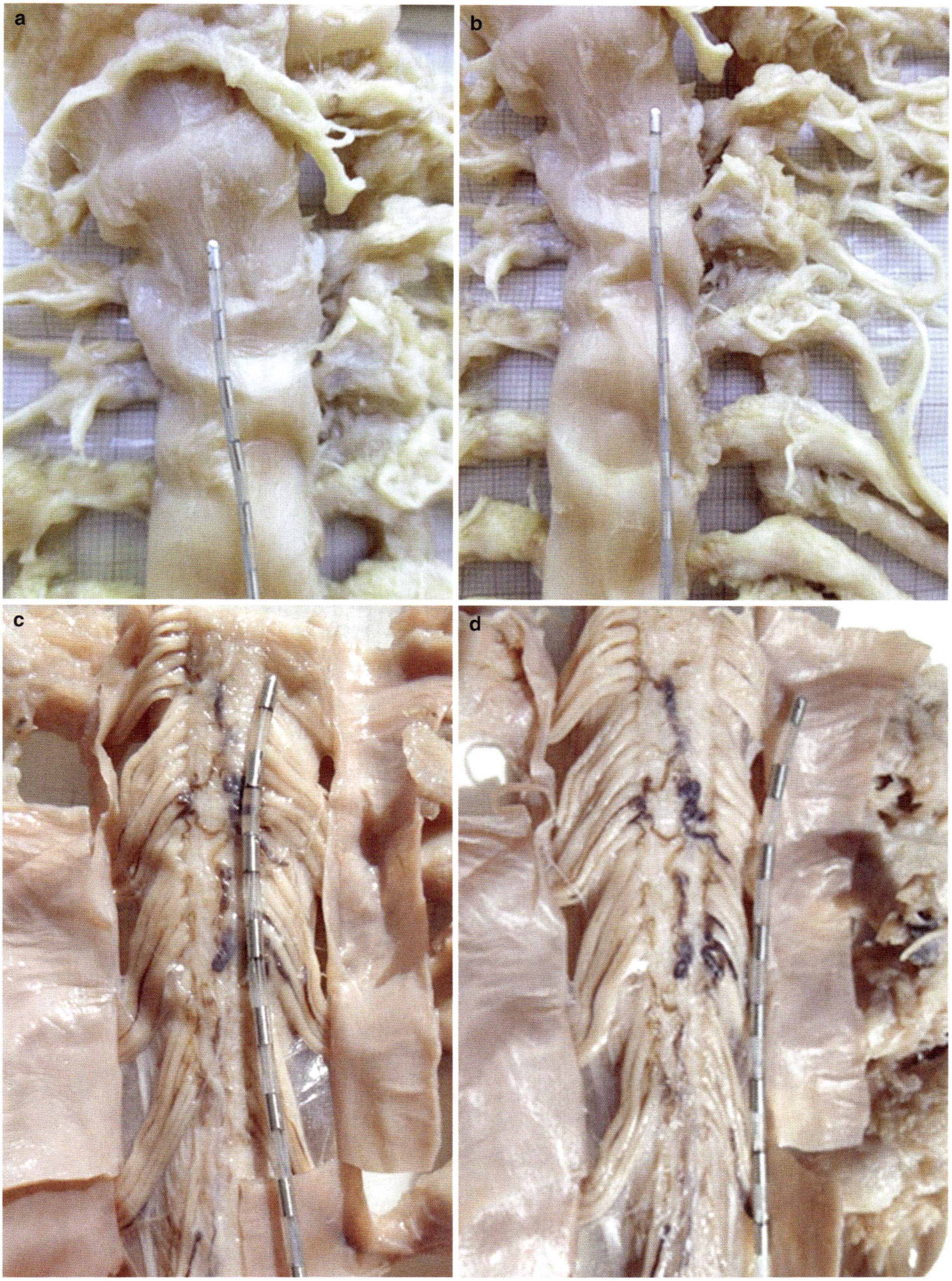

Fig. 40.1 Percutaneous leads located over a human dorsal dural sac at the cervical level. (**a**, **b**) Positioning of percutaneous lead on the dural sac. (**c**, **d**) The dural sac was removed, showing the posterior surface of the spinal cord. The area where the lead lies in direct contact with the spinal cord and rootlets is the site where the electrical field is generated, spreading across the dural sac

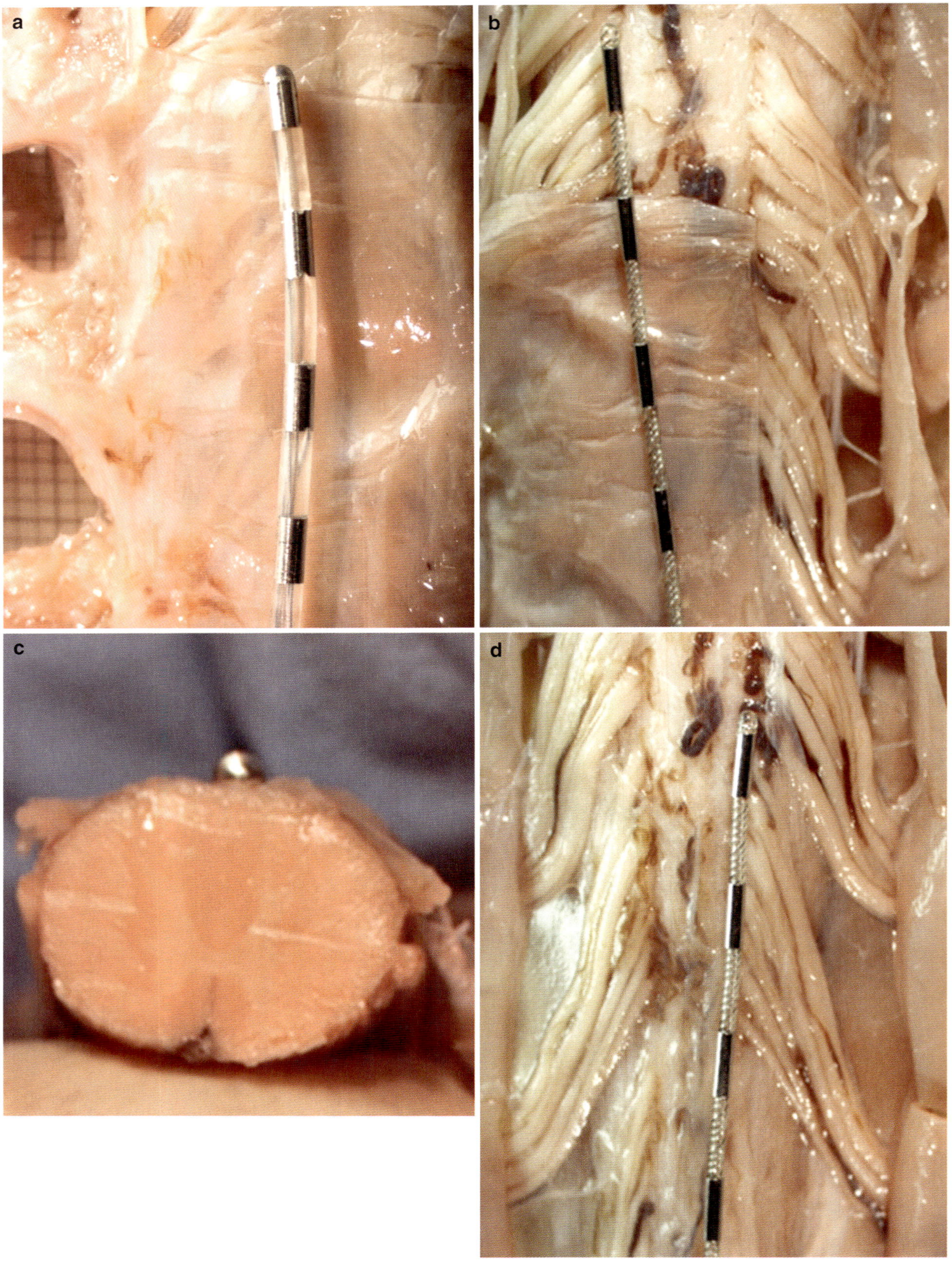

Fig. 40.2 Percutaneous leads located over a human dorsal dural sac at the cervical level. (**a**, **b**) Positioning of percutaneous lead on the dural sac. (**c**) Transversal section of spinal cord with the lead on the posterior surface of the dural sac. (**d**) The dural sac was removed, showing the posterior surface of the spinal cord. The area where the lead lies in direct contact with the spinal cord and rootlets is the area where the electrical field is generated and spread across the dural sac

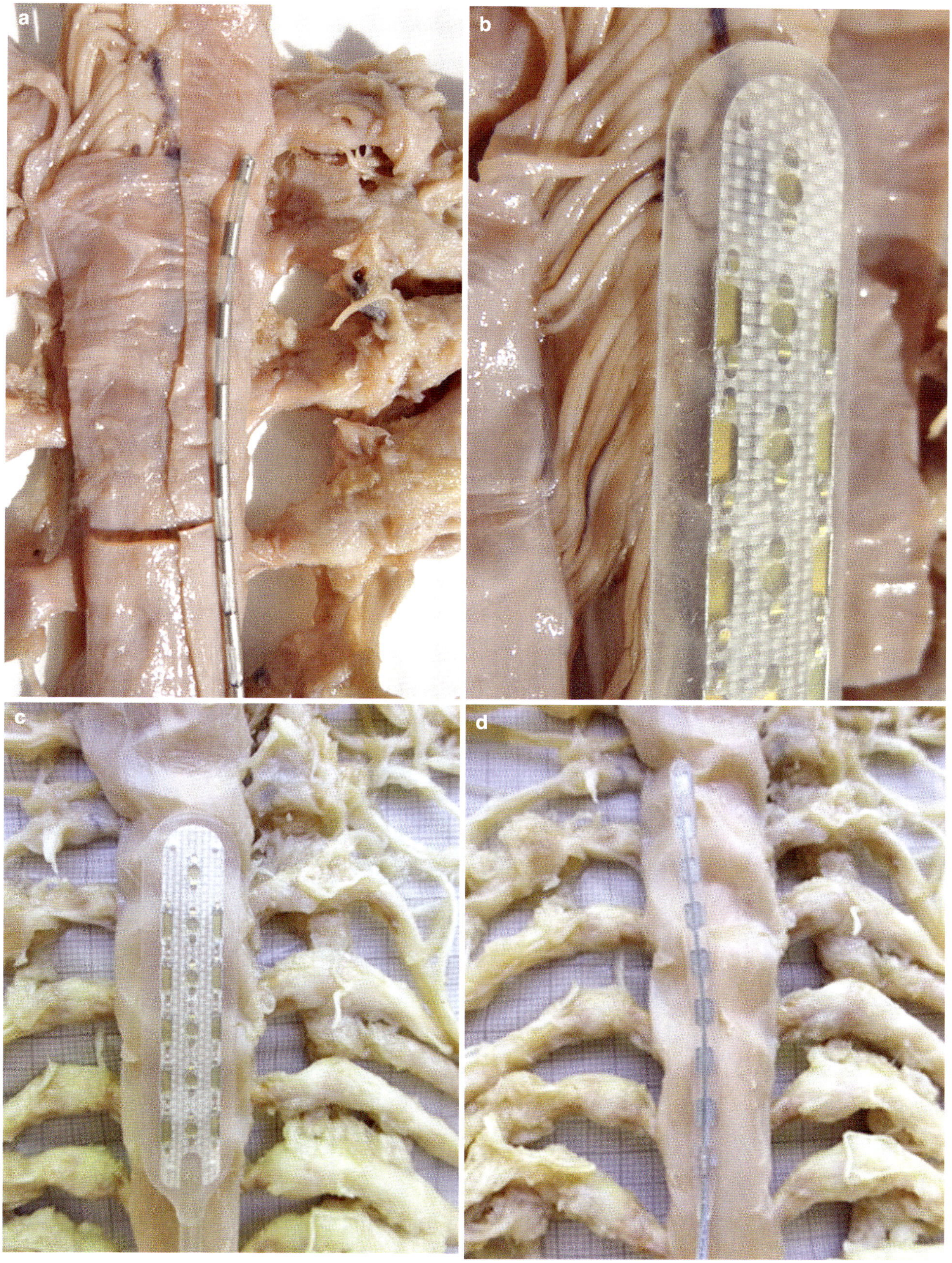

Fig. 40.3 Percutaneous leads and paddle leads located over a human dorsal dural sac at the cervical level. (**a**, **d**) Different designs of percutaneous leads. (**b**, **c**) Different designs of paddle leads. The size of the dural sac and the size occupied by the leads can be identified

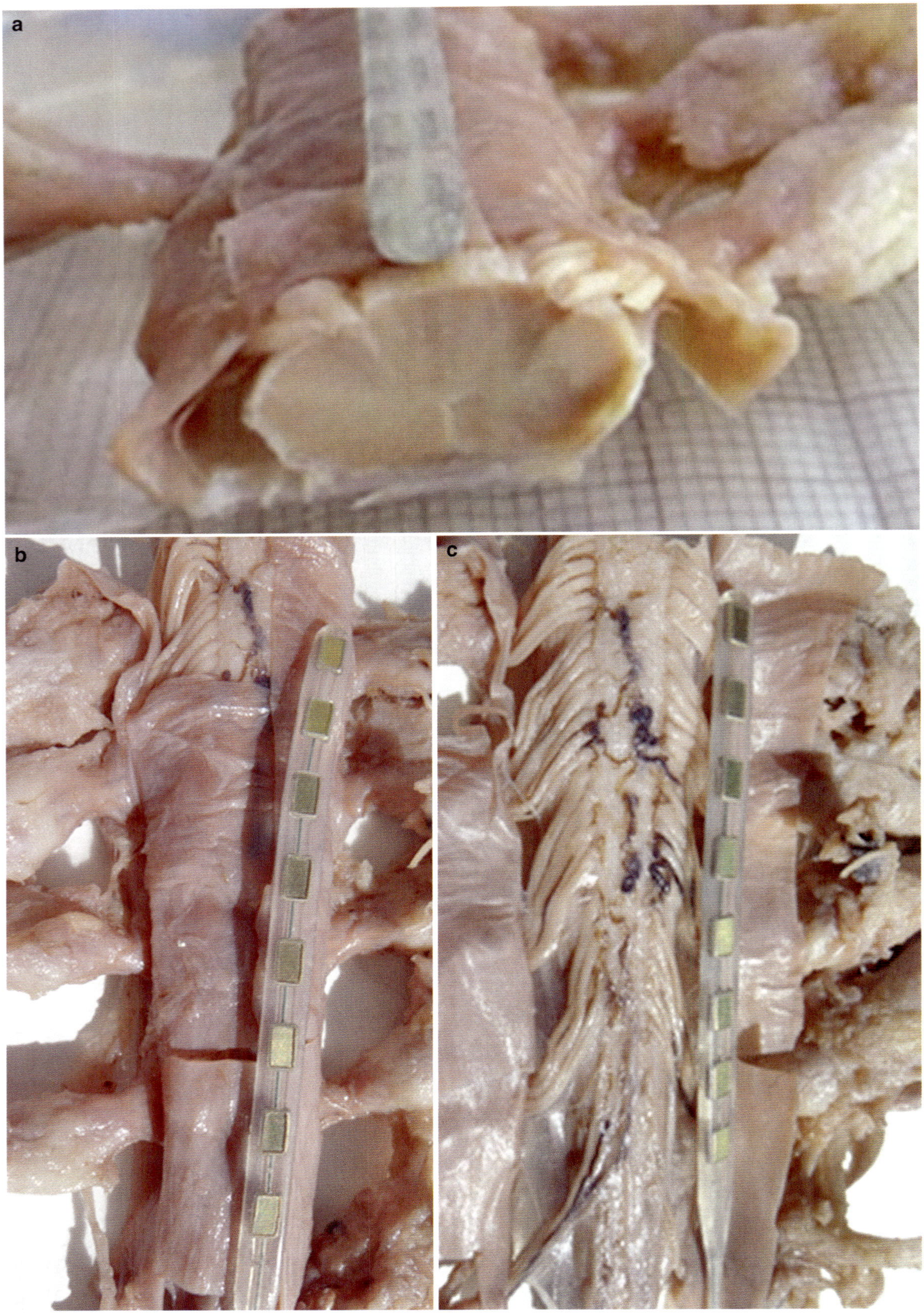

Fig. 40.4 Percutaneous leads and paddle leads located over a human dorsal dural sac at the cervical level. (**a**) Paddle leads of two columns. (**b**, **c**) Percutaneous leads above the dural sac and internal structures affected by the electrical field below the dural sac

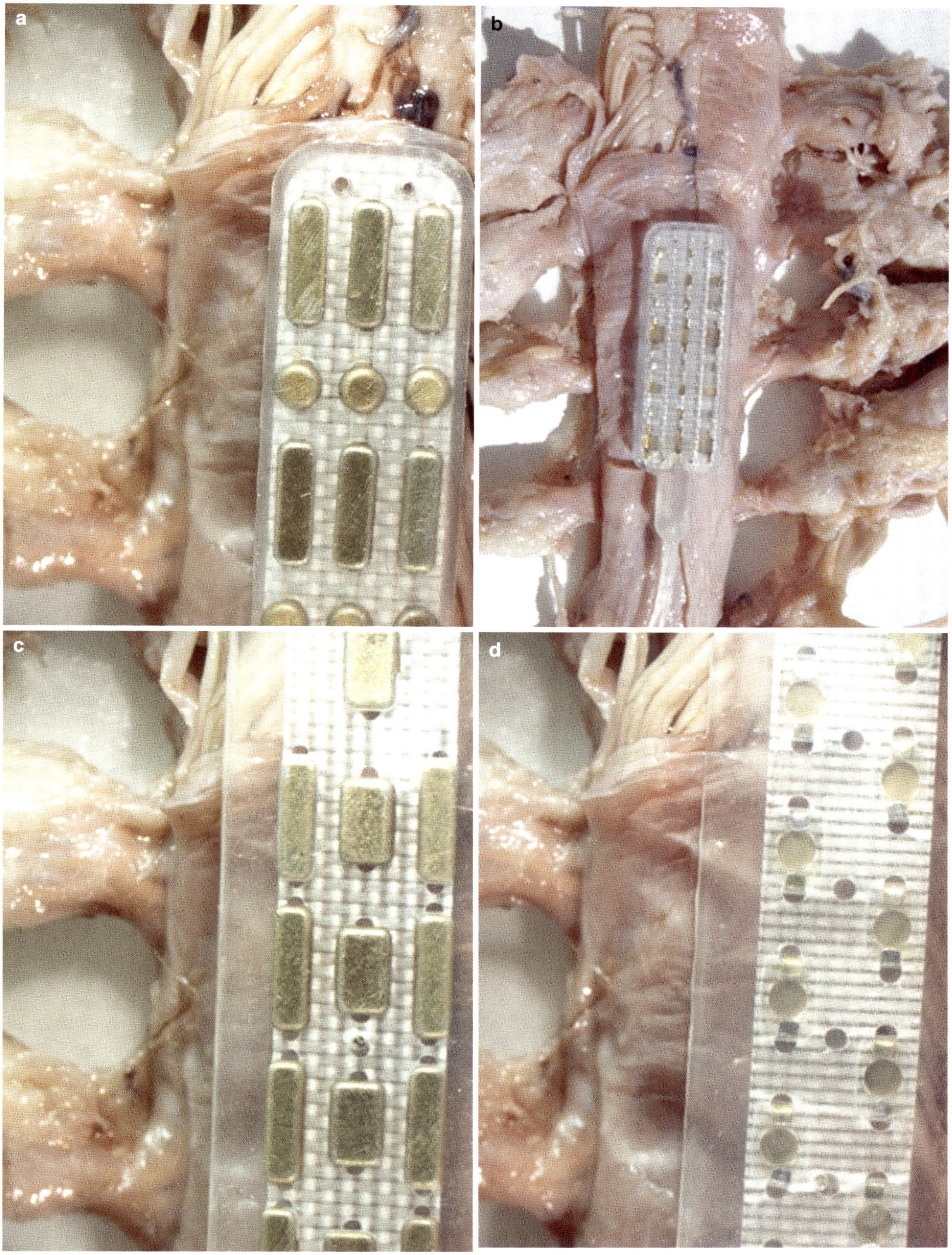

Fig. 40.5 Paddle leads located over a human dorsal dural sac at the cervical level. The size of the dural sac and the size occupied by the leads can be identified. (**a, b**) Same paddle lead, (**a**) The paddle lead is inverted to show the electrodes (**b**) The paddle lead is positioned correctly. (**c, d**) Same paddle lead with another design, (**c**) The paddle lead is inverted to show the electrodes (**d**) The paddle lead is positioned correctly. (From De Andrés et al. [3]; with permission)

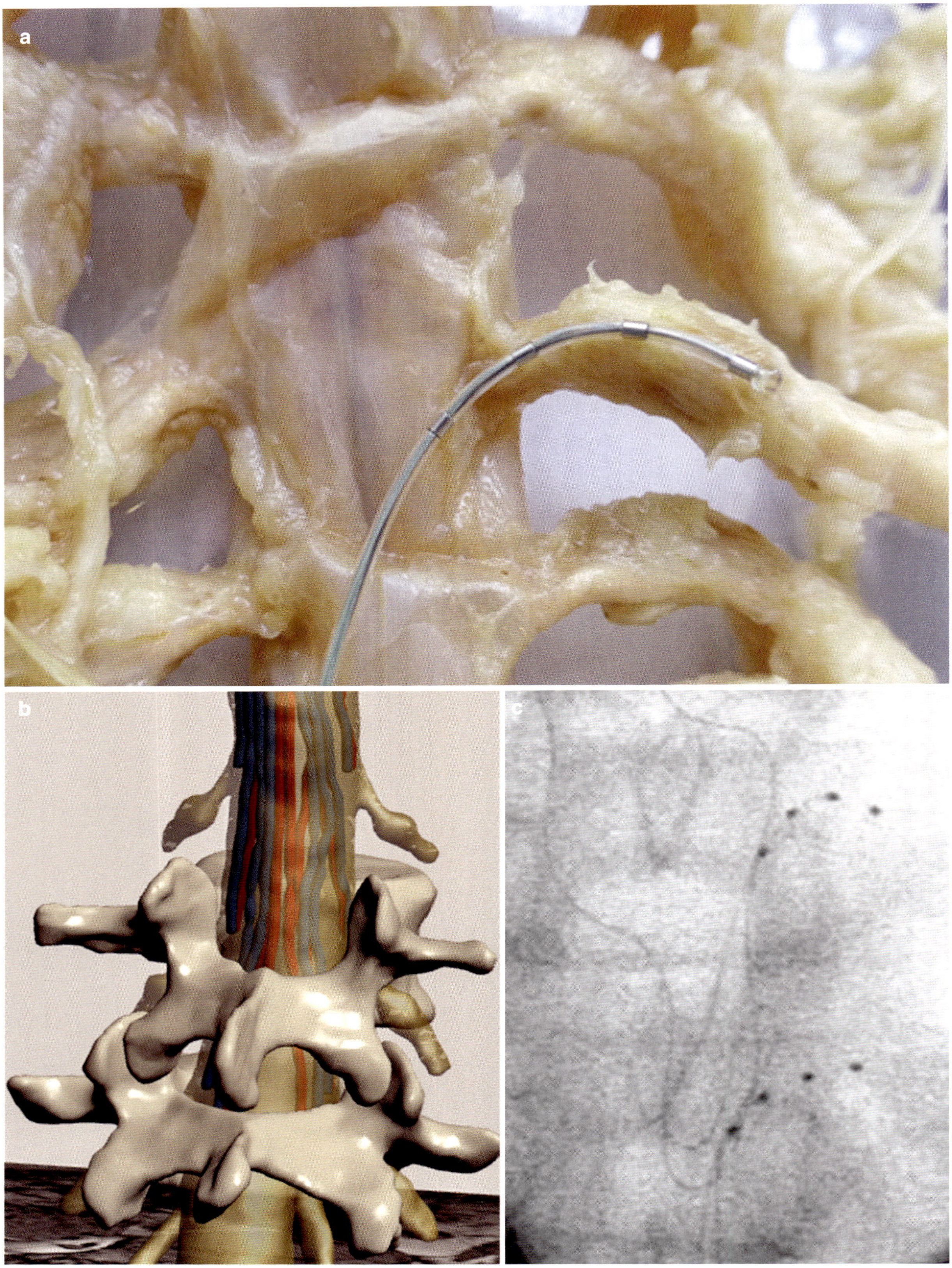

Fig. 40.6 Dorsal root ganglion stimulation at the cervical level. (**a**) The contacts of the percutaneous lead are close to the dorsal root ganglion. (**b**) Three-dimensional (3D) reconstruction from MRI of a patient. (**c**) Radiograph during placement of a percutaneous lead for dorsal root ganglion stimulation

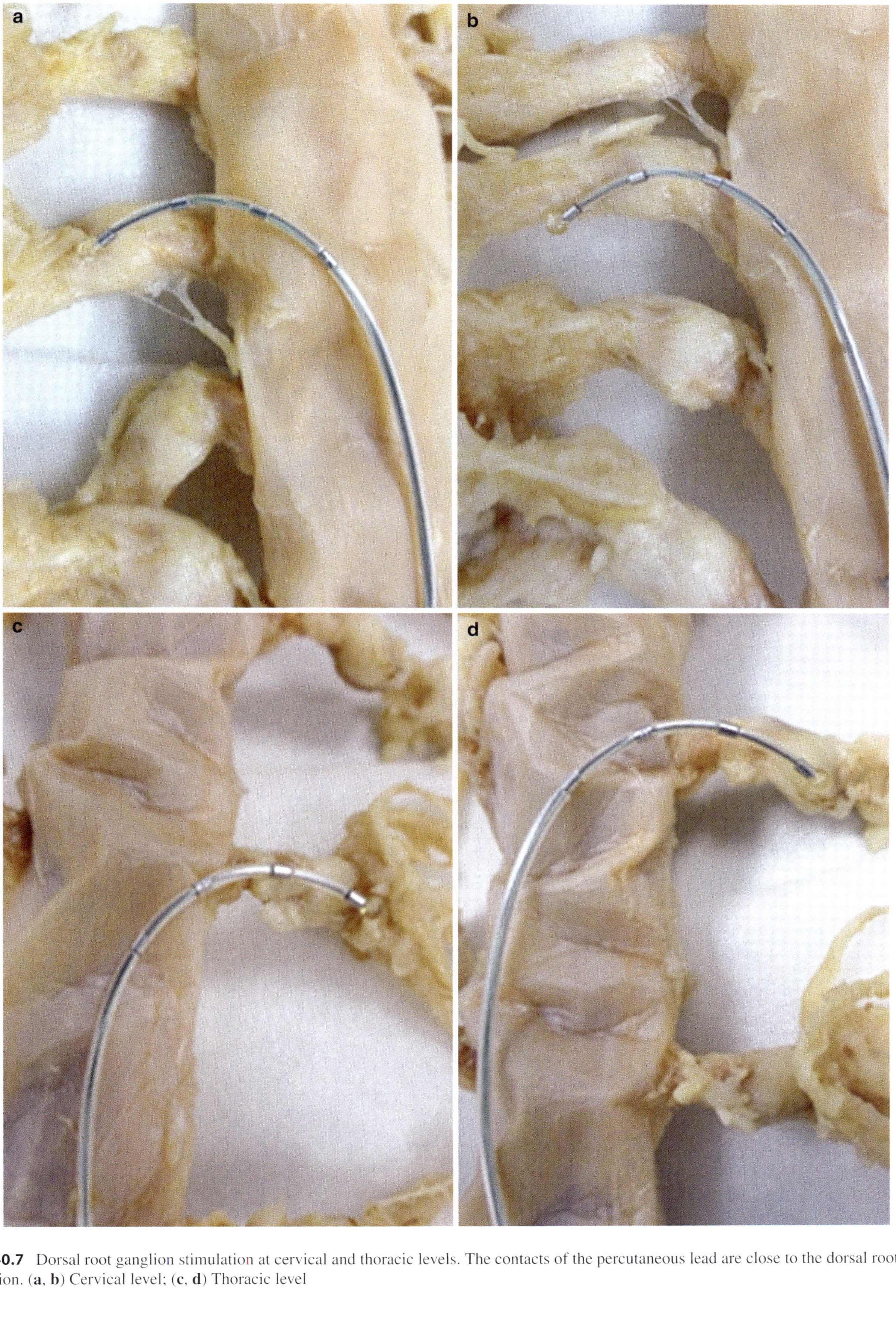

Fig. 40.7 Dorsal root ganglion stimulation at cervical and thoracic levels. The contacts of the percutaneous lead are close to the dorsal root ganglion. (**a**, **b**) Cervical level; (**c**, **d**) Thoracic level

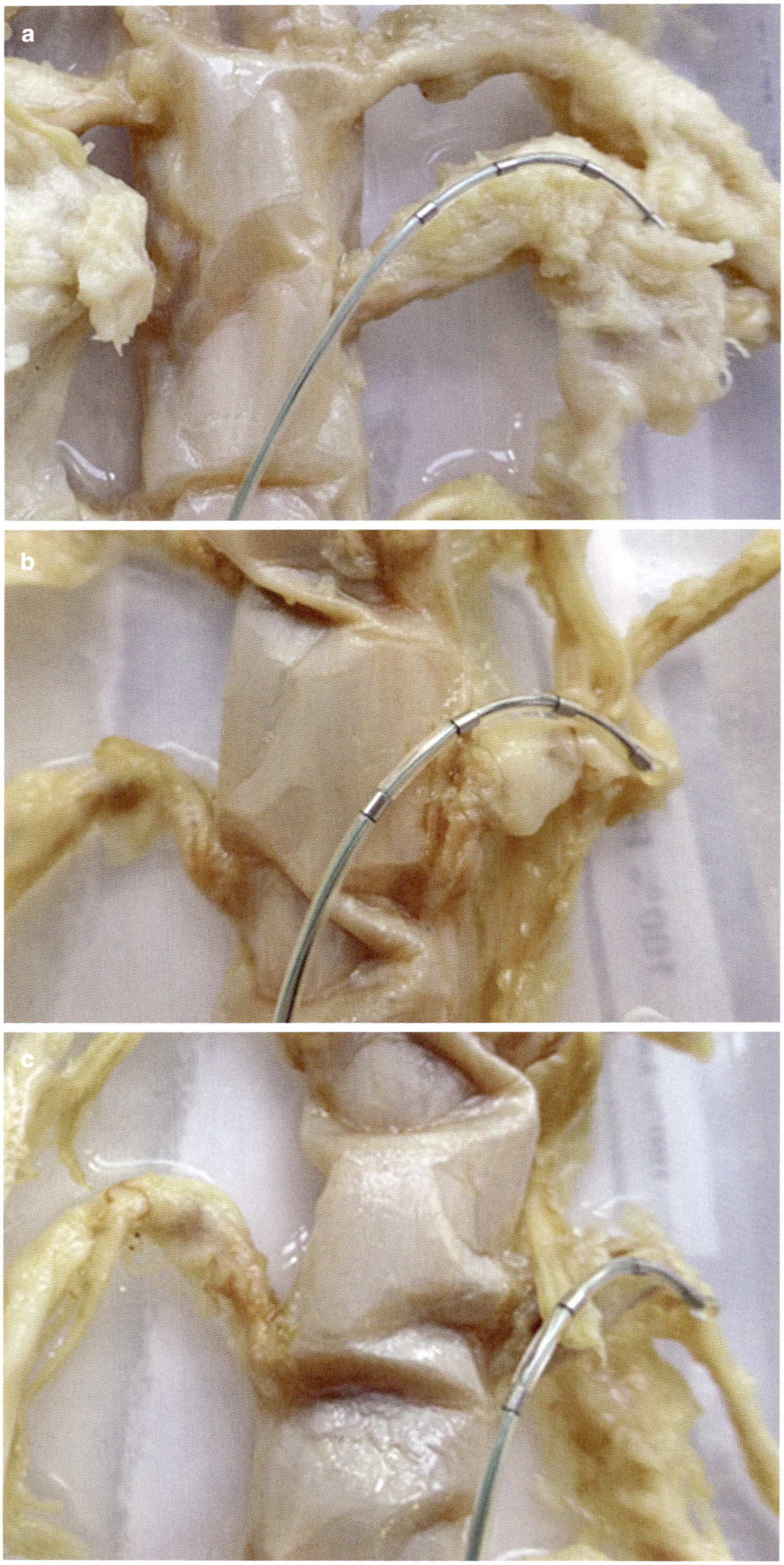

Fig. 40.8 Dorsal root ganglion stimulation at cervical and thoracic levels. The contacts of the percutaneous lead are close to the dorsal root ganglion. (**a**) Cervical level; (**b**, **c**) Thoracic level

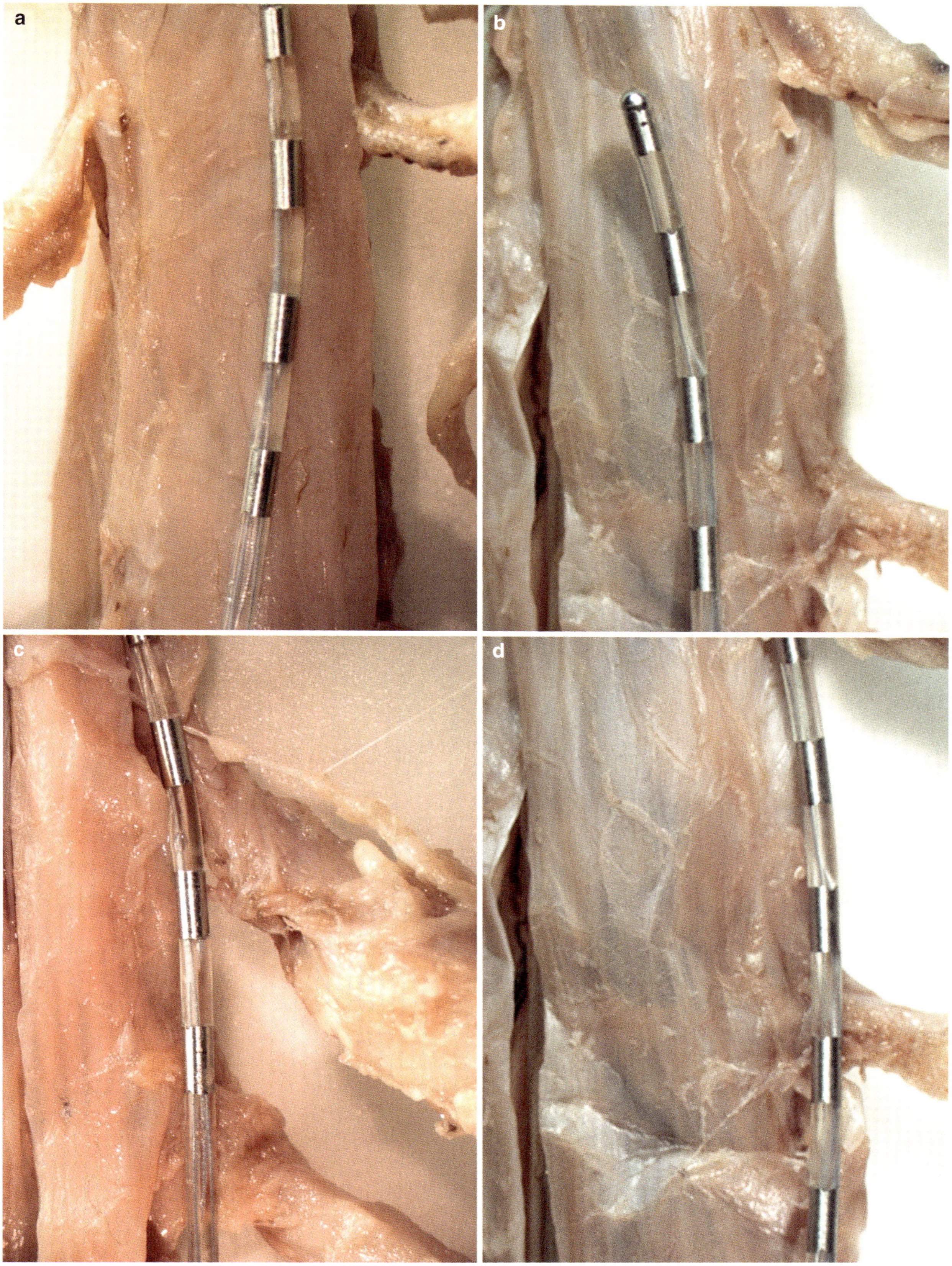

Fig. 40.9 Percutaneous leads located over a human dorsal dural sac at the thoracic level. (**a**, **b**) Positioning of percutaneous lead over the dural sac, showing correct placement. (**c**, **d**) Inadequate placement of percutaneous leads

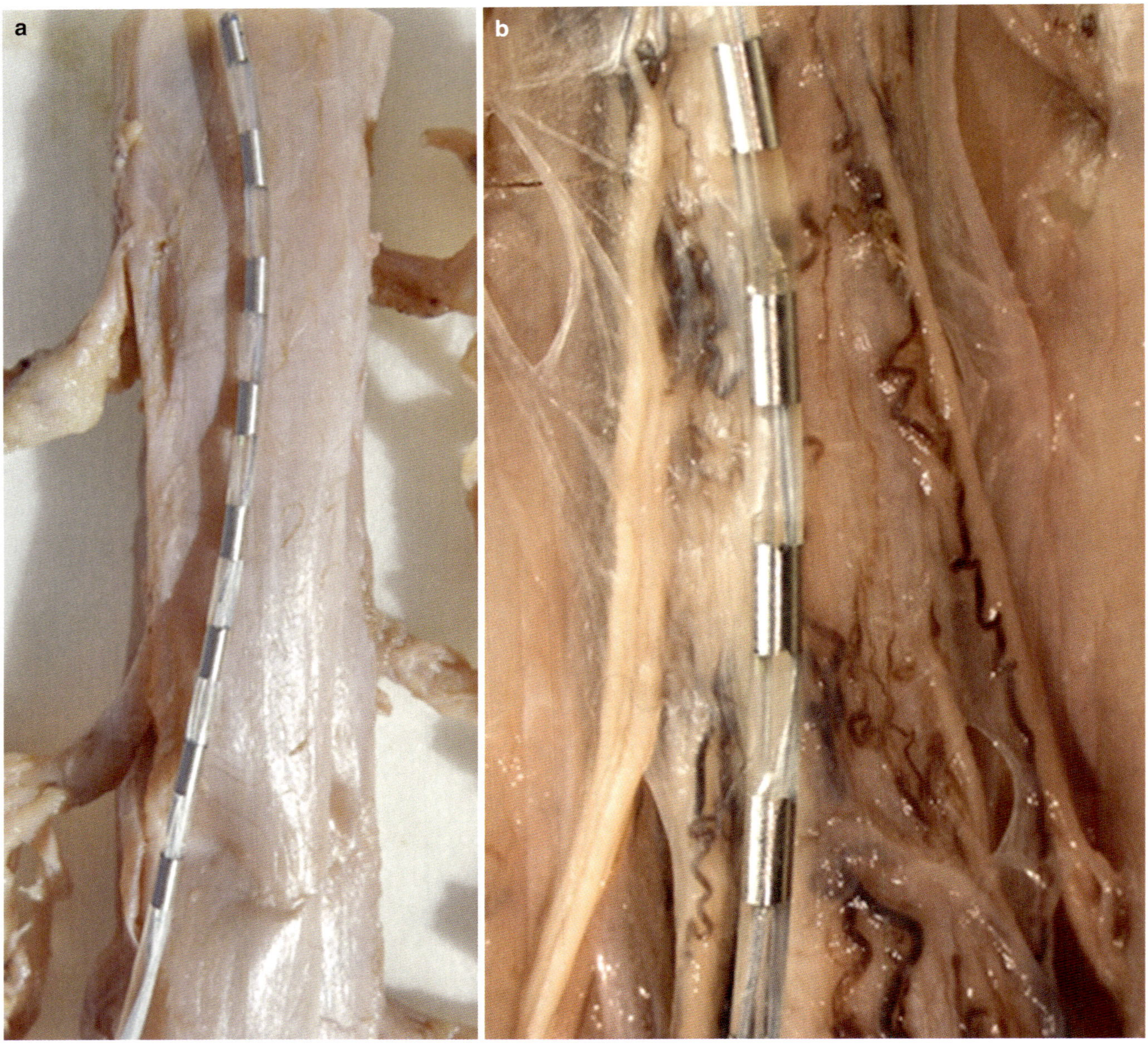

Fig. 40.10 Percutaneous leads with eight contacts located over human dural sac at the thoracic level. (**a**) The lead on the dural sac. (**b**) Internal structures affected by the electrical field

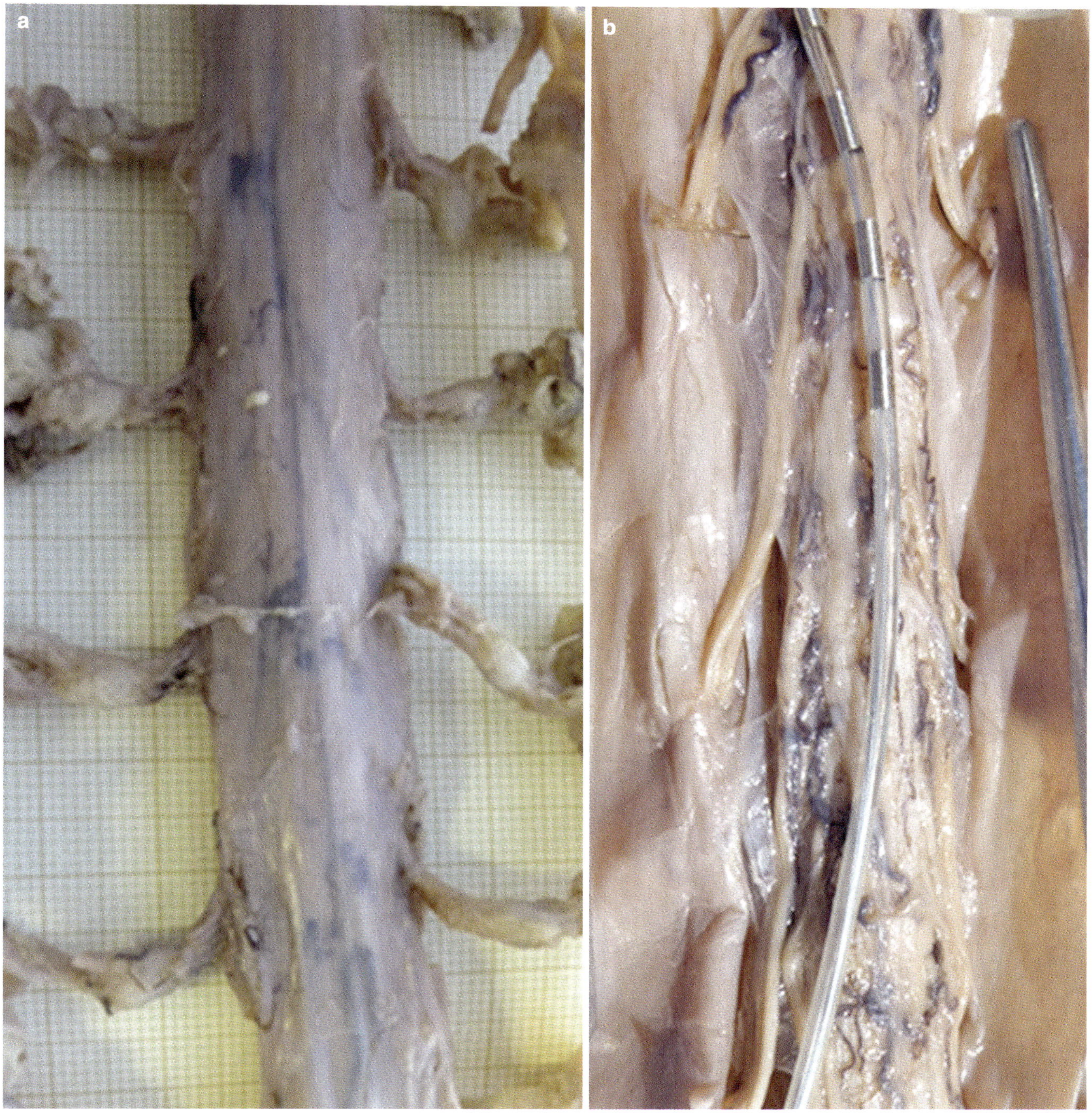

Fig. 40.11 (**a**, **b**) In vitro comparative model. (**a**) External surface of the dural sac at thoracic level (where the leads are commonly placed). (**b**) In vitro model of percutaneous lead with four contacts placed after removing the human dural sac at thoracic level, showing which internal structures are affected by the electrical field

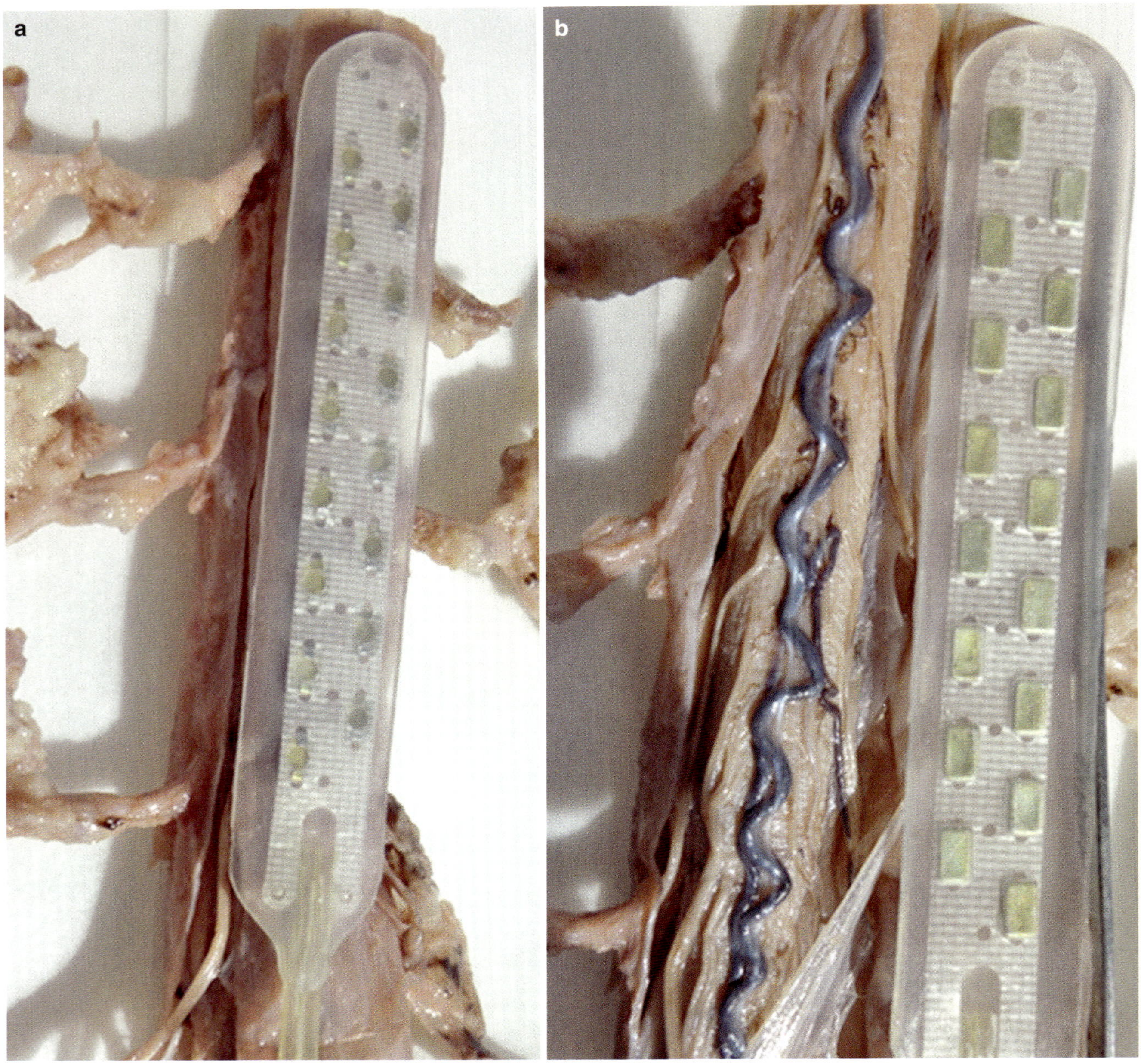

Fig. 40.12 Paddle leads located over a human dorsal dural sac at the thoracic level. Longitudinal views illustrate the sizes of the dural sac as compared with size of the paddle lead. (**a, b**) Same paddle lead, (**a**) The paddle lead is positioned correctly. (**b**) The paddle lead is inverted to show the electrodes (From De Andrés et al. [3]; with permission)

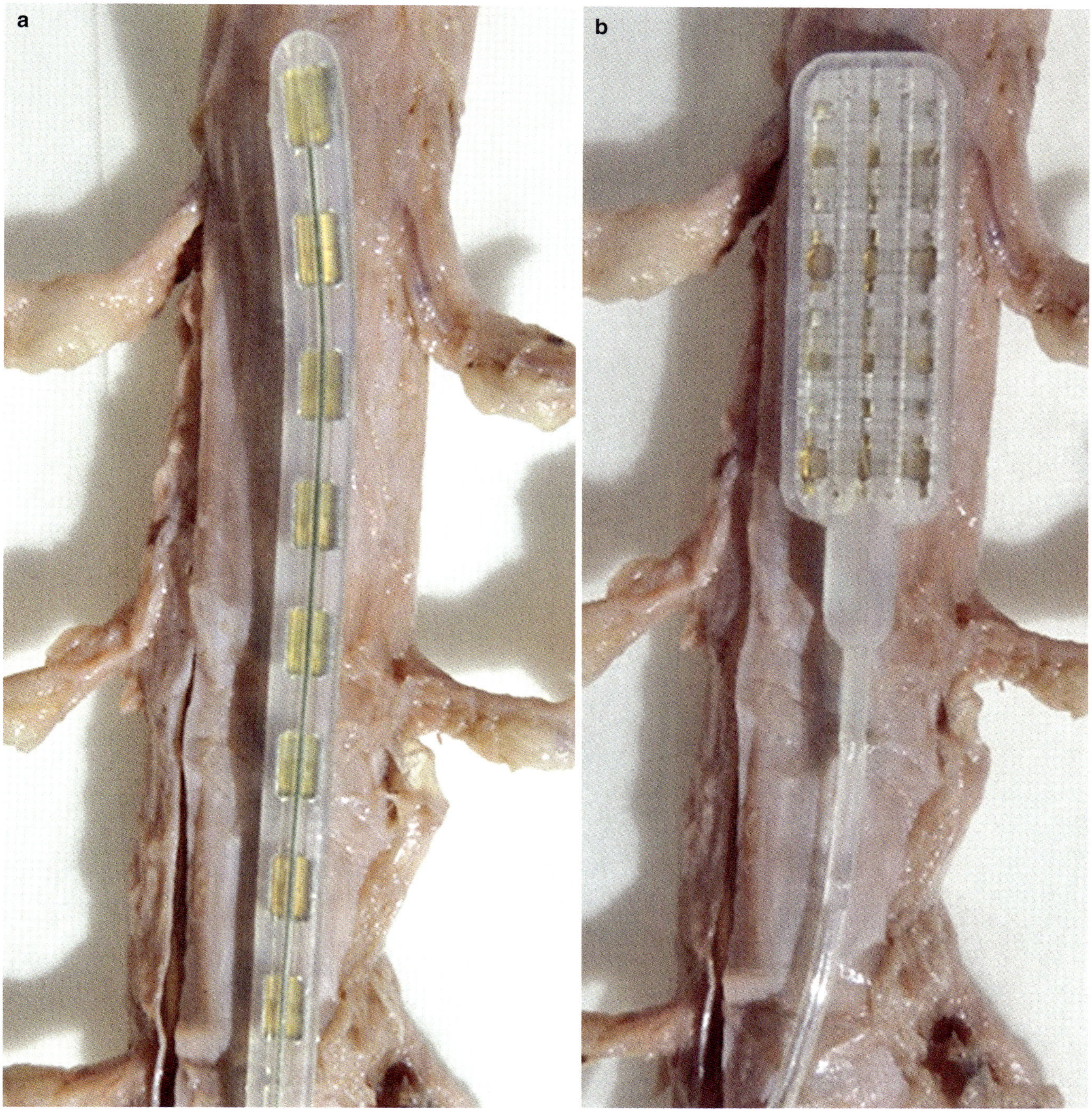

Fig. 40.13 Percutaneous leads and paddle leads located over a human dorsal dural sac at the thoracic level. (**a**) Percutaneous lead. (**b**) Paddle lead. The size of the dural sac and the size occupied by the leads can be identified (From De Andrés et al. [3]; with permission)

Fig. 40.14 Paddle leads located over a human dorsal dural sac at the thoracic level. The size of the dural sac and the size occupied by the paddle lead can be identified. (**a**, **b**) Illustrate the thickness and the width of the paddle lead as compared with the size of the dural sac

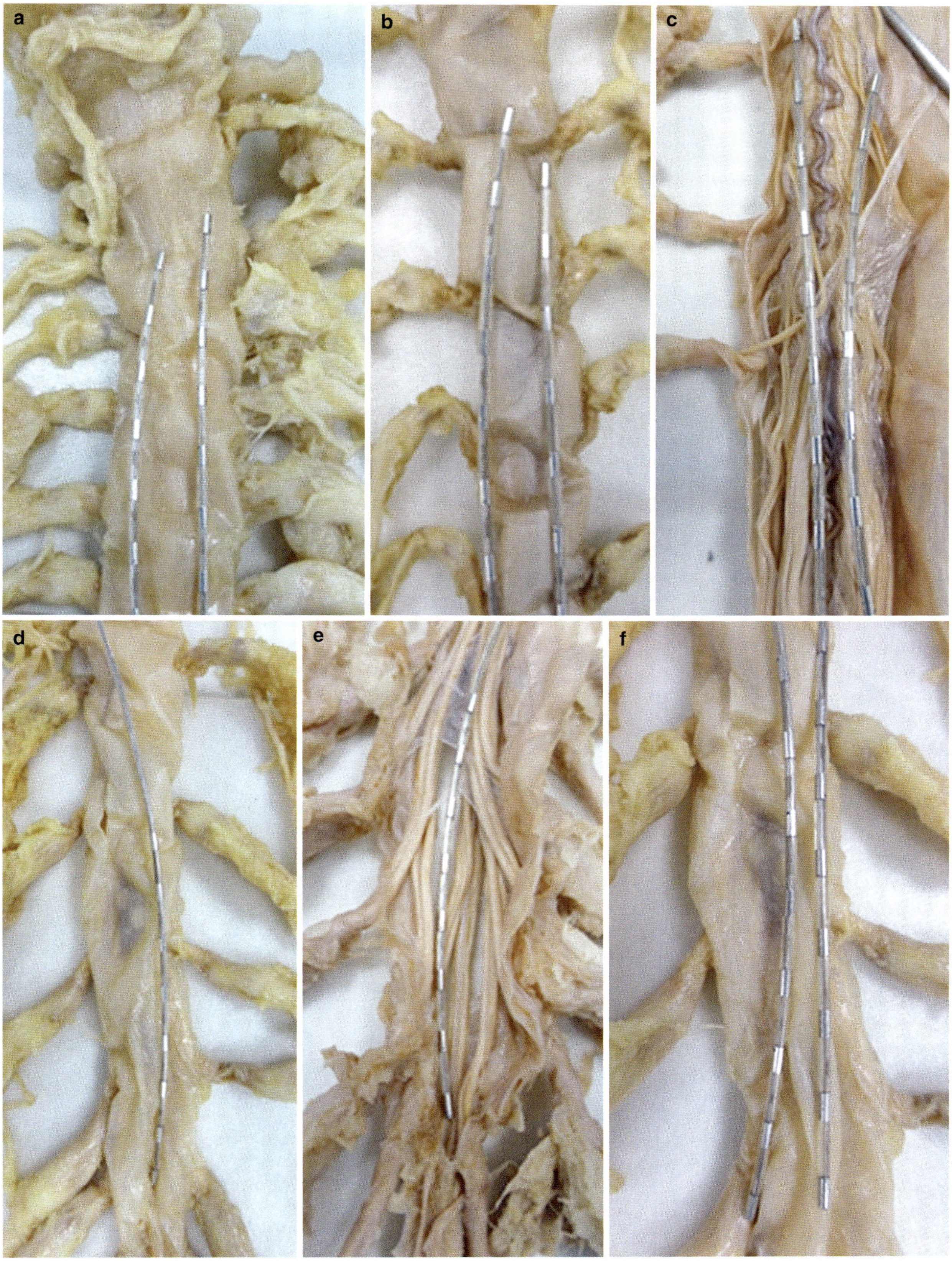

Fig. 40.15 Percutaneous leads with eight contacts arranged in parallel. (**a**) Leads located over a human dural sac at the cervical level. (**b**) Leads located over a human dural sac at the thoracic level. (**c**) Internal structures affected by the electrical field at the thoracic level. (**d**–**f**) Sacral nerve root stimulation

References

1. Holsheimer J. Effectiveness of spinal cord stimulation in the management of chronic pain: analysis of technical drawbacks and solutions. Neurosurgery. 1997;40:990–9.
2. Reina MA, De Andres JA, Hernández JM, Arriazu Navarro R, Durán Mateos EM, Prats-Galino A. Successive changes in extraneural structures from the subarachnoid nerve roots to the peripheral nerve, influencing anesthetic block, and treatment of acute postoperative pain. Eur J Pain Suppl. 2011;5:377–85.
3. De Andrés JA, Reina MA, Hernández JM, Carrera A, Oliva A, Prats-Galino A. Role of spinal anatomical structures for neuromodulation. Reg Anesth Pain Med Suppl. 2011;36:E130–8.
4. De Andrés JA, Reina MA, Machés F, De Sola RG, Oliva A, Prats-Galino A. Epidural fat: considerations for minimally invasive spinal injection and surgical therapies. J Neurosurg Rev. 2012;1:45–53.
5. De Andrés JA, Reina MA, Prats A. Epidural space and regional anaesthesia. Eur J Pain Supplem. 2009;3:55–63.
6. Reina MA, Franco CD, López A, De Andrés JA, van Zundert A. Clinical implications of epidural fat in the spinal canal. A scanning electron microscopic study. Acta Anaesthesiol Belg. 2009;60:7–17.
7. Barolat G. Epidural spinal cord stimulation: anatomical and electrical properties of the intraspinal structures relevant to spinal cord stimulation and clinical correlations. Neuromodulation. 1998;1:63–71.
8. Holsheimer J, Struijk JJ, Tas NR. Effects of electrode geometry and combination on nerve fibre selectivity in spinal cord stimulation. Med Biol Eng Comput. 1995;33:676–82.

Ultrastructure of Dural Lesions Produced in Lumbar Punctures

41

Miguel Angel Reina, Andrés López,
André A.J. van Zundert, and José Antonio De Andrés

During lumbar puncture, the insertion of a single-use, short-beveled spinal needle produces a "tent-like" effect. The dura-arachnoid lesions produced by this Quincke type of needles show a "crescent moon" shape resembling the letter "U" or "V" (similar to the lid of a can), with clean-cut edges [1–8]. As the needle tip advances, the cut fragment is folded inwards, while all dural and arachnoid layers are entirely severed and the edges of the lesion are displaced inwardly. As soon as the spinal needle is withdrawn, the edges of the lesion tend to retract, owing to the viscoelastic properties of the affected dura mater.

Insertion of noncutting, pencil-point needle tips through the dura-arachnoid membrane produces more distension of the dural sac and results in an increased "tent-like" effect of the dural sac thickness (dura mater and arachnoid layer). Pencil-point spinal needles have non-cutting tips and tend to displace the dural sac forwards before piercing the dural lamina during lumbar puncture, producing greater "tent-like effect" because of their lack of sharp beveled tips. Pencil-point tips produce more extensive tearing of tissue margins and folding at the edges around the lesion. The edges of dural lesions caused by pencil-point needles are irregular, with fragmented portions of tissue, owing to the tearing process [1–10].

When epidural Tuohy needles accidentally enter the subarachnoid space, the resulting dura-arachnoid lesions have a "crescent moon" shape, similar to lesions produced by Quincke-tip spinal needles. Lesions caused by epidural Tuohy needles resemble a "U" or "V" letter shape with clean-cut edges, similar to an opening in the lid of a can (Figs. 41.1, 41.2, 41.3, 41.4, 41.5, 41.6, 41.7, 41.8, 41.9, 41.10, 41.11, 41.12, 41.13, 41.14, 41.15, 41.16, 41.17, 41.18, 41.19, 41.20, 41.21, 41.22, 41.23, 41.24, and 41.25).

M.A. Reina, MD, PhD (✉)
Department of Clinical Medical Sciences and Institute of Applied Molecular Medicine, School of Medicine, University of CEU San Pablo, Madrid, Spain

Department of Anesthesiology, Madrid-Montepríncipe University Hospital, Madrid, Spain
e-mail: miguelangel@perticone.e.telefonica.net

A. López, MD
Department of Clinical Medical Sciences, School of Medicine, CEU San Pablo University, Madrid, Spain

Department of Anesthesiology, Madrid-Montepríncipe University Hospital, Madrid-Torrelodones University Hospital, Madrid-Sanchinarro University Hospital, Madrid, Spain
e-mail: alg00009@telefonica.net

A.A.J. van Zundert, MD, PhD, FRCA, EDRA, FANZCA
Discipline of Anesthesiology, The University of Queensland and Royal Brisbane and Women's Hospital, Brisbane, QLD, Australia
e-mail: vanzundertandre@gmail.com

J.A. De Andrés, MD, PhD
Department of Surgical Specialties, School of Medicine, University of Valencia, Valencia, Spain

Department of Anesthesia, Critical Care, and Pain Management, General University Hospital, Valencia, Spain
e-mail: deandres_jos@gva.es

M.A. Reina et al. (eds.), *Atlas of Functional Anatomy for Regional Anesthesia and Pain Medicine: Human Structure, Ultrastructure and 3D Reconstruction Images*, DOI 10.1007/978-3-319-09522-6_41,

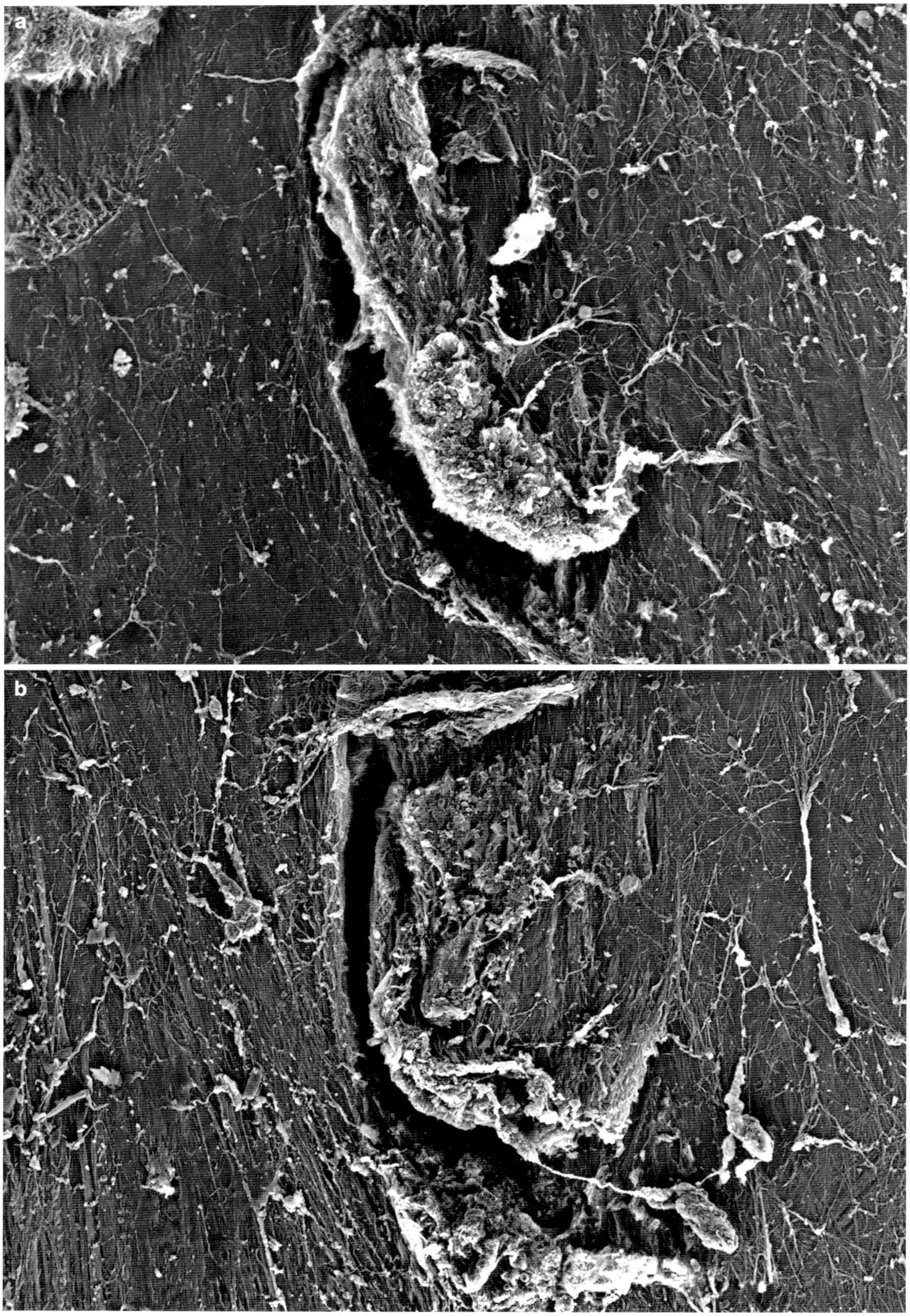

Fig. 41.1 Dura-arachnoid lesion produced by 22G Quincke needle. External surface of human dural sac (external dural laminas). Scanning electron microscopy. Magnification: **a**, ×100; **b**, ×80 (From Reina et al. [9]; with permission)

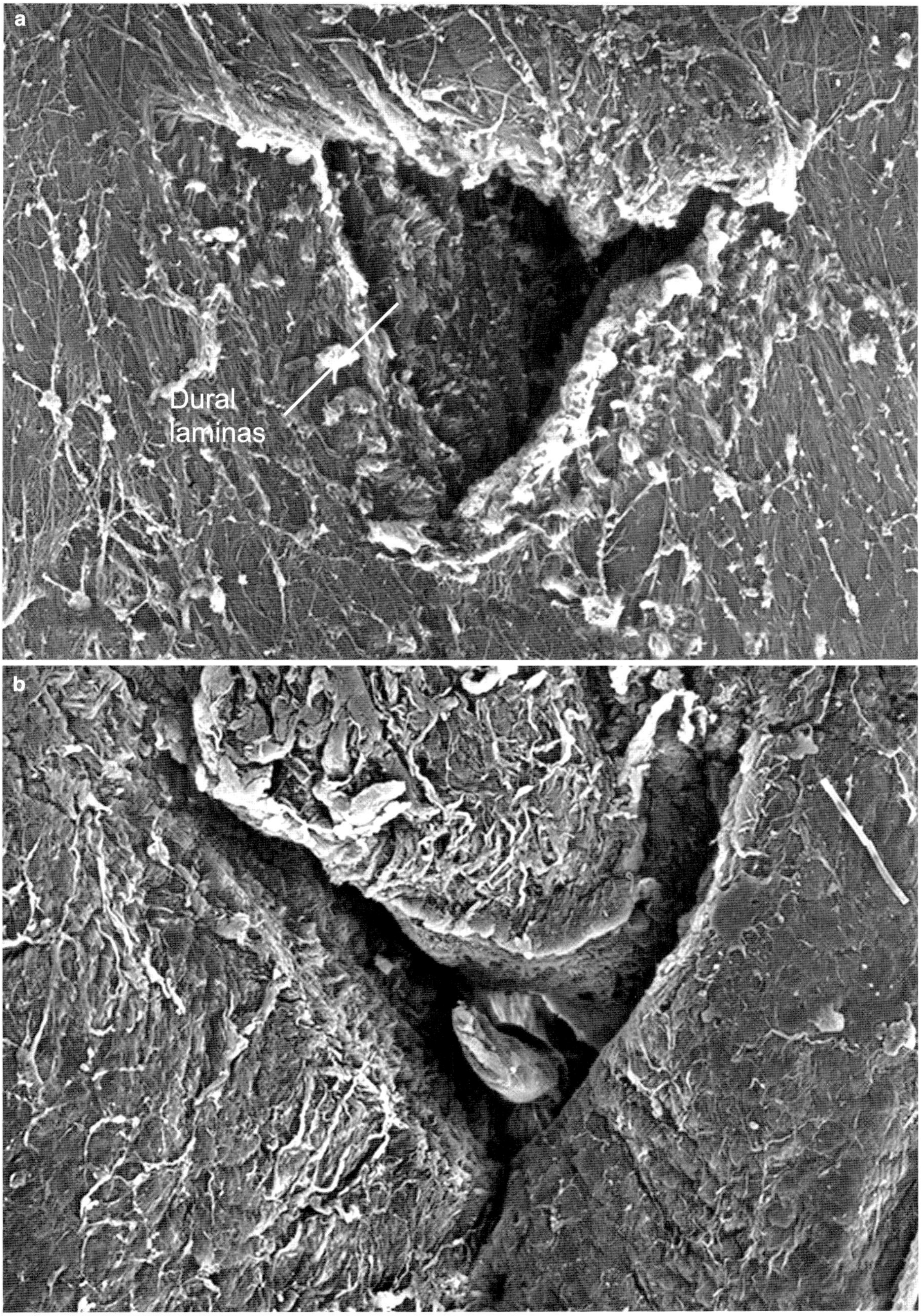

Fig. 41.2 Dura-arachnoid lesion produced by 22G Quincke needle. External surface of human dural sac (external dural laminas). Scanning electron microscopy. Magnification: **a**, ×100; **b**, ×150 (From Reina et al. [9]; with permission)

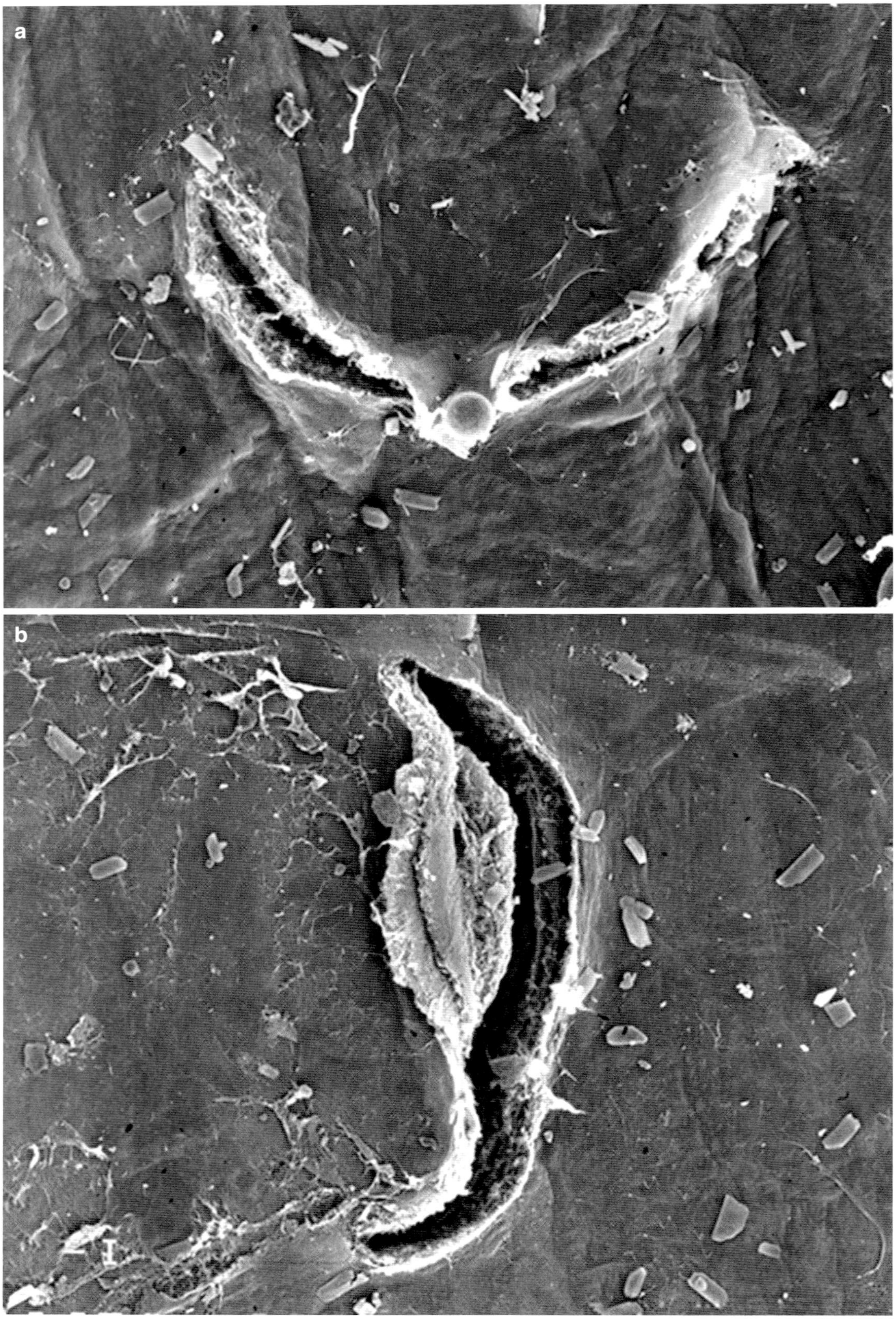

Fig. 41.3 Dura-arachnoid lesion produced by 22G Quincke needle. Internal surface of human dural sac (arachnoid layer). Scanning electron microscopy. Magnification: **a**, ×100; **b**, ×100

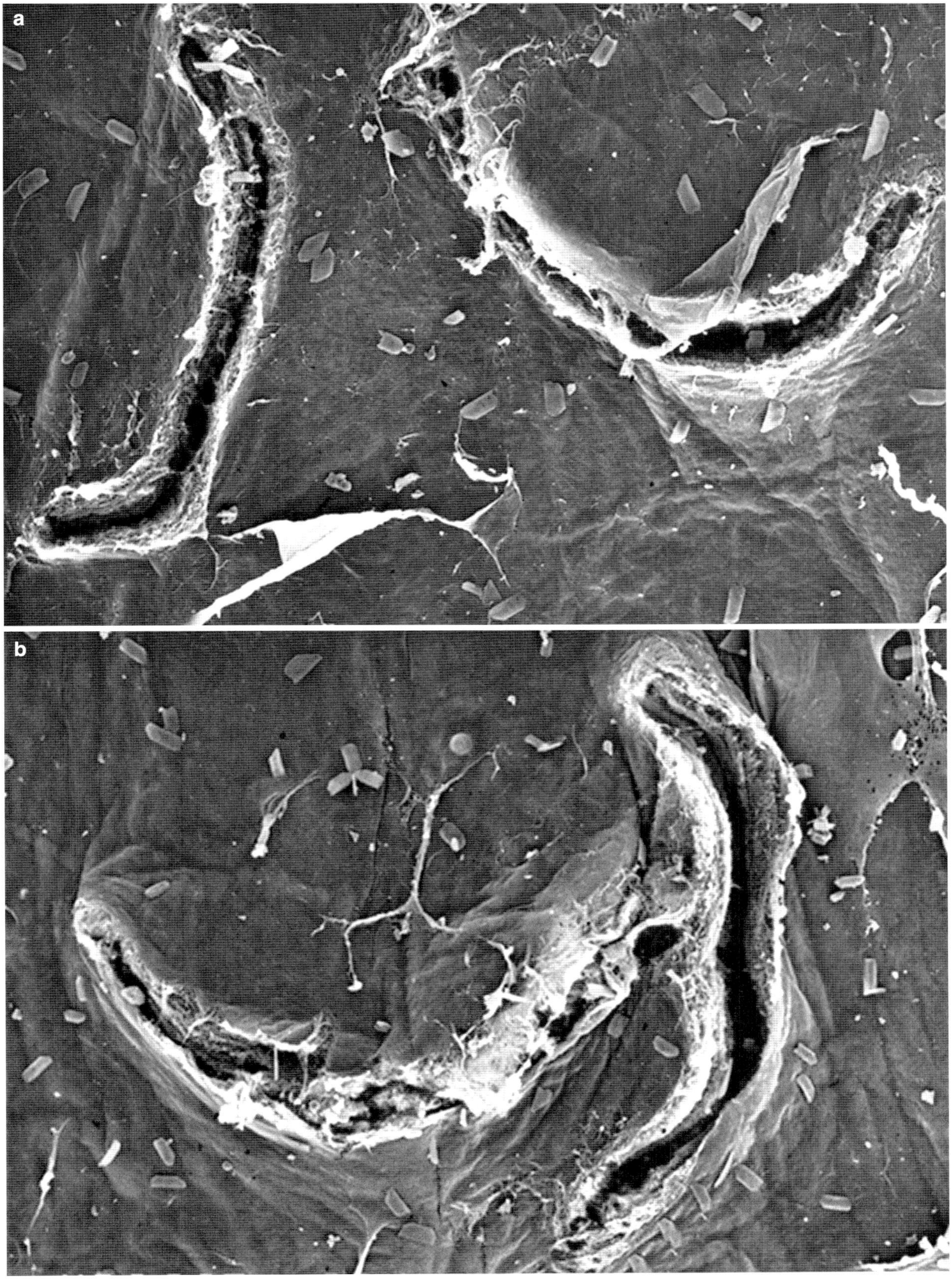

Fig. 41.4 Dura-arachnoid lesion produced by 22G Quincke needle. Internal surface of human dural sac (external dural laminas). Scanning electron microscopy. Magnification: **a**, ×100; **b**, ×100. (With permission (**a** from Reina et al. [10]; with permission. **b** from Reina et al. [3]; with permission))

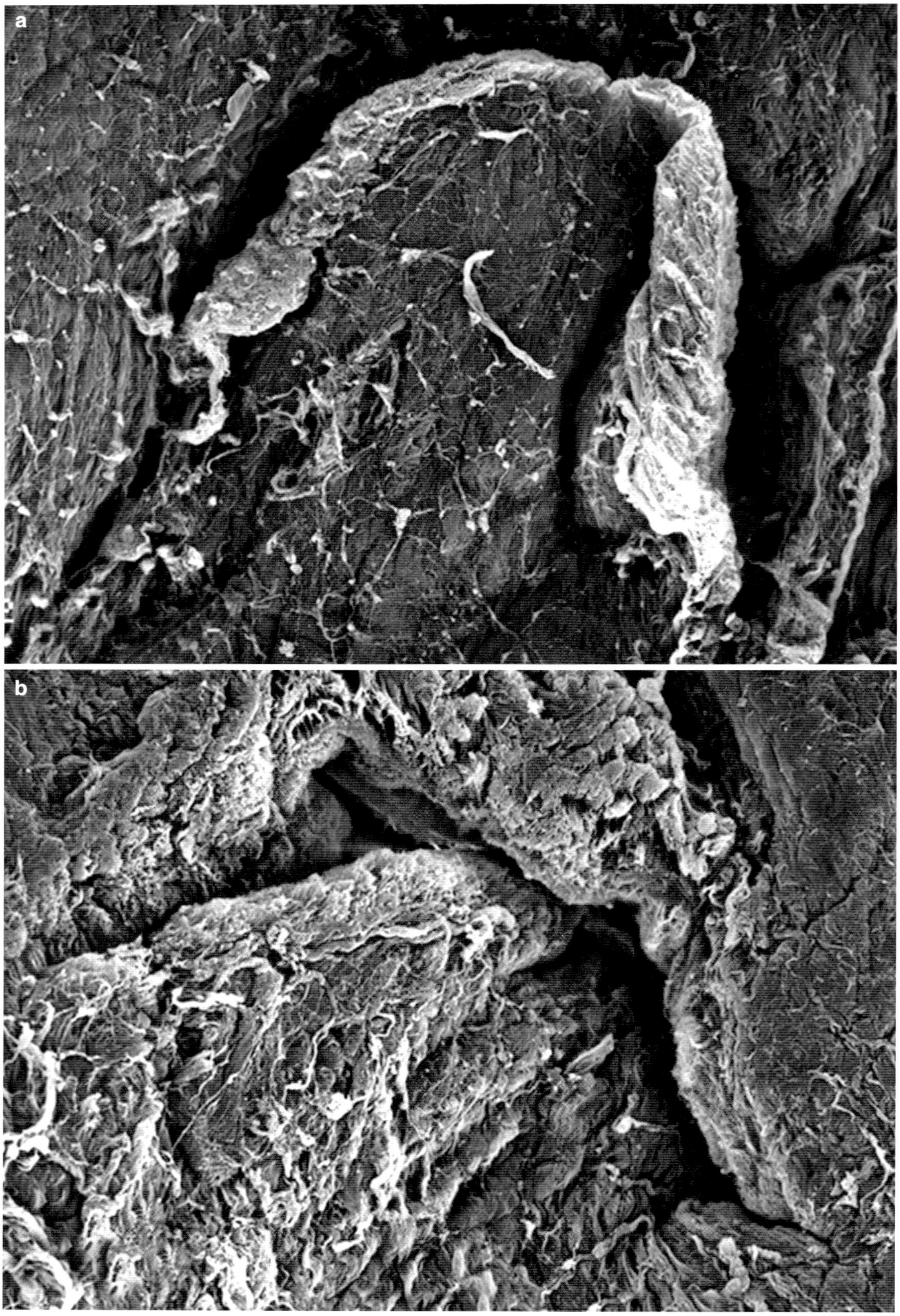

Fig. 41.5 Dura-arachnoid lesion produced by 25G Quincke needle. External surface of human dural sac (external dural laminas). Scanning electron microscopy. Magnification: **a**, ×150; **b**, ×200 (From Reina et al. [9]; with permission)

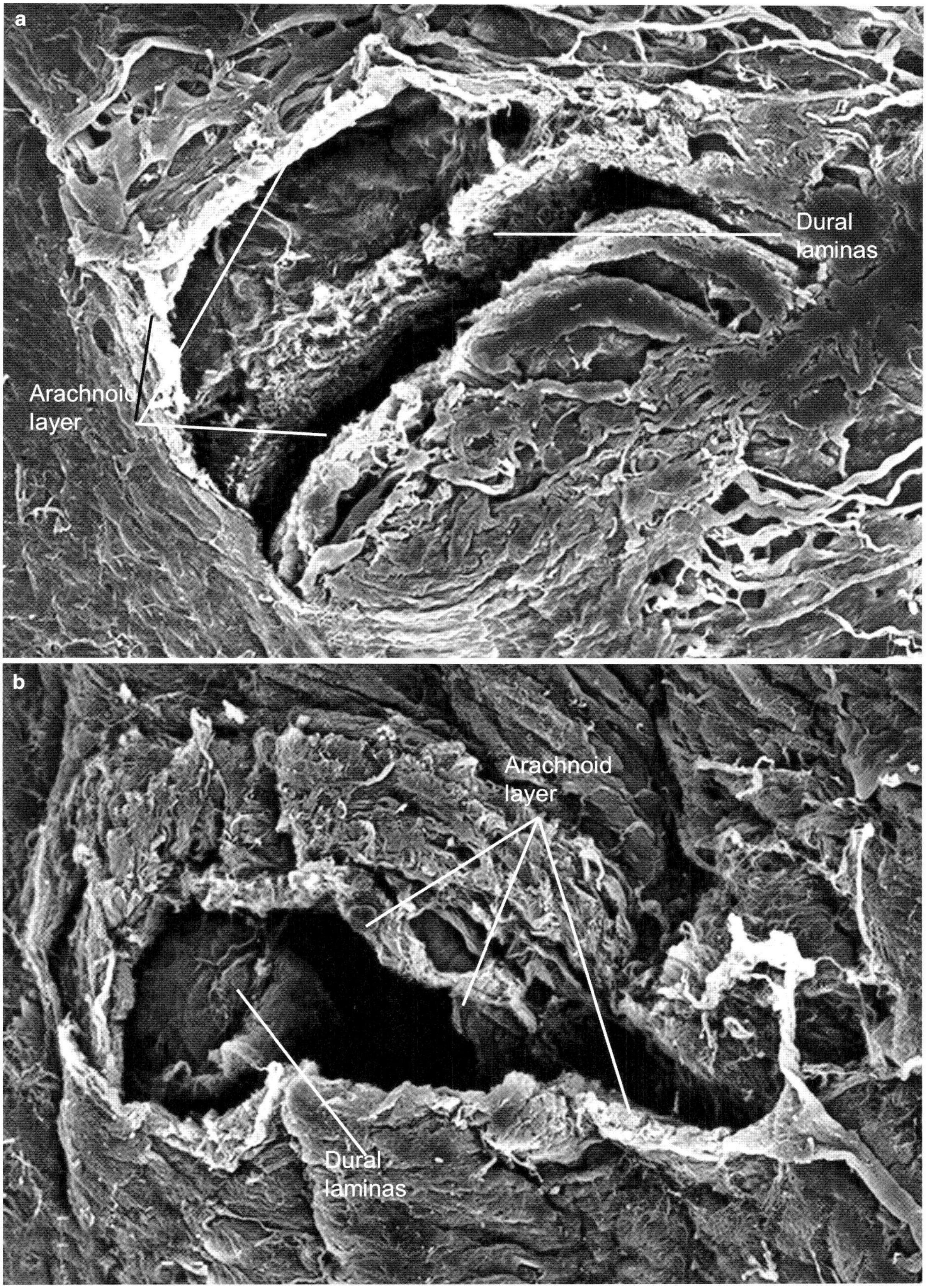

Fig. 41.6 Dura-arachnoid lesion produced by 25G Quincke needle. Internal surface of human dural sac (arachnoid layer). Scanning electron microscopy. Magnification: **a**, ×200; **b**, ×200 (From Reina et al. [3]; with permission)

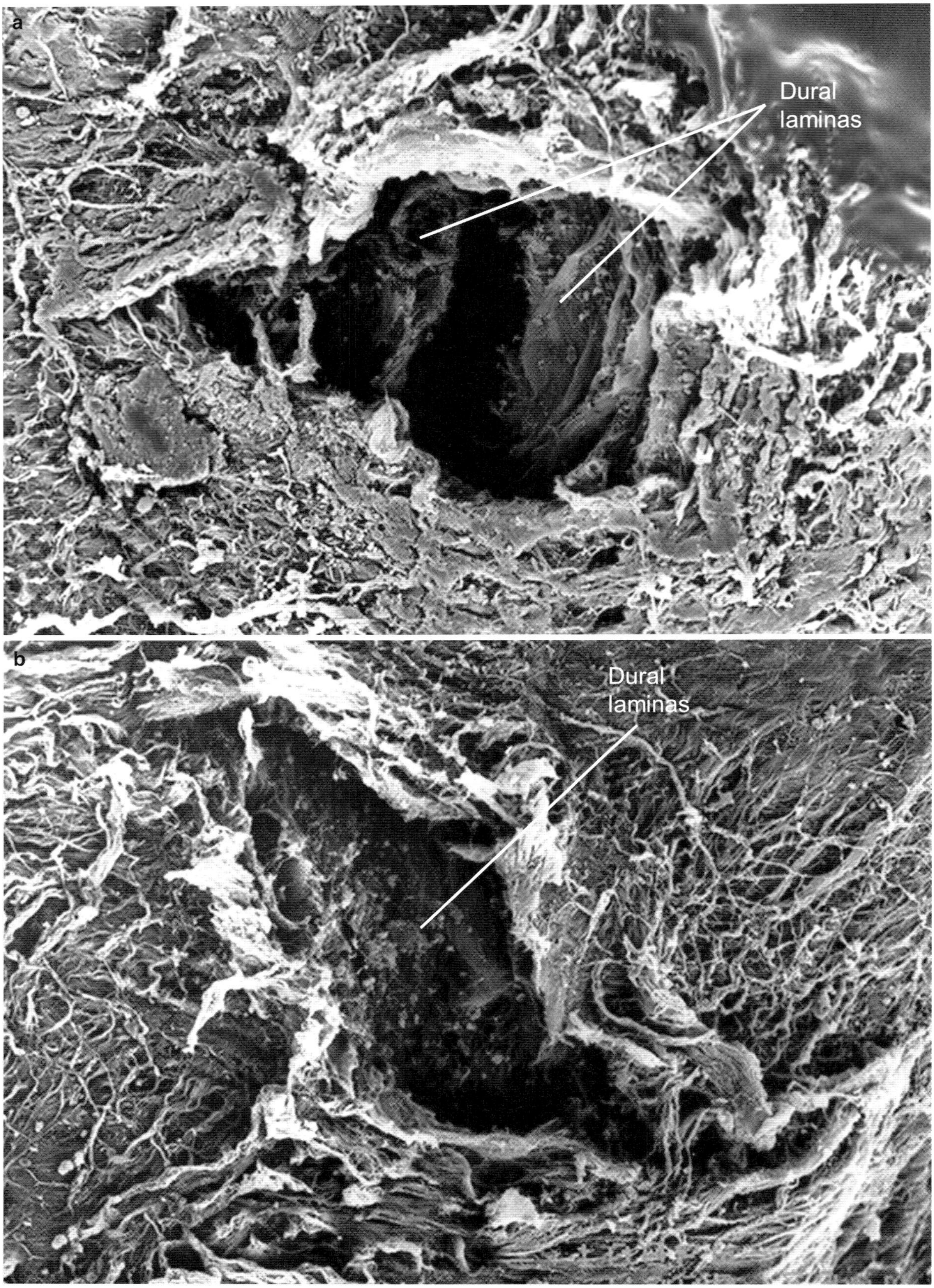

Fig. 41.7 Dura-arachnoid lesion produced by 25G Whitacre needle. External surface of human dural sac (external dural laminas). Scanning electron microscopy. Magnification: **a**, ×200; **b**, ×200 (**a** from Reina et al. [6]; with permission. **b** from Reina et al. [3]; with permission)

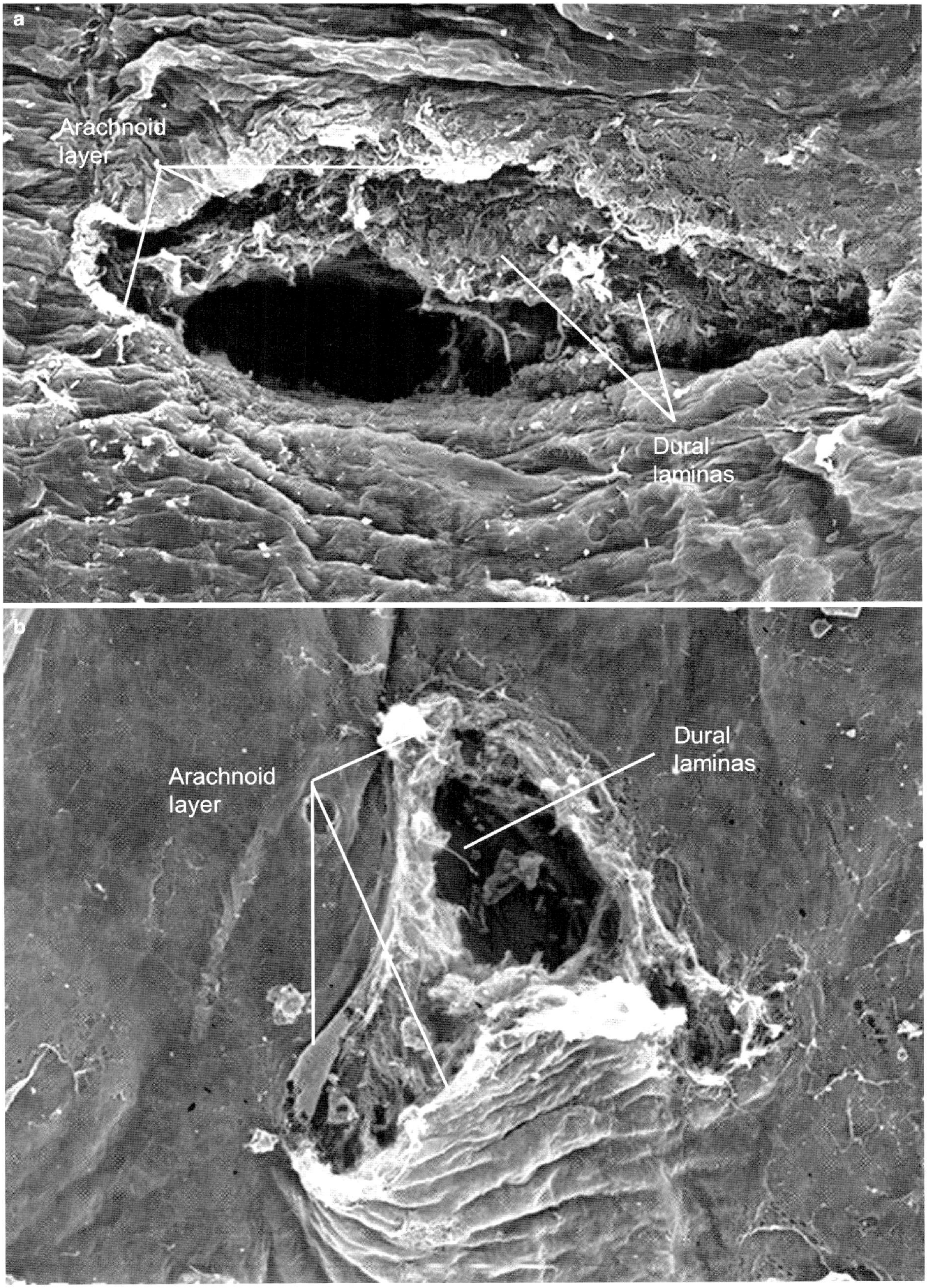

Fig. 41.8 Dura-arachnoid lesion produced by 25G Whitacre needle. Internal surface of human dural sac (arachnoid layer). Scanning electron microscopy. Magnification: **a**, ×200; **b**, ×150 (**a** from Reina et al. [6]; with permission. **b** from Reina et al. [3]; with permission)

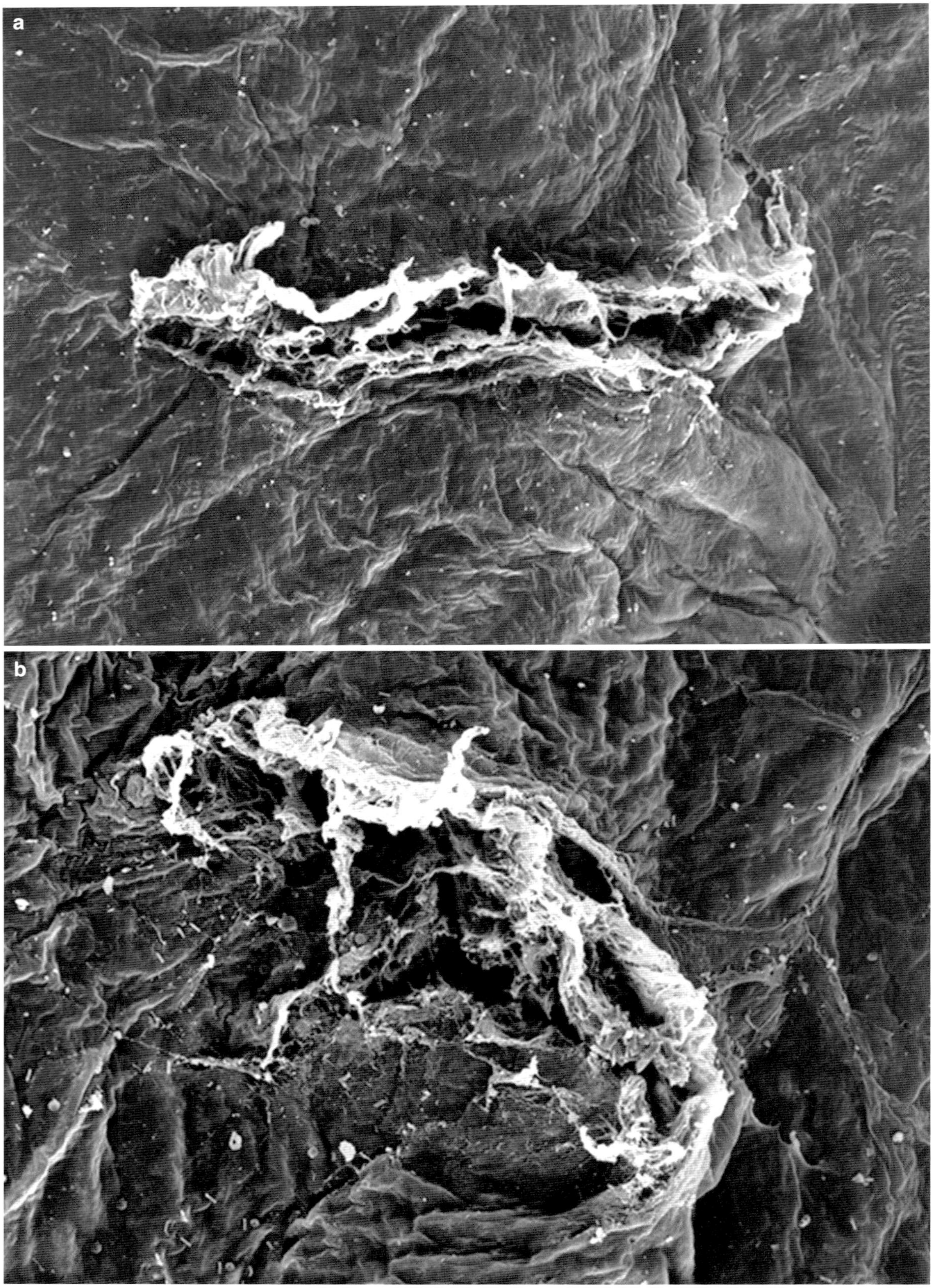

Fig. 41.9 Dura-arachnoid lesion produced by 25G Whitacre needle. Internal surface of human dural sac (arachnoid layer). Scanning electron microscopy. Magnification: **a**, ×100; **b**, ×200 (**a** from Reina et al. [6]; with permission)

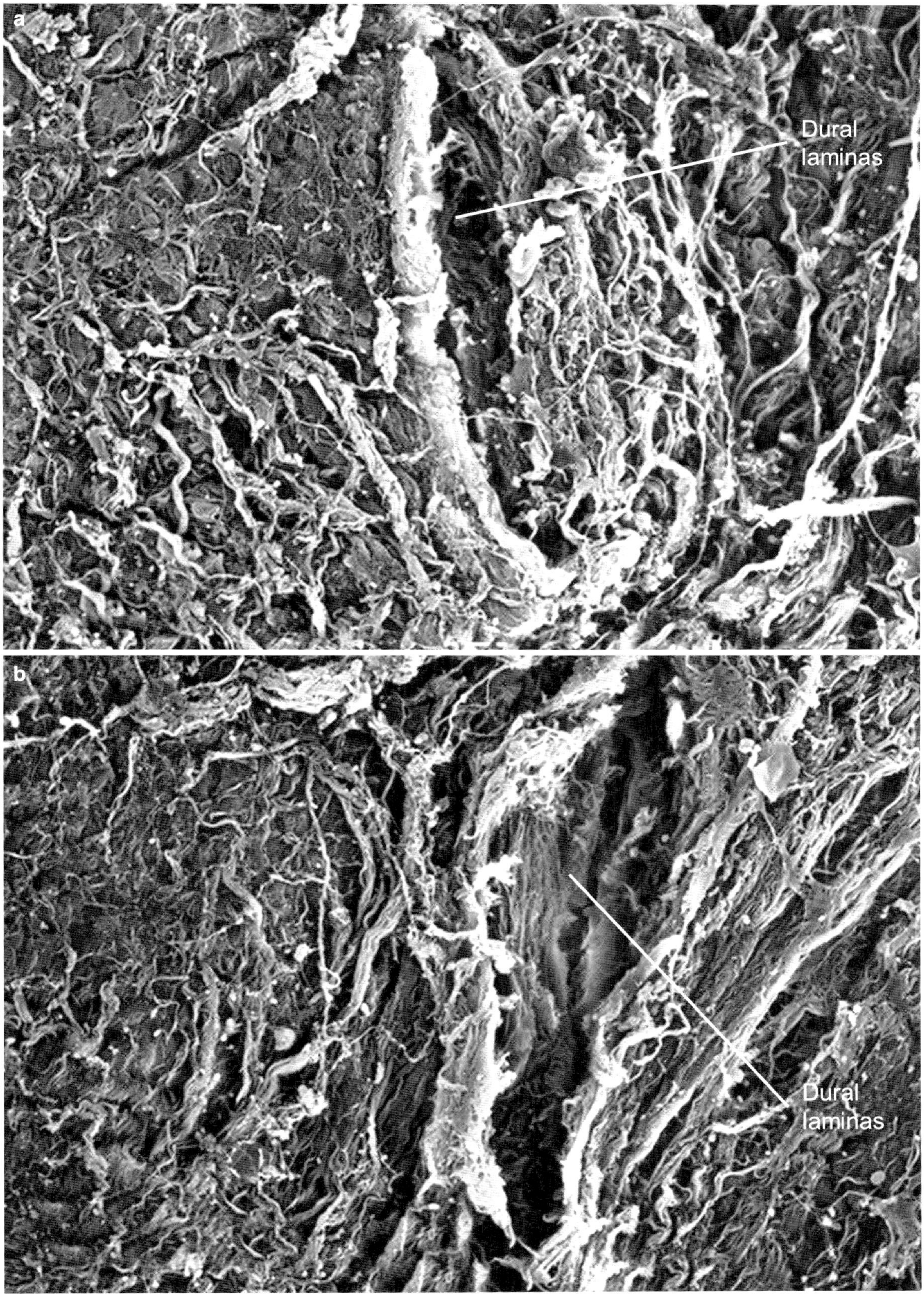

Fig. 41.10 Dura-arachnoid lesion produced by 27G Whitacre needle. External surface of human dural sac (external dural laminas). Scanning electron microscopy. Magnification: **a**, ×200; **b**, ×200

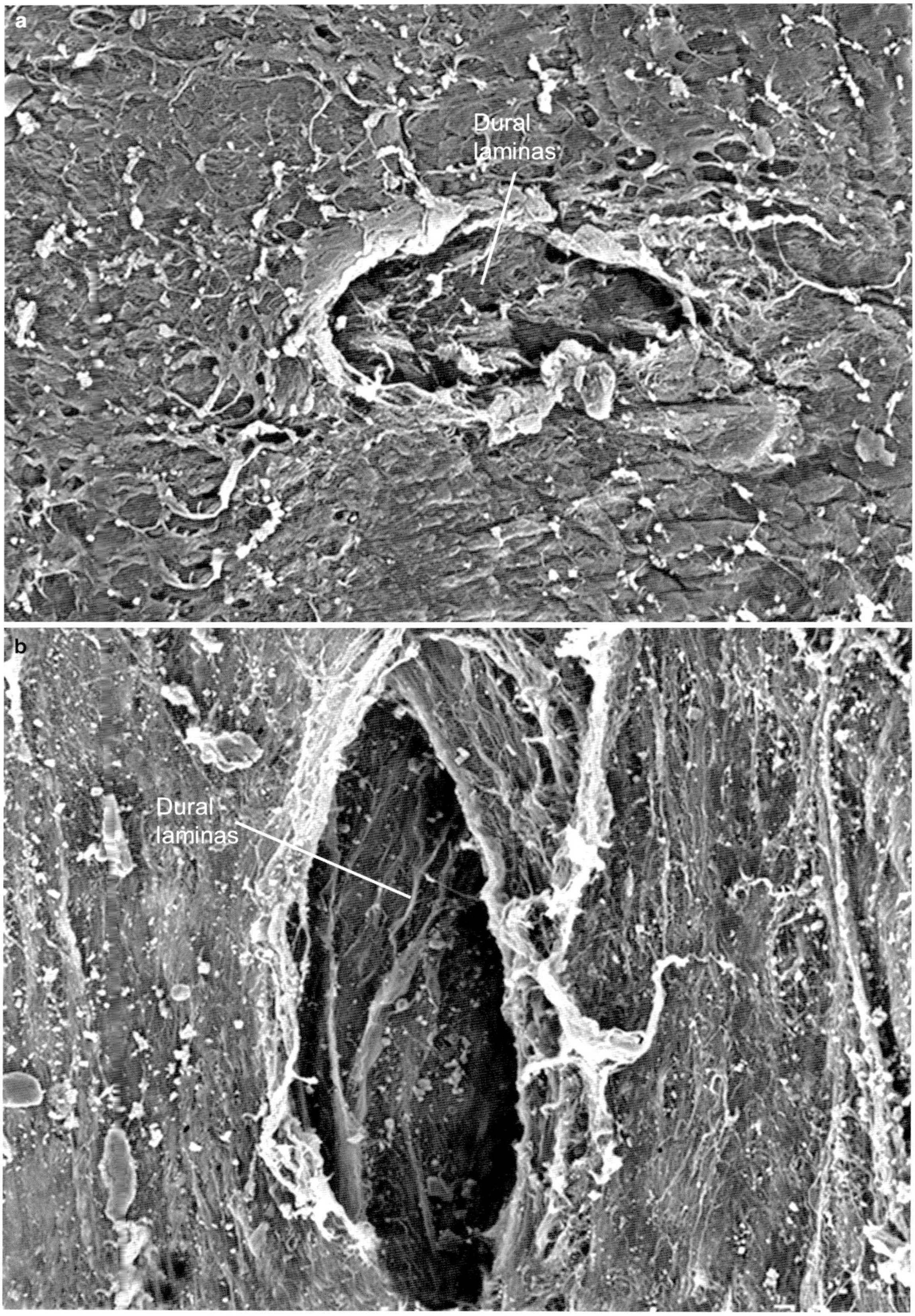

Fig. 41.11 Dura-arachnoid lesion produced by 27G Whitacre needle. External surface of human dural sac (external dural laminas). Scanning electron microscopy. Magnification: **a**, ×200; **b**, ×200 (From Reina et al. [9]; with permission)

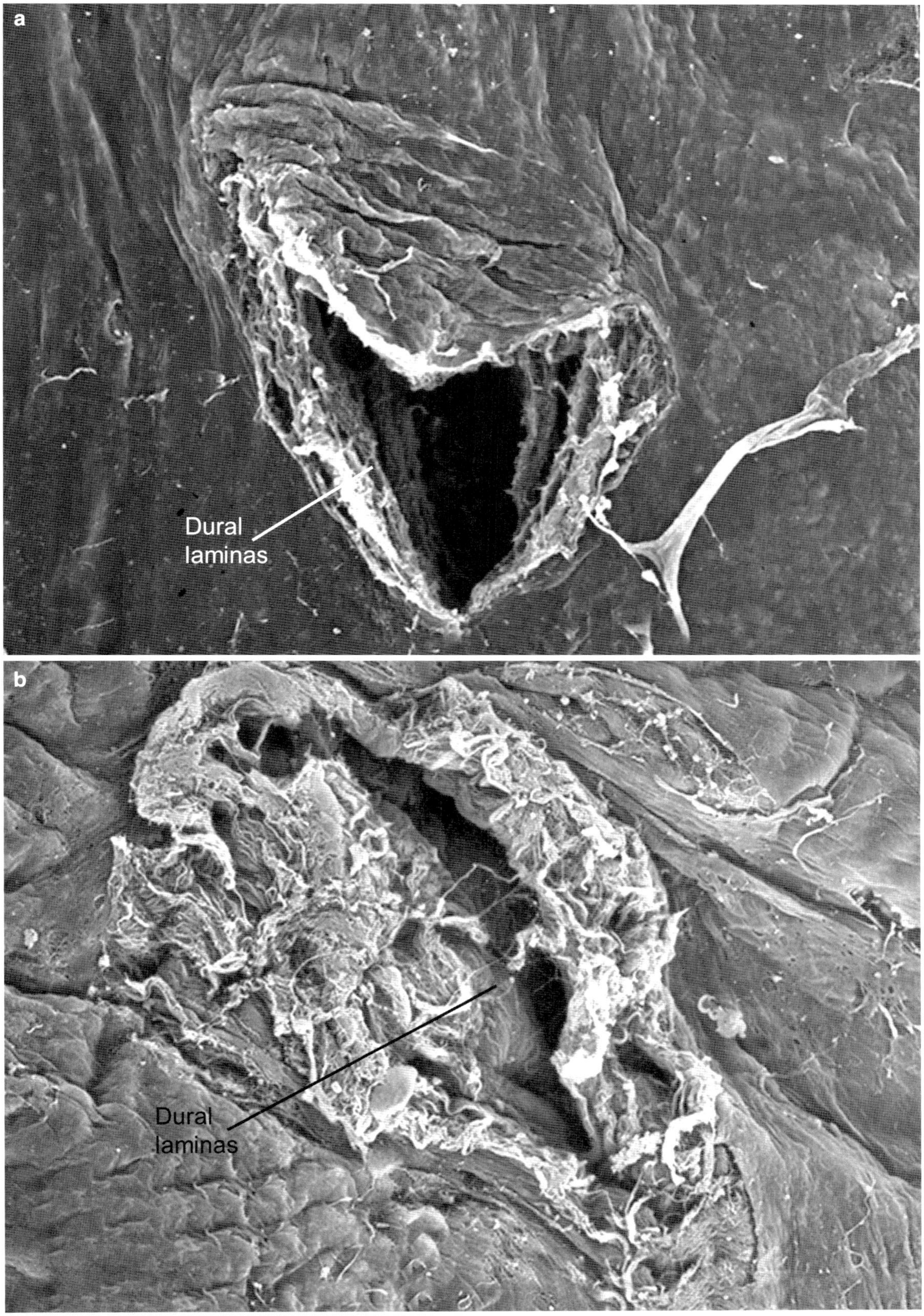

Fig. 41.12 Dura-arachnoid lesion produced by 27G Whitacre needle. Internal surface of human dural sac (arachnoid layer). Scanning electron microscopy. Magnification: **a**, ×200; **b**, ×200 (From Reina et al. [9]; with permission)

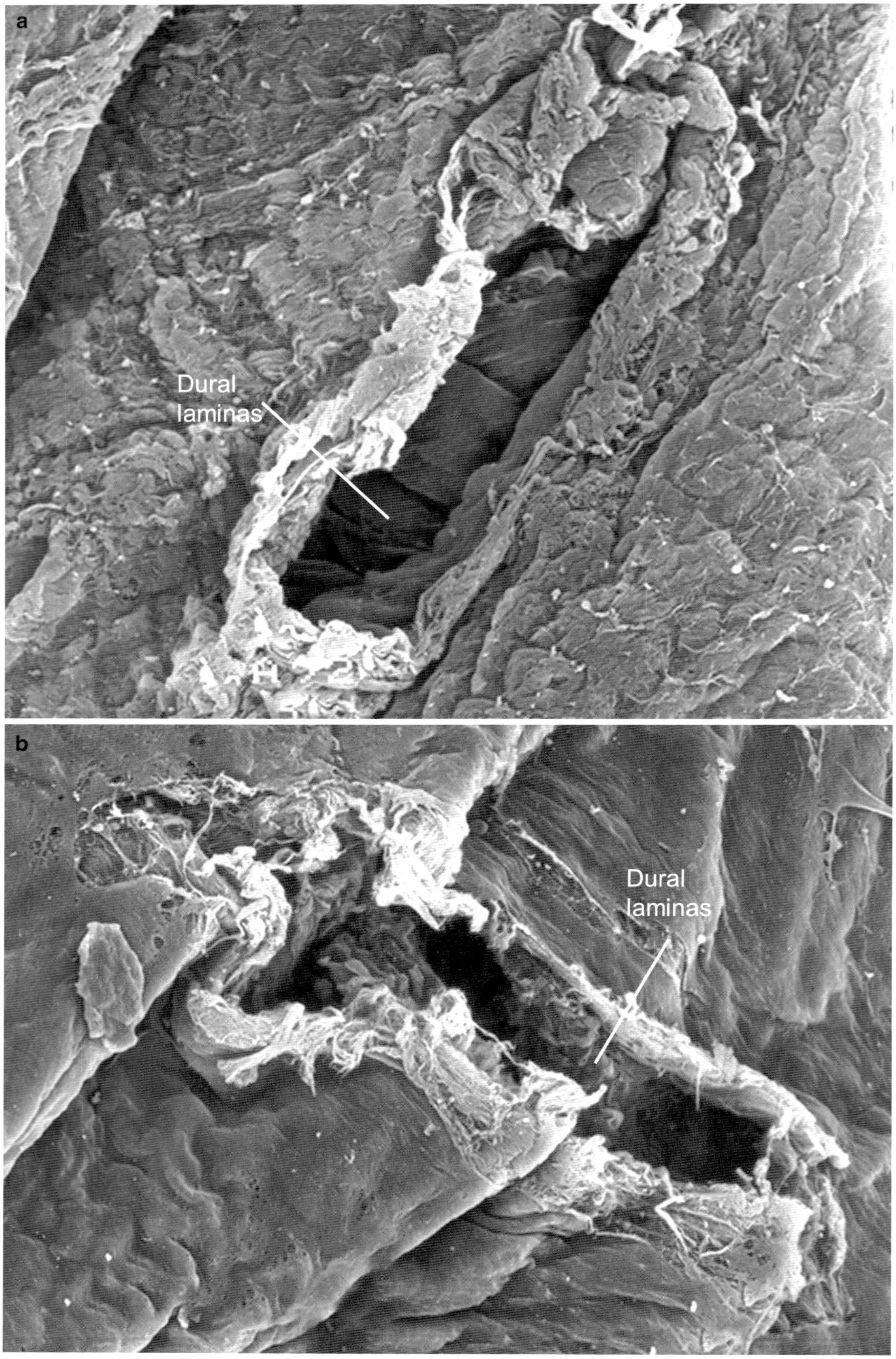

Fig. 41.13 Dura-arachnoid lesion produced by 27G Whitacre needle. Internal surface of human dural sac (arachnoid layer). Scanning electron microscopy. Magnification: **a**, ×200; **b**, ×200

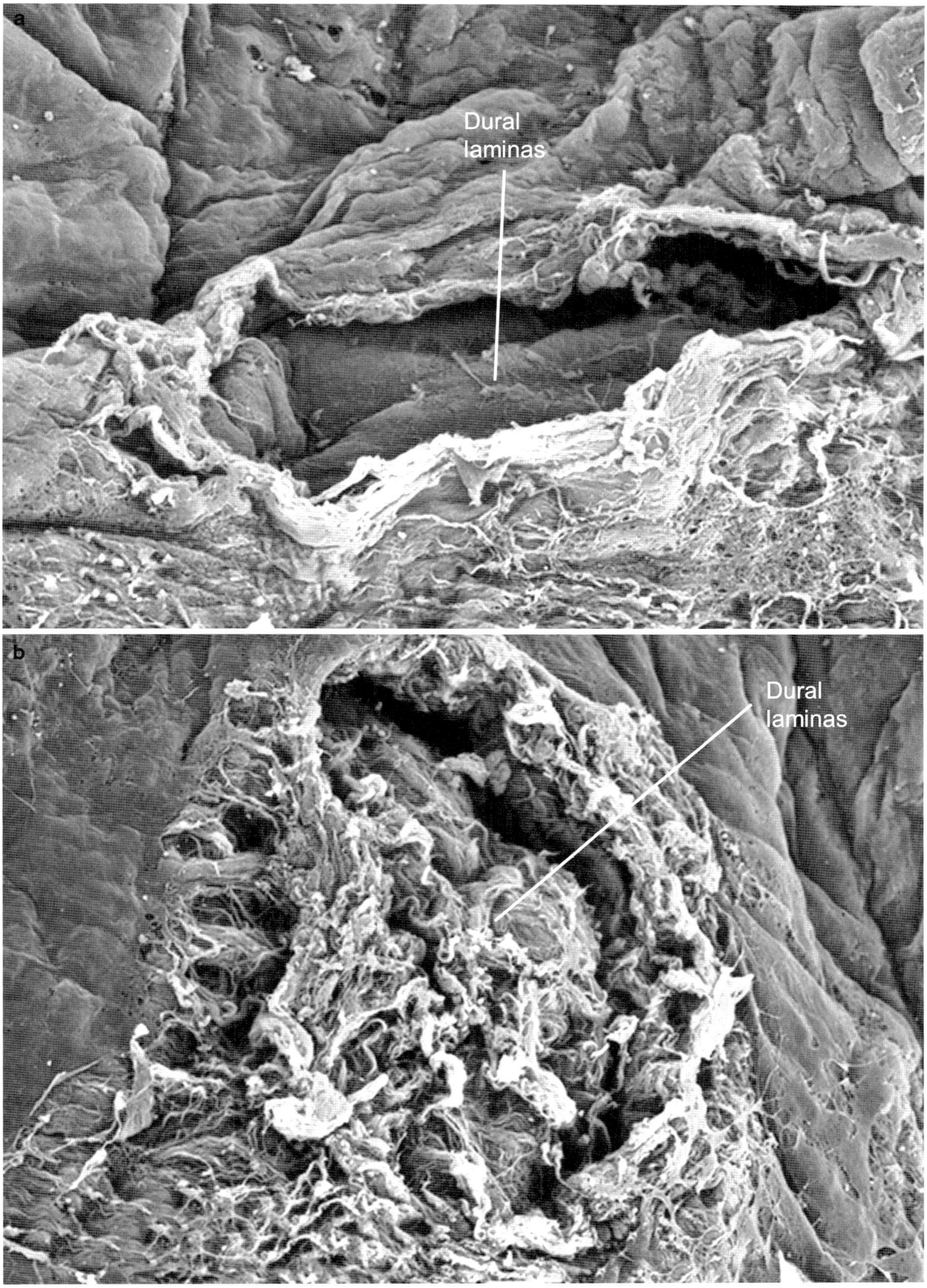

Fig. 41.14 Dura-arachnoid lesion produced by 27G Whitacre needle. Internal surface of human dural sac (arachnoid layer). Scanning electron microscopy. Magnification: **a**, ×200; **b**, ×200

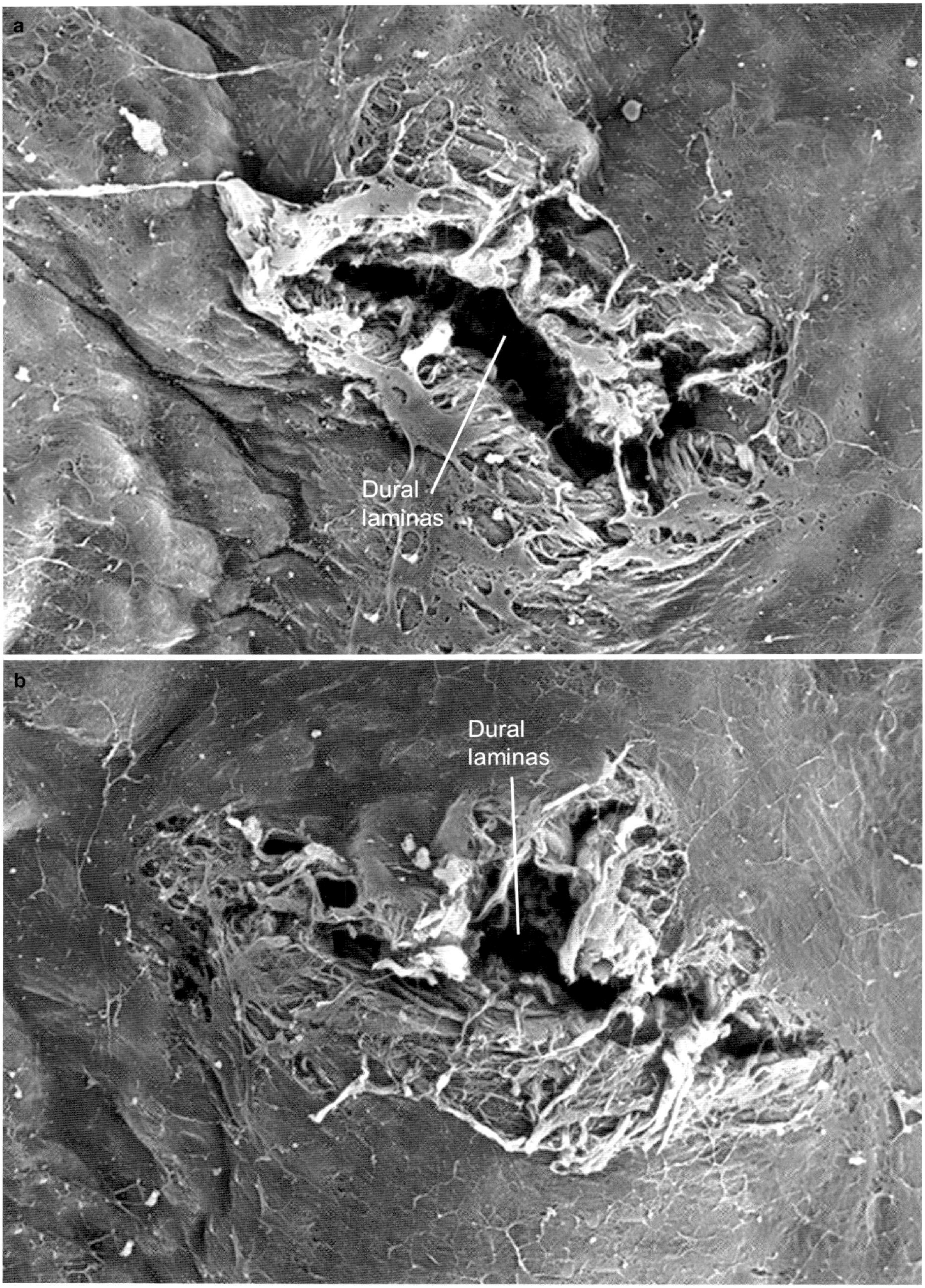

Fig. 41.15 Dura-arachnoid lesion produced by 27G Whitacre needle. Internal surface of human dural sac (arachnoid layer). Scanning electron microscopy. Magnification: **a**, ×200; **b**, ×200 (From Reina et al. [9]; with permission)

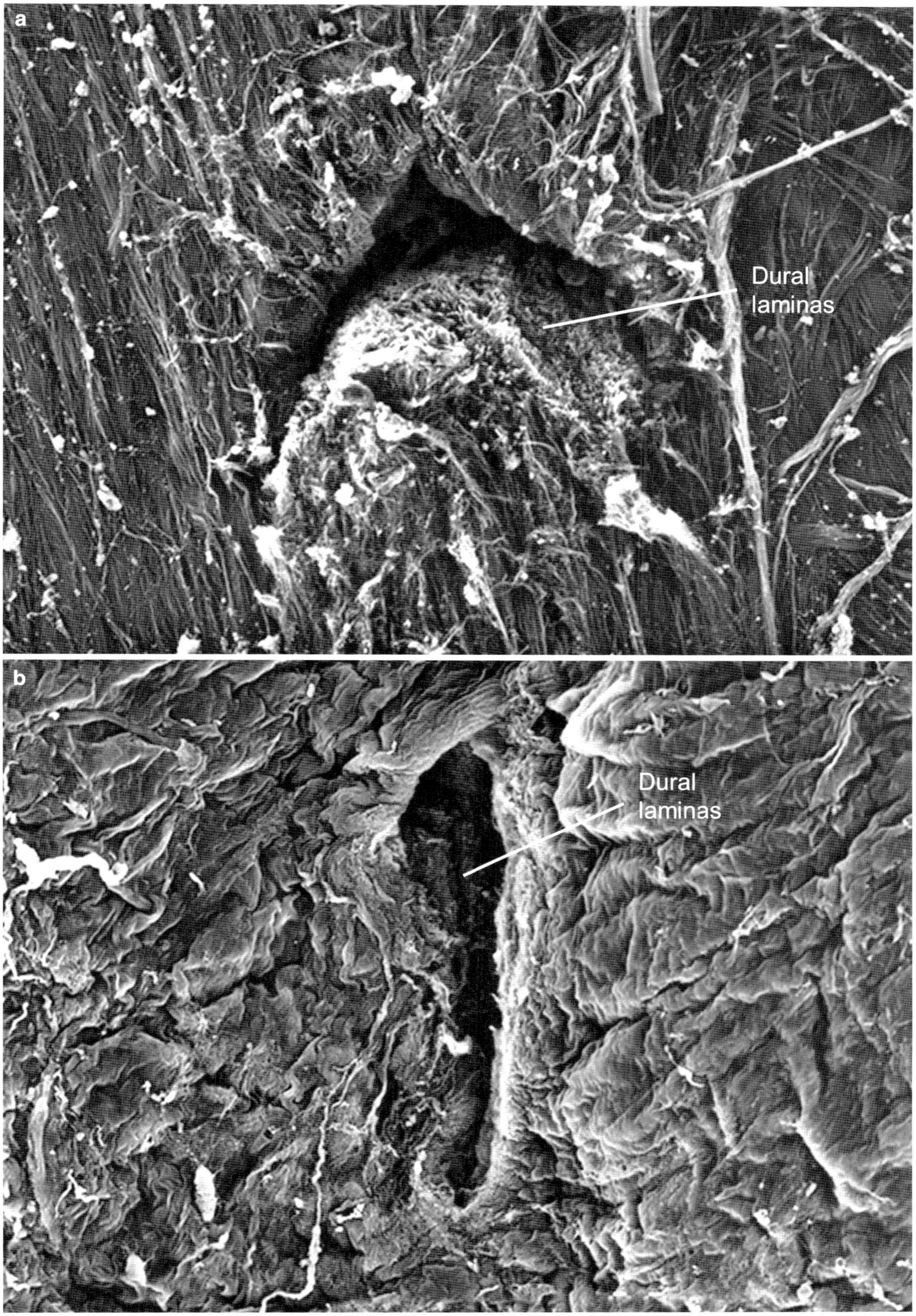

Fig. 41.16 Dura-arachnoid lesion produced by 29G Quincke needle. External surface (external dural laminas). Scanning electron microscopy. Magnification: **a**, ×200; **b**, ×200 (From Reina et al. [9]; with permission)

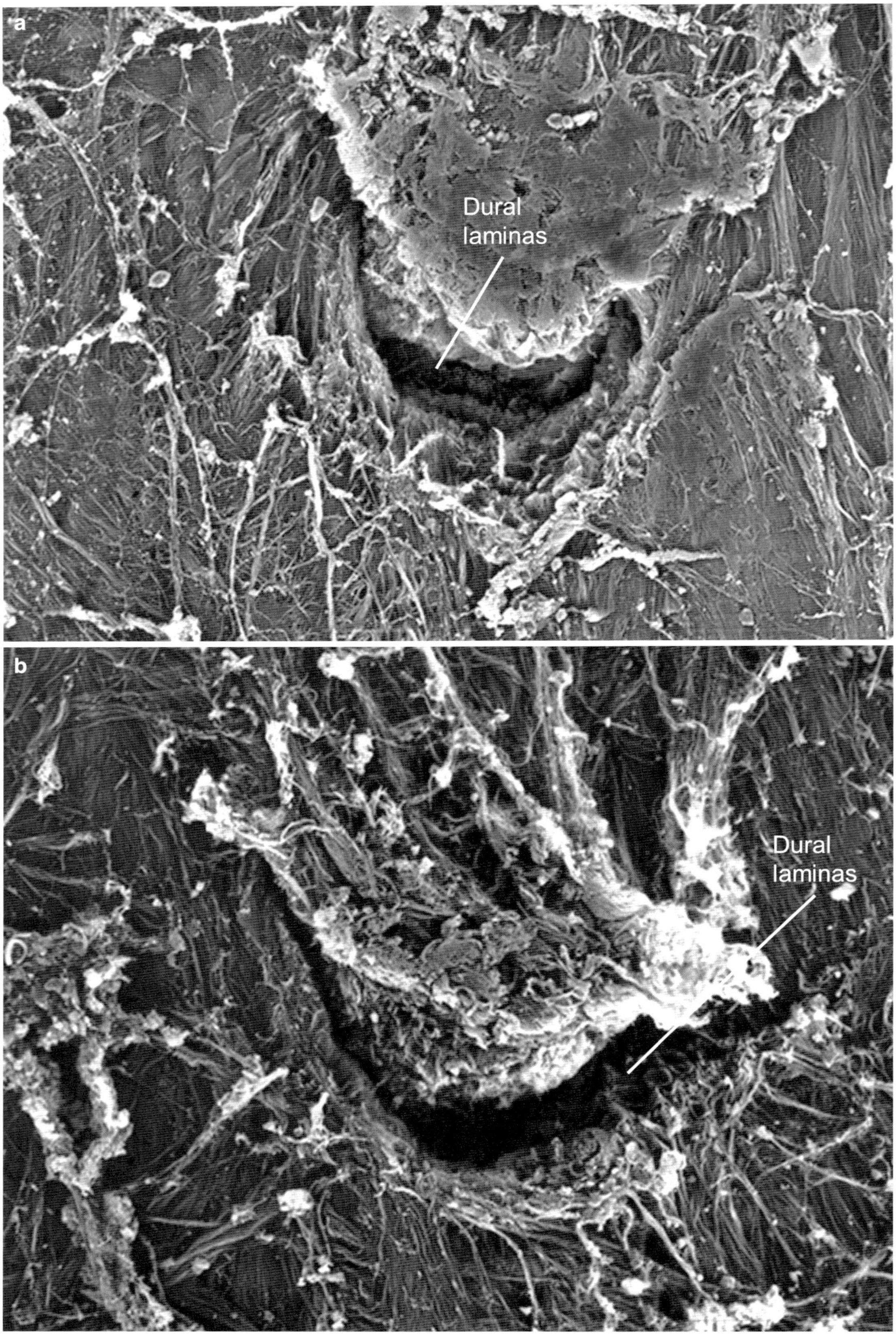

Fig. 41.17 Dura-arachnoid lesion produced by 29G Quincke needle. External surface of human dural sac (external dural laminas). Scanning electron microscopy. Magnification: **a**, ×200; **b**, ×200 (From Reina et al. [9]; with permission)

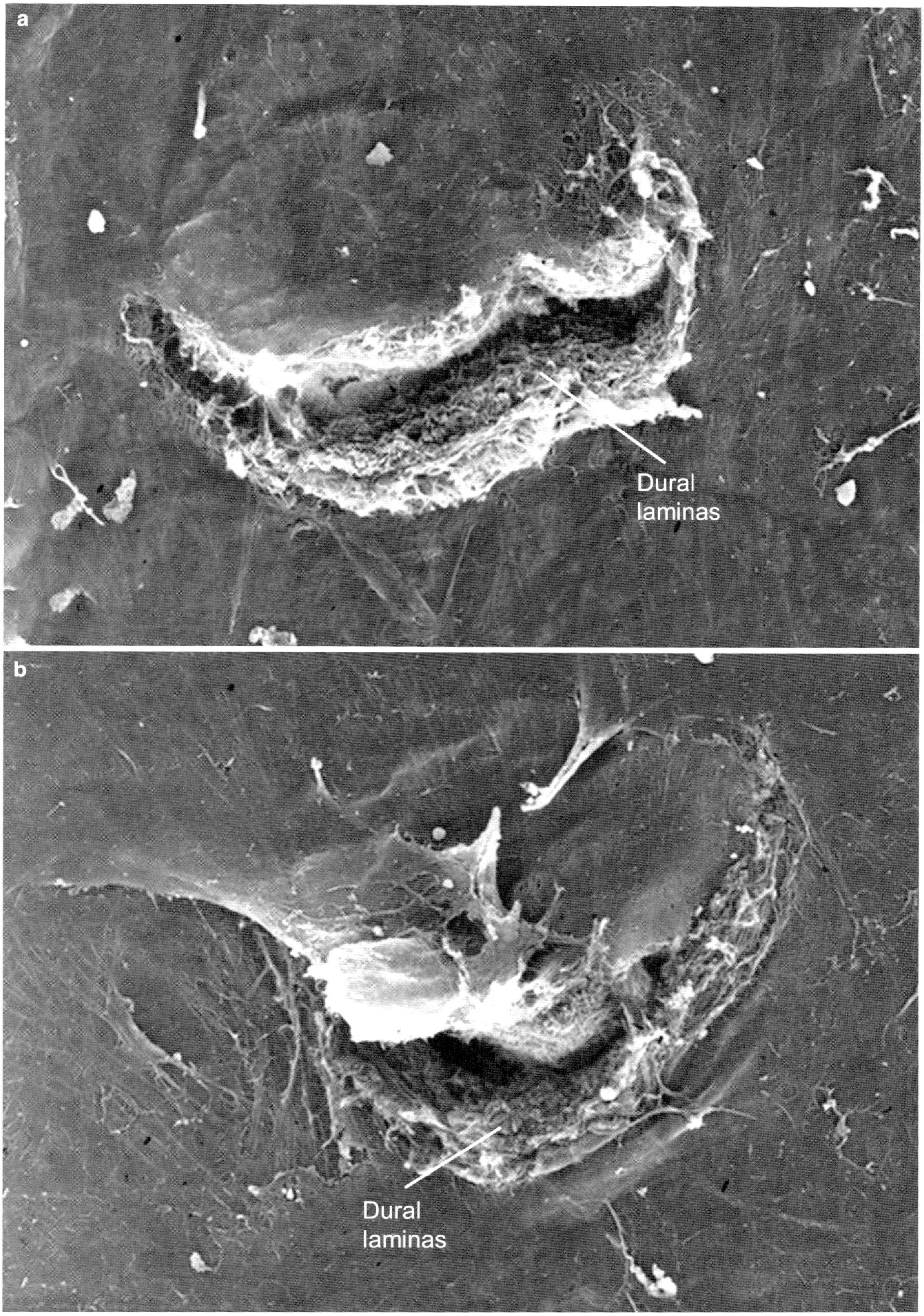

Fig. 41.18 Dura-arachnoid lesion produced by 29G Quincke needle. Internal surface of human dural sac (arachnoid layer). Scanning electron microscopy. Magnification: **a**, ×200; **b**, ×200 (From Reina et al. [9]; with permission)

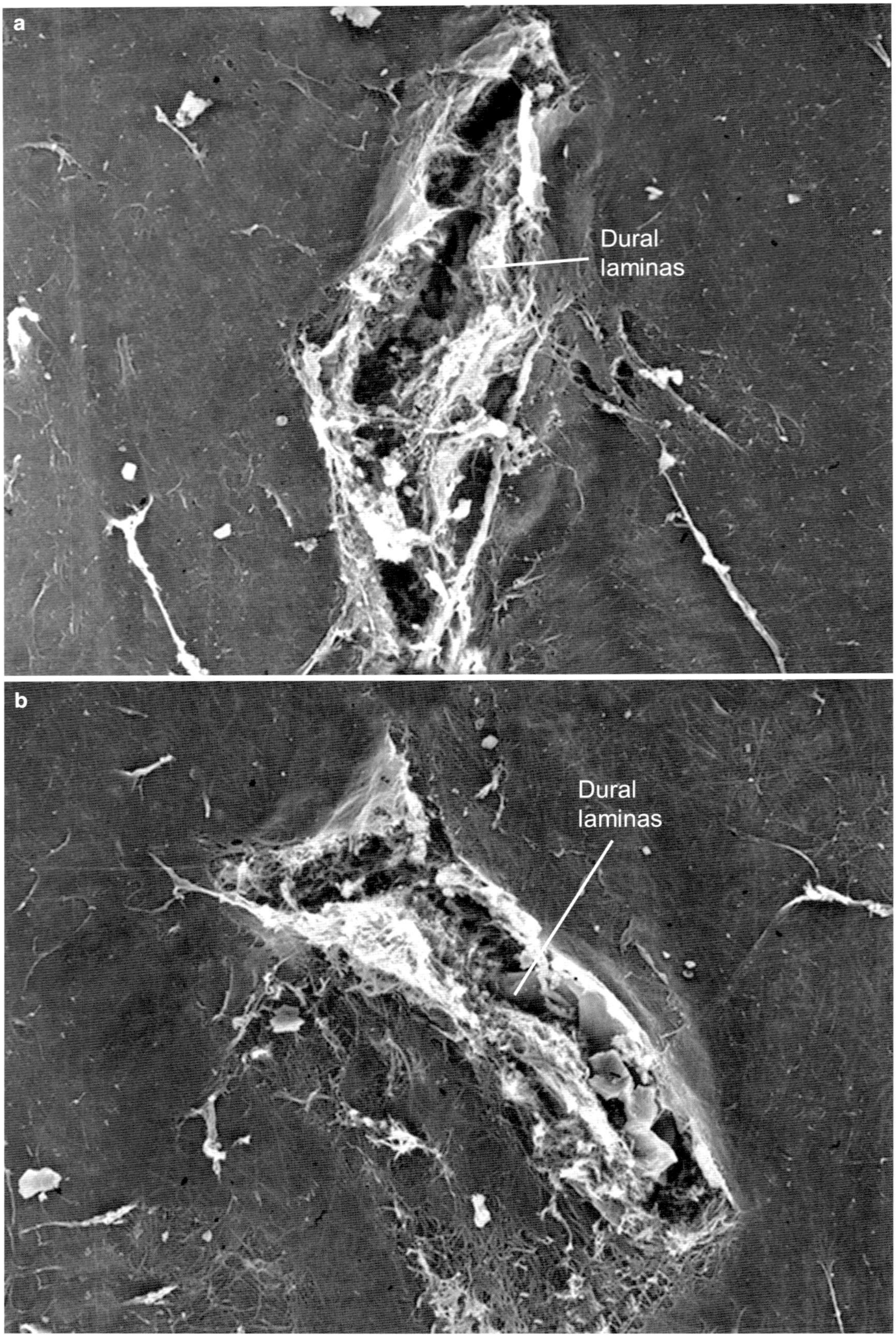

Fig. 41.19 Dura-arachnoid lesion produced by 29G Quincke needle. Internal surface of human dural sac (arachnoid layer). Scanning electron microscopy. Magnification: **a**, ×200; **b**, ×200 (From Reina et al. [9]; with permission)

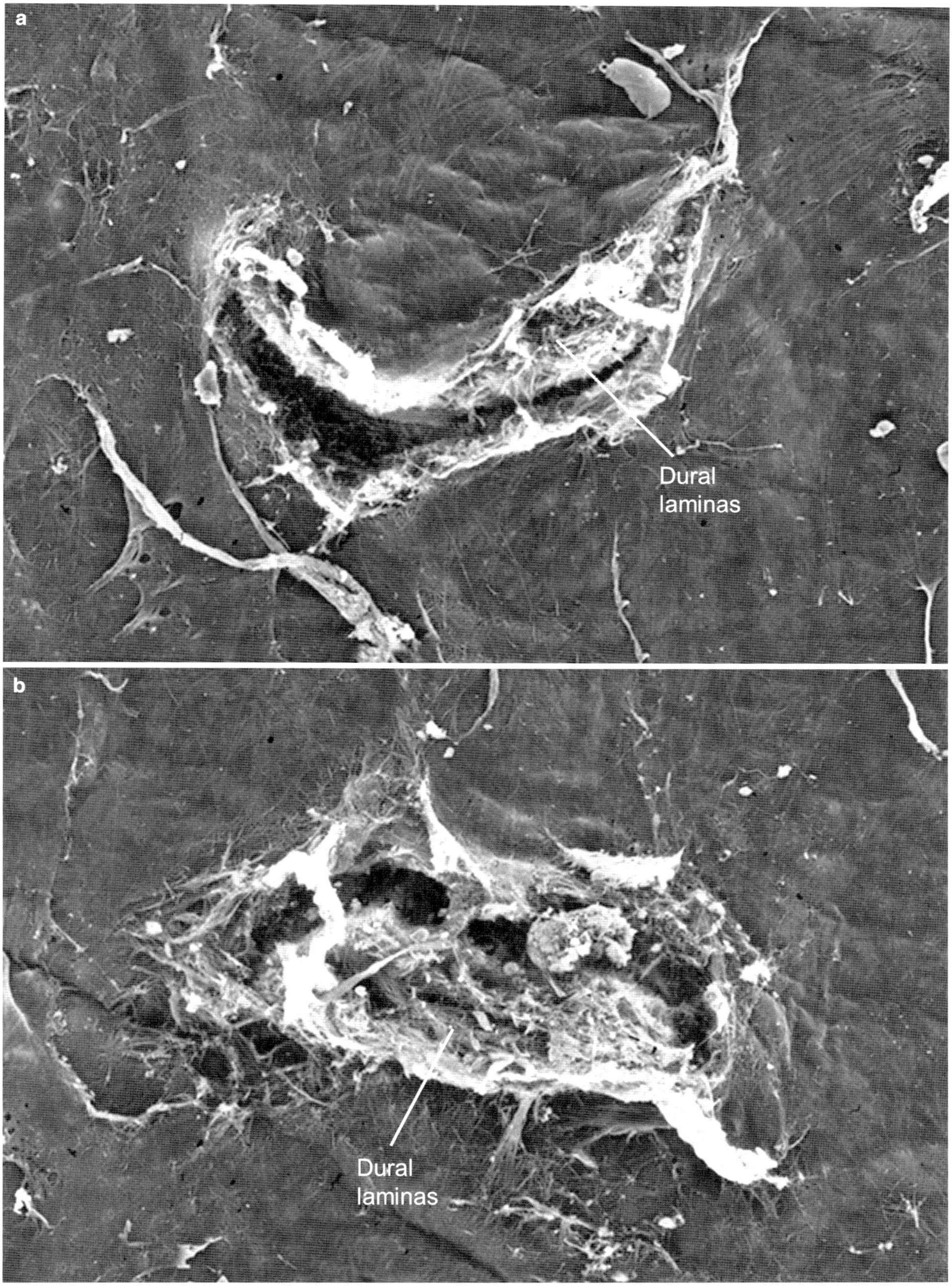

Fig. 41.20 Dura-arachnoid lesion produced by 29G Quincke needle. Internal surface of human dural sac (arachnoid layer). Scanning electron microscopy. Magnification: **a**, ×200; **b**, ×200 (From Reina et al. [9]; with permission)

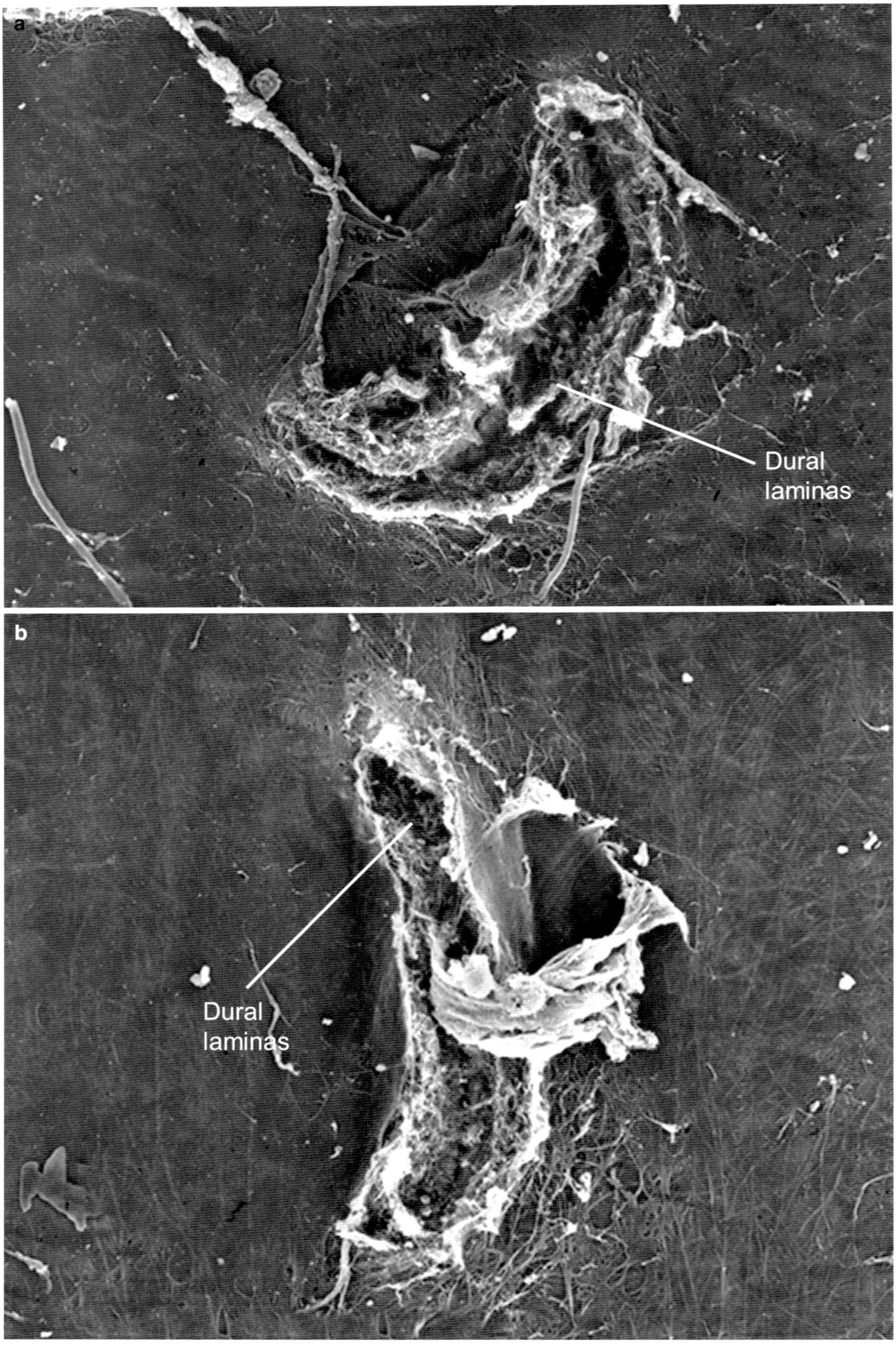

Fig. 41.21 Dura-arachnoid lesion produced by 29G Quincke needle. Internal surface of human dural sac (arachnoid layer). Scanning electron microscopy. Magnification: **a**, ×200; **b**, ×200 (From Reina et al. [9]; with permission)

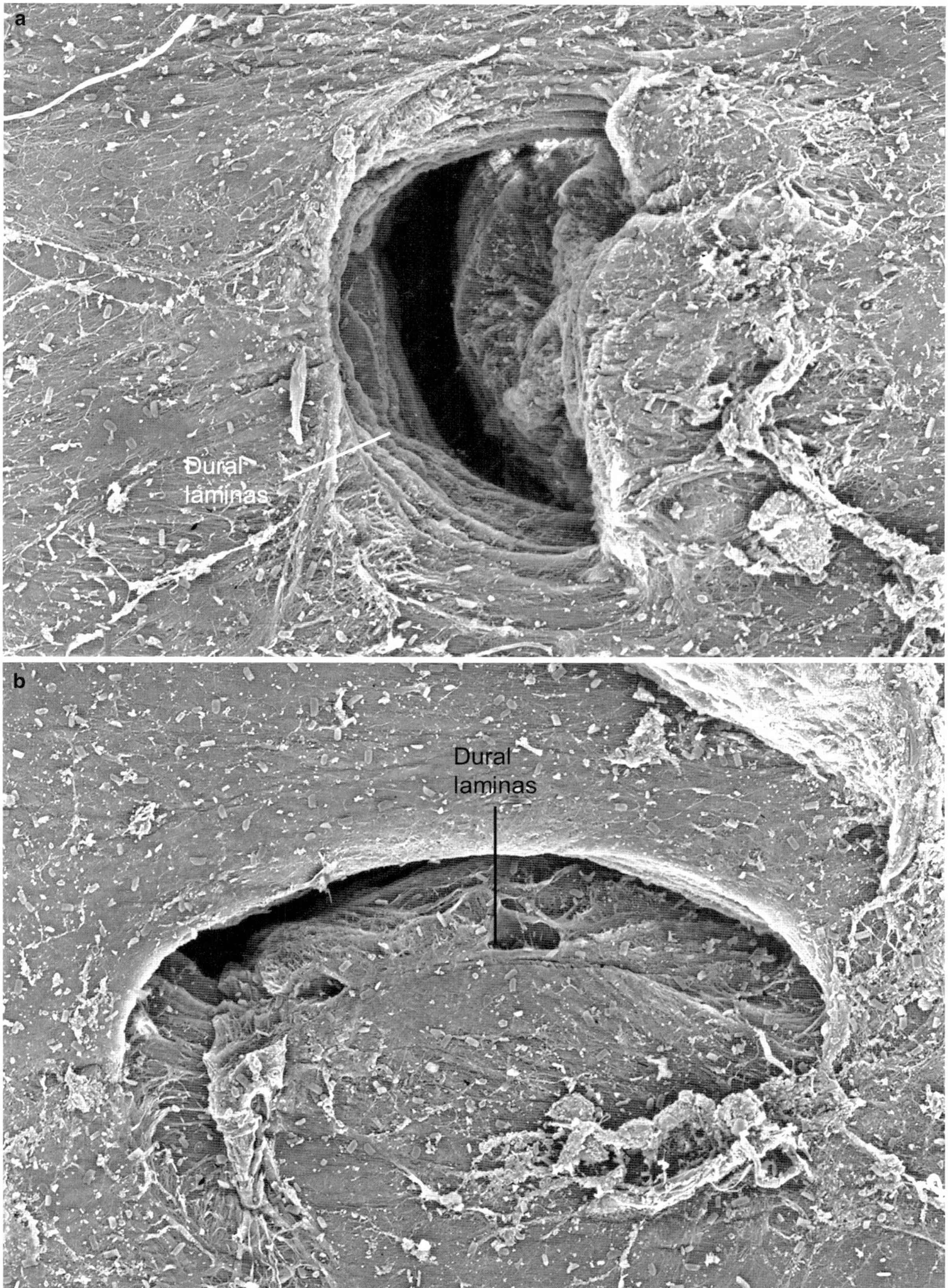

Fig. 41.22 Dura-arachnoid lesion produced by 17G Tuohy needle. External surface of human dural sac (external dural laminas). Scanning electron microscopy. Magnification: **a**, ×50; **b**, ×50 (From Reina et al. [9]; with permission)

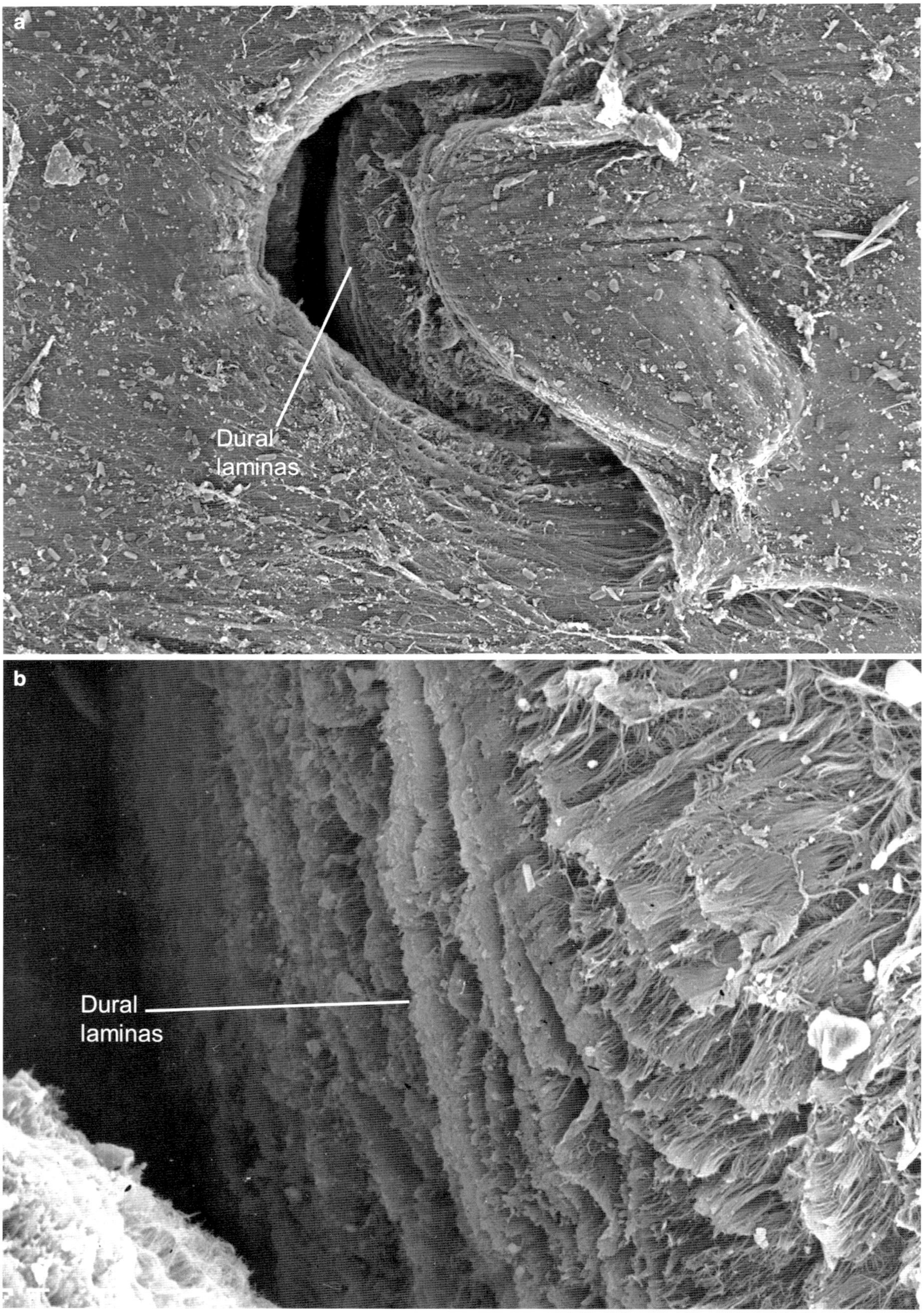

Fig. 41.23 Dura-arachnoid lesion produced by 17G Tuohy needle. External surface of human dural sac (external dural laminas). Scanning electron microscopy. Magnification: **a**, ×50; **b**, ×500 (From Reina et al. [9]; with permission)

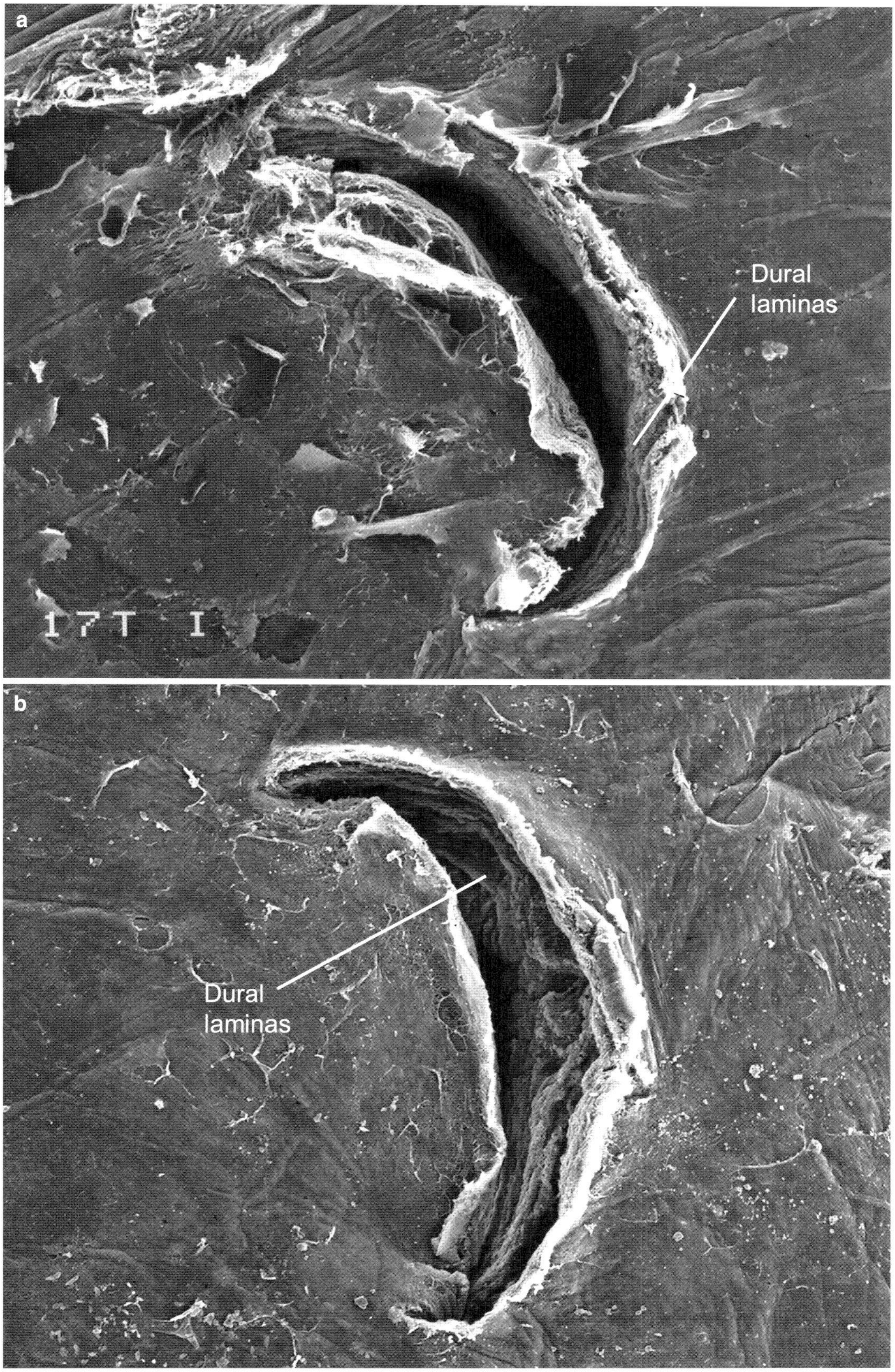

Fig. 41.24 Dura-arachnoid lesion produced by 17G Tuohy needle. Internal surface of human dural sac (arachnoid layer). Scanning electron microscopy. Magnification: **a**, ×50; **b**, ×50

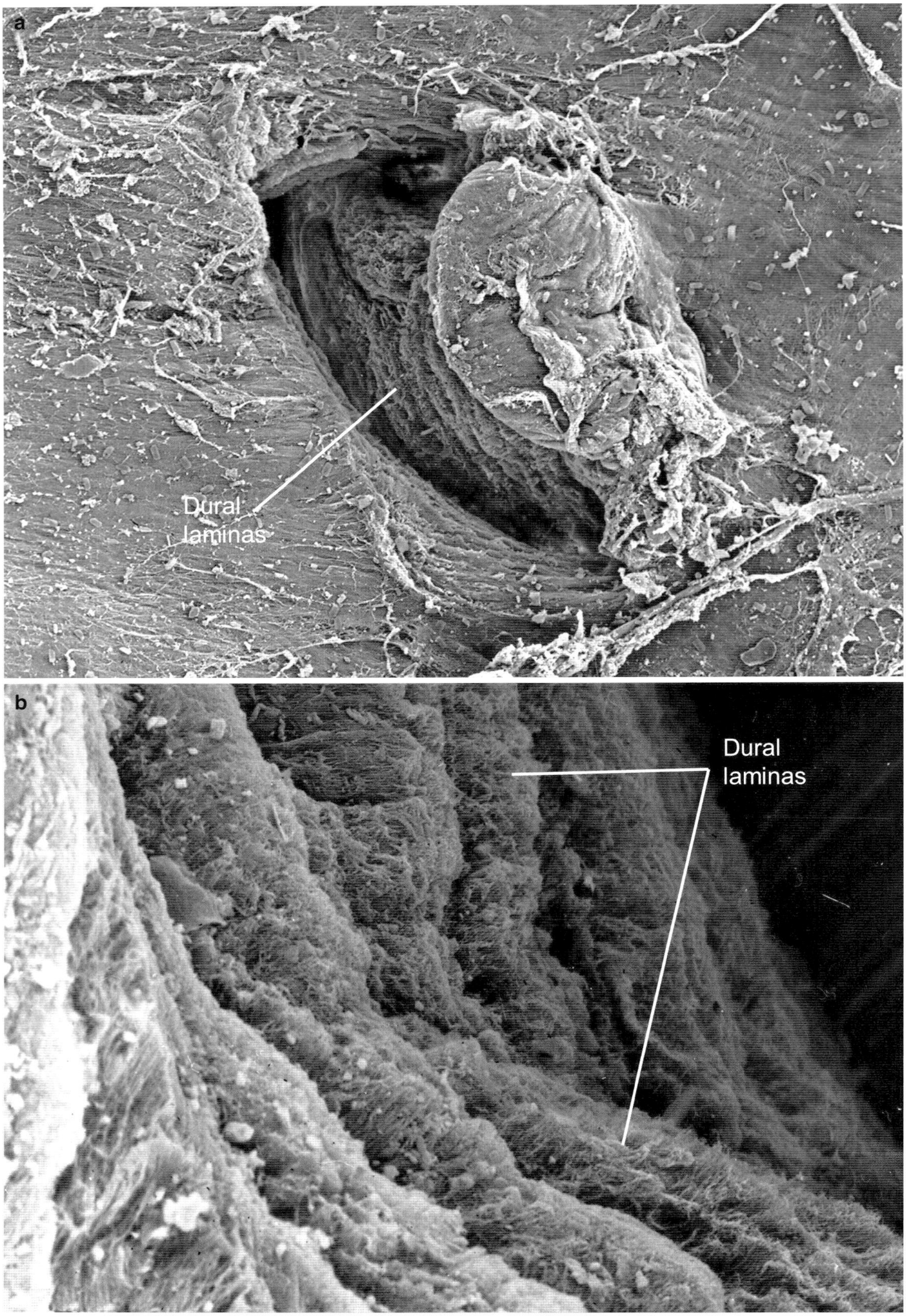

Fig. 41.25 Dura-arachnoid lesion produced by 17G Tuohy needle. External surface of human dural sac (external dural laminas). Scanning electron microscopy. Magnification: **a**, ×50; **b**, ×400 (From Reina et al. [9]; with permission)

References

1. Reina MA, Prats-Galino A, Sola RG, Puigdellívol-Sánchez A, Arriazu Navarro R, De Andrés JA. Structure of the arachnoid layer of the human spinal meninges: a barrier that regulates dural sac permeability. Rev Esp Anestesiol Reanim. 2010;57:486–92.
2. Reina MA, López A, Badorrey V, De Andrés JA, Martín S. Dura-arachnoid lesions produced by 22 gauge Quincke spinal needles during a lumbar puncture. J Neurol Neurosurg Psychiatry. 2004;75:893–7.
3. Reina MA, de Leon-Casasola OA, Lopez A, De Andres J, Martin S, Mora M. An in vitro study of dural lesions produced by 25-gauge Quincke and Whitacre needles evaluated by scanning electron microscopy. Reg Anesth Pain Med. 2000;25:393–402.
4. Dittmann M, Reina MA, López García A. New results in the visualization of the spinal dura mater with scanning electron microscopy. Anaesthesist. 1998;47:409–13.
5. Reina MA, Dittmann M, López Garcia A, van Zundert A. New perspectives in the microscopic structure of human dura mater in the dorsolumbar region. Reg Anesth. 1997;22:161–6.
6. Reina MA, López-García A, de Andrés-Ibáñez JA, Dittmann M, Cascales MR, del Caño MC, et al. Electron microscopy of the lesions produced in the human dura mater by Quincke beveled and Whitacre needles. Rev Esp Anestesiol Reanim. 1997;44:56–61.
7. Reina MA, López-García A, Dittmann M, de Andrés JA. Structural analysis of the thickness of human dura mater with scanning electron microscopy. Rev Esp Anestesiol Reanim. 1996;43:135–7.
8. Reina MA, López-García A, Dittmann M, de Andrés JA. Analysis of the external and internal surface of human dura mater with scanning electron microscopy. Rev Esp Anestesiol Reanim. 1996; 43:130–4.
9. Reina MA, Castedo J, López A. Cefalea pospunción dural. Ultraestructura de las lesiones durales y agujas espinales usadas en las punciones lumbares. Rev Arg Anestesiol. 2008;66:6–26.
10. Reina MA, De Andrés JA, López A. Subarachnoid and epidural anesthesia. In: Raj P, editor. Textbook of regional anesthesia. New York: Churchill Livingstone; 2002. p. 307–24.

Injections of Particulate Steroids for Nerve Root Blockade: Ultrastructural Examination of Complicating Factors

42

Miguel Angel Reina, José Antonio De Andrés, and José María Hernández

Selective nerve root block by a transforaminal approach with radiological control allows injection of an appropriate corticosteroid next to the nerve root, although it has been associated with major neurological complications in a few cases. There have been several reports in which corticosteroid injections performed at the level of the cervical or lumbar spine have led to severe spinal cord ischemic infarction. These have occurred independently of whether the procedure was performed under radiological control, in which the tip of the needle is identified to prevent vascular spreading of the solution injected.

Knowledge of arterial vascularization and of the size of arteries at the intervertebral foraminal level is essential to understanding the possibility of introducing the tip of the needle used. Radicular arteries are medially located inside the intervertebral foramen, close to nerve roots. Some reach the spinal cord; others only reach the nerve roots. Branches of radicular arteries are vertebral arteries, as well as ascending cervical, deep cervical, intercostal, lumbar, and iliolumbar arteries. Huntoon [1] described the importance of the ascending and deep cervical arteries when performing transforaminal cervical injections, because of their critical role in spinal perfusion. The ascending and deep cervical arterial branches enter the external opening of the posterior intervertebral foramen near the target area for transforaminal injections.

At the lumbar level, the artery of Adamkiewicz (arteria radicularis magna) enters the left foramina between T9 and L1 in 75 % of patients, between thoracic vertebrae T7 and T9 in 15 %, and between L1 and L4 in 10 %. The artery usually enters in the superior or middle portion of the neural foramen, slightly ventral and superolateral to the dorsal root ganglion. Injury to the artery of Adamkiewicz can result in devastating ischemia of the lower spinal cord, causing the anterior spinal artery syndrome [1].

In our study, vessels were found next to the nerve root cuff. In an attempt to improve the images, the dural sac was dissected together with nerve root cuffs and plexus roots. During dissection, vessels were kept close to nerve root cuffs and plexus roots. This area is the target region for placement of needles during selective root blocks.

The diameter of these arteries was 1–2 mm at the cervical, thoracic, and lumbar levels near the puncture zone. Commonly used needles with sharp and blunt tips were introduced into these vessels. This study allowed us to verify that arterial injection during a selective root block is possible at the cervical level as well as at the thoracic and lumbar levels.

Steroids with particle sizes larger than 1,000 μm theoretically cannot enter the vessel. Because the solution is usually shaken before it is injected, however, it is possible that the steroid particles coalesce and precipitate to form larger particles after the steroid has entered the blood vessel. Medium-sized particles (51–1,000 μm) can enter and partially occlude the vessels. Smaller particles (10–50 μm) may not be able to occlude the arteries [2], but these smaller particles can occlude arterioles and capillaries, causing tis-

M.A. Reina, MD, PhD (✉)
Department of Clinical Medical Sciences and Institute of Applied Molecular Medicine, School of Medicine, University of CEU San Pablo, Madrid, Spain

Department of Anesthesiology, Madrid-Montepríncipe University Hospital, Madrid, Spain
e-mail: miguelangel@perticone.e.telefonica.net

J.A. De Andrés, MD, PhD
Department of Surgical Specialties, School of Medicine, University of Valencia, Valencia, Spain

Department of Anesthesia, Critical Care, and Pain Management, General University Hospital, Valencia, Spain
e-mail: deandres_jos@gva.es

J.M. Hernández, MD
Department of Anesthesiology, Madrid-Montepríncipe University Hospital, Madrid, Spain

Department of Clinical Medical Sciences, School of Medicine, CEU San Pablo University, Madrid, Spain
e-mail: jhzamora@yahoo.es, jmhernandez@madridsindolor.com

M.A. Reina et al. (eds.), *Atlas of Functional Anatomy for Regional Anesthesia and Pain Medicine: Human Structure, Ultrastructure and 3D Reconstruction Images*, DOI 10.1007/978-3-319-09522-6_42,

sue damage. It is therefore possible that any particulate steroid can occlude the distal portions of the arteries. In epidural steroid injections, the steroids are usually diluted to decrease their concentration, but it has been found that dilution does not decrease the size of the particles, except in compounded betamethasone, in which increasing dilutions with the local anesthetic decreased the proportion of the larger particles [2].

Methylprednisolone, triamcinolone, and betamethasone have been associated with reports of ischemic complications; dexamethasone has not [3]. Dexamethasone sodium phosphate particle size is approximately 10 times smaller than red blood cells, and the particles do not appear to aggregate [3–6]. These characteristics should significantly reduce the risk of embolic infarcts or prevent them from occurring after intra-arterial injection. Triamcinolone acetonide and betamethasone sodium phosphate showed variable particle sizes: some particles were larger than red blood cells, and aggregation of particles was evident. Methylprednisolone acetate showed uniformity in size, and most of the particles were smaller than red blood cells; they were not aggregated, but the particles were densely packed [3–7].

In our study, different particulate steroids were examined under scanning electron microscopy and crystals were identified in normal conditions without dilution and after dilution with local anesthetic. Two solutions were made, one with 1 cc of steroid and 1 cc of lidocaine 1 %, used for transforaminal injection, and another with 1 cc of steroid and 10 cc of bupivacaine 0.25 %, used for epidural administration. Both were filtered through a Millipore filter of 0.2 μm.

In normal conditions, without dilution, triamcinolone particles piled on the filter. Few particles were seen with betamethasone and dexamethasone, and none with methylprednisolone. When diluted with local anesthetic, only triamcinolone appeared in different sizes in the filter: 10–100 μm in a 1:1 dilution with lidocaine 1 %, and 10–60 μm when diluted 1:10 with bupivacaine 0.25 %.

Some authors [8, 9] advocate the use of blunt needles for transforaminal injection to avoid inadvertent arterial injection during cervical transforaminal injection. Unintentional arterial injection has been reported when using other types of needles [10–13]. Although the use of blunt needle tips seems reasonable, there is a lack of clinical evidence that complications will not occur. The tip of a blunt needle may push the wall of an artery rather than pierce it, thus presenting less risk of blood vessel injection, as introduction of 22G sharp needles into arteries is possible (Figs. 42.1, 42.2, 42.3, 42.4, 42.5, 42.6, 42.7, 42.8, 42.9, 42.10, 42.11, 42.12, 42.13, 42.14, 42.15, 42.16, 42.17, 42.18, 42.19, 42.20, 42.21, 42.22, 42.23, 42.24, 42.25, and 42.26).

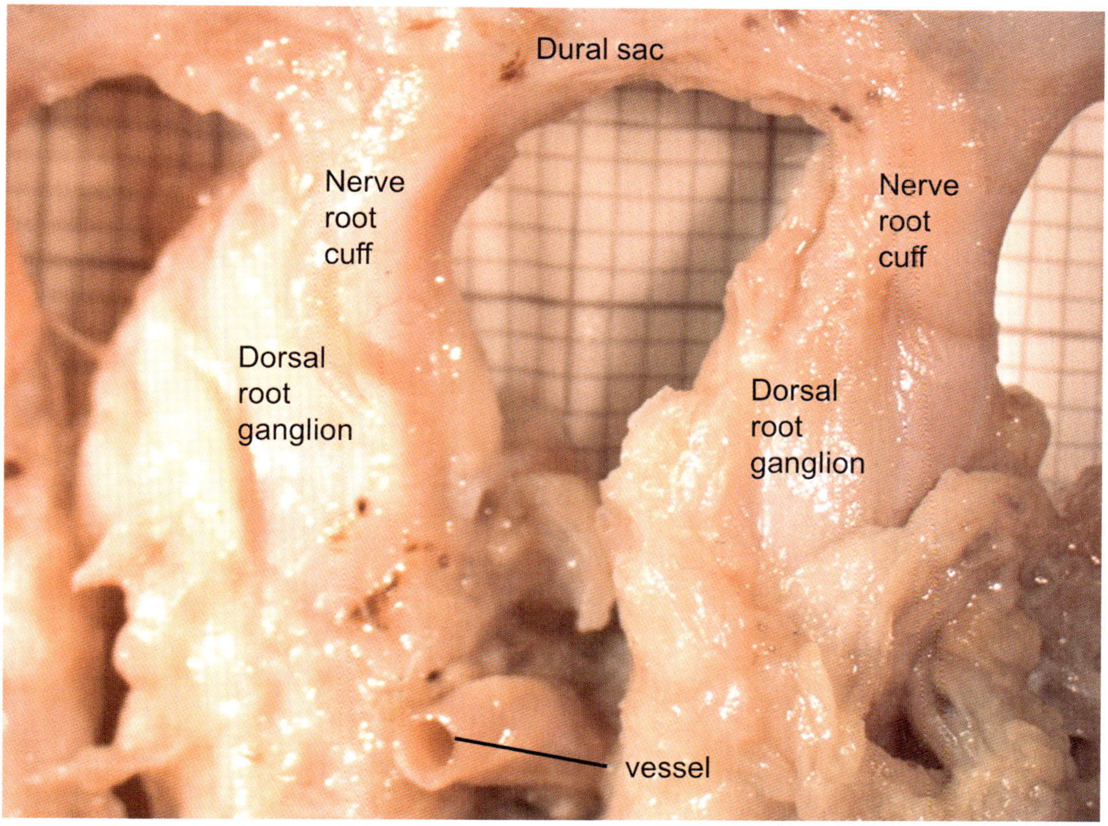

Fig. 42.1 Cervical nerve root cuffs and vessels close to the dorsal root ganglion. Distal to DRG a vessel with 1.5 mm of diameter can be identified

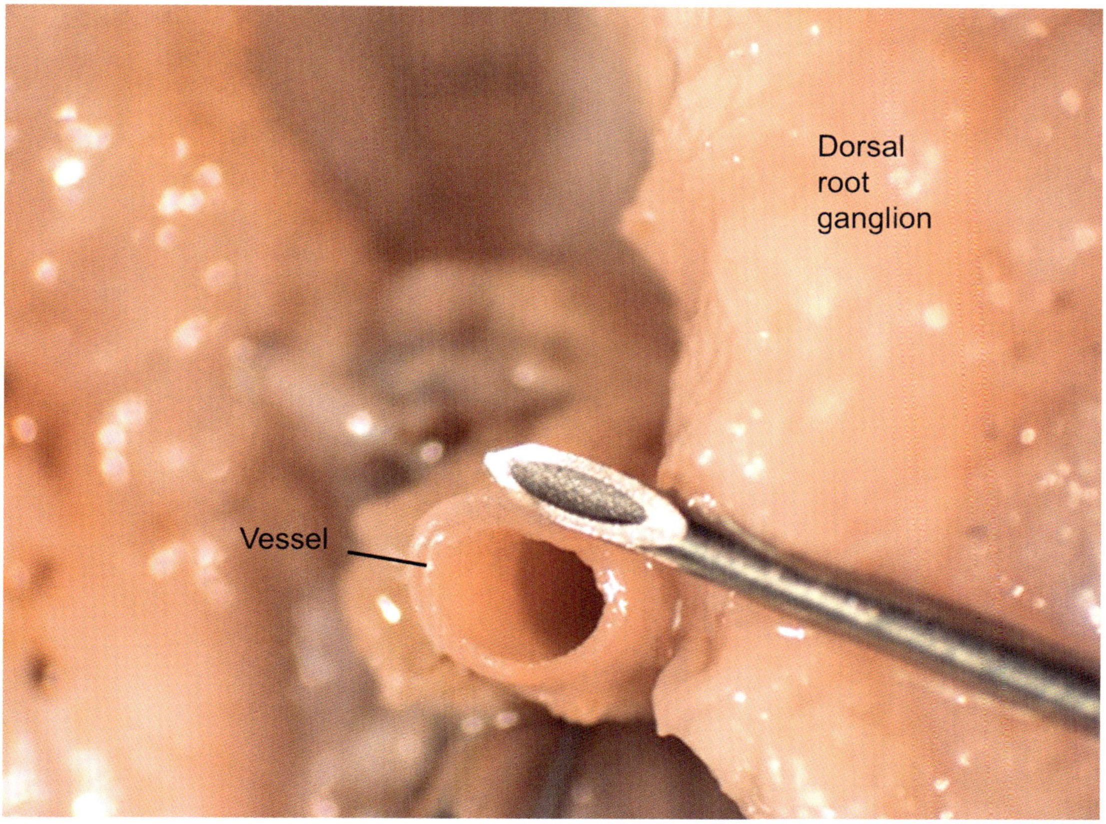

Fig. 42.2 Cervical nerve root cuffs and vessels close to the dorsal root ganglion. We can compare the size of the tip of a 22G Quincke needle with the diameter of the vessel

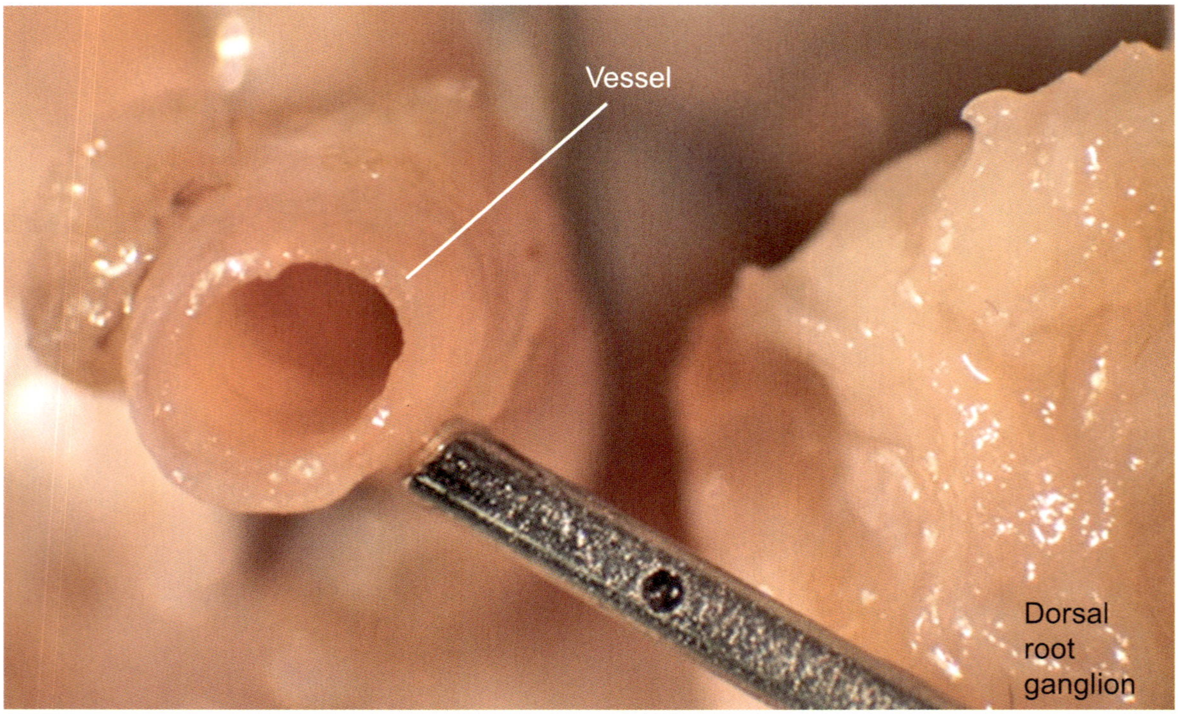

Fig. 42.3 Cervical nerve root cuffs and vessels close to the dorsal root ganglion. We can compare the size of the tip of a 22G blunt needle with the diameter of the vessel

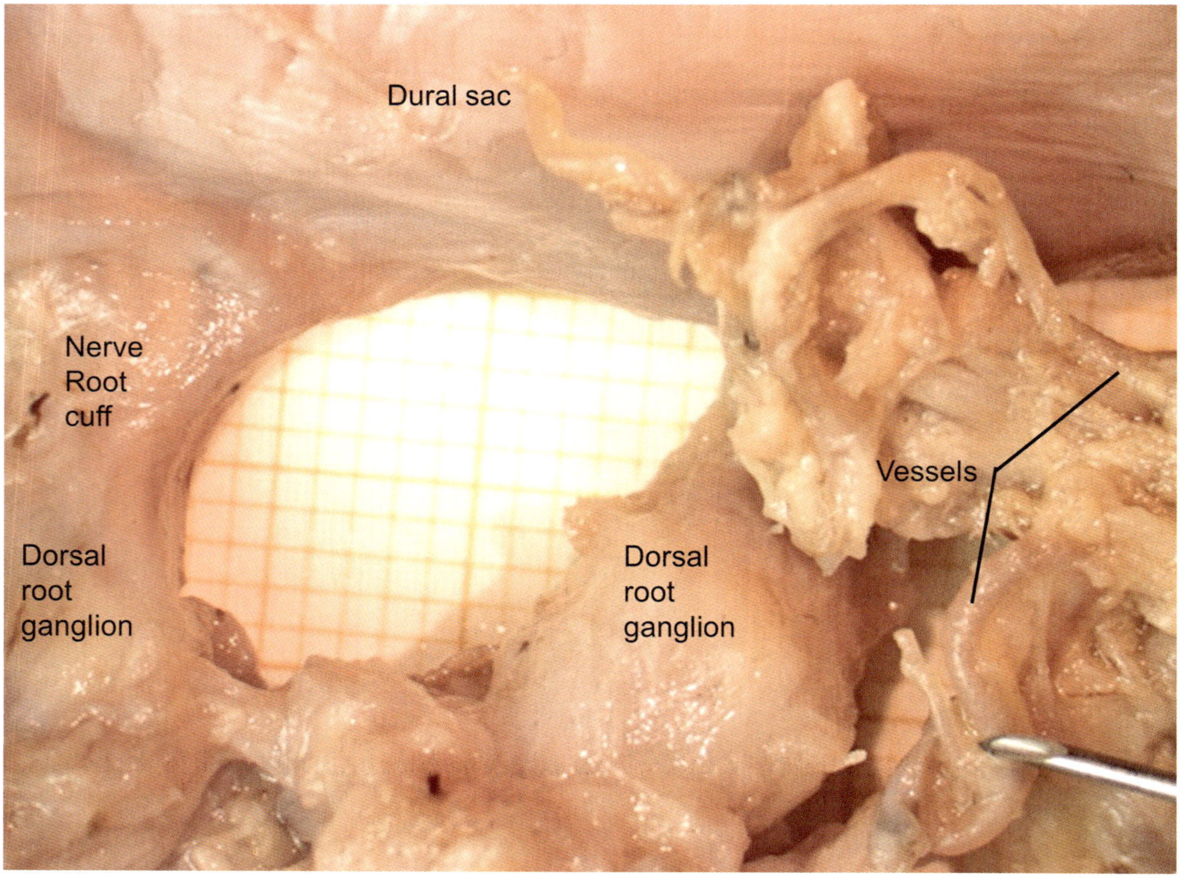

Fig. 42.4 Cervical nerve root cuffs and vessels close to the dorsal root ganglion. We can compare the size of the tip of a 22G Quincke needle and vessel close to DRG

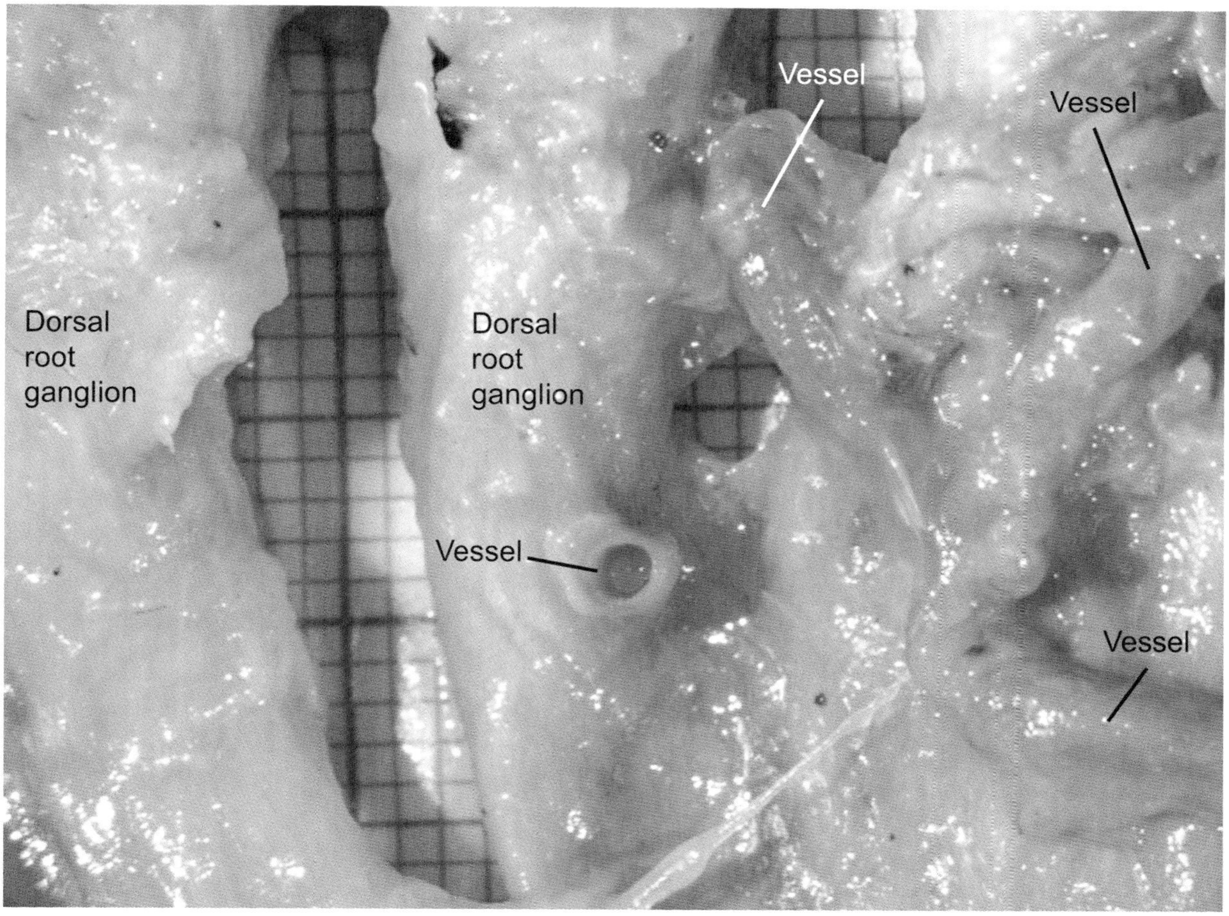

Fig. 42.5 Cervical nerve root cuffs and vessels close to the dorsal root ganglion. Vessel close to DRG can be seen

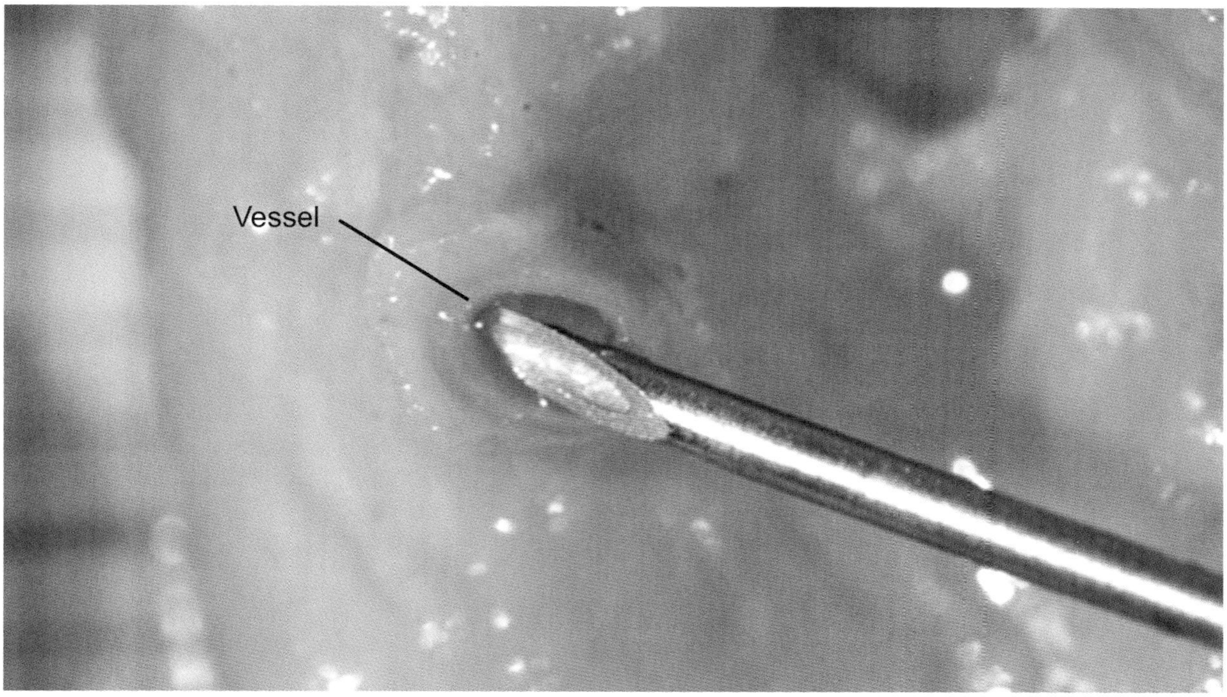

Fig. 42.6 Cervical nerve root cuffs and vessels close to the dorsal root ganglion. The sizes of the tip of a 22G Quincke needle and the diameter of the vessel are identified

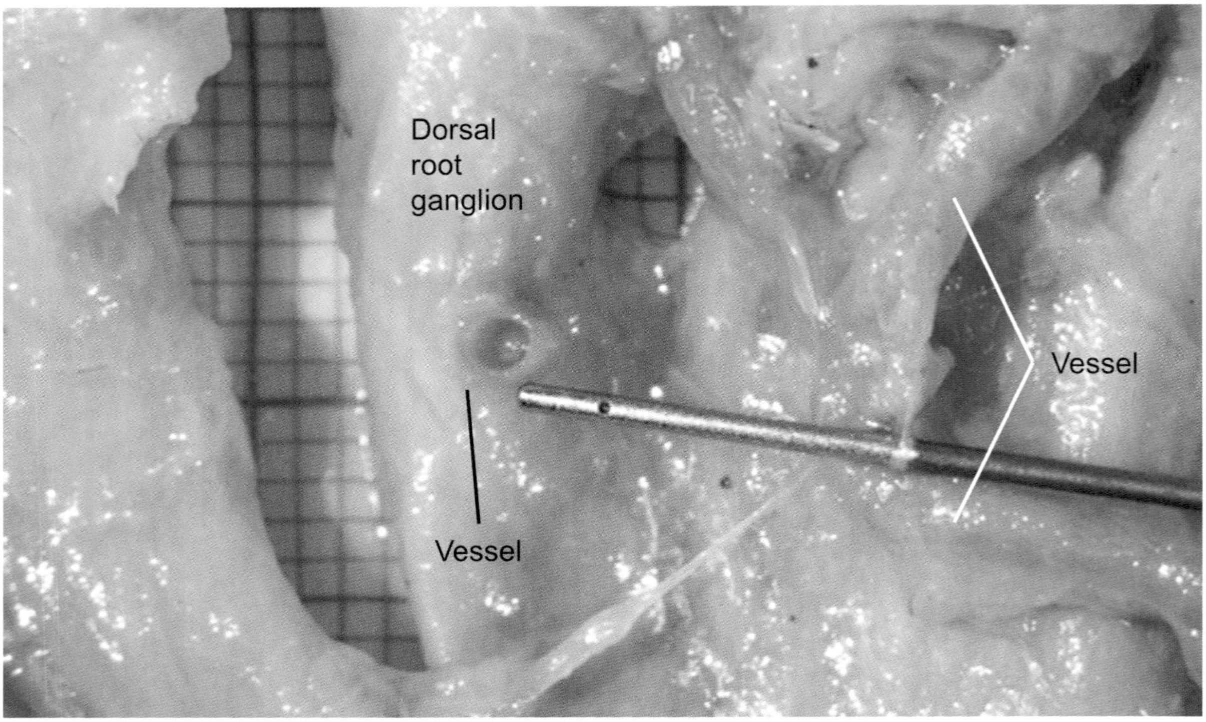

Fig. 42.7 Cervical nerve root cuffs and vessels close to the dorsal root ganglion. The sizes of the tip of a 22G blunt needle and the diameter of the vessel are identified

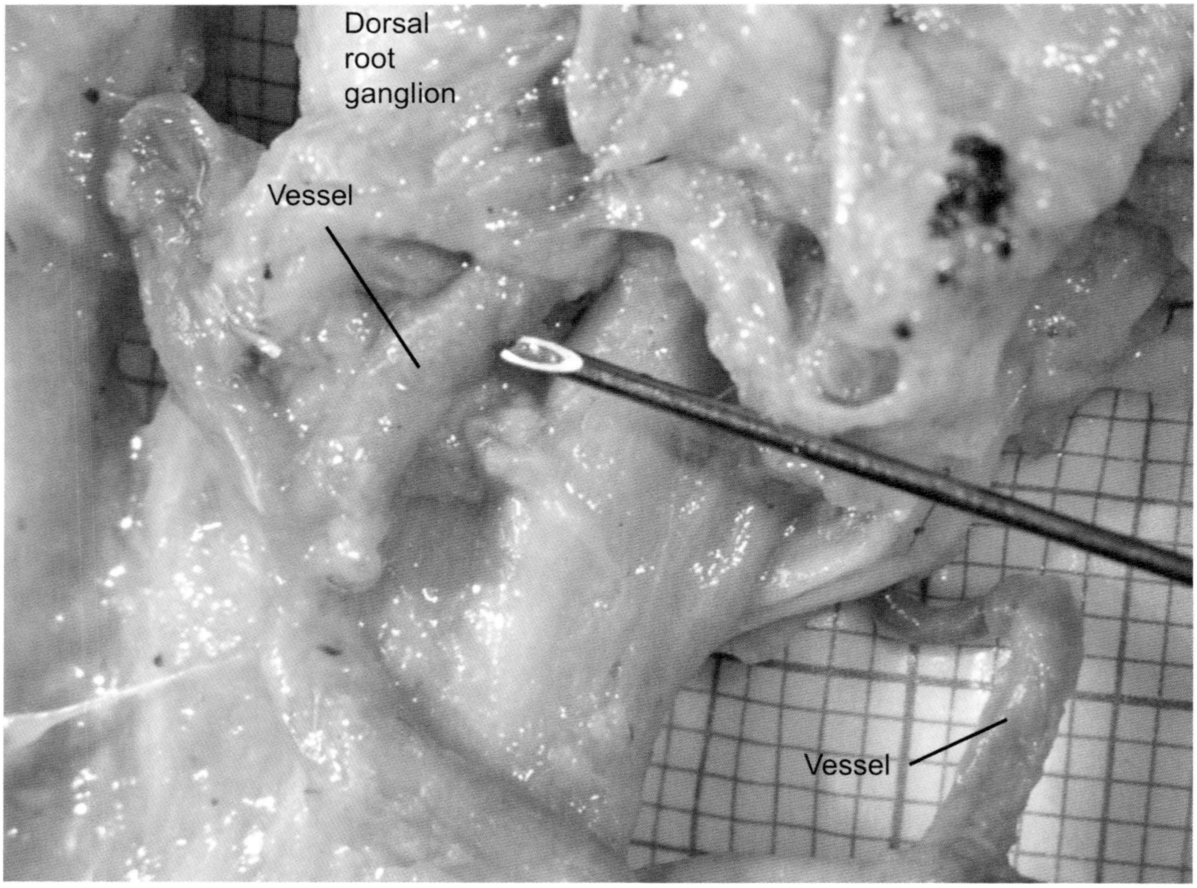

Fig. 42.8 Cervical nerve root cuffs and vessels close to the dorsal root ganglion. The tip of a 22G Quincke needle piercing the wall of the vessel is identified

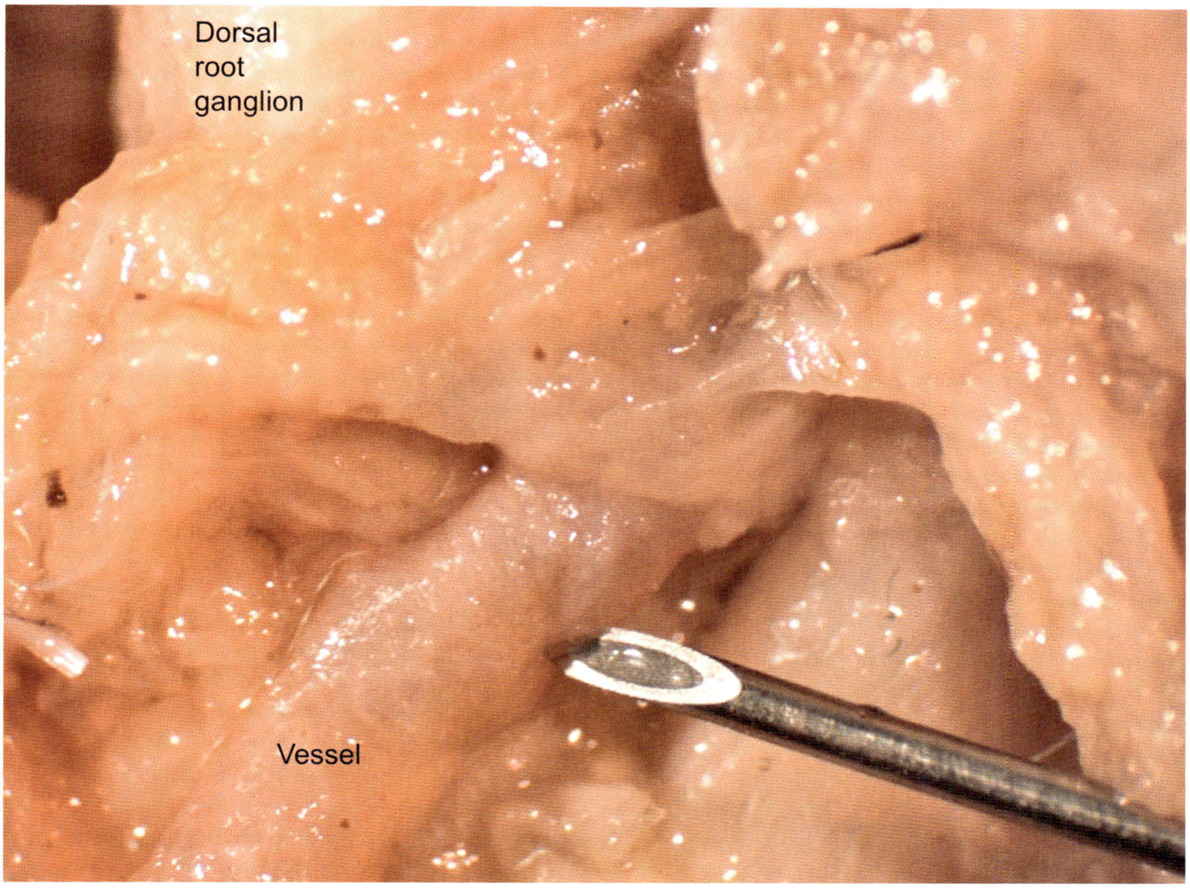

Fig. 42.9 Cervical nerve root cuffs and vessels close to the dorsal root ganglion. The sizes of the tip of the 22G Quincke needle and the diameter of the vessel can be compared

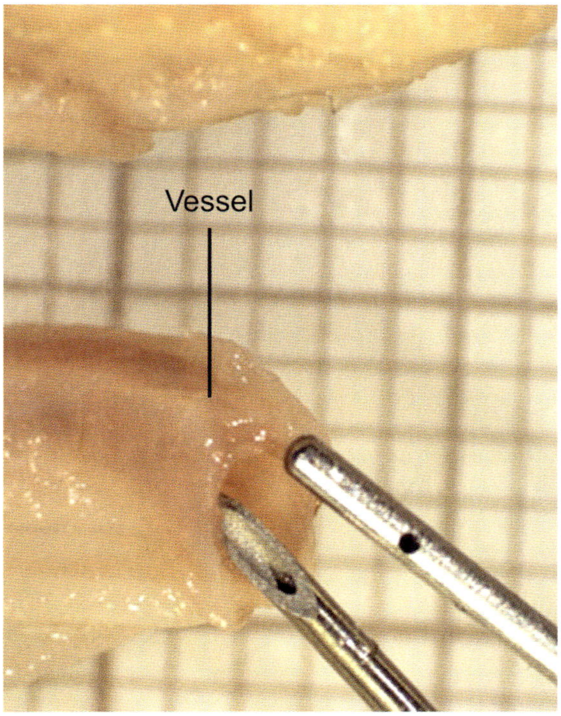

Fig. 42.10 Cervical nerve root cuffs and vessels close to the dorsal root ganglion. The sizes of the tips of a 22G Quincke needle and a 22G blunt needle and the diameter of the vessel can be compared, showing the different distances between their respective tips and orifices (From Reina et al. [4]; with permission)

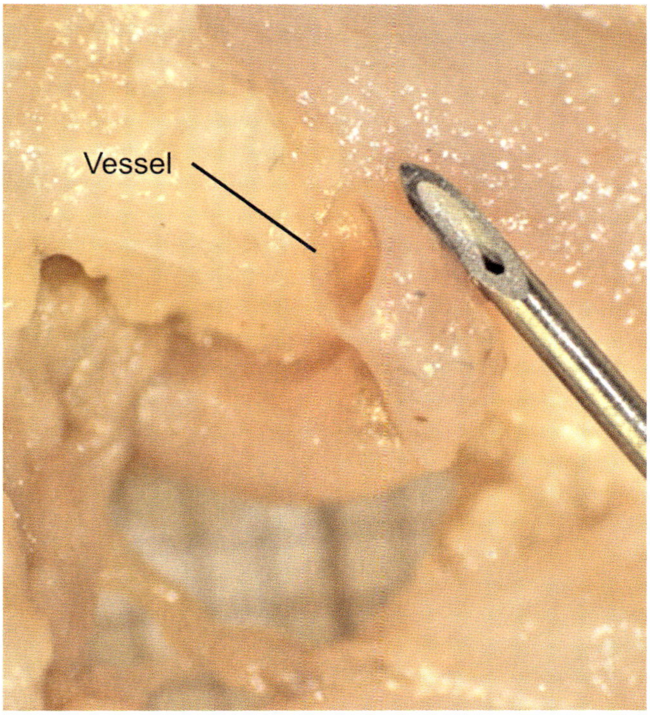

Fig. 42.11 Cervical nerve root cuffs and vessels close to the dorsal root ganglion. The sizes of the tip of a 22G Quincke needle and the diameter of the vessel are identified (From Reina et al. [4]; with permission)

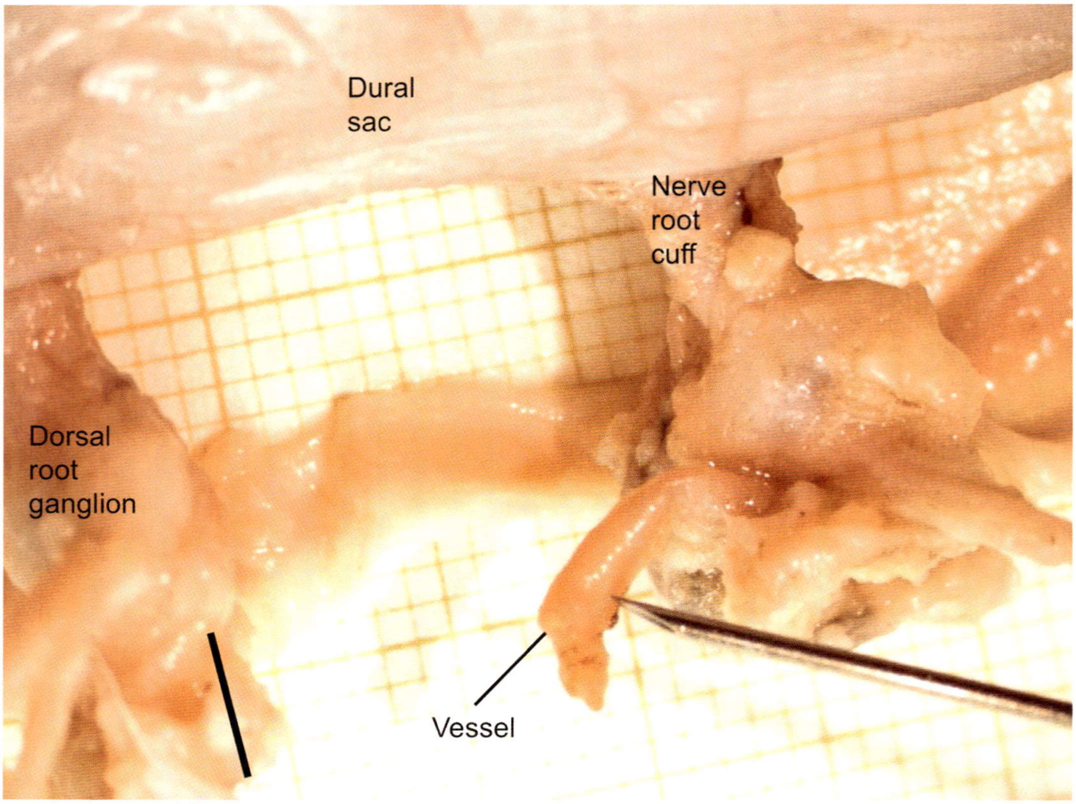

Fig. 42.12 Lumbar nerve root cuffs and vessels close to the dorsal root ganglion. The sizes of the tip of a 22G Quincke needle and the diameter of the vessel are identified

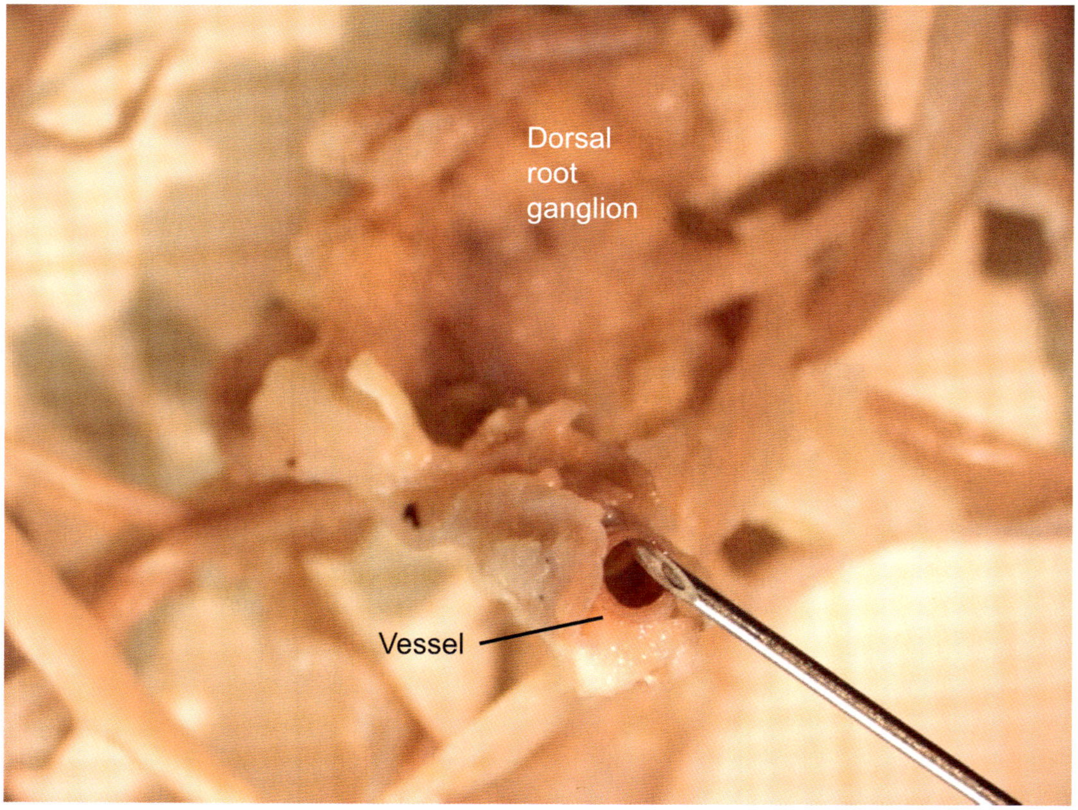

Fig. 42.13 Lumbar nerve root cuffs and vessels close to the dorsal root ganglion. The sizes of the tip of a 22G Quincke needle and the diameter of the vessel are identified

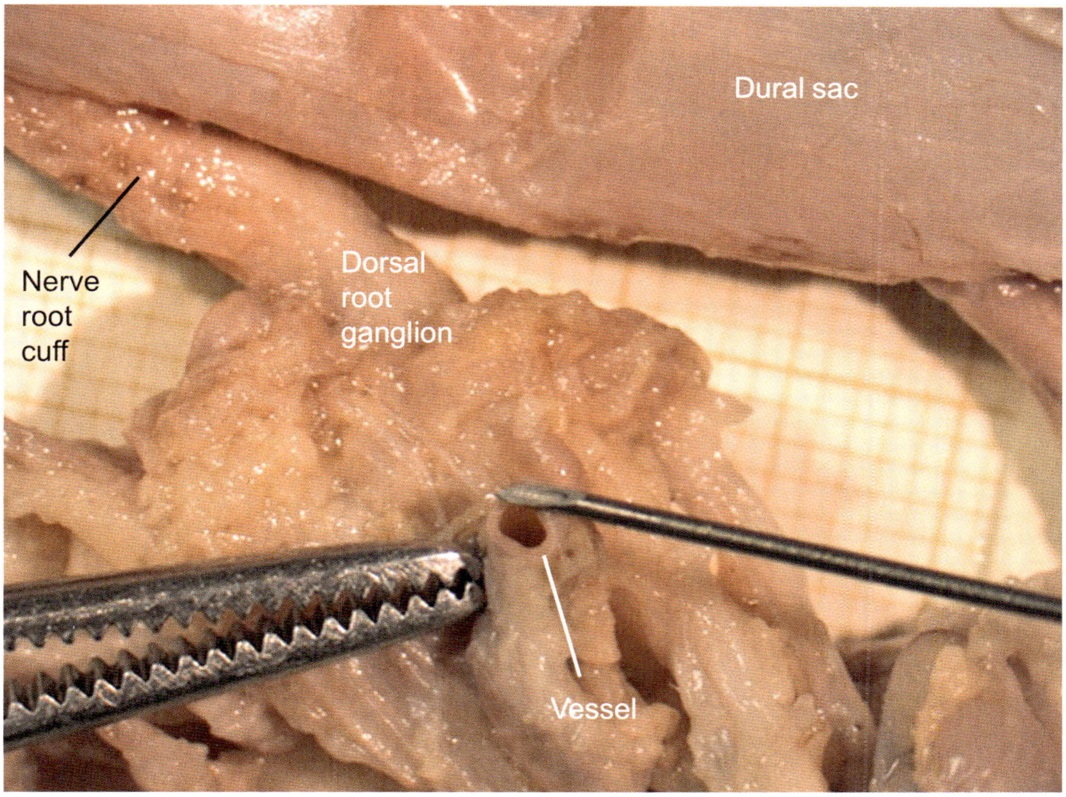

Fig. 42.14 Lumbar nerve root cuffs and vessels close to the dorsal root ganglion. The sizes of the tip of a 22G Quincke needle and the diameter of the vessel are identified

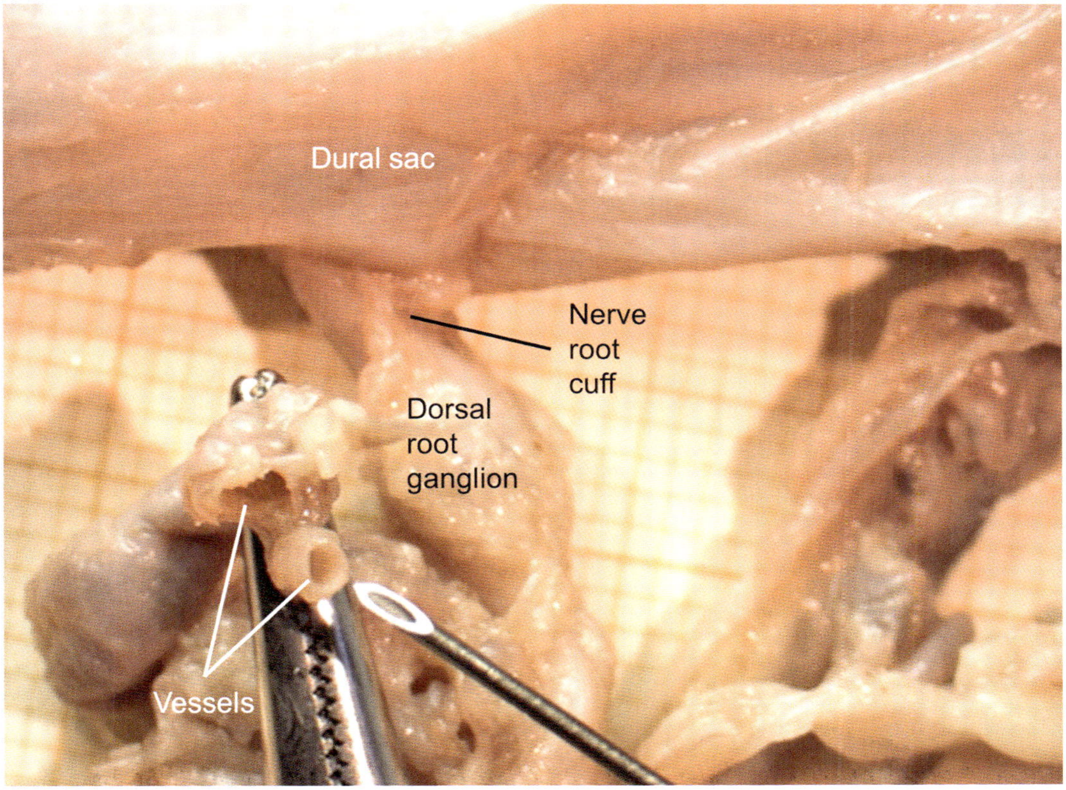

Fig. 42.15 Lumbar nerve root cuffs and vessels close to the dorsal root ganglion. Both artery and vein are clearly identified against the tip of a 22G Quincke needle

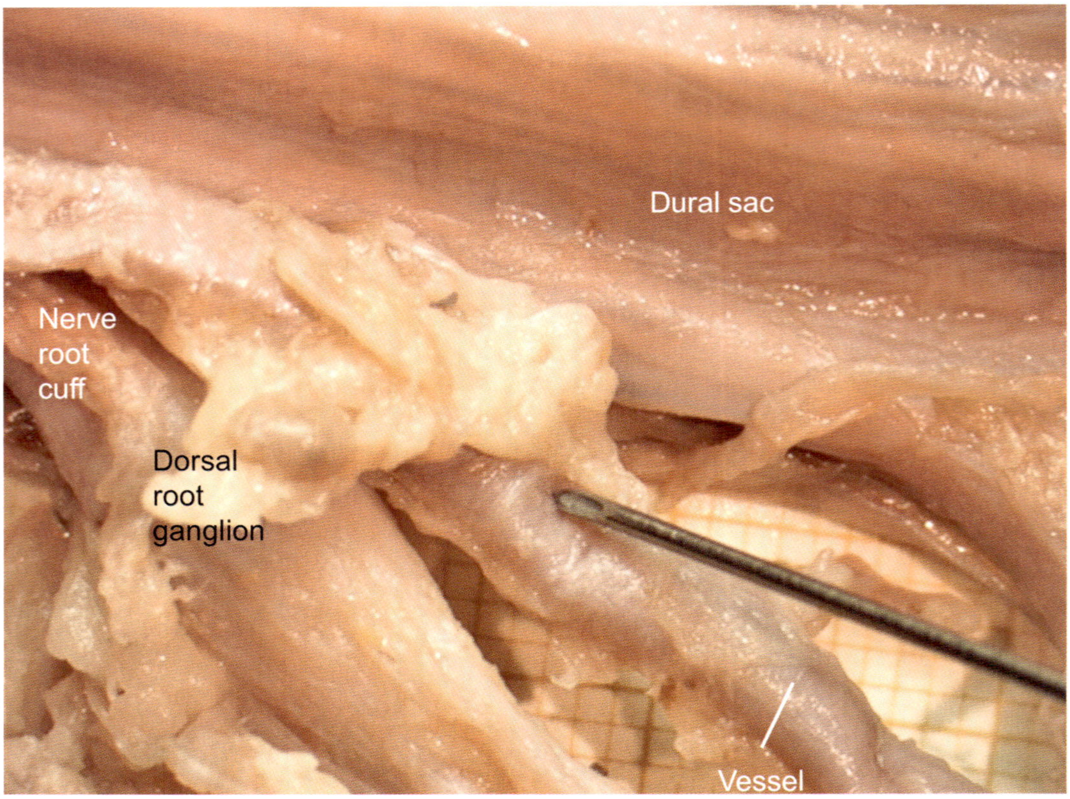

Fig. 42.16 Lumbar nerve root cuffs and vessels close to the dorsal root ganglion. The sizes of the tip of a 22G Quincke needle and the diameter of the vessel are identified

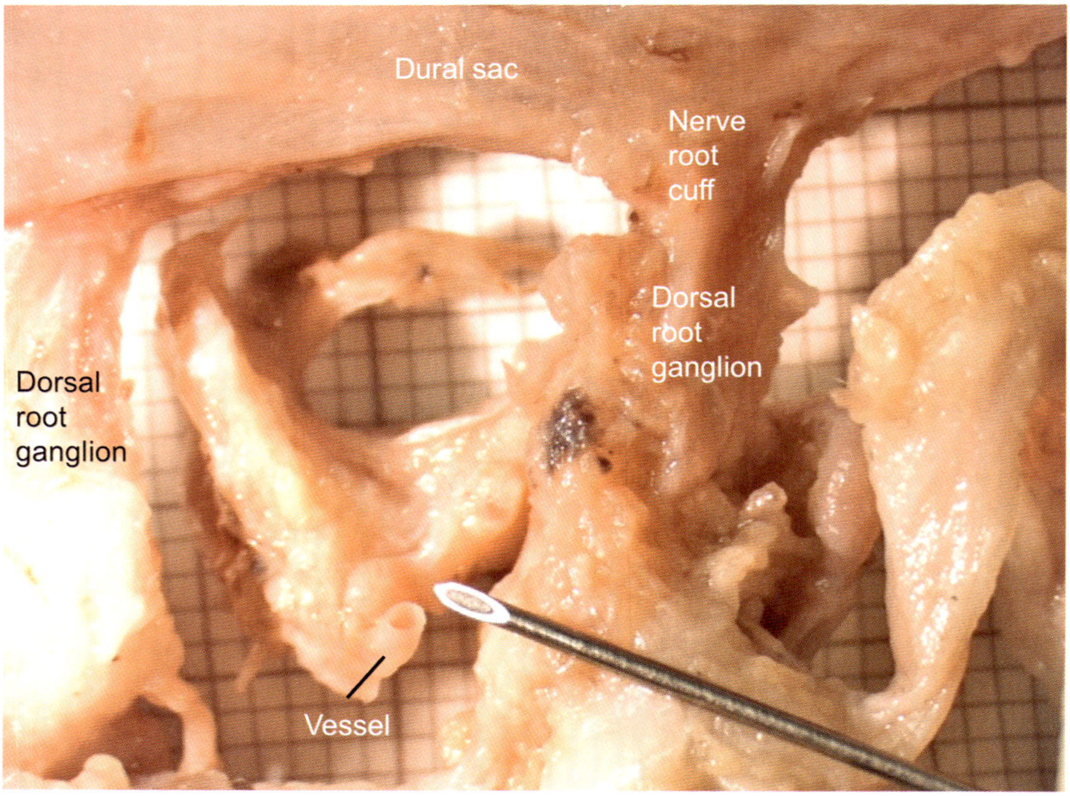

Fig. 42.17 Lumbar nerve root cuffs and vessels close to the dorsal root ganglion. The sizes of the tip of a 22G Quincke needle and the diameter of the vessel are identified

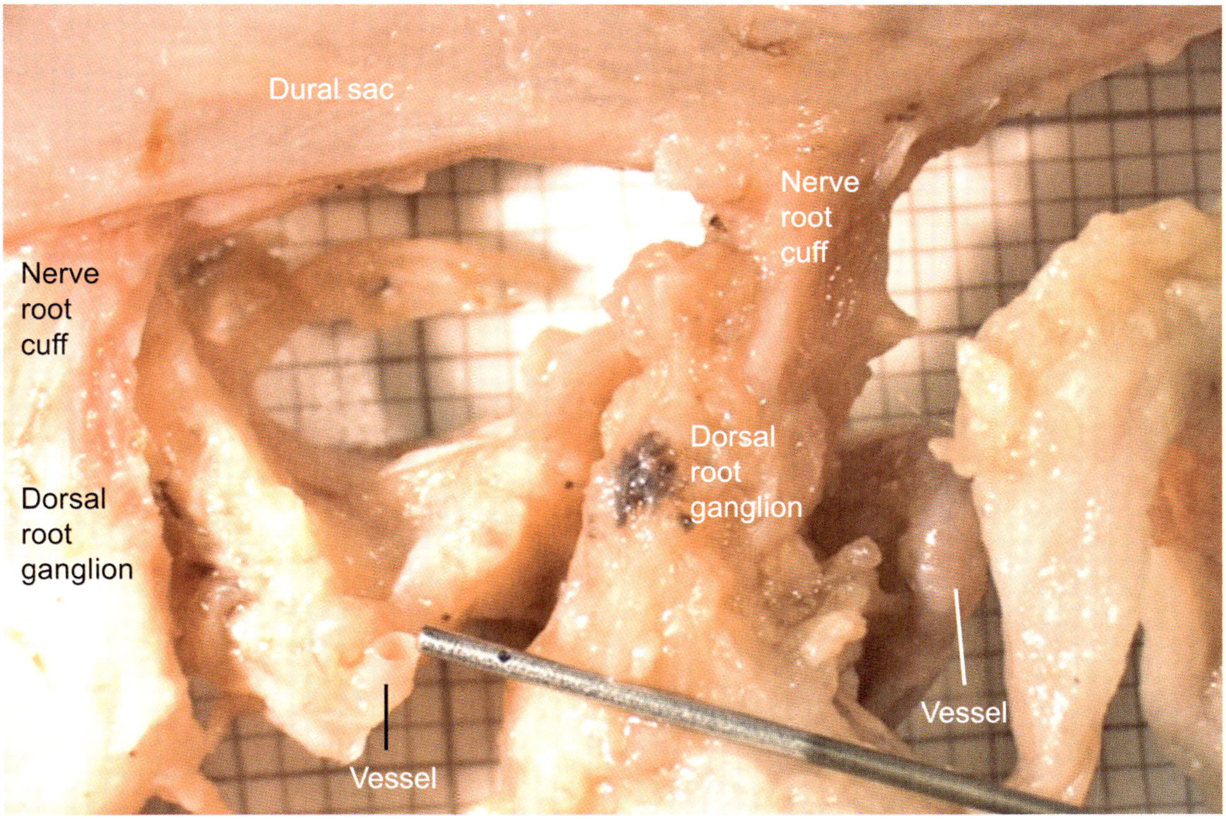

Fig. 42.18 Lumbar nerve root cuffs and vessels close to the dorsal root ganglion. The sizes of the tip of a 22G Quincke needle and the diameter of the vessel are identified

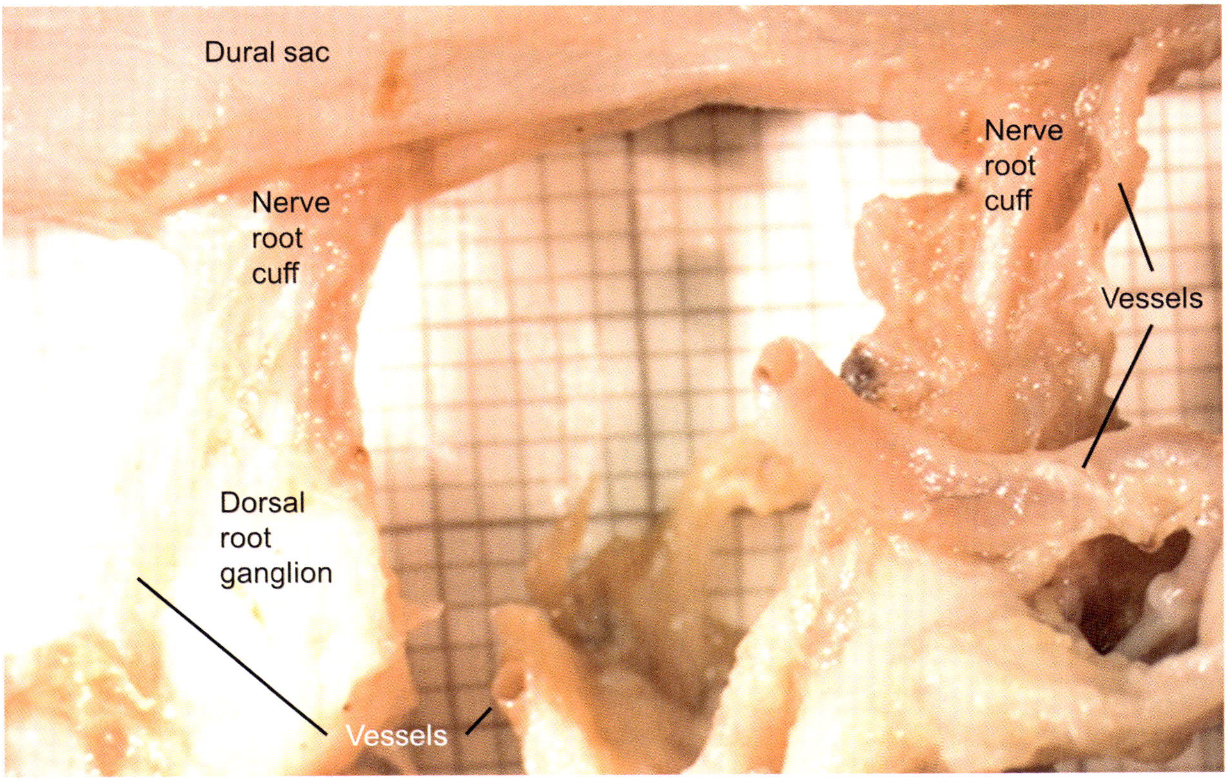

Fig. 42.19 Lumbar nerve root cuffs and vessels close to the dorsal root ganglion. Two contiguous nerve root cuffs exiting from the dural sac. Along the lower nerve root cuff, blood vessels can be identified

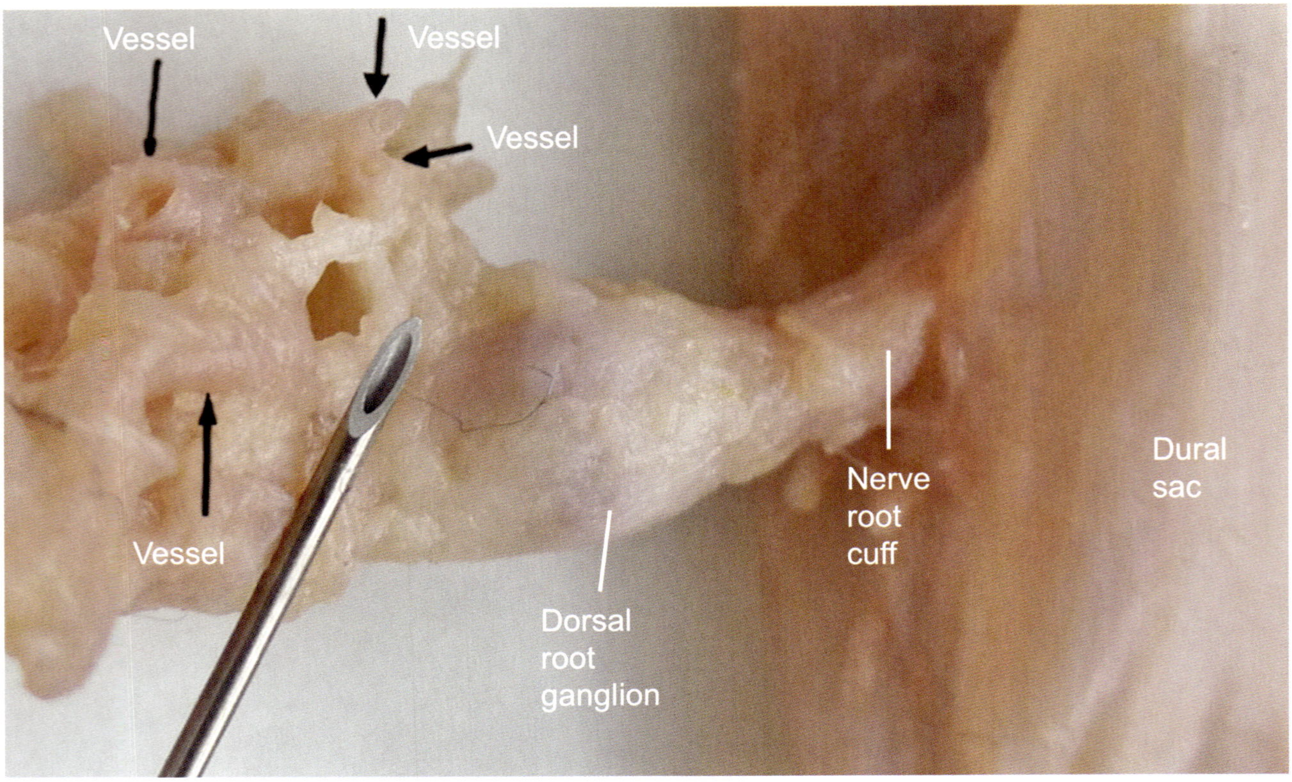

Fig. 42.20 Lumbar nerve root cuffs and vessels close to the dorsal root ganglion. The sizes of the tip of a 22G Quincke needle and the diameter of the vessel are identified (From Reina et al. [4]; with permission)

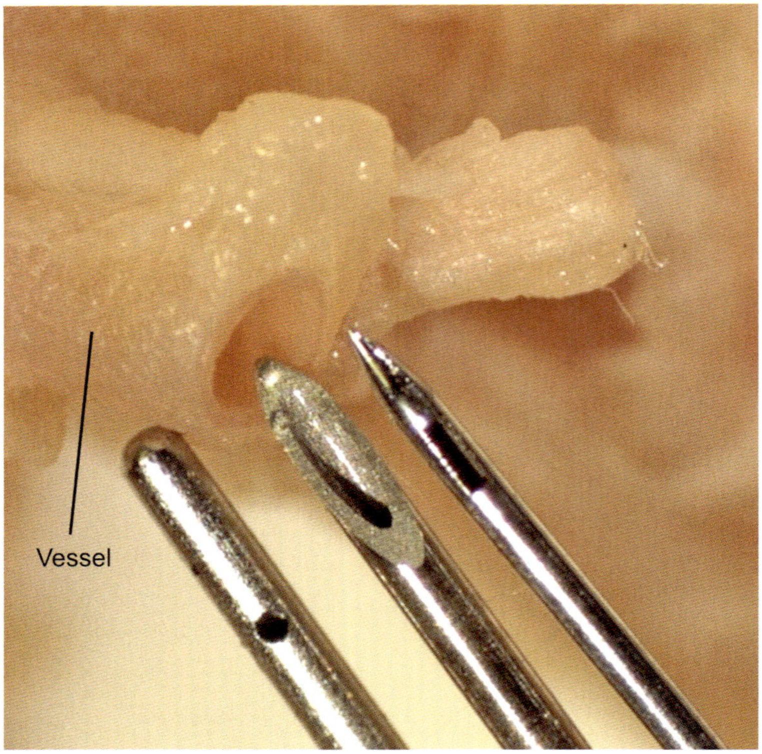

Fig. 42.21 Lumbar nerve root cuffs and vessels close to the dorsal root ganglion. The sizes of the tips of a 22G Quincke needle, 22G blunt needle, and 25G Whitacre needle and the diameter of the vessel are identified (From Reina et al. [4]; with permission)

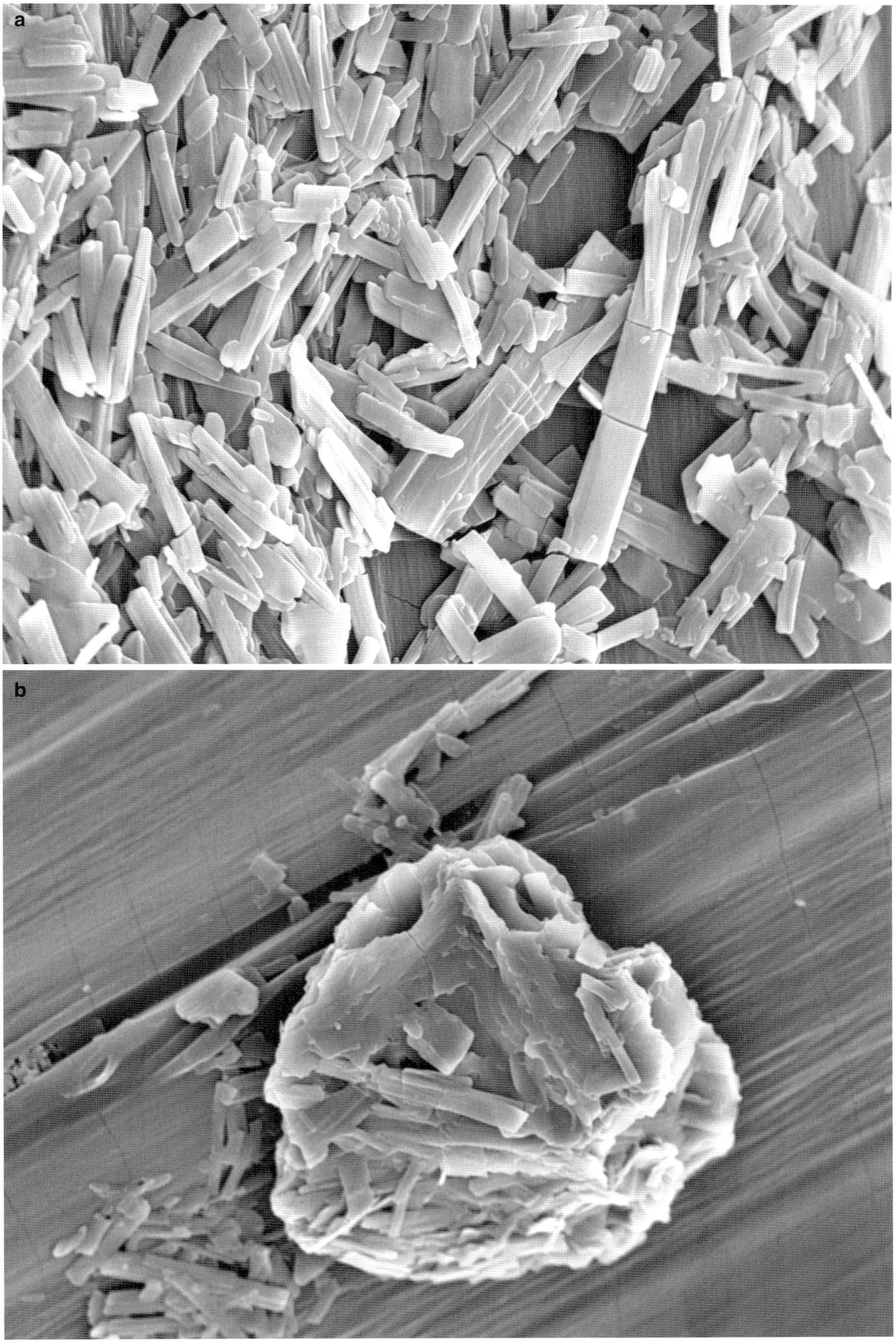

Fig. 42.22 Steroid particles. Betamethasone. Scanning electron microscopy. Magnification: **a**, ×2,000; **b**, ×2,000 (**a** from Reina et al. [5]; with permission)

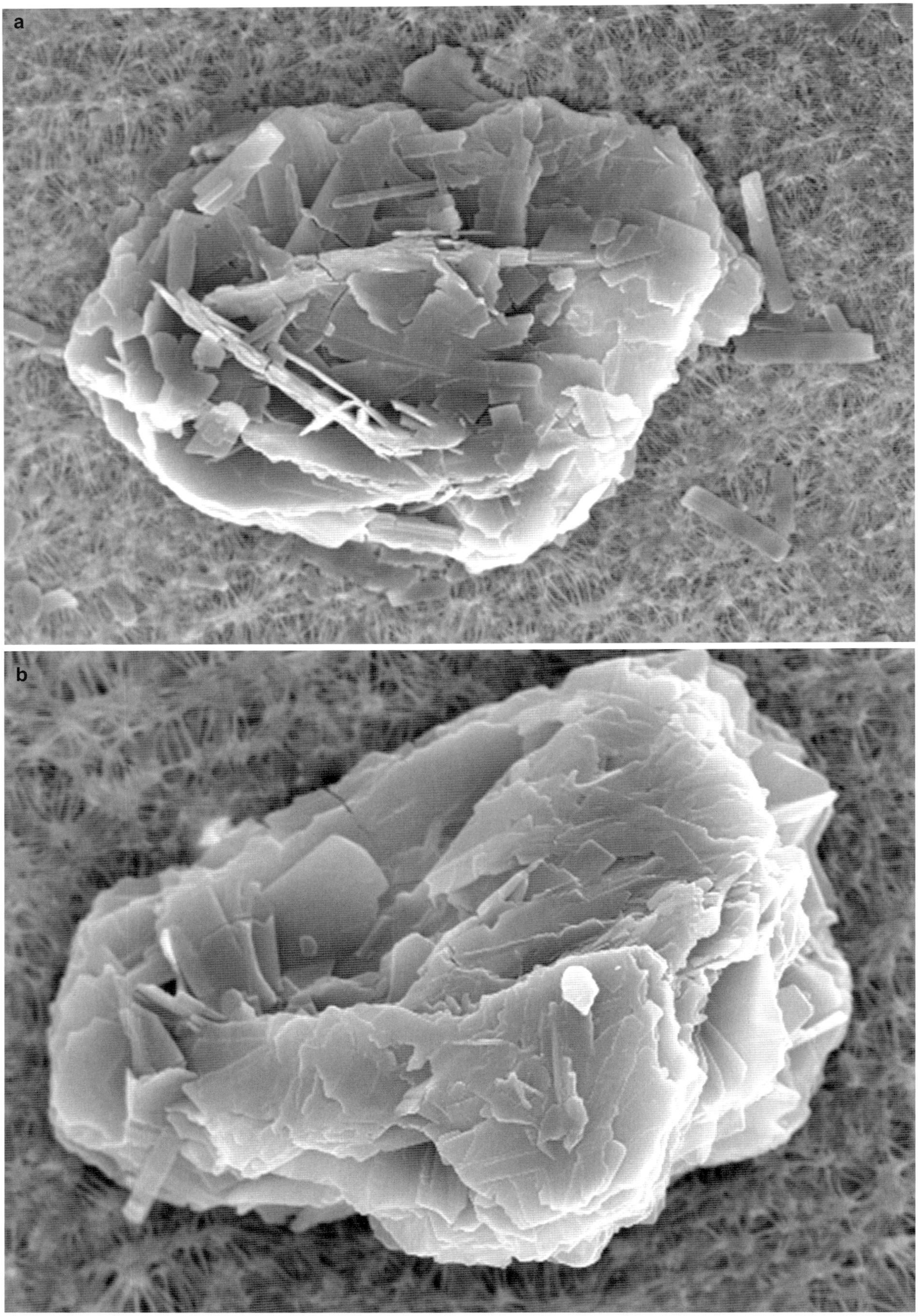

Fig. 42.23 Steroid particles. Betamethasone. Scanning electron microscopy. Magnification: **a**, ×900; **b**, ×900

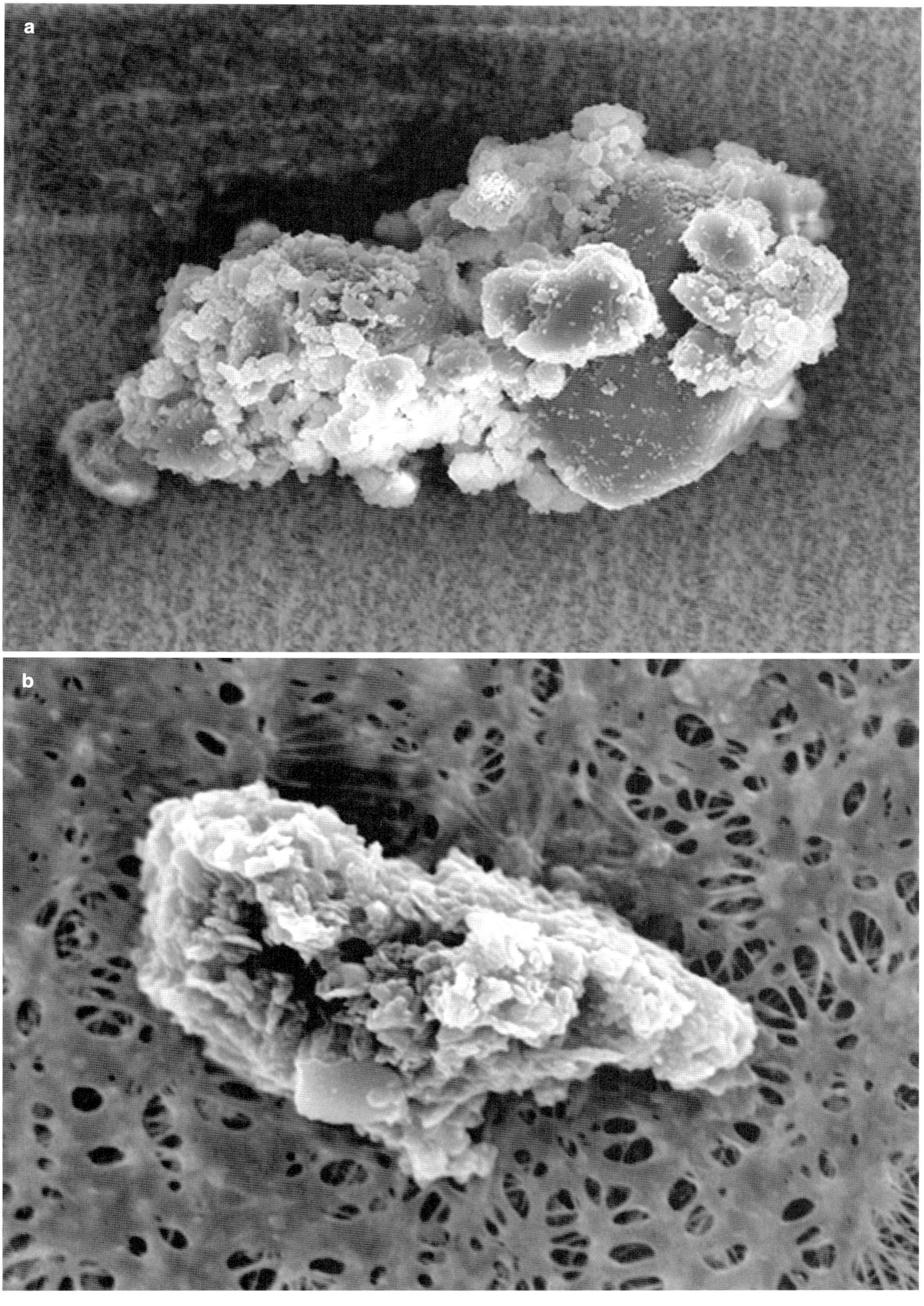

Fig. 42.24 Steroid particles. Dexamethasone. Scanning electron microscopy. Magnification: **a**, ×700; **b**, ×3,000 (**b** from Reina et al. [5]; with permission)

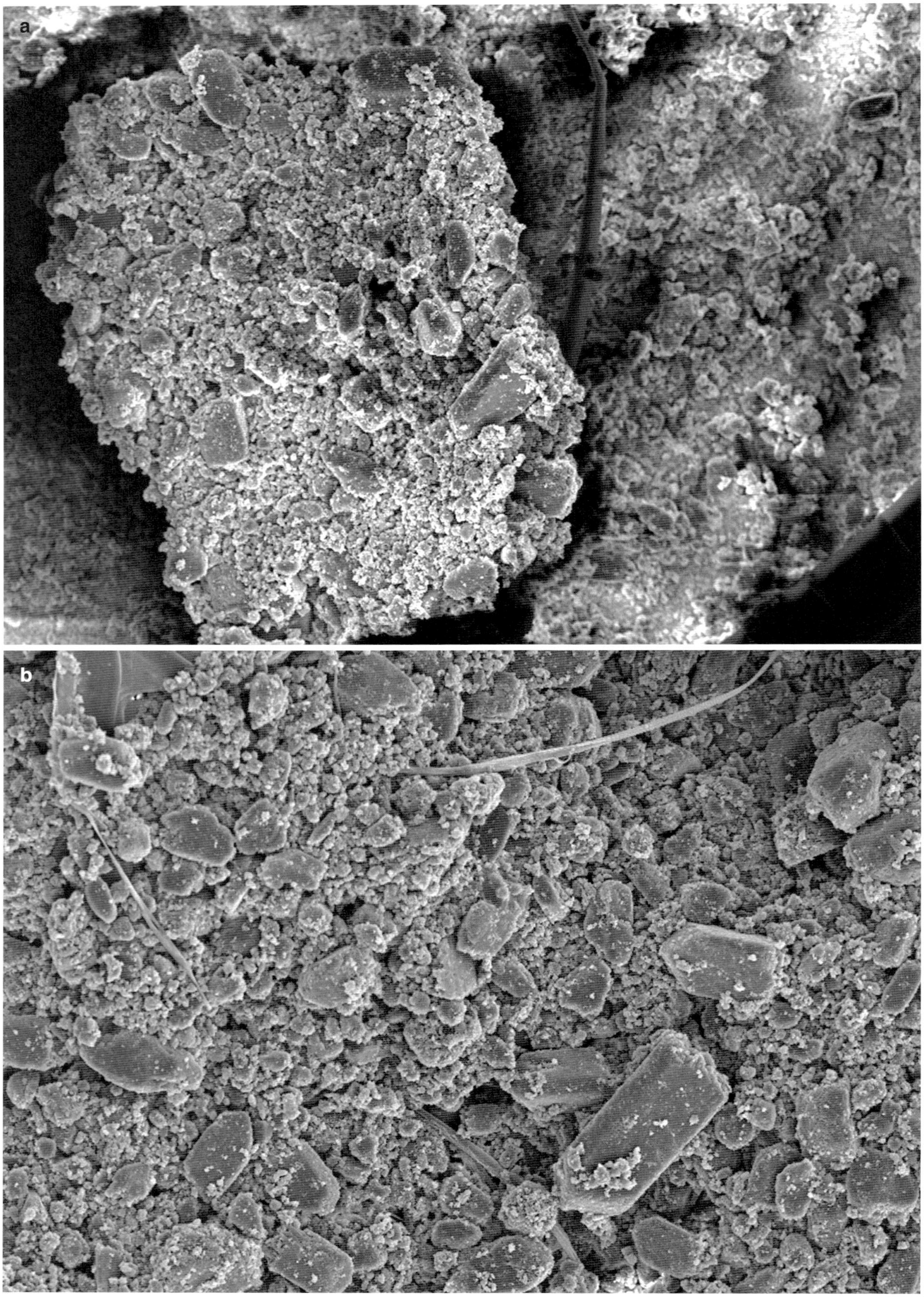

Fig. 42.25 Steroid particles. Triamcinolone. Scanning electron microscopy. Magnification: **a**, ×160; **b**, ×200 (**b** from Reina et al. [5]; with permission)

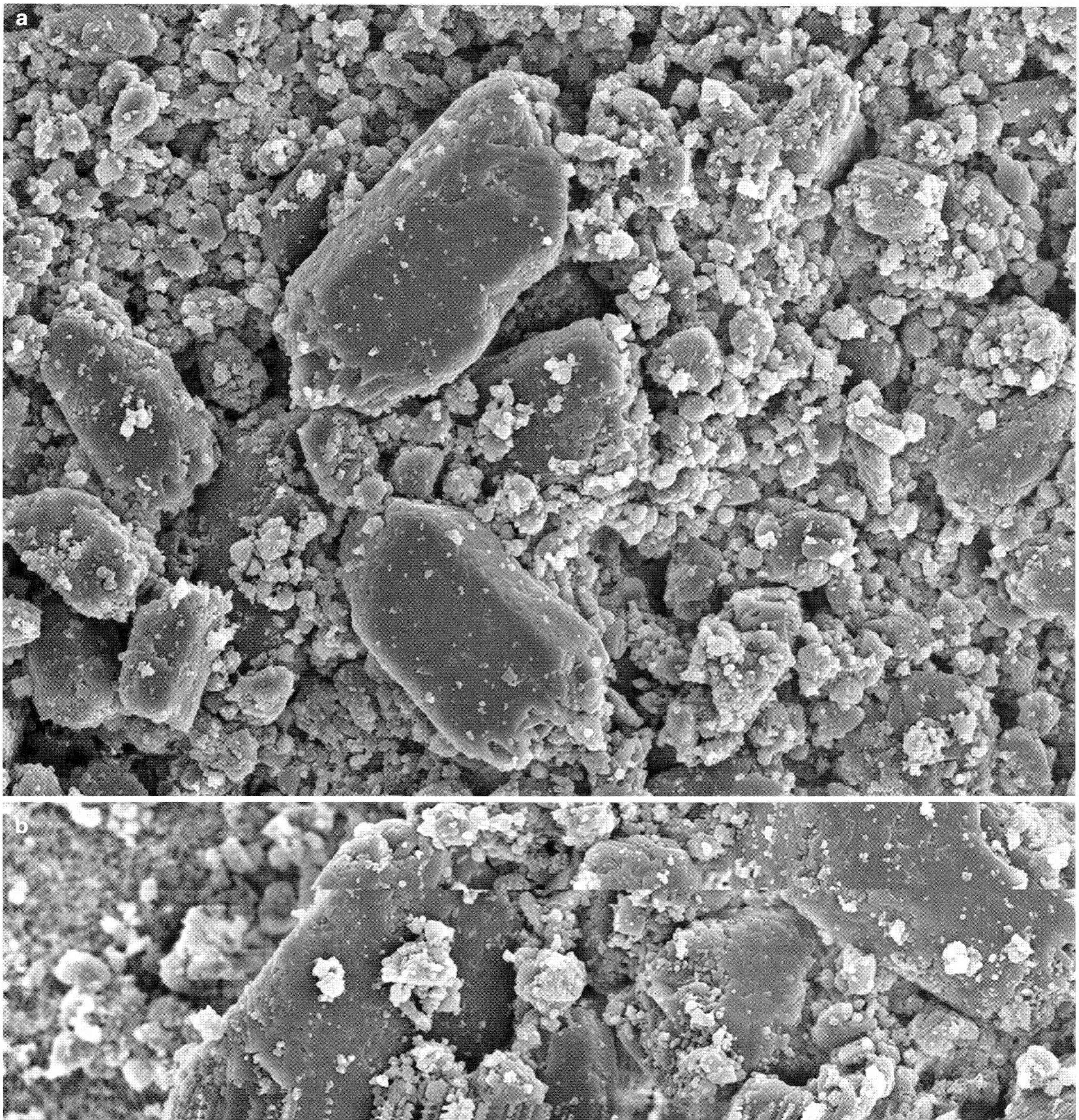

Fig. 42.26 Steroid particles. Triamcinolone. Scanning electron microscopy. Magnification: **a**, ×600; **b**, ×650

References

1. Huntoon MA. Anatomy of the cervical intervertebral foramina: vulnerable arteries and ischemic neurologic injuries after transforaminal epidural injections. Pain. 2005;117:104–11.
2. Benzon HT, Chew TL, McCarthy RJ, Benzon HA, Walega DR. Comparison of the particle sizes of different steroids and the effect of dilution. A review of the relative neurotoxicities of the steroids. Anesthesiology. 2007;106:331–8.
3. Derby R, Lee SH, Date ES, Lee JH, Lee CH. Size and aggregation of corticosteroids used for epidural injections. Pain Med. 2008;9: 227–34.
4. Reina MA, De Andrés JA, Hernández JM, López A, Fernández Domínguez M, Prats-Galino A. Possibility of particulated steroids injection for selective root blocks in the cervical and lumbar regions: anatomical study. Minerva Anestesiol. 2013;79(suppl I): 57–61.
5. Reina MA, De Andrés JA, Hernández JM, Prats-Galino A, Machés F, Peláez J. Transforaminal or translaminar approach for dorsal root ganglion and dorsal nerve root. Anatomical reason for technique decision. Eur J Pain Suppl. 2010;4:287–97.
6. Gazelka HM, Burgher AH, Huntoon MA, Mantilla CB, Hoelzer BC. Determination of the particulate size and aggregation of clonidine and corticosteroids for epidural steroid injection. Pain Physician. 2012;15:87–93.
7. Gharibo C, Koo C, Chung J, Moroz A. Epidural steroid injections: an update on mechanisms of injury and safety. Tech Reg Anesth Pain Manag. 2009;13:266–71.
8. Heavner JE, Racz GB, Jenigiri B, Lehman T, Day MR. Sharp vs blunt needle: a comparative study of penetration of internal structures and bleeding in dogs. Pain Pract. 2003;3:226–31.
9. Raghavendra M, Patel V. Should we cease performing transforaminal injections? Reg Anesth Pain Med. 2005;30:208–9.
10. Helm 2nd S, Jasper JF, Racz GB. Complications of transforaminal epidural injections. Pain Physician. 2003;6:389–90.
11. Smuck M, Tang CT, Fuller BJ. Incidence of simultaneous epidural and vascular injection during cervical transforaminal epidural injections. Spine. 2009;34:E751–5.
12. Smuck M, Yu AJ, Tang CT, Zemper E. Influence of needle type on the incidence of intravascular injection during transforaminal epidural injections: a comparison of short-bevel and long-bevel needles. Spine J. 2010;10:367–71.
13. Smuck M, Leung D. Inadvertent injection of a cervical radicular artery using an atraumatic pencil-point needle. Spine. 2011;36: E220–3.

43 Nerve Root and Types of Needles Used in Transforaminal Injections

José María Hernández, Miguel Angel Reina, and José Antonio De Andrés

The use of transforaminal epidural injections of steroids is common practice in pain medicine, in patients with radicular pain resistant to conventional treatments such as physiotherapy or anti-inflammatory medications. The goal of these injections is to deposit steroids in the proximity of nerve roots, using the epidural or the transforaminal route [1]. The procedure is performed using a translaminar, caudal, or transforaminal approach [2].

The needles most often used in transforaminal epidural injections are of the Quincke type and pencil-point type, and more recently, blunt-tip needles. Our studies have examined different types of blunt needles. The radiofrequency needle (Cosman Medical®, Burlington, MA) has a round distal end that is more pronounced than the Epimed® needle, and its orifice is more distant from the tip [3]. Sharp needles (Quincke type) punctured the nerve easily and offered less resistance than blunt needles [3, 4].

The required length of the needle to introduce the distal orifice inside the nerve root cuff was variable, depending on the type of needle: 1.8 mm for 22G Quincke, 4.1 mm for 22G Epimed®, and 5.7 mm for 20G radiofrequency needles [3]. The differences between the blunt needle tips involve differences in the shape of the distal orifice and in its distance to the tip: 3 mm in the Epimed® needle and 6 mm in the radiofrequency type, which has a smaller orifice, a feature that could be relevant in the prevention of intraneural injection [3].

In contrast to the above-mentioned needles, the distal orifice of sharp needles can be placed closer to the nerve or even introduced inside the nerve root cuff. Once inside the nerve, the distal orifice of the needle could be placed within the fat tissue or inserted in the nerve fibers. The amount of fat found within the nerve root cuffs varies along the root and even among different nerve roots (Figs. 43.1, 43.2, 43.3, 43.4, 43.5, 43.6, 43.7, 43.8, 43.9, 43.10, 43.11, 43.12, 43.13, and 43.14) [5].

J.M. Hernández, MD (✉)
Department of Clinical Medical Sciences, School of Medicine, CEU San Pablo University, Madrid, Spain

Department of Anesthesiology, Madrid-Montepríncipe University Hospital, Madrid, Spain
e-mail: jhzamora@yahoo.es, jmhernandez@madridsindolor.com

M.A. Reina, MD, PhD
Department of Anesthesiology, Madrid-Montepríncipe University Hospital, Madrid, Spain

Department of Clinical Medical Sciences and Institute of Applied Molecular Medicine, School of Medicine, University of CEU San Pablo, Madrid, Spain
e-mail: miguelangel@perticone.e.telefonica.net

J.A. De Andrés, MD, PhD
Department of Surgical Specialties, School of Medicine, University of Valencia, Valencia, Spain

Department of Anesthesia, Critical Care, and Pain Management, General University Hospital, Valencia, Spain
e-mail: deandres_jos@gva.es

M.A. Reina et al. (eds.), *Atlas of Functional Anatomy for Regional Anesthesia and Pain Medicine: Human Structure, Ultrastructure and 3D Reconstruction Images*, DOI 10.1007/978-3-319-09522-6_43,

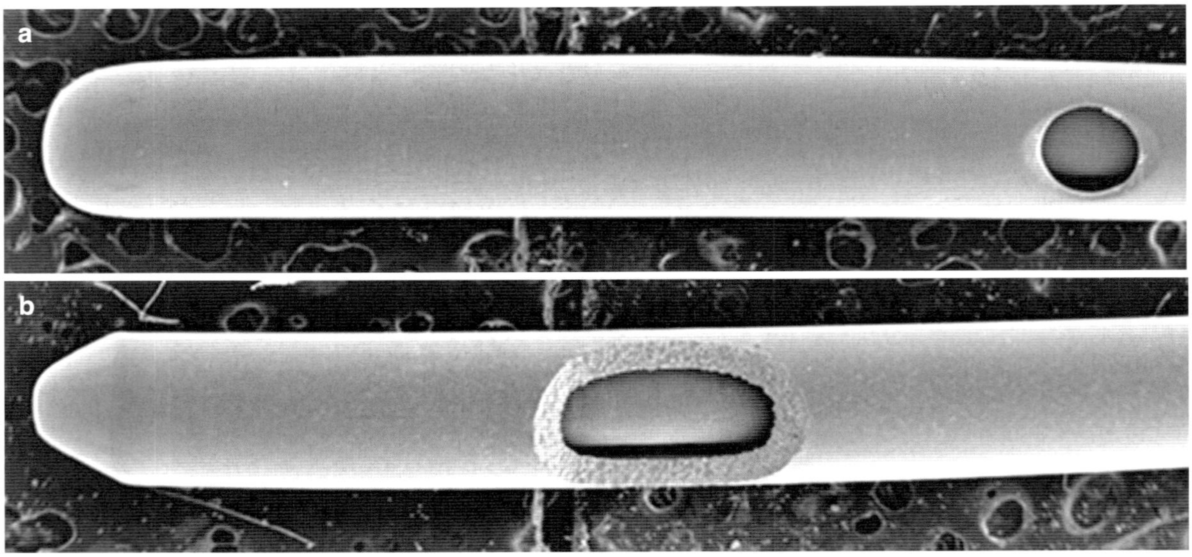

Fig. 43.1 Blunt needles. (**a**) 22G Epimed® needle. (**b**) 20G radiofrequency needle (Cosman Medical®, Burlington, MA). Scanning electron microscopy. Magnification ×18 (From Hernández-Garcia et al. [3]; with permission)

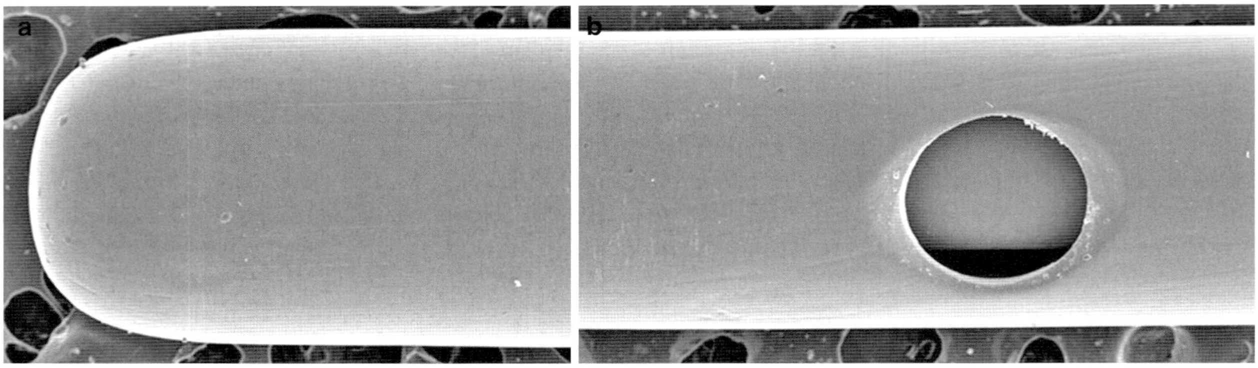

Fig. 43.2 Blunt needles. Details of Epimed® needle. (**a**) Tip. (**b**) Lateral orifice. Scanning electron microscopy. Magnification: **a**, ×80; **b**, ×80 (From Hernández-Garcia et al. [3]; with permission)

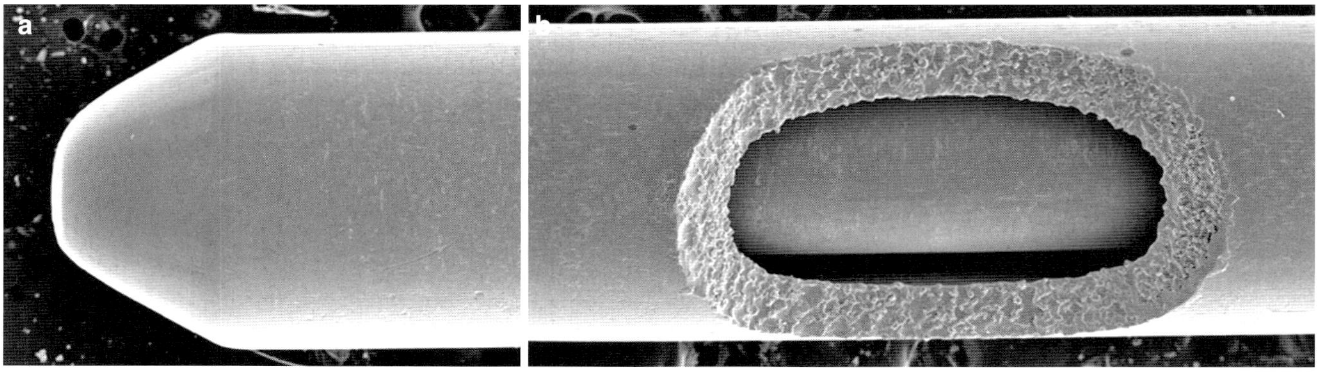

Fig. 43.3 Blunt needles. Details of 20G radiofrequency needle. (**a**) Tip. (**b**) Lateral. Magnification: **a**, ×80; **b**, ×80 (From Hernández-Garcia et al. [3]; with permission)

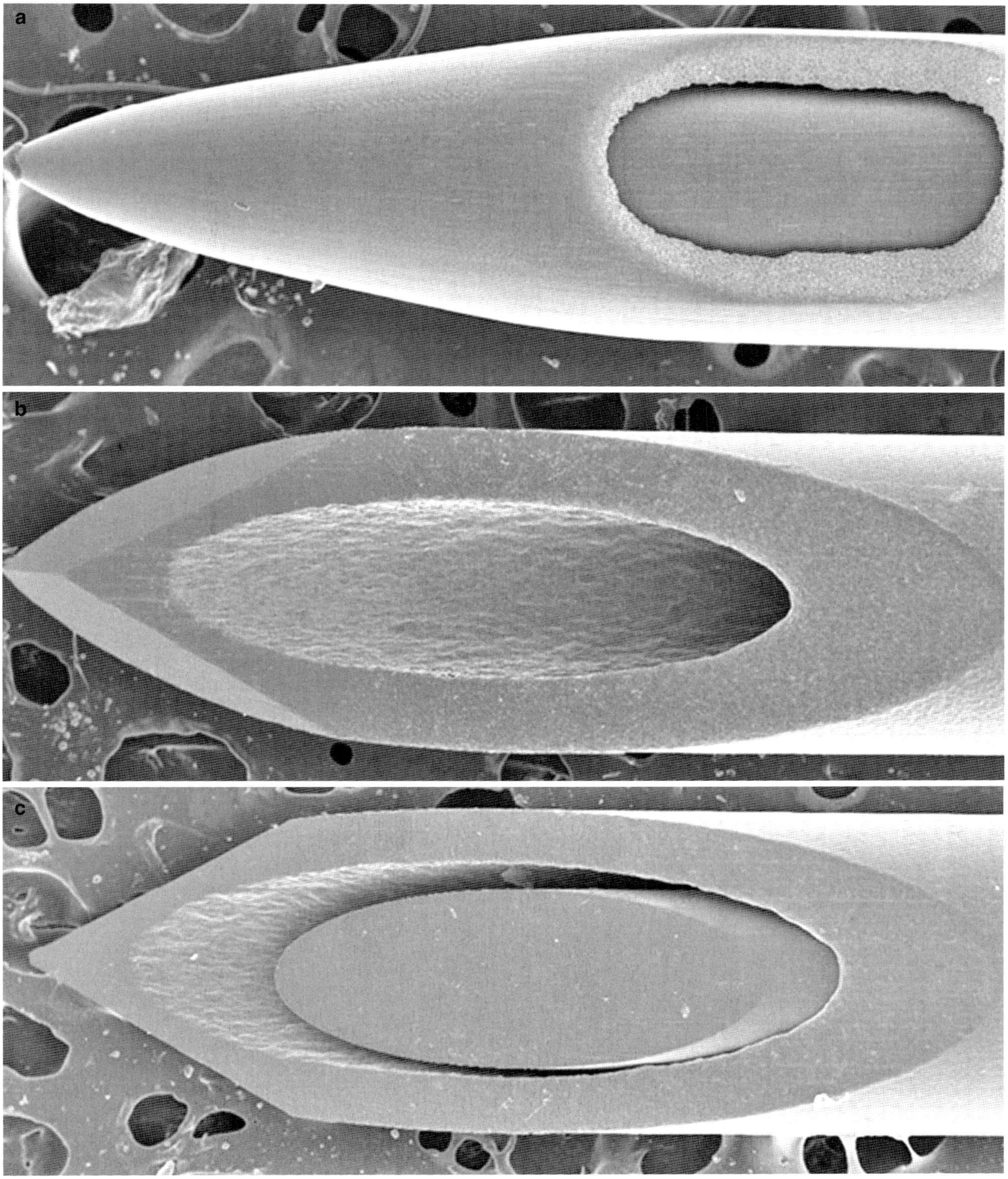

Fig. 43.4 Different designs of needles used in transforaminal injections. (**a**) 25G pencil-point; (**b**) 22G Quincke; (**c**) 22G beveled type. Scanning electron microscopy. Magnification: **a**, ×55; **b**, ×55; **c**, ×55. (**b** from Hernández-Garcia et al. [3]; with permission)

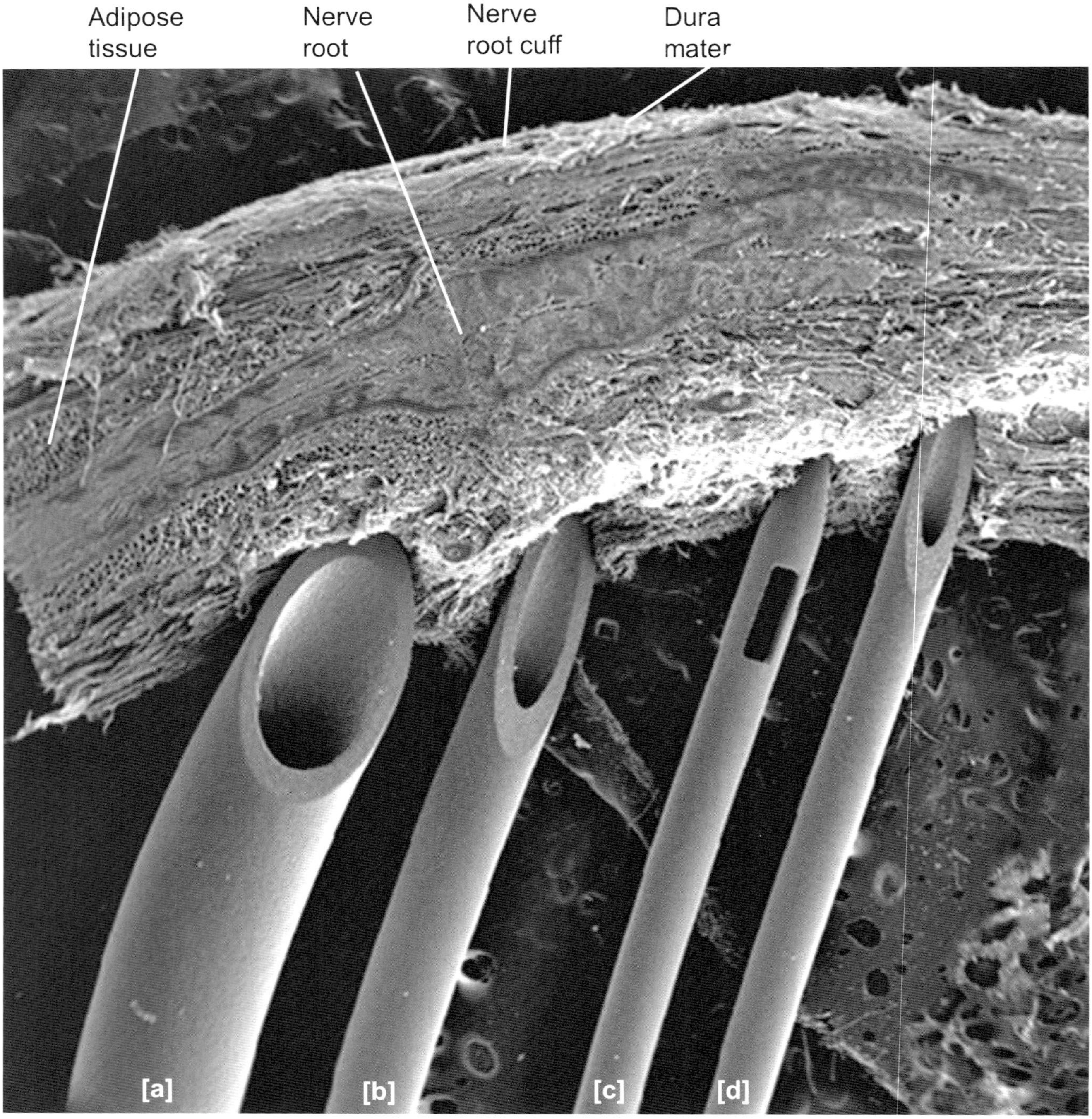

Fig. 43.5 Spinal and epidural needles and nerve root cuff. Different diameters of needles and the size of the nerve root cuff can be identified. (**a**) 18G epidural needle; (**b**) 22G Quincke needle; (**c**) 25G Whitacre needle; (**d**), 25G Quincke needle. Scanning electron microscopy. Magnification ×10

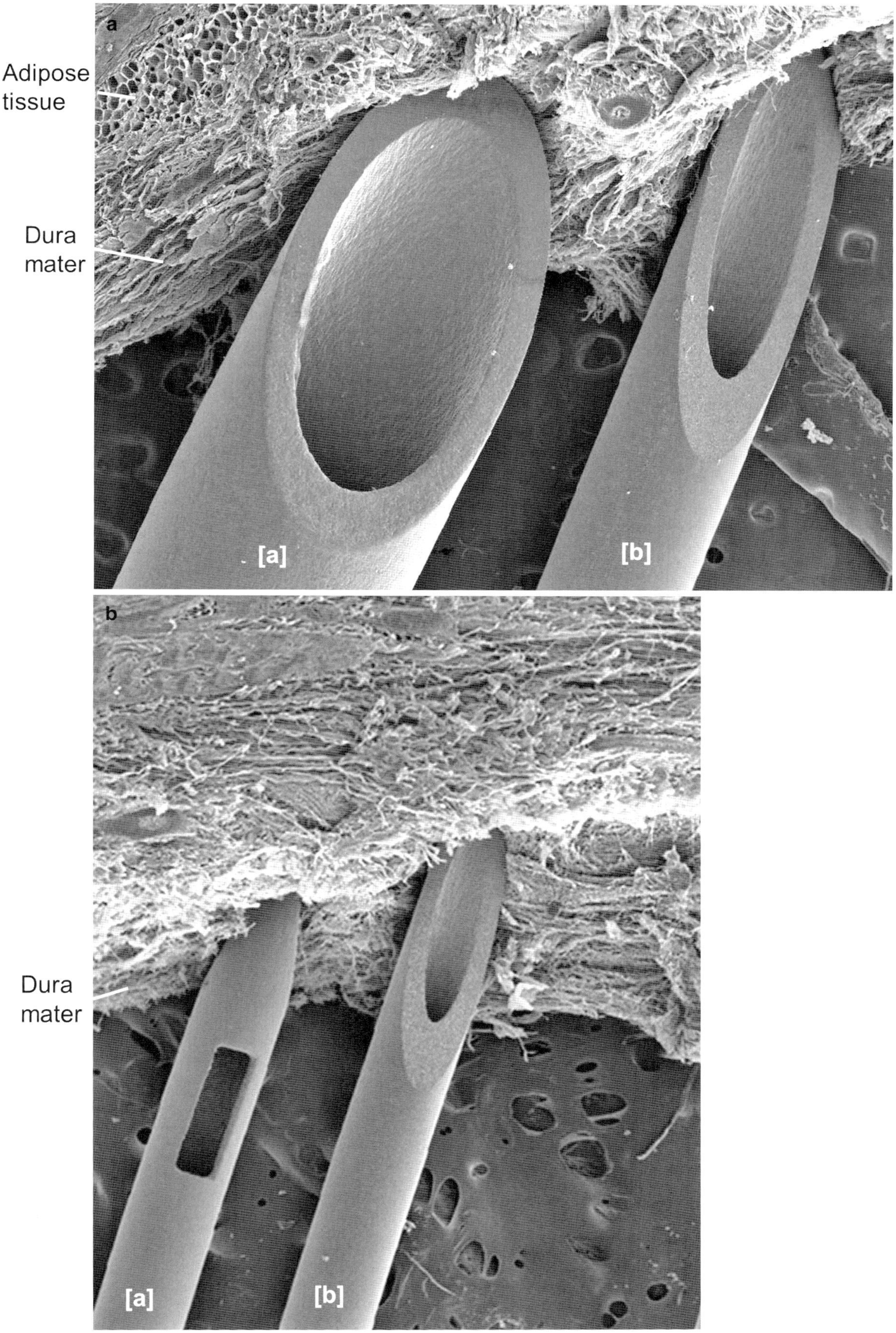

Fig. 43.6 Spinal and epidural needles (details of Fig. 43.5). (**a**) 18G epidural needle (*a*), 22G Quincke needle (*b*). (**b**) 25G Whitacre needle (*a*), 25G Quincke needle (*b*). Scanning electron microscopy. Magnification: **a**, ×30; **b**, ×20

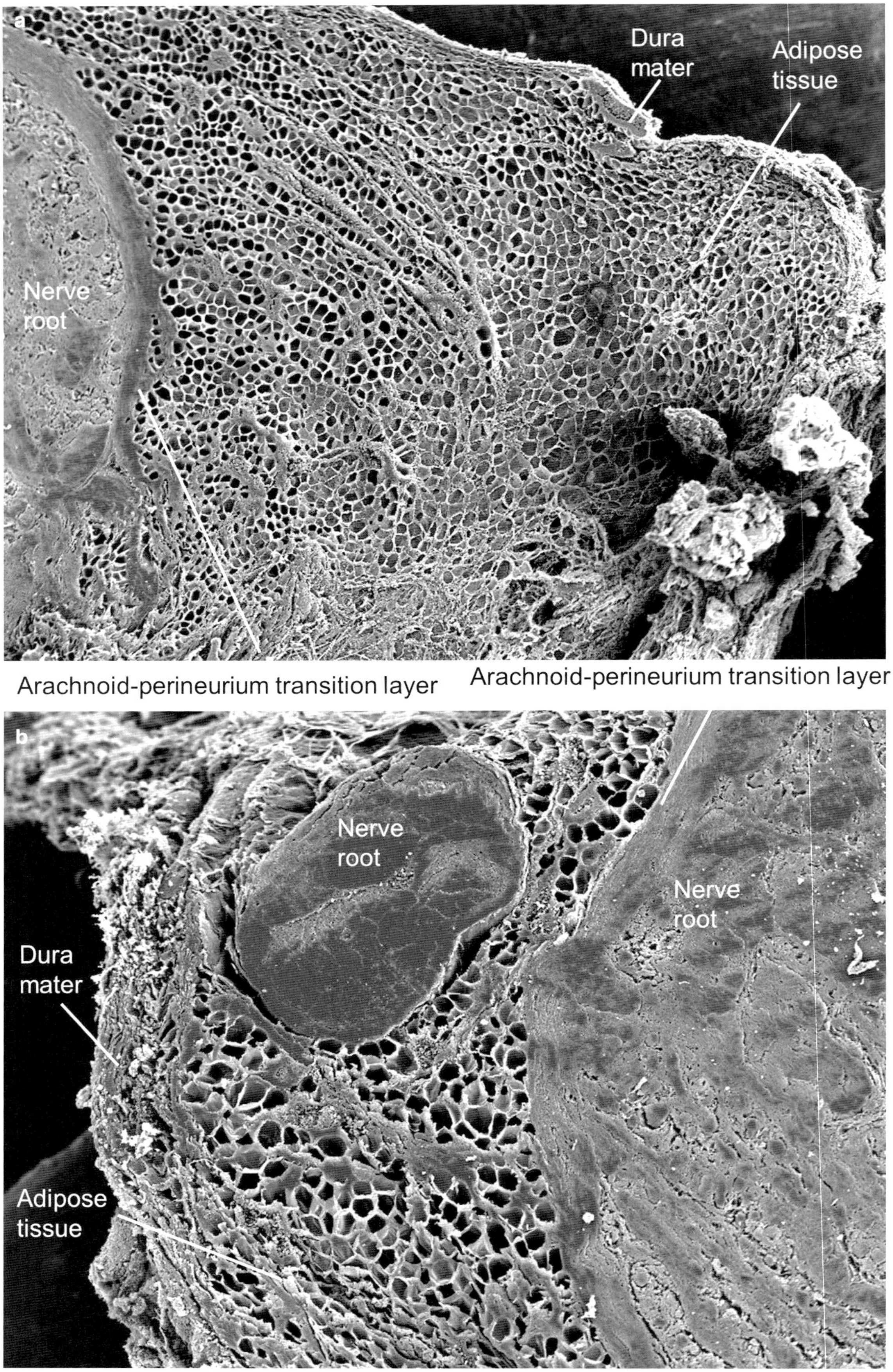

Fig. 43.7 Nerve roots within nerve root cuff. Scanning electron microscopy. Magnification: **a**, ×30; **b**, ×55

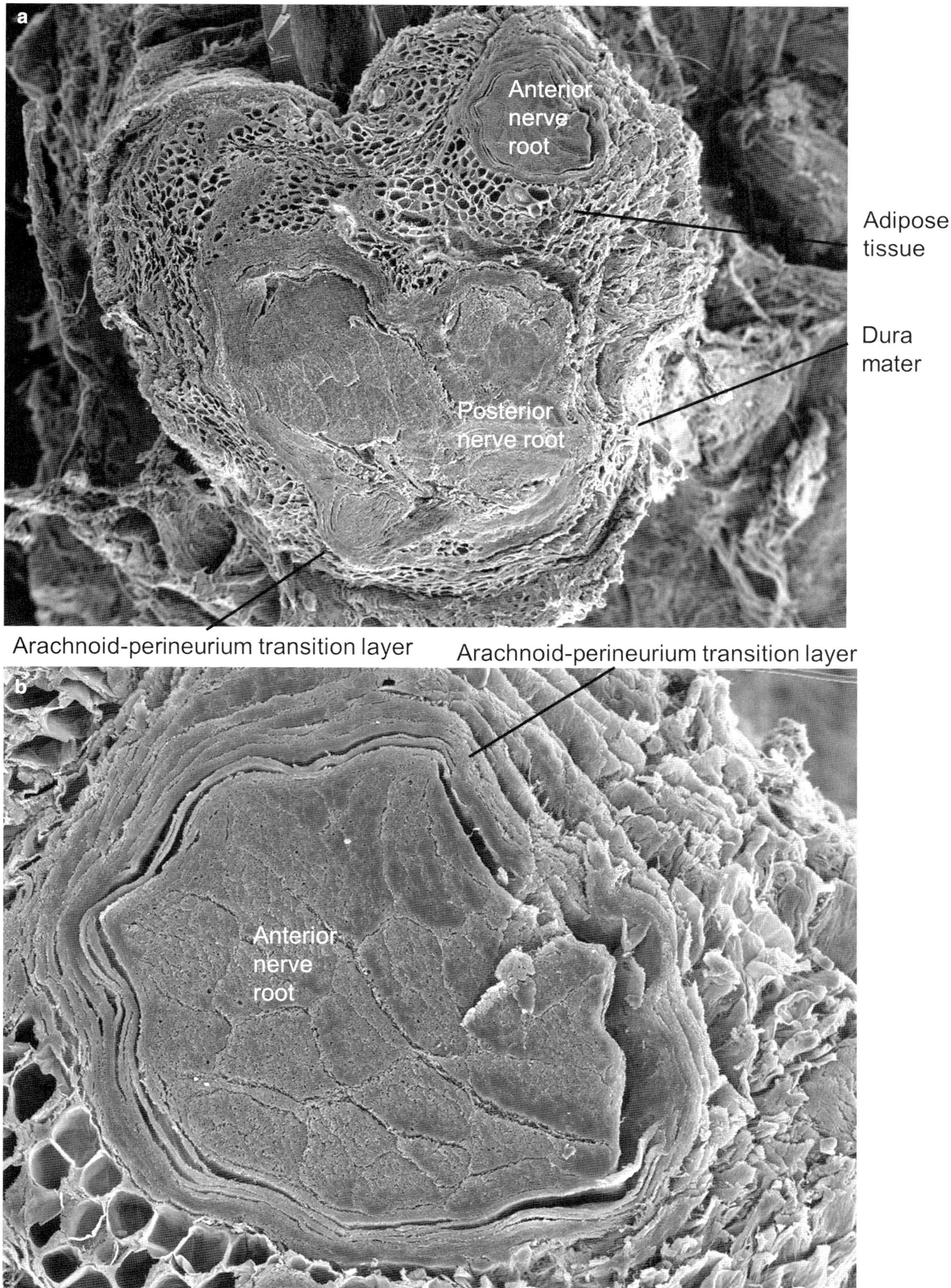

Fig. 43.8 Nerve roots within nerve root cuff. Scanning electron microscopy. Magnification: **a**, ×40; **b**, ×120 (**b** from Hernández-Garcia et al. [3]; with permission)

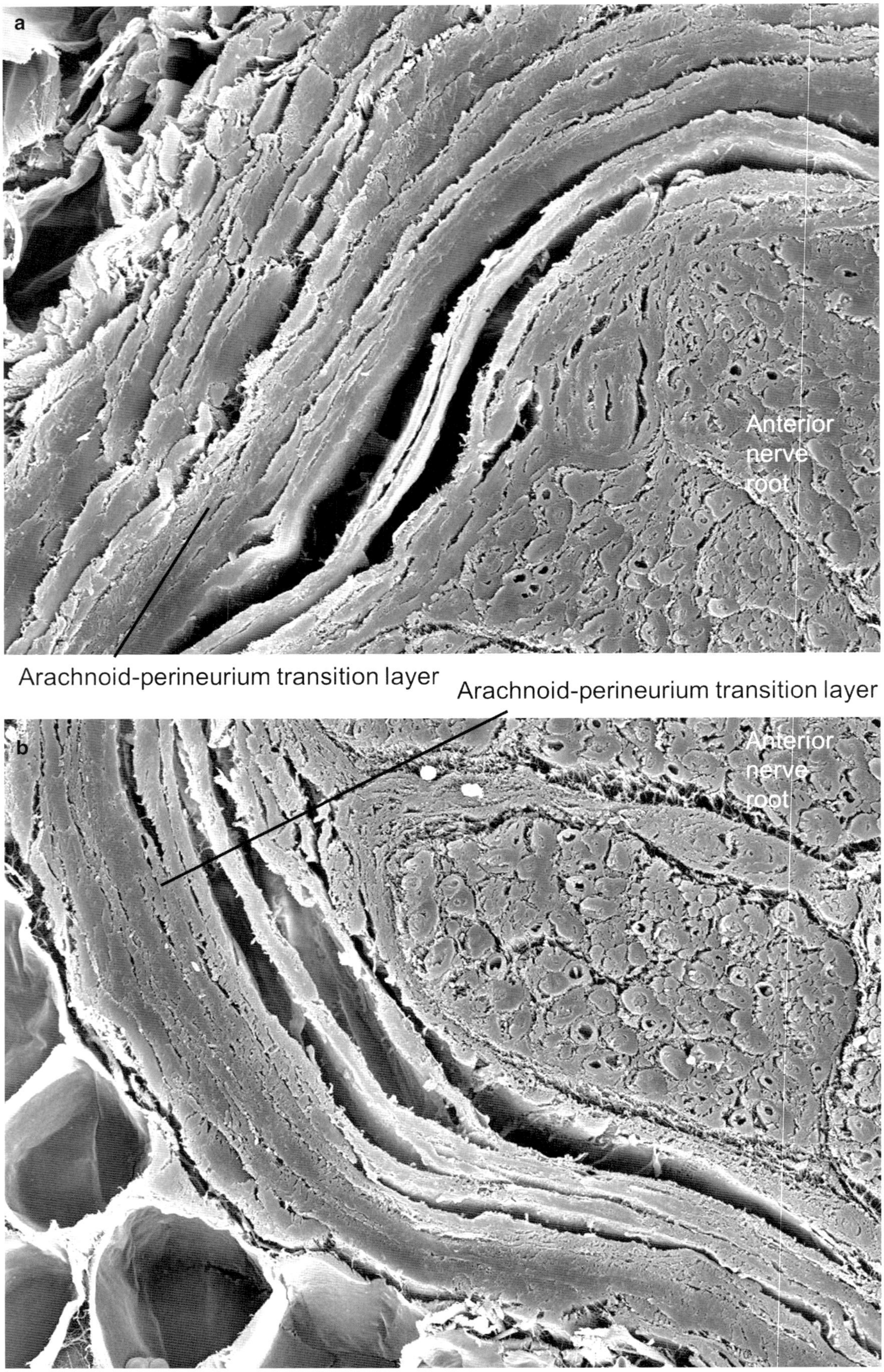

Fig. 43.9 Nerve roots within nerve root cuff (details of Fig. 43.8**b**). Scanning electron microscopy. Magnification: **a**, ×450; **b**, ×450

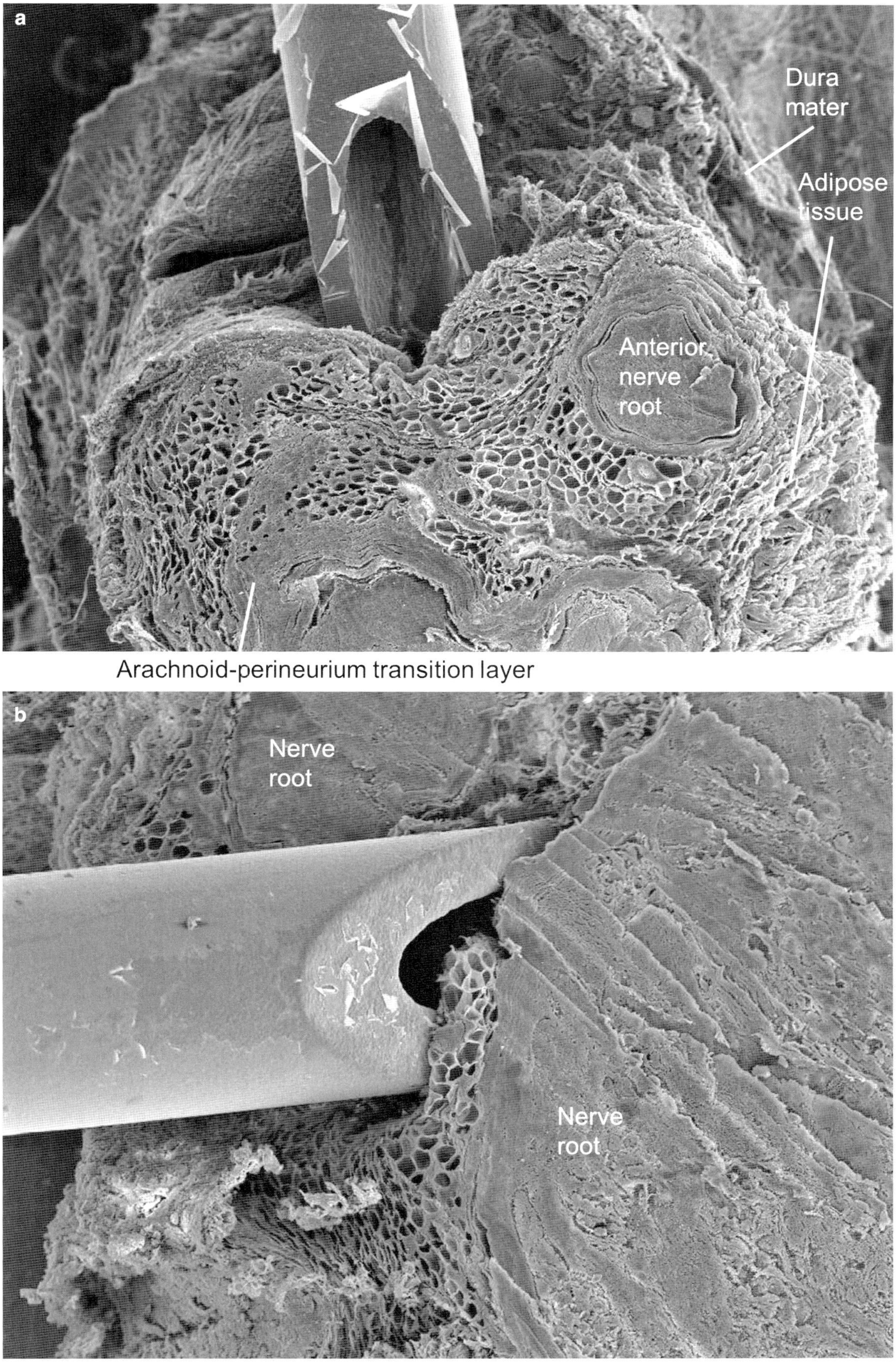

Fig. 43.10 In vitro puncture of human nerve root cuff with 22G Quincke needle. Scanning electron microscopy. Magnification: **a**, ×33; **b**, ×50 (From Hernández-Garcia et al. [3]; with permission)

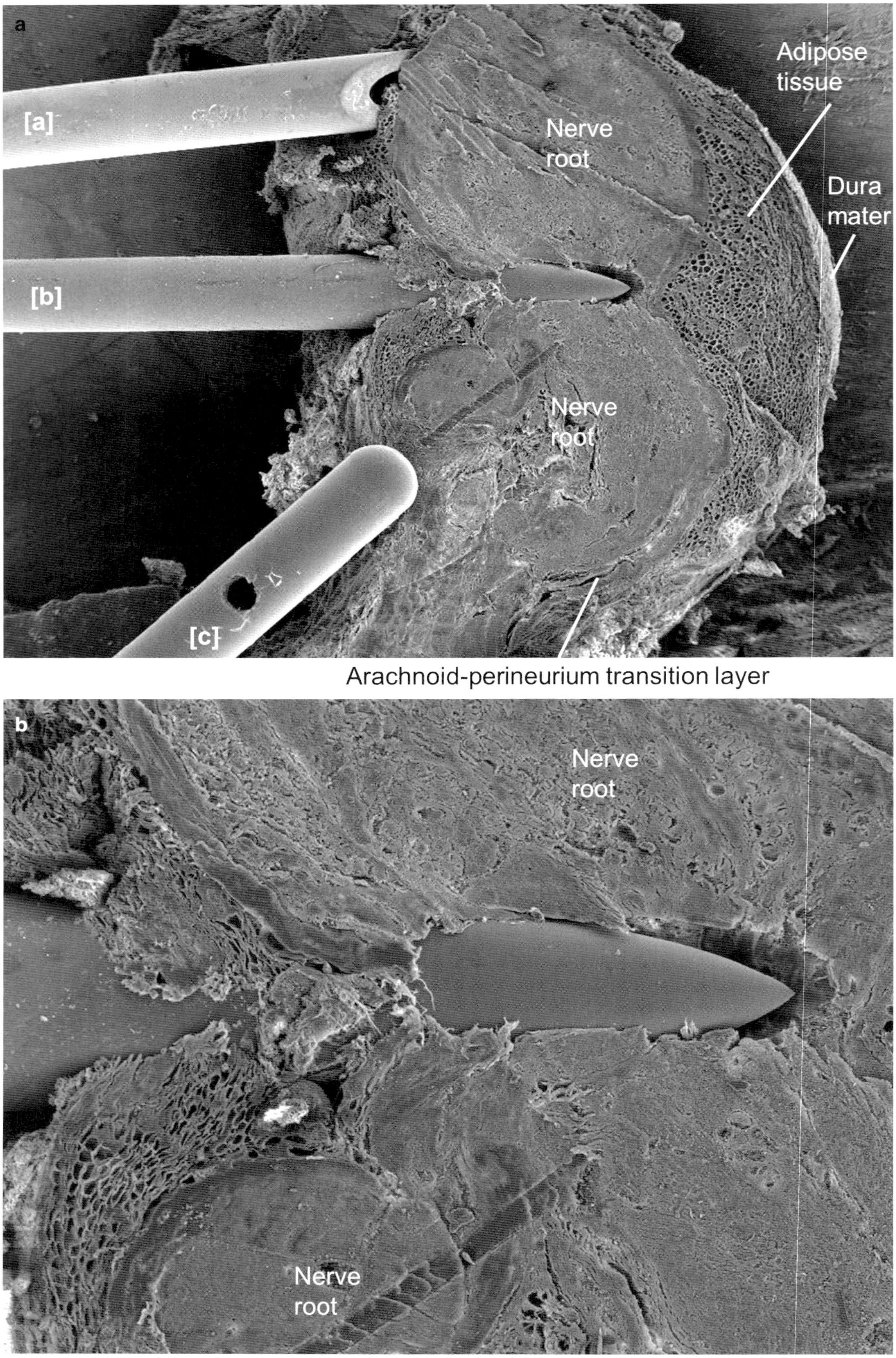

Fig. 43.11 In vitro punctures of nerve root cuff with sharp (**a**, **b**) and blunt Needles (**c**). Scanning electron microscopy. Magnification: a, ×15; b, ×43 (a from Hernández et al. [3]; with permission)

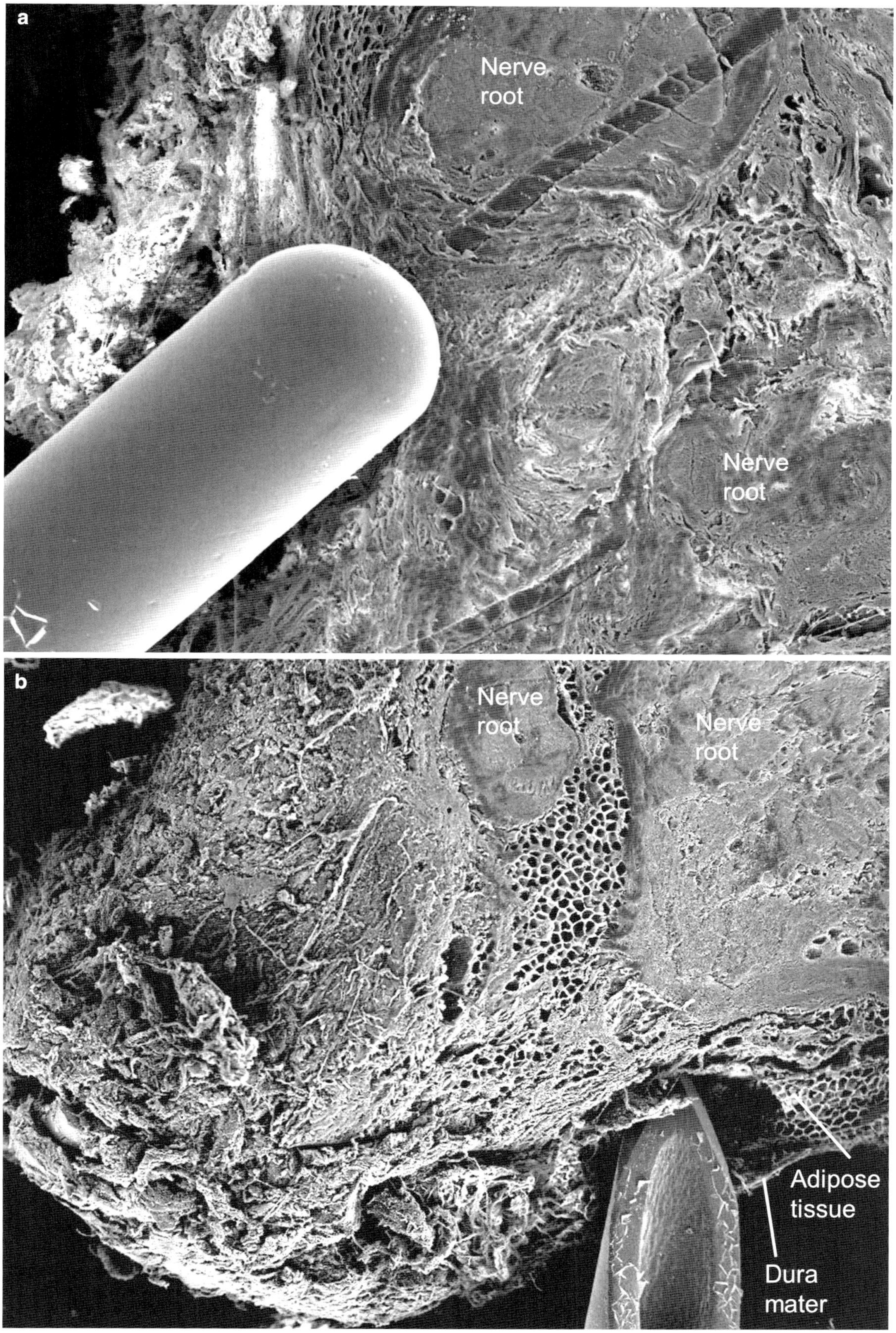

Fig. 43.12 In vitro punctures of nerve root cuff with blunt (**a**) and sharp (**b**) needles. Scanning electron microscopy. Magnification: **a**, ×43; **b**, ×30

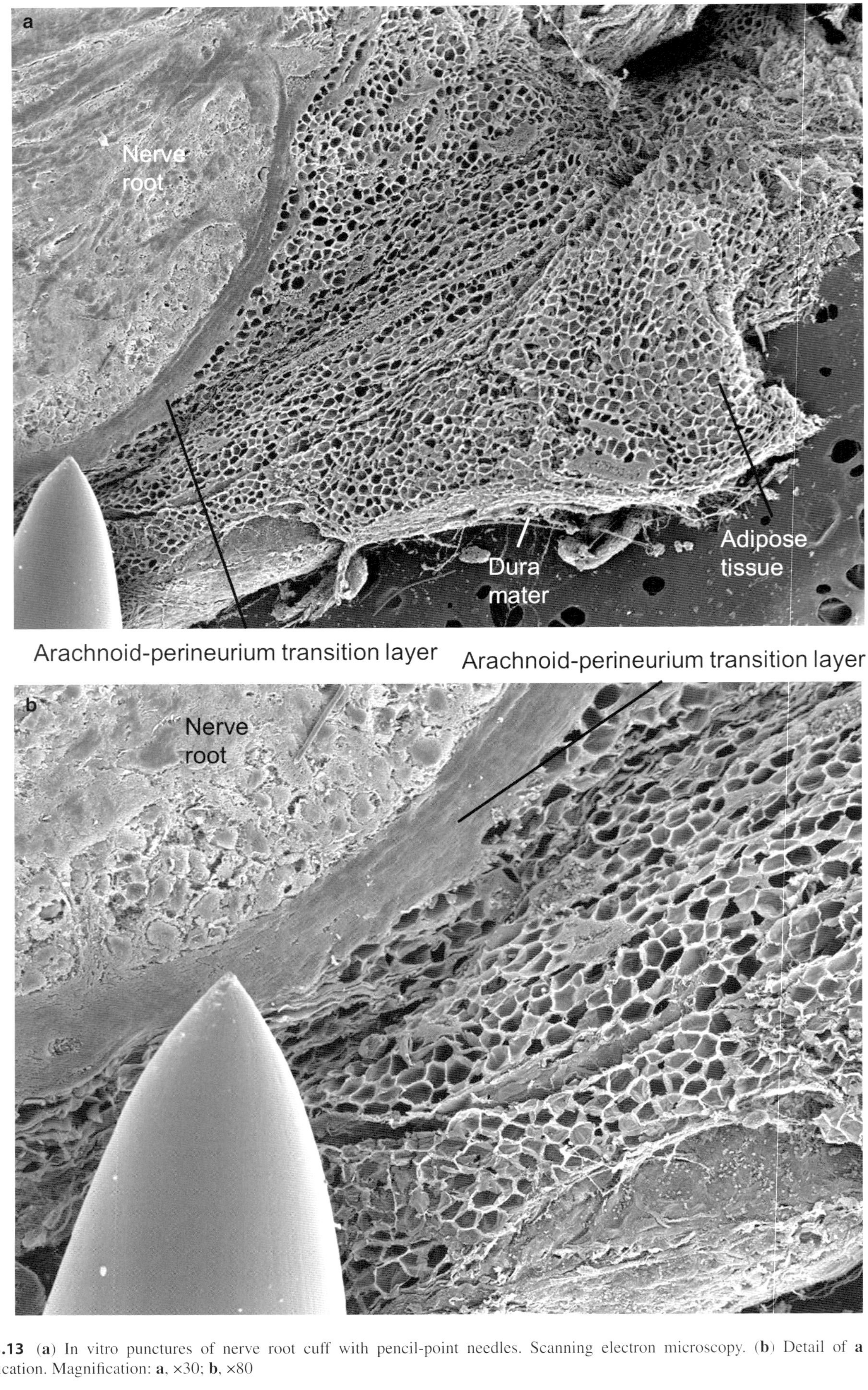

Fig. 43.13 (**a**) In vitro punctures of nerve root cuff with pencil-point needles. Scanning electron microscopy. (**b**) Detail of **a** at higher magnification. Magnification: **a**, ×30; **b**, ×80

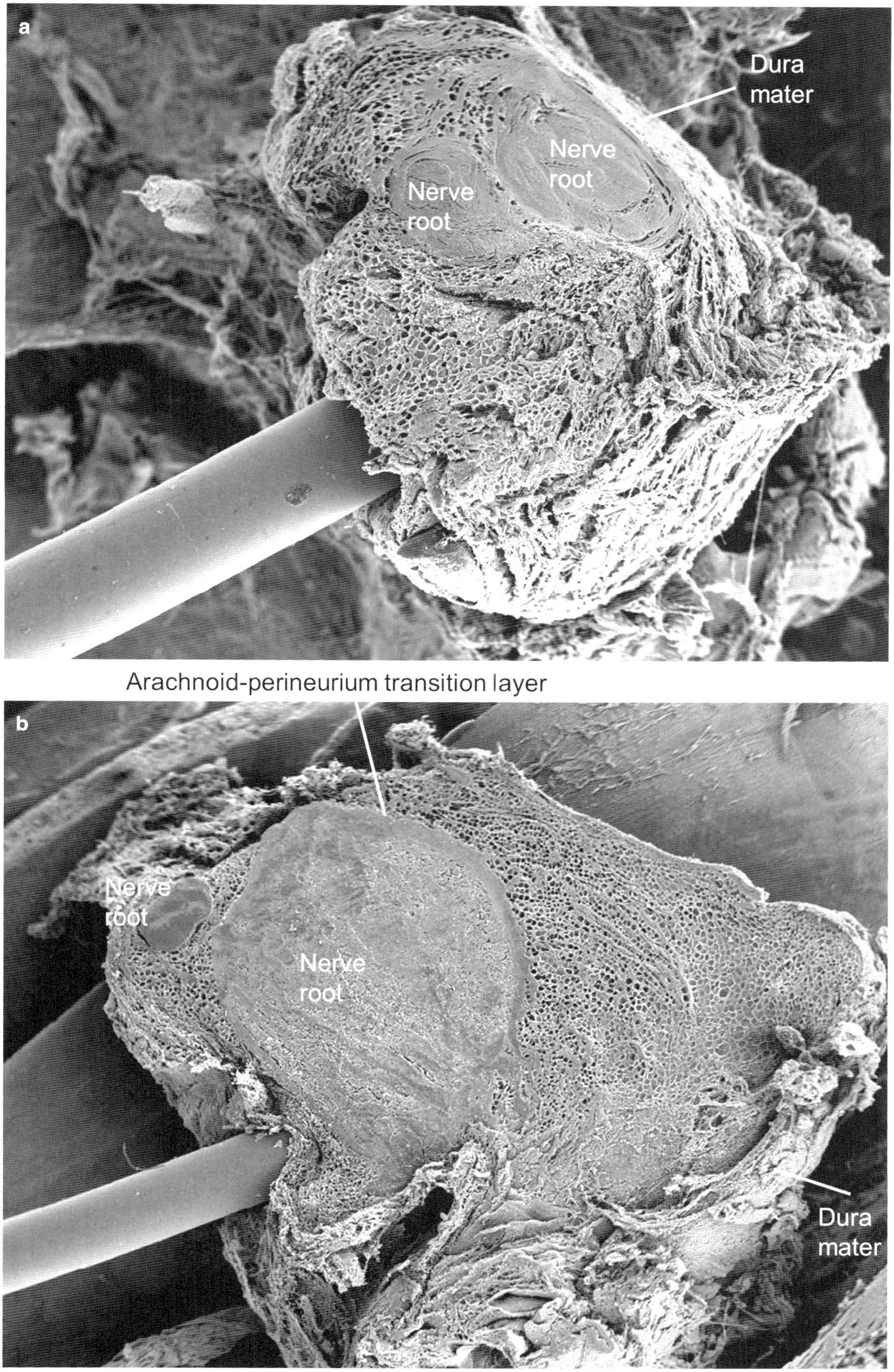

Fig. 43.14 In vitro punctures of nerve root cuff with spinal needles. Scanning electron microscopy. Magnification: **a**, ×22; **b**, ×12

References

1. Baker R, Dreyfuss P, Mercer S, Bogduk N. Cervical transforaminal injection of corticosteroids into a radicular artery: a possible mechanism for spinal cord injury. Pain. 2003;103:211–5.
2. Reina MA, De Andrés JA, Hernández JM, Prats-Galino A, Machés F, Peláez J. Transforaminal or translaminar approach for dorsal root ganglion and dorsal nerve root. Anatomical reason for technique decision. Eur J Pain Suppl. 2010;4:287–97.
3. Hernández-Garcia JM, Reina MA, Prats-Galino A, De Andrés JA. Morphological study of nerve root and types of needle used in transforaminal injections. Reg Anesth Pain Med. 2011;36:278–81.
4. Heavner JE, Racz GB, Jenigiri B, Lehman T, Day MR. Sharp versus blunt needle: a comparative study of penetration of internal structures and bleeding in dogs. Pain Pract. 2003;3:226–31.
5. Reina MA, Villanueva MC, Machés F, Carrera A, López A, De Andrés JA. Ultrastructure of human spinal root cuffs in lumbar spine. Anesth Analg. 2008;106:339–44.

Part III

Materials

44 Needles in Regional Anesthesia

Andrés López and Miguel Angel Reina

Examination of spinal, epidural and neurostimulation needles used in the performance of different anesthetic techniques is relevant to the evaluation of the quality of such needles, as well as of the safety of the techniques and their complications [1–7].

Needles can have imperfections in their tips. The presence of such imperfections may increase the degree of damage to tissues. Examining needles of this type at higher magnifications has the advantage of accurate identification of minute needle-tip defects, which include fissures seen along the surface of the needle tips. These fissures are generated during the production phase of spinal needles, more specifically in the polishing process. The resulting defect causes the needle tip to become more fragile. Examination of needle tips under scanning electron microscopy may be used to improve quality control of the needles used in regional anesthesia, and may help in setting standards to minimize or avoid the presence of imperfections in spinal needles.

There are presently various types of neurostimulation needles. Few of these needles have noninsulated tips, but for some, only a minimal portion of the tip is noninsulated.

Scanning electron microscopy may be a useful tool in the evaluation of quality controls and the establishment of standards for different products used in regional anesthesia. These techniques may help to reduce the presence of imperfections in these products, thus improving the safety and efficacy of regional anesthesia techniques (Figs. 44.1, 44.2, 44.3, 44.4, 44.5, 44.6, 44.7, 44.8, 44.9, 44.10, 44.11, 44.12, 44.13, 44.14, 44.15, 44.16, 44.17, 44.18, 44.19, 44.20, 44.21, and 44.22) [3–8].

A. López, MD (✉)
Department of Clinical Medical Sciences,
School of Medicine, CEU San Pablo University, Madrid, Spain

Department of Anesthesiology, Madrid-Montepríncipe
University Hospital, Madrid-Torrelodones University Hospital,
Madrid-Sanchinarro University Hospital, Madrid, Spain
e-mail: alg00009@telefonica.net

M.A. Reina, MD, PhD
Department of Clinical Medical Sciences and
Institute of Applied Molecular Medicine, School of Medicine,
University of CEU San Pablo, Madrid, Spain

Department of Anesthesiology, Madrid-Montepríncipe
University Hospital, Madrid, Spain
e-mail: miguelangel@perticone.e.telefonica.net

M.A. Reina et al. (eds.), *Atlas of Functional Anatomy for Regional Anesthesia and Pain Medicine: Human Structure, Ultrastructure and 3D Reconstruction Images*, DOI 10.1007/978-3-319-09522-6_44,

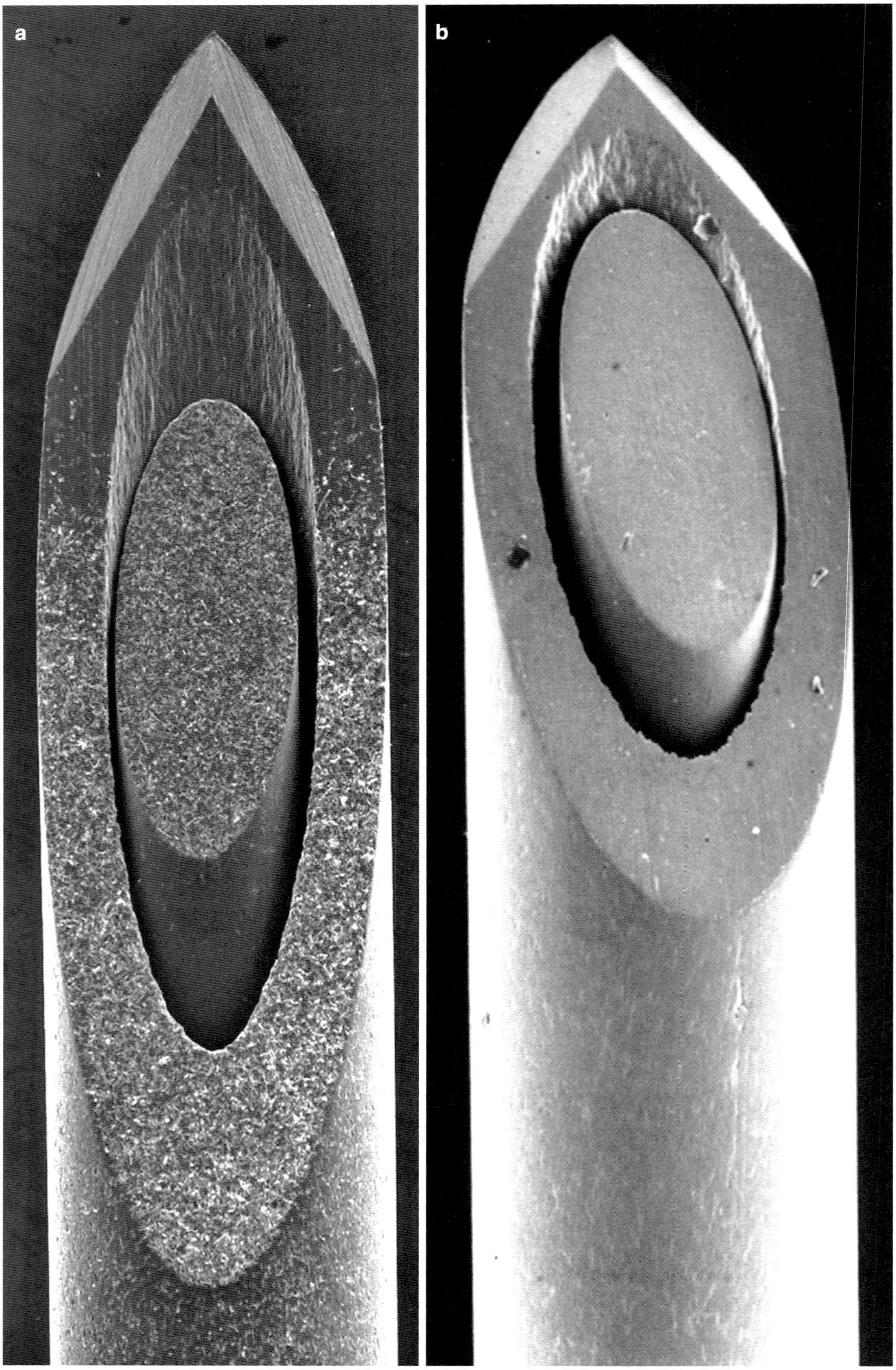

Fig. 44.1 22G Quincke type needles. (**a**, **b**) Needles from different manufacturers. Scanning electron microscopy. Magnification: **a**, ×40; **b**, ×40

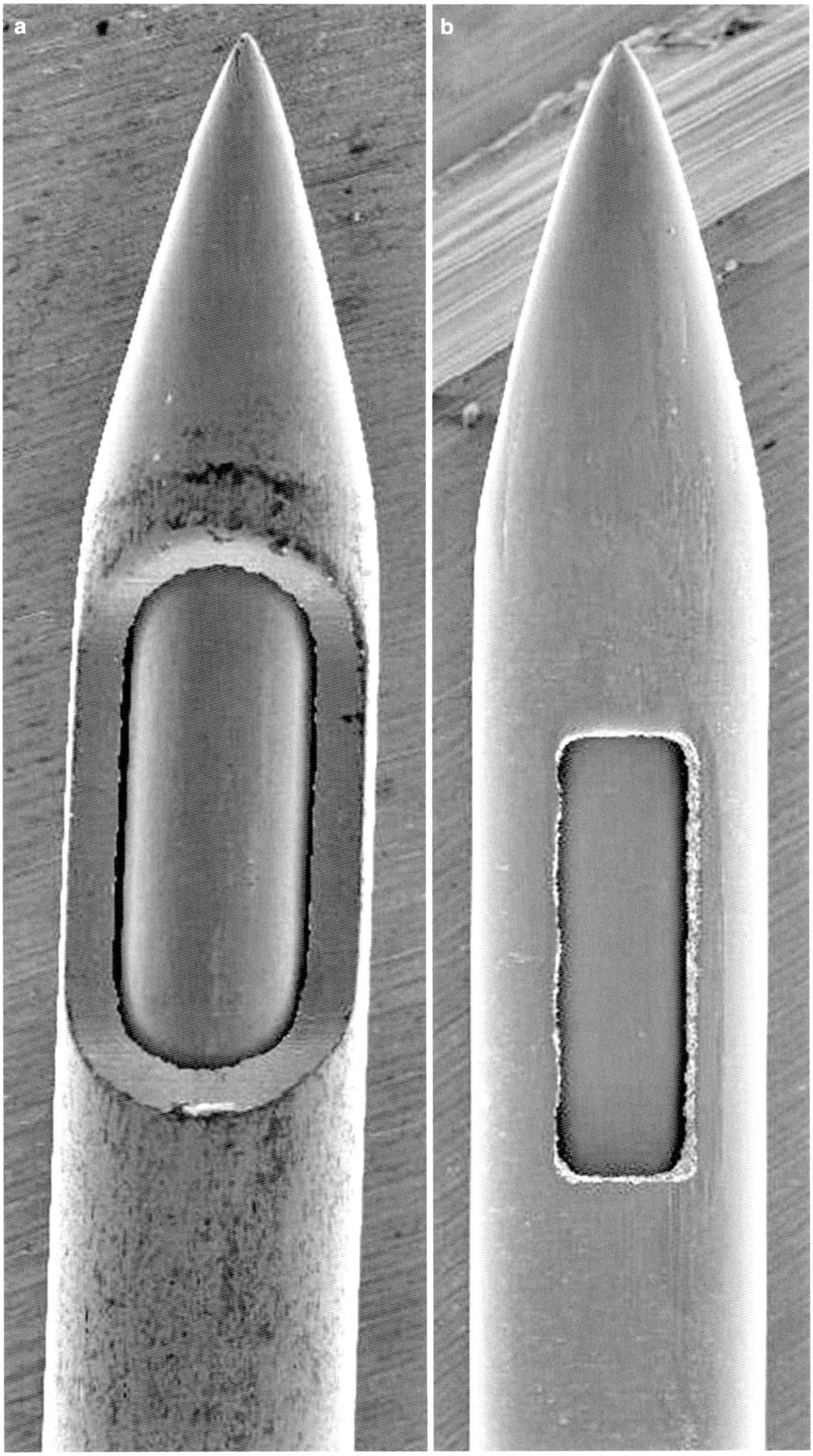

Fig. 44.2 25G Pencil-point type needles. (**a**) Pencil-point type needle. (**b**) Whitacre type needle (scanning electron microscopy). Magnification: **a**, ×35; **b**, ×35

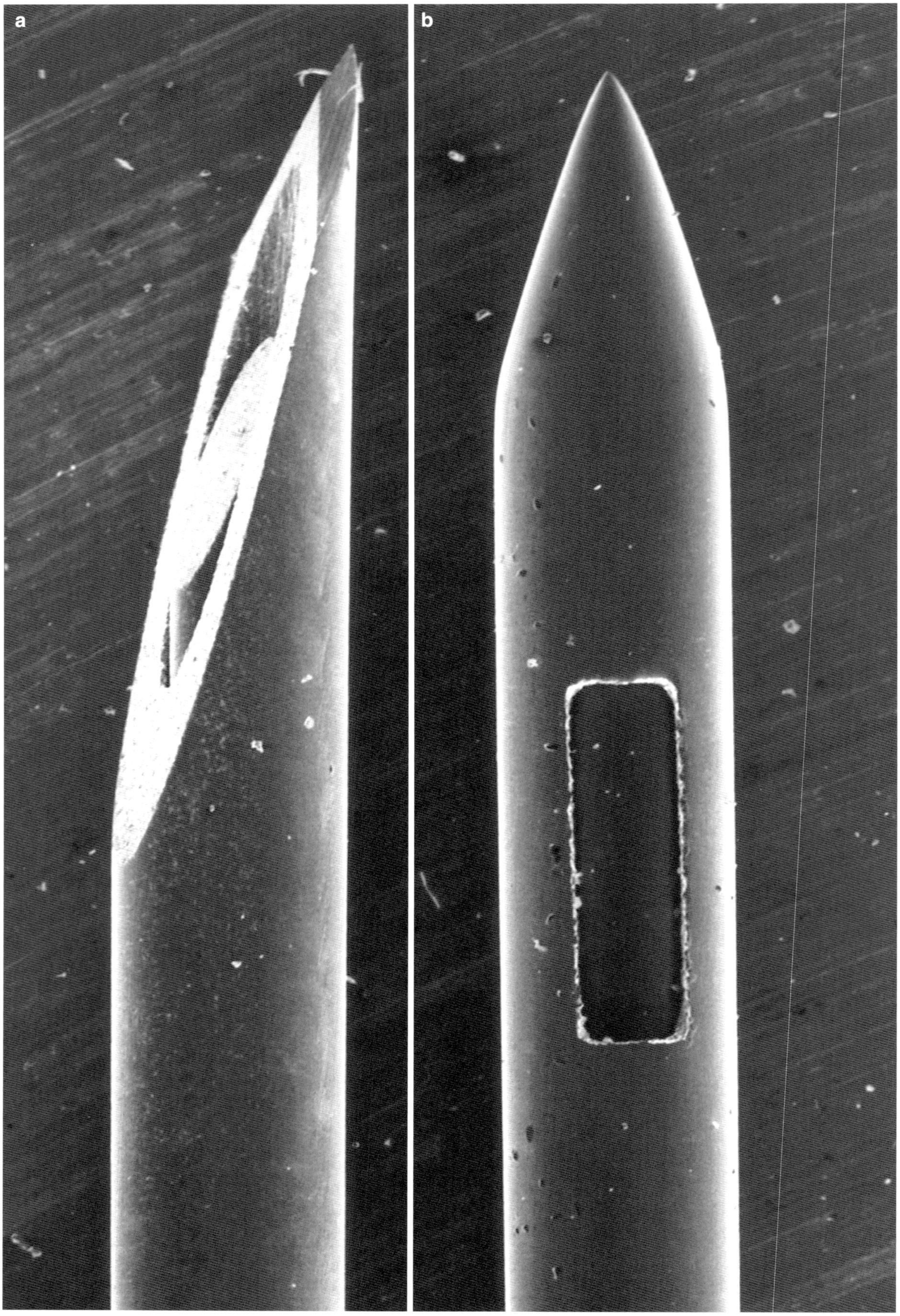

Fig. 44.3 (**a**) 25G Quincke needle. (**b**) 25G Whitacre type needle. Scanning electron microscopy. Magnification: **a**, ×25; **b**, ×25

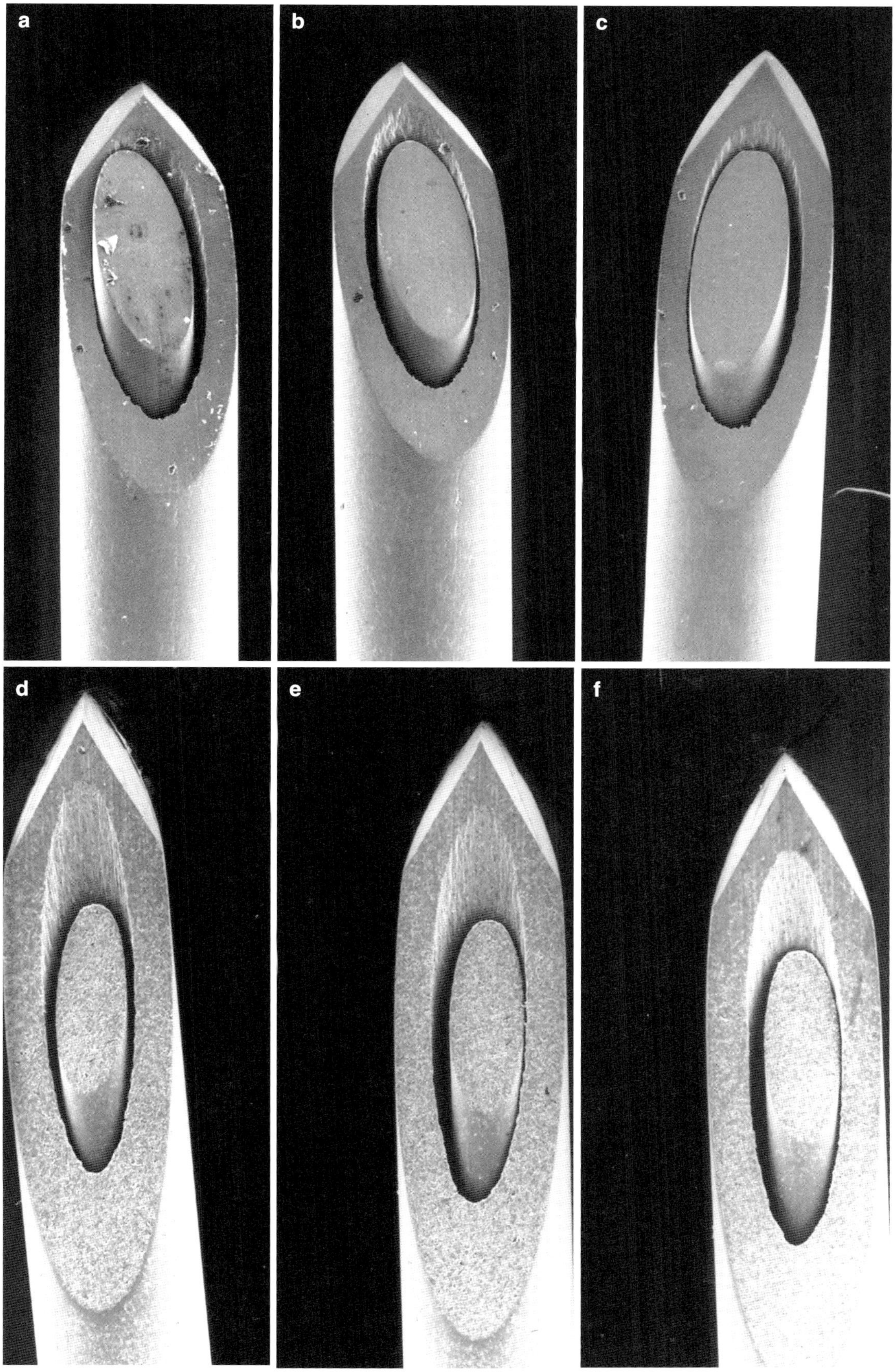

Fig. 44.4 25G Quincke type needles from different manufacturers. (**a–c**) from same manufacturer, (**d-f**) from another manufacturer. Scanning electron microscopy. Magnification: a, b, and c, ×40 (From Reina et al. [5]; with permission)

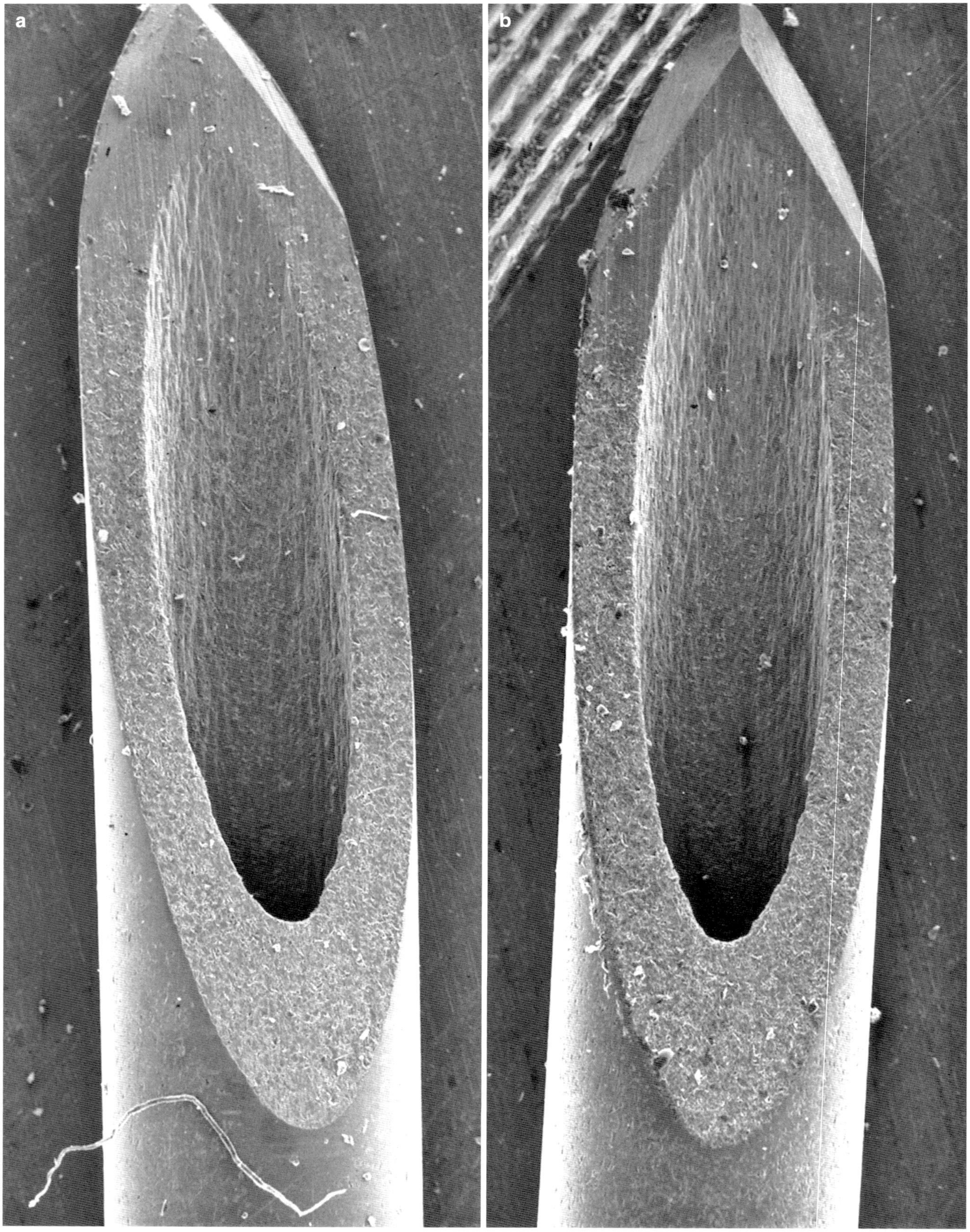

Fig. 44.5 (**a**, **b**) 22G Quincke type needles. Scanning electron microscopy. Magnification: **a**, ×43; **b**, ×43

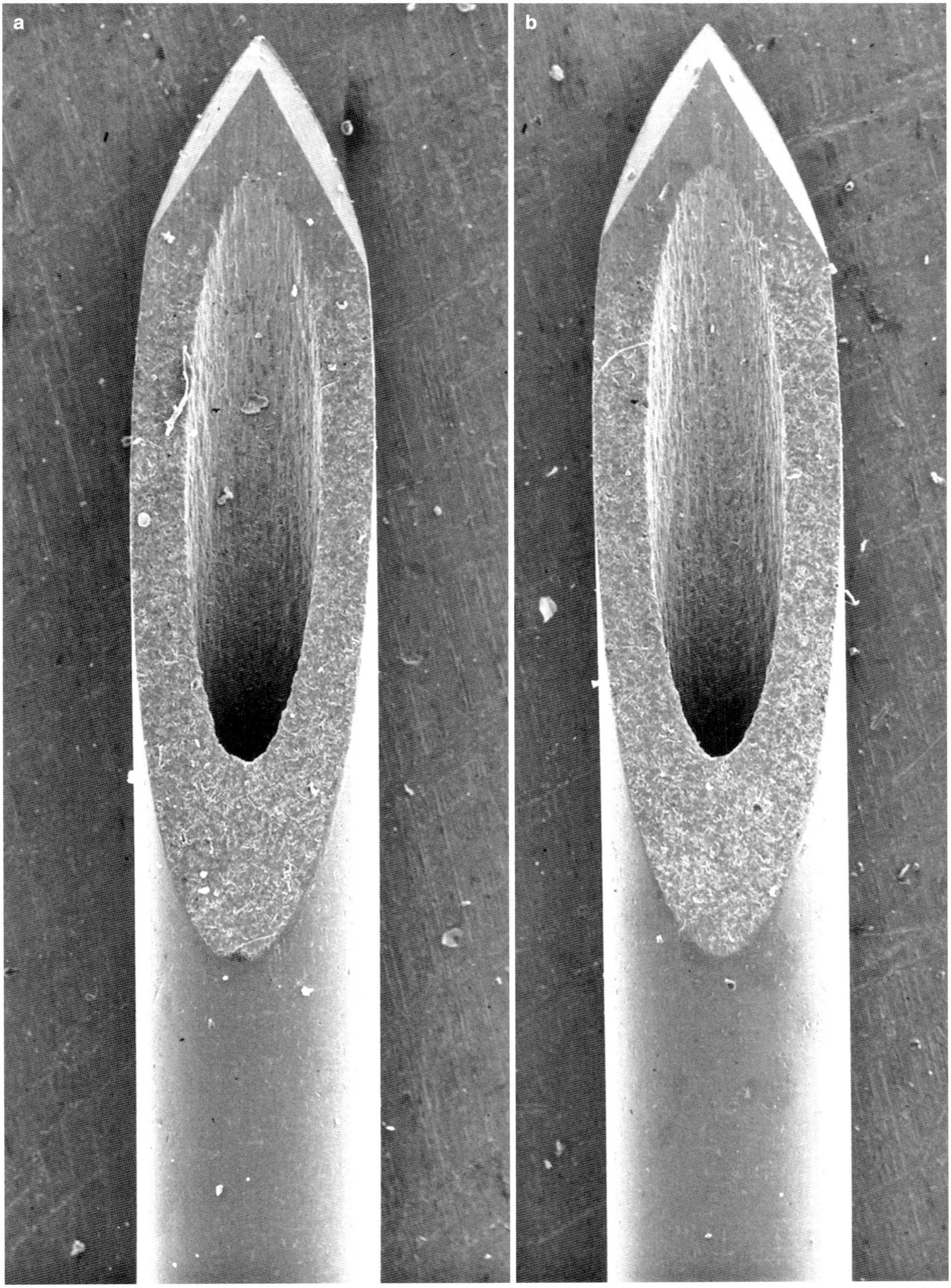

Fig. 44.6 (**a**, **b**) 25G Quincke type needles. Scanning electron microscopy. Magnification: **a**, ×43; **b**, ×43

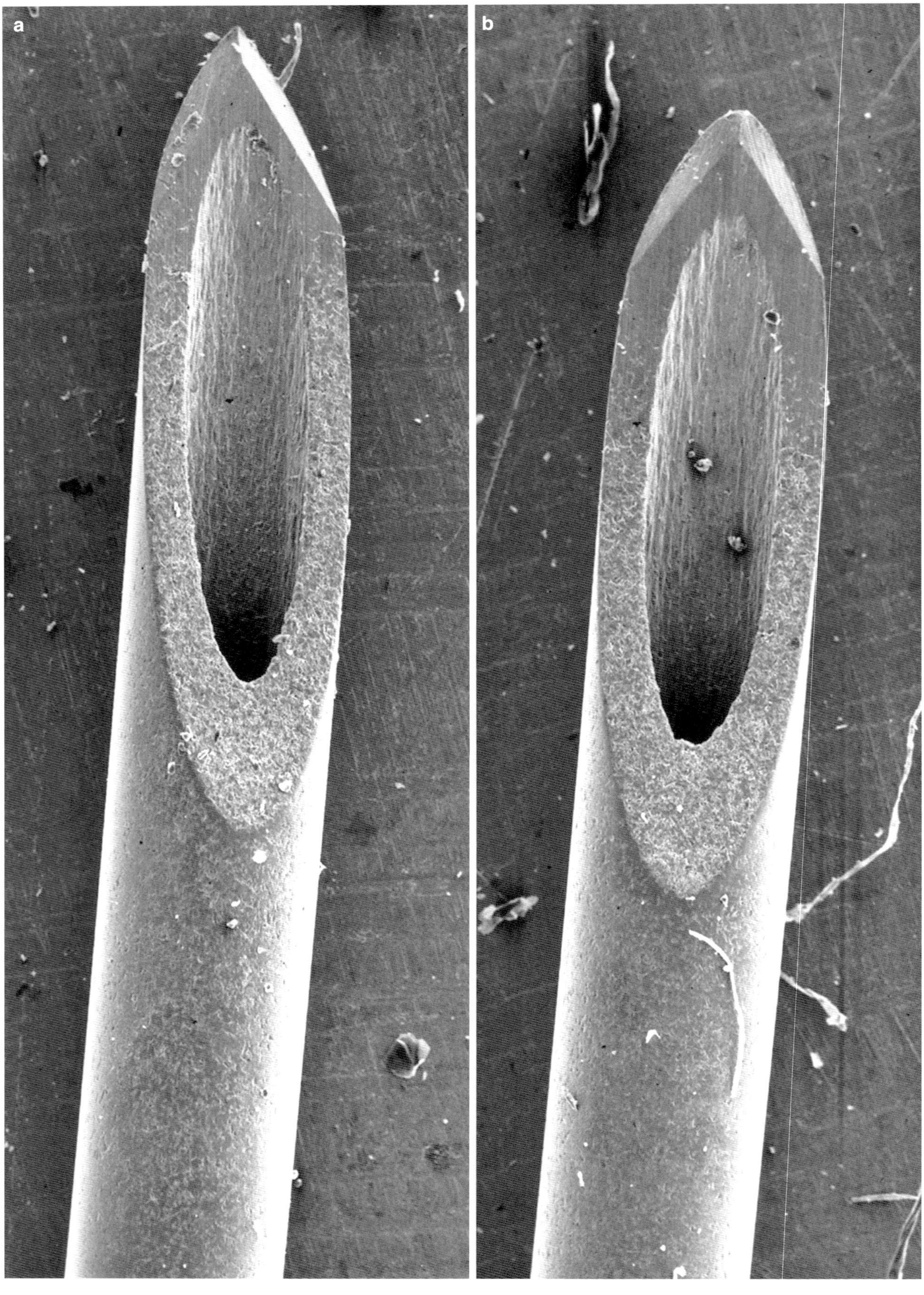

Fig. 44.7 (**a**, **b**) 26G Quincke type needles. Scanning electron microscopy. Magnification: **a**, ×43; **b**, ×43

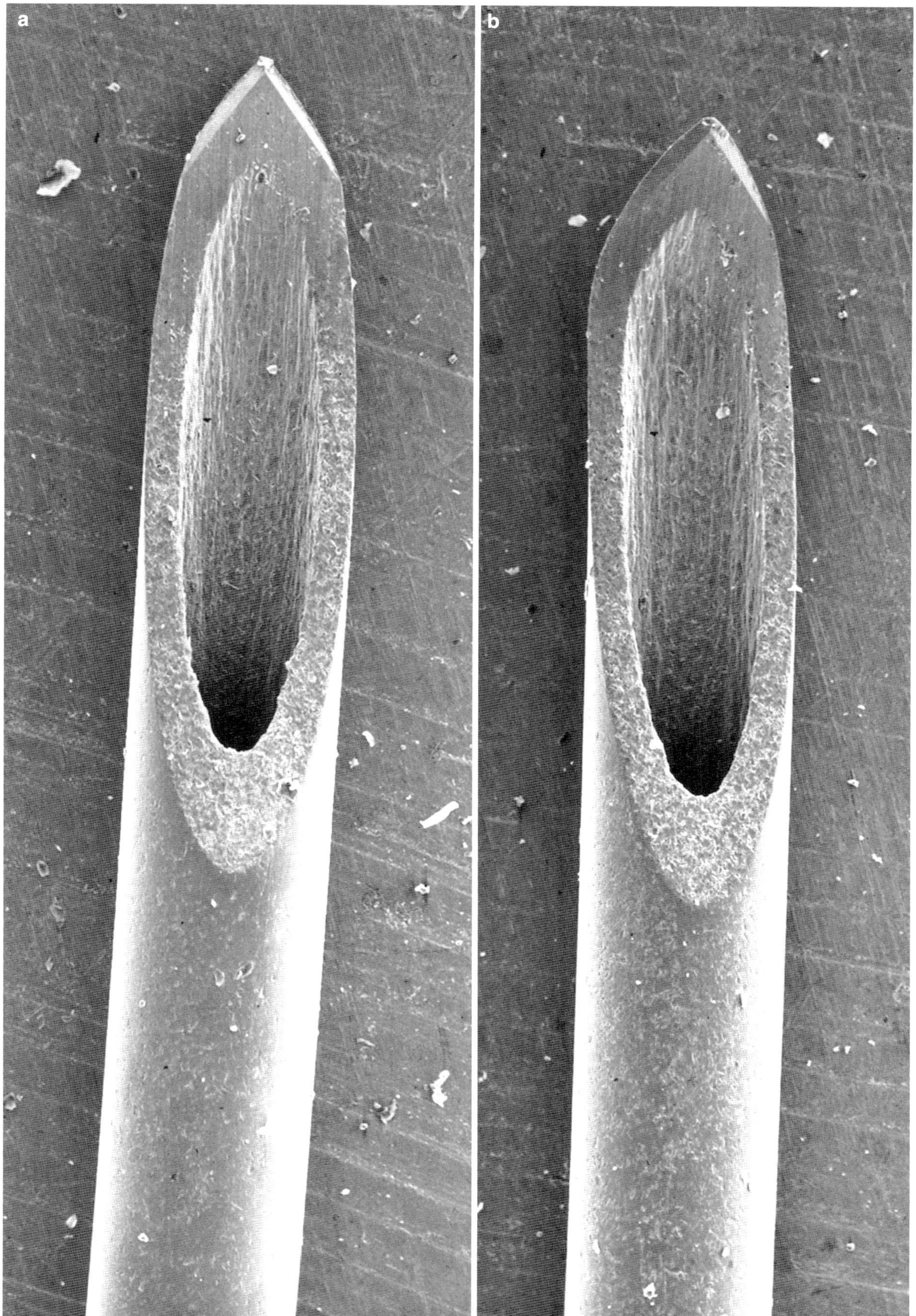

Fig. 44.8 (**a**, **b**) 27G Quincke type needles. Scanning electron microscopy. Magnification: **a**, ×43; **b**, ×43

Fig. 44.9 (**a**, **b**) 29G Quincke type needles. Scanning electron microscopy. Magnification: **a**, ×43; **b**, ×43

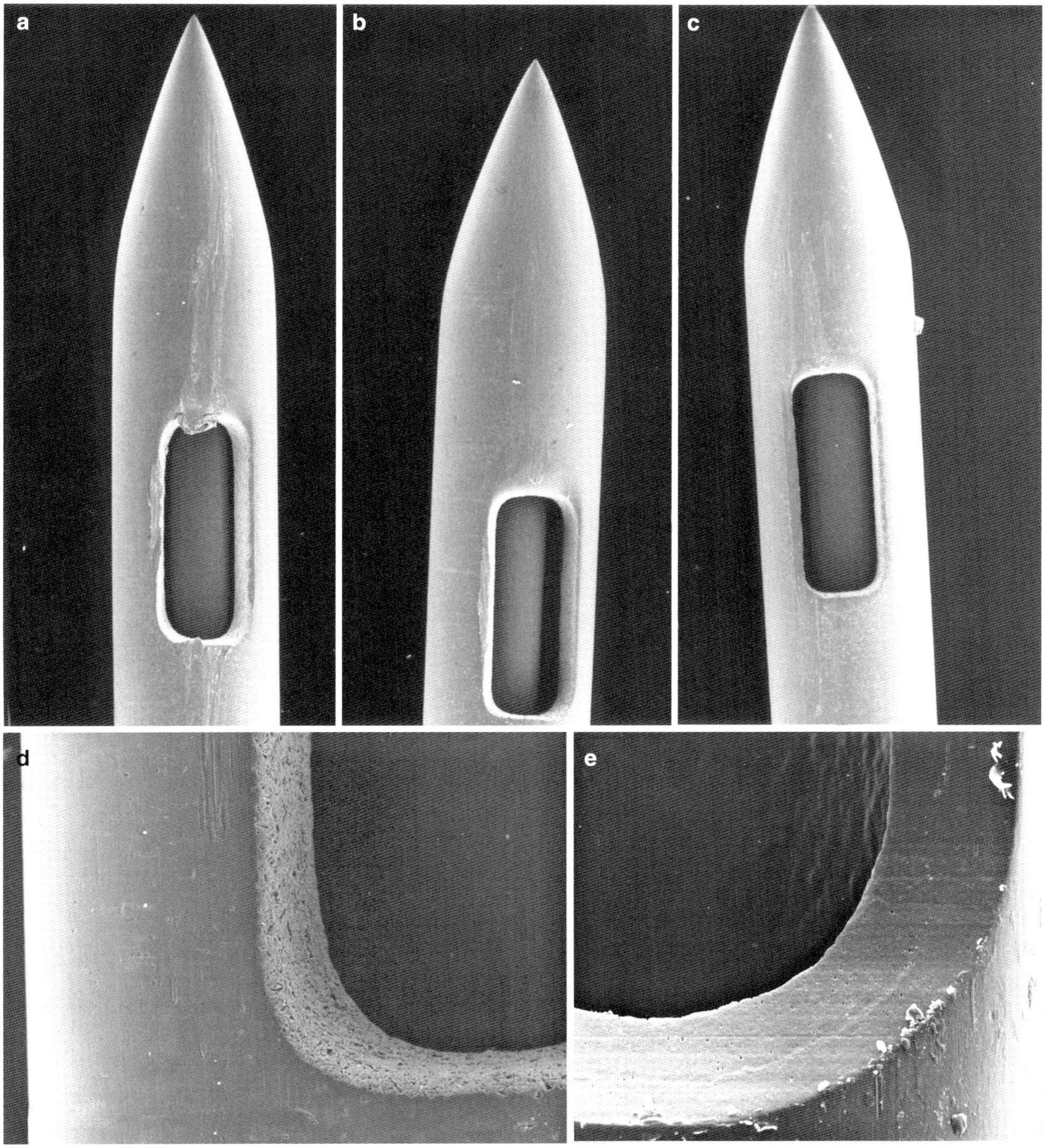

Fig. 44.10 Whitacre type needles. (**a–c**) Needles from same box. (**d**, **e**) Detail of lateral hole. Scanning electron microscopy. Magnification: **a–c**, ×40; **d** and **e**, ×350

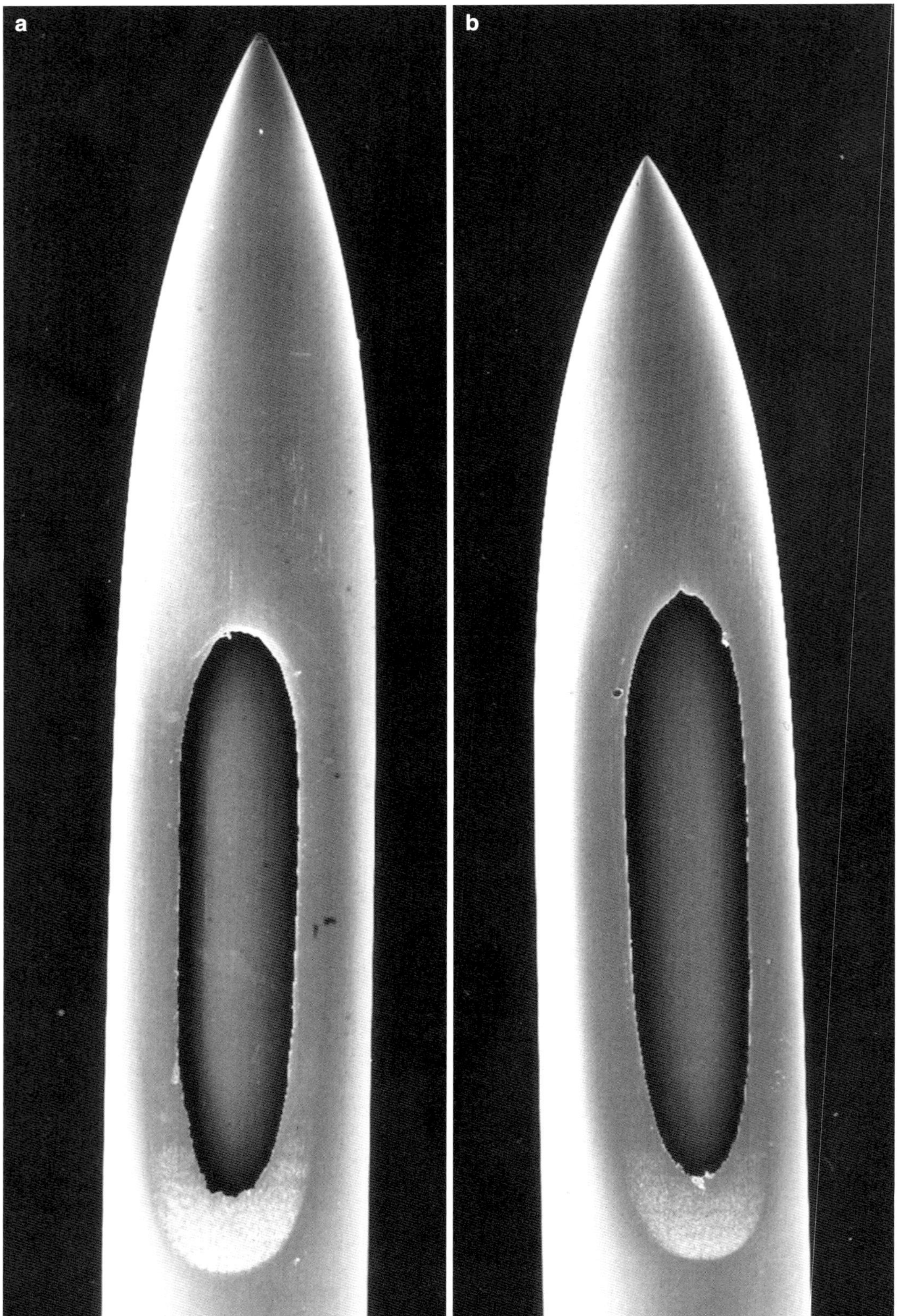

Fig. 44.11 (**a**, **b**) 24G Pencil-point type needles from the same box. Scanning electron microscopy. Magnification: **a** and **b**, ×40

Fig. 44.12 17G Tuohy type needle. Scanning electron microscopy. Magnification ×25

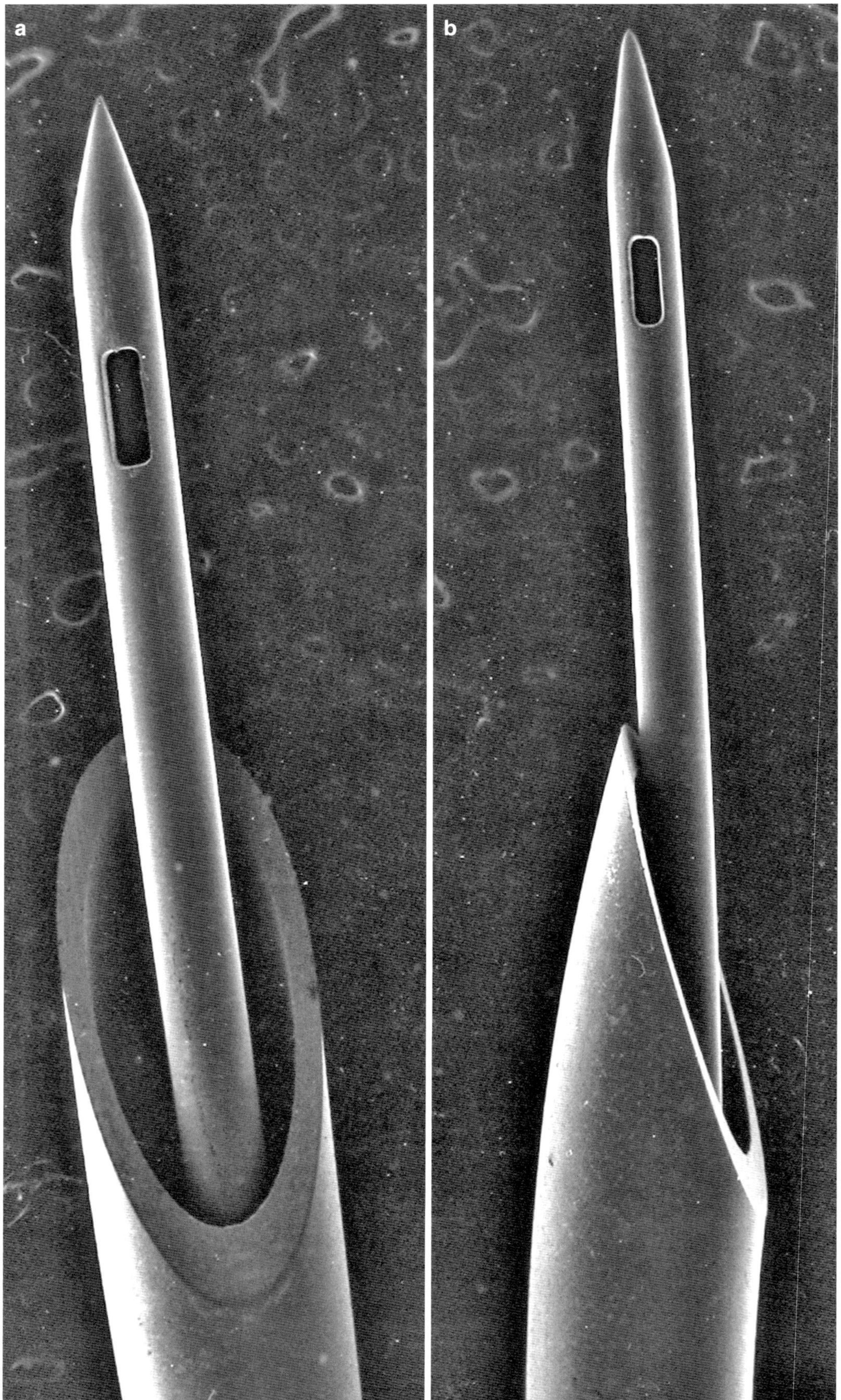

Fig. 44.13 Spinal epidural combined needle. (**a**) 25G Whitacre type needle. (**b**) 27G Whitacre type needle. Scanning electron microscopy. Magnification ×13

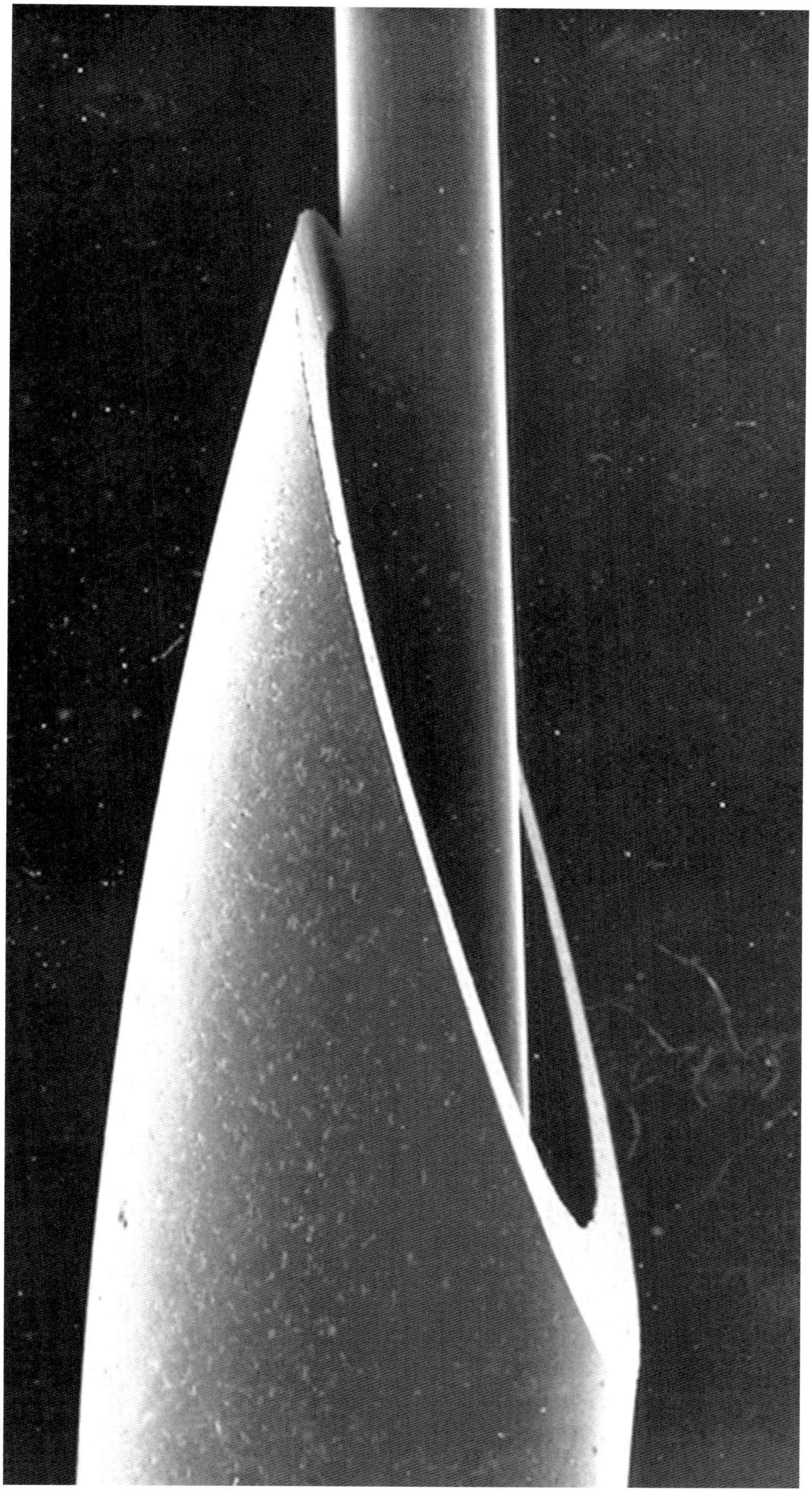

Fig. 44.14 Spinal epidural combined needle (detail of Fig. 44.13). Scanning electron microscopy. Magnification ×30

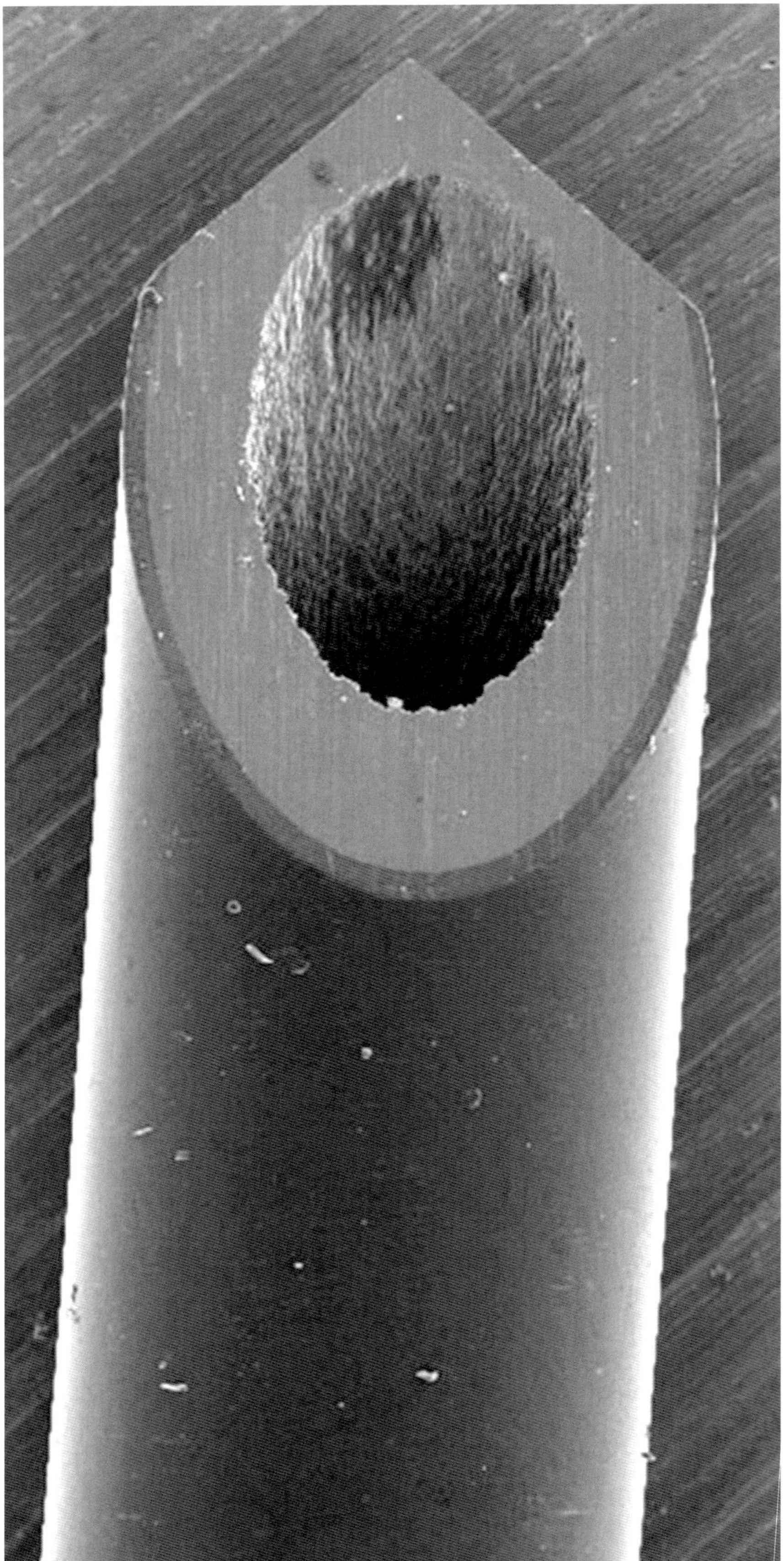

Fig. 44.15 21G Neurostimulation, peripheral needle type A. Scanning electron microscopy. Magnification ×40 (From Reina et al. [8]; with permission)

Fig. 44.16 21G Neurostimulation peripheral needle type A (detail of Fig. 44.15). Scanning electron microscopy. Magnification ×150 (From Reina et al. [8]; with permission)

Fig. 44.17 21G Neurostimulation peripheral needle type A (detail of Fig. 44.15). Scanning electron microscopy. (**a**) Posterior view; (**b**) Anterior view of same needle. Magnification: **a**, ×150; **b**, ×150 (From Reina et al. [8]; with permission)

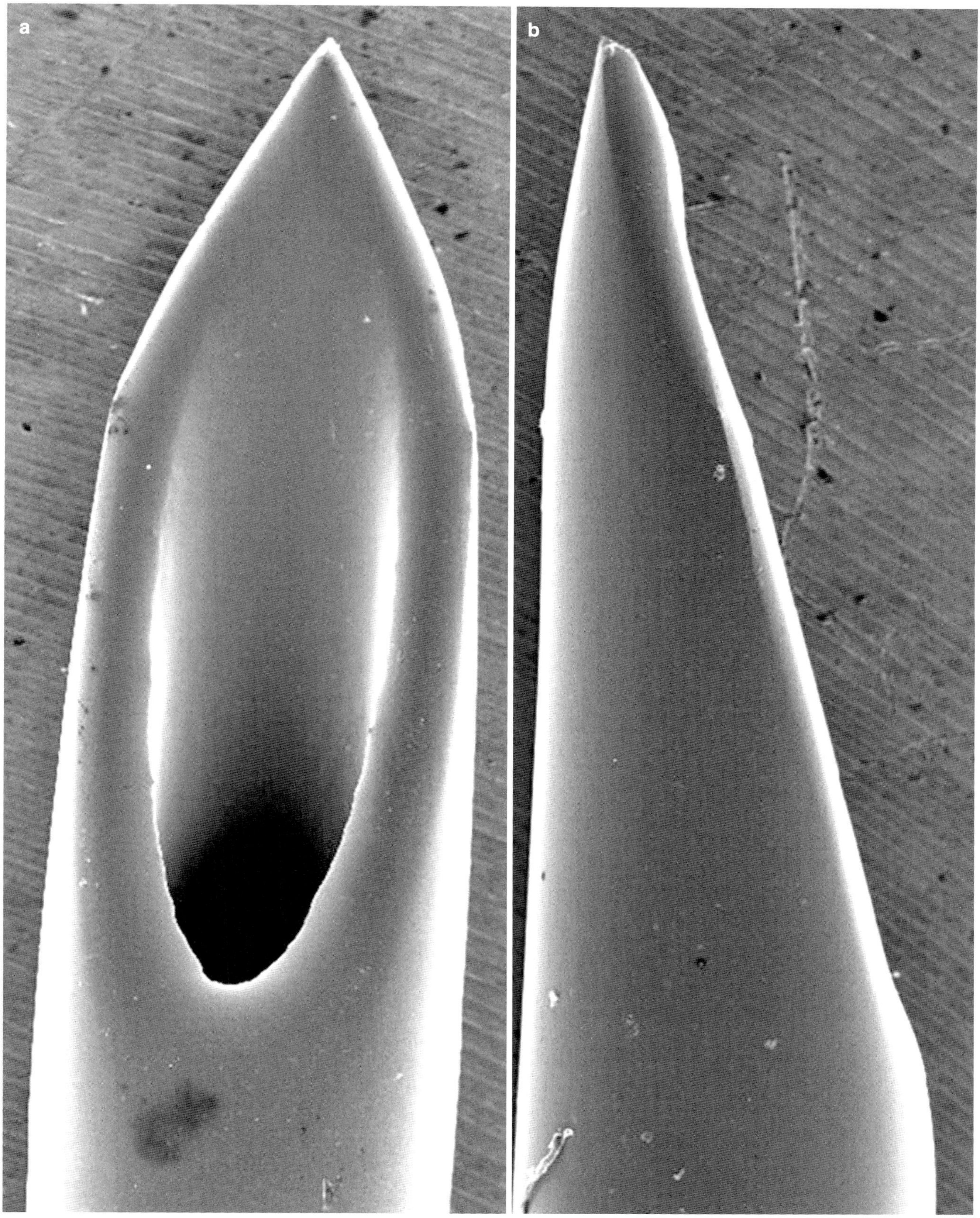

Fig. 44.18 (**a**, **b**) 21G Neurostimulation peripheral needle type D. Scanning electron microscopy. Magnification: **a**, ×40; **b**, ×40

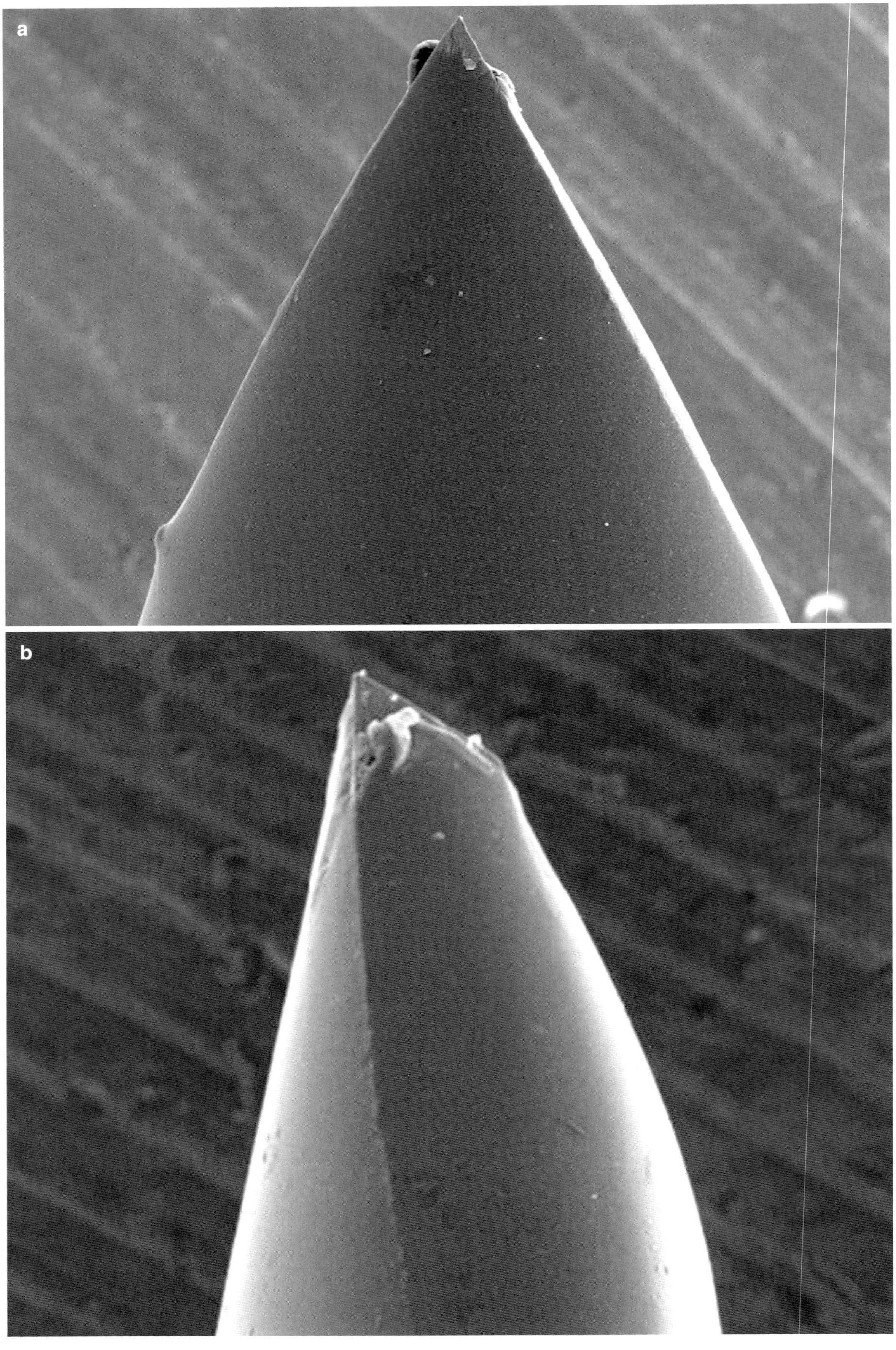

Fig. 44.19 (**a**, **b**) 21G Neurostimulation peripheral needle type D (detail of Fig. 44.18). Scanning electron microscopy. Magnification: **a**, ×150; **b**, ×150

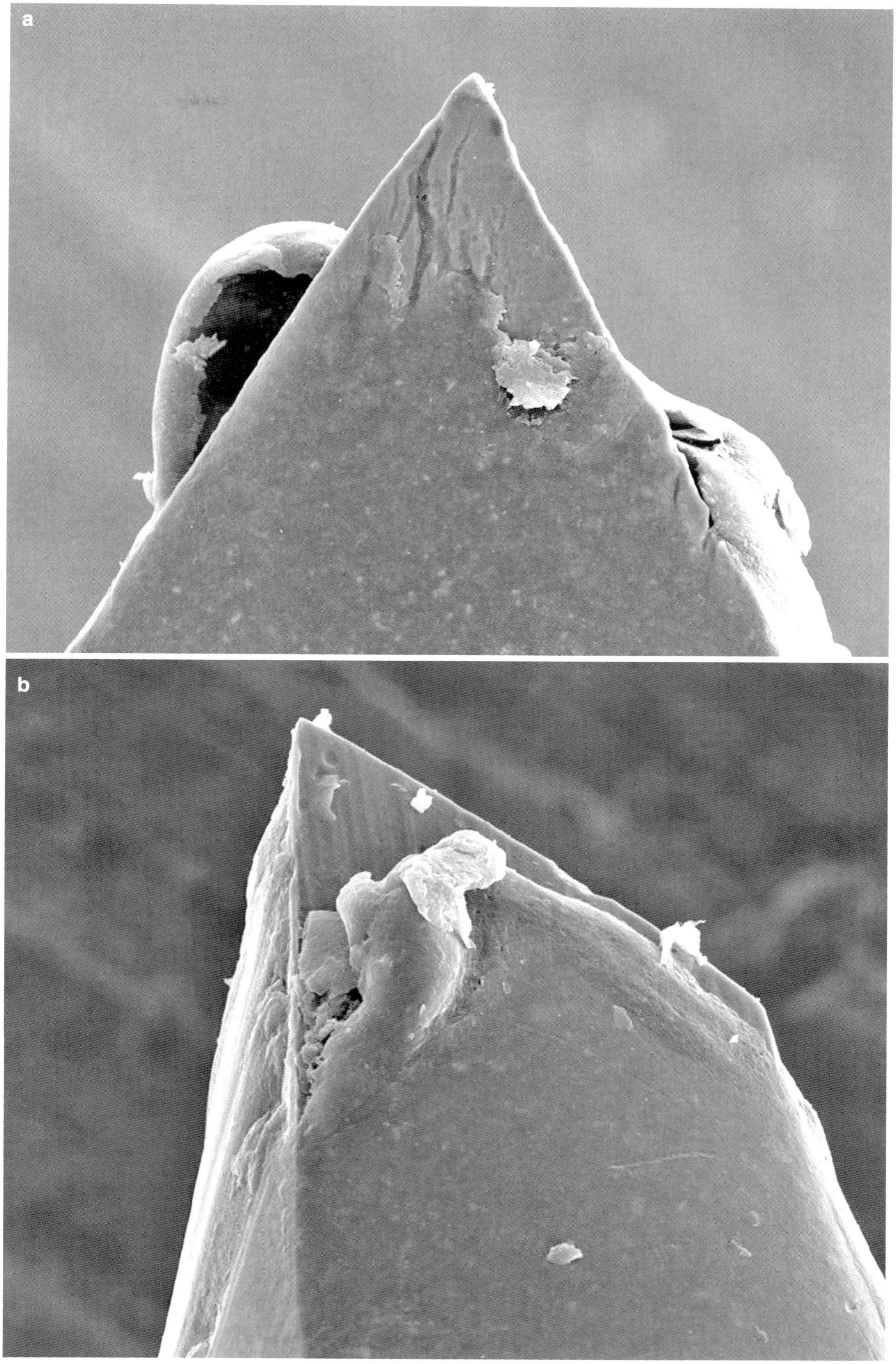

Fig. 44.20 (**a**, **b**) 21G Neurostimulation peripheral needle type D (detail of Fig. 44.18). Scanning electron microscopy. Magnification: **a**, ×1,000; **b**, ×1,000

Fig. 44.21 (**a**) Intramuscular needle. (**b**) Detail of (**a**). Scanning electron microscopy. Magnification: **a**, ×20; **b**, ×150

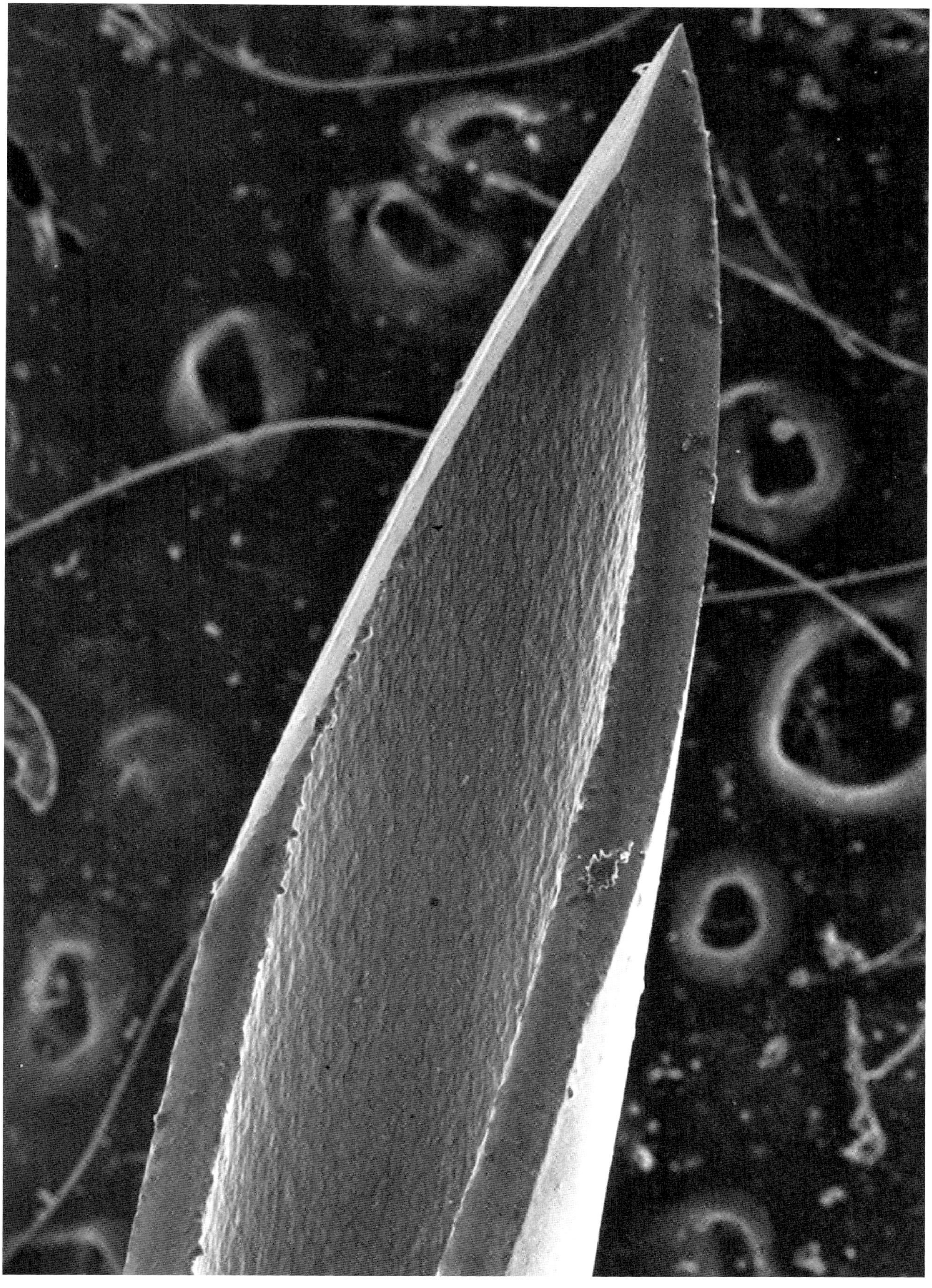

Fig. 44.22 Intramuscular needle. Anterior view of their bevels. Scanning electron microscopy. Magnification ×43

References

1. Reina MA. Contribute of electron microscopy in regional anesthesia. Reg Anesth Pain Med Suppl. 2011;36:E81–5.
2. Reina MA, Lopez A, Badorrey V, De Andres JA, Martín S. Dura-arachnoid lesions produced by 22G Quincke spinal needles during a lumbar puncture. J Neurol Neurosurg Psychiatry. 2004;75:893–7.
3. Reina MA, De León Casasola OA, López A, De Andrés JA, Martín S, Mora M. An in vitro study of dural lesions produced by 25 gauge Quincke and Whitacre needles evaluated by scanning electron microscopy. Reg Anesth Pain Med. 2000;25:393–402.
4. Reina MA, López A, Machés F, De Leon Casasola O, De Andrés JA. Electron microscopy and the expansion of regional anesthesia knowledge. Tech Reg Anesth Pain Management. 2002;6:165–71.
5. Reina MA, Castedo J, López A. Cefalea pospunción dural. Ultraestructura de las lesiones durales y agujas espinales usadas en las punciones lumbares. Rev Arg Anestesiol. 2008;66:6–26.
6. Reina MA, De Andrés JA, López A. Subarachnoid and epidural anesthesia. In: Raj P, editor. Textbook of regional anesthesia. Philadelphia: Churchill Livingstone; 2002. p. 307–24.
7. López A, Reina MA, Machés F, De Leon Casasola O, De Andrés JA, García Trapero J. Electron microscopy in quality control of equipment used in regional anesthesia. Tech Reg Anesth Pain Management. 2002;6:172–9.
8. Reina MA, López A, De Andrés JA, Machés F. Possibility of nerve lesions related to peripheral nerve blocks. A study of the human sciatic nerve using different needles. Rev Esp Anestesiol Reanim. 2003;50:274–83.

45 Catheters in Regional Anesthesia

Miguel Angel Reina, José Antonio De Andrés, and Andrés López

Anesthesiologists use different types of catheters in the performance of continuous regional anesthetic techniques such as epidural, subarachnoid, and paravertebral blocks, as well as peripheral nerve blocks.

Epidural catheters may have a single distal orifice located at the tip of the catheter (open-end catheters) or three lateral orifices located lateral to the tip, which is closed and may be oriented in various positions. The use of subarachnoid catheters specifically designed for continuous regional anesthesia or analgesia is less frequent. In the past, the same type of epidural catheter was placed inside both the epidural and subarachnoid spaces. In the 1980s, special types of microcatheters were manufactured for continuous subarachnoid anesthetic techniques, but production of these types of catheters was discontinued in 1991, after cases of cauda equina syndrome had been reported following the use of this technique. At present, various types of subarachnoid catheters are manufactured. A few of these catheters are indicated for short-term placement (about 2 or 3 days) in surgical interventions and postoperative management of pain. Other catheters are designed with materials suitable for long-term placement lasting several months, such as in the management of chronic pain.

Techniques involving peripheral nerve blockade aim at placing catheters adjacent to the nerve, where bolus doses or continuous perfusions of local anesthetics are administered to produce analgesia during the postoperative period (Figs. 45.1, 45.2, 45.3, 45.4, 45.5, 45.6, 45.7, 45.8, 45.9, 45.10, 45.11, 45.12, 45.13, 45.14, 45.15, 45.16, 45.17, 45.18, and 45.19).

M.A. Reina, MD, PhD (✉)
Department of Clinical Medical Sciences and Institute of Applied Molecular Medicine, School of Medicine, University of CEU San Pablo, Madrid, Spain

Department of Anesthesiology, Madrid-Montepríncipe University Hospital, Madrid, Spain
e-mail: miguelangel@perticone.e.telefonica.net

J.A. De Andrés, MD, PhD
Department of Surgical Specialties, School of Medicine, University of Valencia, Valencia, Spain

Department of Anesthesia, Critical Care, and Pain Management, General University Hospital, Valencia, Spain
e-mail: deandres_jos@gva.es

A. López, MD
Department of Clinical Medical Sciences, School of Medicine, CEU San Pablo University, Madrid, Spain

Department of Anesthesiology, Madrid-Montepríncipe University Hospital, Madrid-Torrelodones University Hospital, Madrid-Sanchinarro University Hospital, Madrid, Spain
e-mail: alg00009@telefonica.net

M.A. Reina et al. (eds.), *Atlas of Functional Anatomy for Regional Anesthesia and Pain Medicine: Human Structure, Ultrastructure and 3D Reconstruction Images*, DOI 10.1007/978-3-319-09522-6_45,

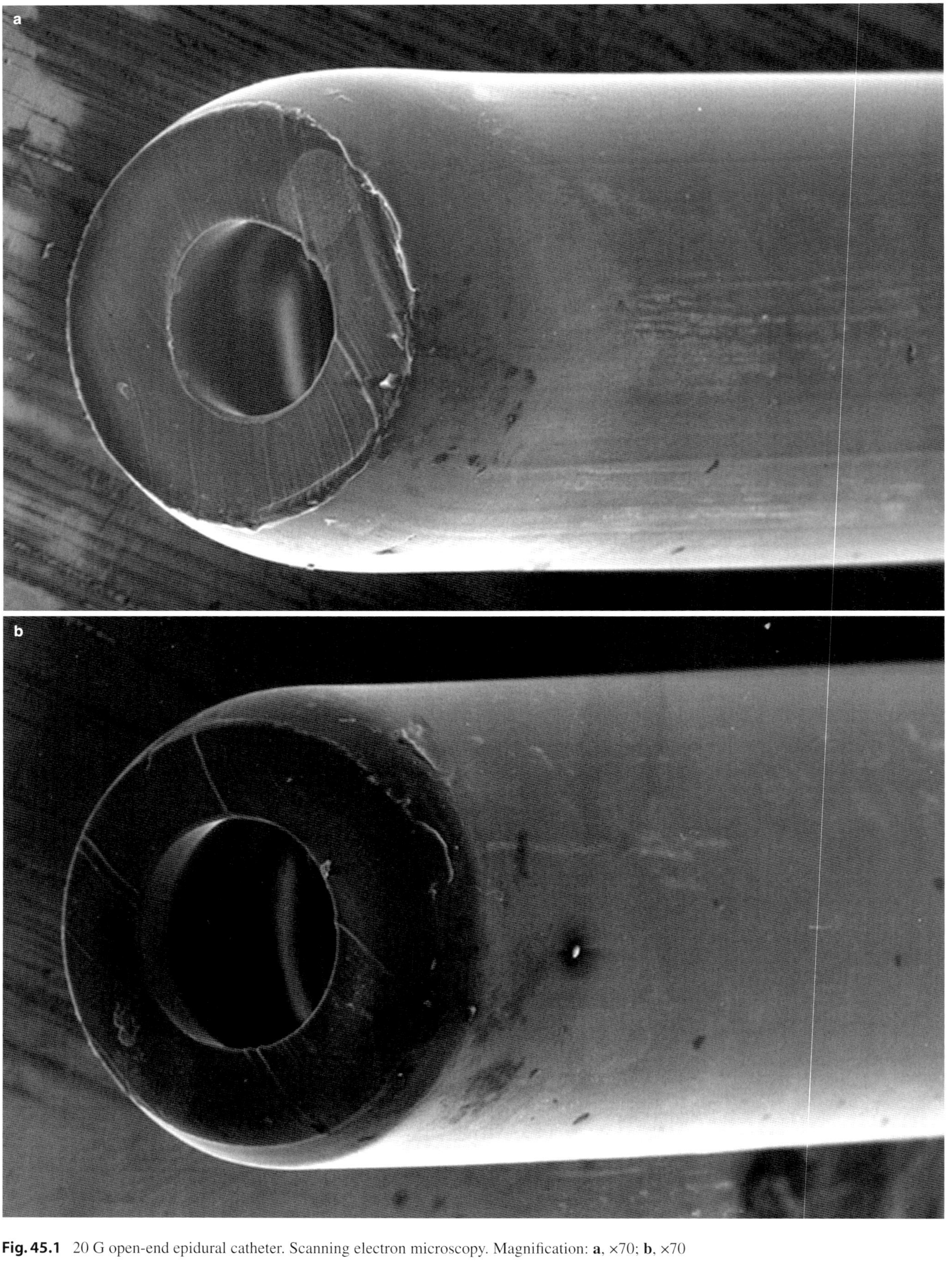

Fig. 45.1 20 G open-end epidural catheter. Scanning electron microscopy. Magnification: **a**, ×70; **b**, ×70

Fig. 45.2 20 G epidural catheter. (**a**) Closed-end tip of catheter. (**b**) Open-end tip of catheter. Scanning electron microscopy. Magnification: **a** and **b**, ×100 (**b** from López et al. [1]; with permission)

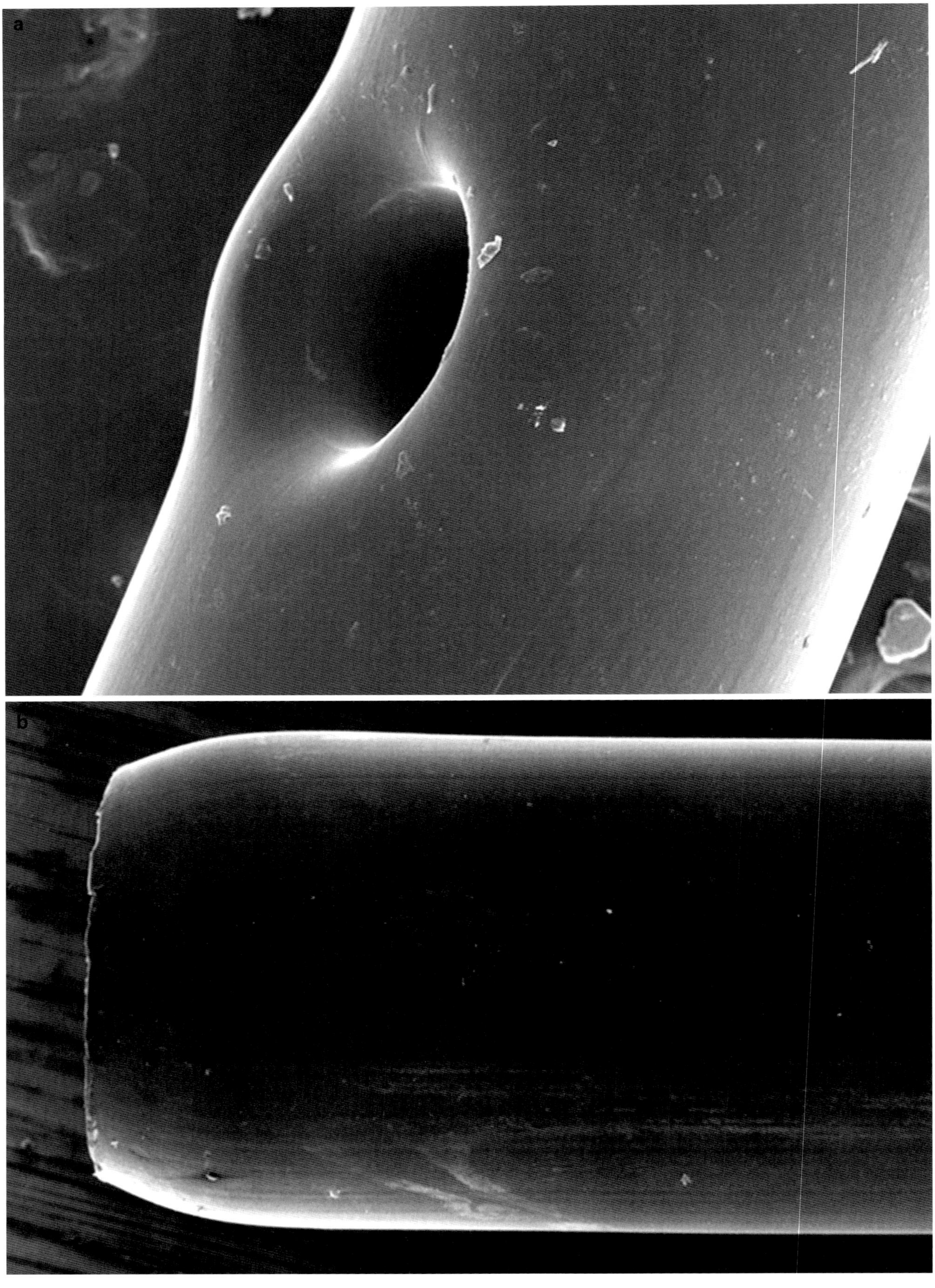

Fig. 45.3 20 G epidural catheter. (**a**) Catheter with orifice. (**b**) Open-end catheter. View of lateral surface. Scanning electron microscopy. Magnification: **a**, ×100; **b**, ×70

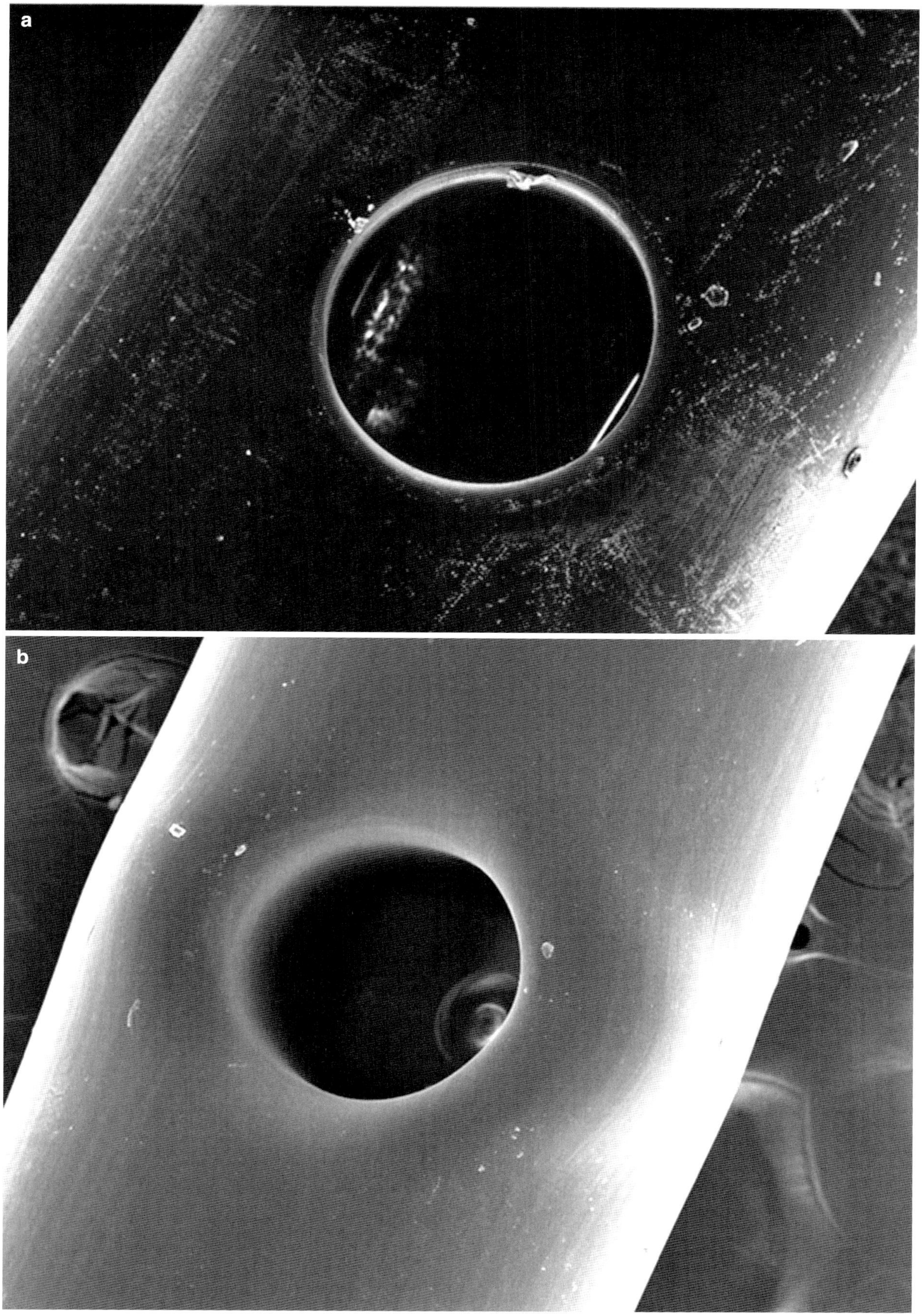

Fig. 45.4 20 G epidural catheter. (**a**, **b**) Catheters with lateral orifices, from different manufacturers. Scanning electron microscopy. Magnification: **a** and **b**, ×100

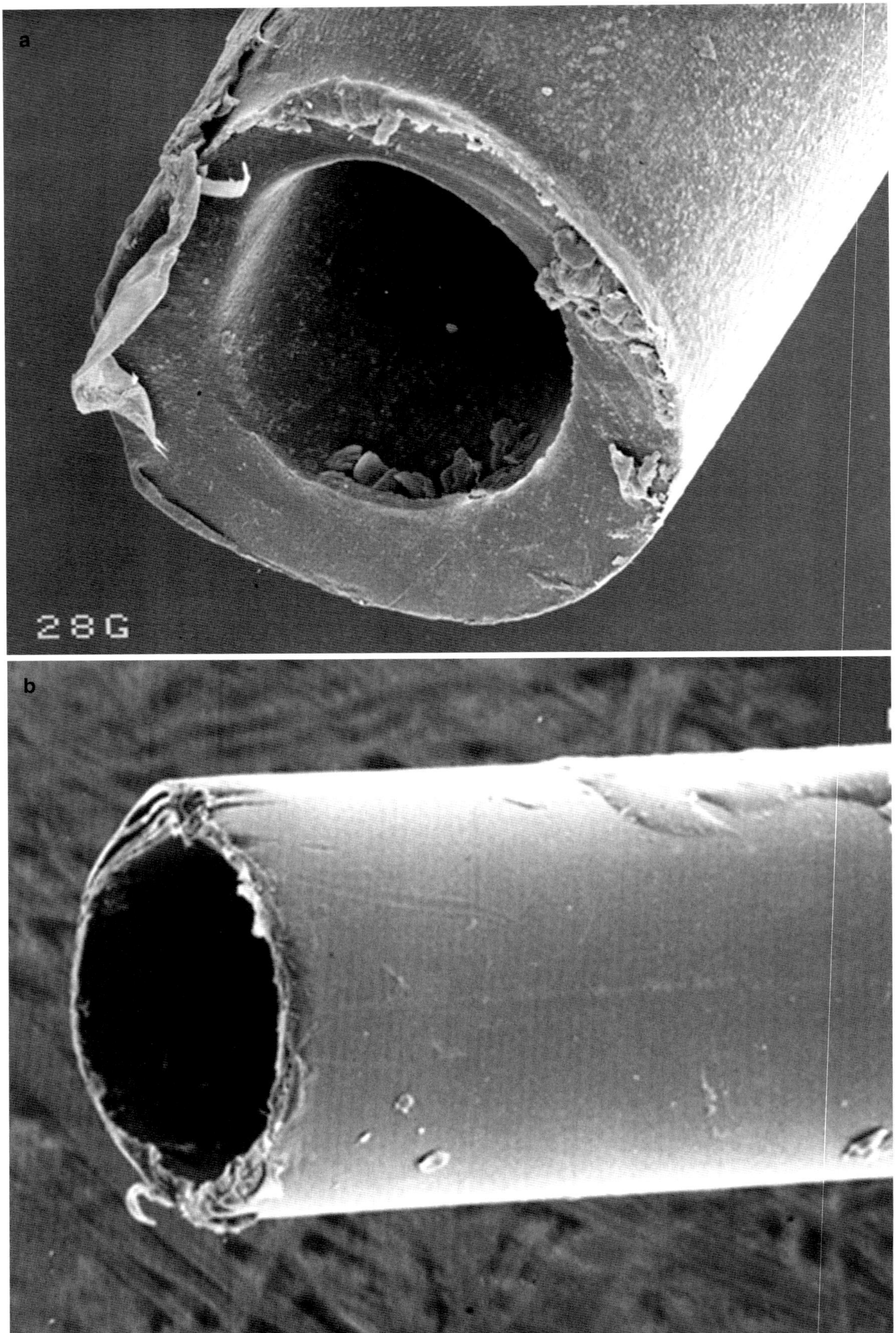

Fig. 45.5 (**a**) 28 G catheter. (**b**) 30 G microcatheter. Scanning electron microscopy. Magnification: **a**, ×200; **b**, ×100 (**b** from López et al. [1]; with permission)

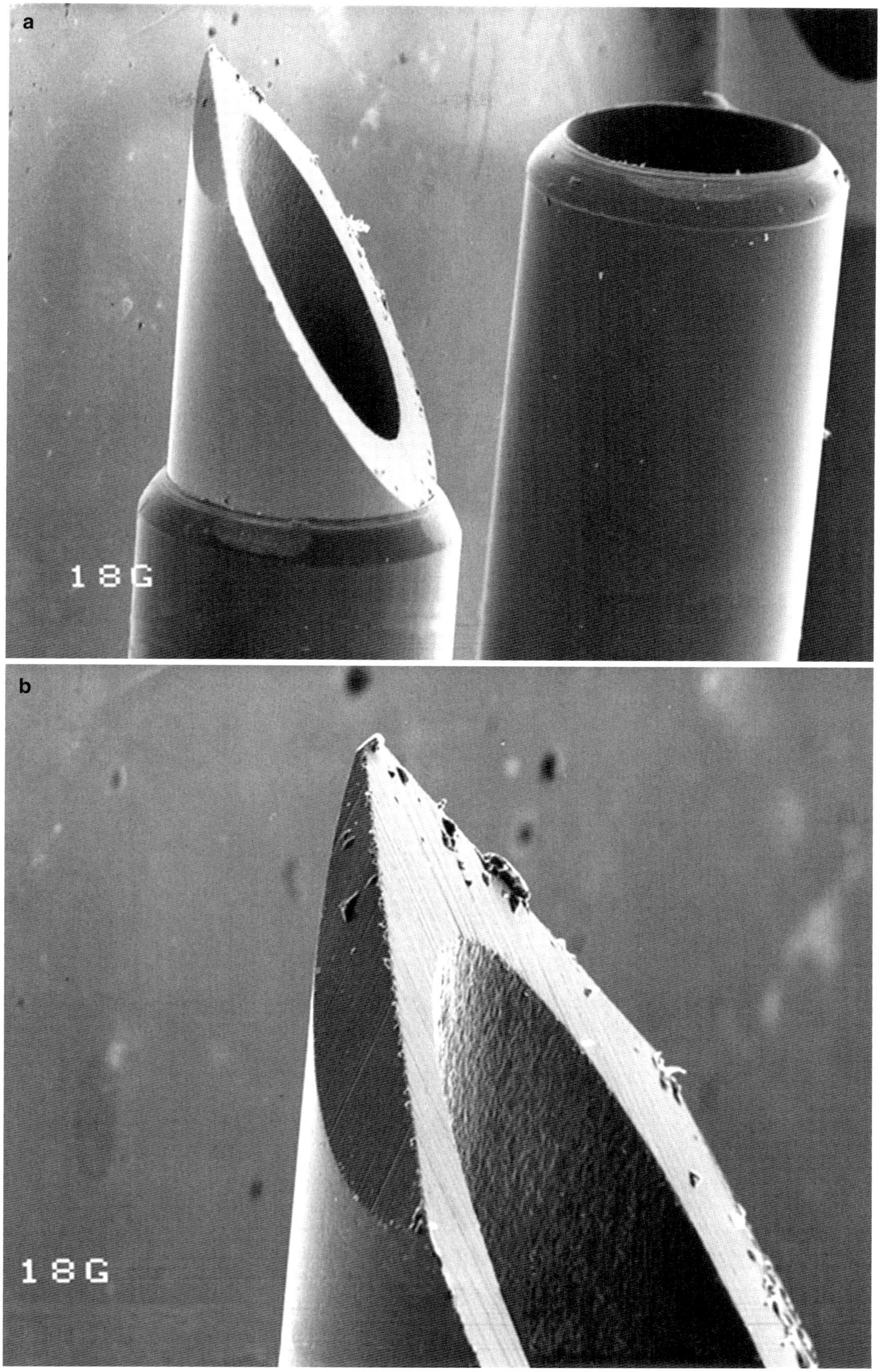

Fig. 45.6 Catheter for peripheral anesthetic blockade. (**a**) Short-bevel neurostimulation needle with its respective catheter. (**b**) Detail of the tip of the needle. Scanning electron microscopy. Magnification: **a**, ×35; **b**, ×100

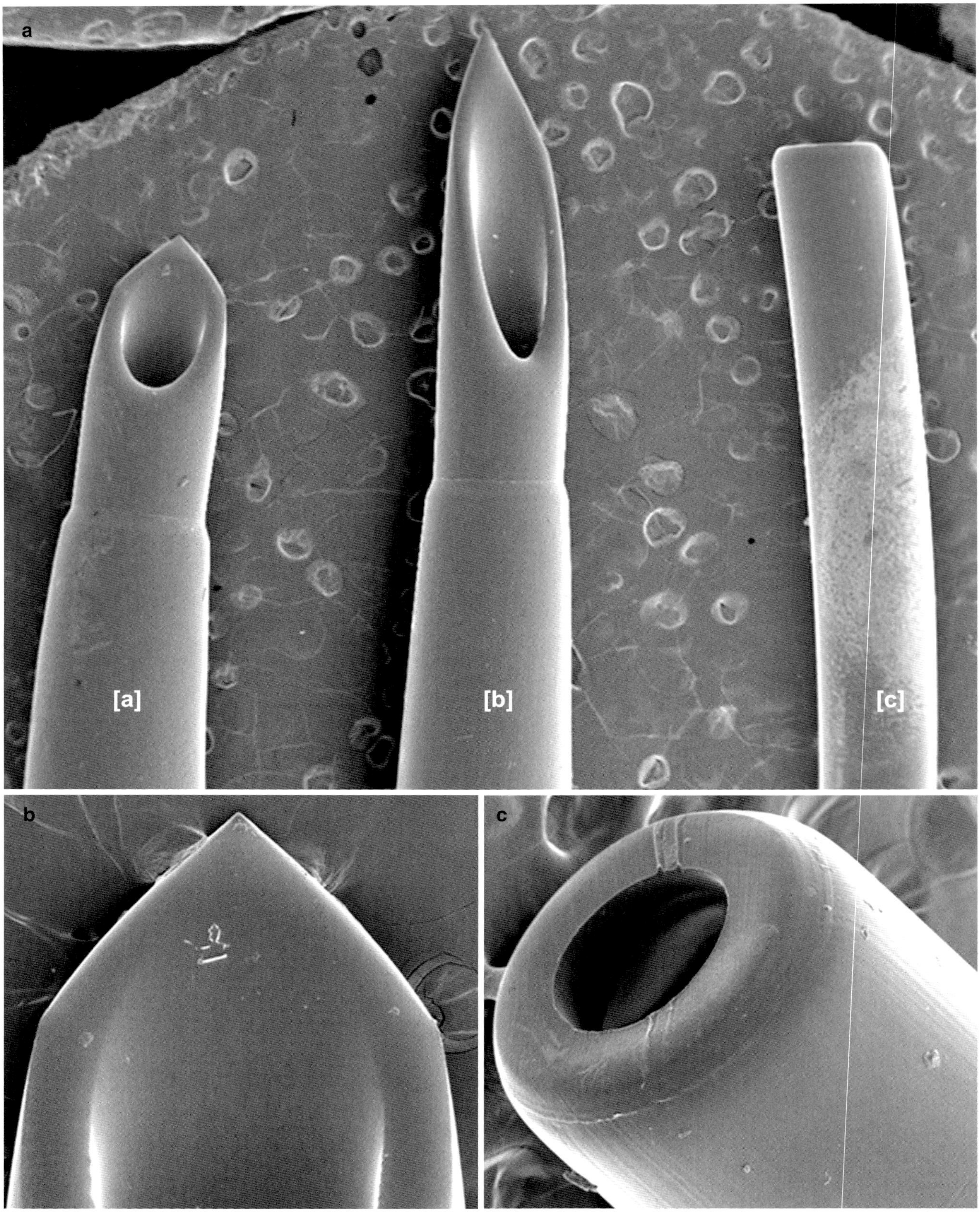

Fig. 45.7 Catheter for peripheral anesthetic blockade. (**a**) Needles with different design, with their respective catheters (*a, b*); the peripheral catheter (*c*) is introduced along catheter a. (**b**) Detail of catheter a at higher magnification. (**c**) Detail of catheter c at higher magnification. Scanning electron microscopy. Magnification: **a**, ×15; **b** and **c**, ×100

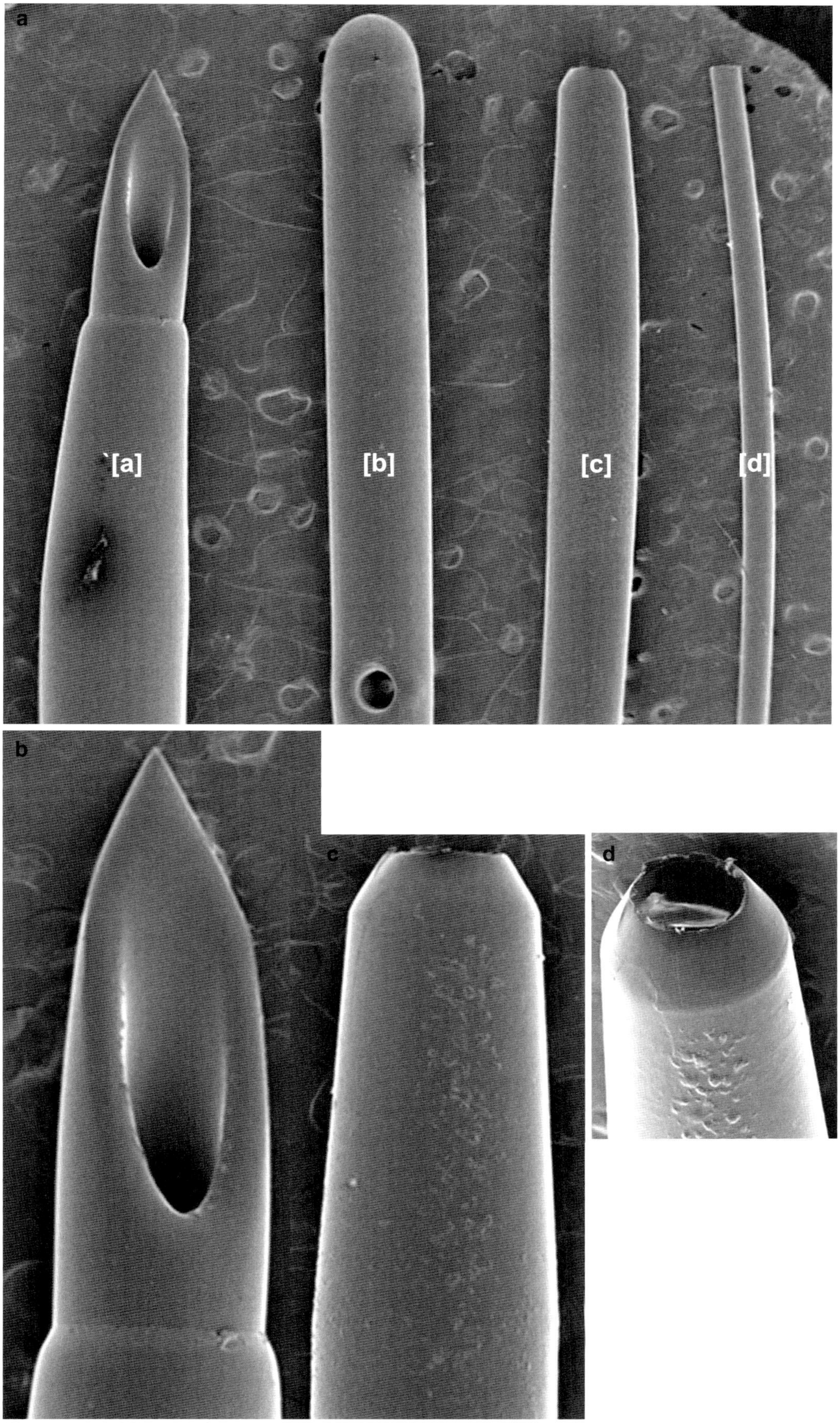

Fig. 45.8 Catheter for peripheral anesthetic blockade. (**a**) Needle with its respective catheter (*a*); epidural catheter (*b*) to compare with peripheral catheter (*c*), which is introduced along catheter a; stylette (*d*) of peripheral catheter *c*. (**b**) Detail at higher magnification of catheter *a*. (**c**, **d**) Detail at higher magnification of stylette *d*. Scanning electron microscopy. Magnification: **a**, ×15; **b**, ×35; **c**, ×100; **d**, ×50

Fig. 45.9 Catheter for peripheral anesthetic blockade. (**a**) Needle used to introduce peripheral catheter. (**b**) Peripheral catheter. (**c**) Detail of **b**. Scanning electron microscopy. Magnification: **a**, ×35; **b**, ×50; **c**, ×100

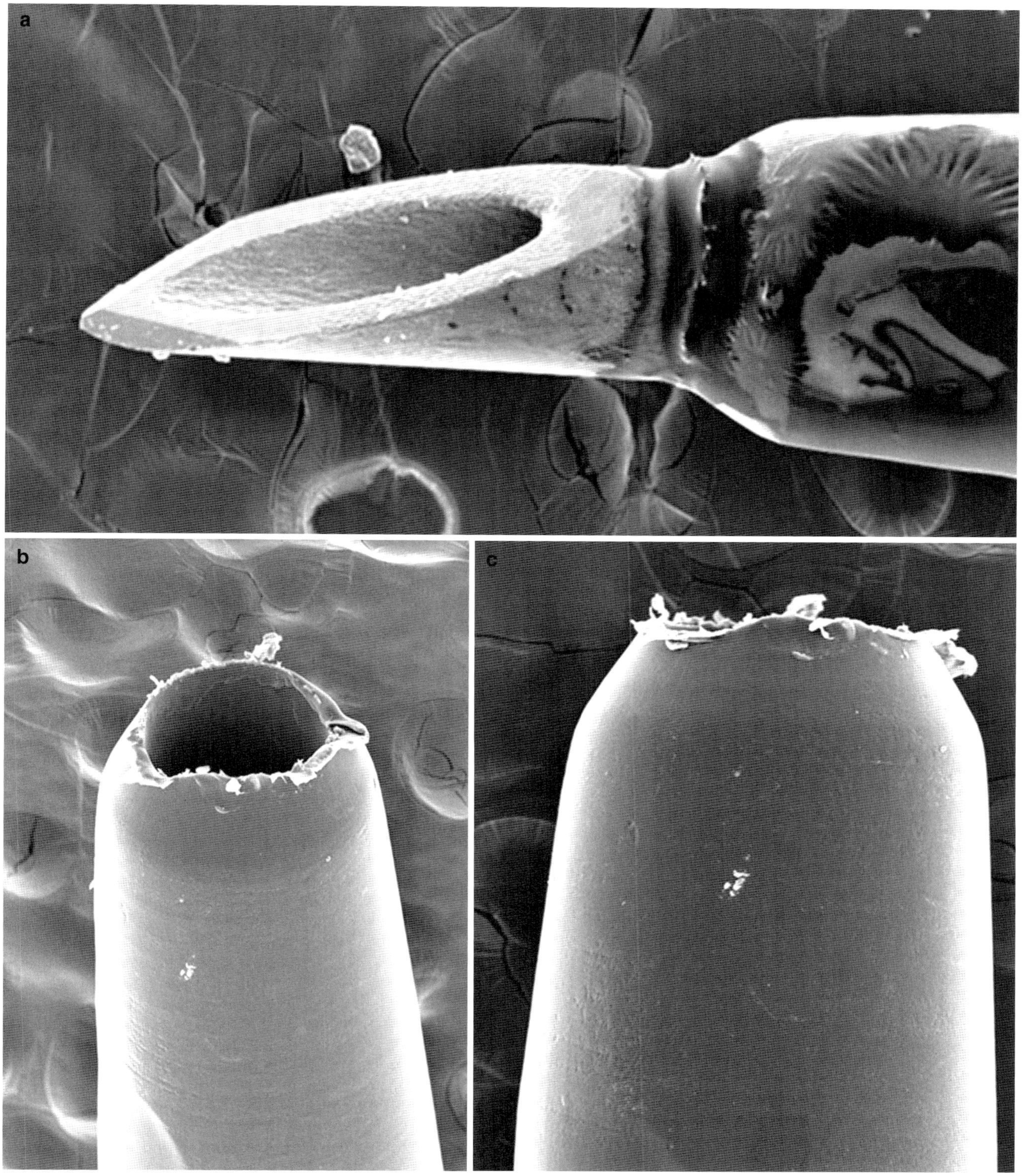

Fig. 45.10 Spinal catheter used in subarachnoid space (short-term use). (**a**) Needle with its respective spinal catheter. (**b**, **c**) Details of the tip of the spinal catheter after the needle is withdrawn. Scanning electron microscopy. Magnification: **a**, ×50; **b** and **c**, ×100

Fig. 45.11 (**a**) Spinal catheter (short-term use) inside subarachnoid space, close to nerve roots. (**b**) Spinal catheter type used in subarachnoid space. Scanning electron microscopy. Magnification: **a** and **b**, ×35

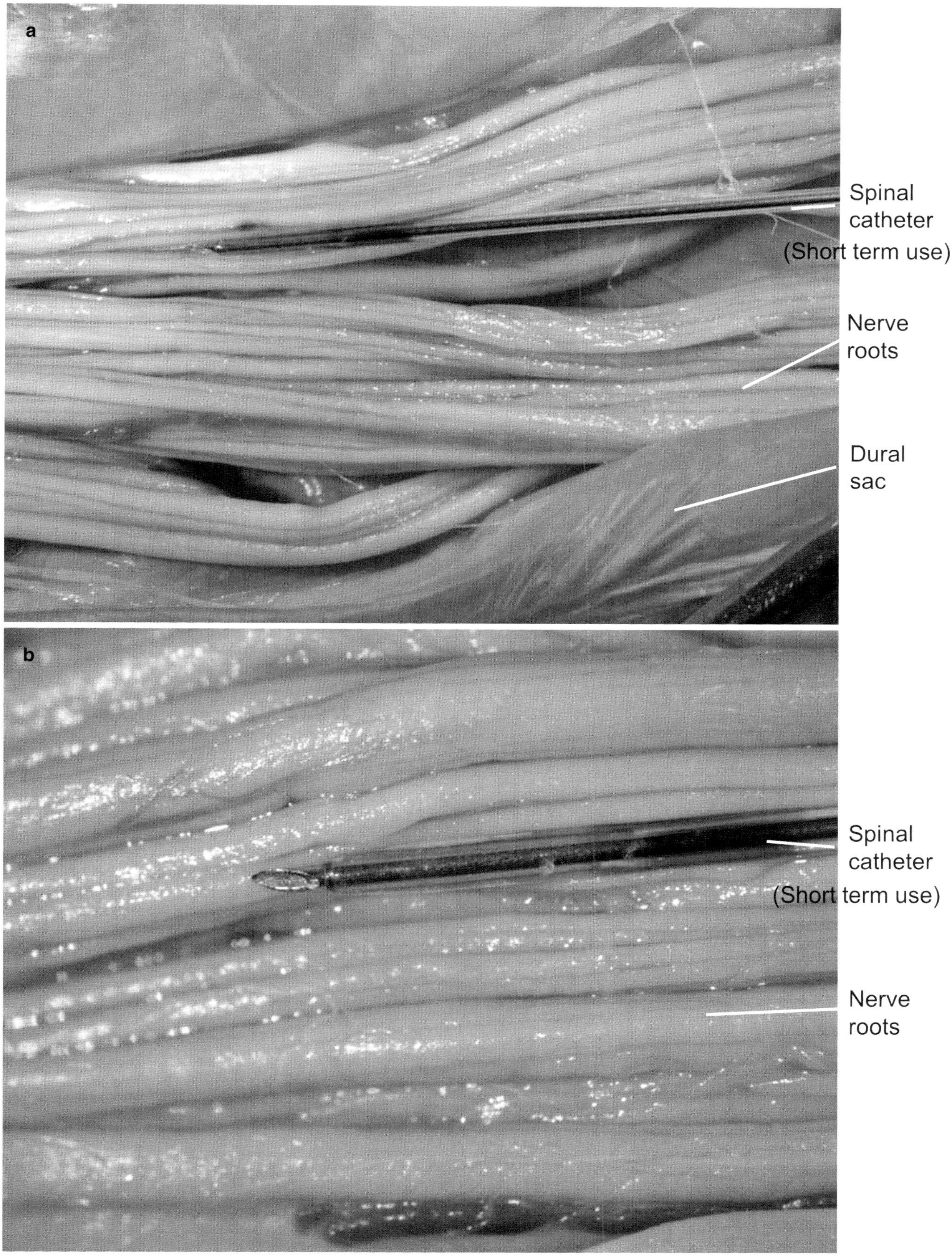

Fig. 45.12 Spinal catheter (short-term use) inside the subarachnoid space. In vitro human anatomic model. (**a**) Spinal catheter inside spinal subarachnoid space adjacent to cauda equina nerve roots. (**b**) Detail at higher magnification

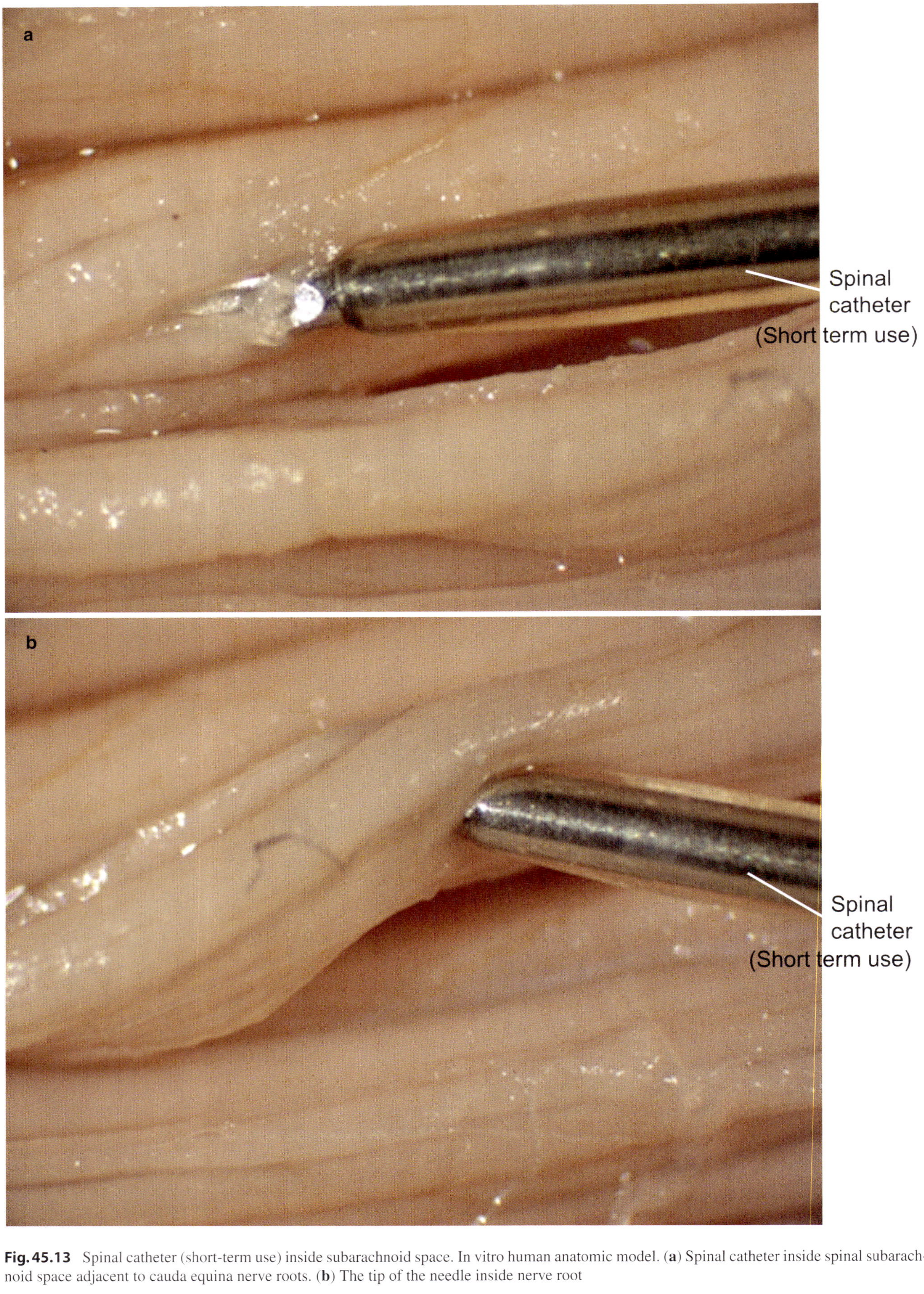

Fig. 45.13 Spinal catheter (short-term use) inside subarachnoid space. In vitro human anatomic model. (**a**) Spinal catheter inside spinal subarachnoid space adjacent to cauda equina nerve roots. (**b**) The tip of the needle inside nerve root

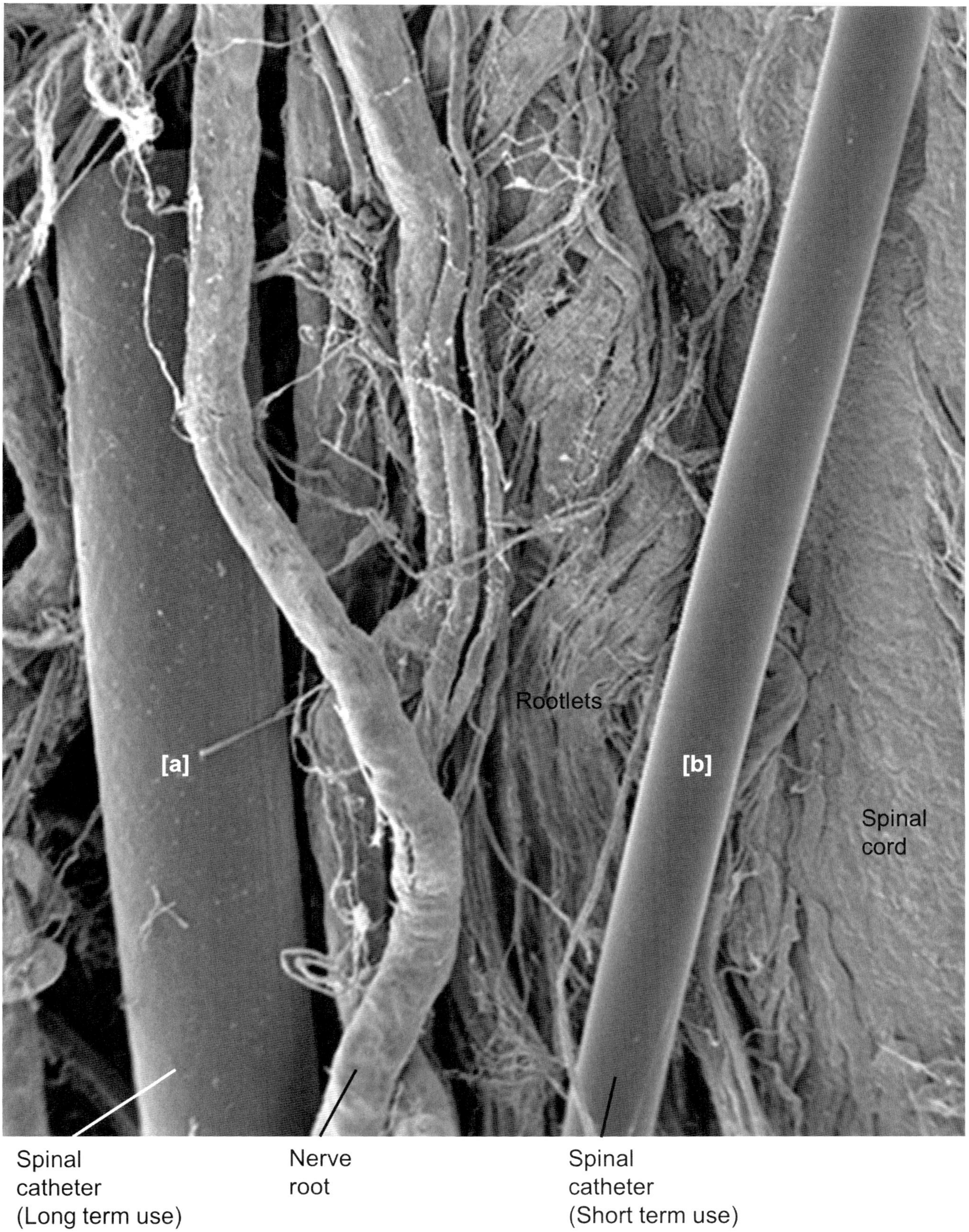

Fig. 45.14 Spinal catheters of the type for long-term use (*a*) and short-term use (*b*) in the subarachnoid space. Scanning electron microscopy. Magnification: ×10

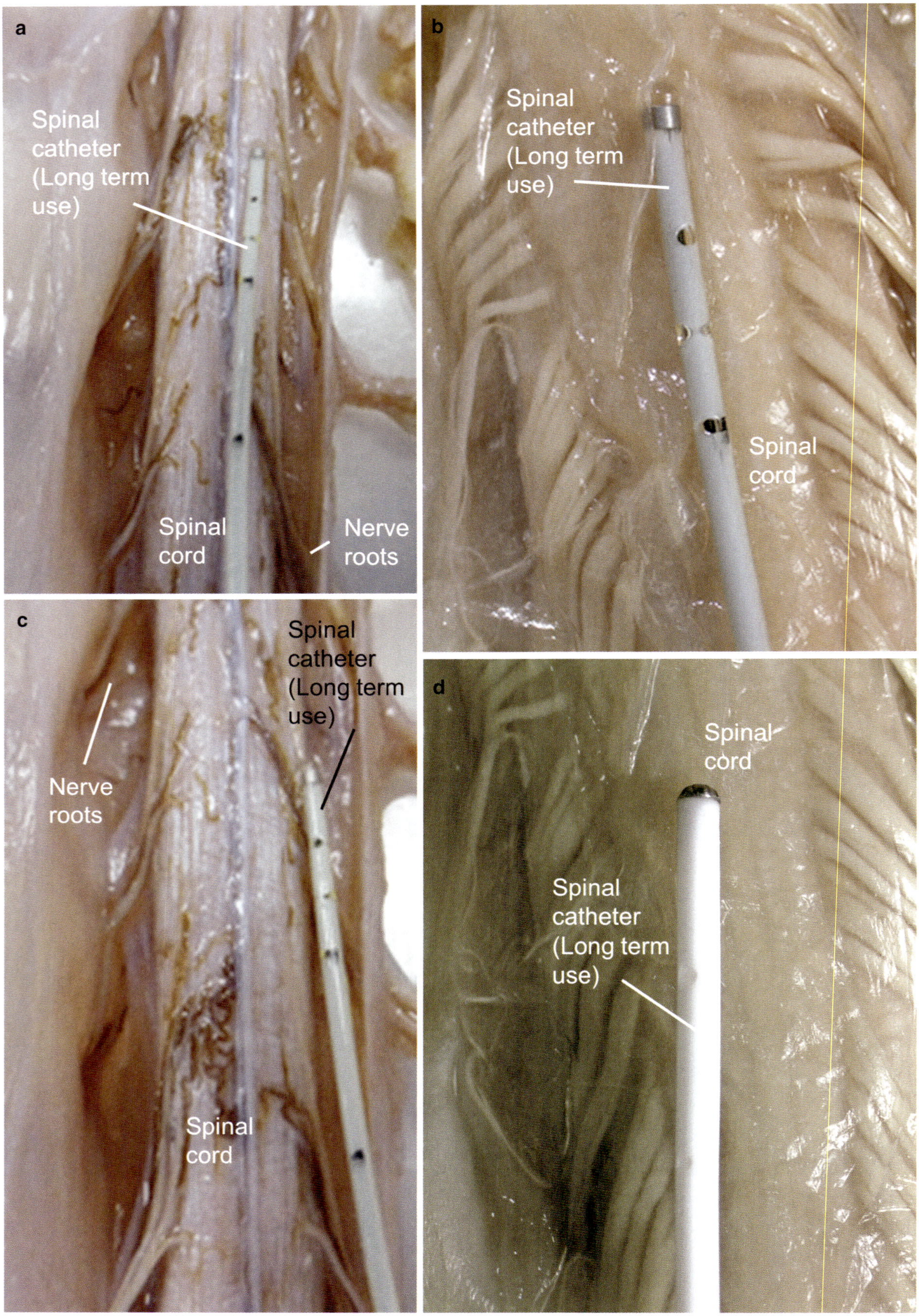

Fig. 45.15 Spinal catheter (long-term use) inside subarachnoid space. In vitro human anatomic model. (**a**, **c**) Spinal catheter inside spinal subarachnoid space adjacent to the spinal cord and nerve roots at the thoracic level. (**b**, **d**) Spinal catheter inside spinal subarachnoid space adjacent to the spinal cord and nerve roots at the cervical level

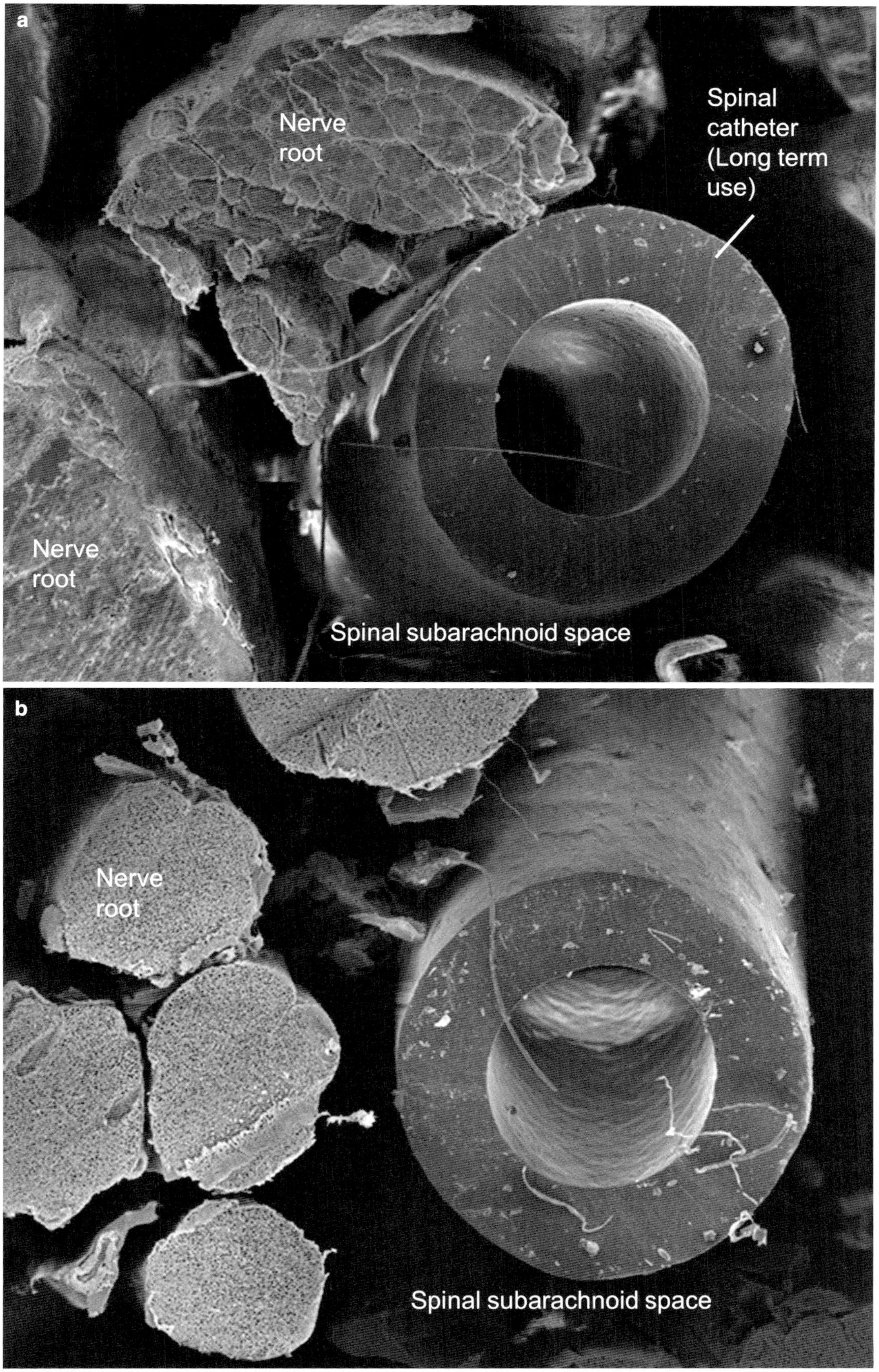

Fig. 45.16 (**a**) Spinal catheter (long-term use) inside subarachnoid space close to nerve roots. (**b**) Spinal catheter of the type used in subarachnoid space. Scanning electron microscopy. Magnification: **a** and **b**, ×35

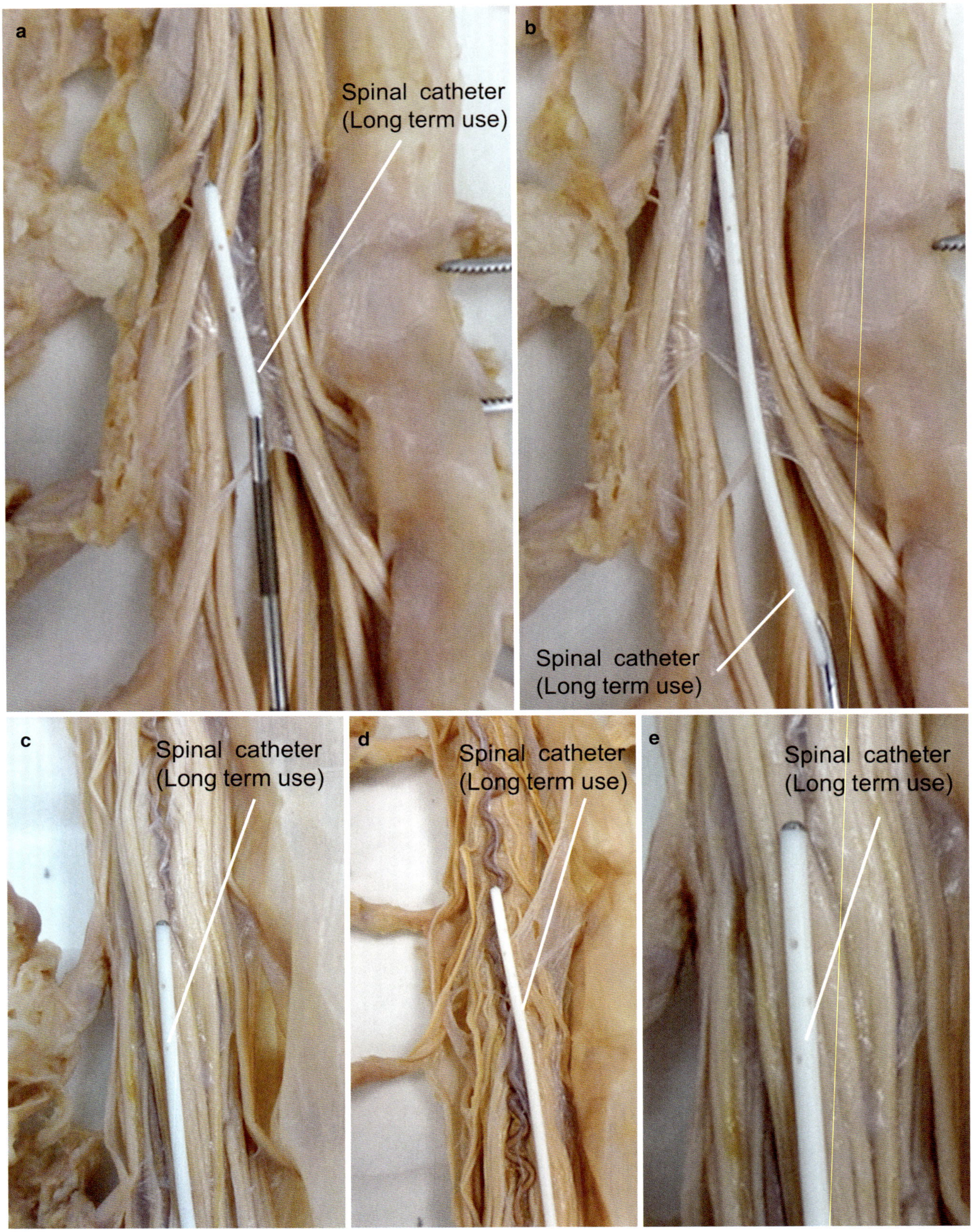

Fig. 45.17 Spinal catheter (long-term use) inside subarachnoid space. In vitro human anatomic model. (**a**, **b**) Spinal catheter inside spinal subarachnoid space adjacent to cauda equina nerve roots at the lumbar level. (**c**–**e**) Spinal catheter inside spinal subarachnoid space adjacent to the spinal cord and nerve roots at the thoracic level

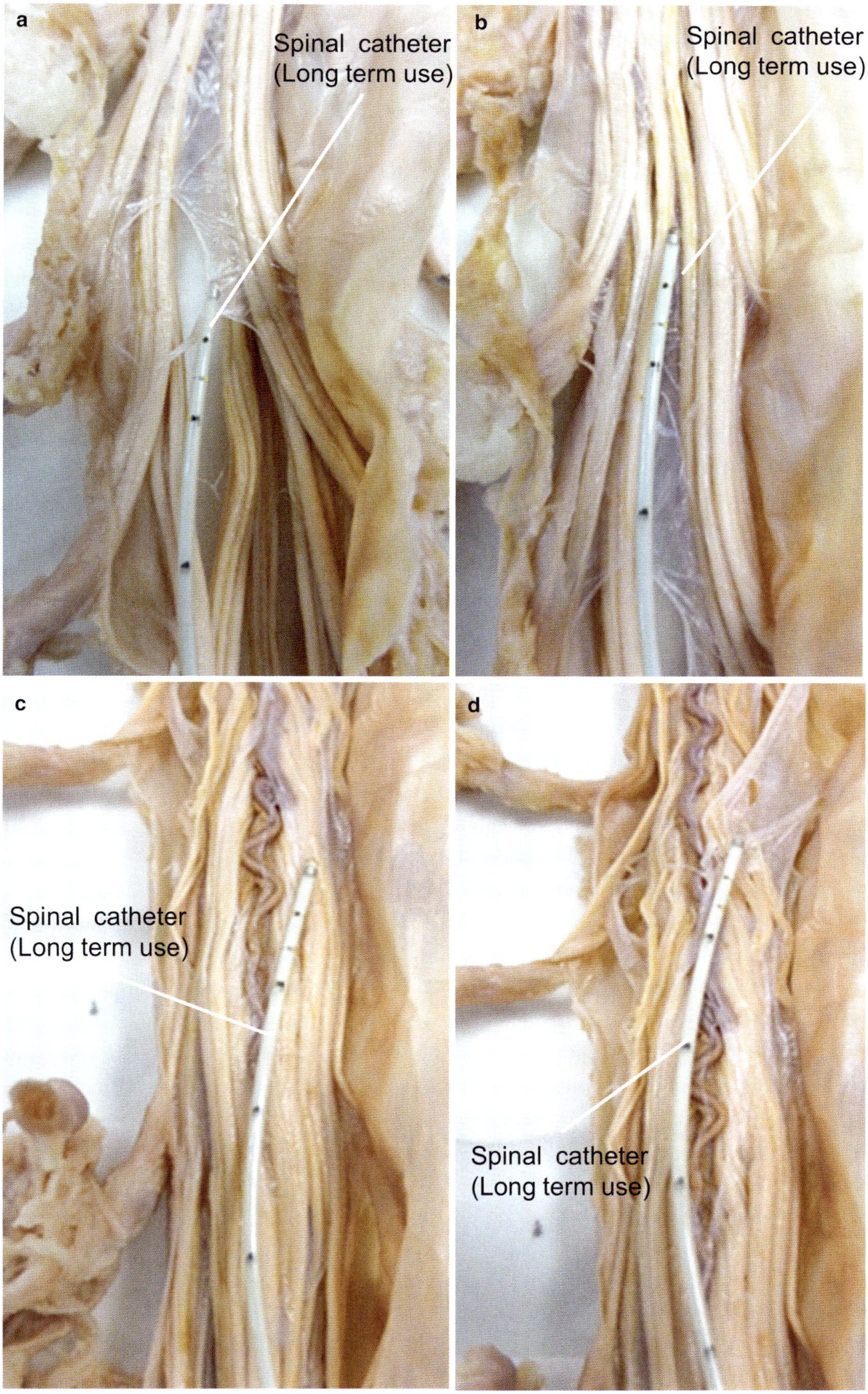

Fig. 45.18 Another spinal catheter type (long-term use) inside subarachnoid space. In vitro human anatomic model. (**a**, **b**) Spinal catheter inside spinal subarachnoid space adjacent to cauda equina nerve roots at the lumbar level. (**c**, **d**) Spinal catheter inside spinal subarachnoid space adjacent to the spinal cord and nerve roots at the thoracic level

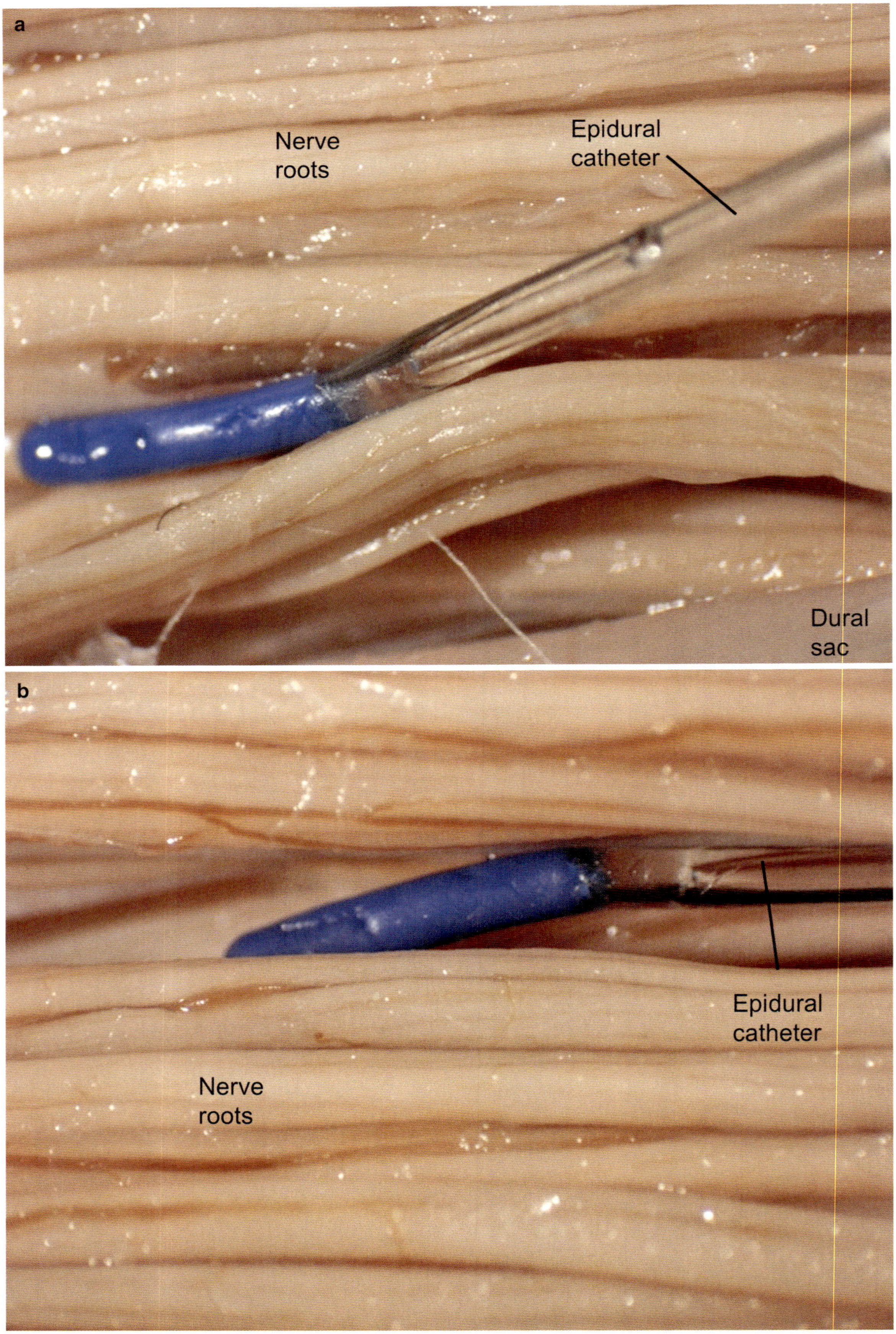

Fig. 45.19 (**a**, **b**), In vitro placement of epidural catheter inside subarachnoid space close to nerve roots. Images illustrate how epidural catheters may be placed after unintentional puncture of the dural sac while performing continuous epidural blockade. Scanning electron microscopy. Magnification: **a** and **b**, ×35

Reference

1. López A, Reina MA, Machés F, De Leon Casasola O, De Andrés JA, García Trapero J. Electron microscopy in quality control of equipment used in regional anesthesia. Tech Reg Anesth Pain Management. 2002;6:172–9.

Epidural Filters and Particles from Surgical Gloves

46

Miguel Angel Reina and Emilse Colman Peyrano

Placement of filters in epidural catheters has been recommended in order to prevent particles and bacteria from reaching the epidural or spinal spaces, as well as peripheral nerves, during the performance of regional or peripheral blockade. The aim of placing filters is to prevent contamination during drug administration, with the subsequent spread of infection along meningeal layers or peripheral nerves. These filters have filtering membranes composed of micropores with variable surface areas that filter particles and bacteria present in substances injected through the epidural catheters. During preparation of local anesthetic infusions, it is possible to introduce contaminant agents such as small particles present in surgical gloves or minute fragments from glass ampoules. Recommended filters for epidural catheters in regional and peripheral blockade include labels such as "Porosity 0.2 μm," referring to the pore size and therefore to the minimum size of particles (including bacteria) that will be retained by the filtering membrane.

In our studies, filtering membranes were observed under scanning electron microscopy (SEM) and micropores were identified and examined. There are numerous micropores across the thickness of the filtering membrane, which effectively filter particles contained in anesthetic solutions. Samples obtained from the filtering membrane of epidural filters underwent a cryofracture process prior to examination under SEM.

Results from the analysis of the surface of the filtering membrane before and after filtering processes showed that in the prefiltration surface of membranes from three different manufacturers, the diameter of pores measured 0.70 (0.66–0.74), 0.45 (0.41–0.49) and 2.07 (2.01–2.15) microns. The pore shapes were different. Examination of the postfiltration surfaces found that the pore diameters of filters from the same three manufacturers were 0.26 (0.25–0.28), 0.26 (0.24–0.28) and 0.46 (0.43–0.49) microns. The thickness of the filters from the various manufacturers measured 130, 118, and 165 μm, with an average number of 140, 220, and 210 filtration planes, respectively.

In regards to particles present in surgical gloves, the particles found on the inner and outer surface of samples from filtering membranes vary in composition and sizes. These particles are noncontaminated. Particles found in the inner surface were related to those used to enable glove application. Particles found in the outer surface (which can be deposited on other surfaces) were related to the manufacturing process of the glove. These particles appear in products with demolding properties, used to facilitate removal of manufactured gloves from their molds.

External glove surfaces carried particles measuring between 3 and 4 μm; their morphology was consistent with calcium carbonate. Over the internal surface, larger particles were identified, measuring about 11–14 μm in diameter, which were shaped differently and of smooth appearance. Analysis of these particles showed traces of magnesium, as well as features consistent with organic molecules. Particles observed on one surface were never observed on the other (Figs. 46.1, 46.2, 46.3, 46.4, 46.5, 46.6, 46.7, 46.8, 46.9, 46.10, 46.11, and 46.12).

M.A. Reina, MD, PhD (✉)
Department of Clinical Medical Sciences
and Institute of Applied Molecular Medicine,
School of Medicine, University of CEU San Pablo, Madrid, Spain

Department of Anesthesiology,
Madrid-Montepríncipe University Hospital, Madrid, Spain
e-mail: miguelangel@perticone.e.telefonica.net

E.C. Peyrano, BS
Research Unit, Anesthesia Reina Colman SLP, Madrid, Spain
e-mail: emilsecolpey@yahoo.es

M.A. Reina et al. (eds.), *Atlas of Functional Anatomy for Regional Anesthesia and Pain Medicine: Human Structure, Ultrastructure and 3D Reconstruction Images*, DOI 10.1007/978-3-319-09522-6_46,

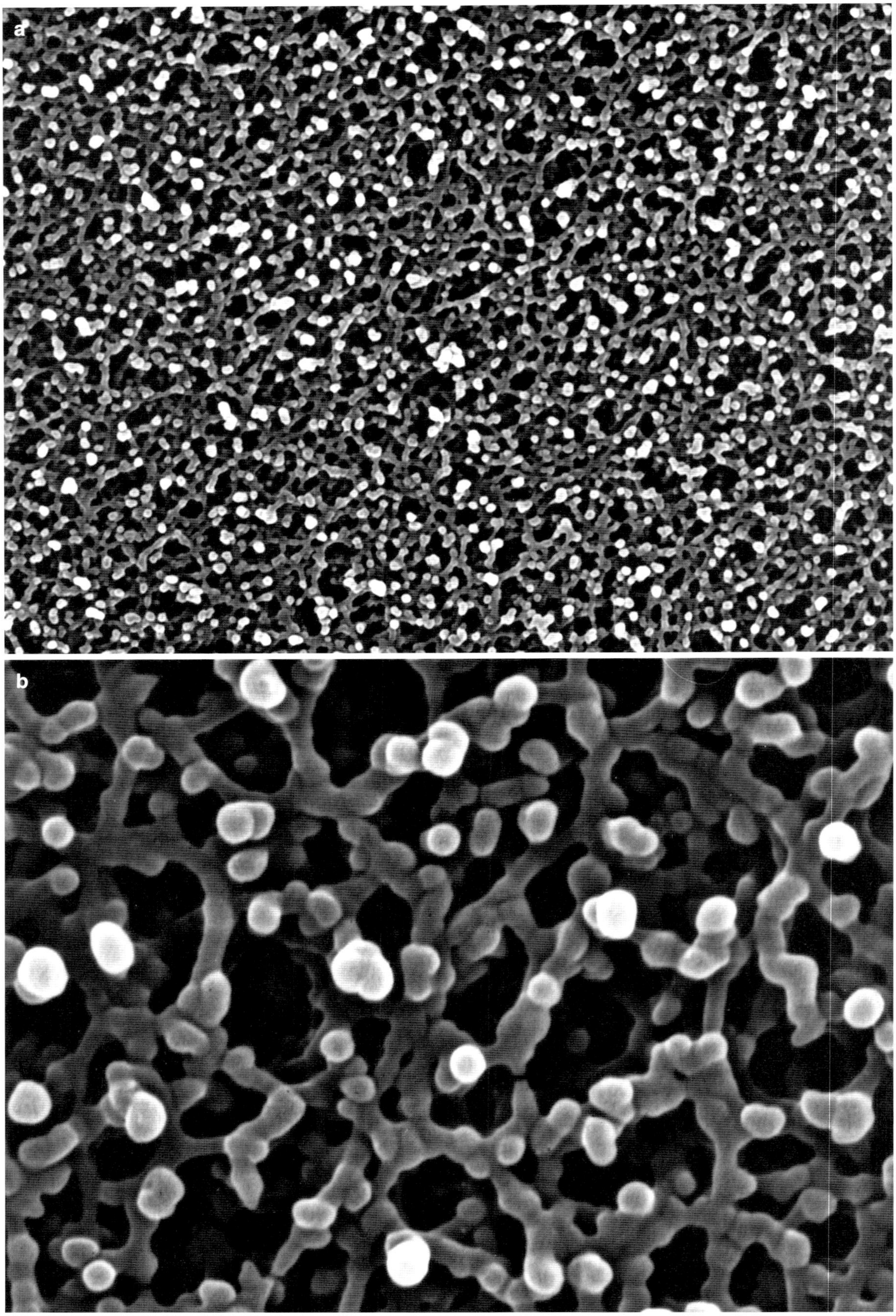

Fig. 46.1 Epidural filter. (**a**) Surface of filtration layer. (**b**) Detail at higher magnification. Scanning electron microscopy. Magnification: **a**, ×2,200; **b**, ×10,000 (From Reina et al. [1]; with permission)

Fig. 46.2 Epidural filter. (**a**) Surface of filtration layer. (**b**) Detail at higher magnification. Scanning electron microscopy. Magnification: **a**, ×1,400; **b**, ×10,000

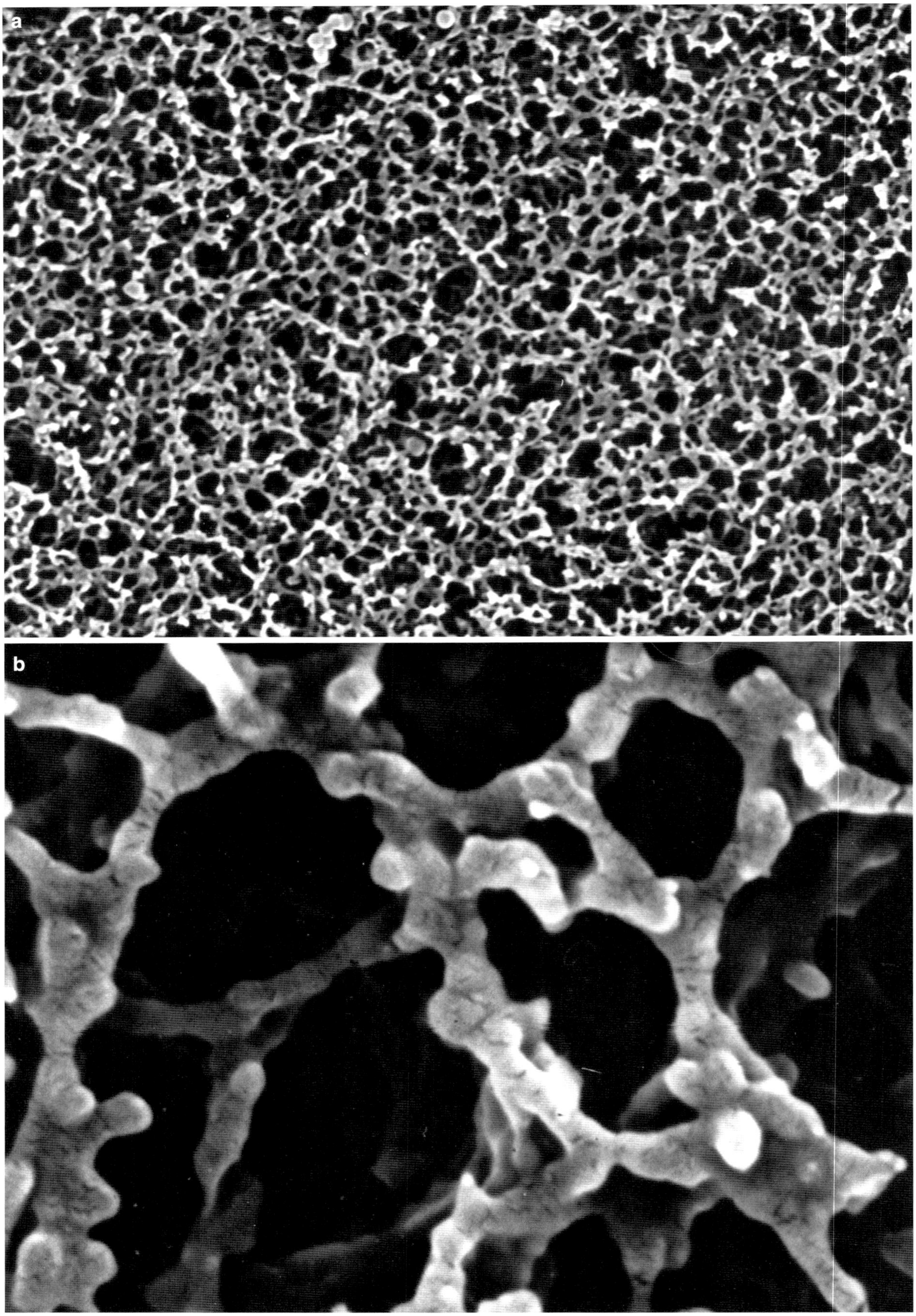

Fig. 46.3 Epidural filter. (**a**), Surface of filtration layer. (**b**) Detail at higher magnification. Scanning electron microscopy. Magnification: **a**, ×2,500; **b**, ×20,000 (From Reina et al. [1]; with permission)

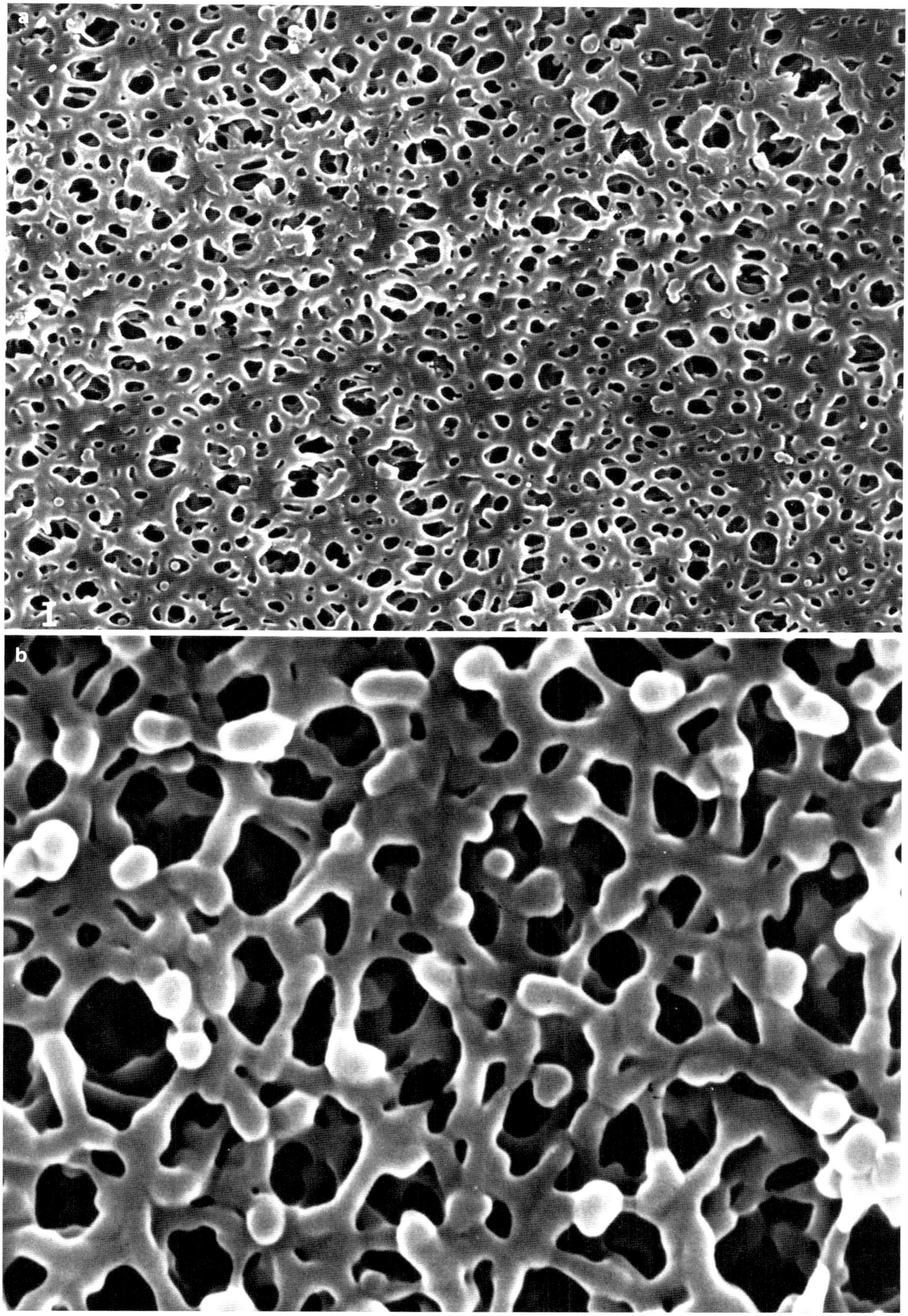

Fig. 46.4 Epidural filter. (**a**) Surface of filtration layer. (**b**) Detail at higher magnification. Scanning electron microscopy. Magnification: **a**, ×3,000; **b**, ×10,000 (From Reina et al. [1]; with permission)

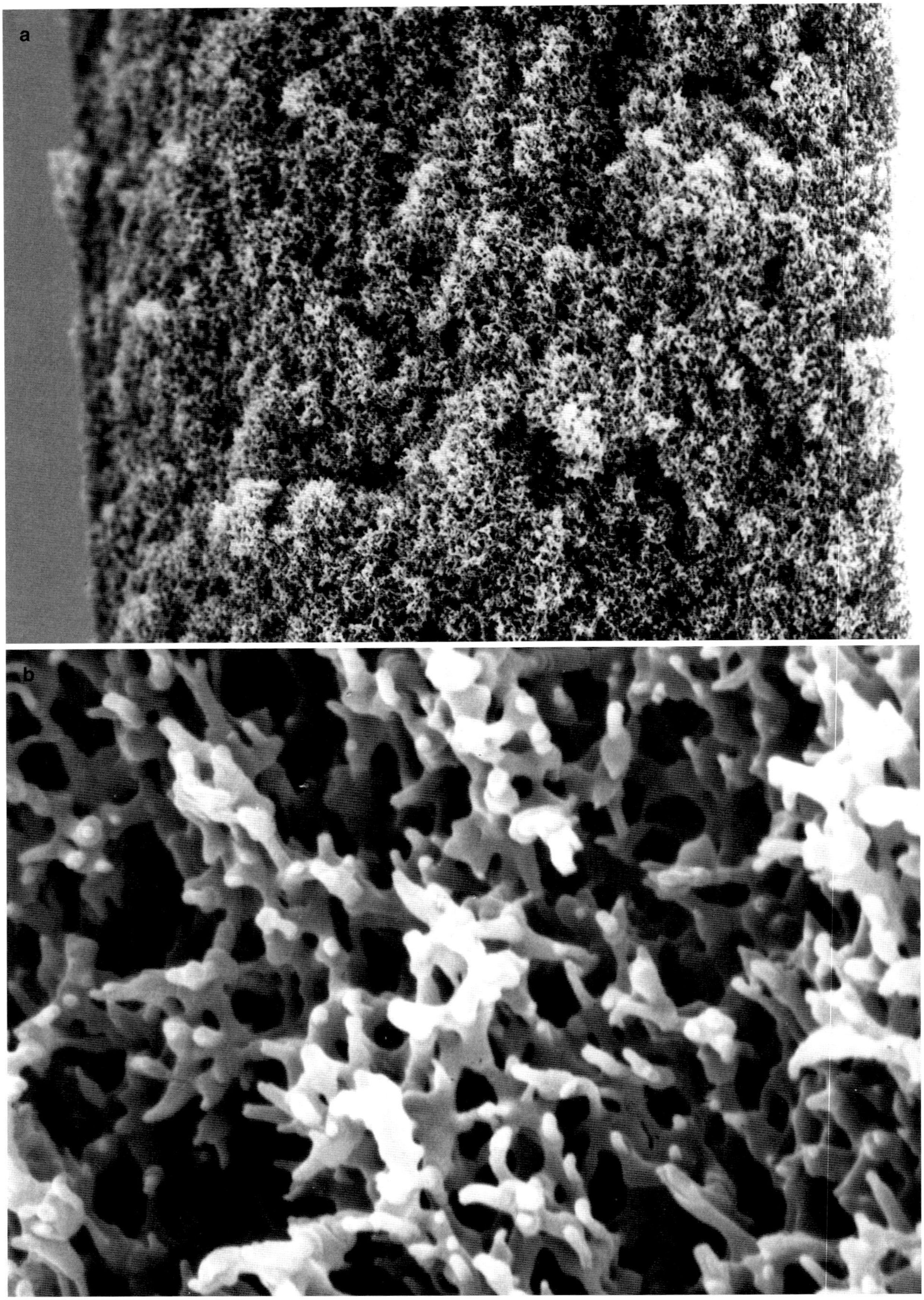

Fig. 46.5 Epidural filter. (**a**) Thickness of filtration layer. (**b**) Detail at higher magnification. Scanning electron microscopy. Magnification **a**: ×650 **b**: ×2,0000

Fig. 46.6 Epidural filter. (**a**) Thickness of filtration layer. (**b**) Detail at higher magnification. Scanning electron microscopy. Magnification: **a**, ×370; **b**, ×10,000

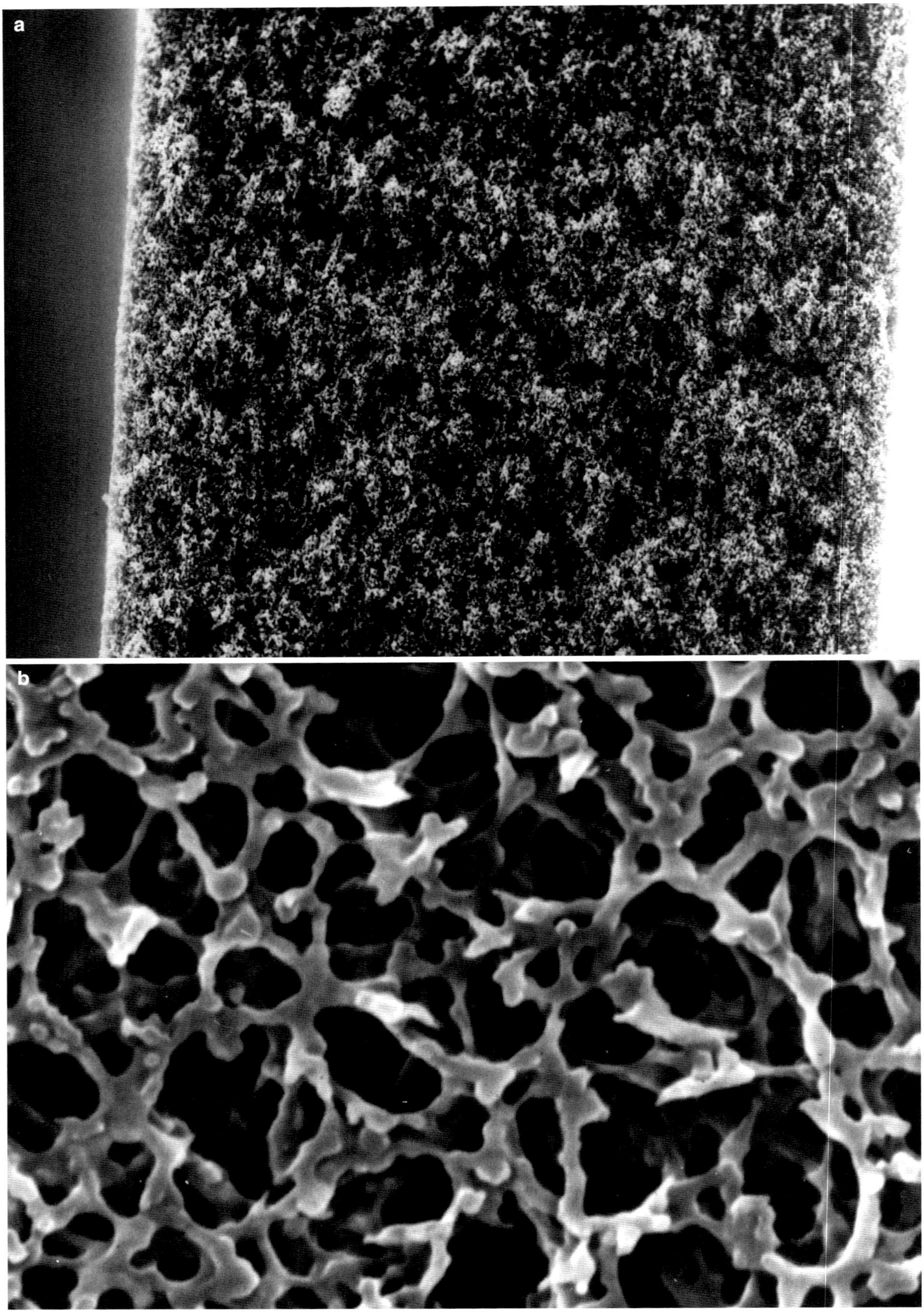

Fig. 46.7 Epidural filter. (**a**) Thickness of filtration layer. (**b**) Detail at higher magnification. Scanning electron microscopy. Magnification: **a**, ×500; **b**, ×10,000 (From Reina et al. [1]; with permission)

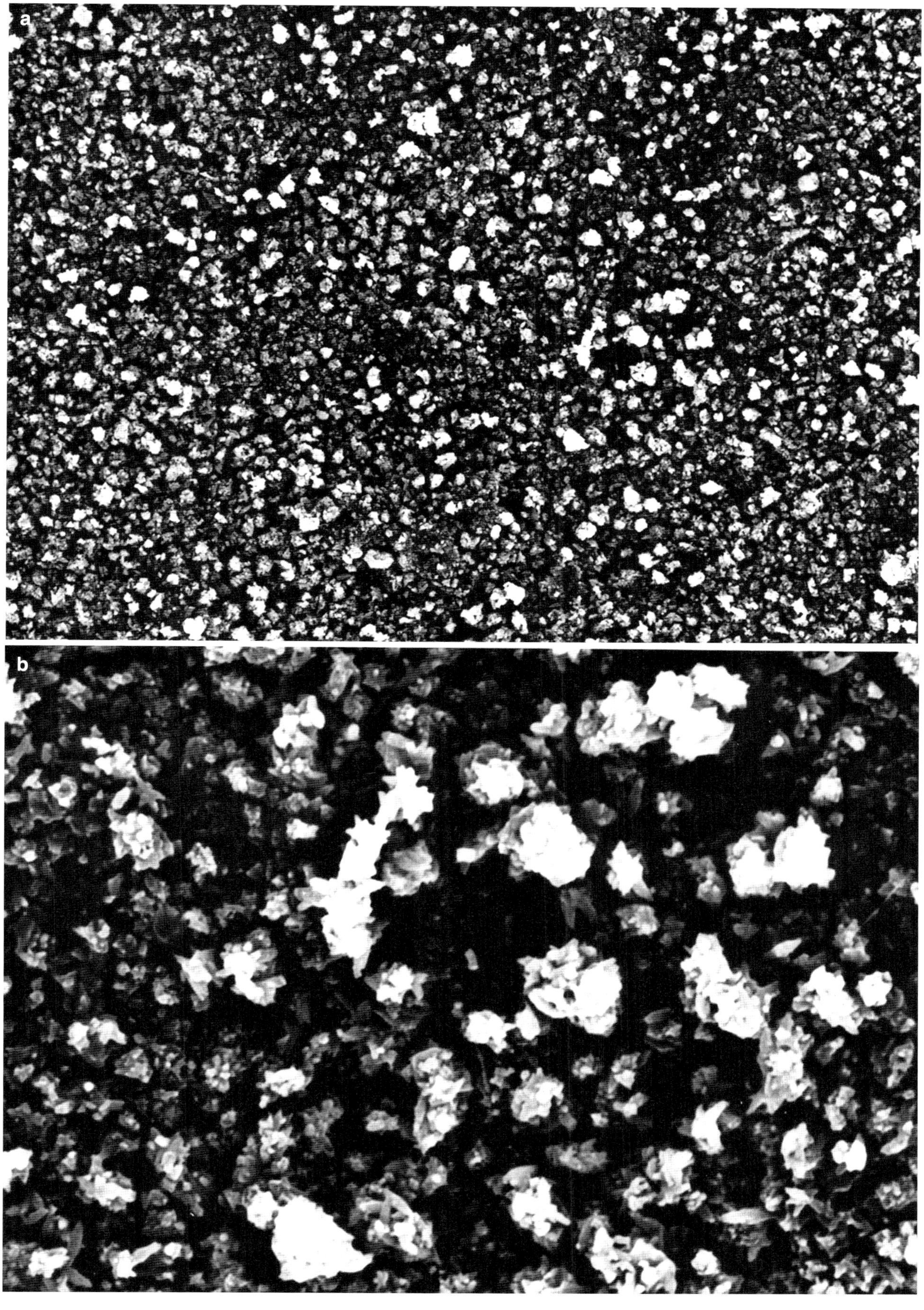

Fig. 46.8 Surgical glove. (**a**) External surface of surgical glove. (**b**) Detail at higher magnification. Scanning electron microscopy. Magnification: **a**, ×500; **b**, ×2,000 (From Reina et al. [2]; with permission.)

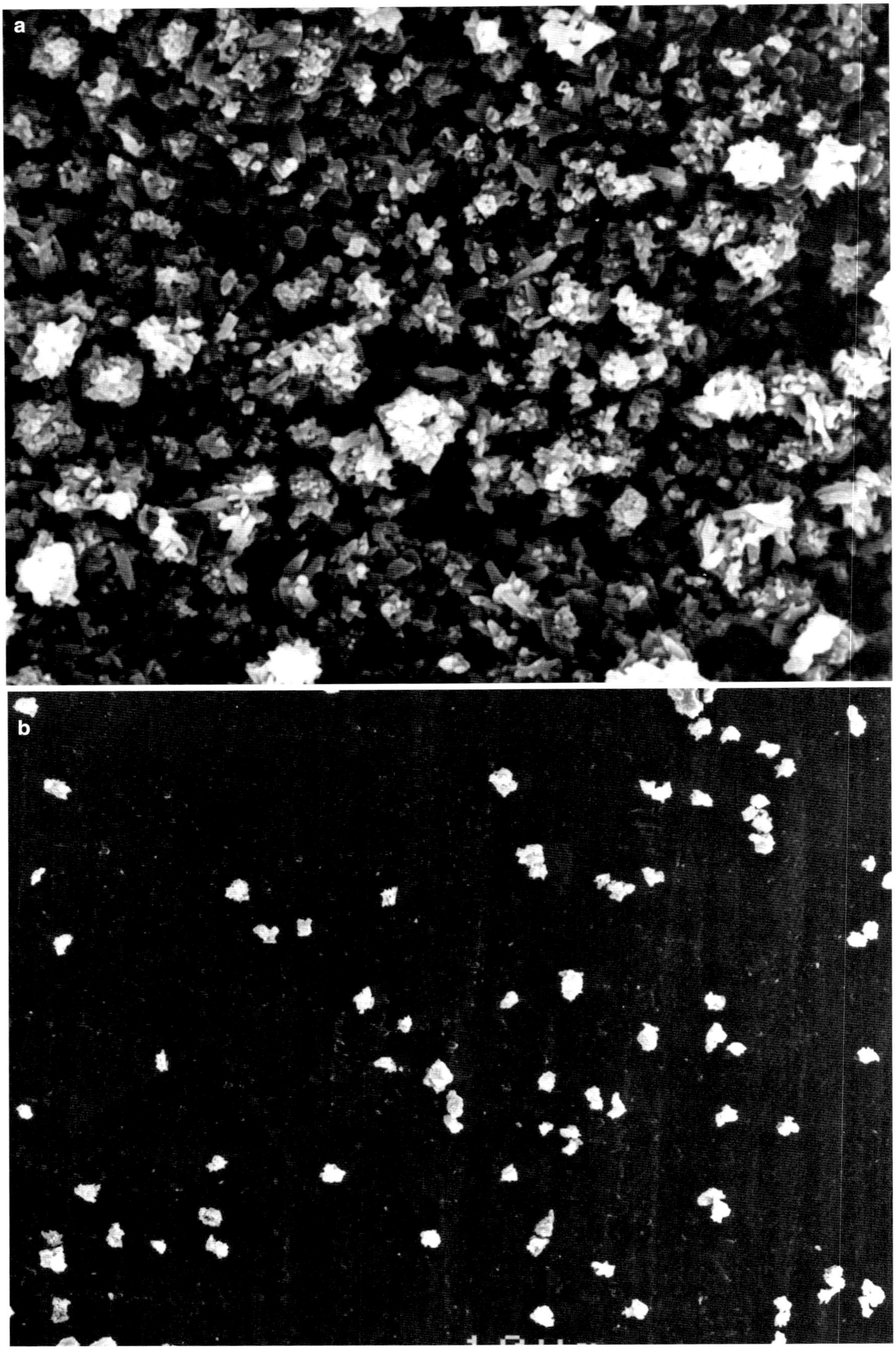

Fig. 46.9 Surgical glove. (**a**) External surface of surgical glove. (**b**) Particles from external surface of surgical glove on epidural catheter surface. Scanning electron microscopy. Magnification: **a**, ×2,000; **b**, ×500 (From Reina et al. [2]; with permission)

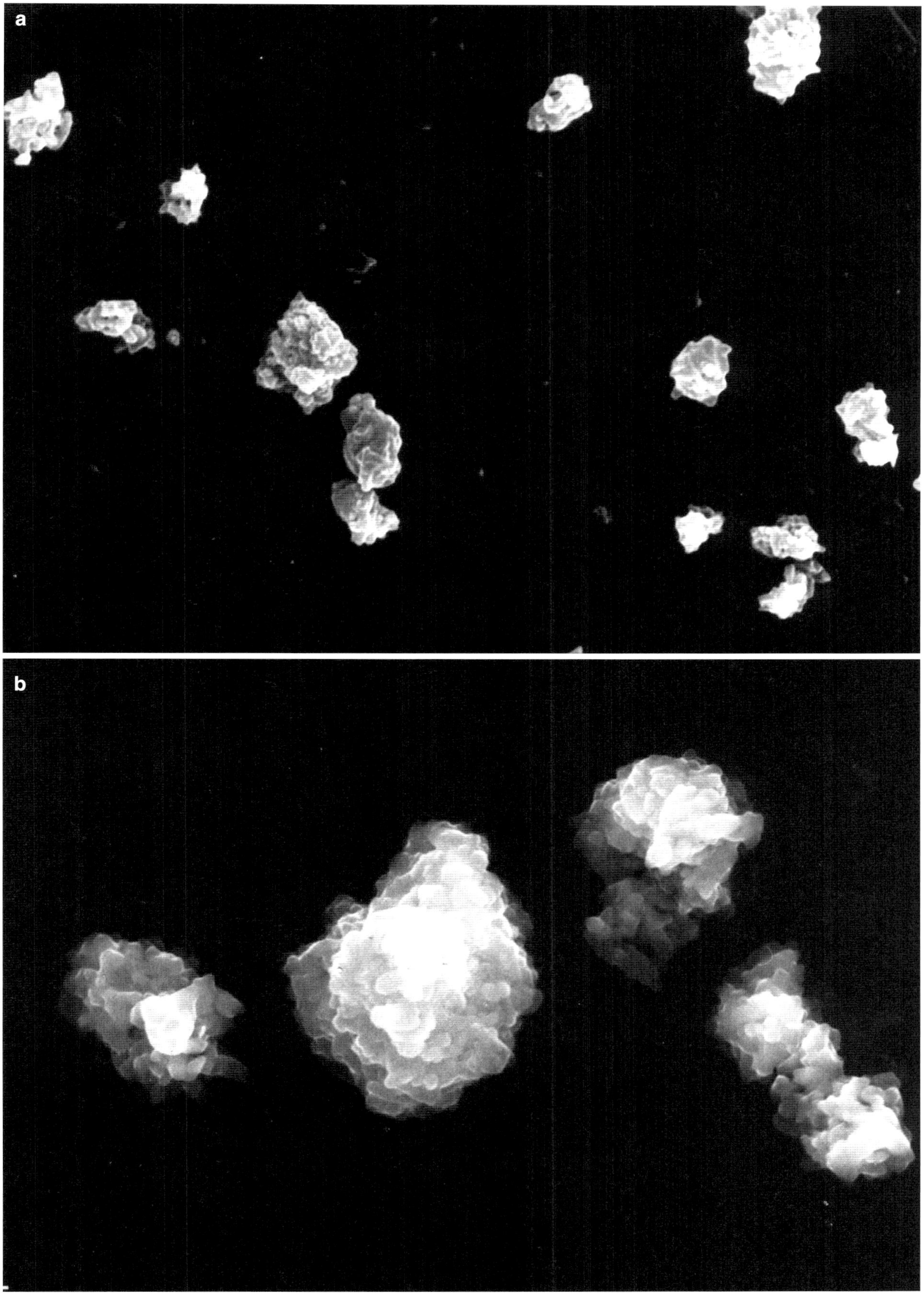

Fig. 46.10 Surgical glove. (**a**, **b**) Particles from external surface of surgical glove on epidural catheter surface. Scanning electron microscopy. Magnification: **a**, ×2000; **b**, ×5,000 (From Reina et al. [2]; with permission)

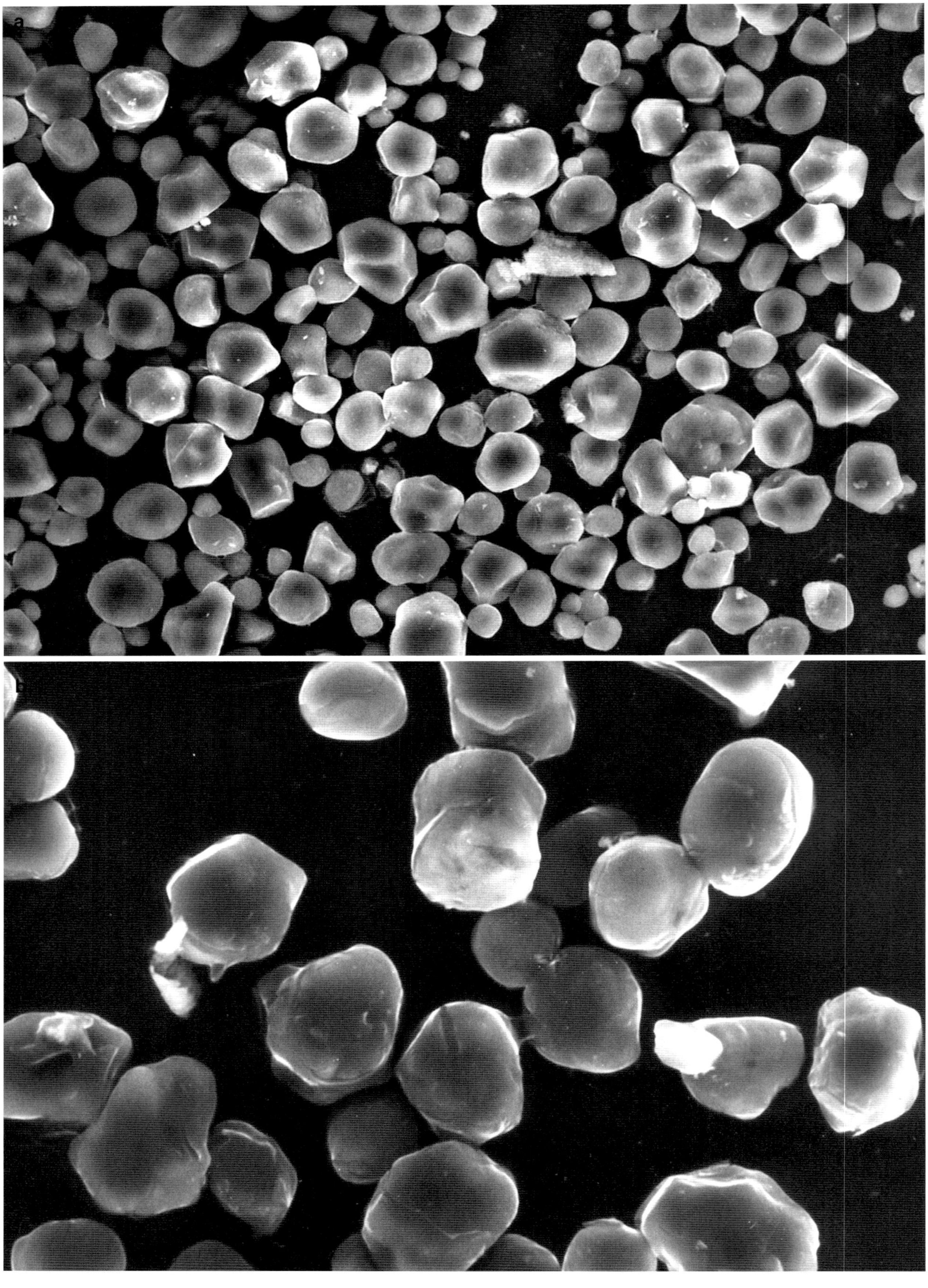

Fig. 46.11 Surgical glove. (**a**) Internal surface of surgical glove. (**b**) Detail at higher magnification. Scanning electron microscopy. Magnification: **a**, ×500; **b**, ×1,000 (From Reina et al. [2]; with permission)

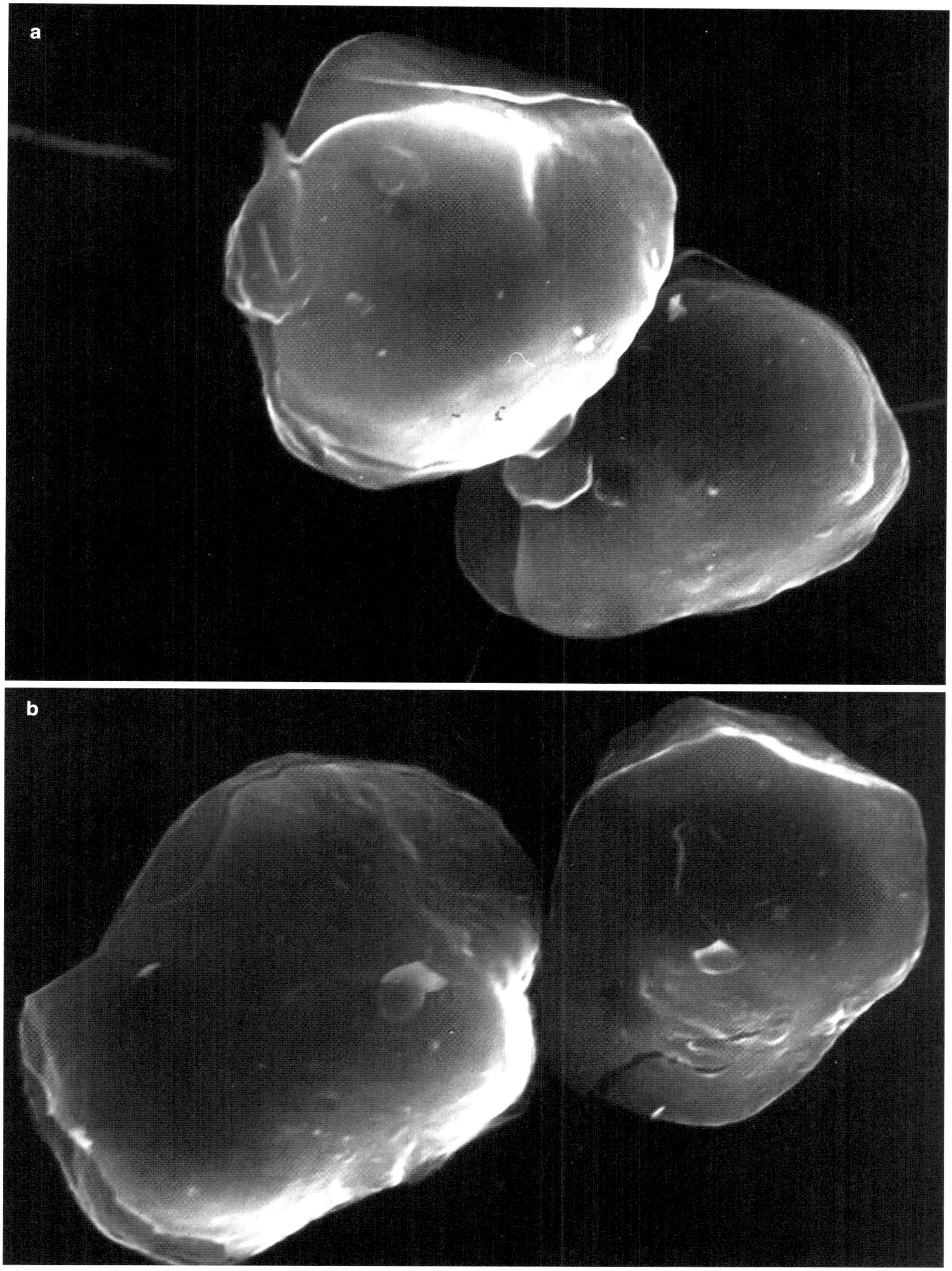

Fig. 46.12 Surgical glove. (**a**, **b**) Detail of particles from internal surface of surgical glove. Scanning electron microscopy. Magnification: **a**, ×3,000; **b**, ×3,000 (From Reina et al. [2]; with permission)

References

1. Reina MA, Català-Puigbó E, López-García A, Amor E, Amador V, Arrizabalaga M, et al. Microscopic characteristics of epidural filter pores. Rev Esp Anestesiol Reanim. 1997;44:352–6.
2. Reina MA, López García A, Aguilar JL, Palacios Martín R. Electron microscopic analysis of particles from surgical gloves and their possible introduction into the epidural space during epidural anesthesia. Rev Esp Anestesiol Reanim. 1999;46:60–6.

Part IV

Research Techniques

47 Three-Dimensional Reconstruction of Spinal Cerebrospinal Fluid, Roots, and Surrounding Structures

Alberto Prats-Galino, Anna Puigdellívol-Sánchez, and Miguel Angel Reina

The chapter describes the process of achieving three-dimensional (3D) reconstruction of cerebrospinal fluid (CSF) and cauda equina roots inside the lumbosacral dural sac, including its surrounding structures.

Lumbosacral magnetic resonance (MR) image scans were taken in the supine position, with a support placed below the popliteal fossa to produce slight knee flexion. T1 and T2 weighted sequences are acquired on a 1.5 T scanner (Phillips Intera, software v1.1) using a 3D fast spin echo. MR images are acquired at 16 bits and exported in DICOM (Digital Imaging and Communication in Medicine) file format. Files are analyzed and processed using Amira v5.2 3D software (Mercury Co., Boston, USA), installed in a Dell Precision graphic workstation.

The following steps were taken: (1) delimitation of the volume of interest (VOI) of the dural sac; (2) validation of the 3D reconstruction of the dural sac and surface model generation (Table 47.1); (3) second segmentation to separate the CSF space and spinal nerve roots; and (4) manual delineation of surrounding structures (vertebrae, intervertebral discs, ligaments, etc.) (Figs. 47.1, 47.2, 47.3, 47.4, 47.5, and 47.6).

A. Prats-Galino, MD, PhD (✉) • A. Puigdellívol-Sánchez, MD, PhD
Laboratory of Surgical NeuroAnatomy (LSNA),
Human Anatomy and Embryology Unit, School of Medicine,
University of Barcelona, Barcelona, Spain
e-mail: aprats@ub.edu; apuigdellivol@ub.edu

M.A. Reina, MD, PhD
Department of Clinical Medical Sciences and Institute of Applied Molecular Medicine, School of Medicine,
University of CEU San Pablo, Madrid, Spain

Department of Anesthesiology, Madrid-Montepríncipe University Hospital, Madrid, Spain
e-mail: miguelangel@perticone.e.telefonica.net

M.A. Reina et al. (eds.), *Atlas of Functional Anatomy for Regional Anesthesia and Pain Medicine: Human Structure, Ultrastructure and 3D Reconstruction Images*, DOI 10.1007/978-3-319-09522-6_47,

Table 47.1 Magnetic resonance settings for three-dimensional reconstruction

Sequence	Acquisition plane	TR/TE, ms	NSA	Slice thickness, mm/gap%	Matrix	Voxel size, mm^3	Sections per block
T1 fast field echo 3D	Axial	32/4.6	1	1.8/0	205 × 256	0.9	160
T2 fast field echo 3D	Axial	7.7/3.8	1	1.3/0	352 × 352	0.65	200

NSA number of signals averaged, *TE* echo time, *TR* repetition time

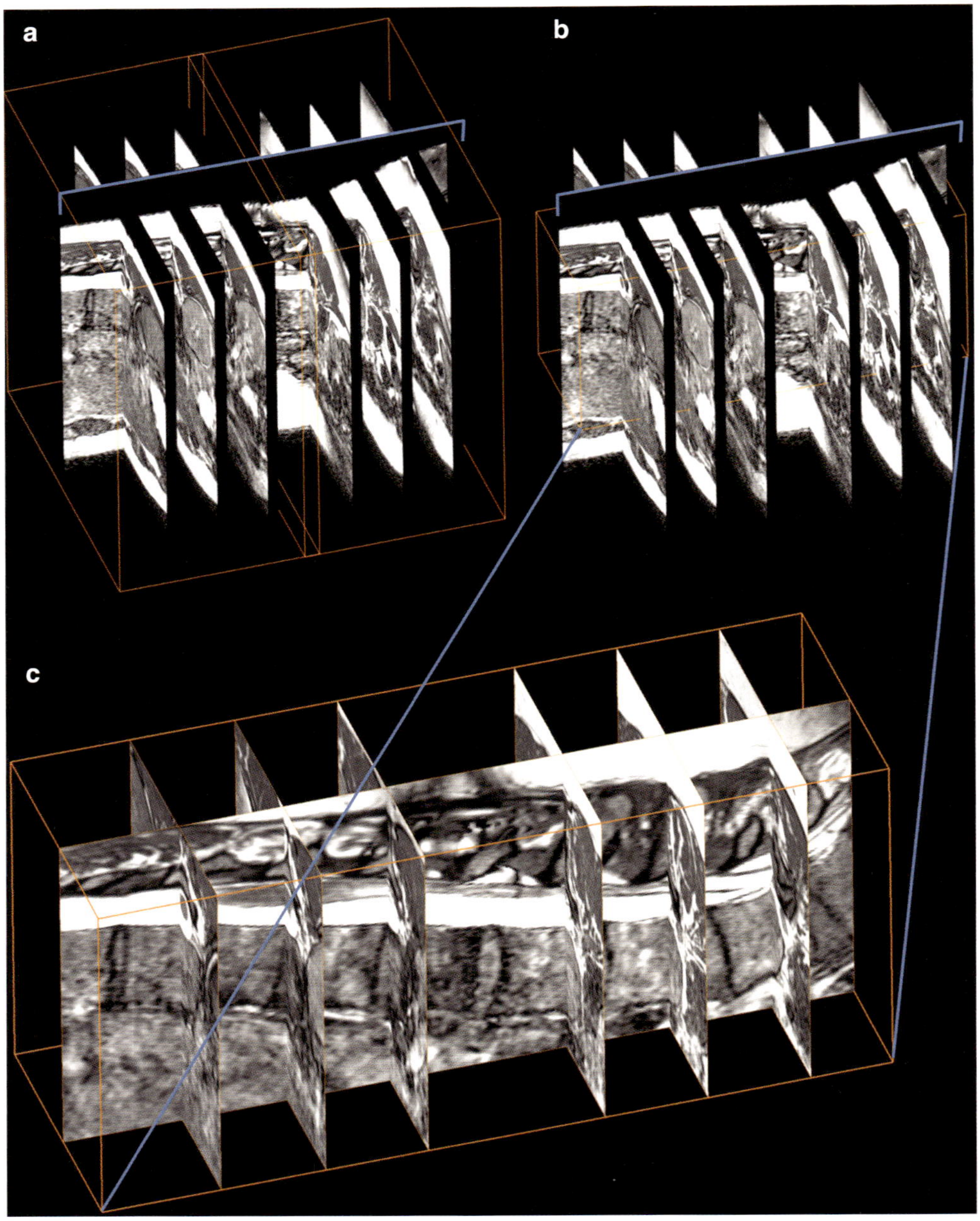

Fig. 47.1 (**a**), Axial sequence acquisitions are grouped into two aligned adjacent blocks of 130 mm, a caudal and a rostral MR block, extending from the lowest end of the dural sac to the lower thoracic vertebrae (T11 or T10, depending on the height of the patient). Adjacent blocks are combined in a single block. (**b**) A smaller volume is selected around the spine to define the limits used to crop the images. (**c**) In the resulting cropped volume, different volumes of interest (VOIs), corresponding to CSF, roots, and surrounding structures (vertebrae, ligaments, etc.) will be delineated using different segmentation tools

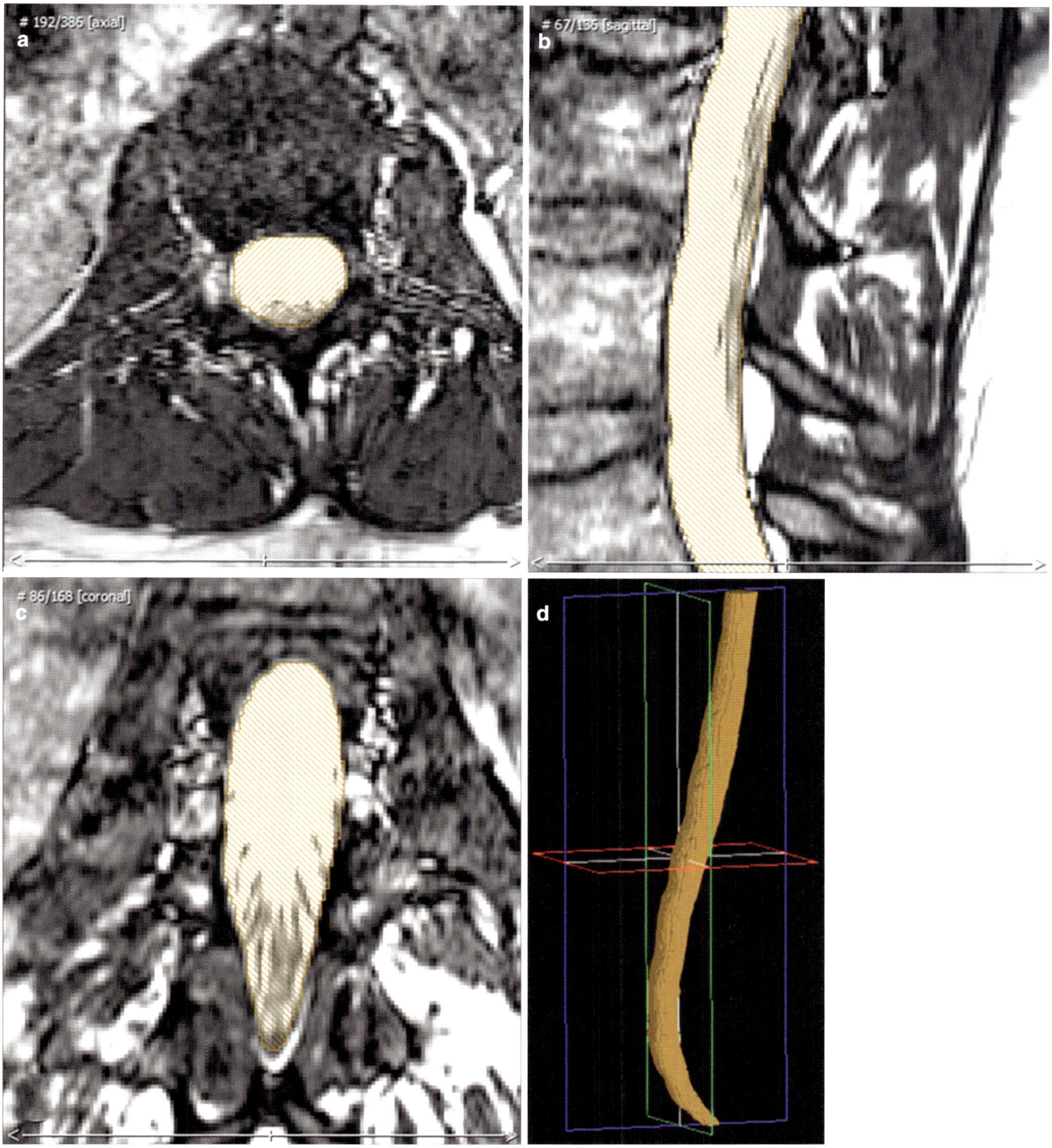

Fig. 47.2 Axial (**a**), sagittal (**b**), and coronal (**c**) MR planes and 3D view (**d**) of the dural sac. Processed images show a gray scale ranging from 0 to 3,000. A threshold segmentation adapted in each case to preferentially select CSF (about 1,000) is applied to T2-weighted MR images. This is followed by interactive segmentation (automatic selection among contiguous voxels with similar gray values), thus separating different tissues enclosed in the dural sac. Manual editing is then performed to delineate the dural sac VOI around the previously selected CSF voxels

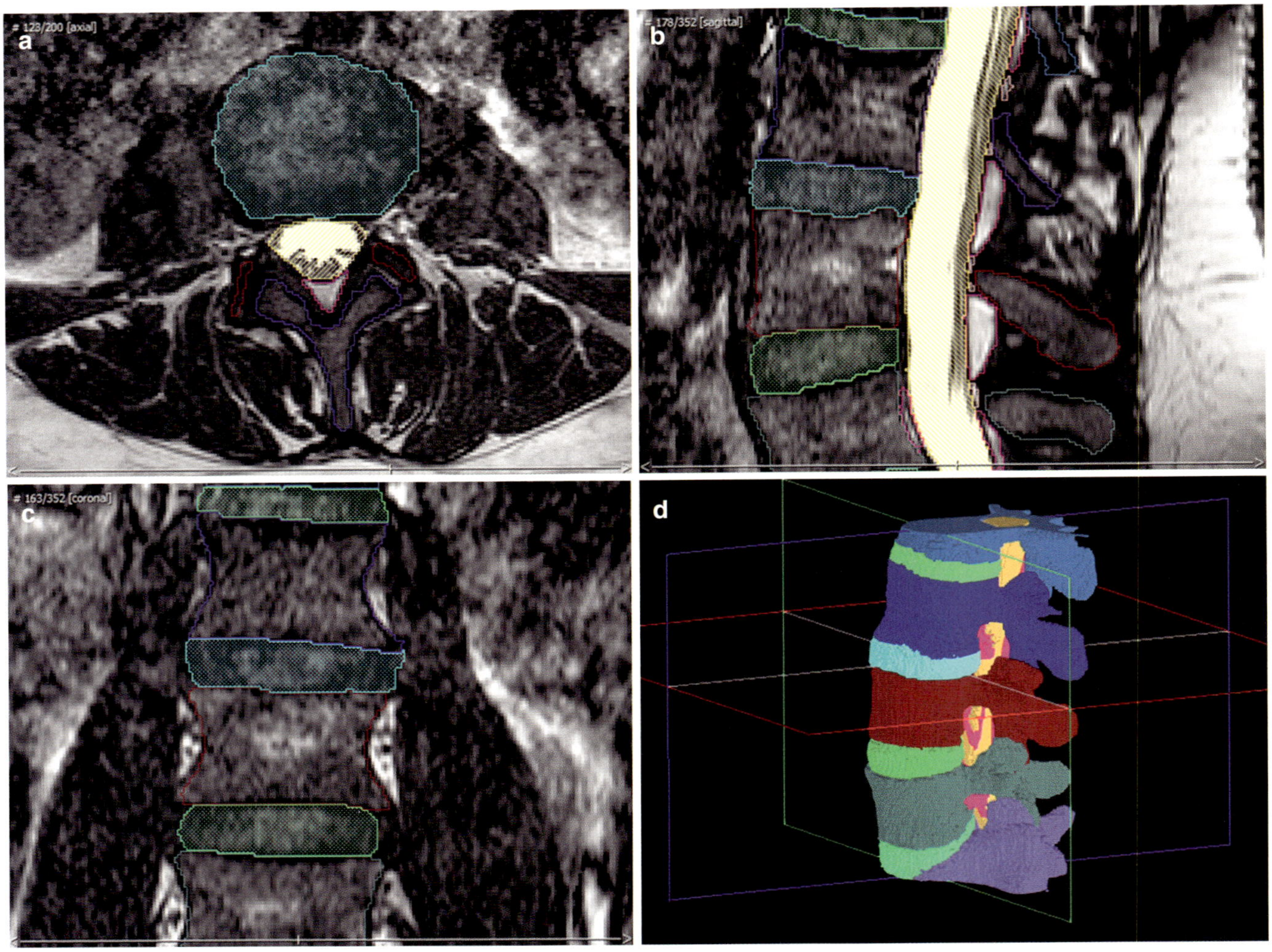

Fig. 47.3 Three-dimensional reconstruction of spinal structures. Manual editing is also required to delineate structures surrounding the spinal canal (vertebrae, intervertebral discs, etc.). Axial (**a**), sagittal (**b**), coronal (**c**), and oblique 3D view (**d**) of different spinal structures

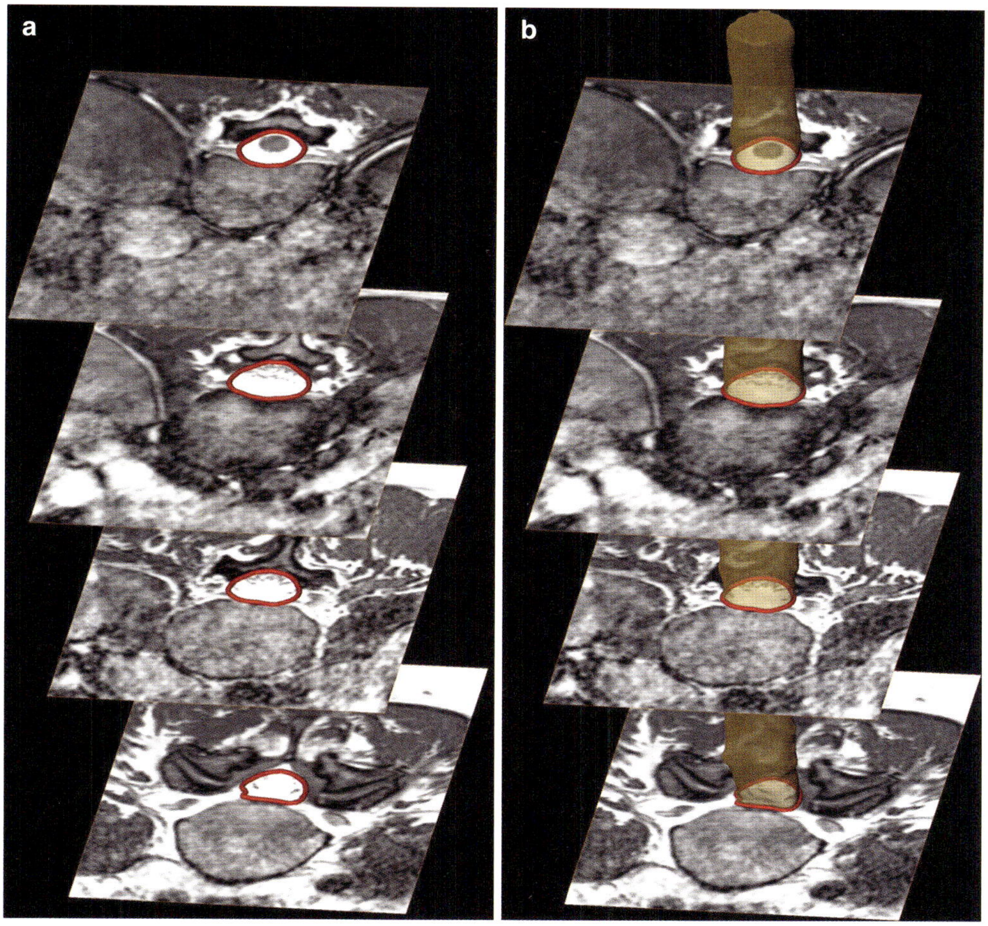

Fig. 47.4 Dural sac VOI validation. Axial planes of the MR image (**a**) and 3D model of the dural sac are intersected (**b**) to identify differences between their respective external contours. If the selected section and the MR image do not visually coincide with the 3D model, the VOI is manually corrected and the procedure is repeated. This validation is also conducted in adjacent structures

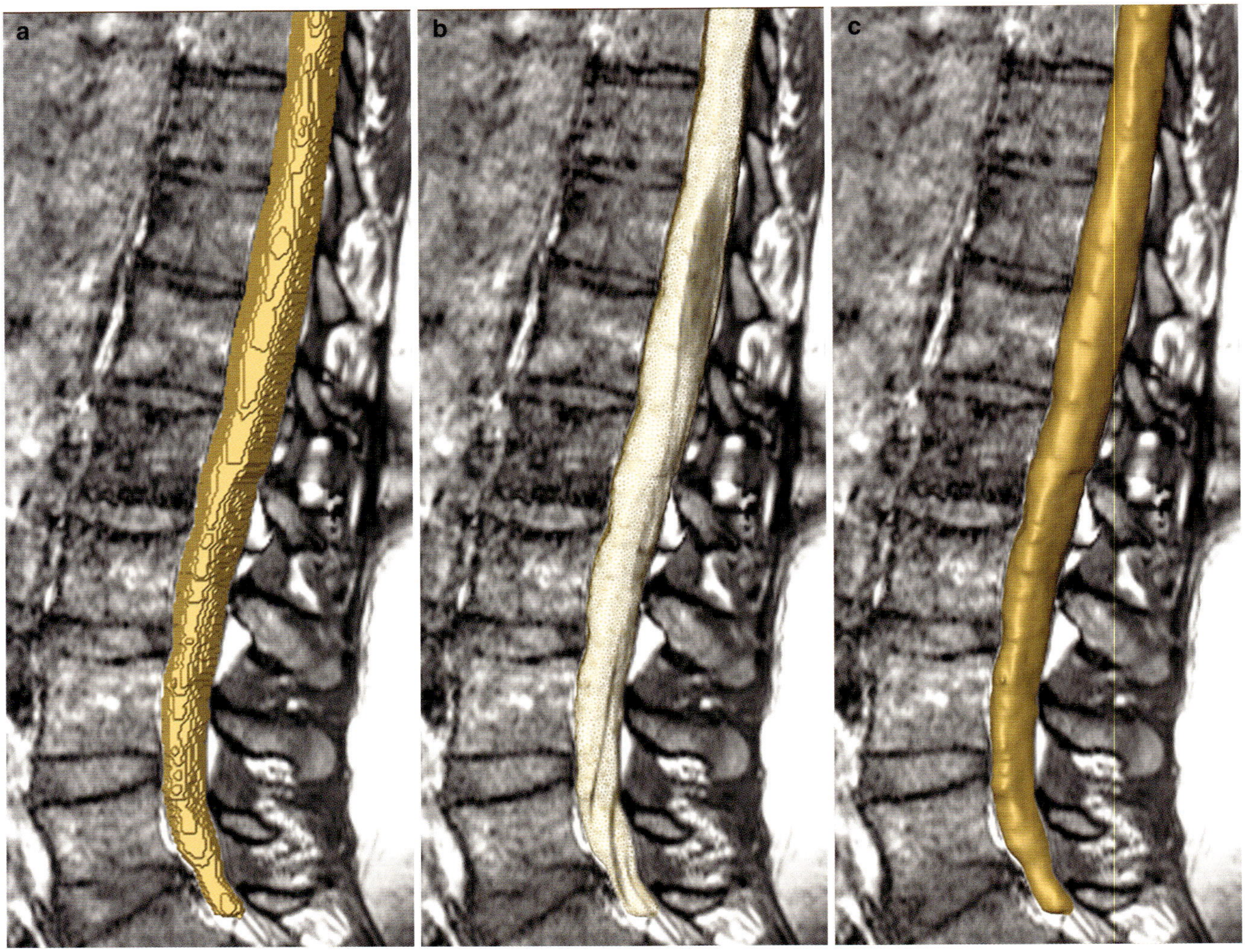

Fig. 47.5 Surface model. Once the dural sac VOI is accepted as correct, a surface model is generated by a marching cubes algorithm (**a**), which is later simplified (**b**), and the resulting model is smoothed (**c**)

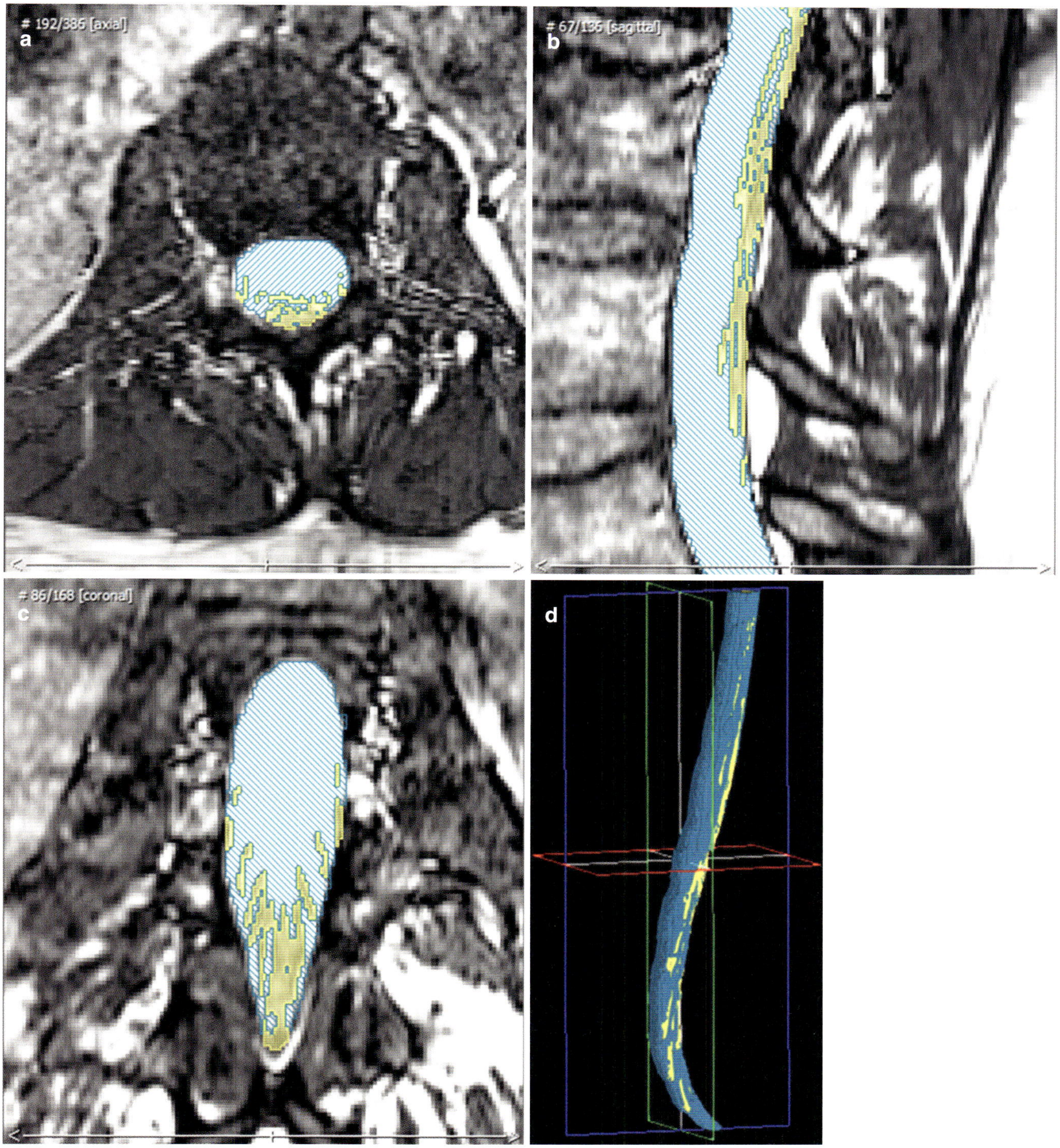

Fig. 47.6 Segmentation of cauda equina, nerve roots, and CSF inside the dural sac VOI. A second segmentation is applied exclusively to the region of the MR image enclosed in the VOI of the dural sac. A segmentation threshold (around 700) is chosen to separate the spinal nerve roots of the cauda equina along the entire portion of the vertebral canal from the CSF inside the dural sac. Axial (**a**), sagittal (**b**), coronal (**c**), and 3D model (**d**) views illustrate the segmentation of cauda equina nerve roots and CSF within the dural sac

Cerebrospinal Fluid and Root Volume Quantification from Magnetic Resonance Images

48

Anna Puigdellívol-Sánchez, Alberto Prats-Galino, and Julio Castedo

Lumbosacral cerebrospinal fluid (CSF) and nerve root volume variation are quantified depending on the threshold selection and according to predefined anatomical criteria [1–10]. High-resolution MR images (T2, 16 bits, voxel size = 0.65 mm^3) are saved as DICOM files and analyzed using 3D analysis software. Criteria for "certain," "likely," and "possible" root and CSF threshold selection is described and applied to the predelineated dural sac volume of interest. When studying CSF, the absence of selected voxels outside the CSF anatomical area is required, and vice versa when selecting roots (Fig. 48.4), combined with a greater or smaller selection of gray voxels in the CSF–root border line (Fig. 48.3). Different thresholds are chosen in a randomly selected slice of the conus medullaris, upper lumbosacral region, and caudal lumbosacral region. Paired Wilcoxon tests are used for comparisons. A standardized block of fifty 3.25-cm slices originating rostrally to the middle dorsal point of the L5-S1 vertebral disk is used to standardize volume comparisons. Significant differences are often found for thresholds within the same structure and criteria in the different anatomical regions. Furthermore, differences are appreciated between thresholds (average max/min values = 2.04) and the resulting volumes (average max/min values = 1.32) in the caudal lumbar region. High variability may be observed between cases within the same criterion (max/min volumes = 1.88 for the "likely CSF" criterion). The selection of different thresholds leads to variations in the volume estimation. However, although max/min threshold differences account for 30 % of the variability in volumes, the interindividual variability may be much higher. The real volume values are expected to be within the range obtained after applying "certain" and "possible" criteria. Real values are probably close to those obtained using the "likely CSF" or "likely root" criteria. If high precision is desired, gray voxels in the anterior dural sac–CSF interface should be assigned manually. Reproducible threshold selection following the criteria proposed may allow future individualized CSF volume estimations prior to intrathecal drug administration (Figs. 48.1, 48.2, 48.3, and 48.4; Tables 48.1 and 48.2).

A. Puigdellívol-Sánchez, MD, PhD (✉) • A. Prats-Galino, MD, PhD
Laboratory of Surgical NeuroAnatomy (LSNA),
Human Anatomy and Embryology Unit,
School of Medicine, University of Barcelona, Barcelona, Spain
e-mail: apuigdellivol@ub.edu; aprats@ub.edu

J. Castedo, MD
Department of Clinical Medical Science,
School of Medicine, CEU San Pablo University, Madrid, Spain

Department of Radiology, Madrid-Torrelodones University
Hospital, Madrid, Spain
e-mail: j.castedo@telefonica.net

M.A. Reina et al. (eds.), *Atlas of Functional Anatomy for Regional Anesthesia and Pain Medicine: Human Structure, Ultrastructure and 3D Reconstruction Images*, DOI 10.1007/978-3-319-09522-6_48,

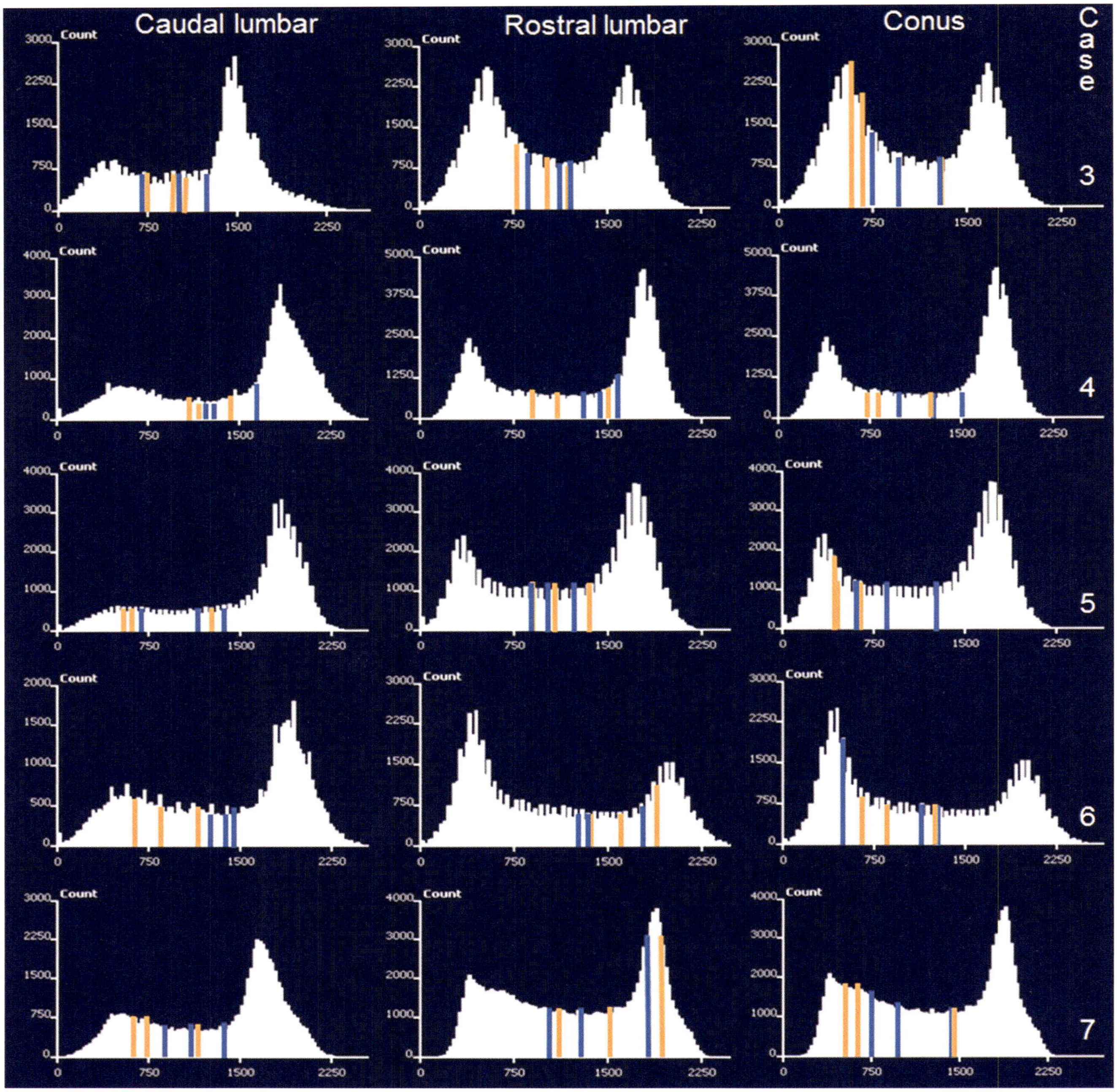

Fig. 48.1 Histogram for gray value ranges from different anatomical regions and cases. The histogram of the dural sac volume of interest (VOI) was generated to determine the gray scale range (about 0–2,300) and its frequency distribution, and to adjust the data window to the maximal gray value prior to threshold selection. Because the distribution was not consistent enough, threshold decision from the histogram visualization itself would not lead to correct anatomical selection of the structures studied. *Orange lines*: nerve root thresholds. *Blue lines*: CSF thresholds

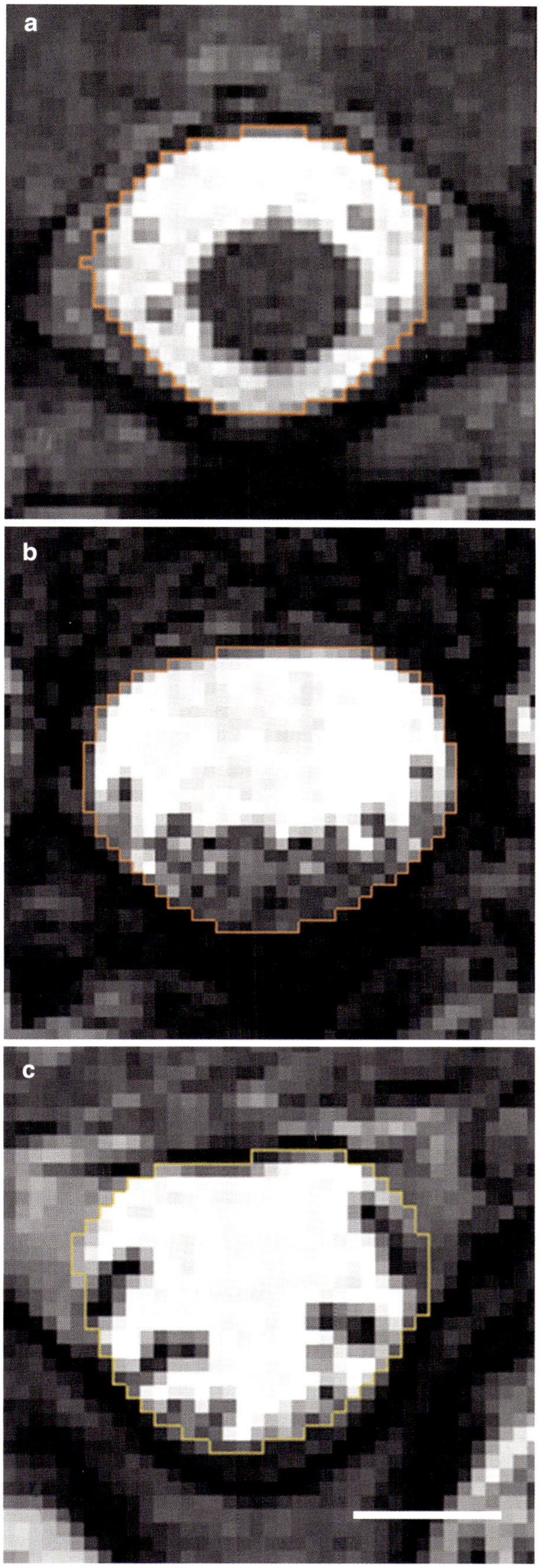

Fig. 48.2 Anatomical regions. (**a**) Caudal lumbar region: cauda equina nerve roots located laterally within the dural sac. (**b**) Rostral lumbar region: cauda equina nerve roots distributed in the dorsal part and CSF in the ventral part of the spinal canal. (**c**) Conus medullaris region. The conus medullaris is present at the center of the dural sac and is surrounded by CSF. Scale bar = 1 cm

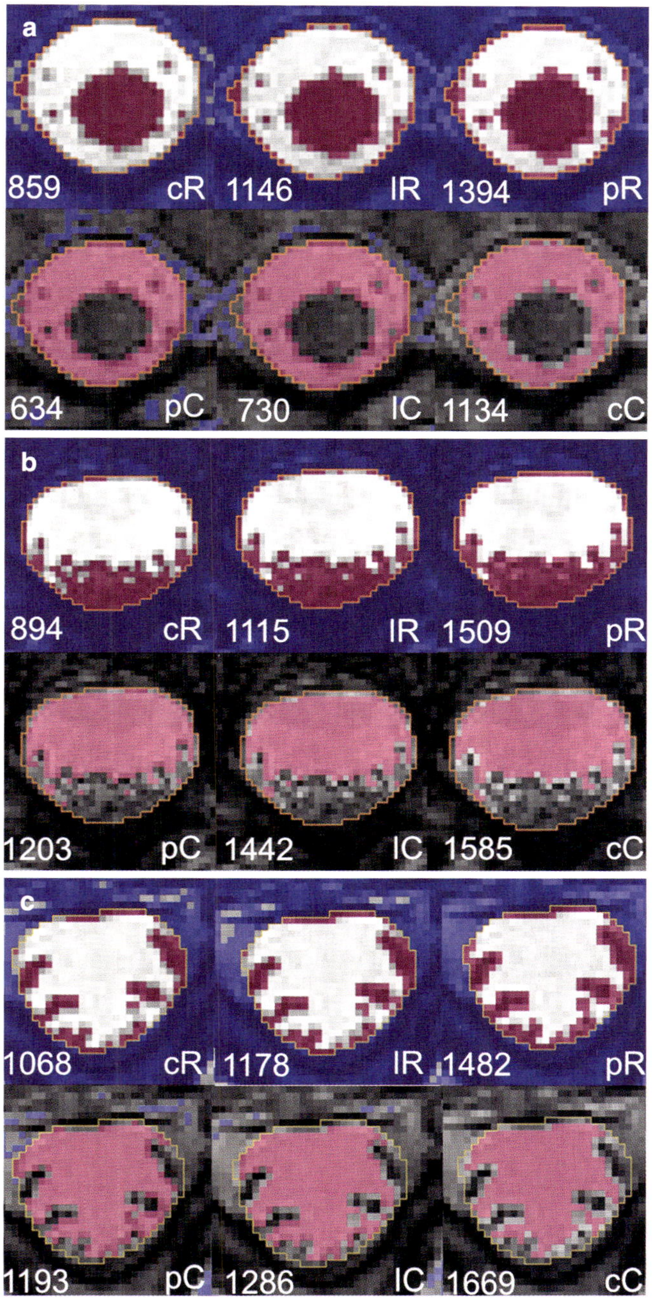

Fig. 48.3 Threshold selection. Nerve root and CSF selected voxels (*magenta*) according to different criteria in the three anatomical regions of one case. Threshold values are indicated in each image. Note the similar appearance of the selected area among images, in spite of great variability in threshold values. Predefined CSF threshold criteria: "Certain CSF" (*cC*): The white voxels are selected, but not the gray voxels around the nerve roots or the dural sac. "Likely CSF" (*lC*): White voxels in the CSF area are selected, as well as some gray voxels close to the roots and dural sac. "Possible CSF" (*pC*): All the white voxels and most of the gray voxels around the cauda equina nerve roots and dural sac are selected. Predefined root threshold criteria: "Certain root" (*cR*): The middle nerve root area is selected, but not adjacent gray voxels. "Likely root" (*lR*): The middle root area is selected, as well as some of the adjacent gray voxels. "Possible root" (*pR*): All the gray voxels in the cauda equina root or conus medullaris area are selected, as are the surrounding gray voxels

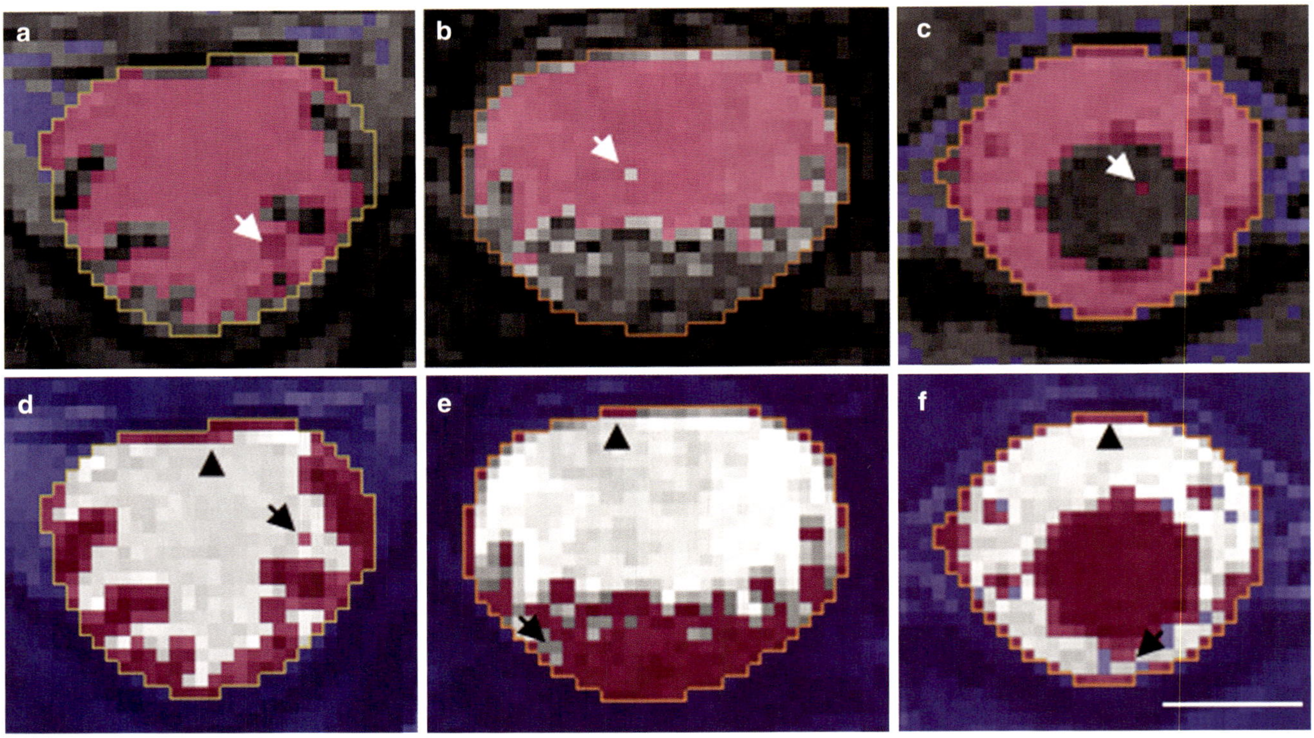

Fig. 48.4 Incorrect threshold values. Threshold values above or below this incorrect threshold value are selected while looking for the ranges of minimal and maximal threshold values ("certain" vs "possible" thresholds). *Magenta voxels*: selected area. In (**a**–**c**) *white arrows* designate wrong CSF voxel selection. (**a**) Gray values with the probable presence of nerve roots are considered CSF. (**b**) Secure CSF voxels are unlabeled. (**c**) Secure conus area is labeled as CSF. In (**d**–**f**) *black arrows* designate wrong nerve root voxel selection including dural sac-CSF interfase (*arrowheads*). (**d**) One secure CSF voxel is selected as nerve root. (**e**) Secure nerve root voxels are unlabeled. (**f**) Secure CSF voxels are considered as conus medullaris and also as dural sac borderline. Scale bar = 1 cm

Table 48.1 Root and CSF threshold values in the different anatomical regions

Region–case	Root			CSF			Max/min
	a. Certain	b. Likely	c. Possible	d. Possible	e. Likely	f. Certain	
1. Conus							
3	547	696	862	684	832	1,128	2.06
4	634	730	1,134	859	1,146	1,394	2.19
5	421	438	649	603	845	1,331	3.16
6	624	864	1,259	496	1,096	1,272	2.56
7	514	615	1,259	717	892	1,221	2.44
Mean	548.00	668.60	1,032.60	671.80	962.20	1,269.20	2.48
SD	87.26	157.12	268.82	134.98	147.73	102.06	0.42
Significance	2.3/b.c	2.3/a.c/e	2/a.b	2.3/e.f	2.3/d.f/b	2/d.e	
Max/min	1.50	1.97	1.93	1.73	1.37	1.23	
2. Rostral lumbar							
3	771	904	1,144	857	1,071	1,161	1.50
4	894	1,115	1,509	1,203	1,442	1,585	1.77
5	868	1,061	1,263	854	988	1,222	1.47
6	1,395	1,597	1,748	1,293	1,375	1,613	1.35
7	1,133	1,568	1,836	1,012	1,294	1,776	1.81
Mean	1,012.20	1,249.00	1,500.00	1,043.80	1,234.00	1,471.40	1.58
SD	252.03	314.32	298.91	199.61	196.11	266.59	0.19
Significance	1/b.c	1.3/a.c	1/a.b	1/e.f	1.3/d.f	1/d.e	
Max/min	1.80	1.76	1.60	1.51	1.45	1.52	
3. Caudal lumbar							
3	756	913	1,095.6	745	932	1,306	1.75
4	1,068	1,178	1,482	1,193	1,286	1,669	1.56
5	519	656	1,267	708	1,148	1,425	2.74
6	698	814	1,055	1,163	1,384	1,440	2.08
7	656	748	1,183	872	958	1,346	2.05
Mean	737.76	861.80	1,216.52	936.36	1,141.76	1,437.36	2.04
SD	203.92	200.12	169.45	229.03	198.08	140.72	0.45
Significance	1/c/	1.2/c/e	/a.b/f	1/e.f/	1.2/d.f/b	/d.e/c	
Max/min	2.05	1.79	1.40	1.68	1.48	1.27	

The variability in threshold selection and volumes were studied in cases 3–7 of Chap. 37; data extending from the sacral region to the conus medullaris are available. Threshold values in the different anatomical regions (conus, rostral and caudal lumbar) according to predefined criteria for selecting nerve roots (a–c) or CSF (d–f). Significant differences (Wilcoxon test) are indicated below SD values: between the same criterion in different regions/between the root or CSF criteria in the same region/between complementary (a–d, b–e, c–f) CSF-root criteria. Max/min threshold values between cases following the same criterion are given below significances; the max/min rate of threshold values for each case are shown in the last column on the right

CSF cerebrospinal fluid, *Max* maximal, *Min* minimal, *SD* standard deviation

Table 48.2 CSF and root resulting volumes from different threshold criteria

Volume/case	According to root thresholds			According to CSF thresholds			Max/min
	a. Certain	b. Likely	c. Possible	d. Possible	e. Likely	f. Certain	
CSF volumes, cm^3							
3	4.20	3.95	3.66	4.23	3.93	3.30	1.27
4	4.22	4.07	3.62	4.05	3.87	3.22	1.31
5	6.41	6.28	5.42	6.22	5.61	5.14	1.24
6	4.00	3.83	3.48	3.31	2.97	2.88	1.39
7	4.91	4.72	3.88	4.47	4.30	3.54	1.38
Mean	4.75	4.57	4.01	4.45	4.14	3.62	1.32
SD	0.99	1.01	0.79	1.07	0.95	0.88	0.06
Max/min	1.60	1.63	1.55	1.87	1.88	1.78	1.72
Root volumes, cm^3							
3	1.22	1.47	1.76	1.19	1.49	2.11	1.77
4	1.30	1.45	1.89	1.47	1.65	2.30	1.77
5	0.41	0.54	1.40	0.60	1.21	1.68	4.07
6	0.64	0.80	1.15	1.33	1.67	1.76	2.75
7	0.67	0.86	1.70	1.11	1.28	2.04	3.03
Mean	0.85	1.02	1.58	1.14	1.46	1.98	2.68
SD	0.39	0.41	0.29	0.33	0.21	0.25	0.96
Max/min	3.15	2.69	1.63	2.42	1.38	1.36	2.11
Significance	b.c/e.f	a.c/e.f	a.b/f	e.f/	d.f/a.b	d.e/a.b.c	

CSF and root volumes resulting from applying caudal lumbar region thresholds in a standardized block of 50 slices ending caudally at the L5-S1 intervertebral disk. Significant differences between them are indicated at the end of the table. Max/min volumes between cases following a specific criterion are shown below SD, while max/min estimations for each case are calculated in the last column on the right. This interindividual variability in volumes is consistent with previous reports of CSF estimations [1–8]

CSF cerebrospinal fluid, *Max* maximal, *Min* minimal, *SD* standard deviation

References

1. Fink BR, Gerlach R, Richards T, Maravilla KR. Improved magnetic resonance (MR) imaging method for measurement of spinal fluid volume in normal subjects: use of fast spin echo pulse sequence. Anesthesiology. 1992;77:A875.
2. Hogan QH, Prost R, Kulier A, Taylor ML, Liu S, Mark L. Magnetic resonance imaging of cerebrospinal fluid volume and the influence of body habitus and abdominal pressure. Anesthesiology. 1996;84:1341–9.
3. Lee RR, Abraham RZ, Quinn CB. Dynamic physiologic changes in lumbar CSF volume quantitatively measured by three-dimensional fast spin-echo MRI. Spine. 2001;26:1172–8.
4. Higuchi H, Hirata J, Adachi Y, Kazama T. Influence of lumbosacral cerebrospinal fluid density, velocity, and volume on extent and duration of plain bupivacaine spinal anesthesia. Anesthesiology. 2004;100:106–14.
5. Sullivan JT, Grouper S, Walker MT, Parrish TB, McCarthy RJ, Wong CA. Lumbosacral cerebrospinal fluid volume in humans using three-dimensional magnetic resonance imaging. Anesth Analg. 2006;103:1306–10.
6. Edsbagge M, Starck G, Zetterberg H, Ziegelitz D, Wikkelso C. Spinal cerebrospinal fluid volume in healthy elderly individuals. Clin Anat. 2011;24:733–40.
7. Puigdellivol-Sánchez A, Prats-Galino A, Reina MA, Machés F, Hernández JM, De Andrés J, et al. Three dimensional magnetic resonance image of structures enclosed in the spinal canal relevant to anesthetists and estimation of the lumbosacral CSF volume. Acta Anaesthesiol Belg. 2011;62:37–45.
8. Prats-Galino A, Reina MA, Puigdellívol-Sánchez A, Juanes Méndez JA, De Andrés JA, Collier CB. Cerebrospinal fluid volume and nerve root vulnerability during lumbar puncture or spinal anaesthesia at different vertebral levels. Anaesth Intensive Care. 2012;40:643–7.
9. Horsfield MA, Sala S, Neema M, Absinta M, Bakshi A, Sormani MP, et al. Rapid semi-automatic segmentation of the spinal cord from magnetic resonance images: application in multiple sclerosis. Neuroimage. 2010;50:446–55.
10. Puigdellívol-Sánchez A, Reina MA, San-Molina J, Escobar JM, Castedo J, Prats-Galino A. Threshold selection criteria for quantification of lumbosacral cerebrospinal fluid and root volumes from MRI. J Neuroimaging. 2014;00:1–6. doi: 10.1111/jon.12150

Scanning Electron Microscopy

49

Ana Vicente Montaña, Alfredo Fernández Larios, and Alfonso Rodríguez Muñoz

The main difference between scanning electron microscopy (SEM) and transmission electron microscopy (TEM) is that SEM provides images of the surface of specimens (as seen in Figs. 49.6, 49.7, 49.8, 49.9, 49.10, 49.11, and 49.12), whereas TEM shows the inner ultrastructure of cells and tissues. Although SEM is not frequently used in the study of ultrastructure, it can be a useful tool, particularly if the information provided is consistent with findings obtained in TEM studies.

A few general considerations are necessary in the preparation of biological specimens for these studies. Every effort must be made to handle biological specimens with extreme care if irreparable damage is to be avoided, certain rules during the processing of specimens in order to achieve the best image possible. These rules are relevant through all phases of the process: collection of specimens, fixation and dehydration, and sample observation.

In relation to the manipulation of specimens, samples must be dissected using scalpel rather than scissors, to avoid the crushing that scissors produce. Tissues taken by diathermy are inadequate, so this method is not recommended for ultrastructure sampling. Whereas an adequately fixed surface is considered essential in SEM, the size of specimens is more significant in TEM techniques. Once the biopsy specimen is obtained from the patient, it must be immediately immersed in an appropriate fixative agent, as minor delays can be detrimental for the electron microscopy study.

Following specimen extraction, fixation of samples must be carried at the earliest possible time in order to preserve the biological structure of tissues. A relevant method for fixing nervous tissue from animal specimens is by means of perfusion, but this method is not suitable for human specimens, even that the ultrastructure of myelin sheaths is highly sensitive to the conditions of perfusion fixation. Glutaraldehyde penetrates slowly and to a depth of only about 2 mm across the surface of samples. Therefore, immersion in modified Karnovsky solution is a more suitable fixation method; the rapid fixative properties of paraformaldehyde, which is formed by small molecules that penetrate tissues quickly, ensure reasonable structure preservation. Fixation intervals depend on the size of the sample and may vary between 2 and 6 h (less in TEM techniques). Specimens are kept immersed for 20 min to ensure adequate pre-fixation, so the samples can be cut into very thin cross-sections in order to achieve full fixation.

Basic SEM techniques are not particularly arduous. Post-fixing of samples with osmium tetroxide 1 % can improve image resolution. At a later stage, samples undergo a process of dehydration; for nerve tissue, acetone is used in preference to ethanol, because an excess of fat can produce artifacts that interfere with the quality of images obtained. After reaching a concentration of about 100 % of dehydrating medium, the drying process is completed by introducing the samples inside a device where, exchange between ethanol and liquid CO_2 occurs. Once critical point is reached liquid CO_2 turns into gas, preserving in this way tissue structure.

Finally, once the sample is dried and mounted on the holder, it must be coated with a conductor substance such as gold or graphite, or it may be double-coated with graphite and gold, or another metal such as palladium. These methods aim to preserve the biological structure of samples and avoid the production of artifacts that interfere with image quality (Figs. 49.1, 49.2, 49.3, 49.4, 49.5, 49.6, 49.7, 49.8, 49.9, 49.10, 49.11, and 49.12).

A. Vicente Montaña, DVM (✉) • A. Fernández Larios, Tech
A. Rodríguez Muñoz, Tech
Scanning Electron Microscopy Unit, ICTS Electron Microscopy National Center, Complutense University, Madrid, Spain
e-mail: amvicent@vet.ucm.es; aflarios@ucm.es; alrodrig@pdi.ucm.es

M.A. Reina et al. (eds.), *Atlas of Functional Anatomy for Regional Anesthesia and Pain Medicine: Human Structure, Ultrastructure and 3D Reconstruction Images*, DOI 10.1007/978-3-319-09522-6_49,

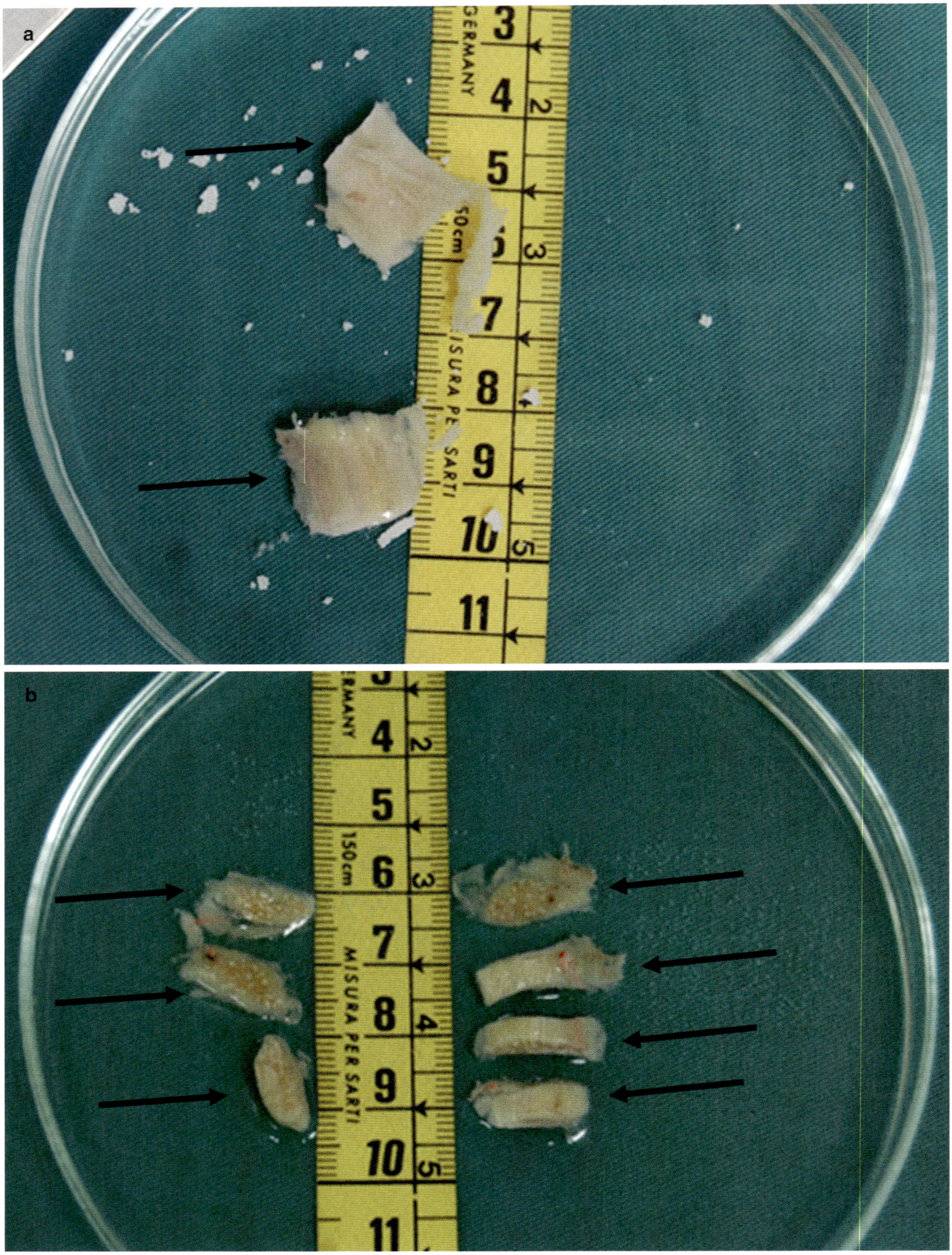

Fig. 49.1 Human sciatic nerve. (**a**) Pieces were immersed in modified Karnovsky fixative for half an hour. Afterwards, samples are cut into smaller pieces. (**b**) The same sample cut into smaller pieces undergoes fixation for a further 2 h to ensure adequate preservation of the tissue architecture. The thickness of each sample measures as little as 5 mm

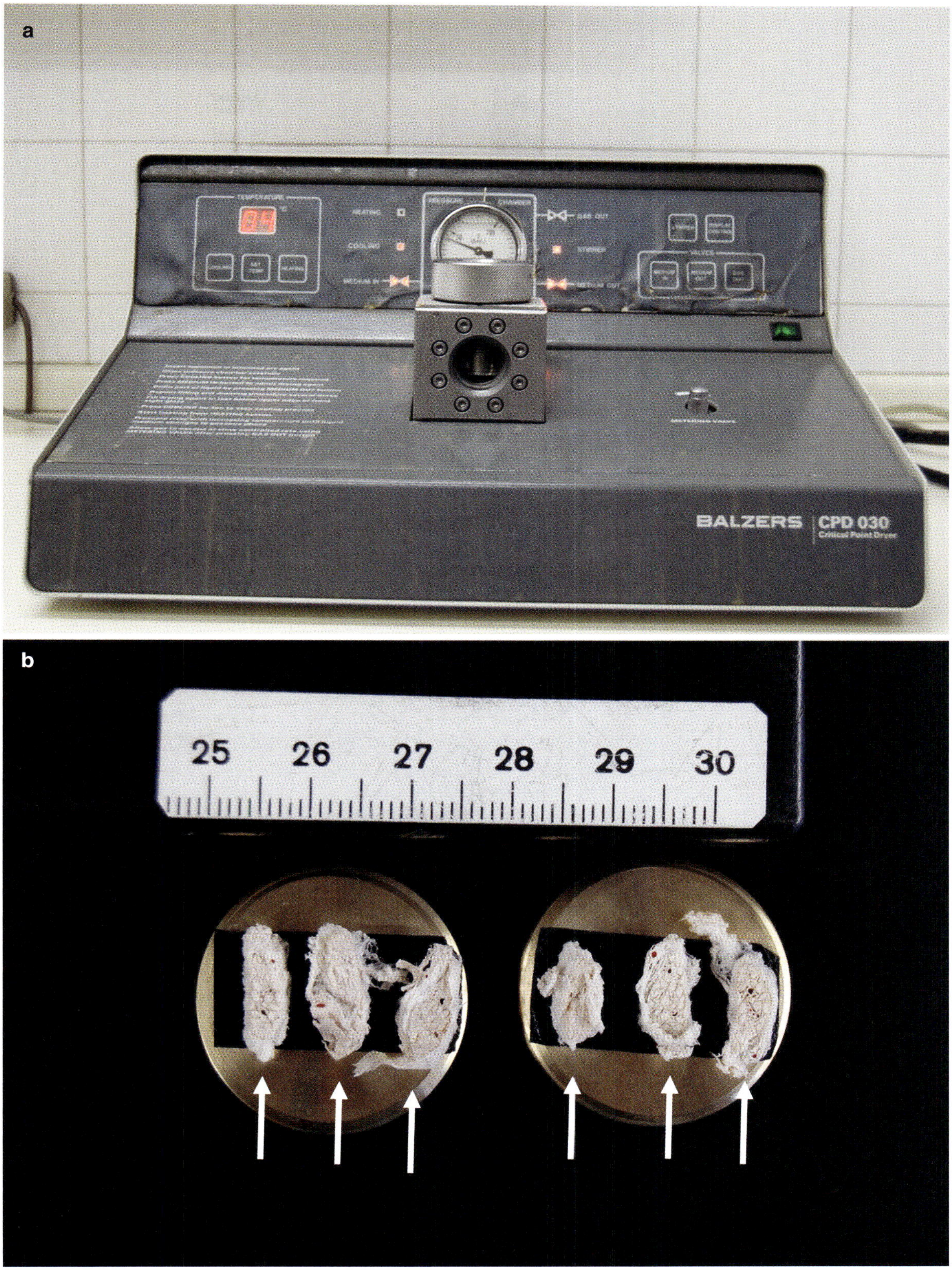

Fig. 49.2 After fixation and dehydration, the sample must pass through a critical point dryer to achieve complete dehydration. (**a**) Dehydrated samples preserve the natural appearance of tissue. Nevertheless, certain tissue retraction is unavoidable. In the critical point dryer, substitution of CO_2 for acetone at low temperature under vacuum conditions takes place. Once the substitution is complete, the temperature is raised to convert liquid CO_2 into gas, maintaining the shape of the tissue. (**b**) Sample subjected to critical point drying. Samples are mounted on microscope holders, on double tapered carbon tapes. Once fixed on the holder, the pieces are covered by gold or by a double coating of carbon and gold

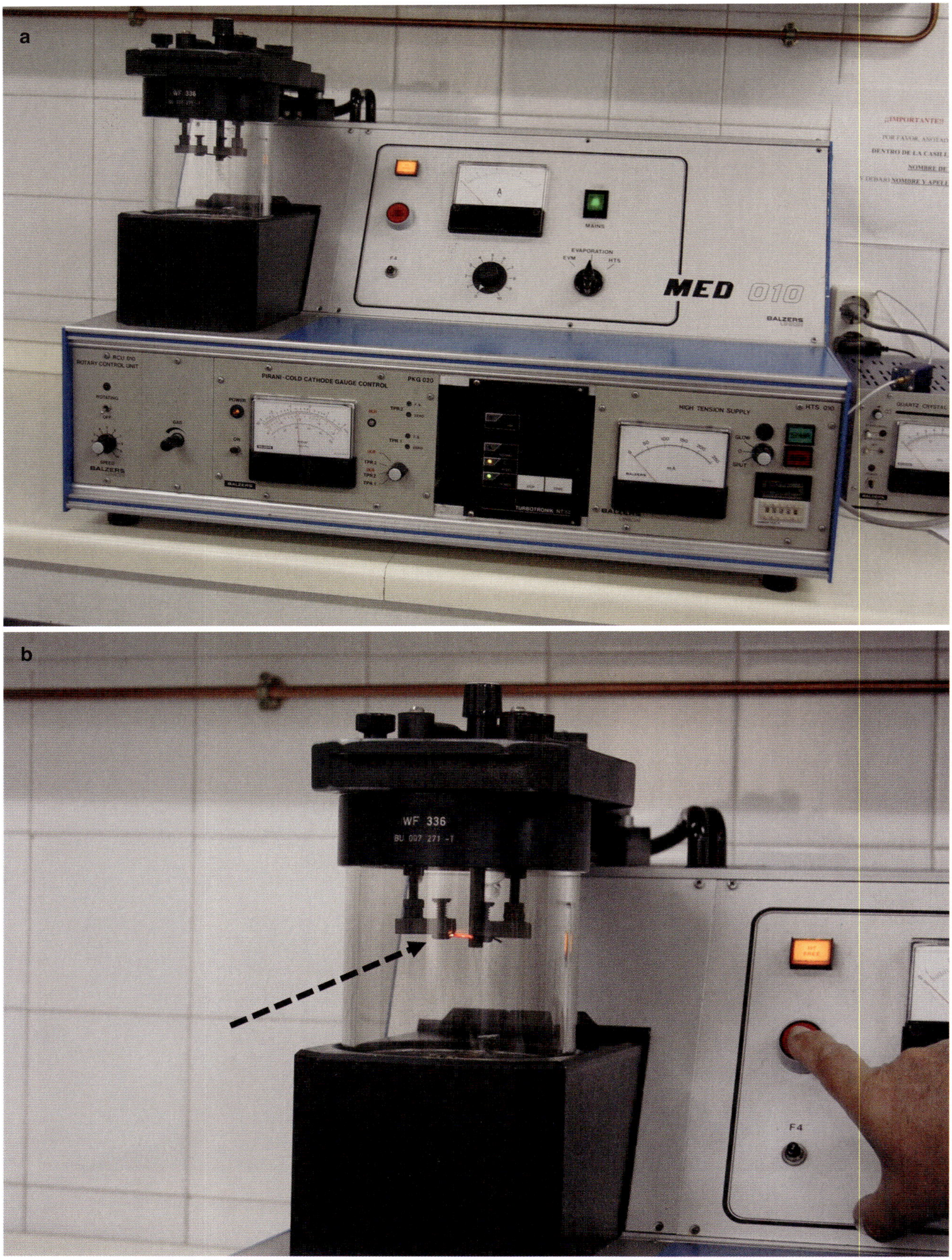

Fig. 49.3 Coal evaporator. (**a**) Samples are covered by graphite to allow conduction. Modern microscopes may not require sample coating. Evaporators may be necessary to remove traces of water from samples. (**b**) Process of sample evaporation in model using graphite thread

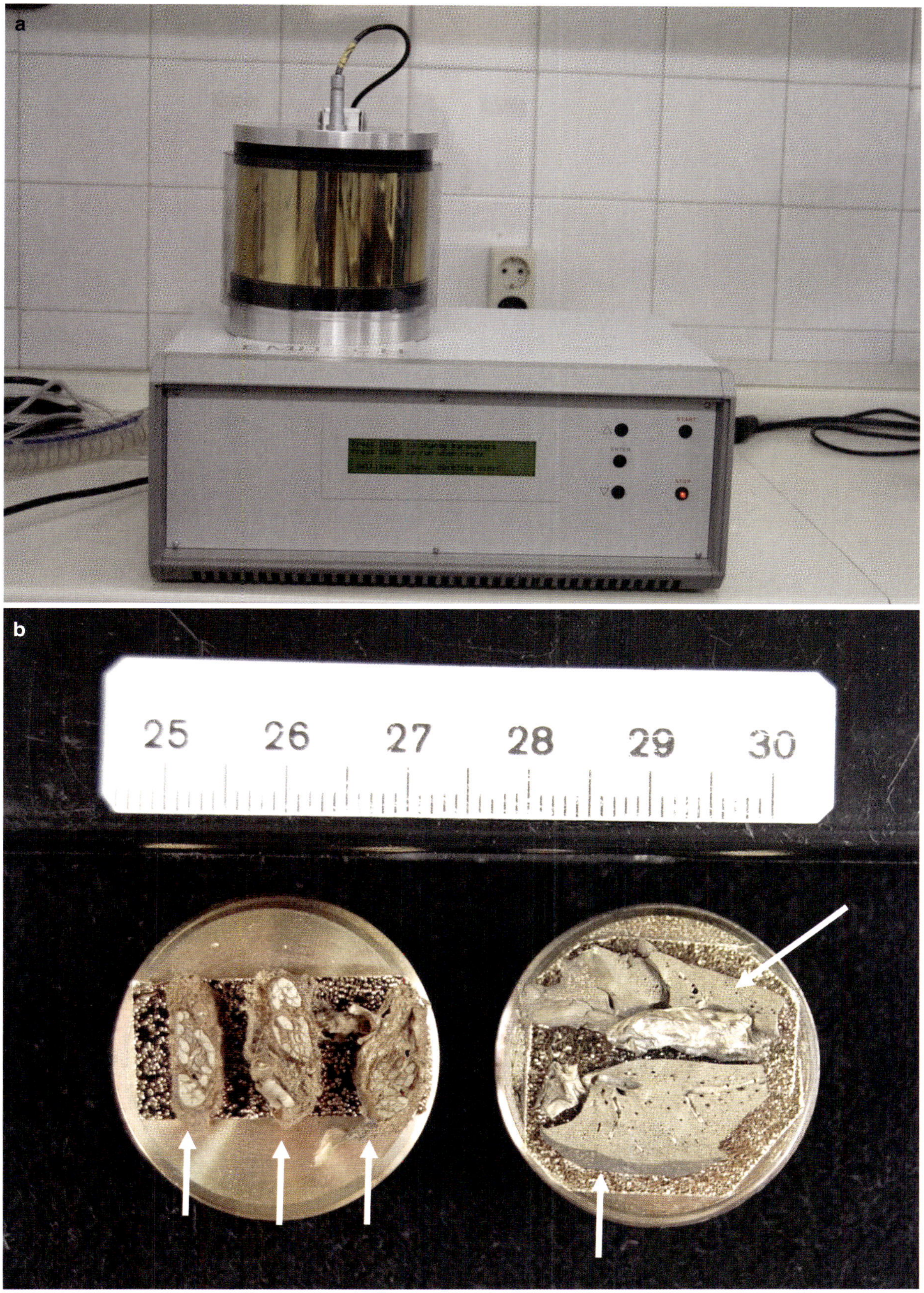

Fig. 49.4 Gold sputter. (**a**) Device used to cover samples with a thin gold layer (10 nm). Graphite evaporators work under vacuum conditions. Different metals (such as palladium or platinum) are used for sample coating in SEM techniques. The overall depth of layers after double coating is about 18 nm (8 nm graphite). (**b**) Detail of samples after graphite and gold coating. *Right*, Rat lung. *Left*, Same human sciatic nerve shown in previous images

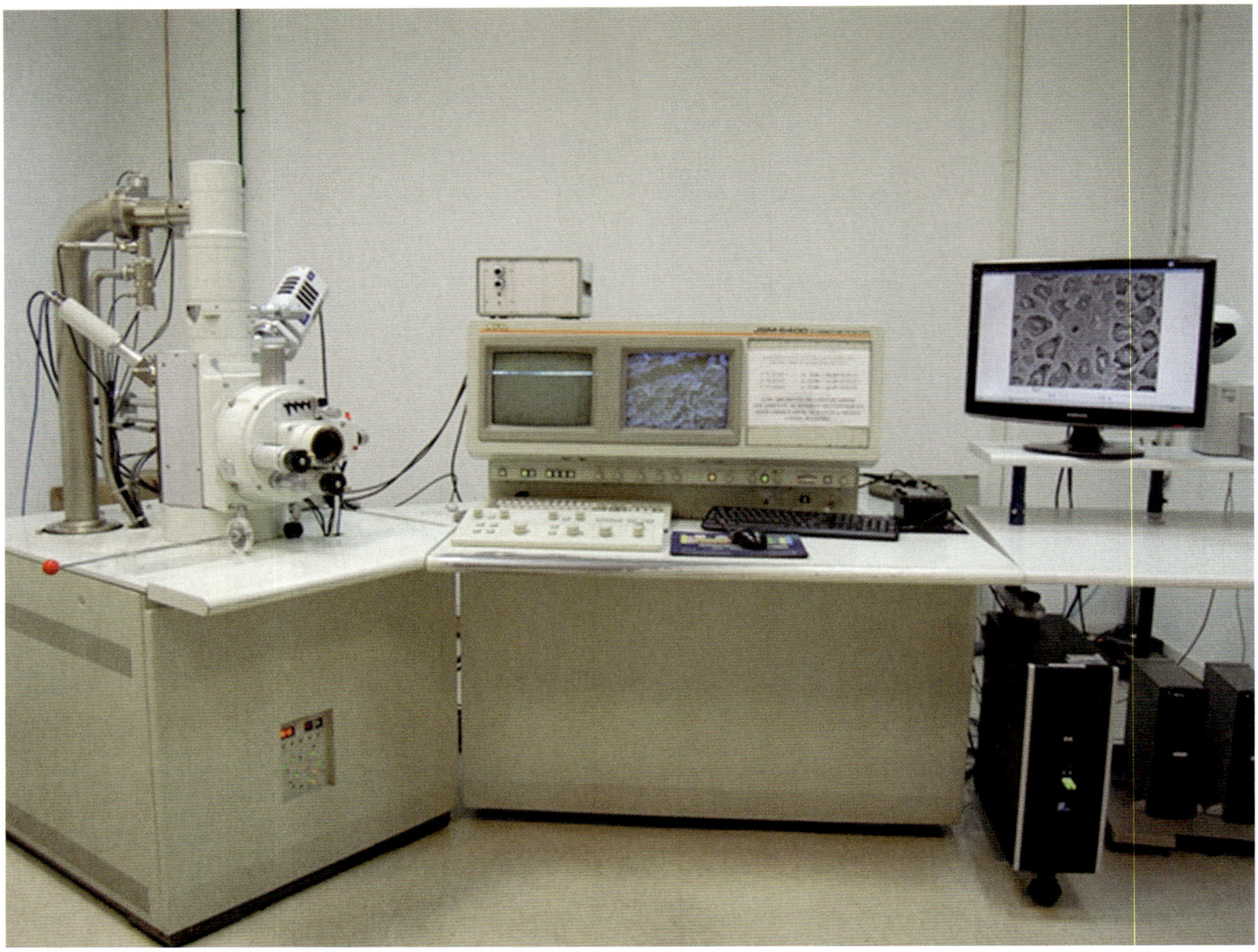

Fig. 49.5 Scanning electron microscope, model JEOL JEM 6400, used in studies of human, animal, and vegetal tissues. The following images have been taken at 20 kilovolts (Kv) and 15 mm working distance (WD). Newer models allow the use of freeze techniques, under reduced kilovoltage or working at lower pressures. Preparation of samples in these models differ slightly, requiring fewer steps

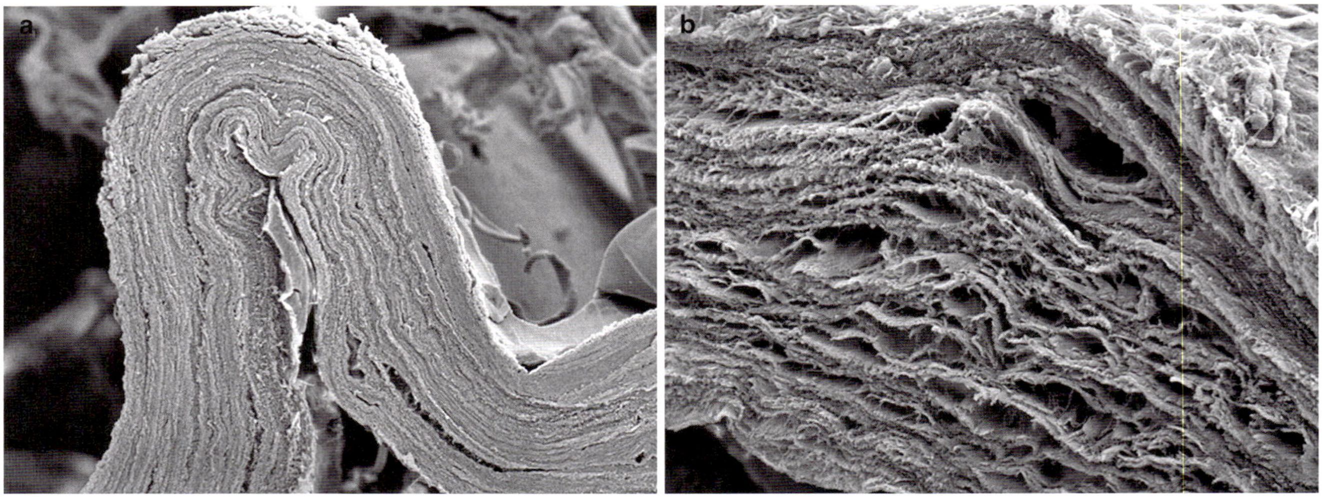

Fig. 49.6 Samples from human dura mater. (**a**) Surface with poor definition of structures. Razor blades are essential to optimize surface cutting before and during fixation, which improves the quality of image details and enables higher magnification. Sample cutting following dehydration in the critical point dryer leads to inaccurate cutting and an uneven surface, severely altering the quality of images. (**b**) Compared with (**a**) all dural laminas from the dura mater are identified. Magnification: (**a**) ×70; (**b**) ×400

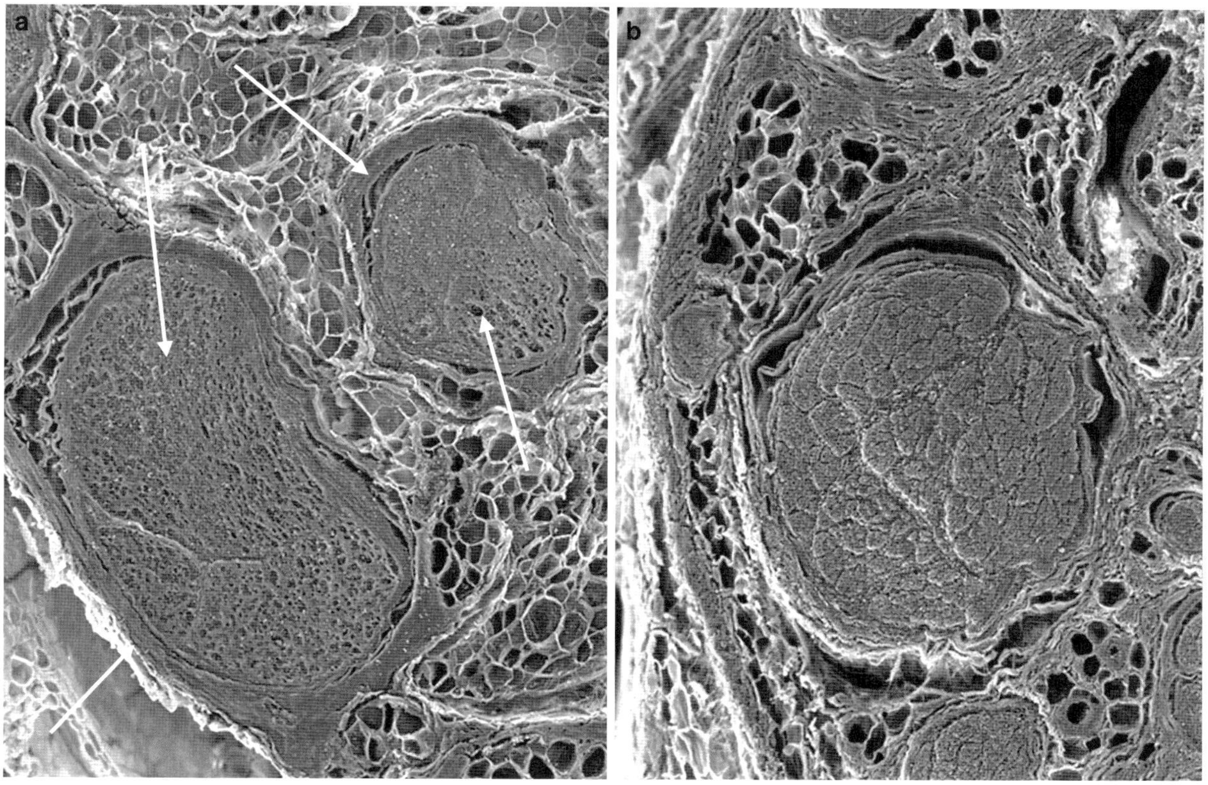

Fig. 49.7 Human sciatic nerve. (**a**) Bright fascicular areas (*arrows*) are an artifact caused by metallization defect. (**b**) This sample maintains its natural appearance, with fat being completely fixed. Magnification: (**a**) ×60; (**b**) ×70

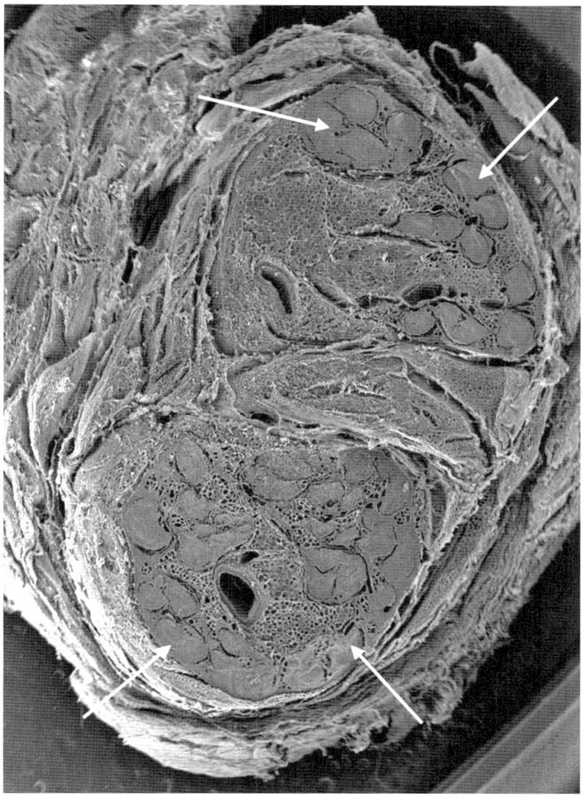

Fig. 49.8 Human sciatic nerve. Bright fascicular area (*arrows*) is an artifact caused by metallization defect. This artifact can be observed in some interfascicular areas. Magnification ×12

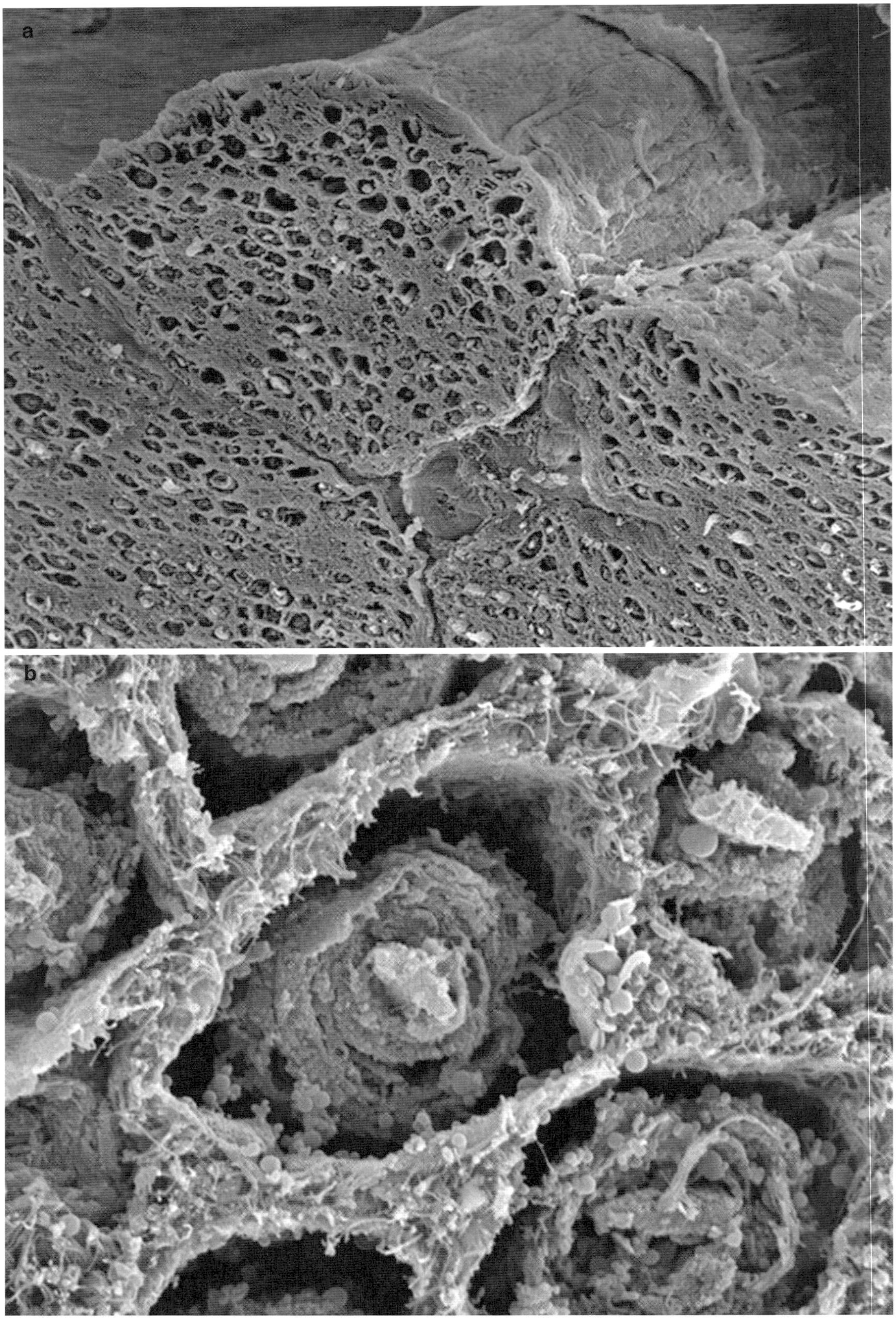

Fig. 49.9 Human spinal nerve roots. (**a**) Detail of its structure. (**b**) Detail of myelinated axon within spinal nerve roots. The appearance of myelinated axons and endoneurium is identical in samples obtained from human peripheral nerves and from human spinal nerve roots. Magnification: (**a**) ×300; (**b**) ×3,500

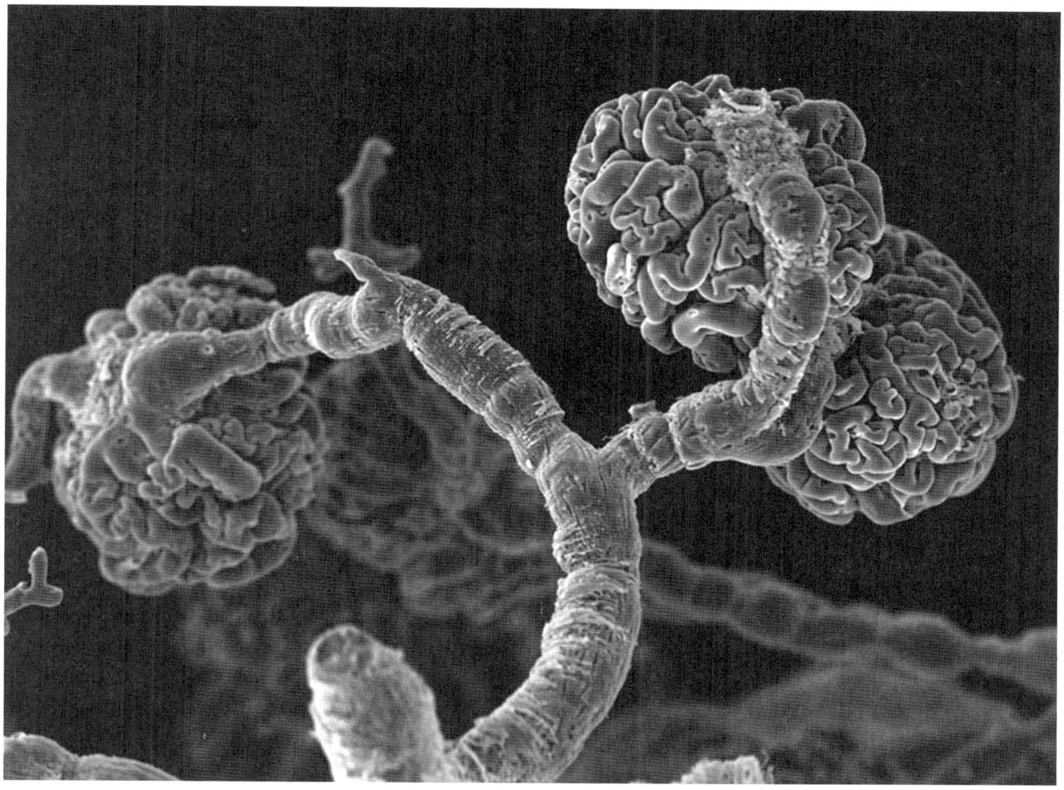

Fig. 49.10 Rat renal glomeruli vascular cast mold. This image is the result of a corrosion vascular cast. After injecting cast in the renal artery, tissue corrosion leaves the vascular mold free. We can see a vascular mold from the vasculature of renal nephrons. The surface of the cast corresponds to the endothelial surface of arteries. Magnification ×150

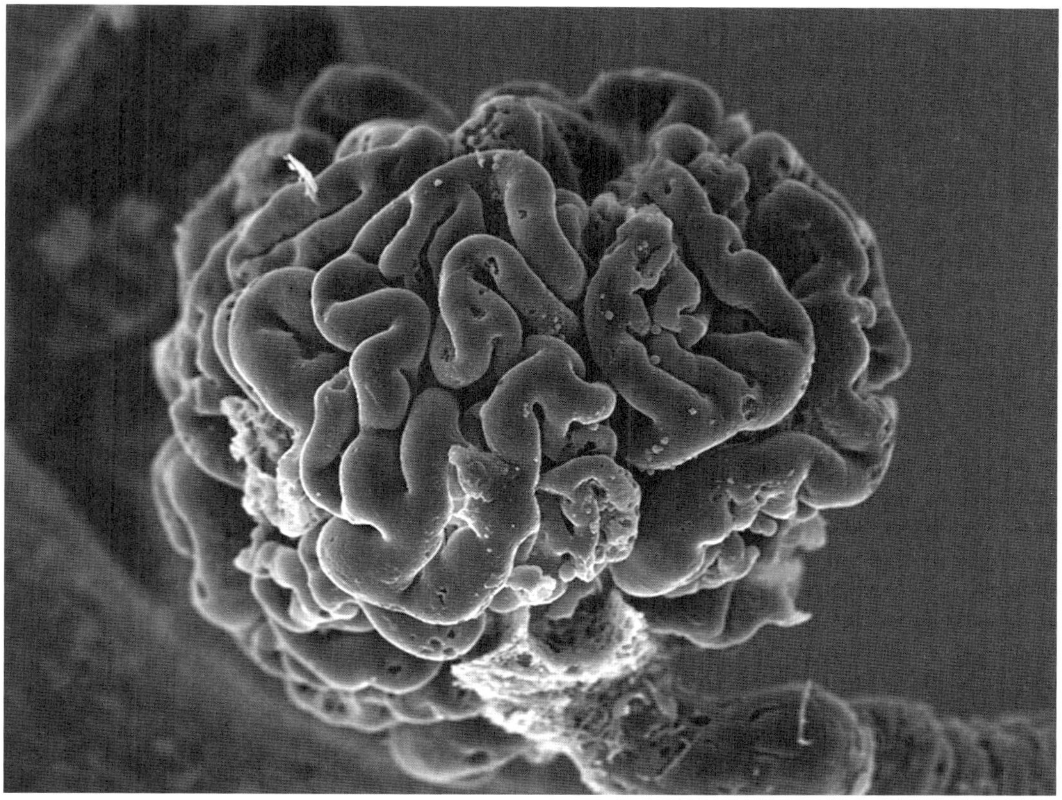

Fig. 49.11 Rat renal glomerulus vascular cast mold (detail of Fig. 49.10). Typical glomerular capillary fenestration. Magnification ×370

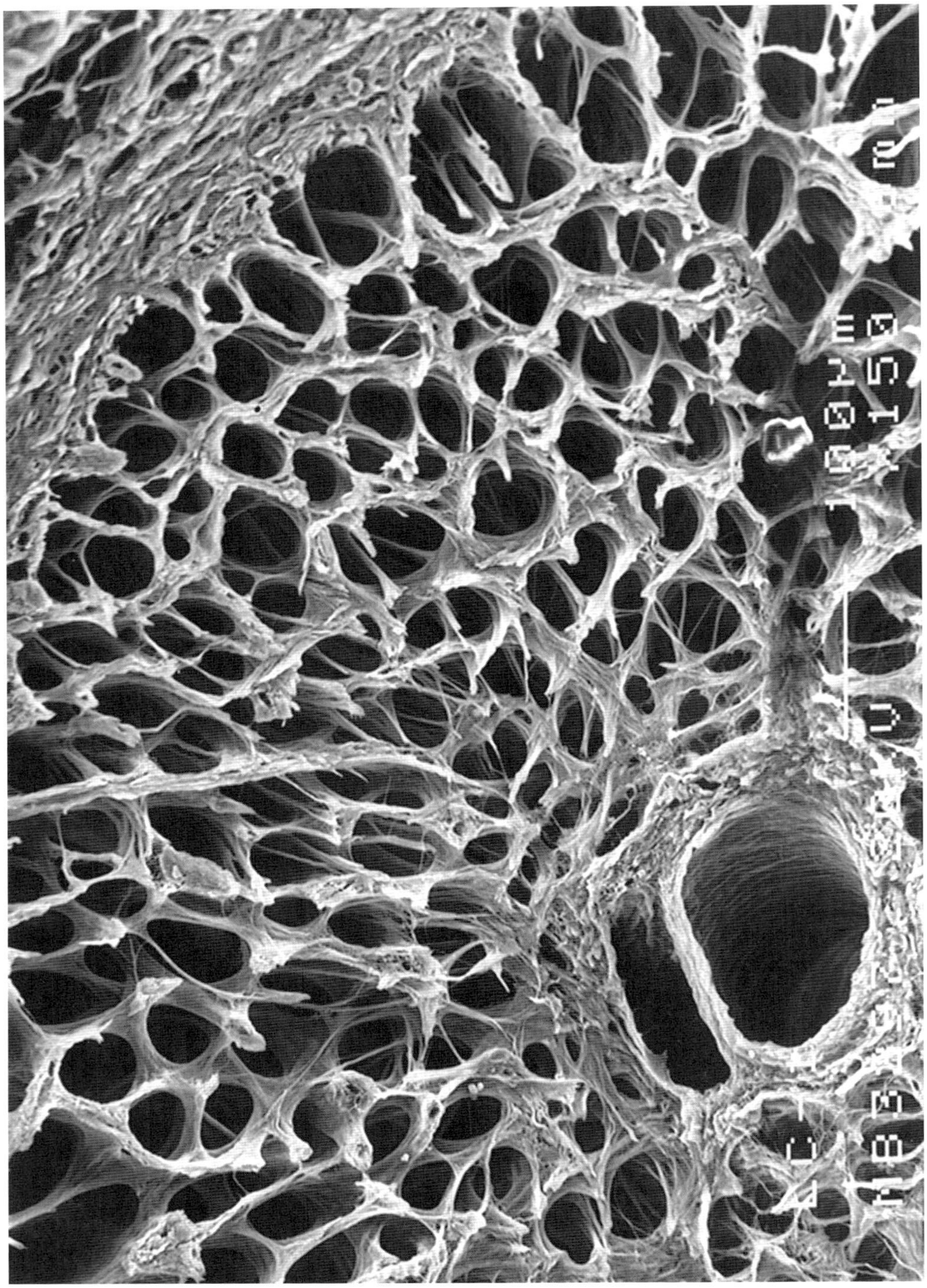

Fig. 49.12 Human optic nerve. Sample prepared using digestion techniques. Detail of collagen fibers that form the skeleton of this nerve. The orifices correspond to spaces previously occupied by the axons. Magnification ×150

Suggested Reading

1. Glauert AM. Practical methods in electron microscopy, vol. 3. Amsterdam: North-Holland/Elsevier; 1987. p. 55–72.
2. Hernández García JM, Reina MA, Prats-Galino A, de Andrés JA. Morphologic study of nerve root and types of needle used in transforaminal injections. Reg Anesth Pain Med. 2011;36:278–81.
3. Reina MA, López A, Machés F, de León Casasola O, de Andrés JA. Electron microscopy and the expansion of regional anesthesia knowledge. Tech Reg Anesth Pain Manag. 2002;6:165–71.
4. Reina MA, Villanueva MC, Machés F, Carrera A, López A, de Andrés JA. The ultrastructure of the human spinal nerve root cuff in the lumbar spine. Anesth Analg. 2008;106:339–44.
5. Robards AW, Wilson AJ. Procedures in electron microscopy. Hoboken: Wiley; 1999.

Transmission Electron Microscopy

50

Maria Luisa García Gil and Agustín Fernández Larios

Transmission electron microscopy (TEM) is a technique whereby a beam of electrons is transmitted through an ultrathin section, interacting with the biological specimen as it passes through. Any method of preparing biological samples for TEM must fulfill four basic requirements:

1. To avoid collapse of the structures in the vacuum of the electron microscope (The effects of water are extremely detrimental to high vacuum systems. Water can be removed using negative staining and tissue fixation, dehydration, and plastic embedding techniques.)
2. To provide a sample sufficiently thin to avoid multiple scattering of electrons
3. To minimize structural alterations resulting from damage by the electron beam
4. To maximize contrast, as biological samples have very low electron density, which results in few scattering events and low-contrast images

These considerations point out the importance of the steps required in sample preparation, in which artifacts can appear that may be erroneously interpreted as part of the ultrastructure of the tissue examined. A number of recommendations are deemed necessary in order to recognize and avoid artifacts. The first and perhaps the most important step in the preparation process is fixation. In this step, tissue is chemically treated to stabilize it and preserve the cellular structure with the minimum alteration of morphology, volume, and spatial relationships. Samples must be dissected, cut into small pieces (about 1–2 mm^3), and quickly immersed in the fixative solution. We use modified Karnovsky fixative—4 % paraformaldehyde and 2.5 % glutaraldehyde in phosphate buffer (0.1 M) [4].

Fixation can produce structural artifacts:

- Osmotic changes may modify the cellular volume, so the pH and osmolarity of the fixation solution must be strictly controlled.
- Redistribution of ions and proteins can occur.
- Elevated temperatures enhance the reaction between glutaraldehyde and cellular reactive groups, extracting cellular constituents through autolysis. For this reason, low temperatures (4 °C) are preferred; these also reduce the shrinkage of mitochondria and other artifacts.

New techniques now preferred for preserving ultrastructure, antigenicity, and immobilization of molecules are ultrarapid cryofixation followed by cryosubstitution or cryoultramicrotomia [3, 4].

After fixation, samples are postfixed in a 1 % osmium tetroxide solution to improve the quality of the fixation, preserving lipids and increasing electron density. To preserve membranes and enhance their contrast potassium ferrocyanide is often added to the osmium solution.

After being fixed, samples must be dehydrated in graded concentrations of ethanol or acetone and finally infiltrated and embedded in epoxy resins. Then an ultramicrotome cuts the samples into ultrathin sections (50–70 nm), which are picked up on grids. Ultrathin sections are stained with heavy metals as uranyl acetate and Reynolds' lead citrate solutions to increase the level of contrast in the final image. In this way, portions of the sample that interact strongly with the metals appear as darker areas.

Finally, it is important to mention one of the most relevant biological applications of TEM: the identification of antigens using immunogold labeling techniques [3, 4]. Immunogold labeling is commonly affected by the problem of nonspecific (background) labeling, however. This background labeling has several causes, each with a suggested solution to reduce or eliminate it:

- Nonspecific charge attraction of the antibody: Add BSA (especially acetylated BSA or normal serum) to approximately 1 % in incubation solutions. Use 1 % Tween in all solutions.

M.L. García Gil, BS, PhD (✉) • A. Fernández Larios, Tech
Transmission Electron Microscopy Unit,
ICTS Electron Microscopy National Center,
Complutense University, Madrid, Spain
e-mail: marisag@pdi.ucm.es; afdez@pdi.ucm.es

M.A. Reina et al. (eds.), *Atlas of Functional Anatomy for Regional Anesthesia and Pain Medicine: Human Structure, Ultrastructure and 3D Reconstruction Images*, DOI 10.1007/978-3-319-09522-6_50,

- Presence of free aldehyde groups in fixed tissue: Float sections on a droplet of 0.1 M glycine or 0.5 M ammonium chloride before incubation with antibodies.
- Inadequate washing between incubations: Wash more thoroughly.
- Primary and secondary gold conjugated antibody concentration too high: Dilute by orders of magnitude (Figs. 50.1, 50.2, 50.3, 50.4, 50.5, 50.6, 50.7, 50.8, 50.9, 50.10, 50.11, 50.12, 50.13, 50.14, and 50.15).

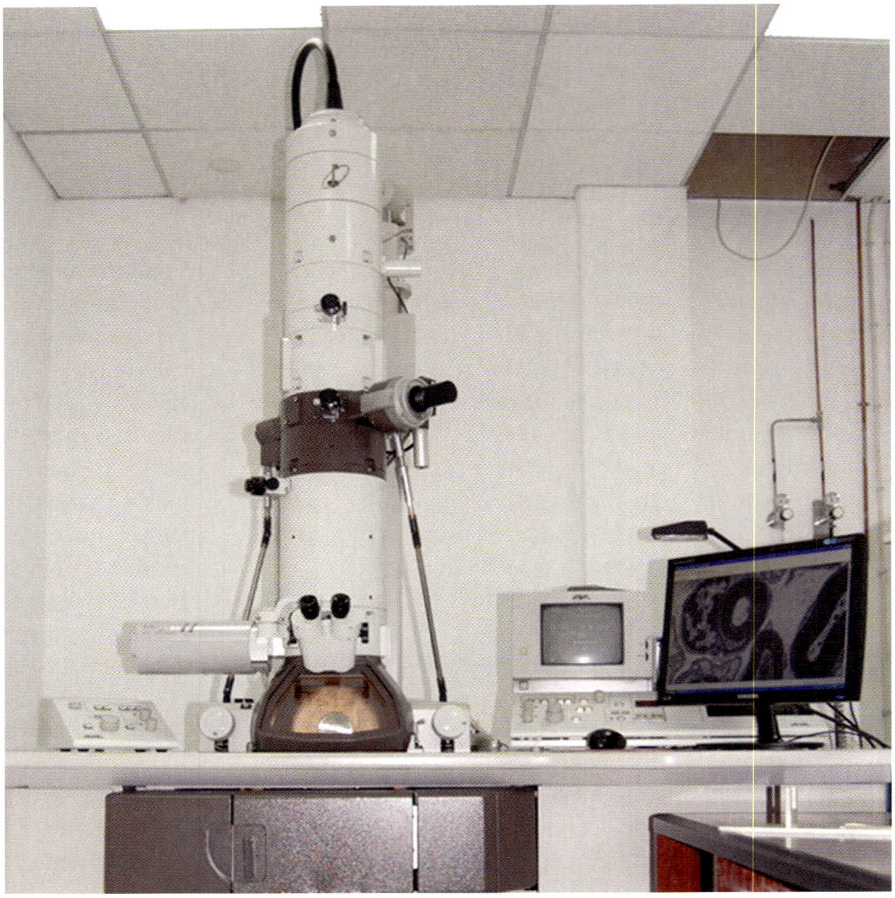

Fig. 50.1 Transmission electron microscope, working at intermediate voltages. It consists of a tungsten filament as an emission source, electromagnetic lenses for beam convergence, a specimen stage to hold the grids, and a fluorescent screen to observe the image

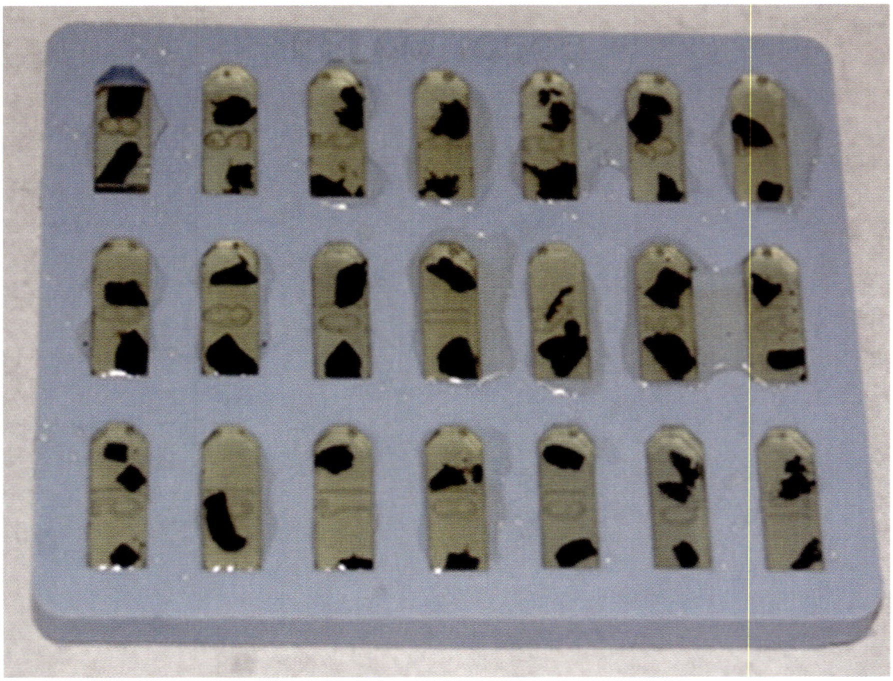

Fig. 50.2 Flat embedding molds. After fixation, postfixation, dehydration, and embedding, the specimen is placed in flat molds. Epoxy resin (Spurr, Epon 812) is poured over the tissue and the molds containing specimens are placed in an oven at 70 °C for polymerization for 2 days. It must be ensured that no air bubbles are trapped within the resin

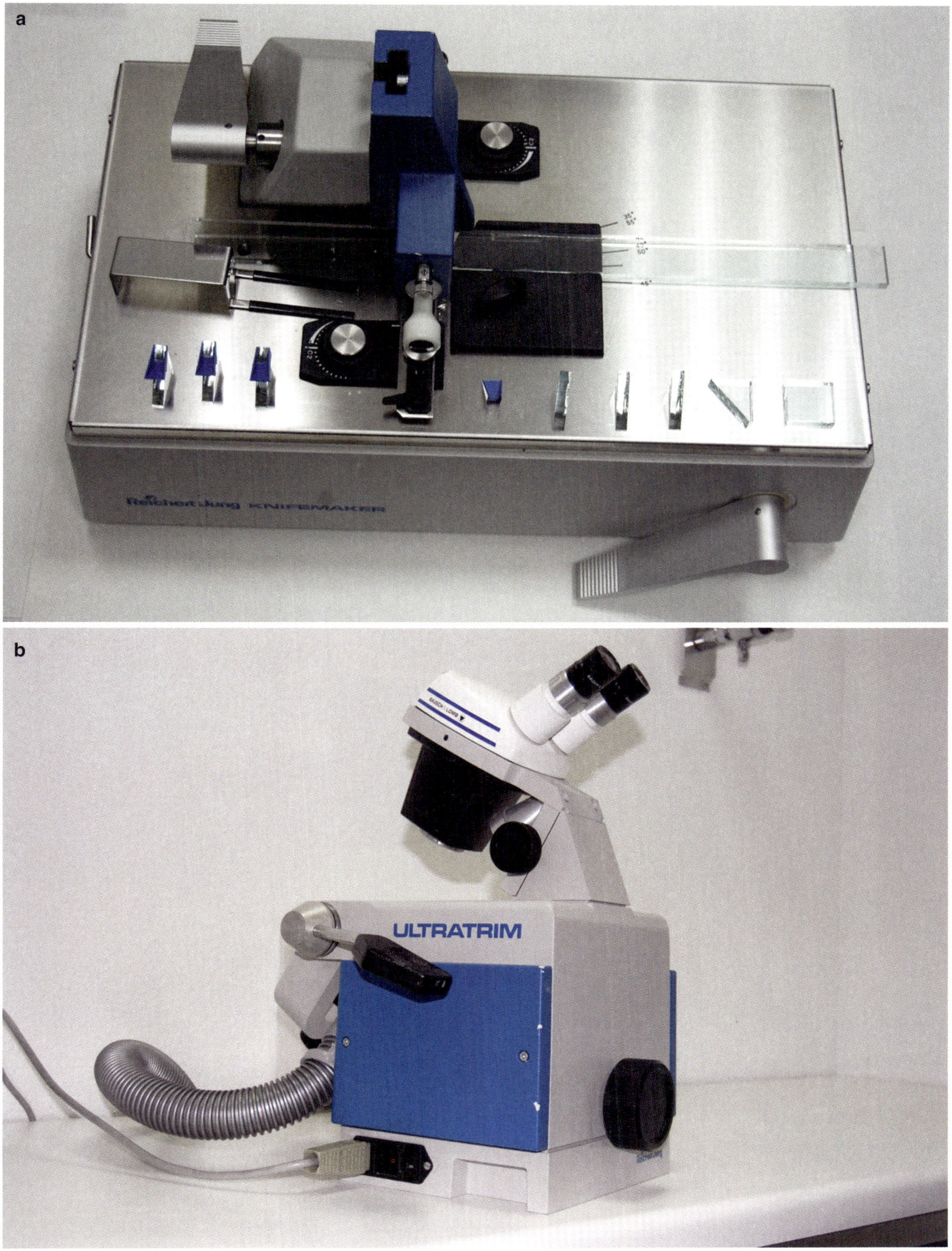

Fig. 50.3 (**a**) Knife preparation; knife maker used to prepare glass knives. Plastic troughs are fixed to the knives with dental wax. (**b**) Block trimming. The block should be trimmed (removing excess resin) to a trapezoid-shaped pyramid, with the section area being as small as possible, to include the part of the specimen of interest

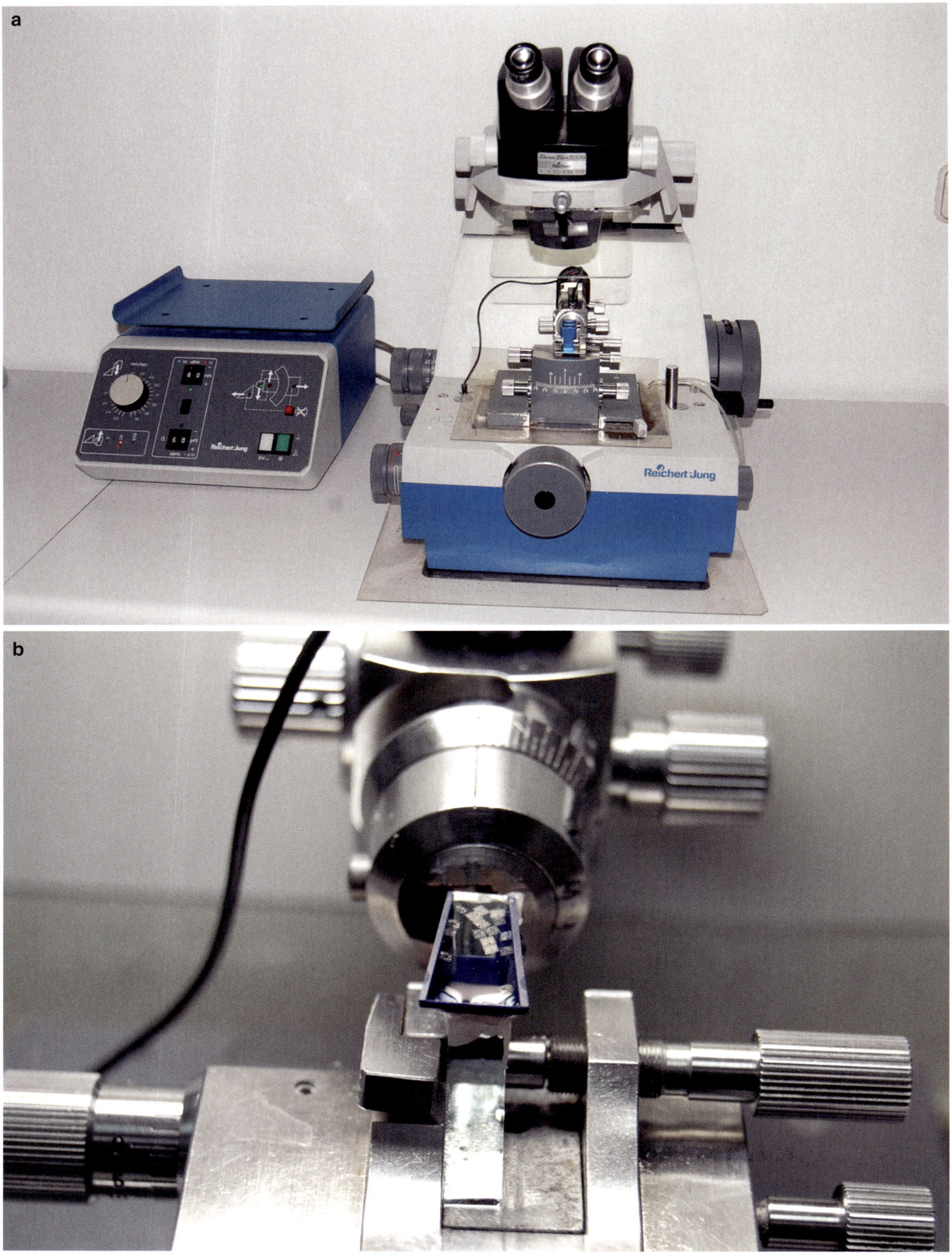

Fig. 50.4 (**a**) Ultramicrotome used to obtain semithin and ultrathin sections. (**b**) Floating semithin sections (0.5–1 μm) are collected on slides and stained with toluidine or Richardson's methylene blue

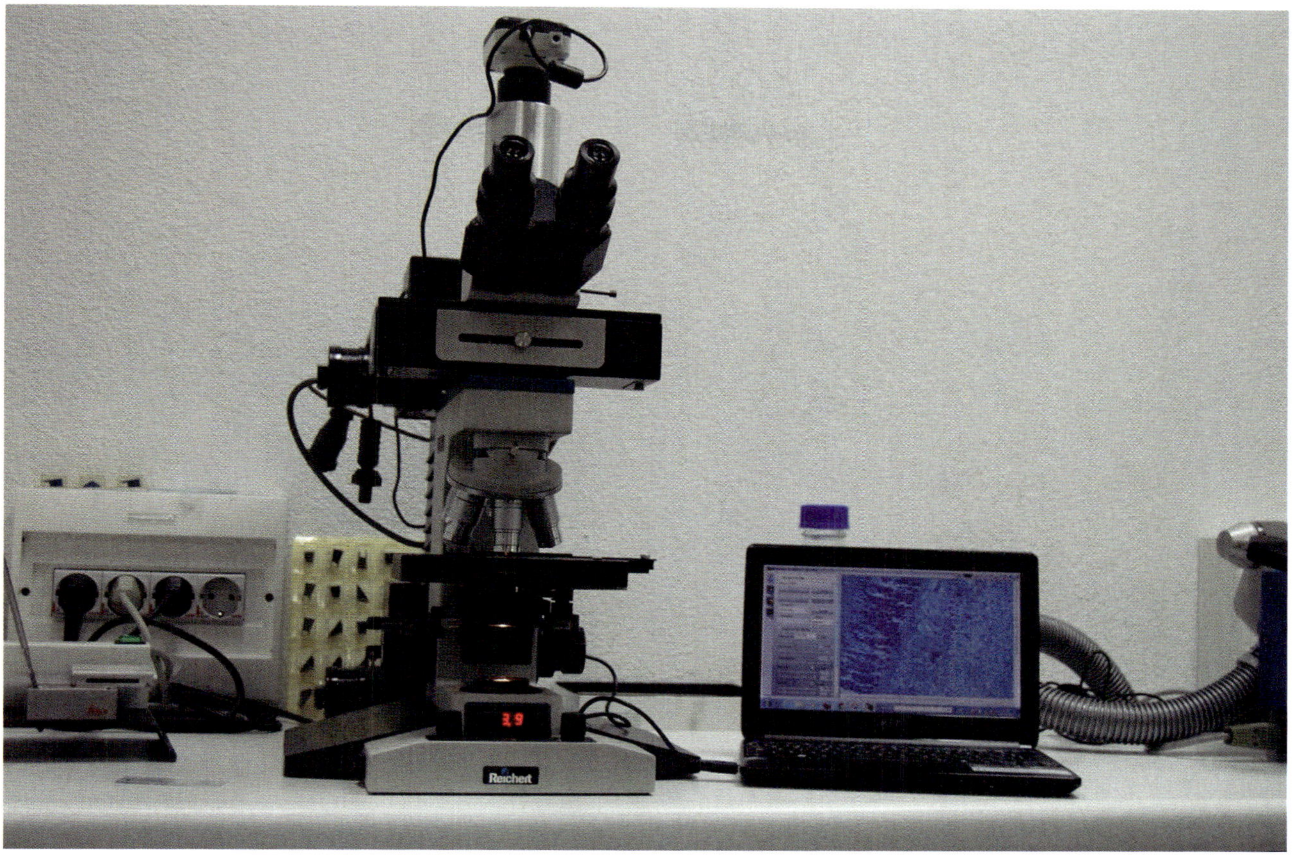

Fig. 50.5 Light microscope used for observation of semithin sections

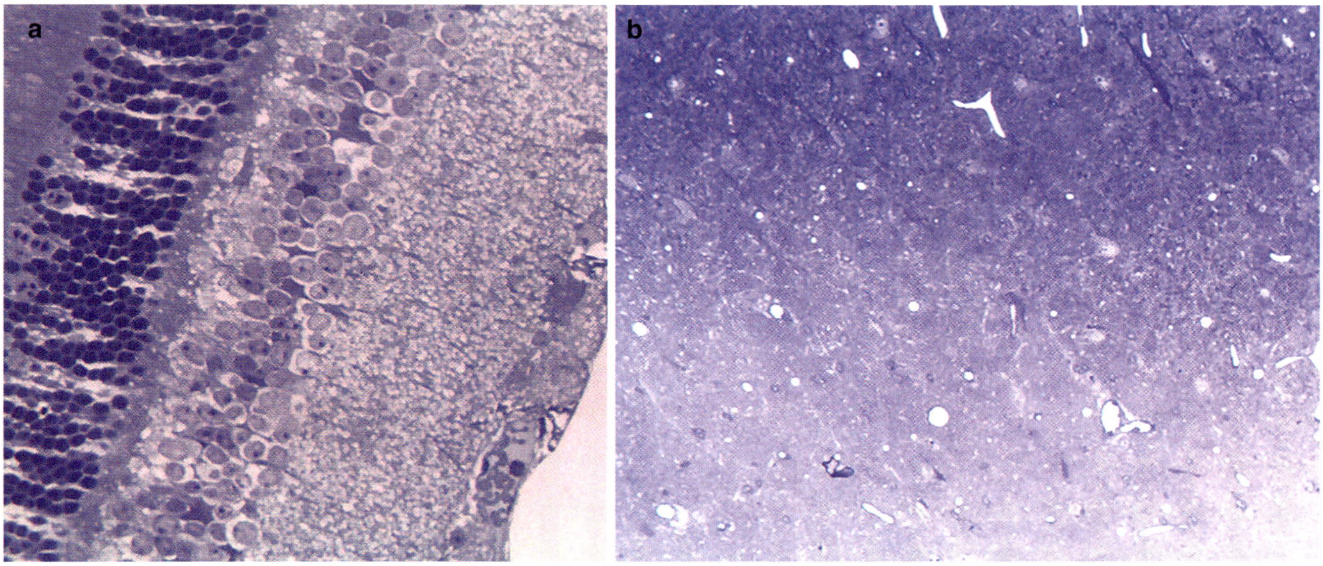

Fig. 50.6 The quality of fixation can be appreciated in semithin sections. (**a**) Well-fixed and postfixed retina semithin section (×1,000). (**b**) In poorly postfixed samples, semithin sections show areas of different staining intensity, which indicates that the osmium tetroxide did not penetrate across the tissue (×400)

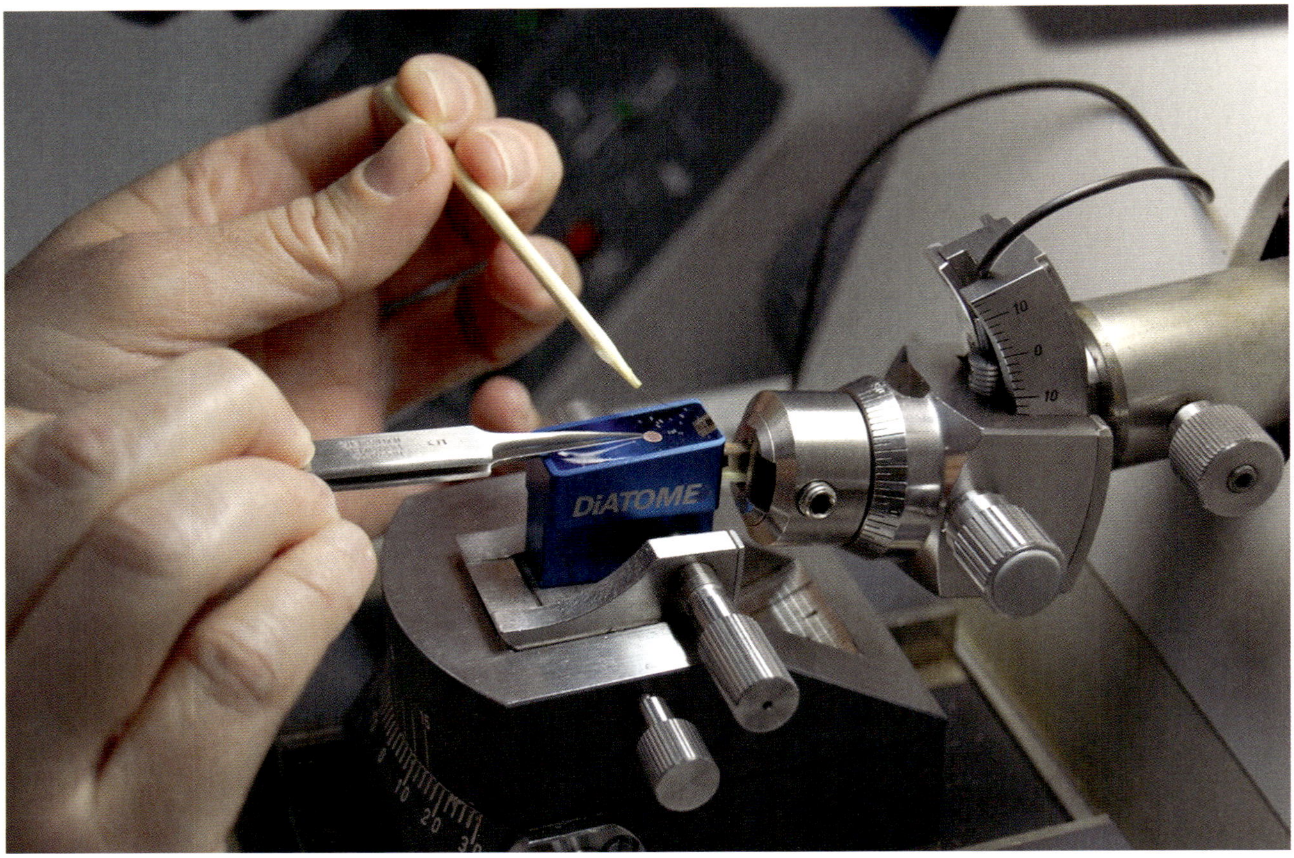

Fig. 50.7 Ultrathin sections (50–70 nm) are cut using a diamond knife and mounted on grids

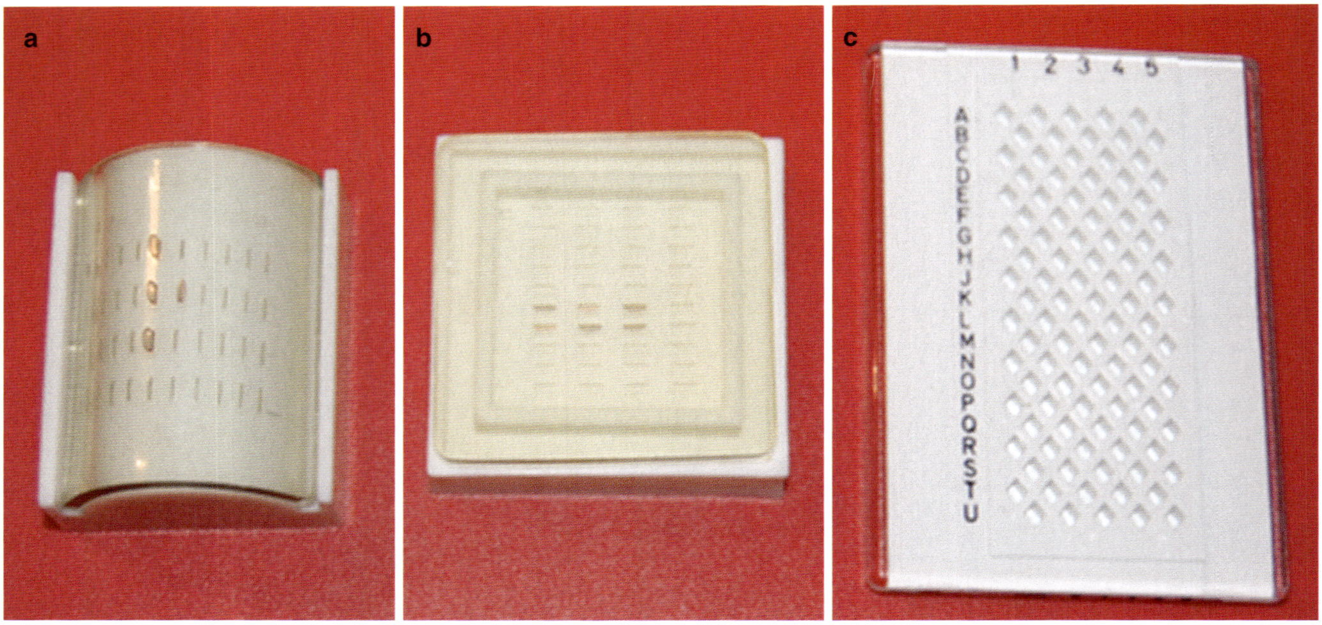

Fig. 50.8 Staining of ultrathin sections using the Hiraoka Staining Kit. (**a**) The grids with the ultrathin sections are inserted into the slits in the plastic plate. (**b**) The plate with the grids is inverted over the dish containing uranyl acetate 1 % and Reynolds' lead citrate successively. (**c**) The grids are stored in the grid box

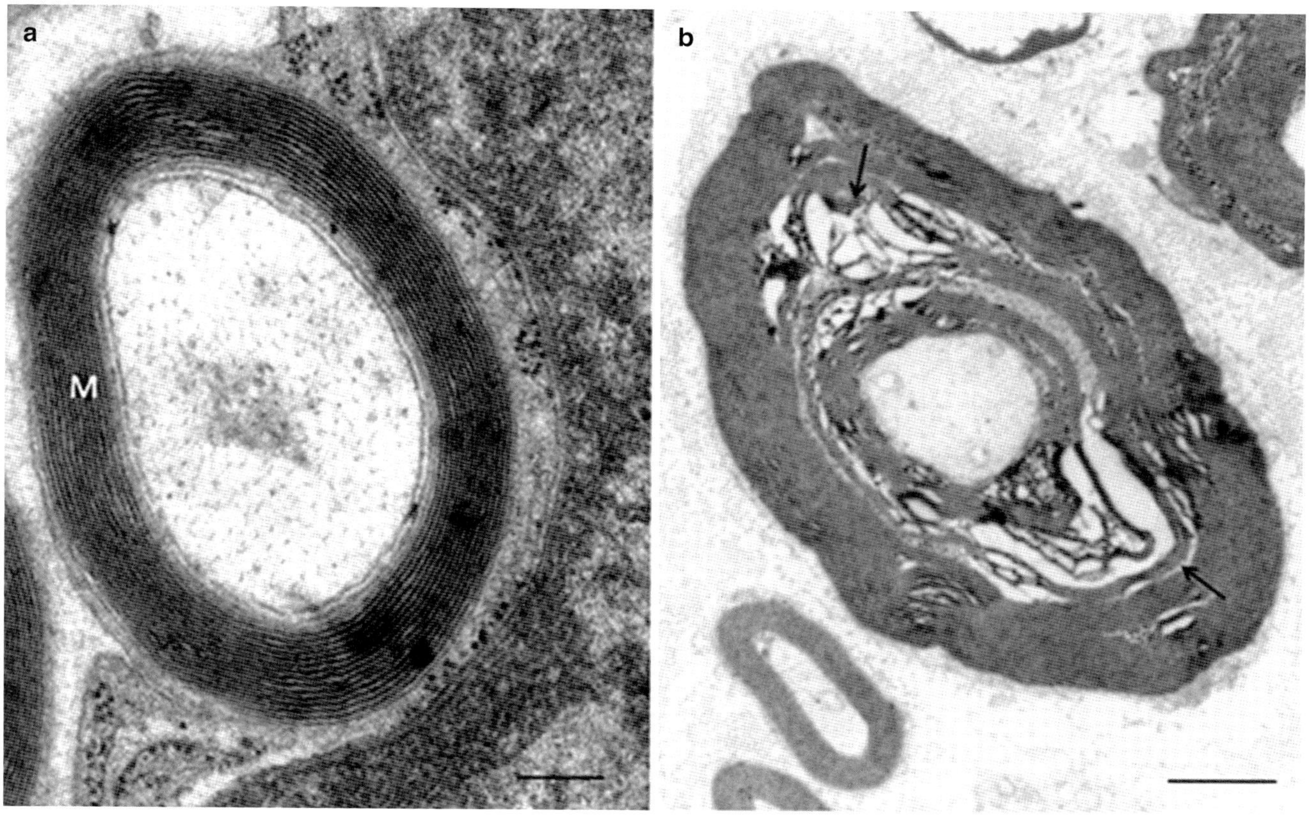

Fig. 50.9 Micrographs showing the effects of poor fixation. (**a**) Well-preserved cross-section of myelinated axons. Multiple layers of phospholipid membrane forming the myelin sheath (M) are well preserved and compacted by glutaraldehyde/osmium tetroxide fixation [2]. (**b**) Artifacts in myelin sheath caused by poor fixation, which affects the myelin compaction

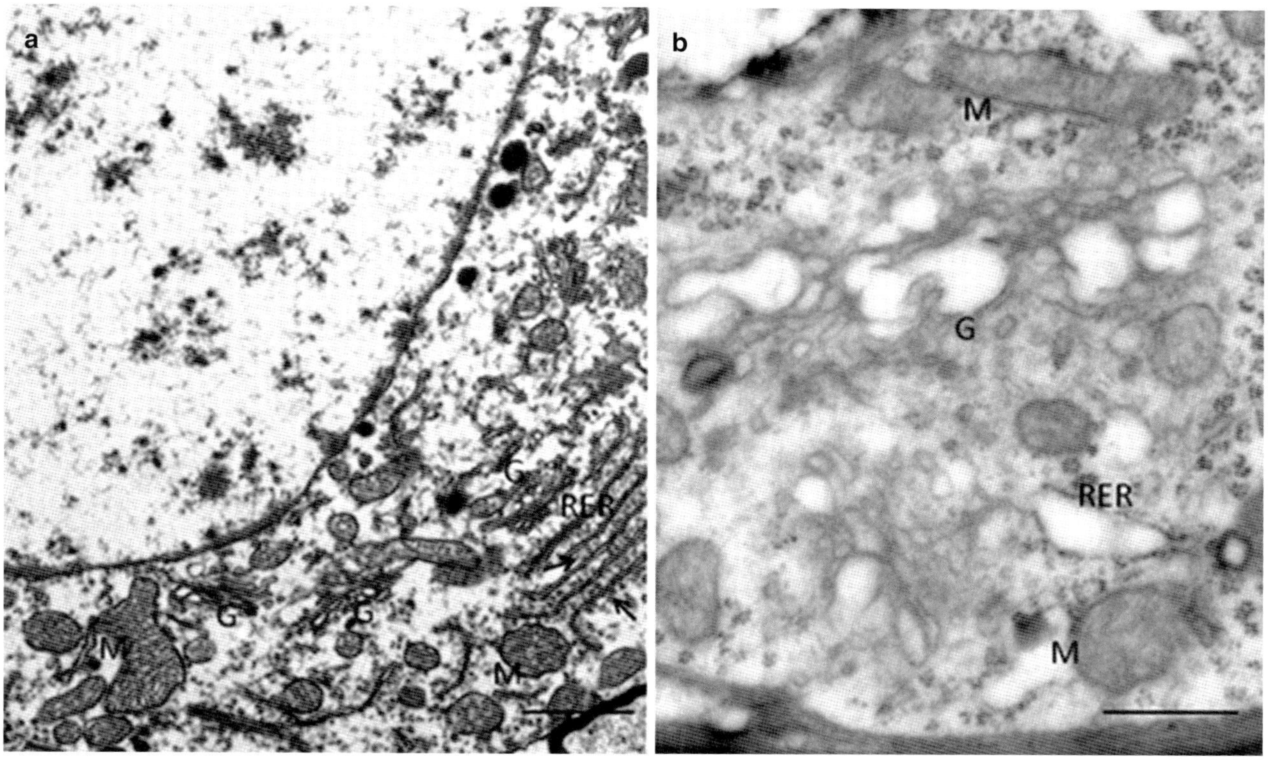

Fig. 50.10 (**a**) Well-fixed motor neuron soma. Mitochondria (M) with detail of cristae, Golgi apparatus (G) in a typical image of a group of four to five Golgi cisternae. Rough endoplasmic reticulum (RER) and network of flattened sacs are clearly visible. Note ribosomes attached to the cytoplasmic side of the membrane [2]. (**b**) Poor fixation of a neuron soma. Notice the lack of detail of the organelles within the cell (blur image). Mitochondria (M) traces of cristae, the ultrastructure of Golgi apparatus (G), and rough endoplasmic reticulum (RER) appear altered and swollen

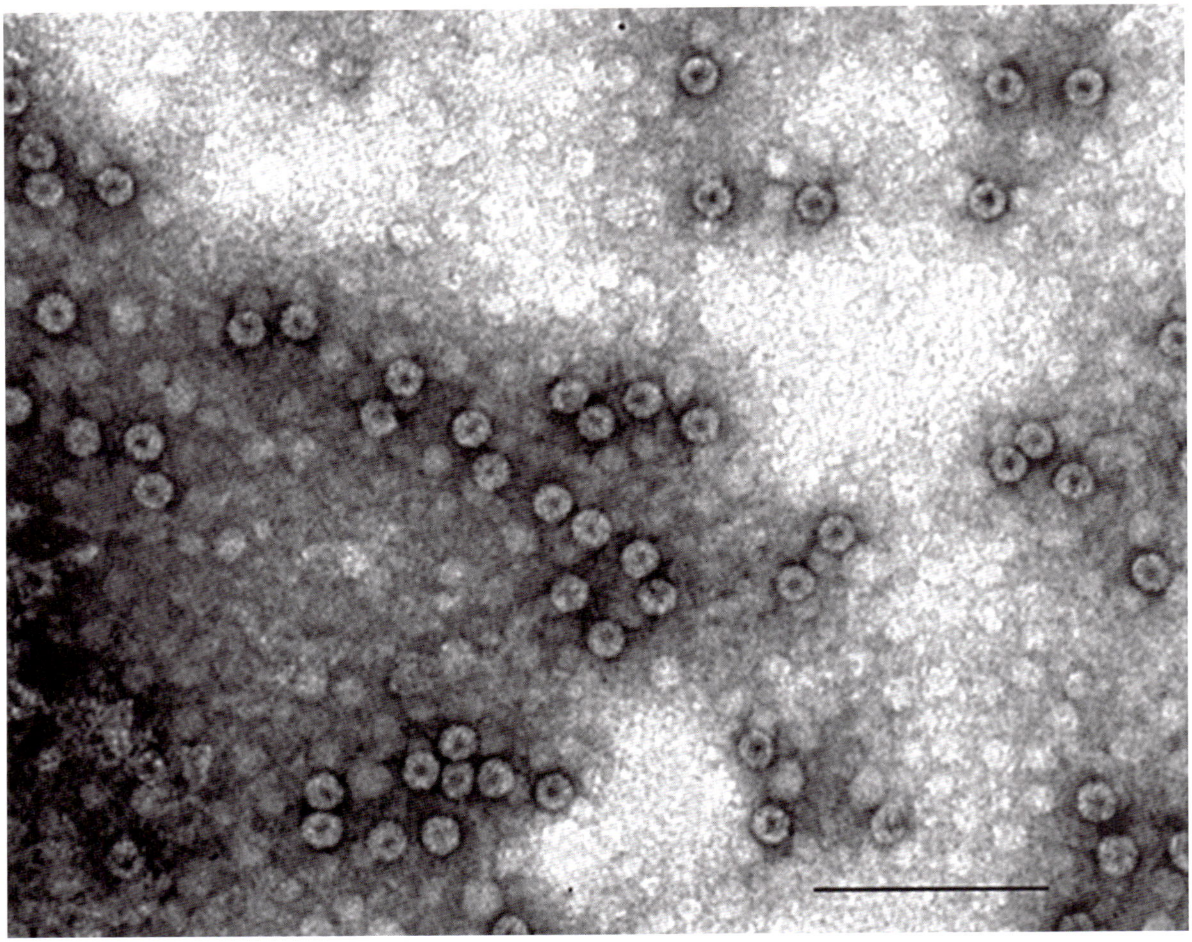

Fig. 50.11 Negative staining techniques are relevant tools enabling examination of isolated organelles, individual macromolecules, bacteria, and viruses by means of electron microscopy. The main purpose of negative staining is to surround or embed the biological sample in heavy metal (uranyl acetate 2 %, phosphotungstic acid [PTA] 1 %), which provides high contrast and good preservation. In the micrograph, viruses are negatively stained with uranyl acetate 2 %. A droplet of a suspension of viruses is placed on the surface of the grid support film, followed by a droplet of stain solution, and then it is dried in air [3, 4].

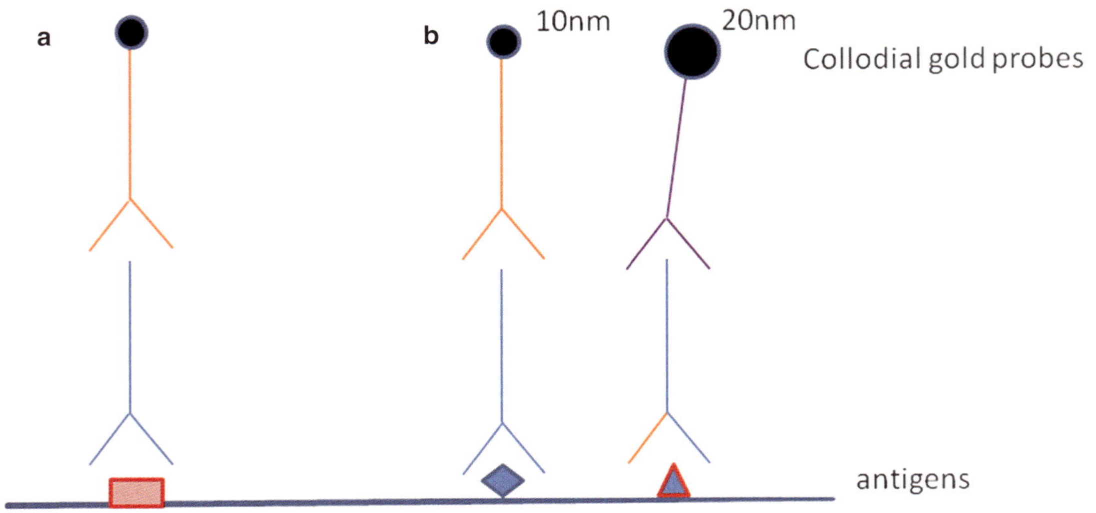

Fig. 50.12 The advantage of immunocytochemistry at the electron microscopy level is the simultaneous detection of antigens and the identification of the cell components where the labeling is observed. Few antigens survive fixation and embedding procedures that utilize osmium tetroxide. It is more beneficial to use a gentler aldehyde fixation (4 % paraformaldehyde and 0.5 % glutaraldehyde), combined with low-temperature embedding, which preserves the antigenicity. To see the labeling in the electron microscope, gold-conjugated secondary antibodies must be used. These secondary antibodies can be purchased in various sizes (10, 15, or 20 nm) to allow for double labeling. (**a**) Specific single gold labeling. (**b**) Specific double gold labeling [4].

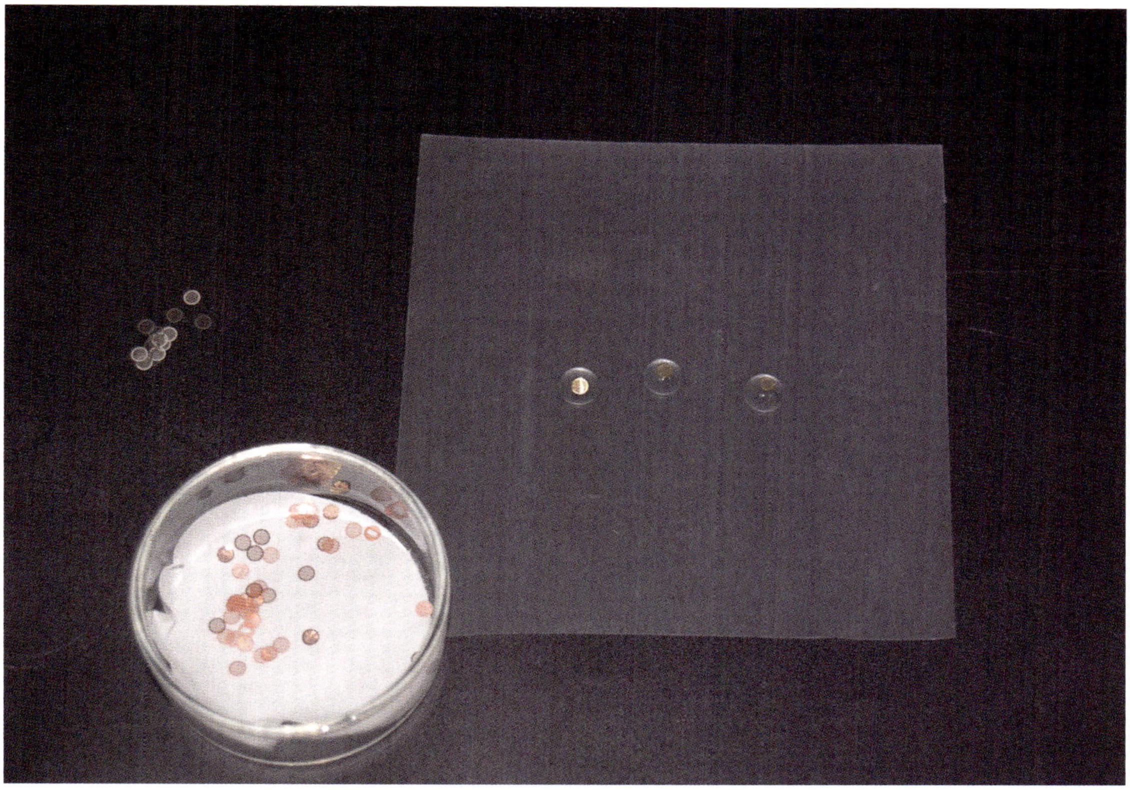

Fig. 50.13 Procedure for post-embedding immunogold labeling. Once the tissue is weakly fixed and properly embedded, it is sectioned by ultramicrotomy. Ultrathin sections are picked up on nickel or gold grids. (Many reagents etch copper grids.) The photograph shows how sections can be immunolabeled by floating the grids on the surface of reagent droplets. Briefly, grids are sequentially floated in blocking solutions (glycine and BSA), the first antibody and the secondary antibody conjugated with colloidal gold. The primary antibody should be about ten times more concentrated than a light-microscopy working dilution; overnight incubation seems better than 2 h at room temperature. A negative control must be included, omitting the first antibody and replacing it with buffer [3, 4].

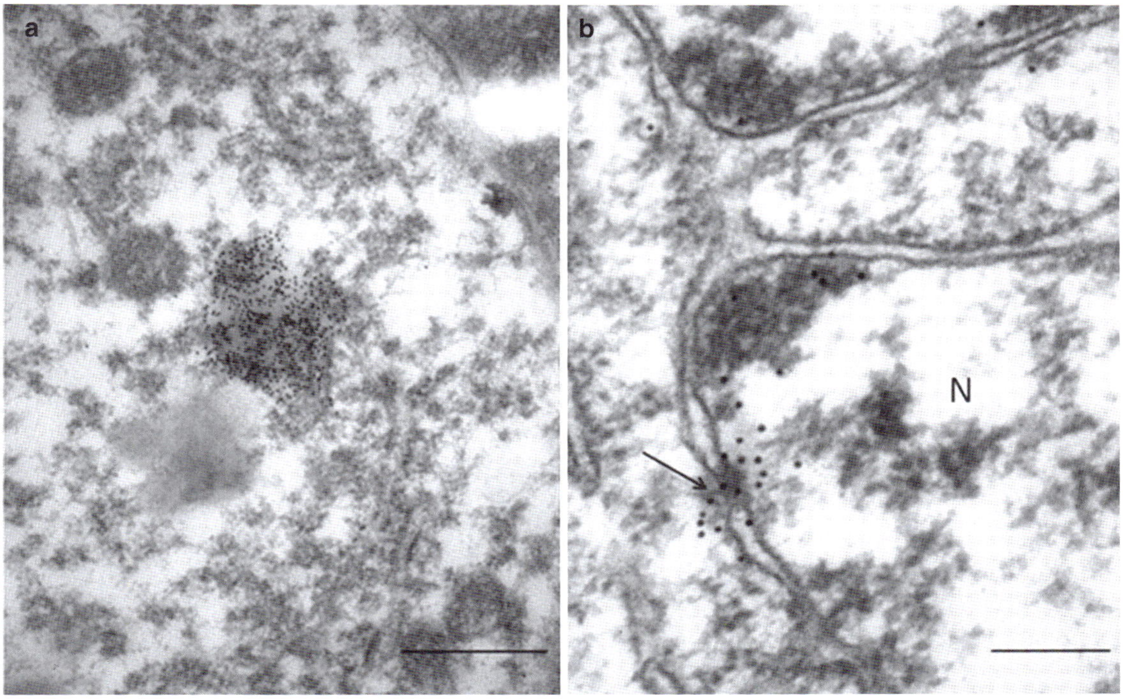

Fig. 50.14 Micrographs showing single immunogold labeling. (**a**) The first antibody labels a cluster of active caspase 3, a protein implicated in cell death, in the motor neuron soma of an aged rat. The secondary antibody is conjugated with 10-nm gold particles (Bar=500 nm) [1]. (**b**) Micrograph showing immunogold labeling of TPD-43, a nuclear RNA-binding protein implicated in neuropathies and neuron degeneration. Note that TDP-43 is being transported through a nuclear pore to nucleus [3, 4].

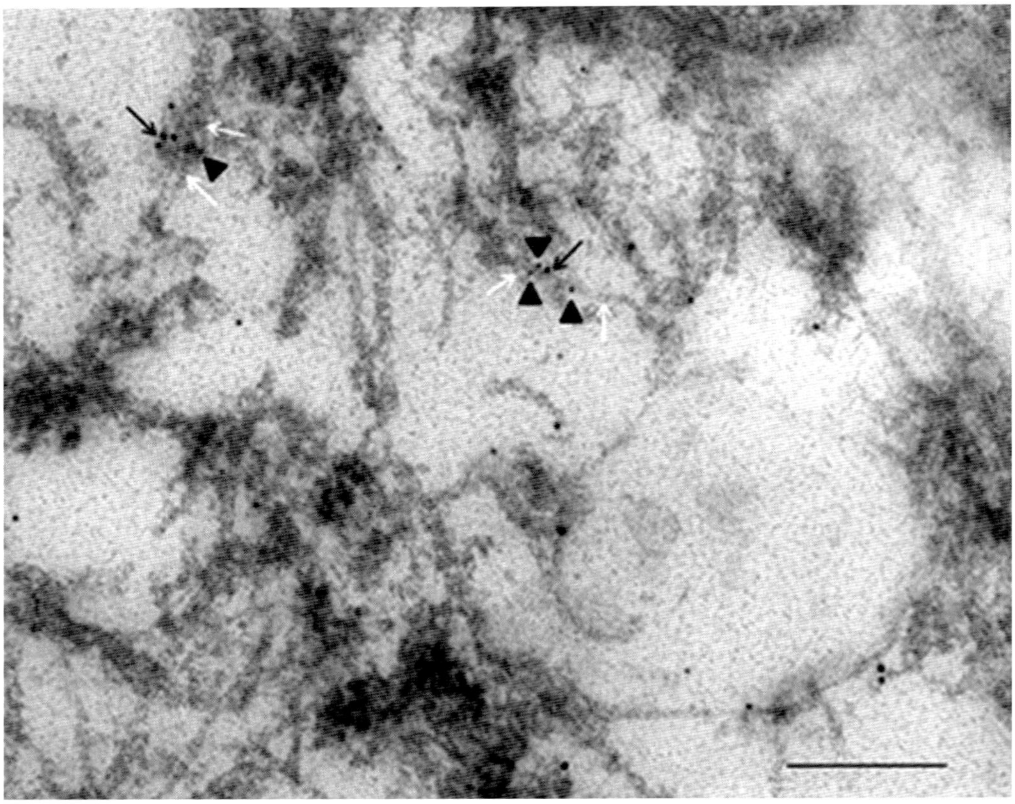

Fig. 50.15 Double immunogold labeling. Spinal cord ultrathin sections from aged rats were collected on nickel grids. The grids were incubated with a mixture of primary antibodies diluted in 1 % BSA-PBS, rabbit anticaspase 3 antibody, and mouse anti-Dynein, which transports cellular substances along microtubules retrogradely. Then the grids were incubated with a mixture of a goat anti–rabbit IgG conjugated to 10-nm gold particles, and a goat anti–mouse IgG conjugated to 5-nm gold particles. Colocalization of dynein and active caspase 3 along axon microtubules suggests that caspase 3 is transported retrogradely in motor neurons of aged rats [1].

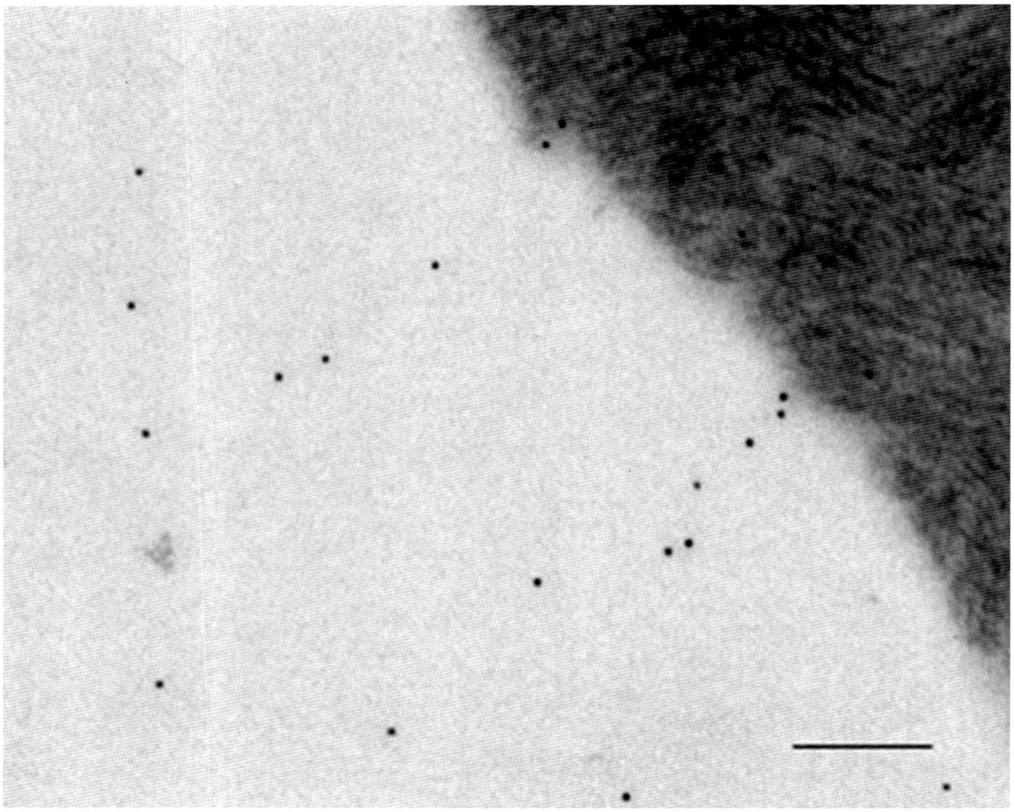

Fig. 50.16 One of the most common problems in immunogold labeling is background (non-specific labeling). The micrograph shows this non-specific labeling where gold particles appear randomly distributed, they even stick on the resin ares where no tissue is found [3, 4].

Suggested Reading

1. García ML, Fernández A, Solas MT. Mitochondria, motor neurons and aging. J Neurol Sci. 2013;330:18–26.
2. Cheville NF. Ultrastructural pathology: the comparative cellular basis of disease. 2nd ed. Ames: Wiley-Blackwell; 2009. p. 3–193.
3. Robards AW, Wilson AJ. Procedures in electron microscopy. Hoboken: Wiley; 1999.
4. Hayat MA. Principles and techniques of electron microscopy: biological applications. 4th ed. Cambridge: Cambridge University Press; 2000. p. 85–138.

Index

M.A. Reina et al. (eds.), *Atlas of Functional Anatomy for Regional Anesthesia and Pain Medicine: Human Structure, Ultrastructure and 3D Reconstruction Images*, DOI 10.1007/978-3-319-09522-6,

Printed in the United States of America